Ion Channels

From Structure to Function

Ion Channels
From Structure to Function

Dr James N. C. Kew

Neurosciences CEDD
GlaxoSmithKline
New Frontiers Science Park, Harlow, UK

Dr Ceri H. Davies

Neurosciences CEDD
GlaxoSmithKline
Medicines Research Centre, Verona, Italy

OXFORD
UNIVERSITY PRESS

OXFORD
UNIVERSITY PRESS

Great Clarendon Street, Oxford OX2 6DP

Oxford University Press is a department of the University of Oxford.
It furthers the University's objective of excellence in research, scholarship,
and education by publishing worldwide in

Oxford New York

Auckland Cape Town Dar es Salaam Hong Kong Karachi
Kuala Lumpur Madrid Melbourne Mexico City Nairobi
New Delhi Shanghai Taipei Toronto

With offices in

Argentina Austria Brazil Chile Czech Republic France Greece
Guatemala Hungary Italy Japan Poland Portugal Singapore
South Korea Switzerland Thailand Turkey Ukraine Vietnam

Oxford is a registered trade mark of Oxford University Press
in the UK and in certain other countries

Published in the United States
by Oxford University Press Inc., New York

British Library Cataloguing in Publication Data

Data available

Library of Congress Cataloging-in-Publication-Data
Ion channels : from structure to function / [edited by] James N.C. Kew
and Ceri Davies. -- 2nd ed.
     p. ; cm.
  Includes bibliographical references and index.
  ISBN 978-0-19-929675-0 (alk. paper)
 1. Ion channels. I. Kew, James N. C. II. Davies, Ceri H.
  [DNLM: 1. Ion Channels--physiology. 2. Biological Transport, Active.
 3. Ion Transport. QU 55.7 I644 2009]
  QH603.I54I57 2009
  571.6'4--dc22
                                        2009020664

Typeset by Cepha Imaging Private Ltd., Bangalore, India
Printed in China by China Translation & Printing Services Ltd

ISBN 978–0–19–929675–0

10 9 8 7 6 5 4 3 2 1

Preface

Since the cloning of the first ion channel (the voltage-gated Na$^+$ channel from the electric eel *Electrophorus electroplax*) over 25 years ago at least 250 structurally distinct ion channel subunits have been identified. Each ion channel allows the rapid passive passage of ions (cations or anions) along electrochemical gradients across poorly permeable lipid bilayers (cell surface or intracellular compartments) at a rate (millisecond time frame) that is at least three orders of magnitude faster than that afforded by pumps or carrier proteins and over ten times faster than diffusion. This unique property allows ion channels to contribute to a wide variety of cellular processes, including electrical excitability, maintenance of membrane potential, hormone/neurotransmitter release, regulation of cell volume, apoptosis and activation of second messenger signalling cascades and of other ion channels, that are central to virtually all physiological processes.

Classically, the ion channel superfamily has been broadly subdivided into: (1) ligand-gated ion channels (LGIC) which comprise channels that are activated by Ca^{2+}, nucleotides and neurotransmitters, and (2) voltage-gated ion channels (VGIC) that are activated by changes in membrane potential. However, this classification is simplistic in that there are (a) many subdivisions that can be made within these two broad classes and (b) there are examples of ion channels that are activated by both ligand and voltage as well as other modalities such as light or mechanical stretch. Not surprisingly, molecular biological approaches have revealed extensive structural diversity within the ion channel superfamily with, for example, primary subunits possessing from between one or two (e.g., intracellular chloride channels or P$_{2X}$ receptors) to 24 (e.g., voltage-gated Ca^{2+} channels) transmembrane domains. Despite the fact that sequencing of the human/rodent genomes is now complete there remain functionally defined conductances for which the molecular identity of the ion channel mediators has not been elucidated. However, for those where the molecular identity and amino acid sequence is available, experimental approaches, such as site-directed mutagenesis in combination with homology modelling and X-ray structural analysis, have identified molecular determinants of aspects of ion channel function such as voltage-dependence, activation, inactivation, deactivation, desensitisation, conductance, ionic selectivity, rectification and ligand-mediated activation/inhibition/modulation. In addition, intracellular regulatory sequence motifs have been discovered on numerous ion channels that enable cytoplasmic factors such as cyclic nucleotides, kinases and phosphatases to modulate their activity both positively and negatively. A further level of complexity is afforded by the fact that most ion channels exist as heteroligomeric assemblies of subunits that are also frequently associated with auxiliary proteins that can have a significant impact on their biophysical and pharmacological properties. Furthermore, it is now recognized that ion channels can exhibit enzymatic activity. Thus, the cystic fibrosis transmembrane conductance regulator (CFTR) Cl$^-$ channel hydrolyzes nucleotide triphosphates.

Looking beyond the characteristics of individual ion channels in isolation there is now a wealth of data on their tissue and subcellular localisation as well as their physiological function at the cellular, organ, and whole animal (including human) levels. By combining this knowledge with that of the biophysical properties of the channels it has been possible to begin to reconcile the aetiology and phenotype of several muscular and neurological diseases (including movement disorders, migraine, and epilepsy) that are caused by inherited mutations in the genes of specific ion channels. The list of these so-called 'channelopathies' is expanding rapidly, as is the phenotypic range associated with mutations within each individual channel (e.g., periodic paralysis, myotonias, malignant hyperthermia). Thus, mutations in different ion channels can result in remarkably similar phenotypes, whereas distinct mutations within the same gene can cause quite distinct disease phenotypes. Furthermore, alterations in ion channel function resulting from autoimmunity as well as acquired disorders of RNA processing can act as risk factors for, or directly precipitate, disease. It should also be noted, however, that there are many central and peripheral diseases that are not directly linked to ion channel mutations that are effectively treated by drugs that specifically modify ion channel activity.

Given that (1) ion channels are central to physiological processes as diverse as diuresis and cognition and (2) the experimental technologies to investigate ion channel function are now well established and widely available, it is certain that these transmembrane proteins will continue to be the focus of intensive interest in both academic and industrial research groups alike. This book was compiled in response to this widespread interest and both as an aid for education in the ion channel field and to facilitate research, in that it reviews the basic principles of ion channel function and provides detailed overviews of the structural, biophysical, physiological and pharmacological aspects of individual ion channels across the ion channel super family.

Contents

List of contributors

Yimy Amarillo
Smilow Program in Neuroscience and Departments of Physiology
and Neuroscience and Department of Biochemistry
New York University School of Medicine
New York
USA

Jorge Arreola
Institute of Physics
University of San Luis Potosí
San Luis Potosí
México

Joel Baumgart
Department of Pharmacology and Neuroscience Graduate
Program
University of Virginia
Charlottesville
Virginia, USA

Cord-Michael Becker
Institut für Biochemie
Emil-Fischer-Zentrum
Universität Erlangen-Nürnberg
Erlangen, Germany

Dietmar Benke
Institute of Pharmacology
University of Zurich
Zurich, Switzerland

Daniel Bertrand
Department of Neuroscience
Medical Faculty
Geneva 4, Switzerland

Jane E Carland
Centre for Neuroscience
Division of Medical Sciences
Ninewells Hospital and Medical School
The University of Dundee
Dundee, United Kingdom

Thomas E Chater
MRC Centre for Synaptic Plasticity
Department of Anatomy
School of Medical Sciences
University Walk
Bristol
United Kingdom

Iain Chessell
NeuroSolutions Ltd
Iconix Park
Cambridge, United Kingdom

Palle Christophersen
NeuroSearch A/S
Ballerup, Denmark

Brian Clark
Smilow Program in Neuroscience and Departments of Physiology
and Neuroscience and Department of Biochemistry
New York University School of Medicine
New York, USA

Michelle A Cooper
Centre for Neuroscience
Division of Medical Sciences
Ninewells Hospital and Medical School
The University of Dundee
Dundee, United Kingdom

Tim Dale
GlaxoSmithKline
New Frontiers Science Park
Harlow, United Kingdom

Sulayman Dib-Hajj
Department of Neurology and Center for Neuroscience and
Regeneration Research
Yale University School of Medicine
New Haven, CT, USA
and Rehabilitation Research Center
VA Connecticut Healthcare System
West Haven, CT, USA

Jean-Marc Fritschy
Institute of Pharmacology
University of Zurich
Zurich, Switzerland

Alasdair Gibb
Research Department of Neuroscience, Physiology & Pharmacology
University College London
London, United Kingdom

Morten Grunnet
NeuroSearch A/S
Pederstrupvej 93
Ballerup, Denmark
and The Danish National Research Foundation Centre for
Cardiac Arrhythmia
University of Copenhagen
Copenhagen, Denmark

Laura Jane Hanley
MRC Centre for Synaptic Plasticity
Department of Anatomy
School of Medical Sciences
University of Bristol
Bristol, United Kingdom

Elizabeth Hartfield
MRC Centre for Synaptic Plasticity
University of Bristol Medical School
Department of Anatomy
Bristol, United Kingdom

Peter G R Hastie
MRC Centre for Synaptic Plasticity
Department of Anatomy
School of Medical Sciences
University of Bristol
Bristol, United Kingdom

Jeremy Henley
MRC Centre for Synaptic Plasticity
Department of Anatomy
School of Medical Sciences
University of Bristol
Bristol, United Kingdom

Hiroshi Hibino
Division of Molecular and Cellular Pharmacology
Department of Pharmacology
Graduate School of Medicine
Osaka University, Osaka, Japan

Jean-Charles Hoda
Department of Neuroscience
Medical Faculty
Geneva, Switzerland

Ronald C Hogg
Department of Neuroscience
Medical Faculty
Geneva, Switzerland

Atsushi Inanobe
Department of Pharmacology II
Graduate School of Medicine
Osaka University
Osaka, Japan

Hilary L Jackson
MRC Centre for Synaptic Plasticity
Department of Anatomy
School of Medical Sciences
University of Bristol
Bristol, United Kingdom

Hyo-Young Jeong
Smilow Program in Neuroscience and Departments of Physiology
and Neuroscience and Department of Biochemistry
New York University School of Medicine
New York, USA

Dawon Kang
Medical Research Center for Neural Dysfunction and
Department of Physiology
Gyeongsang
National University College of Medicine
Jinju, Korea

Stephan Kellenberger
Département de Pharmacologie and Toxicologie
de l'Université
Lausanne, Switzerland

Donghee Kim
Department of Physiology and Biophysics
Rosalind Franklin University of Medicine and Science
The Chicago Medical School
North Chicago, Illinois, USA

Illya Kruglikov
Smilow Program in Neuroscience and Departments of Physiology
and Neuroscience and Department of Biochemistry
New York University School of Medicine
New York, USA

Mira Kuisle
Department of Pharmacology and Neurobiology
Biozentrum
Basel, Switzerland

Yoshihisa Kurachi
Division of Molecular and Cellular Pharmacology
Department of Pharmacology
Graduate School of Medicine
Osaka University
Osaka, Japan

Elaine Kwon
Smilow Program in Neuroscience and Departments of Physiology
and Neuroscience and Department of Biochemistry
New York University School of Medicine
New York, USA

Jeremy J Lambert
Centre for Neuroscience
Division of Medical Sciences
Ninewells Hospital and Medical School
The University of Dundee
Dundee, United Kingdom

Michel Lazdunski
Institut de Pharmacologie Moléculaire et Cellulaire
CNRS-Université de Nice-Sophia Antipolis UMR 6097
Valbonne, France

Marzia Lecchi
Department of Neuroscience
Medical Faculty
Geneva, Switzerland

Stephan Lehnart
Georg August University Medical School
Department of Cardiology & Pulmonology
Center of Molecular Cardiology
Goettingen, Germany

Juan Lerma
Instituto de Neurociencias de Alicante
CSIC-UMH
Sant Joan d'Alacant, Spain

Eric Lingueglia
Institut de Pharmacologie Moléculaire et Cellulaire
CNRS-Université de Nice-Sophia Antipolis UMR 6097
Valbonne, France

Matthew R Livesey
Centre for Neuroscience
Division of Medical Sciences
Ninewells Hospital and Medical School
The University of Dundee
Dundee, United Kingdom

Anita Lüthi
Section of Pharmacology and Neurobiology
Biozentrum
University of Basel
Basel, Switzerland

Jonathon Maffie
Smilow Program in Neuroscience and
Departments of Physiology
and Neuroscience and Department of Biochemistry
New York University School of Medicine
New York, USA

Martin Main
Biological Chemistry
AstraZeneca R&D
Mereside
Cheshire, United Kingdom

Andrew Marks
Department of Physiology and Cellular Biophysics
Columbia University College of Physicians and
Surgeons
New York, USA

Anton D Michel
GlaxoSmithKline
New Frontiers Science Park
Harlow, United Kingdom

Hanns Möhler
Institute of Pharmacology
University of Zurich and Swiss Federal Institute of Technology
(ETH) Zurich
Zurich, Switzerland

Christophe J Moreau
Institut de Biologie Structurale
Laboratoire des Proteines Membranaires IBS/LPM
Grenoble, France

Marcela Nadal
Smilow Program in Neuroscience and Departments of Physiology
and Neuroscience and Department of Biochemistry
New York University School of Medicine
New York, USA

Bernd Nilius
Department of Physiology
Campus Gasthuisberg
Laboratorium voor Fysiologie
KU Leuven
Leuven, Belgium

Søren-Peter Olesen
NeuroSearch A/S
Pederstrupvej 93
Ballerup, Denmark
And The Danish National Research Foundation Centre for
Cardiac Arrhythmia
University of Copenhagen
Copenhagen, Denmark

Grzegorz Owsianik
Department of Physiology
Campus Gasthuisberg
Laboratorium voor Fysiologie
Leuven, Belgium

Gayle Passmore
Department of Pharmacology
University College London
London, United Kingdom

Ana V Paternain
Instituto de Neurociencias de Alicante
CSIC-UMH
Sant Joan d'Alacant, Spain

Randen Patterson
Life Sciences Building Room 230
University Park, PA, USA

Patrica Pérez-Cornejo
School of Medicine
University of San Luis Potosí
San Luis Potosí, México

Edward Perez-Reyes
Department of Pharmacology and Neuroscience Graduate Program
University of Virginia
Charlottesville, Virginia, USA

Matthew Perry
Nora Eccles Harrison Cardiovascular Research
and Training Institute and
Department of Physiology
University of Utah
Salt Lake City, UT, USA

John Peters
Centre for Neuroscience
Division of Medical Sciences
Ninewells Hospital and Medical School
The University of Dundee
Dundee, United Kingdom

Tony Priestley
Endo Pharmaceuticals
100 Endo Boulevard
Chadds Ford
PA, USA

Michael Pusch
Istituto di Biofisica
Consiglio Nazionale delle Ricerche
CNR
Genova, Italy

Juan Pablo Reyes
Institute of Physics
University of San Luis Potosí
San Luis Potosí, Mexico

Rocio Rivera
Instituto de Neurociencias de Alicante
CSIC-UMH
Sant Joan d'Alacant, Spain

Jonathan Robbins
Wolfson Centre for Age Related Diseases
King's College
Guy's Campus
London, United Kingdom

Teresa Rosales-Saavedra
Institute of Physics
University of San Luis Potosí
San Luis Potosí, México

Uwe Rudolph
Laboratory of Genetic Neuropharmacology
McLean Hospital
Harvard Medical School
Belmont, MA USA

Bernardo Rudy
Smilow Program in Neuroscience and Departments of Physiology
and Neuroscience and Department of Biochemistry
New York University School of Medicine
New York, USA

Michael Sanguinetti
Nora Eccles Harrison Cardiovascular Research and
Training Institute and
Department of Physiology
University of Utah
Salt Lake City, UT, USA

Laurent Schild
Département de Pharmacologie and Toxicologie de l'Université
University of Lausanne
Lausanne, Switzerland

Sanja Selak
Instituto de Neurociencias de Alicante
CSIC-UMH
03550 Sant Joan d'Alacant
Spain

Fangxiong Shi
Laboratory of Animal Reproduction
College of Animal Science and Technology
Nanjing Agricultural University
Nanjing, China

Eric Southam
GlaxoSmithKline
New Frontiers Science Park
Harlow, United Kingdom

Dorte Strøbæk
NeuroSearch A/S
Ballerup, Denmark

Andre Terzic
Division of Cardiovascular Diseases
Department of Medicine
Mayo Clinic
Rochester, MN, USA

Derek Trezise
GlaxoSmithKline
New Frontiers Science Park
Harlow, United Kingdom

James Uney
MRC Centre for Synaptic Plasticity
University of Bristol Medical School
Department of Anatomy
Bristol, United Kingdom

Damian van Rossum
Center for Computational Proteomics
Pennsylvannia State University
USA

Carmen Villmann
Institut für Biochemie
Emil-Fischer-Zentrum,
Universität Erlangen-Nürnberg
Erlangen, Germany

Michel Vivaudou
Institut de Biologie Structurale, Laboratoire des Proteines
Membranaires IBS/LPM
Grenoble, France

Thomas Voets
Department of Physiology
Campus Gasthuisberg
Laboratorium voor Fysiologie
Leuven, Belgium

Zhengchao Wang
Laboratory of Animal Reproduction
College of Animal Science and Technology
Nanjing Agricultural University
Nanjing, China

Annette Weil
Neurosciences CEDD
GlaxoSmithKline
New Frontiers Science Park
Harlow, United Kingdom

Edward Zagha
Departments of Physiology and Neuroscience and
Department of Biochemistry
New York University School of Medicine
New York, USA

PART I

Introduction to ion channel structure and function

CHAPTER 1

Ion channels

Principles, terminology and methodology

Derek Trezise, Tim Dale and Martin Main

1.1 Introduction

Ion channels are cellular proteins that conduct the movement of ions from one side of a membrane to the other. The resultant changes in local ion concentrations and electrical field play pivotal roles in physiological processes, as wide ranging as cell to cell communication, cell proliferation and secretion. This chapter provides a brief historical perspective then introduces the basic theory, terminology and generic structural and functional features of ion channels. In addition, an overview of relevant ion channel methodologies is provided. It is hoped to set the scene and to equip the non-specialist reader with sufficient background and understanding to comprehend and enjoy subsequent chapters which provide a more detailed analysis of channel families and individual channels.

1.2 Historical perspective and overview

Our present understanding of ion channel function and structure is the product of extensive, global, multidisciplinary research activities spanning over seven decades. This field continues to grow, as demonstrated by the year on year increase in number of original and review articles published on this topic (currently >7000 per annum, totalling over 95 000). Large Internet search engines yield in excess of 14 million hits when probed with the phrase 'ion channel'.

This research era has been punctuated with a number of landmark discoveries, four of which have thus far culminated in the Nobel prize for Physiology/Medicine or Chemistry. The earliest suggestion of the existence of ion channels was made in the 1950s, largely from the work of Alan Hodgkin and Andrew Huxley at Cambridge University in the UK and Bernard Katz at University College London, also in the UK. Using a sucrose-gap voltage clamp method to measure the current flowing across the membrane of giant squid axon, Hodgkin and Huxley demonstrated that the upstroke of the action potential was caused by an increase in the permeability of the membrane to Na^+, but not K^+ (Hodgkin et al., 1952; Hodgkin and Huxley, 1952a, b, c, d). This pioneering work inspired a generation of biophysicists to unravel the basic principles of ion channel gating, selectivity and modulation, and Hodgkin and Huxley were subsequently awarded the Nobel Prize in 1963. Katz himself went on to become Nobel Laureate in 1970, sharing the

prize with Julius Axelrod and Ulf von Euler, for their achievements in understanding 'the humoral transmitters in nerve terminals and the mechanisms for their storage, release and inactivation'. Katz's contribution pertained to work on the neuromuscular junction in which he demonstrated nicotinic acetylcholine receptor/channel-mediated miniature end-plate potentials, and their critical dependence on synaptic Ca^{2+} concentrations.

Throughout the 1960s and early 1970s a growing body of evidence, albeit indirect, suggested that an individual ion channel would pass a current of a few picoamps when open. Frustratingly, the sensitivity of the best available recording methods of the time was 2–3 orders of magnitude too low to fully resolve these signals. In the mid 1970s, Erwin Neher and Bert Sakmann from the Max-Planck Institute for Biophysical Chemistry in Göttingham, Germany provided a stunning solution with the introduction of the 'patch-clamp' technique (Neher and Sakmann, 1976; Hamill et al., 1981). This method enables the recording of electrical activity in small (<10 μm^2) areas of membrane, with a sensitivity and noise level sufficient to resolve single channel events at the microsecond timescale. Using this approach, Sakmann and Neher were the first to demonstrate unequivocally the presence of discrete ion channels, and to fully resolve single channel events. They went on to apply the patch clamp method, and variants thereof, to mechanistic studies on single nicotinic acetylcholine receptor channels (Fenwick et al., 1982), fusion of secretory vesicles in mast cells (Penner and Neher, 1989), and synaptic transmission processes in the central nervous system. To this day, the patch clamp method remains the gold standard approach for functional analysis of ion channels. With such an impact, it was no surprise when in 1991 the Nobel committee recognized the two for their pioneering work.

Gene cloning, recombinant manipulation and heterologous expression methods began to be applied to ion channel research in earnest in the early 1980s. Allied to patch clamp functional analyses these approaches have proved particularly powerful for elucidating ion channel structure/function. For example, the cloning of the Shaker locus from the fruit fly, Drosophila melanogaster, provided a starting point for the molecular characterization of voltage-gated K^+ channels (Papazian et al., 1987) and DNA sequences encoding mammalian K^+ channels soon followed (see Pongs, 1990 for review). The latter part of the 1990s saw major breakthroughs in the field of structural biology. Using first electron microscopy and then X-ray crystallographic techniques, ion channel structures were resolved for the first time at ultra-high resolution. This field was pioneered by Rod MacKinnon of Rockefeller University, USA whose work focused on the K⁺ channel family. By first overcoming significant technical challenges in protein expression and purification, the Rockefeller lab successfully crystallized several bacterial K⁺ channels, including the 2TM K⁺ channel, KcsA, to reveal three-dimensional structures down to 2Å resolution (Doyle *et al.*, 1998; Gulbis *et al.*, 1999; Roux and MacKinnon, 1999; Gulbis *et al.*, 2000; Jiang and MacKinnon, 2000; Morais-Cabral *et al.*, 2001; Zhou *et al.*, 2001a, b; Jiang *et al.*, 2002a, b). In 2005, the first mammalian voltage-gated K⁺ channel crystal structure, rat Kv1.2, was described (Long *et al.*, 2005a). Questions on assembly, permeation and gating were now accessible as never before. MacKinnon was awarded a share of the Nobel Prize for Chemistry in 2003 for this inspirational work.

This molecular era has now unveiled the entire human ion channel genome, or 'channel-ome'. Overall, there are close to 350 ion channel encoding genes in humans corresponding to 260 putative pore-forming (α) subunits and 90 regulatory subunits. Gene knockout or overexpression experiments in mice and rats have provided valuable insight into the functions of many of these genes, for example *TRPV1* (Caterina *et al.*, 2000). Moreover, clinical investigations into congenital disorders have established causative links between defects (mutations) in ion channel genes and disease, collectively referred to as channelopathies (see Kass 2005 for review). One of the best characterized is the cystic fibrosis transmembrane regulator (CFTR) chloride channel (gene name *ABCC7*), found in lung (and gut) epithelia (Boucher, 2002). Loss of function mutations in CFTR, of which over 1000 are known, cause defective chloride transport across the apical epithelial cell membrane. This culminates in impaired mucociliary clearance, and the resultant thickening of the mucus predisposes the individual to persistent and recurrent bacterial infections and general respiratory malaise. At least 40 different channelopathies have now been described, covering over a dozen ion channel families (see Cannon, 2007 for review).

Human health has benefited directly from this knowledge because channelopathy risk factors and diseases can be readily diagnosed and ion channel pharmacotherapies prescribed. For example, it is now common place to use CFTR genotyping to aid the diagnosis of cystic fibrosis, and to screen genetically at-risk neonates via chorionic villous sampling or amniocentesis. Approximately 180 FDA approved drugs exert their actions via ion channels. The ion channel pharmacopoeia includes analgesics, anticonvulsants and local anaesthetics that block voltage-gated Na⁺ channels (e.g. lidocaine, lamotrigine), inhibitors of ATP-dependent K⁺ channels in pancreatic β-cells to control diabetes (e.g. glibenclamide), and anxiolytic positive modulators of GABA_A receptor channels (e.g. diazepam) to name but a few. Most recently, lubiprostone (Amitiza™), a small molecule activator of ClC-2 chloride channels in gastrointestinal epithelia, was approved by the FDA for idiopathic chronic constipation (approved on 31/01/06, see FDA website: http://www.accessdata.fda.gov/scripts/cder/drugsatfda/). The continued thirst for knowledge and understanding, and the sharp focus of biotechnology and pharmaceutical companies on ion channel targets, should extend this benefit to mankind over the coming decades.

1.3 Basic principles and terminology

Inorganic ions, including Na⁺, K⁺, Ca²⁺, Mg²⁺, Cl⁻ and H⁺, play a key role in cell physiology, mediating a diverse range of processes including neuronal transmission, muscle contraction and cell proliferation. A key aspect of all these processes is the movement of ions between the extracellular environment and the cytoplasm, which impact cell physiology through either a change in membrane potential (e.g. regulation of neuronal firing) or through an alteration in the local concentration of ions inside the cell (e.g. contraction). The plasma membrane that separates extracellular and cytoplasmic compartments is composed of phospholipid, which is impermeable to charged ions. This issue is resolved by the presence of two types of integral membrane protein—ion channels with water-filled pores and ion transporters.

Driving force, ionic gradients and membrane potential

While ion channels provide the means to transport ions across the cell membrane, a useful starting point is to consider the driving forces that lie behind this ionic movement. In turn, to understand driving force we must first examine the terms *membrane potential* and *equilibrium potential*. Most cell types have a shell of negative charge on the cytoplasmic side of the membrane and a shell of positive charge on the extracellular side, i.e. the inside of the cell is negatively charged relative to the outside. This charge separation creates a potential difference or voltage across the membrane, which is called the cellular membrane potential. The origin of the membrane potential is the presence of membrane-impermeant, negatively charged organic molecules (organic acids and proteins) in the cytoplasm. The presence of additional anions in the intracellular compartment favours a greater concentration of cations. The movement of ions continues until equilibrium is reached, where the tendency for ions to move down the concentration gradient into or out of the cell is offset by the electrical gradient, known as the Gibbs–Donnan equilibrium (Hille, 1992).

For simplicity, let us consider a model cell where an equal concentration of K⁺ ions is present on either side of the cell membrane: the membrane exhibits a selective permeability to K⁺ ions (i.e. via K⁺-selective ion channels) and the cytoplasmic compartment contains negatively charged organic molecules. With an equal concentration of K⁺ ions on either side of the membrane, the presence of organic proteins in the cytoplasm generates a net negative charge on the inside of the cell. Thus, there is an *electrostatic gradient* that drives movement of K⁺ ions into the cell (opposite charges attract). As K⁺ ions move into the cell, the electrostatic gradient decreases, but an opposing 'chemical' concentration gradient starts to build up. Eventually, the electrostatic gradient driving K⁺ ions into the cell and the opposing chemical gradient driving K⁺ ions out of the cytoplasm are equal and opposite and equilibrium is reached. The voltage (charge separation) at which there is no net flux of K⁺ ions is called the K⁺ equilibrium potential (E_K) and at room temperature this voltage can be calculated using the Nernst equation (Hille, 1992), and is approximately –80 mV.

Nernst equation

$$E_i \equiv V_{in} - V_{out} = \frac{RT}{zF} \ln \frac{[C]_{out}}{[C]_{in}}$$

where E_i (or E_{ion}) is the equilibrium potential of the ion, R is the gas constant, T is the absolute temperature (273.16 + °C), z is the valence of the ion, F is Faraday's constant, $[C]_{out}$ is concentration of ion outside the cell, $[C]_{in}$ is concentration of ion inside the cell. Simplifying the equation, at a temperature of 20°C, the equilibrium potential can be calculated by multiplying the logarithm base 10 of the ratio of outside to inside concentration by 58 mV. At 37°C the multiplication factor is 62 mV.

In our simple, model system the K^+ equilibrium potential is identical to the resting membrane potential and therefore changes in K^+ conductance have no impact on cellular membrane potential (E_m) provided that a finite K^+ conductance remains. However, in reality many other ionic conductances influence membrane potential and therefore E_K is not identical to E_m. Under these circumstances, the driving force (in volts) for K^+ flux is equal to the difference between equilibrium potential and membrane potential. In most cell types, $E_m > E_K$ under all physiological circumstances and therefore there is a driving force for K^+ efflux from the cell. K^+ efflux through K^+-selective ion channels leads to movement of positive charge out of the cell and this is referred to as an *outward current*. In turn, this outward current increases the degree of charge separation across the membrane and induces *hyperpolarization* of the cell. Conversely, if $E_m < E_K$ there is a driving force for potassium influx, leading to inward movement of positive charge—an *inward current*—and to *depolarization* of the cell. Typically, most mammalian cells have cytoplasmic K^+ concentrations ranging from 100–140 mM compared to 1–10 mM in the extracellular fluid, yielding theoretical E_K values of –60 to –100 mV.

A similar analysis can be carried out for the other major permeating ions—Na^+, Ca^{2+} and Cl^-. Theoretical E_{Na} is typically in the range +40 to +70 mV and E_{Ca} is +60 to +120 mV. These calculations refer to passive distribution of ions across the plasma membrane, however, when one measures ion concentration in the two compartments it is clear that other forces are at work. If E_m typically lies positive to E_K, and there is a driving force for K^+ efflux why doesn't K^+ simply move out of the cell and collapse this driving force? The answer lies in ion transporters. For Na^+ and K^+, the Na^+/K^+ ATPase plays a critical role in maintaining the electrochemical gradient and similar ion transporters exist for Ca^{2+} and Cl^-. E_{Cl} differs widely between cells as intracellular Cl^- concentration can be anywhere between 40–110 mM dependent on active transport. Interestingly, in the central nervous system (CNS) GABA activation of $GABA_A$ receptor Cl^- channels switches from excitatory in the neonate to inhibitory during development (Owens *et al.*, 1996) due to upregulation of a potassium chloride co-transporter KCC2 that lowers the intraneuronal Cl^- concentration below its electrochemical equilibrium (Lee *et al.*, 2005).

Once the contribution of active ion transporters has been taken into account, and knowing the relative ionic permeability and distribution of ions the membrane potential can be calculated according to the Goldman–Hodgkin–Katz (GHK) voltage equation (Goldman, 1943; Hodgkin and Katz, 1949):

$$V_m = \frac{RT}{F} \ln\left(\frac{p_K[K]_o + p_{Na}[Na]_o + p_{Cl}[Cl]_o}{p_K[K]_i + p_{Na}[Na]_i + p_{Cl}[Cl]_i}\right)$$

where R is the gas constant, T is the absolute temperature (273.16 + °C), F is Faraday's constant, p_K, p_{Na} and p_{Cl} are the membrane permeabilites for K^+, Na^+ and Cl^-, respectively. Normally, permeability values are reported as relative permeabilities with p_K having the reference value of one (because in most cells at rest p_K is larger than p_{Na} and p_{Cl}). For a typical neuron at rest, $p_K : p_{Na} : p_{Cl} = 1 : 0.05 : 0.45$.

Ion channel permeability and ion selectivity

To continue the analysis of our simplified model cell: we have defined driving force (DF) as the difference in voltage between membrane potential and the equilibrium potential of the permeating ion: $DF = E_m - E_{ion}$. We can then apply Ohm's law to calculate the current flowing through an open channel:

Voltage = Current × Resistance

When considering a single ion channel, voltage = driving force = $E_m - E_{ion}$. Therefore single channel current = $(E_m - E_{ion})$ / resistance, or in terms more applicable to ion channel biophysics: Current = conductance / $(E_m - E_{ion})$. Single channel conductance (g) is a property of the ion channel protein and can vary from fS to hundreds of pS. At this point it is important to note that even in the lowest conductance channels, current flux is far greater ($>10^6$ ions per second) than can be achieved through simple diffusion (Hille, 1992).

Whilst the major conductance through a channel is often driven by a single ionic species (e.g. Na^+ or Cl^-), the concepts of *relative permeability* and *ion selectivity* are important ones. Different ions pass more or less readily through different ion channel types according to their hydrated size, shape and charge and according to the structure of the pore through which they pass. All channels have a 'selectivity filter' region, several (3–7) angstroms in diameter, which prevents the access of some but not other ion species. At the simplest level, this can be considered as a molecular sieve that cuts off the permeation of larger and inappropriately charged particles. Ion selectivity is however more complex and depends on the balance between the energy required to remove water from the permeating ion and the energy of allowing the partly dehydrated ion to interact with the filter (see Hille 1992, for more details). The relative permeability of two different ions (say K^+ and ion X^+) can be measured by determining the zero current potential with X^+ on the outside and K^+ on the inside of the channel, so called *bi-ionic conditions*. If the two ions have the same valence, and no other permeant ions are present the permeability ratio is given by:

$$E_{rev} = \frac{RT}{zF} \ln\left(\frac{P_A[A]_o}{P_B[B]_i}\right)$$

Using this approach, delayed rectifier K^+ channels have relative permeability sequences of (K^+=1) Tl^+ (2.3) > K^+ (1.0) > Rb^+ (0.9) > NH_4^+ (0.1) > Cs^+ (<0.08) > Li^+ (<0.02) = Na^+ (<0.02).

Ohm's law predicts a linear relationship between driving force (voltage) and current, with the slope equal to channel conductance and the intercept, or *reversal potential* (where current = zero), at the equilibrium potential. In most cases, this is true. However, it is worth pointing out that there are a number of exceptions to this rule. First, a number of ion channel types exhibit the property of *rectification*, where the current-voltage relationship is non-linear. Inward rectifier potassium channels pass far greater inward current at potentials negative compared to potentials positive to E_K, for an equivalent driving force. This property is due to voltage-dependent binding of positively charged Mg^{2+} ions and polyamines (e.g. spermine) at the cytoplasmic face of the channel (Matsuda *et al.*, 1987; Lopatin *et al.*, 1994; Fujiwara and Kubo 2006). Secondly, where the concentration of the permeating ion on one side of the channel is very low, there may be an insufficient number of ions to sustain a linear current. An example of this situation would be calcium-selective ion channels in the voltage-gated calcium channel family (Hille, 1992). Finally, many ion channels exhibit mixed ion selectivity, for example non-selective cations channels, in which case the reversal potential is influenced by the equilibria potentials for several ions.

Ion channel gating

For simplicity, the discussion thus far has focused on open channels. In reality, most ion channels can exist in several conformational states, each of which has a distinct ionic conductance. The process of *gating* describes the movement of channels between these different conformational states. In the following text, we will consider gating between four channel states: closed, open, inactivated, and desensitized state. Different channels are gated by different stimuli, including neurotransmitter binding, temperature, mechanical perturbation etc., but for the purpose of this explanation we will begin with a channel gated by a change in the membrane potential, for example a voltage-gated sodium channel.

In the closed state the pore of the channel exists in a conformation that is impermeable to ions. Depolarization of the membrane evokes a conformational change in the ion channel protein that 'opens' the channel, allowing ions to pass—this is referred to as the 'activation' step and commonly involves movements of positively charged residues in the α-helix S4 and other parts of the channel structure referred to as the voltage sensor (see later). At the level of a single channel, activation is manifest as a binary event, with an almost instantaneous switch from zero to the unitary conductance of the channel. In the face of continued depolarization an 'inactivation' step may then occur, where the channel moves to a third conformation, distinct from the closed state, but equally non-conducting and refractory to subsequent activation. This generally involves cytoplasmic domains of the channel that drop into the inner vestibule and plug it (Ulbricht, 2005). For voltage-gated Na^+ channels this *'fast' inactivation* step occurs within milliseconds of activation, and serves to make Na^+ entry a very transient event. Direct open to closed state transition, known as *deactivation*, can occur but is hard to study for voltage-gated Na^+ channels because inactivation proceeds so swiftly. Inactivated to open state transitions are not energetically favoured, and recovery from inactivation requires a transition through the closed state first to ready the channel for subsequent opening. In reality, detailed kinetic analysis demonstrates several, distinct closed, open and inactivated states in many cases. For example, an additional *'slow' inactivated* state of Na^+ channels involves distinct molecular processes in the outer pore of the channel over a time course of hundreds of milliseconds to several minutes (Elliot and Elliot, 1993).

Conceptually, the state transitions of ligand-gated channels are similar but the conformational changes are driven by the chemical energy of the ligand, rather than the membrane voltage. Taking nicotinic acetylcholine (nAChR) receptors in this case, two molecules of acetylcholine (ACh) are required to bind to extracellular domains on the channel to provide sufficient activation energy to drive the receptor from a closed to open state (Purohit and Auerbach, 2007).

Whilst ACh is bound, the channel oscillates between open and closed states and bursts of current can be observed. Unbinding of ACh from the receptor is slow compared to these state transitions and only occurs when the channel is in its closed state. Interestingly, the length of the bursts varies with the agonist—carbachol or choline produce bursts only half as along as ACh, and this is a product of their different rate constants for opening. In the continued presence of agonist, many channels enter a *desensitized state* where they shut down and become refractory to agonist activation. This bears strong similarities to the voltage-gated channel slow inactivation state. Close scrutiny of nAChR channels indicates that channels can in fact open, albeit with low probability, in the absence of ligand or when only one ACh molecule is bound and thus kinetic allosteric models have been assembled.

Analysis of macroscopic and microscopic currents

The vast majority of electrophysiological experiments in the literature refer to whole-cell recordings (patch-clamp or oocyte), where 'macroscopic' currents are measured from 100s–1000s of ion channels that are expressed in the plasma membrane. As discussed earlier, in most cases single channel ionic currents are binary; either the channel is open and a current is recorded, or the channel is closed and no current flows. In whole cell recordings, population statistics come into play. At any one time, the proportion of channels lying in the closed vs open vs inactivated/desensitized state differs, with the proportion of channels in each state depending upon factors such as voltage, temperature and the presence of ligand. The macroscopic current can be described by a simple equation:

$$\text{Macroscopic current (I)} = i.n.P_o$$

where i is the single channel current (dependent upon conductance and driving force as described earlier), n is the number of channels present in the plasma membrane and P_o is the probability of opening of an individual ion channel. Gating of an ion channel shifts the fraction of channels in each conformational state and therefore modulates P_o.

Kinetic and equilibria analyses of macroscopic currents have provided important insights into ion channels, their structure/function and physiological roles. Fitting single and multi-exponential functions to the activation and decay phases of currents, for example, yields information on rates of state transition and the likely number of conformational states. Current–voltage relationships are used to determine gating thresholds, voltage-sensitivity, reversal potentials and rectification profiles. For voltage-gated channels P_o can be parameterized with a Boltzmann function of the form:

$$P_o = \frac{1}{1 + \exp\left(\dfrac{(V_{1/2} - V)}{k}\right)}$$

where $V_{1/2}$ describes the voltage at which $P_o = 0.5$ and k (mV) the steepness of the voltage dependence, or 'slope factor'. For a non-inactivating K^+ current, gating the channels to different voltages each followed by a step to a common, saturating potential to yield a tail current is a common method for eliminating the driving force (V) component. For inactivating channels (e.g. fast Na^+ and Ca^{2+} channels) it is necessary to transform the peak I-V relationship to a normalized conductance-V relationship first (using Ohms law) to eliminate the driving force. Fitting these data sets to Boltzmann functions then yields the voltage-dependence of activation descriptors, $V_{1/2}$ and k. *Low threshold* channels refers to those which are activated at relatively negative potentials (e.g. T-type Ca^{2+} channels begin to activate at –90 mV and have $V_{1/2}$ values of –70 to –50 mV), whereas a far larger depolarization is required to gate *high threshold* channels (e.g. N-type Ca^{2+} channels $V_{1/2}$ –20 mV). Channels with high voltage sensitivity are those in which only a small change in voltage is required for activation, and k is high. Conversely, low k values indicate weak voltage-dependence.

Similar analyses are used to measure the voltage-dependence of inactivation and deactivation. For inactivation, series of paired voltage steps are employed where the first (variable) step is used to inactivate a fraction of the channels at a given voltage (and time) and the second to monitor the fraction of channels that remain available for gating. Plotting the amplitude of the current from the second step divided by the peak current amplitude as a function of the voltage yields an inactivation plot that can be fitted to Boltzmann parameters. With an understanding of the voltage-dependence of activation and inactivation for a given channel, *window currents* can be identified as those that exist at equilibria at certain potentials where inactivation does not override activation. As one might imagine, window currents are particularly important in setting the resting membrane potential of cells. Paired pulse protocols are also used to derive kinetic measures of state transitions— the time-course of the recovery from inactivation (i.e. inactivated to closed state transition) can be readily determined by varying the inter-pulse interval.

Macroscopic currents can also be analysed in different ways to estimate the density (and number) of channels in the membrane and their single channel conductance. *Noise analysis* derives these parameters by considering the statistical variance in the measured macroscopic current, borne of the stochastic, summated fluctuations in single channel opening and closing (Heinemann and Conti, 1992). 'Stationary' analysis looks at steady state currents (e.g., for nAChR following prolonged acetylcholine application), whilst 'non-stationary' methods utilize the variance throughout the time-course of channel activation and inactivation. For the latter, if we take a representative voltage-gated Na^+ channel as an example, one repeatedly measures the evoked macroscopic current then averages all of the records to produce an ensemble average time-course. This average is then subtracted from each of the individual records to yield a series of 'noisy' traces that fluctuate around zero. If the current variance is then plotted as a function of the total current, a parabolic function is obtained which can be fitted to the equation $\sigma_I^2 = iI - I^2/n$ where σ_I = the variance of the macroscopic current, i = the single channel current, I = the macroscopic current and n = the number of channels. Another approach for voltage-gated channels is to measure gating charges, the tiny membrane currents that immediately precede channel opening and stem from the movements across the membrane of charged residues in the voltage-sensor. With careful measurement, if one knows the gating charge that must be moved across the electric field to open a single channel, the number of channels can be determined from the overall charge movement (Schneider and Chandler, 1973).

Whilst macroscopic analyses are most common, kinetic, gating and permeability parameters can be determined more precisely from 'microscopic' currents, or single channels in excised patches of membrane. Under these conditions, conformational state transitions from non-conducting to conducting forms of a single channel can be monitored in real time. Open channel lifetime refers to the period between when a channel opens and when it next closes, and can be measured directly. For some channels, flickery openings may be observed due to channel subconductance states—these cannot be resolved macroscopically. Channel conductance measurements are made by plotting the current-voltage relationship of a single channel and fitting the points either side of the reversal potential to Ohms Law: G (conductance) = I/V.

From a combination of macroscopic and microscopic analyses it is apparent that in physiological systems channels are typically found at densities ranging from 10–1000 channels per mm^2, and have a single channel conductance of <1 to 300 pS.

Ion channel pharmacology

Ion channels have a rich and interesting pharmacology. Peptides, toxins and small organic molecules alike can modulate ion channel function via a plethora of different sites and mechanisms. The pharmacological terms and mechanisms of modulation that are either unique or highly common to ion channels are introduced here.

State-dependence is a broad descriptor, and captures the idea that compounds exhibit different affinities for different conformations of a channel—the observed modulation depends on the ratios of these affinities and the temporal 'availability' of the states for the drug to bind. In some cases molecules can become trapped in channels extending their access time to binding sites. Classically, local anaesthetic (e.g. lidocaine) and anticonvulsant (e.g. lamotrigine) Na^+ channel blockers preferentially bind to open and inactivated states of the channel, rather than the closed state (Hille, 1992). If a train of depolarizing pulses is applied from a test potential at which all channels are held in the closed state, there is negligible inhibition of the first pulse. As the train proceeds, cumulative increasing block is observed as the drug rapidly binds to previously 'unavailable' open and inactivated state binding sites (from which unbinding is relatively slow). This example of state dependent block is referred to as *use-dependence* and rationalizes the ability of lidocaine and lamotrigine to produce clinical efficacy without undue side effects (see Xie and Hagan, 1998). *Frequency-dependence* refers to when different blocking profiles are observed when channels are gated more or less frequently, and has its root cause in the same principle of preferential binding to transiently exposed channel conformations. In contrast to lidocaine and lamotrigine, the *Fugu* toxin, tetrodotoxin (TTX), does not require channel opening or inactivation per se for binding to occur, and is neither use- nor frequency-dependent. TTX occludes the pore of the channel (Na^+ channel 'site 1') to prevent permeation, and is referred to as a *pore blocker* (Catterall, 2000). MK-801 inhibition of NMDA receptors, picrotoxin block of $GABA_A$ receptor channels and charybdotoxin block of BKCa and ShK block of Kv1.3 channels are other examples of pore block (Etter *et al.*, 1999; Rauer *et al.*, 1999; Meera *et al.*, 2000; Kashiwagi *et al.*, 2002). In some cases, and especially with charged molecules, pore blockers demonstrate *voltage-dependent block*. For example, ruthenium red is a highly charged, polycationic compound that blocks members of the TRPV family in a voltage-dependent manner. Extracellular application of ruthenium red fully blocks TRPV4 at negative test potentials, but produces a greatly reduced block at positive potentials. Mutagenesis data suggest that ruthenium red blocks TRPV4 at a site in the pore, i.e. within the transmembrane electrical field (Voets *et al.*, 2002) and at depolarized potentials the positively charged ruthenium red molecule is driven off its binding site.

As the term suggests, gating modifiers are agents that modulate the channel gating process—the term is most commonly encountered with peptide toxins and voltage-dependent channels. Venoms from a range of species, including spiders, scorpions, cone snails, and snakes, contain toxins that bind to Kv, Na$_v$ and Ca$_v$ channels to either inhibit or promote voltage-dependent gating. Hanatoxin, for example, is a 35 amino acid peptide isolated from the Chilean Rose tarantula venom that binds to the voltage sensor of Kv2.1 channels to stabilize the closed state and produce a depolarizing shift in the voltage-dependence of inactivation (Swartz and MacKinnon, 1995). Tail current analysis also reveals a characteristic acceleration of the open to closed state transition. Mechanistically, the gating modification of Ca$_v$2.1 channels by ω-Agatoxin IVA is very similar (Winterfield and Swartz, 2000). Voltage-gated Na$^+$ channels have five distinct gating modifier sites, which when occupied by neurotoxins produce either slowed inactivation (site 3 and 6), enhanced activation (site 4 and 5) or persistent activation (site 2; Cestele and Catterall, 2000). The diversity and myriad of actions of neurotoxins on channels have proved highly valuable in dissecting the structure/function relationship of gating events. Beyond toxins, capsaicin and menthol have also been shown to act as gating modifiers of the thermo-sensitive TRP channels, TRPV1 and TRPM8, respectively. These agents shift the thermo- and voltage-sensitive gating process into physiological ranges to produce channel activation (Voets et al., 2004).

Channel agonists, openers, activators and positive modulators are all terms used to describe gating modifiers that stimulate, enhance or prolong opening through channels. The terms are not strictly interchangeable but in many cases (some erroneously!) are treated as such. Most precisely, the primary, physiological activating ligands for ligand-gated ion channels are termed agonists, for example acetylcholine for nAChRs and serotonin for 5-HT$_3$ receptors. In other cases, the activation effect is secondary and overtly a modification or amplification of another gating stimulus. RPR260243, for example, a small synthetic molecule, slows the deactivation of voltage-gated hERG K$^+$ channels to prolong channel opening and is best described as an activator (Perry et al., 2007). In the KCNQ (Kv7) potassium channel family a number of activators, for example retigabine and L-364373, have been identified which shift the voltage dependence of channel activation to more hyperpolarized potentials, increase the rate of channel activation, and slow the rate of deactivation (Main et al., 2000; Salata et al., 1998). Intriguingly, another KCNQ channel opener zinc pyrithione, which appears to act at a distinct binding site to retigabine, is able to activate a mutant KCNQ2 (I238A) channel that is non-conducting upon depolarization (Xiong et al., 2008). Another example of gating modification is the calcium-activated potassium channel opener 1-EBIO. This compound binds to the C-terminus of SK (K$_{Ca}$2) and IK (K$_{Ca}$3) K$^+$ channels and activates the channels by modulating the apparent Ca^{2+}-sensitivity of the channel (Pedersen et al., 1999).

The term positive modulator is most commonly encountered for allosteric compound actions on ligand-gated ion channels—examples include cyclothiazide that destabilizes the inactivated state of glutamate-gated AMPA receptor channels to prevent desensitization (Partin et al., 1993), and the allosteric enhancement of the binding of GABA to GABA$_A$ receptor chloride channels observed with benzodiazepines (Olsen, 1982). Negative modulators have the reverse effect, reducing the effect of other gating stimuli. A number of ion channel modulators are thought to act at an auxilliary subunit. Thus, in a recent publication Hendrich et al. (2008) reported that Gabapentin blocks voltage-gated calcium channels by binding to the α2δ subunit and preventing this protein from enhancing trafficking of the channel to the cell surface.

Ion channel topology and structure/function

What is known of the detailed topology and structure of channel families is covered in subsequent chapters—the purpose of this section is to introduce the key elements and highlight the commonalities within the superfamily.

Ion channels are comprised of a single, repeat-structure protein, or more commonly multiple 'subunits' arranged to form a water-filled pore through the plane of the membrane. These protein 'building blocks' are generally termed α subunits if they are the primary pore forming subunits and β or γ if they are secondary, auxiliary ones. Each is made from a chain of 100–2000 amino acids and has a molecular mass of between 10 and 250 kDa. Subunits may be post-translationally modified by glycosylation, phosphorylation and/or linked together by disulfide bonds. Hydrophobic segments of these subunits span the phospholipid membrane as coiled 'α-helices' and are termed *transmembrane domains*. A single α helix is typically 20–27 amino acid residues long and sits at an angle of 10–30° offset to the perpendicular. For some channels, cylindrical 'sheets' are formed, termed β barrels, and nine to thirteen amino acids suffice to span the membrane in this flatter structure. Together the transmembrane domains form the core of the channel and include the pore region.

Voltage-gated and related channels

The transmembrane architecture of all mammalian voltage-gated ion channels has common recognizable features. For Kv channels, a single 300–400 amino acid α-subunit contains 6 α helix transmembrane-domains (TMs), termed S1–S6, with a short membrane re-entrant P-loop between S5–S6. Four subunits associate to form a 24 TM tetramer, which is either *homomeric* (i.e. comprised of four identical subunits) or *heteromeric* (in this case a heterotetramer, in which the subunits are different). A tetramerization domain, termed T1 in the N-terminus is the primary determinant of specific homo- and hetero-multimeric subunit associations that generate functional Kv channels. The overall structure of voltage-gated Na$^+$ (Na$_v$) and Ca^{2+} (Ca$_v$) channels is highly similar to Kvs, but in this case a 300–400 amino acid structural motif is repeated 4 times within the same 1800 amino acid protein to make the 24 TM. These motifs are referred to as *homologous repeat domains*, denoted I, II, III and IV and each domain is the equivalent of a single Kv subunit. Non voltage-gated, inwardly rectifying K$^+$ channels constitute the simplest structural motif in the ion channel protein superfamily. These channels are complexes of four subunits that each has only two transmembrane segments, termed M1 and M2, which are analogous in structure and function to the S5 and S6 segments above. In the twin-pore (2P) K$^+$ channels two of these pore motifs are linked together.

Although the structural basis of the pore and voltage-dependent gating mechanism is not fully understood, an overall view has emerged from a combination of biophysical, mutagenesis, and structural studies. Put simply, the critical components of the conduction pore and ion selectivity filter lie in S5, the P-loop and S6, whilst the S1–S4 TMs, and in particular S4, comprise the voltage-sensor. Side chains of four amino acid residues [Asp-Glu-Lys-Ala (DEKA)], one at the tip of each re-entrant P-loop between the S5 and S6 segments, govern the selectivity of voltage-gated Na$^+$ channels (Heinemann et al., 1992). This pore motif reads EEEE in place of DEKA in voltage-gated Ca^{2+} channels and confers the channel with 1000-fold selectivity for Ca^{2+} over Na$^+$ (Sather and McCleskey, 2003). All K$^+$ channels have the signature sequence of residues, Thr-Val-Gly-Tyr-Gly (TVGYG), in the P loop that project carbonyl oxygen atoms along the 12 Å long, 2.5 Å selectivity filter. These oxygen atoms are ideally positioned to substitute for the water molecules that normally surround each K$^+$ ion and coordinate multiple dehydrated K$^+$ ions during passage (Jiang et al., 2002). Non-specific cation channels that pass Na$^+$ and Ca^{2+}, such as CNG, have similar S5–S6 structures to voltage-gated channels but are less well conserved in the pore region.

The S4 domain is the principal voltage sensor of voltage-gated channels and has 5–7 positively charged amino acids interspersed at regular intervals with hydrophobic ones (Noda et al., 1984). That S4 moves outward and across/within the membrane upon depolarization to initiate opening of the pore is non-controversial (Gandi and Isacoff, 2002). However, the manner of this movement is subject to debate—there is strong evidence for a spiralling, rotation of S4 to move the gating charge across the electric field (reviewed by Gandi and Isacoff, 2002) whilst recent crystallographic evidence suggests an altogether different model where S1–S4 sweeps across the membrane as a paddle (Jiang et al., 2003). These apparently disparate findings have yet to be fully reconciled (Ahern and Horn, 2004). Interestingly, TRP channels also have ordered, positively charged residues in S4 but fewer than are found in classical voltage-gated channels (Clapham, 2003). It has been proposed that TRPs have weak but measurable voltage-dependence that is synergistic with other gating stimuli (e.g. temperature: Voets et al., 2004). HCN channels have up to 10 positively charged residues in S4, more than their voltage-gated counterparts, but are curiously gated by membrane hyperpolarization (Männikkö et al., 2002). For channels gated by intracellular stimuli, such as Ca^{2+}-activated K$^+$ channels and cyclic nucleotide-gated channels the ligand-binding event is translated to channel opening via torsional changes in S6 (Bruening-Wright et al., 2007; Contreras et al., 2008).

Inactivation domains intrinsic to the channel α subunit have also been well characterized. Fast inactivation of voltage-gated K$^+$ channels occurs via physical occlusion of the pore by a cytoplasmic gating particle, the 'ball and chain' model (Armstrong and Bezanilla, 1977). In Shaker K$^+$ channels this inactivation involves a 22 amino acid sequence on the amino (N–) terminus ('the ball') linked to 60 further amino acids ('the chain'), and is referred to as N-type inactivation (Hoshi et al., 1990). The docking site for the ball is between S4 and S5 (Zhou et al., 2001). Similarly, for Na$^+$ channels, a cytoplasmic fast inactivation gate is formed by the highly conserved hydrophobic motif isoleucine–phenylalanine–methionine–threonine (IFMT) within the intracellular linker between domains III and IV (West et al., 1992; Kellenberger et al., 1997). Catterall describes this as a 'hinged lid' model of inactivation (Catterall, 2000). A second, independent and slower form of inactivation, termed C-type inactivation, has a less well-established structural basis but probably involves a gate at the extracellular end of the pore (Hoshi et al., 1991). It is worth noting that not all inactivation can be ascribed to conformational changes in the α subunit

per se—regulatory subunits and phosphorylation are two other key determinants (Wang et al., 1996; Ahern et al., 2005).

Regulatory, sometimes termed 'auxillary', subunits of voltage-gated channels are simpler non pore forming structures that can modify expression, functional properties and subcellular localization of the channel complex. The topology of these subunits varies widely across the voltage-gated superfamily. Voltage-gated Na$^+$ channels have a single family of auxiliary subunits, Na$_V$β1 to Na$_V$β4, that have a single transmembrane segment, a short intracellular C-terminal region, and a large N-terminal extracellular domain that is homologous in structure to a variable chain (V-type) immunoglobulin-like fold. Na$_V$β subunits shift the voltage-dependence of channel gating, enhance cell surface expression, and can interact with cell adhesion molecules, cytoskeletal linker proteins and extracellular matrix components to act as inter- and intracellular communicators (Isom et al., 1994). Min-K related proteins are K$^+$ channel regulatory subunits that are similar in overall transmembrane structure to Na$_V$β, and have been shown to regulate a number of voltage-gated K$^+$ channels including Kv7, K$_v$10 and Kv11 (McCrossan and Abbott, 2004). Other K$^+$ channel regulatory subunits are structurally dissimilar and do not span the membrane. Kv β 1–3 subunits interact with the N-terminal T1 domain on the a chain and form a symmetric tetramer on the intracellular surface (Gulbis et al., 2000). In vertebrates, the N-terminus of the β-subunit serves as an inactivation gate and blocks the pore during sustained channel opening. Likewise, K$^+$ channel interacting proteins, or KChIPs, are Ca^{2+}-sensing proteins (1–4) that bind to Kv4 channels at the T1 domain to modify the properties of the a subunit tetramer (Wang et al., 2007). In this case, the KChIPs provide Ca^{2+}-sensitivity. Voltage–gated Ca^{2+} channel regulatory subunits are somewhat more complex, and comprise of 4 distinct subfamilies: Ca$_V$2α, Ca$_V$2β, Ca$_V$2δ, and Ca$_V$2γ. The Ca$_V$2α and Ca$_V$2δ subunits are encoded by the same gene, whose translation product is proteolytically cleaved and disulfide linked to yield the mature large extracellular α2 subunit glycoprotein and smaller transmembrane disulfide-linked δ subunit. Four Ca$_V$2α2δ genes are known (Davies et al., 2007). Ca$_V$β subunits (1–4) are intracellular proteins, and are superficially similar in structure to Kvβ subunits. They have important regulatory effects on cell surface expression and voltage-dependent gating properties (Richards et al., 2004; Hidalgo and Neely, 2007). There are eight known members of the Ca$_V$2γ subunit gene family, each of which encodes a glycoprotein with four transmembrane domains and intracellular N- and C-termini. The function of the γ subunits is generally less well studied but they are known to play key roles in the assembly and cell surface expression of Ca^{2+} channels (Chen et al., 2007).

Ligand-gated and related channels

Ligand-gated ion channels typically contain three connected structures: an extracellular domain that forms the ligand binding site, a transmembrane cylinder that is the channel itself, and an intracellular region that may be involved in trafficking, localization and regulation by cytoplasmic second messengers.

Nicotinic acetylcholine (nACh), GABA$_A$, glycine and 5-HT$_3$ activated channels together form a family termed the Cys loop channels. These channels are pseudo-symmetrical assemblies of five subunits (pentamers) each 350–500 amino acids in length which contain four hydrophobic transmembrane α helices, termed M1–M4. Their name is derived from a characteristic sequence of 13 amino acids in the extracellular N-terminal region that are flanked by cysteine residues that covalently link to form a closed loop between the binding and channel domains. The intracellular region is largely formed from a loop of approximately 100 amino acids between M3 and M4. Each subunit spans the full length of the channel and contributes to the extracellular, pore and intracellular regions. The extracellular segment is mainly β-sheet, and the intracellular domain is α-helix (Corringer et al., 2000; Sine and Engel, 2006). Homomeric channel complexes can occur but heteromeric assembly is more common: for GABA$_A$ receptors for example a pentamer of two α, two β and a γ subunit arranged counterclockwise around the pore as αβαγβ is prevalent (Sieghart, 2000). The classical nicotinic AChR channel of the Torpedo electric organ is comprised of 2 α, 1 β, 1 γ and a δ subunit (Sine and Engel, 2006). Cys-loop channels can be either anion (e.g. GABA$_A$ and Glycine) or cation (e.g., nACh, 5-HT$_3$ receptor) permeable, depending on the charge and structure of key residues in the M2 region that face into the pore (Wotring et al., 2003). These residues form three rings of charge spanning the entrance and exit to the pore and serve to govern the ion selectivity (Imoto et al. 1988). Agonist binding gates the channel via torsional conformational changes that dilate the pore—the coupling mechanism underlying this phenomenon has been extensively studied for the nAChR for which an X-ray crystal structure of the *Lymnea Stagnalis* acetycholine binding protein has proved most useful (Brejc et al., 2001). At the cytoplasmic end of the channel, the inner vestibule does not open up as a simple axial hole, but rather forms a 'hanging basket' structure. Ions escape the channel to the cytoplasm sideways through five lateral openings in the basket.

The two other major groups of neurotransmitter-gated channels, the glutamate-gated iGluRs, and ATP-gated P2X receptors are structurally distinct from the Cys-loop channels and from each other. iGluRs are formed from hetero-tetramers, which are commonly 'dimers of dimers' (e.g. NR1/N2; Schorge and Colquhoun, 2003). Each subunit has a large extracellular N-terminal domain (NTD) implicated in iGluR oligomerization and modulation and an extracellular S1/S2 ligand binding domain that binds agonists in a Venus fly trap-like mechanism (Stern-Bach et al., 1994; Armstrong et al., 1998). Like the Cys loop channels the channel-forming region is composed of 4 TM regions. The C-terminal domain forms the major intracellular region and is involved in linking the receptor to the membrane scaffold and signal transduction proteins (Hansen et al., 2007). P2X channels are homo- or heterotrimers. Each P2X subunit has only 2-transmembrane spanning regions with intracellular N- and C- termini. The ATP-binding site exists in the structure formed by the extracellular TM1-TM2 loop (Vial et al., 2004).

Miscellaneous
Several ion channel subtypes show structural similarity to transporters. Thus, the epithelial chloride channel CFTR, is a member of the ABC (ATP binding cassette) transporter family of proteins that also includes the multidrug resistance proteins. CFTR is a multi-domain protein that contains two membrane-spanning domains, which form the channel pore, two cytoplasmic nucleotide binding domains (NBD) and a regulatory domain. ATP binding to the two nucleotide binding folds leads to conformational change and to opening of the channel gate. Hydrolysis of one of the ATPs disrupts the NBD1–NBD2 interaction and closes the channel gate (Gadsby et al., 2006).

Another ion channel protein, the sulfonylurea (SUR) subunit of the K$_{ATP}$ potassium channel is also a member of the ABC transporter family. The K$_{ATP}$ channel is a functional complex of two unrelated subunits in a 4:4 stoichiometry: Kir6.2, an inward rectifier potassium channel, forms the pore of the channel and the SUR subunit has a regulatory function, with a role in adenine nucleotide modulation (Inagaki et al., 1995). The SUR subunit also plays a key role in channel trafficking—binding of SUR to Kir6.2 masks ER retention motifs and allows translocation of the channel complex to the cell surface (Zerangue et al., 1999).

The ClC chloride channels, with nine members identified, are an intriguing family of proteins that are ubiquitously expressed in man. ClC proteins play a role at the cell surface, for example controlling excitability in muscle, and also on intracellular organelles, where they play a role in acidification. Early studies on the ClC-0 isoform from electric fish suggested a double-barrelled structure with the channel protein functioning as a homodimer with two ion-conducting pores (Ludewig et al., 1996). This structure was later confirmed when MacKinnon's lab published the X-ray structure of two prokaryotic ClC channels (Dutzler et al., 2002). However, more recently it has been demonstrated that the ClC family contains both genuine chloride channels and also proton-coupled chloride transporters (Miller, 2006).

While the focus of this chapter is on classical ion channel behaviour, namely ion channel permeation across the cell membrane, it is worth highlighting the growing body of data suggesting that some ion channels are multimodal proteins with quite different functional roles (reviewed by Kaczmarek, 2006). For example, the non-selective cation channel TRPM7 has intrinsic protein kinase activity that can phosphorylate both itself and exogenous substrates, and is also essential for channel function (Runnels et al., 2001).

1.4 Ion channel research methods
Voltage-clamp techniques
As introduced earlier, Sakmann and Neher's patch clamp electrophysiology method revolutionized ion channel research and remains the gold standard approach for functional analysis (Neher and Sakmann, 1976; Hamill et al., 1981).

A number of different patch clamp configurations and systems exist (for overview see Cahalan and Neher, 1992). For technical ease, the most common is the *'whole-cell' patch-clamp* mode for measuring macroscopic currents in the entire cell. Conventionally, a glass microelectrode, sometimes referred to as a 'patch pipette', is backfilled with an electrolyte (tip resistance 0.3–3 MΩ) and positioned on the surface of a cell with the aid of a microscope, micromanipulators and gentle suction. Once a high resistance (>1 GΩ) seal is formed, further suction ruptures the membrane beneath the tip of the electrode and allows electrical access to the cell interior. The electrolyte dialyses the cell within minutes so ionic gradients across the membrane can be established as desired. Low noise patch clamp amplifier circuitry is used to monitor membrane potential with reference to a bath ground electrode, inject current to clamp the membrane at fixed voltages and monitor current flow. Voltage steps or ramps, or ligands applied either via the bath or the intracellular solution, can be used to activate channels. Whole-cell currents as small as 10 pA, representing signals from just a few hundred open channels, can be recorded with sub-millisecond temporal resolution. A variant of this whole cell recording method is the *perforated patch clamp* technique in which electrical access to the cell interior is obtained via a membrane permeabilizing agent, typically nystatin or amphoterocin, added to the internal recording solution (Horn and Marty, 1988; Rae *et al.*, 1991). This method produces minimal disruption to the intracellular milieu and generally allows for highly stable recordings.

Three other configurations—*cell-attached, inside-out* and *outside-out patch clamp*, sample channels from a small area ($<10\ \mu m^2$) of the membrane rather than from the entire cell (Neher *et al.*, 1978; Horn and Patlak, 1980). In each case it is possible to resolve single channel events. For cell-attached, the pipette tip is simply positioned on the cell surface and currents arising from the channels directly beneath the tip are measured. This approach is entirely non-invasive, as it leaves the intracellular milieu intact. A major disadvantage is the difficulty in directly applying pharmacological modulators to either side of the membrane, and for this reason this configuration is largely restricted to biophysical analyses. By gently moving the pipette away from the cell, excising the patch inside the pipette tip, it is possible to move to the inside-out recording configuration. With the intracellular surface now exposed to the bath solution, compound addition to the inner facing components of the channel can be readily achieved. Outside-out patches are created by first sealing the pipette tip to the cell and then rupturing the membrane as in whole cell recording. If the pipette is slowly withdrawn from the cell the membrane can seal back over the tip, creating a patch where the extracellular surface is exposed to the bath solution. This is particularly attractive for studying neurotransmitter-gated ion channels or other channels that respond to extracellular ligands.

Minimizing methodological errors in measurement is critical for making high fidelity patch clamp recordings. The relatively large volume of plasma membrane in the whole cell mode can yield capacitance artefacts that distort the ionic currents following voltage steps, as charge is stored and then released when an electric field is applied. The capacitance of small membrane patches is generally so low as to be insignificant. As an aside, membrane capacitance measured in this way has proved to be an extremely useful index of cell surface area in studies on the fusion of secretory vesicles (Neher and Marty, 1982; Penner and Neher, 1989). Non-biological 'leak' currents from the imperfect seal between the glass electrode and membrane can also be confounding. Fortunately, both the undesirable capacitance and leak components are directly proportional to the voltage step amplitude, and can be readily subtracted by scaling signals generated at potentials at which channels are not activated. A third source of error is 'access' (or series) resistance, R_a (Armstrong and Gilly, 1992). The relatively high resistance access path for current injection down the electrode, commonly 3–10 MΩ, gives rise to voltage reading errors which are the product of R_a and the current amplitude. For large currents (>1 nA) and R_a values (10 MΩ) these errors can exceed 10 mV if left uncorrected. Most patch clamp amplifiers have compensation circuitry to minimize the problems of R_a but these are seldom perfect and only 60–80 per cent compensation is usually possible.

The patch clamp paradigm has evolved significantly in the last decade, with the introduction of planar array electrophysiology (Fertig *et al.*, 2002, 2003; Sigworth and Klemic, 2002; Schroeder *et al.*, 2003; see Trezise, 2007 for review). Fertig and co-workers provided the first detailed description of patch clamping cells in suspension by replacing the glass pipette with a planar quartz chip with a 1 mm, 3–5 MΩ aperture as the recording site. A single cell was attracted to the aperture by suction and once a high-resistance seal had formed with the substrate further suction was applied to yield the whole cell recording mode. This advance obviated the need for microscopy and manipulation and opened the way for parallelization and automation. With this approach and the latest system designs, it is now possible to make >3000 recordings in an experimental day, some 100–300-fold faster than conventional patch clamp recording.

The highest throughput system, the IonWorks Quattro marketed by Molecular Devices Corporation, is designed for the industrial drug discovery setting and has some interesting features—the recording substrate is a 384-well consumable in which individual recordings and experiments are made in <15 μL assay volumes. A multichannel electronics head and amplifier circuitry is used to make 48 simultaneous recordings on the plate, and test compounds can be added to assay wells via a fluidics head. Seal resistances are generally lower than conventional patch clamp, and in the region of 100–400 MΩ, and there is no capacity to compensate for stray capacitance and access resistance errors. Experimental design options are constrained somewhat to drug screening applications, and the system is relatively limited for detailed biophysical analyses and the study of certain channels with specific recording challenges (e.g. ultra rapid ligand-gated channels). In a simple but impactful innovation, Finkel *et al.* (2006) recently introduced the 'population patch clamp' method to the IonWorks system whereby recordings can be made from groups (up to 64) rather than individual cells. This allows for currents to be averaged across cells, and increases the precision and statistical power of ion channel bioassays (Dale *et al.*, 2007; John *et al.*, 2007). Other systems of note include the Patch Xpress and QPatch instruments, which provide 16–48 parallel recordings on a chip with GΩ seal resistances and integrated fluidics (http://www.moleculardevices.com/; http://www.sophion.dk/).

Methods for drug application to cells have also evolved to allow the study of ion channels that gate within milliseconds of ligand application, and for analysing the kinetics of drug action. Fenwick *et al.* (1982) introduced a simple drug perfusion system termed the 'U-tube' that comprised a switching valve and a U-shaped large bore gravitational fed tube with a tiny aperture that could be placed near to a cell under patch clamp. Closing the valve forces the drug solution through the hole and onto the cell with very short lag times, as low as 20 ms. A second approach is to use pairs or groups of continuously flowing capillary tubes that can be rapidly positioned over the cell using piezo switches (e.g. Biologic). Most recently, an elegant micro-fluidic device has been introduced (Dynaflow, Cellectricon, Sweden) that flows up to 48 different drug solutions in parallel that are separated via the principles of laminar flow. The cell can be rapidly and precisely moved in and out of any solution flow, and be used almost as a pharmacological barcode reader (Sinclair *et al.*, 2002).

Oocytes from the African Claw frog, *Xenopus Laevis*, provide a highly convenient expression and functional analysis system for recombinant ion channels (Gundersen *et al.*, 1984; see Goldin, 2007 for review). Unlike mammalian cells, *Xenopus* oocytes can be seen with the naked eye (1 mm diameter) and are easy to manipulate and impale with injection needles or recording electrodes. Microinjection methods are used to introduce RNA into the cytoplasm, or cDNA into the nucleus, and after a 6–72 h wait to allow the cell to synthesize and express the encoded protein(s), basic electrophysiological methods can be applied. The most common recording method is two electrode voltage-clamp, where one electrode is used to measure the internal potential of the oocyte and the other is used to inject current. Dependent on the channel type and the amount of RNA injected, currents ranging from 10 nA to 100 μA can be recorded. One limitation of this method is the large membrane capacitance of the oocyte (100–200 nF)—when the voltage is changed sharply (e.g. with a step depolarization) the membrane initially acts as a current sink before the voltage clamp 'settles'. The decay of the capacitance transient can take 1–2 ms, which interferes with measurements of rapid voltage-gated channels such as Na_Vs. To overcome this, other recording configurations have been established, including macro-patches and cut open voltage-clamp. Macro-patches are essentially the same as cell-attached or excised patch recordings, only larger (10 μm diameter), and involve only the channels in the oocyte membrane patch. The capacitance settling time of the smaller volume of membrane is significantly shorter, such that channels with rapid kinetics can be studied.

Oocyte expression systems were once widely used for *expression cloning*, in which pools of RNA/cDNA from a relevant tissue (e.g. rat brain) are injected, characterized for the response of interest and then repeatedly diluted to finally allow the cloning of the novel gene. This approach is now obviated by the availability of cDNA clones encoding ion channels and superior mammalian heterologous expression systems. For structure/function studies, however, the technique remains extremely important, given the ease in which DNA/RNA expression can be titrated and functionally characterized. This is especially true when reconstituting channels comprised of multiple different subunits (e.g. $GABA_A$ and nAChR receptors) or for studies on channel mutations and chimaeras. *Site-directed mutagenesis* is an approach in which changes in the channel subunit DNA sequence are made at specific positions of interest by substituting codons that specify for alternative amino acids (see Stevens *et al.*, 2006 for review). Either naturally or unnaturally occurring amino acids can be introduced in this way (Bennett *et al.*, 2006). If cysteine residues are introduced, for example, their 'accessibility' and hence dynamic positional change (e.g. during gating), can be assessed by electrophysiological analysis of the action of charged hydrophilic sulphydryl agents that covalently bind to cysteines—the so-called *substituted cysteine accessibility method* (SCAM). Similarly, substituted histidine mutants can be probed for accessibility of hydrogen ions based on pH-dependent changes in currents via protonation (Starace and Bezanilla, 2001). Systematic substitution of residues with alanines, to introduce small hydrophobic groups that interact favourably with membrane lipids or hydrophobic regions, is also a common technique for studying gating effects and drug binding sites. In each case, the precision and speed with which the functional consequences of these mutations can be characterized by oocyte (and now in many cases mammalian cell) electrophysiology make this a valuable approach.

As with planar array electrophysiology, oocytes are used for compound screening for drug discovery. Two higher throughput automated recording systems, the Robocyte from Multichannel systems and the OpusXpress from Axon Instruments, have been marketed (Schnizler *et al.*, 2003; Papke, 2006) The Robocyte sequentially processes up to 96 oocytes arrayed on a V-bottomed microtitre plate. Under computer control, the plate is moved to position the oocyte directly below a single head, configured for either injection or recording. Since the oocyte is approached vertically, cytoplasmic or nuclear injections can be achieved by simply changing the depth of the injection. For recording, the two-electrode voltage clamp configuration is used, with semi-automated on line and offline analysis. Test compounds are added either via a valve-controlled gravity perfusion system or a more sophisticated Gilson liquid handler with peristaltic pump. The OpusXpress allows for eight parallel recordings and is comprised of eight independent chambers each with pairs of electrodes and perfusion and ground assemblies. An automated liquid handling system is used to apply compounds from a 96-well plate. There is no provision for automated injection and this must be done offline.

Fluorescence and luminescence probes

For each of the major charge carrying ions in physiology, Ca^{2+}, Na^+, K^+ and Cl^-, ion-sensitive probes, or surrogate measurements thereof, have been developed. Indicators are generally either small molecules or modified proteins that can be loaded into cells and when excited with a light of an appropriate wavelength, and in the presence of the relevant ion, emit a measurable and characteristic light signal. Ion-sensitive dyes differ in their ion specificity, sensitivity, dynamic range, and spatiotemporal resolution.

Molecules such as Ca-Green, Fluo-3 and Fluo-4 (Minta *et al.*, 1989; Gee *et al.*, 2000) are fluorescent Ca^{2+}-chelators based on EGTA. Typically they are loaded into cells as acetoxymethyl (AM) esters which are then hydrolysed by intracellular esterases to the tetra-acid forms which are trapped within cells. A single wavelength excitation at 488 nm for Fluo-3 yields fluorescence increases at 525 nm of up to 100-fold upon Ca^{2+}-binding (K_D~400 nM). Fura-2 is the favoured dye for ratiometric measurements where two excitation wavelengths are used. Cameleons, chimaeric constructs of cyan and yellow variants of green fluorescent protein separated by the Ca^{2+}-binding protein calmodulin, detect Ca^{2+} via fluorescence resonance energy transfer (FRET) (Miyawaki *et al.*, 1997). FRET is the process whereby the emission light from a donor fluorophore acts as the excitation for a second acceptor—only when the donor and acceptor are in close proximity can the FRET occur. In the case of cameleons, the binding of Ca^{2+} to calmodulin causes the protein to pivot around its central axis bringing the two fluorophores together. Aequorin, a photoprotein from the jellyfish *Aequoria Victoria*, (Shimomura *et al.*, 1962) can be transfected into mammalian cells, and generates a flash luminescence in the presence of Ca^{2+} and the reaction co-factor coelenterazine.

Targeting Ca^{2+} sensors to specific regions or organelles of the cell can provide significantly higher spatial resolution, and the study of Ca^{2+} dynamics in the micro- to nano-domain. Both aequorin and cameleons have been genetically targeted to different compartments including mitochondria, nucleus, endoplasmic reticulum and plasma membrane (Mank *et al.*, 2006). Most recently, Roger Tsien and co-workers have developed a new biarsenical Ca^{2+}-indicator, calcium green FlAsH (CaGF), which targets proteins that have been labelled with arsenic binding tetracysteine motifs. Using a tagged α1C aV12 voltage-gated Ca^{2+}-channel expressed in Hela cells it proved possible to resolve real time Ca^{2+} kinetics and domain Ca^{2+} 'hot spots' in the vicinity of the channel (Tour *et al.*, 2007).

Fluorescent indicators for Na^+ and K^+ are based on crown ethers conjugated to fluorophores. For Na^+, the most commonly used are the ratiometric dye SBFI, and the fluorescein-based Sodium Green and CoroNa Red (Minta and Tsien, 1989; Amorino and Fox, 1995). Unlike the Ca^{2+}-dyes, the Na^+-fluorophores have low ion-sensitivity (Na^+ K_D 200 mM CoroNa Red) and this restricts their utility— large (e.g. >10 mM) changes in Na^+ concentration are required to elicit significant changes in fluorescence. Whilst K^+-sensitive dyes are available (e.g. PBFI), for cellular applications the situation is even worse than for Na^+ as intracellular concentrations are basally high, and essentially fixed ($\approx$140 mM). Fluorescent assays for K^+ channels are possible using BTC, a coumarin benzothiazole-based Tl^+ sensor (Weaver et al., 2004; Niswender et al., 2008). Tl^+ readily passes through K^+ channels, and in Cl^- free assay conditions to overcome issues of TlCl insolubility, surrogate assays for K^+ influx or efflux have been assembled. Quinolinium or acrodinium dyes such as SPQ and MQAE have decreased fluorescence when exposed to halides, and can be used to monitor Cl^- fluxes (West and Molloy, 1996). The brightness and signal changes are, however, low and their use is not commonplace. Genetically encoded anion-sensitive fluorescent proteins, such as yellow fluorescent protein (YFP) mutants have recently been introduced and offer greater sensitivity and scope for targeting to cellular organelles (Wachter et al., 1998; Kuner and Augustine, 2000; De la Fuente et al., 2008).

The other major class of optical probes for ion channel research are the 'environment sensors'. Redistribution probes are charged dyes that equilibrate between intracellular hydrophobic sites and the extracellular solution according to the membrane potential (see Gonzalez et al., 2006 for review). Generally, probe fluorescence is low when in aqueous solution and rises on binding to hydrophobic sites such as proteins and membrane. Negatively charged oxonol membrane potential sensors, such as DiBaC4, have been widely used in industrial high throughput screening settings, but have undesirably slow temporal response (seconds to minutes), temperature sensitivity and interference from fluorescent compounds. Next generation oxonol-based assay kits, that include quenching dyes to reduce extracellular background fluorescence, provide improved response times and larger signal to background ratios and are now commercially available (e.g. FMP kit, Molecular Devices). Further benefits can be gained by employing dual fluorophore FRET-based systems. A coumarin FRET donor (e.g. CC1-DMPE) is anchored to the extracellular leaflet of the plasma membrane and interacts with a mobile oxonol-based FRET acceptor (typically a bis-(1,3-dialkythiobarbituric acid) trimethine oxonol DiSBAC) that moves according to the membrane potential. Upon depolarization the oxonol translocates to the positively charged inner surface of the plasma membrane, away from the coumarin, thereby reducing FRET. The ratiometric detection yields fewer experimental artefacts because the signals are independent of optical path length, excitation intensity and the number of cells being detected. Moreover, the subsecond reversible kinetics of the oxonol translocation means that the FRET system can report voltage changes much faster than can be achieved with the conventional redistribution probes. Using an ultra-fast hexyl substituted pentamethine oxonol, $DiSBAC_6$, Huang et al. (2006) were able to follow synchronized action potentials in cell populations evoked by field stimulation with millisecond resolution (see also Djurisic et al., 2004).

Genetically encoded optical probes for voltage-sensing have also been described (Siegel and Isacoff, 1997; see Guerrero and Isacoff, 2001 for review). Green fluorescent protein (GFP) has been fused to the C-terminus of a non-conducting mutant Shaker K^+ channel in such a way that the voltage-dependent channel rearrangements that occur on membrane depolarization result in a measurable spectral change. This approach can yield extremely high temporal resolution. Tagging organic or protein fluorophores to channels has also proved very powerful in dissecting channel conformational rearrangements per se (e.g. Sonnleitner et al., 2002; Pathak et al., 2005). Much of what is known about the pore facing molecular constituents and movements of the S4 domain in the voltage-dependent gating of K^+ channels has been provided by such optical methods (Gandhi and Isacoff, 2005).

Ion flux and ligand binding methods

Ion flux determinations in cells are direct, simple and inexpensive methods for monitoring ion channel function. Radioactive isotopes of the naturally conducting ions, for example $^{45}Ca^{2+}$ and $^{22}Na^+$, have been used as tracers, as have surrogate isotopes such as ^{14}C-guanidinium for Na^+ channels and $^{86}Rb^+$ for K^+ channels. Radioactivity in cell lysates and/or supernatants is quantified by direct Cerenkov counting (e.g. $^{86}Rb^+$) or by liquid scintillation counting. The major limitation of this method is the safety and storage considerations for working with radioisotopes. For these reasons, non-radioactive ion flux measurements using atomic absorption spectrometry (AAS) have gained in popularity (Terstappen, 1999). AAS detects trace amounts of ions in samples by converting them to free ground state atoms in a vapour phase which absorb light of a specific wavelength. Atomization is achieved by spraying the test sample (lysate or supernatant) into the flame of an AA spectrometer, and absorbance is measured with a photomultiplier. This approach is most commonly used for pharmacological studies on K^+ channels using Rb^+ as the marker.

Nature has provided a wealth of highly specific peptide ligands for ion channels which have proved invaluable as pharmacological tools. By attaching radioisotopes to these peptides it is not only possible to probe the distribution of the channel using autoradiography, but also to screen for novel channel modulators using radioligand binding assays. In conjunction with electrophysiology and site-directed mutagenesis, six distinct toxin recognition sites have been mapped on voltage-gated Na^+ channels with radiolabelled peptides (Catterall and Beneski, 1980; Strichartz et al., 1987; Rogers et al., 1996). Numerous small molecules have also been radio-labelled and used for drug screening, especially for ligand-gated ion channels. In heterogeneous competition binding assays, tissue samples, whole cells or membrane preparations are mixed with the radioligand and test compound and then allowed to equilibrate. Channel bound radioligand is separated from free ligand usually by filtration and washing, and quantified as before using scintillation or β-counting. In a screening environment, assay formats that do not require separation or washing, such as scintillation proximity assay (SPA), are preferred. In SPA, scintillant containing beads are coupled to the membrane of interest via wheat germ agglutinin which adheres to glycosyl residues. When mixed with the radioligand (and test compound) binding to the immobilized fraction brings the emitted radiation of the isotope sufficiently close to the scintillant for activation. The short range energy from unbound radioligand is absorbed by the aqueous environment before it reaches the bead. In excess of 100 different ion channel radioligands are commercially available and/or have been described in the literature.

Structural approaches

Compared to soluble proteins such as kinases and nuclear hormone receptors there is comparatively little high resolution information on the protein structure of transmembrane ion channels. There are two major technical challenges: (1) to express and/or purify sufficient channel and (2) to extract this protein into detergent in such a way that it retains a meaningful tertiary structure. Once this has been achieved, three-dimensional images can be reconstructed using either single particle electron microscopy, electron crystallography from two-dimensional protein crystals or X-ray crystallography. Whilst the crystallization step necessary for the latter approaches is in itself challenging, the requirements for ion channels are not significantly different to those of soluble proteins.

To date, approximately 25 different channel structures have been determined by electron microscopy to resolutions of 7–40 Å, and 12 by X-ray crystallography to 2–3 Å (see Bass and Spencer, 2006 for review and the Protein Data Bank http://www.rcsb.pdb.org). Fewer than one in five are native channels, purified from sources rich in protein such as the voltage-gated Na$^+$ channel in the electric organ of *Electrophorus electricus* (Sato *et al.*, 2001) and the nAChR from *Torpedo* Ray (Miyazawa *et al.*, 2003). The remainder are recombinant channels expressed either in mammalian, insect, yeast or bacterial systems. As the primary goal is a high yield of homogeneous protein, mammalian expression systems are generally not favoured since levels of expression are low and intracellular pools of incorrectly processed or folded protein are frequently encountered. Nevertheless, sufficient protein has been obtained using HEK293, COS-7 or BHK cells to support electron microscopic analysis of the TRPC3 cation channel (Mio *et al.*, 2005), Kv4.2/KChIP2 (Kim *et al.*, 2004) and CFTR (Rosenberg *et al.*, 2004). Insect cells, such as Sf9, have shown some worth in the expression of voltage-gated K$^+$ channels (e.g. Li *et al.*, 1994) but have not been widely used. Expression in yeast appears more promising—Parcej and Eckhardt-Strelau (2003) expressed rat Kv1.2 complexed with Kvβ2 in *Pichia Pastoris* at sufficient levels to support electron microscopy work, and this method was adapted by the MacKinnon lab to generate the first mammalian transmembrane high resolution crystal structure (Long *et al.*, 2005a). Heterologous expression in bacteria has proved highly effective for prokaryotic channels. Indeed, all of the prokaryotic ion channel X-ray crystal structures determined thus far were expressed in *Escherichia Coli* (Kuo *et al.*, 2003; Nishida *et al.*, 2007). With eukaryotic channels, however, *E. Coli* expression has proved less fruitful and many mammalian channels are toxic when overexpressed.

Methods for extracting, purifying and crystallizing ion channels are not yet routine, and require significant experimentation in order to succeed. N-Decyl-b-D-maltoside is a commonly employed detergent for extraction from the membrane bilayer, but a wide range of other detergents may need to be tried to establish optimal extraction. Purification approaches are better established and typically involve polyhistidine affinity tagging to a C- or N-terminal domain and immobilized metal chelate chromatography. For crystallization, N,N-diemthyldodecylamine-N-oxide (LDAO), n-decyl-b-D-maltoside, or N-octyl-b-D-maltoside are commonly used detergents. In some cases, it has proved necessary to include an antibody fragment (e.g Fab domain or Fv fragment) in this final step to enhance the polar surface area of the protein and promote the formation of a crystal lattice structure (e.g. KvAP—Jiang *et al.*, 2003).

Concluding remarks

This chapter serves as a generic introduction to the ion channel superfamily highlighting the diversity and complexity of function that is encoded within this family of proteins. The following chapters provide further insights into each of the members of the various subfamilies of channels that the superfamily can be subdivided into.

References

Ahern CA and Horn R (2004). Stirring up controversy with a voltage sensor paddle. *TINS* **27** 303–307.

Ahern CA, Zhang JF, Wookalis MJ *et al.* (2005). Modulation of the cardiac sodium channel NaV1.5 by Fyn, a Src family tyrosine kinase. *Circ Res.* **96**, 991–998.

Amorino GP and Fox MH (1995). Intracellular Na$^+$ measurements using sodium green tetraacetate with flow cytometry. *Cytometry* **21**, 248–256.

Armstrong CM and Bezanilla F (1977). Inactivation of the sodium channel. II. Gating current experiments. *J. Gen. Physiol.* **70**, 567–590.

Armstrong CM and Gilly WF (1992). Access resistance and space clamp problems associated with whole-cell patch clamping. *Methods Enzymol.* **207**, 100–122.

Armstrong N, Sun Y, Chen GQ *et al.* (1998). Structure of a glutamate-receptor ligand-binding core in complex with kainate. *Nature* **395**, 913–917.

Bass RB and Spencer RH (2006). Approaches for ion channel structural studies. In *Expression and analysis of recombinant ion channels.* Eds Clare, Clare JJ and Trezise DJ, Wiley VCH, pp. 213–219.

Bennett PB, Zacharias N, Nicholas JB *et al.* (2006). Unnatural amino acids as probes of ion channel structure-function and pharmacology. In *Expression and analysis of recombinant ion channels* Eds Clare JJ and Trezise DJ, Wiley VCH, pp. 59–77.

Boucher RC (2002). An overview of the pathogenesis of cystic fibrosis lung disease. *Adv Drug Deliv Rev* **54**, 1359–1371.

Brejc K, van Dijk WJ, Klaassen RV *et al.* (2001). Crystal structure of an ACh-binding protein reveals the ligand-binding domain of nicotinic receptors. *Nature* **411**, 269–276.

Bruening-Wright A, Lee WS *et al.* (2007). Evidence for a deep pore activation gate in small conductance Ca^{2+}-activated K$^+$ channels. *J Gen Physiol* **130**, 601–610.

Cahalan M and Neher E (1992). Patch clamp techniques: an overview. *Methods Enzymol.* **207**, 3–14.

Cannon SC (2007). Physiologic principles underlying ion channelopathies. *Neurotherapeutics.* **4**, 174–183.

Caterina MJ, Leffler A, Malmberg AB *et al.* (2000). Impaired nociception and pain sensation in mice lacking the capsaicin receptor. *Science* **288**, 306–313.

Catterall WA and Beneski DA (1980). Interaction of polypeptide neurotoxins with a receptor site associated with voltage-sensitive sodium channels. *J Supramol Struct* **14**, 295–303.

Catterall WA (2000). From ionic currents to molecular mechanisms: the structure and function of voltage-gated sodium channels. *Neuron* **26**, 13–25.

Cestèle S and Catterall WA (2000). Molecular mechanisms of neurotoxin action on voltage-gated sodium channels. *Biochimie* **82**, 883–892.

Chen RS, Deng TC, Garcia T *et al.* (2007). Calcium channel gamma subunits: a functionally diverse protein family. *Cell Biochem Biophys.* **47**, 178–186.

Clapham DE (2003). TRP channels as cellular sensors. *Nature* **426**, 517–524.

Contreras JE, Srikumar D and Holmgren M (2008). Gating at the selectivity filter in cyclic nucleotide-gated channels. *PNAS* **105**, 3310–3314.

Corringer PJ, Le Novère N and Changeux JP (2000). Nicotinic receptors at the amino acid level. *Annu Rev Pharmacol Toxicol.* **40**, 431–458.

Dale TJ, Townsend C, Hollands EC *et al.* (2007). Population patch clamp electrophysiology: a breakthrough technology for ion channel screening. *Mol Biosyst.* **3**, 714–722.

Davies A, Hendrich J, Van Minh AT *et al.* (2007). Functional biology of the $\alpha 2\delta$ subunits of voltage-gated calcium channels. *TIPS* **28**, 220–228.

De La Fuente R, Namkung W, Mills A *et al.* (2008). Small-molecule screen identifies inhibitors of a human intestinal calcium-activated chloride channel. *Mol Pharmacol.* **73**, 758–768.

Djurisic M, Antic S, Chen WR *et al.* (2004). Voltage imaging from dendrites of mitral cells: EPSP attenuation and spike trigger zones. *J Neurosci.* **24**, 6703–6714.

Doyle DA, Morais Cabral J, Pfuetzner RA *et al.* (1998). The structure of the potassium channel: molecular basis of K^+ conduction and selectivity. *Science* **280**, 69–77.

Dutzler R, Campbell EB, Cadene M *et al.* (2002). X-ray structure of a ClC chloride channel at 3.0 Å reveals the molecular basis of anion selectivity. *Nature* **415**, 287–294.

Elliott AA and Elliott JR (1993). Characterization of TTX-sensitive and TTX-resistant sodium currents in small cells from adult rat dorsal root ganglia. *J Physiol. (Lond)* **463**, 39–56.

Etter A, Cully DF, Liu KK *et al.* (1999). Picrotoxin blockade of invertebrate glutamate-gated chloride channels: subunit dependence and evidence for binding within the pore. *J Neurochem* **72**, 318–326.

Fenwick EM, Marty A and Neher E (1982). A patch-clamp study of bovine chromaffin cells and of their sensitivity to acetylcholine. *J Physiol* **331**, 577–597.

Fertig N, Blick RH and Behrends JC (2002). Whole cell patch clamp recording performed on a planar glass chip. *Biophys J.* **82**, 3056–3062.

Fertig N, George M, Klau M *et al.* (2003). Microstructured apertures in planar glass substrates for ion channel research. *Receptors Channels* **9**, 29–40.

Finkel A, Wittel A, Yang N *et al.* (2006). Population patch clamp improves data consistency and success rates in the measurement of ionic currents. *J Biomol Screen.* **11**, 488–496.

Fujiwara Y and Kubo Y (2006). Functional roles of charged amino acid residues on the wall of the cytoplasmic pore of Kir2.1. *J Gen Physiol* **127**, 401–419.

Gadsby DC, Vergani P and Csanády L (2006). The ABC protein turned chloride channel whose failure causes cystic fibrosis. *Nature* **440**, 477–483.

Gandhi CS and Isacoff EY (2002). Molecular models of voltage sensing. *Journal of General Physiology* **120**, 455–463.

Gandhi CS and Isacoff EY (2005). Shedding light on membrane proteins. *TINS* **28**, 472–9.

Gee KR, Brown KA, Chen WN *et al.* (2000). Chemical and physiological characterization of fluo-4 Ca^{2+}-indicator dyes. *Cell Calcium.* **27**, 97–106.

Goldin, A (2007). Expression of ion channels in xenopus oocytes. In *Expression and analysis of recombinant ion channels*. Eds Clare, J.J and Trezise, D.J. Wiley VCH, p1–21.

Goldman DE (1943). Potential, impedence and rectification in membranes. *J Gen Physiol* **27**, 37–60.

Gonzalez JE, Worley J and Van Goor F (2006). Ion channel assays based on ion and voltage-sensitive fluorescent probes. In *Expression and analysis of recombinant ion channels*. Eds Clare JJ and Trezise DJ. Wiley VCH, p187–211.

Guerrero G and Isacoff EY (2001). Genetically encoded optical sensors of neuronal activity and cellular function. *Curr Opin Neurobiol.* **11**, 601–607.

Gulbis JM, Mann S and MacKinnon R (1999). Structure of a voltage-dependent K^+ channel beta subunit. *Cell* **97**, 943–952.

Gulbis JM, Zhou M, Mann S *et al.* (2000). Structure of the cytoplasmic beta subunit-T1 assembly of voltage-dependent K^+ channels. *Science* **289**, 123–127.

Gundersen CB, Miledi R and Parker I (1984). Messenger RNA from human brain induces drug- and voltage-operated channels in Xenopus oocytes. *Nature* **308**, 421–424.

Hamill OP, Marty A, Neher E *et al.* (1981). Improved patch-clamp techniques for high-resolution current recording from cells and cell-free membrane patches. *Pflugers Arch.* **391**, 85–100.

Hansen KB, Yuan H and Traynelis SF (2007). Structural aspects of AMPA receptor activation, desensitization and deactivation. *Curr Opin Neurobiol.* **17**, 281–288.

Heinemann SH and Conti F (1992). Non-stationary fluctuation noise analysis and application to patch clamp recordings. *Methods Enyzmol* **207**, 131–148.

Heinemann SH, Terlau H, Stühmer W *et al.* (1992). Calcium channel characteristics conferred on the sodium channel by single mutations. *Nature* **356**, 441–443.

Hendrich J, Van Minh AT, Heblich F *et al.* (2008). Pharmacological disruption of calcium channel trafficking by the alpha2delta ligand gabapentin. *Proceedings of the National Academy of Science* **105**, 3628–3633.

Hidalgo P and Neely A (2007). Multiplicity of protein interactions and functions of the voltage-gated calcium channel beta-subunit. *Cell Calcium* **42**, 389–396.

Hille, B. (1992). *Ionic channels of excitable membranes.* Sinauer Associates, Sunderland, Mass., 2nd edition.

Hodgkin AL and Huxley AF (1952a). Currents carried by sodium and potassium ions through the membrane of the giant axon of Loligo. *J Physiol* **116**, 449–472.

Hodgkin AL and Huxley AF (1952b). The components of membrane conductance in the giant axon of Loligo. *J Physiol* **116**, 473–496.

Hodgkin AL and Huxley AF (1952c). The dual effect of membrane potential on sodium conductance in the giant axon of Loligo. *J Physiol* **116**, 497–506.

Hodgkin AL and Huxley AF (1952d). A quantitative description of membrane current and its application to conduction and excitation in nerve. *J Physiol* **117**, 500–544.

Hodgkin AL and Katz B (1949). The effect of sodium ions on the electrical activity of the giant axon of the squid. *J Physiol* **108**, 37–77.

Hodgkin AL, Huxley AF and Katz B (1952). Measurement of current–voltage relations in the membrane of the giant axon of Loligo. *J Physiol* **116**, 424–448.

Horn R and Marty A (1988). Muscarinic activation of ionic currents measured by a new whole-cell recording method. *J Gen Physiol* **92**, 145–159.

Horn R and Patlak J (1980). Single channel currents from excised patches of muscle membrane. *PNAS* **77**, 6930–6934.

Hoshi T, Zagotta WN and Aldrich RW (1990). Biophysical and molecular mechanisms of Shaker potassium channel inactivation. *Science* **250**, 533–538.

Hoshi T, Zagotta WN and Aldrich RW (1991). Two types of inactivation in Shaker K^+ channels: effects of alterations in the carboxy-terminal region. *Neuron* **7**, 547–556.

Huang CJ, Harootunian A, Maher MP *et al.* (2006). Characterization of voltage-gated sodium-channel blockers by electrical stimulation and fluorescence detection of membrane potential. *Nat Biotechnol.* **24**, 439–446.

Imoto K, Busch C, Sakmann B *et al.* (1998). Rings of negatively charged amino acids determine the acetylcholine receptor channel conductance. *Nature* **335**, 645–648.

Inagaki N, Tsuura Y, Namba N *et al.* (1995). Cloning and functional characterization of a novel ATP-sensitive potassium channel ubiquitously expressed in rat tissues, including pancreatic islets, pituitary, skeletal muscle, and heart. *J Biol Chem* **270**, 5691–5694.

Isom LL, De Jongh KS and Catterall WA (1994). Auxiliary subunits of voltage-gated ion channels. *Neuron* **12**, 1183–1194.

Jiang Y and MacKinnon R (2000). The barium site in a potassium channel by X-ray crystallography. *J Gen Physiol* **115**, 269–272.

Jiang Y, Lee A, Chen J *et al.* (2002). Crystal structure and mechanism of a calcium-gated potassium channel. *Nature* **417**, 515–522.

Jiang Y, Lee A, Chen J *et al.* (2002). The open pore conformation of potassium channels. *Nature* **417**, 523–526.

Jiang Y, Lee A, Chen J *et al.* (2003). X-ray structure of a voltage-dependent K$^+$ channel. *Nature* **423**, 33–41.

John VH, Dale TJ, Hollands EC *et al.* (2007). Novel 384-well population patch clamp electrophysiology assays for Ca^{2+}-activated K$^+$ channels. *J Biomol Screen* **12**, 50–60.

Kaczmarek L (2006). Non-conducting functions of voltage-gated ion channels. *Nature Reviews Neuroscience* **7**, 761–771.

Kashiwagi K, Masuko T, Nguyen CD *et al.* (2002). Channel blockers acting at N-methyl-D-aspartate receptors: differential effects of mutations in the vestibule and ion channel pore. *Mol Pharmacol* **61**, 533–545.

Kass RS (2005). The channelopathies: novel insights into molecular and genetic mechanisms of human disease. *J Clin Invest* **115**, 1986–1989.

Kellenberger S, West JW, Scheuer T *et al.* (1997). Molecular analysis of the putative inactivation particle in the inactivation gate of brain type IIA Na$^+$ channels. *J. Gen. Physiol.* **109**, 589–605.

Kim LA, Furst J, Gutierrez D *et al.* (2004). Three-dimensional structure of I(to); Kv4.2-KChIP2 ion channels by electron microscopy at 21 Angstrom resolution. *Neuron* **41**, 513–519.

Kuner T and Augustine GJ (2000). A genetically encoded ratiometric indicator for chloride: capturing chloride transients in cultured hippocampal neurons. *Neuron* **27**, 447–459.

Kuo A, Gulbis JM, Antcliff JF *et al.* (2003). Crystal structure of the potassium channel KirBac1.1 in the closed state. *Science* **300**, 1922–1926.

Lee H, Chen CX, Liu YJ *et al.* (2005). KCC2 expression in immature rat cortical neurons is sufficient to switch the polarity of GABA responses. *Eur J Neurosci* **21**, 2593–2599.

Li M, Unwin N, Stauffer KA *et al.* (1994). Images of purified Shaker potassium channels. *Curr Biol* **4**, 110–115.

Long SB, Campbell EB and Mackinnon R (2005a). Crystal structure of a mammalian voltage-dependent Shaker family K$^+$ channel. *Science.* **309**, 897–903.

Long SB, Campbell EB and Mackinnon R (2005b). Voltage sensor of Kv1.2: structural basis of electromechanical coupling. *Science* **309**, 903–908.

Lopatin AN, Makhina EN and Nichols CG (1994). Potassium channel block by cytoplasmic polyamines as the mechanism of intrinsic rectification. *Nature* **372**, 366–369.

Ludewig U, Pusch M and Jentsch TJ (1996). Two physically distinct pores in the dimeric ClC-0 chloride channel. *Nature* **383**, 340–343.

Main MJ, Cryan JE, Dupere JR *et al.* (2000). Modulation of KCNQ2/3 potassium channels by the novel anticonvulsant retigabine. *Mol Pharmacol* **58**, 253–262.

Mank M, Reiff DF, Heim N *et al.* (2006). A FRET-based calcium biosensor with fast signal kinetics and high fluorescence change. *Biophys J* **90**, 1790–1796.

Männikkö R, Elinder F and Larsson HP (2002). Voltage-sensing mechanism is conserved among ion channels gated by opposite voltages. *Nature* **419**, 837–841.

Matsuda HA, Saigusa and Irisawa, H (1987). Ohmic conductance through the inwardly rectifying K channel and blocking by internal Mg^{2+}. *Nature* **325**, 156–159.

McCrossan ZA and Abbott GW (2004). The MinK-related peptides. *Neuropharmacology* **47**, 787–821.

Meera P, Wallner M and Toro L (2000). A neuronal beta subunit (KCNMB4) makes the large conductance, voltage- and Ca^{2+}-activated K$^+$ channel resistant to charybdotoxin and iberiotoxin. *PNAS* **97**, 5562–5567.

Miller C (2006). ClC chloride channels viewed through a transporter lens. *Nature* **440**, 484–489.

Minta A and Tsien RY (1989). Fluorescent indicators for cytosolic sodium. *J Biol Chem* **264**, 19449–19457.

Mio K, Ogura T, Hara Y *et al.* (2005). The non-selective cation-permeable channel TRPC3 is a tetrahedron with a cap on the large cytoplasmic end. *Biochem Biophys Res Commun* **333**, 768–777.

Miyawaki A, Llopis J, Heim R *et al.* (1997). Fluorescent indicators for Ca^{2+} based on green fluorescent proteins and calmodulin. *Nature* **388**, 882–887.

Miyazawa A, Fujiyoshi Y and Unwin N (2003). Structure and gating mechanism of the acetylcholine receptor pore. *Nature* **423**, 949–955.

Morais-Cabral JH, Zhou Y and MacKinnon R (2001). Energetic optimization of ion conduction rate by the K$^+$ selectivity filter. *Nature* **414**, 37–42.

Neher E and Marty A (1982). Discrete changes of cell membrane capacitance observed under conditions of enhanced secretion in bovine adrenal chromaffin cells. *PNAS* **79**, 6712–6716.

Neher E and Sakmann B (1976). Single-channel currents recorded from membrane of denervated frog muscle fibres. *Nature* **260**, 799–802.

Neher E, Sakmann B and Steinbach JH (1978). The extracellular patch clamp: a method for resolving currents through individual open channels in biological membranes. *Pflugers Arch* **375**, 219–228.

Nishida M, Cadene M, Chait BT *et al.* (2007). Crystal structure of a Kir3.1-prokaryotic Kir channel chimera. *EMBO* **26**, 4005–4015.

Niswender CM, Johnson KA, Luo Q *et al.* (2008). A novel assay of Gi/o-linked G protein-coupled receptor coupling to potassium channels provides new insights into the pharmacology of the group III metabotropic glutamate receptors. *Mol Pharmacol* **73**, 1213–1224.

Noda M, Shimizu S, Tanabe T *et al.* (1984). Primary structure of electrophorus electricus sodium channel deduced from cDNA sequence. *Nature* **312**, 121–127.

Olsen RW (1982). Drug interactions at the GABA receptor-ionophore complex. *Annu Rev Pharmacol Toxicol* **22**, 245–277.

Owens DF, Boyce LH, Davis MB *et al.* (1996). Excitatory GABA responses in embryonic and neonatal cortical slices demonstrated by gramicidin perforated-patch recordings and calcium imaging. *J Neurosci* **16**, 6414–6423.

Papazian DM, Schwarz TL, Tempel BL *et al.* (1987). Cloning of genomic and complementary DNA from Shaker, a putative potassium channel gene from Drosophila. *Science* **237**, 749–753.

Papke RL (2006). Estimation of both the potency and efficacy of alpha7 nAChR agonists from single-concentration responses. *Life Sci* **78**, 2812–2819.

Parcej DN and Eckhardt-Strelau L (2003). Structural characterisation of neuronal voltage-sensitive K$^+$ channels heterologously expressed in *Pichia pastoris*. *J Mol Biol* **333**, 103–116.

Partin KM, Patneau DK, Winters CA *et al.* (1993). Selective modulation of desensitization at AMPA versus kainate receptors by cyclothiazide and concanavalin A. *Neuron* **11**, 1069–1182.

Pathak M, Kurtz L, Tombola F *et al.* (2005). The cooperative voltage sensor motion that gates a potassium channel. *J Gen Physiol* **125**, 57–69.

Pedersen KA, Schrøder RL, Skaaning-Jensen B *et al.* (1999). Activation of the human intermediate-conductance Ca^{2+}-activated K$^+$ channel by 1-ethyl-2-benzimidazolinone is strongly Ca^{2+}-dependent. *Biochim Biophys Acta* **1420**, 231–240.

Penner R and Neher E (1989). The patch-clamp technique in the study of secretion. *TINS* **12**, 159–163.

Perry M, Sachse FB and Sanguinetti MC (2007). Structural basis of action for a human ether-a-go-go-related gene 1 potassium channel activator. *PNAS* **104**, 13827–13832.

Pongs O (1990). Structural basis of potassium channel diversity in the nervous system. *J Basic Clin Physiol Pharmacol* **1**, 31–39.

Purohit P and Auerbach A (2007). Acetylcholine receptor gating: movement in the alpha-subunit extracellular domain. *J Gen Physiol* **130**, 569–579.

Rae J, Cooper K, Gates P et al. (1991). Low access resistance perforated patch recordings using amphotericin B. J Neurosci Methods 37, 15–26.

Rauer H, Pennington M, Cahalan M et al. (1999). Structural conservation of the pores of calcium-activated and voltage-gated potassium channels determined by a sea anemone toxin. J Biol Chem 274, 21885–21892.

Richards MW, Butcher AJ and Dolphin AC (2004). Ca^{2+} channel beta-subunits: structural insights AID our understanding. TIPS 25, 626–632.

Rogers JC, Qu Y, Tanada TN et al, (1996). Molecular determinants of high affinity binding of alpha-scorpion toxin and sea anemone toxin in the S3–S4 extracellular loop in domain IV of the Na^+ channel alpha subunit. J Biol Chem 271, 15950–15962.

Rosenberg MF, Kamis AB, Aleksandrov LA et al. (2004). Purification and crystallization of the cystic fibrosis transmembrane conductance regulator (CFTR). J Biol Chem 279, 39051–39057.

Roux B and MacKinnon R (1999). The cavity and pore helices in the KcsA K+ channel: electrostatic stabilization of monovalent cations. Science 285, 100–102.

Runnels LW, Yue L and Clapham DE (2001). TRP-PLIK, a bifunctional protein with kinase and ion channel activities. Science. 291, 1043–1047.

Salata JJ, Jurkiewicz NK, Wang J et al. (1998). A novel benzodiazepine that activates cardiac slow delayed rectifier K+ currents. Mol Pharmacol 54, 220–230.

Sather WA and McCleskey EW (2003). Permeation and selectivity in calcium channels. Ann Rev Physiol. 65, 133–159.

Sato C, Ueno Y, Asai K et al. (2001). The voltage-sensitive sodium channel is a bell-shaped molecule with several cavities. Nature 409, 1047–1051.

Schneider MF and Chandler WK (1973). Voltage dependent charge movement of skeletal muscle: a possible step in excitation-contraction coupling. Nature 242, 244–246.

Schnizler K, Küster M, Methfessel C et al. (2003). The roboocyte: automated cDNA/mRNA injection and subsequent TEVC recording on Xenopus oocytes in 96-well microtiter plates. Receptors Channels 9, 41–48.

Schorge S and Colquhoun DJ (2003). Studies of NMDA receptor function and stoichiometry with truncated and tandem subunits. Neurosci 23, 1151–1158.

Schroeder K, Neagle B, Trezise DJ et al. (2003). Ionworks HT: a new high-throughput electrophysiology measurement platform. J Biomol Screen 8, 50–64.

Shimomura O, Johnson FH and Saiga Y (1962). Extraction, purification and properties of aequorin, a bioluminescent protein from the luminous hydromedusan, Aequorea. J Cell Comp Physiol 59, 223–239.

Siegel MS and Isacoff EY (1997). A genetically encoded optical probe of membrane voltage. Neuron 19, 735–741.

Sieghart W (2000). Unraveling the function of GABA(A) receptor subtypes. Trends Pharmacol Sci. 21, 411–413.

Sigworth FJ and Klemic KG (2002). Patch clamp on a chip. Biophys J 82, 2831–2832.

Sinclair J, Pihl J, Olofsson J et al. (2002). A cell-based bar code reader for high-throughput screening of ion channel-ligand interactions. Anal Chem 74, 6133–6138.

Sine SM and Engel AG (2006). Recent advances in Cys-loop receptor structure and function. Nature 440, 448–455.

Sonnleitner A, Mannuzzu LM, Terakawa S et al. (2002). Structural rearrangements in single ion channels detected optically in living cells. Proceedings of the National Academy of Science 99, 12759–12764.

Starace DM and Bezanilla F (2001). Histidine scanning mutagenesis of basic residues of the S4 segment of the shaker K+ channel. J Gen Physiol 117, 469–490.

Stern-Bach Y, Bettler B, Hartley M et al. (1994). Agonist selectivity of glutamate receptors is specified by two domains structurally related to bacterial amino acid-binding proteins. Neuron 13, 1345–1357.

Stevens L, Powell AJ and Wray D (2006). Molecular biology techniques for structure–function studies of ion channels. In Expression and analysis of recombinant ion channels. Eds Clare, J.J and Trezise, D.J. Wiley VCH, pp. 28–58.

Strichartz G, Rando T and Wang GK (1987). An integrated view of the molecular toxinology of sodium channel gating in excitable cells. Annu Rev Neurosci 10, 237–267.

Swartz KJ and MacKinnon R (1995). An inhibitor of the Kv2.1 potassium channel isolated from the venom of a Chilean tarantula. Neuron 15, 941–949.

Terstappen GC (1999). Functional analysis of native and recombinant ion channels using a high-capacity nonradioactive rubidium efflux assay. Anal Biochem 272, 149–155.

Tour O, Adams SR, Kerr RA et al. (2007). Calcium Green FlAsH as a genetically targeted small-molecule calcium indicator. Nat Chem Biol 3, 423–431.

Trezise DJ (2007). Automated planar array electrophysiology for ion channel research. In Expression and analysis of recombinant ion channels. Eds Clare, J.J and Trezise, D.J. Wiley VCH, pp. 145–164.

Ulbricht W (2005). Sodium channel inactivation: molecular determinants and modulation. Physiol Rev 85, 1271–1301.

Vial C, Roberts JA and Evans RJ (2004). Molecular properties of ATP-gated P2X receptor ion channels. TIPS 25, 487–493.

Voets T, Droogmans G, Wissenbach U et al. (2004). The principle of temperature-dependent gating in cold- and heat-sensitive TRP channels. Nature 430, 748–754.

Voets T, Prenen J, Vriens J et al. (2002). Molecular determinants of permeation through the cation channel TRPV4. J Biol Chem 277, 33704–33710.

Wachter RM, Elsliger MA, Kallio K et al. (1998). Structural basis of spectral shifts in the yellow-emission variants of green fluorescent protein. Structure 6, 1267–1277.

Wang H, Yan Y, Liu Q et al. (2007). Structural basis for modulation of Kv4 K+ channels by auxiliary KChIP subunits. Nature Neuroscience 10, 32–39.

Wang, Z, Kiehn, J, Yang et al. (1996). Comparison of binding and block produced by alternatively spliced Kv 1 subunits. J. Biol. Chem 271, 28311–28317.

Weaver CD, Harden D, Dworetzky SI et al. (2004). A thallium-sensitive, fluorescence-based assay for detecting and characterizing potassium channel modulators in mammalian cells. J Biomol Screen 9, 671–677.

West JW, Patton DE, Scheuer T et al. (1992). A cluster of hydrophobic amino acid residues required for fast Na^+ channel inactivation. Proceedings of the National Academy of Science 89, 10910–10914.

West MR and Molloy CR (1996). A microplate assay measuring chloride ion channel activity. Anal Biochem 241, 51–58.

Winterfield JR and Swartz KJ (2000). A hot spot for the interaction of gating modifier toxins with voltage-dependent ion channels. J Gen Physiol 116, 637–644.

Wotring VE, Miller TS and Weiss DS (2003). Mutations at the GABA receptor selectivity filter: a possible role for effective charges. J Physiol 548, 527–540.

Xie X and Hagan RM (1998). Cellular and molecular actions of lamotrigine: possible mechanisms of efficacy in bipolar disorder. Neuropsychobiology 38, 119–30.

Xiong Q, Sun H, Zhang Y et al. (2008). Combinatorial augmentation of voltage-gated KCNQ potassium channels by chemical openers. Proceedings of the National Academy of Science 105, 3128–3133.

Zerangue N, Schwappach B, Jan YN et al. (1999). A new ER trafficking signal regulates the subunit stoichiometry of plasma membrane K_{ATP} channels. Neuron 22, 537–548.

Zhou M, Morais-Cabral JH, Mann S et al. (2001). Potassium channel receptor site for the inactivation gate and quaternary amine inhibitors. Nature 411, 657–661.

Zhou Y, Morais-Cabral JH, Kaufman A et al. (2001). Chemistry of ion coordination and hydration revealed by a K+ channel-Fab complex at 2.0 A resolution. Nature 414, 43–48.

PART 2

Voltage-gated ion channels

Voltage-gated K⁺ channels

Bernardo Rudy

Contents

Introduction

The ability to establish and maintain an electrical membrane potential is common to all cellular life. In certain specialized cells, rapid modulation of the membrane potential has evolved as a means of coordinating, in time, the spatially distributed cellular processes of fast communication using electrical signals as well as contraction and secretion. The molecular basis of this cellular excitability is provided by a large family of voltage-gated ion channel proteins. Amongst the voltage-gated ion channels, those selective for potassium (K^+) ions are by far the most diverse. In the following chapters we review the molecular and functional diversity of the voltage-gated K^+ channels, paying particular attention to their functional roles in neurons and their relationship to human disease.

Potassium channels dampen excitability in interesting and specific ways

The opening of K^+ channels leads to currents that tend to drive the membrane potential towards the equilibrium potential for K^+, and thus in general are inhibitory. Voltage-gated K^+ channels are activated by membrane depolarization and contribute to many of the non-linear, voltage- and time-dependent electrical properties of neurons. These channels influence and regulate many of the subthreshold properties of neurons, including setting the resting membrane potential and resistance. They contribute to the amplitude and frequency of subthreshold oscillations, thereby influencing the resonant properties of neurons, and to determining the probability of spike generation. In the suprathreshold range of membrane potentials, they are the main determinants of the repolarization of the action potential, governing spike shape and frequency, and have similar functions in other excitable cells, such as all varieties of muscle. In non-excitable cells they contribute to the resting potential and to the regulation of secretion.

Basic organization of the channel complex

Voltage-gated K^+ channels are tetramers of primary, or pore-forming subunits (sometimes called α subunits). With a fourfold symmetry around a central pore, this tetramer forms the infrastructure of the channel and in most cases is sufficient to form a functional channel. In addition, many voltage-gated K^+ channels contain auxiliary proteins (sometimes referred to as β subunits) that have primary sequences not resembling pore-forming subunits and can modify the properties of the channels, often significantly, and in some cases may be essential for the efficient expression of functional channels in the plasma membrane. In addition, many K^+ channel molecular complexes interact with additional proteins such as regulatory enzymes and elements of the cytoskeleton.

Nomenclature

The pore-forming subunits of voltage-gated K^+ channels belong to three major groups of proteins (Figure 2.1, seen here and in the plate section) whose sequence similarity define groups and subgroups that also correspond to functional classes. These groups are:

1. Proteins of the Kv1–Kv6 and Kv8–Kv9 subfamilies, which are components of voltage-gated K^+ channels with fast kinetics. This group is often called the Kv family;

2. Members of the Kv7 subfamily of K^+ channel proteins (often called the KCNQ family); and finally

3. The Kv10–Kv12 subfamilies (also known as the EAG family).

The last two groups are pore-forming subunits of voltage-gated K^+ channels with slow kinetics.

Kv family subunits form the classical voltage-gated K^+ channels that activate relatively quickly upon membrane depolarization.

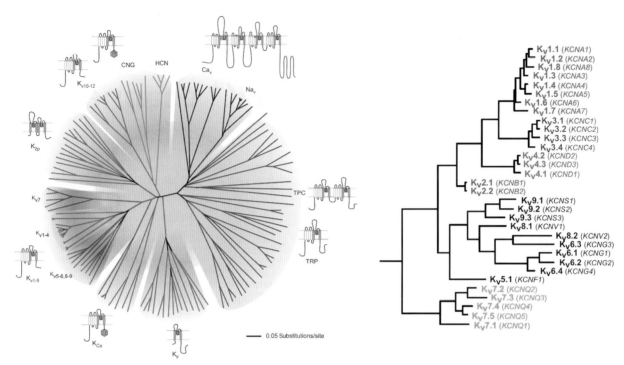

Fig. 2.1 Left: the voltage-gated ion channel superfamily illustrating similarity relationships (based on amino acid sequence) between Kv subunits, subunits of other K[+] channels and other ion channels of the voltage-gated ion channel superfamily. Right: phylogenetic relations of Kv1–Kv9 K[+] channel subunits. The groups discussed in this chapter cluster together based on sequence similarities. From Yu and Catterall (2004). See also the colour plate section.

The subunits of the KCNQ family form the M-type K[+] channels underlying the current known as I_M, a subthreshold operating current of considerable importance in the regulation of neuronal excitability. One member of the EAG family (erg1 or Kv11.1) underlies the cardiac current known as I_{Kr}, a current of major significance in cardiac physiology that was poorly understood prior to the cloning and functional studies of the erg1 subunits (sometimes referred to in the cardiac field as HERG). Moreover, channels of this family had not been identified in neurons or other cells before they were discovered through the molecular–genetic analysis of K[+] channels, and to date, are still insufficiently studied in the central nervous system. Chapter 2.1 focuses on the channels of the Kv family, where we discuss the pore-forming subunits of the Kv family and their auxiliary proteins. The Kv7 (KCNQ) and Kv10–12 (EAG) families are discussed in separate chapters.

2.1.1 The Kv family (Kv1–Kv6 and Kv8–Kv9 subfamilies)

Bernardo Rudy, Jonathon Maffie, Yimy Amarillo, Brian Clark, Hyo-Young Jeong, Illya Kruglikov, Elaine Kwon, Marcela Nadal and Edward Zagha

2.1.1.1 Introduction

In heterologous expression systems, most Kv subunits express as functional homomultimeric channel complexes. However, some

pore-forming subunits (which we call co-assembly pore-forming subunits) (Coetzee *et al.*, 1999) specifically the members of the Kv5–Kv6 and Kv8–Kv9 subfamilies, do not form functional homomultimeric channels. Instead, they must co-assemble with subunits of the Kv2 subfamily for expression of functional channels.

Members of the same subfamily, in the case of the Kv1 and Kv3–Kv4 subfamilies and Kv2 plus Kv5–Kv6 and Kv8–Kv9 subfamilies, can form heteromeric channels. These channels have properties that are intermediary to those of homomeric channels, although some properties of one subunit may dominate in the heteromultimeric channel. Heteromultimerization results in an increase in the number of channels with distinct functional properties that can be generated by Kv subunits[1]. The large number of subunits in some subfamilies (e.g., eight in the case of the Kv1 subfamily), confers an enormous diversity of channels, which can be even larger if one considers other factors, such as interactions with diverse auxiliary proteins or post-translational modifications that can also lead to functional diversification (Coetzee *et al.*, 1999). While it remains to be determined whether much of this diversity is used by neurons or other cells, it should be noted that most Kv proteins are expressed in the nervous system, that there is considerable overlap of expression of multiple Kv subunits of the same subfamily in many types of neurons, and that co-immunoprecipitation studies have demonstrated the existence of heteromeric complexes in native tissue. All of these factors point towards a complex system of molecules capable of fine-tuning neuronal function.

---

[1] The number of different subunit combinations can be calculated from the equation:$[p + (n - 1)]!/[p!(n - 1)!]$, where p is the aggregation number (4) and n is the total number of subunits that can be used to form a channel complex.

Given the subfamily-specific heteromultimerization, each Kv subfamily describes an independent functional unity, and emphasizes the need to discuss each subfamily as a group. As described above, the members of the Kv5–Kv6 and Kv8–Kv9 subfamilies must co-assemble with subunits of the Kv2 subfamily to form functional channels, and are therefore integral parts of Kv2 channels. These proteins are discussed here together with the members of the Kv2 subfamily.

2.1.1.2 Molecular characteristics

Structure of Kv proteins

Kv subunits consist of six membrane spanning-domains (S1–S6) flanked by intracellular N- and C-terminal sequences of variable lengths. There is a loop between the S5 and S6 that starts on the extracellular surface of the membrane, partially enters the membrane, and then re-emerges to the extracellular side known as the P domain. The cytoplasmic N- and C-terminal sequences contain phosphorylation sites, along with sites for other types of post-translational regulation, such as modulation by the redox state of the cell. The P domain contains the 'K$^+$ channel signature sequence'(TVGYG), which is highly conserved not only among Kv proteins, but also among all K$^+$ channel subunits, including those from prokaryotes (Heginbotham *et al.*, 1994; Doyle *et al.*,

1998), contributing to the formation of the K$^+$ selective pore (Figure 2.1.1.1). There is a sequence known as the T1 domain at the N terminus, preceding the S1 helix. T1 domains of members of the same subfamily are very similar and determine subfamily-specific association (Li *et al.*, 1992; Shen and Pfaffinger 1995; Choe *et al.*, 2002).

The fourth membrane-spanning domain (S4) is characterized by the repetition of a motif consisting of two neutral residues (usually hydrophobic except towards the C-terminal end of S4) and one positively charged residue (usually arginine). The number of repetitions of this motif is characteristic of each Kv subfamily: seven in Kv1 subunits; five in Kv2s (as well as Kv5–Kv6 and Kv8–Kv9) and Kv4 subunits and six in Kv3 subunits. This domain is a critical part of the voltage-sensor. The positively charged residues are likely to be the gating charges, i.e. the charges that move in response to changes in membrane potential to open and close the channel's pore (Bezanilla, 2000; Jiang *et al.*, 2003).

The structure of a mammalian Kv channel, a complex consisting of four Kv1.2 subunits and four auxiliary Kvβ2 proteins, has been resolved at 2.9 Å (Long *et al.*, 2005a, b; Figure 2.1.1.1). In the channel structure, four T1 domains, one from each of the four Kv subunits, interact to form a tetrameric assembly at the intracellular membrane surface. This assembly is located directly under the cytoplasmic side of the channel's pore, which communicates

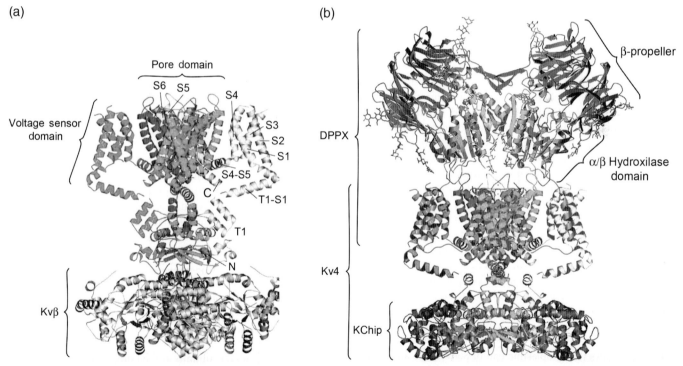

(a)   (b)

Fig. 2.1.1.1 Comparison of a modelled Kv4.3–KChIP1–DPPX channel complex with the Kv1.2–Kvβ2 channel complex. Left: side view of the Kv1.2–Kvβ2 channel complex derived from the crystal structure obtained by MacKinnon and colleagues (Long *et al.* 2005a) based on an image from Wang *et al.* (2007). The four Kv1.2 subunits are in cyan, yellow, pink and green. The four Kvβ2 subunits are in wheat colour. The upper third shows the transmembrane part of the channel. The pore domains of the four Kv1.2 proteins forming an inverted tepee are in the centre flanked by the voltage sensor domains. The middle portion of the figure contains the T1 domains interacting with the KChIP proteins at the bottom. Right: side on view of the K4.3– KChIP1–DPPX channel complex. This figure was obtained by adding molecules of DPPX (also known as DPP6) based on the crystal structure of the extracellular domain obtained by Strop *et al.* (2004) to the model of the Kv4.3–KChIP1 channel complex of Wang *et al.* (2007). The four Kv4.3 subunits are in cyan, yellow, pink and green. The four KChIP1 proteins, also interacting with the channels T1 domains, are in blue. The transmembrane domains of DPPX are shown interacting with the membrane spanning helices of the voltage sensor domains. Only two DPPX molecules are shown to facilitate viewing, but we believe that the channel complex is likely to include four DPPX proteins. Each DPPX subunit includes an α/β hydrolase domain (close to the membrane and a β propeller domain shown above and laterally). The extracellular domains of the DPPX proteins interact at the center forming a dimer. See also the colour plate section.

with the cytoplasm through side portals, or windows, formed by the linkers connecting the T1 domains to the first membrane-spanning helices. These portals allow K^+ ions to flow freely between the pore and the cytoplasm and are large enough to also allow the entry of the N-terminal inactivating polypeptide responsible for N-type inactivation (Kobertz et al., 2000; Lu et al., 2001) (discussed further below). In the crystal structure, under the assembly of T1 domains are the four Kvβ2 subunits, which as described later interact with the T1 domains.

Based largely on these and other structural studies, MacKinnon and colleagues have proposed, that rather than forming a compact protein, each subunit consists basically of two relatively independent domains in the channel complex (Long et al., 2005a, b): a voltage-sensor domain consisting of the S1–S4 helices and a pore domain consisting of the S5 and S6 membrane-spanning helices and the P loop located between S5 and S6, as can be seen in the structures shown in Figure 2.1.1.1. The structure of the pore domain in the mammalian voltage-gated channel is essentially the same as that of channels that are not voltage-dependent and are tetramers of subunits consisting of only two helices (inner and outer, equivalent to the S6 and S5 helices, respectively, and a P loop). The pore-domains of the four Kv subunits create an 'inverted teepee', or cone, with the S6 (or inner helix) facing the inside. The P domains of the four subunits are located in the wider third of the teepee, near the extracellular surface, forming the selectivity filter, which surrounds the narrowest part of the pore. The selectivity filter has the amino acids of the K^+ channel signature sequence, oriented so that they expose their main chain carbonyl oxygen atoms to the inside. These carbonyl oxygen atoms take the place of the oxygen atoms of the water hydrating the K^+ ion allowing the dehydration of the ion and its entry into the narrow pore delimited by the selectivity filter. The carbonyl oxygen atoms are too far apart to properly surround the smaller Na^+ ion, thus explaining K^+ selective permeation.

MacKinnon and colleagues have pointed out that the relative independence of the voltage-sensor and pore domains in the deduced crystal structure is consistent with several recent observations. These include the observation that it is possible to transfer a voltage sensor from a Kv channel to a non-voltage-dependent K^+ channel (Lu et al., 2001) and the recent discovery of a voltage-dependent phosphatase in which an S1–S4 like voltage-sensor domain is coupled to a cytoplasmic phosphatase (Murata et al., 2005). The voltage-sensor domain is connected to the pore domain through helices made by the S4–S5 linker, which makes several amino acid contacts with the S6 or inner helix lining the pore (Long et al., 2005a, b and Figure 2.1.1). The cytoplasmic end of the S6 helix is flexible around a 'hinge' provided by the conserved sequence Pro-X-Pro. These investigators have proposed that movement of the voltage-sensor in response to changes in membrane potential displaces the S4–S5 linker affecting its interactions with the S6 helix. This allows the cytoplasmic end of S6 to move and in turn constrict or dilate the inner opening of the pore.

N- and P/C-type inactivation of Kv channels

Voltage-gated K^+ channels (like most other voltage-gated channels except the HCN pacemaker-channels) activate and open upon membrane depolarization. However, even if the depolarization remains constant, most Kv channels do not remain open, but instead undergo a process known as inactivation (Hodgkin and Huxley, 1952a, b). Depending on the channel type, the rates at which the channels inactivate vary by orders of magnitude.

They are in the range of a few milliseconds in the case of fast inactivating channels mediating transient K^+ currents. There are probably several mechanisms by which Kv channels inactivate; two of which have been extensively characterized: N- and P/C-type inactivation. N-type inactivation is usually fast and is produced by the cytoplasmic N-terminal sequence of some Kv subunits acting as a tethered blocking peptide that enters the channel's pore and occludes it blocking ion permeation (Aldrich, 2001; Hoshi et al., 1990), reminiscent of the 'ball and chain' mechanism proposed by Armstrong and Bezanilla (1977). The Kv pore-forming subunits, Kv1.4, Kv3.3, and Kv3.4, and the Kv1 auxiliary proteins Kvβ1 and Kvβ3 (described below) contain such N-terminal inactivation domains producing N-type inactivation of channels containing these subunits (Ruppersberg et al., 1991; Murrell-Lagnado and Aldrich, 1993; Tseng-Crank et al., 1993; Vega-Saenz de Miera, 1994; Covarrubias et al., 1994; Lee et al., 1996; Rudy et al., 1999; Fernandez et al., 2003).

While the molecular details are not understood, P/C-type inactivation is produced by a time-dependent conformational change that constricts the external mouth of the pore (Hoshi et al., 1991; Liu et al., 1996; Yellen, 1998). Many Kv channels undergo P/C inactivation, but the rates of this can differ by two to three orders of magnitude, from tens of milliseconds to seconds. The rate of P/C inactivation depends on the amino acid sequence of the P domain as well as the S6 domain and can be affected by extracellular ions that can interact with the channel's pore.

Kv1 (Shaker) subfamily

Neuronal Kv1 channels are complexes of Kv1 and Kvβ subunits

There are 8 Kv1 genes (Kv1.1–Kv1.8) (see Table 2.1.1.1). Although not all combinations have been tested, it is assumed that all Kv1 proteins are capable of forming heteromeric channel complexes. Given the overlapping expression of multiple Kv1 proteins in the same cells, it is very likely that native Kv1 channels in many cells are heteromultimeric. In addition, most native Kv1 channel complexes probably include at least one type of Kvβ subunit. As described earlier (Figure 2.1.1.1), the crystallized Kv1.2 channel complex is a tetramer of Kv1.2 pore-forming subunits associated with a tetramer of Kvβ proteins (Long et al., 2005b). We expect that most Kv1 channels in cells will have a similar structure, and, therefore, the properties of native Kv1 channels will depend on the contributions of various Kv1 pore-forming subunits to the heteromultimer and the effects of the specific Kvβ subunits present (see for example Imbrici et al., 2006). Consequently, there is the potential for huge diversity of native Kv1 channels, however, at this point it is extremely difficult to predict channel composition from the properties of the current recorded.

Kinetically, native Kv1 channels can be delayed rectifiers or fast-transient A-type channels, depending on their subunit composition. Among the Kv1 subunits prominently expressed in brain (Kv1.1–Kv1.6)[2], Kv1.4 subunits are the only ones that contain an N-terminal inactivation domain capable of producing fast N-type inactivation. Kv1.4 subunits contribute to the fast inactivation to heteromultimeric channels with a rate that depends on the number of Kv1.4 proteins in the channel complex MacKinnon et al., 1993).

---

2   One of the two alternative spliced Kv1.7 isoforms (Kv1.7L) also has an N-terminal inactivation domain producing channels with fast N-type inactivation (see Kv1.7 table) (Finol-Urdaneta, 2006). However, Kv1.7 is not expressed in brain.

In addition, several Kvβ subunits have N-terminal inactivation domains which give fast N-inactivation to channel complexes (see the section Kvβ below), including to Kv1 channels that do not inactivate in the absence of these proteins.

Many Kv1 channels inactivate slowly via P/C-type inactivation. This inactivation is faster in Kv1.3 compared to other Kv1 channels (time constants of inactivation in the order of hundreds of milliseconds), and is characterized by a slow rate of recovery. Therefore, P/C inactivation of Kv1.3 channels produces pronounced cumulative inactivation during repetitive stimulation (Marom and Levitan, 1994; Grissmer et al., 1994; Bertoli et al., 1996).

Kvβ subunits

Voltage-gated K^+ channels of the Kv1 subfamily associate with Kvβ subunits (Rettig and Heinemann, 1994; Pongs et al., 1999), proteins with a mass of about 40 kD which lack putative transmembrane domains, potential glycosylation sites, or leader sequences, suggesting that they are cytosolic proteins. This has been confirmed by structural analysis (Long et al., 2005a). Kvβ proteins bind 1:1 to the T1 domain located in the N-terminal region of the channel pore-forming subunits, creating channel complexes of eight polypeptides (Long et al., 2005a) (Figure 2.1.1.1; see aslo the section Structure of Kv proteins above). Native Kv1 channels are likely complexes containing four Kv1 pore-forming subunits and four Kvβ associated proteins, as in the case of the structure shown in Figure 2.1.1.

Three genes encode Kvβ proteins. Each of these genes produces several isoforms by alternative splicing in the N-terminal region (Pongs et al., 1999). A conserved core domain is shared by all Kvβ subunits, which has sequence similarity to aldo-keto reductases (McCormack et al., 1994). The Kvβ1 and Kvβ3 subunits also contain an additional, isoform-specific N-terminal sequence resembling the inactivation domain found in Kv1.4 that produces N-type inactivation (Pongs et al., 1999) (see the section N- and C/P-type inactivation of Kv channels above).

Together with their association with Kvβ subunits, most Kv1 proteins are able to interact with scaffolding, membrane-associated putative guanylate kinases (or MAGUKs), which are PDZ domain-containing proteins, such as PSD-95 and SAP97. This association is mediated by the PDZ domains interacting with the C-terminus of Kv1 proteins, which in all members of the subfamily, except for Kv1.8, has the sequence ETDV or ETEV. This association is important for the clustering of the channels in microdomains at the neuronal surface, which can also include other PDZ-binding proteins. Mutations of the Kv1 C-terminal sequence abolish the binding to the MAGUKs and prevent channel clustering (Kim et al., 1995; Kim and Sheng 1996).

Kv2 (Shab) subfamily

There are two known genes in this subfamily in mammals, Kv2.1 and Kv2.2. In rat superior cervical ganglion (SCG) cells, which express both Kv2.1 and Kv2.2 subunits, antibodies to each protein only immunoprecipitated the subunit against which the antibody was targeted, indicating that the two subunits are not associated in the same complex. Similar results were obtained in HEK-293 cells transfected with cDNAs encoding both subunits (Malin and Nerbonne, 2002). Using dominant-negative (DN) constructs, Malin and Nerbonne (2002) also found that a Kv2.1 DN construct

selectively attenuated Kv2.1 currents expressed in HEK-293 cells, while a Kv2.2 DN construct only affected Kv2.2 currents. These results indicate that even in an heterologous expression system Kv2.1 and Kv2.2 are incapable of heteromultimerization. However, Blaine and Ribera (1998) found that a Kv2.2 DN construct suppressed both Kv2.1 and Kv2.2 channels expressed in *Xenopus* oocytes, suggesting that the two proteins can heteromultimerize in this expression system.

Kv3 (Shaw) Subfamily

There are four known Kv3 genes in rodents and humans: Kv3.1, Kv3.2, Kv3.3 and Kv3.4 (reviewed in Rudy et al., 1999, 2001). Homologues of each of these have been identified in other vertebrate species. In contrast to the other Kv subfamilies, the four Kv3 genes undergo extensive alternative splicing that results in the generation of over 10 Kv3 isoforms in mammals (Rudy et al., 1999). Spliced isoforms differ only in their intracellular C-terminal sequence.

Kv4 (Shal) subfamily

The first gene encoding a subunit of A-type channels was cloned from *Drosophila* Shal, (Wei et al., 1990), followed by the mouse homologue mShal or Kv4.1 (Pak et al., 1991). A total of three genes encoding Kv4 subunits—Kv4.1, Kv4.2 and Kv4.3—are now known to be present in mammals (Baldwin et al., 1991; Pak et al., 1991; Salkoff et al., 1992; Serodio et al., 1996). The Kv4.3 gene encodes for two variants via alternative splicing: Kv4.3S and Kv4.3L (short and large).

2.1.1.3 Biophysical characteristics

Kv1 (Shaker) subfamily

Kv1 subunits are components of voltage-gated K^+ channels that tend to activate faster and at lower membrane potentials than Kv channels of other subfamilies. In fact, except for Kv1.8 (better known as KCNA10), Kv1 channels conduct significant current at voltages below spike threshold and hence contribute to the subthreshold properties of neurons. Kv1 channels are also highly sensitive to 4-AP with IC_{50}s for all Kv1 homomeric channels, except Kv1.8 below 1 mM. Their prominent (and in some cases predominant) expression in axons is another important feature of several members of this subfamily, which we address further in later sections.

Kv1.4 channels recover slowly from inactivation

In contrast to A-type K^+ channels of the Kv4 subfamily, which recover fast from inactivation at negative potentials (see the Kv4 subfamily section), K^+ channels containing Kv1.4 proteins recover from inactivation very slowly. The Kvβ subunits do not accelerate the rate of recovery from inactivation. As a result, one would expect cumulative inactivation of these channels during repetitive activity. It has been suggested that cumulative inactivation of these channels mediates activity-dependent spike broadening in hippocampal mossy fibre boutons (MFBs), the large presynaptic terminals of the axons of the granule cells of the dentate gyrus (Geiger and Jonas, 2000). In these terminals the action potential is very brief during low-frequency stimulation but is prolonged up to threefold during high-frequency stimulation. In turn, prolongation of the presynaptic spike leads to an increase in the number of Ca^{2+} ions entering the terminal per action potential and this to potentiation of evoked excitatory postsynaptic currents at MFB-CA3 pyramidal cell synapses.

Kvβ subunits

The Kvβ subunits act as chaperones during channel biosynthesis, which is a key function of these proteins, facilitating proper trafficking to and expression in the cell surface; an effect first described for the interaction of Kvβ2 with Kv1.4. (McCormack *et al.*, 1995; Shi *et al.*, 1996; Nagaya and Papazian, 1997). Additionally, by providing an extrinsic N-type inactivation domain Kvβ1 and Kvβ3 proteins can accelerate the inactivation of Kv1.4-containing channels or induce fast inactivation in otherwise non-inactivating or slowly inactivating Kv1 channels (except for Kv1.3, for which inactivation is not affected by Kvβ1 and Kvβ3 and for Kv1.6, for which inactivation is not affected by Kvβ1) (Rettig *et al.*, 1994; Pongs *et al.*, 1999). Kv1.6 proteins contain an N-type inactivation prevention domain (NIP) that prevents inactivation of channels containing one or more Kv1.6 subunits (Roeper *et al.*, 1998; Pongs *et al.*, 1999). On the other hand, Kvβ2, which lacks an N-terminal inactivation domain, accelerates inactivation only when Kv1.4 subunits form part of the Kv1 channel complex, probably via interactions with the intrinsic Kv1.4 inactivation domain (McCormack *et al.*, 1995; Coetzee *et al.*, 1999). Therefore, the inactivating properties of Kv1 channels are diversified by the Kvβ subunits. In some cases the Kvβ subunits may also shift the voltage dependence of activation in the hyperpolarizing direction (England *et al.*, 1995; Pongs *et al.*, 1999).

The similarity of the amino acid sequence of the Kvβ subunits with aldo-keto reductases continues to be an intriguing feature. The crystal structure of a Kvβ subunit revealed that the protein has a structural fold similar to that of aldo-keto reductases, and that one molecule of the co-factor NADPH (reduced nicotinamide adenine dinucleotide phosphate) was bound to the Kvβ (Gulbis *et al.*, 1999, 2000). A recent study (Weng *et al.*, 2006), provides data suggesting that Kvβ proteins may actually function as redox enzymes. They found that Kvβ2 can use 4-cyanobenzaldehyde as a substrate and reduce it to an alcohol. The catalytic reaction was found to be very slow, suggesting that this might not be the primary function of Kvβ proteins. It has been raised that Kvβs endow Kv1 channels with a sensor module to detect the metabolic state of the cell. This hypothesis rests on the observation that application of the substrate to the cytoplasmic side of the cell influenced the rate of inactivation of Kv1.4–Kvβ2 complexes (Weng *et al.*, 2006; Heinemann and Hoshi, 2006). Unrelated to the Kvβ subunits, KCNA4B is a protein that interacts specifically with KCNA10 (Kv1.8), a Kv1 subunit with a putative cyclic nucleotide-binding domain at the C-terminus, functionally related to both the Kv and the cyclic nucleotide-gated cation channels. KCNA4B increased the KCNA10 current, and increased its sensitivity to activation by cAMP (Tian *et al.*, 2002).

Kv2 (Shab) subfamily

As described below, Kv2 channels can activate over a wide range of voltages, depending on their phosphorylation state and subunit composition. *In vitro* Kv2 subunits have been shown to heteromultimerize with subunits of the Kv5, Kv6, Kv8 and Kv9 subfamilies, which, as mentioned in the introductory section, by themselves do not form conducting channels. These associations affect the voltage-dependence and the kinetics of activation and inactivation

of the channels, as well as their pharmacological properties (Table 2.1.1.1). For example, co-expression of Kv9.1 with Kv2.1 in *Xenopus* oocytes slows the rate of activation, causes a hyperpolarizing shift in the voltage-dependence of activation and inactivation and changes sensitivity to TEA, compared to Kv2.1 subunits alone (Richardson and Kaczmarek, 2000). The molecular composition of channel complexes in native tissues and its relationship to the diversity of I_K in neurons and other cells remains to be investigated.

Kv3 (Shaw) subfamily

Kv3 proteins express voltage-gated channels with unusual properties in heterologous expression systems. In comparison to the currents expressed by other Kv subunits, Kv3 currents become apparent at more depolarized potentials (more positive than −20 to −10 mV) i.e. more depolarized than other known voltage-gated K+ channels by at least 10–20 mV (Coetzee *et al.*, 1999; Table 2.1.1.1).

Another property of the currents mediated by Kv3 channels distinguishing them from other Kv channels is their fast rate of *deactivation* upon repolarization, which is significantly faster than that of other voltage-gated K+ channels. The difference in the deactivation rates of Kv3 and other neuronal voltage-gated K+ channels is approximately an order of magnitude or more; at −70 mV Kv3.1b currents deactivate with a time constant of <1 ms at room temperature (Rudy *et al.*, 1999, 2001). The rightward shifted voltage-dependence and fast deactivation rates of Kv3 channels are likely related properties that probably reflect instability of the channel's open state at voltages near the resting potential (Rudy *et al.*, 1999, 2001). This may be the result of the presence of specific amino acid residues in the S5 transmembrane domain and the S4-S5 linker of Kv3 proteins (Shieh *et al.*, 1997; Kanevsky *et al.*, 1999). Of great interest is the fact that the S4–S5 sequence of Kv3 proteins is 100 per cent identical in teleost fish, birds, frogs, and mammals, including humans. This remarkable degree of conservation suggests that evolution in the vertebrate line has acted to conserve these properties of Kv3 channels, and that these properties might be linked to special physiological roles of these channels.

Kv3 currents activate relatively fast, faster than Kv2, but at a somewhat slower rate than the fastest Kv1 currents. Kv3.1 and Kv3.2 subunits mediate slowly inactivating delayed rectifier type currents, while Kv3.4 proteins produce fast inactivating A-type currents (Rudy *et al.*, 1999). Kv3.3 currents are inactivating in Xenopus oocytes with time constants of the order of tens of milliseconds (Vega-Saenz de Miera *et al.*, 1992). However, in mammalian cell lines used for heterologous expression the inactivation of Kv3.3 currents is slow and variable (Fernandez *et al.*, 2003; Vega-Saenz de Miera *et al.*, 1994; Rudy, 1999; Rae and Shepard, 2000). The reason for, and/or the significance of, these differences is unclear.

Deletion of the first 78 residues of Kv3.3 or the first 28 residues of Kv3.4 produces non-inactivating currents in *Xenopus* oocytes (Vega-Saenz de Miera *et al.*, 1994), suggesting that these proteins contain N-terminal inactivating domains that resemble those producing N-type inactivation of Kv1 channels (see the section N- and P/C-type inactivation of Kv channels). Both Kv3.3 and Kv3.4 have a methionine that can be aligned with the starting methionine of Kv3.1 and Kv3.2 proteins following these N-terminal inactivation domains. This poses the possibility that Kv3.3 proteins produce inactivating, or non-inactivating, currents depending on which methionine is used as translation start site (Vega-Saenz de Miera *et al.*, 1994; Fernandez *et al.*, 2003). Differences in the post-translational modulation of Kv3.3 proteins in oocytes and mammalian cells could also explain the differences in inactivating properties. The observation that the degree of inactivation of Kv3.3 and Kv3.4 currents expressed in heterologous cells is affected by oxidation and PKC-mediated phosphorylation supports this hypothesis (Covarrubias *et al.*, 1994; Vega-Saenz de Miera and Rudy, 1994).

The functional significance of the alternative splicing of Kv3 genes has not been systematically investigated, and in the cases in which it has been investigated, the currents expressed in heterologous expression systems by the isoforms of each Kv3 gene are very similar (Rudy *et al.*, 1999). It is possible that the alternatively spliced C-termini confer isoform-specific regulation by second messenger signalling systems and targeting to distinct neuronal compartments (Ponce *et al.*, 1997; McIntosh *et al.*, 1998; Ozaita *et al.*, 2002; Macica *et al.*, 2003; Deng *et al.*, 2005).

Kv4 (Shal) subfamily

Kv4 channels use novel inactivation mechanisms

Kv4 channels inactivate fast and almost completely, and also recover rapidly from inactivation. This fast recovery from inactivation distinguishes Kv4-mediated currents from other transient K^+ currents and is an essential feature of Kv4 channels, given their role in repetitively firing cardiac and neuronal cells. Compared to most inactivating voltage-gated K^+ channels in which open-state inactivation predominates, Kv4 channels are thought to inactivate preferentially from closed states (Jerng *et al.*, 2004). This implies that modest membrane depolarizations that may not open the channels can render them unable to open in response to a subsequent stronger membrane depolarization, producing cumulative inactivation upon repetitive stimulation (Aldrich *et al.*, 1979; Klemic *et al.*, 1998; Patil *et al.*, 1998). Moreover, the equilibrium between the open state and a pre-open activated state in Kv4 channels does not strongly favour residence in the open state even during depolarization. Therefore, preferential closed-state inactivation produces most of the inactivation that occurs during a depolarizing pulse and hence produces the transient character of the current (Jerng *et al.*, 2004). The molecular mechanisms of inactivation of Kv4 channels remain to be described since they do not seem to correspond to either the N- or the P/C-type inactivation described in N- and C-type inactivation of Kv proteins in the introductory section (reviewed extensively in Jerng *et al.*, 2004).

Kv4 auxilliary subunits

Although the currents mediated by Kv4 channels (particularly those formed by Kv4.2 and Kv4.3 subunits) expressed in heterologous systems resemble the biophysical and pharmacological properties of the fast I_{to} in cardiac myocytes and the native I_{SA} in neurons, they display important functional differences with the native currents (Nadel *et al.*, 2001). The discrepancy between native and heterologously-expressed currents could be explained by post-translational modifications and/or by the existence of auxiliary subunits that modulate the activity of the pore-forming subunits, thus modifying the properties of native channels.

Experiments in *Xenopus* oocytes injected with size fractionated rat brain mRNA found evidence that the channels underlying I_A contain multiple components (Rudy, 1988). The channels expressed in *Xenopus* oocytes by a fraction later shown to have Kv4 mRNAs (Nadel *et al.*, 2001) could be modified by factors encoded

in a fraction of brain mRNA of small size (Rudy, 1988; Chabala *et al.*, 1993; Serodio *et al.*, 1994). Using yeast-two hybrid screens, a family of cytosolic Ca^{2+}-binding EF-hand proteins (200–250 amino acids long), belonging to the neuronal calcium sensor family (NCS) called KChIPs (for Kv channel-interacting proteins) was identified. These proteins bind to the N-terminal region and the T1 domain of Kv4 proteins, facilitate the trafficking, subunit assembly, and surface expression of channel complexes (An *et al.*, 2000; Kuo *et al.*, 2001; Rosati *et al.*, 2001; Holmqvist *et al.*, 2002; Kunjilwar *et al.*, 2004). In addition, KChIPs modify the electrophysiological properties of the channels expressed by Kv4 proteins in heterologous expression systems, which inactivate more slowly and recover from inactivation with faster kinetics in the presence of KChIPs (An *et al.*, 2000; Rosati *et al.*, 2001; Holmqvist *et al.*, 2002). There are four known KChIP genes (*KChIP1–4*). All four exert similar effects on Kv4 channels except for the alternative spliced version *KChIP4a*, which prevents fast inactivation (An *et al.*, 2000; Rosati *et al.*, 2001; Holmqvist *et al.*, 2002).

Although there is sufficient evidence that native Kv4 channels in heart and brain contain KChIPs, the currents expressed by Kv4 proteins in the presence of KChIPs still differ significantly from the native I_{to} in cardiac myocytes (Deschenes and Tomaselli, 2002), or from the I_{SA} recorded in the somata of several neuronal populations that prominently express these subunits (Nadal *et al.*, 2006). As mentioned above, KChIPs slow down the inactivation of Kv4 channels (An *et al.*, 2000; Nakamura *et al.*, 2001), but the native I_{SA} in most neurons inactivates faster, and not slower, than the currents expressed by Kv4.2 and Kv4.3 subunits alone (Nadal *et al.*, 2006). The kinetic differences between the native current in several neuron types where the I_{SA} has been well characterized and Kv4 + KChIP currents are considerable (Nadal *et al.*, 2006). Likewise, the I_{SA} activates at more negative voltages than the currents expressed by Kv4 subunits in the presence or absence of KChIPs in many neurons (Nadal *et al.*, 2006).

The dipeptidyl aminopeptidase-like protein (DPPX, also known as DPP6), a previously identified protein of unknown function (Wada *et al.*, 1992), was identified as an integral membrane glycoprotein associated with Kv4 subunits in native I_A channels (Nadal *et al.*, 2006) using immunoaffinity purification. DPPX increases the rate of inactivation of Kv4 currents when co-expressed with Kv4 subunits in heterologous systems, considerably decreases the time for the currents to reach a maximum (time to peak), and increases the rate of recovery from inactivation. In addition, DPPX produces a very large (28 mV) negative shift in the conductance-voltage relation and a 16 mV negative shift in the voltage-dependence of steady-state inactivation (Nadal *et al.*, 2006). These shifts in voltage-dependence are likely due to effects of DPPX on the Kv4 channel voltage-sensor, since similar shifts are observed on the gating currents recorded from cells expressing Kv4 and DPPX proteins (Dougherty and Covarrubias, 2006). Like KChIPs, DPPX increases the trafficking of Kv4 proteins to the cell surface, contributing to the large increase in current magnitude observed in the presence of this auxiliary subunit. DPPX also increases the single channel conductance of Kv4.3 channels by 50 per cent, a factor that may also contribute to the observed increases in total current magnitude (Rocha *et al.*, 1994).

At least five alternative spliced isoforms of DPPX (DPPX-S, DPPX-L, DPPX-K, DPPX-D, and DPPX-E) exist (Nadal *et al.*, 2006). All of them share the same putative transmembrane domain, a long C-terminal extracellular domain with a hydrolase fold and a β-propeller (Figure 2.1.1.1), but are divergent in the short intracellular amino-terminal sequence (Wada *et al.*, 1992; de Lecca *et al.*, 1994; Kin *et al.*, 2001). All of those isoforms that have been tested have qualitatively similar effects on Kv4 channels expressed in heterologous cells; however, each has a unique pattern of expression in brain (Nadel *et al.*, 2006). In addition, products of a related gene known as DPP10 (which also produces several N-terminal spliced isoforms) were recently shown to have effects on Kv4 channels resembling those of DPPX (Jerng *et al.*, 2004; Ren *et al.*, 2005; Zagha *et al.*, 2005; Takimoto *et al.*, 2006). Like DPPX, DPP10 mainly expresses in brain tissue, however some DPP10 spliced isoforms are also expressed in the adrenal gland and pancreas (Takimoto *et al.*, 2006). KChIP and DPPX proteins are co-precipitated by antibodies to Kv4 proteins, suggesting that the native channel might be a ternary complex of all three types of subunit (Nadal *et al.*, 2003). Consistent with this view, the currents expressed in heterologous cells by Kv4 proteins in the presence of KChIPs and DPPs closely resemble the properties of native channels (Nadal *et al.*, 2003; Jerng *et al.*, 2005). Heterologously expressed channels containing Kv4 proteins and the auxiliary subunits inactivate and recover from inactivation fast as in neuronal I_{SA} channels. Moreover, they activate in the range of voltages seen *in vivo* (Nadal *et al.*, 2003; Jerng *et al.*, 2005). In addition DPP proteins may contribute to the diversity of I_{SA} in neurons. For example, currents containing DPP10 inactivate much faster than currents containing DPPX (Zhaga *et al.*, 2005).

DPPX and DPP10 belong to a family of dipeptidyl-aminopeptidase enzymes that have amino acid replacements in the catalytic active site and lack enzymatic activity. Enzymatic activity is not recovered when the consensus active-site sequence present in functional dipeptidyl proteases is restored by mutagenesis, suggesting other differences that prevent catalysis (Kin *et al.*, 2001). Instead, these DPPs seem to have evolved to function as K⁺ channel associated proteins.

DPPX and DPP10 are related to DPPIV or CD26 (33 per cent identity and ~50 per cent similarity), a cell adhesion protein with dipeptidyl-aminopeptidase activity that has important roles in T cell activation, metabolism of peptide hormones and cell adhesion (reviewed in De Meester *et al.*, 1999; Hildebrandt *et al.*, 2000; Gorrell *et al.*, 2001). The cell adhesion properties of DPPIV proteins are largely mediated by their β propeller, a protein structure that favours protein–protein interactions. It has been suggested that DPPX and DPP10 proteins may confer targeting or cell adhesion properties to Kv4 channels through their homologous extracellular domains because of their similarity with DPPIV (Nadal *et al.*, 2003). This is reminiscent of the cell adhesion properties conferred upon Na⁺ channels by the extracellular immunoglobulin-like domain of their β subunits (Ratcliffe *et al.*, 2001; Isom, 2002).

Interactions of DPPs with the Kv4 voltage sensor and KChIPs with inactivation domains

Mutagenesis studies suggest that the effects of DPPX and DPP10 on Kv4 channels depend on interactions between the transmembrane domain of the DPPs and the channel, possibly the S1 and S2 membrane spanning helices (Zagha *et al.*, 2005; Ren *et al.*, 2005). This is of great interest, given the large effects of DPP proteins on voltage-dependent gating. It is easier to imagine a transmembrane interaction between the pore-forming and the auxiliary subunits if the voltage-sensor is not part of a compact protein, as in the structural model of Kv1 proteins described by MacKinnon and colleagues (Figure 2.1.1.1).

Wang *et al.* (2007) and Pioletti *et al.* (2006) recently described a co-crystal structure of the complex formed between the Kv4.3 N-terminus and KChIP1. The structure shows that each KChIP1 molecule (four in the complex) interacts with two neighbouring Kv4.3 N-termini in a 4:4 manner, forming a cross-shaped octamer. The proximal N-terminal peptide of one Kv4.3 N-terminus is sequestered by its binding to an elongated hydrophobic groove on the surface of KChIP1. At the same time, each KChIP1 binds to the T1 domain of an adjacent Kv4.3 subunit to stabilize the tetrameric Kv4.3 complex. This sequestration of the N-terminus likely explains the slowing of inactivation produced by KChIPs (Pioletti *et al.*, 2006; Wang *et al.*, 2007). The authors generated a model of the Kv4.2–KChip1 complex based on their findings and the Kv1.2 structure. We used this model as a platform for the ternary Kv4–KChIP-DPP model shown in Figure 2.1.1.1.

Modulation of Kv4 channel function by phosphorylation

Modulation of Kv4 channels is not limited to the effects of association with auxiliary proteins. Kv4.2 proteins have been shown to be phosphorylated by extracellular signal-regulated kinase (ERK)-mitogen-activated protein kinase (MAPK) and by PKA; post-translational modifications that modulate current density and voltage dependence (Hoffman *et al.*, 1998; Adams *et al.*, 2000; Anderson *et al.*, 2000; Yuan *et al.*, 2002). Heterologously expressed Kv4.2 channels are also affected by PKC activity. However, it is unclear whether this is mediated through direct PKC phosphorylation of Kv4.2 or through ERK activation (Nakamura *et al.*, 2001). Calcium-calmodulin dependent kinase II (CaMKII) also phosphorylates Kv4.2, a modulation that regulates surface expression of Kv4.2 channels (Varga *et al.*, 2004).

These modulations have considerable physiological significance. Expression in hippocampal neurons of an interfering Rap1 mutant decreased phosphorylation of a membrane-associated pool of p42/44MAPK, impaired cAMP-dependent LTP in the hippocampal Schaffer collateral pathway, decreased complex spike firing, and reduced the p42/44MAPK-mediated phosphorylation of Kv4.2 channels, resulting in impaired spatial memory and context discrimination (Morozov *et al.*, 2003).

Furthermore, it was recently observed that in spinal cord dorsal horn neurons, A-type currents are also modulated by ERKs that are known to mediate central sensitization during inflammatory pain. The excitability of dorsal horn neurons is enhanced and the I_{SA} is absent in Kv4.2 knockout mice. This was accompanied by elevated sensitivity to tactile and thermal stimuli, and the ERK-mediated modulation of excitability and the ERK-dependent forms of pain hypersensitivity were absent in Kv4.2 (–/–) mice (Hu *et al.*, 2006).

2.1.1.4 Cellular and subcellular localization

Kv1 (Shaker) subfamily

Kv1 is widely expressed throughout the body in smooth muscle cells, astrocytes, oligodendrocytes and neurons in organs as diverse as the kidney, heart, skeletal muscle and brain. Table 2.1.1.1 provides more specific details of the expression pattern of each of the Kv1 isoforms.

Kvβ subunits

All Kvβ subunits are expressed mainly in brain (Leicher *et al.*, 1998; Geiger and Jonas, 2000). Using antibodies that may not distinguish among splice isoforms (Rhodes *et al.*, 1997), found that Kvβ1.1 and Kvβ2 are widely distributed in rat brain. Kvβ2 was much more abundant, and a significantly larger portion of the total brain pool of Kv1-containing channel complexes was found to be associated with Kvβ2 rather than with Kvβ1. Kvβ2 codistributes extensively with Kv1.1 and Kv1.2 in all brain regions examined and was strikingly co-localized with these α-subunits in the juxtaparanodal region of nodes of Ranvier as well as in the axons and terminals of cerebellar basket cells. On the other hand, Kvβ1 was found to co-localize extensively with Kv1.1 and Kv1.4 in cortical interneurons, in the hippocampal perforant path and mossy fibre pathways, and in the globus pallidus and substantia nigra.

The distribution of Kvβ3 mRNA was studied in rat brain by *in situ* hybridization and in human brain using region-specific Northern blots. Kvβ3 shows a more restricted expression pattern than Kvβ1 and Kvβ2. In humans it is present mainly in cerebellum. In rat, prominent expression was found in the olfactory bulb, thalamic nuclei, and cerebellum (Heinemann *et al.*, 1995; Leicher *et al.*, 1998). mRNA for KCNA4B another auxillary protein that specifically interacts with Kv1.8 is found in heart, skeletal muscle, kidney, small intestine, and placenta and is notably absent in brain.

Kv2 (Shab) subfamily

The two known members of this subfamily, Kv2.1 and Kv2.2, are both widely expressed in brain tissue (Hwang *et al.*, 1993). Kv2 channels are believed to be present in many, if not all, principal cells and interneurons throughout the brain (Trimmer, 1991; Hwang *et al.*, 1993; Maletic-Savatic *et al.*, 1995; Du *et al.*, 1998). In contrast to other Kv subfamilies, it is possible that the two Kv2 subunits do not form heteromeric channels. For example, immunolocalization studies show that the two Kv2 subunits produce distinct subcellular staining patterns, including in cells that express both proteins (Hwang *et al.*, 1993; Malin and Nerbonne, 2002). Specifically, Kv2.1 channels are found in high-density large clusters in the soma and proximal dendrites, whereas Kv2.2 subunits typically present diffuse localization in somata and neuropil (Trimmer, 1991; Hwang *et al.*, 1993; Du *et al.*, 1998).

Modulation of Kv2.1 channel clustering

We mentioned earlier that Kv2.1 proteins are localized in large clusters in the membranes of somas and proximal dendrites of neurons. Clustering is reproduced in Kv2.1-transfected HEK-293 cells. The clustering mechanisms of these channels have been the subject of intense investigation (Misonou *et al.*, 2005a, b; O'Connell *et al.*, 2006; Mohapatra and Trimmer, 2006). Residues in the intracellular C-terminus are necessary and sufficient for the clustering of Kv2.1 channels (Scannevin *et al.*, 1996; Lim *et al.*, 2000). Furthermore, evidence from immunoelectron microscopy suggests that Kv2.1 clusters are organized in specific subcellular patterns, such as opposite astrocytic processes or inhibitory synapses in cortex and hippocampus (Du *et al.*, 1998) and opposed to cholinergic synapses on spinal motoneurons (Muennich and Fyffe, 2004).

Kv2.1 proteins are extensively phosphorylated in neurons (Misonou *et al.*, 2005a, b), as well as in transfected cultured mammalian cells (Shi *et al.*, 1994) and the phosphorylation state of Kv2.1 subunits influences channel localization and clustering. Large negative shifts in the voltage-dependence of activation and inactivation accompany these changes. Furthermore, phosphorylation, localization, and channel properties are dynamically regulated

by neuronal activity (Misonou *et al.*, 2004a, b). Kv2.1 channels were found to be dephosphorylated relative to control animals, and no longer clustered in pyramidal neuron membranes following kainate-induced seizures in mice (Misonou *et al.*, 2004a, b). Glutamate exposure in cultured hippocampal pyramidal neurons also caused dephosphorylation of Kv2.1 subunits, fragmentation of clusters, and hyperpolarizing shifts in the voltage-dependence of I_K (Misonou *et al.*, 2004a, b). Kv2.1 protein dephosphorylation is mediated by Ca^{2+} influx and the concomitant activation of the Ca^{2+}-dependent phosphatase calcineurin (Misonou *et al.*, 2005a). Following hypoxia and chemical ischemia *in vivo* and *in vitro* (Misonou *et al.*, 2005a, b) and cholinergic stimulation *in vitro* (Mohapatra and Trimmer, 2006), similar modulations of Kv2.1 expression and activity were found. Recently discovered multiple sites of dephosphorylation by calcineurin suggest potential for graded phosphorylation resulting in partial effects (Park *et al.*, 2006). As observed with dephosphorylation, hyperpolarizing shifts of voltage-dependence of activation and inactivation could produce an increase or decrease in I_K. This would decrease or increase membrane excitability, respectively, depending on the resting membrane potential and whether the effect on activation or inactivation is dominant (Misonou *et al.*, 2005a, b). Considering their strong and widespread expression, and the magnitude of the shifts in voltage dependence, dephosphorylation of Kv2.1 subunits may be an effective and pervasive homeostatic mechanism by which neurons are able to alter excitability in response to their chemical environment and neuronal activity.

Kv3 (Shaw) subfamily

Three of the four known Kv3 genes (*Kv3.1–Kv3.3*) are prominently expressed in specific subpopulations of CNS neurons. In contrast, Kv3.4 transcripts are weakly expressed in a few neuronal types in the brain, usually in neurons that also express other Kv3 genes (granule cells of the dentate gyrus in the hippocampal formation, cerebellar Purkinje cells and unidentified neurons in the pontine nucleus (Weiser *et al.*, 1994). However, they are much more abundant in skeletal muscle and sympathetic neurons (Weiser *et al.*, 1994; Veh *et al.*, 1995; Dixon and McKinnon, 1996; Rudy *et al.*, 1999). Despite this, Kv3.4 proteins might be important components of Kv3 channels in some neurons, even when expressed at low levels, since a single Kv3.4 subunit is capable of enabling fast inactivating properties to a Kv3 tetrameric channel (Weiser *et al.*, 1994).

The distribution and developmental regulation of Kv3 gene transcripts and protein products in the rat and mouse CNS has been described in some detail (Perney *et al.*, 1992, Perney and Kaczmarek, 1997; Weiser *et al.*, 1994, 1995; Moreno *et al.*, 1995; Du *et al.*, 1996; Sekirnjak *et al.*, 1997; Chow *et al.*, 1999; Grigg *et al.*, 2000; Atzori *et al.*, 2001; Ozaita *et al.*, 2002; Tansey *et al.*, 2002; Brooke *et al.*, 2004; Dodson and Forsythe, 2004; McDonald and Mascagni, 2006; Chang *et al.*, 2007). Each Kv3 gene exhibits a unique pattern of expression in the CNS, however transcripts of two or more Kv3 genes overlap in many neuronal populations, suggesting a significant potential for heteromultimer formation (Weiser *et al.*, 1995). Co-immunoprecipitation has confirmed heteromultimer formation among Kv3 proteins in native neurons (Chow *et al.*, 1999; Hernandez-Pineda *et al.*, 1999). While Kv3.1–Kv3.3 proteins can be found together in many CNS areas, they are present in selective cell populations. For example, they are found in cortical structures mainly in subpopulations of cortical GABAergic interneurons, including the so-called fast-spiking (FS) interneurons (mainly Kv3.1 and Kv3.2). In different proportions, all three are also expressed in GABAergic cells in other forebrain structures including the globus pallidus, zona incerta, basal forebrain, and GABAergic interneurons (but not medium spiny neurons) in the caudate. Kv3.1 and Kv3.3 are co-expressed in many structures in the brainstem, including all auditory, and other sensory, processing nuclei, some of which also contain Kv3.2. Kv3.1 and to a lesser degree Kv3.3 are expressed in granule cells in the cerebellum and Kv3.3 (but not Kv3.1) in Purkinje cells. All three are present in the deep cerebellar nuclei. Kv3.1 and Kv3.3 are also expressed in several neuronal populations in the spinal cord.

Kv3 proteins are present in somatic and/or axonal membranes. In some instances the axonal expression is dominant. For example, for Kv3.2 in thalamic relay neurons (Moreno et al., 1995) and Kv3.3 and Kv3.4 in hippocampal granule cells (Veh et al., 1995; Chang et al., 2007), the channel proteins are predominantly, if not exclusively, expressed in axons. Kv3.1b proteins, but apparently not other Kv3 subunits, are also expressed in nodes of Ranvier in myelinated axons in the CNS where they may contribute to action potential repolarization during saltatory conduction (Devaux et al., 2003; Chang et al., 2007).

Kv3 channel expression in dendrites is usually limited to the proximal dendrite (Weiser et al., 1995; Perney et al., 1997; Sekirnjak et al., 1997; Chow et al., 1999). However, Kv3.3 proteins are also prominent in distal dendrites in cerebellar Purkinje cells (Martina et al., 2003; McMahon et al., 2004; Chang et al., 2007), resembling the pattern of expression in pyramidal cells of the electrosensory lateral line lobe of weakly electric apteronotid fish (McHaffey et al., 2006). The C-terminal domain of the fish protein appears to target the Kv3.3 channels to dendrites (Deng et al., 2005). This sequence corresponds to the C-terminus of the mouse Kv3.3 spliced version reported by Goldman-Wohl et al. (1994).

Kv4 (Shal) subfamily

Only two of three Kv4 genes, Kv4.2 and Kv4.3, are prominently expressed in the rodent CNS. Kv4.1 is expressed very weakly and in only a few neuronal populations, such as the auxiliary olfactory bulb, granule cells of the olfactory bulb, hippocampus, dentate gyrus, and cerebellar cortex, possibly in Purkinje cells (Serodio and Rudy, 1998; Table 2.1.1.1). It has been suggested that Kv4.1 is more abundant in humans (Isbrandt et al., 2000). In contrast to the Kv1 and Kv3 subfamilies, where multiple subfamily members are expressed in many neuronal populations, Kv4.2 and Kv4.3 mRNAs often display a differential, and sometimes even reciprocal, pattern of expression, although there is overlap of expression in some neuronal populations (Serodio and Rudy, 1998).

2.1.1.5 Pharmacology

Kv1 (Shaker) subfamily

Kv1 channels mediate dendrotoxin-sensitive K$^+$ currents (I$_D$)

Three Kv1 subunits, Kv1.1, Kv1.2, and Kv1.6, express homomultimeric channels that are blocked by the mamba snake toxins dendrotoxin I (DTX-I) and α-dendrotoxin (α-DTX), (reviewed in Harvey and Robertson, 2004). In heteromeric channels, a single Kv1.1, Kv1.2 or Kv1.6 protein is sufficient to confer DTX-sensitivity (Marom and Levitan, 1994). This characteristic does not apply to

other channel blockers such as TEA (Hopkins, 1998). Thus, Kv1 channels are the mediators of the voltage-dependent current known as the D-type current or I$_D$ (Table 2.1.1.1).

Kv2 (Shab) subfamily

Table 2.1.1.1 summarizes the pharmacology of Kv2 family members which are blocked by high mM concentrations of TEA and 4-AP.

Kv3 (Shaw) subfamily

No specific blockers of Kv3 channels are known that could be used to isolate and characterize Kv3 currents in native cells or to investigate the functional consequences of channel blockade. Blood-depressing substance (BDS) toxins from the sea anemone *Anemonia sulcata* were initially reported to specifically block Kv3.4 channels, but were recently shown to block, at similar concentrations, other Kv3 channels (Yeung et al., 2005). Moreover, these toxins are gating modifiers. They shift the voltage-dependence of the channels in the depolarizing direction, tend to produce only a partial channel block at physiological membrane potentials, and modify channel properties. Kv3 currents have been isolated in many studies using the classical K$^+$ channel blockers TEA and 4-AP. Low (<1 mM) TEA has been particularly useful in several preparations. In addition to all Kv3 members, these concentrations of TEA block only a few known K$^+$ channels, including Kv1.1 homomultimeric channels, BK Ca^{2+}-activated K$^+$ channels, and KCNQ2 channels. Specific toxins that block these non-Kv3 TEA-sensitive channels (dendrotoxin, charybdotoxin or iberotoxin and linopirdine, respectively) can be used to inhibit these channels and isolate the Kv3 currents or Kv3-mediated effects. However, the pharmacological isolation of the Kv3 current using this strategy should be used with caution. Low TEA concentrations (1 mM) could block significant amounts of current mediated by channels with a higher IC$_{50}$ for TEA (such as Kv2 or Kv7 channels, for example) if they are very abundant in a particular preparation.

Kv4 (Shal) subfamily

Table 2.1.1.1 outlines the pharmacology of each of the Kv4 family members. All are all sensitive to block by 5-10 mM 4-AP and exhibit differential sensitivity to block by toxins such as the phrix-otoxins and *Buthus martensi* toxin (sBmTX3).

2.1.1.6 Physiology

Kv1 (Shaker) subfamily

Except for Kv1.5, Kv1.7 and Kv1.8, other Kv1 proteins are widely and prominently expressed in neurons. It is likely that most, if not all, neurons in the CNS and PNS express several Kv1 proteins and, as described above, it is these Kv1 channels that are the mediators of the I$_D$ current.

I$_D$ was originally described in CA1 pyramidal neurons as an important subthreshold-operating, slowly-inactivating K$^+$ current that was highly sensitive to 4-AP, was responsible for delayed excitation, and contributed to determining spike threshold and affect other subthreshold properties of these neurons (Storm, 1998). It was later demonstrated that this current is the target of DTX (Wu and Barish, 1992). Given heteromeric association of Kv1 proteins, the widespread expression of Kv1.1, Kv1.2, and Kv1.6 proteins in the nervous system (Table 2.1.1.1), along with the total number of

Kv1 pore-forming subunits and Kvβs, the possible combinations of subunits producing channels containing one of the Kv1 subunits conferring DTX sensitivity is extremely large, resulting in great diversity of I_D with variations in a number properties, including inactivation rates. For example, a channel containing Kv1.1, Kv1.4, and Kvβ1, a combination that has been detected in hippocampus (Rhodes et al., 1997), will produce a fast inactivating DTX-sensitive current. Inactivation would be slower if there is Kvβ2 instead of Kvβ1, and signficantly so if the channels do not contain Kv1.4. On the other hand, Kv1.2 homomultimers activate more slowly than other Kv1.1–Kv1.6 channels, and it is expected that the presence of Kv1.2 subunit would lead to I_D with slower activation kinetics.

Kv1 channels in axons

Although some Kv1 proteins might be expressed in the somatodendritic compartment of neurons, particular attention has been given to the prominent location of several Kv1 channel subunits in axons, where they display unique localization (Figure 2.2.2).

Kv1.1 and Kv1.2 channels have been localized to a distal part of the axon initial segment where they co-localize with ankyrin G and may contribute to regulating action potential initiation (Goldberg and Rudy, 2006; Inda et al., 2006; Van Wart et al., 2007). Kv1 channels are also prominent in presynaptic terminals. In two areas where this has been investigated, the calyx of Held in the auditory brainstem and the terminals of cortical GABAergic interneurons, the Kv1 current does not seem to contribute to the repolarization of the action potential at the terminal (Ishikawa et al., 2003; Goldberg et al., 2005). Instead, Kv1 channels reduce the excitability of the nerve terminal, thereby preventing aberrant transmitter release (Sheng et al., 1994; Dodson et al., 2002, 2003; Dodson and Forsythe, 2004; Ishikawa et al., 2003; Goldberg et al., 2005). This role of Kv1 channels may depend on their localization to the preterminal axon (Dodson et al., 2003; Dodson and Forsythe, 2004).

Kv1.1, Kv1.2 and Kv1.6 are present in the juxtaparanodal region in myelinated axons (Wang et al., 1993; Lai and Jan, 2006; Mert and Rasband, 2006). Whilst their function here is not clear, it has been suggested that they may act to dampen re-entrant excitation (Scherer, 1999). This juxtaparanodal localization is affected by demyelination (Wang et al., 1995; Vabnick et al., 1999; Devaux and Scherer, 2005; Sasaki et al., 2006). Myelin loss results in the exposure and dispersion of these juxtaparanodal K^+ channels, effects which contribute to neurological symptoms (Sinha et al., 2006).

Clarifying the molecular basis for the axonal targeting of these proteins has been the subject of some effort. The T1 domain, located in the N terminus of Kv proteins prior to the first membrane spanning helix, responsible for determining subfamily specific recognition of Kv proteins and hence subfamily-specific heteromultimerization, has been found to be essential for axonal targeting of Kv1 proteins (Gu et al., 2003; Lai and Jan, 2006). The T1 domain is also the major site of association between Kv1 pore-forming and Kvβ subunits. Of these accessory subunits Kvβ2 was recently shown to be responsible for targeting Kv1 channels to the axon. This function depends on their ability to associate with the microtubule (MT) plus-end tracking protein (+TIP) EB1 (Gu et al., 2006).

Kvβ subunits

Mice with Kvβ1 and Kvβ2 gene deletions have been developed. Knockout of Kvβ1 reduced inactivation of A-type current, reduced the slow after-hyperpolarization and caused frequency-dependent spike broadening in hippocampal CA1 pyramidal neurons. It also led to impairments in memory (Geise et al., 1998). Kvβ2-null mice have reduced life spans, occasional seizures, and cold swim-induced tremors similar to those observed in Kv1.1-null mice (McCormack et al., 2002).

Kv2 (Shab) subfamily

Kv2 channels mediate the delayed rectifier current I_K

Kv2.1 and Kv2.2 express delayed rectifier currents which activate more slowly in heterologous expression systems than the delayed rectifiers expressed by Kv1 or Kv3 subunits. The Kv2 currents are also less sensitive to 4-AP and are blocked by millimolar concentrations of TEA (see Table 2.1.1.1). The current view is that Kv2 channels mediate the TEA-sensitive, '4-AP-insensitive' delayed rectifier K^+ current recorded in many types of neurons and typically called I_K (Rudy, 1988). Evidence supporting this comes from studies in several neuronal populations and using different experimental strategies, including hippocampal pyramidal cells (Du et al., 1998; Matina et al., 1998; Murakoshi and Trimmer, 1999) using blocking antibodies, antisense oligonucleotide treatment or pharmacology and RT-PCR, respectively; neocortex (Pal et al., 2003), SCG cells (Malin and Nerbonne, 2002) and globus pallidus (Baranauskas et al., 1999), using pharmacology and RT-PCR; myocytes using pharmacology (Malysz and Huizinga, 1999); and pancreatic β-cells (MacDonald et al., 2001). It is likely that I_K is mediated by Kv2 channels in many, if not all neurons. Overall, the contributions of Kv2.1 subunits to neuronal excitability have been investigated more than those of Kv2.2, and the two may have different functions, but in at least one study (Malin and Nerbonne, 2002) it was shown that both contribute to I_K.

Due to the voltage-dependence and slow kinetics of activation, deactivation, and inactivation of these channels as compared to Kv1, Kv3, and Kv4s (Table 2.1.1.1), they are believed to regulate excitability and Ca^{2+} influx during repetitive firing at higher frequencies, rather than regulating the repolarization of single action potentials (Du et al., 1998). As discussed above, their role in neuronal excitability will depend strongly on the state of phosphorylation of the channels and precise subunit composition due to association with Kv5, Kv6, Kv8 and Kv9 proteins (see Table 2.1.1.1).

Kv3 (Shaw) subfamily

Due to their positively shifted voltage dependence of activation and fast activation/deactivation kinetics, Kv3 channels are activated during the repolarizing phase of the action potential and are quickly deactivated when the spike is terminated. Therefore, Kv3 currents facilitate action potential repolarization more specifically than other voltage-gated K^+ channels that activate more negatively and deactivate more slowly, and therefore will open earlier during the spike and remain active during part of the interspike interval (Erisir et al., 1999; Rudy et al., 1999; Rudy and McBain, 2001; Lien, 2003; Martina, 2007). Hence, when present in sufficient amounts, Kv3 channels dictate action potential duration without compromising action potential threshold, rise time, or magnitude, and more importantly, without contributing current that would increase the duration of the refractory period (Lenz et al., 1994; Vega-Saenz de Miera, 1994; Moreno et al., 1995; Weiser et al., 1995; Massengill et al., 1997; Perney and Kaczmarek, 1997; Sekirnjak et al., 1997; Du et al., 1998; Wang et al., 1998; Erisir et al.,1999; Kanevsky and

Aldrich, 1999; Gurantz *et al.*, 2000; Lau *et al.*, 2000; Vincent *et al.*, 2000; Wigmore and Lacey, 2000; Martina *et al.*, 2003).

Kv3 channels are especially prominent in neurons with the ability to fire sustained trains of action potentials at high frequency or to follow high-frequency stimulation, as in cortical fast-spiking GABAergic interneurons and in neurons in the auditory brainstem. These channels have been shown in several preparations, including cortical fast-spiking GABAergic interneurons, cerebellar Purkinje cells, neurons in the subthalamic nucleus and globus pallidus, and in high-frequency firing neurons in the auditory brainstem to facilitate high-frequency firing. This role is likely the result of their function in spike repolarization: by increasing the rate of spike repolarization and keeping action potentials brief, Kv3 currents reduce the amount of Na^+ channel inactivation that occurs during the action potential and minimize the activation of K^+ channels with slow kinetics. Moreover, recovery from Na^+-channel inactivation and Kv3 channel deactivation is accelerated by the large, fast after hyperpolarization (fAHP) produced by the Kv3 current. By speeding up the recovery of Na^+ channel inactivation and by deactivating fast, Kv3 currents maximize restoring resting conditions (including membrane resistance) and prepare the membrane for a second spike soon after the first, thus minimizing the duration of the refractory period and allowing high-frequency firing (Perney *et al.*, 1992, 1997; Lenz *et al.*, 1994; Weiser *et al.*, 1994, 1995; Du *et al.*, 1996; Sekirnjak *et al.*, 1997; Wang *et al.*, 1998; Chow *et al.*, 1999; Hernandez-Pineda *et al.*, 1999; Rudy *et al.*, 1999, 2001; Atzori *et al.*, 2000; Lau *et al.*, 2000; Wigmore and Lacey, 2000; Macica *et al.*, 2001a, b; Parameshwaren *et al.*, 2001; Tansey *et al.*, 2002; Lien and Jonas, 2003; Ozaita *et al.*, 2004; Fernandez *et al.*, 2005; Kaczmarek *et al.*, 2005).

Correlating with a role in high-frequency firing, Kv3.1 is expressed in a tonotopic gradient within the medial nucleus of the trapezoid body (MNTB) in the auditory brainstem where Kv3.1 levels are highest at the medial end, which corresponds to high auditory frequencies. This pattern is lost in older mice that suffer early cochlear hair cell loss (von Hehn *et al.*, 2004).

There has also been an extensive study of Kv3.3 channels in the electrosensory lateral line lobe of weakly electric apteronotid fish (Mehaffey *et al.*, 2006). These channels regulate burst discharge in pyramidal cells and enable sustained high-frequency firing through their ability to reduce an accumulation of low-threshold potassium current.

Kv3 gene products are also found in neurons that are not known to fire at high frequencies, suggesting that Kv3 channels may have other physiological roles. These functions of Kv3 channels may still be related to their ability to maximize fast action potential repolarization, and the consequences of this function; brief action potentials with quick restoration of membrane properties following activity and limiting calcium entry (Rudy *et al.*, 2001). Kv3 proteins have been found in magnocellular neurosecretory neurons (Shevchenko *et al.*, 2004) and in starburst cells in the retina (Ozaita *et al.*, 2004). Their functions in these cells are yet to be studied. In starburst neurons, which normally do not spike, it has been proposed that Kv3 channels provide a voltage-dependent shunt limiting the amplitude of neuronal depolarization. This might be critical

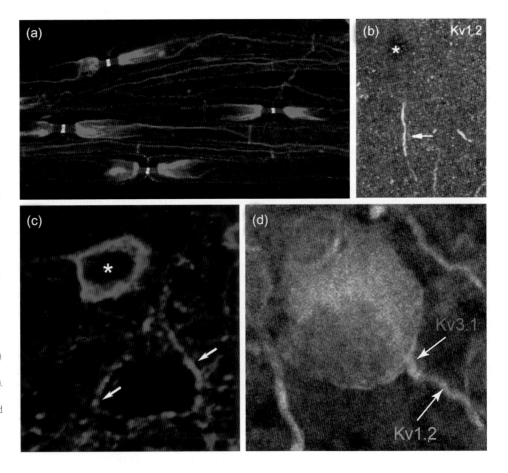

Fig. 2.1.1.2 Neuronal Kv proteins are localized to specific microdomains. (a) Kv1.1 (red) in the juxtaparanodal region of myelinated axons; (green) Na^+ channels at the node of Ranvier; (blue) NCP1 at the paranode (from Bhat *et al.*, 2001). (b) Kv1.2 in the axon initial segment of cortical pyramidal neurons (arrow). The cell body is indicated by an asterisk (from Inda *et al.*, 2006). (c) Kv3.2 in the somatic membrane of a cortical interneuron (asterisk) and the perisomatic (basket) terminals forming synapses on a pyramidal cell (from Chow *et al.*, 1999). (d) Kv3.1 proteins in the calyx of Held terminal of a neuron in the MNTB and Kv1.2 in the pre-terminal axon (from Dodson and Forsythe, 2004). See also the colour plate section.

to the generation of directional selectivity in the retina, believed to be mediated by starburst cells (Ozaita *et al.*, 2004).

Kv3 channels in axons: roles in neurotransmitter release

As discussed earlier, Kv3 channels are prominently expressed in axons and presynaptic terminals of several neuronal populations (see Fig. 2.1.1.2). This distribution suggests roles for Kv3 channels in neurotransmitter release. This function has been investigated in the basket terminals of fast spiking interneurons in the cerebral cortex, in the calyx of Held (the large presynaptic terminals from the ventral cochlear nucleus bushy cells innervating the neurons in the MNTB), in motor nerve terminals, and cerebellar parallel fibre terminals. Kv3 channels in these cells are key determinants of the repolarization of the action potential at the terminal, keeping APs brief, limiting Ca^{2+} influx and hence neurotransmitter release probability, thus contributing to the short-term dynamics of the synapse and the fidelity of synaptic transmission (Ishikawa *et al.*, 2003; Matsukawa *et al.*, 2003; Brooke *et al.*, 2004; Goldberg *et al.*, 2005).

Kv3 channels and CNS function

Gene knockout mice have been developed for three of the four Kv3 genes (*Kv3.1–Kv3.3*); the three Kv3 genes that are prominently expressed in CNS neurons (Ho *et al.*, 1997; Chan, 1997; Lau *et al.*, 2000).

Kv3.1 and Kv3.3 knockout mice show alterations in motor behaviour consistent with their expression in several neuronal populations in brain areas involved in motor control, such as the reticular thalamus, basal ganglia, spinal cord, and cerebellum. The Kv3.1 knockouts have increased locomotor activity and sleep loss (Espinosa *et al.*, 2001, 2004; Joho *et al.*, 2006a, b). Kv3.3 knockouts show impaired gait and reduced motor performance associated with abnormal Purkinje cell discharges and lack harmaline-induced tremor, suggesting defects in the olivocerebellar system (McMahon *et al.*, 2004; Joho *et al.*, 2006). Kv3.1 and Kv3.3 double knockouts suffer from severe ataxia, tremor, and myoclonus (Espinosa *et al.*, 2001; Joho *et al.*, 2006).

Kv3.2 knockout mice have altered EEG rhythms, increased cortical excitability, and enhanced seizure susceptibility perhaps resulting from alterations in GABAergic control in cortical structures (Lau *et al.*, 2000).

Kv3 channels (mainly Kv3.1 and Kv3.3) are prominently expressed in auditory neurons in the mammalian auditory pathway in the brainstem where they regulate firing patterns, and, in particular, the ability to follow high-frequency stimuli, and neurotransmitter release. These are functions believed to be important in preserving the fidelity and timing of firing codes, and as such, contribute to the computation of sound features, such as frequency, intensity, and localization in space (Kaczmarek, 2006).

It has been suggested that Kv3 channels have an important role in modulating circadian neural activity. Itri and colleagues (2005) found that the magnitude of a fast delayed rectifier potassium current, thought to be mediated by Kv3 channels, is under circadian regulation. Blockade of this current prevented the daily rhythm in firing rate of neurons in the SCN.

Kv4 (Shal) subfamily

Kv4 channels underlie fast inactivating or A-type K^+ currents (I_A), including the fast transient outward current I_{to} in cardiac myocytes and most of the subthreshold-operating somato-dendritic A-type K^+ current in mammalian neurons (I_{SA}) (Connor *et al.*, 1971a, b; Serodio *et al.*, 1994, 1996; Fiset *et al.*, 1997; Johns *et al.*, 1997; Nakamura *et al.*, 1997; Barry *et al.*, 1998, Guo *et al.*, 1999; Xu *et al.*, 1999; Malin and Nerbonne, 2000; Shibata *et al.*, 2000; Birnbaum *et al.*, 2004; Jerng *et al.*, 2004a, b; Chen *et al.*, 2006). I_A was first described in molluscan neurons as a fast transient K^+ current that activates at membrane potentials that are below the threshold for Na^+ spike generation (Connor *et al.*, 1971a, b). Similar currents were later found in mammals and other species. These currents are of considerable physiological significance. In the heart, I_{to} plays a role in the initial repolarization of the cardiac action potential, and is considered a target for arrhythmia treatment (Tomaselli and Marban, 1999; Tomaselli and Rose, 2000). In the nervous system, the currents are essential in regulating excitability in many neuronal types. The delay between depolarization and spiking, the rate of action potential repolarization, and the duration of the interspike interval are influenced by the rapid and transient activation of these channels in the subthreshold range of membrane potentials (Connor 1971a; Llinas, 1988; Rudy 1988; Baxter and Byrne, 1991). I_{SA} channels modulate the frequency of repetitive firing through their regulation of the duration of the interspike interval, as illustrated by the recent finding that the pacemaker frequency of individual dopaminergic neurons on which the amount of dopamine release depends correlates with I_{SA} levels, which in turn correlates with the concentration of mRNA products of the Kv4 gene expressed in these cells (Kv4.3) (Liss *et al.*, 2001). Furthermore, as discussed below, the channels underlying the I_{SA} are enriched in the dendrites of several neuronal populations where they regulate action potential back propagation, the integration of synaptic inputs, the establishment and maintenance of long-term potentiation (LTP), and the filtering of fast synaptic potentials (Hoffman *et al.*, 1997; Schoppa and Westbrook, 1999; Johnstone *et al.*, 2000; Watanabe *et al.*, 2002).

Kv4 channels in dendrites

Immunolocalization of Kv4.2 proteins shows that they are most prominently expressed dendrites. This has been well documented both in hippocampal and neocortical pyramidal neurons and in cerebellar granule cells (Sheng *et al.*, 1992; Maletic-Savatic *et al.*, 1995; Rhodes *et al.*, 2004). This is consistent with the observation that there is a gradient of I_{SA} in CA1 pyramidal neurons (Kv4.2 is the predominant Kv4 subunit in these cells [Serodio and Rudy, 1994]), with maximum density in distal dendrites (Hoffman *et al.*, 1998; Johnston *et al.*, 2000, 2003). These dendritic channels have essential roles in regulating dendritic excitability and the backpropagation of action potentials into the dendritic tree, as shown by Johnston and his colleagues (2000, 2003). As a result, the current contributes to the regulation of specific forms of synaptic plasticity reviewed in (Watanabe *et al.*, 2002; Johnston *et al.*, 2003). Utilizing Kv4.2 knockout mice, it was recently confirmed that these functions are mediated by a Kv4.2-dependent I_{SA} (Chen *et al.*, 2006). Kv4.3 immunoreactivity is also strong in dendrites of hippocampal, striatal, and neocortical neurons that express Kv4.3 RNA transcripts (Serodio and Rudy, 1994; Rhodes *et al.*, 2004).

2.1.1.7 Disease relevance

Kv1 (Shaker) subfamily

Mutations in the gene encoding Kv1.1 cause episodic ataxia type 1 and epilepsy

Mutations in the human *KCNA1* gene encoding Kv1.1 proteins have been found to cause a disease with perturbed excitability, episodic ataxia type 1 (EA1) with myokymia, an autosomal-dominant disease affecting central and peripheral nerve function, sometimes associated with partial epilepsy (Browne *et al.*, 1994; Zerr *et al.*, 1998; Zuberi *et al.*, 1999; Lersch *et al.*, 2001). Several mutations causing EA1 have been discovered, all causing reduction in K+ currents as a result of either diminished expression or positive shifts in voltage-dependence. The mutations are dominant and hence affect Kv1 heteromultimeric channels.

Kv1.1 knockout mice have severe, often lethal, epilepsy (Smart *et al.*, 1998), which is consistent with the role of Kv1 channels in regulating neuronal excitability. It is not clear, however, which neurons are responsible for the heightened seizure susceptibility of the mice.

Kv1 channels have also been found as targets of autoimmune disease. Thus, antibodies against several Kv1 channels have been found in various nerve hyperexcitability diseases such as Morvan's syndrome and limbic encephalitis (see Table 2.1.1.1).

Kv2 (Shab) subfamily

Kv2 channels are also implicated in mediating apoptosis in neurons (Pal *et al.*, 2003) and due to their prominent expression and activity in β-cells of the pancreas, are potential therapeutic targets for diabetes (MacDonald *et al.*, 2001).

Kv3 (Shaw) subfamily

Mutations in the human Kv3.3 gene (KCNC3) were recently identified as the cause of spinocerebellar ataxia (SCA13), a disease that can also involve cognitive defects, suggesting cerebellar and extracerebellar dysfunction (Waters *et al.*, 2006). Thus far, two missense mutations have been described: Kv3.3–R420H, with the mutation located in the S4 domain, had no channel activity when expressed alone and a dominant-negative effect when co-expressed with wild-type Kv3.3 proteins. In the second mutation Kv3.3–F448L the activation curve was shifted in the negative direction, and channel deactivation was slow. F488 is a residue in the S5 domain

in a position which is leucine in other Kv channels. It contributes to generating a 'leucine-heptad' or 'leucine-zipper' motif disrupted in Kv3 channels (Vega-Saenz de Miera, 1994). This difference between Kv3 and other Kv channels has been suggested to be important in lowering the stability of the open state and thus governing the voltage-dependence and deactivation kinetics of Kv3 channels (Rettig *et al.*, 1992; Vega-Saenz de Miera *et al.*, 1992, 1994). Therefore, both mutations are likely to perturb the role of Kv3.3 channels in the fast repolarization of Purkinje cells and to change the properties of Purkinje cell outputs.

Kv3.4 channels regulate the resting potential of skeletal muscle

Kv3.4 subunits assemble with MinK-related peptide 2 (MiRP2) to form subthreshold, voltage-gated potassium channels that regulate the resting potential of skeletal muscle cells (Abbott *et al.*, 2001). MiRP2 shifts the voltage dependence of activation of Kv3.4 channels, speeds recovery from inactivation, suppresses cumulative inactivation, and produces an apparent increase in unitary conductance. This suggests that Kv3.4–MiRP2 channels are extremely important for muscle function. An R83H missense mutation in MiRP2 that diminished the effects of MiRP2 on Kv3.4 has been associated with a form of periodic paralysis (Abbott *et al.*, 2001). The R83H variant produces MiRP2–Kv3.4 channels that become very sensitive to physiological changes in intracellular pH (pKa approximately 7.3, consistent with histidine protonation), whereas wild-type channels are largely insensitive. This perhaps explains an observed link between acidosis and episodes of periodic paralysis (Abbott *et al.*, 2006). Moreover, it was recently shown that MiRP2–Kv3.4 function requires phosphorylation of a PKC phosphorylation site at the MiRP2 serine 82 (Abbott *et al.*, 2006).

Kv4 (Shal) subfamily

Genetic elimination of Kv4.2 in dorsal horn neurons results in enhanced sensitivity to tactile and thermal stimuli. Furthermore ERK-dependent forms of pain hypersensitivity are absent in Kv 4.2 knockout mice (Hu *et al.*, 2006). A truncation in Kv4.2 protein was discovered in a patient with temporal lobe epilepsy (Singh *et al.*, 2006) suggesting association of these channels with epilepsy. Moreover, the Kv4 channel associated protein DPP6 (DPPX) has been associated with autism (Marshall *et al.*, 2008) and ALS (Cronin *et al*, 2008) and DPP10 with asthma susceptibility (Allen *et al.*, 2003).

Table 2.1.1.1 Kv K$^+$ channel proteins

Kv1 subfamily						
Channel Subunit	**Molecular Information[1]**	**Electrophysiology Properties in heterologous cells[2]**	**Pharmacology[3]**	**Tissue Expression[4]**	**Associated pathology**	**Channel function**
Kv1.1 **Other names:** HUK1 [1,2] BK1 [3] MK1 [4] RBK1 [5] RCK1 [6] RK1 [7] HBK1 [8]	**Homo sapiens:** Gene symbol: KCNA1 Chr: 12p13 Entrez Gene ID: 3736; Acces: NM_000217 Protein size: 495 aa **Mus musculus:** Gene symbol: Kcna1 Chr: 6 (61.0 cM) Entrez Gene ID: 16485 MGI: 96654 Acces: NM_010595 Protein size: 495aa. **Rattus norvegicus:** Chr: 4q42 Entrez Gene ID: 24520 Acces: NM_173095 Protein size: 495 aa Intronless, single coding exon [14] RNA editing (see electrophysiological properties). **Associated Proteins:** Kvβs. PDZ domain containing proteins such as PSD95 and SAP97 (see text) SNAP25 [248] CASPR2 [56]	**Activation:** [9] V_{on}: -50 mV $V_{0.5}$: -32 mV k= 8.5 mV τ = +++ **Deactivation:** [9] τ = 14 ms **Inactivation:** [9] Very slow **Single Channel conductance:** [9] 10 pS RNA editing profoundly affects channel inactivation conferred by accessory Kvβ subunits [15]	**TEA:** 300 µM [9] **4-AP:** 290 µM [9] **DTX-I:** 20 nM [9] **CTX:** NB [9] **DTX-K** blocks preferentially channels containing Kv1.1 (IC$_{50}$: 0.6 nM) [10] Also blocked by: **HgTx:** 0.12 pM KD [16] **OSK1:** 0.6 nM [17] **ShK:** 16pM [18]	**Tissue:** Strong in brain; weak in atria, less in aorta and skeletal muscle; retina (NB) Smooth muscle (PCR) [6,21,42,76,246,247] **CNS:** RNA: widely expressed throughout rodent CNS [19] Protein: Expressed mainly in axons and presynaptic terminals. Some staining of neuronal somata and proximal dendrites. [20, 21] Juxta-Paranode of myelinated fibers. [22,23] Axon initial segment (AIS) Also expressed in sensory neurons [13,23,24]	**Mouse:** Knockout: Frequent epileptic seizures, short life span [11]. Point mutation V408A: Model of episodic ataxia type-1 (EA1) [12] Spontaneous 11 bp deletion: megencephaly, disturbance in the expression of BDNF and insulin like growth factor binding protein 6 [25-27] **Human:** Kv1 mutations: Episodic Ataxia with myokymia [28-29]; Autoantibodies to several Kv1s found in auto-immune disorders such as Morvan's syndrome and limbic encephalitis; [30,31]	Kv1.1, Kv1.2 and Kv1.6 subunits contribute to the formation of dendrotox n (DTX)-sensitive channels mediating the DTX-sensitive current I$_D$ (see text). **Somata and/or axon initial segment:** I$_D$ regulates neuronal subthreshold excitability (see text). In auditory processing neurons in the MNTB [13] an I$_D$ known as I(LTS) ensures generation of single spikes in response to each EPSP, so that the timing and pattern of AP firing is preserved across the relay synapse (calyx of Held). Tityustoxin and DTX-K suggest I(LTS) channels are Kv1.1/Kv1.2 heteromers. Another I$_D$ known as I(LTR) is probably mediated by Kv1.1/Kv1.6 heteromers. Kv1.1 and Kv1.2 (but not Kv1.6, proteins) are concentrated in the first 20 µm of MNTB axons. **Synaptic terminals:** DTX-sensitive channels regulate excitability preventing reflection of action potentials at axon terminals and aberrant terminal firing and transmitter release (see text) **Juxta-paranodal region of myelinated axons:** see text **Principal neurons of the LSO.** Kv1.1 gradient underlies gradient in firing properties of principal neurons [32] **Mesencephalic nucleus:** 4-AP sensitive K$^+$ currents in Kv1.1 and Kv1.6 immunoreactive cells suppress spike initiation and facilitate axonal spike initiation depending on the origin of the stimulus [2]

1 Unless otherwise stated the molecular information was obtained from the Jackson's laboratory Mouse Genome Informatics (MGI) site: http://www.informatics.jax.org

2 Data is from expression in mammalian transfected cells, and if not available data from expression in *Xenopus* oocytes was used. Values reported at room temperature in physiological K$^+$ concentrations. Von, approximate voltage at which current becomes apparent; $V_{0.5}$, membrane potential at which the conductance is half maximal: k, slope of normalized g-V curve; τ Activation, time constant of activation for fast activating currents; +++ < 5 ms; ++ 5-10 msec at; + >10 msec at + 40 mV; τ deactivation, time constant of deactivation at -60 mV unless otherwise stated; inactivation: $V_{0.5}$ and k refer to potential at which steady-state inactivation g-v curve, respectively; τ, time constant of inactivation at +40 mV for fast inactivating currents; very slow = seconds.

3 Values are IC$_{50}$ unless otherwise stated. NB, not blocked; NA, data not available. Abbreviations: TEA, tetraethylammonium, 4-AP, 4- aminopyridine; DTX-I, dendrotoxin 1; CTX, charybdotoxin.

4 ISH: *In situ* hybridization; NB, northern blot.

Continued

Table 2.1.1.1 (*Continued*) Kv K⁺ channel proteins

Kv1 subfamily

Channel Subunit	Molecular Information	Electrophysiology Properties in heterologous cells	Pharmacology	Tissue Expression	Associated pathology	Channel function
Kv1.2 **Other names:** HuKIV [1,2] MK2 [4] BK2 [33] RCK5 [8,34] RK2 [7] RAK [35] XSha2, [36] NGK1 [37]	**Homo sapiens:** Gene symbol: KCNA2 Chr:1 p13 EntrezGeneID: 3737 Acces #: NM_004974 Protein size: 499aa **Mus musculus:** Gene symbol: Kcna2 Chr: 3 (48.8 cM) MGI:96659 Entrez GeneID: 16490 Acces#: NM_008417 Protein size: 499aa **Rattus norvegicus:** Chr: 2q34 Entrez Gene ID: 25468 Acces: NM_012970 Protein size: 499aa single coding exon **Associated Proteins: Kvβs** see text **PDZ** domain containing proteins (see text) **Small GTP-binding protein RhoA** [195] Tyrosine kinases activated by G protein-coupled receptors phosphorylate and thereby suppress Kv1.2 channels. RhoA is a necessary component in this process and mediates M1 receptor inhibition of Kv1.2 by tyrosine phosphorylation. CASPR2 [56] SNAP25 [14]	**Activation:** [9] V_{on}: -40 mV $V_{0.5}$: 5 mV k= 13 mV τ = + **Deactivation:** [9] τ= 23 ms, **Inactivation:** [9] Very slow. **Single Channel g:** [9] 18pS	**TEA:** 560 mM [14] **4-AP:** 590 µM [9] **DTX I:**17 nM [9] **CTX:**14 nM [9] **Tityustoxin:** preferentially blocks Kv1.2-containing channels (Kd=0.21nM) [38] Also blocked by: **OSK1**, **ShK** (9nM), **HgTx** at concentrations similar to Kv1.1 [16,18,43]	**Tissue:** Vascular smooth muscle cells in rat small cerebral arteries (with Kv1.5) [39]; renal vascular smooth muscle [40,41]; Sensory neurons [42] **CNS:** RNA and proteins: perhaps the most widely expressed Kv1. XXX Kv1 heteromers often contain Kv1.2 [44] anti-Kv1.2 > 1.1 >> 1.6 > 1.4 [45] Protein predominantly in axons and terminals. Pattern highly overlapping with other Kv1s, but not identical. For example in hippocampus [46]; cerebellum [30,47,48] and spinal cord [30,49,50] Proximal dendrites Soma [20,21,24,51]. Juxtaparanodal [21,22,23]. Axon initial segment [65,66]	Autoantibodies to several Kv1s found in autoimmune disorders such as Morvan's syndrome and limbic encephalitis; [30,31]	Subunit that contributes to forming DTX sensitive currents in neuronal somata, axons and terminals (see text; also see Kv1.1). Likely to contribute to I_D with slower kinetics. Maintaining membrane potential and modulating electrical excitability in neurons and muscle Shown to regulate subthreshold excitability in a number of neurons: e.g. striatal medium spiny neurons [250]

	Activation / Deactivation / Inactivation	Pharmacology	Tissue	Function
Kv1.3 **Other names:** HGK5 [52] HLK3 [53] hPCN3 [54] HUKIII {Ramaswami #104; Kamb #33} KV3 [302] RCK3 [259] MK3 [255] **Homo sapiens:** Gene symbol: KCNA3 Chr: 1p13.3 EntrezGeneID: 3738 Acces #: NM_002232 Protein size: 575aa **Mus musculus:** Gene symbol: Kcna3 Chr: 3 (52.3 cM) Entrez GeneID: 16491 MGI: 96660 Acces#: NM_008418 Protein size: 528aa *Rattus norvegicus:* Chr: 2q34 Entrez Gene ID: 29731 Acces: NM_019270 Protein size: 525 aa; **Associated Proteins: Kvβs** (see text). **PDZ domain-containing proteins PSD95, SAP97 & hDLg** (see text); KCNE4 (MirP3) [74] KCNE4 is an inhibitory subunit to Kv1.1 and Kv1.3; β$_1$ integrins [75]	**Activation:** [9] V_{on} = -50 mV $V_{0.5}$ = -26 mV k = 7 mV τ = +++ **Deactivation:** [9] τ = 39 ms **Inactivation:** [14] $V_{0.5}$= -63 mV k= 7.7 mV τ = 250 ms [303] C-type **Single channel g:** [35] 14pS	**TEA:** 10mM [9,58] **4AP:** 190 μM [9, 58] **DTX:** 250nM [9, 58] **CTX:** 3nm [9, 58] Blocked by many Kv1 channel toxins. Most selective are: **ShK-Dap22** 23pM and **ShK(L5)** 69pM, with 100 fold selectivity over Kv1.1 and greater than 250 fold selectivity over other Kv1s [18, 59]; **Margotoxin** 50 pM [72], but see Kv1.6; **ShK** 11pM [18], which also block other Kv1 channels	**Tissue:** Prominently expressed in lymphocytes (type n channel) [58], macrophages [60], microglia [61] Kidney [62] Testis and spermatozoa [63] Osteoclasts [64] **CNS:** [20] Prominent in olfactory bulb, cerebellum, parallel fibres, deep nuclei, Purkinje cell somata; hippocampus, weak, stratum lucidum Mitral and granule cells of the olfactory bulb [73] **Super-smeller mice:** Kv1.3-/- mice have a 1,000- to 10,000-fold lower threshold for detection of odors and an increased ability to discriminate between odorants [67] **Multiple sclerosis:** Blockers of Kv1.3 and KCa3.1, which suppress human T cell activation, considered for MS treatment [58, 68, 69]	**Regulates neuronal excitability, possibly in heteromers with other Kv1s** [67] **Mediates the type n K$^+$ channel in lymphocytes.** Regulates resting potential and is required for lymphocyte activation [57,58]. **Regulates energy homeostasis and body weight** Kv1.3$^{-/-}$ mice weigh significantly less than control littermates [70]. It is also shown to regulate peripheral insulin sensitivity [71] **Microglia** Kv1.3 channels contribute to their ability to kill neurons [61]
Kv1.4 **Other names:** HUKII [2] RK3 [7] HPCN2 [54] HK1 [76] RCK4 [8] RHK1 [77] **Homo sapiens:** Gene symbol: KCNA4 Chr: 11p14 EntrezGeneID: 3739 Acces: NM_002233 Protein size: 653aa **Mus musculus:** Gene symbol: Kcna4 Chr: 2 (61.0cM) Entrez GeneID: 16491 MGI: 96661 Acces#: NM_021275 Protein size: 654aa *Rattus norvegicus:* Chr: 3q33 Entrez Gene ID: 25469 Acces: NM_012971 Protein size: 654aa	**Activation:** [8] V_{on} : -50 mV $V_{0.5}$: -22mV k=-17 mV τ = +++ **Deactivation:** [77] τ = 15 ms **Inactivation:** [77] N-type $V_{0.5}$= -55 mV k= 4.2 mV τ = 30ms τ of recovery = seconds **Single channel g:** [8] 5pS	**TEA:** NB [8] **4-AP:** 0.65 mM [78] **DTX:** NB [8] **CTX:** NB [8] **ShK:** 315pM [18]	**Tissue:** Tissue expression: Heart atria, brain, low levels aorta, skeletal muscle [7] PNS:With TRPV1 and TRPV2 [79-81] altered expression after axotomy [82] sole Kv1 alpha subunit in small diameter neurons nociceptors [42] **CNS:** In hippocampus: [46] mRNA prominent throughout brain, much weaker in cerebellum and brainstem. Strong hippocampus, striatum, thalamus, CX [94] Increased susceptibility to spontaneous seizures in Kv1.4 null mice [83] Patients with myasthenia gravis produce autoantibodies to Kv1.4 [84]	**Function:** **Neurons:** Contributes inactivation to DTX-sensitive K$^+$ channels [24,42,85]. Regulates spike repolarization and spike broadening during repetitive activity in mossy fibre terminals (see text). Neuronal transmission stimulates the PKA-mediated phosphorylation of Kv1.4 channels inhibiting activity [86] **Heart:** Mediate slow component of I$_{to}$ in atria [87]

Continued

Table 2.1.1.1 *(Continued)* Kv K⁺ channel proteins

Kv1 subfamily

Channel Subunit	Molecular Information	Electrophysiology Properties in heterologous cells	Pharmacology	Tissue Expression	Associated pathology	Channel function
	Associated Proteins: Kvβs (see text). **PDZ domain-containing proteins PSD95, SAP97 & hDLg** (see text) [88,89] **Alpha-actinin-2** [90] **KChaP** [91,92] σ-receptor [14,93]			Kv1.4 immunoreactivity In brain is widespread including the cerebral cortex, thalamus, hippocampus, and basal ganglia; low abundance in cerebellum. [94,95] Immunolabeling usually in neuropile, suggesting axonal expression, and, to a lesser extent in cell bodies and proximal dendrite [95,96] In dendrites and spines of neurons in dorsal cochlear nucleus [97]		
Kv1.5 **Other names:** Kv1 [55] HuK II [1] RK4 [7] HK2 76] HpCN1 [54]	**Homo sapiens:** Gene symbol: KCNA5 Chr: 12p13; Entrez GeneID: 3741 Acces: NM_002234 Protein size: 613aa **Mus musculus:** Gene symbol: *Kcna5* Chr: 6 (61.0 cM) MGI: 96662 Access: NM_145983 Protein size: 602aa **Rattus norvegicus:** Chr: 4q42 Acces #: NM_012972 Protein size: 602aa	**Activation:** [9] V_{on}: -50 mV $V_{0.5}$ = -14 mV k = 12 mV τ = +++ **Deactivation:** [9] τ = 23 ms **Inactivation:** [9] very slow **Single channel g:** [9] 8 pS	**TEA:** 330 mM [9]**4-AP:** 270 µM [9]**DTX:** NB [9]**CTX:** NB [9]	**Tissue:** Northern rat: heart> skeletal muscle>brain>lung>kidney [55] Atria~ventricle> Aorta~sm>brain NB [7] Pancreatic β cells [54] Microglia: [98,99] Schwann cells [109]; macrophages [110]. vascular smooth muscle, [111,112] **Brain:** [113], Ab cerebellum: Purkinje cells, DCN [113] Hippocampus: somatodendritic in all principal neurons [114]		4-AP-sensitive Component of IK,slow in mouse ventricle and IKur (Ultrarapid-activating K⁺ current) in human atrium. [100,101] Pharmacological significance potential use in management of atrial fibrillation via blockade of I_{kur} [102]

		Biophysical properties	Pharmacology	Tissue distribution	Comments
	Associated Proteins: Kvβs (see text); **Src tyrosine kinase** [103,104]; see text; **PDZ-containing proteins** see text. **Kchap** [105] **Fyn-** [106] α-actin-2 [90,107] **caveolin** SAP97 [108]				
Kv1.6 **Other names:** HBK2 and RCK2 [34] MK1.6 [115] KV2 [55]	**Homo sapiens:** Gene symbol: KCNA6 Chr: 12p13; Entrez GeneID: 3742 Acces :NM_NM_002235 Protein size: 529 aa **Mus musculus:** Gene symbol: Kcna6 Chr: 6 (61.0 cM) Entrez GeneID: 16494 MGI: 96663 Access: NM_013568 Protein size: 529 aa **Rattus norvegicus:** Chr: 4q42 Entrez GeneID: 64358 Acces: XM_575671 (predicted) Protein size: 530 aa **Associated Proteins:** Kvβs (see text) CASPR2 [56]	**Activation:** [34] V_{on} = -50 mV $V_{0.5}$ = -21 mV k = 8 mV τ = +++ **Deactivation: 20 ms** [116] **Inactivation:** [34] very slow **Single channel g:** [34] 9pS	**TEA:** 1.7-7mM [34,55,117] **4AP:** 0.5-1.5 mM [34,55] **DTX:** 9-20nM [24,55] **CTX:** 1nM [34]; but, >200nM [55] Also blocked by: **ShK** 160pM [18]; **HgTx** [16]; **MgTx** 5nM [72]	**Tissue:** smooth muscle: mouse colon smooth muscle, rat pulmonary arterial smooth muscle, rat mesenteric arterie myocytes [118-120], dorsal root ganglion neurons [82] nodose ganglia [121], oligodendrocytes [122] Astrocytes [11] atrioventricular node [123] **CNS:** Cerebral cortex, Hippocampus cerebellar Purkinje cells (somatodendritic not axonal) and in various olfactory and amygdaloid structures. By in situ and immunostainiong [129] Widely distributed	**Kcna6**^{tm1Lex} Lexicon Genetics Homozygous mutation of this gene results in an increased thermal nociceptive threshold and in females an increase in circulating triglyceride levels. **Cancer cachexia** down-regulated expression in brain from animals with experimental cancer cachexia [124]. Autoantibodies to several Kv1s found in autoimmune disorders such as Morvan's syndrome, neuromyotonia, and limbic encephalitis; [30,1215] Subunit that contributes to forming DTX sensitive currents in neuronal somata, axons and terminals (see text; also see Kv1.1).
Kv1.7	**Homo sapiens:** Gene symbol: KCNA7 Chr: 19q13.3 Entrez GeneID: 3743 Acces: NM_031886 Protein size: 456 aa	**Activation:** [[127] V_{on} = -40 mV $V_{0.5}$ = -20 mV k = 8 mV τ = +++ **Deactivation:** [127] τ = 5 ms **Inactivation:** $V_{0.5}$ = -40mV (mKv1.7L); -21 mV (Kv1.7S)	**TEA:** 80mM [128] **4-AP:** KD 150µM [128] **DTX:** NA **CTX:** NB [128] **ShK:** 11.5nM [18]	**mRNA:** Pulmonary arteries, [130] Heart, skeletal muscle, pancreatic islet cells [127]	Two channel isoforms with different functional characteristics, the long form, Kv1.7L inactivates faster than the short isoform Kv1.7S, predominantly due to an N-type related mechanism, which is impaired in the short form. Kv1.7L, but not mKv1.7S is regulated by the cell's redox state [126] Might be involved in regulation of excitability in response to changes in redox state [126,130]

Continued

Table 2.1.1.1 (*Continued*) Kv K+ channel proteins

Kv1 subfamily

Channel Subunit	Molecular Information	Electrophysiology Properties in heterologous cells	Pharmacology	Tissue Expression	Associated pathology	Channel function
	Mus musculus: Gene symbol: *Kcna7* Chr: 7 (23.0 cM) Entrez GeneID: 3743 MGI: 16495 Two channel isoforms [126]. Access: NM_010596 Protein size: mKv1.7L (532 aa) and a shorter mKv1.7S (489 aa) **Rattus norvegicus:** Chr: 1q22 Entrez GeneID: 365241 Acces: XM_001080729(predicted) Protein size: 457 aa	k=4 mV (mKv1.7L); 7mV (Kv1.7S); τ_{rec}= slow τ=14ms [127] **Single channel g:** [127] 21pS				
Kv1.8 **Other names:** Kv1.10 [131] Kcn1 [132] KCNA10 [133]	**Homo sapiens:** Gene symbol: KCNA10 Chr: 1p13.1 Entrez GeneID: 3744 Acces #: NM_005549 Protein size: 511aa **Mus musculus (predicted):** Gene symbol: Kcna10 Chr: 7 (23.0 cM) Entrez GeneID: 242151 MGI: 3037820 Access: XM_143471 Protein size: 511 aa **Rattus norvegicus :** chr. 2q34 Entrez GeneID: 295360 Acces: XM_227577(predicted) Protein size: 511 aa **Associated Proteins:** KCNA4B: Increases current magnitude and sensitivity to cAMP [136]	**Activation:** [134] V_{on} = -20 mV $V_{0.5}$ = 4 mV k=NA τ=+ **Deactivation: NA** **Inactivation:** NA Very slow **Single channel g : :** [134] 11pS	**TEA:** 50 mM [134] **4-AP:**2 mM [134] **DTX:** NA [134] **CTX:** 200 nM [134]	RNA expressed in kidney, renal blood vessels, heart, brain, aorta [132,135]		KCNA10 may facilitate renal proximal tubular sodium absorption by stabilizing cell membrane potential. Its presence in endothelial and vascular smooth muscle cells suggests that it also regulates vascular tone [133]. Voltage-gated and cyclic nucleotide gated channel (see text).

Kv2 subfamily

Channel Subunit	Molecular information	Electrophysiology	Pharmacology	Expression	Disease and KO	Function
Kv2.1 **Other names:** DRK1 [137]	**Homo sapiens:** Gene symbol: KCNB1 Chr: 20 q13.2 Entrez GeneID: 3745 Acces #: NM_004975 Protein size: 858 aa **Mus musculus:** Gene symbol: Kcnb1 Chr: 2 (97.0cM) Entrez GeneID: 16500 MGI: 96666 Access #: NM_008420 Protein size: 857 aa, **Rattus norvegicus:** Chr: 3 q42 Entrez GeneID: 25736 Acces #: NM_013186 Protein size: 853aa **Associated Proteins:** Kv5-Kv6, Kv8-Kv9 subunits (see text) KChaP (binds to N terminus of Kv2.1), [105] Fyn SH2 domain [106] SNAP25[248]	**Activation:** [138][5] V_{on}= -20 to -30 mV $V_{0.5}$: 1.5-16.5 mV k= 13 τ = + **Deactivation:** [139] τ = 20 ms **Inactivation:** [138] $V_{0.5}$= -25mV k= -6mV τ = very slow τ(recovery)= 1.6s at -90mV[139] **Unitary conductance:** [140] 10 pS Properties are changed by association with Kv5, Kv6, Kv8 and Kv9 subunits. See below.	**TEA:** IC50 5-10mM [137,138,141] 4AP: IC50 17mM [141] Although initially reported as very 4-AP sensitive in oocytes (IC50 0.5 mM; [137]); more recent estimates suggest that is much less 4-AP sensitive [14] **DTX:** NB **CTX:** NB **HaTX:** Kd 100nM (gating modifier) [142]	brain, atria, ventricle, skeletal muscle, olfactory epithelium, retina, kidney [143], SCG (also Kv2.2) [144,145]; spinal motoneurons [146]; lung, p12 cells, pancreatic β cells [135] pulmonary artery [130] In rat SCG neurons, Kv2.1 and Kv2.2 are expressed at the mRNA level 2.1 and 2.2 don't form heteros in HEK or SCG cells Widely expressed [150], but no detailed distribution of Kv2.1 or Kv2.2 mRNA or protein reported **Protein** localized in clusters in somatic and proximal dendritic membrane. Little neuropil staining [153] In pyramidal and interneurons in Cx and hippocampus; apposed to astrocytic processes and in inhibitory postsynaptic sites[154]. In spinal motoneurons apposed to cholinergic synapses [146]		**Kv2 channels mediate** I_K, the TEA-sensitive, 4-AP-insensitive delayed rectifier K+ current recorded in many, if not most types of neurons (see text). Regulate spikes shape and Ca^{2+}-influx during repetitive firing at high frequency in cortical pyramidal neurons [147]. Regulates spike duration in tonic firing sympathetic neurons [144]. Implicated in mediating apoptosis in neurons [148]. Potential therapeutic targets for diabetes due to their prominent expression in pancreatic β-cells [149]
Kv2.2 **Other names:** CDRK (rat) [143]	**Homo sapiens:** Gene symbol: KCNB2 Chr: 8q13.2 Entrez GeneID: 9312 Acces: NM_004770	**Activation:** [151] Von= -20 mV $V_{0.5}$ = 16mV k = 14.5mV τ = +	**TEA:** IC50: 2.6-8 mM [143,152]; **4-AP:** IC50 Close to 10 mM at +10 mV see Fig. 6 in [152]	**Tissue:** Brain, tongue epithelium [143]; SCG [144], sympathetic neurons [145]; GI/mesenteric artery smooth muscle[155]		**Kv2 channels mediate** I_K, the TEA-sensitive, 4-AP-insensitive delayed rectifier K+ current recorded in many, if not most types of neurons (see text).

5 but see also 102.Mohapatra, D.P. and J.S. Trimmer, The Kv2.1 C terminus can autonomously transfer Kv2.1-like phosphorylation-dependent localization, voltage-dependent gating and muscarinic modulation to diverse Kv channels. J Neurosci, 2006. **26**(2): p. 685-95.

Continued

Table 2.1.1.1 (*Continued*) Kv K⁺ channel proteins

Kv2 subfamily

Channel Subunit	Molecular information	Electrophysiology	Pharmacology	Expression	Disease and KO	Function
	Protein size: 911 aa **Mus musculus:** Gene symbol: *Kcnb2* Chr: 1 (1.0 cM) Entrez GeneID: 98741 MGI: 99632 Access: XM_976837 (predicted) Protein size: 907 aa **Rattus norvegicus:** Chr: 5q11 Entrez GeneID: 117105 Acces: NM_054000 Protein size: 802 aa, **Associated Proteins:** Kv5-Kv6, Kv8-Kv9 subunits (see text) mKvβ4 associates with Kv2.2 and enhances expression level [156], KChaP [105]	**Deactivation: NA** **Inactivation:** [151] $V_{0.5}$ =−26mV k =8mV τ = very slow **Single channel conductance:** [152,156] 15pS	**HaTX**- NA **DTX**- NA **CTX**- NB [152] **Quinine** 14 μM [152]	**Brain:** neuropil (axons), maybe in some somas; panoramic protein distribution in brain [150]		Regulate spikes shape and Ca++ influx during repetitive firing at high frequency in cortical pyramidal neurons [147]. Regulates spike duration in tonic firing sympathetic neurons [144]. Implicated in mediating apoptosis in neurons [148]. Potential therapeutic targets for diabetes due to their prominent expression in pancreatic β-cells [149]
Kv5.1 **Other names:** KH1 [157] IK8 [19]	**Homo sapiens:** Gene symbol: KCNF1 Chr: 2p25 Entrez GeneID: 3754 Acces: NM_002236 Protein size: 494aa **Mus musculus:** Gene symbol: Kcnf1 Chr: 12 Entrez GeneID: 382571 MGI: 2687399 Access: NM_201531 Protein size: 493aa **Rattus norvegicus:** Chr: 6q16 Entrez GeneID: 298908 Acces: XM_216678 Protein size: 493 aa	Silent subunit, modifier of Kv2 channels: Decreases current levels, slows down activation, produces negative shift in inactivation, dramatic slowing of deactivation. [139,158]		Rat and human: brain, heart, skeletal muscle, liver [157, 159] Cardiac myocytes [160] **Brain:** limited mRNA expression in rat brain: deep cortical layers, amygdala, medial habenular nucleus [159]		Silent subunit that interacts with Kv2 subunits to modify channel properties
Kv6.1 **Other names:** KH2 [157] K13 [19]	**Homo sapiens:** Gene symbol: KCNG1 Chr: 20q13 Entrez GeneID: 3755 Acces: NM_002237 Protein size: 513aa **Mus musculus:** Gene symbol: Kcng1	Silent subunit, modifier of Kv2 channels Negative shifts in $V_{0.5}$ of activation and inactivation [139,158] Dramatic slowing of deactivation. 158,101],	Reduces Kv2 channel sensitivity to TEA [16]	Human: brain, placenta, skeletal muscle [157] SA cardiac nodal cells [160]		Silent subunit that interacts with Kv2 subunits to modify channel properties.

	Chr: 2 Entrez GeneID: 241794 MGI: 3616086 Access: XM_141545 (predicted) Protein size: 514aa **Rattus norvegicus:** Chr: 3 Entrez GeneID: 296395 Acces: XM_215951 (predicted) Acces: M81784 Protein size: 514aa		
Kv6.2	**Homo sapiens:** Gene symbol: KCNG2 Chr: 18q22-23 Entrez GeneID: 26251 Acces: NM_012283 size: 466aa **Mus musculus:** Gene symbol: Kcng2 Chr: 18 Entrez GeneID: 240444 MGI: 3694646 Access: XM_974521 (predicted) Protein size: 480aa **Rattus norvegicus:** Chr: 18q12.3 Entrez GeneID: 307234 Acces : XM_001058122 Protein size: 480aa Acces #: XM_225718 Protein size: 436 aa	Silent subunit, modifier of Kv2 channels Modifies kinetics and voltage dependence of Kv2.1 channels Confers submicromolar sensitivity to antiarrhythmic drug propafenone [162]	Primarily rat, human heart [162]	Silent subunit that interacts with Kv2 subunits to modify channel properties
Kv6.3 **Other names:** Kv10.1 [163]	**Homo sapiens:** Gene symbol: KCNG3 Chr: 2p21 Entrez GeneID: 170850 Acces: NM_133329 Protein size: 436aa **Mus musculus:** Gene symbol: Kcng3 Chr: 17 (syntenic) Entrez GeneID: 225030 MGI: 2663923 Acces: NM_153512 Protein size: 433aa	zSilent subunit, modifier of Kv2 channels Slows down inactivation [164] and deactivation [165]	Human, rat: brain, testis, thymus, adrenal glad, small intestine, kidney, lung, pancreas, ovary colon [163-165] **Brain:** Rat, widely expressed. mRNA most prominent in cortex, hippocampus, striatum, amygdala [164]	Silent subunit that interacts with Kv2 subunits to modify channel properties

Continued

Table 2.1.1.1 *(Continued)* Kv K⁺ channel proteins

Kv2 subfamily

Channel Subunit	Molecular information	Electrophysiology	Pharmacology	Expression	Disease and KO	Function
	Rattus norvegicus: Chr: 6q12 Entrez GeneID: 171011 Acces: NM_133426 Protein size: 433aa					
Kv6.4 Kv6.3 [163]	**Homo sapiens:** Gene symbol: KCNG4 Chr: 16q24.1 Entrez GeneID: 93107 Acces: NM_133490 Protein size: 519aa **Mus musculus:** Gene symbol: Kcng4 Chr: 8 (syntenic) Entrez GeneID: 66733 MGI: 1913983 Access: NM_025734 Protein size: 506 aa *Rattus norvegicus:* Chr: 19 (q12) Entrez GeneID: 307900 Acces: NM_226524 (predicted) Protein size: 506aa	Silent subunit, modifier of Kv2 channels Negative shifts in $V_{0.5}$ of activation and inactivation, slows deactivation and speeds activation [163]		Human: brain, liver, small intestine, colon [163]		Silent subunit that interacts with Kv2 subunits to modify channel properties
Kv8.1 Other names: Kv2.3 [166]	**Homo sapiens:** Gene symbol: KCNV1 Chr: 8q22.3-24.1 Entrez GeneID: 27012 Acces: NM_014379 NP_055194 Protein size: 500aa **Mus musculus:** Gene symbol: Kcnv1 Chr: 15 (D1) Entrez GeneID: 67498 MGI: 1914748 Access: NM_026200 Protein size: 503 aa **Rattus norvegicus:** Chr: 7q31 Entrez GeneID: 60326 Acces: NM_021697 Protein size: 503aa	Silent subunit, modifier of Kv2 channels Negative shift in $V_{0.5}$ of inactivation, slows activation and inactivation [167]		Hamster brain: mRNA very widespread [168]		Silent subunit that interacts with Kv2 subunits to modify channel properties

	Gene information	Function	Expression	
Kv8.2 **Other names:** Kv11.1 [163]	**Homo sapiens:** Gene symbol: KCNV2 Chr: 9 p24.2 Entrez GeneID: 169522 Acces: NM_133497 Protein size: 545 aa **Mus musculus:** Gene symbol: Kcnv2 Chr: 19 Entrez GeneID: 240595 MGI: 2670981 Access #: NM_183179 Protein size: 562aa **Rattus norvegicus:** Chr: 1 q51 Entrez GeneID: 294065 Acces: XM_220024 (predicted) Protein size: 561aa	Silent subunit, modifier of Kv2 channels. Small negative shift in $V_{0.5}$ of activation and inactivation, slight acceleration of activation [163]	Human: lung, liver, kidney, pancreas, spleen, thymus, prostate, testis, ovary, colon [163]	Silent subunit that interacts with Kv2 subunits to modify channel properties
Kv9.1	**Homo sapiens:** Gene symbol: KCNS1 Chr: 20q12 Entrez GeneID: 3787 Acces: NM_002251 Protein size: 526a **Mus musculus:** Gene symbol: Kcns1 Chr: 2 (H3) Entrez GeneID: 16538 MGI: 1197019 Access: NM_008435 Protein size: 497aa **Rattus norvegicus:** Chr: 3q42 Entrez GeneID: 117023 Acces: NM_053954 Protein size: 497aa	Silent subunit, modifier of Kv2 channels. Reduces currents, negative shift of inactivation. Mild effects on kinetics, except for slowing of deactivation of Kv2.1 and speeding of deactivation of Kv2.2 [158]. Slows down activation of Kv2.1, negative shift in activation and inactivation [169]	Human: brain, prostate, testis by PCR [170] Mouse brain predominantly [158] In mouse brain, mRNA Kv9.1 and Kv9.2 subunits show very similar expression: olfactory bulb, cortex, hippocampus, habenula, basolateral amygdala, cerebellum [158]	Silent subunit that interacts with Kv2 subunits to modify channel properties
Kv9.2	**Homo sapiens:** Gene symbol: KCNS2 Chr: 8q22 Entrez GeneID: 3788 Acces: NM_020697 Protein size: 477aa **Mus musculus:** Gene symbol: Kcns2 Chr: 15 Entrez GeneID: 16539	Silent subunit, modifier of Kv2 channels. Reduces currents, negative shift of inactivation. Mild effects on kinetics, except for slowing of deactivation of Kv2.1 [158]	Mouse brain predominantly [158] Pulmonary artery [130] In mouse brain, mRNA Kv9.1 and Kv9.2 subunits show very similar expression: olfactory bulb, cortex, hippocampus, habenula, basolateral amygdala, cerebellum [158]	Silent subunit that interacts with Kv2 subunits to modify channel properties.

Continued

Table 2.1.1 (Continued) Kv K+ channel proteins

Kv2 subfamily

Channel Subunit	Molecular information	Electrophysiology	Pharmacology	Expression	Disease and KO	Function
	MGI: 1197011 Access: NM_181317 Protein size: 477aa *Rattus norvegicus:* Chr: 7q22 Entrez GeneID: 66022 Acces: NM_023966 Protein size: 477aa					
Kv9.3	**Homo sapiens:** Gene symbol: KCNS3 Chr: 2p24 Entrez GeneID: 3790 Acces: NM_002252 Protein size: 491aa **Mus musculus:** Gene symbol: Kcns3 Chr: 12 (25 cM) Entrez GeneID: 238076 MGI: 1098804 Access: NM_173417 Protein size: 491aa **Rattus norvegicus:** Chr: 6q14 Entrez GeneID: 83588 Acces: NM_031778 Protein size: 491aa	Silent subunit, modifier of Kv2 channels. Negative shifts $V_{0.5}$ activation and inactivation, slow deactivation rates, speed up recovery from inactivation [171]		Widespread expression in human by PCR [170]		Silent subunit that interacts with Kv2 subunits to modify channel properties

Kv3 subfamily

Channel subunit	Molecular information	Electrophysiology properties in heterologous cells[6]	Pharmacology	Expression	Associated pathology	Function
Kv3.1 **Other names:** Kv3.1 [172] NGK2 [37] Kv4 [173]KShIIIB, [174] Raw2, [175]	**Homo sapiens:** Gene symbol: KCNC1 Chr: 11p15 Entrez GeneID: 3746 Two alternative splice versions; Kv3.1a: 511aa, NM_004976; Kv3.1b: EAW 68425, 585 aa. [37,172,173,175,176] **Mus musculus:** Gene symbol: Kcnc1 Chr: 7 23.5 cM Entrez GeneID: 16502 MGI: 96667 Kv31a: Access #: NM_008421 Protein size: 511aa	**Activation:** [9] Von = −10 mV $V_{0.5}$ = 16 mV k = 8.7 mV τ = +++ **Deactivation:** [9] τ = 1.4ms **Inactivation:** slow **Unitary conductance:** [9] 27pS	**TEA:** 0.15-0.2 mM [9,177] **4AP:** .03-6 mM [9,177] **CTX:** NB [9,177] **DTX:** NB [9,177] **BDS I:** 220nM [178] **BDS II:** 750nM [178]	Brain: in situ [179,180]; protein: [181,182,183,184,185] Broadly expressed in brain and spinal cord structures but in specific cell populations, including many (but not only) GABAergic neurons (in somas, axons, and terminals): cerebellar granule cells, globus pallidus, subthalamic nucleus, SNR, reticular thalamic nuclei, cortical and hippocampal interneurons, inferior colliculi, vestibular nuclei, auditory brainstem nuclei	Kv3.1/ mice exhibit impaired motor skills hyperactivity and sleep loss. Kv3.1/Kv3.3 double knockout mice display severe ataxia, tremor, myoclonus, and hypersensitivity to ethanol [187]	Important for spike repolarization and high-frequency firing of brainstem auditory neurons and fast-spiking GABAergic interneurons Reviewed in: [188,189,190] Regulation of Ca2+ influx and neurotransmitter release via controlling spike repolarization and duration in presynaptic terminals [191,192,193]. Channels probably containing Kv3.1 and Kv3.3 regulate release from cerebellar parallel fibre terminals [194] Spike repolarization in nodes of Ranvier in central myelinated axons [196] Type I channel in T lymphocytes [172]

	Electrophysiology	Pharmacology	Expression	Function
Kv3.1b (5' incomplete cDNA): Access #: AK162964 Protein size: 479 aa **Rattus norvegicus:** Chr: 1q22 Entrez GeneID: 25327 Kv3.1b: Acces #: NM_012856 Protein size: 585aa			Dog atrium [186] Molecular basis of canine atrial IKur.d. Type I channel in T lymphocytes [172]	Probably in heteromeric complexes with Kv3.1 important for spike repolarization and high-frequency firing of fast spiking GABAergic interneurons [200,201] Reviewed in: [188,189] Regulation of Ca2+ influx and GABA release via controlling spike repolarization and duration in presynaptic terminals [192] Activity modulated by protein kinase A in heterologous cells and in neurons [202, 203]
Kv3.2 Other names: RKShIIIA [402] Raw1 [175] Kv3.2a [198] rKv3.2b and rKv3.2c [199] **Homo sapiens:** Gene symbol: KCNC2 Chr: 12q14.1 Entrez GeneID: 3747 Kv3.2a: Acces: NM_139136 Protein size: 613aa Kv3.2b: Acces: NM_139137 Protein size: 638aa Kv3.2d?: Acces: AY243473 Protein size: 629aa **Mus musculus:** Gene symbol: Kcnc2 Chr: 10 (62.0 cM) Entrez GeneID: 268345 MGI: 96668 Kv3.2b: Access: BC116290 Protein size: 642aa Kv3.2c: Access: NM_001025581 Protein size: 639 aa **Rattus norvegicus:** Chr: 7q12–22; Entrez GeneID: 246153 Kv3.2a: Acces: NM_139216 Protein size: 613aa Kv3.2b: Acces: NM_139217 Protein size: 638aa Kv3.2c: Acces: partial seq. AAA41820 Protein size: Kv3.2d Acces: S22703 Protein size: 624aa	**Activation:** [188,197] V_{on}= -10mV $V_{0.5}$ = 12mV k = 8.4 mV τ = +++ **Deactivation:** [188] τ= 2.9 ms **Inactivation:** [188] Very slow **Single channel conductance:** [188] 16–20pS	**TEA:** 0.15mM [177] **4AP:** 0.6mM [177] **DTX:** NB [177] **CTX:** NB [177] **BDS I and II:** block at concentrations similar to effects on Kv3.1 channels [178] Blocked by ShK, but at much higher concentrations than Kv1 channels [204]	Tissue expression: mRNA only brain, [179]. Prominently in thalamus. GABAergic interneurons of the neocortex, hippocampus, and caudate; globus pallidus, SNR, sensory nuclei in brainstem Protein in terminal fields of thalamocortical axons. Somata and axons of cortical interneurons and other mRNA-expressing neurons [179,181,183,202]. Islets, [135]; Schwann cells, and Kv3.1b [109]; mesenteric: molecular basis of voltage-dependent delayed rectifier K+ channels in smooth muscle cells from rat tail artery [205]	Kv3.2 null mice show alterations in cortical electroencephalographic patterrs and increased susceptibility to epileptic seizures consistent with an impairment of cortical inhibitory mechanisms [200]

6 Data is from expression in mammalian transfected cells, and if not available data from expression in *Xenopus* oocytes was used. Values reported at room temperature. Von, approximate voltage at which current becomes apparent. Activation rate: +++ < 5 ms at +40 mV; ++ 5-10 msec at +40 mV; + >10 msec at + 40 mV

Continued

Table 2.1.1.1 (*Continued*) Kv K$^+$ channel proteins

Kv3 subfamily

Channel subunit	Molecular information	Electrophysiology properties in heterologous cells	Pharmacology	Expression	Associated pathology	Function
Kv3.3 **Other names:** hKv3.3 mKv3.3 [206] RKShIIID [174] Kv3.3b [207]	**Homo sapiens:** Gene symbol: KCNC3 Chr: 19 q13.3-q13.4 Entrez GeneID: 3748 Acces: NM_004977 Protein size: 757aa[7] **Mus musculus:** Gene symbol: kcnc3 Chr: 7 (23.0 cM) Entrez GeneID: 16504 MGI: 96669 Access: NM_008422[8], 679 aa; S69381[9]; 769aa **Rattus norvegicus:** Chr: 1q22 Entrez GeneID: 117101 Acces: Kv3.3a M84210; 889; Kv3.3b M84211, 679aa; Kv3.3c: AY179604, 758aa; Kv3.3d: NM_053997, 769; Kv3.3e: AAV80432, 729aa.	**Activation:** [208] V_{on}= -10 mV $V_{0.5}$ = 7mV k = 6mV τ = ++ **Deactivation:** [188] τ = 2 ms **Inactivation:** [208][10] $V_{0.5}$ = 5.2mV k = 6.1mV τ = 200ms **Single channel conductance:** [188] 14 pS	**TEA:** 0.14mM [177,208] **4AP:** 1.2mM [177,208] **CTX:** NA **DTX:** NA **BDS** toxins: not yet tested	Expressed predominantly in the CNS. Broadly expressed in somas, axons, and terminals in brain and spinal cord structures but in specific cell types. In many (but not all) GABAergic neurons throughout forebrain and midbrain, and in mossy fiber axons of dentate granule cells in hippocampus. Very strongly expressed in cerebellum, particularly in Purkinje cell somas, axons and dendrites. Expression in brainstem sensory and reticular neurons and motoneurons [179,215]	**Human:** Mutations in Kv3.3 cause spinocerebellar ataxia SCA13 [209] **Mice:** Kv3.3 null mice exhibit cerebellar dysfunction, altered Purkinje cell discharges and olivocerebellar system properties (lack of harmaline-induced tremor [210]; Kv3.1/Kv3.3 double knockout mice display severe ataxia, tremor, myoclonus, and hypersensitivity to ethanol [187]	Regulation of Na+ and Ca2+ spike repolarization and spontaneous firing frequency of Purkinje cells [211-213]. May contribute to Kv3.1-containing channels in auditory brainstem (see Kv3.1) Channels probably containing Kv3.3 and Kv3.4 regulate neurotransmitter release at motor end plates [214] and channels probably containing Kv3.1 and Kv3.3 transmitter release from cerebellar parallel fiber terminals [194]
Kv3.4 [216] **Other names:** Raw3 [216] HKShIIIC [217] mKv3.4 [206]	**Homo sapiens:** Gene symbol: KCNC4 Chr: 1p21 Entrez GeneID: 3749 Kv3.4b Acces: NM_153763 Protein size: 582aa Kv3.4c Acces: NM_001039574 Protein size: 626aa Kv3.4d Acces: NM_004978 Protein size: 635 **Mus musculus:** Gene symbol: Chr: 3 53.0 cM Entrez GeneID: 99738 MGI: 96670 Kv3.4a: Access: NM_145922 Protein size: 628 aa **Rattus norvegicus:** Acces: X62841 Protein size: 625aa	**Activation:** [216-217] V_{on}= -10 mV $V_{0.5}$ = 3.4-19 mV k = 8 mV τ = +++ **Deactivation:** See discussion in [249] **Inactivation** [216,217] $V_{0.5}$ = -15 mV k = 7.2 mV τ = 15 ms Recovery: very slow **Single channel conductance:** [216] 14pS	**TEA:** 0.09-0.3mM [175,177,216] **4AP:** 0.5-0.6mM [175,177,216] **DTX:** NB [216] **CTX:** NB [177] **BDS-I:** 47 nM [218], however [178] report similar sensitivity as Kv3.1 channels	Most prominently in skeletal muscle [49,179,219] In brain: hippocampal granule cells, Purkinje cells, and pontine nuclei [179] parathyroid, prostate, [220] pancreatic acinar cells [127,221]	**Human:** An R83H missense mutation in MiRP2 that diminished the effects of MiRP2 on Kv3.4 has been associated with a form of periodic paralysis [49]	Fast inactivating Kv3 subunit. Together with MirP2 forms low-voltage-activating potassium channels that regulate skeletal muscle resting potential [49]. Channels probably containing Kv3.3 and Kv3.4 regulate transmitter release at motor end plates [214]

Accesory Proteins
MiRP2 forms potassium channels in skeletal muscle with Kv3.4 [49].

Kv4 subfamily

Kv4.1 **Other names:** mShal [222] **Homo sapiens:** Gene symbol: KCND1 Chr: Xp11.23 Entrez GeneID: 3750 Acces: NM_004979 Protein size: 647aa **Mus musculus:** Gene symbol: Kcnd1 Chr: X (2.2 cM) Entrez GeneID: 16506 MGI: 96671 Access: NM_008423 Protein size: 651aa **Rattus norvegicus:** Chr: X q13 Entrez GeneID: 116695 Acces: XM_217601 (predicted) Protein size: 650aa Associated proteins: KChIPs, DPPX and DPP10 (See text).	**Activation:** [223,224] Von= -50mV $V_{0.5}$ = -10mV k = 27 mV **Deactivation:** [223,225] τ = 1 ms **Inactivation:** [223] $V_{0.5}$ =-70mV k =5 mV **Recovery from inactivation:** [223,224] τ = 170ms (-100mV) **Single channel conductance:** [224] 5 pS	**TEA:** NB [222] **4-AP:** 9 mM [222,226] **DTX:** NA **CTX:** NA **Heteropodatoxin:** Not tested [227] **PaTx 1 and 2 (Phrixotoxins):** >300nM [228] **ScTx1** (Stromatoxin) no activity [229,230] **sBmTX3** (*Buthus martensi* toxin): 105 nm [231]	Rat brain: mRNA very weakly expressed in rat [232]. Expressed in human brain [233]; uterus [234]	Kv4 channels underlie the fast transient outward current I_{to} in cardiac myocytes and most of the subthreshold-operating somato-dendritic A-type K^+ current in mammalian neurons (I_{SA}) (see text). However, Kv4.2 and Kv4.3 appear to be the main Kv4 subunits in brain and heart
Kv4.2 **Other names:** RK5 [7] Rat Shal1 [235] **Homo sapiens:** Gene symbol: KCND2 Chr: 7q31 Entrez GeneID: 3751 Acces: NM_012281 Protein size: 630aa **Mus musculus:** Gene symbol: kcnd2 Chr: 6 (7.2 cM) Entrez GeneID: 16508 MGI: 102663 Access: NM_019697 Protein size: 630aa **Rattus norvegicus:** Chr: 4q22	**Activation:** [236] Von= -50mV $V_{0.5}$ = -2.5mV k = 20mV **Deactivation:** [225] τ = 1.5 ms **Inactivation:** [236] $V_{0.5}$ = -65mV k = 6.5 mV **Recovery from inactivation:** [236] τ = 132ms (-110mV) **Single channel conductance:** NA	**TEA:** NB [225] **4-AP:** 5 mM [235] **DTX:** NA **CTX:** NA **HpTX1, 2,3 (Heteropodatoxin)**67-100nM [227,237] **ScTx1:**1.2nM [229,230] **PaTx1:** 1: 5nM, **PaTx2:** 34nM (Phrixotoxins) [228]	**mRNA:** **Brain:** mRNA by ISH: Expressed in selective brain areas, sometimes overlapping with Kv4.3 (e.g. thalamus), but often in areas not expressing Kv4.3 (predominant form in the caudate-putamen, pontine nucleus and several nuclei in the medulla; granule cells in the OB; CA1 pyramidal cells in the hippocampus; reciprocal expression with Kv4.3 in the cerebellar granule cell layer [232]	Kv4 channels underlie the fast transient outward current I_{to} in cardiac myocytes and most of the subthreshold-operating somato-dendritic A-type K^+ current (I_{SA}) in mammalian neurons (see text). These currents regulate repolarization of cardiac action potentials, firing frequency in neurons and dendritic excitability (see text) **Mouse:** The loss of Kv4.2 protein in hippocampal CA1 pyramidal neurons of Kv4.2 produces a near-complete elimination of A-type K^+ currents from apical dendrites, which results in an increase in the amplitude of backpropagating action potentials (APs) and an increase of concomitant Ca^{2+} influx, thus facilitating the induction of long-term potentiation [241]

7 Corresponds to Kv3.3c in rat 22.Rudy, B., A. Chow, D. Lau, et al., *Contributions of Kv3 channels to neuronal excitability.* Ann N Y Acad Sci, 1999. **868**: p. 304-43.

8 Corresponds to Kv33e in rat

9 Corresponds to Kv3.3d in rat

10 Currents are inactivating in Xenopus oocytes but not in mammalian cells (see text and Ref 22.Rudy, B., A. Chow, D. Lau, et al., *Contributions of Kv3 channels to neuronal excitability.* Ann N Y Acad Sci, 1999. **868**: p. 304-43).

Continued

Table 2.1.1.1 *(Continued)* Kv K⁺ channel proteins

Kv4 subfamily

Channel subunit	Molecular information	Electrophysiology properties in heterologous cells	Pharmacology	Expression	Associated pathology	Function
	Entrez GeneID: 65180 Acces: NM_031730 Protein size: 630 aa		Less sensitive to **sBmTX3** than Kv4.1 [231]	**Heart** (rodent ventricular muscle, 238]). **Protein:** Predominantly dendritic expression in hippocampal, striatal and neocortical pyramidal cells [239] and cerebellar granule cells [240]	Genetic elimination of Kv4.2 in dorsal horn neurons increases excitability of dorsal horn neurons, resulting in enhanced sensitivity to tactile and thermal stimuli. Furthermore, ERK-dependent forms of pain hypersensitivity are absent in Kv4.2 (/) mice [242]	
	Associated proteins: KChIPs, DPPX and DPP10 (See text).					
Kv4.3 **Other names:** KShIVB [234,217]	**Homo sapiens:** Gene symbol: KCND3 Chr: 1p13.3 Entrez GeneID: 3752 Two isoforms: Kv4.3S: Acces: NM_172198 Size: 636aa Kv4.3 L: Acces: NM_004980 Size: 655aa **Mus musculus:** Gene symbol: *Kcnd3* Chr: 3 F3 Entrez GeneID: 56543 MGI: 1928743 Two isoforms: Kv4.3S: Acces: NM_001039347 Size: 636aa Kv4.3 L: Acces: NM_019931 Size: 655aa **Rattus norvegicus:** Chr: 2q34 Entrez GeneID: 65195 Two isoforms: Kv4.3S (isoform 2): Acces: NM_031739; Size: 636aa Kv4.3L (isoform 1): Acces: AB003587, Size: 655aa,	**Activation:** [236] $V_{on}=-50mV$ $V_{0.5} = -10mV$ $k = 18\ mV$ **Deactivation:** [224,225] $\tau = 5\ ms$ **Inactivation:** [223,236] $V_{0.5} =-60mV$ $k =6\ mV$ $\tau = 5.2\ ms$ (-70mV) **Recovery from inactivation:** [236] $\tau = 266ms$ (-110mV) **Single channel conductance:** [225] 4 pS	**TEA:** NB [243] **4-AP:** 5-10 mM [243] **DTX:** NA **CTX:** NA **PaTx1:** 28 nM, **PaTx2:** 71 nM (Phrixotoxins) [228] **Heteropodatoxin:** NB [227] **ScTx1** (Stromatoxin) no activity [229,230] *Buthus martensi* toxin (**sBmTX3**) Less sensitive to sBmTX3 than Kv4.1 [231]	**mRNA:** **Brain:** mRNA by ISH: Expressed in selective brain areas, sometimes overlapping with Kv4.2 (eg. thalamus), but often in areas not expressing Kv4.2. Predominant form in substantia nigra pars compacta, restrosplenial cortex, superior colliculus, raphe, and amygdala; periglomerular cells in the OB: interneurons in the hippocampus; in and cerebral cortex; Purkinje cells and cerebellar molecular layer interneurons. In cerebellar granule cells predominant in lateral lobules [232]. **Heart** Kv4.3 main Kv4 gene in canine and human ventricle [245] **Protein:** Predominantly somato-dendritic expression in hippocampal, striatal and neocortical interneurons [239] and cerebellar Purkinje and granule cells [230]		Kv4 channels underlie the fast transient outward current I_{to} in cardiac myocytes and most of the subthreshold-operating somato-dendritic A-type K⁺ current (I_{SA}) in mammalian neurons (see text). These currents regulate repolarization of cardiac action potentials, firing frequency in neurons and dendritic excitability (see text). Kv4.3 is the main Kv4 protein in cortical GABAergic interneurons, cerebellar Purkinje cells and dopaminergic neurons in substantia nigra [232]. The pacemaker frequency of individual dopaminergic neurons, correlates with I_{SA} levels, which in turn correlates with the concentration of Kv4.3 mRNA [244]

Associated proteins:
KChIPs, DPPX and DPP10
(See text).

[1] Ramaswami M *et al.* (1990) *Mol Cell Neurosci* 1, 214–223; [2] Kamb AM *et al.* (1989) *PNAS* 86, 4372–4376; [3] Tempel BL *et al.* (1988) *Nature* 332, 837–839; [4] Chandy KG *et al.* (1990) *Science* 247, 973–975; [5] Christie MJ *et al.* (1989) *Science* 244, 221–224; [6] Baumann A *et al.* (1988) *EMBO J* 7, 2457–2463; [7] Roberds SL *et al.* (1991) *PNAS* 88, 1798–1802; [8] Stuhmer W *et al.* (1989) *EMBO J* 8, 3235–3244; [9] Grissmer S *et al.* (1994) *Mol Pharmacol* 45, 1227–1234; [10] Harvey AL *et al.* (2004) *Curr Med Chem* 11, 3065–3072; [11] Smart SL *et al.* (1998) *Neuron* 20, 809–819; [12] Herson PS *et al.* (2003) *Nat Neurosci* 6, 378–383; [13] Dodson PD *et al.* (2002) *J Neurosci* 22, 6953–6961; [14] Gutman GA *et al.* (2005) *Pharmacol Rev* 57, 473–508; [15] Bhalla T *et al.* (2004) *Nat Struct Mol Biol* 11, 950–956; [16] Koschak A *et al.* (1998) *J Biol Chem* 273, 2639–2644; [17] Mouhat SG *et al.* (2006) *Mol Pharmacol* 69, 354–362; [18] Kalman K *et al.* (1998) *J Biol Chem* 273, 32697–326707; [19] Verma-Kurvari S *et al.* (1997) *Brain Res Mol Brain Res* 46, 54–62; [20] Veh RW *et al.* (1995) *Eur J Neurosci* 7, 2189–2205; [21] Wang H *et al.* (1994) *J Neurosci* 14, 4588–4599; [22] Wang H *et al.* (1993) *Nature* 365, 75–79; [23] Rasband MN *et al.* (2004) *J Neurosci Res* 76, 749–757; [24] Rhodes KJ *et al.* (1995) *J Neurosci* 15, 5360–5371; [25] Diez M *et al.* (2003) *Eur J Neurosci* 18, 3218–3230; [26] Persson AS *et al.* (2005) *BMC Neurosci* 6, 65; [27] Lavebratt C *et al.* (2006) *Neurobiol Dis* 24, 374–383; [28] Browne DL *et al.* (1994) *Nat Genet* 8, 136–140; [29] Jen JC and Baloh RW (2002) *Adv Neurol* 89, 459–461; [30] Newsom-Davis J *et al.* (2003) *Ann N Y Acad Sci* 998, 202–210; [31] Kleopa KA *et al.* (2006) *Brain* 129, 1570–1584; [32] Barnes-Davies M *et al.* (2004) *Eur J Neurosci* 19, 325–333; [33] McKinnon D *et al.* (1989) *J Biol Chem* 264, 8230–8236; [34] Grupe A *et al.* (1990) *EMBO J* 9, 1749–1756; [35] Paulmichl M *et al.* (1991) *PNAS* 88, 7892–7895; [36] Ribera AB and Nguyen DA (1993) *J Neurosci* 13, 4988–4996; [37] Yokoyama S *et al.* (1989) *FEBS Lett* 259, 37–42; [38] Werkman TR *et al.* (1993) *Mol Pharmacol* 44, 430–436; [39] Albarwani S *et al.* (2003) *J Physiol* 551, 751–763; [40] Fergus DJ *et al.* (2003) *Pflugers Arch* 445, 697–704; [41] Kerr PM *et al.* (2001) *Circ Res* 89, 1038–1044; [42] Rasband MN *et al.* (2001) *PNAS* 98, 13373–13378; [43] Mouhat SV *et al.* (2005) *Biochem J* 385, 95–104; [44] Wang FC *et al.* (1999) *Eur J Biochem* 263, 230–237; [45] Scott VE *et al.* (1994) *Biochem* 33, 1617–1623; [46] Monaghan MM *et al.* (2001) *J Neurosci* 21, 5973–5983; [47] Rhodes KJ *et al.* (1997) *J Neurosci* 17, 8246–8258; [48] McNamara NM *et al.* (1996) *Eur J Neurosci* 8, 688–699; [49] Abbott GW *et al.* (2001) *Cell* 104, 217–231; [50] Rasband MN and Trimmer JS (2001) *J Comp Neurol* 429, 166–176; [51] Sheng M *et al.* (1994) *J Neurosci* 14, 2408–2417; [52] Cai YC *et al.* (1992) *DNA Cell Biol* 11, 163–172; [53] Attali B *et al.* (1992) *J Biol Chem* 267 8650–7; [54] Philipson LH *et al.* (1991) *PNAS* 88, 53–7; [55] Swanson R *et al.* (1990) *Neuron* 4, 929–939; [56] Poliak S *et al.* (1999) *Neuron* 24, 1037–47; [57] Grissmer S *et al.* (1990) *PNAS* 87 9411–5; [58] Chandy G *et al.* (2004) *TIPS* 25 280–289; [59] Beeton C *et al.* (2005) *Mol Pharmacol* 67, 1369–1381; [60] Vicente R *et al.* (2006) *J Biol Chem* 281, 37675–37685; [61] Fordyce CB *et al.* (2005) *J Neurosci* 25, 7139–7149; [62] Yao X *et al.* (1996) *J Clin Invest* 97, 2525–2533; [63] Jacob A (2000) *Mol Hum Reprod* 6, 303–313; [64] Komarova *et al.* (2001) *Curr Pharm Des* 7, 637–654; [65] Inda MC *et al.* (2006) *PNAS* 103, 2920–2925; [66] Van Wart A *et al.* (2007) *J Comp Neurol* 500, 339–352; [67] Fadool DA *et al.* (2004) *Neuron* 41, 389–404; [68] Beeton C and Chandy KG (2005) *Neurosci* 11, 550–562; [69] David G *et al.* (1995) *J Physiol* 489, 141–157; [70] Xu J *et al.* (2003) *Hum Mol Genet* 12, 551–559; [71] Xu J *et al.* (2004) *PNAS* 101, 3112–3117; [72] Garcia-Calvo M *et al.* (1993) *J Biol Chem* 268, 18866–18874; [73] Fadool DA and Levitan IB (1998) *J Neurosci* 18, 6126–6137; [74] Grunnet M *et al.* (2003) *Biophys J* 85, 1525–1537; [75] Levite M *et al.* (2000) *J Exp Med* 191, 1167–1176; [76] Tamkun MM *et al.* (1991) *FASEB J* 5, 331–337; [77] Tseng-Crank JC *et al.* (1990) *FEBS Lett* 268, 63–68; [78] Judge SI *et al.* (1999) *Brain Res* 831, 43–54; [79] Binzen U *et al.* (2006) *Neurosci* 142, 527–539; [80] Tanimoto T *et al.* (2005) *Brain Res* 1044, 262–265; [81] Vydyanathan A *et al.* (2005) *J Neurophysiol* 93, 3401–3409; [82] Yang EK *et al.* (2004) *Neurosci* 123, 867–874; [83] London B *et al.* (1998) *J Physiol* 509,171–182; [84] Suzuki S *et al.* (2005) *J Neuroimmunol* 170, 141–149; [85] Guan D *et al.* (2006) *J Physiol* 571, 371–389; [86] Tao Y *et al.* (2005) *J Neurochem* 94, 1512–1522; [87] London B *et al.* (2007) *J Physiol* 578, 115–129; [88] Imamura F *et al.* (2002) *J Biol Chem* 277(5), 3640–3646; [89] Wong W and Schlichter LC (2004) *J Biol Chem* 279, 444–452; [90] Cukovic D *et al.* (2001) *FEBS Lett* 498, 87–92; [91] Kuryshev YA *et al.* (2001) *Am J Physiol* 281, C290–C299; [92] Pourrier M *et al.* (2003) *J Membr Biol* 194, 141–152; [93] Aydar E *et al.* (2002) *Neuron* 34, 399–410; [94] Sheng M *et al.* (1992) *Neuron* 9, 271–284; [95] Lujan R *et al.* (2003) *J Chem Neuroanat* 26, 209–224; [96] Rhodes KJ *et al.* (1996) *J Neurosci* 16, 4846–4860; [97] Juiz JM *et al.* (2000) *Eur J Neurosci* 12, 4345–4356; [98] Pannasch U *et al.* (2006) *Mol Cell Neurosci* 33, 401–411; [99] Kotecha SA and Schlichter LC (1999) *J Neurosci* 19, 10680–10693; [100] Snyders DJ *et al.* (1993) *J Gen Physiol* 101, 513–543; [101] Feng J *et al.* (1997) *Circ Res* 80, 572–579; [102] Brendel J and Peukert S (2003) *Curr Med Chem Cardiovasc Hematol Agents* 1, 273–287; [103] Holmes TC *et al.* (1996) *Science* 274, 2089–2091; [104] Nitabach MN *et al.* (2001) *PNAS* 98, 705–710; [105] Wible BA *et al.* (1998) *J Biol Chem* 273, 11745–11751; [106] Sobko A *et al.* (1998) *EMBO J* 17, 4723–4734; [107] Maruoka ND *et al.* (2000) *FEBS Lett* 473, 188–194; [108] Murata M *et al.* (2001) *Am J Physiol* 281, H2575–H2584; [109] Sobko A *et al.* (1998) *J Neurosci* 18, 10398–10408; [110] Vincent A *et al.* (2000) *J Neurosci* 20, 6087–6094; [111] Cox RH (2005) *Cell Biochem Biophys* 42, 167–195; [112] Archer SL *et al.* (2001) *FASEB J* 15, 1801–1803; [113] Chung YH *et al.* (2001) *Brain Res* 895, 173–177; [114] Maletic-Savatic M *et al.* (1995) *J Neurosci* 15, 3840–3851; [115] Migeon MB *et al.* (1992) *Epilepsy Res Suppl* 9, 173–181; [116] Bertoli A *et al.* (1994) *Eur Biophys J* 23, 379–384; [117] Coetzee WA *Ann N Y Acad Sci* 868, 233–285; [118] Koh SD *et al.* (1999) *J Physiol* 515, 475–487; [119] Yuan XJ *et al.* (1998) *Am J Physiol* 274, L621–L635; [120] Plane F *et al.* (2005) *Circ Res* 96, 216–224; [121] Glazebrook PA *et al.* (2002) *J Physiol* 541, 467–482; [122] Attali B *et al.* (1997) *J Neurosci* 17, 8234–8245; [123] Marionneau C *et al.* (2005) *J Physiol* 562, 223–234; [124] Coma M *et al.* (2003) *FEBS Lett* 536, 45–50; [125] Antozzi C *et al.* (2005) *Neurology* 64, 1290–1293; [126] Finol-Urdaneta RK *et al.* (2006) *J Gen Physiol* 128, 133–145; [127] Kalman K *et al.* (1998) *J Biol Chem* 273, 5851–5857; [128] Bardien-Kruger S *et al.* (2002) *Eur J Hum Genet* 10, 36–43; [129] Couette B *et al.* (2006) *Cardiovasc Hematol Disord Drug Targets* 6, 169–190; [130] Davies AR and Kozlowski RZ (2001) *Lung* 179, 147–161; [131] Fry M *et al.* (2004) *J Neurobiol* 60, 227–235; [132] Yao X *et al.* (1995) *PNAS* 92, 11711–11715; [133] Yao X *et al.* (2002) *J Am Soc Nephrol* 13, 2831–2839; [134] Lang R *et al.* (2000) *Am J Physiol* 278, F1013–F10321; [135] Roe MW *et al.* (1996) *J Biol Chem* 271, 32241–32246; [136] Tian S *et al.* (2002) *Am J Physiol* 283, F142–F2429; [137] Frech GC *et al.* (1989) *Nature* 340, 642–645; [138] Shi G *et al.* (1994) *J Biol Chem* 269, 23204–23211; [139] Kramer JW *et al.* (1998) *Am J Physiol* 274, C1501–C1510; [140] Chapman ML *et al.* (2001) *J Physiol* 530, 21–33; [141] Shieh CC and Kirsch GE (1994) *Biophys J* 67, 2316–23125; [142] Swartz KJ and MacKinnon R (1995) *Neuron* 15, 941–949; [143] Hwang PM *et al.* (1992) *Neuron* 8, 473–481; [144] Malin SA and Nerbonne JM (2002) *J Neurosci* 22, 10094–10105; [145] Dixon JE and McKinnon D (1996) *Eur J Neurosci* 8, 183–191; [146] Muennich EA and Fyffe RE (2004) *J Physiol* 554, 673–685; [147] Du J *et al.* (2000) *J Physiol* 522, 19–31; [148] Pal S *et al.* (2003) *J Neurosci* 23, 4798–4802; [149] MacDonald PE *et al.* (2001) *Mol Endocrinol* 15, 1423–1435; [150] Hwang PM *et al.* (1993) *J Neurosci* 13, 1569–1576; [151] Mohapatra DP and Trimmer JS (2006) *J Neurosci* 26, 685–695; [152] Schmalz F *et al.* (1998) *Am J Physiol* 274, G901–G911; [153] Trimmer JS (1991) *PNAS* 88, 10764–1078; [154] Du J *et al.* (1998) *Neurosci* 84, 37–48; [155] Xu H *et al.* (1999) *J Physiol* 519, 11–21; [156] Fink M *et al.* (1996) *J Biol Chem* 271, 26341–26348; [157] Su K *et al.* (1997) *Biochem Biophys Res Commun* 241, 675–681; [158] Salinas M *et al.* (1997) *J Biol Chem* 272, 24371–243379; [159] Drewe JA *et al.* (1992) *J Neurosci* 12, 538–548; [160] Brahmajothi MV *et al.* (1996) *Circ Res* 78, 1083–1089; [161] Post MA *et al.* (1996) *FEBS Lett* 399, 177–182; [162] Zhu XR *et al.* (1999) *Rec Channels* 6, 337–350; [163] Ottschytsch N *et al.* (2002) *PNAS* 99, 7986–7991; [164] Vega-Saenz de Miera EC (2004) *Brain Res Mol Brain Res* 123, 91–103; [165] Sano Y *et al.* (2002) *FEBS Lett* 512, 230–234; [166] Chiara MD *et al.* (1999) *J Neurosci* 19, 6865–6873; [167] Salinas M *et al.* (1997) *J Biol Chem* 272, 8774–8780; [168] Hugnot, JP *et al.* (1996) *EMBO J* 15, 3322–3331; [169] Richardson FC and Kaczmarek LK (2000) *Hear Res* 147, 21–30; [170] Shepard AR and Rae JL (1999) *Am J Physiol* 277, C412–C424; [171] Kerschensteiner D and Stocker M (1999) *Biophys J* 77, 248–257; [172] Grissmer S *et al.* (1992) *J Biol Chem* 267, 20971–20979; [173] Luneau CJ *et al.* (1991) *PNAS* 88, 3932–3936; [174] Haas M *et al.* (1993) *Mamm Genome* 4, 711–715; [175] Rettig J *et al.* (1992) *EMBO J* 11, 2473–2486; [176] Ried T *et al.* (1993) *Genomics* 15, 405–411; [177] Rudy B (1988) *Neurosci* 25, 729–749; [178] Yeung SY *et al.* (2005) *J Neurosci* 25, 8735–8745; [179] Weiser M *et al.* (1994) *J Neurosci* 14, 949–972; [180] Perney TM *et al.* (1992) *J Neurophysiol* 68, 756–766; [181] Chow A *et al.* (1999) *J Neurosci* 19, 9332–45; [182] Perney TM and Kaczmarek LK (1997) *J Comp Neurol* 386, 178–202; [183] Tansey EP *et al.* (2002) *Hippocampus* 12, 137–148; [184] Weiser M *et al.* (1995) *J Neurosci* 15, 4298–4314; [185] Sekirnjak C *et al.* (1997) *Brain Res* 766, 173–187; [186] Yue L *et al.* (2000) *J Physiol* 527, 467–478; [187] Espinosa F *et al.* (2001) *J Neurosci* 21, 6657–6665; [188] Rudy B *et al.* (1999) *Ann N Y Acad Sci* 868, 304–343; [189] Rudy B and McBain CJ (2001) *TINS* 24, 517–526; [190] Kaczmarek LK *et al.* (2005) *Hear Res* 206, 133–145; [191] Ishikawa T *et al.* (2003) *J Neurosci* 23, 10445–1053; [192] Goldberg EM *et al.* (2005) *J Neurosci* 25, 5230–5235; [193] Dodson PD and Forsythe ID *et al.* (2004) *TINS* 27, 210–217; [194] Matsukawa H *et al.* (2003) *J Neurosci* 23, 7677–7684; [195] Cachero TG *et al.* (1998) *Cell* 93, 1077–1085; [196] Devaux J *et al.* (2003) *J Neurosci* 23, 4509–4518; [197] McCormack *et al.* (1991) *PNAS* 88, 4060; [198] Ponce A *et al.* (1997) *J Membr Biol* 159, 149–159; [199] Luneau C *et al.* (1991) *FEBS Lett* 288, 163–167; [200] Lau D *et al.* (2000) *J Neurosci* 20, 9071–9085; [201] Erisir A *et al.* (1999) *J Neurophysiol* 82, 2476–2489; [202] Moreno H *et al.* (1995) *J Neurosci* 15, 5486–5501; [203] Atzori M *et al.* (2000) *Nat Neurosci* 3, 791–798; [204] Yan L *et al.* (2005) *Mol Pharmacol* 67, 1513–1521; [205] Xu C *et al.* (2000) *Life Sci* 66, 2023–2033; [206] Ghanshani S *et al.* (1992) *Genomics* 12, 190–196; [207] Goldman-Wohl DS *et al.* (1994) *J Neurosci* 14, 511–522; [208] Vega-Saenz de Miera E *et al.* (1992) *Proc Biol Sci* 248, 9–18; [209] Waters MF *et al.* (2006) *Nat Genet* 38, 447–451; [210] McMahon A *et al.* (2004) *Eur J Neurosci* 19, 3317–3327; [211] Martina M *et al.* (2003) *J Neurosci* 23, 5698–5707; [212] Akemann W and Knopfel T B (2006) *J Neurosci* 26, 4602–4612; [213] McKay BE and Turner RW (2004) *Eur J Neurosci* 20, 729–739; [214] Brooke RE *et al.* (2004) *Eur J Neurosci* 20, 3313–3321; [215] Chang SY *et al.* (2007) *J Comp Neurol* 502(6), 953–972; [216] Schroter KH *et al.* (1991) *FEBS Lett* 278, 211–216; [217] Rudy B *et al.* (1991) *J Neurosci Res* 29, 401–412; [218] Diochot S *et al.* (1998) *J Biol Chem* 273, 6744–6749; [219] Vullhorst D *et al.* (1998) *FEBS Lett* 421, 259–262; [220] Riazanski V *et al.* (2001) *J Physiol* 537, 391–406; [221] Gopel SO *et al.* (2000) *J Physiol* 528, 497–507; [222] Pak MD *et al.* (1991) *PNAS* 88, 4386–4390; [223] BeckEJ and Covarrubias M (2001) *Biophys J* 81, 867–883; [224] Beck EJ *et al.* (2002) *J Physiol* 538, 691–706; [225] Jerng HH *et al.* (2004) *Mol Cell Neurosci* 27, 343–369; [226] Jerng HH *et al.* (1999) *J Gen Physiol* 113, 641–660; [227] Sanguinetti MC *et al.* (1997) *Mol Pharmacol* 51, 491–498; [228] Diochot S *et al.* (1999) *Br J Pharmacol* 126, 251–263; [229] Escoubas P *et al.* (2002) *Mol Pharmacol* 62, 48–57; [230] Wang D and Schreurs BG (2006) *Brain Res* 1096, 85–96; [231] Vacher H *et al.* (2006) *Eur J Neurosci* 24, 1325–1340; [232] Pongs O *et al.* (1999) *Ann N Y Acad Sci* 868, 344–355; [233] Suzuki T and Takimoto K (2005) *Am J Physiol* 288, E335–E341; [234] Baldwin TJ *et al.* (1991) *Neuron* 7, 471–483; [235] Nadal MS *et al.* (2001) *J Physiol* 537, 801–809; [236] Nerbonne JM (2000) *J Physiol* 525, 285–298; [237] Dixon JE and McKinnon D (1994) *Circ Res* 75, 252–260; [238] Rhodes KJ *et al.* (2004) *J Neurosci* 24, 7903–7915; [239] Strassle BW *et al.* (2005) *J Comp Neurol* 484, 144–155; [240] Chen X *et al.* (2006) *J Neurosci* 26, 12143–12151; [241] Hu HJ *et al.* (2006) *Neuron* 50, 89–100; [242] Serodio P *et al.* (1996) *J Neurophysiol* 75, 2174–2179; [243] Liss B *et al.* (2001) *EMBO J* 20, 5715–5724; [244] Dixon JE *et al.* (1996) *Circ Res* 79, 659–668; [245] Adda S *et al.* (1996) *J Biol Chem* 271, 13239–13243; [246] Klumpp DJ *et al.* (1991) *Cell Mol Neurobiol* 11, 611–622; [247] Ji *et al.* (2002) *J Biol Chem* 277, 20195–20204; [248] Ruppersberg JP *et al.* (1991) *Nature* 353, 657–660; [249] Shen W *et al.* (2004) *J Neurophysiol* 91, 1337–1349.

References

Abbott GW, Butler MH and Goldstein SA (2006). Phosphorylation and protonation of neighboring MiRP2 sites: function and pathophysiology of MiRP2–Kv3.4 potassium channels in periodic paralysis. *FASEB J* **20(2)**, 293–301.

Abbott GW, Butler MH, Bendahhou S *et al.* (2001). MiRP2 forms potassium channels in skeletal muscle with Kv3.4 and is associated with periodic paralysis. *Cell* **104(2)**, 217–231.

Adams JP, Anderson AE, Varga AW *et al.* (2000). The A-type potassium channel Kv4.2 is a substrate for the mitogen-activated protein kinase ERK. *J Neurochem* **75**, 2277–2287.

Aldrich RW (2001). Fifty years of inactivation. *Nature* **411(6838)**, 643–644.

Aldrich RW Jr, Getting PA and Thompson SH (1979). Inactivation of delayed outward current in molluscan neurone somata. *J Physiol* **291**, 507–530.

Allen M, Heinzmann A, Noguchi E *et al.* (2003). Positional cloning of a novel gene influencing asthma from chromosome 2q14. *Nat Genet* **35**, 258–63.

An WF, Bowlby MR, Betty M *et al.* (2000). Modulation of A-type potassium channels by a family of calcium sensors. *Nature* **403**, 553–556.

Anderson AE, Adams JP, Qian Y *et al.* (2000). Kv4.2 phosphorylation by cyclic AMP-dependent protein kinase. *J Biol Chem* **275**, 5337–5346.

Armstrong CM and Bezanilla F (1977). Inactivation of the sodium channel. II. Gating current experiments. *J Gen Physiol* **70**, 567–590.

Atzori M, Lei S, Evans DI *et al.* (2001). Differential synaptic processing separates stationary from transient inputs to the auditory cortex. *Nat Neurosci* **4**, 1230–1237.

Baldwin TJ, Tsaur ML, Lopez GA *et al.* (1991). Characterization of a mammalian cDNA for an inactivating voltage-sensitive K+ channel. *Neuron* **7**, 471–483.

Baranauskas G, Tkatch T and Surmeier DJ (1999). Delayed rectifier currents in rat globus pallidus neurons are attributable to Kv2.1 and Kv3.1/3.2 K(+) channels. *J Neurosci* **19**, 6394–6404.

Barry DM, Xu H, Schuessler RB *et al.* (1998). Functional knockout of the transient outward current, long-QT syndrome, and cardiac remodeling in mice expressing a dominant-negative Kv4 alpha subunit. *Circ Res* **83**, 560–567.

Baxter DA and Byrne JH (1991). Ionic conductance mechanisms contributing to the electrophysiological properties of neurons. *Curr Opin Neurobiol* **1**, 105–112.

Beeton C, Pennington MW, Wulff H *et al.* (2005). Targeting effector memory T cells with a selective peptide inhibitor of Kv1.3 channels for therapy of autoimmune diseases. *Mol Pharmacol* **67**, 1369–1381.

Bertoli A, Moran O and Conti F (1996). Accumulation of long-lasting inactivation in rat brain K(+)–channels. *Exp Brain Res* **110**, 401–412.

Bezanilla F (2000). The voltage sensor in voltage-dependent ion channels. *Physiol Rev* **80**, 555–592.

Bhat MA, Rios JC, Lu Y *et al.* (2001). Axon–glia interactions and the domain organization of myelinated axons requires neurexin IV/Caspr/Paranodin. *Neuron* **30**, 369–383.

Birnbaum SG, Varga AW, Yuan LL *et al.* (2004). Structure and function of Kv4-family transient potassium channels. *Physiol Rev* **84**, 803–833.

Blaine JT and Ribera AB (1998). Heteromultimeric potassium channels formed by members of the Kv2 subfamily. *J Neurosci* **18**, 9585–9593.

Brooke RE, Moores TS, Morris NP *et al.* (2004). Kv3 voltage-gated potassium channels regulate neurotransmitter release from mouse motor nerve terminals. *Eur J Neurosci* **20**, 3313–3321.

Browne DL, Gancher ST, Nutt JG *et al.* (1994). Episodic ataxia/myokymia syndrome is associated with point mutations in the human potassium channel gene, KCNA1. *Nat Genet* **8**, 136–140.

Chabala LD, Bakry N and Covarrubias M (1993). Low molecular weight poly(A)+ mRNA species encode factors that modulate gating of a non-Shaker A-type K+ channel. *J Gen Physiol* **102**, 713–728.

Chan E (1997). *Regulation and function of Kv3.3*. Ph.D. Thesis. The Rockefeller University.

Chang SY, Zagha E, Kwon ES *et al.* (2007). Distribution of Kv3.3 potassium channel subunits in distinct neuronal populations of mouse brain. *J Comp Neurol* **502(6)**, 953–972.

Chen X, Yuan LL, Zhao C *et al.* (2006). Deletion of Kv4.2 gene eliminates dendritic A-type K+ current and enhances induction of long-term potentiation in hippocampal CA1 pyramidal neurons. *J Neurosci* **26**, 12143–12151.

Choe S, Cushman S, Baker KA *et al.* (2002). Excitability is mediated by the T1 domain of the voltage-gated potassium channel. *Novartis Found Symp* **245**, 169–175; discussion 175–177, 261–264.

Chow A, Erisir A, Farb C *et al.* (1999). K(+) channel expression distinguishes subpopulations of parvalbumin- and somatostatin-containing neocortical interneurons. *J Neurosci* **19**, 9332–9345.

Coetzee WA, Amarillo Y, Chiu J *et al.* (1999). Molecular diversity of K+ channels. *Ann N Y Acad Sci* **868**, 233–285.

Connor JA and Stevens CF (1971a). Inward and delayed outward membrane currents in isolated neural somata under voltage clamp. *J Physiol* **213**, 1–19.

Connor JA and Stevens CF (1971b). Voltage clamp studies of a transient outward membrane current in gastropod neural somata. *J Physiol* **213**, 21–30.

Covarrubias M, Wei A, Salkoff L *et al.* (1994). Elimination of rapid potassium channel inactivation by phosphorylation of the inactivation gate. *Neuron* **13**, 1403–1412.

Cronin S, Berger S, Ding J *et al.* (2008). A genome-wide association study of sporadic ALS in a homogeneous Irish population, *Hum Mol Genet* **17**, 768–74.

De Lecea L, Soriano E, Criado JR *et al.* (1994). Transcripts encoding a neural membrane CD26 peptidase-like protein are stimulated by synaptic activity. *Brain Res Mol Brain Res* **25**, 286–296.

De Meester I, Korom S, Van Damme J *et al.* (1999). CD26, let it cut or cut it down. *Immunol Today* **20**, 367–375.

Deng Q, Rashid AJ, Fernandez FR *et al.* (2005). A C-terminal domain directs Kv3.3 channels to dendrites. *J Neurosci* **25**, 11531–11541.

Deschenes I and Tomaselli GF (2002). Modulation of Kv4.3 current by accessory subunits. *FEBS Lett* **528**, 183–188.

Devaux J and Scherer SS (2005). Altered ion channels in an animal model of Charcot–Marie–Tooth disease type IA. *J Neurosci* **25**, 1470–1480.

Devaux J, Alcaraz G, Grinspan J *et al.* (2003). Kv3.1b is a novel component of CNS nodes. *J Neurosci* **23**, 4509–4518.

Dixon JE and McKinnon D (1994). Quantitative analysis of potassium channel mRNA expression in atrial and ventricular muscle of rats. *Circ Res* **75**, 252–60.

Dodson PD and Forsythe ID (2004). Presynaptic K+ channels: electrifying regulators of synaptic terminal excitability. *TINS* **27**, 210–217.

Dodson PD, Barker MC and Forsythe ID (2002). Two heteromeric Kv1 potassium channels differentially regulate action potential firing. *J Neurosci* **22**, 6953–6961.

Dodson PD, Billups B, Rusznak Z *et al.* (2003). Presynaptic rat Kv1.2 channels suppress synaptic terminal hyperexcitability following action potential invasion. *J Physiol* **550**, 27–33.

Dougherty K and Covarrubias M (2006). A dipeptidyl aminopeptidase-like protein remodels gating charge dynamics in Kv4.2 channels. *J Gen Physiol* **128**, 745–753.

Doyle DA, Morais Cabral J, Pfuetzner RA *et al.* (1998). The structure of the potassium channel: molecular basis of K+ conduction and selectivity. *Science* **280**, 69–77.

Du J, Haak LL, Phillips-Tansey E *et al.* (2000). Frequency-dependent regulation of rat hippocampal somato-dendritic excitability by the K+ channel subunit Kv2.1. *J Physiol* **522**, 19–31.

Du J, Tao-Cheng J-H, Zerfas P *et al.* (1998). The K+ channel, Kv2.1, is apposed to astrocytic processes and is associated with inhibitory postsynaptic membranes in hippocampal and cortical principal neurons and inhibitory interneurons. *Neurosci* **84**, 37–48.

Du J, Zhang L, Weiser M *et al.* (1996). Developmental expression and functional characterization of the potassium-channel subunit Kv3.1b in parvalbumin-containing interneurons of the rat hippocampus. *J Neurosci* **16**, 506–518.

England SK, Uebele VN, Shear H *et al.* (1995). Characterization of a voltage-gated K+ channel beta subunit expressed in human heart. *PNAS* **92**, 6309–6313.

Erisir A, Lau D, Rudy B *et al.* (1999). Function of specific K(+) channels in sustained high-frequency firing of fast-spiking neocortical interneurons. *J Neurophysiol* **82**, 2476–2489.

Espinosa F, Marks G, Heintz N *et al.* (2004). Increased motor drive and sleep loss in mice lacking Kv3-type potassium channels. *Genes Brain Behav* **3**, 90–100.

Espinosa F, McMahon A, Chan E *et al.* (2001). Alcohol hypersensitivity, increased locomotion, and spontaneous myoclonus in mice lacking the potassium channels Kv3.1 and Kv3.3. *J Neurosci* **21**, 6657–6665.

Fernandez FR, Mehaffey WH, Molineux ML *et al.* (2005). High-threshold K+ current increases gain by offsetting a frequency-dependent increase in low-threshold K+ current. *J Neurosci* **25**, 363–371.

Fernandez FR, Morales E, Rashid AJ *et al.* (2003). Inactivation of Kv3.3 potassium channels in heterologous expression systems. *J Biol Chem* **278**, 40890–400898.

Fiset C, Clark RB, Larsen TS *et al.* (1997). Shal-type channels contribute to the Ca2+-independent transient outward K+ current in rat ventricle. *J Physiol* **500**, 51–64.

Geiger JR and Jonas P (2000). Dynamic control of presynaptic Ca(2+) inflow by fast-inactivating K(+) channels in hippocampal mossy fiber boutons. *Neuron* **28**, 927–939.

Giese KP, Storm JF, Reuter D *et al.* (1998). Reduced K+ channel inactivation, spike broadening, and after-hyperpolarization in Kvbeta1.1-deficient mice with impaired learning. *Learn Mem* **5**, 257–273.

Goldberg EM and Rudy B (2006). A DTX-sensitive Kv1 current produces delayed firing at near-threshold potentials in fast-spiking GABAergic interneurons of layer 2/3 mouse barrel cortex. *Soc Neurosci Abstr* 234.15/D53.

Goldberg EM, Watanabe S, Chang SY *et al.* (2005). Specific functions of synaptically localized potassium channels in synaptic transmission at the neocortical GABAergic fast-spiking cell synapse. *J Neurosci* **25**, 5230–5235.

Goldman-Wohl DS, Chan E, Baird D *et al.* (1994). Kv3.3b: a novel Shaw-type potassium channel expressed in terminally differentiated cerebellar Purkinje cells and deep cerebellar nuclei. *J Neurosci* **14**, 511–522.

Gorrell MD, Gysbers V and McCaughan GW (2001). CD26: a multifunctional integral membrane and secreted protein of activated lymphocytes. *Scand J Immunol* **54**, 249–264.

Grigg JJ, Brew HM and Tempel BL (2000). Differential expression of voltage-gated potassium channel genes in auditory nuclei of the mouse brainstem. *Hear Res* **140**(1–2), 77–90.

Grissmer S, Nguyen AN, Aiyar J *et al.* (1994). Pharmacological characterization of five cloned voltage-gated K+ channels, types Kv1.1, 1.2, 1.3, 1.5, and 3.1, stably expressed in mammalian cell lines. *Mol Pharmacol* **45**, 1227–1234.

Gu C, Jan YN and Jan LY (2003). A conserved domain in axonal targeting of Kv1 (Shaker) voltage-gated potassium channels. *Science* **301**, 646–649.

Gu C, Zhou W, Puthenveedu MA *et al.* (2006). The microtubule plus-end tracking protein EB1 is required for Kv1 voltage-gated K+ channel axonal targeting. *Neuron* **52**, 803–816.

Gulbis JM, Mann S and MacKinnon R (1999). Structure of a voltage-dependent K+ channel beta subunit. *Cell* **97** 943–952.

Gulbis JM, Zhou M, Mann S *et al.* (2000). Structure of the cytoplasmic beta subunit-T1 assembly of voltage-dependent K+ channels. *Science* **289**, 123–127.

Guo W, Xu H, London B *et al.* (1999). Molecular basis of transient outward K+ current diversity in mouse ventricular myocytes. *J Physiol* **521**, 587–599.

Gurantz D, Lautermilch NJ, Watt SD *et al.* (2000). Sustained upregulation in embryonic spinal neurons of a Kv3.1 potassium channel gene encoding a delayed rectifier current. *J Neurobiol* **42**, 347–356.

Harvey AL and Robertson B (2004). Dendrotoxins: structure–activity relationships and effects on potassium ion channels. *Curr Med Chem* **11**, 3065–3072.

Heginbotham L, Lu Z, Abramson T *et al.* (1994). Mutations in the K+ channel signature sequence. *Biophys J* **66**, 1061–1067.

Heinemann SH and Hoshi T (2006). Multifunctional potassium channels: electrical switches and redox enzymes, all in one. *Sci STKE* **350**, 33.

Heinemann SH, Rettig J, Wunder F *et al.* (1995). Molecular and functional characterization of a rat brain Kv beta 3 potassium channel subunit. *FEBS Lett*, **377**, 383–389.

Hernandez-Pineda R, Chow A, AmarilloY *et al.* (1999). Kv3.1–Kv3.2 channels underlie a high-voltage-activating component of the delayed rectifier K+ current in projecting neurons from the globus pallidus. *J Neurophysiol* **82**, 512–528.

Hildebrandt M, Rose M, Mayr C *et al.* (2000). Dipeptidyl peptidase IV (DPP IV, CD26) in patients with mental eating disorders. *Adv Exp Med Biol* **477**, 197–204.

Ho CS, Grange RW and Joho RH (1997). Pleiotropic effects of a disrupted K+ channel gene: reduced body weight, impaired motor skill and muscle contraction, but no seizures. *PNAS* **94**, 1533–1538.

Hodgkin AL and Huxley AF (1952a). The dual effect of membrane potential on sodium conductance in the giant axon of Loligo. *J Physiol* **116**, 497–506.

Hodgkin AL and Huxley AF (1952b). A quantitative description of membrane current and its application to conduction and excitation in nerve. *J Physiol* **117**, 500–544.

Hoffman DA and Johnston D (1998). Downregulation of transient K+ channels in dendrites of hippocampal CA1 pyramidal neurons by activation of PKA and PKC. *J Neurosci* **18**, 3521–3528.

Hoffman DA, Magee JC, Colbert CM *et al.* (1997). K+ channel regulation of signal propagation in dendrites of hippocampal pyramidal neurons. *Nature* **387**, 869–875.

Holmqvist MH, Cao, J, Hernandez-Pineda R *et al.* (2002). Elimination of fast inactivation in Kv4 A-type potassium channels by an auxiliary subunit domain. *PNAS* **99**, 1035–1040.

Hopkins WF (1998). Toxin and subunit specificity of blocking affinity of three peptide toxins for heteromultimeric, voltage-gated potassium channels expressed in *Xenopus* oocytes. *JPET* **285**, 1051–1060.

Hoshi T, Zagotta WN and Aldrich RW (1990). Biophysical and molecular mechanisms of Shaker potassium channel inactivation. *Science* **250**(4980), 533–538.

Hoshi T, Zagotta WN and Aldrich RW (1991). Two types of inactivation in Shaker K+ channels: effects of alterations in the carboxy-terminal region. *Neuron* **7**, 547–556.

Hu HJ, Carrasquillo Y, Karim F *et al.* (2006). The kv4.2 potassium channel subunit is required for pain plasticity. *Neuron* **50**, 89–100.

Hwang PM, Fotuhi M, Bredt DS *et al.* (1993). Contrasting immunohistochemical localizations in rat brain of two novel K+ channels of the Shab subfamily. *J Neurosci* **13**, 1569–1576.

Imbrici P, D'Adamo MC, Kullmann DM *et al.* (2006). Episodic ataxia type 1 mutations in the KCNA1 gene impair the fast inactivation properties of the human potassium channels Kv1.4–1.1/Kvbeta1.1 and Kv1.4–1.1/Kvbeta1.2. *Eur J Neurosci* **24**, 3073–3083.

Inda MC, DeFelipe J and Munoz A (2006). Voltage-gated ion channels in the axon initial segment of human cortical pyramidal cells and their relationship with chandelier cells. *PNAS* **103**, 2920–2925.

Isbrandt D, Leicher T, Waldschutz R *et al.* (2000). Gene structures and expression profiles of three human KCND (Kv4) potassium channels mediating A-type currents I(TO) and I(SA). *Genomics* **64**, 144–154.

Ishikawa T, Nakamura Y, Saitoh N *et al.* (2003). Distinct roles of Kv1 and Kv3 potassium channels at the calyx of Held presynaptic terminal. *J Neurosci* **23**, 10445–10453.

Isom LL (2002). The role of sodium channels in cell adhesion. *Front Biosci* **7**, 12–23.

Itri JN, Michel S, Vansteensel MJ *et al.* (2005). Fast delayed rectifier potassium current is required for circadian neural activity. *Nat Neurosci* **8**, 650–656.

Jerng HH, Kunjilwar K and Pfaffinger PJ (2005). Multiprotein assembly of Kv4.2, KChIP3 and DPP10 produces ternary channel complexes with ISA-like properties. *J Physiol* **568**, 767–788.

Jerng HH, Pfaffinger PJ and Covarrubias M (2004). Molecular physiology and modulation of somatodendritic A-type potassium channels. *Mol Cell Neurosci* **27**, 343–369.

Jerng HH, Qian Y and Pfaffinger PJ (2004). Modulation of Kv4.2 channel expression and gating by dipeptidyl peptidase 10 (DPP10). *Biophys J* **87**, 2380–2396.

Jiang Y, Ruta V, Chen J *et al.* (2003). The principle of gating charge movement in a voltage-dependent K$^+$ channel. *Nature* **423**, 42–48.

Johns DC, Nuss HB and Marban E (1997). Suppression of neuronal and cardiac transient outward currents by viral gene transfer of dominant-negative Kv4.2 constructs. *J Biol Chem* **272**, 31598–631503.

Johnston D, Christie BR, Frick A *et al.* (2003). Active dendrites, potassium channels and synaptic plasticity. *Philos Trans R Soc Lond B Biol Sci* **358**, 667–674.

Johnston D, Hoffman DA, Magee JC *et al.* (2000). Dendritic potassium channels in hippocampal pyramidal neurons. *J Physiol* **525**, 75–81.

Joho RH, Marks GA and Espinosa F (2006). Kv3 potassium channels control the duration of different arousal states by distinct stochastic and clock-like mechanisms. *Eur J Neurosci* **23**, 1567–1574.

Joho RH, Street C, Matsushita S *et al.* (2006). Behavioral motor dysfunction in Kv3-type potassium channel-deficient mice. *Genes Brain Behav* **5**, 472–482.

Kaczmarek LK (2006). Policing the ball: a new potassium channel subunit determines inactivation rate. *Neuron* **49**, 642–644.

Kaczmarek LK, Bhattacharjee A, Desai R *et al.* (2005). Regulation of the timing of MNTB neurons by short-term and long-term modulation of potassium channels. *Hear Res* **206**, 133–145.

Kanevsky M and Aldrich RW (1999). Determinants of voltage-dependent gating and open-state stability in the S5 segment of Shaker potassium channels. *J Gen Physiol* **114**, 215–242.

Kim E and Sheng M (1996). Differential K$^+$ channel clustering activity of PSD-95 and SAP97, two related membrane-associated putative guanylate kinases. *Neuropharm* **35**, 993–1000.

Kim E, Niethammer M, Rothschild A *et al.* (1995). Clustering of Shaker-type K$^+$ channels by interaction with a family of membrane-associated guanylate kinases. *Nature* **378**, 85–88.

Kin Y, Misumi Y and Ikehara Y (2001). Biosynthesis and characterization of the brain-specific membrane protein DPPX, a dipeptidyl peptidase IV-related protein. *J Biochem (Tokyo)* **129**, 289–295.

Klemic KG, Shieh CC, Kirsch GE *et al.* (1998). Inactivation of Kv2.1 potassium channels. *Biophys J* **74**, 177917–89.

Kobertz WR, Williams C and Miller C (2000). Hanging gondola structure of the T1 domain in a voltage-gated K(+) channel. *Biochem* **39**, 10347–10352.

Kunjilwar K, Strang C, DeRubeis D *et al.* (2004). KChIP3 rescues the functional expression of Shal channel tetramerization mutants. *J Biol Chem* **279**, 54542–54551.

Kuo HC, Cheng CF, Clark RB *et al.* (2001). A defect in the Kv channel-interacting protein 2 (KChIP2) gene leads to a complete loss of I(to) and confers susceptibility to ventricular tachycardia. *Cell* **107**, 801–813.

Lai HC and Jan LY (2006). The distribution and targeting of neuronal voltage-gated ion channels. *Nat Rev Neurosci* **7**, 548–562.

Lau D, Vega-Saenz de Miera EC, Contreras D *et al.* (2000). Impaired fast-spiking, suppressed cortical inhibition, and increased susceptibility to seizures in mice lacking Kv3.2 K$^+$ channel proteins. *J Neurosci* **20**, 9071–9085.

Lee TE, Philipson LH and Nelson DJ (1996). N-type inactivation in the mammalian Shaker K$^+$ channel Kv1.4. *J Membr Biol* **151**, 225–235.

Leicher T, Bahring R, Isbrandt D *et al.* (1998). Coexpression of the KCNA3B gene product with Kv1.5 leads to a novel A-type potassium channel. *J Biol Chem* **273**, 35095–350101.

Lenz S, Perney TM, Qin Y *et al.* (1994). GABA-ergic interneurons of the striatum express the Shaw-like potassium channel Kv3.1. *Synapse* **18**, 55–66.

Lerche H, Jurkat-Rott K and Lehmann-Horn F (2001). Ion channels and epilepsy. *Am J Med Genet* **106**, 146–159.

Li M, Jan YN and Jan LY (1992). Specification of subunit assembly by the hydrophilic amino-terminal domain of the Shaker potassium channel. *Science* **257**, 1225–1230.

Lien CC and Jonas P (2003). Kv3 potassium conductance is necessary and kinetically optimized for high-frequency action potential generation in hippocampal interneurons. *J Neurosci* **23**, 2058–2068.

Lien CC, Martina M, Schultz JH *et al.* (2002). Gating, modulation and subunit composition of voltage-gated K(+) channels in dendritic inhibitory interneurones of rat hippocampus. *J Physiol* **538**, 405–419.

Lim ST, Antonucci DE, Scannevin RH *et al.* (2000). A novel targeting signal for proximal clustering of the Kv2.1 K$^+$ channel in hippocampal neurons. *Neuron* **25**, 385–397.

Liss B, Franz O, Sewing S *et al.* (2001). Tuning pacemaker frequency of individual dopaminergic neurons by Kv4.3L and KChip3.1 transcription. *EMBO J*, **20**, 5715–5724.

Liu Y, Jurman ME and Yellen G (1996). Dynamic rearrangement of the outer mouth of a K$^+$ channel during gating. *Neuron* **16**, 859–867.

Llinas RR (1988). The intrinsic electrophysiological properties of mammalian neurons: insights into central nervous system function. *Science* **242**, 1654–1664.

Long SB, Campbell EB and Mackinnon R (2005a). Crystal structure of a mammalian voltage-dependent Shaker family K$^+$ channel. *Science* **309**, 897–903.

Long SB, Campbell EB and Mackinnon R (2005b). Voltage sensor of Kv1.2: structural basis of electromechanical coupling. *Science* **309**, 903–908.

Lu Z, Klem AM and Ramu Y (2001). Ion conduction pore is conserved among potassium channels. *Nature* **413**, 809–813.

MacDonald PE, Ha XF, Wang J *et al.* (2001). Members of the Kv1 and Kv2 voltage-dependent K(+) channel families regulate insulin secretion. *Mol Endocrinol* **15**, 1423–1435.

Macica CM and Kaczmarek LK (2001). Casein kinase 2 determines the voltage dependence of the Kv3.1 channel in auditory neurons and transfected cells. *J Neurosci* **21**, 1160–1168.

Macica CM, von Hehn CA, Wang LY *et al.* (2003). Modulation of the kv3.1b potassium channel isoform adjusts the fidelity of the firing pattern of auditory neurons. *J Neurosci* **23**, 1133–1141.

MacKinnon R, Aldrich RW and Lee AW (1993). Functional stoichiometry of Shaker potassium channel inactivation. *Science* **262**, 757–759.

Maletic-Savatic M, Lenn NJ and Trimmer JS (1995). Differential spatiotemporal expression of K$^+$ channel polypeptides in rat hippocampal neurons developing *in situ* and *in vitro*. *J Neurosci* **15**, 3840–3851.

Malin SA and Nerbonne JM (2000). Elimination of the fast transient in superior cervical ganglion neurons with expression of KV4.2W362F: molecular dissection of IA. *J Neurosci* **20**, 5191–5199.

Malin SA and Nerbonne JM (2002). Delayed rectifier K$^+$ currents, IK, are encoded by Kv2 alpha-subunits and regulate tonic firing in mammalian sympathetic neurons. *J Neurosci* **22**, 10094–10105.

Malysz J and Huizinga JD (1999). Searching for intrinsic properties and functions of interstitial cells of Cajal. *Curr Opin Gastroenterol* **15**, 26.

Marom S and Levitan IB (1994). State-dependent inactivation of the Kv3 potassium channel. *Biophys J* **67**, 579–589.

Marshall CR, Noor A, Vincent JB et al. (2008). Structural variation of chromosomes in autism spectrum disorder. Am J Hum Genet 82, 477–88.

Martina M, Schultz JH, Ehmke H et al. (1998). Functional and molecular differences between voltage-gated K+ channels of fast-spiking interneurons and pyramidal neurons of rat hippocampus. J Neurosci 18, 8111–8125.

Martina M, Yao GL and Bean BP (2003). Properties and functional role of voltage-dependent potassium channels in dendrites of rat cerebellar Purkinje neurons. J Neurosci 23, 5698–5707.

Massengill JL, Smith MA, Son DI et al. (1997). Differential expression of K4-AP currents and Kv3.1 potassium channel transcripts in cortical neurons that develop distinct firing phenotypes. J Neurosci 17, 3136–3147.

Matsukawa H, Wolf AM, Matsushita S et al. (2003). Motor dysfunction and altered synaptic transmission at the parallel fiber-Purkinje cell synapse in mice lacking potassium channels Kv3.1 and Kv3.3. J Neurosci 23, 7677–7684.

McCormack K, Connor JX, Zhou L et al. (2002). Genetic analysis of the mammalian K+ channel beta subunit Kvbeta 2 (Kcnab2). J Biol Chem 277, 13219–13228.

McCormack K, McCormack T, Tanouye M et al. (1995). Alternative splicing of the human Shaker K+ channel beta 1 gene and functional expression of the beta 2 gene product. FEBS Lett 370, 32–36.

McCormack T and McCormack K (1994). Shaker K+ channel beta subunits belong to an NAD(P)H-dependent oxidoreductase superfamily. Cell 79, 1133–1135.

McDonald AJ and Mascagni F (2006). Differential expression of Kv3.1b and Kv3.2 potassium channel subunits in interneurons of the basolateral amygdala. Neurosci 138, 537–547.

McIntosh P, Moreno H, Robertson B et al. (1998). Isoform-specific modulation of rat Kv3 potassium channel splice variants. J Physiol 511P, 147.

McMahon A, Fowler SC, Perney TM et al. (2004). Allele-dependent changes of olivocerebellar circuit properties in the absence of the voltage-gated potassium channels Kv3.1 and Kv3.3. Eur J Neurosci 19, 3317–3327.

Mehaffey WH, Fernandez FR, Rashid AJ et al. (2006). Distribution and function of potassium channels in the electrosensory lateral line lobe of weakly electric apteronotid fish. J Comp Physiol A Neuroethol Sens Neural Behav Physiol 192, 637–648.

Mert T (2006). Kv1 channels in signal conduction of myelinated nerve fibers. Rev Neurosci 17, 369–374.

Misonou H and Trimmer JS (2004). Determinants of voltage-gated potassium channel surface expression and localization in Mammalian neurons. Crit Rev Biochem Mol Biol 39, 125–145.

Misonou H, Mohapatra DP and Trimmer JS (2005). Kv2.1: a voltage-gated K+ channel critical to dynamic control of neuronal excitability. Neurotox 26, 743–752.

Misonou H, Mohapatra DP, Menegola M et al. (2005). Calcium- and metabolic state-dependent modulation of the voltage-dependent Kv2.1 channel regulates neuronal excitability in response to ischemia. J Neurosci 25, 11184–11193.

Misonou H, Mohapatra DP, Park EW et al. (2004). Regulation of ion channel localization and phosphorylation by neuronal activity. Nat Neurosci 7, 711–718.

Mohapatra DP and Trimmer JS (2006). The Kv2.1 C terminus can autonomously transfer Kv2.1-like phosphorylation-dependent localization, voltage-dependent gating, and muscarinic modulation to diverse Kv channels. J Neurosci 26, 685–695.

Moreno H, Kentros C, Bueno E et al. (1995). Thalamocortical projections have a K+ channel that is phosphorylated and modulated by cAMP-dependent protein kinase. J Neurosci 15, 5486–5501.

Morozov A, Muzzio IA, Bourtchouladze R et al. (2003). Rap1 couples cAMP signaling to a distinct pool of p42/44MAPK regulating excitability, synaptic plasticity, learning, and memory. Neuron 39, 309–325.

Muennich EA and Fyffe RE (2004). Focal aggregation of voltage-gated, Kv2.1 subunit-containing, potassium channels at synaptic sites in rat spinal motoneurones. J Physiol 554, 673–685.

Murakoshi H and Trimmer JS (1999). Identification of the Kv2.1 K+ channel as a major component of the delayed rectifier K+ current in rat hippocampal neurons. J Neurosci 19, 1728–1735.

Murata Y, Iwasaki H, Sasaki M et al. (2005). Phosphoinositide phosphatase activity coupled to an intrinsic voltage sensor. Nature 435, 1239–43.

Murrell-Lagnado RD and Aldrich RW (1993). Interactions of amino terminal domains of Shaker K channels with a pore blocking site studied with synthetic peptides. J Gen Physiol 102, 949–975.

Nadal MS, Amarillo Y, Vega-Saenz de Miera E et al. (2001). Evidence for the presence of a novel Kv4-mediated A-type K(+) channel-modifying factor. J Physiol 537, 801–809.

Nadal MS, Amarillo Y, Vega-Saenz de Miera E et al. (2006). Differential characterization of three alternative spliced isoforms of DPPX. Brain Res 1094, 1–12.

Nadal MS, Ozaita A, Amarillo Y et al. (2003). The CD26-related dipeptidyl aminopeptidase-like protein DPPX is a critical component of neuronal A-type K+ channels. Neuron 37, 449–461.

Nagaya N and Papazian DM (1997). Potassium channel alpha and beta subunits assemble in the endoplasmic reticulum. J Biol Chem 272, 3022–3027.

Nakamura TY, Coetzee WA, Vega-Saenz De Miera E et al. (1997). Modulation of Kv4 channels, key components of rat ventricular transient outward K+ current, by PKC. Am J Physiol 273, H1775–H1786.

Nakamura TY, Nandi S, Pountney DJ et al. (2001). Different effects of the Ca(2+)-binding protein, KChIP1, on two Kv4 subfamily members, Kv4.1 and Kv4.2. FEBS Lett 499, 205–209.

O'Connell KM, Rolig AS, Whitesell JD et al. (2006). Kv2.1 potassium channels are retained within dynamic cell surface microdomains that are defined by a perimeter fence. J Neurosci 26, 9609–9618.

Ozaita A, Martone ME, Ellisman MH et al. (2002). Differential subcellular localization of the two alternatively spliced isoforms of the Kv3.1 potassium channel subunit in brain. J Neurophysiol 88, 394–408.

Ozaita A, Petit-Jacques J, Volgyi B et al. (2004). A unique role for Kv3 voltage-gated potassium channels in starburst amacrine cell signaling in mouse retina. J Neurosci 24, 7335–7343.

Pak MD, Baker K, Covarrubias M et al. (1991). mShal, a subfamily of A-type K+ channel cloned from mammalian brain. PNAS 88, 4386–4390.

Pal S, Hartnett KA, Nerbonne JM et al. (2003). Mediation of neuronal apoptosis by Kv2.1-encoded potassium channels. J Neurosci 23, 4798–802.

Parameshwaran S, Carr CE and Perney TM (2001). Expression of the Kv3.1 potassium channel in the avian auditory brainstem. J Neurosci 21, 485–494.

Park KS, Mohapatra DP, Misonou H et al. (2006). Graded regulation of the Kv2.1 potassium channel by variable phosphorylation. Science 313, 976–979.

Patil PG, Brody DL and Yue DT (1998). Preferential closed-state inactivation of neuronal calcium channels. Neuron 20, 1027–1038.

Perney TM and Kaczmarek LK (1997). Localization of a high threshold potassium channel in the rat cochlear nucleus. J Comp Neurol 386, 178–202.

Perney TM, Marshall J, Martin KA et al. (1992). Expression of the mRNAs for the Kv3.1 potassium channel gene in the adult and developing rat brain. J Neurophysiol 68, 756–766.

Pioletti M, Findeisen F, Hura GL et al. (2006). Three-dimensional structure of the KChIP1-Kv4.3 T1 complex reveals a cross-shaped octamer. Nat Struct Mol Biol 13, 987–995.

Ponce A, Vega-Saenz de Miera E, Kentros C et al. (1997). K+ channel subunit isoforms with divergent carboxy-terminal sequences carry distinct membrane targeting signals. J Membr Biol 159, 149–159.

Pongs O, Leicher T, Berger M *et al.* (1999). Functional and molecular aspects of voltage-gated K⁺ channel beta subunits. *Ann N Y Acad Sci.* **868**, 344–355.

Rae JL and Shepard AR (2000). Kv3.3 potassium channels in lens epithelium and corneal endothelium. *Exp Eye Res* **70**, 339–348.

Ratcliffe CF, Westenbroek RE, Curtis R *et al.* (2001). Sodium channel beta1 and beta3 subunits associate with neurofascin through their extracellular immunoglobulin-like domain. *J Cell Biol* **154**, 427–434.

Ren X, Hayashi Y, Yoshimura N *et al.* (2005). Transmembrane interaction mediates complex formation between peptidase homologues and Kv4 channels. *Mol Cell Neurosci* **29**, 320–332.

Rettig J, Heinemann SH, Wunder F *et al.* (1994). Inactivation properties of voltage-gated K⁺ channels altered by presence of beta-subunit. *Nature* **369**, 289–294.

Rettig J, Wunder F, Stocker M *et al.* (1992). Characterization of a Shaw-related potassium channel family in rat brain. *EMBO J* **11**, 2473–2486.

Rhodes KJ, Carroll KI, Sung MA *et al.* (2004). KChIPs and Kv4 alpha subunits as integral components of A-type potassium channels in mammalian brain. *J Neurosci* **24**, 7903–7915.

Rhodes KJ, Strassle BW, Monaghan MM *et al.* (1997). Association and colocalization of the Kvbeta1 and Kvbeta2 beta-subunits with Kv1 alpha-subunits in mammalian brain K⁺ channel complexes. *J Neurosci* **17**, 8246–8258.

Richardson FC and Kaczmarek LK (2000). Modification of delayed rectifier potassium currents by the Kv9.1 potassium channel subunit. *Hear Res* **147**, 21–30.

Rocha C, Nadal M, Rudy B *et al.* (1994). Inactivation gating of kv4 K⁺ channels interacting with the dipeptidyl-aminopeptidase-like protein (DPPX) 2780–Pos. *Biophysical Soc Annual Meeting*.

Roeper J, Sewing S, Zhang Y *et al.* (1998). NIP domain prevents N-type inactivation in voltage-gated potassium channels. *Nature* **391**, 390–393.

Rosati B, Pan Z, Lypen S *et al.* (2001). Regulation of KChIP2 potassium channel beta subunit gene expression underlies the gradient of transient outward current in canine and human ventricle. *J Physiol* **533**, 119–125.

Rudy B and McBain CJ (2001). Kv3 channels: voltage-gated K⁺ channels designed for high-frequency repetitive firing. *TINS* **24(9)**, 517–526.

Rudy B, Chow A, Lau D *et al.* (1999). Contributions of Kv3 channels to neuronal excitability. *Ann N Y Acad Sci*, **868**, 304–343.

Rudy B (1988). Diversity and ubiquity of K channels. *Neurosci* **25**, 729–749.

Rudy B (1999). Molecular diversity of ion channels and cell function. *Ann N Y Acad Sci.* **868**, 1–12.

Ruppersberg JP, Frank R, Pongs O *et al.* (1991). Cloned neuronal IK(A) channels reopen during recovery from inactivation. *Nature* **353**, 657–660.

Salkoff L, Baker K, Butler A *et al.* (1992). An essential 'set' of K⁺ channels conserved in flies, mice and humans. *TINS* **15**, 161–166.

Sasaki M, Black JA, Lankford KL *et al.* (2006). Molecular reconstruction of nodes of Ranvier after remyelination by transplanted olfactory ensheathing cells in the demyelinated spinal cord. *J Neurosci* **26**, 1803–1812.

Scannevin RH, Murakoshi H, Rhodes KJ *et al.* (1996). Identification of a cytoplasmic domain important in the polarized expression and clustering of the Kv2.1 K⁺ channel. *J Cell Biol* **135**, 1619–1632.

Scherer SS (1999). Nodes, paranodes, and incisures: from form to function. *Ann N Y Acad Sci* **883**, 131–142.

Schmalz F, Kinsella J, Koh SD *et al.* (1998). Molecular identification of a component of delayed rectifier current in gastrointestinal smooth muscles. *Am J Physiol* **274**, G901–G911.

Schoppa NE and GL Westbrook (1999). Regulation of synaptic timing in the olfactory bulb by an A-type potassium current. *Nat Neurosci* **2(12)**, 1106–1113.

Sekirnjak C, Martone ME, Weiser M *et al.* (1997). Subcellular localization of the K⁺ channel subunit Kv3.1b in selected rat CNS neurons. *Brain Res* **766**, 173–187.

Serodio P and Rudy B (1998). Differential expression of Kv4 K⁺ channel subunits mediating subthreshold transient K⁺ (A-type) currents in rat brain. *J Neurophysiol* **79**, 1081–1091.

Serodio P, Kentros C and Rudy B (1994). Identification of molecular components of A-type channels activating at subthreshold potentials. *J Neurophysiol* **72**, 1516–1529.

Serodio P, Vega-Saenz de Miera E and Rudy B (1996). Cloning of a novel component of A-type K⁺ channels operating at subthreshold potentials with unique expression in heart and brain. *J Neurophysiol* **75**, 2174–2179.

Sheng M, Tsaur ML, Jan YN *et al.* (1992). Subcellular segregation of two A-type K⁺ channel proteins in rat central neurons. *Neuron* **9**, 271–284.

Sheng M, Tsaur ML, Jan YN *et al.* (1994). Contrasting subcellular localization of the Kv1.2 K⁺ channel subunit in different neurons of rat brain. *J Neurosci* **14**, 2408–2417.

Shevchenko T, Teruyama R and Armstrong WE (2004). High-threshold, Kv3-like potassium currents in magnocellular neurosecretory neurons and their role in spike repolarization. *J Neurophysiol* **92**, 3043–3055.

Shi G, Kleinklaus AK, Marrion NV *et al.* (1994). Properties of Kv2.1 K⁺ channels expressed in transfected mammalian cells. *J Biol Chem* **269**, 23204–23211.

Shi G, Nakahira K, Hammond S *et al.* (1996). Beta subunits promote K⁺ channel surface expression through effects early in biosynthesis. *Neuron* **16**, 843–852.

Shibata R, Nakahira K, Shibasaki K *et al.* (2000). A-type K⁺ current mediated by the Kv4 channel regulates the generation of action potential in developing cerebellar granule cells. *J Neurosci* **20**, 4145–4155.

Shieh CC, Klemic GK and Kirsch GE (1997). Role of transmembrane segment S5 on gating of voltage-dependent K⁺ channels. *J Gen Physiol* **109**, 767–778.

Singh B, Ogiwara I, Kanefa M *et al* (2006). A Kv4.2 truncation mutation in a patient with temporal lobe epilepsy. *Neurobiol Dis* **24**, 245–253.

Sinha K, Karimi-Abdolrezaee S, Velumian AA *et al.* (2006). Functional changes in genetically dysmyelinated spinal cord axons of shiverer mice: role of juxtaparanodal Kv1 family K⁺ channels. *J Neurophysiol* **95**, 1683–1695.

Smart SL, Lopantsev V, Zhang CL *et al.* (1998). Deletion of the K(V)1.1 potassium channel causes epilepsy in mice. *Neuron* **20**, 809–819.

Storm JF (1988). Temporal integration by a slowly inactivating K⁺ current in hippocampal neurons. *Nature* **336**, 379–381.

Strop P, Bankovich AJ, Hansen KC *et al.* (2004). Structure of a human A-type potassium channel interacting protein DPPX, a member of the dipeptidyl aminopeptidase family. *J Mol Biol* **343**, 1055–1065.

Takimoto K, Hayashi Y, Ren X *et al.* (2006). Species and tissue differences in the expression of DPPY splicing variants. *Biochem Biophys Res Commun* **348**, 1094–1100.

Tansey EP, Chow A, Rudy B *et al.* (2002). Developmental expression of potassium-channel subunit Kv3.2 within subpopulations of mouse hippocampal inhibitory interneurons. *Hippocampus* **12**, 137–148.

Tian S, Liu W, Wu Y *et al.* (2002). Regulation of the voltage-gated K⁺ channel KCNA10 by KCNA4B, a novel beta-subunit. *Am J Physiol* **283**, F142–F149.

Tomaselli GF and Marban E (1999). Electrophysiological remodeling in hypertrophy and heart failure. *Cardiovasc Res* **42**, 270–283.

Tomaselli GF and Rose J (2000). Molecular aspects of arrhythmias associated with cardiomyopathies. *Curr Opin Cardiol* **15**, 202–208.

Trimmer JS (1991). Immunological identification and characterization of a delayed rectifier K⁺ channel polypeptide in rat brain. *PNAS* **88**, 10764–10768.

Tseng-Crank JC, Yao JA, Berman MF *et al.* (1993). Functional role of the NH2-terminal cytoplasmic domain of a mammalian A-type K channel. *J Gen Physiol* **102**, 1057–10583.

Vabnick I, Trimmer JS, Schwarz TL *et al.* (1999). Dynamic potassium channel distributions during axonal development prevent aberrant firing patterns. *J Neurosci* **19**, 747–758.

Van Wart A, Trimmer JS and Matthews G (2007). Polarized distribution of ion channels within microdomains of the axon initial segment. *J Comp Neurol* **500**, 339–352.

Varga AW, Yuan LL, Anderson AE *et al.* (2004). Calcium-calmodulin-dependent kinase II modulates Kv4.2 channel expression and upregulates neuronal A-type potassium currents. *J Neurosci* **24**, 3643–3654.

Vega-Saenz de Miera E (1994). Shaw-related K⁺ channels in mammals, in *Handbook of membrane channels: molecular and cellular physiology*, C. Peracchia, Ed. Academic Press: San Diego. pp. xix, 591.

Vega-Saenz de Miera E, Moreno H, Fruhling D *et al.* (1992). Cloning of ShIII (Shaw-like) cDNAs encoding a novel high-voltage-activating, TEA-sensitive, type-A K⁺ channel. *Proc Biol Sci* **248**, 9–18.

Vega-Saenz de Miera, E and Rudy B (1994). Modulation of Kv3.3 K⁺ channels by oxidation and phosphorylation. in *Soc. Neurosci. Abstracts*

Veh RW, Lichtinghagen R, Sewing S *et al.* (1995). Immunohistochemical localization of five members of the Kv1 channel subunits: contrasting subcellular locations and neuron-specific co-localizations in rat brain. *Eur J Neurosci* **7**, 2189–2205.

Vincent A, Lautermilch NJ and Spitzer NC (2000). Antisense suppression of potassium channel expression demonstrates its role in maturation of the action potential. *J Neurosci* **20**, 6087–6094.

von Hehn CA, Bhattacharjee A and Kaczmarek LK (2004). Loss of Kv3.1 tonotopicity and alterations in cAMP response element-binding protein signaling in central auditory neurons of hearing impaired mice. *J Neurosci* **24**, 1936–1940.

Wada K, Yokotani N, Hunter C *et al* (1992). Differential expression of two distinct forms of mRNA encoding members of a dipeptidyl aminopeptidase family. *PNAS* **89**, 197–201.

Wang H, Allen ML, Grigg JJ *et al.* (1995). Hypomyelination alters K⁺ channel expression in mouse mutants shiverer and trembler. Neuron 15, 1337–1347.

Wang H, Kunkel DD, Martin TM *et al.* (1993). Heteromultimeric K⁺ channels in terminal and juxtaparanodal regions of neurons. *Nature* **365**, 75–79.

Wang H, Yan Y, Liu Q *et al.* (2007). Structural basis for modulation of Kv4 K⁺ channels by auxiliary KChIP subunits. *Nat Neurosci* **10(1)**, 32–39.

Wang LY, Gan L, Forsythe ID *et al.* (1998). Contribution of the Kv3.1 potassium channel to high-frequency firing in mouse auditory neurones. *J Physiol* **509**, 183–194.

Watanabe S, Hoffman DA, Migliore M *et al.* (2002). Dendritic K⁺ channels contribute to spike-timing dependent long-term potentiation in hippocampal pyramidal neurons. *PNAS* **99**, 8366–8371.

Waters MF, Minassian NA, Stevanin G *et al.* (2006). Mutations in voltage-gated potassium channel KCNC3 cause degenerative and developmental central nervous system phenotypes. *Nat Genet* **38**, 447–451.

Wei A, Covarrubias M, Butler A *et al.* (1990). K⁺ current diversity is produced by an extended gene family conserved in Drosophila and mouse. *Science* **248**, 599–603.

Weiser M, Bueno E, Sekirnjak C *et al.* (1995). The potassium channel subunit KV3.1b is localized to somatic and axonal membranes of specific populations of CNS neurons. *J Neurosci* **15**, 4298–4314.

Weiser M, Vega-Saenz de Miera E, Kentros C *et al.* (1994). Differential expression of Shaw-related K⁺ channels in the rat central nervous system. *J Neurosci* **14**, 949–972.

Weng J, Cao Y, Moss N *et al.* (2006). Modulation of voltage-dependent Shaker family potassium channels by an aldo-keto reductase. *J Biol Chem* **281**, 15194–15200.

Wigmore MA and Lacey MG (2000). A Kv3-like persistent, outwardly rectifying, Cs⁺-permeable, K⁺ current in rat subthalamic nucleus neurones. *J Physiol* **527**, 493–506.

Wu RL and Barish ME (1992). Two pharmacologically and kinetically distinct transient potassium currents in cultured embryonic mouse hippocampal neurons. *J Neurosci* **12**, 2235–2246.

Xu H, Li H and Nerbonne JM (1999). Elimination of the transient outward current and action potential prolongation in mouse atrial myocytes expressing a dominant negative Kv4 alpha subunit. *J Physiol* **519**, 11–21.

Yellen G (1998). Premonitions of ion channel gating. *Nat Struct Biol* **5**, 421.

Yeung SY, Thompson D, Wang Z *et al.* (2005). Modulation of Kv3 subfamily potassium currents by the sea anemone toxin BDS: significance for CNS and biophysical studies. *J Neurosci* **25**, 8735–8745.

Yu FH and Catterall WA (2004).The VGL-chanome: a protein superfamily specialized for electrical signaling and ionic homeostasis. *Sci STKE* **253**, re15.

Yuan LL, Adams JP, Swank M *et al.* (2002). Protein kinase modulation of dendritic K⁺ channels in hippocampus involves a mitogen-activated protein kinase pathway. *J Neurosci* **22**, 4860–4868.

Zagha E, Ozaita A, Chang SY *et al.* (2005). DPP10 modulates Kv4-mediated A-type potassium channels. *J Biol Chem* **280**, 18853–18861.

Zerr P, Adelman JP and Maylie J (1998). Characterization of three episodic ataxia mutations in the human Kv1.1 potassium channel. *FEBS Lett* **431**, 461–464.

Zuberi SM, Eunson LH, Spauschus A *et al.* (1999). A novel mutation in the human voltage-gated potassium channel gene (Kv1.1) associates with episodic ataxia type 1 and sometimes with partial epilepsy. *Brain* **122**, 817–825.

2.1.2 **Kv7 channels**

Jonathan Robbins and Gayle Passmore

2.1.2.1 Introduction

The Kv7 group of voltage-gated potassium channels was first defined in 2002 under the new International Union of Pharmacology nomenclature (Gutman *et al.*, 2003, 2005) and consists of the five members of the KCNQ gene family. Previously known as KCNQ1–5, these channels were renamed in an attempt to standardize the nomenclature for voltage-gated potassium channels. Prior to this Kv7.1 was commonly referred to as KvLQT1 (Barhanin *et al.*, 1996; Wang *et al.*, 1996), owing to its association with a familial form of long QT syndrome (see 2.1.2.7), and the functional current that Kv7.1 partly mediates in the heart was termed I_{Ks}. The other members of the Kv7 channel family are expressed throughout the central and peripheral nervous systems and Kv7.2 + 7.3 heteromers have been identified as the molecular correlates for the M-channel (Wang *et al.*, 1998), which was first functionally described in 1980 (Brown and Adams 1980). More recent research has identified other important roles for the various members of the Kv7 channel family. For example, Kv7.2 has been identified as being responsible for the nodal I_{Ks} current (Devaux *et al.*, 2004; Schwarz *et al.*, 2006), Kv7.4 shows high expression in the auditory system and may contribute to important potassium currents in hair cells of the cochlea (Kharkovets *et al.*, 2000) and Kv7.5 has been shown to contribute to the M-current in a variety of cells, particularly in the brain (Shah *et al.*, 2002; Shen *et al.*, 2005).

Kv7 (KCNQ) genes have been discovered in numerous species including human, mouse, and rat (Table 2.1.2.1) as well as monkey, cow, gerbil, pigeon, zebra fish, pufferfish, dogfish, worm, and fruit fly. Predicted sequences have also been identified in chimpanzee and red jungle fowl.

Table 2.1.2.1 Properties of Kv7 potassium channels

Channel subunit	Biophysical characteristics	Cellular and subcellular localization	Pharmacology (IC$_{50}$/EC$_{50}$)	Physiology	Relevance to disease states
Kv7.1 (KvLQT1) Gene *KCNQ1* Chr. location: mouse 7 rat 1q41 human 11p15.5	Homomer γ (pS) 4.0 [56], 0.7 [79] V0.5 (mV) −24 [12] −6 [56], −8 [53], −23 [32], −12 [43], −28 [1],−18 [69], −4 [51], −22 [70] Slope (mV e-fold G−1) 10 [53], 21 [51], 10 [70], 13 [43] +KCNE1 γ (pS) 15.6 [56], 4.5 [79] V0.5 (mV) 19.6 [12], 20.0 [56], 48 [51], 4 [70], 8 [43] Slope (mV e-fold G−1) 14 [68], 15 [1], 12 [70], 17 [43]	Heart, kidney, rectum pancreas, lung, placenta, stomach, adrenal gland, thyroid gland, colon [2, 34, 43, 72] Cochlea: marginal cells of the stria vascularis [33]	Homomer C293B (μM) 41 [8], 27 [46], 100 [30], 70 [6] TEA (mM) 5 [19] Linopirdine (μM) 8.9 [73], 42 [35], 67 [4] XE991 (μM) 0.8 [73, 74] Retigabine (μM) 100 [66], >100 [30, 78] +KCNE1 C293B (μM) 7 [8], 10 [20], 6 [6], 16 [4] XE991 (μM) 11 [74] +KCNE3 C293B (μM) 4 [6], 3 [47], 0.7 [4] Linopirdine (μM) 40 [35] XE991 (μM) 0.8 [74] 20 [30]	May play a role in renal and GI transport [71] May play a role in smooth muscle contraction [34] +KCNE1 Cardiac I_{Ks} current involved in ventricular repolarization [2, 43] Endolymph cycling in the inner ear [75] +KCNE3 I_{KcAMP} in intestinal crypt cells role in gastric secretion [47]	RWS—autosomal dominant ventricular arrhythmias [12] JLNS – autosomal recessive bilateral deafness and syncopal events associated with ventricular arrhythmias [33] BWS—prenatal overgrowth and predisposition to embryonic tumors [27] LQT1—ventricular arrhythmia due to dysfunction of Kv7.1 subunit gene [63]
Kv7.2 (KQT2) Gene *KCNQ2* Chr. location: mouse 2 rat 3q43 human 20q13.3	Homomer γ (pS) 17.8 [48], 5.8 [54], 6.2 [29] Popen (0 mV) 0.15 [54], 0.17 [29] PRb/PK 0.88 [38] PNH4/PK 0.12 [38] PCs/PK 0.12 [38] V0.5 (mV) −37 [5], −14 [53], −28 [29], −39 [32], −38/−27 [23] Slope (mV e-fold G) +12 [53] +9.9 [29], +6 [32], +10 [23] τd (ms) 29/241 [28] +Kv7.3 γ (pS) 9.0 [54], 8 and 5 [39] Popen (0 mV) 0.30 [54], 0.7 [39] PTl/PK 1.37 [38] PRb/PK 0.71 [38] PNH4/PK 0.12 [38] PCs/PK 0.09 [38] PNa/PK <0.005 [38] V0.5 (mV) −40 [73], −18 [53] −34 [32] −29 [17] Slope (mV e-fold G) 12 [53], 7.9 [32], 7 [73], 10 [17] τd (ms) 47/226 [28]	Brain: hippocampus dentate gyrus neocortex granular layer of cerebellum olfactory bulb caudate putamen [5, 42, 67] F11 cells [24] NG108–15 cells [52] Peripheral ganglia: SCG [52, 55, 73] DRG [37]	Homomer TEA (mM) 0.16 [73], 0.17 [58], 0.13 [76], 0.3 [19], 0.4 [50], 0.13 [9] Linopirdine (μM) 4.8 [73], 3.4 [45], 2.3 [50] XE991 (μM) 0.7 [73], 1.5 [45], 0.3 [50] Retigabine (μM) 2.5 [66], 0.5 [50], 16 [44], 4.1 [78] +Kv7.3 TEA (mM) 3.5 [73], 20 [58], 10.5 [76], 3.8 [19], 1.4 [9] Linopirdine (μM) 4.0 [73], 3.5 [76], 10 [28], 2.7 [45] XE991 (μM) 0.6 [73], 0.9 [45] Retigabine (μM)1.6 [77], 1.9 [66], 1.7 [41]	Homomer? Nodal I_{Ks} a role in axonal saltatory conduction and spike initiation [15, 49] +Kv7.3 M-current ($I_{K(M)}$), role in setting resting membrane potential, spike frequency adaptation, slow cholinergic EPSPs [7]	BFNC1—childhood form of epilepsy, which is self-limiting but increases incidence of generalized epilepsy in later life. Associated with mutations in the gene for Kv7.2 [11, 60, 80] Myokymia— spontaneous involuntary contractions of skeletal muscle fibres [14]
Kv7.3 Gene *KCNQ3* Chr. location: mouse 15 rat 7q33 human 8q24	Homomer γ (pS) 7.3 [48], 9.0 [54], 8.5 [29] Popen (0 mV) 0.59 [54], 0.89 [29] P_{Rb}/P_K 0.77 [38] P_{NH4}/P_K 0.11 [38] P_{Cs}/P_K 0.17 [38] $V_{0.5}$ (mV) −37 [53], −52 [29], −40 [61] Slope (mV e-fold G) 6 [53], 5.7 [29], 11 [61] +Kv7.5 γ (pS) 5.0 (Passmore unpublished) τd (ms) 45/349 [28]	Brain: similar to Kv7.2 but lower levels [13, 18] F11 cells [24] NG108–15 cells [52] Stomach [34] Celiac ganglia, SMG, SCG [73, 55] DRG [37]	Homomer TEA (mM) >30 [19] Linopirdine (μM) 4.8 [40] Retigabine (μM) 0.6 [66], 0.6 [44] +Kv7.4 Linopirdine (μM) 1.4 [45] XE991 (μM) 1.4 [45] +Kv7.5 TEA (mM) 200 [46] Linopirdine (μM) 15 [28], 7.7 [76] Retigabine (μM) 1.4 [77]	+Kv7.5 $I_{K(M)}$ role in setting resting membrane potential, spike frequency adaptation, slow cholinergic EPSP [7]	BFNC2 – childhood form of epilepsy, which is self limiting but increases incidence of generalized epilepsy in later life. Associated with mutations in the gene for Kv7.3 [11, 21]

Table 2.1.2.1 (Continued) Properties of Kv7 potassium channels

Channel subunit	Biophysical characteristics	Cellular and subcellular localization	Pharmacology (IC$_{50}$/EC$_{50}$)	Physiology	Relevance to disease states
Kv7.4 Gene *KCNQ4* Chr. location: mouse 4 rat 5 human 1p34	Homomer γ (pS) 2.1 [29] Popen (0 mV) 0.07 [29] V$_{0.5}$ (mV) −10 [26], −19 [53], −23 [29], −32 [62], −27 [10], −18 [65], −11 [64] Slope (mV e-fold G) 18 [26], 10 [53], 10 [29], 17 [62], 12 [10], 13 [65], 18 [64] τd (ms) 72/365 [10]	Brain: auditory nuclei [25] Cochlea: OHC, IHC [3, 25] Vestibular organ: V1HC [25] Liver, kidney, thymus gland, ovary, lung, retina, spleen, testis [3]	Homomer TEA (mM) 3 [19] Linopirdine (μM) 1.7 [45], 14 [62] XE991 (μM) 1.2 [45], 5.5 [62] Retigabine (μM) 5.2 [66], 1.4 [45] BMS204352 (μM) 2.4 [45]	OHC I$_{Kn}$—resting membrane potential and frequency response [25, 31] IHC I$_{Kn}$—resting membrane potential and [Ca]i [36] V1HC I$_{KL}$ —resting membrane potential and input resistance[25]	DFNA2—autosomal dominate progressive high frequency hearing loss [26]
Kv7.5 Gene *KCNQ5* Chr. location: mouse 1 rat 9 human 6q14	Homomer γ (pS) 2.2 [29] Popen (0mV) 0.17 [29] V$_{0.5}$ (mV) −46 [46], −42 [22], −47 [16], −31 [29] Slope (mV e-fold G) 13.1 [29], 9.0 [46] τd (ms) 51/281 [28], 20/300 [16]	Brain: cerebral cortex [28,22] hippocampus [22, 28] CA1 and CA3 neurons [57] SMSN [59] colon, lung, uterus [22] Sympathetic ganglia [46] Skeletal muscle [28] NG108−15 cells [46]	Homomer TEA (mM) 71 [46] Linopirdine 16 [28], 51 [46] XE991 (μM) 65 [46] Retigabine (μM) 1.7 [22] BMS204352 (μM) 2.4 [16]	I$_{K(M)}$ in some hippocampal and striatal neurons? [57, 59]	None known

BFNC, benign familial neonatal convulsions; BWS, Beckwith–Wiedemann syndrome; C293B, chromanol 293B; Chr., chromosome; DRG, dorsal root ganglia; EPSP, excitatory postsynaptic potential; GI, gastrointestinal; IHC, inner hair cell; JLNS, Jervell and Lange-Nielsen syndrome; LQT1, long QT syndrome type 1; DFNA2, nonsyndromic dominant progressive deafness; OHC, outer hair cell; RWS, Romano–Ward syndrome; SCG, superior cervical ganglia; SMSN, striatal medium spiny neurons; TEA, tetraethylammonium; V$_{0.5}$, half activation potential; V1HC, vestibular type 1 hair cell; γ, single channel conductance; τd, time constant of deactivation.

[1] Abitbol et al. (1999) EMBO J 18, 4137–4148; [2] Barhanin et al. (1996) Nature 384, 78–80; [3] Beisel et al. (2000) Mol Brain Res 82, 137–149; [4] Bett et al. (2006) J Physiol 576(Pt 3), 755–67; [5] Biervert et al. (1998). Science 279, 403–406; [6] Boucherot et al. (2001) J Membr Biol 182, 39–47; [7] Brown (1988) Ion Channels,Vol 1, Plenum Publishing Corporation, New York, pp 55–94; [8] Busch et al. (1997) Br J Pharmacol 122, 187–189; [9] Castaldo et al. (2002) J Neurosci 22, RC(199 1–6; [10] Chambard and Ashmore (2005). Pflugers Arch 450, 34–44; [11] Charlier et al. (1998) Nat Genet 18, 53–55; [12] Chouabe et al. (1997) EMBO Journal **16**, 5472–5479; [13] Cooper et al. (2000) PNAS 97, 4914–4919; [14] Dedek et al (2001) PNAS 98, 12272–12277; [15] Devaux et al. (2004) J Neurosci 24, 1236–1244; [16] Dupuis et al. (2002) Eur J Pharmacol 437, 129–137; [17] Etxeberria et al. (2004) J Neurosci 24, 9146–9152; [18] Geiger et al. (2006) Neurosci Lett 400, 101–104; [19] Hadley et al. (2000) Br J Pharmacol 129, 413–415; [20] Heitzmann et al. (2004) J Physiol 561.2, 547–557; [21] Hirose et al. (2000) Ann Neurol 47, 822–826; [22] Jensen et al. (2005) Mol Brain Res 139, 52–62 (2000); [23] Jow and Wang (2000) Brain Res Mol Brain Res 80, 269–278; [24] Jow et al. (2006) Assay Drug Dev Technol 4, 49–56; [25] Kharkovets et al. (2000) PNAS 97, 4333–4338; [26] Kubisch et al. (1999) Cell 96, 437–446; [27] Lee et al. (2000) Nat Genet 15, 181–185; [28] Lerche et al. (2000) J Biol Chem 275, 22395–22400; [29] Li et al. (2004) J Neurosci 24, 5079–5090; [30] Macvinish et al. (2001) Mol Pharmacol 60, 753–760; [31] Marcotti and Kros, (1999) J Physiol 520.3, 653–660; [32] Nakajo and Kubo (2005) J Physiol 569.1, 59–74; [33] Neyroud et al. (1997) Nat Genet 15, 186–189; [34] Ohya et al. (2002) Am J Physiol Gastrointest Liver Physiol 282, G277–287; [35] Ohya et al. (2003) Circ Res 92, 1016–1023; [36] Oliver et al. (2003) J Neurosci 23, 2141–2149; [37] Passmore et al. (2003) J Neurosci 23, 7227–7236; [38] Prole and Marrion (2004) Biophys J 86, 1454–1469; [39] Prole et al. (2003) J Gen Physiol 122, 775–793; [40] Robbins (2001) Pharmacol Ther 90, 1–19; [41] Rundfeldt and Netzer, (2000) Neurosci Lett 282, 73–76; [42] Sanganich et al. (2001) J Neurosci 21, 4609–4624; [43] Sanguinetti et al. (1996) Nature 384, 80–83; [44] Schenzer et al. (2005) J Neurosci 25, 5051–5060; [45] Schroder et al. (2001) Neuropharmacol 40, 888–898; [46] Schroeder et al. (2000a) J Biol Chem 275, 24089–24095; [47] Schroeder et al. (2000b) Nature 403, 196–199; [48] Schwake et al. (2000) J Biol Chem 275, 13343–13348; [49] Schwarz et al. (2006) J Physiol 573.1, 17–34; [50] Scott et al. (2003) Anal Biochem 319, 251–257; [51] Seebohm et al. (2006) Biophys J 90, 2235–2244; [52] Selyanko et al. (1999) J Neurosci 19, 7742–7756; [53] Selyanko et al. (2000) J Physiol 522.3, 349–355; [54] Selyanko et al. (2001) J Physiol 534.1, 15–24; [55] Selyanko et al. (2002) J Neurosci 22, RC212 1–5; [56] Sesti and Goldstein, (1998) J Gen Physiol 112, 651–663; [57] Shah et al. (2002) J Physiol 544.1, 29–37; [58] Shapiro et al. (2000) J Neurosci 20, 1710–1721; [59] Shen et al. (2005) J Neurosci 25, 7449–7458; [60] Singh et al. (1998) Nat Genet 18, 25–29; [61] Smith et al. (2001) J Neurosci 21, 1096–1103; [62] Sogaard et al. (2001) Am J Physiol Cell Physiol 280, C859–C866; [63] Splawski et al. (1997) N Engl J Med 336, 1562–1567; [64] Strutz-Seebohm et al. (2006) Cell Physiol Biochem 18, 57–66; [65] Su et al. (2006) Biochem Biophys Res Commun 348, 295–300; [66] Tatulian et al. (2001) J Neurosci 21, 5535–5545; [67] Tinel et al. (1998) FEBS Lett 438, 171–176; [68] Toyoda et al. (2006) Biochem Biophys Res Commun 344, 814–820; [69] Tristani-Firouzi and Sanguinetti, (1998) J Physiol 510.1, 37–45; [70] Unsold et al. (2000). Pflugers Arch 441, 368–378; [71] Vallon et al. (2005) PNAS 102, 17864–17869; [72] Wang et al. (1996) Nat. Genet 12, 17–23; [73] Wang et al. (1998) Science 282, 1890–1893; [74] Wang et al. (2000) Mol Pharmacol 57, 1218–1223; [75] Wangemann (2002) Hear Res 165, 1–9; [76] Wickenden et al. (2000) Mol Pharmacol 58, 591–600; [77] Wickenden et al. (2001) Brit J Pharmacol 132, 381–384; [78] Wuttke et al. (2005) Mol Pharmacol 67, 1009–1017; [79] Yang and Sigworth (1998) J Gen Physiol 112, 665–678; [80] Yang et al. (1998) J Biol Chem 273, 19419–19423.

Like other members of the voltage-gated potassium channel superfamily Kv7 channels are probably composed of four subunits, each containing six α-helical transmembrane domains and a single P-region (6TMD-1P; Figure 2.1.2.1). The regularly spaced positive charged amino acids in the fourth transmembrane domain (four in Kv7.1 and 6 in Kv7.2–5) and the potassium signature sequence (TxxTxGYG; Figure 2.1.2.1) in the P-region confirms these proteins as voltage-gated potassium channels. In contrast to other Kv channels, Kv7 channels lack a T1 tetramerization domain and it is the C-terminus instead that is critical for tetramerization (Wehling et al., 2007; Wiener et al., 2008). The C-terminus also determines channel regulation by transduction pathways, surface expression and scaffold protein interaction (Chung et al., 2006; Howard et al., 2007; Wehling et al., 2007; Haitin and Attali, 2008; Wiener et al., 2008). Functional Kv7 channels exist as homomers and as heteromultimers of either two different α subunits, e.g. Kv7.2 and Kv7.3 underlie the M-current, or heteromultimers of an α subunit and a member of their β subunit family, KCNE, e.g. Kv7.1 and KCNE1 underlie I$_{Ks}$.

The Kv7 potassium channel family is generating considerable scientific interest (over 120 papers a year for the last 3 years) partly because mutations in Kv7 channels underlie several human genetic diseases (cardiac dysfunction, epilepsy, and deafness), but also because several of the subunits represent attractive molecular targets for novel anti-epileptic (Main et al., 2000), analgesic (Passmore et al., 2003) anti-migraine (Gribkoff, 2003), anxiolytic (Korsgaard et al., 2005), neuroprotective (Jensen, 2002; Boscia et al., 2006) and antidystonic (Richter et al., 2006) drugs.

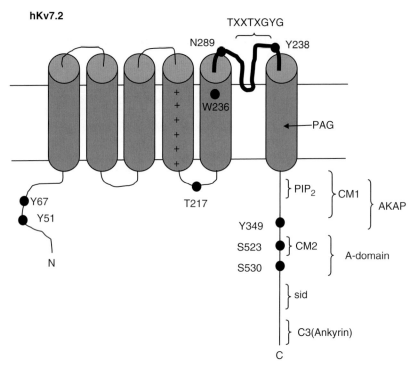

Fig. 2.1.2.1 Schematic diagram of important domains and residues in Kv7 channel subunits (based on human Kv7.2 unless stated). Y51 (Kv7.1), important membrane targeting residue (Jespersen et al., 2005); Y67 and Y349 (Kv7.3), Src tyrosine kinase phosphorylation sites (Li et al., 2004); + − positively charged amino acids in S4 (voltage sensor); T217, putative phosphorylation site (Surti et al., 2005); W236, important residue for retigabine binding (Schenzer et al., 2005; Wuttke et al., 2005); N289 (Kv7.1) putative glycosylation site (Barhanin et al., 1996;Wang et al., 1996); TXXTXGYG, potassium signature sequence in the P-region (Schroeder et al., 1998); Y238, amino acid residue conferring TEA sensitivity (Hadley et al., 2000); PAG, putative gating hinge (Seebohm et al., 2006); PIP2, putative phosphatidylinositol 4,5 bisphosphate binding site (Zhang et al., 2003); CM1 and CM2, calmodulin binding domains 1 and 2 (Yus-Najera et al., 2002); AKAP, A-kinase anchoring protein binding site (Hoshi et al., 2003); S523 and S520, protein kinase C phosphorylation sites (Hoshi et al., 2003); A-domain, possible subunit interaction site (Schwake et al., 2000); sid, subunit interaction domain (Schwake et al., 2003); C3, domain containing an ankyrin-G binding site (Pan et al., 2006).

Previous reviews on the cardiac channel (Kv7.1; Jespersen *et al.*, 2005), the neuronal channels (Kv7.2–7.5; Jentsch, 2000; Rogawski, 2000) and more general reviews of the entire subfamily (Robbins, 2001; Dalby-Brown *et al.*, 2006) have been published.

2.1.2.2 Molecular characterization

The first member of the Kv7 family, Kv7.1, was identified using positional cloning following linkage analysis studies of patients with cardiac long QT syndrome (LQT1). Originally named KvLQT1, the predicted protein exhibited structural properties similar to other voltage-gated K+ channels despite lacking an amino terminal domain (Wang *et al.*, 1996). Several additional KvLQT1 isoforms have since been isolated (Barhanin *et al.*, 1996; Sanguinetti *et al.*, 1996; Lee *et al.*, 1997; Yang *et al.*, 1997; Splawski *et al.*, 1998), including the full-length isoform 1 encoding a protein of 676 amino acid residues (Yang *et al.*, 1997).

Evidence that the Kv7 family consists of multiple genes was first provided in 1998 with the cloning of two novel K+ channel genes KCNQ2 and KCNQ3 (Singh *et al.*, 1998; Charlier *et al.*, 1998; Biervert *et al.*, 1998). These genes encoded proteins of between 825 and 872 amino acid residues and displayed between 60 and 70 per cent homology with Kv7.1 in the transmembrane domain region (Singh *et al.*, 1998; Charlier *et al.*, 1998; Biervert *et al.*, 1998). Alternative splicing of the Kv7.2 gene gives rise to numerous additional isoforms (Nakamura *et al.*, 1998; Tinel *et al.*, 1998; Biervert and Steinlein 1999; Smith *et al.*, 2001; Pan *et al.*, 2001) of which

one class, containing exon 15a, encodes channels exhibiting different kinetics (Pan *et al.*, 2001).

Kv7.4 was first isolated from a human retina cDNA library using a Kv7.3 channel partial cDNA as a probe (Kubisch *et al.*, 1999). The *KCNQ4* gene encodes a protein of 695 amino acid residues with a molecular weight of 77 kDa and shares closest homology (44 per cent) with Kv7.2. Several additional Kv7.4 splice variants, designated KCNQ4_v1 – v4, have also been recently identified (Beisel *et al.*, 2005).

The fifth and final member of the Kv7 channel family, Kv7.5, was first cloned from human brain (Lerche *et al.*, 2000; Schroeder *et al.*, 2000a), although a murine homologue has subsequently been isolated (Jensen *et al.*, 2005). The *KCNQ5* gene encodes a protein of between 897 and 932 amino acids with a molecular weight of approximately 99–102 kDa, sharing closest homology (65 per cent) with Kv7.4. (Lerche *et al.*, 2000; Schroeder *et al.*, 2000a).

2.1.2.3 Biophysical characteristics

Single channel parameters

Kv7.1 homomers show a single channel conductance (γ) of between 0.7 and 4.0 pS which increases when they are co-expressed with their β subunit (KCNE1) to between 4.5 and 15.6 pS (Yang and Sigworth 1998; Sesti and Goldstein, 1998; Pusch, 1998), although there is one study where the converse was seen (Romey *et al.*, 1997). Kv7.2 homomers have a γ of around 6 pS and an open probability (Po) of 0.15–0.17 at 0 mV (Selyanko *et al.*, 2001; Li *et al.*, 2004).

When Kv7.2 is in a heteromer with Kv7.3 the γ increases to around 9 pS and the Po to between 0.3 and 0.7 (Selyanko *et al.*, 2001; Prole and Marrion 2004), although an increase in γ was not seen in one study (Schwake *et al.*, 2000). These increases in γ and Po may partly explain the larger macroscopic current levels reported for Kv7.2+7.3 heteromers over their homomers (Yang *et al.*, 1998), However, a more recent study indicates that the fifteenfold increase in macroscopic current levels following heteromerization of Kv7.2 with 7.3 is a consequence of several complex factors (Etxeberria *et al.*, 2004). Kv7.3 homomers have a γ between 7 and 9 pS and a high Po of between 0.6 and 0.9 (Schwake *et al.*, 2000; Selyanko *et al.*, 2001; Li *et al.*, 2004). In contrast, Kv7.4 and Kv7.5 homomers both have low γ (~2 pS) and Pos of 0.07 and 0.17 at 0 mV, respectively (Li *et al.*, 2004).

Permeability

Consistent with other Kv potassium channels Kv7.1 (with or without KCNE1) and Kv7.2+7.3 heteromeric and homomeric channels have a high permeability to potassium over sodium, in the order of less than one sodium ion to every 200 potassium, and display an Eisenman Type IV permeability series of $Tl^+>K^+>Rb^+>NH_4^+\geq Cs^+>>Na^+$ (Yang and Sigworth 1998; Sesti and Goldstein 1998; Prole and Marrion 2004).

Current kinetics

In general Kv7 channels display characteristics typical of the Kv superfamily with a few important exceptions; namely slow activation and deactivation kinetics, some members of the family display very little or no inactivation (Jensen *et al.*, 2007) and in some cases a relatively negative activation threshold.

The position of the activation curves seem to vary according to the expression system used and the subunit composition of the channel, both in terms of α and β subunits (Table 2.1.2.2). For example Kv7.1 homomers have a half activation potential ($V_{0.5}$) of between −24 and −6 mV, whereas co-expression with KCNE1 leads to a dramatic shift in the activation curve so that the $V_{0.5}$ is more positive (+20 mV; Table 1). Kv7.2 homomers have a $V_{0.5}$ of −39 to −14 mV and a slope of around 10 mV e-fold G^{-1}. These parameters do not change significantly when Kv7.2 is co-expressed with Kv7.3, despite the fact that Kv7.3 homomers have a more negative $V_{0.5}$ and a steeper slope (Table 2.1.2.1).

Kv7.4 homomers, like Kv7.1, have a positive activation curve with a $V_{0.5}$ ranging from −32 to −10 mV (Table 2.1.2.1) but this can be shifted substantially in the negative direction by phosphorylation and prestin binding (Chambard and Ashmore 2005). Half activation potentials for Kv7.5 homomers range from −48 to −31 mV with slopes of 9 to 13 mV e-fold G^{-1} (Table 2.1.2.1).

Kv7 deactivation tail currents are usually best described by the sum of two exponentials, around 50 ms and 300 ms in duration (Table 2.1.2.1). In general, the sum of two exponentials is also used to describe the activation kinetics, although in some cases three have been used (e.g. Schroeder *et al.*, 2000a). A number of Kv7 currents show inactivation such as Kv7.1 homomers (Tristani-Firouzi and Sanguinetti 1998) and Kv7.4 and Kv7.5 currents (Jensen *et al.*, 2007).

2.1.2.4 Localization

Kv7 channels have been detected in a wide variety of tissues including the brain, heart and peripheral nervous system. Kv7.1 channels are predominantly expressed in the heart, whilst Kv7.2, 7.3, 7.4, and 7.5 channels are almost exclusively expressed in the brain and peripheral nervous system.

Early studies reported the expression of Kv7.1 mRNA in the human heart, kidney, lung, placenta, pancreas, adrenal, and thyroid glands and its absence from neuronal tissue, liver, and skeletal muscle (Wang *et al.*, 1996; Barhanin *et al.*, 1996; Sanguinetti *et al.*, 1996). Kv7.1 mRNA has also been detected in the mouse cochlea where its expression is restricted to the stria vascularis (Neyroud *et al.*, 1997).

Immunofluorescence studies have shown prominent expression of Kv7.1 in mouse stomach, small intestine, and colon. In the colon and small intestine, Kv7.1 expression is co-localized with KCNE3 in the basolateral membranes of the crypt cells. In the stomach Kv7.1 is co-expressed with the H^+/K^+-ATPase in the luminal membrane of the acid-secreting parietal cells of gastric glands. KCNE2 has also been detected in a subset of these cells in gastric glands (Dedek and Waldegger, 2001).

Kv7.2 is expressed throughout the central and peripheral nervous systems. Northern blot and *in situ* hybridization studies have shown moderate to high expression of Kv7.2 mRNA in most areas of the brain including the hippocampus, dentate gyrus, neocortex, granular layer of the cerebellum, olfactory bulb, and caudate putamen. Weaker expression is seen in several of the brainstem nuclei including the vestibular nucleus, inferior olive, dorsal cochlear nucleus, pontine nucleus, inferior colliculus, substantia nigra and the red nucleus (Biervert *et al.*, 1998; Tinel *et al.*, 1998; Saganich *et al.*, 2001).

Table 2.1.2.2 Relative current expression for Kv7 alpha subunit homomers and heteromers and Kv7 alpha subunits with KCNE beta subunits

Alpha subunits					
Kv	7.1	7.2	7.3	7.4	7.5
7.1	+	++	+	++	NP
7.2		+	++	++	+
7.3			−	++	++
7.4				+	NP
7.5					+

Beta subunits					
Kv	KCNE1	KCNE2	KCNE3	KCNE4	KCNE5
7.1	++	+*	+*	−	++
7.2	+	+	NP	+	NP
7.3	+	+	NP	NP	NP
7.4	+	+	−	++	+
7.5	NP	NP	NP	+	NP
7.2+7.3	−	+	NP	+	NP

−, reduced current expression; +, current expression; ++, enhanced current expression; NP, not performed; *, loss of time-dependence.

Note: variations may occur in different expression systems.

Data from Grunnet *et al.* (2003); Jesperson *et al.* (2005); Kubisch *et al.* (1999); Schroeder *et al.* (1998); Schroeder *et al.* (2000a); Schwake *et al.* (2000); Schwake *et al.* (2006); Selyanko *et al.* (2000); Strutz-Seebohm *et al.* (2006); Tinel *et al.* (2000); Wang *et al.* (1998); Yang *et al.* (1998).

In general, the expression pattern of Kv7.3 overlaps that of Kv7.2 although Kv7.3 signals are typically weaker and more restricted (Tinel *et al.*, 1998; Schroeder *et al.*, 1998; Saganich *et al.*, 2001). However, some neurons express one of these transcripts in the absence of the other. For example, cerebellar granule cells and Purkinje cells express Kv7.2 but not Kv7.3 whilst the layer IV neurones of the cortex display high Kv7.3 but weak Kv7.2 mRNA expression (Saganich *et al.*, 2001). Kv7.2 mRNA has also been detected in testis and cochlea, although expression levels are low, and Kv7.3 mRNA has been found in the spleen and cochlea (Tinel *et al.*, 1998; Kubisch *et al.*, 1999).

Immunohistochemical analysis has confirmed the overlapping expression of Kv7.2 and Kv7.3 in many neurons of the human cortex and hippocampus. However, Kv7.2 was detected in the absence of Kv7.3 in the axons and/or termini of human hippocampal mossy and granule cells (Cooper *et al.*, 2000). In myelinated axons, Kv7.2 expression has been reported in the axon initial segments and nodes of Ranvier (Devaux *et al.*, 2004).

Kv7.2 and Kv7.3 expression has also been detected in a variety of other neurons including superior cervical ganglion neurons (Wang *et al.*, 1998; Hadley *et al.*, 2003), ventral and dorsal horn neurons (Dedek *et al.*, 2001), dorsal root ganglion neurons (Passmore *et al.*, 2003) and nodose ganglion neurons (Wladyka and Kunze, 2006).

In contrast to the other neuronal Kv7 channels Kv7.4 was initially thought to have a much more limited expression. Significant levels of Kv7.4 transcripts were detected in the cochlea and in brain tissue with fainter expression in the heart and skeletal muscle (Kubisch *et al.*, 1999). However, RT-PCR results from a more recent study suggest that Kv7.4 mRNA expression is much more widespread than originally shown and transcripts have been detected in the liver, thymus, lung, spleen, kidney, ovary, retina, and testicle in addition to the brain, heart, and inner ear (Beisel *et al.*, 2005).

Initial studies suggested that in the cochlea Kv7.4 expression was restricted to the sensory outer hair cells (OHCs) (Kubisch *et al.*, 1999). However, more recent studies indicate that Kv7.4 is also expressed in the inner hair cells (IHCs) (Kharkovets *et al.*, 2000; Beisel *et al.*, 2000; Oliver *et al.*, 2003) and spiral sensory neurons (Beisel *et al.*, 2000).

Immunofluorescence studies from the Jentsch group have shown that the immunofluorescence signal for Kv7.4 is stronger in OHCs at the base of the cochlea than at the apex and that Kv7.4 expression is confined to the basal membrane of OHCs. In contrast, results from a second group indicate a different distribution. Whole mount *in situ* hybridization and immunofluorescence experiments suggest that the highest levels of Kv7.4 expression are in IHCs and spiral ganglion cells at the base of the cochlea and in OHCs at the apex (Beisel *et al.*, 2000, 2005).

In the brain Kv7.4 expression is restricted to certain brainstem nuclei including several nuclei of the central auditory pathway. In addition, intense staining was observed in type I vestibular hair cells at their basal and lateral membrane (Kharkovets *et al.*, 2000). More recent immunofluorescence results suggest Kv7.4 is also expressed in type II vestibular hair cells and in vestibular ganglion neurons (Beisel *et al.*, 2005).

Kv7.5 transcripts have been detected in a wide variety of tissues including the brain and human adult skeletal muscle. Strong expression is seen in the cerebral cortex, occipital pole, frontal and temporal lobes, hippocampus, piriform cortex, entorhinal cortex, pontine medulla, and facial nucleus. Lower levels have been reported in the putamen, cerebellum, lung, colon, uterus, and hypothalamus (Lerche *et al.*, 2000; Schroeder *et al.*, 2000a; Jensen *et al.*, 2005). Kv7.5 transcripts have also been detected in superior cervical ganglia, dorsal root ganglia, rat hepatocytes, human liver, and the NG108-15 mouse neuroblastoma cell line (Schroeder *et al.*, 2000a; Hadley *et al.*, 2003; Passmore *et al.*, 2003; Lan *et al.*, 2004).

Immunocytochemistry has revealed the presence of Kv7.5 protein in cultured hippocampal neurons, human temporal neocortex and hippocampus, superior cervical ganglion (SCG) neurons, dorsal root ganglion (DRG) neurons and nodose ganglia neurons (Shah *et al.*, 2002; Yus-Najera *et al.*, 2003; Hadley *et al.*, 2003; Passmore *et al.*, 2003; Wladyka and Kunze, 2006). In general, the expression pattern of Kv7.5 overlaps that of Kv7.2 and 7.3 suggesting that Kv7.5 might contribute to native M-channels in certain tissues.

2.1.2.5 Pharmacology
Channel blockers

Kv7 currents are sensitive to the pan-potassium channel blockers such as barium ions and tetraethylammonium (TEA). Although TEA is a non-selective K^+-channel blocker it shows some selectivity for the different Kv7 subunits and can therefore be used to probe for the subunit composition of a particular Kv7 mediated current (Hadley *et al.*, 2000, 2003; Schwake *et al.*, 2006). Kv7.2 homomers have the highest sensitivity to TEA (EC_{50} < 1 mM, Table 1), Kv7.2 + 7.3 heteromers and Kv7.4 homomers display an intermediate sensitivity with EC_{50}s between 4–10 mM (Table 2.1.2.1) and Kv7.3 and Kv7.5 homomers have the lowest sensitivity (EC_{50} >30 mM, Table 2.1.2.1). This variation in sensitivity to TEA block correlates well with the nature of the amino acid residue near the P-region (Figure 2.1.2.1); Y283 confers high TEA sensitivity in human Kv7.2 and the presence of a valine in Kv7.1 or threonine in Kv7.3–7.5 reduces this (Hadley *et al.*, 2000).

Kv7 currents are insensitive to the animal toxins dendrotoxin, apamin, and charybdotoxin, and to other potassium channel blockers such as 4-aminopyridine (Yang *et al.*, 1998; Jow and Wang 2000; Passmore *et al.*, 2003).

Kv7.1 blockers include clofilium and chromanol 293B. In general these compounds are less effective on Kv7.2–7.5 channel currents than Kv7.1. Indeed, chromanol 293B is regularly used as a "selective" Kv7.1 channel blocker and blocks Kv7.1 homomers with an IC_{50} of between 27–100 μM (Table 2.1.2.1). When Kv7.1 is expressed with KCNE1 or KCNE3 the sensitivity to chromanol 293B increases to 6–16 μM and 0.7–4 μM, respectively (Table 2.1.2.1; Busch *et al.*, 1997; Schroeder *et al.*, 2000b). This compound has rarely been tested on other Kv7 currents, although it has been reported that 100 μM inhibits Kv7.5 currents by approximately 45 per cent (Lerche *et al.*, 2000).

The most utilized Kv7 channel blockers are linopirdine (DuP996) and its more potent analogue XE991 (Costa and Brown 1997; Zaczek *et al.*, 1998; Wang *et al.*, 2000). Although these compounds have little subunit selectivity (Table 2.1.2.1) they are reasonably good as indicators of the presence of Kv7 currents. However, caution should be used particularly at concentrations above 10 μM for linopirdine and 3 μM for XE991 as it has been demonstrated that they can have effects on other voltage-gated and ligand-gated currents (Lamas *et al.*, 1997; Schnee and Brown 1998; Noda *et al.*, 1998; Gomez-Casati *et al.*, 2004; Wladyka and Kunze, 2006; Elmedyb *et al.*, 2007).

Channel activators

A number of Kv7 channel activators have been synthesized including flupirtine (Dalby-Brown et al., 2006) retigabine (Main et al., 2000; Rundfeldt and Netzer 2000, Wickenden et al., 2000), maxipost (BMS-204352, Dupuis et al., 2002) and WAY-1 (Jow et al., 2006). Unlike some newer compounds these, with the exception of Kv7.1, activate all Kv7 subunits. Channel activators shift the activation curve to more negative potentials resulting in increased current and slowing of the activation and deactivation kinetics. This results in a net hyperpolarization of the cell membrane potential (Rundfeldt and Netzer 2000; Main et al., 2000; Wickenden et al., 2000; Tatulian et al., 2001). The EC_{50} values quoted in Table 2.1.2.1 are derived from various measurements: either the shift in $V_{0.5}$, the increase in current at a given potential or the increase in hyperpolarization. At the single channel level retigabine decreases the longest closed time and increases both open times without altering the single channel conductance (Tatulian and Brown, 2003), the net effect being to increase maximal Po. Chimera and point mutation data have indicated that a tryptophan residue in TMD5 (W236 in human Kv7.2, Figure 2.1.2.1) is critical for retigabine enhancement (Wuttke et al., 2005; Schenzer et al., 2005). This residue is absent from Kv7.1, which is consistent with the lack of effect of retigabine on this subunit.

Retigabine is in Phase III clinical trials as an anti-epileptic drug and has also been shown to display analgesic activity in a variety of animal models of pain (Blackburn-Munro and Jensen, 2003; Passmore et al., 2003; Dost et al., 2004; Nielsen et al., 2004; Hirano et al., 2007). Further, possibly therapeutically useful, actions of Kv7 channel activators (for review see Dalby-Brown et al., 2006) include neuroprotection (Jensen 2002; Mikkelsen 2004; Boscia et al., 2006), anxiolytic (Korsgaard et al., 2005) and micturition control (Streng et al., 2004). A number of recently described channel activators show some subtype selectivity (Bentzen et al., 2006; Wickenden et al., 2008; Xiong et al., 2007) reviewed in Xiong et al. (2008).

2.1.2.6 Physiology

Kv7.1

It is now well established that Kv7.1, together with its β subunit KCNE1, forms the cardiac I_{Ks} current, which is important for repolarization of the cardiac action potential (Barhanin et al., 1996; Sanguinetti et al., 1996: Jespersen et al., 2005). Kv7.1 and KCNE1 heteromers are also thought to play a role in the auditory system, contributing to the control of endolymph homeostasis in the cochlea (Neyroud et al., 1997; Wangemann, 2006). Similar transport roles for Kv7.1, with its alternative β subunits KCNE2 and KCNE3, in other epithelial cells have been suggested including the control of gastric acid secretion and intestinal absorption/secretion (Schroeder et al., 2000b; Vallon et al., 2005). Additional studies have established the presence of Kv7.1 in vascular (Ohya et al., 2003; Yeung and Greenwood 2005; Joshi et al., 2006: Yeung et al., 2007; Mackie et al, 2008) and gastric (Ohya et al., 2002) smooth muscle, indicating a physiological role in controlling smooth muscle contraction.

Kv7.2 and Kv7.3

Kv7.2 and 7.3 heteromultimers are thought to underlie the M-current ($I_{K(M)}$) (Wang et al., 1998), a slowly activating, non-inactivating K$^+$ current that is responsible for damping neuronal excitability (Brown, 1988). The M-current was first described in bullfrog sympathetic neurons (Brown and Adams, 1980) and subsequently in other peripheral ganglia including: rat superior cervical (Constanti and Brown, 1981), thoracolumbar (Norenberg et al., 2000), dorsal root (Passmore et al., 2003), nodose (Wladyka and Kunze, 2006) and guinea pig celiac ganglia (Vanner et al., 1993).

In the central nervous system Kv7.2 and 7.3 play a key role in controlling excitability, particularly in the hippocampus (Cooper et al., 2001; Peters et al., 2005; Yue and Yaari 2004; 2006; Piccinin et al., 2006; Lawrence et al., 2006; Otto et al, 2006; Vervaeke et al.,2006; Yoshida and Alonso., 2007; Peretz et al., 2007; Hu et al., 2007; Chen and Yaari, 2008). Other important brain regions where Kv7 currents may be important include the striatum (Shen et al., 2005) and the mesencephalic area (Koyama and Appel 2006; Hansen et al., 2006, 2007).

Interestingly, Kv7.2 has recently been identified as the nodal slow potassium current (I_{Ks}) and may therefore play an important role in a very wide range of axons. In both peripheral and central neurons Kv7.2 has been detected by immunocytochemistry in the nodal region and initial segments and a physiological role has been confirmed using pharmacological intervention whereby I_{Ks} was inhibited by XE991, linopirdine and low concentrations of TEA (IC_{50} 0.2 mM); results consistent with the presence of Kv7.2 homomers (Devaux et al., 2004; Schwarz et al., 2006).

Kv7.4

Although Kv7.4 has been identified in brain, heart, and skeletal muscle (Kharkovets et al., 2000) it is the role of this subunit in the auditory system that has generated the most interest (Kubisch et al., 1999). Kv7.4 has been identified as the outer hair cell potassium current called I_{Kn} (Housley and Ashmore, 1992; Marcotti and Kros, 1999) and as the analogous current, I_{KL}, in vestibular type 1 hair cells (Chen and Eatock, 2000) and inner hair cells (Oliver et al., 2003). The role of Kv7.4 in the central auditory pathway is yet to be established (Kharkovets et al., 2000).

Kv7.5

Less is known about the specific physiological role of Kv7.5. It is known that in heterologous expression systems it can alone (Lerche et al., 2000; Schroeder et al., 2000a; Dupuis et al., 2002; Jensen et al., 2005) or co-assembled with Kv7.3 (Lerche et al., 2000; Schroeder et al., 2000a; Wickenden et al., 2001) produce an M-like current. In both the hippocampus (Shah et al., 2002) and the striatum (Shen et al., 2005) Kv7.5 subunits may contribute to the M-current. Kv7.5 has been identified in L6E9 myoblast (Roura-Ferrer et al., 2008), Calu-3 airway epithelial (Moser et al., 2008) and A7r5 aortic smooth muscle (Brueggemann et al., 2007) cell lines. However, the mRNA expression of Kv7.5 in cerebral cortex, skeletal muscle, colon, lung, and uterus (Lerche et al., 2000; Schroeder et al., 2000a; Yus-Najera et al., 2003; Jensen et al., 2005), immunohistochemical detection in aortic baroreceptor neurons (Wladyka et al., 2008) and auditory brainstem nuclei Caminos et al., 2007) suggest some unknown physiological roles for this subunit.

Regulation of Kv7 currents

One of the characteristics of voltage-gated Kv7 channels is their regulation by G-protein coupled receptors (GPCRs). In the appropriate heterologous system all of the Kv7 subunits can be inhibited by activation of GPCRs (Selyanko et al., 2000). However, this may not reflect the physiological situation where receptor coupling can vary in different cell types and indeed within a single cell (microdomains). In general, the receptors that couple to the $G_{q/11}$ family of G-proteins are responsible for inhibition of Kv7 currents. Therefore, efforts to elucidate the transduction pathways of modulation have concentrated primarily on the phosphoinositide system. Over the last 20 years there has been considerable controversy about the identity of the second messenger that regulates these channels. However, recently the consensus has formed that phosphatidylinositol 4,5-bisphosphate (PIP_2) may be the primary regulator utilized to mediate GPCR induced inhibition (Suh and Hille 2002; Ford et al., 2003, Zhang et al., 2003, Loussouarn et al., 2003; Suh et al., 2004, Winks et al., 2005; Jensen et al., 2005; Li et al., 2005; Park et al., 2005; Robbins et al., 2006; Suh et al., 2006; Zaika et al., 2006; Hughes et al., 2007). In addition a number of second messengers may regulate the interaction of Kv7 channels with PIP_2 (for review see Delmas and Brown, 2005; Brown et al., 2007),. These regulators include calcium/calmodulin (Wen and Levitan 2002; Yus-Najera et al., 2002; Gamper and Shapiro 2003; Richards et al., 2004; Shahidullah et al., 2005; Gamper et al., 2005; Etxeberria et al., 2008; Bal et al., 2008)—it should be noted that Kv7.1 has recently been shown to require calmodulin for normal function (Shamgar et al., 2006; Ghosh et al., 2006)—phosphorylation by PKC via AKAP (Hoshi et al., 2003; Nakajo and Kubo, 2005), phosphorylation by SRC tyrosine kinase (Gamper et al., 2003; Li et al., 2005), and receptor tyrosine kinase (Jia et al., 2007). Some currents, particularly Kv7.4, may have a more direct regulation by kinase activity (Chambard and Ashmore, 2005). Furthermore, it has been suggested that cyclic ADP-ribose can act as a regulator in some cell types (Higashida et al., 1995, 2000; Bowden et al., 1999). Indeed it has been suggested that some receptors that couple to Kv7 currents will preferentially use one transduction pathway rather than another (Delmas and Brown, 2005; Hermandez et al., 2008).

There are other agents that can modulate the channel directly such as protons (Prole et al., 2003), cysteine-modifiers (Li et al., 2004), oxidation (Gamper et al., 2006), cell volume (Grunnet et al., 2003; Hougaard et al., 2004; Jensen et al., 2005) and ubiquitination (Ekberg et al., 2007). However, the physiological role of these forms of modulation has yet to be established.

2.1.2.7 Pathophysiology and disease

The human Kv7.1 gene maps to chromosome 11p15.5 where there are loci linked with two forms of the long QT syndrome (LQTS): the Romano–Ward syndrome (RWS; Wang et al., 1996) and the Jervell and Lange-Nielsen syndrome (JLNS; Neyroud et al., 1997). A prolongation of the QT interval ('long QT') on the electrocardiogram (ECG) is characteristic of both of these syndromes and indicates abnormal ventricular repolarization, which underlies cardiac arrhythmias such as torsade de pointes and ventricular fibrillation (for review see Keating and Sanguinetti, 2001).

Patients with RWS and JLNS can suffer from an abrupt loss of consciousness (syncope), seizures and sudden death. In patients with JLNS inheritance of homozygous mutations in the Kv7.1 gene not only causes severe arrhythmia but also leads to bilateral congenital deafness (Neyroud et al., 1997).

Numerous mutations in Kv7.1 have been linked with RWS and JLNS. Expression studies have shown that most mutations of Kv7.1 associated with RWS cause a dominant-negative suppression

of Kv7.1 (Chouabe et al., 1997; see also Keating and Sanguinetti 2001) whilst the majority of Kv7.1 mutations associated with JLNS cause a loss of channel function (Wollnik et al., 1997). More recent research indicates that several mutations in the Kv7.1 gene associated with familial long QT syndromes lead to disruption of macromolecular signalling including a reduction in channel affinity for PIP_2 and prevention of Kv7.1 channel regulation by PKA (see Delmas and Brown for review, 2005).

As mentioned earlier (see 2.1.2.6) Kv7.1, together with KCNE1, underlies one of the main repolarizing K^+ currents (I_{Ks}) in the heart. Thus, a reduction in I_{Ks} function (or lack of function) results in abnormal ventricular repolarization. In patients with the more severe form of JLNS, the lack of Kv7.1 also leads to deafness, which probably results from the insufficient production of endolymph and deterioration of the organ of Corti (Neyroud et al., 1997; Rivas and Francis, 2005).

Attempts to develop an animal model of these disorders, by disruption of the Kv7.1 gene, have generated mixed results. Lee and colleagues (2000) did not observe any electrocardiographic abnormalities in their Kv7.1 knockout mice. However, homozygous mice displayed complete deafness and gastric hyperplasia. Mice lacking the Kv7.1 gene exhibited inner ear defects (Lee et al., 2000; Rivas and Francis 2005) including marked atrophy of the stria vascularis, contraction of the endolymphatic compartments and the collapse and adhesion of surrounding membranes. Degeneration of the organ of Corti and spiral ganglion was also observed in homozygous Kv7.1 mutant mice.

In contrast, Casimiro et al., (2001) reported that the phenotype of their Kv7.1 knockout mice resembled patients with JLNS. In addition to suffering with deafness, homozygous mice displayed altered cardiac repolarization. However, defects not found in humans were also observed, including unusual changes in the ECG phenotype and the existence of a severe hyperactivity including circling behaviour. These findings led Casimiro and colleagues (2004) to develop a further mouse model in which point mutations associated with RWS were introduced rather than a complete knockout of the Kv7.1 gene. Mice carrying the A340E mutation displayed normal hearing and a long QT interval on the ECG showing that the long QT in these mouse models is independent of inner ear defects.

Another disorder in which Kv7.1 has been implicated is the Beckwith–Wiedemann syndrome (BWS). BWS is characterized by prenatal overgrowth and cancer and several human genes are thought to be involved. Lee and colleagues (1997) showed that the Kv7.1 gene is disrupted by chromosomal rearrangements in BWS patients. However, the effect of Kv7.1 disruption in BWS was thought to be indirect i.e. the disruption of the Kv7.1 gene disrupts imprinting of other genes on the maternal chromosome. The absence of any features of BWS from Kv7.1 knockout mice supports this notion (Lee et al., 2000).

Kv7.2 and Kv7.3 map to chromosomes 20q13.3 and 8q24, respectively, where there are two known loci that are linked with a rare form of autosomal-dominant idiopathic epilepsy in infants known as benign familial neonatal convulsions (BFNC) (Singh et al., 1998; Charlier et al., 1998; Biervert et al., 1998). BFNC is characterized by brief seizures, which occur frequently and typically begin several days after birth. For most patients, these symptoms spontaneously disappear within a few weeks. However, approximately 10–15 per cent of patients with BFNC develop epilepsy in adulthood (for review see Steinlein, 2001; Cooper and Jan, 2003).

Numerous mutations in the *Kv7.2* gene and several in the *Kv7.3* gene have been linked with BFNC (for review see Lerche *et al.*, 2001). These mutations include structural alterations of the pore region and the C-terminal domain and it was suggested, by analogy with other K$^+$ channel mutations, that these might cause either a loss or gain of function or a dominant negative effect (Stoffel and Jan, 1998). Initial studies by Biervert *et al.*, (1998) found that a mutation in *Kv7.2* linked with BFNC abolished channel function when the mutant channel was expressed alone in *Xenopus* oocytes. When co-expressed at a 1:1 ratio with wild-type *Kv7.2* no dominant negative effect was seen but currents were reduced compared with those recorded from oocytes expressing only wild-type *Kv7.2*. In a further attempt to understand the underlying molecular mechanism of BFNC, Schroeder *et al.*, (1998) utilized a co-expression scheme in which they co-injected various *Kv7.2* or *Kv7.3* mutants with wild-type *Kv7.2* and *7.3*. For example, co-injection of the *Kv7.2* mutant Y284C with wild-type *Kv7.2* and *7.3* at a ratio of 1:1:2 resulted in a 20–30 per cent reduction in currents compared with wild-type *Kv7.2/7.3* heteromeric expression. This small loss of function applied for missense mutations, C-terminal deletions and gene deletions and it was therefore concluded that a 25 per cent loss of heteromeric *Kv7.2/7.3* channel function could cause the neuronal hyperexcitability in BFNC. More recently, Singh *et al.*, (2003) have reported a *Kv7.2* mutation that has a phenotype of neonatal seizures but unlike previously identified mutations displays a dominant-negative mode of action.

Why does a reduced Kv7.2/7.3 current cause seizures in infants, which disappear later in life? The findings of Okada and colleagues (2003) in immature rat hippocampus suggest that the 'age-dependent development and spontaneous remission of BFNC' are a consequence of the interaction between an 'age-dependent reduction in inhibitory Kv7 channel activity' and an 'age-dependent functional switching of the GABAergic-system' in neonates from excitatory to inhibitory. However, the level of Kv7.3 mRNA expression is low in developing rodent brain (and in SCG neurons) and this expression increases with age (Tinel *et al.*, 1998; Hadley *et al.*, 2003). Similar results for Kv7.2 and Kv7.3 expression have been demonstrated in developing human brain (Kanaumi *et al.*, 2008). Furthermore, shorter Kv7.2 splice variants are prominently expressed in immature rodent brain, which disappear in adult brain and are replaced with longer Kv7.2 splice variants (Nakamura *et al.*, 1998; Smith *et al.*, 2001). Expression of a shorter Kv7.2 splice variant in *Xenopus* oocytes did not yield functional currents and co-expression of this shorter splice variant with a longer Kv7.2 variant or with Kv7.3 resulted in a suppression of the potassium conductance (Smith *et al.*, 2001).

Together, these findings might suggest that the overall potassium conductance through Kv7.2/7.3 heteromultimers might be significantly reduced in the developing brain compared with that in adult brain. Thus, it is possible that the upregulation of Kv7 channels and the concurrent expression of other voltage-gated potassium channels during maturation may underlie the remission of BFNC (see Lerche *et al.*, 2001).

Further support for this theory is provided by results from immunohistochemical analyses of Kv7.2 channels in the adult and developing mouse brain, which indicate that the predominantly axonal staining pattern seen in adult mouse hippocampus is not seen before P8 and gradually develops between P11 and P21 (Weber *et al.*, 2006).

Changes in the intensity of staining for Kv7.3 in mouse hippocampus during maturation have also been reported (Geiger *et al.*, 2006), although the apparent stronger expression of Kv7.2 in the axon initial segments and nodes of Ranvier in the hippocampus (Devaux *et al.*, 2004) might indicate that changes in expression of Kv7.2 during maturation may play a more crucial role in the remission of BFNC. Interestingly, recent findings from embryonic Kv7.2 knockout mice suggest that the Kv7.2 subunit may be a critical component of the native M-channel. SCG neurons from embryonic homozygous *Kv7.2* knockout mice lacked functional M-currents and were hyperexcitable (Passmore *et al.*, 2006).

In a few rare cases, patients with BFNC have developed myokymia later in life (Dedek *et al.*, 2001). Characterized by involuntary contractions of skeletal muscle, myokymia has previously been linked with mutations in the *Kv1.1* gene (for review see Lerche *et al.*, 2001). However, one mutation (an arginine to tryptophan substitution in the voltage sensor) in the *Kv7.2* gene also gives rise to myokymia (Dedek *et al.*, 2001). This mutation causes a shift in the activation curve to more depolarized potentials and confers slower channel activation. This could lead to a reduced potassium conductance at the nodes of myelinated axons and may result in the intrinsic motoneuron hyperexcitability thought to underlie myokymia (Dedek *et al.*, 2001; Devaux *et al.*, 2004: Wuttke *et al.*, 2007).

The *Kv7.4* gene maps to chromosome 1p34 to the same region as DFNA2, a locus for autosomal-dominant progressive hearing loss (Kubisch *et al.*, 1999). Several mutations in *Kv7.4* have been identified in families with hearing loss (Kubisch *et al.*, 1999; Coucke *et al.*, 1999; Talebizadeh *et al.*, 1999). Heterologous expression of one of these mutants, *Kv7.4* G285S, in *Xenopus* oocytes revealed a dominant negative suppression of wild-type *Kv7.4* conductance (Kubisch *et al.*, 1999).

Several mechanisms have been proposed to underlie the DFNA2 hearing loss associated with Kv7.4 channel mutations (Jentsch 2000; Beisel *et al.*, 2000). Based on the theory that *Kv7.4* may be responsible for potassium efflux across the basal membrane (see Physiology section; Kubisch *et al.*, 1999), Jentsch (2000) has suggested that the slow progressive hearing loss may be attributed to a slow degeneration of outer hair cells, which may occur as a result of a 'chronic depolarization or potassium overload' caused by the loss of Kv7.4. In addition, Jentsch and colleagues have suggested that a loss of Kv7.4 in the central auditory pathway, where Kv7.4 expression is found, might also contribute to DFNA2 hearing loss (Kharkovets *et al.*, 2000).

Beisel *et al.*, (2000; 2005) have argued that OHC dysfunction cannot fully explain the progressive nature of DFNA2 hearing loss and have suggested that spiral ganglion and/or IHC dysfunction may be the primary cause. A second group of scientists have also suggested that loss of Kv7.4 may lead to DFNA2 hearing loss by way of IHC degeneration (Oliver *et al.*, 2003). Based on their findings from electrophysiological and molecular biological experiments, Oliver *et al.*, (2003) proposed that a lack of Kv7.4 channels could lead to depolarization of the membrane potential of IHCs leading to an increase in intracellular calcium levels. They suggested that a chronic Ca^{2+} overload of IHCs could induce their degeneration resulting in the eventual hearing loss associated with DFNA2.

However, recent studies in mice with altered Kv7.4 channels provide a convincing argument for the hypothesis that DFNA2 deafness is a result of OHC degeneration (Kharkovets *et al.*, 2006). OHCs from *Kv7.4* knockout mice were depolarized by approximately 10–17 mV and it has been suggested that this depolarization may increase Ca^{2+} influx through voltage-gated Ca^{2+} channels and lead to their eventual degeneration. Although a degeneration of IHCs was detected in a few rare cases of older animals, the morphology and *in vitro* function of afferent IHC synapses were reported as being normal.

Currently, no diseases have been linked to mutations in Kv7.5. However, the *Kv7.5* gene has been mapped to chromosome 6q14, where there are two known loci implicated in epileptic diseases including progressive myoclonic epilepsy type 2 and juvenile myoclonic epilepsy (Lerche *et al.*, 2000; Schroeder *et al.*, 2000a). However, mutational analysis studies suggest that the *Kv7.5* gene is an unlikely candidate for BFNC (Kananura *et al.*, 2000). The loci for several retinal diseases also map to this region although as yet there is no functional evidence linking mutations in *Kv7.5* with disease of the retina (Jentsch 2000).

2.1.2.8 Concluding remarks

In the past few years much has been discovered about the Kv7 family. However, there are many important questions still to answer. These include: what are the functional roles for Kv7 channels expressed in tissue such as smooth muscle, skeletal muscle, testis and kidney? Can subtype selective blockers and activators be developed for use as pharmacological probes and therapeutic agents? Furthermore, in light of the discrepancies in activation kinetics (Robbins, 2001) and permeability (Prole and Marrion, 2004), can the native M-current be explained solely by Kv7.2 and 7.3 subunits? With their widespread expression and role in numerous physiological and pathophysiological functions, Kv7 channels are certain to be the subject of much scrutiny in the future.

Acknowledgements

We thank Aban Shuaib for assistance with the reference list. GMP is funded by the MRC from a grant awarded to D.A. Brown and A.H. Dickenson. This article is dedicated to Dr Alex Selyanko, whose work contributed immensely to our understanding of Kv7 channels and who is much missed by the authors of this chapter.

References

Abitbol I, Peretz A, Lerche C *et al.* (1999). Stilbenes and fenamates rescue the loss of I(KS) channel function induced by an LQT5 mutation and other IsK mutants. *EMBO J* **18**, 4137–4148.

Bal M, Zaika O, Martin P *et al.* (2008). Calmodulin binding to M-type K+ channels assayed by TIRF/FRET in living cells. *J Physiol* **586-9**, 2307–2320.

Barhanin J, Lesage F, Guillemare E *et al.* (1996). K(V)LQT1 and IsK (mink) proteins associate to form I(Ks) cardiac potassium current. *Nature* **384**, 78–80.

Beisel KW, Nelson NC, Delimont DC *et al.* (2000). Longitudinal gradients of KCNQ4 expression in spinal ganglion and cochlear hair cells correlate with progressive hearing loss in DFNA2. *Mol Brain Res* **82**, 137–149.

Beisel KW, Rocha-Sanchez SM, Morris KA *et al.* (2005). Differential expression of KCNQ4 in inner hair cells and sensory neurons is the basis of progressive high-frequency hearing loss. *J Neurosci* **25**, 9285–9293.

Bentzen BH, Schmitt N, Calloe K *et al.* (2006). The acrylamide (s)-1 differentially affects Kv7 (KCNQ) potassium channels. *Neuropharmacol* **51**, 1068–1077.

Bett GC, Morales MJ, Beahm DL *et al.* (2006). Ancillary subunits and stimulation frequency determine the potency of chromanol 293B for the KCNQ1 potassium channel. *J Physiol* **576(Pt 3)**, 755–767.

Biervert C and Steinlein OK. (1999). Structural and mutational analysis of KCNQ2, the major gene locus for benign familial neonatal convulsions. *Hum Genet* **104z**, 234–240.

Biervert C, Schroeder BC, Kubisch C *et al.* (1998). A potassium channel mutation in neonatal human epilepsy. *Science* **279**, 403–406.

Blackburn-Munro G and Jensen BS. (2003). The anticonvulsant retigabine attenuates nociceptive behaviours in rat models of persistent and neuropathic pain. *Eur J Pharmacol* **460**, 109–116.

Boscia F, Annunziato L and Taglialatela M. (2006). Retigabine and flupirtine exert neuroprotective actions in organotypic hippocampal cultures. *Neuropharmacol* **51**, 283–294.

Boucherot A, Schreiber R and Kunzelmann K. (2001). Regulation and properties of KCNQ1 (K(V)LQT1) and impact of the cystic fibrosis transmembrane conductance regulator. *J Membr Biol* **182**, 39–47.

Bowden SE, Selyanko AA and Robbins J. (1999). The role of ryanodine receptors in the cyclic ADP ribose modulation of the M-like current in rodent muscarinic receptor-transformed NG108-15 cells. *J Physiol* **519.1**, 23–34.

Brown DA. (1988). M currents. In *Ion channels*, vol. 1, ed. T. Narahashi. Plenum Publishing Corporation, New York, pp. 55–94.

Brown DA and Adams PR. (1980). Muscarinic suppression of a novel voltage-sensitive K+ current in a vertebrate neurone. *Nature* **283**, 673–676.

Brown DA, Hughes SA, Marsh SJ *et al.* (2007). Regulation of M(Kv7.2/7.3) channels in neurons by PIP2 and products of PIP2 hydrolysis: significance for receptor mediated inhibition. *J Physiol* **582(3)**, 917–925.

Brueggemann LI, Moran CJ, Barakat JA *et al.* (2007). Vasopressin stimulates action potential firing by protein kinase C-dependent inhibition of KCNQ5 in A7r5 rat aortic smooth muscle cells. *Am J Physiol Heart Circ Physiol* **292**, H1352–H1363.

Busch AE, Busch GL, Ford E *et al.* (1997). The role of the IsK protein in the specific pharmacological properties of the IKs channel complex. *Br J Pharmacol* **122**, 187–189.

Caminos E, Garcia-Pino E, Martinez-Galan JR *et al.* (2007). The potassium channel KCNQ5/Kv7.5 is localised in synaptic endings of auditory brainstem nuclei of the rat. *J Comp Neurol* **505**, 363–378.

Casimiro MC, Knollmann BC, Ebert SN *et al.* (2001). Targeted disruption of the KCNQ1 gene produces a mouse model of Jervell and Lange-Nielsen Syndrome. *PNAS* **98**, 2526–2531.

Casimiro MC, Knollmann BC, Yamoah EN *et al.* (2004). Targeted point mutagnesis of mouse KCNQ1: phenotypic analysis of mice with point mutation that cause Romano–Ward syndrome in humans. *Genomics* **84**, 555–564.

Castaldo P, del Giudice EM, Coppola G *et al.* (2002). Benign familial neonatal convulsions caused by altered gating of KCNQ2/KCNQ3 potassium channels. *J Neurosci* **22**, RC(199) 1–6.

Chambard JM and Ashmore JF. (2005). Regulation of the voltage-gated potassium channels KCNQ4 in the auditory pathway. *Pflugers Arch* **450**, 34–44.

Charlier C, Singh NA, Ryan SG *et al.* (1998).s A pore mutation in a novel KQT-like potassium channel gene in an idiopathic epilepsy family. *Nat Genet* **18**, 53–55.

Chen JW and Eatock RA. (2000). Major potassium conductance in type I hair cells from rat semicircular canals: characterization and modulation by nitric oxide. *J Neurophysiol* **84**, 139–151.

Chen S and Yaari Y. (2008). Spike Ca2+ influx upmodulates the spike after depolarization and bursting via intracellular inhibition of Kv7/M channels. *J Physiol* **586.5**, 1351–1363.

Chouabe C, Neyroud N, Guicheney P *et al.* (1997). Properties of KvLQT1 K+ channel mutations in Romano–Ward and Jervell and Lange-Nielsen inherited cardiac arrhythmias. *EMBO Journal* **16**, 5472–5479.

Chung HJ, Jan YN and Jan LY. (2006). Polarised axonal surface expression of neuronal KCNQ channels is mediated by multiple signals in the KCNQ2 and KCNQ3 C-terminal domains. *PNAS* **103**, 8870–8875.

Constanti A and Brown DA. (1981). M-currents in voltage–clamped mammalian sympathetic neurons. *Neurosci Lett* **24**, 289–294.

Cooper EC and Jan LY. (2003). M-Channels. Neurological diseases, neuromodulation and drug development. *Arch Neurol* **60**, 496–500.

Cooper EC, Aldape KD, Abosch A *et al.* (2000). Colocalisation and coassembly of two human brain M-type potassium channel subunits that are mutated in epilepsy. *PNAS* **97**, 4914–4919.

Cooper EC, Harrington E, Jan YN *et al.* (2001). M channel KCNQ2 subunits are localized to key sites for control of neuronal network oscillations and synchronization in mouse brain. *J Neurosci* **21**, 9529–9540.

Costa AM and Brown BS. (1997). Inhibition of M-current in cultured rat superior cervical ganglia by linopirdine: mechanism of action studies. *Neuropharmacol* **36**, 1747–1753.

Coucke PJ, Van Hauwe P, Kelley PM *et al.* (1999). Mutations in the KCNQ4 gene are responsible for autosomal dominant deafness in four DFNA2 families. *Hum Mol Genet* **8**, 1321–1328.

Dalby-Brown W, Hansen HH, Korsgaard MP *et al.* (2006). Kv7 channels: function, pharmacology and channel modulators. *Curr Top Med Chem* **6**, 999–1023.

Dedek K and Waldegger S. (2001). Colocalization of KCNQ1/KCNE channel subunits in the mouse gastrointestinal tract. *Pflügers Arch* **442**, 896–902.

Dedek K, Kunath B, Kananura C *et al.* (2001). Myokymia and neonatal epilepsy caused by a mutation in the voltage sensor of the KCNQ2 K+ channel. *PNAS* **98**, 12272–12277.

Delmas P and Brown DA. (2005). Pathways modulating neural KCNQ/M (Kv7) potassium channels. *Nat Rev Neurosci* **6**, 850–862.

Devaux JJ, Kleopa KA, Cooper EC *et al.* (2004). KCNQ2 is a nodal K+ channel. *J Neurosci* **24**, 1236–1244.

Dost R, Rostock A and Rundfeldt C. (2004). The anti-hyperalgesic activity of retigabine is mediated by KCNQ potassium channel activation. *Naunyn-Schmiedebergs Arch Pharmacol* **369**, 382–390.

Dupuis DS, Schr der RL, Jespersen T *et al.* (2002). Activation of KCNQ5 channels stably expressed in HEK293 cells by BMS-204352. *Eur J Pharmacol* **437**, 129–137.

Ekberg J, Schuetz F, Boase NA *et al.* (2007). Regulation of the voltage-gated K+ channels KCNQ2/3 and KCNQ3/5 by ubiquitination. *J Biol Chem* **282**, 12135–12142.

Elmedyb P, Calloe K, Schmitt N *et al.* (2007). Modulation of ERG channels by XE991. *Basic Clin Pharmacol and Toxicol* **100**, 316–322.

Etxeberria A, Aivar P, Rodriguez-Alfaro JA *et al.* (2008). Calmodulin regulates the trafficking of KCNQ2 potassium channels. *FASEB J* **22**, 1135–1143.

Etxeberria A, Santana-Castro I, Regalado MP *et al.* (2004). Three mechanisms underlie KCNQ2/3 heteromeric potassium M-channel potentiation. *J Neurosci* **24**, 9146–9152.

Ford CP, Stemkowski PL, Light PE *et al.* (2003). Experiments to test the role of phosphatidylinositol 4,5-bisphosphate in neurotransmitter-induced M-channel closure in bullfrog sympathetic neurons. *J Neurosci* **23**, 4931–4941.

Gamper N and Shapiro MS. (2003). Calmodulin mediates Ca^{2+}-dependent modulation of M-type K+ channels. *J Gen Physiol* **122**, 17–31.

Gamper N, Li Y and Shapiro MS. (2005). Structural requirements for differential sensitivity of KCNQ K+ channels to modulation by Ca^{2+}/calmodulin. *Mol Biol Cell* **16**, 3538–3551.

Gamper N, Stockand JD and Shapiro MS. (2003). Subunit-specific modulation of KCNQ potassium channels by Src tyrosine kinase. *J Neurosci* **23**, 84–95.

Gamper N, Zaika O, Li Y *et al.* (2006). Oxidative modification of M-type K+ channels as a mechanism of cytoprotective neuronal silencing. *EMBO J* **25**, 4996–5004.

Geiger J, Weber YG, Landwehrmeyer B *et al.* (2006). Immunohistochemical analysis of KCNQ3 potassium channels in mouse brain. *Neurosci Lett* **400**, 101–104.

Ghosh S, Nunziato DA and Pitt GS. (2006). KCNQ1 assembly and function is blocked by long-QT syndrome mutations that disrupt interaction with calmodulin. *Circ Res* **98**, 1048–1054.

Gomez-Casati ME, Katz E, Glowatzki E *et al.* (2004). Linopirdine blocks alpha9alpha10-containing nicotinic cholinergic receptors of cochlear hair cells. *J Assoc Res Otolaryngol* **5**, 261–269.

Gribkoff VK. (2003). The therapeutic potential of neuronal KCNQ channel modulators. *Expert Opin Ther Targets* **7**, 737–748.

Grunnet M, Jespersen T, MacAulay N *et al.* (2003). KCNQ1 channels sense small changes in cell volume. *J Physiol* **549.2**, 419–427.

Gutman GA, Chandy KG, Adelman JP *et al.* (2003). International Union of Pharmacology. XLI. Compendium of voltage-gated ion channels: potassium channels. *Pharmcol Rev* **55**, 583–586.

Gutman GA, Chandy KG, Grissmer S *et al.* (2005). International Union of Pharmacology. LIII. Nomenclature and molecular relationships of voltage-gated potassium channels. *Pharmacol Rev* **57**, 473–508.

Hadley JK, Noda M, Selyanko AA *et al.* (2000). Differential tetraethylammonium sensitivity of KCNQ1-4 potassium channels. *Br J Pharmacol* **129**, 413–415.

Hadley JK, Passmore GM, Tatulian L *et al.* (2003). Stoichiometry of expressed KCNQ2/KCNQ3 potassium channels and subunit composition of native ganglionic M channels deduced from block by tetraethylammonium. *J Neurosci* **23**, 5012–5019.

Haitin Y and Attali B. (2008). The C-terminus of Kv7 channels: a multifunctional module. *J Physiol* **586.7**, 1803–1810.

Hansen HH, Andreasen JT, Weikop P *et al.* (2007). The neuronal KCNQ channel opener retigabine inhibits locomotor activity and reduces forebrain excitatory responses to the psychostimulants cocaine, methyphenidate and phencyclidine. *Eur J Pharmacol.* **570**, 77–88.

Hansen HH, Ebbesen C, Mathiesen C *et al.* (2006). The KCNQ channel opener retigabine inhibits the activity of mesencephalic dopaminergic system of the rat. *JPET* **318**, 1006–1019.

Heitzmann D, Grahammer F, von Hahn T *et al.* (2004). Heteromeric KCNE2/KCNQ1 potassium channels in the luminal membrane of gastric parietal cells. *J Physiol* **561.2**, 547–557.

Hernandez CC, Zaika O, Tolstykh GP *et al.* (2008). Regulation of neuronal KCNQ channels: signalling pathways, structural motifs and functional implications. *J Physiol* **586.7**, 1811–1821.

Higashida H, Brown DA and Robbins J. (2000). Both linopirdine- and WAY123,398-sensitive components of IK(M,ng) are modulated by cyclic ADP ribose in NG108-15 cells. *Pflügers Arch* **441**, 228–234.

Higashida H, Robbins J, Egorova A *et al.* (1995). Nicotinamide-adenine dinucleotide regulates muscarinic receptor-coupled K+ (M) channels in rodent NG108-15 cells. *J Physiol* **482.2**, 317–323.

Hirano K, Kuratani K, Fujiyoshi M *et al.* (2007). Kv7.2–7.5 voltage-gated potassium channel (KCNQ2-5) opener retigabine, reduces capsaicin-induced visceral pain in mice. *Neurosci Lett* **413**, 159–162.

Hirose S, Zenri F, Akiyoshi H *et al.* (2000). A novel mutation of KCNQ3 (c.925T-->C) in a Japanese family with benign familial neonatal convulsions. *Ann Neurol* **47**, 822–826.

Hoshi M, Zhang JS, Omaki M *et al.* (2003). AKAP150 signaling complex promotes suppression of the M-current by muscarinic agonists. *Nat Neurosci* **6**, 564–571.

Hougaard C, Klaerke DA, Hoffmann EK *et al.* (2004). Modulation of KCNQ4 channel activity by changes in cell volume. *Biochem Biophys Acta* **1660**, 1–6.

Housley GD and Ashmore JF. (1992). Ionic currents of outer hair cells isolated from the guinea-pig cochlea. *J Physiol* **448**, 73–98.

Howard RJ, Clark KA, Holton JM *et al.* (2007). Structural insight into KCNQ (Kv7) channel assembly and channelopathy. *Neuron* **53**, 663–675.

Hu H, Veraeke K and Storm JF. (2007). M-channels (Kv7/KCNQ channels) that regulate synaptic integration, excitability, and spike pattern of CA1 pyramidal cells are located in the perisomatic region. *J Neurosci* **27**, 1853–1867.

Hughes S, Marsh SJ, Tinker A *et al.* (2007). PIP2-dependent inhibition of M-type (Kv7.2/7.3) potassium channels: direct on-line assessment of PIP2 depletion by Gq-coupled receptors in single living neurons. *Pflugers Arch* **455**, 115–124.

Jensen BS. (2002). BMS-204352: a potassium channel opener developed for the treatment of stroke. *CNS Drug Rev* **8**, 353–360.

Jensen HS, Call K, Jespersen T *et al.* (2005). The KCNQ5 potassium channel from mouse: a broadly expressed M-current like potassium channel modulated by zinc, pH and volume changes. *Mol Brain Res* **139**, 52–62.

Jensen HS, Grunnet M and Olesen S-P. (2007). Inactivation as a new regulatory mechanism for neuronal Kv7 channels. *Biophys J* **92**, 2747–2756.

Jentsch TJ. (2000). Neuronal KCNQ potassium channels: physiology and role in disease. *Nat Rev Neurosci* **1**, 21–30.

Jespersen T, Grunnet M and Olesen SP. (2005). The KCNQ1 potassium channel: from gene to physiological function. *Physiology (Bethesda)* **20**, 408–416.

Jia Q, Jia Z, Zhao Z *et al.* (2007). activation of epidermal growth factor receptor inhibits KCNQ2/3 current through two distinct pathways: membrane PtdIns(4,5) P2 hydrolysis and channel phosphorylation. *J Neurosci* **27**, 2503–2512.

Joshi S, Balan P and Gurney AM. (2006). Pulmonary vasoconstrictor action of KCNQ potassium channel blockers. *Respir Res* **7**, 31–40.

Jow F and Wang K. (2000). Cloning and functional expression of rKCNQ2 K(+) channel from rat brain. *Brain Res Mol Brain Res* **80**, 269–278.

Jow F, He L, Kramer A *et al.* (2006). Validation of DRG-like F11 cells for evaluation of KCNQ/M-current modulators. *Assay Drug Dev Technol* **4**, 49–56.

Kananura C, Biervert C, Hechenberger M *et al.* (2000). The new voltage-gated potassium channel KCNQ5 and neonatal convulsions. *Neuroreport* **11**, 2063–2067.

Kanaumi T, Takashima S, Iwasaki H *et al.* (2008). Developmental changes in KCNQ2 and KCNQ3 expression in human brain: possible contribution to the age-dependent etiology of benign familial neonatal convulsions. *Brain Develop* **30**, 362–369.

Keating MT and Sanguinetti MC. (2001). Molecular and cellular mechanisms of cardiac arrhtymias. *Cell* **104**, 569–580.

Kharkovets T, Dedek K, Maier H *et al.* (2006). Mice with altered KCNQ4 K+ channels implicate sensory outer hair cells in human progressive deafness. *EMBO J* **25**, 642–652.

Kharkovets T, Hardelin JP, Safieddine S *et al.* (2000). KCNQ4, a K+ channel mutated in a form of dominant deafness, is expressed in the inner ear and the central auditory pathway. *PNAS* **97**, 4333–4338.

Korsgaard MP, Hartz BP, Brown WD *et al.* (2005). Anxiolytic effects of Maxipost (BMS-204352) and retigabine via activation of neuronal Kv7 channels. *JPET* **314**, 282–292.

Koyama S and Appel SB. (2006). Characterization of M-current in ventral tegmental area dopamine neurons. *J Neurophysiol* **96**, 535–543.

Kubisch C, Schroeder BC, Friedrich T *et al.* (1999). KCNQ4, a novel potassium channel expressed in sensory outer hair cells, is mutated in dominant deafness. *Cell* **96**, 437–446.

Lamas JA, Selyanko AA and Brown DA. (1997). Effects of a cognition-enhancer, linopirdine (DuP 996), on M-type potassium currents (IK(M)) and some other voltage- and ligand-gated membrane currents in rat sympathetic neurons. *Eur J Neurosci* **9**, 605–616.

Lan WZ, Abbas H, Lemay AM *et al.* (2004). Electrophysiological and molecular identification of hepatocellular volume-activated K+ channels. *Biochim Biophys Acta* **1668**, 223–233.

Lawrence JJ, Saraga F, Churchill JF *et al.* (2006). Somatodendritic Kv7/KCNQ/M channels control interspike interval in hippocampal interneurons. *J Neurosci* **22**, 12325–12338.

Lee MP, Hu RJ, Johnstone LA *et al.* (1997). Human *KVLQT1* gene shows tissue-specific imprinting and encompasses Beckwith–Wiedemann syndrome chromosomal rearrangements. *Nat Genet* **15**, 181–185.

Lee MP, Ravenel JD, Hu RJ *et al.* (2000). Targeted disruption of the *KVLQT1* gene causes deafness and gastric hyperplasia in mice. *J Clin Invest* **106**, 1447–1455.

Lerche C, Scherer CR, Seebohm G *et al.* (2000). Molecular cloning and functional expression of KCNQ5, a potassium channel subunit that may contribute to M-current diversity. *J Biol Chem* **275**, 22395–22400.

Lerche H, Jurkat-Rott K and Lehmann-Horn F. (2001). Ion channels and epilepsy. *Am J Med Genet* **106**, 146–159.

Li Y, Gamper N and Shapiro MS. (2004). Single-channel analysis of KCNQ K+ channels reveals the mechanism of augmentation by a cysteine-modifying reagent. *J Neurosci* **24**, 5079–5090.

Li Y, Gamper N, Hilgemann DW *et al.* (2005). Regulation of Kv7 (KCNQ) K+ channel open probability by phosphatidylinositol 4,5-bisphosphate. *J Neurosci* **25**, 9825–9835.

Loussouarn G, Park KH, Bellocq C *et al.* (2003). Phosphatidylinositol-4, 5-bisphosphate, PIP2, controls KCNQ1/KCNE1 voltage-gated potassium channels: a functional homology between voltage-gated and inward rectifier K+ channels. *EMBO J* **22**, 5412–5421.

Mackie AR, Brueggemann LI, Henderson KK *et al.* (2008). Vascular KCNQ potassium channels as novel targets for the control of mesenteric artery constriction by vassopressin. *JPET* **325**, 1659–1666.

MacVinish LJ, Guo Y, Dixon AK *et al.* (2001). Xe991 reveals differences in K(+) channels regulating chloride secretion in murine airway and colonic epithelium. *Mol Pharmacol* **60**, 753–760.

Main MJ, Cryan JE, Dupere JR *et al.* (2000). Modulation of KCNQ2/3 potassium channels by the novel anticonvulsant retigabine. *Mol Pharmacol* **58**, 253–262.

Marcotti W and Kros CJ. (1999). Developmental expression of the potassium current Ik,n contributes to maturation of mouse outer hair cells. *J Physiol* **520.3**, 653–660.

Mikkelsen JD. (2004). The KCNQ channel activator retigabine blocks haloperidol-induced c-Fos expression in the striatum of the rat. *Neurosci Lett* **362**, 240–243.

Moser SL, Harron SA, Crack J *et al.* (2008). Multiple KCNQ potassium channel subtypes mediate basal anion secretion from the human airway epithelial cell line Calu-3. *J Membrane Biol* **221**, 153–163.

Nakajo K and Kubo Y. (2005). Protein kinase C shifts the voltage dependence of KCNQ/M channels expressed in Xenopus oocytes. *J Physiol* **569.1**, 59–74.

Nakamura M, Watanabe H, Kubo Y *et al.* (1998). KQT2, a new putative potassium channel family produced by alternative splicing. Isolation, genomic structure and alternative splicing of the putative potassium channels. *Recept Channels* **5**, 255–271.

Neyroud N, Tesson F, Denjoy I *et al.* (1997). A novel mutation in the potassium channel gene KvLQT1 causes the Jervell and Lange-Nielsen cardioauditory syndrome. *Nat Genet* **15**, 186–189.

Nielsen AN, Mathiesen C and Blackburn-Munro G. (2004). Pharmacological characterisation of acid-induced muscle allodynia in rats. *Eur J Pharmacol* **487**, 93–103.

Noda M, Obana M and Akaike N. (1998). Inhibition of M-type K+ current by linopirdine, a neurotransmitter-release enhancer, in NG108-15 neuronal cells and rat cerebral neurons in culture. *Brain Res* **794**, 274–280.

Norenberg W, von Kugelgen I, Meyer A *et al.* (2000). M-type K+ currents in rat cultured thoracolumbar sympathetic neurones and their role in uracil nucleotide-evoked noradrenaline release. *Br J Pharmacol* **129**, 709–723.

Ohya S, Asakura K, Muraki K *et al.* (2002). Molecular and functional characterization of ERG, KCNQ and KCNE subtypes in rat stomach smooth muscle. *Am J Physiol Gastrointest Liver Physiol* **282**, G277–287.

Ohya S, Sergeant GP, Greenwood IA *et al.* (2003). Molecular variants of KCNQ channels expressed in murine portal vein myocytes: a role in delayed rectifier current. *Circ Res* **92**, 1016–1023.

Okada M, Zhu G, Hirose S *et al.* (2003). Age-dependent modulation of hippocampal excitability by KCNQ-channels. *Epilepsy Res* **53**, 81–94.

Oliver D, Knipper M, Derst C *et al.* (2003). Resting potential and submembrane calcium concentration of inner hair cells in the isolated mouse cochlea are set by KCNQ-type potassium channels. *J Neurosci* **23**, 2141–2149.

Otto JF, Yang Y, Frankel WN *et al.* (2006). A spontaneous mutation involving KCNQ2 (Kv7.2) reduces M-current density and spike frequency adaptation in mouse CA1 neurons. *J Neurosci* **26**, 2053–2059.

Pan Z, Kao T, Horvath Z *et al.* (2006). A common ankyrin-G-based mechanism retains KCNQ and NaV channels at electrically active domains of the axon. *J Neurosci* **26**, 2599–2613.

Pan Z, Selyanko AA, Hadley JK *et al.* (2001). Alternative splicing of KCNQ2 potassium channel transcripts contributes to the functional diversity of M-currents. *J Physiol* **531.2**, 347–358.

Park KH, Piron J, Dahimene S *et al.* (2005). Impaired KCNQ1–KCNE1 and phosphatidylinositol-4,5-bisphosphate interaction underlies the long QT syndrome. *Circ Res* **96**, 730–739.

Passmore GM, Robbins J, Abogadie F *et al.* (2006). The KCNQ2 (Kv7.2) gene is required for functional M-channels in embryonic mouse superior cervical ganglion (SCG) neurons. *J Physiol* C106.

Passmore GM, Selyanko AA, Mistry M *et al.* (2003). KCNQ/M currents in sensory neurons: significance for pain therapy. *J Neurosci* **23**, 7227–7236.

Peretz A, Sheinin A, Yue C *et al.* (2007). Pre- and postsynaptic activation of M-channels by a novel opener dampens neuronal firing and transmitter release. *J Neurophysiol* **97**, 283–295.

Peters HC, Hu H, Pongs O *et al.* (2005). Conditional transgenic suppression of M channels in mouse brain reveals functions in neuronal excitability, resonance and behavior. *Nat Neurosci* **8**, 51–60.

Piccinin S, Randall AD and Brown JT. (2006). KCNQ/Kv7 channel regulation of hippocampal gamma-frequency firing in the absence of synaptic transmission. *J Neurophysiol* **95**, 3105–3112.

Prole DL and Marrion NV. (2004). Ionic permeation and conduction properties of neuronal KCNQ2/KCNQ3 potassium channels. *Biophys J* **86**, 1454–1469.

Prole DL, Lima PA and Marrion NV. (2003). Mechanisms underlying modulation of neuronal KCNQ2/KCNQ3 potassium channels by extracellular protons. *J Gen Physiol* **122**, 775–793.

Pusch M. (1998). Increase of the single-channel conductance of KvLQT1 potassium channels induced by the association with mink. *Pflugers Arch* **437**, 172–4.

Richards MC, Heron SE, Spendlove HE *et al.* (2004). Novel mutations in the KCNQ2 gene link epilepsy to a dysfunction of the KCNQ2-calmodulin interaction. *J Med Genet* **41**, e35.

Richter A, Sander SE and Rundfeldt C. (2006). Antidystonic effects of Kv7 (kcnq) channel openers in the dt^SZ mutant, an animal model of primary paroxysmal dystonia. *Brit J Pharmacol* **149**, 747–753.

Rivas A and Francis HW. (2005). Inner ear abnormalities in a Kcnq1 (Kvlqt1) knockout mouse: a model of Jervell and Lange-Nielsen syndrome. *Otol Neurotol* **26**, 415–424.

Robbins J. (2001). KCNQ potassium channels: physiology, pathophysiology and pharmacology. *Pharmacol Ther* **90**, 1–19.

Robbins J, Marsh SJ and Brown DA. (2006). Probing the regulation of M (Kv7) potassium channels in intact neurons with membrane-targeted peptides. *J Neurosci* **26**, 7950–7961.

Rogawski MA. (2000). KCNQ2/KCNQ3 K+ channels and the molecular pathogenesis of epilepsy: implications for therapy. *TINS* **23**, 393–398.

Romey G, Attali B, Chouabe C *et al.* (1997). Molecular mechanism and functional significance of the MinK control of the KvLQT1 channel activity. *J Biol Chem* **272**, 16713–16716.

Roura-Ferrer M, Sole L, Martinez-Marmol R *et al.* (2008). Skeletal muscle Kv7 (KCNQ) channels in myoblast differentiation andv proliferation. *Biochem Biophys Res Commun* doi 10.1016/j.bbrc2008.02.152.

Rundfeldt C and Netzer R. (2000). The novel anticonvulsant retigabine activates M-currents in Chinese hamster ovary cells transfected with human KCNQ2/3 subunits. *Neurosci Lett* **282**, 73–76.

Sanganich MJ, Machado E and Rudy B. (2001). Differential expression of genes encoding subthreshold operating voltage-gated K+ channels in brain. *J Neurosci* **21**, 4609–4624.

Sanguinetti MC, Curran ME, Zou A *et al.* (1996). Coassembly of KvLQT1 and mink (IsK) proteins to form cardiac I_Ks potassium channel. *Nature* **384**, 80–83.

Schenzer A, Friedrich T, Pusch M *et al.* (2005). Molecular determinants of KCNQ (Kv7) K+ channel sensitivity to the anticonvulsant retigabine. *J Neurosci* **25**, 5051–5060.

Schnee ME and Brown BS. (1998). Selectivity of linopirdine (DuP 996), a neurotransmitter release enhancer, in blocking voltage-dependent and calcium-activated potassium currents in hippocampal neurons. *JPET* **286**, 709–717.

Schroder RL, Jespersen T, Christophersen P *et al.* (2001). KCNQ4 channel activation by BMS-204352 and retigabine. *Neuropharmacol* **40**, 888–898.

Schroeder BC, Hechenberger M, Weinreich F *et al.* (2000a). KCNQ5, a novel potassium channel broadly expressed in brain, mediates M-type currents. *J Biol Chem* **275**, 24089–24095.

Schroeder BC, Kubisch C, Stein V *et al.* (1998). Moderate loss of function of cyclic-AMP-modulated KCNQ2/KCNQ3 K+ channels causes epilepsy. *Nature* **396**, 687–690.

Schroeder BC, Waldegger S, Fehr S *et al.* (2000b). A constitutively open potassium channel formed by KCNQ1 and KCNE3. *Nature* **403**, 196–199.

Schwake M, Athanasiadu D, Beimgraben C *et al.* (2006). Structural determinants of M-type KCNQ (Kv7) K+ channel assembly. *J Neurosci* **26**, 3757–3766.

Schwake M, Jentsch TJ and Friedrich T. (2003). A carboxy-terminal domain determines the subunit specificity of KCNQ K+ channel assembly. *EMBO Rep* **4**, 76–81.

Schwake M, Pusch M, Kharkovets T *et al.* (2000). Surface expression and single channel properties of KCNQ2/KCNQ3, M-type K+ channels involved in epilepsy. *J Biol Chem* **275**, 13343–13348.

Schwarz JR, Glassmeier G, Cooper EC *et al.* (2006). KCNQ channels mediate IKs, a slow K+ current regulating excitability in the rat node of Ranvier. *J Physiol* **573.1**, 17–34.

Scott CW, Wilkins DE, Trivedi S *et al.* (2003). A medium-throughput functional assay of KCNQ2 potassium channels using rubidium efflux and atomic absorption spectrometry. *Anal Biochem* **319**, 251–257.

Seebohm G, Strutz-Seebohm N, Ureche ON *et al.* (2006). Differential roles of S6 domain hinges in the gating of KCNQ potassium channels. *Biophys J* **90**, 2235–2244.

Selyanko AA, Delmas P, Hadley JK *et al.* (2002). Dominant-negative subunits reveal potassium channel families that contribute to M-like potassium currents. *J Neurosci* **22**, RC212 1–5.

Selyanko AA, Hadley JK and Brown DA. (2001). Properties of single M-type KCNQ2/KCNQ3 potassium channels expressed in mammalian cells. *J Physiol* **534.1**, 15–24.

Selyanko AA, Hadley JK, Wood IC *et al.* (1999). Two types of K(+) channel subunit, Erg1 and KCNQ2/3, contribute to the M-like current in a mammalian neuronal cell. *J Neurosci* **19**, 7742–7756.

Selyanko AA, Hadley JK, Wood IC *et al.* (2000). Inhibition of KCNQ1-4 potassium channels expressed in mammalian cells via M1 muscarinic acetylcholine receptors. *J Physiol* **522.3**, 349–355.

Sesti F and Goldstein SA. (1998). Single-channel characteristics of wild-type IKs channels formed with mink mutants that cause long QT syndrome. *J Gen Physiol* **112**, 651–663.

Shah MM, Mistry M, Marsh SJ *et al.* (2002). Molecular correlates of the M-current in cultured rat hippocampal neurons. *J Physiol* **544.1**, 29–37.

Shahidullah M, Santarelli LC, Wen H *et al.* (2005). Expression of a calmodulin-binding KCNQ2 potassium channel fragment modulates neuronal M-current and membrane excitability. *PNAS* **102**, 16454–16459.

Shamgar L, Ma L, Schmitt N *et al.* (2006). Calmodulin is essential for cardiac IKS channel gating and assembly: impaired function in long-QT mutations. *Circ Res* **98**, 1055–1063.

Shapiro MS, Roche JP, Kaftan EJ *et al.* (2000). Reconstitution of muscarinic modulation of the KCNQ2/KCNQ3 K(+) channels that underlie the neuronal M current. *J Neurosci* **20**, 1710–1721.

Shen W, Hamilton SE, Nathanson NM *et al.* (2005). Cholinergic suppression of KCNQ channel currents enhances excitability of striatal medium spiny neurons. *J Neurosci* **25**, 7449–7458.

Singh NA, Charlier C, Stauffer D *et al.* (1998). A novel potassium channel gene, *KCNQ2*, is mutated an in inherited epilepsy of newborns. *Nat Genet* **18**, 25–29.

Singh NA, Westenskow P, Charlier C *et al.* (2003). *KCNQ2* and *KCNQ3* potassium channel genes in benign familial neonatal convulsions: expansion of the functional and mutation spectrum. *Brain* **126**, 2726–2737.

Smith JS, Iannotti CA, Dargris P *et al.* (2001). Differential expression of KCNQ2 splice variants: implications to M current function during neuronal development. *J Neurosci* **21**, 1096–1103.

Søgaard R, Ljungstrøm T, Pedersen KA *et al.* (2001). KCNQ4 channels expressed in mammalian cells: functional characteristics and pharmacology. *Am J Physiol Cell Physiol* **280**, C859–C866.

Splawski I, Shen J, Timothy KW *et al.* (1998). Genomic structure of three long QT syndrome genes: *KVLQT1, HERG* and *KCNE1*. *Genomics* **51**, 86–97.

Splawski I, Timothy KW, Vincent GM *et al.* (1997). Molecular basis of the long-QT syndrome associated with deafness. *N Engl J Med* **336**, 1562–1567.

Steinlein OK. (2001). Genes and mutations in idiopathic epilepsy. *Am J Med Genet* **106**, 139–145.

Stoffel M and Jan LY. (1998). Epilepsy genes: excitement traced to potassium channels. *Nat Genet* **18**, 6–8.

Streng T, Christoph T and Andersson KE. (2004). Urodynamic effects of the K+ channel (KCNQ) opener retigabine in freely moving, conscious rats. *J Urol* **172**, 2054–2058.

Strutz-Seebohm N, Seebohm G, Fedorenko O *et al.* (2006). Functional coassembly of KCNQ4 with KCNE-beta-subunits in *Xenopus* oocytes. *Cell Physiol Biochem* **18**, 57–66.

Su CC, Li SY, Yang JJ *et al.* (2006). Studies of the effect of ionomycin on the KCNQ4 channel expressed in *Xenopus* oocytes. *Biochem Biophys Res Commun* **348**, 295–300.

Suh BC and Hille B. (2002). Recovery from muscarinic modulation of M current channels requires phosphatidylinositol 4,5-bisphosphate synthesis. *Neuron* **35**, 507–520.

Suh BC, Horowitz LF, Hirdes W *et al.* (2004). Regulation of KCNQ2/KCNQ3 current by G protein cycling: the kinetics of receptor mediated signaling by Gq. *J Gen Physiol* **123**, 663–683.

Suh BS, Inoue T, Meyer T *et al.* (2006). Rapid chemically induced changes of PtdIns(4,5)P2 gate KCNQ ion channels. *Science* **314**, 1454–1457.

Surti TS, Huang L, Jan YN *et al.* (2005). Identification by mass spectrometry and functional characterization of two phosphorylation sites of KCNQ2/KCNQ3 channels. *PNAS* **102**, 17828–17833.

Talebizadeh Z, Kelley PM, Askew JW *et al.* (1999). Novel mutation in the KCNQ4 gene in a large kindred with dominant progressive hearing loss. *Human Mutat* **14**, 493–501.

Tatulian L and Brown DA. (2003). Effect of the KCNQ potassium channel opener retigabine on single KCNQ2/3 channels expressed in CHO cells. *J Physiol* **549.1**, 57–63.

Tatulian L, Delmas P, Abogadie FC *et al.* (2001). Activation of expressed KCNQ potassium currents and native neuronal M-type potassium currents by the anti-convulsant drug retigabine. *J Neurosci* **21**, 5535–5545.

Tinel N, Diochot S, Lauritzen I *et al.* (2000). M-type KCNQ2-KCNQ3 potassium channels are modulated by the KCNE2 subunit. *FEBS Lett* **480**, 137–141.

Tinel N, Lauritzen I, Chouabe C *et al.* (1998). The KCNQ2 potassium channel: splice variants, functional and developmental expression. Brain localization and comparison with KCNQ3. *FEBS Lett* **438**, 171–176.

Toyoda F, Ueyama H, Ding WG *et al.* (2006). Modulation of functional properties of KCNQ1 channel by association of KCNE1 and KCNE2. *Biochem Biophys Res Commun* **344**, 814–820.

Tristani-Firouzi M and Sanguinetti MC. (1998). Voltage-dependent inactivation of the human K+ channel KvLQT1 is eliminated by association with minimal K+ channel (minK) subunits. *J Physiol* **510.1**, 37–45.

Unsold B, Kerst G, Brousos H *et al.* (2000). KCNE1 reverses the response of the human K+ channel KCNQ1 to cytosolic pH changes and alters its pharmacology and sensitivity to temperature. *Pflugers Arch* **441**, 368–378.

Vallon V, Grahammer F, Volkl H *et al.* (2005). KCNQ1-dependent transport in renal and gastrointestinal epithelia. *PNAS* **102**, 17864–17869.

Vanner S, Evans RJ, Matsumoto SG *et al.* (1993). Potassium currents and their modulation by muscarine and substance P in neuronal cultures from adult guinea pig celiac ganglia. *J Neurophysiol* **69**, 1632–1644.

Vervaeke K, Gu N, Agdestein C *et al.* (2006). Kv7/KCNQ/M-channels in rat glutamatergic hippocampal axons and their role in regulation of excitability and transmitter release. *J Physiol* **576.1**, 235–256.

Wang HS, Brown BS, McKinnon D *et al.* (2000). Molecular basis for differential sensitivity of KCNQ and I(Ks) channels to the cognitive enhancer XE991. *Mol Pharmacol* **57**, 1218–1223.

Wang HS, Pan Z, Shi W *et al.* (1998). KCNQ2 and KCNQ3 potassium channel subunits: molecular correlates of the M-channel. *Science* **282**, 1890–1893.

Wang Q, Curran ME, Splawski I *et al.* (1996). Positional cloning of a novel potassium channel gene: *KVLQT1* mutations cause cardiac arrhythmias. *Nat Genet* **12**, 17–23.

Wangemann P. (2002). K+ cycling and the endocochlear potential. *Hear Res* **165**, 1–9.

Wangemann P. (2006). Supporting sensory transduction: cochlear fluid homeostasis and the endocochlear potential. *J Physiol* **576(Pt 1)**, 11–21.

Weber YG, Geiger J, Kämpchen K *et al.* (2006). Immunohistochemical analysis of KCNQ2 potassium channels in adult and developing mouse brain. *Brain Res* **1077**, 1–6.

Wehling C, Beimgraben C, Gelhaus C *et al.* (2007). Self-assembly of the isolated KCNQ2 subunit interaction domain. *FEBS Lett* **581**, 1594–1598.

Wen H and Levitan IB. (2002). Calmodulin is an auxiliary subunit of KCNQ2/3 potassium channels. *J Neurosci* **22**, 7991–8001.

Wickenden AD, Krajewski JL, London B *et al.* (2008). ICA-27243: a novel, selective KCNQ2/Q3 potassium channel activator. *Mol Pharmacol* **73**, 977–986.

Wickenden AD, Yu W, Zou A *et al.* (2000). Retigabine, a noval anticonvulsant, enhances activation of KCNQ2/KCNQ3 potassium channels. *Mol Pharmacol* **58**, 591–600.

Wickenden AD, Zou A, Wagoner PK *et al.* (2001). Characterization of KCNQ5/3 potassium channels expressed in mammalian cells. *Brit J Pharmacol* **132**, 381–384.

Wiener R, Haitin Y, Shamgar L *et al.* (2008). the KCNQ1 (Kv7.1) COOH terminus, a multitiered scaffold for subunit assembly and protein interaction. *J Biol Chem* **283**, 5815–5830.

Winks JS, Hughes S, Filippov AK *et al.* (2005). Relationship between membrane phosphatidylinositol-4,5-bisphosphate and receptor-mediated inhibition of native neuronal M channels. *J Neurosci* **25**, 3400–3413.

Wladyka CL and Kunze DL. (2006). KCNQ/M-currents contribute to the resting membrane potential in rat visceral sensory neurons. *J Physiol* **575.1**, 175–189.

Wladyka CL, Feng B, Glazebrook PA *et al.* (2008). The KCNQ/M-current modulates arterial baroreceptor function at the sensory terminals in rats. *J Physiol* **586.3**, 795–802.

Wollnik B, Schroeder BC, Kubisch C *et al.* (1997). Pathophysiological mechanisms of dominant and recessive KvLQT1 K$^+$ channel mutations found in inherited cardiac arrhythmias. *Hum Mol Genet* **6**, 1943–1949.

Wuttke TV, Jurkat-Rott K, Paulus W *et al.* (2007). Peripheral nerve hyperexcitability due to dominant-negative KCNQ2 mutations. *Neurol* **69**, 2045–2053.

Wuttke TV, Seebohm G, Bail S *et al.* (2005). The new anticonvulsant retigabine favors voltage-dependent opening of the Kv7.2 (KCNQ2) channel by binding to its activation gate. *Mol Pharmacol* **67**, 1009–1017.

Xiong Q, Gao Z, Wang W *et al.* (2008). Activation of Kv7 (KCNQ) voltage-gated potassium channels by synthetic compounds. *TIPS* **29**, 99–107.

Xiong Q, Sun H and Li M. (2007). Zinc pyrithione-mediated activation of voltage-gated KCNQ potassium channels rescues epileptogenic mutants. *Nature Chem Biol* **3**, 287–296.

Yang WP, Levesque PC, Little WA *et al.* (1997). KvLQT1, a voltage-gated potassium channel responsible for human cardiac arrhythmias. *PNAS* **94**, 4017–4021.

Yang WP, Levesque PC, Little WA *et al.* (1998). Functional expression of two KvLQT-related potassium channels responsible for an inherited idiopathic epilepsy. *J Biol Chem* **273**, 19419–19423.

Yang Y and Sigworth FJ. (1998). Single channel properties of IKs potassium channels. *J Gen Physiol* **112**, 665–678.

Yeung SY and Greenwood IA. (2005). Electrophysiological and functional effects of the KCNQ channel blocker XE991 on murine portal vein smooth muscle cells. *Brit J Pharmacol* **146**, 585–595.

Yeung SYM, Pucovsky V, Moffatt JD *et al.* (2007). Molecular expression and pharmacological identification of a role for Kv7 channels in murine vascular reactivity. *Brit J Pharmacol* **151**, 758–770.

Yoshida M and Alonso A. (2007). Cell-type-specific modulation of intrinsic firing properties and subthreshold membrane oscillations by the m(Kv7)-current in neurons of the entorhinal cortex. *J Neurophysiol* **98**, 2779–2794.

Yue C and Yaari Y. (2004). KCNQ/M channels control spike afterdepolarization and burst generation in hippocampal neurons. *J Neurosci* **24**, 4614–4624.

Yue C and Yaari Y. (2006). Axo-somatic and apical dendritic Kv7/M channels differentially regulate the intrinsic excitability of adult rat CA1 pyramidal cells. *J Neurophysiol* **95**, 3480–3495.

Yus-Nájera E, Muñoz A, Salvador N *et al.* (2003). Localisation of KCNQ5 in the normal and epileptic human temporal neocortex and hippocampal formation. *Neurosci* **120**, 353–364.

Yus-Najera E, Santana-Castro I and Villarroel A. (2002). The identification and characterization of a noncontinuous calmodulin-binding site in noninactivating voltage-dependent KCNQ potassium channels. *J Biol Chem* **277**, 28545–28553.

Zaczek R, Chorvat RJ, Saye JA *et al.* (1998). Two potent neurotransmitter release enhancers 10,10-bis(4pyridinylmethyl)-9(10H)-anthracenone and 10,10-bis(2-fluoro-4-pyridinylmethyl)-9(10H)-anthracenone: comparison to linopirdine. *JPET* **285**, 724–730.

Zaika O, Lara LS, Gamper N *et al.* (2006). Angiotensin II regulates neuronal excitability via phosphatidylinositol 4,5-bisphosphate-dependent modulation of Kv7 (M-type) K$^+$ channels. *J Physiol* **575.1**, 49–67.

Zhang H, Craciun LC, Mirshahi T *et al.* (2003). PIP(2) activates KCNQ channels and its hydrolysis underlies receptor-mediated inhibition of M currents. *Neuron* **37**, 963–975.

2.1.3 **Kv10, Kv11 and Kv12 channels**

Matthew Perry and Michael Sanguinetti

2.1.3.1 Introduction

Kv10, Kv11 and Kv12 channels are encoded by eight different genes, collectively referred to as the eag K$^+$ channel family and designated *KCNHx* for the human forms. The first member of this channel family was cloned from *Drosophila melanogaster* and named *ether-a-go-go* (*eag*). This name refers to the locus for a mutation associated with a leg-shaking phenotype in fruit flies and caused by an increased transmitter release and repetitive firing of motorneurons (Ganetzky and Wu, 1983). Two related channel types were soon discovered in flies, mice, and human tissues by using low stringency screens and degenerate PCR (Warmke and Ganetzky, 1994). These new subfamilies of K$^+$ channels were named eag-related gene (*erg*) and eag-like (*elk*). The amino acid sequences of the channels are about 40 per cent identical between the various members of the three subfamilies.

The Kv10–Kv12 channels share similarities with the typical Kv channels, CNG (cyclic-nucleotide gated) channels and HCN pacemaker channels. Similar to other Kv channels, Kv10–Kv12 channel subunits have six transmembrane α-helical domains (S1–S6) and functional channels are tetramers, formed by co-assembly of four identical or similar subunits. The first four transmembrane domains of each subunit form a voltage-sensing structure, dominated by S4 which contains a basic Arg or Lys residue located in every third position of the α-helix. The S5 and S6 domains of each subunit comprise the pore domain, although X-ray crystal structures have not been determined for any Kv10, Kv11 or Kv12 channel. However, based on sequence similarity it is assumed that the basic topology is very similar to the X-ray crystallographic structures of bacterial KvAP and mammalian Kv1.2 channels (Jiang *et al.*, 2003; Long *et al.*, 2005a, b). Unlike other Kv channels, Kv10–KV12 channel subunits have a Per-Arnt-Sim (PAS) domain located within the cytoplasmic N-terminal region. The function of the PAS domain in these channels is unknown, but deletion or point mutations of the domain in erg1 channels accelerates the rate of channel deactivation (Chen *et al.*, 1999; Morais Cabral *et al.*, 1998), perhaps by disrupting a normal interaction with the S4–S5 linker (Wang *et al.*, 1998). The PAS domain may also be involved in channel phosphorylation (Cayabyab and Schlichter, 2002; Cui *et al.*, 2000) and/or subunit assembly (Paulussen *et al.*, 2002). Similar to CNG and HCN pacemaker channels, Kv10–Kv12 channel subunits have a cyclic nucleotide binding domain (cNBD) located within the cytoplasmic C-terminal region. However, unlike HCN and CNG channels, cAMP has relatively minor effects on the gating of Kv10–Kv12 channels. Binding of cAMP to the cNBD causes a slight increase in the magnitude of eag channel current (Bruggemann *et al.*, 1993), and shifts the voltage dependence of erg1 channel activation by a few mV (Cui *et al.*, 2000).

Eag channels are characterized by very rapid activation and lack of inactivation. Elk channels activate more slowly than eag, and may (elk2) or may not (elk1) exhibit inactivation. Erg channels activate more slowly than eag or elk and rapidly inactivate. In heterologous expression systems, and presumably *in vivo*, functional channels can be formed by co-assembly of subunits from a single subfamily (e.g., elk1 + elk2), but heteromultimers are not formed by subunits from different subfamilies (e.g., elk1 + eag1).

The nomenclature and chromosomal location of the *eag*, *erg* and *elk* genes as well as the major biophysical properties, pharmacology and tissue expression patterns of the encoded channels are summarized in Table 2.1.3.1. The greater emphasis on erg compared to eag or elk channels in this review reflect the lopsided number of published studies on these channel types.

2.1.3.2 Molecular characterization

Kv10 (eag) channels

Ether-a-go-go (eag) channels were first cloned from *Drosophila* (Warmke *et al.*, 1991). In adult fruitflies, the *eag* mutation causes an ether-induced leg shaking (Wu *et al.*, 1983). Two different mammalian eag channel genes (*eag1*, *eag2* that encode Kv10.1, Kv10.2 channel subunits) were subsequently described (Ludwig *et al.*, 1994, 2000; Warmke and Ganetzky, 1994). Two alternatively spliced forms of human eag1 have also been described: eag1a (963 amino acids) and eag1b (990 amino acids). Human and rat eag2 subunits are 988–989 amino acids in length with a predicted molecular weight of 112 kDa. Human eag1 and eag2 channel subunits are 73 per cent identical in amino acid sequence with most differences confined to the C-terminal domain (Ju and Wray, 2002).

In humans, the genes encoding eag1 and eag2 subunits have been designated *KCNH1* and *KCNH5*, and are located on chromosomes 1q32.1–32.3 and 14q24.3, respectively. The coding region for *KCNH1* comprises 11 exons spanning 452 kb of genomic sequence. The coding region for *KCNH5* comprises 11 exons spanning 338 kb of genomic sequence (Ju and Wray, 2002).

Kv11 (erg) channels

The first *ether-a-go-go-related* (erg) gene was cloned from a human hippocampus cDNA library (Warmke and Ganetzky, 1994). The human *erg1* gene (*HERG1* or *KCNH2*) encodes a protein (herg1, Kv11.1) of 1159 amino acids with a predicted molecular mass of 127 kDa. The hydrophobic core (S1 through S6) of the protein shares 49 per cent homology with eag subunits. The originally named erg1 protein is now referred to as erg1a following the discovery in mouse and human heart of an alternatively spliced transcript, named erg1b (Lees-Miller *et al.*, 1997; London *et al.*, 1997). Otherwise identical to erg1a, the erg1b protein has a much shorter and unique N-terminus (36 amino acids for erg1b compared with 376 amino acids for erg1a) that lacks a PAS domain. Erg1b encodes a protein of 819 amino acids with a predicted molecular mass of 94 kDa (Jones *et al.*, 2004). A herg1 C-terminal splice variant, termed herg1$_{USO}$, that may act to modify the expression and gating of herg1a channels, has also been identified (Kupershmidt *et al.*, 1998). *KCNH2* is located on chromosome 7 (7q35-36) and the coding region comprises 16 exons spanning ~34 kb of genomic sequence.

Two additional mammalian *erg* genes (*erg2*, *erg3* that encode Kv11.2 and Kv11.3 channel subunits) have been described (Shi *et al.*, 1997). Human genes encoding erg2 and erg3 subunits were named *KCNH6* and *KCNH7*, respectively. *KCNH6* is located on chromosome 17 (17q23.3) and has a coding region comprising 14 exons spanning 207 kb of genomic sequence. *KCNH7* is located on chromosome 2 (2q24.2–24.3) and the coding region comprises 8 exons spanning 468 kb of genomic sequence. Proteins encoded by *KCNH6* are 994 amino acids in length with a predicted molecular mass of 110 kDa, while those encoded by *KCNH7* are 1196 amino acids in length with a predicted molecular mass of 135 kDa (Bauer *et al.*, 2003). All three erg subunits share a high degree of homology within the core (S1–S6), CNBD and PAS domains. Overall sequence homology with erg1 is 63 per cent for erg2 and 57 per cent for erg3 (Shi *et al.*, 1997).

Two human genes (*KCNE1* and *KCNE2*) encode proteins that have been proposed as ancillary β-subunits for herg1a. Both genes are located on chromosome 21 (21q22.1–22.2) with their coding regions separated by approximately 79 kb. *KCNE1* is comprised of 4 exons spanning ~3.57 kb of genomic sequence, while *KCNE2* comprises 2 exons that spans ~800 bp. *KCNE1* encodes a 129 amino acid protein called MinK (Minimal K channel). *KCNE2* encodes a 123 amino acid protein called MiRP1 (MinK-Related Peptide 1) with a predicted molecular mass of 145 kDa. Both subunits have a single transmembrane domain, but share only 27 per cent amino acid identity (Abbott *et al.*, 1999). When expressed *in vitro*, both MinK and MiRP1 can modulate herg1a channel expression (Bianchi *et al.*, 1999; McDonald *et al.*, 1997). MiRP1 was also reported to alter herg1 channel pharmacology and gating kinetics (Abbott *et al.*, 1999), although the extent of this effect is disputed (Weerapura *et al.*, 2002). In contrast to the striking changes observed when MinK is co-expressed with KCNQ1 to form I$_{Ks}$ channels (Barhanin *et al.*, 1996; Sanguinetti *et al.*, 1996), the physiological relevance of MinK and MiRP1 β-subunits in the formation of I$_{Kr}$ *in vivo* remains controversial, with evidence centred around the effects of MiRP1 mutations that cause long QT syndrome (Abbott *et al.*, 1999).

Kv12 (elk) channels

The first *ether-a-go-go-like* (elk) channel gene was cloned from *Drosophila* (Warmke and Ganetzky, 1994). *Drosophila elk1* encodes a protein (elk1, Kv12.1) of 1284 amino acids and a predicted molecular weight of 141 kDa. Two additional mammalian elk channel genes (*elk2*, *elk3* that encode Kv12.2, Kv12.3 subunits) were subsequently described (Engeland *et al.*, 1998; Miyake *et al.*, 1999). The nomenclature of elk channel subtypes is confusing because three groups reported findings at nearly the same time in 1998–1999. For example, what is now called elk3 (or Kv12.3) was originally named 'elk1' (Engeland *et al.*, 1998) and 'BEC2' (Miyake *et al.*, 1999). Here we adopt the elk terminology suggested by Shi *et al.* (see Table 2.1.3.1) and adopted by the IUPHAR Compendium of voltage-gated ion channels (Gutman *et al.*, 2003). Rat elk1 is 1102 amino acids in length with an estimated mass of 123 kDa and is 41 per cent identical to *Drosophila* elk1 (Shi *et al.*, 1998). Rat elk2 is composed of 1054 amino acids and rat elk3 is 1017 amino acids in length. In humans, the genes for elk1, 2 and 3 have been designated *KCNH8*, *KCNH3* and *KCNH4*, respectively.

In heterologous expression systems, the demonstration of dominant-negative suppression of channel function was used to demonstrate that heteromultimeric elk channels can form by co-assembly of human elk1, elk2 and elk3 subunits (Zou *et al.*, 2003), but it is unknown if such co-assemblies also occur *in vivo*.

Table 2.1.3.1 Properties of eag (KV10), erg (KV11) and elk (KV12) channels

Channel subunit	Biophysical characteristics	Cellular and subcellular localization	Pharmacology	Physiology	Relevance to disease states
Kv10.1 (eag1) gene: KCNH1 chr. location: mouse: 1 H6 rat: 13q27 human: 1q32.1–32.3	Rapidly activating, noninactivating, outward rectifier *Drosophila* eag single channel conductance: 4.9pS [1] hEAG1: fast activation (<100ms) dependent on V_{hold} and $[Mg^{2+}]_o$ V0.5 activation: +10mV (eag1) [2] τ_{Activ} at 0mV: 100ms (10mM $[Mg^{2+}]_i$) and 11 ms (0mM $[Mg^{2+}]_i$) [3]	Transcripts in rat brain high expression in olfactory bulb, piriform cortex, anterior olfactory nucleus, olfactory tubercle, cerebral cortex, hippocampus (CA2, CA3 pyramidal cells), mammary nucleus of the midbrain and granule layer of the cerebellum [4] Moderate expression in CA1 pyramidal cells of the hippocampus, basal ganglia, amygdale, hypothalamus, red nucleus and oculomotor nucleus of the midbrain, and superior olive and facial nucleus of the hindbrain [4] Gliomas [5]	Blockers (IC_{50}): TEA 33mM, quinine 0.7mM, Ba^{2+} (<40% by 1mM) [1] In oocytes: bEAG: Dofetilide 32μM [6] E-4031 162nM [7] Clofilium 0.8nM [8] TBA 1.2μM [8] Cisapride 12μM (dEAG) [9] hEAG1 in HEK cells: Imipramine 2μM [10] Astemizole 200nM [10] hEAG1 in CHO cells: Quinidine 1.4μM [11] Clofilium 255 nM [7] LY97241 4.9nM [7]	Cell cycle regulation and cell proliferation [12, 13]	Not established
Kv10.2 (eag2) gene: KCNH5 chr. location: mouse: 12 C3 rat: 6q24 human: 14q23.1	4-times slower activation than hEAG1 V0.5 activation: −11mV [2]	Transcripts in rat brain (high in olfactory bulb and piriform cortex and one region (layer IV) of the cerebral cortex. Also in the inferior colliculus and nucleus lateral lemniscus of the midbrain. Moderate expression in the thalamus, habenula, and the dorsal and ventral cochlear nucleus of the hindbrain [4] hEAG2 highly expressed in brain>>skeletal muscle>heart, lung, liver>kidney, pancreas [2]	Blockers (IC_{50}): hEAG2 in CHO cells: Quinidine 152nM [11] hEAG2 in HEK cells: LY97241 1.52μM [8]	No information	Not established
Kv11.1 (erg1a) gene: KCNH2 chr. location: mouse: 5 12.0 cM rat: 4q11 human: 7q35–36	In mammalian cells: $Activ_{thresh}$: −50mV [14] Peak I-V: −10mV [15] V0.5 activation: −23mV [14] τ_{Activ} at +20mV: 74ms [14] $\tau_{Deactiv}$ (−60mV to −120mV) range: 344 to 26ms (fast) and 1590 to 238ms (slow) [16] C-type inactivation: $\tau_{Inactiv}$ (−20 to +50mV): 16 to 2ms [17] τ_{Recov} 13ms (−50 mV); 3ms (−120 mV) [16] V0.5 inactivation: −45mV [16] Single channel conductance: 5pS (in 2mM $[K^+]_o$) and 9pS (in 100mM $[K^+]_o$) [18]	Heart: ubiquitously expressed in mammalian cardiac tissues (atrium, ventricle, sino-atrial node and Purkinje fibres) [19–21] Subcellular localization in lateral (sarcolemmal) membranes, intercalated discs, and T-tubules [20,22] Brain (mouse) [23, 24]: High— moderate levels in olfactory bulb, palocortx, basal ganglia (certain nuclei), hippocampus (inc. CA1, CA2, CA3 pyramidal cells), thalamus (inc. reticular nucleus, hypothalamus), midbrain (inc. nucleus lateral lemiscus and red nucleus), vestibular nuclei (inc. lateral vest nucleus, facial nucleus, lateral reticular nucleus), spinal chord and cerebellum (Purkinje cells, deep cerebellar nuclei)	Blockers (IC_{50}) expressed in mammalian cells: dofetilide 12.6nM [35] ibutilide 30nM [37] E-4031 7.7nM [36] terfenadine 56nM [38] astemizole 0.9nM [39] cisapride 44.5nM [38] Toxins (K_D): ErgTx-1 4.4nM [40] BeKm-17.7nM [40] APETx-1 34nM [41] Trafficking blockers (IC_{50}): As_2O_3 1.5μM [42] Pentamidine 7.8μM [43] Activators (EC_{50}): RPR260243 3-10μM [44] NS1643 10.4μM [45]	Encodes α-subunit of I_{Kr} in cardiac myocytes—contributes to action potential repolarization [46] Contributes to the resting membrane potential of several cell types including gastrointestinal smooth muscle cells of eosophagus [47], stomach [48], jejunum [26] and colon [49]. Controls cell excitability and GI motility in smooth muscle [50] Spike-frequency adaptation [50] and firing frequency (e.g, human pancreatic β-cells) [27] Oxygen sensing in carotid body cells? [28]	Cardiac arrhythmia: Long QT syndrome (loss of function mutations) [57] Channel blockers cause acquired (drug-induced) LQTS [46] Short QT syndrome (gain of function mutations) [58] Upregulated in human tumors (proposed roles in cell-cycle control, proliferation, migration, adhesion-dependent signalling and angiogenesis) [59, 60] Proposed role or erg isoforms in neuropathological response to brain ischemia, neurodegeneration and aging phenomena [61]; epileptic seizures [15]

Continued

Table 2.1.3.1 *(Continued)* Properties of eag (KV10), erg (KV11) and elk (KV12) channels

Channel subunit	Biophysical characteristics	Cellular and subcellular localization	Pharmacology	Physiology	Relevance to disease states
		Lower, more variable levels in neocortex, superior and inferior colliculus of midbrain, molecular layer cells and granule cells of cerebellum Embryonic rhomencephalon neurons Other cell types: Lactotrophs [25], gastrointestinal smooth muscle myocytes [26] Pancreatic β-cells [27], carotid body cells [28], several cell types of mouse inner to middle ear [29] Human cell lines including neuroblastomas [30], rhabdomyosarcomas, hematopoietic [31], monoblastic leukemias and colon carcinoma, mammary carcinomas [32] *In vivo* within primary tumours of several tissues including endometrium, colon and lymphocytes [31, 33, 34]		Proposed role in thyrotropin-releasing hormone-induced prolactin secretion of anterior pituitary lactotrophs [16, 51] Cell cycle regulation and neuritogenesis in tumor cells [52, 53] K^+ homeostasis (spatial buffering) in hippocampal astrocytes [54] Proposed role in K^+ regulation across intermediate cells within the cochlear duct. Regulates the endocochlear potential that is vital for hair cell signal transduction [29] Control of neuronal excitability during early parts of spinal network development [55] Firing frequency adaptation and neuronal discharge pattern in mouse cerebellar Purkinje neurons, where it may alter motor control [56]	Skeletal muscle atrophy [62]
Kv11.1 (erg1b) Splice variant of erg1a (lacks 340 NH_2-terminal amino acids)	In *Xenopus* oocytes, mouse erg1b: Activation/deactivation: $Activ_{Thresh}$: −50mV [19] Peak I–V: +20mV [19] $V_{0.5}$ activation: −22mV [19] τ_{Activ} at +20mV: 85ms $\tau_{Deactiv}$ (−20 mV to −90mV) range: 110 to 12ms (fast) and 800 to 100ms (slow) [63]	Heart: atrial and ventricular myocytes from mouse heart. Canine and human heart. Co-localizes with erg1a in T tubules and can associate in human and canine ventricular myocytes *in vivo* Brain (mouse) [24]: Similar levels to erg1a in many regions of the thalamus (including the hypothalamus and amygdala), palocortex, brainstem, hippocampus and cerebellum. More variable expression was also detected in the olfactory bulb, basal ganglia, neocortex, vestibular nuclei and spinal cord Embryonic rhomencephalon neurons	Blockers (IC_{50}): Dofetilide 54nM [19]	May coassemble with erg1a to form I_{Kr} in cardiac myocytes [22]	Not established
Kv11.2 (erg2) gene: *KCNH6* chr. location: mouse: 11E1 rat: 10q32.1 human: 17q23.3	In mammalian cells (CHO): Activation/deactivation: $Activ_{Thresh}$: −30 mV, Peak I–V: +10 mV, $V_{0.5}$ activation: −9 mV, τ_{Activ} at +20mV: 428 ms, $\tau_{Deactiv}$ (−60 mV to −120mV) range: 279 to 35 ms (fast) and 2276 to 539 ms (slow) C-type inactivation: $\tau_{Inactiv}$: (−20 to +50 mV) 16 to 6 ms [15] τ_{Recov}: 18 ms (−50 mV); 5 ms (−120 mV) [16] $V_{0.5}$ inactivation: −51 mV [16]	Brain [24, 61]: Mitral and periglomerular cell layers of rat and mouse olfactory bulb Relatively low levels in hypothalamus, hippocampus, cerebellum, cerebral cortex and brain stem nuclei of rat and mouse Embryonic rhomencephalon neurons Sucellular localization in dendrites of Purkinje cells Retina and prevertebral ganglia in rat [23] Expressed in human retinoblastomas [32]	Blockers (IC_{50}): E-4031 116nM (vs 99nM erg1 in same study) [23] Toxins (K_D): ErgTx-1 210nM (r-erg2 3nM) [40] BeKm-1 7nM [40]	Control of neuronal excitability	Not established

Kv11.3 (erg3) gene: *KCNH7* chr. location: mouse: 2C1.3 rat: 3q21 human: 2q24.2–24.3	Activation/deactivation (CHO cells): $Activ_{Thresh}$: –60 mV, Peak I–V: –25 mV, $V_{0.5}$ activation: –42 mV, τ_{Activ} at +20mV: 17 ms, $\tau_{Deactiv}$ (–60 mV to –120mV) range: 74 to 9ms (fast); 336 to 45ms (slow) C-type inactivation: $\tau_{Inactiv}$ (–20 to +50 mV): 22 to 7ms [15] τ_{Recov} 12 ms (–50mV): 2 ms (–120mV) [16] $V_{0.5}$ inactivation: –44 mV [16]	Brain (mouse) [24]: Similar levels to erg1a in many regions of the thalamus (including the hypothalamus and amygdala), paleocortex, brainstem, hippocampus and cerebellum. More variable expression was also detected in the olfactory bulb, basal ganglia, neocortex, vestibular nuclei and spinal cord Subcellular localization in dendrites of layer V pyramidal neurons Embryonic rhomencephalon neurons Human mammary carcinomas [32]	Blockers (IC_{50}): E-4031: 193nM (vs 99nM erg1 in same study) Toxins (K_D): ErgTx-1 4nM [40] BeKm-1 11.5nM [40]	Control of neuronal excitability	Not established
Kv12.1 (elk1) gene:*KCNH8* chr. location: mouse: 17C rat: 9q11 human: 3q24.3	hELK1 in oocytes [64]: $V_{0.5}$ activation: –62mV τ_{Activ} at +10mV: 45ms (fast); 259ms (slow) $\tau_{Deactiv}$ (–110 to –150mV): 66 to 28ms (fast); 331 to 260ms (slow) No inactivation	oor expression in heart Brain: Rat—high expression in brain (Olfactory bulb and brainstem), sympathetic ganglia and sciatic nerve [65] Downregulated in brain during maturation [66] Human—low expression in cerebral and frontal cortex, hippocampus, caudate nucleus, nucleus accumbens and amygdale. Also in the thalamus, subthalamic nucleus, substantia nigra, pons and spinal cord [64]	Blockers (IC_{50}): Ba^{2+} 0.18mM [64]	Modulation of neural activity [67].	Not established
Kv12.2 (elk2) gene: *KCNH3* chr. location: mouse: 15 rat: 7q36 human: 12q13	Rapid C-type-inactivation m-elk2 in oocytes [67]: $Activ_{Thresh}$: –70mV Peak I–V: +20mV τ_{Activ} at +10mV: 12ms (fast); 83ms (slow) $\tau_{Deactiv}$ (–20mV to –120mV) range: 40 to 100 ms (fast); 220 to 650 ms (slow) $\tau_{Inactiv}$ (+70 to +20mV): 6 to 11ms τ_{Recov} 3 ms (–80mV):16ms (+10mV) hELK2 in CHO cells [68]: $V_{0.5}$ activation: –23mV $V_{0.5}$ inactivation: +10mV	Poor expression in heart Brain: Rat—high expression in cerebellum, hippocampus and brainstem [66] Upregulated in the amygdala, cortex, hippocampus and striatum during maturation [66] Human—high expression in cerebral and frontal cortex, hippocampus, caudate nucleus, nucleus acumbens and amygdala, [17] astrocytoma cells [68]	Blockers (IC_{50}): Cs^+, K_d: 0.68mM at –120mV [68] TEA slows inactivation (IC_{50}: 97mM) [67]	Contributes to the timing of burst duration or to spike frequency adaptation during a long burst of neuronal action potentials [67]	Not established

Continued

Table 2.1.3.1 (*Continued*) Properies of eag (KV10), erg (KV11) and elk (KV12) channels

Channel subunit	Biophysical characteristics	Cellular and subcellular localization	Pharmacology	Physiology	Relevance to disease states
Kv12.3 (elk3) gene: *KCNH4* chr. location: mouse: 11 61.3 cM rat: 10q32.1 human: 17q21.2	No information	Very low expression in the lung. Brain: Rat—high expression in the olfactory bulb and hippocampus [4] Remains relatively constant into adulthood. Human—low expression in cerebral and frontal cortex, hippocampus, caudate nucleus, nucleus acumbens and amygdala [64]	No data	No information	Not established

[1] Bruggemann et al. (1993) *Nature* 365, 445–448; [2] Ju et al. (2002) *FEBS Lett* 524, 204–210; [3] Silverman et al. (2000) *J Gen Physiol* 116, 663–678; [4] Saganich et al. (2001) *J Neurosci* 21, 4609–4624; [5] Patt et al. (2004) *Neurosci Lett* 368, 249–253; [6] Ficker et al. (1998) *Circ Res* 82, 386–395; [7] Gessner et al. (2003) *Br J Pharmacol* 138, 161–171; [8] Gessner et al. (2004) *Mol Pharmacol* 65, 1120–1129; [9] Chen et al. (1999) *J Biol Chem* 274, 10113–10118; [10] Garcia-Ferreiro RE et al. (2004) *J Gen Physiol* 124, 301–317; [11] Schonherr et al. (2002) *FEBS Lett* 514, 204–208; [12] Meyer et al. (1998) *J Physiol* 508 (Pt 1), 49–56; [13] Pardo et al. (1999) *EMBO J* 18, 5540–5547; [14] Wimmers et al. (2002) *Pflugers Arch* 445, 423–430; [15] Sturm et al. (2005) *J Physiol* 564, 329–345; [16] Schledermann et al. (2001) *J Physiol* 532, 143–163; [17] Spector et al. (1996) *Circ Res* 78, 499–503; [18] Kiehn et al. (1996) *Circulation* 94, 2572–2579; [19] Lees-Miller et al. (1997) *Circ Res* 81, 719–726; [20] Pond et al. (2000) *J Biol Chem* 275, 5997–6006; [21] Wymore et al. (1997) *Circ Res* 80, 261–268; [22] Jones et al. (2004) *J Biol Chem* 279, 44690–4464; [23] Shi et al. (1997) *J Neurosci* 17, 9423–9432; [24] Guasti et al. (2005) *J Comp Neurol* 491, 157–174; [25] Bauer et al. (2003) *Pflugers Arch* 445, 589–600; [26] Farrelly et al. (2003) *Am J Physiol* 284, C883–C895; [27] Rosati et al. (2000) *FASEB* 14, 2601–2610; [28] Overholt et al. (2000) *J Neurophysiol* 83, 1150–1157; [29] Nie et al. (2005) *J Neurosci* 25, 8671–8679; [30] Meves et al. (1999) *Br J Pharmacol* 127, 1213–1223; [31] Smith et al. (2002) *J Biol Chem* 277, 18528–18534; [32] Crociani et al. (2003) *J Biol Chem* 278, 2947–2955; [33] Cherubini et al. (2000) *Br J Cancer* 83, 1722–1729; [34] Lastraioli et al. (2004) *Cancer Res.* 64, 606–611; [35] Snyders et al. (1996) *Biophys J* 74, 230–241; [37] Perry et al. (2004) *Mol Pharmacol* 66, 240–249; [38] Rampe et al. (1997) *FEBS Lett* 417, 28–32; [39] Zhou et al. (1999) *Cardiovasc Electrophysiol* 10, 836–843; [40] Restano-Cassulini et al. (2006) *Mol Pharmacol* 69(5), 1673–83; [41] Diochot et al. (2003) *Mol Pharmacol* 64, 59–69; [42] Ficker et al. (2004) *Mol Pharmacol* 66, 33–44; [43] Kuryshev et al. (2005) *JPET* 312, 316–323; [44] Kang et al. (2005) *Mol Pharmacol* 67, 827–836; [45] Hansen et al. (2005) *Mol Pharmacology* 69, 266–277; [46] Sanguinetti et al. (1995) *Cell* 81, 299–307; [47] Akbarali et al. (1999) *Am J Physiol* 277, C1284–1290(1999); [48] Ohya et al. (2002) *Am J Physiol* 282, G277–G287; [49] Shoeb et al. (2003) *Biol Chem* 278, 2503–2514; [50] Chiesa et al. (1997) *J Physiol* 501, 313–318; [51] Kirchberger et al. (2006) *J Physiol* 571, 27–42; [52] Arcangeli et al. (1993) *J Cell Biol* 122, 1131–1143; [53] Arcangeli et al. (1995) *J Physiol* 489, 455–471; [54] Emmi et al. (2000) *J Neurosci* 20, 3915–3925; [55] Furlan et al. (2005) *Neuroscience* 135, 1179–1192; [56] Sacco et al. (2003) *J Neurophysiol* 90, 1817–1828; [57] Curran ME et al. (1995) *Cell* 80, 795–803; [58] Brugada R et al. (2004) *Circulation* 109, 30–35; [59] Arcangeli et al. (1999) *J Neurobiol* 40, 214–225; [60] Bianchi et al. (1998) *Cancer Res* 58, 815–822; [61] Papa et al. (2003) *J Comp Neurol* 466, 119–135; [62] Wang X et al. (2006) *FASEB J* 20, 1531–1533; [63] London et al. (1997) *Circ Res* 81, 870–878; [64] Zou et al. (2003) *Am J Physiol Cell Physiol* 285, C1356–C1366; [65] Shi et al. (1998) *J Physiol* 511, 675–682; [66. Engeland et al. (1998) *J Physiol* 513, 647–654; [67] Trudeau et al. (1999) *Neurosci* 19, 2906–2918; [68] Becchetti et al. (2002) *Eur J Neurosci* 16, 415–428.

2.1.3.3 Biophysical characteristics

Kv10 (eag) channels

When heterologously expressed in *Xenopus* oocytes, *Drosophila* eag channels conduct a rapidly activating, non-inactivating, outward rectifier K^+ current with a single channel conductance of 4.9 pS estimated by nonstationary noise analysis (Bruggemann *et al.*, 1993). The sequence of ion permeability in *Drosophila* eag channels is $K^+>Rb^+>NH_4^+>Cs^+>>Na^+>Li^+$. Surprisingly, the channel was also initially reported to be slightly permeable to Ca^{2+} (Bruggemann *et al.*, 1993).

Two human eag channels have been described, heag1 and heag2 (Ju and Wray, 2002; Schonherr *et al.*, 2002). Heag1 channel currents activate about four times faster than heag2, but both currents reach steady-state values within 100 ms at positive transmembrane potentials (Figure 2.1.3.1a). The voltage dependence of activation

also differs for the two types of eag channels. The half-maximal voltage ($V_{0.5}$) for activation is +10 mV for heag1 and –11 mV for heag2. The slope factor (k) for the conductance-voltage (G–V) curve is steeper for heag1 ($k = 17$ mV) compared to heag2 ($k = 35$ mV) (Ju and Wray, 2002). The structural basis for these differences in the activation process are complex and involve multiple regions of the eag subunit, including the N terminus, S4–S5 linker, S1 and P-S6 domains (Ju and Wray, 2006).

The rate of eag channel activation is highly dependent on the membrane potential preceding a depolarization and $[Mg^{2+}]_o$ (Silverman *et al.*, 2000; Tang *et al.*, 2000). When voltage clamp pulses are applied from a very negative potential, activation of eag currents is slower and more sigmoidal than currents activated from less polarized membrane potentials, a phenomenon known as a Cole–Moore shift. The slower activation associated with a more negative holding potential can be modelled by assuming that

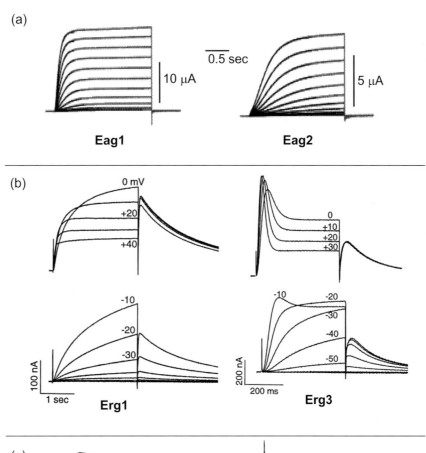

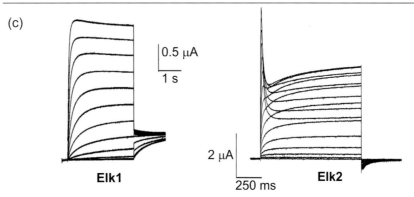

Fig. 2.1.3.1 Representative eag, erg and elk currents. Currents were recorded from oocytes expressing two different types of eag (a), erg (b) and elk (c) channels. Reproduced from Distribution and functional properties of human KCNH8 (Elk1) potassium channels by Anruo Zou *et al.* in the *American Journal of Physiology* 285, © 2003, with permission from The American Physiological Society. Reproduced from Functional analysis of a mouse brain Elk-type K^+ channel by Matthew Trudeau *et al.* in the *Journal of Neuroscience* © 1997, 19 with permission from the Society for Neuroscience Differential expression of genes encoding subthreshold-operating voltage-gated K^+ channels in brain by M J Saganich *et al.* in the *Journal of Neuroscience*, 21 © 2001 with permission from the Society for Neuroscience.

channels must transit through more closed states. Elevation of $[Mg^{2+}]_o$ within a physiologically relevant range also slows the rate of eag activation and shifts the voltage dependence of channel activation to more positive potentials (Silverman *et al.*, 2003, 2004). Mutagenesis studies of *Drosophila* eag revealed that two acidic residues, D278 in S2 and D327 in S3 coordinate Mg^{2+} ions (Silverman *et al.*, 2000). When Mg^{2+} is bound to the channel, it impedes charge pairing between these Asp residues and specific basic residues in S4 that facilitate activation gating. The binding of Mg^{2+} (or other divalent cations such as Ni^{2+} and Mn^{2+}) appears to affect gating transitions at negative transmembrane potentials that can not be detected by measuring gating currents. Optical monitoring of eag channels with a fluorescent probe attached near the S4 voltage sensor has enabled detection of these relatively slow channel transitions that are affected by extracellular divalent cations (Bannister *et al.*, 2005).

Kv11 (erg) channels

All three erg channel isoforms exhibit the same basic inward rectification characteristics, albeit with differences in kinetics and voltage-dependence (Shi *et al.*, 1997). Biophysical properties of herg1a currents have been the most extensively studied, both *in vivo* (as I_{Kr}) and in heterologous expression systems. The left panels of Figure 2.1.3.1b show typical recordings from herg1a channels expressed in *Xenopus* oocytes. The peak currents measured at potentials negative to 0 mV increase with progressive depolarization steps, while currents are progressively decreased when measured at potentials positive to 0 mV. The negative slope conductance at positive potentials and the resulting bell-shaped current-voltage relationship is characteristic of I_{Kr} in native myocytes and heterologously expressed erg channels (Sanguinetti *et al.*, 1995).

Single channel conductance of herg1a is approximately 2 pS with an external $[K^+]$ of 5 mM, but was increased to approximately 9 pS by elevating $[K^+]_e$ to 100 mM (Kiehn *et al.*, 1996). Similar responses to elevated $[K^+]_e$ have been observed for whole cell conductance of erg1 channels expressed in CHO cells, and have been attributed to competitive relief of pore blockage by external Na^+ ions (Mullins *et al.*, 2002).

The threshold voltage and $V_{0.5}$ for activation for heterologously expressed herg1a are approximately −50 mV and −15 mV, respectively (Sanguinetti *et al.*, 1995; Shi *et al.*, 1997; Wang *et al.*, 1997a; Zhou *et al.*, 1998b). Taking into account small variations that depend on the expression system and recording conditions, these values are in agreement with those reported for native cardiac I_{Kr} and the alternatively spliced erg1b isoform (Lees-Miller *et al.*, 1997; London *et al.*, 1997). In comparison, erg2 channels activate at less negative potentials (threshold and $V_{0.5}$ for activation of −30 mV and −3.5 mV, respectively), whereas erg3 channels activate at more negative potentials (threshold and $V_{0.5}$ for activation of −60 mV and −44 mV, respectively) (Shi *et al.*, 1997; Sturm *et al.*, 2005; Wimmers *et al.*, 2002). The activation properties of heteromultimeric channels (combinations of erg1, 2 and 3) are intermediate to their homomeric counterparts (Wimmers *et al.*, 2002).

As illustrated in Figure 2.1.3.1, herg1 channels have a much slower activation rate compared with eag or elk channels. Activation of eag occurs with a time constant of ~10 ms at +10 mV (Silverman *et al.*, 2000), whereas herg1a activation has a time constant closer to 150 ms at the same potential (Wang *et al.*, 1997). The time course of herg1 activation is sigmoidal, indicating that the

channel must pass through multiple closed states prior to opening (Subbiah *et al.*, 2004, 2005; Wang *et al.*, 1997). In contrast to the similarities in voltage dependence, mouse erg1b channels exhibit substantially faster activation than mouse erg1a (Lees-Miller *et al.*, 1997). Notably, the activation rate of native I_{Kr} in cardiac myocytes is intermediate between erg1a and erg1b (Lees-Miller *et al.*, 1997). Activation of erg3 is also significantly faster than erg1 (Figure 2.1.3.1b, right panels), whilst activation of erg2 is slower (Shi *et al.*, 1997). The mechanism that underlies the slow activation of erg1 is not yet fully understood. Fluorescent probes attached to specific Cys residues introduced to the extracellular end of S4 and direct measurement of gating currents identified two kinetic components (fast and slow) associated with S4 movement in response to membrane depolarization (Piper *et al.*, 2003; Smith and Yellen 2002). Similar to eag channels described above, slow movement of the voltage sensor during erg1 activation could reflect the formation of salt bridges between basic residues on the S4 segment and acidic residues on the S1–S3 helices that act to stabilize closed or intermediate states of the channel (Liu *et al.*, 2003; Subbiah *et al.*, 2004; Subbiah *et al.*, 2005). However, the voltage-sensing domains (S1–S4) are highly conserved between erg isoforms despite their significantly different rates of channel activation. Slow activation of herg1 could also reflect coupling between the S4 voltage sensor and the S6 activation gate. Although the coupling mechanism is not yet fully understood, it appears to involve specific interactions between the S4–S5 linker and the S6 domain (Ferrer *et al.*, 2006; Tristani-Firouzi *et al.*, 2002).

Erg1a channel deactivation is a slow and strongly voltage-dependent process that occurs at potentials negative to −20 mV. The time course of deactivation is bi-exponential with time constants that range from approximately 87 ms (at −120 mV) to 350 ms (at −20 mV) for the fast component and from 250 ms to 3 s for the slow component (London *et al.*, 1997; Wimmers *et al.*, 2002). Erg1b deactivates much faster than erg1a with time constants similar to those of native I_{Kr} (London *et al.*, 1997). Channels formed from human erg1b alone are largely retained in the endoplasmic reticulum because of an 'RXR' ER retention signal specific to the N terminus and are thus poorly expressed at the surface membrane (Phartiyal *et al.*, 2008). However, when associated with erg1a subunits, the erg1a/erg1b heteromultimeric channels readily traffic to the plasma membrane.

Erg2 deactivates with a time course similar to erg1a (Restano-Cassulini *et al.*, 2006; Shi *et al.*, 1997; Wimmers *et al.*, 2002) whereas erg3 exhibits faster deactivation (Figure 2.1.3.1b). Slow deactivation may partly reflect the formation of salt bridges that stabilize the voltage sensor in open or intermediate states. However, the substantially faster deactivation observed for erg1b and N-terminal truncated (236–278 aa) herg1 channels suggests a prominent role for the N-terminus in maintaining slow deactivation (London *et al.*, 1997; Spector *et al.*, 1996). In fact, even small deletions, or individual point mutations, in the N-terminal region can also accelerate deactivation (Chen *et al.*, 1999; Morais Cabral *et al.*, 1998). Slow deactivation can be restored in a N-terminal truncated channel by the intracellular application of a peptide corresponding to the initial 16 amino acids ('deactivation domain') of herg1 (Wang *et al.*, 1998). It has been proposed that slow deactivation is mediated by a stabilization of the open state through an interaction between the N-terminal deactivation domain and the S4–S5 linker which couples the voltage sensor to the activation gate (Wang *et al.*,

1998). However, the C-terminal domain of erg may also contribute to these interactions (Aydar and Palmer, 2001).

As noted above, and illustrated in Figure 2.1.3.1b, erg current amplitudes are progressively decreased at positive potentials. This reduction in current is mediated by a rapid and voltage-dependent inactivation process. Repolarization of the membrane to –70 mV generates a fast recovery from inactivation and produces large tail currents which then slowly decay as channels deactivate (Figure 2.1.3.1b). Rate constants for the onset of erg1a inactivation varied between 16 ms (at –20 mV) and 2 ms (at +50 mV), while the $V_{0.5}$ for inactivation measured from a plot of the rectification factor was approximately –49 mV (Sanguinetti et al., 1995; Spector et al., 1996). Thus, erg1 channels inactivate at least 10 times faster than they activate. Rate constants for the recovery from inactivation varied between 5 ms (at –120 mV) and 15 ms (at –50 mV), with a $V_{0.5}$ of –87 mV (Piper et al., 2005). Of the three erg isoforms, erg2 exhibits the strongest, and erg3 the weakest rectification. Weak rectification of erg3 channels represents both a slowing of the onset and a decrease in the steady-state value of inactivation (Schledermann et al., 2001; Shi et al., 1997; Sturm et al., 2005). For erg3, the slower rate of inactivation and faster rate of activation when compared with erg1 reduces the rate of inactivation-to-activation ratio from tenfold (erg1) to twofold (erg3). The consequence of this difference can be observed in the distinct currents produced by erg3 channels at positive potentials when compared with erg1 (Figure 2.1.3.1b). Erg3 also exhibits the fastest rates of recovery from inactivation (Sturm et al., 2005).

Inactivation of erg channels does not require an intact N-terminus (Schonherr and Heinemann 1996; Spector et al., 1996) or the presence of intracellular Mg^{2+} or polyamines (Spector et al., 1996) and thus cannot be caused by an intracellular block or N-type inactivation process previously described for inward rectifier (K$_{ir}$) and Shaker channels, respectively. In contrast, the sensitivity of erg inactivation to tetraethylammonium (TEA), extracellular [K$^+$] and point mutations within the outer pore domain (Ficker et al., 1998; Schonherr and Heinemann 1996; Smith et al., 1996; Wang et al., 1996) are consistent with a C-type inactivation first described for Shaker (Rasmusson et al., 1998). The structural rearrangements that underlie C-type inactivation are not known with certainty, but it has been proposed that a conformational change ('collapse') of the selectivity filter restricts the movement of K$^+$ and makes the channel non-conducting. For all three erg isoforms, an increase in [K$^+$]$_e$ of only 10 mM reduces steady-state inactivation and slows the rate of onset and recovery from inactivation (Sturm et al., 2005). Elevated [K$^+$]$_e$ is likely to increase ion occupancy of the outermost binding site, thereby restricting collapse of the selectivity filter (Kiss and Korn, 1998). Consistent with this mechanism, herg1a channels containing a S631V mutation, which is equivalent to the T449V mutation that removes C-type inactivation in Shaker channels, do not exhibit inactivation (Fan et al., 1999; Herzberg et al., 1998). Furthermore, point mutations of Ser620 or Ser631 in herg1a to their equivalent residues in eag (Thr and Ala, respectively) severely attenuate or remove steady-state inactivation (Ficker et al., 1998; Schonherr and Heinemann 1996).

The mechanism responsible for the intrinsic voltage dependence of erg inactivation is presently unclear. Fluorescence and gating current analysis has revealed a rapid component to voltage sensor movement. However, the voltage dependency of this component does not directly correlate with that of erg1 inactivation (Piper et al., 2003; Smith and Yellen 2002). These inconsistencies may be due to the coupling of the voltage sensor, to inactivation, or to the presence of an as yet unidentified voltage-sensing process. Regardless of the mechanism, the intrinsic voltage dependency, rapid rate of onset and recovery distinguishes erg C-type inactivation from that of Shaker channels.

Kv12 (elk) channels

Elk channels activate faster than erg but slower than eag channels. Elk1 channels do not inactivate, but elk2 channels inactivate rapidly by a C-type mechanism (Figure 2.1.3.1c). The biophysical properties of heterologously expressed mouse elk2 have been determined in Xenopus oocytes (Trudeau et al., 1999) and human elk2 has been characterized in transfected CHO cells (Becchetti et al., 2002). These properties are described below.

In oocytes, the threshold for mouse elk2 channel activation was –70 mV and the current-voltage relationship reached its peak value at +20 mV. The onset of activation was bi-exponential as determined using an envelope of tails protocol. Unlike eag channels, the rate of current activation did not vary with holding potential (no Cole–Moore shift) and the time constants for activation at +10 mV were 12 and 83 ms (Trudeau et al., 1999). Elk2 deactivation was bi-exponential with time constants for the fast and slow components ranging from approximately 40 to 100 ms (fast) and 220 to 650 ms (slow) between –20 and –120 mV. Inactivation and recovery from inactivation of elk2 is very rapid. The time constant for inactivation was 6 to 11 ms between +70 and +20 mV. Time constants for recovery from inactivation vary from 3 ms at –80 mV to 16 ms at +10 mV (Trudeau et al., 1999). Single mutations of specific Ser residues located in the S5-P linker of mouse elk2 can reduce (S475A) or remove (S464T) C-type inactivation (Trudeau et al., 1999). These mutations are equivalent to S631A and S620T that had previously been shown to reduce or eliminate inactivation of herg1a channels. Slowing of the rate of inactivation by external TEA is one of the properties of C-type inactivation (Choi et al., 1991). Accordingly, TEA slows inactivation of mouse elk2 with a half maximal effective concentration of 97 mM (Trudeau et al., 1999).

The $V_{0.5}$ for helk2 activation in CHO cells is –23 mV and the $V_{0.5}$ for C-type inactivation is +10 mV (Becchetti et al., 2002). Both gating processes are weakly voltage-dependent. The slope factor determined from a Boltzmann function was 18 mV for activation and 22 mV for inactivation. The considerable overlap in the voltage dependence of steady-state activation and inactivation for helk2 generates a significant 'window current' where some channels remain in the open state over the entire range of physiologically relevant membrane potentials (Becchetti et al., 2002).

2.1.3.4 Cellular and subcellular localization

Kv10 (eag) channels

Nonradioactive in situ hybridization was used to characterize the distribution of eag transcripts in rat brains (Saganich et al., 2001). Strong expression of eag1 was detected in olfactory bulb, piriform cortex, anterior olfactory nucleus, olfactory tubercle, cerebral cortex, hippocampus (CA2, CA3 pyramidal cells), and the mammary nucleus of the midbrain (Figure 2.1.3.2). Expression of eag1 was also high in granule layer of the cerebellum. Moderate expression of eag1 was detected in CA1 pyramidal cells of the hippocampus, basal ganglia, amygdala, hypothalamus, red nucleus and oculomotor nucleus of the midbrain, and superior olive and facial nucleus of the hindbrain (Saganich *et al.*, 2001). Eag1 is also expressed in gliomas (Patt *et al.*, 2004).

In situ hybridization indicates a different expression pattern for eag2 in rat brain. Like eag1, eag2 expression was also high in olfactory bulb and piriform cortex and one region (layer IV) of the cerebral cortex (Figure 2.1.3.2). However, unlike eag1, high levels of eag2 transcripts were detected in the inferior colliculus and nucleus lateral lemniscus of the midbrain, and moderate expression was found in the thalamus, habenula, and the dorsal and ventral cochlear nucleus of the hindbrain of the rat (Saganich *et al.*, 2001). RT-PCR indicated that human eag2 is highly expressed in several tissues with the following rank order of abundance: brain>>skeletal muscle>heart, lung, liver>kidney, pancreas (Ju and Wray, 2002).

Kv11 (erg) channels

Erg1a mRNA and protein has been detected in all cardiac tissues (atrium, ventricle, sino-atrial node, and Purkinje) from several species including human, dog, rabbit, rat, guinea-pig, mouse, and zebrafish (Langheinrich *et al.*, 2003; Lees-Miller *et al.*, 1997; Pond *et al.*, 2000; Wymore *et al.*, 1997; Zehelein *et al.*, 2001). Western blot analysis identified two erg1a protein bands in human ventricular tissue that correspond to the unglycosylated and maturely glycosylated forms of the full length protein (Jones *et al.*, 2004; Pond *et al.*, 2000). Similar bands were observed in rat and mouse ventricular tissue (Pond *et al.*, 2000). In human and mouse heart, erg1a protein expression was greater in ventricular compared to atrial tissue, whereas the opposite was true in rat (Pond *et al.*, 2000). RT-PCR analysis indicates greater levels of *erg1a* mRNA in canine left ventricular tissue versus right ventricular tissue (Zehelein *et al.*, 2001). In rat and canine ventricular myocytes, immunofluorescence and confocal microscopy demonstrated a subcellular localization of erg1a protein in lateral (sarcolemmal) membranes, intercalated discs, and T-tubules (Jones *et al.*, 2004; Pond *et al.*, 2000).

Erg1b mRNA was abundantly detected in both atrial and ventricular myocytes from mouse heart using RT-PCR analysis (Lees-Miller *et al.*, 1997). Initial western blot studies using first-generation C- or N-anti-erg1 antibodies failed to identify the erg1b-isoform protein band in either mouse or human heart (Pond *et al.*, 2000). However, more recently Jones *et al.* (2004) used an erg1b-specific antibody to identify 94- and 83-kDa bands in both canine and human ventricle. Furthermore, immunocytochemistry and bidirectional coimmunoprecipitation experiments suggest that erg1a and erg1b proteins co-localize in T tubules and can associate in human and canine ventricular myocytes *in vivo* (Jones *et al.*, 2004). In contrast to *erg1*, neither *erg2* nor *erg3* mRNA has been identified in atrial or ventricular cardiac myocytes. Instead both erg2 and erg3 protein appears, at least in rat, to be exclusively located in brain and nervous system (Shi *et al.*, 1997).

As shown in Figure 2.1.3.2, each of the erg isoforms are expressed within the central nervous system of rat, albeit with variations in localization and intensity (Saganich *et al.*, 2001). The specific expression patterns for each erg isoform show some variation between species (Guasti *et al.*, 2005; Papa *et al.*, 2003). Immunohistochemical and non-radioactive *in situ* hybridization (NR-ISH) studies demonstrate that erg1a, erg1b and erg3 are often, although not exclusively, expressed in the same neuronal areas with similar intensities (Guasti *et al.*, 2005). For instance, similar levels of erg1a, erg1b and erg3 protein expression were observed in many regions of the thalamus (including the hypothalamus and amygdala), paleocortex, brainstem, hippocampus, and cerebellum. More variable expression was also detected in the olfactory bulb, basal ganglia, neocortex, vestibular nuclei, and spinal cord. In contrast to the wide range of neuronal expression for erg 1 and erg3, Figure 2.1.3.2 illustrates that erg2 expression occurs at lower levels and appears much more specific. Initial RNase protection studies detected *erg2* mRNA in retina and prevertebral ganglia, but not in rat brain (Shi *et al.*, 1997). However, Saganich and colleagues (2001) used NR-ISH to detect erg2 protein in mitral and

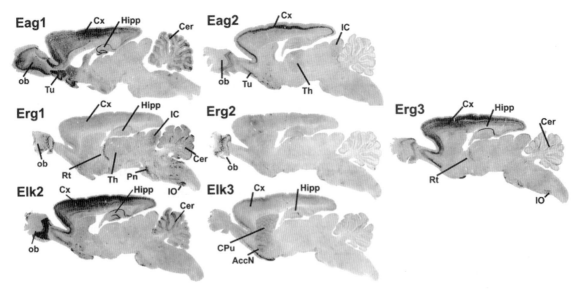

Fig. 2.1.3.2 *In situ* hybridization detection of eag, erg and elk channel transcripts in the rat brain. Non-radioactive *in situ* hybridization for eag, erg and elk channel transcripts in rat brain was performed using DIG-labelled RNA antisense probes. AccN, Accumbens nucleus; CPu, caudate/putamen; Cx, cerebral cortex; Cer, cerebellum; Hipp, hippocampus; IC, inferior colliculus; IO, inferior olive; ob, olfactory bulb; Pn, pontine nucleus; Rt, reticular thalamic nucleus; Th, thalamus; Tu, olfactory tubercle. Reprinted with permission from Saganich *et al.* (2001).

periglomerular cell layers of rat olfactory bulb (see Figure 2.1.3.2). More recently, RT-PCR and NR-ISH studies have further identified *erg2* at relatively low levels in hypothalamus, hippocampus, cerebellum, cerebral cortex, and brainstem nuclei of rat (Papa *et al.*, 2003) and mouse (Guasti *et al.*, 2005). At the subcellular level, erg expression was not limited to the soma of neuronal cells but could also be clearly observed within dendrites of layer V pyramidal neurons (erg3) and Purkinje cells (erg2) of the cerebellum (Guasti *et al.*, 2005).

Interestingly, Guasti *et al.*, (2005) observed that erg2 expression always co-localizes with other erg isoforms. Together with the apparent co-localization of erg1 and erg3, this finding raises the possibility that erg isoforms may form heteromultimeric channels in neuronal cells *in vivo*. Indeed, as in mycoytes, erg1a and erg1b co-immunoprecipitate on Western blots performed on whole brain lysate (Guasti *et al.*, 2005). Although no co-immunoprecipatation of erg1, erg2 and erg3 isoforms has yet been demonstrated *in vivo*, these isoforms can co-assemble to from heteromultimeric channels when over-expressed in CHO cells (Wimmers *et al.*, 2001, 2002). More recently, single-cell RT-PCR was used to reveal differential combinations of erg1a, erg1b, erg2 and erg3 expression in rat embryonic rhomencephalon neurons and these neurons have native erg currents that show the distinct biophysical characteristics of heteromultimeric channels (Hirdes *et al.*, 2005).

Erg1a or its encoded protein has also been identified in a wide range of other tissues including the following cell types: lactotrophs (Bauer *et al.*, 2003), gastrointestinal smooth muscle myocytes (Farrelly *et al.*, 2003), pancreatic β-cells (Rosati *et al.*, 2000) and carotid body cells (Overholt *et al.*, 2000). Immunoreactivity staining identified erg1a in several cell types of mouse inner to middle ear, including the cochlear wall, organ of corti and vestibular tissue (Nie *et al.*, 2005). Erg1a was also found in various human cell lines including neuroblastomas (Meves *et al.*, 1999), rhabdomyosarcomas, hematopoietic (Smith *et al.*, 2002), monoblastic leukemias and colon carcinoma (Crociani *et al.*, 2003). Erg2 was found in human retinoblastomas, while erg1 and erg3 were found in human mammary carcinomas (Crociani *et al.*, 2003). Human erg1 was observed *in vivo* within primary tumours of several tissues including endometrium (Cherubini *et al.*, 2000), colon (Lastraioli *et al.*, 2004) and lymphocytes (Smith *et al.*, 2002).

Kv12 (elk) channels

Northern blot analysis indicates that elk channels are highly expressed in the brain. In addition, elk3 is expressed at very low levels in the lung, and elk1 and elk2 are also poorly expressed in the heart. In the rat, RT-PCR indicates that elk3 is highly expressed in the olfactory bulb and hippocampus, elk2 is most highly expressed in cerebellum, hippocampus and brainstem and elk1 is most highly expressed in the olfactory bulb and brainstem (Engeland *et al.*, 1998). This general pattern of differential expression of elk2 and elk3 (Figure 2.1.3.2) was also detected using nonradioactive *in situ* hybridization of rat brain (Saganich *et al.*, 2001). Using an RNAse protection assay, Shi and colleagues (1998) reported that rat elk1 is undetectable in the brain, but highly expressed in sympathetic ganglia and sciatic nerve. As the brain matures during embryonic development, elk2 mRNA is upregulated, elk1 is downregulated and elk3 remains relatively constant into adulthood (Engeland *et al.*, 1998).

Northern blot analysis of human elk2 indicated expression in the amygdala, cortex, hippocampus and striatum (Miyake *et al.*, 1999).

The expression pattern of human elk channels in the nervous system was also determined by RT-PCR (Zou *et al.*, 2003). Elk2 was the most highly expressed and was most prominent in cerebral and frontal cortex, hippocampus, caudate nucleus, nucleus acumbens and amygdala. Elk1 was expressed at levels about 10-times lower in all these brain tissues as well as in the thalamus, subthalamic nucleus, substantia nigra, pons, and spinal cord. Elk3 was expressed at the lowest levels and was most evident in the same structures as elk2 (Zou *et al.*, 2003). Elk2 channels are also highly expressed in human astrocytoma cells (Becchetti *et al.*, 2002).

2.1.3.5 Pharmacology
Kv10 (eag) channels

Drosophila eag is weakly blocked by TEA (IC_{50}: 33 mM), quinine (IC_{50}: 0.7 mM) and quinidine (IC_{50}: 0.4 mM) and unaffected by 4-aminopyridine at 100 mM (Bruggemann *et al.*, 1993). Cyclic nucleotides (cAMP and cGMP) shift the voltage dependence of *Drosophila* eag channel activation to more negative potentials (Bruggemann *et al.*, 1993).

Eag channels are blocked by several compounds that are better known for their ability to block herg channels, including E-4031, clofilium and haloperidol (Gessner and Heinemann, 2003). In mammalian CHO cells, a tertiary analogue of clofilium, LY97241 blocks heag1 channels with an IC_{50} between 4.9 nM (whole cell recording) and 1.9 nM (inside-out patch) (Gessner and Heinemann 2003). In inside-out patches from *Xenopus* oocytes, clofilium blocks heag1 with an IC_{50} of 0.8 nM (Gessner *et al.*, 2004). Clofilium and LY97241 can only block eag channels after they have opened and can be trapped inside the central cavity of the pore by closure of the activation gate (Gessner *et al.*, 2004). Human eag2 channels are about 150 times less sensitive to block by clofilium when compared to heag1 or herg1 channels. This difference in potency can be accounted for by three specific residues that differ between these channels: a Ser and Val in the pore helix that are Thr and Ile in heag2, and an Ala in the S6 that is a Ser in heag2 channels (Gessner *et al.*, 2004). Heag1 channels are also weakly blocked by quinidine ($IC_{50} = 1.4$ μM) (Schonherr *et al.*, 2002) and Ba^{2+} blocks these channels less than 40 per cent at 1 mM (Shi *et al.*, 1998).

Kv11 (erg) channels

Many common medications reduce I_{Kr} in cardiac myocytes, an effect mediated via block of erg1 channels. A reduction in I_{Kr} delays the repolarization phase and extends the duration of the cardiac ventricular action potential, observed as a prolonged QT interval measured by a body surface electrocardiogram. Pharmacological block of I_{Kr} induces an acquired form of long QT syndrome (LQTS) which increases the risk of patients developing torsades de pointes (TdP) arrhythmia, ventricular fibrillation and sudden death (Sanguinetti and Tristani-Firouzi, 2006). Prolonged QT interval is the intended pharmacological action of cardiac class III antiarrhythmic drugs that block herg1a with high-affinity. These drugs include dofetilide with an IC_{50} of 12.6 nM (Snyders and Chaudhary, 1996) and E-4031 with an IC_{50} of 7.7 nM, (Zhou *et al.*, 1998b). However, several non-cardiac drugs such as the antihistamines terfenadine—56 nM (Rampe *et al.*, 1997) and astemizole—0.9 nM (Zhou *et al.*, 1999) or cisapride, a gastrointestinal prokinetic compound—44.5 nM (Rampe *et al.*, 1997) also block herg1 with high-affinity as an unintended side effect. Although the actual incidence of TdP is rare with non-cardiac medications (1 in 100,000 for cisapride compared to 2–9 per cent for the antiarrhythmic quinidine) it obviously presents an unacceptable risk for the treatment of non-life-threatening disorders (Haverkamp et al., 2000). For this reason, a number of medications including cisapride, sertindole, terfenadine and astemizole have either been withdrawn from the market or their use restricted by drug regulatory agencies.

Blockade of erg channels by drugs occurs from the intracellular side of the channel (Kiehn et al., 1996) and requires opening of the activation gate (Spector et al., 1996) indicating an open channel block mechanism. Once inside the inner cavity of the channel the drug can become trapped when closure of the activation gate is prompted by repolarization of the membrane (Mitcheson et al., 2000). The activation gate is formed by the overlapping S6 helices from each subunit. Drug trapping underlies the slow recovery from block that is a feature of many, but not all, erg blockers. In contrast to other Kv channels that trap only small molecules, herg can accommodate much larger drugs, suggesting that it has a much larger inner cavity. Alanine-scanning mutagenesis of residues within the S6 and pore domain that line the inner cavity identified several residues as potential sites for drug interaction (Mitcheson et al., 2000). In herg1a, these residues are Gly648, Tyr652, Phe656 and Val659 within the S6 domain, as well as Thr623, Ser624 and Val625 located near the base of the pore helix. Mutation of Tyr652 and Phe656, two hydrophobic residues within the S6 domain that are unique to the eag channel family, produced significant decreases in the potency of MK-499, terfenadine and cisapride (Mitcheson et al., 2000). Both residues appear to be crucial interaction sites for nearly all compounds tested so far (Sanguinetti and Mitcheson, 2005), with a few notable exceptions such as fluvoxamine (Milnes et al., 2003). Analysing changes in cisapride and terfenadine potencies following the substitution of Tyr652 and Phe656 with many amino acids suggests an aromatic residue is critical at position 652, while a strongly hydrophobic residue is required at position 656 in herg1a (Fernandez et al., 2004). The effects of polar residue mutations (Thr623 and Ser624) at the base of the pore helix are variable depending on the drug studied. Loss of sensitivity to some compounds may reflect the absence of stabilizing hydrogen bonds between polar groups of the drug and a pore helix residue (Perry et al., 2004, 2006). Thr63 and Ser624 are highly conserved in the Kv family, indicating that these residues are not responsible for the unusual sensitivity of herg1a to open channel block, which is more likely due to the large inner cavity and uniquely positioned hydrophobic residues. Recent studies using herg1 homology models to predict drug interactions broadly support these conclusions (Farid et al., 2006; Osterberg and Aqvist 2005; Rajamani et al., 2005).

Long term exposure to arsenic trioxide (As_2O_3) or pentamidine inhibits I_{Kr} and promotes LQTS through a disruption of channel trafficking to the cell surface (Ficker et al., 2004; Kuryshev et al., 2005). Specifically, As_2O_3 inhibits herg channel–chaperone protein associations (Ficker et al., 2004). These very different mechanisms of reduced herg channel activity add to the difficulty of designing drugs that lack significant QT liability. Current pharmacophore models based on a relatively narrow range of pore blocking compounds do have some predictive value (Cavalli et al., 2002; Ekins et al., 2002; Pearlstein et al., 2003), but a range of screening procedures during the drug development process are still required and even these demonstrate some limitations in accurately predicting QT and TdP liability (Recanatini et al., 2005).

Co-expression of herg1 with MiRP1 (*KCNE2*) has been shown to enhance the rate and potency of block by E4031, with a fast component that is also observed with I_{Kr} block (Abbott et al., 1999). However, similar effects of MiRP1 have not been observed by others using E4031 or other class III antiarrhythmics (Weerapura et al., 2002). Another possible accessory protein termed KCR1 is proposed to reduce the potency of dofetilide and d-sotalol for herg1 block, without affecting kinetics or gating of the channel (Kupershmidt et al., 2003). It remains to be seen if the different kinetics for herg1 block and that of native I_{Kr} can be explained by association with different β-subunits and accessory proteins or whether other factors such as the formation of heteromultimeric channels or the phosphorylation state of the channel may also play a role.

At present there are no open-channel blockers that can select for specific erg isoforms. For instance, E-4031 has similar affinity for all erg isoforms, with IC_{50} values of 99 nM for erg1, 116 nM for erg2 and 193 nM for erg3 (Shi et al., 1997). This lack of specificity is not surprising given that the important residues for drug binding are conserved in all erg isoforms.

Several peptide toxins isolated from scorpions, including BeKm-1 (Korolkova et al., 2001), ErgTx-1 (Pardo-Lopez et al., 2002) and many related peptides from *Centuroides* (Corona et al., 2002) are high affinity blockers of erg1 channels. In addition, a sea anemone toxin, APETx1 has been discovered (Diochot et al., 2003). When measured in CHO cells at −70mV, ErgTx-1 reduces herg1 and herg3 currents with IC_{50} values of approximately 4 to 4.5 nM, but fails to reduce herg2 current up to 210 nM (Restano-Cassulini et al., 2006). The lack of effect on herg2 may be species specific as a high potency effect was observed on rat erg2 (~3 nM). BeKm-1 reduces current generated by all three erg isoforms although with decreased potency for herg2 (77 nM) compared with herg1 (7.7 nM) and herg3 (11.5 nM) (Restano-Cassulini et al., 2006). Despite sharing only 11 per cent amino acid homology, both toxins act in a concentration-dependent (apparent 1:1 stoichiometry), weakly voltage-dependent (potency is reduced with depolarization) and reversible manner (Restano-Cassulini et al., 2006; Zhang et al., 2003). Furthermore, the potency of both toxins is reduced by outer pore mutations and external TEA, increased by extracellular acidification, but insensitive to elevated $[K^+]_e$ or $[K^+]_i$ or lowering of ionic strength (Zhang et al., 2003). Mutation analysis suggests that both toxins bind to the outer mouth of the pore and the S5-P linker (Korolkova et al., 2004; Pardo-Lopez et al., 2002; Zhang et al., 2003). Mutant cycle analysis, using NMR spectroscopy and mutant toxins (Korolkova et al., 2002; Torres et al., 2003), suggest that both toxins bind through predominantly hydrophobic interactions with the outer mouth of the pore and the S5-P linker, but with slightly different orientations such that BeKm-1 may lie deeper into the pore than ErgTx-1 (Korolkova et al., 2004).

Recently, several pharmacological activators of herg1 current have been identified (Hansen et al., 2005; Kang et al., 2005; Zhou et al., 2005). Although the mechanism of action remains unclear, for certain activators a reduction in inactivation due to a shift in the voltage-dependence to more positive potentials has been proposed (Casis et al., 2006). Further studies are required to identify the binding sites and mechanism of action of herg1 activators.

Kv12 (elk) channels

Specific small molecule or toxin blockers of elk channels have not been described. Obviously, the discovery of such drugs or toxins would greatly aid the characterization of the physiology of these channels. Elk channels are unaffected by E-4031 (Shi *et al.*, 1998; Trudeau *et al.*, 1999), a methanesulfonanilide that, as discussed above, blocks erg channels at nM concentrations. Elk channels are also resistant to block by TEA and 4-aminopyridine, but are sensitive to external Ba^{2+} and Cs$^+$. Ba^{2+} (1 mM) blocks elk1 by ~85 per cent (Shi *et al.*, 1998) and elk2 channels are blocked by external Cs$^+$ in a voltage-dependent manner with block increasing at more negative membrane potentials. Elk2 tail currents measured at −120 mV are blocked by Cs$^+$ with a K$_d$ of 0.68 mM.

2.1.3.6 Physiology

Kv10 (eag) channels

Eag channels are highly expressed in several regions of the brain, but their exact physiological role here has not been established. Eag1 has been implicated in cell cycle regulation. It is overexpressed in some cancerous tissues and tumour cell lines and is associated with cell proliferation (Meyer and Heinemann 1998; Pardo *et al.*, 1999). The implication of these findings is discussed below in section 2.1.3.7. Eag1 channels underlie the K$^+$ current known as $I_{K(NI)}$ that contributes to membrane hyperpolarization in human myoblasts during the onset of myoblast differentiation prior to fusion (Bijlenga *et al.*, 1998; Occhiodoro *et al.*, 1998).

The Kvβ subunit *Hyperkinetic* (Hk) was first described as a subunit that interacts with *Shaker* K$^+$ channels, but was subsequently reported to also bind to, and affect the magnitude and kinetics of, eag channels (Wilson *et al.*, 1998). Hk is 48 per cent identical to Kvβ2. Coexpression of *Drosophila* eag with Hk accelerates the rate of activation of eag and enhances its magnitude ~4-fold when measured at +40 mV.

Kv11 (erg) channels

As discussed in section 2.1.3.3, erg isoforms are differentially expressed in cardiac, neuronal, smooth muscle and tumour cells. The exact physiological role for homomeric and heteromeric erg channels is unknown in many of these cell types, particularly in the brain. In contrast, the contribution of erg1a, and possibly erg1b, in maintaining normal electrical activity of the heart has been extensively characterized. Erg channels play an important role in phase 2 (plateau) and phase 3 (repolarization) of action potentials in human atrial and ventricular myocytes. Depolarization initiates fast C-type inactivation of herg1 (I_{Kr}) and suppresses current magnitude during phase 2 of the cardiac action potential (Spector *et al.*, 1996). During phase 3 repolarization, herg1 channels recover from inactivation rapidly, then deactivate relatively slowly. These kinetic properties combine to enhance outward current during the terminal phase of action potential repolarization.

Two lines of evidence suggest that alternatively spliced isoforms of erg1 (full length erg1a and an N-terminal truncated erg1b) form heteromultimeric channels that likely underlie I_{Kr}. First, heterologously expressed erg1a/erg1b heteromultimers exhibit accelerated deactivation kinetics that more closely resemble native I_{Kr} than homomeric erg1a channels (London *et al.*, 1997). As discussed above, the accelerated deactivation is conferred by erg1b that lacks a PAS domain. Secondly, erg1a and erg1b can be co-immunoprecipitated from native human and canine ventricular tissue (Jones *et al.*, 2004). Additionally, MiRP1 can associate with erg1a *in vitro* to alter the expression level of erg channels and accelerate deactivation

(Abbott *et al.*, 1999; McDonald *et al.*, 1997). However, these effects of MiRP1 vary with the heterologous expression system and other unknown factors (Anantharam *et al.*, 2003; Mazhari *et al.*, 2001; Weerapura *et al.*, 2002). Moreover, expression of MiRP1 is limited in the ventricle, suggesting that erg1a-MiRP1 association may be limited to Purkinje cells of the conduction system (Pourrier *et al.*, 2003).

In some cell types that lack inwardly rectifying Kir channels, erg channels can underlie the principle current for maintaining the resting membrane potential (RMP). So-called 'window current', generated by the crossover between the voltage dependence of activation and the steady state inactivation allows for a small but significant erg conductance that maintains the RMP of gastrointestinal smooth muscle cells of the eosophagus (Akbarali *et al.*, 1999), stomach (Ohya *et al.*, 2002), jejunum (Farrelly *et al.*, 2003) and colon (Shoeb *et al.*, 2003). Erg-specific blockers produce a pronounced depolarization of the membrane potential, enhance contractility and induce excitatory bursts in these cells. Thus, $I_{K(ERG)}$ can modulate excitability and GI motility in smooth muscle.

Chiesa and colleagues (Chiesa *et al.*, 1997) proposed that erg can control excitability in neuron-neuroblastoma hybrid cells based on their observation that long trains of action potential spikes act to accumulate outward $I_{K(ERG)}$. Interspike interval during these bursts is too short for erg channels to deactivate which inhibits firing and produces spike-frequency adaptation.

Dofetilide block of $I_{K(ERG)}$ in glomus cells of rabbit carotid body causes an increase in spike frequency of afferent nerve fibres, a response that mimics the effect of hypoxia (Overholt *et al.*, 2000). $I_{K(ERG)}$ is modulated by NOS (Taglialatela *et al.*, 1999) and reactive oxygen species (ROS) (Taglialatela *et al.*, 1997), and it has been suggested that the PAS domain of erg may be involved in oxygen sensing (Schwarz and Bauer 2004). In clonal rat anterior pituitary (GH$_3$/B$_6$) lactotrophs, thyrotropin-releasing hormone differentially reduced erg1a, erg1b, erg2 and erg3-mediated currents leading to changes in membrane excitability and increased prolactin secretion (Schledermann *et al.*, 2001; Kirchberger *et al.*, 2006). Erg-mediated currents regulate firing frequency and insulin secretion in human pancreatic β-cells (Rosati *et al.*, 2000). In addition to its proposed role in spike-frequency adaptation, $I_{K(ERG)}$ in neuroblastoma cells can regulate the cell cycle (Arcangeli *et al.*, 1995) and neuritogenesis (Arcangeli *et al.*, 1993).

In the central nervous system the precise role of $I_{K(ERG)}$ is less clear, although recently some possible functions have been proposed. For example, in cultured and *in situ* hippocampal astrocytes $I_{K(ERG)}$ may be involved in K$^+$ homeostasis by facilitating K$^+$ release in a process known as spatial buffering (Emmi *et al.*, 2000). A similar action has been proposed for $I_{K(ERG)}$ across intermediate cells within the cochlear duct, a process that regulates endocochlear potential that is vital for hair cell signal transduction (Nie *et al.*, 2005). All three erg isoforms are expressed in mouse embryonic spinal cord neurons and ventral horn interneurons when cultured *in vitro*, where they exhibit specific spatiotemporal distribution and are thought to control neuronal excitability during the early period of spinal network development (Furlan *et al.*, 2005). $I_{K(ERG)}$ modulates membrane excitability, firing frequency adaptation and neuronal discharge pattern in mouse cerebellar Purkinje neurons, activities that may play a role in motor control (Sacco *et al.*, 2003).

Kv12 (elk) channels

Based on expression pattern alone, it is almost certain that elk channels play an important role in modulating neural activity. Based on characterization of cloned channels it was proposed that elk2 contributes to the timing of burst duration or to spike frequency adaptation during a long burst of neuronal action potentials (Trudeau *et al.*, 1999). Studies of the physiological roles of these channels in native cells would be greatly aided by the discovery of elk-specific channel blockers.

2.1.3.7 Relevance to disease states

In the human heart, loss of function mutations in *KCNH2* delays ventricular repolarization and causes long QT syndrome type 2 (LQTS2), characterized by prolongation of the QT interval measured by a body surface electrocardiogram (ECG). Delayed repolarization of the ventricle significantly increases the risk of cardiac arrhythmia and ventricular fibrillation. The arrhythmia associated with LQTS is termed torsade de pointes (TdP) and is characterized by sinusoidal twisting of the QRS axis around the isoelectric line of the ECG and can lead to syncope and sudden death. To date, over 200 herg1 mutations have been associated with LQTS2 (Gene connection for the heart website: http://www.fsm.it/cardmoc/). The majority of these mutations disrupt folding or trafficking of herg1 channels to the plasma membrane (Delisle *et al.*, 2004). Less commonly, herg1 mutations alter channel gating (e.g., positive shift in the voltage dependence of activation or negative shift in the voltage dependence of inactivation), cause premature protein truncation or disrupt permeation. Many of these mutations can cause a dominant negative effect to reduce current magnitude to a greater extent than that produced by simple haploinsufficiency. For example, a point mutation within the highly conserved potassium selectivity filter (G638S) produces a subunit that co-assembles with wild-type herg1 subunits to produce heteromultimeric channels that are non-conducting.

Recently, Brugada *et al.*, (2004) identified a point mutation in herg1a (N588K) that causes short QT syndrome. This mutation, located in the S5-Pore linker, shifts the voltage dependence of herg1 inactivation to much more positive potentials, causing a gain of function phenotype. Enhanced herg1 current accelerates ventricular action potential repolarization and shortens the QT interval.

Enhanced expression of erg has been detected in several cancer cell lines (Arcangeli *et al.*, 1999; Bianchi *et al.*, 1998) and primary human tumours (Cherubini *et al.*, 2000; Lastraioli *et al.*, 2004). Many cancer cell lines have a relatively depolarized resting membrane potential (around -40 mV), in order to promote cell-cycle progression, at which erg current would be activated. In neuroblastoma cells, the cell cycle phase can alter the expression and activation of herg1 channels (Arcangeli *et al.*, 1995), with differential effects on herg1a and herg1b isoforms (Crociani *et al.*, 2003). Furthermore, block of herg1 by E4031 or other specific blockers, reduces cell proliferation (Crociani *et al.*, 2003). Other functions for herg channels in tumour cells may involve the regulation of cell migration (Lastraioli *et al.*, 2004), adhesion-dependent signalling (Cherubini *et al.*, 2005; Hofmann *et al.*, 2001) and angiogenesis. Cell migration is essential for cancers with invasive phenotypes such as those in the colon (Lastraioli *et al.*, 2004) and is reduced with specific blockers of herg in a manner that is well correlated with the level of herg expression. This regulatory role may involve interactions with β-integrins, which have been demonstrated for the activation of adhesion-dependent signalling processes in tumour cell lines (Hofmann *et al.*, 2001; Cherubini *et al.*, 2002; Cherubini *et al.*, 2005).

2.1.3.8 Concluding remarks

Erg channels have received a tremendous amount of attention from researchers because of their early recognized role in acquired and inherited cardiac arrhythmias. Future studies will likely reveal the important role of eag and elk channels in numerous neural disorders. Important topics for future research include unraveling the specific roles of these channels in the brain and cancer and the potential of these channels as therapeutic targets.

References

Abbott GW, Sesti F, Splawski I *et al.* (1999). MiRP1 forms I_{Kr} potassium channels with HERG and is associated with cardiac arrhythmia. *Cell* **97**, 175–187.

Akbarali HI, Thatte H, He XD *et al.* (1999). Role of HERG-like K⁺ currents in opossum esophageal circular smooth muscle. *Am J Physiol* **277**, C1284–1290.

Anantharam A, Lewis A, Panaghie G *et al.* (2003). RNA interference reveals that endogenous *Xenopus* MinK-related peptides govern mammalian K⁺ channel function in oocyte expression studies. *J Biol Chem* **278**, 11739–11745.

Arcangeli A, Becchetti A, Mannini A *et al.* (1993). Integrin-mediated neurite outgrowth in neuroblastoma cells depends on the activation of potassium channels. *J Cell Biol* **122**, 1131–1143.

Arcangeli A, Bianchi L, Becchetti A *et al.* (1995). A novel inward-rectifying K⁺ current with a cell-cycle dependence governs the resting potential of mammalian neuroblastoma cells. *J Physiol* **489**, 455–471.

Arcangeli A, Rosati B, Crociani O *et al.* (1999). Modulation of HERG current and herg gene expression during retinoic acid treatment of human neuroblastoma cells: potentiating effects of BDNF. *J Neurobiol* **40**, 214–225.

Aydar E and Palmer C (2001). Functional characterization of the C-terminus of the human ether-a-go-go-related gene K⁺ channel (HERG). *J Physiol* **534**, 1–14.

Bannister JP, Chanda B, Bezanilla F *et al.* (2005). Optical detection of rate-determining ion-modulated conformational changes of the ether-a-go-go K⁺ channel voltage sensor. *PNAS* **102**, 18718–18723.

Barhanin J, Lesage F, Guillemare E *et al.* (1996). KvLQT1 and IsK (minK) proteins associate to form the I_{Ks} cardiac potassium channel. *Nature* **384**, 78–80.

Bauer CK, Wulfsen I, Schafer R *et al.* (2003). HERG K⁺ currents in human prolactin-secreting adenoma cells. *Pflugers Arch* **445**, 589–600.

Becchetti A, De Fusco M, Crociani O *et al.* (2002). The functional properties of the human ether-a-go-go-like (HELK2) K⁺ channel. *Eur J Neurosci* **16**, 415–428.

Bianchi L, Shen Z, Dennis AT *et al.* (1999). Cellular dysfunction of LQT5-minK mutants: abnormalities of I_{Ks}, I_{Kr} and trafficking in long QT syndrome. *Hum Mol Genetics* **8**, 1499–1507.

Bianchi L, Wible B, Arcangeli A *et al.* (1998). herg encodes a K⁺ current highly conserved in tumors of different histogenesis: a selective advantage for cancer cells. *Cancer Res* **58**, 815–822.

Bijlenga P, Occhiodoro T, Liu JH *et al.* (1998). An ether-a-go-go K⁺ current, Ih-eag, contributes to the hyperpolarization of human fusion-competent myoblasts. *J Physiol* **512**, 317–323.

Bruggemann A, Pardo LA, Stuhmer W *et al.* (1993). *Ether-a-go-go* encodes a voltage-gated channel permeable to K⁺ and Ca²⁺ and modulated by cAMP. *Nature* **365**, 445–448.

Casis O, Olesen SP and Sanguinetti MC (2006). Mechanism of action of a novel human ether-a-go-go-related gene channel activator. *Mol Pharmacol* **69**, 658–665.

Cavalli A, Poluzzi E, De Ponti F *et al.* (2002). Toward a pharmacophore for drugs inducing the long QT syndrome: insights from a CoMFA study of HERG K$^+$ channel blockers. *J Med Chem* **45**, 3844–3853.

Cayabyab FS and Schlichter LC (2002). Regulation of an ERG K$^+$ current by Src tyrosine kinase. *J Biol Chem* **277**, 13673–13681.

Chen J, Zou A, Splawski I *et al.* (1999). Long QT syndrome-associated mutations in the Per-Arnt-Sim (PAS) domain of HERG potassium channels accelerate channel deactivation. *J Biol Chem* **274**, 10113–10118.

Cherubini A, Hofmann G, Pillozzi S *et al.* (2005). Human ether-a-go-go-related gene 1 channels are physically linked to beta1 integrins and modulate adhesion-dependent signaling. *Mol Biol Cell* **16**, 2972–2983.

Cherubini A, Pillozzi S, Hofmann G *et al.* (2002). HERG K$^+$ channels and beta1 integrins interact through the assembly of a macromolecular complex. *Ann N Y Acad Sci* **973**, 559–561.

Cherubini A, Taddei GL, Crociani O *et al.* (2000). HERG potassium channels are more frequently expressed in human endometrial cancer as compared to non-cancerous endometrium. *Br J Cancer* **83**, 1722–1729.

Chiesa N, Rosati B, Arcangeli A *et al.* (1997). A novel role for HERG K$^+$ channels: spike-frequency adaptation. *J Physiol* **501**, 313–318.

Choi KL, Aldrich RW and Yellen G (1991). Tetraethylammonium blockade distinguishes two inactivation mechanisms in voltage-activated K$^+$ channels. *PNAS* **88**, 5092–5095.

Corona M, Gurrola GB, Merino E *et al.* (2002). A large number of novel Ergtoxin-like genes and ERG K$^+$-channels blocking peptides from scorpions of the genus *Centruroides*. *FEBS Lett* **532**, 121–126.

Crociani O, Guasti L, Balzi M *et al.* (2003). Cell cycle-dependent expression of HERG1 and HERG1B isoforms in tumor cells. *J Biol Chem* **278**, 2947–2955.

Cui J, Melman Y, Palma E *et al.* (2000). Cyclic AMP regulates the HERG K$^+$ channel by dual pathways. *Curr Biol* **10**, 671–674.

Delisle BP, Anson BD, Rajamani S *et al.* (2004). Biology of cardiac arrhythmias: ion channel protein trafficking. *Circ Res* **94**, 1418–1428.

Diochot S, Loret E, Bruhn T *et al.* (2003). APETx1, a new toxin from the sea anemone *Anthopleura elegantissima*, blocks voltage-gated human ether-a-go-go-related gene potassium channels. *Mol Pharmacol* **64**, 59–69.

Ekins S, Crumb WJ, Sarazan RD *et al.* (2002). Three-dimensional quantitative structure–activity relationship for inhibition of human ether-a-go-go-related gene potassium channel. *JPET* **301**, 427–434.

Emmi A, Wenzel HJ, Schwartzkroin PA *et al.* (2000). Do glia have heart? Expression and functional role for ether-a-go-go currents in hippocampal astrocytes. *J Neurosci* **20**, 3915–3925.

Engeland B, Neu A, Ludwig J *et al.* (1998). Cloning and functional expression of rat ether-a-go-go-like K$^+$ channel genes. *J Physiol* **513**, 647–654.

Fan J-S, Jinag M, Dun W *et al.* (1999). Effects of outer mouth mutations on hERG channel function: a comparison with similar mutations in the Shaker channel. *Biophys J* **76**, 3128–3140.

Farid R, Day T, Friesner RA *et al.* (2006). New insights about HERG blockade obtained from protein modeling, potential energy mapping and docking studies. *Bioorg Med Chem* **14**, 3160–31673.

Farrelly AM, Ro S, Callaghan BP *et al.* (2003). Expression and function of KCNH2 (HERG) in the human jejunum. *Am J Physiol* **284**, G883–G895.

Fernandez D, Ghanta A, Kauffman GW *et al.* (2004). Physicochemical features of the hERG channel drug binding site. *J Biol Chem* **279**, 10120–10127.

Ferrer T, Rupp J, Piper DR *et al.* (2006). The S4–S5 linker directly couples voltage sensor movement to the activation gate in the human ether-a-go-go-related gene (hERG) K$^+$ channel. *J Biol Chem* **281**, 12858–12864.

Ficker E, Jarolimek W, Kiehn J *et al.* (1998). Molecular determinants of dofetilide block of HERG K$^+$ channels. *Circ Res* **82**, 386–395.

Ficker E, Kuryshev YA, Dennis AT *et al.* (2004). Mechanisms of arsenic-induced prolongation of cardiac repolarization. *Mol Pharmacol* **66**, 33–44.

Furlan F, Guasti L, Avossa D *et al.* (2005). Interneurons transiently express the ERG K$^+$ channels during development of mouse spinal networks *in vitro*. *Neuroscience* **135**, 1179–1192.

Ganetzky B and Wu CF (1983). Neurogenetic analysis of potassium currents in Drosophila: synergistic effects on neuromuscular transmission in double mutants. *J Neurogenet* **1**, 17–28.

Gessner G and Heinemann SH (2003). Inhibition of hEAG1 and hERG1 potassium channels by clofilium and its tertiary analogue LY97241. *Br J Pharmacol* **138**, 161–171.

Gessner G, Zacharias M, Bechstedt S *et al.* (2004). Molecular determinants for high-affinity block of human EAG potassium channels by antiarrhythmic agents. *Mol Pharmacol* **65**, 1120–1129.

Guasti L, Cilia E, Crociani O *et al.* (2005). Expression pattern of the ether-a-go-go-related (ERG) family proteins in the adult mouse central nervous system: evidence for coassembly of different subunits. *J Comp Neurol* **491**, 157–174.

Gutman GA, Chandy KG, Adelman JP *et al.* (2003). International Union of Pharmacology. XLI. Compendium of voltage–gated ion channels: potassium channels. *Pharmacol Rev* **55**, 583–586.

Hansen RS, Diness TG, Christ T *et al.* (2005). Activation of hERG potassium channels by the diphenylurea NS1643. *Mol Pharmacology* **69**, 266–277.

Haverkamp W, Breithardt G, Camm AJ *et al.* (2000). The potential for QT prolongation and proarrhythmia by non-antiarrhythmic drugs: clinical and regulatory implications. Report on a policy conference of the European Society of Cardiology. *Eur Heart J* **21**, 1216–1231.

Herzberg IM, Trudeau MC and Robertson GA (1998). Transfer of rapid inactivation and sensitivity to the class III antiarrhythmic drug E-4031 from HERG to M-eag channels. *J Physiol* **511**, 3–14.

Hirdes W, Schweizer M, Schuricht KS *et al.* (2005). Fast erg K$^+$ currents in rat embryonic serotonergic neurones. *J Physiol* **564**, 33–49.

Hofmann G, Bernabei PA, Crociani O *et al.* (2001). HERG K$^+$ channels activation during β_1 integrin-mediated adhesion to fibronectin induces an up-regulation of $\alpha v \beta_3$ integrin in the preosteoclastic leukemia cell line FLG 29.1. *J Biol Chem* **276**, 4923–4931.

Jiang Y, Lee A, Chen J *et al.* (2003). X-ray structure of a voltage-dependent K$^+$ channel. *Nature* **423**, 33–41.

Jones EM, Roti Roti EC, Wang J *et al.* (2004). Cardiac I_{Kr} channels minimally comprise hERG 1a and 1b subunits. *J Biol Chem* **279**, 44690–4464.

Ju M and Wray D (2002). Molecular identification and characterisation of the human eag2 potassium channel. *FEBS Lett* **524**, 204–210.

Ju M and Wray D (2006). Molecular regions responsible for differences in activation between heag channels. *Biochem Biophys Res Commun* **342**, 1088–1097.

Kang J, Chen XL, Wang H *et al.* (2005). Discovery of a small molecule activator of the human ether-a-go-go-related gene (HERG) cardiac K$^+$ channel. *Mol Pharmacol* **67**, 827–836.

Kiehn J, Lacerda A, Wible B *et al.* (1996). Molecular physiology and pharmacology of HERG: single-channel currents and block by dofetilide. *Circulation* **94**, 2572–2579.

Kirchberger NM, Wulfsen I, Schwarz JR *et al.* (2006). Effects of TRH on heteromeric rat erg1a/1b K$^+$ channels are dominated by the rerg1b subunit. *J Physiol* **571**, 27–42.

Kiss L and Korn SJ (1998). Modulation of C-type inactivation by K$^+$ at the potassium channel selectivity filter. *Biophys J* **74**, 1840–1849.

Korolkova YV, Bocharov EV, Angelo K *et al.* (2002). New binding site on common molecular scaffold provides HERG channel specificity of scorpion toxin BeKm-1. *J Biol Chem* **277**, 43104–43109.

Korolkova YV, Kozlov SA, Lipkin AV *et al.* (2001). An ERG channel inhibitor from the scorpion *Buthus eupeus*. *J Biol Chem* **276**, 9868–9876.

Korolkova YV, Tseng GN and Grishin EV (2004). Unique interaction of scorpion toxins with the hERG channel. *J Mol Recognit* **17**, 209–217.

Kupershmidt S, Snyders DJ, Raes A *et al.* (1998). A K⁺ channel splice variant common in human heart lacks a C-terminal domain required for expression of rapidly activating delayed rectifier current. *J Biol Chem* **273**, 27231–27235.

Kupershmidt S, Yang IC, Hayashi K *et al.* (2003). The I_{Kr} drug response is modulated by KCR1 in transfected cardiac and noncardiac cell lines. *FASEB J* **17**, 2263–2265.

Kuryshev YA, Ficker E, Wang L *et al.* (2005). Pentamidine-induced long QT syndrome and block of hERG trafficking. *JPET* **312**, 316–323.

Langheinrich U, Vacun G and Wagner T (2003). Zebrafish embryos express an orthologue of HERG and are sensitive toward a range of QT-prolonging drugs inducing severe arrhythmia. *Toxicol Appl Pharmacol* **193**, 370–382.

Lastraioli E, Guasti L, Crociani O *et al.* (2004). *herg1* gene and HERG1 protein are overexpressed in colorectal cancers and regulate cell invasion of tumor cells. *Cancer Res.* **64**, 606–611.

Lees-Miller JP, Kondo C, Wang L *et al.* (1997). Electrophysiological characterization of an alternatively processed ERG K⁺ channel in mouse and human hearts. *Circ Res* **81**, 719–726.

Liu J, Zhang M, Jiang M *et al.* (2003). Negative charges in the transmembrane domains of the HERG K channel are involved in the activation- and deactivation-gating processes. *J Gen Physiol* **121**, 599–614.

London B, Trudeau MC, Newton KP *et al.* (1997). Two isoforms of the mouse *ether-a-go-go*-related gene coassemble to form channels with properties similar to the rapidly activating component of the cardiac delayed rectifier K⁺ current. *Circ Res* **81**, 870–878.

Long SB, Campbell EB and Mackinnon R (2005a). Crystal structure of a mammalian voltage-dependent Shaker family K⁺ channel. *Science* **309**, 897–903.

Long SB, Campbell EB and Mackinnon R (2005b). Voltage sensor of Kv1.2: structural basis of electromechanical coupling. *Science* **309**, 903–908.

Ludwig J, Terlau H, Wunder F *et al.* (1994). Functional expression of a rat homologue of the voltage gated ether a go-go potassium channel reveals differences in selectivity and activation kinetics between the *Drosophila* channel and its mammalian counterpart. *EMBO J* **13**, 4451–4458.

Ludwig J, Weseloh R, Karschin C *et al.* (2000). Cloning and functional expression of rat eag2, a new member of the ether-a-go-go family of potassium channels and comparison of its distribution with that of eag1. *Mol Cell Neurosci* **16**, 59–70.

Mazhari R, Greenstein JL, Winslow RL *et al.* (2001). Molecular interactions between two long-QT syndrome gene products, HERG and KCNE2, rationalized by *in vitro* and *in silico* analysis. *Circ Res* **89**, 33–38.

McDonald TV, Yu Z, Ming Z *et al.* (1997). A minK-HERG complex regulates the cardiac potassium current I_{Kr}. *Nature* **388**, 289–292.

Meves H, Schwarz JR and Wulfsen I (1999). Separation of M-like current and ERG current in NG108-15 cells. *Br J Pharmacol* **127**, 1213–1223.

Meyer R and Heinemann SH (1998). Characterization of an eag-like potassium channel in human neuroblastoma cells. *J Physiol* **508**(Pt 1), 49–56.

Milnes JT, Crociani O, Arcangeli A *et al.* (2003). Blockade of HERG potassium currents by fluvoxamine: incomplete attenuation by S6 mutations at F656 or Y652. *Br J Pharmacol* **139**, 887–898.

Mitcheson JS, Chen J and Sanguinetti MC (2000). Trapping of a methanesulfonanilide by closure of the HERG potassium channel activation gate. *J Gen Physiol* **115**, 229–240.

Mitcheson JS, Chen J, Lin M *et al.* (2000). A structural basis for drug-induced long QT syndrome. *PNAS* **97**, 12329–12333.

Miyake A, Mochizuki S, Yokoi H *et al.* (1999). New ether-a-go-go K⁺ channel family members localized in human telencephalon. *J Biol Chem* **274**, 25018–25025.

Morais Cabral JH, Lee A, Cohen SL *et al.* (1998). Crystal structure and functional analysis of the HERG potassium channel N terminus: a eukaryotic PAS domain. *Cell* **95**, 649–655.

Mullins FM, Stepanovic SZ, Desai RR *et al.* (2002). Extracellular sodium interacts with the HERG channel at an outer pore site. *J Gen Physiol* **120**, 517–537.

Nie L, Gratton MA, Mu KJ *et al.* (2005). Expression and functional phenotype of mouse ERG K⁺ channels in the inner ear: potential role in K⁺ regulation in the inner ear. *J Neurosci* **25**, 8671–8679.

Occhiodoro T, Bernheim L, Liu JH *et al.* (1998). Cloning of a human ether-a-go-go potassium channel expressed in myoblasts at the onset of fusion. *FEBS Lett* **434**, 177–182.

Ohya S, Asakura K, Muraki K *et al.* (2002). Molecular and functional characterization of ERG, KCNQ and KCNE subtypes in rat stomach smooth muscle. *Am J Physiol* **282**, G277–G287.

Osterberg F and Aqvist J (2005). Exploring blocker binding to a homology model of the open hERG K⁺ channel using docking and molecular dynamics methods. *FEBS Lett* **579**, 2939–2944.

Overholt JL, Ficker E, Yang T *et al.* (2000). HERG-Like potassium current regulates the resting membrane potential in glomus cells of the rabbit carotid body. *J Neurophysiol* **83**, 1150–1157.

Papa M, Boscia F, Canitano A *et al.* (2003). Expression pattern of the ether-a-gogo-related (ERG) K⁺ channel-encoding genes *ERG1*, *ERG2* and *ERG3* in the adult rat central nervous system. *J Comp Neurol* **466**, 119–135.

Pardo LA, del Camino D, Sanchez A *et al.* (1999). Oncogenic potential of EAG K⁺ channels. *EMBO J* **18**, 5540–5547.

Pardo-Lopez L, Zhang M, Liu J *et al.* (2002). Mapping the binding site of a human ether-a-go-go-related gene-specific peptide toxin (ErgTx) to the channel's outer vestibule. *J Biol Chem* **277**, 16403–16411.

Patt S, Preussat K, Beetz C *et al.* (2004). Expression of ether a go-go potassium channels in human gliomas. *Neurosci Lett* **368**, 249–253.

Paulussen A, Raes A, Matthijs G *et al.* (2002). A novel mutation (T65P) in the PAS domain of the human potassium channel HERG results in the long QT syndrome by trafficking deficiency. *J Biol Chem* **277**, 48610–48616.

Pearlstein RA, Vaz RJ, Kang J *et al.* (2003). Characterization of HERG potassium channel inhibition using CoMSiA 3D QSAR and homology modeling approaches. *Bioorg Med Chem Lett* **13**, 1829–1835.

Perry M, de Groot MJ, Helliwell R *et al.* (2004). Structural determinants of HERG channel block by clofilium and ibutilide. *Mol Pharmacol* **66**, 240–249.

Perry MD, Stansfeld PJ, Leaney J *et al.* (2006). Drug binding interactions in the inner cavity of hERG: molecular insights from structure–activity relationships of clofilium and ibutilide analogues. *Mol Pharmacol* **69**, 509–519.

Phartiyal P, Sale H, Jones EM *et al.* (2008). Endoplasmic reticulum retention and rescue by heteromeric assembly regulate human ERG 1a/1b surface channel composition. *J Biol Chem* **283**, 3702–3707.

Piper DR, Hinz WA, Tallurri CK *et al.* (2005). Regional specificity of human ether-a-go-go-related gene channel activation and inactivation gating. *J Biol Chem* **280**, 7206–7217.

Piper DR, Varghese A, Sanguinetti MC *et al.* (2003). Gating currents associated with intramembrane charge displacement in HERG potassium channels. *Proc Natl Acad Sci USA* **100**, 10534–10539.

Pond AL, Scheve BK, Benedict AT *et al.* (2000). Expression of distinct ERG proteins in rat, mouse and human heart. Relation to functional IKr channels. *J Biol Chem* **275**, 5997–6006.

Pourrier M, Zicha S, Ehrlich J, *et al.* (2003). Canine ventricular KCNE2 expression resides predominantly in Purkinje fibers. *Circ Res* **93**, 189–191.

Rajamani R, Tounge BA, Li J *et al.* (2005). A two-state homology model of the hERG K$^+$ channel: application to ligand binding. *Bioorg Med Chem Lett* **15**, 1737–1741.

Rampe D, Roy ML, Dennis A *et al.* (1997). A mechanism for the proarrhythmic effects of cisapride (Propulsid): high affinity blockade of the human cardiac potassium channel HERG. *FEBS Lett* **417**, 28–32.

Rasmusson RL, Morales MJ, Wang S *et al.* (1998). Inactivation of voltage-gated cardiac K$^+$ channels. *Circ Res.* **82**, 739–750.

Recanatini M, Poluzzi E, Masetti M *et al.* (2005). QT prolongation through hERG K$^+$ channel blockade: current knowledge and strategies for the early prediction during drug development. *Med Res Rev* **25**, 133–166.

Restano-Cassulini R, Korolkova YV, Diochot S *et al.* (2006). Species diversity and peptide toxins blocking selectivity of ERG subfamily K$^+$ channels in CNS. *Mol Pharmacol* **69**(**5**),1673–83.

Rosati B, Marchetti P, Crociani O *et al.* (2000). Glucose- and arginine-induced insulin secretion by human pancreatic beta-cells: the role of HERG K$^+$ channels in firing and release. *FASEB J* **14**, 2601–2610.

Sacco T, Bruno A, Wanke E *et al.* (2003). Functional roles of an ERG current isolated in cerebellar Purkinje neurons. *J Neurophysiol* **90**, 1817–1828.

Saganich MJ, Machado E and Rudy B (2001). Differential expression of genes encoding subthreshold-operating voltage-gated K$^+$ channels in brain. *J Neurosci* **21**, 4609–4624.

Sanguinetti MC and Mitcheson JS (2005). Predicting drug-hERG channel interactions that cause acquired long QT syndrome. *Trends Pharmacol Sci* **26**, 119–124.

Sanguinetti MC and Tristani-Firouzi M (2006). hERG potassium channels and cardiac arrhythmia. *Nature* **440**, 463–469.

Sanguinetti MC, Curran ME, Zou A *et al.* (1996). Coassembly of KvLQT1 and minK (IsK) proteins to form cardiac I_{Ks} potassium channel. *Nature* **384**, 80–83.

Sanguinetti MC, Jiang C, Curran ME *et al.* (1995). A mechanistic link between an inherited and an acquired cardiac arrhythmia: *HERG* encodes the I_{Kr} potassium channel. *Cell* **81**, 299–307.

Schledermann W, Wulfsen I, Schwarz JR *et al.* (2001). Modulation of rat erg1, erg2, erg3 and HERG K$^+$ currents by thyrotropin-releasing hormone in anterior pituitary cells via the native signal cascade. *J Physiol* **532**, 143–163.

Schledermann W, Wulfsen I, Schwarz JR *et al.* (2001). Modulation of rat erg1, erg2, erg3 and HERG K$^+$ currents by thyrotropin-releasing hormone in anterior pituitary cells via the native signal cascade. *J Physiol* **532**, 143–163.

Schonherr R and Heinemann SH (1996). Molecular determinants for activation and inactivation of HERG, a human inward rectifier potassium channel. *J Physiol* **493.3**, 635–642.

Schonherr R, Gessner G, Lober K *et al.* (2002). Functional distinction of human EAG1 and EAG2 potassium channels. *FEBS Lett* **514**, 204–208.

Schwarz JR and Bauer CK (2004). Functions of erg K$^+$ channels in excitable cells. *J Cell Mol Med* **8**, 22–30.

Shi W, Wang HS, Pan Z *et al.* (1998). Cloning of a mammalian elk potassium channel gene and EAG mRNA distribution in rat sympathetic ganglia. *J Physiol* **511**, 675–682.

Shi W, Wymore RS, Wang H-S *et al.* (1997). Identification of two nervous system-specific members of the *erg* potassium channel gene family. *J Neurosci* **17**, 9423–9432.

Shoeb F, Malykhina AP and Akbarali HI (2003). Cloning and functional characterization of the smooth muscle ether-a-go-go-related gene K$^+$ channel. Potential role of a conserved amino acid substitution in the S4 region. *J Biol Chem* **278**, 2503–2514.

Silverman WR, Bannister JP and Papazian DM (2004). Binding site in Eag voltage sensor accommodates a variety of ions and is accessible in closed channel. *Biophys J* **87**, 3110–3121.

Silverman WR, Roux B and Papazian DM (2003). Structural basis of two-stage voltage-dependent activation in K$^+$ channels. *PNAS* **100**, 2935–2940.

Silverman WR, Tang CY, Mock AF *et al.* (2000). Mg^{2+} modulates voltage-dependent activation in ether-a-go-go potassium channels by binding between transmembrane segments S2 and S3. *J Gen Physiol* **116**, 663–678.

Smith GA, Tsui HW, Newell EW *et al.* (2002). Functional up-regulation of HERG K$^+$ channels in neoplastic hematopoietic cells. *J Biol Chem* **277**, 18528–18534.

Smith PL and Yellen G (2002). Fast and slow voltage sensor movements in HERG potassium channels. *J Gen Physiol* **119**, 275–293.

Smith PL, Baukrowitz T and Yellen G (1996). The inward rectification mechanism of the HERG cardiac potassium channel. *Nature* **379**, 833–836.

Snyders DJ and Chaudhary A (1996). High affinity open channel block by dofetilide of *HERG* expressed in a human cell line. *Mol Pharmacol* **49**, 949–955.

Spector PS, Curran ME, Keating MT *et al.* (1996). Class III antiarrhythmic drugs block HERG, a human cardiac delayed rectifier K$^+$ channel; open channel block by methanesulfonanilides. *Circ Res* **78**, 499–503.

Spector PS, Curran ME, Zou A *et al.* (1996). Fast inactivation causes rectification of the I_{Kr} channel. *J Gen Physiol* **107**, 611–619.

Sturm P, Wimmers S, Schwarz JR *et al.* (2005). Extracellular potassium effects are conserved within the rat erg K$^+$ channel family. *J Physiol* **564**, 329–345.

Subbiah RN, Clarke CE, Smith DJ *et al.* (2004). Molecular basis of slow activation of the human ether-a-go-go related gene potassium channel. *J Physiol* **558**, 417–431.

Subbiah RN, Kondo M, Campbell TJ *et al.* (2005). Tryptophan scanning mutagenesis of the HERG K$^+$ channel: the S4 domain is loosely packed and likely to be lipid exposed. *J Physiol* **569**, 367–379.

Taglialatela M, Castaldo P, Iossa S *et al.* (1997). Regulation of the human ether-a-gogo related gene (*HERG*) K$^+$ channels by reactive oxygen species. *PNAS* **94**, 11698–11703.

Taglialatela M, Pannaccione A, Iossa S *et al.* (1999). Modulation of the K$^+$ channels encoded by the human ether-a-gogo-related gene-1 (*hERG1*) by nitric oxide. *Mol Pharmacol* **56**, 1298–1308.

Tang CY, Bezanilla F and Papazian DM (2000). Extracellular Mg^{2+} modulates slow gating transitions and the opening of *Drosophila* ether-a-Go-Go potassium channels. *J Gen Physiol* **115**, 319–338.

Torres AM, Bansal PS, Sunde M *et al.* (2003). Structure of the HERG K$^+$ channel S5P extracellular linker: role of an amphipathic alpha-helix in C-type inactivation. *J Biol Chem* **278**, 42136–42148.

Tristani-Firouzi M, Chen J *et al.* (2002). Interactions between S4-S5 linker and S6 transmembrane domain modulate gating of HERG K$^+$ channels. *J Biol Chem* **277**, 18994–9000.

Trudeau MC, Titus SA, Branchaw JL *et al.* (1999). Functional analysis of a mouse brain Elk-type K$^+$ channel. *J Neurosci* **19**, 2906–2918.

Wang J, Trudeau MC, Zappia AM *et al.* (1998). Regulation of deactivation by an amino terminal domain in human ether-a-go-go-related gene potassium channels. *J Gen Physiol* **112**, 637–647.

Wang S, Liu S, Morales MJ *et al.* (1997a). A quantitative analysis of the activation and inactivation kinetics of HERG expressed in *Xenopus* oocytes. *J Physiol* **502**, 45–60.

Wang S, Morales MJ, Liu S *et al.* (1996). Time, voltage and ionic concentration dependence of rectification of h-*erg* expressed in *Xenopus* oocytes. *FEBS Lett* **389**, 167–173.

Wang S, Morales MJ, Liu S *et al.* (1997b). Modulation of HERG affinity for E-4031 by [K$^+$]$_o$ and C-type inactivation. *FEBS Lett* **417**, 43–47.

Warmke J, Drysdale R and Ganetzky B (1991). A distinct potassium channel polypeptide encoded by the *Drosophila eag* locus. *Science* **252**, 1560–1564.

Warmke JW and Ganetzky B (1994). A family of potassium channel genes related to *eag* in *Drosophila* and mammals. *PNAS* **91**, 3438–3442.

Weerapura M, Nattel S, Chartier D *et al.* (2002). A comparison of currents carried by HERG, with and without coexpression of MiRP1 and the native rapid delayed rectifier current. Is MiRP1 the missing link? *J Physiol* **540**, 15–27.

Wilson GF, Wang Z, Chouinard SW *et al.* (1998a). Interaction of the K channel b subunit, *hyperkinetic*, with eag family members. *J Biol Chem* **273**, 6389–6394.

Wimmers S, Bauer CK and Schwarz JR (2002). Biophysical properties of heteromultimeric erg K⁺ channels. *Pflugers Arch* **445**, 423–430.

Wimmers S, Wulfsen I, Bauer CK *et al.* (2001). Erg1, erg2 and erg3 K channel subunits are able to form heteromultimers. *Pflugers Arch.* **441**, 450–455.

Wu CF, Ganetzky B, Haugland FN *et al.* (1983). Potassium currents in *Drosophila*: different components affected by mutations of two genes. *Science* **220**, 1076–1078.

Wymore RS, Gintant GA, Wymore RT *et al.* (1997). Tissue and species distribution of mRNA for the I_{Kr}-like K⁺ channel, *erg*. *Circ Res* **80**, 261–268.

Zehelein J, Zhang W, Koenen M *et al.* (2001). Molecular cloning and expression of cERG, the ether a go-go-related gene from canine myocardium. *Pflugers Arch* **442**, 188–191.

Zhang M, Korolkova YV, Liu J *et al.* (2003). BeKm-1 is a HERG-specific toxin that shares the structure with ChTx but the mechanism of action with ErgTx1. *Biophys J* **84**, 3022–3036.

Zhou J, Augelli-Szafran CE, Bradley JA *et al.* (2005). Novel potent human ether-a-go-go-related gene (hERG) potassium channel enhancers and their *in vitro* antiarrhythmic activity. *Mol Pharmacol* **68**, 876–884.

Zhou W, Cayabyab FS, Pennefather PS *et al.* (1998a). HERG-like K⁺ channels in microglia. *J Gen Physiol* **111**, 781–794.

Zhou Z, Gong Q, Ye B *et al.* (1998b). Properties of HERG channels stably expressed in HEK 293 cells studied at physiological temperature. *Biophys J* **74**, 230–241.

Zhou Z, Vorperian VR, Gong Q *et al.* (1999). Block of HERG potassium channels by the antihistamine astemizole and its metabolites desmethylastemizole and norastemizole. *J Cardiovasc Electrophysiol* **10**, 836–843.

Zou A, Lin Z, Humble M *et al.* (2003). Distribution and functional properties of human KCNH8 (Elk1) potassium channels. *Am J Physiol Cell Physiol* **285**, C1356–C1366.

2.1.4 K_{2P} channels

Dawon Kang and Donghee Kim

2.1.4.1 Introduction

The discovery of K_{2P} channels owes its beginning to the recognition that K⁺ channels possess a well-conserved pore (P) domain that contains the K⁺ channel selectivity sequence, T(I/V)GYG. Searching the genomic database for sequences homologous to the P domain led to the identification and cloning of a yeast K⁺ channel that contained two P domains in tandem (Ketchum *et al.*, 1995). Following this initial discovery, various molecular biological methods have been successfully used to clone K_{2P} channels in fly (Goldstein *et al.*, 1996), nematode (Kunkel *et al.*, 2000; de la Cruz *et al.*, 2003), mammals, and plant (Czempinski *et al.*, 1997; Becker *et al.*, 2004). The potential function of K_{2P} channels was recognized from the observation that a number of them were constitutively active at physiological membrane potentials, and showed behaviour as 'open rectifier' K⁺ conductances when expressed in *Xenopus* oocytes and cloned mammalian cells. Therefore, K_{2P} channels fulfill the role of background (or leak) K⁺ channels whose existence in excitable cells has been known since

the 1960s. Could the K_{2P} channels be the molecular correlates of the leak K⁺ channels found in the native system? This required careful and detailed biophysical and pharmacological analyses of both native leak K⁺ channels and cloned K_{2P} channels expressed in heterologous systems. Such studies have provided good evidence that cloned K_{2P} channels indeed encode leak K⁺ channels in different types of cells including neurons, cardiac myocytes, and smooth muscle cells. Molecular identification of leak K⁺ channels in different cell types continues to be a topic of great interest in many laboratories, and such studies should eventually show which specific K_{2P} channels represent the leak K⁺ channels expressed in different cell types.

If the only property of the K_{2P} channels were its leakiness to K⁺, the interest in studying them would have waned quickly. Fortunately for K_{2P} channels, they have been found to be modulated by a variety of biologically relevant stimuli such as receptor ligands, temperature, lipids, pressure, oxygen tension and volatile anaesthetic agents, indicating that K_{2P} channels are active participants in the regulation of cell excitability and function. Recent studies show that neurotransmitters can activate or inhibit K_{2P} channels via different G proteins. This would explain the earlier observation that certain neurotransmitters enhance neuronal activity by reducing the leak K⁺ current neurotransmitters that stimulate K_{2P} channels would then be expected to depress neuronal activity, but such examples are rare. It is also becoming apparent that K_{2P} channels contribute to diverse cellular functions because of their sensitivity to various chemical and physical stimuli. It is also interesting that many tissues express more than one K_{2P} channel, suggesting that K_{2P} channels provide a tight regulation of cell membrane potential and excitability. The role of K_{2P} channels in various physiological and pathophysiological processes, and uncovering their underlying molecular mechanisms are the focus of many current studies.

2.1.4.2 Molecular characteristics

At present, 15 K_{2P} channel members exist in the mammalian system. These K_{2P} channels can be divided into TWIK, TASK, TREK, THIK, TALK and TRESK subfamilies, as shown in Figure 2.1.4.1. Each K_{2P} channel subunit has two P domains and four transmembrane segments with a short amino terminus and a long carboxyl terminus, with the exception of TRESK that has both short amino and carboxyl termini. As all functional K⁺ channels contain four P domains, K_{2P} channels are probably dimers, and the general structure of the channel is likely to be similar to those of bacterial K⁺ channels whose three dimensional structures have been determined by X-ray crystallography (Doyle *et al.*, 1998; Kuo *et al.*, 2003). The general architecture of the permeation path, the pore α-helix and the selectivity filter of the K_{2P} channels are probably very similar to those of bacterial K⁺ channels. As the second and third transmembrane segments of K_{2P} channels are joined at the cytoplasmic side, the structure at the intracellular entrance will differ from that of the prototypical K⁺ channel. Unlike typical K⁺ channels that have four cytoplasmic termini surrounding the cytoplasmic ion permeation pathway, K_{2P} channels have only two cytoplasmic termini. Therefore, K_{2P} channels may not have a cytoplasmic pore region similar to that of inward rectifier K⁺ channels (Kuo *et al.* 2003). Glycine residues that are believed to help in the bending of the inner helix during channel opening and closing in many inward rectifier K⁺ channels are also present in K_{2P}

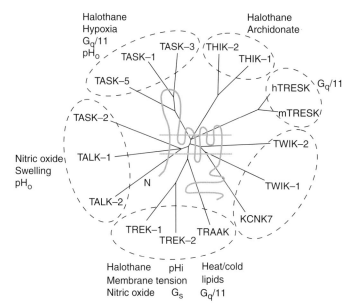

Fig. 2.1.4.1 Membrane topology and family tree of K$_{2P}$ channels. The K$_{2P}$ channel family can be divided into six subgroups based on amino acid identities. Some of the unique properties of each subgroup are listed.

channels, suggesting that the gating mechanisms may be similar. Secondary structure analyses for TASKs and TREKs suggest that the long extracellular loop between the M1 and M2 transmembrane segment consists of two α-helices joined by a short coil, and therefore is likely to be a structured region. This may explain the crucial role of this region in the modulation of channel gating by extracellular ligands as well as stable dimer formation. Plant K$_{2P}$ channels (TPK1-4) possess in their carboxy termini Ca^{2+} binding EF-hand motifs that are not present in functional mammalian K$_{2P}$ channels (Czempinski *et al.*, 1997; Becker *et al.*, 2004). The direct regulation of K$_{2P}$ channels by Ca^{2+} is therefore unique to the TPK family.

Most K$_{2P}$ channels transiently expressed in mammalian cell lines or *Xenopus* oocytes generate large K$^+$ currents. Those that do not produce measurable currents, such as TASK-5 and THIK-2, may instead form proteins that reside in intracellular organelles, although this has not been tested rigorously. At present, it is not clear why they are not functionally expressed in the plasma membrane. As more than one type of K$_{2P}$ channel subunits are expressed in a cell, homodimeric as well as heterodimeric K$_{2P}$ channels may form. For example, TASK-1 and TASK-3 form homomeric and heteromeric channel complex in both heterologous and native systems (Czirjak and Enyedi, 2002; Kang *et al.*, 2004a; Berg *et al.*, 2004; Aller *et al.*, 2005). Heteromeric formation among other members of the K$_{2P}$ family has not yet been reported. Since the heterodimer may possess kinetic properties similar to the homodimer, and there are no selective pharmacological blockers, it is difficult to identify and distinguish the two forms of channel. Immunological, electrophysiological and spectral (such as FRET) analyses would be necessary to show the heteromeric formation of K$_{2P}$ channels. Analysis of genomic sequences shows that most K$_{2P}$ channel genes consist of several exons and introns. Alternatively spliced variants have been identified for several members of the K$_{2P}$ channel family, and there is evidence to suggest that the spliced variants may be expressed in different tissues (Gu *et al.*, 2002). The physiological roles of each functional splice variant have yet to be studied.

One common characteristic of K$_{2P}$ channels is that their low sensitivity to the well-known K$^+$ channel blockers such as tetraethylammonium, 4-aminopyridine, apamin, Cs$^+$ and Ba^{2+}. The lack of specific pharmacological blockers of K$_{2P}$ channels makes it necessary to demonstrate the functional presence of one or more K$_{2P}$ channels in a given cell type by single channel recording. Whole-cell recordings are useful but can lead to misleading conclusions, if the correlation is based on pharmacological criteria alone. The distinct single channel biophysical properties of K$_{2P}$ channels allow a clearer and perhaps an unequivocal determination of whether a native K$^+$ channel is a K$_{2P}$ channel. Such direct recordings of single channels show that members of the K$_{2P}$ channel family are expressed in many types of cells and provide a significant, if not major, contribution to the background K$^+$ current. One good example is the large standing outward background K$^+$ current in cerebellar granules cells (Millar *et al.*, 2000). Single channel recording from these cells shows that the cerebellar background K$^+$ current consists of four types of K$^+$ channels, three of which belong to the K$_{2P}$ channel family (Han *et al.*, 2002). The specific roles of each K$_{2P}$ channel expressed in the cerebellum as well as in other cell types have yet to be determined.

2.1.4.3 Biophysical properties

Studies during the past few years show that K$_{2P}$ channels provide a major part of the background K$^+$ current in a variety of cell types, and contribute to establishing the resting membrane potential. Single channel recordings of functional K$_{2P}$ channels show that each K$_{2P}$ channel possesses unique kinetic properties that may help to identify them in native tissues, as well as distinguish them from other K$^+$ channels. The single channel conductance of K$_{2P}$ channels ranges from ~15 pS (TASK, TRESK) to ~220-pS (TREK), despite the presumed similar architecture of the channel pores. Which molecular structures control the pore architecture and conductance of K$^+$ channels is a topic of great biophysical interest.

TWIK subfamily: TWIK-1 (K$_{2P}$1.1) and TWIK-2 (K$_{2P}$6.1)

TWIK-1 was originally described as a functional background K$^+$ channel (Lesage *et al.*, 1996a). However, a later study was unable to detect any current in heterologous systems (Pountney *et al.*, 1999). Similarly, controversial results were obtained with human TWIK-2 (Chavez *et al.*, 1999; Pountney *et al.*, 1999; Patel *et al.*, 2000). However, the finding that rat TWIK-2 generates ~15 times larger current than human TWIK-2 could explain these discrepant observations. The current recorded with human TWIK-2 was ~200 pA (Patel *et al.*, 2000), just barely larger than the background level observed in non-transfected cells. Nevertheless, the large current recorded with rat TWIK-2 suggests that there must be critical residues present in the rat but not in humans that control functional expression.

Human TWIK-1 may normally be silent because it is sumoylated, as desumoylation activated the cloned channel (Rajan *et al.*, 2005). However, a later study failed to show sumolyation of TWIK-1 and regulation of TWIK-1 by sumoylation (Feliciangeli *et al.*, 2007). Whether sumolyation or other biochemical modifications of TWIK-1 occurs in the native system to regulate the function of TWIK-1 remains unknown. Despite the lack of understanding of TWIK-1 function, its discovery was crucial for identifying additional members of the K$_{2P}$ family. Also, the biochemical evidence for the homodimeric formation of a K_{2P} channel was first demonstrated with TWIK-1 (Lesage *et al.*, 1996b). These studies, together with the mutational and structural evidence that a functional K^+ channel has four P regions, have reinforced the idea that all functional K_{2P} channels are dimers. Hopefully, the controversy surrounding TWIK-1 and TWIK-2 will soon be resolved so that their physiological roles can begin to be studied.

TASK subfamily: TASK-1 (K_{2P}3.1), TASK-3 (K_{2P}9.1) and TASK-5 (K_{2P}15.1)

TASK-1 and TASK-3, but not TASK-5, form functional K^+ channels, and show properties of leak or background K^+ current. In addition, TASK-1 and TASK-3 are inhibited by extracellular acid, hypoxia, and G protein-coupled receptor agonists, and therefore serve as important targets for various stimuli that regulate excitability. Both TASK-1 and TASK-3 are strongly inhibited by extracellular acid such that only about 10 per cent of the current remains following a reduction of pH from 7.3 to 6.3. TASK-1 has a pKa ~7.3 whereas TASK-3 has a pKa ~6.8 (Berg *et al.*, 2004; Kang *et al.*, 2004a). The high sensitivity to proton concentration suggests that TASKs may act as sensors of extracellular pH. Mutagenesis studies show that a histidine residue located next to the K^+ channel selectivity sequence at the external side of the channel pore (GYGH) is the major pH sensor for TASK-3 (Kim *et al.*, 2000; Rajan *et al.*, 2000). For TASK-1, additional histidine and lysine residues located at the extracellular loops may be involved (Morton *et al.*, 2003). Mutations at the outer mouth of TASK-1 that cause loss of ion selectivity also altered the pH sensitivity, suggesting that gating at the selectivity filter is affected (Yuill *et al.*, 2007). TASK-1 and TASK-3 show inwardly rectifying single channel current voltage relationships with single channel conductances of ~16 pS and 32 pS at -40 mV. The channels open and close rapidly and have a mean open time of ~1-2 ms (Kang *et al.*, 2004).

TASK-5 does not form a functional channel at the plasma membrane despite its close homology to TASK-1 and TASK-3 (Ashmole *et al.*, 2001; Karschin *et al.*, 2001; Kim and Gnatenco, 2001). It appears that TASK-5 is expressed in intracellular organelles, based on the expression of GFP-tagged TASK-5. TASK-5 is particularly highly expressed in the human adrenal gland. Whether TASK-5 has a function other than as an ion channel in this organ remains to be seen. The lack of plasma expression of TASK-5 is unlikely to be due to a failure to interact with 14-3-3 protein that is known to promote trafficking of TASK-1 and TASK-3 to the plasma membrane (Rajan *et al.*, 2002). Like TASK-1 and TASK-3, TASK-5 possesses a conserved C-terminal phosphorylation site that also interacts with 14-3-3 proteins. An auxiliary subunit analogous to a β subunit for voltage-gated K^+ channels has not yet been identified for TASKs and may not exist. However, intracellular proteins such as 14-3-3 and p11 interact with TASK-1 and TASK-3 to promote, or inhibit, trafficking of the channels to the surface membrane (Girard *et al.*, 2002; O'Kelly *et al.*, 2002; Rajan *et al.*, 2002; Renigunta *et al.*, 2006).

TREK subfamily: TREK-1 (K_{2P}2.1), TREK-2 (K_{2P}10.1) and TRAAK (K_{2P}4.1)

Native K^+ channels with biophysical and pharmacological properties similar to those of large-conductance TREKs have been described prior to the cloning of their genes. For example, the K^+ channels that have been generally referred to as arachidonic-activated, mechanosensitive K^+ channel (K_{AA} channel) in cardiac myocytes and smooth muscle cells is TREK-1 (Kim *et al.*, 1989; Ordway *et al.*, 1989; Kim 1992; Terrenoire *et al.*, 2001; Xian Tao *et al.*, 2006). In neurons, at least three distinct fatty acid-activated, mechanosensitive K^+ channels have been described and it is now clear that they are functional correlates of the members of the TREK/TRAAK subfamily (Kim *et al.*, 1995; Patel *et al.*, 2001; Kim 2003). One unique feature of TREKs are their high current fluctuations during channel opening, perhaps due to the presence of several subconductance states.

The three most well characterized properties of TREKs are their activation by unsaturated free fatty acids, intracellular acid and pressure applied to the membrane. TREKs are also directly activated by lysophosphatidic acids (Maingret *et al.*, 2000b; Chemin *et al.*, 2005). Structure–function studies show that activation of TREKs by free fatty acids and acidic conditions involves the proximal region of the carboxy terminus for both TREK-1 and TREK-2, but not for TRAAK (Patel *et al.*, 1998; Kim *et al.*, 2001a, b). For example, a mutant in which the TREK-2 carboxy terminus is replaced with that of TASK-3 is insensitive to free fatty acids without loss of its mechanosensitivity. This result is particularly interesting because it shows that activation by free fatty acids and pressure involves distinct mechanisms. It also rules out the idea that lipids such as free fatty acids activate TREKs via membrane deformation, or that membrane deformation activates TREKs via release of free fatty acids. Anionic amphipaths such as trinitrophenol that inserts into the outer leaflet of the lipid bilayer and causes bending of the membrane can elicit activation of TREKs (Patel *et al.*, 2001). With regard to mechanosensitivity, TREKs are mildly sensitive to negative pressure applied to the membrane patch in the cell-attached state, but become much more sensitive in the inside-out state. It is possible that the cytoskeletal proteins are disrupted in the inside-out state, allowing transduction of mechanical energy of the lipid bilayer to the channel gate. However, the role of any cytoskeletal proteins in this process has not been convincingly demonstrated.

Within the proximal carboxyl terminus is a glutamate residue (E306) that may be critical for controlling the channel gating by different activators (Honore *et al.*, 2002). The E306A mutation is fully active at rest whereas the wild type TREKs are nearly closed. Thus, the E306A mutation locks the channel in the open conformation and is not activated further by stretch, acid or arachidonic acid. It has been suggested that protonation of E306 strengthens the interaction between the channel and membrane phospholipids that allows gating by stretch and acid. Exactly how such interactions modulate the sensitivity of TREKs to various stimuli needs to be explored further at the structural level. Interestingly, TREK-1 activity shows voltage-dependence such that there is a preferential opening at depolarized potentials compared to that at hyperpolarized potentials. (Bockenhauer *et al.*, 2001; Maingret *et al.*, 2002). Such voltage-dependence of TREK-1 should play an important role in the modulation of repolarization and firing rate. Phosphorylation of TREK-1 has been reported as the factor that switches the channel from a leak to a voltage-dependent channel, but this is still controversial (Bockenhauer *et al.*, 2001; Maingret *et al.*, 2002).

An interesting observation with TREK-1 and TREK-2 is that they produce channels with different unitary conductance levels when expressed in mammalian cell lines (Kang *et al.*, 2007). The molecular basis for this is unknown, but may be due to switching between different gating modes (Xian Tao *et al.*, 2006), switching between different subconductance levels, different degrees of channel clustering, or an alternative translation initiation of mRNA producing isoforms with different conductance levels. The last mechanism has recently been confirmed, and the studies that the ribosomal translation machinery is able to skip to the second or the third translation-initiation site of TREKs to generate shorter-length isoforms with different unitary conductance levels (Kozak, 2005; Thomas *et al.*, 2008; Simkin *et al.*, 2008).

TALK subfamily: TALK-1 (K_{2P}16.1), TALK-2 (K_{2P}17.1) and TASK-2 (K_{2P}5.1)

TALK-1 and TALK-2 have low single channel conductance (~30 pS) and exhibit rapid kinetics compared to that of TASK-2 (~70 pS) that opens in long bursts. Like TASKs and TREKs, TALKs are active at rest and expected to modulate the resting membrane potential, and perhaps respond to local changes in pH to regulate secretion. In this respect, the three members of the TALK family are sensitive to a wide range of pH_o (6.3–10.3), and have pH-activity curves with different pKa values that are above pH 8 (Girard *et al.*, 2001; Kang and Kim, 2004). This compares with TASK-1 and TASK-3 that are sensitive to pH_o in the 6.0–8.0 range, with pKa at ~6.6 and ~7.3, respectively. TALK-2 is particularly unique, as it has a very low activity at pH 7.3 and is activated strongly by alkali. A lysine residue in TALK-2 near the pore may serve a similar role in pH gating (Niemeyer *et al.*, 2007).

Sensing of pH by TASK-2 was determined to be due to the combined properties of several charged residues at the M1-P1 loop that is usually large in all K_{2P} channels (Morton *et al.*, 2005). A more recent study shows that an arginine residue (R224) near the pore may be the sensor of the pH and affect channel gating via an electrostatic effect (Niemeyer *et al.*, 2007). This contrasts with the single histidine near the K^+ selectivity sequence that confers pH sensitivity to TASK-3. TASK-2 is sensitive not only to pH but also to cell swelling. The molecular basis for volume sensitivity of TASK-2 is not known, as TASK-2 is not a mechanosensitive K^+ channel. Nevertheless, cell swelling caused by hypotonic solution increases, and cell shrinkage caused by hypertonic solution decreases TASK-2 (Niemeyer *et al.*, 2001).

THIK subfamily: THIK-1 (K_{2P}13.1) and THIK-2 (K_{2P}12.1)

THIK-1, but not THIK-2, produces a functional background K^+ current in *Xenopus* oocytes and HEK293 cells. Unlike other K_{2P} channels, THIK-1 exhibits noisy current fluctuations in cell-attached patches, and distinct single channel openings of THIK-1 are not observed in transfected cells. Thus, it is most likely that the single channel conductance of THIK-1 is very small.

TRESK

TRESK is a structurally unique member of the K_{2P} channel family because of its extremely short carboxyl terminus (~13 residues) and a relatively long cytoplasmic loop connecting the second and the third transmembrane segments. Therefore, one might predict that cytoplasmic domains that modulate the gating of TRESK are located within the long loop. Using Xenopus oocytes as an expression system, Czirjak *et al.* (2004) found that TRESK was activated by Ca^{2+}- and calmodulin-dependent protein phosphatase (calcineurin), acting on serine 276 located in the cytoplasmic loop.

In keeping with this interesting finding, receptor agonists that elevate cytosolic Ca^{2+} concentration via G_q also produced activation of TRESK in *Xenopus* oocytes, and S276A mutation (dephosphorylated state) showed high basal activity. Calcineurin was found to interact with TRESK via a specific docking site that is also targeted by immunosuppressive compounds such as cyclosporin A (Czirjak *et al.*, 2006a). TRESK may be the only K_{2P} channel that is upregulated by a rise in cytosolic Ca^{2+} concentration. Whether this signalling pathway leads to activation of TRESK and stabilization of resting membrane potential in the native system remains to be determined, but seems likely if local elevation of $[Ca^{2+}]$ occurs near the channel and calcineurin is also present. Specific protein kinases that phosphorylate TRESK and presumably inhibit channel activity have not yet been reported. The kinetics of TRESK current modulated by calcineurin was affected by 14-3-3 protein, and further studies are needed to address the physiological significance of this interaction.

The single channel kinetics of TRESK expressed in oocytes and mammalian cell lines show very noisy open state when the current is inward. The single channel conductance is ~15 pS and the current–voltage relationship of TRESK is almost linear (Kang *et al.*, 2004b). Such unique single channel behaviour combined with activation by calcineurin should be useful in identifying a native K^+ channel that represents TRESK. A K_{2P} channel named TRESK-2 was initially cloned from mouse testis, and partial sequences of both TRESK and TRESK-2 DNA were identified in human tissues (Kang *et al.*, 2004b). This suggested that two TRESK isoforms exist in human. However, the mouse TRESK-2 sequence is not yet found in the published human genome sequence, and human TRESK is not found in the current rodent genome database. So far, most studies have simply referred to the two as TRESK, as they may simply be orthologues. Although human TRESK and mouse TRESK-2 show nearly identical single channel kinetics, their responses to certain pharmacological agents are not the same (see below). Therefore, the results obtained using human and mouse TRESKs should be interpreted with caution.

2.1.4.4 Cellular and subcellular localization

Studies to date show that K_{2P} channels are expressed in many tissues in human and rodents. Each K_{2P} channel member shows a unique pattern of expression that is specific to each organism examined. Table 2.1.4.1 lists tissues that express K_{2P} channels at relatively high levels, as judged by Northern blot analysis. All such data need to be viewed with some caution, because the functional K^+ current that is recorded can be very small, despite a strong signal from Northern blot analysis. Cell surface protein expression will need to be determined using specific antibodies.

It is not surprising that many cells and tissues express K_{2P} channels, as nearly all types of cells possess a leak K^+ current. What is surprising is the type of K_{2P} channel found in different cells and tissues. Despite our deep understanding of the physiology of various organs and their cells within, it is impossible to predict what members of the K_{2P} channel family would be present in the cells. Even after a specific K_{2P} channel member has been determined to be expressed in a given cell type, it is difficult to see why one K_{2P} channel is expressed and another similar K_{2P} channel is not. Why a cell needs several leak K^+ channels is even more difficult to understand. The most likely explanation is that each K_{2P} channel

Table 2.1.4.1 Nomenclature and properties of K_{2P} channels

Channel subunit	Chromosomal location		Biophysical characteristics		Cellular and subcellular localization		Pharmacology (IC$_{50}$/EC$_{50}$)		Physiology
	Human	Mouse	–60mV	–60 mV	Human	Mouse	Activator	Inhibitor	
K_{2P}1.1 (TWIK-1)	1q42–43	8E2			Brain Heart Kidney	Brain Heart			Sumoylation? Atrial fibrillation Renal ion transport TWIK-1$^{-/-}$ mice has impaired phosphate and water transport
K_{2P}2.1 (TREK-1)	1q32–41	1H6–H5	116	120	Brain Small intestine Ovary	Brain GI smooth Heart	FFA, halothane NO, acid, riluzole mechanosensitive temperature	Lidocaine, norfluoxetine Sipatrigine Hypoxia?	Receptor agonist (Gs,Gq) Neuroprotection Anesthesia TREK-1$^{-/-}$ mice show normal phenotype
K_{2P}3.1 (TASK-1)	2p23	5B1	9	14	Pancreas Placenta Brain Heart, lung	Motor neurons Heart Kidney, lung	Alkali, Halothane	Sa, acid, anandamide? Hypoxia, anarchidonate local anesthetics	Receptor agonists (Gq), apoptosis, Heteromer with TASK-3, tumorigenesis, Oxygen sensor (control of respiration) TASK-1$^{-/-}$ mice show normal phenotype 14-3-3 and p11 regulate expression pulmonary vascular tone
K_{2P}4.1 (TRAAK)	11q13	19A	75	120	Brain Placenta	Brain	Alkali, FFA, heart mechanosensitive riluzole		TRAAK$^{-/-}$ mice show normal phenotype
K_{2P}5.1 (TASK-2)	6p21.2	14A2	34	70	Pancreas Kidney Liver	Kidney Pancreas	Alkali, swelling	Clofilium	Cell swelling, hypoxia K$^+$ transport in kidney TASK-2$^{-/-}$ mice has metabolic acidosis Pancreatic hormone secretion
K_{2P}6.1 (TWIK-2)	19q13.1	17A3.1–A2							
2P7.1 (KCNK7)	11q13	19A	Non functional		Eye Lung Stomach Colon				Contains EF-hand Ca binding motif
K2p9.1 (TASK3)	8q24.2–24.3	15D3	16	38	Brain Cerebellum	Brain Brainstem Motor neurons	Halothaine, alkali	Acid, ruthenium red local anaesthetics	Receptor agonists (Gq), apoptosis, Heteromer with TASK-1. Oxygen sensor (control of respiration) Neuroprotection? 14-3-3 and p11 regulate expression
K_{2P}10.1 (TREK-2)	14q31	12E	50	138	Brain Pancreas Kidney	Cerebellum Pancreas	FFA, temperature mechanosensitive, acid, halothane, riluzole		Receptor agonists (Gs, Gq) K$^+$ transport in astrocyte. Sensory transduction Anaesthesia
K_{2P}12.1 (THIK-2)	2p21	17E4	Non functional		Brain				
K_{2P}13.1 (THIK-1)	14q31–32	12E	Probably very low		Brain Ubiquitous	Arachidonate	Halothane		Oxygen sensor?
K_{2P}15.1 (TASK-5)	20q12–13	2H2	Non functional						
K_{2P}16.1 (TALK-1)	6p21.2	14A2	10	21	Pancreas	Pancreas	Alkali, NO, ROS	Acid	Regulation by PH$_o$
K_{2P}17.1 (TALK-2)	6P21.2		15	33	Pancreas Liver Heart	Pancreas	Alkali, NO, ROS		Pancreatic hormone secretion Regulation by PH$_o$
K_{2P}18.1 (TRESK)	10q25–3	19D2	14	16	Spinal cord	Spinal cord DRG neuron Brain, testis	Ca/CaM (via calcineurin) Halothane	Zn (mouse), Hg Mibefradil Local anaesthetics	Receptor agaonist (Gq) Anaesthesia Sensory transduction

Functionality of TWIK-1 (and maybe TWIK-2) is controversial. NO (nitric oxide); FFA (free fatty acid); ROS (reactive oxygen species). Those K_{2P} channels modulated by phosphorylation should also be acted up by phosphatases. Because functional K_{2P} channels are expressed in many types of cells from different tissues, their cellular localization and physiology are not listed in this table. Because all functional K_{2P} channels act as background K$^+$ channels, they would be involved in setting and stabilizing the resting membrane-potential, in addition to other specific functions. All K_{2P} channels modulated by G$_s$ and G$_i$-linked receptors are expected to be important for receptor-mediated changes in cell excitability. Empty fields indicated that the data is not yet available. Cellular localization of K_{2P} channels is not limited to those tissues listed here. Question marks indicate data that are uncertain and need additional evidence. TRESK-2 reported in mouse may be an orthologue of human TRESK.

possesses unique biophysical properties and is modulated differently by various cellular factors and processes. Therefore, several K_{2P} channels probably work in concert to provide the best level of excitability for a given cell type under different physiological conditions. The answer to these interesting issues will slowly emerge as more information on the properties and cellular localization of the K_{2P} channels becomes available in the future.

2.1.4.5 Pharmacology

K_{2P} channels in general have low sensitivity to the well-known K^+ channel blockers such as tetraethylammonium, 4-aminopyridine, apamin, Cs^+ and Ba^{2+}, and no specific inhibitors have been identified. This has made it difficult to isolate the individual currents produced by K_{2P} channels in a cell and hampered the study of their role in cell physiology. The lack of their sensitivity to K^+ channel blockers has been used as one of the characteristics for identifying K_{2P} channels in the native system, but this is weak evidence for showing that a K_{2P} channel is functionally expressed in the cell. Nevertheless, some studies have identified certain useful organic compounds that can activate or inhibit K_{2P} channels.

TWIK

So far, no selective pharmacological agents have been identified that inhibit or activate TWIK-1 or TWIK-2. As mentioned above, the functionality of TWIKs is still controversial. The finding that rat TWIK-2 produces a large current suggests that TWIK-2 forms a functional channel (Patel et al., 2000). The finding that TWIK-1 knockout mice show impaired ion transport in the kidney suggests that TWIK-1 probably forms functional channels in the native tissues. However, no K^+ channels corresponding to TWIK-1 or TWIK-2 have yet been recorded from any native tissues. Single channel properties of TWIKs have yet to be determined in detail.

TASK

A number of inhibitors of TASK-1 and TASK-3 have been identified using recombinant channels expressed in oocytes and mammalian cell lines. Among them are zinc, copper, anandamide, ruthenium red, Ca^{2+}, and spermine. Unfortunately, none of these are specific for TASK-1 or TASK-3, and their use in identifying the TASKs in native systems should be interpreted with caution. Another important issue is the difference in species-dependent sensitivity of TASKs to an inhibitor, particularly between human and rodent. For example, studies have reported that Zn^{2+} (0.1 mM) has little effect on human TASK-1 but reduces rat TASK-1 by 40 per cent (Leonoudakis et al., 1998; Clarke et al., 2004). Zn^{2+} inhibits human TASK-3 by ~80 per cent but has little effect on TASK-1 and TASK-2, and this is due to a direct interaction of zinc with glutamate and histidine residues at the loop between M1 and M2 segments of TASK-3 (Clarke et al., 2004). Anandamide, an endocannabinoid, has been reported to be a relatively selective inhibitor of human TASK-1, as it does not affect other K_{2P} channels (Maingret et al., 2001). More recent studies, however, show that anandamides inhibit TASK-3 as well (Veale et al., 2007). It seems likely that anandamide inhibits all TASK isoforms, as the amino acids near the proximal C-terminus targeted by anandamide is present in all TASKs. Therefore, one must exercise caution when using any inhibitor of K_{2P} channels, and it would be prudent to test the direct action of a potential inhibitor at the level of the channel itself in the specific experimental system being used by the investigator.

Volatile anesthetic agents such as halothane activate TASK-1 and TASK-3 and this requires the proximal cytoplasmic C-terminal domain (Patel et al., 1999). This is in keeping with the recent finding that TASK-1$^{-/-}$ mice are less susceptible to halothane-induced anesthesia (Linden et al., 2006) TASK-1 and TASK 3 can be distinguished readily using ruthenium red. Ruthenium red is a potent inhibitor of TASK-3 but not TASK-1, due to the presence of negatively charged residues at the M1-P1 loop (Czirjak et al., 2003). TASK-1/TASK-3 heteromers that are expressed in neurons are also inhibited by ruthenium red. Positively charged molecules, such as Ca^{2+}, Mg^{2+}, spermine and ruthenium red, produce an interesting effect on the single channel conductance of TASK-3, but not TASK-1, presumably due to shielding of the negative charges at the M1-P1 loop of TASK-3 (Musset et al., 2006). Removal of external Ca^{2+} or Mg^{2+} produces a large increase in single channel conductance associated with an increase in whole-cell current. However, significant changes in the plasma concentrations of Ca^{2+} and Mg^{2+} are unlikely to occur under physiological conditions. An increase in the concentration of Zn^{2+} may occur during synaptic activity and could potentially reduce the single channel conductance of TASK-3 and lead to depolarization of the neurons involved.

Treprostinil, a stable prostacyclin analogue, was reported to activate TASK-1 in pulmonary smooth muscle cells via the protein kinase A pathway (Olschewski et al., 2006). Further studies are needed to address this issue, because recombinant TASK-1 expressed in mammalian cells and Xenopus oocyte is generally unaffected by cAMP.

TREK/TRAAK

Several organic compounds have been shown to inhibit TREKs. However, nearly all show low potency and poor selectivity. Riluzole, a neuroprotective drug with anticonvulsant and sedative properties, caused a sustained activation of TRAAK, while producing an activation followed by an inhibition of TREK-1 (Duprat et al., 2000). Sipatrigine, another neuroprotective agent, inhibited both TRAAK and TREK-1 (Meadows et al., 2001). Volatile and gaseous anesthetic agents such as halothane, isoflurane, chloroform, xenon, cyclopropane and nitrous oxide activate TREK-1 whereas local anesthetics such as bupivacaine, lidocaine and tetracaine inhibit TREK-1 at clinically relevant concentrations (Patel et al., 1999; Kindler and Yost, 2005).

The Ca^{2+} channel antagonists penfluridol and mibefradil inhibit TREK-1 at low micromolar concentrations, and amlodipine and niguldipine inhibit TREK-2 with IC_{50} values in the 0.5-0.8 μM range (Enyeart et al., 2002; Liu et al., 2007). The antidepressant fluoxetine and its metabolite norfluoxetine inhibit TREK-1 with IC_{50} values of 19 μM and 9 μM, respectively (Kennard et al., 2005). Thus, the block of TREK-1 might produce unwanted side effects such as increased sensitivity to epileptic seizures and ischemia. Caffeic acid derivatives have also been reported to activate TREK-1 (Danthi et al., 2004). Many drugs when used at high concentrations inhibit TREK-1 and these include fenamate, diltiazem, quinidine, caffeine, and theophylline. It is important to recognize that these pharmacological agents are not useful for characterizing and identifying native TREK-1 or studying the role of TREKs in a physiological process due to their low specificity and selectivity.

TALK

Like other K_{2P} channels, TALK-1 and TALK-2 show low sensitivity to K^+ channel blockers such as TEA, 4-AP and Ba^{2+}, and Cs^+. TALK-2 is mildly inhibited by halothane and chloroform (Girard *et al.*, 2001). Clofilium, a potent blocker of the HERG K^+ channel, inhibits TASK-2 with an IC_{50} of 20 μM (Niemeyer *et al.*, 2001).

THIK

THIK-1 produces a large current in *Xenopus* oocytes, and is activated by arachidonic acid (Girard *et al.*, 2001). Thus, TREKs and THIK-1 are the only K_{2P} channels that are activated by arachidonic acid. Unlike TREKs and TASKs, however, THIK-1 is inhibited by halothane.

TRESK

Zinc has been reported to inhibit mouse TRESK but not human TRESK (Czirjak and Enyedi, 2006b). A double mutation of histidine and methionine residues close to the pore region in mouse TRESK to those identical in human TRESK was found to reduce the sensitivity of the mouse TRESK to zinc (Czirjak and Enyedi, 2006b). Amide local anaesthetics are tenfold more potent at inhibiting mouse TRESK than human TRESK (Czirjak and Enyedi, 2006b). Volatile anaesthetics such as halothane and isoflurane are potent activators of human TRESK, but weaker activators of mouse TRESK (Keshavaprasad *et al.*, 2005). The molecular basis for such species-dependent differences in the sensitivity to pharmacological agents is not yet clear. Mercury was also found to be a selective inhibitor of mouse and human TRESK among the K_{2P} channels. Mibefradil, a Ca^{2+} channel blocker, inhibited TRESK by ~60 per cent with a IC_{50} of 2.2 μM (Czirjak and Enyedi, 2006b). These agents, together with the unique single channel properties of cloned TRESK, should aid in identifying this K^+ channel in the native system, but are unlikely to be useful for studying the physiological roles of TRESK.

2.1.4.6 Physiology

Studies so far suggest that K_{2P} channels are involved in the regulation of many physiological processes including synaptic transmission, hormone secretion, cell volume regulation, muscle contraction, respiration, oxygen sensing, cell proliferation and death, renal K^+ homeostasis, and sensory transduction.

TWIK subfamily: TWIK-1 (K_{2P}1.1) and TWIK-2 (K_{2P}6.1)

Although TWIK-1 and TWIK-2 mRNAs are expressed in many regions of the brain and in a number of peripheral tissues, their role as background K^+ channels is still speculative. This is because no K^+ channels considered as TWIK-1 or TWIK-2 have yet been recorded from any native tissues. The possibility that TWIK-1 and TWIK-2 are localized in intracellular organelles needs to be tested to understand their function. Nevertheless, studies in TWIK-1$^{-/-}$ mice suggest that TWIK-1 is involved in K^+/water transport in the kidney and perhaps elsewhere where it is expressed (Nie *et al.*, 2005). Although TWIK-1 is expressed in cardiac myocytes, cerebellar granule neurons and many other tissues, its physiological roles are poorly understood at present.

TASK subfamily: TASK-1 (K_{2P}3.1), TASK-3 (K_{2P}9.1) and TASK-5 (K_{2P}15.1)

In mouse and rat, the expression of TASK-1 mRNA is relatively abundant in the heart (Leonoudakis *et al.*, 1998; Kim *et al.*, 1999). Immunofluorescence studies show that TASK-1 is expressed at the intercalated disks and along the T-tubule network in rat ventricular myocytes (Jones *et al.*, 2002). These findings have led to the speculation that TASK-1 may have a significant role in cardiac electrophysiology. Opening of single TASK-1-like K^+ channels can be recorded from rat ventricular and atrial myocytes (Kim *et al.*, 1999). However, the open probability of TASK-1 is relatively low compared to that of the well-known inwardly rectifying K^+ channel (IRK) at rest, and may not contribute much to the stabilization of resting membrane potential in cardiac myocytes (Kim *et al.*, 1999). Nevertheless, inhibition of TASK-1 by platelet-activating factor via protein kinase C has been reported to cause prolongation of action potential duration in the murine heart (Besana *et al.*, 2004). Therefore, at depolarized potentials, TASK-1 may contribute to early repolarization. TASK-1 has sometimes been referred to as the Kp current that was previously identified in guinea pig ventricular myocytes (Backx and Morban 1993). However, a distinct difference in the single channel kinetics of TASK-1 and Kp channels shows that the two are not the same entity. Also, the presence of the Kp current has not yet been reported from rat or human cardiac myocytes. The level of expression of TASK-1 mRNA in human heart is very weak compared to that in the brain and lung, and functional TASK-1 channels have not yet been recorded in the human cardiac myocytes (Duprat *et al.*, 1997; Vega-Saenz de Miera *et al.*, 2001). Therefore, the role of TASK-1 in human cardiac electrophysiology is difficult to predict and may be insignificant. No particular abnormalities in cardiac function have yet been reported in TASK-1$^{-/-}$ mice (Aller *et al.*, 2005; Linden *et al.*, 2006). Nevertheless, it will be interesting to see whether the action potential duration and cardiac excitability is affected in the TASK-1$^{-/-}$ mice, particularly under ischaemic and hypoxic conditions.

TASK-1 mRNA is highly expressed in the brain, placenta, lung, pancreas, and prostate in humans (Duprat *et al.*, 1997). In the brain, both TASK-1 and TASK-3 mRNAs are abundant in cerebellar granule cells in human as well as rodents. As cerebellar granule cells provide a major excitatory input to the Purkinje cells, the modulation of the standing outward current by various neurotransmitters is an important mechanism for regulating cerebellar output (Millar *et al.*, 2000). Studies show that a large fraction of the standing outward K^+ current is represented by TASK-1 and TASK-3. Functional TASK-1 and TASK-3 currents have been recorded successfully from cultured rat cerebellar granule cells (Brickley *et al.*, 2001; Han *et al.*, 2002). The degree of contribution of TASKs to the standing outward K^+ current may vary depending on the developmental stage, as the levels of their mRNA expression changes (Brickley *et al.*, 2001). The specific roles of TASK-1 and TASK-3 in excitability during development are not yet known, but may involve regulation of resting membrane potential and subtle changes in sensitivity to acid and receptor agonists. In cultured cerebellar granule cells from neonatal rats, TASK-1/TASK-3 heterodimer as well as homodimers were identified (Kang *et al.*, 2004a). TASK-1$^{-/-}$ mice exhibited compromised motor performance, although the standing outward current was not significantly affected due to a shift in subunit composition of TASKs (Aller *et al.*, 2005; Linden *et al.*, 2006). Thus, TASK-1 and TASK-3 are expected to be an important component of the signalling that modulates cerebellar excitability and associated motor function.

In brainstem motor neurons that express high levels of TASK-1 and TASK-3, their high sensitivity to pH (and by inference pCO$_2$) may be an important mechanism for regulation of respiratory neural output (Bayliss *et al.*, 2001; Mulkey *et al.*, 2004). A recent study using TASK-1$^{-/-}$/TASK-3$^{-/-}$ mice shows that these two K$^+$ channels are not involved in central respiratory chemosensitivity (Mulkey *et al.*, 2007). Perhaps another pH$_o$-sensitive K$^+$ channels such as the Type 4 channel expressed in cerebellar granule neurons is involved in the chemosensitivity. TASK-1$^{-/-}$ and TASK-1$^{-/-}$/TASK-3$^{-/-}$ mice exhibit primary hyperaldosteronism, indicating an important role of TASKs in the function of adrenal zona glomerulosa cells (Davies *et al.*, 2008; Heitmann *et al.*, 2008). Deletion of TASKs in these cells removes the background K$^+$ conductance, causing depolarization and overproduction of aldosterone.

The physiological significance of the high acid sensitivity of TASKs is not very clear at this time. It is thought that the release of synaptic vesicles from the nerve terminal is associated with lowering of pH in the synaptic cleft. Perhaps such acidification during synaptic activity inhibits TASKs expressed at the presynaptic or postsynaptic nerve terminal and modulates synaptic transmission. The high sensitivity of TASKs to external acid has been commonly used to provide evidence that a native K$^+$ channel is a functional correlate of a TASK. However, one needs to be cautious because there are other background-like K$^+$ channels (such as the Type 4 channel in cerebellar granule neurons) that are also pH$_o$-sensitive (Han *et al.*, 2002). Both biophysical and pharmacological characteristics must be determined, preferably at the single channel level, to be certain that the native K$^+$ channel is indeed a TASK.

An interesting property of TASK-1 (and probably TASK-3) is its sensitivity to hypoxia (Buckler *et al.*, 2000). Recombinant human TASK-1 expressed in HEK293 cells was inhibited by approximately 20 per cent when pO$_2$ was reduced to less than 40 mmHg (Lewis *et al.*, 2001). The cellular mechanism by which hypoxia inhibits TASK-1 (and presumably TASK-3) is not yet known, but could be mediated by a closely associated enzyme whose activity is sensitive to changes in oxygen concentration, or perhaps via changes in mitochondrial metabolic state. For certain voltage-gated K$^+$ channels such as Kv4.2, its β subunit functions as an oxidoreductase that may sense the redox state of the cell and modulate the channel gating (Pongs *et al.*, 1999). NADPH oxidase and heme-oxygenases are also potential candidates for oxygen sensing (Fu *et al.*, 2000; Kemp 2005). In cerebellar granule cells, the standing outward K$^+$ current was inhibited by hypoxia, consistent with the inhibition of TASKs (Plant *et al.*, 2002). Due to the O$_2$ sensitivity of TASK-1, several studies have focused on Type 1 cells of the carotid body that senses arterial oxygen tension and relays this information to the chemoreceptor afferent nerve fibres that ultimately helps regulate the cardiovascular reflex. Studies show that a TASK-like K$^+$ channel is highly expressed in Type 1 cells, and thus likely to serve as an oxygen sensor (Williams and Buckler, 2004). Another O$_2$-sensing tissue is the airway neuroepithelial body that regulates proper ventilation in the lung (O'Kelly *et al.*, 1999). In these cells, TASK-like K$^+$ channels are also highly expressed, suggesting that TASKs also regulate airway ventilation. Therefore, TASKs probably work in concert with other O$_2$-sensitive ion channels to regulate cell excitability in response to hypoxia. Pulmonary artery smooth muscle cells and brainstem respiratory neurons are other cell types that express TASKs and undergo depolarization in response to hypoxia (Bayliss *et al.*, 2001; Gurney *et al.*, 2003; Olschewski *et al.*, 2006).

An important property of TASKs is that they are inhibitory targets of various neurotransmitters and neuropeptides that act on receptors linked to G$_q$ that stimulates phospholipase C. Thus, in cerebellar granule cells, ACh strongly inhibits the standing outward K$^+$ current via the M3 muscarinic acetylcholine receptor (Millar *et al.*, 2000). In brainstem motor and dorsal vagal neurons that express TASKs, agonists such as serotonin, norepinephrine and substance P were all found to strongly inhibit the acid-sensitive, TASK-like K$^+$ current via specific receptors, leading to an increase in cell excitability (Sirois *et al.*, 2002; Perrier *et al.*, 2003). The neuromodulatory functions of thyrotropin-releasing hormone such as its anti-anaesthetic and general excitatory actions are likely to be via inhibition of background K$^+$ channels such as TASKs and TREKs (Yarbrough, 2003). Such inhibition of background K$^+$ current by receptor agonists is probably a commonly used mechanism to increase cell excitability in various tissues that express TASK-1 and TASK-3.

What is the signalling pathway for inhibition of TASKs by receptor agonists? Although it is known that G$_q$ is involved, conflicting results have been reported on the pathways distal to G$_q$. For example, inhibition of TASK-1 has been attributed to a reduction of phosphatidylinositol-4,5-bisphosphate (PIP$_2$) content in the membrane caused by activation of PLC, to an indirect effect by IP$_3$ or to the action of Gα$_q$ subunit itself (Chemin *et al.*, 2003; Lopes *et al.*, 2005; Chen *et al.*, 2006). It is possible that all of these mechanisms contribute to the inhibition of TASKs, although this seems unlikely. For both TASK-1 and TASK-3, mutations of six carboxyl terminal amino acid residues next to the fourth transmembrane segment was reported to abolish the inhibition by thyrotropin releasing hormone that binds to its G$_q$-coupled receptor (Talley and Bayliss, 2002). Either the mutations removed the binding site for the inhibitory molecule or changed the local conformation of the channel, preventing access to the inhibitory molecule. For other ion channels such as G protein-gated K$^+$ channel (GIRK), inwardly rectifying K$^+$ channel (IRK), M channel (KCNQ2/3) and Ca^{2+} channel, receptor agonists appear to modulate these channels mostly via depletion of PIP$_2$ (Delmas *et al.*, 2005; Suh and Hille, 2005). The general idea is that the negative charges of membrane PIP$_2$ interact with positively charged regions in the cytoplasmic termini located near the membrane interface and stabilize the channel, allowing the channels to be more active. Although the 'PIP$_2$ depletion' hypothesis seems attractive, more definitive evidence is needed to support the mechanism for TASKs. Some of these proposed signalling pathways are shown in Figure 2.1.4.2.

As TASK-1 and TASK-3 are G protein-modulated channels, signalling molecules that modulate the function of the G protein should influence the channel activity. One of these molecules is the RGS (regulator of G protein signalling) protein that is ubiquitously expressed and has GAP (GTPase-activating proteins) activity on G$_{i/o}$ and G$_q$ proteins. As such, certain RGS proteins are expected to regulate the strength and kinetics of effector molecules such as TASK and TREK. One recent study showed that the inhibition of standing outward K$^+$ current in cerebellar granule neurons by a receptor agonist recovers quickly after inhibition if RGS function is normal, but recovers slowly if RGS function is removed (Clark *et al.*, 2006). As the standing outward current consists mostly of K$_{2P}$ channels (TASKs and TREK-2), these findings implicate RGS

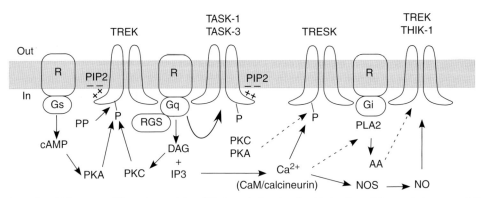

Fig. 2.1.4.2 Signalling pathways by which receptor (R) agonists modulate K2P channels. Agonists that bind to Gs and Gq-coupled receptors inhibit TREK via protein kinase A (PKA) and protein kinase C (PKC), respectively. Ca2+ and depletion of membrane PIP2 have also been suggested to cause inhibition of TREK. Agonists that bind to Gq-coupled receptors inhibit TASK. The reported mechanisms of inhibition include a direct action of Gαq and depletion of membrane PIP2. Agonists that bind to Gq-coupled receptors activate TRESK via dephosphorylation by Ca2+-and calmodulin-dependent phosphatase (calcineurin). Increases in cytosolic Ca2+ concentration may in some cases activate nitric oxide synthase (NOS), and activate TREK-1 via NO. Although not well defined, certain agonists that bind to Gi-coupled receptors coupled to phospholipase A2 (PLA2) may release arachidonic acid (AA) and modulate K2P channels. It is expected that protein phosphatases (PP) also modulate these K2P channels, and that regulator of G protein signalling (RGS) proteins influence K2P channel function by interacting with specific receptors. Solid arrows represent pathways that have been reported in various studies, and dotted arrows represent likely pathways that modulate K2P channel function.

as an important regulator of K_{2P} channel function in the native system.

Glucose-sensing hypothalamic neurons respond to changes in extracellular glucose concentration and provide appropriate information on the energy state of the brain. Recent work shows that inhibition of the neurons by glucose involves activation of TASK-like channel (Burdakov *et al.*, 2006). This could occur via a rise in [ATP] and [PIP_2] that augment TASK activity, but the precise mechanism is not yet known. Therefore, TASK-like channel may be involved in the regulation of wakefulness and metabolism that orexin neurons normally control.

Despite the potentially significant role of TASK-1 in synaptic transmission and other processes discussed above, TASK-1$^{-/-}$ mice appear healthy and show no gross behavioural abnormalities, and have normal size and morphology of their brain and heart (Aller *et al.*, 2005; Linden *et al.*, 2006). Lack of an observable phenotype could be due to upregulation and compensation by other leak K$^+$ channels. More recent work shows that TASK-1$^{-/-}$ mice display enhanced acoustic response, increased sensitivity to thermal nociception in the hotplate test (Linden *et al.*, 2006). In addition, the analgesic effect of WIN55212-2, a cannabinoid receptor agonist, and the anesthetic effects of halothane and isoflurane were also reduced in TASK-1$^{-/-}$ mice compared to those of the wild-type litter mates. These findings suggest that TASK-1 contributes to sensory signal transduction as well as the actions of volatile anaesthetics. Thus, TASK-1$^{-/-}$ mice phenotypes become noticeable when the animal is challenged with stressful environmental conditions.

TREK subfamily: TREK-1 (K_{2P}2.1), TREK-2 (K_{2P}10.1) and TRAAK (K_{2P}4.1)

The behaviour of these K$^+$ channels is somewhat reminiscent of that of TRP ion channels in that TREKs are also sensitive to various sensory-related stimuli such as heat, lipids, pressure, acid, membrane [PIP_2], and receptor agonists (Clapham, 2003). As TREKs are expressed in sensory neurons such as dorsal root and trigeminal ganglion neurons (Matsumoto *et al.*, 2001; Cooper *et al.*, 2004), TREKs presumably work in concert with TRP ion channels in the transduction of sensory and nociceptive signals.

As mentioned above, three of the most well-characterized properties of TREKs are their activation by unsaturated free fatty acids, intracellular acid and pressure applied to the membrane. TREKs are also directly activated by lysophosphatidic acids (Maingret *et al.*, 2000b; Chemin *et al.*, 2005). Thus, if a native K$^+$ channel is easily activated by low micromolar concentrations of free fatty acids, by acid solution of pH$_i$ less than ~6.8 applied to the intracellular side of the membrane, and by negative pressure applied to the membrane in inside-out patch condition, and has single channel kinetic properties similar to those of a TREK, then the native K$^+$ channel is most certainly a TREK. If the cell expresses TREK mRNA, this would help to further establish the identity of the K$^+$ channel. These methods have been used to show that TREK-1, TREK-2 and TRAAK are functionally expressed in various regions in the brain and in peripheral tissues in human and rodents (Medhurst *et al.*, 2001; Talley *et al.*, 2001). What would be the physiological significance of possessing such sensitivities to free fatty acids, membrane tension and acid? It is interesting to note that these three factors are generally associated with metabolically stressed conditions that lead to altered cell volume, pH$_i$ and fatty acid levels. As large changes in cell metabolism probably do not occur under normal physiological conditions, one can only speculate that TREKs perhaps help to finely regulate the background K$^+$ conductance and adjust the excitability of the cell in response to small changes in cell metabolism. In gastrointestinal smooth muscle cells, TREK-1 probably plays an important role in stretch-dependent regulation of muscle tone (Sanders and Koh, 2006). TREK-1$^{-/-}$ mice do not display abnormal phenotype in appearance, body weight and respiration, suggesting that TREK-1 function is not indispensable (Heurteaux *et al.*, 2004). TREK-1$^{-/-}$ mice, however, show altered response to volatile anesthetics and neuronal ischemia, indicating that TREK-1 provides a neuroprotective action (Heurteaux *et al.*, 2004). TREK-1$^{-/-}$ mice show an antidepressant phenotype, indicating that TREK-1 is involved in serotoninergic transmission that affects mood (Heurteaux *et al.*, 2006). Whether the function of the gastrointestinal system that expresses high levels of TREK-1 is affected in TREK-1$^{-/-}$ mice has not yet been reported.

One of the intriguing properties of TREKs is their sensitivity to temperature. TREKs expressed in mammalian cell lines or neuronal TREKs are activated by an increase in temperature in the ~20–45°C range (Maingret *et al.*, 2000a; Kang *et al.*, 2005). Thus, activation by heat is expected to increase background K$^+$ conductance and decrease excitability, whereas cooling is expected to produce the opposite effect. These properties may be important in sensing temperature at the hypothalamus as well as near the skin. TREKs are highly expressed in dorsal root ganglion neurons, but whether TREKs are also expressed at the nerve terminals is not known. TREK mRNA is expressed in a human keratinocyte cell line (HaCaT), suggesting a possible role of TREKs in temperature sensing near the skin (Kang *et al.*, 2007). TRP ion channels (i.e., TRPV1-4, TRPA1 and TRPM8) that sense heat/warm/cold are well known to be functionally expressed in sensory neurons, and presumably work together with TREKs to transduce temperature signals to the brain. The depolarization produced by activation of TRPV1 by capsaicin, for example, would not be opposed by TREKs, but depolarization produced by heat would be opposed, as TREKs may also be activated at the same time. The interactions among different TRP ion channels and TREKs in the modulation of cell excitability and firing frequency have yet to be directly tested in sensory neurons. TREKs may partly account for the desensitization observed following activation of TRP ion channels. It is possible that persistent inhibition of TREKs by substances released during inflammation such as bradykinin, substance P and prostaglandins could be involved in chronic nociception.

Receptor agonists inhibit TREKs via G proteins (Patel *et al.*, 1998; Chemin *et al.*, 2003). Both G$_s$ and G$_q$ mediate inhibition, whereas G$_i$ may cause stimulation of TREKs (Cain *et al.*, 2008). In one study, angiotension II caused inhibition of a non-inactivating K$^+$ current in adrenal glomerulosa cells, and the K$^+$ current was subsequently identified to be TREK-1 by single channel analysis (Enyeart *et al.*, 2005). It seems clear that the mechanism by which an agonist that binds to G$_s$-coupled receptor inhibits TREKs is via PKA-mediated phosphorylation of a serine residue at the proximal carboxy termini of TREKs. However, the mechanism of inhibition by G$_q$ seems complex, as various signalling pathways have been proposed. So far, the proposed mechanism of inhibition of TREKs include degradation of membrane PIP$_2$, a rise in Ca^{2+} and diacylglycerol, PIP$_2$-independent pathway requiring ATP hydrolysis, and protein kinase C-mediated phosphorylation (Chemin *et al.*, 2003; Lopes *et al.*, 2005; Murbartian *et al.*, 2005; Liu *et al.*, 2007). It seems plausible that multiple pathways converge on the channel to cause inhibition, and that difference cell types use distinct pathways. Although inhibition of TREKs by receptor agonists can be demonstrated in transfected cells expressing either TREK-1 or TREK-2, such inhibition has been more difficult to show for native TREKs. The difficulty lies in the fact that isolating the TREK whole-cell current in the native cell is difficult due to lack of specific inhibitors. Since a number of cellular molecules may take part in the signalling pathway for inhibition of TREKs, it will be important to demonstrate inhibition in native cells in the future when specific inhibitors are discovered.

TREKs may be involved in cytoskeletal remodeling that results in the formation of actin- and ezrin-rich membrane protrusions (Lauritzen *et al.*, 2005). The formation of these protrusions was dependent on phosphorylation by PKA and a proton sensor at the C-terminus of TREK-1. The potential regulation of TREK-1 function by cytoskeletal proteins has been suggested previously when the channel activity greatly increased upon patch excision (Patel *et al.*, 2001). Whether the increase in channel activity is due to the disruption of a cytoskeletal protein network that supports the lipid-TREK structure or involves some other actin-mediated processes remains to be determined. Further studies are needed to understand whether and how TREKs might serve as part of the signalling network that induces synaptogenesis. Whether this property of TREK is related to its highly mechanosensitive character is not yet known, but it seems plausible.

TALK subfamily: TALK-1 (K$_{2P}$16.1), TALK-2 (K$_{2P}$17.1) and TASK-2 (K$_{2P}$5.1)

As already mentioned above, members of the TALK family are sensitive to a wide range of pH$_o$ (6.3–10.3). TALK-1 and TALK-2 mRNAs are highly expressed in the human pancreas, particularly in the acinar cells that regulate secretion of pancreatic hormones (Duprat *et al.*, 2005). The significance of the high pH$_o$-sensitivity of TALKs in the native tissues is uncertain at this time. However, TALKs are expected to modulate the resting membrane potential, and perhaps respond to local changes in pH to regulate secretion. Nitric oxide (NO) donor such as sodium nitroprusside was found to produce a large activation of TALK-2 via the cGMP-independent pathway (Duprat *et al.*, 2005). The possibility that NO generated by neurotransmitters that elevate intracellular Ca^{2+} regulates hormone secretion via activation of TALK-2 is yet to be demonstrated. In addition to NO, reactive oxygen radicals and oxidation have been reported to activate TALKs, suggesting that abnormal cell metabolism and conditions that cause cell stress may be associated with activation of background K$^+$ current in pancreatic acinar cells (Duprat *et al.*, 2005). Again, the effects of these molecules have yet to be demonstrated in native systems.

TASK-2 mRNA is highly expressed in the kidney and pancreas. In the pancreas, TASK-2 is expressed in both exocrine glands and Langerhans' islets that secrete insulin. In the kidney, TASK-2 is particularly abundantly expressed in the proximal tubules and collecting duct (Reyes *et al.*, 1998; Barriere *et al.*, 2003). The sensitivity of TASK-2 to both pH and cell volume change may be relevant in renal nephrons that undergo cell swelling and shrinkage due to osmotic challenges. In TASK-2$^{-/-}$ mice, the swelling-activated K$^+$ current was reduced in proximal tubules, and the regulatory volume decrease following swelling was also impaired (Barriere *et al.*, 2003). However, in papillary collecting ducts from TASK-2$^{-/-}$ mice, cell volume regulation was normal, suggesting that the role of TASK-2 may be segment-specific in the kidney. These studies show that TASK-2 that is highly expressed in the nephrons mediates cell volume regulation. In proximal tubular cells, TASK-2 senses the basolateral pH$_o$ that is determined by HCO$_3^-$ transport and regulates the K$^+$ conductance in these cells. The metabolic acidosis exhibited by TASK-2$^{-/-}$ mice was thus attributed to abnormal coupling between TASK-2 and HCO$_3^-$ transport (Warth *et al.*, 2004). Although the properties of cloned TALKs and TASK-2 expressed in mammalian cell lines are generally well known, the native K$^+$ channels that are encoded by their genes have yet to be identified and studied in the kidney and other cell types.

THIK subfamily: THIK-1 (K$_{2P}$13.1) and THIK-2 (K$_{2P}$12.1)

THIK-1 and THIK-2 mRNA is expressed in many peripheral tissues and brain. Interestingly, arachidonic acid activated THIK-1 whereas halothane inhibited it (Rajan *et al.*, 2001). This differs from TREKs that are activated by both arachidonic acid and halothane (Patel *et al.*, 1999). Hypoxia (pO$_2$, ~20 mmHg) inhibited THIK-1 expressed in HEK cells, and it was suggested that this may underlie the O$_2$-sensitive background K$^+$ conductance in the glossopharyngeal nerve and perhaps in other regions in the brain (Campanucci *et al.*, 2003; Campanucci *et al.*, 2005). Interestingly, mitochondrial inhibitors and NADPH oxidase inhibitors were ineffective in inhibiting the hypoxic response of THIK-1, suggesting that the mitochondrial respiratory system may not mediate the response. Clearly, further studies are needed to understand the potential role of THIK-1 in anesthesia and oxygen-sensing. THIK-1 and THIK-2 are also highly expressed in proximal and distal nephrons of the kidney, suggesting that they may be involved in K$^+$ transport in these segments (Theilig *et al.*, 2008). However, native functional THIK K$^+$ channels have yet to be recorded.

TRESK subfamily

TRESK mRNA was initially reported to be expressed exclusively in the human spinal cord. However, further studies show that it is also expressed in many other tissues including brain and testis (Sano *et al.*, 2003; Czirjak *et al.*, 2004; Kang *et al.*, 2004b). Recently, TRESK was found to be one of the background K$^+$ channel in dorsal root ganglion neurons, contributing a significant fraction of the background K$^+$ current (Kang *et al.*, 2006; Dobler *et al.*, 2007). Trigeminal sensory neurons also express TRESK, suggesting that sensory neurons use TRESK to modulate their excitability. As receptor agonists that elevate cytosolic Ca^{2+} may activate TRESK-1 and cause suppression of excitability, TRESK is likely to be an important target of modulation by various neurotransmitters that are known to be involved in sensory transduction. The role of TRESK in any biological function has not yet been studied in native systems.

2.1.4.7 Relevance to disease

The lack of selective pharmacological agents has limited our understanding of the physiology of K$_{2P}$ channels. Therefore, knockout mice are useful for our understanding of the biological roles of K$_{2P}$ channels, particularly in pathological conditions. Knockout mice for TWIK-1, TASK-2, TASK-1, TASK-3 and TREK-2 have been generated so far, and studies using these mice provide interesting and important roles they play in various physiological processes.

TWIK

The controversy surrounding the functionality of TWIK-1 and TWIK-2 poses additional problems in studying these channels. One of the tissues that express high levels of TWIK-1 is kidney where a strong perinuclear immunostaining of TWIK-1 is observed in mouse kidney (Nie *et al.*, 2005). In TWIK-1$^{-/-}$ mice, the renocortical expression of Na$^+$-HPO$_4^-$ co-transporter (NaP$_i$-2a) and internalization of aquaporin-2 in response to vasopressin were reduced. This probably caused the observed impairment of phosphate and water transport in TWIK-1$^{-/-}$ mice. TWIK-1 mRNA is also expressed in the heart, but whether the mechanical or electrophysiological function of the heart is impaired in TWIK-1$^{-/-}$ mice has not yet been reported. A microarray analysis of ion channel mRNA expression showed that TWIK-1, phospholamban, MinK and Mirp2 were up-regulated whereas T-type Ca^{2+} and Kv4.3

channels were down-regulated in patients with valvular heart disease with atrial fibrillation compared with valvular heart disease with sinus rhythm (Gaborit *et al.*, 2005). In another microarray analysis, atrial tissues from patients with chronic atrial fibrillation expressed significantly less TWIK-1 mRNA compared to patients with sinus rhythm, whereas TASK-1, Kv1.5 and Kir2.1 mRNAs were not affected (Ellinghaus *et al.*, 2005). These findings show that one of the ion channels affected by chronic rhythm disturbances is TWIK-1, although the significance of such channel remodeling is not clear. Whether TWIK-1 modulates the trafficking of other ion channels to the plasma membrane and whether the altered expression of TWIK-1 is a consequence of the disease, or cause of it, are not yet known.

TASK

TASK-1 is expressed in pulmonary artery smooth muscle cells, and is believed to be important in the regulation of resting membrane potential and vascular tone (Gurney *et al.*, 2003; Olschewski *et al.*, 2006). Reduction of TASK-1 using siRNA or by hypoxia in pulmonary smooth muscle cells produced a significant depolarization of the resting membrane potential. Although the vascular tension was not directly measured, these findings suggest that TASK-1 is an important regulator of vascular tension by acting as a leak K$^+$ conductance that is sensitive to changes in pH and oxygen tension. TASK-1, along with TREK-1 that is also expressed in vascular smooth muscle and endothelial cells, may therefore regulate blood flow in various vascular beds.

The studies on the potential roles of TASKs in apoptosis and cell protection are interesting and important. In cerebellar granule neurons in culture, blocking K$^+$ efflux via TASK-1 and TASK-3 by low pH$_o$, high K^+_o or ruthenium red greatly reduced cell death by apoptosis (Lauritzen *et al.*, 2003). Overexpressing TASK-1 in rat hippocampal neurons using a viral vector enhanced neuronal cell death while overexpressing the dominant negative mutants of TASK-1 (TASK-1G95E) reduced cell death. These results are in keeping with the previous finding that increased efflux of K$^+$ from the cell is associated with apoptosis (Yu, 2003). Whether the lowering of physiological levels of intracellular K$^+$ concentration is the cause of activation of the mitochondrial apoptotic mechanism is not known, but it is likely to involve many other factors. In contrast to these observations, overexpression of TASK-3 (as well as TASK-1 and TASK-2) in cultured hippocampal slices reduced cell death compared to control, and the apoptosis was also reduced in response to cellular stress induced by low serum and hypoxia (Liu *et al.*, 2005). These conflicting results on the role of TASKs in cell protection could be due to differences in the level of expression of TASKs expressed in the plasma membrane and in resting membrane potentials that determine the actual efflux of K$^+$. A certain basal level of K$^+$ efflux occurs in all cells and is matched by K$^+$ influx via Na$^+$ pump. A slight increase in K$^+$ efflux may be protective by causing hyperpolarization, whereas large K$^+$ efflux may be detrimental to the cell by lowering the bulk concentration of K$^+$ in the cell and causing apoptosis. These issues need to be addressed in future studies by directly measuring K$^+$ flux across the membrane.

Related to the K$^+$ efflux effects of TASKs on apoptosis is their potential role in cancer. In this respect, the TASK-3 gene is overexpressed in a number of human carcinomas (Mu *et al.*, 2003; Pei *et al.*, 2003). The oncogenic potential of TASK-3, as judged by a cell proliferation assay in a partially transformed mouse embryonic fibroblast cells, was reduced by expression of the dominant negative mutant of TASK-3 (TASK-3G95E). This indicates that K$^+$ efflux via TASK-3 is important for cell proliferation. Tumorigenicity due to TASK-3, as assessed by measurement of tumour size after injection of fibroblasts expressing TASK-3 into mice, was also reduced by TASK-3G95E. Block of K$^+$ channels by pharmacological agents inhibited cell proliferation in human lymphocytes and different types of cancer cells (Jensen et al., 1999). Thus, TASK-3 may be one of the K$^+$ channels involved in tumour formation in certain cells. In this respect, most well-designed studies suggest that K$^+$ flux via specific K$^+$ channels during cell growth is critical for proliferation and apoptosis. Whether TASK-3$^{-/-}$ mice show less susceptibility to tumour formation remains to be seen. A recent study shows that TREK-1 expression is markedly increased in prostate cancer tissue, and suggests that TREK-1 is associated with abnormal cell proliferation (Voloshyna et al., 2008). Other K$^+$ channels that are upregulated in a variety of tumours include Kir3.1, Kv1.3, Kv10.1 and Kv11.1. How all of these K$^+$ channels are related to tumour formation is an important topic for future study, as they may prove to be useful targets for anti-cancer therapy.

As mentioned earlier, volatile anesthetic agents such as halothane activate TASK-1 and TASK-3 (Patel et al., 1999) and the observation that TASK-1$^{-/-}$ mice are less sensitive to halothane-induced anaesthesia suggests that these channels contribute to the mechanism(s) underlying the activity of this class of anaesthetics (Linden et al., 2006).

TREK/TRAAK

As mentioned above, the neuroprotective drugs sipatrigine and riluzole affect the function of both TREK-1 and TRAAK (Duprat et al., 2000; Meadows et al., 2001). Although TREK-1$^{-/-}$ mice exhibit normal behaviour and no abnormalities in body weight, respiration, reflex and autonomic function (Heurteaux et al., 2004), they are much more susceptible to kainate-induced epileptic seizures and neuronal ischaemia. Linolenic acid which is known to activate TREK-1 reduced seizure activity in the wild type TREK-1 mouse but not in the TREK-1$^{-/-}$ mouse, providing evidence that TREK-1 helps to protect against neuronal disturbances and ischaemia (Lauritzen et al., 2000). The mechanism by which riluzole and sipatrigine provide neuroprotection is therefore not clear as they can also inhibit TREK-1. The neuroprotective action of linolenic acid may also be present in cerebral vasculature, where free fatty acid activates TREK-1 expressed in smooth muscle and endothelial cells of the cerebral blood vessels in, for example, the basilar and middle cerebral artery (Bryan et al., 2006; Blondeau et al., 2007). By increasing blood flow to critical areas of the brain during ischaemia, for example, linolenic acid may help to reduce or prevent the damage to the brain. Studies in TREK-1$^{-/-}$ mice support the contention that opening of this channel is important for the neuroprotective effect of polyunsaturated free fatty acids. Interestingly, mice lacking TREK-1 are resistant to the vasodilating effects of acetylcholine and bradykinin, due to an impairment of nitric oxide production (Garry et al., 2007). Although these studies suggest that TREK-1 is the primary K$^+$ channel mediating the vasodilating effect of linolenic acid, the large-conductance K$^+$ channels (BK) that are also expressed in the vasculature may also provide vasodilation and neuroprotection by linolenic acid, because BK channels are also activated by polyunsaturated free fatty acids (Kirber et al., 1992). Therefore, it will be necessary to distinguish the roles of the two types of K$^+$ channels.

Whereas volatile and gaseous anaesthetic agents activate TREK-1, local anaesthetics inhibit TREK-1 at clinically relevant concentrations (Patel et al., 1999; Kindler et al., 2003). Direct evidence for the role of TREK-1 in general anaesthesia was obtained from experiments using TREK-1$^{-/-}$ mice in that sensitivity to volatile anaesthetics was markedly reduced in TREK-1$^{-/-}$ mice compared to the wild type. As TREK-1 is expressed in the cortex and other areas of the brain, it makes sense that opening of a background K$^+$ channel such as TREK-1 would decrease neuronal excitability and contribute to anaesthetic-induced hyperpolarization of the cell membrane potential.

Hypoxia has been reported to inhibit not only TASK-1 but also TREK-1 (Miller et al., 2004). Therefore, TREK-1 and probably also TREK-2 may serve as oxygen-sensing K$^+$ channels in the glomus cells of the carotid body that express these channels (Yamamoto and Taniguchi, 2006). Acid- and arachidonate-activated TREK-1 was fully reversed by hypoxia (pO$_2$ = 20 mmHg). Surprisingly, acid and arachidonate failed to activate TREK-1 under hypoxic conditions, and this has raised the question of whether TREK-1 has a neuroprotective role in the brain where the pO$_2$ is lower than that in the arterial blood (Miller et al., 2004). In a subsequent study on this topic, TREK-1 was found to be insensitive to hypoxia, and was activated by arachidonic acid and acid even at an oxygen tension that was less than 4 mmHg (Buckler et al., 2005). These conflicting findings need to be resolved to better understand the role of TREKs as oxygen sensors. Other gases such as NO and CO may activate or inhibit TREKs, and could act as signalling molecules to modulate excitability.

TREKs are highly expressed in dorsal root ganglion (DRG) neurons and may be involved in sensory transduction (Kang and Kim, 2006). The sensitivity of TREKs to membrane tension, acid, various lipids, neurotransmitters, and neuropeptides indicate that TREKs are likely to be involved in pain sensation and sensory transduction. As heat and mechanosensitive TRP ion channels coexist with TREKs in DRG neurons, both channel types probably work in concert to bring about a net change in membrane excitability. These important questions need to be addressed to understand fully the role of TREKs in nociception.

TALK

TALK-1 and TALK-2 are highly expressed in human pancreas, and are expected to be involved in the regulation of cell excitability and hormone secretion. Despite their potential importance, their role in pancreatic physiology remains unknown because of the lack of a specific inhibitor or an activator. Four splice variants of TALK-1 have been identified from human pancreatic tissue and two isoforms form functional channels and two others do not (Han et al., 2003). The two functional splice variants differ markedly in their C-terminus, but show similar biophysical properties. The non-functional isoforms of TALK-1 do not act as dominant negatives when co-expressed with functional isoforms. As the single channel kinetics of the three members of the TALK family are now well described, it should be possible to identify their functional expression in various cell types in the pancreas and allow further study of their function in the native system.

The properties of TASK-2 with respect to oxygen tension and volatile anaesthetics are somewhat similar to those of TASK-1 and

TASK3. TASK-2 is activated by halothane and isoflurane (Gray *et al.*, 2000), and inhibited by local anaesthetics (Kindler *et al.*, 2003). Chronic hypoxia was found to downregulate TASK-2 promoter activity and reduce TASK-2 mRNA expression, causing a depolarization of the membrane potential (Brazier *et al.*, 2005). This finding is particularly relevant to the pathophysiology of the lung and kidney that express TASK-2. Whether or not TASK-2$^{-/-}$ mice show abnormalities in respiratory function has not yet been reported.

THIK

THIK-1 is expressed in many regions of the brain and peripheral tissues. Therefore, its inhibition by a volatile anaesthetic such as halothane should increase cell excitability and produce effects opposite to those produced by the action of halothane on TREKs and TASKs. Hypoxia has been reported to inhibit recombinant THIK-1 as well as THIK-1-like K$^+$ currents in glossopharyngeal neurons (Campanucci *et al.*, 2003). Although THIK-2 protein is expressed in the plasma membrane, it does not form a functional K$^+$ channel, and does not affect the expression of THIK-1. The physiological roles of THIK-2 as well as TASK-5 that do not form functional K$^+$ channels, and understanding why they do not form functional channels are interesting topics for the future.

2.1.4.8 Concluding remarks

Functional K$_{2P}$ channels are expected to provide a significant fraction of the background or leak K$^+$ current in neurons and in many other cell types, and help stabilize the resting membrane potential below the firing threshold. Due to their unique sensitivity to various physiologically relevant factors such as pH, nitric oxide, lipids, anaesthetics, membrane tension, temperature, and receptor agonists, K$_{2P}$ channels are expected to be critical regulators of cell excitability under physiological as well as pathophysiological conditions. Inhibition of the background K$^+$ current provided by K$_{2P}$ channels by various receptor agonists via G protein activation is likely to be an important mechanism for depolarizing the cell and bringing the membrane potential closer to the firing threshold. Studies already suggest that K$_{2P}$ channels are likely to be one of the key components of cellular signalling that affects synaptic transmission, neuroprotection, depression, cerebral blood flow, pulmonary vascular tension, oxygen sensing, gastrointestinal smooth muscle contraction, apoptosis, tumourigenesis and cancer, hormone secretion, and sensory transduction. Thus, some of the K$_{2P}$ channels could be potential therapeutic targets for the treatment of various pathological conditions.

References

Aller MI, Veale EL, Linden AM *et al.* (2005). Modifying the subunit composition of TASK channels alters the modulation of a leak conductance in cerebellar granule neurons. *J Neurosci* **25**, 11455–11467.

Ashmole I, Goodwin PA and Stanfield PR (2001). TASK-5, a novel member of the tandem pore K$^+$ channel family. *Pflugers Arch* **442**, 828–833.

Backx PH and Marban E (1993). Background potassium current active during the plateau of the action potential in guinea pig ventricular myocytes. *Circ Res* **72**, 890–900.

Barriere H, Belfodil H *et al.* (2003). Role of TASK2 potassium channels regarding volume regulation in primary cultures of mouse proximal tubules. *J Gen Physiol* **122**, 177–190.

Bayliss DA, Talley EM, Sirois JE *et al.* (2001). TASK-1 is a highly modulated pH-sensitive 'leak' K$^+$ channel expressed in brainstem respiratory neurons. *Resp Physiol* **129**, 159–174.

Becker D, Geiger D, Dunkel M *et al.* (2004). AtTPK4, an Arabidopsis tandem-pore K$^+$ channel, poised to control the pollen membrane voltage in a pH- and Ca^{2+}-dependent manner. *PNAS* **101**, 15621–15626.

Berg AP, Talley EM, Manger JP *et al.* (2004). Motoneurons express heteromeric TWIK-related acid-sensitive K$^+$ (TASK) channels containing TASK-1 (KCNK3) and TASK-3 (KCNK9) subunits. *J Neurosci* **24**, 6693–6702.

Besana A, Barbuti A, Tateyama MA *et al.* (2004). Activation of protein kinase C epsilon inhibits the two-pore domain K$^+$ channel, TASK-1, inducing repolarization abnormalities in cardiac ventricular myocytes. *J Biol Chem* **279**, 33154–33160.

Blondeau N, Petrault O, Manta S *et al.* (2007) Polyunsaturated fatty acids are cerebral vasodilators via the TREK-1 potassium channel. *Circ Res* **101**, 176–184.

Bockenhauer D, Zilberberg N and Goldstein SA (2001). KCNK2: reversible conversion of a hippocampal potassium leak into a voltage-dependent channel. *Nature Neurosci* **4**, 486–491.

Brazier SP, Mason HS, Bateson AN *et al.* (2005). Cloning of the human TASK-2 (KCNK5) promoter and its regulation by chronic hypoxia. *Biochem Biophys Res Commun* **336**, 1251–1258.

Brickley SG, Revilla V, Cull-Candy SG *et al.* (2001). Adaptive regulation of neuronal excitability by a voltage-independent potassium conductance. *Nature* **409**, 88–92.

Bryan RM Jr, You J, Phillips SC *et al.* (2006) Evidence for two-pore domain potassium channels in rat cerebral arteries. *Am J Physiol Heart Circ Physiol* 291, H770–780.

Buckler KJ and Honore E (2005). The lipid-activated two-pore domain K$^+$ channel TREK-1 is resistant to hypoxia: implication for ischaemic neuroprotection. *J Physiol* **562**, 213–222.

Buckler KJ, Williams BA and Honore E (2000). An oxygen-, acid- and anaesthetic-sensitive TASK-like background potassium channel in rat arterial chemoreceptor cells. *J Physiol* **525** Pt 1, 135–142.

Burdakov D, Jensen LT, Alexopoulos H *et al.* (2006) Tandem-pore K$^+$ channels mediate inhibition of orexin neurons by glucose. *Neuron* 50, 711–722.

Cain SM, Meadows HJ, Dunlop J & Bushell TJ (2008) mGui4 potentiation of K (2P) 2.1 is dependent on C-terminal dephosphorylation. *Mol Cell Neurosci* **37**, 32–9.

Campanucci VA, Brown ST, Hudasek K *et al.* (2005). O2 sensing by recombinant TWIK-related halothane-inhibitable K$^+$ channel-1 background K$^+$ channels heterologously expressed in human embryonic kidney cells. *Neurosci* **135**, 1087–1094.

Campanucci VA, Fearon IM and Nurse CA (2003). A novel O2-sensing mechanism in rat glossopharyngeal neurones mediated by a halothane-inhibitable background K$^+$ conductance. *J Physiol* **548**, 731–743.

Chavez RA, Gray AT, Zhao BB *et al.* (1999). TWIK-2, a new weak inward rectifying member of the tandem pore domain potassium channel family. *J Biol Chem* **274**, 7887–7892.

Chemin J, Girard C, Duprat F *et al.* (2003). Mechanisms underlying excitatory effects of group I metabotropic glutamate receptors via inhibition of 2P domain K$^+$ channels. *EMBO J* **22**, 5403–5411.

Chemin J, Patel A, Duprat F *et al.* (2005). Lysophosphatidic acid-operated K$^+$ channels. *J Biol Chem* **280**, 4415–4421.

Chen X, Talley EM, Patel N *et al.* (2006). Inhibition of a background potassium channel by Gq protein alpha-subunits. *PNAS* **103**, 3422–3427.

Clapham DE (2003). TRP channels as cellular sensors. *Nature* **426**, 517–524.

Clarke CE, Veale EL, Green PJ *et al.* (2004). Selective block of the human 2-P domain potassium channel, TASK-3 and the native leak potassium current, IKSO, by zinc. *J Physiol* **560**, 51–62.

Cooper BY, Johnson RD and Rau KK (2004). Characterization and function of TWIK-related acid sensing K$^+$ channels in a rat nociceptive cell. *Neurosci* **129**, 209–224.

Czempinski K, Zimmermann S, Ehrhardt T *et al.* (1997). New structure and function in plant K$^+$ channels: KCO1, an outward rectifier with a steep Ca^{2+} dependency. *EMBO J* **16**, 2565–2575.

Czirjak G and Enyedi P (2002). Formation of functional heterodimers between the TASK-1 and TASK-3 two-pore domain potassium channel subunits. *J Biol Chem* **277**, 5426–5432.

Czirjak G and Enyedi P (2003). Ruthenium red inhibits TASK-3 potassium channel by interconnecting glutamate 70 of the two subunits. *Mol Pharmacol* **63**, 646–652.

2006a Czirjak G and Enyedi P (2006a). Targeting of calcineurin to an NFAT-like docking site is required for the calcium-dependent activation of the background K$^+$ channel, TRESK. *J Biol Chem* **281**, 14677–14682

Czirjak G and Enyedi P (2006b). Zinc and mercuric ions distinguish TRESK from the other two-pore-domain K$^+$ channels. *Mol Pharmacol* **69**, 1024–1032.

Czirjak G, Toth ZE and Enyedi P (2004). The two-pore domain K$^+$ channel, TRESK, is activated by the cytoplasmic calcium signal through calcineurin. *J Biol Chem* **279**, 18550–18558.

Danthi S, Enyeart JA and Enyeart JJ (2004). Caffeic acid esters activate TREK-1 potassium channels and inhibit depolarization-dependent secretion. *Mol Pharmacol* **65**, 599–610.

Davies LA, Hu C, Guagliardo NA *et al.* (2008) TASK channel deletion in mice causes primary hyperaldosteronism. *Proc Natl Acad Sci USA* 105, 2203–2208.

de la Cruz IP, Levin JZ, Cummins C *et al.* (2003). sup-9, sup-10 and unc-93 may encode components of a two-pore K$^+$ channel that coordinates muscle contraction in *Caenorhabditis elegans*. *J Neurosci* **23**, 9133–9145.

Delmas P, Coste B, Gamper N *et al.* (2005). Phosphoinositide lipid second messengers: new paradigms for calcium channel modulation. *Neuron* **47**, 179–182.

Dobler T, Springauf A, Tovornik S *et al.* (2007) TRESK two-pore-domain K$^+$ channels constitute a significant component of background potassium currents in murine dorsal root ganglion neurones. *J Physiol* 585, 867–879.

Doyle DA, Morais Cabral J, Pfuetzner RA *et al.* (1998). The structure of the potassium channel: molecular basis of K$^+$ conduction and selectivity. *Science* **280**, 69–77.

Duprat F, Lesage F, Fink M *et al.* (1997). TASK, a human background K$^+$ channel to sense external pH variations near physiological pH. *EMBO J* **16**, 5464–5471.

Duprat F, Lesage F, Patel AJ *et al.* (2000). The neuroprotective agent riluzole activates the two P domain K$^+$ channels TREK-1 and TRAAK. *Mol Pharmacol* **57**, 906–912.

Duprat F, Girard C, Jarretou G *et al.* (2005). Pancreatic two P domain K$^+$ channels TALK-1 and TALK-2 are activated by nitric oxide and reactive oxygen species. *J Physiol* **562**, 235–244.

Ellinghaus P *et al.* (2005). Comparing the global mRNA expression profile of human atrial and ventricular myocardium with high-density oligonucleotide arrays. *J Thorac Cardio Surg* **129**, 1383–1390.

Enyeart JJ, Danthi SJ, Liu H *et al.* (2005). Angiotensin II inhibits bTREK-1 K$^+$ channels in adrenocortical cells by separate Ca^{2+}- and ATP hydrolysis-dependent mechanisms. *J Biol Chem* **280**, 30814–30828.

Enyeart JJ, Xu L, Danthi S *et al.* (2002). An ACTH- and ATP-regulated background K$^+$ channel in adrenocortical cells is TREK-1. *J Biol Chem* **277**, 49186–49199.

Feliciangeli S, Bendahhou S, Sandoz G *et al.* (2007) Does sumoylation control K2P1/TWIK1 background K$^+$ channels? *Cell* 130, 563–569.

Fu XW, Wang D, Nurse CA *et al.* (2000). NADPH oxidase is an O2 sensor in airway chemoreceptors: evidence from K$^+$ current modulation in wild-type and oxidase-deficient mice. *PNAS* **97**, 4374–4379.

Gaborit N *et al.* (2005). Human atrial ion channel and transporter subunit gene-expression remodeling associated with valvular heart disease and atrial fibrillation. *Circ* **112**, 471–481.

Garry A, Fromy B, Blondeau N *et al.* (2007) Altered acetylcholine, bradykinin and cutaneous pressure-induced vasodilation in mice

lacking the TREK1 potassium channel: the endothelial link. *EMBO Rep* 8, 354–359.

Girard C *et al.* (2001). Genomic and functional characteristics of novel human pancreatic 2P domain K$^+$ channels. *Biochem Biophys Res Commun* **282**, 249–256.

Girard C *et al.* (2002). p11, an annexin II subunit, an auxiliary protein associated with the background K$^+$ channel, TASK-1. *EMBO J* **21**, 4439–4448.

Goldstein SN, Price LA, Rosenthal DN *et al.* (1996). ORK1, a potassium-selective leak channel with two pore domains cloned from *Drosophila melanogaster* by expression in *Saccharomyces cerevisiae*. *PNAS* **93**, 13256–13261.

Gray AT *et al.* (2000). Volatile anesthetics activate the human tandem pore domain baseline K$^+$ channel KCNK5. *Anesthesiology* **92**, 1722–1730.

Gu W *et al.* (2002). Expression pattern and functional characteristics of two novel splice variants of the two-pore-domain potassium channel TREK-2. *J Physiol* **539**, 657–668.

Gurney AM *et al.* (2003). Two-pore domain K channel, TASK-1, in pulmonary artery smooth muscle cells. *Circ Res* **93**(10), 957–964.

Han J, Kang D and Kim D (2003). Functional properties of four splice variants of a human pancreatic tandem-pore K$^+$ channel, TALK-1. *American J Physiol* **285**, C529–C538.

Han J, Truell J, Gnatenco C *et al.* (2002). Characterization of four types of background potassium channels in rat cerebellar granule neurons. *J Physiol* **542**, 431–444.

Heitmann D *et al.* (2008) Invalidation of TASK1 potassium channels disrupts adrenal gland zonation and mineralocorticoid homeostasis. *EMBO J* **27**(1),179–87.

Heurteaux C *et al.* (2004). TREK-1, a K$^+$ channel involved in neuroprotection and general anesthesia. *EMBO J*, **1–12**.

Heurteaux C, Lucas G and Guy N (2006). Deletion of the background potassium channel TREX-1 results in a depression-resistant phenotype. *Nat Neurosci* **9**, 1134–41.

Honore E, Maingret F, Lazdunski M *et al.* (2002). An intracellular proton sensor commands lipid- and mechano-gating of the K$^+$ channel TREK-1. *EMBO J* **21**, 2968–2976.

Jensen BS, Odum N, Jorgensen NK *et al.* (1999). Inhibition of T cell proliferation by selective block of Ca(2+)-activated K$^+$ channels. *PNAS* **96**, 10917–10921.

Jones SA, Morton MJ, Hunter M *et al.* (2002). Expression of TASK-1, a pH-sensitive twin-pore domain K$^+$ channel, in rat myocytes. *Am J Physiol* **283**, H181–185.

Kang D and Kim D (2006). TREK-2(K2P10.1) and TRESK (K2P18) are major background K$^+$ channels in dorsal root ganglion neurons. *Am J Physiol* **291**(1), C138–46.

Kang D, Choe C and Kim D (2005). Thermosensitivity of the two-pore domain K$^+$ channels TREK-2 and TRAAK. *J Physiol* **564**, 103–116.

Kang D, Choe C, Cavanaugh E *et al.* (2007) Properties of single two-pore domain TREK-2 channels expressed in mammalian cells. *J Physiol* **583**, 57–69.

Kang D, Han J, Talley EM *et al.* (2004a). Functional expression of TASK-1/TASK-3 heteromers in cerebellar granule cells. *J Physiol* **554.1**, 64–77.

Kang D, Mariash E and Kim D (2004b). Functional expression of TRESK-2, a new member of the tandem-pore K$^+$ channel family. *J Biol Chem* **279**, 28063–28070.

Kang D, Kim SH, Hwong EM *et al.* (2007). Expression of thermosensitive two-pore domain K$^+$ channels in human keratinocyte cell line HaCaT. *Exp Dermatol* 16, 1016–1022.

Kang DW and Kim D (2004). Single-channel properties and pH sensitivity of two-pore domain K$^+$ channels of the TALK family. *Biochem Biophys Res Commun* **315**, 836–844.

Karschin C, Wischmeyer E, Preisig-Muller R *et al.* (2001). Expression pattern in brain of TASK-1, TASK-3 and a tandem pore domain K$^+$ channel subunit, TASK-5, associated with the central auditory nervous system. *Mol Cell Neurosci* **18**, 632–648.

Kemp PJ (2005). Hemeoxygenase-2 as an O2 sensor in K$^+$ channel-dependent chemotransduction. *Biochem Biophys Res Commun* **338**, 648–652.

Kennard LE, Chumbley JR, Ranatunga KM *et al.* (2005). Inhibition of the human two-pore domain potassium channel, TREK-1, by fluoxetine and its metabolite norfluoxetine. *Brit J Pharmacol*, **144**, 821–829.

Keshavaprasad B *et al.* (2005). Species-specific differences in response to anesthetics and other modulators by the K2P channel TRESK. *Anesthes Analg* **101**, 1042–1049.

Ketchum KA, Joiner WJ, Sellers AJ *et al.* (1995). A new family of outwardly rectifying potassium channel proteins with two pore domains in tandem. *Nature* **376**, 690–695.

Kim D (1992). A mechanosensitive K$^+$ channel in heart cells: activation by arachidonic acid. *J Gen Physiol* **100**, 1–20.

Kim D (2003). Fatty acid-sensitive two-pore domain K$^+$ channels. *Trends in Pharmacological Science* **24**, 648–654.

Kim D and Clapham DE (1989). Potassium channels in cardiac cells activated by arachidonic acid and phospholipids. *Science* **244**, 1174–1176.

Kim D and Gnatenco C (2001). TASK-5, a new member of the tandem-pore K$^+$ channel family. *Biochem Biophys Res Commun* **284**, 923–930.

Kim D, Sladek CD, Aquado-Velasco C *et al.* (1995). Arachidonic acid activation of a new family of K$^+$ channels in cultured rat neuronal cells. *J Physiol* **484.3**, 643–660.

Kim Y, Bang H and Kim D (1999). TBAK-1 and TASK-1, two-pore K$^+$ channel subunits: kinetic properties and expression in rat heart. *Am J Physiol* **277**, H1669–1678.

Kim Y, Bang H and Kim D (2000). TASK-3, a new member of the tandem pore K$^+$ channel family. *J Biol Chem* **275**, 9340–9347.

Kim Y, Bang H, Gnatenco C *et al.* (2001a). Synergistic interaction and the role of C-terminus in the activation of TRAAK K$^+$ channels by pressure, free fatty acids and alkali. *Pflugers Archiv* **442**, 64–72.

Kim Y, Gnatenco C, Bang H *et al.* (2001b). Localization of TREK-2 K$^+$ channel domains that regulate channel kinetics and sensitivity to pressure, fatty acids and pHi. *Pflugers Arch*, **442**, 952–960.

Kindler CH and Yost CS (2005). Two-pore domain potassium channels: new sites of local anesthetic action and toxicity. *Region Anesthes Pain Med* **30**, 260–274.

Kindler CH *et al.* (2003). Amide local anesthetics potently inhibit the human tandem pore domain background K$^+$ channel TASK-2 (KCNK5). *JPET* **306**, 84–92.

Kirber MT, Ordway RW, Clapp LH *et al.* (1992). Both membrane stretch and fatty acids directly activate large-conductance Ca^{2+}-activated K$^+$ channels in vascular smooth muscle cells. *FEBS Lett* **297**, 24–28.

Kozak M (2005). Regulation of translation via mRNA structure in prokaryotes and eukaryotes. *Gene* **361**, 13–37.

Kunkel MT, Johnstone DB, Thomas JH *et al.* (2000). Mutants of a temperature-sensitive two-P domain potassium channel. *J Neurosci* **20**, 7517–7524.

Kuo A, Gulbis JM, Antcliff JF *et al.* (2003). Crystal structure of the potassium channel KirBac1.1 in the closed state. *Science* **300**, 1922–1926.

Lauritzen I, Blondeau N, Heurteauc C *et al.* (2000). Polyunsaturated fatty acids are potent neuroprotectors. *EMBO J* **19**, 1784–1793.

Lauritzen I, Chemin J, Honere E *et al.* (2003). K$^+$-dependent cerebellar granule neuron apoptosis: Role of TASK leak K$^+$ channels. *J Biol Chem* **278(34)**, 32068–32076.

Lauritzen I, Zanzouri M, Honore E *et al.* (2005). Cross-talk between the mechano-gated K2P channel TREK-1 and the actin cytoskeleton. *EMBO Reports* **6**, 642–648.

Leonoudakis D, Gray AT, Winegar BD *et al.* (1998). An open rectifier potassium channel with two pore domains in tandem cloned from rat cerebellum. *J Neurosci* **18**, 868–877.

Lesage F, Guillemare E, Fink M *et al.* (1996a). TWIK-1, a ubiquitous human weakly inward rectifying K$^+$ channel with a novel structure. *EMBO J* **15**, 1004–1011.

Lesage F, Reyes R, Fink M *et al.* (1996b). Dimerization of TWIK-1 K$^+$ channel subunits via a disulfide bridge. *EMBO J* **15**, 6400–6407.

Lewis A, Hortness ME, Chapman CG *et al.* (2001). Recombinant hTASK1 is an O(2)-sensitive K$^+$ channel. *Biochem Biophys Res Commun* **285**, 1290–1294.

Linden AM, Aller MI, Leppa E *et al.* (2006). The *in vivo* contributions of TASK-1-containing channels to the actions of inhalation anesthetics, the {alpha}2 adrenergic sedative dexmedetomidine and cannabinoid agonists. *JPET* **317**, 615–626.

Liu C, Cotten JF, Schuyler JA *et al.* (2005). Protective effects of TASK-3 (KCNK9) and related 2P K channels during cellular stress. *Brain Res* **1031**, 164–173.

Liu H, Enyeart JA and Enyeart JJ (2007) Potent inhibition of native TREK-1 K$^+$ channels by selected dihydropyridine Ca2$^+$ channel antagonists. *J Pharmacol Exp Ther* **323**, 39–48.

Liu H, Enyeart JA and Enyeart JJ (2007) Angiotensin II inhibits bTREK-1 K$^+$ channels through a PLC- Kinase C- and PIP2-independent pathway requiring ATP hydrolysis. *Am J Physiol* **293**, C682–695.

Lopes CM, Rohacs T, Czirjak G *et al.* (2005). PIP2 hydrolysis underlies agonist-induced inhibition and regulates voltage gating of two-pore domain K$^+$ channels. *J Physiol* **564**, 117–129.

Maingret F *et al.* (2000a). TREK-1 is a heat-activated background K$^+$ channel. *EMBO J* **19**, 2483–2491.

Maingret F, Honore E, Lazdunski M *et al.* (2002). Molecular basis of the voltage-dependent gating of TREK-1, a mechano-sensitive K$^+$ channel. *Biochem Biophys Res Commun* **292**, 339–346.

Maingret F, Patel AJ, Lazdunski M *et al.* (2001). The endocannabinoid anandamide is a direct and selective blocker of the background K$^+$ channel TASK-1. *EMBO J* **20**, 47–54.

Maingret F, Patel AJ, Lesage F *et al.* (2000b). Lysophospholipids open the two-pore domain mechano-gated K$^+$ channels TREK-1 and TRAAK. *J Biol Chem* **275**, 10128–10133.

Matsumoto I, Emori Y, Ninomiya Y *et al.* (2001). A comparative study of three cranial sensory ganglia projecting into the oral cavity: *in situ* hybridization analyses of neurotrophin receptors and thermosensitive cation channels. *Brain Res Mol Brain Res* **93**, 105–112.

Meadows HJ, Chapman CG, Duckworth DM *et al.* (2001). The neuroprotective agent sipatrigine (BW619C89) potently inhibits the human tandem pore-domain K$^+$ channels TREK-1 and TRAAK. *Brain Res* **892**, 94–101.

Medhurst AD, Rennie G, Chapman CG *et al.* (2001). Distribution analysis of human two pore domain potassium channels in tissues of the central nervous system and periphery. *Brain Research. Mol Brain Res* **86**, 101–114.

Millar AJ, Barrat L, Southan AP *et al.* (2000). A functional role for the two-pore domain potassium channel TASK-1 in cerebellar granule neurons. *PNAS* **97**, 3614–3618.

Miller P, Peers C and Kemp PJ (2004). Polymodal regulation of hTREK1 by pH, arachidonic acid and hypoxia: physiological impact in acidosis and alkalosis. *Am J Physiol* **286(2)**, C272–282.

Morton MJ, Abohamed A, Sivaprasadarao A *et al.* (2005). pH sensing in the two-pore domain K$^+$ channel, TASK2. *PNAS* **102**, 16102–16106.

Morton MJ, O'Connell AD, Sivaprasadarao A *et al.* (2003). Determinants of pH sensing in the two-pore domain K$^+$ channels TASK-1 and -2. *Pflugers Archiv* **445**, 577–583.

Mu D, Chen L, Zhag X *et al.* (2003). Genomic amplification and oncogenic properties of the KCNK9 potassium channel gene. *Cancer Cell* **3**, 297–302.

Mulkey DK, Talley OM, Stornetta RL *et al.* (2007). TASK channels determine PH sensitivity in select respiratory neurons but do not contribute to central respiratory chemosensitivity. *J Neurosci* **27**, 14049–14058.

Mulkey DK, Stornetta RL, Weston MC *et al.* (2004). Respiratory control by ventral surface chemoreceptor neurons in rats. *Nature Neurosci* **7**, 1360–1369.

Murbartian J, Lei Q, Sando JJ *et al.* (2005). Sequential phosphorylation mediates receptor- and kinase-induced inhibition of TREK-1 background potassium channels. *J Biol Chem* **280**, 30175–30184.

Musset B, Meuth SG, Liu GX *et al.* (2006). Effects of divalent cations and spermine on the K$^+$ channel TASK-3 and on the outward current in thalamic neurons. *J Physiol* **572(Pt 3)**, 639–57.

Nie X, Amighi I, Kaissling B *et al.* (2005). Expression and insights on function of potassium channel TWIK-1 in mouse kidney. *Pflugers Archiv* **451**, 479–488.

Niemeyer MI, Cid LP, Barros LF *et al.* (2001). Modulation of the two-pore domain acid-sensitive K$^+$ channel TASK-2 (KCNK5) by changes in cell volume. *J Biol Chem* **276**, 43166–43174.

Niemeyer MI, Gonzalez-Nilo FD, Zuniga L *et al.* (2007) Neutralization of a single arginine residue gates open a two-pore domain, alkali-activated K$^+$ channel. *Proc Natl Acad Sci U S A* 104, 666–671.

O'Kelly I, Butler MH, Zilberberg N *et al.* (2002). Forward transport. 14-3-3 binding overcomes retention in endoplasmic reticulum by dibasic signals. *Cell* 111, 577–588.

O'Kelly I, Stephens RH, Peers C *et al.* (1999). Potential identification of the O2-sensitive K$^+$ current in a human neuroepithelial body-derived cell line. *Am J Physiol* **276**, L96–L104.

Olschewski A *et al.* (2006). Impact of TASK-1 in human pulmonary artery smooth muscle cells. *Circ Res* **98(8),** 1072–1080.

Ordway RW, Walsh JV Jr and Singer JJ (1989). Arachidonic acid and other fatty acids directly activate potassium channels in smooth muscle cells. *Science* **244**, 1176–1179.

Patel AJ *et al.* (1998). A mammalian two-pore domain mechano-gated S-like K$^+$ channel. *EMBO J* **17**, 4283–4290.

Patel AJ *et al.* (1999). Inhalational anesthetics activate two-pore-domain background K$^+$ channels. *Nature Neurosci* **2**, 422–426.

Patel AJ *et al.* (2000). TWIK-2, an inactivating 2P domain K$^+$ channel. *J Biol Chem* **275**, 28722–28730.

Patel AJ, Lazdunski M and Honore E (2001). Lipid and mechano-gated 2P domain K$^+$ channels. *Curr Opin Cell Biol* **13**, 422–428.

Pei L *et al.* (2003). Oncogenic potential of TASK3 (Kcnk9) depends on K$^+$ channel function. *PNAS* **100**, 7803–7807.

Perrier JF, Alaburda A and Hounsgaard J (2003). 5-HT1A receptors increase excitability of spinal motoneurons by inhibiting a TASK-1-like K$^+$ current in the adult turtle. *J Physiol* **548**, 485–492.

Plant LD, Kemp PJ, Peers C *et al.* (2002). Hypoxic depolarization of cerebellar granule neurons by specific inhibition of TASK-1. *Stroke* **33**, 2324–2328.

Pokojski S, Busch C, Grgic I *et al.* (2008) TWIK-related two-pore domain potassium channel TREK-1 in carotid endothelium of normotensive and hypertensive mice. *Cardiovasc Res* (to be added).

Pongs O, Leicher T, Berger M *et al.* (1999). Functional and molecular aspects of voltage-gated K$^+$ channel beta subunits. *Annals of the NYAS* **868**, 344–355.

Pountney DJ, Gulkarov I, Vega-Saenz de Hiera E *et al.* (1999). Identification and cloning of TWIK-originated similarity sequence (TOSS): a novel human 2-pore K$^+$ channel principal subunit. *Fed Soc Biochem Mol Biol Letts* **450**, 191–196.

Rajan S, Wischmeyer E, XnLiu G *et al.* (2000). TASK-3, a novel tandem pore domain acid-sensitive K$^+$ channel. An extracellular histidine as pH sensor. *J Biol Chem* **275**, 16650–16657.

Rajan S, Wischmeyer E, Karschin C *et al.* (2001). THIK-1 and THIK-2, a novel subfamily of tandem pore domain K$^+$ channels. *J Biol Chem* **276**, 7302–7311.

Rajan S, Preisig-Muller R, Wischmeyer E *et al.* (2002). Interaction with 14-3-3 proteins promotes functional expression of the potassium channels TASK-1 and TASK-3. *J Physiol* **545**, 13–26.

Rajan S, Plant LD, Rabin ML *et al.* (2005). Sumoylation silences the plasma membrane leak K$^+$ channel K2P1. *Cell* **121**, 37–47.

Renigunta V, Yuan M, Zuzarte M *et al.* (2006). The retention factor p11 confers an endoplasmic reticulum-localization signal to the potassium channel TASK-1. *Traffic* **7**, 168–181.

Reyes R, Duprat F, Lesage F *et al.* (1998). Cloning and expression of a novel pH-sensitive two pore domain K$^+$ channel from human kidney. *J Biol Chem* **273**, 30863–30869.

Sanders KM & Soh SD (2006) Two-pore-domain potassium channels in smooth muscle new components of myogenic regulation. *J Physiol* **570**, 37–43.

Sano Y, Inamura K, Miyake A *et al.* (2003). A novel two-pore domain K$^+$ channel, TRESK, is localized in the spinal cord. *J Biol Chem* **278**: 27406–27412.

Simlein D, Cavanaugh CJ, Kim D. (2008) Control of the single channel conductance of K$_{2p}$10.1 (TREX-2) by the amino-terminus role of alternative translation irritation. *J Physiol* **586**, 5651–5663.

Sirois JE, Lynch C, 3rd and Bayliss DA (2002). Convergent and reciprocal modulation of a leak K$^+$ current and I(h) by an inhalational anaesthetic and neurotransmitters in rat brainstem motoneurones. *J Physiol* **541**, 717–729.

Suh BC and Hille B (2005). Regulation of ion channels by phosphatidylinositol 4,5-bisphosphate. *Curr Opin Neurobiol* **15**, 370–378.

Talley EM and Bayliss DA (2002). Modulation of TASK-1 (Kcnk3) and TASK-3 (Kcnk9) potassium channels: volatile anesthetics and neurotransmitters share a molecular site of action. *J Biol Chem* **277**, 17733–17742.

Talley EM, Solorzano G, Lei Q *et al.* (2001). Cns distribution of members of the two-pore-domain (KCNK) potassium channel family. *J Neurosci* **21**, 7491–7505.

Terrenoire C, Lauritzen I, Lesage F *et al.* (2001). A TREK-1-like potassium channel in atrial cells inhibited by beta-adrenergic stimulation and activated by volatile anesthetics. *Circ Res* **89**, 336–342.

Theilig F, Goranova I, Hirsch JR *et al.* (2008). Cellular localization of THIK-1 (K2P13.1) and THIK-2 (K2P12.1) K channels in the mammalian kidney. *Cell Physiol Biochem* **21**, 63–74.

Thomas D, Plant LD, Wilkens CM *et al* (2008). Alternative translation initiation in rat brain yields K2p2.i potassium channels permeable to sodium. *Neuron* **58**, 859–870.

Veale EL, Buswell R, Clarke CE and Mathie A (2007). Identification of a region in the TASK3 two pore domain potassium channel that is critical for its blockade by methanandamide. *Br J Pharmacol* **152**, 778–786.

Vega-Saenz de Miera E, Lau DH, Zhadina M *et al.* (2001). KT3.2 and KT3.3, two novel human two-pore K$^+$ channels closely related to TASK-1. *J Neurophysiol* **86**, 130–142.

Voloshyna I, Besana A, Castillo M, Matos T *et al* (2008). Trex-1 is a novel molecular target in prostate cancer. *Cancer Res* **68**, 1197–203.

Warth R, Barriere H, Meneton P *et al.* (2004). Proximal renal tubular acidosis in TASK2 K$^+$ channel-deficient mice reveals a mechanism for stabilizing bicarbonate transport. *PNAS* **101**, 8215–8220.

Williams BA and Buckler KJ (2004). Biophysical properties and metabolic regulation of a TASK-like potassium channel in rat carotid body type-1 cells. *Am J Physiol* **286(1)**, L221–230.

Xian Tao L, Dyachenko V, Zuzarte M *et al.* (2006). The stretch-activated potassium channel TREK-1 in rat cardiac ventricular muscle. *Cardiovasc Res* **69**, 86–97.

Yamamoto Y and Taniguchi K (2006). Expression of tandem P domain K$^+$ channel, TREK-1, in the rat carotid body. *J Histochem Cytochem* **54**, 467–472.

Yarbrough GG (2003). Fundamental molecular mechanism of the CNS effects of TRH. *TIPS* **24**, 617–618.

Yu SP (2003). Regulation and critical role of potassium homeostasis in apoptosis. *Prog Neurobiol* **70**, 363–386.

Yuill KH, Stansfeld PJ, Ashmole I (2007). The selectivity, voltage dependence and acid sensitivity of the tandem pore pottassium channel TASK-1: contributions of the pore domains. *Pfuigers Arch* **455**, 333–348.

2.2

Voltage-gated calcium channels

Joel Baumgart and Edward Perez-Reyes

2.2.1 Introduction

Subsequent to their discovery over five decades ago in a crustacean muscle preparation, voltage-gated calcium channels are now recognized for their role at the numerous interfaces between mechanical, electrical, and chemical forms of energy. Native calcium channels do not operate in isolation but work in harmony with other types of ion channels, notably voltage-gated sodium and potassium channels as well as ligand-gated channels. In all cases of signal transduction, the conversion of an electrical signal to a chemical message requires the activity of Ca^{2+} channels. Accordingly, voltage-gated calcium channels have come to be viewed as hallmarks of excitable cells (Hille, 2001). For an interesting review of the early history of voltage-gated Ca^{2+} channels, see Dolphin (2006).

Electrophysiological measurement of Ca^{2+} currents in native cells resolves two groups of Ca^{2+} channels on the basis of their voltage dependence of activation. Low voltage-activated (LVA) channels activate in response to modest depolarizations of the plasma membrane and show rapid voltage-dependent inactivation, while channels requiring larger depolarizations to open are designated high voltage-activated (HVA).

The subunit structure of HVA channels was determined following their purification. Early work on the purification of L-type calcium channels from skeletal muscle revealed a multimeric complex of at least four proteins: α_1, $\alpha_2\delta$, β, and γ subunits in a 1:1:1:1 ratio (Campbell et al., 1988). The γ1 subunit appears to be specific to skeletal muscle channels, as it was not detected in purified preparations of brain N-/P/Q-type channels, which show an α_1-β-$\alpha_2\delta$ structure (Scott et al., 1998; Witcher et al., 1993). In contrast, the subunit structure of LVA channels has yet to be resolved.

Pharmacological advances allowed further subcategorization of the HVA family into L-, N-, P/Q- and R-type channels. Finally, molecular cloning of the voltage-gated calcium channel family members provided the foundation for a taxonomic scheme based on sequence homologies (Ertel et al., 2000). Under this nomenclature voltage-gated Ca^{2+} channels are denoted as 'Ca$_v$', and the ten α_1 subunits are grouped into three families, the Ca$_v$1.x family that encodes L-type channels, the Ca$_v$2.x family that encodes N-, P/Q- and R-type channels, and the Ca$_v$3.x family that encodes T-type channels (Figure 2.2.1). Although less common, α_1 subunits are

sometimes referred to in the α_1X.Y classification scheme (where X = channel family and Y = channel subtype, e.g. $\alpha_1$1.1). The auxiliary subunits are designated Ca$_v\alpha_2\delta$X, Ca$_v\beta$X, and Ca$_v\gamma$X (where X = channel subtype). Interestingly, a comparison of amino acid sequences indicates that an early evolutionary event some 500 million years ago partitioned the ancient calcium channel subunit into the HVA and LVA subfamilies (Perez-Reyes, 2003). There are ten human genes that encode the ion-selective pore-forming α_1 subunit of voltage-gated calcium channels. The α_1 subunit is a large, monomeric protein that consists of four homologous domains. Based on sequence homology, each domain is thought to resemble a voltage-gated K^+ channel, such as Shaker (Long et al., 2005), which possesses six membrane-spanning segments (S1 through S6) and a re-entrant pore loop that dips partially in the plasma membrane between S5 and S6. All voltage-gated channels also contain highly conserved S4 transmembrane segments with positively charged lysine or arginine residues at approximately every third position that act as the predominant voltage sensor.

The molecular cloning of the voltage-gated calcium channels can be divided into three eras based on the techniques employed. As mentioned above, direct sequencing of skeletal muscle proteins initiated the 'skeletal muscle era'. This led to the cloning of the first Ca^{2+} channel subunits (α_{1S}, or Ca$_v$1.1), β1a, and γ1. Screening of cDNA libraries with low stringency oligonucleotide probes designed from these skeletal muscle cDNAs led to the cloning of related L-type channels from heart (α_{1C}, Ca$_v$1.2) and brain (α_{1D}, Ca$_v$1.3), as well as the β2 subunit. The 'brain era' continued with the cloning of the Ca$_v$2 family, which encodes three channels important for synaptic transmission: Ca$_v$2.1 (α1A, P/Q-type channel), Ca$_v$2.2 (α_{1B}, N-type channel), and Ca$_v$2.3 (α_{1E}, R-type channel). The final era has been termed the 'in silico' era as computer-based mining of genetic databases has resulted in the discovery of all the members of the calcium channel gene family. In the α1 family this led to the identification of the Ca$_v$3 family (Perez-Reyes, 1999), which encodes T-type channels, and contains three members: Ca$_v$3.1 (α_{1G}), Ca$_v$3.2 (α_{1H}), and Ca$_v$3.3 (α_{1I}). In the $\alpha_2\delta$ family this led to the identification of three additional members: $\alpha_2\delta$2, $\alpha_2\delta$3, and $\alpha_2\delta$24 (Klugbauer et al., 1999b; Qin et al., 2002). Genetic screens led to the identification of Ca$_v$1.4 and γ2 as the causative genes in congenital stationary night blindness and the *stargazer* mouse model of absence epilepsy, respectively. Although *in silico*

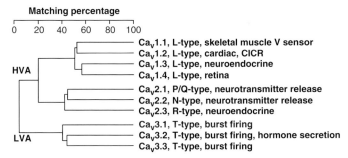

Fig. 2.2.1 Sequence conservation of the Ca$_v$ a$_1$ family. Amino acid sequences of the human channels were aligned using the CLUSTAL algorithm (Higgins and Sharp, 1988) and DNASIS software was then used to create the evolutionary tree. The entire coding sequence was used for the alignment. CICR, calcium-induced calcium release,

cloning has revealed many genes related to γ2, it appears that these proteins are not Ca^{2+} channel subunits (Nicoll *et al.*, 2006), although this remains controversial (Sharp *et al.*, 2001). Genetic screens for a possible tumour suppressor at chromosome 3p21, which is deleted in carcinomas, also led to the cloning of α$_2$δ2 (Gao *et al.*, 2000) and a$_2$d3 (Hanke *et al.*, 2001). The α$_2$δ2 cDNA was also cloned during studies on the locus associated with the absence epilepsy phenotype of the *ducky* mouse (Barclay *et al.*, 2001). With the complete sequencing of the human genome, all the homologs of α$_1$, α$_2$δ, and β subunits have been cloned (Tables 2.2.1).

2.2.2 Ca$_v$1 channels

Introduction

Ca$_v$1.1 channels are the most thoroughly studied subtype of voltage-gated Ca^{2+} channel. Due to its unique expression in skeletal muscle and relatively simple purification, its subunit structure has been definitively characterized as being composed of α$_1$1.1 (formerly α$_{1S}$), α$_2$δ1, β1a, γ1, and this complex has been visualized using electron cryomicroscopy (Wolf *et al.*, 2003; Figure 2.2.2). Although it can function as a voltage-gated Ca^{2+} channel, the primary role for Ca$_v$1.1 is as a voltage-sensor, coupling muscle depolarization to release of Ca^{2+} from the sarcoplasmic reticulum. It provided one of the first examples of a 'channelopathy', an ion channel gene with mutations associated with human disease (Ptacek, 1997). Ca$_v$1.1 binds calcium channel blockers (CCB) such as the dihydropyridines (DHPs) with high affinity but these agents do not block the physiological function of the channel.

Ca$_v$1.2 is arguably the most important and best studied of the L-type channel family. Its role in cardiac contraction was first suggested by the classic studies of Sydney Ringer, who showed that isolated heart required extracellular Ca^{2+} to beat (Ringer, 1883). It has garnered significant attention as the molecular target of many antihypertensive drugs. Ca$_v$1.2 is the predominant channel responsible for Ca^{2+} currents in cardiac and smooth muscle myocytes. Its activity, pharmacology, and regulation have been reviewed extensively (Pelzer *et al.*, 1990; Striessnig, 1999; Lipscombe *et al.*, 2004; Bodi *et al.*, 2005; Godfraind, 2006). Ca$_v$1.2 was the second L-type channel to be defined at the molecular level. Evidence suggests a diversity of function for Ca$_v$1.2 channels, including roles in cardiac and smooth muscle function, action potential propagation, synaptic plasticity, and insulin secretion. Its gene is spliced at

many locations, which alters the resulting channels sensitivity to antihypertensive drugs, and in part explains their tissue selectivity for smooth muscle.

In contrast, much less is known of the biochemistry, pharmacology, and biophysics of Ca$_v$1.3 and Ca$_v$1.4 channels. Ca$_v$1.3 has a broad tissue distribution, and its expression in pancreas and adrenal explains why it is sometimes referred to as a neuroendocrine channel. Although its gene is highly related to the gene encoding Ca$_v$1.2, its biophysical properties are quite different; notably, Ca$_v$1.3 channels open close to the resting membrane potential of most cells as in the case of LVA channels. Ca$_v$1.4 channel expression is almost entirely restricted to the retina, where it plays a role in rod photoreceptor signalling.

Molecular characterization

Ca$_v$1.1

Skeletal muscle L-type Ca^{2+} channels were the first to be defined at the molecular level due to their abundant expression and high affinity binding to a class of antihypertensive drugs known as dihydropyridines (DHPs) (reviewed by Perez-Reyes and Schneider, 1994). T-tubule membranes of skeletal muscle were prelabelled with tritiated dihydropyridines, solubilized, and purified through wheat germ agglutinin and anion exchange chromatography and sucrose gradient centrifugation. The calcium channel activity of purified DHP receptors was measured after reconstitution into lipid bilayers. Purified preparations revealed the presence of at least three proteins in a multimeric receptor complex, a subunit 165–190 kDa in size, designated α, a 55 kDa β subunit, and a 32 kDa γ aubunit. Azidopine photolabelling studies demonstrated that the DHP binding subunit was the 165–190 kDa protein (α). Subsequent findings provided evidence of a second α subunit of ~170 kDa, α$_2$δ, that can be dissociated by reducing agents, decreasing its mobility in SDS-PAGE gels to 142 kDa (α$_2$) and 22–28 kDa (δ). Microsequencing of these proteins allowed for the design of oligonucleotides to screen cDNA libraries. Using these techniques all the subunits of the rabbit skeletal muscle L-type channel were cloned (Tanabe *et al.*, 1987; Ellis *et al.*, 1988; Ruth *et al.*, 1989; Bosse *et al.*, 1990; Jay *et al.*, 1990). The sizes of the purified proteins concur with those deduced from the primary sequence, with the exception of α$_1$, which appears smaller in purified preparations. Only trace amounts of the full length 214 kDa α$_1$ have been detected, a size discrepancy due to either proteolytic processing or artificial degradation of the carboxyl terminus (De Jongh *et al.*, 1991).

Similar Ca$_v$1.1 (α$_{1S}$) cDNAs have now been cloned from many species, including (Swiss-Protein or GenBank accession numbers in parenthesis): human (Q13698), mouse (Q02789), rabbit (P07293), rat (Q02485), chick (O42398), frog (AF037625), carp (M62554), and zebrafish (AY495698). The gene is highly conserved and indeed homologues can be observed as far back as *Paramecia* (Lee, 1998). Therefore, it seems likely to represent the ancestral Ca^{2+} channel gene which subsequently gave rise to the other three members of the Ca$_v$1 family through gene duplication. Since physiologically it now acts as a voltage-sensor, its role could be viewed to have evolved beyond that of a simple Ca^{2+} channel.

Ca$_v$1.2

Attempts to purify the Ca^{2+} channels of cardiac tissue were disadvantaged by a lower number of DHP binding sites in sarcolemmal

Table 2.2.1 Properties of voltage-gated calcium channels

Subunit	Gene	Human chromosomal localization	Knockout mouse phenotype	Cellular and subcellular localization
$Ca_V1.1$, (α_{1S})	CACNA1S	1q32.1	Die at birth due to muscular dysgenesis [1]	Skeletal muscle transverse tubules [2]
$Ca_V1.2$, (α_{1C})	CACNA1C	12p13.3	Embryonic lethal [3]	Cardiac muscle, smooth muscle, and brain [4]; somatodendritic localization in neurons [5]
$Ca_V1.3$, $(\alpha1D)$	CACNA1D	3p14.3	Sinoatrial node dysfunction and deafness [6]	Sensory cells, including photoreceptors and cochlear hair cells [6]; endocrine cells, including pancreatic β-cells, pituitary, adrenal chromaffin cells, and pinealocytes [7]; low density in heart and vascular smooth muscle [8]; somatodendritic localization in neurons [5]
$Ca_V1.4$, (α_{1F})	CACNA1F	Xp11.23		Retinal photoreceptors and bipolar cells [9]
$Ca_V2.1$, (α_{1A})	CACNA1A	19p13	Cerebellar atrophy, muscle spasms, and ataxia; usually die by 3 to 4 weeks postnatal [10, 11]	Neurons, predominantly in presynaptic terminals [12]
$Ca_V2.2$, (α_{1B})	CACNA1B	9q34	Viable, increased mean arterial pressure and other alterations of the sympathetic nervous system [13, 14, 15]	Neurons, predominantly presynaptic terminals [16]
$Ca_V2.3$, (α_{1E})	CACNA1E	1q25–q31	Viable, increased resistance to formalin-induced pain [17, 18], and altered seizure susceptibility [19]	Brain, heart, testes, pituitary; neuronal cell bodies, dendrites, some presynaptic terminals [20]
$Ca_V3.1$, α_{1G}	CACNA1G	17q22	Reduced seizure susceptibility [21], sleep disturbances [22], central control of pain [23]	Brain [24], sinoatrial node [25]; in neurons localized to soma and dendrites [26]
$Ca_V3.2$, (α_{1H})	CACNA1H	16p13.3	Viable, abnormal development of coronary arteries [27]	Kidney, liver, brain [28], Adrenal glomerulosa [29], brain [24], in neurons localized to soma and dendrites [26]
$Ca_V3.3$, (α_{1I})	CACNA1I	22q13.1		Brain [24], in neurons localized to soma and dendrites [26]
$Ca_V\alpha_2d1$ (a_2d1)	CACNA2D1	7q21–q22		Splice variants expressed in a tissue-specific manner in brain, skeletal muscle, heart, and smooth muscle [30]
$Ca_V\alpha_2d2$ (a_2d2)	CACNA2D2	3p21.3	Ducky mice (du/du) exhibit ataxia, paroxysmal dyskinesia, and seizure activity [31]	Brain, heart, testis, pancreas, kidney, liver, lung, prostate, spinal cord [32]
$Ca_V\alpha_2d3$ (a_2d1)	CACNA2D3	3p21.1		Brain, heart, skeletal muscle [33]
$Ca_V\alpha_2d4$ (a_2d1)	CACNA2D4	12p13.3		Heart, pituitary, colon, adrenal gland, skeletal muscle, liver [34]
$Ca_V\beta1$ $(\beta1)$	CACNB1	17q21–q22	Survive to birth but with inability to move and greatly reduced muscle mass [35]	Skeletal muscle (β1a), brain [35, 36]
$Ca_V\beta2$ $(\beta2)$	CACNB2	10p12	Embryonic lethal [37]	Brain, heart, lung, trachea, aorta [38]
$Ca_V\beta3$ $(\beta3)$	CACNB3	12q13	Viable, altered pain response [39]	Brain, smooth muscle, trachea, aorta [40]
$Ca_V\beta4$ $(\beta4)$	CACNB4	2q22–q23	Lethargic mice (lh/lh) exhibit ataxia, lethargy, and seizure activity [41]	Brain [42]
$Ca_V\gamma_1$ (γ_1)	CACNG1	17q24	Healthy appearance and normal behaviour [43, 44]	Skeletal muscle [45]

[1] Chaudhari N (1992) J Biol Chem 267, 25636–25639; [2] Flucher BE et al. (1996) Proc Natl Acad Sci USA 93, 8101–8106; [3] Seisenberger C et al. (2000) J Biol Chem 275, 39193–39199; [4] Splawski I et al. (2004) Cell 119, 19–31; [5] Hell JW et al. (1993) Cell Biol 123, 949–962; [6] Platzer J et al. (2000) Cell 102, 89–97; [7] Takimoto K et al. (1997) J Mol Cell Cardiol 29, 3035–3042; [8] Mangoni ME et al. (2003) Proc Natl Acad Sci USA 100, 5543–5548; [9] McRory JE et al. (2004) J Neurosci 24, 1707–1718; [10] Jun K et al. (1999) Proc Natl Acad Sci USA 96, 15245–15250; [11] Fletcher CF et al. (2001) FASEB J 15, 1288–1290; [12] Westenbroek RE et al. (1998) J Neurosci 18, 6319–6330; [13] Ino M et al. (2001) Proc Natl Acad Sci USA 98, 5323–5328; [14] Mori Y et al. (2002) Trends Cardiovasc Med 12, 1270–275; [15] Beuckmann CT et al. (2003) J Neurosci 23, 6793–6797; [16] Westenbroek RE et al. (1992) Neuron 9, 1099–1115; [17] Saegusa H et al. (2000) Proc Natl Acad Sci USA 97, 6132–6137; [18] Wilson SM et al. (2000) J Neurosci 20, 8566–8571; [19] Weiergräber M et al. (2006) Epilepsia 47, 839–85c; [20] Grabsch H et al. (1999) J Histochem Cytochem 47, 981–994; [21] Kim D et al. (2001) Neuron 31, 35–45; [22] Lee J et al. (2004) J Neurosci 24, 2581–2594; [27] Chen C-C et al. (2003) Science 302, 1416–1418; [28] Cribbs LL et al. (1998) Circ Res 83, 103–109; [29] Schrier AD et al. (2001) Am J Physiol Cell Physiol 280, C265–C272; [30] Angelotti T et al. (1996) FEBS Lett 397, 331–337; [31] Barclay J et al. (2001) J Neurosci 21, 6095–6104; [32] Gao B et al. (2000) European Journal of Neuroscience 24, 2581–2594; [23] Kim D et al. (2003) Science 302, 117–119; [24] Talley EM et al. (1999) J Neurosci 19, 1895–1911; [25] Bohn G et al. (2000) FEBS Lett 481, 73–76; [26] McKay BE et al. (2006) European Journal of Neuroscience 24, 2581–2594; [27] Klugbauer N (1999a) J Biol Chem 275, 12237–12242; [33] Klugbauer N (1999b) J Biol Chem 275, 12237–12242; [33] Klugbauer N (1999b) J Physiol Cell Physiol 280, Gregg RC et al. (1996) Proc Natl Acad Sci USA 93, 13961–13966; [36] Ruth P et al. (1989) Science 245, 1115–1118; [37] Ball SL et al. (2002) Invest Ophthalmol Vis Sci 43, 1595–1603; [38] Perez-Reyes E et al. (1992) J Biol Chem 267, 1792–1797; [39] Murakami M et al. (2002) J Biol Chem 277, 40342–40351; [40] Castellano A et al. (1993) J Biol Chem 268, 3450–3455; [41] Burgess DL et al. (1997) Cell 88, 385–392; [42] Castellano A et al. (1993) J Biol Chem 268, 12359–12366; [43] Ahern CA et al. (2001) BMC Physiol 1, 8; [44] Freise D et al. (2000) J Biol Chem 275, 14476–14481; [45] Bosse E et al. (1990) FEBS Lett 267, 153–156.

Subunit	Pharmacology	Physiology	Relevance to disease states
Ca$_V$1.1, (α_{1S})	Dihydropyridine antagonists (e.g., (+)- isradipine; IC$_{50}$ = 13 nM at -90 mV and 0.15 nM at -65 mV) [1]	Excitation–contraction coupling and Ca^{2+} homeostasis in skeletal muscle [2]	Point mutations cause hypokalemic periodic paralysis and malignant hyperthermia susceptibility in humans and muscular dysgenesis in mice [3, 4]
Ca$_V$1.2, (α_{1C})	Agonists: BayK8644 and FPL64176 [5]; Dihydropyridine antagonists (e.g, isradipine, IC$_{50}$ = 7 nM at -60 mV; nimodipine, IC$_{50}$ = 139 nM at -80 mV) [6]	Excitation–contraction coupling in cardiac or smooth muscle, action potential propagation in sinoatrial and atrioventricular node, synaptic plasticity, hormone (e.g, insulin) secretion [7, 8]	Mutation causes Timothy syndrome [9]
Ca$_V$1.3, (α_{1D})	Agonists: BayK8644 and FPL64176 [5]. Dihydropyridine antagonists (e.g., isradipine, IC$_{50}$ = 30 nM at -50 mV and 300 nM at -90 mV; nimodipine, IC$_{50}$ = 3 μM at -80 mV) [10]	Control of cardiac rhythm, hearing, mood behaviour, and hormone secretion [7, 11, 12]	Mutations have been associated with deafness, sinoatrial and atrioventricular node dysfunction [11]
Ca$_V$1.4, (α_{1F})	Agonist: BayK8644; Dihydropyridine antagonists: nifedipine [13]; isradipine, also blocked by diltiazem and verapamil [14]	Neurotransmitter release in retinal cells [15]	Mutations cause X-linked congenital stationary night blindness type 2 [16, 17]
Ca$_V$2.1, (α_{1A})	ω-conotoxin MVIIC [18]	Neurotransmitter release in CNS neurons as well as neuromuscular junction [19, 20, 21]	Mutations underlie: familial hemiplegic migraine (FHM), episodic ataxia type-2 (EA2), and spinocerebellar ataxia type-6 (SCA6) [22]
Ca$_V$2.2, (α_{1B})	ω-conotoxin GVIA (1–2 μM, irreversible block), ω-conotoxin MVIIA (SNX-111, ziconotide/prialt, ω-conotoxin MVIIC [23]	Neurotransmitter release in central and sympathetic neurons [19], sensation and transmission of pain [24]	Upregulated in neuropathic pain [25]
Ca$_V$2.3, (α_{1E})	SNX-482, Ni^{2+} (IC$_{50}$ = 27 μM), Cd^{2+} (IC$_{50}$ = 0.8 μM), mibefradil (IC$_{50}$ = 0.4 μM) [26], and volatile anesthetics [27]	Neurotransmitter release [20], long-term potentiation [28], insulin release [29]	Mutations associated with type 2 diabetes [30]
Ca$_V$3.1, α_{1G}	Mibefradil, efonidipine, pimozide, kurtoxin [31], ethosuximide [32]	Thalamic oscillations [33]	
Ca$_V$3.2, (α_{1H})	Mibefradil, efonidipine, pimozide, kurtoxin [31], ethosuximide [32]	Sensation of pain [34], muscle development [35], aldosterone secretion [36]	Single nucleotide polymorphisms associated with childhood absence epilepsy [37]
Ca$_V$3.3. (α_{1I})	Mibefradil, efonidipine, pimozide, kurtoxin [31], ethosuximide [32]	Long-lasting burst firing [38]	
Ca$_V$α$_2$δ1 (α$_2$δ1)	Binds gabapentin and pregabalin [39, 40]	Increases current amplitude by stabilizing Ca$_V$α$_1$ at the plasma membrane [41]	Upregulated in models of neuropathic pain [42]
Ca$_V$α$_2$δ2 (α$_2$δ2)	Binds gabapentin [39]	Increases current amplitude [43]	Deleted in lung, breast, and other cancers [44]
Ca$_V$α$_2$δ3 (α$_2$δ1)		Increases current amplitude; modulates voltage-dependence of activation and steady-state inactivation [45]	Deleted in renal cell carcinomas [46]
Ca$_V$α$_2$δ4 (α$_2$δ1)		Increases current amplitude of Ca$_V$1.2 [47]	Mutations associated with autosomal recessive cone dystrophy [48]
Ca$_V$β1 (β1)		Excitation–contraction coupling in skeletal muscle; increases current amplitude; modulates activation and inactivation kinetics; promotes Ca$_V$α$_1$ trafficking to the plasma membrane [49]	
Ca$_V$β2 (β2)		Increases current amplitude; modulates activation and inactivation kinetics; promotes Ca$_V$α$_1$ trafficking to the plasma membrane [49]	
Ca$_V$β3 (β3)		Increases current amplitude; modulates activation and inactivation kinetics; promotes Ca$_V$α$_1$ trafficking to the plasma membrane [49]	
Ca$_V$β4 (β4)		Increases current amplitude; modulates activation and inactivation kinetics; promotes Ca$_V$α$_1$ trafficking to the plasma membrane [49]	Mutations associated with epilepsy [50]
Ca$_V$γ$_1$ (γ1)		Reduces channel activity and modulates activation and inactivation kinetics [51]	

[1] Striessnig J (1999) Cell Physiol Biochem 9, 242–269; [2] Rios E et al. (1992) Annu Rev Physiol 54, 109–133; [3] Striessnig J et al. (2004) Biochem Biophys Res Commun 322, 1341–1346; [4] Chaudhari N (1992) J Biol Chem 267, 25636–25639; [5] Zheng W et al. (1991) Mol Pharmacol 40, 734–741; [6] Peterson BZ et al. (1997) J Biol Chem 272, 18752–18758; [7] Sinnegger-Brauns MJ et al. (2004) J Clin Invest 113, 1430–1439; [8] Schulla V et al. (2003) EMBO J 22, 3844–3854; [9] Splawski I et al. (2004) Cell 119, 19–31; [10] Xu W et al. (2001) J Neurosci 21, 5944–5951; [11] Platzer J et al. (2000) Cell 102, 89–97; [12] Mangoni ME et al. (2006) Circ Res 98, 1422–1430; [13] McRory JE et al. (2004) J Neurosci 24, 1707–1718; [14] Baumann L et al. (2004) Invest Ophthalmol Vis Sci 45, 708–713; [15] Wilkovsky P et al. (1997) J Neurosci 17, 7297–7306; [16] Strom TM et al. (1998) Nat Genet 19, 260–263; [17] Bech-Hansen NT et al. (1998) Nat Genet 19, 264–267; [18] Mintz IM (1994) J Neurosci 14, 2844–2853; [19] Dunlap K et al. (1995) Trends Neurosci 18, 89–98; [20] Wu LG et al. (1999) J Neurosci 19, 726–736; [21] Urbano FJ et al. (2003) Proc Natl Acad Sci USA 100, 3491–3496; [22] Pietrobon D (2002) Mol Neurobiol 25, 31–50; [23] Hillyard DR et al. (1992) Neuron 9, 69–77; [24] Saegusa H et al. (2002) EMBO J 20, 2349–2356; [25] Cizkova D et al. (2002) Exp Brain Res 147, 456–463; [26] Jimenez C et al. (2000) Neuropharmacol 39, 1–10; [27] Nakashima YM et al. (1998) Neuropharmacol 37, 957–972; [28] Dietrich D et al. (2003) Neuron 39, 483–496; [29] Jing X et al. (2005) J Clin Invest 115, 146–154; [30] Li Muller Y et al. (2007) Diabetes 56, 3089–3094; [31] Heady TN et al. (2001) Jpn J Pharmacol 85, 339–350; [32] Gomora JC et al. (2001) Mol Pharmacol 60, 1121–1132; [33] Kim D et al. (2001) Neuron 31, 35–45; [34] Choi S et al. (2006) Genes Brain Behav 6, 425–431; [35] Chen C-C et al. (2003) Science 302, 1416–1418; [36] Rossier MF et al. (1998) Endocr Res 24, 443–447; [37] Chen YC et al. (2003) Ann Neurol 54, 239–243; [38] Kozlov AS et al. (1999) Eur J Neurosci 11, 4149–4158; [39] Klugbauer N et al. (2003) J Bioenerg Biomembr 35, 639–647; [40] Field MJ et al. (2006) Proc Natl Acad Sci USA 103, 17537–17542; [41] Bernstein GM et al. (2007) Cell Calcium 41, 27–40; [42] Li C-Y et al. (2004) J Neurosci 24, 8494–8499; [43] Barclay J et al. (2001) J Neurosci 21, 6095–6104; [44] Gao B et al. (2000) J Biol Chem 275, 12237–12242; [45] Klugbauer N et al. (1999) J Neurosci 21, 684–691; [46] Hanke S et al. (2001) Gene 264, 69–75; [47] Qin N et al. (2002) Mol Pharmacol 62, 485–496; [48] Wycisk KA et al. (2006) Am J Hum Genet 66, 1531–1539; [51] Arikkath J et al. (2003) Curr Opin Neurobiol 13, 298–307.

Fig. 2.2.2 Putative structure of a voltage-gated Ca^{2+} channel including auxiliary subunits. (a) Structure of the skeletal muscle $Ca_v1.1$ channel as deduced using electron cryomicroscopy by Wolf and colleagues (Wolf et al., 2003). Figure reprinted with permission from Elsevier.[1] Antibodies to the subunits were used to localize them in the 3-D reconstructions. The $\alpha_2\delta$ protein protrudes into the cytoplasm, while the β and γ subunits were localized to the cytoplasmic face of the plasma membrane. (b) Hypothetical structure of the channel based on the crystal structure of homologous proteins for $\alpha1$, the Shaker K^+ channel (PDB 2A79); Long et al. (2005) and for $\alpha_2\delta$, the VWF protein CMG2 (PDB 1SHU) Canti et al. (2005); Lacy et al. (2004). The β subunit structure is derived from the crystal structure determined for $\beta2$ in complex with the AID peptide (shown in black, PDB 1TOJ), Van Petegem et al. (2004). The δ protein was modelled with a simple a-helix. Although its structure is unknown, it anchors $\alpha_2\delta$ to the plasma membrane via a disulphide bond. See also the colour plate section.

[1] Reprinted from Journal of Molecular Biology, Volume 332, Wolf, Eberhart, Glossmann, Striessnig and Grigorieff, Visualization of the domain structure of an L-type Ca^{2+} channel using electron cryomicroscopy, pages 171–182, copyright 2003, with permission from Elsevier.

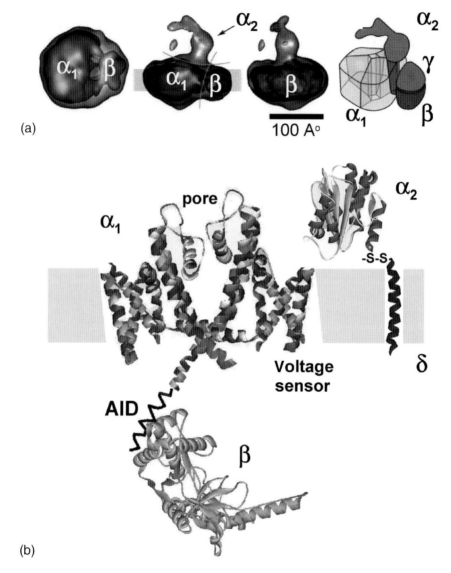

(a)

(b)

membranes from heart relative to skeletal muscle (Gould et al., 1983). Despite this limitation, the channel complex has been purified from various species and shown to contain at least three subunits of the L-type Ca^{2+} channel, α_1, $a_2\delta$, and β subunits (reviewed in Perez-Reyes and Schneider 1994). The size of the purified cardiac α_1 is between 165 and 195 kDa and is much smaller than that deduced from the cDNA clone, which suggests proteolysis. Similar to the skeletal muscle $Ca_v1.1$ (α_{1S}) this truncation occurs at the carboxyl terminus, since antibodies directed against this region only recognize a minor full-length form (242 kDa) in heart and brain preparations (Yoshida et al., 1992; Hell et al., 1994). Recently it was shown that the carboxy terminal fragment migrates to the nucleus in an activity-dependent manner where it acts as a transcription factor (Gomez-Ospina et al., 2006)

The cDNA encoding $Ca_v1.2$ (formerly α_{1C}) was cloned from rabbit heart libraries using low stringency hybridization with skeletal muscle α_1 cDNA as probe (Mikami et al., 1989). The full-length cDNA contains an open reading frame that encodes a protein of 243 kDa that is 66 per cent similar to $Ca_v1.1$. The human cardiac $Ca_v1.2$ (α_{1C}) has also been cloned and its gene, CACNA1C, is extensively

spliced, notably at the amino terminus, IS6, the I-II loop, IIIS2, IVS3, and in the carboxy terminus. A recent study demonstrated that splice variations occur at 19 of its 55 exons (Tang et al., 2004). Many of these variants modulate the biophysical and pharmacological properties of the channel.

$Ca_v1.3$

Three approaches were used to clone the third L-type α_1 subunit ($Ca_v1.3$): low stringency screening of cDNA libraries using $Ca_v1.1$ cDNA as probe (Hui et al., 1991; Williams et al., 1992b), library screening with oligonucleotides based on conserved regions between $Ca_v1.1$ (α_{1S}) and $Ca_v1.2$ (α_{1C}) (Yaney et al., 1992), and reverse transcription polymerase chain reaction (RT-PCR) with degenerate primers to highly conserved regions of $Ca_v1.1$ and 1.2 (Perez-Reyes et al., 1990; Seino et al., 1992). The human $Ca_v1.3$ (α_{1D}) was cloned from pancreatic islet cDNA libraries (Seino et al., 1992).

The deduced amino acid sequence is quite similar to $Ca_v1.2$ (~70 per cent). Remarkably, even the location and sequence of alternatively spliced regions are comparable (Perez-Reyes and Schneider, 1994). Splice variants of the carboxyl terminus have been cloned, of which one is considerably shorter than other L-type $Ca_v\alpha_1$ subunits, with its C terminus comprising only ~30 per cent of the full length protein sequence. Two variants of $Ca_v1.3$ (α_{1D}) have also been detected in immunoprecipitates from brain preparations (Hell *et al.*, 1994). Notably, the larger isoform is upregulated in prostatic tumors (Li HR *et al.*, 2006). Its amino acid sequence is unique in that it begins with seven consecutive methionine (ATG) residues, which appears to act as an enhancer of gene transcription (Kamp *et al.*, 1995).

$Ca_v1.4$

The α_{1F} cDNA ($Ca_v1.4$; *CACNA1F*) was cloned through its association with an X-linked visual disorder known as congenital stationary night blindness type 2 (Bech-Hansen *et al.*, 1998; Strom *et al.*, 1998). Its sequence is approximately 70 per cent identical to the L-type α_{1D}.

Biophysical characteristics

$Ca_v1.1$

Expression of functional Ca^{2+} channels from recombinant $Ca_v1.1$ was achieved by transfecting mouse L cells (Perez-Reyes *et al.*, 1989). Although expression of $\alpha_11.1$ alone can lead to Ca^{2+} currents, the activation kinetics of these currents are extremely slow, which suggested a role for auxiliary subunits and, accordingly, co-expression of $Ca_v\beta1a$ with $\alpha_11.1$ led to a normalization of the current kinetics (Lacerda *et al.*, 1991). This co-expression also led to fivefold increases in high-affinity DHP binding sites, although there was no corresponding increase in the current density. These results demonstrated that $Ca_v\beta$ was a bona fide channel subunit that could alter the biophysical properties of the channel and modify the expression of DHP binding sites. Recombinant channels have also been studied through microinjection of an expression plasmid into cultured dysgenic *mdg* myotubes (Tanabe *et al.*, 1988), which restored both a Ca^{2+} current and skeletal muscle-like excitation-contraction coupling; contraction that occurs even in the absence of external Ca^{2+}. Expression of functional $Ca_v1.1$ channels appears to be dependent on the $Ca_v\beta$ isoform used, both in studies using *Xenopus* oocytes (Ren and Hall, 1997) and the mammalian cell line ts201 (Neuhuber *et al.*, 1998). In all systems the channel requires strong membrane depolarization to open (threshold > -10 mV, $V_{50} \sim 10$ mV). As observed with native skeletal muscle currents (Sipos *et al.*, 1995), recombinant channels have slow kinetics, requiring 100s of ms to reach peak current following activation ($\tau > 50$ ms), and inactivation requiring seconds for complete decay (steady-state current inactivation, $h_\infty \sim -5$ mV) (Catterall *et al.*, 2005). In 110 mM solutions of $BaCl_2$, and in the presence of dihydropyridine agonists to prolong channel opening, skeletal muscle L-type channels display a single channel conductance of 14 pS (Dirksen and Beam, 1995).

$Ca_v1.2$

Injection of $Ca_v1.2$ (α_{1C}) cRNA into *Xenopus* oocytes leads to the appearance of dihydropyridine-sensitive Ba^{2+} currents. In contrast to the stable transfection systems required to express $Ca_v1.1$, oocytes provided an easier system to measure the effects of auxiliary subunits, an approach first used by the Numa group to demonstrate that $Ca_v\alpha_2\delta$ could increase α_{1C} currents approximately twofold (Mikami *et al.*, 1989). In similar studies co-expression with the cloned cardiac $Ca_v1.2$ (α_{1C}) subunit revealed that $Ca_v\beta1a$ not

only conferred an increase in current amplitude but also shifted the voltage-dependence and accelerated the rate of activation (Wei *et al.*, 1991). These effects can be classified as either quantitative, in the case of $Ca_v\beta$s increasing the amount of $Ca_v1.2$-mediated current (by both increasing trafficking and channel opening probability), or qualitative, where $Ca_v\beta$s affect the biophysical properties of the channel.

Calcium current through recombinant $Ca_v1.2$ channels has been measured in many systems, including insect Sf9 cells, COS cells, CHO cells, L cells, and HEK 293 cells. These studies have verified the essential role of $Ca_v\beta$ in increasing the functional expression of DHP binding sites and channel activity. The biophysical properties of the α_1 subunit expressed alone are difficult to measure due to low expression, but currents are typically slower and activate at more depolarized potentials than native L-type currents. When expressed together with an $\alpha_2\delta$ and β subunit the currents are normalized, and channels open after depolarization to above -20 mV with fast kinetics ($\tau = 1$ ms) (Catterall *et al.*, 2005). The $Ca_v1.2$ channel shows considerable overlap of the activation and inactivation curves, i.e. window current range. Its inactivation kinetics are complex, and biexponential fits reveal two slow phases when recorded in Ba^{2+} solutions (τ's of 150 and 1000 ms), and fast inactivation when recorded in Ca^{2+} solutions. Inactivation occurs through two mechanisms, one voltage-dependent (voltage-dependent inactivation, VDI), which accounts for the slow inactivation, and the other Ca^{2+} dependent (Ca^{2+}-dependent inactivation, CDI), which is considerably faster (Cens *et al.*, 2006). The molecular basis for CDI has been extensively studied, and involves calmodulin (CaM) tethered to the carboxyl terminus on an IQ motif (Peterson *et al.*, 1999). This complex has been studied using X-ray crystallography (Van Petegem *et al.*, 2005). Upon binding, Ca^{2+} induces rearrangements in channel structure leading to inactivation, perhaps by interacting with other parts of the channel such as the I-II loop (Soldatov, 2003; Kim *et al.*, 2004).

The single channel conductance of the recombinant $Ca_v1.2$ channels has been estimated at 23 pS in 110 mM $BaCl_2$ (Gondo *et al.*, 1998). Notably, conductance is much smaller in $CaCl_2$ solutions, and native cardiac myocyte channels show an apparent maximum conductance of 5 pS (Guia *et al.*, 2001). The channel pore is composed of two parts: the selectivity filter, which contains the pore loop (P) located between the S5 and S6 segments in each domain, and the walls, which are formed by the S6 segments. At the heart of the selectivity filter are four negatively charged glutamate residues, one from each domain, which form the EEEE locus. Ca^{2+} ions bind relatively tightly and in a coordinated manner at this site, thereby preventing monovalent ion flux (Sather and McCleskey, 2003). Under control conditions the channel has a low probability of opening (P_o) and its infrequent openings are quite brief (mean open time < 1 ms). Phosphorylation of the channel by cAMP-dependent protein kinase and calmodulin-dependent protein kinases increases channel P_o and prolongs channel openings (Yue *et al.*, 1990; Dzhura *et al.*, 2000). More dramatic prolongations of mean open time can be observed with agonists such as Bay K8644 and FPL 64176 (Lacerda and Brown, 1989; Fan *et al.*, 2000).

$Ca_v1.3$

Injection of human brain $Ca_v1.3$, $Ca_v\alpha_2\delta1$ and $Ca_v\beta2$ cRNA into *Xenopus* oocytes produced DHP-sensitive currents (Williams *et al.*, 1992b). This study was also the first to report the surprising finding that injection of $Ca_V\alpha_2\delta$ and $Ca_V\beta$ alone also led to the appearance of non DHP-sensitive currents, suggesting that mammalian subunits can drive the expression of endogenous oocyte channels. Challenging the conventional notion that Ca^{2+} channels can be classified into distinct high and low voltage-activated families, $Ca_V1.3$ and $Ca_V1.4$ were found to activate at lower thresholds than their L-type counterparts $Ca_V1.1$ and $Ca_V1.2$ (Lipscombe *et al.*, 2004). Their midpoint of activation is between –15 and –20 mV, while for inactivation is ~–40 mV (Catterall *et al.*, 2005). Activation kinetics are fast (~1 ms) and inactivation kinetics relatively slow and, as in the case of $Ca_V1.2$, biexponential and regulated by $Ca^{2+}/$ CaM. In hair cells this CaM-mediated fast inactivation is disrupted by expression of a related calcium-binding protein, CaBP4 (Yang *et al.*, 2006). The more modest depolarizations needed for $Ca_V1.3$ activation and its slow inactivation are in keeping with its proposed role in continuous neurotransmitter release in auditory hair cells (Platzer *et al.*, 2000). The single channel properties of $Ca_V1.3$ have yet to be studied.

$Ca_V1.4$

$Ca_V1.4$ channels heterologously expressed in mammalian cell lines cells generated a DHP-sensitive non-inactivating current that was activated at thresholds intermediate to the $Ca_V1.2$ and $Ca_V1.3$ L-type counterparts (Koschak *et al.*, 2003; McRory *et al.*, 2004). The threshold for channel activation was –45 mV, and the midpoint of activation was ~–10 mV. Activation kinetics were fast (~1 ms), while inactivation was extremely slow. Furthermore, the recombinant $Ca_V1.4$ channel exhibited a significant window current (h_∞ ~ –20 mV) and did not display Ca^{2+}-dependent inactivation, consistent with its proposed role in tonic glutamate release from photoreceptors (McRory *et al.*, 2004). The single channel properties of $Ca_V1.4$ have yet to be elucidated.

Localization

Northern blot analysis of RNA from different tissues identified a 6.5 kb transcript of $Ca_V1.1$ in skeletal muscle from rabbit (Biel *et al.*, 1991; Ellis *et al.*, 1988). Expression is almost exclusively limited to skeletal muscle, with minor signals detected in heart, brain, trachea, lung, ileum, stomach, and kidney. $Ca_V1.1$ channels have been localized to the transverse tubules of skeletal muscle, and this targeting is dependent on $Ca_V\beta1a$ (Flucher *et al.*, 2005). Surprisingly, $Ca_V1.1$ is expressed in GABAergic neurons in the caudate nucleus, where it may play a similar physiological role as in muscle, coupling membrane depolarization directly to Ca^{2+} release from intracellular stores (Takahashi *et al.*, 2003).

The $Ca_V1.2$ gene is expressed in many tissues, including heart, bladder, prostate, lung, brain, fibroblasts, kidney, ovary, and pancreas (Biel *et al.*, 1991; Splawski *et al.*, 2004). Subcellularly, $Ca_V1.2$ channels have a preferential somatodendritic distribution in neurons (Hell *et al.*, 1993). They are also localized in high concentrations on the granule-containing portion of pancreatic β-cells (Bokvist *et al.*, 1995).

The $Ca_V1.3$ gene is expressed in brain, endocrine tissues such as pancreas and adrenal, sensory cells such as photoreceptors and hair cells, and in heart, particularly atrial and sinoatrial nodal cells (Catterall *et al.*, 2005). L-type channels in neurons are found principally on the soma and proximal dendrites, suggestive of a role for these channels in regulating gene expression (Hell *et al.*, 1993). Knockout mouse mutants of $Ca_V1.3$ ($Ca_V1.3$ –/–) exhibited

deafness and other abnormalities that indicate a critical role in the development of the cochlea and pancreas (Platzer *et al.*, 2000; Namkung *et al.*, 2001).

$Ca_V1.4$ channels are expressed prominently in retina and function to mediate the release of neurotransmitter from photoreceptor cells (McRory *et al.*, 2004). *In situ* hybridization studies suggest enriched $Ca_V1.4$ mRNA levels in retinal layers with photoreceptors, horizontal, amacrine, and bipolar cells, but not in ganglion cell layers (Bech-Hansen *et al.*, 1998; Strom *et al.*, 1998). Similar results were obtained by immunocytochemistry with affinity purified antibodies (Morgans *et al.*, 2001). Recent findings report a broader distribution, with $Ca_V1.4$ expression extended to adrenal gland, bone marrow, skeletal muscle, spinal cord, spleen, and thymus (McRory *et al.*, 2004).

Pharmacology

The hallmark of Ca_V1 channels is their sensitivity to block by 'calcium channel blockers' of the dihydropyridine class. Classic radioligand binding studies demonstrated that skeletal muscle L-type channels bind dihydropyridines, phenylalkylamines, and benzothiazepines with nanomolar affinity and in an allosteric manner (Glossmann and Striessnig, 1990; Godfraind, 2006). Since drugs of these classes are effective antihypertensive agents (by blocking L-type channels in smooth muscle, i.e. $Ca_V1.2$), their binding sites have been extensively studied. This research was aided by the ability to prepare analogues with photolabile substituents which allowed identification of the regions in the α_1 subunit that form the drug binding pocket (Striessnig *et al.*, 1998). Multiple mutagenesis studies have together established the critical residues involved in conferring sensitivity to both antagonists and agonists (for review see Striessnig *et al.*, 1998). The regions identified included the S6 transmembrane segment in domains III and IV, which is consistent with the drugs' ability to block the ion pore. Notably, the binding regions for the different drug classes overlapped, offering a plausible explanation for their allosteric interactions. The crystal structures of K^+ channels have revealed an inner vestibule in the permeation path, located below the selectivity filter and lined with hydrophobic residues from the S2 segment (Doyle *et al.*, 1998). This S2 segment is structurally similar to the S6 segment of the voltage-gated Ca^{2+} channels (Long *et al.*, 2005), allowing for plausible models of the drug binding pocket (Lipkind and Fozzard, 2003). As noted previously, drug binding to $Ca_V1.1$ channels does not prevent skeletal muscle contraction and is, therefore, not thought to be of therapeutic importance. However, these studies provided important insights into the structure of the $Ca_V1.2$ channel drug binding pocket that is the locus of interaction at which these drugs mediate their antihypertensive, anti-anginal and anti-arrhythmic effects.

Comparatively, the pharmacology of $Ca_V1.3$ and $Ca_V1.4$ channels has not been studied in great detail. Two studies have shown that $Ca_V1.3$ currents were ~tenfold less sensitive to block by DHPs such as isradipine and nimodipine than $Ca_V1.2$ (Koschak *et al.*, 2001; Xu and Lipscombe, 2001). In contrast, radioligand binding of isradipine indicates a similar affinity between these two L-type channels, providing another example of how binding affinity does not necessarily reflect channel block (Koschak *et al.*, 2001). As with other L-type channels, BayK8644 acts as an agonist (Koschak *et al.*, 2001). $Ca_V1.4$ channels are even less sensitive to block by DHPs

(Koschak *et al.*, 2003) and apparently cannot be detected by radiolabelled isradipine (Catterall *et al.*, 2005).

Physiology

Physiological role of Ca$_v$1.1 channels

Ca$_v$1.1 channels contain the highly conserved S4 voltage sensor that senses membrane depolarization and elicits changes in channel conformations. These conformational changes cause both channel opening and rearrangements of the intracellular loops that interact with and regulate channel opening of intracellular Ca^{2+} release channels located in the sarcoplasmic reticulum (SR). Due to their sensitivity to ryanodine, these SR Ca^{2+} release channels are also known as ryanodine receptors (RyRs). Coupling is primarily mediated by direct physical contact of the domain II-III loop of α$_1$1.1 with the release channel (Tanabe *et al.*, 1990) and is stabilized by other intracellular loops (Carbonneau *et al.*, 2005).

The distribution of these two proteins is highly regulated, with four Ca$_v$1.1 molecules (a tetrad) precisely positioned above one RyR (Flucher and Franzini-Armstrong, 1996). Skeletal muscle membranes are richly invaginated, allowing the plasma membrane to reach down into the fibre to contact additional SR in order to form the so-called t-tubule. Ca^{2+} release from the SR triggers muscle contraction and essentially all the Ca^{2+} is pumped back into the SR via Ca-ATPases with little lost to pumps at the plasma membrane (Bers, 2002).

Physiological role of Ca$_v$1.2 channels

Ca$_v$1.2 channels mediate contraction of cardiac and smooth muscle. Ca^{2+} influx through these channels is coupled to Ca^{2+} release from intracellular stores, a phenomenon known as calcium-induced calcium release (CICR) (Somlyo and Somlyo, 1990; Bers, 2002). In pancreatic β cells, Ca$_v$1.2 currents provide the critical Ca^{2+} that triggers vesicle fusion of insulin containing granules. In neurons, Ca$_v$1.2 also appears to play a role in regulating gene transcription, surprisingly not necessarily via alterations in intracellular Ca^{2+}, but by cleavage and translocation of its carboxy terminus to the nucleus where it acts as a transcription factor (Gomez-Ospina *et al.*, 2006).

Physiological role of Ca$_v$1.3 channels

Elucidation of the physiological roles of Ca$_v$1.3 channels was complicated by their co-expression in many tissues with and similar pharmacology to Ca$_v$1.2 channels. One notable exception is auditory hair cells, and transgenic mice lacking Ca$_v$1.3 are deaf (Platzer *et al.*, 2000). A second phenotype of these mice is sinoatrial (SA) node dysfunction (Platzer *et al.*, 2000; Zhang *et al.*, 2002). In keeping with its activation at relatively negative test membrane potentials, isolated SA node cells showed a slower phase 4 depolarization, a role typically ascribed to T-currents (Satoh, 1995). Knockout mice also show decreased development of pancreatic β cells (Namkung *et al.*, 2001).

Physiological role of Ca$_v$1.4 channels

Ca$_v$1.4 is almost exclusively expressed in the retina. Its physiological role has been deduced from mutations that disrupt its function and which lead to myopia and night blindness, which indicates a major role for Ca$_v$1.4 in synaptic transmission between retinal photoreceptors and bipolar cells (Morgans *et al.*, 2005).

Relevance to disease

Mutations in the gene encoding the α$_1$ subunit of Ca$_v$1.1, *CACNA1S*, have been linked to hypokalemic periodic paralysis, (hypoKPP) (Ptacek *et al.*, 1994), thyrotoxic periodic paralysis (Kung *et al.*, 2004), and malignant hyperthermia (MH) (Monnier *et al.*, 1997). Common mutations found in hypoKPP patients are R528H and R1239H, which alter the sequence of the voltage-sensing S4 segment in domains II and IV. Consistent with this localization, Ca^{2+} currents measured from muscle cells isolated from patients showed large shifts in the voltage dependence of inactivation (Sipos *et al.*, 1995). Surprisingly, this result was not reproduced when the mutations were introduced in the recombinant channel (Lapie *et al.*, 1997). The mutation in patients susceptible to MH is R1086H, which was originally observed in a French family (Monnier *et al.*, 1997), and confirmed in ~1 per cent of North American MH individuals (Stewart *et al.*, 2001). R1086 is located in the highly conserved III-IV intracellular loop, and is postulated to disrupt coupling to the RyR. Notably, mutations in skeletal muscle RyR are also associated with malignant hyperthermia (Monnier *et al.*, 2005). The single nucleotide polymorphisms (SNPs) associated with thyrotoxic periodic paralysis occur in untranslated regions and it has been postulated that these may alter the activity of a putative thyroid hormone responsive element (Kung *et al.*, 2004).

Spontaneously arising mutations in the gene encoding Ca$_v$1.2, *CACNA1C*, have been found in patients with Timothy syndrome (Splawski *et al.*, 2004). The mutation leads to the replacement of a highly conserved glycine residue with arginine (G406R) in the IS6 segment, leading to channels with slowed inactivation. Slower inactivation would lead to a longer ventricular action potential with a longer plateau, plausibly explaining the QT prolongation observed in the electrocardiogram of these patients. Timothy syndrome is characterized by malformations in many other organ systems as well (immune deficiency, birth defects such as webbing of the fingers and toes, and hypoglycemia), indicating that Ca$_v$1.2 channels are critical for normal development (Splawski *et al.*, 2004). Consistent with this notion, knockout of the *CACNA1C* gene in mice is lethal (Seisenberger *et al.*, 2000). Selective inactivation in pancreatic β cells using Cre/loxP indicates that Ca$_v$1.2 plays an important role in insulin secretion. Despite these important physiological roles, the long-term use of calcium channel blockers as antihypertensive agents is well tolerated (Ross *et al.*, 2001).

Mutations in the *CACNA1D* gene have not been reported. From the mouse knockout studies, one might predict loss of function mutations may increase the susceptibility to cardiac arrhythmia and deafness. Mutations in *CACNA1F* that disrupt the open reading frame have been linked to X-linked congenital stationary night blindness type 2 (XLCSNB) (Bech-Hansen *et al.*, 1998; Strom *et al.*, 1998). Over 20 mutations have been found to date (for review see McKeown *et al.*, 2006).

Concluding remarks

The skeletal muscle Ca$_v$1.1 channel provided a Rosetta stone for the cloning of the high voltage-activated Ca^{2+} channel family and its auxiliary subunits. Its primary role in muscle is as a voltage sensor and its role in allowing Ca^{2+} entry may simply serve to help refill intracellular stores. It forms a functional complex with the intracellular Ca^{2+} release channel and mutations in either lead to muscle disorders. Ca$_v$1 channels play important roles in sensory transduction, with $Ca_V1.3$ involved in auditory function, and $Ca_V1.4$ involved in vision. The most important drug target of this class of L-type channels is $Ca_V1.2$ due to its role in mediating Ca^{2+} influx, and hence contraction, of vascular smooth muscle myocytes.

2.2.3 Ca_V2 channels

Introduction

Early pharmacological studies identified a subclass of neuronal Ca^{2+} channels that were neither sensitive to L-type antagonists nor gated at potentials as negative as T-type channels that were hence designated 'N-type' channels (Nowycky et al., 1985). Electrophysiological recordings from Purkinje cells of the cerebellum revealed a slowly inactivating calcium current insensitive to DHPs and snail cone toxins (Llinás et al., 1989). The unique pharmacology of these currents expanded the Ca^{2+} channel classification scheme to include 'P-type' for their discovery in Purkinje cells. Subsequently, currents resistant to a cocktail of known blockers were discovered and were named 'R-type' (Randall and Tsien, 1997). Evidence that these were distinct classes of channel were provided by pharmacological studies using toxins from *Conus* fish-hunting snails (e.g. ω-conotoxin GVIA) (Olivera et al., 1994), the funnel-web spider *Agenelopsis aperta* (e.g. ω-agatoxin IVA) (Adams et al., 1993), and the tarantula (e.g. SNX-482), *Hysterocrates gigas* (Newcomb et al., 1998). Cloning of the genes encoding these channels revealed that they were actually closely related and that one gene encoded both P- and Q-types (Bourinet et al., 1999). These channels play important roles in synaptic transmission, coupling Ca^{2+} entry to synaptic vesicle fusion. Their activity is highly regulated and notably they are inhibited by G proteins, which partially explains the analgesic activity of opioids and cannabinoids. Of therapeutic interest, ziconotide (Prialt®), which is a synthetic analog of the ω-conotoxin MVIIA that selectively blocks N-type channels, has been approved for the treatment of neuropathic pain.

Molecular characterization of the Ca_V2 channels

$Ca_V2.1$

The $α_{1A}$ cDNA ($Ca_V2.1$; *CACNA1A*; CaCh 4 or BI) was cloned using low stringency hybridization of rabbit and rat brain cDNA libraries using the skeletal muscle $Ca_V1.1$ cDNA as a probe (Mori et al., 1991; Starr et al., 1991). The sequence of $Ca_V2.1$ ($α_{1A}$) is approximately 40 per cent identical to the L-type $Ca_V1.1$. The highest sequence identity across all voltage-gated Ca^{2+} channel types occurs in their membrane-spanning regions, particularly the residues that make up the voltage sensor (S4) and the pore and its walls (S5-P-S6).

The *CACNA1A* gene is spliced in at least seven regions (Soong et al., 2002). Notably, alternative splicing of the extracellular linker connecting IVS3 to IVS4 alters sensitivity to the funnel-web spider toxin ω-agatoxin IVA and provides the mechanism by which P- and Q-type channels are derived from the same gene (Bourinet et al., 1999). A short form consisting of only domains I and II has been purified and sequenced (Scott et al., 1998), however, its physiological role is unknown. $Ca_V2.1$ channels have been purified by immunoprecipition and shown to contain an $α_2δ$ subunit (Martin-Moutot et al., 1995).

$Ca_V2.2$

The N-type Ca^{2+} channel has been purified from brain using ω-conotoxin-GVIA as a marker and reincorporated into lipid bilayers, where it displayed functional activity (Witcher et al., 1993). Similar to L-type Ca^{2+} channels, the purified complex contains $Ca_Vα_1$ ($α_{1B}$), $Ca_Vα_2δ$, and $Ca_Vβ$ ($β_3$). Subsequent studies revealed that all four $Ca_Vβ$ subunits are associated with N-type channels ($β3>β1>β4>>β2$), although the rank order is developmentally regulated (Vance et al., 1998).

The $α_{1B}$ cDNA ($Ca_V2.2$, *CACNA1B*) was originally isolated by low stringency screening of brain cDNA libraries using the skeletal muscle $Ca_V1.1$ cDNA as probe (Dubel et al., 1992; Williams et al., 1992a). The deduced amino acid sequence of $Ca_V2.2$ is quite similar that of $Ca_V2.1$ (82 per cent sequence identity), with a notable lack of homology between the II-III cytoplasmic loops and carboxyl termini, but is less similar to the L-type $Ca_V1.1$ (43–51 per cent sequence identity). Alternative splicing leading to small insertion and deletion events has been catalogued, some of which confer robust effects on channel properties (Stea et al., 1994). Notably, nociceptors almost exclusively express a particular splice variant that has high activity (longer mean open time) than the other brain variants (Castiglioni et al., 2006). Splicing of the carboxy terminus produces variants that differ in the distribution within neurons (Maximov and Bezprozvanny, 2002).

$Ca_V2.3$

The cDNA encoding the a_1 subunit of $Ca_V2.3$ ($α_{1E}$, *CACNA1E*) was cloned both by screening rat brain cDNA libraries with $α_{1A}$ cDNA as probe (Niidome et al., 1992) and by using PCR with primers based on conserved sequences (Soong et al., 1993; Schneider et al., 1994). As observed in all three Ca_V2 channel genes, alternative splicing creates variants that differ in their II-III loop and carboxyl terminus, while *CACNA1E* also produces N-terminal variants (Pereverzev et al., 1998; Jurkat-Rott and Lehmann-Horn, 2004).

Biophysics

$Ca_V2.1$

The pioneering studies of Mori and coworkers showed that injection of $Ca_V2.1$ cRNA into oocytes led to the appearance of small, but detectable, Ba^{2+} currents (Mori et al., 1991). Co-expression with $Ca_Vγ$ had no effect, while currents were increased after co-expression of $Ca_Vα_2δ$ (threefold), $Ca_Vβ_1$ (20-fold), or the combination of $Ca_Vα_2δ + Ca_Vβ_1$ (over 200-fold to 6.5 μA). These results indicate that there is a synergistic action between the subunits. In addition to increasing surface expression, all four $Ca_Vβ$ subunits are capable of modulating the biophysical properties of the $α_12.1$ subunit (De Waard and Campbell, 1995). As observed with $Ca_V1.2$, co-expression of β subunits shifted the voltage dependence of activation to more negative potentials. $Ca_Vβ$ subunits also modulate kinetics, typically accelerating activation and inactivation, but these effects differ between the βs and their splice variants (Helton and Horne, 2002). Native $Ca_V2.1$ channels and some subunit combinations of the recombinant channel show half-maximal activation at –5 mV, and inactivation in the range of –20 to 0 mV, depending on experimental conditions (Catterall et al., 2003). Activation kinetics are relatively fast (<2 ms) and inactivation is slow (700–1000 ms). Single channel analysis detected a channel with a 16 pS conductance in $BaCl_2$ (Mori et al., 1991). $Ca_V2.1$ channels conduct Ba^{2+} ions more effectively than Ca^{2+} ions (8 pS), but this difference is less than observed with $Ca_V1.2$ (Bourinet et al., 1996).

Ca$_v$2.2

Ca$_v$2.2 channel properties have been studied in many expression systems, including HEK-293 cells, dysgenic myotubes, and *Xenopus* oocytes (Perez-Reyes and Schneider, 1994). The auxiliary subunits α$_2$δ and β regulate α$_1$2.2 channel activity in a similar manner as for α$_1$2.1, increasing current expression and modulating channel biophysics. Currents become measurable around –20 mV, peak around +10 mV and show fast activation kinetics (τ < 3 ms). Inactivation is characterized by relatively fast kinetics (τ < 200 ms), and a relatively negative h$_∞$ (~ –60 mV). The single channel conductance was estimated to be ~15 pS using 100 mM Ba^{2+} as the charge carrier, and ~8 pS using Ca^{2+} (Bourinet *et al.*, 1996).

Ca$_v$2.3

In contrast to the other Ca$_v$1 and Ca$_v$2 channels, injection of a$_1$2.3 cRNA into *Xenopus* oocytes led to the generation of robust (1 μA) Ba^{2+} currents (Schneider *et al.*, 1994). Co-expression with Ca$_v$β$_{1b}$ had no apparent effect on peak currents, but did shift activation and inactivation to more negative potentials (Soong *et al.*, 1993). Subsequent studies revealed the presence of endogenous Ca$_v$β subunits in *Xenopus* oocytes, and antisense oligonucleotide studies supported an interesting model whereby Ca$_v$β subunits regulated channel trafficking of exogenous channels independently of modulating channel biophysics (Tareilus *et al.*, 1997). Ca$_v$2.3 currents activate (~1 ms) and inactivate (>100 ms) relatively quickly. The midpoint of activation varies between clones, but most recombinant Ca$_v$2.3 channels show a V$_{50}$ around 0 mV, and an h$_∞$ around –70 mV. Despite considerable sequence homology in their pores, Ca$_v$2.3 channels differ from other Ca$_v$2 channels in that they conduct Ca^{2+} and Ba^{2+} ions equally well, exhibiting a conductance of ~12 pS (Bourinet *et al.*, 1996).

Prior to the cloning of the Ca$_v$3 T-type channel family, there was considerable debate as to whether they might be encoded by Ca$_v$2.3 (Tsien *et al.*, 1998). The hypothesis was first proposed on the basis of the relatively low voltages at which the recombinant rat channel activated (Soong *et al.*, 1993). The notion was supported by finding that Ca$_v$2.3 channels were similar to (some) native T-channels in their nickel sensitivity (IC$_{50}$ ~ 50 μM) and equal conductance to Ca^{2+} and Ba^{2+} ions (Fox *et al.*, 1987; Zamponi *et al.*, 1996). Biophysical and pharmacological studies revealed that the native counterpart of recombinant Ca$_v$2.3 was an HVA current known as R-type (Randall and Tsien, 1997). This assignment is supported by the loss of R-type currents in Ca$_v$2.3 knockout mice (Sochivko *et al.*, 2002) and the cloning of the Ca$_v$3 family (Perez-Reyes, 2003).

Localization

Northern blot analysis suggests that Ca$_v$2.1 is expressed abundantly in brain, with comparatively less in heart, and none in skeletal muscle, stomach, or whole kidney (Mori *et al.*, 1991). In brain, there is high expression in cerebellum, and moderate expression in hippocampus, olfactory bulb, and spinal cord (Starr *et al.*, 1991; Ludwig *et al.*, 1997). Screening mice with different forms of cerebellar degeneration revealed a lack of Ca$_v$2.1 expression in mice with Purkinje cell degeneration (Mori *et al.*, 1991). Ca$_v$2.1 expression was also detected in the kidney, with one study suggesting localization to distal convoluted tubule (Yu *et al.*, 1992) and another to smooth muscle myocytes lining preglomerular vessels (Hansen *et al.*, 2000).

Localization studies have since identified P/Q-type channels in a variety of neuronal as well as endocrine cells. P-type channels are highly expressed in the aforementioned cerebellar Purkinje cells, among numerous other cells types in the central nervous system, and play the predominant role in neurotransmitter release (Dunlap *et al.*, 1995).

Northern blot analysis identified a 9.3/9.5 kb doublet in rabbit brain (Fujita *et al.*, 1993). *In situ* hybridization to rodent brain slices indicates it has a widespread, but non-uniform, distribution (Coppola *et al.*, 1994; Tanaka *et al.*, 1995). Like P/Q-type channels, N-type channels function in the presynaptic terminal to mediate calcium influx and neurotransmitter release (Westenbroek *et al.*, 1992). Nevertheless, studies using a fluorescein conjugate of ω-conotoxin showed expression of N-type channels is not restricted to presynaptic terminals and channels are also localized on soma and dendrites (Mills *et al.*, 1994). Immunohistochemistry revealed high levels of Ca$_v$2.2 immunoreactivity in presynaptic terminals within lamina I and II of the spinal cord. N-type channels are densely localized in laminae I and II (marginal zone and substantia gelatinosa, respectively) of the grey matter of the dorsal horn of the spinal cord where primary sensory afferents consisting of mainly unmyelinated C- and thinly myelinated Aδ-fibres deliver nociceptive signals, including thermal, mechanical, and inflammatory, from the periphery to the central nervous system (CNS) (Kerr *et al.*, 1988; Gohil *et al.*, 1994). Their high concentration at these central terminals makes N-type calcium channels an attractive target for the development of novel analgesics (McGivern, 2006b).

Although early northern blot studies indicated that Ca$_v$2.3 mRNA was predominantly expressed in brain (Niidome *et al.*, 1992), more recent studies indicate expression in pancreas, kidney, and heart (Vajna *et al.*, 1998; Grabsch *et al.*, 1999). Ca$_v$2.3 distribution in the CNS has been mapped *in situ* hybridization (Soong *et al.*, 1993). Significant expression occurs in the cerebral cortex, hippocampus, cerebellum, and olfactory bulb. Ca$_v$2.3 immunoreactivity is localized predominantly at the soma, but has also been shown in various dendritic regions across structures of the CNS (Yokoyama *et al.*, 1995). For a comprehensive review of Ca$_v$2.3 localization see (Weiergräber *et al.*, 2006b).

Pharmacology

The Ca$_v$2.x channels can be distinguished by peptide toxins isolated from snail and spider venoms. The prototypical P-type channel blocker is ω-agatoxin IVA isolated from funnel-web spider venom. Alternative splicing reduces the sensitivity of Ca$_v$2.1 channels to block by this toxin leading to the notion that there was a separate Q-type channel (Bourinet *et al.*, 1999).

N-type currents are blocked by peptides isolated from cone snails, two examples being ω-conotoxin GVIA and ω-conotoxin MVIIA (SNX-111, Prialt®). N-type channels are attractive drug targets for the development of novel therapeutics to treat neuropathic pain (McGivern, 2006b). The clinical effectiveness of ziconotide provides proof of this principle (Staats *et al.*, 2004). This synthetic conotoxin is administered intrathecally allowing its diffusion into the dorsal horn of the spinal cord where it can block synaptic transmission from nociceptors onto neurons of the spinothalamic tract that transmit the pain signal to the brain (Nelson *et al.*, 2006). Although acting at additional sites in the CNS, both opioids and cannabinoids inhibit N-type channels at this synapse, contributing to their analgesic properties (Pacher *et al.*, 2006). Recently it was proposed that NMED-160 (renamed MK-6721) might be an orally available N-type channel blocker (McGivern, 2006b), however, scientific studies demonstrating its properties have not been published to date.

The pharmacology of $Ca_v2.3$-induced currents is notable in that most Ca^{2+} channel blockers have no effect. R-type currents are partially blocked by amiloride, Ni^{2+}, and Cd^{2+}. However, a novel spider toxin, SNX-482, has been shown to selectively block recombinant $Ca_v2.3$ channels and this compound inhibited R-type currents in some tissues (Zhang *et al.*, 1993). Thus, since these recombinant and native currents share kinetic and pharmacological characteristics, it has been suggested that $Ca_v2.3$ represents an R-type channel (Zhang *et al.*, 1993). In agreement, $Ca_v2.3$ −/− mice lack most, but not all R-type currents (Wilson *et al.*, 2000).

Physiology

The pioneering studies of Katz and Miledi established the importance of Ca^{2+} influx in triggering neurotransmitter release (Katz and Miledi, 1967). It is now known that this Ca^{2+} enters via Ca_v2 channels. Pharmacological studies indicate that $Ca_v2.1$ channels are the predominant player at most synapses, followed by $Ca_v2.2$. $Ca_v2.3$ serves in this capacity more rarely but does play a major role at certain synapses (Olivera *et al.*, 1994; Dunlap *et al.*, 1995; Kamp *et al.*, 2005). In a manner reminiscent to the docking of $Ca_v1.1$ channels with ryanodine receptors, the intracellular loops of Ca_v2 channels interact directly with synaptic vesicle proteins (Catterall, 1999). In particular, the II-III loop has been shown to contain the synaptic protein interaction site, or 'synprint', that interacts with SNARE (soluble N-ethylmaleimide-sensitive fusion attachment protein receptor) proteins such as syntaxin 1, SNAP-25 (synaptosomal-associated protein of 25 kDa), and synaptotagmin. This leads to the docking of vesicles at the plasma membrane, thereby facilitating rapid exocytosis.

The analgesic properties of morphine are partly mediated through its ability to regulate Ca^{2+} influx via Ca_v2 channels and hence synaptic transmission from nociceptors to dorsal horn neurons. Thus, presynaptic G protein-coupled receptors can modulate synaptic transmission via second messengers that modulate Ca_v2 channel activity. Inhibition of channel activity can be mediated by direct binding of the G protein βγ subunits which leads to a shift in the voltage-dependence of channel activation (Bean, 1989). Although many of the molecular details of this interaction have been elucidated it is still not clear how they lead to changes in channel activity (De Waard *et al.*, 2005). It was recently proposed that Gβγ may actually displace $Ca_v\beta$ subunits from the I-II loop, thereby removing their positive influence on channel gating (Sandoz *et al.*, 2004). A second pathway of regulation includes phosphorylation by protein kinases, such as protein kinase C (Maeno-Hikichi *et al.*, 2003). A third pathway involves lipid metabolites (Liu *et al.*, 2004) including endocannabinoids (Guo and Ikeda, 2004).

Studies of transgenic mice lacking functional Ca_v2 genes have provided unexpected insights into their physiological roles. Amongst the most studied mice are those lacking $Ca_v2.3$, which have been generated in at least four academic laboratories (Weiergräber *et al.*, 2006a). $Ca_v2.3$ −/− mice are more susceptible to seizures induced by pentylenetetrazol (PTZ) (Weiergräber *et al.*, 2006a). This study is also notable because the antibodies used to demonstrate $Ca_v2.3$ expression in circuits which are likely to be involved in the seizure generation (e.g. thalamic nucleus reticularis) were shown to be specific, i.e. not to cross-react with any other protein, using the knockout mouse tissue. In addition, $Ca_v2.3$ −/− mice show reduced sensitivity to visceral inflammatory pain, enhanced fear, impaired spatial memory, reduced sperm motility and hyperglyccmia (Weiergräber *et al.*, 2006a). These mice have also been useful for proving the hypothesis that $Ca_v2.3$ underlies R-type Ca^{2+} currents in a variety of neurons. Studies on $Ca_v2.1$ −/− mice have confirmed its important role in cerebellar function as the mice develop dystonia, progressive ataxia and cerebellar atrophy (Jun *et al.*, 1999; Fletcher *et al.*, 2001). $Ca_v2.2$ −/− mice are phenotypically normal but exhibit reduced responses to some painful stimuli (Kim *et al.*, 2001).

Relevance to disease

Of the three Ca_v2 genes, only *CACNA1A* (formerly known as *CACNL1A4*) has been linked to human diseases. Mutations have been found in patients with familial hemiplegic migraine (FHM), episodic ataxia type 2 (EA2), and spinocerebellar ataxia type 6 (SCA6) (Ophoff *et al.*, 1996; Zhuchenko *et al.*, 1997). A large number of mutations have been discovered in EA2 patients including many that would introduce premature chain termination and hence a loss of functional channel expression (McKeown *et al.*, 2006). Many of the FHM mutations are simple amino acid substitutions, however, in some cases these are highly conserved residues involved in channel gating. A knockin mouse model of one of these mutations, R192Q, exhibited increased susceptibility to cortical spreading depression (van den Maagdenberg *et al.*, 2004). The SCA6 mutation is a polyglutamine (CAG) expansion at the carboxyl terminus of $\alpha_12.1$, which leads to the production of a toxic protein and eventually to death of cerebellar Purkinje neurons. Since these neurons constitute the major output of the cerebellum, it follows that their loss, or impaired function, would lead to ataxia. Therefore, ataxia would also be a likely side effect of a non-selective Ca_v2 channel blocker.

Mapping of mutations in juvenile myoclonic epilepsy led to the identification of a protein that interacts with $Ca_v2.3$ channel (Suzuki *et al.*, 2004). The mutated gene, *EFHC1*, encodes a protein with an EF-hand motif. Its overexpression increased recombinant $Ca_v2.3$ currents and induced apoptosis in cultured neurons. Notably, this apoptosis was reduced by the R-type blocker SNX-482. It was also shown that EFHC1 interacts with the carboxy terminus of $Ca_v2.3$. Of the nine SNPs studied, only F229L disrupted the ability of EFHC1 to modulate the $Ca_v2.3$-mediated currents (Suzuki *et al.*, 2004). Further investigation will be required to elucidate how these SNPs contribute to the development of seizures.

Concluding remarks

The isolation of spider and snail toxins that selectively blocked members of the Ca_v2 channel family provided important tools to uncover their critical role in synaptic transmission. More recent discoveries of their physiological roles have been made with transgenic mice. Combined, these techniques have established that Ca_v2 channels are important targets for the development of novel analgesic and anti-epileptic drugs.

2.2.4 Ca_v3 channels
Introduction

The existence of low voltage-activated (LVA) Ca^{2+} channels was first inferred from the classic current clamp studies of Llinás and colleagues on inferior olivary and thalamic neurons (Llinás and Yarom, 1981a, b; Llinás and Jahnsen, 1982). These studies provided the first evidence that a distinct channel mediated low threshold Ca^{2+} spikes and that its intriguing biophysical properties formed the basis for post-anodal exaltation or simply rebound burst firing. Patch clamp recordings firmly established the biophysical differences between LVA and HVA channels (for a comprehensive review of the early history see Perez-Reyes 2003). These early electrophysiological studies established that T-channels were important pacemaker channels in many neurons, including thalamus and nociceptors, but also in peripheral tissues such as heart and adrenal gland. Cloning of the channels led to further insights into their distribution and regulation. There are three T-channel genes: *CACNA1G*, encoding $Ca_v3.1$ (a_{1G}); *CACNA1H*, encoding $Ca_v3.2$ (a_{1H}); and *CACNA1I*, encoding $Ca_v3.3$ (a_{1I}). Of particular interest are their roles in nociception, epilepsy and autism.

Molecular characterization of Ca_v3

The low stringency screening and PCR techniques used so effectively during the 'skeletal muscle' and the 'brain' cloning eras led to the cloning of the family of high voltage-activated subunits, but failed to clone any of the T-channel α_1 subunits. Instead, these genes were discovered during an '*in silico* era', which relied on computer searches of genetic databases, three of which played significant roles: one, the Merck-Washington University EST project, whose goal was to identify all the mRNAs produced by the genome; two, the *C. elegans* genome project; and three, the human genome project (reviewed in Perez-Reyes 2003). The first portion of a human T-type channel to be deposited in the GenBank was an EST fragment derived from brain. The sequence was labelled as being 'similar to a calcium channel'. While searching the Genbank database with DNA fragments many ESTs were noted to be labelled in this manner, which inspired a switch to a text-based search, which had the advantage that one did not have to predict a conserved region. Indeed, the fragment in GenBank represented a poorly conserved region (the first membrane spanning region of repeat III). Further sequencing of this EST clone revealed motifs common to all voltage-gated channels, and more significantly, a pore region with considerable homology to Ca^{2+} channels (Perez-Reyes *et al.*, 1998). Further *in silico* cloning led to the identification of a nearly full-length homologue from *C. elegans* (contained on cosmid C54d2). The 3' end of this protein was then used to screen the GenBank again, leading to the 'cloning' of numerous other brain ESTs. These ESTs were then used to screen cDNA libraries using classical methods, with the end result being the cloning of full-length cDNAs for all three Ca_v3 channels (Cribbs *et al.*, 1998; Perez-Reyes *et al.*, 1998; Lee *et al.*, 1999).

The cloning of the mammalian cDNAs enabled definitive studies into the structure and function of LVA channels, with the first observation being that T-type channel α_1 subunits bear great similarity to other voltage-gated ion channels. As observed in HVA Ca^{2+} and Na^+ channels, these α_1 subunits are large proteins (>200 kDa) containing four repeats of the six transmembrane structure found in voltage-gated K^+ channels. Each repeat contains an S4 voltage sensor, a P loop that forms the selectivity filter, and an S6 segment that lines the inner wall of the channel. Due to the similarities in sequence and structure between voltage-gated K^+, Na^+ and Ca^{2+} channels (Jan and Jan, 1990), it is likely that the structural determinants regulating channel function are also similar. These include

the presence of a helical bundle crossing at the intracellular mouth of the channel or 'top' of the inverted teepee (Doyle *et al.*, 1998), the presence of an internal vestibule where many drugs bind, regulation of channel opening by movements of S6 segments (Jiang *et al.*, 2002) and the involvement of S4 voltage sensors in channel opening (Chanda *et al.*, 2005).

The cDNAs for T-type channels have now been cloned from many species, including human, rat, and mouse, and more T-channel sequences can be deduced from genomic DNA, extending the list to chicken, cow, dog, zebrafish, pufferfish, worm, fruit fly, mosquito, honey bee, snail, and purple urchin. In many cases the deduced sequences are not correctly assembled, particularly at the junction between exons 1 and 2. This is likely due to the presence of minor AT–AC splice junctions that are not detected by the exon prediction algorithm (Wu and Krainer, 1999). Alignments of all Ca_v3 channel proteins from human, dog, cow, rat, and mouse reveal that $Ca_v3.1$ channels are highly conserved, showing 90–95 per cent sequence conservation across these five species, while the $Ca_v3.2$ channels are much less so (70–80 per cent), and the $Ca_v3.3$ channels have an intermediate level of conservation (85–90 per cent). This conservation can be interpreted such that evolution has restricted changes in $Ca_v3.1$ due to its vital physiological roles, such as its contribution to thalamic signalling. In contrast, increased sequence diversity may reflect evolving modulatory roles for $Ca_v3.2$ and $Ca_v3.3$. Sequence identity across the Ca_v3 family is approximately 40 per cent (e.g. comparing $Ca_v3.1$ to $Ca_v3.2$), with the highest level of conservation found in the membrane-spanning regions.

Alternatively spliced variants of all three Ca_v3 genes have been studied in detail (Perez-Reyes and Lory, 2006). The most exhaustive surveys were performed in the Agnew laboratory (Mittman *et al.*, 1999a, b; Emerick *et al.*, 2006; Zhong *et al.*, 2006). Splicing of the III-IV linker is similar in both $Ca_v3.1$ and $Ca_v3.2$ channels and affects their biophysical properties. Both are also spliced in the II-III linker and carboxy terminus. Many splicing events have been reported for $Ca_v3.2$ that are not predicted to allow for the production of functional channels (Gray *et al.*, 2004; Zhong *et al.*, 2006). The significance of these events requires further study. $Ca_v3.3$ splice variants of the I-II loop and carboxy terminus have been studied in detail (Chemin *et al.*, 2001; Murbartián *et al.*, 2004). Notably, the functional effects of these variations are interdependent, suggesting that intracellular loops may interact to modulate channel gating.

The three T-channel genes have been mapped and sequenced. The gene encoding $\alpha 1I$ ($Ca_v3.3$), which resides on chromosome 22q13, was sequenced as part of the Human Genome Project (Dunham *et al.*, 1999) and deposited in GenBank even before the cDNA was isolated (Lee *et al.*, 1999). All three genes are quite large (70–130 kb) and contain numerous exons (35–38). The intron/exon structure of these genes is similar to HVA Ca^{2+} and Na^+ channels, including their use of minor AT-AC introns (Wu and Krainer, 1999).

Biophysical characteristics

Expression of recombinant Ca_v3 channels results in the generation of robust currents that strongly resemble those recorded from native channels. Currents both open and inactivate at potentials near the typical resting membrane potential, recover rapidly from inactivation, and close slowly producing prominent tail currents (Perez-Reyes, 2003). $Ca_v3.1$ and $Ca_v3.2$ currents inactivate (~1.7-fold) faster with Ba^{2+} than with Ca^{2+} as the charge carrier (Klöckner et al., 1999; Klugbauer et al., 1999). Similarly, recording conditions such as choice of charge carrier or length of depolarizing pulse can affect estimates of both deactivation kinetics (Klugbauer et al., 1999b; Kostyuk and Shirokov, 1989; Warre and Randall, 2000) and recovery from inactivation (Warre and Randall, 2000). Therefore, comparisons under a standard set of conditions are most informative. Clearly all three Ca_v3 clones form low voltage-activated channels. Depolarization of the membrane to -70 mV is sufficient to trigger channel opening, and the I-V curves for all three channels peak around -30 mV. Some studies have found that $Ca_v3.3$ currents activate at slightly more positive potentials than either $Ca_v3.1$ or $Ca_v3.2$ (Monteil et al., 2000b; Frazier et al., 2001). Part of this difference can be ascribed to the methods used to estimate the voltage dependence of activation. Of note, activation curves constructed using tail currents that detect both channels show a lower slope and higher midpoint of activation than other methods.

The kinetics of the currents are quite voltage-dependent between -70 and -20 mV, producing a stereotypical crisscrossing pattern of traces (Randall and Tsien, 1997). Therefore, it is useful to compare the kinetics of the three channels at -10 mV or above. $Ca_v3.1$ and $Ca_v3.2$ both activate relatively quickly at this potential ($\tau = 1–2$ ms), and inactivate tenfold slower ($\tau = 11–16$ ms). In stark contrast, $Ca_v3.3$ currents activate and inactivate much slower. This disparity is even greater when comparisons are made using Xenopus oocytes (Lee et al., 1999) for reasons that remain unclear.

Recombinant Ca_v3 channels, like their native counterparts, close slowly producing prominent tail currents. The voltage protocol used to measure tail currents includes a short pulse to open channels, followed by repolarization to a voltage favouring the closed state. Although all three Ca_v3 channels show slow tail currents upon repolarization to -90 mV, $Ca_v3.1$ are the slowest ($t = 3$ ms), while $Ca_v3.3$ are the fastest ($t = 1$ ms). In comparison, HVA channels close tenfold faster.

Despite marked differences in the rate at which they inactivate, the voltage dependence of steady-state inactivation of the three recombinant Ca_v3 channels is remarkably similar. The midpoint of the $h\infty$ curves is -72 mV. Native T-type currents inactivate over a similar range. These studies indicate that T-type channels can transit to inactivated states without passing through open states (Serrano et al., 1999). In fact, during a depolarizing pulse up to 30 per cent of $Ca_v3.3$ channels inactivate without opening (Frazier et al., 2001).

Ca_v3 channels recover rapidly from short-term inactivation. $Ca_v3.1$ channels recover fastest, displaying a mono-exponential recovery with a time constant of approximately 100 ms (Perez-Reyes, 2003). $Ca_v3.3$ channels recover threefold more slowly, while $Ca_v3.2$ channels recover even more slowly (Klöckner et al., 1999). The channels also differ in their recovery from steady-state inactivation, and again, $Ca_v3.2$ was found to recover the slowest (Klöckner et al., 1999). Surprisingly, rates of $Ca_v3.2$ channel recovery from inactivation depend on the length of the inactivating pulse (Uebachs et al., 2006), suggesting the existence of multiple inactivated states, as noted previously for T-type currents from sensory neurons (Bossu and Feltz, 1986). These results also suggest that recovery kinetics can be used to differentiate currents carried by native $Ca_v3.1$ from $Ca_v3.2$ (Satin and Cribbs, 2000). Fast recovery from inactivation is a critical property of native T-type channels that allows them to participate in rebound burst depolarizations.

The ability of T-type channels to open at similar potentials at which they inactivate suggests they might generate a window current under steady-state conditions. All three Ca_v3 channels are predicted to generate window currents supported by less than 1 per cent of the channels. Evidence for window currents has also been obtained in thalamacortical neurons, where they play an important role in signal amplification (Crunelli et al., 2005). Window currents also appear to be important in determining intracellular Ca^{2+} concentrations (Assandri et al., 1999; Chemin et al., 2000; Mariot et al., 2002). Expression of recombinant $Ca_v3.1$ or $Ca_v3.2$ was found to increase basal Ca^{2+} in HEK-293 cells, an increase blocked by appropriate concentrations of either mibefradil or nickel. Expression of $Ca_v3.3$ channels in neuroblastoma cells led to oscillations in intracellular Ca^{2+}, with window currents implicated in mediating this phenomenon (Chevalier et al., 2006). Hormonal regulation of the T-type window current provides cells with another mechanism to regulate intracellular Ca^{2+}. In adrenal glomerulosa cells, angiotensin II increases T-type window currents by selectively shifting activation to more negative voltages (McCarthy et al., 1993; Wolfe et al., 2002), an effect which is mediated by CaMKII phosphorylation.

Another distinguishing property of T-type channels is their lower Ba^{2+} conductance in comparison to HVA channels. Therefore, studies of cloned T-type channels have focused on the amplitude of single channel currents in isotonic barium solutions. All three recombinant channels were found to have small single channel currents, corresponding to slope conductances in the 7–11 pS range. In contrast, $Ca_v1.2$ channels display slope conductances of 20–30 pS, while Ca_v2 channels are in the 13–16 pS range.

Localization

The expression patterns of these genes in peripheral tissues and brain have been studied using northern blots, RNA dot blots, and in situ hybridization. $Ca_v3.1$ mRNA is primarily expressed in human brain but is also found in ovary, placenta, and heart (Chien et al., 1998; Monteil et al., 2000a). A wider distribution was noted in dot blots prepared from human fetal tissues, with strong expression noted in kidney and lung (Monteil et al., 2000a). $Ca_v3.2$ mRNA is primarily expressed in kidney and liver, but also in heart, brain, pancreas, placenta, lung, skeletal muscle, and adrenal cortex (Cribbs et al., 1998; Williams et al., 1999). $Ca_v3.3$ mRNA is almost exclusively expressed in brain (Lee et al., 1999).

The distribution of all three Ca_v3 channel transcripts in rat brain has been studied via in situ hybridization (Talley et al., 1999), and similar patterns of expression have been described using Northern blots of human brain mRNA (Williams et al., 1999; Monteil et al., 2000a, b). The patterns of transcript expression are complementary in many regions, with a majority of brain structures expressing more than one isoform. In fact, some neurons may express all three genes as suggested by the labelling of transcripts in olfactory granule cells and hippocampal pyramidal neurons. Many brain regions showed heavy expression of $Ca_v3.1$ mRNA, including thalamic relay nuclei, olfactory bulb, amygdala, cerebral cortex, hippocampus, hypothalamus, cerebellum, and brain stem. Two separate studies have confirmed this pattern of mRNA expression, and extended the findings by showing a similar distribution of $Ca_v3.1$ protein (Craig et al., 1999; Kase et al., 1999). $Ca_v3.2$ mRNA expression was detected in olfactory bulb, striatum, cerebral cortex, hippocampus, and reticular thalamic nucleus. $Ca_v3.3$ mRNA expression is high in olfactory bulb, striatum, cerebral cortex, hippocampus, reticular nucleus, lateral habenula, and cerebellum. Dorsal root ganglion (DRG) neurons express both $Ca_v3.2$ and 3.3 mRNA (Talley *et al.*, 1999). One study found that this expression was restricted to small and medium-sized neurons (Talley *et al.*, 1999), while a second study reported that $Ca_v3.3$ transcripts were equally abundant in large DRG neurons (Yusaf *et al.*, 2001).

Pharmacology

In contrast to their HVA counterparts, there are no drugs or peptide toxins that selectively block Ca_v3 channels. Although it was first thought to be selective, kurtoxin, a scorpion venom, has subsequently been shown to block both native LVA and HVA currents (Sidach and Mintz, 2002). T-channels are blocked by low micromolar concentrations of a number of drugs from various classes (for comprehensive reviews see Heady *et al.*, 2001; McGivern, 2006a). The most widely used drug is mibefradil, which was briefly on the market as an antihypertensive agent. As is the case for many ion channel blockers, mibefradil has a higher affinity for inactivated states and under appropriate experimental conditions its IC_{50} has been estimated at 70 nM on both native and recombinant T-channels (McDonough and Bean, 1998; Martin *et al.*, 2000). Caution must be exercised in its use as it has also been shown to block a wide range of ion channels in the micromolar range (Heady *et al.*, 2001). This notwithstanding, it has been widely used to implicate T-channels in a variety of physiological processes. Future studies with a truly selective blocker will be required to prove the veracity of these reports.

T-channels were one of the first therapeutic targets to be identified for absence epilepsy (Macdonald and Kelly, 1995). Although drugs such as ethosuximide and valproate block T-currents they are extremely weak blockers, requiring millimolar concentrations. Nevertheless, they do block some of the current at therapeutic concentrations (Gomora *et al.*, 2001). Due to their pacemaker role in the firing of Na^+-dependent action potentials, even a small inhibition may be therapeutically relevant, a concept termed 'pharmacological amplification' (Narahashi, 2000).

Native and recombinant T-channels are also potently blocked by antipsychotic drugs such as pimozide (Orap®) and penfluridol (Enyeart *et al.*, 1987; Santi *et al.*, 2002). The therapeutic mechanism of action of these drugs is primarily mediated by D_2 dopamine receptor antagonism but they are also capable of blocking Ca_v1 and Ca_v2 channels. Accordingly, Santi and coworkers have suggested that T-channel block may contribute to their therapeutic efficacy vs the negative symptoms associated with schizophrenia (Santi *et al.*, 2002).

The antihypertensive drug, efonidipine (Landel®, available in Japan) has also been postulated to act via blockade of T-currents. It blocks both native and recombinant T-channels in the low micromolar range (Masumiya *et al.*, 1998; Lee *et al.*, 2006). In contrast, most antihypertensive drugs of the dihydropyridine class are selective for L-type channels, and only block T-currents at concentrations > 10 µM (Heady *et al.*, 2001). Evidence is accumulating that both T- and L-channels are involved in glomerular filtration, such that block of L channels increases flow to the glomerulus by relaxing afferent arterioles, while block of T relaxes efferent arterioles, and lowers glomerular pressure (Kawabata *et al.*, 1999).

This may explain the renoprotective effects of efonipine (Hayashi *et al.*, 2007). The role of T-channels in aldosterone secretion is well established, and the ability of efodipine to block its secretion may provide additional therapeutic benefit in patients with heart failure (Okayama *et al.*, 2006).

Considerable evidence supports the notion that T-channels are an important target for the treatment of neuropathic pain (McGivern, 2006b; Nelson *et al.*, 2006). Briefly, nociceptors have extremely large T-currents and agents that increase T-currents produce hyperalgesia, while agents that block currents produce analgesia (Nelson *et al.*, 2006). In addition to facilitating pain transmission from the periphery, T-currents have been shown to play a role in central sensitization by inducing long-term potentiation of dorsal horn neuron synaptic transmission (Ikeda *et al.*, 2003). Studies on transgenic knockout mice have also yielded novel insights into the role of T-channels in nociception. As predicted from their prominent expression in nociceptors, $Ca_v3.2$ –/– mice have an attenuated response to pain (Choi *et al.*, 2006). In contrast, $Ca_v3.1$ –/– mice have a prolonged response to painful stimuli, a phenotype attributed to their loss of burst firing of thalamocortical neurons (Kim *et al.*, 2003). In light of these findings, a peripherally active T-channel blocker offers potential as an analgesic mechanism.

Physiology

The expression of T-type channels in various cell types suggests they play a role in diverse physiological functions. The voltage dependence of activation and inactivation provides clues to these roles and places constraints on when these channels will be active. In neurons where the resting membrane potential is in the –90 to –70 mV range, T-type channels can play a secondary pacemaker role: an excitatory postsynaptic potential (EPSP) opens T-type channels and generates a low threshold Ca^{2+} spike, which in turn activates Na^+-dependent action potentials and HVA Ca^{2+} channels. In this manner T-type channels play an important role in the genesis of burst firing. Due to their fast recovery from inactivation, T-type channels can produce a rebound burst in depolarized neurons following an inhibitory postsynaptic potential (IPSP). Electrophysiological recordings indicate that these channels are preferentially localized to dendrites and hence may play a role in signal amplification (Markram and Sakmann, 1994; Magee and Johnston, 1995; Destexhe *et al.*, 1998; Pouille *et al.*, 2000). However, T-type channels have also been suggested to play a role in release of neurotransmitter in the dorsal horn of the spinal cord, retinal bipolar cells, and adrenal chromaffin cells (Carbone *et al.*, 2006).

Large T-type currents have been found in thalamic neurons, where they play an important role in oscillatory behaviour. The thalamus acts as a gateway to the cerebral cortex, and inappropriate oscillations of these circuits, or thalamocortical dysrhythmias, have been implicated in a wide range of neurological disorders (Llinás *et al.*, 2001). T-type currents also appear to play a role in olfaction, vision, and pain reception (Kawai and Miyachi, 2001; Pan *et al.*, 2001; Todorovic *et al.*, 2001).

Calcium influx not only depolarizes the plasma membrane but also acts as a second messenger leading to the activation of a plethora of enzyme and channel activities. T-type channels can cause robust increases in intracellular Ca^{2+}, especially in proximal dendrites (Munsch *et al.*, 1997; Zhou *et al.*, 1997). Calcium and voltage synergistically open Ca^{2+}-activated K^+ channels which contribute to spike repolarization and after hyperpolarizations (Llinás and Yarom, 1981b; Umemiya and Berger, 1994; Wolfart and Roeper, 2002). In addition to transient increases in intracellular Ca^{2+}, T-type window currents may play an important role in slower increases in basal Ca^{2+}. This property appears to be important for hormone secretion from adrenal cortex and pituitary, myoblast fusion (Bijlenga et al., 2000), and possibly smooth muscle contraction (but see Chen et al., 2003).

Relevance to disease

Most of the SNPs that occur in the vicinity of the T-channel genes occur in the intronic, or non-coding regions, which constitute up 90 per cent of the gene. Although changes in intronic sequences can change splicing of exons, such as in the high voltage-activated auxiliary subunit β_4 (Burgess et al., 1997), this has yet to be documented for T-channel genes. Instead, attention has focused on changes in the coding region (missense or non-synonymous SNPs), and in particular, on the possible association of these SNPs with absence epilepsy. Chen and co-workers sequenced the coding regions of *CACNA1G* and *CACNA1H* genes in a large (>100) sample of childhood absence epilepsy (CAE) patients of Han ethnicity (Y Chen et al., 2003a, b). Although no SNPs were found to be associated with epilepsy in the *CACNA1G* gene, there were 12 missense SNPs in *CACNA1H* that were only found in epilepsy patients. Over half the SNPs led to changes in the amino acid sequence of the intracellular loop that connects repeat I to repeat II, two are in extracellular loops, and three are in putative transmembrane loops. Introduction of these mutations into recombinant $Ca_V3.2$ channels alters the time- and/or voltage-dependent properties of the channel (Khosravani et al., 2004; Vitko et al., 2005). A common finding was that mutations increased window currents, and this had a profound effect on simulated neuronal firing (Vitko et al., 2005). However, some mutations decreased simulated firing and others had no significant effect on channel gating. Recent studies indicate that all the mutants tested increased the surface expression of $Ca_V3.2$ channels (Vitko et al., 2007). This provides a potentially unifying mechanism by which SNPs contribute to absence epilepsy, since upregulation of T-currents increases the tendency of thalamic circuits to fire (McCormick and Huguenard, 1992). The SNPs associated with epilepsy are very rare and have not been found in other cohorts (Heron et al., 2004), thus, the finding that a common SNP, R788C, affects channel activity may be of greater population relevance. This SNP is found in about ~10 per cent of the human population and further studies are required to investigate whether it increases seizure susceptibility in these individuals (Vitko et al., 2005).

Due to the important role Ca^{2+} plays as a second messenger, an increase in intracellular Ca^{2+} can trigger a myriad of effects such as changes in neuronal firing, gene expression, and hormone secretion (Chen et al., 1999; Bijlenga et al., 2000; Crunelli et al., 2005). $Ca_V3.2$ channels are expressed in many brain regions, including the reticular nucleus of the thalamus (Talley et al., 1999). This nucleus plays a key role in controlling thalamic rhythms and alterations in these rhythms (termed thalamocortical dysrhythmia) are thought to underlie generalized seizures and other neurological disorders (Llinás et al., 1999), including schizophrenia (Cho et al., 2006). Thus, it is quite possible that SNPs that alter T-channel activity could be one of the predisposing factors that underlie complex polygenic disorders such as absence epilepsy. This hypothesis is further supported by pharmacological studies that show that absence epilepsy treatments such as ethosuximide can block T-channels (Coulter et al., 1989; Gomora et al., 2001), and by animal studies that show increased T-currents and mRNA in rats that exhibit an absence epilepsy phenotype (Tsakiridou et al., 1995; Talley et al., 2000).

Mutations in *CACNA1H* have also been linked to autistic spectrum disorder (ASD) (Splawski et al., 2006). These mutations altered critical amino acids in conserved regions of the channel such as S4 voltage sensors (R212C, R902W) and the pore (W962C). Not surprisingly the mutations dramatically reduced T-currents. It is interesting to note that mutations that increase $Ca_V3.2$ currents are associated with absence epilepsy while those that decrease currents are associated with autism.

Concluding remarks

The cloning of the cDNAs encoding the T-type calcium channel family has led to considerable advances in our understanding of their genomic structure, splicing and how splicing affects channel function. The gene encoding $Ca_V3.1$ is alternatively spliced in many positions, while $Ca_V3.3$ and $Ca_V3.2$ are less so. The $Ca_V3.2$ gene appears to contain more single nucleotide polymorphisms, some of which have been found to be overrepresented in CAE and ASD patients. The expression of $Ca_V3.2$ channels in the reticular nucleus of the thalamus together with the role of T-currents in thalamic oscillations has lead to the hypothesis that SNPs in T-channel genes may contribute to polygenic disorders that are characterized by thalamocortical dysrhythmia (Llinás et al., 1999). Further studies are required to determine if these channels are regulated by auxiliary subunits.

2.2.5 Auxiliary subunits

Introduction

The auxiliary subunits of HVA channels were identified in purified preparations of skeletal muscle L-type channels, a discovery which led to their cloning. Their co-expression with the α_1 subunit was found to be critical not only for the formation of functional channels, but also to confer the electrophysiological and pharmacological properties of native channels. Although their effects on channels are sometimes α_1 specific, a few general observations can be made. $Ca_V\alpha_2\delta$ subunits are important for channel trafficking to the plasma membrane and their co-expression with α_1 subunits increases current density two- to fivefold with little effect on channel biophysics. $Ca_V\beta$ subunits also increase current density ~ fivefold, and this effect is due to both an effect on trafficking and an increase in the probability of channel opening (P_o). Although the skeletal muscle channel contains a γ subunit that is capable of modulating pharmacology and channel inactivation, it remains uncertain whether the brain γ subunits are genuine Ca^{2+} channel subunits or, rather, ancillary α-amino-3-hydroxy-5-methyl-4-isoxazole propionic acid (AMPA) receptor subunits. Ca^{2+} channel auxiliary subunits modulate α_1 structure and thereby pharmacology and the finding that gabapentin binds $\alpha_2\delta$ proteins with high affinity also raised the possibility that they too might represent drug targets.

Molecular characterization

$\alpha_2\delta$

The highly glycosylated $Ca_v\alpha_2\delta$ subunit binds effectively to wheat germ agglutinin affinity columns, a property that, together with the ability of digitonin to solubilize the Ca^{2+} channel as a complex, enabled the purification of $Ca_v1.1$ channels. This purified protein was microsequenced, allowing the subsequent synthesis of oligonucleotides that were used to screen rabbit skeletal muscle cDNA libraries (Ellis *et al.*, 1988). The $Ca_v\delta$ peptide, which in the native state is bound to $Ca_v\alpha_2$ through disulfide bonds, was also purified and sequenced. Comparison of this sequence to the cloned $Ca_v\alpha_2$ demonstrated that these proteins were encoded on the same cDNA (De Jongh *et al.*, 1990). The δ subunit contains a transmembrane region, which anchors the complex to the membrane, while the $\alpha_2\delta$ subunit extends into the extracellular space (Wolf *et al.*, 2003). The α_2 portion of the channel belongs to a large family of proteins with a VWFA motif, named after Von Willebrand factor, and mutagenesis experiments suggest that it folds in a similar manner (Canti *et al.*, 2005).

Human brain $Ca_v\alpha_2\delta$ has also been cloned (Williams *et al.*, 1992b). While most of the amino acid discrepancies between the rabbit skeletal muscle sequence and that of human brain could be attributed to species differences, one region suggests the insertion or deletion of an exon. Three variable regions contribute to the five known $\alpha_2\delta1$ isoforms, all of which display tissue-specific localization (Angelotti and Hofmann, 1996).

The other members of the $Ca_v\alpha_2\delta$ family were cloned using a combination of *in silico* and genetic locus screening. Some patients with small cell and renal cell carcinomas show loss of a large chunk of chromosome 3. This led to an effort to catalogue all the genes within this region, with the hope of identifying a possible tumour suppressor at chromosome 3p21. Instead, it led to the cloning of $\alpha_2\delta2$ (Gao *et al.*, 2000) and $\alpha_2\delta3$ (Hanke *et al.*, 2001). Simultaneously, the search for the mutation that produced the absence epilepsy phenotype of the *ducky* mouse had narrowed the locus to a syntenic region. Based on the recognized involvement of Ca^{2+} channels in epilepsy $\alpha_2\delta2$ was a likely candidate gene and indeed associated mutations were identified within the gene (Barclay *et al.*, 2001). Data mining of the human genome also led to the identification of $\alpha_2\delta2$ (*CACNA2D2*), $\alpha_2\delta3$ (*CACNA2D3*), and $\alpha_2\delta24$ (*CACNA2D4*) (Klugbauer *et al.*, 1999b; Qin *et al.*, 2002). Sequence analysis indicates that $\alpha_2\delta3$ and $\alpha_2\delta4$ are more related to each other than either $\alpha_2\delta1$ or $\alpha_2\delta2$ (Figure 2.2.3).

Two regions in the genetic sequence of $\alpha_2\delta2$ appear to be alternatively spliced. The first, found within the $Ca_v\alpha_2$-encoding region, involves the inclusion or exclusion of eight amino acid residues. The second is found with the $Ca_v\delta$-encoding sequence where three residues are alternatively expressed (Hobom *et al.*, 2000). The $Ca_v\alpha_2\delta2$ isoforms appear to be variably expressed in a number of tissues including brain, kidney, and cardiac tissue, however, all variants are expressed in thyroid cells (Hobom *et al.*, 2000). The functional significance of these variations is unknown. Although there are no known splice variants of $\alpha_2\delta3$, four $\alpha_2\delta4$ variants have been described (Qin *et al.*, 2002).

$\beta1$

The skeletal muscle β ($Ca_v\beta1$) was purified with the skeletal muscle DHP receptor and the cDNA was cloned using oligonucleotides

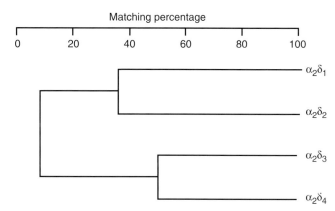

Fig. 2.2.3 Evolutionary tree of the α_2d family. Human sequences corresponding to the four $\alpha_2\delta$ subunits were aligned using CLUSTAL (Higgins and Sharp, 1988) and DNASIS software.

based on the partial protein sequence (Ruth *et al.*, 1989). Screening of rat brain cDNA libraries led to the cloning of both splice variants of the $Ca_v\beta1$ gene and novel $Ca_v\beta$ subtypes (β_2; $Ca_v\beta_2$) (Perez-Reyes *et al.*, 1992). The $Ca_v\beta_1$ gene is alternatively spliced in at least two regions to form three full-length proteins ($Ca_v\beta1a$, $Ca_v\beta1b$, $Ca_v\beta1c$), with a fourth splice variant, β_{1d}, which appears to be a splice error resulting in a jump in the open reading frame and truncation. Despite loss of most of the β protein, β_{1d} retains the ability to modulate $Ca_v1.2$ channels (Cohen *et al.*, 2005). The genomic structure confirms that these alternate forms are derived from splicing of a single gene product.

$\beta2$

As noted above, screening of rat brain cDNA libraries with $Ca_v\beta_{1a}$ cDNA led to cloning of β_2 (Perez-Reyes *et al.*, 1992). A human brain $Ca_v\beta_2$ cDNA was cloned using sera from Lambert–Eaton myasthenic syndrome (LEMS) patients to screen an expression library (Rosenfeld *et al.*, 1993). This was a fortuitous finding since β subunits are intracellular and LEMS sera induces muscle weakness by binding to extracellular epitopes on $\alpha_1 2.1$ (Parsons and Kwok, 2002).

Splice variation of the $Ca_v\beta_2$ gene has been studied in detail (Foell *et al.*, 2004; Takahashi *et al.*, 2003). It is alternatively spliced to generate five variants encoding distinct amino termini, and four variants in the middle of the protein. One of the central splice variants does not restore the proper reading frame, and is predicted to generate a truncated isoform similar to $\beta1d$ (Castellano and Perez-Reyes, 1994). In contrast to $Ca_v\beta_1$, alternative splicing of the carboxyl terminus has not been detected save for one truncated variant found in human cardiac tissue (Harry *et al.*, 2004). The original $\beta2$ isoform cloned, $\beta2a$, is unique in that is can be palmitoylated, which serves as an anchor in the plasma membrane (Chien *et al.*, 1998; Restituito *et al.*, 2000). This anchor alters the way β subunits affect channel activity, dramatically slowing inactivation from the open state (Olcese *et al.*, 1994). This effect may be due to altered mobility of IS6 (Arias *et al.*, 2005) which is thought to form an inner gate in a manner analogous to K^+ channels (Yellen, 2002).

β3

Three approaches were originally used to clone $Ca_v\beta_3$: low stringency screening with the skeletal muscle $Ca_v\beta_{1a}$ cDNA as probe; a PCR-based strategy using primers that matched conserved regions in $Ca_v\beta_1$ and $Ca_v\beta_2$; by using antibodies against $Ca_v\beta_3$ to screen a λ-gtll expression library (Perez-Reyes and Schneider, 1994). The deduced amino acid sequence of the rat $Ca_v\beta_3$ is 98 per cent identical to its human orthologue, while it is only ~70 per cent identical with the other $Ca_v\beta$s (Figure 2.2.4). Most of this similarity is in a central core region, while the carboxy termini are more divergent. In contrast to $Ca_v\beta_1$ and $Ca_v\beta_2$, very little alternative splicing of $Ca_v\beta_3$ has been identified. The $Ca_v\beta_3$ gene expresses a small exon in the central splice site that has a similar sequence to $Ca_v\beta_{1b}$, $Ca_v\beta_{2c}$, and $Ca_v\beta_4$ (consensus sequence AKQKQKQX). Although this exon 6 can be skipped, the reading frame is lost, resulting in a truncated protein ($Ca_v\beta_{3d}$) (Foell et al., 2004).

β4

$Ca_v\beta_4$ was cloned using PCR with primers that matched conserved regions in $Ca_v\beta_1$ and $Ca_v\beta_2$ (Castellano et al., 1993). This cDNA contains an open reading frame that encodes a 58-kDa protein. Its sequence also contains regions that are highly conserved in all the $Ca_v\beta$s, while the carboxyl terminus is more divergent. A phylogenetic tree based on sequence similarities indicates that $Ca_v\beta_4$ is most closely related to the ancestral $Ca_v\beta$ (Figure 2.2.4). Alternatively spliced variants of the amino terminus were recently discovered (Helton and Horne, 2002; Foell et al., 2004). Alternative splicing between exon 2A and 2B has been detected, as well as variants with and without exon 7, yielding a total of four splice variants to date (Foell et al., 2004).

γ1

The $Ca_v\gamma$ protein was purified from skeletal muscle in complex with the DHP receptor. The protein was sequenced and oligonucleotides were made using this sequence information enabling the cloning of its cDNA from skeletal muscle libraries (Bosse et al., 1990). $Ca_v\gamma_1$ was also cloned using antibodies against $Ca_v\gamma$ to screen a λ-gt11 expression library (Jay et al., 1990). The deduced amino acid sequence encodes a protein of 25 kDa, which contains four

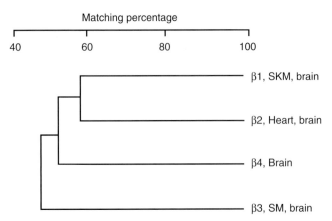

Fig. 2.2.4 Evolutionary tree of the β family. Human sequences corresponding to the four β subunits were aligned using CLUSTAL (Higgins and Sharp, 1988) and DNASIS software. SKM, skeletal muscle; SM, smooth muscle.

putative membrane-spanning regions. Human $Ca_v\gamma$ has been cloned from both cDNA and genomic DNA (Powers et al., 1993).

Almost a decade passed before, serendipitously, a phenotype in a spontaneous mutant mouse line known as *stargazer* led researchers to the discovery of additional $Ca_v\gamma$ subunits. It was shown that the mutant phenotype resulted from a single defective gene whose structure bore strong similarity to that of the skeletal muscle $Ca_v\gamma$ (Letts et al., 1998). This novel sequence was designated as $Ca_v\gamma_2$ with the skeletal muscle subunit as $Ca_v\gamma_1$. Subsequently it was demonstrated that γ2 belongs to a large family of proteins which recent evidence suggests are not in fact Ca^{2+} channel subunits but instead control the targeting of AMPA receptors to the postsynaptic membrane (Nicoll et al., 2006).

Physiology and biophysics

$\alpha_2\delta$ regulation of HVA channel function

Early studies using radioligand binding clearly demonstrated that a_2d1 increased the number of functional receptors (α_1 subunits) two- to fivefold when co-expressed with either $Ca_v1.1$ or $Ca_v2.1$ (Brust et al., 1993; Suh-Kim et al., 1996). In recombinant systems, co-expression of $\alpha_2\delta1$ can increase surface expression of both Ca_v1 and Ca_v2 such that: 1) $\alpha_2\delta1$ increases gating currents of $Ca_v1.2$ (a proxy for the number of channels at the plasma membrane) (Bangalore et al., 1996); and 2) $\alpha_2\delta1$ increases $Ca_v2.3$ current density (Jones et al., 1998). Studies on the trafficking of $Ca_v2.2$ channels reveals that $\alpha_2\delta1$ stabilizes the $\alpha_12.1$ subunit at the plasma membrane by decreasing its rate of degradation (Bernstein and Jones, 2007). Its effects on channel kinetics are less clear, although a role in channel inactivation has been described when co-expressed with $Ca_v1.2$ (Shirokov et al., 1998). One confounding factor is the presence of variable β subunits in the heteromeric channel complex, each of which exert their own effects on the biophysical profiles of the channels. Some evidence suggests the $\alpha_2\delta$ subunit acts synergistically with the β subunit to confer effects on channel properties which include enhanced surface expression (Felix et al., 1997). However, while some effects of the $\alpha_2\delta$ subunit appear to require the β subunit others do not (Klugbauer et al., 1999b). In conclusion, there is no general agreement on the effect of the presence of $\alpha_2\delta$ on the biophysical properties of the Ca_v channel complex (Canti et al., 2003).

β regulation of HVA channel function

In general, all $Ca_v\beta$s affect HVA calcium channel activity in similar ways: they increase functional channel expression by both aiding in trafficking α_1 to the plasma membrane and increasing the probability of channel opening (Perez-Reyes and Schneider, 1994; Arikkath and Campbell, 2003). The most notable difference is that $Ca_v\beta_3$ and $Ca_v\beta_4$ have a more robust effect on inactivation. In contrast, $Ca_v\beta$ interactions with other $Ca_v\alpha$s can have other, sometimes unpredictable, results.

While the mechanisms by which $Ca_v\beta$s modulate the activity of different $Ca_v\alpha_1$s may vary, the interaction site appears to be well conserved. This region was discovered by screening an expression library made from $Ca_v1.1$ cDNA with ^{35}S-labelled $Ca_v\beta_{1b}$ protein (Pragnell et al., 1994). This screening isolated many clones that encoded the domain I to II loop. This $Ca_v\beta$-binding domain was mapped in more detail using site-directed mutagenesis. A recent watershed series of findings revealed the atomic disposition of the α_1 and $Ca_v\beta$ subunit complex which enables insights into the molecular underpinnings of Ca²⁺ channel regulation (Chen *et al.*, 2004; Opatowsky *et al.*, 2004; Van Petegem *et al.*, 2004). X-ray crystallographic analysis demonstrated two well-conserved domains within the Ca$_v$β subunit, a Src homology 3 (SH3) domain comprised of two α helices and five β-strands, and a guanylate kinase (GK) domain consisting of seven α-helices and five β strands. The beta interacting domain (BID) of Ca$_v$β subunits, which was previously thought to bind directly to the α₁ subunits, was shown to contribute to stabilizing the structures of the GK and SH3 domains. The prevailing notion of a BID sequence that is critical to the α1–β interaction was further challenged by the discovery of a Ca$_v$β2 variant that contained none of the BID sequence yet still effectively regulated the Ca$_v$α₁ channel trafficking and kinetics (Harry *et al.*, 2004). Studies with chimeras of HVA and LVA channels indicated that sensitivity to Ca$_v$β regulation could be transferred to the typically unresponsive LVA channels (Arias *et al.*, 2005). This report also presented evidence that β subunits modulate α₁ activity by direct action on IS6. Nevertheless, the chimeric channels only displayed two aspects of β regulation and notably the β subunits' ability to regulate channel P$_o$ was absent. The mechanisms by which β subunits alter channel activity remains a vibrant area of research.

γ modulation of HVA channel function

Expression studies have demonstrated a hyperpolarizing shift in the steady-state inactivation of Ca$_v$1.1 (α$_{1S}$) following co-expression of Ca$_v$γ₁ (Perez-Reyes and Schneider, 1994). These results were confirmed in knockout mice, which are otherwise phenotypically normal, indicating that the γ subunit regulates Ca$_v$1.1 channel function but not its ability to trigger release of Ca²⁺ from the SR (Freise *et al.*, 2000; Ahern *et al.*, 2001).

Auxiliary subunit regulation of LVA channel function

It must be emphasized that the α₁ subunits of T-channels were cloned by homology to α₁ subunits of HVA channels and not by classical methods such as sequential purification, microsequencing, and cloning with oligonucleotides based on the derived protein sequence. This classical approach identified the α₂δ, β, and γ auxiliary subunits of HVA channels from the urified channels, whilst cloning via an *in silico* approach has provided no evidence for T-channel-specific auxiliary subunits. The fact that expression of recombinant Ca$_v$3.1 and Ca$_v$3.2 produced large currents with similar electrophysiological properties to native T-channels suggested that, if there are any T-channel auxiliary subunits, they do not play similar roles as in HVA channels, which are essential for trafficking, regulation of channel P$_o$, and voltage-dependence of gating (for review see Arikkath and Campbell 2003).

Early studies on the effects of recombinant Ca$_v$β subunits suggested that they could regulate endogenous T-like channels in *Xenopus* oocytes (Lacerda *et al.*, 1994). However, these endogenous channels are likely encoded by Ca$_v$2.3 (Lee, 1998). Antisense knockdown of endogenous β subunits in sensory neurons (Lambert *et al.*, 1997) or neuroblastoma cell lines (Leuranguer *et al.*, 1998) showed clear effects on HVA, but not LVA, currents. Similar results were recently obtained with the GTPase Rem2, in sensory neurons, with Rem2 binding and inactivating β, and thereby abolishing β regulation of HVA currents with no effect on LVA currents (Chen *et al.*, 2005). Coexpression of recombinant β subunits with Ca$_v$3.1 led to some modifications in gating parameters when tested in COS cells (Dolphin *et al.*, 1999) but had no effect in HEK-293 cells

(Arias *et al.*, 2005). Using epitope-tagged channels, Dubel and coworkers showed that β subunits could increase the surface expression of Ca$_v$3.x channels, however, they also reported that they were unable to demonstrate high-affinity binding of the two proteins (Dubel *et al.*, 2004). The high-affinity interaction site on HVA α1 subunits that binds β subunits has been well mapped and many of the amino acids that form important contacts between the two proteins (Opatowsky *et al.*, 2004) are not conserved in Ca$_v$3 channels (Perez-Reyes, 2003). Taken together, these studies indicate that HVA β subunits are unlikely to be auxiliary subunits of Ca$_v$3 channels.

Similar co-expression studies using recombinant T-channels and HVA α₂δ subunits have been reported. For example, it was shown that α₂δ1 could increase the surface expression of Ca$_v$3.1 in two distinct expression systems, and the corollary, that Ca$_v$3.1 increased the surface expression of α₂δ1 (Dolphin *et al.*, 1999). This result was confirmed and extended to all three Ca$_v$3 channels using tagged channels and, furthermore, the combination of α₂δ1 plus β had additive effects on Ca$_v$3.x surface expression (Dubel *et al.*, 2004) Studies from the Hofmann group initially reported no effect of α₂δ1 or α₂δ3 (Lacinová *et al.*, 1999) while subsequent investigations did find an effect with α₂δ2 (Hobom *et al.*, 2000). This latter study was the first to report biophysical changes in Ca$_v$3.1 gating when coexpressed with an α₂δ, such that inactivation kinetics were accelerated and the steady-state inactivation curve was shifted. Co-expression of α₂δ2 can also affect the kinetics of the gating current generated by Ca$_v$3.1 (Lacinova and Klugbauer, 2004). Although interesting, these studies do not resolve the issue of whether HVA auxiliary subunits associate with LVA channels *in situ*, leaving open an important avenue for future research.

Localization

Northern blot analysis detects 7- and 8-kb transcripts for Ca$_v$α₂δ mRNA with expression in skeletal muscle, heart, aorta, lung, ileum, and brain (Klugbauer *et al.*, 2003). Ca$_v$α₂δ distribution has also been mapped through the whole rat embryo by *in situ* hybridization (Kim *et al.*, 1992).

Abundant Ca$_v$β₂ transcripts have been detected in brain, heart, and lung (Perez-Reyes and Schneider, 1994). Lower levels of expression can also be detected in kidney. At least three distinct transcripts were detected (6, 4, and 3.5 kb). This heterogeneity is likely due to both alternatively spliced mRNAs and alternative use of polyadenylation signals.

Northern blot analysis indicates that Ca$_v$β₃ is predominantly expressed in brain, with lower amounts detected in heart, aorta, trachea, lung, skeletal muscle, ovary, colon and pancreatic β-cell lines (Perez-Reyes and Schneider, 1994). Ca$_v$β₄ is predominantly expressed in brain, especially in cerebellum (Perez-Reyes and Schneider, 1994). Ca$_v$γ is almost exclusively expressed in skeletal muscle, although similar-sized transcripts have been reported in lung and aorta (Perez-Reyes and Schneider, 1994).

Pharmacology

The Ca$_v$ auxiliary subunits modulate the binding of antagonists to Ca$_v$α₁ subunits, and alter their ability to block currents (Wei *et al.*, 1995; Suh-Kim *et al.*, 1996). Isoform-specific regulation has not been investigated. Studies aimed at elucidating the molecular target of gabapentin (Neurontin®) identified a high affinity binding to C$\alpha_2\delta$1 (Gee *et al.*, 1996). Subsequent studies have shown that it also binds to $\alpha_2\delta$2, but not $\alpha_2\delta$3 or $\alpha_2\delta$4 (Marais *et al.*, 2001; Qin *et al.*, 2002). The related compound, pregabalin (Lyrica®), also binds to the $\alpha_2\delta$1 protein (Belliotti *et al.*, 2005). How this binding might mediate the efficacy of these compounds in treating neuropathic pain and epilepsy remains controversial. Studies of effect on channel function using electrophysiology have yielded conflicting results (Taylor *et al.*, 1998; Maneuf *et al.*, 2006). Recently Luo and collaborators have suggested that these variable results may in part be explained by auxiliary subunit composition since they could observe channel block in DRG neurons isolated from transgenic mice engineered to overexpress $\alpha_2\delta$1 but not in controls (Li CY *et al.*, 2006). Notably, expression of $\alpha_2\delta$1 is upregulated in many animal models of neuropathic pain (Luo *et al.*, 2002). Gabapentin and pregabalin binding to $\alpha_2\delta$1 can be abolished by a point mutation (R217A) and knockin of this mutation into mice abolishes the analgesic properties of these drugs (Field *et al.*, 2006). Therefore, it can be concluded that $\alpha_2\delta$1 is an important drug target for analgesics.

Relevance to disease

As noted in the previous section, $\alpha_2\delta$1 plays an important role in neuropathic pain. Although $\alpha_2\delta$2 and $\alpha_2\delta$3 genes were cloned during a search for a tumour suppressor, there is little evidence that they serve this function *in situ*. Recently, a nonsense mutation in the gene encoding $\alpha_2\delta$4 was linked to autosomal recessive cone dystrophy (Wycisk *et al.*, 2006b). Similarly, a frameshift mutation in the mouse $\alpha_2\delta$4 gene leads to retinopathy (Wycisk *et al.*, 2006a). Mutations in the β4 gene have been found in patients with juvenile myoclonic epilepsy and episodic ataxia type 5 (Escayg *et al.*, 2000).

In mice, mutations in the genes encoding several Ca$_v$ auxiliary subunits have been linked to absence epilepsy phenotypes: mutations in $\alpha_2\delta$2 were found in *ducky* mice; mutations in β4 were found in *lethargic*; and mutations in γ2 were found in *stargazer* (for a comprehensive review of epilepsy associated genes see Noebels 2003). These mutations disrupt expression of the respective auxiliary subunit and, due to their roles in increasing channel activity, lead to a decrease in HVA channel activity and neuronal excitability. This apparent paradox was seemingly resolved by the finding that T-currents are upregulated in these mouse models (Y Zhang *et al.*, 2002). As discussed previously, augmented T-currents would enhance thalamocortical oscillations such as those that occur during absence seizures. As predicted from this hypothesis, crossing of Ca$_v$3.1 –/– mice with either *lethargic* or *stargazer* mice abolished seizures (Song *et al.*, 2004).

Knockout of β3 in mice decreases L-type currents in cardiovascular smooth muscle (Murakami *et al.*, 2003) and N-type currents in DRG neurons (Murakami *et al.*, 2002). As expected from the role of N-type channels in nociceptors, the knockout animals had a reduced response to painful stimuli. Contrary to prediction, β3 –/– mice show enhanced Ca^{2+} oscillations and insulin exocytosis, presumably due to modulation of Ca^{2+} release from intracellular stores (Berggren *et al.*, 2004). Mutations in the human β3 gene have not been reported.

Concluding remarks

Calcium channel auxiliary subunits play crucial roles in trafficking and regulating HVA channel activity. Ca$_v\beta$ subunits also serve important roles in channel regulation by G proteins, and their activity is regulated by phosphorylation and small GTP binding proteins. The determination of the crystal structure of Ca$_v\beta$ subunits has brought new insights into the mechanisms by which they regulate channel activity and with the surging interest in $\alpha_2\delta$ proteins it is likely that elucidation of their structure will follow. Although beyond the scope of the present review, there is a growing list of non-Ca$_v$ subunit proteins that are also capable of binding α_1 subunits and modulating either their activity (e.g. calmodulin) or their subcellular distribution (McKeown *et al.*, 2006). Future studies may also reveal interacting partners for Ca$_v$3 channels.

Acknowledgements
We thank the members of the Perez-Reyes lab and our fellow Ca-channel investigators and the members of their labs at the University of Virginia.

References
Adams ME, Myers RA, Imperial JS *et al.* (1993). Toxityping rat brain calcium channels with omega-toxins from spider and cone snail venoms. *Biochemistry* **32**, 12566–12570.

Ahern CA, Powers PA, Biddlecome GH *et al.* (2001). Modulation of L-type Ca^{2+} current but not activation of Ca2$^+$ release by the gamma$_1$ subunit of the dihydropyridine receptor of skeletal muscle. *BMC Physiol* **1**, 8.

Angelotti T and Hofmann F (1996). Tissue-specific expression of splice variants of the mouse voltage-gated calcium channel α2/δ subunit. *FEBS Lett* **397**, 331–337.

Arias JM, Murbartián J, Vitko I *et al.* (2005). Transfer of β subunit regulation from high to low voltage-gated Ca^{2+} channels. *FEBS Lett* **579**, 3907–3912.

Arikkath J and Campbell KP (2003). Auxiliary subunits: essential components of the voltage-gated calcium channel complex. *Curr Opin Neurobiol* **13**, 298–307.

Assandri R, Egger M, Gassmann M *et al.* (1999). Erythropoietin modulates intracellular calcium in a human neuroblastoma cell line. *J Physiol* **516**, 343–352.

Bangalore R, Mehrke G, Gingrich K *et al.* (1996). Influence of L-type Ca channel α2/δ-subunit on ionic and gating current in transiently transfected HEK 293 cells. *Am J Physiol* **270**, H1521–H1528.

Barclay J, Balaguero N, Mione M *et al.* (2001). Ducky mouse phenotype of epilepsy and ataxia is associated with mutations in the *Cacna2d2* gene and decreased calcium channel current in cerebellar Purkinje cells. *J Neurosci* **21**, 6095–6104.

Bean BP (1989). Neurotransmitter inhibition of neuronal calcium currents by changes in channel voltage dependence. *Nature* **340**, 153–156.

Bech-Hansen NT, Naylor MJ, Maybaum TA *et al.* (1998). Loss-of-function mutations in a calcium-channel alpha1-subunit gene in Xp11.23 cause incomplete X-linked congenital stationary night blindness. *Nat Genet* **19**, 264–267.

Belliotti TR, Capiris T, Ekhato IV *et al.* (2005). Structure–activity relationships of pregabalin and analogues that target the α2δ protein. *J Med Chem* **48**, 2294–2307.

Berggren PO, Yang SN, Murakami M *et al.* (2004). Removal of Ca2$^+$ channel β3 subunit enhances Ca^{2+} oscillation frequency and insulin exocytosis. *Cell* **119**, 273–284.

Bernstein GM and Jones OT (2007). Kinetics of internalization and degradation of N-type voltage-gated calcium channels: role of the $\alpha_2\delta$ subunit. *Cell Calcium* **41**, 27–40.

Bers DM (2002). *Cardiac excitation-contraction coupling*. Kluwer Academic Publishers, Dordrecht.

Biel M, Hullin R, Freundner S et al. (1991). Tissue-specific expression of high-voltage-activated dihydropyridine-sensitive L-type calcium channels. *Eur J Biochem* **200**, 81–88.

Bijlenga P, Liu JH, Espinos E et al. (2000). T-type α1H Ca^{2+} channels are involved in Ca^{2+} signaling during terminal differentiation (fusion) of human myoblasts. *Proc Natl Acad Sci USA* **97**, 7627–7632.

Bodi I, Mikala G, Koch SE et al. (2005). The L-type calcium channel in the heart: the beat goes on. *J Clin Invest* **115**, 3306–3317.

Bokvist K, Eliasson L, Ammala C et al. (1995). Co-localization of L-type Ca^{2+} channels and insulin-containing secretory granules and its significance for the initiation of exocytosis in mouse pancreatic β-cells. *EMBO J* **14**, 50–57.

Bosse E, Regulla S, Biel M et al. (1990). The cDNA and deduced amino acid sequence of the γ subunit of the L-type calcium channel from rabbit skeletal muscle. *FEBS Lett* **267**, 153–156.

Bossu JL, and Feltz A (1986). Inactivation of the low-threshold transient calcium current in rat sensory neurones: evidence for a dual process. *J Physiol* **376**, 341–357.

Bourinet E, Soong TW, Sutton K et al. (1999). Splicing of α$_{1A}$ subunit gene generates phenotypic variants of P- and Q-type calcium channels. *Nat Neurosci* **2**, 407–415.

Bourinet E, Zamponi GW, Stea A et al. (1996). The α$_{1E}$ calcium channel exhibits permeation properties similar to low-voltage-activated calcium channels. *J Neurosci* **16**, 4983–4993.

Brust PF, Simerson S, Mccue AF et al. (1993). Human neuronal voltage-dependent calcium channels: studies on subunit structure and role in channel assembly. *Neuropharmacol* **32**, 1089–1102.

Burgess DL, Jones JM, Meisler MH et al. (1997). Mutation of the Ca^{2+} channel subunit gene *Cchb4* is associated with ataxia and seizures in the lethargic (*lh*) mouse. *Cell* **88**, 385–392.

Campbell KP, Leung AT and Imagawa T (1988). Structural characterization of the nitrendipine receptor of the voltage-dependent Ca^{2+} channel: evidence for a 52,000 dalton subunit. *Journal of Cardiovascular Pharmacology* **12**, S86–90.

Canti C, Davies A and Dolphin AC (2003). Calcium channel α$_2$δ subunits: structure, functions, and target site for drugs. *Curr Neuropharmacol* **1**, 209–217.

Canti C, Nieto-Rostro M, Foucault I et al. (2005). The metal-ion-dependent adhesion site in the Von Willebrand factor-A domain of alpha2delta subunits is key to trafficking voltage-gated Ca2+ channels. *Proc Natl Acad Sci USA* **102**, 11230–11235.

Carbone E, Marcantoni A, Giancippoli A et al. (2006). T-type channels-secretion coupling: evidence for a fast low-threshold exocytosis. *Pflugers Arch* **453**, 373–383.

Carbonneau L, Bhattacharya D, Sheridan DC et al. (2005). Multiple loops of the dihydropyridine receptor pore subunit are required for full-scale excitation–contraction coupling in skeletal muscle. *Biophys J* **89**, 243–255.

Castellano A and Perez-Reyes E (1994). Molecular diversity of Ca^{2+} channel β subunits. *Biochem Soc Trans* **22**, 483–488.

Castellano A, Wei XY, Birnbaumer L et al. (1993). Cloning and expression of a neuronal calcium channel β subunit. *J Biol Chem* **268**, 12359–12366.

Castiglioni AJ, Raingo J and Lipscombe D (2006). Alternative splicing in the C-terminus of Ca$_v$2.2 controls expression and gating of N-type calcium channels. *J Physiol* **576**, 119–134.

Catterall WA (1999). Interactions of presynaptic Ca^{2+} channels and snare proteins in neurotransmitter release. *Ann N Y Acad Sci* **868**, 144–159.

Catterall WA, Perez-Reyes E, Snutch TP et al. (2005). International Union of Pharmacology. XLVIII. Nomenclature and structure-function relationships of voltage-gated calcium channels. *Pharmacol Rev* **57**, 411–425.

Catterall WA, Striessnig J, Snutch TP et al. (2003). International Union of Pharmacology. XL. Compendium of voltage-gated ion channels: calcium channels. *Pharmacol Rev* **55**, 579–581.

Cens T, Rousset M, Leyris JP et al. (2006). Voltage- and calcium-dependent inactivation in high voltage-gated Ca^{2+} channels. *Prog Biophys Mol Biol* **90**, 104–117.

Chanda B, Asamoah OK, Blunck R et al. (2005). Gating charge displacement in voltage-gated ion channels involves limited transmembrane movement. *Nature* **436**, 852–856.

Chemin J, Monteil A, Briquaire C et al. (2000). Overexpression of T-type calcium channels in HEK 293 cells increases intracellular calcium without affecting cellular proliferation. *FEBS Lett* **478**, 166–172.

Chemin J, Monteil A, Dubel S et al. (2001). The α1I T-type calcium channel exhibits faster gating properties when overexpressed in neuroblastoma/glioma NG 108–15 cells. *Eur J Neurosci* **14**, 1678–1686.

Chen C-C, Lamping KG, Nuno DW et al. (2003). Abnormal coronary function in mice deficient in α1H T-type Ca^{2+} channels. *Science* **302**, 1416–1418.

Chen H, Puhl HL, 3rd, Niu SL et al. (2005). Expression of Rem2, an RGK family small GTPase, reduces N-type calcium current without affecting channel surface density. *J Neurosci* **25**, 9762–9772.

Chen XL, Bayliss DA, Fern RJ et al. (1999). A role for T-type Ca^{2+} channels in the synergistic control of aldosterone production by ANG II and K$^+$. *Am J Physiol Renal Physiol* **276**, F674–F683.

Chen Y, Lu J, Zhang Y et al. (2003a). T-type calcium channel gene α$_{1G}$ is not associated with childhood absence epilepsy in the Chinese Han population. *Neurosci Lett* **341**, 29–32.

Chen YC, Lu JJ, Pan H et al. (2003b). Association between genetic variation of CACNA1H and childhood absence epilepsy. *Ann Neurol* **54**, 239–243.

Chen Y-H, Li M-H, Zhang Y et al. (2004). Structural basis of the α1-β subunit interaction of voltage-gated Ca^{2+} channels. *Nature* **429**, 675–680.

Chevalier M, Lory P, Mironneau C et al. (2006). T-type Ca$_v$3.3 calcium channels produce spontaneous low-threshold action potentials and intracellular calcium oscillations. *Eur J Neurosci* **23**, 2321–2329.

Chien AJ, Gao T, Perez-Reyes E et al. (1998). Membrane targeting of L-type calcium channels. Role of palmitoylation in the subcellular localization of the β2a subunit. *J Biol Chem* **273**, 23590–23597.

Cho RY, Konecky RO and Carter CS (2006). Impairments in frontal cortical γ synchrony and cognitive control in schizophrenia. *Proc Natl Acad Sci USA* **103**, 19878–19883.

Choi S, Na HS, Kim J et al. (2006). Attenuated pain responses in mice lacking Ca$_v$3.2 T-type channels. *Genes Brain Behav* **6**, 425–431.

Cohen RM, Foell JD, Balijepalli RC et al. (2005). Unique modulation of L-type Ca2+ channels by short auxiliary β1d subunit present in cardiac muscle. *Am J Physiol Heart Circ Physiol* **288**, H2363–2374.

Coppola T, Waldmann R, Borsotto M et al. (1994). Molecular cloning of a murine N-type calcium channel α1 subunit evidence for isoforms, brain distribution, and chromosomal localization. *FEBS Lett* **338**, 1–5.

Coulter DA, Huguenard JR and Prince DA (1989). Characterization of ethosuximide reduction of low-threshold calcium current in thalamic neurons. *Ann Neurol* **25**, 582–593.

Craig PJ, Beattie RE, Folly EA et al. (1999). Distribution of the voltage-dependent calcium channel α1G subunit mRNA and protein throughout the mature rat brain. *Eur J Neurosci* **11**, 2949–2964.

Cribbs LL, Lee JH, Yang J et al. (1998). Cloning and characterization of α1H from human heart, a member of the T-type Ca^{2+} channel gene family. *Circ Res* **83**, 103–109.

Crunelli V, Toth TI, Cope DW et al. (2005). The 'window' T-type calcium current in brain dynamics of different behavioural states. *J Physiol* **562**, 121–129.

De Jongh KS, Warner C and Catterall WA (1990). Subunits of purified calcium channels. α2 and δ are encoded by the same gene. *J Biol Chem* **265**, 14738–14741.

De Jongh KS, Warner C, Colvin AA et al. (1991). Characterization of the two size forms of the alpha 1 subunit of skeletal muscle L-type calcium channels. *Proc Natl Acad Sci USA* **88**, 10778–10782.

De Waard M and Campbell KP (1995). Subunit regulation of the neuronal α1A Ca²⁺ channel expressed in *Xenopus* oocytes. *J Physiol* **485**, 619–634.

De Waard M, Hering J, Weiss N *et al.* (2005). How do G proteins directly control neuronal Ca2⁺ channel function? *Trends Pharmacol Sci* **26**, 427–436.

Destexhe A, Neubig M, Ulrich D *et al.* (1998). Dendritic low-threshold calcium currents in thalamic relay cells. *J Neurosci* **18**, 3574–3588.

Dirksen RT and Beam KG (1995). Single calcium channel behavior in native skeletal muscle. *J Gen Physiol* **105**, 227–247.

Dolphin AC (2006). A short history of voltage-gated calcium channels. *Br J Pharmacol* **147**, S56–62.

Dolphin AC, Wyatt CN, Richards J *et al.* (1999). The effect of α₂δ and other accessory subunits on expression and properties of the calcium channel α1G. *J Physiol* **519**, 35–45.

Doyle DA, Cabral JM, Pfuetzner RA *et al.* (1998). The structure of the potassium channel: molecular basis of K⁺ conduction and selectivity. *Science* **280**, 69–77.

Dubel SJ, Altier C, Chaumont S *et al.* (2004). Plasma membrane expression of T-type calcium channel α1 subunits is modulated by HVA auxiliary subunits. *J Biol Chem* **279**, 29263–29269.

Dubel SJ, Starr TV, Hell J *et al.* (1992). Molecular cloning of the alpha-1 subunit of an omega-conotoxin-sensitive calcium channel. *Proc Natl Acad Sci USA* **89**, 5058–5062.

Dunham I, Hunt AR, Collins JE *et al.* (1999). The DNA sequence of human chromosome 22. *Nature* **402**, 489–495.

Dunlap K, Luebke JI and Turner TJ (1995). Exocytotic Ca²⁺ channels in mammalian central neurons. *Trends Neurosci* **18**, 89–98.

Dzhura I, Wu Y, Colbran RJ *et al.* (2000). Calmodulin kinase determines calcium-dependent facilitation of L-type calcium channels. *Nat Cell Biol* **2**, 173–177.

Ellis SB, Williams ME, Ways NR *et al.* (1988). Sequence and expression of mRNAs encoding the α1 and α2 subunits of a DHP-sensitive calcium channel. *Science* **241**, 1661–1664.

Emerick MC, Stein R, Kunze R *et al.* (2006). Profiling the array of Ca_v3.1 variants from the human T-type calcium channel gene *CACNA1G*: alternative structures, developmental expression, and biophysical variations. *Proteins* **64**, 320–342.

Enyeart JJ, Sheu SS and Hinkle PM (1987). Pituitary Ca²⁺ channels: blockade by conventional and novel Ca²⁺ antagonists. *Am J Physiol* **253**, C162–170.

Ertel E, Campbell K, Harpold MM *et al.* (2000). Nomenclature of voltage-gated calcium channels. *Neuron* **25**, 533–535.

Escayg A, De Waard M, Lee DD *et al.* (2000). Coding and noncoding variation of the human calcium-channel β₄-subunit gene *CACNB4* in patients with idiopathic generalized epilepsy and episodic ataxia. *Am J Hum Genet* **66**, 1531–1539.

Fan JS, Yuan Y and Palade P (2000). Kinetic effects of FPL 64176 on L-type Ca2⁺ channels in cardiac myocytes. *Naunyn Schmiedebergs Arch Pharmacol* **361**, 465–476.

Felix R, Gurnett CA, De Waard M *et al.* (1997). Dissection of functional domains of the voltage-dependent Ca²⁺ channel α₂δ subunit. *J Neurosci* **17**, 6884–6891.

Field MJ, Cox PJ, Stott E *et al.* (2006). Identification of the α₂δ-1 subunit of voltage-dependent calcium channels as a molecular target for pain mediating the analgesic actions of pregabalin. *Proc Natl Acad Sci USA* **103**, 17537–17542.

Fletcher CF, Tottene A, Lennon VA *et al.* (2001). Dystonia and cerebellar atrophy in Cacna1a null mice lacking P/Q calcium channel activity. *FASEB J* **15**, 1288–1290.

Flucher BE and Franzini-Armstrong C (1996). Formation of junctions involved in excitation–contraction coupling in skeletal and cardiac muscle. *Proc Natl Acad Sci USA* **93**, 8101–8106.

Flucher BE, Obermair GJ, Tuluc P *et al.* (2005). The role of auxiliary dihydropyridine receptor subunits in muscle. *J Muscle Res Cell Motil* **26**, 1–6.

Foell JD, Balijepalli RC, Delisle BP *et al.* (2004). Molecular heterogeneity of calcium channel β-subunits in canine and human heart: evidence for differential subcellular localization. *Physiol Genomics* **17**, 183–200.

Fox AP, Nowycky MC and Tsien RW (1987). Kinetic and pharmacological properties distinguishing three types of calcium currents in chick sensory neurones. *J Physiol* **394**, 149–172.

Frazier CJ, Serrano JR, George EG *et al.* (2001). Gating kinetics of the α1I T-type calcium channel. *J Gen Physiol* **118**, 457–470.

Freise D, Held B, Wissenbach U *et al.* (2000). Absence of the γ subunit of the skeletal muscle dihydropyridine receptor increases L-type Ca²⁺ currents and alters channel inactivation properties. *J Biol Chem* **275**, 14476–14481.

Fujita Y, Mynlieff M, Dirksen RT *et al.* (1993). Primary structure and functional expression of the omega-conotoxin-sensitive N-type calcium channel from rabbit brain. *Neuron* **10**, 585–598.

Gao B, Sekido Y, Maximov A *et al.* (2000). Functional properties of a new voltage-dependent calcium channel α₂δ auxiliary subunit gene (*CACNA2D2*). *J Biol Chem* **275**, 12237–12242.

Gee NS, Brown JP, Dissanayake VU *et al.* (1996). The novel anticonvulsant drug, gabapentin (Neurontin) binds to the α₂δ subunit of a calcium channel. *J Biol Chem* **271**, 5768–5776.

Glossmann H and Striessnig J (1990). Molecular properties of calcium channels. *Rev Physiol Biochem Pharmacol* **114**, 1–105.

Godfraind T (2006). Calcium-channel modulators for cardiovascular disease. *Expert Opin Emerg Drugs* **11**, 49–73.

Gohil K, Bell JR, Ramachandran J *et al.* (1994). Neuroanatomical distribution of receptors for a novel voltage-sensitive calcium-channel antagonist, SNX-230 (omega-conopeptide MVIIC). *Brain Res* **653**, 258–266.

Gomez-Ospina N, Tsuruta F, Barreto-Chang O *et al.* (2006). The C terminus of the L-type voltage-gated calcium channel Ca_v1.2 encodes a transcription factor. *Cell* **127**, 591–606.

Gomora JC, Daud AN, Weiergräber M *et al.* (2001). Block of cloned human T-type calcium channels by succinimide antiepileptic drugs. *Mol Pharmacol* **60**, 1121–1132.

Gondo N, Ono K, Mannen K *et al.* (1998). Four conductance levels of cloned cardiac L-type Ca²⁺ channel α1 and α1/β subunits. *FEBS Lett* **423**, 86–92.

Gould RJ, Murphy KM and Snyder SH (1983). Studies on voltage-operated calcium channels using radioligands. *Cold Spring Harbor Symposia on Quantitative Biology* **1**, 355–362.

Grabsch H, Pereverzev A, Weiergräber M *et al.* (1999). Immunohistochemical detection of α1E voltage-gated Ca²⁺ channel isoforms in cerebellum, INS-1 cells, and neuroendocrine cells of the digestive system. *J Histochem Cytochem* **47**, 981–994.

Gray LS, Perez-Reyes E, Gomora JC *et al.* (2004). The role of voltage gated T-type Ca²⁺ channel isoforms in mediating 'capacitative' Ca²⁺ entry in cancer cells. *Cell Calcium* **36**, 489–497.

Guia A, Stern MD, Lakatta EG *et al.* (2001). Ion concentration-dependence of rat cardiac unitary L-type calcium channel conductance. *Biophys J* **80**, 2742–2750.

Guo J and Ikeda SR (2004). Endocannabinoids modulate N-type calcium channels and G-protein-coupled inwardly rectifying potassium channels via CB1 cannabinoid receptors heterologously expressed in mammalian neurons. *Mol Pharmacol* **65**, 665–674.

Hanke S, Bugert P, Chudek J *et al.* (2001). Cloning a calcium channel α₂δ-3 subunit gene from a putative tumor suppressor gene region at chromosome 3p21.1 in conventional renal cell carcinoma. *Gene* **264**, 69–75.

Hansen PB, Jensen BL, Andreasen D *et al.* (2000). Vascular smooth muscle cells express the α₁ₐ subunit of a P-/Q- type voltage-dependent Ca²⁺ channel, and it is functionally important in renal afferent arterioles. *Circ Res* **87**, 896–902.

Harry JB, Kobrinsky E, Abernethy DR *et al.* (2004). New short splice variants of the human cardiac Ca$_v$β2 subunit: redefining the major functional motifs implemented in modulation of the Ca$_v$1.2 channel. *J Biol Chem* **279**, 46367–46372.

Hayashi K, Wakino S, Sugano N *et al.* (2007). Ca^{2+} channel subtypes and pharmacology in the kidney. *Circ Res* **100**, 342–353.

Heady TN, Gomora JC, Macdonald TL *et al.* (2001). Molecular pharmacology of T-type Ca^{2+} channels. *Jpn J Pharmacol* **85**, 339–350.

Hell JW, Westenbroek RE, Elliott EM *et al.* (1994). Differential phosphorylation, localization, and function of distinct alpha 1 subunits of neuronal calcium channels. Two size forms for class B, C, and D alpha 1 subunits with different COOH-termini. *Ann N Y Acad Sci* **747**, 282–293.

Hell JW, Westenbroek RE, Warner C *et al.* (1993). Identification and differential subcellular localization of the neuronal class C and class D L-type calcium channel alpha 1 subunits. *J Cell Biol* **123**, 949–962.

Helton TD and Horne WA (2002). Alternative splicing of the β4 subunit has α$_1$ subunit subtype-specific effects on Ca^{2+} channel gating. *J Neurosci* **22**, 1573–1582.

Heron SE, Phillips HA, Mulley JC *et al.* (2004). Genetic variation of *CACNA1H* in idiopathic generalized epilepsy. *Ann Neurol* **55**, 595–596.

Higgins DE and Sharp PM (1988). CLUSTAL: a package for performing multiple sequence alignments on a microcomputer. *Gene* **73**, 237–244.

Hille B (2001). *Ionic channels of excitable membranes.* Sinauer Associates, Inc., Sunderland.

Hobom M, Dai S, Marais E *et al.* (2000). Neuronal distribution and functional characterization of the calcium channel α$_2$δ2 subunit. *Eur J Neurosci* **12**, 1217–1226.

Hui A, Ellinor PT, Krizanova O *et al.* (1991). Molecular cloning of multiple subtypes of a novel rat brain isoform of the alpha 1 subunit of the voltage-dependent calcium channel. *Neuron* **7**, 35–44.

Ikeda H, Heinke B, Ruscheweyh R *et al.* (2003). Synaptic plasticity in spinal lamina I projection neurons that mediate hyperalgesia. *Science* **299**, 1237–1240.

Jan LY and Jan YN (1990). A superfamily of ion channels. *Nature* **345**, 672.

Jay SD, Ellis SB, McCue AF *et al.* (1990). Primary structure of the γ subunit of the DHP-sensitive calcium channel from skeletal muscle. *Science* **248**, 490–492.

Jiang Y, Lee A, Chen J *et al.* (2002). The open pore conformation of potassium channels. *Nature* **417**, 523–526.

Jones LP, Wei SK and Yue DT (1998). Mechanism of auxiliary subunit modulation of neuronal α1E calcium channels. *J Gen Physiol* **112**, 125–143.

Jun K, Piedras-Renteria ES, Smith SM *et al.* (1999). Ablation of P/Q-type Ca(2$^+$) channel currents, altered synaptic transmission, and progressive ataxia in mice lacking the alpha(1A)-subunit. *Proc Natl Acad Sci USA* **96**, 15245–15250.

Jurkat-Rott K and Lehmann-Horn F (2004). The impact of splice isoforms on voltage-gated calcium channel alpha1 subunits. *J Physiol* **554**, 609–619.

Kamp MA, Krieger A, Henry M *et al.* (2005). Presynaptic 'Ca2.3-containing' E-type Ca channels share dual roles during neurotransmitter release. *Eur J Neurosci* **21**, 1617–1625.

Kamp TJ, Mitas M, Fields KL *et al.* (1995). Transcriptional regulation of the neuronal L-type calcium channel α1D subunit gene. *Cell Mol Neurobiol* **15**, 307–326.

Kase M, Kakimoto S, Sakuma S *et al.* (1999). Distribution of neurons expressing α1G subunit mRNA of T-type voltage-dependent calcium channel in adult rat central nervous system. *Neurosci Lett* **268**, 77–80.

Katz B and Miledi R (1967). The timing of calcium action during neuromuscular transmission. *J Physiol* **189**, 535–544.

Kawabata M, Ogawa T, Han WH *et al.* (1999). Renal effects of efonidipine hydrochloride, a new calcium antagonist, in spontaneously hypertensive rats with glomerular injury. *Clin Exp Pharmacol Physiol* **26**, 674–679.

Kawai F and Miyachi E (2001). Enhancement by T-type Ca^{2+} currents of odor sensitivity in olfactory receptor cells. *J Neurosci* **21**, RC144.

Kerr LM, Filloux F, Olivera BM *et al.* (1988). Autoradiographic localization of calcium channels with ^{125}I-omega-conotoxin in rat brain. *Eur J Pharmacol* **146**, 181–183.

Khosravani H, Altier C, Simms B *et al.* (2004). Gating effects of mutations in the Ca$_v$3.2 T-type calcium channel associated with childhood absence epilepsy. *J Biol Chem* **279**, 9681–9684.

Kim C, Jun K, Lee T *et al.* (2001). Altered nociceptive response in mice deficient in the α$_{1B}$ subunit of the voltage-dependent calcium channel. *Mol Cell Neurosci* **18**, 235–245.

Kim D, Park D, Choi S *et al.* (2003). Thalamic control of visceral nociception mediated by T-type Ca^{2+} channels. *Science* **302**, 117–119.

Kim HL, Kim H, Lee P *et al.* (1992). Rat brain expresses an alternatively spliced form of the dihydropyridine-sensitive L-type calcium channel alpha 2 subunit. *Proc Natl Acad Sci USA* **89**, 3251–3255.

Kim J, Ghosh S, Nunziato DA *et al.* (2004). Identification of the components controlling inactivation of voltage-gated Ca^{2+} channels. *Neuron* **41**, 745–754.

Klöckner U, Lee JH, Cribbs LL *et al.* (1999). Comparison of the Ca^{2+} currents induced by expression of three cloned α1 subunits, α1G, α1H and α1I, of low-voltage-activated T-type Ca^{2+} channels. *Eur J Neurosci* **11**, 4171–4178.

Klugbauer N, Lacinová L, Marais E *et al.* (1999a). Molecular diversity of the calcium channel α$_2$δ subunit. *J Neurosci* **19**, 684–691.

Klugbauer N, Marais E and Hofmann F (2003). Calcium channel α$_2$δ subunits: differential expression, function, and drug binding. *J Bioenerg Biomembr* **35**, 639–647.

Klugbauer N, Marais E, Lacinová L *et al.* (1999b). A T-type calcium channel from mouse brain. *Pflügers Arch* **437**, 710–715.

Koschak A, Reimer D, Huber I *et al.* (2001). α1D (Cav1.3) subunits can form L-type Ca^{2+} channels activating at negative voltages. *J Biol Chem* **276**, 22100–22106.

Koschak A, Reimer D, Walter D *et al.* (2003). Cav1.4α1 subunits can form slowly inactivating dihydropyridine-sensitive L-type Ca^{2+} channels lacking Ca^{2+}-dependent inactivation. *J Neurosci* **23**, 6041–6049.

Kostyuk PG and Shirokov RE (1989). Deactivation kinetics of different components of calcium inward current in the membrane of mice sensory neurones. *J Physiol* **409**, 343–355.

Kung AW, Lau KS, Fong GC *et al.* (2004). Association of novel single nucleotide polymorphisms in the calcium channel α1 subunit gene (Ca$_v$1.1) and thyrotoxic periodic paralysis. *J Clin Endocrinol Metab* **89**, 1340–1345.

Lacerda AE and Brown AM (1989). Nonmodal gating of cardiac calcium channels as revealed by dihydropyridines. *J Gen Physiol* **93**, 1243–1273.

Lacerda AE, Kim HS, Ruth P *et al.* (1991). Normalization of current kinetics by interaction between the α$_1$ and β subunits of the skeletal muscle dihydropyridine-sensitive Ca^{2+} channel. *Nature* **352**, 527–530.

Lacerda AE, Perez-Reyes E, Wei X, *et al.* (1994). T-type and N-type calcium channels of *Xenopus* oocytes: evidence for specific interactions with β subunits. *Biophys J* **66**, 1833–1843.

Lacinova L and Klugbauer N (2004). Modulation of gating currents of the Ca$_v$3.1 calcium channel by α$_2$δ2a and γ$_5$ subunits. *Arch Biochem Biophys* **425**, 207–213.

Lacinová L, Klugbauer N and Hofmann F (1999). Absence of modulation of the expressed calcium channel α1G subunit by α$_2$δ subunits. *J Physiol* **516**, 639–645.

Lambert RC, Maulet Y, Mouton J, *et al.* (1997). T-type Ca^{2+} current properties are not modified by Ca^{2+} channel β subunit depletion in nodosus ganglion neurons. *J Neurosci* **17**, 6621–6628.

Lapie P, Lory P and Fontaine B (1997). Hypokalemic periodic paralysis: an autosomal dominant muscle disorder caused by mutations in a voltage-gated calcium channel. *Neuromuscul Disord* **7**, 234–240.

Lee J-H (1998). *Cloning and characterization of a novel ion channel*. PhD thesis in Physiology.

Lee JH, Daud AN, Cribbs LL *et al.* (1999). Cloning and expression of a novel member of the low voltage-activated T-type calcium channel family. *J Neurosci* **19**, 1912–1921.

Lee T, Kaku T, Takebayashi S *et al.* (2006). Actions of mibefradil, efonidipine and nifedipine block of recombinant T- and L-type Ca channels with distinct inhibitory mechanisms. *Pharmacology* **78**, 11–20.

Letts VA, Felix R, Biddlecome GH *et al.* (1998). The mouse stargazer gene encodes a neuronal Ca^{2+}-channel γ subunit. *Nat Genet* **19**, 340–347.

Leuranguer V, Bourinet E, Lory P *et al.* (1998). Antisense depletion of β-subunits fails to affect T-type calcium channels properties in a neuroblastoma cell line. *Neuropharmacol* **37**, 701–708.

Li CY, Zhang XL, Matthews EA *et al.* (2006). Calcium channel alpha2delta1 subunit mediates spinal hyperexcitability in pain modulation. *Pain* **125**, 20–34.

Li HR, Wang-Rodriguez J, Nair TM *et al.* (2006). Two-dimensional transcriptome profiling: identification of messenger RNA isoform signatures in prostate cancer from archived paraffin-embedded cancer specimens. *Cancer Res* **66**, 4079–4088.

Lipkind GM and Fozzard HA (2003). Molecular modeling of interactions of dihydropyridines and phenylalkylamines with the inner pore of the L-type Ca^{2+} channel. *Mol Pharmacol* **63**, 499–511.

Lipscombe D, Helton TD and Xu W (2004). L-type calcium channels: the low down. *J Neurophysiol* **92**, 2633–2641.

Liu L, Roberts ML and Rittenhouse AR (2004). Phospholipid metabolism is required for M1 muscarinic inhibition of N-type calcium current in sympathetic neurons. *Eur Biophys J* **33**, 255–264.

Llinás R and Jahnsen H (1982). Electrophysiology of mammalian thalamic neurons in vitro. *Nature* **297**, 406–408.

Llinás R and Yarom Y (1981a). Electrophysiology of mammalian inferior olivary neurones *in vitro*. Different types of voltage-dependent ionic conductances. *J Physiol* **315**, 549–567.

Llinás R and Yarom Y (1981b). Properties and distribution of ionic conductances generating electroresponsiveness of mammalian inferior olivary neurones in vitro. *J. Physiol* **315**, 569–584.

Llinás R, Ribary U, Jeanmonod D *et al.* (2001). Thalamocortical dysrhythmia I. Functional and imaging aspects. *Thalamus and Related Systems* **1**, 237–244.

Llinás R, Sugimori M, Lin JW *et al.* (1989). Blocking and isolation of a calcium channel from neurons in mammals and cephalopods utilizing a toxin fraction (FTX) from funnel-web spider poison. *Proc Natl Acad Sci USA* **86**, 1689–1693.

Llinás RR, Ribary U, Jeanmonod D *et al.* (1999). Thalamocortical dysrhythmia: a neurological and neuropsychiatric syndrome characterized by magnetoencephalography. *Proc Natl Acad Sci USA* **96**, 15222–15227.

Long SB, Campbell EB and Mackinnon R (2005). Crystal structure of a mammalian voltage-dependent Shaker family K^+ channel. *Science* **309**, 897–903.

Ludwig A, Flockerzi V and Hofmann F (1997). Regional expression and cellular localization of the α_1 and β subunit of high voltage-activated calcium channels in rat brain. *J Neurosci* **17**, 1339–1349.

Luo ZD, Calcutt NA, Higuera ES *et al.* (2002). Injury type-specific calcium channel $\alpha_2\delta$-1 subunit up-regulation in rat neuropathic pain models correlates with antiallodynic effects of gabapentin. *J Pharmacol Exp Ther* **303**, 1199–1205.

Macdonald RL and Kelly KM (1995). Antiepileptic drug mechanisms of action. *Epilepsia* **36**, S2–12.

Maeno-Hikichi Y, Chang S, Matsumura K *et al.* (2003). A PKC epsilon-ENH-channel complex specifically modulates N-type Ca2+ channels. *Nat Neurosci* **6**, 468–475.

Magee JC and Johnston D (1995). Synaptic activation of voltage-gated channels in the dendrites of hippocampal pyramidal neurons. *Science* **268**, 301–304.

Maneuf YP, Luo ZD,and Lee K (2006). $\alpha_2\delta$ and the mechanism of action of gabapentin in the treatment of pain. *Semin Cell Dev Biol* **17**, 565–570.

Marais E, Klugbauer N and Hofmann F (2001). Calcium channel $\alpha_2\delta$ subunits-structure and gabapentin binding. *Mol Pharmacol* **59**, 1243–1248.

Mariot P, Vanoverberghe K, Lalevee N *et al.* (2002). Overexpression of an α1H (Ca_v3.2) T-type calcium channel during neuroendocrine differentiation of human prostate cancer cells. *J Biol Chem* **277**, 10824–10833.

Markram H and Sakmann B (1994). Calcium transients in dendrites of neocortical neurons evoked by single subthreshold excitatory postsynaptic potentials via low-voltage-activated calcium channels. *Proc Natl Acad Sci USA* **91**, 5207–5211.

Martin RL, Lee JH, Cribbs LL *et al.* (2000). Mibefradil block of cloned T-type calcium channels. *J Pharmacol Exp Ther* **295**, 302–308.

Martin-Moutot N, Leveque C, Sato K *et al.* (1995). Properties of ω conotoxin MVIIC receptors associated with α1A calcium channel subunits in rat brain. *FEBS Lett* **366**, 21–25.

Masumiya H, Shijuku T, Tanaka H *et al.* (1998). Inhibition of myocardial L- and T-type Ca^{2+} currents by efonidipine: possible mechanism for its chronotropic effect. *Eur J Pharmacol* **349**, 351–357.

Maximov A,and Bezprozvanny I (2002). Synaptic targeting of N-type calcium channels in hippocampal neurons. *J Neurosci* **22**, 6939–6952.

McCarthy RT, Isales C, and Rasmussen H (1993). T-type calcium channels in adrenal glomerulosa cells: GTP-dependent modulation by angiotensin II. *Proc Natl Acad Sci USA* **90**, 3260–3264.

McCormick DA and Huguenard JR (1992). A model of the electrophysiological properties of thalamocortical relay neurons. *J Neurophysiol* **68**, 1384–1400.

McDonough SI and Bean BP (1998). Mibefradil inhibition of T-type calcium channels in cerebellar Purkinje neurons. *Mol Pharmacol* **54**, 1080–1087.

McGivern JG (2006a). Pharmacology and drug discovery for T-type calcium channels. *CNS and Neurological Disorders Drug Targets* **5**, 587–603.

McGivern JG (2006b). Targeting N-type and T-type calcium channels for the treatment of pain. *Drug Discovery Today* **11**, 245–253.

McKeown L, Robinson P and Jones OT (2006). Molecular basis of inherited calcium channelopathies: role of mutations in pore-forming subunits. *Acta Pharmacologica Sinica* **27**, 799–812.

McRory JE, Hamid J, Doering CJ *et al.* (2004). The CACNA1F gene encodes an L-type calcium channel with unique biophysical properties and tissue distribution. *J Neurosci* **24**, 1707–1718.

Mikami A, Imoto K, Tanabe T *et al.* (1989). Primary structure and functional expression of the cardiac dihydropyridine-sensitive calcium channel. *Nature* **340**, 230–233.

Mills L, Niesen C, So A *et al.* (1994). N-type Ca^{2+} channels are located on somata, dendrites, and a subpopulation of dendritic spines on live hippocampal pyramidal neurons. *J Neurosci* **14**, 6815–6824.

Mittman S, Guo J,and Agnew WS (1999a). Structure and alternative splicing of the gene encoding α1G, a human brain T calcium channel α1 subunit. *Neurosci Lett* **274**, 143–146.

Mittman S, Guo J, Emerick MC *et al.* (1999b). Structure and alternative splicing of the gene encoding α1I, a human brain T calcium channel α1 subunit. *Neurosci Lett* **269**, 121–124.

Monnier N, Kozak-Ribbens G, Krivosic-Horber R *et al.* (2005). Correlations between genotype and pharmacological, histological, functional, and clinical phenotypes in malignant hyperthermia susceptibility. *Hum Mutat* **26**, 413–425.

Monnier N, Procaccio V, Stieglitz P *et al.* (1997). Malignant-hyperthermia susceptibility is associated with a mutation of the α1-subunit of the human dihydropyridine-sensitive L-type voltage-dependent calcium-channel receptor in skeletal muscle. *Am J Hum Genet* **60**, 1316–1325.

Monteil A, Chemin J, Bourinet E *et al.* (2000a). Molecular and functional properties of the human α1G subunit that forms T-type calcium channels. *J Biol Chem* **275**, 6090–6100.

Monteil A, Chemin J, Leuranguer V *et al.* (2000b). Specific properties of T-type calcium channels generated by the human α1I subunit. *J Biol Chem* **275**, 16530–16535.

Morgans CW, Bayley PR, Oesch NW *et al.* (2005). Photoreceptor calcium channels: insight from night blindness. *Vis Neurosci* **22**, 561–568.

Morgans CW, Gaughwin P and Maleszka R (2001). Expression of the alpha1F calcium channel subunit by photoreceptors in the rat retina. *Molecular vision* **7**, 202–209.

Mori Y, Friedrich T, Kim MS *et al.* (1991). Primary structure and functional expression from complementary DNA of a brain calcium channel. *Nature* **350**, 398–402.

Munsch T, Budde T and Pape HC (1997). Voltage-activated intracellular calcium transients in thalamic relay cells and interneurons. *Neuroreport* **8**, 2411–2418.

Murakami M, Fleischmann B, De Felipe C *et al.* (2002). Pain perception in mice lacking the β3 subunit of voltage-activated calcium channels. *J Biol Chem* **277**, 40342–40351.

Murakami M, Yamamura H, Suzuki T *et al.* (2003). Modified cardiovascular L-type channels in mice lacking the voltage-dependent Ca^{2+} channel β3 subunit. *J Biol Chem* **278**, 43261–43267.

Murbartián J, Arias JM and Perez-Reyes E (2004). Functional impact of alternative splicing of human T-type $Ca_v3.3$ calcium channels. *J Neurophysiol* **92**, 3399–3407.

Namkung Y, Skrypnyk N, Jeong MJ *et al.* (2001). Requirement for the L-type Ca^{2+} channel α1D subunit in postnatal pancreatic beta cell generation. *J Clin Invest* **108**, 1015–1022.

Narahashi T (2000). Neuroreceptors and ion channels as the basis for drug action: past, present, and future. *J Pharmacol Exp Ther* **294**, 1–26.

Nelson M, Todorovic S and Perez-Reyes E (2006). The role of T-type calcium channels in epilepsy and pain. *Curr Pharm Des* **12**, 2189–2197.

Neuhuber B, Gerster U, Mitterdorfer J *et al.* (1998). Differential effects of Ca2+ channel beta1a and beta2a subunits on complex formation with alpha1S and on current expression in tsA201 cells. *J Biol Chem* **273**, 9110–9118.

Newcomb R, Szoke B, Palma A *et al.* (1998). Selective peptide antagonist of the class E calcium channel from the venom of the tarantula *Hysterocrates gigas*. *Biochemistry* **37**, 15353–15362.

Nicoll RA, Tomita S and Bredt DS (2006). Auxiliary subunits assist AMPA-type glutamate receptors. *Science* **311**, 1253–1256.

Niidome T, Kim MS, Friedrich T *et al.* (1992). Molecular cloning and characterization of a novel calcium channel from rabbit brain. *FEBS Lett* **308**, 7–13.

Noebels JL (2003). The biology of epilepsy genes. *Annu Rev Neurosci* **26**, 599–625.

Nowycky MC, Fox AP and Tsien RW (1985). Three types of neuronal calcium channel with different calcium agonist sensitivity. *Nature* **316**, 440–443.

Okayama S, Imagawa K, Naya N *et al.* (2006). Blocking T-type Ca^{2+} channels with efonidipine decreased plasma aldosterone concentration in healthy volunteers. *Hypertension Research* **29**, 493–497.

Olcese R, Qin N, Schneider T *et al.* (1994). The amino terminus of a calcium channel β subunit sets rates of channel inactivation independently of the subunit's effect on activation. *Neuron* **13**, 1433–1438.

Olivera BM, Miljanich GP, Ramachandran J *et al.* (1994). Calcium channel diversity and neurotransmitter release: the ω-conotoxins and ω-agatoxins. *Annu Rev Biochem* **63**, 823–867.

Opatowsky Y, Chen CC, Campbell KP *et al.* (2004). Structural analysis of the voltage-dependent calcium channel β subunit functional core and its complex with the α1 interaction domain. *Neuron* **42**, 387–399.

Ophoff RA, Terwindt GM, Vergouwe MN *et al.* (1996). Familial hemiplegic migraine and episodic ataxia type-2 are caused by mutations in the Ca^{2+} channel gene *CACNL1A4*. *Cell* **87**, 543–552.

Pacher P, Batkai S and Kunos G (2006). The endocannabinoid system as an emerging target of pharmacotherapy. *Pharmacol Rev* **58**, 389–462.

Pan ZH, Hu HJ, Perring P *et al.* (2001). T-type Ca^{2+} channels mediate neurotransmitter release in retinal bipolar cells. *Neuron* **32**, 89–98.

Parsons KT and Kwok WW (2002). Linear B-cell epitopes in Lambert–Eaton myasthenic syndrome defined by cell-free synthetic peptide binding. *J Neuroimmunol* **126**, 190–195.

Pelzer D, Pelzer S and MacDonald TF (1990). Properties and regulation of calcium channels in muscle cells. *Rev Physiol Biochem Pharmacol* **114**, 108–207.

Pereverzev A, Klöckner U, Henry M *et al.* (1998). Structural diversity of the voltage-dependent Ca^{2+} channel α1E subunit. *Eur J Neurosci* **10**, 916–925.

Perez-Reyes E (1999). Three for T: molecular analysis of the low voltage-activated calcium channel family. *Cell Mol Life Sci* **56**, 660–669.

Perez-Reyes E (2003). Molecular physiology of low-voltage-activated T-type calcium channels. *Physiol Rev* **83**, 117–161.

Perez-Reyes E and Lory P (2006). Molecular biology of T-type calcium channels. *CNS and Neurological Disorders Drug Targets* **5**, 605–609.

Perez-Reyes E and Schneider T (1994). Calcium channels: structure, function, and classification. *Drug Devel Res* **33**, 295–318.

Perez-Reyes E, Castellano A, Kim HS *et al.* (1992). Cloning and expression of a cardiac/brain β subunit of the L-type calcium channel. *J Biol Chem* **267**, 1792–1797.

Perez-Reyes E, Cribbs LL, Daud A *et al.* (1998). Molecular characterization of a neuronal low-voltage-activated T-type calcium channel. *Nature* **391**, 896–900.

Perez-Reyes E, Kim HS, Lacerda AE *et al.* (1989). Induction of calcium currents by the expression of the $α_1$-subunit of the dihydropyridine receptor from skeletal muscle. *Nature* **340**, 233–236.

Perez-Reyes E, Wei XY, Castellano A *et al.* (1990). Molecular diversity of L-type calcium channels. Evidence for alternative splicing of the transcripts of three non-allelic genes. *J Biol Chem* **265**, 20430–20436.

Peterson BZ, DeMaria CD, Adelman JP *et al.* (1999). Calmodulin is the Ca^{2+} sensor for Ca^{2+}-dependent inactivation of L-type calcium channels. *Neuron* **22**, 549–558.

Platzer J, Engel J, Schrott-Fischer A *et al.* (2000). Congenital deafness and sinoatrial node dysfunction in mice lacking class D L-type Ca^{2+} channels. *Cell* **102**, 89–97.

Pouille F, Cavelier P, Desplantez T *et al.* (2000). Dendro-somatic distribution of calcium-mediated electrogenesis in Purkinje cells from rat cerebellar slice cultures. *J Physiol* **527**, 265–282.

Powers PA, Liu S, Hogan K *et al.* (1993). Molecular characterization of the gene encoding the γ subunit of the human skeletal muscle 1,4-dihydropyridine-sensitive Ca2+ channel (CACNLG), cDNA sequence, gene structure, and chromosomal location. *J Biol Chem* **268**, 9275–9279.

Pragnell M, De Waard M, Mori Y *et al.* (1994). Calcium channel β-subunit binds to a conserved motif in the I-II cytoplasmic linker of the $α_1$-subunit. *Nature* **368**, 67–70.

Ptacek LJ (1997). Channelopathies: ion channel disorders of muscle as a paradigm for paroxysmal disorders of the nervous system. *Neuromus Disorders* **7**, 250–255.

Ptacek LJ, Tawil R, Griggs RC *et al.* (1994). Dihydropyridine receptor mutations cause hypokalemic periodic paralysis. *Cell* **77**, 863–868.

Qin N, Yagel S, Momplaisir M-L *et al.* (2002). Molecular cloning and characterization of the human voltage-gated calcium channel $α_2δ$-4 subunit. *Mol Pharmacol* **62**, 485–496.

Randall AD and Tsien RW (1997). Contrasting biophysical and pharmacological properties of T- type and R-type calcium channels. *Neuropharmacol* **36**, 879–893.

Ren D and Hall LM (1997). Functional expression and characterization of skeletal muscle dihydropyridine receptors in *Xenopus* oocytes. *J Biol Chem* **272**, 22393–22396.

Restituito S, Cens T, Barrere C *et al.* (2000). The β2a subunit is a molecular groom for the Ca^{2+} channel inactivation gate. *J Neurosci* **20**, 9046–9052.

Ringer S (1883). A further contribution regarding the influence of the different constituents of the blood on the contraction of the heart. *J Physiol* **4**, 29–42.

Rosenfeld MR, Wong E, Dalmau J *et al.* (1993). Sera from patients with Lambert–Eaton myasthenic syndrome recognize the beta-subunit of Ca^{2+} channel complexes. *Ann N Y Acad Sci* **681**, 408–411.

Ross SD, Akhras KS, Zhang S *et al.* (2001). Discontinuation of antihypertensive drugs due to adverse events: a systematic review and meta-analysis. *Pharmacotherapy* **21**, 940–953.

Ruth P, Rohrkasten A, Biel M *et al.* (1989). Primary structure of the beta subunit of the DHP-sensitive calcium channel from skeletal muscle. *Science* **245**, 1115–1118.

Sandoz G, Lopez-Gonzalez I, Grunwald D *et al.* (2004). Cavβ-subunit displacement is a key step to induce the reluctant state of P/Q calcium channels by direct G protein regulation. *Proc Natl Acad Sci USA* **101**, 6267–6272.

Santi CM, Cayabyab FS, Sutton KG *et al.* (2002). Differential inhibition of T-type calcium channels by neuroleptics. *J Neurosci* **22**, 396–403.

Sather WA and McCleskey EW (2003). Permeation and selectivity in calcium channels. *Annu Rev Physiol* **65**, 133–159.

Satin J and Cribbs LL (2000). Identification of a T-type Ca^{2+} channel isoform in murine atrial myocytes (AT-1 cells). *Circ Res* **86**, 636–642.

Satoh H (1995). Role of T-type Ca^{2+} channel inhibitors in the pacemaker depolarization in rabbit sino-atrial nodal cells. *Gen Pharmacol* **26**, 581–587.

Schneider T, Wei X, Olcese R *et al.* (1994). Molecular analysis and functional expression of the human type E α_1 subunit. *Receptors Channels* **2**, 255–270.

Scott VE, Felix R, Arikkath J *et al.* (1998). Evidence for a 95 kDa short form of the α1A subunit associated with the ω-conotoxin MVIIC receptor of the P/Q-type Ca^{2+} channels. *J Neurosci* **18**, 641–647.

Seino S, Chen L, Seino M *et al.* (1992). Cloning of the α1 subunit of a voltage-dependent calcium channel expressed in pancreatic β cells. *Proc Natl Acad Sci USA* **89**, 584–588.

Seisenberger C, Specht V, Welling A *et al.* (2000). Functional embryonic cardiomyocytes after disruption of the L-type α1C (Ca$_v$1.2) calcium channel gene in the mouse. *J Biol Chem* **275**, 39193–39199.

Serrano JR, Perez-Reyes E and Jones SW (1999). State-dependent inactivation of the α1G T-type calcium channel. *J Gen Physiol* **114**, 185–201.

Sharp AH, Black JL, 3rd, Dubel SJ *et al.* (2001). Biochemical and anatomical evidence for specialized voltage-dependent calcium channel γ isoform expression in the epileptic and ataxic mouse, stargazer. *Neuroscience* **105**, 599–617.

Shirokov R, Ferreira G, Yi J *et al.* (1998). Inactivation of gating currents of L-type calcium channels. Specific role of the α2δ subunit. *J Gen Physiol* **111**, 807–823.

Sidach SS, and Mintz IM (2002). Kurtoxin, a gating modifier of neuronal high- and low-threshold Ca channels. *J Neurosci* **22**, 2023–2034.

Sipos I, Jurkat-Rott K, Harasztosi C *et al.* (1995). Skeletal muscle DHP receptor mutations alter calcium currents in human hypokalaemic periodic paralysis myotubes. *J Physiol* **483**, 299–306.

Sochivko D, Pereverzev A, Smyth N *et al.* (2002). The Ca$_v$2.3 Ca^{2+} channel subunit contributes to R-type Ca^{2+} currents in murine hippocampal and neocortical neurones. *J Physiol* **542**, 699–710.

Soldatov NM (2003). Ca^{2+} channel moving tail: link between Ca^{2+}-induced inactivation and Ca^{2+} signal transduction. *Trends Pharmacol Sci* **24**, 167–171.

Somlyo AP, and Somlyo AV (1990). Flash photolysis studies of excitation-contraction coupling, regulation, and contraction in smooth muscle. *Annu Rev Physiol* **52**, 857–874.

Song I, Kim D, Choi S *et al.* (2004). Role of the α1G T-type calcium channel in spontaneous absence seizures in mutant mice. *J Neurosci* **24**, 5249–5257.

Soong TW, DeMaria CD, Alvania RS *et al.* (2002). Systematic identification of splice variants in human P/Q-type channel $\alpha_1$2.1 subunits: implications for current density and Ca^{2+}-dependent inactivation. *J Neurosci* **22**, 10142–10152.

Soong TW, Stea A, Hodson CD *et al.* (1993). Structure and functional expression of a member of the low-voltage-activated calcium channel family. *Science* **260**, 1133–1136.

Splawski I, Timothy KW, Sharpe LM *et al.* (2004). Ca$_v$1.2 calcium channel dysfunction causes a multisystem disorder including arrhythmia and autism. *Cell* **119**, 19–31.

Splawski I, Yoo DS, Stotz SC *et al.* (2006). *CACNA1H* mutations in autism spectrum disorders. *J Biol Chem* **281**, 22085–22091.

Staats PS, Yearwood T, Charapata SG *et al.* (2004). Intrathecal ziconotide in the treatment of refractory pain in patients with cancer or AIDS: a randomized controlled trial. *JAMA* **291**, 63–70.

Starr TV, Prystay W and Snutch TP (1991). Primary structure of a calcium channel that is highly expressed in the rat cerebellum. *Proc Natl Acad Sci USA* **88**, 5621–5625.

Stea A, Tomlinson WJ, Soong TW *et al.* (1994). Localization and functional properties of a rat brain alpha 1A calcium channel reflect similarities to neuronal Q- and P-type channels. *Proc Natl Acad Sci USA* **91**, 10576–10580.

Stewart S, Hogan K, Rosenberg H *et al.* (2001). Identification of the Arg1086His mutation in the alpha subunit of the voltage-dependent calcium channel (CACNA1S) in a North American family with malignant hyperthermia. *Clin Genet* **59**, 178–184.

Striessnig J (1999). Pharmacology, structure and function of cardiac L-type Ca^{2+} channels. *Cell Physiol Biochem* **9**, 242–269.

Striessnig J, Grabner M, Mitterdorfer J *et al.* (1998). Structural basis of drug binding to L Ca^{2+} channels. *Trends Pharmacol Sci* **19**, 108–115.

Strom TM, Nyakatura G, Apfelstedt-Sylla E *et al.* (1998). An L-type calcium-channel gene mutated in incomplete X-linked congenital stationary night blindness. *Nat Genet* **19**, 260–263.

Suh-Kim H, Wei X, Klos A *et al.* (1996). Reconstitution of the skeletal muscle dihydropyridine receptor. Functional interaction among α_1, β, γ, and α_2δ subunits. *Receptors Channels* **4**, 217–225.

Suzuki T, Delgado-Escueta AV, Aguan K *et al.* (2004). Mutations in EFHC1 cause juvenile myoclonic epilepsy. *Nat Genet* **36**, 842–849.

Takahashi SX, Mittman S and Colecraft HM (2003). Distinctive modulatory effects of five human auxiliary β2 subunit splice variants on L-type calcium channel gating. *Biophys J* **84**, 3007–3021.

Takahashi Y, Jeong SY, Ogata K *et al.* (2003). Human skeletal muscle calcium channel α1S is expressed in the basal ganglia: distinctive expression pattern among L-type Ca^{2+} channels. *Neurosci Res* **45**, 129–137.

Talley EM, Cribbs LL, Lee JH *et al.* (1999). Differential distribution of three members of a gene family encoding low voltage-activated (T-type) calcium channels. *J Neurosci* **19**, 1895–1911.

Talley EM, Solórzano G, Depaulis A *et al.* (2000). Low-voltage-activated calcium channel subunit expression in a genetic model of absence epilepsy in the rat. *Mol Brain Res* **75**, 159–165.

Tanabe T, Beam KG, Adams BA *et al.* (1990). Regions of the skeletal muscle dihydropyridine receptor critical for excitation-contraction coupling. *Nature* **346**, 567–569.

Tanabe T, Beam KG, Powell JA *et al.* (1988). Restoration of excitation–contraction coupling and slow calcium current in dysgenic muscle by dihydropyridine receptor complementary DNA. *Nature* **336**, 134–139.

Tanabe T, Takeshima H, Mikami A *et al.* (1987). Primary structure of the receptor for calcium channel blockers from skeletal muscle. *Nature* **328**, 313–318.

Tanaka O, Sakagami H and Kondo H (1995). Localization of mRNAs of voltage-dependent Ca^{2+}-channels: four subtypes of α_1 and β-subunits in developing and mature brain. *Mol Brain Res* **30**, 1–16.

Tang ZZ, Liang MC, Lu S *et al.* (2004). Transcript scanning reveals novel and extensive splice variations in human L-type voltage-gated calcium channel, Ca$_v$1.2 α1 subunit. *J Biol Chem* **279**, 44335–44343.

Tareilus E, Roux M, Qin N *et al.* (1997). A *Xenopus* oocyte β subunit: evidence for a role in the assembly/expression of voltage-gated calcium channels that is separate from its role as a regulatory subunit. *Proc Natl Acad Sci USA* **94**, 1703–1708.

Taylor CP, Gee NS, Su TZ *et al.* (1998). A summary of mechanistic hypotheses of gabapentin pharmacology. *Epilepsy Res* **29**, 233–249.

Todorovic SM, Jevtovic-Todorovic V, Meyenburg A *et al.* (2001). Redox modulation of T-type calcium channels in rat peripheral nociceptors. *Neuron* **31**, 75–85.

Tsakiridou E, Bertollini L, de Curtis M *et al.* (1995). Selective increase in T-type calcium conductance of reticular thalamic neurons in a rat model of absence epilepsy. *J Neurosci* **15**, 3110–3117.

Tsien RW, Clozel JP and Nargeot J, Eds. (1998). *Low voltage-activated T-type calcium channels.* Chester, Adis International Limited.

Uebachs M, Schaub C, Perez-Reyes E *et al.* (2006). T-type Ca^{2+} channels encode prior neuronal activity as modulated recovery rates. *J Physiol* **571**, 519–536.

Umemiya M, and Berger AJ (1994). Properties and function of low- and high-voltage-activated Ca^{2+} channels in hypoglossal motoneurons. *J Neurosci* **14**, 5652–5660.

Vajna R, Schramm M, Pereverzev A *et al.* (1998). New isoform of the neuronal Ca^{2+} channel α1E subunit in islets of Langerhans and kidney: distribution of voltage-gated Ca^{2+} α1 subunits in cell lines and tissues. *Eur J Biochem* **257**, 274–285.

van den Maagdenberg AM, Pietrobon D, Pizzorusso T *et al.* (2004). A *Cacna1a* knockin migraine mouse model with increased susceptibility to cortical spreading depression. *Neuron* **41**, 701–710.

Van Petegem F, Chatelain FC and Minor DL, Jr (2005). Insights into voltage-gated calcium channel regulation from the structure of the Ca$_v$1.2 IQ domain-Ca^{2+}/calmodulin complex. *Nat Struct Mol Biol* **12**, 1108–1115.

Van Petegem F, Clark KA, Chatelain FC *et al.* (2004). Structure of a complex between a voltage-gated calcium channel β-subunit and an α-subunit domain. *Nature* **429**, 671–675.

Vance CL, Begg CM, Lee WL *et al.* (1998). Differential expression and association of calcium channel α_{1B} and β subunits during rat brain ontogeny. *J Biol Chem* **273**, 14495–14502.

Vitko I, Bidaud I, Arias JM *et al.* (2007). The I-II loop controls plasma membrane expression and gating of Ca$_v$3.2 T-type Ca^{2+} channels: a paradigm for childhood absence epilepsy. *J Neurosci* **27**, 322–330.

Vitko I, Chen Y, Arias JM *et al.* (2005). Functional characterization and neuronal modeling of the effects of childhood absence epilepsy variants of *CACNA1H*, a T-type calcium channel. *J Neurosci* **25**, 4844–4855.

Warre R, and Randall A (2000). Modulation of the deactivation kinetics of a recombinant rat T-type Ca^{2+} channel by prior inactivation. *Neurosci Lett* **293**, 216–220.

Wei X, Pan S, Lang W *et al.* (1995). Molecular determinants of cardiac Ca^{2+} channel pharmacology: subunit requirement for the high affinity and allosteric regulation of dihydropyridine binding. *J Biol Chem* **270**, 27106–27111.

Wei XY, Perez-Reyes E, Lacerda AE *et al.* (1991). Heterologous regulation of the cardiac Ca^{2+} channel α1 subunit by skeletal muscle β and γ subunits. Implications for the structure of cardiac L-type Ca^{2+} channels. *J Biol Chem* **266**, 21943–21947.

Weiergräber M, Henry M, Krieger A *et al.* (2006a). Altered seizure susceptibility in mice lacking the Ca$_v$2.3 E-type Ca^{2+} channel. *Epilepsia* **47**, 839–850.

Weiergräber M, Kamp MA, Radhakrishnan K *et al.* (2006b). The Ca$_v$2.3 voltage-gated calcium channel in epileptogenesis–shedding new light on an enigmatic channel. *Neurosci Biobehav Rev* **30**, 1122–1144.

Westenbroek RE, Hell JW, Warner C *et al.* (1992). Biochemical properties and subcellular distribution of an N-type calcium channel alpha 1 subunit. *Neuron* **9**, 1099–1115.

Williams ME, Brust PF, Feldman DH *et al.* (1992a). Structure and functional expression of an ω-conotoxin-sensitive human N-type calcium channel. *Science* **257**, 389–395.

Williams ME, Feldman DH, McCue AF *et al.* (1992b). Structure and functional expression of α1, α2, and β subunits of a novel human neuronal calcium channel subtype. *Neuron* **8**, 71–84.

Williams ME, Washburn MS, Hans M *et al.* (1999). Structure and functional characterization of a novel human low-voltage activated calcium channel. *J Neurochem* **72**, 791–799.

Wilson SM, Toth PT, Oh SB *et al.* (2000). The status of voltage-dependent calcium channels in α1E knock-out mice. *J Neurosci* **20**, 8566–8571.

Witcher DR, De Waard M, Sakamoto J *et al.* (1993). Subunit identification and reconstitution of the N-type Ca^{2+} channel complex purified from brain. *Science* **261**, 486–489.

Wolf M, Eberhart A, Glossmann H *et al.* (2003). Visualization of the domain structure of an L-type Ca^{2+} channel using electron cryo-microscopy. *J Mol Biol* **332**, 171–182.

Wolfart J, and Roeper J (2002). Selective coupling of T-type calcium channels to SK potassium channels prevents intrinsic bursting in dopaminergic midbrain neurons. *J Neurosci* **22**, 3404–3413.

Wolfe JT, Wang H, Perez-Reyes E *et al.* (2002). Stimulation of recombinant Ca$_v$3.2, T-type, Ca^{2+} channel currents by CaMKIIγ_C. *J Physiol* **538**, 343–355.

Wu Q, and Krainer AR (1999). AT-AC pre-mRNA splicing mechanisms and conservation of minor introns in voltage-gated ion channel genes. *Mol Cell Biol* **19**, 3225–3236.

Wycisk KA, Budde B, Feil S *et al.* (2006a). Structural and functional abnormalities of retinal ribbon synapses due to Cacna2d4 mutation. *Invest Ophthalmol Vis Sci* **47**, 3523–3530.

Wycisk KA, Zeitz C, Feil S *et al.* (2006b). Mutation in the auxiliary calcium-channel subunit CACNA2D4 causes autosomal recessive cone dystrophy. *Am J Hum Genet* **79**, 973–977.

Xu W, and Lipscombe D (2001). Neuronal Ca$_v$1.3α_1 L-type channels activate at relatively hyperpolarized membrane potentials and are incompletely inhibited by dihydropyridines. *J Neurosci* **21**, 5944–5951.

Yaney GC, Wheeler MB, Wei XY *et al.* (1992). Cloning of a novel α1-subunit of the voltage-dependent calcium channel from the β-cell. *Mol Endocrinol* **6**, 2143–2152.

Yang PS, Alseikhan BA, Hiel H *et al.* (2006). Switching of Ca^{2+}-dependent inactivation of Ca$_v$1.3 channels by calcium binding proteins of auditory hair cells. *J Neurosci* **26**, 10677–10689.

Yellen G (2002). The voltage-gated potassium channels and their relatives. *Nature* **419**, 35–42.

Yokoyama CT, Westenbroek RE, Hell JW *et al.* (1995). Biochemical properties and subcellular distribution of the neuronal class E calcium channel alpha 1 subunit. *J Neurosci* **15**, 6419–6432.

Yoshida A, Takahashi M, Nishimura S *et al.* (1992). Cyclic AMP-dependent phosphorylation and regulation of the cardiac dihydropyridine-sensitive Ca channel. *FEBS Lett* **309**, 343–349.

Yu AS, Hebert SC, Brenner BM *et al.* (1992). Molecular characterization and nephron distribution of a family of transcripts encoding the pore-forming subunit of Ca2$^+$ channels in the kidney. *Proc Natl Acad Sci USA* **89**, 10494–10498.

Yue DT, Herzig S, and Marban E (1990). Beta-adrenergic stimulation of calcium channels occurs by potentiation of high-activity gating modes. *Proc Natl Acad Sci USA* **87**, 753–757.

Yusaf SP, Goodman J, Pinnock RD *et al.* (2001). Expression of voltage-gated calcium channel subunits in rat dorsal root ganglion neurons. *Neurosci Lett* **311**, 137–141.

Zamponi GW, Bourinet E, and Snutch TP (1996). Nickel block of a family of neuronal calcium channels: subtype- and subunit-dependent action at multiple sites. *J Mem Biol* **151**, 77–90.

Zhang JF, Randall AD, Ellinor PT *et al.* (1993). Distinctive pharmacology and kinetics of cloned neuronal Ca^{2+} channels and their possible counterparts in mammalian CNS neurons. *Neuropharmacol* **32**, 1075–1088.

Zhang Y, Mori M, Burgess DL *et al.* (2002). Mutations in high-voltage-activated calcium channel genes stimulate low-voltage-activated currents in mouse thalamic relay neurons. *J Neurosci* **22**, 6362–6371.

Zhang Z, Xu Y, Song H *et al.* (2002). Functional roles of Ca$_\mathrm{v}$1.3 (α1D) calcium channel in sinoatrial nodes: insight gained using gene-targeted null mutant mice. *Circ Res* **90**, 981–987.

Zhong XL, Liu JRR, Kyle JW *et al.* (2006). A profile of alternative RNA splicing and transcript variation of *CACNA1H*, a human T-channel gene candidate for idiopathic generalized epilepsies. *Human Molecular Genetics* **15**, 1497 1512.

Zhou Q, Godwin DW, O'Malley DM *et al.* (1997). Visualization of calcium influx through channels that shape the burst and tonic firing modes of thalamic relay cells. *J Neurophysiol* **77**, 2816–2825.

Zhuchenko O, Bailey J, Bonnen P *et al.* (1997). Autosomal dominant cerebellar ataxia (SCA6) associated with small polyglutamine expansions in the alpha 1A-voltage-dependent calcium channel. *Nat Genet* **15**, 62–69.

2.3

Voltage-gated sodium channels

Sulayman Dib-Hajj and Tony Priestley

2.3.1 Introduction

Voltage-gated sodium channels underlie the generation and propagation of action potentials (AP) in excitable cells. They have been identified in bacteria and in all animal phyla (Goldin, 2002) and their conservation reflects a basic function in the life of organisms. They are complex and highly specialized structures and in the case of multicellular organisms, they are critical for transmitting information between, and coordinating activity of, various neural elements and networks. Not surprisingly, mutations in these channels can significantly impact the physiology of cells in which they are expressed and cause a variety of neural and muscular disorders.

Sodium channels in the plasma membrane are closed at rest but undergo structural changes in response to stimuli which permit the channel to cycle through closed, open, inactive and repriming states in order to transmit the response to the initiating stimulus along the particular information axis for that cell (Hille, 2001). External stimuli induce small depolarizations of the membrane which lead to transient sodium channel opening, allowing the passage of sodium ions down their concentration gradient, thus generating an inward current which further depolarizes the membrane. Activated (open) channels inactivate rapidly for most channel isoforms, within milliseconds of opening, and undergo conformational changes to recover from inactivation (reprime) and become available to open upon subsequent depolarization of the membrane; channels are not available to open during their repriming state. These aspects of channel function will be discussed in detail in other sections in this chapter.

In their open conformation, sodium channels are capable of sustaining very high rates of ion flux while maintaining exquisite selectivity for sodium ions. Ion selectivity is critical to the signaling function ascribed to these proteins. To be capable of effectively depolarizing the plasma membrane, they must support a rapid influx of cations but this must exclude calcium ions, to avoid plieotropic effects on other downstream cellular processes, and at the same time avoid a countercurrent of potassium efflux that would undermine depolarization. Selectivity is achieved by tailoring pore biology to the physiochemical properties of hydrated ions (Carrillo-Tripp $et\ al.$, 2006) resulting in a sodium preference of up to ~30-fold over potassium, (Hille, 1972; Favre $et\ al.$, 1996; Kurata $et\ al.$, 1999; Hille, 2001) and even more so over the similarly sized calcium ions (Perez-Garcia $et\ al.$, 1997).

In this chapter we will review more than half a century of research into sodium channel biology. Needless to say, the structural basis for these impressive functional properties has received much attention but the power of molecular biology has yielded a wealth of additional information that revealed these channels interact with, and are modified by, a host of other cellular proteins.

2.3.2 Molecular characterization

Basic structural features

Sodium channels from rodent brain tissue have been purified as heterotrimers of a large pore-forming α-subunit (~260 KDa) and much smaller auxiliary β-subunits (Catterall, 2000; Meadows and Isoms, 2005). The sodium channel family comprises nine functional α-subunit transcripts. Prior to the year 2000, nomenclature reflected the source tissue from which the channel was isolated or its potential function and the preference of the researchers who reported the channel. Thus different names were assigned to the same channel by different groups. A uniform nomenclature based on the International Union of Pharmacology Committee (IUPC) recommendations was adopted in 2000 (Goldin $et\ al.$, 2000): $Na_V1.1$–1.3 replaced brain type I–III; $Na_V1.4$ replaced SkM1 and μ1; $Na_V1.5$ replaced SkM2 and rH1; $Na_V 1.6$ replaced SCN8A, Na6 and PN4; $Na_V 1.7$ replaced hNE-Na, NaS and PN1; $Na_V 1.8$ replaced SNS and PN3; $Na_V 1.9$ replaced NaN, SNS2 and PN5. The adoption of the uniform nomenclature cleared potential confusion in scientific communications and facilitated dissemination of research data.

The nine α-subunit genes, $SCN1A$–$SCN11A$, are located on four chromosomes (Table 2.3.2). Orthologues of individual genes have been identified on autologous regions of rodent chromosomes. $SCN6A$ and $SCN7A$ were assigned to the related channel Nax which was originally thought to be an atypical member of the voltage-gated sodium channel family, but has since been shown not to be voltage-gated but is involved in salt homeostasis (for review, see Noda, 2006). Thus, channels $Na_V1.1$–1.5 are encoded by the genes $SCN1A$–$5A$, $Na_V1.6$–1.9 are encoded by genes $SCN8A$–$11A$; $SCN12A$ is the same as $SCN11A$. The α-subunits are encoded by members of gene families both in invertebrates and vertebrates, suggesting evolutionary gene duplication. Additional diversification has resulted from an accumulation of mutations. Two genes encode α-subunits of sodium channels in invertebrates, but a significant diversity in functional channels is generated by extensive alternative splicing and RNA editing of these channel transcripts (Thackeray and Ganetzhy, 1994; Hanrahan *et al.*, 2000; Liu *et al.*, 2004; Song *et al.*, 2004). However, nine distinct voltage-gated sodium channel genes have been identified in mammals, many of which are also to be found in lower vertebrates. Mammalian voltage-gated sodium channel α-subunits are encoded by up to 26 exons each, with conserved exon–intron boundaries, and several of them have been shown to undergo alternative splicing of coding exons (Sarao, 1991; Gustafson *et al.*, 1993; Belcher *et al.*, 1995; Plummer and Meisler, 1999; Raymond *et al.*, 2004), albeit to a lesser extent than that observed in invertebrate channels, as discussed below. Thus, diversity of sodium channels is reflected by the relatively large gene family and is enhanced by alternative splicing.

Mutually exclusive splicing of two alternative exon 5 (E5N and E5A) has been documented in the genes *SCN1A*, *SCN2A*, *SCN3A* and *SCN9A* (Sarao, 1991; Gustafson *et al.*, 1993; Belcher *et al.*, 1995; Raymond *et al.*, 2004) which are clustered on chromosome 2, and *SCN8A* (Plummer *et al.*, 1997; 1998) which resides on chromosome 15. The conservation of alternative splicing of E5N and E5A in these genes could imply a common lineage prior to the segregation of these genes to different chromosomes. Alternatively, the remaining members of this gene family could have lost this alternative exon since the establishment of gene clusters on four chromosomes.

Alternative splicing of exon 5 which encodes the C-terminal part of S3 and most of S4 in domain I (refer to Figure 2.3.1 and subsequent text for descriptions of these structural elements), and exon 18 which encodes the corresponding segment in domain III, and the usage of an alternative 5′ splice site in exon 11 encoding the C-terminal part of L1 is conserved in several sodium channels. Exon 5 in all members of this family is 92 base pairs (bp) and exists in two forms: E5N is predominantly utilized in embryonic tissues and E5A in adult tissues. While the two alternative exons show many nucleotide differences, they differ only by one or two amino acids because most of the nucleotide substitutions are silent. One amino acid distinguishes the two exons in all five channels: the sixth amino acid from the N-terminus of E5N is Asn but is Asp in E5A. Expression of the two alternative $Na_V1.2$ isoforms in *Xenopus* oocytes did not reveal a functional effect of this substitution (Auld *et al.*, 1990), thus the significance of this alternative splicing remains poorly understood. Alternative splicing of exon 18 which encodes the similar peptide sequence in domain III of *SCN8A* produces a transcript with premature translation termination codon, thus resulting in a two-domain truncated protein (Plummer *et al.*, 1997). The alternative transcript is more abundant than that of normal channel in non-neural tissue, a finding which suggested that alternative splicing of E18 acts as a fail-safe mechanism to prevent the expression of $Na_V1.6$ in non-neural cells (Plummer *et al.*, 1997).

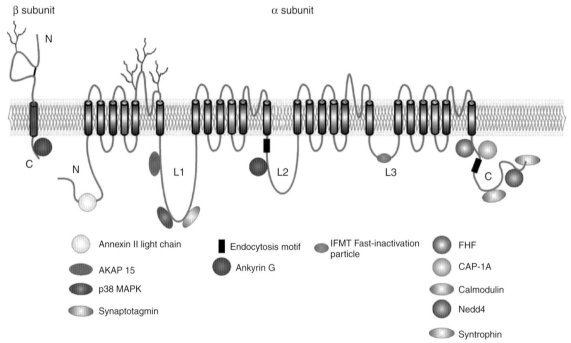

Fig. 2.3.1 Schematic of the predicted structure of α- and β-subunits of voltage-gated sodium channels. The pore-forming α-subunit is a long polypeptide which is organized into four domains (I-IV), each with six transmembrane helices (segments 1–6) that are joined by inta- and extracellular linkers to allow them to weave in and out of the membrane plane. The fourth segment in each domain (S4, shaded red) carries positively charged residues R/K in a repeated motif, and function as a voltage-sensor that changes position upon depolarizing the cell membrane leading to the opening of the Na-selective pore. The N- and C-termini of the α-subunits are cytoplasmic and the domains are joined by cytoplasmic loops (L1–L3). Multiple N-glycosylation sites (represented by violet branches) are identified in the extracellular loops. Consensus sequences for phosphorylation by PKA (cAMP-dependent protein kinase), PKC (calcium-dependent protein kinase), and MAPK (mitogen-activated protein kinase) have been identified, primarily in L1, but also in other cytoplasmic regions. The green oval in L3 represents the blocking particle IFMT which acts as the fast inactivating gate. A variety of cellular proteins have been shown to bind to α-subunits, in isoform-dependent fashion in some cases and regulate channel trafficking to, and stability at the cell membrane, or in modulating the biophysical properties of the channel. The smaller auxiliary β-subunits are type I transmembrane glycoproteins. The larger extracellular domain of β-subunits contains an Ig fold which is necessary for the function of these subunits, and a much smaller cytoplasmic tail. The cytoplasmic tail of β-subunits is phosphorylated by tyrosine kinase and has been shown to bind to AnkyrinG. The structure of β-subunits shows similarity to cell adhesion molecules. See also colour plate section.

Another common alternative splicing event utilizes a different 5′ splice site for intron 11 which leads to an in-frame additional 11 amino acids in the C-terminal end of L1 of $Na_V1.1$, $Na_V1.6$ and $Na_V1.7$ (Raymond et al., 2004). The additional 11 amino acids in $Na_V1.6$ cause the longer channel to recover from inactivation faster than the shorter channel (Dietrich et al., 1998). The effect of the 11 amino acid extension on the behaviour of other channels remains to be investigated.

The α-subunit of sodium channels is a long polypeptide with a range in size from 1700–2000 amino acids for various isoforms, but members of this family are invariant in their basic tertiary structure that comprises 24 transmembrane segments, connected by cytosolic and extracellular linkers, organized into four concatenated, homologous domains (domains I–IV, DI–DIV) each with six transmembrane segments (Catterall, 2000) (Figure 2.3.1). The domains are connected by cytosolic loops (L1–L3) of varying lengths that are less conserved, except for L3, in their primary sequence compared to the transmembrane segments. The variability in loop lengths accounts for most of the variation in the size between α-subunits. The N- and C-termini of α subunits are cytosolic, they too show less sequence variability between isoforms compared to the L1/L2 loop regions.

Sodium channel α-subunits are glycoproteins with a carbohydrate moiety which can account for 5–30 per cent of their mass (Miller et al., 1983; Messner and Catterall, 1985; Cohen and Barchi, 1993). Co-translational glycosylation of the neuronal α-subunit is critical for the association with auxiliary β subunits and the assembly of functional channels in the plasma membrane (Schmidt and Catterall, 1987). Addition of sialic acid to the ends of the complex sugar tree, during maturation of channels, accounts for the bulk of the carbohydrate glycocalyx (Miller et al., 1983; Messner and Catterall 1985). Inhibition of sialic acid addition or enzymatic desialidation does not affect the assembly of resident functional channels at the cell membrane (Schmidt et al., 1987), but does modify the gating properties of such channels (Recio-Pinto et al., 1990; Bennett et al., 1997; Zhang et al., 1999; Tyrrell et al., 2001). Interestingly, in the case of $Na_V 1.9$, the extent of glycosylation of the channel has been shown to be developmentally regulated, with heavier glycosylation on channels in neonatal dorsal root ganglion (DRG) neurons, accompanied by a significant hyperpolarizing shift in steady-state inactivation (Tyrrell et al., 2001).

While the α-subunit alone is necessary and sufficient to produce a functional channel, the β-subunits and a diverse array of other cytosolic channel partners have been shown to regulate various aspects of channel biology (Isom et al., 1994; Cantrell andCatterall 2001; Wood et al., 2004; Abriel and Kass, 2005; Liu et al., 2005). These protein–protein interactions have been shown to impact targeting of the α-subunits to the plasma membrane and to regulate channel stability, thereby influencing current density and membrane excitability. Other channel partners modulate biophysical properties of these channels, either by direct binding or by inducing post-translational modifications such as, for example, phosphorylation. We have expanded on some of these aspects of channel regulation in later sections of this chapter, but a comprehensive review of accessory subunit biology is beyond the scope of this review and interested readers are referred to specialized articles (Meadows and Isom, 2005).

Genetic and biochemical approaches, coupled to electrophysiological studies, have identified clear functional roles for specific primary sequence elements of α-subunits. Sequence analysis has identified a highly conserved primary structure among the different subunit variants making up this family of ion channels. As mentioned previously, that this conservation is more striking in some regions compared to others implies functional significance and this has been confirmed experimentally for many amino acid motifs. The identification of specific mutations in sodium channels underlying neurological and muscular disorders provided a rich source of information on the functional contribution of specific residues to the biology of individual members of the channel family. These disease-specific mutations are discussed later in the chapter.

The sheer size of the sodium channel protein, together with its non-symmetrical structure, has hindered sophisticated crystallographic studies of the sort that has revolutionized our understanding of bacterial and yeast potassium channels recently (Doyle et al., 1998; Jiang et al.., 2002a, b; 2003a; 2004). Nevertheless, there have been several attempts to define the structure in recent years. A 19 Å resolution, three-dimensional structure of the voltage-sensitive sodium channel from the eel *Electrophorus electricus* was determined by cryoelectron microscopy and single-particle image analysis of the solubilized protein (Sato et al., 2001). The channel is depicted as bell-shaped with several inner cavities that are connected to four small holes and eight orifices which are close to the extracellular and cytoplasmic membrane surfaces. The dimensions of the structure are: height of outer surface: 135 Å; side length at the square-shaped bottom: 100 Å; diameter of a spherical top: 65 Å. Another model was created using the known sodium channel residues but coordinated (Lipkind and Fozzard, 2000) to data derived from the crystal structure of the *Streptomyces lividans*, KcsA, channel (Doyle et al., 1998).

The characteristics and features of the various structural blocks of the channels are discussed below. Readers may find it helpful to refer to Figure 2.3.1.

Transmembrane segments

Current models of sodium channels tend to treat each domain as an independent structural element consisting of the transmembrane segments S1–S6 and the extra- and intracellular linkers joining these segments. The functional significance of most of the transmembrane segments has been derived largely from studies of mutant sodium channels, though such analyses have had mixed success in ascribing a specific role to particular segments. Compared to other transmembrane segments, the functional role of S4 and S6 in channel gating and structure is better understood and accepted within the scientific community and is reviewed in detail below (see Activation).

The S4 segments are now universally accepted as the 'voltage-sensors', though it is probably more reasonable to consider the larger S1–S4 region as an integrated voltage-sensor functional unit (Long et al., 2005b)—not an unreasonable proposition given the emerging contributions of multiple adjacent helices that will be discussed in more detail below. The S4 segments are highly unusual in having a striking abundance of basic Arg or Lys residues. These residues are regularly spaced throughout the membrane-spanning S4 segment, in a Arg/Lys-X-X-Arg/Lys motif, where X is typically a hydrophobic residue. This repeated pattern of a positively charged residue every third position, or so, effectively places the charged residues on one side of an amphipathic helix—an arrangement that is, intuitively, well-suited to its proposed function (Catterall, 2000). Though this scheme is common to other voltage-gated channels, sodium channels present additional complexity because the actual number of basic residues and their arrangement varies between the S4s of each of the four domains. The pattern is the same for all sodium channel isoforms except for $Na_V1.9$. Thus, there are 4, 5, 6 and 8 basic residues in the S4 segments of DI, DII, DIII and DIV, respectively; DII/S4 in rodent $Na_V1.9$ has only four basic residues, and DIII/S4 in both human and rodent $Na_V1.9$ has only five basic residues. This asymmetry has posed questions as to the relative contribution of the individual domains to the gating process, this and other aspects of the structure-function relationship are addressed in more detail below (see Biophysical characteristics).

The pore

The pore is thought to be made up of elements from the last two transmembrane segments, S5 and S6, and their connecting extracellular linker, from each of the four domains. The polypeptide chain connecting these two helices displays a structure that is typical to ion channels but unusual in other membrane-resident receptors and proteins. Evidence from multiple sources suggests that this extracellular linker dips into and out of the membrane lipid bilayer (Chiamvimonvat et al., 1996; Yamagishi et al., 2001), this structure is now commonly referred to as the pore-loop, or P-loop.

The model defined by Lipkind and Fozzard (2000) predicts a relatively wide-angle outer vestibule composed of the four α-helix-turn-β-strand motifs provided by each domain. The outer vestibule may act as a funnel to direct ions into the ion conduction channel that is further defined by elements of the S6 segments at the outer surface and by both S5 and S6 segments at the inner regions. These α-helical membrane-spanning structures form an inverted teepee-like arrangement, the outermost parts of the S6 segments form a square array into which the P-loop sequences are docked in a tightly packed configuration (Catterall, 2000; Lipkind and Fozzard, 2000). In defining this model, Lipkind and Fozzard (2000) made use of known multivalent toxin binding sites (discussed in greater detail below) for the vestibule and local anaesthetic binding residues to help constrain the inner water-filled pore. The P-loop not only confers a level of rigidity to the channel structure but the apex of the α-helix-turn-β-strand motif forms an innermost apex or constriction that is known to function as an ion selectivity filter. The dimensions of the constrictive filter region were predicted, long ago, to be around 3×5Å, from experiments that examined the relative permeabilities of various organic cations (Hille, 1971) and these predictions have been largely borne out by subsequent sophisticated molecular modeling data (Lipkind et al., 1994, Lipkind and Fozzard, 2000).

Several elegant studies have revealed critical elements of the selectivity filter. In all of the voltage-gated sodium channels this is known to be comprised of an absolutely conserved DEKA motif provided by DI, II, III and IV, respectively (Heinemann et al., 1992). Each of these amino acid residues are positioned such that they form a ring in three-dimensional space. Though once conceived as being a planar arrangement, recent studies have suggested that each of the individual DEKA residues may, in fact, reside at slightly different depths within the channel (Chiamvimonvat et al., 1996; Yamagishi et al., 2001) but with sufficient proximity to permit a variety of electrostatic interactions. The analogous motif in voltage-gated calcium channels is EEEE and this has provided many clues as to how these different channels achieve their ion selectivities. Substitution of individual DEKA residues, using standard site-directed mutagenesis techniques, has been used to map the structural basis of selectivity (Heinemann et al., 1992; Favre et al., 1996). For example, neutralization of the basic K in a K1237A mutant ($Na_V1.4$ numbering), effectively producing a DEAA locus, led to a channel that was equally permeant to sodium and potassium and also greatly increased calcium permeability. Independent neutralization of the acidic residues (D and E) to create AEKA and DAKA motifs resulted in little, if any, effect, suggesting that K of the DEKA motif is the primary determinant of sodium selectivity. Additional K-substitutions that examined the influence of bulk and charge revealed no clear basis for $Na^+:K^+$ permeability but did suggest that a positive charge at this position negatively influences permeability to calcium (Favre et al., 1996). As might be expected from the corresponding motif in the calcium channel family, substitution of both the K and A residues to create a DEEE mutant channel resulted in a $Na_V1.2$ sodium channel that displayed a preferential selectivity for Ca^{2+} over Na^+ (Heinemann et al., 1992; 1994). It has been suggested that the carbonyl oxygen of the carboxylate side chain of either D or E is required to coordinate Ca^{2+} flux whereas the butylammonium side chain of the critical K residue determines $Na^+:K^+$ selectivity (Favre et al., 1996).

Lipkind and Fozzard (2000) propose that the pore is stabilized in its closed conformation by a strong electrostatic interaction between the ammonium group of K and the carboxylate of E together with a water bridge between the K ammonium and the carboxylate of D. During the process of conduction, Na^+ competes for the ammonium group of K (more efficiently than would a K^+ ion) and pushes K towards D in a way that permits the dehydrated Na^+ ion to coordinate with one of the carboxyl groups from either D or E (the carboxyl group effectively acting as a water surrogates). Favre and colleagues (Favre et al., 1996) appear to favour an even more active role for the ammonium and suggest that this acts as a tethered cationic ligand, capable of neutralizing one of the carboxylic acids and, in the presence of Na^+, creating an overall electrically neutral environment that is energetically favourable for a coordinated metal ion. As an interesting aside, in the case of the voltage-gated calcium channels, a similar state of electroneutrality would result by virtue of the fact that two permeating divalent calcium cations could likely each interact with a pair of glutamate carboxyl groups.

A striking observation to emerge from the sodium channel modelling data is that the conduction process is not likely to be directly analogous to the multi-ion conducting process proposed for the potassium channels. The vestibule is much wider for sodium channels and the constriction shorter. Nevertheless, the same principles of ion dehydration and coordination to carbonyl oxygen atoms is central to the conduction process.

Cytoplasmic elements

N-terminus

A functional role for the N-terminus of sodium channels could be inferred from the presence of disease-causing missense mutations in $Na_V1.1$ and $Na_V1.5$, though the mechanistic bases underlying these channelopathies are not well understood at this time. However, in the case of $Na_V1.8$, a specific protein, annexin II light chain (p11), binds to the N-terminus of the channel causing an increase in the current density in DRG neurons (Okuse et al., 2002). Interestingly, the expression of p11 is increased in cerebellar Purkinje neurons of patients with multiple sclerosis and in experimental murine autoimmune encephalomyelitis (EAE) (Craner *et al.*, 2003a), which are pathological conditions that are known to induce the cryptic expression of Na$_V$1.8 channels (Black *et al.*, 2000).

Loop 1

The loop joining domains I and II (L1) is highly divergent among the sodium channels, both in length and in primary sequence. An important functional role for L1 is supported by the identification of missense mutations in Na$_V$1.1 and Na$_V$1.5. Also, the extension of L1 by 11 amino acids in Na$_V$1.6, which is caused by alternative splicing, as discussed previously, changes recovery from inactivation of the channel. The specific mechanism underlying this role of L1 is not clear at this time. More recently, the N-terminal 50 amino acid residues of human Na$_V$1.5 L1 were shown to bind to a member of the 14-3-3 gene family, and the co-expression of the channel and 14-3-3h induces a hyperpolarizing shift in steady-state inactivation and a delay in recovery from inactivation (Allouis *et al.*, 2006). L1 has also been shown to be the site for direct interaction with synaptotagmin which is regulated by calcium ions, suggesting an important role for this interaction in regulating neuronal excitability (Sampo *et al.*, 2000).

A more specific role for L1 in regulating channel function has been directly linked to phosphorylation of serine residues in L1 of several sodium channels (Cantrell and Catterall, 2001; Chahine *et al.*, 2005; Wittmack *et al.*, 2005). Phosphorylation of several serine residues in L1 in tetrodotoxin- (TTX) sensitive channels by PKA and PKC (Cantrell and Catterall, 2001; Chahine *et al.*, 2005) reduces the current density. Conversely, PKA-mediated phosphorylation of cognate serine residues in L1 of Na$_V$1.5 (Murphy *et al.*, 1996; Zhou *et al.*, 2000; Zhou *et al.*, 2002) and Na$_V$1.8 (Fitzgerald *et al.*, 1999; Vijayaragavan *et al.*, 2004) increases the current density. PKA Phosphorylation of Na$_V$1.2 requires the tethering of the kinase to the channel via an interaction of L1 with the adaptor protein AKAP15 (Cantrell *et al.*, 2002). Phosphorylation of a single serine residue in L1 of Na$_V$1.6 by the mitogen-activated protein kinase (MAPK) p38 reduces the current density without changing gating properties of the channel (Wittmack *et al.*, 2005). The recruitment of activated p38 MAPK to L1 of Na$_V$1.6 and phosphorylation of S553 requires the presence of a MAPK recognition module which includes docking site(s), in addition to the phospho-acceptor proline-directed serine SP dipeptide (Sharrocks *et al.*, 2000) in L1. Putative p38 docking motifs have been identified in L1 of Na$_V$1.6 but their functional significance has not been determined (Wittmack *et al.*, 2005). Studies of p38- and PKA/PKC-mediated modulation of TTX-S channels suggest that a similar outcome of L1 phosphorylation, namely a reduction in current density, could be achieved by different mechanisms.

The p38 MAPK-mediated reduction of Na$_V$1.6 current density, without changes in gating properties, requires a single phosphorylation site (S553) in L1 (Wittmack *et al.*, 2005) unlike the requirement for multiple phosphorylation sites in L1 of TTX-S channels for PKA/PKC to mediate similar reductions in current density (Cantrell and Catterall, 2001; Chahine *et al.*, 2005). The motif PGSP in L1 of Na$_V$1.6 may bind proteins carrying the WW4-type motif (Sudol and Hunter, 2000) but only when the 'S' residue is phosphorylated (Lu *et al.*, 1999; Verdecia *et al.*, 2000; Kato *et al.*, 2002). Thus, the reduction in Na$_V$1.6 current density could be achieved by channel ubiquitination by Nedd4-like protein (E3 ubiquitin ligase carrying the WW4 domain) followed by internalization and degradation.

In contrast, the PKA-mediated reduction in the current density is voltage-dependent (Cantrell and Catterall, 2001). The reduction in current amplitude of TTX-S channels has recently been linked to enhancing slow inactivation of channels (Chen *et al.*, 2006). While the molecular pathway underlying PKA/PKC phosphorylation-mediated alteration of gating of TTX-S channels is not clear, a possible effect of phosphorylated L1 on bending of S6 (the role of S6 bending is discussed further in the section on biophysics of gating) has been suggested as an explanation of the effect on slow inactivation of these channels (Chen *et al.*, 2006).

The reciprocal modulation of TTX-S and TTX-R channels by PKA/PKC might result from opposite effects of phosphorylation of L1 on S6 flexibility in these two types of channels. Based upon the finding of Chen *et al.* (2006), the impairment of slow inactivation of Na$_V$1.8 upon PKA phosphorylation of L1 might lead to the increase in current density, a prediction which could be easily tested. Alternatively, phosphorylation of L1 might interfere with the binding of proteins which normally retard the forward translocation of channels to the cell membrane. Substitution of RRR533–535AAA in human Na$_V$1.5 prevented the PKA-mediated increase in the current density (Zhou *et al.*, 2002). The RXR motif has been shown to function as a retention signal in endoplasmic reticulum (Zerangue *et al.*, 1999). Thus phosphorylation of serine residues close to the RXR motif might change the local structure of the peptide thus releasing channels from the ER. It is not unreasonable to suggest that a similar mechanism could trigger the release of channels from other subcellular compartments, thus freeing the channels to be translocated to the plasma membrane. Additional studies are needed to test these alternative hypotheses.

Na currents can also be regulated by dephosphorylation of serine/threonine residues in L1 causing an increase in current amplitude. It has been shown that different phosphorylated serine residues in L1 of Na$_V$1.2 are preferentially dephosphorylated by protein phosphatases 2A (PP2A) and PP2B (Murphy *et al.*, 1993). The regulation of Na$_V$1.2 current by the concurrent phosphorylation of PKA and PKC sites in L1 of the channel (Cantrell *et al.*, 2002) could be due to inhibition of PP2A/PP2B activities in the presence of activated PKC leading to reduced current, hence neuronal excitability (Kondratyuk and Rossie, 1997). While PKC phosphorylation of serine residues in L1 might change the conformation of the polypeptide to make the PKA sites more accessible to the kinase, it is possible that inhibition of phosphatases could be an alternative mechanism to achieve a similar outcome *in vivo*. It remains to be seen what effect dephosphorylation of serine residues might have on TTX-R channels.

Loop 2

The loop joining domains II and III (L2) is also highly divergent among the sodium channels, and specific residues within L2 have been shown to contribute to regulation of gating properties of sodium channels. Several missense mutations in L2 of Na$_V$1.1 and Na$_V$1.5 have been reported. The missense mutation W1204R in Na$_V$1.1 (Escayg *et al.*, 2001) in a patient with epilepsy changes a tryptophan residue which is conserved in all known vertebrate channels, and has been shown to cause a hyperpolarizing shift of activation and steady-state inactivation of Na$_V$1.1 (Spampanato *et al.*, 2003).

Loop 2 carries molecular determinants which regulate the anchoring of sodium channels at the cell membrane and which target channels to axonal or somatodendritic neuronal compartments. L2 houses a motif (V/API/LAXXES/DD) that binds ankyrin G (Garrido *et al.*, 2003; Lemaillet *et al.*, 2003; Fache *et al.*, 2004; Mohler *et al.*, 2004). The mutant E1053K (the conserved E residue in this motif) in Na$_V$1.5 causes Brugada syndrome by inhibiting the binding of the channel to ankyrin-G thus reducing functional channels on the surface of cardiomyocytes (Mohler *et al.*, 2004). Interestingly, ankyrin G binding to Na$_V$1.6 was shown recently to inhibit the persistent current produced by this channel (Shirahata *et al.*, 2006). Thus, ankyrin G could regulate channel clustering and gating properties.

Extension of the ankyrin G motif in the C-terminal direction identifies a sequence element which is responsible for the targeting of Na$_V$1.6 to axon initial segment (AIS), and presumably to nodes of Ranvier (Garrido *et al.*, 2003; Fache *et al.*, 2004). Additionally, the DII/S6-proximal terminus of L2 of Na$_V$1.2, the channel which is present at immature nodes of Ranvier (Boiko *et al.*, 2001; Kaplan *et al.*, 2001), houses an endocytosis motif which is thought to regulate the preferential removal of this channel from axons, thus leading to their retention in the somatodendritic compartment of neurons (Fache *et al.*, 2004). Naturally occurring substitution of residues in AIS motif in different members of this channel family might account for the absence of these channels from AIS and nodes of Ranvier and selective targeting of Na$_V$1.6 to these specific neuronal compartments.

Loop 3

The loop joining domains III and IV (L3), which is relatively short compared to L1 and L2, is one of the most conserved regions of the channel and fulfils a key function in channel gating (Catterall, 2000). Within milliseconds of channel opening, the peptide isoleucine-phenylalanine-methionine-threonine (IFMT) in L3, acting as a hydrophobic latch on a flexible hinge of proline and glycine residues (Rohl *et al.*, 1999), swings to block the inner pore of the channel thus terminating the inward flow of sodium ions and inactivating the channel (this is covered in greater detail in the section entitled Inactivation and repriming).

An unexpected role for L3 in channel activation was recently reported (Dib-Hajj *et al.*, 2005). Substitution of F1449V at the DIII/S6-proximal end of L3 in Na$_V$1.7 causes a significant hyperpolarizing shift in activation, but only a modest depolarizing shift in steady-state inactivation of the channel. The underlying mechanism for shifting voltage-dependent activation of mutant Na$_V$1.7 is not well understood at this time. However, the proximity of this L3 substitution to the membrane–cytoplasm interface of DIII/S6 suggests the involvement of this S6 segment in the mutant phenotype of the channel.

The L3 of all known sodium channels carries an invariant S residue (S1506 in Na$_V$1.2, for example) which has been shown to be phosphorylated by PKC and, in the case of Na$_V$1.2, causes a reduction in current amplitude and a slowing of fast inactivation (Numann *et al.*, 1991; West *et al.*, 1991). PKC also causes a reduction of the current density, and a significant depolarizing shift in the inactivation of Na$_V$1.8, when expressed in *Xenopus* oocytes (Vijayaragavan *et al.*, 2004). However, and in contrast, activation of PKC in DRG neurons in culture, causes an increase in the current density of the slowly-inactivating TTX-R current (Gold *et al.*, 1998), which is attributable to Na$_V$1.8, and in the persistent TTX-R current which is attributable to Na$_V$1.9 (Baker, 2005). These

discrepancies may reflect additional PKC phosphorylation sites on Na$_V$1.8 and Na$_V$1.9 channels that have yet to be identified. It seems plausible that S1452 of Na$_V$1.8 and S1341 of Na$_V$1.9 (equivalent to 1506 of Na$_V$1.2) might be targets for constitutive PKC phosphorylation and such modulation might contribute to slower inactivation of these TTX-R channels compared to the TTX-S channels.

L3 of sodium channel carries a pair of conserved tyrosine residues at positions 1494–1495, six residues C-terminal to the IFM tripeptide, of Na$_V$1.5 and the corresponding positions in all known vertebrate sodium channels. The double mutant YY1494–1495QQ in Na$_V$1.5 and the equivalent mutation (YY1497–1498QQ) in Na$_V$1.2 slowed macroscopic inactivation of these channels and allowed the channel to recover from inactivation faster than wild-type channels, suggesting an important role in regulating channel fast inactivation (Kellenberger *et al.*, 1997b; Tang *et al.*, 1998). Recently, additional evidence in support of this conclusion was obtained by the demonstration that Y1495 in Na$_V$1.5 is a phosphorylation target for Fyn, a member of the Src family of tyrosine kinases, and that this phosphorylation causes a 10 mV depolarizing shift in steady-state inactivation (Ahern *et al.*, 2005). Paradoxically, phosphorylation of Na$_V$1.2 by tyrosine kinases, at site(s) which remain to be identified, produces the opposite effect of that of Na$_V$1.5, i.e. a hyperpolarizing shift in steady-state inactivation of tyrosine-phosphorylated channels (Ratcliffe *et al.*, 2000); similar results were obtained from the activation of tyrosine kinases on the endogenous TTX-S sodium channels in PC12 cells (Hilborn *et al.*, 1998). The effect, if any, of tyrosine phosphorylation of L3 of Na$_V$1.8 and Na$_V$1.9 remains to be determined.

C-terminus

The C-terminus of sodium channels can be viewed as a membrane-proximal conserved region and a membrane-distal variable region. Both conserved and variable regions carry motifs for binding several different types of cytosolic proteins which regulate channel trafficking, and modulate gating properties of channels (see Inactivation in the section entitled Biophysical characteristics). Mutations in the C-terminus of Na$_V$1.1, Na$_V$1.2, Na$_V$1.4 and Na$_V$1.5 underlie neurological and muscle diseases, confirming an important role of this polypeptide.

Current densities and gating properties of several channels have been shown to be regulated by the binding of channel partners to specific sequence motifs within their C-terminus. Calmodulin (CaM), for example, binds directly to sodium channels via the isoleucine–glutamine (IQ) motif in the C-terminus and has been shown to regulate current density and biophysical properties of these channels (Mori *et al.*, 2000; Deschenes *et al.*, 2002; Tan *et al.*, 2002; Herzog *et al.*, 2003b; Kim *et al.*, 2004; Young *et al.*, 2005; Choi *et al.*, 2006a). Interestingly, CaM binding to the IQ motif in several sodium channel variants has been shown to be Ca^{2+}-independent (Mori *et al.*, 2000; Kim *et al.*, 2004; Choi *et al.*, 2006a). However, published data indicates that Na$_V$1.5 can be regulated by direct binding of Ca^{2+} to putative EF motifs in the acidic-rich, membrane-proximal region of the channel (Tan *et al.*, 2002; Wingo *et al.*, 2004) but this does not exclude a potential interaction between the EF and IQ motifs (Shah *et al.*, 2006). In contrast, Na$_V$1.8 properties did not change in low or high calcium concentrations, but were regulated by binding to apoCaM (Choi *et al.*, 2006a). The demonstration that cell background may change the response of channels to CaM binding (Young and Caldwell, 2005) might explain, at least in part, apparent discrepancies reported in the literature and it underscores the necessity to test such interactions in native neurons whenever possible. Additional studies are certainly needed to reconcile the findings of Ca^{2+}- and CaM-dependent modulation of sodium channels.

Current density and gating properties of several sodium channels are also regulated by binding to members of the fibroblast growth factor family, fibroblast growth factor homologous factors 1–4 (Liu et al., 2001b; Wittmack et al., 2004; Lou et al., 2005; Rush et al., 2006b). Binding of FHF1B to the acidic-rich sequence in the membrane-proximal conserved region of the C-terminus of $Na_V1.5$ shifts the voltage-dependence of inactivation of $Na_V1.5$ in a hyperpolarized direction, similar to the effect of the LQT-3 mutation D1760G (Liu et al., 2003). FHF2A and FHF2B bind to $Na_V1.6$ and increase the current density (Wittmack et al., 2004; Rush et al., 2006b). In the case of FHF2B, this effect is also associated with a depolarizing shift in the voltage-dependence of $Na_V1.6$ steady-state inactivation (Wittmack et al., 2004), but surprisingly, FHF2A increases the frequency-dependent inhibition of this channel (Rush et al., 2006b). Interestingly, a single amino acid substitution in FHF4 (Van Swieten et al., 2003) and ablation of this factor in mice (Wang et al., 2002) cause ataxia and cognitive deficits. Further studies will be necessary to fully understand the complexity of FHF modulation of sodium channel variants.

Sodium channel current density is regulated by an interaction with contactin/F3, a GPI-anchored cell adhesion molecule (Kazarinova-Noyes et al., 2001; Liu et al., 2001a; Shah et al., 2004; Rush et al., 2005a). However, this interaction may not be universal; it has been shown, thus far, to apply to the TTX-S channels $Na_V1.2$ (Kazarinova-Noyes et al., 2001) and $Na_V1.3$ (Shah et al., 2004) in heterologous expression systems, and the TTX-R channels $Na_V1.8$ and $Na_V1.9$ but not $Na_V1.6$ or $Na_V1.7$ in native, uninjured DRG neurons (Rush et al., 2005a). The C-terminus of sodium channels $Na_V1.3$ and $Na_V1.9$ trapped contactin in an affinity purification protocol (Liu et al., 2001a; Shah et al., 2004) and this interaction requires an adaptor protein that bridges the C-terminus of the channel and the mature, GPI-anchored contactin. At least in some cases, the sodium channel β1 subunit appears to serve as the adaptor. For example, β1 has been shown to mediate the interaction of contactin and $Na_V1.2$ (Kazarinova-Noyes et al., 2001) with the C-terminus of the β1 subunit being essential for efficient association (Meadows et al., 2001). Further studies with $Na_V1.1$ have narrowed the point of contact down even further and identified a highly conserved residue, D1866, in the membrane-proximal C-terminus region of the α-subunit as being critical for binding to the β1 subunit (Spampanato et al., 2004). Though contactin fails to co-immunoprecipitate with sodium channels from β1-deficient brain tissue of P17–P19 Scn1b null mice (Chen et al., 2004), there is other evidence that suggests the presence of alternative transmembrane proteins which can function as linkers. For example, the contactin-mediated increase in $Na_V1.3$ current density was not influenced by the co-expression with β1 subunits in HEK 293 mammalian cell line (Shah et al., 2004).

The C-terminus also controls channel density at the cell surface by regulating channel turnover. Clathrin- and ubiquitin-mediated mechanisms have been shown to regulate channel internalization via motifs in the C-termini of sodium channels. While clathrin-mediated internalization of sodium channels might be a universal mechanism, it can be regulated by the interaction of the C-terminus

with other proteins, as will be discussed in the following paragraph. These pathways could play an important role in safeguarding neurons from hyperactivity-induced injury or to establish a polarized distribution of sodium channels in different neuronal compartments.

Clathrin-mediated endocytosis of sodium channels reduces current density at the surface of hyperactive neurons (Dargent et al., 1994; Paillart et al., 1996), or following the infection of DRG neurons with herpes simplex virus (Storey et al., 2002). Clathrin-mediated endocytosis of $Na_V1.2$ has been shown to depend upon a nine amino acid cassette bearing a di-leucine motif, in the membrane-proximal region of the C-terminus of sodium channels (Garrido et al., 2001). The fact that this motif is highly conserved in all mammalian sodium channels suggests that it may regulate internalization of other channels as well as $Na_V1.2$. Interestingly, a D1866Y mutation immediately adjacent to the di-leucine motif, has been identified in $Na_V1.1$ of an Italian lineage with inherited generalized epilepsy (Spampanato et al., 2004). This mutation has been shown to weaken the interaction of $Na_V1.1$ with the β1 subunit (Spampanato et al., 2004) though it may also, conceivably, affect channel internalization. Importantly, the observation serves to underscore the critical role of these protein–protein interactions and highlights the detrimental consequences of their perturbation.

A nine amino acid motif in the C-terminus of $Na_V1.2$ was initially thought to regulate the polarized distribution of this channel in the somatodendritic compartment of neurons by selective removal from axons (Garrido et al., 2001). However, the conservation of these residues in channels that are sorted to different neuronal compartments suggests that this process requires additional factors to ensure specificity of channel distribution. Recently, a novel protein has been identified with properties that are consistent with the notion of isoform-specific factors that can act as linkers between the endocytic complex and sodium channels (Liu et al., 2005). CAP-1A is a modular protein which binds directly to $Na_V1.8$ via a central domain and to clathrin via C-terminal motifs, and whose co-expression with $Na_V1.8$ in DRG neurons reduces the current density of this channel (Liu et al., 2005). Thus, if clathrin-mediated internalization of sodium channels regulates the polarized distribution of sodium channels, it is likely to require additional factors to determine temporal and spatial specificity of channel sorting.

Turnover of several sodium channels, without effect on gating properties, is also regulated by members of the Nedd4 ubiquitin ligase family which bind to L/PPXYXXΨ motif in the C-terminus of channels, where Ψ at the +3 position is a hydrophobic residue (Abriel and Kass, 2005). The canonical motif PXY is present in all but $Na_V1.4$ and $Na_V1.9$ channels. Current densities of $Na_V1.2$, $Na_V1.3$, $Na_V1.5$, $Na_V1.7$ and $Na_V1.8$ channels are reduced upon the co-expression of Nedd4 or Nedd4–2 proteins (Abriel et al., 2000; Fotia et al., 2004; Van Bemmelen et al., 2004; Rougier et al., 2005). Direct binding of Nedd4-like proteins to sodium channels has been confirmed and has been shown to depend upon the PXY motif (Van Bemmelen et al., 2004), and substitution of Y with A or substitution of the hydrophobic residue at the +3 position with a charged amino acid abrogates Nedd4-mediated reduction in current density (Abriel et al., 2000; Fotia et al., 2004; Van Bemmelen et al., 2004; Rougier et al., 2005). The reduction of the current density is caused by the ubiquitination of the channel and its degradation in lysosomal compartments (Van Bemmelen et al., 2004; Rougier et al., 2005). Importantly, specific channels appear to be regulated by distinct members of the Nedd4 family (Fotia et al., 2004; Rougier *et al.*, 2005), suggesting tissue- or cell-specific regulation mediated by selective expression of members of the Nedd4 family.

Syntrophins represent a class of PDZ-adaptors which simultaneously bind to specific sodium channels and other cellular proteins and regulate channel localization and gating (Gee *et al.*, 1998; Schultz *et al.*, 1998; Zhou *et al.*, 2002; Ou *et al.*, 2003). For example, $Na_V1.4$ and $Na_V1.5$ have a characteristic and classical PDZ binding motif, ES/TXV, as their C-terminal tetrapeptide (Songyang *et al.*, 1997). Co-expression of syntrophin $\gamma2$ and $Na_V1.5$ in HEK293 cells shifted voltage-dependence of activation in a depolarized direction and slowed down fast inactivation (Ou *et al.*, 2003). The interactions of $Na_V1.4$ and $Na_V1.5$ with syntrophin appear to directly involve the C-terminal channel tetrapeptide EXSV and the PDZ domain of syntrophin (Gee *et al.*, 1998; Schultz *et al.*, 1998). In other cases, for example the brain-specific channel α-subunit transcripts, the interaction appears to involve the C-terminal syntrophin-unique (SU) domain (Gee *et al.*, 1998). The anchoring of channel α-subunits to cytoskeletal elements via the PDZ-binding motif is thought to hinder the movement of the C-terminus upon channel activation, and thus influence the gating properties of the channels (discussed further in the section Biophysical characteristics—inactivation and repriming').

Intracellular linkers

Studies of mutant channels from human patients with skeletal, cardiac and neurological diseases, and from animal models of neurological disorders support a role for these linkers in regulating activation and inactivation properties of sodium channels. Structure-function studies using site-directed mutagenesis and electrophysiological recordings suggest that S4–S5 linkers can regulate fast inactivation of sodium channels by contributing residues to the docking site of the fast-inactivation particle (Catterall, 2000; Goldin, 2003). Based upon the structure of the potassium channel Kv1.2, the S4–S5 linker could be positioned in close proximity to the C-terminal portion of S6 (Long *et al.*, 2005b) which is bent at a hinge glycine residue (Zhao *et al.*, 2004b). These linkers can regulate activation (Smith and Goldin, 1999; Waxman and Dib-Hajj, 2005) possibly by influencing the movement of the S4 segment which triggers channel activation (Catterall, 2000) or bending of S6 segments which accompanies opening of the pore (Zhao *et al.*, 2004a, b). Also, the effects of these linkers on gating might be dependent on the specific channel carrying the mutation.

Site-directed mutagenesis studies have identified specific residues in DIII (A1329 in $Na_V1.2$) and DIV (N1662 in $Na_V1.2$) S4–S5 linkers which contribute to the receptor site of the IFMT fast inactivation particle (Tang *et al.*, 1996; Lerche *et al.*, 1997; Smith and Goldin, 1997; Filatov *et al.*, 1998; Mcphee *et al.*, 1998; Tang *et al.*, 1998). Analyses of the S4–S5 linkers of DIII and DIV in sodium dodecyl sulfate micelles show these peptides in an alpha helical structure, and confirms the favourable position of residues which were identified by mutagenesis to interact with the fast inactivation gate (Miyamoto *et al.*, 2001). Consistent with the role of this linker in regulating fast inactivation, mutation of the first A, A1152D, in DIII/S4–S5 linker of $Na_V1.4$, which corresponds to A1329 in $Na_V1.2$, disrupts fast inactivation and results in myotonia (Bouhours *et al.*, 2005). Mutations A1330P and A1330T of the second A in DIII/S4–S5 linker of $Na_V1.5$ (corresponds to A1333 in $Na_V1.2$) similarly impairs fast inactivation of the channel causing LQT3 syndrome

(Smits *et al.*, 2005). Thus, the role of this linker in fast inactivation is supported by these naturally occurring mutations.

Mutations in the S4–S5 linkers of DI and DII have also been shown to affect activation and inactivation. Two mutations in the DII/S4–S5 linker of $Na_V1.7$, I848T and L858H cause a hyperpolarized shift in voltage-dependence of activation and slower deactivation (Cummins *et al.*, 2004), while the L858F substitution causes an additional small but significant depolarizing shift in voltage-dependence of fast inactivation (Han *et al.*, 2006). Recent evidence has shown that substitutions of a serine residue in the DI/S4–S5 linker by amino acids with larger side chain groups affect gating properties with S241T and S241L in $Na_V1.7$ causing significant hyperpolarizing shift in the $V_{1/2}$ of activation (Lampert *et al.*, 2006). However, S246L, the corresponding residue in $Na_V1.4$, causes a significant hyperpolarizing shift in voltage-dependence of fast inactivation but has no effect on activation (Tsujino *et al.*, 2003). Thus, it appears that mutations in this region could produce isoform-specific effects. Also, A1317T in DIII/S4–S5 linker in $Na_V1.6$ (corresponds to A1329 of $Na_V1.2$), which causes ataxia in mice (Kohrman *et al.*, 1996), depolarizes voltage-dependence of activation without affecting voltage-dependence or kinetics of fast inactivation of the channel (Smith and Goldin, 1999). Interestingly A1329T substitution in $Na_V1.2$ produced a similar effect on activation of the channel and no effects on inactivation (Kohrman *et al.*, 1996). It is not clear if size or net charge of the side-chain at the conserved A residues of the DIII/S4–S5 linker determines the effect on channel gating characteristics. Whether the observed changes of channel activation are caused by influencing the movement of S4 or bending of S6 upon depolarization of the membrane remains to be determined.

Extracellular linkers

As discussed previously, the extracellular linkers regulate channel function by contributing to the formation of the channel outer vestibule. A single amino acid residue in the S5–S6 linker in DI determines the sensitivity of sodium channels to the neurotoxin TTX with Y or F rendering the channels TTX-sensitive (TTX-S) and A, C or S rendering the channels TTX-resistant (TTX-R) (Catterall, 2000). Single amino acid substitution of C to Y in $Na_V1.5$ (Satin *et al.*, 1992) and S to F in $Na_V1.8$ (Sivilotti *et al.*, 1997) render the channels TTX-S, and substituting Y with S renders $Na_V1.3$ (Cummins *et al.*, 2001), $Na_V1.4$, $Na_V1.6$ and $Na_V1.7$ TTX-R (Herzog *et al.*, 2003a; 2003b). Though TTX is perhaps the best known sodium channel toxin, there are numerous other toxins that interact with these channels and this is an important aspect of sodium channel biology that is discussed in detail in the section entitled Pharmacology. The experimental manipulation of converting TTX-S channels into TTX-R derivatives by single amino acid substitution has permitted studies of these channels in native neurons that do not contain endogenous TTX-R channels or where the endogenous TTX-R currents could be isolated from the recombinant TTX-R current (Cummins *et al.*, 2001; Herzog *et al.*, 2003a, b; Wittmack *et al.*, 2004; 2005).

Natural selection accounts for fixing a TTX-R phenotype on sodium channels in newts (Hirota *et al.*, 1999), certain species of snakes (Geffeney *et al.*, 2002; 2005), soft-shell clams (Bricelj *et al.*, 2005) and puffer fish (Yotsu-Yamashita *et al.*, 2000; Venkatesh *et al.*, 2005). Recently, it was shown that snakes which feed upon newts that have high concentration of TTX in their skin develop resistance to this toxin by amino acid substitutions in the pore-lining sequences from DIV, rendering the channels TTX-R (Geffeney *et al.*, 2005). 'Red tide' refers to the periodic proliferation of algae carrying the neurotoxin saxitoxin (STX), which binds at site 1 (see Pharmacology) as its related toxin TTX (Catterall, 2000). Soft-shell clams in areas which are frequently exposed to 'red tide' proliferation develop resistance to saxitoxin by acquiring a single amino acid substitution of glutamic acid (E) with aspartic acid (D) in the pore-lining sequence from DII, and the introduction of this substitution into Nav1.2 causes 1000–3000-fold reduction in sensitivity to STX and TTX (Bricelj *et al.*, 2005). Similarly, the skeletal muscle sodium channel in puffer fish which accumulate a very high concentration of TTX, carry substitutions in the pore-lining residues in DI (N or C instead of Y or F) or DII (D instead of E, similar to the soft-shell clam channel) which render these channels TTX-R (Yotsu-Yamashita *et al.*, 2000; Venkatesh *et al.*, 2005). The biological relevance of conserving the TTX-S or TTX-R phenotypes in mammals is not clear at this time.

2.3.3 Biophysical characteristics

Activation

An ability to rapidly response to changes in the transmembrane electric field is a fundamental property of all of the voltage-gated sodium channels. The depolarization-driven conformational change that transforms the channel from a closed, non-conducting state to an open, sodium-permeable conformation, is termed 'activation'. In their ground-breaking experiments on giant squid axons, back in the 1950s, Hodgkin and Huxley (1952) predicted that activation gating likely involved the movement of four gating 'particles'. The Hodgkin and Huxley model (H–H model) that best described the data proposed the movement of three particles, their so called '*m* particles', were required for activation. The fourth particle, *h* in their terminology, moved independently of the *m* particles and acted to drive the channel into an inactivated state. Despite the fact that Hodgkin and Huxley were unable to demonstrate charge mobilization experimentally, the basic principles of the H–H model have stood the test of time, though their view that activation and inactivation are separate processes has been repeatedly challenged (for example see Armstrong 1981; Patlak 1991).

Improvements in voltage-clamp technology eventually confirmed the existence of a small capacitive 'gating current' that was detected experimentally when its usual occlusion by bulk ion flux was prevented. Gating current is a direct indication of charge movement prior to channel opening and these experiments, many of which studied sodium channels in giant axon preparations (Armstrong and Bezanilla, 1973; Armstrong and Bezzanilla, 1974; Keynes and Rojas, 1974), have had a major impact on the conceptual understanding of the gating process. The first study of a mammalian channel, rat recombinant $Na_V1.2$, estimated the 'quantal' charge contribution, i.e. that invoked by the 'shot-like' movement of just one gating particle to be around 2.3e- (electron equivalents) (Conti and Stuhmer, 1989). Rationalizing this with the H–H model suggested a total activation gating charge of around 6–7e-. However, more recent studies that have been able to take advantage of site-directed mutagenesis to remove complications introduced by inactivation, have revised these numbers upward and suggest as many as 10 or 12 charges may be needed to account for activation of

voltage-gated potassium (Schoppa *et al.*, 1992; Zagotta *et al.*, 1994) or sodium (Hirschberg *et al.*, 1995) channels.

Publication of the sequence for the first cloned sodium channel (Noda *et al.*, 1984) suggested the probable structural candidates for the H–H gating 'particles' to be the four highly charged, presumably membrane-spanning S4 helices mentioned above (see also Figure 2.3.1). These charged helices are envisioned to translocate within the membrane in response to depolarizing stimuli, a movement which is predicted to induce conformational changes in the channel structure that opens the pore permitting the inflow of sodium ions (Yang and Horn, 1995, 1996; Cha *et al.*, 1999). The voltage-sensor function of the S4 helices has been confirmed experimentally by demonstrating the impact of mutating individual S4 basic residues on the voltage-dependence of channel activation (Stuhmer *et al.*, 1989; Kontis *et al.*, 1997). Combining the power of point mutation with that of chemically modifiable C substitution has led to the widespread use and popularity of the 'substituted-C accessibility method (SCAM)' (Karlin and Akabas, 1998) as a means of addressing tertiary structure. SCAM, when applied to individual S4 basic residues, has helped substantiate the idea of a physical translocation of the S4 regions through the lipid bilayer during activation (Yang and Horn, 1995, Yang *et al.*, 1996). It has also helped clarify the issue of S4 asymmetry in gating. Each of the four S4 segments appears to be involved in activation, though they may not all contribute equally to charge movement (Chen *et al.*, 1996; Kontis *et al.*, 1997).

Several theories have been put forward to explain how S4 voltage sensor movement might be effected, these include 'sliding helix' (Catterall *et al.*, 1986), 'helical screw-like' (Guy and Seetharamulu, 1986) or an even more complex 'propagating helix' models (Guy and Conti, 1990). Each of these hypothetical considerations assumes that a collapse of the transmembrane electric field will force at least some portion of the S4 out of the external plane of the lipid bilayer and there is considerable experimental evidence supporting this notion (Yang and Horn, 1995; Cestele *et al.*, 1998; Cha *et al.*, 1999; Chanda and Bezanilla, 2002; Cestele *et al.*, 2006). As enlightening and compelling as these data are, they raise a number of important questions (Yang *et al.*, 1996) that continue to be debated (Horn, 2004). For example, how is such a high 'concentration' of basic amino acids maintained in an energetically favourable manner, i.e. unsolvated within a hydrophobic lipid environment and without appropriate countercharges? How is the equivalent of 2.5 elementary charges per domain moved across the membrane considering the added burden imposed by intervening hydrophobic residues? Finally, are these theories able to account for the rapid and reversible, highly efficient, gating process given the energetically unfavourable environment? Yang and colleagues suggest that some of the above issues can be explained if the charge movement is restricted to a relatively short contact point within the lipid bilayer (Yang *et al.*, 1996). Their proposal suggests that each S4 sensor moves within a canal in the protein core of its respective domain. The canal, or gating pore, has water-filled crevices or vestibules that act to maintain the high density of charged R/K residues in a solvated, energetically-favorable state. Similar schemes have been proposed to explain gating charge transfer in Shaker potassium channels (Islas and Sigworth, 2001; Asamoah *et al.*, 2003). Charge transfer is suggested to occur across a focused hydrophobic point rather than being a diffuse contact along the length of the sensor (Yang and Horn, 1995; Islas and Sigworth, 2001; Starace and Bezanilla, 2004). Movement of a basic residue through the narrow channel will effectively drag its tethered partners such that successive basic residues are aligned with the focused electric field. Such a model would account for the extraordinary efficiency of gating. Finally, the outward transfer of elementary charge could be effected either by an outward movement of S4 or, conceivably, by an inward movement of the electric field, i.e. an inward collapse of the protein to move the hydrophobic contact point 'down' the S4 sensor (Yang *et al.*, 1996).

Against this backdrop of sustained support, for what might be aptly termed the 'conventional model', an alternative view has emerged more recently, based on the X-ray crystal structure of the KvAP potassium channel, that could hardly be more different in nature (Jiang *et al.*, 2003a; 2003b). This radically different model places the voltage sensor, comprising S4 and the C-terminal half of S3, along the inner plane of the membrane rather than the more usual perpendicular orientation. MacKinnon's group has coined the term 'voltage-sensor paddle' based on appearance. They propose that depolarization-induced gating involves each paddle of four domains carrying their cargo of basic residues through the entire lipid bilayer (Jiang *et al.*, 2003b; Long *et al.*, 2005b). Debate over the relative merits and issues with each of these models continues (Ahern and Horn, 2004; Long *et al.*, 2005b; Cestele *et al.*, 2006; Cohen *et al.*, 2006) and it is likely to persist until crystal structures of a larger number of unmodified voltage-gated channels become available to clarify the picture. A better understanding might be achieved, at least initially, by studying the single domain bacterial sodium channel (Ren *et al.*, 2001).

The work outlined above is impressive and has greatly added to our knowledge about various aspects of channel structure and function. Explaining how the movements of the voltage sensors is translated into a change in pore structure, i.e. the electromechanical coupling that drives the channel from a non-conducting to an ion-permeable state, has presented significant theoretical challenges. Clues to molecular basis of the pore transition were provided recently by the crystal structures of potassium channels, one, KscA, in its closed state (Doyle *et al.*, 1998; Zhou *et al.*, 2001) and two others, MthK (Jiang *et al.*, 2002a; 2002b) and Kv1.2 (Long *et al.*, 2005a) in open configurations. Structural analysis of the open conformations of the MthK and Kv1.2 channels suggested a bending of the pore-lining helices around a critical G residue that is proposed to act as a hinge point with the helix (Jiang *et al.*, 2002a, b; Long *et al.*, 2005a).

A similar situation appears to hold true for the sodium channel family. Sequence alignment reveals an absolute conservation of the putative G hinge, at around the mid point of the S6 segments of DI through DIII (a serine residue is present at the comparable position in DIV/S6), for all of the voltage-gated sodium channel variants (Na$_V$1.1 through Na$_V$1.9). The flexibility conferred by this glycine residue enables the C-terminal portion of S6 to splay outward in the open channel configuration and to straighten in the closed state (Zhao *et al.*, 2004a, b). A G-to-P substitution in the S6 of the bacterial sodium channel causes > 50 mV hyperpolarizing shift in $V_{1/2}$ of activation which is thought to result from increased bending of S6 in the open state of the channel (Zhao *et al.*, 2004a, b). Invariant conservation and site-directed mutagenesis data strongly suggest a common and fundamental role for the S6 glycines in sodium channel gating. It is also interesting to note that proposed G hinge occurs at a level in the sodium channel protein that is close to the plane of the DEKA selectivity filter residues. The close physical juxtapositioning of these structural elements might

be more than just mere coincidence: it might, for example, confer a very efficient coupling of the putative hinging of the S6 elements to the opening of the ion permeation pathway. The fact that S occupies the hinge position in DIV/S6 of all of the Na$_V$1.x variants raises questions as to the comparative flexibility of DIV/S6 during gating and may explain why charge movement associated with DIV/S6 is linked to inactivation.

Inactivation and re-priming

In order to function efficiently as high-fidelity frequency encoders, sodium channel activity must be rapidly terminated and the system reset: these events need to occur on a millisecond timescale. Inactivation describes a time- and voltage-dependent transition of the open and conducting channel into a non-conducting state, a process which has been extensively studied from biophysical, structural, and pharmacological standpoints. Studies of recombinant channels have shown isoform-specific differences in voltage-dependency and in time course of inactivation (Akopian *et al.*, 1996; Chahine *et al.*, 1996; Sangameswaran *et al.*, 1996; Cummins *et al.*, 1999). Though these parameters are readily quantifiable (for examples see Catterall *et al.* 2005), readers are cautioned not to over-interpret the numbers since voltage and time parameters are highly interrelated, and are influenced by the host cell factors in the case of heterologous expression, by the presence or absence of accessory proteins and by post-translational processing (see earlier discussions on these topics). It is probably more helpful, therefore, to appreciate that differences exist rather than to attribute significance to particular numerical values. Arguably, the most striking finding is that the Na$_V$1.8 and Na$_V$1.9, TTX-resistant channels, show slower gating when compared to other family members and potential structural differences that might be responsible for some of the variability in gating profiles have been covered in earlier sections (Akopian *et al.*, 1996; Sangameswaran *et al.*, 1996; Cummins *et al.*, 1999). The striking variability in inactivation seen in recombinant systems is also apparent in certain native neuronal populations, with subtypes of DRG neurons providing, arguably, the best example (Figure 2.3.2). The very similar kinetics shared between specific recombinant channel variants and native DRG sodium currents has been sufficiently compelling to lead to proposals as to the molecular identity of particular native conductances (Rush *et al.*, 1998).

There is a general consensus that favours a process where inactivation is coupled to activation, the biophysical arguments supporting this position have been articulated in detailed reviews (Armstrong, 1981; Patlak, 1991; Sheets and Hanck, 1995; Chen *et al.*, 1996; Mitrovic *et al.*, 1998; Cha *et al.*, 1999; Sheets *et al.*, 1999; Mitrovic *et al.*, 2000; Chanda and Bezanilla, 2002). There are multiple forms of inactivation that are best differentiated by their temporal characteristics. Early observations indicated that sodium channel *fast* inactivation was peptidergic in nature (Armstrong *et al.*, 1973b; Stimers *et al.*, 1985) but provided no clue as to structural details. Over a decade later, and subsequent to publications of the first sodium channel sequences (Noda *et al.*, 1984; 1986; Kayano *et al.*, 1988), additional molecular details began to emerge and pointed to the peptide sequence that links DIII to DIV (see section on intracellular linkers entitled Loop 3). Enzymatic cleavage of the DIII–DIV linker, its deletion by mutagenesis (Stuhmer *et al.*, 1989) or immobilization using epitope-specific antibodies (Vassilev *et al.*, 1988) each effectively modify channel inactivation. As mentioned previously, scrutiny of this linker region, using site-directed mutagenesis, identified

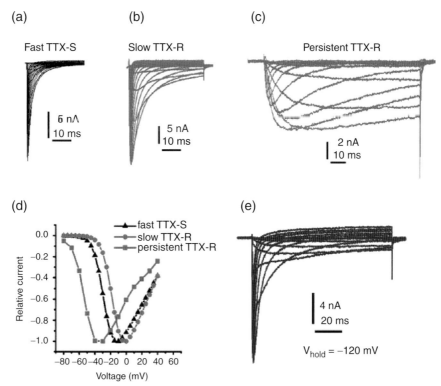

Fig. 2.3.2 Multiple sodium currents are expressed in neurons including small DRG neurons. Family traces of (a) fast-inactivating sodium current which is sensitive to nanomolar concentrations of tetrodotoxin (Fast TTX-S), (b) slowly-inactivating and (c) persistent sodium currents which is resistant to micromolar concentrations of TTX (TTX-R). The amplitudes of the currents in a–c are not normalized. (d) Different current-voltage relationships for fast-inactivating TTX-S (black symbols), slowly-inactivating TTX-R (green symbols) and persistent TTX-R (red symbols) sodium currents recorded from small DRG neurons. (e) Multiple sodium currents are recorded in DRG small (<30 mm diameter) neurons. These currents can be isolated by application of TTX in the bath and by digital subtraction after recording currents in response to depolarizing steps from different holding potentials. Panels a–d are modified, with permission, from Dib-Hajj et al. (2002),[1] and panel E is reproduced, with permission, from Cummins et al. (1999).[2] See also colour plate section.

[1] Reprinted from *Trends in Neurosciences*, Volume 25, S Dib-Hajj, JA Black, TR Cummins, and SG Waxman, NaN/Nav1.9: a sodium channel with unique properties, pages 253–259, Copyright 2002, with permission from Elsevier.

[2] Cummins TR, Dib-Hajj SD, Black JA, Akopian AN, Wood JN, and Waxman SG (1999). A novel persistent tetrodotoxin-resistant sodium current in SNS-null and wild-type small primary sensory neurons. *Journal of Neuroscience* 19, RC43. Reproduced with permission from the Society of Neuroscience, Copyright 1999 by the Society of Neuroscience.

a critical hydrophobic cluster of amino acids, I-F-M-T (commonly referred to as either the IFM or IFMT motif), that appeared to be required for fast inactivation (Patton *et al.*, 1992; West *et al.*, 1992; Kellenberger *et al.*, 1997b). Intracellular application of small peptides containing parts of the IFM motif were shown to mimic inactivation (Eaholtz *et al.*, 1994), suggesting that gating may invoke a movement of IFM such that it occludes ion flux and, as discussed earlier, flexibility of this region may be enhanced by multiple G and P residues (Kellenberger *et al.*, 1997a).

Within the IFM motif, the F residue (e.g. F1489 in $Na_V1.2a$) appears to be particularly important, mutation of this residue alone produces the biggest impact on fast inactivation (West *et al.*, 1992; Nuss *et al.*, 1996); a simple alignment of $Na_V1.x$ sequences reveals the hydrophobic IFM cluster to be absolutely conserved throughout the sodium channel family. However, mutagenesis studies targeting F1485 in $Na_V1.5$, a cardiac-dominant transcript, revealed subtle differences in outcome when compared to $Na_V1.2b$ (Hartmann *et al.*, 1994), suggesting that there are additional structural elements that allow for isoform-dependent tailoring of inactivation parameters. The IFM triad is flanked by a D residue (D1486) on the N-terminus and a T residue (T1491) on the C-terminal end which are invariant in all known voltage-gated sodium channels.

Substitutions of the D1486 do not produce significant effects on channel inactivation (Kellenberger *et al.*, 1997b). In contrast, mutation of T1491 to M (T1491M) in $Na_V1.4$ is a naturally occurring mutation which causes incomplete inactivation of the channel (Cannon, 2006) suggesting a contribution of this residue to the inactivating particle. Studies of additional substitutions of T1491 suggest that this residue binds directly to the receptor of the inactivation particle, and its mutation causes a destabilization of the inactivated state of the channel (Kellenberger *et al.*, 1997b).

A significant number of naturally occurring disease-associated mutations that result in inactivation-deficient gating in $Na_V1.5$ have been identified (Clancy and Rudy, 1999; Wei *et al.*, 1999; Veldkamp *et al.*, 2000; Clancy and Rudy, 2002), the positioning of the mutations suggested an involvement of the C-terminus. Some of the complexity of this region has been discussed previously in the section entitled C-terminus, but the role of this region in fast inactivation represents an additional functionality that appears to vary between channel isoforms (reviewed by Goldin 2003). Studies using sodium channel chimeras, e.g. $Na_V1.2/1.5C$ and $Na_V1.5/1.2C$ (Mantegazza *et al.*, 2001), and $Na_V1.4/1.5C$ and $Na_V1.5/1.4C$ (Deschenes *et al.*, 1998; 2001) have been particularly informative and have revealed a difference in voltage-dependence of inactivation, but not activation, and in the kinetics of inactivation compared to parent channels. The contribution of the C-terminus to fast inactivation has been attributed to sequences in the conserved, membrane-proximal region. The region may act to modulate the interaction of the IFM fast inactivation peptide with its receptor (Cormier et al., 2002; Glaaser et al., 2006) or may converge with the IFM motif to create a stable linker-C-T molecular complex (Motoike et al., 2004). It is also clear, however, that there is yet more to learn because similar studies with Na$_V$1.8 have revealed additional complexity. The voltage-dependence of activation and inactivation of Na$_V$1.8 is significantly more depolarized compared to the other sodium channels, and this channel is characterized by slow inactivation (Akopian et al., 1996; Sangameswaran et al., 1996). In contrast to the chimera channels discussed above, the V$_{1/2}$ of activation of a Na$_V$1.4/1.8C chimera is depolarized by 12 mV compared to parental Na$_V$1.4, i.e. conferring a Na$_V$1.8-like activation profile, while the V$_{1/2}$ of inactivation remains Na$_V$1.4-like (Choi et al., 2004). Paradoxically, a Na$_V$1.8/1.4C chimera did not show a change in either activation or inactivation of the channel. Such complexity may reflect a situation where variable residues in the C-terminus and in its putative interacting partners in other structures of the channel combine to produce the isoform-specific nuances in gating.

The fact that inactivation is quite a stable process, particularly so at depolarized membrane potentials, suggests the presence of a receptor site such that when the DIII–DIV linker-C-T gate closes, it remains closed until other forces come into play to reset the system (i.e. deactivation and repriming, discussed below). Indeed, the inactivation process can be thought of as a binding reaction between the gate and its receptor, with the binding affinity being a function of the activation state such that partially activated channels weakly inactivate and fully activated channels are strongly inactivated (Kuo and Bean, 1994). The ligand-receptor analogy is further supported by the rigid, helical (i.e. structured) and hydrophobic nature of the IFM motif (Rohl et al., 1999) and by the fact that it is known to become inaccessible during inactivation (Kellenberger et al., 1996), implying a translocation to a, presumably hydrophobic, pocket or receptor site. As discussed earlier in the section on intracellular linkers, several site-directed mutagenesis studies have identified potential candidates for the IFM receptor site. These include residues that form the short intracellular loops connecting the S4–S5 segments in DIII (Mcphee et al., 1994; 1995; Smith and Goldin, 1997) and DIV (Mcphee et al., 1998) and residues in DIV/S6 that are also implicated in the local anaesthetic binding site (Bai et al., 2003) and, hence, within the inner vestibule of the channel (discussed in detail below, see pharmacology section). The S4–S5 loop sites are particularly interesting. They are very short and, thus, close to the inner mouth of the channel and they are also likely to experience the movement of the respective S4 voltage sensors during activation–inactivation. This provides an attractive, though speculative, structural basis for the proposed coupling of inactivation to activation and for suggestions that the apparent voltage-dependence of inactivation is derived from that of activation (Sheets and Hanck, 1995; Chen et al., 1996; Mitrovic et al., 1998; Cha et al., 1999; Sheets and Hanck, 1999; Mitrovic et al., 2000; Chanda and Bezanilla, 2002). The availability of crystal structures of the channel, in open and inactivated conformations, would help solidify these hypotheses.

Though this vision of channel gating has many attractions, it would be remiss not to point out that, despite the numerous elegant experiments, there are many subtleties that remain to be explained

and speculations that require experimental support. It was clear to Hodgkin and Huxley that gating was not simply an open and shut case but rather a complex stepwise process with several transitional states (Hodgkin and Huxley, 1952). This tenet has been repeatedly revisited and embellished as models have attempted to account for more and more experimental data and to include additional complexity such as, for example, a path to inactivation from a closed or partially activated channel state (Armstrong and Bezanilla, 1977; Armstrong, 1981; Horn and Vandenberg, 1984; Stimers et al., 1985; Patlak, 1991; Kuo and Bean, 1994; Chanda and Bezanilla, 2002). Such complexity will impact the interpretation of structure–function studies and supporting experimental evidence already exists. For example, the effect on fast inactivation of point mutations within the DIV/S4–S5 linker is dependent upon whether the transition was from open or closed channel conformations (Mcphee et al., 1998). As mentioned earlier, there are also clear indications of additional, kinetically slower inactivation states, including slow and ultraslow transitions that impart further complexity. These states, that are analogous to C-type inactivation in K channels (Yellen, 1998) and involve entirely different structural components of the channel compared to fast inactivation. Rearrangements of the outer vestibule and pore regions have been implicated in slow inactivation states (Balser et al., 1996a; Kambouris et al., 1998; Benitah et al., 1999; Hilber et al., 2002; Xiong et al., 2006), these structures are far removed from the linker and C-T regions discussed above. However, despite their physical separation, it appears as though the various inactivation states are coupled (Rudy, 1978; Featherstone et al., 1996; Nuss et al., 1996; Richmond et al., 1998; Hilber et al., 2002) such that entry into the fast inactivated state reduces the probability of entry into slow inactivation and, moreover, each may be influenced by ion flux through the channel (Benitah et al., 1999). There is also evidence to suggest that there is long-range (i.e. interdomain) coupling between voltage sensor movements (Chanda et al., 2004). These observations suggest extensive allosteric interactions between what were, at least originally, regarded as compartmentalized functional domains.

Numerous cases of further complexity are reported in the literature. Incomplete inactivation and rapid de-inactivation represent just two examples. Incomplete inactivation, when present, results in a low level of persistent sodium channel current flux during sustained depolarizing conditions (see Figure 2.3.2c, for example) and though the residual current amplitude may be small when compared to that flowing initially, it can nevertheless significantly influence excitability (Del Negro et al., 2002; Do and Bean, 2003; Darbon et al., 2004; Do and Bean, 2004; Harvey et al., 2005; Wu et al., 2005; Enomoto et al., 2006; Kuo et al., 2006). The extent of incomplete inactivation can be readily demonstrated as an obvious spike-like tail current elicited upon membrane repolarization. Tail currents reflect sodium current flowing into the cell, driven by the highly favourable electrochemical gradient imposed by hyperpolarization, but they are usually very short-lived because hyperpolarization forces the total channel population into the closed state. De-inactivation is different from incomplete inactivation, it reflects a transition of depolarization-inactivated channels back to an open state, again in response to membrane repolarization, but the tail currents that ensue are much more robust and decay more slowly (Raman et al., 1999, Raman and Bean, 2001). This has also been termed 'resurgent current' and appears to be caused by recovery from an inactivation process that is not only distinct from the DIII–DIV/C-T fast inactivation but, in fact, is likely to compete with the latter and occlude entry into the 'traditional' fast-inactivated state (Raman *et al.*, 2001).

Incomplete inactivation and resurgent current are more than mere biophysical curiosities. These sodium channel properties dictate input–output properties of individual neurons, promoting burst firing and pacemaker-like activity, and hence confer physiological features that are capable of influencing excitability of entire neural networks (Del Negro *et al.*, 2002; Do *et al.*, 2003; Darbon *et al.*, 2004; Do *et al.*, 2004; Harvey *et al.*, 2005; Wu *et al.*, 2005; Enomoto *et al.*, 2006; Kuo *et al.*, 2006). It is not surprising, therefore, that there have been appreciable efforts devoted to understanding the molecular and mechanistic bases of this behaviour. For example, $Na_V1.6$ has been frequently implicated (Burgess *et al.*, 1995; Raman *et al.*, 1997) or shown directly (Cummins *et al.*, 2005) to produce resurgent current in Purkinje cells and DRG neurons, respectively. Though this property was originally attributed, exclusively, to $Na_V1.6$ (Burgess *et al.*, 1995; Raman *et al.*, 1997) more recent evidence suggests other channels may also contribute significantly to this current but the extent of which varies between different neuron types (Do *et al.*, 2004). For example, CA3 neurons which express $Na_V1.6$ channels do not support the production of resurgent current (Grieco *et al.*, 2005), and only half of DRG neurons transfected with $Na_V1.6$ produced the resurgent current (Cummins *et al.*, 2005).

The fact that resurgent currents appear to be more prevalent after strong but brief depolarizations and appear to be correlated to an unusually fast inactivation of the initial peak inward current, provided clues as to potential mechanistic basis for the behaviour. Such a profile might result from a fast open-channel block driven by the strong depolarization that, at the same time, prevents normal inactivation processes and hence enables a resurgent influx of ions as the membrane is repolarized. Studies using outside-out patches from Purkinje cells acutely isolated from mice revealed the transient peak and resurgent tail currents to be highly correlated (Grieco *et al.*, 2002). Treatment with intracellular proteases and alkaline phosphatase essentially abolished resurgent current, suggesting the blocking particle to be proteinaceous and to be sensitive to phosphorylation (Grieco *et al.*, 2002). Recent work has suggested that the blocking 'particle' might be the C-terminal tail of the β4 subunit that may function to occlude the cytoplasmic opening of the channel (Grieco *et al.*, 2005). This would be consistent with the data showing that the blocking element is closely associated with the α-subunit and, indeed, the incidence of detection of robust resurgent current parallels the expression pattern of the β4 subunit (Yu *et al.*, 2003). Inherent instability in the binding of the β4 peptide to the channel would explain resurgent current profile, repolarization presumably relieves the block from channels that are still in an open state and the increasing driving force leads to a substantial flux. The resurgent influx relaxes over time as the channel population is forced into the resting, closed state by membrane repolarization (Raman *et al.*, 2001). The putative role of the β4 subunit in mediating the resurgent current raises the interesting notion that some aspect of $Na_V1.6$ structure predisposes this channel variant to β4-mediated inactivation and this might explain, at least in part, the apparent isoform-specific nature of resurgent current.

Inactivation is a reversible process, the forces that drive the binding of the inactivation gate to its receptor can be driven in reverse by membrane hyperpolarization. Regardless of the particular conformation, fast, slow and ultraslow states are driven back into the closed state when the membrane is sufficiently hyperpolarized. This recovery process has been called a number of terms, including re-activation or de-inactivation but 'repriming' seems to better convey the principles involved. Just as the onset of inactivation ensues after a slight delay (Goldman and Schauf, 1972; Armstrong *et al.*, 1977), so the repriming recovery phase proceeds with a membrane potential-dependent delay and may start out as a deactivation step (i.e. an open but non-conducting state) prior to the reverse movement of the voltage-sensors (Kuo *et al.*, 1994). Following the analogy of a ligand binding scheme, membrane hyperpolarization presumably results in a drastic decrease in affinity of the inactivation gate for it's receptor, a prior deactivation effectively prevents any inward sodium flux that might otherwise occur as a result of the favourable electrochemical gradient imposed by hyperpolarization (Kuo *et al.*, 1994) (but see further considerations below). There have been surprisingly few studies addressing the structural basis for repriming (Ji *et al.*, 1996). There is, however, evidence for isoform-dependent variability in repriming kinetics; $Na_V1.3$ and $Na_V1.8$, for example, are notable for their rapid repriming kinetics (Cummins *et al.*, 1997, 2001). $Na_V1.8$ is particularly interesting in this regard, this channel shows appreciable divergence in the short DIV/S3–S4 loop compared to other family members in which the region is remarkably conserved. Dib-Hajj and colleagues exploited this divergence and showed that a four-residue cassette, SLED, present in $Na_V1.8$ but not in $Na_V1.4$ conferred fast $Na_V1.8$-like repriming when inserted into the corresponding location in $Na_V1.4$ (Dib-Hajj *et al.*, 1997). This is the only study thus far reported hinting at a structural influence on repriming, though it does not explain fast kinetics seen with $Na_V1.3$, implying that there may be additional factors yet to be resolved.

2.3.4 Tissue distribution

Cellular localization

Most sodium channels, $Na_V1.1$–1.3 and $Na_V1.6$–1.9, are expressed in central nervous system (CNS) and/or peripheral nervous system (PNS) neurons and glia, and are thus referred to as 'neuronal' channels, while $Na_V1.4$ and $Na_V1.5$ are expressed primarily in skeletal and cardiac myocytes, respectively (Catterall *et al.*, 2005). These channels are expressed in a tissue- and developmentally specific manner. So far, all neurons have been shown to express multiple sodium channels, while adult skeletal myocytes express $Na_V1.4$ and adult cardiac myocytes express primarily $Na_V1.5$ and, to a much lower extent, few neuronal channels. The electrogenic properties of neurons are influenced by the complement of sodium channels which they express. This view is supported by studies on mutant sodium channels, as will be discussed in a later section.

Sodium channels $Na_V1.1$, 1.2 and 1.3 which are clustered together on human chromosome 2 and the rodents' cognates, have a common lineage which is manifested by a high level of sequence conservation compared to other channels. Despite their common ancestry, the expression of each of these channels is regulated in a different manner. $Na_V1.1$ is expressed in both CNS (Beckh *et al.*, 1989) and DRG neurons (Black *et al.*, 1996) and in cardiac muscle (Maier *et al.*, 2002). In rodents, $Na_V1.1$ expression reaches adult levels within the second and third weeks after birth and appears to be more abundant in the caudal regions of the brain and spinal cord (Beckh *et al.*, 1989; Westenbroek *et al.*, 1989). $Na_V1.2$ reaches adult levels in rostral regions of the brain by one week after birth and lower levels in caudal regions except in cerebellum where it is abundant in adult tissue (Beckh *et al.*, 1989; Westenbroek *et al.*, 1989). $Na_V1.3$ is the dominant channel isoform in embryonic CNS and DRG neurons, but is significantly downregulated after birth (Beckh *et al.*, 1989; Waxman *et al.*, 1994).

$Na_V1.3$ is significantly upregulated following injury to DRG neurons or their axons, or to spinal cord (Waxman *et al.*, 1994; Dib-Hajj *et al.*, 1996; Black *et al.*, 1999a, Dib-Hajj *et al.*, 1999, Hains *et al.*, 2003, 2004, 2005). The elevated levels of $Na_V1.3$ in central neurons following peripheral or central nerve injury is associated with hyperexcitability of these neurons and neuropathic pain behaviour (see a more complete discussion about the role of $Na_V1.3$ in neuropathic pain in the section Sodium channels and disease state). $Na_V1.2$ is expressed at low levels in embryonic DRG neurons which diminish further after birth (Waxman *et al.*, 1994; Chung *et al.*, 2003), and, unlike $Na_V1.3$, is not upregulated in DRG neurons after axotomy (Waxman *et al.*, 1994). Injury-induced upregulation of $Na_V1.3$ can be attenuated by administration of trophic factors, for example NGF and GDNF, to DRG neurons in culture (Black *et al.*, 1997) and *in vivo* (Boucher *et al.*, 2000; Cummins *et al.*, 2000; Leffler *et al.*, 2002).

$Na_V1.4$ is abundant in mature, innervated skeletal muscles, but is absent from CNS tissue (Kallen *et al.*, 1990; Trimmer *et al.*, 1990). While $Na_V1.5$ is the primary channel in cardiac myocytes (Rogart *et al.*, 1989), it is also expressed in immature or denervated skeletal muscles (Kallen *et al.*, 1990; Trimmer *et al.*, 1990). The lack of $Na_V1.4$ in CNS tissues is consistent with the absence of neurological deficits in patients with muscle disorders who carry mutations in $Na_V1.4$. $Na_V1.5$ is also expressed in neurons of the CNS (Hartmann *et al.*, 1999; Wu *et al.*, 2002), in DRG neurons especially during embryonic development (Renganathan *et al.*, 2002), in spinal cord astrocytes (Black *et al.*, 1998) and in T-lymphocytes (Fraser *et al.*, 2004). The contribution of $Na_V1.5$ to neuronal electrogenesis is not clear at this time, and neuronal deficits have not been reported in patients with cardiac arrhythmias caused by mutant $Na_V1.5$ channels (but see additional discussion in the section Disease relevance).

$Na_V1.6$ is expressed in CNS neurons and glia (Burgess *et al.*, 1995; Schaller *et al.*, 1995) and in peripheral sensory neurons (Black *et al.*, 1996; Dietrich *et al.*, 1998). $Na_V1.6$ is also detected by a multiplex reverse transcription-polymerase chain reaction (RT-PCR) assay in sympathetic ganglia, for example superior cervical ganglion neurons (Rush *et al.*, 2006a). $Na_V1.6$ channels have been detected in non-excitable cells, for example microglia and macrophages, where a block of sodium channel activity by treatment with TTX inhibits the phagocytic ability of macrophages (Craner *et al.*, 2005). Expression of $Na_V1.6$ in CNS neurons is detectable after birth and adult levels are achieved by P14 (Schaller *et al.*, 2000). In contrast, $Na_V1.6$ expression is detectable in DRG neurons at E17 (Felts *et al.*, 1997) and is robust at birth (Felts *et al.*, 1997; Chung *et al.*, 2003).

$Na_V1.7$ is expressed in PNS tissues including sensory and sympathetic neurons (Black *et al.*, 1996; Felts *et al.*, 1997; Sangameswaran *et al.*, 1997; Toledo-Aral *et al.*, 1997; Rush *et al.*, 2006a). Within DRG neurons, $Na_V1.7$ is expressed in the majority of nociceptive neurons (Djouhri *et al.*, 2003b). $Na_V1.7$ has been identified in smooth myocytes (Holm *et al.*, 2002; Jo *et al.*, 2004; Saleh *et al.*, 2005), in neuroendocrine cells (Klugbauer *et al.*, 1995; Toledo-Aral *et al.*, 1997), in tumour cells (Diss *et al.*, 2005; Fraser *et al.*,

2005), and in erythrocytes (Hoffman *et al.*, 2004). Nerve growth factor (NGF) has been shown to regulate the expression of $Na_V1.7$ in DRG neurons and PC12 cells (Toledo-Aral *et al.*, 1995; 1997; Gould *et al.*, 2000).

$Na_V1.8$ is expressed in peripheral, sensory neurons of DRG and trigeminal ganglia but not in CNS neurons under normal conditions (Akopian *et al.*, 1996; Sangameswaran *et al.*, 1996). Within DRG neurons, $Na_V1.8$ is expressed in nociceptive neurons (Djouhri *et al.*, 2003a) and in a subpopulation of Aβ neurons (Rizzo *et al.*, 1994, 1995; Black *et al.*, 1996; Amaya *et al.*, 2000). $Na_V1.8$ has been detected in small, peptidergic DRG neurons which are characterized by the presence of NGF high affinity receptor, trkA, and small, non-peptidergic DRG neurons which are characterized by the binding of the lectin IB4 (Fjell *et al.*, 1999; Amaya *et al.*, 2000; Sleeper *et al.*, 2000; Fang *et al.*, 2005; Rush *et al.*, 2005a). NGF and glial-derived neurotrophic factor (GDNF) regulate the expression of $Na_V1.8$ both in culture (Black *et al.*, 1997) and *in vivo* (Dib-Hajj *et al.*, 1998a; Fjell *et al.*, 1999; Boucher *et al.*, 2000; Leffler *et al.*, 2002).

$Na_V1.9$ is expressed primarily within sensory neurons of DRG and trigeminal ganglia, but not in CNS neurons and glia or muscle (Dib-Hajj *et al.*, 1998b; Tate *et al.*, 1998). Within DRG neurons, $Na_V1.9$ is expressed predominantly in nociceptive neurons of small diameter and is detected in nociceptive Aβ neurons (Fang *et al.*, 2002). The distribution of $Na_V1.9$ in small DRG neurons is limited to the non-peptidergic, IB4-positive subpopulation (Fjell *et al.*, 1999; Amaya *et al.*, 2000; Rush *et al.*, 2005a). Consistent with the preferential localization of $Na_V1.9$ in IB4-positive neurons, channel expression is regulated by GDNF (Fjell *et al.*, 1999). While low levels of transcripts were detected by RT-PCR in cerebrum and retina, no detectable $Na_V1.9$ is present in cerebellum or spinal cord, or in satellite or Schwann cells within DRG (Dib-Hajj *et al.*, 1998b). A reported wider distribution of SCN12A, which is identical to human $Na_V1.9$, in CNS tissue (Jeong *et al.*, 2000), and the report of a BDNF-gated $Na_V1.9$ current in CNS neurons (Blum *et al.*, 2002) have yet to be independently reproduced; these findings are inconsistent with the lack of persistent TTX-R Na currents in CNS tissues which casts doubt on the validity of the initial observations.

Subcellular localization

The distribution of sodium channels in the somatodendritic and axonal compartments of myelinated fibers is asymmetric and isoform-specific. However, the distribution of channels is more uniform in cell bodies and unmyelinated axons, for example small diameter DRG neurons. The molecular basis for polarized distribution of sodium channels within neuronal compartments is not well understood despite some evidence that selective endocytosis of channels from one compartment might account for the final distribution of that channel (Garrido *et al.*, 2001).

$Na_V1.1$ appears to be predominantly localized to somatodendritic compartments of myelinated neurons (Westenbroek *et al.*, 1989) and is not detectable along optic nerve axons (Craner *et al.*, 2003b). However, $Na_V1.1$ was shown recently to be clustered at the initial segment of axons in the inner plexiform layer of the retina (Van Wart *et al.*, 2005). $Na_V1.1$ is also present in DRG neurons (Black *et al.*, 1996) but its subcellular distribution has not been well defined.

$Na_V1.2$ is present in cell bodies and non-myelinated axons, or the non-myelinated stretches of axons, for example the proximal regions of optic nerve (Boiko *et al.*, 2001). $Na_V1.2$ is targeted to immature nodes of Ranvier during myelination, but is replaced by $Na_V1.6$ as the node matures and compact myelin is organized (Boiko *et al.*, 2001; Kaplan *et al.*, 2001); a small number of mature nodes retain $Na_V1.2$, but always together with $Na_V1.6$. The molecular basis for the targeting of $Na_V1.2$ channels during maturation of nodes of Ranvier remains incompletely understood. Increased number of $Na_V1.2$-positive nodes of Ranvier has been reported in optic nerve from EAE mice compared to control animals which is accompanied by increased levels of $Na_V1.2$ mRNA and protein in the retinal ganglion cells which give rise to these axons (Craner *et al.*, 2003b); another study also reported $Na_V1.2$ at nodes of Ranvier in remyelinating axons of the optic nerve (Dupree *et al.*, 2005). An increase in the number of nodes with $Na_V1.2$ immunolabelling in remyelinating axons of the optic nerve in EAE animals might reflect a recapitulation of the developmental program of $Na_V1.2$ clustering at immature nodes, or it might suggest an acquired ability of axons to retain $Na_V1.2$ or replace $Na_V1.6$ at mature nodes of Ranvier in remyelinating axons (Waxman *et al.*, 2004).

$Na_V1.3$ is predominantly expressed during pre-myelination stages of development, and it is thought to be distributed in somatodendritic and axonal compartments of embryonic neurons. In rodents, the expression of $Na_V1.3$ is significantly attenuated after birth (Beckh *et al.*, 1989; Westenbroek *et al.*, 1989), while in humans, higher levels of $Na_V1.3$ are detected in adult tissues and the channel is detected in the somatodendritic compartment of myelinated neurons (Whitaker *et al.*, 2001). Following a ligation-induced traumatic injury to DRG or sciatic nerve, or to spinal cord, $Na_V1.3$ expression is upregulated, as discussed earlier. Interestingly, the $Na_V1.3$ channel has been shown to accumulate within axon tips at the neuroma which forms at the ligation site (Black *et al.*, 1999a) where it co-localizes with contactin (Shah *et al.*, 2004), a cell adhesion molecule.

$Na_V1.6$ replaces $Na_V1.2$ as the nodes of Ranvier mature, and becomes the predominant channel isoform at mature nodes in CNS (Boiko *et al.*, 2001; Kaplan *et al.*, 2001) and PNS (Schafer *et al.*, 2006) myelinated axons. $Na_V1.6$ labelling is not detectable in the paranodal regions of myelinated axons. $Na_V1.6$, however, is present along the length of unmyelinated fibres arising from small DRG neurons and has been shown to contribute to conduction in these fibres (Black *et al.*, 2002). The polarized distribution of $Na_V1.6$ to nodes of Ranvier and the replacement of $Na_V1.2$ at these sites are dependent upon the presence of compact myelin. Interestingly, dysmyelination, as seen for example in the Shiverer mutant mouse, leads to a retention of $Na_V1.2$ along the axons and an absence of $Na_V1.6$ from these regions (Boiko *et al.*, 2001). Demyelination of axons in patients with multiple sclerosis (MS) or in mice with experimental allergic encephalitis (EAE) causes a general loss of $Na_V1.6$ staining in the lesion, accompanied by the appearance of linear tracks of $Na_V1.2$ or $Na_V1.6$ immunostaining in a subset of axons in the affected area (Waxman *et al.*, 2004). Remyelination of dorsal column axons by endogenous Schwann cells (Black *et al.*, 2006) or olfactory ensheathing cells (Sasaki *et al.*, 2006) lead to the restoration of $Na_V1.6$ immunolabelling at nodes of Ranvier which is not preceded, or accompanied, by the presence of $Na_V1.2$. A similar finding has also been reported in the sciatic nerve (Schafer *et al.*, 2006). These findings might best be explained by the lack of $Na_V1.2$ expression in DRG neurons which give rise to

the majority of axons in the dorsal columns of the spinal cord and to the sensory axons in the sciatic nerve.

$Na_V1.7$ expression is limited to PNS tissue where this channel is abundant in the soma of small DRG and sympathetic neurons (Toledo-Aral *et al.*, 1997; Rush *et al.*, 2006a). $Na_V1.7$ is also detected along unmyelinated fibres in the sciatic nerve (Rush *et al.*, 2005a), and has been shown to accumulate at neurite endings of DRG neurons in culture (Toledo-Aral *et al.*, 1997). Within DRG neurons, $Na_V1.7$ is expressed in DRG neurons of different sensory modalities, with the majority of nociceptive neurons expressing this channel (Djouhri *et al.*, 2003b).

$Na_V1.8$ is abundant in the soma of small DRG neurons (Amaya *et al.*, 2000; Sleeper *et al.*, 2000), and along unmyelinated fibers innervating cornea (Black and Waxman, 2002) and in the sciatic nerve (Rush *et al.*, 2005a). $Na_V1.8$ is present in a subpopulation of cutaneous afferent medium size DRG neurons (Rizzo *et al.*, 1994) and about 20 per cent of nodes of Ranvier in myelinated axons that innervate tooth pulp show $Na_V1.8$ immunostaining (Henry *et al.*, 2005). $Na_V1.8$ has also been identified at small fibre nerve endings in the cornea (Black and Waxman, 2002) suggesting that it might also be present at nerve endings in the skin.

Overwhelming opinion favours the idea that $Na_V1.8$ is a PNS-specific transcript, though there is an isolated report claiming $Na_V1.8$ to be present at nodes of Ranvier in spinal cord tracts (Arroyo *et al.*, 2002). This latter finding is inconsistent with the lack of $Na_V1.8$ message in normal CNS tissue. It is theoretically possible for a PNS-specific channel to be present at nodes of Ranvier in spinal cord tracts if the signal is localized to fibres in the dorsal columns which are continuous central projections of proprioceptive DRG neurons. However, muscle afferent neurons which are labeled by injecting the tracer fluorogold into peripheral muscle do not produce the slow-inactivating TTX-R current which is produced by $Na_V1.8$ (Rizzo *et al.*, 1994). Additionally, expression of Cre recombinase from the $Na_V1.8$ locus in mouse did not reveal an activity in spinal cord, confirming the absence of normal expression of this channel in CNS tissue (Nassar *et al.*, 2004). Thus, the possibility that $Na_V1.8$ is present at nodes of Ranvier in spinal cord tissue should be viewed with caution and will require independent confirmation.

$Na_V1.9$ is abundant in the soma of small DRG neurons (Nassar *et al.*, 2004), along unmyelinated fibers innervating the cornea (Black and Waxman, 2002) and in the sciatic nerve (Rush *et al.*, 2005a). $Na_V1.9$ is present in a small number of Aβ nociceptive neurons (Fang *et al.*, 2002), which is consistent with the rare detection of this channel at nodes of Ranvier (Fjell *et al.*, 2000). $Na_V1.9$, has been identified at small fibre nerve endings in the cornea (Black and Waxman, 2002) and in skin (Dib-Hajj *et al.*, 2002).

2.3.5 Pharmacology

Naturally occurring sodium channel toxins

The pre-eminence of voltage-gated sodium channels as signalling proteins that are utilized throughout the animal kingdom render them adaptive targets during evolution, either from a protective standpoint or as a predator strategy. The result is a rich collection of neurotoxins that fall into five categories according to their binding sites on the α-subunit of the channel. A comprehensive review of the toxins and their actions is beyond the scope of this text, and interested readers are referred to recent reviews for further details

(Cestele and Catterall, 2000; Blumenthal and Seibert, 2003). The various classes of toxin are listed in Table 2.3.1. These molecules represent not only interesting entities in their own right but have greatly facilitated studies of sodium channel structure and function and have even spurred an independent classification system that has been valuable in defining aspects of sensory nerve physiology.

Site-1

The site-1 toxins comprise the non-peptidic heterocyclic guanidines, TTX and STX, and the peptides μ-conotoxins (μ-CTX). Each of these toxins docks into the outer vestibule-pore-selectivity filter region that is formed from the SS1–SS2 motifs from each of the four domains. Site-directed mutagenesis studies have identified several amino acid residues that are involved in binding TTX, STX and μ-CTX and though there are certainly common contact points, there are also differences suggesting that these molecules occupy an overlapping pharmacophore (Stephan et al., 1994; Dudley et al., 2000). The principal contact points for TTX and STX occupy a relatively large footprint on the channel comprising interactions with residues in two stacked rings that form part of the outer vestibule region (Noda et al., 1989; Terlau et al., 1991) and a critically important interaction with a deeper aromatic residue that forms part of the channel pore (Backx et al., 1992; Satin et al., 1992). The details of each of the interactions have been nicely summarized and involve a combination of multiple ion-pair, salt-bridge and hydrogen-bond interactions between the carboxyl groups of acidic (Asp384 and Glu387; $Na_V1.4$ numbering) residues in DI–S2 and

DII–S2 and the guanidinium moiety and hydroxyl groups of the toxins (Lipkind and Fozzard, 1994). There are additional interactions with Asp1426 and Asp1717 in the SS2 motif of DIII and DIV and a crucial influence of a residue immediately N-terminal to Asp384. In the majority of mammalian sodium channel family members, this residue is Phe but in the cardiac channel ($Na_V1.5$) and the sensory neuron-specific variants ($Na_V1.8$ and 1.9) the Phe is substituted by the non-aromatic amino acids C ($Na_V1.5$) or Ser ($Na_V1.8$ and 1.9). These single-point substitutions result in a major change in potency of TTX block. In the case of $Na_V1.5$, TTX affinity is around 100-fold weaker ($IC_{50} \sim 1\mu M$ vs. ~10nM in the Phe variant channels). The peripheral sensory nerve channels, $Na_V1.8$ and 1.9 are essentially TTX-resistant ($IC_{50}s > 30\mu M$) and, indeed, this differential sensitivity to TTX has been widely used as the basis of a classification into TTX-sensitive and TTX-resistant categories.

The μ-contoxins are peptide snail toxins, they bear no similarity to TTX or STX but they too are very effective, potent sodium channel blockers (Cruz et al., 1985; Ohizumi et al., 1986). The most studied, μ-CTX-GIIIA, is a 22 amino acid with 6 C residues. The C residues are engaged in a specific pattern of three disulfide bonds which provide a conserved scaffold common to this class of toxins which supports loops of variable residues that might account for selective interaction with different targets. Arg-13 is a critical determinant of μ-CTX affinity (~ 11nM for $Na_V1.4$, for example, Chang et al., 1998), the toxin is thought to orient mainly within the channel outer vestibule such that the collar amino acids interact with vestibule residues from each of the four domains with the pendant Arg13 dipping downward toward the selectivity filter (Dudley et al., 2000). The guanidinium of Arg13 is likely to interact electrostatically with the carboxyl groups of Glu758 and Glu403 ($Na_V1.4$ numbering), these residues are each +3 C-terminal to the corresponding selectivity filter residues of DII and DI, respectively, suggesting that even the pendant Arg of the toxin occupies an eccentric location fairly high up in the pore vestibule (Chang et al., 1998). Thus, μ-CTXs appear to share some of the binding interactions with TTX and STX but they occupy an even broader footprint on the channel.

These toxins have not been reported to block the TTX-R $Na_V1.8$ and $Na_V1.9$ channels though they do appear to be able to discriminate between certain TTX-S channel isoforms (Shon et al., 1998; Safo et al., 2000). Interestingly, GIIIA/B μ-conotoxins block rat $Na_V1.4$ with high affinity but human $Na_V1.4$, which shares identical pore residues, is resistant to block. This difference in sensitivity of the two orthologous channels is determined by a natural variant extra-pore residue (S729L) in DII/S5–S6 linker. Substitution of yet another residue, N732, with lysine (N732K), the corresponding residue in $rNa_V1.1a$ and $rNa_V1.7$, reduced the GIIIB sensitivity of rat wild-type $Na_V1.4$ by approximately 20-fold and that of GIIIA by approximately fourfold, demonstrating that GIIIA and GIIIB have distinct interactions with the D2/S5–S6 linker (Cummins et al., 2002). Thus, the isoform and ortholog specificity exhibited by these toxins appears to be mediated through direct interactions with extra-pore residues that either influence the accessibility of the toxin to the binding site within the pore and/or the stability of the toxin-channel complex.

In summary, site-1 toxins presumably act by a simple pore-occlusion mechanism, and they are only weakly voltage-dependent as would be consistent with their primary contact points being outside of the transmembrane potential field.

Table 2.3.1 Sodium channel toxins

Class	Location	Mechanistic basis	Examples
Site 1	Vestibule and pore	Pore occlusion and prevention of ion flux	Tetrodotoxin (TTX); saxitoxin (STX); μ-conotoxin
Site 2	Inner vestibule, cytoplasmic to the selectivity filter	Shift activation in hyperpolarizing direction and delay inactivation	Veratridine; batrachotoxin (BTX); grayanotoxin
Site 3	Multiple contact points involving the extracellular linkers	Influence movement of DIV/S4—delay inactivation	α-scorpion toxins; spider toxins; sea-anemone toxins
Site 4	Bind to a crevice formed by DII/S1–S2 and DII/S3–S4 extracellular linkers, positioned to form additional interactions with externalized DIIS4	'Voltage-sensor trapping', shift activation in hyperpolarized direction	β-scorpion toxins
Site 5	Overlapping site with that of the site-2 toxins	Shift activation in hyperpolarizing direction and delay inactivation	Brevetoxin; ciguatoxin
Site 6	Possible DIIIS3–S4 extracellular linkers, may overlap with site-3 toxins	Yet to be elucidated, in some instances binding is without effect	δ-conotoxins

Site-2 and site-5

These sites are allosterically coupled and will therefore be considered collectively. Site-2 toxins are lipid-soluble molecules that bind within the transmembrane regions of the channel protein and modify the gating process. Examples include plant toxins (e.g. veratridine, aconitine, and grayanotoxin) and a toxin, batrachotoxin (BTX), secreted from the skin of the frog, *Phyllabates aurotaenia*. Each of these molecules are capable of exerting profound effects on channel gating processes, they require channels to be in the open state in order to gain access to their binding site. BTX is extremely potent and repetitive membrane depolarization in the presence of the toxin facilitates binding culminating in a substantial (>30mV) left-shift in activation and a dramatic slowing of both activation and inactivation kinetics (Wang and Wang, 1998). These effects are essentially irreversible and represent a profoundly neurotoxic combination. The plant alkaloids, aconitine and veratridine, though substantially lower in their channel affinity than BTX have similar effects but, additionally, also reduce single channel conductance to around 15 per cent of the former unmodified channel state (Hille, 2001).

Site-directed mutagenesis studies have revealed two regions to be important in defining the pharmacophore of the site-2 toxins. The first implicated region is located in DI/S6 where substitutions at any of three positions, I433, N434 or L437 (Na$_V$1.4 numbering) results in a dramatic loss of BTX potency (Wang and Wang 1998). The second region is located in DIV/S6 where substitutions at either F1579 or N1584 had an equally profound effect on BTX potency (Wang and Wang, 1999). These data suggest that the BTX pharmacophore is formed at the interface between DI/S6 and DIV/S6. Interestingly, the DI–DIV S6 and S5 regions of the protein are thought to contribute to the formation of an inner pore vestibule, i.e. cytoplasmic to the selectivity filter. In addition to being involved in the binding of site-2 neurotoxins, this region is also implicated in the binding of another class of toxins referred to as site-5 and also forms part of the pharmacophore for the local anesthetic-like channel blocking molecules (see later).

The site-5 toxins are the large molecular weight, polycyclic ether, lipid-soluble dinoflagellate brevetoxins (PbTx) and ciguatoxins (CgTx). These potent toxins share a similar binding site, the occupancy of which is able to allosterically potentiate the binding of the site-2 ligand [^{3}H]-BTX (Lombet *et al.*, 1987; Trainer *et al.*, 1991). Site-5 toxins have somewhat similar effects on channel gating to the site-2 toxins, they produce negative shifts in activation and prevent inactivation (Lombet *et al.*, 1987; Trainer *et al.*, 1991; Cestele and Catterall, 2000). Single channel analysis reveals the true range of effects at the molecular level; PbTx-3, for example, elicits a 'flickery' appearance to single channel recordings that reflects the increase in sodium channel open probability, longer open time and shorter shut times mediated by the toxin (Jeglitsch *et al.*, 1998).

Site-3

The site 3 toxins are polypeptides and include the sea anemone toxins (e.g. *Anthopleura* toxins), certain spider toxins (e.g. δ-atracotoxins, μ-agatoxins) and the numerous variants of α-scorpion toxins (reviewed in Blumenthal and Seibert, 2003). These toxins show sufficient commonality in their binding site(s) on the channel to permit a collective assignment but, as might be expected, these diverse toxins display individual, specific requirements and so, once again, the concept is one of an overlapping but non-identical binding domain.

Channel block by site-3 toxins is voltage-sensitive, to a varying extent. Binding affinity (Catterall, 1977; Catterall and Beress, 1978; Cestele *et al.*, 1998) or channel block (Strichartz and Wang, 1986) is reduced by membrane depolarization, and the effect is more prominent for α-scorpion toxins than for sea-anemone toxins (Catterall and Beress, 1978; Wang and Strichartz, 1985). Site-directed mutagenesis has been used to define the molecular determinants of high-affinity site-3 toxin binding to the rat brain channel, Na$_V$1.2a. These studies have indicated that both α-scorpion and sea-anemone toxins bind to an external receptor comprising multiple extracellular loop regions. An electrostatic interaction with acidic amino acids in the DIV/S3–S4 loop, in particular, has been proposed, implicating either E (1613 in Na$_V$1.2a) or D residues (e.g. D1612 in Na$_V$1.5, D1428 in Na$_V$1.4) (Rogers *et al.*, 1996) along with, as-yet, ill-defined contacts with DI/S5–S6 and DIV/S5–S6 loop regions (Tejedor and Catterall, 1988). Certain α-scorpion toxin variants can discriminate between channel isoforms and this has been attributed to variations in toxin affinity that are dictated by the E/D variations in DIV/S3–S4 (Alami *et al.*, 2003). The fact that this site is very close to the voltage-sensors provided a means of rationalizing location with a characteristic action of these toxins, the slowing of channel inactivation. Binding of a site-3 toxin is thought to interfere with the voltage-dependent outward translocation of the DIV voltage-sensor and thereby inhibit fast inactivation. Gating charge experiments using the cardiac channel, Na$_V$1.5, have provided direct evidence of such an interaction (Sheets *et al.*, 1999). However, conformational changes associated with channel gating also reduce the binding affinity of the toxin such that bound toxin rapidly dissociates (Rogers *et al.*, 1996; Chen *et al.*, 2000): more recent data have suggested that slow-inactivated channels may interact differently with site-3 toxins (Gilles *et al.*, 2001).

Site-4

The site-4 designation is reserved for the β-scorpion toxins (e.g. toxins from *Centruroides* and *Tityus* scorpions), these are also gating modifiers (Cahalan, 1975) but they act at a site that is clearly distinct from that of the site-3 group (Jover *et al.*, 1980). The β-scorpion toxins enhance channel activation by producing a hyperpolarizing shift in the voltage-dependent activation (Meves *et al.*, 1982; Wang and Strichartz, 1983; Cestele *et al.*, 1998). Cestele and colleagues have undertaken extensive studies of β-scorpion toxins' interaction with sodium channels that have culminated in a detailed view of the binding site and the functional effects of binding, the key elements of which can be summarized as follows. A crude pharmacophore for the toxin Css IV on Na$_V$1.2a, initially defined using a series of chimera constructs, involves DII/S1–S2 and DII/S3–S4 external loops (Cestele *et al.*, 1998). Subsequent site-directed mutagenesis studies revealed additional interactions involving residues in the DII/S4 voltage sensor (Cestele *et al.*, 2001) and mapped, in detail, the specific residues in the DII external loops (Cestele *et al.*, 2006). Rationalizing all of the mutagenesis data with tertiary structures of a Css IV-related toxin and of the DII/S1–S4 region, each derived from X-ray crystallographic data, has produced a model that suggests the toxin binds to a crevice in the resting-conformation channel protein, formed by the DII–S1/S2, DII–S3/S4 external helical hairpin linkers (Cestele *et al.*, 2006). This strategic position presumably enables the toxin to influence the gating movements of the S4 voltage sensor such that activation and outward movement of the DII–S4 helix exposes additional

S4 binding residues to the already-bound toxin, effectively resulting in this individual sensor being trapped in the activated conformation. Such a voltage-sensor 'trapping' mechanism would account for the consistent findings, mentioned above, showing that the β-scorpion toxins shift the voltage-dependence of activation to the left (i.e. hyperpolarized), such that toxin-bound channels are more easily activated because one of their voltage sensors is already locked in an outward, activated state (Cestele et al., 2006).

Site-6

A distinct binding site for the δ-conotoxins was first proposed over a decade ago (Fainzilber et al., 1994), yet it has received little attention by comparison to other toxin sites. The first examples of this class of toxins to be evaluated, δ-conotoxin-TxVIA, TxIA and TxIB (from Conus textile), presented curious properties. The toxin bound with high affinity to rat brain sodium channels, yet was without any functional effect (Hasson et al., 1993; Shichor et al., 1996). The lack of an effect on rat sodium channel gating was in contrast to a clear action on voltage-dependent steady-state inactivation of sodium channels in Aplysia neurons (Hasson et al., 1993), prompting the authors to coin the term 'silent' binding for the interaction of toxins to the rodent channels (Shichor et al., 1996). More recent studies using a related toxin, δ-conotoxin-SVIE (from Conus stiatus), have revealed an effect on rat $Na_V1.4$ inactivation that is very similar to that of the α-scorpion toxins and appears to involve an interaction with a hydrophobic triad (Tyr-Phe-Val) in the DIII–S3/S4 external linker (Leipold et al., 2005) that partially overlaps with the predicted pharmacophore for the site-3 toxins.

Small molecule pharmacology

Voltage-gated sodium channels are targets for several clinically used drugs that typically fall into three therapeutic classes: the local anaesthetics (e.g. lidocaine, bupivacaine, ropivacaine); anticonvulsants (e.g. phenytoin, carbamazepine, lamotrigine); and class-I anti-arrhythmics (e.g. mexilitine, amiodarone, flecainide). Collectively, these drugs show certain similarities in their channel blocking actions, they generally show a preference for particular channel states and this often confers a use- or frequency-dependency to their profile. At the molecular level, use-dependent block (USB), a term initially coined by Courtney (1975), manifests as a blocking efficiency that is enhanced as a function of the frequency at which the channels are forced to cycle between their closed, open and inactivated conformations, such that increased stimulation frequencies result in greater block. At the network or tissue level, at least in the case of excitable tissues, action potential frequency is a measure of sodium channel activity and these drugs are able to curtail high-frequency firing at concentrations that do little to low-level background activity. This, of course, happens to be a very useful property from a therapeutic perspective. Indeed, were it not for frequency-dependent block, many of these drugs would be regarded more as toxins than useful therapeutics. Not surprisingly, therefore, the characteristics of frequency-dependent block have been described for many sodium channel blocking drugs (Courtney 1975; Courtney et al., 1978; Willow et al., 1985; Balser et al., 1996b; Kuo et al., 1997; Weiser et al., 1999; De Luca et al., 2000; Nuss et al., 2000; Bonifacio et al., 2001; Khodorova et al., 2001).

Theories addressing the mechanistic basis for use-dependent block have their roots in early studies of local anaesthetic (LA) compounds. The first concept, conceived independently by Hille (1977) and Hondeghem and Katzung (1977) and termed the 'modulated receptor hypothesis', suggested that LAs bind to open or inactivated channel conformations with appreciably higher affinity than they do to the closed, so-called 'resting' state. Hille also incorporated elements into the modulated receptor concept that permitted different access modes to the binding site for hydrophilic vs hydrophobic drugs (Hille, 1977). An alternative scheme, termed the 'guarded receptor hypothesis', was proposed several years afterwards, by Starmer and colleagues (Starmer et al., 1984), in which use-dependent block could equally well be explained by assuming a fixed affinity of LA drugs for various channel states but in which access to the binding site was governed by the position of the activation and inactivation gates. Each of these schemes has their merits and it is difficult to completely discount either one, based on the experimental evidence to date. Final judgment may only be forthcoming when we have access to crystal structures of sodium channels in their various gating states and preferably with and without bound drug molecules.

Regardless of the theoretical concepts described above, it was thought, until very recently, that LA-bound channels are effectively stabilized in the inactivated conformation and that the return to the resting state will be governed by the dissociation off-rate for the drug in question. The notion that LA-mediated block of current flux is purely secondary to inactivation-state stabilization has recently been challenged and this is discussed in more detail below. The degree of frequency-dependence or USB is influenced by a number of factors, with contributions imparted by drug binding kinetics and access/egress routes, by channel gating kinetics and membrane voltage. High-affinity blockers dissociate more slowly from the channel, even after a return to negative membrane potentials, and at high-frequency cycles there will be insufficient time to allow unbinding before the next spike-initiating depolarization. In this manner, repetitive cycles lead to a progressive accumulation of drug-bound inactivated channels. This effectively results in an exponential decline in macroscopically available population of channels and ultimately their number is insufficient to support repetitive action potential generation and, hence, propagation fails. The amount of use-dependent block, rate of block and recovery from block are all governed by kinetic parameters imparted by both drug and channel and it is easy to appreciate how drug-blocking profiles can range markedly. The genesis of these concepts, almost 30 years ago, had a significant impact on clinical therapeutics at the time. They firmly established voltage-gated sodium channels as being the clinically relevant target for lidocaine at a time when clinically used doses of the drug were considered insufficient to achieve the plasma levels required to account for any significant channel block.

An incredible research effort, spanning many years and involving many prominent research labs, has been directed toward defining the pharmacophore for the LAs and related drugs. Many of the sodium channel blocking drugs have a common structural motif that comprises, in various guises, a hydrophobic aromatic moiety linked by a variable length alkyl or aryl chain to a hydrophilic tertiary amine, that may be either aliphatic or aromatic in nature. This typical LA structural scaffold is visible in drugs from other therapeutic classes, such as the anticonvulsants and anti-arrhythmics mentioned above but also in certain phenothiazine-based neuroleptics such as chlorpromazine and fluphenazine. All of these drugs are very effective sodium channel blockers (Bolotina *et al.*, 1992; Zhou *et al.*, 2006). It has emerged that representative molecules from each of these drug classes bind within a common pharmacophore in the channel that is defined by structural elements that make up the inner vestibule of the ion-conducting pathway, i.e. cytoplasmic to the selectivity filter. However, the precise details of the binding interactions within this region have evolved over time and have been refined by numerous painstaking site directed mutagenesis studies that have either employed the A-scanning or C-substitution approaches. The complexity and wealth of information provided by this work is well beyond the scope of this chapter and interested readers are referred to a recent excellent and detailed review (Fozzard *et al.*, 2005).

The current view of the LA pharmacophore is that it is the water-filled inner vestibule delineated by the S6 helices from each of the four domains, with the upper margin defined by the DEKA residues comprising the selectivity filter and a lower boundary defined by the activation gate. This proposed site is close to that claimed for the site-2 toxins (Linford *et al.*, 1998), indeed these two sites appear to be allosterically coupled, such that LA-type drugs can influence the binding of $[^3H]$-batrachotoxin (Linford *et al.*, 1998; Riddall *et al.*, 2006). Current conceptions of the LA pharmacophore owe much to the wealth of painstaking site-directed mutagenesis data (Catterall, 1977; Nau and Wang, 2004) and have been enhanced by recent attempts to reconcile this data with modelling scaffolds derived from available crystal structures of the MthK potassium channel (Catterall 2000; Nau and Wang, 2004; Lipkind and Fozzard, 2005). Though it is important to recognize that there may be certain interpretational limitations associated with these data, a consensus appears to be emerging for certain assertions. For example, it is generally agreed that the orientation of the LA molecule within the pharmacophore is such that the polar alkyl amine 'head' region docks in a narrow cavity and faces the selectivity filter with the aromatic 'tail' oriented towards the pore's cytoplasmic opening. A direct physical interaction between drug and selectivity filter seems unlikely (Fozzard *et al.*, 2005) but there is experimental (Sunami *et al.*, 1997) and modelling (Scheib *et al.*, 2006) support for an electrostatic interaction involving the D and E residues of the DEKA ring. As mentioned previously, mutagenesis studies have indicated several residues within three of the four S6 helices (IS6, IIS6 and IVS6) that contribute prominently to LA binding, principally in DI–S6 and DIV–S6, that participate in LA binding. Helix DIII/S6 does not appear to be involved in binding and it is apparent in all of the models to date that this creates an asymmetric pharmacophore. Of all of the implicated residues the F of DIV–S6 (F1764 of $Na_V1.2$; F1579 of $Na_V1.4$) has the largest influence on binding and mutation of this single amino acid residue results in a marked loss in open/inactivated state block by etidocaine or lidocaine (Ragsdale *et al.*, 1994; Ragsdale *et al.*, 1996). Other critical residues ($Na_V1.4$ numbering) are Y1586 (IVS6), L1280 (IIS6) and N418 (IS6). Of all of these residues, only F1579 is sufficiently close to the proposed glycine hinge as to be minimally affected by gating (see gating section) and hence is likely to contribute the most to LA affinity, the remaining residues are likely to move greater distances during gating and these changes in tertiary structure are likely to create more favourable conditions for binding (Lipkind and Fozzard, 2005). Recent studies have suggested that, contrary to the traditional view that LAs act by stabilizing the inactivation gate in its closed conformation, the blocking action may be due primarily to open-channel block in which the bound LA stabilizes part of the voltage-sensing

machinery (Vedantham and Cannon, 1999; Sheets and Hanck 2003). Inactivation gate closure still appears to be promoted by lidocaine in this new model but only as a secondary, allosterically coupled event and, importantly, a repolarization of lidocaine-bound channels provoked a reopening of the inactivation gate yet the channels remained blocked for a significantly longer period, implying fundamentally independent processes (Vedantham and Cannon, 1999). Simple steric hindrance appears unlikely to account for an efficient block of ion flux because the majority of LA drugs are too small to completely occlude the inner pore (Lipkind and Fozzard, 2005), especially in view of the asymmetric nature of the binding contacts. Rationalizing the above data with that derived from modelling of lidocaine docked into a putative open-channel, Lipkind and Fozzard have suggested that ion flux might be occluded by virtue of a charge-screening effect imparted by the LA, the result of which is to counteract the negative field that permits Na^+ ions to traverse the pore (Lipkind and Fozzard, 2005).

Many of the studies described above have used recombinant channel variants, mainly $Na_V1.2$ and $Na_V1.4$ and, to a lesser extent, the cardiac channel $Na_V1.5$. As discussed previously, the cardiac channel is of particular interest due to the fact that it differs in having a non-aromatic C residue immediately C-terminal to selectivity filter D residue (i.e. the D^{+1} position) in the DI pore region, in most other variants this residue is either Y or F. Two other exceptions to this general rule are $Na_V1.8$ and $Na_V1.9$, each of which have a S at the corresponding D^{+1} position. As mentioned previously, this is a critical determinant of sensitivity to TTX but it also appears to confer an external access route (and possibly egress also) for uncharged forms of LAs that is denied to the Y/F channel variants (Qu *et al.*, 1995; Sunami *et al.*, 2000; Lee *et al.*, 2001; Lardin and Lee, 2006). A recent study has confirmed that C substitution in $Na_V1.4$ ($Na_V1.4$-Y401C) permits access to the membrane-impermeable lidocaine derivative, QX-314 (Leffler *et al.*, 2005). However, the same study showed that while $Na_V1.4$-Y401C is sensitive to the external application of QX-314, $Na_V1.4$-Y401S was not, which suggests that the external access of QX-314 through the pore is not simply facilitated by the smaller C residue, compared to Y, since presence of the correspondingly sized residue S at this position within the pore is not sufficient to allow access of the drug. Consistent with this finding, $Na_V1.8$ and 1.9 are not blocked by application of 1 mM QX-314 to DRG neurons (Leffler *et al.*, 2005). Access of LA drugs through the pore, therefore, may depend on additional residues within the vestibule.

2.3.6 Disease relevance

Sodium channelopathies have been implicated in both acquired and inherited neuropathies and in diseases of skeletal and cardiac muscles. Dysregulation of sodium channel expression has been linked to neuropathic pain following trauma injuries, inflammation, metabolic disorders, for example diabetes, or multiple sclerosis, and blocking of sodium channels can ameliorate pain (Waxman 2001; Lai *et al.*, 2004; Wood *et al.*, 2004; Waxman 2006). Also, sodium channel mutations have been shown to underlie excitability disorders of the CNS (epilepsy) (George 2005; Meisler and Kearney, 2005), PNS (inherited erythromelalgia) (Waxman and Dib-Hajj, 2005), cardiac muscle (Viswanathan and Balser, 2004; Moss and Kass, 2005) and skeletal muscle (George, 2005; Cannon, 2006; Venance *et al.*, 2006). These mutations are invariably dominant and in cases of nonsense mutations and deletions demonstrate haploin-sufficiency of $Na_V1.1$, i.e. reduction of channels by 50 per cent is enough to precipitate the mutant phenotype. A functional haploin-sufficiency of $Na_V1.4$ (Cannon, 2006) has been observed in a case of myasthenic syndrome which is caused by mutations that significantly hyperpolarizes steady-state inactivation of the channel (Tsujino et al., 2003). Our current knowledge about the structure of sodium channels and the possible isoform-specific allosteric effect of conserved residues, it is difficult to predict if a particular mutation will result in a hyperexcitability or hypoexcitability phenotype, and as will be discussed later for $Na_V1.7$, the critical role of other sodium channels that may be co-expressed with the mutant channel. The elucidation of the molecular pathophysiology of acquired and inherited sodium channelopathies, however, may identify novel targets for therapeutic intervention.

$Na_V1.1$

Multiple different forms of epilepsy have been associated with mutations in sodium channels $Na_V1.1$ (George 2005; Meisler and Kearney, 2005). Recently, a mutation in $Na_V1.1$ was identified in three families with familial hemiplegic migraine (Dichgans et al., 2005). Mutations in $Na_V1.1$ include both missense substitutions and nonsense mutations which cause truncations of the channel protein. More recently, deletions of exon 21 or of the contiguous exons 21–26 of SCN1A (Mulley et al., 2006), or deletion of a whole copy SCN1A (Suls et al., 2006) have been reported in cases of severe myoclonic epilepsy of infancy which lack missense or nonsense substitutions. Several mutations are predicted to enhance excitability of neurons because they lead to changes in biophysical properties of $Na_V1.1$ in heterologous expression systems which are predicted to lead to neuronal hyperexcitability, for example, reduction in frequency-dependent inhibition of mutant channels (Spampanato et al., 2001; Wallace et al., 2001; Cossette et al., 2003), enhanced recovery from inactivation (Spampanato et al., 2001; Dichgans et al., 2005), impaired fast inactivation and increased persistent current (Claes et al., 2003; Rhodes et al., 2004). One mutation in the C-terminus of $Na_V1.1$ has recently been shown to impair the interaction of $Na_V1.1$ with β1 subunit (Spampanato et al., 2004). Other missense substitutions of $Na_V1.1$ from epilepsy patients cause a reduction in current density and significant depolarization in voltage-dependent activation (Wallace et al., 2001; Lossin et al., 2003; Mantegazza et al., 2005; Barela et al., 2006). Nonsense mutations (Wallace et al., 2001; Ohmori et al., 2002; Fujiwara et al., 2003; Lossin et al., 2003; Sugawara et al., 2003; Fukuma et al., 2004; Rhodes et al., 2004) and deletions (Mulley et al., 2006; Suls et al., 2006) produce non-functional channels.

A genotype–phenotype relationship has emerged from these studies showing that mutations in $Na_V1.1$ which produce non-functional channels, or those that cause truncations of channel protein, underlie severe myoclonic epilepsy of infancy (SMEI), a very aggressive form of epilepsy. These genotypes suggest haploinsufficiency, i.e. 50 per cent of functional channels are simply not enough to maintain normal function (Meisler and Kearney, 2005). The molecular pathophysiology underlying the hyperexcitability of CNS neurons, in those cases of loss of functional $Na_V1.1$ channels, is poorly understood. A better understanding of the role of $Na_V1.1$ in epilepsy awaits functional analysis of the majority of missense mutations, many of which have not been investigated for their effects on the biophysical properties of the channel.

$Na_V1.2$

The SCN2A gene that encodes $Na_V1.2$ is essential, a homozygous null mouse was perinatally lethal, possibly due to neuronal deficiency in brainstem leading to sever hypoxia (Planells-Cases et al., 2000). $Na_V1.2$ has also been shown to carry mutations in patients with epilepsy (Sugawara et al., 2001; Heron et al., 2002; Berkovic et al., 2004; Kamiya et al., 2004; Striano et al., 2006). Epilepsy-causing mutations in $Na_V1.2$, however, are not as numerous as those in $Na_V1.1$, and with one exception (Kamiya et al., 2004), result in milder phenotypes. The nonsense mutation in a patient with SMEI results in a truncation of the channel in the N-terminus which exerts a dominant negative function (Kamiya et al., 2004). However, none of the other sodium channel mutations that are associated with SMEI are located in $Na_V1.2$, suggesting that this channel is not a candidate for haploinsufficiency (Meisler and Kearney, 2005). This conclusion is supported by the finding that a heterozygous mouse lacking one allele of $Na_V1.2$ develops normally, with no evidence of epileptic symptoms or neuronal hyperexcitability (Planells-Cases et al., 2000). Interestingly, though, a mild epilepsy mutation in $Na_V1.2$ may lead to severe symptoms if combined with a mutation in the KCNQ2 potassium channel (Kearney et al., 2006). Thus, the dependence of the phenotype on cell background highlights the role of physiological links between sodium channels and other targets and suggests a plausible explanation for regional manifestation of epileptic forms by the same mutation and for the variability of symptoms among different individuals carrying the same mutation (Noebels, 2003).

A mutation in $Na_V1.2$ has also been linked autism (Weiss et al., 2003). This mutation in the C-terminus of $Na_V1.2$ impairs the binding of a calcium-binding protein calmodulin (CaM) to the channel (Weiss et al., 2003). Details on the modulation of $Na_V1.2$ by CaM are lacking but CaM been shown to modulate biophysical properties of channels and/or channel density, depending on the isoform of the sodium channel that is being tested (Mori et al., 2000; Deschenes et al., 2002; Tan et al., 2002; Herzog et al., 2003b; Kim et al., 2004; Young and Caldwell, 2005; Choi et al., 2006a). The impairment of binding of CaM to $Na_V1.2$ in the above case of autism and of β1 binding to $Na_V1.1$ in a case of epilepsy (Spampanato et al., 2004) suggests caution when attributing any apparent lack of an effect of amino acid mutations in sodium channels expressed in heterologous systems as benign polymorphisms, and highlights the need to evaluate the properties of mutant channels in native neurons.

$Na_V1.3$

$Na_V1.3$ is significantly upregulated following injury to DRG neurons or their axons, or to spinal cord (Waxman et al., 1994; Dib-Hajj et al., 1996; Black et al., 1999a; Dib-Hajj et al., 1999; Hains et al., 2003, 2004, 2005). Unlike $Na_V1.3$, $Na_V1.2$ is expressed at low levels in embryonic DRG neurons, diminishes further after birth (Waxman et al., 1994; Chung et al., 2003) and is not upregulated in DRG neurons after axotomy (Waxman et al., 1994). The upregulation of $Na_V1.3$ following traumatic injury may be either a re-capitulation of an embryonic expression programme or a pathologic response to the injury: additional studies will be required to clarify this issue.

Injury-induced upregulation of $Na_V1.3$ in DRG neurons is influenced, at least in part, by the loss of trophic factors from the skin, and can be attenuated by administration of NGF, or GDNF, to

DRG neurons in culture (Black *et al.*, 1997) or *in vivo* (Boucher *et al.*, 2000; Leffler *et al.*, 2002). In the PNS, $Na_V1.3$ immunostaining has been shown to increase in axons and accumulates at axon tips within the neuroma which forms at the site of a ligation injury (Black *et al.*, 1999a). These same sites have previously been shown to act as ectopic focal generators for spontaneous firing (Matzner and Devor, 1994). Thus, it is likely that the emergence of this channel contributes to the neuronal excitability in primary nociceptive neurons that is associated with neuropathic pain.

$Na_V1.3$ aberrant expression in second and third order nociceptive neurons in the CNS suggests the involvement of this channel in hyperexcitability of these neurons, enlargement of peripheral receptive fields and central pain (Waxman, 2006). This has been reported following peripheral (chronic-constriction injury model) or central nerve injury (spinal cord contusion) and appears to be associated with hyperexcitability of these CNS neurons and with neuropathic pain behaviours. Indeed, these effects have been shown to be attenuated by treatment with an antisense oligonucleotide targeting $Na_V1.3$ (Hains *et al.*, 2003, 2004) and this suggested that the role of $Na_V1.3$ in neuropathic pain extends from primary afferents to CNS neurons in the ascending pain pathway.

However, the role of $Na_V1.3$ in neuropathic pain has been challenged recently by the publication of two recent studies (Lindia *et al.*, 2005; Nassar *et al.*, 2006). First, and surprisingly, mice which lack $Na_V1.3$ globally were normal and fertile, and demonstrated normal neuropathic pain behaviour after nerve injury (Nassar *et al.*, 2006). A potential explanation for these findings is that the role of $Na_V1.3$, in rendering primary and higher order neurons hyperexcitable, is redundant and that there may be compensatory changes such that its function is be replaced by another channel transcript. Other findings argue against such a simple explanation. Lindia *et al.* (2005) have demonstrated an upregulation of $Na_V1.3$ within DRG neurons in rats subjected to a spared nerve injury, and reported that while intrathecal administration of specific antisense (AS) oligonucleotides reduced the levels of $Na_V1.3$ in DRG by 50 per cent, it did not ameliorate the mechanical or cold allodynia typical of this animal model (Lindia *et al.*, 2005). The discordance of these published studies remains unexplained but several factors deserve comment. Studies by Hains and colleagues have reported reduction of $Na_V1.3$ in the second and third order neurons to near uninjured levels, but in their studies the AS reagent was unable to penetrate DRG neurons (Hains *et al.*, 2004, 2005). In contrast, Lindia and colleagues demonstrated an AS-induced downregulation of $Na_V1.3$ within DRG neurons by 50 per cent, but did not look for any effects on second-order neurons (Lindia *et al.*, 2005). The differential distribution of the AS reagents into spinal cord and DRG neurons in the two studies might result from different administration methods and doses. Alternatively, these studies might indicate that 50 per cent reduction of $Na_V1.3$ in inured DRG neurons is not sufficient to ameliorate neuropathic pain. In addition, differences in reagent synthesis, purification, dose, and delivery method could be significant factors which affect the outcome of these studies. None of the studies using anti-$Na_V1.3$ oligonucleotides included an assessment of global gene expression following the treatment which might have revealed off-target effects, both pro- or anti-nociceptive, that might have modified the outcome. Thus, additional investigation is needed to further clarify the role of $Na_V1.3$ in neuropathic pain.

$Na_V1.4$

Nearly 40 missense mutations in $Na_V1.4$ sodium channel underlie skeletal muscle disorders with dominant phenotypes, and, interestingly, can cause a wide spectrum of symptoms ranging from hyperexcitability to hypoexcitability of skeletal myocytes (Lehmann-Horn *et al.*, 2002; George, 2005; Vicart *et al.*, 2005; Cannon, 2006). These mutations can cause hyperexcitability (myotonia, characterized by muscle stiffness), hypoexcitability (periodic paralysis, characterized by muscle weakness) or a combination of the two phenotypes (paramyotnia congenital, characterized by initial muscle stiffness that can be followed by a period of muscle weakness, and which is worsened by repeated activity or cold). The lack of expression of $Na_V1.4$ in CNS tissues is consistent with the absence of neurological deficits in patients with muscle disorders who carry mutations in $Na_V1.4$.

The effects of missense mutations on the biophysical properties of $Na_V1.4$ have been determined primarily by the expression of mutant channels in heterologous systems, for example HEK293 cells, and have demonstrated altered gating properties of the mutant channels. Potassium-aggravated myotonia (PAM) results from mutations that cause a slower rate of $Na_V1.4$ inactivation (Mitrovic *et al.*, 1994). Paramyotnia congenita (PMC) results from mutations that cause an incomplete and slower fast inactivation profile (Yang *et al.*, 1994; Bouhours *et al.*, 2005) and also a slower rate of deactivation (Bendahhou *et al.*, 1999). Hyperkalemic periodic paralysis (HyperPP) mutations produce an incomplete $Na_V1.4$ inactivation leading to a persistent current but without a change in kinetics of inactivation (Cannon *et al.*, 1991, 1993), or to a hyperpolarized shift in activation leading to a 'window' current at voltages between –60 and –50 mV (Cummins *et al.*, 1993). Disruption of slow inactivation (Cummins and Sigworth, 1996; Bendahhou *et al.*, 2002) and, paradoxically, enhancement of slow inactivation (Bendahhou *et al.*, 2000), have been reported in cases of HyperPP. Interestingly, an identical mutation, V1589M, has been associated with PAM (Heine *et al.*, 1993) and PMC (Ferriby *et al.*, 2006) suggesting the influence of other genetic factors in the overall symptomatology of these diseases. The totality of these studies permit the following generalization: mutations that produce persistent or window currents cause a depolarized resting membrane potential inducing resting inactivation of sodium channels and hypoexcitability (Cannon, 2002; Lehmann-Horn *et al.*, 2002), on the other hand, mutations leading to slower fast inactivation cause hyperexcitability (Cannon, 2006).

Another class of muscle disorder, congenital myasthenic syndrome (CMS), has been traditionally linked to mutations in acetylcholine receptor that weaken the endplate potential and impairs the safety factor for neuromuscular transmission (Cannon, 2006). A recent case of CMS, however, presented with normal endplate potential but impaired sodium current activation caused by a mutation in $Na_V1.4$ (Tsujino *et al.*, 2003). The V1442E substitution in $Na_V1.4$ from this patient causes a –33 mV hyperpolarizing shift in $V_{1/2}$ of steady-state inactivation when expressed in HEK 293 cells. This effectively reduced the available channels at the resting membrane potential (-90 mV) of muscles from 98 per cent for wild-type channels to only 13 per cent in the case of the V1442E mutant (Tsujino *et al.*, 2003). Thus, CMS may arise as a functional haploinsufficiency of $Na_V1.4$. It remains to be determined whether reduced levels of $Na_V1.4$ caused by weaker promoter activity, as will be discussed next for $Na_V1.5$, can underlie muscle weakness in cases where mutations in known targets are absent.

Na$_V$1.5

Na$_V$1.5 is the major sodium channel isoform in cardiac myocytes and is critical for normal function of the heart. Not surprisingly, targeted disruption of SCN5A, the gene which encodes Na$_V$1.5, causes intrauterine lethality in mice that are homozygous for the null allele, while heterozygotes survive normally, albeit with cardiac defects (Papadatos et al., 2002). Thus while haploinsufficiency is not lethal, it leads to pathological outcomes. The majority of cardiac pathologies, however, are caused by single amino acid substitutions in Na$_V$1.5 which affect channel gating and lead to either gain- or loss-of-function phenotypes, depending on the specific mutation (Viswanathan and Balser, 2004; George, 2005; Moss and Kass, 2005; Modell and Lehmann, 2006). The discovery that polymorphic sites in SCN5A (Ye et al., 2002; Ackerman et al., 2004; Tan et al., 2005; Poelzing et al., 2006) can modify the functional manifestation of the mutant channels further complicates the Na$_V$1.5 phenotype–genotype relationship.

Conduction deficits are also produced by functional haploinsufficiency resulting from nonsense mutations in SCN5A, and these may exacerbate an effect of separate mutation in the second allele of the gene (Bezzina et al., 2003). Yet other mutations of SCN5A, leading to conduction block, are linked to interactions with cytosolic proteins that do not affect channel gating but, rather, regulate clustering of the channel within the membrane of cardiac myocytes. For example an E1053K mutation impairs the binding of ankyrin G to Na$_V$1.5 channels (Mohler et al., 2004). Transcriptional dysregulation of Na$_V$1.5 has been identified as another cause of conduction block in cardiac myocytes in the absence of known gene mutations. For example, a haplotype of six polymorphic nucleotides was identified in the SCN5A promoter in Asian patients with Brugada symptoms, but not in an ethnically matched control cohort or in a Caucasian control group, and functional analysis using a reporter gene showed reduced activity compared to a control construct with the normal promoter (Bezzina et al., 2006).

More than 100 missense mutations in Na$_V$1.5 are responsible for a spectrum of cardiac pathologies including ventricular arrhythmias (LQT3 and idiopathic ventricular fibrillation, IVF), conduction block, or both (Viswanathan and Balser, 2004; George, 2005; Moss and Kass, 2005; Modell and Lehmann, 2006). Generally speaking, impaired fast inactivation leads to persistent sodium current and a longer depolarization phase of the cardiac action potential in LQT3 cases and enhanced fast inactivation leads to conduction block in the Brugada syndrome and other conduction block pathologies. The severity of each of these mutations varies with the exact location of the amino acid substitution and its impact on the biophysical properties of the channel. However, a more severe phenotype can result from compound heterozygosity, either involving a different mutation in the second allele of SCN5A (Bezzina et al., 2003) or, perhaps, in another LQT gene (Paulussen et al., 2003). There are also cases of a mutant phenotype being rescued by the coexistence of certain polymorphic sites and a known SCN5A mutation (Ye et al., 2002; Ackerman et al., 2004; Tan et al., 2005).

Na$_V$1.5 has also been detected in CNS (Hartmann et al., 1999; Wu et al., 2002), in PNS neurons (Renganathan et al., 2002) and in smooth muscle cells (Holm et al., 2002; Ou et al., 2002; 2003; Strege et al., 2003). Little has been reported on neuronal deficits in patients with SCN5A-related cardiac pathologies, but a recent study (Locke et al., 2006) has reported a higher incidence of gastrointestinal pathological symptoms including abdominal pain in patients with identified SCN5A-related cardiac channelopathy. This finding is consistent with a role for Na$_V$1.5 in regulating the electrogenic properties of smooth muscle cells (Ou et al., 2003), but it is also possible that the gastrointestinal symptoms might arise from treatment of the underlying cardiac pathology.

Na$_V$1.6

Na$_V$1.6 is the major sodium channel isoform at mature nodes of Ranvier in CNS (Boiko et al., 2001; Kaplan et al., 2001) and PNS (Schafer et al., 2006) myelinated axons. Missense mutations of this channel can lead to movement disorders and disruption of both copies of the SCN8A gene in mice (referred to as SCN8A^{med} or SCN8A^{medTg}, depending on the cause of disruption) causes stunted growth which becomes evident 1 week after birth. This results in juvenile lethality in homozygous null mice, while the heterozygotes show normal growth and viability (Meisler et al., 2001). Three ENU-induced missense mutations in the domain-3 pore region of Na$_V$1.6 were homozygous lethal (Buchner et al., 2004). The lethality caused by Na$_V$1.6-null phenotypes has been attributed to a global failure of transmission at neuromuscular junctions. Though not well characterized at the molecular level, the degenerating muscle (dmu) mutation in mice, which causes wasting of skeletal and cardiac muscles, is an autosomal recessive allele of SCN8A (De Repentigny et al., 2001).

Heterozygotes carrying missense mutations of Na$_V$1.6 produce a dominant mutant phenotype while those carrying one disrupted SCN8A allele show normal growth and viability (Meisler et al., 2001). The resulting altered channel properties affect electrogenesis in neurons, similar to mutations of Na$_V$1.1, Na$_V$1.2, Na$_V$1.4, Na$_V$1.5 (see above) and Na$_V$1.7 (see next section), but, unlike Na$_V$1.1 and Na$_V$1.4, Na$_V$1.6 does not manifest haploinsufficiency in mice. Interestingly, Kearney et al. (2002) have shown that a hypomorph, caused by a mutation of the splicing donor site of intron 3 of SCN8A, reduces correct splicing of the transcript by 95 per cent and is juvenile lethal. However, survival is rescued, albeit with motor deficits, by a modifier locus (Sprunger et al., 1999) which encodes a splicing factor allele (Buchner et al., 2003) that doubles the level of correctly spliced Na$_V$1.6 transcripts. Thus a range of 10–50 per cent correctly spliced Na$_V$1.6 transcripts ensures survival and viability.

A screen of SCN8A genomic sequence in 152 human patients with inherited or sporadic ataxia identified an inherited mutation of a 2 bp deletion in exon 24 which leads to the truncation of the channel protein beginning in the pore sequence in domain 4 (Trudeau et al., 2005). In addition to ataxia, patients carrying this mutation have different degrees of cognitive and mental deficits. While these motor and mental deficits may reflect a certain degree of haploinsufficiency (Trudeau et al., 2005), it could also be argued that the truncated channel protein might exert a dominant negative effect, analogous to that of the truncated Na$_V$1.2 channel in a case of SMEI (Kamiya et al., 2004). Another muscle disorder, scapuloperoneal muscular dystrophy (SPMD), which is associated with cardiomyopathy in some patients, has been linked to chromosome 12 (Isozumi et al., 1996; Wilhelmsen et al., 1996). Because of the similarity in symptoms to the mouse mutant dmu and the syntenic localization of SCN8A in the mouse and human genomes near this locus, it has been suggested that SPMD might involve SCN8A (De Repentigny et al., 2001). It remains to be determined, however, whether patients carry SCN8A mutations.

Na$_V$1.6 is also implicated in the pathological outcome of injuries to myelinated axons in acquired CNS neuropathies (Waxman *et al.*, 2004; Waxman 2006). Demyelination of CNS axons in patients with multiple sclerosis or in EAE mice show reduced number of Na$_V$1.6 positive nodes of Ranvier. These patients also have an increased number of nodes which are positive for Na$_V$1.2, and a Na$_V$1.6 staining of long stretches of demyelinated axons (tens of microns), suggesting a dyregulated accumulation of the channel in these compartments Craner *et al.*, 2003b, 2004a, b). The appearance of long stretches of Na$_V$1.6-stained demyelinated axons is associated with the co-expression of the Na$^+$–Ca^{2+} exchanger and axonal degeneration (Craner *et al.*, 2004b). It has been shown previously that the Na$^+$–Ca^{2+} exchanger can act in reverse to promote the influx of Ca^{2+} ions into axons, thereby inducing axonal injury, and that this process is dependent upon the presence of a persistent TTX-S sodium channel (Stys *et al.*, 1991, 1993). Na$_V$1.6 can, in fact, produce such a persistent current even in the absence of an action potential (Rush *et al.*, 2005b) may drive the reverse operation of the Na$^+$–Ca^{2+} exchanger, leading to axonal death (Waxman, 2006). The role of sodium channels in axonal injury in demyelinating diseases is supported by the ability of sodium channel blockers to offer neuroprotection in demyelination animal models (Lo *et al.*, 2002, 2003; Bechtold *et al.*, 2004a, b).

Na$_V$1.7

Animal studies have demonstrated that Na$_V$1.7 is a major target in inherited and acquired painful neuropathies, consistent with its expression in the majority of nociceptive neurons. The embryonic-lethal phenotype of a global SCN9A knockout might reflect a critical role for this channel in sympathetic neurons (Nassar *et al.*, 2004; Waxman and Dib-Hajj, 2005). Na$_V$1.7 expression is elevated in DRG neurons following inflammation of their peripheral receptive fields (Black *et al.*, 2004; Gould *et al.*, 2004). Herpes vector-mediated knockdown of Na$_V$1.7 in primary afferents has been shown to prevent inflammation-induced hyperexcitability in pain-signalling neurons and to attenuate hyperalgesia in a rodent model of pain (Yeomans *et al.*, 2005). A direct role of Na$_V$1.7 in inflammatory pain was inferred from studies of mice lacking Na$_V$1.7 only in a specific subset of DRG neurons which express another voltage-gated sodium channel, the slowly inactivating TTX-R Na$_V$1.8 transcript. This mouse line showed almost complete absence of inflammation-induced pain (Nassar *et al.*, 2004). Taken together, the results of these animal studies suggest that Na$_V$1.7 plays a key role in experimental acquired inflammatory pain.

Early-onset primary erythromelalgia (also known as erythermalgia) is a life-long, autosomal dominant disorder which has been linked to missense mutations in Na$_V$1.7 (Yang *et al.*, 2004; Dib-Hajj *et al.*, 2005; Drenth *et al.*, 2005; Han *et al.*, 2006, Harty *et al.*, 2006, Michiels *et al.*, 2005). Penetrance appears to be 100 per cent for all of these mutations and afflicted individuals suffer painful symptoms that are precipitated by warmth or mild exercise; treatment is empirical and often ineffective (Waxman and Dib-Hajj, 2005). Electrophysiological studies have shown that the most consistent change which has been detected by studies of seven different mutations is a hyperpolarizing shift in the voltage-dependence of activation of mutant channels (Cummins *et al.*, 2004; Dib-Hajj *et al.*, 2005; Choi *et al.*, 2006b; Han *et al.*, 2006; Harty *et al.*, 2006; Lampert *et al.*, 2006). Expression of mutant Na$_V$1.7 channels in DRG neurons has been shown to lower the stimulus threshold for

single action potential and increase the number of action potentials that are fired in response to application of a graded stimulus (Dib-Hajj *et al.*, 2005; Harty *et al.*, 2006; Rush *et al.*, 2006a). This effect has been associated with a depolarization of the resting membrane potential of the neurons in which the mutant channels are expressed (Harty *et al.*, 2006; Rush *et al.*, 2006a). However, depolarization of the resting membrane potential of neurons does not explain all aspects of DRG behaviour (Harty *et al.*, 2006) suggesting the contribution by other changes in the biophysical properties of the channel, for example slower deactivation, faster recovery from inactivation and increase in the current in response to small slow depolarization (ramp current).

The redness of the affected extremities in patients with *SCN9A*-related erythromelalgia can be linked to impaired regulation of blood flow in the affected limbs. This symptom is likely linked to impaired sympathetic neuronal regulation of cutaneous vascularization. In agreement with this hypothesis, the expression of mutant channel L858H in neurons from superior cervical ganglia (SCG) induced hypoexcitability of these sympathetic neurons (Rush *et al.*, 2006a). While mutant channels induced depolarization of the resting membrane potential of these neurons, similar to their effect on DRG neurons, they increased the threshold for single action potential, decreased the amplitude of the action potential overshoot and the number of action potentials that are fired in response to a graded stimulus (Rush *et al.*, 2006a). This dichotomy of response of sensory and sympathetic neurons to mutant Na$_V$1.7 channels is related to the presence of different ionic conductances in the two neuronal types, particularly the expression of Na$_V$1.8 in sensory but not sympathetic neurons because the co-expression of mutant Na$_V$1.7 and Na$_V$1.8 channels rescues firing behaviour of sympathetic neurons (Rush *et al.*, 2006a). The hypoexcitable phenotype of sympathetic neurons may impair cutaneous vasoconstriction contributing to flushing of the affected skin (Choi *et al.*, 2006b).

Recently, another inherited painful syndrome, paroxysmal extreme pain disorder which was previously known as familial rectal pain (Fertleman and Ferrie, 2006) that manifests as severe pain and flushing in the lower body in infants during bowel movement, and which is accompanied at older age by ocular and mandibular pain, has also been linked to eight mutations in Na$_V$1.7. The effect of three of these mutations (I1461T, T1464I and M1627K) on biophysical properties of Na$_V$1.7 were investigated in voltage-clamp studies (Fertleman *et al.*, 2006). These mutants were found to impair fast inactivation and shift the voltage-dependence of steady-state inactivation in a depolarizing direction, both effects which would be expected to enhance the excitability of DRG neurons producing the mutant channels (Fertleman *et al.*, 2006).

Finally, three distinct nonsense, loss-of-function, Na$_V$1.7 mutations with dramatic phenotypic consequences have been identified in three Pakistani consanguineous families. The afflicted individuals are completely unable to perceive any form of noxious pain (Cox *et al.*, 2006). What is particularly interesting in the genotypic analysis of these individuals is that the mutations target distinct regions of the Na$_V$1.7 coding region. One lies in the turret region of domain-2, the second in TM2 of the same domain while the third lies in the intracellular linker connecting domains 1 and 2. Introducing each of these mutations into human wild-type Na$_V$1.7 cDNA produced completely non-functional channels when expressed in HEK293 cells (Cox *et al.*, 2006). This recent finding adds to the growing evidence that $Na_V1.7$ plays a key role in the transduction of painful signals, excessive channel activity leads to hypersensitivity whereas null activity can confer total analgesia.

$Na_V1.8$

$Na_V1.8$ channels contribute most of the current underlying the upstroke of action potential, support repetitive action potentials, and contribute to action potential adaptation in nociceptive DRG neurons (Renganathan *et al.*, 2001; Blair and Bean, 2002, 2003; Tripathi *et al.*, 2006). A key contribution for $Na_V1.8$ in neural electrogenesis was demonstrated by the finding that the channel can rescue the firing properties of SCG neurons that are expressing a mutant $Na_V1.7$ channel (Rush *et al.*, 2006a). Thus, $Na_V1.8$ is a channel which is poised to significantly alter DRG neuron excitability under pathological conditions.

Studies using $Na_V1.8$-null mice provided evidence for an important role of this channel in nociception. These mice demonstrated an attenuated response to noxious mechanical stimuli, but only small deficits in response to noxious thermal stimuli (Akopian *et al.*, 1999). They also showed a delayed development of carrageenan-evoked thermal hyperalgesia (Akopian *et al.*, 1999), an attenuated response to NGF-evoked thermal hyperalgesia (Kerr *et al.*, 2001), and reduced visceral pain responses to inflammatory mediators with referred hyperalgesia (Laird *et al.*, 2002). Recently, it has been shown that $Na_V1.8$-null mice show deficits in visceral pain in a chronic inflammatory pain model induced by infection with the nematode *Nippostrongylus brasiliensis* (Hillsley *et al.*, 2006). This is consistent with finding that $Na_V1.8$ is present in every sensory DRG neuron innervating the colon (Gold *et al.*, 2002). Also, studies of $Na_V1.8$-null mice show that this channel contributes substantially to spontaneous activity in damaged axons or in intact axons that co-mingle with them. The sectioning of the saphenous nerve in mice leads to the formation of neuroma which has been associated with ectopic discharge in 20 per cent of A- and C-fibres in wild-type mice but only in 1 per cent of these fibres in $Na_V1.8$-null mice (Roza *et al.*, 2003). Thus $Na_V1.8$ appears to play a key role in neuropathic and inflammatory painful pathologies.

The contribution of $Na_V1.8$ to hyperexcitability of DRG neuron in axotomy models has been less well defined. Transection of axons causes the loss of $Na_V1.8$ channels within the cell bodies of DRG giving rise to these fibres (Dib-Hajj *et al.*, 1996; Cummins and Waxman, 1997; Sleeper *et al.*, 2000; Decosterd *et al.*, 2002). Similarly, transection of the spinal nerves of L5 and L6 in a spinal nerve ligation (SNL) model causes the loss of $Na_V1.8$ channel in the cell bodies (Gold *et al.*, 2003). These effects are probably the result of an injury-induced loss of trophic factor support from target organs since exogenous NGF or GDNF are able to restore the expression of $Na_V1.8$ both in culture (Black *et al.*, 1997a; Fjell *et al.*, 1999) and *in vivo* (Dib-Hajj *et al.*, 1998a; Cummins *et al.*, 2000; Leffler *et al.*, 2002). Interestingly, transection of the dorsal root (rhizotomy) does not affect the expression of this channel (Sleeper *et al.*, 2000). However, spinal cord transection at T8 leads to decreased expression of $Na_V1.8$ in bladder afferents 28–32 days post injury, a time when the bladder becomes hyperreflexic, but is without effect in afferents which innervate other targets, for example colon (Black *et al.*, 2003). The molecular pathway leading to neuronal hyperexcitability in neurons which have lost $Na_V1.8$ after injury remains poorly understood.

Other injury models, which do not involve physical transection of exons, have been developed to study painful neuropathies. Tying loose ligatures around the sciatic nerve produces what is commonly known as a chronic constriction injury (CCI) that, over time, precipitates a selective axonal transection of large-diameter myelinated axons distal to the ligation site as a consequence of an ensuing Wallerian degeneration (Bennett and Xie, 1988). Two studies using the CCI model reported conflicting results about levels of $Na_V1.8$ in DRG neurons, one study showing the loss of $Na_V1.8$ transcripts and TTX-R currents (Dib-Hajj *et al.*, 1999) and a second showing no change in $Na_V1.8$ transcripts and TTX-R current density but a reduction in $Na_V1.8$-specific immunolabelling within cell bodies and an accumulation of $Na_V1.8$ in the sciatic nerve (Novakovic *et al.*, 1998). Neither of these studies measured changes in identified, un-injured DRG neurons for altered levels of $Na_V1.8$. While the previously mentioned SNL model studies (Gold *et al.*, 2003) revealed no change in $Na_V1.8$ levels within uninjured L4 cell bodies, there was a reported increase in $Na_V1.8$ in the sciatic nerve, presumably from the uninjured L4 axons, and this was accompanied by an increase in a component of the C-fibre compound action potential that was also TTX-R. The apparent elevation of $Na_V1.8$ in uninjured neurons, whose axons mingle with damaged neighbours, may also explain the significantly higher incidence of ectopic firing following saphenous nerve transection in wild-type compared to $Na_V1.8$-null neurons (Roza *et al.*, 2003). The reported attenuation of neuropathic pain behaviour in CCI animal injury models by $Na_V1.8$-specific antisense oligonucleotides (Lai *et al.*, 2002; Gold *et al.*, 2003; Joshi *et al.*, 2006,) or by $Na_V1.8$-specific siRNA (Dong *et al.*, 2007) is also likely caused by targeting $Na_V1.8$ in the spared DRG neurons in this model.

A role for $Na_V1.8$ in inflammatory pain is better understood at the molecular level. For example, though there was no detectable change in the level of $Na_V1.8$ transcript or protein in DRG neurons 48 hrs after complete Freund's adjuvant (CFA) injection into the hindpaw (Okuse *et al.*, 1997), there was an increase in the number of anti-$Na_V1.8$ immunopositive myelinated and unmyelinated axons in sciatic nerve (Coggeshall *et al.*, 2004). Carrageenan-induced inflammation of rat hindpaw has also been reported to increase $Na_V1.8$ current 4 days post-injection (Tanaka *et al.*, 1998). Colitis, induced by injecting trinitrobenzenesulfonic acid into the intestine, also causes an increase in the $Na_V1.8$ current density 10 days post injection (Beyak *et al.*, 2004). The inflammatory mediators PGE2 (Gold *et al.*, 1996) and NGF, together with its second messenger ceramide (Zhang *et al.*, 2002), have also been shown to increase TTX-R current density, attributed to $Na_V1.8$, and this is accompanied by a hyperpolarizing shift in voltage-dependence of activation of the channel. The application of TNF-α to DRG neurons in culture increased the amplitude of the TTX-R current via a p38 MAPK-mediated pathway (Jin and Gereau, 2006). However, treatment of DRG neurons in culture with anisomycin, which directly activates p38 MAPK, caused direct phosphorylation of $Na_V1.8$ and an increase in the current density but without altering the gating properties of the channel (Hudmon *et al.*, 2008). Administration of anti-$Na_V1.8$ antisense oligonucleotides (Khasar *et al.*, 1998) or virally delivered anti $Na_V1.8$ shRNA (Mikami and Yang, 2005) ameliorates pain which is evoked by inflammation of rat hindpaw. Similarly, acetic acid-induced urinary bladder neuropathic pain is attenuated using anti-$Na_V1.8$ antisense oligonucleotide treatment (Yoshimura *et al.*, 2001). Thus, $Na_V1.8$ would appear to be a prime target for development of therapeutic agents for the treatment of neuropathic and inflammatory pain.

Na$_V$1.8 is a peripheral neuron-specific channel which is not normally expressed in the CNS (Akopian *et al.*, 1996; Sangameswaran *et al.*, 1996). However, aberrant expression of Na$_V$1.8 in human Purkinje cells has been observed following demyelination, in patients with MS and also in the EAE mouse model of MS (Black *et al.*, 2000) and the *taiep* mutant rat (Black *et al.*, 1999b). The expression of Na$_V$1.8 in Purkinje cells is maladaptive and changes the firing properties of these neurons (Renganathan *et al.*, 2003). Thus, dysregulation of Na$_V$1.8 could contribute to the ataxia that is associated with demyelinating disorders.

Na$_V$1.9

The very slow kinetics of activation and inactivation of Na$_V$1.9 leads to a persistent current in DRG (Cummins *et al.*, 1999) and myenteric (Rugiero *et al.*, 2003) neurons. This suggests that this channel does not contribute to the upstroke of action potential, but that it may play a role in setting activation thresholds and resting membrane potential (Cummins *et al.*, 1999; Herzog *et al.*, 2001; Baker *et al.*, 2003). Accordingly, the loss of Na$_V$1.9 expression after axotomy (Dib-Hajj *et al.*, 1998b), CCI (Dib-Hajj *et al.*, 1999) and radicular injury to the ganglia (Abe *et al.*, 2002) might depolarize the resting membrane potential and remove resting inactivation of the TTX-S channels in DRG neurons, thereby leading to hyperexcitability (Dib-Hajj *et al.*, 2002). However, DRG neurons from Na$_V$1.9-null mice have been reported to show no difference in the average resting membrane potential compared to control neurons from wild-type mice (Priest *et al.*, 2005). Thus, a more definitive role of this channel in regulating resting membrane potential remains to be verified.

Na$_V$1.9 is preferentially expressed in small-diameter DRG neurons (Dib-Hajj *et al.*, 1998b; Tate *et al.*, 1998), most of which are nociceptive (Fang *et al.*, 2002), suggesting a likely role in pain sensations. However, traumatic injury to DRG neurons, or their axons, causes a downregulation of Na$_V$1.9 (Dib-Hajj *et al.*, 1998b; 1999; Sleeper *et al.*, 2000; Abe *et al.*, 2002; Decosterd *et al.*, 2002) and a reduction in the persistent TTX-R current which is attributable to this channel (Cummins and Waxman, 1997; Cummins *et al.*, 2000). Similar to Na$_V$1.8, a loss of trophic factor support underlies the injury-induced reduction in the current density. However, in contrast, the expression of Na$_V$1.9 is restored by exogenous GDNF but not by NGF, which consistent with the expression of this channel in the non-peptidergic, IB4+ DRG neurons (Fjell *et al.*, 1999; Cummins *et al.*, 2000; Leffler *et al.*, 2002). Mechanical allodynia following partial ligation of the sciatic nerve was not different between groups of wild-type and Na$_V$1.9 null mice (Priest *et al.*, 2005). Thus, the contribution of this channel to mechanical hyperalgesia might be minimal.

A role for Na$_V$1.9 in inflammatory pain is better supported by experimental evidence. Na$_V$1.9 is modulated by inflammatory mediators, for example PGE2, which acts on this channel via a G-protein-coupled mechanism (Rush and Waxman, 2004). Direct activation of G-proteins causes a significant increase in the persistent Na$_V$1.9 current, a subsequent reduction in the threshold for action potential firing and an increase in spontaneous firing (Baker *et al.*, 2003), consistent with predictions from computer simulations (Herzog *et al.*, 2001). In agreement with these findings, both heterozygous and homozygous Na$_V$1.9-knockout mice show deficits in

the late phase of formalin-induced pain, the Na$_V$1.9-null mice also show a reduction in the duration of thermal hyperalgesia following carrageenan or CFA injection into the hindpaw (Priest *et al.*, 2005). An independent strain of Na$_V$1.9-null mouse strain shows similar deficits in several models of inflammation-evoked pain, but not in the spared nerve injury or partial sciatic nerve injury models (Amaya *et al.*, 2006). However, yet another Na$_V$1.9-null mouse strain apparently showed no deficits in visceral pain 20–24 days post-infection with the nematode *Nippostrongylus brasiliensis* (Hillsley *et al.*, 2006). Taken together, this data supports an important role of Nav1.9 in some inflammatory pain conditions.

While the expression of Na$_V$1.9 is preferentially detected in small DRG neurons (Dib-Hajj *et al.*, 1998b; Tate *et al.*, 1998), a subset of larger, nociceptive DRG neurons with Aβ fibres also express this channel (Fang *et al.*, 2002). Diabetic rats show an upregulation of Na$_V$1.9 in DRG neurons 1 week and 8 weeks after treatment with streptozotocin and establishment of elevated blood glucose levels (Craner *et al.*, 2002). Interestingly, small (< 25 μm) and large DRG neurons (> 25 μm) show an increase in Na$_V$1.9 immunostaining. It is noteworthy that Na$_V$1.8 is downregulated while Na$_V$1.3 and Na$_V$1.6 are upregulated under these conditions (Craner *et al.*, 2002).

2.3.7 Concluding remarks

The voltage-gated sodium channels have been known for years to be complex multimeric signalling molecules but the true extent of their complexity continues to unfold as increasingly sophisticated molecular analyses are applied. The biophysical properties of these channels are precisely tailored to their role as high-fidelity transducers and signal propagators. Voltage-dependent gating is a complex process but can be conceptualized as a series of events in which membrane depolarization activates the channel by field-evoked movements in the S4 voltage sensors. Outward movement of the S4 segments of DI, DII and DIII are implicated in channel activation (the '*m*' particles of the H–H formalism), that of DIV-S4, the final sensor to move (the H–H '*h*' particle), initiates inactivation and deactivation and an immobilization of the S4 segments of DIII and DIV. These structural changes are coupled to rearrangements to the structural integrity of the pore-forming elements that briefly enable sodium ion flux and then, after a very short delay, terminate it abruptly. As long as the membrane remains depolarized, the DIII and DIV S4 segments remain immobilized in their outward position and the inner gate remains closed, overall an elegant example of a process of reaction and adaptation to local environmental changes contained entirely within a single molecular entity.

Detailed studies have continued to build upon this general gating theme and it has become clear that channel activity is modulated by interactions with many other cellular molecules and by cytosolic enzymes. These details are proving to be crucially important, not only to our understanding of how these channels operate in their 'normal' environment but, importantly, how their activity is adapted by various pathological and disease states. This may present new opportunities for the development of novel drugs that are capable of selectively targeting particular channel transcripts or counteracting abnormal pathological channel activity. Despite years of intensive research, the identification of truly selective, small molecule drugs has been an elusive goal but it continues to attract appreciable resources—driven by a general acceptance of the potential therapeutic impact of selective pharmacology.

Table 2.3.2 Properties of voltage-gated sodium channels

Channel subunit	Biophysical characteristics	Cellular and subcellular localization	Pharmacology	Physiology	Relevance to disease states
$Na_V1.1$ gene: SCN1A Chr: -h: 2q24	Act ($V_{1/2}$, mV): −33 (h), −20.2 and −21.2 (r, XO, α and $\alpha+\beta1$), −13.7 and −19.6 (r, XO, α and $\alpha+\beta1$) [1–5]. F_{Inact} ($V_{1/2}$, mV): −72mV (h), −27.1 and −39.7 (r, XO, α and $\alpha+\beta1$), −64.9 (h, $\alpha+\beta1$), −37.9 and −41.9 (r, XO, α and $\alpha+\beta1$) [1–5]. $\tau_{inact, fast}$: 0.7ms @ −10mV (h); 8 and 1ms (r, α, and $\alpha+\beta1$, −10mV) [2, 4]. g_{Na}: 17pS (h, $\alpha+\beta1+\beta2$) [5] Ion selectivity: $Na^+>K^+>Ca^{2+}$	Central neurons [6], DRG [7], cardiac myocytes [8]. Expression increases after birth and reaches adult levels by second or third week [6, 9]	Channel blockers: anticonvulsant and local anaesthetics and anti-arrhythmic drugs, site-1 toxins (e.g. TTX, STX) Gating modifiers: site-2, site-3, site-4 and site-6 toxins (see Table 2.3.1) No selective pharmacological tools identified	Action potential initiation	Several missense or nonsense mutations that lead to non-functional channels linked to conditions of hemiplegic migraine and severe myoclonic epilepsy in infants [10–14]
$Na_V1.2$ gene: SCN2A Chr: -h: 2q23–24	Act ($V_{1/2}$, mV): −24mV (h), −21.6 and −24.3 (r, XO, α and $\alpha+\beta1$). F_{Inact} ($V_{1/2}$, mV): −53 (h) [15, 16] $\tau_{inact, fast}$: 8ms @ −24mV [15] g_{Na}: 19pS [17] Ion selectivity: $Na^+>K^+>Ca^{2+}$	Predominantly CNS neurons and largely axonal. Expression is highest in rostral brain and cerebellum, adult levels are attained by 1 week after birth [6, 18]	Channel blockers: anticonvulsant and local anaesthetics and anti-arrhythmic drugs, site-1 toxins (e.g. TTX, STX) Gating modifiers: site-2, site-3, site-4 and site-6 toxins (see Table 2.3.1) No selective pharmacological tools identified	Action potential initiation and axonal conduction	Gene deletion results in perinatal lethality Several loss-of-function mutations associated with epilepsy [19–23] A C-terminal mutation that impairs calmodulin modulation linked to autism [24]
$Na_V1.3$ gene: SCN3A Chr: -h: 2q24	Act ($V_{1/2}$, mV): −23 (h), −9.8 (h, XO), −18.3 (r, XO) [2, 25, 26] F_{Inact} ($V_{1/2}$, mV): −69 (h), −47.3 (h, XO), −24.8 (r, XO) [2, 25, 26] $\tau_{inact, fast}$: 0.8 @ −10mV [2] g_{Na}: 15pS [27] Ion selectivity: $Na^+>K^+>Ca^{2+}$	Predominantly embryonic CNS neurons with expression levels declining significantly, in rat, after birth [6, 18]. Continues to be expressed in adult human brain tissues. Expression induced in rat DRG neurons after peripheral injuries [28, 29]	Channel blockers: anticonvulsant and local anaesthetics and anti-arrhythmic drugs, site-1 toxins (e.g. TTX, STX) Gating modifiers: site-2, site-3, site-4 and site-6 toxins (see Table 2.3.1) No selective pharmacological tools identified	Action potential initiation and axonal conduction	No clinical conditions have been associated with abnormal function of this channel. Rodent studies have suggested a potential role in pain states but this has not been substantiated in human pain conditions
$Na_V1.4$ gene: SCN4A Chr: -h: 17q23–25	Act ($V_{1/2}$, mV): −30 (r, XO), −40 (r, $\alpha+\beta1$) [15] F_{Inact} ($V_{1/2}$, mV): −50.1 (h, XO) [15] $\tau_{inact, fast}$: 0.55ms @ 10mV (h, XO) [15] g_{Na}: 18pS (r) [30] Ion selectivity: $Na^+>K^+>Ca^{2+}$	Abundant in mature innervated skeletal muscles, absent in CNS and DRG neurons [31, 32]	Channel blockers: anticonvulsant and local anaesthetics and anti-arrhythmic drugs, site-1 toxins (e.g. TTX, STX) Gating modifiers: site-2, site-3, site-4 and site-6 toxins (see Table 2.3.1) No selective pharmacological tools identified	Action potential initiation and conduction in skeletal muscle fibres	Large number of missense mutations documented. Skeletal muscle disorders, incl. myotonia, periodic paralysis and paramyotonia congenital linked to $Na_V1.4$ mutations [11, 33–35]. A single case of congenital myasthenic syndrome attributed to a point mutation that results in a substantial hyperpolarizing shift in steady-state inactivation [36]

	Biophysical properties	Expression	Pharmacology	Physiological role	Disease associations
$Na_V1.5$ gene: SCN5A Chr: –h: 3p21	Act ($V_{1/2}$, mV): –47, –56 (h, $\alpha+\beta1$), –27 [15]; F_{inact} ($V_{1/2}$, mV): –84, –100 (h, $\alpha+\beta1$), –61 [15]; $\tau_{inact, fast}$: 1ms @ 0mV [37]; g_{Na}: 19–22pS [38, 39]; Ion selectivity: $Na^+>K^+>Ca^{2+}$	Predominant cardiac sodium channel isoform [40]. Also expressed in denervated skeletal muscles [31, 32] and in certain CNS areas [41, 42], spinal cord astrocytes [43] and T-lymphocytes [44]. Found in a subpopulation of peripheral neurons [45] and in smooth muscle cells [46–48]	Channel blockers: anticonvulsant and local anaesthetics and anti-arrhythmic drugs Insensitive to site-1 toxins (eg. TTX, STX). Gating modifiers: site-2, site-3, site-4 and site-6 toxins (see Table 2.3.1) No selective pharmacological tools identified	Cardiac and neuronal action potential	More than 100 missense mutations in the $Na_V1.5$ gene linked to several cardiac pathologies, incl. arrhythmias (LQT3, idiopathic ventricular fibrillation, conduction block or both [11, 49–51]. More severe cardiac phenotypes resulting from compound heterozygosity reported [52, 53]
$Na_V1.6$ gene: SCN8A Chr: –h: 12q13	Act ($V_{1/2}$, mV): –8.8 (m, XO), –17 (m, XO, $\alpha+\beta1\&\beta2$), –37.7 (r, XO) [15]; F_{inact} ($V_{1/2}$, mV): –55 (m, XO), –51 (m, XO, $\alpha+\beta1\&\beta2$), –97.6 (r, XO) [15]; $\tau_{inact, fast}$: 1–7ms (m and r) @ –30 to +10mV [15]; g_{Na}: Unknown; Ion selectivity: $Na^+>K^+>Ca^{2+}$	Expressed in CNS neurons, glial cells (brain and spinal cord) [54, 55], peripheral sensory neurons [56] and in non-neuronal cells (microglia and macrophages) [57] Particularly prominent at nodes of Ranvier in both central and peripheral axon tracts [58–60]	Channel blockers: anticonvulsant and local anaesthetics and anti-arrhythmic drugs, site-1 toxins (e.g. TTX, STX). Gating modifiers: site-2, and site-3 toxins (see Table 2.3.1) No selective pharmacological tools identified	Action potential initiation and axonal conduction, particularly in myelinated axons. Appears to be involved in the manifestation of resurgent currents that are common to certain neuronal populations (e.g. cerebellar Purkinje cells and some DRG neurons) [61]	A single truncation mutation linked to cases of inherited or sporadic ataxia, patients with this condition may also present with some cognitive or mental deficits [62 Dysregulated nodal expression may be liked to ME [63, 64]
$Na_V1.7$ gene: SCN9A Chr: –h: 2q24	Act ($V_{1/2}$, mV): –31 (r, XO) [15]; F_{inact} ($V_{1/2}$, mV): –78 (r, XO), –60.5 (h), –39.6 (h, $\alpha+\beta1$) [15]; $\tau_{inact, fast}$: 0.46ms @ –30mV (r, XO); g_{Na}: unknown; Ion selectivity: $Na^+>K^+>Ca^{2+}$	Largely a peripherally expressed channel Prominent in sensory (especially nociceptor) and sympathetic neurons [7, 65] Also expressed in smooth muscle myocytes, neuroendocrine cells, tumour cells and erythrocytes [46, 66–71]	Channel blockers: anticonvulsant and local anaesthetics and anti-arrhythmic drugs, site-1 toxins (e.g. TTX, STX). Gating modifiers: site-2 and site-3, toxins (see Table 2.3.1) No selective pharmacological tools identified	Generates ramp currents in response to slow depolarizations, may act as an amplifier of subthreshold depolarizations and thereby facilitate action potential initiation [72]	Several missense mutations that appear to lead to a gain-of-function phenotype, have been associated with early-onset primary erythromelalgia [73–78] Two independent studies have identified several global lineages of inherited loss-of-function mutations that result in pain in what are otherwise apparently normal individuals [79, 80] Three mutations have been linked to paroxysmal extreme pain disorder [81, 82]
$Na_V1.8$ gene: SCN10A Chr: –h: 3p22–24	Act ($V_{1/2}$, mV): 10 (r), –7.7 and –9.4 (r), 10.3 (r, XO, α), 1.6 ($\alpha+\beta1$), 9.5 ($\alpha+\beta2$), 11.5 ($\alpha+\beta3$), 1.7 ($\alpha+\beta1+\beta2+\beta3$); –11 (h); 5.5 (h, $\alpha+\beta1$); 9 (r, $\alpha+\beta1$). [83–88]; F_{inact} ($V_{1/2}$, mV): –34.1 (r); –30.8 (r); –52.9 (r, XO, α), –58.4 ($\alpha+\beta1$), –49 ($\alpha+\beta2$), –47.9 ($\alpha+\beta3$), –54.4 ($\alpha+\beta1+\beta2+\beta3$); –27 (h), –38.3 (h, $\alpha+\beta1$); –40.1 (r, $\alpha+\beta1$) [83–88]; $\tau_{inact, fast}$: 3.8 ms @ 20mV; 14ms @ 10mV [86, 87]; g_{Na}: unknown; Ion selectivity: Na^+	Highly-restricted expression profile. Only found in peripheral sensory and trigeminal neurons [89, 90]. CNS expression is absent except in certain disease states. Sensory neuron expression includes a significant percentage of small-diameter cells but is also found in a subpopulation of larger-diameter $A\beta$ neurons [7, 91–93]	Channel blockers: anticonvulsant and local anaesthetics and anti-arrhythmic drugs. Resistant to site-1 toxins (ie. TTX-R). Selectively blocked by μO-conotoxin [94] and by A-803467 (a substituted carboxamide) [95]	Provide most of the current underlying the upstroke of the action potential in certain sensory neurons [96] Supports repetitive firing and contributes to action potential adaptation in DRG neurons [96–99]	No reported disease associations

Continued

Table 2.3.2 *(Continued)* Properties of voltage-gated sodium channels

Channel Subunit	Biophysical characteristics	Cellular and subcellular localization	Pharmacology	Physiology	Relevance to disease states
Na$_V$1.9 gene: SCN11A Chr: –h: 3p21–24	Detailed studies on the properties of recombinant NaV1.9 have not been published. However, data generated from presumptive native rat NaV1.9 conductances have suggested that this channel has a activation/inactivation voltage range that is appreciably more negative than NaV1.8 and is more comparable to TTX-S channel variants	Exclusively DRG and trigeminal sensory neurons [100, 101]	Resistant to site-1 toxins (i.e. TTX-R). No other pharmacology reported	Slow gating kinetics and persistence of current flux suggests a role in setting activation thresholds and resting membrane potential	No reported disease associations

Values for biophysical characteristics are for recombinant channels only, where: r, rat; m, mouse; h, human. Data generated using mammalian cell expression unless indicated by XO (*Xenopus* oocyte expression). Act, voltage-dependent activation; F_{inact}, voltage-dependent fast inactivation; τ_{inact}, τ_{fast}, time constant for fast inactivation; $g_{Na'}$ single channel conductance.

[1] Barela et al. (2006) *J. Neurosci* 26, 2714–2723; [2] Clare et al. (2000) *Drug Discov Today* 5, 506–520; [3] Mantegazza et al. (2005) *Proc Natl Acad Sci USA* 102, 18177–18182; [4] Spampanato J et al. (2003) *Neuroscience* 116, 37–48; [5] Vanoye CG et al. (2005) *J Gen Physiol* 127, 1–14; [6] Beckh S et al. (1989) *EMBO J* 8, 3611–3616; [7] Black JA et al. (1996) *Brain Res Mol Brain Res* 43, 117–131; [8] Maier SK et al. (2002) *Proc Natl Acad Sci USA* 99, 4073–40788; [9] Westenbroek et al. (1989) *Neuron* 3, 695–704; [10] Dichgans et al. (2005) *Lancet* 366, 371–377; [11] George (2005) *J Clin Invest* 115, 1990–1999; [12] Meisler and Kearney (2005) *J Clin Invest* 115, 2010–2017; [13] Mulley, et al. (2006) *Neurol* 67, 1094–095; [14] Suls A et al. (2006) *Human Mutations* 27, 914–920; [15] Catterall et al. (2002) Voltage-gated sodium channels. In (ed D. Girdlestone) *The IUPHAR Compendium of Voltage-gated Ion Channels*, IUPHAR media, Leeds. pp 9–30; [16] Zhou and Goldin (2004) *Biophys J* 87, 3862–3872; [17] Stuhmer W et al. (1987) *Eur Biophys J* 14, 131–138; [18] Waxman et al. (1994) *J Neurophysiol* 72, 466–470; [19] Berkovic et al. (2004) *Ann Neurol* 55, 550–557; [20] Heron et al. (2002) *Lancet* 360, 851; [21] Kamiya et al. (2004) *J Neurosci* 24, 2690–2698; [22] Striano et al. (2006) *Epilepsia* 47, 218–220; [23] Sugawara et al. (2001) *Proc Natl Acad Sci USA* 98, 6384–6389; [24] Weiss et al. (2003) *Mol Psychiat* 8, 186–194; [25] Thimmapaya et al. (2005) *Eur J Neurosci*, 22, 1–9; [26] Maertens et al. (2006) *Mol Pharmacol* 70, 405–414; [27] Moorman et al. (1990) *Science* 250, 688–691; [28] Dib-Hajj et al. (1999) *Pain* 83, 591–600; [29] Hains et al. (2003) *J Neurosci* 23, 8881–8892; [30] Ukomadu et al. (1992). *Neuron* 8, 663–676; [31] Kallen et al. (1990) *Neuron* 4, 233–242; [32] Trimmer et al. (1990) *Dev Biol* 142, 360–367; [33] Cannon (2006) *Ann Rev Neurosci* 29, 387–415; [34] Lehmann-Horn et al. (2002) *Curr Neurol Neurosci Rep* 2, 61–69; [35] Vicart et al. (2005) *Neurol Sci* 26, 194–202; [36] Tsujino et al. (2003) *Proc Natl Acad Sci USA* 100, 7377–7382; [37] Mantegazza et al. (2001) *Proc Natl Acad Sci USA* 98, 15348–15353; [38] Fozzard and Hanck (1996) *Physiol Rev* 76, 887–926; [39] Gellens et al. (1992) *Proc Natl Acad Sci USA* 89, 554–558; [40] Rogart et al. (1989) *Proc Natl Acad Sci USA* 86, 8170–8174; [41] Hartmann et al. (1999) *Nat Neurosci* 2, 593–595; [42] Wu et al. (2002) *Neuroreport* 13, 2547–2551. [43] Black et al. (1998) *Glia* 23, 200–208; [44] Strege et al. (2003) *Am J Physiol Cell Physiol* 284, C60–66; [49] Modell and Lehmann (2006) *Genet Med* 8, 143–155; [50] Moss and Kass (2005) *J Clin Invest* 115, 2018–2024; [51] Viswanathan and Balser (2004) *Trends Cardiovasc Med* 14, 28–35; [52] Bezzina et al. (2003) *Circ Res* 92, 159–168; [53] Paulussen et al. (2003) *Genet Test* 7, 57–61; [54] Burgess et al. (1995) *Nat Genet* 10, 461–465; [55] Schaller and Caldwell (2000) *Comp Neurol* 420, 84–97; [56] Rush et al. (2006a) *Proc Natl Acad Sci USA* 103, 8245–8250; [57] Craner et al. (2005) *Glia* 49, 220–229; [58] Boiko et al. (2001) *Neuron* 30, 91–104; [59] Kaplan et al. (2001) *Neuron* 30, 105–119; [60] Schafer et al. (2006) *Neuron Glia Biol* 2, 69–79; [61] Cummins et al. (2005) *FEBS Lett* 579, 2166–2170; [62] Trudeau et al. (2005) *J Med Genet* 43, 527–530; [63] Fraser et al., (2005) *Clin Cancer Res* 11, 5381–5389; [64] Waxman et al. (2004) *Proc Natl Acad Sci USA* 101, 12370–12374; [69] Jo et al. (2004) *FEBS Lett* 567 339–343; [70] Klugbauer et al. (1995) *EMBO J* 14, 1084–1090; [71] Diss et al. (2005) *Prostate Cancer Prostatic Dis* 8, 266–273; [67] Fraser et al., (2005) *J Physiol* 568, 155–169; [72] Cummins et al. (2004) *J Neurosci* 24, 8232–8236; [73] Dib-Hajj et al., (2005) *Brain* 128, 1847–5184; [74] Drenth et al. (2005) *Invest Dermatol* 124, 1333–1338; [75] Han et al. (2006) *Ann Neurol* 59, 553–558; [76] Harty et al. (2006) *J Neurosci* 26, 12566–12575; [77] Michiels et al. (2005) *Arch Neurol* 62, 1587–1590; [78] Yang et al. (2004) *J Med Genet* 41, 171–174; [79] Cox et al. (2006) *Nature* 444, 894–898; [80] Goldberg et al. (2007) *Clin Genet* 71, 311–319; [81] Bednarek et al. (2005) *Epileptic Disord* 7, 360–362; [82] Fertleman et al. (2006) *Neuron* 52, 767–774; [83] Dekker et al. (2005) *Eur J Pharmacol* 528, 52–58; [84] John et al. (2004) *Neuropharmacol* 46, 425–438; [85] Leffler (2007) *J Pharmacol Exp Ther* 320, 354–364; [86] Liu et al. (2006) *Assay Drug Dev Technol* 4, 37–48; [87] Vijayaragavan et al. (2004) *Biochem Biophys Res Comm* 319, 531–540; [88] Zhou, al. (2003) *J Physiol* 550, 739–752; [93] Rizzo et al. (1994) *J Neurophysiol* 72, 2796–2815; [94] Ekberg et al. (2006) *Proc Natl Acad Sci USA* 103, 17030–17035; [95] Jarvis et al. (2007) *Proc Natl Acad Sci USA* 104, 8520–8525; [96] Renganathan et al. (2001) *J Neurophysiol* 86, 629–640; [97] Blair and Bean (2002) *J Neurosci* 22, 10277–10290; [98] Blair and Bean (2003) *J Neurosci* 23, 10338–10350; [99] Tripathi et al. (2006) *Neuroscience* 143, 923–938; [100] Dib-Hajj et al. (1998b) *Proc Natl Acad Sci USA* 95, 8963–8968; [101] Tate et al. (1998) *Nat Neurosci* 1, 653–655.

References

Abe M, Kurihara T, Han W *et al.* (2002). Changes in expression of voltage-dependent ion channel subunits in dorsal root ganglia of rats with radicular injury and pain. *Spine* **27**, 1517–1524.

Abriel H and Kass RS (2005). Regulation of the voltage-gated cardiac sodium channel Na$_V$1.5 by interacting proteins. *Trends Cardiovasc Med* **15**, 35–40.

Abriel H, Kamynina E, Horisberger JD *et al.* (2000). Regulation of the cardiac voltage-gated Na$^+$ channel (H1) by the ubiquitin-protein ligase Nedd4. *FEBS Lett* **466**, 377–380.

Ackerman MJ, Splawski I, Makielski JC *et al.* (2004). Spectrum and prevalence of cardiac sodium channel variants among black, white, asian, and hispanic individuals: implications for arrhythmogenic susceptibility and brugada/long QT syndrome genetic testing. *Heart Rhythm* **1**, 600–607.

Ahern CA and Horn R (2004). Stirring up controversy with a voltage sensor paddle. *Trends Neurosci* **27**, 303–307.

Ahern CA, Zhang JF, Wookalis MJ *et al.* (2005). Modulation of the cardiac sodium channel Na$_V$1.5 by Fyn, a Src family tyrosine kinase. *Circ Res* **96**, 991–998.

Akopian AN, Sivilotti L and Wood JN (1996). A tetrodotoxin-resistant voltage-gated sodium channel expressed by sensory neurons. *Nature* **379**, 257–262.

Akopian AN, Souslova V, England S *et al.* (1999). The tetrodotoxin-resistant sodium channel SNS has a specialized function in pain pathways. *Nat Neurosci* **2**, 541–548.

Alami M, Vacher H, Bosmans F *et al.* (2003). Characterization of Amm VIII from *Androctonus mauretanicus mauretanicus*: a new scorpion toxin that discriminates between neuronal and skeletal sodium channels. *Biochem J* **375**, 551–560.

Allouis M, Le Bouffant F, Wilders R *et al.* (2006). 14–3–3 is a regulator of the cardiac voltage-gated sodium channel Na$_V$1.5. *Circ Res* **98**, 1538–1546.

Amaya F, Decosterd I, Samad TA *et al.* (2000). Diversity of expression of the sensory neuron-specific TTX-resistant voltage-gated sodium ion channels SNS and SNS2. *Mol Cell Neurosci* **15**, 331–342.

Amaya F, Wang H, Costigan M *et al.* (2006). The voltage-gated sodium channel Na$_V$1.9 is an effector of peripheral inflammatory pain hypersensitivity. *J Neurosci* **26**, 12852–12860.

Armstrong CM (1981). Sodium channels and gating currents. *Physiol Rev* **61**, 644–683.

Armstrong CM and Bezanilla F (1973). Currents related to movement of the gating particles of the sodium channels. *Nature* **242**, 459–461.

Armstrong CM and Bezanilla F (1974). Charge movement associated with the opening and closing of the activation gates of the na channels. *J Gen Physiol* **63**, 533–552.

Armstrong CM and Bezanilla F (1977). Inactivation of the sodium channel. II. Gating current experiments. *J Gen Physiol* **70**, 567–590.

Armstrong CM, Bezanilla F and Rojas E (1973b). Destruction of sodium conductance inactivation in squid axons perfused with pronase. *J Gen Physiol* **62**, 375–391.

Arroyo EJ, Xu T, Grinspan J *et al.* (2002). Genetic dysmyelination alters the molecular architecture of the nodal region. *J Neurosci* **22**, 1726–1737.

Asamoah OK, Wuskell JP, Loew LM *et al.* (2003). A fluorometric approach to local electric field measurements in a voltage-gated ion channel. *Neuron* **37**, 85–97.

Auld VJ, Goldin AL, Krafte DS *et al.* (1990). A neutral amino acid change in segment IIS4 dramatically alters the gating properties of the voltage-dependent sodium channel. *Proc Natl Acad Sci USA* **87**, 323–327.

Backx PH, Yue DT, Lawrence JH *et al.* (1992). Molecular localization of an ion-binding site within the pore of mammalian sodium channels. *Science* **257**, 248–251.

Bai CX, Glaaser IW, Sawanobori T *et al.* (2003). Involvement of local anesthetic binding sites on IVS6 of sodium channels in fast and slow inactivation. *Neurosci Lett* **337**, 41–45.

Baker MD (2005). Protein kinase C mediates up-regulation of tetrodotoxin-resistant, persistent Na$^+$ current in rat and mouse sensory neurones. *J Physiol* **567**, 851–867.

Baker MD, Chandra SY, Ding Y *et al.* (2003). GTP-induced tetrodotoxin-resistant Na$^+$ current regulates excitability in mouse and rat small diameter sensory neurones. *J Physiol* **548**, 373–382.

Balser JR, Nuss HB, Chiamvimonvat N *et al.* (1996a). External pore residue mediates slow inactivation in mu 1 rat skeletal muscle sodium channels. *Journal of Physiology* **494**, 431–442.

Balser JR, Nuss HB, Romashko DN *et al.* (1996b). Functional consequences of lidocaine binding to slow-inactivated sodium channels. *J Gen Physiol* **107**, 643–658.

Barela AJ, Waddy SP, Lickfett JG *et al.* (2006). An epilepsy mutation in the sodium channel SCN1a that decreases channel excitability. *J Neurosci* **26**, 2714–2723.

Bechtold DA, Kapoor R and Smith KJ (2004a). Axonal protection using flecainide in experimental autoimmune encephalomyelitis. *Ann Neurol* **55**, 607–616.

Bechtold DA, Yue X, Evans RM *et al.* (2004b). Axonal protection in experimental autoimmune neuritis by the sodium channel blocking agent flecainide. *Brain* **128**, 18–28.

Beckh S, Noda M, Lubbert H *et al.* (1989). Differential regulation of three sodium channel messenger RNAs in the rat central nervous system during development. *EMBO J* **8**, 3611–3616.

Belcher SM, Zerillo CA, Levenson R *et al.* (1995). Cloning of a sodium channel alpha subunit from rabbit Schwann cells. *Proc Natl Acad Sci USA* **92**, 11034–11038.

Bendahhou S, Cummins TR, Hahn AF *et al.* (2000). A double mutation in families with periodic paralysis defines new aspects of sodium channel slow inactivation. *J Clin Invest* **106**, 431–438.

Bendahhou S, Cummins TR, Kula RW *et al.* (2002). Impairment of slow inactivation as a common mechanism for periodic paralysis in DIIS4–S5. *Neurology* **58**, 1266–1272.

Bendahhou S, Cummins TR, Tawil R *et al.* (1999). Activation and inactivation of the voltage-gated sodium channel: role of segment S5 revealed by a novel hyperkalaemic periodic paralysis mutation. *J Neurosci* **19**, 4762–4771.

Benitah JP, Chen Z, Balser JR *et al.* (1999). Molecular dynamics of the sodium channel pore vary with gating: interactions between P-segment motions and inactivation. *J Neurosci* **19**, 1577–1585.

Bennett E, Urcan MS, Tinkle SS *et al.* (1997). Contribution of sialic acid to the voltage dependence of sodium channel gating. A possible electrostatic mechanism. *J Gen Physiol* **109**, 327–343.

Bennett GJ and Xie YK (1988). A peripheral mononeuropathy in rat that produces disorders of pain sensation like those seen in man. *Pain* **33**, 87–107.

Berkovic SF, Heron SE, Giordano L *et al.* (2004). Benign familial neonatal-infantile seizures: Characterization of a new sodium channelopathy. *Ann Neurol* **55**, 550–557.

Beyak MJ, Ramji N, Krol KM *et al.* (2004). Two TTX-resistant Na$^+$ currents in mouse colonic dorsal root ganglia neurons and their role in colitis-induced hyperexcitability. *Am J Physiol Gastrointest Liver Physiol* **287**, G845–855.

Bezzina CR, Rook MB, Groenewegen WA *et al.* (2003). Compound heterozygosity for mutations (W156X and R225W) in SCN5a associated with severe cardiac conduction disturbances and degenerative changes in the conduction system. *Circ Res* **92**, 159–168.

Bezzina CR, Shimizu W, Yang P *et al.* (2006). Common sodium channel promoter haplotype in Asian subjects underlies variability in cardiac conduction. *Circulation* **113**, 338–344.

Black JA and Waxman SG (2002b). Molecular identities of two tetrodotoxin-resistant sodium channels in corneal axons. *Exp Eye Res* **75**, 193–199.

Black JA, Cummins TR, Plumpton C *et al.* (1999a). Upregulation of a silent sodium channel after peripheral, but not central, nerve injury in DRG neurons. *J Neurophysiol* **82**, 2776–2785.

Black JA, Cummins TR, Yoshimura N et al. (2003). Tetrodotoxin-resistant sodium channels Na$_V$1.8/SNS and Na$_V$1.9/NAN in afferent neurons innervating urinary bladder in control and spinal cord injured rats. Brain Res **963**, 132–138.

Black JA, Dib-Hajj S, Baker D et al. (2000). Sensory neuron-specific sodium channel SNS is abnormally expressed in the brains of mice with experimental allergic encephalomyelitis and humans with multiple sclerosis. Proc Natl Acad Sci USA **97**, 11598–11602.

Black JA, Dib-Hajj S, Cohen S et al. (1998). Glial cells have heart: rH1 Na$^+$ channel mRNA and protein in spinal cord astrocytes. Glia **23**, 200–208.

Black JA, Dib-Hajj S, McNabola K et al. (1996). Spinal sensory neurons express multiple sodium channel alpha-subunit mRNAs. Mol Brain Res **43**, 117–131.

Black JA, Fjell J, Dib-Hajj S et al. (1999b). Abnormal expression of SNS/PN3 sodium channel in cerebellar Purkinje cells following loss of myelin in the taiep rat. Neuroreport **10**, 913–918.

Black JA, Langworthy K, Hinson AW et al. (1997). NGF has opposing effects on Na$^+$ channel III and SNS gene expression in spinal sensory neurons. Neuroreport **8**, 2331–2335.

Black JA, Liu S, Tanaka M et al. (2004). Changes in the expression of tetrodotoxin-sensitive sodium channels within dorsal root ganglia neurons in inflammatory pain. Pain **108**, 237–247.

Black JA, Renganathan M and Waxman SG (2002a). Sodium channel Na$_V$1.6 is expressed along nonmyelinated axons and it contributes to conduction. Mol Brain Res **105**, 19–28.

Black JA, Waxman SG and Smith KJ (2006). Remyelination of dorsal column axons by endogenous Schwann cells restores the normal pattern of Na$_V$1.6 and Kv1.2 at nodes of Ranvier. Brain **129**, 1319–1329.

Blair NT and Bean BP (2002). Roles of tetrodotoxin (TTX)-sensitive Na$^+$ current, TTX-resistant Na$^+$ current, and Ca^{2+} current in the action potentials of nociceptive sensory neurons. J Neurosci **22**, 10277–10290.

Blair NT and Bean BP (2003). Role of tetrodotoxin-resistant Na$^+$ current slow inactivation in adaptation of action potential firing in small-diameter dorsal root ganglion neurons. J Neurosci **23**, 10338–10350.

Blum R, Kafitz KW and Konnerth A (2002). Neurotrophin-evoked depolarization requires the sodium channel Na$_V$1.9. Nature **419**, 687–693.

Blumenthal K and Seibert A (2003). Voltage-gated sodium channel toxins: poisons, probes, and future promise. Cell Biochem Biophys **38**, 215–238.

Boiko T, Rasband MN, Levinson SR et al. (2001). Compact myelin dictates the differential targeting of two sodium channel isoforms in the same axon. Neuron **30**, 91–104.

Bolotina V, Courtney KR and Khodorov B (1992). Gate-dependent blockade of sodium channels by phenothiazine derivatives: structure–activity relationships. Mol Pharmacol **42**, 423–431.

Bonifacio MJ, Sheridan RD, Parada A et al. (2001). Interaction of the novel anticonvulsant, BIA 2–093, with voltage-gated sodium channels: comparison with carbamazepine. Epilepsia **42**, 600–608.

Boucher TJ, Okuse K, Bennett DL et al. (2000). Potent analgesic effects of GDNF in neuropathic pain states. Science **290**, 124–127.

Bouhours M, Luce S, Sternberg D et al. (2005). A1152D mutation of the Na$^+$ channel causes paramyotonia congenita and emphasizes the role of DIII/S4–S5 linker in fast inactivation. J Physiol **565**, 415–427.

Bricelj VM, Connell L, Konoki K et al. (2005). Sodium channel mutation leading to saxitoxin resistance in clams increases risk of PSP. Nature **434**, 763–767.

Buchner DA, Seburn KL, Frankel WN et al. (2004). Three ENU-induced neurological mutations in the pore loop of sodium channel SCN8a (Na$_V$1.6) and a genetically linked retinal mutation, rd13. Mamm Genome **15**, 344–351.

Buchner DA, Trudeau M, George AL, Jr et al. (2003). High-resolution mapping of the sodium channel modifier Scnm1 on mouse chromosome 3 and identification of a 1.3-kb recombination hot spot. Genomics **82**, 452–459.

Burgess DL, Kohrman DC, Galt J et al. (1995). Mutation of a new sodium channel gene, SCN8a, in the mouse mutant 'motor endplate disease'. Nature Genetics **10**, 461–465.

Cahalan MD (1975). Modification of sodium channel gating in frog myelinated nerve fibres by Centruroides sculpturatus scorpion venom. J Physiol **244**, 511–534.

Cannon SC (2002). An expanding view for the molecular basis of familial periodic paralysis. Neuromuscul Disord **12**, 533–543.

Cannon SC (2006). Pathomechanisms in channelopathies of skeletal muscle and brain. Annu Rev Neurosci **29**, 387–415.

Cannon SC and Strittmatter SM (1993). Functional expression of sodium channel mutations identified in families with periodic paralysis. Neuron **10**, 317–326.

Cannon SC, Brown RH Jr and Corey DP (1991). A sodium channel defect in hyperkalemic periodic paralysis: potassium-induced failure of inactivation. Neuron **6**, 619–626.

Cantrell AR and Catterall WA (2001). Neuromodulation of Na$^+$ channels: an unexpected form of cellular plasticity. Nat Rev Neurosci **2**, 397–407.

Cantrell AR, Tibbs VC, Yu FH et al. (2002). Molecular mechanism of convergent regulation of brain Na$^+$ channels by protein kinase C and protein kinase A anchored to AKAP-15. Mol Cell Neurosci **21**, 63–80.

Carrillo-Tripp M, San-Roman ML, Hernandez-Cobos J et al. (2006). Ion hydration in nanopores and the molecular basis of selectivity. Biophysical Chemistry **124**, 243–250.

Catterall WA (1977). Membrane potential-dependent binding of scorpion toxin to the action potential Na$^+$ ionophore. Studies with a toxin derivative prepared by lactoperoxidase-catalyzed iodination. J Biol Chem **252**, 8660–8668.

Catterall WA (2000). From ionic currents to molecular mechanisms: The structure and function of voltage-gated sodium channels. Neuron **26**, 13–25.

Catterall WA and Beress L (1978). Sea anemone toxin and scorpion toxin share a common receptor site associated with the action potential sodium ionophore. J Biol Chem **253**, 7393–7396.

Catterall WA, Goldin AL and Waxman SG (2005). International union of pharmacology. XlVII. Nomenclature and structure–function relationships of voltage-gated sodium channels. Pharmacol Rev **57**, 397–409.

Catterall WA, Schmidt JW, Messner DJ et al. (1986). Structure and biosynthesis of neuronal sodium channels. Ann N Y Acad Sci **479**, 186–203.

Cestele S and Catterall WA (2000). Molecular mechanisms of neurotoxin action on voltage-gated sodium channels. Biochimie **82**, 883–892.

Cestele S, Qu YS, Rogers JC et al. (1998). Voltage sensor-trapping—enhanced activation of sodium channels by beta-scorpion toxin bound to the S3–S4 loop in domain II. Neuron **21**, 919–931.

Cestele S, Scheuer T, Mantegazza M et al. (2001). Neutralization of gating charges in domain II of the sodium channel alpha subunit enhances voltage-sensor trapping by a beta-scorpion toxin. J Gen Physiol **118**, 291–302.

Cestele S, Yarov-Yarovoy V, Qu Y et al. (2006). Structure and function of the voltage sensor of sodium channels probed by a β-scorpion toxin. J Biol Chem **281**, 21332–21344.

Cha A, Ruben PC, George AL et al. (1999). Voltage sensors in domains III and IV, but not I and II, are immobilized by Na$^+$ channel fast inactivation. Neuron **22**, 73–87.

Chahine M, Deschene I, Chen LQ et al. (1996). Electrophysiological characteristics of cloned skeletal and cardiac muscle sodium channels. Am. J. Physiol. **271**, H498–506.

Chahine M, Ziane R, Vijayaragavan K et al. (2005). Regulation of Na$_V$ channels in sensory neurons. Trends Pharmacol Sci **26**, 496–502.

Chanda B and Bezanilla F (2002). Tracking voltage-dependent conformational changes in skeletal muscle sodium channel during activation. J Gen Physiol **120**, 629–645.

Chanda B, Asamoah OK and Bezanilla F (2004). Coupling interactions between voltage sensors of the sodium channel as revealed by site-specific measurements. *J Gen Physiol* **123**, 217–230.

Chang NS, French RJ, Lipkind GM *et al.* (1998). Predominant interactions between μ-conotoxin ARG-13 and the skeletal muscle Na+ channel localized by mutant cycle analysis. *Biochemistry* **37**, 4407–4419.

Chen C, Westenbroek RE, Xu X *et al.* (2004). Mice lacking sodium channel beta1 subunits display defects in neuronal excitability, sodium channel expression, and nodal architecture. *J Neurosci* **24**, 4030–4042.

Chen H, Gordon D and Heinemann SH (2000). Modulation of cloned skeletal muscle sodium channels by the scorpion toxins Lqh II, Lqh III, and Lqh alphaIT. *Pflugers Arch* **439**, 423–432.

Chen LQ, Santarelli V, Horn R *et al.* (1996). A unique role for the S4 segment of domain 4 in the inactivation of sodium channels. *J. Gen. Physiol.* **108**, 549–556.

Chen Y, Yu FH, Surmeier DJ *et al.* (2006). Neuromodulation of Na+ channel slow inactivation via cAMP-dependent protein kinase and protein kinase C. *Neuron* **49**, 409–420.

Chiamvimonvat N, Perez-Garcia MT, Ranjan R *et al.* (1996). Depth asymmetries of the pore-lining segments of the Na+ channel revealed by cysteine mutagenesis. *Neuron* **16**, 1037–1047.

Choi J, Hudmon A, Waxman S *et al.* (2006a). Calmodulin regulates current density and frequency-dependent inhibition of sodium channel Na$_V$1.8 in DRG neurons. *J Neurophysiol* **96**, 97–108.

Choi JS, Dib-Hajj SD and Waxman SG (2006b). Inherited erythermalgia. Limb pain from an S4 charge-neutral Na channelopathy. *Neurology* **67**, 1563–1567.

Choi JS, Tyrrell L, Waxman SG *et al.* (2004). Functional role of the c-terminus of voltage-gated sodium channel Na$_V$1.8. *FEBS Lett* **572**, 256–260.

Chung JM, Dib-Hajj SD and Lawson SN (2003). Sodium channel subtypes and neuropathic pain. In JO Dostrovsky, DB Carr and M Koltzenberg (eds) *Proceedings of the 10th congress in pain research and management*, vol. 24. IASP Press, Seattle, pp. 99–114.

Claes L, Ceulemans B, Audenaert D *et al.* (2003). De novo SCN1a mutations are a major cause of severe myoclonic epilepsy of infancy. *Hum Mutat* **21**, 615–621.

Clancy CE and Rudy Y (1999). Linking a genetic defect to its cellular phenotype in a cardiac arrhythmia. *Nature* **400**, 566–569.

Clancy CE and Rudy Y (2002). Na+ channel mutation that causes both brugada and long-QT syndrome phenotypes: a simulation study of mechanism. *Circulation* **105**, 1208–1213.

Coggeshall RE, Tate S and Carlton SM (2004). Differential expression of tetrodotoxin-resistant sodium channels Na$_V$1.8 and Na$_V$1.9 in normal and inflamed rats. *Neurosci Lett* **355**, 45–48.

Cohen L, Gilles N, Karbat I *et al.* (2006). Direct evidence that receptor site-4 of sodium channel gating modifiers is not dipped in the phospholipid bilayer of neuronal membranes. *J Biol Chem* **281**, 20673–20679.

Cohen SA and Barchi RL (1993). Voltage-dependent sodium channels. *International Review of Cytology* **137C**, 55–103.

Conti F and Stuhmer W (1989). Quantal charge redistributions accompanying the structural transitions of sodium channels. *Eur Biophys J* **17**, 53–59.

Cormier JW, Rivolta I, Tateyama M *et al.* (2002). Secondary structure of the human cardiac Na+ channel carboxy terminus: evidence for a role of helical structures in modulation of channel inactivation. *J Biol Chem* **277**, 9233–9241.

Cossette P, Loukas A, Lafreniere RG *et al.* (2003). Functional characterization of the D188V mutation in neuronal voltage-gated sodium channel causing generalized epilepsy with febrile seizures plus (GEFS). *Epilepsy Res* **53**, 107–117.

Courtney KR (1975). Mechanism of frequency-dependent inhibition of sodium currents in frog myelinated nerve by the lidocaine derivative GEA. *J Pharmacol Exp Ther* **195**, 225–236.

Courtney KR, Kendig JJ and Cohen EN (1978). Frequency-dependent conduction block: the role of nerve impulse pattern in local anesthetic potency. *Anesthesiology* **48**, 111–117.

Cox JJ, Reimann F, Nicholas AK *et al.* (2006). An SCN9a channelopathy causes congenital inability to experience pain. *Nature* **444**, 894–898.

Craner MJ, Damarjian TG, Liu S *et al.* (2005). Sodium channels contribute to microglia/macrophage activation and function in EAE and MS. *Glia* **49**, 220–229.

Craner MJ, Hains BC, Lo AC *et al.* (2004a). Co-localization of sodium channel Na$_V$1.6 and the sodium-calcium exchanger at sites of axonal injury in the spinal cord in EAE. *Brain* **127**, 294–303.

Craner MJ, Klein JP, Renganathan M *et al.* (2002). Changes of sodium channel expression in experimental painful diabetic neuropathy. *Annals of Neurology* **52**, 786–792.

Craner MJ, Lo AC, Black JA *et al.* (2003b). Abnormal sodium channel distribution in optic nerve axons in a model of inflammatory demyelination. *Brain* **126**, 1552–1561.

Craner MJ, Lo AC, Black JA *et al.* (2003a). Annexin II/p11 is up-regulated in Purkinje cells in EAE and MS. *Neuroreport* **14**, 555–558.

Craner MJ, Newcombe J, Black JA *et al.* (2004b). Molecular changes in neurons in multiple sclerosis: altered axonal expression of nav1.2 and Na$_V$1.6 sodium channels and Na+/Ca2+ exchanger. *Proc Natl Acad Sci USA* **101**, 8168–8173.

Cruz LJ, Gray WR, Olivera BM *et al.* (1985). Conus geographus toxins that discriminate between neuronal and muscle sodium channels. *J Biol Chem* **260**, 9280–9288.

Cummins TR, Aglieco F and Dib-Hajj SD (2002). Critical molecular determinants of voltage-gated sodium channel sensitivity to mu-conotoxins GIIIA/B. *Mol Pharmacol* **61**, 1192–1201.

Cummins TR, Aglieco F, Renganathan M *et al.* (2001). Na$_V$1.3 sodium channels: rapid repriming and slow closed-state inactivation display quantitative differences after expression in a mammalian cell line and in spinal sensory neurons. *J Neurosci* **21**, 5952–5961.

Cummins TR and Sigworth FJ (1996). Impaired slow inactivation in mutant sodium channels. *Biophysical J* **71**, 227–236.

Cummins TR and Waxman SG (1997). Downregulation of tetrodotoxin-resistant sodium currents and upregulation of a rapidly repriming tetrodotoxin-sensitive sodium current in small spinal sensory neurons after nerve injury. *J Neurosci* **17**, 3503–3514.

Cummins TR, Black JA, Dib-Hajj SD *et al.* (2000). Glial-derived neurotrophic factor upregulates expression of functional SNS and NAN sodium channels and their currents in axotomized dorsal root ganglion neurons. *J Neurosci* **20**, 8754–8761.

Cummins TR, Dib-Hajj SD and Waxman SG (2004). Electrophysiological properties of mutant Na$_V$1.7 sodium channels in a painful inherited neuropathy. *J Neurosci* **24**, 8232–8236.

Cummins TR, Dib-Hajj SD, Black JA *et al.* (1999). A novel persistent tetrodotoxin-resistant sodium current in SNS-null and wild-type small primary sensory neurons. *J Neurosci* **19**, RC43.

Cummins TR, Dib-Hajj SD, Herzog RI *et al.* (2005). Na$_V$1.6 channels generate resurgent sodium currents in spinal sensory neurons. *FEBS Lett* **579**, 2166–2170.

Cummins TR, Zhou J, Sigworth FJ *et al.* (1993). Functional consequences of a Na+ channel mutation causing hyperkalemic periodic paralysis. *Neuron* **10**, 667–678.

Darbon P, Yvon C, Legrand JC *et al.* (2004). I$_{NaP}$ underlies intrinsic spiking and rhythm generation in networks of cultured rat spinal cord neurons. *Eur J Neurosci* **20**, 976–988.

Dargent B, Paillart C, Carlier E *et al.* (1994). Sodium channel internalization in developing neurons. *Neuron* **13**, 683–690.

De Luca A, Natuzzi F, Desaphy JF *et al.* (2000). Molecular determinants of mexiletine structure for potent and use-dependent block of skeletal muscle sodium channels. *Mol Pharmacol* **57**, 268–277.

De Repentigny Y, Cote PD, Pool M *et al.* (2001). Pathological and genetic analysis of the degenerating muscle (DMU) mouse: a new allele of SCN8a. *Hum Mol Genet* **10**, 1819–1827.

Decosterd I, Ji RR, Abdi S *et al.* (2002). The pattern of expression of the voltage-gated sodium channels Na$_V$1.8 and Na$_V$1.9 does not change in uninjured primary sensory neurons in experimental neuropathic pain models. *Pain* **96**, 269–277.

Del Negro CA, Koshiya N, Butera RJ Jr *et al.* (2002). Persistent sodium current, membrane properties and bursting behavior of pre-botzinger complex inspiratory neurons *in vitro*. *J Neurophysiol* **88**, 2242–2250.

Deschenes I, Chen L, Kallen RG *et al.* (1998). Electrophysiological study of chimeric sodium channels from heart and skeletal muscle. *J Membr Biol* **164**, 25–34.

Deschenes I, Neyroud N, DiSilvestre D *et al.* (2002). Isoform-specific modulation of voltage-gated Na$^+$ channels by calmodulin. *Circ Res* **90**, E49–57.

Deschenes I, Trottier E and Chahine M (2001). Implication of the c-terminal region of the alpha-subunit of voltage-gated sodium channels in fast inactivation. *J Membr Biol* **183**, 103–114.

Dib-Hajj S, Black JA, Cummins TR *et al.* (2002). NAN/Na$_V$1.9: a sodium channel with unique properties. *Trends Neurosci* **25**, 253–259.

Dib-Hajj S, Black JA, Felts P *et al.* (1996). Down-regulation of transcripts for Na channel alpha-SNS in spinal sensory neurons following axotomy. *Proc Natl Acad Sci USA* **93**, 14950–14954.

Dib-Hajj SD, Black JA, Cummins TR *et al.* (1998a). Rescue of alpha-SNS sodium channel expression in small dorsal root ganglion neurons following axotomy by *in vivo* administration of nerve growth factor. *J Neurophysiol* **79**, 2668–2676.

Dib-Hajj SD, Fjell J, Cummins TR *et al.* (1999). Plasticity of sodium channel expression in DRG neurons in the chronic constriction injury model of neuropathic pain. *Pain* **83**, 591–600.

Dib-Hajj SD, Ishikawa K, Cummins TR *et al.* (1997). Insertion of a SNS-specific tetrapeptide in S3–S4 linker of D4 accelerates recovery from inactivation of skeletal muscle voltage-gated Na channel μ1 in HEK293 cells. *FEBS Lett* **416**, 11–14.

Dib-Hajj SD, Rush AM, Cummins TR *et al.* (2005). Gain-of-function mutation in Na$_V$1.7 in familial erythromelalgia induces bursting of sensory neurons. *Brain* **128**, 1847–1854.

Dib-Hajj SD, Tyrrell L, Black JA *et al.* (1998b). NAN, a novel voltage-gated Na channel, is expressed preferentially in peripheral sensory neurons and down-regulated after axotomy. *Proc Natl Acad Sci USA* **95**, 8963–8968.

Dichgans M, Freilinger T, Eckstein G *et al.* (2005). Mutation in the neuronal voltage-gated sodium channel scn1a in familial hemiplegic migraine. *Lancet* **366**, 371–377.

Dietrich PS, McGivern JG, Delgado SG *et al.* (1998). Functional analysis of a voltage-gated sodium channel and its splice variant from rat dorsal root ganglia. *J Neurochem* **70**, 2262–2272.

Diss JK, Stewart D, Pani F *et al.* (2005). A potential novel marker for human prostate cancer: Voltage-gated sodium channel expression *in vivo*. *Prostate Cancer Prostatic Dis* **8**, 266–273.

Djouhri L, Fang X, Okuse K *et al.* (2003a). The TTX-resistant sodium channel Na$_V$1.8 (SNS/PN3): expression and correlation with membrane properties in rat nociceptive primary afferent neurons. *J Physiol (Lond)* **550**, 739–752.

Djouhri L, Newton R, Levinson SR *et al.* (2003b). Sensory and electrophysiological properties of guinea-pig sensory neurones expressing Na$_V$1.7 (PN1) Na$^+$ channel alpha-subunit protein. *J Physiol (Lond)* **546**, 565–576.

Do MT and Bean BP (2003). Subthreshold sodium currents and pacemaking of subthalamic neurons: modulation by slow inactivation. *Neuron* **39**, 109–120.

Do MT and Bean BP (2004). Sodium currents in subthalamic nucleus neurons from Na$_V$1.6-null mice. *J Neurophysiol* **92**, 726–733.

Dong XW, Goregaoker S, Engler H *et al.* (2007). Small interfering RNA-mediated selective knockdown of NaV1.8 tetrodotoxin-resistant sodium channel reverses mechanical allodynia in neuropathic rats. *Neuroscience* 11, 812–821

Doyle DA, Morais Cabral J, Pfuetzner RA *et al.* (1998). The structure of the potassium channel: molecular basis of K$^+$ conduction and selectivity. *Science* **280**, 69–77.

Drenth JP, Te Morsche RH, Guillet G *et al.* (2005). SCN9a mutations define primary erythermalgia as a neuropathic disorder of voltage-gated sodium channels. *J Invest Dermatol* **124**, 1333–1338.

Dudley SC, Jr., Chang N, Hall J *et al.* (2000). Mu-conotoxin GIIIa interactions with the voltage-gated Na$^+$ channel predict a clockwise arrangement of the domains. *J Gen Physiol* **116**, 679–690.

Dupree JL, Mason JL, Marcus JR *et al.* (2005). Oligodendrocytes assist in the maintenance of sodium channel clusters independent of the myelin sheath. *Neuron Glia Biol* **1**, 1–14.

Eaholtz G, Scheuer T and Catterall WA (1994). Restoration of inactivation and block of open sodium channels by an inactivation gate peptide. *Neuron* **12**, 1041–1048.

Enomoto A, Han JM, Hsiao CF *et al.* (2006). Participation of sodium currents in burst generation and control of membrane excitability in mesencephalic trigeminal neurons. *J Neurosci* **26**, 3412–3422.

Escayg A, Heils A, MacDonald BT *et al.* (2001). A novel SCN1a mutation associated with generalized epilepsy with febrile seizures plus–and prevalence of variants in patients with epilepsy. *Am J Hum Genet* **68**, 866–873.

Fache MP, Moussif A, Fernandes F *et al.* (2004). Endocytotic elimination and domain-selective tethering constitute a potential mechanism of protein segregation at the axonal initial segment. *J Cell Biol* **166**, 571–578.

Fainzilber M, Kofman O, Zlotkin E *et al.* (1994). A new neurotoxin receptor site on sodium channels is identified by a conotoxin that affects sodium channel inactivation in molluscs and acts as an antagonist in rat brain. *J Biol Chem* **269**, 2574–2580.

Fang X, Djouhri L, Black JA *et al.* (2002). The presence and role of the tetrodotoxin-resistant sodium channel Na$_V$1.9 (NAN) in nociceptive primary afferent neurons. *J Neurosci* **22**, 7425–7433.

Fang X, Djouhri L, McMullan S *et al.* (2005). TrkA is expressed in nociceptive neurons and influences electrophysiological properties via Na$_V$1.8 expression in rapidly conducting nociceptors. *J Neurosci* **25**, 4868–4878.

Favre I, Moczydlowski E and Schild L (1996). On the structural basis for ionic selectivity among Na$^+$, K$^+$, and Ca^{2+} in the voltage-gated sodium channel. *Biophys J* **71**, 3110–3125.

Featherstone DE, Richmond JE and Ruben PC (1996). Interaction between fast and slow inactivation in SkM1 sodium channels. *Biophysical Journal* **71**, 3098–3109.

Felts PA, Yokoyama S, Dib-Hajj S *et al.* (1997). Sodium channel alpha-subunit mRNAs I, II, III, NaG, Na6 and hNE (PN1)—different expression patterns in developing rat nervous system. *Mol Brain Res* **45**, 71–82.

Ferriby D, Stojkovic T, Sternberg D *et al.* (2006). A new case of autosomal dominant myotonia associated with the V1589M missense mutation in the muscle sodium channel gene and its phenotypic classification. *Neuromuscular Disorders* **16**, 321–324.

Fertleman CR, Baker MD, Parker KA *et al.* (2006). *SCN9a* mutations in paroxysmal extreme pain disorder: Allelic variants underlie distinct channel defects and phenotypes. *Neuron* **52**, 767–774.

Fertleman CR and Ferrie CD (2006) What's in a name—familial rectal pain syndrome becomes paroxysmal extreme pain disorder. *J Neurol Neurosurg Psychiatry* **77**, 1294-1295.

Filatov GN, Nguyen TP, Kraner SD *et al.* (1998). Inactivation and secondary structure in the D4/S4–5 region of the SkM1 sodium channel. *J Gen Physiol* **111**, 703-715.

Fitzgerald EM, Okuse K, Wood JN *et al.* (1999). cAMP-dependent phosphorylation of the tetrodotoxin-resistant voltage-dependent sodium channel SNS. *J Physiol (Lond)* **516**, 433–446.

Fjell J, Cummins TR, Dib-Hajj SD *et al.* (1999). Differential role of GDNF and NGF in the maintenance of two TTX-resistant sodium channels in adult DRG neurons. *Mol Brain Res* **67**, 267–282.

Fjell J, Hjelmstrom P, Hormuzdiar W *et al.* (2000). Localization of the tetrodotoxin-resistant sodium channel NaN in nociceptors. *Neuroreport* **11**, 199–202.

Fotia AB, Ekberg J, Adams DJ *et al.* (2004). Regulation of neuronal voltage-gated sodium channels by the ubiquitin-protein ligases nedd4 and nedd4–2. *J Biol Chem* **279**, 28930–28935.

Fozzard HA, Lee PJ and Lipkind GM (2005). Mechanism of local anesthetic drug action on voltage-gated sodium channels. *Curr Pharm Des* **11**, 2671–2686.

Fraser SP, Diss JK, Chioni AM *et al.* (2005). Voltage-gated sodium channel expression and potentiation of human breast cancer metastasis. *Clin Cancer Res* **11**, 5381–5389.

Fraser SP, Diss JK, Lloyd LJ *et al.* (2004). T-lymphocyte invasiveness: control by voltage-gated Na$^+$ channel activity. *FEBS Lett* **569**, 191–194.

Fujiwara T, Sugawara T, Mazaki-Miyazaki E *et al.* (2003). Mutations of sodium channel alpha subunit type 1 (SCN1a) in intractable childhood epilepsies with frequent generalized tonic–clonic seizures. *Brain* **126**, 531–546.

Fukuma G, Oguni H, Shirasaka Y *et al.* (2004). Mutations of neuronal voltage-gated Na$^+$ channel alpha 1 subunit gene *SCN1a* in core severe myoclonic epilepsy in infancy (SMEI) and in borderline SMEI (SMEB). *Epilepsia* **45**, 140–148.

Garrido JJ, Fernandes F, Giraud P *et al.* (2001). Identification of an axonal determinant in the c-terminus of the sodium channel Na$_V$1.2. *Embo J* **20**, 5950–5961.

Garrido JJ, Giraud P, Carlier E *et al.* (2003). A targeting motif involved in sodium channel clustering at the axonal initial segment. *Science* **300**, 2091–2094.

Gee SH, Madhavan R, Levinson SR *et al.* (1998). Interaction of muscle and brain sodium channels with multiple members of the syntrophin family of dystrophin-associated proteins. *J Neurosci* **18**, 128–137.

Geffeney S, Brodie ED Jr, Ruben PC *et al.* (2002). Mechanisms of adaptation in a predator-prey arms race: TTX-resistant sodium channels. *Science* **297**, 1336–1339.

Geffeney SL, Fujimoto E, Brodie ED 3rd *et al.* (2005). Evolutionary diversification of TTX-resistant sodium channels in a predator-prey interaction. *Nature* **434**, 759–763.

George AL (2005). Inherited disorders of voltage-gated sodium channels. *J Clin Invest* **115**, 1990–1999.

Gilles N, Leipold E, Chen H *et al.* (2001). Effect of depolarization on binding kinetics of scorpion alpha-toxin highlights conformational changes of rat brain sodium channels. *Biochemistry* **40**, 14576–14584.

Glaaser IW, Bankston JR, Liu H *et al.* (2006). A carboxy terminal hydrophobic interface is critical to sodium channel function: relevance to inherited disorders. *J Biol Chem* **281**, 24015–24023.

Gold MS, Levine JD and Correa AM (1998). Modulation of TTX-R I_{Na} by PKC and PKA and their role in PGE2-induced sensitization of rat sensory neurons *in vitro*. *J Neurosci* **18**, 10345–10355.

Gold MS, Reichling DB, Shuster MJ *et al.* (1996). Hyperalgesic agents increase a tetrodotoxin-resistant Na$^+$ current in nociceptors. *Proc Natl Acad Sci USA* **93**, 1108–1112.

Gold MS, Weinreich D, Kim CS *et al.* (2003). Redistribution of Na$_V$1.8 in uninjured axons enables neuropathic pain. *J Neurosci* **23**, 158–166.

Gold MS, Zhang L, Wrigley DL *et al.* (2002). Prostaglandin E2 modulatesTTX-r I_{Na} in rat colonic sensory neurons. *J Neurophysiol* **88**, 1512–1522.

Goldin AL (2002). Evolution of voltage-gated Na$^+$ channels. *J Exp Biol* **205**, 575–584.

Goldin AL (2003). Mechanisms of sodium channel inactivation. *Curr Opin Neurobiol* **13**, 284–290.

Goldin AL, Barchi RL, Caldwell JH *et al.* (2000). Nomenclature of voltage-gated sodium channels. *Neuron* **28**, 365–368.

Goldman L and Schauf CL (1972). Inactivation of the sodium current in myxicola giant axons. Evidence for coupling to the activation process. *J Gen Physiol* **59**, 659–675.

Gould HJ 3rd, England JD, Soignier RD *et al.* (2004). Ibuprofen blocks changes in Na$_V$1.7 and 1.8 sodium channels associated with complete Freund's adjuvant-induced inflammation in rat. *J Pain* **5**, 270–280.

Gould HJ, 3rd, Gould TN, England JD *et al.* (2000). A possible role for nerve growth factor in the augmentation of sodium channels in models of chronic pain. *Brain Res* **854**, 19–29.

Grieco TM, Afshari FS and Raman IM (2002). A role for phosphorylation in the maintenance of resurgent sodium current in cerebellar Purkinje neurons. *J Neurosci* **22**, 3100–3107.

Grieco TM, Malhotra JD, Chen C *et al.* (2005). Open-channel block by the cytoplasmic tail of sodium channel beta4 as a mechanism for resurgent sodium current. *Neuron* **45**, 233–244.

Gustafson TA, Clevinger EC, O'Neill TJ *et al.* (1993). Mutually exclusive exon splicing of type III brain sodium channel alpha subunit RNA generates developmentally regulated isoforms in rat brain. *J Biol Chem* **268**, 18648–18653.

Guy HR and Conti F (1990). Pursuing the structure and function of voltage-gated channels. *Trends Neurosci* **13**, 201–206.

Guy HR and Seetharamulu P (1986). Molecular model of the action potential sodium channel. *Proc Natl Acad Sci USA* **83**, 508–512.

Hains BC, Klein JP, Saab CY *et al.* (2003). Upregulation of sodium channel Na$_V$1.3 and functional involvement in neuronal hyperexcitability associated with central neuropathic pain after spinal cord injury. *J Neurosci* **23**, 8881–8892.

Hains BC, Saab CY and Waxman SG (2005). Changes in electrophysiological properties and sodium channel Na$_V$1.3 expression in thalamic neurons after spinal cord injury. *Brain* **128**, 2359–2371.

Hains BC, Saab CY, Klein JP *et al.* (2004). Altered sodium channel expression in second-order spinal sensory neurons contributes to pain after peripheral nerve injury. *J Neurosci* **24**, 4832–4839.

Han C, Rush AM, Dib-Hajj SD *et al.* (2006). Sporadic onset of erythermalgia: a gain-of-function mutation in Na$_V$1.7. *Ann Neurol* **59**, 553–558.

Hanrahan CJ, Palladino MJ, Ganetzky B *et al.* (2000). RNA editing of the drosophila para Na$^+$ channel transcript. Evolutionary conservation and developmental regulation. *Genetics* **155**, 1149–1160.

Hartmann HA, Colom LV, Sutherland ML *et al.* (1999). Selective localization of cardiac SCN5a sodium channels in limbic regions of rat brain. *Nat Neurosci* **2**, 593–595.

Hartmann HA, Tiedeman AA, Chen SF *et al.* (1994). Effects of III-IV linker mutations on human heart Na$^+$ channel inactivation gating. *Circ Res* **75**, 114–122.

Harty TP, Dib-Hajj SD, Tyrrell L *et al.* (2006). Na$_V$1.7 mutant A863P in erythromelalgia: effects of altered activation and steady-state inactivation on excitability of nociceptive DRG neurons. *J. Neurosci* **26**, 12566–12575.

Harvey PJ, Grochmal J, Tetzlaff W *et al.* (2005). An investigation into the potential for activity-dependent regeneration of the rubrospinal tract after spinal cord injury. *Eur J Neurosci* **22**, 3025–3035.

Hasson A, Fainzilber M, Gordon D *et al.* (1993). Alteration of sodium currents by new peptide toxins from the venom of a molluscivorous conus snail. *European Journal of Neuroscience* **5**, 56–64.

Heine R, Pika U and Lehmann-Horn F (1993). A novel SCN4a mutation causing myotonia aggravated by cold and potassium. *Hum Mol Genet* **2**, 1349–1353.

Heinemann SH, Schlief T, Mori Y *et al.* (1994). Molecular pore structure of voltage-gated sodium and calcium channels. *Braz J Med Biol Res* **27**, 2781–2802.

Heinemann SH, Terlau H, Stuhmer W *et al.* (1992). Calcium channel characteristics conferred on the sodium channel by single mutations. *Nature* **356**, 441–443.

Henry MA, Sorensen HJ, Johnson LR *et al.* (2005). Localization of the Na$_V$1.8 sodium channel isoform at nodes of Ranvier in normal human radicular tooth pulp. *Neurosci Lett* **380**, 32–36.

Heron S, Crossland K, Andermann E *et al.* (2002). Sodium-channel defects in benign familial neonatal-infantile seizures. *Lancet* **360**, 851.

Herzog RI, Cummins TR and Waxman SG (2001). Persistent TTX-resistant Na$^+$ current affects resting potential and response to depolarization in simulated spinal sensory neurons. *J Neurophysiol* **86**, 1351–1364.

Herzog RI, Cummins TR, Ghassemi F et al. (2003a). Distinct repriming and closed-state inactivation kinetics of Na$_V$1.6 and Na$_V$1.7 sodium channels in mouse spinal sensory neurons. *J Physiol (Lond)* **551**, 741–750.

Herzog RI, Liu C, Waxman SG et al. (2003b). Calmodulin binds to the c terminus of sodium channels Na$_V$1.4 and Na$_V$1.6 and differentially modulates their functional properties. *J Neurosci* **23**, 8261–8270.

Hilber K, Sandtner W, Kudlacek O et al. (2002). Interaction between fast and ultra-slow inactivation in the voltage-gated sodium channel: does the inactivation gate stabilize the channel structure? *J Biol Chem* **277**, 37105–37115.

Hilborn MD, Vaillancourt RR and Rane SG (1998). Growth factor receptor tyrosine kinases acutely regulate neuronal sodium channels through the src signaling pathway. *J Neurosci* **18**, 590–600.

Hille B (1971). The permeability of the sodium channel to organic cations in myelinated nerve. *J Gen Physiol* **58**, 599–619.

Hille B (1972). The permeability of the sodium channel to metal cations in myelinated nerve. *J Gen Physiol* **59**, 637–658.

Hille B (1977). Local anesthetics: hydrophilic and hydrophobic pathways for the drug-receptor reaction. *J Gen Physiol* **69**, 497–515.

Hille B (2001) *Ion channels of excitable membranes.* Sinauer Associates, Inc. Sunderland, MA.

Hillsley K, Lin JH, Stanisz A et al. (2006). Dissecting the role of sodium currents in visceral sensory neurons in a model of chronic hyperexcitability using Na$_V$1.8 and Na$_V$1.9 null mice. *J Physiol* **576**, 257–267.

Hirota K, Kaneko Y, Matsumoto G et al. (1999). Cloning and distribution of a putative tetrodotoxin-resistant Na$^+$ channel in newt retina. *Zoological Science* **16**, 587–594.

Hirschberg B, Rovner A, Lieberman M et al. (1995). Transfer of twelve charges is needed to open skeletal muscle Na$^+$ channels. *Journal of General Physiology* **106**, 1053–1068.

Hodgkin AL and Huxley AF (1952). A quantitative description of membrane current and its application to conduction and excitation in nerve. *J. Physiol. Lond.* **117**, 500–544.

Hoffman JF, Dodson A, Wickrema A et al. (2004). Tetrodotoxin-sensitive Na$^+$ channels and muscarinic and purinergic receptors identified in human erythroid progenitor cells and red blood cell ghosts. *Proc Natl Acad Sci SA* **101**, 12370–12374.

Holm AN, Rich A, Miller SM et al. (2002). Sodium current in human jejunal circular smooth muscle cells. *Gastroenterol* **122**, 178–187.

Hondeghem LM and Katzung BG (1977). Time- and voltage-dependent interactions of antiarrhythmic drugs with cardiac sodium channels. *Biochim Biophys Acta* **472**, 373–398.

Horn R (2004). How S4 segments move charge. Let me count the ways. *J Gen Physiol* **123**, 1–4.

Horn R and Vandenberg CA (1984). Statistical properties of single sodium channels. *J Gen Physiol* **84**, 505–534.

Hudmon A, Choi JS, Tyrrell L et al. (2008) Phosphorylation of sodium channel Nav1.8 by p38 MAPK increases current density in DRG neurons. *J Neurosci*, **28**, 3190–3201.

Islas LD and Sigworth FJ (2001). Electrostatics and the gating pore of shaker potassium channels. *J Gen Physiol* **117**, 69–89.

Isom LL, De Jongh KS and Catterall WA (1994). Auxiliary subunits of voltage-gated ion channels. *Neuron* **12**, 1183–1194.

Isozumi K, DeLong R, Kaplan J et al. (1996). Linkage of scapuloperoneal spinal muscular atrophy to chromosome 12q24.1–q24.31. *Hum Mol Genet* **5**, 1377–1382.

Jeglitsch G, Rein K, Baden DG et al. (1998). Brevetoxin-3 (PbTx-3) and its derivatives modulate single tetrodotoxin-sensitive sodium channels in rat sensory neurons. *Journal of Pharmacology and Experimental Therapeutics* **284**, 516–525.

Jeong SY, Goto J, Hashida H et al. (2000). Identification of a novel human voltage-gated sodium channel alpha subunit gene, SCN12a. *Biochem Biophys Res Comm* **267**, 262–270.

Ji S, George AL Jr, Horn R et al. (1996). Paramyotonia congenita mutations reveal different roles for segments S3 and S4 of domain D4 in hSkS1 sodium channel gating. *Journal of General Physiology* **107**, 183–194.

Jiang QX, Wang DN and MacKinnon R (2004). Electron microscopic analysis of Kvap voltage-dependent K$^+$ channels in an open conformation. *Nature* **430**, 806–810.

Jiang Y, Lee A, Chen J et al. (2002a). Crystal structure and mechanism of a calcium-gated potassium channel. *Nature* **417**, 515–522.

Jiang Y, Lee A, Chen J et al. (2002b). The open pore conformation of potassium channels. *Nature* **417**, 523–526.

Jiang Y, Lee A, Chen J et al. (2003a). X-ray structure of a voltage-dependent K$^+$ channel. *Nature* **423**, 33–41.

Jiang Y, Ruta V, Chen J et al. (2003b). The principle of gating charge movement in a voltage-dependent K$^+$ channel. *Nature* **423**, 42–48.

Jin X and Gereau RW (2006). Acute p38-mediated modulation of tetrodotoxin-resistant sodium channels in mouse sensory neurons by tumor necrosis factor-alpha. *J Neurosci* **26**, 246–255.

Jo T, Nagata T, Iida H et al. (2004). Voltage-gated sodium channel expressed in cultured human smooth muscle cells: involvement of SCN9a. *FEBS Lett* **567**, 339–343.

Joshi SK, Mikusa JP, Hernandez G et al. (2006). Involvement of the TTX-resistant sodium channel Na$_V$1.8 in inflammatory and neuropathic, but not post-operative, pain states. *Pain* **123**, 75–82.

Jover E, Couraud F and Rochat H (1980). Two types of scorpion neurotoxins characterized by their binding to two separate receptor sites on rat brain synaptosomes. *Biochem Biophys Res Comm* **95**, 1607–1614.

Kallen RG, Sheng ZH, Yang J et al. (1990). Primary structure and expression of a sodium channel characteristic of denervated and immature rat skeletal muscle. *Neuron* **4**, 233–242.

Kambouris NG, Hastings LA, Stepanovic S et al. (1998). Mechanistic link between lidocaine block and inactivation probed by outer pore mutations in the rat μ1 skeletal muscle sodium channel. *J Physiol* **512**, 693–705.

Kamiya K, Kaneda M, Sugawara T et al. (2004). A nonsense mutation of the sodium channel gene SCN2a in a patient with intractable epilepsy and mental decline. *J Neurosci* **24**, 2690–2698.

Kaplan MR, Cho M, Ullian EM et al. (2001). Differential control of clustering of the sodium channels Na$_V$1.2 and Na$_V$1.6 at developing CNS nodes of Ranvier. *Neuron* **30**, 105–119.

Karlin A and Akabas MH (1998). Substituted-cysteine accessibility method. *Methods Enzymol* **293**, 123–145.

Kato Y, Ito M, Kawai K et al. (2002). Determinants of ligand specificity in groups I and IV WW domains as studied by surface plasmon resonance and model building. *J Biol Chem* **277**, 10173–10177.

Kayano T, Noda M, Flockerzi V et al. (1988). Primary structure of rat brain sodium channel iii deduced from the cdna sequence. *FEBS Lett* **228**, 187–194.

Kazarinova-Noyes K, Malhotra JD, McEwen DP et al. (2001). Contactin associates with Na$^+$ channels and increases their functional expression. *J Neurosci* **21**, 7517–7525.

Kearney VA, Buchner DA, De Hann G et al. (2002) Molecular and pathological effects of a modifier gene on deficiency of the sodium channel Scn8a (Na(v)1.6). *Human Molecular Genetics* **11**, 2765–2775.

Kearney JA, Yang Y, Beyer B et al. (2006). Severe epilepsy resulting from genetic interaction between SCN2a and KCNQ2. *Human Molecular Genetics* **15**, 1043–1048.

Kellenberger S, Scheuer T and Catterall WA (1996). Movement of the na$^+$ channel inactivation gate during inactivation. *J Biol Chem* **271**, 30971–30979.

Kellenberger S, West JW, Catterall WA et al. (1997a). Molecular analysis of potential hinge residues in the inactivation gate of brain type IIa Na$^+$ channels. *J Gen Physiol* **109**, 607–617.

Kellenberger S, West JW, Scheuer T et al. (1997b). Molecular analysis of the putative inactivation particle in the inactivation gate of brain type IIa Na$^+$ channels. *J Gen Physiol* **109**, 589–605.

Kerr BJ, Souslova V, McMahon SB et al. (2001). A role for the TTX-resistant sodium channel Na$_V$1.8 in NGF-induced hyperalgesia, but not neuropathic pain. Neuroreport **12**, 3077–3080.

Keynes RD and Rojas E (1974). Kinetics and steady-state properties of the charged system controlling sodium conductance in the squid giant axon. J Physiol **239**, 393–434.

Khasar SG, Gold MS and Levine JD (1998). A tetrodotoxin-resistant sodium current mediates inflammatory pain in the rat. Neurosci Letts **256**, 17–20.

Khodorova A, Meissner K, Leeson S et al. (2001). Lidocaine selectively blocks abnormal impulses arising from noninactivating Na channels. Muscle Nerve **24**, 634–647.

Kim J, Ghosh S, Liu H et al. (2004). Calmodulin mediates Ca^{2+} sensitivity of sodium channels. J Biol Chem **279**, 45004–45012.

Klugbauer N, Lacinova L, Flockerzi V et al. (1995). Structure and functional expression of a new member of the tetrodotoxin-sensitive voltage-activated sodium channel family from human neuroendocrine cells. EMBO J **14**, 1084–1090.

Kohrman DC, Smith MR, Goldin AL et al. (1996). A missense mutation in the sodium channel SCN8a is responsible for cerebellar ataxia in the mouse mutant jolting. Journal of Neuroscience **16**, 5993–5999.

Kondratyuk T and Rossie S (1997). Depolarization of rat brain synaptosomes increases phosphorylation of voltage-sensitive sodium channels. J Biol Chem **272**, 16978–16983.

Kontis KJ, Rounaghi A and Goldin AL (1997). Sodium channel activation gating is affected by substitutions of voltage sensor positive charges in all four domains. Journal of General Physiology **110**, 391–401.

Kuo CC and Bean BP (1994). Na$^+$ channels must deactivate to recover from inactivation. Neuron **12**, 819–829.

Kuo CC, Chen RS, Lu L et al. (1997). Carbamazepine inhibition of neuronal Na$^+$ currents: Quantitative distinction from phenytoin and possible therapeutic implications. Mol Pharmacol **51**, 1077–1083.

Kuo JJ, Lee RH, Zhang L et al. (2006). Essential role of the persistent sodium current in spike initiation during slowly rising inputs. J Physiol **574**, 819–834.

Kurata Y, Sato R, Hisatome I et al. (1999). Mechanisms of cation permeation in cardiac sodium channel: description by dynamic pore model. Biophysical J **77**, 1885–1904.

Lai J, Gold MS, Kim CS et al. (2002). Inhibition of neuropathic pain by decreased expression of the tetrodotoxin-resistant sodium channel, Na$_V$1.8. Pain **95**, 143–152.

Lai J, Porreca F, Hunter JC et al. (2004). Voltage-gated sodium channels and hyperalgesia. Annu Rev Pharmacol Toxicol **44**, 371–397.

Laird JM, Souslova V, Wood JN et al. (2002). Deficits in visceral pain and referred hyperalgesia in Na$_V$1.8 (SNS/PN3)-null mice. J Neurosci **22**, 8352–8356.

Lampert A, Dib-Hajj SD, Tyrrell L et al. (2006). Size matters: erythromelalgia mutation S241T in Na$_V$1.7 alters channel gating. J Biol Chem **281**, 36029–36035.

Lardin HA and Lee PJ (2006). The voltage dependence of recovery from use-dependent block by QX-222 separates mechanisms for drug egress in the cardiac sodium channel. Biochem Pharmacol **71**, 1299–1307.

Lee PJ, Sunami A and Fozzard HA (2001). Cardiac-specific external paths for lidocaine, defined by isoform-specific residues, accelerate recovery from use-dependent block. Circ Res **89**, 1014–1021.

Leffler A, Cummins TR, Dib-Hajj SD et al. (2002). GDNF and NGF reverse changes in repriming of TTX-sensitive Na$^+$ currents following axotomy of dorsal root ganglion neurons. J Neurophysiol **88**, 650–658.

Leffler A, Herzog RI, Dib-Hajj SD et al. (2005). Pharmacological properties of neuronal TTX-resistant sodium channels and the role of a critical serine pore residue. Pflugers Arch **451**, 454–463.

Lehmann-Horn F, Jurkat-Rott K and Rudel R (2002). Periodic paralysis: understanding channelopathies. Curr Neurol Neurosci Rep **2**, 61–69.

Leipold E, Hansel A, Olivera BM et al. (2005). Molecular interaction of delta-conotoxins with voltage-gated sodium channels. FEBS Lett **579**, 3881–3884.

Lemaillet G, Walker B and Lambert S (2003). Identification of a conserved ankyrin-binding motif in the family of sodium channel alpha subunits. J Biol Chem **278**, 27333–27339.

Lerche H, Peter W, Fleischhauer R et al. (1997). Role in fast inactivation of the iv/s4–s5 loop of the human muscle na$^+$ channel probed by cysteine mutagenesis. J Physiol **505**, 345–352.

Lindia JA, Kohler MG, Martin WJ et al. (2005). Relationship between sodium channel Na$_V$1.3 expression and neuropathic pain behavior in rats. Pain **117**, 145–153.

Linford NJ, Cantrell AR, Qu Y et al. (1998). Interaction of batrachotoxin with the local anesthetic receptor site in transmembrane segment IVS6 of the voltage-gated sodium channel. Proc Natl Acad Sci USA **95**, 13947–13952.

Lipkind GM and Fozzard HA (1994). A structural model of the tetrodotoxin and saxitoxin binding site of the Na$^+$ channel. Biophys J **66**, 1–13.

Lipkind GM and Fozzard HA (2000). KCSA crystal structure as framework for a molecular model of the Na$^+$ channel pore. Biochemistry **39**, 8161–8170.

Lipkind GM and Fozzard HA (2005). Molecular modeling of local anesthetic drug binding by voltage-gated Na channels. Mol Pharmacol **68**, 1611–1622.

Liu C, Cummins TR, Tyrrell L et al. (2005). Cap-1a is a novel linker that binds clathrin and the voltage-gated sodium channel Na$_V$1.8. Mol Cell Neurosci **28**, 636–649.

Liu C, Dib-Hajj SD and Waxman SG (2001b). Fibroblast growth factor homologous factor 1b binds to the c terminus of the tetrodotoxin-resistant sodium channel rNa$_V$1.9 (NaN). J Biol Chem **276**, 18925–18933.

Liu C, Dib-Hajj SD, Black JA et al. (2001a). Direct interaction with contactin targets voltage-gated sodium channel Na$_V$1.9/NaN to the cell membrane. J Biol Chem **276**, 46553–46561.

Liu C, Dib-Hajj SD, Renganathan M et al. (2003). Modulation of the cardiac sodium channel Na$_V$1.5 by fibroblast growth factor homologous factor 1b. J Biol Chem **278**, 1029–1036.

Liu Z, Song W and Dong K (2004). Persistent tetrodotoxin-sensitive sodium current resulting from U-to-C RNA editing of an insect sodium channel. Proc Natl Acad Sci USA **101**, 11862–11867.

Lo AC, Black JA and Waxman SG (2002). Neuroprotection of axons with phenytoin in experimental allergic encephalomyelitis. Neuroreport **13**, 1909–1912.

Lo AC, Saab CY, Black JA et al. (2003). Phenytoin protects spinal cord axons and preserves axonal conduction and neurological function in a model of neuroinflammation in vivo. J Neurophysiol **90**, 3566–3571.

Locke GR 3rd, Ackerman MJ, Zinsmeister AR et al. (2006). Gastrointestinal symptoms in families of patients with an SCN5a-encoded cardiac channelopathy: evidence of an intestinal channelopathy. Am J Gastroenterol **101**, 1299–1304.

Lombet A, Bidard JN and Lazdunski M (1987). Ciguatoxin and brevetoxins share a common receptor site on the neuronal voltage-dependent Na$^+$ channel. FEBS Lett **219**, 355–359.

Long SB, Campbell EB and Mackinnon R (2005a). Crystal structure of a mammalian voltage-dependent shaker family K$^+$ channel. Science **309**, 897–903.

Long SB, Campbell EB and Mackinnon R (2005b). Voltage sensor of Kv1.2: structural basis of electromechanical coupling. Science **309**, 903–908.

Lossin C, Rhodes TH, Desai RR et al. (2003). Epilepsy-associated dysfunction in the voltage-gated neuronal sodium channel SCN1a. J Neurosci **23**, 11289–11295.

Lou JY, Laezza F, Gerber BR et al. (2005). Fibroblast growth factor 14 is an intracellular modulator of voltage-gated sodium channels. J Physiol **569**, 179–193.

Lu PJ, Zhou XZ, Shen M et al. (1999). Function of WW domains as phosphoserine- or phosphothreonine-binding modules. Science **283**, 1325–1328.

Maier SK, Westenbroek RE, Schenkman KA *et al.* (2002). An unexpected role for brain-type sodium channels in coupling of cell surface depolarization to contraction in the heart. *Proc Natl Acad Sci USA* **99**, 4073–4078.

Mantegazza M, Gambardella A, Rusconi R *et al.* (2005). Identification of a Na$_V$1.1 sodium channel (SCN1a) loss-of-function mutation associated with familial simple febrile seizures. *Proc Natl Acad Sci USA* **102**, 18177–18182.

Mantegazza M, Yu FH, Catterall WA *et al.* (2001). Role of the c-terminal domain in inactivation of brain and cardiac sodium channels. *Proc Natl Acad Sci USA* **98**, 15348–15353.

Matzner O and Devor M (1994). Hyperexcitability at sites of nerve injury depends on voltage-sensitive Na$^+$ channels. *J Neurophysiol* **72**, 349–359.

McPhee JC, Ragsdale DS, Scheuer T *et al.* (1994). A mutation in segment IVS6 disrupts fast inactivation of sodium channels. *Proc Natl Acad Sci USA* **91**, 12346–12350.

McPhee JC, Ragsdale DS, Scheuer T *et al.* (1995). A critical role for transmembrane segment IVS6 of the sodium channel alpha subunit in fast inactivation. *J Biol Chem* **270**, 12025–12034.

McPhee JC, Ragsdale DS, Scheuer T *et al.* (1998). A critical role for the S4–S5 intracellular loop in domain IV of the sodium channel alpha-subunit in fast inactivation. *J Biol Chem* **273**, 1121–1129.

Meadows L, Malhotra JD, Stetzer A *et al.* (2001). The intracellular segment of the sodium channel beta 1 subunit is required for its efficient association with the channel alpha subunit. *J Neurochem* **76**, 1871–1878.

Meadows LS and Isom LL (2005). Sodium channels as macromolecular complexes: implications for inherited arrhythmia syndromes. *Cardiovasc Res* **67**, 448–458.

Meisler MH and Kearney JA (2005). Sodium channel mutations in epilepsy and other neurological disorders. *J Clin Invest* **115**, 2010–2017.

Meisler MH, Kearney J, Escayg A *et al.* (2001). Sodium channels and neurological disease: insights from *SCN8a* mutations in the mouse. *Neuroscientist* **7**, 136–145.

Messner DJ and Catterall WA (1985). The sodium channel from rat brain. Separation and characterization of subunits. *J Biol Chem* **260**, 10597–10604.

Meves H, Rubly N and Watt DD (1982). Effect of toxins isolated from the venom of the scorpion *Centruroides sculpturatus* on the Na currents of the node of Ranvier. *Pflugers Arch* **393**, 56–62.

Michiels JJ, te Morsche RH, Jansen JB *et al.* (2005). Autosomal dominant erythermalgia associated with a novel mutation in the voltage-gated sodium channel alpha subunit Na$_V$1.7. *Arch Neurol* **62**, 1587–1590.

Mikami M and Yang J (2005). Short hairpin RNA-mediated selective knockdown of Na$_V$1.8 tetrodotoxin-resistant voltage-gated sodium channel in dorsal root ganglion neurons. *Anesthesiology* **103**, 828–836.

Miller JA, Agnew WS and Levinson SR (1983). Principal glycopeptide of the tetrodotoxin/saxitoxin binding protein from electrophorus electricus: isolation and partial chemical and physical characterization. *Biochemistry* **22**, 462–470.

Mitrovic N, George AL and Horn R (1998). Independent versus coupled inactivation in sodium channels—role of the domain 2 S4 segment. *J Gen Physiol* **111**, 451–462.

Mitrovic N, George AL Jr and Horn R (2000). Role of domain 4 in sodium channel slow inactivation. *J Gen Physiol* **115**, 707–718.

Mitrovic N, George AL Jr, Heine R *et al.* (1994). K$^+$-aggravated myotonia: destabilization of the inactivated state of the human muscle Na$^+$ channel by the V1589M mutation. *J Physiol* **478**, 395–402.

Miyamoto K, Nakagawa T and Kuroda Y (2001). Solution structures of the cytoplasmic linkers between segments S4 and S5 (S4–S5) in domains III and IV of human brain sodium channels in sds micelles. *J Pept Res* **58**, 193–203.

Modell SM and Lehmann MH (2006). The long QT syndrome family of cardiac ion channelopathies: a huge review. *Genet Med* **8**, 143–155.

Mohler PJ, Rivolta I, Napolitano C *et al.* (2004). Na$_V$1.5 E1053K mutation causing brugada syndrome blocks binding to ankyrin-g and expression of Na$_V$1.5 on the surface of cardiomyocytes. *Proc Natl Acad Sci USA* **101**, 17533–17538.

Mori M, Konno T, Ozawa T *et al.* (2000). Novel interaction of the voltage-dependent sodium channel (VDSC) with calmodulin: does VDSC acquire calmodulin-mediated Ca^{2+}-sensitivity? *Biochemistry* **39**, 1316–1323.

Moss AJ and Kass RS (2005). Long QT syndrome: from channels to cardiac arrhythmias. *J Clin Invest* **115**, 2018–2024.

Motoike HK, Liu H, Glaaser IW *et al.* (2004). The Na$^+$ channel inactivation gate is a molecular complex: a novel role of the COOH-terminal domain. *J Gen Physiol* **123**, 155–165.

Mulley JC, Nelson P, Guerrero S *et al.* (2006). A new molecular mechanism for severe myoclonic epilepsy of infancy: exonic deletions in *scn1a*. *Neurology* **67**, 1094–1095.

Murphy BJ, Rogers J, Perdichizzi AP *et al.* (1996). Camp-dependent phosphorylation of two sites in the alpha subunit of the cardiac sodium channel. *Journal of Biological Chemistry* **271**, 28837–28843.

Murphy BJ, Rossie S, De Jongh KS *et al.* (1993). Identification of the sites of selective phosphorylation and dephosphorylation of the rat brain Na$^+$ channel alpha subunit by cAMP-dependent protein kinase and phosphoprotein phosphatases. *J Biol Chem* **268**, 27355–27362.

Nassar MA, Baker MD, Levato A *et al.* (2006). Nerve injury induces robust allodynia and ectopic discharges in Na$_V$1.3 null mutant mice. *Mol Pain* **2**, 33.

Nassar MA, Stirling LC, Forlani G *et al.* (2004). Nociceptor-specific gene deletion reveals a major role for Na$_V$1.7 (PN1) in acute and inflammatory pain. *Proc Natl Acad Sci USA* **101**, 12706–12711.

Nau C and Wang GK (2004). Interactions of local anesthetics with voltage-gated Na$^+$ channels. *J Membr Biol* **201**, 1–8.

Noda M (2006). The subfornical organ, a specialized sodium channel, and the sensing of sodium levels in the brain. *Neuroscientist* **12**, 80–91.

Noda M, Ikeda T, Suzuki H *et al.* (1986). Expression of functional sodium channels from cloned cDNA. *Nature* **322**, 826–828.

Noda M, Shimizu S, Tanabe T *et al.* (1984). Primary structure of electrophorus electricus sodium channel deduced from cDNA sequence. *Nature* **312**, 121–127.

Noda M, Suzuki H, Numa S *et al.* (1989). A single point mutation confers tetrodotoxin and saxitoxin insensitivity on the sodium channel ii. *FEBS Letters* **259**, 213–216.

Noebels JL (2003). How a sodium channel mutation causes epilepsy. *Epilepsy Curr* **3**, 70–71.

Novakovic SD, Tzoumaka E, McGivern JG *et al.* (1998). Distribution of the tetrodotoxin-resistant sodium channel PN3 in rat sensory neurons in normal and neuropathic conditions. *J Neurosci* **18**, 2174–2187.

Numann R, Catterall WA and Scheuer T (1991). Functional modulation of brain sodium channels by protein kinase c phosphorylation. *Science* **254**, 115–118.

Nuss HB, Balser JR, Orias DW *et al.* (1996). Coupling between fast and slow inactivation revealed by analysis of a point mutation (F1304Q) in mu 1 rat skeletal muscle sodium channels. *J Physiol* **494**, 411–429.

Nuss HB, Kambouris NG, Marban E *et al.* (2000). Isoform-specific lidocaine block of sodium channels explained by differences in gating. *Biophys J* **78**, 200–210.

Ohizumi Y, Nakamura H, Kobayashi J *et al.* (1986). Specific inhibition of [^{3}H] saxitoxin binding to skeletal muscle sodium channels by geographutoxin II, a polypeptide channel blocker. *J Biol Chem* **261**, 6149–6152.

Ohmori I, Ouchida M, Ohtsuka Y *et al.* (2002). Significant correlation of the SCN1a mutations and severe myoclonic epilepsy in infancy. *Biochem Biophys Res Comm* **295**, 17–23.

Okuse K, Chaplan SR, McMahon SB, *et al.* (1997). Regulation of expression of the sensory neuron-specific sodium channel SNS in inflammatory and neuropathic pain. *Mol Cell Neurosci* **10**, 196–207.

Okuse K, Malik-Hall M, Baker MD *et al.* (2002). Annexin II light chain regulates sensory neuron-specific sodium channel expression. *Nature* **417**, 653–656.

Ou Y, Gibbons SJ, Miller SM *et al.* (2002). SCN5a is expressed in human jejunal circular smooth muscle cells. *Neurogastroenterol Motil* **14**, 477–486.

Ou Y, Strege P, Miller SM *et al.* (2003). Syntrophin gamma 2 regulates SCN5a gating by a PDZ domain-mediated interaction. *J Biol Chem* **278**, 1915–1923.

Paillart C, Boudier JL, Boudier JA *et al.* (1996). Activity-induced internalization and rapid degradation of sodium channels in cultured fetal neurons. *J Cell Biol* **134**, 499–509.

Papadatos GA, Wallerstein PM, Head CE *et al.* (2002). From the cover: slowed conduction and ventricular tachycardia after targeted disruption of the cardiac sodium channel gene *scn5a*. *Proc Natl Acad Sci USA* **99**, 6210–6215.

Patlak J (1991). Molecular kinetics of voltage-dependent Na⁺ channels. *Physiol Rev* **71**, 1047–1080.

Patton DE, West JW, Catterall WA *et al.* (1992). Amino acid residues required for fast Na⁺-channel inactivation: charge neutralizations and deletions in the III-IV linker. *Proc Natl Acad Sci USA* **89**, 10905–10909.

Paulussen A, Matthijs G, Gewillig M *et al.* (2003). Mutation analysis in congenital long QT syndrome–a case with missense mutations in KCNQ1 and SCN5a. *Genet Test* **7**, 57–61.

Perez-Garcia MT, Chiamvimonvat N, Ranjan R *et al.* (1997). Mechanisms of sodium/calcium selectivity in sodium channels probed by cysteine mutagenesis and sulfhydryl modification. *Biophys J* **72**, 989–996.

Planells-Cases R, Caprini M, Zhang J *et al.* (2000). Neuronal death and perinatal lethality in voltage-gated sodium channel alpha(II)-deficient mice. *Biophys J* **78**, 2878–2891.

Plummer NW and Meisler MH (1999). Evolution and diversity of mammalian sodium channel genes. *Genomics* **57**, 323–331.

Plummer NW, Galt J, Jones JM *et al.* (1998). Exon organization, coding sequence, physical mapping, and polymorphic intragenic markers for the human neuronal sodium channel gene *SCN8a*. *Genomics* **54**, 287–296.

Plummer NW, McBurney MW and Meisler MH (1997). Alternative splicing of the sodium channel *SCN8a* predicts a truncated two-domain protein in fetal brain and non-neuronal cells. *J Biol Chem* **272**, 24008–24015.

Poelzing S, Forleo C, Samodell M *et al.* (2006). *SCN5a* polymorphism restores trafficking of a brugada syndrome mutation on a separate gene. *Circulation* **114**, 368–376.

Priest BT, Murphy BA, Lindia JA *et al.* (2005). Contribution of the tetrodotoxin-resistant voltage-gated sodium channel Na$_V$1.9 to sensory transmission and nociceptive behavior. *Proc Natl Acad Sci USA* **102**, 9382–9387.

Qu Y, Rogers J, Tanada T *et al.* (1995). Molecular determinants of drug access to the receptor site for antiarrhythmic drugs in the cardiac Na⁺ channel. *Proc Natl Acad Sci USA* **92**, 11839–11843.

Ragsdale DS, McPhee JC, Scheuer T *et al.* (1994). Molecular determinants of state-dependent block of Na⁺ channels by local anesthetics. *Science* **265**, 1724–1728.

Ragsdale DS, McPhee JC, Scheuer T *et al.* (1996). Common molecular determinants of local anesthetic, antiarrhythmic, and anticonvulsant block of voltage-gated Na⁺ channels. *Proc Natl Acad Sci USA* **93**, 9270–9275.

Raman IM and Bean BP (1999). Properties of sodium currents and action potential firing in isolated cerebellar Purkinje neurons. *Ann N Y Acad Sci* **868**, 93–96.

Raman IM and Bean BP (2001). Inactivation and recovery of sodium currents in cerebellar Purkinje neurons: evidence for two mechanisms. *Biophys J* **80**, 729–737.

Raman IM, Sprunger LK, Meisler MH *et al.* (1997). Altered subthreshold sodium currents and disrupted firing patterns in purkinje neurons of *SCN8a* mutant mice. *Neuron* **19**, 881–891.

Ratcliffe CF, Qu Y, McCormick KA *et al.* (2000). A sodium channel signaling complex: modulation by associated receptor protein tyrosine phosphatase beta. *Nat Neurosci* **3**, 437–444.

Raymond CK, Castle J, Garrett-Engele P *et al.* (2004). Expression of alternatively spliced sodium channel alpha-subunit genes: unique splicing patterns are observed in dorsal root ganglia. *J Biol Chem* **279**, 46234–46241.

Recio-Pinto E, Thronhill WB, Duch DS *et al.* (1990). Neuraminidase treatment modifies the function of electroplax sodium channels in planar lipid bilayers. *Neuron* **5**, 675–684.

Ren D, Navarro B, Xu H *et al.* (2001). A prokaryotic voltage-gated sodium channel. *Science* **294**, 2372–2375.

Renganathan M, Cummins TR and Waxman SG (2001). Contribution of Na$_V$1.8 sodium channels to action potential electrogenesis in DRG neurons. *J Neurophysiol* **86**, 629–640.

Renganathan M, Dib-Hajj S and Waxman SG (2002). Na$_V$1.5 underlies the 'third TTX-r sodium current' in rat small DRG neurons. *Brain Res Mol Brain Res* **106**, 70.

Renganathan M, Gelderblom M, Black JA *et al.* (2003). Expression of Na$_V$1.8 sodium channels perturbs the firing patterns of cerebellar Purkinje cells. *Brain Res* **959**, 235–242.

Rhodes TH, Lossin C, Vanoye CG *et al.* (2004). Noninactivating voltage-gated sodium channels in severe myoclonic epilepsy of infancy. *Proc Natl Acad Sci USA* **101**, 11147–11152.

Richmond JE, Featherstone DE *et al.* (1998). Slow inactivation in human cardiac sodium channels. *Biophys J* **74**, 2945–2952.

Riddall DR, Leach MJ and Garthwaite J (2006). A novel drug binding site on voltage-gated sodium channels in rat brain. *Mol Pharmacol* **69**, 278–287.

Rizzo MA, Kocsis JD and Waxman SG (1994). Slow sodium conductances of dorsal root ganglion neurons: intraneuronal homogeneity and interneuronal heterogeneity. *J Neurophysiol* **72**, 2796–2815.

Rizzo MA, Kocsis JD and Waxman SG (1995). Selective loss of slow and enhancement of fast Na⁺ currents in cutaneous afferent dorsal root ganglion neurones following axotomy. *Neurobiol Dis* **2**, 87–96.

Rogart RB, Cribbs LL, Muglia LK *et al.* (1989). Molecular cloning of a putative tetrodotoxin-resistant rat heart Na⁺ channel isoform. *Proc Natl Acad Sci USA* **86**, 8170–8174.

Rogers JC, Qu Y, Tanada TN *et al.* (1996). Molecular determinants of high affinity binding of alpha-scorpion toxin and sea anemone toxin in the S3–S4 extracellular loop in domain IV of the Na⁺ channel alpha subunit. *J Biol Chem* **271**, 15950–15962.

Rohl CA, Boeckman FA, Baker C *et al.* (1999). Solution structure of the sodium channel inactivation gate. *Biochemistry* **38**, 855–861.

Rougier JS, van Bemmelen MX, Bruce MC *et al.* (2005). Molecular determinants of voltage-gated sodium channel regulation by the nedd4/nedd4-like proteins. *Am J Physiol Cell Physiol* **288**, C692–701.

Roza C, Laird JM, Souslova V *et al.* (2003). The tetrodotoxin-resistant Na⁺ channel Na$_V$1.8 is essential for the expression of spontaneous activity in damaged sensory axons of mice. *J Physiol* **550**, 921–926.

Rudy B (1978). Slow inactivation of the sodium conductance in squid giant axons. Pronase resistance. *J Physiol* **283**, 1–21.

Rugiero F, Mistry M, Sage D *et al.* (2003). Selective expression of a persistent tetrodotoxin-resistant Na⁺ current and Na$_V$1.9 subunit in myenteric sensory neurons. *J Neurosci* **23**, 2715–2725.

Rush AM, Brau ME, Elliott AA *et al.* (1998). Electrophysiological properties of sodium current subtypes in small cells from adult dorsal root ganglia. *J Physiol* **511**, 771–789.

Rush AM, Craner MJ, Kageyama T *et al.* (2005a). Contactin regulates the current density and axonal expression of tetrodotoxin-resistant but not tetrodotoxin-sensitive sodium channels in DRG neurons. *Eur J Neurosci* **22**, 39–49.

Rush AM, Dib-Hajj SD and Waxman SG (2005b). Electrophysiological properties of two axonal sodium channels, Na$_V$1.2 and Na$_V$1.6, expressed in mouse spinal sensory neurons. *J Physiol* **564**, 803–815.

Rush AM and Waxman SG (2004) PGE_2 increases the tetrodotoxin-resistant Nav1.9 sodium current in mouse DRG neurons via G-proteins. *Brain Res* **1023**, 264–271.

Rush AM, Dib-Hajj SD, Liu S (2006a). A single sodium channel mutation produces hyper- or hypoexcitability in different types of neurons. *Proc Natl Acad Sci USA* **103**, 8245–8250.

Rush AM, Wittmack EK, Tyrrell L *et al.* (2006b). Differential modulation of sodium channel $Na_V1.6$ by two members of the fibroblast growth factor homologous factor 2 subfamily. *Eur J Neurosci* **23**, 2551–2562.

Safo P, Rosenbaum T, Shcherbatko A *et al.* (2000). Distinction among neuronal subtypes of voltage-activated sodium channels by mu-conotoxin PIIIA. *J Neurosci* **20**, 76–80.

Saleh S, Yeung SY, Prestwich S *et al.* (2005). Electrophysiological and molecular identification of voltage-gated sodium channels in murine vascular myocytes. *J Physiol* **568**, 155–169.

Sampo B, Tricaud N, Leveque C *et al.* (2000). Direct interaction between synaptotagmin and the intracellular loop I-II of neuronal voltage-sensitive sodium channels. *Proc Natl Acad Sci USA* **97**, 3666–3671.

Sangameswaran L, Delgado SG, Fish LM *et al.* (1996). Structure and function of a novel voltage-gated, tetrodoxtoxin-resistant sodium channel specific to sensory neurons. *J Biol Chem* **271**, 5953–5956.

Sangameswaran L, Fish LM, Koch BD *et al.* (1997). A novel tetrodotoxin-sensitive, voltage-gated sodium channel expressed in rat and human dorsal root ganglia. *J Biol Chem* **272**, 14805–14809.

Sarao R, Gupta SK, Auld VJ *et al.* (1991). Developmentally regulated alternative RNA splicing of rat brain sodium channel mRNAs. *Nucl Acid Res* **19**, 5673–5679.

Sasaki M, Black JA, Lankford KL *et al.* (2006). Molecular reconstruction of nodes of Ranvier after remyelination by transplanted olfactory ensheathing cells in the demyelinated spinal cord. *J Neurosci* **26**, 1803–1812.

Satin J, Kyle JW, Chen M *et al.* (1992). A mutant of TTX-resistant cardiac sodium channels with TTX-sensitive properties. *Science* **256**, 1202–1205.

Sato C, Ueno Y, Asai K *et al.* (2001). The voltage-sensitive sodium channel is a bell-shaped molecule with several cavities. *Nature* **409**, 1047–1051.

Schafer DP, Custer AW, Shrager P *et al.* (2006). Early events in node of Ranvier formation during myelination and remyelination in the PNS. *Neuron Glia Biol* **2**, 69–79.

Schaller KL and Caldwell JH (2000). Developmental and regional expression of sodium channel isoform NaCh6 in the rat central nervous system. *J Comp Neurol* **420**, 84–97.

Schaller KL, Krzemien DM, Yarowsky PJ *et al.* (1995). A novel, abundant sodium channel expressed in neurons and glia. *J.Neurosci.* **15**, 3231–3242.

Scheib H, McLay I, Guex N *et al.* (2006). Modeling the pore structure of voltage-gated sodium channels in closed, open, and fast-inactivated conformation reveals details of site 1 toxin and local anesthetic binding. *J Mol Model* **12**, 813–822.

Schmidt JW and Catterall WA (1987). Palmitylation, sulfation, and glycosylation of the alpha subunit of the sodium channel. Role of post-translational modifications in channel assembly. *J Biol Chem* **262**, 13713–13723.

Schoppa NE, McCormack K, Tanouye MA *et al.* (1992). The size of gating charge in wild-type and mutant Shaker potassium channels. *Science* **255**, 1712–1715.

Schultz J, Hoffmuller U, Krause G *et al.* (1998). Specific interactions between the syntrophin PDZ domain and voltage-gated sodium channels. *Nat Struct Biol* **5**, 19–24.

Shah BS, Rush AM, Liu S *et al.* (2004). Contactin associates with sodium channel $Na_V1.3$ in native tissues and increases channel density at the cell surface. *J. Neurosci.* **24**, 7387–7399.

Shah VN, Wingo TL, Weiss KL *et al.* (2006). Calcium-dependent regulation of the voltage-gated sodium channel hH1: intrinsic and extrinsic sensors use a common molecular switch. *Proc Natl Acad Sci USA* **103**, 3592–35927.

Sharrocks AD, Yang SH and Galanis A (2000). Docking domains and substrate-specificity determination for MAP kinases. *Trends Biochem Sci* **25**, 448–453.

Sheets MF and Hanck DA (1995). Voltage-dependent open-state inactivation of cardiac sodium channels: gating current studies with anthopleurin-a toxin. *Journal of General Physiology* **106**, 617–640.

Sheets MF and Hanck DA (2003). Molecular action of lidocaine on the voltage sensors of sodium channels. *J Gen Physiol* **121**, 163–175.

Sheets MF, Kyle JW, Kallen RG *et al.* (1999). The Na channel voltage sensor associated with inactivation is localized to the external charged residues of domain IV, S4. *Biophys J* **77**, 747–757.

Shichor I, Fainzilber M, Pelhate M *et al.* (1996). Interactions of delta-conotoxins with alkaloid neurotoxins reveal differences between the silent and effective binding sites on voltage-sensitive sodium channels. *J Neurochem* **67**, 2451–2460.

Shirahata E, Iwasaki H, Takagi M *et al.* (2006). Ankyrin-g regulates inactivation gating of the neuronal sodium channel, $Na_V1.6$. *J Neurophysiol* **96**, 1347–1357.

Shon KJ, Olivera BM, Watkins M *et al.* (1998). Mu-conotoxin PIIIA, a new peptide for discriminating among tetrodotoxin-sensitive Na channel subtypes. *J Neurosci* **18**, 4473–4481.

Sivilotti L, Okuse K, Akopian AN *et al.* (1997). A single serine residue confers tetrodotoxin insensitivity on the rat sensory-neuron-specific sodium channel SNS. *FEBS Letters* **409**, 49–52.

Sleeper AA, Cummins TR, Dib-Hajj SD *et al.* (2000). Changes in expression of two tetrodotoxin-resistant sodium channels and their currents in dorsal root ganglion neurons after sciatic nerve injury but not rhizotomy. *J Neurosci* **20**, 7279–7289.

Smith MR and Goldin AL (1997). Interaction between the sodium channel inactivation linker and domain III S4-S5. *Biophysical J* **73**, 1885–1895.

Smith MR and Goldin AL (1999). A mutation that causes ataxia shifts the voltage-dependence of the *SCN8a* sodium channel. *Neuroreport.* **10**, 3027–3031.

Smits JP, Veldkamp MW, Bezzina CR *et al.* (2005). Substitution of a conserved alanine in the domain IIIS4-S5 linker of the cardiac sodium channel causes long QT syndrome. *Cardiovasc Res* **67**, 459–466.

Song W, Liu Z, Tan J, Nomura Y *et al.* (2004). RNA editing generates tissue-specific sodium channels with distinct gating properties. *J Biol Chem* **279**, 32554–32561.

Songyang Z, Fanning AS, Fu C *et al.* (1997). Recognition of unique carboxyl-terminal motifs by distinct PDZ domains. *Science* **275**, 73–77.

Spampanato J, Escayg A, Meisler MH *et al.* (2001). Functional effects of two voltage-gated sodium channel mutations that cause generalized epilepsy with febrile seizures plus type 2. *J Neurosci* **21**, 7481–7490.

Spampanato J, Escayg A, Meisler MH *et al.* (2003). Generalized epilepsy with febrile seizures plus type 2 mutation W1204R alters voltage-dependent gating of $Na_V1.1$ sodium channels. *Neuroscience* **116**, 37–48.

Spampanato J, Kearney JA, de Haan G *et al.* (2004). A novel epilepsy mutation in the sodium channel SCN1a identifies a cytoplasmic domain for beta subunit interaction. *J Neurosci* **24**, 10022–10034.

Sprunger LK, Escayg A, Tallaksen-Greene S *et al.* (1999). Dystonia associated with mutation of the neuronal sodium channel SCN8a and identification of the modifier locus SCNM1 on mouse chromosome 3. *Human Molecular Genetics* **8**, 471–479.

Starace DM and Bezanilla F (2004). A proton pore in a potassium channel voltage sensor reveals a focused electric field. *Nature* **427**, 548–553.

Starmer CF, Grant AO and Strauss HC (1984). Mechanisms of use-dependent block of sodium channels in excitable membranes by local anesthetics. *Biophys J* **46**, 15–27.

Stephan MM, Potts JF and Agnew WS (1994). The $\mu1$ skeletal muscle sodium channel: mutation E403Q eliminates sensitivity to tetrodotoxin but not to mu-conotoxins GIIIA and GIIIB. *J Membr Biol* **137**, 1–8.

Stimers JR, Bezanilla F and Taylor RE (1985). Sodium channel activation in the squid giant axon. *Steady state properties. J Gen Physiol* **85**, 65–82.

Storey N, Latchman D and Bevan S (2002). Selective internalization of sodium channels in rat dorsal root ganglion neurons infected with herpes simplex virus-1. *J Cell Biol* **158**, 1251–1262.

Strege PR, Holm AN, Rich A *et al.* (2003). Cytoskeletal modulation of sodium current in human jejunal circular smooth muscle cells. *Am J Physiol Cell Physiol* **284**, C60–66.

Striano P, Bordo L, Lispi ML *et al.* (2006). A novel SCN2a mutation in family with benign familial infantile seizures. *Epilepsia* **47**, 218–220.

Strichartz GR and Wang GK (1986). Rapid voltage-dependent dissociation of scorpion alpha-toxins coupled to Na channel inactivation in amphibian myelinated nerves. *J Gen Physiol* **88**, 413–435.

Stuhmer W, Conti F, Suzuki H *et al.* (1989). Structural parts involved in activation and inactivation of the sodium channel. *Nature* **339**, 597–603.

Stys PK, Sontheimer H, Ransom BR *et al.* (1993). Noninactivating, tetrodotoxin-sensitive Na$^+$ conductance in rat optic nerve axons. *Proc Natl Acad Sci USA* **90**, 6976–6980.

Stys PK, Waxman SG and Ransom BR (1991). Reverse operation of the Na$^+$–Ca^{2+} exchanger mediates Ca^{2+} influx during anoxia in mammalian CNS white matter. *Ann NY Acad Sci* **639**, 328–332.

Sudol M and Hunter T (2000). New wrinkles for an old domain. *Cell* **103**, 1001–1004.

Sugawara T, Tsurubuchi Y, Agarwala KL *et al.* (2001). A missense mutation of the Na$^+$ channel alpha II subunit gene Na$_V$1.2 in a patient with febrile and afebrile seizures causes channel dysfunction. *Proc Natl Acad Sci USA* **98**, 6384–6389.

Sugawara T, Tsurubuchi Y, Fujiwara T *et al.* (2003). Na$_V$1.1 channels with mutations of severe myoclonic epilepsy in infancy display attenuated currents. *Epilepsy Res* **54**, 201–207.

Suls A, Claeys KG, Goossens D *et al.* (2006). Microdeletions involving the SCN1a gene may be common in SCN1a-mutation-negative SMEI patients. *Human Mutations* **27**, 914–920.

Sunami A, Dudley SC and Fozzard HA (1997). Sodium channel selectivity filter regulates antiarrhythmic drug binding. *Proc Natl Acad Sci USA* **94**, 14126–14131.

Sunami A, Glaaser IW and Fozzard HA (2000). A critical residue for isoform difference in tetrodotoxin affinity is a molecular determinant of the external access path for local anesthetics in the cardiac sodium channel. *Proc Natl Acad Sci USA* **97**, 2326–2331.

Tan BH, Valdivia CR, Rok BA *et al.* (2005). Common human SCN5a polymorphisms have altered electrophysiology when expressed in Q1077 splice variants. *Heart Rhythm* **2**, 741–747.

Tan HL, Kupershmidt S, Zhang R *et al.* (2002). A calcium sensor in the sodium channel modulates cardiac excitability. *Nature* **415**, 442–447.

Tanaka M, Cummins TR, Ishikawa K *et al.* (1998). SNS Na$^+$ channel expression increases in dorsal root ganglion neurons in the carrageenan inflammatory pain model. *Neuroreport* **9**, 967–972.

Tang L, Chehab N, Wieland SJ *et al.* (1998). Glutamine substitution at alanine1649 in the S4–S5 cytoplasmic loop of domain 4 removes the voltage sensitivity of fast inactivation in the human heart sodium channel. *J Gen Physiol* **111**, 639–652.

Tang L, Kallen RG and Horn R (1996). Role of an S4–S5 linker in sodium channel inactivation probed by mutagenesis and a peptide blocker. *J Gen Physiol* **108**, 89–104.

Tate S, Benn S, Hick C *et al.* (1998). Two sodium channels contribute to the TTX-r sodium current in primary sensory neurons. *Nat Neurosci* **1**, 653–655.

Tejedor FJ and Catterall WA (1988). Site of covalent attachment of alpha-scorpion toxin derivatives in domain I of the sodium channel alpha subunit. *Proc Natl Acad Sci USA* **85**, 8742–8746.

Terlau H, Heinemann SH, Stuhmer W *et al.* (1991). Mapping the site of block by tetrodotoxin and saxitoxin of sodium channel II. *FEBS Letters* **293**, 93–96.

Thackeray JR and Ganetzky B (1994). Developmentally regulated alternative splicing generates a complex array of *Drosophila* para sodium channel isoforms. *Journal of Neuroscience* **14**, 2569–2578.

Toledo-Aral JJ, Brehm P, Halegoua S *et al.* (1995). A single pulse of nerve growth factor triggers long-term neuronal excitability through sodium channel gene induction. *Neuron* **14**, 607–611.

Toledo-Aral JJ, Moss BL, He ZJ *et al.* (1997). Identification of PN1, a predominant voltage-dependent sodium channel expressed principally in peripheral neurons. *Proc Natl Acad Sci USA* **94**, 1527–1532.

Trainer VL, Thomsen WJ, Catterall WA *et al.* (1991). Photoaffinity labeling of the brevetoxin receptor on sodium channels in rat brain synaptosomes. *Mol Pharmacol* **40**, 988–994.

Trimmer JS, Cooperman SS, Agnew WS *et al.* (1990). Regulation of muscle sodium channel transcripts during development and in response to denervation. *Dev Biol* **142**, 360–367.

Tripathi PK, Trujillo L, Cardenas CA *et al.* (2006). Analysis of the variation in use-dependent inactivation of high-threshold tetrodotoxin-resistant sodium currents recorded from rat sensory neurons. *Neurosci* **143**, 923–938.

Trudeau MM, Dalton JC, Day JW *et al.* (2005). Heterozygosity for a protein truncation mutation of sodium channel SCN8a in a patient with cerebellar atrophy, ataxia and mental retardation. *J Med Genet* **43**, 527–530.

Tsujino A, Maertens C, Ohno K *et al.* (2003). Myasthenic syndrome caused by mutation of the SCN4a sodium channel. *Proc Natl Acad Sci USA* **100**, 7377–7382.

Tyrrell L, Renganathan M, Dib-Hajj SD *et al.* (2001). Glycosylation alters steady-state inactivation of sodium channel Na$_V$1.9/NaN in dorsal root ganglion neurons and is developmentally regulated. *J Neurosci* **21**, 9629–9637.

Van Bemmelen MX, Rougier JS, Gavillet B *et al.* (2004). Cardiac voltage-gated sodium channel Na$_V$1.5 is regulated by nedd4–2 mediated ubiquitination. *Circ Res* **95**, 284–291.

van Swieten JC, Brusse E, de Graaf BM *et al.* (2003). A mutation in the fibroblast growth factor 14 gene is associated with autosomal dominant cerebral ataxia. *Am J Hum Genet* **72**, 191–199.

Van Wart A, Boiko T, Trimmer JS *et al.* (2005). Novel clustering of sodium channel Na$_V$1.1 with ankyrin-G and neurofascin at discrete sites in the inner plexiform layer of the retina. *Mol Cell Neurosci* **28**, 661–673.

Vassilev PM, Scheuer T and Catterall WA (1988). Identification of an intracellular peptide segment involved in sodium channel inactivation. *Science* **241**, 1658–1661.

Vedantham V and Cannon SC (1999). The position of the fast-inactivation gate during lidocaine block of voltage-gated Na$^+$ channels. *J Gen Physiol* **113**, 7–16.

Veldkamp MW, Viswanathan PC, Bezzina C *et al.* (2000). Two distinct congenital arrhythmias evoked by a multidysfunctional Na$^+$ channel. *Circ Res* **86**, E91–97.

Venance SL, Cannon SC, Fialho D *et al.* (2006). The primary periodic paralyses: diagnosis, pathogenesis and treatment. *Brain* **129**, 8–17.

Venkatesh B, Lu SQ, Dandona N *et al.* (2005). Genetic basis of tetrodotoxin resistance in pufferfishes. *Curr Biol* **15**, 2069–2072.

Verdecia MA, Bowman ME, Lu KP *et al.* (2000). Structural basis for phosphoserine-proline recognition by group IV WW domains. *Nat Struct Biol* **7**, 639–643.

Vicart S, Sternberg D, Fontaine B *et al.* (2005). Human skeletal muscle sodium channelopathies. *Neurol Sci* **26**, 194–202.

Vijayaragavan K, Boutjdir M and Chahine M (2004). Modulation of Na$_V$1.7 and Na$_V$1.8 peripheral nerve sodium channels by protein kinase A and protein kinase C. *J Neurophysiol* **91**, 1556–1569.

Viswanathan PC and Balser JR (2004). Inherited sodium channelopathies a continuum of channel dysfunction. *Trends Cardiovasc Med* **14**, 28–35.

Wallace RH, Scheffer IE, Barnett S *et al.* (2001). Neuronal sodium-channel alpha1-subunit mutations in generalized epilepsy with febrile seizures plus. *Am J Hum Genet* **68**, 859–865.

Wang GK and Strichartz G (1985). Kinetic analysis of the action of leiurus scorpion alpha-toxin on ionic currents in myelinated nerve. *J Gen Physiol* **86**, 739–762.

Wang GK and Strichartz GR (1983). Purification and physiological characterization of ncurotoxins from venoms of the scorpions *Centruroides sculpturatus* and *Leiurus quinquestriatus*. *Mol Pharmacol* **23**, 519–533.

Wang Q, Bardgett ME, Wong M *et al.* (2002). Ataxia and paroxysmal dyskinesia in mice lacking axonally transported FGF14. *Neuron* **35**, 25–38.

Wang SY and Wang GK (1998). Point mutations in segment I-S6 render voltage-gated Na⁺ channels resistant to batrachotoxin. *Proc Natl Acad Sci USA* **95**, 2653–2658.

Wang SY and Wang GK (1999). Batrachotoxin-resistant Na⁺ channels derived from point mutations in transmembrane segment D4-S6. *Biophys J* **76**, 3141–3149.

Waxman SG (2001). Transcriptional channelopathies: an emerging class of disorders. *Nat Rev Neurosci* **2**, 652–659.

Waxman SG (2006). Ions, energy and axonal injury: towards a molecular neurology of multiple sclerosis. *Trends Mol Med* **12**, 192–195.

Waxman SG and Dib-Hajj S (2005). Erythermalgia: molecular basis for an inherited pain syndrome. *Trends Mol Med* **11**, 555–562.

Waxman SG and Hains BC (2006). Fire and phantoms after spinal cord injury: Na⁺ channels and central pain. *Trends Neurosci* **29**, 207–215.

Waxman SG, Craner MJ and Black JA (2004). Na⁺ channel expression along axons in multiple sclerosis and its models. *Trends Pharmacol Sci* **25**, 584–591.

Waxman SG, Kocsis JD and Black JA (1994). Type III sodium channel mRNA is expressed in embryonic but not adult spinal sensory neurons, and is reexpressed following axotomy. *J Neurophysiol* **72**, 466–470.

Wei J, Wang DW, Alings M *et al.* (1999). Congenital long-QT syndrome caused by a novel mutation in a conserved acidic domain of the cardiac Na⁺ channel. *Circulation* **99**, 3165–3171.

Weiser T, Qu Y, Catterall WA *et al.* (1999). Differential interaction of r-mexiletine with the local anesthetic receptor site on brain and heart sodium channel alpha-subunits. *Mol Pharmacol* **56**, 1238–1244.

Weiss LA, Escayg A, Kearney JA *et al.* (2003). Sodium channels SCN1a, SCN2a and SCN3a in familial autism. *Mol Psychiat* **8**, 186–194.

West JW, Numann R, Murphy BJ *et al.* (1991). A phosphorylation site in the Na⁺ channel required for modulation by protein kinase C. *Science* **254**, 866–868.

West JW, Patton DE, Scheuer T *et al.* (1992). A cluster of hydrophobic amino acid residues required for fast Na⁺-channel inactivation. *Proc Natl Acad Sci USA* **89**, 10910–10914.

Westenbroek RE, Merrick DK and Catterall WA (1989). Differential subcellular localization of the RI and RII Na⁺ channel subtypes in central neurons. *Neuron* **3**, 695–704.

Whitaker WR, Faull RL, Waldvogel HJ *et al.* (2001). Comparative distribution of voltage-gated sodium channel proteins in human brain. *Brain Res Mol Brain Res* **88**, 37–53.

Wilhelmsen KC, Blake DM, Lynch T *et al.* (1996). Chromosome 12-linked autosomal dominant scapuloperoneal muscular dystrophy. *Ann Neurol* **39**, 507–520.

Willow M, Gonoi T and Catterall WA (1985). Voltage clamp analysis of the inhibitory actions of diphenylhydantoin and carbamazepine on voltage-sensitive sodium channels in neuroblastoma cells. *Mol Pharmacol* **27**, 549–558.

Wingo TL, Shah VN, Anderson ME *et al.* (2004). An EF-hand in the sodium channel couples intracellular calcium to cardiac excitability. *Nat Struct Mol Biol* **11**, 219–225.

Wittmack EK, Rush AM, Craner MJ *et al.* (2004). Fibroblast growth factor homologous factor 2b: Association with Na_V1.6 and selective colocalization at nodes of ranvier of dorsal root axons. *J Neurosci* **24**, 6765–6775.

Wittmack EK, Rush AM, Hudmon A *et al.* (2005). Voltage-gated sodium channel Na_V1.6 is modulated by p38 mitogen-activated protein kinase. *J Neurosci* **25**, 6621–6630.

Wood JN, Boorman JP, Okuse K *et al.* (2004). Voltage-gated sodium channels and pain pathways. *J Neurobiol* **61**, 55–71.

Wu FF, Gordon E, Hoffman EP *et al.* (2005). A c-terminal skeletal muscle sodium channel mutation associated with myotonia disrupts fast inactivation. *J Physiol* **565**, 371–380.

Wu L, Nishiyama K, Hollyfield JG *et al.* (2002). Localization of Na_V1.5 sodium channel protein in the mouse brain. *Neuroreport* **13**, 2547–2551.

Xiong W, Farukhi YZ, Tian Y *et al.* (2006). A conserved ring of charge in mammalian Na⁺ channels: a molecular regulator of the outer pore conformation during slow inactivation. *J Physiol* **576**, 739–754.

Yamagishi T, Li RA, Hsu K *et al.* (2001). Molecular architecture of the voltage-dependent Na channel: functional evidence for alpha helices in the pore. *J Gen Physiol* **118**, 171–182.

Yang N and Horn R (1995). Evidence for voltage-dependent S4 movement in sodium channels. *Neuron* **15**, 213–218.

Yang N, George AL Jr and Horn R (1996). Molecular basis of charge movement in voltage-gated sodium channels. *Neuron* **16**, 113–122.

Yang N, Ji S, Zhou M *et al.* (1994). Sodium channel mutations in paramyotonia congenita exhibit similar biophysical phenotypes *in vitro*. *Proc Natl Acad Sci USA* **91**, 12785–12789.

Yang Y, Wang Y, Li S *et al.* (2004). Mutations in SCN9a, encoding a sodium channel alpha subunit, in patients with primary erythermalgia. *J Med Genet* **41**, 171–174.

Ye B, Valdivia CR, Ackerman MJ *et al.* (2002). A common human *SCN5a* polymorphism modifies expression of an arrhythmia causing mutation. *Physiol Genomics* **12**, 187–193.

Yellen G (1998). The moving parts of voltage-gated ion channels. *Q Rev Biophys* **31**, 239–295.

Yeomans DC, Levinson SR, Peters MC *et al.* (2005). Decrease in inflammatory hyperalgesia by herpes vector-mediated knockdown of Na_V1.7 sodium channels in primary afferents. *Hum Gene Ther* **16**, 271–277.

Yoshimura N, Seki S, Novakovic SD *et al.* (2001). The involvement of the tetrodotoxin-resistant sodium channel Na_V1.8 (PN3/SNS) in a rat model of visceral pain. *J Neurosci* **21**, 8690–8696.

Yotsu-Yamashita M, Nishimori K, Nitanai Y *et al.* (2000). Binding properties of [³H]-PbTx-3 and [³H]-saxitoxin to brain membranes and to skeletal muscle membranes of puffer fish *Fugu pardalis* and the primary structure of a voltage-gated Na⁺ channel alpha-subunit (FMNa1) from skeletal muscle of F. pardalis. *Biochem Biophys Res Comm* **267**, 403–412.

Young KA and Caldwell JH (2005). Modulation of skeletal and cardiac voltage-gated sodium channels by calmodulin. *J Physiol* **565**, 349–370.

Yu FH, Westenbroek RE, Silos-Santiago I *et al.* (2003). Sodium channel beta4, a new disulfide-linked auxiliary subunit with similarity to beta2. *J Neurosci* **23**, 7577–7585.

Zagotta WN, Hoshi T, Dittman J *et al.* (1994). Shaker potassium channel gating II: transitions in the activation pathway. *J Gen Physiol* **103**, 279–319.

Zerangue N, Schwappach B, Jan YN *et al.* (1999). A new er trafficking signal regulates the subunit stoichiometry of plasma membrane K_ATP channels. *Neuron* **22**, 537–548.

Zhang Y, Hartmann HA and Satin J (1999). Glycosylation influences voltage-dependent gating of cardiac and skeletal muscle sodium channels. *J Memb Biol* **171**, 195–207.

Zhang YH, Vasko MR and Nicol GD (2002). Ceramide, a putative second messenger for nerve growth factor, modulates the TTX-resistant Na$^+$ current and delayed rectifier K$^+$ current in rat sensory neurons. *J Physiol* **544**, 385–402.

Zhao Y, Scheuer T and Catterall WA (2004a). Reversed voltage-dependent gating of a bacterial sodium channel with proline substitutions in the S6 transmembrane segment. *Proc Natl Acad Sci USA* **101**, 17873–17878.

Zhao Y, Yarov-Yarovoy V, Scheuer T *et al.* (2004b). A gating hinge in Na$^+$ channels; a molecular switch for electrical signaling. *Neuron* **41**, 859–865.

Zhou J, Shin HG, Yi J, *et al.* (2002). Phosphorylation and putative ER retention signals are required for protein kinase A-mediated potentiation of cardiac sodium current. *Circ Res* **91**, 540–546.

Zhou J, Yi J, Hu N *et al.* (2000). Activation of protein kinase a modulates trafficking of the human cardiac sodium channel in *Xenopus* oocytes. *Circ Res* **87**, 33–38.

Zhou X, Dong XW and Priestley T (2006). The neuroleptic drug, fluphenazine, blocks neuronal voltage-gated sodium channels. *Brain Res* **1106**, 72–81.

Zhou Y, Morais-Cabral JH, Kaufman A *et al.* (2001). Chemistry of ion coordination and hydration revealed by a K$^+$ channel-Fab complex at 2.0 a resolution. *Nature* **414**, 43–48.

2.4

CLC chloride channels and chloride/proton antiporters

Michael Pusch

2.4.1 Introduction

CLC proteins are evolutionary ancient, with members found in all phyla, including bacteria (Jentsch *et al.*, 2002). They are not 'essential' for life, since not all bacterial genomes contain a CLC gene. Nevertheless, all higher organisms exploit the peculiar transport activity of CLC proteins in a variety of physiological settings. The amino acid sequence of CLC proteins bears no resemblance to any other class of membrane proteins, and they constitute a separate class of membrane proteins that is conserved between bacteria and higher organisms. Even though the sequence similarity of bacterial CLC proteins and mammalian CLCs is low overall, several 'signature' sequences leave no doubt about their strict evolutionary and structural relationship (Maduke *et al.*, 2000; Dutzler *et al.*, 2002; Estévez *et al.*, 2003).

The first CLC to be studied was a curious double-barrelled Cl⁻ channel from the electric organ of *Torpedo* (Miller and White 1980), where it serves probably to clamp the non-innervated membrane of the electrocytes to the (hyperpolarized) Cl⁻ equilibrium potential, allowing the build-up of a large transcellular electrical voltage. After the cloning of the *Torpedo* channel, called ClC-0 (Jentsch *et al.*, 1990), and other mammalian homologues, which also proved to be Cl⁻ channels (Steinmeyer *et al.*, 1991b; Thiemann *et al.*, 1992; Saviane *et al.*, 1999; Weinreich and Jentsch 2001), it was implicitly assumed that all CLC proteins were passive, Cl⁻ selective ion channels. In addition, the crystal structures of two bacterial CLC homologues (ClC-ec1 from *E. coli*, and StClC from *S. typhimurium*) (Dutzler *et al.*, 2002, 2003) were initially interpreted in the framework of a Cl⁻ selective ion channel. Thus, it came as a big surprise that the *E. coli* homologue ClC-ec1, whose structure had been determined, is in fact not a Cl⁻ channel at all, but an active chloride–proton (Cl⁻/H⁺) exchange transporter, i.e. a transport protein that moves in a strictly stoichiometric manner two Cl⁻ ions in one direction and one proton in the opposite direction (Accardi and Miller, 2004). This 2Cl⁻: 1 H⁺ antiport is a highly electrogenic process (three elementary charges per transport cycle). This discovery was especially surprising because a Cl⁻/H⁺ antiporter had never before been described in any biological system. Importantly, the antiport mechanism is not restricted to bacteria: several human and plant CLCs were subsequently found to be anion proton exchangers and not Cl⁻ channels (Picollo and Pusch 2005; Scheel *et al.*, 2005; De Angeli *et al.*, 2006). In humans, nine CLC proteins

are encoded by the genome (Jentsch *et al.*, 2005). Four of these are plasma-membrane localized Cl⁻ channels (ClC-1, ClC-2, ClC-Ka, ClC-Kb), whilst the remaining five (ClC-3, ClC-4, ClC-5, ClC-6, ClC-7) are localized to intracellular membranes, where they probably function as Cl⁻/H⁺ antiporters. Human CLCs are associated with several genetic diseases, highlighting their importance, and providing clues to their physiological role (Koch *et al.*, 1992; Lloyd *et al.*, 1996; Simon *et al.*, 1997; Kornak *et al.*, 2001).

2.4.2 Molecular characterization

As already mentioned above, the human genome encodes nine different genes coding for the CLC proteins (see Table 2.4.1). The somewhat odd nomenclature for the CLC-K channels arises from their strong expression in the kidney (Kieferle *et al.*, 1994; Uchida and Sasaki 2005). Functionally, the CLC proteins fall into two classes: ClC-1, ClC-2, ClC-Ka, and ClC-Kb are highly selective Cl⁻ ion channels that are localized in the plasma membrane of various cells. The other CLC proteins reside mostly in intracellular organelles and ClC-4 and ClC-5 and probably also ClC-3, ClC-6, and ClC-7 are secondary active Cl⁻/H⁺ antiporters (Picollo and Pusch 2005; Zifarelli and Pusch 2007). Like the model proteins ClC-0 from *Torpedo* (Ludewig *et al.*, 1996; Middleton *et al.*, 1996) and the bacterial ClC-ec1 (Maduke *et al.*, 1999; Dutzler *et al.*, 2002), all CLC proteins have a homodimeric architecture. Each subunit carries a separate ion translocation pathway, that can work independently from the other subunit (Dutzler *et al.*, 2002). Structurally similar CLC homologues (e.g. ClC-0 and ClC-1) are able to form heterodimers (Lorenz *et al.*, 1996), but the physiological significance of heteromultimerization is largely unexplored. Some evidence exists for physiologically relevant ClC-4/ClC-5 heteromers (Mohammad-Panah *et al.*, 2003). At least three CLC proteins require a small β-subunit for proper membrane expression: ClC-Ka and ClC-Kb are associated with barttin (Estévez *et al.*, 2001), while ClC-7 is associated with ostm1 (Lange *et al.*, 2006). Both associated β-subunits are small transmembrane proteins. These interactions seem to be highly specific, as none of the other CLC proteins appear to interact with either of these small β-subunits. Furthermore, barttin as well as ostm1, do not have other apparent homologues within the human genome (Birkenhäger *et al.*, 2001; Chalhoub *et al.*, 2003).

Table 2.4.1 Overview of the CLC family

Protein (Gene)	Chromos	Function	Ancillary subunit	Localization	Physiology	Disease	Knockout
PLM branch							
ClC-1 (CLCN1)	7q35	Cl⁻ channel	?	Skeletal muscle	Stabilize V_m	Myotonia	Myotonia
ClC-2 (CLCN2)	3q27–q28	Cl⁻ channel	?	Broad	Stabilize V_m and/or epithelial transport	Epilepsy?	Degeneration of testes and retina
ClC-Ka (CLCNKA)	1p36	Cl⁻ channel	Barttin	Kidney; inner ear	Epithelial transport	–	Nephrogenic diabetes insipidus
ClC-Kb (CLCNKB)	1p36	Cl⁻ channel	Barttin	Kidney; inner ear	Epithelial transport	Bartter's syndrome	?
Endosome branch							
ClC-3 (CLCN3)	4q33	Cl⁻/H⁺ antiporter?	?	Neurons; broad	Important for acidification of endosomes and synaptic vesicles?	–	Neurodegeneration
ClC-4 (CLCN4)	Xp22.3	Cl⁻/H⁺ antiporter	?	Broad	Important for acidification of endosomes?	–	?
ClC-5 (CLCN5)	Xp11.23–p11.22	Cl⁻/H⁺ antiporter	?	Kidney	Important for acidification of endosomes?	Dent's disease	Low molecular weight proteinurea, hypercalciuria and other
6/7 branch							
ClC-6 (CLCN6)	1p36	?	?	Nervous system	Important for acidification of late endosomes?	–	Neuronal ceroid lipofuscinosis
ClC-7 (CLCN7)	16p13	?	Ostm1	Osteoclasts; Broad	Important for acidification of lysosomes?	Osteopetrosis	Osteopetrosis and neurodegeneration

PLM, plasma membrane; Chromos., chromosomal localization.

2.4.3 Biophysical characterization

Cl⁻ channel pores are, generally, much less selective among various halide and other small organic or inorganic anions (e.g. NO_3^-) than are cation channels among alkali metal ions. This arises probably from the fact that Cl⁻ is by far the most abundant physiological halide. Thus, there is little evolutionary pressure to select for a specific halide anion. For CFTR and GABA/glycine receptor channels, the bicarbonate permeability seems to play a significant physiological role (Marty and Llano, 2005; Steward et al., 2005). No bicarbonate permeability has yet been reported for CLC channels or transporters. A rather common feature of all CLC channels and transporters studied so far is their preference for Cl⁻ over I⁻, while many other, unrelated Cl⁻ channels show an I⁻>Cl⁻ permeability.

Most CLC channels and transporters are voltage-dependent. In addition, their activity strongly depends on the ambient Cl⁻ concentration and the pH (Pusch, 2004). In fact, the voltage-dependence arises most likely indirectly from a voltage-dependent partial or complete translocation of the substrates (Cl⁻ ions and protons) across the protein (Pusch et al., 1995a; Chen and Miller, 1996; Traverso et al., 2006). For CLC channels, protons are not transported in a substantial manner (Picollo and Pusch, 2005), but they regulate the open probability by protonating residues in the pore (Chen and Chen, 2001; Miller, 2006; Traverso et al., 2006).

For the channel branch, the *Torpedo* ClC-0 channel is the best understood in terms of biophysical mechanisms (Chen 2005). At the single channel level, ClC-0 is characterized by a bursting behaviour (Figure 2.4.1a). Within each burst the two 'protopores' of the

dimeric channel operate independently from one another, with apparent two-state kinetics: open time histograms and closed time histograms are well described by the superposition of two identical pores opening and closing with rate constants α and β, respectively (Hanke and Miller, 1983; Chen and Miller, 1996; Ludewig et al., 1997b). Nevertheless, gating of ClC-0 is not in thermodynamic equilibrium (Richard and Miller, 1990; Chen and Miller, 1996) reflecting the coupling of gating and conduction in CLC channels. The bursts in recordings of single ClC-0 channels are interrupted by long closures, demonstrating the presence of an additional, slow, gating process that shuts off both pores of the double-barrelled channel (Miller and White, 1984) (Figure 2.4.1a). The gating properties of the *Torpedo* ClC-0 provide a framework for the description of the gating of mammalian CLC channels, even though they exhibit quite diverse phenotypes (see Chen, 2005) for a comprehensive overview of ClC-0 fast gating). The fast (protopore) gate of ClC-0 activates at positive voltages with an apparent gating valence ~1 (Pusch et al., 1995a). Opening is favoured by increases in Cl⁻ and also H⁺ concentration, both on the intracellular as well as the extracellular side (Pusch 2004). It has been proposed that the movement of a Cl⁻ ion across the channel pore provides the voltage-dependent step of gating, leading to the observed gating valence of 1 (Pusch et al., 1995a; Chen and Miller 1996). Recently, the possible voltage-dependent movement of protons has also been implicated in the fast gating of ClC-0 (Traverso et al., 2006). The precise mechanism of voltage-dependent gating of ClC-0 remains, however, to be determined. We have even less

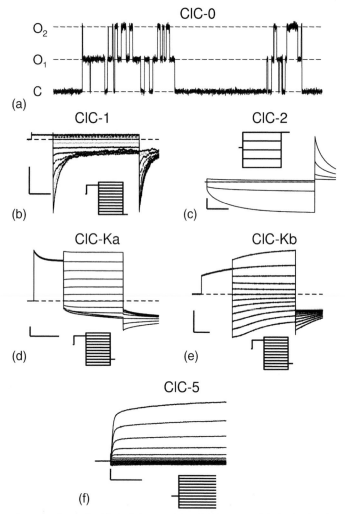

Fig. 2.4.1 Electrophysiological characteristics of CLC proteins. (a) The simulated single channel trace illustrates the characteristic double-barrelled behaviour of ClC-0. (b–e) are typical current traces evoked under voltage-clamp conditions using the respective pulse-protocols shown in the insets. (b) inside-out patch with ClC-1 currents (scale-bars: 50 ms, 50 pA); (c) whole oocyte expressing ClC-2 (scale-bars: 2 s, 2 μA); (d) whole oocyte co-expressing ClC-Ka and barttin (scale-bars: 50 ms, 5 μA); (e) whole oocyte co-expressing ClC-Kb and barttin (scale-bars: 50 ms, 2 μA); (f) transfected cell expressing ClC-5 (scale-bars: 10 ms, 300 pA).

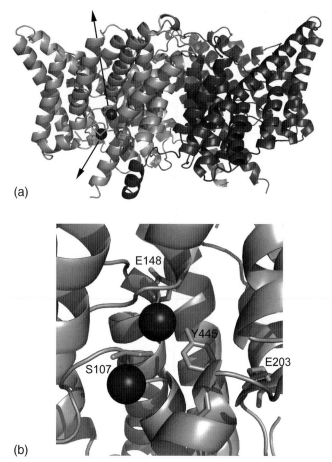

Fig. 2.4.2 Structure of the *E. coli* Cl⁻/H⁺ exchanger, ClC-ec1. The view in (a) is from within the membrane and illustrates the dimeric architecture. One subunit is shown in light grey, while the other subunit is drawn in dark grey. The two Cl⁻ ions of each subunit are shown as dark spheres. The arrows indicate the putative pathways Cl⁻ ions use to exit the protein towards the extracellular side (up) or the intracellular side (down). A short stretch of helix J is shown as a ribbon for clarity. (b) shows a blow-up of the Cl⁻ ion binding region of one subunit. Key residue side-chains are shown as sticks. Several pieces of the protein have been removed for clarity.

mechanistic understanding of the slow (common) gate that acts on both protopores. The slow gate activates at negative voltages with an apparent gating valence around 2, with seconds to minutes kinetics (Ludewig *et al.*, 1997a; Pusch *et al.*, 1997). Also the slow gate strongly depends on the Cl⁻ concentration, being activated at increased Cl⁻ concentrations (Chen and Miller 1996; Pusch *et al.*, 1999). The kinetics of slow gate closure (at positive voltages) shows an extreme temperature dependence with a $Q_{10} \sim 40$ (Pusch *et al.*, 1997). This has led to speculation that the slow gate is associated with a large conformational rearrangement. Consistent with the idea that a rather global conformational change is involved in the slow gate, mutations in many regions alter the slow gate (see Zifarelli and Pusch 2007). The fact that the slow gate is controlled by a conformational change that involves both subunits of the dimeric channel has also been demonstrated using heterodimeric, mutant/wild-type, ClC-0 channels. Mutated subunits in such

constructs can exert a dominant negative effect on the slow gate of the heteromeric channel (Ludewig *et al.*, 1996).

The crystal structure of the bacterial homologue ClC-ec1 marked a breakthrough for the understanding of the biophysical mechanisms that govern CLC function (Dutzler *et al.*, 2002, 2003 Dutzler, 2004) (Figure 2.4.2). ClC-ec1 is a homodimer, with an extensive interface between the two subunits. In each subunit, two bound Cl⁻ ions mark the Cl⁻ transport pathway. The pathways of the two subunits are far from each other (>20 Angstroms), consistent with the independent functioning of the two protopores of ClC-0. The binding sites are denoted by S_{int} for the inner Cl⁻ ion, that is in direct contact with the cytoplasmic solution, and S_{cen}, for the central binding site. The central Cl⁻ ion is completely dehydrated and buried in the protein. It is coordinated by side-chain OH-groups of a conserved serine and a conserved tyrosine residue, and by the backbone amide NH groups of several amino acids (Dutzler *et al.*, 2003) (Figure 2.4.2b). In addition, the Cl⁻ ions are electrostatically stabilized by other parts of the protein (see Zifarelli and Pusch, 2007). The movement of the central Cl⁻ ion towards the extracellular medium is hindered presumably by the charged side-chain of a conserved glutamate residue (Figure 2.4.2b). This glutamate (E148 in ClC-ec1; E166 in ClC-0; E211 in ClC-5) plays a fundamental role for the gating of CLC channels and for the coupled Cl^-/H^+ transport of CLC transporters (Dutzler et al., 2003; Traverso et al., 2003; Accardi and Miller, 2004; Picollo and Pusch, 2005; Scheel et al., 2005). In fact, in the structure of ClC-ec1 in which this glutamate is mutated to uncharged residues, a third Cl^- ion, Cl_{ext}, was crystallographically resolved and found in place of the glutamate side-chain, while the rest of the structure was virtually unchanged (Dutzler et al., 2003). Functionally, the glutamate mutation E148A in ClC-ec1 abolished proton transport, while Cl^- transport was preserved (Accardi and Miller, 2004). In ClC-0 and other CLC channels, mutation of the glutamate abolished voltage and chloride dependent gating (Dutzler et al., 2003; Estévez et al., 2003; Niemeyer et al., 2003; Traverso et al., 2003). Consequently, the movement of the side-chain of the conserved glutamate, probably necessitating its protonation, is considered to be the main event associated with the fast protopore gate of CLC channels. However, the precise mechanisms of gating in CLC channels, or those of the coupled transport in CLC transporter, remain to be identified. For ClC-ec1, another important player has recently been identified: the intracellularly accessible glutamate E203 (Figure 2.4.2b) is essential for coupled Cl^-/H^+ transport (Accardi et al., 2005). As with E148, neutralization of this residue abolishes H^+ transport without impeding Cl^- transport. Based on these observations, Accardi, Miller, and colleagues (2005) proposed that protons enter the transporter from the intracellular side via a route that is distinct from the Cl^- ion pathway, via E203. The corresponding residue has been shown to be involved in H^+ transport also in ClC-4 and ClC-5 (Zdebik et al., 2008).

All mammalian CLCs, and also some bacterial homologues possess a long cytoplasmic C-terminus that is absent in the crystallized ClC-ec1. The C-terminus of mammalian CLCs contains two CBS domains (from cystathionine-β-synthase, the first protein in which these domains were identified). The functional role of the CBS domains, and the rest of the C-terminus, is currently unclear, but these domains have been proposed to be important for protein stability, trafficking and anchoring of other binding partners (see Babini and Pusch, 2004). Several mutations in the C-terminus, within and outside of the CBS domains, have drastic effects on the slow (common) gate of ClC-0 or ClC-1 (Fong et al., 1998; Babini and Pusch, 2004; Estévez et al., 2004). However, these results can not yet be interpreted in a mechanistic manner. Recent fluorescence resonance energy transfer (FRET) measurements are in agreement with a large conformational change of the C-terminus upon slow gating (Bykova et al., 2006).

Recently, isolated CBS domains have been shown to bind ATP and other nucleotides (Scott et al., 2004), and indeed, functional effects of ATP on the ClC-1 channel have been reported (Bennetts et al., 2005, 2007; Tseng et al., 2007). However, in contrast to these studies, our group could not find evidence for a direct regulation of ClC-1 by ATP (Zifarelli and Pusch, 2008). The recent crystal structure of the isolated C-terminus of ClC-0 confirmed the typical CBS1/CBS2 fold for the two CBS-domains of one subunit (Meyer and Dutzler, 2006). No indication of nucleotide binding to the CBS domains was obtained and it is still unclear how the CBS domains are oriented towards the rest of the protein. Interestingly, nucleotides were found to be bound at the CBS1/CBS2 interface in the crystal structure of the ClC-5 C-terminus, and binding studies

indicated an equal affinity or about 100 μM for ATP, ADP, and AMP (Meyer et al., 2007). The possible regulation of CLC function by intracellular nucleotides remains an exciting perspective for the physiology of CLC proteins.

Gating of the muscle channel ClC-1 resembles the fast gate of ClC-0, in that the channel is activated at positive voltages, in a Cl^- and H^+ dependent manner (Steinmeyer et al., 1991b; Rychkov et al., 1996) (Figure 2.4.1b). The single channel conductance of ClC-1 is small (~ 1.5 pS versus 8 pS for ClC-0 (Pusch et al., 1994)) rendering almost impossible a detailed analysis of the gating behaviour at the single channel level. Nevertheless, the available single-channel data (Saviane et al., 1999), as well as other lines of evidence (Accardi and Pusch, 2000; Accardi et al., 2001) indicate that ClC-1 also has a double-barrelled appearance, but that the 'slow' (common) gate is actually only slightly slower than the fast gate, and the two gates are more tightly coupled than in ClC-0. Recently, a further very slow gating transition, that is probably also involved in the common gate, has been identified in ClC-1 (Duffield et al., 2005).

Compared to ClC-1, ClC-2 has an inverted voltage-dependence. It is activated by hyperpolarization with rather slow kinetics (Thiemann et al., 1992) (Figure 2.4.1c). In oocytes, ClC-2 is also strongly activated by low extracellular osmolarity with a considerable delay suggesting the involvement of intracellular mediators (Gründer et al., 1992). Deletions in the N-terminus, as well as in the cytoplasmic loop connecting helices H and I, and in helix I eliminate activation by hypo-osmolarity as well as low pH (Gründer et al., 1992; Jordt and Jentsch, 1997). The dependence of ClC-2 on osmolarity in native systems is not yet fully clear. A property of ClC-2 that is likely to important physiologically is its complex dependence on extracellular pH (Arreola et al., 2002). No detailed single-channel analysis of ClC-2 has been performed, and the assignment of the various kinetic components to the protopore (fast) gate or the common (slow) gate is practically impossible.

ClC-0, ClC-1, and ClC-2 all bear a negatively charged residue at the position corresponding to 148 in the bacterial transporter (see above) in the conserved GK(R)EGP motif. In all three channels, neutralizing this glutamate almost abolishes voltage-, Cl^--, and pH-dependent gating transitions (Dutzler et al., 2003; Estévez et al., 2003; Niemeyer et al., 2003; Traverso et al., 2003). This suggests that gating in all three channels is based on the same fundamental mechanisms, despite the diverse phenotypes. In contrast, the CLC-K channels ClC-Ka and ClC-Kb bear a neutral valine at the corresponding position (sequence GKVGP). Despite the absence of a glutamate, CLC-K channels show a mild voltage-dependence with ClC-Kb being activated at positive voltages and ClC-Ka being activated at negative voltages (Picollo et al., 2004) (Figure 2.4.1d, e). However, introducing a glutamate residue leads to a much more pronounced voltage-dependence (Waldegger and Jentsch, 2000). It is not clear, however, which mechanism underlies the gating relaxations of the wild-type CLC-K channels. Interestingly, CLC-K channel activity depends strongly on the extracellular pH and on extracellular Ca^{2+}, with protons inhibiting and Ca^{2+} ions activating the channels (Estévez et al., 2001). The H^+ dependence of CLC-K channels is different from that of the fast gate of ClC-0. The fast gate of ClC-0 is opened by extracellular protons (Chen and Chen, 2001), presumably by protonating E166 (Dutzler et al., 2003), the 'gating' glutamate. In contrast, CLC-K channels are inhibited by extracellular protons. The Ca^{2+} dependence of CLC-K channels seems to be unique—no Ca^{2+} dependence has been reported for other CLC proteins.

The CLC proteins ClC-3, ClC-4, and ClC-5 share a rather high sequence similarity among each other and have similar biophysical properties in heterologous expression systems. They display an extremely outwardly rectifying current-voltage relationship (Steinmeyer *et al.*, 1995; Friedrich *et al.*, 1999; Li *et al.*, 2000) with no discernable activation or deactivation kinetics (Figure 2.4.1f). While ClC-4 and ClC-5 can be functionally expressed in *Xenopus* oocytes and in cell lines at relatively high levels, ClC-3 yields only very small currents. The dramatic rectification precludes the determination of a true reversal potential. For this reason, it remained undetected for a long time that ClC-4 and ClC-5 (and probably also ClC-3) are actually not passive Cl⁻ channels but secondary active Cl⁻/H⁺ antiporters (Picollo and Pusch, 2005; Scheel *et al.*, 2005) as is the bacterial ClC-ec1(Accardi and Miller, 2004). However, it is nevertheless very difficult to reconcile the rectification of ClC-3–ClC-5 with any possible physiological role, because it is assumed that the strong positive voltages needed for their activation are not reached in the membranes where these proteins are expressed.

No electrophysiological data are available for the intracellular proteins ClC-6 and ClC-7. Thus it is unclear if they are Cl⁻ channels, Cl⁻/H⁺ antiporters, or if they display an as yet undiscovered function. Based on the presence of a glutamate residue at the position corresponding to E203 of the bacterial ClC-ec1, ClC-6 and ClC-7 are probably transporters and not channels, but this remains to be established.

2.4.4 Cellular and subcellular localization

ClC-1

ClC-1 was the first mammalian CLC cloned (Steinmeyer *et al.*, 1991b). It is expressed specifically in skeletal muscle with only faint expression in other tissues (Steinmeyer *et al.*, 1991b). While antibody staining supports a localization in the sarcolemma and not in the t-tubules (Gurnett *et al.*, 1995; Papponen *et al.*, 2005), functional studies predict a dominant localization of gCl in the t-tubules (Palade and Barchi, 1977b; Dutka *et al.*, 2008).

ClC-2

ClC-2 is about 50 per cent identical to ClC-1 and ClC-0 (Thiemann *et al.*, 1992). In contrast to the restricted expression of ClC-1, ClC-2 is broadly expressed in a variety of cell types (Thiemann *et al.*, 1992). Most likely, ClC-2 fulfils its function in the plasma membrane.

ClC-Ka and ClC-Kb

ClC-Ka and ClC-Kb were cloned from kidney where they are highly expressed (Uchida *et al.*, 1993; Kieferle *et al.*, 1994; Uchida and Sasaki 2005). The two proteins are more than 80 per cent identical and the respective genes are close to each other on chromosome 1, suggesting a recent gene duplication. The rodent isoforms are called ClC-K1 and ClC-K2. It is assumed that ClC-K1 is the species homologue of ClC-Ka and ClC-K2 is the species homologue of ClC-Kb. However, it is not clear if this association indeed reflects the different localization, biophysical properties, and the physiological role of the two ClC-K channels. In fact, discordant results have been published regarding the polarized expression of the ClC-K channels in various segments of the nephron (Uchida *et al.*, 1995; Vandewalle *et al.*, 1997). A general agreement exists on the basolateral expression of ClC-Kb in the thick ascending limb of the loop of Henle (TAL).

ClC-3, ClC-4 and ClC-5

ClC-3, ClC-4, and ClC-5 are roughly 70–80 per cent identical to each other and share similar electrophysiological properties in heterologous expression systems (see Biophysical characterization). These proteins are mostly localized in endosomes and ClC-3 is additionally found in synaptic vesicles. While ClC-3 and ClC-4 are rather widely expressed (Kawasaki *et al.*, 1994; van Slegtenhorst *et al.*, 1994), ClC-5 is relatively specific for the kidney and a few other specialized epithelia (Steinmeyer *et al.*, 1995). In the kidney, ClC-5 is predominantly expressed in the proximal tubule but is also found in distal segments (Günther *et al.*, 1998).

ClC-6

ClC-6 is mostly localized intracellularly (Buyse *et al.*, 1998) and, based on mRNA distribution, initial evidence suggested that ClC-6 is rather widely expressed (Brandt and Jentsch, 1995). Using highly specific antibodies and ClC-6 knockout mice, Poët *et al.* (2006) found that ClC-6 is almost exclusively expressed in the nervous system, co-localizing with late endosomal markers.

ClC-7

Like ClC-6, ClC-7 is also widely expressed and bears only a small overall sequence similarity with other CLC proteins (Brandt and Jentsch, 1995). In addition, no functional expression has been obtained for ClC-7, in agreement with its strict intracellular localization in heterologous expression systems (unpublished observation).

2.4.5 Pharmacology

Pharmacological tools of moderate affinity and specificity are only available for ClC-1 and CLC-K channels. These channels will be discussed in more detail below. There are practically no small organic inhibitors known for ClC-2, ClC-3, ClC-4, ClC-5, ClC-6, and ClC-7 that exert effects at concentrations below 100 µM. Several divalent cations have been described to block ClC-1 (Kürz *et al.*, 1997) and ClC-2 (Clark *et al.*, 1998), but these are not very useful from a pharmacological viewpoint. ClC-7 is an interesting potential target to treat osteoporosis, because its activity is necessary for bone resorption. Schaller and colleagues (2004) have identified several compounds that inhibit osteoclastic bone resorption *in vitro* and that also block swelling induced Cl⁻ currents in osteoclasts. It is unclear, however, if these drugs act directly on ClC-7.

The skeletal muscle chloride conductance, gCl, that dominates the background conductance was studied long before the cloning of ClC-1, the channel that mediates gCl. During these early studies two major classes of compounds were identified that inhibit gCl with relatively high affinity. One compound class is exemplified by 9-antracene-carboxylic acid (9-AC, Figure 2.4.3a) that inhibits gCl with an apparent K_D of 11 µM (Palade and Barchi, 1977a). The other class of compounds are derivatives of clofibric acid, exemplified by p-chloro-phenoxy-propionic acid (CPP) (Figure 2.4.3b) that inhibits gCl in a stereoselective manner at micromolar concentrations (Conte-Camerino *et al.*, 1988). The mechanism of inhibition by CPP has been studied in great detail (Pusch *et al.*, 2000, 2001, 2002; Accardi and Pusch, 2003; Estévez *et al.*, 2003; Liantonio *et al.*, 2003).

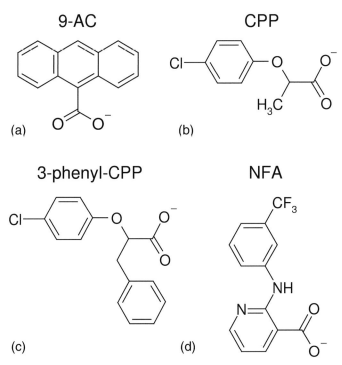

Fig. 2.4.3 Chemical structure of several CLC channel blockers and activators. (a) 9-AC; (b) CPP; (c) 3-phenyl-CPP; (d) NFA.

Most CLC-K blockers, including 3-phenyl-CPP, show a preference for ClC-Ka over ClC-Kb (Picollo et al., 2004). This allowed Picollo et al., (2004) to identify two amino acid residues in helix B as being responsible for the differential drug sensitivity of ClC-Ka and ClC-Kb. The side chains of these amino acids point into the extracellular vestibule of the ion conducting pore, delineating the putative extracellular binding site (Picollo et al., 2004). Recently, benzfuran-derived blockers of ClC-Ka and ClC-Kb with affinity < 10 μM have been reported (Liantonio et al., 2008).

Surprisingly, it was found that niflumic acid (NFA), a member of the fenamates group of non-steroidal anti-inflammatory drugs (NSAID), has a biphasic effect on CLC-K channels: at concentrations below 1 mM, NFA potentiates ClC-Ka currents, and even more strongly ClC-Kb currents, while higher concentrations lead to channel block. In contrast, the related flufenamic acid (FFA) only blocked ClC-Ka (Liantonio et al., 2006). Testing several derivatives of FFA and NFA, it emerged that the rigid co-planar structure of the NFA molecule is essential for the stimulatory effect (Liantonio et al., 2006). The mechanism underlying the potentiation has been studied in some detail (Picollo et al., 2007). However, the relevant binding site(s) are still unknown. Future studies will be necessary to better understand this phenomenon and eventually to develop drugs that stimulate CLC-K channel function in a clinically meaningful way.

2.4.6 Physiology

ClC-1

ClC-1 is found almost exclusively in skeletal muscle where it mediates the large resting Cl⁻ conductance (gCl) that is necessary for an efficient repolarization of the muscle action potential (Jentsch et al., 1995).

ClC-2

The hyperpolarization-activated currents induced after heterologous expression of ClC-2 (see above) resemble currents that have been measured in many cell types (see Jentsch et al., 2002). However, the precise physiological role of this channel is not yet fully clear. Various functions have been ascribed to ClC-2 (see Jentsch et al., 2002), including maintenance of a low neuronal intracellular Cl⁻ concentration (Staley et al., 1996). Based on the phenotype of the knockout mice, it has been proposed that ClC-2 is important for the ionic homeostasis in the extracellular space between tightly associated cells (Bösl et al., 2001). This would also be consistent, for example, with the localization of ClC-2 in astrocytic endfeet (Sik et al., 2000). In epithelial cells, ClC-2 is most likely localized on the basolateral membrane (Zdebik et al., 2004; Peña-Münzenmayer et al., 2005) and is possibly involved in transepithelial transport (Zdebik et al., 2004).

ClC-Ka and ClC-Kb

Although there is still some debate around the specific localization, and therefore the function, of ClC- Ka and ClC-Kb, there is now a consensus opinion on the basolateral expression of ClC-Kb in the thick ascending limb of the loop of Henle (TAL), suggesting a role in the distal reabsorption of NaCl from primary urine through basolateral chloride efflux.

The compound can access its binding site only from the intracellular side and produces a block that is strongly state-dependent, with closed channels having a much higher affinity than open channels. Also 9-AC blocks ClC-1 only when applied from the inside in excised patches (Estévez et al., 2003), but due to its lipophilicity it can cross the membrane quite easily and thus is also effective when applied from the outside in intact cells. The binding sites of both compounds overlap. The site is very close to, and probably partially within the ion conducting pore (Estévez et al., 2003). Most likely, the negatively charged group of these compounds occupies one of the Cl⁻ ion binding sites that have been identified in the crystal structure of ClC-ec1 (Dutzler et al., 2003).

ClC-1 is not an interesting pharmacological target by itself because its inhibition leads to muscle hyperexcitability, generally an undesirable condition. However, its inhibition by drugs, or their metabolites, that are used for other purposes has to be considered and understood. For example, clofibrates are extensively used as lipid-lowering drugs.

In contrast, the kidney and inner ear specific epithelial CLC-K channels are particularly interesting from a pharmacological viewpoint (Fong, 2004). Blockers of CLC-K channels may act as diuretics with different effects from commonly used diuretic drugs, and therefore may be useful in particular pathological circumstances. In addition, activators of CLC-K channels are of potential medical relevance, in treating for example Bartter's syndrome patients who have a residual ClC-Kb channel activity. Recently, several new compounds have been identified that block CLC-K channels from the extracellular side with sub 100 μM inhibition affinity constants (Liantonio et al., 2002, 2004). One of these compounds, 3-phenyl-CPP, is shown in Figure 2.4.3c. These blockers act from the extracellular side, suggesting that the binding site is different from the CPP binding site identified in ClC-1 (Estévez et al., 2003).

ClC-3, ClC-4 and ClC-5

Most insight into the physiological role of ClC-5 and ClC-3 has been gained from knockout studies (Piwon *et al.*, 2000; Wang *et al.*, 2000; Stobrawa *et al.*, 2001). From these studies it has been proposed that ClC-5 is important for the acidification of early endosomes in receptor mediated and liquid phase endocytosis in the proximal tubule (Piwon *et al.*, 2000). In this respect it is believed that ClC-5 provides a Cl$^-$ conductance that neutralizes the electrically polarizing effect of the acidifying vesicular proton pump, allowing an efficient acidification (Piwon *et al.*, 2000). However, two facts are difficult to reconcile with the proposed role of ClC-3, ClC-4, and ClC-5 as shunting anionic conductances. First, recently it has been shown that these proteins are actually not passive Cl$^-$ ion channels, as had been implicitly assumed previously, but secondarily active Cl$^-$/H$^+$ antiporters (Picollo and Pusch, 2005; Scheel *et al.*, 2005). The physiological implication of this finding is still unclear. Furthermore, the biophysical properties of these proteins, i.e. the extreme outward rectification seen in heterologous expression systems, are not easily compatible with a transport function in endosomes, because the endosomal membrane potential is not expected to reach the necessary positive values required for an activation of ClC-5. Clearly, further research is necessary to obtain a satisfactory understanding of the function of these proteins.

ClC-6

Since it cannot yet be functionally expressed in heterologous systems (Brandt and Jentsch, 1995; Buyse *et al.*, 1997) nothing is known about the biophysical properties of ClC-6 and no selective pharmacological tools have been developed to probe its physiological function. However, breakthroughs in the understanding of the physiological role of ClC-6 are likely to be forthcoming with the recent development by Jentsch and colleagues of the ClC-6 knockout mouse (Poët *et al.*, 2006) (see Disease relevance section).

ClC-7

Little is known about the physiological function of ClC-7. However, in osteoclasts ClC-7 co-localizes with the H$^+$-ATPase, and it has been proposed that it acts as a shunt, allowing efficient acidification (Kornak *et al.*, 2001) to dissolve inorganic bone material and to activate proteases that degrade organic material.

2.4.7 Disease relevance

ClC-1—the skeletal muscle Cl$^-$ channel that is mutated in myotonia

Reduction of ClC-1 channel function in skeletal mucle leads to myotonia, i.e. muscle stiffness, a condition in which a single nerve stimulus may elicit a ('myotonic') train of action potentials (Steinmeyer *et al.*, 1991a). Recessive myotonia (Becker-type myotonia) and dominant myotonia (Thomsen's disease) are caused by mutations in the *CLCN1* gene (Koch *et al.*, 1992; Pusch, 2002), while the myotonic phenotype seen in myotonic dystrophy is caused by the interference of transdominant aberrant RNA products with the correct splicing of ClC-1 pre-RNA (Charlet *et al.*, 2002; Mankodi *et al.*, 2002; Berg *et al.*, 2004). Dominant myotonia can often be explained by a dominant alteration of the common gate of the double barrelled ClC-1 channel (Pusch *et al.*, 1995b;

Kubisch *et al.*, 1998; Saviane *et al.*, 1999; Aromataris *et al.*, 2001). Pharmacological treatment of myotonia is aimed at blocking Na$^+$ channels in order to keep the balance between this depolarizing conductance and the reduced repolarizing Cl$^-$ conductance.

ClC-2—a ubiquitous hyperpolarization activated Cl$^-$ channel

In agreement with a role of ClC-2 in neuronal signalling, mutations in the *CLCN2* gene have been linked with epilepsy (Haug *et al.*, 2003). However, some of the initial functional evidence could not be reproduced (Niemeyer *et al.*, 2004), and more solid genetic evidence is probably needed to confirm ClC-2 as a relevant epilepsy gene. A very strong argument against a very critical role of ClC-2 in the central nervous systems is provided by the phenotype of ClC-2 knockout mice. These mice have no epilepsy and no systemic disturbances. They show quite specific, local, cellular degenerations in testes and in the retina that lead to male infertility and blindness (Bösl *et al.*, 2001) and to vacuolation in the white matter of the brain and spinal cord (Blanz *et al.*, 2007).

The highly homologous epithelial channels ClC-Ka and ClC-Kb

Mutations in the gene coding for ClC-Kb lead to Bartter's syndrome (Simon *et al.*, 1997), a nephropathy that is associated with a severe salt loss. In contrast to other plasma membrane CLC channels, ClC-Ka and ClC-Kb are dependent on the interaction with an associated ancillary β-subunit, called barttin. Mutations in barttin cause Bartter's syndrome in conjunction with sensorineural deafness (Birkenhäger *et al.*, 2001). The more severe phenotype caused by mutations in the gene coding for barttin, compared to that caused by mutations in *CLCNKB*, is explained by the fact that both ClC-Ka and ClC-Kb necessitate barttin for an efficient expression at the membrane surface (Estévez *et al.*, 2001; Waldegger *et al.*, 2002). In addition to stabilizing CLC-K channels in the plasma-membrane, barttin has also been suggested to affect the functional properties of ClC-K1, a channel that does not strictly require barttin in heterologous expression systems (Scholl *et al.*, 2006). ClC-Ka, ClC-Kb, and barttin are co-expressed in the basolateral membrane of epithelial cells in the stria vascularis of the inner ear, where they are important for the production of the endolymph (Estévez *et al.*, 2001), explaining the deafness associated with mutations in the barttin gene (Birkenhäger *et al.*, 2001), and in patients that lack both functional ClC-Ka and ClC-Kb coding genes (Schlingmann *et al.*, 2004).

The importance of the ClC-K channels for renal salt and water reabsorption is also suggested by the phenotype of ClC-K1 knockout mice that show an overt nephrogenic diabetes insipidus (Matsumura *et al.*, 1999).

The endosomal transporters ClC-3, ClC-4, and ClC-5

Mutations in the gene coding for ClC-5 lead to Dent's disease (Lloyd *et al.*, 1996). The most characteristic feature of Dent's disease is low molecular weight proteinurea, indicating a primary defect in the endocytosis of small proteins in the proximal tubule of the nephron. Other symptoms include hypercalciuria, nephrocalcinosis, nephrolithiasis (kidney stones) and eventual renal failure (Dent and Friedman, 1964; Wrong *et al.*, 1994; Günther *et al.*, 1998). The role of ClC-5 in acidification of early endosomes in receptor mediated and liquid phase endocytosis in the proximal tubule (Piwon *et al.*, 2000) explains the primary low molecular weight proteinura phenotype of ClC-5 knockout mice that is a common symptom in Dent's disease, because an acidic luminal pH is necessary for proper endosomal function and maturation. The other symptoms are most likely caused by secondary effects (Maritzen *et al.*, 2006). As mentioned above, the simplest hypothesis explaining the ClC-5 loss-of-function phenotype is that the protein provides a Cl⁻ conductance that neutralizes the electrically polarizing effect of the acidifying vesicular proton pump, allowing an efficient acidification (Piwon *et al.*, 2000). The phenotype of the ClC-3 knockout mice has been interpreted along similar lines. These mice show a severe neurodegeneration with a complete loss of the hippocampus (Stobrawa *et al.*, 2001). However, the precise mechanism behind the neurodegeneration is still unclear.

ClC-6, a neuronal, late endosomal channel or transporter

ClC-6(−/−) mice show signs of neuronal ceroid lipofuscinosis (NCL), with swellings of initial axonal segments. However, overall, the phenotype of ClC-6 knockout mice was relatively mild. Even though Poët *et al.* (2006) found no conclusive evidence for an involvement of ClC-6 in a relatively large group of NCL patients, the protein remains a viable candidate for special forms of the disease.

ClC-7, a predominantly lysosomal protein mutated in osteopetrosis

A breakthrough in the understanding of the physiological role of ClC-7 was obtained by the careful analysis of the phenotype of ClC-7 knockout mice (Kornak *et al.*, 2001). These mice display severe osteopetrosis as well as retinal and general neurodegeneration (Kornak *et al.*, 2001; Kasper *et al.*, 2005). The osteopetrosis in ClC-7 knockout mice is caused by an impairment of bone resorption by osteoclasts. In fact, osteoclasts adhere to bone by forming a so-called resorption lacuna. Protons are secreted into the lumen of the lacuna by a V-type H⁺-ATPase and the acidic pH serves to dissolve the inorganic bone material and to activate proteases that degrade organic material. ClC-7 co-localizes with the H⁺-ATPase, and it has been proposed that it acts as a shunt, allowing efficient acidification (Kornak *et al.*, 2001). Mutations in the human gene coding for ClC-7 can cause recessive (Kornak *et al.*, 2001) or dominant (Frattini *et al.*, 2003) osteopetrosis. The molecular mechanism leading to the neurodegeneration in ClC-7 KO mice is, however, unclear (Kasper *et al.*, 2005). Recently, it has been shown that ClC-7 is associated with a small ancillary subunit, called ostm1 (Lange *et al.*, 2006), mutations of which cause osteopetrosis in mice and man (Chalhoub *et al.*, 2003). Ostm1 is needed for the stability of the ClC-7 protein, and the reduction of ClC-7 protein levels is the probable mechanism of how mutations of the OSTM1 gene cause osteopetrosis (Lange *et al.*, 2006).

2.4.8 Concluding remarks

Following the relatively recent cloning of ClC-0, there has been tremendous progress in the CLC channel and transporter field. Their unusual biophysical properties are now beginning to be understood and are believed to arise from the fact that these proteins do not follow the classical distinction of transporters and channels.

In this respect, CLC proteins are structurally unique. They have an unusual 'double-barrelled' architecture and a complex three-dimensional fold. Genetic diseases and gene knockout studies have revealed an unexpected variety of physiological functions. Many challenges remain including better understanding of the molecular mechanisms and the physiological role of the anion/H⁺ antiport of many CLC-proteins and the development of specific and high affinity small molecule ligands, in order to exploit their medical potential.

Acknowledgements

The financial support by Telethon Italy (grant GGP04018) is gratefully acknowledged.

References

Accardi A and Miller C (2004). Secondary active transport mediated by a prokaryotic homologue of ClC Cl⁻ channels. *Nature* **427**, 803–807.

Accardi A and Pusch M (2000). Fast and slow gating relaxations in the muscle chloride channel CLC-1. *J Gen Physiol* **116**, 433–444.

Accardi A and Pusch M (2003). Conformational changes in the pore of CLC-0. *J Gen Physiol* **122**, 277–293.

Accardi A, Ferrera L and Pusch M (2001). Drastic reduction of the slow gate of human muscle chloride channel (ClC-1) by mutation C277S. *J Physiol* **534**, 745–752.

Accardi A, Walden M, Nguitragool W *et al.* (2005). Separate ion pathways in a Cl⁻/H⁺ exchanger. *J Gen Physiol* **126**, 563–570.

Aromataris EC, Rychkov GY, Bennetts B *et al.* (2001). Fast and slow gating of CLC-1: differential effects of 2-(4-chlorophenoxy) propionic acid and dominant negative mutations. *Mol Pharmacol* **60**, 200–208.

Arreola J, Begenisich T and Melvin JE (2002). Conformation-dependent regulation of inward rectifier chloride channel gating by extracellular protons. *J Physiol* **541**, 103–112.

Babini E and Pusch M (2004). A two-holed story: structural secrets about CLC proteins become unraveled? *Physiol* **19**, 293–299.

Bennetts B, Parker MW and Cromer BA (2007). Inhibition of skeletal muscle CLC-1 chloride channels by low intracellular pH and ATP. *J Biol Chem* **282**, 32780–32791.

Bennetts B, Rychkov GY, Ng H-L *et al.* (2005). Cytoplasmic ATP-sensing domains regulate gating of skeletal muscle CLC-1 chloride channels. *J Biol Chem* **280**, 32452–32458.

Berg J, Jiang H, Thornton CA *et al.* (2004). Truncated ClC-1 mRNA in myotonic dystrophy exerts a dominant-negative effect on the Cl current. *Neurology* **63**, 2371–2375.

Birkenhäger R, Otto E, Schurmann MJ *et al.* (2001). Mutation of BSND causes Bartter syndrome with sensorineural deafness and kidney failure. *Nat Genet* **29**, 310–314.

Blanz J, Schweizer M, Auberson M *et al.* (2007). Leukoencephalopathy upon disruption of the chloride channel ClC-2. *J Neurosci* **27**, 6581–6589.

Bösl MR, Stein V, Hübner C *et al.* (2001). Male germ cells and photoreceptors, both dependent on close cell–cell interactions, degenerate upon ClC-2 Cl(-) channel disruption. *EMBO J* **20**, 1289–1299.

Brandt S and Jentsch TJ (1995). ClC-6 and ClC-7 are two novel broadly expressed members of the CLC chloride channel family. *FEBS Lett* **377**, 15–20.

Buyse G, Trouet D, Voets T *et al.* (1998). Evidence for the intracellular location of chloride channel (ClC)-type proteins: co-localization of ClC-6a and ClC-6c with the sarco/endoplasmic-reticulum Ca²⁺ pump SERCA2b. *Biochem J* **330**, 1015–1021.

Buyse G, Voets T, Tytgat J *et al.* (1997). Expression of human pICln and ClC-6 in *Xenopus* oocytes induces an identical endogenous chloride conductance. *J Biol Chem* **272**, 3615–3621.

Bykova EA, Zhang XD, Chen TY *et al.* (2006). Large movement in the C terminus of CLC-0 chloride channel during slow gating. *Nat Struct Mol Biol* **13**,1115–1119.

Chalhoub N, Benachenhou N, Rajapurohitam V *et al.* (2003). Grey-lethal mutation induces severe malignant autosomal recessive osteopetrosis in mouse and human. *Nat Med* **9**, 399–406.

Charlet BN, Savkur RS, Singh G *et al.* (2002). Loss of the muscle-specific chloride channel in type 1 myotonic dystrophy due to misregulated alternative splicing. *Mol Cell* **10**, 45–53.

Chen MF and Chen TY (2001). Different fast-gate regulation by external Cl(-) and H(+) of the muscle-type ClC chloride channels. *J Gen Physiol* **118**, 23–32.

Chen TY (2005). Structure and function of CLC channels. *Ann Rev Physiol* **67**, 809–839.

Chen TY and Miller C (1996). Nonequilibrium gating and voltage dependence of the ClC-0 Cl⁻ channel. *J Gen Physiol* **108**, 237–250.

Clark S, Jordt SE, Jentsch TJ *et al.* (1998). Characterization of the hyperpolarization-activated chloride current in dissociated rat sympathetic neurons. *J Physiol* **506**, 665–678.

Conte-Camerino D, Mambrini M, DeLuca A *et al.* (1988). Enantiomers of clofibric acid analogs have opposite actions on rat skeletal muscle chloride channels. *Pflügers Arch* **413**, 105–107.

De Angeli A, Monachello D, Ephritikhine G *et al.* (2006). The nitrate/proton antiporter AtCLCa mediates nitrate accumulation in plant vacuoles. *Nature* **442**, 939–942.

Dent CE and Friedman M (1964). Hypercalcuric rickets associated with renal tubular damage. *Arch Dis Child* **39**, 240–249.

Duffield MD, Rychkov GY, Bretag AH *et al.* (2005). Zinc inhibits human ClC-1 muscle chloride channel by interacting with its common gating mechanism. *J Physiol* **568**, 5–12.

Dutka TL, Murphy RM, Stephenson DG *et al.* (2008). Chloride conductance in the transverse tubular system of rat skeletal muscle fibres: importance in excitation-contraction coupling and fatigue. *J Physiol* **586**, 875–887.

Dutzler R (2004). The structural basis of ClC chloride channel function. *TINS* **27**, 315–320.

Dutzler R, Campbell EB and MacKinnon R (2003). Gating the selectivity filter in ClC chloride channels. *Science* **300**, 108–112.

Dutzler R, Campbell EB, Cadene M *et al.* (2002). X-ray structure of a ClC chloride channel at 3.0 Å reveals the molecular basis of anion selectivity. *Nature* **415**, 287–294.

Estévez R, Boettger T, Stein V *et al.* (2001). Barttin is a Cl– channel beta-subunit crucial for renal Cl– reabsorption and inner ear K⁺ secretion. *Nature* **414**, 558–561.

Estévez R, Pusch M, Ferrer-Costa C *et al.* (2004). Functional and structural conservation of CBS domains from CLC channels. *J Physiol* **557**, 363–378.

Estévez R, Schroeder BC, Accardi A *et al.* (2003). Conservation of chloride channel structure revealed by an inhibitor binding site in ClC-1. *Neuron* **38**, 47–59.

Fong P (2004). CLC-K channels: if the drug fits, use it. *EMBO Rep* **5**, 565–566.

Fong P, Rehfeldt A and Jentsch TJ (1998). Determinants of slow gating in ClC-0, the voltage-gated chloride channel of *Torpedo marmorata*. *Am J Physiol* **274**, C966–C973.

Frattini A, Pangrazio A, Susani L *et al.* (2003). Chloride channel ClCN7 mutations are responsible for severe recessive, dominant and intermediate osteopetrosis. *J Bone Miner Res* **18**, 1740–1747.

Friedrich T, Breiderhoff T and Jentsch TJ (1999). Mutational analysis demonstrates that ClC-4 and ClC-5 directly mediate plasma membrane currents. *J Biol Chem* **274**, 896–902.

Gründer S, Thiemann A, Pusch M *et al.* (1992). Regions involved in the opening of ClC-2 chloride channel by voltage and cell volume. *Nature* **360**, 759–762.

Günther W, Luchow A, Cluzeaud F *et al.* (1998). ClC-5, the chloride channel mutated in Dent's disease, colocalizes with the proton pump in endocytotically active kidney cells. *PNAS* **95**, 8075–8080.

Gurnett CA, Kahl SD Anderson RD *et al.* (1995). Absence of the skeletal muscle sarcolemma chloride channel ClC-1 in myotonic mice. *J Biol Chem* **270**, 9035–9038.

Hanke W and Miller C (1983). Single chloride channels from *Torpedo* electroplax. Activation by protons. *J Gen Physiol* **82**, 25–45.

Haug K, Warnstedt M, Alekov AK *et al.* (2003). Mutations in *CLCN2* encoding a voltage-gated chloride channel are associated with idiopathic generalized epilepsies. *Nat Genet* **33**, 527–532.

Jentsch TJ, Lorenz C, Pusch M *et al.* (1995). Myotonias due to CLC-1 chloride channel mutations. *Soc Gen Physiol Ser* **50**, 149–159.

Jentsch TJ, Poet M, Fuhrmann JC *et al.* (2005). Physiological functions of CLC Cl channels gleaned from human genetic disease and mouse models. *Ann Rev Physiol* **67**, 779–807.

Jentsch TJ, Stein V, Weinreich F *et al.* (2002). Molecular structure and physiological function of chloride channels. *Physiol Rev* **82**, 503–568.

Jentsch TJ, Steinmeyer K and Schwarz G (1990). Primary structure of *Torpedo marmorata* chloride channel isolated by expression cloning in *Xenopus* oocytes. *Nature* **348**, 510–514.

Jordt SE and Jentsch TJ (1997). Molecular dissection of gating in the ClC-2 chloride channel. *EMBO J* **16**, 1582–1592.

Kasper D, Planells-Cases R, Fuhrmann JC *et al.* (2005). Loss of the chloride channel ClC-7 leads to lysosomal storage disease and neurodegeneration. *EMBO J* **24**, 1079–1091.

Kawasaki M, Uchida S, Monkawa T *et al.* (1994). Cloning and expression of a protein kinase C-regulated chloride channel abundantly expressed in rat brain neuronal cells. *Neuron* **12**, 597–604.

Kieferle S, Fong P, Bens M *et al.* (1994). Two highly homologous members of the ClC chloride channel family in both rat and human kidney. *PNAS* **91**, 6943–6947.

Koch MC, Steinmeyer K, Lorenz C *et al.* (1992). The skeletal muscle chloride channel in dominant and recessive human myotonia. *Science* **257**, 797–800.

Kornak U, Kasper D, Bösl MR *et al.* (2001). Loss of the ClC-7 chloride channel leads to osteopetrosis in mice and man. *Cell* **104**, 205–215.

Kubisch C, Schmidt-Rose T, Fontaine B *et al.* (1998). ClC-1 chloride channel mutations in myotonia congenita: variable penetrance of mutations shifting the voltage dependence. *Hum Mol Genet* **7**, 1753–1760.

Kürz L, Wagner S, George AL Jr *et al.* (1997). Probing the major skeletal muscle chloride channel with Zn2⁺ and other sulfhydryl-reactive compounds. *Pflügers Arch* **433**, 357–363.

Lange PF, Wartosch L, Jentsch TJ *et al.* (2006). ClC-7 requires Ostm1 as a beta-subunit to support bone resorption and lysosomal function. *Nature* **440**, 220–223.

Li X, Shimada K, Showalter LA *et al.* (2000). Biophysical properties of ClC-3 differentiate it from swelling-activated chloride channels in Chinese hamster ovary-K1 cells. *J Biol Chem* **275**, 35994–35998.

Liantonio A, Accardi A, Carbonara G *et al.* (2002). Molecular requisites for drug binding to muscle CLC-1 and renal CLC-K channel revealed by the use of phenoxy-alkyl derivatives of 2-(p- chlorophenoxy) propionic acid. *Mol Pharmacol* **62**, 265–271.

Liantonio A, De Luca A, Pierno S *et al.* (2003). Structural requisites of 2-(p-chlorophenoxy) propionic acid analogues for activity on native rat skeletal muscle chloride conductance and on heterologously expressed CLC-1. *Br J Pharmacol* **139**, 1255–1264.

Liantonio A, Picollo A, Babini E *et al.* (2006). Activation and inhibition of kidney CLC-K chloride channels by fenamates. *Mol Pharmacol* **69**, 165–173.

Liantonio A, Picollo A, Carbonara G et al. (2008). Molecular switch for CLC-K Cl⁻ channel block/activation: optimal pharmacophoric requirements towards high-affinity ligands. *Proc Natl Acad Sci USA* **105**, 1369–1373.

Liantonio A, Pusch M, Picollo A et al. (2004). Investigations of pharmacologic properties of the renal CLC-K1 chloride channel co-expressed with barttin by the use of 2-(p-Chlorophenoxy) propionic acid derivatives and other structurally unrelated chloride channel blockers. *J Am Soc Nephrol* **15**, 13–20.

Lloyd SE, Pearce SH, Fisher SE et al. (1996). A common molecular basis for three inherited kidney stone diseases. *Nature* **379**, 445–449.

Lorenz C, Pusch M and Jentsch TJ (1996). Heteromultimeric CLC chloride channels with novel properties. *PNAS* **93**, 13362–13366.

Ludewig U, Jentsch TJ and Pusch M (1997a). Analysis of a protein region involved in permeation and gating of the voltage-gated *Torpedo* chloride channel ClC-0. *J Physiol* **498**, 691–702.

Ludewig U, Pusch M and Jentsch TJ (1996). Two physically distinct pores in the dimeric ClC-0 chloride channel. *Nature* **383**, 340–343.

Ludewig U, Pusch M and Jentsch TJ (1997b). Independent gating of single pores in CLC-0 chloride channels. *Biophys J* **73**, 789–797.

Maduke M, Miller C and Mindell JA (2000). A decade of CLC chloride channels: structure, mechanism and many unsettled questions. *Annu Rev Biophys Biomol Struct* **29**, 411–438.

Maduke M, Pheasant DJ and Miller C (1999). High-level expression, functional reconstitution and quaternary structure of a prokaryotic ClC-type chloride channel. *J Gen Physiol* **114**, 713–722.

Mankodi A, Takahashi MP, Jiang H et al. (2002). Expanded CUG repeats trigger aberrant splicing of ClC-1 chloride channel pre-mRNA and hyperexcitability of skeletal muscle in myotonic dystrophy. *Mol Cell* **10**, 35–44.

Maritzen T, Rickheit G, Schmitt A et al. (2006). Kidney-specific upregulation of vitamin D3 target genes in ClC-5 KO mice. *Kidney Int* **70**, 79–87.

Marty A and Llano I (2005). Excitatory effects of GABA in established brain networks. *TINS* **28**, 284–289.

Matsumura Y, Uchida S, Kondo Y et al. (1999). Overt nephrogenic diabetes insipidus in mice lacking the CLC-K1 chloride channel. *Nat Genet* **21**, 95–98.

Meyer S and Dutzler R (2006). Crystal structure of the cytoplasmic domain of the chloride channel ClC-0. *Structure* **14**, 299–307.

Meyer S, Savaresi S, Forster IC et al. (2007). Nucleotide recognition by the cytoplasmic domain of the human chloride transporter ClC-5. *Nat Struct Mol Biol* **14**, 60–67.

Middleton RE, Pheasant DJ and Miller C (1996). Homodimeric architecture of a ClC-type chloride ion channel. *Nature* **383**, 337–340.

Miller C (2006). ClC chloride channels viewed through a transporter lens. *Nature* **440**, 484–489.

Miller C and White MM (1980). A voltage-dependent chloride conductance channel from *Torpedo* electroplax membrane. *Ann NY Acad Sci* **341**, 534–551.

Miller C and White MM (1984). Dimeric structure of single chloride channels from *Torpedo* electroplax. *PNAS* **81**, 2772–2775.

Mohammad-Panah R, Harrison R, Dhani S et al. (2003). The chloride channel ClC-4 contributes to endosomal acidification and trafficking. *J Biol Chem* **278**, 29267–29277.

Niemeyer MI, Cid LP, Zúñiga L et al. (2003). A conserved pore-lining glutamate as a voltage- and chloride-dependent gate in the ClC-2 chloride channel. *J Physiol* **553**, 873–879.

Niemeyer MI, Yusef YR, Cornejo I et al. (2004). Functional evaluation of human ClC-2 chloride channel mutations associated with idiopathic generalized epilepsies. *Physiol Genomics* **19**, 74–83.

Palade P and Barchi R (1977a). On the inhibition of muscle membrane chloride conductance by aromatic carboxylic acids. *J Gen Physiol* **69**, 879–896.

Palade PT and Barchi RL (1977b). Characteristics of the chloride conductance in muscle fibers of the rat diaphragm. *J Gen Physiol* **69**, 325–342.

Papponen H, Kaisto T, Myllyla VV et al. (2005). Regulated sarcolemmal localization of the muscle-specific ClC-1 chloride channel. *Exp Neurol* **191**, 163–173.

Peña-Münzenmayer G, Catalán M, Cornejo I et al. (2005). Basolateral localization of native ClC-2 chloride channels in absorptive intestinal epithelial cells and basolateral sorting encoded by a CBS-2 domain di-leucine motif. *J Cell Sci* **118**, 4243–4252.

Picollo A and Pusch M (2005). Chloride/proton antiporter activity of mammalian CLC proteins ClC-4 and ClC-5. *Nature* **436**, 420–423.

Picollo A, Liantonio A, Babini E et al. (2007). Mechanism of interaction of niflumic acid with heterologously expressed kidney CLC-K chloride channels. *J Membr Biol* **216**, 73–82.

Picollo A, Liantonio A, Didonna MP et al. (2004). Molecular determinants of differential pore blocking of kidney CLC-K chloride channels. *EMBO Rep* **5**, 584–589.

Piwon N, Günther W, Schwake M et al. (2000). ClC-5 Cl⁻-channel disruption impairs endocytosis in a mouse model for Dent's disease. *Nature* **408**, 369–373.

Poët M, Kornak U, Schweizer M et al. (2006). Lysosomal storage disease upon disruption of the neuronal chloride transport protein ClC-6. *PNAS* **103**, 13854–13859.

Pusch M (2002). Myotonia caused by mutations in the muscle chloride channel gene CLCN1. *Hum Mutat* **19**, 423–434.

Pusch M (2004). Structural insights into chloride and proton-mediated gating of CLC chloride channels. *Biochemistry* **43**, 1135–1144.

Pusch M, Accardi A, Liantonio A et al. (2001). Mechanism of block of single protopores of the *Torpedo* chloride channel ClC-0 by 2-(p-chlorophenoxy) butyric acid (CPB). *J Gen Physiol* **118**, 45–62.

Pusch M, Accardi A, Liantonio A et al. (2002). Mechanisms of block of muscle type CLC chloride channels. *Mol Membr Biol* **19**, 285–292.

Pusch M, Jordt SE, Stein V et al. (1999). Chloride dependence of hyperpolarization-activated chloride channel gates. *J Physiol* **515**, 341–353.

Pusch M, Liantonio A, Bertorello L et al. (2000). Pharmacological characterization of chloride channels belonging to the ClC family by the use of chiral clofibric acid derivatives. *Mol Pharmacol* **58**, 498–507.

Pusch M, Ludewig U and Jentsch TJ (1997). Temperature dependence of fast and slow gating relaxations of ClC-0 chloride channels. *J Gen Physiol* **109**, 105–116.

Pusch M, Ludewig U, Rehfeldt A et al. (1995a). Gating of the voltage-dependent chloride channel ClC-0 by the permeant anion. *Nature* **373**, 527–531.

Pusch M, Steinmeyer K and Jentsch TJ (1994). Low single channel conductance of the major skeletal muscle chloride channel, ClC-1. *Biophys J* **66**, 149–152.

Pusch M, Steinmeyer K, Koch MC et al. (1995b). Mutations in dominant human myotonia congenita drastically alter the voltage dependence of the ClC-1 chloride channel. *Neuron* **15**, 1455–1463.

Richard EA and Miller C (1990). Steady-state coupling of ion-channel conformations to a transmembrane ion gradient. *Science* **247**, 1208–1210.

Rychkov GY, Pusch M, Astill DS et al. (1996). Concentration and pH dependence of skeletal muscle chloride channel ClC-1. *J Physiol* **497**, 423–435.

Saviane C, Conti F and Pusch M (1999). The muscle chloride channel ClC-1 has a double-barreled appearance that is differentially affected in dominant and recessive myotonia. *J Gen Physiol* **113**, 457–468.

Schaller S, Henriksen K, Sveigaard C et al. (2004). The chloride channel inhibitor NS3736 [corrected] prevents bone resorption in ovariectomized rats without changing bone formation. *J Bone Miner Res* **19**, 1144–1153.

Scheel O, Zdebik AA, Lourdel S *et al.* (2005). Voltage-dependent electrogenic chloride/proton exchange by endosomal CLC proteins. *Nature* **436**, 424–427.

Schlingmann KP, Konrad M, Jeck N *et al.* (2004). Salt wasting and deafness resulting from mutations in two chloride channels. *N Engl J Med* **350**, 1314–1319.

Scholl U, Hebeisen S, Janssen AG *et al.* (2006). Barttin modulates trafficking and function of ClC-K channels. *PNAS* **103**, 11411–11416.

Scott JW, Hawley SA, Green KA *et al.* (2004). CBS domains form energy-sensing modules whose binding of adenosine ligands is disrupted by disease mutations. *J Clin Invest* **113**, 274–284.

Sik A, Smith RL and Freund TF (2000). Distribution of chloride channel-2-immunoreactive neuronal and astrocytic processes in the hippocampus. *Neuroscience* **101**, 51–65.

Simon DB, Bindra RS, Mansfield TA *et al.* (1997). Mutations in the chloride channel gene, *CLCNKB*, cause Bartter's syndrome type III. *Nat Genet* **17**, 171–178.

Staley K, Smith R, Schaack J *et al.* (1996). Alteration of GABA$_A$ receptor function following gene transfer of the CLC-2 chloride channel. *Neuron* **17**, 543–551.

Steinmeyer K, Klocke R, Ortland C *et al.* (1991a). Inactivation of muscle chloride channel by transposon insertion in myotonic mice. *Nature* **354**, 304–308.

Steinmeyer K, Ortland C and Jentsch TJ (1991b). Primary structure and functional expression of a developmentally regulated skeletal muscle chloride channel. *Nature* **354**, 301–304.

Steinmeyer K, Schwappach B, Bens M *et al.* (1995). Cloning and functional expression of rat CLC-5, a chloride channel related to kidney disease. *J Biol Chem* **270**, 31172–31177.

Steward MC, Ishiguro H and Case RM (2005). Mechanisms of bicarbonate secretion in the pancreatic duct. *Annu Rev Physiol* **67**, 377–409.

Stobrawa SM, Breiderhoff T, Takamori S *et al.* (2001). Disruption of ClC-3, a chloride channel expressed on synaptic vesicles, leads to a loss of the hippocampus. *Neuron* **29**, 185–196.

Thiemann A, Gründer S, Pusch M *et al.* (1992). A chloride channel widely expressed in epithelial and non-epithelial cells. *Nature* **356**, 57–60.

Traverso S, Elia L and Pusch M (2003). Gating competence of constitutively open CLC-0 mutants revealed by the interaction with a small organic Inhibitor. *J Gen Physiol* **122**, 295–306.

Traverso S, Zifarelli G, Aiello R *et al.* (2006). Proton sensing of CLC-0 mutant E166D. *J Gen Physiol* **127**, 51–66.

Tseng P-Y, Bennetts B and Chen T-Y (2007). Cytoplasmic ATP inhibition of CLC-1 is enhanced by low pH. *J Gen Physiol* **130**, 217–221.

Uchida S and Sasaki S (2005). Function of chloride channels in the kidney. *Ann Rev Physiol* **67**, 759–778.

Uchida S, Sasaki S, Furukawa T *et al.* (1993). Molecular cloning of a chloride channel that is regulated by dehydration and expressed predominantly in kidney medulla. *J Biol Chem* **268**, 3821–3824.

Uchida S, Sasaki S, Nitta K *et al.* (1995). Localization and functional characterization of rat kidney-specific chloride channel, ClC-K1. *J Clin Invest* **95**, 104–113.

van Slegtenhorst MA, Bassi MT, Borsani G *et al.* (1994). A gene from the Xp22.3 region shares homology with voltage-gated chloride channels. *Hum Mol Genet* **3**, 547–552.

Vandewalle A, Cluzeaud F, Bens M *et al.* (1997). Localization and induction by dehydration of ClC-K chloride channels in the rat kidney. *Am J Physiol* **272**, F678–688.

Waldegger S and Jentsch TJ (2000). Functional and structural analysis of ClC-K chloride channels involved in renal disease. *J Biol Chem* **275**, 24527–24533.

Waldegger S, Jeck N, Barth P *et al.* (2002). Barttin increases surface expression and changes current properties of ClC-K channels. *Pflügers Arch* **444**, 411–418.

Wang SS, Devuyst O, Courtoy PJ *et al.* (2000). Mice lacking renal chloride channel, CLC-5, are a model for Dent's disease, a nephrolithiasis disorder associated with defective receptor-mediated endocytosis. *Hum Mol Genet* **9**, 2937–2945.

Weinreich F and Jentsch TJ (2001). Pores formed by single subunits in mixed dimers of different CLC chloride channels. *J Biol Chem* **276**, 2347–2353.

Wrong OM, Norden AG and Feest TG (1994). Dent's disease; a familial proximal renal tubular syndrome with low-molecular-weight proteinuria, hypercalciuria, nephrocalcinosis, metabolic bone disease, progressive renal failure and a marked male predominance. *QJM* **87**, 473–493.

Zdebik AA, Cuffe JE, Bertog M *et al.* (2004). Additional disruption of the ClC-2 Cl– channel does not exacerbate the cystic fibrosis phenotype of cystic fibrosis transmembrane conductance regulator mouse models. *J Biol Chem* **279**, 22276–22283.

Zdebik AA, Zifarelli G, Bergsdorf EY *et al.* (2008). Determinants of anion-proton coupling in mammalian endosomal CLC proteins. *J Biol Chem* **283**, 4219–4227.

Zifarelli G and Pusch M (2007). CLC chloride channels and transporters: a biophysical and physiological perspective. *Rev Physiol Biochem Pharmacol* **158**, 23–76.

Zifarelli G and Pusch M (2008). The muscle chloride channel ClC-1 is not directly regulated by intracellular ATP. *J Gen Physiol* **131**, 109–116.

2.5

Hyperpolarization-activated cation channels

Mira Kuisle and Anita Lüthi

2.5.1 Introduction

The hyperpolarization-activated cation-non-selective (HCN) currents were identified in the early 1980s as the pacemaker currents driving the autonomous rhythmic discharges of the sinoatrial node (SAN) cells in the heart. These currents have been variously designated as I_h or I_q in the brain ('h' and 'q' stand for 'hyperpolarization' and 'queer', respectively), or I_f for 'funny' current in the heart. This nomenclature reflects the observations that these currents behave oppositely to most other ion currents, i.e. they gate upon membrane hyperpolarization, not depolarization (Figure 2.5.1). Historically, the pacemaking properties have dominated the physiological profile of I_{HCN} for almost 20 years. Recently, the views on the role of I_{HCN} have broadened significantly, largely due to the cloning of the genes for HCN channels. Thus, the function of these channels now covers aspects of synaptic function, dendritic integration, plasticity, learning, and pathological neuronal and cardiac states (see Table 2.5.1). This chapter reviews these clear and straight modern developments of a historically funny and queer current (which we refer to soberly as I_{HCN}).

Strong evidence for multiple involvements of HCN channels in neuronal functions comes from several sources. First, HCN channel proteins show organized expression patterns throughout the brain, including particularly areas involved in cognitive functions, such as learning and memory. Second, they are localized not only in somatic compartments, but also in dendrites, presynaptic zones and axonal elements, thus being ideally placed to be instrumental for synaptic integration and transmission. Third, HCN channels belong to a small subgroup of ion channels whose expression is regulated in an activity-dependent manner. As a consequence, changes in HCN subunit expression accompany plasticity-promoting stimuli. Moreover, aberrant electrical activity, such as that found during epilepsy or cardiopathy, may persistently alter their expression levels, implicating these ion channels in the pathology of excitable systems.

2.5.2 Molecular characterization

Gene products and primary structure of HCN proteins

In the late 1990s the cloning of HCN subunits provided a major breakthrough in the identification of the molecular basis of hyperpolarization-activated cation channels (Santoro et al., 1997; Gauss et al., 1998; Ludwig et al., 1998; Santoro et al., 1998). To date, a family of up to four genes is known for mammals, including mouse (Ludwig et al., 1998; Santoro et al., 1998), rat (Monteggia et al., 2000), rabbit (Ishii et al., 1999; Shi et al., 1999), and human (Ludwig et al., 1999; Seifert et al., 1999; Vaccari et al., 1999) (for review, see Kaupp and Seifert, 2001). The adapted nomenclature for the mammalian gene family is HCN1–HCN4, standing for hyperpolarization-activated cation-nonselective channels. For a detailed list of old and standard nomenclature see Table 1 in Kaupp and Seifert (2001). The chromosomal location of the HCN channel genes has been determined for human HCN2: 19p13.3 (Ludwig et al., 1999; Vaccari et al., 1999) and human HCN4: 15q24–q25 (Seifert et al., 1999). More recently, the location of HCN1 (5p12) and HCN3 (1q21.2) was reported (Jackson et al., 2007).

HCN channel genes were also identified in urochordates, arthropods, and lower vertebrates. Two genes were reported for sea urchin sperm flagella (SpHCN1–2) (Gauss et al., 1998; Galindo et al., 2005), and one each for *Drosophila melanogaster* (DmHCN) (Marx et al., 1999), the silk moth *Heliothis virescens* (HvHCN) (Krieger et al., 1999), the rainbow trout *Oncorhychus mykiss* (Cho et al., 2003), the honeybee *Apis mellifera* (AMIH) (Gisselmann et al., 2004) and the lobster *Panulirus argus* (PAIH) (Gisselmann et al., 2005). A phylogenetic analysis of HCN channel genes via comparative genomics indicates that the genes arose out of a common ancestor, present in urochordates, that underwent three duplications prior to the divergence of mammals (Jackson et al., 2007).

The nucleotide sequence encodes for transcripts of ~780–1200 amino acids, with a predicted primary structure typical for voltage-gated K^+ channels, including six transmembrane-spanning domains S1–S6 that are highly conserved (80–90%) within the gene family, and a positively charged S4 domain. Notably, the pore region between S5 and S6 contains the selectivity filter motif GYG that is a hallmark of voltage-gated K^+-selective channels. However, adjacent amino acids differ for the HCN channel genes, perhaps contributing to the lack of strong K^+ selectivity in I_{HCN}. In addition, all HCN genes show a highly conserved C-terminal cyclic nucleotide-binding domain (CNBD) that is 120 amino acids in length and linked to S6 by an 80 amino acid C-linker region. The CNBD is homologous to those found in cyclic nucleotide-dependent kinases, in the bacterial cAMP binding protein catabolite gene

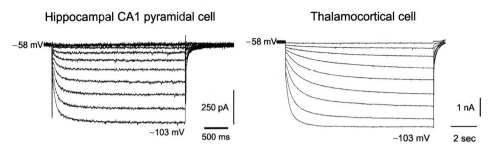

Fig. 2.5.1 Hyperpolarization-activated currents in neurons. Traces were obtained by stepping the membrane voltage to increasingly negative potentials in 5 mV steps (the most negative value reached, –103 mV, is indicated next to the current traces). Note the different timescales over which current activation proceeds in hippocampal pyramidal cells (left) and in thalamocortical cells (right). In hippocampal pyramidal cells, expression of HCN1 and HCN2 predominates, whereas HCN2 and HCN4 are the major channel isoforms in thalamocortical cells.

activator protein, and in cyclic nucleotide-gated ion channels of olfactory sensory and photoreceptor cells. The CNBD shows uniquely positioned amino acids that help explain its selectivity for cAMP over cGMP (Kaupp and Seifert, 2001; Craven and Zagotta, 2006; Flynn *et al.*, 2007). The *HCN* genes are poorly conserved and variable in length within the sequences in the distal N- and C-termini, with the HCN4 protein showing an N-terminus and C-terminus that is longer by ~80–120 and ~300–400 amino acids, respectively, compared to HCN1–HCN3.

Heterologous expression of HCN channels

All four mammalian genes, as well as the invertebrate genes, give rise to hyperpolarization-activated cation currents when expressed heterologously in *Xenopus* oocytes or in mammalian cell lines (Gauss *et al.*, 1998; Ludwig *et al.*, 1998; Santoro *et al.*, 1998; Ludwig *et al.*, 1999; Seifert *et al.*, 1999; Stieber *et al.*, 2005). These currents show characteristic similarities to native I_{HCN} (Figure 2.5.1), in particular with respect to

(a) their activation by membrane hyperpolarization below ~ –60 mV

(b) their complex activation kinetics

(c) their permeability to both Na^+ and K^+ ions

(d) their regulation by cyclic nucleotides (with the exception of HCN3), with a preference for cAMP over cGMP

(e) their lack of inactivation (with the exception of spHCN) and

(f) their blockade by millimolar concentrations of extracellular Cs^+, but not by traditional K^+ current blockers (tetraethylammonium, 4-aminopyridine).

The current generated by HCN1 is the poorly cAMP-sensitive, rapidly activating current isoform, whereas HCN2 and HCN4 give rise to highly cAMP-sensitive currents that activate more slowly. In recordings from intact cells, HCN1-mediated currents showed a half-activation voltage ($V_{1/2}$) around –70 mV, whereas HCN2- and HCN4-mediated currents gated half-maximally around –80 to –90 mV. In cell-free recordings, these values shifted by 40–60 mV in the hyperpolarizing direction for both HCN1 and HCN2 channels (Chen *et al.*, 2001c). A hyperpolarizing shift in $V_{1/2}$ with patch excision was long known for excised patches of cardiac cells (DiFrancesco *et al.*, 1986; DiFrancesco and Mangoni, 1994), but some important determining factors of this run-down were identified just recently (see section on novel modulators in Section 2.5.6.). Activation occurs after an initial lag of tens of ms, and proceeds with a complex time course that is typically best described by a bi-exponential function and accelerates with increased membrane hyperpolarization. At 35°C, around the half-maximal voltage, time constants of activation for HCN1 amount to ~10 ms and ~100 ms, whereas activation of HCN2 and HCN4 is best described by time constants in the order of hundreds of milliseconds and seconds, respectively. In the case of HCN2 and HCN4, cAMP shifts the activation range by ~15–25 mV towards more positive potentials (Ludwig *et al.*, 1999; Chen *et al.*, 2001c, Stieber *et al.*, 2005), whereas this shift amounts to 2–6 mV in the case of HCN1 (Santoro *et al.*, 1998; Chen *et al.*, 2001c; Stieber *et al.*, 2005). The half-maximal concentration for cAMP is < 1 µM (60–800 nM) (Gauss *et al.*, 1998; Ludwig *et al.*, 1998; Chen *et al.*, 2001c; Zagotta *et al.*, 2003), whereas that for cGMP is ~6–8 µM (Ludwig *et al.*, 1998; Zagotta *et al.*, 2003). The HCN3-mediated current is weakly or not modulated by cAMP, yet shows voltage dependence similar to that of HCN2-mediated currents (Mistrík *et al.*, 2005; Stieber *et al.*, 2005).

Formation of heteromers between HCN channel subunits

Evidence for the hetero-oligomerization of HCN subunits was first provided by identifying interactions between the N-termini of HCN1 and HCN2 (Proenza *et al.*, 2002b) and by showing that co-expressed dominant-negative pore mutants of HCN1 inhibited HCN2-mediated currents (Xue *et al.*, 2002). With the exception of HCN2 and HCN3, all dual combinations of channel subunits express and co-localize at the plasma membrane in heteromeric complexes (Much *et al.*, 2003). The most direct evidence for heteromerization is found in (a) the distinct properties of single-channel events arising from co-expressed HCN2 and HCN4 (Michels *et al.*, 2005) and in (b) the demonstration of bioluminescence resonance energy transfer (BRET), requiring apposition of interacting proteins within 10 nm, between co-expressed HCN2 and HCN4 subunits (Whitaker *et al.*, 2007). Co-expression of two of each HCN1, HCN2 and HCN4 produce currents that incorporate properties of both isoforms in ways that do not correspond to those expected from the linear interpolation of homomers (Chen *et al.*, 2001c; Ulens and Tytgat, 2001; Altomare *et al.*, 2003; Michels *et al.*, 2005). The voltage dependence of HCN1–HCN2 heteromeric currents is close to that of HCN2-mediated currents, but currents show a decreased cAMP sensitivity at submaximal cAMP concentrations. This property could have implications for disease, since HCN1 protein is sensitively regulated by abnormal neuronal activity and tends to form heteromers (see Epilepsies section in Section 2.5.7). Co-expression

Table 2.5.1 Overview of the ClC family

Channel subunit	Biophysical characteristics (heterologously expressed channels)	Cellular and sub-cellular localization in brain and heart	Pharmacology	Physiology	Relevance to disease states
HCN1 Gene: *HCN1* chr. location: -human: 5p12 [1]	$V_{0.5\,act}$ ~−70 mV No inactivation Max. cAMP induced shift ~2−6 mV [2−4] Single−channel conductance in Xenopus oocytes and HEK293 cells 1.5−2.5 pS [5−7] Single−channel conductance in CHO cells: 13 pS [8] τ_{act} = tens to hundreds of ms at room temperature (2, 3) Strong voltage hysteresis [9]	Brain: Strong expression in cortex, hippocampus, superior colliculus, cerebellum and brainstem [2, 10−13], olfactory bulb, spinal cord and photoreceptor cells [14−16] Heart: Minor expression in sinus node (<20%), atrial ventricular node and ventricles [17] Subcellular: Expressed in a steep somatodendritic gradient in apical dendrites of cortical neurons [10] and hippocampal CA1 pyramidal neurons [18] Expressed in presynaptic terminals [10, 19].	Sensitive to general HCN blockers, most importantly - extracellular Cs^+ ions in the millimolar range [20, 21] - biadycardiac agents, such as ZD7288 (10−100 μM) [22−26] - Anaesthetic agents (halothane, propofol) [27−31] Some subtype-specific effects reported for halothane [29], propofol [30, 31]	Important in Purkinje cell excitability [32] Important role in the resting membrane potential stabilization [32−34] Important role in the temporal integration of excitatory synaptic input in apical dendrites of hippocampal pyramidal and cortical neurons [33, 34] Limits synaptic plasticity at perforant path afferents into hippocampal CA1 area [35] Acts as a shunting conductance in dendrites [36, 37] Implicated in presynaptic control of neurotransmitter release, together with HCN2 [38] May act as a receptor for sour taste, together with HCN4 [39] Implicated in cerebellar motor learning [32], in spatial learning [35], and in prefrontal cortex working, memory [40]	Upregulated in thalamocortical neurons of animal models or generalized epilepsy [41−43] Downregulated in cortical layer V neurons of an animal model of generalized epilepsy [42, 44] Decreased expression in an animal model of febrile seizures [45, 46] Decreased expression in an animal model of temporal lobe epilepsy [47] Increased expression in chronic epilepsy [48] Increased expression in peripheral nerve injury [49]
HCN2 Gene: *HCN2* chr. location: - human: 19p13. 3 [50] [51]	$V_{0.5\,act}$ ~−80 to −90 mV No inactivation Max. cAMP induced shift ~15−25 mV [3, 4, 50] Half−maximal cAMP conc. <1 μM (60−800 nM) [3, 21, 52, 53] Single−channel conductance: 1.5−2.5 pS [5−7] Single−channel conductance in CHO cells: 3.5 pS [8] τ_{act} hundreds of ms to seconds [3, 50] Generates instantaneous currents [54]	Brain: more uniform throughout the brain, with strong expression in olfactory bulb, thalamus and brainstem [11−13]. Heart: Minor expression in sinus node (<20%), atrial ventricular node and ventricles [17] Subcellular: co-localizes with HCN1 in apical dendrites [13]	Sensitive to general HCN blockers. as HCN1. Some subtype-specific effects reported for halothane [29], propofol [30, 31]	Important role in resting membrane potential and excitability of thalamic neurons [55, 56] Implicated in presynaptic control of neurotransmitter release, together with HCN1 [38] Obligatory role for pacemaker current and sag potential expression in thalamic neurons [55] Plays a role in cardiac pacemaking, in particular its regularity, but is not in involved in regulation of heart rate through adrenergic and muscarinic receptors [55]	Implicated in the regulation of spike-wave discharges [55] Increased expression in an animal model of febrile seizures [45, 46] Decreased expression in an animal model of temporal lobe epilepsy [47] Abnormal expression in heart failure [57, 58] Increased expression in peripheral nerve injury [49]

Continued

Table 2.5.1 (*continued*) Overview of the ClC family

Channel subunit	Biophysical characteristics (heterologously expressed channels)	Cellular and sub-cellular localization in brain and heart	Pharmacology	Physiology	Relevance to disease states
HCN3 Gene: *HCN3* chr. location: -human: 1q21.1 [1]	$V_{0.5\,act}$ ~−80mV No inactivation Insensitive to cAMP or weak hyperpolarizing cAMP effect [4, 59] τ_{act} hundreds of ms to seconds [4, 59]	Brain: generally weakest expression, found in olfactory bulb, hypothalamuus and substania nigra [13]	Sensitive to general HCN blockers extracellular Cs^+ and ZD7288. No subtype-specific effects reported	Function remains unclear.	Function remains unclear.
HCN4 Gene: *HCN4* chr. location: -human: 15q24–q25 [60]	$V_{0.5\,act}$ ~−80 to −100 mV No inactivation Max. cAMP induced shift ~15–2.5 mV [3, 4, 50, 60] Half-maximal cAMP conc. ~1 μM (60–800 nM) [3, 21, 52, 53] Single-channel conductance in CHO cells: 17 pS [8] τ_{act} hundreds of ms to tens of seconds [60]	Brain: Prominent in thalamus, olfactory bulb and hypothalamus [13, 60] Heart: 80% of total HCN mRNA message in sinoatrial node [50, 61, 62], also found in atrial ventricular node and ventricles [17]	Sensitive to general HCN blockers, as HCN1. No subtype-specific effects reported.	Obligatory role for cardiac pacemaker currents and its regulation by cAMP at embryonic stages [63] May act as a receptor for sour taste, together with HCN1 [39]	Mutated in idiopathic forms of sinus node disease [64–66] Syndrome of mental retardaion and of nocturnal frontal lobe epilepsy linked to the genetic locus [67, 68] Abnormal expression at diverse stages of heart failure [57, 58, 69]. Considered as electrical marker for cardiac remodelling during cardiac overload

Table x – Properties of HCN channels

1. Jackson, H.A., C.R. Marshall, and E.A. Accili, 2007. 2. Santoro, B., *et al.*, 1998. 3. Chen, S., J. Wang and S.A. Siegelbaum, 2001. 4. Stieber, J., et al., 2005. 5. Johnson, J.P., Jr. and W.N. Zagotta, 2005. 6. Lyashchenko, A.K., *et al.*, 2007. 7. Dekker, J.P. and G. Yellen, 2006. 8. Michels, G., *et al.*, 2005. 9. Männikkö, R., *et al.*, 2005. 10. Santoro, B., *et al.*, 1997. 11. Moosmang, S., et al., 1999. 12. Monteggia, L.M., *et al.* 2000. 13. Notomi, T. and R. Shigemoto, 2004. 14. Moosmang, S., *et al.*, 2001. 15. Demontis, G.C., *et al.*, 2002. 16. Müller, F., *et al.*, 2003. 17. Marionneau, C., *et al.*, 2005. 18. Lörincz, A., *et al.*, 2002. 19. Luján, R., *et al.*, 2005. 20. DiFrancesco, D., 1982. 21. Gauss, R., R. Seifert, and U.B. Kaupp, 1998. 22. Harris, N.C. and A. Constanti, 1995. 23. Gasparini, S. and D. DiFrancesco, 1997. 24. Williams, S.R., *et al.*, 1997. 25. Lüthi, A., T. Bal, and D.A. McCormick, 1998. 26. Satoh, T.O. and M. Yamada, 2000. 27. Sirois, J.E., *et al.*, 1998. 28. Sirois, J.E., C. Lynch, 3rd, and D.A. Bayliss, 2002. 29. Chen, X., *et al.*, 2005. 30. Cacheaux, L.P., *et al.*, 2005. 31. Chen, X., S. Shu, and D.A. Bayliss, 2005. 32. Nolan, M.F., *et al.*, 2003. 33. Magee, J.C., 1999. 34. Williams, S.R. and G.J. Stuart, 2000. 35. Nolan, M.F., *et al.*, 2004. 36. Berger, T., M.E. Larkum, and H.R. Lüscher, 2001. 37. Fernandez, N., M. Andreasen, and S. Nedergaard, 2002. 38. Aponte, Y., *et al.*, 2006. 39. Stevens, D.R., *et al.*, 2001. 40. Wang, M., *et al.*, 2007. 41. Budde, T., *et al.*, 2005. 42. Strauss, U., *et al.*, 2004. 43. Kuisle, M., *et al.*, 2006. 44. Kole, M.H., A.U. Brauer, and G.J. Stuart, 2007. 45. Brewster, A., *et al.*, 2002. 46. Brewster, A., *et al.*, 2005. 47. Shah, M.M., *et al.*, 2004. 48. Bender, R.A., *et al.*, 2003. 49. Chaplan, S.R., *et al.*, 2003. 50. Ludwig, A., *et al.*, 1999. 51. Vaccari, T., *et al.*, 1999. 52. Ludwig, A., *et al.*, 1998. 53. Zagotta, W.N., et al., 2003. 54. Proenza, C., *et al.*, 2002. 55. Ludwig, A., *et al.*, 2003. 56. Meuth, S.G., *et al.*, 2006. 57. Fernandez-Velasco, M., *et al.*, 2003. 58. Hiramatsu, M., *et al.*, 2002. 59. Mistrik, P., *et al.*, 2005. 60. Seifert, R., *et al.*, 1999. 61. Ishii, T.M., *et al.*, 1999. 62. Shi, W., *et al.*, 1999. 63. Stieber, J., *et al.*, 2003. 64. Milanesi, R., *et al.*, 2006. 65. Ueda, K., *et al.*, 2004. 66. Schulze-Bahr, C., *et al.*, 2003. 67. Mitchell, S.J., *et al.*, 1998. 68. Phillips, H.A., *et al.*, 1998. 69. Borlak, J., anc T. Thum, 2003.

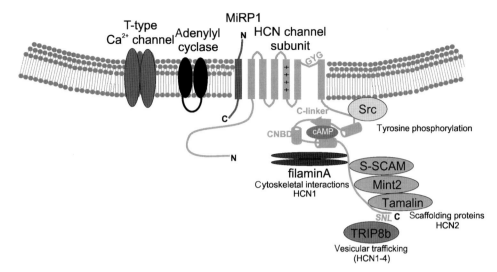

Fig. 2.5.2 Schematic view of accessory and regulatory proteins. The cartoon comprises those proteins for which a close co-localization with HCN channels has been demonstrated functionally (T-type Ca²⁺ channels, adenylyl cyclase, MiRP1) or for which binding to HCN channels has been demonstrated by co-immunoprecipitation from brain tissue (Src, filamin, synaptic scaffolding molecules S-SCAM, Mint2, tamalin) or both co-immunoprecipitation and co-localization in neuronal cells (Trip8b). See also colour plate section.

of HCN1 or HCN2 with HCN4 produces channels with properties approximating to those of HCN4 homomers (Altomare *et al.*, 2003; Michels *et al.*, 2005). Immunoprecipitation and immunohistochemistry studies also suggest the existence of heteromers in embryonic heart (Whitaker *et al.*, 2007), in thalamus (Whitaker *et al.*, 2007) and in the hippocampus (Brewster *et al.*, 2005).

Control of HCN channel expression

Biochemical and functional studies indicate a role for both N-terminal and C-terminal sequences in the assembly and trafficking of functional channels to the cell surface (Proenza *et al.*, 2002b; Tran *et al.*, 2002). A stretch of 52 amino acids (positions 131–182 in HCN2) located N-terminally of the S1 segment is > 90% conserved amongst the HCN isoforms and deletion of this region in HCN2 results in intracellular retention of the protein. In addition, an intact CNBD is required for cell surface expression of HCN2 in CHO cells (Proenza *et al.*, 2002b), although the C-linker is sufficient in *Xenopus* oocytes (Wainger *et al.*, 2001). In contrast to the critical N-terminal sequences, however, not every subunit of a channel needs to contain a CNBD for correct trafficking. Finally, *N*-glycosylation at asparagine 380 of HCN2, lying between S5 and the pore region, is important for subunit expression, although again not required for every channel subunit (Much *et al.*, 2003). The physiological relevance of *N*-glycosylation is underscored by the immunoblot analysis of native subunits, which are about 20 kDa larger than predicted by the primary sequence. Pharmacological removal of *N*-glycosylation restored the predicted molecular weight (Santoro *et al.*, 1997; Much *et al.*, 2003).

Accessory and regulatory proteins of HCN channels

The list of proteins that physically interact with HCN channels and regulate their properties and expression is currently growing (see Figure 2.5.2 for an overview). A widely expressed β subunit for Kv channels, the single transmembrane MinK related protein (MiRP1 or KCNE2) (McCrossan and Abbott, 2004) enhances the amplitude of I_{HCN} and affects activation kinetics, although effects vary depending on the expression systems and HCN isoforms (Yu *et al.*, 2001; Proenza *et al.*, 2002a; Altomare *et al.*, 2003; Decher *et al.*, 2003;

Qu *et al.*, 2004). MiRP1 and HCN2 are highly expressed in SAN, and MiRP1 co-immunoprecipitated with HCN2 in cultured neonatal cardiac myocytes overexpressing both proteins (Qu *et al.*, 2004).

The C-terminal sequence, including the CNBD and the less conserved regions at the distal C-terminal end, mediate important protein–protein interactions. The extreme C-terminal tripeptide (SNL in HCN1, HCN2 and HCN4, ANM in HCN3) tightly binds to the protein TRIP8b (for tetratricopeptide repeat-containing Rab8b interacting protein), which co-localizes with endosomal markers and could be involved in the endocytosis of HCN channels (Santoro *et al.*, 2004). This protein closely matches HCN1 expression in the dendrites of CA1 hippocampal pyramidal and cortical layer V cells and its overexpression reduces the density of native channels. Alternatively spliced isoforms of Trip8b exist, suggesting that the significance of these proteins in controlling HCN channel function may further diversify in the near future. In particular, it will be important to see how the two apparently incongruent observations, co-localization of TRIP8b and HCN1 in a parallel gradient within the apical dendrites on the one hand, but negative regulation of channel expression by TRIP8b on the other hand, can be reconciled. C-terminal sequences also interact with scaffolding proteins that may mediate binding to the cytoskeleton and to postsynaptic elements (Kimura *et al.*, 2004). In the HCN1 protein, a 22-amino acid sequence downstream of the CNBD binds to filamin A, a protein involved in the organization of functional complexes involving receptors, ion channels and cytoskeletal elements (Gravante *et al.*, 2004). In SAN cells, HCN4 protein and components of the β-adrenergic receptor signalling pathway are found in membrane fractions containing structurally specialized portions of the lipid membrane, the caveolae (Barbuti *et al.*, 2004).

Protein kinases, predominantly protein kinase A (PKA) and tyrosine kinases, regulate I_{HCN} properties in a number of cell types (for review see Frère *et al.*, 2004). Additional studies now indicate that kinases may be physically associated with the channels. The C-linker domain, which mediates intersubunit interactions required for channel gating (see Molecular elements controlling cyclic nucleotide regulation section), contains a tyrosine residue (Y476 in

HCN2, Y554 in HCN4) that is phosphorylated by the non-receptor protein tyrosine-kinase Src. In fact, this kinase was used as a bait in the first identification of the *HCN* genes (Santoro *et al.*, 1997) and is now known to bind to the C-terminus (Zong *et al.*, 2005; Arinsburg *et al.*, 2006). Inhibition of Src phosphorylation in HCN2/HCN4-expressing human embryonic kidney (HEK) 293 cells, in SAN cells and in dorsal root ganglia decelerates current activation, suggesting a constitutive regulation of currents by this kinase.

2.5.3 Biophysical properties

The unique biophysical characteristics of I_{HCN} are presented in excellent reviews (Pape, 1996; Kaupp and Seifert, 2001; Robinson and Siegelbaum, 2003; Baruscotti and DiFrancesco, 2004; Cohen and Robinson, 2006; Craven and Zagotta, 2006). Recent research has focused, on the one hand, on identifying additional current properties, such as the single-channel characteristics and instantaneous forms of current activation. On the other hand, the molecular understanding of the events underlying channel gating and its regulation by cyclic nucleotides has advanced. These studies have also demonstrated that HCN channels are composed of a tetrameric arrangement of HCN subunits (Xue *et al.*, 2002; Ulens and Siegelbaum, 2003; Zagotta *et al.*, 2003), in agreement with their evolutionary relationship to the tetrameric voltage-gated K^+ channels. This section briefly describes the main molecular underpinnings of HCN channel properties, and refers to the most relevant reviews for further details and for comparison to additional members of the voltage-gated ion channel family.

Prototypical biophysical properties

The three prototypical biophysical properties of native I_{HCN} that make them unique amongst the family of voltage-gated ionic currents are highlighted here.

First, I_{HCN} typically activates upon membrane hyperpolarization (below ~−60 mV) rather than depolarization, a direction opposite to most voltage-gated ionic currents that are involved in shaping neuronal excitability (Figure 2.5.1). Upon hyperpolarization, the conductance activated is fairly selective for both Na^+ and K^+ ions (permeability ratio Na^+: K^+ = 0.2–0.3) (DiFrancesco, 1981; Wollmuth and Hille, 1992; Pape, 1996; Gauss *et al.*, 1998; Stieber *et al.*, 2005). The current is carried mainly by Na^+ ions at the membrane voltages within its activation range and produces an elevation in the intracellular Na^+ concentration (Knöpfel *et al.*, 1998). In addition to contributing to ion flux, extracellular K^+ ions determine the channel conductance for Na^+ (for a detailed review, see Pape 1996). More recently, a small permeability to Ca^{2+} ions has also been identified for heterologously expressed HCN4 channels (0.6% of the inward current evoked at −120 mV) (Yu *et al.*, 2004) and presynaptically expressed I_{HCN} in crayfish (Zhong *et al.*, 2004). The reversal potential of I_{HCN} lies between −40 to −25 mV, leading to depolarizing currents around resting membrane potentials. The resulting depolarizing drive at subthreshold potentials is the basis for the multifunctionality of I_{HCN}, in particular its pacemaking properties and its control of excitability and synaptic integration.

Second, activation of the current is generally slow, with activation time constants ranging from hundreds of milliseconds to seconds, even at strongly hyperpolarized voltages around −100 mV (for an overview of activation kinetics in different cell types, see Table 2 in Santoro and Tibbs, 1999). The few exceptions include pyramidal neurons from hippocampus and cortex, as well as amygdala and cerebellar neurons, in which activation is complete within tens of milliseconds. Once activated, the current does not inactivate, such that a steadily activated ('standing') I_{HCN} contributes to the resting membrane potential in many neurons, often by opposing the action of tonic outward currents (see HCN currents and resting membrane potentials in section 2.5.6).

Third, I_{HCN} is, in most cases, sensitive to the presence of intracellular cyclic nucleotides. The cyclic nucleotides cAMP and cGMP bind directly to the channels (DiFrancesco and Tortora, 1991; Pedarzani and Storm, 1995), accelerate the kinetics of activation, and shift the voltage dependence of activation towards more depolarized values. In the presence of these ligands, the extent and duration of current activation within the intermediate voltage range is increased (Robinson and Siegelbaum, 2003). This facilitatory effect of cyclic nucleotides can give rise to persistently activated forms of current activation that outlast the presence of available free cAMP (Lüthi and McCormick, 1999; Wang *et al.*, 2002). HCN-currents are also regulated by extra- and intracellular changes in pH. Extracellular pH decreases augment I_{HCN}, whereas intracellular acidification decreases I_{HCN}. The extracellular pH sensitivity has been implicated in the transduction of sour (pH 3–5) stimuli in taste receptor cells (Stevens *et al.*, 2001), whereas the intracellular pH sensitivity may be involved in regulating the discharge of neuronal networks in response to moderate pH changes (<1 unit), such as those occurring during intense neuronal activity (Munsch and Pape, 1999a, b). In HCN2 channels, intracellular pH sensitivity is mediated by direct protonation of a histidine residue (H321) located at the boundary between S4 and the S4–S5 linker (Zong *et al.*, 2001).

Single-channel properties

The first resolution of putative single channel events underlying I_{HCN} was achieved in rabbit SAN (DiFrancesco, 1986) and revealed an unusually low single-channel conductance < 1 pS (DiFrancesco and Mangoni, 1994). A low single-channel conductance (~0.68 pS) was recently determined for I_{HCN} in symmetrical K^+ concentrations in the apical dendrites of cortical layer V pyramidal cells using fluctuation analysis of membrane current noise (Kole *et al.*, 2006). These dendrites express I_{HCN} at up to 550 channels/mm², which is a density about 1000-fold higher than the average channel density on a typical I_{HCN}-expressing cell. This high density renders the low-conductance HCN channels important determinants of membrane current noise in these cells. Single-channel conductances of heterologously expressed channels in symmetrical K^+ concentrations range from 1.5–2.5 pS for HCN1 and HCN2 as quantified in *Xenopus* oocytes using noise analysis (Johnson and Zagotta, 2005; Lyashchenko *et al.*, 2007), and in HEK293 cells using single channel recordings (Dekker and Yellen, 2006) and up to 25–35 pS for channels expressed in CHO cells (Michels *et al.*, 2005). Furthermore, somatic h-channels recorded from dissociated hippocampal neurons show a comparatively high conductance (~10 pS) (Simeone *et al.*, 2005). A high single-channel conductance of ~30 pS has also been reported for the HvHCN channels (Krieger *et al.*, 1999). Recently, cooperative interactions between HCN2 channels recorded in excised patches were observed, evident as bursts of channel openings that are not expected based on the latencies to single-channel openings (Dekker and Yellen, 2006).

Instantaneous HCN-currents

The slowly developing inward current is the predominant manifestation of I_{HCN} activation by voltage. However, a small, voltage-independent leakage current accompanies the gating of HCN2

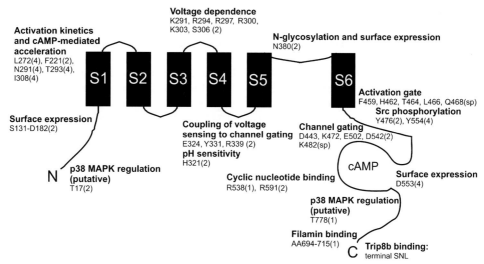

Fig. 2.5.3 Putative transmembrane topology of the HCN channel, and functional roles of the best characterized amino acids and amino acid sequences. The primary structure of the HCN channel, including its N- and C-terminus (N, C), and the six transmembrane-domain sequences S1–S6, are depicted schematically. Position and types of amino acids are indicated below a keyword description of their roles: see text for more details. Numbers and letters indicated in parentheses refer to the HCN channel subunit in which these amino acids were studied. In most cases, these amino acids are conserved in equivalent positions of other HCN isoforms. Amino acids included here are those for which mutations were shown to abrogate or alter their respective functions, or for sequences with which yeast two-hybrid screens were performed to study protein–protein interactions (filamin, Trip8b). The amino acid D553 was found in an inherited form of sinus node disease. The putative phosphorylation sites for p38 MAPK were determined in a screen for consensus sequences, but are not yet validated by mutational analysis. AA stands for amino acids in cases in which the full sequence is not written out.

channels expressed in CHO cells (Proenza *et al.*, 2002a) and of spHCN channels expressed in HEK293 cells (Proenza and Yellen, 2006). This current persisted when mutations were introduced into the channel that abrogated voltage-sensitive current components (Proenza *et al.*, 2002a; Macri *et al.*, 2002). The amplitude of this current correlates with that of the time-dependent current and shows a similar reversal potential (Proenza *et al.*, 2002a; Macri and Accili, 2004), indicating that it is dependent on the expression of HCN2 channels. However, both K^+ and Na^+ ions are independently moving charge carriers (Macri *et al.*, 2002), in contrast to the voltage-gated channel (see Prototypical biophysical properties section). Moreover, the instantaneous current component of HCN2-mediated currents is insensitive to extracellular Cs^+ ions and slightly reduced by intracellular cAMP (Proenza *et al.*, 2002a). Finally, the I_h blocker ZD7288 (see Bradycardiac agents section) more rapidly blocked instantaneous than time-dependent currents. This observation is consistent with a model in which the instantaneous current is generated by an independent channel subgroup that is not in rapid equilibrium with voltage-gating channels (Proenza and Yellen, 2006). To what extent native currents contain instantaneous components remains to be elucidated.

Additional biophysical properties

More recent studies have identified novel properties of expressed channels. Most interestingly, currents carried by spHCN or HCN1 channels show a hysteresis of voltage dependence, evident as a more positive voltage dependence of the current when measured starting from a hyperpolarizing potential (Männikkö *et al.*, 2005). Such hysteresis may contribute to the regularity of action potential discharge, since it helps prolonging the interspike interval, while facilitating repolarization. The hysteresis in voltage dependence was accompanied by a temporally correlated conformational change in the voltage-sensing domain S4 that occurred subsequent to its rapid outward transition during channel gating (Bruening-Wright

and Larsson, 2007). This suggests that slow movements of S4 and/or neighbouring domains, rather than rearrangements within the pore, are responsible for the shift in current voltage dependence. HCN2- and HCN4-mediated currents also show a dependence on extracellular and intracellular Cl^- (Wahl-Schott *et al.*, 2005; Mistrík *et al.*, 2006). Figure 2.5.3 illustrates the amino acids implicated in many of the biophysical characteristics of I_{HCN}.

Molecular elements controlling voltage dependence

The steps leading to channel gating show similarities to the activation of voltage-gated channels and of cyclic nucleotide-gated channels (Chen *et al.*, 2002; Robinson and Siegelbaum, 2003; Horn, 2004; Rosenbaum and Gordon, 2004; Craven and Zagotta, 2006). Thus, the positively charged amino acids in S4 are essential for voltage sensing (Chen *et al.*, 2000; Vaca *et al.*, 2000). Neutralization of each of the first four of these produces negative shifts in current voltage dependence of ~–20 mV/residue. Cysteine-substituted amino acids in S4 show a differential sensitivity to thiol reagents, depending on whether spHCN channels are opened or closed. From these experiments, it could be shown that the S4 domain moves towards the extracellular surface of the membrane upon depolarization, while it moves inward upon hyperpolarization (Männikkö *et al.*, 2002). This direction of S4 translocation is therefore that of the canonical model of charge movement in six-transmembrane-domain voltage-gated ion channels, although the detailed picture of conformational changes differs in HCN channels (Bell *et al.*, 2004; Horn, 2004; Vemana *et al.*, 2004) involving additional conformational changes subsequent to gating charge movement (Männikkö *et al.*, 2005; Elinder *et al.*, 2006; Bruening-Wright and Larsson, 2007) and control by S1 (Ishii *et al.*, 2007). Movement of the S4 domain is dependent on the S3–S4 linker (Tsang *et al.*, 2004) and couples to channel pore opening via amino acids within the S4–S5 linker domain (Chen *et al.*, 2001a; Macri and Accili, 2004). Within this, a requirement for aromatic residues was reported

(Y331 in HCN2, F359 in spHCN) (Chen *et al.*, 2001a; Prole and Yellen, 2006), suggesting that aromatic interactions to inner pore regions are important for transducing voltage changes into gating. One important candidate for such interactions is an activation gate, to date best characterized in spHCN, that localizes to the pore-forming regions of S6 (Figure 2.5.3) and is thought to sterically regulate the access of ions and drugs to the inner cavity of the pore (Liu *et al.*, 1997). Thus, the accessibility of this gate to ions and channel blockers is dependent on channel opening (Shin *et al.*, 2001; Rothberg *et al.*, 2002) and cross-linking of cysteine-substituted amino acids locked the channel in a permanently open or a preferentially closed state (Rothberg *et al.*, 2003). The activation gate also contributes to the peculiar inactivation properties of spHCN channels (Shin *et al.*, 2004).

Altogether, the protein domains essential for voltage sensing and channel gating are conserved between depolarization- and hyperpolarization-activated ion channels (see also Long *et al.* 2005). The similarities in the S4 movements and its coupling to an intracellular activation gate, however, leave open the puzzling question of how channel gating is coupled to hyperpolarizing voltage changes in HCN channels. Progress has been achieved recently by showing that amino acid interactions between the S4–S5 linker and the C-linker could be determinants of how voltage changes are translated oppositely to channel opening. One study identified a residue (R339) within the S4–S5 linker of HCN2 that was required for channel closure and that, via its positively charged side chain, interacted electrostatically with the negatively charged carboxyl group of D443 in the C-linker domain (Decher *et al.*, 2004). The closed state was destabilized if the charge of D443 was reversed, or could be even further stabilized (i.e. voltage-gating is slowed down) once additional carboxyl groups on neighbouring amino acids were introduced. One other study showed that spHCN1-currents became depolarization-activated once F359 in the S4–S5 linker and K482 in the C-linker were forced to interact constitutively via crosslinking (Prole and Yellen, 2006). As such, these studies demonstrate that specific movements of residues in the S4–S5 linker, relative to those of the C-linker, and the disruption or establishment of their interactions, are required for channel closure at positive voltages and hence for the hyperpolarized voltage-gating. Most likely, these interactions are dynamically controlled via the voltage-induced conformational arrangements within S4 on the one hand, and by the cAMP-bound CNBD on the other hand (Prole and Yellen, 2006), thereby helping to explain the unique gating properties of HCN channels.

Molecular elements controlling cyclic nucleotide regulation

The regulation by cyclic nucleotides is a most striking property of HCN channels and affects current properties in a dual manner. First, it shifts the voltage dependence of the current into a more depolarized voltage range. Second, it accelerates activation kinetics (Pape, 1996; Robinson and Siegelbaum, 2003). Numerous studies have reported cAMP-dependent actions on native currents, from which it became clear that cAMP-dependent effects are mediated by direct binding to the channel, rather than by PKA (DiFrancesco and Tortora, 1991; Pedarzani and Storm, 1995) (for review, see Frère *et al.*, 2004). The cAMP-effects are mediated by a cytosolic CNBD within the C-terminus (Barbuti *et al.*, 1999; Wainger *et al.*, 2001; Zagotta *et al.*, 2003). Removal of the CNBD abolishes cyclic nucleotide sensitivity (Wainger *et al.*, 2001), and mutation of a conserved arginine residue within the CNBD (R538 for HCN1, R591 for HCN2) largely abolishes cAMP modulation (Chen *et al.*, 2001c).

The molecular events leading from specific cyclic nucleotide binding to efficacy of channel gating have now been ascribed to a restricted number of amino acid residues (Flynn *et al.*, 2007; Zhou and Siegelbaum, 2007). Thus, the arginine residue identified previously belongs to a group of seven amino acid residues (E582, R591, T592, R632, R635, I636, K638) within the CNBD that mediate high affinity (R591), cAMP selectivity (R635, I636, K638) and gating efficacy (E582, R632) (Zhou and Siegelbaum, 2007). All CNBDs of each subunit must be bound to cAMP to achieve a maximal effect on the voltage dependence of the channel (Ulens and Siegelbaum, 2003).

How can a C-terminally located cytosolic portion of a channel steer voltage-gating and activation kinetics that are determined by the core transmembrane regions of the channel? Although the full picture of the interaction between the C-terminus and the core regions is not yet elaborated, important insights into the cAMP-induced structural arrangements within the CNBD and the C-terminus have been achieved. Cyclic allosteric gating models, containing unliganded and liganded open and closed channel states in a cyclical arrangement, were first used to explain the dual modulation of HCN channels by voltage and ligand (DiFrancesco, 1999; Wang *et al.*, 2001). These studies assumed that the channel subunits undergo a voltage-dependent transition between a closed and an open state, but that the affinity of cAMP for the open state is greater by a factor of ~40–80 compared to the closed state (for review, see Frère *et al.*, 2004). By analysing activation and deactivation kinetics of HCN2-mediated currents at extremely hyperpolarized and depolarized potentials, these allosteric gating models were expanded into eight-state reaction schemes, in which cAMP allosterically modulates the voltage-dependent steps prior to opening (Chen *et al.*, 2007). In addition, cAMP also facilitates a voltage-independent final opening step that is, however, ~10 times faster than the prior steps and only rate-limiting at very hyperpolarized potentials (Chen *et al.*, 2007). These allosteric models capture the essential consequences of some of the mechanistic steps underlying the cAMP-binding to HCN channels. Thus, these models imply that the ligand-free CNBD inhibits both voltage-dependent and -independent steps of channel gating, while cAMP-binding relieves this inhibition and stabilizes the channel in the open state. Indeed, HCN channels with a truncated CNBD activate about as rapidly as an intact channel exposed to a maximal cAMP concentration (Barbuti *et al.*, 1999; Wainger *et al.*, 2001). An essential role for this inhibitory action is played by the C-linker domain that connects the CNBD to S6. The C-linker of each subunit is required for normal cAMP modulation (Ulens and Siegelbaum, 2003) and the formation of salt bridges both between C-linkers and between C-linkers and CNBDs underlies the coupling of ligand binding to channel opening (Craven and Zagotta, 2004; Zhou *et al.*, 2004).

Crystallization and analytical ultracentrifugation of the C-terminal portion of HCN2, including the C-linker and the CNBD, has dramatically shaped the molecular understanding of HCN channel gating. Cyclic nucleotide binding greatly favours the formation of a tetramer of C-terminal fragments, while monomeric and dimeric configurations prevail in the absence of the cyclic nucleotide (Zagotta *et al.*, 2003). These tetramers form a gating ring with a fourfold rotational symmetry, with the major portion of intersubunit interactions mediated by the C-linkers. Elegant support for the tetramerization of CNBDs as a prerequisite for HCN channel gating comes from electrophysiological studies on tandem dimers or tetramers of HCN channels, in which mutated CNBDs were introduced in two subunits lying adjacent or diagonal to each other (Ulens and Siegelbaum, 2003). Cyclic AMP gated those channels more effectively in which functional CNBDs were arranged diagonally to each other. This dependence on the symmetry of CNBD arrangement is best explained in a model whereby a cAMP-induced dimerization of CNBDs is followed by the dimerization of these dimers into a tetramer. Thus, cAMP gating of HCN channels can be visualized as a series of structural rearrangements within the C-terminus, triggered by local arrangements within the C-helix of the CNBD, containing the amino acids interacting with cAMP (Zhou and Siegelbaum, 2007). These lead to the formation of cAMP-induced dimers of CNBDs that promote the interaction between C-linkers and the generation of a tetramer. The tetramer forms a bulky ring-like structure beneath the channel pore that extends by ~5 nm into the cytosol. By an as yet unknown mechanism, the conformational changes induced in the C-linkers are then transmitted to the transmembrane gating domains, including those involved in S4–S5 domains interacting with the C-linker (Prole and Yellen, 2006), thereby facilitating channel opening. This coupling appears to involve the S1–S2 linker and the S2 domain (Stieber et al., 2003b), and additional unknown channel portions (Stieber et al., 2005). The detailed review of Craven and Zagotta (2006) further describes and compares the currently proposed activation mechanisms of HCN channels by cAMP.

Molecular elements controlling isoform-specific properties

Some of the molecular elements underlying the striking differences in basal activation kinetics between HCN1- and HCN4-currents (Ishii et al., 2001) and HCN2- and HCN4-currents were determined (Stieber et al., 2003b). In both cases, exchanging S1 and the S1–S2 linker domain interconverted the basal activation properties of the currents (Ishii et al., 2001; Stieber et al., 2003b). In addition, in HCN1 channels, the C-terminal intracellular portion may contribute to the rapid activation kinetics (Ishii et al., 2001). A particularly important role for the difference between HCN2 and HCN4 is played by the leucine at position 272 within S1, because its mutation to phenylalanine, the respective residue in HCN2, transfers both the basal activation kinetics of HCN2, as well as cAMP-induced changes in current kinetics typical for HCN2. However, when leucine 272 is replaced with methionine, the equivalent amino acid found in HCN1, activation kinetics were not accelerated. This supports the notion that more complex structural differences exist between the rapidly activating HCN1- and the more slowly activating, and structurally tightly related, HCN2- and HCN4-currents.

The potency of modulation by cAMP is strikingly low for HCN1-mediated currents (Santoro et al., 1998), while it is high for HCN2- and HCN4-mediated currents (Ludwig et al., 1999; Seifert et al., 1999). Chimeric channels, composed of domains derived from HCN1 and HCN2 protein, were used to examine the molecular basis for this difference (Wang et al., 2001). The difference in cAMP modulation can be largely attributed to differences in the C-linker domains between the two channel isoforms, whereas differences in basal voltage dependence are determined by the C-terminal portions and by core transmembrane regions.

2.5.4 Cellular and subcellular localization

Molecular expression patterns of HCN channel subunits in brain and heart

All four HCN channel isoforms are expressed in the rodent brain. At the mRNA level, HCN channel subunits show distinct expression patterns that are largely consistent between studies on rat and mouse (Moosmang et al., 1999; Monteggia et al., 2000; Santoro et al., 2000). A detailed map of HCN protein localization at cellular and subcellular levels in the rat brain has been established (Notomi and Shigemoto, 2004). Immunohistochemical expression largely confirmed and further refined mRNA expression studies. Thus, HCN1 expression shows a predominant expression in cortical structures, with distinct expression in dendritic fields of the hippocampus, cerebral cortex and superior colliculus. In addition, Purkinje cells of the cerebellum and brainstem motor nuclei show high HCN1 mRNA levels, although immunoreactivity in Purkinje cells was comparatively modest. In contrast, HCN2 expression is found more uniformly throughout the brain, with strong expression in olfactory bulb, thalamus and brainstem regions. HCN2 expression often overlaps with other HCN channel subunits, most notably with the gradient of HCN1 expression in cortical and hippocampal apical dendrites, but accounts for most of the HCN expression in the nucleus reticularis of the thalamus and the subthalamic nucleus. Notably, a subpopulation of oligodendrocytes, that occupy the perineuronal space throughout the brain, display HCN2-like immunoreactivity. HCN3 generally shows the weakest expression, but is present in olfactory bulb, hypothalamus and the substantia nigra pars compacta. Finally, HCN4 is found in regions in which I_{HCN} functions as a pacemaker, such as the thalamus, but is also present in the olfactory bulb and hypothalamus.

Messenger RNA expressed in the heart is predominantly found for HCN1, HCN2 and HCN4, but isoform expression varies among species and cardiac tissue. The SAN exhibits the highest expression levels of HCN channels, with HCN4 accounting for ~80% of the total HCN message (Ishii et al., 1999; Ludwig et al., 1999; Shi et al., 1999) in all species investigated (rabbit, mouse, dog). Species-dependent differences for HCN2 and HCN1 were seen for the remaining 20% (Moosmang et al., 2001; Marionneau et al., 2005; Zicha et al., 2005). Expression of HCN channel subunits is also detected outside the sinus node region, albeit at lower levels. Both HCN1 and HCN4 message decrease in expression from atrial ventricular node to the ventricles (Marionneau et al., 2005).

Molecular expression patterns of HCN channel subunits in sensory systems and spinal cord

HCN channels are prominently expressed in diverse types of sensory neurons, including neurons of the dorsal root ganglia (Mayer and Westbrook, 1983; Moosmang et al., 2001; Chaplan et al., 2003; Tu et al., 2004) and nodose ganglia (Doan et al., 2004). High levels of HCN1 mRNA have been detected in mouse photoreceptor cells (Moosmang et al., 2001; Demontis et al., 2002; Müller et al., 2003). The cell-type specific distribution of HCN1 channels has been described for the main olfactory bulb (Holderith et al., 2003), and for spinal cord and medulla oblongata (Milligan et al., 2006).

Correlation of expression patterns with native current properties

The expression patterns of HCN subunits are remarkably predictive, at least in a semi-quantitative manner, for some of the basic properties of expressed currents, in particular their activation kinetics and their cAMP sensitivity. Typically, rapidly activating I_{HCN} is found in HCN1-expressing tissue, while slow currents predominate in regions with HCN2 and HCN4 (Santoro et al., 2000). This correlation has been established more quantitatively for HCN1. Current properties correlate with HCN1 mRNA in single cells (Franz *et al.*, 2000) and gradients of current amplitudes in dendrites of cortical layer V pyramidal neurons are proportional to gradients of HCN1 subunit expression (Lörincz *et al.*, 2002; Kole *et al.*, 2006).

Notably, in the nucleus laminaris of the chick, HCN1 is expressed in a gradient, along which current properties gradually change correspondingly (Yamada *et al.*, 2005). Studies in animal models of epilepsy indicate that slight imbalances in HCN1 expression, be it up- or downregulation, alter current properties accordingly in different cell types (Strauss *et al.*, 2004; Budde *et al.*, 2005; Kuisle *et al.*, 2006). Finally, developmental studies show that increased expression of HCN1 leads to an acceleration of current kinetics (Vasilyev and Barish, 2002). In addition, in ventricular myocytes a decreased expression of HCN1 and HCN4 mRNA is paralleled by a decrease in I_{HCN} (Cerbai *et al.*, 1999b; Shi *et al.*, 1999; Yasui *et al.*, 2001).

There are also a number of cases in which a correlation between the native current properties and expressed isoforms could only be partially achieved. In SAN, native current properties are not fully congruent with those of HCN1–HCN4 heteromers (Altomare *et al.*, 2003), but currents generated by HCN2–HCN4 heteromers fit well with the properties of atrial I_{HCN}. In nucleus reticularis cells, HCN2 is expressed strongly, but current amplitudes are comparatively small (Santoro *et al.*, 2000; Rateau and Ropert, 2006).

Subcellular localization

Two prominent cases point to a highly regulated subcellular expression of HCN channels. The first is the gradient of HCN1 and HCN2 expression of pyramidal cells in layer V cortical neurons and in hippocampal CA1 neurons (Santoro *et al.*, 1997; Lörincz *et al.*, 2002). Immunogold labelling shows that this labelling is predominantly dendritic, and largely absent from glutamatergic synapses (Notomi and Shigemoto, 2004). This increase in HCN channel density correlates with an increasing I_{HCN} amplitude (Magee, 1998; Williams and Stuart, 2000; Kole *et al.*, 2006) and is of great relevance in the dendritic computational properties of these cells (see Physiology section). How the subcellular trafficking of HCN1 protein gives rise to this steep gradient is unclear, but the first protein believed to regulate channel expression has now been identified (Santoro *et al.*, 2004) (see Accessory and regulatory proteins of HCN channels section). Labelling in the pyramidal cell layer is mostly due to presynaptic basket cell terminals.

Recordings from presynaptic terminals revealed the presence of I_{HCN}. These include chick ciliary ganglion neurons (Fletcher and Chiappinelli, 1992), crayfish neuromuscular junction (Beaumont and Zucker, 2000), cerebellar basket cells (Southan *et al.*, 2000) and the calyx of Held in auditory brainstem (Cuttle *et al.*, 2001). HCN1 protein is expressed in basket cell terminals (Santoro *et al.*, 1997), and co-localizes with GAD65 and synaptophysin, suggesting that HCN1-currents are involved in inhibitory synaptic transmission (Luján *et al.*, 2005). Indeed, in some neurons, pharmacologically reducing I_{HCN} with ZD7288 affects GABAergic transmission in a manner consistent with a presynaptic mechanism (Southan *et al.*, 2000; Aponte *et al.*, 2006; Boyes *et al.*, 2007). Presynaptic expression of HCN channel protein was also found in retinal bipolar cells and in photoreceptors (Müller *et al.*, 2003).

HCN channels may be localized in discrete microdomains of the cellular membrane, in conjunction with other ion channels and regulatory molecules. HCN4 channels are found in membrane fractions containing structural proteins of lipid rafts ('caveolae') in HEK and SAN cells, and disruption of lipid rafts altered their current-voltage dependence (Barbuti *et al.*, 2004). Additionally, HCN channels appear to be co-localized with T-type, low-threshold Ca^{2+} channels in dendrites of thalamocortical neurons (Stuart and Williams, 2000), perhaps explaining the regulation of I_{HCN} by Ca^{2+} in these cells (Lüthi and McCormick, 1998b).

Developmental expression patterns

Differential up- and downregulation of individual HCN channel subunits occurs during development (for review see Frère *et al.*, 2004), as shown in most detail for hippocampal CA1 and CA3 principal cells and interneurons (Bender *et al.*, 2001; Vasilyev and Barish, 2002; Bender *et al.*, 2005; Brewster *et al.*, 2006; Surges *et al.*, 2006). The temporal profile of HCN channel expression and current density matches that of synchronized electrical activity appearing during circuit maturation (Bender *et al.*, 2005, 2006). Activation rates increase with development, in agreement with a strong increase of HCN1 channel expression in comparison to HCN2 and HCN4 channels. Presynaptic expression of HCN protein is also regulated developmentally. Thus, in projections of entorhinal cortex layer II cells to the molecular layer of the hippocampal dentate gyrus, robust presynaptic expression and function of HCN1-containing currents is found up until ~P25 in rats, but is absent at P90, although mRNA and protein at somatic levels remain unchanged (Bender *et al.*, 2007).

In embryonic ventricular myocytes, an initially large I_{HCN} gradually decays until adulthood is achieved (Robinson *et al.*, 1997; Cerbai *et al.*, 1999b), and is accompanied perinatally by the loss of the ability to generate spontaneous activity (Cerbai *et al.*, 1999b, Shi *et al.*, 1999; Yasui *et al.*, 2001). This decrement is paralleled by a gradually reduced expression of HCN1 and HCN4 mRNA, until HCN2 becomes predominant in adult cardiac myocytes (Shi *et al.*, 1999). The dominance of HCN2 over HCN4 is further pronounced in the aged SAN, although both HCN2 and HCN4 transcription weakens markedly during ageing (Huang *et al.*, 2007).

2.5.5 Pharmacology

The sensitivity of I_{HCN} to millimolar concentrations of extracellular Cs^+, and its relative insensitivity to Ba^{2+}, was crucial for its original isolation as a cationic current in cardiac cells (Fain *et al.*, 1978; Bader *et al.*, 1979; DiFrancesco, 1981; Wollmuth and Hille, 1992). The mechanism of action of these ions involves binding within a multi-ion conducting pore that competes with the fully permeant ions Na^+ and K^+ (Pape, 1996). The pharmacological profile of I_{HCN} has diversified in recent years. In addition to bradycardiac agents, volatile and general anaesthetics interfere with HCN channel gating, while the anti-epileptic agent lamotrigine potentiates hippocampal I_{HCN}. However, a common characteristic of all I_{HCN} blockers, including the most recently described ones, is the lack of full selectivity for HCN channels. Furthermore, the pharmacology of I_{HCN} awaits the development of isoform-specific compounds that would allow the characterization of subunit-specific roles in discrete brain regions.

Extracellular cations

Millimolar concentrations of Cs^+ ions, and, to a much weaker extent, Rb^+ ions, are fast blockers of I_{HCN}, acting from the extracellular side of the channel and penetrating deep into the pore to bind to an inner 'blocking site' (DiFrancesco, 1982; Gauss *et al.*, 1998). The block is voltage-dependent, being greater at hyperpolarized potentials, and vanishing at more depolarized potentials (for review, see Pape, 1996).

The disadvantage of the experimental use of Cs^+ is its limited specificity: it also blocks neuronal K^+ channels (Constanti and Galvan, 1983) and interferes with K^+ uptake in glial cells (Janigro et al., 1997).

Extracellular Ba^{2+} ions, principally known as blockers of inward rectifier K^+ currents, fairly strongly (~55%) reduce I_{HCN} in photoreceptor cells (Wollmuth, 1995). In contrast, Ba^{2+} effects are minor for hippocampal I_{HCN} and for heterologously expressed HCN1 and HCN2 channels (~20–40%) (van Welie et al., 2005). In general, the use of Ba^{2+} ions is well established to distinguish (even small) I_{HCN} components from large inward rectifier currents (see for example Rateau and Ropert, 2006) and to isolate cyclic-nucleotide-dependent current modulation via G_i-coupled neurotransmitter receptors that also target inward rectifier K^+ currents (Frère and Lüthi, 2004).

Bradycardiac agents

The important role of HCN channels for pacing in the heart is strongly supported by a class of heart rate-reducing agents that powerfully block the cardiac isoform of I_{HCN} (for review see Baruscotti et al., 2005; DiFrancesco, 2005). These agents are termed bradycardiacs, because they selectively slow heart rate without interfering with other cardiovascular functions. Alinindine (ST567), was the first member of the family but its mechanism of action is not limited to I_{HCN} (Snyders and Van Bogaert, 1987).

Today the most widely used bradycardiac agent in I_{HCN} pharmacology is ZD7288 (originally named ICI D7288) (Briggs et al., 1994). ZD7288 blocks I_{HCN} in neurons and photoreceptor cells at concentrations between 10–100 µM (Harris and Constanti, 1995; Gasparini and DiFrancesco, 1997; Williams et al., 1997; Lüthi et al., 1998; Satoh and Yamada, 2000). Additionally, ZD7288 induces a negative shift of the I_{HCN} activation curve at a concentration of 0.3 µM in vitro in SAN cells (BoSmith et al., 1993). Distinctive characteristics of ZD7288 actions on whole-cell currents involve (a) its slow kinetics, ranging in the order of minutes, (b) its poor reversibility, (c) and its apparent lack of use dependence in moderate voltage ranges, although its actions are relieved by strong hyperpolarizations. These properties are consistent with the localization of ZD7288 binding in the vicinity of the intracellular portions of the channel pore (Shin et al., 2001). In spite of the remarkable selectivity of ZD7288 for HCN channels, it should be considered that ZD7288 induces a down-regulation of synaptic transmission that is independent of I_{HCN} (Chevaleyre and Castillo, 2002) and could involve a direct action on ionotropic glutamate receptors (Chen, 2004). This renders the use of ZD7288 particularly delicate when assessing potential roles for I_{HCN} in synaptic transmission (Mellor et al., 2002).

In addition to ZD7288, zatebradine (UL-FS 49) (Pape, 1994) and its more potent derivative cilobradine (DK AH 268) (Pape, 1994; Van Bogaert and Pittoors, 2003) block I_{HCN} in a concentration range of 10–100 µM. In contrast to ZD7288, however, these agents are 'open channel blockers' and their effects are strictly use-dependent.

The member of the family of bradycardiac agents with greatest specificity for cardiac I_{HCN} is ivabradine (Corlentor®). It is clinically used to reduce heart rate in the treatment of stable angina pectoris (Bucchi et al., 2002, 2006). Like zatebradine, ivabradine blocks I_{HCN} in a use-dependent manner from the intracellular side in a low micromolar concentration range. Unlike zatebradine, the action of ivabradine is dependent on ion flow through the channel pore (Bucchi et al., 2002).

Currently available studies on bradycardiac agents reveal no specificity for HCN channel isoforms, although the mechanisms of block may differ for HCN1 and HCN4 (Bucchi et al., 2006; Stieber et al., 2006).

Anaesthetic agents

Volatile anaesthetics have been predominantly known to act as powerful modulators of the two pore domain K^+ channels (for review see Franks and Lieb, 1988). However, studies on both native and cloned HCN channels have revealed a strong inhibitory action of volatile anaesthetics at clinically relevant concentrations (Sirois et al., 1998, 2002; Chen et al., 2005b). These studies reveal the first evidence for an isoform-specific blockade of HCN channels. Thus, halothane (Fluothane®) primarily shifts the voltage dependence of HCN1-currents towards a more hyperpolarized range, while reducing the maximal conductance of HCN2-currents (Chen et al., 2005b). These effects correlate strongly with the molecular make-up of the two channel isoforms: the CNBD and the C-linker of HCN1 channels mediate the halothane-induced shift in voltage dependence, while core transmembrane domains of HCN2 mediate the inhibition of conductance. The differential actions of halothane on isoforms are lost when cAMP is bound, implying a role for the CNBD, and its inhibitory effects on channel gating, in the isoform specificity of halothane.

In addition to its modulatory action on GABAergic inhibition (Jurd et al., 2003) and on small conductance (SK)-type K^+ channels (Ying and Goldstein, 2005), the general anaesthetic propofol (Diprivan®) inhibits I_{HCN}. At clinically relevant concentrations of 3–5 µM, propofol induced strong negative shifts of about 10 mV in the voltage dependence of activation of I_{HCN} in both cortical pyramidal neurons (Chen et al., 2005a) and thalamocortical cells (Ying et al., 2006). It also inhibited HCN channels in an isoform-specific manner, with the most pronounced actions on HCN1-currents, and more moderate effects on HCN2- and HCN4-currents (Cacheaux et al., 2005; Chen et al., 2005a). Propofol mediates its action via binding to the membrane-bound channel core and by lowering open probability (Lyashchenko et al., 2007). Given the strong expression of HCN2 and HCN4 protein in thalamus, it remains unclear which native channel isoform in thalamocortical cells confers high propofol sensitivity. Interestingly, propofol also induces a decrement of HCN2 and HCN4 protein expression 3–24 hours after a single propofol injection (Ying et al., 2006).

Anti-epileptic agents

Commonly used anti-epileptic drugs are powerful potentiators of HCN channels (Poolos et al., 2002; Berger et al., 2003; Surges et al., 2003). In hippocampal principal cells and cortical layer V pyramidal cells in vitro, lamotrigine (Lamictal®) acutely causes a several fold upregulation of dendritic I_{HCN} amplitude around resting membrane potentials and an 11 mV positive shift in current voltage-dependence (Poolos et al., 2002; Berger and Lüscher, 2004) at concentrations of 50–100 µM. This enhancement of the dendritic current may attenuate neuronal excitability, either via a decrease in dendritic input resistance or by reducing temporal summation of synaptic inputs (see HCN-currents in dendritic integration section). However, additional effects of lamotrigine on voltage-gated Na^+ channels may explain its neuronal-specific actions (Berger and Lüscher, 2004). A potentiating action has also been reported for gabapentin (Neurontin®) (Surges et al., 2003).

Additional blockers

An inhibitory effect on HCN1-mediated currents (Gill et al., 2004), and of hippocampal I_{HCN} (Ray et al., 2003), has been reported for capsazepine, a blocker of vanilloid receptors, with an IC_{50} of 8 µM. QX-314, a quaternary derivative of lidocaine and best known as an intracellular blocker of voltage-activated Na$^+$ channels, completely blocked I$_{HCN}$ in CA1 pyramidal cells (Perkins and Wong, 1995). Furthermore, the analgesic loperamide, a μ-opioid receptor agonist, exerts a use- and voltage-dependent block on I$_{HCN}$ in dorsal root ganglia neurons independently of μ-opioid receptor activation (Vasilyev et al., 2007). Additionally, it has been proposed that the tyrosine kinase inhibitor genistein (Shibata et al., 1999; Altomare et al., 2006) and the α$_2$-receptor agonist clonidine (Parkis and Berger, 1997), an antihypertensive and antisympathetic drug, might directly inhibit I$_{HCN}$. Indeed, in mice deficient in all three α$_2$-receptor subtypes, clonidine exerts bradycardiac actions on acutely isolated atrial myocytes that are occluded by Cs$^+$ ions, and clonidine blocks heterologous HCN-currents, with an IC$_{50}$ ~10 μM for HCN2 and HCN4, and ~40 μM for HCN1 (Knaus et al., 2007).

2.5.6 Physiology

Previous reviews have excellently summarized the role of HCN-currents in neuronal and cardiac functions (Pape, 1996; Kaupp and Seifert, 2001; Robinson and Siegelbaum, 2003; Baruscotti and DiFrancesco, 2004; Stieber et al., 2004; Cohen and Robinson, 2006). This chapter reports mostly on the novel insights into long-recognized functions of I$_{HCN}$, and then describes some of the most recently established functions of this current.

Physiological and behavioural deficiencies in HCN-subunit knockout mice

Animals deficient in HCN-channel subunits illustrate the diverse involvement of HCN channels in neuronal and cardiac functions (Herrmann et al., 2007). The lack of the HCN1 gene in Purkinje and basket cells of cerebellum causes selective deficits in the training of repetitive movements that involve the phasic excitation and inhibition of cerebellar Purkinje cells (Nolan et al., 2003). In these cells, the HCN1-deficiency retards the resumption of action potential discharge after hyperpolarization, since any voltage deviation induced by transient inhibition is greater than normal and repolarization is slowed. Thus, by allowing a rapid response to oscillatory inputs, I$_{HCN}$ may facilitate the coincidence detection of pre- and postsynaptic activity that is required for synaptic plasticity of afferents to Purkinje cells. In contrast, the lack of HCN1 in cortical structures facilitates hippocampal-dependent spatial learning, potentiates long-term synaptic plasticity at some hippocampal afferents, and enhances subthreshold neuronal responses to inputs in the theta frequency range (Nolan et al., 2004). Deleting the HCN2 gene produces striking deficiencies in thalamic and cardiac cells. The resting membrane potential of thalamic neurons is strongly hyperpolarized, the propensity to discharge in bursts increased, and synchronous thalamocortical activity typical for generalized spike-and-wave discharges recorded in the EEG. HCN2-deficient mice show cardiac arrhythmia, while sympathetic stimulation remains intact (Ludwig et al., 2003). Finally, HCN4-deficient mice die at embryonic stages; show a strongly reduced heart rate and insensitivity to sympathetic stimuli. Cardiomyocytes from these animals are virtually devoid of hyperpolarization-activated currents and pacemaker-like action potentials (Stieber et al., 2003a). Taken together, the deficits observed in HCN-channel subunit deficient animals illustrate, and further strengthen, the long-recognized importance of HCN-currents for some of the cornerstones of neuronal excitability and of pacemaker functions. However, they also point to an unexpected involvement of I$_{HCN}$ in synaptic integration and long-term synaptic plasticity (Nolan et al., 2003, 2004).

HCN-currents and pacemaking

The discovery that the diastolic phase of the cardiac action potential is carried predominantly by a cationic hyperpolarization-activated current has set the ground for the unique profile of I$_{HCN}$ as a pacemaker current (Brown et al., 1979; DiFrancesco, 1981). Detailed reviews on the mechanisms of pacemaking in cardiac and neuronal preparations are found in Lüthi and McCormick (1998a), Robinson and Siegelbaum (2003), Frère et al. (2004), Baruscotti et al. (2005) and Cohen and Robinson (2006). Neuronal pacemaker functions attributed to I$_{HCN}$ figure prominently in rhythms studied extensively in in vitro neuronal preparations, including sleep-related oscillations in thalamocortical cells and circuits (for review, see Lüthi and McCormick, 1998a) γ-oscillations in hippocampus (Fisahn et al., 2002), synchronized oscillations in inferior olive (Bal and McCormick, 1997), and subthreshold oscillations in entorhinal cortex (Dickson et al., 2000). In autonomously firing neurons, a kinetically precisely tuned I$_{HCN}$ sets the regularity and timing of action potential discharge (Chan et al., 2004).

Importantly, I$_{HCN}$ does not serve as a pacemaker solely by virtue of its voltage-dependent activation at hyperpolarized potentials. In this respect, the dual allosteric gating by cAMP and voltage also allows for a slow, persistent form of current enhancement that may underlie the recurrence of synchronous oscillations at timescales of once every 10–30 s (Lüthi and McCormick, 1998a; Wang et al., 2002).

Most recent advances concerning the role of HCN channels in pacemaker mechanisms include the molecular identification of the HCN channel isoforms for distinct types and aspects of cardiac and neuronal pacemaking. In cardiac myocytes, overexpression and dominant-negative suppression of HCN2 and HCN4 increase or suppress cardiac beating, respectively, both in vitro and in vivo (for review, see Cohen and Robinson, 2006). Genetic deletion of HCN2 reduces SAN currents by 30%, whereas eliminating HCN4 in embryonic myocytes almost abolished the current. Consistent with these effects, heart rate was more variable, but still present in HCN2-deficient animals, whereas the HCN4-deficient animals are embryonically lethal due to a failure of heart beating.

The fact that HCN2- and HCN4-mediated currents are largely responsible for the diastolic phase of the cardiac rhythm has prompted the idea that HCN genes could be used to recreate a pacemaker for the arrhythmic or failing heart. The initial experiments with viral or cell-based delivery of HCN genes into cardiac cells have proven remarkably successful (for review, see Robinson et al., 2006). Thus, within three days after injection of an HCN2-containing adenovirus in the canine atrium or in the Purkinje fibres, substantial current was expressed that promoted escape rhythms driving the heart after sinus node arrest (Qu et al., 2003; Plotnikov et al., 2004).

HCN-currents and resting membrane potentials

By virtue of its 10–15% activation around resting membrane potential and its non-inactivating properties, standing HCN-currents exert a depolarizing action on neuronal resting properties in many different cell types (Pape, 1996). Activation of I$_{HCN}$ around rest stabilizes membrane potential against hyperpolarization (Ludwig et al., 2003; Nolan et al., 2003), while its deactivation antagonizes membrane depolarization (Magee, 1999). The importance of shortening transient hyperpolarization to reinstate cellular excitability is perhaps best demonstrated in bulbospinal neurons involved in rhythmic breathing movements (Dekin, 1993) and in Purkinje cells that are phasically inhibited during motor activity (Nolan *et al.*, 2003). Around rest, the depolarizing effect of I_{HCN} is cancelled by hyperpolarizing currents which abolish net current flow at the resting membrane potential. In thalamocortical neurons, HCN2-currents and TASK3-currents counterbalance each other (Meuth *et al.*, 2006). Consequently, although both these channel types are highly sensitive to extracellular acidification, thalamocortical cellular resting membrane potential is relatively unaffected by drops in pH. Conversely, volatile anaesthetics, which inhibit HCN and potentiate TASK currents, dramatically hyperpolarize the resting membrane potential (Sirois *et al.*, 2002). Interactions between HCN channels and K_{ir}/K_{leak} channels are prominent in pyramidal cell dendrites and maintain I_{HCN} activation sufficiently high to allow for its control of temporal summation (Day *et al.*, 2005). A standing I_{HCN} is also crucial for stabilizing the ongoing action potential discharge in Purkinje neurons (Williams *et al.*, 2002). Primary sensory neurons exploit the pH and temperature sensitivity of HCN currents to generate sensory receptor potentials (Stevens *et al.*, 2001; Viana *et al.*, 2002).

Regulation by cAMP and novel modulators

A canonical way of I_{HCN} regulation is via the direct binding to cyclic nucleotides, independently of the action of PKA (DiFrancesco and Tortora, 1991; Pedarzani and Storm, 1995). Since this discovery, it has been shown that many G-protein-coupled neurotransmitters, endogenous neuropeptides and inflammatory agents regulate I_{HCN} via altering intracellular cAMP and cGMP levels (for an overview see Pape (1996) and Frère *et al.* (2004)). These compounds principally mediate their cAMP-dependent actions by shifting the activation curve towards more depolarized potentials, while leaving maximal current conductance unaltered. The regulation by cAMP is bidirectional, in that both increases and decreases in standing cAMP levels, mediated by G_s- and $G_{i/o}$-coupled neurotransmitter receptors, respectively, can be detected (for an overview, see Frère *et al.*, 2004). In entorhinal cortex, the dopaminergic decrease in neuronal excitability is mediated, at least in part, by cAMP-dependent regulation of HCN channels present in apical dendrites (Rosenkranz and Johnston, 2006). A number of effects of neurotransmitters on I_{HCN} appear, however, unrelated to cAMP. In hippocampus, muscarinic receptor agonists augment I_{HCN} by increasing its maximal conductance, independently of cAMP (Colino and Halliwell, 1993; Fisahn *et al.*, 2002), and serotonin augments I_{HCN} by affecting both maximal conductance and activation curve, an effect that is not fully mimicked by cAMP (Gasparini and DiFrancesco, 1999; Bickmeyer *et al.*, 2002). Furthermore, a μ-opioid receptor agonist augments the maximum conductance of I_{HCN} in brainstem neurons via mobilization of intracellular Ca^{2+}, without shifting the I_{HCN} activation curve (Pan, 2003). In hypothalamic paraventricular neurons, neuromedin-U, best known for its effect on gut smooth muscle, positively shifted the voltage dependence of I_{HCN}, likely via stimulation of phospholipase C pathways (Qiu *et al.*, 2003). Finally, the sleep-promoting peptide cortistatin augments the maximal conductance of I_{HCN} by a cAMP-independent mechanism (Schweitzer *et al.*, 2003). From these available data, it is reasonable to assume that additional modulators of HCN channels exist. Some of these were thought to be found in the second messenger pathways involving mobilization of intracellular Ca^{2+} and phosphoinositide pathways, although Ca^{2+} has no direct effect on at least the thalamic isoforms of I_{HCN} (Budde *et al.* 1997,

Lüthi and McCormick, 1999; Fan *et al.*, 2005). Indeed, a number of recent studies demonstrate a powerful regulation of I_{HCN} by the phosphorylated products of the membrane lipid phosphatidyl inositol, the so-called phosphoinositides. These are widely used signalling molecules that are present in particular concentrations in plasma membranes and that control ion channel function (Suh and Hille, 2005) in addition to exerting important signalling roles in communication and trafficking between cytosol and membrane compartments (Di Paolo and De Camilli, 2006). The polyanionic lipid phosphatidylinositol 4,5-bisphosphate (PIP_2) shifts the activation range of cloned HCN2-currents, as well as of cardiac and neuronal currents, more positively by ~20–30 mV (Pian *et al.*, 2006; Zolles *et al.*, 2006). This effect requires the negatively charged phosphates on the inositol rings, but is largely independent of cAMP-dependent channel regulation. Interestingly, exposure to PIP_2 largely prevents the hyperpolarizing shift ('run-down') of HCN-current voltage dependence traditionally found in excised patches (DiFrancesco *et al.*, 1986; Chen *et al.*, 2001c). This can be explained by the observation that endogenous phosphoinositide levels, maintained by opposing actions of corresponding kinases and phosphatases, are sufficiently high to exert a tonic depolarizing action on the voltage dependence of native currents (Pian *et al.*, 2006; Zolles *et al.*, 2006). Further acidic lipids such as the PIP_2 precursor phosphatidic acid, as well as arachidonic acid, also positively shift current voltage dependence (Fogle *et al.*, 2007). In addition, p38 mitogen-associated protein kinase (MAPK) was identified as a powerful modulator of the voltage dependence of hippocampal I_{HCN} (Poolos *et al.*, 2006). Inhibition of p38 MAPK negatively shifted the activation curve of hippocampal I_{HCN} by ~25 mV, whereas activation of this kinase induced an 11 mV positive shift. These authors proposed that phosphorylation of T778 on HCN1 (Figure 2.5.3), located distal to the CNBD, underlies the MAPK actions, but the recent findings of Fogle *et al.* (2007) now point to a possible involvement of arachidonic acid. The existence of further consensus phosphorylation sites for additional kinases (Poolos *et al.*, 2006) (Figure 2.5.3), and the diverse types of tyrosine kinase modulation of heterologously expressed (see for example Arinsburg *et al.*, 2006) and native isoforms of I_{HCN} (Thoby-Brisson *et al.*, 2003) strongly suggests that elucidation of the pathways controlling HCN channel phosphorylation may soon lead to additional insights into the physiological roles of this current.

HCN-currents in dendritic integration

The active properties of dendrites play an important role in how synaptic input is shaped and integrated. A notable feature of dendritic HCN1 is its expression along a steep gradient in the apical dendrites of CA1 pyramidal neurons and layer V cortical neurons, with channel density increasing by over 60-fold towards the distal end in the latter, and about 7-fold in the former (see Correlation of expression patterns with native current properties in section 2.5.4). A standing I_{HCN} contributes up to 11 mV to the resting membrane potential of apical dendrites (Williams and Stuart, 2000). When this current is deactivated partially, as may occur through small subthreshold depolarization, it is rapid enough to accelerate the decay of these deviations from the resting potential. The deactivation of I_{HCN} is boosted during repetitive depolarizations and produces a net membrane hyperpolarization that effectively impairs the temporal summation of postsynaptic excitatory potentials, generated by presynaptic firing up to about 100 Hz. These effects are most pronounced for inputs at distal dendrites, where HCN channel density is highest. Notably, the density of HCN channels appears tuned such that it exactly compensates the incrementing electrotonic filtering of the tapering dendritic cables, thereby rendering the temporal summation essentially independent of dendritic location (Magee, 1999; Williams and Stuart, 2000). By a similar mechanism, dendritic I_{HCN} also antagonizes the summation of back-propagating action potentials into cortical layer V dendrites (Berger *et al.*, 2003). The acceleration, and hence sharpening of EPSPs, could be particularly important in auditory processing, in which high temporal precision in the detection of inputs is required (Yamada *et al.*, 2005). The high HCN channel density, however, also imposes a shunting conductance on the hippocampal (Berger *et al.*, 2001) and cortical (Fernandez *et al.*, 2002) neuronal dendrites, and produces a disproportionate attenuation of the amplitude of distal inputs. A higher density of AMPA-type glutamate receptors at distal dendrites reinstates the site independence of EPSP amplitude (Andrasfalvy and Magee, 2001).

The rapid deactivation and reactivation of I_{HCN} shapes the responsiveness of dendrites to oscillatory input, imparting to them a resonance at frequencies in the theta range (~5 Hz) (Magee, 2001; Hu *et al.*, 2002). Computational models show that stochastic fluctuations of HCN channels, present at high densities in cortical dendrites, are strong enough to help regulate the precision of action potential discharge to periodic inputs (Kole *et al.*, 2006).

A recent study suggests that I_{HCN}, expressed on distal dendrites of pyramidal neurons in prefrontal cortex, may play a role in prefrontal cortex-specific working memory function (Wang *et al.*, 2007). Working memory is known to be impaired when cAMP levels are elevated and noradrenergic receptors of the α2A-type strengthen working memory via suppression of cAMP synthesis. When I_{HCN} was inhibited or HCN1 mRNA knocked down, working memory was enhanced and suppressive effects of cAMP-elevating agents were reduced. Based on the co-localization of α2A-receptor and HCN1 protein at perisynaptic locations within dendritic spines, the authors propose that the memory-promoting effects of noradrenergic input may be mediated via downregulation of cAMP-HCN1 signalling, thereby decreasing a shunting action of opened HCN channels on excitatory synaptic inputs.

HCN-currents in synaptic plasticity

A novel and exciting theme in HCN channel physiology is the recognition that these channels are members of the growing family of ion channels the expression of which is rapidly regulated by neuronal activity. Thus, I_{HCN} is an important determinant of how intrinsic excitability can be modified, thereby influencing how synaptic input is transformed into axonal output (Zhang and Linden, 2003). Modifications in excitability, accompanied by changes in I_{HCN}, have been reported from different groups. HCN 1-containing channels reduce the amount of synaptic plasticity that can be induced at the synapses made by the perforant path onto distal dendrites in the hippocampal CA1 area (Nolan *et al.*, 2004). In the CA1 area, HCN-mediated currents were rapidly (within tens of minutes) enhanced several-fold following exposure to pressure-applied glutamate (van Welie *et al.*, 2004), following synaptic release of glutamate via a theta-burst pairing protocol, or following brief increases in extracellular K^+ levels (Fan *et al.*, 2005). In younger animals, input-specific modifications of I_{HCN}, as well as local changes in dendritic integration, were reported (Wang *et al.*,

2003). Long-term plasticity was accompanied by an enhanced expression of HCN1 protein, while HCN2 expression remained unchanged. Entry of Ca^{2+} was required for this effect, with NMDA receptors as the principal source, and CaMKII and the protein synthesis machinery as likely mediators. The increased I_{HCN} decreases neuronal excitability persistently and could therefore be involved in the dendritic mechanisms controlling synaptic plasticity. Presynaptically expressed I_{HCN} has also been implicated in mediating presynaptic forms of synaptic plasticity, both in mammalian hippocampus (Mellor *et al.*, 2002; but see Chevaleyre and Castillo 2002), as well as in invertebrates (Beaumont and Zucker, 2000; Beaumont *et al.*, 2002).

HCN-currents in presynaptic neurotransmitter release

In one of the original cloning studies, it was realized that HCN1 protein was strongly expressed in axon terminals of cerebellar basket cells (Santoro *et al.*, 1997). In these terminals (Southan *et al.*, 2000), as well as in the giant presynaptic terminals of the rat auditory pathway (Cuttle *et al.*, 2001) and the avian ciliary ganglion (Fletcher and Chiappinelli, 1992), substantial I_{HCN} is expressed. However, so far, evidence for an involvement of this current in presynaptic neurotransmitter release in these terminals is lacking, while presynaptic I_{HCN} in dentate gyrus basket cells (Aponte *et al.*, 2006) and in crustacean terminals was proposed to facilitate neurotransmitter release (Beaumont and Zucker, 2000). In cerebellar basket cells, as well as in stratum oriens interneurons of hippocampus (Lupica *et al.*, 2001), blocking I_{HCN} reduces spontaneous GABAergic transmission, consistent with I_{HCN} controlling somatodendritic excitability of these neurons and axonal propagation. More recently, it was shown that blocking I_{HCN} increases the threshold for axonal action potential initiation (Hiramatsu *et al.*, 2002; Aponte *et al.*, 2006), probably because the membrane hyperpolarization generated by the electrogenic Na^+/K^+ ATPase (Pachucki *et al.*, 1999) is no longer antagonized by I_{HCN}.

2.5.7 Relevance to disease states

HCN channels belong to the few ionic channels known to be sensitively altered at the transcriptional level in response to brief periods of aberrant electrical activity (Waxman, 2001). The ensuing abnormal functional expression of HCN channels may produce maladaptive increases in neuronal excitability, such as those typical of epilepsy and neuropathic pain. In the heart, ventricular myocytes show changes in HCN subunit transcription in response to cardiac stress, which may contribute to the arrhythmias that are seen in cardiac disease.

Human diseases related to the genetic locus of HCN channels

Three patients with idiopathic sinus node disease, characterized by marked bradycardia and life-threatening cardiac arrhythmia, were found to carry a mutation in one of the two *HCN4* genes. In two cases, the mutation co-segregated with the disease in the pedigree of the family in a dominant fashion (Ueda *et al.*, 2004; Milanesi *et al.*, 2006). Mutated genes either encode a channel subunit with a deleted CNBD (Schulze-Bahr *et al.*, 2003) or cause poor expression (Ueda *et al.*, 2004). In the third case, mutated channels show a negatively shifted activation range (Milanesi *et al.*, 2006). Mutated channel subunits displayed dominant-negative effects on the expression of the wild-type channels. Additional human diseases were linked to the chromosomal locus 15q24, such as a syndrome of mental retardation (Mitchell *et al.*, 1998), and an autosomal dominant nocturnal frontal-lobe epilepsy (Phillips *et al.*, 1998), although the involvement of mutated *HCN* genes remains to be demonstrated. As discussed later in this section in animal models, abnormal mRNA expression for HCN channels is found in cardiac hypertrophy and failure as well as in a number of neuronal pathologies. Human myocytes from patients with end-stage heart failure show an ~three-fold increase in HCN4 mRNA (Borlak and Thum, 2003) and an augmented current density (Hoppe *et al.*, 1998; Cerbai *et al.*, 2001).

Epilepsies

Changes in HCN expression were found in a developmental form of epilepsy, in temporal lobe epilepsy and in generalized epilepsy (Santoro and Baram, 2003; Poolos, 2004). In animal models of developmental and temporal lobe epilepsy, HCN channel expression is disturbed following single seizure episodes and these changes may persist for weeks, implicating them in the development of chronic epilepsy. Animal models of generalized epilepsy show abnormal HCN expression prior to the onset of the seizures that persists unaltered in the chronic phase, suggesting a genetically determined defect in HCN channel function that accompanies epilepsy. Thus, HCN channels appear to be involved in both inherited, as well as in acquired forms of epilepsy, and their rapid and high susceptibility to single phases of hyperexcitation could render them important causative factors in the processes of epileptogenesis.

Generalized epilepsies

Perhaps the most extensive characterization of HCN channel expression and properties is now available for models of absence epilepsy, a generalized form of absence epilepsy characterized by spike-and-wave discharges (SWDs) in the electroencephalogram (Waxman, 2001; Meeren *et al.*, 2005). These discharges arise from a hypersynchronous oscillation in reverberating thalamocortical loops, and both thalamic as well as cortical mechanisms underlie these pathological oscillations. Remarkably, the two independent rat models of absence epilepsy, the Wistar Albino Glaxo rats, bred in Rijswijk (WAG/Rij) (Strauss *et al.*, 2004; Budde *et al.*, 2005) and the Genetic Absence Epilepsy Rats from Strasbourg (GAERS) (Kuisle *et al.*, 2006) as well as the stargazer mouse model (Di Pasquale *et al.*, 1997), show altered I_{HCN} properties. Thalamocortical neurons of both WAG/Rij and GAERS show an increased expression of the cAMP-insensitive HCN1 channel isoform that is accompanied by a weakened cAMP sensitivity of HCN channels. The enhanced HCN1 expression found during juvenile, epilepsy-free stages, persists throughout chronic epilepsy, and may lead to an enhanced heteromer formation with HCN2, thereby reducing the sensitivity of the channels to cAMP transients typical for normal sleep-related oscillations (Kuisle *et al.*, 2006). Conversely, in cortex of the WAG/Rij model, HCN1 expression was downregulated in layers II/III and V, producing an increase in the excitability of these neurons (Strauss *et al.*, 2004; Kole *et al.*, 2007). The loss of HCN1 protein abrogates the steep somatodendritic gradient of I_{HCN} in cortical layer V cells, producing enhanced somatodendritic coupling and facilitated dendritic Ca^{2+} electrogenesis (Kole *et al.*, 2007). This renders cortical layer V cells primed for high-frequency

firing during whisking behaviour, a finding that is consistent with the recent identification of a cortical focus of seizure initiation in deep layers of GAERS somatosensory cortex (Polack *et al.*, 2007). Conversely, the stargazer mouse model shows an enhanced current amplitude in layer V principal neurons (Di Pasquale *et al.*, 1997).

Febrile seizures

Seizures induced by fever are the most common type of seizure in the developing brain and affect up to 5% of children younger than five years. In young rats, febrile seizures can be induced by briefly exposing the brain to a single (~20 min) period of hyperthermia. Such animals reliably (> 98%) show seizures and an increased tendency to develop epilepsy in adulthood (Toth *et al.*, 1998), suggesting that early life seizures may predispose to later epileptic susceptibility (Walker and Kullmann, 1999; Baram *et al.*, 2002). Febrile seizures are accompanied by a number of persistent modifications of neuronal excitability in hippocampal neurons, amongst which changes in HCN mRNA expression figure prominently (Brewster *et al.*, 2002, 2005). A single experimental febrile seizure induced a persistent decrease in HCN1 mRNA and an increase in HCN2 mRNA in hippocampal CA1 pyramidal cells (Brewster *et al.*, 2002, 2005), that were accompanied by a depolarizing shift in the current activation curve (Chen *et al.*, 2001b). These altered expression patterns were not found when animals were given anti-epileptic drugs prior to induction of hyperthermia, indicating that seizure activity, and not the elevated brain temperature, triggered changes in subunit expression. Notably, the changes in HCN mRNA expression levels persisted for up to 3 months, suggesting a long-lasting perturbation of the developmental programme of HCN channel expression. In addition to the altered expression, HCN1 and HCN2 proteins increasingly associated in heteromers following seizures, further promoting the functional expression of I_{HCN} with altered biophysical properties (Brewster *et al.*, 2005).

Temporal lobe epilepsy

The kainate model of temporal lobe epilepsy is produced by a single injection of the convulsant agent kainic acid, which leads to chronic epilepsy after a latency period of a few weeks (White, 2002). During this latent phase, which lacks behaviourally overt epileptic activity, HCN1 and HCN2 proteins in entorhinal cortex, the major component of the temporoammonic pathway into the hippocampus, are markedly down regulated within 24 h, and current amplitude is reduced for up to 1 week (Shah *et al.*, 2004). This downregulation of I_{HCN} causes enhanced excitability of entorhinal cortex layer III neurons and increased temporal summation of excitatory input. It is widely believed that such modifications may represent a first step in the processes leading from the latent, seizure-free episode to chronic epilepsy.

In contrast, during advanced stages of chronic temporal lobe epilepsies, the expression of HCN protein in surviving neurons is altered in a manner that suggests a neuroprotective effect. Thus, in surviving dentate granule cells of sclerotic human hippocampus, HCN1 is markedly increased compared to the non-sclerotic case. Such current enhancements in surviving cells may help to counteract the excessive excitation and the associated excitotoxic cell death (Bender *et al.*, 2003).

These currently available studies document that, in at least some instances, an altered HCN subunit expression in neurons may be a defining characteristic of neuronal networks prone to develop an epileptic phenotype. The detailed patterns of modified channel expression are determined not only by the type of seizures, but also by the cell type and the developmental stage. In addition, HCN subunit expression, in particular HCN1 and HCN2, shows a high vulnerability to single periods of enhanced neuronal activity. The influences during developmental stages appear to be particularly multifactorial, given that single febrile seizures produce persistent current enhancements (Chen et al., 2001b), but can be influenced by neonatal behavioural experience (Schridde et al., 2006). The mechanisms translating seizures into altered HCN channel expression remain, so far, unexplored, but could range from acute influences, such as synaptic activity (see for example van Welie et al. 2004) to long-term, chronic modulation of channel expression, for example via hormones and inflammatory processes (see for example Armoundas et al., 2001; Vasilyev and Barish 2004).

Cardiopathies

Cardiac myocytes undergo electrical and structural remodelling to adapt to external stressful factors, such as pressure overload (e.g. hypertension), inflammation and infarction (for review, see Armoundas et al., 2001). A number of cardiovascular diseases are accompanied by abnormal expression of HCN mRNA, notably in areas in which this channel is poorly expressed under healthy conditions. Most prominently, cardiac ventricular myocytes, which are not normally involved in cardiac pacemaking, show increases in I_{HCN} magnitude or shifts in voltage dependence, rendering them potentially rhythmogenic (Cerbai and Mugelli, 2006).

An unexpected presence of I_{HCN} was first observed in ventricular myocytes of spontaneously hypertensive rats that was accompanied by a small diastolic depolarization (Cerbai et al., 1994). The degree of rat myocardial hypertrophy is positively correlated with an increase in the density of I_{HCN}. This correlation indicates that the mechanisms underlying cardiac hypertrophy, such as alteration in action potential properties and response to cellular stress, steer HCN subunit expression (Cerbai et al., 1996). At the molecular level, an upregulation of the HCN2 and HCN4 mRNA (Hiramatsu et al., 2002; Fernandez-Velasco et al., 2003) and HCN2 protein expression (Fernandez-Velasco et al., 2006) the predominant isoforms underlying ventricular I_{HCN} (Shi et al., 1999), have been described. Again, the change in expression levels was most pronounced in those cardiac regions that had highest pressure overload (Fernandez-Velasco et al., 2003; Sartiani et al., 2006). These changes are fairly specific to HCN proteins amongst the family of currents involved in cardiac diastolic potentials (Fernandez-Velasco et al., 2006). Cardiac remodelling also alters β-adrenergic receptor mediated regulation of HCN currents in ventricular myocytes (Cerbai et al., 1999a; Sartiani et al., 2006). Qualitatively similar observations have also been reported for the failing heart in humans, including larger current densities and a positive correlation between the severity of hypertrophy and increased current density (Hoppe et al., 1998; Cerbai et al., 2001). Additionally, gene microarray analysis points to an upregulated *HCN4* gene expression in the failing human ventricle (Borlak and Thum, 2003).

Signalling pathways switching on/off HCN channel expression in non-pacemaker cells may involve the renin-angiotensin system. Chronic administration of the type I angiotensin receptor-blocker losartan to old hypertensive rats not only reduces cardiac hypertrophy, but also reverses I_{HCN} upregulation and overexpression of

HCN2 and HCN4 mRNA (Cerbai et al., 2000; Hiramatsu et al., 2002). G-protein coupled receptors also seem to play a role in HCN channel expression. The β2-adrenergic receptor overexpressing mice show a five times larger ventricular I_{HCN} than normal animals, and a preferential upregulation of HCN4 compared to HCN2 (Graf et al., 2001). Furthermore, HCN mRNA is correlated to thyroid hormone and/or thyroid hormone receptor level (Pachucki et al., 1999; Gloss et al., 2001).

HCN channel dysregulation not only plays a role in ventricular cells, but also in the pacing regions of the heart, further increasing the risk of arrhythmic impulses. For example, in the rabbit failing heart, SAN cell automaticity is impaired by down regulation of I_{HCN} (Verkerk et al., 2003), and a reduced expression of HCN2 and HCN4 subunits was described in a dog model of congestive heart failure (Zicha et al., 2005).

Taken together HCN expression patterns in cardiac myocytes appear strongly correlated with the degree of cardiac overload and are thus electrical markers of cardiac remodelling. Given the similarities in the expression profile of fetal and hypertrophied myocytes, it has been speculated that cardiac hypertrophy involves a recapitulation of gene expression patterns typical for neonatal stages and a re-entry of cells into a juvenile programme.

Injuries

Peripheral nerve injury is often accompanied by syndromes of neuropathic pain such as allodynia (strong painful sensation evoked by light mechanical stimuli) and spontaneous painful sensations. Sensory pathways and a misrepresentation of sensory information are primarily thought to contribute to the clinical symptoms of neuropathic pain. This manifests as a hyperexcitability of dorsal root ganglion cells bodies that give rise to large, myelinated Aβ/δ-fibres, which are normally not involved in the transmission of pain (Shir and Seltzer, 1990; Chaplan et al., 2003). Rat models of spinal cord injury, such as axotomy (Black et al., 1999), chronic constriction (Dib-Hajj et al., 1999) or ligation of spinal nerves (Kim et al., 2001) show that an altered expression of several voltage-gated Na$^+$ channel subunits contributes to the persistence of neuronal firing in injured cells (for review see Waxman, 2001). However, administration of ZD7288 reduced allodynia in rat models of neuropathic pain (Chaplan et al., 2003; Dalle and Eisenach, 2005; Lee et al., 2005) and reversed the spontaneous discharges in injured large myelinated fibres (Chaplan et al., 2003). In this latter model of spinal nerve ligation (Chaplan et al., 2003) as well as in a model of chronic compression (Yao et al., 2003), the maximal I_{HCN} density was enhanced 1.5–2-fold compared to control, with variable effects on voltage dependence and kinetics. These findings establish I_{HCN} upregulation, resulting from nerve injury, as an essential factor leading to neuropathic pain. The molecular identity of the HCN channels that contribute to these changes remains to be determined but appears to involve a decrease in the amount of HCN1 and HCN2 mRNA and protein in the case of nerve ligation (Chaplan et al., 2003).

Besides neuronal injury, lesions in excitatory input can also cause an altered expression of HCN channels. Functional changes of I_{HCN} and HCN channel expression are described for a lesion of the enthorhinal cortex, where a decreased expression of HCN1 channels was accompanied by a negative shift of I_h activation and a faster kinetic in several neuronal cell types of the hippocampus (Bräuer et al., 2001). These changes were partly reversed following reactive sprouting and replacement of entorhinal input by septal and associational afferents.

HCN channel activity is also modified after other forms of insults, such as hypoxia (Erdemli and Crunelli, 1998) and inflammation (Ingram and Williams, 1996; Linden et al., 2003).

2.5.8 Concluding remarks

The HCN channels have seen a paradigm change in their physiological standing within the voltage-gated ion channel family. Funny and queer in earlier times, these channels have evolved to some of the best characterized ion channels today. Particularly remarkable within this ion channel family is that comparatively minor differences amongst channel isoforms, such as their kinetics, voltage dependence and cAMP sensitivity, implicate them in the most diverse neuronal functions. Two areas are particularly interesting. First, HCN channel expression in hippocampus is regulated rapidly by glutamatergic excitatory activity. This implicates these channels in the major control pathways regulating neuronal excitability, plasticity and homeostasis. It is thus reasonable to expect that, in the near future, a major pathway in the HCN channel field will consist of the elucidation of the activity-dependent co-assembly, trafficking and regulation of HCN channels in excitatory neuronal networks. Second, accruing evidence indicates that dysregulated HCN channel function and abnormal expression is related to cardiopathies and central nervous system diseases, and some of the rules by which this happens are currently being established. On the one hand, dysfunction of HCN channels in the thalamus precedes the onset of epilepsy in rat models, suggesting that a common genetically, or developmentally, determined defect in I_{HCN} may predispose to epilepsy. On the other hand, brief periods of epileptic activity or neuronal injury persistently alter hippocampal I_{HCN}, pointing to a high susceptibility of the mechanisms controlling HCN channel expression to unbalanced activity. Elucidating the genetic and activity-dependent mechanisms underlying these disturbances will prove crucial for therapeutic approaches for neuronal and cardiac diseases, including those aiming at the de novo creation of biological pacemakers as a substitute for electronic pacemakers.

In conclusion, by merging the voltage- and ligand-gating modules into a single ion channel, it is probably fair to speculate that nature has generated a masterpiece of flexibility that it may not have previously anticipated.

Acknowledgements

We thank Drs B Santoro and T Baram for discussion. AL's research is supported by the Swiss National Science Foundation (No. 3100A0–103655), the Jubiläumsstiftung der Schweiz. Mobiliarversicherungsgesellschaft and the Fonds zur Förderung von Lehre und Forschung.

References

Altomare C, Terragni B, Brioschi C et al. (2003). Heteromeric HCN1–HCN4 channels: a comparison with native pacemaker channels from the rabbit sinoatrial node. J Physiol 549, 347–359.
Altomare C, Tognati A, Bescond J et al. (2006). Direct inhibition of the pacemaker (I_f) current in rabbit sinoatrial node cells by genistein. Br J Pharmacol 147, 36–44.
Andrasfalvy BK and Magee JC (2001). Distance-dependent increase in AMPA receptor number in the dendrites of adult hippocampal CA1 pyramidal neurons. J Neurosci 21, 9151–9159.
Aponte Y, Lien C, Reisinger CE et al. (2006). Hyperpolarization-activated cation channels in fast-spiking interneurons of rat hippocampus. J Physiol 574, 229–243.
Arinsburg SS, Cohen IS and Yu HG (2006). Constitutively active Src tyrosine kinase changes gating of HCN4 channels through direct binding to the channel proteins. J Cardiovasc Pharmacol 47, 578–586.
Armoundas AA, Wu R, Juang G et al. (2001). Electrical and structural remodeling of the failing ventricle. Pharmacol Ther 92, 213–230.
Bader CR, Macleish PR and Schwartz EA (1979). A voltage-clamp study of the light response in solitary rods of the tiger salamander. J Physiol 296, 1–26.
Bal T and McCormick DA (1997). Synchronized oscillations in the inferior olive are controlled by the hyperpolarization-activated cation current I_h. J Neurophysiol 77, 3145–3156.
Baram TZ, Eghbal-Ahmadi M and Bender RA (2002). Is neuronal death required for seizure-induced epileptogenesis in the immature brain? Prog Brain Res 135, 365–375.
Barbuti A, Baruscotti M, Altomare C et al. (1999). Action of internal pronase on the f-channel kinetics in the rabbit SA node. J Physiol 520, 737–744.
Barbuti A, Gravante B, Riolfo M et al. (2004). Localization of pacemaker channels in lipid rafts regulates channel kinetics. Circ Res 94, 1325–1331.
Baruscotti M and DiFrancesco D (2004). Pacemaker channels. Ann N Y Acad Sci 1015, 111–121.
Baruscotti M, Bucchi A and DiFrancesco D (2005). Physiology and pharmacology of the cardiac pacemaker ("funny") current. Pharmacol Ther 107, 59–79.
Beaumont V and Zucker RS (2000). Enhancement of synaptic transmission by cyclic AMP modulation of presynaptic I_h channels. Nat Neurosci 3, 133–141.
Beaumont V, Zhong N, Froemke RC et al. (2002). Temporal synaptic tagging by I_h activation and actin: involvement in long-term facilitation and cAMP-induced synaptic enhancement. Neuron 33, 601–613.
Bell DC, Yao H, Saenger RC et al. (2004) Changes in local S4 environment provide a voltage-sensing mechanism for mammalian hyperpolarization-activated HCN channels. J Gen Physiol 123, 5–19.
Bender RA, Brewster A, Santoro B et al. (2001). Differential and age-dependent expression of hyperpolarization-activated, cyclic nucleotide-gated cation channel isoforms 1–4 suggests evolving roles in the developing rat hippocampus. Neurosci 106, 689–698.
Bender RA, Galindo R, Mameli M et al. (2005). Synchronized network activity in developing rat hippocampus involves regional hyperpolarization-activated cyclic nucleotide-gated (HCN) channel function. Eur J Neurosci 22, 2669–2674.
Bender RA, Kirschstein T, Kretz O et al. (2007). Localization of HCN1 channels to presynaptic compartments: novel plasticity that may contribute to hippocampal maturation. J Neurosci 27, 4697–4706.
Bender RA, Soleymani SV, Brewster AL et al. (2003). Enhanced expression of a specific hyperpolarization-activated cyclic nucleotide-gated cation channel (HCN) in surviving dentate gyrus granule cells of human and experimental epileptic hippocampus. J Neurosci 23, 6826–6836.
Berger T and Lüscher HR (2004). Associative somatodendritic interaction in layer V pyramidal neurons is not affected by the antiepileptic drug lamotrigine. Eur J Neurosci 20, 1688–1693.
Berger T, Larkum ME and Lüscher HR (2001). High I_h channel density in the distal apical dendrite of layer V pyramidal cells increases bidirectional attenuation of EPSPs. J Neurophysiol 85, 855–868.
Berger T, Senn W and Lüscher HR (2003). Hyperpolarization-activated current I_h disconnects somatic and dendritic spike initiation zones in layer V pyramidal neurons. J Neurophysiol 90, 2428–2437.

Bickmeyer U, Heine M, Manzke T *et al.* (2002). Differential modulation of I_h by 5-HT receptors in mouse CA1 hippocampal neurons. *Eur J Neurosci* **16**, 209–218.

Black JA, Cummins TR, Plumpton C *et al.* (1999). Upregulation of a silent sodium channel after peripheral, but not central, nerve injury in DRG neurons. *J Neurophysiol* **82**, 2776–2785.

Borlak J and Thum T (2003). Hallmarks of ion channel gene expression in end-stage heart failure. *FASEB J* **17**, 1592–1608.

Bosmith RE, Briggs I and Sturgess NC (1993). Inhibitory actions of ZENECA ZD7288 on whole-cell hyperpolarization activated inward current (I_f) in guinea-pig dissociated sinoatrial node cells. *Br J Pharmacol* **110**, 343–349.

Boyes J, Bolam JP, Shigemoto R *et al.* (2007). Functional presynaptic HCN channels in the rat globus pallidus. *Eur J Neurosci* **25**, 2081–2092.

Bräuer AU, Savaskan NE, Kole MHP *et al.* (2001). Molecular and functional analysis of hyperpolarization-activated pacemaker channels in the hippocampus after entorhinal cortex lesion. *FASEB J* **15**, 2689–2701.

Brewster A, Bender RA, Chen Y *et al.* (2002). Developmental febrile seizures modulate hippocampal gene expression of hyperpolarization-activated channels in an isoform- and cell-specific manner. *J Neurosci* **22**, 4591–4599.

Brewster A, Bernard JA, Gall CM *et al.* (2005). Formation of heteromeric hyperpolarization-activated cyclic nucleotide-gated (HCN) channels in the hippocampus is regulated by developmental seizures. *Neurobiol Disorders* **19**, 200–207.

Brewster AL, Chen Y, Bender RA *et al.* (2006). Quantitative analysis and subcellular distribution of mRNA and protein expression of the hyperpolarization-activated cyclic nucleotide-gated channels throughout development in rat hippocampus. *Cereb Cortex* **17**, 702–712.

Briggs I, Bosmith RE and Heapy CG (1994). Effects of Zeneca ZD7288 in comparison with alinidine and UL-FS 49 on guinea pig sinoatrial node and ventricular action potentials. *J Cardiovasc Pharmacol* **24**, 380–387.

Brown HF, DiFrancesco D and Noble SJ (1979). How does adrenaline accelerate the heart? *Nature* **280**, 235–236.

Bruening-Wright A and Larsson HP (2007). Slow conformational changes of the voltage sensor during the mode shift in hyperpolarization-activated cyclic-nucleotide-gated channels. *J Neurosci* **27**, 270–278.

Bucchi A, Baruscotti M and DiFrancesco D (2002). Current-dependent block of rabbit sino-atrial node I_f channels by ivabradine. *J Gen Physiol* **120**, 1–13.

Bucchi A, Tognati A, Milanesi R *et al.* (2006). Properties of ivabradine-induced block of HCN1 and HCN4 channels. *J Physiol* **572**, 335–346.

Budde T, Biella G, Munsch T *et al.* (1997). Lack of regulation by intracellular Ca^{2+} of the hyperpolarization-activated cation current in rat thalamic neurones. *J Physiol* **503**, 79–85.

Budde T, Caputi L, Kanyshkova T *et al.* (2005). Impaired regulation of thalamic pacemaker channels through an imbalance of subunit expression in absence epilepsy. *J Neurosci* **25**, 9871–9882.

Cacheaux LP, Topf N, Tibbs GR *et al.* (2005). Impairment of hyperpolarization-activated, cyclic nucleotide-gated channel function by the intravenous general anesthetic propofol. *J Pharmacol Exp Ther* **315**, 517–525.

Cerbai E and Mugelli A (2006). I_f in non-pacemaker cells: role and pharmacological implications. *Pharmacol Res* **53**, 416–423.

Cerbai E, Barbieri M and Mugelli A (1994). Characterization of the hyperpolarization-activated current, I_f, in ventricular myocytes isolated from hypertensive rats. *J Physiol* **481**, 585–591.

Cerbai E, Barbieri M and Mugelli A (1996). Occurrence and properties of the hyperpolarization-activated current I_f in ventricular myocytes from normotensive and hypertensive rats during aging. *Circ* **94**, 1674–1681.

Cerbai E, Crucitti A, Sartiani L *et al.* (2000). Long-term treatment of spontaneously hypertensive rats with losartan and electrophysiological remodeling of cardiac myocytes. *Cardiovasc Res* **45**, 388–396.

Cerbai E, Pino R, Rodriguez ML *et al.* (1999a). Modulation of the pacemaker current I_f by β-adrenoceptor subtypes in ventricular myocytes isolated from hypertensive and normotensive rats. *Cardiovasc Res* **42**, 121–129.

Cerbai E, Pino R, Sartiani L *et al.* (1999b). Influence of postnatal-development on I_f occurrence and properties in neonatal rat ventricular myocytes. *Cardiovasc Res* **42**, 416–423.

Cerbai E, Sartiani L, DE Paoli P *et al.* (2001). The properties of the pacemaker current I_F in human ventricular myocytes are modulated by cardiac disease. *J Mol Cell Cardiol* **33**, 441–448.

Chan CS, Shigemoto R, Mercer JN *et al.* (2004). HCN2 and HCN1 channels govern the regularity of autonomous pacemaking and synaptic resetting in globus pallidus neurons. *J Neurosci* **24**, 9921–9932.

Chaplan SR, Guo H-Q, Lee DH *et al.* (2003). Neuronal hyperpolarization-activated pacemaker channels drive neuropathic pain. *J Neurosci* **23**, 1169–1178.

Chen C (2004). ZD7288 inhibits postsynaptic glutamate receptor-mediated responses at hippocampal perforant path-granule cell synapses. *Eur J Neurosci* **19**, 643–649.

Chen J, Mitcheson JS, Lin M *et al.* (2000). Functional roles of charged residues in the putative voltage sensor of the HCN2 pacemaker channel. *J Biol Chem* **275**, 36465–36471.

Chen J, Mitcheson JS, Tristani-Firouzi M *et al.* (2001a). The S4–S5 linker couples voltage sensing and activation of pacemaker channels. *PNAS* **98**, 11277–11282.

Chen J, Piper DR and Sanguinetti MC (2002). Voltage sensing and activation gating of HCN pacemaker channels. *Trends Cardiovasc Med*, **12**, 42–45.

Chen K, Aradi I, Thon N *et al.* (2001b). Persistently modified h-channels after complex febrile seizures convert the seizure-induced enhancement of inhibition to hyperexcitability. *Nat Med* **7**, 331–337.

Chen S, Wang J and Siegelbaum SA (2001c). Properties of hyperpolarization-activated pacemaker current defined by coassembly of HCN1 and HCN2 subunits and basal modulation by cyclic nucleotide. *J Gen Physiol* **117**, 491–504.

Chen S, Wang J, Zhou L *et al.* (2007). Voltage sensor movement and cAMP binding allosterically regulate an inherently voltage-independent closed-open transition in HCN channels. *J Gen Physiol* **129**, 175–188.

Chen X, Shu S and Bayliss DA (2005a). Suppression of I_h contributes to propofol-induced inhibition of mouse cortical pyramidal neurons. *J Neurophysiol* **94**, 3872–3883.

Chen X, Sirois JE, Lei Q *et al.* (2005b). HCN subunit-specific and cAMP-modulated effects of anesthetics on neuronal pacemaker currents. *J Neurosci* **25**, 5803–5814.

Chevaleyre V and Castillo PE (2002). Assessing the role of I_h channels in synaptic transmission and mossy fiber LTP. *PNAS* **99**, 9538–9543.

Cho WJ, Drescher MJ, Hatfield JS *et al.* (2003). Hyperpolarization-activated, cyclic AMP-gated, HCN1-like cation channel: the primary, full-length HCN isoform expressed in a saccular hair-cell layer. *Neurosci* **118**, 525–534.

Cohen IS and Robinson RB (2006). Pacemaker current and automatic rhythms: toward a molecular understanding. *Handb Exp Pharmacol* 41–71.

Colino A and Halliwell JV (1993). Carbachol potentiates Q current and activates a calcium-dependent non-specific conductance in rat hippocampus *in vitro*. *Eur J Neurosci* **5**, 1198–1209.

Constanti A and Galvan M (1983). Fast inward-rectifying current accounts for anomalous rectification in olfactory cortex neurons. *J Physiol* **335**, 153–178.

Craven KB and Zagotta WN (2004). Salt bridges and gating in the COOH-terminal region of HCN2 and CNGA1 channels. *J Gen Physiol* **124**, 663–677.

Craven KB and Zagotta WN (2006). CNG and HCN channels: two peas, one pod. *Annu Rev Physiol* **68**, 375–401.

Cuttle MF, Rusznak Z, Wong AY *et al.* (2001). Modulation of a presynaptic hyperpolarization-activated cationic current (I_h) at an excitatory synaptic terminal in the rat auditory brainstem. *J Physiol* **534**, 733–744.

Dalle C and Eisenach JC (2005). Peripheral block of the hyperpolarization-activated cation current (I_h) reduces mechanical allodynia in animal models of postoperative and neuropathic pain. *Reg Anesth Pain Med* **30**, 243–248.

Day M, Carr DB, Ulrich S *et al.* (2005). Dendritic excitability of mouse frontal cortex pyramidal neurons is shaped by the interaction among HCN, Kir2, and K_{leak} channels. *J Neurosci* **25**, 8776–8787.

Decher N, Bundis F, Vajna R *et al.* (2003). KCNE2 modulates current amplitudes and activation kinetics of HCN4: influence of KCNE family members on HCN4 currents. *Pflügers Arch* **446**, 633–640.

Decher N, Chen J and Sanguinetti MC (2004). Voltage-dependent gating of hyperpolarization-activated, cyclic nucleotide-gated pacemaker channels: molecular coupling between the S4–S5 and C-linkers. *J Biol Chem* **279**, 13859–13865.

Dekin MS (1993). Inward rectification and its effects on the repetitive firing properties of bulbospinal neurons located in the ventral part of the nucleus tractus solitarius. *J Neurophysiol* **70**, 590–601.

Dekker JP and Yellen G (2006). Cooperative gating between single HCN pacemaker channels. *J Gen Physiol* **128**, 561–567.

Demontis GC, Moroni A, Gravante B *et al.* (2002). Functional characterization and subcellular localisation of HCN1 channels in rabbit retinal rod photoreceptors. *J Physiol* **542**, 89–97.

Di Paolo G and De Camilli P (2006). Phosphoinositides in cell regulation and membrane dynamics. *Nature* **443**, 651–657.

Di Pasquale E, Keegan KD and Noebels JL (1997). Increased excitability and inward rectification in layer V cortical pyramidal neurons in the epileptic mutant mouse *Stargazer*. *J Neurophysiol* **77**, 621–631.

Dib-Hajj SD, Fjell J, Cummins TR *et al.* (1999). Plasticity of sodium channel expression in DRG neurons in the chronic constriction model of neuropathic pain. *Pain* **83**, 591–600.

Dickson CT, Magistretti J, Shalinsky MH *et al.* (2000). Properties and role of I_h in the pacing of subthreshold oscillations in entorhinal cortex layer II neurons. *J Neurophysiol* **83**, 2562–2579.

DiFrancesco D (1981). A study of the ionic nature of the pace-maker current in calf Purkinje fibres. *J Physiol* **314**, 377–393.

DiFrancesco D (1982). Block and activation of the pace-maker channel in calf Purkinje fibres: effects of potassium, caesium and rubidium. *J Physiol* **329**, 485–507.

DiFrancesco D (1986). Characterization of single pacemaker channels in cardiac sino-atrial node cells. *Nature* **324**, 470–473.

DiFrancesco D (1999). Dual allosteric modulation of pacemaker (f) channels by cAMP and voltage in rabbit SA node. *J Physiol* **515**, 367–376.

DiFrancesco D (2005). Cardiac pacemaker I_f current and its inhibition by heart rate-reducing agents. *Curr Med Res Opin* **21**, 1115–1122.

DiFrancesco D and Mangoni M (1994). Modulation of single hyperpolarization-activated channels (i_f) by cAMP in the rabbit sino-atrial node. *J Physiol* **474**, 473–482.

DiFrancesco D and Tortora P (1991). Direct activation of cardiac pacemaker channels by intracellular cyclic AMP. *Nature* **351**, 145–147.

DiFrancesco D, Ferroni A, Mazzanti M *et al.* (1986). Properties of the hyperpolarizing-activated current (i_f) in cells isolated from the rabbit sino-atrial node. *J Physiol* **377**, 61–88.

Doan TN, Stephans K, Ramirez AN *et al.* (2004). Differential distribution and function of hyperpolarization-activated channels in sensory neurons and mechanosensitive fibers. *J Neurosci* **24**, 3335–3343.

Elinder F, Männikkö R, Pandey S *et al.* (2006) Mode shifts in the voltage gating of the mouse and human HCN2 and HCN4 channels. *J Physiol* **575**, 417–431.

Erdemli G and Crunelli V (1998). Response of thalamocortical neurons to hypoxia: a whole-cell patch-clamp study. *J Neurosci* **18**, 5212–24.

Fain GL, Quandt FN, Bastian BL *et al.* (1978). Contribution of a caesium-sensitive conductance increase to the rod photoresponse. *Nature* **272**, 466–469.

Fan Y, Fricker D, Brager DH *et al.* (2005). Activity-dependent decrease of excitability in rat hippocampal neurons through increases in I_h. *Nat Neurosci* **8**, 1542–1551.

Fernandez N, Andreasen M and Nedergaard S (2002). Influence of the hyperpolarization-activated cation current, I_h, on the electrotonic properties of the distal apical dendrites of hippocampal CA1 pyramidal neurones. *Brain Res* **930**, 42–52.

Fernandez-Velasco M, Goren N, Benito G *et al.* (2003). Regional distribution of hyperpolarization-activated current (I_f) and hyperpolarization-activated cyclic nucleotide-gated channel mRNA expression in ventricular cells from control and hypertrophied rat hearts. *J Physiol* **553**, 395–405.

Fernandez-Velasco M, Ruiz-Hurtado G and Delgado C (2006). I_{K1} and I_f in ventricular myocytes isolated from control and hypertrophied rat hearts. *Pflügers Arch* **452**, 146–154.

Fisahn A, Yamada M, Duttaroy A *et al.* (2002). Muscarinic induction of hippocampal gamma oscillations requires coupling of the M1 receptor to two mixed cation currents. *Neuron* **33**, 615–624.

Fletcher GH and Chiappinelli VA (1992). An inward rectifier is present in presynaptic nerve terminals in the chick ciliary ganglion. *Brain Res* **575**, 103–112.

Flynn GE, Black KD, Islas LD *et al.* (2007). Structure and rearrangements in the carboxy-terminal region of SpIH channels. *Structure* **15**, 671–682.

Fogle KJ, Lyashchenko AK, Turbendian HK *et al.* (2007). HCN pacemaker channel activation is controlled by acidic lipids downstream of diacylglycerol kinase and phospholipase A2. *J Neurosci* **27**, 2802–2814.

Franks NP and Lieb WR (1988). Volatile general anaesthetics activate a novel neuronal K^+ current. *Nature* **333**, 662–4.

Franz O, Liss B, Neu A *et al.* (2000). Single-cell mRNA expression of HCN1 correlates with a fast gating phenotype of hyperpolarization-activated cyclic nucleotide-gated ion channels (I_h) in central neurons. *Eur J Neurosci* **12**, 2685–2693.

Frère SGA and Lüthi A (2004). Pacemaker channels in mouse thalamocortical neurons are regulated by distinct pathways of cAMP synthesis. *J Physiol* **554**, 111–125.

Frère SGA, Kuisle M and Lüthi A (2004). Regulation of recombinant and native hyperpolarization-activated cation channels. *Mol Neurobiol* **30**, 279–306.

Galindo BE, Neill AT and Vacquier VD (2005). A new hyperpolarization-activated, cyclic nucleotide-gated channel from sea urchin sperm flagella. *Biochem Biophys Res Commun* **334**, 96–101.

Gasparini S and DiFrancesco D (1997). Action of the hyperpolarization-activated current (I_h) blocker ZD 7288 in hippocampal CA1 neurons. *Pflügers Arch* **435**, 99–106.

Gasparini S and DiFrancesco D (1999). Action of serotonin on the hyperpolarization-activated cation current (I_h) in rat CA1 hippocampal neurons. *Eur J Neurosci* **11**, 3093–3100.

Gauss R, Seifert R and Kaupp UB (1998). Molecular identification of a hyperpolarization-activated channel in sea urchin sperm. *Nature* **393**, 583–587.

Gill CH, Randall A, Bates SA *et al.* (2004). Characterization of the human HCN1 channel and its inhibition by capsazepine. *Br J Pharmacol* **143**, 411–421.

Gisselmann G, Marx T, Bobkov Y *et al.* (2005). Molecular and functional characterization of an I_h-channel from lobster olfactory receptor neurons. *Eur J Neurosci* **21**, 1635–1647.

Gisselmann G, Wetzel CH, Warnstedt M *et al.* (2004). Functional characterization of I_h-channel splice variants from *Apis mellifera*. *FEBS Lett* **575**, 99–104.

Gloss B, Trost S, Bluhm W *et al.* (2001). Cardiac ion channel expression and contractile function in mice with deletion of thyroid hormone receptor α or β. *Endocrinology* **142**, 544–550.

Graf EM, Heubach JF and Ravens U (2001). The hyperpolarization-activated current I$_f$ in ventricular myocytes of non-transgenic and β2-adrenoceptor overexpressing mice. *Naunyn Schmiedebergs Arch Pharmacol* **364**, 131–139.

Gravante B, Barbuti A, Milanesi R*et al.* (2004). Interaction of the pacemaker channel HCN1 with filamin A. *J Biol Chem* **279**, 43847–43853.

Harris NC and Constanti A (1995). Mechanism of block by ZD 7288 of the hyperpolarization-activated inward rectifying current in guinea pig substantia nigra neurons *in vitro*. *J Neurophysiol* **74**, 2366–2378.

Herrmann S, Stieber J and Ludwig A (2007). Pathophysiology of HCN channels. *Pflügers Arch* **454**, 517–522.

Hiramatsu M, Furukawa T, Sawanobori T *et al.* (2002). Ion channel remodeling in cardiac hypertrophy is prevented by blood pressure reduction without affecting heart weight increase in rats with abdominal aortic banding. *J Cardiovasc Pharm* **39**, 866–874.

Holderith NB, Shigemoto R and Nusser Z (2003). Cell type-dependent expression of HCN1 in the main olfactory bulb. *Eur J Neurosci* **18**, 344–354.

Hoppe UC, Jansen E, Südkamp M *et al.* (1998). Hyperpolarization-activated inward current in ventricular myocytes from normal and failing human hearts. *Circ* **97**, 55–65.

Horn R (2004), How S4 segments move charge. Let me count the ways. *J Gen Physiol* **123**, 1–4.

Hu H, Vervaeke K and Storm JF (2002). Two forms of electrical resonance at theta frequencies, generated by M-current, h-current and persistent Na$^+$ current in rat hippocampal pyramidal cells. *J Physiol* **545**, 783–805.

Huang X, Yang P, Du Y *et al.* (2007). Age-related down-regulation of HCN channels in rat sinoatrial node. *Basic Res Cardiol* **102**, 429–435.

Ingram SL and Williams JT (1996). Modulation of the hyperpolarization-activated current (I$_h$) by cyclic nucleotides in guinea-pig primary afferent neurons. *J Physiol* **492**, 97–106.

Ishii TM, Nakashima N and Ohmori H (2007). Tryptophan-scanning mutagenesis in the S1 domain of mammalian HCN channel reveals residues critical for voltage-gated activation. *J Physiol* **579**, 291–301.

Ishii TM, Takano M and Ohmori H (2001). Determinants of activation kinetics in mammalian hyperpolarization-activated cation channels. *J Physiol* **537**, 93–100.

Ishii TM, Takano M, Xie LH *et al.* (1999). Molecular characterization of the hyperpolarization-activated cation channel in rabbit heart sinoatrial node. *J Biol Chem* **274**, 12835–12839.

Jackson HA, Marshall CR and Accili EA (2007). Evolution and structural diversification of hyperpolarization-activated cyclic nucleotide-gated channel genes. *Physiol Genomics* **29**, 231–245.

Janigro D, Gasparini S, D'Ambrosio R *et al.* (1997). Reduction of K$^+$ uptake in glia prevents long-term depression maintenance and causes epileptiform activity. *J Neurosci* **17**, 2813–2824.

Johnson JP Jr and Zagotta WN (2005). The carboxyl-terminal region of cyclic nucleotide-modulated channels is a gating ring, not a permeation path. *PNAS* **102**, 2742–2747.

Jurd R, Arras M, Lambert S *et al.* (2003). General anesthetic actions *in vivo* strongly attenuated by a point mutation in the GABA$_A$ receptor β3 subunit. *FASEB J* **17**, 250–252.

Kaupp UB and Seifert R (2001). Molecular diversity of pacemaker ion channels. *Annu Rev Physiol* **63**, 235–257.

Kim CH, Oh Y, Chung JM *et al.* (2001). The changes in expression of three subtypes of TTX sensitive sodium channels in sensory neurons after spinal nerve ligation. *Mol Brain Res* **95**, 153–161.

Kimura K, Kitano J, Nakajima Y *et al.* (2004). Hyperpolarization-activated, cyclic nucleotide-gated HCN2 cation channel forms a protein assembly with multiple neuronal scaffold proteins in distinct modes of protein–protein interaction. *Genes to Cells* **9**, 631–640.

Knaus A, Zong X, Beetz N *et al.* (2007). Direct inhibition of cardiac hyperpolarization-activated cyclic nucleotide-gated pacemaker channels by clonidine. *Circ* **115**, 872–880.

Knöpfel T, Guatteo E, Bernardi G *et al.* (1998). Hyperpolarization induces a rise in intracellular sodium concentration in dopamine cells of the substantia nigra pars compacta. *Eur J Neurosci* **10**, 1926–1929.

Kole MH, Braüer AU and Stuart GJ (2007). Inherited cortical HCN1 channel loss amplifies dendritic calcium electrogenesis and burst firing in a rat absence epilepsy model. *J Physiol* **578**, 507–525.

Kole MH, Hallermann S and Stuart GJ (2006). Single I$_h$ channels in pyramidal neuron dendrites: properties, distribution, and impact on action potential output. *J Neurosci* **26**, 1677–1687.

Krieger J, Strobel J, Vogl A *et al.* (1999). Identification of a cyclic nucleotide- and voltage-activated ion channel from insect antennae. *Insect Biochem Mol Biol* **29**, 255–267.

Kuisle M, Wanaverbecq N, Brewster AL *et al.* (2006). Functional stabilization of weakened thalamic pacemaker channel regulation in absence epilepsy. *J Physiol* **575**, 83–100.

Lee DH, Chang L, Sorkin LS *et al.* (2005). Hyperpolarization-activated, cation-nonselective, cyclic nucleotide-modulated channel blockade alleviates mechanical allodynia and suppresses ectopic discharge in spinal nerve ligated rats. *J Pain* **6**, 417–424.

Linden DR, Sharkey KA and Mawe GM (2003). Enhanced excitability of myenteric AH neurones in the inflamed guinea-pig distal colon. *J Physiol* **547**, 589–601.

Liu Y, Holmgren M, Jurman ME *et al.* (1997). Gated access to the pore of a voltage-dependent K$^+$ channel. *Neuron* **19**, 175–184.

Long SB, Campbell EB and MacKinnon R (2005). Voltage sensor of Kv1.2: structural basis of electromechanical coupling. *Science* **309**, 903–908.

Lörincz A, Notomi T, Tamas G *et al.* (2002). Polarized and compartment-dependent distribution of HCN1 in pyramidal cell dendrites. *Nat Neurosci* **5**, 1185–1193.

Ludwig A, Budde T, Stieber J *et al.* (2003). Absence epilepsy and sinus dysrhythmia in mice lacking the pacemaker channel HCN2. *EMBO J* **22**, 216–224.

Ludwig A, Zong H, Stieber J *et al.* (1999). Two pacemaker channels from human heart with profoundly different activation kinetics. *EMBO J* **18**, 2323–2329.

Ludwig A, Zong X, Jeglitsch M *et al.* (1998). A family of hyperpolarization-activated mammalian cation channels. *Nature* **393**, 587–591.

Luján R, Albasanz JL, Shigemoto R *et al.* (2005). Preferential localization of the hyperpolarization-activated cyclic nucleotide-gated cation channel subunit HCN1 in basket cell terminals of the rat cerebellum. *Eur J Neurosci* **21**, 2073–2082.

Lupica CR, Bell JA, Hoffman AF *et al.* (2001). Contribution of the hyperpolarization-activated current (I$_h$) to membrane potential and GABA release in hippocampal interneurons. *J Neurophysiol* **86**, 261–268.

Lüthi A and McCormick DA (1998a). H-current: properties of a neuronal and network pacemaker. *Neuron* **21**, 9–12.

Lüthi A and McCormick DA (1998b). Periodicity of thalamic synchronized oscillations: the role of Ca^{2+}-mediated upregulation of I$_h$. *Neuron* **20**, 553–563.

Lüthi A and McCormick DA (1999). Modulation of a pacemaker current through Ca^{2+}-induced stimulation of cAMP production. *Nat Neurosci* **2**, 634–641.

Lüthi A, Bal T and McCormick DA (1998). Periodicity of thalamic spindle waves is abolished by ZD7288, a blocker of I$_h$. *J Neurophysiol* **79**, 3284–3289.

Lyashchenko AK, Redd KJ, Yang J *et al.* (2007). Propofol inhibits HCN1 pacemaker channels by selective association with the closed states of the membrane embedded channel core. *J Physiol* **583**, 37–56.

Macri V, Proenza C, Agranovich E et al. (2002) Separable gating mechanisms in a mammalian pacemaker channel. J Biol Chem 277, 35939–35946.

Macri VS and Accili EA (2004). Structural elements of instantaneous and slow gating in hyperpolarization-activated cyclic nucleotide-gated channels. J Biol Chem 279, 16832–16846.

Magee JC (1998). Dendritic hyperpolarization-activated currents modify the integrative properties of hippocampal CA1 pyramidal neurons. J Neurosci 18, 7613–7624.

Magee JC (1999). Dendritic I_h normalizes temporal summation in hippocampal CA1 neurons. Nat Neurosci, 2, 508–514.

Magee JC (2001). Dendritic mechanisms of phase precession in hippocampal CA1 pyramidal neurons. J Neurophysiol 86, 528–532.

Männikkö R, Elinder F and Larsson HP (2002). Voltage-sensing mechanism is conserved among ion channels gated by opposite voltages. Nature 419, 837–841.

Männikkö R, Pandey S, Larsson HP et al. (2005). Hysteresis in the voltage dependence of HCN channels: conversion between two modes affects pacemaker properties. J Gen Physiol 125, 305–326.

Marionneau C, Couette B, Liu J et al. (2005). Specific pattern of ionic channel gene expression associated with pacemaker activity in the mouse heart. J Physiol 562, 223–234.

Marx T, Gisselmann G, Stortkuhl KF et al. (1999). Molecular cloning of a putative voltage- and cyclic nucleotide-gated ion channel present in the antennae and eyes of Drosophila melanogaster. Invert Neurosci 4, 55–63.

Mayer ML and Westbrook GL (1983). A voltage-clamp analysis of inward (anomalous) rectification in mouse spinal sensory ganglion neurones. J Physiol 340, 19–45.

McCrossan ZA and Abbott GW (2004). The MinK-related peptides. Neuropharmacol 47, 787–821.

Meeren H, Van Luijtelaar G, Lopes Da Silva F et al. (2005). Evolving concepts on the pathophysiology of absence seizures: the cortical focus theory. Arch Neurol 62, 371–376.

Mellor J, Nicoll RA and Schmitz D (2002). Mediation of hippocampal mossy fiber long-term potentiation by presynaptic I_h channels. Science 295, 143–147.

Meuth SG, Kanyshkova T, Meuth P et al. (2006). Membrane resting potential of thalamocortical relay neurons is shaped by the interaction among TASK3 and HCN2 channels. J Neurophysiol 96, 1517–1529.

Michels G, Er F, Khan I et al. (2005). Single-channel properties support a potential contribution of hyperpolarization-activated cyclic nucleotide-gated channels and I_f to cardiac arrhythmias. Circ 111, 399–404.

Milanesi R, Baruscotti M, Gnecchi-Ruscone T et al. (2006). Familial sinus bradycardia associated with a mutation in the cardiac pacemaker channel. N Engl J Med 354, 151–157.

Milligan CJ, Edwards IJ and Deuchars J (2006). HCN1 ion channel immunoreactivity in spinal cord and medulla oblongata. Brain Res 1081, 79–91.

Mistrík P, Mader R, Michalakis S et al. (2005). The murine HCN3 gene encodes a hyperpolarization-activated cation channel with slow kinetics and unique response to cyclic nucleotides. J Biol Chem 280, 27056–27061.

Mistrík P, Pfeifer A and Biel M (2006). The enhancement of HCN channel instantaneous current facilitated by slow deactivation is regulated by intracellular chloride concentration. Pflügers Arch 452, 718–727.

Mitchell SJ, McHale DP, Campbell DA et al. (1998). A syndrome of severe mental retardation, spasticity, and tapetoretinal degeneration linked to chromosome 15q24. Am J Human Genet 62, 1070–1076.

Monteggia LM, Eisch AJ, Tang MD et al. (2000). Cloning and localization of the hyperpolarization-activated cyclic nucleotide-gated channel family in rat brain. Brain Res Mol Brain Res 81, 129–139.

Moosmang S, Biel M, Hofmann F et al. (1999). Differential distribution of four hyperpolarization-activated cation channels in mouse brain. Biol Chem 380, 975–980.

Moosmang S, Stieber J, Zong X et al. (2001). Cellular expression and functional characterization of four hyperpolarization-activated pacemaker channels in cardiac and neuronal tissues. Eur J Biochem 268, 1646–1652.

Much B, Wahl-Schott C, Zong X et al. (2003). Role of subunit heteromerization and N-linked glycosylation in the formation of functional hyperpolarization-activated cyclic nucleotide-gated channels. J Biol Chem 278, 43781–43786.

Müller F, Scholten A, Ivanova E et al. (2003). HCN channels are expressed differentially in retinal bipolar cells and concentrated at synaptic terminals. Eur J Neurosci 17, 2084–2096.

Munsch T and Pape HC (1999a). Modulation of the hyperpolarization-activated cation current of rat thalamic relay neurones by intracellular pH. J Physiol 519, 493–504.

Munsch T and Pape HC (1999b). Upregulation of the hyperpolarization-activated cation current in rat thalamic relay neurones by acetazolamide. J Physiol 519, 505–514.

Nolan MF, Malleret G, Dudman JT et al. (2004). A behavioral role for dendritic integration: HCN1 channels constrain spatial memory and plasticity at inputs to distal dendrites of CA1 pyramidal neurons. Cell 119, 719–732.

Nolan MF, Malleret G, Lee KH et al. (2003). The hyperpolarization-activated HCN1 channel is important for motor learning and neuronal integration by cerebellar Purkinje cells. Cell 115, 551–564.

Notomi T and Shigemoto R (2004). Immunohistochemical localization of I_h channel subunits, HCN1–4, in the rat brain. J Comp Neurol 471, 241–276.

Pachucki J, Burmeister LA and Larsen PR (1999). Thyroid hormone regulates hyperpolarization-activated cyclic nucleotide-gated channel (HCN2) mRNA in the rat heart. Circ Res 85, 498–503.

Pan ZZ (2003). κ-opioid receptor-mediated enhancement of the hyperpolarization-activated current (I_h) through mobilization of intracellular calcium in rat nucleus raphe magnus. J Physiol 548, 765–775.

Pape HC (1994). Specific bradycardiac agents block the hyperpolarization-activated cation current in central neurons. Neurosci 59, 363–373.

Pape HC (1996). Queer current and pacemaker: the hyperpolarization-activated cation current in neurons. Annu Rev Physiol 58, 299–327.

Parkis MA and Berger AJ (1997). Clonidine reduces hyperpolarization-activated inward current (I_h) in rat hypoglossal motoneurons. Brain Res 769, 108–118.

Pedarzani P and Storm JF (1995). Protein kinase A-independent modulation of ion channels in the brain by cyclic AMP. PNAS 92, 11716–11720.

Perkins KL and Wong RK (1995). Intracellular QX-314 blocks the hyperpolarization-activated inward current I_q in hippocampal CA1 pyramidal cells. J Neurophysiol 73, 911–915.

Phillips HA, Scheffer IE, Crossland KM et al. (1998). Autosomal dominant nocturnal frontal-lobe epilepsy: genetic heterogeneity and evidence for a second locus at 15q24. Am J Human Genet 63, 1108–1116.

Pian P, Bucchi A, Robinson RB et al. (2006). Regulation of gating and rundown of HCN hyperpolarization-activated channels by exogenous and endogenous PIP_2. J Gen Physiol 128, 593–604.

Plotnikov AN, Sosunov EA, Qu J et al. (2004). Biological pacemaker implanted in canine left bundle branch provides ventricular escape rhythms that have physiologically acceptable rates. Circ 109, 506–512.

Polack PO, Guillemain I, Hu E et al. (2007). Deep layer somatosensory cortical neurons initiate spike-and-wave discharges in a genetic model of absence seizures. J Neurosci 27, 6590–6599.

Poolos NP (2004). The yin and yang of the h-Channel and its role in epilepsy. Epilepsy Curr 4, 3–6.

Poolos NP, Bullis JB and Roth MK (2006). Modulation of h-channels in hippocampal pyramidal neurons by p38 mitogen-activated protein kinase. *J Neurosci* **26**, 7995–8003.

Poolos NP, Migliore M and Johnston D (2002). Pharmacological upregulation of h-channels reduces the excitability of pyramidal neuron dendrites. *Nat Neurosci* **5**, 767–774.

Proenza C and Yellen G (2006). Distinct populations of HCN pacemaker channels produce voltage-dependent and voltage-independent currents. *J Gen Physiol* **127**, 183–190.

Proenza C, Angoli D, Agranovich E et al. (2002a). Pacemaker channels produce an instantaneous current. *J Biol Chem* **277**, 5101–5109.

Proenza C, Tran N, Angoli D et al. (2002b). Different roles for the cyclic nucleotide binding domain and amino terminus in assembly and expression of hyperpolarization-activated, cyclic nucleotide-gated channels. *J Biol Chem* **277**, 29634–29642.

Prole DL and Yellen G (2006). Reversal of HCN channel voltage dependence via bridging of the S4–S5 linker and post-S6. *J Gen Physiol* **128**, 273–282.

Qiu DL, Chu CP, Shirasaka T et al. (2003). Neuromedin U depolarizes rat hypothalamic paraventricular nucleus neurons *in vitro* by enhancing I_H channel activity. *J Neurophysiol* **90**, 843–850.

Qu J, Kryukova Y, Potapova IA et al. (2004). MiRP1 modulates HCN2 channel expression and gating in cardiac myocytes. *J Biol Chem* **279**, 43497–43502.

Qu J, Plotnikov AN, Danilo P Jr et al. (2003). Expression and function of a biological pacemaker in canine heart. *Circ* **107**, 1106–1109.

Rateau Y and Ropert N (2006). Expression of a functional hyperpolarization-activated current (I_h) in the mouse nucleus reticularis thalami. *J Neurophysiol* **95**, 3073–3085.

Ray AM, Benham CD, Roberts JC et al. (2003). Capsazepine protects against neuronal injury caused by oxygen glucose deprivation by inhibiting I_h. *J Neurosci* **23**, 10146–10153.

Robinson RB and Siegelbaum SA (2003). Hyperpolarization-activated cation currents: from molecules to physiological function. *Annu Rev Physiol* **65**, 453–480.

Robinson RB, Brink PR, Cohen IS et al. (2006). I_f and the biological pacemaker. *Pharmacol Res* **53**, 407–415.

Robinson RB, Yu H, Chang F et al. (1997). Developmental change in the voltage-dependence of the pacemaker current, i_f, in rat ventricle cells. *Pflügers Arch* **433**, 533–535.

Rosenbaum T and Gordon SE (2004). Quickening the pace: looking into the heart of HCN channels. *Neuron* **42**, 193–196.

Rosenkranz JA and Johnston D (2006). Dopaminergic regulation of neuronal excitability through modulation of I_h in layer V entorhinal cortex. *J Neurosci* **26**, 3229–3244.

Rothberg BS, Shin KS and Yellen G (2003). Movements near the gate of a hyperpolarization-activated cation channel. *J Gen Physiol* **122**, 501–510.

Rothberg BS, Shin KS, Phale PS et al. (2002). Voltage-controlled gating at the intracellular entrance to a hyperpolarization-activated cation channel. *J Gen Physiol* **119**, 83–91.

Santoro B and Baram TZ (2003). The multiple personalities of h-channels. *TINS* **26**, 550–554.

Santoro B and Tibbs GR (1999). The HCN gene family: molecular basis of the hyperpolarization-activated pacemaker channels. *Ann NY Acad Sci* **868**, 741–764.

Santoro B, Chan S, Lüthi A et al. (2000). Molecular and functional heterogeneity of hyperpolarization-activated pacemaker channels in the mouse CNS. *J Neurosci* **20**, 5264–5275.

Santoro B, Grant SG, Bartsch D et al. (1997). Interactive cloning with the SH3 domain of *N*-src identifies a new brain specific ion channel protein, with homology to eag and cyclic nucleotide-gated channels. *PNAS* **94**, 14815–14820.

Santoro B, Liu DT, Yao H et al. (1998). Identification of a gene encoding a hyperpolarization-activated pacemaker channel of brain. *Cell* **93**, 717–729.

Santoro B, Wainger BJ and Siegelbaum SA (2004). Regulation of HCN channel surface expression by a novel C-terminal protein–protein interaction. *J Neurosci* **24**, 10750–10762.

Sartiani L, De Paoli P, Stillitano F et al. (2006). Functional remodeling in post-myocardial infarcted rats: focus on β-adrenoceptor subtypes. *J Mol Cell Cardiol* **40**, 258–266.

Satoh TO and Yamada M (2000). A bradycardiac agent ZD7288 blocks the hyperpolarization-activated current (I_h) in retinal rod photoreceptors. *Neuropharmacol* **39**, 1284–1291.

Schridde U, Strauss U, Braüer AU et al. (2006). Environmental manipulations early in development alter seizure activity, I_h and HCN1 protein expression later in life. *Eur J Neurosci* **23**, 3346–3358.

Schulze-Bahr C, Neu A, Friederich P et al. (2003). Pacemaker channel dysfunction in a patient with sinus node disease. *J Clin Invest* **111**, 1537–1545.

Schweitzer P, Madamba SG and Siggins GR (2003). The sleep-modulating peptide cortistatin augments the h-current in hippocampal neurons. *J Neurosci* **23**, 10884–10891.

Seifert R, Scholten A, Gauss R et al. (1999). Molecular characterization of a slowly gating human hyperpolarization-activated channel predominantly expressed in thalamus, heart, and testis. *PNAS* **96**, 9391–9396.

Shah MM, Anderson AE, Leung V et al. (2004). Seizure-induced plasticity of h channels in entorhinal cortical layer III pyramidal neurons. *Neuron* **44**, 495–508.

Shi W, Wymore R, Yu H et al. (1999). Distribution and prevalence of hyperpolarization-activated cation channel (HCN) mRNA expression in cardiac tissues. *Circ Res* **85**, e1–e6.

Shibata S, Ono K and Iijima T (1999). Inhibition by genistein of the hyperpolarization-activated cation current in porcine sino-atrial node cells. *Br J Pharmacol* **128**, 1284–1290.

Shin KS, Maertens C, Proenza C et al. (2004). Inactivation in HCN channels results from reclosure of the activation gate: desensitization to voltage. *Neuron* **41**, 737–744.

Shin KS, Rothberg BS and Yellen G (2001). Blocker state dependence and trapping in hyperpolarization-activated cation channels: evidence for an intracellular activation gate. *J Gen Physiol* **117**, 91–101.

Shir Y and Seltzer Z (1990). A-fibers mediate mechanical hyperesthesia and allodynia and C-fibers mediate thermal hyperalgesia in a new model of causalgiform pain disorders in rats. *Neurosci Lett* **115**, 62–67.

Simeone TA, Rho JM and Baram TZ (2005). Single channel properties of hyperpolarization-activated cation currents in acutely dissociated rat hippocampal neurones. *J Physiol* **568**, 371–380.

Sirois JE, Lynch C 3rd and Bayliss DA (2002). Convergent and reciprocal modulation of a leak K$^+$ current and I_h by an inhalational anaesthetic and neurotransmitters in rat brainstem motoneurones. *J Physiol* **541**, 717–729.

Sirois JE, Pancrazio JJ III CL and Bayliss DA (1998). Multiple ionic mechanisms mediate inhibition of rat motoneurones by inhalation anaesthetics. *J Physiol* **512**, 851–862.

Snyders DJ and Van Bogaert PP (1987). Alinidine modifies the pacemaker current in sheep Purkinje fibers. *Pflügers Arch* **410**, 83–91.

Southan AP, Morris NP, Stephens GJ et al. (2000). Hyperpolarization-activated currents in presynaptic terminals of mouse cerebellar basket cells. *J Physiol* **526**, 91–97.

Stevens DR, Seifert R, Bufe B et al. (2001). Hyperpolarization-activated channels HCN1 and HCN4 mediate responses to sour stimuli. *Nature* **413**, 631–635.

Stieber J, Herrmann S, Feil S et al. (2003a). The hyperpolarization-activated channel HCN4 is required for the generation of pacemaker action potentials in the embryonic heart. *PNAS* **100**, 15235–15240.

Stieber J, Hofmann F and Ludwig A (2004). Pacemaker channels and sinus node arrhythmia. *Trends Cardiovasc Med* **14**, 23–28.

Stieber J, Stöckl G, Herrmann S et al. (2005). Functional expression of the human HCN3 channel. *J Biol Chem* **280**, 34635–34643.

Stieber J, Thomer A, Much B et al. (2003b). Molecular basis for the different activation kinetics of the pacemaker channels HCN2 and HCN4. *J Biol Chem* **278**, 33672–33680.

Stieber J, Wieland K, Stockl G et al. (2006). Bradycardic and proarrhythmic properties of sinus node inhibitors. *Mol Pharmacol* **69**, 1328–1337.

Strauss U, Kole MH, Brauer AU et al. (2001). An impaired neocortical I_h is associated with enhanced excitability and absence epilepsy. *Eur J Neurosci* **19**, 3048–3058.

Stuart G and Williams S (2000). Co-localization of I_H and I_T channels in dendrites of thalamocortical neurons. *Soc Neurosci Abstr* **30**, 610.9.

Suh BC and Hille B (2005). Regulation of ion channels by phosphatidylinositol 4,5-bisphosphate. *Curr Opin Neurobiol* **15**, 370–378.

Surges R, Brewster AL, Bender RA et al. (2006). Regulated expression of HCN channels and cAMP levels shape the properties of the h current in developing rat hippocampus. *Eur J Neurosci* **24**, 94–104.

Surges R, Freiman TM and Feuerstein TJ (2003). Gabapentin increases the hyperpolarization-activated cation current I_h in rat CA1 pyramidal cells. *Epilepsia* **44**, 150–156.

Thoby-Brisson M, Cauli B, Champagnat J et al. (2003). Expression of functional tyrosine kinase B receptors by rhythmically active respiratory neurons in the pre-Bötzinger complex of neonatal mice. *J Neurosci* **23**, 7685–7689.

Toth Z, Yan XX, Haftoglou S et al. (1998). Seizure-induced neuronal injury: vulnerability to febrile seizures in an immature rat model. *J Neurosci* **18**, 4285–4294.

Tran N, Proenza C, Macri V et al. (2002). A conserved domain in the NH2 terminus important for assembly and functional expression of pacemaker channels. *J Biol Chem* **277**, 43588–43592.

Tsang SY, Lesso H and Li RA (2004). Dissecting the structural and functional roles of the S3–S4 linker of pacemaker (hyperpolarization-activated cyclic nucleotide-modulated) channels by systematic length alterations. *J Biol Chem* **279**, 43752–43759.

Tu H, Deng L, Sun Q et al. (2004). Hyperpolarization-activated, cyclic nucleotide-gated cation channels: roles in the differential electrophysiological properties of rat primary afferent neurons. *J Neurosci Res* **76**, 713–722.

Ueda K, Nakamura K, Hayashi T et al. (2004). Functional characterization of a trafficking-defective HCN4 mutation, D553N, associated with cardiac arrhythmia. *J Biol Chem* **279**, 27194–27198.

Ulens C and Siegelbaum SA (2003). Regulation of hyperpolarization-activated HCN channels by cAMP through a gating switch in binding domain symmetry. *Neuron* **40**, 959–970.

Ulens C and Tytgat J (2001). Functional heteromerization of HCN1 and HCN2 pacemaker channels. *J Biol Chem* **276**, 6069–6072.

Vaca L, Stieber J, Zong X et al. (2000). Mutations in the S4 domain of a pacemaker channel alter its voltage dependence. *FEBS Lett* **479**, 35–40.

Vaccari T, Moroni A, Rocchi M et al. (1999). The human gene coding for HCN2, a pacemaker channel of the heart. *Biochim Biophys Acta* **1446**, 419–425.

Van Bogaert PP and Pittoors F (2003). Use-dependent blockade of cardiac pacemaker current (I_f) by cilobradine and zatebradine. *Eur J Pharmacol* **478**, 161–171.

Van Welie I, Van Hooft JA and Wadman WJ (2004). Homeostatic scaling of neuronal excitability by synaptic modulation of somatic hyperpolarization-activated I_h channels. *PNAS* **101**, 5123–5128.

Van Welie I, Wadman WJ and Van Hooft JA (2005). Low affinity block of native and cloned hyperpolarization-activated I_h channels by Ba^{2+} ions. *Eur J Pharmacol* **507**, 15–20.

Vasilyev DV and Barish ME (2002). Postnatal development of the hyperpolarization-activated excitatory current I_h in mouse hippocampal pyramidal neurons. *J Neurosci* **22**, 8992–9004.

Vasilyev DV and Barish ME (2004). Regulation of the hyperpolarization-activated cationic current I_h in mouse hippocampal pyramidal neurones by vitronectin, a component of extracellular matrix. *J Physiol* **560**, 659–675.

Vasilyev DV, Shan Q, Lee Y et al. (2007). Direct inhibition of I_h by analgesic loperamide in rat DRG neurons. *J Neurophysiol* **97**, 3713–3721.

Vemana S, Pandey S and Larsson HP (2004). S4 movement in a mammalian HCN channel. *J Gen Physiol* **123**, 21–32.

Verkerk AO, Wilders R, Coronel R et al. (2003). Ionic remodeling of sinoatrial node cells by heart failure. *Circ* **108**, 760–766.

Viana F, De La Pena E and Belmonte C (2002). Specificity of cold thermotransduction is determined by differential ionic channel expression. *Nat Neurosci* **5**, 254–260.

Wahl-Schott C, Baumann L, Zong X et al. (2005). An arginine residue in the pore region is a key determinant of chloride dependence in cardiac pacemaker channels. *J Biol Chem* **280**, 13694–13700.

Wainger BJ, Degennaro M, Santoro B et al. (2001). Molecular mechanism of cAMP modulation of HCN pacemaker channels. *Nature* **411**, 805–810.

Walker MC and Kullmann DM (1999). Febrile convulsions: a 'benign' condition? *Nature Med* **5**, 871–872.

Wang J, Chen S and Siegelbaum SA (2001). Regulation of the hyperpolarization-activated HCN channel gating and cAMP modulation due to interactions of COOH terminus and core transmembrane regions. *J Gen Physiol* **118**, 237–250.

Wang J, Chen S, Nolan MF et al. (2002). Activity-dependent regulation of HCN pacemaker channels by cyclic AMP: signaling through dynamic allosteric coupling. *Neuron* **36**, 451–461.

Wang M, Ramos BP, Paspalas CD et al. (2007). α2A-adrenoceptors strengthen working memory networks by inhibiting cAMP-HCN channel signaling in prefrontal cortex. *Cell* **129**, 397–410.

Wang Z, Xu NL, Wu CP et al. (2003). Bidirectional changes in spatial dendritic integration accompanying long-term synaptic modifications. *Neuron* **37**, 463–472.

Waxman SG (2001). Transcriptional channelopathies: an emerging class of disorders. *Nat Rev Neurosci* **2**, 652–659.

Whitaker GM, Angoli D, Nazzari H et al. (2007). HCN2 and HCN4 isoforms self-assemble and co-assemble with equal preference to form functional pacemaker channels. *J Biol Chem* **282**, 22900–22909.

White HS (2002). Animal models of epileptogenesis. *Neurology* **59**, S7–S14.

Williams SR and Stuart GJ (2000). Site independence of EPSP time course is mediated by dendritic I_h in neocortical pyramidal neurons. *J Neurophysiol* **83**, 3177–3182.

Williams SR, Christensen SR, Stuart GJ et al. (2002). Membrane potential bistability is controlled by the hyperpolarization-activated current I_H in rat cerebellar Purkinje neurons *in vitro*. *J Physiol* **539**, 469–483.

Williams SR, Turner JP, Hughes SW et al. (1997). On the nature of anomalous rectification in thalamocortical neurones of the cat ventrobasal thalamus *in vitro*. *J Physiol* **505**, 727–747.

Wollmuth LP (1995). Multiple ion binding sites in I_h channels of rod photoreceptors from tiger salamanders. *Pflügers Arch* **430**, 34–43.

Wollmuth LP and Hille B (1992). Ionic selectivity of I_h channels of rod photoreceptors in tiger salamanders. *J Gen Physiol* **100**, 749–765.

Xue T, Marban E and Li RA (2002). Dominant-negative suppression of HCN1- and HCN2-encoded pacemaker currents by an engineered HCN1 construct: insights into structure–function relationships and multimerization. *Circ Res* **90**, 1267–1273.

Yamada R, Kuba H, Ishii TM et al. (2005). Hyperpolarization-activated cyclic nucleotide-gated cation channels regulate auditory coincidence detection in nucleus laminaris of the chick. *J Neurosci* **25**, 8867–8877.

Yao H, Donnelly DF, Ma C et al. (2003). Upregulation of the hyperpolarization-activated cation current after chronic compression of the dorsal root ganglion. *J Neurosci* **23**, 2069–2074.

Yasui K, Liu W, Opthof T et al. (2001). I_f current and spontaneous activity in mouse embryonic ventricular myocytes. *Circ Res* **88**, 536–542.

Ying SW, Abbas SY, Harrison NL *et al.* (2006). Propofol block of I_h contributes to the suppression of neuronal excitability and rhythmic burst firing in thalamocortical neurons. *Eur J Neurosci* **23**, 465–480.

Ying SW and Goldstein PA (2005). Propofol-block of SK channels in reticular thalamic neurons enhances GABAergic inhibition in relay neurons. *J Neurophysiol* **93**, 1935–1948.

Yu H, Wu J, Potapova I *et al.* (2001). MinK-related peptide 1: A β subunit for the HCN ion channel subunit family enhances expression and speeds activation. *Circ Res* **88**, E84–E87.

Yu,X, Duan KL, Shang CF *et al.* (2004). Calcium influx through hyperpolarization-activated cation channels (I_h channels) contributes to activity-evoked neuronal secretion. *PNAS* **101**, 1051–1056.

Zagotta WN, Olivier NB, Black KD *et al.* (2003). Structural basis for modulation and agonist specificity of HCN pacemaker channels. *Nature* **425**, 200–205.

Zhang W and Linden DJ (2003). The other side of the engram: experience-driven changes in neuronal intrinsic excitability. *Nat Rev Neurosci* **4**, 885–900.

Zhong N, Beaumont V and Zucker RS (2004). Calcium influx through HCN channels does not contribute to cAMP-enhanced transmission. *J Neurophysiol* **92**, 644–647.

Zhou L and Siegelbaum SA (2007). Gating of HCN channels by cyclic nucleotides: residue contacts that underlie ligand binding, selectivity, and efficacy. *Structure* **15**, 655–670.

Zhou L, Olivier NB, Yao H *et al.* (2004). A conserved tripeptide in CNG and HCN channels regulates ligand gating by controlling C-terminal oligomerization. *Neuron* **44**, 823–834.

Zicha S, Fernandez-Velasco M, Lonardo G *et al.* (2005). Sinus node dysfunction and hyperpolarization-activated (HCN) channel subunit remodeling in a canine heart failure model. *Cardiovasc Res* **66**, 472–481.

Zolles G, Klöcker N, Wenzel D *et al.* (2006). Pacemaking by HCN channels requires interaction with phosphoinositides. *Neuron* **52**, 1027–1036.

Zong X, Eckert C, Yuan H *et al.* (2005). A novel mechanism of modulation of hyperpolarization-activated cyclic nucleotide-gated channels by src kinase. *J Biol Chem* **280**, 34224–34232.

Zong X, Stieber J, Ludwig A *et al.* (2001). A single histidine residue determines the pH sensitivity of the pacemaker channel HCN2. *J Biol Chem* **276**, 6313–6319.

PART 3

Extracellular ligand-gated ion channels

Extracellular
ligand-gated ion
channels

3.1

Cys loop receptors

Contents

3.1.1 Nicotinic acetylcholine receptors

Marzia Lecchi, Jean-Charles Hoda,
Ronald Hogg and Daniel Bertrand

3.1.1.1 Introduction

Ligand-gated ion channels are membrane proteins that are specialized in the fast conversion of a chemical into an electrical signal. Vertebrate ligand-gated ion channels can be divided into subfamilies and classified according to their endogenous ligands or structural features.

The nicotinic acetylcholine receptors (nAChRs) belong to the family of cys-loop receptors which possess four transmembrane domains. Substantial progress has been made during the past 10 years in advancing our understanding of this protein family. In this chapter we will focus our attention on nAChR physiology, pharmacology and pathophysiology.

3.1.1.2 Structure

Our knowledge about the structure of the nAChRs originated from studies conducted on muscle receptors with an important contribution from data obtained from the electric organ of *Torpedo* rays. This specialized muscle-derived organ is extremely rich in receptors and enabled the first observation of the receptor structure using negative staining and electron microscopy (Cartaud *et al.*, 1973). Proteins isolated from the *Torpedo* electric organ allowed the determination of the molecular weight of the receptor complex (~270 kDa) and revealed that the complex contains both the ligand binding site and the channel domain (Raftery *et al.*, 1980, Popot and Changeux, 1984). Unwin and collaborators took advantage of the pseudo crystalline structure of nAChRs in microtubules reconstituted from *Torpedo* electric organ and, using highresolution cryoelectron microscopy, were able to resolve the muscle nAChR structure at 4 Å (Miyazawa *et al.*, 2003; Unwin, 2005). Although such resolution is not sufficient to resolve atomic positions, these data confirmed numerous results from biochemical studies. Both approaches suggest that nAChRs result from the assembly of five subunits that span the membrane and are arranged in a ring-like manner with a water-filled ionic pore through which ions flow when the receptor is activated (Dani, 1986). The ligand binding sites are formed by the extracellular domains of two adjacent subunits. Muscle nAChRs possess two acetylcholine binding sites which display distinct properties reflecting their amino acid composition (Changeux, 1990; Galzi *et al.*, 1991a; Taylor *et al.*, 1994; Taylor *et al.*, 2000). Figure 3.1.1.1 illustrates the structure of the nAChR inserted into the membrane and highlights the importance of the extracellular domain that faces the synaptic space, the transmembrane domains and the intracellular domains.

Protein sequence deduced from the *Torpedo* receptor constituted an important tool for the cloning, first of muscle receptor subunits, and later by low-stringency hybridization of neuronal nAChRs (Noda *et al.*, 1983; Patrick *et al.*, 1983; Ballivet *et al.*, 1988; Monteggia *et al.*, 1995; for review see Devillers-Thiery *et al.*, 1993). To date 17 genes encoding the muscle and neuronal nAChR subunits have been identified in vertebrates and a consensus nomenclature was adopted (Lukas *et al.*, 1999). Table 3.1.1.1 summarizes the gene nomenclature, human chromosomal localizations of

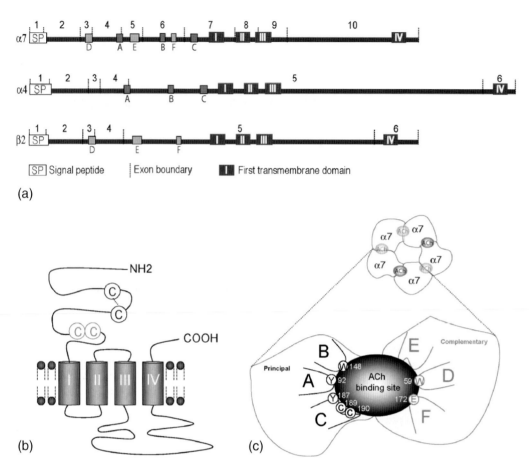

Fig. 3.1.1.1 Structure of the nAChRs. (a) Typical cDNA nAChR sequences comprise approximately 2000 base pairs that correspond to a coding sequence of about 600 residues. This panel represents a linear scheme of α7, α4 and β2 subunit sequences, respectively. Drawings reflect the coding sequence length. Positions of the signal peptide, exon boundaries, and transmembrane segments are indicated. Major loops of the ligand binding pocket are indicated by capital letters. (b) Schematic drawing of a subunit inserted in the membrane with the NH₂ and COOH termini facing the extracellular space. The cys-loop and the adjacent cysteines typical of an α subunit are indicated. (c) Representation of the ligand binding site located at the interface of two adjacent subunits. The major loops of the principal and complementary components corresponding to panel A are schematized (for review see Grutter and Changeux, 2001). Numbering corresponds to the chick α7 nAChR (Galzi *et al.*, 1991a, b).

neuronal nAChRs, and genetically modified animal models in which they have been studied. Muscle subunits include the α1, β1, γ, δ and ε subunits, whereas all other subunits, α2–α10 and β2–β4, belong to the neuronal nAChR family. A large N-terminal domain that contains the typical disulphide bridge of the cys-loop ligand-gated ion channels characterizes all subunits. The presence of two vicinal cysteines that form another disulphide bridge in the N-terminal domain distinguishes the α subunits. Each subunit presents four hydrophobic regions that are thought to span the membrane—transmembrane (TM) regions 1–4—a large intracellular domain formed by the residues between TM3 and TM4 and a short C-terminal tail. Both the N and C termini face the extracellular domain.

The N-terminal domain and the acetylcholine-binding site

The hydrophobic nature of the transmembrane domain has so far prevented the crystallization of ligand-gated ion channels, therefore, different strategies were followed to obtain more precise data about the ligand binding site. Photoaffinity labelling with ligands structurally related to acetylcholine allowed the determination of amino acids which interact with acetylcholine (Galzi *et al.*, 1991a; Corringer *et al.*, 2000). The role of these residues was later confirmed using site-directed mutagenesis combined with functional investigations (Galzi *et al.*, 1991b; Sine *et al.*, 1994). Finally, models based on the knowledge of the protein sequences and analogies

with other known protein structures gave a first set of plausible structures that were in agreement with experimental data and a putative three-dimensional organization of the N-terminal domain (Le Novere *et al.*, 1999, 2002).

However, a major breakthrough in our understanding of the ligand binding domain came with the identification of a soluble protein which binds acetylcholine with high affinity and is secreted by glial cells in the snail nervous system (Smit *et al.*, 2001). Cloning of the DNA that encodes this acetylcholine binding protein (AChBP) revealed a striking similarity with the N-terminal domain of the nAChRs and, moreover, the AChBP is assembled from five monomers (Brejc *et al.*, 2001). Crystallization of this water-soluble protein offered for the first time a three-dimensional structure that resembles the structure of the N-terminal domain of the nAChRs. Indeed, numerous studies have since then confirmed the similarity between the AChBP and the extracellular domain of the nAChRs and have allowed, by structural homology, the prediction of the three-dimensional organization of the nAChR ligand binding site (Sixma and Smit, 2003; Celie *et al.*, 2004; Sine *et al.*, 2004). Thus, the N-terminal domain is mostly formed by folded βsheets and the ligand binds at the interface between two adjacent subunits. The principal component of the binding site is formed by three loops of the α subunit (A,B,C) and the complementary component by three loops present on the adjacent subunit (D, E, F) (Figure 3.1.1.1). In the case of homomeric receptors, such as α7 receptors, each subunit contributes both to the principal and to the complementary component.

Table 3.1.1.1 Neuronal nAChRs with their corresponding gene nomenclature and human chromosomal localization are indicated. The two last columns indicate the respective reported genetically engineered mice models including both knockout and knockin mutations

Neuronal subunit	Gene	Human chromosomal localization	Knockout mice	Knockin mice
α2	CHRNA2	8p21.2 [1]	–	
α3	CHRNA3	15q24.3 [2]	Survive to birth but with impaired growth and increased mortality before and after weaning. Develop megacystis and mydriasis [3]	
α4	CHRNA4	20q13.33 [4]	Survive with reduced antinociception [5, 6]	L9'S and L9'A: higher sensitivity to seizures induced by nicotine and altered sleep–wake cycle [7, 8–10]. S252F with or without L264: model of ADNFLE [11]
α5	CHRNA5	15q24.3 [2]	Healthy appearance and normal behaviour [12]. Reduced sensitivity to nicotine-induced seizures [13]. Higher susceptibility to experimental colitis [14]	
α6	CHRNA6	8p11.21 [15]	Viable with no major physical or neurological defects [16]	
α7	CHRNA7	15q13.1 [17]	Viable and anatomically normal [14]. Reduction in sustained attention and olfactory working memory [18]. Baroreflex impairment [19]. Impaired sperm motility [20]	L250T: die within 2–24 h of birth. Extensive apoptotic cell death throughout the somatosensory cortex [21]
α9	CHRNA9	4p14–15.1 [22]	Survive with impairment in efferent cochlear innervation [23]	
α10	CHRNA10	11p15.5 [24]	Breed normally without overt behavioural phenotype [25]	
β2	CHRNB2	1q23.1 [26]	Survive [27] with reduced visual acuity [28, 29] and a modified sleep organization [30]. Impaired spatial learning in aged mice [31]. Lack of reinforcing properties of nicotine [32]	
β3	CHRNB3	8p11.21 [33]	Survive with altered locomotor activity [34]	
β4	CHRNB4	15q24.3 [2]	Survive with altered autonomic functions [12]. Highly resistant to nicotine-induced seizures [13]	

[1] Wood S et al. (1995) Somat Cell Mol Genet 21, 147–150; [2] Raimondi E et al. (1992) Genomics 12, 849–850; [3] Xu W et al. (1999) Proc Natl Acad Sci USA 96, 5746–5751; [4] Steinlein O et al. (1994) Genomics 22, 493–495; [5] Marubio LM et al. (1999) Nature 398, 805–810; [6] Ross SA et al. (2000) J Neurosci 20, 6431–6441; [7] Labarca C et al. (2001) Proc Natl Acad Sci USA 98, 2786–2791; [8] Fonck C et al. (2005) J Neurosci 25, 11396–11411; [9] Fonck C et al. (2003) J Neurosci 23, 2582–2590; [10] Orb S et al. (2004) Physiol Genomics 18, 299–307; [11] Glykys J et al. (2005) Soc Neurosci Abstr 964.968; [12] Wang N et al. (2003) Mol Pharmacol 63, 574–580; [13] Salas R et al. (2004) Neuropharmacology 47, 401–407; [14] Orr-Urtreger A et al. (1997) J Neurosci 17, 9165–9171; [15] Ebihara M et al. (2002) Gene 298, 101–108; [16] Champtiaux N et al. (2002) J Neurosci 22, 1208–1217; [17] Chini B et al. (1994) Genomics 19, 379–381; [18] Young JW et al. (2007) Eur Neuropsychopharmacol 17, 145–155; [19] Franceschini D et al. (2000) Behav Brain Res 113, 3–10; [20] Bray C et al. (2005) Biol Reprod 73, 807–814; [21] Orr-Urtreger A et al. (2000) J Neurochem 74, 2154–2166; [22] Lustig LR et al. (2002) Cytogenet Genome Res 98, 154–159; [23] Vetter DE et al. (1999) Neuron 23, 93–103; [24] Lustig LR et al. (2001) Genomics 73, 272–283; [25] Vetter D et al. (2005) Soc Neurosci Abstr 954.920; [26] Rempel N et al. (1998) Hum Genet 103, 645–653; [27] Picciotto MR et al. (1995) Nature 374, 65–67; [28] Grubb MS et al. (2003) Neuron 40, 1161–1172; [29] Rossi FM et al. (2001) Proc Natl Acad Sci USA 98, 6453–6458; [30] Lena C et al. (2004) J Neurosci 24, 5711–5718; [31] Zoli M et al. (1999) Embo J 18, 1235–1244; [32] Picciotto MR et al. (1998) Nature 391, 173–177; [33] Koyama K et al. (1994) Genomics 21, 460–461; [34] Cui C et al. (2003) J Neurosci 23, 11045–11053.

This suggests that in homomeric receptors five identical ligand binding sites may be present in the extracellular domain of the receptor.

The channel domain

The structural organization of the receptor is such that the water-filled pore through which ions can flow is localized in the centre of the nAChR. Biochemical studies using labelled open channel blockers which enter and block the ionic pore first suggested that the channel is lined by the second transmembrane segment of the nAChRs and that it must be coiled in a α-helix like manner (Giraudat et al., 1989). Site-directed mutagenesis of the charged residues that were presumed to be at the inner mouth or at the outer mouth of the channel of the muscle receptor confirmed the prevalent role of TM2 (Konno et al., 1991; Mishina et al., 1991). Since then, a large number of studies have confirmed that TM2 from each of the subunits lines the ionic pore and has an α-helix structure. Site-directed mutagenesis and functional studies revealed the critical role of the uncharged leucine residue (L247 in the chick α7) that is located in the

lower part of the pore and is important for the closing of the channel during desensitization (Revah et al., 1991, Bertrand et al., 1993b). This residue is highly conserved throughout the family of ligand-gated ion channels and was suggested to correspond to the constriction of the channel observed in the cryoelectron microscopy images (Unwin, 2005).

The high degree of conservation of the TM2 protein sequences observed between different animal species and even between cationic and anionic receptors suggests that receptors require a precise organization of the channel domain for proper functioning. Thus, different functionalities, such as ionic selectivity, may depend upon a few residues without structural changes of the channel domain. In agreement with such hypothesis, and as discussed below, it was observed that substitution of three residues in TM2 segment of the α7 nAChR is sufficient to convert the ionic selectivity of this receptor from cationic to anionic (Galzi et al., 1992; Corringer et al., 1999). Although it is beyond the scope of this chapter to discuss invertebrate receptors in detail, it is interesting to note that this structural feature is also present in invertebrate nAChRs such as those isolated from molluscs and that ionic selectivity was successfully predicted based on the protein sequence of TM2 (van Nierop et al., 2005, 2006).

The TM1, TM3 and TM4 transmembrane domains

The TM1, TM3 and TM4 transmembrane domains have received less attention than TM2 and our current knowledge about the role of these protein segments is rather limited. Nonetheless, it was shown that TM1 plays an important role in determining the capacity of the receptor to form homo-oligomers (Vicente-Agullo et al., 1996). Furthermore, as the agonist binding sites are present on truncated nAChR subunits that were comprised only of the N-terminal domain and TM1, it seems that this transmembrane segment plays an important role in receptor assembly and insertion into the membrane (Person et al., 2005).

The relative contribution of a given protein segment to receptor function is difficult to predict. The main reason for this difficulty resides in the fact that characteristics of the receptor are not dictated by a short segment of amino acids but by the ensemble of the protein complex. This is illustrated by the fact that a mutation in TM4 modifies the muscle receptor sensitivity to steroids, although this site may not correspond to a point of interaction between the receptor and the steroid molecules (Bouzat, 1996).

The M2–M3 extracellular loop

Although rather short, the extracellular M2–M3 domain plays an important role in the function of these four transmembrane domain ligand-gated ion channels. The first indications of the importance of this short segment of the receptor came with the observation that the genetically transmissible form of hyperekplexia is associated with a mutation of a lysine residue in this region of the structurally related glycine receptor (Shiang et al., 1993). Mutations of this short extracellular loop in the nAChRs were proposed to modify the coupling between ligand binding and receptor gating (Campos-Caro et al., 1996). More recently, it was proposed that in another cys-loop ligand-gated ion channel, the Gamma aminobutyric acid$_A$ (GABA$_A$) receptor, residues from the M2–M3 segment interact with amino acids from the ligand binding site and form bridges that could participate in the coupling between binding and receptor activation (Kash et al., 2003). Interestingly, chimeric receptors formed by fusing the AChBP with the transmembrane and C-terminal end of the 5HT$_3$ receptor yielded functional channels only when residues in the M2–M3 loop were substituted (Bouzat et al., 2004). Although debate still exists about the state and functionality of such fusion proteins (Grutter et al., 2005), these results suggest that the short extracellular segment located between TM2 and TM3 plays a crucial role in the receptor function and may couple the structural changes observed during ligand binding with the TM2 rotation that is thought to take place while the channel is opening.

The M3–M4 intracellular loop

The TM3 and TM4 transmembrane segments delimit a large amino acid sequence that protrudes into the cytosol. This sequence, which displays a large degree of variability between subunits, probably plays important roles in the interaction with intracellular components as well as in receptor function.

For the muscle receptor it was proposed that this intracellular segment interacts with the anchoring molecule rapsyn that is thought to anchor the receptor at the neuromuscular junction (Huh and Fuhrer, 2002; Zhu et al., 2006). Although no specific anchoring molecule has been identified to date for the neuronal nAChRs, several candidate proteins have been suggested.

In addition to its putative anchoring role, the M3–M4 segment was shown to play a determinant role in processes of receptor targeting, phosphorylation and ion conduction (Williams et al., 1998; Fenster et al., 1999; Williams et al., 1999; Valor et al., 2002; Charpantier et al., 2005; Cho et al., 2005; Peters et al., 2005). For example, the targeting process that allows a cell to insert nAChRs at an appropriate cellular localization was shown to depend strongly on the amino acid sequence of the M3–M4 segment (Williams et al., 1998; Williams et al., 1999). Exchange of the α7 nAChR M3–M4 segment with that of either α3 or α5 resulted in the clustering of the receptors at a distinct synaptic localization without marked modification of receptor function. The presence of highly conserved serine-threonine phosphorylation consensus sequences in this M3–M4 intracellular segment suggested that nAChRs could be regulated by phosphorylation. In agreement with this hypothesis it was recently shown that genistein, a kinase inhibitor, which causes protein dephosphorylation, increases the acetylcholine-evoked current of the α7 receptor (Charpantier et al., 2005; Cho et al., 2005).

While numerous studies have shown that ion flow through ligand-gated ion channels is limited by diffusion in the channel pore, it is widely agreed that the shape and size of both the outer and inner mouth of the channel can influence ion permeation. Recent experiments on the M3–M4 segment support this hypothesis and demonstrate that exchange of charged residues in the M3–M4 segment of the 5HT$_3$ receptor with those found in the nAChRs resulted in major modifications of the single channel properties (Peters et al. 2005). Thus, it was proposed that α-helices in the M3–M4 segment form a funnel at the inner mouth of the channel domain.

The C-terminus

The short C-terminal sequence comprises a few amino acid residues facing the extracellular domain. Determined unequivocally by different techniques, including antibody labelling, flag targeting, etc., the localization of this segment in the extracellular domain further confirms the topology of the four transmembrane domain nAChRs. Despite its short length this segment was shown to play a determinant role in the steroid modulation of α4β2 nAChRs (Paradiso et al., 2001; Curtis et al., 2002). Additionally, it was suggested that this short terminal sequence might participate in receptor assembly or in the interaction with the N-terminal domain of the adjacent subunit.

3.1.1.3 Biophysical properties

In this section we will describe the biophysical properties of nAChRs and discuss them with reference to the structural determinants that govern receptor function (Table 3.1.1.2). Because of its size and relative tractability for investigation, the neuromuscular junction was used extensively as a model of synaptic transmission and much important knowledge was gained from such studies. For example, the basic principles of synaptic transmission and quantal release were established from electrophysiological recordings of endplate receptors (Katz, 1966).

Studies conducted at ganglionic receptors highlighted the differences between muscle and neuronal nAChRs (Ascher *et al.*, 1979). However, major advances in our understanding of this second class of nAChRs only followed the cloning, sequencing and expression of neuronal nAChR genes (Boulter *et al.*, 1986; Ballivet *et al.*, 1988).

Current–voltage relationship and ion selectivity

An important characteristic of any ion channel is its ionic selectivity and voltage dependency. To determine which ions flow through a particular receptor it is necessary to record the so-called current–voltage (I–V) relationship. In practice, the amount of current evoked by an agonist test pulse is measured at different holding potentials. For the muscle receptor the I–V relationship is almost ohmic and reverses near 0 mV (Dani, 1986). The muscle receptors are permeable to cations with permeability ranking in the following order pCs > pK > pNa >> pCa (Adams *et al.*, 1980). These studies highlighted the low permeability ratio for divalent cations such as calcium and suggested that the muscle nAChR ion pore displays a diameter of approximately 5–6 Å. Activity of a single receptor complex was first demonstrated at the neuromuscular junction and revealed that opening of a receptor caused a sharp transition of the current that remained steady until the receptor channel closed (Sakmann *et al.*, 1980). The endplate single channel has a conductance of about 40 pS. Single channel activity is well described by a two-state stochastic model in which the channel is either closed or open (Colquhoun and Rang, 1973; Colquhoun and Sakmann, 1983; Gardner *et al.*, 1984). Plot of the single channel current as a function of the membrane voltage yields an I–V relationship that is linear and reverses at potential of approximately 0 mV, resembling macroscopic recordings. Single channel measurements also provided further evidence that the TM2 segment lines the ionic pore. For example, substitution of the charged residue at the putative outer mouth or inner mouth of the channel caused a modification of the single channel conductance (Konno *et al.*, 1991).

The neuronal nAChRs exhibit significantly different properties relative to muscle receptors. First, all neuronal nAChRs display a non-linear IV relationship with an inward rectification that is manifested as a progressive reduction of the conductance as the membrane is depolarized (Bertrand *et al.*, 1990; Fieber and Adams 1991; Sands and Barish, 1992; Elgoyhen *et al.*, 1994; Forster and Bertrand 1995; Buisson *et al.*, 1996). Rectification of the α7 nAChR is such that almost no current can be detected at positive membrane potentials. A typical example of rectification recorded at the α4β2 and α7 nAChRs is presented in Figure 3.1.1.2a.

Inward rectification could be attributed to a lower probability of channel opening, channel blockade or reduction of the single channel conductance. Site-directed mutagenesis experiments carried out at the α7 nAChRs highlighted the role of the charged glutamate residue at the beginning of the TM2 segment that is thought to form the inner mouth of the channel (Forster and Bertrand, 1995). The inward rectification of heteromeric nAChRs probably depends upon comparable mechanisms as mutation of the corresponding amino acid residues also suppressed the I–V nonlinearity (Haghighi and Cooper, 2000). Single channel measurements conducted at neuronal nAChRs yielded currents of a few

pA with properties resembling those of the endplate receptors. However, in contrast with the macroscopic currents recorded at α4β2 nAChRs, single channel currents display a linear relationship versus voltage with a conductance of about 46 pS and a reversal potential of approximately 7 mV (Buisson *et al.*, 1996). Comparable findings were obtained for different neuronal nAChR compositions, indicating that the inward rectification observed at the macroscopic level does not reflect a modification of the channel conductance but rather a reduction of the opening probability that is compatible with channel block by intracellular molecules.

Despite the high degree of homology observed between the TM2 segments of the different nAChR subunits there are minor sequence differences in this critical protein region. Interestingly in this respect, the receptor subtypes have different channel conductances ranging from 5 pS to 76 pS. Although it would be tempting to correlate the amplitude of the single channel current with receptor subunit composition, variability in measurement of native receptor currents prevents definitive conclusions. Studies of single channel kinetics of muscle receptors have elegantly illustrated that the probability of opening of the receptor is increased in the presence of acetylcholine. While the same principles probably hold for the neuronal nAChR, a progressive decline of channel activity has thus far prevented kinetic analysis (Buisson *et al.*, 1996).

The ionic selectivity of the neuronal nAChRs differs markedly from the muscle receptors in their higher permeability to divalent cations (Sands and Barish, 1991; Vernino *et al.*, 1992; Seguela *et al.*, 1993; Bertrand *et al.*, 1993a; Fucile, 2004; Fucile *et al.*, 2005). With a permeability ratio pCa/pNa of approximately 10–20 the α7 homomeric nAChR is even more permeable to calcium than the *N*-methyl-D-aspartate (NMDA) receptor (Bertrand *et al.*, 1993a). Similarly, the α9α10 receptor which is expressed mainly in the outer hair cells of the inner ear has a high calcium permeability that is thought to play an important physiological role in these cells (Elgoyhen *et al.*, 2001; Sgard *et al.*, 2002). The high calcium permeability of the neuronal nAChRs critically depends upon the structure and composition of the channel domain. Mutation of a single uncharged leucine residue (L254 or L255) in the TM2 segment or a glutamate at the inner mouth of the channel was shown to reduce calcium permeability to a negligible value (Bertrand *et al.* 1993a). This illustrates that calcium permeability depends upon a complex mechanism, which is determined by both the diameter and amino acid residue lining of the pore.

Sequence alignment of the TM2 segments from cationic (nAChRs, 5-hydroxytryptamine₃ [5-HT₃]) and anionic (GABA_A, glycine) ligand-gated channels reveals striking homologies between these ligand-gated ion channels. A consistent difference is, however, observed in the lower section of the channel domain with the presence of an extra proline residue in anionic receptors. Based on this observation it was suggested that the selectivity filter could be located in this area of the channel. As first demonstrated at the chick α7 nAChR, conversion of ionic selectivity from cationic to anionic was obtained by insertion of an extra residue at the inner mouth of the channel (Galzi *et al.*, 1992). Numerous studies have subsequently confirmed these initial results and indicate that the ionic selectivity filter lies in the narrowest part of the channel at the inner mouth (Corringer *et al.*, 1999; Keramidas *et al.*, 2000; Gunthorpe and Lummis, 2001).

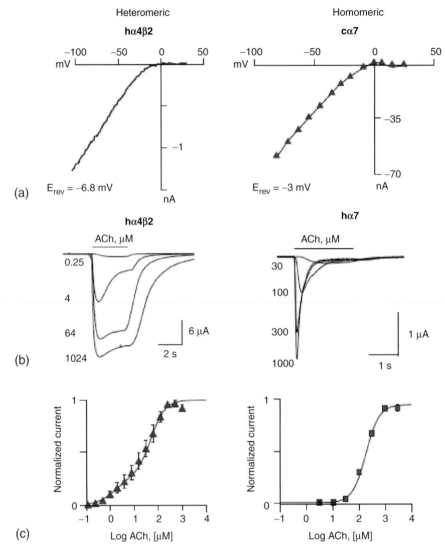

Fig. 3.1.1.2 Biophysical properties of neuronal nAChRs. (a) I–V relationships for AChevoked currents from α4β2 and α7 nAChRs. Left panel, I–V relationship recorded in a K177 cell expressing the human α4β2 nAChR (Buisson et al., 1996[1]). Right panel, I–V relationship of the chick α7 nAChR recorded in a *Xenopus* oocyte (Bertrand et al., 1993a[2]). To avoid contamination by calciumactivated conductances recordings were performed in presence of a calciumchelating agent. (b) Typical currents evoked by a series of ACh concentrations for the human α4β2 (left) and α7 (right) nAChRs, expressed in *Xenopus* oocytes, are superimposed. Note the difference in time course of the AChevoked response with relatively fast desensitization of the α7 receptors. (c) Agonist concentration–response curves for the human α4β2 (left) and α7 (right) nAChRs determined by reporting the peak of the AChevoked current, normalized to unity for saturating ACh concentration, as a function of the logarithm of the agonist concentration. Bars represent the standard error obtained for 6 and 28 cells for the α4β2 and α7 nAChRs, respectively. Curves through the data points are the best fit obtained with a dual Hill equation for α4β2 (left) and with a single Hill equation for α7 (right) nAChRs (Buisson and Bertrand, 2001;[3] Charpantier et al., 2005[4]). Equations and values for the best fit are indicated in these publications.

[1] Adapted with permission from Buisson B, Gopalakrishnan M, Arneric SP, Sullivan JP, and Bertrand D (1996). Human α4β2 neuronal nicotinic acetylcholine receptor in HEK 293 cells: a patch-clamp study. *J Neurosci* **16**, 7880–7891. Copyright, 1996, Society for Neuroscience.

[2] Adapted with permission from Bertrand D, Galzi JL, Devillers-Thiéry A, Bertrand S and Changeux JP (1993). Mutations at two distinct sites within the channel domain M2 alter calcium permeability of neuronal α7 nicotinic receptor. *Proc Natl Acad Sci USA* **90**, 6971–6975. Copyright 1993, National Academy of Sciences USA.

[3] Adapted with permission from Buisson B and Bertrand D (2001). Chronic exposure to nicotine upregulates the human (alpha)4(beta)2 nicotinic acetylcholine receptor function. *J Neurosci* **21**, 1819–1829. Copyright 2001, Society for Neuroscience.

[4] Adapted with permission from Charpantier E, Wiesner A, Huh KH, Ogier R, Hoda JC, Allaman G, Raggenbass M, Feuerbach D, Bertrand D and Fuhrer C (2005). Alpha7 neuronal nicotinic acetylcholine receptors are negatively regulated by tyrosine phosphorylation and Src-family kinases. *J Neurosci* **25**, 9836–9849. Copyright, 2005, Society for Neuroscience.

Ligand binding and activation

Understanding of the composition of the receptors highlighted the diversity between muscle, ganglionic and brain neuronal nAChRs. The constitution of the ligand binding site at the interface between two adjacent subunits suggests that the affinity and properties must be governed by the amino acid composition of these two subunits.

Indeed, receptors composed of α4β2 subunits have different properties to those formed by α4β4 or by α3β2 (Couturier et al., 1990a; Hernandez et al., 1995; Chavez-Noriega et al., 1997). Moreover, because neuronal nAChRs may result from more complex assembly with different α and β subunits, the corresponding number of possible configurations of functional receptors is large. When considering the ligand binding site, an important point is the number of putative binding sites present on a given receptor and how many sites need to bind acetylcholine for receptor activation. Results obtained at the muscle receptor clearly indicate that this receptor subtype possesses two acetylcholine binding sites that are, in the embryonic receptors, at the αγ and αδ interface and at the αδ and αε in the adult form of the receptor (Changeux, 1990). In the putatively simplest and, in the evolutionary context, probably oldest receptor subtype, a single subunit can constitute a functional receptor. This is best exemplified by the α7 receptor that is composed, both in recombinant and native receptors, of five identical subunits (Couturier et al., 1990a; Seguela et al., 1993; Drisdel and Green, 2000). Thus, it follows that an α7 receptor complex must include five ligand binding sites. While the number of binding sites that must be occupied to activate the α7 receptor still remains to be determined, the existence of five identical binding sites for competitive inhibitors was suggested (Palma et al., 1996). Analysis of the decay time of the inhibition caused by exposure to a fixed concentration of the competitive inhibitor methyllycaconitine (MLA) indicates that a bimolecular process (single exponential) adequately describes this inhibition and, thus, that binding of a single MLA molecule is sufficient to block the α7 receptor complex. In contrast, recovery from inhibition displays an S-shaped time course that is not compatible with a bimolecular process. These data can be adequately described by a model in which the α7 receptor harbours five identical MLA binding sites and that opening of the receptor will be observed only if all sites are free of the competitive inhibitor (Palma et al., 1996).

Receptor stoichiometry

Different strategies have been employed in attempt to resolve the stoichiometry of heteromeric receptors such as α4β2. The reporter approach consists of modifying one subunit to allow its identification and quantification. A reporter mutation, which causes a detectable modification of the receptor properties, is employed. First employed at α4β2 recombinant receptors with a mutation that changes the single channel conductance, this technique revealed the stoichiometry of these neuronal nAChRs as probably composed of two α4 and three β2 subunits (Cooper et al., 1991). Subsequently, introduction of a gain of function mutation was used to illustrate that the β3 subunit, which is otherwise silent, participates in the assembly of functional nAChRs (Boorman et al., 2003). Other approaches have included the concatenation of subunits in a precise assembly and assessment of their functional properties in comparison to native receptors (Zhou et al., 2003). While all these studies confirm that a single receptor complex results from the assembly of five subunits, the stoichiometry of the native receptor subtypes still remains to be determined.

Allosteric model

Several models have been proposed to describe the physiological and pharmacological properties of the muscle and neuronal nAChRs. These models can be divided in two categories with on one hand the induced fit model, such as initially described by Koshland et al. (1966), and on the other hand the allosteric model (Monod et al., 1965). In this section we shall briefly review the allosteric model and why it is an appropriate model for ligand-gated ion channels.

Introduced in 1965 by Monod, Wyman, and Changeux, the allosteric concept is based on the hypothesis that proteins resulting from assembly of monomers can exist in multiple conformations and that presence of a compound can stabilize the protein complex in a particular conformation (Monod et al., 1965). The equilibrium between different states and the probabilities of a transition depend upon the isomerization coefficients that correspond to the energy barriers between the different states. This concept was extended to ligand-gated ion channels by considering that binding of a ligand stabilizes the receptor in a particular state (for review see Changeux and Edelstein, 2005). An important aspect that distinguishes the allosteric model from the induced fit model is that in the allosteric model spontaneous transitions may occur in the absence of ligand. In agreement with this definition, spontaneous openings were observed at the single channel level of the neuromuscular junction receptor (Jackson, 1994). Another important feature of the allosteric model is that effects of a ligand are determined by the state that it has stabilized. Thus, it follows that agonists are ligands that preferentially stabilize the active open state whereas competitive antagonists are ligands that preferentially stabilize a closed state. Interestingly, mutation of a single amino acid (L247T) in TM2 of the α7 receptor causes major modifications of its properties that are best explained using the allosteric model (Revah et al., 1991; Bertrand et al., 1992). To explain the pleiotropic effects caused by this single point mutation which affects the sensitivity to acetylcholine, desensitization, and converts competitive antagonists such as dihydro-β-erythroidine (DHβE) into agonists, it was proposed that the L247T mutation renders an otherwise closed desensitized state conducting (Revah et al., 1991). This single hypothesis readily explains all properties of the L247T mutant together with its unconventional pharmacology (Bertrand et al., 1992). This new open state presents a lower isomerization coefficient causing spontaneous opening of the receptor that can be observed in the absence of agonist. Competitive antagonists such as MLA, that are thought to stabilize the resting state, inhibit this spontaneous opening whereas DHβE stabilizes this open state of the receptor and exhibits agonist activity (Bertrand et al. 1997).

The allosteric model can describe other receptor properties, such as desensitization or cooperativity, and is the best-suited model to explain how substances that are neither competitive agonists nor antagonists can modulate receptor function. Positive modulators are substances that increase receptor function whereas negative modulators decrease receptor function. Such modulators, also known as allosteric effectors, can bind anywhere on the protein at a site distinct from the ligand binding site but still affect the receptor function. As reviewed by Galzi and collaborators allosteric modulators can have multiple effects depending on their actions on ligand binding affinity and receptor isomerization coefficients (Galzi et al., 1996b). Modulators that reduce the isomerization coefficient between the resting and active states are positive allosteric modulators whereas an increase of the isomerization coefficient will be observed for negative allosteric modulators. Positive and negative modulator activity mediated by small chemical molecules, steroids or even polypeptides, has been observed at both muscle and neuronal nAChRs (reviewed in Buisson and Bertrand 1998a; Hogg et al. 2005). Studies of the positive allosteric modulation mediated by extracellular calcium revealed that calcium probably binds at a site located in the vicinity of the acetylcholinebinding site. This was best illustrated by the fact that grafting the short amino acid sequence thought to mediate calcium binding from the neuronal α7 nAChR to the 5HT_3 receptor is sufficient to confer sensitivity to extracellular calcium to this otherwise insensitive receptor (Galzi *et al.*, 1996a). The observation that the high affinity nicotinic receptor α4β2 is modulated by steroids and neurosteroids could explain gender differences of brain function or sensitivity to nicotine addiction. The allosteric nature of this modulation is exemplified by the finding that the C-terminal end of the α4 subunit plays a crucial role in the response to 17βoestradiol and that small differences observed between the human and rat sequences are sufficient to confer or suppress the steroid response (Paradiso *et al.*, 2001; Curtis *et al.*, 2002). Allosteric modulators that affect the isomerization coefficient can also produce major changes of the physiological and pharmacological properties of the receptor. For example ivermectin, which is a positive allosteric modulator at the α7 nAChR, increases the apparent affinity of acetylcholine, the maximal evoked current, decay time of the response and transforms a partial agonist into a full agonist (Krause *et al.*, 1998).

3.1.1.4 Pharmacology

The neurotransmitter acetylcholine physiologically activates two classes of receptors that were classified according to their sensitivity to natural compounds into muscarinic and nicotinic receptors. Muscarine extracted from fungi was shown to stimulate G-protein-coupled receptors while the plant alkaloid nicotine was first recognized as a potent agonist at the ligand-gated ion channel activated by acetylcholine and was later used to identify nAChRs. Since then, numerous natural substances and synthetic compounds acting at the nAChRs have been identified (Table 3.1.1.2).

Agonists

Agonists are compounds that cause activation of the receptors and evoke a physiological response that can be evaluated using different techniques. Because agonists bind at the interface between two adjacent subunits, a full agonist at one receptor subtype may be partial agonist or have no activity at a closely related nAChR composed of different subunits. Agonists selective for a given receptor subtype can therefore be envisaged and are thought to offer potential therapeutic utility. For example several compounds displaying selectivity for the α7 nAChR have been recently reported (e.g. PNU 282987, SSR-180711, AR-R 17779). Substances that selectively activate the heteromeric receptors have also been found with a good example provided by the frog toxin epibatidine that activates heteromeric receptors such as the α3β4 or α4β2 in the nM range whereas it activates α7 receptors in the μM range.

The capacity of an agonist to activate the receptor is best characterized by the agonist concentration–response curve that is the amplitude of the evoked response plotted against the logarithm of the agonist concentration (Figure 3.1.1.2b,c). When plotted on a semi-logarithmic scale, these curves are typically Sshaped and can generally be fitted by the empirical Hill equation. The curve is characterized by three values, the half activation (EC_{50}) a measure of agonist efficiency, the Hill coefficient (nH) that is an index of cooperativity, and the amplitude of the maximal response (A). As reviewed by Changeux and Edelstein (2005) these empirical factors which are widely used in the literature find equivalent significance in the allosteric model. While many nAChR concentration–response profiles are adequately described by a single Hill equation, properties of the α4β2 nAChRs require a more complex model. The acetylcholine concentration–response curve measured

at this receptor subtype is best fitted by a dual Hill equation with a high- and a low-affinity profile (Covernton and Connolly, 2000; Buisson and Bertrand, 2001, 2002; Vallejo *et al.*, 2005). This suggests either the existence of two receptor subpopulations that may have a different α versus β stoichiometry or that receptors may have different conformations displaying distinct affinity profiles.

At a given receptor subtype agonists evoke responses of different amplitudes even at saturating concentration when it is assumed that all receptors are occupied. To describe their lower efficacy these compounds are termed partial agonists. Because agonists and partial agonists enter the same ligand binding pocket an interaction that depends upon their relative affinity is expected. The concomitant application of a partial agonist and a full agonist causes a reduction of the amplitude of the response that is otherwise elicited by the full agonist alone.

Although agonists provide a reasonable way to classify receptor subtypes, caution must be exercised in drawing conclusions. For example, whilst muscarine and nicotine have been used historically to discriminate between cholinergic G-protein-coupled receptors and ligand-gated ion channels, cloning and expression of the nAChR α9 and subsequently α10 subunits has cast doubt over this clear-cut pharmacology (Elgoyhen *et al.*, 1994, 2001; Sgard *et al.*, 2002).

Antagonists

Antagonists are molecules that inhibit receptor function and can be described as competitive or non-competitive, according to their mode of action. Because competitive antagonists are compounds that occupy the ligand binding pocket their inhibition can be overcome by increasing the agonist concentration. The natural compound DHβE is a typical competitive antagonist acting at both the muscle and neuronal nAChRs (Bertrand *et al.*, 1990; Alkondon and Albuquerque, 1995). Competitive antagonists interact with amino acids that compose the ligand binding pocket and can exhibit selectivity for the different receptor subtypes. This is illustrated by the high affinity of the natural alkaloid MLA which inhibits α7 nAChRs in the pM–nM range, whereas it acts only in the μM range at the α4β2 nAChRs (Wonnacott *et al.*, 1993) whilst DHβE blocks α4β2 nAChRs in the nM range and α7 nAChRs only in the μM range (Alkondon *et al.*, 1992). Concentration–response data can be fitted by the Hill equation and antagonists can be quantified by their IC_{50} (half inhibition) and Hill coefficient (nH, for the cooperativity). Polypeptides, such as toxins found in venom, can display very high affinity for nAChRs as particularly well illustrated by the toxin isolated from the snake *Bungarus Multicintus*. The muscle receptor displays a very high affinity for the α-bungarotoxin with an almost irreversible binding (Blanchard *et al.*, 1979; Oswald and Changeux, 1982). Models based on the AChBP structure for the binding of α-bungarotoxin at the muscle receptors are compatible with the hypothesis that, like acetylcholine, this toxin binds at the interface between the α and its adjacent subunit (Fruchart-Gaillard *et al.*, 2002). While α-bungarotoxin was first used to discriminate between muscle and ganglionic receptors the observation that brain tissue displayed a high affinity for this toxin raised some doubt about its specificity (Clarke *et al.*, 1985; Clarke 1992). Cloning and expression of the neuronal α7 nAChR that displayed a high affinity for the α-bungarotoxin clarified this issue and it was later recognized that α7 nAChRs were highly expressed in the mammalian brain (Couturier *et al.*, 1990a; Seguela *et al.*, 1993, Drisdel and Green, 2000; Tribollet *et al.*, 2004).

Non-competitive antagonists represent a distinct class as the magnitude of their action is independent of the agonist concentration. This suggests that they act at a site that is different from the ligand binding site. Non-competitive antagonists are further divided in two classes: open channel blockers (OCBs); and substances that do not interact with the ionic pore. Since most OCBs are charged molecules and they enter the pore to block the channel by steric hindrance, OCBs generally display marked voltage-dependency (Buisson and Bertrand, 1998b). Because OCBs can enter the channel only when the receptors are activated they display a use dependency that is a blockade proportional to the time in which the channels are open. Therefore, amplitude of the block must be related to the duration of the agonist exposure. As OCBs interact with the amino acids of the highly conserved pore domain, it can be hypothesized that they should display poor selectivity amongst receptor subtypes. Indeed, a poor selectivity for OCBs was observed within the nAChRs family for hexamethonium with a slightly higher affinity for the neuronal than the muscle receptors (Ascher et al., 1979; Bertrand et al., 1990). Notably, an OCB such as MK-801 can block receptors as different as the neuronal nAChRs and the NMDA receptor subtype of glutamate receptor family (Amador and Dani, 1991; Buisson and Bertrand, 1998b).

As discussed, pharmacological studies have provided evidence of receptor diversity and, based on the differences between ligand binding properties, it was proposed that ligands that are specific for a given receptor subtype could be used for the localization of these receptors in tissue. Brain slices exposed to radiolabelled αbungarotoxin, epibatidine, MLA or nicotine indeed display different patterns of labelling for these compounds (Jones and Wonnacott, 2004; Tribollet et al., 2004). Moreover, in certain cases specific brain pathways can be revealed by analysing the displacement of a given labelled molecule by another agonist (Whiteaker et al., 2000). Although these studies provide important clues about receptor distribution and the putative role of these nAChRs in brain function, caution must be taken in their interpretation. Similarly, radiolabelled compounds are often employed in biochemical experiments to examine how a given substance displaces the reference marker. Interestingly, values obtained from such binding experiments can differ markedly from those derived through functional techniques. The allosteric model provides a simple explanation for such discrepancy. According to this theory, the receptor can bind a compound in any of its states and the desensitized and resting states might display a higher affinity than the active state. Because labelling experiments including autoradiography and biochemical experiments employ long exposure of the receptors to the ligand in order to achieve equilibrium, these experiments preferentially assay the properties of the desensitized or resting states. Furthermore, binding experiments are often conducted with compounds that are competitive antagonists such as αbungarotoxin, and MLA which precludes a direct correlation with functional studies.

3.1.1.5 Physiology

In 1857 Claude Bernard proposed that a chemical substance mediates transmission between the motor nerve and the muscle fibres. The French physiologist was able to show that a plant-derived poison used by South American tribesmen reversibly blocks the nerveevoked contractions of frog muscle without affecting electrically evoked contractions. This observation led him to propose that a chemical substance was released from the nerve and that the poison, today known as curare, blocked its effect. This first evidence of a chemical synapse attracted great interest from physiologists and pharmacologists. The identification of the neurotransmitter acetylcholine by Otto Loewi represented another important step and it was subsequently demonstrated that the motor nerve releases this neurotransmitter molecule into the synaptic cleft of the neuromuscular junction.

Subcellular distribution of nAChRs

It is widely recognized that nAChRs can be somatic, postsynaptic, presynaptic, perisynaptic, and axonal (Lena et al., 1993; Fabian-Fine et al., 2001; Jones and Wonnacott, 2004). Hence, the physiological role of nAChRs in the peripheral nervous system (PNS) and central nervous system (CNS) will depend upon multiple factors including the nAChR subtype(s) and their subcellular localization (Table 3.1.1.2).

Since nAChRs are cationic receptors, the action of postsynaptic receptors is easy to understand as they are expected to depolarize the postsynaptic neuron and thereby mediate synaptic transmission. Surprisingly, such activity of postsynaptic neuronal nAChRs has been functionally detected in only very limited cases (Frazier et al., 1998, Hefft et al., 1999).

The physiological role of presynaptic receptors depends upon the subunit expressed in the synaptic bouton. Activation of α4β2 or α3β2 nAChRs by acetylcholine will cause a depolarization of the presynaptic bouton and activate voltage-gated calcium channels with the subsequent calcium influx modulating the release of the neurotransmitter from the presynaptic bouton (Radcliffe and Dani, 1998). A pharmacological dissection of such effects has been carried out by application of calcium channel blockers. In contrast, a direct effect of acetylcholine on intracellular calcium concentration has been observed for presynaptic boutons that express the homomeric α7 receptors (Dani, 2001). The calcium permeability of the α7 receptors is sufficient to influence neurotransmitter release directly. The importance of the modulation of neurotransmitter release is exemplified in the mesolimbic system where nicotine has been shown to induce dopamine release and it was proposed that nicotine addiction depends on this feedback mechanism (Dani and De Biasi, 2001; Dani, 2003; Wonnacott et al., 2005).

Evidence for somatic receptors comes from electrophysiological recordings performed either in isolated cells or in brain slices (Zaninetti et al., 2000; Dani, 2001; Zaninetti et al., 2002). Acetylcholine or nicotine application evokes an inward current in many cells, when monitored in voltageclamp, or a depolarization when recorded in current-clamp. Although this demonstrates that many neurons express nAChRs on their soma the role of these receptors can be complex. The observation that on some neurons activation of these receptors can provoke action potential firing suggests a role in the regulation of neuronal excitability. However, the absence of clear synaptic contacts and in most cases of excitatory postsynaptic potentials lead to the conclusion that these nAChRs have a modulatory role rather than participating directly in fast signal transmission (Dani, 2001; Hogg et al., 2003). It was proposed that the depolarization caused by nAChR activation brings the membrane potential of the neuron closer to its threshold value and therefore modulates the firing activity. Recent findings in vertebrate neurons have revealed that variation of the somatic potential modulates synaptic transmission (Alle and Geiger, 2006; Shu et al., 2006). This suggests that activation of nAChRs that results in the depolarization of the neuron soma could enhance synaptic transmission and therefore modulate network activity.

Activation of neuronal nAChRs expressed on axons can produce distinct effects depending upon their precise cellular localization. Activation of receptors expressed in the axon terminal can cause a localized depolarization of the axon segment and prevent the propagation of the action potential (Dani, 2001; Hogg et al., 2003; Zhang et al., 2004). Blockade of the axon potential propagation results in a silencing of the corresponding synapses. Receptors expressed on axon branches could play a comparable role, depolarizing short axonal segments and gating action potentials in specific branches of the axon tree.

Neuromuscular transmission

The physiological role of the neuromuscular junction is to amplify the electrical signal that propagates along the motor nerve to trigger an action potential in the muscle fibres. Moreover, every afferent motor neuron action potential must trigger an action potential in the muscle fibre. This implies the existence of a large safety factor both at the level of neurotransmitter release and that of nAChR function. Because action potential firing frequency in the motor nerve can be as high as 80 Hz, neuromuscular endplate currents have a fast decay time and exhibit little or no desensitization. Multiple processes exist to support this high-frequency signalling at the neuromuscular endplate. For example, it is known that neuronal nAChRs are expressed by the motor nerve and insert into the presynaptic process of the neuromuscular junction. These receptors have been shown to play a role in sustained neurotransmission by facilitating neurotransmitter release such that their blockade causes a reduction of the endplate current during trains of stimuli (Lingle and Steinbach, 1988).

The neuromuscular system displays important regenerative characteristics that allow both muscle fibres and motor nerves to regenerate following injury (Witzemann et al., 1989; Koenen et al., 2005). Muscle fibres initially regenerate in the absence of innervation, a condition that is similar to that observed during muscle development. During this transition phase, muscle fibres express the $\alpha1,\beta1,\gamma,\delta$ receptor subunit combination with receptors dispersed on the fibre surface rather than the high-density clustering observed in the mature endplate. Functional properties of the γ containing receptors are distinct from those of ε containing receptors and it is thought that their longer channel opening time and increased acetylcholine sensitivity facilitates acetylcholine detection, promoting the regeneration of synaptic connections (Koenen et al., 2005). The differential expression of the γ and ε subunits provides an example of how gene expression and receptor subunit composition can be dynamically regulated. In addition to this switch in subunit expression the presence of $\alpha7$ containing receptors was observed at the sarcolemma. The unique pharmacological properties of these receptors with respect to the $\alpha7$ homomeric receptors suggests, however, that either a special form of $\alpha7$ is expressed by the cell during this regeneration period or that $\alpha7$ associates with another subunit (Fischer et al., 1999; Tsuneki et al., 2003).

The peripheral nervous system

Synaptic transmission in the peripheral nervous system is largely mediated by acetylcholine through activation of nAChRs (Conroy

and Berg, 1995). Although the exact subunit composition of nAChRs in ganglia remains unknown, it is generally accepted that they contain $\alpha3$ and $\beta4$ subunits and that $\alpha5$ contributes to a fraction of the receptors. Recombinant $\alpha3\beta4$ nAChRs display a high sensitivity to acetylcholine and relatively little desensitization (Couturier et al., 1990b; Chavez-Noriega et al., 1997). These characteristics allow nAChRs to efficiently mediate synaptic transmission at the sustained discharge rates of preganglionic fibres. A closer examination of the parasympathetic synapse reveals a complex postsynaptic organization with specific localization of receptor subtypes (Williams et al., 1998; Berg and Conroy, 2002). In addition, the expression of $\alpha7$ subunits suggests that these homomeric nAChRs may also participate in synaptic transmission. Indeed, a significant contribution of $\alpha7$ receptors has been demonstrated in the ciliary ganglion (Berg and Conroy, 2002).

Both immunoprecipitation and physiological results suggest that different ganglia express specific receptor subtypes that confer distinct physiological and pharmacological profiles. For example, intracardiac ganglia that are known to express nAChRs present an unusual pharmacological profile with sensitivity to αbungarotoxin but fast reversibility of the blockade (Cuevas and Berg, 1998). Cloning and sequencing of the genes expressed in the mouse intracardiac ganglia revealed the expression of a distinct form of $\alpha7$ receptor that results from alternate splicing (Severance et al., 2004). These data illustrate a potential diversity of ganglionic nAChR function and await confirmatory experiments from other animal species and humans. Ganglionic-type receptors are also expressed in specific glands such as the adrenomedula, submandibular glands etc. where they control activity of the gland. Adrenomedula chromaffin cells receive cholinergic innervation and it is well established in different animal species that these cells express at least the $\alpha3$, $\beta4$, $\alpha5$ and $\alpha7$ subunits (Garcia-Guzman et al., 1995; Campos-Caro et al., 1997; Wenger et al., 1997; Tachikawa et al., 2001). Acetylcholine released by the nerve terminals evokes a response of both $\alpha7$-containing and $\alpha3\beta4$-containing receptors that provokes the release of mineralocorticoids (Inoue and Kuriyama, 1991; Di Angelantonio et al., 2003).

The broad expression of nAChRs throughout the PNS and their involvement in neurotransmission in systems as different as the muscle endplate, signalling in ganglia, control of gland secretion or functions of the urinary system, reveals the complexity and diversity of the role of nAChRs throughout the body.

The central nervous system

Behavioural experiments have demonstrated memory enhancing effects of cholinergic agonists while antagonists cause a loss of cognitive functions (Simpson, 1978; Levin, 2002; Van Kampen et al., 2004). The macroscopic localization of the nAChRs in the CNS provides important information about the role of these ligand-gated ion channels in brain function. Labelling of brain slices with radiolabelled αbungarotoxin and epibatidine reveals the distribution of $\alpha7$-containing versus heteromeric nAChRs (Clarke et al., 1985; Tribollet et al., 2004). These studies revealed that nAChRs are widely distributed both in central brain areas and in the neocortex but that the pattern of expression of receptor subtypes differs markedly. Whilst distribution of nAChRs could differ between rodents and higher mammals, comparable results were obtained in monkey brain using iodinated MLA (Kulak et al.,

2006). Further information about receptor distribution has been provided by *in situ* hybridization using subunit-specific mRNA labelling and immunoprecipitation (Zoli *et al.*, 1995; Gotti *et al.*, 1997; Charpantier *et al.*, 1998; Sgard *et al.*, 1999).

Post-mortem labelling of human brain confirmed both the wide overall distribution of nAChRs throughout the CNS and specific patterns of expression of the individual subtypes (Wevers *et al.*, 1994; Banerjee *et al.*, 2000; Wevers *et al.*, 2000; Perry *et al.*, 2001; Graham *et al.*, 2002). More recently the development of A85380, an α4β2-specific ligand that is suitable for positron emission tomography (PET), provided the first *in vivo* images of nAChR distribution in humans (Kimes *et al.*, 2003; Gallezot *et al.*, 2005). These studies confirmed a high level of expression of α4β2 nAChRs in the thalamus and more diffuse labelling in the cortex. Use of these techniques in longterm studies will allow the quantification of receptor expression in the course of ageing or in the progression of neurological diseases.

Physiological studies carried out in central brain areas such as the mesolimbic system, ventral tegmental area, or hypothalamus suggest that nAChRs have multiple roles and that they can modulate network activity (Mansvelder and McGehee, 2000; Zaninetti *et al.*, 2000; Hatton and Yang, 2002). Distinct mechanisms of neuronal modulation are possible according to nAChR subtype and localization in a given neuron. For example stimulation of nAChRs expressed on the cell soma can cause depolarization of the neuron and trigger action potential firing. However, activation of presynaptic receptors on the synaptic bouton can influence neurotransmitter release either by causing the depolarization of the presynaptic terminal and the activation of voltage-gated calcium channels or by direct calcium influx (Lena *et al.*, 1993; Mcgehee and Role, 1996; Dani *et al.*, 2000). Such modulatory mechanisms are though to play an important role in the mesolimbic dopaminergic system and may be critical in the development of nicotine addiction (Pidoplichko *et al.*, 2004, for reviews see Dani and De Biasi, 2001; Laviolette and van der Kooy, 2004).

Projections from the central brain areas, such as the thalamus, to cortical structures exhibit high levels of nAChR expression. However, in spite of this abundance of nAChRs, our understanding of the role that these receptors play remains limited. While the presence and function of nAChRs is easily revealed by electrophysiological recordings carried out in brain slices, these studies have produced little evidence for a role of nAChRs in mediating fast neurotransmission (Frazier *et al.*, 1998; Hefft *et al.*, 1999; Hogg *et al.*, 2003). In agreement with histological data many studies have, however, confirmed the presence and function of different receptor subtypes that are MLA- and/or DHβE-sensitive in cortical structures (Inoue and Kuriyama, 1991; Alkondon and Albuquerque, 1995; Ji and Dani, 2000; Hurst *et al.*, 2005; Rousseau *et al.*, 2005; Mansvelder *et al.*, 2006). The lack of characterized cholinergic synapses in these regions suggests, however, a diffuse transmission mediated by acetylcholine and supports the hypothesis that nAChRs have a modulatory function rather than a role in fast neurotransmission (Descarries *et al.*, 1997).

The early expression of nAChRs in the CNS during development suggests that these receptors may play a role in the establishment of synaptic transmission in neuronal networks (Agulhon *et al.*, 1998, 1999). Indeed, it has been shown that nicotine exposure activates immature connections in the developing hippocampus (Maggi *et al.*, 2003, 2004). Developmental regulation of synaptic activity

has also been observed in the visual and auditory cortex (Aramakis *et al.*, 2000).

nAChRs in retina

Both labelling and immunoprecipitation studies with the retina revealed that this specialized neuronal tissue expresses high levels of multiple nAChR subtypes (Whiting *et al.*, 1991; Keyser *et al.*, 1993; Kaneda *et al.*, 1995; Champtiaux *et al.*, 2003; Moretti *et al.*, 2004; Gotti *et al.*, 2005). The axons of retinal ganglion cells make up the optic nerve that connects the retina to the thalamus. In embryonic stages these cells express nAChRs even before forming connections with the thalamic nuclei (Matter *et al.*, 1992, 1995). Furthermore, retinal development and proper synaptic organization strongly depend upon spontaneous waves of activity requiring the activation of cholinergic synaptic transmission (Feller *et al.*, 1996).

Electrophysiological recordings of isolated retinal ganglion cells have confirmed the presence of functional nAChRs on these specialized neurons (Lipton *et al.*, 1987; Aizenman *et al.*, 1990). Subsequently, multiple nAChR subtypes have been characterized in isolated retina and have been shown to be developmentally regulated (Lecchi *et al.*, 2004; Reed *et al.*, 2004; Lecchi *et al.*, 2005; Marritt *et al.*, 2005; Strang *et al.*, 2005). Although much further investigation is necessary to enable a thorough understanding of the nAChRs in the retina, this work highlights the relevance of these ligand-gated ion channels in sensory functions.

The specific innervation of the inner ear

Transduction of the mechanical energy in sound waves into neuronal activity is mediated by the cochlear inner hair cells that contact synaptic endings of the auditory nerve. The inner ear contains, however, another group of ciliary cells, the outer hair cells, that participate in the tuning and amplification of the mechanical waves in the organ of corti. These cells receive efferent stimuli mediated by cholinergic fibres. Because outer hair cells display a peculiar pharmacological profile that appears mixed between muscarinic and nicotinic receptors, it was thought that these cells probably express a distinct ligand-gated ion channel. Cloning and expression of the α9 subunit was the first step in the resolution of this issue. Subsequently, the contribution of α9-containing nAChRs has been confirmed by numerous studies including genetically engineered mice in which the gene encoding for this subunit was silenced (Vetter *et al.*, 1999). While expression of the α9 subunit was sufficient to obtain functional receptors, the amplitude of the currents and difficulty of expression indicated that additional subunit(s) must be expressed by outer hair cells. The identification both in rat and human of the α10 subunit and the finding that it coassembles with the α9 subunit suggests that cochlear nAChRs are heteromeric (Elgoyhen *et al.*, 1994, 2001; Sgard *et al.*, 2002; Lustig, 2006). These nAChRs display a high calcium permeability resembling that observed for α7 receptors (Katz *et al.*, 2000; Sgard *et al.*, 2002). Physiologically, the calcium influx entering outer hair cells during nAChR activation is sufficient to activate calcium-sensitive potassium channels. This mechanism is thought to create a signal amplification and later triggers the physiological response in outer hair cells. Examination of α9 and α10 expression reveals that these subunits are expressed in very few brain areas and their role in these particular structures remains to be determined (Sgard *et al.*, 2002).

Expression of nAChRs in non-neuronal cells

Mapping of nAChRspecific mRNAs or labelling with nAChR specific ligands indicated that these receptors are expressed by many non-neuronal cells (for review see Gahring and Rogers, 2005). Given the common embryonic origin between neurons and epidermis, it is not surprising to observe that keratinocytes express neuronal nAChRs (Nguyen *et al.*, 2000; Arredondo *et al.*, 2003; Grando, 2006). Expression of nAChRs in glial cells has also been reported (Gahring *et al.*, 2004; Shytle *et al.*, 2004). Moreover, stimulation of α7 containing receptors in astrocytes caused a significant increase in intracellular calcium (Sharma and Vijayaraghavan, 2001). More unexpectedly, nAChR expression has been detected in adipose tissue and also in the immune system, where cholinergic transmission and nAChR activity has also been demonstrated (Blanchet *et al.*, 2005; Metz and Tracey, 2005; Saeed *et al.*, 2005; Ulloa, 2005). These results potentially explain the antiinflammatory effects of nicotine and nAChRspecific agonists. The discovery of cholinergic transmission in non-neuronal cells has further stimulated research into these ligand-gated ion channels.

Functional upregulation

Post-mortem ligand binding studies performed on control brain subjects and brains from smokers revealed a clear increase in radiolabelled nicotine binding in smokers' brains (Perry *et al.*, 1999). This increase in binding was termed upregulation. Numerous studies have since replicated these results both *in vivo* and *in vitro*.

To date, two hypotheses have been formulated to account for receptor upregulation: (1) the increase in nicotine binding corresponds to an increase in number of nAChRs expressed in the cell membrane (Peng *et al.*, 1994; Bencherif *et al.*, 1995; Flores *et al.*, 1997; Peng *et al.*, 1997; Corringer *et al.*, 2006) and (2) increase in the nicotine binding reflects a change in the receptor affinity for the ligand (Vallejo *et al.*, 2005). Functional studies showed that nicotine upregulation causes an increase of the current amplitude accompanied by an increase in apparent acetylcholine affinity and reduced desensitization (Buisson and Bertrand, 2001, 2002; Sallette *et al.*, 2005; Vallejo *et al.*, 2005).

3.1.1.6 Pathology

nAChR-associated diseases range from cognitive dysfunction to impaired motor control (Figure 3.1.1.3) and illustrate the importance of nAChRs in numerous pathological states (Table 3.1.1.2).

Myasthenia gravis, affecting about 14/100,000 of the population, is an impairment of the neuromuscular junction nAChRs. The large majority of myasthenia gravis cases are caused by an autoimmune reaction (Kalden, 1975; Simpson, 1978; Hohlfeld and Wekerle, 1981; Lindstrom, 2000). For reasons that remain uncertain, the immune system of these patients produces antibodies directed against the neuromuscular junction nAChRs. Binding of these antibodies can result in different effects depending upon the epitope at which they are targeted. While some myasthenia gravis antibodies directly interfere with receptor function,

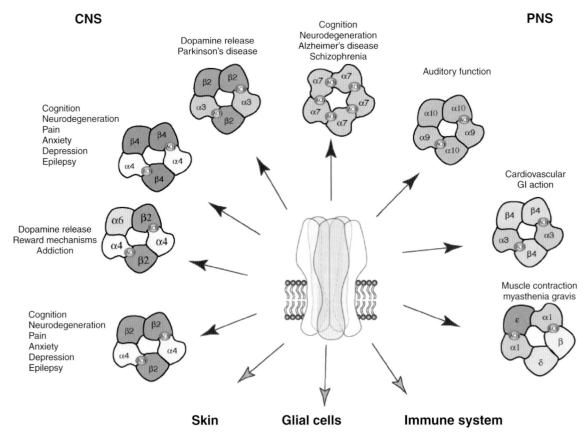

Fig. 3.1.1.3 Roles of nAChRs in brain function and disease. Schematic representation of putative native nAChR composition with their related brain functions. Known diseases associated with nAChR dysfunction are also indicated. Centre, schematic representation of the nAChR receptor inserted into the lipid membrane. Grey arrows and text in the lower part of the figure illustrate major known non-neuronal tissues that express neuronal nAChRs.

most of them increase receptor internalization. This causes a reduction in the number of receptors in the muscle plasma membrane and therefore attenuation of the synaptic transmission. When the acetylcholineevoked current does not reach the threshold for the action potential firing in the muscle fibre, a paralysis is observed. While the use of anticholinesterase agents is generally sufficient for patients' symptom relief, some cases necessitate blood treatment to reduce the level of antibodies. A distinct, but rarer form of myasthenia gravis is genetically transmissible (Ohno and Engel, 2003, 2005a, 2005b). These patients harbour a mutation in one of the genes encoding the neuromuscular junction nAChRs that causes a dysfunction of the receptor.

Although no autoimmune disease has been characterized to date for the neuronal nAChR receptor, impairments comparable to that observed in myasthenia gravis cannot be excluded and could be a cause of pathologies. Progress in the understanding of our genes and their variation with the sequencing of the human genome, and studies of single nucleotide polymorphisms (SNPs) has furthered understanding of the link between genetics and pathology. As it is well recognized that genetic variability can account for many factors that distinguish one person from another, it is expected that variation or polymorphisms may be present at nAChR encoding genes and that they may be the cause of some pathologies, such as myasthenia gravis. The finding of an association between an $\alpha 4$ nAChR subunit polymorphism and a rare form of genetically transmissible nocturnal epilepsy represents compelling evidence for the determinant role of nAChRs in brain function (Steinlein et al., 1995). This genetic study carried out on a very large Australian family revealed that mutation of a single amino acid in CHRNA4, the gene encoding the $\alpha 4$ nAChR subunit, is associated with epileptic seizures. The autosomal dominant mode of transmission, frontal lobe seizure initiation and the preferential occurrence of seizures during sleep gave rise to the acronym ADNFLE (autosomal dominant nocturnal frontal lobe epilepsy) that is currently used as reference for this form of epilepsy. Since then, several distinct mutations have been identified in patients suffering from this particular form of epilepsy (Steinlein, 2004). As the major brain high-nicotine-affinity nAChRs are heteromers composed of $\alpha 4$ and $\beta 2$ subunits, it was first hypothesized and later demonstrated that mutations in CHRNB2, the gene that encodes for the $\beta 2$ subunit, are also associated with ADNFLE (De Fusco et al., 2000; Phillips et al., 2001; Bertrand et al., 2005). Expression of the receptors harbouring the mutations found in ADNFLE patients in a heterologous system allowed functional studies to be performed. While each mutation causes slightly different modifications of the receptor properties they all present a common trait, which is an increase in acetylcholine sensitivity (De Fusco et al., 2000; Itier and Bertrand, 2002; Bertrand et al., 2005). It was therefore proposed that seizures triggered in the frontal lobe must be caused by this increase in acetylcholine sensitivity that induces an inbalance between excitation and inhibition (Hogg and Bertrand, 2004).

The high prevalence of smoking in psychiatric patients has prompted much speculation about the possible role of nicotine as a selfmedication used by patients to relieve their psychiatric symptoms (Leonard et al., 1998). This is perhaps best exemplified in the case of schizophrenia which exhibits a prevalence of approximately 1% in the general population with a high proportion of affected individuals also smoking (Leonard et al. 1996). Furthermore it was observed that auditory gating, as assayed by P50 inhibition, is markedly impaired in many schizophrenic patients and that nicotine administration partially restores this deficit (Adler et al. 1998; De Luca et al. 2004). An initial genetic study carried out in schizophrenic patients exhibiting such auditory gating deficits suggested an association of the phenotype with polymorphisms at the CHRNA7 locus, the gene encoding the $\alpha 7$ subunit (Freedman et al., 1995, 1997; Leonard et al., 2002). Although subsequent studies failed to confirm this association, the $\alpha 7$ gene remains a focus of current study. Selective agonists for $\alpha 7$ nAChRs have been identified and have been shown to restore impaired auditory gating in animal models. Similarly, the $\alpha 7$ positive allosteric modulator PNU120596 was shown to improve amphetamine-induced auditory gating deficits in rodents (Hurst et al., 2005). Most recently a positive outcome was reported in a study with a small cohort of schizophrenic patients utilizing an $\alpha 7$ selective agonist, suggesting therapeutic potential for ligands at this target (Olincy et al., 2006).

Alzheimer's and Parkinson's disease are progressive neurodegenerative disorders where cholinergic dysfunction has been associated with the observed cognitive impairment (Banerjee et al., 2000; D'Andrea and Nagele, 2006). Alzheimer's disease is characterized by the accumulation of β amyloid peptides within the brain and several studies have reported an interaction between the β amyloid 142 peptide and the $\alpha 7$ nAChR, although difficulties in replicating these results are of concern (Liu et al., 2001; Lamb et al., 2005; Pym et al., 2005; Spencer et al., 2006). Moreover, post-mortem studies have not revealed any correlation between β amyloid levels and nAChR expression. A correlation between the prevalence of Parkinson's disease and smoking suggested a putative neuroprotective role of nicotine in the development of this neurological disorder (Newhouse et al., 1997; Court et al., 2000; Costa et al., 2001). Notably, the demonstration of a rather specific pattern of expression of the $\alpha 6$ nAChR subunit in brain areas that contain the dopaminergic neurons which progressively degenerate in Parkinson's disease suggests a possible role of the $\alpha 6$ subunit in the disease (Quik et al., 2004).

3.1.1.7 Concluding remarks

nAChRs were first identified at the neuromuscular junction where they mediate synaptic transmission between the nerve terminal and muscle fibres and these receptors have been utilized extensively as a model of ligand-gated ion channel function. Cloning and sequencing of genes that encode for multiple forms of nAChRs that are widely expressed both in the PNS and CNS, as well as by non-neuronal cells, and the development of receptor subtype selective ligands have presented new opportunities to explore the role of these receptors. Whilst nAChRs have been implicated both in disease pathophysiology and as therapeutic targets, many questions regarding their functional roles remain unanswered and further studies are needed to gain a better understanding of their roles in brain function and dysfunction.

Table 3.1.2 The nicotinic acetylcholine receptors (nAChRs)

Receptor subtype	Biophysical characteristics	Cellular and subcellular localization	Pharmacology (heterologous expression)	Physiology	Relevance to disease states
Muscle nAChR α1β1γδ	*Mouse:* P_{Ca}/P_{Na} 2.1 [1] τ_{des} 53 ms and 459 ms at 100 μM ACh [2] *Calf:* g 40 pS [3]	Fetal muscle and denervated adult muscle [3, 4] Torpedo electric organs [3] Myoid cells of the thymus [5] Extraocular muscles [6, 7]	Agonists EC_{50} ACh 5.6 μM *t* [8] Nicotine 250 μM *t* [8] +Epibatidine 1.6 μM *t* [8]; 7.3 μM *h/t* [8] −Epibatidine 3.9 μM *t* [8]; 15.9 μM *h/t* [8] Antagonists IC_{50} α-Bgt 100 nM complete and irreversible block *t* [9]	Mediate signal transmission from the nerve to the muscle to induce contraction [10]	Reduced function in fetal arthrogryposis multiplex congenita [11] Impaired function and altered expression in autoimmune form of myasthenia gravis [7, 12–17]
Muscle nAChR α1β1εδ	*Mouse:* P_{Ca}/P_{Na} 4.2 [1] *Calf:* g 60 pS [3]	Adult endplates [3] Thymic epithelial cells [18]	Antagonists IC_{50}: d-TC 40 nM *m* [19, 20]	Mediate signal transmission from the nerve to the muscle to induce contraction [10]	Associated with congenital form of myasthenia gravis [21–30]
α4β2	*Human:* g 46 pS [31] τ_{des} 120 and 749 ms at 300 μM ACh [32] *Rat:* g 12, 22, 34 pS [33, 34] *Chick:* P_{Ca}/P_{Na} 2.9 [1] g 20 pS [35]; 18 and 30 pS [36] Rectification of macroscopic currents [37]	Habenula neurons, interpeduncular nuclei, lateral spiriform nucleus [38, 39] Cerebral cortex and hippocampus [40–42] Dopaminergic neurons in striatum, substantia nigra and VTA [43–45] Retina [46, 47] Sympathetic ganglia [39] Nociceptive dorsal root ganglion neurons [48] Myeloma, hybridoma cells and B lymphocytes [49, 50]	Agonists EC_{50} ACh high affinity 0.7–16 μM and low affinity 68–74 μM *h* [32, 51]; high-affinity 0.3 μM and low-affinity 58.3 μM *r* [52]; 0.3–0.77 μM *c* [8, 37, 53, 54] Nicotine high-affinity 0.3–2.4 μM and low-affinity 14.5–18 μM *h* [32, 51] ±Epibatidine 4 nM *c* [8] +Epibatidine 21 nM *r* [55] −Epibatidine 23 nM *r* [55] Cytisine high-affinity 4.5 nM and low-affinity 2.1 μM *h* [56] DMPP 1.9–20 μM *h* [32, 57, 58] Choline 3.8 μM *h* [57] Antagonists IC_{50}: DHβE 80–110 nM *h* [31, 58]; 370 nM *r* [59] Mecamylamine 0.47 μM *h* [57] MLA 1.5 μM *h* [31] d-TC 3.16–62 μM *h* [57, 60] Hexametonium 6.83–11 μM *h* [57, 60]	Modulation of GABA release in hippocampus, cerebral cortex [40, 41] Regulation of dopamine and glutamate release in the striatum [45, 61] Excitatory signals mediated by postsynaptic receptors in dopaminergic neurons in the substantia nigra [62, 63] Modulation of dopaminergic cell activity in VTA [44] Regulation of dopamine release from striatal synaptosomes [64] Participation in regulating myeloma and hybridoma cell proliferation and function [49] Implication in the preimmune status and in the regulation of antibody response [50]	Reduced expression in the temporal cortex of individuals with Alzheimer's disease [65] Altered nicotine binding sites in Parkinson's, Lewy body dementia and Alzheimer's disease [66] Target of nicotine therapy in patients with Alzheimer's disease [67] Mutations that modify receptor function associated with autosomal dominant nocturnal frontal lobe epilepsy (ADNFLE) [68–80] Single point mutation associated with benign familial neonatal convulsion epilepsy [81] Neuroprotection against β-amyloid cytotoxicity [82] Possible association of receptor upregulation with nicotine addiction [83] Possible targets of carbamate pesticide toxicology [84]
α4β2α6		Transcripts for α6 in: locus coeruleus, ventral tegmental area, substantia nigra, reticular thalamic nucleus, supramammillary nucleus, mesencephalic V nucleus, medial habenula, interpeduncular nucleus, olfactory bulb, trigeminal ganglion, retina [85–87] Transcripts for α6 in nasal mucosa [88] Immunohistochemical localization in: substantia nigra, ventral tegmental area, locus coeruleus, medial habenula, dopaminergic neurons of midbrain [89], striatum [90]	Agonists EC_{50}: ACh 2.6 μM *h* [91] Nicotine 1.3 μM *h* [91]	Possible involvement in modulation of dopaminergic transmission [89] Nicotine-induced dopamine release [90] Suggested involvement of α6-containing receptors in the modulation of locomotor behaviour and reward [85]	Reduced transcripts in substantia nigra after induced nigrostriatal degeneration [92] α6-containing receptors as possible target of neurodegenerative disorders linked to the nigrostriatal system, such as Parkinson's disease [93]

α4β4	*Chick:* P_{Ca}/P_{Na} 3 [1] Rectification of macroscopic currents [37]	Dopaminergic neurons [94] Retina [95], [46] Transcripts in: hippocampus, habenula, interpeduncular nuclei [39]	Agonists EC_{50}: ACh 4.8 μMc [37]; 19.68 μM h [58] Nicotine 5 μM h [58] Cytisine high-affinity 3.6 nM and low-affinity 1.1 μM h [56] DMPP 18.71 μM h [58] Antagonists IC_{50}: DHβE 10 nM h [58]; 190 nM r [59] d-TC 0.21 μM h [58]	Decreased expression by interleukin-1β [96]	Possible targets of carbamate pesticide toxicology [84] Possible site of interaction for the effects of cocaine in the nervous system [97]
α3β2	*Human:* g 26 pS [98] τ_{act} 0.23 s, $\tau_{1/2des}$ 1.1 s [99] *Rat:* P_{Ca}/P_{Na} 1.5 [100] 15.4, 5.1 pS [33] Rectification of macroscopic currents [37]	Striatal synaptosomes [101] Neuromuscular junction of the rat diaphragm [102] Noradrenergic axons in neocortex and hippocampus [103, 104] Retina [46]	Agonists EC_{50}: ACh 26, 28, 443 μM h [8, 58, 105]; 4.1, 4.4, 5.6 μM c [8, 37, 106]; 30, 44 μM r [107, 108] Nicotine 6.7, 6.8 μM h [8, 105]; 200 μM c [8] +Epibatidine 22 nM h [8]; 17 nM c [8] −Epibatidine 0.13 μM h [8]; 42 nM c [8] Cytisine 67, 71 μM h [58, 99] DMPP 2.1, 56 μM h [58, 99] Antagonists IC_{50}: α-CTX MII 0.5 nM r [109] DHβE 1.62 μM h [58]; 410 nM r [59] d-TC 2.41 μM h [58] Mecamylamine 2.8 μM r [107] Hexamethonium 8.3 μM r [107]	Modulation of dopamine release from striatal synaptosomes [101, 110] Mediation of the facilitation of ACh release at the neuromuscular junction of the rat diaphragm [102] Regulation of noradrenaline release in hippocampus and neocortex [103, 104]	Possible targets of carbamate pesticide toxicology [84]
α3β4	*Human:* g 31 pS [98] τ_{act} 0.77 s, $\tau_{1/2des}$ 1.8 s [99] *Chick:* P_{Ca}/P_{Na} 4.6 [1] τ_{des} 0.8 s, 22 s at 100 μM ACh [37] Rectification of macroscopic currents [37]	Autonomic ganglion neurons [111] Autonomic ganglia that control bladder function [112] Hippocampal synaptosomes [101] Sensory neurons in trigeminal ganglia [113] Inhibitory presynaptic terminals in substantia gelatinosa of the spinal cord [114] Chromaffin cells [115–117] Retina [46, 47] Bronchial epithelial and endothelial cells [118] Medial habenula neurons [119]	Agonists EC_{50}: ACh 160, 163, 203 μM h [8, 58, 105]; 52, 207 μM, high-affinity 39 μM and low-affinity 2.9 mM r [52, 107, 108]; 53, 158 μM c [8, 37] Nicotine 80.3, 106, 110 μM h [8, 58, 105]; 65 μM r [120]; 48, 410 μM c [8, 106] +Epibatidine 73 nM h [8]; 21 nM c [8]; 36 nM r [55] −Epibatidine 21 nM h [8]; 9 nM c [8]; 19 nM r [55] Cytisine 72, 76 μM h [58] Gerzanich (1998); 134 μM r [121] DMPP 10, 18.7 μM h [58, 99]; 92 μM r [121] Antagonists IC_{50}: Mecamylamine 140 nM r [107] Hexamethonium 310 nM r [107] α-CTX AuIB 750 nM r [101] d-TC 2.24 μM h [58] α-CTX IMI 2.5 μM b [115] DHβE 13.77 μM h [58]; 23.1 μM r [59]	Modulation of norepinephrine release from hippocampal synaptosomes [101] Proposed control of inhibitory activity in substantia gelatinosa of the spinal cord [114] Association with catecholamine secretion from chromaffin cells [116]	Possible involvement in opioid withdrawal [122] Possible targets of carbamate pesticide toxicology [84]

Continued

Table 3.1.2 (*Continued*) The nicotinic acetylcholine receptors (nAChRs)

Receptor subtype	Biophysical characteristics	Cellular and subcellular localization	Pharmacology (heterologous expression)	Physiology	Relevance to disease states
$\alpha3\beta4\alpha5$	*Human:* g 36 pS [98] τ_{act} 0.43 s, $\tau_{1/2des}$ 0.7 s [99] Rectification of macroscopic currents [99]	Autonomic ganglion neurons [111] Chick ciliary ganglion neurons [123] Human neuroblastoma cells [105, 124] Chromaffin cells [115–117]	Agonists EC_{50}: ACh 122 μM h [105] Nicotine 105 μM h [105] Cytisine 20 μM h [99] DMPP 20 μM h [99] Antagonists IC_{50}: α-CTX IMI 2.5 μM b [115]	Association with catecholamine secretion from chromaffin cells [116]	
$\alpha7$	*Rat:* P_{Ca}/P_{Na} 20 [125] *Chick:* P_{Ca}/P_{Na} 10 [126] g 45 pS [127] τ_{act} 0.05 s at 100 μM ACh [128] Rectification of macroscopic currents [129]	Hippocampus [40, 130–137, 138] Thalamus [133] Magnocellular neurons of the hypothalamic supraoptic and paraventricular nuclei [139, 140] Neurons from olfactory bulb [141, 142] Cerebral cortex [41] Ventral tegmental area (VTA) [143] Vagal motoneurons [144] Ciliary ganglion neurons [145–150], superior cervical ganglion neurons [151], sympathetic neurons [152, 153], cardiac parasympathetic neurons [154, 155] Nociceptive dorsal root ganglion neurons [48] Hippocampal [156] and cortical astrocytes [157] Retina [158] Transcripts in: amygdala, brainstem, Purkinje cells in the cerebellum, spinal cord [125] Lymphocytes [50, 159], macrophages [160], myeloma and hybridoma cells [49] Keratinocytes [161, 162] Bronchial epithelial and endothelial cells [118], lung cells [163] and small-cell lung carcinoma [164] Vascular endothelial cells [165, 166] Chromaffin cells [115] Denervated muscles [167]	Agonists EC_{50}: ACh 79, 155 μM h [8, 168]; 150, 316 μM r [108, 169]; 115–120 μM c [8, 127, 129, 170, 171] Nicotine 49, 83 μM h [168, 172]; 25 μM r [169]; 10, 18 μM c, [8, 170, 171] +Epibatidine 1.2 μM h [8]; 2.5 μM r [55]; 2 μM c [8] −Epibatidine 1.1 μM h, [8]; 2.03 μM r [55]; 2.2 μM c [8] Cytisine 71.4 μM h [58]; 14.9 μM c [171] DMPP 95 μM h [168] Choline 1.6 mM r [142] Antagonists IC_{50}: MLA 0.025 nM c [173] α-Bgt 0.73 nM c [129] a-CTX IMI 220 nM r [174] d-TC 3.1 μM h [58]; 0.27 μM c [175] DHβE 1.6 μM c [170]; 19.6 μM h [58]	Excitatory signals mediated by postsynaptic receptors in hippocampal CA1 interneurons [137] and in dopaminergic neurons in the substantia nigra [62, 63] Modulation of GABAergic synaptic transmission in the hippocampus [138] and cerebral cortex [41] Modulation of calcium signalling in CNS [156] Regulation of glutamate [176] and noradrenaline release [177] in the hippocampus Lymphocyte development [159] Regulation of myeloma and hybridoma cell proliferation and function [49] Implication in the preimmune status and in the regulation of antibody response [50] Regulation of inflammatory response [160, 178, 179] Control of cutaneous homeostasis: keratinocyte migration [161, 180], apoptosis [162] Angiogenic effects [165, 166]	Proposed to be involved in neuronal development [181] Decreased expression associated with schizophrenia [182, 183, 184] Associated with auditory sensory gating deficit in schizophrenia [185, 186] Decreased expression associated with Alzheimer's disease [187, 188] Associated with β-amyloid peptide in Alzheimer's disease [189]

α9α10	Human: P_{Ca}/P_{Na} 14.4 [190] Rat: P_{Ca}/P_{Na} 9 [191] g 76 pS [192] Rectification of macroscopic currents [190]	Outer hair cells of the cochlea [193] Dorsal root ganglion neurons [194] α10 subunit in nociceptive dorsal root ganglion neurons [48] Pituitary gland ([195], [190]) Lymphocytes [196] α9 subunit in keratinocytes [197]	Mediate synaptic transmission between olivocochlear fibers and outer hair cells of the cochlea [193]	Agonists EC_{50}: ACh 27 μM h [190]; 10–14 μM r [191, 192] Choline 519 μM h [190]; 538 μM r [192] Antagonists IC_{50}: Nicotine 3.9, 4.6 μM r [192, 193] α-Bgt 14 nM h [190] d-TC 730 nM h [190]; 110 nM r [193]

b, bovine; c, chick; h, human; m, mouse; r, rat; t, Torpedo; α-Bgt, alpha-bungarotoxin; ACh, acetylcholine; MLA, methyllycaconitine; DHβE, dihydrobeta-erithroidine; d-TC. D-tubocuraine; α-CTX, alpha-conotoxin; DMPP, dimethylphenyl-piperazinium-iodide.

[1] Ragozzino D et al. (1998) J Physiol 507, 749–757; [2] Naranjo D et al. (1993) Science 260, 1811–1814; [3] Mishina M et al. (1986) Nature 321, 406–411; [4] Fumagalli G et al. (1990) Neuron 4, 563–569; [5] Kao I et al. (1977) Science 195, 74–75; [6] Horton RM et al. (1993) Neurology 43, 983–986; [7] Lindstrom JM (2000) Muscle Nerve 23, 453–477; [8] Gerzanich V et al. (1995) Mol Pharmacol 48, 774–782; [9] Changeux JP et al. (1970) Proc Natl Acad Sci USA 67, 124–1247; [10] Hille B (1992) Ionic Channels of Excitable Membranes. Sinauer Associates Inc, Sunderland, Massachusetts; [11] Vincent A et al. (1995) Immunol Commun 5, 323–344; [13] Lennon VA et al. (1976) Ann N Y Acad Sci 274, 283–299. [14] Appel SH et al. (1982) Am J Physiol 243, E31–36; [15] Ashizawa T et al. (1982) Ann Neurol 11, 22–27; [16] Clementi F et al. (1985) Eur J Cell Biol 37, 220–228; [17] Lennon VA et al. (1985) Proc Natl Acad Sci USA 82, 8805–8809; [18] Wakkach A et al. (1996) J Immunol 157, 3752–3760; [19] Kopta C et al. (1994) J Neurosci 14, 3922–3933; [20] Steinbach JH et al. (1995) Neurosci 15, 230–240; [21] Ohno K et al. (1995) Proc Natl Acad Sci USA 92, 758–762; [22] Sine SM et al. (1995) Neuron 15, 229–239; [23] Engel AG et al. (1996) Ann Neurol 40, 810–817; [24] Engel AG et al. (1996) Hum Mol Genet 5, 1217–1227; [25] Gomez CM et al. (1996) Ann Neurol 39, 712–723; [26] Ohno K et al. (1996) Neuron 17, 157–170; [27] Milone M et al. (1997) Neurosci 17, 5651–5665; [28] Croxen R et al. (2002) Neurology 59, 162–168; [29] Kraner S et al. (2002) Neurogenetics 4, 87–91; [30] Shelley C et al. (2005) J Physiol 564, 377–396; [31] Buisson B et al. (1996) J Neurosci 16, 7880–7891; [32] Nelson ME et al. (2003) Mol Pharmacol 63, 332–341; [33] Papke RL et al. (1989) Neuron 3, 589–596; [34] Charnet P et al. (1992) J Physiol 450, 375–394; [35] Balliver M et al. (1988) Neuron 1, 847–852; [36] Pereira EF et al. (1994) J Pharmacol Exp Ther 270, 768–778; [37] Couturier S et al. (1990) Biol Chem 265, 17560–17567; [38] Brussaard AB et al. (1994) Pflugers Arch 429, 27–43; [39] McGehee DS et al. (1995) Annu Rev Physiol 57, 521–546; [40] Alkondon M et al. (1988) Neuron 1, 1847–852; [41] Albuquerque EX et al. (2000) Behav Brain Res 113, 131–141; [42] Christophe E et al. (2002) J Neurophysiol 88, 1318–1327; [43] Zoli M et al. (2002) J Neurosci 22, 8785–8789; [44] Chen Y et al. (2003) Neuropharmacology 45, 334–344; [45] Skok MV et al. (2003) Mol Pharmacol 64, 885–889; [50] Skok M et al. (2005) Eur J Pharmacol 517, 246–251; [51] Buisson B et al. (2001) FEBS Lett 450, 273–279; [56] Houlihan LM et al. (2001) J Neurochem 78, 1029–1043; [57] Eaton JB et al. (2003) Mol Pharmacol 64, 1283–1294; [58] Chavez-Noriega LE et al. (1997) J Pharmacol Exp Ther 280, 346–356; [59] Harvey SC et al. (1996) J Neurochem 67, 1953–1959; [60] Buisson B et al. (1998) Mol Pharmacol 53, 555–563; [61] Matsubayashi H et al. (2001) Jpn J Pharmacol 86, 429–43; [62] Matsubayashi H et al. (2004) Brain Res 1005, 1–8; [63] Matsubayashi H et al. (2004) Brain Res Mol Brain Res 129, 1–7; [64] Soliakov L et al. (1995) Neuropharmacology 34, 1535–1541; [65] Warpman U et al. (1991) Neuroreport 6, 2419–2423; [66] Perry EK et al. (1995) Neuroscience 64, 385–395; [67] Baldinger SL et al. (1995) Ann Pharmacother 29, 314–315; [68] Steinlein OK et al. (1995) Nat Genet 11, 201–203; [69] Weiland S et al. (1996) FEBS Lett 398, 91–96; [70] Kuryatov A et al. (1997) J Neurosci 17, 9035–9047; [71] Steinlein OK et al. (1997) Hum Mol Genet 6, 943–947; [72] Bertrand S et al. (1998) Br J Pharmacol 125, 751–760; [73] Figl A et al. (1998) J Physiol 513, 655–670; [74] Hirose S et al. (1999) Neurology 53, 1749–1753; [75] De Fusco M et al. (2000) Nat Genet 26, 275–276; [76] Phillips HA et al. (2001) Am J Hum Genet 68, 225–231; [77] Matsushima N et al. (2002) Epilepsy Res 48, 181–186; [78] Rodrigues-Pinguet N et al. (2003) J Physiol 550, 11–26; [79] Bertrand D et al. (2005) Neurobiol Dis 20, 799–804; [80] Rodrigues-Pinguet NO et al. (2005) Mol Pharmacol 68, 487–501; [81] Beck C et al. (1994) Neurobiol Dis 1, 95–99; [82] Kihara T et al. (1998) Brain Res 792, 331–334; [83] Buisson B et al. (2002) Trends Pharmacol Sci 23, 130–136; [84] Smulders CJ et al. (2003) Toxicol Appl Pharmacol 193, 139–146; [85] Le Novere N et al. (1996) Eur J Neurosci 8, 2428–2439; [86] Keiger CJ et al. (2000) Biochem Pharmacol 59, 233–240; [87] Vailati S et al. (1999) Mol Pharmacol 56, 11–19; [88] Keiger CJ et al. (2003) Ann Otol Rhinol Laryngol 112, 77–84; [89] Goldner FM et al. (1997) Neuroreport 8, 2739–2742; [90] Champtiaux N et al. (2003) J Neurosci 23, 7820–7829; [91] Kuryatov A et al. (2000) Eur J Pharmacol 393, 223–30; [93] Quik M et al. (2006) J Pharmacol Exp Ther 316, 481–489; [94] Charpantier E et al. (1998) Neuroreport 9, 3097–3101; [95] Barabino B et al. (2001) Mol Pharmacol 59, 1410–1417; [96] Gahring LC et al. (2005) J Neuroimmunol 166, 88–101; [97] Fran-is et al. (2003) Synapse 49, 77–88; [103] Sershen H et al. (1997) Neuroscience 77, 121–130; [104] Amtage F et al. (2004) Brain Res Bull 62, 413–423; [105] Wang F et al. (1996) J Biol Chem 271, 17656–17665; [106] Hussy N et al. (1994) J Neurophysiol 72, 1317–1326; [107] Cachelin AB et al. (1995) Pflugers Arch 429, 449–451; [108] Papke RL et al. (1997) Br J Pharmacol 120, 429–438; [109] Cartier GE et al. (1996) J Biol Chem 271, 7522–7528; [110] Kulak JM et al. (1997) J Neurosci 17, 5263–5273; [111] Lukas RJ et al. (1999) Pharmacol Rev 51, 397–401; [112] De Biasi M et al. (2000) Eur J Pharmacol 393, 137–140; [113] Flores CM et al. (1996) J Neurosci 16, 7892–7901; [114] Takeda D et al. (2003) Pain 101, 13–23; [115] Broxton NM et al. (1999) J Neurochem 72, 1656–1662; [116] Tachikawa E et al. (2001) Neurosci Lett 312, 161–164; [117] Di Angelantonio S et al. (2003) Eur J Neurosci 17, 2313–2322; [118] Wang Y et al. (2001) Mol Pharmacol 60, 1201–1209; [119] Quick MW et al. (1999) Neuropharmacology 38, 769–783; [120] Fenster CP et al. (1997) J Neurosci 17, 5747–5759; [121] Wong ET, et al. (1995) Brain Res Mol Brain Res 28, 101–109; [122] Taraschenko OD et al. (2005) Eur J Pharmacol 513, 207–218; [123] Conroy WG et al. (1995) J Biol Chem 270, 4424–4443; [124] Lukas RJ (1993) Pharmacol Exp Ther 265, 294–302; [125] Seguela P et al. (1993) J Neurosci 13, 596–604; [126] Bertrand D et al. (1993) Proc Natl Acad Sci USA 90, 6971–6975; [127] Revah F et al. (1991) Nature 353, 846–849; [128] Devillers-Thiery A et al. (1992) Neuroreport 3, 1001–1004; [129] Couturier S et al. (1990) Neuron 5, 847–856; [130] Alkondon M et al. (1993) J Pharmacol Exp Ther 265, 1455–1473; [131] Alkondon M et al. (1994) J Pharmacol Exp Ther 271, 494–506; [132] Albuquerque EX et al. (1995) Ann N Y Acad Sci 757, 48–72; [133] Alkondon M et al. (1997) J Pharmacol Exp Ther 283, 1396–1411; [134] Hefft S et al. (1999) J Physiol 515, 769–776; [135] Frazier CJ et al. (1998) J Neurosci 18, 1187–1195; [137] Alkondon M et al. (1998) Brain Res 810, 257–263; [138] Alkondon M et al. (2000) Eur J Pharmacol 393, 59–67; [139] Zaninetti M et al. (2000) Neuroscience 95, 319–323; [140] Zaninetti M et al. (2002) Neuroscience 110, 287–299; [141] Alkondon M et al. (1994) Neurosci Lett 176, 152–156; [142] Alkondon M et al. (1997) Eur J Neurosci 9, 2734–2742; [143] Pidoplichko VI et al. (1997) Nature 390, 401–404; [144] Zaninetti M et al. (1999) Eur J Neurosci 11, 2737–2748; [145] Chiappinelli VA et al. (1978) Proc Natl Acad Sci USA 75, 2999–3003; [146] Jacob MH et al. (1983) J Neurosci 3, 260–271; [147] Loring RH et al. (1985) Neuroscience 14, 645–660; [148] Horch HL et al. (1995) Neuron 15, 7778–7795; [149] Ullian EM et al. (1997) J Neurosci 17, 7210–7219; [150] Blumenthal EM et al. (1999) J Neurophysiol 81, 111–120; [151] Cuevas J et al. (2000) J Physiol 525, 735–746; [152] Listerud M et al. (1991) Science 254, 1518–1521; [153] Yu CR et al. (1998) J Physiol 509, 651–665; [154] Cuevas J et al. (1998) J Neurosci 18, 10335–10344; [155] Bibevski S et al. (2000) J Neurosci 20, 5076–5082; [156] Sharma G et al. (2001) Proc Natl Acad Sci USA 98, 4148–4153; [157] Okawa H et al. (2005) J Neurosci Res 79, 535–544; [158] Gotti C et al. (1997) Eur J Neurosci 9, 1201–1211; [159] Skok M et al. (2006) J Neuroimmunol 171, 86–98; [160] Wang H et al. (2003) Nature 421, 384–388; [161] Zia S et al. (2000) Pharmacol Exp Ther 293, 973–981; [162] Arredondo J et al. (2003) Life Sci 72, 2063–206- ; [163] Plummer HK et al. (2005) Respir Res 6, 29; [164] Codignola A et al. (1996) Brain Res Mol Brain Res 129, 1–7; [165] Heeschen C et al. (2006) Life Sci 78, 1863–1870; [167] Tsuneki H et al. (2003) J Physiol 547, 165–179; [168] Gopalakrishnan M et al. (1995) Eur J Pharmacol 290, 237–246; [169] Puchacz E et al. (1994) FEBS Lett 342, 286–290; [171] Galzi JL et al. (1991) FEBS Lett 294, 198–202; [171] Neurosci Lett 146, 87–90, [172] Briggs CA et al. (1995) Neuropharmacology 34, 583–590; [173] Palma E et al. (1996) J Physiol 491, 151–161; [174] Johnson DS et al. (1995) Mol Pharmacol 48, 194–199; [175] Gotti C et al. (1994) Eur J Neurosci 6, 1281–1291; [176] Kanno T et al. (2005) Biochem Biophys Res Commun 338, 742–747; [177] Barik J et al. (2006) Mol Pharmacol 69, 618–628; [178] Floto RA et al. (2003) Lancet 361, 1069–1070; [179] de Jonge WJ et al. (2005) Nat Immunol 6, 844–851; [180] Chernyavsky AI et al. (2004) J Cell Sci 117, 5665–5679; [181] Tagulhon C et al. (1999) Neuroreport 10, 2223–2227; [182] Freedman R et al. (1995) Biol Psychiatry 38, 22–33; [183] Freedman R et al. (2000) J Chem Neuroanat 20, 299–306; [184] Olincy A et al. (2006) Arch Gen Psychiatry 63, 630–638; [185] Freedman R et al. (1997) Proc Natl Acad Sci USA 94, 587–592; [186] Adler LE et al. (1998) Schizophr Bull 24, 189–202; [187] Hellstrom-Lindahl E et al. (1999) Brain Res Mol Brain Res 66, 94–103; [188] Guan ZZ et al. (2000) J Neurochem 74, 237–243; [189] Wang HY t al. (2000) J Biol Chem 275, 5626–5632; [190] Sgard F et al. (2002) Mol Pharmacol 61, 150–159; [191] Weisstaub N et al. (2002) Hear Res 167, 122–135; [192] Plazas PV et al. (2005) Br J Pharmacol 145, 963–974; [193] Elgoyhen AB et al. (2001) Proc Natl Acad Sci USA 98, 3501–3506; [194] Lips KS et al. (2002) Neuroscience 115, 1–5; [195] Elgoyhen AB et al. (1994) Cell 79, 705–715; [196] Peng H et al. (2004) Life Sci 76, 263–280; [197] Nguyen VT et al. (2000) Am J Pathol 157, 1377–1391.

References

Adams DJ, Dwyer TM and Hille B (1980). The permeability of endplate channels to monovalent and divalent metal cations. *J Gen Physiol* 75, 493–510.

Adler LE, Olincy A, Waldo M *et al.* (1998). Schizophrenia, sensory gating, and nicotinic receptors. *Schizophr Bull* 24, 189–202.

Agulhon C, Abitbol M, Bertrand D *et al.* (1999). Localization of mRNA for CHRNA7 in human fetal brain. *Neuroreport* 10, 2223–2227.

Agulhon C, Charnay Y, Vallet P *et al.* (1998). Distribution of mRNA for the alpha4 subunit of the nicotinic acetylcholine receptor in the human fetal brain. *Brain Res Mol Brain Res* 58, 123–131.

Aizenman E, Loring RH and Lipton SA (1990). Blockade of nicotinic responses in rat retinal ganglion cells by neuronal bungarotoxin. *Brain Res* 517, 209–214.

Alkondon M and Albuquerque EX (1995). Diversity of nicotinic acetylcholine receptors in rat hippocampal neurons.3. Agonist actions of the novel alkaloid epibatidine and analysis of type II current. *J Pharmacol Exp Ther* 274, 771–782.

Alkondon M, Pereira EF, Wonnacott S *et al.* (1992). Blockade of nicotinic currents in hippocampal neurons defines methyllycaconitine as a potent and specific receptor antagonist. *Mol Pharmacol* 41, 802–808.

Alle H and Geiger JR (2006). Combined analog and action potential coding in hippocampal mossy fibers. *Science* 311, 1290–1293.

Amador M and Dani JA (1991). MK-801 inhibition of nicotinic acetylcholine receptor channels. *Synapse* 7, 207–215.

Aramakis VB, Hsieh CY, Leslie FM *et al.* (2000). A critical period for nicotine-induced disruption of synaptic development in rat auditory cortex. *J Neurosci* 20, 6106–6116.

Arredondo J, Nguyen VT, Chernyavsky AI *et al.* (2003). Functional role of alpha7 nicotinic receptor in physiological control of cutaneous homeostasis. *Life Sci* 72, 2063–2067.

Ascher P, Large WA and Rang HP (1979). Two modes of action of ganglionic blocking drugs. *Br J Pharmacol* 66, 79P.

Ballivet M, Nef P, Couturier S *et al.* (1988). Electrophysiology of a chick neuronal nicotinic acetylcholine receptor expressed in xenopus oocytes after cDNA injection. *Neuron* 1, 847–852.

Banerjee C, Nyengaard JR, Wevers A *et al.* (2000). Cellular expression of alpha7 nicotinic acetylcholine receptor protein in the temporal cortex in Alzheimer's and Parkinson's disease—a stereological approach. *Neurobiol Dis* 7, 666–672.

Bencherif M, Fowler K, Lukas RJ *et al.* (1995). Mechanisms of up-regulation of neuronal nicotinic acetylcholine receptors in clonal cell lines and primary cultures of fetal rat brain. *J Pharmacol Exp Ther* 275, 987–994.

Berg DK and Conroy WG (2002). Nicotinic alpha 7 receptors: synaptic options and downstream signaling in neurons. *J Neurobiol* 53, 512–523.

Bertrand D, Ballivet M and Rungger D (1990). Activation and blocking of neuronal nicotinic acetylcholine receptor reconstituted in *Xenopus* oocytes. *Proc Natl Acad Sci USA* 87, 1993–1997.

Bertrand D, Devillers TA, Revah F *et al.* (1992). Unconventional pharmacology of a neuronal nicotinic receptor mutated in the channel domain. *Proc Natl Acad Sci USA* 89, 1261–1265.

Bertrand D, Elmslie F, Hughes E *et al.* (2005). The CHRNB2 mutation I312M is associated with epilepsy and distinct memory deficits. *Neurobiol Dis* 20, 799–804.

Bertrand D, Galzi JL, Devillers-Thiéry A *et al.* (1993a). Mutations at two distinct sites within the channel domain M2 alter calcium permeability of neuronal α7 nicotinic receptor. *Proc Natl Acad Sci USA* 90, 6971–6975.

Bertrand D, Galzi JL, Devillers-Thiéry A *et al.* (1993b). Stratification of the channel domain in neurotransmitter receptors. *Current Opinion in Neurobiology* 5, 688–693.

Bertrand S, Devillersthiery A, Palma E *et al.* (1997). Paradoxical allosteric effects of competitive inhibitors on neuronal alpha 7 nicotinic receptor mutants. *Neuroreport* 8, 3591–3596.

Blanchard SG, Quast U, Reed K *et al.* (1979). Interaction of ^{125}I-alpha-bungarotoxin with acetylcholine receptor from *Torpedo californica*. *Biochemistry* 18, 1875–1889.

Blanchet MR, Israel-Assayag E and Cormier Y (2005). Modulation of airway inflammation and resistance in mice by a nicotinic receptor agonist. *Eur Respir J* 26, 21–27.

Boorman JP, Beato M, Groot-Kormelink PJ *et al.* (2003). The effects of beta 3 subunit incorporation on the pharmacology and single channel properties of oocyte-expressed human alpha3beta4 neuronal nicotinic receptors. *J Biol Chem* 278, 44033–44040.

Boulter J, Evans K, Goldman D *et al.* (1986). Isolation of a cDNA clone coding for a possible neural nicotinic acetylcholine receptor alpha-subunit. *Nature* 319, 368–374.

Bouzat C (1996). Ephedrine blocks wild-type and long-lived mutant acetylcholine receptor channels. *Neuroreport* 8, 317–321.

Bouzat C, Gumilar F, Spitzmaul G *et al.* (2004). Coupling of agonist binding to channel gating in an ACh-binding protein linked to an ion channel. *Nature* 430, 896–900.

Brejc K, van Dijk WJ, Klaassen RV *et al.* (2001). Crystal structure of an ACh-binding protein reveals the ligand-binding domain of nicotinic receptors. *Nature* 411, 269–276.

Buisson B and Bertrand D (1998a). Allosteric modulation of neuronal nicotinic acetylcholine receptors. *J Physiol Paris* 92, 89–100.

Buisson B and Bertrand D (1998b). Open-channel blockers at the human α4β2 neuronal nicotinic acetylcholine receptor. *Mol Pharmacol* 53, 555–563.

Buisson B and Bertrand D (2001). Chronic exposure to nicotine upregulates the human (alpha)4(beta)2 nicotinic acetylcholine receptor function. *J Neurosci* 21, 1819–1829.

Buisson B and Bertrand D (2002). Nicotine addiction: the possible role of functional upregulation. *Trends Pharmacol Sci* 23, 130–136.

Buisson B, Gopalakrishnan M, Arneric SP *et al.* (1996). Human α4β2 neuronal nicotinic acetylcholine receptor in HEK 293 cells: a patch-clamp study. *J Neurosci* 16, 7880–7891.

Campos-Caro A, Sala S, Ballesta JJ *et al.* (1996). A single residue in the M2–M3 loop is a major determinant of coupling between binding and gating in neuronal nicotinic receptors. *Proc Natl Acad Sci USA* 93, 6118–6123.

Campos-Caro A, Smillie FI, Dominguez del Toro E *et al.* (1997). Neuronal nicotinic acetylcholine receptors on bovine chromaffin cells: cloning, expression and genomic organization of receptor subunits. *J Neurochem* 68, 488–497.

Cartaud J, Benedetti EL, Cohen JB *et al.* (1973). Presence of a lattice structure in membrane fragments rich in nicotinic receptor protein from the electric organ of *Torpedo marmorata*. *FEBS Lett* 33, 109–113.

Celie PH, van Rossum-Fikkert SE, van Dijk WJ *et al.* (2004). Nicotine and carbamylcholine binding to nicotinic acetylcholine receptors as studied in AChBP crystal structures. *Neuron* 41, 907–914.

Champtiaux N, Gotti C, Cordero-Erausquin M *et al.* (2003). Subunit composition of functional nicotinic receptors in dopaminergic neurons investigated with knock-out mice. *J Neurosci* 23, 7820–7829.

Changeux JP (1990). Functional architecture and dynamics of the nicotinic acetylcholine receptor: an allosteric ligand-gated ion channel. In JP Changeux, RR Llinàs, D Purves and FE Bloom, eds, *Fidia Research Foundation Neuroscience Award Lectures*. Raven Press Ltd, New York, pp. 21–168.

Changeux JP and Edelstein SJ (2005). Allosteric mechanisms of signal transduction. *Science* 308, 1424–1428.

Charpantier E, Barneoud P, Moser P *et al.*F (1998). Nicotinic acetylcholine subunit mRNA expression in dopaminergic neurons of the rat substantia nigra and ventral tegmental area. *Neuroreport* 9, 3097–3101.

Charpantier E, Wiesner A, Huh KH *et al.* (2005). Alpha 7 neuronal nicotinic acetylcholine receptors are negatively regulated by tyrosine phosphorylation and Src-family kinases. *J Neurosci* 25, 9836–9849.

Chavez-Noriega LE, Crona JH, Washburn MS *et al.* (1997). Pharmacological characterization of recombinant human neuronal nicotinic acetylcholine receptors h alpha 2 beta 2, h alpha 2 beta 4, h alpha 3 beta 2, h alpha 3 beta 4, h alpha 4 beta 2, h alpha 4 beta 4 and h alpha 7 expressed in *Xenopus* oocytes. *J Pharmacol Exp Ther* **280**, 346–356.

Cho CH, Song W, Leitzell K *et al.* (2005). Rapid upregulation of alpha7 nicotinic acetylcholine receptors by tyrosine dephosphorylation. *J Neurosci* **25**, 3712–3723.

Clarke PB (1992). The fall and rise of neuronal alpha-bungarotoxin binding proteins. *Trends Pharmacol Sci* **13**, 407–413.

Clarke PBS, Schwartz RD, Paul SM *et al.* (1985). Nicotinic binding in rat brain: autoradiographic comparison of [3H]acetylcholine, [3H] nicotine and [125I]-alpha-bungarotoxin. *J Neurosci* **5**, 1307–1315.

Colquhoun D and Rang HP (1973). *The relation between classical and cooperative models for drug action in drug receptors*. Macmillan, London.

Colquhoun D and Sakmann B (1983). Burst of openings in transmitter-activated ion channels. In B Sakmann and E Neher, eds, *Single channel recording*. Plenum Press, New Yok, pp. 345–364.

Conroy WG and Berg DK (1995). Neurons can maintain multiple classes of nicotinic acetylcholine receptors distinguished by different subunit compositions. *J Biol Chem* **270**, 4424–4431.

Cooper E, Couturier S and Ballivet M (1991). Pentameric structure and subunit stoichiometry of a neuronal nicotinic acetylcholine receptor. *Nature* **350**, 235–238.

Corringer PJ, Bertrand S, Galzi JL *et al.* (1999). Mutational analysis of the charge selectivity filter of the alpha7 nicotinic acetylcholine receptor. *Neuron* **22**, 831–843.

Corringer PJ, Le Novere N and Changeux JP (2000). Nicotinic receptors at the amino acid level. *Annu Rev Pharmacol Toxicol* **40**, 431–458.

Corringer PJ, Sallette J and Changeux JP (2006). Nicotine enhances intracellular nicotinic receptor maturation: a novel mechanism of neural plasticity? *J Physiol Paris* **99**, 162–171.

Costa G, Abin-Carriquiry JA and Dajas F (2001). Nicotine prevents striatal dopamine loss produced by 6-hydroxydopamine lesion in the substantia nigra. *Brain Res* **888**, 336–342.

Court JA, Piggott MA, Lloyd S *et al.* (2000). Nicotine binding in human striatum: elevation in schizophrenia and reductions in dementia with Lewy bodies, Parkinson's disease and Alzheimer's disease and in relation to neuroleptic medication. *Neuroscience* **98**, 79–87.

Couturier S, Bertrand D, Matter JM *et al.* (1990a). A neuronal nicotinic acetylcholine receptor subunit (alpha 7) is developmentally regulated and forms a homo-oligomeric channel blocked by alpha-BTX. *Neuron* **5**, 847–856.

Couturier S, Erkman L, Valera S J *et al.* (1990b). Alpha 5, alpha 3 and non-alpha 3. Three clustered avian genes encoding neuronal nicotinic acetylcholine receptor-related subunits. *J Biol Chem* **265**, 17560–17567.

Covernton PJ and Connolly JG (2000). Multiple components in the agonist concentration–response relationships of neuronal nicotinic acetylcholine receptors. *J Neurosci Methods* **96**, 63–70.

Cuevas J and Berg DK (1998). Mammalian nicotinic receptors with alpha7 subunits that slowly desensitize and rapidly recover from alpha-bungarotoxin blockade. *J Neurosci* **18**, 10335–10344.

Curtis L, Buisson B, Bertrand S *et al.* (2002). Potentiation of human alpha4beta2 neuronal nicotinic acetylcholine receptor by estradiol. *Mol Pharmacol* **61**, 127–135.

D'Andrea MR and Nagele RG (2006). Targeting the alpha 7 nicotinic acetylcholine receptor to reduce amyloid accumulation in Alzheimer's disease pyramidal neurons. *Curr Pharm Des* **12**, 677–684.

Dani JA (1986). Ion-channel entrances influence permeation. Net charge, size, shape and binding considerations. *Biophys J* **49**, 607–618.

Dani JA (2001). Overview of nicotinic receptors and their roles in the central nervous system. *Biol Psychiatry* **49**, 166–174.

Dani JA (2003). Roles of dopamine signaling in nicotine addiction. *Mol Psychiatry* **8**, 255–256.

Dani JA and De Biasi M (2001). Cellular mechanisms of nicotine addiction. *Pharmacol Biochem Behav* **70**, 439–446.

Dani JA, Radcliffe KA and Pidoplichko VI (2000). Variations in desensitization of nicotinic acetylcholine receptors from hippocampus and midbrain dopamine areas. *Eur J Pharmacol* **393**, 31–38.

De Fusco M, Becchetti A, Patrignani A *et al.* (2000). The nicotinic receptor beta 2 subunit is mutant in nocturnal frontal lobe epilepsy. *Nat Genet* **26**, 275–276.

De Luca V, Wong AH, Muller DJ *et al.* (2004). Evidence of association between smoking and alpha7 nicotinic receptor subunit gene in schizophrenia patients. *Neuropsychopharmacology* **29**, 1522–1526.

Descarries L, Gisiger V and Steriade M (1997). Diffuse transmission by acetylcholine in the CNS. *Prog Neurobiol* **53**, 603–625.

Devillers-Thiery A, Galzi JL, Eisele JL *et al.* (1993). Functional architecture of the nicotinic acetylcholine receptor: a prototype of ligand-gated ion channels. *J Membr Biol* **136**, 97–112.

Di Angelantonio S, Matteoni C, Fabbretti E *et al.* (2003). Molecular biology and electrophysiology of neuronal nicotinic receptors of rat chromaffin cells. *Eur J Neurosci* **17**, 2313–2322.

Drisdel RC and Green WN (2000). Neuronal alpha-bungarotoxin receptors are alpha7 subunit homomers. *J Neurosci* **20**, 133–139.

Elgoyhen AB, Johnson DS, Boulter J *et al.* (1994). Alpha 9: an acetylcholine receptor with novel pharmacological properties expressed in rat cochlear hair cells. *Cell* **79**, 705–715.

Elgoyhen AB, Vetter DE, Katz E *et al.* (2001). alpha10: a determinant of nicotinic cholinergic receptor function in mammalian vestibular and cochlear mechanosensory hair cells. *Proc Natl Acad Sci USA* **98**, 3501–3506.

Fabian-Fine R, Skehel P, Errington ML *et al.* (2001). Ultrastructural distribution of the alpha7 nicotinic acetylcholine receptor subunit in rat hippocampus. *J Neurosci* **21**, 7993–8003.

Feller MB, Wellis DP, Stellwagen D *et al.* (1996). Requirement for cholinergic synaptic transmission in the propagation of spontaneous retinal waves. *Science* **272**, 1182–1187.

Fenster CP, Beckman ML, Parker JC *et al.* (1999). Regulation of alpha4beta2 nicotinic receptor desensitization by calcium and protein kinase C. *Mol Pharmacol* **55**, 432–443.

Fieber LA and Adams DJ (1991). Acetylcholine-evoked currents in cultured neurones dissociated from rat parasympathetic cardiac ganglia. *J Physiol* **434**, 215–237.

Fischer U, Reinhardt S, Albuquerque EX *et al.* (1999). Expression of functional alpha7 nicotinic acetylcholine receptor during mammalian muscle development and denervation. *Eur J Neurosci* **11**, 2856–2864.

Flores CM, Davila-Garcia MI, Ulrich YM *et al.* (1997). Differential regulation of neuronal nicotinic receptor binding sites following chronic nicotine administration. *J Neurochem* **69**, 2216–2219.

Forster I and Bertrand D (1995). Inward rectification of neuronal nicotinic acetylcholine receptors investigated by using the homomeric alpha 7 receptor. *Proc Biol Sci* **260**, 139–148.

Frazier CJ, Buhler AV, Weiner JL *et al.* (1998). Synaptic potentials mediated via alpha-bungarotoxin-sensitive nicotinic acetylcholine receptors in rat hippocampal interneurons. *J Neurosci* **18**, 8228–8235.

Freedman R, Coon H, Mylesworsley M *et al.* (1997). Linkage of a neurophysiological deficit in schizophrenia to a chromosome 15 locus. *Proc Natl Acad Sci USA* **94**, 587–592.

Freedman R, Hall M, Adler LE *et al.* (1995). Evidence in postmortem brain tissue for decreased numbers of hippocampal nicotinic receptors in schizophrenia. *Biol Psychiatry* **38**, 22–33.

Fruchart-Gaillard C, Gilquin B, Antil-Delbeke S *et al.* (2002). Experimentally based model of a complex between a snake toxin and the alpha 7 nicotinic receptor. *Proc Natl Acad Sci USA* **99**, 3216–3221.

Fucile S (2004). Ca^{2+} permeability of nicotinic acetylcholine receptors. *Cell Calcium* **35**, 1–8.

Fucile S, Sucapane A and Eusebi F (2005). Ca^{2+} permeability of nicotinic acetylcholine receptors from rat dorsal root ganglion neurones. *J Physiol* **565**, 219–228.

Gahring LC and Rogers SW (2005). Neuronal nicotinic acetylcholine receptor expression and function on nonneuronal cells. *Aaps J* **7**, E885–894.

Gahring LC, Persiyanov K and Rogers SW (2004). Neuronal and astrocyte expression of nicotinic receptor subunit beta4 in the adult mouse brain. *J Comp Neurol* **468**, 322–333.

Gallezot JD, Bottlaender M, Gregoire MC et al. (2005). *In vivo* imaging of human cerebral nicotinic acetylcholine receptors with 2-18F-fluoro-A-85380 and PET. *J Nucl Med* **46**, 240–247.

Galzi JL, Bertrand D, Devillers TA et al. (1991a). Functional significance of aromatic amino acids from three peptide loops of the alpha 7 neuronal nicotinic receptor site investigated by site-directed mutagenesis. *FEBS Lett* **294**, 198–202.

Galzi JL, Bertrand S, Corringer PJ et al. (1996a). Identification of calcium binding sites that regulate potentiation of a neuronal nicotinic acetylcholine receptor. *EMBO J* **15**, 5824–5832.

Galzi JL, Devillers TA, Hussy N et al. (1992). Mutations in the channel domain of a neuronal nicotinic receptor convert ion selectivity from cationic to anionic. *Nature* **359**, 500–505.

Galzi JL, Edelstein SJ and Changeux JP (1996b). The multiple phenotypes of allosteric receptor mutants. *Proc Natl Acad Sci USA* **93**, 1853–1858.

Galzi JL, Revah F, Bouet F et al. (1991b). Allosteric transitions of the acetylcholine receptor probed at the amino acid level with a photolabile cholinergic ligand. *Proc Natl Acad Sci USA* **88**, 5051–5055.

Garcia-Guzman M, Sala F, Sala S et al. (1995). alpha-Bungarotoxin-sensitive nicotinic receptors on bovine chromaffin cells: molecular cloning, functional expression and alternative splicing of the alpha 7 subunit. *Eur J Neurosci* **7**, 647–655.

Gardner P, Ogden DC and Colquhoun D (1984). Conductances of single ion channels opened by nicotinic agonists are indistinguishable. *Nature* **309**, 160–162

Giraudat J, Galzi JL, Revah F et al. (1989). The noncompetitive blocker chlorpromazine labels segment MII but not segment MI on the nicotinic acetylcholine receptor alpha-subunit. *FEBS Lett* **253**, 190–198.

Gotti C, Fornasari D and Clementi F (1997). Human neuronal nicotinic receptors. *Prog Neurobiol* **53**, 199–237.

Gotti C, Moretti M, Zanardi A et al. (2005). Heterogeneity and selective targeting of neuronal nicotinic acetylcholine receptor (nAChR) subtypes expressed on retinal afferents of the superior colliculus and lateral geniculate nucleus: identification of a new native nAChR subtype alpha3beta2(alpha5 or beta3) enriched in retinocollicular afferents. *Mol Pharmacol* **68**, 1162–1171.

Graham A, Court JA, Martin-Ruiz CM et al. (2002). Immunohistochemical localisation of nicotinic acetylcholine receptor subunits in human cerebellum. *Neuroscience* **113**, 493–507.

Grando SA (2006). Cholinergic control of epidermal cohesion. *Exp Dermatol* **15**, 265–282.

Grutter T and Changeux JP (2001). Nicotinic receptors in wonderland. *Trends Biochem Sci* **26**, 459–63.

Grutter T, Prado de Carvalho L, Virginie D et al. (2005). A chimera encoding the fusion of an acetylcholine-binding protein to an ion channel is stabilized in a state close to the desensitized form of ligand-gated ion channels. *C R Biol* **328**, 223–234.

Gunthorpe MJ and Lummis SC (2001). Conversion of the ion selectivity of the 5-HT(3a) receptor from cationic to anionic reveals a conserved feature of the ligand-gated ion channel superfamily. *J Biol Chem* **276**, 10977–10983.

Haghighi AP and Cooper E (2000). A molecular link between inward rectification and calcium permeability of neuronal nicotinic acetylcholine alpha3beta4 and alpha4beta2 receptors. *J Neurosci* **20**, 529–541.

Hatton GI and Yang QZ (2002). Synaptic potentials mediated by alpha 7 nicotinic acetylcholine receptors in supraoptic nucleus. *J Neurosci* **22**, 29–37.

Hefft S, Hulo S, Bertrand D et al. (1999). Synaptic transmission at nicotinic acetylcholine receptors in rat hippocampal organotypic cultures and slices. *J Physiol* **515**, 769–776.

Hernandez MC, Erkman L, Matter SL et al. (1995). Characterization of the nicotinic acetylcholine receptor beta 3 gene. Its regulation within the avian nervous system is effected by a promoter 143 base pairs in length. *J Biol Chem* **270**, 3224–3233.

Hogg RC and Bertrand D (2004). Neuronal nicotinic receptors and epilepsy, from genes to possible therapeutic compounds. *Bioorg Med Chem Lett* **14**, 1859–1861.

Hogg RC, Buisson B and Bertrand D (2005). Allosteric modulation of ligand-gated ion channels. *Biochem Pharmacol* **70**, 1267–1276.

Hogg RC, Raggenbass M and Bertrand D (2003). Nicotinic acetylcholine receptors: from structure to brain function. *Rev Physiol Biochem Pharmacol* **147**, 1–46.

Hohlfeld R and Wekerle H (1981). Myasthenia gravis. Prototype of an anti-receptor autoaggression disease. *MMW Munch Med Wochenschr* **123**, 1207–1211.

Huh KH and Fuhrer C (2002). Clustering of nicotinic acetylcholine receptors: from the neuromuscular junction to interneuronal synapses. *Mol Neurobiol* **25**, 79–112.

Hurst RS, Hajos M, Raggenbass M et al. (2005). A novel positive allosteric modulator of the alpha7 neuronal nicotinic acetylcholine receptor: *in vitro* and *in vivo* characterization. *J Neurosci* **25**, 4396–4405.

Inoue M and Kuriyama H (1991). Properties of the nicotinic-receptor-activated current in adrenal chromaffin cells of the guinea-pig. *Pflugers Arch* **419**, 13–20.

Itier V and Bertrand D (2002). Mutations of the neuronal nicotinic acetylcholine receptors and their association with ADNFLE. *Neurophysiol Clin* **32**, 99–107.

Jackson MB (1994). Single channel currents in the nicotinic acetylcholine receptor: a direct demonstration of allosteric transitions. *Trends Biochem Sci* **19**, 396–399.

Ji D and Dani JA (2000). Inhibition and disinhibition of pyramidal neurons by activation of nicotinic receptors on hippocampal interneurons. *J Neurophysiol* **83**, 2682–2690.

Jones IW and Wonnacott S (2004). Precise localization of alpha7 nicotinic acetylcholine receptors on glutamatergic axon terminals in the rat ventral tegmental area. *J Neurosci* **24**, 11244–11252.

Kalden JR (1975). Auto-immune diseases of the skeletal muscle-system. Immune-pathogenetic mechanisms in myasthenia gravis and polymyositis. *Immun Infekt* **3**, 100–115.

Kaneda M, Hashimoto M and Kaneko A (1995). Neuronal nicotinic acetylcholine receptors of ganglion cells in the cat retina. *Jpn J Physiol* **45**, 491–508.

Kash TL, Jenkins A, Kelley JC et al. (2003). Coupling of agonist binding to channel gating in the GABA(A) receptor. *Nature* **421**, 272–275.

Katz B (1966). *Nerve muscle and synapse*. Mc Graw and Hill, New York.

Katz E, Verbitsky M, Rothlin CV et al. (2000). High calcium permeability and calcium block of the alpha9 nicotinic acetylcholine receptor. *Hear Res* **141**, 117–128.

Keramidas A, Moorhouse AJ, French CR et al. (2000). M2 pore mutations convert the glycine receptor channel from being anion- to cation-selective. *Biophys J* **79**, 247–259.

Keyser KT, Britto LRG, Schoepfer R et al. (1993). Three subtypes of alpha-bungarotoxin-sensitive nicotinic acetylcholine receptors are expressed in chick retina. *J Neurosci* **13**, 442–454.

Kimes AS, Horti AG, London ED et al. (2003). 2-[18F]F-A-85380: PET imaging of brain nicotinic acetylcholine receptors and whole body distribution in humans. *FASEB J* **17**, 1331–1333.

Koenen M, Peter C, Villarroel A et al. (2005). Acetylcholine receptor channel subtype directs the innervation pattern of skeletal muscle. *EMBO Rep* **6**, 570–576.

Konno T, Bush C, Von Kitzing E et al. (1991). Rings of anionic amino acids as structural determinants of ion selectivity in the acetylcholine receptor. *Proc R Soc Lond B* **244**, 69–79.

Koshland DE, Jr., Nemethy G and Filmer D (1966). Comparison of experimental binding data and theoretical models in proteins containing subunits. *Biochemistry* **5**, 365–385.

Krause RM, Buisson B, Bertrand S *et al.* (1998). Ivermectin: a positive allosteric effector of the alpha7 neuronal nicotinic acetylcholine receptor. *Mol Pharmacol* **53**, 283–294.

Kulak JM, Ivy Carroll F and Schneider JS (2006). [I]Iodomethyllycaconitine binds to alpha7 nicotinic acetylcholine receptors in monkey brain. *Eur J Neurosci* **23**, 2604–2610.

Lamb PW, Melton MA and Yakel JL (2005). Inhibition of neuronal nicotinic acetylcholine receptor channels expressed in *Xenopus* oocytes by beta-amyloid1-42 peptide. *J Mol Neurosci* **27**, 13–21.

Laviolette SR and van der Kooy D (2004). The neurobiology of nicotine addiction: bridging the gap from molecules to behaviour. *Nat Rev Neurosci* **5**, 55–65.

Le Novere N, Corringer PJ and Changeux JP (1999). Improved secondary structure predictions for a nicotinic receptor subunit: incorporation of solvent accessibility and experimental data into a two-dimensional representation. *Biophys J* **76**, 2329–2345.

Le Novere N, Grutter T and Changeux JP (2002). Models of the extracellular domain of the nicotinic receptors and of agonist- and Ca²⁺-binding sites. *Proc Natl Acad Sci USA* **99**, 3210–3215.

Lecchi M, Marguerat A, Ionescu A *et al.* (2004). Ganglion cells from chick retina display multiple functional nAChR subtypes. *Neuroreport* **15**, 307–311.

Lecchi M, McIntosh JM, Bertrand S *et al.* (2005). Functional properties of neuronal nicotinic acetylcholine receptors in the chick retina during development. *Eur J Neurosci* **21**, 3182–3188.

Lena C, Changeux JP and Mulle C (1993). Evidence for 'preterminal' nicotinic receptors on GABAergic axons in the rat interpeduncular nucleus. *J Neurosci* **13**, 2680–2688.

Leonard S, Adams C, Breese CR *et al.* (1996). Nicotinic receptor function in schizophrenia. *Schizophrenia Bull* **22**, 431–445.

Leonard S, Gault J, Adams C *et al.* (1998). Nicotinic receptors, smoking and schizophrenia. *Restor Neurol Neurosci* **12**, 195–201.

Leonard S, Gault J, Hopkins J *et al.* (2002). Association of promoter variants in the alpha7 nicotinic acetylcholine receptor subunit gene with an inhibitory deficit found in schizophrenia. *Arch Gen Psychiatry* **59**, 1085–1096.

Levin ED (2002). Nicotinic receptor subtypes and cognitive function. *J Neurobiol* **53**, 633–640.

Lindstrom JM (2000). Acetylcholine receptors and myasthenia. *Muscle Nerve* **23**, 453–477.

Lingle CJ and Steinbach JH (1988). Neuromuscular blocking agents. *Int Anesthesiol Clin* **26**, 288–301.

Lipton SA, Aizenman E and Loring RH (1987). Neural nicotinic acetylcholine responses in solitary mammalian retinal ganglion cells. *Pflügers Arch* **410**, 37–43.

Liu Q, Kawai H and Berg DK (2001). beta -Amyloid peptide blocks the response of alpha 7-containing nicotinic receptors on hippocampal neurons. *Proc Natl Acad Sci USA* **98**, 4734–4739.

Lukas RJ, Changeux JP, Le Novere N *et al.* (1999). International Union of Pharmacology. XX. Current status of the nomenclature for nicotinic acetylcholine receptors and their subunits. *Pharmacol Rev* **51**, 397–401.

Lustig LR (2006). Nicotinic acetylcholine receptor structure and function in the efferent auditory system. *Anat Rec A Discov Mol Cell Evol Biol* **288**, 424–434.

Maggi L, Le Magueresse C, Changeux JP *et al.* (2003). Nicotine activates immature 'silent' connections in the developing hippocampus. *Proc Natl Acad Sci USA* **100**, 2059–2064.

Maggi L, Sola E, Minneci F *et al.* (2004). Persistent decrease in synaptic efficacy induced by nicotine at Schaffer collateral-CA1 synapses in the immature rat hippocampus. *J Physiol* **559**, 863–874.

Mansvelder HD and McGehee DS (2000). Long-term potentiation of excitatory inputs to brain reward areas by nicotine. *Neuron* **27**, 349–357.

Mansvelder HD, van Aerde KI, Couey JJ *et al.* (2006). Nicotinic modulation of neuronal networks: from receptors to cognition. *Psychopharmacology* **184**, 292–305.

Marritt AM, Cox BC, Yasuda RP *et al.* (2005). Nicotinic cholinergic receptors in the rat retina: simple and mixed heteromeric subtypes. *Mol Pharmacol* **68**, 1656–1668.

Matter JM, Mattersadzinski L and Ballivet M (1995). Activity of the beta 3 nicotinic receptor promoter is a marker of neuron fate determination during retina development. *J Neurosci* **15**, 5919–5928.

Matter SL, Hernandez MC, Roztocil T *et al.* (1992). Neuronal specificity of the alpha 7 nicotinic acetylcholine receptor promoter develops during morphogenesis of the central nervous system. *EMBO J* **11**, 4529–4538.

Mcgehee DS and Role LW (1996). Presynaptic ionotropic receptors. *Curr Opin Neurobiol* **6**, 342–349.

Metz CN and Tracey KJ (2005). It takes nerve to dampen inflammation. *Nat Immunol* **6**, 756–757.

Mishina M, Sakimura K, Mori H *et al.* (1991). A single amino acid residue determines the Ca⁺⁺ permeability of AMPA-selective glutamate receptor channels. *Biochem Biophys Res Comm* **180**, 813–821.

Miyazawa A, Fujiyoshi Y and Unwin N (2003). Structure and gating mechanism of the acetylcholine receptor pore. *Nature* **423**, 949–955.

Monod J, Wyman J and Changeux JP (1965). On the nature of allosteric transitions: a plausible model. *J Mol Biol* **12**, 88–118.

Monteggia LM, Gopalakrishnan M, Touma E *et al.* (1995). Cloning and transient expression of genes encoding the human alpha 4 and beta 2 neuronal nicotinic acetylcholine receptor (nAChR) subunits. *Gene* **155**, 189–193.

Moretti M, Vailati S, Zoli M *et al.* (2004). Nicotinic acetylcholine receptor subtypes expression during rat retina development and their regulation by visual experience. *Mol Pharmacol* **66**, 85–96.

Newhouse PA, Potter A and Levin ED (1997). Nicotinic system involvement in Alzheimer's and Parkinson's diseases. Implications for therapeutics. *Drugs Aging* **11**, 206–228.

Nguyen VT, Ndoye A and Grando SA (2000). Novel human alpha9 acetylcholine receptor regulating keratinocyte adhesion is targeted by *Pemphigus vulgaris* autoimmunity. *Am J Pathol* **157**, 1377–1391.

Noda M, Furutani Y, Takahashi H *et al.* (1983). Cloning and sequence analysis of calf cDNA and human genomic DNA encoding alpha-subunit precursor of muscle acetylcholine receptor. *Nature* **305**, 818–823.

Ohno K and Engel AG (2003). Congenital myasthenic syndromes: gene mutations. *Neuromuscul Disord* **13**, 854–857.

Ohno K and Engel AG (2005a). Gene symbol: CHRNE. Disease: endplate acetylcholine receptor deficiency. *Hum Genet* **117**, 302.

Ohno K and Engel AG (2005b). Splicing abnormalities in congenital myasthenic syndromes. *Acta Myol* **24**, 50–54.

Olincy A, Harris JG, Johnson LL *et al.* (2006). Proof-of-concept trial of an alpha7 nicotinic agonist in schizophrenia. *Arch Gen Psychiatry* **63**, 630–638.

Oswald RE and Changeux JP (1982). Crosslinking of alpha-bungarotoxin to the acetylcholine receptor from *Torpedo marmorata* by ultraviolet light irradiation. *FEBS Lett* **139**, 225–229.

Palma E, Bertrand S, Binzoni T *et al.* (1996). Neuronal nicotinic alpha 7 receptor expressed in *Xenopus* oocytes presents five putative binding sites for methyllycaconitine. *J Physiol* **491**, 151–161.

Paradiso K, Zhang J and Steinbach JH (2001). The C terminus of the human nicotinic alpha4beta2 receptor forms a binding site required for potentiation by an estrogenic steroid. *J Neurosci* **21**, 6561–6568.

Patrick J, Ballivet M, Boas L *et al.* (1983). Molecular cloning of the acetylcholine receptor. *Cold Spring Harb Symp Quant Biol* **48**, 71–78.

Peng X, Gerzanich V, Anand R *et al.* (1997). Chronic nicotine treatment up-regulates alpha3 and alpha7 acetylcholine receptor subtypes expressed by the human neuroblastoma cell line SH-SY5Y. *Mol Pharmacol* **51**, 776–784.

Peng X, Gerzanich V, Anand R *et al.* (1994). Nicotine-induced increase in neuronal nicotinic receptors results from a decrease in the rate of receptor turnover. *Mol Pharmacol* **46**, 523–530.

Perry DC, Davila-Garcia MI, Stockmeier CA *et al.* (1999). Increased nicotinic receptors in brains from smokers: membrane binding and autoradiography studies. *J Pharmacol Exp Ther* **289**, 1545–1552.

Perry EK, Martin-Ruiz CM and Court JA (2001). Nicotinic receptor subtypes in human brain related to aging and dementia. *Alcohol* **24**, 63–68.

Person AM, Bills KL, Liu H *et al.* (2005). Extracellular domain nicotinic acetylcholine receptors formed by alpha4 and beta2 subunits. *J Biol Chem* **280**, 39990–40002.

Peters JA, Hales TG and Lambert JJ (2005). Molecular determinants of single-channel conductance and ion selectivity in the Cys-loop family: insights from the 5-HT3 receptor. *Trends Pharmacol Sci* **26**, 587–594.

Phillips HA, Favre I, Kirkpatrick M *et al.* (2001). CHRNB2 is the second acetylcholine receptor subunit associated with autosomal dominant nocturnal frontal lobe epilepsy. *Am J Hum Genet* **68**, 225–231.

Pidoplichko VI, Noguchi J, Areola OO *et al.* (2004). Nicotinic cholinergic synaptic mechanisms in the ventral tegmental area contribute to nicotine addiction. *Learn Mem* **11**, 60–69.

Popot JL and Changeux JP (1984). Nicotinic receptor of acetylcholine: structure of an oligomeric integral membrane protein. *Physiol Rev* **64**, 1162–1239.

Pym L, Kemp M, Raymond-Delpech V *et al.* (2005). Subtype-specific actions of beta-amyloid peptides on recombinant human neuronal nicotinic acetylcholine receptors (alpha7, alpha4beta2, alpha3beta4) expressed in *Xenopus laevis* oocytes. *Br J Pharmacol* **146**, 964–971.

Quik M, Bordia T, Forno L *et al.* (2004). Loss of alpha-conotoxinMII- and A85380-sensitive nicotinic receptors in Parkinson's disease striatum. *J Neurochem* **88**, 668–679.

Radcliffe KA and Dani JA (1998). Nicotinic stimulation produces multiple forms of increased glutamatergic synaptic transmission. *J Neurosci* **18**, 7075–7083.

Raftery MA, Hunkapiller M, Strader CD *et al.* (1980). Acetylcholine receptor: complex of homologous subunits. *Science* **208**, 1454, 1457.

Reed BT, Keyser KT and Amthor FR (2004). MLA-sensitive cholinergic receptors involved in the detection of complex moving stimuli in retina. *Vis Neurosci* **21**, 861–872.

Revah F, Bertrand D, Galzi JL *et al.* (1991). Mutations in the channel domain alter desensitization of a neuronal nicotinic receptor. *Nature* **353**, 846–849.

Rousseau SJ, Jones IW, Pullar IA *et al.* (2005). Presynaptic alpha7 and non-alpha7 nicotinic acetylcholine receptors modulate [3H]d-aspartate release from rat frontal cortex *in vitro*. *Neuropharmacology* **49**, 59–72.

Saeed RW, Varma S, Peng-Nemeroff T *et al.* (2005). Cholinergic stimulation blocks endothelial cell activation and leukocyte recruitment during inflammation. *J Exp Med* **201**, 1113–1123.

Sakmann B, Patlak J and Neher E (1980). Single acetylcholine-activated channels show burst-kinetics in presence of desensitizing concentrations of agonist. *Nature* **286**, 71–73.

Sallette J, Pons S, Devillers-Thiery A *et al.* (2005). Nicotine upregulates its own receptors through enhanced intracellular maturation. *Neuron* **46**, 595–607.

Sands SB and Barish ME (1991). Calcium permeability of neuronal nicotinic channels in PC12 cells. *Brain Res* **560**, 38–42.

Sands SB and Barish ME (1992). Neuronal nicotinic acetylcholine receptor currents in phaeochromocytoma (PC12) cells: dual mechanisms of rectification. *J Physiol* **447**, 467–487.

Seguela P, Wadiche J, Dineley MK, *et al.* (1993). Molecular cloning, functional properties and distribution of rat brain alpha 7: a nicotinic cation channel highly permeable to calcium. *J Neurosci* **13**, 596–604.

Severance EG, Zhang H, Cruz Y *et al.* (2004). The alpha7 nicotinic acetylcholine receptor subunit exists in two isoforms that contribute to functional ligand-gated ion channels. *Mol Pharmacol* **66**, 420–429.

Sgard F, Charpantier E, Barneoud P *et al.* (1999). Nicotinic receptor subunit mRNA expression in dopaminergic neurons of the rat brain. *Ann N Y Acad Sci* **868**, 633–635.

Sgard F, Charpantier E, Bertrand S *et al.* (2002). A novel human nicotinic receptor subunit, alpha10, that confers functionality to the alpha9-subunit. *Mol Pharmacol* **61**, 150–159.

Sharma G and Vijayaraghavan S (2001). Nicotinic cholinergic signaling in hippocampal astrocytes involves calcium-induced calcium release from intracellular stores. *Proc Natl Acad Sci USA* **98**, 4148–4153.

Shiang R, Ryan S, Zhu Y-Z, Hanhn AF *et al.* (1993). Mutations in the α1 subunit if the inhibitory glycine receptor cause the dominant neurologic disorder, hyperekplexia. *Nature Genetics* **5**, 351–357.

Shu Y, Hasenstaub A, Duque A *et al.* (2006). Modulation of intracortical synaptic potentials by presynaptic somatic membrane potential. *Nature* **441**, 761–765.

Shytle RD, Mori T, Townsend K *et al.* (2004). Cholinergic modulation of microglial activation by alpha 7 nicotinic receptors. *J Neurochem* **89**, 337–343.

Simpson JA (1978). Myasthenia gravis: a personal view of pathogenesis and mechanism, part 2. *Muscle Nerve* **1**, 151–156.

Sine SM, Quiram P, Papanikolaou F *et al.* (1994). Conserved tyrosines in the alpha subunit of the nicotinic acetylcholine receptor stabilize quaternary ammonium groups of agonists and curariform antagonists. *J Biol Chem* **269**, 8808–8816.

Sine SM, Wang HL and Gao F (2004). Toward atomic-scale understanding of ligand recognition in the muscle nicotinic receptor. *Curr Med Chem* **11**, 559–567.

Sixma TK and Smit AB (2003). Acetylcholine binding protein (AChBP): a secreted glial protein that provides a high-resolution model for the extracellular domain of pentameric ligand-gated ion channels. *Annu Rev Biophys Biomol Struct* **32**, 311–334.

Smit AB, Syed NI, Schaap D *et al.* (2001). A glia-derived acetylcholine-binding protein that modulates synaptic transmission. *Nature* **411**, 261–268.

Spencer JP, Weil A, Hill K *et al.* (2006). Transgenic mice over-expressing human beta-amyloid have functional nicotinic alpha 7 receptors. *Neuroscience* **137**, 795–805.

Steinlein OK (2004). Genetic mechanisms that underlie epilepsy. *Nat Rev Neurosci* **5**, 400–408.

Steinlein OK, Mulley JC, Propping P *et al.* (1995). A missense mutation in the neuronal nicotinic acetylcholine receptor alpha 4 subunit is associated with autosomal dominant nocturnal frontal lobe epilepsy. *Nat Genet* **11**, 201–203.

Strang CE, Andison ME, Amthor FR *et al.* (2005). Rabbit retinal ganglion cells express functional alpha7 nicotinic acetylcholine receptors. *Am J Physiol Cell Physiol* **289**, C644–655.

Tachikawa E, Mizuma K, Kudo K *et al.* (2001). Characterization of the functional subunit combination of nicotinic acetylcholine receptors in bovine adrenal chromaffin cells. *Neurosci Lett* **312**, 161–164.

Taylor P, Malanz S, Molles BE *et al.* (2000). Subunit interface selective toxins as probes of nicotinic acetylcholine receptor structure. *Pflugers Arch* **440**, R115–117.

Taylor P, Radic Z, Kreienkamp HJ *et al.* (1994). Expression and ligand specificity of acetylcholinesterase and the nicotinic receptor: a tale of two cholinergic sites. *Biochem Soc Trans* **22**, 740–745.

Tribollet E, Bertrand D, Marguerat A *et al.* (2004). Comparative distribution of nicotinic receptor subtypes during development, adulthood and aging: an autoradiographic study in the rat brain. *Neuroscience* **124**, 405–420.

Tsuneki H, Salas R and Dani JA (2003). Mouse muscle denervation increases expression of an alpha7 nicotinic receptor with unusual pharmacology. *J Physiol* **547**, 169–179.

Ulloa L (2005). The vagus nerve and the nicotinic anti-inflammatory pathway. *Nat Rev Drug Discov* **4**, 673–684.

Unwin N (2005). Refined structure of the nicotinic acetylcholine receptor at 4A resolution. *J Mol Biol* **346**, 967–989.

Vallejo YF, Buisson B, Bertrand D *et al.* (2005). Chronic nicotine exposure upregulates nicotinic receptors by a novel mechanism. *J Neurosci* **25**, 5563–5572.

Valor LM, Mulet J, Sala F *et al.* (2002). Role of the large cytoplasmic loop of the alpha 7 neuronal nicotinic acetylcholine receptor subunit in receptor expression and function. *Biochemistry* **41**, 7931–7938.

Van Kampen M, Selbach K, Schneider R *et al.* (2004). AR-R 17779 improves social recognition in rats by activation of nicotinic alpha7 receptors. *Psychopharmacology* **172**, 375–383.

van Nierop P, Bertrand S, Munno DW *et al.* (2006). Identification and functional expression of a family of nicotinic acetylcholine receptor subunits in the central nervous system of the mollusc *Lymnaea stagnalis*. *J Biol Chem* **281**, 1680–1691.

van Nierop P, Keramidas A, Bertrand S *et al.* (2005). Identification of molluscan nicotinic acetylcholine receptor (nAChR) subunits involved in formation of cation- and anion-selective nAChRs. *J Neurosci* **25**, 10617–10626.

Vernino S, Amador M, Luetje CW *et al.* (1992). Calcium modulation and high calcium permeability of neuronal nicotinic acetylcholine receptors. *Neuron* **8**, 127–134.

Vetter DE, Liberman MC, Mann J *et al.* (1999). Role of alpha9 nicotinic ACh receptor subunits in the development and function of cochlear efferent innervation. *Neuron* **23**, 93–103.

Vicente-Agullo F, Rovira JC, Campos-Caro A *et al.* (1996). Acetylcholine receptor subunit homomer formation requires compatibility between amino acid residues of the M1 and M2 transmembrane segments. *FEBS Lett* **399**, 83–86.

Wenger BW, Bryant DL, Boyd RT *et al.* (1997). Evidence for spare nicotinic acetylcholine receptors and a beta 4 subunit in bovine adrenal chromaffin cells: studies using bromoacetylcholine, epibatidine, cytisine and mAb35. *J Pharmacol Exp Ther* **281**, 905–913.

Wevers A, Burghaus L, Moser N *et al.* (2000). Expression of nicotinic acetylcholine receptors in Alzheimer's disease: postmortem investigations and experimental approaches. *Behav Brain Res* **113**, 207–215.

Wevers A, Jeske A, Lobron C *et al.* (1994). Cellular distribution of nicotinic acetylcholine receptor subunit mRNAs in the human cerebral cortex as revealed by non-isotopic *in situ* hybridization. *Brain Res Mol Brain Res* **25**, 122–128.

Whiteaker P, Jimenez M, McIntosh JM *et al.* (2000). Identification of a novel nicotinic binding site in mouse brain using [(125)I]-epibatidine. *Br J Pharmacol* **131**, 729–739.

Whiting PJ, Schoepfer R, Conroy WG *et al.* (1991). Expression of nicotinic acetylcholine receptor subtypes in brain and retina. *Brain Res Mol Brain Res* **10**, 61–70.

Williams BM, Temburni MK, Bertrand S *et al.* (1999). The long cytoplasmic loop of the alpha 3 subunit targets specific nAChR subtypes to synapses on neurons *in vivo*. *Ann N Y Acad Sci* **868**, 640–644.

Williams BM, Temburni MK, Levey MS *et al.* (1998). The long internal loop of the alpha 3 subunit targets nAChRs to subdomains within individual synapses on neurons *in vivo*. *Nat Neurosci* **1**, 557–562.

Witzemann V, Barg B, Criado M *et al.* (1989). Developmental regulation of five subunit-specific mRNAs encoding acetylcholine receptor subtypes in rat muscle. *FEBS Lett* **242**, 419–424.

Wonnacott S, Albuquerque EX and Bertrand D (1993). Methyllycaconitine: a new probe that discriminates between nicotinic acetylcholine receptor subclasses. *Methods Neurosci* **12**, 263–275.

Wonnacott S, Sidhpura N and Balfour DJ (2005). Nicotine: from molecular mechanisms to behaviour. *Curr Opin Pharmacol* **5**, 53–59.

Zaninetti M, Blanchet C, Tribollet E *et al.* (2000). Magnocellular neurons of the rat supraoptic nucleus are endowed with functional nicotinic acetylcholine receptors. *Neuroscience* **95**, 319–323.

Zaninetti M, Tribollet E, Bertrand D *et al.* (2002). Nicotinic cholinergic activation of magnocellular neurons of the hypothalamic paraventricular nucleus. *Neuroscience* **110**, 287–299.

Zhang CL, Verbny Y, Malek SA *et al.* (2004). Nicotinic acetylcholine receptors in mouse and rat optic nerves. *J Neurophysiol* **91**, 1025–1035.

Zhou Y, Nelson ME, Kuryatov A *et al.* (2003). Human alpha4beta2 acetylcholine receptors formed from linked subunits. *J Neurosci* **23**, 9004–9015.

Zhu D, Xiong WC and Mei L (2006). Lipid rafts serve as a signaling platform for nicotinic acetylcholine receptor clustering. *J Neurosci* **26**, 4841–4851.

Zoli M, Le Novere N, Hill JA *et al.* (1995). Developmental regulation of nicotinic ACh receptor subunit mRNAs in the rat central and peripheral nervous systems. *J Neurosci* **15**, 1912–1939.

3.1.2 5-HT$_3$ receptors

John Peters, Michelle Cooper,
Matthew Livesey, Jane Carland
and Jeremy Lambert

3.1.2.1 Introduction

The initial description of the 5-hydroxytryptamine type-3 (5-HT$_3$) receptor dates back to work performed over 50 years ago upon the guinea pig isolated ileum preparation. Utilizing a classical pharmacological approach, Gaddum and Picarelli (1957) established that the contractile effect of exogenously applied 5-HT upon the ileum involved the activation of 'two kinds of tryptamine receptor'. They further surmised that one of these (the M receptor in their terminology; reclassified as the 5-HT$_3$ receptor by Bradley *et al.* [1986]) 'are probably in the nervous tissue of the intestine'. Some 32 years later the application of the patch-clamp technique to neurons isolated from the submucous plexus of the guinea pig, and neuronal clonal cell lines, provided unambiguous evidence that the 5-HT$_3$ receptor is a cation-selective transmitter-gated ion channel (Derkach *et al.*, 1989; Lambert *et al.*, 1989).

Shortly thereafter, the cloning of the first 5-HT$_3$ receptor subunit, originally termed 5-HT$_{3A}$ (Maricq *et al.*, 1991), assigned the channel to the Cys-loop family of transmitter-gated ion channels that includes the cation-selective nicotinic acetylcholine (nACh) receptors, the anion-selective gamma aminobutyric acid-type-A (GABA$_A$) and strychnine-sensitive glycine receptors and a less well characterized cation-selective Zn^{2+}-activated ion channel (Peters *et al.*, 1997; 2005; Reeves and Lummis, 2002; Davies *et al.*, 2003; Lester *et al.*, 2004; Houtani *et al.*, 2005). Throughout the remainder of this Chapter, we employ the revised nomenclature for ligand-gated ion channels issued by The International Union of Basic and Clinical Pharmacology Committee on Receptor Nomenclature and Drug Classification (NC-IUPHAR) which recommends that subscripted text should not be used in subunit names, but may be retained specifically for the 5-HT$_3$ and GABA$_A$ receptors (see Collingridge *et al.* 2009 for full details).

In common with all members the Cys-loop family, 5-HT$_3$ receptors are a pentameric assembly of subunits (Boess *et al.*, 1995; Green *et al.*, 1995; Barerra *et al.*, 2005) that pseudo-symmetrically surround a central, water-filled, ion channel. The individual subunits comprise a large extracellular N-terminal domain (ECD) that harbours the ligand binding site, four transmembrane domains (M1–M4) connected by intracellular (M1–M2 and M3–M4) and extracellular (M2–M3) loops, of which the M3–M4 linker is the most extensive, and an extracellular C-terminus (Reeves and Lummis, 2002; Karlin, 2002; Lester *et al.*, 2004; Peters *et al.*, 2005; Unwin, 2005; Sine and Engel, 2006; Figure 3.1.2.1a). M2 lines the ion channel (Figure 3.1.2.1b), fringed by M1, M3 and M4 that partition it from the membrane lipid (Miyazawa *et al.*, 2003). However, not all of these domains are indispensable for channel function. A Cys-loop receptor homolog (Glvi) from the cyanobacterium *Gloeobacter violaceus* lacks a large intracellular M3–M4 linker and also structural features within the ECD that are conserved in eukaryotic Cys-loop receptor subunits, yet retains function as a proton gated, cation-selective, ion channel (Bocquet *et al.*, 2007). Indeed, as discussed in section 3.1.2.3, it has proved possible to completely replace the large intracellular loop of the mouse 5-HT3A receptor subunit with the heptapeptide M3–M4 linker of Glvi without loss of function (Jansen *et al.*, 2008).

3.1.2.2 Molecular characterization

The 5-HT3A receptor subunit (Maricq *et al.*, 1991) was initially isolated via an expression cloning strategy utilizing a murine hybridoma cell line (NCB-20) known to express functional 5-HT$_3$ receptors at high density (Peters and Lambert, 1989; McKernan, 1992). This subunit assembles into functional homopentamers in a variety of heterologous expression systems. Subsequently, 5-HT3A subunit species orthologues have been cloned from rat brain (Miyake *et al.*, 1995; Akuzawa *et al.*, 1996) and superior cervical ganglia (Isenberg *et al.*, 1993), human amygdala (Miyake *et al.*, 1995), hippocampus (Belelli *et al.*, 1995), and colon (Lankiewicz *et al.*, 1998), guinea pig small intestine (Lankiewicz *et al.*, 1998), ferret colon (Mochizuki *et al.*, 2000), and dog brain (Jensen *et al.*, 2006). The predicted polypeptides range between 478 and 490 amino acids in length, including a predicted signal sequence of 23 residues. The degree of sequence identity (excluding the signal peptide) between the human 5-HT3A subunit and each of the other species orthologues is high (83–87 per cent), the most conspicuous region of variability being, as for other Cys-loop receptors, within the large intracellular M3–M4 linker. Notably, the channel lining M2 segment and the M2–M3 linker, which are involved in channel gating (Lummis *et al.*, 2005), demonstrate absolute conservation between species. In addition, only a single, conservative (i.e. Glu → Asn), amino acid substitution is apparent between the human and all other subunit orthologues at the − 4′ position within the M1–M2 linker that contributes to the intracellular entrance to the pore region (Figure 3.1.2.1b).

The human *HTR3A* gene maps to chromosome 11 at locus 11q23.1 and comprises 9 exons and 8 introns that span approximately 14.5 KBa from the start to stop codon (Weiss *et al.*, 1995; Bruss *et al.*, 2000). The presence of two splice acceptor sites

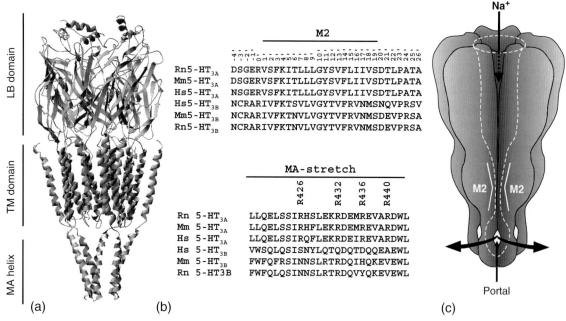

Fig. 3.1.2.1 Structural features of 5-HT3 receptor subunits that influence single channel conductance and ion selectivity. (a) Homology model of the wild-type 5-HT$_3$ receptor constructed using the *Torpedo marmorata* model as a template. The model depicts (i) the extracellular ligand binding (LB) domain; (ii) the transmembrane (TM) spanning domains, of which M2 forms the lining of the ion channel, and (iii) the MA-helix domains that are postulated to frame intracellular portals that form conduits for ion flow into and out of the intracellular vestibule of the channel. (b) Sequence alignment of the M2 domain and adjacent sequences (upper) and MA-helix (lower) of 5-HT3A and 5-HT3B subunit species orthologues. The numbering of the amino acid residues within the M2 domain and flanking sequences corresponds to that conventionally employed for cation-selective members of the Cys-loop family (e.g. Peters *et al.*, 2005) whereas that of the MA-helix corresponds to the human orthologue of the 5-HT3A receptor subunit (Kelley *et al.*, 2003). The abbreviations are: Hs, human; Mm, mouse; Rn, rat. (c) Schematic representation of the proposed revised mode of ion permeation. In this scheme, the intracellular vestibule contains five parallel conduction pathways ('portals') that are situated at the interface of adjacent subunits of the receptor. Elements of the MA-stretch contribute to the frames of each portal.

in exon 9 of the gene encoding mouse, rat and guinea pig, but not human, dog or ferret, 5-HT3A subunits, gives rise to iso-forms termed 5-HT3A(a) ('long') and 5-HT3A(b) ('short') (Hope *et al.*, 1993; Isenberg *et al.*, 1993; Werner *et al.*, 1994; Belelli *et al.*, 1995; Lankiewicz *et al.*, 1998; Mochizuki *et al.*, 2000; Jensen *et al.*, 2006). The functional significance of such alternative splicing in which six (mouse and guinea pig), or five (rat), residues are omitted from the M3-M4 linker in the 5-HT3A(b) variant is unclear, although differences in the efficacy of certain 5-HT$_3$ receptor agonists at the two isoforms have been described and agents that activate protein kinase A (PKA) and protein kinase C (PKC) differentially modulate the two splice variants (Downie *et al.*, 1994; van Hooft *et al.*, 1997; Niemeyer and Lummis, 1999; Hubbard *et al.*, 2000). The human canonical isoform corresponds to the mouse 5-HT3A(b) subunit (Belelli *et al.*, 1995; Miyake *et al.*, 1995). Novel truncated (238 amino acids in length with only a single transmembrane [M1] domain) and 'long' (containing a 32 amino acid insertion between M2 and M3) splice variants of the human 5-HT3A subunit, unofficially denoted 5-HT3AT and 5-HT3AL, respectively, have also been reported (Bruss *et al.*, 2000). Unsurprisingly, neither forms a functional receptor when expressed alone. However, upon co-expression with canonical 5-HT3A subunits, receptors with modified properties are gener-ated. Whether such receptors occur in nature is presently a matter for conjecture, but co-expression of the relevant mRNA species within the amygdala and hippocampus has been documented (Bruss *et al.*, 2000).

The human 5-HT3A subunit is, in common with other mem-bers of the Cys-loop family, subject to post-translational modifica-tions that are functionally important. *N*-linked glycosylation of the subunit occurs at four locations within the N-terminal ECD and suppression of this process at any one of three of these (i.e. N104, N170 and N186; numbering includes the signal peptide) essentially abolishes cell surface expression and the formation of a high-af-finity ligand binding site (Monk *et al.*, 2004). Glycosylation of the homologous residues within the mouse 5-HT3A subunit affects either receptor assembly, or plasma membrane targeting (Quirk *et al.*, 2004). Phosphorylation of the intracellular M3–M4 linker of the guinea pig 5-HT3A subunit, possibly via PKA, occurs at S409, but a functional correlate of this specific modification has not been reported (Lankiewicz *et al.*, 2000). The function of native 5-HT$_3$ receptors is modulated by agents that increase the activity of PKA (Yakel *et al.*, 1991), PKC (Zhang *et al.*, 1995) or by kinases such as casein kinase II introduced into the cell interior (Jones and Yakel, 2003). However, although the M3–M4 loop of the 5-HT3A subu-nit possesses the appropriate consensus sequences for such kinases, there is no proof that their action is direct. Indeed, potentiation of responses mediated by the mouse 5-HT$_3$A receptor by activators of PKC persists when all PKC consensus sequences are mutated from the M3–M4 loop (Sun *et al.*, 2003). Instead, enhancement of mac-roscopic current responses evoked by 5-HT results from increased cell surface expression of the receptor due to structural rearrange-ments of F-actin (Sun *et al.*, 2003) with which the receptor clusters (see section 3.1.2.4).

A second member of the 5-HT$_3$ receptor subunit family (i.e. 5-HT3B), with 41 per cent amino acid sequence identity to the human 5-HT$_{3A}$ subunit, was isolated in 1999 by the screening of human genomic sequence data (Davies *et al.*, 1999; Dubin *et al.*, 1999). Species orthologues of the 5-HT3B subunit that have a rela-

tively low (73 per cent) degree of identity to the human sequence, but a conserved gene structure, have been cloned from the mouse and rat (Hanna *et al.*, 2000). The human *HTR3B* gene locus is immediately upstream to that of *HTR3A* (11q23.1) suggesting that a local duplication event generated these two members of the family (Davies *et al.*, 1999). The human 5-HT3B subunit was initially reported to comprise 441 amino acids (Davies *et al.*, 1999; Dubin *et al.*, 1999), but this was subsequently amended to 436 following the cloning of rat and mouse 5-HT3B subunit orthologues and a revised prediction of the sequence at which translation is initiated (Hanna *et al.*, 2000). More recent data suggests that tissue-specific alternative promoters in the *HTR3B* gene give rise to distinct transcription start sites in the intestine (the canonical form) versus brain. The brain transcripts (unof-ficially termed BT-1 and BT-2) predict translation products that differ from the canonical form only in the signal peptide (BT-1), or more dramatically by lacking a substantial portion of the N-terminal sequence (BT-2) (Tzvetkov *et al.*, 2007). Whether such proteins are synthesized within the brain remains to be demonstrated.

The introduction of cDNA encoding the 5-HT3B subunit into either mammalian cells, or oocytes, unlike cDNA coding for the 5-HT3A subunit, does not result in functional receptors nor, indeed, specific 5-HT$_3$ receptor binding sites (Davies *et al.*, 1999; Dubin *et al.*, 1999; Hanna *et al.*, 2000; Boyd *et al.*, 2002, Holbrook *et al.*, 2008). Boyd *et al.* (2002) reported retention of the 5-HT3B receptor subunit within the endoplasmic reticulum (ER), but a recent study has presented evidence for cell surface expression in the absence of the 5-HT3A subunit (Holbrook *et al.*, 2008). When co-expressed with the 5-HT3A subunit, 5-HT3B subunit is able to form functional heteromeric 5-HT$_3$AB receptors with distinct biophysical characteristics as described in section 3.1.2.3 (Davies *et al.*, 1999; Dubin *et al.*, 1999; Boyd *et al.*, 2002; Reeves and Lummis, 2006). Atomic force microscopy applied to epitope-tagged receptor subunits suggests that the heteromeric 5-HT$_3$AB receptor is composed, *in vitro*, of two 5-HT3A and three 5-HT3B subunits arranged as a rosette with the order B-B-A-B-A (Barrera *et al.*, 2005).

Additional putative human 5-HT$_3$ receptor subunit genes, occurring as a close cluster on human chromosome 3 at 3q27.1, provisionally termed *HTR3C* (aka *HTR3C3*), *HTR3D*, *HTR3E* (aka *HTR3C1long*), *HTR3C2* and *HTR3C4* have been described (Karnovsky *et al.*, 2003; Niesler *et al.*, 2003). *HTR3C2* and *HTR3C4* are considered to be pseudogenes in man (Karnovsky *et al.*, 2003). The 5-HT3D subunit predicted by Niesler *et al.* (2003) contains only 279 amino acid residues, has no signal peptide, and lacks most of the ECD including the signature Cys–Cys loop. The predicted products of the putative *HTR3C* and *HTR3E* genes, which com-prise 447 and 471 amino acid residues, respectively, share 36 per cent and 39 per cent identity with the human 5-HT3A subunit (Karnovsky *et al.*, 2003). A splice variant of the 5-HT3E subunit, unofficially termed 5-HT3Ea (aka 5-HT3C1short), differs from the former in lacking 15 N-terminal amino acid residues and present-ing a completely different signal peptide sequence (Karnovsky *et al.*, 2003; Neisler *et al.*, 2007). Unlike the 5-HT3D subunit, the 5-HT3C (5-HT3C3) and 5-HT3E (5-HT3C1) subunits conserve the essential structural features of Cys-loop receptor subunits. Nonetheless, one report indicates that, in common with the 5-HT3D subunit, they do not traffic to the cell surface, or form a ligand binding site, when expressed alone in HEK-293 cells (Niesler et al., 2007). However, a second report suggests that all three subunits are capable of cell surface expression when expressed singularly, but lack function (Holbrook et al., 2008). The reason for this discrepancy is uncertain, but may relate to the fact that Boyd et al. (2002) and Niesler et al. (2007) incorporated epitope tags to map subunit distribution in the extreme N-terminal domain whereas Holbrook et al. (2008) attached tags to the C-terminus via a short, flexible, polyglycine spacer. Co-immunoprecipitation of the 5-HT3C, 5-HT3D or 5-HT3E subunits with the 5-HT3A subunit has been demonstrated by Niesler et al. (2007). Co-expression of the 5-HT3A subunit with either the 5-HT3C, 5-HT3D, or 5-HT3E subunits does not result in receptors with biophysical, or pharmacological, characteristics that differ substantially from those of homomeric $5-HT_3A$ receptors (Niesler et al., 2007; Holbrook et al. 2008, see section 3.1.2.5). Curiously, the 5-HT3C, 5-HT3D and 5-HT3E subunits appear to be absent from the rodent genome, although they are present the dog, cow, pig, rabbit, ferret, and chicken genomes (Karnovsky et al., 2003; Holbrook et al., 2008).

3.1.2.3 Biophysical characteristics

Native $5-HT_3$ receptors are cation-selective ion channels that very weakly select between monovalent metals in the sequence $Cs^+>K^+>Li^+>Na^+>Rb^+$, but essentially exclude anions (Derkach et al., 1989; Lambert et al., 1989; Yakel et al., 1990; Yang, 1990; Furukawa et al., 1992; Yang et al., 1992; Peters et al., 1993; Jackson and Yakel, 1995). The relative permeability of monovalent organic cations decreases as a function of their mean geometric diameter and from such measurements the diameter of the most constricted region of the $5-HT_3$ receptor channel pore in the open state has been estimated to be approximately 7.6 Å in N18 neuroblastoma cells (Yang, 1990) and 8.2 Å in rabbit nodose ganglion neurons (Malone et al., 1994).

Numerous reports indicate that the biophysical properties of the ion channel integral to native $5-HT_3$ receptors are heterogeneous. This is exemplified by estimates of single channel conductance made by direct observation of channel activity in excised membrane patches, or inferred from fluctuation analysis of macroscopic current responses recorded from whole-cells under voltage-clamp. Recordings from membrane patches excised from guinea pig submucous and myenteric plexus neurons yield well resolved single channel events with conductances of 9–15 pS (Derkach et al., 1989) and 17 pS (Zhou and Galligan, 1999), respectively. Direct observation of channel activity elicited by $5-HT_3$ receptor activation in rat and mouse superior cervical ganglion neurons, hippocampal neurons of the same species, and rabbit nodose ganglion neurons provides comparable estimates of conductance (9–19 pS) (Yang et al., 1992; Peters et al., 1993; Hussy et al., 1994; Jones and Surprenant, 1994). By contrast, discrete channel events are not resolved in equivalent recordings made from murine N1E-115 (Lambert et al., 1989; Hussy et al., 1994; Barann et al., 1997), or N18 (Yang, 1990) neuroblastoma cells. Indeed, fluctuation analysis indicates that the conductance of the $5-HT_3$ receptor channel of clonal cell lines is far too small (0.3–0.6 pS) (Lambert et al., 1989; Yang, 1990; Hussy et al., 1994) to permit direct resolution of single channel events (but see Shao et al., 1991 and van Hooft and Vijverberg, 1995). Thus, based on their divergent conductances, at least two broad populations of $5-HT_3$ receptor-channel could exist in nature (Fletcher

and Barnes, 1998). Interestingly, fluctuation analysis applied to rat and mouse superior cervical ganglion neurons yields an estimate of channel conductance (i.e. 2–3 pS) much smaller than that seen in single channel recording in the same cell type (Yang et al., 1992; Hussy et al., 1994). A similar value, inferred by fluctuation analysis, has been reported for rat petrosal ganglion neurons (Zhong et al., 1999). It is conceivable that such cells express both low (0.3–0.6 pS) and relatively high conductance (~10 pS) ion channels and that fluctuation analysis provides a weighted mean estimate of the two (Yang et al., 1992; Hussy et al., 1994). The presence of both 5-HT3A and 5-HT3B subunit transcripts in many superior cervical ganglion neurons of the rat supports this possibility (Morales and Wang, 2002).

Diversity in the channel properties of native $5-HT_3$ receptors extends to the conduction of divalent cations such as Ca^{2+}, for which varying estimates of relative permeability have been reported (Yakel et al., 1990; Yang, 1990; Yang et al., 1992; Jackson and Yakel, 1995; Fletcher and Barnes, 1998; van Hooft and Wadman, 2003). In addition, although there is a consensus that extracellular Ca^{2+} and Mg^{2+} reduce macroscopic currents mediated by $5-HT_3$ receptors, blockade can be either voltage-independent, as found for N1E 115 and N18 neuroblastoma cells (Peters et al., 1988; Yang, 1990), or voltage-dependent, as consistently reported for hippocampal basket cells and interneurons (Kawa, 1994; McMahon and Kauer, 1997; van Hooft and Wadman, 2003). In this context it is interesting that the homomeric $5-HT_3A$ receptor, which shares the low single channel conductance of $5-HT_3$ receptors expressed by neuroblastoma cells is also, when expressed in mammalian cell hosts, blocked by extracellular Ca^{2+} in a voltage-independent manner (Gill et al., 1995; Brown et al., 1998; Mochizuki et al., 1999a; Niemeyer and Lummis, 2001; Hu and Lovinger, 2005). However Noam et al. (2008) recently reported that dialysis of the cell interior with phosphate, or phosphate-containing compounds, confers a voltage-dependent blockade upon the human $5-HT_3A$ receptor by Ca^{2+}. This effect appears to involve positively changed arginine residues with the large M3–M4 loop that additonally exert a profound influence upon single channel conductance (Kelley et al., 2003; see below). Neutralization, or inversion, of the negative charge of the extracellular ring present at position 20′ of M2 (Figure 3.1.2.1b) of the mouse $5-HT_3A$ receptor, a position predicted to experience only a small fraction of the transmembrane field, completely abolished blockade of macroscopic currents by extracellular Ca^{2+} consistent with the voltage-independent nature of the block reported in many studies of the wild-type receptor (Hu and Lovinger, 2005). However, neutralization of the extracellular ring of a high conductance variant of the human $5-HT_3A$ receptor (see below) did not alleviate voltage-independent blockade of single channel conductance by Ca^{2+} (Livesey et al., 2008).

The pattern of rectification of whole-cell and single channel currents evoked by the activation of native $5-HT_3$ receptors provides a further indication of heterogeneity (Fletcher and Barnes, 1998). For example, the single channel conductance of the receptor expressed by murine N1E-115 cells increases markedly with membrane hyperpolarization (inward rectification) as does that of the more efficiently conducting channels present in rat and mouse superior cervical ganglion neurons (Yang et al., 1992; Hussy et al., 1994). By contrast, little, or no, rectification occurs in the case of single channel currents recorded from guinea pig myenteric and submucous-plexus neurons, or rabbit nodose ganglion neurons (Derkach et al., 1989; Peters et al., 1993; Zhou and Galligan, 1999).

Collectively, the weight of biophysical evidence indicates 5-HT$_3$ receptors to be heterogeneous. Unlike pharmacological heterogeneity (see section 3.1.2.5), the differences in ion channel function cannot be explained by interspecies variation. Below, we consider to what extent recombinant homomeric receptors composed of 5-HT3A receptor subunits and heteromeric receptors assembled from both 5-HT3A and 5-HT3B subunits can explain the heterogeneity observed for the 5-HT$_3$ receptors endogenous to neurons. Recent studies in which the structural determinants of single channel conductance and ion selectivity have been explored will also be discussed.

Homomeric 5-HT$_3$A and heteromeric 5-HT$_3$AB receptors are both cation-selective ion channels (Maricq et al., 1991; Brown et al., 1998; Davies et al., 1999; Mochizuki et al., 1999a; Thompson and Lummis, 2003), but important differences exist between them with respect to ion permeation and conduction. Most significantly, the single channel conductance of homomeric 5-HT$_3$A receptors is unusually small (<1 pS) (Hussy et al., 1994; Werner et al., 1994; Gill et al., 1995; Brown et al., 1998; Davies et al., 1999; Mochizuki et al., 1999a; Gunthorpe et al., 2000; Kelley et al., 2003b; Hu et al., 2006; Sun et al., 2008) and must be inferred by fluctuation analysis. By contrast, the much larger single channel conductance of the human 5-HT$_3$AB receptor, which has a slope conductance of 16 pS in outside-out membrane patches, readily permits direct resolution of single channel events (Davies et al., 1999). An even larger single channel slope conductance of 25 pS has been reported from cell-attached patch recordings (Krzywkowski et al., 2008). In addition, 5-HT$_3$AB receptors demonstrate substantially reduced permeability to Ca^{2+} and unlike 5-HT$_3$A receptors are impermeant to Mg^{2+} and, at least for the human receptor, do not display the profile of marked inward rectification that is a characteristic of whole-cell and single channel currents mediated by 5-HT$_3$A receptors expressed in mammalian cell hosts (Hussy et al., 1994; Brown et al., 1998; Davies et al., 1999).

As a generalization, the biophysical properties of 5-HT$_3$A receptors most closely resemble those of 5-HT$_3$ receptors endogenous to neuronal clonal cell lines such as N1E-115 (Lambert et al., 1989; Hussy et al., 1994) and N18 (Yang, 1990) neuroblastoma cells. The commonalities include (i) a sub-picosiemen single channel conductance; (ii) depression of single channel conductance by extracellular Ca^{2+}; (iii) similar relative permeabilities to Ca^{2+} and Mg^{2+} and (iv) inward rectification at the single channel level (Hussy et al., 1994; Brown et al., 1998; Davies et al., 1999). Thus, it is tempting to speculate that such receptors are assembled as homomeric complexes of the 5-HT3A subunit. However, such a simple interpretation is complicated by the fact that the neuroblastoma cell lines studied clearly express transcripts that encode both the 5-HT3A and 5-HT3B subunits (Hanna et al., 2000; Stewart et al., 2003).

It is axiomatic that the channel lining M2 domain of Cys-loop receptor subunits is a major determinant of ion selectivity and single channel conductance (Karlin, 2002; Keramidas et al., 2004; Peters et al., 2005; Sine and Engel, 2006). The importance of M2 in the cation selectivity of the murine 5-HT$_3$A receptor has been established by neutralisation of the intermediate ring of negative charges (Glu) at the -1′ position by replacement with alanine, in concert with the insertion of proline between Glu-1′Ala and Gly-2′ and the exchange of the 13′ valine residue by threonine (Figure 3.1.2.1b). Receptors assembled from such mutant subunits are anion-selective ($P_{Cl}/P_{Na}=12.3$), but have a single channel conductance that is

reduced to a value even lower than that (0.4 pS) (Gill et al., 1995) of the wild-type homomeric receptor (Gunthorpe and Lummis, 2001). The Glu-1′ residue in particular is a crucial component of the ion selectivity filter, because replacement of this residue by alanine in the mouse 5-HT$_3$A receptor abolishes the monovalent ion selectivity of the channel ($P_{Cl}/P_{Na} = 1.12$) (Thompson and Lummis, 2003). Interestingly, 5-HT$_3$AB receptors that contain three copies of the 5-HT3B subunit, wherein alanine occupies the −1′ position (Figure 3.1.2.1b), are cation-selective, although their relative permeability to divalent cations in comparison to 5-HT$_3$A receptors is reduced (Davies et al., 1999). Evidently, two negative charges within the intermediate ring are sufficient to preserve cation selectivity. The relative permeability of the human 5-HT$_3$A receptor to Ca^{2+} is also reduced by mutation of Glu20′ to alanine (Livesey et al., 2008). Consistent with the reduced relative permeability of the human 5-HT$_3$AB, versus human 5-HT$_3$A, receptor, to Ca^{2+}, the 5-HT3B subunit presents an asparagine, rather than glutamate, residue at the 20′ position (Figure 3.1.2.1b). The inclusion of the 5-HT3B subunit within the heteromeric receptor thus reduces the negative charge of the extracellular ring and may reduce the local accumulation of Ca^{2+}. Consistent with such an interpretation, mutation of Glu20′ of a high-conductance variant of the 5-HT$_3$A receptor (see below) to alanine suppresses inward, but not outward, single channel currents (Livesey et al., 2008).

Studies employing the substituted cysteine accessibility method (SCAM) strongly suggest that the M2 domain of the 5-HT3A subunit adopts an α-helical conformation between residues 1′ and 19′ (Reeves et al., 2001; Panicker et al., 2002). By contrast, the M1–M2 linker, which contributes to the intracellular vestibule of the channel, probably exists as an extended loop structure (Reeves et al., 2001; Panicker et al., 2002).

The primary amino acid sequence of the 5-HT3A subunit within M2 and its extra- and intracellular flanking regions shows strong homology to those of nACh receptor subunits and, moreover, conservation at the −4′ (cytoplasmic), −1′ (intermediate), 2′ (polar) and 20′ (extracellular) rings that are well established and important determinants of single channel conductance in nACh receptors (Imoto et al., 1988; Corringer et al., 2000). Paradoxically, the homologous domain of the human 5-HT3B subunit presents features that appear less conducive to cation conduction, yet its inclusion within heteromeric receptors enhances single channel conductance by nearly 40-fold (Davies et al., 1999). Notably, the −1′ and 20′ positions are occupied by non-polar (alanine) and polar (asparagine) residues, respectively, reducing the net negative charge of the intermediate and extracellular rings and, most unusually, valine replaces the hydroxyl-bearing residue (serine, or threonine) commonly occurring at the 2′ position in cation-selective Cys-loop receptor subunits (Figure 3.1.2.1b). This conundrum was resolved by the finding that an intracellular domain of the 5-HT3A subunit, rather than M2, sets an upper limit to the rate of ion translocation through the 5-HT$_3$A receptor. Specifically, an amphipathic helix (the MA-stretch) (Finer-Moore and Stroud, 1984) within the M3–M4 loop proximal to M4 presents positively charged arginine residues (R432, R436 and R440 in the human sequence) that are unique to 5-HT3A subunit species orthologues within all subunits of the cation-selective members of the Cys-loop family (Kelley et al., 2003b; Peters et al., 2005; Figure 3.1.2.1b). Replacement of the arginine residues by their counterparts within the human 5-HT3B subunit (i.e. alanine, aspartate and glutamine), to generate the 5-HT$_3$A(QDA) receptor construct, is sufficient to enhance single channel conductance of the 5-HT$_3$A receptor by at least 37-fold (Kelley *et al.*, 2003b; Reeves *et al.*, 2005; Hales *et al.*, 2006). The individual mutations R436D and R440A cause nine- and five-fold increases in single channel conductance, respectively, whereas the R432Q mutation has no substantial effect (Hales *et al.*, 2006). Replacement of R427 (equivalent to R440 of the human sequence) of the mouse 5-HT$_3$A receptor by glycine, increases single channel conductance by approximately two-fold (Hu *et al.*, 2006). Consistent with an important role of the large intracellular loop in regulating ion flux, its complete substitution in the mouse 5-HT$_3$A receptor by the heptapeptide M3–M4 linker of the prokaryotic proton-gated ion channel, Glvi, increases single channel conductance to 43 pS. However, such an exchange does not cause any measurable change in cation versus anion selectivity, confirming a dominant role for M2 and bracketing sequences in this process (Jansen *et al.*, 2008). Nonetheless, the MA-stretch contributes to discrimination between mono- and divalent cations, since the 5-HT$_3$A(QDA) construct has a greater relative permeability to Ca^{2+} than the wild-type receptor (Livesey *et al.*, 2008).

The α-helical MA-stretch of the α4β2 nACh receptor has also been found to influence single channel conductance via α4 and β2 subunit residues homologous to R432 and R436 of the human 5-HT$_3$A receptor (Hales *et al.*, 2006). The recent data that implicates the MA-stretch as a component of the ion permeation pathway can be incorporated into the traditional model of conduction by postulating that the transmembrane pore, lined by five M2 segments, funnels into an intracellular vestibule that is perforated by five narrow openings ('portals') that form obligate pathways for ions fluxing to and from the ion channel (Figure 3.1.2.1c). Ultrastructural evidence for the existence of such portals has been provided by cryoelectron microscopy and image reconstruction applied to tubular crystals of the nACh receptor of *Torpedo marmorata* (Miyazawa *et al.*, 1999; Unwin, 2005). Such images reveal that a curved α-helical MA-stretch projects from the base of each subunit and these collectively converge on the central axis of the receptor in the manner of an inverted tee-pee. The dimensions of the portals, which reside between adjacent MA-stretches, are estimated to be no greater than the diameter of a permeating ion, including its first hydration shell (Miyazawa *et al.*, 1999; Unwin, 2005). Thus, the charge of the residues that frame the portals should impact upon ionic movements and additionally, in the manner of a 'selectivity filter', influence the local concentration of ions within the vestibule (Unwin, 2005). Thus, it is likely that the anomalously low conductance of the 5-HT$_3$A receptor is attributable to an unfavourable charge distribution over the portals, or at least in proximity to them. R432, R436 and R440 are predicted to lie on the same stripe of an α-helix, potentially providing a powerful repulsive electrostatic influence upon cation flow that might be exacerbated by the large side chain volume of the arginine residue. Direct evidence for both steric and electrostatic influences of the residue occupying the 436 position has recently been obtained by studying a substituted cysteine residue covalently modified by neutral, positive and negative methanethiosulphonate (MTS) reagents (Deeb *et al.*, 2007). Presumably, five identical portals exist in the inner vestibule of the 5-HT$_3$A receptor and it seems likely that the influence of the 5-HT$_3$B subunit upon single channel conductance results not from its M2 sequence, but the formation 5-HT3A/5-HT3B subunit interfaces that provide permeation pathway(s) more conducive to cation conduction.

The unusually low subpicosiemen single channel conductance of the homomeric 5-HT$_3$A receptor has, in one study, been attributed to blockade of the channel by extracellular divalent cations (Hapfelmeier *et al.*, 2002a). However, in other studies (Brown *et al.*, 1998) extracellular Ca^{2+} and Mg^{2+} at physiological concentrations have been found by fluctuation analysis to exert only a modest depressant effect upon the single channel conductance of the human 5-HT$_3$A receptor, comparable to that reported for the nACh receptor of skeletal muscle (Decker and Dani, 1990). Indeed, employing the human 5-HT$_3$A(QDA) receptor construct that permits direct resolution of single channel events, we find Ca^{2+} (0.1–30 mM) to depress single channel conductance in a voltage-independent manner, but even with Ca^{2+} as the sole charge carrier, single channel currents are clearly evident (Livesey *et al.*, 2008). Earlier studies upon 5-HT$_3$ receptors native to the neuroblastoma cell line N1E-115 have additionally suggested that single channel conductance can be regulated by phosphorylation (van Hooft and Vijverberg, 1995). In view of the now established influence of the M3–M4 loop upon channel conductance (Kelley *et al.*, 2003b; Hales *et al.*, 2006; Deeb *et al.*, 2007; Jansen *et al.*, 2008), a mechanism by which phosphorylation might regulate this biophysical property is not difficult to envisage and might readily be tested by mutagenesis and complementary pharmacological approaches.

In addition to augmenting single channel conductance, the 5-HT3B subunit has a pronounced effect upon the kinetics of agonist-evoked macroscopic current responses. Inclusion of the 5-HT3B subunit within heteromeric receptors greatly accelerates the rate of onset of desensitization of current responses to 5-HT (Dubin *et al.*, 1999; Hapfelmeier *et al.*, 2002b; Stewart *et al.*, 2003; Hu and Peoples, 2008). Co-expression of the 5-HT3B subunit also increases the rate at which recovery from desensitization occurs following the termination of agonist application (Hapfelmeier *et al.*, 2002b). A recent study (Hu and Peoples, 2008) has proposed that unlike 5-HT$_3$A receptors, 5-HT$_3$AB receptors display spontaneous channel openings. Such an assertion is based on the observation that a competitive antagonist, bemesetron (MDL72222), and Zn^{2+} (see section 3.1.2.5) elicit outwardly directed macroscopic currents, at negative holding potentials, in HEK-293 cells expressing the 5-HT$_3$AB receptor. However, such currents are extremely small compared to the inwardly directed currents evoked by 5-HT, indicating that either the probability of spontaneous openings is very low, or that spontaneous events have a relatively brief open time, or small unitary conductance.

The biophysical properties of recombinant 5-HT$_3$ receptors are summarized in Table 3.1.2.1.

3.1.2.4 Cellular and subcellular localization

The distribution of the 5-HT3A subunit within neuronal and non-neuronal cells has been assessed by the use of antibodies raised against elements of the subunit sequence, or constructs tagged with epitopes, or fluorescent labels. In HEK-293 cells expressing the mouse 5-HT3A subunit tagged with yellow fluorescent protein (YFP) at the extreme N-terminus (i.e. YFP-5-HT3A), live imaging localized the subunit to the plasma membrane, intracellular vesicles and, to a lesser extent, the endoplasmic reticulum (ER) (Grailhe *et al.*, 2004). Such a distribution is similar to that reported in fixed intact and permeabilised HEK-293, or COS-7, cells in which the localization of the heterologously expressed mouse 5-HT3A subunit

Table 3.1.2.1 Properties of 5-HT$_3$ receptors

Channel subunit	Biophysical characteristics †	Cellular and subcellular localisation	Pharmacology ‡	Physiology *	Relevance to disease
5-HT3A Gene: *HTR3A* Chromosome location: mouse: 9 A5.3 rat: 8q23 human: 11q23.1	Functional as a homomer [1] Monovalent permeability series: Li$^+$ = Cs$^+$ = Rb$^+$ K$^+$ = Na$^+$ [2] Relative permeability to divalents: P$_{Ca}$/P$_{Cs}$ = 1.0–1.08, P$_{Mg}$/P$_{Cs}$ = 0.41–0.61 [3, 4]. Single channel conductance: 0.3–1.0 pS [2–4] Single channel conductance reduced by extracellular divalent cations [3]. Inwardly rectifying single channel current-voltage relationship [3]	Brain (area postrema, nucleus tractus solitarius, nucleus of the spinal tract of the trigeminal nerve (Sp5), dorsal motor nucleus of the vagal nerve, hippocampus (pyramidal cell layer of CA1 to CA3 and granule cell layer of dentate gyrus), amygdala, frontal, piriform and entorhinal cortices, reticular and paraventricular thalamic nuclei, olivar nuclei, olfactory bulb, nucleus accumbens, putamen, caudate [5–13] Dorsal and ventral horn of spinal cord [6, 10, 11]. Vagus nerve [9] Neurons of the myenteric and submucus plexus and interstitial cells of Cajal [14] Nerve fibres of the urinary bladder [15]	Antagonists (pK$_D$): [^{3}H]-ramosetron 9.8 [16], [^{3}H]-granisetron 8.9 [17, 18], [^{3}H]-GR65630 8.6 [19] Antagonists (pK$_i$): S-zacopride 9.0 [17], granisetron 8.6–8.8 [16, 18], tropisetron 8.5–8.8 [16, 19], azasetron 8.5 [17], R-zacopride 7.9 [17], ondansetron 7.8–8.3 [16–18] Agonists (pK$_i$): quipazine 9.5 [17], m-chlorophenylbiguanide 6.6–7.2 [16–19], 5-HT 6.5–6.9 [17, 18, 19], 2-methyl-5-HT 6.0–6.7 [16, 18, 19], 1-phenylbiguanide 4.7–5.6 [16–19] Agonists (pEC$_{50}$): m-chlorophenylbiguanide 5.4–5.7 [4, 16, 19, 20], 5-HT 5.5–5.9 [4, 16, 19–22], 2-methyl-5-HT 5.3 – 5.6 [4, 16, 19, 20], 1-phenylbiguanide [19, 20] Channel blockers: diltiazem pK$_i$ = 5.0 (at –60 mV) [23], picrotoxin pIC$_{50}$ = 4.4–4.5 [24, 25]	Null mutant mice demonstrate no gross deficits [26–28] Postsynaptic receptors mediate fast excitatory synaptic transmission in cortex [29, 30, 31], amygdala [32] and myenteric neurons [33] Presynaptic receptors evoke/facilitate release of GABA from central interneurons [34, 35] Contributes to persistent nociception [28, 36]	Polymorphisms possibly associated with bipolar disorder [37] and personality trait of lower harm avoidance in women [38] Inappropriate activation in chemo- and radiotherapy evokes nausea and vomiting (relative involvement of homomeric and heteromeric receptors unknown) [39]
5-HT3B Gene: *HTR3B* Chromosome location mouse: 9 B rat: 8q23 human: 11q23.1	No functional expression as a homomer [4, 22]. Expresses as a heteromer with the 5-HT3A subunit (tabulated below)	Brain (cerebral cortex including occipital, frontal, and temporal regions, amygdala, caudate nucleus, hippocampus and thalamus) [4, 22, 40–43], but expression in brain is disputed [44] Neurons of the myenteric and submucous plexus and mucosal epithelial cells of intestinal crypts [45] Superior cervical, nodose and dorsal root ganglion neurons [44]	Does not form a binding site as a homomer [46]. Down regulation of mRNA encoding the 5-HT3B subunit as a consequence of the knockout of SERT is associated with an increase in the EC$_{50}$ for 5-HT in myenteric neurons [47]	Requires co-expression with the 5-HT3A receptor subunit for function [4, 22]	Polymorphisms possibly protective in bipolar disorder and major depression [38, 48]
5-HT3C Gene: *HTR3C* (aka *HTR3C*) Chromosome location: human: 3q27.1	No functional expression as a homomer [49]	mRNA detected by RT-PCR in brain, colon, intestine, lung muscle, and stomach [50]	Does not form a binding site as a homomer [49]	Not characterised	None known
5-HT3D Gene: *HTR3D* Chromosome location: human: 3q27.1	No functional expression as a homomer [49]	Possibly a pseudogene [50, 51]	Does not form a binding site as a homomer [49]	Not characterised	None known

Continued

Table 3.1.2.1 (*Continued*) Properties of 5-HT$_3$ receptors

Channel subunit	Biophysical characteristics†	Cellular and subcellular localisation	Pharmacology†	Physiology*	Relevance to disease
5-HT3E Gene: *HTR3E* (aka *HTR3C1*) Chromosome location: human: 3q27.1	No functional expression as a homomer [49]	mRNA detected by RT-PCR in colon, intestine, spinal cord [50, 51]	Does not form a binding site as a homomer [49]	Not characterised	None known
(5-HT3A)$_2$(5-HT3B)$_3$§ (composition determined for recombinant receptor *in vitro* [52])	Monovalent permeability series: Cs$^+$ = K$^+$ = Na$^+$ [4] Relative permeability to divalents: P_{Ca}/P_{Cs} = 0.62, P_{Mg}/P_{Cs} = 0 [4] Single channel conductance: 16.0 pS (outside-out patch) [4], 30 pS (cell attached patch) [48] Linear single channel current-voltage relationship [4]	Potentially at cellular locations of 5-HT3A and 5-HT3B subunit co-expression (e.g. submucus plexus neurons and nodose, superior cervical and dorsal root ganglion cells) [45, 44]	Antagonists (pK$_D$): [^{3}H]-granisetron 8.8 [17] Antagonists (pK$_i$): S-zacopride 8.8 [17], tropisetron 8.5–8.8 [17], azasetron 8.4 [17], R-zacopride 7.7 [17], ondansetron 7.8 [17] Agonists (pK$_i$): quipazine 9.0 [17], m-chlorophenylbiguanide 7.0 [17], 5-HT 6.0 [6], 1-phenylbiguanide 4.9 [17] *Agonists* (pEC$_{50}$) m-chlorophenylbiguanide 5.7 [4], 5-HT 4.8–5.3 [4, 21, 48], 2-methyl-5-HT 4.9 [4] *Channel blockers:* picrotoxin pIC$_{50}$ = 2.9 [24]		

* Currently available data do not allow specific physiological functions to be ascribed to homomeric 5-HT$_3$A versus heteromeric 5-HT$_3$AB receptors.

† Data refer to the human orthologues of the 5-HT$_3$ receptor. Rodent heteromeric receptors assembled from 5-HT3A and 5-HT3B subunits display a smaller single channel conductance than the human orthologues and mediate single channel currents that rectify inwardly. [53]

‡ Data for agonist and antagonist compounds refers to the human orthologues of the 5-HT$_3$ receptor. Prominent species differences exist, particularly for agonists of the arylbiguandide class [54]. Data for channel blockers refers to the mouse 5-HT$_3$ receptor [23, 24, 25].

§The stoichiometry is that proposed for heterologously expressed heteromeric receptors [52]. The stoichiometry of the heteromeric receptor native to neuron's is not known.

[1] Maricq A et al. (1991) Science, 254, 432–437; [2] Mochizuki S et al. (1999) Amino acids 17, 243–255; [3] Brown AM et al. (1998) J Physiol 507, 653–665; [4] Davies PA et al. (1999) Nature 397, 359–363; [5] Bufton KE et al. (1993) Neuropharmacology 32, 1325–1331; [6] Doucet E et al. (2000) Neuroscience 95, 881–892; [7] Hewlett WA et al. (1999) J Pharmacol Exp Ther 288, 221–231; [8] Huang J et al. (2004) Brain Res 1028, 156–169; [9] Marazitti D et al. (2001) Neurochem Res 26, 187–190; [10] Miquel MC et al. (2002) Eur J Neurosci 15, 449–457; [11] Morales M et al. (1998) J Comp Neurol 402, 385–401; [12] Parker RM et al. (1996) J Neurol Sci 144, 119–127; [13] Tecott LH et al. (1993) Proc Natl Acad Sci USA 90, 1430–1434; [14] Glatzle J et al. (2002) Gastroenterology 123, 217–226; [15] Bhattacharya A et al. (2004) J Neurosci 24, 5537–5548; [16] Miyake A et al. (1995) Mol Pharmacol 48, 407–416; [17] Brady CA et al. (2001) Neuropharmacology 41, 282–284; [18] Hope AG et al. (1996) Br J Pharmacol 118, 1237–1245; [19] Lankiewicz S et al. (1998) Mol Pharmacol 53, 202–212; [20] Belelli D et al. (1995) Mol Pharmacol 48, 1054–1062; [21] Solt K et al. (2005) J Pharmacol Exp Ther 315, 771–776; [22] Dubin AE et al. (1999) J Biol Chem 274, 30799–30810; [23] Gunthorpe MJ and Lummis SCR (1999) J Physiol 519, 713–722; [24] Das P, and Dillon GH (2003) Brain Res Mol Brain Res 119, 207–212; [25] Das P and Dillon GH (2005) J Pharmacol Exp Ther 314, 320–328; [26] Bhatnagar S et al. (2004) Behav Brain Res 153, 527–535; [27] Kelley SP et al. (2003) Eur J Pharmacol 461, 19–25; [28] Zeitz KP et al. (2002) J Neurosci 22, 1010–1019; [29] Chameau P et al. (2006) Soc Neurosci Abs 125.10–A22; [30] Férézou I et al. (2002) J Neurosci 22, 7389–7397; [31] Roerig B et al. (1997) J Neurosci 17, 8353–8362; [32] Sugita S et al. (1992) Neuron 8, 199–203; [33] Zhou X and Galligan JJ (1999) J Physiol 529, 373–383; [36] Kayser V et al. (2007) Pain 130, 235–248; [37] Niesler B et al. (2001) Pharmacogenetics 11, 471–475; [38] Krzywkowski K et al. (2008) Proc Natl Acad Sci USA 105, 722–727; [49] Niesler B et al. (2007) Mol Pharmacol 72, 8–17; [50] Niesler B et al. (2003) Gene 310, 101–111; [51] Karnovsky AM et al. (2003) Gene 319, 137–148; [52] Barrera NP et al. (2005) Proc Natl Acad Sci USA 102, 12595–12600; [53] Hanna MC et al. (2000) J Neurochem 75, 240–247; [54] Mair ID et al. (1998) Br J Pharmacol 124, 1667–1674.

was studied using a polyconal antibody (pAb120) directed to the N-terminal domain of the subunit (Spier *et al.*, 1999), or a polyclonal antibody raised against a hexapeptide sequence within the large intracellular loop (Doucet *et al.*, 2000). Elegant fluorescence imaging studies performed in real time upon HEK-293 cells transfected to express the 5-HT3A(b) subunit, tagged with enhanced cyan fluorescent protein (CYP), indicate that such a distribution represents newly synthesized proteins within the ER that traffic *via* the Golgi cisternae and microtubule-associated vesicle-like structures to the plasma membrane where they subsequently cluster in actin-rich domains (Ilegems *et al.*, 2004). Similarly, at the plasma membrane of COS-7, CHO, HEK-293 cells and hippocampal neurons in culture, mouse 5-HT3A(b) receptor subunits occur in clusters in structures enriched with F-actin, such as lamellipodia and microspikes (Emerit *et al.*, 2002; Grailhe *et al.*, 2004). The 5-HT$_3$ receptor native to NG108-15 cells presents a similar membrane topology and associates with F-actin (Emerit *et al.*, 2002). In hippocampal neurons in culture, transfected to express the mouse 5-HT3(a) subunit, the receptor is highly localized to dendrites and is targeted to the tips of filopodia and dendritic spines (Grailhe *et al.*, 2004). In addition, the mouse 5-HT$_3$A receptor has been observed to co-localize with the light chain (LC1) of microtubule-associated protein 1B (MAP1B) in clusters at adhesion and branching spots and central growth cones of the neurites of hippocampal neurons in culture (Sun *et al.*, 2008). It is postulated that LC1 interacts with the distal half of the M3–M4 linker of the 5-HT3A receptor subunit. Interestingly, such an interaction appears, in HEK-293 cells, to result in reduced cell surface expression of 5-HT3A receptors and an acceleration of the kinetics of desensitization onset of 5-HT-evoked currents (Sun *et al.*, 2008).

Within the central nervous system (CNS), the 5-HT$_3$ receptor resides in both pre- and postsynaptic locations (Chameau and van Hooft, 2006). Immunohistochemistry and immunocytochemistry using antibodies directed against the 5-HT3A subunit (Doucet *et al.*, 2000; Miquel *et al.*, 2002; Huang *et al.*, 2004) and autoradiography with tritiated, or iodinated, 5-HT$_3$ receptor antagonists (which bind to the 5-HT3A, but not 5-HT3B, subunit) (Parker *et al.*, 1996a; Hewlett *et al.*, 1999; Marazitti *et al.*, 2001) indicate a high density of the receptor within the caudal medulla (i.e. nucleus tractus solitarius, motor nucleus of the vagus nerve, area postrema and spinal nucleus of the trigemminal nerve [p5]), where they appear to be largely, although not exclusively, presynaptically located in association with nerve terminals and axonal profiles (Pratt and Bowery, 1989; Leslie *et al.*, 1990; Glaum *et al.*, 1992; Wang *et al.*, 1998; Doucet *et al.*, 2000; Huang *et al.*, 2004). A predominantly presynaptic location in such regions is consistent with the paucity of mRNA encoding the 5-HT3A subunit, as probed by *in situ* hybridization (Tecott *et al.*, 1993; Fonseca *et al.*, 2001). Electrophysiological recordings conducted *in vivo*, or *in vitro*, indicate that 5-HT$_3$ receptors facilitate the release of glutamate onto dorsal vagal preganglionic, nucleus tractus solitarius and area postrema neurons via a presynaptic site (Wang *et al.*, 1998; Funahashi *et al.*, 2004; Jeggo *et al.*, 2005; Wan and Browning, 2008). A study of nucleus tractus solitarius neurons *in vitro* presents a more complex scenario, wherein the activation of 5-HT$_3$ receptors increased the frequency of both spontaneous excitatory and inhibitory synaptic currents in addition to exerting a direct postsynaptic excitation (Glaum *et al.*, 1992). Due to its association with the vomiting reflex, the dorsal vagal complex represents a central site

at which 5-HT$_3$ receptor antagonists might exert their anti-emetic action (Reynolds *et al.*, 1995).

Within the rat spinal cord, autoradiography performed with radiolabelled 5-HT$_3$ receptor antagonists reveals expression of the receptor within the superficial layers of the dorsal horn, largely in association with the central terminals of primary sensory afferent neurons (Hamon *et al.*, 1989; Kidd *et al.*, 1993). The pattern of immunostaining with antibodies directed against the 5-HT3A subunit confirms this distribution but, together with hybridization signals obtained with riboprobes for the 5-HT3A subunit mRNA, additionally indicates that some intrinsic spinal neurons within the dorsal and ventral horn also express the receptor (Tecott *et al.*, 1993; Kia *et al.*, 1995; Morales *et al.*, 1998). Indeed, a recent study has demonstrated the expression of 5-HT3A subunit mRNA in a subpopulation of GABAergic and enkephalinergic neurons of the dorsal horn of the spinal cord of the mouse (Huang *et al.*, 2008).

In comparison to the brainstem, the 5-HT3A receptor subunit is expressed at considerably lower levels in the forebrain (reviewed by Barnes and Sharpe, 1999). Immunohistochemistry reveals the subunit to be present in the hippocampus (pyramidal cell layer of CA1 to CA3 and granule cell layer of dentate gyrus), amygdala, frontal, piriform, and entorhinal cortices, olfactory regions and the reticular and paraventricular thalamic nuclei (Morales *et al.*, 1996b; Miquel *et al.*, 2002). A broadly similar distribution is apparent from studies employing tritiated and iodinated 5-HT$_3$ receptor antagonists (Bufton *et al.*, 1993; Hewlett *et al.*, 1999), or riboprobes directed against mRNA encoding the 5-HT3A receptor subunit (Tecott *et al.*, 1993; Morales *et al.*, 1996b). It is notable that the relative abundance of the 5-HT$_3$ receptor within such regions varies between rodents and humans (Barnes and Sharpe, 1999).

As assessed by immunocytochemistry and *in situ* hybridization, the 5-HT3A receptor subunit in cortical and hippocampal regions is overwhelmingly associated with a subpopulation of GABA-ergic interneurons that frequently express the calcium binding protein calbindin, but rarely parvalbumin, and contain the peptide cholecystokinin (CCK), but rarely somatostatin (Morales *et al.*, 1996a, b; Morales and Bloom, 1997). An elegant combination of electrophysiological, single-cell reverse transcription and multiplex PCR techniques has confirmed the selective expression of 5-HT3A, but not 5-HT3B, subunit mRNA in GABA-ergic interneurons containing CCK and vasoactive intestinal peptide (VIP). Importantly, the same studies reveal that 5-HT$_3$ receptors mediate fast excitatory postsynaptic currents in response to local electrical stimulation (Férézou *et al.*, 2002; see below). In addition, a high degree of co-expression of the cannabinoid CB1 receptor and the 5-HT3A receptor subunit mRNA has been observed in hippocampal and dentate gyrus interneurons by the use of double *in situ* hybdridization histochemistry (Morales and Backman, 2002). 5-HT3A subunit and CB1 receptor transcripts also frequently coexist in GABA-ergic interneurons of the cortex and amygdala (Morales *et al.*, 2004). Such observations raise the possibility that 5-HT$_3$ and CB1 receptors mediate opposing effects upon GABA-ergic neurotransmission, namely facilitation (see section 3.1.2.6) and inhibition, respectively (Morales and Backman, 2002; Morales *et al.*, 2004). In this respect, it is interesting that the endogenous cannabinoid receptor agonist, anandamide, also acts as an inhibitor of the 5-HT$_3$A homomeric receptor, independent of CB1 receptor agonism (Oz *et al.*, 2002). The mechanism of inhibition is complex involving an acceleration of the onset of 5-HT-evoked current desensitization which itself is influenced by the density of receptor expression at the cell surface (Xiong *et al.*, 2008). Inhibition by anandamide is, thus, inversely correlated with the level of expression of the 5-HT$_3$A receptor, at least in heterologous expression systems.

The expression of the 5-HT3B subunit within the CNS has been a matter of considerable debate (van Hooft and Yakel, 2003). Davies *et al.* (1999), employing Northern blot analysis, initially reported that 5-HT3B subunit mRNA is detectable in human caudate nucleus, hippocampus and thalamus, and particularly the amygdala. In agreement, polymerase chain reaction (PCR)-based detection methods applied to human brain have detected 5-HT3B subunit mRNA in regions that include the occipital, frontal, and temporal cortices, amygdala, hippocampus, and caudate nucleus (Dubin *et al.*, 1999; Brady *et al.*, 2007; Tzvetkov *et al.*, 2007; Holbrook *et al.*, 2008). Additionally, polyclonal antibodies directed against the 5-HT3B subunit label subpopulations of neurons within the rat and human hippocampus (Monk *et al.*, 2001; Reeves and Lummis, 2006; Brady *et al.*, 2007; Doucet *et al.*, 2007). The single channel conductance of 5-HT$_3$ receptors within murine hippocampal cultures (~9 pS) (Jones and Surprenant, 1994) is also consistent with a heteromeric 5-HT$_3$AB receptor (Hanna *et al.*, 2000), and estimates of conductance made from rat hippocampal slices using fluctuation analysis (~4 pS) (Sudweeks *et al.*, 2002) might most parsimoniously be explained by a mixed population of 5-HT$_3$A and 5-HT$_3$AB receptors. However, such an interpretation is inconsistent with the reported absence of mRNA encoding the 5-HT3B subunit in rat hippocampal neurons (or neocortical neurons, Férézou *et al.*, 2002) as assessed by single cell reverse transcription (RT)-PCR (Sudweeks *et al.*, 2002). Nonetheless, Sudweeks *et al.* (2002) and Férézou *et al.* (2002) did detect 5-HT3B subunit mRNA in rat whole brain samples. However, Morales and Wang (2002) did not detect mRNA encoding the 5-HT3B subunit in rat whole brain, or discrete brain regions, using either RT-PCR, or *in situ* hybridization histochemistry, although such transcripts were readily detected in the same study in a variety of peripheral ganglia (see below). The reasons for the discrepant reports concerning the central expression of the 5-HT3B subunit are unknown, but it is pertinent to note that the evidence for a centrally located 5-HT3B subunit (other than that potentially present on the centrally projecting terminals of peripheral neurons; see below) rests to a significant degree upon the specificity of the polyclonal antibodies directed against it.

5-HT3C subunit mRNA is present in the brain (Niesler *et al.*, 2003) and peripherally is prominent in the dorsal root ganglia (Holbrook *et al.*, 2008). The distribution of the 5-HT3D and 5-HT3E subunit transcripts, as judged by RT-PCR, is the subject of conflicting reports. Niesler *et al.* (2003) originally found both to be confined to the periphery but central expression, particularly of the 5-HT3E subunit transcript, has subsequently been claimed (Holbrook *et al.*, 2008).

The distribution of the 5-HT$_3$ receptor within the peripheral nervous system (PNS) is widespread, including post-ganglionic neurons of the sympathetic and parasympathetic divisions of the autonomic nervous system, neurons of the myenteric- and submucous-plexuses of the enteric nervous system and primary somatic and visceral afferents (Fozard, 1984b; Wallis, 1989; Peters *et al.*, 1991; Gershon, 1999; 2000). At a minority of neurons within the myenteric plexus, 5-HT$_3$ receptors contribute to the fast excitatory postsynaptic potential (EPSP) evoked by electrical stimulation of interganglionic connectives (Zhou and Galligan, 1999). The paucity

of synaptic responses as opposed to the high incidence of neurons that express 5-HT$_3$ receptors is likely to be due to the rather limited number (1 per cent of total) of serotonergic neurons within the myenteric plexus (Gershon, 2000). More widespread excitation of myenteric neurons (and certain extrinsic sensory fibres) mediated by 5-HT$_3$ receptors is likely to occur under pathophysiological conditions, wherein the secretion of 5-HT from enterochromaffin cells is abnormally enhanced (Gershon, 2000).

In contrast to the controversy that has surrounded the central expression of the 5-HT3B subunit, there is ample evidence from *in situ* hybridization histochemistry that peripheral neurons, including dorsal root, superior cervical and nodose ganglion cells, have the capacity to express both the 5-HT3A and 5-HT3B subunits, or solely the 5-HT3A subunit (Morales *et al.*, 2001; Morales and Wang, 2002). Immunocytochemistry also supports co-expression of 5-HT3A and 5-HT3B subunits in trigeminal ganglion neurons (Doucet *et al.*, 2007). Accordingly, the single channel conductance of 5-HT$_3$ receptors native to rat superior cervical and rabbit nodose ganglion neurons are compatible with the expression of a population of 5-HT$_3$AB receptors (Yang *et al.*, 1992; Peters *et al.*, 1994), whereas the electrophysiological data obtained from dorsal root ganglion neurons are suggestive of the expression of homomeric constructs (Robertson and Bevan, 1991). Interestingly, dorsal root and nodose ganglion neurons projecting to the superficial layers of the spinal cord and nucleus tractus solitarius possess mRNA encoding both the 5-HT3A and 5-HT3B subunit, or only the 5-HT3A subunit. Hence, the centrally located presynaptic terminals of such neurons may express either 5-HT$_3$A, or 5-HT$_3$AB receptors (Morales and Wang, 2002), a feature that may be functionally relevant in view of the differing permeabilities of such receptors to Ca^{2+} (Davies *et al.*, 1999). The potential for diversity in the composition of 5-HT$_3$ receptors expressed by dorsal root ganglion (DRG) neurons has recently been expanded by the detection of relatively abundant levels of mRNA transcripts encoding the 5-HT3C, 5-HT3D and 5-HT3E subunits in such cells (Holbrook *et al.*, 2008).

Immunohistochemical studies suggest that neurons of the human submucous plexus express 5-HT3A and 5-HT3B subunits (Michel *et al.*, 2005) and a subpopulation of mouse myenteric neurons in culture is immunoreactive for the 5-HT3A subunit and contains mRNA encoding the 5-HT3B subunit (Liu *et al.*, 2002). The expression of the latter is consistent with the single channel properties of 5-HT$_3$ receptors endogenous to guinea pig submucous plexus neurons (Derkach *et al.*, 1989).

The distribution of 5-HT$_3$ receptor subunits in the CNS and PNS is summarized in Table 3.1.2.1.

3.1.2.5 Pharmacology

Responses to 5-HT mediated by 5-HT$_3$ receptors are readily isolated from those elicited by the activation of G-protein-coupled 5-HT receptors by the use of a variety of selective and competitively acting antagonist compounds. Prototypical agents are bemesetron (MDL72222, Fozard, 1984b) and tropisetron (ICS205–930, Richardson *et al.*, 1985), developed from the structures of cocaine and 5-HT, respectively. Additional antagonists that are of clinical interest, or frequently used as experimental compounds, include: ondansetron (GR38082, Butler *et al.*, 1988); zacopride (AHR1190B, Smith *et al.*, 1988); dolasetron (MDL73147, Boeijinga *et al.*, 1992); granisetron (BRL43694, Sanger and Nelson, 1991); azasetron

(Y-25130, Sato *et al.*, 1992) cilansetron (KC9946, van Wijngaarden *et al.*, 1993); palonosetron (RS 25259–197, Wong *et al.*, 1995); ramosetron (YM060, Akuzawa *et al.*, 1996) and alosetron (GR68755, Clayton *et al.*, 1999). Such agents typically bind to both native and recombinant 5-HT$_3$ receptors with high affinity (see Thompson and Lummis 2006 for a quantitative summary) and a high degree of selectivity. However, it should be noted that the substituted benzamide, zacopride, also activates 5-HT$_4$ receptors. In addition, at concentrations higher than those necessary to produce 5-HT$_3$ receptor blockade, tropisetron acts as an antagonist of 5-HT$_4$ receptors (Gershon, 1999). The *in vitro* pharmacology of 5-HT$_3$ receptor antagonists has been reviewed in detail elsewhere (Peters *et al.*, 1994; Costall and Naylor, 2004; Thompson and Lummis, 2006), but an issue worthy of mention is that antagonism, although competitive in radioligand binding studies conducted under equilibrium conditions (Hoyer *et al.*, 1989), frequently appears to be non-competitive in functional studies, especially for agents with high binding affinity. This apparent contradiction is readily explained by the slow rate of dissociation of high-affinity antagonists from the receptor complex. The latter precludes the development of equilibrium between agonist and antagonist binding before the onset of receptor desensitization truncates functional responses (Peters *et al.*, 1994). Extrapolating to the situation *in vivo*, one would predict that antagonism of 5-HT$_3$ receptors might be 'surmounted' principally by activation of receptors not occupied by antagonist, rather than a decrease in fractional receptor occupancy by the drug.

Selective activation of 5-HT$_3$ receptors can be achieved, with varying degrees, by compounds that include 2-methyl-5-HT (Richardson *et al.*, 1985), SR57227 (4-amino-(6-chloro-2-pyridyl)-1-piperidine hydrochloride), which crosses the blood–brain barrier (Bachy *et al.*, 1993), and a range of arylbiguanides, the most extensively studied of which are 1-phenylbiguanide (1-PBG) and its 3-chloro-substituted derivative, *meta*-chlorophenylbiguanide (mCPBG) (Morain *et al.*, 1994; Dukat *et al.*, 1996). The potency and efficacy of individual compounds relative to 5-HT differs markedly between experimental preparations. Some of this variability is probably artefactual and may relate to differences in the degree of 'receptor reserve' inherent in some experimental paradigms. However, intra-species variation in the pharmacological properties of the 5-HT$_3$ receptor is also a major influence (Butler *et al.*, 1990; Newberry *et al.*, 1991; Peters *et al.*, 1994; Miyake *et al.*, 1995; Lankiewicz *et al.*, 1998; Mair *et al.*, 1998).

In addition to agents acting at the recognition site for 5-HT, or a site that overlaps with the latter, numerous chemical classes have been documented to modulate receptor activity via allosteric mechanisms. These include: simple alcohols (e.g. ethanol); halogenated alcohols (e.g. trichloroethanol, the active metabolite of the sedative, chloral hydrate); intravenous- (e.g. propofol, ketamine) and volatile- (e.g. chloroform, halothane, isoflurane) general anaesthetics; gonadal steroids; structurally distinct antidepressants and antipsychotics; antagonists of L-type Ca^{2+} channels (e.g. diltiazem) and di- and trivalent cations. The actions of these diverse agents, which are beyond the scope of the present review, have been discussed in detail elsewhere (Parker *et al.*, 1996b; Wetzel *et al.*, 1998; Lovinger, 1999; Rupprecht *et al.*, 2001; Eisensamer *et al.*, 2005; Thompson and Lummis, 2006).

Homomeric receptors composed of 5-HT3A subunits present a pharmacological profile that reflects the species from which they were cloned (Hope *et al.*, 1993; Downie *et al.*, 1994; Hussy *et al.*, 1994; Belelli *et al.*, 1995; Miyake *et al.*, 1995; Akuzawa *et al.*, 1996; Lankiewicz *et al.*, 1998; Mair *et al.*, 1998; Hope *et al.*, 1999). Indeed, when comparisons within a species are made, the ligand binding profiles of the cloned and native receptors of a given species are essentially identical (Akuzawa *et al.*, 1996; Hope *et al.*, 1996; Mair *et al.*, 1998). Hence, the species-dependent pharmacology described for native 5-HT$_3$ receptors (Butler *et al.*, 1990; Newberry *et al.*, 1991) applies to homomeric 5-HT$_3$A receptors also. In several instances this has proved instructive in defining regions of the N-terminal ECD that contribute to ligand binding. Sequence alignments between the highly conserved species homologues of 5-HT3A subunit reveal differences in primary amino acid sequence that could contribute to species-dependent pharmacology. This approach has led to the identification of amino acids that influence the apparent affinity of (+)-tubocurarine (Hope *et al.*, 1999) and mCBPG (Mochizuki *et al.*, 1999b), although none of these individually are crucial to binding.

However, a far greater insight into the residues within the ECD of the homomeric 5-HT$_3$A receptor that contribute to ligand binding has been obtained by a combination of: (i) methodical mutagenesis of aromatic and acidic amino acids that might be anticipated to bond with specific moieties of 5-HT$_3$ receptor ligands (e.g. Boess *et al.*, 1997; Hope *et al.*, 1999; Yan *et al.*, 1999; Spier and Lummis, 2000; Venkataraman *et al.*, 2002a, b; Price and Lummis, 2004; Suryanarayanan *et al.*, 2005; Sullivan *et al.*, 2006; Thompson *et al.*, 2005, 2006), (ii) the application of unnatural amino acid mutagenesis to probe specific chemical interactions (Beene *et al.*, 2004) and (iii) homology modelling based on the crystal structure of the acetylcholine binding protein of *Lymnaea stagnalis* (Brejc *et al.*, 2001), a water soluble homolog of the ECD of the nACh receptor (Reeves *et al.*, 2003; Thompson *et al.*, 2005; Joshi *et al.*, 2006; Sullivan *et al.*, 2006). The picture that emerges is, that as for other members of the Cys-loop family of transmitter-gated ion channels, ligand binding at the 5-HT$_3$ receptor occurs at interfaces composed of discontinuous stretches of amino acid residues (binding domains; termed A, B, C, D, E, and F) contributed by the N-terminal ECD of adjacent subunits within the pentamer (Thompson and Lummis, 2006; Barnes *et al.*, 2009). In homomeric receptors, such as the 5-HT$_3$A receptor, each subunit contributes both 'principal' (domains A, B and C) and 'complementary' (domains D, E and F) binding components that lie on opposite faces of the protein (Thompson and Lummis, 2006, Barnes *et al.*, 2009). Specific residues within these binding domains that influence the binding of agonist and antagonists are beyond the scope of this overview, but are reviewed in detail by Thompson and Lummis (2006).

The 5-HT3B subunit has a pronounced effect upon the biophysics of the 5-HT$_3$ receptor (Davies *et al.*, 1999; Dubin *et al.*, 1999; Hanna *et al.*, 2000; Hapfelmeier *et al.*, 2002b), but only a subtle effect upon agonist and competitive antagonist pharmacology. This is despite significant differences between 5-HT3A and 5-HT3B subunits in the specific amino acids that are thought to contribute to the ligand binding pocket (Thompson and Lummis, 2006). In our own studies (Low *et al.*, 2001), the rank order of antagonist affinities assessed from competition assays utilizing [^{3}H]granisetron was identical at human homomeric 5-HT$_3$A and heteromeric 5-HT$_3$ AB receptors (Figure 3.1.2.2). However, all agonist compounds evaluated demonstrated a small, but statistically significant, reduction in apparent affinity at 5-HT$_3$AB versus 5-HT$_3$A receptors

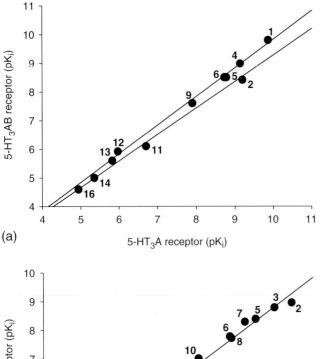

(a)

(b)

Fig. 3.1.2.2 Pharmacological profile of human recombinant 5-HT₃A- and 5-HT₃AB receptors expressed in HEK-293 cells. (a) and (b) Correlation between the potencies (pK_i values) of ligands competing for specific [³H]-granisetron binding at homomeric (5-HT₃A, abcissa) *versus* hetero-oligomeric (5-HT₃AB, ordinate) receptors. In (a) (from Low *et al.*, 2001[1]) separate lines were fitted to the agonist (i.e. compounds 2, 9, 10, 11, 14 and 16; see below) and antagonist data sets using least squares regression analysis whereas in (b) (constructed from data provided by Professor NM Barnes, University of Birmingham) no such distinction was made. In either case, the data emphasize that the apparent affinity of agonists and the affinity of antagonists in binding assays differs little between human 5-HT₃A and 5-HT₃AB receptors. The compounds investigated were: 1. alosetron; 2. quipazine; 3. (S)-zacopride; 4. (R,S)-zacopride; 5. azasetron (Y25130); 6. ondansetron; 7. BRL46470; 8. (R)-zacopride; 9. 2,5-dichlorophenylbiguanide; 10 m-chlorophenylbiguanide; 11. 5-HT; 12. metoclopramide; 13. (+)-tubocurarine; 14. 1-phenylbiguanide; 15 cocaine; 16. tryptamine; 17. morphine.

[1] Reproduced from Low PB, Lambert JJ, and Peters JA (2001). A comparative study of the pharmacological properties of homo-oligomeric and hetero-oligomeric human recombinant 5-hydroxytryptamine type-3 (5-HT₃) receptors. *British Journal of Pharmacology* **133**, 144P.

(Low *et al.*, 2001) (Figure 3.1.2.2a). Brady *et al.* (2001) reached the same conclusion regarding the affinities of antagonists, but only 5-HT was highlighted as an agonist with lower apparent affinity for the 5-HT₃AB receptor (Figure 3.1.2.2b). In agreement with the results obtained in radioligand binding assays, electrophysiological studies consistently demonstrate the agonist potency of 5-HT to be

reduced at 5-HT₃AB versus 5-HT₃A receptors (Davies *et al.*, 1999; Dubin *et al.*, 1999; Hanna *et al.*, 2000; Hapfelmeier *et al.*, 2002b; Solt *et al.*, 2005). It should be noted that differences in efficacy may impinge upon the apparent affinity of agonists measured in either binding, or functional, assays (Colquhoun, 1998).

In contrast to their lack of discrimination in radioligand binding assays, (+)– tubocurarine (Davies *et al.*, 1999; Hanna *et al.*, 2000) and azasetron (Y-25130) (Dubin *et al.*, 1999) are reported to preferentially antagonize homomeric 5-HT₃A receptors in functional assays. It is possible that, in addition to interacting with the ligand binding domain, (+)– tubocurarine and azasetron may exert other mechanisms of antagonism (e.g. channel block) that underlie their differential effect in functional assays. This possibility is underscored by the fact that picrotoxin, better known as a blocker of GABA_A receptors, discriminates between 5-HT₃A and 5-HT₃AB receptors of murine origin, most likely via a channel-blocking mechanism (Das and Dillon, 2003, 2005). In addition, 5-HT₃A and 5-HT₃AB receptors differ in the potentiation of their function by chloroform, halothane and small volume n-alcohols (Solt *et al.*, 2005; Rüsch *et al.*, 2007). It has been claimed that 5-methoxyindole discriminates between 5-HT₃A and 5-IIT₃AB receptors, by acting as a partial agonist at the former and as a 'protean' agonist (i.e. initially activating and subsequently reducing spontaneous activity) at the latter (Hu and Peoples, 2008). Such observations are surprising, since the structurally related compound, 5-methoxytrytamine, by virtue of its lack of activity at 5-HT₃ receptors, was once used as a tool to distinguish the 5-HT₃ receptor from G-protein-coupled 5-HT receptors (Bradley *et al.*, 1986).

The pharmacological properties of homomeric 5-HT₃A and heteromeric 5-HT₃AB receptors are summarized in Table 3.1.2.1.

3.1.2.6 Physiology

The precise physiological role(s) of the 5-HT₃ receptor have proved difficult to define, but recent studies of transgenic mice have revealed some potential functions. Targeted disruption of the *Htr3a* gene does not produce an obvious phenotype; null mutant animals are healthy, fertile, eat and drink normally, have similar body weight and overall appearance to wild-type siblings, and display no gross deficits in motor ability (Zeitz *et al.*, 2002; Kelley *et al.*, 2003a; Bhatnagar *et al.*, 2004a). However, more detailed investigation reveals that persistent, but not acute, nociception in response to tissue injury is blunted in null mutant animals (Zeitz *et al.*, 2002; Kayser *et al.*, 2007). A comparison of the latter with wild-type mice indicates that the 5-HT₃ receptor mediates activation of a novel subset of myelinated and unmyelinated nociceptive afferents that do not express the pro-inflammatory neuropeptide substance P, explaining why injury-associated edema is unaffected in the null mutant animals (Zeitz *et al.*, 2002). Such results confirm the pronociceptive action of the 5-HT₃A receptor previously suggested from pharmacological studies (e.g. Oyama *et al.*, 1996; Green *et al.*, 2000) and add to the evidence that central serotonergic circuits facilitate nociceptive synaptic transmission via spinal 5-HT₃ receptors.

The analysis of 5-HT3A subunit knockout mice has also suggested an involvement of the receptor in the function of the hypothalamo–pituitary–adrenal axis and behaviours related to anxiety and depression, although the phenotypes are complex and can be influenced by gender (Kelley *et al.*, 2003a; Bhatnagar *et al.*, 2004a, b;

Bhatnagar and Vining, 2004). The potential role(s) of the 5-HT$_3$ receptor in higher order phenomema, such as addictive behaviours (see below), learning, memory, and attention, has been inferred from a mouse model in which the 5-HT3A receptor subunit was selectively overexpressed in the forebrain (Engel *et al.*, 1998; Engel and Allan, 1999; Harrell and Allan, 2003). In brief, enhanced expression of the receptor is reported to improve learning and attention that is dependent upon the hippocampus (Harrell and Allan, 2003). However, the latter assertion would appear to contradict earlier behavioural observations in which 5-HT$_3$ receptor antagonists were found to enhance cognition (reviewed by Costall and Naylor, 2004) and also appears inconsistent with the ability of such drugs to facilitate long-term potentiation in the hippocampus (Staubli and Xu, 1995).

In contrast to the phenotype of the 5-HT3A subunit null mutant mouse, animals engineered to express a mutant 5-HT3A receptor subunit (via a Va113′ Ser mutation within the M2 domain) display pronounced uropathic changes that ultimately result in premature death (Bhattacharya *et al.*, 2004). *In vitro*, the mutation confers agonist hypersensitivity and constitutive activity to 5-HT$_3$A and 5-HT$_3$AB receptors, respectively. It would appear that excessive 5-HT$_3$ receptor activity is associated with excitotoxic neuronal cell death and changes in the function of the bladder that resemble those of urinary bladder outlet obstruction (Bhattacharya *et al.*, 2004).

At the cellular level, activation of neuronal 5-HT$_3$ receptors elicits a rapidly activating and desensitizing cation current with a reversal potential close to zero millivolts. Synaptically released 5-HT in brain slices elicits a 5-HT$_3$ receptor-mediated current that contributes to fast excitatory synaptic transmission in the rat lateral (but not basolateral) amygdala (Sugita *et al.*, 1992; Koyama *et al.*, 2000), ferret developing visual cortex (Roerig *et al.*, 1997) and rat neocortical GABA-ergic interneurons that contain CCK and VIP (Férézou *et al.*, 2002). The temporal characteristics of the 5-HT$_3$ receptor-mediated evoked excitatory postsynaptic current (eEPSC) recorded from the latter are intriguing, having a relatively brief latency to peak (1.6 ms), large amplitude (−240 pA) and fast and slow decay time constants of 0.7 ms and 4.8 ms, respectively (Férézou *et al.*, 2002). Such characteristics are strikingly dissimilar from the far more slowly rising and decaying 5-HT$_3$ receptor-mediated eEPSCs recorded from visual cortex neurons (Roerig *et al.*, 1997), or neurons of the lateral amygdala. Moreover, they differ greatly from the much slower kinetics of activation and deactivation (or desensitization) found for heterologously expressed 5-HT$_3$A receptors in response to rapid agonist applications that mimic synaptic delivery (Mott *et al.*, 2001). Expression of the 5-HT3B subunit cannot explain this discrepancy, because it is undetectable in neocortical interneurons by single cell RT-PCR (Férézou *et al.*, 2002). It is possible that the kinetic properties of the 5-HT3A receptor subunit in a native, neuronal, environment differ from those in non-neuronal, heterologous expression systems, perhaps as a consequence of cell-specific processing, or post-translational modifications.

In agreement with immunocytochemical and *in situ* hydridization studies (see above), electrophysiological recordings from single neurons in rat brain slices have documented the presence of functional 5-HT$_3$ receptors upon subsets of inhibitory interneurons in the CA1 area (Ropert and Guy, 1991; McMahon and Kauer, 1997) and dentate gyrus (Kawa, 1994) of the hippocampus and layer I of the cerebral cortex (Zhou and Hablitz, 1999). Activation of the receptor at such sites by exogenously applied agonist elicits a large increase in the frequency of GABA$_A$ receptor-mediated spontaneous inhibitory postsynaptic currents (sIPSCs) recorded from the postsynaptic targets of these interneurons (Ropert and Guy, 1991; Kawa, 1994; Zhou and Hablitz, 1999; Turner *et al.*, 2004). Similar findings have been obtained using single neurons of the rat basolateral amygdala and CA1 neurons of the hippocampus isolated mechanically to preserve functional presynaptic nerve terminals (Koyama *et al.*, 2000; Katsurabayashi *et al.*, 2003). In such studies, the activation of 5-HT$_3$ receptors increased the frequency of miniature inhibitory postsynaptic currents (mIPSCs) elicited by the vesicular release of GABA. Although facilitation was dependent upon the presence of extracellular Ca^{2+}, it was not prevented by blockade of presynaptic voltage-activated Ca^{2+} channels with Cd^{2+}, suggesting that Ca^{2+} influx occurs via activated 5-HT$_3$ receptor-channel complexes (Koyama *et al.*, 2000). Indeed, 5-HT$_3$ receptor-mediated influx of Ca^{2+} into single isolated nerve terminals (synaptosomes) has been demonstrated using confocal microscopy and Ca^{2+}-imaging techniques (Nichols and Mollard, 1996; Rondé and Nichols, 1998). Accordingly, an inward Ca^{2+} flux evoked by 5-HT$_3$ receptor activation facilitates the release of GABA from hippocampal synaptosomes (Turner *et al.*, 2004). Neurochemical studies measuring the release of either endogenous neurotransmitter, or radiolabelled tracers, suggest that 5-HT$_3$ receptor activation additionally modulates transcellular signalling mediated by ACh, dopamine, noradrenaline, CCK and 5-HT itself. Such actions, and their potential relevance to pathological conditions that include anxiety, schizophrenia, drug addiction, and cognitive disorders, are discussed in detail by Barnes and Sharp (1999) and Costall and Naylor (2004).

3.1.2.7 Relevance to disease states

Numerous polymorphisms of the *HTR3A* and *HTR3B* genes have been described and some linkage studies are suggestive of a role for 5-HT$_3$ receptors in psychiatric disorders (reviewed by Krzywkowski, 2006).

Psychiatric disorders, such as schizophrenia and bipolar disorder, segregate with cytogenetic abnormalities involving a region on chromosome 11 that contains the *HTR3A* gene (Weiss *et al.*, 1995). Niesler *et al.* (2001b) identified five polymorphisms within the *HTR3A* gene, but none of these could be associated with either schizophrenia, or bipolar disorder. Similarly, two missense mutations (R344H and P391R), found in separate schizophrenic patients, were initially thought to occur with an allelic frequency that is far too low to play any major role in the aetiology of the disorder (Niesler *et al.*, 2001b). However, the genotyping of additional ethnic populations indicates that the R344H variant occurs with a much higher minor allele frequency (MAF) of 0.038 to 0.108. Although the functional properties of this variant are unperturbed, its total abundance and level of cell surface expression are reduced in comparison to wild-type receptors when expressed in tsA-201 cells (Krzywkowski *et al.*, 2007). Similarly, although the P391R mutation is functionally silent (Kurzwelly *et al.*, 2004), it too is associated with reduced cell surface expression (Krzywkowski *et al.*, 2007). Additional non-synonymous polymorphisms of the 5-HT3A subunit include the A33T and M257I variants, which result in reduced levels of cell surface expression in comparison to the wild-type 5-HT$_3$A receptor, and the S253N variant which does not compromise plasma membrane expression. The S253N and M257I substitutions, both of which occur in M1, cause a profound reduction in the maximal response to 5-HT in Ca^{2+} imaging and fluorescent membrane potential assays in comparison to the wild-type receptor, suggesting that they impair signal transduction (Krzywkowski et al., 2007).

A relatively common polymorphism within the regulatory region of the *HTR3A* gene (i.e. –42C>T, aka C178T), which results in increased translation of a luciferase reporter construct (Niesler et al., 2001a), is associated with bipolar disorder (in Caucasians), and the personality trait of lower harm avoidance in women (reviewed by Krzywkowski, 2006). Intriguingly, functional magnetic resonance imaging of normal subjects performing a face recognition task indicates that the less common C/T allele, versus C/C allele, is associated with a reduced level of neuronal activation in the right amygdala and prefrontal cortex during the task (Iidaka et al., 2005). It has been speculated that the cellular basis of this effect is enhanced 5-HT$_3$ receptor expression on inhibitory GABA-ergic interneurons (Iidaka et al., 2005).

The very common Y129S polymorphism encoded by the *HTR3B* gene (MAF = 0.433) occurs at a significantly reduced frequency in female patients suffering from major depression and also patients with bipolar disorder, consistent with a protective influence of the variant allele (Krzywkowski, 2006). Intriguingly, the Y129S variant is a gain of function mutation. When assembled into heteromeric receptors with the wild-type 5-HT3A subunit, the 5-HT3B(Y129S) variant confers an increased maximal response to 5-HT, decreased desensitization and deactivation kinetics and a seven-fold increase in mean channel open time in comparison to wild-type 5-HT$_3$AB receptors (Krzywkowski et al., 2008; Walstab et al., 2008). An intermediate effect upon the maximal response to 5-HT in Ca^{2+} imaging and fluorescent membrane potential assays is evident for receptors assembled from a mixture of wild-type 5-HT3A, wild-type 5-HT3B and 5-HT3B(Y129S) subunits. Hence, the Y129S variant might significantly affect signalling via the 5-HT$_3$ receptor in heterozygous, as well as homozygous, individuals (Krzywkowski et al., 2008).

An –100–102 AAG deletion variant in the promoter region of the *HTR3B* gene has been associated with an increased frequency of vomiting caused by chemotherapy in patients receiving 5-HT$_3$ receptor antagonists as anti-emetic therapy (Tremblay et al., 2003). Interestingly, the same mutation is also associated with an increased incidence of nausea in response to paroxetine, a selective serotonin reuptake inhibitor (Tanaka et al., 2008). Furthermore, the –100–102 AAG deletion variant is underrepresented in a sample of patients suffering from bipolar depression (Frank et al., 2004b), but occurs at significantly higher frequency in treatment-resistant schizophrenic (TRS) patients versus non-TRS patients (Ji et al., 2008). The –100–102 AAG deletion has recently been shown to increase the promoter activity of the *HTR3B* gene *in vitro* (Ji et al., 2008; Meineke et al., 2008).

5-HT$_3$ receptors located within the gastrointestinal tract are important targets in the treatment of certain forms of emesis and irritable bowel syndrome. Pathological activation of 5-HT$_3$ receptors, via efflux of 5-HT from enterochromaffin cells within the intestinal mucosa, occurs in response to anti-cancer chemo- or radiotherapy. The subsequent excitation of chemosensitive vagal afferent fibres stimulates the vomiting centre of the brainstem to elicit often severe emesis accompanied by extremely unpleasant sensations and nausea (Reynolds et al., 1995; Hillsley and Grundy, 1998; Gershon, 1999, 2000). The amelioration of such effects

in cancer sufferers is a major indication of competitively acting 5-HT$_3$ receptor antagonists, such as ondansetron, granisetron, tropisetron, dolasetron, ramosetron and the more recently introduced long-acting compound, palonosetron (Rubenstein et al., 2006). Benefit most likely derives, at least in part, from blockade of receptors present upon the peripheral terminals of vagal afferents, with receptors located within the dorsal vagal complex presenting an additional target (Minami et al., 2003). Unfortunately, nausea and vomiting are not adequately controlled by 5-HT$_3$ receptor antagonists in a substantial proportion of patients undergoing chemotherapy. To date, there is little evidence to suggest that polymorphisms within the *HTR3A*, or *HTR3B* genes, contribute substantially to such therapeutic failures (Tremblay et al., 2003; Kaiser et al., 2004). 5-HT$_3$ receptor antagonists are also effective in the prevention of postoperative nausea and vomiting (Gan, 2005), but lack efficacy against other causes of emesis such as abnormal motion, or agents that increase dopaminergic neurotransmission.

Within the gastrointestinal tract 5-HT$_3$ receptors mediate signalling to the CNS, as noted above, and are additionally involved in absorption/secretion and motility. In animals, 5-HT$_3$ receptor antagonists, such as alosetron and cilansetron, reduce visceral sensitivity, absorption/secretion and motility and, in man, slow orocecal and colonic transit times and increase colonic compliance (reviewed by Camilleri et al., 2006). Alosetron is effective in the treatment of severe diarrhoea-predominant irritable bowel syndrome (IBS-D) in females, but can produce a low incidence of unwanted effects, such as constipation with serious complications and ischaemic colitis (Chang et al., 2006).

Pre-clinical behavioural studies conducted in the 1980s and 1990s provided evidence for the involvement of 5-HT$_3$ receptors in conditions including anxiety, schizophrenia, cognitive disorders, and drug addiction and withdrawal (Costall and Naylor, 2004). Unfortunately, subsequent clinical trials with 5-HT$_3$ receptor antagonists in man did not confirm the efficacy of such agents in most psychiatric conditions evaluated. However, ondansetron has been shown to be superior to placebo in the treatment of adult patients with early-onset alcoholism, an effect associated with reduced craving (Johnson et al., 2000, 2002). A preliminary study additionally suggests ondansetron to be effective in reducing alcohol consumption in adolescents with alcohol dependence (Dawes et al., 2005). Somewhat paradoxically, mice overexpressing the 5-HT3A receptor subunit self-administer less ethanol than controls (Engel et al., 1998), but this might be explained by the observation that the transgenic animals are more sensitive to low doses of ethanol (Engel and Allan, 1999). Experiments performed on 5-HT3A subunit knockout mice clearly indicate that 5-HT$_3$ receptor antagonists reduce alcohol consumption specifically by interacting with the 5-HT$_3$ receptor (Hodge et al., 2004).

Selective 5-HT$_3$ receptor antagonists possess efficacy in a variety of additional conditions, suggesting the direct, or indirect, involvement of 5-HT$_3$ receptor activation in their pathogenesis, or symptomatic manifestations. Pruritus, which is the most common adverse effect of intrathecal morphine administration for postoperative pain relief, is in some studies reported to be reduced in incidence and severity by the prophylatic use of ondansetron, or dolasetron (e.g. Latrou et al., 2005), but others find no benefit (e.g. Sarvela et al., 2006). Similarly, the effectiveness of 5-HT$_3$ receptor antagonists in cholestatic pruritus, or pruritus accompanying haemodialysis, is controversial. The efficacy of 5-HT$_3$ receptor antagonists in reducing pain in rheumatic diseases, such as fibromyalgia, has recently been reviewed by Muller *et al.* (2006). Polymorphisms of the *HTR3A* and *HTR3B* genes do not appear to be linked to this condition (Frank *et al.*, 2004a).

3.1.2.8 Concluding remarks

The biophysical and pharmacological properties of native and recombinant 5-HT₃A and 5-HT₃AB receptors are now well established. An important area for future investigation will be the verification, or otherwise, of the inclusion 5-HT3C, 5-HT3D, or 5-HT3E subunits within 5-HT₃ receptors *in vivo*. In addition, the controversy surrounding the expression of the 5-HT3B subunit within central neurons remains to be conclusively resolved, although much evidence now suggests this to be the case. The recent generation of a transgenic mouse, in which 5-HT₃ receptor expressing neurons have been tagged with enhanced green fluorescent protein (Chameau *et al.*, 2006), will greatly facilitate electrophysiological studies of fast synaptic transmission mediated by 5-HT₃ receptors within the CNS, reports of which are presently very limited in number. In addition, the 5-HT₃ receptor will continue to provide an excellent model system from which to explore the properties of the Cys-loop family of transmitter-gated ion channels as a whole (Reeves and Lummis, 2002; Peters *et al.*, 2005; Barnes *et al.*, 2009).

Acknowledgements

We are grateful to the Wellcome Trust, Tenovus Scotland and the Anonymous Trust for supporting 5-HT₃ receptor research conducted in the laboratories of JAP and JJL. MRL is supported by a Biotechnology and Biological Sciences Research Council Case award in conjunction with Eli Lilly.

References

Akuzawa S, Miyake A, Miyata K *et al.* (1996). Comparison of [³H]YM060 binding to native and cloned rat 5-HT₃ receptors. *Eur J Pharmacol* **296**, 227–230.

Bachy A, Heaulme M, Giudice A *et al.* (1993). SR57227A: a potent and selective agonist at central and peripheral 5-HT₃ receptors *in vitro* and *in vivo*. *Eur J Pharmacol* **237**, 299–309.

Barann M, Göthert M, Bönisch H *et al.* (1997). 5-HT₃ receptors in outside-out membrane patches of N1E-115 neuroblastoma cells: basic properties and effects of pentobarbital. *Neuropharmacology* **36**, 655–664.

Barnes NM and Sharp T (1999). A review of central 5-HT receptors and their function. *Neuropharmacology* **38**, 1083–1152.

Barnes NM, Hales TG, Lummis SC *et al.* (2009). The 5-HT₃ receptor—the relationship between structure and function. *Neuropharmacology* **56**, 273–284.

Barrera NP, Herbert P, Henderson RM *et al.* (2005). Atomic force microscopy reveals the stoichiometry and subunit arrangement of 5-HT₃ receptors. *Proc Natl Acad Sci USA* **102**, 12595–12600.

Beene DL, Price KL, Lester HA *et al.* (2004). Tyrosine residues that control binding and gating in the 5-hydroxytryptamine₃ receptor revealed by unnatural amino acid mutagenesis. *J Neurosci* **24**, 9097–9104.

Belelli D, Balcarek JM, Hope AG (1995). Cloning and functional expression of a human 5-hydroxytryptamine type 3A subunit. *Mol Pharmacol* **48**, 1054–1062.

Bhatnagar S and Vining C (2004). Pituitary–adrenal activity in acute and chronically stressed male and female mice lacking the 5-HT-3A receptor. *Stress* **7**, 251–256.

Bhatnagar S, Nowak N, Babich L (2004a). Deletion of the 5-HT₃ receptor differentially affects behavior of males and females in the Porsolt forced swim and defensive withdrawal tests. *Behav Brain Res* **153**, 527–535.

Bhatnagar S, Sun LM, Raber J (2004b). Changes in anxiety-related behaviors and hypothalamic–pituitary–adrenal activity in mice lacking the 5-HT-3A receptor. *Physiol Behav* **81**, 545–555.

Bhattacharya A, Dang H, Zhu QM *et al.* (2004). Uropathic observations in mice expressing a constitutively active point mutation in the 5-HT₃A receptor subunit. *J Neurosci* **24**, 5537–5548.

Bocquet N, Prado de Carvahlo L, Cartaud J *et al.* (2007). A prokaryotic proton-gated ion channel from the nicotinic acetylcholine receptor family. *Nature* **445**, 116–119.

Boeijinga PH, Galvan M, Baron BM *et al.* (1992). Characterization of the novel 5-HT₃ antagonists MDL 73147EF (dolasetron mesilate) and MDL 74156 in NG108–15 neuroblastoma x glioma cells. *Eur J Pharmacol* **219**, 9–13.

Boess FG, Beroukhim R and Martin IL (1995). Ultrastructure of the 5-hydroxytryptamine₃ receptor. *J Neurochem* **64**, 1401–1405.

Boess FG, Steward LJ, Steele JA *et al.* (1997). Analysis of the ligand-binding site of the 5-HT₃ receptor using site-directed mutagenesis: importance of glutamate 106. *Neuropharmacology* **36**, 637–647.

Boyd GW, Low PB, Dunlop JI *et al.* (2002). Assembly and cell surface expression of homomeric and heteromeric 5-HT₃ receptors: the role of oligomerization and chaperone proteins. *Mol Cell Neurosci* **21**, 38–50.

Bradley PB, Engel G, Feniuk W *et al.* (1986). Proposals for the classification and nomenclature of function receptors for 5-hydroxytryptamine. *Neuropharmacology* **25**, 563–567.

Brady CA, Dover TJ, Massoura AN *et al.* (2007). Identification of 5-HT₃A and 5-HT₃B receptor subunits in human hippocampus. *Neuropharmacology* **52**, 1284–1290.

Brady CA, Stanford IM, Ali I *et al.* (2001). Pharmacological comparison of human homomeric 5-HT₃A versus heteromeric 5-HT₃A/5-HT₃B receptors. *Neuropharmacology* **41**, 282–284.

Brejc K, van Dijk WJ, Klaassen RV *et al.* (2001). Crystal structure of an ACh-binding protein reveals the ligand-binding domain of nicotinic receptors. *Nature* **411**, 269–276.

Brown AM, Hope AG, Lambert JJ *et al.* (1998). Ion permeation and conduction in a human recombinant 5-HT₃ receptor subunit (h5-HT₃A). *J Physiol* **507**, 653–665.

Bruss M, Barann M, Hayer-Zillgen M *et al.* (2000). Modified 5-HT₃A receptor function by co-expression of alternatively spliced human 5-HT3A receptor isoforms. *Naunyn-Schmiedeberg's Arch Pharmacol* **362**, 392–401.

Bufton KE, Steward LJ, Barber PC *et al.* (1993). Distribution and characterization of the [³H]granisetron-labelled 5-HT₃ receptor in the human forebrain. *Neuropharmacology* **32**, 1325–1331.

Butler A, Elswood CJ, Burridge J *et al.* (1990). The pharmacological characterization of 5-HT₃ receptors in three isolated preparations derived from guinea pig tissues. *Br J Pharmacol* **101**, 591–598.

Butler A, Hill JM, Ireland SJ *et al.* (1988). Pharmacological properties of GR38032F, a novel antagonist at 5-HT₃ receptors. *Br J Pharmacol* **94**, 397–412.

Camilleri M, Bueno L, de Ponti F *et al.* (2006). Pharmacological and pharmacokinetic aspects of functional gastrointestinal disorders. *Gastroenterology* **130**, 1421–1434.

Chameau P and van Hooft JA (2006). Serotonin 5-HT₃ receptors in the central nervous system. *Cell Tissue Res* **326**, 573–581.

Chameau P, Inta DI, Vitalis T *et al.* (2006). Serotonergic control of postnatal cortical development by 5-HT₃ receptors. *Soc Neurosci Abs* **125**.10/A22.

Chang L, Chey WD, Harris L *et al.* (2006). Incidence of ischemic colitis and serious complications of constipation among patients using alosetron: systematic review of clinical trials and post-marketing surveillance data. *Am J Gastroenterol* **101**, 1069–1079.

Clayton NM, Sargent R, Butler A *et al.* (1999). The pharmacological properties of the novel selective 5-HT₃ receptor antagonist, alosetron and its effects on normal and perturbed small intestinal transit in the fasted rat. *Neurogastroenterol Motil* **11**, 207–217.

Collingridge GL, Olsen RW, Peters J *et al.* (2009). A nomenclature for ligand-gated ion channels. *Neuropharmacology* **56**, 2–5.

Colquhoun D (1998). Binding, gating, affinity and efficacy: the interpretation of structure–activity relationships for agonists and the effects of mutating receptors. *Br J Pharmacol* **125**, 924–947.

Corringer PJ, Le Novere N and Changeux JP (2000). Nicotinic receptors at the amino acid level. *Annu Rev Pharmacol Toxicol* **40**, 431–458.

Costall B and Naylor RJ (2004). 5-HT$_3$ receptors. *Curr Drug Targets CNS Neurol Disord* **3**, 27–37.

Das P and Dillon GH (2003). The 5-HT$_{3B}$ subunit confers reduced sensitivity to picrotoxin when co-expressed with the 5-HT$_{3A}$ receptor. *Brain Res Mol Brain Res* **119**, 207–212.

Das P and Dillon GH (2005). Molecular determinants of picrotoxin inhibition of 5-hydroxytryptamine type 3 receptors. *J Pharmacol Exp Ther* **314**, 320–328.

Davies PA, Pistis M, Hanna MC *et al.* (1999). The 5-HT$_{3B}$ subunit is a major determinant of serotonin receptor function. *Nature* **397**, 359–363.

Davies PA, Wang W, Hales TG *et al.* (2003). A novel class of ligand-gated ion channel is activated by Zn^{2+}. *J Biol Chem* **278**, 712–717.

Dawes MA, Johnson BA, Ma JZ *et al.* (2005). Reductions in and relations between 'craving' and drinking in a prospective, open-label trial of ondansetron in adolescents with alcohol dependence. *Addict Behav* **30**, 1630–1637.

Decker ER and Dani JA (1990). Calcium permeability of the nicotinic acetylcholine receptor: the single channel calcium influx is significant. *J Neurosci* **10**, 3413–3420.

Deeb TZ, Carland JE, Cooper MA *et al.* (2007). Dynamic modification of a mutant cytoplasmic cysteine residue modulates the conductance of the human 5-HT$_{3A}$ receptor. *J Biol Chem* **282**, 6172–6182.

Derkach V, Surprenant A and North RA (1989). 5-HT$_3$ receptors are membrane ion channels. *Nature* **339**, 706–709.

Doucet E, Latrémolière A, Darmon M *et al.* (2007). Immunolabelling of the 5-HT$_{3B}$ receptor subunit in the central and peripheral nervous systems in rodents. *Eur J Neurosci* **26**, 355–366.

Doucet E, Miquel MC, Nosjean A *et al.* (2000). Immunolabeling of the rat central nervous system with antibodies partially selective of the short form of the 5-HT$_3$ receptor. *Neuroscience* **95**, 881–892.

Downie DL, Hope AG, Lambert JJ *et al.* (1994). Pharmacological characterization of the apparent splice variants of the murine 5-HT$_3$R-A subunit expressed in *Xenopus laevis* oocytes. *Neuropharmacology* **33**, 473–482.

Dubin AE, Huvar R, D'Andrea MR *et al.* (1999). The pharmacological and function characteristics of the serotonin 5-HT$_{3A}$ receptor are specifically modified by a 5-HT$_{3B}$ receptor subunit. *J Biol Chem* **274**, 30799–30810.

Dukat M, Abdel-Rahman AA, Ismaiel AM *et al.* (1996). Structure–activity relationships for the binding of arylpiperazines and arylbiguanides at 5-HT$_3$ receptors. *J Med Chem* **39**, 4017–4026.

Eisensamer B, Uhr M, Meyr S *et al.* (2005). Antidepressants and antipsychotic drugs colocalize with 5-HT$_3$ receptors in raft-like domains. *J Neurosci* **25**, 10198–10206.

Emerit MB, Doucet E, Darmon M *et al.* (2002). Native and cloned 5-HT$_{3A}$ (S) receptors are anchored to F-actin in clonal cells and neurons. *Mol Cell Neurosci* **20**, 110–124.

Emerit MB, Doucet E, Latrémolière M *et al.* (2006). Immunolabelling of central and peripheral rodent nervous systems with antibodies directed against the 5-HT$_{3B}$ subunit of the 5-HT$_3$ receptor. *Soc Neurosci Abs* **625**, C72.

Engel SR and Allan AM (1999). 5-HT$_3$ receptor over-expression enhances ethanol sensitivity in mice. *Psychopharmacology* **144**, 411–415.

Engel SR, Lyons CR and Allan AM (1998). 5-HT$_3$ receptor over-expression decreases ethanol self-administration in transgenic mice. *Psychopharmacology* **140**, 243–248.

Férézou I, Cauli B, Hill EL *et al.* (2002). 5-HT$_3$ receptors mediate serotonergic fast synaptic excitation of neocortical vasoactive interstinal peptide/cholecytokinin interneurons. *J Neurosci* **22**, 7389–7397.

Finer-Moore J and Stroud RM (1984). Amphipathic analysis and possible formation of the ion channel in an acetylcholine receptor. *Proc Natl Acad Sci USA* **81**, 155–159.

Fletcher S and Barnes NM (1998). Desperately seeking subunits: are native 5-HT$_3$ receptors really homomeric complexes? *Trends Pharmacol Sci* **19**, 212–215.

Fonseca MI, Ni TG, Dunning DD *et al.* (2001). Distribution of serotonin 2A, 2C and 3 receptor mRNA in spinal cord and medulla oblongata. *Brain Res Mol Brain Res* **89**, 11–19.

Fozard JR (1984a). MDL 72222: a potent and highly selective antagonist at neuronal 5-hydroxytryptamine receptors. *Naunyn Schmiedeberg's Arch Pharmacol* **326**, 36–44.

Fozard JR (1984b). Neuronal 5-HT receptors in the periphery. *Neuropharmacology* **23**, 1473–1486.

Frank B, Niesler B, Bondy B *et al.* (2004a). Mutational analysis of serotonin receptor genes: *HTR3A* and *HTR3B* in fibromyalgia patients. *Clin Rheumatol* **23**, 338–344.

Frank B, Niesler B, Nöthen MM *et al.* (2004b). Investigation of the human serotonin receptor gene HTR3B in bipolar affective and schizophrenic patients. *Am J Med Genet B Neuropsychiatr Genet* **131**, 1–5.

Funahashi M, Mitoh Y and Matsuo R (2004). Activation of presynaptic 5-HT$_3$ receptors facilitates glutamatergic synaptic inputs to area postrema neurons in rat brain slices. *Methods Find Exp Clin Pharmacol* **26**, 615–622.

Furukawa K, Akaike N, Onodera H *et al.* (1992). Expression of 5-HT$_3$ receptors in PC12 cells treated with NGF and 8-Br-c-AMP. *J Neurophysiol* **67**, 812–819.

Gaddum JH and Picarelli ZP (1957). Two kinds of tryptamine receptor. *Br J Pharmacol* **9**, 323–328.

Gan TJ (2005). Selective serotonin 5-HT$_3$ receptor antagonists for postoperative nausea and vomiting: are they all the same? *CNS Drugs* **19**, 225–238.

Gershon MD (1999). Roles played by 5-hydroxytryptamine in the physiology of the bowel. *Aliment Pharmacol Ther* **13** (Suppl 2), 15–30.

Gershon MD (2000). 5-HT (serotonin) physiology and related drugs. *Curr Opin Gastroenterol* **16**, 113–1120.

Gill CH, Peters JA and Lambert JJ (1995). An electrophysiological investigation of the properties of a murine recombinant 5-HT$_3$ receptor stably expressed in HEK 293 cells. *Br J Pharmacol* **114**, 1211–1221.

Glatzle J, Sternini C, Robin C *et al.* (2002). Expression of 5-HT$_3$ receptors in the rat gastrointestinal tract. *Gastroenterology* **123**, 217–226.

Glaum SR, Brookes PA, Spyer KM *et al.* (1992). 5-Hydroxytryptamine-3 receptors modulate synaptic activity in the rat nucleus tractus solitarius *in vitro*. *Brain Res* **589**, 62–68.

Grailhe R, de Carvalho LP, Paas T *et al.* (2004). Distinct subcellular targeting of fluorescent nicotinic a3b4 and serotoninergic 5-HT$_{3A}$ receptors in hippocampal neurons. *Eur J Neurosci* **19**, 855–862.

Green GM, Scarth J and Dickenson A (2000). An excitatory role for 5-HT in spinal inflammatory nociceptive transmission; state-dependent actions via dorsal horn 5-HT$_3$ receptors in the anaesthetized rat. *Pain* **89**, 81–88.

Green T, Stauffer KA and Lummis SCR (1995). Expression of recombinant homo-oligomeric 5-hydroxytryptamine$_3$ receptors provides new insights into their maturation and structure. *J Biol Chem* **270**, 6056–6061.

Gunthorpe MJ and Lummis SCR (1999). Diltiazem causes open channel block of recombinant 5-HT$_3$ receptors. *J Physiol* **519**, 713–722.

Gunthorpe MJ and Lummis SCR (2001). Conversion of the ion selectivity of the 5-HT$_{3A}$ receptor from cationic to anionic reveals a conserved feature of the ligand-gated ion channel superfamily. *J Biol Chem* **276**, 10977–10983.

Gunthorpe MJ, Peters JA, Gill CH *et al.* (2000). The 4' lysine in the putative channel lining domain affects desensitization but not the single-channel conductance of recombinant homomeric 5-HT$_{3A}$ receptors. *J Physiol* **522**, 187–198.

Hales TG, Dunlop JI, Deeb TZ *et al.* (2006). Common determinants of single channel conductance within the large cytoplasmic loop of 5-hydroxytryptamine type 3 and a4b2 nicotinic acetylcholine receptors. *J Biol Chem* **281**, 8062–8071.

Hamon M, Gallissot MC, Menard F *et al.* (1989). 5-HT$_3$ receptor binding sites are on capsaicin-sensitive fibres in the rat spinal cord. *Eur J Pharmacol* **164**, 315–322.

Hanna MC, Davies PA, Hales TG *et al.* (2000). Evidence for expression of heteromeric serotonin 5-HT$_3$ receptors in rodents. *J Neurochem* **75**, 240–247.

Hapfelmeier G, Haseneder R, Lampadius K *et al.* (2002a). Cloned human and murine serotonin$_{3A}$ receptors expressed in human embryonic kidney 293 cells display different single-channel kinetics. *Neurosci Lett* **335**, 44–48.

Hapfelmeier G, Tredt C, Hasender R *et al.* (2002b). Co-expression of the 5-HT$_{3B}$ serotonin receptor subunit alters the biophysics of the 5-HT$_3$ receptor. *Biophys J* **84**, 1720–17233.

Harrell AV and Allen AM (2003). Improvements in hippocampal-dependent learning and decremental attention in 5-HT$_3$ receptor overexpressing mice. *Learn Mem* **10**, 410–419.

Hewlett WA, Trivedi BL, Zhang ZJ *et al.* (1999). Characterization of (S)-des-4-amino-3-[^{125}I]iodozacopride ([^{125}I]DAIZAC), a selective high-affinity radioligand for 5-hydroxytryptamine$_3$ receptors. *J Pharmacol Exp Ther* **288**, 221–231.

Hillsley K and Grundy D (1998). Sensitivity to 5-hydroxytryptamine in different afferent subpopulations within the mesenteric nerves supplying the rat jejunum. *J Physiol* **509**, 717–728.

Hodge CW, Kelley SP, Bratt AM *et al.* (2004). 5-HT$_{3A}$ receptor subunit is required for 5-HT$_3$ antagonist-induced reductions in alcohol drinking. *Neuropsychopharmacology* **29**, 1807–1813.

Holbrook JD, Gill CH, Zebda N *et al.* (2009). Characterisation of 5-HT$_{3C}$, 5-HT$_{3D}$ and 5-HT$_{3E}$ receptor subunits: evolution distribution and function. *J Neurochem* **108**, 384–396

Hope AG, Belelli D, Mair ID *et al.* (1999). Molecular determinants of (+)-tubocurarine binding at recombinant 5-hydroxytryptamine$_{3A}$ receptor subunits. *Mol Pharmacol* **55**, 1037–1043.

Hope AG, Downie, DL, Sutherland L *et al.* (1993). Cloning and functional expression of an apparent splice variant of the murine 5-HT$_3$ receptor A subunit. *Eur J Pharmacol* **245**, 187–192.

Hope AG, Peters JA, Brown AM *et al.* (1996). Characterization of a human 5-hydroxytryptamine$_3$ receptor type A (h5-HT$_3$R-AS) subunit stably expressed in HEK 293 cells. *Br J Pharmacol* **118**, 1237–1245.

Houtani T, Munemoto Y, Kase M *et al.* (2005). Cloning and expression of ligand-gated ion-channel receptor L2 in central nervous system. *Biochem Biophys Res Commun* **335**, 277–285.

Hoyer D, Neijt HC and Karpf A (1989). Competitive interaction of agonists and antagonists with 5-HT$_3$ recognition sites in membranes of neuroblastoma cells labelled with [^{3}H]ICS 205–930. *J Receptor Res* **9**, 65–79.

Hu XQ and Lovinger DM (2005). Role of aspartate 298 in mouse 5-HT$_{3A}$ receptor gating and modulation by extracellular Ca^{2+}. *J Physiol* **568**, 381–396.

Hu XQ and Peoples RW (2008). The 5-HT$_{3B}$ subunit confers spontaneous channel opening and altered ligand properties of the 5-HT$_3$ receptor. *J Biol Chem* **283**, 6826–6831.

Hu XQ, Sun H, Peoples RW *et al.* (2006). An interaction involving an arginine residue in the cytoplasmic domain of the 5-HT$_{3A}$ receptor contributes to receptor desensitization mechanism. *J Biol Chem* **281**, 21781–21788.

Huang J, Spier AD and Pickel VM (2004). 5-HT$_{3A}$ receptor subunits in the rat medial nucleus of the solitary tract: subcellular distribution and relation to the serotonin transporter. *Brain Res* **1028**,156–169.

Huang J, Wang YY, Wang W *et al.* (2008). 5-HT$_{3A}$ receptor subunit is expressed in a subpopulation of GABAergic and enkephalinergic neurons in the mouse dorsal spinal cord. *Neurosci Lett* **441**, 1–6.

Hubbard PC, Thompson AJ and Lummis SCR (2000). Functional differences between splice variants of the murine 5-HT$_{3A}$ receptor: possible role for phosphorylation. *Brain Res Mol Brain Res* **81**, 101–108.

Hussy N, Lukas W and Jones KA (1994). Functional properties of a cloned 5-hydroxytryptamine ionotropic receptor subunit: comparison with native mouse receptors. *J Physiol* **481**, 311–323.

Iiatrou CA, Dragoumanis CK, Vogiatzaki TD *et al.* (2005). Prophylactic intravenous ondansetron and dolasetron in intrathecal morphine-induced pruritus: a randomized, double-blinded, placebo-controlled study. *Anesth Analg* **101**, 1516–1520.

Iidaka T, Ozaki N, Matsumoto A *et al.* (2005). A variant C178T in the regulatory region of the serotonin receptor gene *HTR3A* modulates neural activation in the human amygdala. *J Neurosci* **25**, 6460–6466.

Ilegems E, Pick HM, Deluz C *et al.* (2004). Noninvasive imaging of 5-HT$_3$ receptor trafficking in live cells: from biosynthesis to endocytosis. *J Biol Chem* **279**, 53346–53352.

Imoto K, Busch C, Sakmann B *et al.* (1988). Rings of negatively charged amino acids determine the acetylcholine receptor channel conductance. *Nature* **335**, 645–648.

Isenberg KE, Ukhun SG, Holstad S *et al.* (1993). Partial cDNA cloning and NGF regulation of a rat 5-HT$_3$ receptor subunit. *NeuroReport* **5**, 121–124.

Jackson MB and Yakel JL (1995). The 5-HT$_3$ receptor channel. *Annu Rev Physiol* **57**, 447–468.

Jansen M, Bali M and Akabas MH (2008). Modular design of Cys-loop ligand-gated ion channels: functional 5-HT$_3$ and GABA r1 receptors lacking the large cytoplasmic M3M4 loop. *J Gen Physiol* **131**, 137–146.

Jeggo RD, Kellett DO, Wang Y *et al.* (2005). The role of central 5-HT$_3$ receptors in vagal reflex inputs to neurones in the nucleus tractus solitarius of anaesthetized rats. *J Physiol* **566**, 939–953.

Jensen TN, Nielsen J, Frederiksen K *et al.* (2006). Molecular cloning and pharmacological characterization of serotonin 5-HT$_{3A}$ receptor subtype in dog. *Eur J Pharmacol* **538**, 23–31.

Ji X, Takahashi N, Branko A *et al.* (2008). An association between serotonin receptor 3B gene (*HTR3B*) and treatment-resistant schizophrenia (TRS) in a Japanese population. *Nagoya J Med Sci* **70**, 11–17.

Johnson BA, Ait-Daoud N and Prihoda TJ (2000). Combining ondansetron and naltrexone effectively treats biologically predisposed alcoholics: from hypotheses to preliminary clinical evidence. *Alcohol Clin Exp Res* **24**, 737–742.

Johnson BA, Roache JD, Ait-Daoud N *et al.* (2002). Ondansetron reduces the craving of biologically predisposed alcoholics. *Psychopharmacology* **160**, 408–413.

Jones KA and Surprenant A (1994). Single channel properties of the 5-HT$_3$ subtype of serotonin receptor in primary cultures of rodent hippocampus. *Neurosci Lett* **174**, 133–136.

Jones S and Yakel JL (2003). Casein kinase II (protein kinase ck2) regulates serotonin 5-HT$_3$ receptor channel function in NG108–15 cells. *Neuroscience* **119**, 629–634.

Joshi PR, Suryanarayanan A, Hazai E *et al.* (2006). Interactions of granisetron with an agonist-free 5-HT$_{3A}$ receptor model. *Biochemistry* **45**, 1099–1105.

Kaiser R, Trembley PB, Sezer O *et al.* (2004). Investigation of the association between *5-HT3A* receptor gene polymorphisms and efficiency of antiemetic treatment with 5-HT$_3$ receptor antagonists. *Pharmacogenetics* **14**, 271–278.

Karlin A (2002). Emerging structure of the nicotinic acetylcholine receptors. *Nat Rev Neurosci* **3**, 102–114.

Karnovsky AM, Gotow LF, McKinley DD *et al.* (2003). A cluster of novel serotonin receptor 3-like genes on human chromosome 3. *Gene* **319**, 137–148.

Katsurabayashi S, Kubota H, Tokutomi N *et al.* (2003). A distinct distribution of functional presynaptic 5-HT receptor subtypes on GABAergic nerve terminals projecting to single hippocampal CA1 pyramidal neurons. *Neuropharmacology* **44**, 1022–1030.

Kawa K (1994). Distribution and functional properties of 5-HT$_3$ receptors in the rat hippocampal dentate gyrus: a patch-clamp study. *J Neurophysiol* **71**, 1935–1947.

Kayser V, Elfassi IE, Aubel B *et al.* (2007). Mechanical, thermal and formalin-induced nociception is differentially altered in 5-HT$_{1A}$–/–, 5-HT$_{1B}$–/–, 5-HT$_{2A}$–/–, 5-HT$_{3A}$–/– and 5-HTT–/– knock-out male mice. *Pain* **130**, 235–248.

Kelley SP, Bratt AM and Hodge CW (2003a). Targeted gene deletion of the 5-HT$_{3A}$ receptor subunit produces an anxiolytic phenotype in mice. *Eur J Pharmacol* **461**, 19–25.

Kelley SP, Dunlop JI, Kirkness EF *et al.* (2003b). A cytoplasmic region determines single-channel conductance in 5-HT$_3$ receptors. *Nature* **424**, 321–324.

Keramidas A, Moorhouse AJ, Schofield PR *et al.* (2004). Ligand-gated ion channels: mechanisms underlying ion selectivity. *Prog Biophys Mol Biol* **86**, 161–204.

Kia HK, Miquel MC, McKernan RM *et al.* (1995). Localization of 5-HT$_3$ receptors in the rat spinal cord: immunohistochemistry and *in situ* hybridization. *Neuroreport* **6**, 257–261.

Kidd EJ, Laporte AM, Langlois X *et al.* (1993). 5-HT$_3$ receptors in the rat central nervous system are mainly located on nerve fibres and terminals. *Brain Res* **612**, 289–298.

Koyama S, Matsumoto N, Kubo C *et al.* (2000). Presynaptic 5-HT$_3$ receptor-mediated modulation of synaptic GABA release in the mechanically dissociated rat amygdala neurons. *J Physiol* **529**, 373–383.

Krzywkowski K (2006). Do polymorphisms in human 5-HT$_3$ genes contribute to pathological phenotypes? *Biochem Soc Trans* **34**, 872–876.

Krzywkowski K, Davies PA, Feinberg-Zadek PL *et al.* (2008). High-frequency *HTR3B* variant associated with major depression dramatically augments the signaling of the human 5-HT$_{3AB}$ receptor. *Proc Natl Acad Sci USA* **105**, 722–727.

Krzywkowski K, Jensen AA, Connolly CN *et al.* (2007). Naturally occurring variations in the human 5-HT$_{3A}$ gene profoundly impact 5-HT$_3$ receptor function and expression. *Pharmacogenet Genomics* **17**, 255–266.

Kurzwelly D, Barann M, Kostanian A *et al.* (2004). Pharmacological and electrophysiological properties of the naturally occurring Pro391Arg variant of the human 5-HT$_{3A}$ receptor. *Pharmacogenetics* **14**, 165–172.

Lambert JJ, Peters JA, Hales TG *et al.* (1989). The properties of 5-HT$_3$ receptors in clonal cell lines studied by patch-clamp techniques. *Br J Pharmacol* **97**, 27–40.

Lankiewicz S, Huser MB, Heumann R *et al.* (2000). Phosphorylation of the 5-hydroxytryptamine$_3$ (5-HT$_3$) receptor expressed in HEK293 cells. *Receptors Channels* **7**, 9–15.

Lankiewicz S, Lobitz N, Wetzel CHR *et al.* (1998). Molecular cloning, functional expression and pharmacological characterization of 5-hydroxytryptamine$_3$ receptor cDNA and its splice variants from the guinea pig. *Mol Pharmacol* **53**, 202–212.

Leslie RA, Reynolds DJM andrews PLR *et al.* (1990). Evidence for presynaptic 5-hydroxytryptamine$_3$ recognition sites on vagal afferent terminals in the brainstem of the ferret. *Neuroscience* **38**, 667–673.

Lester HA, Dibas MI, Dahan DS *et al.* (2004). Cys-loop receptors: new twists and turns. *Trends Neurosci* **27**, 329–336.

Liu MT, Rayport S, Jiang Y *et al.* (2002). Expression and function of 5-HT$_3$ receptors in the enteric neurons of mice lacking the serotonin transporter. *Am J Physiol Gastrointest Liver Physiol* **283**, G1398-G1411.

Livesey MR, Cooper MA, Deeb TZ *et al.* (2008). Structural determinants of Ca^{2+} permeability and conduction in the human 5-hydroxytryptamine type 3A receptor. *J Biol Chem* **283**, 19301–19313.

Lovinger DM (1999). 5-HT$_3$ receptors and the neural actions of alcohols: an increasingly exciting topic. *Neurochem Int* **53**, 125–130.

Low PB, Lambert JJ and Peters JA (2001). A comparative study of the pharmacological properties of homo-oligomeric and hetero-oligomeric human recombinant 5-hydroxytrytamine type-3 (5-HT$_3$) receptors. *Br J Pharmacol* **133**,144P.

Lummis SCR, Beene DL, Lee LW *et al.* (2005). Cis-trans isomerization at a proline opens the pore of a neurotransmitter-gated ion channel. *Nature* **438**, 248–252.

Mair ID, Lambert JJ, Yang J *et al.* (1998). Pharmacological characterization of a rat 5-hydroxytryptamine type$_3$ receptor subunit (r5-HT$_{3A(b)}$) expressed in *Xenopus laevis* oocytes. *Br J Pharmacol* **124**, 1667–1674.

Malone HM, Peters JA and Lambert JJ (1994). The permeability of 5-HT$_3$ receptors of rabbit nodose ganglion neurones to organic and monovalent cations. *J Physiol* **475.P**, 151P.

Marazziti D, Betti L, Giannaccini G *et al.* (2001). Distribution of [^{3}H]GR65630 binding in human brain postmortem. *Neurochem Res* **26**, 187–190.

Maricq AV, Peterson AS, Brake AJ *et al.* (1991). Primary structure and functional expression of the 5-HT$_3$ receptor, a serotonin-gated ion channel. *Science* **254**, 432–437.

McKernan RM (1992). Biochemical properties of the 5-HT$_3$ receptor. In M Hamon, ed., *Central and peripheral 5-HT$_3$ receptors*. Academic Press, London UK, pp. 89–102.

McMahon L and Kauer JA (1997). Hippocampal interneurones are excited via serotonin-gated ion channels. *J Neurophysiol* **78**, 2493–2502.

Meineke C, Tzvetkov MV, Bokelmann K *et al.* (2008).Functional characterization of a –100_–102delAAG deletion-insertion polymorphism in the promoter region of the *HTR3B* gene. *Pharmacogenet Genomics* **18**, 219–30

Michel K, Zeller F, Langer R *et al.* (2005). Serotonin excites neurons in the human submucous plexus via 5-HT$_3$ receptors. *Gastroenterology* **128**, 1317–1326.

Minami M, Endo T, Hirafugi M *et al.* (2003). Pharmacological aspects of anticancer drug-induced emesis with emphasis on serotonin release and vagal nerve activity. *Pharmacol Ther* **99**, 149–165.

Miquel MC, Emerit MB, Nosjean A *et al.* (2002). Differential subcellular localization of the 5-HT3-As receptor subunit in the rat central nervous system. *Eur J Neurosci* **15**, 449–457.

Miyake A, Mochizuki S, Takemoto Y *et al.* (1995). Molecular cloning of human 5-hydroxytryptamine$_3$ receptor: heterogeneity in distribution and function among species. *Mol Pharmacol* **48**, 407–416.

Miyazawa A, Fujiyoshi Y and Unwin N (2003). Structure and gating mechanism of the acetylcholine receptor pore. *Nature* **423**, 949–955.

Miyazawa A, Fujiyoshi Y, Stowell M *et al.* (1999). Nicotinic acetylcholine receptor at 4.6 A resolution: transverse tunnels in the channel wall. *J Mol Biol* **288**, 765–786.

Mochizuki S, Miyake A and Furuichi K (1999a). Ion permeation properties of a cloned human 5-HT$_3$ receptor transiently expressed in HEK 293 cells. *Amino Acids* **17**, 243–255.

Mochizuki S, Miyake A and Furuichi K (1999b). Identification of a domain affecting agonist potency of meta-chlorophenylbiguanide in 5-HT$_3$ receptors. *Eur J Pharmacol* **369**, 125–132.

Mochizuki S, Watanabe T, Miyake A *et al.* (2000). Cloning, expression and characterization of ferret 5-HT$_3$ receptor subunit. *Eur J Pharmacol* **399**, 97–106.

Monk SA, Desai K, Brady CA *et al.* (2001). Generation of a selective 5-HT$_{3B}$ subunit-recognising polyclonal antibody; identification of immunoreactive cells in rat hippocampus. *Neuropharmacology* **41**, 1013–1016.

Monk SA, Williams JM, Hope AG *et al.* (2004). Identification and importance of N-glycosylation of the human 5-hydroxytryptamine$_{3A}$ receptor subunit. *Biochem Pharmacol* **68**, 1787–1796.

Morain P, Abraham C, Portevin B *et al.* (1994). Biguanide derivatives: agonist pharmacology at 5-hydroxytryptamine type 3 receptors *in vitro*. *Mol Pharmacol* **46**, 732–742.

Morales M and Backman C (2002). Coexistence of serotonin 3 (5-HT$_3$) and CB1 cannabinoid receptors in interneurons of hippocampus and dentate gyrus. *Hippocampus* **12**, 756–764.

Morales M and Bloom FE (1997). The 5-HT$_3$ receptor is present in different subpopulations of GABAergic neurons in the rat telencephalon. *J Neurosci* **17**, 3157–3167.

Morales M and Wang S-D (2002). Differential composition of 5-hydroxytryptamine$_3$ receptors synthesised in the rat CNS and peripheral nervous system. *J Neurosci* **22**, 6732–6741.

Morales M, Battenberg E and Bloom FE (1998). Distribution of neurons expressing immunoreactivity for the 5HT$_3$ receptor subtype in the rat brain and spinal cord. *J Comp Neurol* **402**, 385–401.

Morales M, Battenberg E, de Lecea L *et al.* (1996a). The type 3 serotonin receptor is expressed in a subpopulation of GABAergic neurons in the rat neocortex and hippocampus. *Brain Res* **731**, 199–202.

Morales M, Battenberg E, de Lecea L *et al.* (1996b). Cellular and subcellular immunolocalization of the type 3 serotonin receptor in the rat central nervous system. *Brain Res Mol Brain Res* **36**, 251–260.

Morales M, McCollum N and Kirkness EF (2001). 5-HT$_3$-receptor subunits A and B are co-expressed in neurons of the dorsal root ganglion. *J Comp Neurol* **438**, 163–172.

Morales M, Wang SD, Diaz-Ruiz O *et al.* (2004). Cannabinoid CB1 receptor and serotonin 3 receptor subunit A (5-HT$_{3A}$) are co-expressed in GABA neurons in the rat telencephalon. *J Comp Neurol*, **468**, 205–216.

Mott DD, Erreger K, Banke TG *et al.* (2001). Open probability of homomeric murine 5-HT$_{3A}$ serotonin receptors depends on subunit occupancy. *J Physiol* **535**, 427–443.

Muller W, Fiebich BL and Stratz T (2006). New treatment options using 5-HT$_3$ receptor antagonists in rheumatic diseases. *Curr Top Med Chem* **6**, 2035–2042.

Newberry NR, Cheshire SH and Gilbert MJ (1991). Evidence that the 5-HT$_3$ receptors of the rat, mouse and guinea pig superior cervical ganglion may be different. *Br J Pharmacol* **102**, 615–620.

Nichols RA and Mollard P (1996). Direct observation of serotonin 5-HT$_3$ receptor-induced increases in calcium levels in individual brain nerve terminals. *J Neurochem* **67**, 581–592.

Niemeyer M-I and Lummis SCR (1999). Different efficacy of specific agonists at 5-HT$_3$ receptor splice variants: the role of the extra six amino acid segment. *Br J Pharmacol* **123**, 661–666.

Niemeyer M-I and Lummis SCR (2001). The role of the agonist binding site in Ca^{2+} inhibition of the recombinant 5-HT$_{3A}$ receptor. *Eur J Pharmacol* **428**, 153–161.

Niesler B, Flohr T, Nöthen MM *et al.* (2001a). Association between the 5′ UTR variant C178T of the serotonin receptor gene *HTR3A* and bipolar affective disorder. *Pharmacogenetics* **11**, 471–475.

Niesler B, Frank B, Kapeller J *et al.* (2003). Cloning, physical mapping and expression analysis of the human 5-HT$_3$ serotonin receptor-like genes *HTR3C, HTR3D* and *HTR3E. Gene* **310**, 101–111.

Niesler B, Walstab J, Combrink S *et al.* (2007). Characterization of the novel human serotonin receptor subunits 5-HT$_{3C}$, 5- HT$_{3D}$ and 5-HT$_{3E}$. *Mol. Pharmacol* **72**, 8–17

Niesler B, Weiss B, Fischer C *et al.* (2001b). Serotonin receptor gene *HTR3A* variants in schizophrenic and bipolar affective patients. *Pharmacogenetics* **11**, 21–27.

Noam Y, Wadman WJ, van Hooft JA (2008). On the voltage-dependent Ca^{2+} block of serotonin 5-HT$_3$ receptors: a critical role of intracellular phosphates. *J Physiol* **586**, 3629–3638.

Oyama T, Ueda M, Kuraishi Y *et al.* (1996). Dual effect of serotonin on formalin-induced nociception in the rat spinal cord. *Neurosci Res* **25**, 129–135.

Oz M, Zhang L and Morales M (2002). Endogenous cannabinoid, anandamide, acts as a noncompetitive inhibitor on 5-HT$_3$ receptor-mediated responses in *Xenopus* oocytes. *Synapse* **46**, 150–156.

Panicker S, Cruz H, Arrabit C *et al.* (2002). Evidence for a centrally located gate in the pore of a serotonin-gated ion channel. *J Neurosci* **22**, 1629–1639.

Parker RM, Barnes JM, Ge J *et al.* (1996a). Autoradiographic distribution of [^{3}H]-(S)-zacopride-labelled 5-HT$_3$ receptors in human brain. *J Neurol Sci* **144**, 119–127.

Parker RM, Bentley KR and Barnes NM (1996b). Allosteric modulation of 5-HT$_3$ receptors: focus on alcohols and anaesthetic agents. *Trends Pharmacol Sci* **17**, 95–99.

Peters JA and Lambert JJ (1989). Electrophysiology of 5-HT$_3$ receptors in neuronal cell lines. *Trends Pharmacol Sci* **10**, 172–175.

Peters JA, Hales TG and Lambert JJ (1988). Divalent cations modulate 5-HT$_3$ receptor-induced currents in N1E-115 neuroblastoma cells. *Eur J Pharmacol* **151**, 491–495.

Peters JA, Hales TG and Lambert JJ (2005). Molecular determinants of single-channel conductance and ion selectivity in the Cys-loop family: insights from the 5-HT$_3$ receptor. *Trends Pharmacol Sci* **26**, 587–594.

Peters JA, Hope AG, Sutherland L *et al.* (1997). Recombinant 5-hydroxytryptamine$_3$ receptors. In MJ Browne, ed., *Recombinant cell surface receptors—targets for therapeutic intervention*. CRC Press, Boca Raton, pp. 119–154.

Peters JA, Lambert JJ and Malone HM (1991). Physiological and pharmacological aspects of 5-HT$_3$ receptor function. In TW Stone, ed., *Aspects of synaptic transmission: LTP, galanin, opioids, autonomic and 5-HT*. Taylor and Francis, London, pp. 283–313.

Peters JA, Lambert JJ and Malone HM (1994). Electrophysiological studies of 5-HT$_3$ receptors, In FD King, BJ Jones and GJ Sanger, eds, *5-Hydroxytryptamine-3 receptor antagonists*. CRC Press, Boca Raton, pp. 116–153.

Peters JA, Malone HM and Lambert JJ (1993). An electrophysiological investigation of the properties of 5-HT$_3$ receptors of rabbit nodose ganglion neurones in culture. *Br J Pharmacol* **110**, 665–676.

Pratt GD and Bowery NG (1989). The 5-HT$_3$ receptor ligand [^{3}H]BRL 43694 binds to presynaptic sites in the nucleus tractus solitarius of the rat. *Neuropharmacology* **28**, 1367–1376.

Price KL and Lummis SCR (2004). The role of tyrosine residues in the extracellular domain of the 5-hydroxytryptamine$_3$ receptor. *J Biol Chem* **279**, 23294–23301.

Quirk PL, Rao S, Roth BL *et al.* (2004). Three putative N-glycosylation sites within the murine 5-HT$_{3A}$ receptor sequence affect plasma membrane targeting, ligand-binding and calcium influx in heterologous mammalian cells. *J Neurosci Res* **77**, 498–506.

Reeves DC and Lummis SCR (2002). The molecular basis of the structure and function of the 5-HT$_3$ receptor: a model ligand-gated ion channel. *Mol Membr Biol* **9**, 11–26.

Reeves DC and Lummis SCR (2006). Detection of human and rodent 5-HT$_{3B}$ receptor subunits by anti-peptide polyclonal antibodies. *BMC Neurosci* **7**, 27.

Reeves DC, Goren EN, Akabas MH *et al.* (2001). Structural and electrostatic properties of the 5-HT$_3$ receptor pore revealed by substituted cysteine accessibility mutagenesis. *J Biol Chem* **276**, 42035–42042.

Reeves DC, Jansen M, Bali M *et al.* (2005). A role for the β1– β2 loop in the gating of 5-HT$_3$ receptors. *J Neurosci* **25**, 9358–9366.

Reeves DC, Sayed MF, Chau PL *et al.* (2003). Prediction of 5-HT$_3$ receptor agonist-binding residues using homology modeling. *Biophys J* **84**, 2338–2344.

Reynolds DJM andrews PLR, Davis CJ eds (1995). *Serotonin and the scientific basis of anti-emetic therapy*. Oxford Clinical Communications, Oxford.

Richardson BP, Engel G, Donatsch P *et al.* (1985). Identification of serotonin M-receptor subtypes and their specific blockade by a new class of drugs. *Nature* **316**, 126–131.

Robertson B and Bevan S (1991). Properties of 5-hydroxytryptamine$_3$ receptor-gated currents in adult rat dorsal root ganglion neurones. *Br J Pharmacol* **102**, 272–276.

Roerig B, Nelson DA and Katz LC (1997). Fast signaling by nicotinic acetylcholine and serotonin 5-HT$_3$ receptors in developing visual cortex. *J Neurosci* **17**, 8353–8362.

Rondé P and Nichols RA (1998). High calcium permeability of serotonin 5-HT$_3$ receptors on presynaptic nerve terminals from rat striatum. *J Neurochem* **70**, 1094–1103.

Ropert D and Guy N (1991). Serotonin facilitates GABAergic transmission in the CA1 region of the rat hippocampus *in vitro. J Physiol* **441**, 121–136.

Rubenstein EB, Slusher BS, Rojas C *et al.* (2006). New approaches to chemotherapy-induced nausea and vomiting: from neuropharmacology to clinical investigations. *Cancer J* **12**, 341–347.

Rupprecht R, di Michele F, Hermann B *et al.* (2001). Neuroactive steroids: molecular mechanisms of action and implications for neuropsychopharmacology. *Brain Res Brain Res Rev* **37**, 59–67.

Rüsch D, Musset B, Wulf H *et al.* (2007). Subunit-dependent modulation of the 5-hydroxytryptamine type 3 receptor open-close equilibrium by n-alcohols. *J Pharmacol Exp Ther* **321**, 1069–1074.

Sanger GJ and Nelson DR (1991). Selective and functional 5-hydroxytryptamine$_3$ receptor antagonism by BRL43694 (granisetron). *Eur J Pharmacol* **159**, 113–124.

Sarvela PJ, Halonen PM, Soikkeli AI *et al.* (2006). Ondansetron and tropisetron do not prevent intraspinal morphine and fentanyl-induced pruritus in elective Caesarean delivery. *Acta Anaesthesiol Scand* **50**, 239–244.

Sato N, Sakamori M, Hage K *et al.* (1992). Antagonistic activity of Y-25130 on 5-HT$_3$ receptors. *Jpn J Pharmacol* **59**, 443–448.

Shao XM, Yakel JL and Jackson MB (1991). Differentiation of NG108–15 cells alters single channel conductance and desensitisation kinetics of the 5-HT$_3$ receptor. *J Neurophysiol* **65**, 630–638.

Sine SM and Engel AG (2006). Recent advances in Cys-loop receptor structure and function. *Nature* **440**, 448–455.

Smith WW, Sancillo LF, Owera-Atepo JB *et al.* (1988). Zacopride, a potent 5-HT$_3$ antagonist. *J Pharm Pharmacol* **40**, 301–302.

Solt K, Stevens RJ, Davies PA *et al.* (2005). General anesthetic-induced channel gating enhancement of 5-hydroxytryptamine type 3 receptors depends on receptor subunit composition. *J Pharmacol Exp Ther* **315**, 771–776.

Spier AD and Lummis SCR (2000). The role of tryptophan residues in the 5-hydroxytryptamine$_3$ receptor ligand-binding domain. *J Biol Chem* **275**, 5620–5625.

Spier AD, Wotherspoon G, Nayak SV *et al.* (1999). Antibodies against the extracellular domain of the 5-HT$_3$ receptor label both native and recombinant receptors. *Brain Res Mol Brain Res* **67**, 221–230.

Staubli U and Xu FB (1995). Effects of 5-HT$_3$ receptor antagonism on hippocampal theta rhythm, memory and LTP induction in the freely moving rat. *J Neurosci* **15**, 2445–2452.

Stewart A, Davies PA, Kirkness EF *et al.* (2003). Introduction of the 5-HT$_{3B}$ subunit alters the function properties of 5-HT$_3$ receptors native to neuroblastoma cells. *Neuropharmacology* **44**, 214–223.

Sudweeks SN, van Hooft JA and Yakel JL (2002). Serotonin 5-HT$_3$ receptors in rat CA1 hippocampal interneurons: functional and molecular characterization. *J Physiol* **544**, 715–726.

Sugita S, Shen K-Z and North RA (1992). 5-Hydroxytryptamine is a fast excitatory transmitter at 5-HT$_3$ receptors in rat lateral amygdala. *Neuron* **8**, 199–203.

Sullivan NL, Thompson AJ, Price KL *et al.* (2006). Defining the roles of Asn-128, Glu-129 and Phe-130 in loop A of the 5-HT$_3$ receptor. *Mol Membr Biol* **23**, 442–451.

Sun H, Hu XQ, Emerit MB *et al.* (2008). Modulation of 5-HT$_3$ receptor desensitization by the light chain of microtubule-associated protein 1B expressed in HEK 293 cells. *J Physiol* **586**, 751–762.

Sun H, Hu XQ, Moradel EM *et al.* (2003). Modulation of 5-HT$_3$ receptor-mediated response and trafficking by activation of protein kinase C. *J Biol Chem* **278**, 34150–34157.

Suryanarayanan A, Joshi PR, Bikádi Z *et al.* (2005). The loop C region of the murine 5-HT$_{3A}$ receptor contributes to the differential actions of 5-hydroxytryptamine and m-chlorophenylbiguanide. *Biochemistry* **44**, 9140–9149.

Tanaka M, Kobayashi D, Murakami Y *et al.* (2008). Genetic polymorphisms in the 5-hydroxytryptamine type 3B receptor gene and paroxetine-induced nausea. *Int J Neuropsychopharmacol*, **11**, 261–267.

Tecott LH, Maricq AV and Julius D (1993). Nervous system distribution of the serotonin 5-HT3 receptor mRNA. *Proc Natl Acad Sci USA* **90**,1430–1434.

Thompson AJ and Lummis SCR (2003). A single ring of charged amino acids at one end of the pore can control ion selectivity in the 5-HT$_3$ receptor. *Br J Pharmacol* **140**, 359–365.

Thompson AJ and Lummis SCR (2006). 5-HT$_3$ receptors. *Curr Pharm Des* **12**, 3615–3630.

Thompson AJ, Padgett CL and Lummis SCR (2006). Mutagenesis and molecular modeling reveal the importance of the 5-HT$_3$ receptor F-loop. *J Biol Chem* **281**, 16576–16582.

Thompson AJ, Price KL, Reeves DC *et al.* (2005). Locating an antagonist in the 5-HT$_3$ receptor binding site using modeling and radioligand-binding. *J Biol Chem* **280**, 20476–20482.

Tremblay RB, Kaiser R, Sezer O *et al.* (2003). Variations in the 5-hydroxytryptamine type 3B receptor gene as predictors of the efficacy of antiemetic treatment in cancer patients. *J Clin Oncol* **21**, 2147–2155.

Turner TJ, Mokler DJ and Luebke JI (2004). Calcium influx through presynaptic 5-HT$_3$ receptors facilitates GABA release in the hippocampus: *in vitro* slice and synaptosome studies. *Neuroscience* **129**, 703–718.

Tzvetkov MV, Meineke C, Oetjen E *et al.* (2007). Tissue-specific alternative promoters of the serotonin receptor gene *HTR3B* in human brain and intestine. *Gene* **386**, 52–62.

Unwin N (2005). Refined structure of the nicotinic acetylcholine receptor at 4Å resolution. *J Mol Biol* **346**, 967–989.

van Hooft JA and Vijverberg HPM (1995). Phosphorylation controls conductance of 5-HT$_3$ receptor ligand-gated ion channels. *Receptors Channels* **3**, 7–12.

van Hooft JA and Wadman W (2003). Ca^{2+} ions block and permeate serotonin 5-HT$_3$ receptor channels in rat hippocampal interneurons. *J Neurophysiol* **89**, 1864–1869.

van Hooft JA and Yakel JL (2003). 5-HT$_3$ receptors in the CNS: 3B or not 3B. *Trends Pharmacol Sci* **24**, 157–160.

van Hooft JA, Kreinkamp AP and Vijverberg HPM (1997). Native serotonin 5-HT$_3$ receptors expressed in *Xenopus laevis* oocytes differ from homopentameric receptors. *J Neurochem* **69**, 1316–1321.

van Wijngaarden I, Hamminga D, van Hes R *et al.* (1993). Development of high-affinity 5-HT$_3$ receptor antagonists. Structure–affinity relationships of novel 1,7-annelated indole derivatives. *J Med Chem* **36**, 3693–3699.

Venkataraman P, Joshi P, Venkatachalan SP *et al.* (2002b). Identification of critical residues in loop E in the 5-HT$_{3A}$R binding site. *BMC Biochem* **3**, 15.

Venkataraman P, Joshi P, Venkatachalan SP *et al.* (2002a). Functional group interactions of a 5-HT$_3$R antagonist. *BMC Biochem* **3**, 16.

Wallis DI (1989). Interaction of 5-hydroxytryptamine with autonomic and sensory neurones. In JR Fozard, ed., *Peripheral actions of 5-hydroxytryptamine*. Oxford University Press, Oxford, pp. 220–246.

Walstab J, Hammer C, Bönisch H *et al.* (2008). Naturally occurring variants in the *HTR3B* gene significantly alter properties of human heteromeric 5-hydroxytryptamine-3A/B receptors. *Pharmacogenet Genomics* **18**, 793–802.

Wan S and Browning KN (2008). Glucose increases synaptic transmission from vagal afferent central nerve terminals via modulation of 5-HT$_3$ receptors. *Am J Physiol Gastrointest Liver Physiol* **295**, G1050–1057.

Wang Y, Ramage AG and Jordan D (1998). Presynaptic 5-HT$_3$ receptors evoke an excitatory response in dorsal vagal preganglionic neurones in anaesthetized rats. *J Physiol* **509.3**, 683–694.

Weiss B, Mertz A, Schrock E *et al.* (1995). Assignment of a human homolog of the mouse *Htr3* receptor gene to chromosome 11q23.1–q23.2. *Genomics* **29**, 304–305.

Werner P, Kawashima E, Reid J *et al.* (1994). Organization of the mouse 5-HT$_3$ receptor gene and functional expression of two splice variants. *Brain Res Mol Brain Res* **26**, 233–241.

Wetzel CHR, Hermann B, Behl C *et al.* (1998). Functional antagonism of gonal steroids at the 5-hydroxytryptamine type 3 receptor. *Mol Endocrinol* **12**, 1441–1451.

Wong EH, Clark R and Leung E *et al.* (1995). The interaction of RS 25259–197, a potent and selective antagonist, with 5-HT$_3$ receptors, *in vitro*. *Br J Pharmacol* **114**, 851–859.

Xiong W, Hosoi M, Koo BN *et al.* (2008). Anandamide inhibition of 5-HT$_{3A}$ receptors varies with receptor density and desensitization. *Mol Pharmacol* **73**, 314–322

Yakel JL, Shao XM and Jackson MB (1990). The selectivity of the channel coupled to the 5-HT$_3$ receptor. *Brain Res* **533**, 46–52.

Yakel JL, Shao XM and Jackson MB (1991). Activation and desensitization of the 5-HT$_3$ receptor in a rat glioma x mouse neuroblastoma hybrid cell. *J Physiol* **436**, 293–308.

Yan D, Schulte MK, Bloom KE *et al.* (1999). Structural features of the ligand-binding domain of the serotonin 5HT$_3$ receptor. *J Biol Chem* **274**, 5537–5541.

Yang J (1990). Ion permeation through 5-HT-gated channels in neuroblastoma N18 cells. *J Gen Physiol* **96**, 1177–1198.

Yang J, Mathie A and Hille B (1992). 5-HT$_3$ receptor channels in dissociated rat superior cervical ganglion neurones. *J Physiol* **448**, 237–256.

Zeitz KP, Guy N, Malmberg AB *et al.* (2002). The 5-HT$_3$ subtype of serotonin receptor contributes to nociceptive processing via a novel subset of myelinated and unmyelinated nociceptors. *J Neurosci* **22**, 1010–1019.

Zhang L, Oz M and Weight FF (1995). Potentiation of 5-HT$_3$ receptor-mediated responses by protein kinase C activation. *Neuroreport* **6**, 1464–1468.

Zhong H, Zhang M and Nurse CA (1999). Electrophysiological characterization of 5-HT receptors on rat petrosal neurons in dissociated cell culture. *Brain Res* **816**, 544–553.

Zhou F-M and Hablitz JJ (1999). Activation of serotonin receptors modulates synaptic transmission in rat cerebral cortex. *J Neurophysiol* **82**, 2989–2999.

Zhou X and Galligan JJ (1999). Synaptic activation and properties of 5-hydroxytryptamine$_3$ receptors in myenteric neurons of guinea pig intestine. *J Pharmacol Exp Ther* **290**, 803–810.

3.1.3 **Glycine receptors**

Carmen Villmann and Cord-Michael Becker

3.1.3.1 Introduction

In the central nervous system (CNS), fast synaptic inhibition is primarily mediated by activation of glycine receptors (GlyRs) and GABA$_{A/C}$ (gamma aminobutyric acid) receptors. Both receptor families are members of the Cys-loop superfamily of ligand-gated ion channels. They share a similar transmembrane topology as well as other structural and functional properties. Presently, four GlyR

α-subunits are known encoded by different genes, as well as a single β-subunit. The receptor channel complex exists as a rosette-like heteropentamer of five individual subunits, most likely comprising 2α and 3β subunits. Considerable progress has been made in identifying domains on GlyR subunits important for lig- and binding, oligomerization and trafficking during protein synthesis, including exit from the endoplasmatic reticulum, transit to the Golgi apparatus, protein–protein interactions and synaptic clustering. GlyR clusters are stabilized by the scaffolding protein gephyrin in the plasma membrane that also contributes to the plasticity of inhibitory synapses. The opening of the GlyR ion channel pore follows a ligand-induced transition process. As evident from mutagenesis studies, α- and β-subunits are involved in ligand binding at a heterodimeric ligand binding interface, however, the role of at least one existing β-β interface still remains to be elucidated. Degradation of membrane bound GlyRs is regulated by ubiquitination and followed by proteolysis via the endocytotic pathway.

The focus of this chapter is on GlyR receptor heterogeneity, ligand binding domains, physiology, assembly, and trafficking to the synaptic compartments.

3.1.3.2 Molecular characterization

The strychnine-sensitive glycine receptor is an important mediator of synaptic inhibition in brainstem and vertebrate spinal cord. GlyRs exist as a heterogenous group of pentameric ligand-gated chloride channels. GlyRs comprise two major subunits, the α and the β subunit (Figure 3.1.3.1) (Becker, 1995). In the human, three functional α subunit genes have been identified (α1–α3), while there are four different α-subunits known (α1–α4) in other species. In contrast, there is only one β-subunit gene. Alternative splicing leads to further diversity among the glycine receptor subunits and splice variants have been described for the α1, α2, α3, and β subunits (Kuhse *et al.*, 1991; Malosio *et al.*, 1991a; Nikolic *et al.*, 1998; Oertel *et al.*, 2007).

GlyRs belong to the superfamily of Cys-loop recetors, in addition to the nicotinic acetylcholine receptors (nAChR), the 5-HT$_3$ receptor, and GABA$_{A/C}$ receptors (Lynch, 2004; Betz and Laube, 2006). The superfamily of Cys-loop receptors is characterized by a paradigmatic N-terminal cysteine bridge (between C138 and C152). The topology of this superfamily is characterized by four transmembrane domains (TMs) connected by intra- or extracellular loop structures (Figures 3.1.3.1 and 3.1.3.2a). While TM2 forms the interior of the ion channel, the N- and the C-termini are extracellular. In contrast to a short C-terminus, the N-terminus is a long highly structured domain, which exhibits a high homology to the acetylcholine binding protein (AChBP) of the pond snail *Lymnea stagnalis* whose structure was solved by X-ray crystallography with a resolution of 2.7 Å (Figure 3.1.3.1) (Brejc *et al.*, 2001). The AChBP protein displays an immunoglobulin-like protein fold and is characterized by a short α-helical domain in the far N-terminus that is followed by 10 β-sheets. The X-ray data show that the ligand binding cavity is situated at the interface between adjacent subunits (Brejc *et al.*, 2001; Unwin, 2005). This inter-subunit localization of the ligand binding site differs from the glutamate receptor family where the ligand binding cavity is formed by two regions of the same subunit (Villmann and Becker, 2007). In contrast to the β-subunit, all of the α-subunits are able to form functional

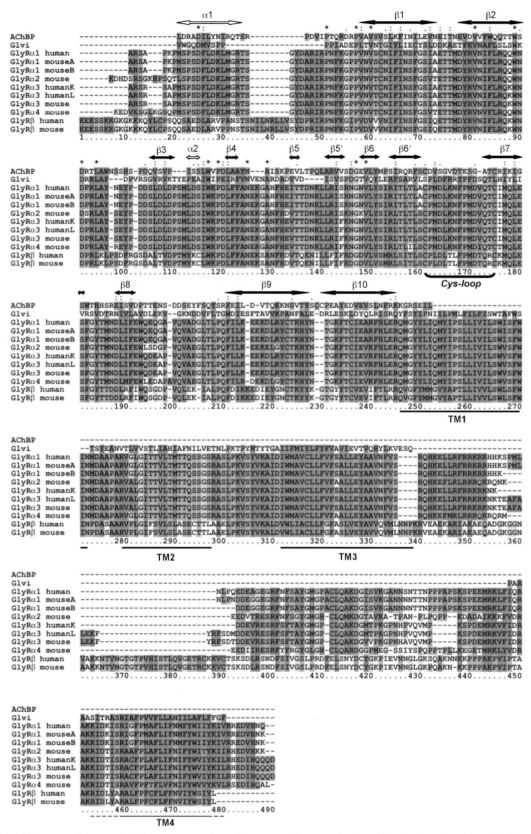

Fig. 3.1.3.1 Amino acid sequence alignment of Cys-loop receptor variants. The sequence alignment of various GlyR subunit variants from either human (human GlyR α1 NP_00162.2; human GlyRα3K NP_001036008, human GlyRα3L NP_006520.2, human GlyRβ P48167) or mouse (mouse GlyRα1 NP_065238, the variants A and B refer to two GlyRα1 splice variants described by Malosio *et al.*, [1991]; mouse GlyRα2 NP_906272.1; mouse GlyRα3 NP_536686.2; mouse GlyRα4 CAA53468; mouse GlyRβ NP_034428) together with the acetylcholine binding protein (AchBP) isolated from *Lymnea stagnalis* (2BG9) and an homologous prokaryotic proton-gated ion channel isolated from the cyanobacterium *Gloeobacter violaceus* (Glvi). The sequence shown represents the mature proteins. The characteristic Cys-loop of this protein family is indicated. Transmembrane domains (TM) are marked below the sequence by black bars. According to the X-ray structure of the AchBP the α-helices of the N-terminal domain are shown with open arrows, β-sheets 1 to 10 are represented by solid arrows. Small and hydrophobic amino acids (AVMILW) are shown in blue, acidic residues (DE) are magenta, basic residues (KR) are red, hydroxyl + amine residues (QSTN) are shown with a green background, FYH are dark blue, prolines are yellow, cysteines are light red and glycine residues are light orange. The alignment was performed using clustalW2 (http://www.ebi.ac.uk/Tools/clustalw2/index.html). See also colour plate section.

homomeric receptor complexes when expressed in recombinant systems (Table 3.1.3.1). In heteromeric GlyR protein complexes from native CNS tissue, the α-subunits are indispensable for functionality. However, although originally described as a key feature of the α-subunits, Grudzinska *et al.* recently demonstrated that the β-subunit contributes to the formation of the ligand binding cavity at the heteromeric subunit interface (Schmieden and Betz, 1995; Breitinger *et al.*, 2004a; Grudzinska *et al.*, 2005). Key residues important for glycine binding include oppositely charged amino acids located at the + and − sides of the GlyR interface (R65 and E157 in the α-subunit, which are conserved in the β-subunit) (Grudzinska *et al.*, 2005). Using recombinant tandem constructs joining the subunits in the order of α1β or βα1, analysis of the GlyR complex stochiometry revealed that at least one β–β interface

exists. This might provide a unique pharmacological target site of the GlyR family (Figure 3.1.3.2b).

Developmental regulation

During development, the GlyR variant prevailing in the embryonic and neonatal spinal cord, $GlyR_N$, is replaced by the adult isoform $GlyR_A$, which comprizes α1 and β subunits. The neonatal isoform, $GlyR_N$, is a homopentamer formed by five α2 subunits whose transcription is downregulated after birth. The developmental shift from the embryonic receptor complex to an adult receptor form starts around P8 and is almost complete at P21 (Becker *et al.*, 1988). Reciprocal to the α2 subunit expression, the expression of α1 and α3 subunits is upregulated (Malosio *et al.*, 1991b; Harvey *et al.*, 2004) during postnatal development of rodent spinal cord. As a

Table 3.1.3.1 Physiological and pharmacological properties of recombinantly expressed homomeric and heteromeric GlyR receptor complexes

Receptor subtype	Agonist pharmacology	Antagonist pharmacology	Single channel conductances	References
α1 human	*HEK293 cells:* EC50 glycine 18 ± 2 μM, EC50 β-alanine 52 ± 4 μM, EC50 taurine 153 ± 43 μM; *oocytes:* EC50 glycine 290 ± 20 μM, EC50 taurine 3600 ± 160 μM	*HEK293 cells:* IC_{50} strychnine 130 nM, IC_{50} picrotoxinin 9 μM, IC_{50} CTB 2.6 ± 0.7 μM; *oocytes:* IC_{50} strychnine 36 ± 10 nM	86 pS,* 64 pS, 46 pS, 30 pS, 18 pS,[†]	Grenningloh *et al.* (1990b); Pribilla *et al.* (1992); Bormann *et al.* (1993); Langosch *et al.* (1994); Rundström *et al.* (1994); Lynch *et al.* (1997); Rajendra *et al.* (1997); Grudzinska *et al.* (2005)
α1 rat	*Oocytes:* EC_{50} glycine 260 ± 20 μM, EC_{50} β-alanine 730 ± 70 μM, EC_{50} taurine 2200 ± 400 μM	*Ooyctes:* IC_{50} strychnine 37 ± 7 nM	75 pS,* 59 pS, 43 pS, 25 pS, 15 pS,[†]	Grenningloh *et al.* (1990b); Kuhse *et al.* (1990); Takahashi *et al.* (1992)
α1human/β human	*HEK293 cells:* EC_{50} glycine 74 ± 7 μM; *oocytes:* EC_{50} glycine 188 ± 14 μM;	*Oocytes:* IC_{50} strychnine 32 ± 8 nM	44 pS,* 29 pS, 20 pS,[†]	Bormann *et al.* (1993); Handford *et al.* (1996); Lewis *et al.* (1998); Grudzinska *et al.* (2005)
α1human/β rat	*HEK293 cells:* EC_{50} glycine 48 ± 17 μM; *oocytes:* EC_{50} glycine 380 ± 70 μM	*HEK293 cells:* IC_{50} strychnine 110 nM, IC_{50} picrotoxinin >1000 μM, IC_{50} CTB 1.3 ± 1.0 μM		Pribilla *et al.* (1992); Bormann *et al.* (1993); Langosch *et al.* (1994); Rundström *et al.* (1994)
α2 human	*HEK293 cells:* EC_{50} glycine 112 ± 23 μM; *oocytes:* EC_{50} glycine 310 ± 10 μM, EC_{50} β-alanine 3200 ± 700 μM, EC_{50} taurine 6000 ± 300 μM;	*Oocytes:* IC_{50} strychnine 32 ± 3 nM; *HEK293 cells:* IC_{50} strychnine 180 nM, IC_{50} picrotoxinin 6 μM, IC_{50} CTB >>20 μM	111 pS,* 91 pS, 66 pS, 48 pS, 36 pS[†] 23 pS[†]	Grenningloh *et al.* (1990b); Pribilla *et al.* (1992); Bormann *et al.* (1993); Rundström *et al.* (1994); Wick *et al.* (1999)
α2 rat	*Oocytes:* EC_{50} glycine 290 ± 20 μM, EC_{50} β-alanine 2800 ± 420 μM, EC_{50} taurine 3700 ± 530 μM	*Oocytes:* IC_{50} strychnine 50 ± 7 nM	88 pS,* 72 pS,* 42 pS, 24 pS	Kuhse *et al.* (1990); Schmieden *et al.* (1992); Takahashi *et al.* (1992)
α2 human/β rat	*HEK293 cells:* EC_{50} glycine 95 ± 34 μM	*HEK293 cells:* IC_{50} strychnine 120 nM, IC_{50} picrotoxinin 300 μM, IC_{50} CTB 2.8 ± 2.5 μM	112 pS, 80 pS,[†] 54 pS,* 36 pS	Pribilla *et al.* (1992); Bormann *et al.* (1993); Rundström *et al.* (1994)
α3 human	*HEK293 cells:* α3K EC_{50} glycine 64 ± 14 μM, α3L EC_{50} glycine 54 ± 12 μM	*HEK293 cells:* IC_{50} strychnine 200 nM		Nikolic *et al.* (1998)
α3 rat	*HEK293 cells:* EC_{50} glycine 60 μM, *transfected hippocampal neurons:* EC_{50} glycine 70 ± 9 μM	*HEK293 cells:* IC_{50} picrotoxinin 5 μM, *transfected hippocampal neurons:* IC_{50} strychnine 123 ± 28 nM	105 pS,* 85 pS. 62 pS, 42 pS, 30 pS, 20 pS[†]	Pribilla *et al.* (1992); Bormann *et al.* (1993); Meier *et al.* (2005)
α3 rat/β rat		*HEK293 cells:* IC_{50} strychnine 150 nM, IC_{50} picrotoxinin >1000 μM	48 pS,* 34 pS, 23 pS	Pribilla *et al.* (1992); Bormann *et al.* (1993)
α4 mouse	*HEK293 cells:* EC_{50} glycine 120 μM, EC_{50} β-alanine 500 μM, EC_{50} taurine 1120 μM	*HEK293 cells:* IC_{50} strychnine 30 nM, IC_{50} nipecotic acid 0.7 mM		Harvey *et al.* (2000)

* main state conductance; [†] state of lowest frequency; CTB, cyanotriphenylborate.

consequence, homomeric α2 complexes are replaced by heteromeric adult receptor complexes (GlyR$_A$), yielding a 2α:3β stoichiometry of GlyR subunits. This may apply to either α1/β or α3/β combinations (Breitinger and Becker, 1998; Grudzinska et al., 2005). The β subunit is a ubiquitous polypeptide that has been detected inside as well as outside the central nervous system (Oertel et al., 2007). Transcripts of the β-subunit were found very early in development displaying signals around E14 (Grenningloh et al., 1990a; Malosio et al., 1991b), and exist in at least two major splice variants (Oertel et al., 2007).

Topology of transmembrane domains

The characteristic topology of the Cys-loop superfamily receptors is of four transmembrane domains which are connected by one short intracellular, one short extracellular, and one large intracellular loop which exhibits the highest diversity both amongst the different GlyRs and with other Cys-loop receptor subunits (Figures 3.1.3.1 and 3.1.3.2a). While TM2 forms the ion channel pore, the N- and the C-termini are localized extracellularly. Transmembrane domains are structurally important elements for the overall folding of the protein: most GlyR transmembrane domains are confined by the positively charged amino acid arginine delineating the intracellular side. Mutagenesis studies have shown that arginine residues bordering the intracellular faces of transmembrane segments represent determinants essential for correct folding and membrane integration. The aliphatic part of the arginine residues is thought to favour localization in the hydrophobic environment of the bilayer, whereas the positively charged amino- and guanidino-groups might interact with negatively

charged phosphate groups of the membrane (de Planque et al., 1999). The import of the nascent polypeptide chain may be viewed as a 'snorkelling' process where positively charged amino acids appear to be important determinants for a correct transmembrane topology. Besides folding, transmembrane domains are defined interaction sites: Transmembrane domain 2 forms the ion channel determining the Cl⁻ conductance of the receptor family. As recognised in several cases of the human hereditary motor disorder, hyperekplexia, point mutations in the *GLRAI* gene, mainly localised within the ion channel pore or adjacent loop structures, result in disturbances of receptor physiology or assembly (Lynch, 2004).

The first conclusive structural information for the superfamily of Cys-loop receptors was obtained from a paradigmatic study on the X-ray structure of the acetylcholine binding protein from *Lymnea stagnalis*, implying an immunoglobulin-like fold (Brejc et al., 2001). Recently, the X-ray structure of an nAChR α1 subunit construct bound to α-bungarotoxin was solved with a resolution of 1.94 Å (Dellisanti et al., 2007) as well as that of a prokaryotic pentameric ligand-gated ion channel from the bacterium *Erwinia chrysanthemi* (ELIC) (Hilf and Dutzler, 2008). A homopentameric organization was also demonstrated for another prokaryotic proton-gated ion channel isolated from the cyanobacterium *Gloeobacter violaceus* (Bocquet et al., 2007).

To date, a more detailed investigation of the structure of transmembrane compartments of eukaryotic Cys-loop channels was possible only by electron microscopy. Using this technique on intact postsynaptic membranes from the electric organ of the *Torpedo* ray, similar structural data on the N-terminus were observed for the

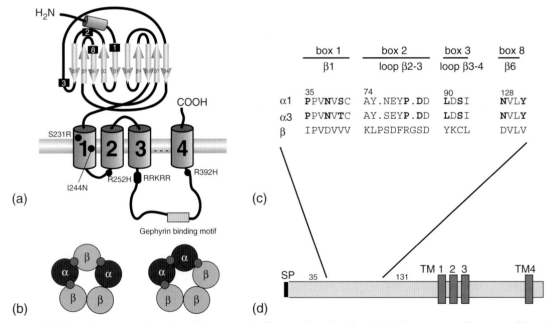

Fig. 3.1.3.2 Topology of the glycine receptor. A single subunit (a) is composed of four TMs (cyan barrels 1–4), which are connected by two small loops (TM1-2, TM2-3), and a long intracellular loop (TM3-4). In homology to the acetylcholine binding protein, the N-terminus shows a well-organized immunoglobulin-like fold starting with a short α-helical stretch (cyan barrel) followed by 10 β-sheets (yellow arrows). The C-terminus is short and extracellular. All residues or domains important for assembly are coloured in violet. N-terminal assembly boxes 1, 2, 3, and 8 are shown with rectangles. Violet circles correspond to single point mutations described for recessive cases of hyperekplexia (S231R, I244N, R252H, R392H) resulting in disturbed assembly behaviour. Another motif important for correct membrane integration is the RRKRR motif in the TM3-4 loop (GlyRα1 and GlyRα3). The gephyrin-binding motif within the TM3-4 loop of the GlyR β-subunit only is shown in light blue. (b) GlyRs form heteropentameric ion channels consisting of α- (blue circles) and β-subunits (yellow circles) with a 2α:3β stoichiometry thereby creating homo- (α–α) or heteromeric (α–β) ligand binding interfaces for glycine (red circles), and at least one β–β interface. (c and d) Bar diagram of a single subunit (d) starting with the signal peptide (SP) with the four TMs illustrated as blue boxes. In (c) the N-terminal assembly boxes important for homo-oligomerization localized between amino acid positions 35–131 are shown (boxes 1, 2, 3, and 8) with key amino acids determinants highlighted in bold. See also colour plate section.

nicotinic acetylcholine receptor (Unwin, 2005). To mimic the synaptic release of acetylcholine a rapid spray-freezing technique was employed to trap the open channel form. The analysis suggested that the mechanism of channel opening includes a ligand-induced rotation of the inner β-sheets towards the pore-lining helices, causing them to switch cooperatively to an alternative conformation. This change in conformation is thought to destabilize the gate and to open the channel pore (Unwin, 2003, 2005). A wealth of structural data has recently been gathered, producing an insight into structure–function relationships that are more detailed than previous studies, furnishing our understanding of the functional properties of this ion channel superfamily. However, the structure of the intracellular TM3–4 loop region remains unsolved.

3.1.3.3 Physiology and pharmacology

Following ligand binding, a conformational change leads to opening of the intrinsic ion channel. The open–closed state transition of the ion channel is termed 'gating'. For the hyperekplexia mutants GlyR α1(K276E), and GlyR α1(Q266H) located within the TM2-3 extracellular loop (Table 3.1.3.2) and TM2 (Lewis et al., 1998; Moorhouse et al., 1999), an impaired channel gating was reported with a reduced duration of channel openings. Mutations within the TM1-2 intracellular loop of the glycine receptor α1 subunit resulted in both an altered receptor desensitization (Breitinger et al., 2001; Breitinger and Becker, 2002a) and ion permeation (Lee et al., 2003). Similar effects on gating have been observed with GABA$_A$ receptor and nicotinic acetylcholine receptor mutants (Akk and Steinbach, 2000; Wotring et al., 2003).

Besides glycine, the amino acids taurine and β-alanine are able to induce channel opening with most variants of the inhibitory receptor complex (Table 3.1.3.1) (Laube et al., 2002). A comparison of the agonist activities of α- and β-amino acids revealed that β-amino acids act as partial agonists (Schmieden and Betz, 1995). At low concentrations of β-amino acids, glycine responses are inhibited whereas the β-amino acids elicit significant responses in their own right at high concentrations. Strychnine is a classical antagonist of the GlyR that causes hypertonic motor disturbances and increased muscle tone (Becker et al., 1992). Further selective GlyR antagonists have emerged from studies on quinolinic acid derivatives (Schmieden et al., 1996). With regard to allosteric modulation of the GlyR complex, there are only a few agents known, with most of those described lacking specificity. Anaesthetics and alcohols have been shown not only to potentiate GABA$_A$ receptors but also GlyRs. The potency of n-alcohols at recombinant GlyRs increases with alkyl chain length (Mascia et al., 1996). The amino acid residue S267 in the α1 subunit plays a unique role in mediating the effects of ethanol (Ye et al., 1998; Mascia et al., 2000). Interestingly, a case of hyperekplexia has been reported carrying the mutation S267N which results in a reduction in agonist affinity and efficacy at the GlyR and has the potential to modify therapeutic drug responses in affected patients (Becker et al., 2008).

Volatile halogenated hydrocarbons, including halothane, enflurane, isoflurane, methoxyflurane, and seroflurane, exert a potentiating effect on glycine-gated currents when glycine is applied at low concentrations (Downie et al., 1996). In experiments utilizing *Xenopus* oocytes and transgenic mice (O'Shea et al., 2004), propofol was found to potentiate GlyR responses and thereby reversed the functional and behavioural effects of hyperekplexia. Steroids constitute another class of modulatory agents that act on GlyRs as well as GABA$_A$ receptors. Steroid inhibition of GlyRs may be either fully or partially competitive, suggesting the existence of heterogenous binding sites. Moreover, a potentiating effect of some steroids has also been described, which is dependent on subunit composition (Laube et al., 2002; Maksay et al., 2002).

The divalent cation Zn^{2+} acts as an important modulator of glycinergic neurotransmission (Hirzel et al., 2006). Interestingly, in addition to acting on inhibitory neurotransmitter receptors, Zn^{2+} also modulates current responses mediated by excitatory receptors (Erreger and Traynelis, 2005; Rachline et al., 2005). The physiological relevance of zinc modulation of GlyRs is supported by studies in transgenic mice with a knock-in mutation, D80A, in the *Glra1* gene that results in an elimination of the potentiating effect of Zn^{2+}. *Glra1*(D80A) homozygous mice developed a neuromotor phenotype. Physiological characterization of these mice confirmed that the potentiation by Zn^{2+} was completely abolished, arguing

Table 3.1.3.2 Mutant alleles in human *hyperekplexia*

GlyRα1 residue	Mode of inheritance	Functional effect	References
S231R	Recessive	Reduced glycine affinity, low surface expression	Humeny et al., 2002
I244N/A	Recessive	Reduced glycine affinity, decreased I$_{max}$ increased desensitization rate	Rees et al., 1994
P250T	Dominant	Reduced channel conductance, increased desensitization rate	Saul et al., 1999
R252H/R392H	Recessive	Almost absent from cell surface	Vergouwe et al., 1999, Rea et al., 2002
V260M	Dominant	Increased EC$_{50}$ for agonists	Del Giudice et al., 2001, Castaldo et al., 2004
Q266H	Dominant	Reduced glycine sensitivity, reduced open probability	Milani et al., 1996, Moorhouse et al., 1999
S267N	Dominant	Reduced glycine sensitivity, no sensitivity to ethanol	Becker et al., 2008
R271Q/L	Dominant	Reduced glycine sensitivity, reduced single channel conductance	Ryan et al., 1992, Shiang et al., 1993, Rajendra et al., 1994
K276E	Dominant	Reduced glycine sensitivity, reduced open probability	Elmslie et al., 1996, Lynch et al., 1997
Y279C	Dominant	Reduced glycine sensitivity, reduced I$_{max}$ values	Shiang et al., 1995, Lynch et al., 1997
Deletion exon 1–6	Recessive	Non-functional	Brune et al., 1996, Becker et al., 2006

for an important role of endogenous Zn^{2+} in the regulation of glycinergic neurotransmission (Hirzel *et al.*, 2006).

Channel blocking agents

Generated by hydrolysis of the plant alkaloid picrotoxin, the agent picrotoxinin acts as an antagonist of $GABA_A$ receptors. It has been also used to distinguish between homo- and heteromeric GlyR receptors: receptors composed of either $\alpha1/\beta$ or $\alpha3/\beta$ are more resistant to picrotoxinin relative to $\alpha1$ or $\alpha3$ homomers which are efficiently blocked (Table 3.1.3.1) (Pribilla *et al.*, 1992). Recently, ginkgolide B was described to non-competitively block glycine-induced currents of heteromeric GlyRs with a 20-fold higher affinity for heteromeric than homomeric receptor complexes (Kondratskaya *et al.*, 2005). A discrimination of different α-homomers has become possible by use of the compounds cyanotriphenylborate (CTB), and possibly, cyclothiazide (Zhang *et al.*, 2008). Cyanotriphenylborate is a bulky molecule that acts as a use-dependent inhibitor of $\alpha1$-containing receptors whereas $\alpha2$-containing receptors are largely unaffected (Rundström *et al.*, 1994).

Biosynthetic assembly and trafficking of GlyRs

During the last decade, our knowledge of the assembly and trafficking of GlyRs has progressively increased. Through site-directed mutagenesis of the N-terminus, several critical assembly cassettes of GlyRs were determined (Figures 3.1.3.2c and 3.1.3.2d). In addition, single amino acids adjacent to transmembrane domains, as well as in the TM3-4 loop, contribute to correct assembly of GlyRs within the plasma membrane (Figure 3.1.3.2) (Griffon *et al.*, 1999; Rea *et al.*, 2002; Sadtler *et al.*, 2003). Following oligomerization of the ion channel complex, the clustering of GlyR surface receptors is a highly dynamic process mediated by a number of interacting proteins. The TM3-4 loop of the GlyR β-subunit harbours a binding motif for the postsynaptic anchoring protein gephyrin (Figures 3.1.3.2a and 3.1.3.3) (Kirsch *et al.*, 1993; Kneussel and Betz, 2000b). Interaction between the β subunit and gephyrin mediates clustering of the GlyR receptor complex at the postsynaptic membrane (Feng *et al.*, 1998). Synaptic clustering, however, does not imply a rigid or inflexible complex: recently, it was demonstrated that synaptic GlyRs (Meier *et al.*, 2001; Dahan *et al.*, 2003), as well as α-amino-3-hydroxy-5-methyl-4-isoxazole propionic acid (AMPA) (Takahashi *et al.*, 2003) and N-methyl-D-aspartate (NMDA) receptors (Groc *et al.*, 2004), are highly mobile and are exchanged through lateral diffusion between distinct synapses. With a novel imaging technique utilizing semiconductor quantum dots, Triller and Choquet have shown that glycine receptors are mobile both inside and outside the synaptic complexes (Triller and Choquet, 2005). Direct visualization has changed our view of the organization of the neuronal membrane such that lateral diffusion of receptors is now recognized as a key parameter in the regulation of synaptic function and plasticity (Dahan *et al.*, 2003; Hanus *et al.*, 2006).

During biogenesis, post-translational protein assembly is a multistep process, which includes both folding reactions and the post-translational modifications that follow synthesis of the protein and its insertion into the endoplasmic reticulum (ER) membrane. Within each subfamily of the ligand-gated ion channels only certain combinations of subunits are oligomerized and targeted towards the plasma membrane (Figures 3.1.3.2 and 3.1.3.3). Studies on nicotinic acetylcholine receptors using fluorescently labelled proteins introduced into the TM3-4 loop, allowed a study of the subunit composition and regulation of AChR at the plasma membrane without disturbing functional properties of the subunit (Nashmi *et al.*, 2003; Drenan *et al.*, 2008). Similarly, $GABA_A$ receptor assembly occurs within the ER and involves interactions with chaperone molecules, e.g. calnexin or immunoglobulin heavy-chain-binding protein. For $GABA_A$ receptors, only specific subunit combinations produce functional surface receptors, with a fixed stoichiometry of 2α, 2β and 1γ (Sieghart *et al.*, 1999; Baumann *et al.*, 2001; Moss and Smart, 2001; Bollan *et al.*, 2003; Stephenson, 2006). In contrast, other subunit combinations are thought to be retained within the ER and degraded. Thus, receptor assembly occurs by defined pathways to limit the diversity of $GABA_A$ receptors. The key to understanding the control of receptor diversity was the identification of assembly signals capable of distinguishing between other subunit partners. An amino acid stretch (MEYTIDVFFRQS corresponding to $\alpha1$) localized in the $\beta2$-sheet of the N-terminal domain was shown to be an important assembly signal and regulator of $GABA_A$ receptor diversity. Replacing this motif in the $\alpha1$ subunit with the sequence of the $GABA_C$ receptor polypeptide $\rho1$ abolished surface expression of $\alpha1$. Furthermore, it has been shown that this motif plays a role in promoting assembly of $\alpha1$ with β-subunits (Taylor *et al.*, 2000).

In the GlyR, several assembly cassettes have been identified within the N-terminal domain (Figure 3.1.3.2). In addition, there are residues at the ends of transmembrane domains, which also play a role either as topogenic signals or signals for transport from the ER to golgi compartments. A chimeric protein consisting of the N-terminal domain from the β-subunit and the C-terminal domain from an $\alpha1$-subunit fails to form homooligomeric channels. Replacement of defined residues within the so-called assembly boxes 1, 2, 3, and 8 restored homooligomeric GlyR formation (Pro35, Asn38, Ser40 from box 1; Pro79 from box 2; Leu90 and Ser92 from box 3; Asn125 and Tyr128 from box 8; Figure 3.1.3.2) (Kuhse *et al.*, 1993; Griffon *et al.*, 1999). Some of the mutants failed to exit the ER. In contrast, they were transported towards the membrane when complexed with wild-type $\alpha1$ subunits. These data were supported by glycosidase studies, which determined an Endo H resistance only in the presence of $\alpha1$-wild-type. This observation argues for a requirement of glycosylation and oligomerization before ER exit (Griffon *et al.*, 1999) and in fact N-glycosylation seems to be a prerequisite for exit from the ER. Corresponding to the structure of the acetylcholine binding protein, the residues responsible for assembly lie within the first and the sixth β-sheet and in loops $\beta2$–3 and $\beta3$–4, the latter with a short α-helical stretch (Brejc *et al.*, 2001). ER retention consensus motifs, 'RRRRR' or 'RXR', which play roles both in ER retention and in retrieval of glutamate receptors and ATP-sensitive potassium channels, have not been described for GlyRs (Scott *et al.*, 2001; Ma and Jan, 2002; Coussen and Mulle, 2006; Nasu-Nishimura *et al.*, 2006). A motif present in the TM3-4 loop, harbouring the basic residues 'RRKRR' appears to be involved in trafficking of glycine receptors to the plasma membrane (Figures 3.1.3.1 and 3.1.3.2) (Sadtler *et al.*, 2003). Mutating two of these residues resulted in an ER translocation and improper integration into the plasma membrane.

Arginine residues localized either at the intra- or extracellular faces of transmembrane domains are important determinants of GlyR function. Mutations of Arg271 within the TM2-3 loop at the extracellular face of the human $\alpha1$-subunit polypeptide underly a variant of hyperekplexia which is characterized by a modified physiology of the ion channel (Maksay *et al.*, 2002; 2008). In contrast, arginine mutations at the intracellular face of the plasma membrane are associated with disturbances in protein assembly. The importance of Arg252 (Figure 3.1.3.2a) for correct biogenesis has become evident from a recombinant study, where ER retention resulted when the arginine was replaced by either a leucine or a glutamine. In a case of compound heterozygosity of hyperekplexia, two arginines at the intracellular face of TM2 (R252) and TM4 (R392) are affected (Vergouwe *et al.*, 1999). Upon recombinant expression, both of the mutant receptor variants were virtually absent from the cell surface, consistent with a lack of Cl⁻ channel formation. The importance of these cytoplasmic arginine residues for membrane integration suggests that their positive charges act as topogenic signals during protein biogenesis.

Studies on the isolated N-terminal domain of the human α1-subunit (α1 residues 1–219) revealed a membrane association indistinguishable from that of full-length wild-type α1 polypeptides. These data suggest that the N-terminus harbours not only amino acid residues important for oligomerization, but also for membrane association of the N-terminal domain, despite the absence of any recognized canonical transmembrane domains (Breitinger *et al.*, 2004b).

Synaptic clustering of the glycine receptor: role of gephyrin and downstream interacting proteins

Following assembly of the ion channel complex proper, the clustering of GlyR surface receptors is modulated by binding of the anchor protein gephyrin to the β-subunit of the GlyR complex. The binding motif for gephyrin consists of a stretch of 18 amino acids localized within the TM3-4 loop of the GlyR β-subunit (Kirsch *et al.*, 1993; Kneussel and Betz, 2000b). Analysis of tagged α1, α2, and an α1 subunit harbouring portions of the gephyrin-binding motif derived from the β polypeptide revealed different subcellular distribution patterns in heterologous cells and neurons. In neurons, GlyR clusters composed of either α1 or α2 polypeptides, are formed independently, without interaction with gephyrin. In contrast, when expressed in COS7 cells, receptor clustering depends on the presence of the β-gephyrin binding motif (Meier *et al.*, 2000).

Through assembly and disassembly, conformational changes of gephyrin are thought to regulate dynamic receptor movements. Gephyrin is a highly structured protein resulting from the fusion of two anchestral genes. Homologous genes exist in *E. coli* encoding the enzymes involved in the biosynthesis of the molybdenum cofactor (Stallmeyer *et al.*, 1995). On the basis of the homology of the gephyrin E-domain to the MoeA protein from *E. coli*, whose crystal structure was solved, a dimeric gephyrin E was postulated (Schrag *et al.*, 2001; Xiang *et al.*, 2001). The bacterial enzymes possess a high homology to the G- and E-domains of gephyrin, which are joined by a highly flexible linker region of 170 amino acid residues in length. Recently, the X-ray structure of the G-domain of gephyrin showed the formation of interacting homotrimers (Sola *et al.*, 2001). Among the different gephyrin domains, binding to the GlyR β subunit is conveyed by E-domains, which harbour both high- and low-affinity binding sites for the β polypeptide (Schrader *et al.*, 2004). Based on these observations, Sola *et al.* proposed a model for gephyrin scaffolding at the synapse (Sola *et al.*, 2004): The trimeric appearance of G-domains of gephyrin, as observed with isolated domain constructs (Sola *et al.*, 2001), may prevent

dimerization via its E-domains. Accordingly, an as yet unknown signal may trigger a conformational change of gephyrin and initiate the dimerization of E-domains from different trimers, respectively. Dimeric gephyrin E may dock onto β-subunits derived from two different GlyR complexes or onto two β-subunits of the same GlyR complex. Thus, E-domain dimerization may lead to the formation of a submembraneous hexagonal lattice and thereby stabilize GlyR clusters. If this holds true, conformational changes of gephyrin could regulate dynamic receptor movements at the synapse, and thus synaptic plasticity (Sola *et al.*, 2004; Saiyed *et al.*, 2007). Meier *et al.* showed that clustering is a two-step process, where a phase of gephyrin-independent formation of cellular GlyR clusters precedes that of gephyrin-mediated postsynaptic accumulation of clusters (Figure 3.1.3.3) (Meier *et al.*, 2000). These studies characterize the dynamic oligomerization behaviour of gephyrin as an important modulator of postsynaptic plasticity.

The interactions of the various gephyrin domains with GlyRβ predict a complex hexagonal lattice that clusters the receptor complex to the cytoskeleton. By virtue of these arrangements, gephyrin may interact with its binding partners tubulin, collybistin, profilin, and RAFT1 (Figure 3.1.3.3) (Feng *et al.*, 1998; Stallmeyer *et al.*, 1999; Kneussel and Betz, 2000a). Among these, collybistin is important for the oligomerization step of gephyrin (Kins *et al.*, 2000). The brain specific guanine nucleotide exchange factor collybistin plays an essential role at selected GABAergic synapses (Papadopoulos *et al.*, 2007). Indeed, collybistin deficiency leads to a loss of gephyrin-dependent GABA$_A$ receptor subtypes from postsynaptic sites of specific brain regions, but not of GlyRs. At the behavioural level, collybistin null mutant mice exhibit increased anxiety and impaired spatial learning. The protein profilin has been shown to link the gephyrin/GlyR complexes to the microfilament system (Giesemann *et al.*, 2003) and dynein acts as an important mediator in this interplay (Figure 3.1.3.3) (Fuhrmann *et al.*, 2002; Maas *et al.*, 2006). Blockade of dynein and dynein–gephyrin interactions, interfered with the retrograde recruitment of gephyrin from dendrites of cultured hippocampal neurons. In this model, the GlyR-gephyrin-dynein triple complex contributes to the dynamics of postsynaptic GlyRs and underlies the regulation of synaptic strength.

Cycling and mobility of GlyRs

Membrane receptors and submembrane proteins are subject to fast and constant protein turnover, processes which underlie synaptogenesis and synaptic plasticity. Studies on receptor trafficking have long relied on immunocytochemistry. Initial experiments analysing the transport route of newly synthesized GlyRs in transfected neurons had made use of thrombin-cleavable extracellular tags (Rosenberg *et al.*, 2001). Subsequently, to study receptor movements, latex beads coupled via antibodies to the N-terminal domains of different receptors, e.g. glycine receptors, metabotropic glutamate receptors, and AMPA receptors were used (Meier *et al.*, 2001; Borgdorff and Choquet, 2002; Serge *et al.*, 2002). Until recently, however, direct visualization of receptor movements at synaptic sites was limited by the size of latex beads employed as markers (200–500 nm). Real time imaging of receptor movements within the lipid bilayer and during transport by cytoskeletal elements has become possible with the introduction of inorganic semiconductor quantum dots (QDs). These tags are fluorescent probes of 5–10 nm that possess a higher photostability than fluorophores

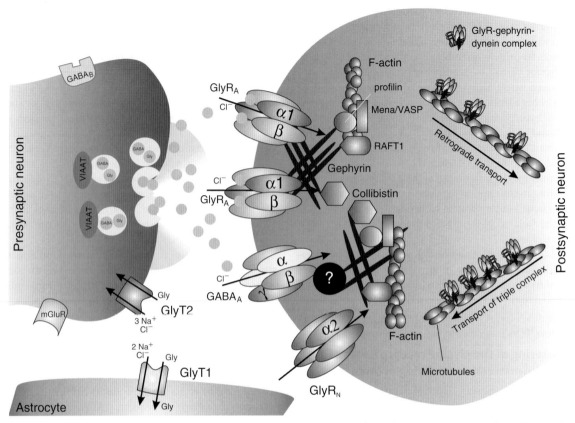

Fig. 3.1.3.3   The inhibitory synapse. GlyRs are localized at the postsynaptic membrane. The neonatal isoform of the receptor complex is formed by $\alpha 2$ subunits (GlyR$_N$, violet pentamers). Around day 14 a developmental switch leads to an upregulation of the adult heteromeric isoform of $2\alpha/3\beta$ subunit (GlyR$_A$, green pentamers) and replaces GlyR$_N$. GABA$_A$ receptors are also found at the postsynaptic site (illustrated as $2\alpha/2\beta/\gamma$ subunits). Upon excitation of the presynaptic neuron, the ligand glycine (orange circles) is released from the presynaptic terminal in a Ca^{2+}-dependent manner. Following binding to the GlyR at the postsynaptic terminal, Cl$^-$ conductance is increased. Within the presynaptic terminal synthesized glycine as well as GABA (light green circles) is stored in small synaptic vesicles (yellow circles) by the vesicular inhibitory amino acid transporter (VIATT, red). The termination of glycinergic transmission is mediated by the Na$^+$- and Cl$^-$-dependent transporters GlyT1 and GlyT2, which are responsible for glycine reuptake. In contrast to GlyT2, which is localized at the presynaptic membrane, GlyT1 is at the glial plasma membrane. In glial cells, glycine is degraded in mitochondria via the glycine cleavage system. At the postsynaptic synapse the GlyR is clustered via gephyrin (blue ellipse) that binds selectively to the β-subunit (light green ellipses). Direct binding of gephyrin to GABA$_A$ receptors has not been demonstrated, however, current data suggest that an as yet unknown factor (black circle, ?) link GABA$_A$ receptors and gephyrin at the postsynaptic site. Gephyrin domains form a lattice structure and anchor the receptor complexes to the cytoskeleton. Several proteins have been identified to be involved in this interaction. One of these proteins is RAFT1 (violet), which is a translational regulator. Collibistin (red hexagon) is involved as a GDP/GTP exchange factor thereby regulating actin dynamics via Rho-type GTPases. RAFT1 as well as collibistin also interact with GABA$_A$ receptors. The gephyrin scaffold is attached to microfilaments mediated by the same proteins for both receptor classes. Profilin (light brown circles) is an actin-binding protein involved in cytoskeleton dynamics. Mena (mammalian enabled)/VASP (Vasodilator-stimulated phosphoprotein) (green rectangle) facilitates actin polymerization. Parallel F-actin strands (grey ellipses) twist around each other in a helical formation, giving rise to microfilaments of the cytoskeleton. Moreover, gephyrin also plays a role in the transport of GlyR complexes towards the membrane (triple complex—GlyR/gephyrin/dynein) and in retrograde transport. These processes are mediated by microtubules (grey ellipses). Here, gephyrin links the GlyR complex with the microtubule-dependent dynein motor complex (see triple complex). See also colour plate section.

(Dahan *et al.*, 2003; Triller and Choquet, 2005). Single QD tracking was used to study the exchange of GlyRs between synaptic and extrasynaptic membranes, as well as the rapid lateral dynamics of GlyRs, AMPARs, and NMDARs. The latter is a parameter important for the numbers of receptors at synapses (Dahan *et al.*, 2003; Tardin *et al.*, 2003; Groc *et al.*, 2004). In the fluid mosaic model, transmembrane proteins, such as GlyRs, that integrate in the cell surface membrane, are thought to freely diffuse within its phospholipid bilayer. In native cell membranes, however, the lateral mobility is restricted by regulated interactions with submembrane proteins, e.g. gephyrin. Furthermore, GlyR dynamics at synaptic and extrasynaptic sites are modulated by microtubules and F-actin. Depolymerizing agents, e.g. latrunculin and nocodazole which

disrupt the cytoskeleton, decrease GlyR numbers and gephyrin at synapses. Concomitantly, a disruption of the cytoskeleton resulted in increased GlyR exchange between synaptic and extrasynaptic compartments as well as decreased receptor dwell times (Charrier *et al.*, 2006). From these studies, it is becoming apparent that the number of receptors at the synapse is regulated by an equilibrium between lateral diffusion within the membrane and a transient anchoring by scaffolding proteins (Ehrensperger *et al.*, 2007).

Receptor turnover/degradation

Synaptic receptor turnover is regulated by variants of gephyrin. The gephyrin variant harboring cassette 5, which prevents G-domain trimerization, is not able to stabilize GlyR clusters at the cell surface.

Moreover, it depletes synapses of both GlyR and full-length gephy-rin (Bedet *et al.*, 2006). Ubiquitin-mediated proteolysis is also a critical component in protein turnover (Hegde, 2004). Targeted protein internalization and degradation via the ubiquitin–proteas-ome pathway plays a vital role in synaptic plasticity (Cremona *et al.*, 2003), including neurotransmitter receptor retrieval by endo-cytosis and degradation (Büttner *et al.*, 2001; Christianson and Green, 2004). Intracellular ubiquitin molecules are able to covalently attach themselves to specific lysine chains on the cyto-plasmatic protein surface. The TM3-4 loop of the GlyR α1 subunit contains 13 lysine residues that serve as attack points for ubiquiti-nation. Ubiquitination precipitates the internalization and degra-dation of α1 GlyRs expressed at the cell surface (Büttner *et al.*, 2001). Following ubiquitination, the receptor molecule is nicked into a 35 kDa fragment, which has been shown to be glycosylated, and a 13 kDa C-terminal fragment. Cleavage of ubiquinated α1 subunits has been attributed to an endocytotic pathway and is inhibited by concanamycin, an inhibitor of the vesicular H$^+$-ATPase, which blocks the acidification of endosomes and lyso-somes. These results were confirmed by the complete inhibition of cleavage with the serine protease inhibitor, phenylmethylsulfonyl fluoride, which does not raise lysosomal pH values (Büttner *et al.*, 2001). Dysfunctional ubiquitination in the nervous system has been implicated in Parkinson's, Huntington's, and Alzheimer's as well as other neurological diseases (Ehlers, 2004).

3.1.3.4 Relevance to disease state
Human hyperekplexia: stiff baby syndrome

Mutant alleles of the GlyR α1 subunit gene (*GLRA1; OMIM# 138491*) give rise to hypertonic motor disorders in humans (Table 3.1.3.2) (Brune *et al.*, 1996; Humeny *et al.*, 2002; Breitinger *et al.*, 2004b; Sola *et al.*, 2004) and mice (Table 3.1.3.3) (Buckwalter *et al.*, 1994; Kingsmore *et al.*, 1994; Mülhardt *et al.*, 1994; Saul *et al.*, 1994). Characteristic symptoms of the human neurological disorder, hyperekplexia (startle disease, stiff baby syndrome, STHE; OMIM# 149400), include muscle stiffness in the neonate, as well as exagger-ated startle responses to unexpected acoustic and tactile stimuli that persist into adulthood (Tijssen *et al.*, 1995). Dominant forms of startle disease (STHE) have been linked to *GLRA1* mutations that result in amino acid substitutions within transmembrane region 2. This transmembrane region lines the inner wall of the ion channel pore, or the adjacent extra- and intracellular regions (Becker *et al.*, 2000; Breitinger and Becker, 2002b; Lynch, 2004). In addition to missense alleles, a null allele of *GLRA1* resulting in recessive hyperekplexia has been characterized as a deletion of exons 1–7 (Brune *et al.*, 1996; Becker *et al.*, 2006). Recessive traits of the disease are associated with the amino acid exchanges I244N, S231R, and a case of compound heterozygosity R252H/R392H (Rees *et al.*, 1994; Vergouwe *et al.*, 1999; Humeny *et al.*, 2002). Recombinant expression of these receptors resulted in a complete

Table 3.1.3.3 Pathological alleles of the mouse *Glra* and *Glrb* genes

Neuronal subunit	Gene	Chromosomal localization	Mutation	Pathological mechanism	Phenotype	Symptoms	References
α1	Glra1	11 B1.3	Spasmodic	Point mutation A52S, decrease in glycine affinity	Mild	Stiff posture, rapid tremor, impaired righting reflexes	Lane *et al.*, 1987 Buckwalter *et al.*, 1993 Ryan *et al.*, 1994, Plappert *et al.*, 2001
α1	Glra1	11 B1.3	Oscillator	Microdeletion, frameshift, premature truncation of the protein, loss of GlyR function	Lethal at 3 weeks of age	Massive tremor, hypertonia	Buckwalter *et al.*, 1994 Kling *et al.*, 1997
α1	Glra1	11 B1.3	Knockin	D80A	Mild	Increased muscular tone and inducible tremor, fertile, normal lifespan	Hirzel *et al.*, 2006
α1	Glra1	11 B1.3	Knockin	S267Q	Lethal at 3 weeks of age	Seizures, hindfeet clenching, increase in the acoustic startle responses	Findlay *et al.*, 2003
α2	Glra2	X F5	Knockout	Glra2$^{-/-}$	No	Normal weight and life expectancy and fertile, glycinergic sIPSCs of narrow-field amacrine cells had slow kinetics	Derry and Barnard, 1991 Young-Pearse *et al.*, 2006 Weiss *et al.*, 2008
α3	Glra3	8 B2	Knockout	Glra3$^{-/-}$	No	No notable postural abnormality, tremor or startle response, lack of pain sensitization	Kingsmore *et al.* 1994 Harvey *et al.*, 2004
β	Glrb	3E–3F1	Spastic	Insertion of LINE1 element into intron 6, abberant splicing, excess of truncated transcripts, decrease of full-length transcripts, <10 %	Mild	Hypertonia, myoclonia, pronounced startle responses, impaired righting reflexes	Becker *et al.*, 1986 Kingsmore *et al.*, 1994 Mülhardt *et al.*, 1994

loss of functional GlyRs or pronounced decreases in affinities for the receptor agonists glycine, β-alanine, and taurine. Given the pronounced reduction in I_{max}, recessive mutants are thought primarily to result in alterations of receptor numbers due to changes in assembly or degradation, rather than ion channel properties. Additional genes associated with hereditary hyperekplexia include the glycine receptor β-subunit gene (*GLRB*), the gephyrin gene (*GPHRN*), and the gene encoding glycine transporter subtype 2 (*SLC6A5*) (Gomeza *et al.*, 2003; Rees *et al.*, 2006).

Mouse models to study the pathophysiology of hyperekplexia

Glycine receptor defects associated with hypertonic motor disorders are also known in mice (Table 3.1.3.3). Indeed, mouse models are useful tools to study the pathological mechanisms underlying hyperekplexia ('startle disease'). The recessive mouse mutants *spasmodic*, *oscillator* and *spastic* harbour pathological alleles of the *Glra* and *Glrb* genes. In the glycine receptor β subunit mutant *spastic*, the intronic insertion of a LINE1 element is associated with abberant splicing. As a consequence, the content of full-length β transcript and the number of GlyRs in the central nervous system is highly reduced (Kingsmore *et al.*, 1994; Mülhardt *et al.*, 1994). The allelic variants *spasmodic* and *oscillator* affect the α1-subunit of the GlyR. *Spasmodic* displays the mildest phenotype, correlated with a reduced affinity for glycine due to a point mutation in the far N-terminal region (A52S) (Buckwalter *et al.*, 1994; Saul *et al.*, 1994). In contrast, the *oscillator* mutation is lethal. *Oscillator* mice are not able to produce functional receptor complexes. Lethality around postnatal day 21 matches the developmental shift from α2 to α1 containing receptors with downregulation of α2 and upregulation of α1 GlyRs after birth. The *oscillator* mutation within the α1 subunit gene sequence is a 7 bp deletion resulting in a frameshift. Depending on use of an alternative splice site in the TM3-4 loop, two different transcripts result (Buckwalter *et al.*, 1994; Saul *et al.*, 1994). When translated, neither transcript is able to generate functional ion channels. Rather, mutant polypeptides are likely to be degraded within the ER compartment. In contrast to the human null allele, where exons 1 to 7 are deleted, there is no mechanism to compensate for the complete loss of functional GlyRs in *oscillator* mice. Thus, *oscillator* is an excellent model for interaction studies with regard to assembly and GlyR complex formation.

3.1.3.5 Conclusion

While the inhibitory glycine receptor serves as an interesting model for molecular physiology and pathology of synaptic ion channels, its pharmacology is characterized by a paucity of agonist ligands. Further structural studies may pave the way for novel ligand design and enable the therapeutic targeting of the GlyR.

Acknowledgements:

Work in the authors laboratory is supported by Deutsche Forschungsgemeinschaft, European Union, and Fonds der Chemischen Industrie. Drs Christoph Kluck, Heike Meiselbach and Nima Melzer are gratefully acknowledged for helpful discussions.

References

Akk G and Steinbach JH (2000). Structural elements near the C-terminus are responsible for changes in nicotinic receptor gating kinetics following patch excision. *J Physiol* **527**, 405–417.

Baumann SW, Baur R and Sigel E (2001). Subunit arrangement of gamma-aminobutyric acid type A receptors. *J Biol Chem* **276**, 36275–36280.

Becker CM (1995). Review: glycine receptors: molecular heterogeneity and implications for disease. *Neuroscientist* **1**, 130–141.

Becker CM, Hermans-Borgmeyer I, Schmitt B *et al.* (1986). The glycine receptor deficiency of the mutant mouse spastic: evidence for normal glycine receptor structure and localization. *J Neurosci* **6**, 1358–1364.

Becker CM, Hoch W and Betz H (1988). Glycine receptor heterogeneity in rat spinal cord during postnatal development. *Embo J* **7**, 3717–3726.

Becker CM, Schmieden V, Tarroni P *et al.* (1992). Isoform-selective deficit of glycine receptors in the mouse mutant spastic. *Neuron* **8**, 283–289.

Becker K, Breitinger HG and Becker CM (2000). The inhibitory glycine receptor as a model of hereditary channelopathies. In F. Lehmann-Horn and K. Jurkat-Rott, eds., *Channelopathies*. Amsterdam, Elsevier, pp. 199–222.

Becker K, Breitinger HG, Humeny A *et al.* (2008). The novel hyperekplexia allele GLRA1(S267N) affects the ethanol site of the glycine receptor. *Eur J Hum Genet* **16**, 223–228.

Becker K, Hohoff C, Schmitt B *et al.* (2006). Identification of the microdeletion breakpoint in a GLRA1null allele of Turkish hyperekplexia patients. *Hum Mutat* **27**, 1061–1062.

Bedet C, Bruusgaard JC, Vergo S *et al.* (2006). Regulation of gephyrin assembly and glycine receptor synaptic stability. *J Biol Chem* **281**, 30046–30056.

Betz H and Laube B. (2006). Glycine receptors: recent insights into their structural organization and functional diversity. *J Neurochem* **97**, 1600–1610.

Bocquet N, Prado de Carvalho L, Cartaud J *et al.* (2007). A prokaryotic proton-gated ion channel from the nicotinic acetylcholine receptor family. *Nature* **445**, 116–119.

Bollan K, King D, Robertson LA *et al.* (2003). GABA(A) receptor composition is determined by distinct assembly signals within alpha and beta subunits. *J Biol Chem* **278**, 4747–4755.

Borgdorff AJ and Choquet D (2002). Regulation of AMPA receptor lateral movements. *Nature* **417**, 649–653.

Bormann J, Rundström N, Betz H *et al.* (1993). Residues within transmembrane segment M2 determine chloride conductance of glycine receptor homo- and hetero-oligomers. *Embo J* **12**, 3729–3737.

Breitinger HG and Becker CM (1998). The inhibitory glycine receptor: prospects for a therapeutic orphan? *Curr Pharm Des* **4**, 315–334.

Breitinger HG and Becker CM (2002a). Statistical coassembly of glycine receptor alpha1 wildtype and the hyperekplexia mutant alpha1(P250T) in HEK 293 cells: impaired channel function is not dominant in the recombinant system. *Neurosci Lett* **331**, 21–24.

Breitinger HG and Becker CM (2002b). The inhibitory glycine receptor—simple views of a complicated channel. *Chembiochem* **3**, 1042–1052.

Breitinger HG, Lanig H, Vohwinkel C *et al.* (2004a). Molecular dynamics simulation links conformation of a pore-flanking region to hyperekplexia-related dysfunction of the inhibitory glycine receptor. *Chem Biol* **11**, 1339–1350.

Breitinger HG, Villmann C, Becker K *et al.* (2001). Opposing effects of molecular volume and charge at the hyperekplexia site alpha 1(P250) govern glycine receptor activation and desensitization. *J Biol Chem* **276**, 29657–29663.

Breitinger U, Breitinger HG, Bauer F *et al.* (2004b). Conserved high-affinity ligand-binding and membrane association in the native and refolded extracellular domain of the human glycine receptor alpha1-subunit. *J Biol Chem* **279**, 1627–1636.

Brejc K, van Dijk WJ, Klaassen RV *et al.* (2001). Crystal structure of an ACh-binding protein reveals the ligand-binding domain of nicotinic receptors. *Nature* **411**, 269–276.

Brune W, Weber RG, Saul B, von Knebel Doeberitz M *et al.* (1996). A GLRA1 null mutation in recessive hyperekplexia challenges the functional role of glycine receptors. *Am J Hum Genet* **58**, 989–997.

Buckwalter MS, Cook SA, Davisson MT *et al.* (1994). A frameshift mutation in the mouse alpha 1 glycine receptor gene (Glra1) results in progressive neurological symptoms and juvenile death. *Hum Mol Genet* **3**, 2025–2030.

Buckwalter MS, Testa CM, Noebels JL *et al.* (1993). Genetic mapping and evaluation of candidate genes for spasmodic, a neurological mouse mutation with abnormal startle response. *Genomics* **17**, 279–286.

Büttner C, Sadtler S, Leyendecker A *et al.* (2001). Ubiquitination precedes internalization and proteolytic cleavage of plasma membrane-bound glycine receptors. *J Biol Chem* **276**, 42978–42985.

Castaldo P, Stefanoni P, Miceli F *et al.* (2004). A novel hyperekplexia-causing mutation in the pre-transmembrane segment 1 of the human glycine receptor alpha1 subunit reduces membrane expression and impairs gating by agonists. *J Biol Chem* **279**, 25598–25604.

Charrier C, Ehrensperger MV, Dahan M *et al.* (2006). Cytoskeleton regulation of glycine receptor number at synapses and diffusion in the plasma membrane. *J Neurosci* **26**, 8502–8511.

Christianson JC and Green WN (2004). Regulation of nicotinic receptor expression by the ubiquitin–proteasome system. *Embo J* **23**, 4156–4165.

Coussen F and Mulle C (2006). Kainate receptor-interacting proteins and membrane trafficking. *Biochem Soc Trans* **34**, 927–930.

Cremona O, Collesi C and Raiteri E (2003). Protein ubiquitylation and synaptic function. *Ann N Y Acad Sci* **998**, 33–40.

Dahan M, Levi S, Luccardini C *et al.* (2003). Diffusion dynamics of glycine receptors revealed by single-quantum dot tracking. *Science* **302**, 442–445.

de Planque MR, Kruijtzer JA, Liskamp RM *et al.* (1999). Different membrane anchoring positions of tryptophan and lysine in synthetic transmembrane alpha-helical peptides. *J Biol Chem* **274**, 20839–20846.

del Giudice EM, Coppola G, Bellini G *et al.* (2001). A mutation (V260M) in the middle of the M2 pore-lining domain of the glycine receptor causes hereditary hyperekplexia. *Eur J Hum Genet* **9**, 873–876.

Dellisanti CD, Yao Y, Stroud JC *et al.* (2007). Crystal structure of the extracellular domain of nAChR alpha1 bound to alpha-bungarotoxin at 1.94 A resolution. *Nat Neurosci* **10**, 953–962.

Derry JM and Barnard PJ (1991). Mapping of the glycine receptor alpha 2-subunit gene and the GABAA alpha 3-subunit gene on the mouse X chromosome. *Genomics* **10**, 593–597.

Downie DL, Hall AC, Lieb WR *et al.* (1996). Effects of inhalational general anaesthetics on native glycine receptors in rat medullary neurones and recombinant glycine receptors in Xenopus oocytes. *Br J Pharmacol* **118**, 493–502.

Drenan RM, Nashmi R, Imoukhuede PI *et al.* (2008). Subcellular trafficking, pentameric assembly and subunit stoichiometry of neuronal nicotinic ACh receptors containing fluorescently-labeled {alpha}6 and {beta}3 subunits. *Mol Pharmacol* **73**, 27–41.

Ehlers MD (2004). Deconstructing the axon: Wallerian degeneration and the ubiquitin–proteasome system. *Trends Neurosci* **27**, 3–6.

Ehrensperger MV, Hanus C, Vannier C *et al.* (2007). Multiple association states between glycine receptors and gephyrin identified by SPT analysis. *Biophys J* **92**, 3706–3718.

Elmslie FV, Hutchings SM, Spencer V *et al.* (1996). Analysis of GLRA1 in hereditary and sporadic hyperekplexia: a novel mutation in a family cosegregating for hyperekplexia and spastic paraparesis. *J Med Genet* **33**, 435–436.

Erreger K and Traynelis SF (2005). Allosteric interaction between zinc and glutamate binding domains on NR2A causes desensitization of NMDA receptors. *J Physiol* **569**, 381–393.

Feng G, Tintrup H, Kirsch J *et al.* (1998). Dual requirement for gephyrin in glycine receptor clustering and molybdoenzyme activity. *Science* **282**, 1321–1324.

Findlay GS, Phelan R, Roberts MT *et al.* (2003). Glycine receptor knock-in mice and hyperekplexia-like phenotypes: comparisons with the null mutant. *J Neurosci* **23**, 8051–8059.

Fuhrmann JC, Kins S, Rostaing P *et al.* (2002). Gephyrin interacts with dynein light chains 1 and 2, components of motor protein complexes. *J Neurosci* **22**, 5393–5402.

Giesemann T, Schwarz G, Nawrotzki R *et al.* (2003). Complex formation between the postsynaptic scaffolding protein gephyrin, profilin, and Mena: a possible link to the microfilament system. *J Neurosci* **23**, 8330–8339.

Gomeza J, Ohno K and Betz H (2003). Glycine transporter isoforms in the mammalian central nervous system: structures, functions and therapeutic promises. *Curr Opin Drug Discov Devel* **6**, 675–682.

Grenningloh G, Pribilla I, Prior P *et al.* (1990a). Cloning and expression of the 58 kd beta subunit of the inhibitory glycine receptor. *Neuron* **4**, 963–970.

Grenningloh G, Schmieden V, Schofield PR *et al.* (1990b). Alpha subunit variants of the human glycine receptor: primary structures, functional expression and chromosomal localization of the corresponding genes. *Embo J* **9**, 771–776.

Griffon N, Büttner C, Nicke A *et al.* (1999). Molecular determinants of glycine receptor subunit assembly. *Embo J* **18**, 4711–4721.

Groc L, Heine M, Cognet L *et al.* (2004). Differential activity-dependent regulation of the lateral mobilities of AMPA and NMDA receptors. *Nat Neurosci* **7**, 695–696.

Grudzinska J, Schemm R, Haeger S *et al.* (2005). The beta subunit determines the ligand-binding properties of synaptic glycine receptors. *Neuron* **45**, 727–739.

Handford CA, Lynch JW, Baker E *et al.* (1996). The human glycine receptor beta subunit: primary structure, functional characterisation and chromosomal localisation of the human and murine genes. *Brain Res Mol Brain Res* **35**, 211–219.

Hanus C, Ehrensperger MV and Triller A (2006). Activity-dependent movements of postsynaptic scaffolds at inhibitory synapses. *J Neurosci* **26**, 4586–4595.

Harvey RJ, Schmieden V, Von Holst *et al.* (2000). Glycine receptors containing the alpha4 subunit in the embryonic sympathetic nervous system, spinal cord and male genital ridge. *Eur J Neurosci* **12**, 994–1001.

Harvey, RJ, Depner UB, Wässle H *et al.* (2004). GlyR alpha3: an essential target for spinal PGE2-mediated inflammatory pain sensitization. *Science* **304**, 884–887.

Hegde AN (2004). Ubiquitin–proteasome-mediated local protein degradation and synaptic plasticity. *Prog Neurobiol* **73**, 311–357.

Hilf RJ and Dutzler R (2008). X-ray structure of a prokaryotic pentameric ligand-gated ion channel. *Nature* **452**, 375–379.

Hirzel K, Müller U, Latal AT *et al.* (2006). Hyperekplexia phenotype of glycine receptor alpha1 subunit mutant mice identifies $Zn(2^+)$ as an essential endogenous modulator of glycinergic neurotransmission. *Neuron* **52**, 679–690.

Humeny A, Bonk T, Becker K *et al.* (2002). A novel recessive hyperekplexia allele GLRA1 (S231R): genotyping by MALDI-TOF mass spectrometry and functional characterisation as a determinant of cellular glycine receptor trafficking. *Eur J Hum Genet* **10**, 188–196.

Kingsmore SF, Giros B, Suh D *et al.* (1994). Glycine receptor beta-subunit gene mutation in spastic mouse associated with LINE-1 element insertion. *Nat Genet* **7**, 136–141.

Kins S, Betz H and Kirsch J (2000). Collybistin, a newly identified brain-specific GEF, induces submembrane clustering of gephyrin. *Nat Neurosci* **3**, 22–29.

Kirsch J, Wolters I, Triller A *et al.* (1993). Gephyrin antisense oligonucleotides prevent glycine receptor clustering in spinal neurons. *Nature* **366**, 745–748.

Kling C, Koch M, Saul B *et al.* (1997). The frameshift mutation oscillator (Glra1(spd-ot)) produces a complete loss of glycine receptor alpha1-polypeptide in mouse central nervous system. *Neuroscience* **78**, 411–417.

Kneussel M and Betz H (2000a). Clustering of inhibitory neurotransmitter receptors at developing postsynaptic sites: the membrane activation model. *Trends Neurosci* **23**, 429–435.

Kneussel M and Betz H (2000b). Receptors, gephyrin and gephyrin-associated proteins: novel insights into the assembly of inhibitory postsynaptic membrane specializations. *J Physiol* **525**, 1–9.

Kondratskaya EL, Betz H, Krishtal OA *et al.* (2005). The beta subunit increases the ginkgolide B sensitivity of inhibitory glycine receptors. *Neuropharmacology* **49**, 945–951.

Kuhse J, Kuryatov A, Maulet Y et al. (1991). Alternative splicing generates two isoforms of the alpha 2 subunit of the inhibitory glycine receptor. FEBS Lett 283, 73–77.

Kuhse J, Laube B, Magalei D et al. (1993). Assembly of the inhibitory glycine receptor: identification of amino acid sequence motifs governing subunit stoichiometry. Neuron 11, 1049–1056.

Kuhse J, Schmieden V and Betz H (1990). A single amino acid exchange alters the pharmacology of neonatal rat glycine receptor subunit. Neuron 5, 867–873.

Lane PW, Ganser AL, Kerner AL et al. (1987). Spasmodic, a mutation on chromosome 11 in the mouse. J Hered 78, 353–356.

Langosch D, Herbold A, Schmieden V et al. (1993). Importance of Arg-219 for correct biogenesis of alpha 1 homooligomeric glycine receptors. FEBS Lett 336, 540–544.

Langosch D, Laube B, Rundström N et al. (1994). Decreased agonist affinity and chloride conductance of mutant glycine receptors associated with human hereditary hyperekplexia. Embo J 13, 4223–4228.

Laube B, Maksay G, Schemm R et al. (2002). Modulation of glycine receptor function: a novel approach for therapeutic intervention at inhibitory synapses? Trends Pharmacol Sci 23, 519–527.

Lee DJ, Keramidas A, Moorhouse AJ et al. (2003). The contribution of proline 250 (P-2') to pore diameter and ion selectivity in the human glycine receptor channel. Neurosci Lett 351, 196–200.

Lewis TM, Sivilotti LG, Colquhoun D et al. (1998). Properties of human glycine receptors containing the hyperekplexia mutation alpha1(K276E), expressed in Xenopus oocytes. J Physiol 507, 25–40.

Lynch JW (2004). Molecular structure and function of the glycine receptor chloride channel. Physiol Rev 84, 1051–1095.

Lynch JW, Rajendra S, Pierce KD et al. (1997). Identification of intracellular and extracellular domains mediating signal transduction in the inhibitory glycine receptor chloride channel. Embo J 16, 110–120.

Ma D and Jan LY (2002). ER transport signals and trafficking of potassium channels and receptors. Curr Opin Neurobiol 12, 287–292.

Maas C, Tagnaouti N, Loebrich S et al. (2006). Neuronal cotransport of glycine receptor and the scaffold protein gephyrin. J Cell Biol 172, 441–451.

Maksay G, Biro T and Laube B (2002). Hyperekplexia mutation of glycine receptors: decreased gating efficacy with altered binding thermodynamics. Biochem Pharmacol 64, 285–288.

Maksay G, Biro T, Laube B et al. (2008). Hyperekplexia mutation R271L of alpha(1) glycine receptors potentiates allosteric interactions of nortropeines, propofol and glycine with [(3)H]strychnine binding. Neurochem Int 52, 235–40.

Malosio ML, Grenningloh G, Kuhse J et al. (1991a). Alternative splicing generates two variants of the alpha 1 subunit of the inhibitory glycine receptor. J Biol Chem 266, 2048–2053.

Malosio ML, Marqueze-Pouey B, Kuhse J et al. (1991b). Widespread expression of glycine receptor subunit mRNAs in the adult and developing rat brain. Embo J 10, 2401–2409.

Mascia MP, Gong DH, Eger EI et al. (2000). The anesthetic potency of propanol and butanol versus propanethiol and butanethiol in alpha1 wild-type and alpha1(S267Q) glycine receptors. Anesth Analg 91, 1289–1293.

Mascia MP, Machu TK and Harris RA (1996). Enhancement of homomeric glycine receptor function by long-chain alcohols and anaesthetics. Br J Pharmacol 119, 1331–1336.

Meier J, Meunier-Durmort C, Forest C et al. (2000). Formation of glycine receptor clusters and their accumulation at synapses. J Cell Sci 113 (Pt 15), 2783–2795.

Meier J, Vannier C, Serge A et al. (2001). Fast and reversible trapping of surface glycine receptors by gephyrin. Nat Neurosci 4, 253–260.

Meier JC, Henneberger C, Melnick I et al. (2005). RNA editing produces glycine receptor alpha3(P185L), resulting in high agonist potency. Nat Neurosci 8, 736–744.

Milani N, Dalpra, L, del Prete A et al. (1996). A novel mutation (Gln266->His) in the alpha 1 subunit of the inhibitory glycine-receptor gene (GLRA1) in hereditary hyperekplexia. Am J Hum Genet 58, 420–422.

Moorhouse AJ, Jacques P, Barry PH et al. (1999). The startle disease mutation Q266H, in the second transmembrane domain of the human glycine receptor, impairs channel gating. Mol Pharmacol 55, 386–395.

Moss SJ and Smart TG (2001). Constructing inhibitory synapses. Nat Rev Neurosci 2, 240–250.

Mülhardt C, Fischer M, Gass P et al. (1994). The spastic mouse: aberrant splicing of glycine receptor beta subunit mRNA caused by intronic insertion of L1 element. Neuron 13, 1003–1015.

Nashmi R, Dickinson ME, McKinney S et al. (2003). Assembly of alpha4beta2 nicotinic acetylcholine receptors assessed with functional fluorescently labeled subunits: effects of localization, trafficking, and nicotine-induced upregulation in clonal mammalian cells and in cultured midbrain neurons. J Neurosci 23, 11554–11567.

Nasu-Nishimura Y, Hurtado D, Braud S et al. (2006). Identification of an endoplasmic reticulum-retention motif in an intracellular loop of the kainate receptor subunit KA2. J Neurosci 26, 7014–7021.

Nikolic Z, Laube B, Weber RG et al. (1998). The human glycine receptor subunit alpha3. Glra3 gene structure, chromosomal localization, and functional characterization of alternative transcripts. J Biol Chem 273, 19708–19714.

Oertel J, Villmann C, Kettenmann H et al. (2007). A novel glycine receptor beta subunit splice variant predicts an unorthodox transmembrane topology. Assembly into heteromeric receptor complexes. J Biol Chem 282, 2798–2807.

O'Shea SM, Becker L, Weiher H et al. (2004). Propofol restores the function of 'hyperekplexic' mutant glycine receptors in Xenopus oocytes and mice. J Neurosci 24, 2322–2327.

Papadopoulos T, Korte M, Eulenburg V et al. (2007). Impaired GABAergic transmission and altered hippocampal synaptic plasticity in collybistin-deficient mice. Embo J 26, 3888–3899.

Plappert CF, Pilz PK, Becker K et al. (2001). Increased sensitization of acoustic startle response in spasmodic mice with a mutation of the glycine receptor alpha1-subunit gene. Behav Brain Res 121, 57–67.

Pribilla I, Takagi T, Langosch D et al. (1992). The atypical M2 segment of the beta subunit confers picrotoxinin resistance to inhibitory glycine receptor channels. Embo J 11, 4305–4311.

Rachline J, Perin-Dureau F, Le Goff A et al. (2005). The micromolar zinc-binding domain on the NMDA receptor subunit NR2B. J Neurosci 25, 308–317.

Rajendra S, Lynch JW and Schofield PR (1997). The glycine receptor. Pharmacol Ther 73, 121–146.

Rajendra S, Lynch JW, Pierce KD et al. (1994). Startle disease mutations reduce the agonist sensitivity of the human inhibitory glycine receptor. J Biol Chem 269, 18739–18742.

Rea R, Tijssen MA, Herd C et al. (2002). Functional characterization of compound heterozygosity for GlyRalpha1 mutations in the startle disease hyperekplexia. Eur J Neurosci 16, 186–196.

Rees MI, Andrew M, Jawad S et al. (1994). Evidence for recessive as well as dominant forms of startle disease (hyperekplexia) caused by mutations in the alpha 1 subunit of the inhibitory glycine receptor. Hum Mol Genet 3, 2175–2179.

Rees MI, Harvey K, Pearce BR et al. (2006). Mutations in the gene encoding GlyT2 (SLC6A5) define a presynaptic component of human startle disease. Nat Genet 38, 801–806.

Rosenberg M, Meier J, Triller A et al. (2001). Dynamics of glycine receptor insertion in the neuronal plasma membrane. J Neurosci 21, 5036–5044.

Rundström N, Schmieden V, Betz H et al. (1994). Cyanotriphenylborate: subtype-specific blocker of glycine receptor chloride channels. Proc Natl Acad Sci USA 91, 8950–8954.

Ryan SG, Buckwalter MS, Lynch JW et al. (1994). A missense mutation in the gene encoding the alpha 1 subunit of the inhibitory glycine receptor in the spasmodic mouse. Nat Genet 7, 131–135.

Ryan SG, Dixon MJ, Nigro MA *et al.* (1992). Genetic and radiation hybrid mapping of the hyperekplexia region on chromosome 5q. *Am J Hum Genet* **51**, 1334–1343.

Sadtler S, Laube B, Lashub A *et al.* (2003). A basic cluster determines topology of the cytoplasmic M3–M4 loop of the glycine receptor alpha1 subunit. *J Biol Chem* **278**, 16782–16790.

Saiyed T, Paarmann I, Schmitt B *et al.* (2007). Molecular basis of gephyrin clustering at inhibitory synapses: role of G- and E-domain interactions. *J Biol Chem* **282**, 5625–5632.

Saul B, Kuner T, Sobetzko D *et al.* (1999). Novel GLRA1 missense mutation (P250T) in dominant hyperekplexia defines an intracellular determinant of glycine receptor channel gating. *J Neurosci* **19**, 869–877.

Saul B, Schmieden V, Kling C *et al.* (1994). Point mutation of glycine receptor alpha 1 subunit in the spasmodic mouse affects agonist responses. *FEBS Lett* **350**, 71–76.

Schmieden V and Betz H (1995). Pharmacology of the inhibitory glycine receptor: agonist and antagonist actions of amino acids and piperidine carboxylic acid compounds. *Mol Pharmacol* **48**, 919–927.

Schmieden V, Jezequel S and Betz H (1996). Novel antagonists of the inhibitory glycine receptor derived from quinolinic acid compounds. *Mol Pharmacol* **50**, 1200–1206.

Schmieden V, Kuhse J and Betz H (1992). Agonist pharmacology of }neonatal and adult glycine receptor alpha subunits: identification of amino acid residues involved in taurine activation. *Embo J* **11**, 2025–2032.

Schrader N, Kim EY, Winking J *et al.* (2004). Biochemical characterization of the high-affinity binding between the glycine receptor and gephyrin. *J Biol Chem* **279**, 18733–18741.

Schrag JD, Huang W, Sivaraman J *et al.* (2001). The crystal structure of *Escherichia coli* MoeA, a protein from the molybdopterin synthesis pathway. *J Mol Biol* **310**, 419–431.

Scott DB, Blanpied TA, Swanson GT *et al.* (2001). An NMDA receptor ER retention signal regulated by phosphorylation and alternative splicing. *J Neurosci* **21**, 3063–3072.

Serge A, Fourgeaud L, Hemar A *et al.* (2002). Receptor activation and homer differentially control the lateral mobility of metabotropic glutamate receptor 5 in the neuronal membrane. *J Neurosci* **22**, 3910–3920.

Shiang R, Ryan SG, Zhu YZ *et al.* (1995). Mutational analysis of familial and sporadic hyperekplexia. *Ann Neurol* **38**, 85–91.

Shiang R, Ryan SG, Zhu YZ *et al.* (1993). Mutations in the alpha 1 subunit of the inhibitory glycine receptor cause the dominant neurologic disorder, hyperekplexia. *Nat Genet* **5**, 351–358.

Sieghart W, Fuchs K, Tretter V *et al.* (1999). Structure and subunit composition of GABA(A) receptors. *Neurochem Int* **34**, 379–385.

Sola M, Bavro VN, Timmins J *et al.* (2004). Structural basis of dynamic glycine receptor clustering by gephyrin. *Embo J* **23**, 2510–2519.

Sola M, Kneussel M, Heck IS *et al.* (2001). X-ray crystal structure of the trimeric N-terminal domain of gephyrin. *J Biol Chem* **276**, 25294–25301.

Stallmeyer B, Nerlich A, Schiemann J *et al.* (1995). Molybdenum co-factor biosynthesis: the *Arabidopsis thaliana* cDNA cnx1 encodes a multifunctional two-domain protein homologous to a mammalian neuroprotein, the insect protein Cinnamon and three *Escherichia coli* proteins. *Plant J* **8**, 751–762.

Stallmeyer B, Schwarz G, Schulze J *et al.* (1999). The neurotransmitter receptor-anchoring protein gephyrin reconstitutes molybdenum cofactor biosynthesis in bacteria, plants, and mammalian cells. *Proc Natl Acad Sci USA* **96**, 1333–1338.

Stephenson FA (2006). Structure and trafficking of NMDA and GABAA receptors. *Biochem Soc Trans* **34**, 877–881.

Takahashi T, Momiyama A, Hirai K *et al.* (1992). Functional correlation of fetal and adult forms of glycine receptors with developmental changes in inhibitory synaptic receptor channels. *Neuron* **9**, 1155–1161.

Takahashi T, Svoboda K and Malinow R (2003). Experience strengthening transmission by driving AMPA receptors into synapses. *Science* **299**, 1585–1588.

Tardin C, Cognet L, Bats C *et al.* (2003). Direct imaging of lateral movements of AMPA receptors inside synapses. *Embo J* **22**, 4656–4665.

Taylor PM, Connolly CN, Kittler JT *et al.* (2000). Identification of residues within GABA(A) receptor alpha subunits that mediate specific assembly with receptor beta subunits. *J Neurosci* **20**, 1297–1306.

Tijssen MA, Shiang R, van Deutekom J *et al.* (1995). Molecular genetic reevaluation of the Dutch hyperekplexia family. *Arch Neurol* **52**, 578–582.

Triller A and Choquet D (2005). Surface trafficking of receptors between synaptic and extrasynaptic membranes: and yet they do move! *Trends Neurosci* **28**, 133–139.

Unwin N (2003). Structure and action of the nicotinic acetylcholine receptor explored by electron microscopy. *FEBS Lett* **555**, 91–95.

Unwin N (2005). Refined structure of the nicotinic acetylcholine receptor at 4A resolution. *J Mol Biol* **346**, 967–989.

Vergouwe MN, Tijssen MA, Peters AC *et al.* (1999). Hyperekplexia phenotype due to compound heterozygosity for GLRA1 gene mutations. *Ann Neurol* **46**, 634–638.

Villmann C and Becker CM (2007). On the hypes and falls in neuroprotection: targeting the NMDA receptor. *Neuroscientist* **13**, 594–615.

Weiss J, O'Sullivan GA, Heinze L *et al.* (2008). Glycinergic input of small-field amacrine cells in the retinas of wild-type and glycine receptor deficient mice. *Mol Cell Neurosci* **37**, 40–55.

Wick MJ, Bleck V, Whatley VJ *et al.* (1999). Stable expression of human glycine alpha1 and alpha2 homomeric receptors in mouse L(tk-) cells. *J Neurosci Methods* **87**, 97–103.

Wotring, VE, Miller TS and Weiss DS (2003). Mutations at the GABA receptor selectivity filter: a possible role for effective charges. *J Physiol* **548**, 527–540.

Xiang S, Nichols J, Rajagopalan Kv *et al.* (2001). The crystal structure of *Escherichia coli* MoeA and its relationship to the multifunctional protein gephyrin. *Structure* **9**, 299–310.

Ye Q, Koltchine VV, Mihic SJ *et al.* (1998). Enhancement of glycine receptor function by ethanol is inversely correlated with molecular volume at position alpha267. *J Biol Chem* **273**, 3314–3319.

Young-Pearse TL, Ivic L, Kriegstein AR *et al.* (2006). Characterization of mice with targeted deletion of glycine receptor alpha 2. *Mol Cell Biol* **26**, 5728–5734.

Zhang XB, Sun GC, Liu LY *et al.* (2008). Alpha2 subunit specificity of cyclothiazide inhibition on glycine receptors. *Mol Pharmacol* **73**, 1195–1202.

3.1.4 GABA$_A$ receptors

Hanns Möhler, Dietmar Benke, Uwe Rudolph and Jean-Marc Fritschy

3.1.4.1 Introduction

Neuronal inhibition in the brain is mainly mediated by gamma aminobutyric acid (GABA)ergic neurons which are largely local, functionally highly diverse and morphologically strikingly different. Inhibitory interneurons play a key role in regulating the activity pattern of neuronal circuits by controlling spike timing and sculpting neuronal rhythms. For instance, GABAergic interneurons are essential for neuronal oscillations in the β and γ bands of the electroencephalogram (EEG) which are considered to underlie cognitive functions such as object perception, selective attention, working memory, and consciousness. GABAergic neurons are also crucial for the formation and retrieval of cell assemblies required for encoding, consolidating and recalling information. The pronounced functional heterogeneity of GABAergic neurons requires a comparative heterogeneity of GABA$_A$ receptors to match the signal transduction kinetics with the functional needs of the respective circuits. More than a dozen genes provide the subunit infrastructure. GABA$_A$ receptor subtypes therefore vary in their subunit architecture, display distinct cellular expression patterns and subcellular localization, differ in their ability to undergo covalent modification and vary in their kinetic properties. Study of the GABA$_A$ receptor subtypes therefore provides multiple avenues into the analysis of behaviour and its modulation by drugs (Mody and Pearce, 2004; Möhler et al., 2005; Rudolph and Möhler, 2006; Möhler, 2007.)

3.1.4.2 Subunit architecture of GABA$_A$ receptors

GABA$_A$ receptors are heteropentameric chloride channels of the Cys-loop ligand-gated ion channel superfamily. Molecular cloning revealed a pronounced heterogeneity of the GABA$_A$ receptor based on numerous subunits (Sieghart, 1995; Barnard et al., 1998a; Barnard, 2001), which are divided by their sequence homology into three main classes (α1–6, β1–3, γ1–3) (Figure 3.1.4.1) and several more specialized forms (δ, ε, π, θ, ρ1–3). GABA receptors consisting exclusively of either ρ_1, ρ_2 or ρ_3 subunits were originally termed GABA$_C$ receptors, largely based on their homomeric structure (Bormann, 2000). However, since homomeric ionotropic GABA receptors operate as GABA-gated chloride channels the homomeric receptors are not considered to be a distinct class of receptors. The term "GABA$_C$ receptors" is not part of the International Union of basic and clinical pharmacology (IUPHAR) GABA receptor nomenclature (Barnard et al., 1998b).

All subunits exhibit a similar topology with a large extracellular N-terminal domain (~200 amino acids), four α-helical transmembrane segments (M1–M4), a large intracellular loop connecting transmembrane segments 3 and 4 and a short extracellular C-terminal sequence (Figure 3.1.4.2). Within the extracellular N-terminal domain all subunits contain a 15 amino acid-long disulfide-linked loop, which is characteristic for all members of the Cys-loop superfamily. The Cys-loop receptors are considered to fold into a similar structure (Galzi and Changeux, 1994). While the extracellular domain harbours the ligand binding sites, the intracellular loop is amenable to regulation by phosphorylation and receptor associated proteins (Sigel, 2002; Ernst et al., 2003; Kittler and Moss, 2003; Lüscher and Keller, 2004). Studies on recombinant GABA$_A$ receptors showed that fully functional receptors require the assembly of α, β and γ subunits in a stoichiometry of 2α, 2β and 1γ subunit (Im et al., 1995; Chang et al., 1996; Tretter et al., 1997; Farrar et al., 1999; Baumann et al., 2001), although, in a minority of receptors, the γ subunit can be replaced by the δ-subunit (Table 3.1.4.1).

Although there is considerable information on the contribution of specific amino acids to functional domains of GABA$_A$ receptors as derived from site-directed mutagenesis and substituted cysteine accessibility methods (for details see Akabas, 2004) no high-resolution structure is available today. The recent identification of the X-ray structure of a soluble acetylcholine binding protein (AChBP) from the snail *Lymnaea stagnalis* provided an initial template for the extracellular receptor domain (Brejc et al., 2001). Its low sequence identity (15–18 per cent), however, permitted only a preliminary estimate of the putative extracellular GABA$_A$ receptor structure. A refined electron microscopic structure of the nicotinic acetylcholine receptor (nAChR) from electric fish at 4.0 Å resolution (Unwin, 2005) (Figure 3.1.4.2) shed further light onto a

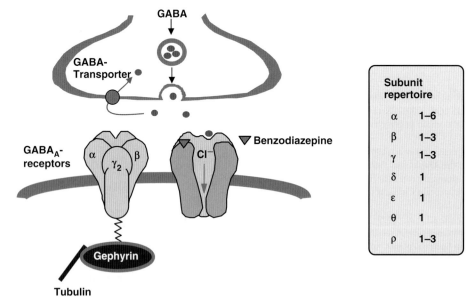

Fig. 3.1.4.1 Scheme of GABAergic synapse depicting major elements of signal transduction. The GABA$_A$ receptors are heteromeric membrane proteins which are linked by an as yet unknown mechanism to the synaptic anchoring protein gephyrin and the cytoskeleton. The binding sites for GABA and benzodiazepines are located at the interface of α/β and α/γ2 subunits, respectively. Synaptic GABA$_A$ receptors mediate phasic inhibition providing a rapid point-to-point communication for synaptic integration and control of rhythmic network activities. Extrasynaptic GABA$_A$ receptors (not shown) are activated from synaptic spillover or non-vesicular release of GABA. By mediating tonic inhibition they provide a maintenance level of reduction in neuronal excitability (Mody and Pearce, 2004; Möhler et al., 2005). See also colour plate section.

Table 3.1.4.1 GABA$_A$ receptor subtypes

Receptor subtype	Major subunit combination	Cellular and subcellular localization * [1, 2, 3, 4]	Pharmacology and behavioural activity[†‡§]		
α_1 GABA$_A$ receptor	$\alpha_1\beta_2\gamma_2$	Major receptor subtype (60%) Synaptic and extrasynaptic Present in most brain areas	Benzodiazepine-sensitive receptor. Mediates sedation and most of the anticonvulsant activity of classical benzodiazepines. Ligands of the benzodiazepine site with preferential affinity for α_1GABA$_A$ receptors include:		
			Zolpidem [5]	Hypnotic	
			Zaleplone [5]	Hypnotic	
			Indiplone [6]	Hypnotic	
α_2 GABA$_A$ receptor	$\alpha_2\beta_3\gamma_2$	Minor receptor subtype (15–20%) Synaptic localization, largely on soma and axon initial segment of principal cells, e.g. in cerebral cortex and hippocampus Present in central nucleus of amygdala [7]	Benzodiazepine-sensitive receptor. Mediates anxiolytic activity. Ligands of the benzodiazepine site with anxiolytic activity (although not selective for α_2GABA$_A$ receptors) include:		
			L-838 417 [8]	Anxiolytic	Comparable affinity at α_1, α_2, α_3, α_5 subtype Partial agonist at α_2, α_3, α_5 (not α_1) subtype
			Ocinaplon [9]	Anxiolytic	Comparable affinity at α_1, α_2, α_3, α_5 subtypes Partial agonist at α_2, α_3, α_5 subtypes, nearly full agonist at α_1
			SL 651 498 [10]	Anxiolytic	Agonist at α_2, α_3, partial agonist at α_1 and α_5 subtypes
			TPA 023 [11]	Anxiolytic	Partial agonist at α_2, α_3 subtypes, antagonist at α_1, α_5 subtypes
α_3 GABA$_A$ receptor	$\alpha_3\beta_n\gamma_2$	Minor receptor subtype	Benzodiazepine-sensitive receptor. Mediates anxiolytic activity at high receptor occupancy [12]. Ligands include:		
			TP 003 [12]	Anxiolytic	Partial agonist at α_3 subtype
			ELB 139 [13]	Anxiolytic	Receptor selectivity profile uncertain
			α_3 IA [14]	Anxiogenic	Weak inverse agonist at α_3
α_5 GABA$_A$ receptor	$\alpha_5\beta_{1,3}\gamma_2$	Less than 5% of GABA$_A$ receptors Extrasynaptic Largely restricted to hippocampus (pyramidal cells), cerebral cortex and olfactory bulb	Benzodiazepine-sensitive receptor. Mediates modulation of temporal and spatial memory:		
			L-655 708 [15–18]	Memory-enhancer, anxiogenic	Partial inverse agonist with preferential affinity for α_5 GABA$_A$ receptors. Not proconvulsant
α_4 GABA$_A$ receptor	$\alpha_4\beta_n\delta$, $\alpha_4\beta_n\gamma$	Less than 5% of receptors Extrasynaptic	Benzodiazepine-sensitive receptors. Ligands acting at sites other than the benzodiazepine site include (not selective for α_4):		
			Ethanol at low concentration [19, 20]	Anxiolytic, sedative	High sensitivity ($\geq$ 3 mM) at $\alpha_4(\alpha_6)\beta_3\delta$ Medium sensitivity ($\geq$ 30 mM) at $\alpha_4(\alpha_6)\beta_2\delta$ Low sensitivity ($\geq$ 100 mM) at $\alpha_4(\alpha_6)\beta_3\gamma_2$
			Gaboxadol [21]	Hypnotic	Partial agonist at α_1, α_3 subtypes, full agonist at α_5 and super agonist at $\alpha_4\beta_3\delta$ receptors[†]
			Neurosteroids (e.g. 3α,5α-THDOC) [22]	Anxiolytic Sedative Anaesthetic	High sensitivity at δ-containing subtypes and at α_1, α_3 receptors in combination with β_1
α_6 GABA$_A$ receptor	$\alpha_6\beta_n\delta$	Small population, only in cerebellum, extrasynaptic	Benzodiazepine-insensitive receptors. Modulators acting at sites other than the benzodiazepine site include, although not selective for α_6 GABA$_A$ receptors:		
			Ethanol [19, 20]	Anxiolytic, sedative	High sensitivity ($\geq$ 3 mM) at $\alpha_6(\alpha_4)\beta_3\delta^¶$ Medium sensitivity ($\geq$ 30 mM) at $\alpha_6(\alpha_4)\beta_2\delta^¶$ Low sensitivity ($\geq$ 100 mM) at $\alpha_6(\alpha_4)\beta_3\gamma_2$
	$\alpha_6\beta_{2,3}\gamma_2$	Less than 5% of receptors, only in cerebellum, synaptic	Neurosteroids [22] (e.g. 3α,5α THDOC)	Anxiolytic, sedative, anaesthetic	High sensitivity at δ-containing subtypes[‡] and at α_1, α_3 receptors in combination with β_1

Continued

Table 3.1.4.1 *(Continued)* GABA$_A$ receptor subtypes

Receptor subtype	Major subunit combination	Cellular and subcellular localization * [1, 2, 3, 4]	Pharmacology and behavioural activity [†‡§]		
β$_3$ GABA$_A$ receptors	α$_2$β$_3$γ$_2$ α$_3$β$_3$γ$_2$	See α$_2$, α$_3$ GABA$_A$ receptors	Intravenous anaesthetics (etomidate, propofol) [23]	Sedative, anaesthetic	Act on receptor subtypes containing β$_3$ subunit i.e. mainly α$_2$ and α$_3$ subtypes. At anaesthetic doses, the ligands can activate the receptors in the absence of GABA
δ GABA$_A$ receptor	α$_4$β$_n$δ α$_6$β$_n$δ	see α$_4$ and α$_6$ GABA$_A$ receptors	See α$_4$ and α$_6$ GABA$_A$ receptors		

*The % values are estimates taking total brain GABA$_A$ receptors as 100%.

†The term benzodiazepine refers to diazepam and structurally related ligands in clinical use.

‡Classical partial agonists which do not differentiate between GABA$_A$ receptor subtypes such as Bretazenil (Haefely *et al.*, 1990) or Pagoclone (Atack *et al.*, 2006b) are not considered in this review.

§Data should be treated with caution as properties of recombinant receptors that are expressed in foreign host cells might not give an accurate reflection of their native counterparts.

¶GABA is a weak partial agonist on δ-containing receptors, which largely explains the strong modulatory response of ligands acting on δ-containing receptors (Bianchi and MacDonald, 2003).

THDOC, 3α, 5α-tetrahydrodeoxycorticosterone.

[1] Möhler *et al.* (2002) *J Pharm Exp Ther* 300, 2–8; [2] Fritschy and Brünig (2003) *Pharmacol and Therapeutics* 98, 299–323; [3] Wallner *et al.* (2006) *Proc Natl Acad Sci USA* 103, 8540–8545; [4] Möhler (2007) *J Neurochem* 102, 1–12; [5] Dämgen and Lüdens (1999) *Neurosci Res Comm* 25, 139–148; [6] Foster *et al* (2004) *J Pharmacol Exp Ther* 311, 547–559; [7] Marowsky *et al.* (2004) *Eur J Neurosci* 20, 1281–1289; [8] McKernan *et al.* (2000) *Nature Neurosci* 3, 587–592; [9] Lippa *et al.* (2005) *Proc Natl Acad Sci USA* 102, 7380–7385; [10] Griebel *et al.* (2003) *CNS Drug Rev* 9, 3–20; [11] Atack *et al.* (2006) *J Pharm Exp Ther* 316, 410–422; [12] Dias *et al.* (2005) *J Neurosci* 25, 10682–10688; [13] Langen *et al.* (2005) *J Pharmacol Exp Ther* 314, 717–724; [14] Atack *et al.* (2005) *Br J Pharmacol* 144, 357–366; [15] Sternfeld *et al.* (2004) *J Med Chem* 47, 2176–2179; [16] Chambers *et al.* (2004) *J Med Chem* 47, 5829–5832; [17] Navarro *et al.* (2002) *Prog Neuropsychopharmacol Biol Psychiatry* 26, 1389–1392; [18] Navarro *et al.* (2004) *Aggress Behav* 30, 319–325; [19] Wallner *et al.* (2003) *Proc Natl Acad Sci USA* 100, 15218–15223; [20] Wallner *et al.* (2006) *Proc Natl Acad Sci USA* 103, 8540–8545; [21] Storustovu and Ebert (2003) *Eur J Pharmacol* 467, 49–56; [22] Belelli and Lambert (2005) *Nat Rev Neurosci* 6, 565–575; [23] Rudolph and Antkowiak (2004) *Nat Rev Neurosci* 5, 709–720.

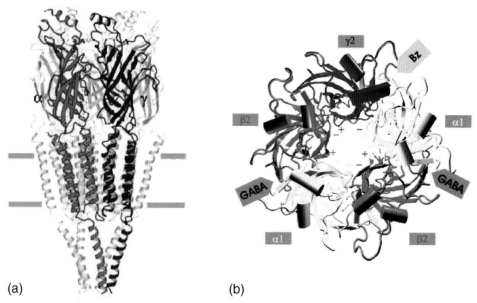

(a)   (b)

Fig. 3.1.4.2 The nicotinic acetylcholine receptor and an acetylcholine binding protein (AChBP) as a blueprint of the GABA$_A$ receptor structure. (a) Ribbon representation of the nicotinic acetylcholine receptor (nAChR) from the *Torpedo* electric organ with 4Å resolution (Unwin, 2005).[1] Only the front two subunits are highlighted (α, red; γ, blue). The lines indicate the limits of the cell membrane. The GABA$_A$ receptor is assumed to share the main structural motifs with the nAChR. (b) Model of the GABA$_A$ receptor extracellular domain as viewed from the synaptic cleft. The crystal structure of the soluble AChBP was used as template for comparative modeling (Ernst *et al.*, 2003; 2005).[2] Since the sequence identity between the snail AChBP and the vertebrate Cys-loop extracellular receptor domains is only 15–30 per cent the structural details are only a first approximation. See also colour plate section.

[1] Reprinted from *Journal of Molecular Biology*, volume 346, N Unwin, Refined structure of the nicotinic acetylcholine receptor at 4A resolution, pages 967–989, Copyright 2005, with permission from Elsevier.

[2] Reprinted from *Neuroscience*, volume 119, M Ernst, D Brauchart, S Boresch and W Sieghart, Comparative modeling of GABA$_A$ receptors: limits, insights, future developments, pages 933–943, Copyright 2003, with permission from Elsevier.

putative structure of GABA$_A$ receptors. Modelling the structure on the basis of the ACh-binding protein, the nAChR and the information known on residues involved in the formation of binding sites and subunit interfaces, resulted in only one possible absolute subunit arrangement, γ-α-β-α-β (Cromer *et al.*, 2002; Trudell, 2002; Ernst *et al.*, 2003; Trudell and Bertaccini, 2004; Ernst *et al.*, 2005) (Figure 3.1.4.2). This configuration was experimentally confirmed by co-expression experiments of concatenated GABA$_A$ receptor subunits (Baumann *et al.*, 2002).

The emerging structure of the GABA$_A$ receptor comprises an extracellular domain of mainly β-sheets, a channel domain of α-helices and a cytoplasmic domain of α-helices (Figure 3.1.4.2). The structure leaves considerable space for solvent-accessible cavities which, at least in part, are candidates for distinct ligand binding sites (Ernst *et al.*, 2005). Depending on the model, these pockets form a continuum from the extracellular domain into the transmembrane domain and, in case of the nAChR, have been proposed to provide space for the conformational transition from the closed to the open state upon agonist binding (Unwin, 2003; Ernst *et al.*, 2005).

3.1.4.3 Structure of GABA and benzodiazepine binding sites

The GABA binding site is located at the interface between the α and β subunits since amino acids of both subunits have been identified by site-directed mutagenesis to be important for ligand binding, including αPhe64, αArg66, αSer68, αArg119, αIle120, βTyr62 βTyr97, βLeu99, βTyr157, βThr160, βThr202, βSer204, βTyr205, βArg207 and βSer209 (for review see Sigel, 2002; Smith and Olsen, 1995; Akabas, 2004). The putative binding site for GABA was proposed to be accessible from the outside of the channel (Cromer *et al.*, 2002), a view which was questioned based on the low sequence identity with the AChBP (Ernst *et al.*, 2003). In line with the information gained from mutational analyses, the modelling studies suggest the binding pocket is surrounded by a box of aromatic residues (βTyr157, βPhe200, βTyr205 and αPhe64), which are possibly involved in agonist binding. The antagonist binding pocket is considered to overlap with the agonist pocket and extend further into the solvent-accessible cavity, which is in line with the larger size of GABA$_A$ receptor competitive antagonists compared to agonists (Ernst *et al.*, 2003).

The putative benzodiazepine binding site is located at the α/γ subunit interface as amino acids of both subunits are important for ligand binding, including αHis101, αTyr159, αTyr162, αGly200, αSer205, αThr206, αTyr209, γPhe77, γAla79, γThr81 and γMet130 (reviewed by Sigel, 2002; Akabas, 2004; Rudolph and Möhler, 2004). Like the GABA binding site, the benzodiazepine binding pocket appears to be surrounded by a box of aromatic residues (αTyr159, αTyr 210 and γPhe 77) with the fourth aromatic residue being replaced by αSer205. Based on the amino acid analogy, the benzodiazepine site may have evolved from a GABA binding site (Galzi and Changeux, 1994). Various other drug binding sites which are structurally less well-defined include the binding sites for barbiturates, neurosteroids, ethanol, picrotoxinin, mefenanic acid, Zn^{2+}, and others (Table 3.1.4.1). In summary, the presently available data on the GABA$_A$ receptor structure offer only a first glimpse for an understanding the three steps of chemical-to-electrical signal transduction: neurotransmitter binding, communication between the binding site and the channel, opening and closing of the barrier and, finally, the modulation of these steps by drugs.

3.1.4.4 Synaptic versus extrasynaptic receptors

Although GABA$_A$ receptors mediate most of the fast synaptic inhibition in the central nervous system (CNS), a large fraction of the receptor population is located extrasynaptically on dendrites and somata. These receptors mediate tonic inhibition and are under increasing scrutiny with respect to their role in modulating network activity (Mody and Pearce, 2004; Semyanov *et al.*, 2004; Farrant and Nusser, 2005), notably in cerebellum, thalamus, cerebral cortex, hippocampus, and dentate gyrus (Stell and Mody, 2002; Rossi *et al.*, 2003; Caraiscos *et al.*, 2004; Belelli *et al.*, 2005; Cope *et al.*, 2005; Jia *et al.*, 2005). They are present at a lower density than postsynaptic receptors (Somogyi *et al.*, 1989; Nusser *et al.*, 1995a, b). The functional and pharmacological relevance of extrasynaptic GABA$_A$ receptors is best documented for those containing the δ subunit (Table 3.1.4.1). The discovery of selective modulators of these GABA$_A$ receptors, including GABA agonists such as gaboxadol and muscimol (Drasbek and Jensen, 2005; Storustovu and Ebert, 2006), and neurosteroids (Belelli and Herd, 2003; Stell *et al.*, 2003; Belelli *et al.*, 2005) pointed to possible novel therapeutic potential, notably for sleep disorders, neuropathic pain, epilepsy, and premenstrual dysphoric syndrome (Krogsgaard-Larsen *et al.*, 2004) (Table 3.4.1.1).

Two subunits, γ2 and δ, appear to be major determinants of post- and extrasynaptic GABA$_A$ receptors, respectively. *In vitro* and *in vivo* studies demonstrated that the γ2 subunit is required for postsynaptic clustering of GABA$_A$ receptors and gephyrin (Essrich *et al.*, 1998; Lüscher and Fritschy, 2001; Schweizer *et al.*, 2003), a cytoskeletal protein selectively concentrated in inhibitory synapses in the CNS (Triller *et al.*, 1985; Sassoè-Pognetto and Fritschy, 2000). Multiple GABA$_A$ receptor subtypes are clustered at postsynaptic sites in defined neuronal populations (α1-, α2-, α3-, and, in part, α5-GABA$_A$ receptors). A major additional property of the γ2 subunit is to confer diazepam sensitivity to receptor complexes containing these α subunit variants. A conserved histidine residue (in position 101 of the α1 subunit) is an additional determinant of diazepam binding affinity (Wieland *et al.*, 1992). It is important to note, however, that GABA$_A$ receptors containing the γ2 subunit are not only confined to postsynaptic sites. For instance, most α5-GABA$_A$ receptors are extrasynaptic (Crestani *et al.*, 2002) and contribute to tonic inhibition which can be modulated by diazepam (Caraiscos *et al.*, 2004; Glykys and Mody, 2006; Prenosil *et al.*, 2006). Consequently, postsynaptic and extrasynaptic receptors formed with the γ2 subunit are diazepam-sensitive and contribute to the pharmacological profile of classical benzodiazepine-site agonists.

It is not known why α5-GABA$_A$ receptors are preferentially located extrasynaptically despite the fact that they contain a γ2 subunit. However, a selective interaction with activated radixin, an actin-binding protein, has been demonstrated recently which is required for anchoring of α5-GABA$_A$ receptors to the actin cytoskeleton (Loebrich *et al.*, 2006). Ironically, it has not been established how postsynaptic GABA$_A$ receptors associated with gephyrin are linked to the cytoskeleton. Gephyrin has been shown to bind to tubulin, but recent work suggests that this interaction might be more relevant for gephyrin transport along microtubules.

The interdependence of $GABA_A$ receptors and gephyrin for post-synaptic clustering was underscored by the demonstration that mutations in the GDP–GTP exchange factor collybistin impaired clustering of gephyrin, leading to a mislocalization of the $\gamma2$ subunit (Harvey et al., 2004). Interestingly, palmitoylation of the $\gamma2$ subunit by GODZ (Golgi apparatus specific protein with a DHHC zinc finger domain), a Golgi-specific palmitoyltransferase belonging to the superfamily of DHHC (asparate-histidine-histidine-cysteine)-cysteine-rich domain (DHHC-CRD) polytopic membrane proteins, is required for clustering of $GABA_A$ receptors (Keller et al., 2004; Rathenberg et al., 2004), implicating covalent modification of the receptor in synaptic targeting and possibly interaction with gephyrin. In line with this conclusion, distinct domains in the fourth transmembrane segment and intracellular loop of the $\gamma2$ subunit ensure clustering of $GABA_A$ receptors and recruitment of gephyrin to postsynaptic receptor clusters, respectively (Alldred et al., 2005). In addition, a direct binding of $GABA_A$ receptor α_2 subunits to gephyrin has been demonstrated (Tretter et al., 2008).

Receptors containing the δ subunit, in contrast to the $\gamma2$ subunit, appear to be excluded from postsynaptic sites, as demonstrated by immunoelectron microscopy (Nusser et al., 1998; Wei et al., 2003). The δ subunit is associated with the $\alpha4$ subunit, notably in thalamus and dentate gyrus (Peng et al., 2002; Sun et al., 2004), or the $\alpha6$ subunit in cerebellar granule cells (Jones et al., 1997). These receptors are diazepam-insensitive (Kapur and Macdonald, 1996; Mäkelä et al., 1997). The importance of these associations was uncovered in mutant mice. Thus, the δ subunit protein disappears from cerebellar granule cells of $\alpha6$-null mice (Jones et al., 1997) whereas the $\alpha4$ subunit protein was markedly reduced in thalamus and dentate gyrus of δ-null mice (Peng et al., 2002). However, the $\alpha4$ subunit can contribute to functional $GABA_A$ receptors without δ, most likely associated with the $\gamma2$ subunit. Interestingly, there is also evidence in a model of ethanol withdrawal for a translocation of $\alpha4$-$GABA_A$ receptors from extrasynaptic to postsynaptic location in the hippocampus (Hanchar et al., 2005). According to this model, ethanol dependence includes an altered balance between synaptic and extrasynaptic $GABA_A$ receptors containing the $\alpha1$ and $\alpha4$ subunit, suggesting that ligands which preferentially activate the latter receptors, such as gaboxadol, might be suitable for the treatment of ethanol withdrawal symptoms (Liang et al., 2006).

Studies from knockout mice also provided indirect evidence for a mutual exclusion of the δ and $\gamma2$ subunit from $GABA_A$ receptor complexes. Thus, in $\alpha1$-null mice, there is a complete loss of clustered $GABA_A$ receptors and gephyrin in neurons of the ventrobasal complex of the thalamus, accompanied by an increased expression of extrasynaptic $\alpha4$-$GABA_A$ receptors in the same cells (Kralic et al., 2006). Despite the absence of $\alpha1$ subunit, no evidence was found for association of the $\gamma2$ subunit with these receptors. Conversely, in δ-null mice, tonic, but not phasic, inhibition is reduced in thalamus and dentate gyrus (Mihalek et al., 1999; Spigelman et al., 2002; Porcello et al., 2003), despite an increased amount of $\gamma2$ subunit in all brain regions where the δ subunit is normally abundant, including the cerebellum (Tretter et al., 2001; Korpi et al., 2002; Peng et al., 2002).

Functionally, the role of extrasynaptic receptors has been analysed in detail in dentate gyrus granule cells and CA1 pyramidal cells. In whole-cell patch clamp experiments, tonic inhibition is evidenced by the shift in baseline holding current occurring upon application of a $GABA_A$ receptor antagonist. In dentate gyrus granule cells,

which express high amounts of $\alpha4\beta x\delta$ receptors, tonic inhibition is diazepam-insensitive. However, it carries a larger overall current than phasic inhibition, owing to the high affinity and slow desensitization kinetics of these receptors (Nusser and Mody, 2002; Mtchedlishvili and Kapur, 2006). In CA1 pyramidal cells, which express lower levels of $\alpha4$ subunit and virtually no δ subunit, both diazepam-insensitive and sensitive tonic inhibitory currents occur (Scimemi et al., 2005; Glykys and Mody, 2006; Prenosil et al., 2006; Rudolph and Möhler, 2006). A systematic analysis of mutant knockin mice (Möhler et al., 2002) carrying diazepam-insensitive $GABA_A$ receptors revealed the selective involvement of $\alpha5$-$GABA_A$ receptors to the latter form of tonic inhibition (Prenosil et al., 2006). Interestingly, the same receptors also contributed to the generation of slow phasic inhibitory postsynaptic potentials (IPSPs), characterized by high amplitude and slow kinetics (Pearce, 1993), which occur at low frequency in recordings of spontaneous inhibitory activity in the CA1 region. Slow IPSPs are likely to arise from the stimulation of synaptic and perisynaptic receptors, the latter containing the $\alpha5$ subunit in CA1 pyramidal cells.

Ectopic expression of the $\alpha6$ subunit under the control of the Thy-1.2 promoter has been used to assess the functional and pharmacological significance of enhanced tonic inhibition. These transgenic mice overexpress $\alpha1\alpha6\beta\gamma2$-$GABA_A$ receptors and exhibit a fivefold increase in tonic inhibition in CA1 pyramidal cells (Wisden et al., 2002). Behaviourally, these mice are essentially normal but are more sensitive than wild-type to the convulsant effects of $GABA_A$ receptor antagonists (Sinkkonen et al., 2004), suggesting an imbalance between phasic and tonic inhibition, with an overall decrease in GABAergic synaptic strength.

3.1.4.5 Plasticity of $GABA_A$ receptors

Three principal mechanisms contribute to the dynamic regulation of $GABA_A$ receptor function, which is essential for fine-tuning of neuronal networks and the generation of rhythmic activities:

1. Regulation of $GABA_A$ receptor trafficking, synaptic clustering, and cell-surface mobility;

2. Regulation of receptor function by chemical modification, with phosphorylation being one of the major covalent modifiers. Increasing evidence indicates that chemical modification affects receptor trafficking and cell surface expression, as well as intrinsic functions of the ligand-gated ion channel (Hinkle and Macdonald, 2003; Kittler and Moss, 2003);

3. Regulation of subunit expression, at the transcriptional and translational level (Steiger and Russek, 2004) determining the abundance and subunit profile of $GABA_A$ receptors in a given cell type or brain region.

This third mechanism appears to be of particular relevance for physiological alterations of network function, such as occurs upon hormonal fluctuations during the ovarian cycle (Brussaard and Herbison, 2000; Maguire et al., 2005), and for pathophysiological changes underlying chronic disorders such as epilepsy (Peng et al., 2004; Li et al., 2006) or drug abuse (Wei et al., 2004; Liang et al., 2006). Experimentally, changes in $GABA_A$ receptor subunit repertoire can be induced in mutant mice lacking a given $GABA_A$ receptor subunit, allowing the dissection of mechanisms regulating the subunit repertoire of defined neuron types (Vicini and Ortinski, 2004; Kralic et al., 2006; Ogris et al., 2006).

Plasticity of GABAergic synapses is also ensured by modifications affecting presynaptic function (regulation of GABA synthesis, storage and reuptake, modulation of GABA release probability, etc.) and GABA$_B$ receptors, but these aspects are outside the scope of the present review.

Regulation of cell surface expression

GABA$_A$ receptor internalization has emerged as a major mechanism of short- and long-term plasticity of GABAergic synapses. Like α-amino-3-hydroxy-5-methyl-4-isoxazole propionic acid (AMPA) receptors (Man et al., 2000), internalization of GABA$_A$ receptors is mediated by clathrin-coated vesicle endocytosis (Herring et al., 2003; van Rijnsoever et al., 2005) and a direct binding of the clathrin AP-2 adaptor complex to GABA$_A$ receptors has been demonstrated (Kittler et al., 2005). However, unlike AMPA receptors, it is not triggered by agonist exposure but is regulated by phosphorylation (see next section). In addition, several tyrosine kinase receptor ligands, such as tumour necrosis factor (TNF)-α, insulin, or brain-derived neurotrophic factor (BDNF) also modulate GABA$_A$ receptor cell surface expression by regulating rate of internalization and/or membrane insertion (Wan et al., 1997; Brünig et al., 2001; Jovanovic et al., 2004; Gilbert et al., 2006). An intriguing observation was made upon immunocytochemical analysis of GABA$_A$ receptor internalization, demonstrating that internalized GABA$_A$ receptors are targeted towards subsynaptic clusters where they are colocalized with gephyrin (van Rijnsoever et al., 2005), suggesting that a continuous cycle between cell surface, synaptic receptors and a subsynaptic pool might serve as a basis for short-term regulation of GABAergic synapse function. Along these lines, GABA$_A$ receptor cell surface mobility, reflecting diffusion within the plasma membrane, has been suggested as another mechanism to supply receptors to inhibitory synapses (Thomas et al., 2005). This process is regulated by interaction with gephyrin (Jacob et al., 2005), which might favour postsynaptic clustering by reducing the lateral mobility of GABA$_A$ receptors.

Finally, there is a growing list of intracellular proteins identified as interacting with GABA$_A$ receptors which regulate receptor trafficking and cell surface expression (Lüscher and Keller, 2004; Kneussel, 2005). The majority of proteins identified, including NSF (N-ethylmaleimide-sensitive factor) (Goto et al., 2005), GRIF1 (GABA$_A$ receptor interacting factor) (Brickley et al., 2005), and GRIP1 (glutamate receptor interacting protein 1) (Charych et al., 2004) contribute to GABA$_A$ receptor trafficking and transport. Interestingly, all these proteins interact with β subunit variants, suggesting that a major role for these subunits might be differential regulation of GABA$_A$ receptor trafficking and synaptic targeting.

Chemical modification

Several protein kinases have been shown to positively or negatively modulate the function of GABA$_A$ receptors, in part depending on their subunit composition (Krishek et al., 1994; Moss et al., 1995; McDonald et al., 1998; Brandon et al., 2002a; Kittler and Moss, 2003). The β subunits appear to be a major target of Protein kinase A (PKA)-, Protein kinase C (PKC)-, and tyrosine kinase-mediated phosphorylation (Brandon et al., 2002b; Boehm et al., 2004; Terunuma et al., 2004; Kanematsu et al., 2006; Marutha Ravindran and Ticku, 2006). Phosphorylation influences the kinetic properties and rate of desensitization of GABA$_A$ receptors (Hinkle and Macdonald, 2003), but also their sensitivity to allosteric modulators,

including benzodiazepines, neurosteroids and ethanol, and regulates cell-surface expression and internalization (Harney et al., 2003; Song and Messing, 2005). Importantly, several regulators of protein kinases or phosphatases have been shown to bind to the β2 or β3 subunit (Brandon et al., 2002b; 2003; Terunuma et al., 2004; Kittler et al., 2005; Kanematsu et al., 2006). Inhibition of this binding impairs phosphorylation (or phosphatase activity). According to these studies, increased phosphorylation of the β2/3 subunits reduces GABA$_A$ receptor internalization and the accompanying receptor rundown observed electrophysiologically. In particular, the well-characterized effects of BDNF on GABA$_A$ receptor internalization are mediated by phospholipase C-related inactive protein-1 (PRIP-1), which binds to phosphatases and β subunits, thereby reducing their degree of phosphorylation (Kanematsu et al., 2006). Phosphorylation-dependent endocytosis of GABA$_A$ receptors can be stimulated by activation of G-protein-coupled receptors, leading to a crosstalk among neurotransmitter systems to regulate neuronal excitability, as demonstrated in the nucleus accumbens (Chen et al., 2006).

Modulation of GABA$_A$ receptor sensitivity to allosteric modulators, such as neurosteroids, also appears as a mechanism for short-term GABA$_A$ receptor plasticity regulated by phosphorylation, as shown in the hippocampus (Harney et al., 2003) and in the supraoptic nucleus. In the latter case, oxytocin increased PKC-dependent GABA$_A$ receptor phosphorylation, thereby rendering the receptors insensitive to the neurosteroid allopregnanolone, accounting in part for the plasticity of oxytocin neuron function shortly after parturition (Koksma et al., 2003). In cerebellum, neurosteroids prolong GABA$_A$-mediated IPSCs in cells expressing the δ subunit (Fodor et al., 2005). Contrary to oxytocinergic neurons, this effect depends on stimulation of PKC, as demonstrated pharmacologically in recordings from δ-null mice (Vicini et al., 2002).

Change of GABA$_A$ receptor subunit expression and composition

Numerous alterations in GABA$_A$ receptor subunit mRNA expression in response to a physiological, pharmacological, or pathological stimulus have been reported in various experimental models or clinical conditions (Steiger and Russek, 2004). It should be emphasized, however, that many of these studies did not use quantitative methods and did not address the issue of whether gene transcription or mRNA half-life were changed. Furthermore, the relevance of these changes for the function of GABA$_A$ receptors is difficult to assess due to the major impact of receptor localization, targeting, and regulation of function, in addition to the mere abundance of a given subunit mRNA. Indeed, there are very few studies which have demonstrated a direct correlation between altered GABA$_A$ receptor subunit gene transcription and a corresponding functional alteration.

Even more striking is the fact that major changes in the abundance of subunit proteins have been reported in mutant mice carrying targeted deletion of a GABA$_A$ receptor subunit, which were not accompanied by a change in gene expression. Thus, a complete loss of δ subunit protein occurs in the cerebellum of α6-null mice, without change of δ subunit mRNA expression (Jones et al., 1997; Nusser et al., 1999). Conversely, a several-fold increase in α3 subunit protein was reported in cerebellum of α1-null mice (Kralic et al., 2006; Ogris et al., 2006), again without a corresponding change in mRNA expression. While these findings might be interpreted as evidence that $GABA_A$ receptor subunit mRNAs are present in vast surplus in most cell types, this conclusion cannot be generalized. For instance, a partial deficit in $\gamma2$ subunit protein was observed in mutant mice carrying a single copy of the $\gamma2$ subunit gene (Crestani *et al.*, 1999). Most interestingly, the deficit was region-specific, being largest in CA1 pyramidal cells, leading to an anxiety phenotype reminiscent of human generalized anxiety disorder.

A better understanding of $GABA_A$ receptor subunit gene promoter regulation is nevertheless required, in particular since these genes occur in clusters distributed on distinct chromosomes (Russek, 1999; Barnard, 2001). These clusters correspond only in part to the most abundant $GABA_A$ receptor subtypes found *in vivo*, indicating that complex regulatory mechanisms ensure proper transcription and translation of all subunits that form the $GABA_A$ receptor subtype repertoire of each cell type in the CNS. In this respect, the discovery that α subunit variants are not interchangeable functionally underscores the necessity of a strict, cell-specific regulation of gene expression.

Changes in $GABA_A$ receptor subunit composition, underlying changes in efficacy of GABAergic transmission, occur physiologically under the influence of hormonal fluctuations during the ovarian cycle. Of particular relevance for premenstrual dysphoric syndrome are the reversible changes in δ subunit expression (and therefore extrasynaptic receptors in the hippocampus) occurring during the ovarian cycle, leading to a decrease in seizure susceptibility in late diestrus (Maguire *et al.*, 2005). These alterations can be reproduced pharmacologically by neurosteroid treatment and affect $GABA_A$ receptor function (Sundstrom-Poromaa *et al.*, 2002; Shen *et al.*, 2005).

Alterations in $GABA_A$ receptor subunit composition are also well documented in several chronic diseases, in particular epilepsy (Peng *et al.*, 2004; Gilby *et al.*, 2005; Nishimura *et al.*, 2005; Roberts *et al.*, 2005; Li *et al.*, 2006). These recent studies extend previous work by demonstrating a major contribution of extrasynaptic $GABA_A$ receptors to the changes in inhibitory function that might underlie epileptogenesis and occurrence of chronic recurrent seizures.

3.1.4.6 Pharmacology and functional roles of $GABA_A$ receptor subtypes

Receptors containing the α_1, α_2, α_3 or α_5 subunit in combination with any of the β-subunits and the γ_2-subunit are most prevalent in the brain, with the major subtype being $\alpha_1\beta_2\gamma_2$ (Table 3.4.1.1). These receptors are sensitive to benzodiazepine modulation. In contrast, receptors containing the α_4 or α_6 subunit and/or the δ-subunit are insensitive to classical benzodiazepines. The pharmacological relevance of $GABA_A$ receptor subtypes for the spectrum of benzodiazepine effects was recently identified based on a genetic approach and the development of ligands which differentiate by efficacy and/or affinity between receptor subtypes (Table 3.1.4.1) (Rudolph *et al.*, 1999; Löw *et al.*, 2000; McKernan *et al.*, 2000; Whiting *et al.*, 2000; Rudolph *et al.*, 2001; Crestani *et al.*, 2002; Möhler, 2002; Möhler *et al.*, 2002; Whiting, 2003; Möhler, 2007; Rudolph and Möhler, 2006). Experimentally, the $GABA_A$ receptor subtypes were rendered diazepam-insensitive by replacing a conserved histidine residue with an arginine residue in the respective α subunit gene—α_1(H101R), α_2(H101R), α_3(H126R) and α_5(H105R)—(Rudolph *et al.* 1999; Löw *et al.*, 2000; Crestani *et al.*, 2002). This strategy permitted the allocation of the benzodiazepine

drug actions to the α_1-, α_2-, α_3- and α_5- $GABA_A$-receptor subtypes (Rudolph *et al.*, 2001; Crestani *et al.*, 2002). In addition, it implicated the neuronal networks expressing the particular receptor in mediating the corresponding drug actions.

Receptor subtype mediating for sedation

Among α_1-, α_2- and α_3-point-mutated mice only the α_1(H101R) mutants were resistant to the depression of motor activity by diazepam (Rudolph *et al.* 1999; Crestani *et al.*, 2000; Löw *et al.*, 2000; McKernan *et al.*, 2000). This effect was specific for ligands of the benzodiazepine site since pentobarbital or a neurosteroid remained as effective in α_1(H101R) mice as in wild-type mice in inducing sedation. Clearly, sedation is linked to α_1-$GABA_A$ receptors and differs neurobiologically from the anxiolytic action of benzodiazepines. Ligands with preferential affinity for α_1-$GABA_A$ receptors comprise common hypnotic drugs (Table 3.1.4.1).

Receptor subtype mediating protection against seizures

The anticonvulsant activity of diazepam, assessed by its protection against pentyleneterazole-induced tonic convulsions, was strongly reduced in α_1(H101R) mice compared to wild-type animals (Rudolph *et al.* 1999). Sodium phenobarbital remained fully effective as anticonvulsant in α_1(H101R) mice. Thus, the anticonvulsant activity of benzodiazepines is partially but not fully mediated by α_1-$GABA_A$ receptors. The anticonvulsant action of zolpidem is exclusively mediated by α_1-$GABA_A$ receptors, since its anticonvulsant action and the zolpidem-induced changes in EEG are completely absent in α_1 (H101R) mice (Crestani *et al.*, 2000; Kopp *et al.*, 2004).

Receptors subtype mediating anxiolysis

New strategies for the development of daytime anxiolytics which are devoid of drowsiness and sedation are of high priority. Experimentally, the anxiolytic-like action of diazepam is due to the modulation of α_2-$GABA_A$ receptors as shown by the lack of tranquillizing action of diazepam in α_2(H101R) mice (elevated plus maze; light/dark choice test) (Löw *et al.*, 2000). The α_2-$GABA_A$ receptors, which comprise only about 15 per cent of all diazepam-sensitive $GABA_A$ receptors, are mainly expressed in the amygdala and in principal cells of the cerebral cortex and the hippocampus with particularly high densities on their axon initial segments (Nusser *et al.* 1996a; Fritschy *et al.*, 1998a, b). Thus, the inhibition of the output of these principal neurons appears to be a major mechanism of anxiolysis. In keeping with this notion, the ligand L-838417 with partial efficacy at α_2, α_3 and α_5 but not on $\alpha_1$$GABA_A$ receptors was anxiolytic in wild-type rats (McKernan *et al.*, 2000) (Table 3.1.4.1).

It had previously been postulated that the anxiolytic action of diazepam is based on the dampening of the reticular activating system which is mainly represented by noradrenergic and serotonergic neurons of the brain stem. These neurons express preponderantly α_3-$GABA_A$ receptors. The analysis of the α_3-point-mutated mice—α_3(H126R)—indicated that the anxiolytic effect of benzodiazepine drugs was unaffected (Löw *et al.*, 2000). The reticular activating system, therefore, does not appear to be a major contributor to anxiolysis. Nevertheless, at least at high receptor occupancy, the α_3-selective ligand TP003 showed anxiolytic activity (Dias *et al.*, 2005) (Table 3.1.4.1). However, while TP003 displayed high selectivity at recombinant α_3-$GABA_A$ receptors, its specificity for native α_3-$GABA_A$ receptors *in vivo* remains to be verified.

Receptor subtype mediating myorelaxation

The muscle relaxant effect of diazepam is largely mediated by α_2-GABA$_A$ receptors, as shown by the failure of diazepam to induce changes in muscle tone in the α_2-point mutated mouse line (Crestani et al., 2001). α_2-GABA$_A$ receptors in the spinal cord, notably in the superficial layer of the dorsal horn and in motor neurons (Bohlhalter et al., 1996) are most likely implicated in this effect. The muscle relaxant effect requires considerably higher doses of diazepam than its anxiolytic-like activity which is mediated by α_2GABA$_A$ receptors located in the limbic system (see above). It was only at very high doses of diazepam that α_3- and α_5-GABA$_A$ receptors were also implicated in mediating myorelaxation (Crestani et al., 2001; Crestani et al., 2002) (Table 3.1.4.1).

Receptor subtypes mediating effects on associative learning and memory

The acquisition of spatial and temporal memory is associated with excitatory synaptic plasticity involving hippocampal NMDA receptors (Morris et al., 1986; 1989; Davis et al., 1992; McHugh et al., 1996; Tsien et al., 1996; Tang et al., 1999; Huerta et al., 2000; Nakazawa et al., 2002). A counter regulatory GABAergic control involves the α_5-GABA$_A$ receptors, located extrasynaptically on hippocampal pyramidal cells (Fritschy et al., 1998b, Crestani et al., 2002).

In α_5(H105R) mice, the content of α_5GABA$_A$ receptors was reduced by 30–40 per cent exclusively in the hippocampus (Crestani et al., 2002). This is presumably due to an effect of the mutation on receptor assembly or insertion. There was no indication for adaptive changes of other GABA$_A$ receptors expressed in the same pyramidal cells. Behaviourally, the partial deficit of hippocampal α_5GABA$_A$ receptors resulted in an improved performance in trace fear conditioning, a hippocampus-dependent memory task (Crestani et al., 2002). These results pointed to a role of α_5GABA$_A$ receptors in the function of temporal memory. When the α_5-GABA$_A$ receptors were deleted in the entire brain by targeting the α_5 subunit gene (Collinson et al., 2002; Whiting, 2003) a significantly improved performance in a water maze model of spatial learning was observed. In addition, the amplitude of hippocampal IPSCs was decreased and the paired-pulse facilitation of field EPSP amplitudes was enhanced. These data strongly suggest that α_5-GABA$_A$ receptors play a crucial role in cognitive processes of hippocampal learning and memory. A partial inverse agonist acting at α_5GABA$_A$ receptors enhanced the performance of wild-type rats in the water maze test (Chambers et al., 2004).

Receptors mediating effects on sensorimotor processing

A deficit in GABAergic inhibitory control is one of the major hypotheses underlying the symptomatology of schizophrenia (Lewis et al., 2005). A potential contribution of GABA$_A$ receptor subtypes was therefore investigated with regard to the overactivity of the dopaminergic system, considered to be a major factor in schizophrenia pathogenesis. The dopaminergic system is under GABAergic inhibitory control mainly via α_3-GABA$_A$ receptors (Fritschy and Möhler, 1995; Pirker et al., 2000). In mice lacking the α_3 subunit gene no adaptive changes in the expression of α_1, α_2 and α_5 subunits was observed (Studer et al., 2006) and anxiety-related behaviour was normal (Yee et al., 2005). However, the mice displayed a marked deficit in prepulse inhibition of the acoustic startle reflex, pointing to a deficit in sensorimotor information processing (Yee et al., 2005). This deficit in prepulse inhibition was normalized by administration of the antipsychotic dopamine D2 receptor antagonist haloperidol, suggesting that the phenotype is caused by hyperdopaminergia (Yee et al., 2005).

The hippocampus is also believed to play an important role in the modulation of prepulse inhibition. In the α_5(H105R) point-mutated mice (see above) prepulse inhibition was attenuated concomitant with an increase in spontaneous locomotor activity in a novel open field (Hauser et al., 2005), although increased locomotor activity was not apparent in an earlier study (Crestani et al., 2002). Thus, the α_5-GABA$_A$ receptors which are located extrasynaptically and are thought to mediate tonic inhibition (Caraiscos et al., 2004; Scimemi et al., 2005; Glykus and Mody, 2006; Prenosil et al., 2006) are important regulators of the expression of prepulse inhibition. Attenuation of prepulse inhibition is a frequent phenotype of psychiatric conditions including schizophrenia. These results suggest that α_3- and/or α_5-selective agonists may constitute an effective treatment for sensorimotor gating deficits in various psychiatric conditions.

Receptors mediating anaesthetic activity

In the quest for neuronal correlates of consciousness (Koch, 2004), the clarification of the mechanism of action of anaesthetic drugs is an important strategy. Various molecular targets have been invoked in mediating the clinical effects of general anaesthetics (Franks and Lieb, 2000; Campagna et al., 2003; Rudolph and Antkowiak, 2004). Recently, β_3-containing GABA$_A$ receptors were found to mediate in full the immobilizing action of etomidate and propofol (Table 3.1.4.1) (Jurd et al., 2003), such that they were abolished in mice carrying a point mutation in the β_3 subunit. Thus, β_3 containing GABA$_A$ receptors are a major control element for anaesthesia (Table 3.1.4.1). These receptors also mediate, although only to a limited degree, the immobilizing action of enflurane, isoflurane, and halothane (Jurd et al., 2003; Lambert et al., 2005; Liao et al., 2005). The hypnotic action of etomidate and propofol, assessed by the loss of righting reflex, was partly mediated via β_3 containing GABA$_A$ receptors (Jurd et al., 2003), although this was apparently not the case for the volatile anaesthetics (Jurd et al., 2003; Lambert et al., 2005). In fact, the hypnotic action of etomidate was found to be mainly mediated by β_2-containing GABA$_A$ receptors (Reynolds et al., 2003). The respiratory depressant action of etomidate and propofol is also mediated by β_3-containing GABA$_A$ receptors, while their heart rate depressant action and to a large part their hypothermic action are mediated by other targets (Zeller et al., 2005; Cirone et al., 2004). Thus, a β_3-selective agent would be predicted to be immobilizing and respiratory depressant, but largely lack the heart rate depressant and hypothermic actions of etomidate and propofol.

Pentobarbital is widely used as general anaesthetic. Its immobilizing action was found to be fully mediated via β_3-containing GABA$_A$ receptors since this effect was absent in β_3(N265M) point mutated mice (Zeller et al., 2007a). Its hypnotic action was only partially mediated via this receptor subtype (Zeller et al., 2007a). Surprisingly, and in contrast to etomidate and propofol, the respiratory depressant effect of pentobarbital was independent of β_3 containing GABA$_A$ receptors and is attributed to other yet unidentified targets (Zeller et al., 2007a). Isoflurane decreased core body temperature and heart rate to a small degree via β_3 containing GABA$_A$ receptors (Zeller et al., 2007b).

Abuse and dependence liability

Drug abuse, the inappropriate use of substances, is a complex phenomenon. A key property with regard to abuse liability for a given compound is the degree to which it has reinforcing properties, often assessed by self-administration procedures in primates. Another major determinant for abuse liability is the occurrence of physical dependence with repeated administration, i.e. the emergence of withdrawal symptoms on cessation of chronic drug treatment. Tolerance to some or all of the effect of a drug often accompanies the development of physical dependence. Finally, a key component of abuse liability is the subjective, or interoceptive drug effect, which is often assessed by drug discrimination procedures. There is a general consensus that certain subpopulations of patients are at greater risk for inappropriate use of benzodiazepines. These include poly-drug abusers and individuals with a history of alcohol abuse. The observation that benzodiazepine-type drugs do not have reinforcing effects in normal healthy subjects, suggests that poly-drug abusers and alcoholics are likely to administer these compounds as a result of particular characteristics of these patient populations (Rowlett *et al.*, 2007; Licata and Rowlett, 2008).

With regard to the roles of the GABA$_A$ receptor subtypes in mediating the reinforcing effects of benzodiazepines, no clear answer is yet available. In general, benzodiazepine drugs have relatively modest reinforcing effects. An intriguing exception is zolpidem (and zaleplon) which exhibit effects exceeding those of classical benzodiazepines (Rowlett *et al.*, 2005, 2007). These findings suggested that α_1GABA$_A$ receptors may be important substrates for driving self-administration. Similarly, the partial agonist SL 651498 (Table 3.1.4.1) partially substituted for triazolam, an effect that was blocked by an α_1 preferring beta-carboline, implicating α_1 GABA$_A$ receptors in the subjective effect of SL 651498 (Licata *et al.*, 2005). Furthermore, L-838 417, a partial agonist acting on α_2, α_3, α_5 GABA$_A$ receptors, was inactive in a drug discrimination assay in primates. Finally, no sign of dependence liability was detected for TPA 023, a partial α_2,α_3 receptor agonist (Ator, 2005; Atack *et al.*, 2006). These results clearly point to a role of α_1 GABA$_A$ receptors, which are the most prevalent subtype in the brain, in mediating the reinforcing properties of benzodiazepines (Ator, 2005). However, despite its lack of efficacy at α_1GABA$_A$ receptors (McKernan *et al.*, 2000), L-838 417 was self-administered by primates, although to a much lower degree than zolpidem or classical benzodiazepines (Rowlett *et al.*, 2005). This finding suggested that α_1 GABA$_A$ receptors may not exclusively mediate the reinforcing properties of benzodiazepines. Clearly, any definitive conclusion as to the relative roles of GABA$_A$ receptor subtypes in mediating the reinforcng properties of benzodiazepines awaits demonstration of the efficacy and affinity of the ligands at native receptor subtypes, preferably human. In conclusion, at present, there is no simple, unequivocal answer to the question of which GABA$_A$ subtypes mediate the reinforcing effect of benzodiazepine-type drugs, although, α_1-GABA$_A$ receptors remain the prime candidates.

References

Akabas MH (2004). GABA$_A$ receptor structure-function studies: a reexamination in light of new acetylcholine receptor structures. *Int Rev Neurobiol* **62**, 1–43.

Alldred MJ, Mulder-Rosi J, Lingenfelter SE *et al.* (2005). Distinct γ2 subunit domains mediate clustering and synaptic function of postsynaptic GABA$_A$ receptors and gephyrin. *J Neurosci* **19**, 594–603.

Atack JR, Hutson PH, Collinson N *et al.* (2005). Anxiogenic properties of an inverse agonist selective for α_3 subunit-containing GABA$_A$ receptors. *Br J Pharmacol* **144**, 357–366.

Atack JR, Pike A, Marshall G *et al.* (2006b). The in vivo properties of pagoclone in rat are most likely mediated by 5'-hydroxy pagoclone. *Neuropharmacol* **50**, 677–689.

Atack JR, Wafford K, Tye SJ *et al.* (2006a). TPA023, an agonist selective for α_2- and α_3-containing GABA$_A$ receptors, is a non-sedating anxiolytic in rodents and primates. *J Pharm Exp Ther* **316**, 410–422.

Ator, NA (2005). Contribution of GABA$_A$ receptor subtype selectivity to abuse liability and dependence potential of pharmacological treatments for anxiety and sleep disorders. *CNS Spectr* **10**, 33.

Barnard EA (2001). The molecular architecture of GABA$_A$ receptors. In H Mohler, ed., *Pharmacology of GABA and glycine neurotransmission*. Berlin, Springer-Verlag, pp. 79–100.

Barnard EA, Skolnick P, Olsen RW *et al.* (1998a). Subtypes of g-aminobutyric acid$_A$ receptors: classification on the bases of subunit structure and receptor function. *Pharmacol Reviews* **50**, 291–313.

Barnard EA, Skolnick P, Olsen RW *et al.* (1998b). International Union of Pharmacology. XV. Subtypes af γ-aminobutyric acidA receptors: classification of the basis of subunit structure and receptor function. *Pharmacol Rev* **50**, 291–313.

Baumann SW, Baur R and Sigel E (2001). Subunit arrangement of γ-aminobutyric acid type A receptors. *J Biol Chem* **276**, 36275–36280.

Baumann SW, Baur R and Sigel E (2002). Forced subunit assembly in 1β2γ2 GABA$_A$ receptors: insight into the absolute arrangement. *J Bio Chem* **277**, 46020–46025.

Belelli D and Herd MB (2003). The contraceptive agent Provera enhances GABA$_A$ receptor-mediated inhibitory neurotransmission in the rat hippocampus: evidence for endogenous neurosteroids? *J Neurosci* **23**, 10013–10020.

Belelli D and Lambert JJ (2005). Neurosteroids: endogenous regulators of the GABA(A) receptor. *Nat Rev Neurosci* **6**, 565–575.

Belelli D, Peden DR, Rosahl TW *et al.* (2005). Extrasynaptic GABA$_A$ receptors of thalamocortical neurons: a molecular target for hypnotics. *J Neurosci* **25**, 11513–11520.

Bianchi MT and McDonald RL (2003). Neurosteroids shift partial agonist activation of GABA(A) receptor channels from low- to high-efficacy gating patterns. *J Neurosci* **23**, 10934–10943.

Boehm SL, Peden L, Harris RA *et al.* (2004). Deletion of the fyn-kinase gene alters sensitivity to GABAergic drugs: dependence on β 2/ β 3 GABA$_A$ receptor subunits. *J Pharmacol Exp Therap* **309**, 1154–1159.

Bohlhalter S, Weinmann O, Möhler H *et al.* (1996). Laminar compartmentalization of GABA$_A$-receptor subtypes in the spinal cord: an immunohistochemical study. *J Neurosci* **16**, 283–297.

Bormann J (2000). The ABC of GABA receptors. *Trends Pharmacol Sci* **21**, 16–19.

Brandon NJ, Jovanovic JN and Moss SJ (2002a). Multiple roles of protein kinases in the modulation of γ-aminobutyric acid$_A$ receptor function and cell surface expression. *Pharmacol Ther* **94**, 113–122.

Brandon NJ, Jovanovic JN, Colledge M *et al.* (2003). A-kinase anchoring protein 79/150 facilitates the phosphorylation of GABA$_A$ receptors by cAMP-dependent protein kinase via selective interaction with receptor β subunits. *Mol Cell Neurosci* **22**, 87–97.

Brandon NJ, Jovanovic JN, Smart TG *et al.* (2002b). Receptor for activated C kinase-1 facilitates protein kinase C-dependent phosphorylation and functional modulation of GABA$_A$ receptors with the activation of G-protein-coupled receptors. *J Neurosci* **22**, 6353–6361.

Brejc K, van Dijk WJ, Klaassen RV *et al.* (2001). Crystal structure of an ACh-binding protein reveals the ligand-binding domain of nicotinic receptors. *Nature* **411**, 269–276.

Brickley K, Smith MJ, Beck M *et al.* (2005). GRIF-1 and OIP106, members of a novel gene family of coiled-coil domain proteins: association *in vivo* and *in vitro* with kinesin. *J Biol Chem* **280**, 14723–14732.

Brünig I, Penschuck S, Berninger B *et al.* (2001). BDNF reduces miniature inhibitory postsynaptic currents by rapid down-regulation of GABA$_A$ receptor surface expression. *Eur J Neurosci* **13**, 1320–1328.

Brussaard AB and Herbison AE (2000). Long-term plasticity of postsynaptic GABA$_A$-receptor function in the adult brain: insights from the oxytocin neurone. *Trends Neurosci* **23**, 190–195.

Campagna JA, Miller KW and Forman SA (2003). Mechanisms of actions of inhaled anesthetics. *N Engl J Med* **348**, 2110–2124.

Caraiscos VB, Elliott EM, You-Ten KE *et al.* (2004). Tonic inhibition in mouse hippocampal CA1 pyramidal neurons is mediated by α5 subunit-containing γ-aminobutyric acid type A receptors. *Proc Natl Acad Sci USA* **101**, 3662–3667.

Chambers MS, Atack JR, Carling RW *et al.* (2004). An orally bioavailable, functionally selective inverse agonist at the benzodiazepine site of GABA$_A$ α5 receptors with cognition-enhancing properties. *J Med Chem* **47**, 5829–5832.

Chang Y, Wang R, Barot S *et al.* (1996). Stoichiometry of a recombinant GABA$_A$ receptor. *J Neurosci* **16**, 5415–5424.

Charych EI, Yu W, Miralles CP *et al.* (2004). The brefeldin A-inhibited GDP/GTP exchange factor 2, a protein involved in vesicular trafficking, interacts with the β subunits of the GABA$_A$ receptors. *J Neurochem* **90**, 173–189.

Chen G, Kittler JT, Moss SJ *et al.* (2006). Dopamine D3 receptors regulate GABA$_A$ receptor function through a phospho-dependent endocytosis mechanism in nucleus accumbens. *J Neurosci* **26**, 2513–2521.

Cirone J, Rosahl TW, Reynolds DS *et al.* (2004). Gamma-aminobutyric acid type A receptor β2 subunit mediates the hypothermic effect of etomidate in mice. *Anesthesiology* **100**, 1438–1445.

Collinson N, Kuenzi FM, Jarolimek W *et al.* (2002). Enhanced learning and memory and altered GABAergic synaptic transmission in mice lacking the α$_5$ subunit of the GABA$_A$ receptor. *J Neurosci* **22**, 5572–5580.

Cope DW, Hughes SW and Crunelli V (2005). GABA$_A$ receptor-mediated tonic inhibition in thalamic neurons. *J Neurosci* **25**, 11533–11563.

Crestani F, Keist R, Fritschy JM *et al.* (2002). Trace fear conditioning involves hippocampal α5 GABA$_A$ receptors. *Proc Natl Acad Sci USA* **99**, 8980–8985.

Crestani F, Lorez M, Baer K *et al.* (1999). Decreased GABA$_A$-receptor clustering results in enhanced anxiety and a bias for threat cues. *Nature Neurosci* **2**, 833–839.

Crestani F, Löw K, Keist R *et al.* (2001). Molecular targets for the myorelaxant action of diazepam. *Mo Pharmacol* **59**, 442–445.

Crestani F, Martin JR, Möhler H *et al.* (2000). Mechanism of action of the hypnotic zolpidem in vivo. *B J Pharmacol* **131**, 1251–1254.

Cromer BA, Morton CJ and Parker MW (2002). Anxiety over GABA$_A$ receptor structure relieved by AChBP. *Trends Biochem Sci* **27**, 280–287.

Dämgen K and Lüddens H (1999). Zaleplon diaplays a selecitvity to recombinant GABA$_A$ receptors different from zolpidem, zopiclone and benzodiazepines. *Neurosci Res Comm* **25**, 139–148.

Davis S, Butcher SP and Morris RGM (1992). The NMDA receptor antagonist D-2-amino-5-phosphonopentanoate (D-AP5) impairs spatial learning and LTP *in vivo* at intracerebral concentrations comparable to those that block LTP in *vitro*. *J Neurosci* **12**, 21–34.

Dias R, Sheppard WF, Fradley RL *et al.* (2005). Evidence for a significant role of alpha3-containing GABAA receptors in mediating the anxiolytic effects of benzodiazepines. *J Neurosci* **25**, 10682–10688.

Drasbek KR and Jensen K (2005). THIP, a hypnotic and antinociceptive drug, enhances an extrasynaptic GABA$_A$ receptor-mediated conductance in mouse neocortex. *Cereb Cortex* **8**, 1134–1141.

Ernst M, Brauchart D, Boresch S *et al.* (2003). Comparative modeling of GABA$_A$ receptors: limits, insights, future developments. *Neuroscience* **119**, 933–943.

Ernst M, Bruckner S, Boresch S *et al.* (2005) Comparative models of GABA$_A$ receptor extracellular and transmembrane domains: important insights in pharmacology and function. *Mol Pharmacol* **68**, 1291–1300.

Essrich C, Lorez M, Benson JA, Fritschy JM *et al.* (1998). Postsynaptic clustering of major GABA$_A$ receptor subtypes requires the γ2 subunit and gephyrin. *Nature Neurosci* **1**, 563–571.

Farrant M and Nusser Z (2005). Variations on an inhibitory theme: phasic and tonic activation of GABA$_A$ receptors. *Nature Rev Neurosci* **6**, 215–229.

Farrar SJ, Whiting PJ, Bonnert TP *et al.* (1999). Stoichiometry of a ligand-gated ion channel determined by fluorescence energy transfer. *J Biol Chem* **274**, 10100–10104.

Fodor L, Biro T and Maksay G (2005). Nanomolar allopregnanolone potentiates rat cerebellar GABA$_A$ receptors. *Neurosci Lett* **383**, 127–130.

Foster AC, Pelleymounter MA, Cullen MJ *et al.* (2004). *In vivo* pharmacological characterization of indiplon, a novel pyrazolopyrimidine sedative-hypnotic. *J Pharmacol Exp Ther* **311**, 547–559.

Franks, NP and Lieb WR (2000). The role of NMDA receptors in consciousness: what can we learn from anaesthetic mechanisms? In T Metzinger, ed., *Neural correlates of consciousness: empirical and conceptual questions*. MIT Press, Cambridge, MA, pp. 265–269.

Fritschy JM and Brünig I (2003). Formation and plasticity of GABAergic synapses: physiological mechanisms and pathophysiological implications. *Pharmacol and Therapeutics* **98**, 299–323.

Fritschy JM and Möhler H (1995). GABA$_A$ receptor heterogeneity in the adult rat brain: differential regional and cellular distribution of seven major subunits. *J Comp Neuro* **359**, 154–194.

Fritschy JM, Johnson DK, Möhler H *et al.* (1998b). Independent assembly and subcellular targeting of GABA$_A$ receptor subtypes demonstrated in mouse hippocampal and olfactory neurons *in vivo*. *Neurosci Lett* **249**, 99–102.

Fritschy JM, Weinmann O, Wenzel A *et al.* (1998a). Synapse-specific localization of NMDA and GABA$_A$ receptor subunits revealed by antigen-retrieval immunohistochemistry. *J Comp Neurol* **390**, 194–210.

Galzi IJ and Changeux JP (1994). Neurotransmitter-gated ion channels as unconventional allosteric proteins. *Curr Opin Struct Biol* **4**, 554–565.

Gilbert SL, Zhang L, Forster ML *et al.* (2006). Trak1 mutation disrupts GABA$_A$ receptor homeostasis in hypertonic mice. *Nature Genetics* **38**, 245–250.

Gilby KL, Da Silva EA and McIntyre DC (2005). Differential GABA$_A$ subunit expression following status epilepticus in seizure-prone and seizure-resistant rats: a putative mechanism for refractory drug response. *Epilepsia* **46**, 3–9.

Glykys J and Mody I (2006). Hippocampal network hyperactivity after selective reduction of tonic inhibition in GABA$_A$ receptor α5 subunit-deficient mice. *J Neurophysiol* **95**, 2796–2807.

Goto H, Terenuma M, Kanematsu T *et al.* (2005). Direct interaction of N-ethylmaleimide-sensitve factor with GABA$_A$ receptor β subunits. *Mol Cell Neurosci* **2**, 197–206.

Griebel G, Perrault G, Simiand J *et al.* (2003). SL651498, a GABA$_A$ receptor agonist with subtype-selective efficacy, as a potential treatment for generalized anxiety disorder and muscle spasms. *CNS Drug Rev* **9**, 3–20.

Haefely W, Martin JR and Schoch P (1990). Novel anxiolytics that act as partial agonists at benzodiazepine receptors. *Trends Pharmacol Sci* **11**, 452–456.

Hanchar HJ, Dodson PD, Olsen RW *et al.* (2005). Alcohol-induced motor impairment caused by increased extrasynaptic GABA$_A$ receptors activity. *Nature Neurosci* **8**, 339–345.

Harney SC, Frenguelli BG and Lambert JJ (2003). Phosphorylation influences neurosteroid modulation of synaptic GABA$_A$ receptors in rat CA1 and dentate gyrus neurons. *Neuropharmacol* **45**, 873–883.

Harvey K, Duguid IC, Alldred MJ *et al.* (2004). The GDP–GTP exchange factor collybistin: an essential determinant of neuronal gephyrin clustering. *J Neurosci* **24**, 5816–5826.

Hauser J, Rudolph U, Keist R *et al.* (2005). Hippocampal α$_5$ subunit containing GABA$_A$ receptors modulate expression of prepulse inhibition. *Mol Psychiatry* **10**, 201–207.

Herring D, Huang RQ, Singh M *et al.* (2003). Constitutive GABA$_A$ receptor endocytosis is dynamin-mediated and dependent on a dileucine AP2 adaptin-binding motif within the β2 subunit of the receptor. *J Biol Chem* **278**, 24046–24052.

Hinkle DJ and Macdonald RL (2003). β subunit phosporylation selectivity increases fast desensitization and prolongs deactivation of α1 β1 and γ2$_L$ and α1 β3 γ2$_L$ GABA$_A$ receptor currents. *J Neurosci* **23**, 11698–11710.

Huerta PT, Sun LD, Wilson MA *et al.* (2000). Formation of temporal memory requires NMDA receptors within CA1 pyramidal neurons. *Neuron* **25**, 473–480.

Im WB, Pregenzer JF, Binder JA *et al.* (1995). Chloride channel expression with the tandem construct of α6–β2 GABA$_A$ receptor subunit requires a monomeric subunit of α6 or γ2. *J Biol Chem* **270**, 26063–26066.

Jacob TC, Bogdanov YD, Magnus C *et al.* (2005). Gephyrin regulates the cell surface dynamics of synaptic GABA$_A$ receptor. *J Neurosci* **25**, 10469–10478.

Jia F, Pignataro L, Schofield CM *et al.* (2005). An extrasynaptic GABA$_A$ receptor mediates tonic inhibition in thalamic VB neurons. *J Neurophysiol* **94**, 4491–4501.

Jones A, Korpi ER, McKernan RM *et al.* (1997). Ligand-gated ion channel subunit partnerships: GABA$_A$ receptor α6 subunit gene inactivation inhibits δ subunit expression. *J Neurosci* **17**, 1350–1362.

Jovanovic JN, Thomas P, Kittler JT *et al.* (2004). Brain-derived neurotrophic factor modulates fast synaptic inhibition by regulating GABA$_A$ receptor phosphorylation, activity and cell-surface stability. *J Neurosci* **24**, 522–530.

Jurd R, Arras M, Lambert S *et al.* (2003). General anesthetic actions *in vivo* strongly attenuated by a point mutation in the GABA(A) receptor β3 subunit. *Faseb J* **17**, 250–252.

Kanematsu T, Yasunaga A, Mizoguchi Y *et al.* (2006). Modulation of GABA$_A$ receptor phosphorylation and membrane trafficking by phospholipase C-related inactive protein/protein phosphatase 1 and 2A signaling complex underlying BDNF-dependent regulation of GABAergic inhibition. *J Biol Chem* **281**, 22180–22189.

Kapur J and Macdonald RL (1996). Pharmacological properties of γ-aminobutyric acid$_A$ receptors from acutely dissociated rat dentate granule cells. *Mol Pharmacol* **50**, 458–466.

Keller CA, Yuan X, Panzanelli P *et al.* (2004). The γ2 subunit of GABA$_A$ receptors is a substrate for palmitoylation by GODZ. *J Neurosci* **24**, 5881–5891.

Kittler JT and Moss SJ (2003). Modulation of GABA$_A$ receptor activity by phosphorylation and receptor trafficking: implications for the efficacy of synaptic inhibition. *Curr Opin Neurobiol* **13**, 341–347.

Kittler JT, Chen G, Honing S *et al.* (2005). Phospho-dependent binding of the clathrin AP2 adaptor complex to GABA$_A$ receptors regulates the efficacy of inhibitory synaptic transmission. *Proc Natl Acad Sci USA* **102**, 14871–14876.

Kneussel M (2005). Postsynaptic scaffold proteins at non-synaptic sites. The role of postsynaptic scaffold proteins in motor–protein-receptor complexes. *EMBO Rep* **6**, 22–27.

Koch, C (2004). *The quest of consciousness, a neurobiological approach.* Roberts and Company, Englewood.

Koksma JJ, Van Kesteren RE, Rosahl TW *et al.* (2003). Oxytoxin regulates neurosteroid modulation of GABA$_A$ receptors in supraoptic nucleus around parturition. *J Neurosci* **23**, 788–797.

Kopp C, Rudolph U and Tobler I (2004a). Sleep EEG changes after zolpidem in mice. *Neuroreport* **15**, 2299–2302.

Korpi ER, Mihalek RM, Sinkkonen ST *et al.* (2002). Altered receptor subtypes in the forebrain of GABA$_A$ receptor δ subunit-deficient mice: recruitment of γ2 subunits. *Neuroscience* **109**, 733–743.

Kralic JE, Sidler C, Parpan F *et al.* (2006). Compensatory alteration of inhibitory synaptic circuits in thalamus and cerebellum of GABA$_A$ receptor α1 subunit knockout mice. *J Comp Neurol* **495**, 408–421.

Krishek BJ, Xie X, Blackstone C *et al.* (1994). Regulation of GABA$_A$ receptor function by protein kinase C phosphorylation. *Neuron* **12**, 1081–1095.

Krogsgaard-Larsen P, Frolund B, Liljefors T *et al.* (2004). GABA$_A$ agonists and partial agonists: THIP (gaboxadol) as a non-opioid analgesic and a novel type of hypnotic. *Biochem Pharmacol* **68**, 1573–1580.

Lambert S, Arras M, Vogt KE *et al.* (2005). Isoflurane-induced surgical tolerance mediated only in part by beta3-containing GABA(A) receptors. *Eur J Pharmacol* **516**, 23–27.

Langen B, Egerland U, Bernoster K *et al.* (2005). Characterization in rats of the anxiolytic potential of ELB139 [1-(4-chlorophenyl)-4-piperidin-1-yl-1,5-dihydro-imidazol-2-on], a new agonist at the benzodiazepine binding site of the GABA$_A$ receptor. *J Pharmacol Exp Ther* **314**, 717–724.

Lewis DA, Hashimoto T and Volk DW (2005). Cortical inhibitory neurons and schizophrenia. *Nature Rev Neurosci* **6**, 312–324.

Li H, Wu J, Huguenard JR *et al.* (2006). Selective changes in thalamic and cortical GABA$_A$ receptor subunits in a model of acquired absence epilepsy in the rat. *Neuropharmacol* **51**,121–128.

Liang J, Zhang N, Cagetti E *et al.* (2006). Chronic intermittent ethanol-induced switch of ethanol actions from extrasynaptic to synaptic hippocampal GABA$_A$ receptors. *J Neurosci* **26**, 1749–1758.

Liao M, Sonner JM, Jurd R *et al.* (2005). β3-containing γ-aminobutyric acidA receptors are not major targets for the amnesic and immobilizing actions of isoflurane. *Anesth Analg* **101**, 412–418.

Licata SC and Rowlett JK (2008). Abuse and dependence liability of benzodiazepine-type drugs: GABA$_A$ receptor modulation and beyond. *Pharmacol Biochem Behav* (in press).

Licata SC, Platt DM, Cook JM *et al.* (2005). Contribution of GABAA receptor subtypes to the anxiolytic-like, motor and discriminative stimulus effects of benzodiazepines: studies with the functionally selective ligand SL651498. *J Pharmacol Exp Ther* **313**, 1118–1125.

Lippa A, Czobor P, Stark J *et al.* (2005). Selective anxiolysis produced by ocinaplon, a GABA(A) receptor modulator. *Proc Natl Acad Sci USA* **102**, 7380–7385.

Loebrich S, Bahring R, Katsuno T *et al.* (2006). Activated radixin is essential for GABA$_A$ receptor α5 subunit anchoring at the actin cytoskeleton. *EMBO J* **25**, 987–999.

Löw K, Crestani F, Keist R *et al.* (2000). Molecular and neuronal substrate for the selective attenuation of anxiety. *Science* **290**, 131–134.

Lüscher B and Fritschy JM *et al.* (2001). Subcellular localization and regulation of GABA$_A$ receptors and associated proteins. *Int Rev Neurobiol* **48**, 31–64.

Lüscher B and Keller CA (2004). Regulation of GABA$_A$ receptor trafficking, channel activity and functional plasticity of inhibitory synapses. *Pharmacol Ther* **102**, 195–221.

Maguire JL, Stell BM, Rafizadeh M *et al.* (2005). Ovarian cycle-linked changes in GABA$_A$ receptors mediating tonic inhibition alter seizure susceptibility and anxiety. *Nature Neurosci* **8**, 797–804.

Mäkelä R, Uusi-Oukari M, Homanics GE *et al.* (1997). Cerebellar γ-aminobutyric acid type A receptors: pharmacological subtypes revealed by mutant mouse lines. *Mol Pharmacol* **52**, 380–388.

Man HY, Lin J, Ju WH *et al.* (2000). Regulation of AMPA receptor-mediated synaptic transmission by clathrin-dependent receptor internalization. *Neuron* **25**, 649–662.

Marowsky A, Fritschy JM and Vogt KE (2004). Functional mapping of GABA$_A$ receptor subtypes in the amygdala. *Eur J Neurosci* **20**, 1281–1289.

Marutha Ravindran CR and Ticku MK (2006). Tyrosine kinase phosphorylation of GABA$_A$ receptor subunits following chronic ethanol exposure of cultured cortical neurons of mice. *Brain Res* **1086**, 35–41.

McDonald BJ, Amato A, Connolly CN *et al.* (1998). Adjacent phosphorylation sites on GABA$_A$ receptor β subunits determine regulation by cAMP-dependent protein kinase. *Nature Neurosci* **1**, 23–28.

McHugh TJ, Blum KI, Tsien JZ *et al.* (1996) Impaired hippocampal representation of space in CA1-specific NMDAR1 knockout mice. *Cell* **87**, 1339–1349.

McKernan RM, Rosahl TW, Reynolds DS *et al.* (2000). Sedative but not anxiolytic properties of benzodiazepines are mediated by the GABA$_A$ receptor α$_1$ subtype. *Nature Neurosci* **3**, 587–592.

Mihalek RM, Banerjee PK, Korpi ER *et al.* (1999). Attenuated sensitivity to neuroactive steroids in γ-aminobutyrate type A receptor δ subunit knockout mice. *Proc Natl Acad Sci USA* **96**, 12905–12910.

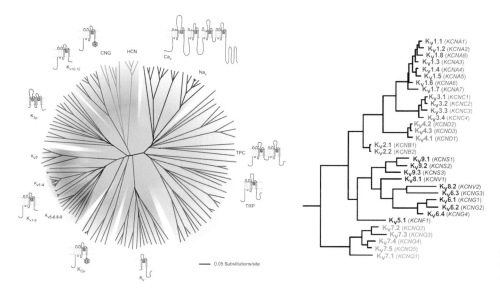

Kᵥ1.1 (KCNA1)
Kᵥ1.2 (KCNA2)
Kᵥ1.8 (KCNA8)
Kᵥ1.3 (KCNA3)
Kᵥ1.4 (KCNA4)
Kᵥ1.5 (KCNA5)
Kᵥ1.6 (KCNA6)
Kᵥ1.7 (KCNA7)
Kᵥ3.1 (KCNC1)
Kᵥ3.2 (KCNC2)
Kᵥ3.3 (KCNC3)
Kᵥ3.4 (KCNC4)
Kᵥ4.2 (KCND2)
Kᵥ4.3 (KCND3)
Kᵥ4.1 (KCND1)
Kᵥ2.1 (KCNB1)
Kᵥ2.2 (KCNB2)
Kᵥ9.1 (KCNS1)
Kᵥ9.2 (KCNS2)
Kᵥ9.3 (KCNS3)
Kᵥ8.1 (KCNV1)
Kᵥ8.2 (KCNV2)
Kᵥ6.3 (KCNG3)
Kᵥ6.1 (KCNG1)
Kᵥ6.2 (KCNG2)
Kᵥ6.4 (KCNG4)
Kᵥ5.1 (KCNF1)
Kᵥ7.2 (KCNQ2)
Kᵥ7.3 (KCNQ3)
Kᵥ7.4 (KCNQ4)
Kᵥ7.5 (KCNQ5)
Kᵥ7.1 (KCNQ1)

Fig. 2.1 The voltage-gated ion channel superfamily illustrating similarity between voltage gated potassium channels and other family members.

Fig. 2.1.1.1 Comparison of a modelled Kv4.3–KChIP1–DPPX channel complex with the Kv1.2–Kvβ2 channel complex. Left: side view of the Kv1.2–Kvβ2 channel complex. The four Kv1.2 subunits are in cyan, yellow, pink and green. The four Kvβ2 subunits are in wheat colour. The upper third shows the transmembrane part of the channel. The middle portion of the figure contains the T1 domains interacting with the KChIP proteins at the bottom. Right: side on view of the K4.3–KChIP1–DPPX channel complex. The four Kv4.3 subunits are in cyan, yellow, pink and green. The four KChIP1 proteins, also interacting with the channels T1 domains, are in blue. Only two of the likely four DPPX molecules in the channel are shown.

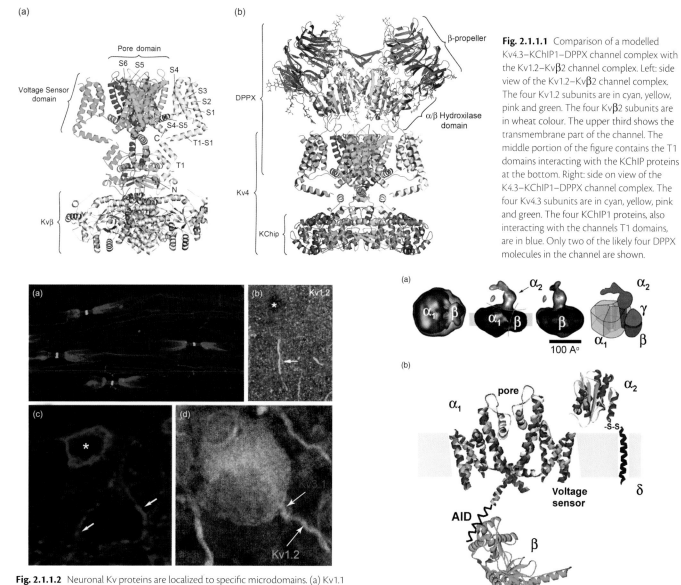

Fig. 2.1.1.2 Neuronal Kv proteins are localized to specific microdomains. (a) Kv1.1 (red) in the juxtaparanodal region of myelinated axons; (green) Na⁺ channels at the node of Ranvier; (blue) NCP1 at the paranode. (b) Kv1.2 in the axon initial segment of cortical pyramidal neurons (arrow). * = cell body. (c) Kv3.2 in the somatic membrane of a cortical interneuron (*) and the perisomatic (basket) terminals innervating pyramidal cells (d) Kv3.1 in the calyx of Held terminal and Kv1.2 in the pre-terminal axon.

Fig. 2.2.2 Putative structure of a voltage-gated Ca²⁺ channel including auxiliary subunits. (a) Structure of the skeletal muscle Caᵥ1.1 channel as deduced using electron cryomicroscopy. (b) Hypothetical structure of the channel based on the crystal structure of homologous proteins.

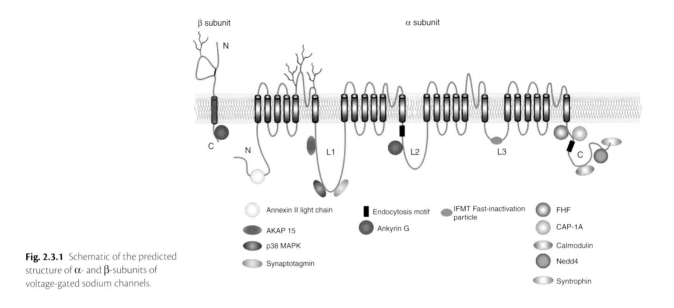

β subunit   α subunit

N   N

C   N   L1   L2   L3   C

⬤ Annexin II light chain   ■ Endocytosis motif   ⬤ IFMT Fast-inactivation particle   ⬤ FHF

⬤ AKAP 15   ⬤ Ankyrin G   ⬤ CAP-1A

⬤ p38 MAPK   ⬤ Calmodulin

⬤ Synaptotagmin   ⬤ Nedd4

                                                                                       ⬤ Syntrophin

Fig. 2.3.1 Schematic of the predicted structure of α- and β-subunits of voltage-gated sodium channels.

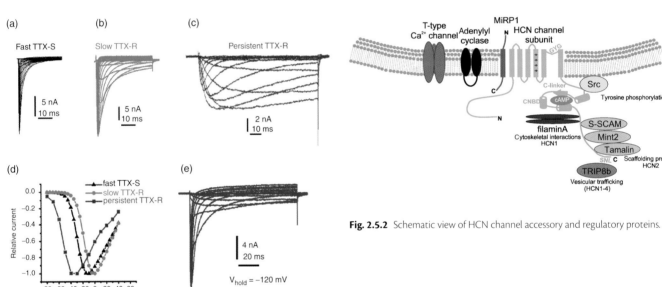

(a) Fast TTX-S   (b) Slow TTX-R   (c) Persistent TTX-R

5 nA / 10 ms   5 nA / 10 ms   2 nA / 10 ms

(d)   (e)

▲ fast TTX-S
● slow TTX-R
■ persistent TTX-R

Relative current

Voltage (mV)

4 nA / 20 ms

$V_{hold} = -120$ mV

Fig. 2.3.2 Multiple sodium currents are expressed in neurons including small DRG neurons. Family traces of (a) fast-inactivating, (b) slowly-inactivating and (c) persistent sodium currents and (d) their current-voltage relationship. (e) Multiple sodium currents are recorded in DRG small (<30 mm diameter) neurons.

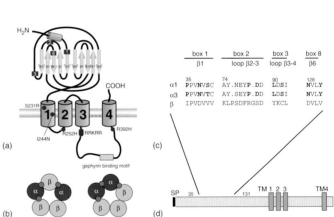

Fig. 2.5.2 Schematic view of HCN channel accessory and regulatory proteins.

H₂N   COOH

S231R   I244N   R252H RRKRR R392H

gephyrin binding motif

(a)

(b)

(c)

	box 1 β1	box 2 loop β2-3	box 3 loop β3-4	box 8 β6
α1	35 PPVNVSC	74 AY.NEYP.DD	90 LDSI	128 NVLY
α3	PPVNVTC	AY.SEYP.DD	LDSI	NVLY
β	IPVDVVV	KLPSDFRGSD	YKCL	DVLV

(d)

SP 35   131   TM 1 2 3   TM4

Fig. 3.1.3.2 Topology of the glycine receptor. A single subunit (a) is composed of 4 TMs (cyan barrels 1–4), which are connected by two small loops (TM1–2, TM2–3), and a long intracellular loop (TM3–4). (b) GlyRs form heteropentameric ion channels consisting of α- (blue circles) and β-subunits (yellow circles) with a 2α:3β stoichiometry. (c and d) Bar diagram of a single subunit (d), illustrating in (c) the N-terminal assembly boxes important for homo-oligomerization.

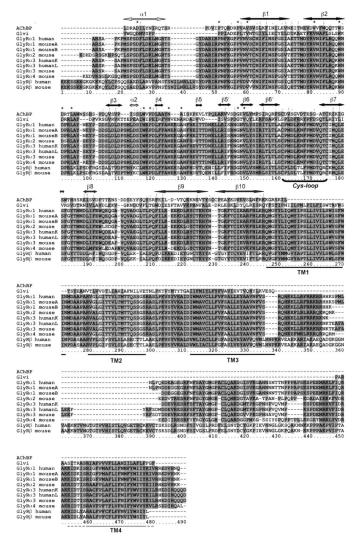

Fig. 3.1.3.1 Amino acid sequence alignment of Cys-loop receptor variants.

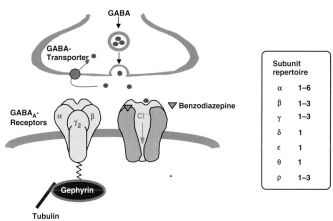

Fig. 3.1.4.1 Scheme of GABAergic synapse depicting major elements of signal transduction.

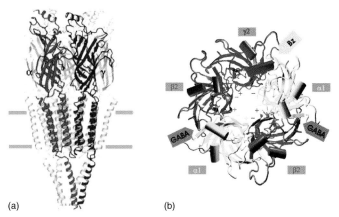

(a)   (b)

Fig. 3.1.4.2 The nicotinic acetylcholine receptor and an acetylcholine binding protein (AChBP) as a blueprint of the GABAA receptor structure. (a) Ribbon representation of the nicotinic acetylcholine receptor (nAChR). (b) Model of the GABA$_A$ receptor extracellular domain as viewed from the synaptic cleft.

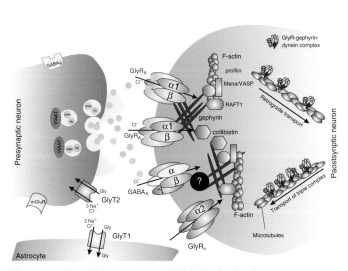

Fig. 3.1.3.3 The inhibitory synapse, with GlyRs localized at the postsynaptic membrane.

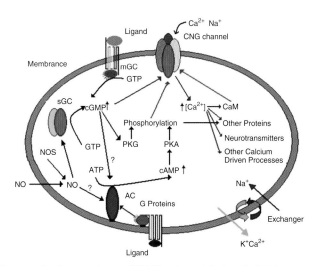

Fig. 4.2.2 Regulatory pathways of CNG ion channels in the cell. CNG ion channels are opened by cGMP generated by membrane bound guanylyl cyclase (mGC), and\or NO activation of soluble guanylyl cyclase (sGC). Ca^{2+} entering through the channel initiates numerous calcium-driven processes including Ca^{2+}/CaM-mediated negative feedback. CNG channel activity is also regulated by cGMP and cAMP-dependent pathways.

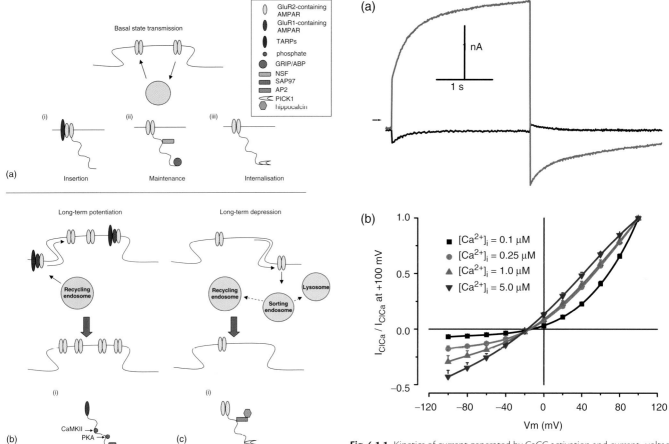

Fig. 3.2.1.2 AMPAR trafficking in synaptic plasticity. (a) During basal conditions, AMPARs undergo constitutive cycles of endocytosis from and reinsertion into the synaptic membrane. (b) In LTP, GluR1-containing receptors are inserted into the membrane and are later replaced by GluR2/3 receptors. (c) There is a net loss of AMPARs at the synaptic membrane during LTD.

Fig. 4.1.1 Kinetics of current generated by CaCC activation and current–voltage relationships at different $[Ca^{2+}]_i$. (a) $I_{Cl(Ca)}$ traces at +100 (red) and −100 mV (black) from a parotid acinar cell with $[Ca^{2+}]_i$ = 250 nM held at −50 mV. Arrow indicates 0 current level. (b) I–V curves obtained with different free $[Ca^{2+}]$ levels.

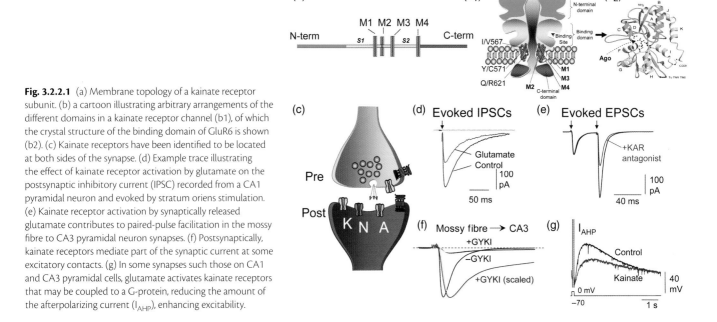

Fig. 3.2.2.1 (a) Membrane topology of a kainate receptor subunit. (b) a cartoon illustrating arbitrary arrangements of the different domains in a kainate receptor channel (b1), of which the crystal structure of the binding domain of GluR6 is shown (b2). (c) Kainate receptors have been identified to be located at both sides of the synapse. (d) Example trace illustrating the effect of kainate receptor activation by glutamate on the postsynaptic inhibitory current (IPSC) recorded from a CA1 pyramidal neuron and evoked by stratum oriens stimulation. (e) Kainate receptor activation by synaptically released glutamate contributes to paired-pulse facilitation in the mossy fibre to CA3 pyramidal neuron synapses. (f) Postsynaptically, kainate receptors mediate part of the synaptic current at some excitatory contacts. (g) In some synapses such those on CA1 and CA3 pyramidal cells, glutamate activates kainate receptors that may be coupled to a G-protein, reducing the amount of the afterpolarizing current (I_{AHP}), enhancing excitability.

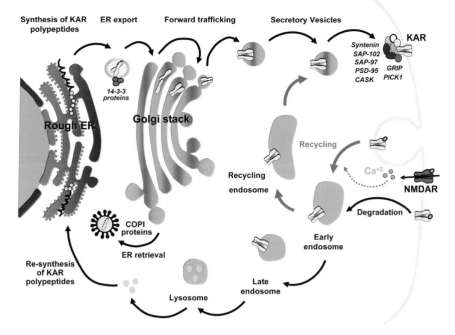

Fig. 3.2.2.2 Intracellular trafficking of kainate receptors mediated by the interactions with receptor-binding proteins.

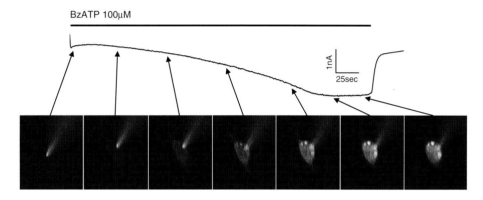

Fig. 3.3.1.1 Whole-cell current elicited by BzATP (100 μM) in a single HEK-293 cell expressing hP2X$_7$.

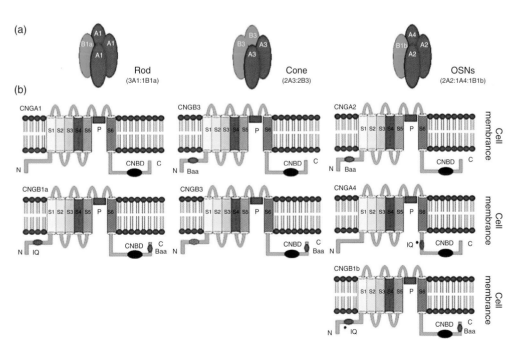

Fig. 4.2.1 Subunits and stoichiometry of CNG ion channels in photoreceptors and OSNs. (a) Subunit composition of rod, cone and olfactory CNG channels.
(b) Topological models of CNG channel subunits of rod, cone and OSNs. A site in CNGB3 lacking a critical signature residue of an IQ sequence is depicted in grey.

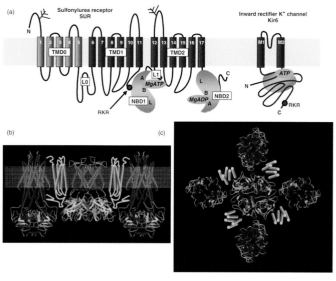

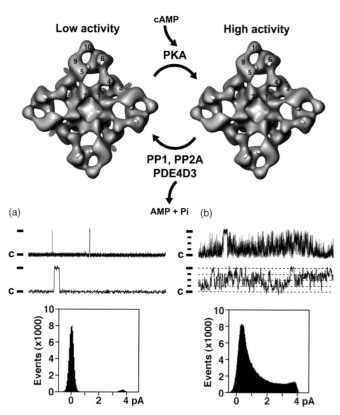

Fig. 4.4.3.1 Secondary structure and molecular models of the 2 subunits of the K_{ATP} channel. (a) Kir6.x has a large cytoplasmic domain harbouring an inhibitory ATP binding site formed by the proximal C-terminus of one subunit and N-terminus of its neighbour. In SUR nucleotide binding domains (NBD) 1 and 2 probably function as a dimer with nucleotide sites made up of Walker A and B of one NBD and linker L of the other. (b (side view) and c (extracellular view)) Homology model of a Kir6.2 tetramer based on the KirBac1.1 structure (centre) and model of Sav1866 together with a cartoon representation of TMD 0 and L0 in yellow. (α-helices are red; β-sheets are blue; ATP is green).

Fig. 4.5.2 RyR2 function is regulated by cAMP-dependent PKA phosphorylation.

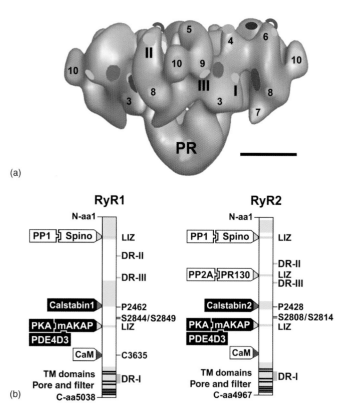

Fig. 4.5.1 The homotetrameric RyR channel complex. (a) Side view of the RyR2 channel comprised of four identical subunits by cryo-EM surface reconstruction. (b) Schematic of the peptide sequence of RyR1 and RyR2 showing linear organization and putative C-terminal transmembrane (8-TM) pore domain.

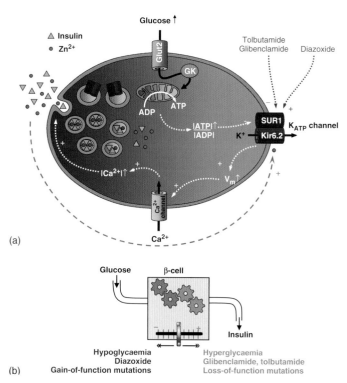

Fig. 4.4.3.4 K_{ATP} channels and insulin secretion in pancreatic β-cells. In pancreatic β-cells, K_{ATP} channel blockers like sulfonylureas mimic high glucose and upregulate secretion while openers like diazoxide down regulate secretion.

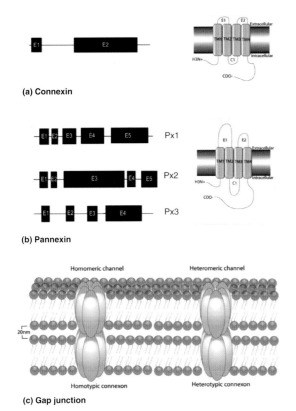

(a) Connexin

(b) Pannexin

(c) Gap junction

Fig. 5.2.1 Schematic representation of connexin and pannexin genes (left hand schematics in a and b) and proteins. Connexin assembly into gap junctions (c).

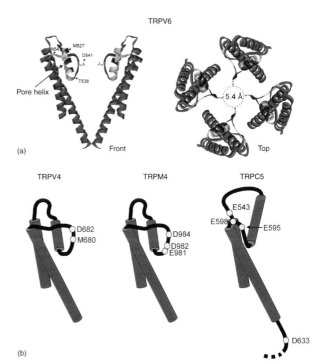

(a)

(b)

Fig. 5.1.2 Predicted topologies of the pore region of TRPV, TRPM and TRPC channels. (a) Structural model of the TRPV6 pore region (left = side-on, right = top-down views). At the narrowest point, the pore is formed by the acidic side chain of Asp541 (orange). Blue residues = TM5 and TM6. Amino acids that were subjected to SCAM analysis (residues P526 to N547) are coloured: red for rapid reactivity, yellow for slow reactivity with Ag+ (< 5x10⁶ M⁻¹s⁻¹) and green for no reactivity. Grey residues denote cysteine substitution induced loss of function. (b) Schematic representation of crucial residues in the putative selectivity filters of TRPV4, TRPM4 and TRPC5 channels.

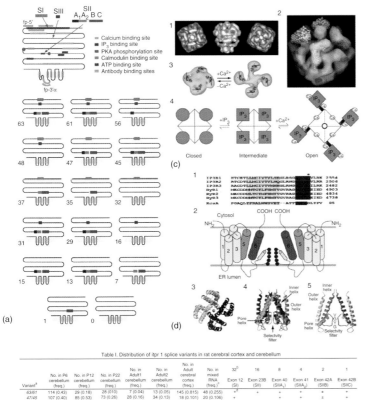

Fig. 4.6.1 (a) Schematic representation of naturally occurring rat brain *itpr1* splice variants (numbered in Table 4.6.1). (b) Developmental expression of IP₃R splice variants. (c) EM derived 3D reconstructions of the type 1 IP₃R. (1) Highest resolution in the absence of InsP₃ and Ca²⁺. (2) Modelling of IP₃R core in the InsP₃-free state into the density map of the whole receptor. (3) Global structural change resulting from the presence of Ca²⁺ (IP₃R core in blue; InsP₃ in red) (4) Possible 3-state model for receptor activation. (d) Structure of the IP₃R channel domain and its pore. (1) Sequence alignment of IP₃Rs and RyRs at the pore helix (yellow) and the selectivity filter (red). (2) Model of the channel domain for the two IP₃R monomers based on the KcsA structure. Helix 5 and 6 correspond to outer and inner helices of KcsA. (3) Structure of the KcsA potassium channel comprising four monomers (green, blue, red and yellow) viewed from the extracellular side. (4) Side view of KcsA channel. (5) Side view of the two KcsA monomers (outer helix (light grey), inner helix (dark grey), pore helix (yellow) and selectivity filter (red)).

Table I. Distribution of itpr 1 splice variants in rat cerebral cortex and cerebellum

Variant[a]	No. in P6 cerebellum (freq.)	No. in P12 cerebellum (freq.)	No. in P22 cerebellum (freq.)	No. in Adult1 cerebellum (freq.)	No. in Adult2 cerebellum (freq.)	No. in Adult cerebral cortex (freq.)	No. in Adult mixed RNA (freq.)[c]	32[b] Exon 12 (SI)	16 Exon 23B (SII)	8 Exon 40 (SIIA₁)	4 Exon 41 (SIIA₂)	2 Exon 42A (SIIB)	1 Exon 42B (SIIC)
63:61	114 (0.43)	29 (0.18)	28 (010)	7 (0.04)	13 (0.05)	145 (0.815)	48 (0.255)	+[e]	+				
47:45	107 (0.40)	85 (0.53)	73 (0.26)	28 (0.16)	34 (0.13)	18 (0.101)	20 (0.106)	+		+			
31:29	8 (0.03)	5 (0.03)	34 (013)	39 (0.22)	62 (0.24)	5 (0.028)	29 (0.154)		+	+			
15/13	11 (0.04)	32 (0.20)	138 (0.50)	101 (0.56)	145 (0.56)	0	90 (0.479)			+			
0	3 (0.01)	1 (0.006)	1 (0.006)	1 (0.006)	3 (0.01)	4 (0.022)	0						
48	0	1 (0.006)	1 (0.004)	1 (0.006)	0	3 (0.017)	0		+				
32	16 (0.06)	6 (0.04)	0	0	1 (0.004)	0	0	+					
16	1 (0.004)	0	1 (0.004)	0	0	3 (0.017)	1 (0.005)	+	+				
35	1 (0.004)	0	0	1 (0.006)	0	0	0					+	+
1	0	0	0	0	0	1 (0.006)	0	+					
56	0	1 (0.006)	0	0	0	0	0	+	+				+
7	0	1 (0.006)	0	0	0	0	0	+				+	
37	2 (0.008)	0	0	0	0	0	0	+				+	+
Total clones	263	162	277	180	258	179	188						

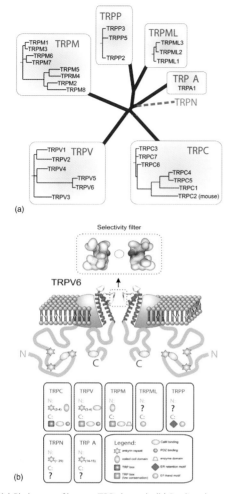

(a)

(b)

Fig. 5.1.1 (a) Phylogeny of human TRP channels. (b) Predicted structural topology of TRPV6. The structure of TM1–4 is currently unknown. Structural motifs found in the N- and C-terminal tails of the different TRP subfamilies are provided for each subfamily of TRP channels.

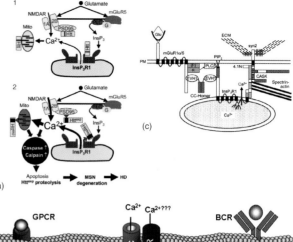

(a)

(c)

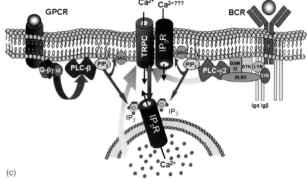

(c)

Fig. 4.6.2 (a) Huntington expression causes cytosolic and mitochondrial Ca^{2+} overload and apoptosis of striatal medium spiny neurons through activation of NR1A/NR2B NMDA and mGlu5 receptors. (b) Hypothetical marcomolecular signalling complex linking mGluR1/5 to syndecan-2 (syn2) through type 1 IP_3R. (c) **Multiple roles of** IP_3R in regulating intracellular Ca^{2+} levels following activation of G protein-coupled and tyrosine kinase-coupled receptors in B cells.

| (1) Receptor binding (cell membrane) - Glutamate binds to the metabotropic glutamate receptor (mGluR) activating PLC to generate $InsP_3$. This $InsP_3$ can activate IP_3R in the ER and IP_3R which is physically coupled to the mGluR receptor by Homer. | (2) Channel activation (cell membrane) - $InsP_3$ generated by PLC can bind to IP_3R either resident or juxtaposed to the plasma membrane. This could activate Ca^{2+} influx through cell membrane IP_3R, or via physical coupling to TRP channels. | (3) Enzyme activation (cell membrane) - Increased Ca^{2+} levels can activate $PLC\delta$ to generate more $InsP_3$. Ca^{2+} can also activate Ca^{2+} sensitive adenylyl cyclases which generate cAMP. cAMP stimulates PKA phosphorylation of the IP3R, leading to more Ca^{2+} release. |

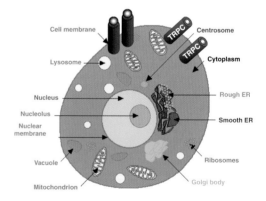

| (4) Transcription (cytosol/nucleus) Ca^{2+} generated in response to mGluR stimulation activates transcription factors which translocate to the nucleus initiating gene transcription. Gene transcription can be further regulated by nuclear IP_3Rs. | (5) Translation (cytosol/ER/golgi/ secretory vesicles) After transcription, mRNA is translated into proteins which are trafficked through the ER, golgi, and secretory vesicles, all of which can contain and are regulated by IP_3R. | (6) Excitotoxicity (mitochondria /ER/ cytosol) If the glutamate levels remain high, excitotoxicity occurs inducing apoptosis via the IP_3R overloading the mitochondria with Ca^{2+} This causes the release of cytochrome C from the mitochondrion which locks the ER resident IP_3Rs in an open conformation which poisons the mitochondria with Ca^{2+}, depletes ER Ca^{2+}, and creates high cytosolic Ca^{2+} which increases apoptotic protein activity. |

Fig. 4.6.3 Multiple areas of influence of the IP_3R within the mGluR signalling pathway.

Mody I and Pearce RA (2004). Diversity of inhibitory neurotransmission through GABA_A receptors. *Trends Neurosci* **27**, 569–575.

Möhler H (2002). Pathophysiological aspects of diversity in neuronal inhibition: a new benzodiazepine pharmacology. *Dialogues in Clinical Neuroscience* **4**, 261–269.

Möhler H (2007). Molecular regulation of cognitive functions and developmental plasticity: impact of GABA_A receptors. *J Neurochem* **102**,1–12.

Möhler H, Fritschy JM and Rudolph U (2002). A new benzodiazepine pharmacology. *J Pharm Exp Ther* **300**, 2–8.

Möhler H, Fritschy JM, Vogt K *et al.* (2005). Pathophysiology and pharmacology of GABA_A receptors. *Hanb Exp Pharmacol* **169**, 225–247.

Morris RGM (1989). Synaptic plasticity and learning: selective impairment of learning in rats and blockade of long-term potentiation *in vivo* by the N-methyl-D-aspartate receptor antagonist AP5. *J Neurosci* **9**, 3040–3057.

Morris RGM anderson A, Lynch GS *et al.* (1986). Selective impairment of learning and blockade of long-term potentiation by an N-methyl-D-aspartate receptor antagonist AP5. *Nature* **319**, 774–776.

Moss SJ, Gorrie GH, Amato A *et al.* (1995). Modulation of GABA_A receptors by tyrosine phosphorylation. *Nature* **377**, 344–348.

Mtchedlishvili Z and Kapur J (2006). High-affinity, slowly desensitizing GABA_A receptors mediate tonic inhibition in hippocampal dentate granule cells. *Mol Pharmacol* **69**, 564–575.

Nakazawa K, Quirk MC, Chitwood RA *et al.* (2002). Requirement for hippocampal CA3 NMDA receptors in associative memory recall. *Science* **297**, 211–218.

Navarro JF, Buron E and Martin-Lopez M (2002). Anxiogenic-like activity of L-655,708, a selective ligand for the benzodiazepine site of GABA(A) receptors which contain the α5 subunit, in the elevated plus-maze test. *Prog Neuropsychopharmacol Biol Psychiatry* **26**, 1389–1392.

Navarro JF, Buron E and Martin-Lopez M (2004). Behavioral profile of L-655,708, a selective ligand for the benzodiazepine site of GABA_A receptors which contain the α_5 subunit in social encounters between male mice. *Aggress Behav* **30**, 319–325.

Nishimura T, Schwarzer C, Gasser E *et al.* (2005). Altered expression of GABA_A and GABA_B receptor subunit mRNAs in the hippocampus after kindling and electrically induced status epilepticus. *Neurosci* **134**, 691–704.

Nusser Z, Ahmad Z, Tretter V *et al.* (1999). Alterations in the expression of GABA_A receptor subunits in cerebellar granule cells after the disruption of the α6 subunit gene. *Eur J Neurosci* **11**, 1685–1697.

Nusser Z and Mody I (2002). Selective modulation of tonic and phasic inhibitions in dentate gyrus granule cells. *J Neurophysiol* **87**, 2624–2628.

Nusser Z, Roberts JDB, Baude A *et al.* (1995a). Relative densities of synaptic and extrasynaptic GABA_A receptors on cerebellar granule cells as determined by a quantitative immunogold method. *J Neurosci* **15**, 2948–2960.

Nusser Z, Roberts JDB, Baude A *et al.* (1995b). Immunocytochemical localization of the α1 and β2/3 subunits of the GABA_A receptor in relation to specific GABAergic synapses in the dentate gyrus. *Eur J Neurosci* **7**, 630–646.

Nusser Z, Sieghart W and Somogyi P (1998). Segregation of different GABA_A receptors to synaptic and extrasynaptic membranes of cerebellar granule cells. *J Neurosci* **18**, 1693–1703.

Nusser Z, Sieghart W, Benke D *et al.* (1996a). Differential synaptic localization of two major g-aminobutyric acid type A receptor a subunits on hippocampal pyramidal cells. *Proc Natl Acad Sci USA* **93**, 11939–11944.

Ogris W, Lehner R, Fuchs K *et al.* (2006). Investigation of the abundance and subunit composition of GABA_A receptor subtypes in the cerebellum of α1-subunit-deficient mice. *J Neurochem* **96**, 136–147.

Pearce RA (1993). Physiological evidence for two distinct GABA_A responses in rat hippocampus. *Neuron* **10**, 189–200.

Peng Z, Hauer B, Mihalek RM *et al.* (2002). GABA_A receptor changes in δ subunit-deficient mice: altered expression of α4 and γ2 subunits in the forebrain. *J Comp Neurol* **446**, 179–197.

Peng ZC, Huang CS, Stell BM *et al.* (2004). Altered expression of the δ subunit of the GABA_A receptor in a mouse model of temporal lobe epilepsy. *J Neurosci* **24**, 8629–8639.

Pirker S, Schwarzer C, Wieselthaler A *et al.* (2000). GABA_A receptors: immunocytochemical distribution of 13 subunits in the adult rat brain. *Neurosci* **101**, 815–850.

Porcello DM, Huntsman MM, Mihalek RM *et al.* (2003). Intact synaptic GABAergic inhibition and altered neurosteroid modulation of thalamic relay neurons in mice lacking δ subunit. *J Neurophysiol* **89**, 1378–1386.

Prenosil GA, Schneider Gasser EM, Rudolph U *et al.* (2006). Specific subtypes of GABA_A receptors mediate phasic and tonic forms of inhibition in hippocampal pyramidal neurons. *J. Neurophysiol.* **96**, 846–857.

Rathenberg J, Kittler JT and Moss SJ (2004). Palmitoylation regulates the clustering and cell surface stability of GABA_A receptors. *Mol Cell Neurosci* **26**, 251–257.

Reynolds DS, Rosahl TW, Cirone J *et al.* (2003). Sedation and anesthesia mediated by distinct GABA(A) receptor isoforms. *J Neurosci* **23**, 8608–8617.

Roberts DS, Raol YH, Bandyopadhyay S *et al.* (2005). Egr3 stimulation of GABRA4 promoter activity as a mechanism for seizure-induced upregulation of GABA_A receptor α4 subunit expression. *Proc Natl Acad Sci USA* **102**, 11894–11899.

Rossi DJ, Hamann M and Attwell D (2003). Multiple modes of GABAergic inhibition of rat cerebellar granule cells. *J Physiol* **548**, 97–110.

Rowlett JK, Duke AN and Platt DM (2007). Abuse and dependence liability of GABA_A receptor modulators. In SJ Enna and H Möhler, eds, *The GABA receptors* Humana Press, Totowa, NJ, pp. 143–168.

Rowlett JK, Platt DM, Lelas S *et al.* (2005). Different GABA_A receptors subtypes mediate the anxiolytic, abuse-related and motor effects of benzodiazepine-like drugs in primates. *Proc Natl Acad Sci USA* **102**, 915–920.

Rudolph U and Antkowiak B (2004). Molecular and neuronal substrates for general anaesthetics. *Nat Rev Neurosci* **5**, 709–720.

Rudolph U and Möhler H (2004). Analysis of GABA_A receptor function and dissection of the pharmacology of benzodiazepines and general anesthetics though mouse genetics. *Ann Rev Toxicol Pharmacol* **44**, 475–498.

Rudolph U and Möhler H (2006). GABA-based therapeutic approaches: GABA_A receptor subtype functions. *Curr Opin Pharmacol* **6**, 18–23.

Rudolph U, Crestani F and Möhler H (2001). GABA_A receptor subtypes: dissecting their pharmacological functions. *Trends Pharmacol Sci* **22**, 188–194.

Rudolph U, Crestani F, Benke D *et al.* (1999). Benzodiazepine actions mediated by specific γ-aminobutyric acid_A receptor subtypes. *Nature* **401**, 796–800.

Russek SJ (1999). Evolution of GABA_A receptor diversity in the human genome. *Gene* **227**, 213–222.

Sassoè-Pognetto M and Fritschy JM (2000). Gephyrin, a major postsynaptic protein of GABAergic synapses. *Eur J Neurosci* **7**, 2205–2210.

Schweizer C, Balsiger S, Bluethmann H *et al.* (2003). The γ2 subunit of GABA_A receptors is required for maintenance of receptors at mature synapses. *Mol Cell Neurosci* **24**, 442–450.

Scimemi A, Semyanov A, Sperk G *et al.* (2005). Multiple and plastic receptors medate tonic GABA_A receptor currents in the hippocampus. *J Neurosci* **25**, 10016–10024.

Semyanov A, Walker MC, Kullmann DM *et al.* (2004). Tonically active GABA_A receptors: modulating gain and maintaining the tone. *Trends Neurosci* **27**, 262–269.

Shen H, Gong QH, Yuan M *et al.* (2005). Short-term steroid treatment increases δ GABA_A receptor subunit expression in rat CA1 hippocampus: pharmacological and behavioral effects. *Neuropharmacol* **49**, 573–586.

Sieghart W (1995). Structure and pharmacology of gamma-aminobutyric acid_A receptor subtypes. *Pharmacol Rev* **47**, 181–234.

Sigel E (2002). Mapping of the benzodiazepine recognition site on GABA_A receptors. *Curr Top Med Chem* **2**, 833–839.

Sinkkonen ST, Vekovischeva OY, Möykkynen T *et al.* (2004). Behavioural correlates of an altered balance between synaptic and extrasynaptic GABA_Aergic inhibition in a mouse model. *Eur J Neurosci* **20**, 2168–2178.

Smith GB and Olsen RW (1995). Functional domains of GABA_A receptors. *Trends Pharmacol Sci* **16**, 162–168.

Somogyi P, Takagi H, Richards JG *et al.* (1989). Subcellular localization of benzodiazepine/GABA_A receptors in the cerebellum of rat, cat and monkey using monoclonal antibodies. *J Neurosci* **9**, 2197–2209.

Song M and Messing RO (2005). Protein kinase C regulation of GABA_A receptors. *Cell Mol Life Sci* **62**, 119–127.

Spigelman I, Li Z, Banerjee PK *et al.* (2002). Behavior and physiology of mice lacking the GABA_A receptor δ subunit. *Epilepsia* **43**, 3–8.

Steiger JL and Russek SJ (2004). GABA_A receptors: building the bridge between subunit mRNAs, their promoters and cognate transcription factors. *Pharmacol Ther* **101**, 259–281.

Stell BM and Mody I (2002). Receptors with different affinities mediate phasic and tonic GABA_A conductances in hippocampal neurons. *J Neurosci* **22**, 1–5.

Stell BM, Brickley SG, Tang CY *et al.* (2003). Neuroactive steroids reduce neuronal excitability by selectively enhancing tonic inhibition mediated by δ subunit-containing GABA_A receptors. *Proc Natl Acad Sci USA* **100**, 14439–14444.

Sternfeld F, Carling RW, Jelley RA *et al.* (2004) Selective, orally active γ-aminobutyric acidA α5 receptor inverse agonists as cognition enhancers. *J Med Chem* **47**, 2176–2179.

Storustovu S and Ebert B (2003). Gaboxadol: *in vitro* interaction studies with benzodiazepines and ethanol suggest functional selectivity. *Eur J Pharmacol* **467**, 49–56.

Storustovu SI and Ebert B (2006). Pharmacological characterization of agonists at δ-containing GABA_A receptors: functional selectivity for extrasynaptic receptors is dependent on the absence of γ2. *J Pharmacol Exp Therap* **316**, 1351–1359.

Studer R, von Boehmer L, Haenggi T *et al.* (2006). Alteration of GABAergic synapses and gephyrin clusters in the thalamic reticular nucleus of GABA_A receptor α3 subunit-null mice. *Eur J Neurosci* **24**,1307–15.

Sun C, Sieghart W and Kapur J (2004). Distribution of α1, α4, γ2 and δ subunits of GABA_A receptors in hippocampal granule cells. *Brain Res* **1029**, 207–216.

Sundstrom-Poromaa I, Smith AD, Gong QH *et al.* (2002). Hormonally regulated α4β2δ GABA_A receptors are a target for alcohol. *Nature Neurosci* **5**, 721–722.

Tang YP, Shimizu E, Dube GR *et al.* (1999). Genetic enhancement of learning and memory in mice. *Nature* **401**, 63–69.

Terunuma M, Jang IS, Ha SH *et al.* (2004). GABA_A receptor phospho-dependent modulation is regulated by phospholipase C-related inactive protein type1, a novel protein phospatase 1 anchoring protein. *J Neurosci* **24**, 7074–7084.

Thomas P, Mortensen M, Hosie AM *et al.* (2005). Dynamic mobility of functional GABA_A receptors at inhibitory synapses. *Nature Neurosci* **8**, 889–897.

Tretter V, Ehya N, Fuchs K *et al.* (1997). Stoichiometry and assembly of a recombinant GABA_A receptor subtype. *J Neurosci* **17**, 2728–2737.

Tretter V, Hauer B, Nusser Z *et al.* (2001). Targeted disruption of the GABA_A receptor δ subunit gene leads to an up-regulation of γ2 subunit-containing receptors in cerebellar granule cells. *J Biol Chem* **276**, 10532–10538.

Tretter V, Jacob TJ, Mukherjee J *et al.* (2008). The clustering of GABA_A receptor subtypes at inhibitory synapses is facilitated via direct binding of receptor α_2 subunits to gephyrin. *J Neurosci* **28**, 1356–1365.

Triller A, Cluzeaud F, Pfeiffer F *et al.* (1985). Distribution of glycine receptors at central synapses: an immunoelectron microscopy study. *J Cell Biol* **101**, 683–688.

Trudell J (2002). Unique assignment of inter-subunit association in GABA_A α1β3γ2 receptors determined by molecular modeling. *Biochim Biophys Acta* **1565**, 91.

Trudell JR and Bertaccini E (2004). Comparative modeling of a GABA_A α1 receptor using three crystal structures as templates. *J Mol Graph Model* **23**, 39–49.

Tsien JZ, Huerta PT and Tonegawa S (1996). The essential role of hippocampal CA1 NMDA receptor-dependent synaptic plasticity in spatial memory. *Cell* **87**, 1327–1338.

Unwin N (2003). Structure and action of the nicotinic acetylcholine receptor explored by electron microscopy. *FEBS Lett* **555**, 91–95.

Unwin N (2005). Refined structure of the nicotinic acetylcholine receptor at 4A resolution. *J Mol Biol* **346**, 967–989.

van Rijnsoever C, Sidler C and Fritschy JM (2005). Internalized GABA_A-receptor subunits are transferred to an intracellular pool associated with the postsynaptic density. *Eur J Neurosci* **21**, 327–338.

Vicini S and Ortinski P (2004). Genetic manipulations of GABA_A receptor in mice make inhibition exciting. *Pharmacol Ther* **103**, 109–120.

Vicini S, Losi G and Homanics GE (2002). GABA_A receptor δ subunit deletion prevents neurosteroid modulation of inhibitory synaptic currents in cerebellar neurons. *Neuropharmacol* **43**, 646–650.

Wallner M, Hanchar HJ and Olsen RW (2003). Ethanol enhances α4 β3 δ and α6 β3 δ γ-aminobutyric acid type A receptors at low concentration known to affect humans. *Proc Natl Acad Sci USA* **100**, 15218–15223.

Wallner M, Hanchar HJ and Olsen RW (2006). Low-dose alcohol actions on α_4β_3δ GABA_A receptors are reversed by the behavioural alcohol antagonist Ro 15–4513. *Proc Natl Acad Sci USA* **103**, 8540–8545.

Wan Q, Xiong ZG, Man HY *et al.* (1997). Recruitment of functional GABA_A receptors to postsynaptic domains by insulin. *Nature* **388**, 686–690.

Wei ZW, Faria LC and Mody I (2004). Low ethanol concentrations selectively augment the tonic inhibition mediated by δ subunit-containing GABA_A receptors in hippocampal neurons. *J Neurosci* **24**, 8379–8382.

Wei ZW, Zhang N, Peng ZC *et al.* (2003). Perisynaptic localization of δ subunit-containing GABA_A receptors and their activation by GABA spillover in the mouse dentate gyrus. *J Neurosci* **23**, 10650–10661.

Whiting P, Wafford KA and McKernan RM (2000). Pharmacologic subtypes of GABA_A receptors based on subunit composition. In Martin DL and Olsen RW., eds, *GABA in the nervous system: the view at fifty years*. Lippincott, Philadelphia, pp. 113–126.

Whiting PJ (2003). GABA_A receptor subtypes in the brain: a paradigm for CNS drug discovery? *Drug Discovery Today* **8**, 445–450.

Wieland HA, Luddens H and Seeburg PH (1992). A single histidine in GABA_A receptors is essential for benzodiazepine agonist binding. *J Biol Chem* **267**, 1426–1429.

Wisden W, Cope D, Klausberger T *et al.* (2002). Ectopic expression of the GABA_A receptor α6 subunit in hippocampal pyramidal neurons produces extrasynaptic receptors and an increased tonic inhibition. *Neuropharmacol* **43**, 530–549.

Yee BK, Keist R, von Boehmer L *et al.* (2005). A schizophrenia-related sensorimotor deficit links α_3-containing GABA_A receptors to a dopamine hyperfunction. *Proc Natl Acad Sci USA* **102**, 17154–17159.

Zeller A, Arras M, Jurd R *et al.* (2007a). Identification of a molecular target mediating the general anaesthetic actions of pentobarbital. *Mol Pharmacol* **71**, 852–859.

Zeller A, Arras M, Jurd R *et al.* (2007b). Mapping the contribution of β_3-containing GABA_A receptors to volatile and intravenous general anesthetic actions. *BMC Pharmacology* **7**, 2–14.

Zeller A, Arras M, Lazaris A *et al.* (2005). Distinct molecular targets for the central respiratory and cardiac actions of the general anesthetics etomidate and propofol. *Faseb J* **12**, 1677–1679.

3.2

Glutamate receptors

Contents

3.2.1 AMPA receptors

Laura Jane Hanley, Hilary Jackson, Thomas
Chater, Peter Hastie and Jeremy Henley

3.2.1.1 Introduction

α-amino-3-hydroxy-5-methyl-4-isoxazole propionic acid (AMPA) receptors are found principally in the CNS where they perform a crucial role in neuronal signalling by mediating the fast component of excitatory glutamatergic neurotransmission. AMPA receptors (AMPARs) have a critical role in the expression of synaptic plasticity, which is thought to form the cellular basis of learning and memory. Long-term potentiation (LTP) involves the activity-dependent recruitment of AMPARs to the synapse and a concurrent increase in AMPA-mediated transmission. A decrease in functional AMPARs at the synapse characterizes long-term depression (LTD). Given their importance, AMPARs are stringently regulated and the mechanisms of this regulation have been, and continue to be, the subject of intense study. Furthermore, there is considerable potential for AMPARs and associated proteins as targets for therapeutic intervention in neuronal dysfunction and disease. Here we review the currently available information and highlight recent progress made and outstanding questions that remain to be answered.

3.2.1.2 Molecular characterization of AMPA receptor subunits

AMPARs are tetrameric proteins (Rosenmund *et al.*, 1998; Safferling *et al.*, 2001) that are responsible for most fast excitatory synaptic transmission in the central nervous system (Stern *et al.*, 1992; Jones and Westbrook, 1996). The subunits available to contribute to the AMPA tetramer are GluR1–4 (or GluRA-D in the alternative nomenclature; Keinänen *et al.*, 1990) (see Table 3.2.1.1 for a summary of subunit characteristics). GluR1, initially identified as a kainate receptor subunit, was eventually shown to have higher affinity for glutamate and AMPA (Hollmann *et al.*, 1989). The other subunits were subsequently cloned using sequence data from GluR1 (Boulter *et al.*, 1990; Nakanishi *et al.*, 1990). The subunits are similar in size (around 900 amino acids), with sequence homology of approximately 70 per cent (Bettler *et al.*, 1990; Keinänen *et al.*, 1990; Nakanishi *et al.*, 1990;). They have been postulated to resemble the K^+ channel in structure (Kuner *et al.*, 2003), and in common with other ionotropic glutamate receptors, their structure comprises four domains that penetrate the plasma membrane (Figure 3.2.1.1). Membrane domains 1, 3, and 4 (M1, 3, and 4) traverse the plasma membrane, but M2, entering the membrane from the intracellular side, forms a re-entrant loop (Hollmann *et al.*, 1994; Kuner *et al.*, 1996). The M2 loops of each subunit in the tetramer coalesce to form the receptor's ion channel.

RNA editing

Although just four AMPAR subunits are encoded in the genome, many functional variants are found within the cell. Several mechanisms contribute to the functional diversity of proteins over and above what is encoded by the genome. One of the most significant in AMPARs is RNA editing. Substitution of a single nucleotide in RNA may lead to encoding of a completely different amino acid. The function of the resulting protein may be altered and this change is known as RNA editing (Bass, 2002). RNA editing is a 1-step hydrolytic deamination of adenosine to inosine at a particular point in an RNA sequence (Polson *et al.*, 1991). This deamination is catalysed by adenosine deaminases that act on RNA (ADARs) (Bass and Weintraub, 1988; Wagner *et al.*, 1989), and occurs before intron removal from double-stranded RNA (Higuchi *et al.*, 1993).

Table 3.2.1.1 Characterization of the AMPA receptor subunits

Channel subunit	Biophysical characteristics	Cellular and subcellular localization	Pharmacology	Physiology	Relevance to disease states
GluR1 (GluRA, GluA1[*]) [1, 2] Gene: *GRIA1* Chromosome location: mouse:11 rat: 10 human: 5q33 Alternatively spliced flip/flop in 38AA sequence in extracellular loop between M3 and 4	GluR2 lacking receptors are inwardly rectifying permeable to calcium have a higher unitary conductance [3] Reversal potential Ca^{2+}/Cs^+: 13.0 mV [4]. t_{desens}: flip 3.4 ms, flop 3.7 ms [5]	Widespread in CNS: parallel and climbing fibres of the cerebellum, Bergmann glia [6, 7] pyramidal neurons of hippocampus [8] non-pyramidal neurons of parietal cortex and parietal cells [8] Found pre- and postsynaptically in the neuron and throughout the soma and dendrites [9, 10]	EC_{50}: S-glutamate, 51 µM; (RS)-AMPA, 53 µM; kainate, 23 µM [11] IC_{50}: GYKI52466, 18 µM; LY293606, 24 µM; LY300168, 6 µM [12] Sensitive to polyamine block. IC_{50}: PhTx, 2.8 µM [13] Potentiation by 100 µM cyclothiazide: flip, 103-fold; flop, 22 fold [14]	Dominant subunit in LTP Phosphorylated by CaMKII and PKA [15, 16] Interacts with TARPS, SAP97 and protein 4.1 [17–19] Palmitoylated [20]	Increased synaptic GluR1 in hyperalgesia [21]
GluR2 (GluRB, GluA2[*]) [22, 23, 24, 25, 26, 27] Gene: *GRIA2* Chromosome location: mouse: 3 rat: 2q33 human: 4q32–33 Q/R editing at critical site in M2 loop determines calcium permeability of pore Alternatively spliced flip/flop in 38AA sequence in extracellular loop between M3 and 4. R/G editing may occur at a codon preceding the splice site Alternative splicing determines C-terminus length. Long form resembles GluR1, short form resembles GluR3	R edited GluR2 at Q/R site renders receptor calcium impermeable Dominant subunit in conferring ion channel characteristics in heteromeric receptor assemblies Reversal potential Ca^{2+}/Cs^+: GluR2R −50.8 mV, GluR2Q 10.3 mV [5]. t_{desens} GluR2Q: flip, 5.89 ms; flop, 1.18 ms [28] t_{deact} GluR2Q: flip, 0.62 ms; flop, 0.54 ms [28] Conductance: GluR2Rflip Glutamate, 0.36 pS; AMPA, 0.21 pS; GluR2Rflop kainite, <0.3 pS [29]	Ubiquitous in CNS especially: principal neurons of telecephalon pyramidal cells of hippocampus Purkinje cells of cerebellum [8, 9] Found pre- and postsynaptically in the neuron and throughout the soma and dendrites [9, 10]	EC_{50} (GluR2Q): S-glutamate, 190 µM; (RS)-AMPA, 67 µM; kainite, 380 µM [11] Insensitive to polyamine block IC_{50}: PhTx, 270 µM [13]. Does not affect GluR1 sensitivity to cyclothiazide [14]	Drives constitutive receptor cycling and LTD Phosphorylated by PKC Interacts with TARPS, NSF, AP2, GRIP/ABP and PICK [17, 30–33] Palmitoylated [20] Usually assembled with GluR1 or GluR3	ALS and ischaemia: deficient Q/R editing results in an increase in calcium permeable receptors at the membrane and excitotoxicity [34] decreased mRNA in ischaemia [35] overexpression can attenuate ischaemic injury [36] Alzheimer's Disease: decreased levels in presence of soluble Aβ [37] decreased levels in early stages of disease [38]
GluR3 (GluRC, GluA3[*]) [22–27] Gene: *GRIA3* Chromosome location: mouse: X rat: Xq11 human: Xq25–26 Alternatively spliced flip/flop in 38AA sequence in extracellular loop between M3 and 4. R/G editing may occur at a codon preceding the splice site	t_{desens}: flip, 4.8 ms; flop, 1.4 ms [5]	Ubiquitous in CNS especially: principal neurons of telecephalon pyramidal cells of hippocampus Purkinje cells of cerebellum [8, 9] Found pre- and postsynaptically in the neuron and throughout the soma and dendrites [8, 10] Usually forms a heteromer with GluR2	EC_{50}: S-glutamate, 52 µM; (RS)-AMPA, 37 µM; kainate, 40 µM [11] Potentiation by 100 µM cyclothiazide: flip, 69-fold [14]	Usually assembled with GluR2 Important in constitutive cycling Palmitoylated [20] Interacts with many of the proteins also interacting with GluR2	Alzheimer's Disease: decreased levels in intermediate stages of disease [38]
GluR4 (GluRD, GluA4[*]) [22, 23, 24, 25, 26, 27] Gene: *GRIA4* Chromosome location: mouse: 9	Reversal potential Ca^{2+}/Cs^+: 12.7 mV [4]. t_{desens}: flip, 3.6 ms; flop, 0.9 ms [5]	Widespread in CNS: parallel and climbing fibres of the cerebellum, Bergmann glia	EC_{50}: S-glutamate, 20 µM; (RS)-AMPA, 16 µM; kainate, 55 µM [11]	Dominant subunit in developmental plasticity, plays a similar role to GluR1	Altered levels following injury to neurons of spinal cord [41, 42]

Table 3.2.1.1 *(Continued)* Characterization of the AMPA receptor subunits

Channel subunit	Biophysical characteristics	Cellular and subcellular localization	Pharmacology	Physiology	Relevance to disease states
rat: 8q11 human: 11q22–23 Alternatively spliced flip/ flop in 38AA sequence in extracellular loop between M3 and 4. R/G editing may occur at a codon preceding the splice site. Alternative splicing determines C-terminus length. Long form resembles GluR1, short form resembles GluR3	Conductance: flip— glutamate, 7/16/27 pS (i.e. subconductance states), kainite, 2.5 pS; flop— kainite, 4.0 pS [29]	pyramidal neurons of hippocampus [9] In hippocampus levels decrease in adulthood [39]	IC$_{50}$: GYKI52466, 23 μM; LY293606, 28 μM; LY300168, 3 μM Potentiation by 100 μM cyclothiazide: flip 171-fold	Phosphorylation by PKA necessary and sufficient for LTP in young animals [40] Palmitoylated [20]	

TARPS, transmembrane AMPA receptor regulatory proteins; ABP, AMPAR binding protein; AP2, Adaptor protein 2; ALS, Amyotrophic lateral sclerosis.

* The classification of glutamate receptor subunits has recently been re-addressed by NC-IUPHAR and the revised recommended nomenclature is shown here. See Alexander SPH, Mathie A, and Peters JA (2008). Guide to receptors and channels (GRAC), third edn. *Br J Pharmacol* 153, S1–S209.)

[1] Hollmann M et al. (1989) *Nature* 342, 643–648; [2] Sommer B et al. (1990) *Science* 249, 1580–1585; [3] Bochet P et al. (1994) *Neuron* 12, 383–388; [4] Burnashev N et al. (1992) *Neuron* 8, 189–198; [5] Mosbacher J et al. (1994) *Science* 266, 1059–1062; [6] Baude A et al. (1994) *J Neurosci* 14, 2830–2843; [7] Martin D et al. (1993) *Eur J Pharmacol* 250, 473–476; [8] Petralia RS et al. (1997) *J Comp Neurol* 385, 456–476; [9] Petralia RS et al. (1992) *J Comp Neurol* 318, 329–354; [10] Pittaluga A et al. (2006) *Neuropharmacology* 50, 286–296; [11] Strange M et al. (2006) *Comb Chem High Throughput Screen* 9, 147–158; [12] Bleakman D et al. (1996) *Mol Pharmacol* 49, 581–585; [13] Brackley PT et al. (1993) *J Pharmacol Exp Ther* 266, 1573–1580; [14] Partin KM et al. (1996) *J Neurosci* 16, 6634–6647; [15] Mammen AI et al. (1997) *J Biol Chem* 272, 32528–32533; [16] Roche KW et al. (1996) *Neuron* 16, 1179–1188; [17] Cuadra AE et al. (2004) *J Neurosci* 25, 7491–7502; [18] Leonard AS et al. (1998) *J Biol Chem* 273, 19518–19524; [19] Shen L et al. (2000) *J Neurosci* 20, 7932–7940; [20] Hayashi T et al. (2005) *Neuron* 47, 709–723; [21] Galan A et al. (2005) *Pain* 112, 315–323; [22] Boulter J et al. (1990) *Science* 249, 1033–1037; [23] Nakanishi N et al. (1990) *Neuron* 5, 569–581; [24] Sommer et al. (1991) *Cell* 67, 11–19; [25] Lomeli M et al. (1994) *Science* 266, 1709–1713; [26] Kohler M et al. (1994) *J Biol Chem* 269, 17367–17370; [27] Gallo V et al. (1992) *J Neurosci* 12, 1010–1023; [28] Koike M et al. (2000) *J Neurosci* 15, 2166–2174; [29] Swanson GT et al. (1997) *J Neurosci* 17, 58–69; [30] Nishimune A et al. (1998) *Neuron* 21, 87–97; [31] Lee SH et al. (2002) *Neuron* 36, 661–674; [32] Dong H et al. (1997) *Nature* 20, 279–284; [33] Dev KK et al. (1999) *Neuropharmacology* 38, 635–644; [34] Peng PL et al. (2006) *Neuron* 49, 719–733; [35] Gorter JA et al. (1997) *J Neurosci* 17, 6179–6188; [36] Liu S et al. (2004) *Neuron* 43, 43–55; [37] Rosseli F et al. (2005) *J Neurosci* 25, 11061–11070; [38] Carter TL et al. (2004) *Exp Neurol* 187, 299–309; [39] Boehm J et al. (2005) *Biochem Soc Trans* 33, 1354–1356; [40] Esteban JA et al. (2003) *Nat Neurosci* 6, 136–143; [41] Mennini T et al. (2002) *J Neurosci Res* 70, 553–560; [42] Brown K et al. (2004) *Brain Res Dev Brain Res* 152, 61–68.

ADARs require a specific intronic sequence in the RNA, the editing complimentary site (ECS), to direct them to the correct sequence of double-stranded RNA (dsRNA; Seeburg et al., 1998). The ADAR then unwinds this sequence and deamination takes place. The deamination of adenosine changes the resultant protein because the translation mechanisms recognize inosine as guanosine (Basilio et al., 1962), potentially leading to a codon change.

The M2 loop of the AMPAR subunit contains a crucial editing site. While GluR1, 3, and 4 subunits are not edited and contain the amino acid glutamine (Q) in the critical position (marked*, Figure 3.2.1.1), almost 100 per cent of GluR2 subunits in the mammalian brain contain arginine (R) rather than glutamine at the editing site (Sommer et al., 1991; Seeburg et al., 2001). Q/R editing results from an RNA sequence reading CAG (encoding glutamine) being edited to CIG by ADAR2. It is then translated as CGG at the ribosome, thus encoding arginine (Sommer et al., 1991; Higuchi et al., 1993). This gives the edited GluR2 subunit a substantial influence over the AMPAR's permeability to calcium: indeed it is the only subunit to have such an effect (Hollmann et al., 1990; Hume et al., 1991; Verdoorn et al., 1991; Burnashev et al, 1992). Influx of Ca^{2+} ions modulates intracellular processes while the ionic charge also affects the cell's electrical properties. The influence of GluR2 is significant because although under normal conditions Ca^{2+} is tightly regulated, uncontrolled increases in cellular Ca^{2+} can lead to excitotoxicity, a condition important physiologically and in disease (Choi 1992; Slemmer et al., 2005). The ion pore of the AMPAR is formed from the M2 of each of the four constituent subunits

(Keinänen et al., 1990; Hume et al., 1991; Verdoorn et al., 1991). Without edited GluR2, each subunit in the pore contains uncharged glutamine at the critical point. This has no effect on the ability of charged ions, crucially, Ca^{2+}, to pass through the pore and into the cell. However, if the tetramer contains edited GluR2, the positively charged arginine at the critical Q/R point prevents the passage of Ca^{2+} through the receptor pore (Hume et al., 1991; Burnashev et al., 1992). The presence of edited GluR2 in the tetramer transforms the usual inward rectifying current (signifying a correlation between depolarization of the membrane and Ca^{2+} entry through the pore) into a linear current–voltage relationship (Boulter et al., 1990; Hume et al., 1991; Verdoorn et al., 1991). Interestingly, this is in direct contrast to editing in kainate receptors, where a conformational difference means that Q/R editing in M2 of GluR6 leads to an increase in Ca^{2+} permeability (Köhler et al., 1993).

In addition to channel gating, Q/R editing also appears to determine the trafficking of AMPAR subunits to the cell membrane (Greger et al., 2002). GluR1 tetramerizes and is trafficked directly to the plasma membrane, while GluR2 remains in the endoplasmic reticulum (ER) in a stable pool of dimerized subunits (Greger et al., 2003). GluR2 is maintained in the ER as a result of sequence motifs in the C-terminus and the edited arginine residue at the Q/R editing site. If Q takes the place of R, GluR2 is swiftly released from its pool in the ER and is expressed at the surface in a similar manner to GluR1. Accordingly, when R is replaced by Q in GluR1's normally unedited Q/R site, the subunit is held in the ER in the same way as edited GluR2. These findings suggest that GluR2 is held in

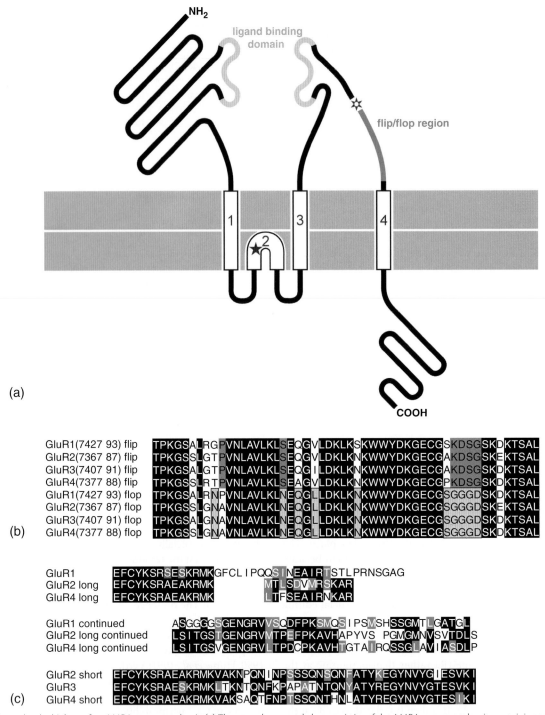

Fig. 3.2.1.1 The molecular biology of an AMPA receptor subunit. (a) The general structural characteristics of the AMPA receptor subunit, containing an extracellular N-terminus, four membrane domains, and an intracellular C-terminus. There are two RNA editing sites: Q/R (filled star) in the M2 loop, and R/G (unfilled star) in the extracellular loop preceding flip/flop. (b) The peptide sequence for each flip/flop variant of GluR1–4. (c) The peptide sequence for each C-terminal variant. Key to the peptide sequences: black with white type, homologous between isoforms; dark grey with black type, homologous in the flip isoform; light grey with black outline, homologous in the flop isoform; light grey with white type, isoforms share character.

the ER until it forms a complex with GluR1 or GluR3, which may mask an ER retention signal and allow forward trafficking (Greger *et al.*, 2002).

Alongside editing at the Q/R site in GluR2, editing also occurs at a residue in the extracellular loop of GluR2, 3, and 4 (marked* in Figure 3.2.1.1 (Lomeli *et al.*, 1994). Immediately preceding, and

intimately related to, the flip/flop alternative splicing sequence (see below), another adenosine to inosine change in a single codon (AGA to IGA/GGA) leads to a functionally significant protein sequence alteration, this time arginine (R) to glycine (G). This R/G editing has a significant effect on the recovery rate of the receptor from desensitization. AMPARs that contain R/G edited subunits recover from desensitization more quickly than those that do not (Krampfl et al., 2002). Interestingly, GluR1 lacks the necessary intron for RNA editing at this site (Lomeli et al., 1994; Seeburg, 1996). The relationship between R/G editing and alternative splicing is explored later in this chapter.

Alternative splicing

Alternative splicing in AMPAR genes allows certain exons to appear interchangeably in the translated proteins (Early et al., 1980; Stetefeld and Ruegg, 2005). Like RNA editing, this allows greater functional diversity amongst AMPARs than is coded in the genome. Two major sites in GluR1–4 undergo alternative splicing: flip/flop, which influences receptor desensitization and resensitization; and in GluR2 and GluR4, the C-terminus, which directs trafficking, targeting, and assembly of the subunits.

Each of the four AMPAR subunits contains the flip/flop region which is a 115 base pair sequence that exists in one of two versions (Keinänen et al., 1990; Sommer et al., 1990). Situated in the extracellular loop between M3 and M4, these 115 base pairs form a 38 amino acid sequence that differs in only 9 to 11 amino acids (Figure 3.2.1.1) (Sommer et al., 1990; Partin et al., 1994), and is unique to mammals (Chen et al., 2006). In murine GRIA1–4, the genes encoding GluR1–4, flip and flop are situated on sequential exons, separated by a 900 base-pair intron (Sommer et al., 1990).

Despite the sequence similarity of flip and flop, in GluR2 to 4 they have significant functional differences in their desensitization kinetics (Sommer et al., 1990; Mosbacher et al., 1994; Koike et al., 2000). Almost all ligand-gated ionotropic receptors lose sensitivity to agonist after a period of time, despite the continued presence of an agonist. Although all subunits desensitize at different rates, AMPARs that contain GluR2–4 flop desensitize far more quickly than those made up from GluR2–4 flip (Mosbacher et al., 1994; Koike et al., 2000): the desensitization rates are similar in GluR1 flip and flop). These differences in desensitization kinetics are thought to be due to only three amino acids in the flip/flop sequence (Quirk et al., 2004).

In addition to faster desensitization rates related to the flop cassette, subunits expressing flip show greater inward currents in response to activation at a given membrane potential (Sommer et al., 1990; Koike et al., 2000). GluR1–4 flip subunits also recover from desensitization much more quickly than subunits containing flop, allowing faster reactivation of receptors containing these subunits (Koike et al., 2000). Additionally, GluR2 flip, which resensitizes more quickly than GluR1 flip/flop or GluR2 flop, dominates any receptor in which it is combined (Mosbacher et al., 1994; Schlesinger et al., 2005). In heteromeric receptors, resensitization is dominated by flip-containing subunits (Partin et al., 1994; Schlesinger et al., 2005) and experiments using recombinantly expressed GluR1 and 2 flip and flop indicate that heteromers are preferentially expressed over homomers (Mansour et al., 2001; Brorson et al., 2004).

Processing of pre-mRNA to remove introns is carried out by a spliceosome. This massive complex forms from a collection of many small nuclear ribonucleoprotein particles (snRNPs) and non-snRNPs (Jurica and Moore, 2003). The spliceosome recognizes the 5′ and 3′ splice sites and catalyses the transesterification reactions that cause exon splicing. The precise mechanism for mutually exclusive selection of flip rather than flop (or vice versa) is currently unknown but may occur in one of a number of ways,

often related to conformational compatibility between the particular spliceosome and the RNA (Ast, 2004; Graveley, 2005).

Due to increased inward current, slower desensitization, and quicker resensitization (Sommer et al., 1990; Mosbacher et al., 1994; Schlesinger et al., 2005) it has been hypothesized that insertion of flip to an AMPAR at the synapse would lead to increased response of that synapse to glutamate. Consequently, the receptor would be more likely to allow Ca^{2+} entry, either directly or indirectly as a result of activation of voltage-gated channels, exposing the cell to an increased risk of excitotoxicity (Weiss et al., 1990; Schlesinger et al., 2005).

In most brain regions, notably the cortex, hippocampus, and caudate nucleus, flop levels are low until postnatal day (p) 8, whereupon they increase to adult levels by p14. Subunits containing the flip cassette are expressed at consistent levels throughout development but flop subunits dominate in the mature animal. The increase in AMPARs that contain flop allows a mature response to receptor activation due to the quicker desensitization rates (and consequently lower level of cellular Ca^{2+} influx) of the flop isoform (Sommer et al., 1990; Monyer et al., 1991). It has been postulated that flip dominance that lasted into maturity would lead to excitotoxicity.

Flip/flop alternative splicing and Q/R editing are critically interlinked. When Q/R editing in the GluR2 subunit is eliminated, there is a concomitant lack of splicing in the subunit (Higuchi et al., 2000). In addition, R/G editing changes the biophysical characteristics of the receptor as it has a key effect on flip/flop splicing (Bratt and Öhman, 2003).

Post-transcriptional modification and neurological disorders

AMPARs have a significant role in neurodegenerative disease. Because the effects of R/G editing and flip/flop alternative splicing are so intertwined, it is difficult to disentangle the effects of these two mechanisms in health and disease. As described above, the fast rate of desensitization associated with GluR1–4 flop is associated with mature synapses in healthy neurons. It is increasingly clear that excitotoxicity associated with a range of neurological disorders, including Parkinson's and Alzheimer's disease may be AMPAR-mediated, and modulation of flip/flop expression is a potential target for neuroprotective treatment (Jayakar and Dikshit, 2004).

Other post-transcriptional modifications also have an impact on health and disease. In much of the healthy brain almost 100 per cent of GluR2 is edited at the Q/R site (Seeburg et al., 2001). However, in certain regions of the brain and under some physiological conditions, underediting has a substantial effect on function. In mice, a subset of striatal and cortical neurons containing NADPH-diaphorase (a nitric oxide synthase) are particularly susceptible to calcium-dependent excitotoxicity brought on by experimental application of AMPA and kainate (Weiss et al., 1994). In these cells, just 46 per cent of the GluR2 subunits exhibit the arginine residue at the Q/R editing site. The large number of unedited receptors renders the cells more permeable to Ca^{2+} on activation of AMPARs, and therefore vulnerable to excitotoxicity (Kim DY et al., 2001).

C-terminal splicing

GluR2 and GluR4 express either long or short alternatively spliced C-termini (Gallo et al., 1992; Köhler et al., 1994). When long, these isoforms are in essence homologous to the C-terminus in GluR1, and when short, they are largely homologous to the GluR3 C-terminus (Figure 3.2.1.1) (Braithwaite *et al.*, 2000). Expression of the isoforms differs depending on the brain region and subunit: 90 per cent of murine GluR2 transcripts encode the short C-terminus (Köhler *et al.*, 1994). GluR4-long is preferentially expressed in Bergman glial cells while GluR4-short is preferentially expressed in cerebellar granule cells (Gallo *et al.*, 1992). These differences are functionally significant. Like the *N*-methyl-D-aspartate receptor (NMDAR) subunit, the C-terminus of the GluR1–4 subunits is an important site for interaction with intracellular proteins. As described in more detail below, this region interacts with PDZ-containing proteins including protein interacting with C-Kinase (PICK), glutamate receptor interacting protein (GRIP), and synapse associated protein 97 (SAP97) (Dong *et al.*, 1997; Leonard *et al.*, 1998; Dev *et al.*, 1999; Sheng and Sala, 2001), and with proteins that do not contain PDZ domains such as N-ethylmaleimide sensitive fusion protein (NSF) and α and β soluble NSF attachment proteins (SNAPS) (Henley *et al.*, 1997; Osten *et al.*, 1998; Braithwaite *et al.*, 2000). As suggested by the C-terminus-related expression patterns of the subunits, the differing C-terminal interactions with these intracellular proteins regulate receptor clustering, trafficking, and targeting (Xia *et al.*, 1999; Ruberti and Dotti, 2000; Cuadra *et al.*, 2004; Lu and Ziff, 2005).

3.2.1.3 Biophysical characteristics

Ion conductance

AMPARs act as non-selective cation channels, permeable to Na^+ and K^+. Receptors lacking the Q/R edited GluR2 subunit are also permeable to Ca^{2+} and have a higher unitary conductance (Hume *et al.*, 1991; Bochet *et al.*, 1994). GluR2-lacking receptors are inwardly rectifying due to polyamine block of the channel at positive voltages; the presence of the arginine residue in the channel pore prevents this block in GluR2-containing channels (Iino *et al.*, 1994).

Kinetics

AMPARs are known to be activated much faster than NMDARs with measurement of rise times determined within the limits of resolution of the apparatus as at least as fast as 1.2 ms (Trussell and Fischbach, 1989). AMPARs are the fastest desensitizing ligand-gated ion channels but the contribution of desensitization to EPSC decay is not clear (Glavinovic, 2002). Outside-out patches from CA1 hippocampal synapses have a decay constant of 2.3 ± 0.7 ms and a time constant of desensitization of 9.3 ± 2.8 ms in response to brief exposure to glutamate (Colquhoun *et al.*, 1992). This suggests that, at synapses where glutamate is efficiently removed from the synapse by the vesicular glutamate transporter (VGLUT), receptor deactivation rather than desensitization contributes most greatly to the decay of AMPAR-mediated EPSCs (Takahashi, 2005). Where glutamate is less efficiently removed, such as at chick spinal neurons, rates of EPSC decay are equivalent to rates of AMPAR desensitization (Trussell and Fischbach, 1989). Further complexity is added by a process termed 'developmental speeding' in which decay constants for the currents are reduced during early postnatal development from P7 to P14 as the contribution of desensitization decreases (Koike-Tani *et al.*, 2005).

Mechanism of channel gating

The mechanism by which agonist binding to AMPARs is translated into channel opening has been well studied and is thought to be a template for other members of the ionotropic glutamate receptor (iGluR) group (Mayer, 2006). The ligand-binding domain of each subunit is formed from the membrane proximal region of the N-terminus (S1) and the extracellular linker region between TM2 and TM3. Chimeric receptors in which this domain was swapped between AMPARs and kainate receptors show that agonist selectivity is determined by these domains (Stern-Bach *et al.*, 1994). Fusing these two domains with a short linker peptide allowed the bacterial expression and crystallization of this ligand binding pocket. This has led to the production of high resolution (<2Å) images of the GluR2 binding pocket in complex with a number of ligands. These have shown that closure of the binding domain clamshell structures is promoted by agonist binding. The agonist binds first to the upper lobe (D1) which is fixed in place via a dimer interface with the D1 lobe of the neighbouring subunit, providing torque for an upward movement of the lower lobe, D2 (Horning and Mayer, 2004). The degree of lobe closure determines agonist efficacy with glutamate causing a movement of ~21°.

Binding of two or more agonist molecules to the receptor complex results in channel opening (Clements and Westbrook, 1994). It has been proposed that the upward movement of the D2 lobe is transferred to the TM3 domains to cause channel opening (Stern-Bach *et al.*, 2004). As these domains are kinked, the movement is amplified, increasing the diameter of the pore. The binding of a greater number of agonist molecules is thought to cause movement of more of the TM3 domains, explaining the existence of multiple AMPAR conductance states (Rosenmund *et al.*, 1998). However, the difficulty of crystallizing the hydrophobic transmembrane domains (TMDs) has so far prevented confirmation of this model.

The cyclothiazide binding site and residues determined by flip/flop variation lie within the dimer interface (Sun *et al.*, 2002). Introduction of the mutation L483Y to the bacterially expressed binding domain construct increases the dissociation constant (Kd) of dimer formation 10^5-fold. This mutation blocks desensitization of GluR2 homomers suggesting that desensitization of the agonist bound AMPAR is due to destabilization of the D1–D1 dimer interface. These data support the proposed model for channel gating as this destabilization would relieve the tension required to transfer force to the TMDs, allowing them to rotate into the lower-energy closed conformation. The effect of this relaxation can be quite strikingly seen by comparing the structures of agonist-bound and apoAMPARs (Nakagawa *et al.*, 2005a).

3.2.1.4 AMPA receptor pharmacology

Agonist

Early studies involving glutamate receptor agonists identified NMDA receptor and non-NMDAR (i.e. AMPAR and kainate receptor) channels. The endogenous ligand, L-glutamate (also known as (S)-glutamate), acts more potently at NMDARs than at non-NMDARs with EC_{50} values of 2.3 mM and 480 mM, respectively, recorded in embryonic mouse hippocampus (McBain and Mayer, 1994). Early studies used quisqualate to activate the non-NMDARs although this drug also activates mGluRs (Murphy and Miller, 1989). The development of AMPA allowed these receptors to be activated more specifically, although functional studies show that this drug also activates kainate receptors (Krogsgaard-Larsen et al., 1980; Seeburg, 1993). Similarly, kainate also activates AMPARs but here acts as a partial agonist with reduced receptor desensitization (Sommer et al., 1990). The partial agonist effects of kainate are due to steric hindrance between the isopropyl group of the ligand and the residue Leu 650 in the ligand binding pocket of the receptor, preventing full closure of the binding pocket (Armstrong et al., 2003). This model for partial agonist action was confirmed by the use of the 5-substituted Willardiines: larger atomic substitutions lead to lesser degrees of binding pocket closure and weaker efficacy (Jin et al., 2003).

Competitive antagonists

A widely used group of competitive antagonists are the quinoxalinediones. The most widely used members of this group include 6-cyano-7-nitroquinoxaline-2,3-dione (CNQX) and 6,7-dinitro-quinoxaline-2,3-dione (DNQX) (Drejer and Honoré, 1988). 2,3-Dihydroxy-6-nitro-7-sulfamoyl-benzo(F)quinoxaline (NBQX) is more potent than CNQX with IC_{50} values of 60 nM and 400 nM, respectively, and displaces [^{3}H]AMPA with 30-fold selectivity over [^{3}H]kainate (Sheardown et al., 1990; Stein et al., 1992). A recent addition to this group, 6-azido-7-nitro-1,4-dihydroquinoxaline-2,3-dione (ANQX; IC_{50}, 1 µM) promises to be a useful research tool as ultraviolet light causes the reactive azide group of the ligand to bind irreversibly to the receptor, albeit with a tenfold reduction in potency (Chambers et al., 2004; Adesnik et al., 2005). A subset of this class of compounds contain an imidazole group and include the compounds YM90K and YM872, which are notable due to their high water solubility, raising the possibility that they could potentially be useful therapeutic lead compounds (Ohmori et al., 1994; Kohara et al., 1998).

Other classes include tetrazole substituted decahydroisoquinolones such as LY293558, a potent antagonist (IC_{50}, 0.45 µM) that is highly selective for GluR1–4 over GluR6 (Ki 2–4 µM vs >1mM) (Ornstein et al., 1993; Bleakman et al., 1995). The isoxazole compound (S)-2-amino-3-[5-tert-butyl-3-(phosphonomethoxy)-4-isoxazolyl]propionic acid (ATPO) acts as an antagonist at AMPARs but is a partial agonist at GluR5-containing kainate receptors (Wahl et al., 1998). Barbiturates such as pentobarbitone are partially selective for GluR2-containing over GluR2-lacking receptors (IC_{50} 200 µM vs 1–2 mM) (Taverna et al., 1994).

Non-competitive antagonists

The first non-competitive antagonist to be described was the 2,3-benzodiazepine GYKI52466 which blocked AMPA and kainate responses in rat hippocampal neurons with IC_{50} values of 11 µM and 7.5 µM, respectively, (Donevan and Rogawski, 1993; Solyom and Tarnawa, 2002). The most powerful of this class of antagonists is GYKI53655, also known as LY300168, which exhibits IC_{50} values as low as 1.5 µM measured at native receptors (Bleakman et al., 1996). Although inhibition is not dependent on the state of the channel, the presence of cyclothiazide (discussed below) causes a tenfold increase in IC_{50} value (Donevan and Rogawski, 1993; Balannik et al., 2005). However, the binding sites for these drugs are distinct and GYKI53655 acts with equal potency at flip and flop variants (Partin and Mayer, 1996). GYKI53655 binds in the S1–M1 and S2–M4 linker regions and it is likely that the change in potency

in the presence of cyclothiazide reflects the conformational changes in this region during channel gating (Balannik et al., 2005). This class of antagonist has an IC_{50} of >200 µM at kainate receptors and this selectivity for AMPARs has made it a very useful pharmacological tool (Wilding and Huettner, 1995).

Other classes of noncompetitve antagonists include phthalazine derivatives and quinazolinones. 3 µM dihydrophthalazine (SYM2207) inhibits cortical neuron responses to 50 µM kainate to 89 per cent and has an IC_{50} value of 1.8 µM. The quinazolinone CP-465,022 is extremely potent with an IC_{50} of 36 nM, which represents a marked improvement over the lead compound piriqualone (IC_{50} = 0.5 µM) (Welch et al., 2001). IC_{50}s of 25nM have been measured in rat cortical neurons making this compound ~100-fold more potent than GYKI53655 (Lazarro et al., 2002). CP-465,022 completely inhibits kainate-induced AMPAR responses in cortical neurons at 3.2 µM (Lazzaro et al., 2002). As with 2,3-benzodiazepines, the quinazolinones bind with higher affinity to the desensitized state of the receptor and at a similar binding site, although mutations in the S1–M1 region have a greater effect on CP-465,022 whilst those in the S2–M4 region have a greater effect on GYKI53655 (Balannik et al., 2005).

Channel blockers

A number of polyamine-based toxins are known to block AMPAR responses. This type of antagonism is reversible, non-competitive and partly voltage-dependent (Brackley et al., 1993). Joro spider toxin from *Nephila clavata* and its synthetic analogue 1-napthyl-acetylspermine (naspm) block AMPAR responses with an IC_{50} of 40 nM (Blaschke et al., 1993). Interestingly, this blockade is specific to Ca^{2+}-permeable, GluR2-lacking receptors and is therefore selective for type II neurons which do not express this subunit (Iino et al., 1996). The presence of the GluR2 subunit shifts the IC_{50} of philanthotoxin from 2.8 µM to 270 µM (Brackley et al., 1993). Mutation of the residue Glu 590 in the second transmembrane domain to arginine completely abolishes inhibition by naspm (Blaschke et al., 1993). This mutation also abolishes Ca^{2+} permeability and confers a linear rectification curve on the receptor. These data support the view that polyamine-based toxins block AMPAR responses by entering the channel pore.

The more negative the membrane potential, the more rapid the onset of block by the positively charged polyamines as they are attracted to the pore (Priestley et al., 1989). These toxins are therefore use-dependent blockers of the open channel. Synthetic argiotoxin 636 has no effect on subsequent kainate responses if applied in the absence of agonist: however, if applied while channels are active, inhibition constants in the submicromolar range are measured (Herlitze et al., 1993). Variations in the polyamine group of the toxins confer different properties, for example philanthotoxin 433, the synthetic analogue of the toxin from the *Philanthus triangulum* wasp, is more readily removed from the channel than Joro spider toxin, removal of which requires application of agonist under depolarizing conditions (Jones and Lodge, 1991; Iino et al., 1996).

Modulators

One of the most widely used compounds in studying AMPAR function is cyclothiazide, a benzothiadiazine drug. AMPAR and kainite receptor responses can be easily dissected through their potentiation in the presence of cyclothiazide and concavalin A, respectively (Partin *et al.*, 1993; Lynch, 2004). 100 µM cyclothiazide causes an 8- to 12-fold potentiation of responses to 100 µM kainate and a ~100-fold potentiation of responses to 300 µM glutamate. This difference may be a reflection of the fact that AMPAR responses to kainate are relatively non-desensitizing. It is thought that cyclothiazide promotes dimerization between ligand binding domains, preventing desensitization (Sun *et al.*, 2002). Indeed, cyclothiazide potentiates responses from flip receptors to a greater extent than the faster desensitizing flop receptors (Partin *et al.*, 1994). In contrast, cyclothiazide has no effect on receptor deactivation (Yamada and Tang, 1993). Other benzothiadizides include diazoxide and idra-21 (Yamada and Rothman, 1992; Thompson *et al.*, 1995).

The nootropic agents aniracetam and idebenone also potentiate AMPAR responses but are less potent than cyclothiazide (Tsuzuki *et al.*, 1992; Nakamura *et al.*, 1994). AMPAkines, benzamide compounds such as CX614, also block desensitization but unlike cyclothiazide these drugs also slow receptor deactivation (Arai *et al.*, 2000). These compounds share one of the cyclothiazide binding sites but are more potent at flop receptor isoforms. The mechanism of AMPAkine action varies, with CX516 accelerating channel opening and CX546 slowing channel closing (Arai *et al.*, 2002). 4-[2-(phenylsulfonylamino)thio]-difluor-phenoxyacetamide, commonly referred to as PEPA, is also more potent at flop receptors, potentiating these 50-fold in contrast to a threefold potentiation of flip receptors (Sekiguchi *et al.*, 1997). A series of biarylpropylsulphonomides including LY392098 and LY404187 have been identified as AMPAR potentiators that are an order of magnitude more potent than cyclothiazide (Ornstein *et al.*, 2000; Miu *et al.*, 2001).

High concentrations (>3 mM) of thiocyanate ions also potentiate AMPAR responses although lower concentrations are inhibitory (Donevan and Rogawski, 1998). The opposite is true of Zn^{2+} ions that potentiate AMPAR and kainate receptor responses at low concentrations (<10 mM) but are inhibitory for these currents at higher concentrations responses (Rassendren *et al.*, 1990). This effect is most pronounced when GluR3 is present (Dreixler and Leonard, 1994).

3.2.1.5 Cellular and subcellular localization

AMPARs are the most abundant excitatory receptor of the central nervous system. It is important to note that although AMPAR expression is largely restricted to the central nervous system, they can also be found in the sensory nerve terminals of the skin and islet cells of the pancreas (Inagaki *et al* 1995; Weaver *et al.*, 1996). This distribution has implications for receptor function in pain-processing pathways and also in the aberrant insulin secretion symptomatic of diabetes. Within the CNS AMPARs are expressed in the motor neurons of the spinal cord and in many regions of the brain including the olfactory bulb, septal nuclei, amygdala, hippocampus, interpendicular nucleus, cerebellum (Petralia and Wenthold, 1992), and cochlear nucleus (Hunter *et al.*, 1993). The distribution of subunits across these structures varies greatly. The immunostaining pattern for the GluR2/3 subunit is strong and ubiquitous whereas GluR1 staining is dense in some areas but weak in others (Petralia and Wenthold, 1992). The distribution of GluR4 is more diverse than GluR1 but is not as widespread as Glu2/3 (Petralia and Wenthold, 1992).

The GluR2 subunit is present in the vast majority of principle neurons of the telecephalon but is absent from principle neurons of the brain stem and spinal cord and also from interneurons (Petralia *et al*, 1997). GluR2 is predominant throughout the soma and dendrites of pyramidal neurons in the hippocampus and Purkinje cells in the cerebellum. All immunostained GluR2 is co-localized with immunoreactivity to GluR1 or GluR2/3/4 indicating that the majority of AMPARs contain the GluR2 subunit. The GluR1 and GluR4 subunits are found in parallel and climbing fibre synapses of the cerebellum (Baude *et al.*, 1994), surrounding Bergmann glia (Martin *et al.*, 1993), and non-pyramidal neurons (but not granule cells), of the parietal cortex and olfactory bulb. GluR1 is also strongly expressed in hippocampal pyramidal cells. It should be noted that some astrocytes in these regions also express AMPAR subunits (Janssens and Lesage, 2001).

The subcellular location of AMPARs within the neuron has until recently been considered exclusively postsynaptic. Initially, strong AMPAR staining could only be observed in the postsynaptic density (Petralia *et al.*, 1992), but its immunoreactivity has subsequently been identified in the soma, dendrites and spines of principal neurons.

As discussed below, AMPARs are constantly moving between 'intracellular pools' and the synaptic membrane in response to changes in neuronal activity. Consequently, the exact location of a given AMPAR varies over time. Intracellular pools of edited GluR2 are retained in the ER whereas GluR1 heteromers are released for transport to the surface (Greger *et al.*, 2002; 2003) During basal state transmission, there are few GluR1-containing receptors at the synapse while 40 per cent of the total population of GluR2 heteromers are found at the cell membrane. During LTP the number of surface GluR1-containing receptors is transiently increased and then replaced by GluR2 heteromers, thus increasing the overall number of AMPARs at the synapse. During LTD there is a net loss of AMPARs, specifically those containing GluR2, from the synaptic membrane. The internalized receptors pass quickly through early endosomes and depending on the prior activity of the neuron, they are sorted to either lysosomes for degradation or recycling endosomes for reinsertion into the membrane (Ehlers, 2000; Lee *et al.*, 2004).

Interestingly, it is now apparent that the presynaptic membrane is an additional site of AMPAR expression. These receptors also undergo cycles of insertion and removal from the membrane and are thought to regulate neurotransmitter release (Pittaluga *et al.*, 2006).

3.2.1.6 Physiology: AMPARs in synaptic plasticity

Synapses are considered 'plastic' because of their ability to alter the strength of communication between neurons. Some of the first evidence to suggest AMPARs are dynamic and that their trafficking is an integral part of synaptic plasticity came from the identification of silent synapses; glutamatergic synapses that do not show any AMPAR-mediated transmission at resting membrane potential (Isaac *et al.*, 1995; Liao *et al.*, 1995), but display AMPAR EPSCs following LTP induction which are maintained when the membrane is returned to resting potential. The interpretation of these findings was that LTP caused the insertion of AMPARs into the synaptic membrane resulting in an 'unsilencing' of the synapse.

This phenomenon is also thought to be important in the processes involved in the development of cortical circuits (Isaac, 2003).

Following the identification of silent synapses it is now widely accepted that during LTP, AMPARs are inserted into the synaptic membrane by exocytosis. AMPAR subunit isoforms with long C-termini drive this process with GluR1-containing heteromers dominant in the adult brain and GluR4 during development. During NMDAR-dependent LTP in the mature hippocampus, GluR1/GluR2-containing receptors are trafficked from recycling endosomes to the synapse (Park et al., 2004). However, it is currently unclear whether the receptors are inserted into the extrasynaptic membrane and diffuse laterally to the synaptic membrane or if they are inserted directly into the synapse. The GluR1-containing heteromers are then replaced by GluR2/R3 complexes to maintain the increased AMPAR response (Shi et al., 2001) but without increased calcium conductance. In order for GluR2/3 to be inserted in the correct synapse, it has been proposed that 'slot' proteins are added to the membrane alongside GluR1 heteromers. PSD-95 and stargazin (see below) have been proposed as potential slot proteins (Schnell et al., 2002) and the phosphorylation of the GluR1 C-terminus has been suggested to act as a 'tag' to direct incoming GluR2/R3-containing receptors (Shi et al., 2001).

AMPARs can also be removed from the membrane in response to activity. Initial studies showed that brief application of glutamate to hippocampal cultures resulted in a decrease in the number of receptors that could be detected immunocytochemically (Lissin et al., 1999) and electrophysiologically (Carroll et al., 1999). This activity-dependent removal of AMPARs is mediated by clathrin-dependent endocytosis (Beattie et al., 2000; Man et al., 2000). GluR2 interacts with the β-adaptin subunit of the clathrin adapter protein AP2 and is thought to be the dominant subunit in LTD (Lee et al., 2002). The β-adaptin subunit also interacts with the calcium sensor hippocalcin, a protein thought to play a role in coupling changes in intracellular calcium to endocytotic mechanisms (Palmer et al., 2005). AMPAR endocytosis is also reliant on calcium-dependent intracellular signalling cascades (Daniel et al 1998). The situation is highly complex and it seems that the destination of the internalized receptors depends on the history of synapse activity and the method of stimulation (Ehlers, 2000; Lee et al., 2004).

It is now also clear that AMPARs undergo rapid cycles of membrane exocytosis and endocytosis during basal levels of activity (Nishimune et al., 1998; Luscher et al., 1999; Shi et al., 2001) that are tightly regulated to maintain a stable level of activity. The short C-termini isoforms of GluR2 and GluR3 subunits drive this constitutive recycling (Passafaro et al., 2001), and the cycling is dependent on the interaction between GluR2 and NSF (see below).

There are several mechanistic variations of LTP and LTD that are specific to the brain region in which they occur. NMDAR-dependent LTP and LTD in the CA1 region of the hippocampus have been the most extensively researched and will be the focus here. In this type of plasticity, NMDAR activation results in the calcium-dependent insertion or removal of AMPARs via posttranslational modification of the subunits such as phosphorylation and palmitoylation, and the interaction of the subunits with PDZ and non-PDZ domain-containing proteins (see Figure 3.2.1.2).

Auxiliary subunits

Recently much attention has been paid to the prospect that AMPARs may be associated with auxiliary subunits that modulate many aspects of their function. These subunits belong to a family known as TARPs (transmembrane AMPAR regulatory proteins) (Tomita et al., 2003). They interact strongly and directly with all the subunits of AMPARs, possibly via their extracellular domain (Cuadra et al., 2004). The most studied TARP, stargazin/γ2, modulates AMPAR channel properties directly by increasing single channel conductance, slowing receptor desensitization and deactivation (Tomita et al., 2003), possibly by stabilizing receptor conformation (Priel et al., 2005) and by increasing the affinity of the AMPAR for glutamate. All these modifications contribute to increasing the AMPAR response, making stargazin a likely candidate protein for regulating LTP.

Further to a direct effect on AMPAR gating mechanisms, expression of stargazin has also been shown to increase the number of AMPARs at the synapse (Tomita et al., 2003; Vandenberghe et al., 2005). The AMPAR–TARP interaction is crucial for export of AMPARs from the ER (Tomita et al., 2003) and trafficking through the early biosynthetic pathway (Vandenberghe et al., 2005). Stargazin is then involved in a two-stage process that controls synaptic delivery of AMPARs. It first helps to recruit AMPARs to the plasma membrane and then assists their targeting to the synapse (Chen et al., 2000; Tomita et al., 2005b). Stargazin associates with PSD-95 (Schnell et al., 2002), a synaptic scaffolding protein (El-Husseini et al., 2000) thought to facilitate this process (Ehrlich and Malinow, 2004).

In support of these findings, mice with a targeted deletion of the hippocampal stargazin homologue g8 gene display decreased basal AMPAR-mediated transmission and a deficit in LTP (Rouach et al., 2005). LTD was not affected indicating that TARPs are not directly required for the endocytosis of receptors. Moreover, it has been demonstrated that dissociation of TARPs from AMPARs is required for receptor internalization (Tomita et al., 2004). Phosphorylation may be a crucial factor in determining whether TARPs remain bound to AMPARs. Regions of the stargazin C-terminus have been shown to be phosphorylated by Ca^{2+}-calmodulin-dependent protein kinase II (CaMKII) and protein kinase C (PKC) (Tomita et al., 2005b). Phosphorylation site mutants, which could not be phosphorylated, were trafficked efficiently to the membrane but LTP could not be induced. Conversely, mutants mimicking phosphorylation did not display LTD. These findings suggest that phosphorylation prevents internalization of the receptors and that dephosphorylation involving PP1 and PP2A is required for the TARP–AMPAR interaction to be disrupted (Tomita et al., 2005), as a consequence of which, AMPARs are free to move to extrasynaptic sites for internalization and/or the AMPAR current is reduced.

Overall, it is clear that TARPs play an important role in the constitutive cycling of AMPARs and in their activity-dependent incorporation into the synaptic membrane mediated through regulation of receptor properties and increasing receptor abundance at the membrane and preventing receptor removal.

Phosphorylation

Phosphorylation of AMPAR subunits is an important posttranslational modification essential to synaptic plasticity. Receptor phosphorylation regulates channel properties and protein trafficking via protein interactions of receptors (Palmer et al., 2005; Swope et al., 1999).

The GluR1 C-terminal domain can be phosphorylated at position Ser831 by both CaMKII (Mammen et al., 1997; Roche et al.,

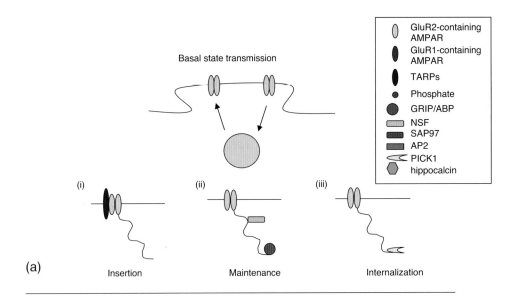

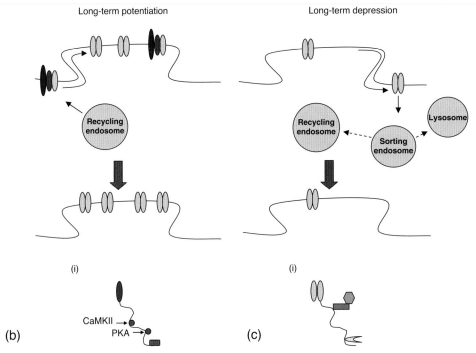

Fig. 3.2.1.2 AMPAR trafficking in synaptic plasticity. (a) During basal conditions, AMPARs undergo constitutive cycles of endocytosis from and reinsertion into the synaptic membrane that are dependent on protein–protein interactions: (i) TARPs are involved in the insertion of AMPARs; (ii) NSF and GRIP/ABP are thought to anchor GluR2-containing AMPARs at the synapse; and (iii) the interaction of PICK and GluR2 may be important in constitutive endocytosis. (b) In LTP, GluR1-containing receptors are inserted into the membrane and are later replaced by GluR2/3 receptors to maintain the increase in synaptic receptor number, (i) phosphorylation of GluR1 and interaction of GluR1 with SAP97 are important in this process. (c) There is a net loss of AMPARs at the synaptic membrane during LTD via clatherin-dependent endocytosis, (i) the interaction of GluR2 with PICK1 and AP2 is essential for this process. See also colour plate section.

1996) and PKC, and at Ser845 by protein kinase A (PKA). The effects of phosphorylation are dependent on the history of activity in each synapse (Palmer *et al.*, 2005).

CaMKII, activated by calcium influx through NMDARs (Lisman *et al.*, 1997; Poncer *et al.*, 2002), is considered to be a mediator of LTP and to play a key role in synaptic unsilencing (Liao *et al.*, 2001). LTP induction increases CaMKII-mediated phosphorylation of the GluR1 subunit (Barria *et al.*, 1997) and consequently enhances synaptic response (Yakel *et al.*, 1995) by increasing single channel conductance (Benke *et al.*, 1998; Derkach *et al.*, 1999; Poncer *et al.*, 2002). Overexpression of activated CaMKII combined with LTP induction increases the presence of GluR1 at the synapse (Shi *et al.*, 2001). It is unclear if this results from a direct effect of CaMKII on the subunit (Shi *et al.*, 2001) or is dependent on an interaction between GluR1 and a PDZ domain-containing protein (see below; Hayashi *et al.*, 2000).

LTP induction also activates PKA, a kinase that phosphorylates Ser 845 of GluR1 (Roche *et al.*, 1996; Banke *et al.*, 2000). The PDZ domain-containing protein SAP97 (synapse-associated protein of 97kDa) and AKAP79 (A kinase anchoring protein 79) form a complex, facilitating the interaction of PKA and GluR1 (Colledge et al., 2000). Like CaMKII, PKA phosphorylation also results in an increased AMPAR response but does so via an increase in the open probability of the receptor (Banke et al., 2000). PKA-mediated phosphorylation also accompanies insertion of AMPARs at the synaptic membrane (Ehlers, 2000). Indeed, mutation of Ser 845 on GluR1 prevents its delivery to the synapse (Shi et al., 2001) and PKA-mediated phosphorylation is critical for LTP during development (Yasuda et al., 2003), however, it is not sufficient for LTP in adult animals. CaMKII activation is also required for increased GluR1 insertion in adults and so consequently, the phosphorylation of PKA acts as a 'gating' mechanism to determine the number of receptors that are available for synaptic incorporation (Esteban et al., 2003). The importance of phosphorylation at these two sites has been emphasized by studies using a genetically modified mouse with mutations at Ser831 and Ser845 (Lee et al., 2003). It was not possible to induce NMDAR-dependent LTP in these mice and LTD was deficient.

The GluR2 subunit is also phosphorylated, most notably at Ser880 by PKC. Phosphorylation at this site results in reduced synaptic AMPARs (Seidenman et al., 2003) by differentially affecting the interaction of the GluR2 CT with PDZ domain-containing proteins GRIP, ABP and PICK (see below) and stimulating subsequent internalization of the receptors.

The dephosphorylation of GluR1 by protein phosphatases is important in LTD induction. NMDAR activation can initiate dephosphorylation involving the phosphatases PP1 and calcineurin (PP2A). If calcineurin is inhibited then AMPAR internalization stimulated by both AMPA and NMDA receptor activation is blocked (Beattie et al, 2000; Ehlers, 2000; Lin et al., 2000). The involvement of PP1 is controversial, with some evidence suggesting that PP1 inhibition blocks AMPA endocytosis (Ehlers, 2000) whereas other laboratories have shown inhibition causes an enhancement of internalization (Beattie et al., 2000, Lin et al., 2000). Further work is needed to establish the exact role of these phosphatases.

Palmitoylation

It has recently been shown that all of the AMPAR subunits can be palmitoylated (Hayashi et al., 2005). Palmitoylation is a post-translational modification whereby a 16-carbon fatty acid is attached to specific cysteine residues in certain proteins and enhances their membrane targeting and association (for review see Smotrys and Linder, 2004). The AMPAR subunits are palmitoylated at a cysteine residue in the pore forming TMD2, and in the proximal C-terminal domain. Using overexpression of mutant GluR1 and GluR2 subunits that could not be palmitoylated, it was revealed that palmitoylation of the residue at TMD2 leads to decreased trafficking of AMPARs to the surface and accumulation of the AMPARs in the Golgi (Hayashi et al., 2005). The palmitoylacetyl transferase GODZ is involved in the palmitoylation of TMD2. Palmitoylation of the C-terminal residue is not important during basal state transmission but is necessary for AMPAR or NMDAR-dependent receptor internalization. It was concluded that palmitoylation prevents the binding of GluR1 to protein 4.1N and leads to internalization (Hayashi et al., 2005).

Protein–protein interactions

The AMPAR subunits are known to bind to many proteins that play an important role in their cycling. Each of the subunits contains a C-terminal PDZ binding domain which mediates many of these protein–protein interactions.

Type II PDZ proteins

GRIP1 and ABP (AMPAR-binding protein) are two highly homologous proteins that exist in many isoforms, the most studied of which are 7 PDZ domain containing proteins that have palmitoylated and unpalmitoylated splice variants. All isoforms interact with the PDZ binding domain of GluR2. GRIP/ABP is thought to anchor receptors at the synapse during basal transmission (Osten et al., 2000) and at intracellular membranes during LTD (Daw et al., 2000). Studies on ABP have shown that the palmitoylated variant is targeted to spines consistent with a role for palmitate in membrane targeting (DeSouza et al., 2002) whereas the unpalmitoylated form is targeted intracellularly (Fu et al., 2003), indicating that ABP could anchor both at the surface and intracellularly depending on the palmitoylation state of the protein

GRIP and ABP are often considered collectively, although the finding that GRIP1 interacts with KIF5, a member of the kinesin heavy transport chain (Setou et al., 2002), suggests that GRIP1 may have additional roles in AMPAR trafficking. GRIP1 has also been implicated in recycling as a result of its interaction with NEEP21 and syntaxin13 (Steiner et al., 2005).

The constitutive and activity dependent internalization of AMPARs is dependent on PICK1 (Xia et al., 2000; Kim et al., 2001). PICK1 contains a PDZ domain that binds to both the GluR2 C-terminus (Dev et al., 1999) and the activated form of PKC (Perez et al., 2001). Phosphorylation of GluR2 at Ser880 by PKC prevents GRIP/ABP binding (Chung et al., 2000), permits the binding of PICK (see below) and facilitates internalization (Seidenman et al., 2003).

PICK1 also contains a BAR (bin/amphiphysin/Rvs) domain, a site of interaction with GRIP/ABP (Lu and Ziff, 2005), but perhaps more relevantly, a region capable of detecting membrane curvature (Peter et al., 2004). It is known that overexpressed PICK1 is targeted to synapses and results in a decrease of synaptic GluR2 (Perez et al., 2001), thus, an interesting hypothesis is that PICK1 is targeted to areas of membrane invagination and assists in endocytosis. PICK1 has been identified as a calcium-sensing protein (Hanley and Henley, 2005), thus providing a mechanism whereby changes in calcium influx resulting from NMDAR activation can be transduced directly into AMPAR internalization. Furthermore, it has been shown that in vivo deletion of PICK1 or the GluR2 PDZ binding domain prevents the induction of LTD (Steinberg et al., 2006) and confirms the importance of PICK in activity-dependent receptor internalization.

Type I PDZ proteins

GluR1 also contains a PDZ binding motif that binds to the second PDZ domain of SAP97 which interacts with GluR1 at the synapse (Leonard et al., 1998) where it may help to anchor GluR1-containing AMPARs (Valschanoff et al., 2000). SAP97 is able to multimerize with itself and other proteins of a similar class, a characteristic consistent with other anchoring proteins (Lee et al., 2002). When SAP97 is overexpressed, the presence of functional

AMPARs at the membrane is increased (Rumbaugh *et al.*, 2003) and LTP is occluded (Nakagawa *et al.*, 2004). These data support a role for SAP97 in the maintenance of synaptic GluR1. There is also a proposed role for CaMKII in the synaptic targeting of SAP97 (Mauceri *et al.*, 2004).

Non-PDZ proteins

NSF interacts with the GluR2 region spanning Lys844–Gln853 (Nishimune *et al.*, 1998; Osten *et al.*, 1998; Song *et al.*, 1998). Some uncertainty surrounds the role of NSF in AMPAR trafficking, much of which is results from initial studies that used the peptide pep2m to block the interaction of NSF and endogenous GluR2. Subsequent studies revealed that the clathrin adaptor protein AP2, a protein involved in endocytosis, also binds to GluR2 at an overlapping site with NSF and so would also be disrupted by pep2m (Lee *et al.*, 2002). Thus, it is difficult to determine what mechanisms are responsible for the decreased response of AMPARs and occlusion of LTD (Luscher *et al.*, 1999; Luthi *et al.*, 1999) that is seen with pep2m infusion (Noel *et al.*, 1999). Nonetheless, NSF appears to stabilize GluR2 at the membrane during constitutive recycling. It may do this by preventing the interaction of GluR2 with AP2 and subsequent internalization via clathrin-dependent vesicles and/or by preventing interaction with PICK1 (Hanley *et al.*, 2002). A complex of NSF, GluR2 and SNAPs was initially identified (Osten *et al.*, 1998) which has subsequently been shown to associate with PICK1 (Hanley *et al.*, 2002). The interaction of GluR2 and PICK can be disrupted by NSF to inhibit internalization. NSF is also implicated in activity-dependent cycling, as when NSF interactions are blocked, receptors that are internalized in response to AMPAR stimulation are aberrantly targeted to lysosomes instead of recycling endosomes (Lee *et al.*, 2004).

Protein 4.1N belongs to a family of cytoskeletal proteins. It provides a link between AMPARs and the actin cytoskeleton via its interaction with a region in the C-terminus of GluR1 (Shen *et al.*, 2000). Surface expression of AMPARs is dependent on the GluR1–4.1N interaction (Hayashi *et al.*, 2005).

3.2.1.7 AMPARs and disease

Given the knowledge that AMPARs mediate fast excitatory transmission and synaptic plasticity, attention has naturally been directed to how AMPARs may be affected in neurodegenerative disease. Furthermore, the theory that synaptic plasticity is the molecular correlate of learning and memory, and the fundamental role that AMPARs play in this process, positions these receptors as a prime target for cognition-enhancing drugs.

A common mechanism thought to contribute to many neurodegenerative diseases is excitotoxicity. The overactivation of excitatory glutamate receptors gives rise to abnormally high levels of intracellular calcium. This triggers release of cell-damaging reactive oxygen species (ROS) and apoptotic mediators. AMPARs are believed to play a crucial role in this pathological process. In normal physiological conditions, the subunit composition and movements of AMPARs are strictly controlled. In the hippocampus, the majority of receptors at the synapse are rendered calcium impermeable by the presence of the GluR2 subunit and calcium permeable GluR1-containing AMPARs are transiently inserted into the synaptic membrane in response to induction of LTP. Disrupted regulation of edited GluR2 would result in altered AMPAR composition at the synaptic membrane and consequently a change

in receptor calcium conductance would occur that could contribute to the toxic calcium levels seen in degenerating neurons.

In amyotropic lateral sclerosis (ALS) there is a progressive and selective death of motor neurons (MNs). The total levels of GluR2 mRNA in ALS cases compared to normal controls are identical (Kawahara *et al.*, 2003). However, there is a decrease in the editing efficiency at the Q/R site of GluR2 in ALS that is a result of reduced ADAR2 enzyme activity (Kawahara *et al.*, 2004). Consequently, 56 per cent of GluR2 is not edited and a greater number of calcium-permeable AMPARs are assembled (Kwak and Kawahara, 2005). In normal cells, MNs have higher levels of edited GluR2 containing receptors at the membrane than found in the hippocampus, but despite this they do not buffer calcium well. Excess calcium is taken up by mitochondria, promoting the release of ROS and activation of apoptotic pathways. It is considered that ROS may disrupt the functioning of glutamate transporters in astrocytes (Rao *et al.*, 2003). This in turn may contribute to a pathogenic cycle of increased extracellular glutamate, AMPAR activation, intracellular calcium and further toxicity. It has also been proposed that a decrease in the flop variants of GluR1–4 in motor neurons could contribute to the vulnerability of these cells to AMPAR-mediated excitotoxicity (Kawahara and Kwak, 2005).

The mechanism underlying neuronal injury resulting from ischaemia may be similar to that involved in ALS under certain conditions, whereby there is a localized deficiency in the ADAR enzyme in CA1 hippocampal neurons (Peng *et al.*, 2006). This deficit is accompanied by a loss of GluR2 mRNA and an increased contribution of calcium-permeable AMPARs to receptor-mediated currents in the CA1 following an ischaemic episode (Gorter *et al.*, 1997). Thus, as in ALS the decrease in the GluR2 subunit is likely to result in increased calcium conductance via calcium-permeable AMPA receptors. In support of this hypothesis, the overexpression of calcium-impermeable GluR2 *in vivo* can attenuate ischaemic injury (Liu *et al.*, 2004).

Additionally, studies on tissue from patients with malignant glioma (who display an increased incidence of epileptic seizures) show a 300-fold increase in unedited GluR2 (Maas *et al.*, 2001), which can also be attributed to deficits in ADAR2 activity.

The increased level of calcium-permeable AMPARs at the synaptic membrane of vulnerable neurons in ALS and ischaemia is also thought to result in an increased AMPAR-mediated Zn^{2+} conductance. Zn^{2+} is thought to impair mitochondrial function more potently than excess Ca^{2+} and can induce PARP (poly(ADP-ribose) polymerase) activation and downstream enzyme activation resulting in the generation of ROS (Sensi and Jeng, 2004). In an *in vitro* slice model of ischaemia, Zn^{2+} chelators and blockers of Ca^{2+}-permeable ion channels which reduce the levels of intracellular Zn^{2+} and Ca^{2+} attenuate neuronal damage (Yin *et al.*, 2002). The positive effect of channel blockers was also apparent if the drugs were applied hours to days after an ischaemic insult (Noh *et al.*, 2005), indicating that drugs with this mode of action might offer therapeutic potential.

Similar to expression of mutant non-editable GluR2, ablation of ADAR2 expression (Brusa *et al.*, 1995; Higuchi *et al.*, 2000; Seeburg *et al.*, 2001) or disruption of editing mechanisms results in seizures and premature death in animal models (Feldmeyer *et al.*, 1999). The lethal phenotype resulting from ADAR2 ablation can be rescued by generating a Q to R point mutation at the Q/R editing site, thus rendering editing obsolete (Higuchi *et al.*, 2000).

Mechanisms that protect some cells from AMPAR-mediated excitotoxicity are beginning to emerge. For example, prolonged synaptic activity in cerebellar stellate neurons causes substitution of GluR2-lacking receptors at these synapses with GluR2-containing ones, thus decreasing receptor-mediated Ca^{2+} influx (Liu and Cull-Candy, 2000).

Q/R editing appears to be preserved in the vulnerable brain regions involved in other neurodegenerative conditions such as Alzheimer's and Parkinson's diseases. However, this does not exclude a role for excitotoxicity in these disorders or other changes in AMPAR physiology. For example, Alzheimer's disease (AD), characterized pathologically by the presence of amyloid plaques and neurofibrillary tangles, and behaviourally by memory loss, is considered to be a consequence of synaptic malfunction. It has recently been hypothesized that AD is a result of a downscaling mechanism to maintain the overall activity of the cell. This theory is supported by observations from double knockin transgenic mice carrying human mutations in the genes for amyloid precursor protein and presenilin-1 which exhibit age-related decreases in AMPAR-mediated currents and deficits in LTP and LTD (Chang et al., 2006). In addition, application of amyloid β peptide in vitro results in deficits of fast axonal transmission (Hiruma et al., 2003) and LTP (Freir et al., 2001) which has been proposed to result from mechanisms including disruption of activity-dependent CaMKII phosphorylation (Zhao et al., 2004), AMPAR dysfunction (Szegedi et al., 2005) and increased AMPAR proteolysis by caspases (Chan et al.,1999). Soluble Aβ is also targeted to synapses where it is thought to cause an NMDAR activity-dependent degradation of PSD95 that results in a loss of GluR2 from the synaptic membrane (Roselli et al., 2005).

Analysis of post-mortem brain tissue of patients with AD correlated to their scores on the Braak AD assessment scale before death, revealed that in the vulnerable pyramidal cells of the subiculum, the GluR2 subunit is decreased in early stages of the disease followed by a decrease in GluR3 in intermediate stages. GluR1 levels are unchanged throughout the disease course (Carter et al., 2004). The trend for an early decrease in GluR2 has been supported by an observed decrease in the subunit before the formation of neurofibrillary tangles (Ikonomovic et al., 1997).

In addition to their potential involvement in neurodegenerative disorders, AMPARs have become a focus of research in the mechanisms underlying neuropathic pain. Analogous to their role in synaptic plasticity under normal physiological conditions, evidence suggests that AMPARs are involved in activity-dependent changes in the synapse in response to pain stimuli. Pathways involved in mediating chronic pain are thought to exhibit altered glutamatergic transmission (Gebhart et al., 2004; for review Woolf and Salter, 2000). The dorsal spinal cord has a high density of AMPARs that are permeable to calcium both pre- and postsynaptically (Lu et al., 2002). A comprehensive study showed that GluR1 knockout mice exhibited decreased AMPAR currents and receptor-mediated calcium influx, accompanied by a loss of nociceptive plasticity. Conversely, evidence of increased behavioural responses to painful stimuli could be observed in GluR2 knockout mice, accompanied by increased receptor-mediated calcium flux (Hartman et al., 2004). It was concluded that similar to LTP, increased GluR1 incorporation, or otherwise increased activity of calcium-permeable receptors, is required at the synapse for rapid sensitization in response to pain stimuli. This increase could be due to modulation

of RNA editing mechanisms such as those described in ALS and ischaemia, rapid synthesis of GluR1 in the dendrite, or changes in other factors regulating receptor function such as phosphorylation or interaction with other proteins (Hartman et al., 2004). Indeed, recent studies have shown that induced hyperalgaesia in animal models results in increased synaptic GluR1 and the activation of CaMKII. Additionally, it has been suggested that the application of Brefeldin-A prevents GluR1 trafficking and referred hyperalgaesia by blocking exocytosis (Galan et al., 2004) and that bradykinin increases the AMPAR EPSC and decreases pain threshold (Wang et al., 2004).

AMPARs are also implicated in other diseases including Parkinson's and Huntington's and schizophrenia and recently they have been implicated in addiction and reward mechanisms (Sutton et al., 2003; Thomas and Malenka, 2003). The full extent of AMPARs involvement in disease is beyond the scope of this review. However, we have highlighted the areas in which significant progress has been in made in elucidating the mechanisms by which AMPARs might contribute to certain degenerative disorders and consequently how they might represent potential therapeutic targets. Facilitation of AMPAR function, for example using positive allosteric modulatory ligands such as those described above, may also represent an attractive therapeutic strategy in certain disorders. Clearly the tight regulation of AMPAR number and function plays a critical role both in normal physiology and in disease pathophysiology and as such, further research into the mechanisms underlying these regulatory processes and their dysfunction is of great interest.

3.2.1.8 Concluding remarks

The biophysical characteristics of AMPARs, particularly those conferred by post-transcriptional modification, enable them to perform a critical role in the CNS, where they are tightly regulated and modulated by phosphorylation, trafficking, and interacting proteins. Changes in AMPAR number and properties at the synapse constitute key mechanisms underlying synaptic plasticity. Increased understanding of the role of AMPARs, both in normal physiology and in disease states, together with the ongoing development of novel pharmacological tools renders them attractive therapeutic targets for a number of disorders. While research in this field to date has generated a vast wealth of understanding, there remains enormous potential for future discovery.

References

Adesnik H, Nicoll RA and England PM (2005). Photoinactivation of native AMPA receptors reveals their real-time trafficking. *Neuron* **48**, 977–985.

Arai AC, Kessler M, Rogers G et al. (2000). Effects of the potent ampakine CX614 on hippocampal and recombinant AMPA receptors: interactions with cyclothiazide and GYKI 52466. *Mol Pharmacol* **58**, 802–813.

Arai AC, Xia YF, Rogers G et al. (2002) Benzamide-type AMPA receptor modulators form two subfamilies with distinct modes of action. *J Pharmacol Exp. Ther* **303**, 1075–1085.

Armstrong N, Mayer M and Gouaux E (2003). Tuning activation of the AMPA-sensitive GluR2 ion channel by genetic adjustment of agonist-induced conformational changes. *Proc Nat Acad Sci* USA **100**, 5736–5741.

Ast G (2004). How did alternative splicing evolve? *Nat Rev Genet* **5**, 773–782.

Balannik V, Menniti FS, Paternain AV et al. (2005). Molecular mechanism of AMPA receptor noncompetitive antagonism. *Neuron* **48**, 279–288.

Banke TG, Bowie D, Lee H *et al.* (2000). Control of GluR1 AMPA receptor function by cAMP-dependent protein kinase. *J Neurosci* **20**, 89–102.

Barria A, Muller D, Derkach V *et al.* (1997). Regulatory phosphorylation of AMPA-type glutamate receptors by CaM-KII during long-term potentiation. *Science* **276**, 2042–2045.

Basilio C, Wahba AJ, Lengyel P *et al.* (1962). Synthetic polynucleotides and the amino acid code. *Proc Nat Acad Sci USA* **48**, 613–616.

Bass BL (2002). RNA editing by adenosine deaminases that act on RNA. *Annu Rev Biochem* **71**, 817–846.

Bass BL and Weintraub H (1988). An unwinding activity that covalently modifies its double-stranded RNA substrate. *Cell* **55**, 1089–1098.

Baude A, Molnar E, Latawiec D *et al.* (1994). Synaptic and nonsynaptic localization of the GluR1 subunit of the AMPA-type excitatory amino acid receptor in the rat cerebellum. *J Neurosci* **14**, 2830–2843.

Beattie EC, Carroll RC, Yu X *et al.* (2000). Regulation of AMPA receptor endocytosis by a signaling mechanism shared with LTD. *Nat Neurosci* **3**, 1291–1300.

Benke TA, Lüthi A, Isaac JT *et al.* (1998). Modulation of AMPA receptor unitary conductance by synaptic activity. *Nature* **393**, 793–797.

Bettler B, Boulter J, Hermans-Borgmeyer I *et al.* (1990). Cloning of a novel glutamate receptor subunit, GluR5: expression in the nervous system during development. *Neuron* **5**, 583–595.

Blaschke M, Keller BU, Rivosecchi R *et al.* (1993). A single amino acid determines the subunit-specific spider toxin block of α-amino-3-hydroxy-5-methyl-4-isoxazoleproprionate/kainate receptor channels. *Proc Nat Acad Sci USA* **90**, 6528–6532.

Bleakman D, Ballyk BA, Schoepp DD *et al.* (1996). Activity of 2,3-benzodiazepines at native rat and recombinant human glutamate receptors *in vitro*: stereospecificity and selectivity profiles. *Neuropharmacology* **35**, 1689–1702.

Bleakman D, Pearson K, Harnan S *et al.* (1995). Effects of LY293558 and NBQX on glutamate receptor responses in rat cerebellar Purkinje neurones and HEK293 cells expressing the human GluR6 glutamate receptor. *Br J Pharmacol* **115**, 112.

Bochet P, Audinat E, Lambolez B *et al.* (1994). Subunit composition at the single-cell level explains functional properties of a glutamate-gated channel. *Neuron* **12**, 383–388.

Boulter J, Hollmann M, O'shea-Greenfield A *et al.* (1990). Molecular cloning and functional expression of glutamate receptor subunit genes. *Science* **249**, 1033–1037.

Brackley PTH, Bell DR, Choi S-K *et al.* (1993) Selective antagonism of native and cloned kainate and NMDA receptors by polyamine-containing toxins. *J Pharmacol Exp Ther* **266**, 1573–1580.

Braithwaite SP, Meyer G and Henley JM (2000). Interactions between AMPA receptors and intracellular proteins. *Neuropharmacology* **39**, 919–930.

Bratt E and Öhman M (2003). Coordination of editing and splicing of glutamate receptor pre-mRNA. *RNA* **9**, 309–318.

Brorson JR, Li D and Suzuki T (2004). Selective expression of heteromeric AMPA receptors driven by flip-flop differences. *J Neurosci* **24**, 3461–3470.

Brusa R, Zimmermann F, Koh DS *et al.* (1995). Early-onset epilepsy and postnatal lethality associated with an editing-deficient GluR-B allele in mice. *Science* **270**, 1677–1680.

Burnashev N, Monyer H, Seeburg PH *et al.* (1992). Divalent ion permeability of AMPA receptor channels is dominated by the edited form of a single subunit. *Neuron* **8**, 189–198.

Carroll RC, Beattie EC, Xia H *et al.* (1999). Dynamin-dependent endocytosis of ionotropic glutamate receptors. *Proc Nat Acad Sci USA* **96**, 14112–14117.

Carter TL, Rissman RA, Mishizen-Eberz AJ *et al.* (2004). Differential preservation of AMPA receptor subunits in the hippocampi of Alzheimer's disease patients according to Braak stage. *Exp Neurol* **187**, 299–309.

Chambers JJ, Gouda H, Young DM *et al.* (2004). Photochemically knocking out glutamate receptors *in vivo. J Am Chem Soc* **126**, 13886–13887.

Chan SL, Griffin WS and Mattson MP (1999). Evidence for caspase-mediated cleavage of AMPA receptor subunits in neuronal apoptosis and Alzheimer's disease. *J Neurosci Res* **57**, 315–323.

Chang EH, Savage MJ, Flood DG *et al.* (2006). AMPA receptor downscaling at the onset of Alzheimer's disease pathology in double knockin mice. *Proc Nat Acad Sci USA* **103**, 3410–3415.

Chen L, Chetkovich DM, Petralia RS *et al.* (2000). Stargazin regulates synaptic targeting of AMPA receptors by two distinct mechanisms. *Nature* **408**, 936–943.

Chen YC, Lin WH, Tzeng DW *et al.* (2006). The mutually exclusive flip and flop exons of AMPA receptor genes were derived from an intragenic duplication in the vertebrate lineage. *J Mol Evol* **62**, 121–131.

Choi DW (1992). Excitotoxic cell death. *J Neurobiol* **23**, 1261–1276.

Chung HJ, Xia J, Scannevin RH *et al.* (2000). Phosphorylation of the AMPA receptor subunit GluR2 differentially regulates its interaction with PDZ domain-containing proteins. *J Neurosci* **20**, 7258–7267.

Clements JD and Westbrook GL (1994) Kinetics of AP5 dissociation from NMDA receptors: evidence for two identical cooperative binding sites. *J Neurophysiol* **71**, 2566–2569.

Colledge M, Dean RA, Scott GK *et al.* (2000). Targeting of PKA to glutamate receptors through a MAGUK-AKAP complex. *Neuron* **27**, 107–119.

Colquhoun D, Jonas P and Sakmann B (1992) Action of brief pulses of glutamate on AMPA/kainate receptors in patches from different neurones of rat hippocampal slices. *J Physiol* **458**, 261–287.

Cuadra AE, Kuo SH, Kawasaki Y *et al.* (2004). AMPA receptor synaptic targeting regulated by stargazin interactions with the Golgi-resident PDZ protein nPIST. *J Neurosci* **24**, 7491–7502.

Daniel H, Levenes C and Crepel F (1998). Cellular mechanisms of cerebellar LTD. *Trends Neurosci* **21**, 401–407.

Daw MI, Chittajallu R, Bortolotto ZA *et al.* (2000). PDZ proteins interacting with C-terminal GluR2/3 are involved in a PKC-dependent regulation of AMPA receptors at hippocampal synapses. *Neuron* **28**, 873–886.

Derkach V, Barria A and Soderling TR (1999). Ca2+/calmodulin-kinase II enhances channel conductance of alpha-amino-3-hydroxy-5-methyl-4-isoxazolepropionate type glutamate receptors. *Proc Nat Acad Sci USA* **96**, 3269–3274.

DeSouza S, Fu J, States BA *et al.* (2002). Differential palmitoylation directs the AMPA receptor-binding protein ABP to spines or to intracellular clusters. *J Neurosci* **22**, 3493–3503.

Dev KK, Nishimune A, Henley JM *et al.* (1999). The protein kinase C alpha binding protein PICK1 interactis with short but not long form alternative splice variants of AMPA receptor subunits. *Neuropharmacology* **38**, 635–644.

Donevan SD and Rogawski MA (1993). GYKI 52466, a 2,3-benzodiazepine, is a highly selective, noncompetitive antagonist of AMPA/kainate receptor responses. *Neuron* **10**, 51–59.

Donevan SD and Rogawski MA (1998). Allosteric regulation of alpha-amino-3-hyroxy-5-methyl-4-isoaxazole-proprionate receptors by thiocyanate and cyclothiazide at a common modulatory site distinct from that of 2,3-benzodiazepines. *Neuroscience* **87**, 615–629.

Dong H, O'Brien RJ, Fung ET *et al.* (1997). GRIP: a synaptic PDZ domain containing protein that interacts with AMPA receptors. *Nature* **386**, 279–284.

Dreixler JC and Leonard JP (1994). Subunit-specific enhancement of glutamate receptor responses by zinc. *Brain Res Mol Brain Res* **22**, 144–150.

Drejer J and Honoré T (1988). New quinoxalinediones show potent antagonism of quisqualate responses in cultured mouse cortical neurons. *Neurosci Lett* **87**, 104–108.

Early P, Rogers J, Davis M *et al.* (1980). Two mRNAs can be produced from a single immunoglobulin *mu* gene by alternative RNA processing pathways. *Cell* **20**, 313–319.

Ehlers MD (2000). Reinsertion or degradation of AMPA receptors determined by activity-dependent endocytic sorting. *Neuron* **28**, 511–525.

Ehrlich I and Malinow R (2004). Postsynaptic density 95 controls AMPA receptor incorporation during long-term potentiation and experience-driven synaptic plasticity. *J Neurosci* **24**, 916–27.

El-Husseini AE, Schnell E, Chetkovich DM *et al.* (2000). PSD-95 involvement in maturation of excitatory synapses. *Science* **290**, 1364–1368.

Esteban JA, Shi SH, Wilson C *et al.* (2003). PKA phosphorylation of AMPA receptor subunits controls synaptic trafficking underlying plasticity. *Nat Neurosci* **6**, 136–143.

Feldmeyer D, Kask K, Brusa R *et al.* (1999). Neurological dysfunctions in mice expressing different levels of the Q/R site-unedited AMPAR subunit GluR-B. *Nat Neurosci* **2**, 57–64.

Freir DB, Holscher C and Herron CE (2001). Blockade of long-term potentiation by beta-amyloid peptides in the CA1 region of the rat hippocampus *in vivo*. *J Neurophysiol* **85**, 708–713.

Fu J, deSouza S and Ziff EB (2003). Intracellular membrane targeting and suppression of Ser880 phosphorylation of glutamate receptor 2 by the linker I-set II domain of AMPA receptor-binding protein. *J Neurosci* **23**, 7592–7601.

Galan A, Laird JM and Cervero F (2004). *In vivo* recruitment by painful stimuli of AMPA receptor subunits to the plasma membrane of spinal cord neurons. *Pain* **112**, 315–323.

Gallo V, Upson LM, Hayes WP *et al.* (1992). Molecular cloning and development analysis of a new glutamate receptor subunit isoform in cerebellum. *J Neurosci* **12**, 1010–1023.

Gebhart GF (2004). Descending modulation of pain. *Neurosci Biobehav Rev* **27**, 729–737.

Glavinovic MI (2002) Mechanisms shaping fast excitatory postsynaptic currents in the central nervous system. *Neural Comput* **14**, 1–19

Gorter JA, Petrozzino JJ, Aronica EM *et al.* (1997). Global ischemia induces downregulation of Glur2 mRNA and increases AMPA receptor-mediated Ca2+ influx in hippocampal CA1 neurons of gerbil. *J Neurosci* **17**, 6179–6188.

Graveley BR (2005). Mutually exclusive splicing of the insect Dscam pre-mRNA directed by competing intronic RNA secondary structures. *Cell* **123**, 65–73.

Greger IH, Khatri L and Ziff EB (2002). RNA editing at arg607 controls AMPA receptor exit from the endoplasmic reticulum. *Neuron* **34**, 759–772.

Greger IH, Khatri L, Kong X *et al.* (2003). AMPA receptor tetramerization is mediated by Q/R editing. *Neuron* **40**, 763–774.

Hanley JG and Henley JM (2005). PICK1 is a calcium-sensor for NMDA-induced AMPA receptor trafficking. *EMBO J* **24**, 3266–3278.

Hanley JG, Khatri L, Hanson PI *et al.* (2002). NSF ATPase and alpha-/beta-SNAPs disassemble the AMPA receptor-PICK1 complex. *Neuron* **34**, 53–67.

Hartmann B, Ahmadi S, Heppenstall PA *et al.* (2004). The AMPA receptor subunits GluR-A and GluR-B reciprocally modulate spinal synaptic plasticity and inflammatory pain. *Neuron* **44**, 637–650.

Hayashi T, Rumbaugh G and Huganir RL (2005). Differential regulation of AMPA receptor subunit trafficking by palmitoylation of two distinct sites. *Neuron* **47**, 709–723.

Hayashi Y, Shi SH, Esteban JA *et al.* (2000). Driving AMPA receptors into synapses by LTP and CaMKII: requirement for GluR1 and PDZ domain interaction. *Science* **287**, 2262–2267.

Henley JM, Nishimune A, Nash SR *et al.* (1997). Use of the two-hybrid system to find novel proteins that interact with AMPA receptor subunits. *Biochem Soc Trans* **25**, 838–842.

Herlitze S, Raditsch M, Ruppersberg JP *et al.* (1993). Argiotoxin detects molecular differences in AMPA receptor channels. *Neuron* **10**, 1131–1140.

Higuchi M, Maas S, Single FN *et al.* (2000). Point mutation in an AMPA receptor gene rescues lethality in mice deficient in the RNA-editing enzyme ADAR2. *Nature* **406**, 78–81.

Higuchi M, Single FN, Köhler M *et al.* (1993). RNA editing of AMPA receptor subunit GluR-B: a base-paired intron–exon structure determines position and efficiency. *Cell* **75**, 1361–1370.

Hiruma H, Katakura T, Takahashi S *et al.* (2003). Glutamate and amyloid beta-protein rapidly inhibit fast axonal transport in cultured rat hippocampal neurons by different mechanisms. *J Neurosci* **23**, 8967–8977.

Hollmann M, Maron C and Heinemann S (1994). N-glycosylation site tagging suggests a three trasmembrane domain topology for the glutamate receptor GluR1. *Neuron* **13**, 1331–1343.

Hollmann M, O'shea-Greenfield A, Rogers SW *et al.* (1989). Cloning by functional expression of a member of the glutamate receptor family. *Nature* **342**, 643–648.

Hollmann M, Rogers SW, O'shea-Greenfield A *et al.* (1990). Glutamate receptor GluR-K1: structure, function and expression in the brain. *Cold Spring Harb Symp Quant Biol* **55**, 41–55.

Horning MS and Mayer ML (2004) Regulation of AMPA receptor gating by ligand binding core dimers. *Neuron* **41**, 379–388.

Hume RI, Dingledine R and Heinemann SF (1991). Identification of a site in glutamate receptor subunits that controls calcium permeability. *Science* **253**, 1028–1031.

Hunter C, Petralia RS, Vu T *et al.* (1993). Expression of AMPA-selective glutamate receptor subunits in morphologically defined neurons of the mammalian cochlear nucleus. *J Neurosci* **13**, 1932–1946.

Iino M, Koike M, Isa T *et al.* (1996). Voltage-dependent blockage of Ca^{2+}-permeable AMPA receptors by joro spider toxin in cultured rat hippocampal neurones. *J Physiol* **496**, 431–437.

Iino M, Mochizuki S and Ozawa S (1994). Relationship between calcium permeability and rectification properties of AMPA receptors in cultured rat hippocampal neurons. *Neurosci Lett* **173**, 14–16.

Ikonomovic MD, Mizukami K, Davies P *et al.* (1997). The loss of GluR2(3) immunoreactivity precedes neurofibrillary tangle formation in the entorhinal cortex and hippocampus of Alzheimer brains. *J Neuropathol Exp Neurol* **56**, 1018–1027.

Inagaki N, Kuromi H, Gonoi T *et al.* (1995). Expression and role of ionotropic glutamate receptors in pancreatic islet cells. *FASEB J* **9**, 686–691.

Isaac JT (2003). Postsynaptic silent synapses: evidence and mechanisms. *Neuropharmacology* **45**, 450–460.

Isaac JT, Nicoll RA and Malenka RC (1995). Evidence for silent synapses: implications for the expression of LTP. *Neuron* **15**, 427–434.

Janssens N and Lesage AS (2001). Glutamate receptor subunit expression in primary neuronal and secondary glial cultures. *J Neurochem* **77**, 1457–1474.

Jayakar SS and Dikshit M (2004). AMPA receptor regulation mechanisms: future target for safer neuroprotective drugs. *Int J Neurosci* **114**, 695–734.

Jin R, Banke TG, Mayer ML *et al.* (2003). Structural basis for partial agonist action at ionotropic glutamate receptors. *Nat Neurosci* **6**, 803–810.

Jones MG and Lodge D (1991). Comparison of some arthropod toxins and toxin fragments as antagonists of excitatory amino acid-induced excitation of rat spinal neurones. *Eur J Pharmacol* **204**, 203–209.

Jones MV and Westbrook GL (1996). The impact of receptor desensitization on fast synaptic transmission. *Trends Neurosci* **19**, 96–101.

Jurica MS and Moore MJ (2003). Pre-mRNA splicing: awash in a sea of proteins. *Mol Cell* **12**, 5–14.

Kawahara Y and Kwak S (2005). Excitotoxicity and ALS: what is unique about the AMPA receptors expressed on spinal motor neurons? *Amyotroph Lateral Scler Other Motor Neuron Disord* **6**, 131–144.

Kawahara Y, Ito K, Sun H *et al.* (2004). Glutamate receptors: RNA editing and death of motor neurons. *Nature* **427**, 801.

Kawahara Y, Kwak S, Sun H *et al.* (2003). Human spinal motor neurons express low relative abundance of GluR2 mRNA: an implication for excitotoxicity in ALS. *J Neurochem* **85**, 680–689.

Keinänen K, Wisden W, Sommer B *et al.* (1990). A family of AMPA-selective glutamate receptors. *Science* **249**, 556–560.

Kim CH, Chung HJ, Lee HK *et al.* (2001). Interaction of the AMPA receptor subunit GluR2/3 with PDZ domains regulates hippocampal long-term depression. *Proc Nat Acad Sci USA* **98**, 11725–11730.

Kim DY, Kim SH, Choi HB *et al.* (2001). High abundance of GluR1 mRNA and reduced Q/R editing of GluR2 mRNA in individual NADPH-diaphorase neurons. *Mol Cell Neurosci* **17**, 1025–1033.

Kohara A, Okada M, Tsutsumi R *et al.* (1998). *In-vitro* characterization of YM872, a selective, potent and highly water-soluble alpha-amino-3-hydroxy-5-methylisoxazole-4-propionate receptor antagonist. *J Pharm Pharmacol* **50**, 795–801.

Köhler M, Burnashev N, Sakmann B *et al.* (1993). Determinants of Ca^{2+} permeability in both TM1 and TM2 of high-affinity kainate receptor channels: diversity by RNA editing. *Neuron* **10**, 491–500.

Köhler M, Kornau HC and Seeburg PH (1994). The organization of the gene for the functionally dominant AMPA receptor subunit GluR-B. *J Biol Chem* **269**, 17367–17370.

Koike M, Tsukada S, Tsuzuki K *et al.* (2000). Regulation of kinetic properties of GluR2 AMPA receptor channels by alternative splicing. *J Neurosci* **20**, 2166–2174.

Koike-Tani M, Saitoh N and Takahashi T (2005) Mechanisms underlying developmental speeding in AMPA-EPSC decay time at the calyx of Held. *J Neurosci* **25**, 199–207.

Krampfl K, Schlesinger F, Zörner A *et al.* J (2002). Control of kinetic properties of GluR2 flop AMPA-type channels: impact of R/G nuclear editing. *Eur J Neurosci* **15**, 51–62.

Krogsgaard-Larsen P, Honore T, Hansen JJ *et al.* (1980). New class of glutamate agonist structurally related to ibotenic acid. *Nature* **284**, 64–66.

Kuner T, Seeburg PH and Guy HR (2003). A common architecture for K$^+$ channels and ionotropic glutamate receptors? *Trends Neurosci* **26**, 27–32.

Kuner T, Wollmuth LP, Karlin A *et al.* (1996). Structure of the NMDA receptor channel M2 segment inferred from the accessibility of substituted cysteines. *Neuron* **17**, 343–352.

Kwak S and Kawahara Y (2005). Deficient RNA editing of GluR2 and neuronal death in amyotropic lateral sclerosis. *J Mol Med* **83**, 110–120.

Lazzaro JT, Paternain AV, Lerma J *et al.* (2002). Functional characterisation of CP-465,022, a selective, noncompetitive AMPA receptor antagonist. *Neuropharmacology* **42**, 143–153.

Lee HK, Takamiya K, Han JS *et al.* (2003). Phosphorylation of the AMPA receptor GluR1 subunit is required for synaptic plasticity and retention of spatial memory. *Cell* **112**, 631–643.

Lee SH, Liu L, Wang YT *et al.* (2002). Clathrin adaptor AP2 and NSF interact with overlapping sites of GluR2 and play distinct roles in AMPA receptor trafficking and hippocampal LTD. *Neuron* **36**, 661–674.

Lee SH, Simonetta A and Sheng M (2004). Subunit rules governing the sorting of internalized AMPA receptors in hippocampal neurons. *Neuron* **43**, 221–236.

Leonard AS, Davare MA, Horne MC *et al.* (1998). SAP97 is associated with the AMPA receptor GluR1 subunit. *J Biol Chem* **273**, 19518–19524.

Liao D, Hessler NA and Malinow R (1995). Activation of postsynaptically silent synapses during pairing-induced LTP in CA1 region of hippocampal slice. *Nature* **375**, 400–404.

Liao D, Scannevin RH and Huganir R (2001). Activation of silent synapses by rapid activity-dependent synaptic recruitment of AMPA receptors. *J Neurosci* **21**, 6008–6017.

Lin JW, Ju W, Foster K *et al.* (2000). Distinct molecular mechanisms and divergent endocytotic pathways of AMPA receptor internalization. *Nat Neurosci* **3**, 1282–1290.

Lisman J, Malenka RC, Nicoll RA *et al.* (1997). Learning mechanisms: the case for CaM-KII. *Science* **276**, 2001–2002.

Lissin DV, Carroll RC, Nicoll RA *et al.* (1999) Rapid, activation-induced redistribution of ionotropic glutamate receptors in cultured hippocampal neurons. *J Neurosci* **19**, 1263–1272.

Liu S, Lau L, Wei J *et al.* (2004). Expression of Ca(2+)-permeable AMPA receptor channels primes cell death in transient forebrain ischemia. *Neuron* **43**, 43–55.

Liu SQ and Cull-Candy SG (2000). Synaptic activity at calcium-permeable AMPA receptors induces a switch in receptor subtype. *Nature* **405**, 454–458.

Lomeli H, Mosbacher J, Melcher T *et al.* (1994). Control of kinetic properties of AMPA receptor channels by nuclear RNA editing. *Science* **266**, 1709–1713.

Lu CR, Hwang SJ, Phend KD *et al.* (2002). Primary afferent terminals in spinal cord express presynaptic AMPA receptors. *J Neurosci* **22**, 9522–9529.

Lu W and Ziff EB (2005). PICK1 interacts with ABP/GRIP to regulate AMPA receptor trafficking. *Neuron* **47**, 407–421.

Lu W, Man H, Ju W *et al.* (2001a). Activation of synaptic NMDA receptors induces membrane insertion of new AMPA receptors and LTP in cultured hippocampal neurons. *Neuron* **29**, 243–254.

Lu X, Wyszynski M, Sheng M *et al.* (2001b). Proteolysis of glutamate receptor-interacting protein by calpain in rat brain: implications for synaptic plasticity. *J Neurochem* **77**, 1553–1560.

Luscher C, Xia H, Beattie EC *et al.* (1999). Role of AMPA receptor cycling in synaptic transmission and plasticity. *Neuron* **24**, 649–658.

Lüthi A, Chittajallu R, Duprat F *et al.* (1999). Hippocampal LTD expression involves a pool of AMPARs regulated by the NSF-GluR2 interaction. *Neuron* **24**, 389–399.

Lynch G (2004). AMPA receptor modulators as cognitive enhancers. *Curr Op Pharmacol* **4**, 4–11.

Maas S, Patt S, Schrey M *et al.* (2001). Underediting of glutamate receptor GluR-B mRNA in malignant gliomas. *Proc Nat Acad Sci USA* **98**, 14687–14692.

Mammen AL, Kameyama K, Roche KW *et al.* (1997). Phosphorylation of the alpha-amino-3-hydroxy-5-methylisoxazole4-propionic acid receptor GluR1 subunit by calcium/calmodulin-dependent kinase II. *J Biol Chem* **272**, 32528–32533.

Man HY, Lin JW, Ju WH *et al.* (2000). Regulation of AMPA receptor-mediated synaptic transmission by clathrin-dependent receptor internalization. *Neuron* **25**, 649–662.

Mansour M, Nagarajan N, Nehring R *et al.* (2001). Heteromeric AMPA receptors assemble with a preferred subunit stoichiometry and spatial arrangement. *Neuron* **32**, 841–853.

Martin LJ, Blackstone CD, Levey AI *et al.* (1993). AMPA glutamate receptor subunits are differentially distributed in rat brain. *Neuroscience* **53**, 327–358.

Mauceri D, Cattabeni F, Di Luca M *et al.* (2004). Calcium/calmodulin-dependent protein kinase II phosphorylation drives synapse-associated protein 97 into spines. *J Biol Chem* **279**, 23813–23821.

Mayer ML (2006) Glutamate receptors at atomic resolution. *Nature* **440**, 456–462

McBain CJ and Mayer ML (1994). N-methyl-D-aspartic acid receptor structure and function. *Physiol Rev* **74**, 723–760.

Miu P, Jarvie KR, Radhakrishnan V *et al.* (2001). Novel AMPA receptor potentiators LY392098 and LY404187: effects on recombinant human AMPA receptors *in vitro*. *Neuropharmacology* **40**, 976–983.

Monyer H, Seeburg PH and Wisden W (1991). Glutamate-operated channels: developmentally early and mature forms arise by alternative splicing. *Neuron* **6**, 799–810.

Mosbacher J, Schoepfer R, Monyer H *et al.* (1994). A molecular determinant for submillisecond desensitization in glutamate receptors. *Science* **266**, 1059–1062.

Murphy SN and Miller RJ (1989) Two distinct quisqualate receptors regulate Ca^{2+} in homeostasis in hippocampal neurous in vitro. *Mol Pharmacol* **35**, 671–680.

Nakagawa T, Cheng Y, Ramm E *et al.* (2005a) Structure and different conformational states of native AMPA receptor complexes. *Nature* **433**, 545–549.

Nakagawa T, Feliu-Mojer MI, Wulf P *et al.* (2005b). Generation of lentiviral transgenic rats expressing glutamate receptor interacting protein 1 (GRIP1) in brain, spinal cord and testis. *J Neurosci Methods* **152**, 1–9.

Nakagawa T, Futai K, Lashnel HA *et al.* (2004) Quaternary structure, protein dynamics, and synaptic function of SAP97 controlled by C27 domain interactions. *Neuron* **44**, 453–467.

Nakamura S, Kaneko S and Satoh M (1994). Potentiation of alpha-amino-3-hydroxy-5-methyl-4-isoxazoleproprionic acid (AMPA)-selective glutamate receptor function by a nootropic drug, idebenone. *Biol Pharm Bull* **17**, 70–73.

Nakanishi N, Shneider NA and Axel R (1990). A family of glutamate receptor genes: evidence for the formation of heteromultimeric receptors with distinct channel properties. *Neuron* **5**, 569–581.

Nishimune A, Isaac JT, Molnar E *et al.* (1998). NSF binding to GluR2 regulates synaptic transmission. *Neuron* **21**, 87–97.

Noel J, Ralph GS, Pickard L *et al.* (1999). Surface expression of AMPA receptors in hippocampal neurons is regulated by an NSF-dependent mechanism. *Neuron* **23**, 365–376.

Noh KM, Yokota H, Mashiko T *et al.* (2005). Blockade of calcium-permeable AMPA receptors protects hippocampal neurons against global ischemia-induced death. *Proc Nat Acad Sci USA* **102**, 12230–12235.

Ohmori J, Sakamoto S, Kubota H *et al.* (1994). 6-(1H-imidazol-1-yl)-7-nitro-2,3(1H,4H)-quinoxalinedione hydrochloride (YM90K) and related compounds: structure–activity relationships for the AMPA-type non-NMDA receptor. *J Med Chem* **37**, 467–475.

Ornstein PL, Arnold MB, Augenstein NK *et al.* (1993). 3SR,4aRS,6RS, 8aRS-6-(2-(1H-tetrazol-5-yl)ethyl)-decahydroisoquinoline-3carboxylic acid: a structurally novel systemically active, competitive AMPA receptor antagonist. *J Med Chem* **36**, 2046–2048.

Ornstein PL, Zimmerman DM, Arnold MB *et al.* (2000). Biarylpropylsulfonamides as novel, potent potentiators of 2-amino-3-(5-methyl-3-hydroxyisoxazol-4-yl)- propanoic acid (AMPA) receptors. *J Med Chem* **43**, 4354–4358.

Osten P, Khatri L, Perez JL *et al.* (2000). Mutagenesis reveals a role for ABP/GRIP binding to GluR2 in synaptic surface accumulation of the AMPA receptor. *Neuron* **27**, 313–325.

Osten P, Srivastava S, Inman GJ *et al.* (1998). The AMPA receptor GluR2 C-terminus can mediate a reversible, ATP-dependent interaction with NSF and alpha- and beta-SNAPs. *Neuron* **21**, 99–110.

Palmer CL, Lim W, Hastie PGR *et al.* (2005). Hippocalcin functions as a calcium sensor in hippocampal LTD. *Neuron* **47**, 487–494.

Park M, Penick EC, Edwards JG *et al.* (2004). Recycling endosomes supply AMPA receptors for LTP. *Science* **305**, 1972–1975.

Partin KM and Mayer ML (1996). Negative allosteric modulation of wild-type and mutant AMPA receptors by GYKI 53655. *Mol Pharmacol* **49**, 142–148.

Partin KM, Patneau DK and Mayer ML (1994). Cyclothiazide differentially modulates desensitization of AMPA receptor splice variants. *Mol Pharmacol* **46**, 129–138.

Partin KM, Patneau DK, Winters CA *et al.* (1993). Selective modulation of desensitization at AMPA versus kainate receptors by cyclothiazide and concavalin A. *Neuron* **11**, 1069–1082.

Passafaro M, Piech V and Sheng M (2001). Subunit-specific temporal and spatial patterns of AMPA receptor exocytosis in hippocampal neurons. *Nat Neurosci* **4**, 917–926.

Pelletier JC, Hesson DP, Jones KA *et al.* (1995). Substituted 1,2-dihydrophthalazines: potent, selective and noncompetitive inhibitors of the AMPA receptor. *J Med Chem* **39**, 343–346.

Pelletier JC, Hesson DP, Jones KA *et al.* (1996) Substituted 1,2-dihydrophthalazines: potent, selective and noncompetitive inhibitors of the AMPA receptor. *J. Med Chem* **39**, 343–346.

Peng PL, Zhong X, Tu W *et al.* (2006). ADAR2-dependent RNA editing of AMPA receptor subunit GluR2 determines vulnerability of neurons in forebrain ischemia. *Neuron* **49**, 719–733.

Perez JL, Khatri L, Chang C *et al.* (2001). PICK1 targets activated protein kinase Calpha to AMPA receptor clusters in spines of hippocampal neurons and reduces surface levels of the AMPA-type glutamate receptor subunit 2. *J Neurosci* **21**, 5417–5428.

Peter BJ, Kent HM, Mills IG *et al.* (2004). BAR domains as sensors of membrane curvature: the amphiphysin BAR structure. *Science* **303**, 495–499.

Petralia RS and Wenthold RJ (1992). Light and electron immunocytochemical localization of AMPA-selective glutamate receptors in the rat brain. *J Comp Neurol* **318**, 329–354.

Petralia RS, Wang YX, Mayat E *et al.* (1997). Glutamate receptor subunit 2-selective antibody shows a differential distribution of calcium-impermeable AMPA receptors among populations of neurons. *J Comp Neurol* **385**, 456–476.

Pittaluga A, Feligioni M, Longordo F *et al.* (2006). Trafficking of presynaptic AMPA receptors mediating neurotransmitter release: neuronal selectivity and relationships with sensitivity to cyclothiazide. *Neuropharmacology* **50**, 286–296.

Polson AG, Crain PF, Pomerantz SC *et al.* (1991). The mechanism of adenosine to inosine conversion by the double-stranded RNA unwinding/modifying activity: a high-performance liquid chromatography-mass spectrometry analysis. *Biochemistry* **30**, 11507–11514.

Poncer JC, Esteban JA and Malinow R (2002). Multiple mechanisms for the potentiation of AMPA receptor-mediated transmission by alpha-Ca2+/calmodulin-dependent protein kinase II. *J Neurosci* **22**, 4406–4411.

Priel A, Kolleker A, Ayalon G *et al.* (2005). Stargazin reduces desensitization and slows deactivation of the AMPA-type glutamate receptors. *J Neurosci* **25**, 2682–2686.

Priestley T, Woodruff GN and Kemp JA (1989). Antagonism of responses to excitatory amino acids on rat cortical neurones by the spider toxin, argiotoxin 636. *Br J Pharmacol* **97**, 1315–1323.

Quirk JC, Siuda ER and Nisenbaum ES (2004). Molecular determinants responsible for differences in desensitization kinetics of AMPA receptor splice variants. *J Neurosci* **24**, 11416–11420.

Rao SD, Yin HZ and Weiss JH (2003). Disruption of glial glutamate transport by reactive oxygen species produced in motor neurons. *J Neurosci* **23**, 2627–2633.

Rassendren FA, Lory P, Pin JP *et al.* (1990) Zinc has opposite effects on NMDA and non-NMDA receptors expressed in Xenopus oocytes. *Neuron* **4**, 733–740.

Roche KW, O'Brien RJ, Mammen AL *et al.* (1996). Characterization of multiple phosphorylation sites on the AMPA receptor GluR1 subunit. *Neuron* **16**, 1179–1188.

Roselli F, Tirard M, Lu J *et al.* (2005). Soluble beta-amyloid1-40 induces NMDA-dependent degradation of postsynaptic density-95 at glutamatergic synapses. *J Neurosci* **25**, 11061–11070.

Rosenmund C, Stern-Bach Y and Stevens CF (1998) The tetrameric structure of a glutamate receptor channel. *Science* **280**, 1596–1599

Rouach N, Byrd K, Petralia RS *et al.* (2005). TARP gamma-8 controls hippocampal AMPA receptor number, distribution and synaptic plasticity. *Nat Neurosci* **8**, 1525–1533.

Ruberti F and Dotti CG (2000). Involvement of the proximal C terminus of the AMPA receptor subunit GluR1 in dendritic sorting. *J Neurosci* **20**, RC78.

Rumbaugh G, Sia GM, Garner CC *et al.* (2003). Synapse-associated protein-97 isoform-specific regulation of surface AMPA receptors and synaptic function in cultured neurons. *J Neurosci* **23**, 4567–4576.

Safferling M, Tichelaar W, Kümmerle G *et al.* (2001). First images of a glutamate receptor ion channel: oligomeric state and molecular dimensions of GluRB homomers. *Biochemistry* **40**, 13948–13953.

Schlesinger F, Tammena D, Krampfl K *et al.* (2005). Desensitization and resensitization are independently regulated in human recombinant GluR subunit coassemblies. *Synapse* **55**, 176–182.

Schnell E, Sizemore M, Karimzadegan S *et al.* (2002). Direct interactions between PSD-95 and stargazin control synaptic AMPA receptor number. *Proc Natl Acad Sci USA* **99**, 13902–13907.

Seeburg PH (1993). The TINS/TiPS Lecture. The molecular biology of mammalian glutamate receptor channels. *Trends Neurosci* **16**, 359–365.

Seeburg PH (1996). The role of RNA editing in controlling glutamate receptor channel properties. *J Neurochem* **66**, 1–5.

Seeburg PH, Higuchi M and Sprengel R (1998). RNA editing of brain glutamate receptor channels: mechanism and physiology. *Brain Res Brain Res Rev* **26**, 217–229.

Seeburg PH, Single F, Kuner T *et al.* (2001). Genetic manipulation of key determinants of ion flow in glutamate receptor ion channels in the mouse. *Brain Res* **907**, 233–243.

Seidenman KJ, Steinberg JP, Huganir R *et al.* (2003). Glutamate receptor subunit 2 serine 880 phosphorylation modulates synaptic transmission and mediates plasticity in CA1 pyramidal cells. *J Neurosci* **23**, 9220–9228.

Sekiguchi M, Fleck MW, Mayer ML *et al.* (1997). A novel allosteric potentiator of AMPA receptors: 4-[2-(phenylsulfonylamino) ethylthio]-2,6-difluoro-phenoxyacetamide. *J Neurosci* **17**, 5760–5771.

Sensi SL and Jeng JM (2004). Rethinking the excitotoxic ionic milieu: the emerging role of Zn(2+) in ischemic neuronal injury. *Curr Mol Med* **4**, 87–111.

Setou M, Seog DH, Tanaka Y *et al.* (2002). Glutamate-receptor-interacting protein GRIP1 directly steers kinesin to dendrites. *Nature* **417**, 83–87.

Sheardown MJ, Nielsen EO, Hansen AJ *et al.* (1990). 2,3-Dihydroxy-6-nitro-7-sulfamoyl-benzo(F) quinoxaline: a neuroprotectant for cerebral ischemia. *Science* **247**, 571–574.

Shen L, Liang F, Walensky LD *et al.* (2000). Regulation of AMPA receptor GluR1 subunit surface expression by a 4.1N-linked actin cytoskeletal association. *J Neurosci* **20**, 7932–7940.

Sheng M and Sala C (2001). PDZ domains and the organization of supramolecular complexes. *Annu Rev Neurosci* **24**, 1–29.

Shi S, Hayashi Y, Esteban JA *et al.* (2001). Subunit-specific rules governing AMPA receptor trafficking to synapses in hippocampal pyramidal neurons. *Cell* **105**, 331–343.

Slemmer JE, De Zeeuw CI and Weber JT (2005). Don't get too excited: mechanisms of glutamate-mediated Purkinje cell death. *Prog Brain Res* **148**, 367–390.

Smotrys JE and Linder ME (2004). Palmitoylation of intracellular signaling proteins: regulation and function. *Annu Rev Biochem* **73**, 559–587.

Solyom S and Tarnawa I (2002). Non-competitive AMPA antagonists of 2,3-benzodiazepine type. *Curr Pharm Design* **8**, 913–939.

Sommer B, Keinanen K, Verdoorn TA *et al.* (1990). Flip and flop: a cell-specific functional switch in glutamate-operated channels of the CNS. *Science* **249**, 1580–1585.

Sommer B, Köhler M, Sprengel R *et al.* (1991). RNA editing in brain controls a determinant of ion flow in glutamate-gated channels. *Cell* **67**, 11–19.

Song I, Kamboj S, Xia J *et al.* (1998). Interaction of the N-ethylmaleimide-sensitive factor with AMPA receptors. *Neuron* **21**, 393–400.

Stein E, Cox JA, Seeburg PH *et al.* (1992). Complex pharmacological properties of recombinant α-amino-3-hydroxy-5-methyl-4-isoxazoleproprionate receptor subtypes. *Mol Pharmacol* **42**, 864–871.

Steinberg JP, Takamiya K, Shen Y *et al.* (2006). Targeted *in vivo* mutations of the AMPA receptor subunit GluR2 and its interacting protein PICK1 eliminate cerebellar long-term depression. *Neuron* **49**, 845–860.

Steiner P, Alberi S, Kulangara K *et al.* (2005). Interactions between NEEP21, GRIP1 and GluR2 regulate sorting and recycling of the glutamate receptor subunit GluR2. *EMBO J* **24**, 2873–2884.

Stern P, Edwards FA and Sakmann B (1992). Fast and slow components of unitary EPSCs on stellate cells elicited by focal stimulation in slices of rat visual cortex. *J Physiol* **449**, 247–278.

Stern-Bach Y, Bettler B, Hartley M *et al.* (1994) Agonist selectivity of glutamate receptors is specified by two domains structurally related to bacterial amino acid-binding proteins. *Neuron* **13**, 1345–1357.

Stetefeld J and Ruegg MA (2005). Structural and functional diversity generated by alternative mRNA splicing. *Trends Biochem Sci* **30**, 515–521.

Strange M, Brauner-Osborne H and Jensen AA (2006). Functional characterisation of homomeric ionotropic glutamate receptors GluR1–GluR6 in a fluorescence-based high-throughput screening assay. *Com Chem High Throughput Screen* **9**, 147–158.

Sun Y, Olson R, Horning M *et al.* (2002) Mechanism of glutamate receptor desensitisation. *Nature* **417**, 245–253.

Sutton MA, Schmidt EF, Choi KH *et al.* (2003). Extinction-induced upregulation in AMPA receptors reduces cocaine-seeking behaviour. *Nature* **421**, 70–75.

Swope SL, Moss SJ, Raymond LA *et al.* (1999). Regulation of ligand-gated ion channels by protein phosphorylation. *Adv Second Messenger Phosphoprotein Res* **33**, 49–78.

Szegedi V, Juhasz G, Budai D *et al.* (2005). Divergent effects of Abeta1-42 on ionotropic glutamate receptor-mediated responses in CA1 neurons *in vivo*. *Brain Res* **1062**, 120–126.

Tachibana M, Wenthold RJ, Morioka H *et al.* (1994). Light and electron microscopic immunocytochemical localization of AMPA-selective glutamate receptors in the rat spinal cord. *J Comp Neurol* **344**, 431–454.

Takahashi T (2005). Postsynaptic receptor mechanisms underlying developmental speeding of synaptic transmission. *Neurosci Res* **53**, 229–240.

Taverna FA, Cameron BR, Hampson DL *et al.* (1994). Sensitivity of AMPA receptors to pentobarbital. *Eur J Pharmacol* **267**, R3–5.

Thomas MJ and Malenka ZC (2003) Synaptic plasticity in the mesolimbic dopamine system. *Philos Trans R Soc Lond B Biol Sci* **358**, 815–819.

Thompson DM, Guidotti A, DiBella M *et al.* (1995). 7-chloro-3-methyl-3,4-dihydro-2H-1,2,4-benzothiadiazine S,S-dioxide (IDRA-21), a congener of aniracetam, potently abates pharmacologically induced cognitive impairments in patas monkeys. *Proc Nat Acad Sci USA* **92**, 7667–7671.

Tomita S, Adesnik H, Sekiguchi M *et al.* (2005a). Stargazin modulates AMPA receptor gating and trafficking by distinct domains. *Nature* **435**, 1052–1058.

Tomita S, Chen L, Kawasaki Y *et al.* (2003). Functional studies and distribution define a family of transmembrane AMPA receptor regulatory proteins. *J Cell Biol* **161**, 805–816.

Tomita S, Fukata M, Nicoll RA *et al.* (2004). Dynamic interaction of stargazin-like TARPs with cycling AMPA receptors at synapses. *Science* **303**, 1508–1511.

Tomita S, Stein V, Stocker TJ *et al.* (2005b). Bidirectional synaptic plasticity regulated by phosphorylation of stargazin-like TARPs. *Neuron* **45**, 269–277.

Trussell LO and Fischbach GD (1989). Glutamate receptor desensitisation and its role in synaptic transmission. *Neuron* **3**, 209–218.

Tsuzuki K, Takeuchi T and Ozawa S (1992). Agonist- and subunit-dependent potentiation of glutamate receptors by a nootropic drug aniracetam. *Mol Brain Res* **16**, 105–110.

Valtschanoff JG, Burette A, Davare MA *et al.* (2000). SAP97 concentrates at the postsynaptic density in cerebral cortex. *Eur J Neurosci* **12**, 3605–3614.

Vandenberghe W, Nicoll RA and Bredt DS (2005). Interaction with the unfolded protein response reveals a role for stargazin in biosynthetic AMPA receptor transport. *J. Neurosci* **25**, 1095–1102.

Verdoorn TA, Burnashev N, Monyer H *et al.* (1991). Structural determinants of ion flow through recombinant glutamate receptor channels. *Science* **252**, 1715–1718.

Wagner RW, Smith JE, Cooperman BS *et al.* (1989). A double-stranded RNA unwinding activity introduces structural alterations by means of adenosine to inosine conversions in mammalian cells and *Xenopus* eggs. *Proc Nat Acad Sci USA* **86**, 2647–2651.

Wahl P, Frandsen A, Madsen U *et al.* (1998). Pharmacology and toxicology of ATOA, an AMPA receptor antagonist and a partial agonist at GluR5 receptors. *Neuropharmacology* **37**, 1205–1210.

Wang H, Kohno T, Amaya F *et al.* (2005). Bradykinin produces pain hypersensitivity by potentiating spinal cord glutamatergic synaptic transmission. *J Neurosci* **25**, 7986–7892.

Weaver CD, Yao TL, Powers AC *et al.* (1996). Differential expression of glutamate receptor subtypes in rat pancreatic islets. *J Biol Chem* **271**, 12977–12984.

Weiss JH, Hartley DM, Koh J et al. (1990). The calcium channel blocker nifedipine attenuates slow excitatory amino acid neurotoxicity. *Science* **247**, 1474–1477.

Weiss JH, Turetsky D, Wilke G et al. (1994). AMPA/kainate receptor-mediated damage to NADPH-diaphorase-containing neurons is Ca2+-dependent. *Neurosci Lett* **167**, 93–96.

Welch WM, Ewing FE, Huang FS et al. (2001). Atropisomeric quinazolin-4-one derivatives are potent noncompetitive α-amino-3-hydroxy-5-methyl-4-isoxazoleproprionic acid (AMPA) receptor antagonists. *Bioorg Med Chem Lett* **11**, 177–181.

Wilding TJ and Huettner JE (1995). Differential antagonism of α-amino-3-hydroxy-5-methyl-4-isoxazoleproprionic acid-preferring and kainate-preffering receptors by 2,3-benzodiazepines. *Mol Pharmacol* **47**, 582–587.

Woolf CJ and Salter MW (2000). Differential expression of glutamate receptor subtypes in rat pancreatic islets. *Science* **288**, 1765–1769.

Xia J, Zhang X, Staudinger J et al. (1999). Clustering of AMPA receptors by the synaptic PDZ domain-containing protein PICK1. *Neuron* **22**, 179–187.

Yakel JL, Vissavajjhala P, Derkach VA et al. (1995). Identification of a Ca2+/calmodulin-dependent protein kinase II regulatory phosphorylation site in non-N-methyl-D-aspartate glutamate receptors. *Proc Nat Acad Sci USA* **92**, 1376–1380.

Yamada KA and Rothman SM (1992). Diazoxide blocks glutamate desensitization and prolongs excitatory postsynaptic currents in rat hippocampal neurons. *J Physiol* **458**, 409–423.

Yamada KA and Tang CM (1993). Benzothiadiazides inhibit rapid glutamate receptor desensitisation and enhance glutamatergic synaptic currents. *J Neurosci* **13**, 3904–3915.

Yasuda H, Barth AL, Stellwagen D et al. (2003). A developmental switch in the signaling cascades for LTP induction. *Nat Neurosci.* **6**, 15–16.

Yin HZ, Sensi SL, Ogoshi F et al. (2002). Blockade of Ca2+-permeable AMPA/kainate channels decreases oxygen–glucose deprivation-induced Zn2+ accumulation and neuronal loss in hippocampal pyramidal neurons. *J Neurosci* **22**, 1273–1279.

Zhao D, Watson JB and Xie CW (2004). Amyloid beta prevents activation of calcium/calmodulin-dependent protein kinase II and AMPA receptor phosphorylation during hippocampal long-term potentiation. *J Neurophysiol* **92**, 2853–2858.

3.2.2 Kainate receptors

Sanja Selak, Rocio Rivera, Ana Paternain and Juan Lerma

3.2.2.1 Introduction

Glutamate is an excitatory substance to the central nervous system (CNS), a fact known since the 1950s (Hayashi, 1956). We now know that glutamate plays a key role in many physiological processes in addition to acting as a neurotransmitter. Indeed, it is involved in the ontogeny of the CNS (Rauschecker and Hahn, 1987; Brewer and Cotman, 1989), in long-lasting plastic phenomena (Collingridge and Bliss, 1999) as well as in neuronal death-associated pathologies.

Glutamate targets receptor proteins embedded in the plasma membrane of neurons and glia which have been divided into two families: metabotropic glutamate receptors (mGluRs) and ionotropic glutamate receptors (iGluRs), depending on the signalling mechanism elicited upon glutamate binding. Pharmacological, biophysical and molecular data support the existence of three

different iGluR families (Hollman and Heinemann, 1994), named after the agonists that activate them with somewhat more specificity. N-methyl-D-aspartate (NMDA), α-amino-3-hydroxy-5-methyl-4-isoxazole propionic acid (AMPA) and kainate receptors (KARs) are among the most studied molecules in the nervous system. As AMPA receptors (AMPARs) and KARs share many pharmacological agonist and antagonist molecules, for many years they were pooled together as non-NMDA, as opposed to NMDA receptors (NMDARs). Subsequently the cloning of the glutamate receptor subunits (Nakanishi, 1992; Seeburg, 1993; Hollman and Heinemann, 1994) supported the initial pharmacological characterization. However, the discovery that 2,3-benzodiazepines, such as GYKI 53655 (LY300268) selectively antagonized AMPARs but not KARs (Paternain et al., 1995; Wilding and Huettner, 1995) helped to uncover the true KARs in neurons. Following a search for their functional significance, an overwhelming amount of data demonstrates that KARs are located at both sides of the synapse. Pre- and postsynaptic KARs perform different tasks. Thus, postsynaptic KARs co-localize with NMDARs and AMPARs, mediating synaptic transmission at certain CNS synapses. At the presynaptic side, they modulate transmitter release at both excitatory and inhibitory synapses. The specific subunit composition of KARs confers the receptor with distinct biophysical properties and roles which translates into efficient control of synaptic physiology. Further understanding of the specific roles of KARs at the synapse is now feasible due to the generation of mice that are deficient for KAR subunits (Mulle et. al., 1998; Contractor et al., 2001, 2003). The generation of antibodies against the different KAR subunits has become an important research goal, albeit uncompleted as yet. Finally, one striking advance in the field is the resolution at an atomic level of the ligand-binding cores for individual glutamate receptors, including KARs. Hence, definition of the molecular biology of KAR subunits has represented a real step forward in the study of these receptors, establishing the foundations to better understand their physiology.

3.2.2.2 Molecular characterization of the kainate receptor family

KARs are integral transmembrane proteins that co-assemble into tetramers to give rise to glutamate-gated ion channels. Each monomer carries its own ligand binding site and contributes to the lumen of the channel with a specific hydrophobic amino acid sequence (M2), giving rise to a hairpin-like structure that lies within the membrane. The rest of the subunit is arranged so that the N-terminal domain lies towards the extracellular side followed by the first transmembrane segment (M1). This is followed by the M2 region and then two consecutive transmembrane domains (M3 and M4) linked by a large extracellular loop. The C-terminal region lies inside the cell. The agonist binding site is formed by the so-called S1 and S2 segments (Stern-Bach et al., 1994) which correspond to the amino acid stretches located N-terminal to M1 (S1) and forming the loop connecting the transmembrane segments M3 and M4 on the extracellular side (see Lerma, 2003 for review) (Figure 3.2.2.1).

The diversity of kainate receptors

Five different kainate subunits have been cloned to date: GluR5 (Bettler et al., 1990), GluR6 (Egebjerg et al., 1991), GluR7 (Bettler et al., 1992), KA1 (Werner et al., 1991) and KA2 (Herb et al., 1992) and, based on the sequence of the human genome, it seems that there are no additional members of this family. GluR5–7 share a 75–80 per cent homology and can co-assemble as homomers or heteromers to give rise to functional receptors (Cui and Mayer, 1999; Paternain *et al.*, 2000). On the other hand, KA1 and KA2, which share a 45 per cent homology and bind glutamate, must co-assemble with one of the GluR5–7 subunits to form functional receptors, conferring the channel with a different affinity for agonists.

As in other GluRs (glutamate receptors), KAR subunits undergo alternative splicing (Bettler *et al.*, 1990; Sommer, 1992; Gregor *et al.*, 1993; Schiffer *et al.*, 1997). The GluR5 subunit has an alternatively spliced exon in the N-terminal domain such that GluR5–1 contains a 15 aa (amino acid) insert between residues 402–416 whilst GluR5–2 lacks this insert. GluR5 also presents different splice variants at the C-terminus including GluR5–2a with a premature stop codon that generates a truncated C terminus of 16 aa and GluR5–2b whose C-terminal domain is 49 aa longer than GluR5–2a. GluR5–2c is similar to GluR5–2b but with a 29 aa insert

starting from the end of the 16 aa sequence. Finally, GluR5–2d has a completely different C-terminus that is not homologous to the C termini of the other variants which follows the initial 16 aa starting sequence.

In the case of GluR6, two splice variants C-terminal splice variants have been identified: GluR6 and GluR6–2, in which the first 889 aa are conserved, but with each variant carrying a different C-terminal tail. In the case of GluR7, two variants, GluR7a and GluR7b, have been discovered to date. In this case, the insertion of 40 nucleotides out-of-frame in the carboxy-terminal sequence of GluR7a leads to the production of a shorter GluR7b, which bears no significant homology in this region to any of the known iGluRs. No splice variants of the KA1 and KA2 subunits have been identified to date. The changes at the C-terminus of the KAR subunits do not affect their physiological properties but instead, they seem to have a major influence on the intracellular trafficking and targeting at the synapse which is mediated by different receptor-interacting proteins that also contribute to their functional roles. In fact, there

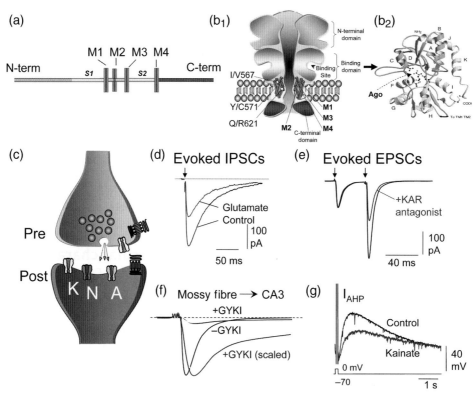

Fig. 3.2.2.1 (a) Membrane topology of a kainate receptor subunit. Kainate receptors are integral proteins that co-assemble into tetramers. Each monomer carries its own ligand-binding site, formed by segments S1 and S2, and contributes to the lumen of the channel with the hydrophobic amino acid sequence M2. The M2 segment is thought to dip into the membrane forming a structure akin to a hairpin. The subunit is arranged so that the N-terminal domain lies towards the extracellular side, whilst the C-terminal region lies inside the cell. (b) a cartoon illustrating arbitrary arrangements of the different domains in a kainate receptor channel (b1), of which the crystal structure of the binding domain of GluR6 is shown (b2) (from Mayer 2005).[1] The highlighted residues correspond to sites susceptible to editing in the mRNA coding for GluR5 (Q/R site) of GluR6 (all three sites) subunits. (c) Kainate receptors have been identified to be located at both sides of the synapse. Presynaptically, they can control the release of both gamma-amino-butyric acid (GABA) and glutamate (K, kainate receptor; N, NMDA receptor; A, AMPA receptor). (d) Example trace illustrating the effect of kainate receptor activation by glutamate on the postsynaptic inhibitory current (IPSC) recorded from a CA1 pyramidal neuron and evoked by stratum oriens stimulation. (e) Kainate receptor activation by synaptically released glutamate contributes to paired-pulse facilitation in the mossy fibre to CA3 pyramidal neuron synapses. In the presence of kainate receptor antagonists, the facilitation is partially abolished. (f) Postsynaptically, kainate receptors mediate part of the synaptic current at some excitatory contacts, where they generate a small and long-lasting current which is apparent after AMPA receptor blockade by GYKI53655. (g) In some synapses such those on CA1 and CA3 pyramidal cells, glutamate activates kainate receptors that, like those on GABAergic terminals, may be coupled to a G-protein, reducing the amount of the afterpolarizing current (I_{AHP}), enhancing excitability. See also colour plate section.

[1] Reprinted from *Neuron*, volume 45, ML Mayer, Crystal structures of the GluR5 and GluR6 ligand binding cores: molecular mechanisms underlying kainate receptor selectivity, pages 539–552, Copyright 2005, with permission from Elsevier.

also are key sites for allosteric modulation by protein kinases and phosphatases (Schiffer *et al.*, 1997; Jaskolski *et al.*, 2004; Coussen *et al.*, 2005) although, to date, very little is known regarding the molecular organization of this region.

In addition to this diversity of subunits, one has to consider the fact that the mRNA of GluR5 and GluR6 undergo editing at the so-called Q/R site of the M2 domain (Sommer *et al.*, 1991; Bernard and Khrestchatisky, 1994; Seeburg, 1996) (Figure 3.2.2.1). As in the case of AMPA receptor subunits, the glutamine-to-arginine substitution in both GluR5 and GluR6 subunits decreases the permeability to calcium (Sommer *et al.*, 1991). Moreover, it provokes a change in the rectification properties of the functional channel (Bettler and Mulle, 1995; Huettner, 2003; Lerma, 2003). GluR6 contains two additional sites in the first transmembrane domain (M1) that are susceptible to editing, such that I567 and Y571 can be edited to V and C, respectively. While the role of these M1 editing sites remains unclear, editing within the M2 domain of GluR6 kainate receptors has recently been shown to exert a significant effect on synaptic physiology (Vissel *et al.*, 2001).

Kainate receptor expression in the central nervous system

KARs are expressed widely throughout the entire nervous system with distinct, overlapping patterns. Most of our knowledge of KARs distribution comes from *in situ* hybridization analysis (e.g. Wisden and Seeburg, 1993; Bahn *et al.*, 1994; Bischoff *et al.*, 1997). Due to the lack of specific antibodies for the KAR subunits the precise subcellular distribution of most subunits is lacking. GluR5 is present in dorsal root ganglion neurons (DRG), in subicular neurons, in the septal nuclei, in the piriform and cingulate cortices, in Purkinje cells of the cerebellum and in hippocampal interneurons. GluR6, is expressed in cerebellar granule cells, dentate gyrus, striatum and in the CA1 and CA3 pyramidal cells of the hippocampus. GluR7 mRNA is expressed at low levels throughout the brain but particularly in the deep layers of the cerebral cortex, in the striatum and in the inhibitory neurons of the molecular layer of the cerebellum. KA1 is almost exclusively restricted to the CA3 region, although it is also expressed at lower levels in the dentate gyrus, amygdala and entorhinal cortex. KA2 is found in essentially every part of the nervous system. Although already present in the embryo, expression of KAR subunits becomes prominent during the late embryonic and early postnatal period (Bahn *et al.*, 1994). Expression of GluR5 mRNA peaks between postnatal day (PN) 0 and PN5 after which its expression falls, reaching adult levels by PN12. Likewise, the GluR6 and KA1 expression patterns observed within the hippocampal formation in the adult begin to emerge at around PN0 (GluR6) and PN12 (KA1).

To achieve a full and clear understanding of the synaptic function of KARs, determination of the precise synaptic localization of each subunit is of great importance. Although specific antibodies for GluR5/6/7 have not yet been identified, those for GluR6/7, KA1 and KA2 are starting to render promising data. Thus, KA2 has been detected on parallel fibres of cerebellar granule cells, whereas GluR6/7 immunoreactivity has been identified in a large population of terminals that form axospinal and axodendritic asymmetric synapses in the monkey striatum (Petralia *et al.*, 1994; Charara *et al.*, 1999; Kieval *et al.*, 2001).

Kainate receptors at the atomic level

Crystallography studies along with genetic approaches have enabled the generation of models of high-resolution crystal structures for the ligand-binding cores of individual iGluR subunits. These studies have revealed the crystal structure of the GluR5 and GluR6 ligand binding cores complexed with glutamate and other agonists, such as kainate and quisqualate, at a resolution of 1.65 (GluR6) and 2.1 Å (GluR5) (Mayer, 2005). As in the case of AMPARs and NMDARs, the structure of the KAR binding domain revealed a two-domain closed shell motif linked by β strands (Figure 3.2.2.1). Thus, a common molecular structure for iGluRs has been suggested (Mayer, 2006) whereby iGluRs are likely to be formed by the fusion of three genetically discrete segments which seem to have derived from a bacterial ancestor. The proposed model is derived from the solved crystal structures of both the N-terminal domain (ATD) of mGluRs and the agonist binding domain of KARs. To date, the structure of the glutamate receptor ion channel at the atomic resolution has not been elucidated and, based on the remarkable sequence homology that it exhibits with bacterial potassium channels, the KcsA potassium channel has been used as a template for the iGluR pore region. The definitive solution of the structure of the glutamate receptor channel remains a major goal.

3.2.2.3 Biophysical properties of kainate receptors

KARs are cationic ion channels whose kinetics properties have been studied mostly in heterologous expression systems. Under these conditions, stationary noise analyses have revealed:

1 GluR5(Q) and GluR6(Q) receptors have single-channel conductance in the picosiemens (pS) range with reported values of 2.9 and 5.4, respectively (Swanson *et al.*, 1996).

2 Editing of the Q/R site drastically alters single channel conductance such that the edited forms (R) of GluR5 and GluR6 have unitary conductance in the femtosiemens (fS) scale (Swanson *et al.*, 1996). For instance, the single channel conductance of GluR6(R) estimated from non-stationary variance analysis in excised outside-out patches is 17 fS (Traynelis and Wahl, 1997).

3 The combination of either GluR5 or GluR6 with the KA2 subunit produces heteromeric receptors with two- to threefold larger conductances. However, GluR6(Q)/KA2 present single-channel events which are quiet similar to homomeric GluR6 receptors (7.1 pS) (Swanson *et al.*, 1996).

4 Three subconductance levels could be directly resolved from outside-out patches excised from HEK cells expressing the unedited (Q) form of GluR5 (5/9/14 pS) and GluR6 (8/15/25 pS) as homomeric channels or with KA2 in the heteromeric configuration (Swanson, 1996). It has been reported that the mean value of single channel conductance depends on how many agonist molecules bind to the receptor (Rosemund *et al.*, 1998) provided that the receptor does not enter into the desensitized state. Therefore, the three conductance levels observed for KARs at low concentrations of a partially desensitizing agonist are consistent with a tetrameric conformation of KARs and binding of two, three or four agonist molecules for receptor activation. This may be the origin of the discrepancy between the mean single-channel conductance calculated from stationary analysis (0.22 pS and 5.4 pS for GluR6 (R) and GluR6 (Q), respectively; Swanson *et al.*, 1996) and those calculated by non-stationary analysis at saturating concentration of glutamate (0.4 pS and 17 pS, respectively) (Traynelis and Wahl, 1997).

There are just a few studies analysing single channel conductance of native KARs. In DRG cells the unitary conductance estimated by whole-cell current fluctuations was 2–4 pS. However, openings of 8 pS and 15–18 pS were occasionally observed (Huettner, 1990) and agree well with those reported for recombinant receptors composed of GluR5(Q) and/or GluR5(Q)/KA2. KARs channels expressed by cerebellar granule cells in culture exhibit properties that are mostly similar to those observed for heteromeric channels formed from GluR6(R) and KA2 (conductances of approximately 1 pS; Pemberton et al., 1998). Interestingly, unitary conductance of kainate-type channels is significantly higher in proliferating (4 pS) than post-migratory granule cerebellar cells (Smith et al., 1999).

Activation and desensitization of kainate receptors

One of the most characteristic features of KARs is the rapid desensitization that they undergo in the continuous presence of an agonist. The time course of the current decay follows a single exponential curve (Lerma et al. 1993; Paternain et al., 1998, 2000), although double exponential decays have also been described (Lerma et al., 1993; Ruano et al., 1995; Wilding and Huettner, 1997). This characteristic phenomenon has been studied extensively both in native and recombinant KARs. As a result, there are compendiums of time constants measured by different authors depending on the kainate receptor and cell type analysed. The common view is that speed of desensitization is very rapid. For this reason there are two necessary considerations when analysing desensitization kinetics. First, if binding of the agonist must occur prior to desensitization, its rate may affect the desensitization onset kinetics when estimated from the current decay upon agonist perfusion. Therefore, a slow binding process could constitute a source of error when measuring desensitization. Second, the speed of the current relaxation falls in the range of resolution of rapid perfusion systems, raising the possibility that solution changes could affect the desensitization rate measured. Indeed, the time constant of desensitization in both native and recombinant receptors appears to be concentration-dependent, decreasing as agonist concentration increases (Heckmann et al., 1996; Paternain et al., 1998). One way to overcome these problems has been to measure the time constant of current decay at very high concentration of agonist, i.e. when the binding rate is no longer the limiting step. Another aspect to consider is that as ligand concentration increases the rate of desensitization, the asymptotic value should be the value of the real desensitization rate. Taking this into account, it has been determined that GluR6 homomers and native KARs expressed by cultured hippocampal neurons have a similar desensitization time constant, irrespective of the agonist used (11–13 ms) (Paternain et al., 1998). The desensitization rate estimated from recombinant GluR6 in outside-out patches falls within the same range (5–8 ms; Heckmann et al., 1996; Traynelis and Wahl, 1997). In some cases, however, the decay rate is better fitted by the sum of two exponentials, the faster component being predominant. This seems to be particularly relevant in DRG cells and with recombinant GluR5 receptors (Huettner, 1990; Sommer et al., 1992; Wong et al., 1994; Sahara et al., 1997).

In contrast, recovery from the desensitized state of KARs proceeds slowly in comparison to other glutamate receptors and largely depends on the agonist used. For example, a full recovery of glutamate- and kainite-induced responses in the hippocampal neurons takes 15 s and 1 min after the initial pulse, respectively

(Paternain et al., 1998). Recombinant receptors, however, show additional variability in their recovery times depending on the subunit composition. For instance, GluR5 homomeric receptors recover from (S)-5-iodowillardiine-induced desensitization with a time course of minutes (2.5 min) (Swanson et al., 1998). The incorporation of the KA2 subunit within this receptor leads to faster recovery with a time constant of approximately 12 s.

In conclusion, the profound differences in the time scale between desensitization and recovery imply that the equilibrium between both states is displaced towards desensitization. Thus, KARs must spend long periods of time in the desensitized state, which can be considered as an absorbing state. This characteristic makes it difficult to understand how synaptic KARs behave (see below). In addition, native KARs do not have a high affinity for glutamate ($EC_{50} = 330$ μM as determined in cultured neurons). However, desensitization occurs at concentrations two orders of magnitude lower which do not even elicit activation of the channel ($IC_{50} = 2.8$ μM). A similar situation has been observed for recombinant GluR6 receptors and when kainate was used as an agonist in cultured hippocampal neurons ($EC_{50} = 22$ μM, $IC_{50} = 0.31$ μM). The activation and desensitization curves for KARs overlap over a range of agonist concentrations at which the channel is open and at the same time does not evoke the complete desensitization of KARs (i.e. they present a sort of window current). For glutamate, the calculated window presents a maximum at around 100 μM. This property may be physiologically relevant as it could explain the fact that modulation of gamma aminobutyric acid (GABA) release by presynaptic KARs (see below) follows a bell-shaped dose–response curve (Rodriguez-Moreno et al., 1997; Paternain et al., 1998). These observations imply that KARs must be very sensitive to resting levels of glutamate and that a large fraction of channels are inactive before they can be activated by a bolus of synaptic glutamate.

3.2.2.4 Subcellular localization of kainate receptors

The molecular diversity of KARs determines their unique physiological and pharmacological properties (Huettner, 2003; Lerma, 2003). There are a few reliable immunolabelling studies combined with electron microscopy that indicate that KARs are expressed in both pre- and postsynaptic nerve terminals. In retina, GluR6/7 and KA2 subunits are selectively distributed in a subset of postsynaptic domains (Brandstatter et al., 1997). At mossy fibre synapse in the hippocampus, KA1 and KA2 subunits are present in presynaptic boutons and vesicular structures resembling endosomes and trafficking vesicles (Darstein et al., 2003). However KA1 and KA2 immunoreactivity in most synapses is predominantly postsynaptic. Interestingly, more than 70 per cent of GluR6/7 and KA2 labelling is intracellular (Kieval et al., 2001; Darstein et al., 2003), and more than 60 per cent of the plasma membrane-bound KARs are extrasynaptic (Kieval et al., 2001). Such localization suggests that KARs play special roles in synaptic transmission apart from a classical ionotropic channel function. Pharmacological studies also suggest that KARs are located extrasynaptically in somatodendritic compartments (Bureau et al., 1999; Chergui et al., 2000) and that they are not involved in the conventional excitatory neurotransmission (Frerking et al., 1998). Moreover, the KAR distribution pattern resembles the distribution of G-protein-coupled metabotropic receptors which are predominantly expressed intracellularly or at extrasynaptic sites (Pasquini *et al.*, 1992; Hanson and Smith, 1999). Recent electrophysiological studies support this observation. Kainate receptors on presynaptic GABAergic terminals reduce transmitter release by a G-protein-mediated activation of phospholipase C and protein kinase C (PKC) (Rodriguez-Moreno *et al.*, 1997) and similar metabotropic mechanisms of action have been documented in several other studies (Ziegra *et al.*, 1992; Cunha *et al.*, 1997, 1999, 2000; Lauri *et al.*, 2005, 2006). However, it remains to be determined whether these metabotropic actions of kainate involve a direct binding of specific G-proteins to KARs or if they are induced by an indirect G-protein coupling mechanism.

3.2.2.5 Membrane delivery and trafficking of kainate receptors

KARs play crucial roles in modifying the strength of excitatory synapses and their cell-surface expression is tightly controlled by the mechanisms involving selective subunit assembly within the endoplasmic reticulum (ER), intracellular trafficking, membrane targeting and recycling in an activity-dependent manner (Figure 3.2.2.2). Therefore, the number and the type of functional KARs expressed at a particular synapse are determined not only by the cell or tissue type, but also by the local intracellular environment and extracellular stimuli which ultimately determine the fate of each subunit.

One of the mechanisms recently discovered to regulate subunit assembly and intracellular trafficking of KARs involves interactions with chaperone proteins that bind specific domains encoded in the C-termini of each subunit. The GluR6a subunit contains a forward trafficking motif which favours plasma membrane expression; consequently GluR6a subunits can be highly expressed as homomeric KARs on the plasma membrane even in the absence of other KAR subunits (Jaskolski *et al.*, 2004; Yan, 2004). On the contrary, GluR5c and KA2 subunits contain an arginine-rich motif which serves as an ER retention signal, so in order to exit the ER and

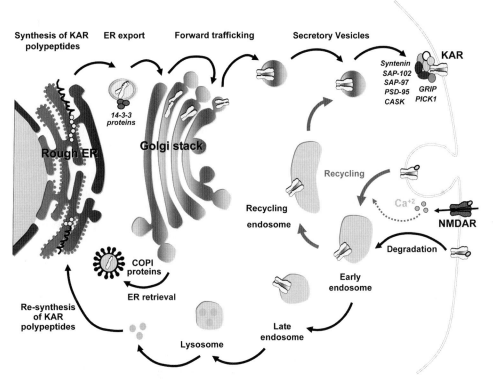

Fig. 3.2.2.2 Intracellular trafficking of kainate receptors mediated by the interactions with receptor-binding proteins. Various protein binding motifs encoded in the KAR C-termini have been identified. Upon mRNA translation the newly synthesized kainate receptor polypeptides traffic from the endoplasmic reticulum (ER) to the golgi. This trafficking is regulated by the interactions with 14-3-3 and COPI proteins which bind to the KAR C-terminal sequences. An ER export signal (forward trafficking motif) is located within the C-terminus of GluR6a [870-KCQRRLKHKPQ-880], whereas ER retention (retrieval) motifs are encoded within the C-termini of GluR5b [910-QRRTQRKETVA-920], GluR5c [872-KSSRLRFYFRN-882] and KA2 [861-SRRRRRP-867]. COPI proteins recognize and bind the ER retention motifs thereby retrieving the receptor to the ER. Co-assembly of the GluR subunits into homomeric or heteromeric receptors masks the retention motif and favours forward trafficking mediated by 14-3-3 proteins. KA1 and KA2 subunits are retained in the ER unless they are assembled with the GluR subunits. Mature and properly assembled KARs are exported from the Golgi and trafficked to the plasma membrane (PM) inside secretory vesicles. KARs at the PM are not static and are continuously cycled in and out by exo- and endocytosis in an activity-dependent fashion regulated by the receptor-interacting proteins. The last four C-terminal amino acids of GluR6a [ETMA] and GluR5b/c [ETVA] bind the PDZ domain-containing proteins: PSD-95/SAP 90, SAP-102, SAP-97 and CASK which regulate receptor clustering at the PM target sites, whereas GRIP1 and PICK1 play a role in the KAR exo- and endocytosis. Endocytosis is additionally regulated by the endocytic motifs, such as dileucine motif within the KA2 C-terminus [907-RLLN-910] which may be involved in the interactions with various endocytic proteins. In hippocampal neurons, KAR activation in the absence of NMDAR activity induces GluR6 KAR endocytosis followed by the transport to late endosomes and lysosomes where the receptors are degraded to small peptides and amino acids which can be reused for protein synthesis. KAR receptor activation coupled with NMDAR activity leads to the activation of alternative calcium-dependent signalling pathways favouring GluR6 receptor recycling. See also colour plate section.

reach the plasma membrane they have to be assembled with other KAR subunits (Ren *et al.*, 2003; Jaskolski *et al.*, 2004). GluR5a and GluR5b are targeted to plasma membrane although at low levels. This lower expression level is thought to be due to the presence of an ER retention signal which is partially masked in the assembled homomeric receptors (Ren *et al.*, 2003; Jaskolski *et al.*, 2005). It has been recently shown that the ER retention by the polyarginine motif is mediated by its interactions with coatomer protein complex I (COPI), which regulates retrograde trafficking of misfolded or unassembled proteins from the golgi to the ER (Vivithanaporn *et al.*, 2006). KA2 assembly with GluR6a greatly reduces the association between KA2 and COPI, while simultaneously increasing the interactions with 14-3-3 proteins, which mediate forward trafficking of integral membrane proteins (Vivithanaporn *et al.*, 2006). However, the KA2 C-terminal domain involved in the interactions with 14-3-3 proteins remains to be determined. In addition to these motifs encoded in the C-termini of the KAR subunits, N-terminal domains have been also found to play an important role in the KAR trafficking and surface expression (Fleck *et al.*, 2003; Mah *et al.*, 2005; Valluru *et al.*, 2005). The formation of the glutamate binding site has been proposed to control forward trafficking of the intracellular KARs.

After the successful ER exit and the membrane targeting by forward trafficking, KARs do not remain steadily expressed at their functional sites. The surface expression of KARs is dynamically regulated by rapid activity-dependent endocytic sorting, upon which the receptors are either recycled back to the plasma membrane or transported to late endosomes and lysosomes where they are eventually degraded (Figure 3.2.2.2).

Although much work is needed to verify these aspects of KAR regulation, recycling and membrane sorting they seem to be regulated by specific stimuli and receptor-interacting proteins. It has been reported that in the cultured hippocampal neurons exogenous kainate or glutamate application induces protein kinase C (PKC)-dependent KAR internalization ultimately leading to degradation. Stimulation of NMDARs rather than KARs favours receptor recycling back to the cell surface via early and recycling endosomes (Martin and Henley, 2004). This stimulus-dependent trafficking is promoted by the intracellular signalling machinery involving numerous proteins, some of which have been identified recently.

Kainate receptor subunits and their splice variants display a high degree of sequence homology, and the various subunit isoforms are distinguished by the distinct sequences in their C-termini (see Figure 3.2.2.1). This suggests that the differential regulation of KAR trafficking may be due to the interactions of the signalling molecules with the specific C-terminal sequences of each receptor subunit. Consequently, a number of studies have attempted to identify the proteins that interact with particular C-terminal domains. Various approaches have been undertaken, including yeast-two-hybrid (Y2H) screening, co-immunoprecipitation, peptide-binding and glutathione *S*-transferase (GST)-pull down assays. The protein interactions identified by these techniques were subsequently tested for their effects on receptor trafficking by various functional studies, including electrophysiology and *in vitro* membrane trafficking studies in polarized and non-polarized cells.

A number of PDZ-containing proteins have been reported to interact with the short C-terminal sequence motifs or internal PDZ motifs of KARs. PDZ domains are modular protein interaction domains that bind in a sequence-specific fashion to C-terminal or internal short peptides that fold in a beta-finger (Sheng and Sala, 2001). PDZ-containing proteins are typically involved in the assembly of macromolecular complexes that perform localized signalling functions at particular subcellular locations. In Y2H and GST-pull down screens, four PDZ proteins, PICK1, GRIP, PSD95 and syntenin, were detected which interact with the extreme C-termini of two splice variants of GluR5 (GluR5b and GluR5c) and GluR6 (Hirbec *et al.*, 2003). Other studies have used similar approaches to identify additional PDZ-binding proteins: SAP102 (102-kDa synapse-associated protein) and SAP97 (97-kDa synapse-associated protein) (Garcia *et al.*, 1998), as well as CASK (calmodulin-associated serine/threonine kinase) (Coussen *et al.*, 2002). Deletion of the PDZ-domain interaction sites does not affect trafficking of KARs from the ER to the plasma membrane (Coussen *et al.*, 2002; Ren *et al.*, 2003). However, disrupting PDZ-domain interactions by infusing inhibitory peptides alters kainate receptor-mediated excitatory postsynaptic currents (EPSCs) (Hirbec *et al.*, 2003).

Interestingly, PICK1 and GRIP have been previously identified as proteins interacting with the AMPA receptor subunit GluR2 (Dev *et al.*, 1999; Xia *et al.*, 1999; Osten *et al.*, 2000). PICK1 and GRIP can bind both AMPAR and KARs because both receptors contain PDZ interaction domains and are likely to be expressed at the same synapses. Therefore, it is surprising that these two proteins simultaneously differentially regulate AMPAR and KAR activity. For example, blocking PICK1 activity by small inhibitory peptides increases the number of GluR2-containing AMPARs, presumably reflected by an increase in the $EPSC_{AMPA}$ (Perez *et al.*, 2001; Terashima *et al.*, 2004), while at the same time causing a gradual decrease in the KAR-mediated EPSCs (Hirbec *et al.*, 2003). This mechanism of differential modulation of AMPA and KARs has been suggested to have important physiological consequences for the dynamic properties of synapses which are strongly affected by the relative proportions of these two receptor types (Collingridge and Isaac, 2003). Nevertheless, these results should be taken with caution since many aspects of the AMPAR and KAR modulation by PICK1 and GRIP remain to be deciphered. For example, it is presently unclear whether the effects on the receptor surface expression are due to the stabilization of the receptor on the plasma membrane or inside at intracellular compartments (Figure 3.2.2.2).

3.2.2.6 Pharmacology

The ionotropic glutamate receptors were named after the agonists that activate them preferentially. A clear pharmacological boundary between NMDARs and the other iGluRs has been recognized for some time. On the other hand, due to the fact that kainate not only activates KARs but also AMPARs at relatively low doses, the two subfamilies were often pooled together. The discovery of 2,3-benzodiazepines, such as GYKI 53655 (LY300268), that specifically antagonize AMPARs but not KARs, permitted the elucidation of the roles of the two glutamate subfamilies in synaptic transmission (Paternain *et al.*, 1995; Wilding and Huettner, 1995). Subsequently newer drugs that have been shown to act specifically on the different KARs subunits have also become essential research tools in the field.

Agonists

Kainate receptors are activated ($EC_{50} = 6-23\ \mu M$) and desensitized by kainate and partially desensitized by domoate ($EC_{50} = \sim 1\ \mu M$ and $\sim 30\ \mu M$ for native and recombinant receptors, respectively) (Huettner, 1990; Lerma et al., 1993). ATPA (5-tert-butyl-4-isoxazolepropionic acid), a substituted analogue of AMPA; domoate; (S)-5-iodowillardiine, a diastereomer of 4-methylglutamate; 2S-4R-methyl-glutamate (SYM2081) and LY339434 have been identified as kainate receptor-selective agonists. Each of these drugs has different effects on the KAR depending upon its subunit composition. Thus, ATPA and (S)-5-iodowillardiine show some selectivity for GluR5-containing ($EC_{50} = \sim 1\ \mu M$ and $0.14\ \mu M$ in DRGs) over GluR6 or GluR7-containing receptors. Similarly it is more potent at KARs than AMPARs ($EC_{50} = \sim 350\ \mu M$). AMPA, ATPA and SYM2081 can also, to a lesser degree, activate GluR6/KA2 heteromeric receptors (ATPA $EC_{50} = \sim 20\ \mu M$ for) (Paternain et al., 2000; Alt et al., 2004). SYM2081 fully desensitizes KARs with a half-maximal value of about 10 nM, which is a 100 times less than the EC_{50} for the activation of recombinant GluR6 containing receptors (Jones et al., 1997) and, thus, it has been used as a functional antagonist of KARs. When GluR5, GluR6 and GluR7 form functional heteromeric receptors with KA1 or KA2, ion channels with distinct properties arise. The inclusion of KAR subunits in GluR6 or GluR7 heteromeric receptors makes them sensitive to AMPA (Schiffer et al., 1997), whereas heteromeric GluR5/KA2 receptors exhibit faster desensitization kinetics than homomeric GluR5 channels. One striking fact regarding KA1 and KA2 subunits is that they have a high affinity for kainate. However, although one would, thus, expect a higher kainate affinity for heteromeric receptors that include these subunits, the inclusion of KA subunits into functional receptors actually renders them less sensitive to the agonist (Howe, 1996).

Antagonists

KARs and AMPARs initially proved difficult to distinguish from one another with antagonist ligands. 6-Cyano-7-nitroquinoxaline (CNQX), the prototypic non-NMDA receptor antagonist, can be used to antagonize KARs once AMPARs have been selectively blocked by GYKI 53665, or its active isomer LY303070, which has an IC_{50} for AMPARs of about 1 μM (Paternain et al., 1995; Wilding and Huettner, 1995). NBQX (2,3-dihydroxy-6-nitro-7-sulphamoyl-benzo (F) quinoxaline) has also been shown to selectively antagonize AMPARs over KARs in the hippocampus at low concentrations (1 μM) (Mulle et al., 2000). Pharmacological tools are now available that selectively antagonize specific KARs subunits. Thus, it was first shown that NS-102 blocks recombinant GluR6 and KARs at a concentration 20 times lower than required to antagonize AMPARs (Verdoorn et al., 1994). However, since the IC_{50} values of this antagonist at native KARs and AMPARs in cultured hippocampal neurons are similar (2.2 and 4.1 μM, respectively) and also that it is only partially soluble in water, its utility in physiological experiments is limited (Paternain et al., 1996). LY293558, a competitive antagonist of AMPARs ($IC_{50} = 0.5\ \mu M$) also blocks GluR5-mediated responses ($IC_{50} = 2.5\ \mu M$) with no effect on GluR6 homomers (Bleakman et al., 1996) ($IC_{50} > 300\ \mu M$). A related derivative, LY294486 has a significantly higher affinity for GluR5 than for AMPARs ($IC_{50}s = 4\ \mu M$ and 30–100 μM, respectively) and lacks any activity at GluR6 or GluR7 homomeric channels

(Clarke et al., 1997). LY382884 exhibits even higher selectivity for GluR5-containing receptors over the other KAR subunits and, more importantly, over AMPARs (Simmons et al., 1998; Bortolotto et al., 1999). KARs are also subject to blockade by lanthanides (Huettner et al., 1998) but the utility of these compounds in physiological experiments is limited. Recently, the willardiine derivative UBP 296 has been reported to be a potent and selective antagonist of GluR5-containing receptors. A series of new willardiine derivatives are now being synthesized and tested but are not yet commercially available (see More et al., 2004).

Plant lectins, such as Concanavalin A (ConA), have been widely used in the study of KARs. ConA binds to a series of N-glycosylated residues present on this receptor (Everts et al., 1997; 1999) causing a reduction of the fast desensitization observed upon ligand binding (Huettner 1990; Partin et al., 1993; Wong and Mayer 1993). This compound is active at native and recombinant receptors, although its potency at GluR7 is lower compared to the other subunits (Schmitt et al., 1996). Unfortunately, the utility of ConA is restricted for several reasons. On the one hand, ConA binds non-specifically to sugars in cell membranes, rendering it of no use when working with brain slices. Secondly, ConA increases the agonist affinity of KARs by one or two orders of magnitude (Jones et al., 1997; Paternain et al., 1998) leading to confusing biophysical and pharmacological results in terms of the channel properties. In fact, it has recently been shown with macroscopic responses, that application of ConA results in a shift in the relative contribution of different open states of the channel (Bowie et al., 2003). Finally, ConA has been observed to modulate other subtypes of glutamate receptors such that it has been reported to increase currents through homomeric GluR1 channels and NMDAR receptors (Everts et al., 1999).

3.2.2.7 Kainate receptor physiology

The detailed study of KARs physiology has been feasible since the discovery of pharmacological tools which enable the separation of the KAR and AMPAR-mediated components of a given response. In 1995, the demonstration that the 2,3-benzodiazepines, GYKI 53655 (LY300268 or the active isomer LY303070), antagonize AMPARs but not KARs (Wilding and Huettner, 1995; Paternain et al., 1995) was the starting point in the study of KARs from a physiological perspective. This work has also been advanced with the generation of mice deficient for the subunits that conform KARs (Mulle et al., 1998; Contractor et al., 2003) and the development of several compounds specific for KAR subunits.

It is now clear that KARs are present pre- and postsynaptically and that they have different roles in synaptic transmission depending upon their localization at the synapse. At the postsynaptic level KARs, along with the AMPARs and NMDARs, carry part of the current charge of the synaptic response. Presynaptically, KARs play a role in the regulation of transmitter release, at both excitatory and inhibitory synapses. Furthermore, KARs have been shown to function in a metabotropic fashion by activating phospholipase C and PKC assisted by a G protein. This relatively recently recognized non-canonical signalling mechanism illustrates how much remains to be learnt of the physiological roles of the KARs in glutamate signalling. Finally, KARs are believed to play a significant role in synaptic plasticity and may have a fundamental role in epileptogenesis due to their control of network excitability.

Kainate receptors support excitatory synaptic transmission

The availability of GYKI 53655, a selective antagonist of AMPARs, was essential for experiments aimed at determining that synaptically released glutamate mediates the activation of postsynaptic KARs in several brain areas. Although CA1 hippocampal pyramidal neurons express KAR subunits, no EPSC mediated by these receptors could be recorded in the presence of AMPAR and NMDAR antagonist (Lerma et al., 1997). However, further experiments performed in CA3 pyramidal neurons revealed the presence of postsynaptic KARs in these synapses (Castillo et al., 1997; Vignes and Collindridge, 1997). Indeed, the CA3 region of the hippocampus where pyramidal neurons receive excitatory input from mossy fibres (MFs) was the first place where synaptically activated KARs were identified. Repetitive stimulation of the MF, in the presence of antagonists of AMPARs and NMDARs (GYKI 53655 and APV), evoked KAR-mediated responses with slow kinetics and these responses were blocked by CNQX (Castillo et al., 1997; Vignes and Collindridge, 1997) (Figure 3.2.2.1f). These effects are absent in mice lacking the GluR6 subunit, suggesting that GluR6-containing receptors are involved in mediating transmission at these synapses (Mulle, 1998). Immunogold and biochemical studies (Darstein et al., 2003) together with the use of KA2-deficient mice (Contractor et al., 2003) provided evidence for GluR6/KA2 heteromers as the major postsynaptic receptor at MF synapses. Interestingly, in CA3 pyramidal cells, stimulation of the associational commissural fibres fails to elicit a KAR-mediated EPSC (Castillo et al., 1997). Furthermore, local application of kainate along the dendritic tree reveals that postsynaptic KARs are restricted to the proximal dendrites opposed to MF terminals (Castillo et al., 1997).

The interneurons present in the CA1 area of the hippocampus also contain a population of postsynaptic KARs at synapses made by Schaffer collaterals (Cossart et al., 1998; Frerking et al., 1998). Here, a single stimulus elicits a slow synaptic response. Other brain regions where postsynaptic KARs have been identified include synapses on cerebellar golgi neurons (Bureau et al., 2000), Purkinje cells (Huang et al., 2004), in both pyramidal cells and interneurons of the neocortex (Kidd and Isaac, 1999; Ali 2003; Eder et al., 2003; Wu et al., 2005), the spinal cord (Li et al., 1999), amygdala (Li and Rogawski, 1998) and off bipolar cells in the retina (DeVries and Schwartz, 1999).

Recently, it has been shown that activation of postsynaptic KARs enhances CA1 and CA3 pyramidal cell excitability by inhibiting the potassium current that underlies the slow after-hyperpolarization (I_{sAHP}) (Melyan et al., 2002; 2004; Fisahn et al., 2004, 2005). This action of kainate was mimicked by domoate but not by ATPA, and it was blocked by CNQX but not by GYKI 53655. Further analysis suggested that this metabotropic mechanism may be mediated by KARs containing KA2 subunits (Ruiz et al., 2005) (Figure 3.2.2g).

Slow kinetic properties of postsynaptic kainate receptors

Synaptic responses mediated by KARs share two features: the synaptic current is small compared to that mediated by AMPARs (about 10 per cent of the total peak current) and has significantly slower deactivation kinetics (Figure 3.2.1f). The long-lasting current is in principle difficult to reconcile with the properties of recombinant and native KARs. Therefore, it was initially suggested that KARs could be extra-synaptic rather than being located at the

postsynaptic density. However, there is experimental evidence showing quantal activation of KARs by glutamate release at both MF synapses and Schaffer collaterals synapses onto CA1 interneurons (Cossart et al., 2002), which rules out such a possibility. The explanation for the differences in kinetics with recombinant KARs may be that the composition of the synaptic KARs differs from any combination of subunits so far analysed recombinantly. Postsynaptic KARs could contain additional unidentified subunits or bind auxiliary proteins that influence the response kinetics. There are indications that coupling of recombinant KARs to scaffolding and trafficking proteins could alter the kinetic properties of the responses (Garcia et al., 1998; Coussen et al., 2002; Bowie et al., 2003; Hirbec et al., 2003; Coussen et al., 2005). The functional significance of the relatively small amplitude and slow kinetics of KAR-mediated currents might be related to neuronal integrative properties. KARs and AMPARs both cause neuronal activation but encode different types of information. While AMPARs would provide transient excitation, the KAR component of synaptic current may contribute a substantial tonic depolarization that brings the neuron close to the firing threshold in such a way that a single afferent input can be highly effective (Frerking and Ohliger-Frerking, 2002). Such a slow, small amplitude EPSC as mediated by KARs offers the possibility of integrating excitatory inputs over a wide time window.

Kainate receptors modulate transmitter release

A number of early experiments carried out more than 20 years ago indicated that KARs might have a presynaptic localization. For example, autoradiographic studies revealed a band of high-affinity kainate binding sites along the stratum lucidum in the CA3 region of the hippocampus. In addition, when the granule cells were destroyed, there was a parallel reduction in the number of binding sites (Represa et al., 1987) and a decrease in the susceptibility to the kainate-induced toxicity (Debonnel et al., 1989) and epileptogenic activity (Gaiarsa et al., 1994). Nowadays there is compelling evidence that KARs are present in both excitatory and inhibitory terminals where they can modulate transmitter release.

Excitatory synapses

It has been found that nanomolar concentrations of exogenous kainate enhance glutamate release at CA3 mossy fibre buttons (Figure 3.2.2.1e), whereas higher concentrations depress release (Contractor et al., 2000; Lauri et al., 2001a; Schmitz et al., 2001a). Both phenomena are reproduced by synaptic activity illustrating that KARs in MFs work as autoreceptors that sense synaptically released glutamate. Thus, on one hand synaptic activation of presynaptic KARs at the mossy fibre synapse induced frequency-dependent facilitation (Contractor et al., 2001; Lauri et al., 2001a; Schmitz et al., 2001), while increased activity produces frequency-dependent depression. In addition, the induction of long-term potentiation (LTP) seems to be influenced by the activation of KARs at these synapses (Bortolotto et al., 1999; Contractor et al., 2001; Schmitz et al., 2003).

Dentate granular cells express mRNA for GluR6, GluR7, KA1 and KA2 subunits. However, the exact subunit composition of presynaptic KARs on mossy fibre terminals remains unclear. Pharmacological studies using LY382884 have suggested a role for GluR5 containing KARs in short-term as well as long-term plasticity at mossy fibre synapses (Bortolotto et al., 1999; Lauri et al., 2001b,

2003). This result is a matter of dispute, as the GluR5 subunit is in fact poorly expressed in the granule cells that form the presynaptic afferents at MF synapses (Bahn et al., 1994; Paternain et al., 2000). Other studies using knockout (KO) mice have concluded that LTP at MF synapses was reduced but not abolished in GluR6-deficient mice, whereas GluR5-deficient mice showed a normal LTP at these synapses (Contractor et al., 2001; Schmitz et al., 2003; Breustedt and Schmitz, 2004). Clearly, more investigation is needed to understand which subunits form the KARs at MF synapses which mediate long-lasting changes in release properties.

Several groups have suggested that facilitation of glutamate release is due to the depolarizing effect of KAR activation on the presynaptic terminal. This would be enough to facilitate release at low level of activation (Kamiya et al., 2002), although a Ca^{2+}-induced Ca^{2+} release from the intracellular stores has been claimed to be required for this phenomenon to occur (Lauri et al., 2003). On the other hand, KAR-mediated large depolarizations of synaptic terminals could inactivate the generation of action potentials leading to a failure in release. However, there is data showing that inhibition but not facilitation of release by presynaptic KARs depends on a G-protein-dependent mechanism, independent of their ion channel activity (Lauri et al., 2005). Therefore, it is possible that two different KARs systems co-localize in presynaptic terminals with different signalling pathways such that the threshold for activating one or other KAR could determine the physiological response.

Besides the modulation of glutamate release by KARs in the MF pathway, it has also been found that presynaptic KARs regulate excitatory synapses made by Schaffer collaterals onto CA1 pyramidal cells. Application of kainate elicits a dose-dependent decrease in glutamate release from rat hippocampal synaptosomes and also depresses glutamatergic synaptic transmission due to the inhibition of Ca^{+2} influx into presynaptic terminals (Chittajallu et al, 1996; Kamiya and Ozawa, 1998). The inhibition of glutamate release by KARs in these synapses seems to depend on Gi/Go proteins (Frerking et al., 2001), which might inhibit Ca^{2+} channels (Rozas et al., 2003).

Presynaptic kainate autoreceptors are also involved in regulating thalamocortical synapses in early stages of postnatal development. These receptors induce depression of glutamate release at high frequencies of stimulation (>50 Hz) which is inhibited by LY382884. Interestingly, such frequencies of afferent activity are associated with whisker activation. This presynaptic KAR-mediated mechanism is lost by the end of the first postnatal week and, thus, may be important for the maturation of sensory processing (Kidd et al., 2002).

Finally, it has been reported that synaptically activated KARs presynaptically either facilitate or depress transmission at parallel fibre synapses in the cerebellar cortex. Low-frequency stimulation of parallel fibres (five stimuli, 10 Hz) facilitates synapses onto both stellate cells and Purkinje cells, whereas high-frequency stimulation (five stimuli, 100 Hz) depresses stellate cell synapses but continues to facilitate Purkinje cell synapses (Delaney and Jahr, 2002). The difference in the sensitivity of the two parallel fibre synapses has not been explained and might rely on subtle differences in receptor composition, density or precise location.

Inhibitory synapses

KARs have also been implicated in regulating the evoked release of GABA at interneuron-pyramidal cell connections in the hippocampus

(Clarke et al., 1997; Rodriguez-Moreno et al., 1997; 2000; Rodriguez-Moreno and Lerma, 1998; Maingret et al., 2005). Glutamate seems to spill over from adjacent terminals activating presynaptic KARs at GABA terminals (Min et al., 1999). Kainate also induces a marked increase in the firing rate of the interneurons (Fisher and Alger, 1984), which is due to the presence of KARs in their somatodendritic compartment. Thus, KARs might strongly regulate CA1 GABAergic circuitry through two distinct and opposing mechanisms. Whereas activation of somatodendritic KARs increases the activity of GABA interneurons (Cossart et al., 1998; Frerking et al., 1998; 1999; Jiang et al., 2001), KARs located in the presynaptic terminals reduce transmitter release, a mechanism that involves G protein and protein kinase C (PKC) activation (Rodriguez-Moreno et al., 1997, Rodriguez-Moreno and Lerma, 1998) (Figure 3.2.2.1d). More recent experiments have provided evidence indicating that these two effects are mediated by receptor populations that are functionally, molecularly, and pharmacologically different (Rodriguez-Moreno et al., 2000; Christensen et al., 2004; Maingret et al., 2005).

Role of kainate receptors in synaptic development

Although it is well known that the expression of KARs subunits is developmentally regulated (Bahn et al., 1994), their role in the plasticity and development of synaptic circuits is only now beginning to be explored. For example, developing thalamocortical synapses express postsynaptic KARs as well as AMPARs that do not co-localize at the same synapse in a number of cases. Coinciding with the critical period for experience-dependent plasticity (early in development), the contribution of KARs to synaptic transmission decreases. In fact, induction of LTP causes a rapid reduction in the number of synapses containing KARs. This result provides direct evidence that postsynaptic KARs are modified in an activity-dependent manner during development (Kidd and Isaac, 1999). Recent evidence also supports the notion that presynaptic KARs are involved in regulating the dynamic properties of the hippocampal synapses during maturation (Lauri et al., 2005, 2006). In addition, KARs may regulate the refinement of neural circuits and the growth and stabilization of axonal and dendritic structures. The role of neuronal activity in the regulation of pre- and postsynaptic motility is controversial, but new data have provided evidence that axonal motility is bidirectionally regulated by neural activity through KARs (Tashiro et al., 2003). Whereas low KAR activity stimulates motility by a mechanism that involves depolarization and Ca^{2+} influx, high KAR activity blocks motility via a metabotropic mechanism mediated by G-protein activation. These results suggest a role for non-canonical signalling of KARs (see below) in synapse formation. These opposing effects are potentially mediated by different types of KARs although further work is required to identify which subunits are involved.

Canonical and non-canonical signalling of kainate receptors

As mentioned above, an emerging characteristic of KARs and other ionotropic glutamate receptors is that signalling might occur through ion flux (canonical) and alternatively by their coupling to G-proteins (non-canonical). The first evidence of the G-protein-mediated signalling by KARs was discovered in the hippocampus (Rodriguez-Moreno et al., 1998) and was later confirmed by other authors (Cunha et al., 2000; Frerking et al., 2001; Tashiro et al., 2003; Fisahn et al., 2005; Lauri et al., 2005). This mechanism profoundly affects

Table 3.2.2.1 Properties of kainate receptors

Receptor subunit	Gene	Chromosomal location			Cellular and subcellular localization	Pharmacology	Knockout mice
		Mouse	Rat	Human			
GluR5 (GluK1*)	GRIK1	16 (58.0 cM)	11q22	21q22.11	Dorsal root ganglion neurons, subicular neurons, septal nuclei, piriform and cingulated cortices, Purkinje cells, hippocampal interneurons	Agonists: Glutamate, kainate, domoate, AMPA [1], ATPA [2], (S)-5-I-willardiine [3], SYM 2081 [4–6] Antagonists: LY293558 and LY382884 [7], LY294486 [8], SYM 2081 [4]	GluR5−/− [9]
GluR6 (GluK2*)	GRIK2	10 (27.0 cM)	20q13	6q16.3–q21	Cerebellar granular cells, dentate gyrus and CA3 region of the hippocampus and the striatum	Agonists: Glutamate, kainate, domoate [10], SYM 2081 [4]; ATPA at GluR5/GluR6 heteromers [11] Antagonists: NBQX [7], SYM 2081 [4]	GluR6−/− [12]
GluR7 (GluK3*)	GRIK3	4 (57.2 cM)	5q36	1p34–p33	Low levels in the brain	Agonists: Glutamate, kainate [13]	GluR7−/− [14]
KA1 (GluK4*)	GRIK4	9 (23.0 cM)	8q22	11q22.3	CA3 region of hippocampus and lower levels in the dentate gyrus, amygdala and entorhinal cortex	Agonists: Glutamate, kainate, AMPA, domoate and ATPA (heteromeric GluR6/KA1,GluR5/KA1and GluR7/KA1) [13, 15]	
KA2 (GluK5*)	GRIK5	7 (6.5 cM)	1q21	19q13.2	Throughout the CNS	Agonists: Glutamate, AMPA and ATPA (heteromeric GluR5/KA2 and Glu6/KA2) [11, 15]	KA2−/− [16]

*The classification of glutamate receptor subunits has recently been re-addressed by NC-IUPHAR and the revised recommended nomenclature is shown here. See Alexander SPH, Mathie A and Peters JA (2008). Guide to receptors and channels (GRAC), 3rd edn. *Br J Pharmacol* 153, S1–S209.)

[1] Sommer B et al. (1992) EMBO J 11, 1651–1656; [2] Clarke VR et al. (1997) Nature 389, 599–603; [3] Swanson GT et al. (1998) Mol Pharmacol 53, 942–949; [4] Jones KA et al. (1997) Neuropharmacology 36, 853–863; [5] Wilding TJ, and Huettner JE (1997) Mol Pharmacol 47, 582–587; [6] Zhou LM et al. (1997) J Pharmacol Exp Ther 280, 422–427; [7] Bleackman D et al. (1996) Neuropharmacology 35, 1689–1702; [8] Simmons RM et al. (1998) Neuropharmacology 37, 25–36; [9] Contractor A et al. (2000) J Neurosci 22, 8269–78; [10] Lerma J et al. (1993) Proc Natl Acad Sci USA 90, 11688–11692; [11] Paternain AV et al. (2000) J Neurosci 20, 196–205; [12] Mulle C et al. (1998) Nature 392, 601–605; [13] Schiffer HH et al. (1997) Neuron 19, 1141–1146; [14] Pinheiro PS et al. (2007) Proc Natl Acad Sci USA 29, 12181–12186; [15] Herb A et al. (1992) Neuron 8, 775–785; [16] Contractor A et al. (2003) J Neurosci 23, 422–429.

neuronal excitability, as it results in the inhibition of transmitter release. In cultured dorsal root ganglion cells, that express GluR5 and KA2 subunits, activation of KARs produced a release of Ca^{+2} from intracellular stores in a G-protein-dependent manner. The subsequent activation of PKC inhibits N-type Ca^{2+} channels (Rozas et al., 2003). This non-canonical signalling is independent of ion channel activity since it is present under conditions in which ion channel flux is minimized and kainate does not induce any ionotropic response. However, both pathways depend on a common ionotropic subunit, GluR5 (Rozas et al., 2003). The mechanism by which the ion channel couples to the G protein is unresolved but it seems likely that an accessory or linked protein participates, although none of the KARs interacting proteins identified to date appear to perform this role.

3.2.2.8 Relevance to disease states

KARs have been implicated in several brain disorders, the best established being involvement in the excitatory imbalances linked to epilepsy. The best studied system in this regard has been the animal model of temporal lobe epilepsy. In this model kainate injections cause epileptic-like seizures and electrical discharges in the hippocampus that propagate to other limbic structures (Ben-Ari and Cossart, 2000). The evidence for KAR involvement in this phenomenon came from the study of GluR6−/− mice which are resistant to epileptic seizures induced by kainate injection (Mulle et al., 1998). However, this effect was evident only at low doses of the agonist. In addition, KAR antagonists have been shown to prevent epileptic seizures in chemical models of epilepsy both *in vitro* and *in vivo* (Smolders et al., 2002).

The acute epileptogenic effect of kainate has been suggested to result from inhibition of GABA release (Rodriguez-Moreno et al., 1997). However, this mechanism does not explain the chronic epilepsy generated months after kainate treatment. It is well-known that both in this model of temporal lobe epilepsy and in human patients a large number of aberrant synapses are formed following sprouting of glutamatergic fibres. Recently it has been demonstrated that the KAR-mediated synaptic activity constitutes a substantial component of the excitatory transmission at these functional aberrant synapses (Epztein et al. 2005). Therefore, the formation of the KAR-driven aberrant synapses might participate in the pathogenesis of temporal lobe epilepsy. In addition, changes in the KAR mRNA expression and editing have been also documented in the patients with temporal lobe epilepsy (Mathern et al.,

1998; Kortenbruck et al., 2001). These and other data suggest that KAR-targeted anti-epileptic strategies might be of therapeutic value.

In addition to their role in epilepsy, KARs have received attention in other pathophysiological states. KARs are present at spinal afferent synapses and their blockade has been demonstrated to have analgesic actions (Simmons et al., 1998; Liu et al., 2005). This observation raises the possibility that KAR antagonists could be used in the treatment of chronic pain. Excitotoxicity has been also proposed to be mediated by KARs under certain circumstances as kainate administration causes an increase in intracellular Ca^{2+} and activates Ca^{2+}-dependent signalling pathways that lead to neuronal apoptosis (Nadler et al., 1978). However, the extent to which such KAR-dependent mechanisms operate in pathological states is not well understood.

A few human genetic and genomic studies have suggested a possible pathological role of KARs in diseases such as Down's syndrome, schizophrenia and amyotropic lateral sclerosis (Lerma, 2003). The involvement of KARs in the pathology of these diseases has been suggested based on disease-associated changes in the KAR mRNA expression and disease association with variants in the genome mapping to chromosomal regions carrying the KAR genes (Table 3.2.2.1). However, to date, there is no clear demonstration of a role of KARs in the pathophysiology in these diseases.

3.2.2.9 Concluding remarks

In contrast to the other iGluRs, KARs have been generally elusive to study. The discovery of the selective AMPAR antagonist, GYKI53655, enabled a number of experiments which have contributed to an understanding of the role of KARs in physiological processes. Indeed, it seems that the activity of KARs can control excitability of brain networks. The availability of mice deficient for KAR subunits or carrying modified KARs has also been enabling in deciphering their physiological roles. Nowadays, it is generally accepted that KARs can be located postsynaptically, where they can mediate part of the synaptic current at particular synapses, and presynaptically, where they can modulate the release of both glutamate and GABA in a bidirectional manner. Interestingly, KARs can signal in two ways: they are ion channels but they can also couple to G-proteins to trigger a second messenger cascade. These two modes of signalling have been referred to as a canonical and non-canonical pathways These discovery raises the issue of how an ion channel is able to activate a G-protein and, accordingly, much current work is concentrated on the identification of proteins interacting with KARs, which could determine not only their correct targeting but also their signalling capabilities.

A number of issues remain to be determined for KARs. For instance, the synaptic expression of KARs most likely depends on many as yet unidentified proteins playing important roles in receptor trafficking and membrane targeting. It is not known how different KAR subunits are targeted differentially to axons and dendrites and what determines the specific subunit composition of the receptors at a particular synapse. Novel research approaches are needed to investigate the large numbers of receptor interacting proteins and to define the macromolecular signalling complexes. Potential pitfalls of currently available screening assays for the identification of the interacting proteins include the specificity of the antibodies used for pull-down assays, the need for a

relatively large quantity of the protein in the analysed sample and the difficulty in detecting transient and labile protein interactions, which may have a potentially crucial role in determining the ultimate fate of a particular receptor subunit. Innovative techniques, such as protein microarrays and microchips, may help answer some of the questions and it is expected that scientific progress will rapidly lead to better understanding of KAR function in physiology and pathology. Future studies would be greatly facilitated by the development of novel means for high-throughput screening of the unstable protein interactions, simultaneous studies of multiple protein–protein interactions and the analysis of biological specimens which are only available in minute quantities. On the other hand, high-resolution structural analysis is already revealing the atomic structure of KARs subunits. This information will be crucial for enabling the rational design of specific agonists and antagonists. Importantly, the role of KARs in brain diseases remains largely unclear and remains to be elucidated with the possible exception of epilepsy, where KARs seem to play an important role in the regulation of network excitability.

Acknowledgements

Work in the authors' lab was supported by grants from the Spanish Ministry BFI2003–00161 and BFI-2006–07138. The lab belongs to the CONSOLIDER-Ingenio 2010 Programme (ref. CSD2007–0023). SS was a fellow of the Program of Foreign Doctors and Technologists in Spain (MCYT) and currently is an IP3 Program CSIC Research Fellow.

References

Ali AB (2003). Involvement of post-synaptic kainate receptors during synaptic transmission between unitary connections in rat neocortex. Eur J Neurosci 17, 2344–2350.

Alt A, Weiss B, Ogden AM et al. (2004) Pharmacological characterization of glutamatergic agonists and antagonists at recombinant human homomeric and heteromeric kainate receptors in vitro. Neuropharmacology 46, 793–806.

Bahn S, Volk B and Wisden W (1994). Kainate receptor gene expression in the developing rat brain. J Neurosci 14, 5525–5547.

Ben-Ari Y and Cossart R (2000) Kainate, a double agent that generates seizures: two decades of progress. Trends Neurosci 23, 580–587.

Bernard A and Khrestchatisky M (1994) Assessing the extent of RNA editing in the TMII regions of GluR5 and GluR6 kainate receptors during rat brain development. J Neurochem 62, 2057–2060.

Bettler B, Egebjerg V, Scharma G et al. (1992). Cloning of a putative glutamate receptor: a low affinity kainate-binding subunit. Neuron 8, 257–265.

Bettler B and Mulle C (1995). Review: neurotransmitter receptors. II. AMPA and kainate receptors. Neuropharmacology 34, 123–139.

Bettler B, Boulter J, Hermans-Borgmeyer I et al. (1990) Cloning of a novel glutamate receptor subunit, GluR5: expression in the nervous system during development. Neuron 5, 583–595.

Bischoff S, Barhanin J, Bettler B et al. (1997). Spatial distribution of kainate receptor subunit mRNA in the mouse basal ganglia and ventral mesencephalon. J Comp Neurol 379, 541–562.

Bleakman D, Ballyk BA, Schoepp DD et al. (1996). Activity of 2,3-benzodiazepines at native rat and recombinant human glutamate receptors in vitro: stereospecificity and selectivity profiles. Neuropharmacology 35, 1689–1702.

Borgmeyer U, Hollman M and Heinemann S (1990). Cloning of a novel glutamate receptor subunit, GluR5: expression in the nervous system during development. Neuron 5, 583–595.

Bortolotto ZA, Clarke VR, Delany CM et al. (1999). Kainate receptors are involved in synaptic plasticity. *Nature* **402**, 297–301.

Bowie D, Garcia EP, Marshall J et al. (2003). Allosteric regulation and spatial distribution of kainate receptors bound to ancillary proteins. *J Physiol* **547**, 373–385.

Brandstatter JH, Koulen P and Wassle H (1997). Selective synaptic distribution of kainate receptor subunits in the two plexiform layers of the rat retina. *J Neurosci* **17**, 9298–9307.

Breustedt J and Schmitz D (2004). Assessing the role of GLUK5 and GLUK6 at hippocampal mossy fiber synapses. *J Neurosci* **24**, 10093–10098.

Brewer GJ and Cotman CW (1989). NMDA receptor regulation of neuronal morphology in cultured hippocampal neurons. *Neurosci Lett* **99**, 268–273.

Bureau I, Bischoff S, Heinemann SF et al. (1999). Kainate receptor-mediated responses in the CA1 field of wild-type and GluR6-deficient mice. *J Neurosci* **19**, 653–663.

Bureau I, Dieudonne S, Coussen F et al. (2000). Kainate receptor-mediated synaptic currents in cerebellar golgi cells are not shaped by diffusion of glutamate. *Proc Natl Acad Sci USA* **97**, 6838–6843.

Castillo PE, Malenka RC and Nicoll RA (1997). Kainate receptors mediate a show postsynaptic current in hippocampal CA3 neurons. *Nature* **388**, 182–186.

Charara A, Blankstein E and Smith Y (1999). Presynaptic kainate receptors in the monkey striatum. *Neuroscience* **91**, 1195–1200.

Chergui K, Bouron A, Normand E et al. (2000). Functional GluR6 kainate receptors in the striatum: indirect downregulation of synaptic transmission. *J Neurosci* **20**, 2175–2182.

Chittajallu R, Vignes M, Dev KK et al. (1996). Regulation of glutamate release by presynaptic kainate receptors in the hippocampus. *Nature* **379**, 78–81.

Christensen JK, Paternain AV, Selak S et al. (2004). A mosaic of functional kainate receptors in hippocampal interneurons. *J Neurosci* **24**, 8986–8993.

Clarke VR, Ballyk BA, Hoo KH et al. (1997). A hippocampal GluR5 kainate receptor regulating inhibitory synaptic transmission. *Nature* **389**, 599–603.

Collingridge GL and Bliss TV (1995). Memories of NMDA receptors and LTP. *Trends Neurosci* **18**, 54–56.

Collingridge GL and Isaac JT (2003). Functional roles of protein interactions with AMPA and kainate receptors. *Neurosci Res* **47**, 3–15.

Contractor A, Sailer AW, Darstein M et al. (2003). Loss of kainate receptor-mediated heterosynaptic facilitation of mossyfiber synapses in KA2–/– mice. *J Neurosci* **23**, 422–429.

Contractor A, Swanson G and Heinemann SF (2001). Kainate receptors are involved in short- and long-term plasticity at mossy fiber synapses in the hippocampus. *Neuron* **29**, 209–216.

Contractor A, Swanson GT, Sailer A et al. (2000). Identification of the kainate receptor subunits underlying modulation of excitatory synaptic transmission in the CA3 region of the hippocampus. *J Neurosci* **20**, 8269–8278.

Cossart R, Epsztein J, Tyzio R et al. (2002). Quantal release of glutamate generates pure kainate and mixed AMPA/kainate EPSCs in hippocampal neurons. *Neuron* **35**, 147–159.

Cossart R, Esclapez M, Hirsch JC et al. (1998). GluR5 kainate receptor activation in interneurons increases tonic inhibition of pyramidal cells. *Nat Neurosci* **1**, 470–478.

Coussen F, Normand E, Marchal C et al. (2002). Recruitment of the kainate receptor subunit glutamate receptor 6 by cadherin/catenin complexes. *J Neurosci* **22**, 6426–6436.

Coussen F, Perrais D, Jaskolski F et al. (2005). Co-assembly of two GluR6 kainate receptor splice variants within a functional protein complex. *Neuron* **47**, 555–566.

Cui C and Mayer ML (1999). Heteromeric kainate receptors formed by the coassembly of GluR5, GluR6 and GluR7. *J Neurosci* **19**, 8281–8291.

Cunha RA, Constantino MD and Ribeiro JA (1997) Inhibition of [3H] gamma-aminobutyric acid release by kainate receptor activation in rat hippocampal synaptosomes. *Eur J Pharmacol* **323**, 167–172.

Cunha RA, Malva JO and Ribeiro JA (2000). Pertussis toxin prevents presynaptic inhibition by kainate receptors of rat hippocampal [(3)H] GABA release. *FEBS Lett* **469**, 159–162.

Cunha RA, Malva JO and Ribeiro JA (1999). Kainate receptors coupled to G(i)/G(o) proteins in the rat hippocampus. *Mol Pharmacol* **56**, 429–433.

Darstein M, Petralia RS, Swanson GT et al. (2003). Distribution of kainate receptor subunits at hippocampal mossy fiber synapses. *J Neurosci* **23**, 8013–8019.

Debonnel G, Weiss M and De Montigny C (1989). Reduced neuroexcitatory effect of domoic acid following mossy fiber denervation of the rat dorsal hippocampus: further evidence that toxicity of domoic acid involves kainate receptor activation. *Can J Physiol Pharmacol* **67**, 904–908.

Delaney AJ and Jahr CE (2002). Kainate receptors differentially regulate release at two parallel fiber synapses. *Neuron* **36**, 475–482.

Dev KK, Nishimune A, Henley JM et al. (1999). The protein kinase C alpha binding protein PICK1 interacts with short but not long form alternative splice variants of AMPA receptor subunits. *Neuropharmacology* **38**, 635–644.

DeVries SH and Schwartz EA (1999). Kainate receptors mediate synaptic transmission between cones and 'Off' bipolar cells in a mammalian retina. *Nature* **397**, 157–160.

Eder M, Becker K, Rammes G et al. (2003). Distribution and properties of functional postsynaptic ainate receptors on neocortical layer V pyramidal neurons. *J Neurosci* **23**, 6660–6670.

Egebjerg J, Bettler B, Hermans-Borgmeyer I et al. (1991). Cloning of a cDNA for a glutamate receptor subunit activated by kainate but not AMPA. *Nature*, 351745–351748.

Epztein J, Represa A, Jorquera I et al. (2005). Recurrent mossy fibers establish aberrant kainate receptor-operated synapses on granule cells from epileptic rats. *J Neurosci* **25**, 8229–8239.

Everts I, Petroski R, Kizelsztein P et al. (1999). Lectin-induced inhibition of desensitization of the kainate receptor GluR6 depends on the activation state and can be mediated by a single native or ectopic N-linked carbohydrate side chain. *J Neurosci* **19**, 916–927.

Everts I, Villmann C and Hollmann M (1997). N-Glycosylation is not a prerequisite for glutamate receptor function but is essential for lectin modulation. *Mol Pharmacol* **52**, 861–873.

Fisahn A, Contractor A, Traub RD et al. (2004). Distinct roles for the kainate receptor subunits GluR5 and GluR6 in kainate-induced hippocampal gamma oscillations. *J Neurosci* **24**, 9658–9668.

Fisahn A, Heinemann SF and McBain CJ (2005). The kainate receptor subunit GluR6 mediates metabotropic regulation of the slow and medium AHP currents in mouse hippocampal neurones. *J Physiol* **562**, 199–203.

Fisher RS and Alger BE (1984). Electrophysiological mechanisms of kainic acid-induced epileptiform activity in the rat hippocampal slice. *J Neurosci* **4**, 1312–1323.

Fleck MW, Cornell E and Mah SJ (2003). Amino-acid residues involved in glutamate receptor 6 kainate receptor gating and desensitization. *J Neurosci* **23**, 1219–1227.

Frerking M and Ohliger-Frerking P (2002). AMPA receptors and kainate receptors encode different features of afferent activity. *J Neurosci* **22**, 7434–7443.

Frerking M, Malenka R and Nicoll RA (1998). Synaptic activation of KARs on hippocampal interneurons. *Nature Neuroscience* **1**, 479–486.

Frerking M, Petersen CC and Nicoll RA (1999). Mechanisms underlying kainate receptor-mediated disinhibition in the hippocampus. *Proc Natl Acad Sci USA* **96**, 12917–12922.

Frerking M, Schmitz D, Zhou Q et al. (2001). Kainate receptors depress excitatory synaptic transmission at CA3->CA1 synapses in the hippocampus via a direct presynaptic action. *J Neurosci* **21**, 2958–2966.

Gaiarsa JL, Zagrean L and Ben-Ari Y (1994). Neonatal irradiation prevents the formation of hippocampal mossy fibers and the epileptic action of kainate on rat CA3 pyramidal neurons. *J Neurophysiol* **71**, 204–215.

Garcia EP, Mehta S, Blair LA *et al.* (1998). SAP90 binds and clusters kainate receptors causing incomplete desensitization. *Neuron* 21, 727–739.

Gregor P, O'Hara BF, Yang X *et al.* (1993). Expression and novel subunit isoforms of glutamate receptor sgenes GluR5 and GluR6. *Neuroreport* 4, 1343–1346.

Hanson JE and Smith Y (1999). Group I metabotropic glutamate receptors at GABAergic synapses in monkeys. *J Neurosci* 19, 6488–6496.

Hayashi T (1956). Chemical physiology of excitation in muscle and nerve. Tokyo, Nakayama Shoten Ltd.

Heckmann M, Bufler J, Franke C *et al.* (1996). Kinetics of homomeric GluR6 glutamate receptor channels. *Biophys J* 71, 1743–1750.

Herb A, Burnashev N, Werner P *et al.* (1992). The KA-2 subunit of excitatory amino acid receptors shows widespread expression in brain and forms ion channels with distantly related subunits. *Neuron* 8, 775–785.

Hirbec H, Francis JC, Lauri SE *et al.* (2003). Rapid and differential regulation of AMPA and kainate receptors at hippocampal mossy fibre synapses by PICK1 and GRIP. *Neuron* 37, 625–638.

Hollman M and Heinemann S (1994). Cloned glutamate receptors. *Ann Rev Neurosci* 17, 31–108.

Howe JR (1996). Homomeric and heteromeric ion channels formed from the kainate-type subunits GluR6 and KA2 have very small, but different, unitary conductances. *J Neurophysiol* 76, 510–519.

Huang YH, Dykes-Hoberg M, Tanaka K *et al.* (2004). Climbing fiber activation of EAAT4 transporters and kainate receptors in cerebellar Purkinje cells. *J Neurosci* 24, 103–111.

Huettner JE (1990). Glutamate receptor channels in rat DRG neurons-activation by kainate and quisqualate and blockade of desensitization by Con-A. *Neuron* 5, 255–266.

Huettner JE (2003). Kainate receptors and synaptic transmission. *Prog Neurobiol* 70, 387–407.

Huettner JE, Stack E and Wilding TJ (1998). Antagonism of neuronal kainate receptors by lanthanum and gadolinium. *Neuropharmacology* 37, 1239–1247.

Jaskolski F, Coussen F and Mulle C (2005). Subcellular localization and trafficking of kainate receptors. *Trends Pharmacol Sci* 26, 20–26.

Jaskolski F, Coussen F, Nagarajan N *et al.* (2004). Subunit composition and alternative splicing regulate membrane delivery of kainate receptors. *J Neurosci* 24, 2506–2515.

Jiang L, Xu J, Nedergaard M *et al.* (2001). A kainate receptor increases the efficacy of GABAergic synapses. *Neuron* 30, 503–513.

Jones KA, Wilding TJ, Huettner KE *et al.* (1997). Desensitization of kainate receptors by kainate, glutamate and diastereomers of 4-methylglutamate. *Neuropharmacology* 36, 853–863.

Kamiya H and Ozawa S (1998). Kainate receptor-mediated inhibition of presynaptic Ca^{2+} influx and EPSP in area CA1 of the rat hippocampus. *J Physiol* 509, 833–845.

Kamiya H, Ozawa S and Manabe T (2002). Kainate receptor-dependent short-term plasticity of presynaptic Ca^{2+} influx at the hippocampal mossy fiber synapses. *J Neurosci* 22, 9237–9243.

Kidd FL and Isaac JT (1999). Developmental and activity-dependent regulation of kainate receptors at thalamocortical synapses. *Nature* 400, 569–573.

Kidd FL, Coumis U, Collingridge GL *et al.* (2002). A presynaptic kainate receptor is involved in regulating the dynamic properties of thalamocortical synapses during development. *Neuron* 34, 635–646.

Kieval JZ, Hubert GW, Charara A *et al.* (2001). Subcellular and subsynaptic localization of presynaptic and postsynaptic kainate receptor subunits in the monkey striatum. *J Neurosci* 21, 8746–8757.

Kortenbruck G, Berger E, Speckmann EJ *et al.* (2001). RNA editing at the Q/R site for the glutamate receptor subunits GLUR2, GLUR5 and GLUR6 in hippocampus and temporal cortex from epileptic patients. *Neurobiol Dis* 8, 459–468.

Lauri SE, Bortolotto ZA, Bleakman D *et al.* (2001b). A critical role of a facilitatory presynaptic kainate receptor in mossy fiber LTP. *Neuron* 32, 697–709.

Lauri SE, Bortolotto ZA, Nistico R *et al.* (2003). A role for Ca^{2+} stores in kainate receptor-dependent synaptic facilitation and LTP at mossy fiber synapses in the hippocampus. *Neuron* 39, 327–341.

Lauri SE, Delany C, J Clarke VR *et al.* (2001a). Synaptic activation of a presynaptic kainate receptor facilitates AMPA receptor-mediated synaptic transmission at hippocampal mossy fibre synapses. *Neuropharmacology* 41, 907–915.

Lauri SE, Segerstrale M, Vesikansa A *et al.* (2005). Endogenous activation of kainate receptors regulates glutamate release and network activity in the developing hippocampus. *J Neurosci* 25, 4473–4484.

Lauri SE, Vesikansa A, Segerstrale M *et al.* (2006). Functional maturation of CA1 synapses involves activity-dependent loss of tonic kainate receptor-mediated inhibition of glutamate release. *Neuron* 50, 415–429.

Lerma J (1998). Kainate reveals its targets. *Neuron* 19, 1155–1158.

Lerma J (2003) Roles and rules of kainate receptors in synaptic transmission. *Nature Rev Neurosci* 4, 481–495.

Lerma J, Morales M, Vicente MA *et al.* (1997). Glutamate receptors of the kainate type and synaptic transmission. *Trends Neurosci* 20, 9–12.

Lerma J, Paternain AV, Naranjo JR *et al.* (1993). Functional kainate-selective glutamate receptors in cultured hippocampal neurons. *Proc Natl Acad Sci USA* 90, 11688–11692.

Lerma J, Paternain AV, Rodriguez-Moreno A *et al.* (2001). Molecular physiology of kainate receptors. *Physiol Rev* 81, 971–998.

Li H and Rogawski MA (1998). GluR5 kainate receptor mediated synaptic transmission in rat basolateral amygdala *in vitro*. *Neuropharmacology* 37, 1279–1286.

Li P, Wilding TJ, Kim SJ *et al.* (1999). Kainate receptor-mediated sensory synaptic transmission in mammalian spinal cord. *Nature* 397, 161–164.

Liu QS, Pu L and Poo MM (2005). Repeated cocaine exposure *in vivo* facilitates LTP induction in midbrain dopamine neurons. *Nature* 437, 1027–1031.

Mah SJ, Cornell E, Mitchell NA *et al.* (2005). Glutamate receptor trafficking: endoplasmic reticulum quality control involves ligand binding and receptor function. *J Neurosci* 25, 2215–2225.

Maingret F, Lauri SE, Taira T *et al.* (2005). Profound regulation of neonatal CA1 rat hippocampal GABAergic transmission by functionally distinct kainate receptor populations. *J Physiol* 567, 131–142.

Martin S and Henley JM (2004). Activity-dependent endocytic sorting of kainate receptors to recycling or degradation pathways. *EMBO J* 23, 4749–4759.

Mathern GW, Pretorius JK, Kornblum HI *et al* (1998). Altered hippocampal kainate-receptor mRNA levels in temporal lobe epilepsy patients. *Neurobiol Dis* 5, 151–176.

Mayer ML (2005). Glutamate receptor ion channels. *Curr Opin Neurobiol* 15, 282–288.

Mayer ML (2006). Glutamate receptors at atomic resolution. *Nature* 440, 456–462.

Melyan Z, Lancaster B and Wheal HV (2004). Metabotropic regulation of intrinsic excitability by synaptic activation of kainate receptors. *J Neurosci* 24, 4530–4534.

Melyan Z, Wheal HV and Lancaster B (2002). Metabotropic-mediated kainate receptor regulation of IsAHP and excitability in pyramidal cells. *Neuron* 34, 107–114.

Min MY, Melyan Z and Kullmann DM (1999). Synaptically released glutamate reduces gamma-aminobutyric acid (GABA)ergic inhibition in the hippocampus via kainate receptors. *Proc Natl Acad Sci USA* 96, 9932–9937.

More JC, Nistico R, Dolman NP *et al.* (2004). Characterisation of UBP296: a novel, potent and selective kainate receptor antagonist. *Neuropharmacology* 47, 46–64.

Mulle C, Sailer A, Perez-Otano I *et al.* (1998). Altered synaptic physiology and reduced susceptibility to kainate-induced seizures in GluR6-deficient mice. *Nature* 392, 601–605.

Mulle C, Sailer A, Swanson GT *et al.* (2000). Subunit composition of kainate receptors in hippocampal interneurons. *Neuron* **28**, 475–284.

Nadler JV, Perry BW and Cotman CW (1978). Intraventricular kainic acid preferentially destroys hippocampal pyramidal cells. *Nature* **271**, 676–677.

Nakanishi S (1992). Molecular diversity of glutamate receptors and implications for brain function. *Science* **258**, 597–603.

Osten P, Khatri L, Perez JL *et al.* (2000). Mutagenesis reveals a role for ABP/GRIP binding to GluR2 in synaptic surface accumulation of the AMPA receptor. *Neuron* **27**, 313–325.

Partin KM, Patneau DK, Winters CA *et al.* (1993). Selective modulation of desensitization at AMPA versus kainate receptors by cyclothiazide and concanavalin A. *Neuron* **11**, 1069–1082.

Pasquini F, Bochet P, Garbay-Jaureguiberry C *et al.* (1992). Electron microscopic localization of photo affinity labelled delta opioid receptors in the neostriatum of the rat. *J Complement Neurol* **326**, 229–244.

Paternain AV, Herrera MT, Nieto MA *et al.* (2000). GluR5 and GluR6 kainate receptor subunits coexist in hippocampal neurons and coassemble to form functional receptors. *J Neurosci* **20**, 196–205.

Paternain AV, Morales M and Lerma J (1995). Selective antagonism of AMPA receptor unmasks kainate receptor-mediated responses in hippocampal neurons. *Neuron* **14**, 185–189.

Paternain AV, Rodriguez-Moreno A, Villarroel A *et al.* (1998). Activation and desensitization properties of native and recombinant kainate receptors. *Neuropharmacology* **37**, 1249–1259.

Paternain AV, Vicente MA, Nielsen E *et al.* (1996). Comparative antagonism of kainate-activated AMPA and kainate receptors in hippocampal neurons. *Eur J Neurosci* **8**, 2129–2136.

Pemberton KE, Belcher SM, Ripellino JA *et al.* (1998). High-affinity kainate-type ion channels in rat cerebellar granule cells. *J Physiol* **510**, 401–420.

Perez JL, Khatri L, Chang C *et al.* (2001) PICK1 targets activated protein kinase C alpha to AMPA receptor clusters in spines of hippocampal neurons and reduces surface levels of the AMPA-type glutamate receptor subunit 2. *J Neurosci* **21**, 5417–5428.

Petralia RS, Wang YX and Wenthold RJ (1994) Histological and ultrastructural localization of the kainate receptor subunits KA2 and GluR6/7, in the rat central nervous system using selective antipeptide antibodies. *J Comp Neurol* **349**, 85–110.

Rauschecker JP and Hahn S (1987). Ketamine-xylazine anaesthesia blocks consolidation of ocular dominance changes in kitten visual cortex. *Nature* **12**, 183–185.

Ren Z, Riley NJ, Garcia EP *et al.* (2003) Multiple trafficking signals regulate kainate receptor KA2 subunit surface expression. *J Neurosci* **23**, 6608–6616.

Represa A Tremblay E and Ben-Ari Y (1987). Kainate binding sites in the hippocampal mossy fibers: localization and plasticity. *Neuroscience* **20**, 739–748.

Rodriguez-Moreno A and Lerma J (1998). Kainate receptor modulation of GABA release involves a metabotropic function. *Neuron* **20**, 1211–1218.

Rodriguez-Moreno A, Herreras O *et al.* (1997). Kainate receptors presynaptically downregulate GABAergic inhibition in the rat hippocampus. *Neuron* **19**, 893–901.

Rodriguez-Moreno A, López-Garcia VC and Lerma J (2000) Two populations of kainate receptors with separate signalling mechanisms in hippocampal interneurons. *Proc Natl Acad Sci USA* **97**, 1293–1298.

Rosenmund C, Stern-Bach Y and Stevens CF (1998). The tetrameric structure of a glutamate receptor channel. *Science* **280**, 1596–1599.

Rozas JL, Paternain AV and Lerma J (2003). Noncanonical signaling by ionotropic kainate receptors. *Neuron* **39**, 543–553.

Ruano D, Lambolez B, Rossier J *et al.* (1995). Kainate receptor subunits expressed in single cultured hippocampal neurons: molecular and functional variants by RNA editing. *Neuron* **14**, 1009–1017.

Ruiz A, Sachidhanandam S, Utvik JK *et al.* (2005). Distinct subunits in heteromeric kainate receptors mediate ionotropic and metabotropic function at hippocampal mossy fiber synapses. *J Neurosci* **25**, 11710–11718.

Sahara Y, Noro N, Lida Y *et al.* (1997). Glutamate receptor subunits GluR5 and KA-2 are coexpressed in rat trigeminal ganglion neurons. *J Neurosci* **17**, 6611–6620.

Schiffer HH, Swanson GT and Heinemann SF (1997). Rat GluR7 and a carbosy-terminal splice variant, GluR7b, are functional kainate receptor subunits with a low sensitivity to glutamate. *Neuron* **19**, 1141–1146.

Schmitt J, Dux E, Gissel C *et al.* (1996). Regional analysis of developmental changes in the extent of GluR6 mRNA editing in rat brain. *Brain Res Dev Brain Res* **91**, 153–157.

Schmitz D, Mellor J and Nicoll RA (2001). Presynaptic kainate receptor mediation of frequency facilitation at hippocampal mossy fiber synapses. *Science* **291**, 1972–1976.

Schmitz D, Mellor J, Breustedt J *et al.* (2003). Presynaptic kainate receptors impart an associative property to hippocampal mossy fiber long-term potentiation. *Nat Neurosci* **6**, 1058–1063.

Seeburg PH (1993). The molecular biology of mammalian glutamate receptor channels. *Trends Neurosci* **16**, 359–365.

Seeburg PH (1996). The role of RNA editing in controlling glutamate receptor channel properties. *J Neurochem* **66**, 1–5.

Sheng M and Sala C (2001). PDZ domains and the organization of supramolecular complexes. *Annu Rev Neurosci* **24**, 1–29.

Simmons RM, Li DL, Hoo KH *et al.* (1998). Kainate GluR5 receptor subtype mediates the nociceptive response to formalin in the rat. *Neuropharmacology* **37**, 25–36.

Smith TC, Wang LY and Howe JR (1999). Distinct kainate receptor phenotypes in immature and nature mouse cerebellar granule cells. *J Physiol* **517**, 51–58.

Smolders I, Bortolotto ZA, Clarke VR *et al.* (2002). Antagonists of GLU(K5)-containing kainate receptors prevent pilocarpine-induced limbic seizures. *Nat Neurosci* **5**, 796–804.

Sommer B, Burnashev N, Verdoorn TA *et al.* (1992). A glutamate receptor with high affinity for domoate and kainate. *EMBO J* **11**, 1651–1656.

Sommer B, Kohler M, Sprengel R *et al.* (1991). RNA editing in brain controls a determinant of ion flow in glutamate-gated channels. *Cell* **67**, 11–19.

Stern-Bach Y, Bettler B, Hartley M *et al.* (1994). Agonist selectivity of glutamate receptors is specified by two domains structurally related to bacterial amino acid-binding proteins. *Neuron* **13**, 1345–1357.

Swanson GT, Feldmeyer D, Kaneda M *et al.* (1996). Effect of RNA editing and subunit co-assembly single-channel properties of recombinant kainate receptors. *J Physiol (Lond)* **492**, 129–142.

Swanson GT, Green T and Heinemann SF (1998). Kainate receptors exhibit differential sensitivities to (S)-5-iodowillardiine. *Mol Pharmacol* **53**, 942–949.

Tashiro A, Dunaevsky A, Blazeski R *et al.* (2003). Bidirectional regulation of hippocampal mossy fiber filopodial motility by kainate receptors: a two-step model of synaptogenesis. *Neuron* **38**, 773–784.

Terashima A, Cotton L, Dev KK *et al.* (2004). Regulation of synaptic strength and AMPA receptor subunit composition by PICK1. *J Neurosci* **24**, 5381–5390.

Traynelis SF and Wahl P (1997). Control of rat GluR6 glutamate receptor open probability by protein kinase A and calcineurin. *J Physiol (Lond)* **503**, 513–531.

Valluru L, Xu J, Zhu Y *et al.* (2005). Ligand binding is a critical requirement for plasma membrane expression of heteromeric kainite receptors. *J Biol Chem* **280**, 6085–6093.

Verdoorn TA, Johansen TH, Drejer J *et al.* (1994). Selective block of recombinant GluR6 receptors by NS-102, a novel non-NMDA receptor antagonist. *Eur J Pharmacol* **269**, 43–49.

Vignes M and Collingridge GL (1997). The synaptic activation of kainate receptors. *Nature* **388**, 179–182.

Vissel B, Royle GA, Christie BR *et al.* (2001). The role of RNA editing of kainate receptors in synaptic plasticity and seizures. *Neuron* **29**, 217–227.

Vivithanaporn P, Yan S and Swanson GT (2006). Intracellular trafficking of KA2 kainate receptors mediated by interactions with coatomer protein complex I (COPI) and 14-3-3 chaperone systems. *J Biol Chem* **281**, 15475–15484.

Wang LY, Taverna FA Huang XP *et al.* (1993). Phosphorylation and modulation of a kainate receptor (GluR6) by cAMP-dependent protein kinase. *Science* **259**, 1173–1175.

Werner P, Voight M, Keinänen K *et al.* (1991). Cloning of a putative high-affinity kainate receptor expressed predominantly in hippocampal CA3 cells. *Nature* **351**, 742–744.

Wilding TJ and Huettner JE (1995). Differential antagonism of α-amino-3-hydroxy-5–4-isoxazolepropionic acid-preferring and kainate-preferring receptors by 2,3-benzodiazepines. *Mol Pharmacol* **47**, 582–587.

Wilding TJ and Huettner JE (1996). Antagonist pharmacology of kainate- and alpha-amino-3-hydroxy-5-methyl-4-isoxazolepropionic acid-preferring receptors. *Mol Pharmacol* **49**, 540–546.

Wilding TJ and Huettner JE (1997). Activation and desensitization of hippocampal kainate receptors. *J Neurosci* **17**, 2713–2721.

Wisden W and Seeburg PH (1993) A complex mosaic of high-affinity kainate receptors ion rat brain. *J Neurosci* **13**, 3582–3598.

Wong LA and Mayer ML (1993). Differential modulation by cyclothiazide and concanavalin A of desensitization at native alpha-amino-3-hydroxy-5-methyl-4-isoxazolepropionic acid- and kainate-preferring glutamate receptors. *Mol Pharmacol* **44**, 504–510.

Wong LA, Mayer ML, Jane DE *et al.* (1994). Willardiines differentiate agonist binding sites for kainate- versus AMPA-preferring glutamate receptors in DRG and hippocampal neurons. *J Neurosci* **14**, 3881–3897.

Wu LJ, Zhao MG, Toyoda H *et al.* (2005). Kainate receptor-mediated synaptic transmission in the adult anterior cingulate cortex. *J Neurophysiol* **94**, 1805–1813.

Xia J, Zhang X, Staudinger J *et al.* (1999). Clustering of AMPA receptors by the synaptic PDZ domain-containing protein PICK1. *Neuron* **22**, 179–187.

Yan S, Sanders JM, Xu J *et al.* (2004). A C-terminal determinant of GluR6 kainate receptor trafficking. *J Neurosci* **24**, 679–691.

Ziegra CJ, Willard JM and Oswald RE (1992). Coupling of a purified goldfish brain kainate receptor with a pertussis toxin-sensitive G protein. *Proc Natl Acad Sci USA* **89**, 4134–4138.

3.2.3 NMDA receptors

Alasdair Gibb

3.2.3.1 Introduction

N-methyl-D-aspartate (NMDA) receptors are one class of ionotropic glutamate receptor (Dingledine *et al.*, 1999) that were originally characterized by their sensitivity to the selective agonist NMDA (reviewed by Watkins and Jane, 2006). Among the ligand-gated ion channel receptors, NMDA receptors are unique in a number ways. They are the only receptor requiring binding of two distinct ligands, glycine (or D-serine) and glutamate (Johnson and Ascher, 1987; Kleckner and Dingledine, 1988), before receptor activation can occur leading to a slowly activating and deactivating post-synaptic current (compared to the much faster α-amino-3-hydroxy-5-methyl-4-isoxazole propionic acid [AMPA] receptors) that carries significant calcium into the cell (Mayer and Westbrook, 1987; Forsythe and Westbrook, 1988). Their essential role in most forms of long-term potentiation (LTP) (Collingridge *et al.*, 1983; Bliss and Collingridge, 1993) and their importance in excitotoxic cell death in the brain (Choi, 1992) has fuelled interest in their properties over the past 25 years. Their exquisite sensitivity to voltage-dependent block by magnesium ions (Nowak *et al.*, 1984; Mayer *et al.*, 1984) endows them with the ability to act as 'coincidence detectors' in the brain: only when glutamate release from a presynaptic terminal is followed closely by postsynaptic depolarization (hence relieving the magnesium block), will significant current carried by Na^+ and Ca^{2+} ions flow through the NMDA receptor channels (Nevian and Sakmann, 2004) leading to activation of a number of calcium-dependent signalling pathways located in the postsynaptic density (Kennedy, 2000; Kennedy *et al.*, 2005). This calcium-dependent signalling is a two-edged sword: in normal physiology it is the key event in initiating changes in synaptic strength (Malenka, 1991; Malenka and Nicoll, 1993; Bliss and Collingridge, 1993; Collingridge, 2003; Lisman, 2003) that likely underlie the ability of the brain to learn from experience, while in pathologies such as stroke and anoxia, excessive NMDA receptor activation kills neurons (reviewed by Szatkowski and Attwell, 1994; Arundine and Tymianski, 2004). NMDA receptors have therefore been the focus of intense investigation over the past 25 years and this chapter aims to summarize some of the key results from this research effort.

3.2.3.2 Molecular characterization of NMDA receptors

NMDA receptors are hetero-oligomeric membrane proteins with each receptor composed of four subunits. Three subunit families have been identified: NR1, NR2 (NR2A, 2B, 2C, 2D) and NR3 (NR3A, 3B) (Table 3.2.3.1). The most obvious structural differences between these subunits lies in their C-termini, with NR2A and NR2B having the longest while NR1 and NR3B have the shortest. Each subunit has the transmembrane topology (Figure 3.2.3.1) of an extracellular N-terminus, three membrane-spanning regions (M1, M3 and M4) and a long intracellular C-terminus. The M2 region forms a re-entrant loop from the inside of the membrane that lines the major part of the channel and determines the key channel properties of Ca^{2+} permeability and Mg^{2+} block (Burnashev *et al.*, 1992; Wollmouth *et al.*, 1998).

The N-terminal domain is an important site of drug action and of tonic inhibition by protons (NR1) or zinc (NR2A) (Paoletti *et al.*, 2000; Erreger and Traynelis, 2005). The agonist binding site is formed from a 'clam-shell' arrangement of the S1 region that is N-terminal to M1 and the S2 region formed by the M3–M4 extracellular loop (Furukawa *et al.*, 2005, reviewed by Chen and Wyllie, 2006). The intracellular C-terminal domain is a region rich in regulatory features containing serine/threonine and tyrosine kinase phosphorylation sites (Swope *et al.*, 1999), and binding sites for calmodulin (Ehlers *et al.*, 1996), alpha-actinin and yotiao (Wyszynski *et al.*, 1997; Lin *et al.*, 1998) and PSD-95 family proteins (Kornau *et al.*, 1995; Wenthold *et al.*, 2003; Chung *et al.*, 2004; Kim and Sheng, 2004).

Table 3.2.3.1

Receptor subunit	Biophysical characteristics	Cellular and subcellular localization	Pharmacology	Physiology	Relevance to disease states
NR1 (GluN1*) [1, 2, 3, 4, 5, 6] Gene: *GRIN1* Chromosome location: mouse: 2 12.0 cM rat: 3p13 human: 9q34.3 22 exons, ~31kb, alternative splicing of exons 5, 21 and 22, creates 8 splice variants [7–9]	Obligatory subunit to form functional NMDA receptors with NR2 subunits [2, 10] and/or NR3 subunits [11, 12] providing glycine binding site [13–16] Splice variants determine spermine [17], proton and calmodulin sensitivity [18, 19], regulate ER retention of NR1 [20, 21] and modulate NMDA receptor kinetic properties [17–19, 22]	Expressed throughout CNS. Brain (high in cortex, striatum, hippocampus, olfactory bulb, cerebellum, brainstem, spinal cord) [1, 2, 23–26] Spinal cord neurons [23, 24] Trigeminal, vagal and dorsal root ganglia (small and medium nociceptive sensory neurons) [27–29] Taste receptor cells [30] Retina [31] Cochlea and vestibular ganglia [32] Bone [33, 34]	For NR1 S1S2 fusion protein [16, 35] K_i: glycine 26 µM K_i: D-serine 7 µM K_i: D-cycloserine 241 µM K_i: 5,7-DCK 0.5 µM K_i: CNQX 6.3 µM	Essential in order to form functional NMDA receptors Mice lacking ζ (NR1) die soon after birth showing impairment of suckling response [36] and disrupted whisker-related patterning in the trigeminal nucleus [37]. Mice with ζ (NR1) mutation N598R giving calcium-impermeable and Mg²⁺-insensitive NMDA receptors fail to develop normal trigeminal patterning and die soon after birth [38, 39]	Upregulated phosphorylation in rat dorsal horn neurons in chronic pain [40–42] Neuronal death associated with brain ischaemia [43] Potential role in schizophrenia [44]
NR2A (GluN2A*) [2, 5, 10] Gene: *GRIN2A* Chromosome location: mouse: 16 3.4 cM rat: 10q11 human: 16p13.2 14 exons	$P_{Ca}/P_{Na} = 4$ [45] g_{Na+} (1 mM Ca) 50pS /40pS [46] τ_{act} = 13 ms [23, 47] τ_{deact} = 40–120 ms [17, 23, 47, 48] With NR1–1a isoform [26–30]	Brain (high in cortex, striatum, hippocampus, olfactory bulb, brainstem, spinal cord) [2, 10, 23, 49, 50] Cerebellar granule cells (>P10) [26] Spinal cord neurons [24, 51] Dorsal root ganglia (not expressed) [28] Retina [31] Cochlea and vestibular ganglia [32]	With NR1–1a EC_{50}: glutamate 2–4 µM [52, 53] EC_{50}: glycine 2 µM [52] EC_{50}: NMDA 9 µM [54] K_i: APS 0.3 uM [55], K_B 1–2 µM [53, 54] IC_{50}: Zn²⁺ 6 nM [56] IC_{50}: (±)ketamine 3 µM [57] IC_{50}: (+)MK-801 7–15 nM [57, 58] IC_{50}: memantine 4 µM [57] K_B: NVP-AAM077 15 nM [59]	Mice lacking the e1 (NR2A) subunit show apparently normal growth and mating behaviour, but an enhanced startle response, deficiencies in social interaction, and reduced hippocampal long-term potentiation (LTP) and spatial learning [60]. Gene deletion of e1 (NR2A) prevents developmental speeding of NMDA EPSCs in cerebellar granule cells [61]	Upregulated by inflammation in rat DRG [62] Potential role in schizophrenia: polymorphisms [63]
NR2B (GluN2B*) [2, 10, 54] Gene: *GRIN2B* Chromosome location: mouse: 6 64.5 cM rat: 4q43 human: 12p12 14 exons Splice variant	$P_{Ca}/P_{Na} = 4$ [45, 64] g_{Na+} (1 mM Ca) 50pS /40pS [46] τ_{act} 14 ms [65] τ_{deact} 180–400 ms (with NR1–1a) [20, 29, 30, 32] With NR1–1a isoform [17, 45, 46, 48, 65]	Brain (high in cortex, striatum, hippocampus, olfactory bulb, brainstem, spinal cord) [2, 10, 23, 25, 50] Cerebellar granule cells (<P10) [23, 26] Spinal cord neurons [24] Retina [31] Cochlea and vestibular ganglia [32] Trigeminal ganglion and dorsal root ganglion [28, 29] Heart [66]	With NR1–1a EC_{50}: glutamate 1–4 µM [52, 65, 67] EC_{50}: glycine 0.3 – 0.5 µM [52, 67] EC_{50}: NMDA 7 µM [54]. K_i: APS 0.5 µM [55], K_B 4 µM [54] IC_{50}: Zn²⁺ 490 nM [56] IC_{50}: (±)ketamine 0.9 µM [57] IC_{50}: (+)MK-801 9 nM [57] IC_{50}: memantine 12 µM [57] Potentiated by Mg²⁺ (pH 7.2) K_{Mg} 2 mM [68] K_B:NVP-AAM07778nM [59] NR2B selective ligands: IC_{50}: ifenprodil, 167–340 nM [69, 70] IC_{50}: Ro 25–6981, [71] (with NR1–2a) 9nM IC_{50}: CP-101,606, 13–62 nM [70]	Mice lacking the e2 (NR2B) subunit die soon after birth showing impairment of suckling response, impairment of trigeminal neuronal pattern formation and impaired hippocampal long-term depression (LTD) [72] Overexpression of NR2B enhances learning and memory [73] and affects behavioural responses to inflammatory pain [62]	Huntington's disease: enhanced NR2B receptor currents [74] Upregulated in forebrain following peripheral inflammation [75] Altered expression following methamphetamine administration [76] Potential role in schizophrenia: polymorphisms [77]

NR2C (GluN2C*) [2, 10, 78] Gene: GRIN2C Chromosome location: mouse: 11 78.0 cM rat: 10q32.2 human: 17q25 14 exons 4 splice variants [79]	P_{Ca}/P_{Na} = 2.5 [64] g_{Na+} (1 mM Ca) 35pS /18pS [46] τ_{act} 8–14 ms [2, 79] τ_{deact} 280–380 ms [23, 48] With NR1–1a isoform [46–48, 79]	Brain (high in cerebellum, found in olfactory bulb, subthalamic nucleus, dentate granule cells, hippocampal formation, habenula, basolateral amygdaloid nuclei, corpus callosum, brainstem and cerebellar granule cells (>P10) [2, 20, 39, 68, 69] Spinal cord neurons [37] Retina [40] Cochlea and vestibular ganglia [47] Heart, skeletal muscle, pancreas [68]	With NR1–1a EC_{50}: glutamate 0.7 μM [52] EC_{50}: glycine 0.2 – 0.6 μM [52, 78] EC_{50}: NMDA 22 μM [78] K_i: AP5 = 1.6μM [55] IC_{50}: Zn^{2+} 14 μM [56] IC_{50}: (±)ketamine 1.7 μM [57] IC_{50}: (+)MK-801 = 24 nM [57] IC_{50}: memantine 0.6 μM [57]	Mice lacking the e3 (NR2C) subunit show apparently normal development and behaviour [81]. Knockout of both e1 (NR2A) and e3 (NR2C) results in loss of spontaneous and evoked EPSCs in cerebellar granule cells, and disturbance of motor coordination [82]
NR2D (GluN2D*) [10, 23] Gene: GRIN2D 7 23.5 cM Chromosome location: mouse: 7 23.5 cM rat: 1q22 human: 19q13.1–qter Splice variants	g_{Na+} (1 mM Ca) 36pS /18pS [47, 83] τ_{act} 50 ms [47] τ_{deact} 2000–4000 ms [23, 47, 48] With NR1–1a isoform [47, 48]	Brain (high in midbrain, embryonic and neonatal hippocampal formation, cerebellum and brainstem) [10, 23, 50]. Cerebellar Purkinje cells (<P7) [23, 26] Spinal cord neurons [24] Dorsal root ganglia [28] Cochlea and vestibular ganglia [32] Bone [34]	With NR1–1a EC_{50}: glycine 0.08 μM [84, 85] EC_{50}: glutamate 0.5 μM [84, 85] EC_{50}: NMDA 3.7 μM [84] EC_{50}: aspartate 3.3 μM [84] K_i: AP5 3.7 μM [55] IC_{50}: (±)ketamine 2.4 μM [57] IC_{50}: (+)MK-801 38 nM [57] IC_{50}: memantine 0.82 μM [57]	Mice lacking the e4 (NR2D) subunit show apparently normal growth and mating behaviour, but reduced spontaneous behavioural activity [86], lower sensitivity to stress and altered monoaminergic neuronal function [87]
NR3A (GluN3A*) [11, 88–90] Gene: GRIN3A Chromosome location: mouse: 4 B1 rat:4q11 human: 9q31.1 9 EXONS Two splice variants in rat [91, 92] but not in human [90]	P_{Ca}/P_{Na} = 0.6 (with NR1–1a and NR2A) [12] g_{Na+} (1 mM Ca) (with NR1–1a and NR2A) 28pS [12] Forms a glycine-activated cation channel when expressed only with NR1 in oocytes [93]	Brain (cortex, striatum, hippocampus, olfactory bulb, brainstem, spinal cord) [88, 89] Retina [94]	With NR1–1a EC_{50}: glycine 5–7 μM [93, 95] NR3A S1S2 fusion protein forms a high affinity binding site for glycine [35] K_i: glycine 40 nM K_i: D-serine 643 nM K_i: D-cycloserine 277 μM K_i: 5,7-DCK 647 μM K_i: CNQX 2.5 μM	Mice lacking NR3A have increased spine density in cortex and enhanced NMDA responses [11] No obvious developmental or behavioural effects in knockout mice [11]

(Continued)

Table 3.2.3.1 (continued)

Receptor subunit	Biophysical characteristics	Cellular and subcellular localization	Pharmacology	Physiology	Relevance to disease states
NR3B (GluN3B*) [90, 95, 96] Gene: GRIN3B Chromosome location: mouse: 10 C1 rat: 7q11 human: 19p13.3	$P_{Ca}/P_{Na} = 2$ (with NR1–1a and NR2A) [97, 98] g_{Na+} (1 mM Ca) (with NR1–1a and NR2A) 28pS [12] g_{Na+} (with NR1–1a 37pS/12pS) [93] Forms a glycine-activated cation channel when expressed with NR1 [93]	Brain (particularly hippocampus, cerebellum, midbrain, medulla, corpus collosum) and spinal cord motor neurons [90, 96–99]	With NR1–1a and NR2B EC$_{50}$: glutamate 1.6 μM [100] EC$_{50}$: glycine 0.18 μM [100],6 μM [95] No effect on Mg^{2+} sensitivity IC$_{50}$: ketamine 1.8 μM [100]	Functional role not yet established	

*The classification of glutamate receptor subunits has recently been re-addressed by NC-IUPHAR and the revised recommended nomenclature is shown here. See Alexander SPH, Mathie A, and Peters JA (2008). Guide to receptors and channels (GRAC), 3rd edn. Br J Pharmacol 153, S1–S209.

1 Moriyoshi K et al. (1991) Nature 354, 31–37; 2 Monyer H et al. (1992) Science 256, 1217–1221; 3 Karp SJ et al. (1993) J Biol Chem 268, 3728–3733; 4 Planells-Cases R et al. (1993) Proc Natl Acad Sci USA 90, 5057–5061; 5 Le Bourdelles B et al. (1994) J Neurochem 62, 2091–2098; 6 Zimmer M et al. (1995) Gene 159, 219–223; 7 Sugihara H et al. (1992) Biochem Biophys Res Commun 185, 826–832; 8 Durand GM et al. (1992) Proc Natl Acad Sci USA 89, 9359 –9363; 9 Nakanishi N et al. (1992) Neuron 12, 1291–1300; 14 Wafford K et al. (1995) Mol Pharmacol 47, 374 –380; 15 Laube B et al. (1998) J Neurosci 18, 2954–2961; 16 Furukawa H and Gouaux E (2003) EMBO J 22, 1–13; 17 Rumbaugh G et al. (2000) J Neurophysiol 83, 1300–1306; 18 Ehlers MD et al. (1996) Cell 84, 745–755; 19 Traynelis SF et al. (1995) Science 268, 873–876; 20 Standley S et al. (2000) Neuron 28, 887–89; 21 Pérez-Otaño I and Ehlers MD (2005) Trends Neurosci 28, 229–238; 22 Rycroft BK and Gibb AJ (2002) J Neurosci 22, 8860–8868; 23 Monyer H et al. (1994) Neuron 12, 529–540; 24 Watanabe M et al. (1994) J Comp Neurol 343, 513–519; 27 Watanabe M et al. (1994) Neurosci Lett 165, 183–186; 28 Marvizon JC et al. (2002) J Comp Neurol 446, 325–341; 29 Li B et al. (2002) Nat Neurosci 5, 833–834; 30 Caicedo A et al. (2004) Chem Senses 29, 463–471; 31 Watanabe M et al. (1994) Brain Res 634, 328–332; 32 Niedzielski AS and Wenthold RJ (1995) J Neurosci 15, 2338–2353; 33 Chenu C et al. (1998) Bone 22, 295–299; 34 Patton AJ et al. (1998) Bone 22, 645–649; 35 Yao Y and Mayer ML (2006) J Neurosci 26, 4559–4566. 36 Forrest D et al. (1994) Cell 76, 427–437; 38 Single FN et al. (2000) J Neurosci 20, 2558–2566; 39 Rudhard Y et al. (2003) J Neurosci 23, 2323–2332; 40 Gao X et al. (2005) Pain 116, 62–72; 41 Caudle RM et al. (2005) Mol Pain 1, 25; 42 Kim HW et al. (2006) Br J Pharmacol 148, 490–498; 43 Arundine M and Tymianski M (2004) Cell Mol Life Sci 61, 657–668; 44 Javitt DC (2007) Int Rev Neurobiol 78, 69–108; 45 Schneggenburger R (1996) Biophys J 70, 2165–2174; 46 Stern P et al. (1992) Proc Biol Sci 250, 271–277; 47 Wyllie DJ et al. (1998) J Physiol 510, 1–18; 48 Vicini S et al. (1998) J Neurophysiol 79, 555–566; 49 Goebel DJ and Poosch MS (1999) Molecular Brain Research 69, 164–170; 50 Paarmann I et al. (2000) J Neurochem 74, 1335–1345; 51 Tolle TR et al. (1993) J Neurosci 13, 5009–5028; 52 Kutsuwada T et al. (1992) Nature 358, 36–41; 53 Anson LC et al. (1998) J Neurosci 18, 581–589; 54 Priestley T et al. (1995) Mol Pharmacol 48, 841–848; 55 Morley RM et al. (2005) J Med Chem 48, 2627–2637; 56 Paoletti P et al. (1997) Neurosci 17, 5711–5725; 57 Dravid SM et al. (2007) J Physiol 581, 107–128; 58 Chazot PL et al. (1994) Biol Chem 269, 24403–24409; 59 Frizelle PA et al. (2006) Mol Pharmacol 70, 1022–1032; 60 Sakimura K et al. (1995) Nature 373, 151–155; 61 Fu Z et al. (2005) Physiol 563, 867–881; 62 Wei F et al. (2001) Nat Neurosci 4, 164–169; 63 Itokawa M et al. (2003) Pharmacogenetics 13, 271–278; 64 Burnashev N et al. (1995) J Physiol 485, 403–418; 65 Banke TJ and Traynelis SF (2003) Nat Neurosci 6, 144–145; 66 Seeber S et al. (2004) J Biol Chem 279, 21062–21068; 67 Laube B et al. (1997) Neuron 18, 493–503; 68 Paoletti P et al. (1995) Neuron 15, 1109–1120; 69 Williams K (1993) Mol Pharmacol 44, 851–859; 70 Mott DD et al. (1998) Nat Neurosci 1, 659–667; 71 Fischer G et al. (1997) J Pharmacol Exp Ther 283, 1285–1292; 72 Kutsuwada T et al. (1996) Neuron 16, 333–344; 73 Tang YP et al. (1999) Nature 401, 63–69; 74 Li L et al. (2004) J Neurophysiol 92, 2738–2746; 75 Wu LJ et al. (2005) J Neurosci 25, 11107–11116; 76 Yamamoto H et al. (2006) Ann N Y Acad Sci 1074, 97–103; 77 Martucci L et al. (2006) Schizophr Res 84, 214–221; 78 Daggett LP et al. (1998) J Neurochem 71, 1953–1968; 79 Chen BS et al. (2006) J Biol Chem 281, 16583–16590; 80 Lin YJ et al. (1996) Brain Res Mol Brain Res 43, 57–64; 81 Ebralidze AK et al. (1996) J Neurosci 16, 5014–5025; 82 Kadotani H et al. (1996) J Neurosci 16, 7859–7867; 83 Wyllie DJ et al. (1996) Proc Biol Sci 263, 1079–1086; 84 Chen PE et al. (2004) J Physiol 558, 45–58; 85 Ikeda K et al. (1992) FEBS Lett 313, 34–38; 86 Ikeda K et al. (1995) Brain Res Mol Brain Res 33, 61–71; 87 Miyamoto Y et al. (2002) J Neurosci 22, 2335–2342; 88 Sucher NJ et al. (1995) J Neurosci 15, 6509–6520; 89 Ciabarra AM et al. (1995) J Neurosci 15, 6498–6508; 90 Andersson O et al. (2001) Genomics 78, 178–184; 91 Sun L et al. (1998) FEBS Lett 441, 392–396; 92 Pérez-Otaño I et al. (1998) J Neurosci 21, RC185; 97 Matsuda K et al. (2002) Brain Res Mol Brain Res 100, 43–52; 98 Matsuda K et al. (2003) J Neurosci 23, 10064–10073; 99 Fukaya M et al. (2005) Eur J Neurosci 21, 1432–1436; 100 Yamakura T et al. (2005) Anesth Analg 100, 1687–1692.

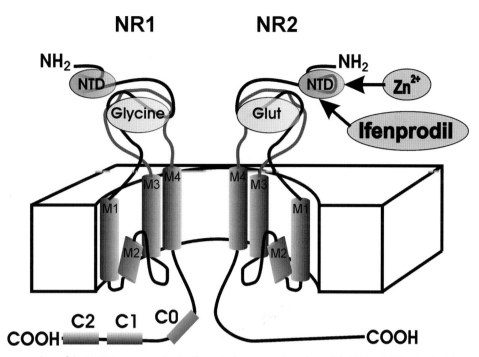

Fig. 3.2.3.1 Transmembrane topology of the NMDA receptor subunits. Shown is the proposed topology of the NR1 and NR2 subunits. It is assumed that NR3 subunits will follow a similar architecture. Each subunit has an intracellular C-terminus and an extracellular N-terminus. The N-terminal domains (NTD) form the binding site for polyamines in NR1 and for modulatory ligands such as Zn^{2+} (NR2A) and ifenprodil (NR2B). Conserved regions in the protein between the NTD and M1 and in the extracellular loop between M3 and M4 form the ligand-binding sites (a glycine site for NR1 and NR3, a glutamate site for NR2). Three transmembrane domains, M1, M3, M4, and the pore-loop formed by M2 form the selectivity filter of the channel and the outer and inner channel vestibules. The C-terminal region of NR1 and NR2 subunits bind both kinases and phosphatases, and structural proteins such as PSD-95 and alpha-actinin. The NR1 subunit C-terminal region contains two alternatively spliced domains, C1 and C2, and an invariant domain, C0, that are particularly important targets for calmodulin binding (C0 and C1) and PKA and PKC phosphorylation (C1) while PDZ-containing proteins can bind to C2. The C-terminus of NR2 subunits contains sites for AP-2 binding as well as tyrosine kinase phosphorylation sites.

Genes and splice variants

The NR1 subunit is the product of a single gene (*GRIN1*) containing 22 exons with the possibility of alternative splicing of exons 5 (the N1 cassette), 21 and 22 (the C1 and C2 cassettes) where N1 and C1 can be present or absent and C2 can alternate with C2′, resulting in eight different splice variants: NR1–1a (presence of C1 and C2), NR1–1b (N1, C1, C2), NR1–2a (C2), NR1–2b (N1C2), NR1–3a (C1C2′), NR1–3b (N1C1C2′), NR1–4a (C2′), and NR1–4b (N1C2′). NR1–1a and NR1–2a are most widely distributed in the brain (Laurie and Seeburg, 1994) but in some areas, expression is developmentally regulated. In the cerebellum, NR1–1b increases with age, becoming the predominant form in adult granule cells (Prybylowski *et al.*, 2000) while NR1–1a is widely distributed at all postnatal ages. It should be noted that the presence of exon 5 in NR1–1b increases the deactivation rate approximately threefold for NR2B receptors (Rumbaugh *et al.*, 2000) and so is expected to contribute to speeding of the synaptic current decay with development as well as affecting the proton sensitivity of the receptor. The presence or absence of C1, splices in or out one of the calmodulin binding sites (Ehlers *et al.*, 1996) that regulate the NMDA receptor channel open time and open probability (Ehlers *et al.*, 1996; Rycroft and Gibb, 2002) and the interaction with alpha-actinin (Wyszynski *et al.*, 1997; Rycroft and Gibb, 2004b), yotiao, PKC or PKA, protein phosphatase 1 (PP1) or calcineurin (PP2a) providing some of the mechanisms for regulating the trafficking and activity of NMDA receptors (Wyszynski *et al.*, 1997; Lin *et al.*, 1998; Wenthold

et al., 2003; Rycroft and Gibb, 2004a; Pérez-Otaño and Ehlers, 2005). The presence of N1 (exon 5) prevents the normal proton inhibition observed with the most common NR1 subunit in the brain (NR1–1a) and removes potentiation by extracellular polyamines (Traynelis *et al.*, 1995). Apart from NR2A (Table 3.2.3.1) each of the NR2 and NR3 subunits have possible splice variants, although the functional significance of these is not well defined. Four isoforms of the human NR2C subunit have been identified (Daggett *et al.*, 1998) with 89 per cent sequence identity to the rat NR2C. In functional studies human NR2C-1 and NR2C-2 expressed together with NR1 gave agonist responses while NR2C-3 and NR2C-4 did not.

Subunit composition

The NMDAR subunits (NR1, NR2A, NR2B, NR2C, NR2D, NR3A, and NR3B) form functional NMDARs that are found in neurons and glia throughout the brain and spinal cord (Hollmann and Heinemann, 1994; Dingledine *et al.*, 1999; Cull-Candy *et al.*, 2001; Wenthold *et al.*, 2003; Karadottir *et al.*, 2005; Salter and Fern, 2005). There are intriguing and in some cases critically important variations in spatial and temporal subunit expression patterns that give some insights into the functional importance of NMDA receptor properties in the brain. *In situ* hybridization and protein immuno-histochemistry have shown that NR2 subunit expression shows dramatic spatial and temporal changes in expression during development that are likely to be important to the maturation of neural networks in the mature brain. The receptor subunit composition is only known with confidence for relatively few synapses in the brain (reviewed by Cull-Candy and Lieskiewicz, 2004). The subunit composition of recently discovered NMDA receptors in oligodendrocytes (Karadotir et al., 2005; Salter and Fern, 2005) and astrocytes (Lalo et al., 2006) is as yet unclear, although their pharmacology and low Mg^{2+} sensitivity suggest they are likely to be triheteromers containing both NR2C or 2D and NR3A or 3B subunits.

The relatively simple architecture and accessibility of the cerebellum has meant that considerable progress has been made towards identifying the subunit composition of receptors expressed at different developmental stages in both principal and interneurons, mainly by a combination of careful biophysical measurement and receptor pharmacology (reviewed by Cull-Candy et al., 2001; Cull-Candy and Leskiewicz, 2004). In cerebellar granule cells, NR2B subunit mRNA and protein expression begin before birth and then decrease during the second postnatal week. Meanwhile, NR2A and NR2C subunit expression begins a few days after birth (Watanabe et al., 1994c) during migration of granule cells to the internal granular layer (Farrant et al., 1994; Wang et al., 1995; Takahashi et al., 1996). Recombinant NR1/NR2B and NR1/NR2C subunits have slower deactivation kinetics than NR1/NR2A receptors (Monyer et al., 1994; Vicini et al., 1998) and these data correlate with the faster decay of the NMDA EPSC decay after the second postnatal week (Cathala et al., 2000; Rumbaugh and Vicini, 1999) followed by a slowing from P21 as NR2C subunit expression increases (Cathala et al., 2000). In granule cells from NR2A knockout mice, these changes in NMDA EPSC time course are not observed (Fu et al., 2005). Thus, it seems likely that the mature receptors are triheteromeric NR1/NR2A/NR2C receptors.

In contrast, cerebellar golgi cells express NR2B and NR2D, and interestingly, the NR1/NR2B/NR2D-containing triheteromeric receptors seem to be located only extra-synaptically (Brickley et al., 2003; Cull-Candy and Leskiewicz, 2004). Purkinje cells, on the other hand, are unusual in being one of the few neurons in the brain expressing almost no NMDA receptors in the adult while expressing NR2D receptors during the first postnatal week (Momiyama et al., 1996; Cull-Candy and Lieskiewicz, 2004).

In hippocampus, cortex and striatum electrophysiological recordings have been combined with the use of NR2B subunit selective antagonists such as ifenprodil and immunostaining with subunit-selective antibodies to show that early in development there are mainly NR2B and NR2D-containing receptors, which then change to a mixture of NR2A and NR2B-containing receptors around the first to second postnatal weeks in rodents. There is the possibility that many of these receptors may be triheteromeric NR1/NR2A/NR2B receptors (Sheng et al., 1994; Chazot and Stephenson, 1997) that have a sensitivity to ifenprodil similar to NR2B diheteromers but decreased maximum effect reflected in a slower drug onset and faster offset (Tovar and Westbrook, 1999; Hatton and Paoletti, 2005) compared to the NR2B diheteromer. Because ifenprodil acts by enhancing the H$^+$ sensitivity of the receptor (Mott et al., 1998), the use of ifenprodil to investigate changes in native receptor expression during development could be complicated in some regions by concomitant developmental changes in the expression of the NR1 subunit exon 5.

In midbrain dopamine neurons there is both biochemical (Dunah et al., 1998) and electrophysiological evidence (Jones

and Gibb, 2005) for expression of NR1/NR2B/NR2D triheteromeric receptors at extra-synaptic sites, but the identity of synaptic NMDA receptors is not yet known. These data contrast with the thalamus where there is expression of both NR2B and NR2D subunits but no evidence from co-immunoprecipitation experiments to suggest the presence of NR1/NR2B/NR2D triheteromers (Dunah et al., 1998). In the midbrain, brainstem and spinal cord, NR2B and NR2D expression continues into adulthood with varying levels of NR2A and NR2C (Akazawa et al., 1994; Monyer et al., 1994). There is thus considerable potential for increased receptor functional diversity if these subunits come together in triheteromeric receptors, but in many instances subunit expression at the single cell level is not yet known and a lack of good antibodies directed at the extracellular N-terminus of subunits, or subunit-selective antagonists (other than NR2B-selective drugs such as ifenprodil) (Williams, 1993) has severely hampered identification of receptor subtypes at the single cell level.

The NR2 and NR3 subunits are all products of distinct genes (Grin2a–2d, Grin3a, Grin3b). The NR2 subunits provide the main source of functional diversity among NMDA receptors, while NR3 subunits seem to have rather more specialized roles giving a reduced channel calcium permeability when co-expressed with NR1 and NR2A (Pérez-Otaño et al., 2001; Matsuda et al., 2002; 2003) and additional PACSIN1-dependent regulation of receptor trafficking (Pérez-Otaño et al., 2006). NR3 subunits can also assemble with NR1 subunits to form glycine-activated NR1/NR3 excitatory receptors (Chatterton et al., 2002) that have decreased calcium permeability and magnesium sensitivity compared to NR2 subunit-containing receptors. A physiological role for these receptors is not yet defined.

Subunit stoichiometry

The precise subunit stoichiometry is not known for most native NMDA receptors, but based on studies using heterologous expression systems most receptors are likely to be composed of two glycine-binding NR1 subunits (Behe et al., 1995; Laube et al., 1998) and two glutamate-binding NR2 subunits (Laube et al.,1997; Anson et al., 1998; Ulbrich and Isacoff, 2007) arranged around a central ion channel. Using crystals of the fused extracellular domains of NR1 and NR2A, Furukawa et al. (2005) have suggested that the NMDA receptor is assembled from two NR1/NR2 heterodimers while expression studies using NR1 and NR2A subunits spliced together (Schorge and Colquhoun, 2003), suggest that the receptor is assembled as a pair of NR1/NR2 and NR2/NR1 heterodimers so that the two NR1 subunits are |adjacent and facing a pair of NR2 subunits.

The stoichiometry of receptors containing one or more glycine-binding NR3 subunits (Yao and Mayer, 2006) in combination with NR1 and NR2 subunits is uncertain (Pérez-Otaño et al., 2001). It is likely that the NR3 subunits take the place of NR2 in assembly of NR1/NR3 receptors (Ulbrich and Isacoff, 2007) and hence by analogy, the stoichiometry of the NR1/NR2A/NR3A triheteromer (Pérez-Otaño et al., 2001) is likely to be two copies of NR1 and one copy of each of NR2A and NR3A. This means the NR1/NR2A/NR3A triheteromer likely has reduced maximum inhibition by zinc (Hatton and Paoletti, 2005) and assuming that the two NR1 subunits are adjacent to each other (Furukawa et al., 2005) this means the NR2A subunit has a different environment compared to that of an NR2A diheteromer. Despite these likely structural differences, the single channel currents of NR1/NR2A/NR3A triheteromeric receptors have similar kinetic behaviour to those of NR1/NR2A diheteromers (Pérez-Otaño et al., 2001). The glycine-activated NR1/NR3 receptors (Chattereton et al., 2002; Madry et al., 2007) have been shown using GFP-tagged bleaching to have a tetrameric 'two plus two' stoichiometry (Ulbrich and Isacoff, 2007). By analogy, NR1/NR2B/NR2D triheteromeric receptors are likely have a stoichiometry of $NR1_2$, NR2B, NR2D.

3.2.3.3 Biophysical characteristics

The key biophysical properties of NMDA receptors are their slow activation and deactivation kinetics, which underlie the slow NMDA receptor-mediated synaptic current, their permeability to calcium and voltage-dependent block by magnesium.

Synaptic current time course

At excitatory synapses throughout the nervous system, NMDA receptors mediate the slow component of synaptic transmission which at room temperature rises slowly to peak about 20 ms after initiation and then decays with a time constant of about 100 ms. Under normal conditions, the slow time course is not dependent on glutamate diffusion, but is determined by the kinetic properties of the receptor. Elegant evidence for this comes from 'concentration jump' experiments (Lester et al., 1990) at synapses in hippocampal neuron cultures and on outside-out patches. At synapses in culture, rapid application of the competitive antagonist AP5 during the synaptic current fails to block the current demonstrating that glutamate is bound to the receptor, throughout the synaptic current (Lester et al., 1990). When brief (1 ms duration) applications of glutamate are made to outside-out patches, the resulting current was found to have the same time course as the synaptic current, indicating that the receptor kinetics determine the synaptic current time course.

The synaptic current time course depends on the receptor subunit composition and in the cortex is slower (decay time constant ~250 ms) early in development (Carmignoto and Vicini, 1992; Hestrin, 1992) when NR2B subunits predominate, and becomes faster (decay time constant ~80 ms) as NR2A expression rises (Flint et al., 1997; Lu et al., 2001), around the same time as the end of the critical period for synaptic plasticity in somatosensory cortex. However, it should be noted that the critical period ends normally in NR2A knockout mice so that it seems the change in NMDA receptor kinetics associated with NR2A expression is not the determining factor in closing the critical period (Lu et al., 2001). In brain regions such as hippocampus, somatosensory (Flint et al., 1997; Lu et al., 2001) and visual cortex (Nase et al., 1999), NR2B subunits are expressed early in development, with a rise in NR2A subunit expression occurring in rodents around 1–2 weeks of age. In midbrain, brainstem and spinal cord, developmental changes in expression patterns are less clear, perhaps in part reflecting the very diverse nature of the neuronal populations in these areas. In cerebellar granule cells, the switch from slow NR2B receptors (decay ~250 ms) to a mixture of NR2A and NR2C (Rumbaugh and Vicini, 1999; Lu et al., 2006) gives, interestingly, a deactivation (~80 ms) not significantly different to NR2A alone in NR2C knockout mice (Lu et al., 2006). Synaptic currents with the very slow deactivation expected of diheteromeric

NR2D receptors (Monyer et al., 1994; Wyllie et al., 1998) have not been observed (see e.g. Misra et al., 2000).

NMDA receptor desensitization

Desensitization of NMDA receptors has been found to involve more than one mechanism and depends on receptor subunit composition, with desensitization being evident on a timescale of seconds for NR2A or NR2B receptors and not apparent for NR2C and NR2D (Krupp et al. 1998; Wyllie et al., 1998). Two distinct forms of desensitization have been described. In one form negative cooperativity between the affinity of glutamate and glycine binding sites results in a weakening of glycine affinity upon glutamate binding such that at subsaturating concentrations of glycine, the response to prolonged glutamate application wanes along a time course of hundreds of milliseconds dictated by the unbinding rate for glycine (Mayer et al., 1989; Vycklicky et al., 1990; Nahum-Levy et al., 2001). Whether this form of desensitization is significant during synaptic transmission may depend on the local activity of glycine transporters (Berger et al., 1998). A fast component of desensitization of NR2A receptors is due to a change in proton inhibition as a result of an increased Zn^{2+} affinity following glutamate binding and so is not observed in the presence of Zn^{2+} chelators such as TPEN or EDTA (Erreger and Traynelis, 2005).

A second form of desensitization, calcium-dependent inactivation, depends on calcium influx through the NMDA channel (Mayer et al., 1989) initiating activation of calmodulin (Ehlers et al., 1996; Krupp et al., 1999) and other calcium-sensitive proteins and subsequent inhibition of receptor function, perhaps by dissociation of the linkage between receptor and cytoskeleton (Rosenmund and Westbrook, 1993; Legendre et al., 1993; Wyszynski et al., 1997; Krupp et al., 1999) that may increase receptor internalization. This form of desensitization occurs with a time constant of 1–2 seconds and contributes to synaptic depression of the NMDA EPSC (Tong et al., 1995). For reasons not yet understood, glycine-sensitive desensitization is not observed when receptors are isolated in membrane patches where a glycine-insensitive desensitization is observed (Sather et al., 1990; Lester and Jahr, 1992). Thus the rate and extent of NMDA desensitization at any particular synapse will depend not only on subunit composition but on cellular factors such as intracellular calcium buffering and hence, because the NMDA receptor channel is calcium permeable, will be influenced by receptor activity.

Calcium permeability and magnesium block of NMDA receptors

The steep voltage-dependence of Mg^{2+} block of the channel underlies the crucial role that it plays in imparting NMDA receptors with the property of 'coincidence detectors' in the nervous system. This property probably underlies the Hebbian behaviour of excitatory synapses in the brain and can, in principle, allow networks of neurons to adapt their behaviour according to experience, in effect, allowing the nervous system to 'learn' from experience. Typically, around 10 per cent of the current through the NMDA receptor channel is carried by Ca^{2+} ions in 1 mM external calcium (Schneggenburger, 1996) while Mg^{2+} block is so effective that around resting potential for most neurons (~ −60 mV) more than

90 per cent of channels are blocked. This steep voltage-dependent Mg^{2+} block and Ca^{2+} permeability meant that early studies of synaptic potentials often could not detect an NMDA receptor-mediated synaptic potential in response to single low-frequency stimulation. Subsequently it was demonstrated clearly (Dale and Roberts, 1985; Herron et al., 1986) that there was an NMDA EPSP and that this summated dramatically during repetitive stimulation as each successive response in a train of stimuli briefly relieved the Mg^{2+} block in precisely the way expected for the role of NMDA receptors in frequency-dependent synaptic plasticity (Herron et al., 1986; Collingridge et al., 1988).

Selectivity of the channel for Mg^{2+} block and Ca^{2+} permeability depends on an asparagine residue located in the apex of the M2 segment (Figure 3.2.3.1) of the NR1 subunit (position 598 for NR1) and the second of a pair of asparagines in the NR2 subunit (596 for NR2A) (Wollmuth et al., 1998). These positions are equivalent to the Q/R (glutamine-arginine) site of AMPA and kainate receptors. Structural predictions suggest this asparagine residue should be located about halfway through the channel (Wollmuth et al., 1998), but the steep voltage-dependence of Mg^{2+} block infers that the Mg^{2+} ion senses about 90 per cent of the membrane electric field at its binding site. This apparent paradox was resolved with the discovery by Antonov and Johnson (1999) of the effect of occupancy of permeant ion binding sites in the channel on the voltage-dependence of Mg^{2+} block. Antonov and Johnson (1999) demonstrated that taking this effect into account places the Mg^{2+} ion binding site at a much shallower position (~50 per cent) in the membrane electric field, consistent with structural predictions made from cysteine scanning in the pore region and point mutation of the equivalent asparagine residues in NR1 and NR2 (Wollmuth et al. 1998).

NMDA receptor activation mechanisms

NMDA receptors generate a slow synaptic current that peaks in about 20 ms and decays (depending on subunit composition) (Table 3.2.3.1) over a time course of 100 ms or more at room temperature (reviewed by Cull-Candy and Leszkiewicz, 2004; Erreger et al., 2004; Popescu and Auerbach, 2004). Assuming a $Q_{10} \sim 3$, this translates into a rise time around 4 ms and decay time constant of 25 ms at 37°C (Feldmeyer et al., 2002). The receptor activation process is remarkable: two small amino acids collide with the receptor protein and following binding, the protein undergoes conformational changes resulting in the opening of a channel across the membrane. The channel then opens and closes repeatedly during the period while glutamate and glycine are bound (Gibb and Colquhoun, 1992) until finally the channel closes and the glutamate or glycine unbind to end the activation. The slow time course of the synaptic current was elegantly shown to be determined by the slow kinetics of the receptor (Lester et al., 1990) and is crucial for the role of NMDA receptors as coincidence detectors of pre- and postsynaptic activity (Markram et al., 1997; Navian and Sakmann, 2004). The time course of both synaptic currents and the response of NMDA receptors to brief or prolonged applications of glutamate was successfully described with a model incorporating the sequential binding of two molecules of glutamate followed by either channel opening, or entry into a relatively long-lived closed ('desensitized') state (Lester and Jahr, 1992) (Figure 3.2.3.2). However, this mechanism was not based on knowledge of the receptor structure, and could not explain the

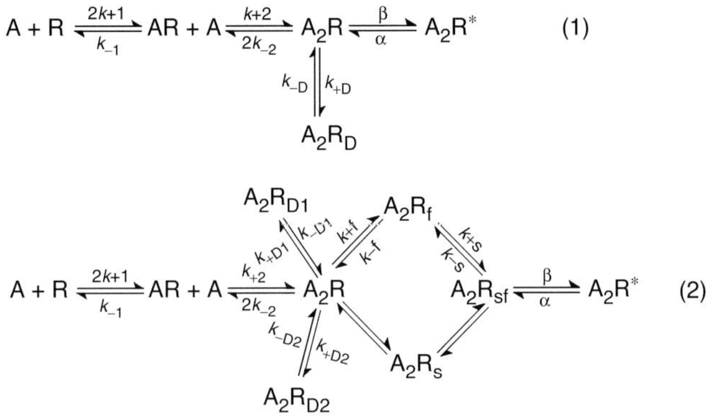

$$A + R \xrightleftharpoons[k_{-1}]{2k+1} AR + A \xrightleftharpoons[2k_{-2}]{k+2} A_2R \xrightleftharpoons[\alpha]{\beta} A_2R^* \qquad (1)$$

Fig. 3.2.3.2 Mechanism (1) shows NMDA receptor activation as described by Lester and Jahr (1992). The mechanism describes two sequential agonist binding reactions, followed by channel opening, or desensitization. Mechanism (2) shows the receptor activation as described by Banke and Traynelis (2003). In this mechanism two new states, A_2R_f and A_2R_s, have been added to reflect the fast gating transitions of the NR1 subunits (k_{+f} and k_{-f}) and slower NR2 subunits (k_{+s} and k_{-s}) which can occur in either order to arrive at the A_2R_{sf} state that is then followed by channel opening.

complicated channel opening and closing activity observed in single channel recordings. Over the past 15 years considerable progress has been made in investigating the structural basis of NMDA receptor function and the kinetic behaviour of the receptor has been described in much more detail (reviewed by Erreger et al., 2004; Popescu and Auerbach, 2004). In particular, the use of partial agonists at the NR1 site (full agonist: glycine, partial agonists: HA-966 or the antibiotic D-cycloserine) and at the glutamate site of the NR2 subunit (full agonist: glutamate, partial agonists: NMDA or the neurotoxin, quinolinic acid) (Banke and Traynelis, 2003) or homoquinolinate (Erreger et al., 2005a) has refined the mechanisms used to describe the receptor activation by allowing identification of kinetic steps that can be associated with pre-gating conformational changes in either the NR1 or NR2 subunits of the receptor. In addition, two directly interconverting open states allow a more accurate description of the channel open times (Popescu et al., 2004; Schorge et al., 2005). Using a combination of single channel recordings and recordings of macroscopic responses, while activating receptors with a combination of brief (1–4 ms long) glutamate applications (mimicking synaptic transmission) and long glutamate applications (3 seconds to assess how receptor desensitization impinges on receptor activation), enough kinetic information can be obtained to identify conformational changes in the model that can be related to the NR1 or NR2 receptor subunits. In addition, by changing the rate constants in the scheme, the same mechanism can be used to describe the behaviour of NR2A or NR2B receptors (Erreger et al., 2005b). These results give new insights into the sluggish behaviour of the NMDA receptor. They suggest that glutamate binding (k_{+1}) is rapid, while unbinding (k_{-1}) is relatively slow, but it is the subsequent conformational changes in the receptor protein, as the NR1 and NR2 subunits adjust to the rearrangements resulting from closing of the clam shell-like domains of the agonist binding sites (Furukawa et al., 2005; Mayer, 2006), that take time, before the channel can finally open (state A_2R^*). Although each opening is relatively short (~3.0 ms, Gibb and Colquhoun, 1992), the channel can open and close many times during the activation, with gaps between openings ranging from 100 ms to 100 ms (Gibb and Colquhoun, 1992). During these structural rearrangements, the receptor may slip into one or more desensitized states (state A_2R_{D1}, A_2R_{D2}) but on exiting these states, may continue to open and close for a further burst of activity. Addition of the extra pre-gating steps allows the behaviour of the receptor in both single channel recordings and macroscopic recordings (e.g. synaptic currents or following a glutamate concentration jump) to be accurately described in a wide range of situations (Banke and Traynelis, 2003; Popescu et al., 2004; Erreger et al., 2005b; Schorge et al., 2005; Wyllie et al., 2006). The idea that structural features of the receptor can be related to different aspects of the channel gating mechanism as suggested in this mechanism by Banke and Traynelis (2003) is a major advance in being able to relate structure to function because it is the beginnings of providing a physical basis for the receptor behaviour (Figure 3.2.3.2). That different domains of each subunit may contribute discrete functional domains is already recognized (Mayer, 2006) and a major challenge is to understand how these different domains (N-terminal, ligand binding, channel, C-terminal) communicate. In the future, molecular dynamics simulations that currently can only run for tens of nanoseconds, will begin to predict protein conformational changes on the micro and millisecond time scale of channel gating and combined with X-ray crystallographic structural data and high resolution single channel data will produce a unified picture of receptor function.

3.2.3.4 Cellular and subcellular localization

NMDA receptors are expressed both at synapses and in the extra-synaptic membrane. It has been proposed (Hardingham and Bading, 2003) that synaptic receptors activate signalling pathways that promote neuron survival, while extrasynaptic receptors dominate in excitotoxicity (Soriano and Harding, 2007). The mechanisms controlling their subcellular localization have been a subject of intense study in recent years (reviewed by Wenthold et al., 2003; Pérez-Otaño and Ehlers, 2005; Cognet et al., 2006; Lau and Zukin, 2007).

The mechanisms controlling NMDA receptor subunit gene expression in different cells and at different developmental stages are not yet fully understood. NR1 subunit mRNAs and protein are expressed throughout the brain and spinal cord (Laurie and Seeburg, 1994; Watanabe et al., 1994a). They are also expressed in small and medium nociceptive sensory neurons in the trigeminal, vagal and dorsal root ganglia (Watanabe et al., 1994e; Marvizon et al. 2002; L Li et al. 2004), in taste receptor cells (Caicedo et al., 2004), in retina (Watanabe et al., 1994d), in cochlea and vestibular ganglia (Niedzielski and Wenthold, 1995) and in bone (Chenu et al., 1998; Patton et al., 1998). In general NR2B and NR2D subunits are expressed early in development. From the end of the first postnatal week, NR2A expression is found in association with NR2B, particularly in hippocampus and cortex. In the cerebellum, NR2C expression in association with NR2A appears from the end of the first postnatal week in rodents and continues into adulthood. In contrast, in the midbrain, brainstem and spinal cord, NR2B and NR2D expression continue into adulthood with varying levels of NR2A and NR2C (Akazawa et al., 1994; Monyer et al., 1994; Watanabe et al., 1994a; 1994b; Dunah et al., 1998; Goebel and Poosch, 1999).

There is increasing evidence that in some neurons, synaptic and extra-synaptic receptors may have different subunit composition, or there may be different proportions of NMDA receptor subtypes in synaptic and extra-synaptic membrane and that these variations are under homeostatic control (see e.g. Wenthold et al., 2003; Pérez-Otaño and Ehlers, 2005; Cognet et al., 2006). NMDA receptors are stabilized at synapses by interaction of their intracellular C-terminal domains with postsynaptic density proteins such as alpha-actinin and members of the PSD-95 class of MAGUK proteins. The mechanisms regulating the trafficking of surface (Cognet et al., 2006) or intracellular receptors (Wenthold et al., 2003) to the synapse are complex and not yet fully elucidated. Trafficking of NMDA receptors to the cell surface can be enhanced by PKC activation via SNARE-dependent exocytosis (Lan et al., 2001). Internalization of NMDA receptors (Roche et al., 2001) is subunit-dependent (Lavezzari et al., 2004) and mediated by association of receptors with the clathrin adaptor AP-2 (Nong et al., 2003; Prybylowski et al., 2005), is regulated by tyrosine phosphorylation, and can be promoted by mGluR activation (Snyder et al., 2001) and ligand binding (Nong et al., 2003). In the striatum Dunah et al. (2004) have shown that deletion of the gene for the protein tyrosine kinase, Fyn, inhibits dopamine D1 receptor-induced enhancement of the abundance of NR1, NR2A and NR2B subunits in the synaptosomal membrane fraction of striatal homogenates (Dunah et al., 2004), while Lee et al. (2002) demonstrated inhibition of NMDA responses by a direct protein–protein interaction between the dopamine D1 receptor and NR2A subunit C-termini (reviewed by Cepeda and Levine, 2006). It is interesting to speculate whether other more subtle effects initiated by activation or inhibition of G-protein-coupled receptors in the brain may have as a component of their action, alteration of the density or subtypes of synaptic NMDA receptors, leading then to changes in NMDA receptor signalling and synaptic strength (Lau and Zukin, 2007).

The most well-established developmental change is the increasing expression of NR2A subunits during the first postnatal weeks in rodents. This will have clearly established functional effects. First, a decrease in the time course of the NMDA receptor-mediated synaptic current, for example in cortical synapses and at synapses on cerebellar granule cells (Carmignoto andVicini, 1992; Hestrin, 1992; Flint et al. 1997; Cathala et al., 2000). Secondly, a decrease in glycine sensitivity: this may mean that in some areas of the brain the size of NMDA synaptic currents is regulated by the local activity of glycine transporters (Schell, 2004) such as in brainstem (Berger et al., 1998; Lim et al., 2004) but not in others such as the cerebellum (Billups and Attwell, 2003). Thirdly, increased sensitivity to Zn^{2+} inhibition (Paoletti et al., 1997; Erreger and Traynelis, 2005; Hatton and Paoletti, 2005): this means that in some areas of the brain the size of NMDA currents will be tonically regulated by the background level of Zn^{2+}.

NMDA receptors, or NMDA receptor subunits have been found to be expressed in a number of unexpected cell types (Table 3.2.3.1). The NR1 and NR2D subunits are expressed in bone cells (Patton et al., 1988; Chenu et al., 1998), taste receptor cells express NR1 (Caicedo et al., 2004), while NR2B and NR2C are found in the heart (Lin et al., 1996; Seeber et al., 2004) and NR2C in the pancreas and skeletal muscle (Lin et al., 1996). Possibly these proteins have evolved distinct physiological roles in these cells, for example as chaperones in the ER, or have a structural role (Seeber et al., 2004), unrelated to glutamate signalling.

3.2.3.5 Pharmacology

NMDA receptors were originally defined by their sensitivity to the synthetic agonist, N-methyl-D-aspartate (reviewed by Watkins and Jane, 2006) and they have subsequently been shown to have a rich pharmacology. They are now more often defined and identified in physiological systems by their sensitivity to the glutamate site antagonist AP5, the glycine site antagonist, 7-Cl-kynurenate, or by the use of channel blocking drugs like MK-801 (dizocilpine) or the anaesthetic ketamine.

A central role for proton inhibition in NMDA receptor pharmacology

As discussed above (section 3.2.3.2), NMDA receptors are inhibited by protons, with an IC_{50} around pH 7.3; $[H^+] = 60$ nM (Traynelis and Cull-Candy, 1990, Traynelis et al., 1995; Banke et al., 2005). Proton block is central to the action of drugs such as ifenprodil that bind to the amino terminal domain of the NR2B subunit (Mott et al., 1998; Kew and Kemp, 2005) and also to the inhibition of NR2A receptors by zinc ($IC_{50} \sim 10$nM), which binds to the N-terminal domain of NR2A (Paoletti et al., 1997, 2000; Hatton and Paoletti, 2005). Protons inhibit NMDA receptors by preventing channel opening while ifenprodil and zinc act to increase the sensitivity of the receptor to proton block (Mott et al., 1998; Erreger and Traynelis, 2005). A positive cooperativity between the glutamate binding domain and the N-terminal domain of NR2A subunits results in a zinc-sensitive desensitization on applying glutamate to NR2A receptors at a constant pH and zinc concentration, as glutamate binding increases the receptor affinity for zinc, which in turn increases the receptor sensitivity to proton block (Erreger and Traynelis, 2005).

Receptor subtype-selective drugs

Appart from the selective block of NR2B receptors by ifenprodil and other phenylethanolamines, there are relatively few drugs with significant selectivity between NMDA receptor subtypes (Table 3.2.3.1). For example the glutamate site antagonist AP5 and the channel blocking drugs, MK-801 and ketamine, show only two- to threefold higher affinity for NR2A and NR2B, compared to NR2C and NR2D (Table 3.2.3.1). The low-affinity channel blocking drug, memantine (Johnson and Kotermanski, 2006; Lipton, 2006) has two- to threefold higher affinity for NR2C and NR2D compared to NR2A and NR2B receptors (Dravid et al., 2007). Drug selectivity for NR2A over NR2B has been difficult to achieve with NVP-AAM077 showing only fivefold selectivity (NR2A, 15 nM and NR2B, 78 nM) (Frizelle et al., 2006). Progress has only recently been made in developing agonist drugs (Erreger et al., 2007) and antagonist drugs with improved selectivity for NR2C and NR2D receptors (Morley et al., 2005). The NR3A S1S2 fusion protein displays high affinity for glycine and D-serine but much lower affinity for 5,7-dichlorokynurenic acid (647 μM) compared to NR1 (0.5 μM).

Non-competitive antagonists

Among non-competitive antagonists of NMDA receptors, the phenylethanolamines like ifenprodil (Williams, 1993; Mott et al., 1998) and the ion channel blocking drugs like ketamine and MK-801 have received most attention. Ifenprodil and other ligands that bind to the N-terminal domain are voltage-independent but

activity-dependent and mediate an increase in glutamate potency (Kew et al., 1996; Mott et al., 1998). In contrast, ketamine and MK-801 are highly voltage-dependent blockers that show a trapping block mechanism with use-dependence of the block and high-affinity binding that may be an important factor in the unacceptable side effects of these drugs in humans (Kemp and McKernan, 2002). The low-affinity channel-blocking drug memantine was recently approved for treatment of dementia and shows a significantly improved profile that has been suggested to be due to having a faster unblocking rate that reduces use-dependence of the block (Lipton, 2006).

3.2.3.6 Physiology

A key physiological role of NMDA receptors was established during the 1980s following the discovery that their activation was essential for the initiation of certain forms of LTP (Collingridge et al., 1983). Detailed investigation of excitatory synaptic transmission at many synapses throughout the brain and spinal cord have further established the role of NMDA receptors in normal synaptic transmission and in the integration of synaptic activity as a result of the calcium permeability and voltage-dependent Mg^{2+} block of the NMDA receptor channel, as discussed above.

Considerable interest has also focused on attempting to identify the physiological roles of the different NMDA receptor subtypes, principally receptors defined by their NR2 subunit type. These investigations have been hampered by a lack of subtype-specific antagonists, except in the case of NR2B receptors where ifenprodil and similar drugs can be used. An alternative to the use of pharmacological tools has been the use of molecular genetic techniques to selectively disrupt gene expression or target receptor function by mutagenesis. While this approach gives the advantage that expression of a receptor subunit can be switched off or a mutant subunit switched on throughout development, or at a particular time point in development, or sometimes in a particular cell type, the disadvantage is that it is rarely possible to block receptor function, and then recover it so acutely as can be achieved with selective pharmacology. Nevertheless, early studies demonstrating the importance of NMDA receptor function for correct maturation of the visual system (Kleinschmidt et al., 1987; Bear et al., 1990) gave clear evidence of the importance of NMDA receptors in the development and maturation of the brain. It was subsequently shown that mice lacking the NR1 subunit (ζ1) die soon after birth, showing impairment of their suckling reflex (Forrest et al., 1994) and disruption of whisker-related patterning in the trigeminal nucleus (Y Li et al., 1994), although in general the nervous sytem had developed normally. More specifically, mice with a targeted mutation in NR1 (N598R) giving calcium impermeable and Mg^{2+} insensitive NMDA receptors fail to develop normal trigeminal nucleus patterning and die soon after birth (Rudhard et al., 2003) illustrating the critical importance of these two functional properties to the physiological role of NMDA receptors.

Targeted gene deletion of the NR2A subunit (ε1) has demonstrated that these subunits are not essential for survival with animals showing apparently normal growth and mating behaviour, but an enhanced startle response, deficiencies in social interaction, and reduced hippocampal LTP (Sakimura et al., 1995). Deletion of NR2A also abolishes the developmental speeding of NMDA receptor-mediated EPSC kinetics in cortex and cerebellum but despite this, in NR2A knockout mice, the critical period for synaptic plasticity in somatosensory cortex ends normally indicating that the change in NMDA receptor kinetics associated with NR2A expression is not the determining factor in closing the critical period (Lu *et al.*, 2001). In contrast to NR2A, mice lacking the NR2B subunit die soon after birth, showing impairment in suckling response, impaired trigeminal neuron pattern formation and impaired hippocampal long-term depression (LTD) (Kutsuwada *et al.*, 1996). The NR2B knockout thus has a similar phenotype to the NR1 knockout, confirming the essential role of these two subunits in providing normal NMDA receptor function in the neonatal nervous system. Interestingly, overexpression of NR2B enhances learning and memory in behavioural tests (Tang *et al.*, 1999) and also enhances the behavioural response to inflammatory pain (Wei *et al.*, 2001).

The observation that the NR2A knockout lacks LTP (Sakimura et al., 1995) while the NR2B knockout lack LTD (Kutsuwada et al., 1996) may have led to the idea that LTP is supported by NR2A receptors and LTD by NR2B receptors. Indeed the lower open probability and slower activation of NR2B receptors will generate a longer, slower calcium signal that might selectively support LTD, while the faster kinetics and higher open probability of NR2A receptors might selectively support LTP (Erreger *et al.*, 2005b). This ideas is consistent with the observation that it is generally easier to evoke LTP in older animals (where NR2A expression is higher) and generally easier to evoke LTD in younger animals (where NR2B expression is higher). However, attributing LTP to NR2A receptors and LTD to NR2B receptors ignores the fact that a proportion of receptors in the cortex and hippocampus are likely to be NR1/NR2A/NR2B triheteromers and it is unclear how will these receptors influence synaptic plasticity. At present this area is still very much open to debate (Neyton and Paoletti, 2006).

Both NR2C (ε3) and NR2D (ε4) knockout mice show apparently normal development and growth (Ikeda *et al.*, 1995; Ebralidze *et al.*, 1996) although the NR2D knockout shows reduced spontaneous behavioural activity (Ikeda *et al.*, 1995), lower sensitivity to stress and altered monoaminergic neuronal function (Miyamoto *et al.*, 2001, 2002). These results perhaps point to an important role for NR2D receptors which have been shown to be expressed in midbrain dopamine nuclei in the adult (Dunah *et al.*, 1998). Mice with the double knockout of NR2A and NR2C show a loss of both spontaneous and evoked EPSCs in cerebellar granule cells (as expected if the mature receptor is a triheteromer of NR1/NR2A/NR2C or a mixed NR1/NR2A and NR1/NR2C population) and disturbed motor coordination (Kadotani *et al.*, 1996).

3.2.3.7 Relevance to disease states

NMDA receptors have been implicated in a wide range of disease states and this fact is one of the main drivers to develop receptor antagonists that may be clinically useful (Kemp and McKernan, 2002). Their importance in excitotoxic cell death (Choi, 1992) has been clearly demonstrated in pathologies such as stroke and anoxia where excessive NMDA receptor activation kills neurons (Szatkowski and Attwell, 1994; Arundine and Tymianski, 2004). Equally provoking are reports noting changes in NMDA receptor expression or properties that may be either a consequence or a contributing factor in a disease, or animal model of a disease. Considerable interest has focused on the role of NMDA receptors

in pain processes, particularly in the sensitization to painful stimuli that may be part of the development of chronic pain (Woolf and Salter, 2000). In the spinal cord dorsal horn there is an upregulation of NR1 subunit phosphorylation in response to formalin injection, inflammation or chronic pain stimuli (Caudle *et al.*, 2005; Gao et al. 2005; Kim *et al.*, 2006). Similarly, an increased tyrosine phosphorylation of the NR2B subunit is observed in the spinal cord following development of inflammatory hyperalgesia (Guo *et al.*, 2002). Overexpression of NR2B subunit in the forebrain results in increased sensitivity to inflammatory pain (Wei *et al.*, 2001). It is not yet clear how these changes relate to the disease state, although it may be hypothesized that they affect the neural circuits processing sensory information by alterations in synaptic strength in sensory pathways. In the rat amygdala there is a PKA-dependent enhancement of pain-related synaptic plasticity that may reflect part of the aversive nature of pain (Bird *et al.*, 2005).

There is considerable evidence supporting the glutamate hypofunction hypothesis of schizophrenia (Coyle *et al.*, 2003; Moghaddam, 2003; Pilowsky *et al.*, 2006; Javitt, 2007). This is supported by evidence from genetic linkage studies showing polymorphisms of the NR1 and NR2B subunit gene are associated with schizophrenia and bipolar disorder (Mundo *et al.*, 2003; Martucci *et al.*, 2006) and also with obsessive–compulsive disorder (Arnold *et al.*, 2004). It is intriguing that phosphorylation of receptor subunits is particularly associated with changes in the trafficking and surface expression of NMDA receptors and that this could have an influence on other psychiatric disease states (Lau and Zukin, 2006). For example, methamphetamine treatment alters the expression of the NR2B subunit in mice forebrain (Yamamoto *et al.*, 2006).

3.2.3.8 Concluding remarks

The material in this chapter has centred around the structure and function of NMDA receptors and the effects of drugs at this receptor class. In particular, the aim has been to integrate current knowledge in these areas to allow an overview to be made of NMDA receptor properties and function in the nervous system. In so far as understanding receptor properties is at the heart of advancing understanding of drug action, this approach is likely to continue to be a useful part of pharmacology. This is particularly so for studies in the central nervous system where understanding the properties of receptor subtypes is an essential aspect of assessing the therapeutic potential of subtype-selective drugs.

Acknowledgements

Supported by the Wellcome Trust and the BBSRC.

References

Akazawa C, Shigemoto R, Bessho Y *et al.* (1994). Differential expression of five N-methyl-D-aspartate receptor subunit mRNAs in the cerebellum of developing and adult rats. *J Comp Neurol* **347**, 150–160.

Andersson O, Stenqvist A, Attersand A *et al.* (2001). Nucleotide sequence, genomic organization, and chromosomal localization of genes encoding the human NMDA receptor subunits NR3A and NR3B. *Genomics* **78**, 178–184.

Anson LC, Chen PE, Wyllie DJ *et al.* (1998). Identification of amino acid residues of the NR2A subunit that control glutamate potency in recombinant NR1/NR2A NMDA receptors. *J Neurosci* **18**, 581–589.

Antonov SM and Johnson JW (1999). Permeant ion regulation of N-methyl-D-aspartate receptor channel block by Mg(2+). Proc *Natl Acad Sci USA* **96**, 14571–14576.

Antonov SM, Gmiro VE and Johnson JW (1998). Binding sites for permeant ions in the channel of NMDA receptors and their effects on channel block. *Nat Neurosci* **1**, 451–461.

Arnold PD, Rosenberg DR, Mundo E *et al.* (2004). Association of a glutamate (NMDA) subunit receptor gene (*GRIN2B*) with obsessive–compulsive disorder: a preliminary study. *Psychopharmacology (Berl)* **174**, 530–538.

Arundine M and Tymianski M (2004). Molecular mechanisms of glutamate-dependent neurodegeneration in ischemia and traumatic brain injury. *Cell Mol Life Sci* **61**, 657–668.

Banke TG and Traynelis SF (2003). Activation of NR1/NR2B NMDA receptors. *Nat Neurosci* **6**, 144–145.

Banke TG, Dravid SM and Traynelis SF (2005). Protons trap NR1/NR2B NMDA receptors in a nonconducting state. *J Neurosci* **25**, 42–51.

Bear MF, Kleinschmidt A, Gu QA *et al.* (1990). Disruption of experience-dependent synaptic modifications in striate cortex by infusion of an NMDA receptor antagonist. *J Neurosci* **10**, 909–925.

Béhé P, Stern P, Wyllie DJ, Nassar M, Schoepfer R and Colquhoun D (1995). Determination of NMDA NRI subunit copy number in rccombinant NMDA receptors. *Proc Biol Sci* **262**, 205–213.

Berger AJ, Dieudonne S and Ascher P (1998). Glycine uptake governs glycine site occupancy at NMDA receptors of excitatory synapses. *J Neurophysiol* **80**, 3336–3340.

Billups D and Attwell D (2003). Active release of glycine or D-serine saturates the glycine site of NMDA receptors at the cerebellar mossy fibre to granule cell synapse. *Eur J Neurosci* **18**, 2975–2980.

Bird GC, Lash LL, Han JS *et al.* (2005). Protein kinase A-dependent enhanced NMDA receptor function in pain-related synaptic plasticity in rat amygdala neurones. *J Physiol* **564**, 907–921.

Bliss TV and Collingridge GL (1993). A synaptic model of memory: long-term potentiation in the hippocampus. *Nature* **361**, 31–39.

Brickley SG, Misra C, Mok MH, Mishina M and Cull-Candy SG (2003). NR2B and NR2D subunits coassemble in cerebellar Golgi cells to form a distinct NMDA receptor subtype restricted to extrasynaptic sites. *J Neurosci* **23**, 4958–4966.

Burnashev N, Schoepfer R, Monyer H *et al.* (1992). Control by asparagine residues of calcium permeability and magnesium blockade in the NMDA receptor. *Science* **257**, 1415–1419.

Burnashev N, Zhou Z, Neher E *et al.* (1995). Fractional calcium currents through recombinant GluR channels of the NMDA, AMPA and kainate receptor subtypes. *J Physiol* **485**, 403–418.

Caicedo A, Zucchi B, Pereira E *et al.* (2004). Rat gustatory neurons in the geniculate ganglion express glutamate receptor subunits. *Chem Senses* **29**, 463–471.

Carmignoto G and Vicini S (1992). Activity-dependent decrease in NMDA receptor responses during development of the visual cortex. *Science* **258**, 1007–1011.

Cathala L, Misra C and Cull-Candy SG (2000). Developmental profile of the changing properties of NMDA receptors at cerebellar mossy fiber-granule cell synapses. *J Neurosci* **20**, 5899–5905.

Caudle RM, Perez FM, Del Valle-Pinero AY *et al.* (2005). Spinal cord NR1 serine phosphorylation and NR2B subunit suppression following peripheral inflammation. *Mol Pain* **1**, 25.

Cepeda C and Levine MS (2006). Where do you think you are going? The NMDA-D1 receptor trap. *Sci STKE* pe20.

Chatterton JE, Awobuluyi M, Premkumar LS *et al.* (2002). Excitatory glycine receptors containing the NR3 family of NMDA receptor subunits. *Nature* **415**, 793–798.

Chazot PL and Stephenson FA (1997). Molecular dissection of native mammalian forebrain NMDA receptors containing the NR1 C2 exon: direct demonstration of NMDA receptors comprising NR1, NR2A, and NR2B subunits within the same complex. *J Neurochem* **69**, 2138–2144.

Chazot PL, Coleman SK, Cik M *et al.* (1994). Molecular characterization of N-methyl-D-aspartate receptors expressed in mammalian cells yields evidence for the coexistence of three subunit types within a discrete receptor molecule. *J Biol Chem* **269**, 24403–24409.

Chen BS, Braud S, Badger JD *et al.* (2006). Regulation of NR1/NR2C N-methyl-D-aspartate (NMDA) receptors by phosphorylation. *J Biol Chem* **281**, 16583–16590.

Chen PE and Wyllie DJ (2006). Pharmacological insights obtained from structure–function studies of ionotropic glutamate receptors. *Br J Pharmacol* **147**, 839–853.

Chen PE, Johnston AR, Mok MH *et al.* (2004). Influence of a threonine residue in the S2 ligand binding domain in determining agonist potency and deactivation rate of recombinant NR1a/NR2D NMDA receptors. *J Physiol* **558**, 45–58.

Chenu C, Serre CM, Raynal C *et al.* (1998). Glutamate receptors are expressed by bone cells and are involved in bone resorption. *Bone* **22**, 295–299.

Choi DW (1992). Excitotoxic cell death. *J Neurobiol* **23**, 1261–1276.

Chung HJ, Huang YH, Lau L-F *et al.* (2004). Regulation of the NMDA receptor complex and trafficking by activity-dependent phosphorylation of the NR2B subunit PDZ ligand. *J Neurosci* **24**, 10248–10259.

Ciabarra AM, Sullivan JM, Gahn LG *et al.* (1995). Cloning and characterization of x-1: a developmentally regulated member of a novel class of the ionotropic glutamate receptor family. *J Neurosci* **15**, 6498–6508.

Cognet L, Groc L, Lounis B *et al.* (2006). Multiple routes for glutamate receptor trafficking: surface diffusion and membrane traffick cooperate to bring receptors to synapses. *Science STKE* **2006**, pe13.

Collingridge GL (2003). The induction of N-methyl-D-aspartate receptor-dependent long-term potentiation. *Phil Trans Roy Soc Lond B* **358**, 635–641.

Collingridge GL, Herron CE and Lester RA (1988). Frequency-dependent N-methyl-D-aspartate receptor-mediated synaptic transmission in rat hippocampus. *J Physiol* **399**, 301–312.

Collingridge GL, Kehl SJ and McLennan H (1983). Excitatory amino acids in synaptic transmission in the Schaffer collateral–commissural pathway of the rat hippocampus. *J Physiol* **334**, 33–46.

Coyle JT, Tsai G and Goff D (2003). Converging evidence of NMDA receptor hypofunction in the pathophysiology of schizophrenia. *Ann NY Acad Sci* **1003**, 318–327.

Cull-Candy S, Brickley S and Farrant M (2001). NMDA receptor subunits: diversity, development and disease. *Curr Opin Neurobiol* **11(3)**, 327–335.

Cull-Candy SG and Leszkiewicz DN (2004). Role of distinct NMDA receptor subtypes at central synapses. *Science STKE* **2004**, re16.

Daggett LP, Johnson EC, Varney MA *et al.* (1998). The human N-methyl-D-aspartate receptor 2C subunit: genomic analysis, distribution in human brain, and functional expression. *J Neurochem* **71**, 1953–1968.

Dale N and Roberts A (1985). Dual-component amino-acid-mediated synaptic potentials: excitatory drive for swimming in *Xenopus* embryos. *J Physiol* **363**, 35–59.

Das S, Sasaki YF, Rothe T *et al.* (1998). Increased NMDA current and spine density in mice lacking the NMDA receptor subunit NR3A. *Nature* **393**, 377–381.

Dingledine R, Borges K, Bowie D *et al.* (1999). The glutamate receptor ion channels. *Pharmacol Rev* **51**, 7–61.

Dravid SM, Erreger K, Yuan H *et al.* (2007). Subunit-specific mechanisms and proton sensitivity of NMDA receptor channel block. *J Physiol* **581**, 107–128.

Dunah AW, Luo J, Wang Y-H *et al.* (1998). Subunit composition of N-Methyl-D-aspartate receptors in the central nervous system that contain the NR2D subunit. *Mol Pharm* **53**, 429–437.

Dunah AW, Sirianni AC, Fienberg AA *et al.* (2004) Dopamine D1-dependent trafficking of striatal N-methyl-D-aspartate glutamate receptors requires Fyn protein tyrosine kinase but not DARPP-32. *Mol Pharmacol* **65**, 121–129.

Durand GM, Gregor P, Zheng X *et al.* (1992). Cloning of an apparent splice variant of the rat N-methyl-Daspartate receptor NMDAR1 with altered sensitivity to polyamines and activators of protein kinase C. *Proc Natl Acad Sci USA* **89**, 9359–9363.

Ebralidze AK, Rossi DJ, Tonegawa S *et al.* (1996). Modification of NMDA receptor channels and synaptic transmission by targeted disruption of the NR2C gene. *J Neurosci* **16**, 5014–5025.

Ehlers MD, Zhang S, Bernhadt JP *et al.* (1996). Inactivation of NMDA receptors by direct interaction of calmodulin with the NR1 subunit. *Cell* **84**, 745–55.

Erreger K and Traynelis SF (2005). Allosteric interaction between zinc and glutamate binding domains on NR2A causes desensitization of NMDA receptors. *J Physiol* **569**, 381–393.

Erreger K, Chen PE, Wyllie DJ *et al.* (2004). Glutamate receptor gating. *Crit Rev Neurobiol* **16**, 187–224.

Erreger K, Dravid SM, Banke TG *et al.* (2005b). Subunit-specific gating controls rat NR1/NR2A and NR1/NR2B NMDA channel kinetics and synaptic signalling profiles. *J Physiol* **563**, 345–358.

Erreger K, Geballe MT, Dravid SM *et al.* (2005a). Mechanism of partial agonism at NMDA receptors for a conformationally restricted glutamate analog. *J Neurosci* **25(34)**, 7858–7866.

Erreger K, Geballe MT, Kristensen AS *et al.* (2007). Subunit-specific agonist activity at NR2A, NR2B, NR2C, and NR2D containing N-methyl-D-aspartate glutamate receptors. *Mol Pharmacol* **72**, 907–920.

Farrant M, Feldmeyer D, Takahashi T *et al.* (1994). NMDA-receptor channel diversity in the developing cerebellum. *Nature* **368**, 335–339.

Feldmeyer D, Lübke J, Silver RA and Sakmann B (2002). Synaptic connections between layer 4 spiny neurone-layer 2/3 pyramidal cell pairs in juvenile rat barrel cortex: physiology and anatomy of interlaminar signalling within a cortical column, *J Physiol* **538**, 803–822.

Fischer G, Mutel V, Trube G *et al.* (1997). Ro 25–6981, a highly potent and selective blocker of N-methyl-D-aspartate receptors containing the NR2B subunit. Characterization *in vitro. J Pharmacol Exp Ther* **283**, 1285–1292.

Flint AC, Maisch US, Weishaupt JH, Kriegstein AR and Monyer H (1997). NR2A subunit expression shortens NMDA receptor synaptic currents in developing neocortex. *J Neurosci* **17**, 2469–2476.

Forrest D, Yuzaki M, Soares HD *et al.* (1994). Targeted disruption of NMDA receptor 1 gene abolishes NMDA response and results in neonatal death. *Neuron* **13**, 325–338.

Forsythe ID and Westbrook GL (1988). Slow excitatory postsynaptic currents mediated by N-methyl-D-aspartate receptors on cultured mouse central neurones. *J Physiol* **396**, 515–533.

Frizelle PA, Chen PE and Wyllie DJ (2006). Equilibrium constants for (R)-[(S)-1-(4-bromo-phenyl)-ethylamino]-(2,3-dioxo-1,2,3,4-tetrahydroquinoxalin-5-yl)-methyl]-phosphonic acid (NVP-AAM077) acting at recombinant NR1/NR2A and NR1/NR2B N-methyl-D-aspartate receptors: implications for studies of synaptic transmission. *Mol Pharmacol* **70(3)**, 1022–1032.

Fu Z, Logan SM and Vicini S (2005). Deletion of the NR2A subunit prevents developmental changes of NMDA-mEPSCs in cultured mouse cerebellar granule neurones. *J Physiol* **563**, 867–881.

Fukaya M, Hayashi Y and Watanabe M (2005). NR2 to NR3B subunit switchover of NMDA receptors in early postnatal motoneurons. *Eur J Neurosci* **21**, 1432–1436.

Furukawa H and Gouaux E (2003). Mechanisms of activation, inhibition and specificity: crystal structures of NR1 ligand-binding core. *EMBO J* **22**, 1–13.

Furukawa H, Singh SK, Mancusso R *et al.* (2005). Subunit arrangement and function in NMDA receptors. *Nature* **438**, 185–192.

Gao X, Kim HK, Chung JM *et al.* (2005). Enhancement of NMDA receptor phosphorylation of the spinal dorsal horn and nucleus gracilis neurons in neuropathic rats. *Pain* **116**, 62–72.

Gibb AJ (2006). Glutamate unbinding reveals new insights into NMDA receptor activation. *J Physiol* **574**, 329.

Gibb AJ and Colquhoun D (1992). Activation of N-methyl-D-aspartate receptors by L-glutamate in cells dissociated from adult rat hippocampus. *J Physiol* **456**, 143–179.

Goebel DJ and Poosch MS (1999). NMDA receptor subunit gene expression in the rat brain: a quantitative analysis of endogenous mRNA levels of NR1Com, NR2A, NR2B, NR2C, NR2D and NR3A. *Molecular Brain Research* **69**, 164–170.

Guo W, Zou S, Guan Y *et al.* (2002). Tyrosine phosphorylation of the NR2B subunit of the NMDA receptor in the spinal cord during the development and maintenance of inflammatory hyperalgesia. *J Neurosci* **22**, 6208–6217.

Hatton CJ and Paoletti P (2005). Modulation of triheteromeric NMDA receptors by N-terminal domain ligands. *Neuron* **46**, 261–274.

Herron CE, Lester RAJ, Coan EJ *et al.* (1986). Frequency-dependent involvement of NMDA receptors in the hippocampus: a novel synaptic mechanism. *Nature* **322**, 265–267.

Hestrin S (1992). Developmental regulation of NMDA receptor-mediated synaptic currents at a central synapse. *Nature* **357**, 686–689.

Hollmann M and Heinemann S (1994). Cloned glutamate receptors. *Annu Rev Neurosci* **17**, 31–108.

Ikeda K, Araki K, Takayama C *et al.* (1995). Reduced spontaneous activity of mice defective in the epsilon 4 subunit of the NMDA receptor channel. Brain *Res Mol Brain Res* **33**, 61–71.

Ikeda K, Nagasawa M, Mori H *et al.* (1992). Cloning and expression of the epsilon 4 subunit of the NMDA receptor channel. *FEBS Lett* **313**, 34–38.

Ishii T, Moriyoshi K, Sugihara H *et al.* (1993). Molecular characterization of the family of the *N*-methyl-D-aspartate receptor subunits. *J Biol Chem* **268**, 2836–2843.

Itokawa M, Yamada K, Yoshitsugu K *et al.* (2003). A microsatellite repeat in the promoter of the N-methyl-D-aspartate receptor 2A subunit (*GRIN2A*) gene suppresses transcriptional activity and correlates with chronic outcome in schizophrenia. *Pharmacogenetics* **13**, 271–278.

Javitt DC (2007). Glutamate and schizophrenia: phencyclidine, N-methyl-d-aspartate receptors, and dopamine–glutamate interactions. *Int Rev Neurobiol* **78**, 69–108.

Johnson JW (2003). Acid tests of N-methyl-D-aspartate receptor gating basics. *Mol Pharmacol* **63**, 1199–1201.

Johnson JW and Ascher P (1987). Glycine potentiates the NMDA response in cultured mouse brain neurons. *Nature* **325**, 529–531.

Johnson JW and Kotermanski SE (2006). Mechanism of action of memantine. *Curr Opin Pharmacol* **6**, 61–67.

Jones S and Gibb AJ (2005). Functional NR2B- and NR2D-containing NMDA receptor channels in rat substantia nigra dopaminergic neurones. *J Physiol* **569**, 209–221.

Kadotani H, Hirano T, Masugi M, Nakamura K, Nakao K, Katsuki M and Nakanishi S (1996). Motor discoordination results from combined gene disruption of the NMDA receptor NR2A and NR2C subunits, but not from single disruption of the NR2A or NR2C subunit. *J Neurosci* **16**, 7859–7867.

Karadottir R, Cavelier P, Bergersen LH *et al.* (2005). NMDA receptors are expressed in oligodendrocytes and activated in ischaemia. *Nature* **438**, 1162–1166.

Karp SJ, Masu M, Eki T *et al.* (1993). Molecular cloning and chromosomal localization of the key subunit of the human N-methyl-D-aspartate receptor. *J Biol Chem* **268**, 3728–3733.

Kemp JA and McKernan RM (2002). NMDA receptor pathways as drug targets. *Nat Neurosci* **5**(Suppl), 1039–1042.

Kennedy MB (2000). Signal-processing machines at the post-synaptic density. *Science* **290**, 750–754.

Kennedy MB, Beale HC, Carlisle HJ *et al.* (2005). Integration of biochemical signalling in spines. *Nat Rev Neurosci* **6**, 423–434.

Kew JN and Kemp JA (2005). Ionotropic and metabotropic glutamate receptor structure and pharmacology. *Psychopharmacology* **179**, 4–29.

Kew JN, Trube G and Kemp JA (1996). A novel mechanism of activity-dependent NMDA receptor antagonism describes the effect of ifenprodil in rat cultured cortical neurones. *J Physiol* **497**, 761–772.

Kim E and Sheng M (2004). PDZ domain proteins of synapses. *Nat Rev Neurosci* **5**, 771–781.

Kim HW, Kwon YB, Roh DH *et al.* (2006). Intrathecal treatment with sigma 1 receptor antagonists reduces formalin-induced phosphorylation of NMDA receptor subunit 1 and the second phase of formalin test in mice. *Br J Pharmacol* **148**, 490–498.

Kleckner NW and Dingledine R (1988). Requirement for glycine in activation of NMDA receptors expressed in *Xenopus* oocytes. *Science* **241**, 835–837.

Kleinschmidt A, Bear MF and Singer W (1987). Blockade of 'NMDA' receptors disrupts experience-dependent plasticity of kitten striate cortex. *Science* **238**, 355–358.

Kornau HC, Schenker LT, Kennedy MB *et al.* (1995). Domain interaction between NMDA receptor subunits and the postsynaptic density protein PSD-95. *Science* **269**, 1737–1740.

Krupp JJ, Vissel B, Heinemann SF and Westbrook GL (1998). N-terminal domains in the NR2 subunit control desensitization of NMDA receptors. *Neuron* **20**, 317–327.

Krupp JJ, Vissel B, Thomas CG, Heinemann SF and Westbrook GL (1999). Interactions of calmodulin and alpha-actinin with the NR1 subunit modulate Ca2+-dependent inactivation of NMDA receptors. *J Neurosci* **19**, 1165-1178.

Kuryatov A, Laube B, Betz H *et al.* (1994) Mutational analysis of the glycine-binding site of the NMDA receptor: structural similarity with bacterial amino acid-binding proteins. *Neuron* **12**, 1291–1300.

Kutsuwada T, Kashiwabuchi N, Mori H *et al.* (1992). Molecular diversity of the NMDA receptor channel. *Nature* **358**, 36–41.

Kutsuwada T, Sakimura K, Manabe T *et al.* (1996). Impairment of suckling response, trigeminal neuronal pattern formation, and hippocampal LTD in NMDA receptor epsilon 2 subunit mutant mice. *Neuron* **16**, 333–344.

Lalo U, Pankratov Y, Kirchhoff F, North RA and Verkhratsky A (2006). NMDA receptors mediate neuron-to-glia signaling in mouse cortical astrocytes. *J Neurosci* **26**, 2673–2683.

Lan JY, Skeberdis VA, Jover T et al. (2001). Protein Kinase C modulates NMDA receptor trafficking and gating. *Nat Neurosci* **4**, 382–390.

Lau CG and Zukin RS (2007). NMDA receptor trafficking in synaptic plasticity and neuropsychiatric disorders. *Nat Rev Neurosci* **8**, 413–426.

Laube B, Hirai H, Sturgess M *et al.* (1997). Molecular determinants of agonist discrimination by NMDA receptor subunits: analysis of the glutamate binding site on the NR2B subunit. *Neuron* **18**, 493–503.

Laube B, Kuhse J and Betz H (1998). Evidence for a tetrameric structure of recombinant NMDA receptors. *J Neurosci* **18**, 2954–2961.

Laurie DJ and Seeburg PH (1994). Ligand affinities at recombinant N-methyl-D-aspartate receptors depend on subunit composition. *Eur J Pharmacol* **268**, 335–345.

Laurie DJ and Seeburg PH (1994). Regional and developmental heterogeneity in splicing of the rat brain NMDAR1 mRNA. *J Neurosci* **14**, 3180–3194.

Lavezzari G, McCallum J, Dewey CM *et al.* (2004). Subunit-specific regulation of NMDA receptor endocytosis. *J Neurosci* **24**, 6383–6391.

Le Bourdelles B, Wafford KA, Kemp JA *et al.* (1994). Cloning, functional coexpression, and pharmacological characterisation of human cDNAs encoding NMDA receptor NR1 and NR2A subunits. *J Neurochem* **62**, 2091–2098.

Lee FJ, Xue S, Pei L *et al.* (2002) Dual regulation of NMDA receptor functions by direct protein–protein interactions with the dopamine D1 receptor. *Cell* **111**, 219–302.

Legendre P, Rosenmund C and Westbrook GL (1993). Inactivation of NMDA channels in cultured hippocampal neurons by intracellular calcium. *J Neurosci* **13**, 674–684.

Lester RA, Clements JD, Westbrook GL *et al.* (1990). Channel kinetics determine the time course of NMDA receptor-mediated synaptic currents. *Nature* **346**, 565–567.

Lester RA and Jahr CE (1992). NMDA channel behavior depends on agonist affinity. *J Neurosci* **12**, 635–643.

Li B, Chen N, Luo T, Otsu Y *et al.* (2002). Differential regulation of synaptic and extra-synaptic NMDA receptors. *Nat Neurosci* **5**, 833–834.

Li L, Murphy TH, Hayden MR *et al.* (2004). Enhanced striatal NR2B-containing N-methyl-D-aspartate receptor-mediated synaptic currents in a mouse model of Huntington disease. *J Neurophysiol* **92(5)**, 2738–2746.

Li Y, Erzurumlu RS, Chen C *et al.* (1994). Whisker-related neuronal patterns fail to develop in the trigeminal brainstem nuclei of NMDAR1 knockout mice. *Cell* **76(3)**, 427–437.

Lim R, Hoang P and Berger AJ (2004). Blockade of glycine transporter-1 (GLYT-1) potentiates NMDA receptor-mediated synaptic transmission in hypoglossal motorneurons. *J Neurophysiol* **92(4)**, 2530–2537.

Lin JW, Wyszynski M, Madhavan R *et al.* (1998). Yotiao, a novel protein of neuromuscular junction and brain that interacts with specific splice variants of NMDA receptor subunit NR1. *J Neurosci* **18**, 2017–2027.

Lin YJ, Bovetto S, Carver JM *et al.* (1996). Cloning of the cDNA for the human NMDA receptor NR2C subunit and its expression in the central nervous system and periphery. *Brain Res Mol Brain Res* **43**, 57–64.

Lipton S (2006). Paradigm shift in neuroprotection by NMDA receptor blockade: memantine and beyond. *Nature Reviews in Drug Discovery* **5**, 163–170.

Lisman J (2003). Long-term potentiation: outstanding questions and attempted synthesis. *Phil Trans R Soc B* **358**, 829–842.

Li-Smerin Y, Levitan ES and Johnson JW (2001). Free intracellular Mg(2+) concentration and inhibition of NMDA responses in cultured rat neurons. *J Physiol* **533**, 729–43.

Liu XY, Chu XP, Mao LM *et al.* (2006). Modulation of D2R–NR2B interactions in response to cocaine. *Neuron* **52(5)**, 897–909.

Lu C, Fu Z, Karavanov I *et al.* (2006). NMDA receptor subtypes at autaptic synapses of cerebellar granule neurons. *J Neurophysiol* **96**, 2282–2294.

Lu HC, Gonzalez E and Crair MC (2001). Barrel cortex critical period plasticity is independent of changes in NMDA receptor subunit composition. *Neuron* **32**, 619–634.

Madry C, Mesic I, Bartholomaus I *et al.* (2007). Principal role of NR3 subunits in NR1/NR3 excitatory glycine receptor function. *Biochem Biophys Res Commun* **354**, 102–108.

Magleby KL (2004). Modal gating of NMDA receptors. *Trends Neurosci* **5**, 231–233.

Malenka RC (1991). The role of postsynaptic calcium in the induction of long-term potentiation. *Mol Neurobiol* **5(2–4)**, 289–295.

Malenka RC and Nicoll RA (1993). NMDA-receptor-dependent synaptic plasticity: multiple forms and mechanisms. *Trends Neurosci* **16(12)**, 521–527.

Mandich P (1994). Mapping of the human NMDAR2B receptor subunit gene (GRIN2B) to chromosome 12 p12. *Genomics* **22**, 216–218.

Markram H, Lubke J, Frotscher M *et al.* (1997). Regulation of synaptic efficacy by coincidence of postsynaptic APs and EPSPs. *Science* **275**, 213–215.

Martucci L, Wong AH, De Luca V *et al.* (2006). N-methyl-D-aspartate receptor NR2B subunit gene *GRIN2B* in schizophrenia and bipolar disorder: polymorphisms and mRNA levels. *Schizophr Res* **84**, 214–221.

Marvizon JC, McRoberts JA, Ennes HS *et al.* (2002). Two N-methyl-D-aspartate receptors in rat dorsal root ganglia with different subunit composition and localization. *J Comp Neurol* **446**, 325–341.

Matsuda K, Fletcher M, Kamiya Y *et al.* (2003). Specific assembly with the NMDA receptor 3B subunit controls surface expression and calcium permeability of NMDA receptors. *J Neurosci* **23(31)**, 10064–10073.

Matsuda K, Kamiya Y, Matsuda S *et al.* (2002). Cloning and characterization of a novel NMDA receptor subunit NR3B: a dominant subunit that reduces calcium permeability. *Brain Res Mol Brain Res* **100**, 43–52.

Mayer ML (2006). Glutamate receptors at atomic resolution. *Nature* **440**, 456–462.

Mayer ML and Westbrook GL (1987). Permeation and block of N-methyl-D-aspartic acid receptor channels by divalent cations in mouse cultured central neurones. *J Physiol* **394**, 501–527.

Mayer ML, Westbrook GL and Guthrie PB (1984). Voltage-dependent block by Mg2+ of NMDA responses in spinal cord neurones. *Nature* **309**, 261–263.

Mayer ML, Vyklicky L Jr and Clements J (1989). Regulation of NMDA receptor desensitization in mouse hippocampal neurons by glycine. *Nature* **338**, 425–427.

Misra C, Brickley SG, Wyllie DJA *et al.* (2000). Slow deactivation kinetics of NMDA receptors containing NR1 and NR2D subunits in rat cerebellar Purkinje cells. *J Physiol* **525**, 299–305.

Miyamoto Y, Yamada K, Noda Y *et al.* (2001). Hyperfunction of dopaminergic and serotonergic neuronal systems in mice lacking the NMDA receptor epsilon1 subunit. *J Neurosci* **21**, 750–757.

Miyamoto Y, Yamada K, Noda Y *et al.* (2002). Lower sensitivity to stress and altered monoaminergic neuronal function in mice lacking the NMDA receptor epsilon 4 subunit. *J Neurosci* **22**, 2335–2342.

Moghaddam B (2003). Bringing order to the glutamate chaos in schizophrenia. *Neuron* **40**, 881–884.

Momiyama A, Feldmeyer D and Cull-Candy SG (1996). Identification of a native low-conductance NMDA channel with reduced sensitivity to Mg2+ in rat central neurones. *J Physiol* **494**, 479–492.

Monyer H, Burnashev N, Laurie DJ *et al.* (1994). Developmental and regional expression in the rat brain and functional properties of four NMDA receptors. *Neuron* **12**, 529–540.

Monyer H, Sprengel R, Schoepfer R *et al.* (1992). Heteromeric NMDA receptors: molecular and functional distinction of subtypes. *Science* **256**, 1217–1221.

Moriyoshi K, Masu M, Ishii T *et al.* (1991). Molecular cloning and characterization of the rat NMDA receptor. *Nature* **354**, 31–37.

Morley RM, Tse H-W, Feng B *et al.* (2005). Synthesis and pharmacology of N1-substituted piperazine-2,3-dicarboxylic acid derivatives acting as NMDA receptor antagonists. *J Med Chem* **48**, 2627–2637.

Mott DD, Doherty JJ, Zhang S *et al.* (1998). Phenylethanolamines inhibit NMDA receptors by enhancing proton inhibition. *Nat Neurosci* **1(8)**, 659–667.

Mundo E, Tharmalingham S, Neves-Pereira M *et al.* (2003). Evidence that the N-methyl-D-aspartate subunit 1 receptor gene (*GRIN1*) confers susceptibility to bipolar disorder. *Mol Psychiatry* **8**, 241–245.

Nahum-Levy R, Lipinski D, Shavit S and Benveniste M (2001). Desensitization of NMDA receptor channels is modulated by glutamate agonists. *Biophys J.* **80**, 2152–2166.

Nakanishi N, Axel R and Shneider NA (1992). Alternative splicing generates functionally distinct N-methyl-D-aspartate receptors. *Proc Natl Acad Sci USA* **89**, 8552–8556.

Nase G, Weishaupt J, Stern P, Singer W and Monyer H (1999). Genetic and epigenetic regulation of NMDA receptor expression in the rat visual cortex. *Eur J Neurosci* **11**, 4320–4326.

Nevian T and Sakmann B (2004). Single spine Ca2+ signals evoked by coincident EPSPs and backpropagating action potentials in spiny stellate cells of layer 4 in the juvenile rat somatosensory barrel cortex. *J Neurosci* **24**, 1689–1699.

Neyton J and Paoletti P (2006). Relating NMDA receptor function to receptor subunit composition: limitations of the pharmacological approach. *J Neurosci* **26**, 1331–1333.

Niedzielski AS and Wenthold RJ (1995). Expression of AMPA, kainate, and NMDA receptor subunits in cochlear and vestibular ganglia. *J Neurosci* **15**, 2338–2353.

Nishi M, Hinds H, Lu HP *et al.* (2001). Motorneuron-specific expression of NR3B, a novel NMDA-type glutamate receptor subunit that works in a dominant-negative manner. *J Neurosci* **21**, RC185.

Nong Y, Huang YQ, Ju W *et al.* (2003). Glycine binding primes NMDA receptor internalization. *Nature* **422**, 302–307.

Nowak L, Bregestovski P, Ascher P *et al.* (1984). Magnesium gates glutamate-activated channels in mouse central neurones. *Nature* **307**, 462–465.

Paarmann I, Frermann D, Keller BU *et al.* (2000). Expression of 15 glutamate receptor subunits and various splice variants in tissue slices and single neurons of brainstem nuclei and potential functional implications. *J Neurochem* **74**, 1335–1345.

Paoletti P, Ascher P and Neyton J (1997). High-affinity zinc inhibition of NMDA NR1–NR2A receptors. *J Neurosci* **17**, 5711–5725.

Paoletti P, Neyton J and Ascher P (1995). Glycine-independent and subunit-specific potentiation of NMDA responses by extracellular Mg2+. *Neuron* **15**, 1109–1120.

Paoletti P, Perin-Dureau F, Fayyazuddin A *et al.* (2000). Molecular organization of a zinc binding N-terminal modulatory domain in a NMDA receptor subunit. *Neuron* **28**, 911–925.

Patton AJ, Genever PG, Birch MA *et al.* (1998). Expression of an N-methyl-D-aspartate-type receptor by human and rat osteoblasts and osteoclasts suggests a novel glutamate signaling pathway in bone. *Bone* **22**, 645–649.

Pérez-Otaño I and Ehlers MD (2005). Homeostatic plasticity and NMDA receptor trafficking. *Trends in the Neurosciences* **28**, 229–238.

Pérez-Otaño I, Contractor A, Schulteis CT *et al.* (1998). NMDAR3A-2, a novel splice variant of the NMDA receptor subunit formerly known as L-1. *Soc Neurosci Abstr* **24**, 1087.

Pérez-Otaño I, Luján R, Tavalin SJ *et al.* (2006). Endocytosis and synaptic removal of NR3A-containing NMDA receptors by PACSIN1/syndapin1. *Nat Neurosci* **9**, 611–621.

Pérez-Otaño I, Schulteis CT, Contractor A *et al.* (2001). Assembly with the NR1 subunit is required for surface expression of NR3A-containing NMDA receptors. *J Neurosci* **21**, 1228–1237.

Pilowsky LS, Bressan RA, Stone JM *et al.* (2006). First *in vivo* evidence of an NMDA receptor deficit in medication-free schizophrenic patients. *Mol Psychiatry* **11**, 118–119.

Planells-Cases R, Sun W, Ferrer-Montiel AV *et al.* (1993). Molecular cloning, functional expression, and pharmacological characterization of an N-methyl-D-aspartate receptor subunit from human brain. *Proc Natl Acad Sci USA* **90**, 5057–5061.

Popescu G and Auerbach A (2003). Modal gating of NMDA receptors and the shape of their synaptic response. *Nat Neurosci* **5**, 476–483.

Popescu G and Auerbach A (2004).The NMDA receptor gating machine: lessons from single channels. *Neuroscientist* **10(3)**, 192–198.

Popescu G, Robert A, Howe JR *et al.* (2004). Reaction mechanism determines NMDA receptor response to repetitive stimulation. *Nature* **430**, 790–793.

Priestley T, Laughton P, Myers J *et al.* (1995). Pharmacological properties of recombinant human N-methyl-D-aspartate receptors comprising NR1a/NR2A and NR1a/NR2B subunit assemblies expressed in permanently transfected mouse fibroblast cells. *Mol Pharmacol* **48**, 841–848.

Prybylowski K, Chang K, Sans N *et al.* (2005). The synaptic localization of NR2B-containing NMDA receptors is controlled by interactions with PDZ proteins and AP-2. *Neuron* **47(6)**, 845–857.

Prybylowski K, Rumbaugh G, Wolfe BB and Vicini S (2000). Increased exon 5 expression alters extrasynaptic NMDA receptors in cerebellar neurons. *J Neurochem* **75**, 1140–1146.

Roche KW, Standley S, McCallum J *et al.* (2001). Molecular determinants of NMDA receptor internalization. *Nat Neurosci* **4**, 794–802.

Rosenmund C and Westbrook GL (1993). Calcium-induced actin depolymerization reduces NMDA channel activity. *Neuron* **10**, 805–314.

Rudhard Y, Kneussel M, Nassar MA *et al.* (2003). Absence of whisker-related pattern formation in mice with NMDA receptors lacking coincidence detection properties and calcium signaling. *J Neurosci* **23**, 2323–2332.

Rumbaugh G, Prybylowski K, Wang JF *et al.* (2000). Exon 5 and spermine regulate deactivation of NMDA receptor subtypes. *J. Neurophysiol* **83**, 1300–1306.

Rumbaugh G and Vicini S (1999). Distinct synaptic and extrasynaptic NMDA receptors in developing cerebellar granule neurons. *J Neurosci* **19**, 10603–10610.

Rycroft BK and Gibb AJ (2002). Direct effects of calmodulin on NMDA receptor single-channel gating in rat hippocampal granule cells. *J Neurosci* **22**, 8860–8868.

Rycroft BK and Gibb AJ (2004a). Inhibitory interactions of calcineurin (phosphatase 2B) and calmodulin on rat hippocampal NMDA receptors. *Neuropharmacology* **47**, 505–514.

Rycroft BK and Gibb AJ (2004b). Regulation of single NMDA receptor channel activity by alpha-actinin and calmodulin in rat hippocampal granule cells. *J Physiol* **557**, 795–808.

Sakimura K, Kutsuwada T, Ito I et al. (1995). Reduced hippocampal LTP and spatial learning in mice lacking NMDA receptor epsilon 1 subunit. *Nature* **373**, 151–155.

Salter MG and Fern R (2005). NMDA receptors are expressed in developing oligodendrocyte processes and mediate injury. *Nature* **438**, 1167–1171.

Sather W, Johnson JW, Henderson G and Ascher P (1990). Glycine-insensitive desensitization of NMDA responses in cultured mouse embryonic neurons. *Neuron* **4**, 725–731.

Schell MJ (2004). The N-methyl D-aspartate receptor glycine site and D-serine metabolism: an evolutionary perspective. *Phil Trans Roy Soc Lond B* **359**, 943–964.

Schorge S and Colquhoun D (2003). Studies of NMDA receptor function and stoichiometry with truncated and tandem subunits. *J Neurosci* **23**, 1151–1158.

Schneggenburger R (1996). Simultaneous measurement of Ca2+ influx and reversal potentials in recombinant N-methyl-D-aspartate receptor channels. *Biophys J* **70**, 2165–2174.

Schorge S, Elenes S and Colquhoun D (2005). Maximum likelihood fitting of single channel NMDA activity with a mechanism composed of independent dimers of subunits. *J Physiol* **569**, 395–418.

Seeber S, Humeny A, Herkert M et al. (2004). Formation of molecular complexes by N-methyl-D-aspartate receptor subunit NR2B and ryanodine receptor 2 in neonatal rat myocard. *J Biol Chem* **279**, 21062–21068.

Sheng M, Cummings J, Roldan LA, Jan YN and Jan LY (1994). Changing subunit composition of heteromeric NMDA receptors during development of rat cortex. *Nature* **368**, 144–147.

Single FN, Rozov A, Burnashev N et al. (2000). Dysfunctions in mice by NMDA receptor point mutations NR1(N598Q) and NR1(N598R). *J Neurosci* **20**, 2558–2566.

Snyder EM, Nong Y, Almeida CG et al. (2005). Regulation of NMDA receptor trafficking by amyloid-beta. *Nat Neurosci* **8**(8), 1051–1058.

Snyder EM, Ohilpot BD, Huber KM et al. (2001). Internalization of ionotropic glutamate receptors in response to mGluR activation. *Nat Neurosci* **4**, 1079–1085.

Soriano FX and Hardingham GE (2007). Compartmentalized NMDA receptor signalling to survival and death. *J Physiol* **584**, 381–387.

Standley S, Roche KW, McCallum J et al. (2000). PDZ domain suppression of an ER retention signal in NMDA receptor NR1 splice variants. *Neuron* **28**, 887–898.

Stern P, Béhé P, Schoepfer R et al. (1992). Single-channel conductances of NMDA receptors expressed from cloned cDNAs: comparison with native receptors. *Proc Biol Sci* **250**, 271–277.

Suchanek B, Seeburg PH and Sprengel R (1995). Gene structure of the murine N-methyl-D-aspartate receptor subunit NR2C. *J Biol Chem* **270**, 41–44.

Sucher NJ, Akbarian S, Chi CL et al. (1995). Developmental and regional expression pattern of a novel NMDA receptor-like subunit (NMDAR-L) in the rodent brain. *J Neurosci* **15**, 6509–6520.

Sucher NJ, Kohler K, Tenneti L et al. (2003). N-methyl-D-aspartate receptor subunit NR3A in the retina: developmental expression, cellular localization, and functional aspects. *Invest Ophthalmol Vis Sci* **44**(10), 4451–4456.

Sugihara H, Moriyoshi K, Ishii T et al. (1992). Structures and properties of seven isoforms of the NMDA receptor generated by alternative splicing. *Biochem Biophys Res Commun* **185**, 826–832.

Sun L, Margolis FL, Shipley MT et al. (1998). Identification of a long variant of mRNA encoding the NR3 subunit of the NMDA receptor: its regional distribution and developmental expression in the rat. *FEBS Lett* **441**, 392–396.

Swope SL, Moss SJ, Raymond LA et al. (1999). Regulation of ligand-gated ion channels by protein phosphorylation. *Adv Second Messenger Phosphoprotein Res* **33**, 49–78.

Szatkowski M and Attwell D (1994). Triggering and execution of neuronal death in brain ischaemia: two phases of glutamate release by different mechanisms. *Trends Neurosci* **17**, 359–365.

Tang YP, Shimizu E, Dube GR et al. (1999). Genetic enhancement of learning and memory in mice. *Nature* **401**, 63–69.

Tolle TR, Berthele A, Zieglgansberger W et al. (1993). The differential expression of 16 NMDA and non-NMDA receptor subunits in the rat spinal cord and in periaqueductal gray. *J Neurosci* **13**, 5009–5028.

Traynelis SF and Cull-Candy SG (1990). Proton inhibition of N-methyl-D-aspartate receptors in cerebellar neurons. *Nature* **345**, 347–350.

Traynelis SF, Hartley M and Heinemann SF (1995). Control of proton sensitivity of the NMDA receptor by RNA splicing and polyamines. *Science* **268**, 873–876.

Ulbrich MH and Isacoff EY (2007). Subunit counting in membrane-bound proteins. *Nature Methods* **4**(4), 319–321.

Vicini S, Wang JF, Li JH et al. (1998). Functional and pharmacological differences between recombinant N-methyl-D-aspartate receptors. *J Neurophysiol* **79**, 555–566.

Vyklicky L Jr, Benveniste M and Mayer ML (1990). Modulation of N-methyl-D-aspartic acid receptor desensitization by glycine in mouse cultured hippocampal neurones. *J Physiol* **428**, 313–331.

Wafford KA, Kathoria M, Bain CJ et al. (1995). Identification of amino acids in the N-methyl-D-aspartate receptor NR1 subunit that contribute to the glycine binding site. *Mol Pharmacol* **47**, 374–380.

Watanabe M, Mishina M and Inoue Y (1994a). Distinct spatiotemporal distributions of the N-methyl-D-aspartate receptor channel subunit mRNAs in the mouse cervical cord. *J Comp Neurol* **345**, 314–319.

Watanabe M, Mishina M and Inoue Y (1994b). Distinct distributions of five NMDA receptor channel subunit mRNAs in the brainstem. *J Comp Neurol* **343**, 520–531.

Watanabe M, Mishina M and Inoue Y (1994c). Distinct spatiotemporal expressions of five NMDA receptor channel subunit mRNAs in the cerebellum. *J Comp Neurol* **343**, 513–519.

Watanabe M, Mishina M and Inoue Y (1994d). Differential distributions of the NMDA receptor channel subunit mRNAs in the mouse retina. *Brain Res* **634**, 328–332.

Watanabe M, Mishina M and Inoue Y (1994e). Distinct gene expression of the N-methyl-D-aspartate receptor channel subunit in peripheral neurons of the mouse sensory ganglia and adrenal gland. *Neurosci Lett* **165**, 183–186.

Watkins JC and Jane DE (2006). The glutamate story. *Br J Pharmacol* **147**(Suppl 1), S100–108.

Wei F, Wang GD, Kerchner GA et al. (2001). Genetic enhancement of inflammatory pain by forebrain NR2B overexpression. *Nat Neurosci* **4**, 164–169.

Wenthold RJ, Prybylowski K, Standley S et al. (2003). Trafficking of NMDA receptors. *Annu Rev Pharmacol Toxicol* **43**, 335–358.

Williams K (1993). Ifenprodil discriminates subtypes of the N-methyl-D-aspartate receptor: selectivity and mechanisms at recombinant heteromeric receptors. *Mol Pharmacol* **44**(4), 851–859.

Wollmuth LP, Kuner T and Sakmann B (1998). Adjacent asparagines in the NR2-subunit of the NMDA receptor channel control the voltage-dependent block by extracellular Mg^2+. *J Physiol* **506**, 13–32.

Woolf CJ and Salter MW (2000). Neuronal plasticity: increasing the gain in pain. *Science* **288**, 1765–1769.

Wu LJ, Toyoda H, Zhao MG et al. (2005). Upregulation of forebrain NMDA NR2B receptors contributes to behavioral sensitization after inflammation. *J Neurosci* **25**, 11107–11116.

Wyllie DJ, Behe P and Colquhoun D (1998). Single-channel activations and concentration jumps: comparison of recombinant NR1a/NR2A and NR1a/NR2D NMDA receptors. *J Physiol* **510**, 1–18.

Wyllie DJ, Béhé P, Nassar M *et al.* (1996). Single-channel currents from recombinant NMDA NR1a/NR2D receptors expressed in *Xenopus* oocytes. *Proc Biol Sci* **263**, 1079–1086.

Wyllie DJ, Johnston AR, Lipscombe D *et al.* (2006). Single-channel analysis of a point mutation of a conserved serine residue in the S2 ligand-binding domain of the NR2A NMDA receptor subunit. *J Physiol* **574**, 477–489.

Wyszynski M, Lin J, Rao A *et al.* (1997). Competitive binding of α-actinin and calmodulin to the NMDA receptor. *Nature* **385**, 439–442.

Yamakura T and Shimoji K (1999). Subunit and site-specific pharmacology of the NMDA receptor channel. *Prog Neurobiol* **59**, 279–298.

Yamakura T, Askalany AR, Petrenko AB *et al.* (2005). The NR3B subunit does not alter the anesthetic sensitivities of recombinant N-methyl-D-aspartate receptors. *Anesth Analg* **100**, 1687–1692.

Yamamoto H, Imai K, Kamegaya E *et al.* (2006). Repeated methamphetamine administration alters expression of the NMDA receptor channel epsilon2 subunit and kinesins in the mouse brain. *Ann N Y Acad Sci* **1074**, 97–103.

Yao Y and Mayer ML (2006). Characterization of a soluble ligand binding domain of the NMDA receptor regulatory subunit NR3A. *J Neurosci* **26**, 4559–4566.

Zimmer M, Fink TM, Franke Y *et al.* (1995). Cloning and structure of the gene encoding the human N-methyl-D-aspartate receptor (NMDAR1). *Gene* **159**, 219–223.

3.3

ATP receptors

Contents

3.3.1 P2X receptors

Iain Chessell and Anton Michel

3.3.1.1 Introduction

Adenosine triphosphate (ATP) is a ubiquitous molecule, and in recent years its well-known role as a source of intracellular energy has been supplemented with a vast body of research describing an important role as a transmitter substance. ATP interacts with two main families of receptors: metabotropic P2Y receptors, and ionotropic P2X receptors. This chapter focuses on the structure and function of the latter family.

Seven P2X receptor genes have been identified following isolation of the first cDNAs encoding P2X receptor subunits in 1994, although the existence of ATP-gated ion channels was extensively described and reviewed before this date (e.g. Burnstock, 1972; Burnstock and Kennedy, 1985). When heterologously expressed, all seven of the cDNAs encode ATP-gated non-selective cation channels. However, the properties of these channels in terms of pharmacology and biophysics differ considerably between the receptors, and some of these subunits are also capable of forming heteromultimers with other members of the P2X receptor family which confers further different properties. When functionally activated, all members of the family, whether alone or in combination with a different subunit, are capable of producing cell depolarization with physiological or pathophysiological consequences which depend on cellular localization and environment.

3.3.1.2 Molecular properties of P2X receptors

The coding sequences of the seven P2X receptor genes contain up to 13 exons. Their chromosomal localization is described in Table 3.3.1.1. There are two gene subunit 'pairs'; $P2X_4$ and $P2X_7$ are localized near the tip of chromosome 12, and are separated by less than 24 kb (similarly in the mouse these genes are separated by <27kb on chromosome 5), and share closely related amino acid sequences. $P2X_1$ and $P2X_5$ genes are also close together on chromosome 13, with the remaining genes localized on different chromosomes. All seven genes share a common structure, and many splice variants have been described. P2X receptor subunits are 384 to 595 amino acids long and each has two hydrophobic regions which are of sufficient length to cross the plasma membrane; extrapolation of the conformational motifs of these subunits predicts both N and C termini of the subunits to be intracellular. The presence of multiple motifs for N-linked glycosylation in the putative extracellular domain supports this hypothesis (Newbolt et al., 1998; Torres et al., 1998b).

The stoichiometry of P2X subunit assembly has not been fully elucidated, but it is assumed that, in order to form a pore large enough to pass mono- and divalent cations, oligomeric assembly of the subunits is a requirement. Initial studies using concatameric assembly and cross-linking suggests that these receptors are composed either of trimers, or assemblies of trimers (Nicke et al., 1998; Stoop et al., 1999). More recent studies suggest that trimeric architecture is the functional hallmark of both homomeric and heteromeric P2X receptors (Aschrafi et al., 2004), and study of cross-linked $P2X_2$ receptors using atomic force microscopy further confirms these proposals (Barrera et al., 2005).

Functional expression studies have provided evidence for heteromultimeric assembly of different P2X receptor subunits. Lewis et al. (1995) demonstrated functional co-assembly of $P2X_2$ and $P2X_3$, following observation of a discrepancy between native ATP-induced responses in rat dorsal root ganglion (DRG) cells, and those observed for the then known receptor cDNAs following heterologous expression. Subsequently Torres et al. (1999a) followed up findings from various groups describing the ability of subunit pairs to heteropolymerize, with a study using co-immunoprecipitation to determine subunit pair specificities. Table 3.3.1.2 summarizes these findings (Le et al., 1998; Surprenant et al., 2000; Aschrafi et al., 2004; Nicke et al., 2005), together with findings from functional studies. While these studies have clearly demonstrated the capability for co-assembly of different subunits to produce a receptor phenotype which differs from either of the respective contributing partners, most of these studies have been performed using heterologous expression techniques. Critical questions thus remain regarding the subunit identity of many native-tissue P2X receptor-mediated effects (e.g. see Chessell *et al.*, 1997a; Sansum *et al.*, 1998).

3.3.1.3 Biophysical characteristics

All of the P2X receptor subunits capable of forming homomeric or heteromeric assemblies which are functionally active produce non-selective cation channels which are gated by ATP and other agonists. The P2X receptors are permeable to sodium, potassium and calcium; $P2X_2$, $P2X_4$, $P2X_5$ and particularly $P2X_7$ are able to develop a further open state which is permeable to much larger ions (Egan *et al.*, 2006). However, there are key differences in permeation properties of each of the receptors and desensitization and inactivation profiles vary greatly. These properties of each subunit or assembly, together with any pertinent information on splice variants or identification of molecular motifs which confer specific features, are discussed below.

P2X₁ receptors

$P2X_1$ receptors were first isolated from a cDNA library from rat vas deferens, long known to express functional ionotropic receptors for ATP (Valera *et al.*, 1994). Human and mouse orthologues have subsequently been identified and a number of splice variants of the $P2X_1$ receptor have been described (North, 2002). One of these, involving a 17aa deletion from exon 6, produces a receptor with preferential sensitivity to adenosine diphosphate (ADP) (Greco *et al.*, 2001, but see Oury *et al.*, 2002). In addition a single amino acid deletion (leucine) in the platelet $P2X_1$ was identified in a patient with a severe bleeding disorder and this led to identification of a non-functional, dominant negative $P2X_1$ receptor when expressed recombinantly (Oury *et al.*, 2000).

The agonist binding site of the $P2X_1$ receptor has not been fully elucidated. However, studies where conserved positively charged amino acids in the extracellular loop of the $P2X_1$ were mutated suggest that the ATP binding domain (but not the binding domain for the antagonist, suramin) lies close to the channel vestibule. Thus, mutation of Lys-68, Lys-70, Arg-292 and Lys-309, which are close to the predicted transmembrane domains of the receptor, resulted in significant changes in ATP potency (Ennion *et al.*, 2000).

The homomeric $P2X_1$ receptor is a typical non-selective cation channel which shows little permeation selectivity for Na^+ over K^+ (Evans *et al.*, 1996). The receptor also has relatively high permeability to Ca^{2+} (Table 3.3.1.1), and is not markedly inhibited by the presence of high concentrations of Ca^{2+}, in contrast to some of the other P2X receptors. On activation, the kinetics of channel opening are rapid (10–90 per cent, rise time 7 ms using 10 μM ATP), but the receptor desensitizes very rapidly (Valera *et al.*, 1994)—within the P2X receptor family, $P2X_1$ and $P2X_3$ receptors show the most marked and rapid desensitization. Desensitization is less marked at lower agonist concentrations ($<EC_{50}$), and it is under these conditions that most characterization of the receptor pharmacology has been made, however, even under these conditions agonist inter-application times must be 15 min or more in order to obtain reproducible responses.

P2X₂ receptors

The rat $P2X_2$ receptor was isolated from a cDNA library made from differentiated pheochromocytoma (PC12) cells (Brake *et al.*, 1994).

Several splice variants of the $P2X_2$ receptor have been described, which confer altered functional properties with respect to agonist and antagonist sensitivity as well as desensitization properties (Brandle *et al.*, 1997; Simon *et al.*, 1997; Chen *et al.*, 2000).

Alanine scanning mutagenesis has been utilized to attempt to identify the ATP binding site of the $P2X_2$ receptor (Jiang *et al.*, 2006), and has identified a region proximal to the first transmembrane domain containing two lysine residues which are critical for the action of ATP. These data together with that from methanesulfonate attachment experiments suggest that residues 67, 69 and 71 are close to the binding pocket for ATP.

The permeation characteristics of $P2X_2$ are similar to those of $P2X_1$, with the exception that the $P2X_2$ receptor is less permeable to Ca^{2+} (Table 3.3.1.1). Determination of Ca^{2+} permeability is complicated by fast divalent cation block of the channel (Ding and Sachs, 2000), with rank order of inhibitory activity in keeping with ionic radii. However, in contrast activity of $P2X_2$ is potentiated by Zn^{2+}, an effect which recent studies suggest is conferred by an intersubunit binding site comprising His120 and His213 on adjacent subunits (Nagaya *et al.*, 2005).

The biophysical properties of $P2X_2$ differ from those of $P2X_1$ in several important ways. First, $P2X_2$ desensitizes much more slowly than $P2X_1$, such that currents evoked by ATP decline little during agonist applications of a few seconds (Brake *et al.*, 1994). Secondly, desensitization has been demonstrated to be regulated by phosphorylation in the NH_2 terminus (Boué-Grabot *et al.*, 2000) and to differ between splice variants (Brandle *et al.*, 1997). Furthermore valine at position 370 has been shown to critically regulate desensitization rate (Smith *et al.*, 1999). In addition to Ca^{2+} block of the channel, this ion can also influence desensitization (Ding and Sachs, 2000). The $P2X_2$ receptor also exhibits strong inward rectification and although the underlying mechanisms mediating this are unclear, it does not appear to involve voltage-dependent divalent cation block, as rectification persists in divalent cation-free solutions (Ding and Sachs, 1999).

$P2X_2$ receptors also undergo changes in permeability to cations following prolonged activation and become permeable to larger ions such as N-methyl-D-glucamine (NMDG) and the fluorescent DNA binding dye YO-PRO-1 (molecular weight [MW] ~375) which normally do not permeate through $P2X_2$ channels (Virginio *et al.*, 1999). These changes in permeability are similar to those described at $P2X_4$, $P2X_5$, and $P2X_7$ receptors and it seems likely that they are caused by conformational changes in the receptor (Fisher *et al.*, 2004).

P2X₃ receptors

The $P2X_3$ receptor, and the properties of its heteromeric combination with $P2X_2$ receptors, are amongst the most studied of the P2X family (see below for properties of $P2X_{2/3}$). $P2X_3$ cDNAs were first isolated from rat DRG libraries (Chen *et al.*, 1995; Lewis *et al.*, 1995), and it was originally proposed that these receptors were restricted in localization to sensory ganglia. Subsequent reports described isolation of cDNAs from heart (Garcia-Guzman *et al.*, 1997b).

Homomeric $P2X_3$ receptors are cation selective with similar permeability for Ca^{2+} and Na+ (Virginio *et al.*, 1998b). Marked inward rectification is apparent in most expression systems (Chen *et al.*, 1995). $P2X_3$ receptors desensitize as rapidly as $P2X_1$ receptors although application of low agonist concentrations (e.g. <100

Table 3.3.1.1 Properies of P2X receptors

Channel subunit	Biophysical characteristics	Cellular and subcellular localization	Pharmacology	Physiology	Relevance to disease states
P2X$_1$ Chromosomal localization: 17p13.3 (h) 11 40.0 cM (m) 10q24 (r)	Rise time (10 μM ATP): 7ms Reversal potential (Erev):~0 mV Inward rectification: marked Relative permeability (P)$_{Ca2+}$:P$_{Na+}$: 4.8 (hP2X$_1$;3.9) Chord conductance (−140 to −80 mV): 19 picoSeimens (pS) Desensitization: very rapid (ms) [1, 2]	Smooth muscle, platelets, CNS tissue (cerebellum, pituitary), vas deferens, spinal cord, sensory ganglia [3, 4]	ATP EC$_{50}$: ~0.6 μM Agonist potency: BzATP≥2meSATP = ATP> αβmeATP>ATPγS = ADPβS>ADP Antagonist potency: NF449>IP$_5$> TNP-ATP> PPADS = suramin P2X$_{1/4}$: ATP>αβmeATP (agonists), some sensitivity to TNP-ATP (antagonist) P2X$_{1/5}$: ATP = 2meSATP≥αβmeATP (agonists), some sensitivity to TNP-ATP (partial agonist), suramin and PPADS (antagonist) [1, 5, 6, 7, 8, 9, 10, 11, 12]	Sympathetic co-transmitter with noradrenaline [13], mediating muscle contraction Control of bladder contractility [14] Possible role in tonic control of vascular smooth muscle [15]. Activation of platelets leading to prothrombotic phenotype [16] Regulation of renal function [17] Possible role of P2X$_{1/5}$ in mediating excitatory junction potentials in submucosal arterioles [10]	Possible role in male infertility; KO mice fertility markedly reduced [18] Thromboses [16, 19]
P2X$_2$ Chromosomal localization: 12q24.33 (h) 5F (m) 12q16 (r)	Erev:~ −5 mV Inward rectification: marked Conductance: 30 pS at − 100 mV P$_{Ca2+}$:P$_{Na+}$: 2.8 Desensitization: slow (s) Fast block by divalent cations [2, 11, 20]	Smooth muscle, CNS tissue, retina, chromoffin cells, GI and bladder tissue, autonomic and sensory ganglia [3, 21, 22]	ATP EC$_{50}$: ~10 μM (5-60 μM) Agonist potency: BzATP(human)>ATP> BzATP(rat) = 2meSATP = ATPγS>[αβmeATP] Agonist responses potentiated by zinc. Antagonist potency: RB2>PPADS = BBG = TNP-ATP = suramin [5, 6, 11, 20, 23]	Possible role in myenteric neurons [24] Possible role in carotid body function [25] Other physiological roles described as part of heteromeric combination with P2X$_3$	
P2X$_3$ Chromosomal localization: 11q12 (h) 2D (m) 3q24 (r)	P2X$_3$ homomers: Activation time constant (τ$_{on}$): 1.9 ms 2 component fitted decay time constant (τ$_d$)$_1$: 86 ms, τ$_{d2}$: 1.3 s Erev:~0 mV Rectification: minimal P$_{Ca2+}$:P$_{Na+}$: 1.2 Desensitization: rapid (10–100 ms) P2X$_{2/3}$ heteromers: τ$_{on}$: 27 ms τ$_d$: 14 ms PCa^{2+}:Na$^+$: 1.3 Rectification: minimal Desensitization: slow (s) [2, 26, 27]	Restricted localization to sensory neurons and ganglia [26, 28]	P2X$_3$ homomers: ATP EC$_{50}$: ~0.8 μM Agonist potency: BzATP>2meSATP≥ATP = αβmeATP = ATPγS>D-βγmeATP>[ADP] Antagonist potency: TNP-ATP>A-317491> PPADS>suramin = RB2 [6, 26, 28–30]	P2X$_3$ homomers: Initiation of ATP-induced action potentials in sensory nerves [31–34] Control of bladder function [34] Note that it is not possible to conclusively separate physiological role of P2X$_3$ homomers from P2X$_{2/3}$ heteromers	Chronic inflammatory and neuropathic pain; P2X$_3$/P2X$_{2/3}$ antagonists reverse these pain types, and P2X$_3$ −/− animals have impaired pain behaviours [34, 35] Bladder hyper-reactivity, neurogenic detrusor overactivity [36]
P2X$_4$ Chromosomal localization: 12q24.32 (h) 5 65.0 cM (m) 12q16 (r)	Rise time (30 μM ATP): 1–8 ms τ$_d$: 17.6 s Erev:~1 mV Rectification: minimal P$_{Ca2+}$:P$_{Na+}$: 4.2 Desensitization: slow (s) [2, 37]	Extensive distribution throughout the CNS, microglia, epithelia of ducted glands, smooth muscle of bladder, GI tract, uterus, arteries, fat cells [38]	ATP EC$_{50}$: ~10 μM Agonist potency: ATP = BzATP>2meSATP> αβmeATP>ATPγS (αβmeATP active at human and mouse, but not rat P2X$_4$ receptors) Antagonist potency: All plC$_{50}$<5. Ivermectin and several antagonists potentiate responses P2X$_{4/6}$ rank order agonist potency: 2meSATP = ATP≥αβmeATP [6, 11, 37, 39–41]	Modulation of release of cytokines, BDNF from microglia [42, 43] Control of haemodynamic responses [44, 45]	Upregulated in microglia following nerve injury, role in control of pain processing [42, 43] Potential to be involved in aberrant control of blood pressure in response to volume changes [44]

P2X5 Chromosomal localization: 17p13.3 (h) 11B5 (m) 10q24 (r)	τ_{on}: 410 ms τ_d: 0.2 s Erev: ~ −0.5 mV Rectification: minimal P_{Ca2+}:P_{Na+}: 1.5 Desensitization: slow (s); voltage dependent P2X5 receptor naturally truncated; data taken from chimaeric assembly of human receptors [2, 46]	Proliferating cells of the skin, GI tissue (myenteric and submucosal plexuses), bladder, thymus, spinal cord, skeletal muscle satellite cells [3, 47, 48]	ATP EC50: 0.4–14 μM Agonist potency: ATPγS = ATP = BzATP = αβmeATP(rat)>αβmeATP(human) Antagonist potency: PPADS>TNP-TP = BBG >suramin>RB2 [46, 49]	Truncated receptor; physiology not elucidated [46]. Possible role of P2X1/5 in mediating excitatory junction potentials in submucosal arterioles [10]
P2X6 Chromosomal localization: 22q11.21 (h) 16A1 (m) 11q23 (r)	Erev: ~ −5 mV Inward rectification: slight Desensitization: slow (s) Properties of P2X6 receptors in fully glycosylated form [50]	Overlaps with distribution of P2X4 [49]	ATP EC50: ~0.5 μM Agonist potency: ATP>>αβmeATP Antagonist potency: TNP-ATP>[PPADS> suramin] [49, 50]	Potentially contributes to responses to αβmeATP observed in native tissues when present in fully glycosylated form; depolarization of medial habenula, medial vestibular and locus coeruleus neurons [51, 52, 53]
P2X7 Chromosomal localization: 12q24.31 (h) 12q16 (r)	Rise time: variable Erev: ~ −2 mV Rectification: none Ionic permeability dependent on status or pore dilation Conductance dependent on status of pore dilation (up to 65 nS) Desensitization: absent, current amplitude increases with prolonged agonist application (pore dilation) [54–56]	Cells of haemopoetic origin: immune cells, microglia, macrophages, lymphocytes, mast cells, bone [57–59]	ATP EC50: >1 mM in physiological solutions but 100-700 μM in low divalent cation solutions. Agonist potency: BzATP>ATP>[2meSATP = αβmeATP = ATPγS = ADPβS] Antagonist potency: KN62(human)>BBG> PPADS = suramin>KN62(rat) [6, 11, 60, 61, 62]	Release of mature IL-1β [63, 64] Lysis of T-lymphocytes [65, 66] Activation of PLD [67] Shedding of L-selectin [68] Role in inflammatory and neuropathic pain [69] May play a role in arthritides [70] Loss of function mutation association with chronic lymphocytic leukaemia [71] Loss of function mutation may be associated with impaired host defence [72]

IP5, diinosine pentaphosphate; 2meSATP, 2-methylthio-ATP; PLD, phospholipase-D.

Agonists denoted by square parentheses are classified as inactive (pEC50<4).

[1] Valera S et al. (1994) Nature 371, 516–519; [2] Egan TM et al. (2006) Pflügers Arch 452, 501–512; [3] Burnstock G et al. (2004) Int Rev Cytology 240, 31–304; [4] Ashour F et al. (2006) Neurosci Lett 397, 120–125; [5] Evans RJ et al. (1995) Mol Pharmacol 48, 178–183; [6] Bianchi BR et al. (1999) Eur J Pharmacol 376, 127–138; [7] Abe K et al. (1991) Neurosci Lett 125, 172–174; [8] Lê KT et al. (1999) Biol Chem 274, 15415–15419; [9] North RA et al. (2000) Ann Rev Pharmacol Toxicol 40, 563–580; [10] Surprenant A et al. (2000) Autonom Nerv Sys, 81, 249–263; [11] North RA (2002) Physiol Rev 82, 1013–1067; [12] Nicke A et al. (2005) J Neurochem 92, 925–933; [13] Burnstock G (1972) Pharmacol Rev 24, 509–581; [14] Rapp DE et al. (2005) Eur Urol 48, 303–308; [15] Cario-Toumaniantz C et al. (1998b) Circ Res 83, 196–203; [16] Oury C et al. (2003) Blood 101, 3969–3976; [17] Inscho EW et al. (2003) Clin Invest 112, 1895–1905; [18] Dunn PM (2000) Current Biology 10, R305–R307; [19] Mulryan K et al. (2000) Nature 403, 86–89; [20] Brake AJ et al. (1994) Nature 371, 519–523; [21] Studeny S et al. (2005) Am J Physiol 289, R1155–R1168; [22] Puthssery T et al. (2006) J Comp Neurol 496, 595–609; [23] Lynch KJ et al. (1999) Mol Pharmacol 56, 1171–1181; [24] Ren J et al. (2003) J Physiol 552.3, 809–821; [25] Rong W et al. (2003) J Neurosci 10, 11315–11321; [26] Lewis C et al. (1995) Nature 377, 432–435; [27] Radford K et al. (1997) J Neurosci 17, 6529–6533; [28] Chen C–C et al. (1995) Nature 377, 428–431; [29] Garcia-Guzman M et al. (1997) Mol Brain Res 47, 59–66; [30] Rae MG et al. (1998) Br J Pharmacol 122, 176–180; [31] Dowd E et al. (1997) Br J Pharmacol 122, 286P; [32] Kirkup AJ et al. (1999) J Physiol 520, 551–563; [33] Hilliges M et al. (2002) Pain 98, 59–68; [34] Cockayne DA et al. (2000) Nature 407, 1011–1015; [35] McGaraughy S et al. (2003) Br J Pharmacol 140, 1381–1388; [36] Brady CM et al. (2004) Eur Urol 46, 247–253; [37] Buell G et al. (1996) EMBO J 15, 55–62; [38] Bo X et al. (2003b) Cell Tissue Res 313, 159–165; [39] Garcia-Guzman M et al. (1997) Mol Pharmacol 51, 109–118; [40] Khakh BS et al. (1999) J Neurosci 19, 7289–7299; [41] Jones CA et al. (2000) Br J Pharmacol 129, 388–394; [42] Tsuda M et al. (2003) Nature 424, 778–783; [43] Coull JAM et al. (2005) Nature 438, 1017–1021; [44] Yamamoto K et al. (2006) Nature Med 12, 133–137; [45] Shen J-B et al. (2006) FASEB Journal 20, 277–284; [46] Bo X et al. (2003a) Mol Pharmacol 63, 1407–1416; [47] Ryten M et al. (2002) J Cell Biol 158, 345–355; [48] Ruan HZ et al. (2005) Cell Tissue Res 319, 191–200; [49] Collo G et al. (1996) J Neurosci 16, 2495–2507; [50] Jones CA et al. (2004) Mol Pharmacol 65, 979–985; [51] Edwards FA et al. (1992) Nature 359, 144–147; [52] Chessell IP et al. (1997) Neuroscience 77, 783–791; [53] Sansum AJ et al. (1998) Neuropharmacology 37, 875–885; [54] Surprenant A et al. (1996) Science 272, 735–738; [55] Rassendren F et al. (1997) J Biol Chem 272, 5482–5486; [56] Chessell IP et al. (1997) Br J Pharmacol 121, 1429–1437; [57] Collo G et al. (1997) Neuropharmacology 36, 1277–1283; [58] Buell G et al. (1998) Blood 92, 3521–3528; [59] Sim JA et al. (2004) J Neurosci 24, 6307–6314; [60] Gargett CE et al. (1997) Br J Pharmacol 120, 1483–1490; [61] Chessell IP et al. (1998) Br J Pharmacol 124, 1314–1320; [62] Baraldi PG et al. (2004) Curr Top Med Chem 4, 1707–1717; [63] Perregaux DG et al. (1994) J Biol Chem 269, 15195–15203; [64] Solle M et al. (2001) J Biol Chem 276, 125–132; [65] Blanchard DK et al. (1991) J Immunol 147, 2579–2585; [66] Blanchard DK et al. (1995) J Cell Biochem 57, 452–464; [67] Gargett CE et al. (1997) Cell Biochem 57, 452–464; [67] Jamieson GP et al. (1996) J Cell Physiol 166, 637–642; [69] Chessell IP et al. (2005) Pain 114, 386–396; [70] Labasi JM et al. (2002) J Immunol 168, 6436–6445; [71] Wiley JS et al. (2002) Lancet 359, 1114–1119; [72] Saunders BM et al. (2003) J Immunol 171, 5442–5446.

Table 3.3.1.2 Summary of theoretical (T) and functionally observed (+) heteromeric assembly of P2X receptor subunits

	P2X$_1$	P2X$_2$	P2X$_3$	P2X$_4$	P2X$_5$	P2X$_6$	P2X$_7$
P2X$_1$	+	T	T	+	+	T	-
P2X$_2$		+	+	–	T	T	-
P2X$_3$			+	–	T	–	-
P2X$_4$				+	T	+	-
P2X$_5$					+	T	-
P2X$_6$						+	-
P2X$_7$							T

Lewis *et al.* (1995); Le *et al.* (1998); Torres *et al.* (1999); Surprenant *et al.* (2000); Nicke *et al.* (2005); Aschrafi *et al.* (2004).

nM ATP), elicits inward currents which are sustained for several seconds. However, higher concentrations induce profound and long-lasting desensitization which often confounds pharmacological analysis (Chen *et al.*, 1995; Lewis *et al.*, 1995).

Recovery rates from desensitization are agonist-dependent with recovery from desensitization being more rapid for low-potency agonists such as αβ—methylene-adenosine 5′-triphosphate(αβmeATP) or adenosine 5′-(3-thio) triphosphate (ATPγS), than for higher potency agonists such as ATP (Sokolova *et al.*, 2004; Pratt *et al.*, 2005). The relationship between potency and rate of recovery from desensitization suggests that dissociation of agonist from the receptor may underlie the differences in recovery rate. Recovery from desensitization can also be accelerated by calcium ions (Cook *et al.*, 1998).

Inflammatory mediators, such as substance P and bradykinin, enhance P2X$_3$ receptor responses (Paukert *et al.*, 2001), possibly due to effects on the phosphorylation status of intracellular serine and threonine residues. P2X$_3$ receptor currents may also be increased by extracellular nucleotides (UTP, GTP and ATP) through activation of ectoprotein kinase C (Stanchev *et al.*, 2006).

P2X$_4$ receptors

P2X$_4$ cDNAs were isolated by a number of groups from ganglia, brain and endocrine tissue (Buell *et al.*, 1996b; Seguela *et al.*, 1996; Soto *et al.*, 1996b; Wang *et al.*, 1996). Subsequently, P2X$_4$ receptor cDNAs have been identified from other species (Garcia-Guzman *et al.*, 1997a; Townsend-Nicholson *et al.*, 1999; Diaz-Hernandez *et al.*, 2002). Splice variants of the human and mouse P2X$_4$ receptors have also been identified and characterized (Dhulipala *et al.*, 1998; Townsend-Nicholson *et al.*, 1999), with the human splice variant cDNA containing a substitution of the first 90 amino acids with an alternative 35 amino acid coding region. However, only the full-length cDNA gives rise to functional currents when expressed in *Xenopus* oocytes (Dhulipala *et al.*, 1998). The murine P2X$_4$ splice variant lacks a 27 amino acid region in the putative extracellular domain, and does form, an albeit poorly functional, channel. This splice variant may interact with the full-length receptor to produce a functional channel with reduced affinity for ATP (Townsend-Nicholson *et al.*, 1999).

The P2X$_4$ receptor is classified as a slowly desensitizing receptor, with a relatively high permeability to Ca^{2+}, which does not exhibit robust inward rectification (Buell *et al.*, 1996b). Application of ~EC$_{50}$ concentrations of ATP activates a sustained current which

does not wane appreciably over several seconds; some desensitization is observed at higher ATP concentrations. Recent studies suggest that for the human P2X$_4$ receptor, two positions within the C-terminal tail are critical for controlling the rate of desensitization. Thus, alanine scanning mutagenesis revealed that Lys^{373} and Tyr^{374} play a key role in the regulation of desensitization rate (Fountain and North, 2006). The P2X$_4$ receptor appears to be susceptible to long-lasting inactivation or run-down, such that in oocytes, a period of 10–15 min is required to obtain reproducible responses (Bo *et al.*, 1995). Run-down has also been described for P2X$_4$ receptor expressed heterologously in mammalian systems (Jones *et al.*, 2000). The process of run-down appears unrelated to the desensitization process described above. It can be prevented by recording from contiguous, electrically coupled rafts of cells (Miller *et al.*, 1998) or by using a perforated patch approach (Fountain and North, 2006) suggesting that loss of a diffusible intracellular component is responsible.

Finally, prolonged activation of P2X$_4$ receptors leads to an increase in permeability to large molecules such as NMDG and YO-PRO-1 similar to that occurring at P2X$_2$, P2X$_5$, and P2X$_7$ receptors (Khakh *et al.*, 1999a; Virginio *et al.*, 1999).

P2X$_5$ receptors

The first P2X$_5$ cDNA described was isolated from rat sympathetic ganglia and heart libraries (Collo *et al.*, 1996; Garcia-Guzman *et al.*, 1996). Since then, multiple P2X$_5$ cDNAs have been described from various species, which largely result in membrane-expressed receptors, albeit with limited or difficult to detect function. The human P2X$_5$ cDNA, however, first isolated from brain (Lê *et al.*, 1997), is a truncated receptor with 62 per cent identity to the rat orthologue, and as such displays only a single transmembrane domain. When expressed alone, it does not form detectable channels, but chimeric combination with the rat P2X$_5$ receptor (Lê *et al.*, 1997) yields a functional channel.

Heterologous expression of rat or mouse P2X$_5$ produces channels which are activated by ATP, but evoked currents are very small, and have not been extensively characterized (Collo *et al.*, 1996; Garcia-Guzman *et al.*, 1996). However, construction of the above chimera (Lê *et al.*, 1997), or construction of a 'full-length' human P2X$_5$ receptor by incorporation of a sequence corresponding to exon 10 where a consensus sequence for splicing is present (Bo *et al.*, 2003a), both give rise to fully functional P2X$_5$ channels. The former gave rise to currents which desensitized rapidly at high

ATP concentrations, and did not readily recover from desensitization even with extensive washout periods. The latter gave rise to responses which persisted during short periods of agonist applications (<2 s), but declined with longer application. The currents showed little rectification and studies of the permeation properties of the channel revealed high permeability to Ca^{2+}.

The $P2X_5$ receptor has several atypical biophysical properties. The rate of opening is slower than at $P2X_2$ receptors. Responses are modified by membrane potential such that current deactivation after agonist removal is faster at −60 than at +30 mV holding potential (Bo *et al.*, 2003a) and desensitization is considerable at −60 mV but minimal at +30 or +40 mV (Ruppelt *et al.*, 2001; Bo *et al.*, 2003a). Reducing extracellular calcium concentration reduces receptor desensitization (Bo *et al.*, 2003a) and run-down of responses (Wildman *et al.*, 2002). The $P2X_5$ receptor undergoes the same changes in permeability to NMDG and YO-PRO-1 observed at $P2X_2$, $P2X_4$ and $P2X_7$ receptors. At the human $P2X_5$ receptor, permeability to NMDG is observed immediately after agonist application, which contrasts with the more slowly developing increase observed with the other P2X receptors. Finally, a most unusual feature of the $P2X_5$ receptor is its high permeability to chloride ions (Ruppelt *et al.*, 2001; Bo *et al.*, 2003a). This is not observed with any other P2X receptors, including those receptors that also undergo increases in permeability to large ions.

$P2X_6$ receptors

The rat $P2X_6$ receptor was isolated from superior cervical ganglia cDNA and from rat brain (Collo *et al.*, 1996; Soto *et al.*, 1996a). When heterologously expressed in mammalian systems, $P2X_6$ was originally described as being similar to $P2X_4$ (Collo *et al.*, 1996), but receptor expression was highly inefficient, such that only a small fraction of transfects displayed any functional responses.

More recent studies suggest that functional expression of the $P2X_6$ receptor is determined by glycosylation status (Jones *et al.*, 2004). Here, dilution cloning of stably transfected HEK-293 cells revealed a population expressing a slowly desensitizing, antagonist sensitive current. This receptor was also sensitive to the $P2X_1$ and $P2X_3$ receptor agonist, αβmeATP, which may account for responses in tissues which apparently lack receptors traditionally recognized to be sensitive to this agonist. These studies also revealed that non-functional full length $P2X_6$ receptors are expressed at the cell membrane, and further glycolsylation is required to confer a functional phenotype. Studies by Ormond *et al.* (2006) have also shown that trafficking of the $P2X_6$ receptor through the endoplasmic reticulum is inhibited by an uncharged region of the N-terminal; removal of this region, or addition of charge results in homotrimeric assembly, glycosylation, and delivery to the plasma membrane.

$P2X_7$ receptors

The $P2X_7$ was functionally identified and studied in the guise of the cytolytic P2Z receptor (Gordon, 1986; Burnstock, 1990) long before its formal identification from a rat brain cDNA library (Surprenant *et al.*, 1996). The human (Rassendren *et al.*, 1997) and mouse (Chessell *et al.*, 1998b) $P2X_7$ receptors were identified shortly thereafter.

Multiple polymorphisms have been described for human $P2X_7$ receptors and five loss of function polymorphisms have been described. Two are in the C-terminal sequence (bases A1513C and T1729A). A third is in the putative ATP binding site (G946A). The fourth is an intronic polymorphism at the exon-1/intron-1 boundary, and the fifth is in the C-terminus (C1096G). A gain of function polymorphism has been reported for a T489C mutation (Cabrini *et al.*, 2005) although this was not confirmed by Denlinger *et al.* (2005) who instead suggested that an A1405G polymorphism may cause gain of function. These findings and corresponding amino acid changes are detailed in Shemon *et al.* (2006) and Denlinger *et al.* (2006). Many of the polymorphisms are present at high frequency (>0.1) and more than one polymorphisms can exist in the same individual. Indeed, the presence of multiple polymorphisms may exacerbate loss of function (Fernando *et al.*, 2005). It should be noted that loss of function for most polymorphisms has been assessed in pore-formation assays and there is no information on their effects on cytokine release. Polymorphisms also exist in mice with the P451L (Adriouch *et al.*, 2002) and S342F (Guerra *et al.*, 2003) amino acid changes causing loss of function.

Multiple splice variants of the human $P2X_7$ receptor ($P2X_{7a-j}$) have been identified (Cheewatrakoolpong *et al.*, 2005; Feng *et al.*, 2006). These lead to receptors truncated at the N- or C-termini and/or missing exons. Most of these are non-functional or have reduced function but are expressed in spleen, lung, and brain, particularly those truncated at the C-terminus (Cheewtrakoolpong *et al.*, 2005). The $P2X_{7-j}$ splice variant, truncated close to TM2, is not functional but is detected in several types of cancer cell and may block wild-type $P2X_7$ receptor function (Feng *et al.*, 2006).

The $P2X_7$ receptor mediates non-rectifying, non-desensitizing currents. This receptor also demonstrates strikingly different properties to the other P2X family members, notably dramatic changes in current time course and amplitude. Put most simply, at high agonist concentrations, sustained or repeated activation of the receptor yields an apparent pore dilation phenomenon which is reflected by either dramatic growth of current amplitude, by very sustained tail currents, or by both (Chessell *et al.*, 1997b; Chessell *et al.*, 2001). This pore dilation phenomenon is associated with increasing permeability of the channel assembly to NMDG and YO-PRO1 (Figure 3.3.1.1) and this may occur through pannexin channels (Pelegrin and Surprenant, 2006). It is notable that the changes in ionic permeability at $P2X_7$ receptors can be observed on a macroscopic level and detected without the complex ionic substitution experiments required to detect 'pore-dilation' at the $P2X_2$ and $P2X_4$ receptors. A further complex feature of the $P2X_7$ receptor is the very slow closure time (5–20 s) of the channel after agonist removal. This is very pronounced at rat receptors (Surprenant *et al.*, 1996), modest at human $P2X_7$ receptors where it is manifest as a prolonged slowly declining tail current (Rassendren *et al.*, 1997), and absent at mouse receptors (Chessell *et al.*, 1997b). Channel closure times appear related to agonist potency and may reflect slow dissociation of agonist from the receptor (Hibell *et al.*, 2001a), although such a slow dissociation rate is unusual given that agonist potency at $P2X_7$ receptors is low (1–100 μM).

The protein sequence of the $P2X_7$ receptor is 35–45 per cent identical to the other six P2X receptors, and differs most strikingly in its much longer intracellular COOH terminal. Studies suggest that this long C-terminal tail contributes to the unusual properties of this subunit, as truncations abolish transition into the 'large-pore' form of the receptor (Smart *et al.*, 2003). However, given that this pore dilation phenomenon has also been described

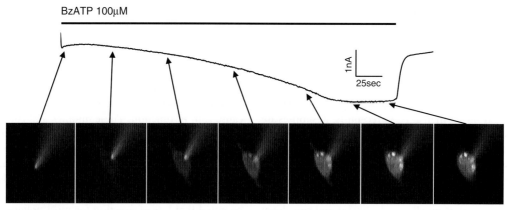

BzATP 100µM

1nA

25sec

Fig. 3.3.1.1 Whole-cell current elicited by BzATP (100 µM) in a single HEK-293 cell expressing hP2X$_7$. The low divalent cation-containing extracellular solution included 2 µM YO-PRO-1, a dye which fluoresces when bound to DNA. The time course demonstrates the formation of a YO-PRO-1 permeable large pore with increasing time of agonist application. See also colour plate section.

for P2X$_2$ and P2X$_4$ receptors (Khakh et al., 1999a; Virginio et al., 1999), the long C-terminal tail is clearly not the only determinant. More recently, studies suggest that there are two independent pore phenomena; one allowing increases in permeability to large cations such as N-methyl-D-glucamine, and another allowing permeability of DNA binding dyes (Jiang et al., 2005). Moreover, convincing evidence suggest that pannexin-1 mediates the P2X$_7$-receptor-activated dye-uptake pathway (Pelegrin and Surprenant, 2006). The role of pannexin-1 in dye uptake and pore-dilation at P2X$_2$, P2X$_4$ and P2X$_5$ receptors is not known.

In addition to causing changes in permeability, prolonged or repeated exposure to agonist results in increases in agonist potency at recombinant (Chessell et al., 1997b; Hibell et al., 2001a) and native (Chakfe et al., 2002) P2X$_7$ receptors. The mechanism underlying these changes is not known but phosphorylation of P2X$_7$ receptors can also change agonist potency (Adinolfi et al., 2003).

Not only are the biophysics of the P2X$_7$ receptor complex but the range of cell responses initiated following receptor activation is considerable. Thus, P2X$_7$ activation has been shown to activate mitogen-activated protein kinase (MAPK), extracellular regulated kinase (ERK) 1/2, phospholipase A2, nuclear factor of activated T lymphocytes, and to lead to membrane blebbing, microvesicle shedding, cytokine release and necrotic and apoptotic cell death (North, 2002).

Heteromeric P2X receptors

Whilst biochemical evidence for co-assembly of different P2X receptor subunits (see Table 3.3.1.2) has been generated using a variety of techniques: functional effects have been shown only for P2X$_{1/4}$, P2X$_{1/5}$, P2X$_{2/3}$ P2X$_{2/6}$ and P2X$_{4/6}$.

P2X$_{1/4}$ heteromeric receptors were observed following co-purification of these subunits, in a trimeric assembly, from a *Xenopus* expression system (Nicke et al., 2005). Functionally, P2X$_{1/4}$ receptors show a kinetic profile similar to P2X$_4$ homomeric receptors, with desensitization rate an order of magnitude lower than that of homomeric P2X$_1$ receptor (Nicke et al., 2005).

P2X$_{1/5}$ receptor heteromers produce slowly desensitizing currents which were not observed with either subunit expressed alone (Lê et al., 1999). As with P2X$_{1/4}$, the most striking differences were observed with respect to pharmacology (see below). Co-expression of P2X$_{1/5}$ in a mammalian system revealed a similar phenotype; here the authors also characterized Ca^{2+} permeability, which was

less than that observed for the other P2X receptors (Surprenant et al., 2000). The phenotype of heteromeric P2X$_{1/5}$ receptors has also been observed in native tissue systems (Surprenant et al., 2000).

The P2X$_{2/3}$ hetero-oligomeric receptor appears to share its pharmacological profile with homomeric P2X$_3$, but some of the biophysical properties resemble homomeric P2X$_2$. Thus, P2X$_{2/3}$ receptors desensitize slowly, and activation produces inward currents which are sustained for several seconds even with EC$_{90}$ concentrations of ATP (Lewis et al., 1995). Functional responses for P2X$_{2/3}$ are readily observed in both heterologous and native systems.

The phenotype of P2X$_{4/6}$ closely resembles that described for homomeric P2X$_6$ receptors when adequately glycosylated to produce a functional, membrane-expressed channel (Lê et al., 1998; Jones et al., 2004). Thus, the assembly produces a slowly desensitizing receptor which has a distinct pharmacology to P2X$_4$, but a similar pharmacology to P2X$_6$. In contrast P2X$_{2/6}$ receptors more closely resemble P2X$_2$ receptors (King et al., 2000).

P2X receptor interacting proteins

As with many ion channels and receptors it is becoming increasingly evident that P2X receptors co-assemble with or interact with multiple membrane proteins. In the case of the P2X$_2$ receptor there is now considerable evidence to suggest interactions with nicotinic (Khakh et al., 2005), gamma aminobutyric acid$_A$ (GABA$_A$) (Boue-Grabot et al., 2004) and 5-hydroxytryptamine (5-HT) receptors (Barajas-Lopez et al., 2002) and that this can lead to crosstalk of responses. The P2X$_2$ receptor can also directly interact with Fe65, a brain beta-amyloid associated protein, and this association inhibits the ability of the P2X$_2$ receptor to change its permeability (Masin et al., 2006).

P2X$_4$ receptors interact with adaptor protein 2 and this may regulate their cell-surface expression and recycling (Royle et al., 2002). Finally, P2X$_7$ receptors interact with a host of proteins (Kim et al., 2001), and this may account for the diverse range of responses elicited by P2X$_7$ receptor activation.

3.3.1.4 Localization of P2X receptors

The distribution of P2X$_1$ receptors is broad. They have been localized at the mRNA or protein level to smooth muscle, platelets, central nervous system (CNS) tissue; particularly cerebellum, pituitary gland and vas deferens, and spinal cord dorsal horn neurons (Burnstock and Knight, 2004; see also Ashour et al., 2006). In keeping with the majority of the P2X receptors, P2X$_1$ has also been found in sensory ganglia (Brake et al., 1994).

P2X$_2$ receptors have been localized to smooth muscle, CNS tissue, cochlea, retina, chromoffin cells, gastrointestinal (GI) and bladder tissue, and autonomic and sensory ganglia (Housley, 1998; Burnstock and Knight,2004; Studeny et al., 2005; Puthssery and Fletcher, 2006).

The P2X$_3$ receptor is largely localized to sensory nerve tissue, which includes sensory nerve terminals, and associated ganglia (Chen et al., 1995; Lewis et al., 1995). Of the P2X receptors described, P2X$_3$ appears to exhibit the most restricted localization. The receptor is found principally in peripherin-containing cells (Chen et al., 1995), and expression in sensory neurons is limited to neurons which express isolectin B4 binding sites (Bradbury et al., 1998). Co-expression of P2X$_3$ with TRPV1 in some sensory neurons has increased the level of interest in this receptor as a potential pain target (Guo et al., 1999).

The P2X$_4$ receptor is extensively distributed throughout mammalian tissues. As well as recent studies demonstrating functional P2X$_4$ receptors on microglial cells (Tsuda et al., 2003), use of a monoclonal antibody has shown distribution of P2X$_4$ throughout the central and peripheral nervous systems, epithelia of ducted glands and airways, smooth muscle of bladder, GI tract, uterus, arteries, and fat cells (Bo et al., 2003b).

The P2X$_5$ receptor has been localized to proliferating cells in the skin, GI tissue, bladder, thymus, and spinal cord (Burnstock and Knight, 2004), and skeletal muscle satellite cells (Ryten et al., 2002). Within the GI tract, immunoreactivity is distributed in the myenteric and submucosal plexuses (Ruan and Burnstock, 2005).

The distribution of P2X$_6$ receptors overlaps heavily with P2X$_4$ (Collo et al., 1996). In addition to distribution in the brain, P2X$_6$ is also found in motor neurons of the spinal cord (Burnstock and Knight, 2004). Within the kidney, P2X$_6$, together with P2X$_4$, is found throughout the renal tubule epithelium (Turner et al., 2003).

Distribution of P2X$_7$ receptors has been the subject of some debate. When first identified, it was concluded that P2X$_7$ is expressed only on cells of immune and haemopoetic origin (Collo et al., 1997), and is not present in neurons, although it is unequivocally expressed in microglia and other neuroglia. Studies with a monoclonal antibody confirmed localization to human lymphoid tissue (Buell et al., 1998). With respect to localization on neuronal cells, the field has been confounded by the use of antibodies which appear to lack specificity for the P2X$_7$ receptor, and recent studies suggest that P2X$_7$ protein is not expressed on neuronal cells (Sim et al., 2004). The P2X$_7$ receptor has also been functionally identified in bone cells (Ke et al., 2003).

3.3.1.5 Pharmacology

The complexities of studying P2X receptors are considerable and are worth considering before describing their pharmacological characterization. Most P2X agonists are nucleotide triphosphates which can be degraded by ecto-enzymes (Zimmerman, 2000). This can confound agonist potency determination and the metabolites generated may also activate other P2X, P2Y, and adenosine receptors.

This problem is not restricted to ATP as αβmeATP, ATPγS and 2′-3′-o-(4-benzoylbenzoyl)-adenosine 5′-triphosphate (BzATP) can be metabolized (Cascalheira and Sebastiao 1992; Kukley et al., 2004). ATP and BzATP can even be broken down by ectoATPases present in tissue culture serum while BzATP binds to serum albumen with resultant reduction in apparent potency (Cowen et al., 1991; Michel et al., 2001). Metabolism is more likely to occur in native tissues (Evans and Kennedy, 1994), although Rettinger and Schmaltzing (2004) have shown that even in electrophysiology experiments with recombinant receptors, oocyte follicle cells, known to express ectoATPase (Ziganshin et al., 1995), could reduce apparent ATP potency at P2X$_1$ receptors 100-fold.

Nucleotide triphosphates also activate ecto-protein kinases (Redegeld et al., 1999) and this may affect P2X$_3$ receptor function (Stanchev et al., 2006). In addition, nucleotides can be interconverted by nucleoside diphosphokinase (NDPK) which can generate GTPγS from exogenous ATPγS and endogenous GDP or produce UTP from ATP and endogenous UDP (Blevins et al., 1996; Harden et al., 1997).

A further complexity is accurate determination of agonist potency for rapidly desensitizing receptors where measurement of equilibrium responses may be impossible due to desensitization (e.g. P2X$_1$) (Rettinger and Schmaltzing, 2004). Similar problems may occur for receptors prone to run-down (see P2X$_1$, P2X$_4$ and P2X$_5$) or P2X$_7$ receptors, where continued activation increases agonist potency. The now well-characterized ability of cells to release various nucleotides in responses to diverse stimuli (osmotic and shear stress) can further complicate studies. The nucleotides released may serve as substrates for NDPK or activate P2 receptors (Harden et al., 1997, Lazarowski et al., 2000). Indeed, ATP released from cells has been shown to mediate desensitization of P2X$_1$ and P2X$_{1/5}$ receptors (Buell et al., 1996a; Surprenant et al., 2000).

Agonist potency can also be affected by ions present in physiological solutions. Sodium and chloride ions affect agonist potency at P2X$_7$ receptors while divalent cations bind to nucleotides and affect the form of nucleotide present which may, thus, affect the potency of agonists at P2 receptors (Gordon, 1986). Calcium and magnesium also activate ectoATPases and, as with zinc, copper and protons, can change agonist potency at several P2X receptor types. High concentrations of nucleotides chelate calcium ions and reduce its concentration in physiological solutions.

Agonist potency varies according to technique and is generally higher in calcium influx studies (Bianchi et al., 1999) than in electrophysiology studies. Thus, BzATP is a potent P2X$_1$ receptor agonist in calcium influx studies (EC$_{50}$~2 nM) (Bianchi et al., 1999) but is much weaker (EC$_{50}$~700 nM) (Evans et al., 1995; Wildman et al., 2002) in electrophysiology studies. The majority of the agonist potency estimates quoted in the following sections are from electrophysiology studies.

Studies using antagonists are hampered by limited characterization of antagonist activity, poor specificity of compounds and often marked species- and technique- dependent differences in antagonist effect. Few studies have calculated pA$_2$ or pK$_B$ values, and use of IC$_{50}$ values may contribute to the often considerable variations in antagonist potency reported. Many studies on P2X receptors use suramin and pyridoxalphosphate-6-azophenyl-2′-4′-disulphonate (PPADS) as antagonists but the specificity of suramin is limited (Voogd et al., 1993), with non-P2X effects often observed at lower concentrations than affect P2X receptors (e.g. IC$_{50}$ ~10 nM for protein kinase A) (Okabe *et al.*, 2006). PPADS interacts with multiple P2X and P2Y receptors subtypes (Burnstock and Knight, 2004) and at high concentrations may have non-specific actions (Shehnaz *et al.*, 2000). Many coloured dyes such as reactive blue 2 (RB2, probably the same molecule as cibacron blue) and brilliant blue G (BBG, also known as coomassie brilliant blue) block P2X receptors but also affect ecto-nucleotidases and other ATP-dependent enzymes complicating their use (Bultmann *et al.*, 1999), although this may be an issue with BBG.

A diverse range of compounds have been found to interact with the P2X receptors. The anion transport inhibitor 4,4′-diisothiocyanatostilbene-2,2′-disulfonate (DIDS) (Bültmann and Starke., 1994), hexamethylene amiloride (HMA) (Wiley *et al.*, 1992), d-tubocurarine (Nakazawa *et al.*, 1991), decavanadate (Michel *et al.*, 2006), phenol red and phenolphthalein (King *et al.*, 2005) block P2X receptors. While many of these compounds are not useful for receptor characterization there are clear implications when they are utilized for other purposes or are present in culture media.

Species differences in antagonist sensitivity have been well documented with $P2X_7$ receptors, and also occur at $P2X_3$ and $P2X_4$ receptors (see below). Finally, there are differences in antagonist potency and effects between laboratories (see $P2X_4$) and even between techniques, e.g. in electrophysiology studies, calmidazolium is a potent inhibitor of $P2X_7$ receptors but has no effect when dye accumulation is measured (Virginio *et al.*, 1997).

The pharmacological properties of the various P2X receptor types are described below and summarized in Table 3.3.1.1. More detailed information on key antagonists is presented in Table 3.3.1.3.

$P2X_1$ receptors

The $P2X_1$ receptor is the prototypic 'αβmeATP-sensitive' receptor and is activated by most P2X agonists. D-βγmeATP has higher potency at $P2X_1$ receptors (Evans *et al.*, 1995) than at $P2X_3$ receptors (Chen *et al.*, 1995) and may be of value in differentiating between these receptors (North, 2002), although D-βγmeATP also activates $P2X_5$ receptors (Wildman *et al.*, 2002). L-βγmeATP is equipotent to D-βγmeATP at $P2X_1$ receptors (Evans *et al.*, 1995) but inactive on rat $P2X_3$ receptors in DRG (Rae *et al.*, 1998) and may be a more selective $P2X_1$ receptor agonist.

The desensitizing properties of αβmeATP have been used as a technique for 'blocking' $P2X_1$ receptors (Kasakov and Burnstock, 1982) and Rettinger and Schmaltzing (2004) have also shown that very low concentrations of ATP can mediate $P2X_1$ receptor desensitization ($EC_{50} = 3$nM).

In addition to the compounds illustrated in Tables 3.3.1.1 and 3.3.1.3, numerous dyes (Bultmann *et al.*, 1999), DIDS ($IC_{50} = 2\,\mu M$) (Bultmann and Starke., 1994), several PPADS analogues (cyclic pyridoxine-α4,5-monophosphate-6-azo-phenyl-2′,5′disulphonate (MRS2220) (Jacobsen *et al.*, 1998); pyridoxal-5′-phosphate-6-(2′-naphthylazo-6′-nitro-4′,8′-disulfonate) (PPNDS), (Lambrecht *et al.* 2000), and several suramin analogues (8,8′-[carbonylbis(imino-4,1-phenylenecarbonylimino-4,1-ph enylenecarbonylimino)) bis(1,3,5-naphthalenetrisulfonic acid] (NF279), (Rettinger *et al.*, 2000); 4,4′,4″,4‴-(carbonylbis(imino-5,1,3-benzenetriylbis (carbonylimino)))tetrakis-benzene-1,3-disulfonic acid (NF449), Hulsmann *et al.*, 2003) block $P2X_1$ receptors. NF449 is the most potent $P2X_1$ receptor antagonist ($pA_2 = 10.7$) with 1000-fold selectivity verses most P2X and several P2Y receptors, although its association rate is slow and this may contribute to discrepancies between potency estimates ($P2X_1$ $IC_{50} = 490$ nM) (Kassack *et al.*, 2004). NF449 also has high affinity for $G_s\alpha$ ($IC_{50} = 490$ nM) (Hohenegger *et al.*, 1998) and a protein kinase A isoform ($IC_{50} = 70$ nM) (Okabe *et al.*, 2006).

Ions do not greatly facilitate characterization of $P2X_1$ receptors although zinc inhibits $P2X_1$ responses and protons produce modest decreases in agonist potency (Stoop *et al.*, 1997, Wildman *et al.*, 1999).

Overall, $P2X_1$ receptors can be defined in terms of their sensitivity to low micromolar concentrations of αβmeATP, rapid desensitization, inhibition by zinc ions and blockade by selective $P2X_1$ receptor antagonists (NF449).

$P2X_{1/4}$ receptors

The $P2X_{1/4}$ receptor exhbits some of the pharmacological properties of the $P2X_1$ receptor (Nicke *et al.*, 2005). αβmeATP produced a biphasic dose–response curve (DRC) and had higher efficacy than at $P2X_4$ receptors while effects of 500 nM trinitrophenyl-ATP (TNP-ATP) and 10 μM suramin were similar to those observed at $P2X_1$ receptors.

Table 3.3.1.3 Summary of activity of commonly used antagonists at P2X receptor subtypes

	P2X1	P2X1/4	P2X1/5	P2X2	P2X2/3	P2X3	P2X4*	P2X5**	P2X6	P2X7
Suramin	1 μM	~ $P2X_1$	1 μM	10–33 μM	1 μM	3–100 μM	>30 μM	1–4 μM	>100 μM	1–300 μM
PPADS	1 μM	–	1 μM	1–3 μM	0.1–1 μM	1–5 μM	>30 μM	0.2–3 μM	>22 μM	0.05–45 μM
TNP-ATP	6 nM	~ $P2X_1$	64–720 nM	2 μM	2–7 nM	0.9 nM	15 μM	0.4–3 μM†	0.8 μM	>30 μM
RB2	10 μM	–	–	0.4 μM	–		>50 μM	18 μM	–	40 μM
BBG	>5 μM	–	>10 μM	1.4 μM	>10 μM	>10 μM	>10 μM	530 nM	–	10–400 nM

* Several antagonists potentiate responses. ** Human $P2X_5$ full length receptor constructed to include exon 10. † Partial agonist in 1 study (North, 2002).

P2X1, Evans *et al.* (1995); Bultmann *et al.* (1999). P2X1/4, Nicke *et al.* (2005); P2X1/5, Torres *et al.* (1998a); Haines *et al.* (1999); Lê *et al.* (1999); Surprenant *et al.* (2000); P2X2, Brake *et al.* (1994); Evans *et al.* (1995); King *et al.* (1997, 2000); Lynch *et al.* (1999); North (2002); P2X3 and P2X2/3, Chen *et al.*(1995); Lewis *et al.* (1995); Garcia-Guzman *et al.* (1997b); North (2002); Spelta *et al.* (2002); Jarvis *et al.* (2003); P2X4 and P2X4/6, Bo *et al.* (1995); Buell *et al.* (1996b); Seguela *et al.* (1996); Soto *et al.* (1996b); Lê *et al.* (1998); Khakh *et al.* (1999b); Townsend-Nicholson *et al.* (1999); Jones *et al.* (2004); P2X5, Collo *et al.* (1996); Wildman *et al.* (2002); Bo *et al.* (2003a); P2X6, Collo *et al.* (1996); Jones *et al.* (2004); P2X7, Soltoff *et al.* (1989); Surprenant *et al.* (1996); Rassendren *et al.* (1997); Chessell *et al.* (1998a; 1998b); Bianchi *et al.* (1999); Hibell *et al.* (2001b); Baraldi *et al.* (2004); Suramin and PPADS, Bianchi *et al.* (1999); Trinitrophenyl-ATP (TNP-ATP), Virginio *et al.* (1998a); Brilliant blue G (BBG), Jiang *et al.* (2000).

P2X$_{1/5}$ receptors

ATP potency at P2X$_{1/5}$ receptor differs between studies (EC$_{50}$ = 55 nM-5 μM) but the slope of the DRC consistently appears shallower than at P2X$_1$, receptors possibly reflecting the slower desensitization of P2X$_{1/5}$ receptors (Torres et al., 1998a; Lê et al., 1998; Haines et al., 1999; Surprenant et al., 2000). 2meSATP is a full agonist while αβmeATP is 2–30-fold less potent than ATP and a partial agonist in some studies. Suramin and PPADS are potent antagonists but TNP-ATP possesses relatively low potency and was a partial agonist in one study (Surprenant et al., 2000).

P2X$_2$ receptors

P2X$_2$ receptors are activated by 3–10-fold higher concentrations of ATP than activate P2X$_1$ or P2X$_3$ receptors. At rat P2X$_2$ receptors, BzATP is a partial agonist (EC$_{50}$ = 23 μM), with threefold lower potency than ATP. However, at human receptors it is a full agonist (EC$_{50}$ = 0.4 μM) and 2.5-fold more potent than ATP (Lynch et al., 1999). A defining feature of the P2X$_2$ receptor is its universal insensitivity to αβmeATP (EC$_{50}$ ~ 1 mM) (Spelta et al., 2002) and βγmeATP.

There are no P2X$_2$ receptor selective antagonists. Most antagonists block P2X$_2$ receptors although TNP-ATP and NF449 do so with low affinity (Table 3.3.1.3). Suramin potency is affected by pH, being 130-fold more potent at pH 6.5 than at 7.5 (King et al., 1997). d-Tubocurarine (1-1000 μM) blocked recombinant and native P2X$_2$ receptors in some studies (Nakazawa et al., 1991; Brake et al., 1994; but see Evans et al., 1995). Calcium inhibits P2X$_2$ receptors with an IC$_{50}$ of 5 mM (Evans et al., 1995) which is lower than its IC$_{50}$ at P2X$_1$, P2X$_3$ and P2X$_4$ receptors, although similar to that at the P2X$_7$ receptor (Virginio et al., 1997).

Zinc and copper differentiate P2X$_2$ from P2X$_4$ receptors as both ions potentiate responses at P2X$_2$ receptors but only zinc potentiates responses at P2X$_4$ receptors (Xiong et al., 1999). Protons potentiate P2X$_2$ receptor-mediated responses and ATP potency at P2X$_2$ receptors increases as pH is reduced from 7.4 to 5. This contrasts with effects of pH at all other P2X receptors with the exception of P2X$_{2/3}$ (Stoop et al., 1997).

Overall, the P2X$_2$ receptor can be defined in terms of its insensitivity to αβmeATP, low potency for antagonists such as TNP-ATP and N449, potentiation by zinc, copper and protons and by its biophysical properties, including its slow desensitization rate.

P2X$_{2/6}$ receptors

A single report of the P2X$_{2/6}$ heteromer demonstrated some slight differences from P2X$_2$ receptors but responses to ATP (EC$_{50}$ = 32 μM) were still potentiated by zinc and protons (King et al., 2000). Furthermore, as observed at P2X$_2$ receptors, suramin was less potent at pH 7.5 (IC$_{50}$ = 6 μM) than at pH 6.5 (IC$_{50}$ = 13 and 1610 nM). Inhibition curves were biphasic at pH 6.5, perhaps suggesting some modulation by the P2X$_6$ subunit.

P2X$_3$ receptors

The rank order of potency is similar to that at P2X$_1$ receptors except for the aforementioned low potency of D- and L-βγmeATP (Table 3.3.1.1). CTP is a full P2X$_3$ receptor agonist although 30-fold weaker than ATP.

PPADS potently blocks P2X$_3$ receptors but its effects are slow in onset and offset (Spelta et al., 2002). Suramin is an antagonist of rat P2X$_3$ receptors (IC$_{50}$ = 0.8 μM; Bianchi et al., 1999) but is weaker at human receptors (IC$_{50}$>20 μM, Garcia-Guzman et al., 1997b; Bianchi et al., 1999). Cibacron blue potentiates responses at human P2X$_3$ receptors (Alexander et al., 1999) while estradiol blocks P2X$_3$ receptors in mouse DRG (IC$_{50}$ = 27 nM) but this is an indirect effect via cell surface ER-alpha receptors (Chaban et al., 2003).

TNP-ATP (Virginio et al., 1998a) is a potent and selective antagonist of P2X$_3$ and P2X$_{2/3}$ receptors although it possesses high potency at P2X$_1$ receptors (IC$_{50}$ = 6 nM) and there is no detailed data on its stability in vitro or in vivo. In a manner analogous to that described at the P2X$_1$ receptor, several agonists have been shown to possess unexpectedly high potency to desensitize P2X$_3$ receptors (McDonald et al., 2002; Sokolova et al., 2004; Pratt et al., 2005). More recently, 5-({[3-phenoxybenzyl][(1S)-1,2,3,4-tetrahydro-1-naphthalenyl]amino}carbonyl)-1,2,4-benzenetricarboxylic acid (A-317491) has been identified as a non-nucleotide high-affinity antagonist of P2X$_3$ (K$_i$ = 22 nM) and P2X$_{2/3}$ (K$_i$ = 92 nM) receptors with >100-fold selectivity against other P2X receptors (Jarvis et al., 2003).

Zinc produces variable effects at P2X$_3$ receptors, potentiating or slightly inhibiting responses depending on concentration applied and time of exposure (Wildman et al., 1999; North, 2002).

Overall, P2X$_3$ receptors can be defined on the basis of their αβmeATP sensitivity, insensitivity to D- and L-βγmeATP, rapid desensitization and high sensitivity to TNP-ATP and A-317491.

P2X$_{2/3}$ receptors

The agonist sensitivity of this heteromer is similar to the P2X$_3$ receptor with a few subtle differences in terms of absolute potency of compounds (Bianchi et al., 1999). The potency of A-317491 and TNP-ATP at rat P2X$_{2/3}$ receptors is slightly lower than at P2X$_3$ receptors (Virginio et al., 1998a; Jarvis et al., 2003). A-317491 has higher potency at human than at rat P2X$_{2/3}$ receptors (Jarvis et al., 2003).

The P2X$_{2/3}$ receptor is differentiated from the other αβmeATP-sensitive receptors (P2X$_1$ and P2X$_3$) by virtue of its much slower rate of desensitization and by the effects of protons, which potentiate responses at P2X$_{2/3}$ receptors (Li et al., 1996), but inhibit at P2X$_1$ and P2X$_3$ receptors (Stoop et al., 1999; Wildman et al., 1999).

P2X$_4$ receptors

Rank order of agonist potency at P2X$_4$ receptors varies between studies, although ATP usually possesses higher potency/efficacy than 2meSATP. αβmeATP possesses very low efficacy at rat P2X$_4$ receptors but is a partial agonist at human and mouse receptors (EC$_{50}$ = 7–19 μM) (Jones et al., 2004). ATP potency varies slightly between the species orthologues (EC$_{50}$~1–10 μM), but not in any consistent manner (Townsend-Nicholson et al., 1999; Jones et al., 2000). Agonist potency and efficacy at P2X$_4$ receptors may be affected by channel density. Thus, P2X$_4$ receptors can be internalized (Royle et al., 2002) and Toulme et al. (2006) have demonstrated that ATP potency and αβmeATP efficacy can be increased by mutations that prevent receptor recycling and increase cell-surface expression. Furthermore, ivermectin increases ATP potency and potentiates responses to αβmeATP (Khakh et al., 1999b), and this may reflect effects of ivermectin on cell surface channel density (Toulme et al., 2006).

There are no selective P2X$_4$ receptor antagonists and the receptor is usually defined in terms of its insensitivity to antagonists. Initial studies suggested P2X$_4$ receptors were insensitive to PPADS because they lacked a lysine at position 249 that was present in P2X$_1$ and P2X$_2$ receptors (Buell et al., 1996b). This may be an over-simplification as human and mouse orthologues lack lysine at position 249 but are moderately sensitive to PPADS (IC$_{50}$ = 10–30 μM; Jones et al., 2000; Garcia-Guzman et al., 1997a). Garcia-Guzman et al. (1997a) found that suramin weakly blocked human P2X$_4$ receptors (IC$_{50}$ = 178 μM) and that lysine at position 78 of the human receptor contributed to the species difference. However, other studies failed to confirm effects of suramin at the human P2X$_4$ receptor (Bianchi et al., 1999; Jones et al., 2000). Further discrepancies in antagonist effects at P2X$_4$ receptors have been noted. Suramin and PPADS potentiated responses at mouse P2X$_4$ receptors (Townsend-Nicholson et al., 1999) but this was not observed in other studies. Similarly, cibacron blue potentiated responses at rat receptors in some studies (Miller et al., 1998) but inhibited responses in others (Garcia-Guzman et al., 1997a). There is no obvious explanation for these discrepancies, although the receptor has complex run-down and desensitization properties (Fountain and North, 2006) which may be a factor.

It is also noteworthy that facilitation of responses at P2X$_4$ receptors can be consistently induced by the anti-parasitic agent, ivermectin; the proposed mechanism of action is to reduce channel desensitization and reduce the rate of deactivation (Khakh et al., 1999b; Priel and Sildeberg, 2004) although effects on receptor density may also contribute (Toulme et al., 2006).

As discussed above (see P2X$_2$), effects of copper and zinc may differentiate P2X$_2$ and P2X$_4$ receptors since zinc, but not copper, potentiates P2X$_4$ responses (Xiong et al., 1999).

Overall, P2X$_4$ receptors can be defined on the basis of their biophysical properties, the relatively low potency and efficacy of αβmeATP, low sensitivity to P2X receptor antagonists, inhibition of responses by protons and enhancement of responses by ivermectin and zinc but not by copper.

P2X$_{4/6}$ receptors

The major feature of the rat P2X$_{4/6}$ receptor is its greater sensitivity to 2meSATP and αβmeATP when compared to P2X$_4$ receptors (Lê et al., 1998; Khakh et al., 1999b). In addition, RB2 potentiated responses at P2X$_4$ but not at P2X$_{4/6}$ receptors.

P2X$_5$ receptors

Due to the lack of a cloned native full-length receptor, studies on human P2X$_5$ receptors used a receptor constructed according to the predicted full length sequence. Furthermore, in some studies extracellular calcium concentration was manipulated in order to obtain reproducible responses. P2X$_5$ receptors are activated by most P2X agonists (Table 3.3.1.1). BzATP (EC$_{50}$ = 1.3 μM) and D-βγmeATP (EC$_{50}$~12 μM) are relatively potent, but low efficacy, agonists while GTP is a weak full agonist. αβmeATP activated rat and human P2X$_5$ receptors (Wildman et al., 2002; Bo et al., 2003a), but was inactive or had low efficacy in other studies (Collo et al., 1996; Torres et al., 1998a).

P2X$_5$ receptors are blocked by most antagonists (Table 3.3.1.3) except diinosine pentaphosphate (IP$_5$I, IC$_{50}$>30 μM) (Wildman et al., 2002). Protons inhibit P2X$_5$ responses while zinc potentiates responses at low doses but inhibits at higher doses (Wildman et al.,

2002). Calcium effects are complex: absence of calcium prevents run-down of responses, but responses are increased if calcium is added shortly before agonist addition (Wildman et al., 2002).

P2X$_6$ receptors

The P2X$_6$ receptor is poorly expressed and has only been characterized in two studies. In one study ATP, 2meSATP and ATPγS were equipotent (EC$_{50}$~10 μM) and αβmeatp inactive (Collo et al., 1996). However, in a subsequent study (Jones et al., 2004), ATP and αβmeATP were equipotent (EC$_{50}$~0.5 μM). Interestingly, αβmeATP was more effective on P2X$_{4/6}$ receptors than on P2X$_4$ receptors (Lê et al., 1998; Khakh et al., 1999b) although it is not an agonist at P2X$_{2/6}$ receptors (King et al., 2000).

P2X$_7$ receptors

The defining pharmacological feature of the P2X$_7$ receptor is the low potency of ATP and greater potency for BzATP coupled with lack of effect of all other agonists except at very high concentrations and/or after prolonged activation of the receptor (North, 2002). However, even these properties may be an over-simplification as ATP activates high- and low-affinity sites in cells expressing recombinant human P2X$_7$ receptors (Klapperstuck et al., 2000). In addition, agonist potency increases after repeated receptor activation (Chessell et al., 1997b). There are also marked species differences in potency of ATP and BzATP (potency at rat > human >> mouse) (Chessell et al., 1998b). Finally, NAD can activate mouse P2X$_7$ receptors but this is indirect via ADP-ribosylation (Seman et al., 2003).

In many studies suramin and PPADS are weak (IC$_{50}$>50 μM) or inactive P2X$_7$ receptor antagonists (Surprenant et al., 1996; Rassendren et al., 1997; Bianchi et al., 1999) but their potency at P2X$_7$ receptors is greatly increased when they are pre-incubated or studied in different buffers (Chessell et al., 1998a; Hibell et al., 2001b). Periodate-oxidized ATP (OxATP) is the archetypal P2X$_7$ antagonist (Murgia et al., 1993) although its utility is clearly limited as it also blocks P2X$_1$ and P2X$_2$ receptors (Evans et al., 1995) and has non-specific effects (Beigi et al., 2003). KN62 is a calcium/calmodulin-dependent kinase II inhibitor (IC$_{50}$~0.9 μM) but at lower concentrations it selectively blocks human P2X$_7$ receptors (IC$_{50}$~10–100 nM) (Gargett and Wiley, 1997). KN62 exhibits marked species selectivity (Humphreys et al., 1998), possessing low potency (IC$_{50}$>10 μM) at rat receptors and intermediate potency at mouse receptors (IC$_{50}$~200-400 nM). It is likely that KN62 is an allosteric inhibitor of P2X$_7$ receptors as its effects on human and mouse receptors are saturable (Humphreys et al., 1998; Michel et al., 2000). Several more potent analogues of KN62 have been synthesized (Baraldi et al., 2004).

Brilliant blue G (BBG) (Jiang et al., 2000; Hibell et al., 2001b) is a potent antagonist of rat (IC$_{50}$ = 10 nM) and mouse (IC$_{50}$ = 80 nM) P2X$_7$ receptors and exhibits 300–1000 selectivity for rat P2X$_7$ receptors over rat P2X$_{1-4}$ receptors. However, in some studies BBG potency at rat P2X$_7$ receptors is much lower and varied depending upon assay conditions (IC$_{50}$ 100–1000 nM) (Soltoff et al., 1989; Hibell et al., 2001b). It is also a weak antagonist of human P2X$_7$ receptors (IC$_{50}$~300nM).

Several other compounds can block P2X$_7$ receptors including a human P2X$_7$ receptor monoclonal antibody (Buell et al., 1998), HMA (Wiley et al., 1992), DIDS (Soltoff et al., 1993) and

17-beta-oestrodiol (Cario-Toumaniantz *et al.*, 1998a). The gap junction blocker, mefloquine, was reported to block P2X$_7$ receptors (IC$_{50}$ = 2.5 nM; Suadicani *et al.*, 2006) but this has not been confirmed (Pelegrin and Surprenant, 2006). HMA and tenidap also potentiate responses at mouse receptors (Sanz *et al.*, 1998; Hibell *et al.*, 2001b). More recently, a number of high-affinity, selective, P2X$_7$-receptor antagonists have been described in patents (Baraldi *et al.*, 2004). Although there is little detailed information on their properties, these compounds should offer great potential for characterizing P2X$_7$ receptors.

P2X$_7$ receptor function is inhibited by protons, by many divalent cations including, calcium, magnesium, copper and zinc (Virginio *et al.*, 1997) and by sodium and chloride ions (Wiley *et al.*, 1992; Michel *et al.*, 1999). Ionic effects on ligand potency are more marked than at other P2X receptors. These ions probably affect P2X$_7$ receptor function in an allosteric manner rather than through chelating ATP, although definitive proof is awaited (Virginio *et al.*, 1997).

Overall, P2X$_7$ receptors are best described by low sensitivity to ATP and BzAP, particularly in solutions with normal divalent cation concentrations, lack of desensitization, pronounced ability to enable fluorescent dyes to enter cells and blockade by ions (sodium, chloride, calcium, magnesium, copper) and antagonists such as BBG (1–400 nM, rat and mouse) and KN62 (10–100 nM, human).

Receptor binding studies

There have been only a few successful attempts to label P2X receptors and in several of the studies the pharmacological properties are complex and show marked differences from data derived from functional studies. [^{3}H]-αβmeATP was initially used to label high-affinity P2X$_1$ receptors in vas deferens and bladder (Bo and Burnstock, 1990). However, subsequent studies identified non P2X binding sites in other tissues (Michel *et al.*, 1995) limiting its utility. Radio-labelled ATP and ATPγS have been successfully used to label P2X$_{1-4}$ receptors (see Michel *et al.*, 1997) but only when expressed at very high density in recombinant systems or in native tissues (bladder, vas deferens). All studies with nucleotides require careful characterization to avoid artefacts due to nucleotide metabolism/interconversion and are hampered by high levels of endogenous non-specific binding sites.

More recently [^{3}H]-A-317491 was introduced as the first non-nucleotide radioligand and used to label recombinant P2X$_{2/3}$ and P2X$_3$ receptors (Jarvis *et al.*, 2004). A tritiated KN62 analogue was has been used to label P2X$_7$ receptors (Romagnoli *et al.*, 2004).

In binding studies on P2X$_{1-4}$ receptors antagonist affinity estimates are similar to those from functional studies but agonist affinity is 100–1000-fold higher than in functional studies. It has been suggested that this represents binding to a desensitized state of the receptor (Michel *et al.*, 1997). Consistent with this Rettinger and Schmaltzing (2003) demonstrated a high affinity desensitizing activity of ATP (IC$_{50}$ = 3 nM) at P2X$_1$ receptors. Similarly, many agonists desensitize P2X$_3$ receptors at lower doses than activate the receptors (McDonald *et al.*, 2002; Sokolova *et al.*, 2004; Pratt *et al.*, 2005; Ford *et al.*, 2005).

3.3.1.6 Physiology and relevance to disease states

The physiological properties, and relevance to disease states, of the defined P2X receptors has been the subject of intense scrutiny since the identification of the first cDNA encoding these receptors.

However, as discussed briefly above, the properties of heterologously expressed P2X cDNAs seldom resemble native ATP responses observed in tissues, and coupled with the lack of available pharmacologically selective tools with which to dissect receptor identity, unequivocal attribution of a specific physiological function to a certain receptor subtype remains difficult. Nonetheless, for some of the subunits, there is a clearly described role, and the advent of transgenic manipulation has aided identification of at least some of the important functions of these receptors. The current thinking on the role of P2X receptors is discussed below; notably few functions have been well described for the heteromeric combinations with the exception of P2X$_{2/3}$.

P2X$_1$ receptors

Although the P2X$_1$ receptor has a fairly widespread distribution, its best described functions derive from studies on smooth muscle and platelets. The original tissue for which ATP was proposed to be a sympathetic transmitter was vas deferens (Burnstock, 1972). Studies using mice from which the P2X$_1$ receptor is genetically deleted show much reduced or absent contractile responses in the vas deferens following nerve stimulation (Mulryan *et al.*, 2000), and male fertility is reduced by ~90 per cent. This has led to suggestions that P2X$_1$ antagonists may be useful in developing non-hormonal male contraceptive drugs (Dunn, 2000). These data do conclusively establish that the P2X$_1$ receptor is an essential component of the vas deferens smooth muscle receptor, but do not unequivocally establish that homomeric P2X$_1$ receptors are functionally expressed as the ATP receptor in all smooth muscle systems. A role for the P2X$_1$ has also been described in control of bladder contractility (Rapp *et al.*, 2005).

Localization of P2X$_1$ to vascular smooth muscle has also led to research into the role of this receptor in control of blood flow and blood pressure. In human saphenous vein, ATP and analogues mediate an contractile response compatible with actions at P2X$_1$ receptors (Cario-Toumaniantz *et al.*, 1998b), and in kidney, preglomerular vasoconstriction is mediated via the action of ATP on P2X$_1$ (Inscho *et al.*, 2003). Studies with P2X$_1$-deficient mice confirm that the P2X$_1$ receptor underlies the artery smooth muscle P2 receptor contractile response (Vial and Evans, 2002). P2X1 receptors may also contribute to the Bezold–Jarisch reflex following intravenous administration of αβmeATP in rats (McQueen et al., 1998).

Several studies have shown that the P2X$_1$ receptor is responsible for the rapid nucleotide-evoked Ca^{2+} influx in platelets (MacKenzie *et al.*, 1996; Vial *et al.*, 1997). The consequences of P2X$_1$ activation on these cells is a transient but substantial shape change which is associated with platelet activation (Mahaut-Smith *et al.*, 2004). However, P2X$_1$ knockout animals do not appear to display spontaneous thromboses or changes in bleeding time, which suggests that the receptor does not play a major role in basic haemostatic function, with the caveat that some form of developmental compensation may have occurred in the knockout setting (Mulryan *et al.*, 2000). Overexpression of the P2X$_1$ receptor does lead to a prothrombotic phenotype, particularly where shear-stress triggered platelet activation may underlie thromboboses (Oury *et al.*, 2003). Overall, these studies demonstrate that P2X$_1$ receptors contribute to platelet function *in vivo*.

P2X$_2$ receptors

The physiological functions of P2X$_2$ are not clear; most studies in which P2X$_2$ is localized and studied functionally focus on the contribution of P2X$_2$ to the heteromeric combination with P2X$_3$. Indeed, genetic deletion of P2X$_2$ converts sustained responses to ATP in nodose neurons to transient, rapidly desensitizing responses, and a similar finding was also observed when recording from DRG neurons of wild-type and P2X$_2^{-/-}$ animals (Cockayne et al., 2005). However, examination of myenteric neurons from these P2X$_2^{-/-}$ animals revealed a role for presumed homomeric P2X$_2$ receptors as in P2X$_2$ deleted mice ATP responses in S-type neurons (interneurons and motorneurons in myenteric plexus) were abolished, but responses in AH (likely to be intrinsic sensory neurons) neurons were unaffected (Ren et al., 2003). In addition, P2X$_2^{-/-}$, but not P2X$_3^{-/-}$ mice display markedly attenuated ventilatory response to hypoxia, indicating that P2X$_2$ receptors may play a pivotal role in carotid body function (Rong et al., 2003). Finally, there is a considerable body of evidence demonstrating the presence of P2X$_2$ receptors in the ear (Housley, 1998) and these receptors may regulate sensitivity to sound (Zhao et al., 2005).

P2X$_3$ and P2X$_{2/3}$ receptors

Research regarding the physiological role of P2X$_2$ and P2X$_{2/3}$ receptors is dominated by investigations into the sensory role of these subunits. Since Bleehan and Keele (1977) described algogenic actions of ATP on the human blister base preparation, it has been recognized that receptors for ATP must play a major role in sensory trafficking. More recently it has been demonstrated that injection of ATP can elicit firing of the peroneal nerve in humans (Hilliges et al., 2000).

In in vivo studies in rats, αβmeATP elicits nocifensive responses (Bland-Ward and Humphrey,1997), and this work was followed by demonstration that the fibres activated are predominantly classified as C-mechanoheat nociceptors (Hamilton et al., 2001). In addition to activation of skin afferents, αβmeATP also activates afferents innervating the joint (Dowd et al., 1998) and mesenteric afferents (Kirkup et al., 1999). Two further pieces of evidence firmly link P2X$_3$ with pain sensation and signalling. First, mice in which the P2X$_3$ receptor is deleted have reduced pain behaviours in response to injection of ATP or formalin (Cockayne et al., 2000). Secondly, novel antagonists directed at P2X$_3$/P2X$_{2/3}$ receptors inhibit pain responses in both inflammatory and neuropathic models (McGaraughty et al., 2003). In the latter study, the antagonist produced dose-related antinociception following either intrathecal or intraplantar injection in a model of inflammatory pain, but in a model of neuropathic pain, the molecule was not effective when administered by the intraplantar route.

In all of these studies, it is not possible to determine whether afferent activation, or inhibition by genetic or pharmacological means, is via inhibition of homomeric P2X$_3$ receptor, or the heteromeric combination of P2X$_{2/3}$. It is clear that deletion of P2X$_3$ reduces both pain behaviours, and ATP-elicited currents in sensory neurons, but these effects may be mediated by the resultant inability to form P2X$_{2/3}$ heteromeric complexes. Nonetheless, it is clear that P2X$_3$ and P2X$_{2/3}$ receptors are important targets for modulation of pain, and a number of centres are working to bring antagonists at these receptors forward as novel clinical analgesics.

A second important role for the P2X$_3$ receptor is in control of bladder function. Thus, P2X$_3$-null mice exhibit a marked urinary bladder hyporeflexia, characterized by reduced voiding frequency and an increased bladder size (Cockayne et al., 2000). While the P2X$_3$ receptor is localized to bladder afferents, importantly,

deletion of the P2X$_3$ receptor did not have any effect on neuronal innervation density (Cockayne et al., 2000). Interestingly, in a study of patients with neurogenic detrusor overactivity, increased P2X$_3$ immunoreactivity was observed in suburothelial bladder (Brady et al., 2004), which further reinforces the potential importance of this receptor in bladder function.

Finally, P2X$_{2/3}$ and P2X$_3$ receptors may play a role in taste since these receptors have been detected in taste-sensitive nerves and P2X$_2$ and P2X$_3$ double knockout mice have a marked disturbance in their sense of taste (Finger et al., 2005).

P2X$_4$ receptors

An important role for P2X$_4$ receptors has been observed only quite recently. Following studies examining the analgesic efficacy of TNP-ATP administered intrathecally following nerve injury, it was noted that the antagonist, PPADS, had no effect, even at quite high doses. This led to re-examination of the receptor responsible for the TNP-ATP-induced analgesia, and further study has revealed that the P2X$_4$ receptor plays an important role in the spinal cord in mediating these effects (Tsuda et al., 2003). P2X$_4$ receptors are upregulated in ipsilateral spinal cord following nerve injury, as well as following spinal cord injury (Schwab et al., 2006), and cellular localization studies suggests that these receptors are localized to microglial cells in the cord (Tsuda et al., 2003); intrathecal treatment with P2X$_4$ antisense oligonucleotides suppressed allodynia following nerve injury.

One mechanism which may underlie the effects of P2X$_4$ activation in these pain states is via microglial release of brain-derived neurotrophic factor (BDNF). Thus, ATP-treated microglia release BDNF and intrathecal administration of ATP-treated microglia induces allodynia which can be blocked by preventing the release or blocking the action of BDNF (Coull et al., 2005). The allodynia is accompanied by a switch in the anion gradient in spinal neurons which reverses the effects of the normally inhibitory transmitter, GABA, to cause depolarization, rather than hyperpolarization of spinal lamina I neurons, thus facilitating pain transmission rather than inhibiting it. In support of the analgesic role of P2X$_4$, the author's group has examined P2X$_4^{-/-}$ animals for a pain phenotype following intraplantar injection of Freund's complete adjuvant (FCA), and found the P2X$_4^{-/-}$ animals to be resistant to development of hyperalgesia when compared to P2X$_4^{+/+}$ animals.

The P2X$_4$ receptor also appears to play an important role in haemodynamic responses. P2X$_4^{-/-}$ mice do not display normal endothelial cell responses to changes in blood flow, such as Ca^{2+} influx and production of nitric oxide, and vessel dilation induced by blood volume changes is also impaired. In these animals resting blood pressure was higher and it appears that P2X$_4$ is essential for blood vessel remodelling such that no decrease in vessel size was observed to chronic decreases in blood flow (Yamamoto et al., 2006). In support of a critical haemodynamic role, Shen and colleagues (2006) have shown that in mice overexpressing the P2X$_4$ receptor, inward currents in cardiac ventricular myocytes in response to 2meSATP were greater in magnitude than in cells from wild-type animals (Shen et al., 2006).

Taken together, these data suggest a critical role for the P2X$_4$ receptor in pain processing which may render this an important target for development of novel analgesics. However, the cardiovascular phenotype of this receptor may preclude development of a drug with an acceptable safety profile.

P2X$_5$ and P2X$_{1/5}$ receptors

Few functions have been ascribed to the P2X$_5$ receptor, largely because the native receptor appears to be a truncated non-functional form (Bo et al., 2003a). However, a subset of individuals with a single nucleotide polymorphism at a 3′ splice site will make full length P2X$_5$; identification of such individuals and subsequent characterization of the potential roles of full length P2X$_5$ is awaited with interest. In in vitro studies, it has been proposed that P2X$_5$ in combination with P2X$_1$ may mediate excitatory junction potentials in a guinea pig submucosal arteriole preparation (Surprenant et al., 2000).

P2X$_6$ receptors

The novel phenotype of P2X$_6$ receptors revealed following adequate glycosylsation to enable functional responses (Jones et al., 2004) makes it tempting to speculate that this receptor may be responsible for αβmeATP-sensitive responses observed in some native neuronal preparations, where responses due to the presence of P2X1 or P2X3 can be largely ruled out. Thus, responses in the medial habenula (Edwards et al., 1992), the medial vestibular nucleus (Chessell et al., 1997a), and locus coeruleus (Sansum et al., 1998) are sensitive to αβmeATP, and these brain regions do not express receptors which are conventionally responsive to this agonist. However, fully glycolsylated P2X6 receptors appear to be relatively insensitive to the antagonist, suramin, whereas functional responses recorded in brain nuclei do show sensitivity to this antagonist. In the case of the P2 receptor in the locus coeruleus and medial vestibular nucleus, responses were observed to αβmeADP, to which the receptor described by Jones et al. (2004) is insensitive. Thus, the identity of the functional CNS receptor described in these studies remains elusive, although P2X6 may contribute to these responses.

P2X$_7$ receptors

The roles for the P2X$_7$ receptor are largely defined by two key properties: (1) the ability to form large cytolytic pores and (2) the ability to regulate release of mature, biologically active interleukin-1β (IL-1β). Given the localization of P2X$_7$ receptors to cells of immune origin and its role in mediating release of inflammatory cytokines, it is perhaps not surprising that much of the research has focused on role in inflammatory conditions and, more recently, in pain. A role for the P2X$_7$ receptor has also been described in bone (see below). It is important to note that the P2X$_7$ receptor is capable of triggering multiple and divergent signalling cascades which include regulation of cytokines; these include actions as diverse as activation of phospholipase-D (Gargett et al., 1996) to shedding of L-selectin (Jamieson et al., 1996).

The P2X$_7$ receptor, when originally known as P2Z, was implicated in the lysis of macrophages and antigen-presenting cells by cytotoxic T lymphocytes, as well as mitogenic stimulation of T cells and the formation of multinucleated giant cells (Blanchard et al., 1991, 1995; Falzoni et al., 1995; Di Virgilio et al., 1998). The large pore-forming ability of the receptor was also utilized to permeabilize mast cells (Cockcroft and Gomperts, 1979). Given these roles, it has been speculated that the P2X$_7$ receptor may play a pivotal role in host defence. However, P2X$_7$$^{-/-}$ animals control mycobacterium tuberculosis infection in lungs as effectively as wild-type mice, suggesting that perhaps the P2X$_7$ receptor is not required to mount

an efficient immune response (Myers et al., 2005). However, in vitro, it has also been shown that macrophages with a loss of function P2X$_7$ mutation do show functional deficits in ATP-mediated apoptosis and mycobacterium killing, so further investigation is required (Saunders et al., 2003; Shemon et al., 2006). The same loss of function polymorphisms also appeared to be associated with enhanced susceptibility to chronic lymphocytic leukaemia (Wiley et al., 2002) but this has not been subsequently confirmed (see Paneesha et al., 2006).

More recently, a critical role for the P2X$_7$ receptor in regulating the release of mature IL-1β has been described. Perregaux and Gabel (1994) first observed this release of IL-1β from macrophages following treatment with ATP, and studies using P2X$_7$$^{-/-}$ mice provided unequivocal confirmation that the P2X$_7$ receptor is indeed the pivotal transducer (Solle et al., 2001). The mechanism of action of P2X$_7$ in mediating this effect remains unclear; interleukin-1 converting enzyme (ICE) (caspase-1) is activated following activation of P2X$_7$ (Ferrari et al., 2000). It has been shown more recently that activation of P2X$_7$ results in rapid formation of microvesicles containing IL-β, which may represent a major secretory pathway for such leaderless secretory proteins.

The functional consequences of regulation of the release of IL-1β by P2X$_7$ are numerous. This cytokine elicits signalling cascades which are linked to regulation of a host of inflammatory mediators, including IL-6, cyclooxygenase-2, nitric oxide synthase, tumour necrosis factor-α, and matrix metalloproteinases (see Solle et al., 2001; Chessell et al., 2005). In a model of arthritis, using monoclonal anti-collagen antibodies and LPS challenge, P2X$_7$$^{-/-}$ animals were markedly less affected than the wild-type animals with respect to development of inflamed paws, lesions of the cartilage, collagen degradation and loss of proteoglycan content (Labasi et al., 2002). The phenotype of P2X$_7$$^{-/-}$ animals is also unusual with respect to pain; P2X$_7$$^{-/-}$ animals fail to develop hypersensitivity in either inflammatory or neuropathic pain models, and also produce less IL-10 in response to inflammatory insult, suggesting yet another pathway which may be regulated by P2X$_7$ (Chessell et al., 2005).

The role of P2X$_7$ in bone formation and resorption is complex. Ke et al. (2003) found that P2X$_7$$^{-/-}$ animals have reduced cortical bone, reduced periosteal bone formation, and increased trabecular bone resorption in the femur, and Gartland et al. (2003a) showed that blockade of the human P2X$_7$ receptor with a monoclonal antibody (Buell et al., 1998) inhibited the fusion of osteoclast precursors to form multinucleated osteoclasts. However, the same group also showed that P2X$_7$$^{-/-}$ animals maintain the ability to form such multinucleated cells (Gartland et al., 2003b), perhaps indicating redundancy in this pathway.

Finally, although the presence of P2X$_7$ receptors in neurons remains controversial recent studies have suggested a genetic association between the P2X$_7$ receptor and both major depressive disorder and bipolar affective disorder (Barden et al., 2006). Since a loss of function polymorphism (A1513C) showed a strong association with these disorders, P2X$_7$ receptors may play a hitherto unexpected role in brain function.

3.3.1.7 Concluding remarks

Since the first cDNA was identified in 1994, a major research effort has been mobilized to understand the functions and roles of the P2X receptor family. Some of the fundamental questions have either been answered, or sufficient data exists to make a reasoned interpretation regarding the most likely scenario. Thus, it is reasonably likely that P2X subunits assemble with a trimeric architecture to form functional receptors; the binding site for ATP has been partially identified; seemingly the majority of heteromeric combinations have been identified; and there is a good understanding of the gross localization of each of the subunits.

Despite this, major questions remain unresolved. Whilst the phenotype of heterologously expressed receptors is clear, these seldom resemble the functional characteristics of endogenous P2X receptors; expression patterns help make a best guess at which subunits may contribute to endogenous functional effects, but the exact identity of these functional receptors remains, in the majority of cases, a mystery. The pharmacology of the P2X receptors remains a critical challenge—antagonists with some specificity for the $P2X_3$/ $P2X_{2/3}$ and $P2X_7$ receptors are beginning to appear (McGaraughty *et al.*, 2003; Baraldi *et al.*, 2004), but antagonists at the other receptors, along with selective agonist molecules are desperately needed to help unravel the pleiotropic actions of this receptor family.

One major advance of recent years is in understanding the role of P2X receptors in sensory processing. Here, there is a surfeit of riches: the $P2X_3$/$P2X_{2/3}$, $P2X_4$ and $P2X_7$ receptors all seem to be critically involved in either the induction or maintenance of pain, as evidenced by antagonist, antisense oligonucleotide knockdown, and knockout studies. The mechanisms by which these three receptor groups convey their responses do seem to be different; $P2X_3$/ $P2X_{2/3}$ receptors are very much associated with sensory afferent signal trafficking, and antagonists at these receptors may prevent the sensory signal from being transduced at the level of the sensory afferent. $P2X_4$ and $P2X_7$ receptors, however, appear to modulate pain signalling by interrupting the sensitization process, either in the periphery, or in the spinal cord, and it remains to be seen how much overlap of molecular mechanisms exists; certainly in the case of modulation of IL-1β, only $P2X_7$ has been implicated.

Overall, there is little doubt that ultimately the P2X receptor family will prove to be a tractable target for the development of new drugs, be they analgesics, immune modulators, treatments for bone remodelling disorders, or treatments for bladder dysfunction. The field awaits development of truly selective tools with which to probe function and provide validation for the development of new medicines.

References

Adinolfi E, Kim M, Young MT *et al.* (2003). Tyrosine phosphorylation of HSP90 within the P2X7 receptor complex negatively regulates P2X7 receptors. *J Biol Chem* **278**, 37344–37351.

Adriouch S, Dox C, Welge V *et al.* (2002). Cutting edge: a natural P451L mutation in the cytoplasmic domain impairs the function of the mouse P2X7 receptor. *J Immunol* **169**, 4108–4112.

Alexander K, Niforatos W, Bianchi B *et al.* (1999). Modulation and accelerated resensitization of human P2X(3) receptors by cibacron blue. *J Pharmacol Exp Ther* **291**, 1135–1142.

Aschrafi A, Sadtler S, Niculescu C *et al.* (2004). Trimeric architecture of homomeric P2X2 and heteromeric P2X1+2 receptor subtypes. *J Mol Biol* **342**, 333–343.

Ashour F, Atterbury-Thomas M, Deuchars J *et al.* (2006). An evaluation of antibody detection of the P2X1 receptor subunit in the CNS of wild type and P2X1-knockout mice. *Neurosci Lett* **397**, 120–125.

Barajas-Lopez C, Montano LM and Espinosa-Luna R (2002). Inhibitory interactions between 5-HT3 and P2X channels in submucosal neurons. *Am J Physiol* **283**, G1238–G1248.

Baraldi PG, Di Virgilio F and Romagnoli R (2004). Agonists and antagonists acting at P2X7 receptor. *Curr Top Med Chem* **4**, 1707–1717.

Barden N, Harvey M, Gagne B *et al.* (2006). Analysis of single nucleotide polymorphisms in genes in the chromosome 12Q24.31 region points to P2RX7 as a susceptibility gene to bipolar affective disorder. *Am J Med Genet Neuropsych Gen* **4**, 374–382.

Barrera NP, Ormond SJ, Henderson RM *et al.* (2005). Atomic force microscopy imaging demonstrates that P2X2 receptors are trimers but that P2X6 receptors do not oligomerize. *J Biol Chem* **280**, 10759–10765.

Beigi RD, Kertesy SB, Aquilina G *et al.* (2003). Oxidized ATP (oATP) attenuates proinflammatory signaling via P2 receptor-independent mechanisms. *Br J Pharmacol* **140**, 507–519.

Bianchi BR, Lynch KJ, Touma E *et al.* (1999). Pharmacological characterization of recombinant human and rat P2X receptor subtypes. *Eur J Pharmacol* **376**, 127–138.

Blanchard DK, Hoffman SL and Djeu JY (1995). Inhibition of extracellular ATP-mediated lysis of human macrophages by calmodulin antagonists. *J Cell Biochem* **57**, 452–464.

Blanchard DK, McMillen S and Djeu JY (1991). IFN-gamma enhances sensitivity of human macrophages to extracellular ATP-mediated lysis. *J Immunol* **147**, 2579–2585.

Bland-Ward PA and Humphrey PPA (1997). Acute nociception mediated by hindpaw P2X receptor activation in the rat. *Br J Pharmacol* **122**, 365–371.

Bleehan T and Keele CA (1977). Observations on the algogenic actions of adenosine compounds on the human blister base preparation. *Pain* **3**, 367–377.

Blevins GT, Vandewesterlo EMA, Logsdon CD *et al.* (1996). Nucleotides regulate the binding affinity of the recombinant type a cholecystokinin receptor in cho k1 cells. *Reg Peptides* **61**, 87–93.

Bo X, Jiang LH, Wilson HL *et al.* (2003a). Pharmacological and biophysical properties of the human P2X5 receptor. *Mol Pharmacol* **63**, 1407–1416.

Bo X, Kim M, Nori SL *et al.* (2003b). Tissue distribution of P2X4 receptors studied with an ectodomain antibody. *Cell Tiss Res* **313**, 159–165.

Bo X, Zhang Y, Nassar M *et al.* (1995). A P2X purinoceptor cDNA conferring a novel pharmacological profile. *FEBS Lett* **375**, 129–133.

Bo XN and Burnstock G (1990). High- and low-affinity binding sites for [3H]-alpha, beta- methylene ATP in rat urinary bladder membranes. *Br J Pharmacol* **101**, 291–296.

Boué-Grabot E, Archambault V and Séguéla P (2000). A protein kinase C site highly conserved in P2X subunits controls the desensitisation kinetics of P2X2 ATP-gated channels. *J Biol Chem* **275**, 10190–10195.

Boue-Grabot E, Toulme E, Emerit MB *et al.* (2004). Subunit-specific coupling between gamma-aminobutyric acid type A and P2X2 receptor channels. *J Biol Chem* **279**, 52517–52525.

Bradbury EJ, Burnstock G and McMahon SB (1998). The expression of P2X3 purinoceptors in sensory neurons. Effects of axotomy and glial-derived neurotrophic factor. *Mol Cell Neurosci* **12**, 256–268.

Brady CM, Apostolidis A, Yiangou Y *et al.* (2004). P2X3-immunoreactive nerve fibres in neurogenic detrusor overactivity and the effect of intravesical resiniferatoxin. *Eur Urol* **46**, 247–253.

Brake AJ, Wagenbach MJ and Julius D (1994). New structural motif for ligand-gated ion channels defined by an ionotropic ATP receptor. *Nature* **371**, 519–523.

Brandle U, Spielmanns P, Osteroth R *et al.* (1997). Desensitization of the $P2X_2$ receptor controlled by alternative splicing. *FEBS Lett* **404**, 294–298.

Buell G, Chessell IP, Michel AD *et al.* (1998). Blockade of human $P2X_7$ receptor function with a monoclonal antibody. *Blood* **92**, 3521–3528.

Buell G, Lewis C, Collo G *et al.* (1996b). An antagonist-insensitive P_{2X} receptor expressed in epithelia and brain. *EMBO J* **15**, 55–62.

Buell G, Michel AD, Lewis C *et al.* (1996a). P2x(1) receptor activation in hl60 cells. *Blood* **87**, 2659–2664.

Bultmann R and Starke K (1994). Blockade by 4, 4′-diisothiocyanatostilbene-2,2′-disulphonate (DIDS) of P2X-purinoceptors in rat vas deferens. *Br J Pharmacol* **112**, 690–694.

Bultmann R, Trendelenburg M, Tuluc F *et al.* (1999). Concomitant blockade of P2X-receptors and ecto-nucleotidases by P2-receptor antagonists: functional consequences in rat vas deferens. *Naunyn-Schmiedebergs. Arch Pharmacol* **359**, 339–344.

Burnstock G (1972). Purinergic nerves. *Pharmacol Rev* **24**, 509–581.

Burnstock G (1990). Overview: purinergic mechanisms. *Ann N Y Acad Sci* **603**, 1–18.

Burnstock G and Kennedy C (1985). Is there a basis for distinguishing two types of P2 purinoceptor? *Gen Pharmacol* **16**, 433–440.

Burnstock G and Knight GE (2004). Cellular distribution and functions of P2 receptor subtypes in different systems. *Int Rev Cytology* **240**, 31–304.

Cabrini G, Falzoni S, Forchap SL *et al.* (2005). A His-155 to Tyr polymorphism confers gain-of-function to the human P2X7 receptor of human leukemic lymphocytes. *J Immunol* **175**, 82–89.

Cario-Toumaniantz C, Loirand G, Ferrier L *et al.* (1998a). Non-genomic inhibition of human P2X$_7$ purinceptor by 17β-oestradiol. *J Physiol* **508.3**, 659–666.

Cario-Toumaniantz C, Loirand G, Ladoux A *et al.* (1998b). P2X7 receptor activation-induced contraction and lysis in human saphenous vein smooth muscle. *Circ Res* **83**, 196–203.

Cascalheira JF and Sebastiao AM (1992). Adenine nucleotide analogues, including gamma-phosphate- substituted analogues, are metabolised extracellularly in innervated frog sartorius muscle. *Eur J Pharmacol* **222**, 49–59.

Chaban VV, Mayer EA, Ennes HS *et al.* (2003). Estradiol inhibits atp-induced intracellular calcium concentration increase in dorsal root ganglia neurons. *Neuroscience* **118**, 941–948.

Chakfe Y, Seguin R, Antel JP *et al.* (2002). ADP and AMP induce interleukin-1beta release from microglial cells through activation of ATP-primed P2X7 receptor channels. *J Neurosci* **22**, 3061–3069.

Cheewatrakoolpong B, Gilchrest H, Anthes JC *et al.* (2005). Identification and characterization of splice variants of the human P2X7 ATP channel. *Biochem Biophys Res Commun* **332**, 17–27.

Chen C, Parker MS, Barnes AP *et al.* (2000). Functional expression of three P2X2 receptor splice variants from guinea-pig cochlea. *J Neurophysiol* **83**, 1502–1509.

Chen C-C, Akopian AN, Sivilotti L *et al.* (1995). A P2X purinoceptor expressed by a subset of sensory neurons. *Nature* **377**, 428–431.

Chessell IP, Grahames CBA, Michel AD *et al.* (2001). Dynamics of P2X$_7$ receptor pore dilation: pharmacological and functional consequences. *Drug Dev Res* **53**, 60–65.

Chessell IP, Hatcher J, Bountra C *et al.* (2005). Disruption of the P2X7 purinoceptor gene abolishes chronic inflammatory and neuropathic pain. *Pain* **114**, 386–396.

Chessell IP, Michel AD and Humphrey PPA (1997a). Functional evidence for multiple purinoceptor subtypes in the rat medial vestibular nucleus. *Neuroscience* **77**, 783–791.

Chessell IP, Michel AD and Humphrey PPA (1997b). Properties of the pore-forming P2X$_7$ purinoceptor in mouse NTW8 microglial cells. *Br J Pharmacol* **121**, 1429–1437.

Chessell IP, Michel AD and Humphrey PPA (1998a). Effects of antagonists at the human recombinant P2X$_7$ receptor. *Br J Pharmacol* **124**, 1314–1320.

Chessell IP, Simon J, Hibell AD *et al.* (1998b). Cloning and functional characterisation of the mouse P2X$_7$ receptor. *FEBS Lett* **439**, 26–30.

Cockayne DA, Dunn PM, Zhong Y *et al.* (2005). P2X2 knockout mice and P2X2P2X3 double knockout mice reveal a role for the P2X2 receptor subunit in mediating multiple sensory effects of ATP. *J Physiol* **567.2**, 621–639.

Cockayne DA, Hamilton SG, Zhu Q-M *et al.* (2000). Urinary bladder hyporeflexia and reduced pain-related behaviour in P2X3-deficient mice. *Nature* **407**, 1011–1015.

Cockcroft S and Gomperts BD (1979). ATP induces nucleotide permeability in rat mast cells. *Nature* **279**, 541–542.

Collo G, Neidhart S, Kawashima E *et al.* (1997). Tissue distribution of the P2X$_7$ receptor. *Neuropharmacology* **36**, 1277–1283.

Collo G, North RA, Kawashima E *et al.* (1996). Cloning of P2X5 and P2X6 receptors and the distribution and properties of an extended family of ATP-gated ion channels. *J Neurosci* **16**, 2495–2507.

Cook SP, Rodland KD and McCleskey EW (1998). A memory for extracellular Ca2+ by speeding recovery of P2X receptors from desensitization. *J Neurosci* **18**, 9238–9244.

Coull JAM, Beggs S, Boudreau D *et al.* (2005). BDNF from microglia causes the shift in neuronal anion gradient underlying neuropathic pain. *Nature* **438**, 1017–1021.

Cowen DS, Berger M, Nuttle L *et al.* (1991). Chronic treatment with P2-purinergic receptor agonists induces phenotypic modulation of the HL-60 and U937 human myelogenous leukemia cell lines. *J Leuk Biol* **50**, 109–122.

Denlinger LC, Angelini G, Schell K *et al.* (2005). Detection of human P2X7 nucleotide receptor polymorphisms by a novel monocyte pore assay predictive of alterations in lipopolysaccharide-induced cytokine production. *J Immunol* **174**, 4424–4431.

Denlinger LC, Coursin DB, Schell K *et al.* (2006). Human P2X7 pore function predicts allele linkage disequilibrium. *Clin Chem* **52**, 995–1004.

Dhulipala PDK, Wang Y-X and Kotlikoff MI (1998). The human P2X$_4$ receptor gene is alternatively spliced. *Gene* **207**, 259–266.

Di Virgilio F, Chiozzi P, Falzoni S *et al.* (1998). Cytolytic P2X purinoceptors. *Cell Death Differ* **5**, 191–199.

Diaz-Hernandez M, Cox JA, Migita K *et al.* (2002). Cloning and characterization of two novel zebrafish P2X receptor subunits. *Biochem Biophys Res Comm* **295**, 849–853.

Ding S and Sachs C (1999). Single channel properties of P2X2 purinoceptors. *J Gen Physiol* **113**, 695–720.

Ding S and Sachs F (2000). Inactivation of P2X2 purinoceptors by divalent cations. *J Physiol* **522**, 199–214.

Dowd E, McQueen DS, Chessell IP *et al.* (1998). P2X receptor-mediated excitation of nociceptive afferents in the normal and arthritic rat knee joint. *Br J Pharmacol* **125**, 341–346.

Dunn PM (2000). Fertility. Purinergic receptors and the male contraceptive pill. *Current Biology* **10**, R305–R307.

Edwards FA, Gibb AJ and Colquhoun D (1992). ATP receptor-mediated synaptic currents in the central nervous system. *Nature* **359**, 144–147.

Egan TM, Samways DSK and Li Z (2006). Biophysics of P2X receptors. *Pflügers Arch* **452**, 501–512.

Ennion S, Hagan S and Evans RJ (2000). The role of positively charged amino acids in ATP recognition by human P2X1 receptors. *J Biol Chem* **275**, 29361–29367.

Evans RJ and Kennedy C (1994). Characterization of P2-purinoceptors in the smooth muscle of the rat tail artery: a comparison between contractile and electrophysiological responses. *Br J Pharmacol* **113**, 853–860.

Evans RJ, Lewis C, Buell G *et al.* (1995). Pharmacological characterization of heterologously expressed ATP-gated cation channels. *Mol Pharmacol* **48**, 178–183.

Evans RJ, Lewis C, Virginio C *et al.* (1996). Ionic permeability of and divalent cation effect on, two ATP-gated cation channels (P2X receptors) expressed in mammalian cells. *J Physiol* **497.2**, 413–422.

Falzoni S, Munerati M, Ferrari D *et al.* (1995). The purinergic P2Z receptor of human macrophage cells. *J Clin Invest* **95**, 1207–1211.

Feng YH, Li X, Wang L *et al.* (2006). A truncated P2X7 receptor variant (P2X7-j) endogenously expressed in cervical cancer cells antagonizes the full-length P2X7 receptor through hetero-oligomerization. *J Biol Chem* **281**, 17228–17237.

Fernando SL, Saunders BM, Sluyter R *et al.* (2005). Gene dosage determines the negative effects of polymorphic alleles of the P2X7 receptor on adenosine triphosphate-mediated killing of mycobacteria by human macrophages. *J Infect Dis* **192**, 149–155.

Ferrari D, Los M, Bauer MK *et al.* (2000). P2Z purinoreceptor ligation induces activation of caspases with distinct roles in apoptotc and necrotic alterations of cell death. *FEBS Lett* **447**, 71–75.

Finger TE, Danilova V, Barrows J *et al.* (2005). ATP signaling is crucial for communication from taste buds to gustatory nerves. *Science* **310**, 1495–1499.

Fisher JA, Girdler G and Khakh BS (2004). Time-resolved measurement of state-specific P2X2 ion channel cytosolic gating motions. *J Neurosci* **24**, 10475–10487.

Ford KK, Matchett M, Krause JE *et al.* (2005). The P2X3 antagonist P1, P5-di[inosine-5′] pentaphosphate binds to the desensitized state of the receptor in rat dorsal root ganglion neurons. *J Pharmacol Exp Ther* **315**, 405–413.

Fountain SJ and North RA (2006). A C-terminal lysine that controls human P2X4 receptor desensitisation. *J Biol Chem* **281**, 15044–15049.

Garcia-Guzman M, Soto F, Gomez-Hernandez JM *et al.* (1997a). Characterization of recombinant human P2X₄ receptor reveals pharmacological differences to the rat homologue. *Mol Pharmacol* **51**, 109–118.

Garcia-Guzman M, Soto F, Laube B *et al.* (1996). Molecular cloning and functional expression of a novel rat heart P2X purinoceptor. *FEBS Lett* **388**, 123–127.

Garcia-Guzman M, Stuhmer W and Soto F (1997b). Molecular characterization and pharmacological properties of the human p2x(3) purinoceptor. *Mol Brain Res* **47**, 59–66.

Gargett CE and Wiley JS (1997). The isoquinoline derivative KN-62 a potent anyagonist of the P2Z-receptor of human lymphocytes. *Br J Pharmacol* **120**, 1483–1490.

Gargett CE, Cornish EJ and Wiley JS (1996). Phospholipase D activation by P2Z-purinoceptor agonists in human lymphocytes is dependent on bivalent cation influx. *Biochem J* **313**, 529–535.

Gartland A, Buckley KA, Bowler WB *et al.* (2003b). Blockade of the pore-forming P2X7 receptor inhibits formation of multinucleated human osteoclasts *in vitro*. *Calcified Tiss Int* **73**, 361–369.

Gartland A, Buckley KA, Hipskind RA *et al.* (2003a). P2X₇ receptor-deficient mice maintain the ability to form multinucleated osteoclasts *in vivo* and *in vitro*. Crit Rev Euk Gene Expr **13**, 243–253.

Gordon JL (1986). Extracellular ATP. Effects, sources and fate. *Biochem J* **233**, 309–319.

Greco NJ, Tonon G, Chen W *et al.* (2001). Novel structurally altered P2X1 receptor is preferentially activated by adenosine diphosphate in platelets and megakaryocytic cells. *Blood* **98**, 100–107.

Guerra AN, Fisette PL, Pfeiffer ZA *et al.* (2003). Purinergic receptor regulation of LPS-induced signaling and pathophysiology. *J Endotox Res* **9**, 256–263.

Guo A, Vulchanova L, Wang J *et al.* (1999). Immunocytochemical localization of the vanilloid receptor 1 (VR1): relationship to neuropeptides, the P2X₃ purinoceptor and IB4 binding sites. *Eur J Neurosci* **11**, 946–958.

Haines WR, Torres GE, Voigt MM *et al.* (1999). Properties of the novel ATP-gated ionotropic receptor composed of the P2X(1) and P2X(5) isoforms. *Mol Pharmacol* **56**, 720–727.

Hamilton SG, McMahon SB and Lewin GR (2001). Selective activation of nociceptors by P2X receptor agonists in normal and inflamed rat skin. *J Physiol* **534**, 437–445.

Harden TK, Lazarowski ER and Boucher RC (1997). Release, metabolism and interconversion of adenine and uridine nucleotides: implications for G protein-coupled P2 receptor agonist selectivity. *Trends Pharmacol Sci* **18**, 43–46.

Hibell AD, Kidd EJ, Chessell IP *et al.* (2000). Apparent species differences in the kinetic properties of P2X₇ receptors. *Br J Pharmacol* **130**, 167–173.

Hibell AD, Thompson KM, Simon J *et al.* (2001a). Species- and agonist-dependent differences in the deactivation-kinetics of P2X7 receptors. *Naunyn-Schmiedebergs Arch Pharmacol* **363**, 639–648.

Hibell AD, Thompson KM, Xing M *et al.* (2001b). Complexities of measuring antagonist potency at P2X(7) receptor orthologs. *J Pharmacol Exp Ther* **296**, 947–957.

Hilliges M, Weidner C, Schmeltz M *et al.* (2000). ATP respones in human C nociceptors. *Pain* **98**, 59–68.

Hohenegger M, Waldhoer M, Beindl W *et al.* (1998). Gsalpha-selective G protein antagonists. *Proc Natl Acad Sci USA* **95**, 346–351.

Housley GD (1998). Extracellular nucleotide signaling in the inner ear. *Mol Neurobiol* **16**, 21–48.

Hulsmann M, Nickel P, Kassack M *et al.* (2003). NF449, a novel picomolar potency antagonist at human P2X1 receptors. *Eur J Pharmacol* **470**, 1–7.

Humphreys BD, Virginio C, Surprenant A *et al.* (1998). Isolquinolines as antagonists of the P2X₇ nucleotide receptor: high slectivity for the human versus the rat receptor homologues. *Mol Pharmacol* **54**, 22–32.

Inscho EW, Cook AK, Imig JD *et al.* (2003). Physiological role for P2X1 receptors in renal microvascular autoregulatory behaviour. *J Clin Invest* **112**, 1895–1905.

Jacobson KA, Kim YC, Wildman SS *et al.* (1998). A pyridoxine cyclic phosphate and its 6-azoaryl derivative selectively potentiate and antagonize activation of P2X1 receptors. *J Med Chem* **41**, 2201–2206.

Jamicson GP, Snook MB, Thurlow PJ *et al.* (1996). Extracellular ATP causes loss of L-selectin from human lymphocytes via occupancy of P2Z purinoceptors. *J Cell Physiol* **166**, 637–642.

Jarvis MF, Bianchi B, Uchic JT *et al.* (2004). [3H]A-317491, a novel high-affinity non-nucleotide antagonist that specifically labels human P2X2/3 and P2X3 receptors. *J Pharmacol Exp Ther* **310**, 407–416.

Jarvis MF, Burgard EC, McGaraughty S *et al.* (2003). A-317491, a novel potent and selective non-nucleotide antagonist of P2X3 and P2X2/3 receptors, reduces chronic inflammatory and neuropathic pain in the rat. *Proc Natl Acad Sci USA* **99**, 17179–17184.

Jiang LH, MacKenzie AB, North RA *et al.* (2000). Brilliant blue G selectively blocks ATP-gated rat P2X(7) receptors. *Mol Pharmacol* **58**, 82–88.

Jiang LH, Rassendren F, MacKenzie A *et al.* (2005). N-methyl-D-glucamine and propridium dyes utilize different permeation pathways at rat P2X7 receptors. *Am J Physiol* C1295–C1302.

Jiang LH, Rassendren F, Surprenant A *et al.* (2006). Identification of amino acid residues contributing to the ATP-binding site of a purinergic P2X receptor. *J Biol Chem* **275**, 34190–34196.

Jones CA, Chessell IP, Simon J *et al.* (2000). Functional characterisation of the P2X₄ receptor orthologues. *Br J Pharmacol* **129**, 388–394.

Jones CA, Vial C, Sellers LA *et al.* (2004). Functional regulation of P2X₆ receptors by *N*-linked glycosylation: identification of a novel alpha beta-methylene ATP-sensitive phenotype. *Mol Pharmacol* **65**, 979–985.

Kasakov L and Burnstock G (1982). The use of the slowly degradable analog, alpha, beta-methylene ATP, to produce desensitisation of the P2-purinoceptor: effect on non-adrenergic, non-cholinergic responses of the guinea-pig urinary bladder. *Eur J Pharmacol* **86**, 291–294.

Kassack MU, Braun K, Ganso M *et al.* (2004). Structure–activity relationships of analogues of NF449 confirm NF449 as the most potent and selective known P2X1 receptor antagonist. *Eur J Med Chem* **39**, 345–357.

Ke HZ, Hong Q, Weidema F *et al.* (2003). Deletion of the P2X7 nucleotide receptor reveals its regulatory roles in bone formation and resorption. *Mol Endocrinology* **17**, 1356–1367.

Khakh BS, Bao XR, Labarca C *et al.* (1999a). Neuronal P2X transmitter-gated cation channels change their ion selectivity in seconds. *Nature Neurosci* **2**, 322–330.

Khakh BS, Fisher JA, Nashmi R *et al.* (2005). An angstrom scale interaction between plasma membrane ATP-gated P2X2 and alpha4beta2 nicotinic channels measured with fluorescence resonance energy transfer and total internal reflection fluorescence microscopy. *J Neurosci* **25**, 6911–6920.

Khakh BS, Proctor WR, Dunwiddie TV *et al.* (1999b). Allosteric control of gating and kinetics at P2X$_4$ receptor channels. *J Neurosci* **19**, 7289–7299.

Kim M, Jiang LH, Wilson HL *et al.* (2001). Proteomic and functional evidence for a P2X7 receptor signalling complex. *EMBO J* **20**, 6347–6358.

King BF, Liu M, Townsend-Nicholson A *et al.* (2005). Antagonism of ATP responses at P2X receptor subtypes by the pH indicator dye, phenol red. *Br J Pharmacol* **145**, 313–322.

King BF, Townsend-Nicholson A, Wildman SS *et al.* (2000). Coexpression of rat P2X2 and P2X6 subunits in *Xenopus* oocytes. *J Neurosci* **20**, 4871–4877.

King BF, Wildman SS, Ziganshina LE *et al.* (1997). Effects of extracellular ph on agonism and antagonism at a recombinant P2X2 receptor. *Br J Pharmacol* **121**, 1445–1453.

Kirkup AJ, Booth CE, Chessell IP *et al.* (1999). Excitatory effect of P2X receptor activation on mesenteric afferent nerves in the anaesthetised rat. *J Physiol* **520**, 551–563.

Klapperstuck M, Buttner C, Schmalzing G *et al.* (2001). Functional evidence of distinct ATP activation sites at the human P2X(7) receptor. *JPhysiol* **534**, 25–35.

Kukley M, Stausberg P, Adelmann G *et al.* (2004). Ecto-nucleotidases and nucleoside transporters mediate activation of adenosine receptors on hippocampal mossy fibers by P2X7 receptor agonist 2′-3′-O-(4-benzoylbenzoyl)-ATP. *J Neurosci* **24**, 7128–7139.

Labasi JM, Petrushova N, Donovan C *et al.* (2002). Absence of the P2X$_7$ receptor alters leukocyte function and attenuates an inflammatory response. *J Immunol* **168**, 6436–6445.

Lambrecht G, Rettinger J, Baumert HG *et al.* (2000). The novel pyridoxal-5′-phosphate derivative PPNDS potently antagonizes activation of P2X(1) receptors. *Eur J Pharmacol* **387**, R19–R21.

Lazarowski ER, Boucher RC and Harden TK (2000). Constitutive release of ATP and evidence for major contribution of ecto-nucleotide pyrophosphatase and nucleoside diphosphokinase to extracellular nucleotide concentrations. *J Biol Chem* **275**, 31061–31068.

Lê K-T, Babinski K and Seguela P (1998). Central P2X$_4$ and P2X$_6$ channel subunits coassemble into a novel heteromeric ATP receptor. *J Neurosci* **18**, 7152–7159.

Lê KT, Grabot E-B, Archambault V *et al.* (1999). Functional and biochemical evidence for heteromeric ATP-gated channels composed of P2X1 and P2X5 subunits. *J Biol Chem* **274**, 15415–15419.

Lê KT, Paquet M, Nouel D *et al.* (1997). Primary structure and expression of a naturally truncated P2X ATP receptor subunit from brain and immune system. *FEBS Lett* **418**, 195–199.

Lewis C, Neidhart S, Holy C *et al.* (1995). Coexpression of P2X$_2$ and P2X$_3$ receptor subunits can account for ATP-gated currents in sensory neurons. *Nature* **377**, 432–435.

Li CY, Peoples RW and Weight FF (1996). Proton potentiation of atp-gated ion channel responses to atp and zn2+ in rat nodose ganglion neurons. *J Neurophysiol* **76**, 3048–3058.

Lynch KJ, Touma E, Niforatos W *et al.* (1999). Molecular and functional characterization of human P2X$_2$ receptors. *Mol Pharmacol* **56**, 1171–1181.

MacKenzie AB, Mahaut-Smith MP and Sage SO (1996). Activation of receptor-operated cation channels via P2X1 not P2T purinoceptors in human platelets. *J Biol Chem* **271**, 2879–2881.

Mahaut-Smith MP, Tolhurst G and Evans RJ (2004). Emerging roles for P2X1 receptors in platelet activation. *Platelets* **15**, 131–144.

Masin M, Kerschensteiner D, Dumke K *et al.* (2006). Fe65 interacts with P2X2 subunits at excitatory synapses and modulates receptor function. *J Biol Chem* **281**, 4100–4108.

McDonald HA, Chu KL, Bianchi BR *et al.* (2002). Potent desensitization of human P2X3 receptors by diadenosine polyphosphates. *Eur J Pharmacol* **435**, 135–142.

McGaraughty S, Wismer CT, Zhu CZ *et al.* (2003). Effects of A-317491, a novel and selective P2X3/P2X2/3 receptor antagonist, on neuropathic, inflammatory and chemogenic nociception following intrathecal and intraplantar administration. *Br J Pharmacol* **140**, 1381–1388.

McQueen DS, Bond SM, Moores C *et al.* (1998). Activation of P2X receptors for adenosine evokes cardiorespiratory reflexes in anaesthetized rats. *J Physiol* **507**, 843–855.

Michel AD, Chau NM, Fan TP *et al.* (1995). Evidence that [3H]-alpha,beta-methylene ATP may label an endothelial-derived cell line 5′-nucleotidase with high affinity. *Br J Pharmacol* **115**, 767–774.

Michel AD, Chessell IP and Humphrey PPA (1999). Ionic effects on human recombinant P2X$_7$ receptor function. *Naunyn-Schmiedebergs Arch Pharmacol* **359**, 102–109.

Michel AD, Kaur R, Chessell IP *et al.* (2000). Antagonist effects on human P2X(7) receptor-mediated cellular accumulation of YO-PRO-1. *Br J Pharmacol* **130**, 513–520.

Michel AD, Miller KJ, Lundstrom K *et al.* (1997). Radiolabeling of the rat p2x4 purinoceptor: evidence for allosteric interactions of purinoceptor antagonists and monovalent cations with p2x purinoceptors. *Mol Pharmacol* **51**, 524–532.

Michel AD, Xing M and Humphrey PP (2001). Serum constituents can affect 2′-and 3′-O-(4-benzoylbenzoyl)-ATP potency at P2X(7) receptors. *Br J Pharmacol* **132**, 1501–1508.

Michel AD, Xing M, Thompson KM *et al.* (2006). Decavanadate, a P2X receptor antagonist and its use to study ligand interactions with P2X7 receptors. *Eur J Pharmacol* **534**, 19–29.

Miller KJ, Michel AD, Chessell IP *et al.* (1998). Cibacron blue allosterically modulates the rat P2X4 receptor. *Neuropharmacol* **37**, 1579–1586.

Mulryan K, Gitterman DP, Lewis CJ *et al.* (2000). Reduced vas deferens contraction and male infertility in mice lacking P2X1 receptors. *Nature* **403**, 86–89.

Murgia M, Hanau S, Pizzo P *et al.* (1993). Oxidized ATP. An irreversible inhibitor of the macrophage purinergic P2Z receptor. *J Biol Chem* **268**, 8199–8203.

Myers AJ, Eilertson B, Fulton SA *et al.* (2005). The purinergic P2X7 receptor is not required for control of pulmonary *Mycobacterium tuberculosis* infection. *Infect Immunity* **73**, 3192–3195.

Nagaya N, Tittle RK, Saar N *et al.* (2005). An intersubunit zinc binding site in rat P2X2 receptors. *J Biol Chem* **280**, 25982–25993.

Nakazawa K, Inoue K, Fujimori K *et al.* (1991). Effects of ATP antagonists on purinoceptor-operated inward currents in rat phaeochromocytoma cells. *Pflugers Arch* **418**, 214–219.

Newbolt A, Stoop R, Virginio C *et al.* (1998). Membrane topology of an ATP-gated ion channel (P2X receptor). *J Biol Chem* **273**, 15177–15182.

Nicke A, Baumert HG, Rettinger J *et al.* (1998). P2X$_1$ and P2X$_3$ receptors form stable trimers: a novel structural motif of ligand-gated ion channels. *EMBO J* **17**, 3016–3028.

Nicke A, Kerschensteiner D and Soto F (2005). Biochemical and functional evidence for heteromeric assembly of P2X1 and P2X4 subunits. *J Neurochem* **92**, 925–933.

North RA (2002). Molecular physiology of P2X receptors. *Physiol Rev* **82**, 1013–1067.

North RA and Surprenant A (2000). Pharmacology of cloned P2X receptors. *Ann Rev Pharmacol Toxicol* **40**, 563–580.

Okabe M, Enomoto M, Maeda H *et al.* (2006). Biochemical characterization of suramin as a selective inhibitor for the PKA-mediated phosphorylation of HBV core protein *in vitro*. *Biolog Pharmaceut Bull* **29**, 1810–1814.

Ormond SJ, Barrera NP, Qureshi OS *et al.* (2006). An uncharged region within the N terminus of the P2X6 receptor inhibits its assembly and exit from the endoplasmic reticulum. *Mol Pharmacol* **69**, 1692–1700.

Oury C, Kuijpers MJE, Toth-Zsamboki E *et al.* (2003). Overexpression of the platelet P2X1 ion channel in transgenic mice generates a novel prothrombotic phenotype. *Blood* **101**, 3969–3976.

Oury C, Toth-Zsamboki E, Van Geet C *et al.* (2000). A natural dominant negative P2X1 receptor due to deletion of a single amino acid residue. *J Biol Chem* **275**, 22611–22614.

Oury C, Toth-Zsamboki E, Vermylen J *et al.* (2002). Does the P(2X1del) variant lacking 17 amino acids in its extracellular domain represent a relevant functional ion channel in platelets? *Blood* **99**, 2275–2277.

Paneesha S, Starczynski J, Pepper C *et al.* (2006). The P2X7 receptor gene polymorphism 1513 A–>C has no effect on clinical prognostic markers and survival in multiple myeloma. *Leukemia Lymphoma* **47**, 281–284.

Paukert M, Osteroth R, Geisler HS *et al.* (2001). Inflammatory mediators potentiate ATP-gated channels through the P2X(3) subunit. *J Biol Chem* **276**, 21077–21082.

Pelegrin P and Surprenant A (2006). Pannexin-1 mediates large pore formation and interleukin-1β release by the ATP-gated P2X7 receptor. *EMBO J* **25**, 5071–5082.

Perregaux DG and Gabel CA (1994). Interleukin-1β maturation and release in response to ATP and nigericin. *J Biol Chem* **269**, 15195–15203.

Pratt EB, Brink TS, Bergson P *et al.* (2005). Use-dependent inhibition of P2X3 receptors by nanomolar agonist. *J Neurosci* **25**, 7359–7365.

Priel A and Silberberg SD (2004). Mechanism of ivermectin facilitation of human P2X4 receptor channels. *J Gen Physiol* **123**, 281–293.

Puthssery T and Fletcher EL (2006). P2X2 receptors on ganglion and amacrine cells in cone pathways of the rat retina. *J Comp Neurol* **496**, 595–609.

Radford K, Virginio C, Surprenant A *et al.* (1997). Baculovirus expression provides direct evidence for heteromeric assembly of P2X$_2$ and P2X$_3$ receptors. *J Neurosci* **17**, 6529–6533.

Rae MG, Rowan EG and Kennedy C (1998). Pharmacological properties of P2X3-receptors present in neurones of the rat dorsal root ganglia. *Br J Pharmacol* **124**, 176–180.

Rapp DE, Lyon MB, Bales GT *et al.* (2005). A role for the P2X receptor in urinary tract physiology and in the pathophysiology of urinary dysfunction. *Eur Urol* **48**, 303–308.

Rassendren F, Buell G, Virginio C *et al.* (1997). The permeabilizing ATP receptor (P2X$_7$): Cloning and expression of a human cDNA. *J Biol Chem* **272**, 5482–5486.

Redegeld FA, Caldwell CC and Sitkovsky MV (1999). Ecto-protein kinases: ecto-domain phosphorylation as a novel target for pharmacological manipulation? *Trends Pharmacol Sci* **20**, 453–459.

Ren J, Bian X, DeVries M *et al.* (2003). P2X2 subunits contribute to fast synaptic excitation in myenteric neurones of the mouse small intestine. *J Physiol* **552.3**, 809–821.

Rettinger J and Schmalzing G (2003). Activation and desensitization of the recombinant P2X1 receptor at nanomolar ATP concentrations. *J GenPhysiol* **121**, 451–461.

Rettinger J and Schmalzing G (2004). Desensitization masks nanomolar potency of ATP for the P2X1 receptor. *J Biol Chem* **279**, 6426–6433.

Rettinger J, Schmalzing G, Damer S *et al.* (2000). The suramin analogue NF279 is a novel and potent antagonist selective for the P2X(1) receptor. *Neuropharmacology* **39**, 2044–2053.

Romagnoli R, Baraldi PG, Pavani MG *et al.* (2004). Synthesis, radiolabeling and preliminary biological evaluation of [3H]-1-[(S)-N,O-bis-(isoquinolinesulfonyl)-N-methyl-tyrosyl]-4-(o-tolyl)-piperazine, a potent antagonist radioligand for the P2X7 receptor. *Bioorg Med Chem Letts* **14**, 5709–5712.

Rong W, Gourine AV, Cockayne DA *et al.* (2003). Pivotal role of nucleotide P2X2 receptor subunit of the ATP-gated ion channel mediating ventilatory responses to hypoxia. *J Neurosci* **10**, 11315–11321.

Royle SJ, Bobanovic LK and Murrell-Lagnado RD (2002). Identification of a non-canonical tyrosine-based endocytic motif in an ionotropic receptor. *J Biol Chem* **277**, 35378–35385.

Ruan HZ and Burnstock G (2005). The distribution of P2X5 purinergic receptors in the enteric nervous system of mouse. *Cell Tiss Res* **319**, 191–200.

Ruppelt A, Ma W, Borchardt K *et al.* (2001). Genomic structure, developmental distribution and functional properties of the chicken P2X(5) receptor. *J Neurochem* **77**, 1256–1265.

Ryten M, Dunn PM, Neary JT *et al.* (2002). ATP regulates differentiation of mammalian skeletal muscle by activation of a P2X5 receptor on satellite cells. *J Cell Biol* **158**, 345–355.

Sansum AJ, Chessell IP, Hicks GA *et al.* (1998). Evidence that P2X purinoceptors mediate the excitatory effects of αβ methylene-ADP in rat locus coeruleus neurones. *Neuropharmacology* **37**, 875–885.

Sanz JM, Chiozzi P and Di Virgilio F (1998). Tenidap enhances P2Z/P2X7 receptor signalling in macrophages. *Eur J Pharmacol* **355**, 235–244.

Saunders BM, Fernando SL, Sluyter R *et al.* (2003). A loss-of-function polymorphism in the human P2X7 receptor aboloishes ATP-mediated killing of mycobacterium. *J Immunol* **171**, 5442–5446.

Schwab JM, Guo L and Schluesener HJ (2006). Spinal cord injury induces early and persistent lesional P2X4 receptor expression. *J Neuroimmunol* **163**, 186–189.

Seguela P, Haghighi A, Soghomonian J-J *et al.* (1996). A novel P$_{2X}$ ATP receptor ion channel with widespread distribution in the brain. *J Neurosci* **16**, 448–455.

Seman M, Adriouch S, Scheuplein F *et al.* (2003). NAD-induced T cell death: ADP-ribosylation of cell surface proteins by ART2 activates the cytolytic P2X7 purinoceptor. *Immunity* **19**, 571–582.

Shehnaz D, Torres B, Balboa MA *et al.* (2000). Pyridoxal-phosphate-6-azophenyl-2′,4′-disulfonate (PPADS), a putative P2Y(1) receptor antagonist, blocks signaling at a site distal to the receptor in Madin–Darby canine kidney-D(1) cells. *J Pharmacol Exp Ther* **292**, 346–350.

Shemon AN, Sluyter R, Fernando SL *et al.* (2006). A Thr357 to Ser polymorphism in homozygous and compound heterozygous subjects causes absent or reduced P2X7 function and impairs ATP-induced mycobacterial killing by macrophages. *J Biol Chem* **281**, 2079–2086.

Shen J-B, Pappano AJ and Liang BT (2006). Extracellular ATP-stimulated current in wild-type and P2X4 receptor transgenic mouse ventricular myocytes: implications for a cardiac physiologic role of P2X4 receptors. *FASEB Journal* **20**, 277–284.

Sim JA, Young MT, Sung H-Y *et al.* (2004). Reanalysis of P2X7 receptor expression in rodent brain. *J Neurosci* **24**, 6307–6314.

Simon J, Kidd EJ, Smith FM *et al.* (1997). Localization and functional expression of splice variants of the P2X$_2$ receptor. *Mol Pharmacol* **52**, 237–248.

Smart ML, Gu B, Panchal RG *et al.* (2003). P2X7 receptor cell surface expression and cytolytic pore formation are regulated by a distal c-terminal region. *J Biol Chem* **278**, 8853–8860.

Smith FM, Humphrey PP and Murrell-Lagnado RD (1999). Identification of amino acids within the P2X2 receptor C-terminus that regulate desensitization. *J Physiol* **520**, 91–99.

Sokolova E, Skorinkin A, Fabbretti E *et al.* (2004). Agonist-dependence of recovery from desensitization of P2X(3) receptors provides a novel and sensitive approach for their rapid up or downregulation. *Br J Pharmacol* **141**, 1048–1058.

Solle M, Labasi J, Perregaux DG *et al.* (2001). Altered cytokine production in mice lacking P2X$_7$ receptors. *J Biol Chem* **276**, 125–132.

Soltoff SP, McMillian MK and Talamo BR (1989). Coomassie brilliant blue G is a more potent antagonist of P2 purinergic responses than reactive blue 2 (cibacron blue 3GA) in rat parotid acinar cells. *Biochem Biophys Res Commun* **165**, 1279–1285.

Soltoff SP, McMillian MK, Talamo BR *et al.* (1993). Blockade of ATP binding site of P2 purinoceptors in rat parotid acinar cells by isothiocyanate compounds. *Biochem Pharmacol* **45**, 1936–1940.

Soto F, Garciaguzman M, Gomezhernandez JM *et al.* (1996b). P2x(4)—an atp-activated ionotropic receptor cloned from rat brain. *Proc Natl Acad Sci USA* **93**, 3684–3688.

Soto F, Garcia-Guzman M, Karschin C *et al.* (1996a). Cloning and tissue distribution of a novel P2X receptor from rat brain. *Biochem Biophys Res Comm* **223**, 456–460.

Spelta V, Jiang LH, Surprenant A *et al.* (2002). Kinetics of antagonist actions at rat P2X2/3 heteromeric receptors. *Br J Pharmacol* **135**, 1524–1530.

Stanchev D, Flehmig G, Gerevich Z *et al.* (2006). Decrease of current responses at human recombinant P2X3 receptors after substitution by Asp of Ser/Thr residues in protein kinase C phosphorylation sites of their ecto-domains. *Neurosci Lett* **393**, 78–83.

Stoop R, Surprenant A and North RA (1997). Different sensitivities to pH of ATP-induced currents at four cloned P2X receptors. *J Neurophysiol* **78**, 1837–1840.

Stoop R, Thomas S, Rassendren F *et al.* (1999). Contribution of individual subunits to the multimeric P2X(2) receptor: estimates based on methanethiosulfonate block at T336C. *Mol Pharmacol* **56**, 973–981.

Studeny S, Torabi A and Vizzard MA (2005). P2X2 and P2X3 receptor expression in postnatal and adult urinary bladder and lumbrosacral spinal cord. *Am J Physiol* **289**, R1155–R1168.

Suadicani SO, Brosnan CF and Scemes E (2006). P2X7 receptors mediate ATP release and amplification of astrocytic intercellular Ca2+ signaling. *J Neurosci* **26**, 1378–1385.

Surprenant A, Rassendren F, Kawashima E *et al.* (1996). The cytolytic P_{2Z} receptor for extracellular ATP identified as a P_{2X} receptor ($P2X_7$). *Science* **272**, 735–738.

Surprenant A, Schneider DA, Wilson HL *et al.* (2000). Functional properties of heteromeric P2X1/5 receptors expressed in HEK cells and excitatory junction potentials in guinea pig submucosal arterioles. *J Autonom Nerv Sys* **81**, 249–263.

Torres GE, Egan TM and Voigt M (1998b). Topological analysis of the ATP-gated ionotropic P2X2 receptor subunit. *FEBS Lett* **425**, 19–23.

Torres GE, Egan TM and Voigt MM (1999). Hetero-oligomeric assembly of P2X receptor subunits. Specificities exist with regard to possible partners. *J Biol Chem* **274**, 6653–6659.

Torres GE, Haines WR, Egan TM *et al.* (1998a). Co-expression of $P2X_1$ and $P2X_5$ receptor subunits reveals a novel ATP-gated ion channel. *Mol Pharmacol* **54**, 989–993.

Toulme E, Soto F, Garret M *et al.* (2006). Functional properties of internalization-deficient P2X4 receptors reveal a novel mechanism of ligand-gated channel facilitation by ivermectin. *Mol Pharmacol* **69**, 576–587.

Townsend-Nicholson A, King BF, Wildman SS *et al.* (1999). Molecular cloning, functional characterization and possible cooperativity between the murine $P2X_4$ and $P2X_{4a}$ receptors. *Mol Brain Res* **64**, 246–254.

Tsuda M, Shigemoto-Mogami Y, Koizumi S *et al.* (2003). P2X4 receptors induced in spinal microglia gate tactile allodynia after nerve injury. *Nature* **424**, 778–783.

Turner CM, Vonend O, Chen C *et al.* (2003). The pattern of distribution of selected ATP-sensitive P2 receptor subtypes in normal rat kidney: an immunohistological study. *Cell Tiss Organs* **175**, 105–117.

Valera S, Hussy N, Evans RJ *et al.* (1994). A new class of ligand-gated ion channel defined by P_{2X} receptor for extracellular ATP. *Nature* **371**, 516–519.

Vial C and Evans RJ (2002). P2X1 receptor-deficient mice establish the native P2X receptor and a P2Y6-like receptor in arteries. *Mol Pharmacol* **62**, 1438–1445.

Vial C, Hechler B, Leon C *et al.* (1997). Presence of P2X1 purinoceptors in human platelets and megakaryoblastic cell lines. *Thromb Haemost* **78**, 1500–1504.

Virginio C, Church D, North RA *et al.* (1997). Effects of divalent cations, protons and calmidazolium at the rat $P2X_7$ receptor. *Neuropharmacology* **36**, 1285–1294.

Virginio C, MacKenzie A, Rassendren FA *et al.* (1999). Pore dilation of neuronal P2X receptor channels. *Nature Neurosci* **2**, 315–321.

Virginio C, North RA and Surprenant A (1998b). Calcium permeability and block at homomeric and heteromeric P2X2 and P2X3 receptors and receptors in rat nodose neurones. *J Physiol* **510**, 27–25.

Virginio C, Robertson G, Surprenant A *et al.* (1998a). Trinitrophenyl-substituted nucleotides are potent antagonists selective for $P2X_1$, $P2X_3$ and heteromeric $P2X_{2/3}$ receptors. *Mol Pharmacol* **53**, 969–973.

Voogd TE, Vansterkenburg EL, Wilting J *et al.* (1993). Recent research on the biological activity of suramin. *Pharmacol Rev* **45**, 177–203.

Wang CZ, Namba N, Gonoi T *et al.* (1996). Cloning and pharmacological characterization of a fourth p2x receptor subtype widely expressed in brain and peripheral tissues including various endocrine tissues. *Biochem Biophys Res Commun* **220**, 196–202.

Wildman SS, Brown SG, Rahman M *et al.* (2002). Sensitization by extracellular Ca(2+) of rat P2X(5) receptor and its pharmacological properties compared with rat P2X(1). *Mol Pharmacol* **62**, 957–966.

Wildman SS, King BF and Burnstock G (1999). Modulatory activity of extracellular H+ and Zn2+ on ATP-responses at rP2X1 and rP2X3 receptors. *Br J Pharmacol* **128**, 486–492.

Wiley JS, Chen R, Wiley MJ *et al.* (1992). The ATP4- receptor-operated ion channel of human lymphocytes: inhibition of ion fluxes by amiloride analogs and by extracellular sodium ions. *Arch Biochem Biophys* **292**, 411–418.

Wiley JS, Dao-Ung LP, Gu B *et al.* (2002). A loss-of-function polymorphic mutation in the cytolytic P2X7 receptor gene and chronic lymphocytic leukaemia: a molecular study. *Lancet* **359**, 1114–1119.

Xiong K, Peoples RW, Montgomery JP *et al.* (1999). Differential modulation by copper and zinc of P2X2 and P2X4 receptor function. *J Neurophysiol* **81**, 2088–2094.

Yamamoto K, Sokabe T, Matsumoto T *et al.* (2006). Impaired flow-dependent control of vascular tone and remodeling in P2X4-deficient mice. *Nature Med* **12**, 133–137.

Zhao HB, Yu N and Fleming CR (2005). Gap junctional hemichannel-mediated ATP release and hearing controls in the inner ear. *Proc Nat Acad Sci USA* **102**, 18724–18729.

Ziganshin AU, Ziganshina LE, King BE *et al.* (1995). Characteristics of ecto-ATPase of *Xenopus* oocytes and the inhibitory actions of suramin on ATP breakdown. *Pflugers Arch* **429**, 412–418.

Zimmermann H (2000). Extracellular metabolism of ATP and other nucleotides. *Naunyn-Schmiedebergs Arch Pharmacol* **362**, 299–309.

3.4

Others

Contents

3.4.1 Acid-sensing ion channels

Eric Lingueglia and Michel Lazdunski

3.4.1.1 Introduction

Acid-sensing ion channels (ASICs) are neuronal homo- or hetero-multimeric voltage-insensitive cationic channels activated by extracellular acidification, which can trigger membrane depolarization in response to local acidosis (for reviews, see Krishtal, 2003; Kress and Waldmann, 2006; Wemmie et al., 2006; Lingueglia, 2007). They are mostly Na^+ selective and belong to the epithelial amiloride-sensitive Na^+ channel and degenerin (ENaC/DEG) family of ion channels. ASICs desensitize rapidly and the properties of most family members, and especially those expressed in the CNS (ASIC1a, -2a, -2b), make these channels sensitive to dynamic pH fluctuations. However, activated ASIC3-containing channels carry a sustained current that does not inactivate while the medium remains acidic. ASIC3, like ASIC1b, is specifically localized in sensory neurons whereas ASIC1a, -2a and -2b are found in both the peripheral and central nervous systems (PNS and CNS), where ASIC-like currents have been detected in almost all neurons. The physiological function of these channels in the nervous system remains incompletely understood. In recent years, knockout mice have helped to illustrate a role of ASICs in many physiological and pathophysiological processes such as hippocampal long-term potentiation, learning and memory, acquired fear-related behaviour, brain ischaemia, autoimmune inflammation of the CNS and a number of different sensory processes such as nociception, visual transduction, sour taste perception, hearing, and mechanoperception. ASIC1a appears to be an important subunit in the CNS that is modulated by ASIC2 and is required for normal synaptic plasticity in hippocampal neurons, contributes to the neural mechanisms underlying fear conditioning in the amygdala and participates in the central modulation of pain in the spinal cord. ASIC1a has the unique property among ASICs to provide a voltage-independent pathway for Ca^{2+} entry into neurons and has been associated with neuronal death in brain ischaemia and with axonal degeneration in a mouse model of multiple sclerosis. On the other hand, ASIC3 is the most important ASIC in the PNS where it participates in the modulation of pain perception and has been proposed to sense painful tissue acidosis that occurs for instance in ischemic, damaged or inflamed tissue where ASICs are upregulated. ASIC3 and ASIC2a have also been implicated in cutaneous mechanical sensitivity and almost all ASICs contribute to visceral mechanotransduction, but it remains to be determined if these channels have the ability to transduce mechanical signals, either directly or indirectly. Some functions attributed to ASICs might therefore not (or not only) rely on their activation by protons, and the existence of other stimuli remains an open question.

The first family member, ASIC1a, was identified in 1997 (Waldmann et al., 1997b), although ASIC2a was cloned earlier in 1996 (Price et al., 1996; Waldmann et al., 1996) but was not identified as a proton-gated channel at that time. The epithelial amiloride-sensitive Na^+ channel and degenerin (ENaC/DEG) family of ion channels (Figure 3.4.1.1; Kellenberger and Schild, 2002), to which the ASICs belong, was named based on the first two members identified, the epithelial Na^+ channel involved in taste perception and Na^+ homeostasis in mammals (Canessa et al., 1993; Lingueglia et al., 1993) and the neuronal and muscular degenerins of the nematode *Caenorhabditis elegans* involved in mechanoperception (Driscoll and Chalfie, 1991; Huang and Chalfie, 1994; Liu et al., 1996; Syntichaki and Tavernarakis, 2004). This family also comprises channels in *Drosophila melanogaster* including the *Drosophila* gonad-specific amiloride-sensitive Na^+ channel (dGNaC1 or RPK) (Adams et al., 1998a; Darboux et al., 1998a) suggested to participate in gametogenesis and/or early embryonic development, a putative stretch-activated channel (dmdNaC1 or PPK1) (Adams et al., 1998a; Darboux et al., 1998b) involved in proprioception (Ainsley et al., 2003), genes expressed in the tracheal system where they may be involved in liquid clearance from the airways (PPK4, -7, -10, -11, -12, -13, -14, -19, and -28)

(Liu *et al.*, 2003a), genes involved in salt taste (*PPK11* and *PPK19*) (Liu *et al.*, 2003b) and a gene necessary for response to female pheromones by *Drosophila* males (*PPK25*) (Lin *et al.*, 2005). The ENaC/DEG family also comprises a sodium channel expressed in mammalian brain, liver and intestine whose mode of activation and physiological roles remain unknown (BLINaC) (Sakai *et al.*, 1999; Schaefer *et al.*, 2000). Of particular interest is the significant structure and sequence homology of ASICs with the FMRFamide-gated Na^+ channel (FaNaC), which is an ionotropic peptide-gated receptor identified from invertebrate nervous system (Lingueglia *et al.*, 1995; 2006).

3.4.1.2 Molecular characterization of the ASIC channel family

Four different genes encoding seven ASIC isoforms have been described to date in mammals (512–563 amino acid residues long with 45–80% amino acid identity between rat isoforms, Table 3.4.1.1): the splice variants ASIC1a (also named ASIC, ASICα or BNaC2α) (Waldmann *et al.*, 1997b), ASIC1b (also named ASICβ or BNaC2β) (Chen *et al.*, 1998; Bassler *et al.*, 2001) and ASIC1b2 (also named ASICβ2) (Ugawa *et al.*, 2001); the splice variants ASIC2a (also named MDEG1, BNaC1α or BNC1a) (Price *et al.*, 1996; Waldmann *et al.*, 1996; Garcia-Anoveros *et al.*, 1997; Champigny *et al.*, 1998) and ASIC2b (also named MDEG2 or BNaC1β) (Lingueglia *et al.*, 1997); ASIC3 (also named DRASIC or TNaC) (Waldmann *et al.*, 1997a; de Weille *et al.*, 1998; Ishibashi and Marumo, 1998; Babinski *et al.*, 1999); and ASIC4 (also named SPASIC) (Akopian *et al.*, 2000; Grunder *et al.*, 2000). Some ASICs are particularly well conserved between rat or mouse and human, suggesting important functions in the mammalian nervous system. Human and mouse ASIC1a and ASIC2a for instance share 98% and 99% amino acid identity, respectively. The chromosomal localizations of ASIC1, ASIC2, ASIC3 and ASIC4 in mouse, rat, and human are listed in Table 3.4.1.2. ASICs have also been characterized from ascidian (Coric *et al.*, 2008), toadfish, lamprey, shark (Coric *et al.*, 2003, 2005), and zebrafish (Paukert *et al.*, 2004b; Chen *et al.*, 2007).

ASICs share the overall structure of members of the ENaC/DEG family, i.e., two transmembrane domains surrounding a large extracellular loop containing conserved cysteines (Figure 3.4.1.1) (Saugstad *et al.*, 2004). The NH_2- and COOH-terminal domains are therefore cytoplasmic. Functional ASICs are homomeric or heteromeric channels. They form trimers, as shown from the determination of the tridimensional structure of a chicken ASIC1 deletion mutant (Jasti *et al.*, 2007).

ASIC channels are assembled in the membrane of neurons as complexes containing accessory proteins. These essential ASIC

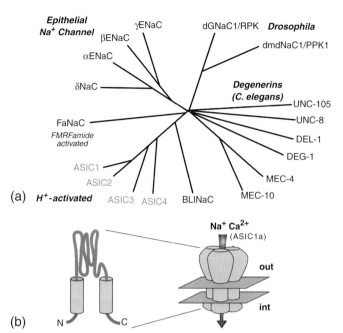

Fig. 3.4.1.1 ASICs belong to the ENaC/DEG family of ion channels. (a) Dendrogram illustrating the amiloride-sensitive Na^+ channel and degenerin (ENaC/DEG) family comprising members in different species with functions often related to sensory processing. The epithelial Na^+ (ENaC), which is comprised of α, β, γ and possibly δ subunits is involved in taste perception and Na^+ homeostasis in mammals. The *C. elegans* neuronal and muscular degenerins are involved in mechanoperception and *Drosophila* dmdNaC1/PPK1 has been implicated in proprioception, while dGNaC1/RPK was suggested to participate in gametogenesis and/or early embryonic development. It should be noted that there are several additional members of the ENaC/DEG family in *C. elegans* and *Drosophila*, which are not shown here. ASICs share significant structure and sequence homology with a sodium channel subunit of unknown function (BLINaC), which is expressed in mammalian brain, liver, and intestine. Perhaps more interestingly, ASICs are also related to the ionotropic peptide receptor FaNaC, which is gated by FMRFamide and has been identified from the invertebrate nervous system. (b) ASICs as other members of the ENaC/DEG family possess two hydrophobic transmembrane regions flanking a large extracellular domain with many conserved cysteines (representing more than 60% of the protein) and relatively short intracellular NH_2- and COOH-termini. Functional ASICs are trimers, as indicated by the recent determination of the tridimensional structure of a chicken ASIC1 deletion mutant… ASICs predominantly conduct Na^+ but ASIC1a is also substantially permeable to Ca^{2+}, providing a voltage-independent pathway for Ca^{2+} entry into neurons. Note that mammalian ASIC4 is not activated by protons.

Table 3.4.1.1 The percentage of amino acid identity between rat ASIC isoforms

	ASIC1b	ASIC2a	ASIC2b	ASIC3	ASIC4
ASIC1a	80	68	60	54	50
ASIC1b		62	60	53	50
ASIC2a			78	51	47
ASIC2b				49	45
ASIC3					47

partners are important both for control of ASIC function and subcellular targeting. ASICs contain PDZ-binding motifs at their C-termini and interact with several PDZ-containing proteins. PICK1 (protein interacting with C-kinase-1) associates with ASIC1 and ASIC2 C-terminal regions through its single PDZ domain (Duggan *et al.*, 2002; Hruska-Hageman *et al.*, 2002). PICK1 co-localizes with ASICs in the nervous system and modifies the cellular distribution of ASIC2a in heterologous expression systems. The PICK1 association with ASIC1 is downregulated by cAMP-dependent protein kinase A (PKA), suggesting a mechanism to control the cellular distribution of ASIC1 (Leonard *et al.*, 2003). In addition, PICK1 association potentiates the regulation of ASIC2a by protein kinase C (PKC) phosphorylation (Baron *et al.*, 2002a) and has been proposed to participate in the PKC regulation of ASIC3 through

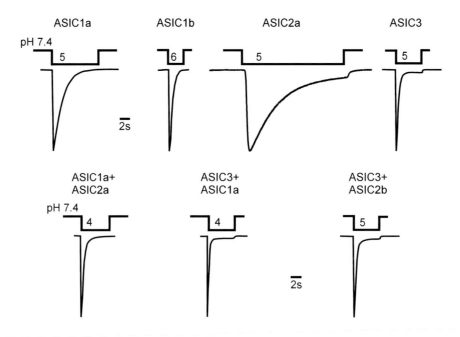

Fig. 3.4.1.2 Typical current traces of homomeric and heteromeric ASIC channels. Typical inward currents recorded from mammalian cells expressing recombinant ASIC channels. Channel activity is induced by a fast decrease in the extracellular pH from 7.4 to the value indicated above each current trace. The sensitivity to external H+, the activation and inactivation kinetics and the pharmacology vary according to the ASIC subtypes and the subunit composition of the channel complex (see Tables 3.4.1.2 and 3.4.1.3). ASIC3-containing channels have an additional sustained component that does not inactivate while the pH remains acidic and is smaller than the transient component.

its association with the ASIC2b subunit (Deval *et al.*, 2004). The PDZ-mediated interaction of PICK1 with parkin promotes PICK1 mono-ubiquitination and inhibits the PKC-induced, PICK1-dependent potentiation of ASIC2a currents (Joch *et al.*, 2007). The PDZ-containing proteins CIPP (channel-interacting PDZ domain protein) (Anzai *et al.*, 2002), NHERF-1 (Na+/H+ exchanger regulatory factor-1) (Deval *et al.*, 2006), and both PSD-95 and Lin-7b (Hruska-Hageman *et al.*, 2004), interact with the ASIC3 C-terminus and control surface expression of the channel. In addition, NHERF-1 considerably increases the sustained current generated by ASIC3 and also modifies ASIC3 subcellular localization in heterologous expression systems, probably through its association with the cortical cytoskeleton (Deval *et al.*, 2006). The annexin II light chain P11 interacts with the N-terminus of ASIC1a and enhances the cell-surface expression of the channel without altering its properties (Donier *et al.*, 2005). The integral membrane protein stomatin has also been shown to associate with ASIC1a, ASIC2a and ASIC3 and to potently reduce ASIC3 currents and to accelerate the rate of desensitization of the ASIC2a, ASIC1a+3 and ASIC2a+3 channels (Price *et al.*, 2004b). More recently, the stomatin-like protein 3 (SLP3), which is indispensable for the function of a subset of cutaneous mechanoreceptors in mouse, has been shown to associate in vitro with ASIC2a, ASIC2b and ASIC3 and to inhibit the endogenous ASIC current in HEK-293 cells. In addition, SLP3 mutant sensory neurons display larger sustained proton-gated currents compared to wild-type neurons (Wetzel *et al.*, 2007). An association between CaMKII and ASIC1a has been described in brain, which is enhanced by ischemia and leads to an increase of the ASIC1a current by CaMKII phosphorylation (Gao *et al.*, 2005). An interaction between ASICs and the high-conductance (BK) Ca2+- and voltage-activated K+channel has been described, which results in an inhibition of BK channel current with acidosis disrupting the interaction and relieving the inhibition (Petroff *et al.*, 2008). ASIC1a and ASIC2a associate with A kinase-anchoring protein 150 (AKAP150) and the protein phosphatase calcineurin (Chai *et al.*, 2007). A positive modulation of ASICs by AKAP150 in association with protein kinase A has been suggested, while calcineurin-dependent dephosphorylation seems to inhibit ASICs. A positive modulation of the ASIC1a+2a heteromeric channel by the cystic fibrosis transmembrane regulator (CFTR) has also been reported in vitro (Ji *et al.*, 2002) but it is not clear whether this effect relies on a direct physical interaction between the two channels. An interaction between ASIC3 and CFTR has also been described in airway epithelial cells (Su *et al.*, 2006).

3.4.1.3 Biophysical characteristics

To date, the only known activators of ASICs are extracellular protons, although activation by other stimuli cannot be ruled out at this time. ASIC1b2, ASIC2b and ASIC4 do not form functional proton-activated channels when expressed alone, however ASIC1b2 and ASIC2b form hetreromeric complexes with other ASICs when co-expressed, yielding receptors with distinct properties (Lingueglia *et al.*, 1997; Ugawa *et al.*, 2001). ASIC4 does not appear to co-assemble with the other ASICs and its function remains unclear. The sensitivity to external H+, the activation, inactivation and reactivation kinetics and the pharmacology vary according to the ASIC subtypes and the subunit composition of the channel complex, with $pH_{0.5}$ ranging from 4.0–6.8 and activation thresholds close to pH 7.0 (Tables 3.4.1.2 and 3.4.1.3). As a result most ASICs are activated by pH variations in the physiological range. Protons trigger a transient inward current that desensitizes rapidly (within seconds; see Figure 3.4.1.2). ASICs show no voltage-dependence and the desensitization kinetics and recovery from desensitization depend on channel subtypes. Interestingly, incorporation of the ASIC2a subunit into ASIC1a or ASIC3 channel complexes significantly accelerates the recovery from desensitization ($\tau_{recovery}$ at pH 7.4 ~12 sec and ~0.5 sec for homomeric ASIC1a and ASIC3, respectively versus ~0.6 sec and ~0.08 sec for heteromeric ASIC1a+2a and ASIC3+2a, respectively; Tables 3.4.1.2 and 3.4.1.3). The strong desensitization and the pH dependence of inactivation make ASICs very sensitive to pre-existing pH. This means that a small shift of the resting pH may have important effects on the ASIC response to a rapid drop in pH. In addition, prolonged or slow acidification of

Table 3.4.1.2 Properties of ASIC channels

Channel subunit	Biophysical characteristics	Cellular and subcellular localization	Pharmacology	Physiology	Relevance to disease states
ASIC1a (ASIC, ASICα, BNaC2α) [1, 2] Gene: ACCN2 Chromosomal location: mouse: 15F3 rat: 7q36 human: 12q12	pH0.5 act ~6.2–6.8 [1, 3–8] pH 0.5 inact ~7.2–7.4 [3, 5, 6, 7, 8] P_{Na}/P_K ~5.5–13 [1, 3, 9–11] P_{Na}/P_{Ca} ~2.5–18.5 [1, 3, 9] g_{Na}^+ 14.3pS [1] τ_{act} 5.8–13.7ms (pH 6) [3, 9] τ_{inact}* 1.1–4s (pH 6) [3, 4, 6, 8, 9, 12, 13] τ_{recov} 5.3–13s (pH 7.4) [3–5, 8]	Brain (high in cortex, striatum, hippocampus, olfactory bulb, amygdala and cerebellum, probably postsynaptic in hippocampal neurons) [1, 2, 14–16] Spinal cord neurons [8, 17–20] Trigeminal, vagal and dorsal root ganglia (small and medium nociceptive sensory neurons) [19, 21–24] Taste receptor cells [25] Lung epithelial cells [26] Retina [27–29] Bone [30] Immune cells [31] Vascular smooth muscle cells [32]	IC50: amiloride 10 μM (pH 6.0) [1], Pb2+ ~3.7 μM [33], Ni2+ ~0.58 mM (pH5.0) [34], Zn2+ ~7 nM [35]; Ca2+ ~1.3 mM (pH 6), 3.9–9.2 mM (pH 5.5–5) [1, 10, 36]; ibuprofen/flurbiprofen ~ 350 μM [21]; PcTx1 ~0.9nM [13]; EC50: FMRFa ~33 μM [11]; Ca2+ ~2.0 mM [5, 6]; Ba2+ ~2.9 mM; Mg2+ ~2.0 mM [5] Enhanced by lactate [37], spermine [5], arachidonic acid [38], NO [39], glutathione and DTT [40]	Proposed role in synaptic transmission Provides a voltage-independent pathway for Ca2+ entry into neurons. In the CNS, knockout mice exhibit impaired hippocampal long-term potentiation and defects in spatial learning, eye-blink conditioning and fear conditioning [15, 16, 41] Influences visceral but not cutaneous mechanoreceptor function [24, 42] Knockout mice present no apparent sensory or motor deficit [15, 16] Role in visual transduction in mouse [29] Central modulation of pain through the opioid system [43] Assembles in heteromeric receptors with at least ASIC1b, –2a, –2b and –3	Upregulated by inflammation in rat DRG [21, 44] and dorsal horn neurons [18, 20] Participates in the spinal cord in central sensitization associated with inflammatory pain [20] No role in acid-induced mechanical hyperalgesia in mouse [45] Proposed to contribute to neuronal death associated with brain ischemia [46, 47] Possible implication in anxiety disorders in human, based on its role in the fear circuit in mouse Downregulated in a rat model of epilepsy [48] Downregulated in a rat model of neuropathic pain [49] Detected in high-grade human glioma cells derived from tumor resections [50, 51] Axonal degeneration in autoimmune inflammation of the central nervous system in a mouse model of multiple sclerosis [31]
ASIC1b (ASICβ, BNaC2β) [9, 19] Splice variant of ASIC1a (first 218 amino acids including TM1 and a part of the extracellular loop are different)	pH0.5 act ~5.1–6.2 [4, 5, 7, 19, 52, 53] pH0.5 inact ~7.0 [5, 7] P_{Na}/P_K ~14.0 [9, 19] $P_{Na} > P_{Ca}$ τ_{act} 9.9 ms (pH 6) [9] τ_{inact} 0.9–1.7 s (pH 6) [3, 4, 9, 52], ~0.8s (pH 4.7) [7] τ_{recov} 4.4–7.7 s (pH 7.4) [3–5]	Dorsal root ganglia (small diameter and large diameter neurons) [19, 23] Cochlear hair cells (stereocilia) [54] Taste receptor cells [55] Carotid body [56] Immune cells [31]	IC50: amiloride 21–23 μM (pH 5.5) [19, 53]; Pb2+ ~1.5 μM [33]; Ca2+>10 mM (pH 4.7)[36] EC50: Ca2+ ~2.0 mM (conditioning pH 7.1) [5] Potentiated by spermine [5] Potentiated by NO [39] and hypotonicity [57] PcTx1 promotes its opening [58]	Proposed role in pain perception. Assembles in heteromeric receptors with at least ASIC1a, –1b2, –2a and –3	Upregulated by inflammation in rat DRG [21, 44]
ASIC1b2 (ASICβ2) [53] Splice variant of ASIC1b (114 amino acid deletion just after the TM1)	Not activated by extracellular pH	Trigeminal ganglion and dorsal root ganglion [53]		Function remains unclear Assembles in heteromeric receptors with ASIC1b to modify its pH dependence [53]	
ASIC2a (MDEG1, BNaC1α, BNC1a) [2, 59–61] Gene: ACCN1 Chromosomal location: mouse: 11 rat: 10q26 human: 17q11.2–q12	pH0.5 act ~4.1–5.0 [4, 6, 52, 61–63] pH0.5 inact ~6.0–6.2 [6, 8] P_{Na}/P_K ~10 [61] P_{Ca}/P_K ~0.5 [61] g_{Na}^+ 10.4–13.4 pS [61] τ_{inact} 1.4 s (pH 4.5), 2.8 s (pH 4), 3.3–5.5 s (pH5) [4, 6, 8, 12, 52] τ_{recov} 0.6–0.9 s (pH 7.4) [4, 8]	Brain (high in olfactory bulb, neo and allocortical regions, dentate granule cells, pyramidal cells of CA1–CA3 subfields of the hippocampal formation, habenula, basolateral amygdaloid nuclei, and the Purkinje and granule cells of the cerebellum) [2, 12, 14, 59, 60, 64] Dendrites and soma of Purkinje cells [65] and pyramidal neurons [64]	IC50: amiloride 28 μM (pH 4.0)[61], Cd2+~0.99 mM (pH5.0) [34] EC50 Zn2+ ~ 120 μM [62] Potentiated by nitric oxide (NO) [39]	Modulation of ASIC1a current in central neurons [74] Role in visual transduction in mouse [70] No [75] or little [67] alteration in cutaneous mechanical sensitivity in knockout mice Some influence on visceral mechanoreceptor function [24]	Protect from light-induced retinal degeneration in mouse [70] Upregulated in rat brain during ischemia [78] Upregulated by peripheral inflammation in rat dorsal horn neurons [18] Detected in high-grade human glioma cells derived from tumour resections [50, 51] Dampen acid-evoked afferent input from the stomach to the brain stem [79]

Continued

Table 3.4.1.2 (Continued) Properties of ASIC channels

Channel subunit	Biophysical characteristics	Cellular and subcellular localization	Pharmacology	Physiology	Relevance to disease states
		Spinal cord neurons [8, 18]; Dorsal root ganglia (small and medium nociceptive neurons; large mechanosensory neurons) [22, 66, 67], specialized cutaneous mechanosensory structures including Meissner corpuscles, Merkel nerve endings, palisades of lanceolate nerve endings [67, 68]; Nodose ganglia [24]; Spiral ganglion in the cochlea [69]; Retina (both neurons and photoreceptors) [27, 28, 70]; Taste receptor cells (rat only) [25, 71, 72]; Bone [30]; Low level in carotid body [56]; Vascular smooth muscle cells [32, 73]		Contributes to suprathreshold hearing functions [69]; Proposed to be a mammalian sour taste receptor [71, 72] but not confirmed in knockout mice [76]; Upregulated by PKC [77]; Assembles in heteromeric receptors with at least ASIC1a, -1b, -2b and -3	
ASIC2b (MDEG2, BNaC1β) [12] Splice variant of ASIC2a (first 235 amino acids including TM1 and a part of the extracellular loop are different)	Not activated by extracellular pH	Brain (high in olfactory bulb, neo- and allo-cortical regions, hippocampal formation, habenula, basolateral amygdaloid nuclei, and cerebellum) [12]; Spinal cord neurons [8, 18]; Dorsal root ganglia [12]; Taste receptor cells [72]; Retina (both neurons and photoreceptors) [27, 70]		Modulation of ASIC2a and ASIC3 [12]; Participates in the PKC regulation of ASIC3 [80]; Assembles in heteromeric receptors with at least ASIC1a, -2a and -3	Upregulated by inflammation in rat DRG [21, 44]; Downregulated in a rat model of epilepsy [48]
ASIC3 (DRASIC, TNaC1) [81-84] Gene: ACCN3 Chromosomal location: mouse: 5A3 rat: 4q11 human: 7q35	$pH_{0.5\,act}$ (peak) -6.2-6.7 [3, 4, 7, 52, 82, 83]; $pH_{0.5\,act}$ (sustained) ~3.5 (rat) [83], ~4.3 (human) [82]; $pH_{0.5\,inact}$ ~7-7.1 [3, 7]; P_{Na}/P_K (rat) ~4.5-13.5 (peak~sustained); P_{Na}/P_K (human) ≥ 10 (peak)/1.62 (sustained) [3, 82, 83]; P_{Na}/P_{Ca} >100 [3]; g_{Na+} 12.6-15 pS [82, 83]; τ_{act} <5 ms (pH 6) [3]; τ_{inact} 0.3-0.4 s (pH 6-6.4) [3, 4, 52, 85], ~0.7 s (pH 4.7) [7]; τ_{recov} 0.4-0.6 s (pH 7.4) [3, 4]	Dorsal root ganglia (small and medium nociceptive neurons; large mechanosensory neurons) [21-23, 45, 86], specialized cutaneous mechanosensory structures including Meissner corpuscles, Merkel nerve endings, palisades of lanceolate nerve endings [87]; Jugular and nodose ganglia [24, 88]; Spinal cord [89, 90] although not detected in [8, 18]; Testis [84]; Taste receptor cells [25]; Lung epithelial cells [26, 91]; Inner ear [92]; Retina [27]; Bone [30, 93]; Carotid body [56]; Immune cells [31]; Vascular smooth muscle cells [32]	IC_{50}: amiloride (peak current) 16 μM (human, pH 4.0), 63 μM (rat, pH4.0) [82, 83]; Gd^{3+} ~40 μM [94]; salicylic acid (sust. current) ~260 μM; aspirin/diclofenac (sust. current) ~92 μM [21]; APETx2 ~63 nM [85]; 34 % block by 10 mM Ca^{2+} [3], inhibited by Cd^{3+} [34]; EC_{50} FMRFa ~49 μM [95]; Enhanced by lactate [37]; Sustained current enhanced by 200 μM (pH 4) [83], or 100 μM (pH 7) [96] amiloride; Potentiated by arachidonic acid [38] and NO [39]	Modulation of pain sensation (knockout and transgenic mice) [87, 97, 98]; Sensor of acidic and primary inflammatory pain [110]; Sensing of tissue acidosis in muscle [45, 87]; Alterations in cutaneous mechanical sensitivity in knockout mice [87]; Contributes to acid sensation in gastroesophageal afferents and plays a minor role in responses to distending stimuli [99]; Affects visceral mechanoreceptor function [24, 100]; No obvious developmental alterations of DRG in knockout mice [97]; Potentiated by PKC [80]; Assembles in heteromeric receptors with at least ASIC1a, -1b, -2a and -2b	Upregulated by inflammation in rat DRG [21, 44] and in human intestine [101]; Contributes to sensitization but not activation of colon afferent fibres by acidic inflammatory environment [100]; Suggested to be the sensor of myocardial acidity that triggers angina during cardiac ischemia [3, 102]; Participates in the development of chronic mechanical, but not heat, hyperalgesia after muscle insult [45, 87, 103]; Enhanced muscle fatigue in ASIC3 knockout mice, but only in males [104]; Male ASIC3 knockout mice display lower plasma levels of testosterone [104]; Involved in the secondary, but not the primary, mechanical hyperalgesia produced by joint inflammation in mice [105]; Role in the acid hyperresponsiveness associated with experimental gastritis [79]

ASIC4 (SPASIC) [90, 106, 107] Gene: ACCN4 Chromosomal location: mouse: 1C4 rat: 9q33 human: 2q35	Not activated by extracellular pH (mammalian isoform)	Brain (including cerebellum, hippocampus, cortex, striatum and globus pallidus, inferior and superior calliculi, the amygdala, the magnocellular preoptic nucleus, islands of Calleja, and olfactory tubercule) and spinal cord [90, 106] Pituitary gland [106] Inner ear [106] Low level in DRG [90] Retina [27, 28] Immune cells [31]	Function remains unknown Associates with ASIC1a [109]	mRNA slightly increased in a rat neuropathic pain model [108] Not associated with paroxysmal dystonic choreoathetosis [107]

* τ_{inact} of ASIC1a is strongly pH-dependent.

[1] Waldmann R et al. (1997b) Nature 386, 173–177; [2] Garcia-Anoveros J et al. (1997) Proc Natl Acad Sci USA 94, 1459–1464; [3] Sutherland SP et al. (2001) Proc Natl Acad Sci USA 98, 711–716; [4] Benson CJ et al. (2002) Proc Natl Acad Sci U S A 99, 2338–2343; [5] Babini E et al. (2002) J Biol Chem 277, 41597–41603; [6] de Weille J et al. (2001b) Brain Res 900, 277–281; [7] Chen X et al. (2006) Neuropharmacology 50, 964–974; [8] Baron A et al. (2008) J Neurosci 28, 1498–1508; [9] Bassler EL et al. (2001) J Biol Chem 276, 33782–33787; [10] Zhang P et al. (2002) J Gen Physiol 120, 553–566; [11] Askwith CC et al. (2000) Neuron 26, 133–141; [12] Lingueglia E et al. (1997) J Biol. Chem. 272, 29778–29783; [13] Escoubas P et al. (2000) J Biol Chem 275, 25116–25121; [14] Bassilana F et al. (1997) J Biol Chem. 272, 28819–28822; [15] Wemmie JA et al. (2002) Neuron 34, 463–477; [16] Wemmie JA et al. (2003) J Neurosci 23, 5496–5502; [17] Alvarez de la Rosa D et al. (2003) J Physiol 546, 77–87; [18] Wu LJ et al. (2004) J Biol Chem 279, 43716–43724; [19] Chen CC et al. (1998) Proc Natl Acad Sci U S A 95, 10240–10245; [20] Duan B et al. (2007) J Neurosci 27, 11139–11148; [21] Voilley N et al. (2001) J Neurosci 21, 8026–8033; [22] Alvarez de la Rosa D et al. (2002) Proc Natl Acad Sci USA 99, 2326–2331; [23] Ugawa S et al. (2005) Brain Res Mol Brain Res 136, 125–133; [24] Page A et al. (2005) Gut 54, 1408–1415; [25] Richter TA et al. (2004) J Neurosci 24, 4088–4091; [26] Agopyan N et al. (2003) Toxicol Appl Pharmacol 192, 21–35; [27] Lilley S et al. (2004) J Neurosci 24, 1013–1022; [28] Brockway LM et al. (2002) Am J Physiol Cell Physiol 283, C126–134; [29] Ettaiche M et al. (2006) J Neurosci 26, 5800–5809; [30] Jahr H et al. (2005) Biochem Biophys Res Commun 337, 349–354; [31] Friese MA et al. (2007) Nat Med 13, 1483–1489; [32] Grifoni SC et al. (2008) Microvasc Res 75, 202–210; [33] Wang W et al. (2006) J Biol Chem 281, 2497–2505; [34] Staruschenko A et al. (2006) J Neurobiol 67, 97–107; [35] Chu XP et al. (2004) J Neurosci 24, 8678–8689; [36] Paukert M et al. (2004) J Gen Physiol 124, 383–394; [37] Immke DC et al. (2001) Nat Neurosci 4, 869–870; [38] Smith ES et al. (2007) Neuroscience 145, 686–698; [39] Cadiou H et al. (2007) J Neurosci 27, 13251–13260; [40] Cho JH et al. (2007) Am J Physiol Cell Physiol 292, C2161–2174; [41] Wemmie JA et al. (2004) Proc Natl Acad Sci USA 101, 3621–3626; [42] Page AJ et al. (2004) Gastroenterology 127, 1739–1747; [43] Mazzuca M et al. (2007) Nat Neurosci 10, 943–945; [44] Mamet J et al. (2002) J Neurosci 22, 10662–10670; [45] Sluka KA et al. (2003) Pain 106, 229–239; [46] Xiong ZG et al. (2004) Cell 118, 687–698; [47] Gao J et al. (2005) Neuron 48, 635–646; [48] Biagini G et al. (2001) Neurobiol Dis 8, 45–58; [49] Poirot O et al. (2006) J Physiol 576, 215–234; [50] Bubien JK et al. (1999) Am J Physiol 276, C1405–1410; [51] Berdiev BK et al. (2003) J Biol Chem 278, 15023–15034; [52] Hesselager M et al. (2004) J Biol Chem 279, 11006–11015; [53] Ugawa S et al. (2001) Neuroreport 12, 2865–2869; [54] Ugawa S et al. (2006) Neuroreport 17, 1235–1239; [55] Liu L et al. (2001) Brain Res 923, 58–70; [56] Tan ZY et al. (2007) Circ Res 101, 1009–1019; [57] Ugawa S et al. (2008) Biochem Biophys Res Commun 367, 530–534; [58] Chen X et al. (2006) J Gen Physiol 127, 267–276; [59] Waldmann R et al. (1996) J Biol Chem 271, 10433–10436; [60] Price MP et al. (1996) J Biol Chem 271, 7879–7882; [61] Champigny G et al. (1998) J Biol Chem 273, 15418–15422; [62] Baron A et al. (2001) J Biol Chem 276, 35361–35367; [63] Coscoy S et al. (1999) J Biol Chem 274, 10129–10132; [64] Duggan A et al. (2002) J Biol Chem 277, 5203–5208; [65] Zha XM et al. (2006) Proc Natl Acad Sci USA 103, 16556–16561; [66] Mcliwrath SL et al. (2005) Neuroscience 131, 499–511; [67] Price MP et al. (2000) Nature 407, 1007–1011; [68] Garcia-Anoveros J et al. (2001) J Neurosci 21, 2678–2686; [69] Peng BG et al. (2004) J Neurosci 24, 10167–10175; [70] Ettaiche M et al. (2004) J Neurosci 24, 1005–1012; [71] Ugawa S et al. (1998) Nature 395, 555–556; [72] Ugawa S et al. (2003) J Neurosci 23, 3616–3622; [73] Grifoni SD et al. (2008) Am J Physiol Heart Circ Physiol; [74] Askwith CC et al. (2004) J Physiol 558, 659–669; [76] Kinnamon SC et al. (2000) Internat Symp Olfact Taste 13, 80; [77] Baron A et al. (2002) J Biol Chem 277, 50463–50468; [78] Johnson MB et al. (2001) Cereb Blood Flow Metab 21, 734–740; [79] Wultsch T et al. (2008) Pain 134, 245–253; [80] Deval E et al. (2004) J Biol Chem 279, 19531–19539; [81] Babinski K et al. (1999) J Neurochem 72, 51–57; [82] de Weille JR et al. (1998) FEBS Lett 433, 257–260; [83] Waldmann R et al. (1997) J. Biol. Chem. 272, 20975–20978; [84] Ishibashi K et al. (1998) Biochem Biophys Res Commun 245, 589–593; [85] Diochot S et al. (2004) Embo J 23, 1516–1525; [86] Molliver DC et al. (2005) Mol Pain 1, 35; [87] Price MP et al. (2001) Neuron 32, 1071–1083; [88] Fukuda T et al. (2006) Brain Res 1081, 150–155; [89] Hruska-Hageman AM et al. (2004) J Biol Chem 279, 46962–46968; [90] Akopian AN et al. (2000) Neuroreport 11, 2217–2222; [91] Su X et al. (2006) Biol Chem 281, 36960–36968; [92] Hildebrand MS et al. (2004) Hear Res 190, 149–160; [93] Uchiyama Y et al. (2007) J Bone Miner Res 22, 1996–2006; [94] Babinski K et al. (2000) J Biol Chem 275, 28519–28525; [95] Deval E et al. (2003) Neuropharmacology 44, 662–671; [96] Yagi J et al. (2006) Circ Res 99, 501–509; [97] Chen CC et al. (2002) Proc Natl Acad Sci USA 99, 8992–8997; [98] Mogil JS et al. (2005) J Neurosci 25, 9893–9901; [99] Bielefeldt K et al. (2008) Am J Physiol Gastrointest Liver Physiol 294, G130–138; [100] Jones RC, 3rd et al. (2005) J Neurosci 25, 10981–10989; [101] Yiangou Y et al. (2001) Eur J Gastroenterol Hepatol 13, 891–896; [102] Benson CJ et al. (2001) Ann N Y Acad Sci 940, 96–109; [103] Sluka KA et al. (2006) Pain 129, 102–112; [104] Burnes LA et al. (2008) Am J Physiol Regul Integr Comp Physiol 294, R1347–1355; [105] Ikeuchi M et al. (2008) Pain doi:10.1016/j.pain.2008.01.020; [106] Grunder S et al. (2000) Neuroreport 11, 1607–1611; [107] Grunder S et al. (2001) Eur J Hum Genet 9, 672–676; [108] Macdonald R et al. (2001) Brain Res Mol Brain Res 90, 48–56; [109] Donier E et al. (2008) Eur J Neurosci 28, 74–86; [110] Deval E et al. (2008) EMBO J 27, 3047–3055.

Table 3.4.1.3 Biophysical characteristics and pharmacology of heteromeric ASIC channels

Subunit combination	ASIC1a+ ASIC1b	ASIC1a+ ASIC2a	ASIC1a+ ASIC2b	ASIC1a+ ASIC3	ASIC1b+ ASIC1b2	ASIC1b+ ASIC2a	ASIC1b+ ASIC3	ASIC2a+ ASIC2b	ASIC2a+ ASIC3	ASIC2b+ ASIC3
Biophysical characteristics	$pH_{0.5\ act}$ ~6.0 [1] τ_{inact} 1.3 s (pH 6)[1]	$pH_{0.5\ act}$ ~4.8–6.1 (1:1 ratio) [1–5]; ~6.1 (2:1 ratio) and ~5.1 (1:2 ratio) [5] $pH_{0.5\ inact}$ ~6.8 [4]; ~6.5 (2:1 ratio) [5] p_{Na}/p_K ~7.2 [2] p_{Na}/p_{Ca} ~36 [2] g_{Na}+ 10 pS [2] τ_{inact} 0.41s (pH 5.5), 0.9s (pH 6), 1.2 s (pH 5) [1, 3, 4]; 0.8 s (2:1 ratio), 2 s (1:1 ratio) and 3.5 s (1:2 ratio) (pH5.0) [5] τ_{recov} 0.63 s (pH 7.4) [3]; 0.5 s (2:1 ratio) (pH 7.4) [5]	$pH_{0.5\ act}$ ~6.2 [1] τ_{inact} 1 s (pH 6.2) [1]	$pH_{0.5\ act}$ ~6.3–6.6 [1, 3, 6] $pH_{0.5\ inact}$ ~7.1 [6] τ_{inact} 0.14– 0.26 s (pH 6) [1, 3, 7, 8], ~0.5 s (pH 4.7) [6] τ_{recov} 0.64 s (pH 7.4) [3]	$pH_{0.5\ act}$ ~4.4 [9]	$pH_{0.5\ act}$ ~4.9 [1] τ_{inact} 1.0 s (pH 5) [1]	$pH_{0.5\ act}$ ~6.0–6.2 [1, 6] $pH_{0.5\ inact}$ ~6.8 [6] τ_{inact} 0.23–0.35 s (pH 6) [1, 7], ~0.5s [6]	$pH_{0.5\ act}$ ~3.9–4.8 [1, 10] $p_{Na}>p_K$ g_{Na}+ 13.7 pS [11] τ_{inact} 3.5 s (pH 4.8) [1]	$pH_{0.5\ act}$ ~4.3–6.1 [1, 3, 12] $p_{Na}>p_K$ τ_{inact} 0.2– 0.26 s (pH 6) [1, 3] τ_{recov} 0.08 s (pH 7.4) [3]	$pH_{0.5\ act}$ ~6.5 (peak) [1, 11] $p_{Na}>>p_K$ (peak) p_{Na}~p_K (sustained) [11] τ_{inact} 0.23 s (pH 6.5) [1]
Pharmacology		IC_{50}: amiloride 20μM (pH 4.0) [2]; Pb^{2+} ~4.9 μM [13]; Zn^{2+} ~10 nM [14], inhibited by Ni^{2+} and Cd^{2+} [15] EC_{50} Zn^{2+} ~ 111 μM [16]	IC_{50} Pb^{2+} ~2.8μM [13]	IC_{50} APETx2 ~2μM [7] Inhibited by Cd^{2+} [15]	IC_{50} amiloride 17.2 μM (pH 5.5) [9]		IC_{50}: amiloride 112μM [6]; APETx2 ~0.9 μM [7]	IC_{50} amiloride 4.5 μM (pH 5.5), 22 μM (pH 4.0) [10]	IC_{50} Gd^{3+} ~40 μM [12] Inhibited by Cd^{2+} [15] Potentiated by Zn^{2+} [16] EC_{50} FMRFa ~11 μM and NPFF ~2μM [17]	IC_{50}: APETx2 ~117 nM [7] Inhibited by salicylic acid and diclofenac [18]

[1] Hesselager M et al. (2004) J Biol Chem 279, 11006–11015; [2] Bassilana F et al. (1997) J Biol Chem. 272, 28819–28822; [3] Benson CJ et al. (2002) Proc Natl Acad Sci U S A 99, 2338–2343; [4] de Weille J et al. (2001) Brain Res 900, 277–281; [5] Baron A et al. (2008) J Neurosci 28, 1498–1508; [6] Chen X et al. (2006) Neuropharmacology 50, 964–974; [7] Diochot S et al. (2004) EMBO J 23, 1516–1525; [8] Escoubas P et al. (2000) J Biol Chem 275, 25116–25121; [9] Ugawa S et al. (2001) Neuroreport 12, 2865–2869; [10] Ugawa S et al. (2003) J Neurosci 23, 3616–3622; [11] Lingueglia E et al. (1997) J. Biol. Chem. 272, 29778–29783; [12] Babinski K et al. (2000) J Biol Chem 275, 28519–28525; [13] Wang W et al. (2006) J Biol Chem 281, 2497–2505; [14] Chu XP et al. (2004) J Neurosci 24, 8678–8689; [15] Staruschenko A et al. (2006) J Neurobiol 67, 97–107; [16] Baron A et al. (2001) J Biol Chem 276, 35361–35367; [17] Catarsi S et al. (2001) Neuropharmacology 41, 592–600; [18] Voilley N et al. (2001) J Neurosci 21, 8026–8033.

the medium may result in inactivation of most of the transient response, although ASIC3-containing channels have an additional slowly activating, sustained current component that does not inactivate while the pH remains acidic (Waldmann et al., 1997a; de Weille et al., 1998; Babinski et al., 1999). For moderate extracellular acidification (between pH 7.3 and 6.7), this sustained current results from the window of overlap between inactivation and activation of the channel (Yagi et al., 2006), while the sustained current activated by lower pH seems to involve a different mechanism.

ASICs are low conductance channels that predominantly conduct Na^+ (gNa ~ 10–15 pS) (Bassilana et al., 1997; Lingueglia et al., 1997; Waldmann et al., 1997a, b; Champigny et al., 1998; de Weille et al., 1998). ASIC1a is also substantially permeable to Ca^{2+} (Waldmann et al., 1997b; Bassler et al., 2001; Gunthorpe et al., 2001; Sutherland et al., 2001) and, thus, provides a voltage-independent pathway for Ca^{2+} entry into neurons (Yermolaieva et al., 2004). ASICs are regulated by extracellular Ca^{2+} with a dual inhibitory and stimulatory effect (Waldmann et al., 1997b; Berdiev et al., 2001; de Weille and Bassilana, 2001; Sutherland et al., 2001; Zhang and Canessa, 2001; Babini et al., 2002; Zhang and Canessa, 2002; Paukert et al., 2004a). It has been proposed for ASIC3 that the Ca^{2+} block is achieved by ions binding with high affinity at a site on the extracellular side of the open pore. Accordingly, H^+ ions could activate the channel by displacing Ca^{2+} from that site (Immke and McCleskey, 2003), i.e., without significant conformational change. However, the channel block by Ca^{2+} and interaction with H^+-gating is likely to be more complex as suggested for ASIC1a (Paukert et al., 2004a; Zhang et al., 2006) and in fact it seems most likely that ASICs do undergo conformational changes upon gating (Chen et al., 2006a; Zhang et al., 2006).

3.4.1.4 Cellular and subcellular localization

ASICs are largely expressed in neurons but have also been detected in testis (hASIC3; Ishibashi and Marumo, 1998; Babinski et al., 1999); pituitary gland (ASIC4; Grunder et al., 2000); taste receptor cells (ASIC1–3; Ugawa et al., 1998; Liu and Simon, 2001; Ugawa et al., 2003; Richter et al., 2004); photoreceptors and retinal cells (ASIC1–3; Brockway et al., 2002; Ettaiche et al., 2004; Lilley et al., 2004; Ettaiche et al., 2006); cochlear hair cells (ASIC1b; Ugawa et al., 2006), lung epithelial cells (ASIC1a and -3; Agopyan et al., 2003; Su et al., 2006); vascular smooth muscle cells (ASIC1–3; Drummond et al., 2008; Gannon et al., 2008; Grifoni et al., 2008a, b); immune cells (ASIC1, -3 and -4; Friese et al., 2007); and bone (ASIC1–3; Jahr et al., 2005; Uchiyama et al., 2007). ASIC1 and ASIC3 were also shown to be expressed in rat carotid body (Tan et al., 2007). ASICs are present in primary sensory neurons of the trigeminal, vagal, and dorsal root ganglia (DRG) (Lingueglia et al., 1997; Waldmann et al., 1997a, b; Chen et al., 1998; Olson et al., 1998; Price et al., 2000; Garcia-Anoveros et al., 2001; Price et al., 2001; Ugawa et al., 2001; Voilley et al., 2001; Alvarez de la Rosa et al., 2002; Benson et al., 2002; Sluka et al., 2003; Peng et al., 2004; Schicho et al., 2004; McIlwrath et al., 2005; Molliver et al., 2005; Page et al., 2005; Ugawa et al., 2005; Christianson et al., 2006; Holzer, 2006; Jiang et al., 2006; Hughes et al., 2007; Page et al., 2007). ASIC1, ASIC2 and ASIC3 are significantly expressed in the small and medium nociceptive sensory neurons, i.e. the nociceptors that are able to detect noxious chemical, thermal and high threshold mechanical stimuli, but ASIC2 and ASIC3 have also been detected in large neurons that mostly correspond to low-threshold mechanoreceptors (Olson et al., 1998; Price et al., 2000; Garcia-Anoveros et al., 2001; Price et al., 2001; Voilley et al., 2001; Alvarez de la Rosa et al., 2002; Molliver et al., 2005; Ugawa et al., 2005). ASIC2 gene expression was shown to be controlled by BDNF in subsets of medium and large diameter sensory neurons (McIlwrath et al., 2005). ASIC1, ASIC2a and ASIC3 proteins are present in the soma and in the peripheral nerve endings of DRG neurons (Olson et al., 1998; Price et al., 2000; Garcia-Anoveros et al., 2001; Price et al., 2001; Alvarez de la Rosa et al., 2002). However, ASIC1a has also been detected on central projections in the dorsal horn of the spinal cord (Olson et al., 1998), but these data should be interpreted cautiously due to the incomplete characterization of the antibodies used. ASIC2a and ASIC3 are expressed in mechanoreceptors, including specialized cutaneous mechanosensory structures such as Meissner corpuscles, Merkel nerve endings, and palisades of lanceolate nerve endings surrounding the hair shaft (Price et al., 2000; Garcia-Anoveros et al., 2001; Price et al., 2001). ASIC3 knockout mice have no obvious developmental alterations of the DRG (Chen et al., 2002).

ASICs are also expressed in central neurons (Price et al., 1996; Waldmann et al., 1996; Bassilana et al., 1997; Garcia-Anoveros et al., 1997; Lingueglia et al., 1997; Waldmann et al., 1997b; Chen et al., 1998; Akopian et al., 2000; Babinski et al., 2000; Grunder et al., 2000; Duggan et al., 2002; Wemmie et al., 2002; Alvarez de la Rosa et al., 2003; Wemmie et al., 2003; Askwith et al., 2004; Wu et al., 2004), including spinal cord neurons (Chen et al., 1998; Babinski et al., 1999; Akopian et al., 2000; Grunder et al., 2000; Alvarez de la Rosa et al., 2003; Hruska-Hageman et al., 2004; Wu et al., 2004; Duan et al., 2007; Baron et al., 2008). ASICs are widely distributed in the mammalian brain, with a large expression in the cortex, striatum,

hippocampus, olfactory bulb, amygdala and cerebellum (Waldmann and Lazdunski, 1998; Wemmie et al., 2003) (Table 3.4.1.2). ASIC2a has been located in dendrites and soma in Purkinje neurons of cerebellum and in some pyramidal neurons of brain cortex and hippocampus (Duggan et al., 2002). Transfected ASIC1a in rat hippocampal neurons shows a dendritic and synaptic pattern of localization (Hruska-Hageman et al., 2002; Wemmie et al., 2002) and ASIC1 is detected in synaptosome-containing brain fractions (Wemmie et al., 2002; Alvarez de la Rosa et al., 2003). Paired-pulse facilitation (an index of presynaptic activity and neurotransmitter release probability) is normal in hippocampal slices from ASIC1 knockout mice (Wemmie et al., 2002). A study using multiple approaches and ASIC1 knockout mice shows that ASIC1 protein is preferentially distributed in brain regions with strong excitatory synaptic input (Wemmie et al., 2003). Together, these data suggest the presence of ASIC1 and probably ASIC2a in the synaptic membranes of neurons, presumably located postsynaptically in the hippocampus. In agreement, ASIC1a has been reported to be localized in dendritic spines in hippocampal slices and to affect their density (Zha et al., 2006).

ASIC-like currents have been reported in acutely dissociated or cultured sensory neurons (Krishtal and Pidoplichko, 1981; Davies et al., 1988; Kovalchuk et al., 1990; Bevan and Yeats, 1991; Akaike and Ueno, 1994; Benson et al., 1999; Escoubas et al., 2000; Liu and Simon, 2000; Petruska et al., 2000; Sutherland et al., 2001; Voilley et al., 2001; Alvarez de la Rosa et al., 2002; Petruska et al., 2002; Xie et al., 2002; Dirajlal et al., 2003; Deval et al., 2004; Diochot et al., 2004; Liu et al., 2004; Peng et al., 2004; Connor et al., 2005; Sugiura et al., 2005; Gu and Lee, 2006; Leffler et al., 2006), in neurons of the CNS (Vrublevskii et al., 1985; Grantyn and Lux, 1988; Ueno et al., 1992; Akaike and Ueno, 1994; Li et al., 1997, Varming, 1999; Escoubas et al., 2000; Allen and Attwell, 2002; Baron et al., 2002b; Bolshakov et al., 2002; Wemmie et al., 2003; Gao et al., 2004; Lilley et al., 2004; Roza et al., 2004; Vukicevic and Kellenberger, 2004; Weng et al., 2004; Wu et al., 2004; Ettaiche et al., 2006), in oligodendrocytes (Sontheimer et al., 1989), in carotid body glomus cells (Tan et al., 2007) and in taste receptor cells (Lin et al., 2002). It is notable that ASIC-like currents are detected in almost all central neurons. Electrophysiological (Benson et al., 1999; Sutherland et al., 2001; Benson et al., 2002), pharmacological (Escoubas et al., 2000; Baron et al., 2002b; Diochot et al., 2004) and genetic (Price et al., 2001; Benson et al., 2002; Wemmie et al., 2002; Xie et al., 2002; Wemmie et al., 2003; Drew et al., 2004; Peng et al., 2004; Roza et al., 2004) evidence clearly supports the participation of ASICs in these native proton-activated cation currents, although a significant part of the sustained response to low pH can be attributed to the capsaicin receptor TRPV1 in sensory neurons (Caterina et al., 1997; 2000; Davis et al., 2000). Expression of different ASIC isoforms and/or different combinations of subunits accounts for the heterogeneous properties of H^+-gated currents in neurons (Bassilana et al., 1997; Lingueglia et al., 1997; Babinski et al., 2000; Sutherland et al., 2001; Benson et al., 2002; Askwith et al., 2004; Hesselager et al., 2004).

3.4.1.5 Pharmacology

ASICs are reversibly blocked by amiloride and its derivatives benzamil and ethylisopropylamiloride (EIPA) at micromolar concentration levels (Table 3.4.1.2). A paradoxical enhancing effect of amiloride has also been described on the sustained component of the ASIC3-mediated current (Waldmann *et al.*, 1997a; Yagi *et al.*, 2006) and, likewise, on currents mediated by an ASIC2a mutant (Adams *et al.*, 1999). ASICs are also directly inhibited by therapeutic concentrations of non-steroidal anti-inflammatory drugs (NSAIDs) (IC$_{50}$ ~92–350 μM) (Voilley *et al.*, 2001). ASIC channels in rat cerebellar Purkinje cells (Allen and Attwell, 2002) and ASIC3 containing channels expressed in *Xenopus* oocytes (Babinski *et al.*, 2000) are blocked by Gd^{3+} (IC$_{50}$ ~40 μM). ASIC currents in spinal dorsal horn and hippocampal CA1 neurons and ASIC1- or ASIC3-containing channels in CHO cells are inhibited by Pb^{2+} (IC$_{50}$ ~1.5–5 μM) (Wang *et al.*, 2006). External Ni^{2+} and to a lower extent Cd^{2+} also inhibit ASIC currents in hippocampal neurons (IC$_{50}$ ~1.6 mM for Ni^{2+}) as well as ASIC1a, 1a+2a (Ni^{2+}), ASIC2a, ASIC3, ASIC1a+2a, 1a+3 and 2a+3 (Cd^{2+}) in CHO cells (Staruschenko *et al.*, 2006). Extracellular Zn^{2+} potentiates the acid activation of homomeric and heteromeric ASIC2a-containing channels (EC$_{50}$ ~ 110–120 μM) (Baron *et al.*, 2001) and the native ASIC currents in rat hippocampal neurons (Baron *et al.*, 2002b). On the other hand, high-affinity zinc inhibition of ASIC1a-mediated currents has also been described both in mouse cultured cortical neurons and in recombinant expression systems (IC$_{50}$ ~ 7–10 nM) (Chu *et al.*, 2004). Zn^{2+}, which can be co-released with glutamate during synaptic activity, is thus an important modulator of ASIC channels in the central nervous system.

All these compounds have a relatively low specificity and more specific blockers of ASICs have, thus, been developed. Two toxins that specifically and efficiently block ASICs *in vitro* and *in vivo* have been isolated (Diochot *et al.*, 2007): the psalmotoxin 1 (PcTx1) from tarantula blocks ASIC1a homomeric channels (IC$_{50}$ ~ 0.9 nM) (Escoubas *et al.*, 2000, 2003; Salinas *et al.*, 2006) and the sea anemone toxin APETx2 blocks homomeric ASIC3 (IC$_{50}$ ~ 63 nM) and ASIC3-containing heteromeric channels (IC$_{50}$ ~ 0.1–2 μM) (Diochot *et al.*, 2004; Chagot *et al.*, 2005). PcTx1 also interacts with ASIC1b but does not inhibit the channel and rather promotes its opening (Chen *et al.*, 2006a). More recently, the compound A-317567 was described as a non-amiloride blocker of native ASIC currents in DRG neurons (IC$_{50}$ between 2 and 30 μM) and was shown to be more potent than amiloride in rat *in vitro* and *in vivo* models of inflammatory and postoperative pain (Dube *et al.*, 2005).

Organic compounds released during ischaemia, such as lactate and arachidonic acid, are also able to potentiate ASIC activity in sensory neurons that innervate the heart (Immke and McCleskey, 2001) and in Purkinje cells and DRG neurons (Allen and Attwell 2002; Smith *et al.*, 2007). The Purkinje cell ASIC current is also potentiated by cell swelling (Allen and Attwell, 2002), and ASIC-like currents in the sensory ganglia and hippocampal neurons are potentiated by the reducing agent dithiothreitol (DTT) and by glutathione (GSH), but inhibited by oxidizing reagents (Andrey *et al.*, 2005; Chu *et al.*, 2006; Cho and Askwith, 2007). Glutathione and DTT also increase ASIC1a currents expressed in CHO cells (Cho and Askwith, 2007). Nitric oxide (NO) potentiates both ASIC currents in DRG neurons and proton-gated currents mediated by recombinant ASIC1a, ASIC1b, ASIC2a and ASIC3 (Cadiou *et al.*, 2007). Cold temperature also affects recombinant ASICs as well as the proton-gated current in sensory neurons by slowing desensitization (Askwith *et al.*, 2001). ASIC1a and ASIC1b can be irreversibly modulated by extracellular serine proteases through proteolytic degradation (Poirot *et al.*, 2004; Vukicevic *et al.*, 2006). Proteases

are present in the CNS under physiological and pathological conditions as well as in the PNS, where they are released, for example, during inflammation.

ASICs are related to the invertebrate peptide ionotropic receptor FaNaC, which is directly activated by FMRFamide (Lingueglia *et al.*, 1995; 2006). FMRFamide and structurally related peptides, such as FRRFamide and neuropeptide FF (NPFF, FLFQPQRFamide), are not able to activate ASIC channels but potentiate H$^+$-gated currents of heterologously expressed ASIC1 and ASIC3, but not ASIC2a (EC$_{50}$ ~ 10–50 μM; threshold ~ 1 μM) (Askwith *et al.*, 2000; Catarsi *et al.*, 2001; Deval *et al.*, 2003; Chen *et al.*, 2006b). The neuropeptides increase the peak amplitude and/or slow inactivation of the H$^+$-gated current, inducing or increasing the sustained phase during acidification, and appear to act directly on the channel (Askwith *et al.*, 2000; Deval *et al.*, 2003; Xie *et al.*, 2003). FMRFamide, FRRFamide, and to a lower extent NPFF, neuropeptide SF (NPSF, SLAAPQRFamide), RFamide, QRFamide (an inactive metabolite of NPFF), RFRP-1 (VPHSAANLPLRFamide) and RFRP-2 (SHFPSLPQRFamide) also affect the H$^+$-gated current of cultured sensory neurons (Askwith *et al.*, 2000; Xie *et al.*, 2002; Deval *et al.*, 2003; Xie *et al.*, 2003; Ostrovskaya *et al.*, 2004), hippocampal (Askwith *et al.*, 2004) and Purkinje cells (Allen and Attwell, 2002). On the other hand, the opioid Met enkephalin-RF (YGGFMRF), which is structurally related to FMRFamide but lacks the amide group, has no effect (Askwith *et al.*, 2000; Allen and Attwell, 2002; Lingueglia *et al.*, 2006). Studies using knockout mice have demonstrated that ASIC3 plays a major role in mediating the sensory response to FMRFamide and FRRFamide, whilst the ASIC1 contribution to the effect seems to be very modest (Xie *et al.*, 2002, 2003), consistent with the lower potency of FMRFamide and related peptides in modulating ASIC1a compared with ASIC3. The peptides alter steady-state desensitization of ASIC1a (Sherwood and Askwith, 2008). ASIC2 did not play a major role in mediating the DRG current response to the peptides (Xie *et al.*, 2003). However, the presence of ASIC2 subunits enhanced peptide modulation of ASIC1a-containing channels in CHO cells and in hippocampal neurons (Askwith *et al.*, 2004). ASIC2a is, thus, capable of increasing the response to peptides when present in heteromeric channels containing ASIC1a or ASIC3 subunits (Catarsi *et al.*, 2001; Askwith *et al.*, 2004). The modulation of ASICs by RFamide neuropeptides is strongly pH dependent (half-maximal effect achieved at pH 5.6–6.0); (Ostrovskaya *et al.*, 2004) and requires FMRFamide addition at pH 7.4, i.e., when the channel is closed (Askwith *et al.*, 2000). Interestingly, once the channel has been activated by acid, the effect of FMRFamide is retained until the pH is returned to 7.4, even if the peptide is removed from the bath just before acidification (Askwith *et al.*, 2000). The binding of FMRFamide to ASIC1b+ASIC3 heteromers has been shown to be competitive with the binding of extracellular calcium (Chen *et al.*, 2006b). NPFF, NPAF and NPSF are mainly expressed in the CNS, with high levels in the spinal cord (Panula *et al.*, 1999; Vilim *et al.*, 1999), where they can be released (Zhu *et al.*, 1992; Devillers and Simonnet, 1994; Devillers *et al.*, 1995). NPFF expression is increased in the spinal cord during chronic inflammation (Kontinen *et al.*, 1997; Vilim *et al.*, 1999; Yang and Iadarola, 2003). NPFF has also been detected in small and intermediate-sized DRG neurons (Allard *et al.*, 1999). The potentiation of mammalian ASIC activity by endogenous RFamide neuropeptides has therefore been proposed to play a role in the response to noxious acidosis in the peripheral and central nervous system. This modulation may be complex because of the diversity of ASIC channels, the diversity of FMRFamide-related peptides and the diversity of their effects on ASICs. In addition, the effects and mechanism of action of these peptides on ASICs *in vivo* remains to be clarified.

3.4.1.6 Physiology
CNS

In central neurons in culture, ASIC activation following acidification of the extracellular medium can induce action potential firing (Varming, 1999; Baron et al., 2002b; Lilley et al., 2004; Vukicevic and Kellenberger, 2004; Wu et al., 2004). Studies with ASIC knockout mice have demonstrated the importance of these channels in brain function. Disrupting the ASIC1 gene eliminates the H^+-evoked currents in hippocampal neurons (Wemmie et al., 2002) and in the amygdala (Wemmie et al., 2003), illustrating the importance of ASIC1 in brain. ASIC1 knockout mice exhibit impaired hippocampal long-term potentiation and defects in hippocampus-dependent spatial learning, cerebellum-dependent eye-blink conditioning and amygdala-dependent fear conditioning (Wemmie et al., 2002, 2003). Conversely, overexpression of ASIC1a in transgenic mice facilitates fear conditioning (Wemmie et al., 2004). ASIC1a has therefore been proposed to participate in synaptic transmission, although no component that could be attributed to the activity of ASICs has been detected so far (Alvarez de la Rosa et al., 2003; Krishtal, 2003). A proton pump acidifies synaptic vesicles and provides the electrochemical gradient for neurotransmitter uptake. When vesicles fuse with the plasma membrane, they release protons into the synaptic cleft which can affect synaptic ion channels, as proposed for the presynaptic Ca^{2+} channel mediating vesicle release in mammalian cone photoreceptors (Barnes et al., 1993; DeVries, 2001). Similarly, H^+ released by chemical synaptic transmission might activate synaptic ASICs. A functional interaction between ASIC1a and N-methyl-D-aspartate (NMDA) receptors in hippocampal neurons has been suggested (Wemmie et al., 2002). Postsynaptic ASIC-generated depolarization may facilitate release of the Mg^{2+} block of the NMDA receptor (Wemmie et al., 2002; Krishtal, 2003). A functional link between the NMDA receptor- Ca^{2+}-calmodulin-dependent protein kinase II (CaMKII) cascade and ASIC1a has recently been described, such that the ASIC1a current is increased by CaMKII phosphorylation, and this is thought to contribute to acidotoxicity during ischaemia, where excessive glutamate release and acidosis occur (Gao et al., 2005). During the metabolic acidosis accompanying experimental stroke, ASIC1a, which provides a voltage-independent pathway for Ca^{2+} entry into neurons (Yermolaieva et al., 2004), has been proposed to contribute to neuronal death associated with brain ischemia (Xiong et al., 2004; Gao et al., 2005). Ca^{2+}-mediated CaMKII signalling has also been proposed to participate in the positive modulation of dendritic spine density by ASIC1a reported in hippocampal slice preparations. Acid evokes an ASIC1a-dependent elevation of Ca^{2+} concentration within spines and ASIC1a modulates CaMKII phosphorylation (Zha et al., 2006). Interestingly, this effect on spines does not seem to be related to the impaired learning observed in ASIC1 knockout mice since these animals do not have reduced spine density. Thus, the function of ASICs in the central nervous system most likely relies on their ability to sense extracellular pH fluctuations. Activity-dependent pH fluctuations in the extracellular

domain of the CNS typically accompany many processes (Kaila and Ransom, 1998; Chesler, 2003) although the extent of these pH change remains generally poorly understood. Alternatively, ASICs in the central nervous system might be activated by a stimulus other than protons, such as as yet undiscovered extracellular ligand(s), which might include neuropeptides based on ASIC similarities with the FMRFamide-gated sodium channel FaNaC (Lingueglia et al., 1995, 2006).

ASIC1a and ASIC2 are expressed in spinal second-order sensory neurons, where they are upregulated by peripheral inflammation, and may therefore contribute to the processing of noxious stimuli and participate in the generation of hyperalgesia and allodynia in persistent pain states (Wu et al., 2004; Duan et al., 2007; Baron et al., 2008). ASIC1a has been recently implicated in central sensitization induced by inflammation (Duan et al., 2007). ASIC1a, ASIC2, ASIC3, and ASIC4 have also been detected in rodent retina (Brockway et al., 2002; Ettaiche et al., 2004; Lilley et al., 2004; Ettaiche et al., 2006), and ASIC2, which is present both in neurons and photoreceptors, is a negative modulator of rod phototransduction (Ettaiche et al., 2004), while ASIC1a is a positive modulator of cone phototransduction and adaptation (Ettaiche et al., 2006). Inactivation of the ASIC2 gene in mouse also sensitizes the retina toward light-induced degeneration (Ettaiche et al., 2004).

ASIC1a participates in the central modulation of pain. Intrathecal or intracerebroventricular injections of the PcTx1 toxin that blocks homomeric ASIC1a has a potent analgesic effect in rodent (Mazzuca et al., 2007). Knock-down of the channel after intrathecal injection of antisense oligonucleotides has a similar effect. Blocking ASIC1a results in an activation of the endogenous enkephalin pathway and in increased levels of Met-enkephalin in the cerebrospinal fluid.

Finally, ASICs are expressed in the CNS early in development both in mammals (before embryonic day 7 and day 11 for mouse ASIC2a and ASIC1a, respectively, (Garcia-Anoveros et al., 1997; Alvarez de la Rosa et al., 2003), and before embryonic day 17 for rat ASIC2a (Waldmann et al., 1996) and in zebrafish (between 24 and 48 hours postfertilization, (Paukert et al., 2004b) and preliminary data with triple knockout mice lacking ASIC1, ASIC2 and ASIC3 suggest a possible role for these channels in neurodevelopment (Price et al., 2004a).

PNS

A decrease in pH has been associated with non-adapting pain in human volunteers (Steen et al., 1995) and this cutaneous acid-induced pain appears to be largely mediated by ASIC channels, especially at moderate pH (Ugawa et al., 2002; Jones et al., 2004; McMahon and Jones, 2004). In rat, activation of slowly conducting sensory fibres (C-type) by acid is inhibited by amiloride (Yudin et al., 2004). A decrease in extracellular pH is associated with conditions such as ischaemia, inflammation, tumours or lesions. ASICs have therefore been proposed to be sensors of acidic pH on peripheral terminals of primary sensory neurons and to participate in the perception of pain that accompanies tissue acidosis. Consistent with this role, ASICs are able to excite primary sensory neurons and cause their depolarization and the initiation of action potentials upon extracellular acidification (Mamet et al., 2002; Deval et al., 2003). Interestingly, action potentials were only spontaneously triggered during the peak phase and not during the sustained depolarization due to the plateau phase of ASIC3-like currents in DRG neurons (Mamet et al., 2002; Deval et al., 2003). However, as previously mentioned, the fast transient component of ASIC currents is probably rapidly inactivated when pH decreases slowly, which is likely to happen during the onset of a tissue acidosis. This suggests that in addition to a direct effect on neuronal firing, ASIC channels and protons provide a way of modulating nociceptor excitability. In good agreement, the ASIC3-induced sustained depolarization increases the excitability of sensory neurons in response to another depolarizing stimulus during weak extracellular acidification while it decreases excitability at more acidic pH, probably through the inactivation of voltage-dependent sodium channels mediated by the stronger depolarization (Deval *et al.*, 2003). ASIC3 thus appears to play an important role in the perception of non-adapting pain caused by acids and for fine tuning of nociceptive neuron excitability. In agreement, gene inactivation and transgenic expression of a dominant-negative subunit in mice have revealed a role for ASIC3 in modulating pain sensation (Price *et al.*, 2001; Chen *et al.*, 2002; Mogil *et al.*, 2005), with a particular emphasis on stimuli of moderate to high intensity (Chen *et al.*, 2002). ASIC3 has been implicated in the process of detection of tissue acidosis in muscle (Price *et al.*, 2001; Sluka *et al.*, 2003) and contributes to secondary mechanical hyperalgesia produced by both muscle (Sluka *et al.*, 2006) and joint (Ikeuchi *et al.*, 2008) inflammation. ASIC3 has been proposed to play an important role in inflammation (Voilley *et al.*, 2001; Yiangou *et al.*, 2001; Mamet *et al.*, 2002; Voilley, 2004; Jones *et al.*, 2005) and has been suggested to be the major sensor of painful acidification in cardiac ischemia (Benson and Sutherland, 2001; Sutherland *et al.*, 2001). However, the available experimental data do not allow definitive conclusions concerning the roles of ASIC3 in nociception. For instance, mutant mice that lack ASIC3 display increased sensitivity to some painful stimuli (Price *et al.*, 2001; Chen *et al.*, 2002; Mogil *et al.*, 2005). However, it is important to note that interpretation of such data from mouse models could be complicated by factors including compensatory mechanisms or species differences in expression of ASIC channels. Transient proton-induced currents in nociceptors innervating the skin have been shown for instance to be less frequent and to have a smaller current density in mouse compared to rat (Leffler *et al.*, 2006).

Conversely, ASIC1 does not participate in acid-induced mechanical hyperalgesia in muscle (Sluka *et al.*, 2003) and the behavioural responses to mechanical and thermal stimuli are unaffected in knockout mice (Page *et al.*, 2004). Subcutaneous injections of the toxin PcTx1 in wild-type animals have no effect in acute pain models (Mazzuca *et al.*, 2007) or in a model of inflammatory pain (Duan *et al.*, 2007), suggesting that homomeric ASIC1a does not significantly contribute to pain at the periphery. Furthermore, ASIC1 knockout mice do not exhibit sensory or motor deficits (Wemmie *et al.*, 2002, 2003).

Mice with targeted deletion of ASIC2 and ASIC3 were reported to display subtle alterations in cutaneous mechanical sensitivity (Price *et al.*, 2000, 2001; Welsh *et al.*, 2002), whilst such a phenotype was not evident in ASIC1 knockouts (Page *et al.*, 2004). ASIC1, ASIC2 and ASIC3 also contribute to visceral mechanotransduction (Page *et al.*, 2004, 2005; Jones *et al.*, 2005, 2007; Page *et al.*, 2007). ASIC2a and ASIC3 are expressed in mechanoreceptors (Price *et al.*, 2000; Garcia-Anoveros *et al.*, 2001; Price *et al.*, 2001) and it was suggested that ASICs may participate in mechanotransduction, as proposed for the degenerins of *C. elegans*, possibly by contributing to mechanosensitive ion channels. The *C. elegans* degenerins

MEC-4 and MEC-10 are expressed in mechanosensory neurons where they are required for normal touch sensation (Driscoll and Chalfie, 1991; Huang and Chalfie, 1994). Family members that are required for coordinated movement have also been identified in motor neurons (UNC-8, DEL-1) (Tavernarakis *et al.*, 1997) and muscle (UNC-105) (Liu *et al.*, 1996). Recently, *in vivo* imaging and recording from the worm touch receptor neurons have convincingly demonstrated that MEC-4 transduces mechanical signals (Suzuki *et al.*, 2003; O'Hagan *et al.*, 2005). ASICs may fulfil a similar role in mammals but experimental evidence for a mechanical gating of these channels is still lacking and in fact some data do not support a role as direct peripheral mechanosensors (Drew *et al.*, 2004; Roza *et al.*, 2004). ASIC1, ASIC2, and ASIC3 are also expressed in the mouse inner ear (Hildebrand *et al.*, 2004; Peng *et al.*, 2004), however, mutant mice exhibit no significant hearing loss (ASIC2 knockout mice (Peng *et al.*, 2004; Roza *et al.*, 2004) and ASIC3 knockout mice at 2 months (Hildebrand *et al.*, 2004)), except in the mild-to-moderate range for ASIC3 knockout mice at 4 months (Hildebrand *et al.*, 2004). However, ASIC2, which is mainly expressed in spiral ganglion neurons in the adult cochlea, does contribute to suprathreshold hearing functions (Peng *et al.*, 2004).

Other potential roles

ASICs are also expressed in rat and mouse taste receptor cells (Ugawa *et al.*, 1998; Lin *et al.*, 2002; Ugawa *et al.*, 2003; Richter *et al.*, 2004; Shimada *et al.*, 2006) and ASIC2 has been proposed to be a mammalian sour taste receptor (Ugawa *et al.*, 1998, 2003). However, ASIC2 knockout mice respond normally to acid taste stimuli (Kinnamon *et al.*, 2000) and taste cells from these mice respond normally to acid stimuli (Richter *et al.*, 2004). In addition, no ASIC2 subunit has been detected in mouse taste buds (Richter *et al.*, 2004), suggesting that ASICs probably do not act as general mammalian sour receptors.

ASIC1 and ASIC3 are expressed in rat carotid body and extracellular acidosis evokes ASIC-like inward current in glomus cells, suggesting a contribution of ASICs to chemotransduction of low pH by carotid body chemoreceptors (Tan *et al.*, 2007). A role for ASICs in cellular migration has also been proposed in both smooth muscle cells (Grifoni *et al.*, 2008a) and in glioma cells (Vila-Carriles *et al.*, 2006).

Recently, an H^+-activated current, probably mediated by ASIC1a, has been recorded from a human skeletal muscle cell line (Gitterman *et al.*, 2005). Interestingly, the UNC-105 degenerin of *C. elegans* is expressed in muscle (Liu *et al.*, 1996; Garcia-Anoveros *et al.*, 1998) and an amiloride-sensitive H^+-gated Na^+ channel has been recorded in *C. elegans* body wall muscle cells (Jospin and Allard, 2004). However, expression of ASICs in vertebrate muscle has not been clearly documented to date and, thus, the presence and the role of ASICs in muscle cells remains to be established.

3.4.1.7 Relevance to disease states

No inherited human disease caused by mutations in ASICs has been identified to date. However, expression of ASIC subunits in the PNS and CNS is modified in several pathological situations and ASICs have been directly implicated in processes such as ischaemia and inflammation. ASIC2a is upregulated in brain during ischaemia (Johnson *et al.*, 2001) and activation of ASIC1a during the metabolic acidosis accompanying experimental stroke has been proposed to contribute to the neuronal death associated with brain ischaemia (Xiong et al., 2004; Gao et al., 2005). ASIC1a has also been proposed to contribute to axonal degeneration in autoimmune inflammation of the CNS in a mouse model of multiple sclerosis, where a significant tissue acidosis occurs (Friese et al., 2007). ASIC1a is also involved in synaptic plasticity and contributes to learning and memory (Wemmie et al., 2002) and may, therefore, represent a novel pharmacological target for modulating excitatory neurotransmission. ASIC3 modulates pain sensation in sensory neurons (Price et al., 2001; Chen et al., 2002) and inhibition of ASIC3 may, thus, be of therapeutic benefit (Dube et al., 2005). In cardiac afferents, ASIC3 has been proposed to be the sensor of myocardial acidity that triggers angina during cardiac ischaemia (Benson and Sutherland, 2001; Sutherland et al., 2001). ASIC3 also plays an important role in the generation of chronic mechanical hyperalgesia and central sensitization which is observed in a mouse model of non-inflammatory muscular pain following repeated acid injection into the muscle (Sluka et al., 2003), as well as in the development of cutaneous mechanical, but not heat, hyperalgesia induced by muscle inflammation (Sluka et al., 2006). Enhanced muscle fatigue also occurs in ASIC3 knockout mice, but only in males which display lower plasma levels of testosterone (Burnes et al., 2008). ASICs also play a role in the metabolic component of the exercise pressor reflex (Hayes et al., 2007). Organic compounds released during ischaemia in heart, brain and muscle, such as lactate and arachidonic acid, are able to potentiate ASICs in sensory neurons that innervate the heart (Immke and McCleskey, 2001; Smith et al., 2007) and in Purkinje cells (Allen and Attwell, 2002). The expression of ASICs is induced by chronic inflammation in sensory neurons (Voilley et al., 2001; Yiangou et al., 2001; Mamet et al., 2002, 2003) and ASICs are regulated by nerve growth factor (NGF) at the transcriptional level (Mamet et al., 2002, 2003). This may contribute, in association with PKC-mediated potentiation of ASIC3 (Deval et al., 2004) and positive regulation by arachidonic acid (Smith et al., 2007) and nitric oxide (Cadiou et al., 2007) to the peripheral sensitization of nociceptors, as proposed in colon afferent fibres (Jones et al., 2005). Peripheral inflammation also induced ASIC expression in second-order sensory neurons of the spinal cord (Wu et al., 2004; Duan et al., 2007) and ASIC1a has been proposed to participate in the central sensitization associated with prolonged pain (Duan et al., 2007). It is interesting to note that the FMRFamide-related peptide NPFF, which directly modulates ASIC1- and ASIC3-containing channels, is similarly increased in the spinal cord during inflammation (Kontinen et al., 1997; Vilim et al., 1999; Yang and Iadarola, 2003). ASIC-like currents have been recorded in gastric sensory neurons (Sugiura et al., 2005) and ASIC3 has been suggested to contribute to acid sensation in gastroesophageal afferents and to play a minor role in responses to distending stimuli (Bielefeldt and Davis, 2008). ASIC3 also plays a role in the acid hyperresponsiveness associated with experimental gastritis while ASIC2 dampens acid-evoked afferent input from the stomach to the brainstem (Wultsch et al., 2008). Sensory neurons that innervate bone and joints express ASICs. They could participate in inflammatory arthritides and bone cancer pain via their capacity to detect osteoclast- and tumor-induced tissue acidosis (Mantyh et al., 2002). In agreement, ASIC3 has been involved in the second-, but not the primary, mechanical hyperalgesia produced by joint inflammation in mice (Ikeuchi et al., 2008). ASIC-like currents

have been recorded from rat vagal pulmonary sensory neurons (Gu and Lee, 2006) and ASIC3 is expressed in rat vagal and glossopharyngeal sensory ganglia (Fukuda et al., 2006), as well as in spinal afferent neurons projecting to the rat lung and pleura (Groth et al., 2006). ASICs have been proposed to play a role in the effect of acidosis on airway basal tone and responsiveness in the guinea pig (Faisy et al., 2007). ASICs could therefore participate in responses to airway acidification, such as cough and bronchoconstriction (Ricciardolo, 2001; Kollarik and Undem, 2002; Canning et al., 2006), associated with both physiological (e.g. exercise) and pathophysiological (e.g. chronic obstructive pulmonary disease) conditions. Expression of ASICs has not been extensively evaluated in neuropathic pain models, however, a downregulation of ASIC-like currents associated with a substantial decrease of ASIC1a mRNA levels has been described (Poirot et al., 2006), while ASIC4 mRNA levels seem to be slightly increased (Macdonald et al., 2001). ASIC1a and ASIC2b expression is downregulated in an animal model of epilepsy (Biagini et al., 2001) and ASIC1 and ASIC2 transcripts have been detected in high-grade human glioma cells derived from tumour resections (Bubien et al., 1999; Berdiev et al., 2003). Knockout mice have revealed a role for ASIC2 in light-induced retinal degeneration (Ettaiche et al., 2004). Similarly, transgenic mice have provided evidence for a role of ASIC1a in the fear circuit (Wemmie et al., 2003, 2004; Coryell et al., 2007) raising the interesting possibility of the involvement of ASIC channels in anxiety disorders in man. ASIC1a and ASIC2 are constitutively activated by gain-of-function mutations at a particular position just before the second transmembrane region (substitution of an alanine or a glycine for a bulkier amino acid, such as valine, phenylalanine or modified cysteine) (Waldmann et al., 1996; Bassilana et al., 1997; Adams et al., 1998b). Interestingly, similar mutations in *C. elegans* degenerins were previously found to induce neurodegeneration (Driscoll and Chalfie, 1991) and ASIC2a mutant channels can also mediate cell death (Waldmann et al., 1996). However, such mutations have not been associated with human disease to date. Finally, the broad and early expression of ASICs during the development of the CNS suggests a possible role for these channels in neurodevelopment and related disorders (Price et al., 2004a), although knockout of the individual ASICs has no drastic consequences on animal development and viability.

3.4.1.8 Concluding remarks

The ASIC channel family has been relatively recently identified in comparison with many of the other ligand-gated ion channel families and the structure and function of these channels remain only partially understood. An important step forward came from the recent determination of the three-dimensional structure of chicken ASIC1. However, important unresolved issues remain to be elucidated, including the structure of the intracellular domains, the gating mechanisms, and the possibility that the large extracellular loop could bind ligands other than protons. ASICs are broadly expressed in the nervous system but they are also present in some non-neuronal cells and it is probable that novel roles outside the nervous system, possibly in relation to their ability to sense extracellular pH, will be assigned to these channels. The multigenic nature of the ASIC family also makes the analysis of knockout mouse models of multiple members of the family (i.e. double, treble knockouts) necessary to further clarify and expand our understanding of their repertoire of functions. It is particularly noteworthy that almost all CNS neurons express these channels, suggesting that they serve a core functional role. However, although knockout mice have highlighted the importance of ASIC1a in synaptic plasticity and fear conditioning as well as in several pathological conditions, our overall knowledge of their physiological roles in the CNS remains limited. Protons are the only known activators of these channels and it is reasonable to assume that ASIC functions in brain are related to their capacity to behave as extracellular pH sensors. In this regard, an increasing body of evidence suggests a role for proton-mediated signalling in the CNS, including in the retina, and pH fluctuations occur in the brain during normal physiological processes as well as in pathological situations such as epilepsy, ischaemia and inflammation where ASICs seem to play an important role. However, little is known about the amplitude of these physiological pH fluctuations, especially in relevant microdomains such as the synaptic cleft, and an important remaining question is to understand how ASICs contribute to neuronal function within the physiological range of pH. Extracellular acidification also occurs in peripheral tissues in conditions such as ischaemia, inflammation, tumours or injury and has clearly been associated with pain, in part through the activation of ASIC channels in sensory neurons. However, the precise mechanisms underlying ASIC function in nociception remain to be clarified. The proposed role of ASICs in sensory processes such as mechanoperception may be unrelated to the detection of extracellular pH. This raises the interesting possibility of the existence of activating stimuli other than protons. Consistent with this suggestion, mammalian ASIC4, whose function remains unclear, is not activated by protons. ASIC similarities with the FMRFamide-gated sodium channel FaNaC and their susceptibility to modulation by FMRFamide-related peptides raises the possibility of mammalian peptides acting as direct agonists for these channels. The participation of ASICs in mechanosensory function is now well established but its molecular basis remains largely unknown. ASICs on their own are not directly activated by mechanical force but in certain cellular contexts ASICs might be able to transduce mechanical signals, either directly or indirectly. Genetic data in *C. elegans* have clearly shown that several proteins on both sides of the membrane are able to interact with the degenerin channels and participate in mechanotransduction (Syntichaki and Tavernarakis, 2004). In this regard, gaining data on the ASIC-associated protein networks in the membrane of neurons, including proteins that determine their localization, association to the cytoskeleton, surface expression, and function could prove to be of importance. Indeed, determination of the mechanisms regulating ASIC expression and function is of clear importance considering the participation of these channels in highly regulated processes such as long-term potentiation and nociception.

Acknowledgements

We thank Drs E Deval and A Baron for helpful suggestions and careful reading of the manuscript, as well as the Association Française contre les Myopathies (AFM), the Agence Nationale de la Recherche (ANR), the Association pour la Recherche sur le Cancer (ARC), the Institut National du Cancer (INCa) and the Institut UPSA de la Douleur (IUD) for financial support.

References

Adams CM anderson MG, Motto DG et al. (1998a). Ripped pocket and pickpocket, novel *Drosophila* DEG/ENaC subunits expressed in early development and in mechanosensory neurons. *J Cell Biol* **140**, 143–152.

Adams CM, Snyder PM and Welsh MJ (1999). Paradoxical stimulation of a DEG/ENaC channel by amiloride. *J Biol Chem* **274**, 15500–15504.

Adams CM, Snyder PM, Price MP et al. (1998b). Protons activate brain Na$^+$ channel 1 by inducing a conformational change that exposes a residue associated with neurodegeneration. *J Biol Chem* **273**, 30204–30207.

Agopyan N, Bhatti T, Yu S et al. (2003). Vanilloid receptor activation by 2- and 10-microm particles induces responses leading to apoptosis in human airway epithelial cells. *Toxicol Appl Pharmacol* **192**, 21–35.

Ainsley JA, Pettus JM, Bosenko D et al. (2003). Enhanced locomotion caused by loss of the *Drosophila* DEG/ENaC protein Pickpocket1. *Curr Biol* **13**, 1557–1563.

Akaike N and Ueno S (1994). Proton-induced current in neuronal cells. *Prog Neurobiol* **43**, 73–83.

Akopian AN, Chen CC, Ding Y et al. (2000). A new member of the acid-sensing ion channel family. *Neuroreport* **11**, 2217–2222.

Allard M, Rousselot P, Lombard MC et al. (1999). Evidence for neuropeptide FF (FLFQRFamide) in rat dorsal root ganglia. *Peptides* **20**, 327–333.

Allen NJ and Attwell D (2002). Modulation of ASIC channels in rat cerebellar Purkinje neurons by ischaemia-related signals. *J Physiol* **543**, 521–529.

Alvarez de la Rosa D, Krueger SR, Kolar A et al. (2003). Distribution, subcellular localization and ontogeny of ASIC1 in the mammalian central nervous system. *J Physiol* **546**, 77–87.

Alvarez de la Rosa D, Zhang P, Shao D et al. (2002). Functional implications of the localization and activity of acid-sensitive channels in rat peripheral nervous system. *Proc Natl Acad Sci USA* **99**, 2326–2331.

Andrey F, Tsintsadze T, Volkova T et al. (2005). Acid-sensing ionic channels: modulation by redox reagents. *Biochim Biophys Acta* **1745**, 1–6.

Anzai N, Deval E, Schaefer L et al. (2002). The multivalent PDZ domain-containing protein CIPP is a partner of acid-sensing ion channel 3 in sensory neurons. *J Biol Chem* **277**, 16655–16661.

Askwith CC, Benson CJ, Welsh MJ et al. (2001). DEG/ENaC ion channels involved in sensory transduction are modulated by cold temperature. *Proc Natl Acad Sci USA* **98**, 6459–6463.

Askwith CC, Cheng C, Ikuma M et al. (2000). Neuropeptide FF and FMRFamide potentiate acid-evoked currents from sensory neurons and proton-gated DEG/ENaC channels. *Neuron* **26**, 133–141.

Askwith CC, Wemmie JA, Price MP et al. (2004). Acid-sensing ion channel 2 (ASIC2) modulates ASIC1 H$^+$-activated currents in hippocampal neurons. *J Biol Chem* **279**, 18296–18305.

Babini E, Paukert M, Geisler HS et al. (2002). Alternative splicing and interaction with di- and polyvalent cations control the dynamic range of acid-sensing ion channel 1 (ASIC1). *J Biol Chem* **277**, 41597–41603.

Babinski K, Catarsi S, Biagini G et al. (2000). Mammalian ASIC2a and ASIC3 subunits co-assemble into heteromeric proton-gated channels sensitive to Gd^{3+}. *J Biol Chem* **275**, 28519–28525.

Babinski K, Le KT and Seguela P (1999). Molecular cloning and regional distribution of a human proton receptor subunit with biphasic functional properties. *J Neurochem* **72**, 51–57.

Barnes S, Merchant V and Mahmud F (1993). Modulation of transmission gain by protons at the photoreceptor output synapse. *Proc Natl Acad Sci USA* **90**, 10081–10085.

Baron A, Deval E, Salinas M et al. (2002a). Protein kinase C stimulates the acid-sensing ion channel ASIC2a via the PDZ domain-containing protein PICK1. *J Biol Chem* **277**, 50463–50468.

Baron A, Schaefer L, Lingueglia E *et al.* (2001). Zn^{2+} and H^+ are coactivators of acid-sensing ion channels. *J Biol Chem* **276**, 35361–35367.

Baron A, Voilley N, Lazdunski M *et al.* (2008). Acid sensing ion channels in dorsal spinal cord neurons. *J Neurosci* **28**, 1498–1508.

Baron A, Waldmann R and Lazdunski M (2002b). ASIC-like, proton-activated currents in rat hippocampal neurons. *J Physiol* **539**, 485–494.

Bassilana F, Champigny G, Waldmann R *et al.* (1997). The acid-sensitive ionic channel subunit ASIC and the mammalian degenerin MDEG form a heteromultimeric H^+-gated Na^+ channel with novel properties. *J Biol Chem.* **272**, 28819–28822.

Bassler EL, Ngo-Anh TJ, Geisler HS *et al.* (2001). Molecular and functional characterization of acid-sensing ion channel (ASIC) 1b. *J Biol Chem* **276**, 33782–33787.

Benson CJ and Sutherland SP (2001). Toward an understanding of the molecules that sense myocardial ischemia. *Ann N Y Acad Sci* **940**, 96–109.

Benson CJ, Eckert SP and McCleskey EW (1999). Acid-evoked currents in cardiac sensory neurons: a possible mediator of myocardial ischemic sensation. *Circ Res* **84**, 921–928.

Benson CJ, Xie J, Wemmie JA *et al.* (2002). Heteromultimers of DEG/ENaC subunits form H^+-gated channels in mouse sensory neurons. *Proc Natl Acad Sci U S A* **99**, 2338–2343.

Berdiev BK, Mapstone TB, Markert JM *et al.* (2001). pH alterations 'reset' Ca^{2+} sensitivity of brain Na^+ channel 2, a degenerin/epithelial Na^+ ion channel, in planar lipid bilayers. *J Biol Chem* **276**, 38755–38761.

Berdiev BK, Xia J, McLean LA *et al.* (2003). Acid-sensing ion channels in malignant gliomas. *J Biol Chem* **278**, 15023–15034.

Bevan S and Yeats J (1991). Protons activate a cation conductance in a sub-population of rat dorsal root ganglion neurones. *J Physiol* **433**, 145–161.

Biagini G, Babinski K, Avoli M *et al.* (2001). Regional and subunit-specific downregulation of acid-sensing ion channels in the pilocarpine model of epilepsy. *Neurobiol Dis* **8**, 45–58.

Bielefeldt K and Davis BM (2008). Differential effects of ASIC3 and TRPV1 deletion on gastroesophageal sensation in mice. *Am J Physiol Gastrointest Liver Physiol* **294**, G130–138.

Bolshakov KV, Essin KV, Buldakova SL *et al.* (2002). Characterization of acid-sensitive ion channels in freshly isolated rat brain neurons. *Neuroscience* **110**, 723–730.

Brockway LM, Zhou ZH, Bubien JK *et al.* (2002). Rabbit retinal neurons and glia express a variety of ENaC/DEG subunits. *Am J Physiol Cell Physiol* **283**, C126–134.

Bubien JK, Keeton DA, Fuller CM *et al.* (1999). Malignant human gliomas express an amiloride-sensitive Na^+ conductance. *Am J Physiol* **276**, C1405–1410.

Burnes LA, Kolker SJ, Danielson JF *et al.* (2008). Enhanced muscle fatigue occurs in male but not female ASIC3–/– mice. *Am J Physiol Regul Integr Comp Physiol* **294**, R1347–1355.

Cadiou H, Studer M, Jones NG *et al.* (2007). Modulation of acid-sensing ion channel activity by nitric oxide. *J Neurosci* **27**, 13251–13260.

Canessa CM, Horisberger JD and Rossier BC (1993). Epithelial sodium channel related to proteins involved in neurodegeneration. *Nature* **361**, 467–470.

Canning BJ, Farmer DG and Mori N (2006). Mechanistic studies of acid evoked coughing in anesthetized guinea pigs. *Am J Physiol Regul Integr Comp Physiol* **291**, R454–463.

Catarsi S, Babinski K and Seguela P (2001). Selective modulation of heteromeric ASIC proton-gated channels by neuropeptide FF. *Neuropharmacology* **41**, 592–600.

Caterina MJ, Leffler A, Malmberg AB *et al.* (2000). Impaired nociception and pain sensation in mice lacking the capsaicin receptor. *Science* **288**, 306–313.

Caterina MJ, Schumacher MA, Tominaga M *et al.* (1997). The capsaicin receptor: a heat-activated ion channel in the pain pathway. *Nature* **389**, 816–824.

Chagot B, Escoubas P, Diochot S *et al.* (2005). Solution structure of APETx2, a specific peptide inhibitor of ASIC3 proton-gated channels. *Protein Sci* **14**, 2003–2010.

Chai S, Li M, Lan J *et al.* (2007). A kinase-anchoring protein 150 and calcineurin are involved in regulation of acid-sensing ion channels ASIC1a and ASIC2a. *J Biol Chem* **282**, 22668–22677.

Champigny G, Voilley N, Waldmann R *et al.* (1998). Mutations causing neurodegeneration in *Caenorhabditis elegans* drastically alter the pH sensitivity and inactivation of the mammalian H^+-gated Na^+ channel MDEG1. *J Biol Chem* **273**, 15418–15422.

Chen CC, England S, Akopian AN *et al.* (1998). A sensory neuron-specific, proton-gated ion channel. *Proc Natl Acad Sci USA* **95**, 10240–10245.

Chen CC, Zimmer A, Sun WH *et al.* (2002). A role for ASIC3 in the modulation of high-intensity pain stimuli. *Proc Natl Acad Sci USA* **99**, 8992–8997.

Chen X, Kalbacher H and Grunder S (2006a). Interaction of acid-sensing ion channel (ASIC) 1 with the tarantula toxin psalmotoxin 1 is state-dependent. *J Gen Physiol* **127**, 267–276.

Chen X, Paukert M, Kadurin I *et al.* (2006b). Strong modulation by RFamide neuropeptides of the ASIC1b/3 heteromer in competition with extracellular calcium. *Neuropharmacology* **50**, 964–974.

Chen X, Polleichtner G, Kadurin I *et al.* (2007). Zebrafish acid-sensing ion channel (ASIC) 4, characterization of homo- and heteromeric channels and identification of regions important for activation by H^+. *J Biol Chem* **282**, 30406–30413.

Chesler M (2003). Regulation and modulation of pH in the brain. *Physiol Rev* **83**, 1183–1221.

Cho JH and Askwith CC (2007). Potentiation of acid-sensing ion channels by sulfhydryl compounds. *Am J Physiol Cell Physiol* **292**, C2161–2174.

Christianson JA, Traub RJ and Davis BM (2006). Differences in spinal distribution and neurochemical phenotype of colonic afferents in mouse and rat. *J Comp Neurol* **494**, 246–259.

Chu XP, Close N, Saugstad JA *et al.* (2006). ASIC1a-specific modulation of acid-sensing ion channels in mouse cortical neurons by redox reagents. *J Neurosci* **26**, 5329–5339.

Chu XP, Wemmie JA, Wang WZ *et al.* (2004). Subunit-dependent high-affinity zinc inhibition of acid-sensing ion channels. *J Neurosci* **24**, 8678–8689.

Connor M, Naves LA and McCleskey EW (2005). Contrasting phenotypes of putative proprioceptive and nociceptive trigeminal neurons innervating jaw muscle in rat. *Mol Pain* **1**, 31.

Coric T, Passamaneck YJ, Zhang P *et al.* (2008). Simple chordates exhibit a proton-independent function of acid-sensing ion channels. *Faseb J.* DOI 10.1096/fj.07–100313

Coric T, Zhang P, Todorovic N *et al.* (2003). The extracellular domain determines the kinetics of desensitization in acid-sensitive ion channel 1. *J Biol Chem* **278**, 45240–45247.

Coric T, Zheng D, Gerstein M *et al.* (2005). Proton sensitivity of ASIC1 appeared with the rise of fishes by changes of residues in the region that follows TM1 in the ectodomain of the channel. *J Physiol* **568**, 725–735.

Coryell MW, Ziemann AE, Westmoreland PJ *et al.* (2007). Targeting ASIC1a reduces innate fear and alters neuronal activity in the fear circuit. *Biol Psychiatry* **62**, 1140–1148.

Darboux I, Lingueglia E, Champigny G *et al.* (1998a). dGNaC1, a gonad-specific amiloride-sensitive Na^+ channel. *J Biol Chem* **273**, 9424–9429.

Darboux I, Lingueglia E, Pauron D *et al.* (1998b). A new member of the amiloride-sensitive sodium channel family in *Drosophila melanogaster* peripheral nervous system. *Biochem Biophys Res Commun* **246**, 210–216.

Davies NW, Lux HD and Morad M (1988). Site and mechanism of activation of proton-induced sodium current in chick dorsal root ganglion neurones. *J Physiol* **400**, 159–187.

Davis JB, Gray J, Gunthorpe MJ *et al.* (2000). Vanilloid receptor-1 is essential for inflammatory thermal hyperalgesia. *Nature* **405**, 183–187.

de Weille J and Bassilana F (2001). Dependence of the acid-sensitive ion channel, ASIC1a, on extracellular Ca(2$^+$) ions. *Brain Res* **900**, 277–281.

de Weille JR, Bassilana F, Lazdunski M *et al.* (1998). Identification, functional expression and chromosomal localisation of a sustained human proton-gated cation channel. *FEBS Lett* **433**, 257–260.

Deval E, Baron A, Lingueglia E *et al.* (2003). Effects of neuropeptide SF and related peptides on acid-sensing ion channel 3 and sensory neuron excitability. *Neuropharmacology* **44**, 662–671.

Deval E, Friend V, Thirant C *et al.* (2006). Regulation of sensory neuron-specific acid-sensing ion channel 3 by the adaptor protein na$^+$/h$^+$ exchanger regulatory factor-1. *J Biol Chem* **281**, 1796–1807.

Deval E, Salinas M, Baron A *et al.* (2004). ASIC2b-dependent regulation of ASIC3, an essential acid-sensing ion channel subunit in sensory neurons via the partner protein PICK-1. *J Biol Chem* **279**, 19531–19539.

Devillers JP and Simonnet G (1994). Modulation of neuropeptide FF release from rat spinal cord slices by glutamate. Involvement of NMDA receptors. *Eur J Pharmacol* **271**, 185–192.

Devillers JP, Labrouche SA, Castes E *et al.* (1995). Release of neuropeptide FF, an anti-opioid peptide, in rat spinal cord slices is voltage- and Ca(2$^+$)-sensitive: possible involvement of P-type Ca^{2+} channels. *J Neurochem* **64**, 1567–1575.

DeVries SH (2001). Exocytosed protons feedback to suppress the Ca^{2+} current in mammalian cone photoreceptors. *Neuron* **32**, 1107–1117.

Diochot S, Baron A, Rash LD *et al.* (2004). A new sea anemone peptide, APETx2, inhibits ASIC3, a major acid-sensitive channel in sensory neurons. *Embo J* **23**, 1516–1525.

Diochot S, Salinas M, Baron A *et al.* (2007). Peptides inhibitors of acid-sensing ion channels. *Toxicon* **49**, 271–284.

Dirajlal S, Pauers LE and Stucky CL (2003). Differential response properties of IB(4)-positive and -negative unmyelinated sensory neurons to protons and capsaicin. *J Neurophysiol* **89**, 513–524.

Donier E, Rugiero F, Okuse K *et al.* (2005). Annexin ii light chain p11 promotes functional expression of acid-sensing ion channel ASIC1a. *J Biol Chem* **280**, 38666–38672.

Drew LJ, Rohrer DK, Price MP *et al.* (2004). Acid-sensing ion channels ASIC2 and ASIC3 do not contribute to mechanically activated currents in mammalian sensory neurones. *J Physiol* **556**, 691–710.

Driscoll M and Chalfie M (1991). The mec-4 gene is a member of a family of *Caenorhabditis elegans* genes that can mutate to induce neuronal degeneration. *Nature* **349**, 588–593.

Drummond HA, Grifoni SC and Jernigan NL (2008). A new trick for an old dogma: ENaC proteins as mechanotransducers in vascular smooth muscle. *Physiology (Bethesda)* **23**, 23–31.

Duan B, Wu LJ, Yu YQ *et al.* (2007). Upregulation of acid-sensing ion channel ASIC1a in spinal dorsal horn neurons contributes to inflammatory pain hypersensitivity. *J Neurosci* **27**, 11139–11148.

Dube GR, Lehto SG, Breese NM *et al.* (2005). Electrophysiological and *in vivo* characterization of A-317567, a novel blocker of acid-sensing ion channels. *Pain* **117**, 88–96.

Duggan A, Garcia-Anoveros J and Corey DP (2002). The PDZ domain protein PICK1 and the sodium channel BNaC1 interact and localize at mechanosensory terminals of dorsal root ganglion neurons and dendrites of central neurons. *J Biol Chem* **277**, 5203–5208.

Escoubas P, Bernard C, Lambeau G *et al.* (2003). Recombinant production and solution structure of PcTx1, the specific peptide inhibitor of ASIC1a proton-gated cation channels. *Protein Sci* **12**, 1332–1343.

Escoubas P, De Weille JR, Lecoq A *et al.* (2000). Isolation of a tarantula toxin specific for a class of proton-gated Na$^+$ channels. *J Biol Chem* **275**, 25116–25121.

Ettaiche M, Deval E, Cougnon M *et al.* (2006). Silencing acid-sensing ion channel 1a alters cone-mediated retinal function. *J Neurosci* **26**, 5800–5809.

Ettaiche M, Guy N, Hofman P *et al.* (2004). Acid-sensing ion channel 2 is important for retinal function and protects against light-induced retinal degeneration. *J Neurosci* **24**, 1005–1012.

Faisy C, Planquette B, Naline E *et al.* (2007). Acid-induced modulation of airway basal tone and contractility: role of acid-sensing ion channels (ASICs) and TRPV1 receptor. *Life Sci* **81**, 1094–1102.

Friese MA, Craner MJ, Etzensperger R *et al.* (2007). Acid-sensing ion channel-1 contributes to axonal degeneration in autoimmune inflammation of the central nervous system. *Nat Med* **13**, 1483–1489.

Fukuda T, Ichikawa H, Terayama R *et al.* (2006). ASIC3-immunoreactive neurons in the rat vagal and glossopharyngeal sensory ganglia. *Brain Res* **1081**, 150–155.

Gannon KP, Vanlandingham LG, Jernigan NL *et al.* (2008). Impaired pressure-induced constriction in mouse middle cerebral arteries of ASIC2 knockout mice. *Am J Physiol Heart Circ Physiol* **294**, H1793–1803.

Gao J, Duan B, Wang DG *et al.* (2005). Coupling between NMDA receptor and acid-sensing ion channel contributes to ischemic neuronal death. *Neuron* **48**, 635–646.

Gao J, Wu LJ, Xu L *et al.* (2004). Properties of the proton-evoked currents and their modulation by Ca^{2+} and Zn^{2+} in the acutely dissociated hippocampus CA1 neurons. *Brain Res* **1017**, 197–207.

Garcia-Anoveros J, Derfler B, Neville-Golden J *et al.* (1997). BNaC1 and BNaC2 constitute a new family of human neuronal sodium channels related to degenerins and epithelial sodium channels. *Proc Natl Acad Sci U S A* **94**, 1459–1464.

Garcia-Anoveros J, Garcia JA, Liu JD *et al.* (1998). The nematode degenerin UNC-105 forms ion channels that are activated by degeneration- or hypercontraction-causing mutations. *Neuron* **20**, 1231–1241.

Garcia-Anoveros J, Samad TA, Zuvela-Jelaska L *et al.* (2001). Transport and localization of the DEG/ENaC ion channel BNaC1alpha to peripheral mechanosensory terminals of dorsal root ganglia neurons. *J Neurosci* **21**, 2678–2686.

Gitterman DP, Wilson J and Randall AD (2005). Functional properties and pharmacological inhibition of ASIC channels in the human SJ-RH30 skeletal muscle cell line. *J Physiol* **562**, 759–769.

Grantyn R and Lux HD (1988). Similarity and mutual exclusion of NMDA- and proton-activated transient Na$^+$-currents in rat tectal neurons. *Neurosci Lett* **89**, 198–203.

Grifoni SC, Jernigan NL, Hamilton G *et al.* (2008a). ASIC proteins regulate smooth muscle cell migration. *Microvasc Res* **75**, 202–210.

Grifoni SD, McKey SE and Drummond HA (2008b). Hsc70 regulates cell surface ASIC2 expression and vascular smooth muscle cell migration. *Am J Physiol Heart Circ Physiol.* DOI 10.1152/ajpheart.01271.2007.

Groth M, Helbig T, Grau V *et al.* (2006). Spinal afferent neurons projecting to the rat lung and pleura express acid-sensitive channels. *Respir Res* **7**, 96.

Grunder S, Geissler HS, Bassler EL *et al.* (2000). A new member of acid-sensing ion channels from pituitary gland. *Neuroreport* **11**, 1607–1611.

Gu Q and Lee LY (2006). Characterization of acid-signaling in rat vagal pulmonary sensory neurons. *Am J Physiol Lung Cell Mol Physiol* **291**, L58–65.

Gunthorpe MJ, Smith GD, Davis JB *et al.* (2001). Characterisation of a human acid-sensing ion channel (hASIC1a) endogenously expressed in HEK293 cells. *Pflugers Arch* **442**, 668–674.

Hayes SG, Kindig AE and Kaufman MP (2007). Blockade of acid-sensing ion channels attenuates the exercise pressor reflex in cats. *J Physiol* **581**, 1271–1282.

Hesselager M, Timmermann DB and Ahring PK (2004). pH Dependency and desensitization kinetics of heterologously expressed combinations of acid-sensing ion channel subunits. *J Biol Chem* **279**, 11006–11015.

Hildebrand MS, de Silva MG, Klockars T *et al.* (2004). Characterisation of DRASIC in the mouse inner ear. *Hear Res* **190**, 149–160.

Holzer P (2006). Taste receptors in the gastrointestinal tract. V. Acid sensing in the gastrointestinal tract. *Am J Physiol Gastrointest Liver Physiol* **292**, G699–705.

Hruska-Hageman AM, Benson CJ, Leonard AS *et al.* (2004). PSD-95 and Lin-7b interact with acid-sensing ion channel-3 and have opposite effects on H^+- gated current. *J Biol Chem* **279**, 46962–46968.

Hruska-Hageman AM, Wemmie JA, Price MP *et al.* (2002). Interaction of the synaptic protein PICK1 (protein interacting with C kinase 1) with the non-voltage gated sodium channels BNC1 (brain Na^+ channel 1) and ASIC (acid-sensing ion channel). *Biochem J* **361**, 443–450.

Huang M and Chalfie M (1994). Gene interactions affecting mechanosensory transduction in *Caenorhabditis elegans*. *Nature* **367**, 467–470.

Hughes PA, Brierley SM, Young RL *et al.* (2007). Localization and comparative analysis of acid-sensing ion channel (ASIC1, 2 and 3) mRNA expression in mouse colonic sensory neurons within thoracolumbar dorsal root ganglia. *J Comp Neurol* **500**, 863–875.

Ikeuchi M, Kolker SJ, Burnes LA *et al.* (2008). Role of ASIC3 in the primary and secondary hyperalgesia produced by joint inflammation in mice. *Pain*. DOI 10.1016/j.pain.2008.01.0.

Immke DC and McCleskey EW (2001). Lactate enhances the acid-sensing Na^+ channel on ischemia-sensing neurons. *Nat Neurosci* **4**, 869–870.

Immke DC and McCleskey EW (2003). Protons open acid-sensing ion channels by catalyzing relief of Ca^{2+} blockade. *Neuron* **37**, 75–84.

Ishibashi K and Marumo F (1998). Molecular cloning of a DEG/ENaC sodium channel cDNA from human testis. *Biochem Biophys Res Commun* **245**, 589–593.

Jahr H, van Driel M, van Osch GJ *et al.* (2005). Identification of acid-sensing ion channels in bone. *Biochem Biophys Res Commun* **337**, 349–354.

Jasti J, Furukawa H, Gonzales EB *et al.* (2007). Structure of acid-sensing ion channel 1 at 1.9 A resolution and low pH. *Nature* **449**, 316–323.

Ji HL, Jovov B, Fu J *et al.* (2002). Up-regulation of acid-gated $Na^{(+)}$ channels (ASICs) by cystic fibrosis transmembrane conductance regulator co-expression in *Xenopus* oocytes. *J Biol Chem* **277**, 8395–8405.

Jiang N, Rau KK, Johnson RD *et al.* (2006). Proton sensitivity, Ca^{2+} permeability and molecular basis of ASIC channels expressed in glabrous and hairy skin afferents. *J Neurophysiol* **95**, 2466–2478.

Joch M, Ase AR, Chen CX *et al.* (2007). Parkin-mediated monoubiquitination of the PDZ protein PICK1 regulates the activity of acid-sensing ion channels. *Mol Biol Cell* **18**, 3105–3118.

Johnson MB, Jin K, Minami M *et al.* (2001). Global ischemia induces expression of acid-sensing ion channel 2a in rat brain. *J Cereb Blood Flow Metab* **21**, 734–740.

Jones NG, Slater R, Cadiou H *et al.* (2004). Acid-induced pain and its modulation in humans. *J Neurosci* **24**, 10974–10979.

Jones RC 3rd, Otsuka E, Wagstrom E *et al.* (2007). Short-term sensitization of colon mechanoreceptors is associated with long-term hypersensitivity to colon distention in the mouse. *Gastroenterology* **133**, 184–194.

Jones RC 3rd, Xu L and Gebhart GF (2005). The mechanosensitivity of mouse colon afferent fibers and their sensitization by inflammatory mediators require transient receptor potential vanilloid 1 and acid-sensing ion channel 3. *J Neurosci* **25**, 10981–10989.

Jospin M and Allard B (2004). An amiloride-sensitive H^+-gated Na^+ channel in *Caenorhabditis elegans* body wall muscle cells. *J Physiol* **559**, 715–720.

Kaila K and Ransom BR (1998). *pH and brain function*. New York, Wiley-Liss.

Kellenberger S and Schild L (2002). Epithelial sodium channel/degenerin family of ion channels: a variety of functions for a shared structure. *Physiol Rev* **82**, 735–767.

Kinnamon SC, Price MP, Stone LM *et al.* (2000). The acid-sensing ion channel BNC1 is not required for sour taste transduction. *Internat Symp Olfact Taste* **13**, 80.

Kollarik M and Undem BJ (2002). Mechanisms of acid-induced activation of airway afferent nerve fibres in guinea-pig. *J Physiol* **543**, 591–600.

Kontinen VK, Aarnisalo AA, Idanpaan-Heikkila JJ *et al.* (1997). Neuropeptide FF in the rat spinal cord during carrageenan inflammation. *Peptides* **18**, 287–292.

Kovalchuk YN, Krishtal OA and Nowycky MC (1990). The proton-activated inward current of rat sensory neurons includes a calcium component. *Neurosci Lett* **115**, 237–242.

Kress M and Waldmann R (2006). Acid-sensing ionic channels. *Curr Top Membr* **57**, 241–276.

Krishtal O (2003). The ASICs: signaling molecules? Modulators? *Trends Neurosci* **26**, 477–483.

Krishtal OA and Pidoplichko VI (1981). A receptor for protons in the membrane of sensory neurons may participate in nociception. *Neuroscience* **6**, 2599–2601.

Leffler A, Monter B and Koltzenburg M (2006). The role of the capsaicin receptor TRPV1 and acid-sensing ion channels (ASICS) in proton sensitivity of subpopulations of primary nociceptive neurons in rats and mice. *Neuroscience* **139**, 699–709.

Leonard AS, Yermolaieva O, Hruska-Hageman A *et al.* (2003). cAMP-dependent protein kinase phosphorylation of the acid-sensing ion channel-1 regulates its binding to the protein interacting with C-kinase-1. *Proc Natl Acad Sci USA* **100**, 2029–2034.

Li YX, Schaffner AE, Li HR *et al.* (1997). Proton-induced cation current in embryonic rat spinal cord neurons changes ion dependency over time in vitro. *Brain Res Dev Brain Res* **102**, 261–266.

Lilley S, LeTissier P and Robbins J (2004). The discovery and characterization of a proton-gated sodium current in rat retinal ganglion cells. *J Neurosci* **24**, 1013–1022.

Lin H, Mann KJ, Starostina E *et al.* (2005). A *Drosophila* DEG/ENaC channel subunit is required for male response to female pheromones. *Proc Natl Acad Sci USA* **102**, 12831–12836.

Lin W, Ogura T and Kinnamon SC (2002). Acid-activated cation currents in rat vallate taste receptor cells. *J Neurophysiol* **88**, 133–141.

Linguelia E (2007). Acid-sensing ion channels in sensory perception. *J Biol Chem* **282**, 17325–17329.

Linguelia E, Champigny G, Lazdunski M *et al.* (1995). Cloning of the amiloride-sensitive FMRFamide peptide-gated sodium channel. *Nature* **378**, 730–733.

Linguelia E, de Weille JR, Bassilana F *et al.* (1997). A modulatory subunit of acid-sensing ion channels in brain and dorsal root ganglion cells. *J Biol Chem* **272**, 29778–29783.

Linguelia E, Deval E and Lazdunski M (2006). FMRFamide-gated sodium channel and ASIC channels: a new class of ionotropic receptors for FMRFamide and related peptides. *Peptides* **27**, 1138–1152.

Linguelia E, Voilley N, Waldmann R *et al.* (1993). Expression cloning of an epithelial amiloride-sensitive Na^+ channel. A new channel type with homologies to Caenorhabditis elegans degenerins. *FEBS Lett* **318**, 95–99.

Liu J, Schrank B and Waterston RH (1996). Interaction between a putative mechanosensory membrane channel and a collagen. *Science* **273**, 361–364.

Liu L and Simon SA (2000). Capsaicin, acid and heat-evoked currents in rat trigeminal ganglion neurons: relationship to functional VR1 receptors. *Physiol Behav* **69**, 363–378.

Liu L and Simon SA (2001). Acidic stimuli activates two distinct pathways in taste receptor cells from rat fungiform papillae. *Brain Res* **923**, 58–70.

Liu L, Johnson WA and Welsh MJ (2003a). *Drosophila* DEG/ENaC pickpocket genes are expressed in the tracheal system, where they may be involved in liquid clearance. *Proc Natl Acad Sci USA* **100**, 2128–2133.

Liu L, Leonard AS, Motto DG et al. (2003b). Contribution of *Drosophila DEG/ENaC* genes to salt taste. *Neuron* **39**, 133–146.

Liu M, Willmott NJ, Michael GJ et al. (2004). Differential pH and capsaicin responses of *Griffonia simplicifolia* IB4 (IB4)-positive and IB4-negative small sensory neurons. *Neuroscience* **127**, 659–672.

Macdonald R, Bingham S, Bond BC et al. (2001). Determination of changes in mRNA expression in a rat model of neuropathic pain by Taqman quantitative RT-PCR. *Brain Res Mol Brain Res* **90**, 48–56.

Mamet J, Baron A, Lazdunski M et al. (2002). Proinflammatory mediators, stimulators of sensory neuron excitability via the expression of acid-sensing ion channels. *J Neurosci* **22**, 10662–10670.

Mamet J, Lazdunski M and Voilley N (2003). How nerve growth factor drives physiological and inflammatory expressions of acid-sensing ion channel 3 in sensory neurons. *J Biol Chem* **278**, 48907–48913.

Mantyh PW, Clohisy DR, Koltzenburg M et al. (2002). Molecular mechanisms of cancer pain. *Nat Rev Cancer* **2**, 201–209.

Mazzuca M, Heurteaux C, Alloui A et al. (2007). A tarantula peptide against pain via ASIC1a channels and opioid mechanisms. *Nat Neurosci* **10**, 943–945.

McIlwrath SL, Hu J, Anirudhan G et al. (2005). The sensory mechanotransduction ion channel ASIC2 (acid sensitive ion channel 2) is regulated by neurotrophin availability. *Neuroscience* **131**, 499–511.

McMahon SB and Jones NG (2004). Plasticity of pain signaling: role of neurotrophic factors exemplified by acid-induced pain. *J Neurobiol* **61**, 72–87.

Mogil JS, Breese NM, Witty MF et al. (2005). Transgenic expression of a dominant-negative ASIC3 subunit leads to increased sensitivity to mechanical and inflammatory stimuli. *J Neurosci* **25**, 9893–9901.

Molliver DC, Immke DC, Fierro L et al. (2005). ASIC3, an acid-sensing ion channel, is expressed in metaboreceptive sensory neurons. *Mol Pain* **1**, 35.

O'Hagan R, Chalfie M and Goodman MB (2005). The MEC-4 DEG/ENaC channel of *Caenorhabditis elegans* touch receptor neurons transduces mechanical signals. *Nat Neurosci* **8**, 43–50.

Olson TH, Riedl MS, Vulchanova L et al. (1998). An acid-sensing ion channel (ASIC) localizes to small primary afferent neurons in rats. *Neuroreport* **9**, 1109–1113.

Ostrovskaya O, Moroz L and Krishtal O (2004). Modulatory action of RFamide-related peptides on acid-sensing ionic channels is pH dependent: the role of arginine. *J Neurochem* **91**, 252–255.

Page AJ, Brierley SM, Martin CM et al. (2007). Acid-sensing ion channels 2 and 3 are required for inhibition of visceral nociceptors by benzamil. *Pain* **133**, 150–160.

Page AJ, Brierley SM, Martin CM et al. (2004). The ion channel ASIC1 contributes to visceral but not cutaneous mechanoreceptor function. *Gastroenterology* **127**, 1739–1747.

Page AJ, Brierley SM, Martin CM et al. (2005). Different contributions of ASIC channels 1a, 2 and 3 in gastrointestinal mechanosensory function. *Gut* **54**, 1408–1415.

Panula P, Kalso E, Nieminen M et al. A (1999). Neuropeptide FF and modulation of pain. *Brain Res* **848**, 191–196.

Paukert M, Babini E, Pusch M et al. (2004a). Identification of the Ca^{2+} blocking site of acid-sensing ion channel (ASIC) 1: implications for channel gating. *J Gen Physiol* **124**, 383–394.

Paukert M, Sidi S, Russell C et al. (2004b). A family of acid-sensing ion channels from the zebrafish: widespread expression in the central nervous system suggests a conserved role in neuronal communication. *J Biol Chem* **279**, 18783–18791.

Peng BG, Ahmad S, Chen S et al. (2004). Acid-sensing ion channel 2 contributes a major component to acid-evoked excitatory responses in spiral ganglion neurons and plays a role in noise susceptibility of mice. *J Neurosci* **24**, 10167–10175.

Petroff EY, Price MP, Snitsarev V et al. (2008). Acid-sensing ion channels interact with and inhibit BK K^+ channels. *Proc Natl Acad Sci USA* **105**, 3140–3144.

Petruska JC, Napaporn J, Johnson RD et al. (2002). Chemical responsiveness and histochemical phenotype of electrophysiologically classified cells of the adult rat dorsal root ganglion. *Neuroscience* **115**, 15–30.

Petruska JC, Napaporn J, Johnson RD et al. (2000). Subclassified acutely dissociated cells of rat DRG: histochemistry and patterns of capsaicin-, proton- and ATP-activated currents. *J Neurophysiol* **84**, 2365–2379.

Poirot O, Berta T, Decosterd I et al. (2006). Distinct ASIC currents are expressed in rat putative nociceptors and are modulated by nerve injury. *J Physiol* **576**, 215–234.

Poirot O, Vukicevic M, Boesch A et al. (2004). Selective regulation of acid-sensing ion channel 1 by serine proteases. *J Biol Chem* **279**, 38448–38457.

Price MP, Gong H, Schnizler MK et al. (2004a). Mice missing multiple acid-sensing ion channel subunits show severe deficits in cognitive, social and motor behaviors. *Society for Neuroscience Abstracts*, No. 679.6.

Price MP, Lewin GR, McIlwrath SL et al. (2000). The mammalian sodium channel BNC1 is required for normal touch sensation. *Nature* **407**, 1007–1011.

Price MP, McIlwrath SL, Xie J et al. (2001). The DRASIC cation channel contributes to the detection of cutaneous touch and acid stimuli in mice. *Neuron* **32**, 1071–1083.

Price MP, Snyder PM and Welsh MJ (1996). Cloning and expression of a novel human brain Na^+ channel. *J Biol Chem* **271**, 7879–7882.

Price MP, Thompson RJ, Eshcol JO et al. (2004b). Stomatin modulates gating of acid-sensing ion channels. *J Biol Chem* **279**, 53886–53891.

Ricciardolo FL (2001). Mechanisms of citric acid-induced bronchoconstriction. *Am J Med* **111** Suppl 8A, 18S-24S.

Richter TA, Dvoryanchikov GA, Roper SD et al. (2004). Acid-sensing ion channel-2 is not necessary for sour taste in mice. *J Neurosci* **24**, 4088–4091.

Roza C, Puel JL, Kress M et al. (2004). Knockout of the ASIC2 channel in mice does not impair cutaneous mechanosensation, visceral mechanonociception and hearing. *J Physiol* **558**, 659–669.

Sakai H, Lingueglia E, Champigny G et al. (1999). Cloning and functional expression of a novel degenerin-like Na^+ channel gene in mammals. *J Physiol* **519 Pt 2**, 323–333.

Salinas M, Rash LD, Baron A et al. (2006). The receptor site of the spider toxin PcTx1 on the proton-gated cation channel ASIC1a. *J Physiol* **570**, 339–354.

Saugstad JA, Roberts JA, Dong J et al. (2004). Analysis of the membrane topology of the acid-sensing ion channel 2a. *J Biol Chem* **279**, 55514–55519.

Schaefer L, Sakai H, Mattei M et al. (2000). Molecular cloning, functional expression and chromosomal localization of an amiloride-sensitive $Na(^+)$ channel from human small intestine. *FEBS Lett* **471**, 205–210.

Schicho R, Florian W, Liebmann I et al. (2004). Increased expression of TRPV1 receptor in dorsal root ganglia by acid insult of the rat gastric mucosa. *Eur J Neurosci* **19**, 1811–1818.

Sherwood TW and Askwith CC (2008). Endogenous arginine-phenylalanine-amide-related peptides alter steady-state desensitization of ASIC1a. *J Biol Chem* **283**, 1818–1830.

Shimada S, Ueda T, Ishida Y et al. (2006). Acid-sensing ion channels in taste buds. *Arch Histol Cytol* **69**, 227–231.

Sluka KA, Price MP, Breese NM et al. (2003). Chronic hyperalgesia induced by repeated acid injections in muscle is abolished by the loss of ASIC3, but not ASIC1. *Pain* **106**, 229–239.

Sluka KA, Radhakrishnan R, Benson CJ et al. (2006). ASIC3 in muscle mediates mechanical, but not heat, hyperalgesia associated with muscle inflammation. *Pain* **129**, 102–112.

Smith ES, Cadiou H and McNaughton PA (2007). Arachidonic acid potentiates acid-sensing ion channels in rat sensory neurons by a direct action. *Neuroscience* **145**, 686–698.

Sontheimer H, Perouansky M, Hoppe D *et al.* (1989). Glial cells of the oligodendrocyte lineage express proton-activated Na$^+$ channels. *J Neurosci Res* **24**, 496–500.

Staruschenko A, Dorofeeva NA, Bolshakov KV *et al.* (2007). Subunit-dependent cadmium and nickel inhibition of acid-sensing ion channels. *J Neurobiol* **67**, 97–107.

Steen KH, Issberner U and Reeh PW (1995). Pain due to experimental acidosis in human skin: evidence for non-adapting nociceptor excitation. *Neurosci Lett* **199**, 29–32.

Su X, Li Q, Shrestha K *et al.* (2006). Interregulation of proton-gated Na(+) channel 3 and cystic fibrosis transmembrane conductance regulator. *J Biol Chem* **281**, 36960–36968.

Sugiura T, Dang K, Lamb K *et al.* (2005). Acid-sensing properties in rat gastric sensory neurons from normal and ulcerated stomach. *J Neurosci* **25**, 2617–2627.

Sutherland SP, Benson CJ, Adelman JP *et al.* (2001). Acid-sensing ion channel 3 matches the acid-gated current in cardiac ischemia-sensing neurons. *Proc Natl Acad Sci USA* **98**, 711–716.

Suzuki H, Kerr R, Bianchi L *et al.* (2003). *In vivo* imaging of *C. elegans* mechanosensory neurons demonstrates a specific role for the MEC-4 channel in the process of gentle touch sensation. *Neuron* **39**, 1005–1017.

Syntichaki P and Tavernarakis N (2004). Genetic models of mechanotransduction: the nematode *Caenorhabditis elegans*. *Physiol Rev* **84**, 1097–1153.

Tan ZY, Lu Y, Whiteis CA *et al.* (2007). Acid-sensing ion channels contribute to transduction of extracellular acidosis in rat carotid body glomus cells. *Circ Res* **101**, 1009–1019.

Tavernarakis N, Shreffler W, Wang S *et al.* (1997). unc-8, a DEG/ENaC family member, encodes a subunit of a candidate mechanically gated channel that modulates *C. elegans* locomotion. *Neuron* **18**, 107–119.

Uchiyama Y, Cheng CC, Danielson KG *et al.* (2007). Expression of acid-sensing ion channel 3 (ASIC3) in nucleus pulposus cells of the intervertebral disc is regulated by p75NTR and ERK signaling. *J Bone Miner Res* **22**, 1996–2006.

Ueno S, Nakaye T and Akaike N (1992). Proton-induced sodium current in freshly dissociated hypothalamic neurones of the rat. *J Physiol* **447**, 309–327.

Ugawa S, Inagaki A, Yamamura H *et al.* (2006). Acid-sensing ion channel-1b in the stereocilia of mammalian cochlear hair cells. *Neuroreport* **17**, 1235–1239.

Ugawa S, Minami Y, Guo W *et al.* (1998). Receptor that leaves a sour taste in the mouth. *Nature* **395**, 555–556.

Ugawa S, Ueda T, Ishida Y *et al.* (2002). Amiloride-blockable acid-sensing ion channels are leading acid sensors expressed in human nociceptors. *J Clin Invest* **110**, 1185–1190.

Ugawa S, Ueda T, Takahashi E *et al.* (2001). Cloning and functional expression of ASIC-beta2, a splice variant of ASIC-beta. *Neuroreport* **12**, 2865–2869.

Ugawa S, Ueda T, Yamamura H *et al.* (2005). *In situ* hybridization evidence for the coexistence of ASIC and TRPV1 within rat single sensory neurons. *Brain Res Mol Brain Res* **136**, 125–133.

Ugawa S, Yamamoto T, Ueda T *et al.* (2003). Amiloride-insensitive currents of the acid-sensing ion channel-2a (ASIC2a)/ASIC2b heteromeric sour-taste receptor channel. *J Neurosci* **23**, 3616–3622.

Varming T (1999). Proton-gated ion channels in cultured mouse cortical neurons. *Neuropharmacology* **38**, 1875–1881.

Vila-Carriles WH, Kovacs GG, Jovov B *et al.* (2006). Surface expression of ASIC2 inhibits the amiloride-sensitive current and migration of glioma cells. *J Biol Chem* **281**, 19220–19232.

Vilim FS, Aarnisalo AA, Nieminen ML *et al.* (1999). Gene for pain modulatory neuropeptide NPFF: induction in spinal cord by noxious stimuli. *Mol Pharmacol* **55**, 804–811.

Voilley N (2004). Acid-sensing ion channels (ASICs): new targets for the analgesic effects of non-steroid anti-inflammatory drugs (NSAIDs). *Curr Drug Targets Inflamm Allergy* **3**, 71–79.

Voilley N, de Weille J, Mamet J *et al.* (2001). Nonsteroid anti-inflammatory drugs inhibit both the activity and the inflammation-induced expression of acid-sensing ion channels in nociceptors. *J Neurosci* **21**, 8026–8033.

Vrublevskii SV, Kryshtal OA, Osipchuk IV *et al.* (1985). Proton-activated sodium conduction in the brain neurons of the rat. *Dokl Akad Nauk SSSR* **284**, 990–993.

Vukicevic M and Kellenberger S (2004). Modulatory effects of acid-sensing ion channels on action potential generation in hippocampal neurons. *Am J Physiol Cell Physiol* **287**, C682–690.

Vukicevic M, Weder G, Boillat A *et al.* (2006). Trypsin cleaves acid-sensing ion channel 1a in a domain that is critical for channel gating. *J Biol Chem* **281**, 714–722.

Waldmann R and Lazdunski M (1998). H(+)-gated cation channels: neuronal acid sensors in the NaC/DEG family of ion channels. *Curr Opin Neurobiol* **8**, 418–424.

Waldmann R, Bassilana F, De Weille JR *et al.* (1997a). Molecular cloning of a non-inactivating proton-gated Na$^+$ channel specific for sensory neurons. *J. Biol. Chem.* **272**, 20975–20978.

Waldmann R, Champigny G, Bassilana F *et al.* (1997b). A proton-gated cation channel involved in acid-sensing. *Nature* **386**, 173–177.

Waldmann R, Champigny G, Voilley N *et al.* (1996). The mammalian degenerin MDEG, an amiloride-sensitive cation channel activated by mutations causing neurodegeneration in *Caenorhabditis elegans*. *J Biol Chem* **271**, 10433–10436.

Wang W, Duan B, Xu H *et al.* (2006). Calcium-permeable acid-sensing ion channel is a molecular target of the neurotoxic metal ion lead. *J Biol Chem* **281**, 2497–2505.

Welsh MJ, Price MP and Xie J (2002). Biochemical basis of touch perception: mechanosensory function of degenerin/epithelial Na$^+$ channels. *J Biol Chem* **277**, 2369–2372.

Wemmie JA, Askwith CC, Lamani E *et al.* (2003). Acid-sensing ion channel 1 is localized in brain regions with high synaptic density and contributes to fear conditioning. *J Neurosci* **23**, 5496–5502.

Wemmie JA, Chen J, Askwith CC *et al.* (2002). The acid-activated ion channel ASIC contributes to synaptic plasticity, learning and memory. *Neuron* **34**, 463–477.

Wemmie JA, Coryell MW, Askwith CC *et al.* (2004). Overexpression of acid-sensing ion channel 1a in transgenic mice increases acquired fear-related behavior. *Proc Natl Acad Sci U S A* **101**, 3621–3626.

Wemmie JA, Price MP and Welsh MJ (2006). Acid-sensing ion channels: advances, questions and therapeutic opportunities. *Trends Neurosci* **29**, 578–586.

Weng XC, Zheng JQ, Gai XD *et al.* (2004). Two types of acid-sensing ion channel currents in rat hippocampal neurons. *Neurosci Res* **50**, 493–499.

Wetzel C, Hu J, Riethmacher D *et al.* (2007). A stomatin-domain protein essential for touch sensation in the mouse. *Nature* **445**, 206–209.

Wu LJ, Duan B, Mei YD *et al.* (2004). Characterization of acid-sensing ion channels in dorsal horn neurons of rat spinal cord. *J Biol Chem* **279**, 43716–43724.

Wultsch T, Painsipp E, Shahbazian A *et al.* (2008). Deletion of the acid-sensing ion channel ASIC3 prevents gastritis-induced acid hyperresponsiveness of the stomach–brainstem axis. *Pain* **134**, 245–253.

Xie J, Price MP, Berger AL *et al.* (2002). DRASIC contributes to pH-gated currents in large dorsal root ganglion sensory neurons by forming heteromultimeric channels. *J Neurophysiol* **87**, 2835–2843.

Xie J, Price MP, Wemmie JA *et al.* (2003). ASIC3 and ASIC1 mediate FMRFamide-related peptide enhancement of H$^+$-gated currents in cultured dorsal root ganglion neurons. *J Neurophysiol* **89**, 2459–2465.

Xiong ZG, Zhu XM, Chu XP *et al.* (2004). Neuroprotection in ischemia: blocking calcium-permeable acid-sensing ion channels. *Cell* **118**, 687–698.

Yagi J, Wenk HN, Naves LA *et al.* (2006). Sustained currents through ASIC3 ion channels at the modest pH changes that occur during myocardial ischemia. *Circ Res* **99**, 501–509.

Yang HY and Iadarola MJ (2003). Activation of spinal neuropeptide FF and the neuropeptide FF receptor 2 during inflammatory hyperalgesia in rats. *Neuroscience* **118**, 179–187.

Yermolaieva O, Leonard AS, Schnizler MK *et al.* (2004). Extracellular acidosis increases neuronal cell calcium by activating acid-sensing ion channel 1a. *Proc Natl Acad Sci USA* **101**, 6752–6757.

Yiangou Y, Facer P, Smith JA *et al.* (2001). Increased acid-sensing ion channel ASIC-3 in inflamed human intestine. *Eur J Gastroenterol Hepatol* **13**, 891–896.

Yudin YK, Tamarova ZA, Ostrovskaya OI *et al.* (2004). RFa-related peptides are algogenic: evidence *in vitro* and *in vivo*. *Eur J Neurosci* **20**, 1419–1423.

Zha XM, Wemmie JA, Green SH *et al.* (2006). Acid-sensing ion channel 1a is a postsynaptic proton receptor that affects the density of dendritic spines. *Proc Natl Acad Sci USA* **103**, 16556–16561.

Zhang P and Canessa CM (2001). Single-channel properties of recombinant acid-sensitive ion channels formed by the subunits ASIC2 and ASIC3 from dorsal root ganglion neurons expressed in *Xenopus* oocytes. *J Gen Physiol* **117**, 563–572.

Zhang P and Canessa CM (2002). Single channel properties of rat acid-sensitive ion channel-1alpha, -2a and -3 expressed in *Xenopus* oocytes. *J Gen Physiol* **120**, 553–566.

Zhang P, Sigworth FJ and Canessa CM (2006). Gating of acid-sensitive ion channel-1: release of Ca^{2+} block vs. allosteric mechanism. *J Gen Physiol* **127**, 109–117.

Zhu J, Jhamandas K and Yang HY (1992). Release of neuropeptide FF (FLFQPQRF-NH2) from rat spinal cord. *Brain Res* **592**, 326–332.

Table 3.4.2.1 Functional properties of epithelial sodium channels

	g (pS)	PNa+/PK+	IC$_{50}$ amiloride (µM)	Reference
Highly selective Na⁺ channels				
Mammalian CCD	4.9	>10	0.07	Palmer and Frindt (1986)
Alveolar type II cells	4	>10	0.09	Voilley *et al.* (1994)
Toad urinary bladder	4.5	>1000	0.36	Frings *et al.* (1988)
A6 cell line	4.8	>20	-	Hamilton and Eaton (1985)
ENaC channel (expressed in *Xenopus* oocyte)				
αβγ	5	>50	<0.1	Canessa *et al.* (1994b)
Nonselective Na⁺ channels				
Microvessels	23	1.2	10	Vigne *et al.* (1989)
Thyroid cells	2.6	1.0	0.15	Verrier *et al.* (1989)
IMCD cells (primary culture)	28	1.0	-	Light *et al.* (1988)

Classification of different epithelial sodium channels in native tissues, cultured cells, and heterologous expression system, based on functional characteristics such as the single channel conductance with Na⁺ ion as charge carrier (g (pS)), the Na⁺ over K⁺ ionic selectivity (PNa⁺/PK⁺), and to inhibition by amiloride (IC$_{50}$).

3.4.2 **Epithelial sodium channels**

Stephan Kellenberger and Laurent Schild

3.4.2.1 Introduction

In salt-reabsorbing epithelia such as the distal nephron, the colon, salivary and sweat glands or the respiratory epithelium, the epithelial sodium channels are components of the apical membrane allowing entry of Na⁺ ions into the cell by electrodiffusion without a direct coupling with the flow of other solutes and without the direct need of metabolic energy. To achieve a vectorial transepithelial Na⁺ transport, intracellular Na⁺ ions are then extruded across the basolateral membrane by the Na-K/ATPase against an electrochemical gradient (Figure 3.4.2.3). Different sodium channels have been described and characterized in these epithelia, based on their single channel properties and their sensitivity to the channel blocker amiloride (for review see Garty and Palmer, 1997). These epithelial Na channels can be divided in two main categories: first, the highly selective, small conductance sodium channels with a high affinity for the channel blocker amiloride, and second, the moderately to non-selective channels with usually a large single channel conductance and/or lower affinity for amiloride (Table 3.4.2.1). So far, only the amiloride-sensitive, highly selective epithelial Na⁺

channel (ENaC) has been characterized at the molecular level. Furthermore, the physiological importance of ENaC in whole body Na⁺ homeostasis and in epithelial Na⁺ transport in many organs is beginning to be unravelled using genetic approaches (Lifton, 1996). This chapter deals with the structure, function, physiology and pathophysiology of ENaC.

In kidney, colon and sweat glands, ENaC-mediated transepithelial Na⁺ transport helps to adjust Na⁺ excretion in order to maintain Na⁺ homeostasis. In lung and airways, ENaC activity is important for alveolar liquid clearance and regulation of mucous fluidity. In the taste buds of the tongue, ENaC is likely involved in salt tasting, whereas ENaC expression in the eye and inner ear may help to control the ionic composition of the aqueous humour and the endolymph, respectively (for review see Kellenberger and Schild, 2002).

ENaC belongs to the recently discovered ENaC/degenerin gene family that encodes Na⁺ channels involved in various cell functions in metazoans (Kellenberger and Schild, 2002) (Figure 3.4.2.1). The subfamilies include the degenerins (UNC-105, UNC-8, DEL-1, DEG-1, MEC-10, MEC-4), which are part of a mechanosensory complex in the nematode *C. elegans*, the FMRFamide peptide-gated Na⁺ channel FaNaC present in snails, the *Drosophila* channels RPK/dGNaC1 and PPK/dmdNaC1, and the mammalian subfamilies ENaC and acid-sensing ion channels (ASICs).

3.4.2.2 Molecular characterization

Primary structure, membrane topology and genomic organization

ENaC was cloned by functional expression. Like other ion channels such as K⁺ channels or ligand-gated channels, the functional ENaC

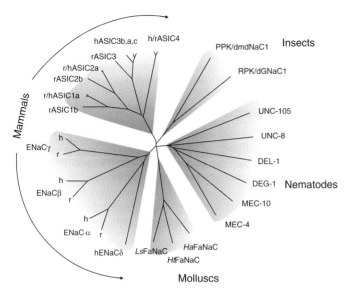

Fig. 3.4.2.1 Phylogenetic tree of the ENaC/DEG family, showing its organization into subfamilies of related sequences. ENaC/DEG proteins of invertebrates can be divided into three groups: (1) the degenerins from *C. elegans*, (2) the *drosophila* channels RPK/dGNaC1 and PPK/dmdNaC1, and (3) FaNaC, the FMRFamide-gated ion channel expressed in molluscs. The channels from vertebrates are divided into two groups, ENaC and ASICs (acid-sensing ion channels). r, rat; h, human; PPK, pickpocket; RPK, ripped pocket; HaFaNaC, Helix aspersa FaNaC; HtFaNaC, Helisoma trivolvis FaNaC; LsFaNaC, Lymnaea stagnalis FaNaC.

channels are formed by heteromeric assembly of homologous channel subunits (Canessa *et al.*, 1993, 1994b). The ENaC subunits αβγ have been identified in amphibian and mammalian Na⁺-absorbing epithelia such as the distal colon, the distal nephron, and the respiratory epithelium, while α δ ENaC subunit is found in humans only, with a maximal expression at the mRNA level in testis, ovary, pancreas, and brain (Waldmann *et al.*, 1995). In *Xenopus laevis*, an ε subunit has been reported, whose sequence is equally distant from α and δ subunits (Babini *et al.*, 2003). Each ENaC subunits contain 640–670 amino acids, and show 27–37% sequence identity among each other, with the highest homology between α and δ, and between β and γ (both 37%). The homology between human and rat ENaC orthologues is ~85%. The homology between ENaC subunits and subunits of other ENaC/degenerin subfamilies is 15–20%.

All ENaC subunits share the same membrane topology with intracellular N- and C-termini, two transmembrane segments and a large extracellular loop, that makes up ~70 % of the total protein mass (Figure 3.4.2.2a). (Canessa *et al.*, 1994a; Renard *et al.*, 1994). This topology has been confirmed by the crystal structure of ASIC.

The human α *ENaC* gene covers 17 kb on chromosome 12p13. The human β and δ ENaC genes are located on chromosome 16p in very close proximity, suggesting that the two genes arose from gene duplication. The three *ENaC* genes α β and γ share a high degree of conservation in their genomic organization (Chow *et al.*, 1999; Saxena *et al.*, 1998; Thomas *et al.*, 1996) with the presence of 13 exons.

Multimeric channels and subunit stoichiometry

The initial cloning experiments demonstrated that the three *ENaC* subunits αβγ are required for maximal expression of ENaC activity in a heterologous expression system such as the *Xenopus* oocytes

(Canessa *et al.*, 1994b). Expression of the a ENaC subunit alone or the αβ or αγ combinations of ENaC subunits generate 10- to 100-fold smaller magnitude of amiloride-sensitive Na⁺ current compared to the co-expression of the three αβγ subunits (Canessa *et al.*, 1994b). The expression of homomeric α, β or γ ENaC subunits alone is very inefficient in generating functional ENaC channels, indicating that in native tissue ENaC is a heteromultimeric channel. This is further supported by the fact that functional ENaC resulting from the co-expression of αβγ ENaC has the same functional properties as the highly selective epithelial Na channels found in the distal nephron (Palmer and Frindt, 1986), indicating that the highly selective epithelial Na⁺ channel of the distal nephron is formed by ENaC α, β and γ subunits.

Different approaches have been used to determine the subunit stoichiometry of functional ENaC present at the cell surface. Such experiments were consistent with a fixed subunit stoichiometry that requires the three ENaC αβγ for channel assembly (Firsov *et al.*, 1998). Several studies concluded that ENaC is a tetramer, formed by two α and one β and one γ subunit (Firsov *et al.*, 1998; Kosari *et al.*, 1998; Dijkink *et al.*, 2002; Anantharam and Palmer, 2007). Other studies has reported a 8–9 subunit stoichiometry for the functional ENaC channel (Eskandari *et al.*, 1999; Snyder *et al.*, 1998), making the ENaC stoichiometry still a matter of debate. The recently published crystal structure of the related ASIC1 channel (Jasti *et al.*, 2007) showed that the functional ASIC channel is composed of three subunits. It is therefore highly likely that the functional ENaC also is a trimer. However, the functional data from ENaC expression in *Xenopus* oocytes are consistent with the presence of two α subunits in the functional ENaC channels. This observation and the finding that the three αβγ ENaC subunits are required for maximal channel activity are difficult to reconcile with a heterotrimeric structure.

3.4.2.3 Biophysical characteristics

The highly Na⁺ selective amiloride-sensitive ENaC was first characterized functionally in tight epithelia, such as frog skin and toad urinary bladder, using noise analysis (Lindemann and Van Driessche, 1977; Frings *et al.*, 1988; Palmer *et al.*, 1980). Later, ENaC activity at the single-channel level was resolved in the rat renal cortical collecting duct (RCCD) and in an amphibian kidney cell line (A6 cells), using the patch-clamp technique (Palmer and Frindt, 1986; Hamilton and Eaton, 1985). The biophysical and pharmacological characteristics of the highly selective sodium channel in native tissues closely match those of the cloned ENaC made of the three αβγ subunits when expressed in *Xenopus* oocytes (Canessa *et al.*, 1994b). The ENaC current was found to be highly selective for the small inorganic ions Na⁺ and Li⁺, and the Na⁺/K⁺ selectivity ratio was estimated to be > 100 in toad bladder, mammalian CCD and in oocytes expressing α, β and γ ENaC. In contrast to some members of the ENaC/DEG channel family, ENaC is not permeable to divalent cations.

In the CCD and in *Xenopus* oocytes expressing α, β and γ ENaC, the unitary conductance of the epithelial Na⁺ channel measured in the presence of 140 mM Na⁺ is 5 pS, and 9–10 pS in the presence of 100–150 mM Li⁺. Single-channel recordings in the CCD and in *Xenopus* oocytes expressing α, β and γ ENaC show slow kinetics with open and closed dwell times in the range of seconds at room temperature. Systematic analysis of ENaC gating kinetics showed that the open probability of the channel (P_o) varies from <0.1 to >0.9, and that active ENaC channels can be divided into two groups, one with low and one with high P_o (Palmer and Frindt, 1996). This variation in gating may reflect different gating modes. ENaC gating involves specific structural domains in extracellular and intracellular parts of the channel protein and is modulated by intra- and extracellular factors. Interestingly, ENaC formed by α and β subunits alone is almost always open with P_o close to 1, but the functional relevance of heterodimeric αβ ENaC channels remains uncertain (Fyfe and Canessa, 1998).

Modulators

As mentioned before, ENaC channels at the cell surface are constitutively active and the half life of the open channel at the plasma membrane has been estimated to be between 15 minutes and 1 hour (May et al., 1997; Hanwell et al., 2002; de la Rosa et al., 2002). Channel opening is triggered neither by changes in the membrane voltage nor by external ligands, however, different factors that modulate ENaC activity at the cell surface have been reported.

External and internal Na+ concentration

Inhibition of ENaC by high extracellular Na+ concentrations has been termed Na+ self-inhibition. This phenomenon was initially demonstrated in amphibian epithelia as a fast (i.e. order of seconds) decrease in the transepithelial short circuit current upon a rapid increase in extracellular Na+ concentration (reviewed in Garty and Palmer, 1997) and could also be shown in Xenopus oocytes expressing ENaC channels (Fuchs et al., 1977; Chraibi and Horisberger, 2002). The time course for ENaC self-inhibition is too slow to represent Na+ ions entering the channel pore and too fast to be related to changes in the intracellular Na+ concentration. This phenomenon is an intrinsic property of the channel and strongly suggests the presence of an allosteric binding site for Na+ ions in the extracellular loop that downregulates the channel. The apparent Na+ affinity of this site was estimated to be ≥100 mM (Chraibi and Horisberger, 2002). A number of agents that increase the ENaC-mediated transport rate of Na+ seem to act, at least in part, by relieving the channel self-inhibition: they include sulfhydryl reagents, detergents, trypsin and other proteases such as furin (Chraibi and Horisberger, 2002; Sheng et al., 2004; Sheng et al., 2006).

The term 'feedback regulation' defines a process that controls the rate of Na+ entry into the epithelial cells by intracellular factors to maintain an intracellular Na+ homeostasis. This phenomenon was first proposed by MacRobbie and Ussing (1961), who noted that cells of Na+ -absorbing epithelium, such as the frog skin expressing an apical amiloride-sensitive Na+ channel, did not swell when Na+ extrusion by the Na+/K+-ATPase at the basolateral cell was blocked. They proposed that the rise in [Na+]$_i$ upon inhibition of the Na/K-ATPase downregulates Na+ permeability of the apical membrane to prevent the cell overload with Na+ ions. The presence of this feedback regulation that inhibits ENaC when Na+ enters the cell has since been confirmed in CCT cells and in Xenopus oocytes expressing ENaC (Silver et al., 1993; Anantharam et al., 2006). Feedback inhibition is thought to serve to maintain intracellular ion homeostasis and to allow homogenous reabsorption of Na+ along the distal nephron. More recent studies indicate that feedback inhibition relies on reducing both channel open probability and channel density at the cell surface to limit Na+ entry when [Na+]$_i$ is high (Kellenberger et al., 1998). These studies also provide evidence that feedback inhibition is reduced in ENaC containing mutations associated with a severe form of hereditary hypertension, known as Liddle's syndrome. It is not clear whether this feedback regulation involves a direct or indirect effect of intracellular Na+ ions on ENaC (Abriel and Horisberger, 1999).

Membrane voltage

ENaC activity exhibits a slight voltage-dependence. In contrast to the effects of membrane potential on voltage-gated Na+ channels in excitable cells, hyperpolarization of the cell membrane slightly activates epithelial Na+ channels, while depolarization deactivates them. The response to voltage is weak compared with classical voltage-gated channels, with P_o increasing by ~2 % per mV (Palmer and Frindt, 1996). This depolarization-induced deactivation of ENaC may represent a feedback regulation of the channel to prevent excessive entry of Na+ ions into the cell.

pH

As the maintenance of cell pH depends, in part, on the gradient of Na+ from outside to inside the cell, an elevation of the intracellular Na+ concentration could lead to acidification of the cytoplasm. It was shown in ENaC-expressing epithelia, such as frog skin, A6 cells, rat CCD or in Xenopus oocytes, that ENaC activity is inhibited by intracellular acidification (Palmer and Frindt, 1987; Abriel and Horisberger, 1999). In tight epithelia, intracellular pH regulates Na+ permeability within a narrow pH range, with a maximal permeability for Na+ ions at pH > 7.5 and almost complete inhibition at pH ≤ 6.4. Single-channel analysis indicates that lowering of the intracellular pH reduces channel open probability.

Extracellular pH does not affect αβγ ENaC activity, however, two recent studies suggested that extracellular acidification increases the δβγ ENaC current in Xenopus oocytes by several fold (Ji et al., 2004). The physiological relevance of these regulatory processes of ENaC activity for transepithelial Na+ transport remains to be precisely established.

Proteases

Extracellular serine proteases, such as trypsin and elastase, as well as the membrane-bound 'channel-activating proteases' 1–3 (CAP1–3), have been shown to increase ENaC currents by several-fold in heterologous expression systems (Vallet et al., 1997; Chraibi et al., 1998; Caldwell et al., 2005; Caldwell et al., 2004). Aprotinin, an inhibitor of serine proteases, can block this effect. The activating effect on ENaC function is not dependent on a change in [Ca]$_i$ and is not modified by GTPγS, suggesting that this effect is not mediated by activation of a G-protein-coupled receptor. The effect of proteases does not influence the ENaC protein density at the oocyte surface nor its single-channel conductance, indicating that proteases act by acutely increasing the channel open probability. In the A6 kidney cell line from Xenopus laevis, Na+ transport could not be further activated by trypsin, suggesting that ENaC is likely constitutively activated through the presence of endogenously expressed proteases. Consistent with such a mechanism, the basal ENaC activity in A6 cells and in several other epithelial cell lines, such as mouse mpk-CCDC14 cells and human bronchial epithelial cells, is inhibited by 50–90% after addition of the protease inhibitor aprotinin or its analogue BAY39–9437 (Vallet et al., 1997; Vuagniaux et al., 2000, 2002; Adebamiro et al., 2005).

Site-directed mutagenesis of the catalytic triad of CAP proteins was performed to investigate whether the catalytic activity is required for the ENaC current increase. These experiments showed that the catalytic activity was required for CAP2 and CAP3, but that CAP1 with deleted catalytic triad could still activate ENaC (Vallet et al., 2002; Andreasen et al., 2006). Recent evidence supports a furin-dependent processing and cleavage of α and γ ENaC subunits during channel maturation in CHO cells. Furin is a protease that resides mainly in the trans-golgi network. Site-directed mutations of consensus sites for furin cleavage on ENaC subunits and the use of furin-specific inhibitors indicate that the furin-dependent proteolysis of ENaC subunits is required for channel activation (Hughey et al., 2004). Proteolytic cleavage of ENaC subunits during channel maturation and processing is likely to be important in the physiological response of increased ENaC activity in the distal nephron, under conditions of salt deprivation and/or increased plasma aldosterone secretion (Ergonul et al., 2006).

Functional domains of ENaC

Regions of conserved amino acid sequences in ENaC/DEG family members are likely to represent structural elements important for channel function. A few sequences are completely conserved among the ENaC/DEG family members; including a HG motif (*His-Gly*) located in the N-terminal cytoplasmic domain in close proximity to the M1 segment, a FPxxTxC signature sequence following M1 (post-M1) and completely conserved residues in the transmembrane M2 segment (Figure 3.4.2.2). The large extracellular loop contains conserved cysteine-rich domains in which *Cys* residues are involved in disulfide bridges (Firsov, 1999) likely to maintain the tertiary structure of the large extracellular loop.

ENaC subunits contain conserved proline-rich motifs in the cytoplasmic C-terminus that are unique to the ENaC subfamily. These motifs have the consensus sequence PPPXYXXL (PY motif) and are involved in the binding interaction of ENaC subunits and the cytosolic ubiquitin ligase Nedd4 (see below).

The ion permeation pathway

In the crystal structure of ASIC1, the large extracellular domain exhibits a dense packing of the three subunits without evidence for a central ion pathway. However, lateral fenestrations located extracellularly above the plane of the membrane might provide an access for ions to the transmembrane pore of the channel.

Functional analyses of the ionic selectivity properties of ENaC indicate that the channel discriminates among monovalent cations based on their size, with the divalent cations being excluded from the permeation pathway. Therefore, the selectivity filter on ENaC likely constitutes the narrowest part of the pore, allowing only small monovalent cations such as Na^+ and Li^+ ions to pass through the channel. Mutagenesis studies indicate that the ENaC selectivity filter involves a stretch of three amino acids, G/S x S (x being any amino acid residue) that is located according to the ASIC1 structure deep in the second transmembrane M2 segment (Kellenberger et al., 1999, 2001). Of the residues constituting the selectivity filter, the effects of amino acid substitutions are most dramatic for the third selectivity filter position of the α subunit, the conserved Ser589. Mutations of this residue allow larger ions such as K^+, Rb^+, Cs^+, NH_4^+ and divalent cations to pass through the channel, consistent with an enlargement of the pore at the selectivity filter.

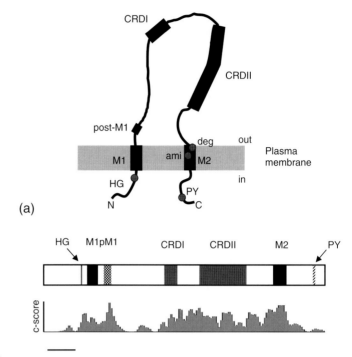

(a)

(b)   100 amino acid residues

Fig. 3.4.2.2 Primary structure of the homologous ENaC subunits. (a) Membrane topology of an ENaC subunit with 2 transmembrane segments (M1 and M2), the intracellular HisGly (HG) and ProLys (PY) motifs; the extracellular loop contains the Cys-rich domains CRDI and II, the degenerin site (deg), the amiloride binding site (ami). (b) Top, linear representation of ENaC subunits with M1 and M2 (black), showing functional and structural domains. The lower part of panel (b) illustrates the conservation of the amino acid sequence across the ENaC/DEG family. The height of the grey column at a given position is proportional to the degree of conservation (obtained from column score of an alignment of ~50 ENaC/DEG family members in Clustal X, using a sliding window of 10 aminos acids).

From our knowledge of the function and three-dimensional structure of the potassium channel KcsA, a very close contact between the permeant ions and the residues lining the channel pore in the selectivity filter is essential to achieve a high ion selectivity. This close contact critically depends on the distances between the permeating ion and the carbonyl oxygen atoms of the amino acid main chains lining the channel pore. By analogy, alterations in the ENaC pore geometry by mutations in the selectivity filter are expected to perturb the tight interactions between the permeant ions Na^+ or Li^+ and the pore leading to changes in ion conduction and channel conductance. Indeed, mutations in α β or γ subunits at the first and third position of the selectivity filter reduced Li^+ and/or Na^+ single-channel conductance and the αS588I mutation (second position) increased the Na^+ single-channel conductance. These alterations in single-channel conductance are associated with changes in channel affinity for the permeant ion (Kellenberger et al., 2001). For instance, a decrease in the Li^+ single-channel conductance in ENaC containing mutations at the first selectivity filter position in the β or γ subunit (βG529, γS541) are associated with an increase in the channel affinity for Li^+, resulting in a lower rate of dissociation from its binding site. The effects of the mutations in the conserved G/SxS sequence on ion selectivity and unitary conductance indicate that this region of the channel pore is essential for the ion recognition and for its permeation through the channel.

In the ASIC1 structure, the transmembrane domain has an hourglass-like shape and the interior of the pore is defined primarily by residues from M2, although M1 also lines portions of the pore. The ASIC1 structure was obtained from desensitized channels that were thus not in the active, conducting conformation. Therefore it is currently not clear whether the orientation of the transmembrane helices corresponds to that of the actively conducting ASIC. The selectivity filter sequence GxS lies deep in the transmembrane segment, almost at the intracellular end of M2. Functional work with ASICs has suggested a possible contribution of the M1 segment and cytoplasmic N-terminal parts to the internal part of the channel pore. The analysis of an ASIC2b/2a chimera identified residues in the cytoplasmic N-terminus (I19, F20, T25), whose mutation changed the Na^+/K^+ permeability ratio up to sevenfold (Coscoy et al., 1999). The same region was shown to play a role in the selectivity of ASIC1 towards divalent cations (Bassler et al., 2001). As the crystal structure of ASIC1 that was determined from a truncated channel starts with S40, it does not provide any structural information about this N-terminal domain involved in ion selectivity. A screening of N-terminal cytoplasmic residues in the proximity of the M1 of ASIC 1a and of some residues of M1 showed inhibition by sulfhydryl reagents of engineered cys residues at four sites, A22C, I33C, F34C, and R43C. The R43C residue, which is located just at the cytoplasmic end of the M1 segment, was accessible to sulfhydryl reagents from the cytoplasmic as well as from the extracellular side, indicating that it contributes to the channel pore (Pfister et al., 2006). In summary, structural information from ASIC1 together with experimental evidence from ENaC and other ENaC/degenerin family members indicates that in addition to the GxS sequence in M2, part of the intracellular N-terminus and of the M1 segment also participate in the formation of the ENaC pore.

Gating domains

As discussed above, ENaC is a constitutively active channel, but many factors influence channel gating, which varies between different gating modes characterized by high or low P_o. Two functional domains that control ENaC activity have been identified on the channel protein, the degenerin site in the extracellular pore entry and the HG-gating domain in the cytoplasmic N-terminus.

The mutation of a conserved Ala residue in the homolog of ENaC in *C. elegans* (degenerin channels) causes degeneration of touch receptor cells with morphological features that are consistent with an abnormal cation leak into the cell (Driscoll and Chalfie, 1991). The corresponding residues in ENaC subunits are small residues (Ala, Ser or Gly) located seven amino acid residues upstream of the amiloride binding site (i.e. αS576, βS518, γS530). It has been shown for most of the ENaC/degenerin family members that this conserved residue plays an important role in channel gating. Mutation of this residue to larger residues, or mutation to cys and subsequent modification of the engineered cys residue by sulfhydryl reagents increase the ENaC current by several-fold and induces a sustained current in ASICs (Snyder et al., 2000; Kellenberger et al., 2002; Champigny et al., 1998). At the single channel level, these modifications result in long channel openings, consistent with alterations in conformational changes during channel gating.

A loss-of-function mutation in the highly conserved HG motif in the N-terminus of β ENaC was found to be associated with the syndrome pseudohypoaldosteronism type 1 (PHA-I) that results in abnormal renal sodium losses and the inability for the kidney to retain sodium (Chang et al., 1996). When expressed in *Xenopus* oocytes, the βG37S mutation causing PHA-I, as well as mutations of the corresponding gly residues in α or β ENaC subunits, result in an important reduction in ENaC P_o (Grunder et al., 1997; Grunder et al., 1999). ENaC surface expression is not affected by these mutations, suggesting a role in channel gating. Mutations of the conserved neighbouring residues of the homologous Gly residue in α ENaC (G95), H94 and R98, resulted in a similar inhibition. The His-Gly motif is located in an intracellular linker between a putative hydrophobic a helix of the N-terminus and the M1 segment. Furthermore, this region of the N-terminus in the αβγ ENaC subunits is rich in cys residues. These residues are responsible for the high ENaC sensitivity to inhibition by a variety of intracellular sulfhydryl reagents, including methanethiosulfonates, metal divalent cations, and oxidizing agents (Kellenberger et al., 2005). These reagents inhibit ENaC activity from the cytosolic side by inducing long and slowly reversible channel closures.

3.4.2.4 Cellular and subcellular localization

In the kidney, ENaC is the major pathway for sodium absorption in the distal part of the nephron where Na^+ transport is under the control of the mineralocorticoid hormone aldosterone (the aldosterone-sensitive distal nephron or ASDN). ENaC is present in the apical membrane (urinary pole) of principal cells that mediates sodium absorption in the ASDN, whereas the intercalated cells secreting protons or bicarbonate in the ASDN do not express ENaC at the apical membrane. Other channel proteins such as the potassium channel ROMK or aquaporins (AQP2) are present at the apical membrane of principal cells, mediating K^+ secretion or water reabsorption.

Morphological and functional studies on rodent and human kidneys (Reilly and Ellison, 2000; Bachmann et al., 1999) indicate that at least three successive tubule portions, i.e. the late portion of the distal convoluted tubule (DCT), the connecting tubule (CNT), and the collecting duct (CD), contribute to the ASDN. Although these segments have distinct structural and functional features (Loffing and Kaissling, 2003), they have in common the expression of ENaC, the mineralocorticoid receptor (MR) and the 11-β hydroxysteroid dehydrogenase, type II (11-βHSD2) proteins (Bachmann et al., 1999). The former binds aldosterone, but also glucocorticoids (cortisol) and the latter confers mineralocorticoid-selectivity to the MR by rapid metabolism of circulating glucocorticoids into metabolites unable to bind and activate the MR.

The subcellular localization of ENaC subunits along the axis of the ASDN changes drastically with the elevation of plasma aldosterone levels in response to changes in sodium diet (Masilamani et al., 1999; Loffing et al., 2000). In rodents kept under a high-sodium diet that lowers plasma aldosterone levels, ENaC subunits are barely detectable at the luminal membrane; the α ENaC subunit shows very low levels of intracellular expression, whereas the β and γ ENaC subunits are detected exclusively in an intracellular compartment that remains to be identified. On a standard sodium diet, ENaC subunits are traceable at the luminal membrane of late DCT and early CNT. However, in further downstream segments (i.e. late CNT and CD), β and γ ENaC subunits in particular remain almost exclusively localized at intracellular sites. Under a low dietary sodium intake with high plasma aldosterone levels, ENaC subunits become detectable in the luminal membrane along the late DCT, CNT and CD (Loffing et al., 2000). Nevertheless, the axial gradient for apical ENaC still prevails and the apical localization of ENaC subunits is more prominent in early ASDN than in late ASDN. This immunohistochemically traceable axial gradient of apical ENaC localization is corroborated by measuring the number of active channels in ASDN segments isolated from rats, using patch-clamp recording, and of amiloride-sensitive currents measured in microperfused rat and rabbit tubules (Costanzo, 1984; Tomita *et al.*, 1985; Frindt and Palmer, 2003; Almeida and Burg, 1982). Together, these studies show that ENaC-mediated Na$^+$ absorption is highest in the CNT and decreases from the initial to the late part of the collecting duct (Frindt and Palmer, 2003). Thus, the aldosterone-dependent adaptation of renal sodium excretion to dietary sodium intake occurs predominantly in the early ASDN, while ENaC channels in the late ASDN get recruited only under extreme salt deprivation and high plasma aldosterone levels.

Like the ASDN, the distal colon is an aldosterone-responsive tight epithelium mediating Na$^+$ absorption, and ENaC is also expressed at the luminal membrane of the absorbing epithelial cells, with a lower expression level in the crypt colonocytes or in goblet cells (Coric *et al.*, 2004). Low salt diet associated with high plasma levels of aldosterone results in the appearance of the αβγ ENaC subunits at the apical membrane and in the subapical space. The respiratory epithelium also transports sodium ions but in an aldosterone-independent manner. Messenger RNAs encoding the three αβγ ENaC subunits are found in the respiratory epithelium and in lung, with an increasing expression from the trachea to the terminal bronchioles, suggesting a higher Na$^+$ transport capacity in the distal airways (Burch *et al.*, 1995; Talbot *et al.*, 1999; Farman *et al.*, 1997). ENaC subunits are found at the mRNA level in salivary glands, in taste cells of the fungiform papillae and in hair follicles.

In the principal cells of the ASDN, the trafficking of the αβγ ENaC subunits from intracellular sites (endothelial reticulum) to the cell surface is relatively inefficient under conditions of low plasma levels of aldosterone associated with a high salt diet, and/or an extracellular volume expansion. Insertion of ENaC at the plasma membrane is highly increased within 3 hours of aldosterone secretion (Loffing *et al.*, 2000, 2001). The cellular mechanisms involved in ENaC trafficking between its intracellular pool and the membrane pool are only partially understood. Recent evidence indicates that ENaC maturation and processing in response to an increase in plasma aldosterone involve a proteolytic cleavage and a processing of the glycosylation of ENaC subunits, that occur rapidly enough to account for the daily regulation of Na$^+$ absorption in the ASDN (Ergonul *et al.*, 2006). Insertion of active ENaC at the apical membrane may involve interactions with proteins such as syntaxins and proteins of the SNARE complex. In addition, cAMP controls this exocytotic process (Condliffe *et al.*, 2003; Butterworth *et al.*, 2005). ENaC stability at the apical cell surface is controlled by the ubiquitin-protein ligase Nedd4–2 that binds to conserved proline-rich motifs (PPPxYxxL sequence) in the C-terminus of β and γ ENaC (Kamynina *et al.*, 2001; Snyder *et al.*, 2004). Binding of Nedd4–2 reduces the expression and the activity of ENaC at the cell surface by promoting ubiquitylation of the channel, endocytosis, and subsequent degradation by the proteasome and the lysosome. The tyrosine residue of the PY motif sequence (xPPxY) is in addition part of a consensus internalization signal (NPxY) found in receptors that are removed from the plasma membrane via clathrin-coated pits (Shimkets *et al.*, 1997). Evidence supporting the role of the clathrin-coated pit pathway for ENaC endocytosis

comes from the ENaC retention at the cell surface obtained with the co-expression of a dominant-negative dynamin mutant.

3.4.2.5 Pharmacology

In the 1960s, a period of intense research directed towards the synthesis of new diuretics, Merck Sharp and Dohme sought a diuretic that elicits an enhanced urinary Na$^+$ excretion and a diminished K$^+$ output to prevent the hypokalemia usually associated with thiazide diuretics (hydrochlorothiazide) or loop diuretics, such as furosemide. Out of 25 000 compounds screened in a rat assay, the pyrazine compound amiloride was shown to be efficient in promoting natriuresis with minimal K$^+$ loss by the kidney.

The pyrazinoylguanidine derivative amiloride, and triamterene, a pteridine compound, both decrease Na$^+$ absorption and K$^+$ secretion in the ASDN. Amiloride and its derivative benzamil are high-affinity blockers of ENaC, inhibiting channel activity at submicromolar concentrations, with inhibitory half-maximal concentrations (IC$_{50}$) of 0.1 µM and 0.01 µM, respectively. Triamterene has a lower affinity for ENaC inhibition with an IC$_{50}$ of 4.5 mM (Kellenberger *et al.*, 2003). Amiloride and triamterene are clinically used as K$^+$-sparing diuretics. Recent development of amiloride derivatives for the potential use in cystic fibrosis yielded a substance with an up to 100-fold higher potency at ENaC (Hirsh *et al.*, 2008).

Early work on the effect of amiloride on electrogenic Na$^+$ absorption in epithelia showed that amiloride interacts competitively with the permeant sodium or Li$^+$ ions, suggesting that amiloride acts as an ENaC pore blocker to prevent the flow of permeant Na$^+$ ions through the channel pore (Palmer, 1984).

The identification of the primary structure of ENaC, together with site-directed mutagenesis, allowed the identification of the amiloride binding site on the channel protein. Mutation of amino acid residues at corresponding positions at the centre of M2 of three αβγ ENaC subunits decreases the channel affinity for amiloride up to 1000-fold (Schild *et al.*, 1997). These mutations affect similarly ENaC affinity for benzamil and triamterene, consistent with a unique binding site for these blockers and a similar mechanism of action of these blockers (Kellenberger *et al.*, 2003). The decrease in apparent affinity of amiloride and benzamil for these ENaC mutants is essentially due to an unstable binding interaction between the ligand and its receptor, as shown by a faster dissociation rate of benzamil and amiloride.

The binding site for amiloride and its analogues is located quite deep in the transmembrane segment, in close vicinity to the ENaC selectivity filter where the permeant ions intimately interact with the residues lining the channel pore (see above). This location of the amiloride binding site in the external channel pore close to the selectivity filter can account for the competitive interaction of the pore blocker with the permeant ions.

Note that amiloride binds to and inhibits other Na$^+$ channels or transporters such as the Na$^+$-H$^+$ antiporter, however, with a lower affinity than for ENaC.

3.4.2.6 Physiology

In the kidney, ENaC mediates Na$^+$ absorption in the distal nephron where Na$^+$ transport is tightly regulated by hormones such as aldosterone and vasopressin. This fine tuning of Na$^+$ absorption is essential for maintaining a balance between the daily sodium intake and renal sodium excretion. Extracellular Na$^+$ is the major osmolite of the extracellular fluid (~2400 mEq) and therefore determines the magnitude of extracellular volume. The kidney filters approximately 25 000 mEq/day of Na^+ from the plasma and approximately 90–95 % of the filtered Na^+ is reabsorbed before reaching the ASDN. The challenge of the ASDN is to adapt the amount of Na^+ excreted in the urine with the daily salt intake, which is on average between 120 and 150 mEq /day in Western countries. When Na^+ intake exceeds renal Na^+ excretion (positive balance of Na^+), the osmotically active Na^+ ions that are retained expand the extracellular volume and decrease the secretion of aldosterone, which in turns prevents Na^+ absorption in the ASDN. In contrast, when renal Na^+ output exceeds daily Na^+ intake (negative balance of Na^+), the body Na^+ loss shrinks the extracellular volume, the activation of the renin–angiotensin cascade stimulates aldosterone secretion, and Na^+ absorption is enhanced in the ASDN. This constant adaptation of the ENaC-mediated Na^+ absorption by aldosterone in the ASDN allows the maintenance of the Na^+ balance and extracellular fluid volume.

Aldosterone promotes the insertion of active ENaC in the luminal membrane of principal cells in the ASDN, allowing the entry of Na^+ ions from the tubule lumen (urine) into the cell along a favourable electrochemical gradient (Figure 3.4.2.3) (Loffing et al., 2001). The appearance of newly synthesized ENaC at the apical membrane is supported by recordings of ENaC currents by patch clamp in rat CCT principal cells (Frindt et al., 2001; Frindt et al., 2002). The half-life of ENaC at the cell surface is approximately 1 hour (Hanwell et al., 2002; May et al., 1997). Intracellular Na^+ ions are extruded across the basolateral membrane by the Na-K/

ATPase. The electrogenic entry of Na^+ ions through ENaC channels from the tubule lumen depolarizes the luminal membrane, and increases the electrochemical driving force for K^+ secretion into the tubule lumen. Thus, in the ASDN, Na^+ absorption is linked to K^+ secretion, a phenomenon which is best illustrated by the anti-kaliuretic effect of diuretics such as amiloride or triamterene that primarily target and inhibit ENaC channels, and indirectly decrease K^+ secretion.

Moreover, sodium transport through ENaC may osmotically drive transepithelial water transport via vasopressin-dependent apical water channels (i.e. aquaporin-2). As such, ENaC in the ASDN plays a pivotal role in the final adjustment of renal sodium, potassium and water excretion. The importance of ENaC for sodium, potassium, and fluid homeostasis is emphasized by the observation that gain-of-function mutations, or loss-of-function mutations, of ENaC lead, respectively, to extracellular volume expansion (Liddle's syndrome or pseudohyperaldosteronism), or to renal salt-wasting syndromes (pseudohypoaldosteronism type I) associated with alterations in potassium homeostasis (Rossier et al., 2002; Oh and Warnock, 2000) (see below).

The first step of the aldosterone action leading to the recruitment of active ENaC at the cell surface is the binding of the hormone to the cytosolic mineralocorticoid receptor (MR receptor) (Figure 3.4.2.3). The hormone-receptor complex is then translocated to the nucleus where transcriptional events occur, including the increase in ENaC mRNAs, together with other aldosterone-induced transcripts (AITs), or alternatively to a decrease in expression of transcripts, i.e. the aldosterone-repressed transcripts (ARTs)

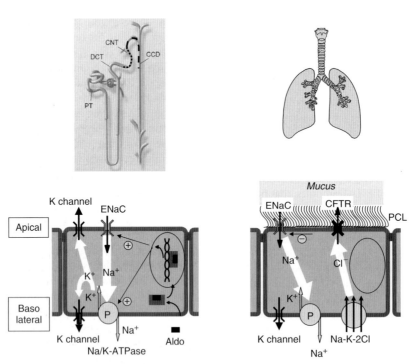

Fig. 3.4.2.3 ENaC in the kidney and in the lungs. Upper left: anatomy of the nephron with the proximal tubule (PT), distal convoluted tubule (DCT), connecting tubule (CNT) and the cortical collecting duct (CCD); sites of ENaC expression along the nephron are indicated by a dashed and an interrupted line (early/late ASDN). Lower left: scheme illustrating ion transport across an epithelial cell (principal cell) lining the CNT and CCD; ENaC at the apical membrane (urinary pole) allows entry of Na^+ ions into the cell which are then extruded across the basolateral membrane (blood side) by the Na^+/K^+-ATPase. The apical K^+ channel (ROMK) is responsible for K^+ secretion. Na^+ absorption and is under the control of aldosterone which binds to the cytosolic mineralocorticoid receptor (MR). Upper and lower right: schematic view of bronchial tree and illustration of an airway epithelium cell lined at the apical pole by the periciliary liquid (PCL) and a layer of mucus. The height of the PCL results from a balance between ENaC mediated Na^+ absorption and CFTR (cystic fibrosis transmembrane regulator) mediated Cl^- secretion. Chloride transport across the basolateral membrane is achieved by a Na^+-K^+-$2Cl^-$ cotransporter.

(Robert-Nicoud *et al.*, 2001). The AITs and ARTs are key elements for ENaC activation at the apical membrane and the upregulation of the Na$^+$-K$^+$/ATPase at the basolateral membrane underlying the stimulation of a transepithelial Na$^+$ flux (Verrey *et al.*, 2001). Aldosterone rapidly increases the synthesis in principal cells of the mouse CCD, of a serine/threonine kinase called serum- and glucocorticoid-regulated kinase (SGK) (Chen *et al.*, 1999; Naray *et al.*, 1999). The aldosterone-induced increase in SGK correlates nicely with the marked redistribution of ENaC from a cytoplasmic pool to the apical membrane (Loffing *et al.*, 2001). This increase in ENaC channel density at the cell surface is thought to contribute essentially to the early response to aldosterone which increases Na$^+$ transport within 1–3 hours of stimulation by the hormone. Alternative signalling pathways for aldosterone action on ENaC likely exist since SGK knockout mice show little impairment in Na$^+$ handling by the kidney and in extracellular fluid homeostasis (Wulff *et al.*, 2002).

Vasopressin increases both water and Na$^+$ absorption in the distal nephron (Hawk *et al.*, 1996; Nielsen *et al.*, 1995). Binding of vasopressin to vasopressin receptors (V2 receptors) at the basolateral membrane of principal cells causes a redistribution of aquaporins (AQP-2) from an intracellular compartment to the apical membrane, leading to an increase in cell membrane water permeability. In parallel, the stimulation of the V2 receptors increases ENaC activity at the apical membrane that accounts for the enhanced Na$^+$ absorption (Morris and Schafer, 2002). Both vasopressin effects on AQP-2 and ENaC are mediated by cAMP and can be mimicked by cAMP analogues or forskolin (Auberson *et al.*, 2003; van Balkom *et al.*, 2004). Beside this signalling pathway, vasopressin also induces the expression of a gene network that might be involved in the long-term regulation of Na$^+$ absorption (Robert-Nicoud *et al.*, 2001). Expression of genes such as VIP32 or RAMP3 is enhanced, but their precise role in the modulation of ENaC activity remains to be established. All of these cellular and molecular mechanisms contribute to the Na$^+$ and water homeostasis as well as the control of the extracellular volume.

The airway epithelia absorb Na$^+$ ions via an amiloride-sensitive electrogenic transport (Figure 3.14.3). In adult rats and humans the α, β, γ ENaC subunits are highly expressed in small- and medium-sized airways (Farman *et al.*, 1997; Talbot *et al.*, 1999; Burch *et al.*, 1995). Apical Na$^+$ transport by ENaC is important for the maintenance of the composition and the volume of the airway surface liquid (ASL), which forms a thin layer of liquid (the periciliary liquid layer—PCL) that covers the airway epithelium (Boucher, 2004). The depth of the PCL is controlled by opposing transport processes, the ENaC-mediated Na$^+$ absorption, and the CFTR-dependent Cl secretion (Figure 3.4.2.3). The PCL homeostasis, together with the beating cilia at the surface of the airway epithelium, is important for the clearance of particles and microbes that enter the airways. In a transgenic mouse model overexpressing the β ENaC subunit in the airway epithelium, Na$^+$ absorption was increased in the lungs, and the mouse exhibited a phenotype similar to cystic fibrosis with a reduced depth of PCL, mucus stasis and chronic inflammation (Mall *et al.*, 2004).

At birth, the amiloride-sensitive electrogenic Na$^+$ transport is important to clear the liquid that fills the alveoli and the airways of the fetal lung. Messenger RNAs for α, β, γ ENaC become detectable in the mouse fetal lung, at around days 15–17 of gestation and expression of ENaC subunits (mainly α and γ ENaC) sharply increases in the late fetal and early postnatal life when the lung

turns from a secretory to an absorptive organ (Talbot *et al.*, 1999). The physiological role of ENaC in lung liquid balance was clearly demonstrated in mice in which the αENaC gene was inactivated by homologous recombination (Hummler *et al.*, 1996). These αENaC knockout mice die soon after birth from a respiratory failure, due to a severe defect in the clearance of the fetal liquid that fills the lungs, associated with severe plasma electrolyte imbalance. These studies suggest that at birth α ENaC in the mouse fetal lung is important for Na$^+$ absorption. Disruption of the β or γ ENaC gene loci results in slower clearance of fetal lung liquid at birth but does not severely affect blood gas parameters. The β or γ ENaC knockout mice die slightly later than the α knockout, from severe electrolyte imbalance, namely hyperkalemia due to deficient renal K$^+$ secretion (Barker *et al.*, 1998; McDonald *et al.*, 1999). Thus, in contrast to the kidney, the Na$^+$ transport in the lung can be maintained efficiently by only two functional ENaC genes, i.e. combination of α and β, or α and γ ENaC.

The colon is a tight epithelium absorbing Na$^+$ by an electrogenic transport, sensitive to amiloride and stimulated by aldosterone (Epple *et al.*, 2000). Initially the α, β and γ ENaC cDNAs were isolated from a rat distal colon cDNA library, because of the high level of channel expression in this tissue, suggesting that the distal colon is an important site for Na$^+$ reabsorption. The contribution of the distal colon to the maintenance of the Na$^+$ homeostasis is not clear, and differences may exist between species.

3.4.2.7 Relevance to disease states

Renal salt transport and hypertension

A multitude of genetic, environmental, and regulatory factors determine blood pressure in individuals. The autonomic nervous system protects against major rapid changes in blood pressure, whereas long-term regulation of blood pressure within hours or days is achieved by the renin–angiotensin–aldosterone system controlling Na$^+$ excretion by the kidney and consequently extracellular fluid volume (Guyton, 1992). Normally, when blood pressure rises, the kidneys excrete more Na$^+$ and fluid than the amount ingested. Consequently, the extracellular fluid volume decreases, as well as the cardiac output, and blood pressure drops. Hypertension starts to develop when higher blood pressure is required to maintain a balance between ingestion and excretion of Na$^+$ and fluid by the kidneys.

Genetic studies of rare Mendelian syndromes associating elevated blood pressure with changes in extracellular volume and in electrolyte balance have identified important genes involved in the maintenance of Na$^+$ balance and extracellular fluid volume (Lifton, 1996). Remarkably, these genes whose mutations cause rare Mendelian forms of hypertension encode key enzymes, receptors or channels in the renin–angiotensin–aldosterone cascade and in its target kidney epithelium, the ASDN. The hereditary forms of hypertension include the glucocorticoid-remediable aldosteronism (GRA), the apparent mineralocorticoid excess (AME) and pseudoaldosteronism (or Liddle syndrome) all increasing Na$^+$ absorption in the aldosterone-sensitive distal nephron (for review, see Lifton *et al.*, 2001).

Pseudoaldosteronism, described by G. Liddle in the early 1950s (Liddle syndrome) associates low renin, low aldosterone, hypertension, hypokalemia, and metabolic alkalosis (Liddle *et al.*, 1963). Pharmacological investigations of the patients suffering from Liddle syndrome revealed that triamterene, a K$^+$-sparing diuretic, causes a significant increase in Na$^+$ excretion and a decrease in K$^+$ excretion, and, when associated with a low Na$^+$ diet, the drug normalized blood pressure. The mineralocorticoid receptor antagonist, spironolactone, was without effects on the renal handling of Na$^+$ and K$^+$, and also on blood pressure. GW Liddle concluded that the primary defect in these patients lies in the renal tubular transporters for Na$^+$ and K$^+$ ions.

Except for one case, all the mutations associated with Liddle's syndrome delete, or modify, the sequence of a conserved proline-rich motif xPPxY (x being any amino acid, P proline and Y tyrosine) in the C-terminus of the β and γ ENaC subunits (Shimkets et al., 1994; Hansson et al., 1995a, b). The tyrosine residue of the PY motif sequence (xPPxY) is part of a consensus internalization signal (NPxY) found in receptors removed from the plasma membrane via clathrin-coated pits (Shimkets et al., 1997). Furthermore, proline-rich motifs such as the PY motif are involved in specific protein–protein interactions. The ubiquitin protein ligase Nedd4–2 catalyses the final steps in the ubiquitylation cascade; in particular the attachment of ubiquitin moieties onto lysine residues of the N-terminus of ENaC subunits (Staub et al., 1996; Kamynina et al., 2001). Ubiquitylation of ENaC targets the channel for degradation by the lysosomal and proteasomal pathway. The mutations or deletions of the PY motif in the C-terminus of β or γ ENaC subunits result in a retention of active ENaC at the cell surface, an increase in Na$^+$ absorption in the distal nephron and a subsequent elevation of blood pressure.

These studies provide direct evidence that ENaC is important in the control of blood pressure.

ENaC and salt loosing nephropathy

The type 1 pseudohypoaldosteronism (PHA-1) is characterized in the first week of life by severe dehydration, hyponatraemia and hyperkalaemia resistant to mineralocorticoids, due to the inability of the kidney to retain sodium in the distal part of the nephron (Chang et al., 1996). Two clinically distinct forms of PHA-1 have been described, an autosomal recessive form affecting multiple organs (systemic form), including kidneys, colon, salivary glands, and sweat ducts, and an autosomal dominant form, characterized by salt loss in the kidneys only (Hanukoglu, 1991; Bonny and Rossier, 2002). Patients with a systemic form of PHA-1 fail to absorb liquid from the airway surface, resulting in an increase in the volume and the clearance of the airway surface liquid, and present frequent lower respiratory tract infections. The systemic form is caused by homozygous mutations in α, β, γ ENaC subunits (Chang et al., 1996). The autosomal dominant form of PHA-1 is caused by nonsense or frameshift mutations of the mineralocorticoid receptors and is consistent with salt loss restricted to the kidneys (Geller et al., 1998).

3.4.2.8 Concluding remarks

During the last two decades, the epithelial sodium channel (ENaC) was functionally characterized at the single channel level and its primary structure identified by expression cloning, revealing a novel ion channel family. Genetic studies have demonstrated the important role of ENaC in the adaptation of Na$^+$ excretion by the kidney to maintain the extracellular fluid volume and blood pressure within physiological ranges. In lung, ENaC plays a key role in the regulation of the airway surface liquid composition and volume. Many unanswered questions still remain to be addressed regarding the regulation, trafficking and structure of ENaC in its native cellular environment.

References

Abriel H and Horisberger JD (1999). Feedback inhibition of rat amiloride-sensitive epithelial sodium channels expressed in Xenopus laevis oocytes. J Physiol 516(1), 31–43.

Adebamiro A, Cheng Y, Johnson JP et al. (2005). Endogenous protease activation of ENaC: effect of serine protease inhibition on ENaC single channel properties. J Gen Physiol 126(4), 339–352.

Almeida AJ and Burg MB (1982). Sodium transport in the rabbit connecting tubule. Am J Physiol 243(4), F330–F334.

Anantharam A, Tian Y and Palmer LG (2006). Open probability of the epithelial sodium channel is regulated by intracellular sodium. J Physiol 574(2), 333–347.

Anantharam, A and Palmer LG (2007). Determination of epithelial Na$^+$ channel subunit stoichiometry from single-channel conductances. J Gen Physiol 130, 55–70.

Andreasen D, Vuagniaux G, Fowler-Jaeger N et al. (2006). Activation of epithelial sodium channels by mouse channel activating proteases (mCAP) expressed in Xenopus oocytes requires catalytic activity of mCAP3 and mCAP2 but not mCAP1. J Am Soc Nephrol 17(4), 968–976.

Auberson M, Hoffmann-Pochon N, Vandewalle A et al. (2003). Epithelial Na$^+$ channel mutants causing Liddle's syndrome retain ability to respond to aldosterone and vasopressin. Am J Physiol 285(3), F459–F471.

Babini E, Geisler HS, Siba M et al. (2003). A new subunit of the epithelial Na$^+$ channel identifies regions involved in Na$^+$ self-inhibition. J Biol Chem 278(31), 28418–28426.

Bachmann S, Bostanjoglo M, Schmitt R et al. (1999). Sodium transport-related proteins in the mammalian distal nephron—distribution, ontogeny and functional aspects. Anat Embryol 200(5), 447–468.

Barker PM, Nguyen MS, Gatzy JT et al. (1998). Role of g-ENaC subunit in lung liquid clearance and electrolyte balance in newborn mice. J Clin Investig 102(8), 1634–1640.

Bassler EL, Ngo-Anh TJ, Geisler HS et al. (2001). Molecular and functional characterization of acid-sensing ion channel (ASIC) 1b. J Biol Chem 276(36), 33782–33787.

Bonny O and Rossier BC (2002). Disturbances of Na/K balance: pseudohypoaldosteronism revisited. J Am Soc Nephrol 13(9), 2399–2414.

Boucher RC (2004). New concepts of the pathogenesis of cystic fibrosis lung disease. Eur Respir J 23(1), 146–158.

Burch LH, Talbot CR, Knowles MR et al. (1995). Relative expression of the human epithelial Na$^+$ channel subunits in normal and cystic fibrosis airways. Am J Physiol 269, C511–C518.

Butterworth MB, Frizzell RA, Johnson JP et al. (2005). PKA-dependent ENaC trafficking requires the SNARE-binding protein complexin. Am J Physiol 289(5), F969–F977.

Caldwell RA, Boucher RC and Stutts MJ (2004). Serine protease activation of near-silent epithelial Na$^+$ channels. Am J Physiol 286(1), C190–C194.

Caldwell RA, Boucher RC and Stutts MJ (2005). Neutrophil elastase activates near-silent epithelial Na$^+$ channels and increases airway epithelial Na$^+$ transport. Am J Physiol 288(5), L813–L819.

Canessa CM, Horisberger J-D and Rossier BC (1993). Epithelial sodium channel related to proteins involved in neurodegeneration. Nature 361, 467–470.

Canessa CM, Merillat A-M and Rossier BC (1994). Membrane topology of the epithelial sodium channel in intact cells. Am J Physiol 267, C1682–C1690.

Canessa CM, Schild L, Buell G et al.C (1994). Amiloride-sensitive epithelial Na$^+$ channel is made of three homologous subunits. Nature 367, 463–467.

Champigny G, Voilley N, Waldmann R *et al.* (1998). Mutations causing neurodegeneration in *Caenorhabditis elegans* drastically alter the pH sensitivity and inactivation of the mammalian H⁺-gated Na⁺ channel MDEG1. *J Biol Chem* **273**(**25**), 15418–15422.

Chang SS, Gründer S, Hanukoglu A *et al.* (1996). Mutations in subunits of the epithelial sodium channel cause salt wasting with hyperkalaemic acidosis, pseudohypoaldosteronism type 1. *Nature Genet* **12**, 248–253.

Chen SY, Bhargava A, Mastroberardino L *et al.* (1999). Epithelial sodium channel regulated by aldosterone-induced protein sgk. *PNAS* **96**(**5**), 2514–2519.

Chow YH, Wang Y, Plumb J *et al.* (1999). Hormonal regulation and genomic organization of the human amiloride-sensitive epithelial sodium channel alpha subunit gene. *Pediat Res* **46**(**2**), 208–214.

Chraibi A and Horisberger AD (2002). Na self-inhibition of human epithelial Na channel: temperature dependence and effect of extracellular proteases. *J Gen Physiol* **120**(**2**), 133–145.

Chraibi A, Vallet V, Firsov D *et al.* (1998). Protease modulation of the activity of the epithelial sodium channel expressed in *Xenopus* oocytes. *J Gen Physiol* **111**(**1**), 127–138.

Condliffe SB, Carattino MD, Frizzell RA *et al.* (2003). Syntaxin 1A regulates ENaC via domain-specific interactions. *J Biol Chem* **278**(**15**), 12796–12804.

Coric T, Hernandez N, Varez de la RD *et al.* (2004). Expression of ENaC and serum- and glucocorticoid-induced kinase 1 in the rat intestinal epithelium. *Am J Physiol* **286**(**4**), G663–G670.

Coscoy S, de Weille JR, Lingueglia E *et al.* (1999). The pre-transmembrane 1 domain of acid-sensing ion channels participates in the ion pore. *J Biol Chem* **274**(**15**), 10129–10132.

Costanzo LA (1984). Comparison of calcium and sodium transport in early and late rat distal tubules: effect of amiloride. *Am J Physiol* **246**, F937-F945.

de la Rosa DA, Li H and Canessa CM (2002). Effects of aldosterone on biosynthesis, traffic and functional expression of epithelial sodium channels in A6 cells. *J Gen Physiol* **119**(**5**), 427–442.

Dijkink L, Hartog A, Van Os CH *et al.* (2002). The epithelial sodium channel (ENaC) is intracellularly located as a tetramer. *Pflugers Archiv-Eur J Physiol* **444**(**4**), 549–555.

Driscoll M and Chalfie M (1991). The *mec-4* gene is a member of a family of *Caenorhabditis elegans* genes that can mutate to induce neuronal degeneration. *Nature* **349**, 588–593.

Epple HJ, Amasheh S, Mankertz J *et al.* (2000). Early aldosterone effect in distal colon by transcriptional regulation of ENaC subunits. *Am J Physiol* **278**(**5**), G718–G724.

Ergonul Z, Frindt G and Palmer LG (2006). Regulation of maturation and processing of ENaC subunits in the rat kidney. *Am J Physiol* **291** (**3**), F683–F693.

Eskandari, S, Snyder PM, Kreman M *et al.* (1999). Number of subunits comprising the epithelial sodium channel. *J Biol Chem* **274**(**38**), 27281–27286.

Farman N, Talbot CR, Boucher R *et al.* (1997). Noncoordinated expression of α, β and γ subunit mRNAs of epithelial Na⁺ channel along rat respiratory tract. *Am J Physiol* **272**(**1**), C131–141.

Firsov D, Gautschi I, Merillat AM *et al.* (1998). The heterotetrameric architecture of the epithelial sodium channel (ENaC). *EMBO J* **17**(**2**), 344–352.

Frindt G and Palmer LG (2004). Na channels in the rat connecting tubule. *Am J Physiol* **286**(**4**), F669–674.

Frindt G, Masilamani S, Knepper MA *et al.* (2001). Activation of epithelial Na channels during short-term Na deprivation. *Am J Physiol* **280**(**1**), F112–F118.

Frindt G, McNair T, Dahlmann A *et al.* (2002). Epithelial Na channels and short-term renal response to salt deprivation. *Am J Physiol* **283**(**4**), F717–F726.

Frings S, Purves RD and Macknight, ADC (1988). Single-channel recordings from the apical membrane of the toad urinary bladder epithelial cell. *J Membr Biol* **106**, 157–172.

Fuchs W, Larson EH and Lindemann B (1977). Current voltage curve of sodium channels and concentration dependence of sodium permeability in frog skin. *J Physiol* **267**, 137–166.

Fyfe GK and Canessa CM (1998). Subunit composition determines the single channel kinetics of the epithelial sodium channel. *J Gen Physiol* **112**(**4**), 423–432.

Garty H and Palmer LG (1997). Epithelial sodium channels—function, structure and regulation. *Physiol Rev* **77**(**2**), 359–396.

Geller DS, Rodriguezsoriano J, Boado AV *et al.* (1998). Mutations in the mineralocorticoid receptor gene cause autosomal-dominant pseudohypoaldosteronism type 1. *Nature Genet* **19**(**3**), 279–281.

Grunder S, Firsov D, Chang SS *et al.* (1997). A mutation causing pseudohypoaldosteronism type 1 identifies a conserved glycine that is involved in the gating of the epithelial sodium channel. *EMBO J* **16**(**5**), 899–907.

Grunder S, Jaeger NF, Gautschi I *et al.* (1999). Identification of a highly conserved sequence at the N-terminus of the epithelial Na⁺ channel alpha subunit involved in gating. *Pflugers Arch – Eur J Physiol* **438**(**5**), 709–715.

Guyton AC (1992). Kidneys and fluids in pressure regulation. Small volume but large pressure changes. *Hypertension* **19**(**1**), I2-I8.

Hamilton KL and Eaton DC (1985). Single-channel recordings from amiloride-sensitive epithelial sodium channel. *Am J Physiol* **249**, C200–C207.

Hansson JH, Nelson-Williams C, Suzuki H *et al.* (1995a). Hypertension caused by a truncated epithelial sodium channel gamma subunit: genetic heterogeneity of Liddle syndrome. *Nature Genet* **11**, 76–82.

Hansson JH, Schild L, Lu Y *et al.* (1995b). A *de novo* missense mutation of the β subunit of the epithelial sodium channel causes hypertension and Liddle syndrome, identifying a proline-rich segment critical for regulation of channel activity. *PNAS* **92**, 11495–11499.

Hanukoglu A (1991). Type I pseudohypoaldosteronism includes two clinically and genetically distinct entities with either renal or multiple target organ defects. *J Clin Endocrinol Metab* **73**, 936–944.

Hanwell D, Ishikawa T, Saleki R *et al.* (2002). Trafficking and cell surface stability of the epithelial Na⁺ channel expressed in epithelial Madin–Darby canine kidney cells. *J Biol Chem* **277**(**12**), 9772–9779.

Hawk CT, Li L and Schafer JA (1996). AVP and aldosterone at physiological concentrations have synergistic effects on Na⁺ transport in rat CCD. *Kidney Int* **57**, S35–41.

Hirsh AJ, Zhang J, Zamurs A *et al.* (2008). Pharmacological properties of N-(3,5-diamino-6-chloropyrazine-2-carbonyl)-N'-4-[4-(2,3-dihydroxypropoxy) phenyl]butyl-guanidine methanesulfonate (552–02), a novel epithelial sodium channel blocker with potential clinical efficacy for cystic fibrosis lung disease. *JPET* **325**, 77–88.

Hughey RP, Bruns JB, Kinlough CL *et al.* (2004). Epithelial sodium channels are activated by furin-dependent proteolysis. *J Biol Chem* **279**(**18**), 18111–18114.

Hummler E, Barker P, Gatzy J *et al.* (1996). Early death due to defective neonatal lung liquid clearance in aENaC-deficient mice. *Nature Genet* **12**, 325–328.

Jasti J, Furukawa H, Gonzales EB *et al.* (2007). Structure of acid-sensing ion channel 1 at 1.9 A resolution and low pH. *Nature* **449**, 316–323.

Ji HL, Bishop LTR, Anderson SJ *et al.* (2004). The role of pre-H2 domains of alpha- and delta-epithelial Na⁺ channels in ion permeation, conductance and amiloride sensitivity. *J Biol Chem* **279**(**9**), 8428–8440.

Kamynina E, Debonneville C, Bens M *et al.* (2001). A novel mouse Nedd4 protein suppresses the activity of the epithelial Na⁺ channel. *FASEB J* **15**(**1**), 204–214.

Kellenberger S and Schild L (2002). Epithelial sodium channel/degenerin family of ion channels: a variety of functions for a shared structure. *Physiol Rev* **82** (**3**), 735–767.

Kellenberger S, Auberson M, Gautschi I et al. (2001). Permeability properties of ENaC selectivity filter mutants. J Gen Physiol 118(6), 679–692.

Kellenberger S, Gautschi I and Schild L (1999). A single point mutation in the pore region of the epithelial Na+ channel changes ion selectivity by modifying molecular sieving. PNAS 96(7), 4170–4175.

Kellenberger S, Gautschi I and Schild L (2003). Mutations in the epithelial Na+ channel ENaC outer pore disrupt amiloride block by increasing its dissociation rate. Mol Pharmacol 64(4), 848–856.

Kellenberger S, Gautschi I and Schild, L (2002). An external site controls closing of the epithelial Na+ channel ENaC. J Physiol 543(2), 413–424.

Kellenberger S, Gautschi I, Pfister Y et al. (2005). Intracellular thiol-mediated modulation of epithelial sodium channel activity. J Biol Chem 280 (9), 7739–7747.

Kellenberger S, Gautschi I, Rossier BC et al. (1998). Mutations causing Liddle syndrome reduce sodium-dependent downregulation of the epithelial sodium channel in the Xenopus oocyte expression system. J Clin Invest 101, 2741–2750.

Kosari F, Sheng SH, Li JQ et al. (1998). Subunit stoichiometry of the epithelial sodium channel. J Biol Chem 273(22), 13469–13474.

Liddle GW, Bledsoe T and Coppage WS (1963). A familial renal disorder simulating primary aldosteronism but with negligible aldosterone secretion. Trans Assoc Am Physicians 76, 199–213.

Lifton RP (1996). Molecular genetics of human blood pressure variation. Science 272, 676–680.

Lifton RP, Gharavi AG and Geller DS (2001). Molecular mechanisms of human hypertension. Cell 104(4), 545–556.

Light, DB, McCann FV, Keller TM et al. (1988). Amiloride-sensitive cation channel in apical membrane of inner medullary collecting duct. Am J Phsiol 255, F278–F286.

Lindemann B and Van Driessche W (1977). Sodium-specific membrane channels of frog skin are pores: current fluctuations reveal high turnover. Science 195, 292–294.

Loffing J and Kaissling B (2003). Sodium and calcium transport pathways along the mammalian distal nephron: from rabbit to human. Am J Physiol 284(4), F628–F643.

Loffing J, Pietri L, Aregger F et al. (2000). Differential subcellular localization of ENaC subunits in mouse kidney in response to high- and low-Na diets. Am J Physiol 279(2), F252–F258.

Loffing J, Zecevic M, Feraille E et al. (2001). Aldosterone induces rapid apical translocation of ENaC in early portion of renal collecting system: possible role of SGK. Am J Physiol 280(4), F675–F682.

Mall M, Grubb BR, Harkema JR et al. (2004). Increased airway epithelial Na+ absorption produces cystic fibrosis-like lung disease in mice. Nat Med 10(5), 487–493.

Masilamani S, Kim GH, Mitchell C et al. (1999). Aldosterone-mediated regulation of ENaC alpha, beta and gamma subunit proteins in rat kidney. J Clin Investig 104(7), R19–R23.

May A, Puoti A, Gaeggeler HP et al. (1997). Early effect of aldosterone on the rate of synthesis of the epithelial sodium channel alpha subunit in A6 renal cells. J Am Soc Nephrol 8(12), 1813–1822.

McDonald, FJ, Yang BL, Hrstka RF et al. (1999). Disruption of the β subunit of the epithelial Na+ channel in mice: hyperkalemia and neonatal death associated with a pseudohypoaldosteronism phenotype. PNAS 96(4), 1727–1731.

Morris RG and Schafer JA (2002). cAMP increases density of ENaC subunits in the apical membrane of MDCK cells in direct proportion to amiloride-sensitive Na(+) transport. J Gen Physiol 120(1), 71–85.

Naray FT, Canessa C, Cleaveland ES et al. (1999). SgK is an aldosterone-induced kinase in the renal collecting duct—effects on epithelial Na+ channels. J Biol Chem 274(24), 16973–16978.

Nielsen S, Chou C-L, Marples D et al. (1995). Vasopressin increases water permeability of kidney collecting duct by inducing translocation of aquaporin-CD water channels to plasma membrane. PNAS 92, 1013–1017.

Oh YS and Warnock DG (2000). Disorders of the epithelial Na(+) channel in Liddle's syndrome and autosomal recessive pseudohypoaldosteronism type 1. Exp Nephrol 8(6), 320–325.

Palmer LG (1984). Voltage-dependent block by amiloride and other monovalent cations of apical Na channels in the toad urinary bladder. J Membr Biol 80, 153–165.

Palmer LG and Frindt G (1986). Amiloride-sensitive Na channels from the apical membrane of the rat cortical collecting tubule. PNAS 83, 2727–2770.

Palmer LG and Frindt G (1987). Effects of cell Ca and pH on Na channels from rat cortical collecting tubule. Am J Physiol 253, F333–F339.

Palmer LG and Frindt G (1996). Gating of Na channels in the rat cortical collecting tubule: effects of voltage and membrane stretch. J Gen Physiol 107, 35–45.

Palmer LG, Edelman IS and Lindemann B (1980). Current-voltage analysis of apical sodium transport in toad urinary bladder: effects of inhibitors of transport and metabolism. J Membr Biol 57, 59–71.

Pfister Y, Gautschi I, Takeda AN et al. (2006). A gating mutation in the internal pore of ASIC1a. J Biol Chem 281(17), 11787–11791.

Reilly RF and Ellison DH (2000). Mammalian distal tubule: physiology, pathophysiology and molecular anatomy. Physiol Rev 80(1), 277–313.

Renard S, Lingueglia E, Voilley N et al. (1994). Biochemical analysis of the membrane topology of the amiloride-sensitive Na+ channel. J Biol Chem 269, 12981–12986.

Robert-Nicoud M, Flahaut M, Elalouf JM et al. (2001). Transcriptome of a mouse kidney cortical collecting duct cell line: effects of aldosterone and vasopressin. PNAS 98(5), 2712–2716.

Rossier BC, Pradervand S, Schild L et al. (2002). Epithelial sodium channel and the control of sodium balance: interaction between genetic and environmental factors. Annu Rev Physiol 64, 877–897.

Saxena A, Hanukoglu I, Strautnieks SS et al. (1998). Gene structure of the human amiloride-sensitive epithelial sodium channel β subunit. Biochem Biophys Res Commun 252(1), 208–213.

Schild L, Schneeberger E, Gautschi I et al. (1997). Identification of amino acid residues in the α, β, γ subunits of the epithelial sodium channel (ENaC) involved in amiloride block and ion permeation. J Gen Physiol 109, 15–26.

Sheng S, Carattino MD, Bruns JB et al. (2006). Furin cleavage activates the epithelial Na+ channel by relieving Na+ self-inhibition. Am J Physiol 290(6), F1488–F1496.

Sheng S, Perry CJ and Kleyman TR (2004). Extracellular Zn2+ activates epithelial Na+ channels by eliminating Na+ self-inhibition. J Biol Chem 279(30), 31687–31696.

Shimkets RA, Lifton RP and Canessa CM (1997). The activity of the epithelial sodium channel is regulated by clathrin-mediated endocytosis. J Biol Chem 272(41), 25537–25541.

Shimkets RA, Warnock DG, Bositis CM et al. (1994). Liddle's syndrome: heritable human hypertension caused by mutations in the β subunit of the epithelial sodium channel. Cell 79, 407–414.

Silver RB, Frindt G, Windhager EE et al. (1993). Feedback regulation of Na channels in rat CCT. I. Effects of inhibition of Na pump. Am J Physiol 264, F557–F564.

Snyder PM, Bucher DB and Olson DR (2000). Gating induces a conformational change in the outer vestibule of ENaC. J Gen Physiol 116(6), 781–790.

Snyder PM, Cheng C, Prince LS et al. (1998). Electrophysiological and biochemical evidence that deg/enac cation channels are composed of nine subunits. J Biol Chem 273(2), 681–684.

Snyder PM, Steines JC and Olson DR (2004). Relative contribution of Nedd4 and Nedd4–2 to ENaC regulation in epithelia determined by RNA interference. J Biol Chem 279(6), 5042–5046.

Staub O, Dho S, Henry PC et al. (1996). WW domains of Nedd4 bind to the proline-rich PY motifs in the epithelial Na+ channel deleted in Liddle's syndrome. EMBO J 15, 2371–2380.

Talbot CL, Bosworth DG, Briley EL *et al.* (1999). Quantitation and localization of ENaC subunit expression in fetal, newborn and adult mouse lung. *Am J Resp Cell Mol Biol* **20**(**3**), 398–406.

Thomas CP, Doggett NA, Fisher R and Stokes JB (1996). Genomic organization and the 5′ flanking region of the gamma subunit of the human amiloride-sensitive epithelial sodium channel. *J Biol Chem* **271**, 26062–26066.

Tomita K, Pisano JJ and Knepper MA (1985). Control of sodium and potassium transport in the cortical collecting duct of the rat. Effects of bradykinin, vasopressin and deoxycorticosterone. *J Clin Invest* **76**(**1**), 132–136.

Vallet V, Chraibi A, Gaeggeler HP *et al.* (1997). An epithelial serine protease activates the amiloride-sensitive sodium channel. *Nature* **389**(**6651**), 607–610.

Vallet VR, Pfister C, Loffing J *et al.* (2002). Cell-surface expression of the channel activating protease xCAP-1 is required for activation of ENaC in the *Xenopus* oocyte. *J Am Soc Nephrol* **13**(**3**), 588–594.

van Balkom BW, Graat MP, van Raak M *et al.* (2004). Role of cytoplasmic termini in sorting and shuttling of the aquaporin-2 water channel. *Am J Physiol* **286**(**2**), C372–C379.

Verrey F, Hummler E, Schild L *et al.* (2001). Control of Na$^+$ transport by aldosterone. In DW Seldin and G Giebisch, eds, *The Kidney*, 3rd edn. Philadelphia, Lippincott Williams and Wilkins, pp. 1441–1472.

Verrier B, Champigny G, Barbry P *et al.* (1989). Identification and properties of a novel type of Na$^+$-permeable amiloride-sensitive channel in thyroid cells. *Eur J Biochem* **183**, 499–505.

Vigne P, Champigny G, Marsault R *et al.* (1989). A new type of amiloride-sensitive cationic channel in endothelial cells of brain microvessels. *J Biol Chem* **264**, 7663–7668.

Voilley N, Lingueglia E, Champigny G *et al.* (1994). The lung amiloride-sensitive Na$^+$ channel: biophysical properties, pharmacology, ontogenesis and molecular cloning. *PNAS* **91**, 247–251.

Vuagniaux G, Vallet V, Jaeger NF *et al.* (2002). Synergistic activation of ENaC by three membrane-bound channel-activating serine proteases (mCAP1, mCAP2 and mCAP3) and serum- and glucocorticoid-regulated kinase (Sgk1) in *Xenopus* oocytes. *J Gen Physiol* **120**(**2**), 191–201.

Vuagniaux G, Vallet V, Jaeger NF *et al.* (2000). Activation of the amiloride-sensitive epithelial sodium channel by the serine protease mCAP1 expressed in a mouse cortical collecting duct cell line. *J Am Soc Nephrol* **11**(**5**), 828–834.

Waldmann R, Champigny G, Bassilana F *et al.* (1995). Molecular cloning and functional expression of a novel amiloride-sensitive Na$^+$ channel. *J Biol Chem* **270**, 27411–27414.

Wulff P, Vallon V, Huang DY *et al.* (2002). Impaired renal Na$^+$ retention in the sgk1-knockout mouse. *J Clin Investig* **110**(**9**), 1263–1268.

PART 4

Intracellular ligand-gated ion channels

PART 4

Intracellular ligand-
gated ion channels

4.1

Chloride channels activated by intracellular ligands

Jorge Arreola, Juan Pablo Reyes, Teresa Rosales-Saavedra and Patricia Pérez-Cornejo

4.1.1 Introduction

The discovery of ion channels gated by extracellular ligands was made by Bernard Katz and José del Castillo while working with the motor endplate (del Castillo and Katz, 1955). Years later Katz and Miledi reported that the noisiness of the membrane potential increased when acetylcholine was applied and interpreted this noise as the result of opening and closing of individual channels (Katz and Miledi, 1972). Since then, several other channels activated by external ligands have been characterized (see other sections of this book). In some cases, detailed structural data is available making possible a comprehensive understanding of the structure–function relationship for these channels. In contrast, the discovery of plasma membrane channels gated by intracellular ligands has been less prolific.

Classically, a ligand is regarded as an organic molecule capable of interacting with binding site(s) located in biologically active macromolecule(s). Thus, according to this definition, neurotransmitters are among the most widely recognized extracellular ligands, which can activate both metabotropic and ionotropic receptors. This definition, however, can be extended to include inorganic ions such as Ca^{2+}, which bind to both extracellular and intracellular binding sites. Extracellular Ca^{2+} activates receptors such as the well-characterized Ca^{2+}-sensing receptors (Brown et al., 1998) and Ca^{2+}-inactivated Cl^- channels (Weber, 1999, 2002). Activation of Ca^{2+}-sensing receptors by extracellular Ca^{2+} results in activation of a signal transduction cascade (Brown, 2000; Brown and MacLeod, 2001) typical of metabotropic receptors. On the other hand, subtle changes in $[Ca^{2+}]_i$ (i.e. from 100 to 1000 nM) result in activation of Cl^- and K^+ channels. Direct binding of Ca^{2+} gates Ca^{2+}-activated K^+ channels (i.e. ionotropic mechanism) by binding to structural Ca^{2+}-binding motifs already characterized in cloned channels (Bao et al., 2004; Zeng et al., 2005). Thus, Ca^{2+} can be regarded as an inorganic ligand that activates both metabotropic and ionotropic receptors, paralleling the action of neurotransmitters. Hence, Ca^{2+}-activated Cl^- channels (CaCCs) and Ca^{2+}-activated K^+ channels are considered ion channels gated by an intracellular ligand.

CaCCs were first described in Xenopus oocytes (Miledi, 1982) and rod photoreceptors of salamander retina (Bader et al., 1982); subsequently CaCCs were identified in many other cell types where they play important roles in several physiological processes.

In this chapter we will review some known functional and molecular characteristics of CaCCs, a prototype of Cl^- channels activated by intracellular ligand.

4.1.2 Molecular characterization

The presence of multiple types of CaCCs is suggested by the existence of different single-channel conductances as well as different mechanisms of activation. This idea has not been well established because the molecular identity of CaCCs remains unsettled. The recognition of the gene product(s) responsible for the expression of CaCCs in native tissue remains a priority in order to understand the role of CaCCs in normal physiology and disease. Traditional cloning strategies have not been successful in the case of CaCCs. For example, expression cloning in Xenopus oocytes is compromised as oocytes express endogenous CaCCs. Likewise, approaches based in protein purification have failed owing to the lack of specific and high affinity ligands. Finally, none of the Cl^- channels cloned so far (i.e. cystic fibrosis transmembrane conductance regulator (CFTR), ligand-gated anion channels such as $GABA_A$ and glycine receptors, ClC chloride channels) has properties resembling those of CaCCs, thus rendering homology screening ineffective. In spite of these difficulties, a few molecular candidates for CaCCs have been proposed (Table 4.1.1).

Calcium-activated chloride channels (CLCA) family

The first CLCA gene, named bCLCA1, was cloned from a bovine tracheal cDNA expression library (Cunningham et al., 1995). Subsequently, several related genes from different mammalian species and tissues have been cloned and analysed. These clones include one additional isoform from bovine, four human (hCLCA1–4), two porcine (pCLCA1–2), and six murine clones (mCLCA1–6) (Fuller and Benos 2000; Elble et al., 2002; Beckley et al., 2004; Evans et al., 2004). Human genes have been mapped to the same site on chromosome 1p31–p22 and show a high degree of homology in size, sequence, and predicted structure of gene products but diverge significantly in their tissue expression pattern.

Sequence analysis of bCLCA1, mCLCA1 and bCLCA2 has resulted in ambiguous predictions of one or two transmembrane domains (Elble et al., 1997; Romio et al., 1999) or up to four putative transmembrane domains (Cunningham et al., 1995). In contrast,

Table 4.1.1 Putative human cloned calcium-activated chloride channels

Family	Members	Chromosome location	Ca²⁺-gated channel formation	I–V	Ca²⁺ EC₅₀ (µM)	References
CLCA	hCLCA-1 (NM_001285)	1p31–p22	Uncertain			[1–6]
	hCLCA-2 (NM_006536)	1p31–p22	Uncertain			[7–12]
	hCLCA-3 (NM_004921)	1p31–p22	Uncertain			[11, 13, 14]
	hCLCA-4 (NM_012128)	1p31–p22	Uncertain			[11, 15–17]
Bestrophins	hBest-1 (VMD2) (NM_004183)	11q13	Yes	~Linear	?	[18–22]
	hBest-2 (VMD2L1) (NM_017682)	19p13.13	Yes	Linear	0.21	[23, 24]
	hBest-3 (VMD2L2) (NM_153274)	1p33–p32.3	Yes	Inwardly rectifying	?	[23, 25]
	hBest-4 (VMD2L3) (NM_152439)	12q14.2–q15	Yes	Linear	0.23	[23, 25, 26]
Tweety	Tweety homolog 1 (*Drosophila*) [homo sapiens] TTYH1 (NM_001005367)	19q13.4	No			[27, 28]
	Tweety homolog 2 (*Drosophila*) [homo sapiens] TTYH2 (NM_032646)	17q24	Yes			[27, 29]
	Tweety homolog 3 (*Drosophila*) [homo sapiens] TTYH3 (NM_025250)	7p22	Yes	Linear	~2	[27]
ClC3	CLCN3 [homo sapiens] (NM_001829)	4q33	Uncertain	Slightly outward rectifying		[30–32]

[1] Gibson A et al. (2005) J Biol Chem 280, 27205–27212; [2] Ritzka M et al. (2004) Hum Genet 115, 483–491; [3] Loewen ME et al. (2002) Biophys Res Commun 298, 531–536; [4] Greenwood IA et al. (2002) J Biol Chem 277, 22119–22122; [5] Hume JR et al. (2000) Physiol Rev 80, 31–81; [6] Gruber AD et al. (1998) Genomics 54, 200–214; [7] Connon CJ et al. (2005) Acta Histochem 106, 421–425; [8] Abdel–Ghany M et al. (2001) J Biol Chem 276, 25438–25446; [9] Koegel H and Alzheimer C (2001); FASEB J 15, 145–154; [10] Itoh R et al. (2000) Curr Eye Res 21, 918–925; [11] Pauli BU et al. (2000) Clin Exp Pharmacol Physiol 27, 901–905; [12] Gruber AD et al. (1999) Am J Physiol 276, C1261–C1270; [13] Gruber AD and Pauli BU (1999) Genome 42, 1030–1032; [14] Gruber AD and Pauli BU (1998) Biochim Biophys Acta 1444, 418–423; [15] Ritzka M et al. (2004) Hum Genet 115, 483–491; [16] Clark HF et al. (2003) Genome Res 13, 2265–2270; [17] Agnel M et al. (1999) FEBS Lett 455, 295–301; [18] Tsunenari T et al. (2003) J Biol Chem 278, 41114–41125; [19] Marmorstein LY et al. (2002) J Biol Chem 277, 30591–30597; [20] Sun H et al. (2002) Proc Natl Acad Sci USA 99, 4008–4013; [21] Marmorstein AD et al. (2000) Proc Natl Acad Sci USA 97, 12758–12763; [22] Petrukhin K et al. (1998) Nat Genet 19, 241–247; [23] Marquardt A et al. (1998) Hum Mol Genet 7, 1517–1525; [24] Qu Z et al. (2003) J Biol Chem 278, 49563–49572; [25] Stohr H et al. (2002) Eur J Hum Genet 10, 281–284; [26] Tsunenari T et al. (2006) J Gen Physiol 127, 749–754; [27] Suzuki M and Mizuno A (2004) J Biol Chem 279, 22461–22468; [28] Campbell HD et al. (2000) Genomics 68, 89–92; [29] Rae FK et al. (2001) Genomics 77, 200–207; [30] Yeung CH et al. (2005) Biol Reprod 73, 1057–1063; [31] Huang P et al. (2001) J Bio Chem 276, 20093–20100; [32] Taine L et al. (1998) Human Genet 102, 178–181.

biochemical analysis of the *hCLCA2* gene product suggests five transmembrane domains (Gruber et al., 1999). Three of these domains are located in the amino terminus and two are located in the carboxyl terminus, within an 8-kDa-cleavage product.

An increasing body of evidence strongly suggests that CLCA genes might not be accountable for CaCCs expression since the Ca²⁺-dependent Cl⁻ currents recorded from cells transfected with CLCA cDNAs show weak sensitivity to Ca²⁺ and no membrane voltage (Vm) dependence (Elble et al., 1997; Gandhi et al., 1998; Gruber et al., 1998; Pauli et al., 2000). In addition, some cells display CaCC activity although they lack CLCA expression (Papassotiriou et al., 2001). Moreover, CLCA proteins are homologous to adhesion proteins and at least one of them (hCLCA3) is

secreted (Gruber and Pauli, 1998; Pauli et al., 2000). Even though CLCA expression does not duplicate native CaCC current there is the possibility that some or all CLCA proteins may be modulators of endogenous Cl⁻ channels (Loewen et al., 2003; 2004; Gibson et al., 2005; for an excellent review see Loewen and Forsyth, 2005). Table 4.1.1 summarizes the main characteristics of CLCA, Bestrophin, Tweety and ClC-3 genes.

The Bestrophin family

In the retina, a form of macular degeneration called Best vitelliform macular dystrophy has been associated with mutations in the *Bestrophin 1 (Best-1)* gene (Petrukhin et al., 1998). Other genes that belong to the Bestrophin family have been cloned from different species including humans. In humans the Bestrophin family is composed of four members, *hBest 1–4* (Table 4.1.1). Pairwise comparison of Bestrophin sequences indicate that the first 360 amino acids of the N-terminus share between 56–66% identity, even though the C-terminus shows minimal homology (Tsunenari *et al.*, 2003). Human Bestrophins are widely expressed in excitable and non-excitable tissues. High levels of Best-2 mRNA are found in retinal pigmented epithelium (RPE), liver, and spleen, whereas moderate levels are detected in lung and retina and only modest amounts are found in brain, heart, and gut (Leblanc *et al.*, 2005). Moreover, human Bestrophins have few phosphorylation consensus sequences that are conserved. In contrast, *Xenopus* Bestrophins (XBest) contain multiple protein kinase A, casein kinase, and protein kinase C consensus sequences (Qu *et al.*, 2003).

Sequence analyses of Bestrophins suggest the presence of four (Bakall *et al.*, 1999; Sun *et al.*, 2002), five (Stohr *et al.*, 2000), or six (Qu *et al.*, 2003) transmembrane domains suggesting that Bestrophins are integral membrane proteins. There is strong evidence that indicates Bestrophins function as Ca^{2+}-activated Cl^- channels (Table 4.1.1) when expressed heterologously (Sun *et al.*, 2002; Qu *et al.*, 2003; Tsunenari *et al.*, 2003). Ca^{2+} appears to activate the Bestrophin Cl^- channel without participation of diffusible messengers or protein phosphorylation (Tsunenari *et al.*, 2006). Additionally, Bestrophin channels exhibit a generic lyotropic anion-selectivity sequence as well as block by niflumic acid, akin to CaCCs behaviour (Qu and Hartzell, 2004; for comparison see Arreola and Perez-Cornejo, 2006). These functional similarities suggest that *Best* genes may encode CaCCs. Major progress in demonstrating that Bestrophins form a Cl^- channel pore was recently achieved by showing that mutations within the putative TM2 domain of mBest-2 alter anion selectivity and conductance, both of which are properties determined by the pore (Tsunenari *et al.*, 2003; Qu and Hartzell, 2004; Qu *et al.*, 2006). Besides these similarities Bestrophins and CaCCs also show important differences. For example, CaCCs exhibit time- and Vm-dependent kinetics along with outward rectification in their steady-state I–V curve while current traces from hBest-1 or hBest-2-transfected cells are nearly flat within a broad Vm range and display roughly linear I–V curves (Table 4.1.1). Moreover, hBest-3-transfected cells display I-V curves with marked inward rectification, instead of the outward rectification expected for CaCCs (Table 4.1.1). Finally, hBest-4-transfected cells show a nearly linear I–V curve and time- and Vm-dependent currents (Table 4.1.1) that look like CaCC currents with slow kinetics (Tsunenari *et al.*, 2003; 2006). Strong evidence supporting the idea that bestrophins are components of native CaCCs at the plasma membrane was reported recently (Chien *et al.*, 2006). Using interfering RNA (RNAi) against dbest1 and dbest2 in the *Drosophila* S2 cell line that express four bestrophins (dbest1–4) and have an endogenous CaC current, it was shown that the CaCC current is abolished. The endogenous CaCC current was mimicked by expression of dbest1 in HEK cells. Furthermore, the rectification and relative permeability of the current were altered by site-specific mutations of this gene. Single channel analysis of the S2 bestrophin currents revealed an ~2-pS channel with fast gating kinetics and linear current–voltage relationship. A similar channel was observed in CHO cells transfected with dbest1, but no such channel was seen in S2 cells treated with RNAi to dbest1. In contrast, Tsunenari *et al.* (2006) described macroscopic CaCC currents in inside-out patches excised from cells expressing hBest4.

However, these channels activated with a time constant of ~15 s at saturating internal Ca^{2+} concentrations (10 μM). Although these activation kinetics are rather slow when compared to native CaCC currents, still is several-fold faster than the current reported by Chien *et al.*, (2006) for dbest1 which requires ~2 min for complete activation. In contrast currents generated by mBest2 expressed in HEK cells are activated very quickly after breaking the seal (<10 s). The mechanism of this slow activation is unknown, but is unlikely to be due to direct binding of Ca^{2+} to the channel.

Notably, in heterologous expression systems a large fraction of the Bestrophins is located in intracellular organelles, raising the possibility that Bestrophins are Cl^- channels located both in intracellular organelles as well as plasma membrane (Qu *et al.*, 2003; Tsunenari *et al.*, 2003). Further evidence against Best-1 as a molecular candidate for CaCC is the presence of normal CaCC activity in RPE cells from mBest-1 knockout mice (Marmorstein *et al.*, 2006), although these results do not rule out the possibility that upregulation of mBest-2 expression could compensate for the lack of mBest-1. Interestingly, mouse RPE cells are unlike *Xenopus* RPE cells because they do not express CaCC with classical kinetics (Hartzell and Qu, 2003; Marmorstein *et al.*, 2006). Thus, the effect of mBest-1 knockout remains to be tested in a tissue endogenously expressing CaCC to demonstrate the effect of mBest-1 ablation on classical CaCC and if Bestrophins only comprised the pore subunit of CaCCs, then the differences in biophysical properties could be explained assuming that additional unidentified subunits are needed to recreate the endogenous CaCC response.

Novel CaCCs?

Suzuki and Mizuno (2004) reported that the human genes *hTTYH2* and *hTTYH3*, whose homologues are known as *Tweety* in *Drosophila*, are Ca^{2+}-regulated maxi-Cl^- channels that show a conductance of ~260 pS (Table 4.1.1), while the related gene, hTTYH1 is a volume-sensitive large conductance Cl^- channel, which is not regulated by Ca^{2+}. Members of the hTTYH family encode Cl^- channels that consist of five or six transmembrane segments. *hTTYH1* is expressed mainly in the brain, hTTYH2 is predominantly expressed in brain and testis and is expressed to a lower extent in heart, ovary, spleen and peripheral blood leukocytes; this protein may play a role in kidney tumorigenesis since expression of this gene is upregulated in 81% of cell samples taken from renal carcinoma (Rae *et al.*, 2001). hTTYH3 is expressed in heart, brain and skeletal muscle (Suzuki, 2006), tissues known to express CaCCs, but interestingly it is not expressed in salivary glands where endogenous CaCCs have been extensively studied (Melvin *et al.*, 2002). Like native Ca^{2+} activated Cl^- channels, the hTTYH3 channel showed complex gating kinetics and voltage-dependent inactivation, and was dependent on micromolar intracellular Ca^{2+} concentration (Suzuki, 2006).

hTTYH2 and hTTYH3 may also represent genes coding for maxi Cl^- channels regulated by Ca^{2+} in spinal neurons and lactotrophs (Hussy, 1992; Fahmi *et al.*, 1995).

Interestingly, ClC-3, a member of the ClC family that encompasses Vm-dependent Cl^- channels and H^+/Cl^- antiporters, can be activated by CaMKII (i.e. indirectly gated by Ca^{2+}) (Huang *et al.*, 2001; Robinson *et al.*, 2004). The robust endogenous CaMKII-activated Cl^- current of arterial smooth muscle cells is absent in ClC-3 knockout mice (Robinson *et al.*, 2004). However, CaCCs are still present in salivary acinar cells from ClC-3 knockout mice (Arreola *et al.*, 2002). As we shall see, it is very likely that different types of CaCCs exist, some of them directly activated by Ca^{2+} and others indirectly activated via CaMKII or cGMP; ClC-3 channels might belong to the latter category.

4.1.3 Biophysical characterization

Most of what we know about the biophysical properties of intracellular ligand-gated Cl$^-$ channels comes from studies of endogenous CaCCs in different tissues. Thus, in this section we summarize some of the known biophysical properties of CaCCs and when possible we compare their properties to those of cloned channels.

Whole-cell currents and current-voltage relationship

Functional properties of CaCCs including Vm-dependence, permeability and pharmacology have been assessed using the whole-cell configuration of the patch clamp technique. Figure 4.1.1a illustrates raw currents generated by CaCCs activation (I_{ClCa}) recorded at +100 and –100 mV from an acinar cell dialysed with a solution containing ~250 nM free Ca^{2+}. At +100 mV a slowly activating Cl$^-$ current (red trace) is observed, which reaches a maximum after about 1.5 s. At the end of the pulse, when the membrane potential is hyperpolarized I_{ClCa} displays a rapidly decaying

tail current. In contrast, at –100 mV (black trace) little current is observed. This outward rectification is due to the Vm-dependence of Ca^{2+} binding to a site in the channel: since at positive voltages Ca^{2+}-affinity is higher than at negative voltages, a larger fraction of channels are open at positive voltages. When $[Ca^{2+}]_i$ is elevated to >5 μM, the current activates nearly instantaneously and loses its time dependency. Figure 4.1.1b illustrates I-V curves obtained from cells dialysed with different $[Ca^{2+}]_i$. As the ligand concentration increases, the I-V relationship changes from outward rectification to nearly linear, indicating that CaCCs are nearly fully opened at high Ca^{2+} concentrations. This pattern of Ca^{2+} and Vm dependence has been observed in practically all cells where CaCCs have been studied. It has also been demonstrated that the macroscopic conductance increases in a non-linear fashion with the external Cl$^-$ concentration. In rat parotid acinar cells (Figure 4.1.2) and *Xenopus laevis* oocytes (Qu and Hartzell, 2000) the half maximum saturation is reached with 59 and 73 mM Cl$^-$, respectively.

Activation by intracellular Ca^{2+}

Based on the interaction with Ca^{2+}, at least three different mechanisms of CaCC activation can be distinguished (Hartzell *et al.*, 2005). The simplest mechanism is direct gating, that is, Ca^{2+} binds to the protein forming the channel and induces molecular rearrangements that lead to opening of the pore (Arreola *et al.*, 1996; Kuruma and Hartzell, 2000). A second type of mechanism is indirect gating of CaCCs by Ca^{2+}, via activation of a Ca^{2+} and calmodulin-dependent protein kinase (CaMKII), which phosphorylates the channel and induces pore opening (Nishimoto *et al.*, 1991; Wagner *et al.*, 1991; Worrell and Frizzell, 1991; Schumann *et al.*, 1993). A third mechanism of activation requires both Ca^{2+} and cGMP and has been described in smooth muscle cells (Matchkov *et al.*, 2004, 2005).

Evidence for direct gating comes from studies in several cellular preparations. For example, inside-out patches isolated from hepatocytes and *Xenopus* oocytes display currents that activate when

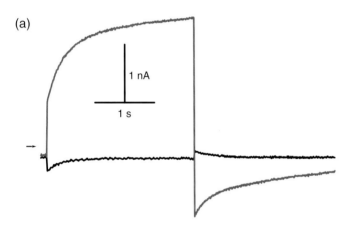

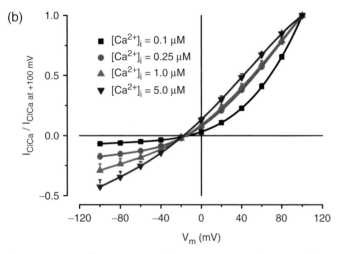

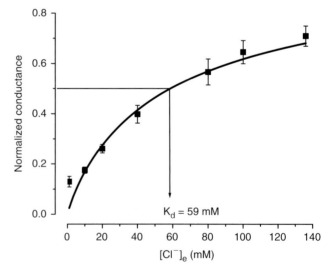

Fig. 4.1.1 Kinetics of current generated by CaCC activation and current–voltage relationships at different $[Ca^{2+}]_i$. (a) Raw $I_{Cl(Ca)}$ traces recorded at +100 (red) and –100 mV (black) from a parotid acinar cell dialysed with 250 nM free $[Ca^{2+}]_i$ and held at –50 mV. Arrow indicates 0 current level. (b) I-V curves obtained from cells dialysed with solutions containing the indicated free $[Ca^{2+}]$. See also colour plate section.

Fig. 4.1.2 Cl$^-$ dependency of CaCC conductance. Conductance was determined as external [Cl$^-$] was progressively increased. The continuous line is a fit with a bimolecular reaction scheme. The EC_{50} obtained was 59 mM. Conductance was estimated from whole-cell currents measured in isolated mouse parotid acinar cells dialysed with 250 nM Ca^{2+} as described in Arreola *et al.* (1996).

Ca^{2+} is repeatedly applied in the absence of ATP (Koumi et al., 1994; Kuruma and Hartzell, 2000). Most likely Ca^{2+} binds directly to the channel because in inside-out patches cytosolic molecules such as kinases or other cellular elements are lost. Furthermore, CaCCs are rapidly activated by photorelease of Ca^{2+} inside a cell in the absence of ATP, thus supporting the idea that phosphorylation plays little or no role in activation (Giovanucci et al., 2002). Finally, in rat parotid acinar cells, CaCCs activation is independent of calmodulin and CaMKII since their inhibition fails to prevent such activation (Arreola et al., 1998).

Curves of apparent open probabilities constructed as a function of intracellular $[Ca^{2+}]$ at different voltages have been used to determine the Ca^{2+} sensitivity of the channels. In general, the EC_{50} for channel activation is around 200–400 nM at positive voltages but increases to about 500–600 nM at negative potentials (Arreola et al., 1996). This behaviour partially accounts for the outwardly rectifying steady-state I-V relationship as previously mentioned (Figure 4.1.1b). In addition, the Vm dependency of Ca^{2+} binding suggests that Ca^{2+} may directly interact with the channel protein during activation. Dose–response curves show that the Hill coefficient is 2–4 suggesting that more than one Ca^{2+} ion interacts with the channel (Arreola et al., 1996; Frings et al., 2000; Nilius and Droogmans, 2002; Hartzell et al., 2005). At subsaturating $[Ca^{2+}]_i$ the macroscopic conductance increases in a sigmoidal fashion when Vm becomes more positive, indicating that the open probability of the channel is also enhanced by depolarization (Arreola et al., 1996; Kuomi et al., 1994; Nilius et al., 1997a; Kuruma and Hartzell, 2000).

Mathematical models that describe the direct activation of the channel by Ca^{2+} (see Figure 4.1.3) have been constructed incorporating both the Ca^{2+} and Vm dependency of open probability as well as channel kinetics (Arreola et al., 1996; Kuruma and Hartzell, 2000). In such models CaCCs are closed when $[Ca^{2+}]_i$ is low and the potential is hyperpolarized. As $[Ca^{2+}]_i$ is increased, CaCCs transit between several closed states before reaching the open state(s). The models successfully describe the kinetics and Vm-dependence of activation; however, most of them fail to describe tail currents and the step(s) that confers the Vm-dependence of channel opening. Certainly, further detailed analytical studies are required to build a single model for CaCCs activation. To do this, single channel data will be required to determine the number of closed and open states, rate constants of transition between states and Ca^{2+}- and Vm-dependence of rate constants.

CaCC activation is also dependent on the permeant anion. Those anions with higher permeability than Cl^- (SCN^-, NO_3^-, or I^-) speed up activation and slow down deactivation by a factor similar to the anion permeability ratio. This effect is not dependent on the channel affinity for Ca^{2+} (Greenwood and Large, 1999; Perez-Cornejo et al., 2004). Thus, it is possible that permeant anions increase channel opening by coupling the process of channel gating to the permeation mechanism.

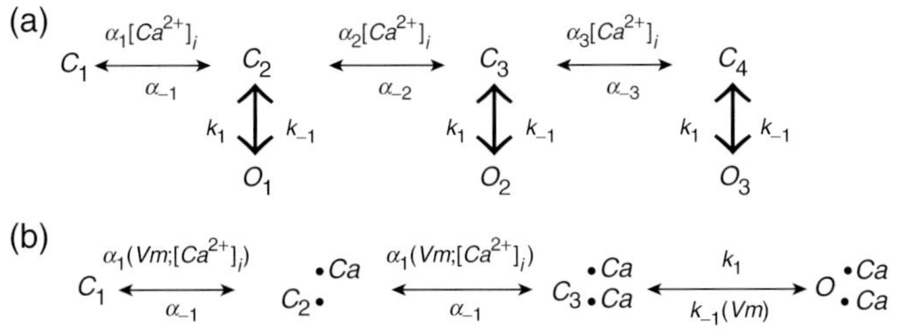

Evidence of indirect gating by Ca^{2+} was gathered using inhibitors of CaMKII. In preparations like airway epithelial cells, neutrophils and Jurkat T lymphocytes (Nishimoto et al., 1991; Schumann et al., 1993; Wagner et al., 1991; Worrell and Frizzell, 1991) these drugs block activation of CaCCs, suggesting that activation requires calmodulin and CaMKII. Accordingly, dialysis of activated CaMKII enzyme into airway epithelial cells (Wagner et al., 1991) or application of this enzyme to excised inside-out patches of Jurkat T lymphocytes (Nishimoto et al., 1991) cause CaCC activation.

Finally, a third gating mechanism was revealed in rat mesenteric artery smooth muscle cells. Two types of CaCCs are observed in these cells, one is activated directly by Ca^{2+} while the other is activated by Ca^{2+} in a cyclic guanosine monophosphate (cGMP)-dependent manner (Piper and Large, 2003; Matchkov et al., 2004). These two types of channels can be distinguished by their distinct sensitivities to blockers. cGMP-dependent CaCCs are blocked by Zn^{2+} but not by niflumic acid (NFA) whilst the 'classical' CaCCs are blocked by NFA but not Zn^{2+} (Matchkov et al., 2004). In addition, cGMP-dependent CaCCs are potentiated by calmodulin but unaffected by CaMKII blockade (Piper and Large, 2004). It is thought that cGMP acts via a cGMP-dependent protein kinase (Matchkov et al., 2004).

Ca^{2+} sensitivity of heterologously expressed Bestrophins has been determined. xBest-1, xBest-2, mBest-2 and hBest-4 are activated at +100 mV with an EC_{50} between 220–240 nM (Qu et al., 2003; 2004; Tsunenari et al., 2006). These values are within the physiological range and are similar to those determined for endogenous CaCCs (Kuomi et al., 1994; Arreola et al., 1996; Nilius et al., 1997b; Kuruma and Hartzell, 2000; Nilius and Droogmans, 2002).

Single-channel conductance

At the single channel level, CaCCs do not seem to be a homogeneous population. Diverse single-channel conductances (from 1 to 50 ps) have been reported in studies using different species and tissues. CaCCs with low conductance, between 1–3 pS, were described in Xenopus oocytes (Takahashi et al., 1987), cardiac myocytes (Collier et al., 1996), and arterial smooth muscle cells (Klockner, 1993; Piper and Large, 2003). In endothelial cells (Nilius et al., 1997a; Nilius and Droogmans, 2002), acinar cells from rat submandibular gland (Martin, 1993) and hepatocytes (Koumi et al., 1994) CaCCs have conductances of 3, 5 and 8 pS, respectively. Biliary (Schlenker and Fitz, 1996) and colon cell lines (Morris and Frizzell, 1993) express CaCCs with a 15 pS conductance. Single channel recordings of cGMP-dependent CaCCs from rat mesenteric artery smooth muscle cells show substate conductances of 15, 35 and 55 pS (Piper and Large, 2003). Intermediate conductance CaCCs, 40–50 pS, have been described in Xenopus spinal neurons (Hussy, 1992), vascular smooth muscle cells (Piper and Large, 2003), Jurkat T cells (Nishimoto et al., 1991), and airway epithelial cells (Frizzell et al., 1986). An even larger conductance, a maxi-CaCC of 310 pS was reported in Xenopus spinal neurons (Hussy, 1992). This myriad of single channel conductances as well as the different mechanisms of CaCC activation indicate the presence of multiple types of CaCCs.

Fig. 4.1.3 Two kinetic models of CaCC activation. (a) Kuruma and Hartzell's (2000) model with four closed and three open states. (b) Alternative model with three closed states and one open state from Arreola et al. (1996). Closed and open states are indicated as C and O respectively; CaCCs transition rate constants are indicated by αs and ks.

Ion permeability

Under physiological conditions, Cl^- is the main current carrier, but this does not mean that CaCCs are highly selective for Cl^-. In fact, anions with larger ionic radii, such as SCN^- or ClO^{4-} are considerably more permeant than Cl^-. CaCCs also show some cation permeability, for example the reported P_{Na}/P_{Cl} in *Xenopus* oocytes is 0.1 (Qu and Hartzell, 2000). Thus, in terms of anion permeation these channels must be regarded as non-selective. Most CaCCs studied, including those present in *Xenopus* oocytes (Qu and Hartzell, 2000), lachrymal (Evans and Marty, 1986), and parotid glands (Perez-Cornejo et al., 2004) have the same rank order of anion permeability: $SCN^->I^->Br^->Cl^->F^-$. However, there are exceptions and CaCCs in olfactory receptor neurons have a permeability sequence of $Cl^->F^->I^->Br^-$ (Hallani et al., 1998).

Permeant anions have complex effects on CaCC behaviour. For example, anions with high permeability ratios like $C(CN)_3^-$ and SCN^- become trapped in the pore and in fact block the channel in a Vm-dependent manner (Qu and Hartzell, 2000). This property has been extensively characterized in *Xenopus* oocytes where external SCN^- blocks CaCCs with a Kd at 0 mV of 1.7 mM (Qu and Hartzell, 2000). Additionally, highly permeant anions facilitate CaCC activation and slow down deactivation, which would increase channel open time. The degree of facilitation parallels permeation (Perez-Cornejo et al., 2004). This may suggest that CaCC gating is coupled to permeation, as has been shown in ClC channels.

The anion selectivity sequence displayed by endogenous CaCCs is reproduced by Bestrophins (Qu and Hartzell, 2004; for a comparison of CaCCs vs Bestrophins see Arreola and Perez-Cornejo, 2006). For example, HEK 293 cells transfected with mBest-2 and studied using whole-cell patch in the presence of saturating $[Ca^{2+}]_i$, display CaCCs with the following anion selectivity: $SCN>I>Br>Cl>F$. Likewise, mBest-2 displays other characteristics present in CaCCs. mBest-2 is blocked by SCN^- with a Kd of 9.8 mM but in a Vm-insensitive manner (Qu and Hartzell, 2004). However, introducing the mutation S79T in mBest-2 decreases SCN^- affinity substantially and causes the block to become Vm-dependent (Qu and Hartzell, 2004).

4.1.4 Cellular and subcellular localization

CaCCs are expressed in different cell types including brown fat adipocytes (Pappone and Lee, 1995), Ehrlich ascites (Papassotiriou et al., 2001), endothelial cells (Nilius et al., 1997a), epithelial cells (Kidd and Thorn, 2000), hepatocytes (Koumi et al., 1994), insulin-secreting beta cells (Kozak and Logothetis, 1997), lymphocytes (Nishimoto et al., 1991), mast cells (Matthews et al., 1989), cardiac (Hirayama et al., 2002), skeletal (Kourie et al., 1996) and smooth muscle cells (Large and Wang, 1996), neurons (Owen et al., 1984; Sanchez-Vives and Gallego, 1994), neutrophils (Schumann et al., 1993), oocytes and eggs (Machaca et al., 2002), renal cells (Rubera et al., 2000); secretory glands (Melvin et al., 2005), sensory receptors (Thoreson and Burkhardt, 1991; Lowe and Gold, 1993; Taylor and Roper, 1994), Sertoli cells (Lalevee and Joffre, 1999) and *Vicia faba* guard cells (Schroeder and Hagiwara, 1989, Table 1) (reviewed by Hartzell et al., 2005 and Fuller, 2002).

A local increase in $[Ca^{2+}]_i$ near the apical membrane of secretory epithelial cells activates CaCCs although the same manipulation on the basolateral membrane has no effect (Tan et al., 1992). This indicates that CaCCs expression is mainly restricted to the apical membrane (Tan et al., 1992; Giovanucci et al., 2002). CaCCs

co-localize to the apical membrane with CFTR, a Cl^- channel defective in cystic fibrosis (CF) patients (Quinton, 1983; Anderson and Welsh, 1991; Gabriel et al., 2000). Co-expression of CFTR and CaCC results in down regulation of the CaCC conductance, possibly by direct protein–protein interaction (Kunzelmann, 1997; Wei et al., 2001; Perez-Cornejo and Arreola, 2004). Thus, CaCCs could be suitable candidates to rescue fluid secretion in CF patients since CaCC current increases in the absence of CFTR. However, this possibility remains untested due to the lack of specific activators for CaCCs.

The Ca^{2+}-gated Cl^- channel, Best-1, is localized in the basolateral membrane of RPE cells from macaque and porcine (Marmorstein et al., 2000), where it might play a major role in Cl^- fluxes during the dark–light cycle.

In skeletal muscle cells, large conductance CaCCs have been identified in the membrane of the sarcoplasmic reticulum (Dulhunty and Laver, 2002). Although the physiological role of these intracellular channels is unknown, it has been suggested that they could act as counterion carriers during Ca^{2+} release. More data is needed to understand the role of CaCCs in intracellular organelles.

4.1.5 Pharmacology

To date, no specific ligands have been found to selectively inhibit or enhance CaCC activity. Such ligands are essential in order to assay the physiological role of these proteins. The most common blockers for native CaCCs are in general non-selective and high concentrations are needed to see any effect (see Hartzell et al. 2005 for a comprehensive list of different drugs and concentrations used to block CaCC activity). Despite this, niflumic and flufenamic acids have been employed as 'specific blockers' of CaCCs and at concentrations of 10 μM these drugs can inhibit CaCCs from *Xenopus* oocytes (Qu and Hartzell, 2001). However, NFA also blocks Ca^{2+}, K^+ and volume-regulated channels and can even enhance CaCCs current in smooth muscle at negative Vm (Reinsprecht et al., 1995; Wang et al., 1997; Xu et al., 1997; Doughty et al., 1998; Piper et al., 2002).

Less effective blockers of CaCCs include tamoxifen, 4,4'-di-isothiocyanatostilbene-2,2'-disulphonic acid (DIDS), 4-aceta-mido-4'-isothiocyanatostilbene-2,2'-disulfonic acid (SITS), 5-nitro-2-(3-phenylpropylamino) benzoic acid (NPPB), anthracene-9-carboxylic acid (A9C), diphenylamine-2-carboxylic acid (DPC), fluoxetine and mefloquine (Frings et al., 2000; Maertens et al., 2000). The small peptide chlorotoxin appears to specifically block CaCCs from rat astrocytoma cells (Dalton et al., 2003) but is ineffective when used on CaCCs from secretory epithelia (Maertens et al., 2000a).

Some blockers including DIDS, SITS, A9C and NFA exert their effect in a Vm-dependent manner (Qu and Hartzell, 2001; Qu et al., 2003) which indicates that these molecules are likely to penetrate into the pore, plug it and thus inhibit ion permeation. As a result, these Vm-sensitive blockers have been used as molecular probes to obtain information about the size and geometry of the pore. Recently, a detailed analysis of the correlation between blocker structure and CaCCs blockade has been undertaken to create a model of pore architecture (Qu and Hartzell, 2001). According to this model, the pore of CaCC has a large opening (0.6×0.94 nm) with an elliptical cone shape that opens into the extracellular space.

4.1.6 Physiology

Physiological functions of CaCCs

CaCCs serve many important functions in the cells where they are expressed. Table 4.1.2 summarizes the cell distribution of CaCCs as well as possible functional roles. In most cells, CaCC activation results in Cl$^-$ exit and subsequent depolarization of the cell membrane (Hartzell *et al.*, 2005). This Cl$^-$ efflux and the depolarization that accompanies it have a critical role in epithelial secretion (Quinton, 1983; Wagner *et al.*, 1991; Grubb and Gabriel, 1997), membrane excitability in cardiac muscle and neurons (Zygmunt and Gibbons, 1992; Kawano *et al.*, 1995; Frings *et al.*, 2000), olfactory transduction (Frings *et al.*, 2000), regulation of vascular tone (Large and Wang, 1996) and modulation of photoreceptor light responses (Kaneko *et al.*, 2006). In addition, membrane depolarization in *Xenopus* oocytes somehow prevents fusion of additional sperm during fertilization (Cross and Elinson, 1980). Unfortunately, the role of CaCCs in other cells is unknown or unsettled. This is the case for cells like brown fat adipocytes, Ehrlich ascites, endothelial cells, hepatocytes, leech neurons, lymphocytes, mast cells, neutrophils, Sertoli cells, and *Vicia faba* guard cells. Excellent reviews on the physiological roles of CaCCs have been published elsewhere (see Large and Wang, 1996; Kotlikoff and Wang, 1998; Frings *et al.*, 2000; Fuller, 2002; Eggermont, 2004; Hartzell *et al.*, 2005; Leblanc *et al.*, 2005).

Regulation

CaCC activity is highly regulated in several ways, including Ca^{2+}-dependent enzymes (CaMKII, protein phosphatases, and annexins), inositol 3,4,5,6-tetrakisphosphate (IP$_4$), H$^+$, cGMP, G-proteins and other regulatory proteins such as CFTR. The mechanisms that underlie CaCC regulation are largely unknown. However, understanding these regulatory mechanisms will shed light on the role of CaCCs under normal and abnormal conditions.

CaMKII and IP$_4$

CaMKII can activate CaCC in T84 cells and neutrophils (Worrell and Frizzell, 1991; Schumann *et al.*, 1993), however, in arterial and tracheal smooth muscle CaMKII inhibits CaCC (Greenwood *et al.*, 2001; Ledoux *et al.*, 2003; Leblanc *et al.*, 2005). In CFPAC-1 and T84 colonic carcinoma cells the CaMKII effect is modulated by IP$_4$ (Vajanaphanich *et al.*, 1994; Ismailov *et al.*, 1996; Xie *et al.*, 1996; Nilius *et al.*, 1998; Xie *et al.*, 1998) via channel dephosphorylation because the IP$_4$ effect is prevented by inhibiting phosphatase activity.

Table 4.1.2 Physiological roles of CaCCs in different cell types

Cell type	Influence on membrane potential	Physiological role	References
Acinar cells of secretory glands	Depolarization	Fluid secretion associated with Cl$^-$ efflux	[1, 2]
Cardiac myocytes	Repolarization of action potentials	Likely to be important for cardiac rhythm	[3, 4]
Endothelial cells	Depolarization	May control membrane potential, vectorial transport, modulation of agonist-induced Ca^{2+} signals and cell proliferation	[5, 6]
Epithelial cells	Depolarization	Fluid secretion associated with Cl$^-$ efflux. Regulation of surface liquid in airway epithelium	[7–9]
Frog eggs	Depolarization	Fast block of polyspermy.	[10–12]
Neurons: autonomic system and dorsal root ganglion	Depolarization and generation of after-depolarizing potentials	Regulation of neuronal firing activity	[13–15]
Neurons: spinal cord	Repolarization and decrease of neuronal excitability (E$_{Cl}$ ~60 mV)	Regulation of neuronal firing activity	[16, 17]
Renal cells	Depolarization	Likely to contribute to urine composition through Cl$^-$ secretion in the distal tubule	[18, 19]
Sensory receptor cells	Depends on cell type. For example: depolarization in olfactory neurons and hyperpolarization in taste cells from *Necturus*	Amplification of receptor current in olfactory neurons. Probable participation in taste adaptation and lateral inhibition in the retina	[20–25]
Smooth muscle cells	Depolarization	Promote opening of Vm-dependent Ca^{2+} channels, this in turn augments muscle contraction. Probable participation in penile erection (corpus cavernosum smooth muscle cells)	[26–31]

[1] Evans MG *et al.* (1986) *J Physiol* 378, 437–460; [2] Melvin JE *et al.* (2005) *Ann Rev Physiol* 67, 445–469; [3] Hiraoka M *et al.* (1998) *Cardiovascular Res* 40, 23–33; [4] Hirayama Y *et al.* (2002) *Jpn J Physiol* 52, 293–300; [5] Nilius B *et al.* (1997) *J Physiol* 498, 381–396; [6] Zhong N *et al.* (2000) *Acta Pharmacol Sinica* 21, 215–220; [7] Kidd JF and Thorn P (2000) *Ann Rev Physiol* 62, 493–513; [8] Kidd JF and Thorn P (2001) *Pflugers Arch* 441, 489–497; [9] Tarran R *et al.* (2002) *J Gen Physiol* 120, 407–418; [10] Webb DJ and Nuccitelli R (1985) *Dev Biol* 107, 395–406; [11] Machaca K *et al.* (2002) *Calcium-activated chloride channels* pp. 3–39; [12] Glahn D and Nuccitelli R (2003) *Dev Growth Differentiation* 45, 187–197; [13] Mayer ML (1985) *J Physiol* 364, 217–239; [14] Sanchez-Vives MV and Gallego R (1993) *J Physiol* 471, 801–815; [15] Sanchez-Vives MV and Gallego R (1994) *J Physiol* 475, 391–400; [16] Owen DG *et al.* (1984) *Nature* 311, 567–570; [17] Owen DG *et al.* (1986) *J Neurophysiol* 55, 1115–1135; [18] Bidet M *et al.* (1996) *Am J Physiol* 271, F940–F950; [19] Rubera I *et al.* (2000) *Am J Physiol* 279, F102–F111; [20] Maricq AV and Korenbrot JI (1988) *Neuron* 1, 503–515; [21] Thoreson WB and Burkhardt DA (1991) *J Neurophysiol* 65, 96–110; [22] Lowe G and Gold GH (1993) *Nature* 366, 283–286; [23] Kurahashi T and Yau K-W (1994) *J Gen Physiol* 115, 59–80; [24] Taylor R and Roper S (1994), *J Neurophysiol* 72, 475–478; [25] Reisert J *et al.* (2003) *J Gen Physiol* 122, 349–364; [26] Pacaud P *et al.* (1991) *Br J Pharmacol* 104, 1000–1006; [27] Large WA and Wang Q (1996) *Am J Physiol* 271, C435–C454; [28] Yuan XJ (1997) *Am J Physiol* 272, L959–L968; [29] Lamb FS and Barna TJ (1998) *Am J Physiol* 275, H151–H160; [30] Karkanis T *et al.* (2003) *J Appl Physiol* 94, 301–313; [31] Craven M *et al.* (2004) *J Physiol* 556, 495–506.

This later type of regulation could serve to adjust CaCC activation during electrical activity. Other Ca^{2+}-dependent enzymes such as calcineurin and alkaline phosphatase act also as positive regulators of CaCC (Marunaka and Eaton, 1990; Wang and Kotlikoff, 1997).

Calmodulin

Calmodulin decreases CaCC sensitivity to intracellular Ca^{2+} in *Odora* cells. Hindering the Ca^{2+} binding sites 1, 2 and 4 located in EF-hands of calmodulin decreases Ca^{2+} sensitivity of CaCCs by approximately twofold (Kaneko *et al.*, 2006). In sensory neurons, CaCC regulation by calmodulin could be critical to amplify sensory stimuli.

Ca^{2+}-dependent K^+ channel regulators

NFA, a CaCC blocker, is able to increase the activity of Ca^{2+}-dependent K^+ channels (Ottolia and Toro, 1994). Furthermore, the putative Ca^{2+}-gated Cl^- channel CLCA appears to be modulated by the beta 1 subunit of Ca^{2+}-dependent K channels when both proteins are co-expressed in HEK 293 cells (Greenwood *et al.*, 2002). These observations suggest that CaCC and the large conductance Ca^{2+}- activated K^+ channels (BKCa channels) may have some structural resemblance, however, additional experiments are needed to test this hypothesis.

Annexins

Phospholipids- and annexins inhibit CaCCs (Chan *et al.*, 1994; Kaetzel *et al.*, 1994; Jorgensen *et al.*, 1997). CaCCs from *Xenopus* oocytes are inhibited quite potently (IC_{50} near 50 nM) by annexins II, III and V isolated from Ehrlich ascites cells. However, porcine and bovine annexins type II and V do not inhibit CaCC. Block by annexins is potentiated by IP_4 (Xie *et al.*, 1996).

G-proteins

Application of GTPγS to inside-out patches isolated from submandibular acinar cells induces the appearance of small-conductance CaCCs (Martin, 1993) which may suggest that CaCCs are regulated by G-proteins. New data indicate that CaCCs can be indirectly regulated by IP_3 through activation of a PLC β3-like enzyme, following G-protein stimulation with ginsenosides or GTP-γ-S (Choi *et al.*, 2001; Kaibara *et al.*, 2001; Kilic and Fitz 2002).

CFTR

CFTR, a Cl^- channel defective in cystic fibrosis disease, regulates several ion channels, including CaCCs (Kunzelmann *et al.*, 2000; Kunzelmann, 2001; Tarran *et al.*, 2002). Expression of CFTR in either bovine pulmonary artery endothelial (CPAE) cells (Wei *et al.*, 2001) or *Xenopus* oocytes (Kunzelmann *et al.*, 1997) reduces endogenous CaCC current. Similarly, CaCC current obtained from mouse parotid acinar cells isolated from CFTR knockout mice is larger than the current obtained from WT mice (Perez-Cornejo and Arreola 2004). Moreover, expression of CFTR lacking part of the R-domain does not produce CaCC current attenuation (Wei *et al.*, 2001). Thus, it has been proposed that an interaction between the C-terminal part of the R-domain and CaCC could be the underlying mechanism for this modulation.

Protons

Most of the Cl^- channels characterized to date are sensitive to changes in internal pH, external pH, or both. In *Xenopus* oocytes, extracellular alkalinization decreases CaCC current at positive Vm (Qu and Hartzell, 2000). In contrast, intracellular acidification has little effect, even though the inward current can be blocked by intracellular alkalinization (Qu and Hartzell, 2000). Conversely, intracellular acidification inhibits CaCCs in acinar cells from lachrymal and parotid glands, and T84 cells (Arreola *et al.*, 1995; Park and Brown, 1995). Despite the proton sensitivity displayed by CaCCs, the mechanism that underlies proton regulation is unknown. A partial explanation could be that intracellular protons compete for Ca^{2+} binding sites on the channel, thus decreasing its open probability.

4.1.7 Relevance to disease state

CaCC activity may be linked in some way(s) to arrhythmogenesis or its prevention since CaCC activation contributes to the transient outward current that repolarizes the cardiac myocyte (Zygmunt, 1997, 1998). This current is abnormal in German shepherd dogs with inherited arrhythmias which are prone to cardiac sudden death (Freeman *et al.*, 1997). It is possible that changes in the distribution or the properties of CaCCs could account for the cardiac affection to some extent. On the other hand, CaCCs have been proposed to play an anti-arrhythmogenic role during acidosis induced by ischemia or other pathologic events. Extracellular acidification enhances CaCC activity favouring repolarization, which in turn prevents early depolarizations and other pathological cardiac activities (Hirayama *et al.*, 2002). Accordingly, blocking CaCCs with DIDS during external acidosis results in a 13% increment of pathological activity (Hirayama *et al.*, 2002). In contrast, CaCC block by 9-AC and SITS results in cardioprotection against ischaemia–reperfusion damage (Tanaka *et al.*, 1996). CaCCs might be a factor in arrhythmogenic activity due to its contribution to a transient cardiac inward current linked to arrhythmogenesis (Han and Ferrier, 1992; 1996; Zygmunt *et al.*, 1998).

Another example of a disease linked to CaCC activity comes from Bestrophins rather than classical CaCCs. To date more than 85 mutations in the *hBest-1* gene have been associated with early macular degeneration, a disease that causes blindness (Hartzell *et al.*, 2005). These mutations are found within the first 300 residues of the Best-1 protein. Some of these amino acids are located in what appears to be the transmembrane domain 2 that forms part of the pore (Qu *et al.*, 2006). Linking hBest-1 abnormal function to macular degeneration is not straightforward. In fact, it has been recently shown that hBest-1 regulates a Vm-gated Ca^{2+} channel (Rosenthal *et al.*, 2006) and that the light peak of the electro-oculogram (whose affectation is diagnostic of macular degeneration) is diminished by nimodipine (a Ca^{2+} channel blocker) or ablation of β_4, a Ca^{2+} channel subunit (Marmorstein *et al.*, 2006). The light peak, initially thought to be generated mainly by Best-1 activity, does not change following ablation of Best-1 in mice. Moreover, no change was seen in the whole-cell current recorded from Best-1 knockout mice RPE cells dialysed with high Ca^{2+} solutions (Marmorstein *et al.*, 2006). Thus, the electro-oculogram light peak has a component mediated by Ca^{2+} channel activity that seems to be regulated by Best-1 but it may not be a good diagnostic of Best1-associated macular dystrophy.

4.1.8 Concluding remarks

CaCCs are presented here as the prototype of Cl^- channels activated by intracellular ligand. An increase in intracellular Ca^{2+} ions, resulting from Ca^{2+} influx and/or release from intracellular stores, can interact by itself or together with other molecules to activate CaCCs. So far, there is evidence of at least three distinctive mechanisms of channel activation: the first and best characterized is through a direct interaction between Ca^{2+} and the channel; the second mechanism involves activation of various kinases by Ca^{2+} and subsequent activation of the channel; the third and most recently reported mechanism is the activation of the channel by Ca^{2+} in the presence of cGMP. Several ions as well as proteins can modulate CaCC activity; these include protons, permeant anions, CaMKII, IP_4, annexins, calmodulin, G-proteins, phosphatases and CFTR. Finally, agonists that mobilize Ca^{2+} dynamically also regulate CaCCs.

The presence of multiple mechanisms of activation suggests the existence of more than one type of CaCC; this idea finds further support in the range of values reported for single channel conductance. In analogy with Ca^{2+}-activated K^+ channels the diversity in single channel conductance could mean that more than one gene encodes CaCCs. Accordingly, several putative genes have been reported to encode these Cl^- channels but the CaCC molecular identity remains unclear. In search of candidate genes the families of CLCA, Bestrophins and Tweety proteins have been characterized. Bestrophins are until now the best studied proteins for which there is evidence indicating that they form a Ca^{2+}-sensitive pore that replicates some of the functional properties of CaCCs.

Many physiological roles of CaCCs are dependent on depolarization or the osmotic water drag induced by Cl^- exit. Unfortunately, many other physiological functions have not been critically evaluated because of the lack of specific blockers and CaCCs clones. Thus, a research area that needs to be aggressively developed is the search for drugs that will specifically bind to CaCCs. Advances in this field would also help us understand the contribution of this channel to disease states and potentially enable the development of novel therapeutics.

Note added in proof

Recently three different groups have shown that the TMEM16 family of 'transmembrane proteins with unknown function' are a major subunit of Ca^{2+}-dependent chloride channels. The TMEM16 family was identified through multiple approaches including expression cloning (Schroeder et al., 2008), searches in public domain databases (Yang et al., 2008) and through analysis of IL-4-upregulated membrane proteins using global gene expression analysis (Caputo et al., 2008). In all cases expression of TMEM16A (also known as anoctamin 1 or ANO1) or TMEM16B, cloned from Xenopus oocytes, mouse or airway epithelial cells, resulted in the appearance of anion-selective currents that were activated by increases in intracellular Ca^{2+} stimulated via Ca^{2+}-mobilizing receptors, IP_3 or Ca^{2+} ionophores. TMEM16 currents were inhibited by known blockers of CaCC and by siRNA-mediated knockdown of TMEM16 protein. The current displayed the classical time-dependence at positive voltages with large tail currents upon repolarization. Single channel conductance was 8.3 pS and the Ca^{2+} EC_{50} values were 2.6 and 0.4 μM at −60 and +60 mV, respectively. As expected, TMEM16 was found in mammary and salivary glands, lung, pancreas, kidney, retina, dorsal root ganglion, prostate, large intestine, trachea, uterus and, vomeronasal organ. An initial attempt to assay a functional role of TMEM16 in saliva production was conducted using mice treated with siRNA. In these animals pilocarpine-induced saliva production was significantly reduced. All functional properties and physiological responses are consistent with TMEM16 being the classical CaCC.

Acknowledgements

The authors research is supported in part by NIH grants DE-09692 and DE13539, Fogarty International Center grant R03TW006429 (JE Melvin), P01-HL18208 (R Waugh), CONACyT-Mexico grants 42561 (J. Arreola) and 45895 (P Pérez-Cornejo) and The Academy of Sciences for the Developing World (TWAS) grant 04–459RG/BIO/LA (P Perez-Cornejo).

References

Anderson MP and Welsh MJ (1991). Calcium and cAMP activate different chloride channels in the apical membrane of normal and cystic fibrosis epithelia. *Proc Natl Acad Sci USA* **88**, 6003–6007.

Arreola J and Pérez-Cornejo P (2006). Functional properties of Ca^{2+}-dependent Cl^- channels and bestrophins: do they correlate? In E Bittar and M Pusch, eds, *Chloride movements across cellular membranes. Advances in Molecular and Cell Biology series.* San Diego, California, USA. Elsevier, pp. 181–198.

Arreola J, Begenisich T, Nehrke K et al. (2002). Secretion and cell volume regulation by salivary acinar cells from mice lacking expression of the Clcn3 Cl^- channel gene. *J Physiol* **545**, 207–216.

Arreola J, Melvin JE and Begenisich T (1995). Inhibition of Ca^{2+}-dependent Cl^- channels from secretory epithelial cell by low internal pH. *J Membr Biol* **147**, 95–104.

Arreola J, Melvin JE and Begenisich T (1996). Activation of calcium-dependent chloride channels in rat parotid acinar cells. *J Gen Physiol* **108**, 35–48.

Arreola J, Melvin JE and Begenisich T (1998). Differences in regulation of Ca2+- activated Cl^- channels in colonic and parotid secretory cells. *Am J Physiol* **274**, C161–C166.

Bader CR, Bertrand D and Schwartz EA (1982). Voltage-activated and calcium-activated currents studied in solitary rod inner segments from the salamander retina. *J. Physiol*, **331**, 253–284.

Bakall B, Marknell T, Ingvast S et al. (1999). The mutation spectrum of the bestrophin protein—functional implications. *Hum Genet* **104**, 383–389.

Bao L, Kaldany C, Holmstrand EC et al. (2004). Mapping the BK_{Ca} channel's 'Ca2+ bowl': side-chains essential for Ca^{2+} sensing. *J Gen Physiol* **123**, 475–489.

Beckley JR, Pauli BU and Elble RC (2004). Re-expression of detachment-inducible chloride channel mCLCA5 suppresses growth of metastatic breast cancer cells. *J Biol Chem* **279**, 41634–41641.

Brown EM (2000). G protein-coupled, extracellular Ca2+ (Ca^{2+}_o)-sensing receptor enables Ca^{2+}_o to function as a versatile extracellular first messenger. *Cell Biochem Biophys* **33**, 63–95.

Brown EM and MacLeod RJ (2001). Extracellular calcium sensing and extracellular calcium signaling. *Physiol Rev* **81**, 239–297.

Brown EM, Pollak M and Hebert SC (1998). The extracellular calcium-sensing receptor: its role in health and disease. *Annu Rev Med* **49**, 15–29.

Caputo A, Caci E, Ferrera L et al. (2008). TMEM16A, A membrane protein associated with calcium-dependent chloride channel activity. *Science* **322**, 590–594.

Chan HC, Kaetzel MA, Gotter AL et al. (1994). Annexin IV inhibits calmodulin-dependent protein kinase II-activated chloride conductance: a novel mechanism for ion channel regulation. *J Biol Chem* **269**, 32464–32468.

Chien L-T, Zhang Z-R and Hartzell HC (2006). Single Cl^- channels activated by Ca^{2+} in Drosophila S2 cells are mediated by bestrophins. *J Gen Physiol* **128**, 247–259.

Choi S, Rho SH, Jung SY *et al.* (2001). A novel activation of Ca^{2+}-activated Cl^- channel in *Xenopus* oocytes by ginseng saponins: evidence for the involvement of phospholipase C and intracellular Ca^{2+} mobilization. *Br J Pharmacol* **132**, 641–648.

Collier ML, Levesque PC, Kenyon JL *et al.* (1996). Unitary Cl^- channels activated by cytoplasmic Ca^{2+} in canine ventricular myocytes. *Circ Res* **78**, 936–944.

Cross NL and Elinson RP (1980). A fast block to polyspermy in frogs mediated by changes in the membrane potential. *Dev Biol* **75**, 187–198.

Cunningham SA, Awayda MS, Bubien JK *et al.* (1995). Cloning of an epithelial chloride channel from bovine trachea. *J Biol Chem* **270**, 31016–31026.

Dalton S, Gerzanich V, Chen M *et al.* (2003). Chlorotoxin-sensitive Ca^{2+}-activated Cl^- channel in type R2 reactive astrocytes from adult rat brain. *Glia* **42**, 325–339.

del Castillo J and Katz B (1955). On the localization of acetylcholine receptors. *J Physiol* **128**, 157–181.

Doughty JM, Miller AL and Langton PD (1998). Non-specificity of chloride channel blockers in rat cerebral arteries: block of the L-type calcium channel. *J Physiol* **507**, 433–439.

Dulhunty AF and Laver DR (2002). A Ca^{2+}-activated anion channel in the sarcoplasmic reticulum of skeletal muscle. In CM Fuller, ed., *Current topics in membranes 53. Calcium-activated chloride channels*. San Diego, Academic Press, pp. 59–80.

Eggermont J (2004). Calcium-activated chloride channels: (un)known, (un) loved? *Proc Am Thor Soc* **1**, 22–27.

Elble RC, Ji G, Nehrke K *et al.* (2002). Molecular and functional characterization of a murine calcium-activated chloride channel expressed in smooth muscle. *J Biol Chem* **277**, 18586–18591.

Elble RC, Widom J, Gruber AD *et al.* (1997). Cloning and characterization of lung endothelial cell adhesion molecule—I suggest it is an endothelial chloride channel. *J Biol Chem* **272**, 27853–27861.

Evans MG and Marty A (1986). Calcium dependent chloride currents in isolated cells from rat lacrimal glands. *J Physiol* **378**, 437–460.

Evans MG, Marty A, Tan YP *et al.* (1986). Blockage of Ca^{2+}-activated Cl^- conductance by furosemide in rat lacrimal glands. *Pflugers Arch* **406**, 65–68.

Evans SR, Thoreson WB and Beck CL (2004). Molecular and functional analyses of two new CLCA family members from mouse eye and intestine. *J Biol Chem* **279**, 41792–41800.

Fahmi M, Garcia L, Taupignon A *et al.* (1995). Recording of a large conductance chloride channel in normal rat lactotrophs. *Am J Physiol* **269**, E969-E976.

Freeman LC, Pacioretty LM, Moise NS *et al.* (1997). Decreased density of Ito in left ventricular myocytes from German shepherd dogs with inherited arrhythmias. *J Cardiovasc Electrophysiol* **8**, 872–883.

Frings S, Reuter D and Kleene SJ (2000). Neuronal Ca^{2+}-activated Cl^- channels: homing in on an elusive channel species. *Prog Neurobiol* **60**, 247–289.

Frizzell RA, Rechkemmer G and Shoemaker RL (1986). Altered regulation of airway epithelial cell chloride channels in cystic fibrosis. *Science* **233**, 558–560.

Fuller CM (2002). Calcium-activated chloride channels. *Current Topics in Membranes 53*. San Diego, Academic Press.

Fuller CM and Benos DJ (2000). Electrophysiological characteristics of the Ca^{2+}-activated Cl^- channel family of anion transport proteins. *Clin Exp Pharmacol Physiol* **27**, 906–910.

Gabriel SE, Makhlina M, Martsen E *et al.* (2000). Permeabilization via the P2X7 purinoreceptor reveals the presence of a Ca^{2+}-activated Cl^- conductance in the apical membrane of murine tracheal epithelial cells. *J Biol Chem* **275**, 35028–35033.

Gandhi R., Elble RC, Gruber AD *et al.* (1998). Molecular and functional characterization of a calcium-sensitive chloride channel from mouse lung. *J Biol Chem* **273**, 32096–32101.

Gibson A, Lewis AP, Affleck K *et al.* (2005). hClCA1 and mClCA3 are secreted non-integral membrane proteins and therefore are not ion channels. *J Biol Chem* **280**, 27205–27212.

Giovannucci DR, Bruce JIE, Straub SV *et al.* (2002). Cytosolic Ca^{2+} and Ca^{2+}-activated Cl^- current dynamics: insights from two functionally distinct mouse exocrine cells. *J Physiol* **540**, 469–484.

Greenwood IA and Large WA (1999). Modulation of the decay of Ca^{2+}-activated Cl– currents in rabbit portal vein smooth muscle cells by external anions. *J Physiol* **516**, 365–376

Greenwood IA, Ledoux J and Leblanc N (2001). Differential regulation of Ca^{2+}-activated Cl^- currents in rabbit arterial and portal vein smooth muscle cells by Ca^{2+}-calmodulin-dependent kinase. *J Physiol* **534**, 395–408.

Greenwood IA, Miller LJ, Ohya S *et al.* (2002). The large conductance potassium channel beta-subunit can interact with and modulate the functional properties of a calcium-activated chloride channel, CLCA1. *J Biol Chem* **277**, 22119–22122.

Grubb BR and Gabriel SE (1997). Intestinal physiology and pathology in gene-targeted mouse models of cystic fibrosis. *Am J Physiol* **273**, G258-G266.

Gruber AD and Pauli BU (1998). Molecular cloning and biochemical characterization of a truncated, secreted member of the human family of Ca^{2+}-activated Cl channels. *Biochim Biophys Acta* **1444**, 418–423.

Gruber AD and Pauli BU (1999). Clustering of the human CLCA gene family on the short arm of chromosome 1 (1p22–31). *Genome* **42**, 1030–1032.

Gruber AD, Elble RC, Ji HL *et al.* (1998). Genomic cloning, molecular characterization and functional analysis of human CLCA1, the first human member of the family of Ca^{2+}- activated Cl^- channel proteins. *Genomics* **54**, 200–214.

Gruber AD, Schreur KD, Ji HL *et al.* (1999). Molecular cloning and transmembrane structure of hCLCA2 from human lung, trachea and mammary gland. *Am J Physiol* **276**, C1261–C1270.

Hallani M, Lynch JW and Barry PH (1998). Characterization of calcium-activated chloride channels in patches excised from the dendritic knob of mammalian olfactory receptor neurons. *J Memb Biol* **161**, 163–171.

Han X and Ferrier GR (1992). Ionic mechanisms of transient inward current in the absence of Na^+-Ca^{2+} exchange in rabbit cardiac Purkinje fibres. *J Physiol* **456**, 19–38.

Han X and Ferrier GR (1996). Transient inward current is conducted through two types of channels in cardiac Purkinje fibres. *J Mol Cell Cardiol* **28**, 2069–2084.

Hartzell C, Putzier I and Arreola J (2005). Calcium-activated chroride channels. *Annu Rev Physiol* **67**, 719–758.

Hartzell HC and Qu Z (2003). Chloride currents in acutely isolated *Xenopus* retinal pigment epithelial cells. *J Physiol* **549**, 453–469.

Hirayama Y, Kuruma A, Hiraoka M *et al.* (2002). Calcium-activated Cl^- current is enhanced by acidosis and contributes to the shortening of action potential duration in rabbit ventricular myocytes. *Jpn J Physiol* **52**, 293–300.

Huang P, Liu J, Di A *et al.* (2001). Regulation of human CLC-3 channels by multifunctional Ca^{2+}/calmodulin-dependent protein kinase. *J Biol Chem* **276**, 20093–20100.

Hussy N (1992). Calcium-activated chloride channels in cultures embryonic *Xenopus* spinal neurons. *J Neurophysiol* **68**, 2042–2050.

Ismailov II, Fuller CM, Berdiev BK *et al.* (1996). A biologic function for an 'orphan' messenger: D-myo-inositol 3,4,5,6-tetrakisphosphate selectively blocks epithelial calcium-activated chloride channels. *Proc Natl Acad Sci USA* **93**, 10505–10509.

Jorgensen AJ, Bennekou P, Eskesen K *et al.* (1997). Annexins from Ehrlich ascites inhibit the calcium-activated chloride current in *Xenopus laevis* oocytes. *Eur J Physiol* **434**, 261–266.

Kaetzel MA, Chan HC, Dubinsky WP *et al.* (1994). A role for annexin IV in epithelial cell function. Inhibition of calcium-activated chloride conductance. *J Biol Chem* **269**, 5297–5302.

Kaibara M, Nagase Y, Murasaki O et al. (2001). GTPgS-induced Ca^{2+} activated Cl$^-$ currents: its stable induction by G$_q$ alpha overexpression in *Xenopus* oocytes. *Jpn J Pharmacol* **86**, 244–247.

Kaneko H, Mohrlen F and Frings S (2006). Calmodulin contributes to gating control in olfactory calcium-activated chloride channels. *J Gen Physiol* **127**, 737–748.

Katz B and Miledi R (1972). The statistical nature of the acetycholine potential and its molecular components. *J Physiol* **224**, 665–699.

Kawano S, Hirayama Y and Hiraoka M (1995). Activation mechanism of Ca^{2+} sensitive transient outward current in rabbit ventricular myocytes. *J Physiol* **486**, 593–604.

Kidd JF and Thorn P (2000). Intracellular Ca^{2+} and Cl$^-$ channel activation in secretory cells. *Ann Rev Physiol* **62**, 493–513.

Kilic G and Fitz JG (2002). Heterotrimeric G-proteins activate Cl$^-$ channels through stimulation of a cyclooxygenase-dependent pathway in a model liver cell line. *J Biol Chem* **277**, 11721–11727.

Klockner U (1993). Intracellular calcium ions activate a low-conductance chloride channel in smooth-muscle cells isolated from human mesenteric artery. *Pflugers Arch.* **424**, 231–237.

Kotlikoff MI and Wang YX (1998). Calcium release and calcium-activated chloride channels in airway smooth muscle cells. *Am J Respir Crit Care Med* **158**, S109–114.

Koumi S, Sato R and Aramaki T (1994). Characterization of the calcium-activated chloride channel in isolated guinea-pig hepatocytes. *J Gen Physiol* **104**, 357–373.

Kourie JI, Laver DR, Ahern GP et al. (1996). A calcium-activated chloride channel in sarcoplasmic reticulum vesicles from rabbit skeletal muscle. *Am J Physiol*, **270**, C1675–C1686.

Kozak JA and Logothetis DE (1997). A calcium-dependent chloride current in insulin-secreting beta TC-3 cells. *Pflugers Arch* **433**, 679–690.

Kunzelmann K (2001). CFTR: interacting with everything? *News Physiol Sci* **16**, 167–170.

Kunzelmann K, Mall M, Briel M et al. (1997). The cystic fibrosis transmembrane conductance regulator attenuates the endogenous Ca^{2+} activated Cl– conductance of *Xenopus* oocytes. *Pflügers Arch* **435**, 178–81.

Kunzelmann K, Schreiber R, Nitschke R et al. (2000). Control of epithelial Na+ conductance by the cystic fibrosis transmembrane conductance regulator, *Pflügers Arch* **440**, 193–201.

Kuruma A and Hartzell HC (2000). Bimodal control of a Ca^{2+}-activated Cl$^-$ channel by different Ca^{2+} signals. *J Gen Physiol* **115**, 59–80.

Lalevee N and Joffre M (1999). Inhibition by cAMP of calcium-activated chloride currents in cultured Sertoli cells from immature testis. *J Memb Biol* **169**, 167–174.

Large WA and Wang Q (1996). Characteristics and physiological role of the Ca^{2+}-activated Cl$^-$ conductance in smooth muscle. *Am J Physiol* **271**, C435–C454.

Leblanc N, Ledoux J, Saleh S et al. (2005). Regulation of calcium-activated chloride channels in smooth muscle cells: a complex picture is emerging. *Can J Physiol Pharmacol* **83**, 541–556.

Ledoux J, Greenwood I, Villeneuve LR et al. (2003). Modulation of Ca^{2+}-dependent Cl$^-$ channels by calcineurin in rabbit coronary arterial myocytes. *J Physiol* **552**, 701–714.

Loewen ME, Bekar LK, Gabriel SE et al. (2002). pCLCA1 becomes a cAMP-dependent chloride conductance mediator in Caco-2 cells biochem. *Biophys Res Commun* **298**, 531–536.

Loewen ME, Bekar LK,Walz W et al. (2004). pCLCA1 lacks inherent chloride channel activity in an epithelial colon carcinoma cell line. *Am J Physiol Gastrointest Liver Physiol* **287**, G33–G41.

Loewen ME and Forsyth EW (2005). Structure and function of CLCA proteins. *Physiol Rev* **85**, 1061–1092.

Loewen ME, Smith NK, Hamilton DL et al. (2003). CLCA protein and chloride transport in canine retinal pigment epithelium. *Am J Physiol Cell Physiol* **285**, C1314–C1321.

Lowe G and Gold GH (1993). Nonlinear amplification by calcium-dependent chloride channels in olfactory receptor cells. *Nature* **366**, 283–286.

Machaca K, Qu Z, Kuruma A et al. (2002). The endogenous calcium-activated chloride channel in Xenopus oocytes: a physiologically and biophysically rich model system. In CM Fuller, ed., *Current topics in membranes 53. Calcium-activated chloride channels.* San Diego, Academic Press, pp. 3–39.

Maertens C, Wei L, Droogmans G et al. (2000). Inhibition of volume-regulated and calcium-activated chloride channels by the antimalarial mefloquine. *J Pharm Exp Ther* **295**, 29–36.

Maertens C, Wei L, Tytgat J et al. (2000a). Chlorotoxin does not inhibit volume-regulated, calcium-activated and cyclic AMP-activated chloride channels. *Br J Pharmacol* **129**, 791–801.

Marmorstein AD, Marmorstein LY, Rayborn M et al. (2000). K Bestrophin, the product of the Best vitelliform macular dystrophy gene (*VMD2*), localizes to the basolateral plasma membrane of the retinal pigment epithelium. *Proc Natl Acad Sci USA* **97**, 12758–12763.

Marmorstein LY, McLaughlin PJ, Stanton JB et al. (2002). Bestrophin interacts physically and functionally with protein phosphatase 2A. *J Biol Chem* **277**, 30591–30597.

Marmorstein LY, Wu J, McLaughlin P et al. (2006). The light peak of the electroretinogram is dependent on voltage-gated calcium channels and antagonized by bestrophin (Best-1). *J Gen Physiol* **127**, 577–589.

Martin DK (1993). Small conductance chloride channels in acinar cells from the rat mandibular salivary gland are directly controlled by a G-protein. *Biochem Biophys Res Commun* 192, 1266–1273.

Marunaka Y and Eaton DC (1990). Effects of insulin and phosphatase on a Ca^{2+}-dependent Cl$^-$ channel in a distal nephron cell line (A6). *J Gen Physiol* **95**, 773–789.

Matchkov VV, Aalkjaer C and Nilsson H (2004). A cyclic GMP-dependent calcium-activated chloride current in smooth-muscle cells from rat mesenteric resistance arteries. *J Gen Physiol* **123**, 121–134.

Matchkov VV, Aalkjaer C and Nilsson H (2005). Distribution of cGMP-dependent and cGMP-independent Ca^{2+}-activated Cl$^-$ conductances in smooth muscle cells from different vascular beds and colon. *Pflugers Arch* **451**, 371–379.

Matthews G, Neher E and Penner R (1989). Chloride conductance activated by external agonists and internal messengers in rat peritoneal mast cells. *J Physiol* **418**, 131–144.

Melvin JE, Arreola J, Nehrke K et al. (2002). Ca^{2+}-activated cl$^-$ currents in salivary and lacrimal glands. *Current Topics in Membranes* **53**, 209–230.

Melvin JE, Yule D, Shuttleworth TJ et al. (2005). Regulation of fluid and electrolyte secretion in salivary gland cells. *Ann Rev Physiol* **67**, 445–469.

Miledi R (1982). A calcium-dependent transient outward current in *Xenopus laevis* oocytes. *Proc Roy Soc B* **315**, 491–497.

Morris AP and Frizzell RA (1993). Ca^{2+}-dependent Cl$^-$ channels in undifferentiated human colonic cells (HT-29). II. Regulation and rundown. *Am J Physiol Cell Physiol* **264**, C977–C985.

Nilius B and Droogmas G (2002). Calcium-activated chloride channels vascular endothelial cells. In CM Fuller, ed., *Calcium-activated chloride channels. Current Topics in Membranes* 53. San Diego, Academic Press, pp. 327–344.

Nilius B, Prenen J, Szucs G et al. (1997a). Calcium-activated chloride channels in bovine pulmonary artery endothelial cells. *J Physiol* **498**, 381–396.

Nilius B, Prenen J, Voets T et al. (1998). Inhibition by inositoltetrakisphosphates of calcium- and volume-activated Cl$^-$ currents in macrovascular endothelial cells. *Pflugers Archiv* **435**, 637–644.

Nilius B, Prenen J, Voets T et al. (1997b). Kinetic and pharmacological properties of the calcium-activated chloride current in macrovascular endothelial cells. *Cell Calcium* **22**, 53–63.

Nishimoto I, Wagner J, Schulman H et al. (1991). Regulation of Cl$^-$ channels by multifunctional CaM kinase. *Neuron* **6**, 547–555.

Ottolia M and Toro L (1994). Potentiation of large conductance KCa channels by niflumic, flufenamic and mefenamic acids. *Biophys J* **67**, 2272–2279.

Owen DG, Segal M and Barker JL (1984). A Ca-dependent Cl⁻ conductance in cultured mouse spinal neurones. *Nature* **311**, 567–570.

Papassotiriou J, Eggermont J, Droogmans G et al. (2001). Ca²⁺-activated Cl⁻ channels in Ehrlich ascites tumor cells are distinct from mCLCA1, 2 and 3. *Pflugers Arch* **442**, 273–279.

Pappone PA and Lee SC (1995). Alpha-adrenergic stimulation activates a calcium-sensitive chloride current in brown fat cells. *J Gen Physiol* **106**, 231–258.

Park K and Brown PD (1995). Intracellular pH modulates the activity of chloride channels in isolated lacrimal gland acinar cells. *Am J Physiol* **268**, C647-C650.

Pauli BU, Abdel-Ghany M, Cheng HC et al. (2000). Molecular characteristics and functional diversity of CLCA family members. *Clin Exp Pharmacol Physiol* **27**, 901–905.

Perez-Cornejo P and Arreola J (2004). Regulation of Ca²⁺-activated chloride channels by cAMP and CFTR in parotid acinar cells. *Biochem Biophys Res Comm* **316**, 612–617.

Perez-Cornejo P, De Santiago JA and Arreola J (2004). Permeant anions control gating of calcium-dependent chloride channels. *J Membr Biol* **198**, 125–133.

Petrukhin K, Koisti MJ, Bakall B et al. (1998). Identification of the gene responsible for Best macular dystrophy. *Nat Genet* **19**, 241–247.

Piper AS and Large WA (2003). Multiple conductance states of single Ca²⁺-activated Cl⁻ channels in rabbit pulmonary artery smooth muscle cells. *J Physiol* **547**, 181–196.

Piper AS and Large WA (2004). Direct effect of Ca²⁺-calmodulin on cGMP-activated Ca²⁺-dependent Cl⁻ channels in rat mesenteric artery myocytes. *J Physiol* **559**, 449–457.

Piper AS, Greenwood IA and Large WA (2002). Dual effect of blocking agents on Ca²⁺-activated Cl⁻ currents in rabbit pulmonary artery smooth muscle cells. *J Physiol* **539**, 119–131.

Qu Z and Hartzell HC (2000). Anion permeation in Ca²⁺-activated Cl⁻ channels. *J Gen Physiol* **116**, 825–844.

Qu Z and Hartzell HC (2001). Functional geometry of the permeation pathway of Ca²⁺ activated Cl⁻ channels inferred from analysis of voltage-dependent block. *J Biol Chem* **276**, 18423–18429.

Qu Z and Hartzell HC (2004). Determinants of anion permeation in the second transmembrane domain of the mouse bestrophin-2 chloride channel. *J Gen Physiol* **124**, 371–382.

Qu Z, Chien LT, Cui Y et al. (2006). The anion-selective pore of the bestrophins, a family of chloride channels associated with retinal degeneration. *J Neurosci* **26**, 5411–5419.

Qu Z, Fischmeister R and Hartzell HC (2004). Mouse bestrophin-2 is a bona fide Cl⁻ channel: identification of a residue important in anion binding and conduction. *J Gen Physiol* **123**, 327–340.

Qu Z, Wei RW, Mann W et al. (2003). Two bestrophins cloned from *Xenopus laevis* oocytes express Ca²⁺- activated Cl⁻ currents. *J Biol Chem* **278**, 49563–49572.

Quinton PM (1983). Chloride impermeability in cystic fibrosis. *Nature* **301**, 421–422.

Rae FK, Hooper JD, Eyre HJ et al. (2001). TTYH2, a human homologue of the *Drosophila melanogaster* gene *tweety* is located on 17q24 and upregulated in renal cell carcinoma. *Genomics* **77**, 200–207.

Reinsprecht M, Rohn MH, Spadinger RJ et al. (1995). Blockade of capacitive Ca²⁺ influx by Cl⁻ channel blockers inhibits secretion from rat mucosal-type mast cells. *Mol Pharmacol* **47**, 1014–1020.

Robinson NC, Huang P, Kaetzel MA et al. (2004). Identification of an N-terminal amino acid of the CLC-3 chloride channel critical in phosphorylation-dependent activation of a CaMKII-activated chloride current. *J Physiol* 556, 353–368.

Romio L, Musante L, Cinti R et al. (1999). Characterization of a murine gene homologous to the bovine CaCC chloride channel. *Gene.* **228**, 181–188.

Rosenthal R, Bakall B, Kinnick T et al. (2006). Expression of bestrophin-1, the product of the *VMD2* gene, modulates voltage-dependent Ca²⁺ channels in retinal pigment epithelial cells. *FASEB J* **20**, 178–180.

Rubera I, Tauc M, Bidet M et al. (2000). Extracellular ATP increases [Ca²⁺]ᵢ in distal tubule cells. II. Activation of a Ca²⁺-dependent Cl⁻ conductance. *Am J Physiol* **279**, F102–F111.

Sánchez-Vives MV and Gallego R (1994). Calcium-dependent chloride current induced by axotomy in rat sympathetic neurons. *J Physiol* **475**, 391–400.

Schlenker T and Fitz JG (1996). Ca²⁺- activated Cl⁻ channels in human biliary cell line: regulation by Ca²⁺/calmodulindependent protein kinase. *Am J Physiol* **271**, G304–G310.

Schroeder BC, Cheng T, Jan YN et al. (2008). Expression cloning of TMEM16A as a calcium-activated chloride channel subunit. *Cell* **134**,1019–1029.

Schroeder JI and Hagiwara S (1989). Cytosolic calcium regulates ion channels in the plasma membrane of *Vicia faba* guard cells. *Nature*, **338**, 427–430.

Schumann MA, Gardner P and Raffin TA (1993). Recombinant human tumor necrosis factor alpha induces calcium oscillation and calcium-activated chloride current in human neutrophils: the role of calcium/calmodulin-dependent protein kinase. *J Biol Chem* **268**, 2134–2140.

Stohr H, Marquardt A, Nanda I et al. (2002). Three novel human *VMD2*-like genes are members of the evolutionary highly conserved RFP-TM family. *Eur J Hum Genet* **10**, 281–284.

Stohr H, Marquardt A, White K et al. (2000). cDNA cloning and genomic structure of a novel gene (C11orf9) localized to chromosome 11q12 q13.1 which encodes a highly conserved, potential membrane-associated protein. *Cytogenet Cell Genet* **88**, 211–216.

Sun H, Tsunenari T, Yau K-W et al. (2002). The vitelliform macular dystrophy protein defines a new family of chloride channels. *Proc Natl Acad Sci USA* **99**, 4008–4013.

Suzuki M (2006). The *Drosophila tweety* family: molecular candidates for large-conductance Ca²⁺-activated Cl⁻ channels. *Exp Physiol* **91**, 141–147.

Suzuki M and Mizuno A (2004). A novel human cl⁻ channel family related to *Drosophila* flightless locus. *J Biol Chem* **279**, 22461–22468.

Takahashi T, Neher E and Sakmann B (1987). Rat brain serotonin receptors in *Xenopus* oocytes are coupled by intracellular calcium to endogenous channels. *Proc Natl Acad Sci USA* **84**, 5063–5067.

Tan YP, Marty A and Trautmann A (1992). High density of Ca²⁺-dependent K⁺ and Cl⁻ channels on the luminal membrane of lacrimal acinar cells. *Proc Natl Acad Sci USA* **89**, 11229–11233.

Tanaka H, Matsui S, Kawanishi T et al. (1996). Use of chloride blockers: a novel approach for cardioprotection against ischemia–reperfusion damage. *J Pharmacol Exp Ther* **278**, 854–861.

Tarran R, Loewen ME, Paradiso AM et al. (2002). Regulation of murine airway surface liquid volume by CFTR and Ca²⁺-activated Cl⁻ conductances. *J Gen Physiol* **120**, 407–418.

Taylor R and Roper S (1994). Ca²⁺-dependent Cl⁻ conductance in taste cells from *Necturus*. *J Neurophysiol* **72**, 475–478.

Thoreson WB and Burkhardt DA (1991). Ionic influences on the prolonged depolarization of turtle cones *in situ*. *J Neurophysiol* **65**, 96–110.

Tsunenari T, Nathans J and Yau K (2006). Ca²⁺ activated Cl⁻ current from human bestrophin-4 in excised membrane patches. *J Gen Physiol* **127**, 749–754.

Tsunenari T, Sun H, Williams J et al. (2003). Structure-function analysis of the bestrophin family of anion channels. *J Biol Chem* **278**, 41114–41125.

Vajanaphanich M, Schultz C, Rudolf MT et al. (1994). Long-term uncoupling of chloride secretion from intracellular calcium levels by Ins(3,4,5,6)P4. *Nature* **371**, 711–714.

Wagner JA, Cozens AL, Schulman H et al. (1991). Activation of chloride channels in normal and cystic fibrosis airway epithelial cells by multifunctional calcium/calmodulin-dependent protein kinase. *Nature* **349**, 793–796.

Wang HS, Dixon JE and McKinnon D (1997). Unexpected and differential effects of Cl⁻ channel blockers on the Kv4.3 and Kv4.2 K⁺ channels. Implications for the study of the I(to2) current. *Circ Res* **81**, 711–718.

Wang YX and Kotlikoff MI (1997). Inactivation of calcium-activated chloride channels in smooth muscle by calcium/calmodulin-dependent protein kinase. *Proc Natl Acad Sci USA* **94**, 14918–14923.

Weber WM (1999). Ion currents of *Xenopus laevis* oocytes: state of the art. *Biochim Biophys Acta* **1421**, 213–233.

Weber WM (2002). Ca^{2+}-inactivated Cl⁻ channels in *Xenopus laevis* oocytes. In CM Fuller, ed., *Calcium-activated chloride channels: Current topics in membranes*, 53. San Diego, Academic Press, pp. 41–55.

Wei L, Vankeerberghen A, Cuppens H *et al.* (2001). The C-terminal part of the R-domain, but not the PDZ binding motif, of CFTR is involved in interaction with Ca^{2+}-activated Cl⁻ channels. *Pflugers Arch.* **442**, 280–285.

Worrell RT and Frizzell RA (1991). CaMKII mediates stimulation of chloride conductance by calcium in T84 cells. *Am J Physiol* **260**, C877-C882.

Xie W, Kaetzel MA, Bruzik KS *et al.* (1996). Inositol 3,4,5,6-tetrakisphosphate inhibits the calmodulin-dependent protein kinase II-activated chloride conductance in T84 colonic epithelial cells. *J Biol Chem* **271**, 14092–14097.

Xie W, Solomons KR, Freeman S *et al.* (1998). Regulation of Ca^{2+}-dependent Cl⁻ conductance in a human colonic epithelial cell line (T_{84}): cross-talk between $Ins(3,4,5,6)P_4$ and protein phosphates. *J Physiol* **510**, 661–673.

Xu WX, Kim SJ, So I *et al.* (1997). Volume-sensitive chloride current activated by hyposmotic swelling in antral gastric myocytes of the guinea-pig. *Pflugers Arch*, **435**, 9–19.

Yang YD, Cho H, Koo JY *et al.* (2008). TMEM16A confers receptor-activated calcium-dependent chloride conductance. *Nature* **455**, 1210–1215.

Zeng XH, Xia XM and Lingle CJ (2005). Divalent cation sensitivity of BK channel activation supports the existence of three distinct binding sites. *J Gen Physiol* **125**, 273–286.

Zygmunt AC and Gibbons WR (1992). Properties of the calcium-activated chloride channel in heart. *J Gen Physiol* **99**, 391–414.

Zygmunt AC, Goodrow RJ and Weigel CM (1998). I_{NaCa} and $I_{Cl(Ca)}$ contribute to isoproterenol-induced after hyperpolarizations in midmyocardial cells. *Am J Physiol* **275**, H979–H992.

Zygmunt AC, Robitelle DC and Eddlestone GT (1997). Ito1 dictates behavior of $I_{Cl(Ca)}$ during early repolarization of canine ventricle. *Am J Physiol* **273**, H1096–H1106.

Cyclic nucleotide-gated ion channels

Zhengchao Wang and Fangxiong Shi

4.2.1 Introduction

Cyclic nucleotide-gated ion channels (CNGs) are non-selective cation channels, which are opened by the direct binding of cyclic nucleotides (cAMP and cGMP). CNG channels were initially discovered in the plasma membrane of the outer segment of the vertebrate rod photoreceptors and then observed in cone photoreceptors, olfactory sensory neurons (OSNs), and other neuronal and non-neuronal cell types (Kaupp, 1991; Distler *et al.*, 1994; Menin, 1995; Friedmann, 2000; Kaupp and Seifert, 2002; Bradley *et al.*, 2005; Giorgetti *et al.*, 2005; Pifferi *et al.*, 2006; Wang *et al.*, 2006a). With the recent molecular cloning and functional expression of CNG channels, their study has sparked much progress. The functions of CNG channels, such as molecular structures, physiological and pathological roles, regulatory mechanisms, and gating mechanisms, have been firmly established in retinal photoreceptors and in OSNs (Kaupp, 1991; Menin, 1995; Kaupp and Seifert, 2002; Weitz *et al.*, 2002; Giorgetti and Carloni, 2003; Zhong *et al.*, 2003; Zheng and Zagotta, 2004; Peng *et al.*, 2004; Bradley *et al.*, 2005; Giorgetti *et al.*, 2005; Pifferi *et al.*, 2006; Wang *et al.*, 2006a, 2007).

Although CNG channel activity shows very little voltage dependence, CNG channels belong to the superfamily of voltage-gated ion channels. Like their cousins the voltage-gated K$^+$ channels, CNG channels form heterotetrameric complexes consisting of two or three different types of subunits. In vertebrates, there are at least six CNG subunit genes, *CNGA1*, *CNGA2*, *CNGA3*, *CNGA4*, *CNGB1*, and *CNGB3*, with homology between families ranging from 30–70 per cent (Kaupp and Seifert, 2002; Bradley *et al.*, 2005; Wang, 2006). These subunits are generically designated as principle subunits (or CNGA subunits) and modulatory subunits (or CBGB subunits). Important functional features of these channels, like ligand sensitivity and selectivity, ion permeation, and gating, are determined by the subunit composition and stoichiometry of the respective channel complex (Kaupp, 1991; Distler *et al.*, 1994; Menin, 1995; Friedmann, 2000; Kaupp and Seifert, 2002; Weitz *et al.*, 2002; Zhong *et al.*, 2003; Giorgetti *et al.*, 2003, 2005; Peng *et al.*, 2004; Zheng *et al.*, 2004; Bradley *et al.*, 2005; Pifferi *et al.*, 2006; Wang *et al.*, 2006, 2007).

CNG channels discriminate between alkali ions poorly and even pass divalent cations, in particular Ca^{2+}. Calcium entry through CNG channels is important for both excitation and adaptation of sensory cells. To permeate the channel, the ions must bind to a site inside the channel pore. The dwell time at this binding site is significantly longer for Ca^{2+} than for monovalent cations. As a result, Ca^{2+} blocks the current of the more permeant Na$^+$, which is crucially important for the channel's function and underlies, for example, the ability of rod photoreceptors to detect single photons and to adapt to steady illumination (Kaupp, 1991; Menin, 1995; Pifferi *et al.*, 2006).

CNG channels can serve as a molecular switch that faithfully tracks the cAMP or cGMP levels in a cell, since they do not desensitize in the continuous presence of the ligand, unlike ligand-gated neurotransmitter receptors. Although CNG channels do not desensitize, CNG channel activity is nonetheless modulated by Ca^{2+}/calmodulin and phosphorylation. Other factors may also be involved in channel regulation, including the influence of co-expressed CNGB subunits (CNGB1 or CNGB3) (Kaupp and Seifert, 2002; Bradley *et al.*, 2005; Wang *et al.*, 2006a).

Targeted disruption of CNG channel genes has been used to demonstrate that mutations in CNG channels can give rise to retinal degeneration and colour blindness, such as various forms of complete and incomplete achromatopsia, in humans especially.

4.2.2 Molecular characterization

Gene diversity, nomenclature and character of CNG subunits

CNG channels belong to a heterogeneous gene superfamily of ion channels that share a common transmembrane topology and pore structure. In vertebrates, there are at least six *CNG* subunit genes encoding CNG proteins (Kaupp and Seifert, 2002; Bradley *et al.*, 2005; Wang *et al.*, 2006). The National Centre for Biotechnology Information (NCBI) database contains the sequences of six different human genes encoding CNG channels. These genes have been identified from cloning of human cDNA or other vertebrate orthologues including bovine, mouse, rabbit, rat, and chick. The *Drosophila* and *C. elegans* genomes harbour four and six different *CNG* channel genes, respectively. The mammalian CNG channel subunits fall into two different subfamilies, CNGA and CNGB (Kaupp and Seifert, 2002; Bradley *et al.*, 2005; Wang *et al.*, 2006a).

CNGAs are defined as functional subunits of channels because they are functional when four different CNGAs are assembled.

In contrast, CNGBs are defined as regulatory subunits of channels because CNGBs are non-functional if four different CNGBs are assembled together. However, this functional definition is not always consistent with the phylogenetic assignment of subunit type, as is the case for CNGA4 (see below). Although certain types of CNG subunits can form functional homomultimeric channels when expressed in a heterologous system *in vivo*, they appear to be constructed as heterotetramers with 2–3 different types of subunits per channel complex (Kaupp, 1991; Distler *et al.*, 1994; Menin, 1995; Friedmann, 2000; Kaupp and Seifert, 2002; Weitz *et al.*, 2002; Zhong *et al.*, 2003; Giorgetti and Carloni, 2003, 2005; Peng *et al.*, 2004; Zheng and Zagotta, 2004; Bradley *et al.*, 2005; Pifferi *et al.*, 2006; Wang *et al.*, 2006a, 2007).

Nomenclature of CNG channels

Subunits that were initially cloned from either rod, or cone, or OSN have been named according to their cellular origin. However, CNG channels form heteromultimers, and subunits are not specific for rod, cone, and OSN but are expressed in other cells as well. Therefore, classification of channel subunits based on the tissue from which they were originally identified has become inappropriate. Thus, a new nomenclature has been formulated based on the following two guidelines (Kaupp and Seifert, 2002; Wang *et al.*, 2006a; Bradley *et al.*, 2005).

First, based on sequence comparisons, CNG channel genes fall into different subfamilies. In mammals, two gene subfamilies can be distinguished. Homologous members of these two subfamilies are found in the genomes of species as distant as human, *C. elegans*, and *Drosophila*. The encoded polypeptides of the two subfamilies are referred to as CNGA and CNGB subunits. Members of a subfamily are numbered CNGA1, CNGA2, etc.

Second, a designation used in the past for one gene cannot be used for another gene even if the former usage has been discontinued. This rule prevented the numbering of CNGA and CNGB subunits based on functional relatedness or functional context rather than on the order of molecular cloning or baptizing. Moreover, this rule created gaps in the numbering system. For example, one subunit of olfactory CNG channels has previously been thought to be a B subunit (formerly CNGB2). This subunit is now designated CNGA4. As a consequence, the designation CNGB2 is excluded from future use, and one of the other B subunits cannot be renamed but must keep their original designation (CNGB3).

Composition and stoichiometry of CNG channels

It is now generally believed that CNG channels in various tissues form heteromeric complexes consisting of two or more distinct subunits, because CNGA and CNGB subunits from nematode to human co-assemble functionally with each other (Kaupp and Seifert, 2002; Weitz *et al.*, 2002; Giorgetti and Carloni, 2003; Zhong *et al.*, 2003; Peng *et al.*, 2004; Zheng and Zagotta, 2004; Bradley *et al.*, 2005; Giorgetti *et al.*, 2005; Wang *et al.*, 2006a, 2007). As such, a large number of distinct channels can, in principle, be combinatorially generated from the six subunit types. Furthermore, several splice variants of the CNGA3 and CNGB1 subunits have been identified (Oda *et al.*, 1997).

Understanding the nature of subunit interactions and their contribution to channel gating requires knowledge of subunit stoichiometry and arrangement as heteromultimers. The cell type-specific expression of CNG genes determines the make up of the sensory transduction channels. Functional and biochemical assays together with other approaches including fluorescently tagged subunits and fluorescence resonance energy transfer (FRET) have helped us to understand the native assembly of CNG channels (Kaupp and Seifert, 2002; Weitz *et al*, 2002; Giorgetti *et al.*, 2003; Zhong *et al.*, 2003; Peng *et al.*, 2004; Zheng *et al.*, 2004; Bradley *et al.*, 2005; Giorgetti *et al.*, 2005; Wang *et al.*, 2006a, 2007). The stoichiometry of native heteromeric CNG channels in olfactory, rod, and cone cells has been worked out. Thus, in olfactory receptor neurons, CNGs require three types of subunits, CNGA2, CNGA4 and CNGB1b, to exhibit properties necessary for olfactory transduction (Kaupp, 1991; Menin, 1995; Pifferi *et al.*, 2006). In a photoreceptor, rod CNG channels are made up of CNGA1 and CNGB1, whereas cone CNG channels are comprised of CNGA3 and CNGB3 subunits (Kaupp, 1991; Menin, 1995; Pifferi *et al.*, 2006). However, information in other tissues is very limited. Many studies have suggested that the spatial arrangement of subunits within the heterotetramer has a profound effect on channel functions (Kaupp, 1991; Distler *et al.*, 1994; Menin, 1995; Friedmann, 2000; Kaupp and Seifert, 2002; Weitz *et al.*, 2002; Zhong *et al.*, 2003; Giorgetti *et al.*, 2003, 2005; Peng *et al.*, 2004; Zheng *et al.*, 2004; Bradley *et al.*, 2005; Pifferi *et al.*, 2006; Wang *et al.*, 2006, 2007). This is a burgeoning area of research with a number of groups examining the stoichiometry and arrangement of subunits by determining the effects of mutant subunits (Brown *et al.*, 2006; Wang *et al.*, 2006; Biel and Michalakis, 2007) on channel properties, often using new biochemical approaches and direct structural measurements such as single particle electron microscopy.

4.2.3 Biophysical characteristics

Although activated by a ligand and nearly voltage independent, the transmembrane topology, sequence similarity, and tetrameric assembly of CNG channels place them within the superfamily of voltage-gated ion channels (Kaupp and Seifert, 2002; Bradley *et al.*, 2005; Giorgetti *et al.*, 2005; Wang *et al.*, 2006, 2007). Therefore, CNG channels, like the voltage-dependent K^+ channel family, form as tetramers with each subunit containing six putative transmembrane segments (S1–S6), a charged S4 region, a P region between S5 and S6, and a cytoplasmic amino and carboxyl terminus (Kaupp and Seifert, 2002; Bradley *et al.*, 2005; Giorgetti *et al.*, 2005; Wang *et al.*, 2006a, 2007). The major structural sequence motifs and functional domains of CNG channels are depicted in Figure 4.2.1 and are discussed below.

Transmembrane topology

The model for the membrane topology of CNG channel subunits is illustrated in Figure 4.2.1. The core structural unit consists of six membrane-spanning segments, designated S1–S6, followed by a cNMP (cyclic nucleotide monophosphate) binding domain near the COOH terminus. A pore region is located between S5 and S6. The S4 segment in CNG channels resembles the voltage-sensor motif found in the S4 segment of voltage-gated K^+, Na^+ and Ca^{2+} channels. These three structures are also characteristic features of voltage-gated channels (Kaupp and Seifert, 2002; Bradley *et al.*, 2005; Giorgetti *et al.*, 2005; Wang *et al.*, 2006a, 2007).

However, certain subunit regions are important for gating: one is the intracellular cyclic nucleotide-binding domain (CNBD), another is an iris-like structure formed by a loop which connects the

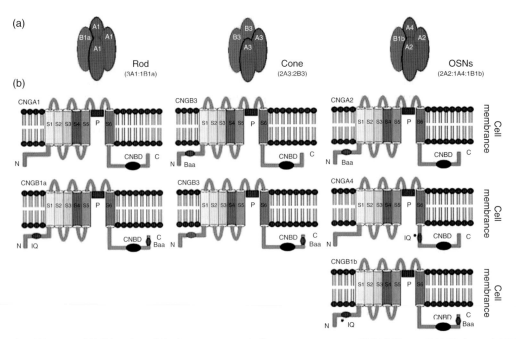

Fig. 4.2.1 Subunits and stoichiometry of CNG ion channels in photoreceptors and olfactory sensory neurons (OSNs) (Kaupp, 1991; Distler *et al.*, 1994; Menin, 1995; Friedmann, 2000; Kaupp and Seifert, 2002; Weitz *et al.*, 2002; Zhong *et al.*, 2003; Peng *et al.*, 2004; Zheng *et al.*, 2004; Bradley *et al.*, 2005; Giorgetti *et al.*, 2005; Pifferi *et al.*, 2006; Wang *et al.*, 2006a; Wang *et al.*, 2007). (a) Subunit composition of rod, cone and olfactory CNG channels. (b) Topological models of the CNG channel subunits of rod, cone and OSNs. Six transmembrane domains (TMDs) are indicated by S1–S6, respectively, the pore loop indicated P is located between S5 and S6. Both N- and C-termini are intracellular and contain functional regions for channel regulation: the cyclic nucleotide-binding sites (CNBD, black), the calmodulin-binding site of the calcium independent 'IQ-type' and calcium-dependent 'Baa-type' (blue and purple, respectively). A site in CNGB3 lacking a critical signature residue of an IQ sequence is depicted in grey. Two IQ-type sites indicated by asterisks may have important functions in the regulation of channels. See also colour plate section.

S5 and S6 segments, and a third is the helix bundle (Matulef *et al.*, 1999; Liu and Siegelbaum, 2000; Zheng and Zagotta, 2000; Johnson and Zagotta, 2001; Kwon *et al.*, 2002; Rosenbaum and Gordon, 2002; Simoes *et al.*, 2002; Hua and Gordon, 2005; Qu *et al.*, 2006). The S6 region of the bovine rod CNG channel (CNGA1) shares a sequence similarity with both the bacterial potassium channel from *Streptomyces lividans* (KcsA) and the voltage-gated potassium channel (Shaker B). On the basis of sequence alignment between CNGA1 and KcsA, a homology model was constructed to represent the putative pore structure of CNG channels (Simoes *et al.*, 2002). Notably, the CNG pore model is different from KcsA in the selectivity filter, as CNG channels lack the GYG signature sequence of potassium-selective channels (Liu *et al.*, 2000; Kwon *et al.*, 2002; Qu *et al.*, 2006). Outside the selectivity filter, the S6 α-helices enter the membrane at an angle to form a helix bundle on the intracellular side (Matulef *et al.*, 1999; Johnson and Zagotta, 2001; Simoes *et al.*, 2002; Rosenbaum and Gordon, 2002). As in KcsA, this inverted teepee structure forms the inner vestibule of the channel.

Cyclic nucleotide-binding domain (CNBD)

The CNBD is an intracellular domain in the C-terminal region, and shares sequence similarity with other cyclic nucleotide-binding proteins, including cGMP- and cAMP-dependent protein kinases (PKG and PKA, respectively) and the *Escherichia coli* catabolite gene activator protein (CAP) (Matulef *et al.*, 1999; Zheng and Zagotta, 2000; Johnson and Zegotta, 2001). This domain is thought to contain a β-roll and two α-helices, designated as the B and C helices.

The binding of the ligand to the CNBD is followed by an allosteric conformational change coupled to the opening of the pore

(Matulef *et al.*, 1999; Zheng and Zagotta, 2000; Johnson and Zagotta, 2001; Biskup *et al.*, 2007). Various cyclic nucleotides, including the full agonist cGMP, the partial agonist cAMP and inosine 3′, 5′-cyclic monophosphate (cIMP), can bind to the CNBDs of bovine CNGA1 channels. To probe the conformational changes that occur in CNBD during channel gating, the substituted cysteine accessibility method (SCAM) has been used. It was found that the residue C505, in the β-roll, is more accessible in un-liganded channels than in liganded channels, whereas the residue G597C, in the C helix, is more accessible in closed channels than in open channels (Matulef *et al.*, 1999; Zheng and Zagotta, 2000; Johnson and Zegotta, 2001). These observations led to a molecular mechanism for channel activation in which the ligand initially binds to the β-roll. This is followed by an allosteric opening transition involving the relative movement of the C helix towards the β-roll. When the ligand initially binds to the β-roll, the bound ligand will stabilize a relative movement of the C helix towards the β-roll in each subunit, which pulls the C helices away from each other. When disulfide bonds are formed between C helices primarily in closed channels, they will inhibit the relative movement of the C helices toward the β-rolls, and the channel closes (Matulef *et al.*, 1999; Zheng and Zagotta, 2000; Johnson and Zegotta, 2001; Biskup *et al.*, 2007).

C-linker

In CNG channels, binding of cGMP or cAMP drives a conformational change that leads to the opening of an ion-conducting pore. The region implicated in the coupling of ligand binding to the opening of the pore is the C-linker.

Cross-linking of endogenous cysteines has been used to investigate inter-region proximity, and an individual amino acid (C481)

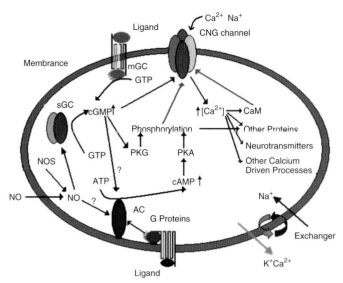

Fig. 4.2.2 Regulatory pathways of CNG ion channels in the cell (Distler *et al.*, 1994; Friedmann, 2000; Kaupp and Seifert, 2002; Bradley *et al.*, 2005; Giorgetti *et al.*, 2005; Wang *et al.*, 2006a, b, 2007). Note that CNG ion channels are opened by the direct binding of the cyclic nucleotide, cGMP. The concentration of cyclic GMP (cGMP) is increased mainly from transferring GTP by two pathways, one is activation of membrane bound guanylyl cyclase (mGC), and another is activation of soluble guanylyl cyclase (sGC) by nitric oxide (NO). CNG ion channels are opened through binding cGMP, then, Ca^{2+} enters through opened CNG ion channels and increases the intracellular Ca^{2+} level. The increased Ca^{2+} can interact with CaM, neurotransmitters, other functional proteins or other calcium-driven processes. The intracellular cyclic AMP (cAMP) concentration is increased mainly from transferred ATP by G protein-linked pathways, which stimulates the activity of adenylyl cyclase (AC). The activity of adenylyl cyclase can be regulated by intracellular cGMP and NO. The regulation of CNG ion channels has two major pathways, one is the Ca^{2+}/CaM-mediated negative feedback, the other is through cGMP and cAMP-dependent pathways. The Na^+–K^+–Ca^{2+} exchanger plays critical roles in limiting the change in intracellular Ca^{2+} concentration. See also colour plate section.

in each C-linker region of two neighbouring subunits has been shown to form a disulfide bond. Furthermore, a disulfide bond between a cysteine residue at position 35 of the N-terminal region (C35) and a cysteine residue at position 481 of the C-linker region (C481) can be formed either within a subunit, or between subunits, using tandem dimers (Kwon *et al.*, 2002; Simoes *et al.*, 2002; Rosenbaum and Gordon, 2002). Formation of this disulfide bond alters channel functions, suggesting that N-terminal and C-linker regions lie in close proximity in the tertiary structure. The characteristics of C481 in the CNGA1 C-linker region allow it to be a reporter of disulfide bond formation. Since it is modified preferentially when the channels are in the open state, disulfide bonds that are involved decrease the free energy of the open state relative to the closed state (Kwon *et al.*, 2002; Simoes *et al.*, 2002; Rosenbaum and Gordon, 2002). Therefore, opening of the channels in a C481 disulfide bond is more favourable.

Although rare conformations of a protein can certainly be trapped by disulfide bond formation, the propensity of C linker regions to participate in the C481–C481 disulfide bond suggests that the proximity between C-linker regions may be prevalent in functional channels (Matulef *et al.*, 1999; Zheng *et al.*, 2000; Johnson and Zagotta, 2001; Rosenbaum and Gordon, 2002).

Formation of a disulfide bond between cysteines in neighbouring subunits indicates that the regions containing these cysteines lie in close proximity in functional channels. In fact, C481 is positioned just a few amino acids away from the CNBD.

S4 region

The S4 region of all CNG channels comprises a charged sequence motif that is reminiscent of the 'voltage-sensing' S4 segment of voltage-gated channels. This segment, in voltage-gated channels, has been proposed to serve as the sensor of membrane voltage that on depolarization moves towards the extracellular surface and thus opens the channel (Kaupp and Seifert, 2002; Bradley *et al.*, 2005; Wang *et al.*, 2006a). The S4 segment of K^+ channels is characterized by five to seven positively charged residues (Arg or Lys) at every third position interspersed with predominantly hydrophobic residues. The core of the S4 segment in CNG channels displays significant sequence similarity to the respective region of K^+ channels, although it includes only three or four regularly spaced Arg or Lys residues. However, the functional significance of the S4 segment is still elusive, as CNG channels show very little voltage dependence. In CNG channels the S4 charges might be 'locked' in a fixed conformation by hydrophilic residues in the S3–S4 loop, and gating of these channels, despite the S4 motif, is solely promoted by the binding of cyclic nucleotides (Kaupp and Seifert, 2002; Bradley *et al.*, 2005; Wang *et al.*, 2006). Clearly, more work is needed to understand the role of the S4 segment of CNG channels.

S6 region

To probe the architecture of the CNG channel pore, Simoes and his colleagues (2002) took advantage of the unique biochemistry of cysteine and substituted the amino acid into S6 sites of a cysteine-free CNG1 channel. For those positions that are further away from the pore, modification is expected to have little effect on permeation, since the positive charge is likely to be buried behind the helix away from the permeation pathway. Formation of a spontaneous disulfide bond at S399C in the absence of cyclic nucleotide suggests that these residues are in close proximity in the closed state (Matulef *et al.*, 1999; Zheng and Zagotta, 2000; Johnson and Zagotta, 2001). To better understand the open state, disulfide bond formation was investigated by maintaining S399C channels in saturating cGMP (Matulef *et al.*, 1999; Zheng and Zagotta, 2000; Johnson and Zagotta, 2001). State-dependent formation of a spontaneous disulfide bond at S399C indicated that a conformational change in the helix bundle was associated with channel gating. This suggests that the smoke hole may widen as the channel opens.

Since a conformational change occurs in the helix bundle that is associated with gating, it is necessary to know whether the helix bundle acts as an activation gate for CNG channels (Samoes *et al.*, 2002). Cysteine residues lining the inner vestibule are protected from internal methanethiosulfonate (MTS) ethyltrimethylammonium (MTSET) modification in the closed state. MTSEA and Ag^+ can pass into the inner vestibule, even when CNG channels are closed, implying that the cytoplasmic ends of the S6 helices are in close proximity to each other. Opening is sufficient to permit the passage of small cations. This implies that the gate is beyond the helix bundle and movement of the S6 helices is coupled with conformational changes in the selectivity filter (Liu and Siegelbaum, 2000).

P region

The P region connects the S5 and S6 segments, forming a loop that extends towards the central axis of the channel (Liu and Siegelbaum, 2000; Kwon *et al.*, 2002; Qu *et al.*, 2006). Residues in the loop between the S5 and S6 transmembrane segments are the major determinants of CNG channel ion permeation properties. This loop was proposed to form the blades of an iris-like structure, which constricts the large diameter pore. Although residues in the S4–S5 loop and in the S5 and S6 transmembrane segments also affect ion conduction, the P region is the major determinant of ion selectivity in CNG channels (Giorgetti *et al.*, 2005). Moreover, these regions determine the pore diameter of CNG channels.

To investigate the structure and the role of the P region in gating, the accessibility of 11 cysteine-substituted P region residues to small, charged sulfhydryl reagents, methanethiosulfonate (MTS) derivatives, MTS-ethylammonium (MTSEA) and MTS-ethylsulfonate (MTSES) as well as Ag$^+$, applied to the inside-out configuration of membrane patches in the open and closed states of the bovine retinal CNG channel have been determined (Liu and Siegelbaum 2000; Kwon *et al.*, 2002; Qu *et al.*, 2006). The center of the iris, possibly involving residues T16–E19, forms a short, narrow region of the pore that serves as the selectivity filter. Since the channel must contain a gate to prevent ion permeation in the closed state, the P region itself, in addition to forming the ion selectivity filter, must function as the channel gate, the structure of which changes when the channel opens.

Gating rearrangement

Recordings of site-specific fluorescence have shown great promise in understanding conformational changes in signalling proteins. This approach, together with patch-clamp recording from inside-out patches, allows direct monitoring of the gating movements of CNGs (Matulef *et al.*, 1999; Johnson and Zagotta, 2001; Zheng and Zagotta, 2004). Fluorescence signals are reliably observed when a fluorophore is covalently attached to a site between CNBD and the pore. Iodide, an anionic quencher, has a higher quenching efficiency in the channel's closed state. Thallium ions, a cationic quencher, have a higher quenching efficiency in the open state. The state and charge dependence of quenching suggests movements of

charged or dipolar residues near the fluorophore during CNG channel activation (Matulef *et al.*, 1999; Zheng and Zagotta, 2000; Johnson and Zagotta, 2001). Therefore, CNG gating involves a concerted, or cooperative, conformational change of all four subunits, and the stability of the opening transition depends largely on the number and type of cyclic nucleotides bound to CNG channels (Matulef *et al.*, 1999; Zheng and Zagotta 2000; Johnson and Zagotta, 2001; Kaupp and Seifert, 2002; Weitz *et al.*, 2002; Zhong *et al.*, 2003; Zheng and Zagotta 2004; Peng *et al.*, 2004; Bradley *et al.*, 2005; Giorgetti *et al.*, 2005; Wang *et al.*, 2006a, 2007; Biskup *et al.*, 2007).

4.2.4 Localization

Subunit localization in neuronal cells and tissues

CNG channels were initially discovered in the plasma membrane of the outer segment of the vertebrate rod photoreceptors and then subsequently observed in cone photoreceptors, olfactory sensory neurons (OSNs), and other neuronal and non-neuronal cell types (Kaupp, 1991; Distler *et al.*, 1994; Menin, 1995; Friedmann, 2000; Kaupp and Seifert, 2002; Pifferi *et al.*, 2006; Wang *et al.*, 2006a, b). CNGs play critical roles in phototransduction and olfactory transduction, which can convert stimuli into electrical signals that are processed as visual or olfactory information in visual and olfactory sensory cells (Kaupp, 1991; Menin, 1995; Pifferi *et al.*, 2006).

CNG subunit composition in rod, cone and OSNs has been elucidated in detail. Recently, localization of CNG channel subunits in the brain has also been examined by various techniques (Table 4.2.1). In particular, extensive PCR amplification and *in situ* hybridization with different primers or cDNA probes, amplification protocols, and hybridization conditions, suggest that the levels of mRNAs encoding CNG channel subunits in the brain are very low. That said, there are many reports of CNGA1 expression in the brain with intense staining being observed in the hippocampus and cerebellum and weaker staining in the cortex. Notably, the expression of CNGA1, CNGA2, and CNGA3 subunits in the cortex is developmentally regulated in a specific temporospatial pattern (Samanta and Barnstaple, 1999). For example, the level of CNGA1 transcript in rat cortical areas is high at birth and gradually decays with time until it becomes practically non-detectable at

Table 4.2.1 Cell and tissue localizations of cyclic nucleotide-gated ion channel subunit

Subunit	Sensory cells or neuronal tissues	Non-sensory cells or non-neuronal tissues
CNGA1	Rod photoreceptors, retinal ganglion cells, cortex, hippocampus, pinealocytes, cerebellum, anterior pituitary, GnRH-secreting cell line (GT1)	Heart, kidney, liver, lung, testis, spleen, pancreas, keratinocyte, muscle, vascular endothelium, M-1 cell line
CNGA2	OSNs, cortex, hippocampus, cerebellum, olfactory bulb, thalamic and hypothalamic nuclei	Heart, lung
CNGA3	Cone photoreceptors, subpopulation OSNs, embryonic chick brain, pinealocytes, cortex	Heart, kidney, lung, testis, colon
CNGA4	OSNs, hippocampus, olfactory bulb, GnRH-secreting cell line (GT1)	
CNGB1	Rod photoreceptors or retina, OSNs, GnRH-secreting cell line (GT1)	Testis
CNGB3	Retina	Testis

Various CNG subunits may exist in some cells and tissues as determined by RT-PCR, Northern blot, *in situ* hybridization, Western Blot or IHC.

Ahmad *et al.* (1990) *Biochem Biophys Res Commun* 173, 463–470; Biel *et al.* (1994) *Proc Natl Acad Sci USA* 91, 3505–3509; Cho *et al.* (2005) *Mol Cells* 19, 149–154; Ding *et al.* (1997) *Am J Physiol* 272, C1335–1344; Distler *et al.* (1994) *Neuropharm* 33, 1275–1282; Friedmann (2000) *Cell Calcium* 27,127–138; Kaupp (1991) *TINS* 14, 150–157; Kaupp and Seifert (2002) *Physiol Rev* 82, 769–824; Kruse *et al.*, (2006) *Neurosci Lett* 404, 202–207; Menin (1995) *Biophys Chem* 55, 185–196; Podda *et al.* (2005) *Neuroreport* 16, 1939–1943; Ruiz *et al.* (1996) *J Mol Cell Cardiol* 28, 1453–1461; Samanta *et al.* (1999) *Cereb Cortex* 9, 340–347; Tetreault *et al.* (2006) *Biochem Biophys Res Commun* 348, 441–449; Wang *et al.* (2006a, b) *Prog Vet Med* 27, 1–4; Weyand *et al.* (1994) *Nature* 368, 859–863; Wiesner *et al.* (1998) *J Cell Biol* 142, 473–484.

postnatal day 55, whereas a similar developmental regulation was not observed in the hippocampal region (Samanta and Barnstaple, 1999).

Subunit localization in non-neuronal cells and tissues

The expression of CNG channel subunits in non neuronal tissue has also been studied by various techniques (Table 4.2.1). Biel and his group (1994) demonstrated expression of CNGA3 transcripts in renal cortex and renal medulla and a weak signal in cardiac atrium and ventricle by Northern blot analysis. Ahmad and colleagues (1990) detected the presence of CNGA1 in heart and kidney of rat. The CNGA1 transcript was also detected in keratinocytes (Oda et al., 1997). Ruiz and colleagues (1996) used RNase protection assay to detect the A2 transcript in mouse heart and brain. Ding and colleagues (1997) studied the expression pattern of CNGA1 in rat using an RNAase protection assay and suggested the presence of CNGA1 transcripts in retina, brain, lung, and spleen; by in situ hybridization they suggested the presence of CNGA1 transcripts in alveolar tissue of lung, the endothelial layer of aorta, thymus, and spleen and by RT-PCR they suggested CNGA1 fragment expression in retina, brain, lung, spleen, liver, heart, testis, and kidney. The expression pattern of other CNG subunits has been also studied by Reverse Transcription Polymerase Chain Reaction (RT-PCR) with mixed results. For example, CNGA2 seems to be expressed in the heart of rabbit, but not in rat and cattle, whereas CNGA3 was detected in bovine heart, but not in rat and rabbit (Distler et al., 1994). These findings may suggest that CNG subunits express in cell-, tissue-, and species-specific manners.

Compared with other non-neuronal tissues, the experimental evidence for CNG subunit expression in testis and spermatozoa is strong. In testis, CNGA3, CNGB1 and CNGB3 mRNA were all detected by cDNA cloning and Northern blot analysis (Biel et al., 1994; Weyand et al., 1994; Wiesner et al., 1998). CNGA3 and CNGB1 subunits have been localized immunohistochemically to the flagellum of ejaculated sperm and precursor cells in cross-sections of seminiferous tubules using polyclonal antibodies. Cloning has also identified one short and three long transcripts of the CNGB1 subunit in cauda epididymal and ejaculated sperm (Wiesner et al., 1998).

In summary, although many studies demonstrate the presence of CNG channel polypeptides in non-sensory cells, further studies in this field will be required to understand the expression and functions of CNG channels in these cells and tissues, especially as in the few studies in which cyclic nucleotide-sensitive currents have been recorded the properties of these do not match those of the well-characterized CNG channels in photoreceptors and OSNs. Furthermore, targeted disruption of the CNGA2 and CNGA3 genes in mice revealed no obvious phenotypes other than anosmia and loss of cone function, respectively (Biel et al., 1999; Brunet et al., 1996).

4.2.5 Pharmacology

CNG channels play a central role in vision and olfaction, generating the electrical responses to light in photoreceptors and to odourants in olfactory receptors. Whilst these channels have been detected in many other tissues their functions here are largely unclear. The use of gene knockouts and other methods have yielded

some information, but there is a pressing need for potent and specific pharmacological agents directed at CNG channels to help unravel their physiological roles (Kaupp and Seifert, 2002; Bradley et al., 2005; Wang et al., 2006a; Brown et al., 2006; Biel and Michalakis, 2007). The pharmacology of CNG channels is beginning to be understood with the identification of first generation inhibitor compounds, which can block CNG channels, although not with very high affinity. In this section, we discuss these compounds and their potential for treating certain forms of retinal degeneration.

CNG channel blockers include L-cis-diltiazem, pimozide, amiloride and its derivative, tetracain, polyamine, H-8 ([N-2-(methylamino) ethyl-5-isoquinoline-sulfonamide] (a PKA/PKG inhibitor), W-7 (a calmodulin inhibitor), LY83583 ([6-(phenylamino)-5,8-quinolin-edione]), D-cis enantiomer of diltiazem, and pseudechetoxin. Most inhibitors block CNG channels at micromolar concentrations.

The most specific blocker amongst these inhibitors is L-cis diltiazem which has been studied most extensively, and which blocks CNG channels in a voltage-dependent manner at micromolar concentration (Brown et al., 2006; Wang et al., 2006; Biel et al., 2007). L-cis diltiazem exerts its effect from the cytoplasmic face of the channel; extracellular application is much less effective. In the fish, L-cis-diltiazem blocks the rod CNG channel at tenfold lower concentrations than the cone CNG channel. The D-cis enantiomer of diltiazem, that is used therapeutically as a blocker of the L-type calcium channel, is much less effective than the L-cis enantiomer in blocking CNG channels. Notably, high-affinity binding of L-cis diltiazem is only observed with heteromeric CNG channels containing the CNGB1 subunit (Brown et al., 2006; Wang et al., 2006a; Biel and Michalakis, 2007).

The most potent blocking agent for CNG channels is pseudechetoxin (Wang et al., 2006a, b; Biel and Michalakis, 2007; Brown et al., 2006), which inhibits the homomeric CNGA2 channel with a Ki of 5 nM and the homomeric CNGA1 channel with a Ki of 100 nM. Pseudechetoxin is several orders of magnitude less effective in blocking the heteromeric channels.

CNG channels are also moderately sensitive to block by some other inhibitors of the L-type calcium channel (e.g., nifedipine), the local anaesthetic tetracaine and calmodulin antagonists (Brown et al., 2006; Wang et al., 2006a, b; Biel and Michalakis, 2007). Interestingly, LY83583 blocks both soluble guanylate cyclase and some CNG channels at similar concentrations to those at which it blocks CNG channels. H-8 has been widely used as a non-specific cyclic nucleotide-dependent protein kinase inhibitor, which can also block CNG channels at significantly higher concentrations than needed to inhibit protein kinases.

Cyclic nucleotide analogues can be both activators and competitive inhibitors. Thus, Strassmaier and Karpen (2007) have shown that when combined with 8-parachlorophenylthio derivatization, the resulting cGMP analogue is more potent than cGMP itself, suggesting the N7 and N1 positions of cGMP could be targets for modification in the design of novel CNG channel agonists. Furthermore, another study has shown that changes in ligand sensitivity of CNGA3+CNGB3 channels were prevented by inhibition of phosphatidylinositol 3-kinase (PI3-kinase) using wortmannin or 2-(4-morpholinyl)-8-phenyl-1(4H)-benzopyran-4-one hydrochloride (LY294002). This suggests that phospholipid metabolism can regulate the channels (Bright et al., 2007).

4.2.6 Physiology

The physiological functions of CNG channels in photoreceptors and OSNs have been firmly established beyond reasonable doubt. Although CNG channels also exist in other neurons and non-neuronal tissues, their specific functions are yet to be rigorously determined. In this section, we will discuss the physiological functions of CNG channels mainly in retinal photoreceptors, OSNs, taste receptor cells, spermatozoa, and some other non-neuronal cells (Ahmad *et al.*, 1990; Kaupp, 1991; Pittler *et al.*, 1992; Dister *et al.*, 1994; Weyand *et al.*, 1994; Biel *et al.*, 1994; Menin, 1995; Ruiz-Avila *et al.*, 1995; Brunet *et al.*, 1996; Ruiz *et al.*, 1996; Ding *et al.*, 1997; Wiesner *et al.*, 1998; Biel *et al.*, 1999; Samanta and Barnstable, 1999; Xu *et al.*, 1999; Junor *et al.*, 1999, 2000; Friedmann, 2000; Vitalis *et al.*, 2000; Wu *et al.*, 2000; Kaupp and Seifert, 2002; Bradley *et al*, 2005; Podda *et al.*, 2005; Cho *et al.*, 2005; Shi *et al.*, 2005; Brady *et al.*, 2006; Chen *et al.*, 2006; Tetreault *et al.*, 2006; Kruse *et al.*, 2006; Strünker *et al.*, 2006; Pifferi *et al.*, 2006; Flannery *et al.*, 2006; Wang *et al.*, 2006a, b; Boccaccio and Menini, 2007).

Physiological functions of CNG channels in retinal photoreceptors

Transactivation of vision is consecutive in the visual system. Photoreceptors respond to light either with a depolarization or hyperpolarization in the vertebrate, via use of cGMP-signallling pathways which depending on the direction of activation/inhibition either open or close CNG channels.

Rods respond to a light stimulus with a brief hyperpolarization by closing CNG channels in the surface membrane of the outer segment (Kaupp, 1991; Menin, 1995; Pifferi *et al.*, 2006). In the dark, CNG channels are activated by the binding of cGMP, allowing a steady cationic current to flow into the outer segment. Light triggers a sequence of enzymatic reactions that leads to the hydrolysis of cGMP. When CNG channels close, the inward current ceases and the cell hyperpolarizes. The enzyme cascade comprises the photopigment rhodopsin, the G protein transducin, and a phosphodiesterase (PDE). Light stimulation decreases the cytoplasmic Ca^{2+} concentration ($[Ca^{2+}]_i$), which initiates the recovery from the light response by enhancing the synthesis of new cGMP molecules and adjusting the sensitivity of the transduction machinery; a process known as light adaptation (Kaupp and Seifert, 2002; Bradley *et al.*, 2005; Wang *et al.*, 2006). The CNG channels are crucially important for the control of $[Ca^{2+}]_i$, because they provide the only source for Ca^{2+} influx into the outer segment. Ca^{2+} entry through open CNG channels is balanced by Ca^{2+} extrusion through a Na^+/Ca^{2+}–K^+ exchange mechanism (Kaupp, 1991; Distler *et al.*, 1994; Menin, 1995; Friedmann, 2000; Kaupp and Seifert, 2002; Bradley *et al.*, 2005; Pifferi *et al.*, 2006; Wang *et al.*, 2006). In the light, when CNG channels close but the exchanger continues to clear Ca^{2+} from the cytosol, the balance between Ca^{2+} entry and Ca^{2+} extrusion is disturbed. The resulting decline in $[Ca^{2+}]_i$ provides a negative feedback mechanism which controls at least three biochemical processes. First, the activity of guanylyl cyclase (GC) that synthesizes cGMP is stimulated as Ca^{2+} levels decrease. Second, the lifetime of active PDE is shortened through the phosphorylation of light-activated rhodopsin by rhodopsin kinase. Finally, the ligand sensitivity of the CNG channel increases as $[Ca^{2+}]_i$ decreases. All three reactions, to varying degrees, help to restore the dark state

and to adjust the light sensitivity of the cell (Kaupp and Seifert, 2002; Bradley *et al.*, 2005; Wang *et al.*, 2006).

In cones a similar transduction scheme exists. Thus, fundamentally similar events underlie phototransduction in rods and cones, and the two photoreceptor types utilize similar protein isoforms of the enzyme cascade. However, the light sensitivity of cones is 30- to 100-fold lower than that of rods, and cones adapt over a wider range of light intensities than rods (Kaupp, 1991; Menin, 1995; Pifferi *et al.*, 2006), suggesting that differences in Ca^{2+} homeostasis underlie the distinct light sensitivity and adaptation range of the two photoreceptor types. Several observations demonstrate that the cGMP sensitivity, its modulation by $[Ca^{2+}]_i$, and the Ca^{2+} permeation are profoundly different in CNG channels of rods and cones, supporting the notion that the CNG channel is a pivotal determinant of the dynamics of Ca^{2+} homeostasis in vertebrate photoreceptor cells (Kaupp and Seifert, 2002; Bradley *et al.*, 2005; Wang *et al.*, 2006a).

Physiological functions of CNG channels in OSNs

OSNs are embedded in the olfactory epithelium lining the cavity of the nose, which house all the molecular components to register odourants, to amplify the signal through a series of enzymatic reactions and to generate the electrical response. Quite remarkably, a cousin of the retinal CNG channels takes centre stage in odourant signalling. Specifically, the vast majority of OSNs respond to brief pulses of odourants with a transient receptor current by opening cAMP-gated channels in the ciliary membrane (Kaupp and Seifert, 2002; Brady *et al.*, 2006; Chen *et al.*, 2006; Flannery *et al.*, 2006; Wang *et al.*, 2006a; Boccaccio and Menini, 2007). Like rods, OSNs are exquisitely sensitive, and they can respond to stimulation by a few odourant molecules with a high sensitivity and a rich selectivity. For example, humans are probably able to discriminate between more than 10,000 or so different odourous compounds.

The binding of odourants to their cognate receptors in the membrane of chemosensitive cilia first activates a G protein (G_{olf}) and then an adenylyl cyclase (ACIII). The ensuing rise in the concentration of cAMP opens CNG channels and thereby produces a depolarization of the cell membrane. Like its retinal cousins, the CNG channel in OSNs is highly Ca^{2+} permeable and channel activation causes a rapid increase of $[Ca^{2+}]_i$ (Brady *et al.*, 2006; Chen *et al.*, 2006; Flannery *et al.*, 2006; Boccaccio and Menini, 2007). The odourant-stimulated rise of $[Ca^{2+}]_i$ plays an important role in both excitation and odour adaptation. The increase of $[Ca^{2+}]_i$ serves as a delayed negative feedback signal that reduces the cAMP sensitivity of the CNG channels and stimulates cAMP hydrolysis by a Ca^{2+}-dependent ciliary form of PDE (PDE1C2) (Brady *et al.*, 2006; Chen *et al.*, 2006; Flannery *et al.*, 2006; Boccaccio and Menini, 2007). Both processes, channel desensitization and PDE activation, are controlled by Ca^{2+}/CaM (Kaupp and Seifert, 2002; Bradley *et al.*, 2005; Wang *et al.*, 2006). The increase in $[Ca^{2+}]_i$ serves also as a feedforward signal that enhances the depolarizing response by activating Ca^{2+}-dependent Cl^- channels, which in fact carry a large fraction of the receptor current (Kaupp and Seifert, 2002; Bradley *et al.*, 2005; Wang *et al.*, 2006a).

In contrast to the outer segment of photoreceptors, where the light-sensitive current is solely carried by CNG channels, the receptor current originating in chemosensitive cilia has two ionic components: an inward cationic component mediated by CNG channels, followed by an inward anionic component mediated by Ca^{2+}-activated Cl^- channels (Wang et al., 2006a; Brady et al., 2006; Chen et al., 2006; Flannery et al., 2006; Boccaccio and Menini, 2007). In addition to cAMP, other signalling molecules have been implicated in odourant transduction, in particular the two gaseous messengers NO and carbon monoxide (CO). Furthermore, a cGMP-signalling pathway that targets a highly cGMP-selective CNG channel has also been identified in a small subset of OSNs. In contrast, components that furnish the prototypical cAMP-signalling pathway are absent in these cells. These observations provide compelling evidence that cGMP serves as the principal messenger for chemosensory signalling in some OSNs (Kaupp and Seifert, 2002; Brady et al., 2006; Chen et al., 2006; Flannery et al., 2006; Wang et al., 2006a, b; Boccaccio and Menini, 2007), suggesting the functions of CNG channels in OSNs are more complicated than those in photoreceptors.

Physiological functions of CNG channels in taste receptor cells

CNG channels play a special role in taste sensation. Thus, it is generally believed that after food enters the mouth, it interacts with taste buds, taste cells and their receptors in the lingua epidermis. Tastants then bind to G protein-coupled receptors, activate taste-related protein-coupled receptors, and stimulate CNG channels in the taste tube. This leads to depolarization of the cell membrane potential, resulting in a nerve impulse that travels to the cerebral cortex. Thus, CNG channels play a vital role in the taste bud, and taste signal transduction is achieved through regulation of cyclic nucleotide second messengers. The cone CNGA3 subunit was cloned from taste buds in the rat (Ruiz-Avila et al., 1995), and immunohistochemistry (IHC) has confirmed the presence of CNGA3 protein around taste cells. Furthermore, degeneration of taste buds through the cutting of the lingua throat nerve results in the loss of CNG channel immunoreactivity.

Physiological functions of CNG channels in spermatozoa

Cyclic nucleotides, cAMP and cGMP, are key elements of cellular signalling in spermatozoa, mediating several cellular responses, including acrosomal exocytosis, swimming behaviour, and chemo-attraction (Weyand et al., 1994; Wiesner et al., 1998; Biel and Michalakis, 2007). The testicular expression of several CNG channel subunits (CNGA1, CNGA3, CNGB1, and CNGB3) has been confirmed by cloning of cDNA from testis libraries as well as by Northern blot analysis and IHC (Biel et al., 1994; Weyand et al., 1994; Wiesner et al., 1998; Biel and Michalakis, 2007; Shi et al., 2005). Antibodies specific for the CNGA3 and CNGB1 subunits labeled the flagellum of mature spermatozoa and precursor cells in cross-sections of seminiferous tubules (Wiesner et al., 1998). Heterologous expression of the CNGA3 subunit cloned from testis produces channels that are cGMP sensitive and cGMP selective (Weyand et al., 1998). These CNG channels therefore might be involved in a cGMP-stimulated Ca^{2+} influx into intact spermatozoa (Wiesner et al., 1998). CNG channel activity is also detected in small vesicles that might have been derived from cytoplasmic droplets and in patches excised from osmotically swollen spermatozoa. Cyclic nucleotide-mediated Ca^{2+} influx into spermatozoa has been studied by confocal laser scanning microscopy (Wiesner et al., 1998).

The Ca^{2+} influx depends on the presence of extracellular Ca^{2+} and is greatly reduced at high extracellular Mg^{2+} concentrations. Since knockout mice lacking the CNGA3 subunit are fertile (Biel et al., 1999), the functional role of CNG channels in spermatozoa is still unclear. However, CNG channels might be involved in some aspect of motility control or, more specifically, in chemotactic swimming behaviour or possibly in the process of capacitation or acrosomal exocytosis.

In addition, many reports have suggested that CNG channels are involved in the maturation of granulosa cells during ovarian follicular development (Wang et al., 2006a). Vitalis et al., (2000) also report the identification by PCR of transcripts for the CNG channel subunits CNGA2, CNGA4, and CNGB1 in a neuronal cell line (GT1) secreting the gonadotropin releasing hormone (GnRH) (Vitalis et al., 2000).

Physiological functions of CNG channels in non-neuronal cells

CNG channels have been reported to exist in several non-neuronal cells, but their physiological roles have not been well-defined. Xu and colleagues (1999) report the expression of transcripts coding for the CNG channels from rod photoreceptors (CNGA1 and CNGB1 subunits) in a human alveolar cell line (A549). CNG channels have also been suggested to be involved in liquid homeostasis in the lung of 6-month-old sheep but not in 6-week-old sheep on the basis of a pharmacological study using dichlorobenzil, amiloride, and pimozide, which block CNG channels amongst other channels (Junor et al., 1999, 2000). Wu and colleagues (2000) have identified CNG channels in endothelial cells of pulmonary artery by recording current–voltage relationships in the whole-cell configuration under various conditions. They suggest that these currents may be carried by a CNG channel comprising the CNGA2 subunit. CNG channels have also been identified in cells of the renal inner medullary collecting duct (IMCD cells). These CNG channels display 99 per cent nucleotide sequence identity to the respective cDNA of the CNGA1 subunit cloned from the mouse retina (Pittler et al., 1992), suggesting that the differences between the retinal and renal cDNA sequence may account for functional differences of these channels.

4.2.7 Relevance to disease states

CNG channels are important cellular switches that mediate influx of Na^+ and Ca^{2+} in response to increases in the intracellular concentration of cAMP and cGMP. In photoreceptors and olfactory receptor neurons, these channels serve as end stage targets for cGMP and cAMP signalling pathways that are initiated by the absorption of photons and the binding of odourants, respectively. CNG channels have also been found in other types of neurons and in non-excitable cells. However, in most of these cells, the physiological role of CNG channels has yet to be determined. Recently, mutations in human CNG channel genes leading to inherited diseases (so-called channelopathies) have been functionally characterized. Moreover, mouse knockout models have been generated in an attempt to define the role of CNG channel proteins in vivo. In this section, we summarize recent insights into the physiological and pathophysiological role of CNG channels that have emerged from genetic studies in mice and humans (Brunet et al., 1996; Biel et al., 1999; Trudeau and Zagotta, 2002; Patel et al., 2005;

Michalakis *et al* 2005; Liu and Varnum 2005; Hüttl *et al.*, 2005; Brown *et al.*, 2006; Michalakis *et al.*, 2006; Wang *et al.*, 2006a; Biel and Michalakis, 2007).

Mutations in the *CNGA1* gene cause a rare autosomal recessive form of retinitis pigmentosa (RP) (Wang *et al.*, 2006; Brown *et al.*, 2006; Biel *et al.*, 2007), a genetically heterogeneous group of diseases that are characterized by a progressive degeneration of the rod and cone photoreceptors, ultimately leading to blindness. Photoreceptor degeneration is due to the paucity, or entire lack, of the rod CNG channel. Specifically, three of the five subunit alleles identified are null mutants and the other two alleles encode channels that mostly fail to reach the plasma membrane when heterologously expressed in HEK293 cells. The function of CNG channels has also been studied by targeted disruption of genes encoding the CNGA2, the CNGA3 and the CNGA4 subunit (Brunet *et al.*, 1996; Biel *et al.*, 1999). *CNGA2* $^{-/-}$ mice lacking functional cAMP-sensitive olfactory channels exhibit no detectable responses to odourants, i.e. they suffer from general anosmia (Brunet *et al.*, 1996). The disruption of the *CNGA2* gene also has biochemical and morphological consequences (Wang *et al.*, 2006a; Brown *et al.*, 2006; Biel *et al.*, 2007). The olfactory epithelium in *CNGA2* knockout mice is thinner and shows lower expression of an olfactory marker protein than in wild-type mice. Moreover, recording from hippocampal slices of *CNGA2* $^{-/-}$ mice, Patel and colleagues have shown that on weaker more physiological theta-burst stimulation, the initial amplitude of LTP is smaller and the post-titanic response decays faster in knockout mice compared with wild-type mice (Trudeau and Zagotta, 2002; Hüttl *et al.*, 2005; Liu and Varnum, 2005; Michalakis *et al.*, 2005, 2006; Patel *et al.*, 2005), suggesting that CNG channels play a significant role in long-term potentiation (LTP) in the hippocampus. Deletion of the CNGA3 subunit produced mice lacking any cone-mediated photoresponses, whereas the rod pathway is completely intact (Trudeau and Zagotta, 2002; Patel *et al.*, 2005; Michalakis *et al.*, 2005; Liu and Varnum, 2005; Hüttl *et al.*, 2005; Michalakis *et al.*, 2006). *CNGA3* knockout mice are also fertile. Kohl and collaborators have identified 46 mutations in the *CNGA3* gene in families originating from Germany, Norway, and the United States of America. Most mutations (39 of 46) represent amino acid substitutions (Wang *et al.*, 2006; Brown *et al.*, 2006; Biel and Michalakis, 2007). Four mutations (R277C, R283W, R436W, and F547L) account for almost 42 per cent of all mutant *CNGA3* alleles. The affected residues are located in the NH2-terminal domain near S1, in the S4 motif, the linker region, and the cGMP-binding site.

Another form of autosomal recessive RP is caused by mutation of the *CNGB1* gene encoding the CNGB1 subunit of the rod CNG channel (Hüttl *et al.*, 2005; Michalakis *et al.*, 2006). Mutations in either the *CNGB3* or the *CNGA3* genes, encoding CNGB3 or CNGA3 subunits of the cone photoreceptors, respectively, cause achromatopsia (or total colour blindness) (Patel *et al.*, 2005; Michalakis *et al.*, 2005; Liu and Varnum, 2005), a rare autosomal recessive disorder characterized by the total loss of colour discrimination, by photophobia, nystagmus, and severely reduced visual acuity. In the *CNGB3* gene encoding the B3 subunit, three missense mutations have been identified in the Pingelapese islanders of Micronesia (Trudeau and Zagotta, 2002; Liu *et al.*, 2005; Hüttl *et al.*, 2005). Six different mutations have been identified in caucasian families originating from Germany, Italy, and the United States of America, including a missense mutation, two stop-codon mutations, a 1– and

a 8–bp deletion, and a putative splice-site mutation of intron 13. None of the mutations has been functionally characterized. The disruption of cone photoreceptor function in achromatopsia, however, suggests that these mutations cause a total or at least severe impairment of CNG channel functions (Wang *et al* 2006a; Brown *et al.*, 2006; Biel and Michalakis, 2007).

Given the reportedly widespread expression of various CNG channel subunits in neuronal and non-neuronal tissues, it is surprising that in CNGA2- and CNGA3-deficient mice no phenotypic alterations were reported other than loss of smell and vision, respectively, and that patients suffering from either RP or achromatopsia associated with defects in the respective CNGA1, CNGB1, or CNGA3 channel subunits display no other phenotypic abnormalities.

4.2.8 Concluding remarks

Since the discovery of CNG channels in rods by Fesenko and colleagues in 1985 tremendous advances have been made. It has become clear that these channels are present in many systems, from sensory cells to non-sensory cells and plant cells. Molecular and physiological studies have already revealed that CNG channels are heteromultimers that contain at least two kinds of subunits. However, the physiological function of CNG channels in cells other than sensory neurons is still ill-defined.

CNGs are non-selective cation channels, which are very important in phototransduction and olfactory transduction as well as in neuronal path finding and synaptic plasticity. Mutations in CNG genes cause retinal degeneration, colour blindness, and many other diseases (Trudeau and Zagotta, 2002; Patel *et al.*, 2005; Michalakis *et al.*, 2005; Liu and Varnum, 2005; Hüttl *et al.*, 2005; Wang *et al.*, 2006a; Brown *et al.*, 2006; Michalakis *et al.*, 2006; Biel and Michalakis, 2007). In particular, mutations in the A and B subunits of the CNG channel expressed in human cones cause various forms of complete and incomplete achromatopsia. At the same time, the mutant channels provide a rich source of proteins with which to study the functional significance of individual amino acid residues. Further studies on the molecular mechanisms of CNG channel gating would be helpful in the prevention and treatment of channelopathies related to CNG channels.

CNG channels serve as a link between second messengers, membrane potential, ion transport and many aspects of cell physiology. Future studies will no doubt reveal new locations and functions of CNG channels, as well as new information on their functional modulation.

Acknowledgements

The financial support by National Natural Science Foundation of China (No.30571335 and 30771553) is gratefully acknowledged.

References

Ahmad I, Redmond LJ and Barnstable CJ (1990). Developmental and tissue-specific expression of the rod photoreceptor cGMP-gated ion channel gene. *Biochem Biophys Res Commun* **173**, 463–470.

Biel M and Michalakis S (2007). Function and dysfunction of CNG channels: insights from channelopathies and mouse models. *Mol Neurobiol* **35**, 266–277.

Biel M, Seeliger M, Pfeifer A *et al.* (1999). Selective loss of cone function in mice lacking the cyclic nucleotide-gated channel CNG3. *Proc Natl Acad Sci USA* **96**, 7553–7557.

Biel M, Zong X, Distler M *et al.* (1994). Another member of the cyclic nucleotide-gated channel family, expressed in testis, kidney, and heart. *Proc Natl Acad Sci USA* **91**, 3505–3509.

Biskup C, Kusch J, Schulz E *et al.* (2007). Relating ligand binding to activation gating in CNGA2 channels. *Nature* **446**, 440–443.

Boccaccio A and Menini A (2007).Temporal development of cyclic nucleotide-gated and Ca^{2+}-activated Cl– currents in isolated mouse olfactory sensory neurons. *J Neurophysiol* **98**, 153–160.

Bradley J, Reisert J and Frings S (2005). Regulation of cyclic nucleotide-gated channels. *Curr Opin Neurobiol* 15, 343–349.

Brady JD, Rich ED, Martens JR *et al.* (2006). Interplay between PIP3 and calmodulin regulation of olfactory cyclic nucleotide-gated channels. *Proc Natl Acad Sci USA* **103**, 15635–1540.

Bright SR, Rich ED and Varnum MD (2007). Regulation of human cone cyclic nucleotide-gated channels by endogenous phospholipids and exogenously applied phosphatidylinositol 3,4,5-trisphosphate. *Mol Pharmacol* **71**, 176–183.

Brown RL, Strassmaier T, Brady JD *et al.* (2006). The pharmacology of cyclic nucleotide-gated channels: emerging from the darkness. *Curr Pharm Des* **12**, 3597–3613.

Brunet LJ, Gold GH and Ngai J (1996). General anosmia caused by a targeted disruption of the mouse olfactory cyclic nucleotide-gated cation channel. *Neuron* **17**, 681–93.

Chen TY, Takeuchi H and Kurahashi T (2006). Odourant inhibition of the olfactory cyclic nucleotide-gated channel with a native molecular assembly. *J Gen Physiol* **128**, 365–371.

Cho SW, Cho JH, Song HO *et al.* (2005). Identification and characterization of a putative cyclic nucleotide-gated channel, CNG-1, in *C. elegans*. *Mol Cells* **19**, 149–154.

Ding C, Potter ED, Qiu W *et al.* (1997). Cloning and widespread distribution of the rat rod-type cyclic nucleotide-gated cation channel. *Am J Physiol* **272**, C1335–1344.

Distler M, Biel M, Flockerzi V *et al.* (1994). Expression of cyclic nucleotide-gated cation channels in non-sensory tissues and cells. *Neuropharm* **33**, 1275–1282.

Fesenko EE, Kolesnikov SS and Lyubarsky AL (1985). Induction by cyclic GMP of cationic conductance in plasma membrane of retinal rod outer segment. *Nature* **313**, 310–313.

Flannery RJ, French DA and Kleene SJ (2006). Clustering of cyclic-nucleotide-gated channels in olfactory cilia. *Biophys J* **91**, 179–188.

Friedmann NK (2000). Cyclic nucleotide-gated channels in nonsensory organs. *Cell Calcium* **27**, 127–138.

Giorgetti A and Carloni P (2003). Molecular modelling of ion chanels: structural predictions. *Curr Opin Chem Biol* 7, 150–156.

Giorgetti A, Nair AV, Codega P *et al.* (2005). Structural basis of gating of CNG channels. *FEBS Lett* **579**, 1968–1972.

Hua L and Gordon SE (2005). Functional interactions between A' helices in the C-linker of open CNG channels. *J Gen Physiol* **125**, 335–344.

Hüttl S, Michalakis S, Seeliger M *et al.* (2005). Impaired channel targeting and retinal degeneration in mice lacking the cyclic nucleotide-gated channel subunit CNGB1. *J Neurosci* **25**, 130–138.

Johnson J and Zagotta WN (2001). Rotation movement during cyclic nucleotide-gated channel opening. *Nature* **412**, 917–921.

Junor RW, Benjamin AR, Alexandrou D *et al.* (1999). A novel role for cyclic nucleotide-gated cation channels in lung liquid homeostasis in sheep. *J Physiol* **520**, 255–260.

Junor RW, Benjamin AR, Alexandrou D *et al.* (2000). Lack of a role for cyclic nucleotide-gated cation channels in lung liquid absorption in fetal sheep. *J Physiol* **523**, 493–502.

Kaupp UB (1991). The cyclic nucleotide-gated channels of vertebrate photoreceptors and olfactory epithelium. *TINS* **14**, 150–157.

Kaupp UB and Seifert R (2002). Cyclic nucleotide-gated ion channels. *Physiol Rev* **82**, 769–824.

Kruse LS, Sandholdt NT, Gammeltoft S *et al.* (2006). Phosphodiesterase 3 and 5 and cyclic nucleotide-gated ion channel expression in rat trigeminovascular system. *Neurosci Lett* **404**, 202–207.

Kwon RJ, Ha TS, Kim W *et al.* (2002). Binding symmetry of extracellular divalent cations to conduction pore studied using tandem dimers of a CNG channel. *Biochem Biophys Res Commun* **29**, 478–485.

Liu C and Varnum MD (2005). Functional consequences of progressive cone dystrophy-associated mutations in the human cone photoreceptor cyclic nucleotide-gated channel CNGA3 subunit. *Am J Physiol Cell Physiol* **289**, C187–198.

Liu J and Siegelbaum SA (2000). Change of pore helix conformational state upon opening of cyclic nucleotide-gated channels. *Neuron* **28**, 899–909.

Matulef K, Flynn GE and Zagotta WN (1999). Molecular rearrangements in the ligand-binding domain of cyclic nucleotide-gated channels. *Neuron* **24**, 443–452.

Menin A (1995). Cyclic nucleotide-gated channels in visual and olfactory transduction. *Biophys Chem* **55**, 185–196.

Michalakis S, Geiger H, Haverkamp S *et al.* (2005). Impaired opsin targeting and cone photoreceptor migration in the retina of mice lacking the cyclic nucleotide-gated channel CNGA3. *Invest Ophthalmol Vis Sci* **46**, 1516–1524.

Michalakis S, Reisert J, Geiger H *et al.* (2006).Loss of CNGB1 protein leads to olfactory dysfunction and subciliary cyclic nucleotide-gated channel trapping. *J Biol Chem* **281**, 35156–35166.

Oda Y, Timpe LC, McKenzie RC *et al.* (1997). Alternatively spliced forms of the cGMP-gated channel in human keratinocytes. *FEBS Lett* **414**, 140–145.

Patel KA, Bartoli KM, Fandino RA *et al.* (2005).Transmembrane S1 mutations in CNGA3 from achromatopsia 2 patients cause loss of function and impaired cellular trafficking of the cone CNG channel. *Invest Ophthalmol Vis Sci* **46**, 2282–2290.

Peng CH, Rich ED and Vamum MD (2004). Subunit configuration of heteromeric cone cyclic nucleotide-gated channels. *Neuron* **42**, 401–410.

Pifferi S, Boccaccio A and Menini A (2006). Cyclic nucleotide-gated ion channels in sensory transduction. *FEBS Lett* **580**, 2853–2859.

Pittler SJ, Lee AK, Altherr MR *et al.* (1992). Primary structure and chromosomal localization of human and mouse rod photoreceptor cGMP-gated cation channel. *J Biol Chem* **267**, 6257–6262.

Podda MV, Marcocci ME, Del Carlo B *et al.* (2005). Expression of cyclic nucleotide-gated channels in the rat medial vestibular nucleus. *Neuroreport* **16**, 1939–1943.

Qu W, Moorhouse AJ, Chandra M *et al.* (2006). A single P-loop glutamate point mutation to either lysine or arginine switches the cation-anion selectivity of the CNGA2 channel. *J Gen Physiol* **127**, 375–389.

Rosenbaum T and Gordon SE (2002). Dissecting intersubunit contacts in cyclic nucleotide-gated ion channels. *Neuron* **33**, 703–713.

Ruiz ML, London B and Nadal-Ginard B (1996). Cloning and characterization of an olfactory cyclic nucleotide-gated channel expressed in mouse heart. *J Mol Cell Cardiol* **28**, 1453–1461.

Ruiz-Avila L, McLaughlin SK, Wildman D *et al.* (1995). Coupling of bitter receptor to phosphodiesterase through transducin in taste receptor cells. *Nature* **376**, 80–85.

Samanta Roy DR and Barnstable CJ (1999). Temporal and spatial pattern of expression of cyclic nucleotide-gated channels in developing rat visual cortex. *Cereb Cortex* **9**, 340–347.

Shi F, Perez E, Wang T *et al.* (2005). Stage- and cell-specific expression of soluble guanylyl cyclase alpha and beta subunits, cGMP-dependent protein kinase I alpha and beta, and cyclic nucleotide-gated channel subunit 1 in the rat testis. *J Androl* **26**, 258–263.

Simoes M, Garneau L, Klein H *et al.* (2002). Cysteine mutagenesis and computer modeling of the S6 region of an intermediate conductance IKCa channels. *J Gen Physiol* **120**, 99–116.

Strassmaier T and Karpen JW (2007). Novel N7- and N1-substituted cGMP derivatives are potent activators of cyclic nucleotide-gated channels. *J Med Chem* **50**, 4186–4194.

Strünker T, Weyand I, Bönigk W *et al.* (2006). A K$^+$-selective cGMP-gated ion channel controls chemosensation of sperm. *Nat Cell Biol* **8**, 1149–1154.

Tetreault ML, Henry D, Horrigan DM *et al.* (2006). Characterization of a novel cyclic nucleotide-gated channel from zebrafish brain. *Biochem Biophys Res Commun* **348**, 441–449.

Trudeau MC and Zagotta WN (2002). An intersubunit interaction regulates trafficking of rod cyclic nucleotide-gated channels and is disrupted in an inherited form of blindness. *Neuron* **34**, 197–207.

Vitalis EA, Costantin JL, Tsai PS *et al.* (2000). Role of the cAMP signaling pathway in the regulation of gonadotropin-releasing hormone secretion in GT1 cells. *Proc Natl Acad Sci USA* **97**, 1861–1866.

Wang ZC, Huang RH, Pan LM *et al.* (2006a). Molecular structures, physiological roles and regulatory mechanism of cyclic nucleotide-gated ion channels. *Chin J Biochem Mol Biol* **22**, 282–288.

Wang ZC, Jiang YQ, Lu LZ *et al.* (2007). Molecular mechanisms of cyclic nucleotide-gated ion channel gating. *J of Genetics and Genomics* **34**, 477–485.

Wang ZC, Luo JB, Pang US *et al.* (2006b). Involvement of cyclic nucleotide-gated ion channel in disease pathogenesis. *Prog Vet Med* **27**, 1–4.

Wang ZC and Shi FX (2007). Phosphodiesterase 4 and compartmentalization of cyclic AMP signaling. *Chin Sci Bull* **52**, 34–46.

Weitz D, Ficek N, Kremmer E *et al.* (2002). Subunit stoichiometry of the CNG channel of rod photoreceptors. *Neuron* **36**, 881–889.

Weyand I, Godde M, Frings S *et al.* (1994). Cloning and functional expression of a cyclic-nucleotide-gated channel from mammalian sperm. *Nature* **368**, 859–863.

Wiesner B, Weiner J, Middendorff R *et al.* (1998). Cyclic nucleotide-gated channels on the flagellum control Ca2+ entry into sperm. *J Cell Biol* **142**, 473–484.

Wu S, Moore TM, Brough GH *et al.* (2000). Cyclic nucleotide-gated channels mediate membrane depolarization following activation of store-operated calcium entry in endothelial cells. *J Biol Chem* **275**, 18887–18896.

Xu W, Leung S, Wright J *et al.* (1999).Expression of cyclic nucleotide-gated cation channels in airway epithelial cells. *J Membr Biol* **171**, 117–126.

Zheng J, Zagotta WN (2000). Gating rearrangements in cyclic nucleotide-gated channels revealed by patch-clamp fluorometry. *Neuron* **28**, 369–374.

Zheng J, Zagotta WN (2004). Stoichiometry and assembly of olfactory cyclic nucleotide-gated channels. *Neuron* **42**, 411–421.

Zhong II, Lai J, Yau KW (2003). Selective heteromeric assembly of cyclic nucleotide-gated channels. *Proc Natl Acad Sci USA* **100**, 5509–5513.

4.3

K$_{Ca}$1—K$_{Ca}$5 families

Morten Grunnet, Dorte Strøbæk, Søren-Peter Olesen and Palle Christophersen

4.3.1 Introduction

The activity of Ca^{2+}-activated K$^+$ channels (K$_{Ca}$) depends on the intracellular Ca^{2+} concentration ([Ca^{2+}]$_i$) and thus constitutes a link between this important second messenger system and the electrical activity of cells. All mammalian cell types express K$_{Ca}$ channels and the membrane hyperpolarizations they induce feed negatively or positively back to [Ca^{2+}]$_i$ depending on whether the Ca^{2+} entry pathways are voltage-dependent (Ca$_v$ or N-methyl-D-aspartate [NMDA] receptors), as is often the case in excitable cells, or voltage-independent (Ca^{2+}-release-activated Ca^{2+} channels [CRAC]), as in many non-excitable cells. K$_{Ca}$ channels have classically been distinguished according to biophysical and pharmacological characteristics as small conductance (SK) channels being sensitive to apamin, intermediate conductance (IK) channels being sensitive to clotrimazole, and big conductance (BK) channels being sensitive to iberiotoxin. With their molecular cloning it has become clear that the SK channels (three members, K$_{Ca}$2.1–2.3) and the IK channel (one member, K$_{Ca}$3.1) are closely related families and constitute one group at the molecular level. The BK channel is very different from SK/IK channels and belongs to a special Ca^{2+}-activated group of intracellular ion-gated K$^+$ channels also representing OH$^-$, Na$^+$ and Cl$^-$-activated channels (K$_{Ca}$1.1 [BK]; K$_{Ca}$5.1; K$_{Ca}$4.1–4.2 [slack and slick]). For explanation of the IUPHAR nomenclature on Ca^{2+}-activated K$^+$ channels, see Wei *et al.* (2005).

The first unequivocal description of Ca^{2+}-activated K$^+$ transport from cells was made by Gárdos (1958). Studying the accelerating effect of glycolytic poisons on K$^+$ loss from erythrocytes, he showed that this effect required the presence of Ca^{2+}. Although the ion channel nature of this pathway was subesquently questioned and final proof was not forthcoming until the development of patch-clamp technology (Grygorczyk and Schwarz, 1983), the erythrocyte K$_{Ca}$ channel was dubbed 'the Gárdos channel'. The intermediate conductance K$_{Ca}$ channel, cloned in 1997 by Ishii and colleagues (Ishii *et al.*, 1997b), was shown to have physiological, biophysical and pharmacological properties very similar to those described for the Gárdos channel. The final proof, closing this loop of 50 years of scientific effort, was provided by Hoffman and co-workers who demonstrated that the only detectable mRNA transcript encoding K$_{Ca}$ channels in purified reticulocytes, was the IK mRNA (Hoffman *et al.*, 2003). The notion that different types of closely related Ca^{2+}-activated channels exist arose following the observation of channels with smaller single channel conductances than the Gárdos channel, as well as sensitivity to the neurotoxin apamin that had no effect in erythrocytes. These channels were identified when the three subtypes of SK channels were finally cloned from brain tissue (Köhler *et al.*, 1996). For a review on molecular and functional aspects of SK channels, see Stocker (2004).

The initial electrophysiological characterization of BK channels was reported about 15 years before the BK gene was cloned. In neuronal recordings from snails a Ca^{2+}- and voltage-gated potassium current was described (Heyer and Lux, 1976) and a potassium channel with these properties was identified shortly afterwards in mammals (Pallotta *et al.*, 1981). A breakthrough towards cloning of the BK channel came in 1986, with the demonstration of a mutation in the genome of the fruitfly *Drosophila melanogaster* that eliminated a Ca^{2+}- and voltage-gated potassium current (Elkins *et al.*, 1986). This led to the cloning of the BK channel 5 years later and it was initially named 'slowpoke' after the characteristic jerky movements of the fly (Atkinson *et al.*, 1991). The BK channel was subsequently identified in mammalian species including human (Adelman *et al.*, 1992; Butler *et al.*, 1993; Dworetzky *et al.*, 1994).

SK and IK channels are classic K$^+$ channels with 6 transmembrane (TM) regions, 1 pore loop (P), and with both N- and C-termini positioned at the cytoplasmic side of the membrane. The functional channels are composed of four subunits arranged around a central pore lined with P and 5-TM amino acid residues, and with the potassium channel signature sequence (GYG) as the central selectivity filter for K$^+$. Calmodulin (CaM), constitutively associated with a calmodulin binding domain (CaMBD) in the C-terminal (Xia *et al.*, 1998; Fanger *et al.*, 1999), is the binding site for Ca^{2+} that drives activation and deactivation of these channels (Schumacher *et al.*, 2001).

Initially, it was believed that BK channel subunits were composed of the classical six TM segments with a bulky C-terminus which could constitute up to four additional TM segments (Knaus *et al.*, 1995). It now seems apparent that there are seven TM domains (numbered 0–6), with the N-terminus located extracellularly and the C-terminus intracellularly (Wallner *et al.*, 1996; Meera *et al.*, 1997). As for SK and IK channels, four BK α-subunits create the pore-forming functional channel and a tetramerization domain has been located slightly C-terminal to TM6 (Quirk and Reinhart, 2001). In contrast to SK/IK channels, activation of the BK channel occurs by binding of intracellular Ca^{2+} directly to the C-terminal (Schreiber and Salkoff, 1997). The BK channel is in addition voltage-dependent and TM4, which is essential as voltage-sensor in Kv channels, is also in BK channels characterized by positively charged residues at every third position, each separated by two hydrophobic residues (Atkinson *et al.*, 1991).

The remaining three members of the K_{Ca} family are $K_{Ca}4.1$, $K_{Ca}4.2$ and $K_{Ca}5.1$. $K_{Ca}4.X$ was originally cloned from the nematode *Caenorhabditis elegans* and these channels are regulated by Ca^{2+} and Cl^- (Yuan *et al.*, 2000). In mammals two closely related genes denoted $K_{Ca}4.1$ and $K_{Ca}4.2$ have been identified, and when expressed, these channels are synergistically gated by Na^+ and Cl^-. $K_{Ca}4.1$ and $K_{Ca}4.2$ are six TM channels lacking the positively charged residues in the fourth TM domain and, thus, they exhibit a poor voltage-sensitivity (Yuan *et al.*, 2003; Salkoff *et al.*, 2006). Finally, $K_{Ca}5.1$ was cloned from mammalian spermatocytes. Its membrane topology is identical to BK channels (seven TM domains) with a voltage-sensor in TM4. The calcium binding site is lacking and the channel is thus regulated by voltage and is insensitive to internal Ca^{2+}. Instead, a prominent sensitivity to regulation by pH has been demonstrated (Schreiber *et al.*, 1998).

For completeness, in addition to reviewing all cloned members of the SK and BK families of K^+ channels (including the members not activated by Ca^{2+}), this chapter also considers the hitherto molecularly undefined Ca^{2+}-dependent K^+ current underlying the slow after-hyperpolarization in some neurons. For a summary of the basic features of the cloned K_{Ca} families, see Figure 4.3.1.

4.3.2 Gene structure

The SK/IK channel proteins differ greatly in lengths (for the human clones: hSK1, 543 aa; hSK2, 579 aa; hSK3, 736 aa; and hIK, 427 aa) in particular due to variation at their N-termini. The SK1–3 channels exhibit high homology, whereas they are only about ~40 per cent homologous with the IK channel. The genes for the human, mouse and rat SK1–3 and IK channels have been cloned and are positioned (within the human genome) at chromosomes 19p13.1, 5q22.3, 1q21.3, and 19q13.2, respectively. The intron/exon structure of SK/IK channels is complex with the SK1/IK channel having nine and the SK2/3 channels eight exons with intron/exon junctions in the coding regions nearly identically positioned (Ghansani *et al.*, 2000). The conserved gene structure of SK/IK channels across species indicates a common evolutionary origin of this group of channels. SK1 has a unique three-amino acid exon in the CaMBD. Alternative splicing is documented for hSK1 with four gene products capable of CaM binding expressed at significant levels in the brain (Litt *et al.*, 1999; Shmukler *et al.*, 2001). SK2 exists in two forms, SK2-S corresponding to the original clone and an N-terminally extended isoform called SK2-L, that can co-assemble with SK2S and SK3 (Strassmaier *et al.*, 2005). Alternative splicing of SK3 between TM5 and the pore region yields channels with altered blocker pharmacology (Wittekindt *et al.*, 2004), whereas N-terminally truncated, non-functional transcripts can be dominant negative suppressors of channel expression (Tomita *et al.*, 2003). SK channels tend to have repetitive motives in their N-terminals with SK1 carrying a polyglutamate, and SK3 a polyglutamine, repeat sequence.

Only one gene coding for the BK α-subunit has been described with the human isoform positioned at chromosome 10q22.3. The original clone consisted of 1184 amino acids (Atkinson *et al.*, 1991) but at least ten different alternative splice sites are known. These are all located

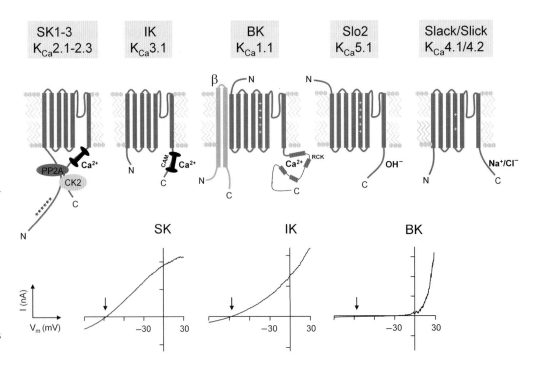

Fig. 4.3.1 Basic characteristics of K_{Ca} channels. (Upper panel) Overall membrane topology of the K_{Ca} families. CAM, calmodulin; PP2A, protein phosphatase 2A; CK2, casein kinase 2; RCK, regulator of K^+ conductance homology domain; β, BK β-subunit (β1–β4); ****** Poly amino acid repeats. (Lower panel) Representative whole-cell I–V relationships of cloned human SK3, IK, and BK channels, respectively, expressed in HEK293 cells. The arrows indicate the K^+ equilibrium potential, E_K. $[Ca^{2+}]$ in pipette solutions was buffered at 400 nM.

in the C-terminal part of the protein distal to TM6. Ca^{2+} and phosphorylation regulate the isoforms differentially. Functional diversity through alternative splicing was identified by Adelman *et al.* (1992) and later characterized in detail by others (reviewed by Shipston 2001).

Within the K_{Ca} channel families, BK is the only channel for which specific β-subunits have been described. The β1 subunit was cloned from bovine tracheal smooth muscle in 1994 (Knaus *et al.*, 1994) and recently three more β-subunits named β2–4 have been cloned. The four different β-subunits (β1–4) all have two TM domains with the N- and C-termini located intracellularly. The β3-subunit has been described by a number of laboratories and exists in four different splice variants named β3a–d (Wallner *et al.*, 1999; Xia *et al.*, 1999).

4.3.3 Distribution

SK and BK channels are distributed in the central nervous system (CNS) and in peripheral tissues, whereas the IK channel is mainly expressed peripherally. The distribution of K_{Ca} channels has traditionally been evaluated functionally through the identification of K_{Ca} currents sensitive to specific blockers, as well as by quantitative receptor autoradiography. With the cloning of the various subtypes more detailed information on subtype expression and subcellular localization has been obtained from *in situ* hybridization and immunohistochemistry studies.

SK channels exhibit widespread but distinct expression patterns in the brains of mice and rats. The overall picture is that SK1 and SK2 co-express in neocortical and hippocampal regions, whereas SK3 is confined to evolutionary older brain regions such as the basal ganglia and thalamus (Stocker and Pedarzani, 2000; Sailer *et al.*, 2002, 2004). The strong expression of SK3 in the habenula, in monoaminergic neurons of the ventral tegmental area (VTA) and substantia nigra pars compacta (SNc) (dopaminergic) as well as in the raphe (serotoninergic) and locus coeruleus (noradrenergic) is noteworthy (Tacconi *et al.*, 2001; Sarpal *et al.*, 2004). At the subcellular level SK1 and SK2 immunoreactivity have been associated with soma and dendrites, whereas SK3 also appears to be localized presynaptically (Obermair *et al.*, 2003; Sailer *et al.*, 2004). In the periphery SK channels are found in vascular endothelium, liver, adrenal chromaffin cells, skeletal, and smooth muscle cells and in sensory nerves (Bahia *et al.*, 2005; Mongan *et al.*, 2005). The general distribution pattern also applies to human tissue where SK1 has the narrowest distribution, being largely restricted to neuronal tissue. SK3 is highly expressed in many peripheral tissues including vascular and genital tissues (Chen *et al.*, 2004).

The IK channel is prominently expressed in epithelia, endothelia, smooth muscle cells, and blood cells: erythrocytes, lymphocytes, and various inflammatory cells (Ishii *et al.*, 1997b; Logsdon *et al.*, 1997; Jensen *et al.*, 1998). As an exception to the general peripheral distribution, the IK channel is expressed in CNS microglia, where they participate in activation and induction of NO production (Kaushal *et al.*, 2007). In human tissue the IK channel was found in placenta, prostate, salivary glands, and trachea (Chen *et al.*, 2004). In the intestinal wall IK immunoreactivity was confined to inflammatory cells and enteric neurons, a finding that corresponds to the established 'IK-pharmacology' of the Ca^{2+}-dependent AHP in rodent enteric neurons (Nguyen *et al.*, 2007).

Neuronal expression of BK channels is prominent in substantia nigra, outer layers of the neocortex, in pallidal areas, in the suprachiasmatic nucleus, striatum, habenula, the olfactory bulb, sensory fibres, caudate putamen, thalamus, cerebellar Purkinje cell layers, and in the granule and pyramidal cell layers of the hippocampal formation (Knaus *et al.*, 1996; Grunnet and Kaufmann 2004; Meredith *et al.*, 2006). At the subcellular level BK immunoreactivity has been associated both with presynaptic terminals, soma, and dendrites.

BK channels are also widely expressed in peripheral tissues, such as in parts of the kidney epithelia including the visceral and parietal layer of Bowman's capsule, medullary thick ascending limb, thin limb of Henle's loop, distal tubulus, and cortical collecting tubule (Frindt and Palmer, 1987; Lu *et al.*, 1990; Grunnet *et al.*, 2005). BK channels are also found in the epithelia of distal colon, airways, and lens (Rae and Shepard, 1998; Tamaoki *et al.*, 1998; Grunnet *et al.*, 1999; Hay-Schmidt *et al.*, 2003). The channels are strongly expressed in smooth muscle cells and some endothelial cells (O'Neill and Steinberg, 1995; Holland *et al.*, 1996; Herrera *et al.*, 2001). BK channels can be targeted to different subcellular compartments, as exemplified by selective expression at the apical or basolateral membranes of various epithelia (Hay-Schmidt *et al.*, 2003), or mitochondrial membranes (Xu *et al.*, 2002).

The distribution of $K_{Ca}4.1$ (Slack) and $K_{Ca}4.2$ (Slick) has not been investigated to the same degree as for BK channels. $K_{Ca}4.1$ is found primarily in brain with the highest expression in mitral cells of the olfactory bulb, but is also expressed in kidney and testis. $K_{Ca}4.2$ is more widely distributed and can be detected in brain, heart, skeletal muscle, kidney, testis, lung, and liver (Joiner *et al.*, 1998; Yuan *et al.*, 2003). $K_{Ca}5.1$ has to date only been identified in spermatocytes and mature spermatozoa (Schreiber *et al.*, 1998; Salkoff *et al.*, 2006).

4.3.4 Biophysical characteristics

Even though K_{Ca} channels share the basic K^+ selectivity they differ in other open channel properties, such as rectification and single channel conductance, and even more at the level of gating, where they only share the ability to be activated by various intracellular inorganic ions.

SK and IK channels are closed at 'resting' intracellular Ca^{2+} and their gating is mediated via Ca^{2+}-binding to CaM that is constitutively bound to a CaMBD in the C-terminus (Xia *et al.*, 1998; Fanger *et al.*, 1999). It is likely that four CaM molecules, each potentially binding two Ca^{2+}, account for channel activation (Schumacher *et al.*, 2001). More generally SK channels (SK2 and probably SK3) are part of intracellular macromolecular protein complexes, consisting of CaM, the N- and C-termini, a protein kinase (casein kinase 2, CK2), a phosphatase (protein phosphatase 2A, PP2A) and scaffolding proteins. This complex dynamically controls the phosphorylation state of CaM and thereby the Ca^{2+}-sensitivity of SK2 channels (Bildl *et al.*, 2004), a mechanism that imposes state-dependent regulation of the Ca^{2+}-sensitivity (Allen *et al.*, 2007). It has recently been demonstrated that CK2 phosphorylation and reduced Ca^{2+} sensitivity of SK2 channels can be induced by stimulation of noradrenergic receptors (Maingret *et al.*, 2008). N- and C- terminal interactions have also been demonstrated for SK3 (Frei *et al.*, 2006). All SK subtypes exhibit identical steady-state activation curves for Ca^{2+} (Xia et al., 1998) that follow steep ($n_H = 3–5$) Hill-type equations with EC_{50} values around 300 nM. Single channel recordings at SK2 channels using inside-out patches showed that the maximal probability of opening (Po) attained at high $[Ca^{2+}]_i$ was ~0.8 (Hirschberg et al., 1998). The rate constant for activation of SK channels by fast (msec) Ca^{2+} steps is proportional to the $[Ca^{2+}]_i$ (47 $\mu M^{-1}s^{-1}$), whereas the deactivation rate is independent of $[Ca^{2+}]_i$ (13 s^{-1}). Despite this appealing macroscopic simplicity a kinetic scheme of at least four closed and two open states was needed to model single channel kinetics (Hirschberg et al., 1998). The Ca^{2+} activation curve for the IK channel is less steep ($n_H = 1.7$), but with a similar EC_{50} value (Ishii et al., 1997b) indicating that differences in details of Ca^{2+} binding/gating exist even between SK and IK channels. Multiple closed and open states are also discerned for the IK channel (Bennekou and Christophersen, 1990; Syme et al., 2000). SK and IK channels can be activated by certain Ca^{2+} 'analogues' such as Sr^{2+}, Cd^{2+}, Pb^{2+}, and Co^{2+}. It is noteworthy that the BK channel is not activated by Cd^{2+}, which makes this ion selective for activation SK/IK channels (Leinders et al., 1992). The gate of SK and IK channels has been localized to a position central in the membrane close to the selectivity filter (Bruening-Wright

et al., 2002, 2007; Klein et al., 2007) and this may also apply to BK channels (Wilkens and Aldrich, 2006). This is in contrast to Kv channels that open and close at the internally located TM6 bundle crossing (the inverted tepee model).

SK channels have a single channel conductance of 8–10 pS (0 mV, symmetrical 150 mM K^+), whereas the corresponding value for IK is 25–40 pS (Christophersen, 1991; Ishii et al., 1997b). Both SK and (less pronounced) the IK channel exhibit inward rectifying IV curves in symmetrical K^+ solutions, an effect which largely vanishes at physiological ion gradients (Figure 4.3.1), clearly distinguishing it from the inward rectification of Kir channels. SK and IK channels are blocked by impermeable, monovalent K^+ analogues, but also by divalent ions, and voltage-dependent block by intracellular Ca^{2+} and Mg^{2+} have been suggested to account for the inward rectification of SK channels (Soh and Park, 2001). From macroscopic flux experiments with erythrocytes, the Gárdos channel was found to exhibit classic multi-ion conduction, showing single-file diffusion (n = 2.7) (Vestergaard-Bogind et al., 1985) and voltage-dependent block (z > 1) by intracellular Na^+ (Stampe and Vestergaard-Bogind, 1989). As expected from the canonical K^+ channel sequence (GYG) in the SK and IK channel selectivity

Fig. 4.3.2 Physiology (mAHP and sAHP) and pharmacology of K_{Ca} channels. (a) Left panel, Ca^{2+}-activation curves for SK channels in the presence of varied concentrations of NS309, a positive modulator of SK/IK channels. Right panel, the $EC_{50}(Ca^{2+})$ values plotted vs [NS309] derived from recordings from inside-out patches pulled from SK3 expressing HEK293 cells. Reproduced from Hougaard et al. (2007).[1] (b) Augmentation of apamin sensitive spike frequency adaptation by NS309 (10 μM) following current injection to a CA1 neuron recorded from a rat hippocampal slice. (c) Specific NS309-mediated augmentation of the apamin-sensitive mAHP in a CA1 neuron. Reproduced from Pedarzani et al. (2005).[2]

[1] Reproduced from Hougaard C, Eriksen BL, Jørgensen S, Johansen TH, Dyhring T, Madsen LS, Strøbaek D and Christophersen P (2007). Selective positive modulation of the SK3 and SK2 subtypes of small conductance Ca^{2+}-activated K^+ channels. *British Journal of Pharmacology* **151**, 655–666.

[2] Reproduced from Pedarzani P, McCutcheon JE, Rogge G, Jensen BS, Christophersen P, Hougaard C, Strøbaek D, and Stocker M (2005). Specific enhancement of SK channel activity selectively potentiates the after-hyperpolarizing current I_{AHP} and modulates the firing properties of hippocampal pyramidal neurons. *Journal of Biological Chemistry* **280**, 41404–41411.

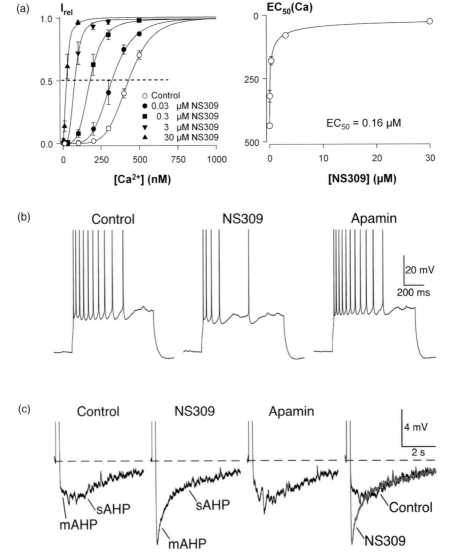

filter, they exhibit the classic selectivity sequence for monovalent cations ($Tl^+ > K^+ > Rb^+ > NH_4^+ > Cs^+ \approx Na^+ \approx Li^+$), which, for the alkali metals, corresponds to the Eisenman selectivity sequence IV or V (Christophersen, 1991; Park, 1994). However, some selectivity properties are surprising for a K^+ channel. Shin *et al.* (2005) reported that cloned SK2 and IK channels exhibit significant Cs^+, Na^+, and Li^+ permeabilities ($P_X/P_K \approx 0.1–0.2$) that were at least one order of magnitude higher than that found for BK in the same study. For the IK channel these results contrast with the permeability ratios calculated from tracer fluxes ($P_K/P_{Na} > 1000$) with human erythrocytes under maximal Gárdos channel activation and also with single channel recordings where only currents carried by Tl^+, K^+, Rb^+, and NH_4^+ were measured (Christophersen, 1991). A complicating factor in the comparison of single channel currents and macroscopic currents is that permeating ions themselves strongly influence gating properties (Bennekou and Christophersen, 1990; Jensen *et al.*, 1998) and stability of the IK channel (Hoffman *et al.*, 2003). The relation between permeation and gating may still be worth investigating in light of the proposed close apposition of the selectivity filter and gating structures in SK channels (Bruening-Wright *et al.*, 2002, 2007).

The mechanism for the Ca^{2+}-sensitivity for BK channels is more speculative, since no Ca^{2+} binding consensus motives are found in the α-subunit and Ca^{2+} binding auxiliary subunits have not been demonstrated. A potential binding site containing negative charges has been identified in the C-terminal domain and has been termed the 'calcium bowl' (Wei *et al.*, 1994; Schreiber and Salkoff, 1997). The Ca^{2+} bowl hypothesis has been challenged by experiments demonstrating that the Ca^{2+}- and voltage-regulation of BK channels is largely unaffected even if the entire C-terminus is deleted (Piskorowski and Aldrich, 2002). This finding, however, has not yet been confirmed by other laboratories. Other than the Ca^{2+} bowl, BK channels have two separate domains in the C-terminus called RCK1 and RCK2, that are also believed to participate in Ca^{2+}-dependent gating of BK channels; the RCK2 domain in particular is responsible for the Mg^{2+}-modulation of the channels (Shi and Cui, 2001; Xia *et al.*, 2002).

As for Kv channels the exact mechanism by which changes in membrane potential are converted to movement of the TM segments is still a matter of debate. It is clear, however, that voltage-mediated activation takes place in a Ca^{2+}-independent manner (Cox *et al.*, 1997; Vergara *et al.*, 1998). Taking the voltage-dependency, the three putative binding sites for Ca^{2+}, and the stoichiometry of a functional BK channel into consideration the gating mechanism becomes complex and a complete model of BK channel gating consists of 1250 states (Salkoff *et al.*, 2006). Variations in Ca^{2+}-sensitivity of skeletal muscle BK channels have been reported ranging from $EC_{50} = 300$ nM to 30 µM (Latorre, 1989). In addition to the complex gating, this phenomenon might be explained by the phosphorylation state of the channel. BK channels are affected by various kinases: protein kinase A (PKA), protein kinase C (PKC), protein kinase G (PKG) and Ca^{2+}-CaM-dependent protein kinase II (CaMKII) (Zhou *et al.*, 2001; Salkoff *et al.*, 2006). The general PKA phosphorylation site is S869 (in some publications S899), phosphorylation of which normally results in an increase in activity (Nara *et al.*, 1998) possibly due to increased sensitivity to Ca^{2+} (Reinhart *et al.*, 1991; Klaerke *et al.*, 1996). However, to add to the complexity, different splice variants of the BK channel are differently affected by phosphorylation (Tian *et al.*, 2001a), which may

explain why BK channels in some neuroendocrine cells are inhibited after PKA phosphorylation.

BK channels have the largest single-channel conductance among K^+ channels with values for native channels ranging between 126–250 pS when recorded in a 150 mM symmetrical potassium solution (Adelman *et al.*, 1992; Vergara *et al.*, 1998) where the single channel IV relationship is linear. In asymmetric transmembrane K^+ gradients the IV curve becomes slightly outward rectifying. The unique high single-channel conductance is due to structural elements apart from the central selectivity filter: (1) A ring of eight negatively charged amino acids at both the inner and outer mouth of the pore increases the local K^+ concentration around the selectivity filter, thereby increasing the single-channel conductance (Brelidze *et al.*, 2003; Nimigean *et al.*, 2003); (2) the architecture of the inner pore where a large cavity ensures assess to cytoplasmic potassium ions (Brelidze and Magleby, 2005); (3) an outer vestibule composed of a large mouth with a number of additional negative charges, which has been mapped with the scorpion peptide charybdotoxin (Miller *et al.*, 1985). Despite the very specialized conduction pathway of BK channels the classical K^+ channel selectivity pattern is preserved (Soh and Park, 2001) and there is evidence of multi-ion conduction (i.e. anomalous mole fraction behaviour) (Eisenman *et al.*, 1984).

The presence of β subunits has a large impact on the biophysical properties of BK channels. When β1-subunits are expressed together with the BK α-subunit in *Xenopus laevis* oocytes, a dramatic shift to the left in the voltage-dependent activation curve occurs (Meera *et al.*, 1996; Tseng-Crank *et al.*, 1996) suggesting that the presence of β1 either increases Ca^{2+}-sensitivity, voltage-sensitivity or facilitates the coupling between voltage/Ca^{2+} sensing and channel opening. Knocking out the β1 gene reduces the Ca^{2+}-sensitivity of the native BK channel and knockout mice show increased contractility of arterial, bladder and airway smooth muscle (Brenner *et al.*, 2000b; Petkov *et al.*, 2001; Semenov *et al.*, 2006). Expression of the β2-subunit changes the kinetic properties of BK channels by mediating a rapid and complete inactivation of the current and by shifting the voltage-dependence towards more negative potentials. The N-terminal region of β2 is responsible for mediating channel inactivation and works like the 'ball and chain' mechanism of some Kv channels and Kvβ subunits (Wallner *et al.*, 1999; Xia *et al.*, 1999). The β3a–c isoforms convey a partial inhibition of BK current while this is not the case for the β3d. None of the β3 subunits changes the voltage-dependence for activation of the channel (Brenner *et al.*, 2000a; Uebele *et al.*, 2000; Xia *et al.*, 2000). Expression of the β4 subunit results in a slowing of activation and deactivation kinetics. In addition, β4 affects BK channel Ca^{2+}-sensitivity in a manner different from all other β-subunits, namely by decreasing the Ca^{2+}-sensitivity at low $[Ca^{2+}]$, but increasing it at high $[Ca^{2+}]$ (Brenner *et al.*, 2000a). Accordingly, the specific tissue expression of the β-subunits plays a role in shaping BK channel properties to a given function and as such the β-subunits appear to exert a relatively larger physiological impact than that of alternative splicing of the α-subunit (Xia *et al.*, 2000).

BK channel gating is also influenced by a number of factors such as pH, membrane stretch, phosphorylation, direct hormone binding, and volume regulation (Kume *et al.*, 1989; Nara *et al.*, 1998; Kawahara *et al.*, 1991). BK channels are probably not sensitive to volume changes *per se*, but are rather mechano- and shear stress-sensitive (Taniguchi and Imai, 1998; Grunnet *et al.*, 2002).

Finally, it is worth mentioning the functional coupling of BK channels to a number of membrane proteins. These include NMDA receptors (Isaacson and Murphy, 2001); $Na^+/2Cl^-/K^+$ co-transporters (O'Neill and Steinberg, 1995); the cystic fibrosis transmembrane regulator (CFTR) (Sorensen *et al.*, 2001), β-adrenergic receptors (Kume *et al.*, 1989), opiate and cannabinoid receptors (Sade *et al.*, 2006) and ligand-gated P_{2U} receptors (Wu *et al.*, 1998). Physical coupling between BK channels and voltage-gated Ca^{2+} channels of the L-type, P/Q-type and N-type, has also been described, (Grunnet and Kaufmann, 2004a; Berkefeld *et al.*, 2006) (Figure 4.3.3).

4.3.5 Pharmacology

Pharmacology of K_{Ca} channels has traditionally been centered on various peptide toxin blockers but now also includes small organic molecules that either inhibit by a pore-blocking mechanism or act as positive or inhibitory modifiers of the gating processes.

Blockers/negative modulators

SK channels are specifically blocked by apamin, an 18 amino acid basic peptide from bee and wasp venoms. Apamin physically occludes the pore from the outside, interacting with specific acidic

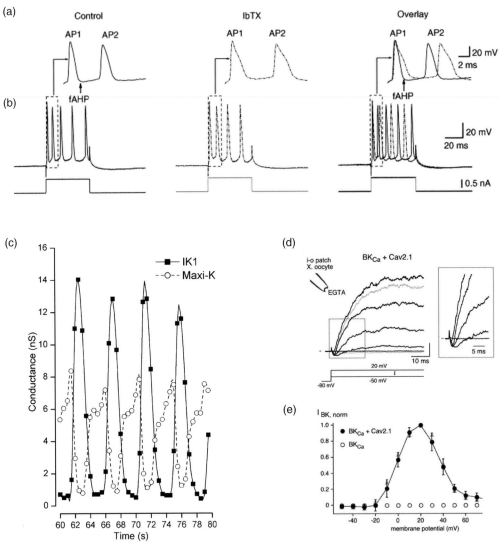

Fig. 4.3.3 K_{Ca} channel activation, regulation and cellular impact depends critically on microdomains. (a) and (b), the effect of inhibiting BK channels (fAHP) in a high-frequency firing CA1 neuron. (a) Action potentials in the absence and presence of the specific BK channel blocker iberiotoxin (notice action potential broadening). (b) Illustrates the contra-intuitive observation that inhibition of BK channels can decrease the firing frequency firing, probably due to inefficient re-priming of Na_V channels. Reproduced from Gu *et al.* (2007).[3] (c) Reciprocal membrane-delimited regulation of IK and BK channel activity in parotid acinar cells, probably due to direct IK-BK channel interactions. Reproduced from Thompson and Begenisich (2006).[4] (d) and (e), Activation of cloned BK channels in macropatches is linked to specific Ca^{2+} influx via co-localized voltage-gated Ca^{2+} channels ($Ca_V2.1$). Reproduced from Berkefeld *et al.* (2006).[5]

[3] Reproduced with permission from Wiley-Blackwell Publishing, Gu N, Vervaeke K and Storm JF (2007) BK potassium channels facilitate high-frequency firing and cause early spike frequency adaptation in rat CA1 hippocampal pyramidal cells. *Journal of Physiology* **580**, 859–882.

[4] Reproduced from *The Journal of General Physiology*, 2006, **127**, 159–169. Copyright 2006, The Rockefeller University Press.

[5] From Berkefeld H, Sailer CA, Bildl W, Rohde V, Thumfart JO, Eble S, Klugbauer N, Reisinger E, Bischofberger J, Oliver D, Knaus HG, Schulte U and Fakler B (2006). BK$_{Ca}$-Cav channel complexes mediate rapid and localized Ca^{2+}-activated K$^+$ signaling. *Science* **314**, 615–620. Reprinted with permission from AAAS.

and hydrophobic amino acids (Asp^{341} and Asn^{368}) in the pore vestibule (Ishii *et al.*, 1997a) as well as a non-pore amino acid (Thr^{216}) in the linker between TM 3 and 4 (Nolting *et al.*, 2006). Apamin exerts distinct blocker affinity for the SK subtypes with the selectivity sequence: SK2>SK3>SK1 (Table 4.3.1). However, it is noteworthy that a variable fraction of SK channels may become apamin- (and bicuculline) insensitive in expression systems (Khawaled *et al.*, 1999; Dale *et al.*, 2002), a phenomenon that is currently unexplained. Several scorpion toxins also block SK channels: scyllatoxin/leiurotoxin (Chicchi *et al.*, 1988) and tamapin (Pedarzani *et al.*, 2002) are high-affinity SK channel blockers that exhibit overlapping binding sites with apamin. An artificial peptide LeiDap^7, based on the structure of scyllatoxin, exhibits improved selectivity for SK2 over SK3/SK1 (Shakkottai *et al.*, 2001). A splice variant of SK3 with a 15 amino acid insert between TM5 and P is insensitive to block by apamin and scyllatoxin (Wittekindt *et al.*, 2004).

Small organic molecules carrying one or two positive charges block SK channels and displace radioactive apamin by interaction with the external blocker binding site. The anti-cholinergic substances d-tubocurarine and dequalinium (Gater *et al.*, 1985; Castle *et al.*, 1993), as well as the quaternized analogues of bicuculline (Johnson and Seutin, 1997; Khawaled *et al.*, 1999; Strøbæk *et al.*, 2000; Grunnet *et al.*, 2001b), methyl-laudanosine and methyl-noscapine (Scuvee-Moreau *et al.*, 2004) are blockers with high-to-low µM affinity for SK channels. High-affinity, small organic blockers (IC$_{50}$ below 1 nM) such as UCL1684 (Campos Rosa *et al.*, 2000) and UCL1848 (Shah and Haylett, 2000) have been synthezized. In general, small-molecule pore blockers are non-selective or exhibit a weak selectivity profile akin to that of apamin. Recently, an SK inhibitory compound (NS8593) has been described, that does not displace radioactive apamin and causes a shift in the SK channel activation curve towards higher [Ca^{2+}]$_i$ (Strøbæk *et al.*, 2006). NS8593 interacts with the gating process of SK channels (but not IK) at a site distinct from the external blocker site and represents a novel functional and chemical class of SK channel inhibitors.

IK channels are blocked by several scorpion toxins (charybdotoxin, maurotoxin, margatoxin, sticholydactylatoxin) (Brugnara *et al.*, 1995; Cai *et al.*, 1998; Castle *et al.*, 2003), but none act exclusively on IK channels. However, a mutated peptide, Glu^{32}-charybdotoxin, exhibits improved selectivity for IK channels over Kv1.3 (Rauer *et al.*, 2000). The classic small-molecule blocker of IK channels is the antimycotic drug clotrimazole, originally found to block the Gárdos channel (Alvarez *et al.*, 1992). Analogues of clotrimazole have been synthezized in order to improve potency and selectivity. TRAM-34 was profiled as an immune suppressant (Wulff *et al.*, 2000) and ICA-17043 (Senicapoc) was developed for the treatment of sickle cell anaemia (Stocker *et al.*, 2003). The clotrimazole-like IK blockers are pore-blockers acting from the cytoplasmic side, coordinating with hydrophobic amino acid residues (Thr^{250} and Val^{275}) supposedly positioned close to the selectivity filter (Wulff *et al.*, 2001). Another class of small-molecule IK blockers is certain dihydropyridine Ca^{2+} antagonists exemplified by nitrendipine (Ellory *et al.*, 1992). The dihydropyridine binding site on IK channels is unknown, but is distinct from the clotrimazole site (Wulff *et al.*, 2001).

The first peptide inhibitor to be identified for BK channels was charybdotoxin (ChTX) from the scorpion *Leiurus quinquestriatus*. This is a peptide toxin of 37 amino acids which mediates

a high-affinity inhibition of BK channels (IC$_{50}$ 1–5 nM) that has proven useful in the characterization of BK channels (Miller *et al.*, 1985), although its specificity is not very high. Subsequently, other peptide inhibitors were identified such as iberiotoxin (IbTX), limbatustoxin (LbTX), and kaliotoxin-1 (1KTX) (Galvez *et al.*, 1990; Crest *et al.*, 1992). IbTX is selective and also exists as a genetically modified toxin called IbTX-D19Y/Y36F that can be radioactively labelled without losing its specificity and affinity for BK channels (Koschak *et al.*, 1997; Grunnet *et al.*, 1999). A number of non-peptide compounds also inhibit BK channels. These include Ba^{2+}, tetraethylammonium (TEA) and organic modulators such as paxilline.

Activators/positive modulators

The prototype activator of IK/SK channels is 1-EBIO, which activates IK channels more potently than SK channels (Table 4.3.1). Analogues of 1-EBIO, such as DCEBIO (Singh *et al.*, 2001) have been synthezized and some drugs with related chemical structures (chlorzoxazone, zoxazolamine and riluzole) also possess IK/SK channel activation properties that may contribute to their therapeutic profiles (Syme *et al.*, 2000; Grunnet *et al.*, 2001b). NS309, a high-affinity compound, also belongs to this category of small, inflexible, heteroaromatic structures (Strøbæk *et al.*, 2004; Pedarzani *et al.*, 2005). Disregarding their different potencies, the above-mentioned compounds all possess the same selectivity sequence as 1-EBIO, meaning that they are ~10 times more potent at IK than at SK channels, between which they do not distinguish. Recently, however, CyPPA, a compound belonging to a completely different chemical class, was found to be a selective SK3/SK2 channel activator exerting no effects on the SK1 and IK channels (Hougaard *et al.*, 2007). Some controversy exists regarding the mode of action of the IK/SK activators. Based on single channel recordings, chlorzoxazone was reported to enhance the maximally attainable open state propability (Po) of hIK and rSK2 channels expressed in *Xenopus* oocytes, without much effect on Ca^{2+}-sensitivity (Syme *et al.*, 2000). However, Pedersen *et al.* (1999) found evidence for a shift in the Ca^{2+}-sensitivity of cloned IK channels, a mode of action that clearly accounts for the activation of SK channels by 1-EBIO (Pedarzani *et al.*, 2001), NS309 and CyPPA (Hougaard *et al.*, 2007) when tested in multichannel inside-out patches. 1-EBIO interacts with SK channels at the C-terminal (Pedarzani *et al.*, 2001) close to the CaMBD and, consistent with this, 1-EBIO also activates the apamin-insensitive splice variant of SK3 (Wittekindt *et al.*, 2004). Using a fast application technique Pederzani *et al.* (2001) showed that 1-EBIO does not influence the on-rate for Ca^{2+}, rather it decreases the off-rate, thereby strongly prolonging the deactivation time constant upon Ca^{2+} removal. This observation is potentially important since the duration of the SK-mediated AHP is often determined by the rate of decay of the SK tail current, rather than the rate of decrease in intracellular Ca^{2+} (Oliver *et al.*, 2000).

For BK channels several organic compounds acting as activators have been identified (Nardi and Olesen, 2008). These are dihydrosoyasaponin I (DHS-1, McManus *et al.*, 1993), NS004 (Olesen *et al.*, 1994a), NS1619 (Olesen *et al.*, 1994b), NS1608 (Strøbæk *et al.*, 1996), NS11021 (Bentzen *et al.*, 2007) and Maxipost (Cheney *et al.*, 2001). The most potent and selective of these openers is DHS-1, which requires the presence of the β1 subunit to exert its action, in contrast to the other compounds which act directly on the α-subunit. The mechanism of action of the NS compounds is an increased

Table 4.3.1 Properties of K_{Ca} channels

Channel subunit	Biophysical characteristics	Cellular and subcellular localization	Pharmacology	Physiology	Relevance to disease states
SK1 ($K_{Ca}2.1$) Gene: *KCNN1* Chromosomal location: mouse: 8 rat: 16p14 human: 19p13.1 9 exons 4 isoforms detected in human brain N-terminal glutamate repeat [1, 2] rSK1 does not express recombinantly as a homomer Possibly creates heteromeric channels with SK2 [3]	Open channel properties: Conductance: 9.2 pS (at 0 mV, symmetrical K) [4] Inward rectifying (voltage-dependent block by divalent cations) Permeability sequence: $K^+>Cs^+(0.19)>Li^+(<0.02) = Na^+(<0.02)$ [5] Gating properties: $[Ca^{2+}]_{0.5} = 0.3–0.7\ \mu M$; $n_H = 3.8–4.5$ [4,6,7,8] Fast activation/deactivation by $[Ca^{2+}]$ steps [8] Constitutively associated with CaM [8] Po independent of voltage	Rodents: Neocortex and hippocampus [9–11] Primarily in soma and dendrites [10] Human: Almost exclusively in neurons, including enteric neurons [12]	Blockers: Apamin (3–8 nM) [5, 13, 14, 15] Tamapin (42 nM) [16] Leiuro-/scyllatoxin (80–325 nM) [15, 17] Lei-Dab7 (6 mM) [17] UCL1684 (762 pM) [15] UCL1848 (1 nM) [14] Dequalinium (0.4 μM) [14, 15] d-tubocurarine (~25 mM) [13, 14, 15] Fluoxetine (9 μM) [18] Cyproheptadine (33 μM) [5] Bicuculline salts (1.7–15 μM) [15, 19] N-methyl-laudanosine (1.2 μM) [20] N-methyl-noscapine (5.9 μM) [20] TEA (5–14 mM) [13, 21] Negative modulator: NS8593 (0.42 μM at 0.5 mM Ca^{2+}) [22] Positive modulators: 1-EBIO (630 μM) [5, 23] NS309 (~0.3–0.6 μM) [C. Hougaard; unpublished] Also activated by DCEBIO; chlorzoxazone; zoxazolamine; and riluzole at concentrations similar to SK2 and SK3		
SK2 ($K_{Ca}2.2$) Gene: *KCNN2* Chromosomal location: mouse: 18 rat: 18q11 human: 5q22.3 8 exons Two isoforms: SK2S and SK2L (with N-terminal cysteine-rich extension) SK2L can form heteromers with SK2S and SK3 [24]	Open channel properties: Conductance: 10 pS (0 mV, symmetrical K) [4, 25] Inward rectifying (voltage-dependent block by divalent cations) [26] Permeability sequence: $K^+>Rb^+(0.8)>NH_4^+(0.19)-Cs^+(0.19)>Li^+(0.14)>Na^+(0.12)$ [27] Gating properties: $[Ca^{2+}]_{0.5} = 0.3–0.7\ \mu M$; $n_H = 2.5–6$ [4, 6–8, 13, 25, 26, 28] Fast activation/deactivation by $[Ca^{2+}]$ steps [8] $k_{on,Ca} = 47\ \mu M^{-1}s^{-1}$ and $k_{off,Ca} = 13\ s^{-1}$ [25] Constitutively associated with CaM [8, 29] Po independent of voltage [25]	Rodents: Neocortex and hippocampus [9–11] Primarily soma and dendrites [9, 10, 30] Inner ear [31] Adrenal glands (mRNA) [4] Human: Adrenal gland, brain, prostate, bladder, liver, heart [12]	Blockers: Apamin (27–95 pM) [3, 4, 13, 15] Tamapin (24 pM) [16] Leiuro-/scyllatoxin (200–287 pM) [15, 17] LeiDab7 (5.5 nM) [17] PO5 (22 nM), Tsk (80 nM), Pi1-NH2 (100 nM) and Maurotoxin (1 μM) [17] UCL1848 (110–240 pM) [3, 32, 33] UCL1684 (280–364 pM) [15, 32] Dequalinium (162 nM) [15] d-tubocurarine (2.4–17 μM) [4, 15, 21] Bicuculline salts (1–25 μM) [15, 19] N-methyl-laudanosine (0.8 μM) [20] N-methyl-noscapine (5.6 μM) [20] Fluoxetine (7 μM) [18] Negative modulator: NS8593 (0.60 μM at 0.5 μM Ca^{2+}) [22] Positive modulators: 1-EBIO (0.5–1 mM) [7, 23, 34, 35] DCEBIO (27 μM) [35] Chlorzoxazone (27–960 μM) [28, 34] Zoxazolamine (696 μM) [34] NS309 (0.62 μM) [35] CyPPA (14 μM) [6] Also activated by riluzole [36]	Protein kinase (CK2), phosphatase (PP2A) and scaffolding proteins regulate Ca^{2+}-sensitivity [37] mAHP in CA1 and possibly other neurons [38] Keeps excitatory dendritic potentials short [39] Negative feedback on NMDA-mediated responses in CA1 [30] Hippocampal-dependent learning and memory [40, 41] Inhibitory cholinergic synapses in inner ear [31, 42] Regulation of excitability in gabaergic SNr neurons [43] Regulation of excitability and firing pattern in cerebellar purkinje neurons [44]	Head trauma-induced epilepsia [45] Cerebellar Ataxia [46, 47]

	Open channel properties	Tissue distribution	Blockers	Physiological function	Associated diseases
SK3 ($K_{Ca}2.3$) Gene: KCNN3 Chromosomal location: mouse: 3 rat: 2q34 human: 1q21.3 8 exons, N-terminal poly-glutamine repeats [6,8] Apamin-insensitive and dominant-negative isoforms [48, 49]	Open channel properties: Conductance: 10 pS (0 mV, symmetrical K) [50] Inward rectifying (voltage-dependent block by divalent cations) Permeability sequence: K+>Rb+(0.79)>Cs+(0.17) [49] Gating properties: [Ca²⁺]$_{0.5}$ = 0.3–0.4 µM; n_H = 5–6.3 [6,8] Fast activation/deactivation by [Ca²⁺] steps [8] Constitutively associated with CaM [8] Po independent of voltage	Rodents: Basal ganglia, thalamus, hypothalamus, habenula, ventral tegmental area, substantia nigra compacta, raphe, and locus coeruleus [9–11, 51, 52] Soma and dendrites as well as varicose fibres [10] Superior cervical ganglion neurons [33] Human: Brain, skeletal and smooth muscle, spleen, thymus, adrenal, thyroid, prostate, kidney, testis [49] Smooth muscle, vascular endothelium, genital tissues [12]	Blockers: Apamin (0.2–4 nM) [22, 33, 49, 53] Tamapin (1.7 nM) [16] Leiuro-/scyllatoxin (1.1–8.3 nM) [17, 33, 49] LeiDab⁷ (2.5 µM) [46] PO5 (25 nM), Tsk (197 nM), Pi1-NH2 (250 nM), Pi1-OH (330 nM) [17] Bicuculline salts (6 µM) [54] N-methyl-laudanosine (1.8 µM) [20] N-methyl-noscapine (3.9 µM) [20] UCL1684 (1.8–9.5 nM) [22, 32, 33] UCL1848 (2.1 nM) [33] Dequalinium (136 nM) [22] d-tubocurarine (33 µM) [49] Fluoxetine (17 µM) [18] TEA (2.2 mM) [49] Negative modulator: NS8593 (0.47 µM at 0.3 µM Ca²⁺, 0.73 µM at 0.5 µM Ca²⁺, 14 µM at 10 µM Ca²⁺) [22] Positive modulators: 1-EBIO (0.17–1 mM) [6, 23, 49] DCEBIO (16 µM) [6] NS309 (0.3 µM) [6] CyPPA (5.6 µM) [6] Riluzole activates from 3 µM [54] [EC₅₀ = 16 µM] [C. Hougaard, unpublished] Also activated by chlorzoxazone and zoxazolamine.	Ca²⁺-dependent AHP in DA neurons [55] Maintains regular firing pattern in DA neurons [56–58] Knockout phenotypes reveal importance for: Bladder stability [59] Vascular tone [60] Uterine contractability [61] N- and C-terminal interactions [62]	Schizophrenia? [63] Anorexia nervosa [64] Migraine [65] Myotonic muscular dystrophy [66, 67]
IK ($K_{Ca}3.1$, Gárdos channel) Gene: KCNN4 Chromosomal location: mouse: 7 rat: 1q21 human: 19q13.2 9 exons	Open channel properties: Conductance: 30–43 pS (0 mV, symmetrical K) [68–72] Permeability sequence: Selectivity follows Eisenman series IV or V: Tl+>K+>Rb+(0.71)>NH₄⁺(0.12)>Cs⁺(<0.02); Li⁺(<0.02); Na⁺(<0.001) [73] K+= Rb+(0.96)NH₄⁺(0.17)Cs⁺(0.07)>>Na⁺(<0.0042) [68] K+= Rb+(1.0)Cs⁺(0.096)>>Na⁺, Li⁺, NMDG (<0.02) [70] K+>Rb+(0.97)>NH₄⁺(0.21)>Cs⁺(0.17)>Li⁺(0.11)>Na⁺(0.10) [27] Multi-ion conduction (single file diffusion) [74] Gating properties: [Ca²⁺]$_{0.5}$ = 0.27–0.7 µM; n_H = 1.7–3.9 [28, 68, 69, 71, 72]	Epithelia, endothelia, proliferating smooth muscle cells, erythrocytes, lymphocytes and various inflammatory cells, Placenta, prostate, salivary glands [69, 70] Brain microglia [78] Trachea [12] Enteric nervous system [79]	Blockers: Charybdotoxin (2–28 nM) [68–70, 72, 76, 80] Glu³²-charybdotoxin (33 nM) [80] Maurotoxin (1 nM) [81] Margatoxin (459 nM) [70] Stichodactylatoxin (30 nM) [70] Clotrimazole (25–387 nM) [69, 70, 72, 76, 82, 83, 84] TRAM-34 (20–25 nM) [84] ICA-17043 (11 nM) [83] 1-[(2-hlorophenyl)diphenylmethyl]-1,2,3-triazole (12 nM) and methyl 4-[4-chloro-3-(trifluoromethyl) phenyl]-6-methyl-3-oxo-1,4,7-tetrahydroisobenzofuran-5-carboxylate (13 µM) [85] Nitrendipine (0.03–0.9 µM) [70, 84, 86] Nimodipine (1 µM) [84] Nifedipine (3–4 µM) [84, 86] Tioconazole (0.3 µM) [82] Miconazole (0.8–1.5 µM) [70, 82] Econazole (1.8–10 µM) [70, 82, 84] Ketoconazole (30 µM) [84] Cetiedil (79 µM) [70]	Volume regulation in red blood cells? Upregulation in activated lymphocytes. Proliferation of lymphocytes? [68, 89, 90] Epithelial transport [91] NO production in microglia [79] Head trauma [85] Regulation of intestinal peristaltics [79]	Diarrhoea [91] Experimental allergic encephalitis [92] Sickle cell disease [83, 93, 94]

Continued

Table 4.3.1 (*Continued*) Properties of K_{Ca} channels

	Properties	Tissue distribution	Pharmacology	Physiological function	Remarks
	Constitutively associated with CaM [75, 76] Po independent of voltage [68, 70, 72] Dependence on extracellular K$^+$ [77]		TEA (30–40 mM) [68, 72] Positive modulators: 1-EBIO (28–210 μM) [7, 23, 28, 70, 87] DCEBIO (19 μM) [87] Chlorzoxazone (98 μM) [28] NS309 (~0.01 μM) [88] also activated by zoxazolamine, riluzole, and methylxanthines		Gain-of-function in human result in epilepsy and paroxysmal movement disorder [130]
BK (K_{Ca}1.1, Slo, maxi K) Gene: *KCNMA1* Chromosomal location: mouse: 14 rat: 15p16 human: 10q22.3 >10 exons Several isoforms with different Ca^{2+} sensitivity and sensitivity to phosphorylation have been cloned [95]	Open channel properties Conductance: 126–250 pS (symmetrical K) [95, 96] Permeability sequence: Tl$^+$(1.3)>K$^+$(1.0)> Rb$^+$(0.7) >NH$_4$$^+$(0.1) [97] K$^+$>Rb$^+$(0.75)>NH$_4$$^+$(0.13)>Cs$^+$(0.10)> Li$^+$(0.01) = Na$^+$(0.01) [27] Multi-ion conduction (anomalous mole fraction behaviour) [97] Gating properties Ca^{2+} – and voltage-activated [98] [Ca^{2+}]$_{0.5}$ = 0.3–30 μM, dependent on voltage [99] Putative mode of Ca^{2+} activation by binding to a Ca^{2+} bowl in the C-terminal region [100]	Substantia nigra, outer layers of the neocortex, pallidal areas, striatum, habenula, olfactory bulb, sensory fibres, caudate putamen, thalamus, cerebellar Purkinje cell layers, granule and pyramidal cell layers of the hippocampal formation [101–103] Presynaptic terminals, soma, and dendrites [101, 104] Epithelia [105–107] Smooth muscle cells and some endothelial cells [108–110] Heart mitochondria [111]	Blockers: Charybdotoxin (1–5 nM) Iberiotoxin (250 pM) [112] Paxilline (1.9 nM) TEA (0.14 mM) Activators: dihydrosoyasaponin (10 nM) [113] NS004 (10 μM) [114] NS1608 and NS1619 (3 μM) [115, 116] Minimal active concentrations are stated for the BK channel openers	The four different β-subunits might modify Ca^{2+}-sensitivity, inactivation properties and gating kinetics [117–120] Knockout mice exhibit ataxia and high-frequency hearing loss [121, 122] fAHP of neurons [104, 123] Prevention of exaggerated transmitter release [101, 124] Circadian timing [103] Oxygen-sensing in dopaminergic glomus cells of the carotid body [125] Dilatation of penile arteries [126] Regulation of blood pressure[127] Airway tone [128] Protection against heart ischemia [111] Transepithelial transport [129]	
Slack (K_{Ca}4.1, slo2.2) Gene: *KCNT1* Chromosomal location: mouse: 2 rat: 3p13 human: 9q34	Regulated by intracellular Na$^+$ and Cl$^-$ [131]	Primarily in brain with highest expression in mitral cells. Also found in kidney and testis [131–133]		Fine tuning of excitability [98, 134]	
Slick (K_{Ca}4.2, slo2.1) Gene: *KCNT2* Chromosomal location: mouse: 1 rat: 13q13 human: 1q31.3	Regulated by intracellular Na$^+$ and Cl$^-$ [131]	Brain, heart, skeletal muscle, kidney, testis, lung, liver [131, 133, 135]		Fine tuning of excitability [98, 134]	

$K_{Ca}5.1$ (slo3)	Regulated by intracellular pH [136]	Spermatozytes [98]
Gene: KCNU1	Voltage-sensitive [131]	
Chromosomal location:		
mouse: 8		
rat: 15p16		
human: 8p11.2		

[1] Litt et al. (1999) Cytogenet Cell Genet 86, 70–73; [2] Shmukler et al. (2001) Biochem Biophys Acta 1518, 36–46; [3] Benton et al. (2003) J Physiol 553, 13–19; [4] Köhler et al. (1996) Science 273, 1709–1714; [5] Dale et al. (2002) Naunyn Schmiedebergs Arch Pharmacol 366, 470–477; [6] Hougaard et al. (2007) Br J Pharmacol 151, 655–665; [7] Pedarzani et al. (2001) Biol Chem 276, 9762–9769; [8] Xia et al. (1998) Nature 395, 503–507; [9] Sailer et al. (2002) J Neurosci 22, 9698–9707; [10] Sailer et al. (2004) Mol Cell Neurosci 26, 458–469; [11] Stocker and Pedarzani (2000) Mol Cell Neurosci 15, 476–493; [12] Chen et al. (2004) Naunyn Schmiedebergs Arch Pharmacol 369, 602–615; [13] Nolting et al. (2006) J Biol Chem 282, 3478–3486; [14] Shah and Haylett (2000) Br J Pharmacol 129, 627–630; [15] Strobaek et al. (2000) Br J Pharmacol 129, 991–999; [16] Pedarzani et al. (2002) Biol Chem 277, 46101–46109; [17] Shakkottai et al. (2001) J Biol Chem 276, 43145–43151; [18] Terstappen et al. (2003) Neurosci Lett 346, 85–88; [19] Khawaled et al. (1999) Pflugers Archiv 438, 314–321; [20] Scuvée-Moreau et al. (2004) Br J Pharmacol 143, 753–764; [21] Ishii et al. (1997) J Biol Chem 272, 23195–23200; [22] Strobaek et al. (2006) Mol Pharmacol 70, 1–12; [23] Lappin et al. (2005) Neurosci Lett 346, 37–46; [24] Strassmaier et al. (2005) J Biol Chem 280, 21231–21236; [25] Hirschberg et al. (1998) J Gen Physiol 111, 565–581; [26] Soh and Park (2001) Biophys J 80, 2207–2215; [27] Shin et al. (2005) Biophys J 89, 3111–3119; [28] Syme et al. (2000) Am J Cell Physiol 278, C570–581; [29] Schumacher et al. (2001) Nature 410, 1120–1124; [30] Ngo-Anh et al. (2005) Nat Neurosci 8, 642–649; [31] Nie et al. (2004) J Neurophysiol 91, 1536–1544; [32] Fanger et al. (2001) J Biol Chem 276, 12249–12256; [33] Hosseini et al. (2001) J Physiol 535, 323–334; [34] Cao et al. (2001) Pharmacol Exp Ther 296, 683–689; [35] Pedarzani et al. (2005) Biol Chem 280, 41404–41411; [36] Cao et al. (2002) Eur J Pharmacol 449, 47–54; [37] Bildl et al. (2004) Neuron 43, 847–858; [38] Bond et al. (2004) J Neurosci 24, 5301–5306; [39] Cai et al. (2004) Neuron 44, 351–364; [40] Deschaux and Bizot (2005) Neursci lett 386, 5–8; [41] Hammond et al. (2006) J Neurosci 26, 1844–1853; [42] Oliver et al. (2004) Neuron 26, 595–601; [43] Yanovsky et al. (2005) Neuroscience 136, 1027–1036; [44] Cingolani et al. (2002) J Neurosci 22, 4456–4467; [45] Cai et al. (2007) J Neurosci 27, 59–68; [46] Shakkottai et al. (2004) Clin Invest 113, 582–590; [47] Walter et al. (2006) Nat Neurosci 9, 389–397; [48] Tomita et al. (2003) Mol Psychiatry 8, 524–535; [49] Wittekindt et al. (2004) Mol Pharmacol 65, 788–801; [50] Barford et al. (2001) Am J Physiol Cell Physiol 280, C836–842; [51] Tacconi et al. (2001) Neuroscience 102, 209–215; [53] Grunnet et al. (2001) Pflugers Archiv 441, 544–550; [54] Grunnet et al. (2001) Neuropharmacol 40, 879–887; [55] Wolfart et al. (2001) J Neurosci 21, 3443–3456; [56] Ji and Shepard (2006) Neuroscience 140, 623–633; [57] Waroux et al. (2005) Eur J Neurosci 22, 3113–3121; [58] Wolfart and Roeper (2001) J Neurosci 22, 3404–3413; [59] Herrera et al. (2003) J Physiol 551, 893–903; [60] Taylor et al. (2003) Circ Res 93, 124–131; [61] Brown et al. (2007) Am J Physiol Cell Physiol 292, C832–840; [62] Frei et al. (2006) Cell Physiol Biochem 18, 165–176; [63] Chandy et al. (1998) Mol Psychiatry 3, 32–37; [64] Koronyo-Hamaoui et al. (2007) Mol Psychiatry 7, 82–85; [65] Mössner et al. (2005) Headache 45, 132–136; [66] Behrens et al. (1994) Muscle Nerve 17, 1264–1270; [67] Neelands et al. (2001) J Physiol 536, 397–407; [68] Ishii et al. (1997) Proc Natl Acad Sci USA 94,11651–11656; [70] Jensen et al. (1998) Am J Physiol 275, C848–856; [71] Joiner et al. (1997) Proc Natl Acad Sci 94, 11013–11018; [72] Logsdon et al. (1997) Biol Chem 272, 32723–32726; [73] Christophersen (1991) J Membr Biol 119, 75–83; [74] Vestergaard-Bogind et al. (1985) J Membr Biol 95, 121–130; [75] Fanger et al. (1999) Biol Chem 274, 5746–5754; [76] Khanna et al. (1999) Biol Chem 274, 14838–14849; [77] Hoffman et al. (2003) Proc Natl Acad Sci 100, 7366–7371; [78] Kaushal et al. (2007) J Neurosci 27, 234–244; [79] Nguyen et al. (2007) J Neurophysiol 97, 2024–2031; [80] Rauer et al. (2000) Proc Natl Acad Sci 97, 8151–8156; [82] Alvarez et al. (1992) Biol Chem 267, 11789–11793; [83] Stocker et al. (2003) Blood 101, 2412–2418; [84] Wulff et al. (2000) Proc Natl Acad Sci 97, 8151–8156; [85] Mauler et al. (2004) Eur J Neurosci 20, 1761–1768; [86] Ellory et al. (1992) FEBS 296, 219–221; [87] Singh et al. (2001) Pharmacol Exp Ther 296, 600–611; [88] Strobaek et al. (2004) Biochim Biophys Acta 1665, 1–5; [89] Ghanshani et al. (2000) Biol Chem 275, 37137–37149; [90] Wulff et al. (2004) Immunol 173, 776–786; [91] Rufo et al. (1996) Clin Invest 98, 2066–2075; [92] Reich et al. (2005) Eur J Immunol 35, 1027–1036; [93] Ataga et al. (2006) Pharmacotherapy 26, 1557–1564; [94] Brugnara et al. (1993) Clin Invest 92, 520–526; [95] Adelman et al. (1992) Neuron 9, 209–216; [96] Vergara et al. (1998) Cur Opin Neurobiol 8, 321–329; [97] Eisenman et al. (1984) Biophys J 50, 1025–1034; [98] Salkoff et al. (2006) Nat Rev Neurosci 7, 921–931; [99] Latorre (1989) Acta Physiol Scand Suppl 582, 13; [100] Schreiber and Salkoff (1997) Biophys J 73, 1355–1363; [101] Knaus et al. (1996) J Neurosci 16, 955–963; [102] Grunnet et al. (2004) Biol Chem 279, 36445–36453; [103] Meredith et al. (2006) Nat Neurosci 9, 1041–1049; [104] Lancaster and Nicoll (1987) J Physiol 389, 187–203; [105] Frindt and Palmer (1987) Am J Physiol 252, F458–467; [106] Grunnet et al. (2005) Biochim Biophys Acta 1714, 114–124; [107] Lu et al. (2002) Biol Chem 265, 16190–16194; [108] Herrera et al. (2001) Am J Physiol Cell Physiol 280, C481–490; [109] Holland et al. (1996) Br J Pharmacol 117, 119–129; [110] O'Neill and Steinberg (1995) Am J Physiol 269, C267–274; [111] Xu et al. (2002) Science 298, 1029–1033; [112] Galvez et al. (1990) Biol Chem 265, 11083–11090; [113] McManus et al. (1993) Biochemistry 32, 6128–6133; [114] Olesen et al. (1994) Eur J Pharmacol 251, 53–59; [116] Strobaek et al. (1996) Neuropharmacology 35, 903–914; [117] Brenner et al. (2000) J Biol Chem 275, 6453–6461; [118] Knaus et al. (1994) J Biol Chem 269, 17274–17278; [119] Wallner et al. (1999) Proc Natl Acad Sci 96, 4137–4142; [120] Xia et al. (1999) J Neurosci 19, 5255–5264; [121] Ruttiger et al. (2004) Proc Natl Acad Sci 101, 12922–12927; [122] Sausbier et al. (2004) Proc Natl Acad Sci 101, 9474–9478; [123] Viana et al. (1993) J Neurophysiol 69, 2150–2163; [124] Hu et al. (2001) J Neurosci 21, 9585–9597; [125] Williams et al. (2004) Science 306, 2093–2097; [126] Werner et al. (2005) J Physiol 567, 545–556; [127] Brenner et al. (2000) Nature 407, 870–876; [128] Benoit et al. (2001) Am J Physiol Lung Cell Mol Physiol 280, L965–973; [129] Klaerke et al. (1996) J Membr Biol 151, 11–18; [130] Du et al. (2005) Nat Genet 37, 733–738; [131] Yuan et al. (2003) Neuron 37, 765–773; [132] Bhattacharjee et al. (2002) J Comp Neurol 454, 241–254; [133] Joiner et al. (1998) Nat Neurosci 1, 462–469; [134] Santi et al. (2006) J Neurosci 26, 5059–5068; [135] Bhattacharjee et al. (2005) J Comp Neurol 484, 80–92; [136] Schreiber et al. (1998) J Biol Chem 273, 3509–3516.

Po at all potentials which is accomplished by an increased open dwell time and decreased closed times (Olesen *et al.*, 1994b). The compounds left-shift the Ca^{2+}/voltage activation curves at all $[Ca^{2+}]_i$ and also work in the absence of internal Ca^{2+} (Strøbæk *et al.*, 1996).

4.3.6 Physiology

An important function of Ca^{2+}-activated K^+ channels in neurons is to generate after-hyperpolarizations (AHPs) following individual or trains of action potentials. Three different Ca^{2+}-dependent AHPs of distinct durations are observed in many neurons with hippocampal CA1 neurons providing a particularly well studied case: (1) A fast TEA- and iberiotoxin-sensitive AHP (fAHP, 3–10 ms) contributes actively to repolarization of the action potential and is carried by BK channels (Storm 1987) (Figure 4.3.3); (2) an apamin-sensitive AHP of intermediate duration (mAHP, 100–200 ms) contributes to early spike frequency adaptation and is carried by SK channels (Stocker *et al.*, 1999) (Figure 4.3.2); (3) a slow AHP (sAHP, 1–2 s) (Lancaster and Nicoll, 1987) that is inhibited by metabotropic pathways via specific G-proteins (Krause *et al.*, 2002) and which mediates late spike frequency adaptation and long hyperpolarizing responses between trains of impulses (Figure 4.3.2). The channel(s) mediating the sAHP is currently unknown. SK1 was originally suggested as a candidate due to initial findings of a very low apamin-sensitivity of that subtype (Ishii *et al.*, 1997a) and, furthermore, 1-EBIO increased I_{sAHP} which suggested that an SK channel might be responsible (Pedarzani *et al.*, 2001). However, more recent studies have clearly demonstrated that the apamin-sensitivity of hSK1 (rSK1 does not express as a functional homomer) is in the low nM range (Shah and Haylett, 2000; Strøbæk *et al.*, 2000; Grunnet *et al.*, 2001a) and that the much more potent SK/IK activator NS309 did not affect I_{sAHP} (Pedarzani *et al.*, 2005). This last finding also speaks against the (remote) possibility that IK could be involved in the generation of sAHP, even though clotrimazole and certain analogues of this drug are quite potent inhibitors (Shah *et al.*, 2006).

In addition to their role in CA1 neurons SK channels also mediate the apamin-sensitive I_{AHP} and early spike frequency adaptation in pyramidal neurons from amygdala and neocortex. SK1 and SK2 are co-expressed in these areas and may form functional heteromers (Benton *et al.*, 2003). Overexpression of SK1/SK2 channels in slices of the somatosensory cortex (layer V) selectively increased I_{AHP}, whereas a dominant negative transgene (SK3–1B) selectively abolished it (neither of the treatments changed sI_{AHP}) (Villalobos *et al.*, 2004). However, studies from SK knockout mice show that the SK2 channel is the sole SK subtype mediating I_{AHP} in hippocampal CA1 neurons, again without effects on sI_{AHP} (Bond *et al.*, 2004). The efficient activation of SK channels by Ca_v channels during action potentials may suggest co-localization as suggested for hippocampal SK channels (Bowden *et al.*, 2001).

In autonomously firing neurons such as dopaminergic SNc neurons or serotonergic raphe neurons, SK channels are cyclically activated (Wolfart and Roeper, 2002) and are pivotal in maintaining regular firing. SK3 has been strongly implicated as the SK subtype critical for pacemaking control in SNc dopaminergic cells (Wolfart *et al.*, 2001). Application of apamin or small molecule blockers to these neurons cause irregular firing or even burst firing in rat midbrain slices *in vitro* (Shepard and Bunney, 1988;

Johnson and Seutin, 1997) or *in vivo* in anaesthetized rats (Waroux *et al.*, 2005; Ji and Shepard, 2006). Likewise, irregular firing is also induced in raphe neurons upon application of blockers (Rouchet *et al.*, 2008). In accordance with the robust expression of SK3 in monoaminergic neurons the SK3 conditional knockout mouse was recently described to exert altered serotonergic and dopaminergic neurotransmission and reduced behavioural despair following knock-down of the SK3 gene (Jacobsen *et al.*, 2008). In GABAergic SNr neurons SK2 channels regulate excitability by the Ca_v-linked post spike mAHP, but also contribute to the resting membrane potential by Ca^{2+} release from ryanodine receptors (Yanovsky *et al.*, 2005). In cerebellar Purkinje cells, pacing precision is clearly under control of SK2 in young rats (Cingolani *et al.*, 2002), whereas other mechanisms take over in adults concomitant with decreased SK2 expression. In adult mice, however, SK channels apparently still play a significant role in maintaining the stability of pacemaking (Walter *et al.*, 2006).

In addition to being activated via Ca_v channels during action potentials, SK channels can also be activated by ionotropic receptor activation during excitatory synaptic input. SK channels thereby contribute to synaptic integration (Cai *et al.*, 2004) in various neurons (Bond *et al.*, 2005). Functional downregulation of dendritic SK2 channels in CA1 neurons after partial deafferentation has recently been shown to cause prolonged dendritic plateau potentials and may contribute to head trauma-induced epilepsia (Cai *et al.*, 2007). The SK2 channels expressed in dendritic spines are colocalized with NMDA receptors (Ngo-Anh *et al.*, 2005) and exert negative feedback on excitatory stimulation via the hyperpolarization-mediated Mg^{2+} block (Hammond *et al.*, 2006). During induction of long-term potentiation (LTP) SK2 channels are internalized from dendritic spines, thereby facilitating LTP induction (Lin *et al*, 2008) and influencing learning and memory (Brosh *et al.*, 2007). Plasticity in the dendritic NMDA/SK system may furthermore be subject to exogenous metabotropic regulation by sigma-1 receptors (Martina *et al.*, 2007). A general nootropic effect of SK channels is also indicated by reports of age-related overexpression of SK3 in the hippocampus inducing cognitive deficits (Blank *et al.*, 2003) and of apamin improving learning in rats (Deschaux and Bizot, 2005). In outer hair cells of the inner ear, SK2 functionally couples with Ca^{2+}-permeable nicotinic α-9 and α-10 receptors, thereby creating inhibitory cholinergic synapses (Oliver *et al.*, 2000; Nie *et al.*, 2004). Surprisingly, SK2 knockout eliminates not only the inhibitory K^+ current, but also the excitatory cholinergic signal in hair cells (Kong *et al.*, 2008).

SK channels are expressed in both endothelia and smooth muscles and are therefore inferred to play an important role in the regulation of contractility and tone of smooth muscle tissue. Analysis of overexpressing/conditional knockouts of the SK3 and SK2 channel has confirmed this and a number of characteristic phenotypes related to endothelial/smooth muscle function such as bladder instability (Herrera *et al.*, 2003; Thorneloe *et al.*, 2008), vascular tone (Taylor *et al.*, 2003), and uterine contractility (Brown *et al.*, 2007) have been reported.

The physiological role of IK channels in human red blood cells remains speculative. A role in erythrocyte volume regulation has been suggested, although the Ca^{2+} entry pathway(s) are difficult to demonstrate in healthy erythrocytes under physiological conditions (Baunbaek and Bennekou, 2008). IK channels are expressed in low numbers on naive circulating T- and B-lymphocytes, whereas the expression level increases dramatically upon mitogenic stimulation *in vitro* (Grissmer *et al.*, 1993; Ghansani *et al.*, 2000; Wulff *et al.*, 2004). IK channels may play a role in concert with the voltage-dependent Kv1.3 in stabilizing a negative membrane potential driving Ca^{2+} influx via CRAC and thereby in the proliferation and cytokine production of T-cells (Grissmer *et al.*, 1993; Jensen *et al.*, 1999). Although the importance of IK channels in autoimmune and reactive diseases is presently unclear, a pivotal role is suggested from the reported effects of selective IK blockers in some animal models of multiple sclerosis (Reich *et al.*, 2005) and head trauma (Mauler *et al.*, 2004). Recently IK channels have been shown to translocate to the immunological synapse of immune cells upon antigen presentation (Nicolaou *et al.*, 2007), a feature shared with CD3, Kv1.3 and Orai-1. In addition, IK channels expressed in the intestinal epithelium may contribute to the basolateral K^+ conductances which hyperpolarize the epithelial cells and thereby maintain the driving force for luminal Cl^- and transcellular Na^+ secretion (Rufo *et al.*, 1996).

An apparently healthy IK channel knockout mouse line has been generated (Begenisich *et al.*, 2004) and at the cellular level, induced K_{Ca} transport and volume regulation of erythrocytes were completely abolished, as was the T-lymphocyte regulatory volume decrease (RVD). Saliva secretion was unaffected, although complete and selective knockdown of IK currents from parotis epithelia cells was obtained. Recently, the IK knockout has been shown to lack the Ca^{2+}-activated K^+ conductance of the intestinal epithelium and to be deficient in colonic fluid secretion (Flores *et al.*, 2007). Surprisingly no effects on haematology or immune responses have been noted to date.

The physiological significance of the BK channels in various tissues remains a fairly open question. Despite their nearly ubiquitous expression BK channels are only functional under specific intracellular conditions. At resting intracellular $[Ca^{2+}]$ of 100 nM the channel is silent at −80 mV and activation requires voltage steps to very positive membrane potentials. BK currents of 1 nA are often measured in smooth muscle cells at +100 mV, but since the current drops by tenfold for each 20 mV hyperpolarization, the BK current is about a billion-fold smaller at a membrane potential of −80 mV. Thus, the BK channels are without significance unless they are activated by an increased Ca^{2+} level, depolarization, phosphorylation, or by specific drugs rendering them the dominating K^+ conductance.

In neurons BK channels are present both presynaptically and in soma/dendrites. The function of presynaptic BK channels appears to be to prevent exaggerated transmitter release following high excitability of the neurons (Knaus *et al.*, 1996; Hu *et al.*, 2001), whereas BK channels with somatodendritic localization contribute to the repolarization of action potentials (fAHP) (Lancaster *et al.*, 1987; Viana *et al.*, 1993). The overall effect of BK channels in the CNS is addressed in BK α-subunit knockout mice, which exhibit a number of disorders including ataxia and high-frequency hearing loss (Ruttiger *et al.*, 2004; Sausbier *et al.*, 2004) supporting the notion that a physiological function of BK channels is to prevent hyperactivity.

A unique BK channel function occurs in oxygen-sensing dopaminergic glomus cells of the carotid body. Here BK channels react to changes in oxygen pressure via hemoxygenase-2 that is a distinct part of the BK channel complex in these cells (Williams *et al.*, 2004). Another special function is undertaken in the cochlea where inactivating BK channels are located to the inner hair cells

(Pyott *et al.*, 2004; Hafidi *et al.*, 2005). BK channels are also involved in the regulation of circadian behavioural rhythms controlled by the suprachiasmatic nucleus and daily variations in BK channel expression are essential for pacemaker output from this nucleus (Meredith *et al.*, 2006).

Another interesting function of BK channels is their role in endothelial cells and smooth muscle cells. The phenotypes of α and β subunit knockout mice indicate that the BK channel complex plays a role in normal regulation of the blood pressure (Brenner *et al.*, 2000b; Sausbier *et al.*, 2004) and other studies suggest that hypertensive rats have fewer vascular BK channels than controls (Amberg and Santana, 2005). The channels may thus be involved in the development of essential hypertension and have a particular function in coronary perfusion (Miura *et al.*, 1999; 2001). Vascular tone is controlled by a complex mechanism of opposing events mediated by intracellular Ca^{2+}. Vasoconstriction is achieved by global influx of Ca^{2+} through L-type Ca^{2+} channels. However, the Ca^{2+} influx also stimulates intracellular ryanodine receptors resulting in local Ca^{2+} increases, known as Ca^{2+} sparks, that activate BK channels, resulting in hyperpolarization and vasodilation. The BK β1 subunit is essential in this process as is increases the Ca^{2+}-sensitivity of the BK channel (Nelson and Quayle, 1995; Brenner *et al.*, 2000b). BK channels may be the targets for regulation by nitric oxide (NO) through the cGMP/PKG pathway, and therefore may play a role in NO-mediated vasodilation in certain tissues such as the lung (Elmedal *et al.*, 2005). Tissue-specific vasodilation has been demonstrated in BK α-subunit knockout mice that showed erectile dysfunction, indicating that BK channels are necessary for dilation of penile arteries (Werner *et al.*, 2005). BK α and β1-subunits are expressed in the urinary bladder smooth muscle cells and BK channel activity profoundly modulates bladder contraction (Ohya *et al.*, 2000; Petkov *et al.*, 2001). Further, BK channels are key regulators of airway tone. In guinea pig they hyperpolarize smooth muscle cells and relax airways precontracted with acetylcholine (Benoit *et al.*, 2001). Human airway smooth muscle cells express numerous BK channels which are sensitive to internal Ca^{2+} mobilization (Snetkow and Ward, 1999).

In the heart, cellular damage following a coronary artery obstruction can be diminished by transient, brief, obstructions before the pathological challenge. This phenomenon is known as preconditioning and depends on the cardiac mitochondria. BK channels are present in the mitochondria inner membrane and their activation has a preconditioning effect. The exact mechanism is unknown (Xu *et al.*, 2002; Ohya *et al.*, 2005; Sato *et al.*, 2005, Bentzen *et al.*, 2009) but it is possible that BK channels depolarize the inner membrane, thereby uncoupling the electron transport chain resulting in reduced production of ATP and harmful superoxides. Finally, a role for BK channels in epithelia may be to provide a potassium conductance pivotal for transepithelial transport. BK channels are present in cells responsible for Na^+ absorption and could be involved in this process (Buttefield *et al.*, 1997).

The functions of $K_{Ca}4.1$, $K_{Ca}4.2$ and $K_{Ca}5.1$ remain speculative. $K_{Ca}4.1$ and $K_{Ca}4.2$ are strongly regulated through activation of G-protein-coupled receptors. Both channels are modulated by activation of $G_{\alpha q}$ receptors, but interestingly the two channels are regulated in opposite manners by neurotransmitters. $K_{Ca}4.1$ and $K_{Ca}4.2$ channels could thus participate in fine tuning the general excitability of neurons through their relative expression levels in different brain regions (Salkoff *et al.*, 2006; Santi *et al.*, 2006).

4.3.7 Relevance to disease states

Mutations in potassium channels including loss- and gain-of-function mutations of Kir2 and many Kv families and their accessory subunits are known to cause a broad spectrum of monogenetic diseases. With a few possible exceptions no such disease-causing mutations have been described for K_{Ca} channels hitherto. It is important to note, however, that K_{Ca} channels could represent effective drug targets without the channels playing a causative role in disease pathogenesis.

Epidemiological studies have pointed towards a possible link between excessively prolonged glutamine stretches of SK3 and various diseases. The original observation reported was an association with schizophrenia (Chandy et al., 1998), however, the majority of follow-up studies have failed to replicate this finding. Recently, the SK3 poly-Q polymorphism was associated with a genetic predisposition for anorexia nervosa, acting additively with an NMDA receptor polymorphism also reported to be associated with this disorder (Koronyo-Hamaoui et al., 2007). A preponderance of extended poly-Q sequences in migraine patients has also been reported (Mössner et al., 2005).

Dominant-negative N-terminal truncated splice variants of SK3, that are able to suppress the expression of all SK isoforms, as well as IK (but not BK) (Kolski-Andreaco et al., 2004), have been isolated from humans (Tomita et al., 2003). Expression of one of these (SK3–1B) causes induction of life-long cerebellar ataxia in transgenic mice (Shakkottai et al., 2004). Whether truncated SK3 expression may be a factor in the development of psychiatric diseases is unknown, although one of these mutants was isolated from a schizophrenic patient. Cerebellar ataxia type I results from loss-of-function mutations in P-type Ca_v channels, which make cerebellar Purkinje cell firing less precise due to inefficient cyclic activation of SK2 channels. Using the ataxic tottering mice that carry a similar mutation as the patients, Walter et al. (2006) showed that local cerebellar injections of the SK activator 1-EBIO partially corrected the ataxic behavior. Recently, the *frissonnant* mice, a phenotype characterized by tremors and locomotor deficits, has been shown to be caused by a deletion mutation in the SK2 gene (Szatanik et al., 2008). Future studies are needed in order to reveal whether SK2 is generally involved in motor coordination and motor disorders of humans. Hopf et al. (2007) showed that chronic ethanol administration to rats increased the excitability of VTA neurons by reducing the apamin-sensitive afterhyperpolarization, a finding which may suggest a general role of SK3 in addiction.

In the periphery, skeletal muscle T-tubule clustering of SK3 channels has paradoxally been demonstrated to lower the threshold for AP generation due to local increased extracellular K^+ causing hyperexcitability of denervated muscles (Behrens et al., 1994). This may be an element in the etiology of myotonic muscular dystrophy (Neelands et al., 2001).

Although not the causative event for sickle cell anaemia (a hemoglobinopathy) the human erythrocyte IK channel is pivotal in the chain of events leading to erythrocyte dehydration and acceleration of the sickling cascade that eventually leads to vascular occlusion, painful crises and tissue damage in these patients (Brugnara et al., 1993). Cetiedil, and recently Senicapoc (Icagen), have shown efficacy in normalizing blood parameters in a transgenic sickle cell model (Stocker et al., 2003) and Senicapoc has been in

phase III development for the treatment of sickle cell disease (Ataga et al., 2006). IK channels may play a pivotal role in the establishment of the proliferative phenotype of vascular smooth muscle cells that preceedes the formation of atherosclerotic lesions in patients and may constitute a target for treatment of atherosclerosis (Toyama et al., 2008).

BK channels are thought to play a significant role in the feedback loop attenuating cellular signalling in conditions such as brain and cardiac ischemia, epilepsy, airway hyperactivity, arterial hypertension, and other states with a concurrent increased internal Ca^{2+}-mobilization and membrane depolarization (Garcia and Kaczorowski, 2005). However, a human BK channel gain-of-function mutation identified in subjects suffering from an increased excitability exemplified by generalized epilepsy and paroxysmal dyskinesia (probably due to accelerated release of Na^+ channel inactivation), has challenged this simplified picture of BK channel function (Du et al., 2005). Nevertheless, the first human disease targeted therapeutically with a BK channel opener was cerebral stroke. The studies were discontinued in phase III due to lack of efficacy (BMS204352) (Cheney et al., 2001). Two BK channel-opening compounds are currently in clinical development for the treatment of hyperactive bladder (TA-1702 from Tanabe/GSK) and COPD/asthma (Andolast from Rottapharm), respectively.

For reviews on the possible role of K_{Ca} channels as targets for drug development, see Jensen et al. (2001), Liegeois et al. (2003), Blank et al. (2004), Garcia and Kaczorowski (2005), and Wulff et al. (2007).

4.3.8 Concluding remarks

K_{Ca} channels have been a focus for scientific investigation for half a century (since Gárdos, 1958) and there is no indication of a decline in interest in these channels. Extensive recent literature covers K_{Ca} function at multiple levels including molecular structure, cell biology, pharmacology, animal behaviour, and epidemiological investigations of possible disease-linked genetic variants.

An important area of investigation at the molecular level is the further clarification of the organization of K_{Ca} channels, probably in a cell-specific manner, in functional complexes with β-subunits, enzymes, and other ion channels, and to delineate the structure, the flow of information, and the degree of regulation within these complexes. Specific apposition of Ca^{2+}-activated K^+ channels with Ca^{2+} sources, Ca^{2+}-mediated gating, regulation of Ca^{2+}-sensitivity by ß-subunits or by phosphorylation, and direct protein/protein interactions have already been documented (Figure 4.3.3). A further challenge is to define the position of these 'molecular networks' in the larger cellular cascades thereby understanding, for example, under which conditions SK channel-associated calmodulin becomes phosphorylated and thereby less Ca^{2+}-sensitive. The understanding of how the context or micro-milieu of a given ion channel impacts on the overall outcome of cellular activity has direct clinical significance as shown by the epilepsy putatively caused by a gain-of-function in BK channels, or by the increased muscle excitability in muscular dystrophy patients due to overexpression of SK3.

It is conceivable that the potential of K_{Ca} channels as drug targets will grow further with the accumulating understanding of the functions they exert in specific tissues, while possibly being devoid of function in other tissues despite their presence.

References

Adelman JP, Shen KZ, Kavanaugh MP *et al.* (1992). Calcium-activated potassium channels expressed from cloned complementary DNAs. *Neuron* **9**, 209–216.

Allen D, Fakler B, Maylie J *et al.* (2007). Organization and regulation of small conductance Ca^{2+}-activated K^+ channel multiprotein complexes. *J Neurosci* **27**, 2369–2376.

Alvarez J, Montero M and Garcia-Sancho J (1992). High affinity inhibition of Ca^{2+}-dependent K^+ channels by cytochrome P-450 inhibitors. *J Biol Chem* **267**, 11789–11793.

Amberg GC and Santana LF (2003). Downregulation of the BK channel beta1 subunit in genetic hypertension. *Circ Res* **93**, 965–971.

Ataga KI, Orringer EP, Styles L *et al.* (2006). Dose-escalation study of ICA-17043 in patients with sickle cell disease. *Pharmacotherapy* **26**, 1557–1564.

Atkinson NS, Robertson GA and Ganetzky B (1991). A component of calcium-activated potassium channels encoded by the *Drosophila* slo locus. *Science* **253**, 551–555.

Bahia PK, Suzuki R, Benton DC *et al.* (2005). A functional role for small-conductance calcium-activated potassium channels in sensory pathways including nociceptive processes. *J Neurosci* **25**, 3489–3498.

Barfod ET, Moore AL and Lidofsky SD (2001). Cloning and functional expression of a liver isoform of the small conductance Ca^{2+}-activated K^+ channel SK3. *Am J Physiol Cell Physiol* **280**, C836–842.

Baunbaek M and Bennekou P (2008) Evidence for a random entry of Ca^{2+} into human red cells. *Bioelectrochemistry* **73**, 145–150.

Begenisich T, Nakamoto T, Ovitt CE *et al.* (2004). Physiological roles of the intermediate conductance, Ca^{2+}-activated K channel, Kcnn4. *J Biol Chem* **279**, 47681–47687.

Behrens MI, Jalil P, Serani A *et al.* (1994). Possible role of apamin-sensitive K^+ channels in myotonic dystrophy. *Muscle Nerve* **17**, 1264–1270.

Bennekou P and Christophersen P (1990). The gating of human red cell Ca^{2+}-activated K^+-channels is strongly affected by the permeant cation species. *Biochim Biophys Acta* **1030**, 183–187.

Benoit C, Renaudon B, Salvail D *et al.* (2001). EETs relax airway smooth muscle via an EpDHF effect: BK(Ca) channel activation and hyperpolarization. *Am J Physiol Lung Cell Mol Physiol* **280**, L965–973.

Benton DC, Monaghan AS, Hosseini R *et al.* (2003). Small conductance Ca^{2+}-activated K^+ channels formed by the expression of rat SK1 and SK2 genes in HEK 293 cells. *J Physiol* **553**, 13–19.

Bentzen BH, Nardi A, Calloe K *et al.* (2007). The small molecule NS11021 is a potent and specific activator of Ca^{2+}-activated big-conductance K^+ channels. *Mol Pharmacol* **72**,1033–1044.

Bentzen BH, Osadchii O, Jespersen T *et al.* (2009). Activation of big conductance Ca^{2+}-activated K^+ channels (BK) protects the heart against ischemia–reperfusion injury. *Pflügers Arch Eur J Physiol* **475**, 978–988.

Berkefeld H, Sailer CA, Bildl W *et al.* (2006). BKCa-Cav channel complexes mediate rapid and localized Ca^{2+}-activated K^+ signaling. *Science* **314**, 615–620.

Bhattacharjee A, Gal L and Kaczmarek LK (2002). Localization of the Slack potassium channel in the rat central nervous system. *J Comp Neurol* **454**, 241–254.

Bhattacharjee A, von Hehn CA, Mei X *et al.* (2005). Localization of the Na^+-activated K^+ channel Slick in the rat central nervous system. *J Comp Neurol* **484**, 80–92.

Bildl W, Strassmaier T, Thurm H *et al.* (2004). Protein kinase CK2 is coassembled with small conductance Ca^{2+}-activated K^+ channels and regulates channel gating. *Neuron* **43**, 847–858.

Blank T, Nijholt I, Kye MJ *et al.* (2004). Small conductance Ca^{2+}-activated K^+ channels as targets of CNS drug development. *Curr Drug Targets CNS Neurol Disord* **3**, 161–167.

Blank T, Nijholt I, Kye MJ *et al.* (2003). Small-conductance, Ca^{2+}-activated K^+ channel SK3 generates age-related memory and LTP deficits. *Nat Neurosci* **6**, 911–912.

Bond CT, Herson PS, Strassmaier T *et al.* (2004). Small conductance Ca^{2+}-activated K^+ channel knock-out mice reveal the identity of calcium-dependent afterhyperpolarization currents. *J Neurosci* **24**, 5301–5306.

Bond CT, Maylie J and Adelman JP (2005). SK channels in excitability, pacemaking and synaptic integration. *Curr Opin Neurobiol* **15**, 305–311.

Bowden SE, Fletcher S, Loane DJ *et al.* (2001). Somatic colocalization of rat SK1 and D class (Ca(v)1.2) L-type calcium channels in rat CA1 hippocampal pyramidal neurons. *J Neurosci* **21**, RC175.1–6.

Brelidze TI and Magleby KL (2005). Probing the geometry of the inner vestibule of BK channels with sugars. *J Gen Physiol* **126**, 105–121.

Brelidze TI, Niu X and Magleby KL (2003). A ring of eight conserved negatively charged amino acids doubles the conductance of BK channels and prevents inward rectification. *Proc Natl Acad Sci* **100**, 9017–9022.

Brenner R, Jegla TJ, Wickenden A *et al.* (2000a). Cloning and functional characterization of novel large conductance calcium-activated potassium channel beta subunits, hKCNMB3 and hKCNMB4. *J Biol Chem* **275**, 6453–6461.

Brenner R, Peréz GJ, Bonev AD *et al.* (2000b). Vasoregulation by the beta1 subunit of the calcium-activated potassium channel. *Nature* **407**, 870–876.

Brosh I, Rosenblum K and Barkai E (2007). Learning-induced modulation of SK channels-mediated effect on synaptic transmission. *Eur J Neurosci* **26**, 3253–3260.

Brown A, Cornwell T, Korniyenko I *et al.* (2007). Myometrial expression of small conductance Ca^{2+}-activated K^+ channels depresses phasic uterine contraction. *Am J Physiol Cell Physiol* **292**, C832-C840.

Bruening-Wright A, Lee WS, Adelman JP *et al.* (2007). Evidence for a deep pore activation gate in small conductance Ca^{2+}-activated K^+ channels. *J Gen Physiol* **130**, 601–610.

Bruening-Wright A, Schumacher MA, Adelman JP *et al.* (2002). Localization of the activation gate for small conductance Ca^{2+}-activated K^+ channels. *J Neurosci* **22**, 6499–6506.

Brugnara C, Armsby CC, De Franceschi L *et al.* (1995). Ca^{2+}-activated K^+ channels of human and rabbit erythrocytes display distinctive patterns of inhibition by venom peptide toxins. *J Membr Biol* **147**, 71–82.

Brugnara C, de Franceschi L and Alper SL (1993). Inhibition of Ca^{2+}-dependent K^+ transport and cell dehydration in sickle erythrocytes by clotrimazole and other imidazole derivatives. *J Clin Invest* **92**, 520–526.

Butler A, Tsunoda S, McCobb DP *et al.* (1993). mSlo, a complex mouse gene encoding 'maxi' calcium-activated potassium channels. *Science* **261**, 221–224.

Buttefield I, Warhurst G, Jones MN *et al.* (1997). Characterization of apical potassium channels induced in rat distal colon during potassium adaptation. *J Physiol* **501**, 537–547.

Cai S, Garneau L and Sauve R (1998). Single-channel characterization of the pharmacological properties of the K(Ca2+) channel of intermediate conductance in bovine aortic endothelial cells. *J Membr Biol* **163**, 147–158.

Cai X, Liang CW, Muralidharan S *et al.* (2004). Unique roles of SK and Kv4.2 potassium channels in dendritic integration. *Neuron* **44**, 351–364.

Cai X, Wei DS, Gallagher SE *et al.* (2007). Hyperexcitability of distal dendrites in hippocampal pyramidal cells after chronic partial deafferentation. *J Neurosci* **27**, 59–68.

Campos Rosa J, Galanakis D, Piergentili A *et al.* (2000). Synthesis, molecular modeling and pharmacological testing of bis-quinolinium cyclophanes: potent, non-peptidic blockers of the apamin-sensitive Ca^{2+}-activated K^+ channel. *J Med Chem* **43**, 420–431.

Cao Y, Dreixler JC, Roizen JD *et al.* (2001). Modulation of recombinant small-conductance Ca^{2+}-activated K^+ channels by the muscle relaxant chlorzoxazone and structurally related compounds. *J Pharmacol Exp Ther* **296**, 683–689.

Cao YJ, Dreixler JC, Coucy JJ *et al.* (2002). Modulation of recombinant and native neuronal SK channels by the neuroprotective drug riluzole. *Eur J Pharmacol* **449**, 47–54.

Castle NA, Haylett DG, Morgan JM *et al.* (1993). Dequalinium: a potent inhibitor of apamin-sensitive K^+ channels in hepatocytes and of nicotinic responses in skeletal muscle. *Eur J Pharmacol* **236**, 201–207.

Castle NA, London DO, Creech C *et al.* (2003). Maurotoxin: a potent inhibitor of intermediate conductance Ca^{2+}-activated potassium channels. *Mol Pharmacol* **63**, 409–418.

Chandy KG, Fantino E, Wittekindt O *et al.* (1998). Isolation of a novel potassium channel gene hSKCa3 containing a polymorphic CAG repeat: a candidate for schizophrenia and bipolar disorder? *Mol Psychiatry* **3**, 32–37.

Chen MX, Gorman SA, Benson B *et al.* (2004). Small and intermediate conductance Ca^{2+}-activated K^+ channels confer distinctive patterns of distribution in human tissues and differential cellular localisation in the colon and corpus cavernosum. *Naunyn Schmiedebergs Arch Pharmacol* **369**, 602–615.

Cheney JA, Weisser JD, Bareyre FM *et al.* (2001). The maxi-K channel opener BMS-204352 attenuates regional cerebral edema and neurologic motor impairment after experimental brain injury. *J Cereb Blood Flow Metab* **21**, 396–403.

Chicchi GG, Gimenez-Gallego G, Ber E *et al.* (1988). Purification and characterization of a unique, potent inhibitor of apamin binding from *Leiurus quinquestriatus hebraeus* venom. *J Biol Chem* **263**, 10192–10197.

Christophersen P (1991). Ca^{2+}-activated K^+ channel from human erythrocyte membranes: single channel rectification and selectivity. *J Membr Biol* **119**, 75–83.

Cingolani LA, Gymnopoulos M, Boccaccio A *et al.* (2002). Developmental regulation of small-conductance Ca^{2+}-activated K^+ channel expression and function in rat Purkinje neurons. *J Neurosci* **22**, 4456–4467.

Cox DH, Gui J and Aldrich RW (1997). Allosteric gating of a large conductance Ca^{2+}-activated K^+ channel. *J Gen Physiol* **110**, 257–281.

Crest M, Jacquet G, Gola M *et al.* (1992). Kaliotoxin, a novel peptidyl inhibitor of neuronal BK-type Ca^{2+}-activated K^+ channels characterized from *Androctonus mauretanicus mauretanicus* venom. *J Biol Chem* **267**, 1640–1647.

Dale TJ, Cryan JE, Chen MX *et al.* (2002). Partial apamin sensitivity of human small conductance Ca^{2+}-activated K^+ channels stably expressed in Chinese hamster ovary cells. *Naunyn Schmiedeberg Arch Pharmacol* **366**, 470–477.

Deschaux O and Bizot JC (2005). Apamin produces selective improvements of learning in rats. *Neurosci Lett* **386**, 5–8.

Du W, Bautista JF, Yang H *et al.* (2005). Calcium-sensitive potassium channelopathy in human epilepsy and paroxysmal movement disorder. *Nat Genet* **37**, 733–738.

Dworetzky SI, Trojnacki JT and Gribkoff VK (1994). Cloning and expression of a human large-conductance calcium-activated potassium channel. *Brain Res Mol Brain Res* **27**, 189–193.

Eisenman G, Latorre R and Miller C (1984). Multi-ion conduction and selectivity in the high-conductance Ca^{2+}-activated K^+ channel from skeletal muscle. *Biophys J* **50**, 1025–1034.

Elkins T, Ganetzky B and WU CF (1986). A *Drosophila* mutation that eliminates a calcium-dependent potassium current. *Proc Natl Acad Sci* **83**, 8415–8419.

Ellory JC, Kirk K, Culliford SJ *et al.* (1992). Nitrendipine is a potent inhibitor of the Ca^{2+}-activated K^+ channel of human erythrocytes. *FEBS* **296**, 219–221.

Elmedal B, Mulvany MJ and Simonsen U (2005). Dual impact of a nitric oxide donor, GEA 3175, in human pulmonary smooth muscle. *Eur J Pharmacol* **516**, 78–84.

Fanger CM, Ghanshani S, Logsdon NJ *et al.* (1999). Calmodulin mediates calcium-dependent activation of the intermediate conductance KCa channel, IKCa1. *J Biol Chem* **274**, 5746–5754.

Fanger CM, Rauer H, Neben AL *et al.* (2001). Calcium-activated potassium channels sustain calcium signaling in T lymphocytes. Selective blockers and manipulated channel expression levels. *J Biol Chem* **276**, 12249–12256.

Flores CA, Melvin JE, Figueroa CD *et al.* (2007). Abolition of Ca^{2+}-mediated intestinal anion secretion and increased stool dehydration in mice lacking the intermediate conductance Ca^{2+}-dependent K^+ channel Kcnn4. *J Physiol* **583**, 705–717.

Frei E, Spindler I, Grissmer S *et al.* (2006). Interactions of N-terminal and C-terminal parts of the small conductance Ca^{2+} activated K^+ channel, hSK3. *Cell Physiol Biochem* **18**, 165–176.

Frindt G and Palmer LG (1987). Ca-activated K channels in apical membrane of mammalian CCT and their role in K secretion. *Am J Physiol* **252**, F458–467.

Galvez A, Gimenez-Gallego G, Reuben JP *et al.* (1990). Purification and characterization of a unique, potent, peptidyl probe for the high conductance calcium-activated potassium channel from venom of the scorpion *Buthus tamulus*. *J Biol Chem* **265**, 11083–11090.

Garcia ML and Kaczorowski GJ (2005). Potassium channels as targets for therapeutic intervention. *Sci STKE* **20**, pe46.

Gárdos G (1958) The function of calcium in the potassium permeability of human erythrocytes. *Biochim Biophys Acta* **30**, 653–654.

Gater PR, Haylett DG and Jenkinson DH (1985). Neuromuscular blocking agents inhibit receptor-mediated increases in the potassium permeability of intestinal smooth muscle. *Br J Pharmacol* **86**, 861–868.

Ghanshani S, Wulff H, Miller MJ *et al.* (2000). Up-regulation of the IKCa1 potassium channel during T-cell activation. Molecular mechanism and functional consequences. *J Biol Chem* **275**, 37137–37149.

Grissmer S, Nguyen AN and Cahalan MD (1993). Calcium-activated potassium channels in resting and activated human T lymphocytes. Expression levels, calcium dependence, ion selectivity and pharmacology. *J Gen Physiol* **102**, 601–630.

Grunnet M and Kaufmann W (2004a). Coassembly of big conductance Ca^{2+}-activated K^+ channels and L-type voltage-gated Ca^{2+} channels in rat brain. *J Biol Chem* **279**, 36445–36453.

Grunnet M, Hay-Schmidt A and Klaerke DA (2005). Quantification and distribution of big conductance Ca^{2+}-activated K^+ channels in kidney epithelia. *Biochim Biophys Acta* **1714**, 114–124.

Grunnet M, Jensen BS, Olesen SP *et al.* (2001a). Apamin interacts with all subtypes of cloned small-conductance Ca^{2+}-activated K^+ channels. *Pflugers Arch* **441**, 544–550.

Grunnet M, Jespersen T, Angelo K *et al.* (2001b). Pharmacological modulation of SK3 channels. *Neuropharmacology* **40**, 879–887.

Grunnet M, Jespersen TJ and Perrier JF (2004b). 5-HT1A receptors modulate small-conductance Ca^{2+}-activated K^+ channels. *J Neurosci Res* **78**, 845–854.

Grunnet M, Knaus HG, Solander C *et al.* (1999). Quantification and distribution of Ca^{2+}-activated maxi K channels in rabbit distal colon. *Am J Physiol* **277**, G22–30.

Grunnet M, MacAulay N, Jorgensen NK *et al.* (2002). Regulation of cloned, Ca^{2+}-activated K^+ channels by cell volume changes. *Pflugers Arch* **444**, 167–177.

Grygorczyk R and Schwarz W (1983). Properties of the Ca^{2+}-activated K^+ conductance of human red cells as revealed by the patch-clamp technique. *Cell Calcium* **4**, 499–510.

Gu N, Vervaeke K and Storm JF (2007). BK potassium channels facilitate high-frequency firing and cause early spike frequency adaptation in rat CA1 hippocampal pyramidal cells. *J Physiol* **580**, 859–882.

Hafidi A, Beurg M and Dulon D (2005). Localization and developmental expression of BK channels in mammalian cochlear hair cells. *Neuroscience* **130**, 475–84.

Hammond RS, Bond CT, Strassmaier T *et al.* (2006). Small-conductance Ca^{2+}-activated K^+ channel type 2 (SK2) modulates hippocampal learning, memory and synaptic plasticity. *J Neurosci* **26**, 1844–1853.

Hay-Schmidt A, Grunnet M, Abrahamsen SL *et al.* (2003). Localization of Ca^{2+}-activated big-conductance K^+ channels in rabbit distal colon. *Pflugers Arch* **446**, 61–68.

Herrera GM, Heppner TJ and Nelson MT (2001). Voltage dependence of the coupling of Ca^{2+} sparks to BK(Ca) channels in urinary bladder smooth muscle. *Am J Physiol Cell Physiol* **280**, C481–490.

Herrera GM, Pozo MJ, Zvara P *et al.* (2003). Urinary bladder instability induced by selective suppression of the murine small conductance calcium-activated potassium (SK3) channel. *J Physiol* **551**, 893–903.

Heyer CB and Lux HD (1976). Control of the delayed outward potassium currents in bursting pace-maker neurones of the snail, *Helix pomatia*. *J Physiol* **262**, 349–382.

Hirschberg B, Maylie J, Adelman JP *et al.* (1998). Gating of recombinant small-conductance Ca^{2+}-activated K^+ channels by calcium. *J Gen Physiol* **111**, 565–581.

Hoffman JF, Joiner W, Nehrke K *et al.* (2003). The hSK4 (KCNN4) isoform is the Ca^{2+}-activated K^+ channel (Gardos channel) in human red blood cells. *Proc Natl Acad Sci* **100**, 7366–7371.

Holland M, Langton PD, Standen NB *et al.* (1996). Effects of the BKCa channel activator, NS1619, on rat cerebral artery smooth muscle. *Br J Pharmacol* **117**, 119–129.

Hopf FW, Martin M, Chen BT *et al.* (2007). Withdrawal from intermittent ethanol exposure increases probability of burst firing in VTA neurons *in vitro*. *J Neurophysiol* **98**, 2297–2310.

Hosseini R, Benton DC, Dunn PM *et al.* (2001). SK3 is an important component of K^+ channels mediating the afterhyperpolarization in cultured rat SCG neurons. *J Physiol* **535**, 323–334.

Hougaard C, Eriksen BL, Jørgensen S *et al.* (2007). Selective positive modulation of the SK3 and SK2 subtypes of small conductance Ca^{2+}-activated K^+ channels. *Br J Pharmacol* **151**, 655–665.

Hu H, Vervaeke K and Storm JF (2001). Presynaptic Ca^{2+}-activated K^+ channels in glutamatergic hippocampal terminals and their role in spike repolarization and regulation of transmitter release. *J Neurosci* **21**, 9585–9597.

Isaacson JS and Murphy GJ (2001). Glutamate-mediated extrasynaptic inhibition: direct coupling of NMDA receptors to Ca^{2+}-activated K^+ channels. *Neuron* **31**, 1027–1034.

Ishii TM, Maylie J and Adelman JP (1997a). Determinants of apamin and d-tubocurarine block in SK potassium channels. *J Biol Chem* **272**, 23195–23200.

Ishii TM, Silvia C, Hirschberg B *et al.* (1997b). A human intermediate conductance calcium-activated potassium channel. *Proc Natl Acad Sci USA* **94**, 11651–11656.

Jacobsen JP, Weikop P, Hansen HH *et al.* (2008). SK3 K^+ channel-deficient mice have enhanced dopamine and serotonin release and altered emotional behaviors. *Genes Brain Behav* **7**, 836–848.

Jensen BS, Odum N, Jorgensen NK *et al.* (1999). Inhibition of T cell proliferation by selective block of Ca^{2+}-activated K^+ channels. *Proc Natl Acad Sci* **96**, 10917–10921.

Jensen BS, Strobaek D, Christophersen P *et al.* (1998). Characterization of the cloned human intermediate-conductance Ca^{2+}-activated K^+ channel. *Am J Physiol* **275**, C848-C856.

Jensen BS, Strobaek D, Olesen SP *et al.* (2001). The Ca^{2+}-activated K^+ channel of intermediate conductance: a molecular target for novel treatments? *Curr Drug Targets* **2**, 401–422.

Ji H and Shepard PD (2006). SK Ca^{2+}-activated K^+ channel ligands alter the firing pattern of dopamine-containing neurons *in vivo*. *Neuroscience* **140**, 623–633.

Johnson SW and Seutin V (1997). Bicuculline methiodide potentiates NMDA-dependent burst firing in rat dopamine neurons by blocking apamin-sensitive Ca^{2+}-activated K^+ currents. *Neurosci Lett* **231**, 13–16.

Joiner WJ, Tang MD, Wang LY *et al.* (1998). Formation of intermediate-conductance calcium-activated potassium channels by interaction of Slack and Slo subunits. *Nat Neurosci* **1**, 462–469.

Joiner WJ, Wang LY, Tang MD *et al.* (1997). hSK4, a member of a novel subfamily of calcium-activated potassium channels. *Proc Natl Acad Sci* **94**, 11013–11018.

Kaushal V, Koeberle PD, Wang Y *et al.* (2007). The Ca^{2+}-activated K^+ channel KCNN4/KCa3.1 contributes to microglia activation and nitric oxide-dependent neurodegeneration. *J Neurosci* **27**, 234–244.

Kawahara K, Ogawa A and Suzuki M (1991). Hyposmotic activation of Ca-activated K channels in cultured rabbit kidney proximal tubule cells. *Am J Physiol* **260**, F27–33.

Khanna R, Chang MC, Joiner WJ *et al.* (1999). hSK4/hIK1, a calmodulin-binding KCa channel in human T lymphocytes. Roles in proliferation and volume regulation. *J Biol Chem* **274**, 14838–14849.

Khawaled R, Bruening-Wright A, Adelman JP *et al.* (1999). Bicuculline block of small-conductance calcium-activated potassium channels. *Pflugers Arch* **438**, 314–321.

Klaerke DA, Wiener H, Zeuthen T *et al.* (1996) Regulation of Ca^{2+} activated K^+ channel from rabbit distal colon epithelium by phosphorylation and dephosphorylation. *J Membr Biol* **151**, 11–18.

Klein H, Garneau L, Banderali U *et al.* (2007). Structural determinants of the closed KCa3.1 channel pore in relation to channel gating: results from a substituted cysteine accessibility analysis. *J Gen Physiol* **129**, 299–315.

Knaus HG, Eberhart A, Koch RO *et al.* (1995). Characterization of tissue-expressed alpha subunits of the high conductance Ca^{2+}-activated K^+ channel. *J Biol Chem* **270**, 22434–22439.

Knaus HG, Folander K, Garcia-Calvo M *et al.* (1994). Primary sequence and immunological characterization of beta-subunit of high conductance Ca^{2+}-activated K^+ channel from smooth muscle. *J Biol Chem* **269**, 17274–17278.

Knaus HG, Schwarzer C, Koch RO *et al.* (1996). Distribution of high-conductance Ca^{2+}-activated K^+ channels in rat brain: targeting to axons and nerve terminals. *J Neurosci* **16**, 955–963.

Köhler M, Hirschberg B, Bond CT *et al.* (1996). Small-conductance, calcium-activated potassium channels from mammalian brain. *Science* **273**, 1709–1714.

Kolski-Andreaco A, Tomita H, Shakkottai VG *et al.* (2004). SK3-1C, a dominant-negative suppressor of SKCa and IKCa channels. *J Biol Chem* **279**, 6893–6904.

Kong JH, Adelman JP and Fuchs P (2008). Expression of the SK2 calcium-activated potassium channel is required for cholinergic function in mouse cochlear hair cells. *J Physiol* **586**, 5471–5485.

Koronyo-Hamaoui M, Danziger Y, Frisch A *et al.* (2007). Association between anorexia nervosa and the hSKCa3 gene: a family-based and case control study. *Mol Psychiatry* **7**, 82–85.

Koschak A, Koch RO and Liu J (1997). [125I]Iberiotoxin-D19Y/Y36F, the first selective, high-specific activity radioligand for high-conductance calcium-activated potassium channels. *Biochemistry* **36**, 1943–1952.

Krause M, Offermanns S, Stocker M *et al.* (2002). Functional specificity of G alpha q and G alpha 11 in the cholinergic and glutamatergic modulation of potassium currents and excitability in hippocampal neurons. *J Neurosci* **22**, 666–673.

Kume H, Takai A, Tokuno H *et al.* (1989). Regulation of Ca^{2+}-dependent K^+-channel activity in tracheal myocytes by phosphorylation. *Nature* **341**, 152–154.

Lancaster B and Nicoll RA (1987). Properties of two calcium-activated hyperpolarizations in rat hippocampal neurones. *J Physiol* **389**, 187–203.

Lappin SC, Dale TJ, Brown JT *et al.* (2005). Activation of SK channels inhibits epileptiform bursting in hippocampal CA3 neurons. *Brain Res* **1065**, 37–46.

Latorre R (1989). Ion channel modulation by divalent cations. *Acta Physiol Scand Suppl* **582**,13.

Leinders T, van Kleef RG and Vijverberg HP (1992).Divalent cations activate small- (SK) and large-conductance (BK) channels in mouse neuroblastoma cells: selective activation of SK channels by cadmium. *Pflugers Arch* **422**, 217–22.

Liegeois JF, Mercier F, Graulich A *et al.* (2003). Modulation of small conductance calcium-activated potassium (SK) channels: a new challenge in medicinal chemistry. *Curr Med Chem* **10**, 625–647.

Lin MT, Luján R, Watanabe M *et al.* (2008). SK2 channel plasticity contributes to LTP at Schaffer collateral-CA1 synapses. *Nat Neurosci* **11**, 170–177.

Litt M, LaMorticella D, Bond CT *et al.* (1999). Gene structure and chromosome mapping of the human small-conductance calcium-activated potassium channel SK1 gene (KCNN1). *Cytogenet Cell Genet* **86**, 70–73.

Logsdon NJ, Kang J, Togo JA *et al.* (1997). A novel gene, hKCa4, encodes the calcium-activated potassium channel in human T lymphocytes. *J Biol Chem* **272**, 32723–32726.

Lu L, Montrose-Rafizadeh M and Guggino WB (1990). Ca^{2+}-activated K^+ channels from rabbit kidney medullary thick ascending limb cells expressed in *Xenopus* oocytes. *J Biol Chem* **265**, 16190–16194.

Maingret F, Coste B, Hao J *et al.* (2008). Neurotransmitter modulation of small-conductance Ca^{2+}-activated K^+ channels by regulation of Ca^{2+} gating. *Neuron* **59**, 439–449.

Martina M, Turcotte ME, Halman S *et al.* (2007). The sigma-1 receptor modulates NMDA receptor synaptic transmission and plasticity via SK channels in rat hippocampus. *J Physiol* **578**, 143–157.

Mauler F, Hinz V, Horváth E *et al.* (2004). Selective intermediate-/small-conductance calcium-activated potassium channel (KCNN4) blockers are potent and effective therapeutics in experimental brain oedema and traumatic brain injury caused by acute subdural haematoma. *Eur J Neurosci* **20**, 1761–1768.

McManus OB, Harris GH, Giangiacomo KM *et al.* (1993). An activator of calcium-dependent potassium channels isolated from a medicinal herb. *Biochemistry* **32**, 6128–6133.

Meera P, Wallner M, Jiang Z *et al.* (1996). A calcium switch for the functional coupling between alpha (hslo) and beta subunits (KV,Ca beta) of maxi K channels. *FEBS Lett* **382**, 84–88.

Meera P, Wallner M, Song M *et al.* (1997). Large conductance voltage- and calcium-dependent K^+ channel, a distinct member of voltage-dependent ion channels with seven N-terminal transmembrane segments (S0–S6), an extracellular N terminus and an intracellular (S9–S10) C terminus. *Proc Natl Acad Sci* **94**, 14066–14071.

Meredith AL, Wiler SW, Miller BH *et al.* (2006). BK calcium-activated potassium channels regulate circadian behavioral rhythms and pacemaker output. *Nat Neurosci* **9**, 1041–1049.

Miller C, Moczydlowski E, Latorre R *et al.* (1985). Charybdotoxin, a protein inhibitor of single Ca^{2+}-activated K^+ channels from mammalian skeletal muscle. *Nature* **313**, 316–318.

Miura H, Liu Y and Gutterman DD (1999). Human coronary arteriolar dilation to bradykinin depends on membrane hyperpolarization: contribution of nitric oxide and Ca^{2+}-activated K^+ channels. *Circulation* **99**, 3132–3138.

Miura H, Wachtel RE, Liu Y *et al.* (2001). Flow-induced dilation of human coronary arterioles: important role of Ca^{2+}-activated K^+ channels. *Circulation* **103**, 1992–1998.

Mongan LC, Hill MJ, Chen MX *et al.* (2005). The distribution of small and intermediate conductance calcium-activated potassium channels in the rat sensory nervous system. *Neuroscience* **131**, 161–175.

Mössner R, Weichselbaum A, Marziniak M *et al.* (2005). A highly polymorphic poly glutamine stretch in the potassium channel KCNN3 in migraine. *Headache* **45**, 132–136.

Nara M, Dhulipala PD, Wang YX *et al.* (1998). Reconstitution of beta-adrenergic modulation of large conductance, calcium-activated potassium (maxi-K) channels in *Xenopus* oocytes. *J Biol Chem* **273**, 14920–14924.

Nardi A and Olesen SP (2008). BK channel modulators: a comprehensive overview. *Curr Med Chem* **15**, 1126–1146.

Neelands TR, Herson PS, Jacobson D *et al.* (2001). Small-conductance calcium-activated potassium currents in mouse hyperexcitable denervated skeletal muscle. *J Physiol* **536**, 397–407.

Nelson MT and Quayle JM (1995). Physiological roles and properties of potassium channels in arterial smooth muscle. *Am J Physiol* **268**, C799–822.

Ngo-Anh TJ, Bloodgood BL, Lin M *et al.* (2005). SK channels and NMDA receptors form a Ca^{2+}-mediated feedback loop in dendritic spines. *Nat Neurosci* **8**, 642–649.

Nguyen TV, Matsuyama H, Baell J *et al.* (2007). Effects of compounds that influence IK (KCNN4) channels on afterhyperpolarizing potentials and determination of IK channel sequence, in guinea pig enteric neurons. *J Neurophysiol* **97**, 2024–2031.

Nicolaou SA, Neumeier L, Peng Y *et al.* (2007). The Ca^{2+}-activated K^+ channel KCa3.1 compartmentalizes in the immunological synapse of human T lymphocytes. *Am J Physiol Cell Physiol* **292**, C1431–1439.

Nie L, Song H, Chen MF *et al.* (2004). Cloning and expression of a small-conductance Ca^{2+}-activated K^+ channel from the mouse cochlea: coexpression with alpha9/alpha10 acetylcholine receptors. *J Neurophysiol* **91**, 1536–1544.

Nimigean CM, Chappie JS and Miller C (2003). Electrostatic tuning of ion conductance in potassium channels. *Biochemistry* **42**, 9263–9268.

Nolting A, Ferraro T, D'hoedt D *et al.* (2006). An amino acid outside the pore region influences apamin sensitivity in small conductance Ca^{2+}-activated K^+ channels. *J Biol Chem* **282**, 3478–3486.

O'Neill WC and Steinberg DF (1995). Functional coupling of Na^+-K^+-2Cl- cotransport and Ca^{2+}-dependent K^+ channels in vascular endothelial cells. *Am J Physiol* **269**, C267–274.

Obermair GJ, Kaufmann WA, Knaus HG *et al.* (2003). The small conductance Ca^{2+}-activated K^+ channel SK3 is localized in nerve terminals of excitatory synapses of cultured mouse hippocampal neurons. *Eur J Neurosci* **17**, 721–731.

Ohya S, Kuwata Y, Sakamoto K *et al.* (2005). Cardioprotective effects of estradiol include the activation of large-conductance Ca^{2+}-activated K^+ channels in cardiac mitochondria. *Am J Physiol* **289**, H1635–H1642.

Ohya S, Tanaka M, Watanabe M *et al.* (2000). Diverse expression of delayed rectifier K^+ channel subtype transcripts in several types of smooth muscles of the rat. *J Smooth Muscle Res* **36**, 101–115.

Olesen SP, Munch E, Moldt P *et al.* (1994b). Selective activation of Ca^{2+}-dependent K^+ channels by novel benzimidazolone. *Eur J Pharmacol* **251**, 53–59.

Olesen SP, Munch E, Watjen F *et al.* (1994a). NS004—an activator of Ca^{2+}-dependent K^+ channels in cerebellar granule cells. *Neuroreport* **5**, 1001–1004.

Oliver D, Klöcker N, Schuck J *et al.* (2000). Gating of Ca^{2+}-activated K^+ channels controls fast inhibitory synaptic transmission at auditory outer hair cells. *Neuron* **26**, 595–601.

Pallotta BS, Magleby KL and Barrett JN (1981). Single channel recordings of Ca^{2+}-activated K^+ currents in rat muscle cell culture. *Nature* **293**, 471–474.

Park YB (1994). Ion selectivity and gating of small conductance Ca^{2+}-activated K^+ channels in cultured rat adrenal chromaffin cells. *J Physiol* **481**, 555–570.

Pedarzani P, D'hoedt D, Doorty KB *et al.* (2002). Tamapin, a venom peptide from the Indian red scorpion (*Mesobuthus tamulus*) that targets small conductance Ca^{2+}-activated K^+ channels and afterhyperpolarization currents in central neurons. *J Biol Chem* **277**, 46101–46109.

Pedarzani P, McCutcheon JE, Rogge G *et al.* (2005). Specific enhancement of SK channel activity selectively potentiates the afterhyperpolarizing current I (AHP) and modulates the firing properties of hippocampal pyramidal neurons. *J Biol Chem* **280**, 41404–41411.

Pedarzani P, Mosbacher J, Rivard A *et al.* (2001). Control of electrical activity in central neurons by modulating the gating of small conductance Ca^{2+}-activated K^+ channels. *J Biol Chem* **276**, 9762–9769.

Pedersen KA, Schrøder RL, Skaaning-Jensen B et al. (1999). Activation of the human intermediate-conductance Ca^{2+}-activated K$^+$ channel by 1-ethyl-2-benzimidazolinone is strongly Ca^{2+}-dependent. *Biochim Biophys Acta* **1420**, 331–340.

Petkov GV, Bonev AD, Heppner TJ et al. (2001). Beta1-subunit of the Ca^{2+}-activated K$^+$ channel regulates contractile activity of mouse urinary bladder smooth muscle. *J Physiol* **537**, 443–452.

Piskorowski R and Aldrich RW (2002). Calcium activation of BK(Ca) potassium channels lacking the calcium bowl and RCK domains. *Nature* **420**, 499–502.

Pyott SJ, Glowatzki E, Trimmer JS et al. (2004). Extrasynaptic localization of inactivating calcium-activated potassium channels in mouse inner hair cells. *J Neurosci* **24**, 9469–9474.

Quirk JC and Reinhart PH (2001). Identification of a novel tetramerization domain in large conductance K(ca) channels. *Neuron* **32**, 13–23.

Rae JL and Shepard AR (1998). Molecular biology and electrophysiology of calcium-activated potassium channels from lens epithelium. *Curr Eye Res* **17**, 264–275.

Rauer H, Lanigan MD, Pennington MW et al. (2000). Structure-guided transformation of charybdotoxin yields an analog that selectively targets Ca^{2+}-activated over voltage-gated K$^+$ channels. *J Biol Chem* **275**, 1201–1208.

Reich EP, Cui L, Yang L et al. (2005). Blocking ion channel KCNN4 alleviates the symptoms of experimental autoimmune encephalomyelitis in mice. *Eur J Immunol* **35**, 1027–1036.

Reinhart PH, Chung S, Martin BL et al. (1991). Modulation of calcium-activated potassium channels from rat brain by protein kinase A and phosphatase 2A. *J Neurosci* **11**, 1627–1635.

Rouchet N, Waroux O, Lamy C et al. (2008). SK channel blockade promotes burst firing in dorsal raphe serotonergic neurons. *Eur J Neurosci* **28**, 1108–1115.

Rufo PA, Jiang L, Moe SJ et al. (1996). The antifungal antibiotic, clotrimazole, inhibits Cl$^-$ secretion by polarized monolayers of human colonic epithelial cells. *J Clin Invest* **98**, 2066–2075.

Rüttiger L, Sausbier M, Zimmermann U et al. (2004). Deletion of the Ca^{2+}-activated potassium (BK) alpha-subunit but not the BKbeta1-subunit leads to progressive hearing loss. *Proc Natl Acad Sci* **101**, 12922–12927.

Sade H, Muraki K, Ohya S et al. (2006). Activation of large-conductance, Ca^{2+}-activated K$^+$ channels by cannabinoids. *Am J Physiol Cell Physiol* **290**, C77–86.

Sailer CA, Hu H, Kaufmann WA et al. (2002). Regional differences in distribution and functional expression of small-conductance Ca^{2+}-activated K$^+$ channels in rat brain. *J Neurosci* **22**, 9698–9707.

Sailer CA, Kaufmann WA, Marksteiner J et al. (2004). Comparative immunohistochemical distribution of three small-conductance Ca^{2+}-activated potassium channel subunits, SK1, SK2 and SK3 in mouse brain. *Mol Cell Neurosci* **26**, 458–469.

Salkoff L, Butler A, Ferreira G et al. (2006). High-conductance potassium channels of the SLO family. *Nat Rev Neurosci* **7**, 921–931.

Santi CM, Ferreira G, Yang B et al. (2006). Opposite regulation of Slick and Slack K$^+$ channels by neuromodulators. *J Neurosci* **26**, 5059–5068.

Sarpal D, Koenig JI, Adelman JP et al. (2004). Regional distribution of SK3 mRNA-containing neurons in the adult and adolescent rat ventral midbrain and their relationship to dopamine-containing cells. *Synapse* **53**, 104–113.

Sato A, Terata K, Miura H et al. (2005). Mechanism of vasodilation to adenosine in coronary arterioles from patients with heart disease. *Am J Physiol Heart Circ Physiol* **288**, H1633–1640.

Sausbier M, Hu H, Arntz C et al. (2004). Cerebellar ataxia and Purkinje cell dysfunction caused by Ca^{2+}-activated K$^+$ channel deficiency. *Proc Natl Acad Sci* **101**, 9474–9478.

Schreiber M and Salkoff L (1997). A novel calcium-sensing domain in the BK channel. *Biophys J* **73**, 1355–1363.

Schreiber M, Wei A, Yuan A et al. (1998). Slo3, a novel pH-sensitive K$^+$ channel from mammalian spermatocytes. *J Biol Chem* **273**, 3509–3516.

Schumacher MA, Rivard AF, Bachinger HP et al. (2001). Structure of the gating domain of a Ca2+-activated K$^+$ channel complexed with Ca^{2+}/calmodulin. *Nature* **410**, 1120–1124.

Scuvée-Moreau J, Boland A, Graulich A et al. (2004). Electrophysiological characterization of the SK channel blockers methyl-laudanosine and methyl-noscapine in cell lines and rat brain slices. *Br J Pharmacol* **143**, 753–764.

Semenov I, Wang B, Herlihy JT et al. (2006). BK channel α1-subunit regulation of calcium handling and constriction in tracheal smooth muscle. *Am J Physiol Lung Cell Mol Physiol* **291**, L802–L810.

Shah M and Haylett DG (2000). The pharmacology of hSK1 Ca^{2+}-activated K$^+$ channels expressed in mammalian cell lines. *Br J Pharmacol* **129**, 627–630.

Shah MM, Javadzadeh-Tabatabaie M, Benton DC et al. (2006). Enhancement of hippocampal pyramidal cell excitability by the novel selective slow-afterhyperpolarization channel blocker 3-(triphenyl-methyl-aminomethyl)pyridine (UCL2077). *Mol Pharmacol* **70**, 1494–1502.

Shakkottai VG, Chou CH, Oddo S et al. (2004). Enhanced neuronal excitability in the absence of neurodegeneration induces cerebellar ataxia. *J Clin Invest* **113**, 582–590.

Shakkottai VG, Regaya I, Wulff H et al. (2001). Design and characterization of a highly selective peptide inhibitor of the small conductance calcium-activated K$^+$ channel, SkCa2. *J Biol Chem* **276**, 43145–43151.

Shepard PD and Bunney BS (1988). Effects of apamin on the discharge properties of putative dopamine-containing neurons *in vitro*. *Brain Res* **463**, 380–384.

Shi J and Cui J (2001). Intracellular Mg^{2+} enhances the function of BK-type Ca^{2+}-activated K$^+$ channels. *J Gen Physiol* **118**, 589–606.

Shin N, Soh H, Chang S et al. (2005). Sodium permeability of a cloned small-conductance calcium-activated potassium channel. *Biophys J* **89**, 3111–3119.

Shipston MJ (2001). Alternative splicing of potassium channels: a dynamic switch of cellular excitability. *Trends Cell Biol* **11**, 353–358.

Shmukler BE, Bond CT, Wilhelm S et al. (2001). Structure and complex transcription pattern of the mouse SK1 K_{Ca} channel gene, KCNN1. *Biochim Biophys Acta* **1518**, 36–46.

Singh S, Syme CA, Singh AK et al. (2001). Benzimidazolone activators of chloride secretion: potential therapeutics for cystic fibrosis and chronic obstructive pulmonary disease. *J Pharmacol Exp Ther* **296**, 600–611.

Snetkov VA and Ward JP (1999). Ion currents in smooth muscle cells from human small bronchioles: presence of an inward rectifier K$^+$ current and three types of large conductance K$^+$ channels. *Exp Physiol* **84**, 835–846.

Soh H and Park CS (2001). Inwardly rectifying current-voltage relationship of small-conductance Ca^{2+}-activated K$^+$ channels rendered by intracellular divalent cation blockade. *Biophys J* **80**, 2207–2215.

Sorensen JB, Nielsen MS, Gudme CN et al. (2001). Maxi K$^+$ channels co-localised with CFTR in the apical membrane of an exocrine gland acinus: possible involvement in secretion. *Pflugers Arch* **442**, 1–11.

Stampe P and Vestergaard-Bogind B (1989) Ca^{2+}-activated K$^+$ conductance of the human red cell membrane: voltage-dependent Na+ block of outward-going currents. *J Membr Biol* **112**, 9–14.

Stocker JW, De Franceschi L, McNaughton-Smith GA et al. (2003) ICA-17043, a novel Gardos channel blocker, prevents sickled red blood cell dehydration in vitro and in vivo in SAD mice. *Blood* **101**, 2412–2418.

Stocker M and Pedarzani P (2000). Differential distribution of three Ca^{2+}-activated K$^+$ channel subunits, SK1, SK2 and SK3, in the adult rat central nervous system. *Mol Cell Neurosci* **15**, 476–493.

Stocker M, Krause M and Pedarzani P (1999). An apamin-sensitive Ca^{2+}-activated K$^+$ current in hippocampal pyramidal neurons. *Proc Natl Acad Sci* **96**, 4662–4667.

Stocker, M (2004). Ca^{2+}-activated K$^+$ channels: Molecular determinants and function of the SK family. *Nat Rev Neurosci* **5**, 758–770.

Storm JF (1987). Action potential repolarization and a fast afterhyperpolarization in rat hippocampal pyramidal cells. *J Physiol* **385**, 733–759.

Strassmaier T, Bond CT, Sailer CA *et al.* (2005). A novel isoform of SK2 assembles with other SK subunits in mouse brain. *J Biol Chem* **280**, 21231–21236.

Strøbaek D, Christophersen P, Holm NR *et al.* (1996). Modulation of the Ca-dependent K channels, *hslo*, by the substituted dipehnylurea NS1608, paxilline and internal Ca. *Neuropharmacology* **35**, 903–914.

Strøbaek D, Hougaard C, Johansen TH *et al.* (2006). Inhibitory gating modulation of small conductance Ca^{2+}-activated K$^+$ channels by the synthetic compound *(R)-N-*(Benzimidazol-2-yl)-1,2,3,4-tetrahydro-1-naphtylamine (NS8593) reduces afterhyperpolarizing current in hippocampal CA1 neurons. *Mol Pharmacol* **70**, 1–12.

Strøbaek D, Jørgensen TD, Christophersen P *et al.* (2000). Pharmacological characterization of small-conductance Ca^{2+}-activated K$^+$ channels stably expressed in HEK 293 cells. *Br J Pharmacol* **129**, 991–999.

Strøbaek D, Teuber L, Jørgensen TD *et al.* (2004). Activation of human IK and SK Ca^{2+}-activated K$^+$ channels by NS309 (6,7-dichloro-1H-indole-2,3-dione 3-oxime). *Biochim Biophys Acta* **1665**, 1–5.

Syme CA, Gerlach AC, Singh AK *et al.* (2000). Pharmacological activation of cloned intermediate- and small-conductance Ca^{2+}-activated K$^+$ channels. *Am J Physiol Cell Physiol* **278**, C570–581.

Szatanik M, Vibert N, Vassias I *et al.* (2008) Behavioral effects of a deletion in *Kcnn2*, the gene encoding the SK2 subunit of small-conductance Ca^{2+}-activated K$^+$ channels. *Neurogenetics* **9**, 237–248.

Tacconi S, Carletti R, Bunnemann B *et al.* (2001). Distribution of the messenger RNA for the small conductance calcium-activated potassium channel SK3 in the adult rat brain and correlation with immunoreactivity. *Neuroscience* **102**, 209–215.

Tamaoki J, Isono K, Kondo M *et al.* (1998). A human bronchial epithelial cell line releases arginine vasopressin: involvement of Ca^{2+} -activated K$^+$ channels. *Regul Pept* **74**, 91–95.

Taniguchi J and Imai M (1998). Flow-dependent activation of maxi K$^+$ channels in apical membrane of rabbit connecting tubule. *J Membr Biol* **164**, 35–45.

Taylor MS, Bonev AD, Gross TP *et al.* (2003). Altered expression of small-conductance Ca^{2+}-activated K$^+$ (SK3) channels modulates arterial tone and blood pressure. *Circ Res* **93**, 124–131.

Terstappen GC, Pellacani A, Aldegheri L *et al.* (2003). The antidepressant fluoxetine blocks the human small conductance calcium-activated potassium channels SK1, SK2 and SK3. *Neurosci Lett* **346**, 85–88.

Thompson J and Begenisich T (2006). Membrane-delimited inhibition of maxi-K channel activity by the intermediate conductance Ca^{2+}-activated K channel. *J Gen Physiol* **127**, 159–169.

Thorneloe KS, Knorn AM, Doetsch PE *et al.* (2008). Small conductance, Ca^{2+}- activated K$^+$ channel 2 (SK2) is the key functional component of SK channels in mouse urinary bladder. *Am J Physiol Regul Integr Comp Physiol* **294**, R1737-R1743.

Tian L, Duncan RR, Hammond MS *et al.* (2001a). Alternative splicing switches potassium channel sensitivity to protein phosphorylation. *J Biol Chem* **276**, 7717–7720.

Tomita H, Shakkottai VG, Gutman GA *et al.* (2003). Novel truncated isoform of SK3 potassium channel is a potent dominant-negative regulator of SK currents: implications in schizophrenia. *Mol Psychiatry* **8**, 524–535.

Toyama K, Wulff H, Chandy KG *et al.* (2008). The intermediate-conductance calcium-activated potassium channel KCa3.1 contributes to atherogenesis in mice and humans. *J Clin Invest* **118**, 3025–3037.

Tseng-Crank J, Godinot N, Johansen TE *et al.* (1996). Cloning, expression and distribution of a Ca^{2+}-activated K$^+$ channel beta-subunit from human brain. *Proc Natl Acad Sci USA* **93**, 9200–9205.

Uebele VN, Lagrutta A, Wade T *et al.* (2000). Cloning and functional expression of two families of beta-subunits of the large conductance calcium-activated K$^+$ channel. *J Biol Chem* **275**, 23211–23218.

Vergara C, Latorre R, Marrion NV *et al.* (1998). Calcium-activated potassium channels. *Curr Opin Neurobiol* **8**, 321–329.

Vestergaard-Bogind B, Stampe P *et al.* (1985). Single-file diffusion through the Ca^{2+}-activated K$^+$ channel of human red cells. *J Membr Biol* **95**, 121–130.

Viana F, Bayliss DA and Berger AJ (1993). Multiple potassium conductances and their role in action potential repolarization and repetitive firing behavior of neonatal rat hypoglossal motoneurons. *J Neurophysiol* **69**, 2150–2163.

Villalobos C, Shakkottai VG, Chandy KG *et al.* (2004). SKCa channels mediate the medium but not the slow calcium-activated afterhyperpolarization in cortical neurons. *J Neurosci* **24**, 3537–3542.

Wallner M, Meera P and Toro L (1996). Determinant for beta-subunit regulation in high-conductance voltage-activated and Ca^{2+}-sensitive K$^+$ channels: an additional transmembrane region at the N terminus. *Proc Natl Acad Sci* **93**, 14922–14927.

Wallner M, Meera P and Toro L (1999). Molecular basis of fast inactivation in voltage and Ca^{2+}-activated K$^+$ channels: a transmembrane beta-subunit homolog. *Proc Natl Acad Sci* **96**, 4137–4142.

Walter JT, Alvina K, Womack MD *et al.* (2006). Decreases in the precision of Purkinje cell pacemaking cause cerebellar dysfunction and ataxia. *Nat Neurosci* **9**, 389–397.

Waroux O, Massotte L, Alleva L *et al.* (2005). SK channels control the firing pattern of midbrain dopaminergic neurons *in vivo. Eur J Neurosci* **22**, 3111–3121.

Wei AD, Gutman GA, Aldrich R *et al.* (2005). International Union of Pharmacology. LII. Nomenclature and molecular relationships of calcium-activated potassium channels. *Pharmacol Rev* **57**, 463–472.

Werner ME, Zvara P, Meredith AL *et al.* (2005). Erectile dysfunction in mice lacking the large-conductance calcium-activated potassium (BK) channel. *J Physiol* **567**, 545–556.

Wilkens CM and Aldrich RW (2006). State-independent block of BK channels by an intracellular quaternary ammonium. *J Gen Physiol* **128**, 347–364.

Williams SE, Wootton P, Mason HS *et al.* (2004). Hemoxygenase-2 is an oxygen sensor for a calcium-sensitive potassium channel. *Science* **306**, 2093–2097.

Wittekindt OH, Visan V, Tomita H *et al.* (2004). An apamin- and scyllatoxin-insensitive isoform of the human SK3 channel. *Mol Pharmacol* **65**, 788–801.

Wolfart J and Roeper J (2002). Selective coupling of T-type calcium channels to SK potassium channels prevents intrinsic bursting in dopaminergic midbrain neurons. *J Neurosci* **22**, 3404–3413.

Wolfart J, Neuhoff H, Franz O *et al.* (2001). Differential expression of the small-conductance, calcium-activated potassium channel SK3 is critical for pacemaker control in dopaminergic midbrain neurons. *J Neurosci* **21**, 3443–3456.

Wu WL, So SC, Sun YP *et al.* (1998). Functional expression of P2U receptors in rat spermatogenic cells: dual modulation of a Ca^{2+}-activated K$^+$ channel. *Biochem Biophys Res Commun* **248**, 728–732

Wulff H, Gutman GA, Cahalan MD *et al.* (2001). Delineation of the clotrimazole/TRAM-34 binding site on the intermediate conductance calcium-activated potassium channel, IKCa1. *J Biol Chem* **276**, 32040–32045.

Wulff H, Knaus HG, Pennington M *et al.* (2004). K$^+$ channel expression during B cell differentiation: implications for immunomodulation and autoimmunity. *J Immunol* **173**, 776–786.

Wulff H, Kolski-Andreaco A, Sankaranarayanan A *et al.* (2007). Modulators of small- and intermediate-conductance calcium-activated potassium channels and their therapeutic indications. *Curr Med Chem* **14**, 1437–1457.

Wulff H, Miller MJ, Hansel W *et al.* (2000). Design of a potent and selective inhibitor of the intermediate-conductance Ca^{2+}-activated K$^+$ channel, IKCa1: a potential immunosuppressant. *Proc Natl Acad Sci* **97**, 8151–8156.

Xia XM, Ding JP and Lingle CJ (1999). Molecular basis for the inactivation of Ca^{2+}- and voltage-dependent BK channels in adrenal chromaffin cells and rat insulinoma tumor cells. *J Neurosci* **19**, 5255–5264.

Xia XM, Ding JP, Zeng XH *et al.* (2000). Rectification and rapid activation at low Ca^{2+} of Ca^{2+}-activated, voltage-dependent BK currents: consequences of rapid inactivation by a novel beta subunit. *J Neurosci* **20**, 4890–4903.

Xia XM, Fakler B, Rivard A *et al.* (1998). Mechanism of calcium gating in small-conductance calcium-activated potassium channels. *Nature* **395**, 503–507.

Xia XM, Zeng X and Lingle CJ (2002). Multiple regulatory sites in large-conductance calcium-activated potassium channels. *Nature* **418**, 880–884.

Xu W, Liu Y, Wang S *et al.* (2002). Cytoprotective role of Ca^{2+}- activated K^+ channels in the cardiac inner mitochondrial membrane. *Science* **298**, 1029–1033.

Yanovsky Y, Zhang W and Misgeld U (2005). Two pathways for the activation of small-conductance potassium channels in neurons of substantia nigra pars reticulata. *Neuroscience* **136**, 1027–1036.

Yoshihara S, Morimoto H, Yamada Y *et al.* (2004). Cannabinoid receptor agonists inhibit sensory nerve activation in guinea pig airways. *Am J Respir Crit Care Med* **170**, 941–946.

Yuan A, Dourado M, Butler A *et al.* (2000). SLO-2, a K^+ channel with an unusual Cl^- dependence. *Nat Neurosci* **3**, 771–779.

Yuan A, Santi CM, Wei A *et al.* (2003). The sodium-activated potassium channel is encoded by a member of the Slo gene family. *Neuron* **37**, 765–773.

Zhou XB, Arntz C, Kamm S *et al.* (2001). A molecular switch for specific stimulation of the BKCa channel by cGMP and cAMP kinase. *J Biol Chem* **276**, 43239–43245.

4.4

Kir family

Contents

4.4.1   Kir1, Kir2, Kir4, Kir5 and Kir7 families

Hiroshi Hibino and Yoshihisa Kurachi

4.4.1.1 Introduction

The inwardly rectifying K$^+$ (Kir) channel was first identified by electrophysiology in skeletal muscle, and was designated as an anomalous rectifier K$^+$ current (Katz, 1949) because current flows more readily in the inward direction than the outward direction, which is a property opposite to that of the voltage-gated K$^+$ (Kv) channel current in squid giant axon (Hodgikin *et al.*, 1952). The properties of this type of K$^+$ channel do not follow classical Hodgkin–Huxley kinetics (Hagiwara, 1983); that is, its kinetics depend not only on the membrane potential (E_m) but also on the equilibrium potential of K$^+$ (E_K). The channel conductance increases as the concentration of extracellular K$^+$ ($[K^+]_o$) is augmented. Under physiological conditions, rectification of Kir channels is sufficiently strong that these channels generate a large K$^+$-conductance at potentials negative to E_K but allow very little current flow at potentials positive to E_K (Noble, 1965). Such asymmetry in the current–voltage relationship results from block of outward-flowing current by intracellular polyamines and Mg^{2+} (Matsuda *et al.*, 1987; Lopatin *et al.*, 1994). Due to these properties, Kir channels play central roles in the maintenance of the resting membrane potential (E_{rmp}), in regulation of the action potential duration and in receptor-dependent inhibition of cellular

excitability (Hagiwara and Takahashi, 1974; Miyazaki *et al.*, 1974; Sakmann and Trube, 1984). Theoretically, cells expressing high levels of Kir channels are expected to have a hyperpolarized E_{rmp} close to E_K and may not be excited spontaneously. Following excitation, these cells may exhibit action potentials with a long-lasting plateau when the expression of voltage-gated outward K$^+$ channels is relatively low. Kir channels are expressed in a variety of cells such as cardiac myocytes (McAllister and Noble, 1966; Rougier *et al.*, 1968; Beeler and Reuter, 1970; Kurachi, 1985), neurons (Gahwiler and Brown, 1985; North *et al.*, 1987; Lacey *et al.*, 1988; Williams *et al.*, 1988; Brown *et al.*, 1990; Takahashi, 1990), blood cells (McKinney and Gallin, 1988; Lewis *et al.*, 1991), osteoclasts (Sims and Dixon, 1989), endothelial cells (Silver and DeCoursey, 1990), glial cells (Kuffler and Nicholls, 1966; Newman, 1984), and epithelial cells (Greger *et al.*, 1990). Accordingly, Kir channels are important in orchestrating the passive and active electrical properties of many cells within the body. After the identification of classical Kir channels, a number of K$^+$ channels, including G protein-gated K$^+$ channels (K$_G$) (Sakmann *et al.*, 1983; Kurachi, 1995) and glial K$^+$ channels (Kuffler and Nicholls, 1966; Newman, 1984), have been found to share the property of inward rectification. K$_G$ channels are activated via pertussis toxin-sensitive G proteins by a variety of neurotransmitters (Kurachi *et al.*, 1986a, b, c; North, 1989; Brown, 1990; Oh *et al.*, 1995). Recent molecular biological studies have further disclosed that the subunit encoding the pore of ATP-sensitive K$^+$ (K$_{ATP}$) channels, which is opened by a decrease in intracellular ATP (Noma, 1983), also belongs to the same family of inwardly rectifying K$^+$ channels (Inagaki *et al.*, 1995, 1996; Isomoto *et al.*, 1996). These functional Kir channels closely associate with various cell signalling and metabolic systems. As such, Kir channels are important for both receptor- and metabolism-dependent regulation of electrical activity of cells.

Molecular, structural biology and biochemical studies have revealed significant diversity of Kir channel protein structure, function, biophysics, physiological regulation, and pharmacology. Furthermore, extensive genetic analyses have revealed a number of diseases that are caused by mutations in Kir channels. In this chapter, we will discuss these topics, focusing specifically on the molecular and functional properties as well as pathophysiological relevance of both classical Kir channels (Kir2.x) and K$^+$-transport channels (Kir1.x, 4.x, 5.x, 7.x). Latter channels are expressed mainly in glial and epithelial cells where their major functional role is in transcellular K^+-transport, referred to as 'K^+-buffering' in glia, or 'K^+-recycling'; a role that is responsible for maintaining the normal activity of ion transporters.

4.4.1.2 Molecular characterization

Kir channel genes were first identified in 1993 by expression-cloning technique. The ATP-dependent Kir channel ROMK1/Kir1.1 (Ho *et al.*, 1993) and the classical Kir channel IRK1/Kir2.1 (Kubo *et al.*, 1993) were cloned from the outer medulla of rat kidney and a mouse macrophage cell line, respectively. These channel subunits possess a common structural motif of two putative membrane-spanning domains (M1 and M2) and one pore-forming region (H5) with cytoplasmic N- and C-terminal domains in their primary structure; a structure conserved in all types of Kir channel subunit. The H5 region acts as an 'ion-selectivity filter' that allows only K^+ to pass through the channel pore (Heginbotham *et al.*, 1994). As with other K^+ channels, all Kir channels have the signature sequence TXGY(F)G, which is a major determinant of K^+-selectivity (Bichet *et al.*, 2003). Biophysical relationships between the channels and their structures are described in detail later. Notably, Kir channels lack an S4 voltage sensor region, which is conserved not only in voltage-gated K^+ (Kv) channels but also in other voltage-gated ion channels such as Na^+ and Ca^{2+} channels. As a result, Kir channels are active at all membrane potentials. Their inward rectification is mediated by the blocking action of intracellular substances such as Mg^{2+} and polyamines on the outward-flowing K^+ current. Thus, the apparent kinetics of Kir channels depend on the voltage-shift from E_K but not on the membrane potential itself.

Recent molecular cloning techniques have identified fifteen Kir subunit genes, which can be classified into four subfamilies: (i) classical Kir channels (Kir2.x), (ii) G-protein gated Kir channels (Kir3.x), (iii) K^+-transport channels (Kir1.x, Kir4.x, Kir5.x, and Kir7.x) and (iv) ATP-sensitive K^+ channels (Kir6.x). Table 4.4.1.1 summarizes the nomenclature and chromosomal localization of these subunits (Kubo *et al.*, 2005). A phylogenetic tree of Kir channel-subunits is shown in Figure 4.4.1.1.

Like Kv channels, functional Kir channels are composed of tetrameric complex of subunits (Glowatzki *et al.*, 1995; Yang *et al.*, 1995b).

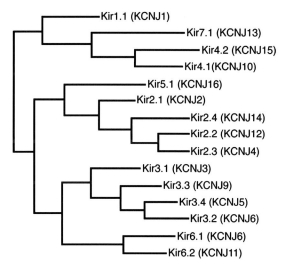

Fig. 4.4.1.1 Phylogenetic tree of Kir channel-subunits. Amino acid sequence alignment and phylogenetic analysis of the 15 known subunits of human Kir channels.

Both homomeric and heteromeric combinations of multimers of Kir-channel subunits have been identified. Such multimerization generally occurs between members of the same subfamily rather than between members of different subfamilies. For example, Kir2.1 can be assembled with Kir2.2 (see the section Classical inwardly rectifying K^+ channels (Kir2.x)), and Kir3.1 forms a heteromeric complex with Kir3.2, Kir3.3 or Kir3.4 (see Chapter 4.4.2 on Kir3 channels). An exception is Kir4.1 which can assemble with Kir5.1 that belongs to another subfamily (see the sections on K^+ transport channels—Kir1.x, Kir4.x, Kir5.x, Kir7.x). Heteromeric assemblies confer distinct properties on the channel, which further contributes to the unique functions of these channels in different cell types. Details of such properties and their relevance to physiological roles will be discussed in the following sections.

Classical inwardly rectifying K^+ channels (Kir2.x)

The channels referred to as 'classical Kir channels' show strong inward-rectification and are constitutively active. As previously mentioned, Kir channels were first identified in skeletal muscle (Katz, 1949). Similar types of channels were subsequently found and further characterized in frog skeletal muscle, starfish, tunicate eggs, cardiac myocytes, and neurons (McAllister and Noble, 1966; Adrian, 1969; Hille and Schwarz, 1978; Hagiwara and Jaffe, 1979). The classical Kir channels are not directly regulated by neurotransmitters.

So far, four Kir subunits have been identified as members of the classical Kir channel subfamily. They are Kir2.1(IRK1)/KCNJ2 (Kubo *et al.*, 1993), Kir2.2(IRK2)/KCNJ12 (Koyama *et al.*, 1994; Takahashi *et al.*, 1994), Kir2.3(IRK3, BIR11)/KCNJ4 (Bond *et al.*, 1994; Morishige *et al.*, 1994; Bredt *et al.*, 1995), and Kir2.4(IRK4)/KCNJ14 (Topert *et al.*, 1998). Chromosomal localization of each subunit is shown in Table 4.4.1.1. The amino acid sequence of mouse brain Kir2.1 shares 70, 61 and 63 per cent identity with Kir2.2, Kir2.3 and Kir2.4, respectively. The sequences are most highly conserved in the M1, M2, and H5 regions.

The question as to whether or not the Kir2 subfamily can form functional heterotetramers has been a highly debated issue for a number of years. Recent studies have revealed that hetero-tetramers

Table 4.4.1.1 Chromosomal localization of Kir channel subunits in various species

Names		Chromosomal localization		
Protein	Gene	Human	Rat	Mouse
Kir1.1	KCNJ1	chr. 11q24	chr. 8q21	chr. 9A
Kir2.1	KCNJ2	chr. 17q23.1–24.2	chr. 10q32.1	chr. 11E2, 11, 68.0 cm
Kir2.2	KCNJ12	chr. 17p11.1	chr. 10q22	chr. 11, 34.15 cm
Kir2.3	KCNJ4	chr. 22q13.10	chr. 7q34	chr. 15, 46.7 cm
Kir2.4	KCNJ14	chr. 19q13.1–13.3	chr. 1q22	chr. 7
Kir4.1	KCNJ10	chr. 1q22-q2	chr. 13q24	chr. 1, 93.5 cm
Kir4.2	KCNJ15	chr. 21q22.2	chr. 11q.11	chr. 16, 69.1 cm
Kir5.1	KCNJ16	chr. 17q23.1–24.2	chr. 10q32.1	chr. 11, 71.0 cm
Kir7.1	KCNJ13	chr. 2p37	chr. 9q35	NE

chr, chromosome; cm, centimorgans; NE, not examined.

do exist both *in vitro* and *in vivo*. In this respect, *in vitro* electrophysiological experiments have now clarified that each of Kir2.1, Kir2.2, and Kir2.3 can assemble with any one of the other subunits and, in each respective case, the heteromer so formed exhibits different properties from that of the homomer (Preisig-Muller *et al.*, 2002). Example heteromeric assemblies include Kir2.1/2.2 and Kir2.1/2.3 channels that are expressed in cardiac myocytes (see the sections on Cellular and subcellular localization and Physiological roles of classical Kir channels) and Kir2.4/ Kir2.1 which forms in the *Xenopus* oocyte heterologous expression system as well as in the brain (Schram *et al.*, 2002).

K⁺-transport channels (Kir1.x, Kir4.x, Kir5.x, Kir7.x)

Kir1.1/ROMK1

The first Kir subunit gene to be identified was Kir1.1 (Ho *et al.*, 1993), which was originally called 'rat outer medullary K⁺ channel', ROMK1. Six alternative splicing isoforms have been cloned so far (Kir1.1a–f/ROMK1–6) (Shuck *et al.*, 1994; Boim *et al.*, 1995; Doupnik *et al.*, 1995; Kondo *et al.*, 1996). Chromosomal localization of Kir1.1 is described in Table 4.4.1.1. Like other Kir channels, the Kir1.1 subunit seems to form as tetramer when generating functional channels (Leng *et al.*, 2006). Heteromeric assemblies of Kir1.1 with different other subfamilies have not yet been reported, suggesting that Kir1.1 exists as a homomer *in vivo*.

Kir4.x and Kir5.1

The Kir4.1 gene was discovered by several groups and its product has multiple names such as BIR10 (Bond *et al.*, 1994), K_{AB}-2 (Takumi *et al.*, 1995), BIRK-1 (Bredt *et al.*, 1995), and Kir1.2 (Shuck *et al.*, 1997). The salmon homologue of the mammalian Kir4.1 is referred to as Kir4.3 (Kubo *et al.*, 1996). Kir5.1 was first identified in 1994 and called BIR9 (Bond *et al.*, 1994). The chromosomal localization of these subunits is described in Table 4.4.1.1. The deduced amino-acid sequence revealed that Kir4.1 has 53, 43 and 43 per cent identity with Kir1.1, Kir2.1, and Kir3.1, respectively, whereas Kir5.1 shares 39, 50 and 40 per cent identity with these subunits. Kir4.1, when expressed alone in a heterologous expression system, forms a tetramer and elicits a K⁺-current (Pessia *et al.*, 1996). In native tissues such as brain, the channel also exists as a tetramer (Hibino *et al.*, 2004a; Kaiser *et al.*, 2006). Although Kir5.1, which is expressed alone in heterologous expression system, is non-functional, its co-expression with Kir4.1 forms a functional channel whose properties are different from the Kir4.1-homomer (see the section on Biophysical characteristics of Kir4.x and Kir5.1) (Bond *et al.*, 1994; Pessia *et al.*, 1996; Tanemoto *et al.*, 2000; Tucker *et al.*, 2000). The Kir4.1/5.1-heteromer also exists as a tetramer (Pessia *et al.*, 1996), and is detected in some tissues including kidney and brain (see the section on Cellular and subcellular localization of Kir4.x and Kir5.1).

Kir4.2 (Kir1.3) was first isolated from a human kidney cDNA library (Shuck *et al.*, 1997). It was also identified in mouse liver (Pearson *et al.*, 1999). Kir4.2 has 62 per cent identity to human Kir4.1. Although Kir4.1 has a Walker type-A ATP-binding cassette in its carboxyl terminus (Takumi *et al.*, 1995), Kir4.2 does not possess this feature (Pearson *et al.*, 1999). Kir4.2 forms a functional channel when expressed alone in heterologous expression systems

(Pearson *et al.*, 1999; Pessia *et al.*, 2001). In addition, Kir4.2 can also be assembled with Kir5.1 to generate a channel with novel characteristics (Pearson *et al.*, 1999; Pessia *et al.*, 2001) (see the section on Biophysical characteristics of Kir4.x and Kir5.1).

Kir7.1

Kir7.1 was cloned in 1998 by three different groups (Doring *et al.*, 1998; Krapivinsky *et al.*, 1998; Partiseti *et al.*, 1998). Its sequence is quite distinct from those of other Kir channels in that it is only ~38 per cent identical to the closest Kir relative, Kir4.2. To date, only one isoform has been isolated in this subfamily. Chromosomal localization of Kir7.1 is described in Table 4.4.1.1. By analogy to other Kir channels, Kir7.1 is expected to function as a tetramer. No observations have been published to support heteromeric assembly of Kir7.1 with other subunits.

4.4.1.3 Biophysical characteristics

Classical inwardly rectifying K⁺ channels (Kir2.x)

The Kir2.x subunits heterologously expressed in *Xenopus* oocytes and cultured cells generate strong inwardly rectifying K⁺ currents. Furthermore, as summarized in Table 4.4.1.2 channels composed of various homo- and heteromultimers of Kir2.x subunits exhibit unique single channel conductances (Kubo *et al.*, 1993; Morishige *et al.*, 1994; Takahashi *et al.*, 1994; Topert *et al.*, 1998; Liu *et al.*, 2001; Preisig-Muller *et al.*, 2002). The steady-state open probability P_o of Kir2.1 decreases with hyperpolarization, whereas those of Kir2.1 and Kir2.3 remain constant. Furthermore, a remarkable feature of Kir2.3 is its activation by intracellular or extracellular alkalization ($pK_a = 6.76$ and 7.4, respectively) (Coulter *et al.*, 1995; Zhu *et al.*, 1999; Qu *et al.*, 2000). Extracellular alkalization also enhances the Kir2.4 channel current with a pK_a of 7.14 (Hughes *et al.*, 2000). The determinant of the pH-sensitivity is a single histidine residue (H117 in Kir2.3) in the M1 to H5 linker region (Coulter *et al.*, 1995).

Inward rectification, the hallmark of Kir channels, is caused by intracellular ions such as Mg^{2+} (Matsuda *et al.*, 1987; Vandenberg, 1987) and polyamines (Lopatin *et al.*, 1994; Yamada and Kurachi, 1995). In Kir channels, inward rectification can be classified into three types, i.e. strong, intermediate and weak. Whereas Kir2.x and Kir3.x elicit strongly rectifying K⁺ currents, Kir1.x and Kir6.x channels exhibit weak rectification. On the other hand, Kir4.x forms channels that generate a K⁺ current with intermediate rectification. Electrophysiological studies and mutational analyses of the kinetics of Mg^{2+} and polyamine block revealed that Kir channels have more than one binding site for these ions. One binding site was identified as being located in the second transmembrane helix (M2 region). Kir2.1, which exhibits very strong rectification, harbours a negatively charged residue, D172, at this position. On the other hand, the weakly rectifying Kir1.1 possesses an uncharged residue N171 at the corresponding site. When N171 was mutated to D171 in Kir1.1, the mutant exhibited increased affinity for Mg^{2+} and thus strong rectification (Lu and MacKinnon, 1994; Stanfield *et al.*, 1994; Wible *et al.*, 1994; Yang *et al.*, 1995a). Another site of importance is S165 (Kir2.1) in the M2 region. This residue plays a crucial role in Mg^{2+} but not polyamine block (Fujiwara and Kubo, 2002). Moreover, site-directed mutagenesis (SDM) analyses identified a negatively charged amino acid (glutamate) at two different positions (E224 and E229 for Kir2.1) in the carboxyl-terminus

Table 4.4.1.2 Comparison of biophysical properties of cloned classical Kir channels, Kir2.x

Channel name	Unitary conductance	Ba²⁺ sensitivity (IC₅₀)	References
Kir2.1	23 ~ 30.6–pS	3.2 µM (60 mM [K⁺]ₒ in XO) 8 µM (96 mM [K⁺]ₒ in XO)	Kubo et al. (1993); Topert et al. (1998); Liu et al. (2001); Preisig-Müller et al. (2002)
Kir2.2	34.2 ~ 40–	0.5 µM (60 mM [K⁺]ₒ in XO) 6 µM (96 mM [K⁺]ₒ in XO)	(Takahashi et al. (1994); Topert et al. (1998); Liu et al. (2001); Preisig-Muller et al. (2002)
Kir2.3	13 ~ 14.2–pS	10.3 µM (60 mM [K⁺]ₒ in XO)	(Morishige et al. (1994); Liu et al. (2001); Preisig-Muller et al. (2002)
Kir2.4	15–pS	390 µM (96 mM [K⁺]ₒ in XO)	Topert et al. (1998)
Kir2.1-Kir2.2 (Concatamers)	30–pS	0.68 µM (60 mM [K⁺]ₒ in XO)	Preisig-Muller et al. (2002)
Kir2.1-Kir2.3 (Concatamers)	28.1–pS	3.39 µM (60 mM [K⁺]ₒ in XO)	Preisig-Muller et al. (2002)
Kir2.2-Kir2.3 (Concatamers)	NE	1.73 µM (60 mM [K⁺]ₒ in XO)	Preisig-Muller et al. (2002)
Kir2.1+Kir2.2 (Co-expression)	NE	0.64 µM (60 mM [K⁺]ₒ in XO)	Preisig-Muller et al. (2002)
Kir2.1+Kir2.3 (Co-expression)	NE	6.32 µM (60 mM [K⁺]ₒ in XO)	Preisig-Muller et al. (2002)
Kir2.2+Kir2.3 (Co-expression)	NE	1.94 µM (60 mM [K⁺]ₒ in XO)	Preisig-Muller et al. (2002)

NE, not examined; XO, *Xenopus* oocytes

of cytoplasmic domains that was critically involved in Mg^{2+} and polyamine sensitivity (Taglialatela et al., 1994, 1995; Yang et al., 1995a; Kubo and Murata, 2001). Further SDM and substituted cysteine accessibility experiments have demonstrated that these residues might directly interact with Mg^{2+} and polyamines (Lu et al., 1999; Minor et al., 1999). In support of this idea, the crystal structures of KirBac and Kir3.1 disclosed that the glutamate side chain faced towards the centre of the conduction pathway, forming rings of negatively charged residues creating a complementary electrostatic match for the binding of positively charged polyamines (Nishida and MacKinnon, 2002; Kuo et al., 2003).

Many types of Kir channels are activated by a membrane-anchored phospholipid, phosphatidylinositol 4,5-bisphosphate (PIP_2) (Huang et al., 1998; Hilgemann et al., 2001; Takano and

Kuratomi, 2003). In patch-clamp analyses, when membranes expressing Kir channels are excised, channel activity is gradually suppressed. Furthermore, channel activity was restored by intracellular application of Mg-ATP; an effect proposed to be mediated by replenishment of PIP_2 via lipid kinases (Hilgemann and Ball, 1996). Mutation analyses revealed that PIP_2 directly interacts with positively charged residues at the carboxyl terminus of the channels (Huang et al., 1998; Lopes et al., 2002). It was shown that a PIP_2 antibody inhibits the activity of Kir channels with the following order of potency: Kir3.1/3.4 ≈ Kir3.1 > Kir2.1 ≈ Kir1.1 (Huang et al., 1998). This result may indicate that the affinity for PIP_2 of Kir3.1/3,4 or Kir3,1 is low and that of Kir2.1 and Kir1.1 is high. Thus, it has been suggested that the constitutively activity of classical Kir2.x channels is a result of their high affinity for PIP_2. Furthermore, it was reported that one of the mutations in Kir2.1 identified in Andersen's syndrome reduces the channel-PIP_2 interaction (Lopes et al., 2002) (see the sections on Physiological roles and Pathophysiological relevance of classical Kir channels).

K⁺-transport channels (Kir1.x, Kir4.x, Kir5.x, Kir7.x)

Kir1.1/ROMK1

When expressed in *Xenopus* oocytes, Kir1.1 exhibits weak inward rectification, a single channel conductance of ~35-pS, and high open probability over a wide range of E_m (Ho et al., 1993; Boim et al., 1995; Chepilko et al., 1995). The deduced amino acid sequence of Kir1.1 reveals that it possesses a Walker type-A ATP-binding cassette at its carboxyl-terminal region. However, the functional importance of this cassette has not yet been established (Ho et al., 1993; Bond et al., 1994; Choe et al., 1998).

Intracellular but not extracellular changes in pH modulate the channel activity of Kir1.1. Acidification closes the channel with a pK_a of ~6.5 (Tsai et al., 1995; Doi et al., 1996; Fakler et al., 1996; Choe et al., 1997; McNicholas et al., 1998). Mutation analyses demonstrated that residue K80 close to the M1 region at the amino-terminus is crucial for the pH sensitivity of Kir1.1 (Fakler et al., 1996; Choe et al., 1997; McNicholas et al., 1998; Schulte et al., 1999). Further studies revealed that R41 in the amino terminus, K80, and R311 in the carboxyl terminus appear to establish electrostatic intramolecular interactions, which are essential for the sensitivity of Kir1.1 to intracellular pH (pH_i) (Schulte et al., 1999). Indeed, each of the mutations R41Q and R311Q greatly shift the pH_i-sensitivity to an alkaline pH range, which results in the loss of the channel's activity within the physiological pH_i range. The relationship between mutations found in a genetic renal disease, Bartter's syndrome, and altered pH_i-sensitivity will be described later in the section Pathophysiological relevance of Kir1.1 channels.

The channel activity and number of active Kir1.1 channels on the cell surface are regulated by serine-threonine kinases (e.g., PKA and SGK), tyrosine kinases and phosphatases. The maintenance of channel activity requires a PKA-dependent phosphorylation process (McNicholas et al., 1994). Kir1.1 has three PKA-phosphorylation sites (S44 in the amino terminus, S219 and S313 in the carboxyl terminus of Kir1.1a). In Kir1.1b, mutation of any of these single sites suppresses the whole cell K⁺-current by ~35–40 per cent in *Xenopus* oocytes by reducing the channel open probability without affecting single channel conductance. Two or three mutations abolish Kir1.1b channel activity altogether (Xu et al., 1996). Phosphorylation of the aforementioned residues augments cell surface expression of the Kir1.1 protein. In addition, a recent study has strongly suggested that PKA activates Kir1.1 by enhancing the interaction between the channel and PIP_2 (Liou et al., 1999).

S44 is also phosphorylated by serum-glucocorticoid-regulated kinase SGK1, which results in enhancement of the Kir1.1 current probably due to an increase in the surface expression level of the channel (Yoo et al., 2003). Since aldosterone stimulates SGK-1 transcription, the activity of Kir1.1, like epithelial Na^+ channels, could be regulated by this hormone. However, another study reported that cell surface expression and current density of Kir1.1 were not altered by co-expression of SGK-1 alone but were augmented by that of SGK-1 and a scaffolding protein, Na^+, H^+-exchanger regulatory factor-2 (NHERF-2) (Yun et al., 2002). The mechanism of augmentation is unknown, because the authors also demonstrated that NHERF-2 itself, when co-expressed with Kir1.1, had little effect on channel-function. NHERF-2, which binds to multiple ion transport systems, was shown to be directly associated with Kir1.1 by an in vitro biochemical assay and co-localized with the channel at the apical membrane of renal epithelial cells (Shenolikar et al., 2004; Yoo et al., 2004). This suggests that Kir1.1-activity may be regulated by a combination of SGK-1 and NHERF-2 in vivo as well.

A recent study further clarified the mechanism by which cell surface expression and channel activity of Kir1.1 is increased by S44 phosphorylation. Kir1.1 has an endoplasmic reticulum (ER) retention signal, RXR (Ma et al., 2001), in its carboxyl-terminal region and a certain proportion of channels are not expressed at the cell surface. It is likely that S44 phosphorylation suppresses this ER retention signal, which results in augmentation of delivery of Kir1.1 to the cell surface (O'Connell et al., 2005).

Two serine residues in Kir1.1a, S4 and S201, are reported to be phosphorylated by PKC (Lin et al., 2002). Although PKC-phosphorylation promoted surface expression of Kir1.1a in oocytes and HEK cells (Lin et al., 2002), this modification has been known to inhibit 35 pS K^+ channels composed of Kir1.1 observed in CCD cells (Wang and Giebisch, 1991) (see the section Physiological roles of Kir1.1). It has recently been reported that PKC-induced suppression of Kir1.1 is due to the reduction of membrane PIP_2, a phospholipid essential for the channel's activation, albeit by an unknown mechanism (Zeng et al., 2003).

Moreover, the function of Kir1.1 is controlled by other serine-threonine kinases, such as WNK4 and WNK1. Both kinases reduced Kir1.1 current by inhibiting the recruitment of Kir1.1 to the plasma membrane through distinct mechanisms (Kahle et al., 2003; Wade et al., 2006). Kir1.1 is retrieved from the cell surface by a clathrin-dependent mechanism because of its internalization motif, NPXY, at the carboxyl terminal region (Zeng et al., 2002). The effect of WNK4 is attributed to an enhancement of clathrin-dependent endocytosis but not to kinase activity (Kahle et al., 2003). On the other hand, WNK1 seems to inhibit cell surface expression of Kir1.1 by phosphorylating certain residues within the protein (Wade et al., 2006). Since mutations in either the WNK1 or WNK4 gene associate with pseudohypoaldosteronism type II (PHAII: Gordon's syndrome; an autosomal dominant form of hypertension with hyperkalaemia (Wilson et al., 2001)), abnormal modulation of Kir1.1 function by the mutated WNKs is physiologically linked with its phenotypes. Indeed, some mutations of WNK4, which were found as a source of the disease, largely increased its inhibitory effect on Kir1.1 (Zeng et al., 2002).

Finally, K22 in the amino-terminus of Kir1.1 is mono-ubiquitinated (Lin et al., 2005). Substitution of K22 for R greatly enhanced the K^+ current mediated by Kir1.1 by increasing its cell surface expression. Therefore, ubiquitination acts as a negative regulation system for Kir1.1 function. Altogether, cell surface expression and function of Kir1.1 is dynamically controlled by a combination of physiological events such as phosphorylation, clathrin-dependent endocytosis, and mono-ubiquitination.

Kir4.x and Kir5.1

The single channel conductance of the Kir4.1-homomer and Kir4.1/5.1-heteromer is variable, possibly because these channels possess multiple subconductance states. Whereas the conductance of the former is ~20–40 pS, that of the latter is ~40–60 pS (Takumi et al., 1995; Pessia et al., 1996; Tanemoto et al., 2000; Yang et al., 2000; Lourdel et al., 2002). The Kir4.1 homomer exhibits intermediate inward rectification, but the rectification profile of the Kir4.1/5.1 heteromer is much stronger. The proximal region of the carboxyl terminus of Kir4.1 plays a crucial role in the assembly with Kir5.1 and Kir4.1 itself (Konstas et al., 2003). In particular, in Kir4.1, the E177 residue in the proximal region of the carboxyl terminus just under the M2 region is essential for its interaction with Kir5.1 (Konstas et al., 2003).

An even more remarkable difference between homomeric Kir4.1 and heteromeric Kir4.1/5.1 is the sensitivity of these channels to changes in pH_i. In the physiological range of pH_i of 6.5–8.0, the channel activity of the Kir4.1 homomer is slightly inhibited by acidification with a pK_a of ~6 (Bond et al., 1994; Pessia et al., 1996; Tanemoto et al., 2000; Yang et al., 2000). In contrast, over the same pH_i range, the activity of the heteromeric channels is dramatically suppressed by a small intracellular acidification and enhanced by alkalization with a pK_a of ~7.5 (Bond et al., 1994; Pessia et al., 1996; Tanemoto et al., 2000; Yang et al., 2000). Several residues of Kir4.1 such as E158 in the M2 region (Xu et al., 2000), K67 in the cytoplasmic amino-terminal region (Yang et al., 2000), and H190 in the carboxyl terminal region (Casamassima et al., 2003) are reported to be critically involved in pH_i-sensitivity of not only homomeric Kir4.1 but also heteromeric Kir4.1/5.1. The residues responsible for this property in Kir5.1 have not been identified.

Kir4.2 displays different biophysical behaviour. When the Kir4.2 subunit was expressed alone in Xenopus oocytes, its single channel conductance was measured to be ~25 pS. Although the amino acid sequence of Kir4.2 is similar to that of Kir4.1, the former is more sensitive to a change in pH_i in the physiological range than the latter (Kir4.2; $pK_a = 7.1$). Therefore, little change in pH_i-sensitivity was observed even when Kir4.2 was coexpressed with Kir5.1. The heteromeric channel exhibited larger unitary conductance of ~54 pS than the Kir4.2-homomer (Pessia et al., 2001). Interestingly, the mutant K66M in the amino terminal region of Kir4.2 almost abrogated its pH_i-sensitivity (Pessia et al., 2001). It was also shown that a titratable lysine residue in the relatives of Kir4.2, i.e. K80 in Kir1.1 and K67 in Kir4.1, was commonly crucial for achieving their pH_i-sensitivity. However, it is impossible to explain the difference of the pH_i-sensitivity between Kir4.1 and Kir4.2 solely on the role of the lysine residue and it is the intracellular carboxyl terminal part of these subunits that appears to be responsible for this difference (Pessia et al., 2001). However, the crucial residues in this region have not yet been identified.

Kir7.1

Kir7.1, when expressed in heterologous expression systems, yields inwardly rectifying K$^+$ currents but exhibits very unique properties. First, the single channel conductance of Kir7.1 is extremely small: ~50 fS (Krapivinsky et al., 1998). Second, the sensitivity of the channel to Ba^{2+} and Cs$^+$, common blockers of Kir channels, is very low with IC$_{50}$ values of ~1 mM and ~10 mM, respectively, which are ~ 10 times higher than the IC$_{50}$ values for other Kir channels (Krapivinsky et al., 1998). Third, the inward rectification property of this channel is independent of [K$^+$]$_o$ (Doring et al., 1998; Krapivinsky et al., 1998). The residue responsible for these unusual characteristics of the pore is M125, which is R in all other Kir subunits. Replacement of this residue with the conserved R dramatically increased the single channel conductance of the channel by ~20 fold and Ba^{2+} sensitivity by ~tenfold as well as allowed the channel to exhibit the rectification profile of other Kir channels (Doring et al., 1998; Krapivinsky et al., 1998). Since R148 in Kir2.1 is located in the same position as M125 in Kir7.1 it is reported to be crucially involved in single channel conductance and sensitivity to blockers (Sabirov et al., 1997). This residue may be commonly important for establishing the pore structure in all of the Kir channels.

4.4.1.4 Cellular and subcellular localization

Classical inwardly rectifying K$^+$ channels (Kir2.x)

A number of studies have shown that Kir2.x subunits are differentially expressed in many tissues such as heart, endothelial cells, brain, skeletal muscle, and kidney. Each subunit displays a distinct localization pattern in these various tissues.

Heart

In heart, analyses of mRNA transcripts by in situ hybridization histochemistry, RNase protection assay, and RT-PCR showed that cardiomyocytes express Kir2.1, Kir2.2, and Kir2.3 (Raab-Graham et al., 1994; Wible et al., 1995; Brahmajothi et al., 1996; Wang et al., 1998; Liu et al., 2001). In contrast, Kir2.4 is restricted to neuronal cells (Liu et al., 2001). These observations suggested that multimers among Kir2.1, Kir2.2, and Kir2.3 subunits could generate the I_{K1} current, a major Kir conductance in cardiomyocytes. Within this subfamily, Kir2.1 was considered to be the core subunit forming the I_{K1} current because levels of its transcript correlated strongly with the amount of current observed in electrophysiological experiments (Brahmajothi et al., 1996; see Physiological roles of classical Kir channels section).

Blood vessels

The major constituents of vasculature are endothelial and smooth muscle cells. Electrophysiological analyses have demonstrated that both types of cells express classical Kir channels (Nilius and Droogmans, 2001; Adams and Hill, 2004). Whereas Kir2.1 and Kir2.2 proteins are expressed in aortic endothelial cells, patch clamp analysis identified Kir2.2 as the dominant channel (Fang et al., 2005). In vascular smooth muscle cells, Kir2.1 transcripts but not Kir2.2 or Kir2.3 were identified (Bradley et al., 1999). That Kir2.1 is the dominant channel in these cell types was further confirmed by functional analysis of Kir2.1 knockout mice (Zaritsky et al., 2000) (see Physiological roles of classical Kir channels section).

Brain

Classical Kir channels are abundantly and differentially expressed in brain. Kir2.1 is expressed diffusely and weakly in the whole brain, Kir2.2 moderately throughout the forebrain and strongly in the cerebellum, Kir2.3 mainly in forebrain, and Kir2.4 in the cranial nerve motor nuclei in the midbrain, pons, and medulla (Horio et al., 1996; Karschin et al., 1996; Topert et al., 1998; Inanobe et al., 2002; Pruss et al., 2005). Their distribution is restricted to neurons and immunohistochemical analyses revealed that the majority of Kir2.x-proteins are located in somata and dendrites of these cells (Inanobe et al., 2002; Pruss et al., 2005). For example, in mouse olfactory bulb, Kir2.3 is specifically expressed at the postsynaptic membrane of dendritic spines of granule cells, which receive mostly excitatory synaptic inputs (Inanobe et al., 2002). Putative physiological roles of these channels in neurons will be discussed in Physiological roles of classical Kir channels section.

Skeletal muscle

All Kir2.x subunits are reported to be expressed in skeletal muscle at the mRNA level (Kubo et al., 1993; Perier et al., 1994; Raab-Graham et al., 1994; Takahashi et al., 1994; Topert et al., 1998).

Kidney

In kidney, two classical Kir subunits, Kir2.1 and Kir2.3, have been identified as being expressed at the protein level (Le Maout et al., 1997; Welling, 1997; Leichtle et al., 2004). Kir2.1 specifically distributes to juxtaglomerular cells but is not present in either epithelial cells or glomeruli (Leichtle et al., 2004). In contrast, Kir2.3 is found in the basolateral membrane of epithelial cells in collecting ducts (Le Maout et al., 1997; Welling, 1997).

Endoplasmic reticulum (ER) export signal

It has been reported that classical Kir channels possess an amino acid sequence responsible for recruitment of the channel from the endoplasmic reticulum (ER) to the cell surface. This ER export signal is reported to be the amino-acid sequence FCYENE, which is present in the cytoplasmic carboxyl terminus and conserved in all subunits of the Kir2.x subfamily (Ma et al., 2001). Disruption of this motif by mutagenesis causes retention of Kir2.1 in the cytosol. It is of note that Kir1.1 has different ER export signals (VLS and EXD) in its carboxyl terminal region (Ma et al., 2001). It is therefore likely that the surface expression of channels from each Kir subfamily is regulated by these specific signals.

K$^+$-transport channels (Kir1.x, Kir4.x, Kir5.x, Kir7.x)

Kir1.1/ROMK1

Kir1.1 is expressed in renal epithelial cells such as the thick ascending limb of the loop of Henle (TAL), distal convoluted tubule (DCT), and cortical collecting duct (CCD) (Lee and Hebert, 1995; Fakler et al., 1996; Hebert et al., 2005). Its isoforms (Kir1.1a-c) are differentially distributed along the nephron (Figure 4.4.1.2) (Boim et al., 1995) and play respective roles in each segment (see Physiological roles of Kir1.1 section). Immunohistochemical examination revealed that Kir1.1 is specifically localized at the apical but not at the basolateral membrane of the epithelial cells (Xu et al., 1997; Kohda et al., 1998). Furthermore, in situ hybridization demonstrated that the Kir1.1 transcript was expressed in neurons of the cortex and hippocampus but not in other areas

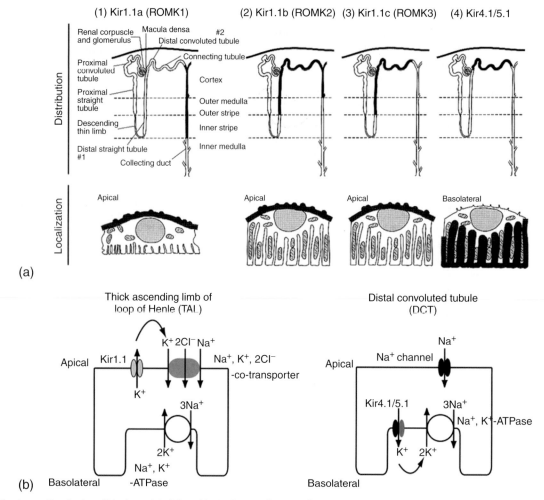

Fig. 4.4.1.2 Distribution and localization of Kir channels in kidney. (a) Distribution of Kir1.1 isoforms and Kir4.1/5.1 in the kidney. Note that each Kir1.1 isoform is differentially expressed in unique portions of renal epithelia (upper panels). #1: The distal straight tubule is also called the thick ascending limb of loop of Henle (TAL). Lower panels show the membrane domains where each channel subunit is expressed. Whereas Kir1.1 isoforms are expressed at the apical membrane of renal epithelia, the Kir4.1/5.1 heteromer is expressed at the basolateral membrane. (b) Localization of Kir channels in renal epithelia, and ion transport systems that are functionally coupled with these channels. In TAL cells (left), Kir1.1 expressed at the apical membrane is co-localized with the $Na^+/K^+/2Cl^-$ co-transporter. In distal convolute tubule (DCT) cells (right), the heteromeric Kir4.1/5.1 channel is co-localized with the Na^+/K^+-ATPase at the basolateral membrane. These K^+ channels supply K^+ to the extracellular K^+ site of the K^+ transporters and sustain their activity; a process referred to as 'K^+-recycling' action.

(Kenna *et al.*, 1994). The physiological function of Kir1.1 in the brain is still unknown.

Kir4.x and Kir5.1

Kir4.1 and Kir5.1 are expressed in epithelial and glial cells of several organs. In each tissue, the channel proteins exhibit unique distribution profiles which map on to their specific physiological roles (see Pharmacology of Kir4.x and Kir5.1 section).

Epithelial tissues in kidney, stomach, and ear

Immunohistochemical studies have demonstrated that both Kir4.1 and Kir5.1 are located at the basolateral membrane of epithelial cells in DCT (Ito *et al.*, 1996; Tanemoto *et al.*, 2000; Tucker *et al.*, 2000) (Figure 4.4.1.2). Patch clamp analysis of renal epithelia and biochemical experiments identified that these subunits formed a functional heteromeric channel at that area (Tanemoto *et al.*, 2000; Lourdel *et al.*, 2002). The Kir4.2/5.1 channel is also proposed to participate in the basolateral K^+ channel (Lourdel *et al.*, 2002).

In the stomach, Kir4.1 is specifically distributed to the apical membrane of parietal cells (Fujita *et al.*, 2002). Careful analysis revealed that the channel was located at the microvilli of the apical membrane surface but not in the tubulovesicles. No transcript of the Kir5.1-gene was detected, indicating that the Kir4.1 homomer dominates in gastric parietal cells.

The distribution pattern of Kir4.1 and Kir5.1 in the cochlea of the inner ear is unique in comparison with other epithelial tissues. Both are expressed in the cochlear lateral wall. However, whereas Kir4.1 is located in epithelial tissue stria vascularis (Hibino *et al.*, 1997; Ando and Takeuchi, 1999), Kir5.1 is located in the spiral ligament that contains several types of fibrocytes (Hibino *et al.*, 2004b). Ultra-structural analysis suggests that, in stria vascularis, Kir4.1-expression is restricted to the apical site of intermediate cells (Ando and Takeuchi, 1999; Hibino *et al.*, 2004b). Furthermore, Kir4.1 is expressed in both supporting cells beneath hair cells and satellite cells surrounding the spiral ganglion neurons, and Kir5.1 is expressed in the ganglion neurons (Hibino *et al.*, 1997;

Hibino *et al.*, 1999; Hibino *et al.*, 2004b). These observations strongly suggest that both subunits exist as homomers in the cochlea.

Glial cells in central nervous system and retina

In the central nervous system (CNS) and retina, Kir4.1 and Kir5.1 subunits are specifically expressed in astroglial cells. They form different multimers, which exhibit unique distribution patterns in distinct types of cells.

Kir4.1 widely distributes in astrocytes of the CNS in both brain and spinal cord (Takumi *et al.*, 1995; Poopalasundaram *et al.*, 2000; Higashi *et al.*, 2001; Li *et al.*, 2001; Kaiser *et al.*, 2006; Neusch *et al.*, 2006; Olsen *et al.*, 2006). Kir5.1 is also expressed in brain astrocytes (Hibino *et al.*, 2004a; Benfenati *et al.*, 2006). Both subunits are present in Müller cells of the retina (Ishii *et al.*, 1997; Ishii *et al.*, 2003; Nagelhus *et al.*, 2004).

Recent studies report differential distribution of Kir4.1 and Kir5.1 multimers in brain astrocytes and retinal Müller cells. In the brain, their distribution pattern can be classified into two categories; specifically, in the same astrocyte, both the Kir4.1 homomer and Kir4.1/5.1 heteromer simultaneously occur or only the Kir4.1 homomer is expressed. The former case is observed in neocortex and the latter is detectable in hippocampus and thalamus (Hibino *et al.*, 2004a). In cortical astrocytes, both heteromeric and homomeric channels are detected in the astrocytic processes surrounding synapses (perisynaptic processes), whereas only the heteromer was present at the end feet attached pia and processes surrounding blood vessels (perivascular processes) (Figure 4.4.1.3). Retinal Müller cells express the Kir4.1 homomer at the perivascular processes and the end feet facing vitreous humor and the Kir4.1/5.1 heteromer at the perisynaptic processes (Ishii *et al.*, 2003) (Figure 4.4.1.3).

In the spinal cord, Kir4.1 is reported to be expressed in oligodendrocytes during early postnatal development (Neusch *et al.*, 2001) but disappears when animals become mature (Neusch *et al.*, 2006).

Kir7.1

According to the results of Northern blot analyses, Kir7.1 is expressed in brain, kidney, testis, lung, thyroid gland, retina, stomach and intestine (Doring *et al.*, 1998; Nakamura *et al.*, 1999; Shimura *et al.*, 2001). *In situ* hybridization and immunohistochemistry have been used to study the cellular and subcellular localization of Kir7.1 in detail. In the brain, Kir7.1 is specifically expressed in the secretory epithelial cells of the choroid plexus (Doring *et al.*, 1998). Its expression is restricted to the apical membrane of the epithelial cells (Nakamura *et al.*, 1999). Furthermore, in the retina, the channel is expressed at the apical membrane of the pigment epithelia (Kusaka *et al.*, 2001; Shimura *et al.*, 2001). On the other hand, Kir7.1-immunoreactivity was detected at the basolateral membrane of thyroid follicular cells and of renal epithelia in the distal convoluted tubule, proximal tubule, and collecting duct (Nakamura *et al.*, 1999; Ookata *et al.*, 2000; Derst *et al.*, 2001).

4.4.1.5 Pharmacology

Classical inwardly rectifying K+ channels (Kir2.x)

Specific blockers and activators for classical Kir channels are not known. Tetraethylammonium and 4-aminopyridine, inhibitors for Kv channels, have little effect on Kir channels (Hagiwara *et al.*, 1976; Oonuma *et al.*, 2002). Ba^{2+} and Cs^+ effectively block the majority of Kir channels including Kir2.x. Although high concentration of Ba^{2+} can also block Kv channels, this cation is relatively specific for Kir channels at micromolar concentrations (Quayle *et al.*, 1993; Franchini *et al.*, 2004). These cations are often used to investigate the physiological roles of classical Kir channels in native cells and tissues.

Externally applied Ba^{2+} and Cs^+ suppress Kir currents in an E_m-dependent manner. Specifically, they inhibit Kir channels more strongly as the membrane is hyperpolarized (Hagiwara *et al.*, 1976; Hagiwara *et al.*, 1978; French and Shoukimas, 1985; Oliver *et al.*, 1998). In addition, the blocking effect of Cs^+ and Ba^{2+} decreases substantially as $[K^+]_o$ increases (Hagiwara *et al.*, 1976; Hagiwara *et al.*, 1978). However, there are a few different aspects in the behaviour of the two blockers. First, following a voltage step, the approach to steady-state blocking is much faster for Cs^+ than Ba^{2+}. Second, the dissociation constant for Cs^+ is dependent on $[K^+]_o$ (Hagiwara *et al.*, 1976), but that for Ba^{2+} is independent of this factor (Hagiwara *et al.*, 1978).

It is of note that each of the homomeric Kir2.x channels differs substantially in their sensitivity to Ba^{2+} and Cs^+ (Table 4.4.1.2). In particular, Kir2.4 was much less sensitive to these blockers than other subunits (K_i values for Ba^{2+} and Cs^+ = 390 μM and ~8 mM for Kir2.4, respectively, and ~8 μM and 420 μM for Kir2.1)

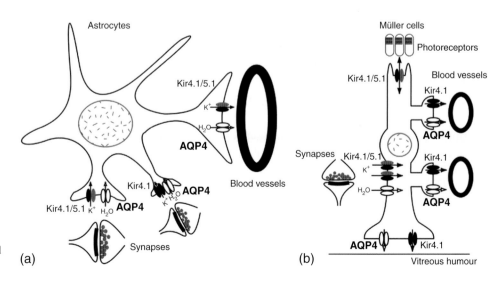

Fig. 4.4.1.3 Localization of astroglial K+ channels. (a) Brain astrocytes. The astrocytes harbour and distribute homomeric Kir4.1 and heteromeric Kir4.1/5.1 at their unique membrane elements: both the homomer and the heteromer are present at the perisynaptic processes and the heteromer alone at the perivascular processes. (b) Retinal Müller cells. These cells express the heteromeric Kir4.1/5.1 at perisynaptic sites and the homomeric Kir4.1 at perivascular processes and end feet. Note that in the astrocytes and Müller cells, the Kir channels coexist at the same membrane domains with the astroglial water channel, AQP4 (a and b).

(Topert *et al.*, 1998) (Table 2). In addition, when K^+-currents generated by co-expression of different Kir2.x subunits or expression of concatamers (tandem) were analyzed, it was found that their Ba^{2+}-sensitivity is clearly distinct from that of homomeric Kir2.x channels (Table 2) (Preisig-Muller *et al.*, 2002). These findings have been taken as proof of heteromeric assembly of Kir subunits.

K^+-transport channels (Kir1.x, Kir4.x, Kir5.x, Kir7.x)

Kir1.1/ROMK1

In addition to the non-specific blockers of Kir channels, Ba^{2+} and Cs^+, two inhibitors obtained from venoms, δ-dendrotoxin (K_d: 150 nM) and teriapin, are reported to inhibit Kir1.1 channels (Imredy *et al.*, 1998; Jin and Lu, 1998; Makino *et al.*, 2004). Both appear to bind the external vestibule of the channel pore. The blocking effect of teriapin is sensitive to pH such that alkaline pH diminishes the dissociation constant for the interaction of the inhibitor with Kir1.1 (Ramu *et al.*, 2001).

Kir4.x and Kir5.1

Ba^{2+} effectively inhibits Kir4.1- and Kir4.2-containing channels; almost completely blocking them at ~ 1 mM (Takumi *et al.*, 1995; Tanemoto *et al.*, 2000; Konstas *et al.*, 2003). Kir4.1 and Kir4.2 are also inhibited by Cs^+ at concentrations of several hundred mM (Takumi *et al.*, 1995; Konstas *et al.*, 2003). In addition to these well-known Kir inhibitors, it has recently been found that tricyclic antidepressants (TCAs) such as nortriptyline, amitriptyline, desipramine, and imipramine block Kir4.1 channels (Su *et al.*, 2006). The inhibition of Kir4.1 by nortriptyline depends on the voltage difference from E_K, with greater potency at more positive potentials (IC_{50} = ~16 μM at +30 mV, ~38 μM at –110 mV in 40 mM $[K^+]_o$). In addition, another group of antidepressants, selective serotonin reuptake inhibitors (SSRIs) such as fluoxetine, sertraline and fluvoxamine, are also found to block Kir4.1 current in a voltage-independent manner (IC_{50} of fluoxetine: 15.2 μM) (Ohno *et al.*, 2007). Interestingly, these drug barely affected the K^+ current elicited by the neuronal Kir2.1 channel. It is conceivable that the block of astroglial Kir channels by TCAs and SSRIs may be involved in the therapeutic and/or adverse actions of the drugs.

Kir7.1

Neither specific blockers nor activators are available to modulate Kir7.1 activity. However, the channel has the unique property amongst Kir channels of being unusually insensitive to block by Ba^{2+} and Cs^+ in that these cations display IC_{50} values ~10 fold lower for the Kir7.1 channel than they do for other Kir channels.

4.4.1.6 Physiology

Classical inwardly rectifying K^+ channels (Kir2.x)

Heart

The classical Kir conductance is abundantly expressed in cardiac myocytes including Purkinje fibres as well as ventricular and atrial tissues (McAllister and Noble, 1966; Rougier *et al.*, 1968; Beeler and Reuter, 1970; Kurachi, 1985) but not in nodal cells (Noma *et al.*, 1984). This current is the dominant component of the resting conductance of these tissues, and is defined as a time-independent background K^+ current, I_{K1}. This current exhibits a large inward current at membrane potentials below E_K and a relatively large outward conductance at membrane potentials just above E_K. This is essential to stabilize the resting potential of cardiac myocytes near E_K. Due to inwardly rectifying property of I_{K1}, as the membrane is depolarized the I_{K1} current decreases progressively generating the so-called negative slope of I_{K1}. Effectively, the current becomes practically zero at the plateau potential of the cardiac action potential. The lack of outward conductance at plateau-level potentials prevents excess K^+-efflux during the action potential plateau, resulting in the maintenance of depolarization. When repolarization is initiated by activation of the Kv channel at plateau potential, relatively large outward currents pass through the negative slope region of the I_{K1} channel, which accelerates the final stage of repolarization. Hence, I_{K1} is critically involved in determining the shape of the cardiac action potential, namely (i) setting the resting potential, (ii) keeping the plateau phase, and (iii) inducing rapid repolarization. The single-channel conductance of I_{K1} is ~40 pS (Kurachi, 1985).

Precise molecular constituents that form I_{K1} channel have recently been identified by the analysis of Kir2.1- and Kir2.2 knockout mice. Whereas Kir2.2 knockout mice showed ~50 per cent reduction in the I_{K1} current, no detectable current was observed in Kir2.1 knockout mice (Zaritsky *et al.*, 2001). Therefore, it has been suggested that a large population of Kir2.2 subunits assemble with Kir2.1 to form I_{K1} current. In rabbit ventricular myocytes, expression of the dominant negative form of either Kir2.1 or Kir2.2 suppressed ~70 per cent of the I_{K1} current, supporting the idea that I_{K1} is generated by a heteromeric Kir2.1/2.2 channel (Zobel *et al.*, 2003). The cardiac phenotypes observed in these knockout mice will be discussed in the section Pathophysiological relevance of classical Kir channels. A heteromeric assembly of Kir2.1/2.3 was also observed in guinea pig cardiomyocytes (Preisig-Muller *et al.*, 2002).

Blood vessels

Previous physiological experiments suggested that classical Kir channels expressed in vascular smooth muscle cells might contribute to vasodilation in response to an increase in extracellular K^+ concentration $[K^+]_o$ (Edwards *et al.*, 1988; McCarron and Halpern, 1990a, b; Nelson *et al.*, 1995; Knot *et al.*, 1996). Although elevated $[K^+]_o$ usually depolarizes smooth muscle cells and is expected to constrict blood vessels, a mild increase in $[K^+]_o$ from 6 to 15 mM is reported to hyperpolarize E_m and dilate cerebral and coronary arteries perhaps by the following mechanism (McCarron and Halpern, 1990a; Nelson *et al.*, 1995; Knot *et al.*, 1996). The resting E_m is known to be ~–45 mV, and an elevation of $[K^+]_o$ to 15 mM hyperpolarizes the E_m to ~–60 mV by increasing the Kir conductance (Nelson *et al.*, 1995; Knot *et al.*, 1996). This hyperpolarization closes voltage-gated Ca^{2+} channels and therefore reduces $[Ca^{2+}]_i$, which results in vasodilation (Knot and Nelson, 1998). Consistently, Ba^{2+} blocked both K^+-induced hyperpolarization and vasodilation of coronary and cerebral arteries (Knot *et al.*, 1996), and the cerebral artery isolated from Kir2.1 knockout mice was not dilated when it was exposed to an extracellular solution containing 15 mM K^+ (Zaritsky *et al.*, 2000). In the cerebral arteries of brain, it was reported that $[K^+]_o$ around the smooth muscle cells was elevated due to K^+-secretion from the end feet of astrocytes (Filosa *et al.*, 2006). It is of interest that this local augmentation of $[K^+]_o$ occurred when neurons were stimulated. Therefore, this system may couple neuronal activity to control of local blood flow in the brain (Filosa *et al.*, 2006).

Kir channels are considered to be the most prominent channels in vascular endothelial cells (Nilius et al., 1993; Nilius and Droogmans, 2001). Functional expression of classical Kir channels sets the resting E_m of endothelial cells to a negative potential, which provides the driving force for Ca^{2+} influx through Ca^{2+}-permeable channels. Indeed, inhibition of endothelial Kir by Ba^{2+} blocked flow-induced Ca^{2+} influx and vasodilatation (Wellman and Bevan, 1995; von Beckerath et al., 1996; Kwan et al., 2003). Since an increase in $[Ca^{2+}]_i$ activates enzymes such as NO synthase and phospholipase A2 and thus induces the secretion of vasoactive factors, control of membrane potential by classical Kir channels is one of the key regulatory systems for vascular tone. In this respect it would be of interest to examine the endothelial function of Kir2.1 knockout mice.

Brain

In the central nervous system, Kir currents are detected in neuronal cells such as hippocampal neurons (Brown et al., 1990) and neonatal rat spinal motor neurons (Takahashi, 1990). The currents, which are generated by various tetramers of Kir2.x subunits (see the section Cellular and subcellular localization of classical Kir channels), are considered to be critically involved in the maintenance of the E_{rmp} and regulation of the excitability of the neurons. In support of this idea, application of Ba^{2+}, a blocker of Kir channels, to an isolated neuron caused depolarization and initiation of action potential firing (Day et al., 2005).

Kir2.1 and Kir2.2 knockout mice have already been generated, but no obvious phenotypes related to neuronal abnormality have been observed (Zaritsky et al., 2000). Since in many cases neurons express multiple Kir channel subunits (Pruss et al., 2005), loss of one Kir2.x protein could be substituted for by another subunit(s) in the mutant mice.

As mentioned above in Cellular and subcellular localization of classical Kir channels, Kir2.3 is specifically located in the dendritic spines of granule cells (Inanobe et al., 2002). Therefore, Kir2.3 may determine the activity of various ionotropic glutamate receptors by keeping the E_{rmp} of the spines negative, which could regulate excitatory synaptic transmission.

Postsynaptic localization of Kir2.x subunits may be further controlled by mechanisms such as protein–protein interactions. The PDZ domain containing proteins (PDZ-proteins) are well-known to interact with various molecules such as glutamate receptors and Kv channels and target them to postsynaptic sites (Cohen et al., 1996; Sheng, 1996; Craven and Bredt, 1998; Nehring et al., 2000; Kim and Sheng, 2004). The interaction occurs between the PDZ domains and consensus motifs at the C-terminal end of the receptors and channels. Since Kir2.x subunits possess a class I PDZ domain recognition sequence (X-S/T-X-V/I), MAGUK family proteins are strong candidates for determining postsynaptic localization of Kir subunits. Recent biochemical analyses advocate this idea as follows: (i) Kir2.1 can interact with PSD-95 and PSD-93/chapsyn110 (Cohen et al., 1996; Nehring et al., 2000; Leyland and Dart, 2004), (ii) Kir2.2 can associate with SAP97, CASK, Velis, and Mint-1 (Leonoudakis et al., 2004a, b) and (iii) Kir2.3 can bind to PSD-95, PSD-93/chapsyn110, and SAP97 (Cohen et al., 1996; Nehring et al., 2000; Inanobe et al., 2002; Leonoudakis et al., 2004a). Notably, association of PSD-95 causes reduction of the single-channel conductance of Kir2.3 (Nehring et al., 2000). Since these PDZ-proteins simultaneously gather different types

of signalling molecules such as NO-synthase (Brenman et al., 1996a, b), SynGAP (Kim et al., 1998) and A-kinase anchoring proteins (Colledge et al., 2000), the PDZ-proteins may not only determine the postsynaptic localization of Kir2.x channels but also control their function. However, the majority of such functional regulation in vivo via protein–protein interaction has yet to be demonstrated.

Interestingly, in Schwann cells surrounding peripheral nerve fibres, a Kir current was recorded using patch-clamp recording (Wilson and Chiu, 1990). Immunohistochemical analysis revealed that Kir2.1 and Kir2.3 were specifically expressed in the microvilli of Schwann cells at the nodes of Ranvier (Mi et al., 1996). Since the villi are facing towards the axon, these Kir channels may maintain a constant extracellular K^+ concentration ($[K^+]_o$) by picking up excess K^+ elicited from an excited axon. This function is similar to the 'K$^+$-buffering' action of astrocytic K^+ channels (see K^+ transport channels [Kir1.x, Kir4.x, Kir5.x, Kir7.x]).

Skeletal muscle

A major determinant of the resting E_m in skeletal muscle is a Cl^- conductance (Hoffman, 1995). Besides this, functional expression of Kir channels participates in setting the resting E_m by shifting it in a hyperpolarized direction. The importance of Kir2.1 in muscle function was highlighted by genetic analysis of Andersen's syndrome, which is induced by a 'loss of function' mutation of Kir2.1 (Plaster et al., 2001). This disease is accompanied by periodic paralysis, which is caused by abnormal muscle relaxation resulting from cellular hyperexcitability (see the section Pathophysiological relevance of classical Kir channels). In this particular case, reduction of the Kir2.1-conductance should depolarize the resting E_m, which would inactivate Na^+ channels and thus make them unavailable for initiation and propagation of action potentials.

Furthermore, functional expression of Kir2.1 seems necessary not only for differentiation of myoblasts (Konig et al., 2004) but also for the fusion of mononucleated myoblasts to a multinucleated skeletal muscle fibre (Fischer-Lougheed et al., 2001). Both events are Ca^{2+}-dependent processes and essential for skeletal muscle development, growth, and repair. During development, the E_{rmp} of myoblasts gradually hyperpolarizes from ~-10 mV to ~-70 mV (Liu et al., 1998), due to the developmental increase in Kir2.1 expression (Fischer-Lougheed et al., 2001). This Kir2.1-induced hyperpolarization sets E_m in a range where Ca^{2+} can enter the myoblasts through Ca^{2+}-permeable channels, which promotes the differentiation and fusion of myoblasts. In addition, differentiation is at least in part mediated by expression and activity of myogenic transcription factors, such as myogenin and myocyte enhancer factor-2. Interestingly, their expression is triggered by Kir2.1-induced hyperpolarization (Konig et al., 2004). Although abnormality in morphology of skeletal muscles is not obvious in either Andersen's syndrome patients or Kir2.1-null mice, the 'slender' build of this patient group could be caused by the slight impairment of muscle development (Jongsma and Wilders, 2001).

K^+-transport channels (Kir1.x, Kir4.x, Kir5.x, Kir7.x)

Kir1.1/ROMK1

The ionic concentration and the volume in the extracellular fluids are finely controlled by the kidney. To perform this task, various types of ion transport systems and water channels are expressed and functionally coupled together at certain membrane domains, i.e. apical and basolateral membrane of renal epithelial cells (Wang et al., 1997; Hebert et al., 2005). In urine and blood, K$^+$ channels are crucial in regulating not only K$^+$ concentration but also the concentrations of other ions such as Na$^+$ and Cl$^-$ (Hebert et al., 2005). Patch-clamp analysis, immunolabelling, and knockout mouse studies revealed that Kir1.1 was the source of 35 pS and 70 pS K$^+$ channels expressed at the apical membrane of TAL cells as well as that 35 pS K$^+$ channels in CCD cells (Lorenz et al., 2002; Lu et al., 2002; Hebert et al., 2005).

Renal TAL cells

TAL cells in the kidney reabsorb ~25 per cent of filtered Na$^+$ from urine mainly by Na$^+$,K$^+$,2Cl$^-$cotransporter (NKCC) located at their apical membrane. K$^+$ channels comprising Kir1.1 supply K$^+$ to the extracellular K$^+$ site of NKCC and maintain the activity of this transporter (Wang et al., 1997; Hebert et al., 2005) (Figure 4.4.1.2). The channels also facilitate Cl$^-$ absorption, via the transporter, from the urine and thus promote uptake of NaCl. This function of Kir1.1 is called 'K$^+$-recycling'. Furthermore, Kir1.1 hyperpolarizes the membrane potential not only by secreting K$^+$ but also accelerating Cl$^-$ exit from the basolateral membrane via a Cl$^-$ channel (Hebert et al., 2005). It establishes 'the lumen positive transepithelial potential' that is the main driving force for paracellular Na$^+$, Ca^{2+}, and Mg^{2+} transport (Bleich et al., 1990; Greger et al., 1990). These functions were confirmed in knockout mice, which exhibited an impairment of renal NaCl absorption and exhibited 'Bartter's syndrome'-like phenotypes (Lorenz et al., 2002; Lu et al., 2002) (see Pathophysiological relevance of Kir1.1 channels.

Renal CCD cells

The CCD in the kidney plays a central role in secretion of K$^+$ into the urine (Giebisch and Wang, 1996; Giebisch, 1998). K$^+$ secretion occurs in principal cells of the CCD via 35 pS K$^+$ channels (Frindt and Palmer, 1987, 1989). The 35 pS channel is encoded probably by Kir1.1, because Kir1.1 knockout mice did not exhibit this K$^+$ current in principal cells (Lu et al., 2002).

Kir4.x and Kir5.1

Epithelial tissues in kidney, stomach and ear

The DCT plays an important role in reabsorption of Na$^+$ (Hebert et al., 2005). This is probably mediated by epithelial Na$^+$ channels at the apical membrane (Costanzo, 1984), whose function is further coupled with the activity of the basolateral Na$^+$/K$^+$-ATPase. K$^+$ channels at the basolateral membrane of DCT epithelial cells are key players in this process, because they supply K$^+$ to the extracellular K$^+$ site of the Na$^+$, K$^+$-ATPase to maintain its activity, in a so-called 'K$^+$-recycling' action (Figure 4.4.1.2). Electrophysiologically, a ~48–60 pS K$^+$ channel has been identified at the basolateral membrane of the rabbit distal tubule (Taniguchi et al., 1989). In mouse distal tubule cells, a Kir channel with unitary conductance of ~37–40 pS dominates and is inhibited by intracellular acidification with a pK_a of ~7.6 (Lourdel et al., 2002). Molecular biological, biochemical, and immunohistochemical studies confirmed that this channel is composed of heteromeric Kir4.1/5.1 and/or Kir4.2/5.1 (Tanemoto et al., 2000; Tucker et al., 2000; Lourdel et al., 2002). The pH$_i$-sensitivity of these heteromeric channels strongly suggests that they act as intracellular pH sensors and are crucial for pH-dependent regulation of ionic transport in the kidney.

In gastric parietal cells, the Kir4.1-homomer is colocalized with a H$^+$/K$^+$-ATPase at the apical membrane. This pump secretes protons by exchanging them for K$^+$. Thus, as in renal epithelial cells, Kir4.1 is necessary to maintain the activity of the H$^+$/K$^+$-ATPase by providing K$^+$ for the pump. Consistent with this idea, application of Ba^{2+} to the parietal cells diminished their proton-secretion (Fujita et al., 2002).

The cochlea of the inner ear is filled with two extracelluar fluids, i.e. perilymph and endolymph. The ionic composition of perilymph is almost identical to regular extracellular fluid. Surprisingly, the endolymph contains ~150 mM K$^+$ and possesses a highly positive potential of ~+80 mV in respect to blood and perilymph; a potential referred to as the 'endocochlear potential' (Von Bekesy, 1951, 1952; Hibino and Kurachi, 2006). This unique ionic and voltage environment is essential for hearing function. It has been proposed that high K$^+$ and the positive potential in endolymph are maintained by K$^+$ circulation from perilymph to endolymph through the cochlear lateral wall. The lateral wall comprises two components, i.e. stria vascularis and spiral ligament, which are critical for K$^+$ circulation. It was previously reported that application of Ba^{2+} to stria vascularis dramatically suppressed the endocochlear potential (Marcus et al., 1985). Since Kir4.1 is the only subunit expressed in stria vascularis it is believed that the Kir4.1 homomer in stria vascularis is essential for formation of the positive potential and K$^+$ circulation (Hibino et al., 1997; Ando and Takeuchi, 1999; Takeuchi et al., 2000; Nin et al., 2008). In support of this concept, Kir4.1 knockout mice are deaf and exhibit an ~50 per cent reduction of [K$^+$] in endolymph and an endocochlear potential of almost 0 mV (Marcus et al., 2002). Since perilymphatic perfusion of Ba^{2+} slightly increased the endocochlear potential, it is possible that the Kir5.1 homomer expressed in the ligament could negatively regulate the K$^+$ circulation (Marcus, 1984; Hibino et al., 2004b).

Glial cells of central nervous system and retina

Brain astrocytes and retinal Müller cells project their processes not only to synapses and soma of neurons but also to blood vessels, pia matter, and vitreous humour (Figure 4.4.1.3). These astroglial cells play various roles in the control of synaptic functions. One of the most important tasks of astrocytes is to maintain the ionic and osmotic environment in the extracellular space. They conduct a large inwardly rectifying K$^+$ current involved in transporting K$^+$ from regions of high extracellular K$^+$ ([K$^+$]$_o$), which is elicited by synaptic excitation, to those of low [K$^+$]$_o$ (Kuffler and Nicholls, 1966; Newman, 1984). This polarized transport is referred to as 'spatial buffering of K$^+$' in the brain, and 'K$^+$-siphoning' in retinal Müller cells. As the excess [K$^+$]$_o$ would interfere with normal signalling of neurons by depolarizing them continuously (Orkand et al., 1966; Newman, 1984), its rapid clearance by astrocytes is essential for the proper function of synapses. To achieve the 'K$^+$ buffering' function, astroglial cells express Kir channels at specific membrane domains.

As mentioned in the section Cellular and subcellular localization of Kir4.x and Kir5.1, in cortical astrocytes of the brain, both heteromeric Kir4.1/5.1 and homomeric Kir4.1 channels are located in the perisynaptic processes, whereas only the heteromeric channel is expressed at perivascular processes (Figure 4.4.1.3). Therefore, the cortical astrocytes take up K$^+$ by either the heteromer or the homomer and excrete K$^+$ by the heteromer. On the other hand, the astrocytes of thalamus and hippocampus, where abundant synapses dominate, express only the Kir4.1 homomer. In the case of retina, Müller cells harbour Kir4.1/5.1 at their perisynaptic processes and

Kir4.1 at their perivascular processes (Figure 4.4.1.3), suggesting that the former takes up K^+ and the latter secretes K^+. The physiological significance of the difference of these profiles among the organs, tissues, and regions is largely unknown. However, it may offer unique characteristics to each cell as follows. The major difference of the Kir4.1 homomer and Kir4.1/5.1 heteromer is their sensitivity to pH_i (see Biophysical characteristics of Kir4.x and Kir5.1 section). Astrocytes express various ion transport systems including a Na^+/HCO_3^- co-transporter. This transporter, when activated, induces intracellular alkalinization and membrane hyperpolarization, because it takes up one Na^+ and ~2–3 HCO_3^- ions (Anderson and Swanson, 2000; Blaustein et al., 2002). Excess extracellular K^+ elicited by synaptic activation accelerates the transporter by depolarizing the astrocytic membrane. The concomitant alkalinization would enhance the activity of heteromeric Kir4.1/5.1 channels within the physiological pH_i range, and could facilitate the uptake of K^+ by the astrocytes. In contrast, astrocytes in the thalamus and hippocampus, which mainly express the Kir4.1 homomer, express little Na^+/HCO_3^- co-transporter (Schmitt et al., 2000). The physiological role of the heterogeneity of Kir channels expressed at the perisynaptic processes in the astrocytes is unclear. Since the brain has complicated and sophisticated neuronal networks, the heterogeneity could contribute to the effective and proper tuning of synaptic function.

It is of interest that these Kir channels are co-localized with the water channel AQP4 at the perivascular processes of brain astrocytes and retinal Müller cells (Nagelhus et al., 1999; Ishii et al., 2003; Hibino et al., 2004a; Nagelhus et al., 2004) (Figure 4.4.1.3). K^+ transport simultaneously accompanies the flow of anions such as Cl^-, which produces an osmotic gradient across the astrocytic plasma membrane and thus facilitates water flux. The co-localization between the Kir and AQP4 channels may be critical for establishing coupling between K^+ and water fluxes. Deletion of AQP4-protein slows the rate of astrocytic K^+-uptake in mice (Padmawar et al., 2005), which reflects a linkage between K^+ and water transport. Our recent study has further identified detergent-resistant microdomains (DRMs) of astrocytic membrane play key roles in control of localization and activity of Kir4.1 and AQP4 (Hibino and Kurachi, 2007). Biochemical approaches show that both channel proteins were abundant in non-caveolar DRMs in the brain and in HEK293T cells when exogenously expressed. In the HEK cells, depletion of membrane cholesterol by methyl-β-cyclodextrin resulted in loss of Kir4.1 association with DRMs as well as its channel activity but did not affect either the DRM-distribution or the function of AQP4. Therefore, astroglial non-caveolar DRMs may be made up of at least two distinct compartments, cholesterol-dependent and cholesterol-independent microdomains, which respectively express Kir4.1 and AQP4 and regulate their function. Moreover, immunolabelling showed that Kir4.1-DRM and AQP4-DRM, which occurred in distinct compartments, were occasionally distributed in close proximity on the astrocytic membrane. Such spatial organization of channels may be involved in the concomitant transport of K^+ and water in the astrocytes.

Kir4.1 is also functionally coupled to glutamate transporters. Knockdown and knockout of Kir4.1-gene resulted in impairment of glutamate-uptake by astrocytes (Kucheryavykh et al., 2007; Djukic et al., 2007). Astrocytic glutamate transporters, when conveying the glutamate, depolarize the membrane potential. Thus, under physiological conditions, expression of Kir4.1, which hyperpolarizes the

membrane, would facilitate the glutamate influx from the extracellular space into the astrocytes by the transporters.

The mechanism that determines the specific localization of Kir channels is, at least in part, identified. Mutant mice lacking dystrophin (called 'mdx^{3Cv} mice'), which is a constituent of the dystrophin-associated protein complex (DAPC), lose Kir4.1 at perivascular processes and end feet of Müller cells (Connors and Kofuji, 2002). Therefore, DAPC may determine the localization of Kir channels at these particular membrane domains. On the other hand, in mdx^{3Cv} mice, the expression of Kir4.1 at the perisynaptic membrane surface was unaffected and still clearly detectable, suggesting that other mechanisms beyond DAPC traffic Kir channels to this area.

Kir7.1

The physiological functions of Kir7.1 are still largely unknown. However, the specific localization of Kir7.1 in various epithelial cells suggests its importance in cellular ion transport mechanisms.

The epithelial cells of the choroid plexus and the retinal pigment epithelia have a unique polarity that is opposite to usual types of epithelial cells. These two epithelial cells are known to express a Na^+/K^+-ATPase at the apical membrane (DiBona and Mills, 1979; Masuzawa et al., 1981; Caldwell and McLaughlin, 1984; Masuzawa et al., 1985; Ernst et al., 1986; Okami et al., 1990). In contrast, thyroid follicular cells and renal epithelial cells harbour the pump at their basolateral membrane (Nakamura et al., 1999; Hebert et al., 2005). Clearly, Kir7.1 is co-localized with Na^+/K^+-ATPase in these epithelial cells. This suggests that Kir7.1 maintains the activity of the Na^+/K^+-ATPase by supplying K^+ to the extracelllular K^+ site of the pump; a function which resembles the 'K^+-recycling' action of Kir1.1 and Kir4.1/5.1 channels (see Kir1.1/ROMK1, Kir4.x and Kir5.1 sections).

4.4.1.7 Pathophysiology and disease

Classical inwardly rectifying K^+ channels (Kir2.x)

Several years ago, Kir2.1 and Kir2.2 knockout mice were generated and their phenotypes analysed. As already mentioned, in the heart, whereas Kir2.2 knockout mice showed ~50 per cent reduction in the I_{K1} current, no detectable current was observed in Kir2.1 knockout mice (Zaritsky et al., 2001). These findings imply that Kir2.1 is a major constituent of the I_{K1} current and, although Kir2.2 also contributes to the formation of the current, it does so by assembling with Kir2.1 (Zaritsky et al., 2001). The majority of ventricular myocytes isolated from wild-type mice were quiescent and their resting E_m was ~-72 mV. In contrast, most of the cells from Kir2.1 knockout mice frequently showed spontaneous rhythmical action potential firing, which did not enable the authors to measure a true E_{rmp}. In addition, Kir2.1 knockout myocytes exhibited significantly broader action potentials than wild-type myocytes. Such phenotypes were not observed in Kir2.2 knockout mice. On the other hand, the electrocardiograms of Kir2.1 knockout mice showed neither ectopic beats nor re-entry arrhythmias, indicating that their cardiac abnormalities did not disrupt the sinus pacing of the heart. Surprisingly, Kir2.1 knockout mice had consistently slower heart rates, which arose indirectly from the influence of disrupting the channel outside of the heart. Similar phenotypes in heart physiology were also detected in transgenic mice that overexpressed a dominant negative form of Kir2.1 (McLerie and Lopatin, 2003).

Furthermore, blood vessels, such as cerebral arteries, of Kir2.1 knockout mice do not dilate in response to a modest increase in extracellular K^+. This seems to be due to the lack of the channel in vascular smooth muscle. On the other hand, neither Kir2.2 nor Kir2.3 transcripts were detected in vascular smooth muscle cells (Bradley et al., 1999), and blood vessels in Kir2.2 knockout mice were normally dilated in response to high $[K^+]_o$-stimulation (Zaritsky et al., 2000). These findings highlight an essential role of Kir2.1 in the regulation of vascular tone.

Andersen's syndrome, which is accompanied by cardiac arrhythmias reminiscent of long QT syndrome, periodic paralysis, and dysmorphic bone structure of face and fingers (Tawil et al., 1994), was first identified as a genetic channelopathy associated with classical Kir channels. The syndrome was an autosomal dominant disorder and found to be caused by mutations in the KCNJ2 gene, which encodes the Kir2.1 subunit (Plaster et al., 2001). In vitro experiments clearly showed that some of the mutations had dominant negative effects on the K^+-current. The mutations of Kir2.1 observed in Andersen's syndrome abolished, or decreased, an interaction of the channel with PIP_2 (Lopes et al., 2002), a molecule generally essential for activating all of the Kir channels (Huang et al., 1998) (see the section Biophysical characteristics of classical Kir channels). Therefore, the symptoms of Andersen's syndrome are caused by the following mechanisms: reduction of Kir2.1 function prolongs the plateau phase of the action potential and makes the E_{rmp} reduced and unstable, which may trigger arrhythmias. In skeletal muscle, when Kir2.1 expression decreases, the E_{rmp} would be depolarized, which could lead to inhibition of Na^+ channel activation. This depolarization-induced block could further result in complete inactivation of Na^+ channels, which would induce the paralysis observed in Andersen's syndrome. Abnormal bone structure of the disease would be caused by dysfunction of osteoclasts. Low extracellular pH in the extracellular matrix is critical for proper degradation of the bone by osteoclasts. Low pH is maintained by H^+-secretion via an ATP-driven proton pump. Since the H^+-secretion is achieved in exchange for K^+ through Kir channels, disruption of these channels would cause dysfunction of osteoclasts and the induction of severe deformity. Several different mutations in various regions of Kir2.1 have been attributed to variable levels of phenotypic expression in Andersen's syndrome.

The Kir2.1 knockout mice can partially mimic the symptoms of Andersen's syndrome. The mice exhibit the narrow maxilla and complete cleft of the secondary palate (Zaritsky et al., 2000). These phenotypes may correspond to the facial dysmorphology observed in patients. The isolated ventricular myocytes of the knockout mice showed longer action potentials and prolonged QT interval. However, no arrhythmia was observed in the mutant mice.

Interestingly, recent studies have reported that the function of classical Kir channels is modulated by the cholesterol content of the membrane. Application of cholesterol suppressed the Kir2.1 current density expressed in CHO cells and native classical Kir channels in bovine and human aortic endothelial cells (Romanenko et al., 2002, 2004). More importantly, analysis of hypercholesterolemic pigs revealed that an increase in plasma cholesterol levels strongly reduced endothelial Kir currents and depolarized the E_{rmp} (Fang et al., 2006). These observations suggest that suppression of classical Kir currents is a key factor not only in hypercholesterolemia-induced endothelial dysfunction but also in various vascular diseases.

K^+-transport channels (Kir1.x, Kir4.x, Kir5.x, Kir7.x)

Kir1.1/ROMK1

The physiological importance of Kir1.1 is highlighted by the phenotypes of the Kir1.1 knockout mouse and its 'loss of function disease', Bartter's syndrome. Bartter's syndrome (Bartter et al., 1962) is an autosomal recessive renal tubulopathy that is characterized by hypokalaemic metabolic alkalosis, renal salt wasting, hyperreninemia, and hyperaldosteronism (Guay-Woodford, 1995; Asteria, 1997; Karolyi et al., 1998; Rodriguez-Soriano, 1998; Peters et al., 2002). Genetic analyses revealed that type II of this disease was caused by various mutations in the Kir1.1 gene that impaired its function (Simon et al., 1996; Feldmann et al., 1998; Vollmer et al., 1998; Hebert et al., 2005). In TAL cells, abrogation of Kir1.1-function greatly diminishes K^+ supply to NKCC that is co-localized with the channel and induces inactivation of the transporter. In addition, the lumen positive transepithelial voltage that drives the ~50 per cent reabsorption of Na^+ paracellularly is lost (Bleich et al., 1990; Greger et al., 1990). The Bartter's mutations of the Kir1.1 gene cause alterations in PKA-phosphorylation, pH sensing, channel gating, proteolytic processing, or sorting to the apical membrane of the channel (Derst et al., 1997, 1998; Schwalbe et al., 1998; Flagg et al., 1999; Schulte et al., 1999; Jeck et al., 2001; Flagg et al., 2002; Lopes et al., 2002; Starremans et al., 2002; Peters et al., 2003). In particular, several mutations were found to be located close to R41, K80, and R311 residues, which are crucial for the pH_i sensitivity of wild-type Kir1.1. These mutations shift the pK_a to the alkaline range, which results in the loss of the channel's activity over a physiological pH_i range (Schulte et al., 1999).

Ablation of the Kir1.1-gene in mice causes a renal Na^+, Cl^-, and K^+ wasting phenotype (Lorenz et al., 2002; Lu et al., 2002), which is consistent with a crucial role of this channel in salt absorption in TAL. The mice also mimicked the phenotypes of Bartter's syndrome. Patch-clamp analyses showed that, in TAL cells, both 35 pS and 70 pS channels at the apical membrane disappeared, indicating that Kir1.1 forms the two different channels in the kidney (Lu et al., 2002, 2004). In addition, K^+ channels in CCD cells were not observed in the knockout mice (Lu et al., 2002). Like Bartter's syndrome patients, knockout mice showed hypokalemia and had an excess K^+ excretion in their urine, even though the K^+ channels in the TAL and CCD cells are expected to secrete K^+ into the urine. Its mechanism is still unclear but it may be due to diminished reabsorption of K^+ in TAL cells via NKCC (Hropot et al., 1985).

Kir4.x and Kir5.1

The phenotypes of Kir4.1 knockout mice suggest that Kir4.1-containing channels are relevant to a variety of pathophysiological conditions. First, light-evoked electroretinogram of Kir4.1 knockout mice revealed that their slow PIII response, which is generated by K^+ flux via Müller cells, was totally absent (Kofuji et al., 2000). This observation indicates that deletion of Kir4.1 causes a dysfunction of the retina by impairing the K^+ buffering action of Müller cells. Second, Kir4.1 knockout mice displayed marked motor impairment (Neusch et al., 2001). The cellular basis of this phenotype appears to be hypomyelination in the spinal cord, accompanied by severe spongiform vacuolation, axonal swellings, and degeneration (Neusch et al., 2001). Accordingly, Kir4.1, which is expressed in oligodendrocytes only in early postnatal development (see the section Biophysical characteristics of Kir4.x and Kir5.1), is involved in the development, differentiation and myelination of oligodendrocytes. It also suggests that the knockout mice could be useful as a model of various neuronal diseases. Third, as mentioned in Physiological roles of Kir4.x and Kir5.1, Kir4.1 knockout mice are deaf because of a lack of the endocochlear potential and loss of endolymphatic K$^+$. This finding highlights the importance of this channel in cochlear K$^+$-circulation and hearing function. In addition, in mutant mice that lack the gene for the anion transporter SLC26A4, Kir4.1 protein disappeared from cochlear stria vascularis (Wangemann et al., 2004). Since these mice are considered to be a good model for Pendred syndrome which displays thyroid goitre and deafness, the loss of strial Kir4.1 could possibly represent a point of origin for the hearing impairment in this disease.

Two studies reported that mutations of the Kir4.1 gene might be involved in epilepsy and seizure. The missense variation (T262S) was found to be a candidate for causing the seizure susceptibility of DBA/2 mice (Ferraro et al., 2004). In addition, linkage analysis identified a mutation of the human Kir4.1 gene, R271C, that might be associated with generalized seizures in humans (Buono et al., 2004). However, these mutations of Kir4.1, when expressed alone or together with Kir5.1 in a heterologous system, elicited significant K$^+$ currents with properties almost identical to wild-type channels (Shang et al., 2005).

Finally, genetic analysis suggests that the Kir4.2 gene is located close to the locus of the Down syndrome chromosome region-1 (DCR1); the chromosome 21 trisomy that causes dysmorphic features, hypotonia, and psychomotor delay (Gosset et al., 1997). This observation suggests a possible linkage between dysfunction of Kir4.2 and the disease.

Kir7.1

No disease that is linked to mutations in the Kir7.1 gene has been reported and no disease has yet been identified from phenotypic evaluation of the Kir7.1 knockout mouse. However, very interestingly, a recent study revealed that irregular pigment pattern is observed in so-called jaguar/ovelix Zebrafish when the Kir7.1 gene is mutated (Iwashita et al., 2006). The mutations were found to occur in either the pore-region or in the M2 helix. The electrophysiological assays demonstrated that all of the mutations inhibited Kir7.1 currents. Furthermore, Kir7.1 is expressed in melanophores and cells expressing mutated Kir7.1 could not correctly respond to the melanosome dispersion signal derived from neurons. These findings suggest that Kir7.1 is critically involved in migration of melanophores in Zebrafish.

References

Adams DJ and Hill MA (2004). Potassium channels and membrane potential in the modulation of intracellular calcium in vascular endothelial cells. J Cardiovasc Electrophysiol 15, 598–610.

Adrian RH (1969). Rectification in muscle membrane. Prog Biophys Mol Biol 19, 339–369.

Anderson CM and Swanson RA (2000). Astrocyte glutamate transport, review of properties, regulation and physiological functions. Glia 32, 1–14.

Ando M, Takeuchi S (1999). Immunological identification of an inward rectifier K$^+$ channel (Kir4.1) in the intermediate cell (melanocyte) of the cochlear stria vascularis of gerbils and rats. Cell Tissue Res 298, 179–183.

Asteria C (1997). Molecular basis of Bartter's syndrome, new insights into the correlation between genotype and phenotype. Eur J Endocrinol 137, 613–615.

Bartter FC, Pronove P, Gill JR Jr et al. (1962). Hyperplasia of the juxtaglomerular complex with hyperaldosteronism and hypokalemic alkalosis. A new syndrome. Am J Med 33, 811–828.

Beeler GW Jr and Reuter H (1970). Voltage clamp experiments on ventricular myocarial fibres. J Physiol 207, 165–190.

Benfenati V, Caprini M, Nobile M et al. (2006). Guanosine promotes the up-regulation of inward rectifier potassium current mediated by Kir4.1 in cultured rat cortical astrocytes. J Neurochem 98, 430–445.

Bichet D, Haass FA and Jan LY (2003). Merging functional studies with structures of inward-rectifier K$^+$ channels. Nat Rev Neurosci 4, 957–967.

Blaustein MP, Juhaszova M, Golovina VA et al. (2002). Na/Ca exchanger and PMCA localization in neurons and astrocytes, functional implications. Ann NY Acad Sci 976, 356–366.

Bleich M, Schlatter E and Greger R (1990). The luminal K$^+$ channel of the thick ascending limb of Henle's loop. Pflugers Arch 415, 449–460.

Boim MA, Ho K, Shuck ME et al. (1995). ROMK inwardly rectifying ATP-sensitive K$^+$ channel. II. Cloning and distribution of alternative forms. Am J Physiol 268, F1132–1140.

Bond CT, Pessia M, Xia XM et al. (1994). Cloning and expression of a family of inward rectifier potassium channels. Receptors Channels 2, 183–191.

Bradley KK, Jaggar JH, Bonev AD et al. (1999). Kir2.1 encodes the inward rectifier potassium channel in rat arterial smooth muscle cells. J Physiol 515(Pt 3), 639–651.

Brahmajothi MV, Morales MJ, Liu S et al. (1996). In situ hybridization reveals extensive diversity of K$^+$ channel mRNA in isolated ferret cardiac myocytes. Circ Res 78, 1083–1089.

Bredt DS, Wang TL, Cohen NA et al. (1995). Cloning and expression of two brain-specific inwardly rectifying potassium channels. PNAS 92, 6753–6757.

Brenman JE, Chao DS, Gee SH et al. (1996b). Interaction of nitric oxide synthase with the postsynaptic density protein PSD-95 and alpha1-syntrophin mediated by PDZ domains. Cell 84, 757–767.

Brenman JE, Christopherson KS, Craven SE et al. (1996a). Cloning and characterization of postsynaptic density 93, a nitric oxide synthase interacting protein. J Neurosci 16, 7407–7415.

Brown DA (1990). G-proteins and potassium currents in neurons. Annu Rev Physiol 52, 215–242.

Brown DA, Gahwiler BH, Griffith WH et al. (1990). Membrane currents in hippocampal neurons. Prog Brain Res 83, 141–160.

Buono RJ, Lohoff FW, Sander T et al. (2004). Association between variation in the human KCNJ10 potassium ion channel gene and seizure susceptibility. Epilepsy Res 58, 175–183.

Caldwell RB and McLaughlin BJ (1984). Redistribution of Na-K-ATPase in the dystrophic rat retinal pigment epithelium. J Neurocytol 13, 895–910.

Casamassima M, D'Adamo MC, Pessia M et al. (2003). Identification of a heteromeric interaction that influences the rectification, gating and pH sensitivity of Kir4.1/Kir5.1 potassium channels. J Biol Chem 278, 43533–43540.

Chepilko S, Zhou H, Sackin H et al. (1995). Permeation and gating properties of a cloned renal K$^+$ channel. Am J Physiol 268, C389–401.

Choe H, Sackin H and Palmer LG (1998). Permeation and gating of an inwardly rectifying potassium channel. Evidence for a variable energy well. J Gen Physiol 112, 433–446.

Choe H, Zhou H, Palmer LG et al. (1997). A conserved cytoplasmic region of ROMK modulates pH sensitivity, conductance and gating. Am J Physiol 273, F516–529.

Cohen NA, Brenman JE, Snyder SH et al. (1996). Binding of the inward rectifier K$^+$ channel Kir 2.3 to PSD-95 is regulated by protein kinase A phosphorylation. Neuron 17, 759–767.

Colledge M, Dean RA, Scott GK et al. (2000). Targeting of PKA to glutamate receptors through a MAGUK-AKAP complex. Neuron 27, 107–119.

Connors NC and Kofuji P (2002). Dystrophin Dp71 is critical for the clustered localization of potassium channels in retinal glial cells. *J Neurosci* **22**, 4321–4327.

Costanzo LS (1984). Comparison of calcium and sodium transport in early and late rat distal tubules, effect of amiloride. *Am J Physiol* **246**, F937–945.

Coulter KL, Perier F, Radeke CM *et al.* (1995). Identification and molecular localization of a pH-sensing domain for the inward rectifier potassium channel HIR. *Neuron* **15**, 1157–1168.

Craven SE and Bredt DS (1998). PDZ proteins organize synaptic signaling pathways. *Cell* **93**, 495–498.

Day M, Carr DB, Ulrich S *et al.* (2005). Dendritic excitability of mouse frontal cortex pyramidal neurons is shaped by the interaction among HCN, Kir2 and Kleak channels. *J Neurosci* **25**, 8776–8787.

Derst C, Hirsch JR, Preisig-Muller R *et al.* (2001). Cellular localization of the potassium channel Kir7.1 in guinea pig and human kidney. *Kidney Int* **59**, 2197–2205.

Derst C, Konrad M, Kockerling A *et al.* (1997). Mutations in the ROMK gene in antenatal Bartter syndrome are associated with impaired K+ channel function. *Biochem Biophys Res Commun* **230**, 641–645.

Derst C, Wischmeyer E, Preisig-Muller R *et al.* (1998). A hyperprostaglandin E syndrome mutation in Kir1.1 (renal outer medullary potassium). channels reveals a crucial residue for channel function in Kir1.3 channels. *J Biol Chem* **273**, 23884–23891.

DiBona DR and Mills JW (1979). Distribution of Na+-pump sites in transporting epithelia. *Fed Proc* **38**, 134–143.

Djukic B, Casper KB, Philpot BD *et al.* (2007). Conditional knock-out of Kir4.1 leads to glial membrane depolarization, inhibition of potassium and glutamate uptake, and enhanced short-term synaptic potentiation. *J Neurosci* **17**, 11354–11365.

Doi T, Fakler B, Schultz JH *et al.* (1996). Extracellular K+ and intracellular pH allosterically regulate renal Kir1.1 channels. *J Biol Chem* **271**, 17261–17266.

Doring F, Derst C, Wischmeyer E *et al.* (1998). The epithelial inward rectifier channel Kir7.1 displays unusual K+ permeation properties. *J Neurosci* **18**, 8625–8636.

Doupnik CA, Davidson N and Lester HA (1995). The inward rectifier potassium channel family. *Curr Opin Neurobiol* **5**, 268–277.

Edwards FR, Hirst GD and Silverberg GD (1988). Inward rectification in rat cerebral arterioles; involvement of potassium ions in autoregulation. *J Physiol* **404**, 455–466.

Ernst SA, Palacios JR 2nd and Siegel GJ (1986). Immunocytochemical localization of Na+, K+-ATPase catalytic polypeptide in mouse choroid plexus. *J Histochem Cytochem* **34**, 189–195.

Fakler B, Schultz JH, Yang J *et al.* (1996). Identification of a titratable lysine residue that determines sensitivity of kidney potassium channels (ROMK) to intracellular pH. *EMBO J* **15**, 4093–4099.

Fang Y, Mohler ER 3rd, Hsieh E *et al.* (2006). Hypercholesterolemia suppresses inwardly rectifying K+ channels in aortic endothelium *in vitro* and *in vivo*. *Circ Res* **98**, 1064–1071.

Fang Y, Schram G, Romanenko VG *et al.* (2005). Functional expression of Kir2.x in human aortic endothelial cells, the dominant role of Kir2.2. *Am J Physiol Cell Physiol* **289**, C1134–1144.

Feldmann D, Alessandri JL and Deschenes G (1998). Large deletion of the 5′ end of the ROMK1 gene causes antenatal Bartter syndrome. *J Am Soc Nephrol* **9**, 2357–2359.

Ferraro TN, Golden GT, Smith GG *et al.* (2004). Fine mapping of a seizure susceptibility locus on mouse Chromosome 1, nomination of Kcnj10 as a causative gene. *Mamm Genome* **15**, 239–251.

Filosa JA, Bonev AD, Straub SV *et al.* (2006). Local potassium signaling couples neuronal activity to vasodilation in the brain. *Nat Neurosci* **9**(11), 1397–1403.

Fischer-Lougheed J, Liu JH, Espinos E *et al.* (2001). Human myoblast fusion requires expression of functional inward rectifier Kir2.1 channels. *J Cell Biol* **153**, 677–686.

Flagg TP, Tate M, Merot J *et al.* (1999). A mutation linked with Bartter's syndrome locks Kir 1.1a (ROMK1) channels in a closed state. *J Gen Physiol* **114**, 685–700.

Flagg TP, Yoo D, Sciortino CM *et al.* (2002). Molecular mechanism of a COOH-terminal gating determinant in the ROMK channel revealed by a Bartter's disease mutation. *J Physiol* **544**, 351–362.

Franchini L, Levi G and Visentin S (2004). Inwardly rectifying K+ channels influence Ca2+ entry due to nucleotide receptor activation in microglia. *Cell Calcium* **35**, 449–459.

French RJ and Shoukimas JJ (1985). An ion's view of the potassium channel. The structure of the permeation pathway as sensed by a variety of blocking ions. *J Gen Physiol* **85**, 669–698.

Frindt G and Palmer LG (1987). Ca-activated K channels in apical membrane of mammalian CCT and their role in K secretion. *Am J Physiol* **252**, F458–467.

Frindt G and Palmer LG (1989). Low-conductance K channels in apical membrane of rat cortical collecting tubule. *Am J Physiol* **256**, F143–151.

Fujita A, Horio Y, Higashi K *et al.* (2002). Specific localization of an inwardly rectifying K+ channel, Kir4.1, at the apical membrane of rat gastric parietal cells; its possible involvement in K+ recycling for the H+-K+-pump. *J Physiol* **540**, 85–92.

Fujiwara Y and Kubo Y (2002). Ser165 in the second transmembrane region of the Kir2.1 channel determines its susceptibility to blockade by intracellular Mg2+. *J Gen Physiol* **120**, 677–693.

Gahwiler BH and Brown DA (1985). GABA_B-receptor-activated K+ current in voltage-clamped CA3 pyramidal cells in hippocampal cultures. *PNAS* **82**, 1558–1562.

Giebisch G (1998). Renal potassium transport, mechanisms and regulation. *Am J Physiol* **274**, F817–833.

Giebisch G and Wang W (1996). Potassium transport, from clearance to channels and pumps. *Kidney Int* **49**, 1624–1631.

Glowatzki E, Fakler G, Brandle U *et al.* (1995). Subunit-dependent assembly of inward-rectifier K+ channels. *Proc Biol Sci* **261**, 251–261.

Gosset P, Ghezala GA, Korn B *et al.* (1997). A new inward rectifier potassium channel gene (KCNJ15). localized on chromosome 21 in the Down syndrome chromosome region 1 (DCR1). *Genomics* **44**, 237–241.

Greger R, Bleich M and Schlatter E (1990). Ion channels in the thick ascending limb of Henle's loop. *Ren Physiol Biochem* **13**, 37–50.

Guay-Woodford LM (1995). Molecular insights into the pathogenesis of inherited renal tubular disorders. *Curr Opin Nephrol Hypertens* **4**, 121–129.

Hagiwara S (1983). Anomalous rectification in membrane potential-dependent ion channels in cell membrane. *Phylogenet Dev Approaches*, 65–74.

Hagiwara S and Jaffe LA (1979). Electrical properties of egg cell membranes. *Annu Rev Biophys Bioeng* **8**, 385–416.

Hagiwara S and Takahashi K (1974). The anomalous rectification and cation selectivity of the membrane of a starfish egg cell. *J Membr Biol* **18**, 61–80.

Hagiwara S, Miyazaki S and Rosenthal NP (1976). Potassium current and the effect of cesium on this current during anomalous rectification of the egg cell membrane of a starfish. *J Gen Physiol* **67**, 621–638.

Hagiwara S, Miyazaki S, Moody W *et al.* (1978). Blocking effects of barium and hydrogen ions on the potassium current during anomalous rectification in the starfish egg. *J Physiol* **279**, 167–185.

Hebert SC, Desir G, Giebisch G *et al.* (2005). Molecular diversity and regulation of renal potassium channels. *Physiol Rev* **85**, 319–371.

Heginbotham L, Lu Z, Abramson T *et al.* (1994). Mutations in the K+ channel signature sequence. *Biophys J* **66**, 1061–1067.

Hibino H and Kurachi Y (2006). Molecular and physiological bases of the K+ circulation in the mammalian inner ear. *Physiology (Bethesda)* **21**, 336–345.

Hibino H, Fujita A, Iwai K *et al.* (2004a). Differential assembly of inwardly rectifying K+ channel subunits, Kir4.1 and Kir5.1, in brain astrocytes. *J Biol Chem* **279**, 44065–44073.

Hibino H, Higashi-Shingai K, Fujita A *et al.* (2004b). Expression of an inwardly rectifying K⁺ channel, Kir5.1, in specific types of fibrocytes in the cochlear lateral wall suggests its functional importance in the establishment of endocochlear potential. *Eur J Neurosci* **19**, 76–84.

Hibino H, Horio Y, Fuijta A *et al.* (1999). Expression of an inwardly rectifying K⁺ channel, Kir4.1, in the satellite cells of rat cochlear ganglia. *Am J Physiol; Cell Physiol* **277**, C638–C644.

Hibino H, Horio Y, Inanobe A *et al.* (1997). An ATP-dependent inwardly rectifying potassium channel, K$_{AB}$-2 (Kir4. 1), in cochlear stria vascularis of inner ear, its specific subcellular localization and correlation with the formation of endocochlear potential. *J Neurosci* **17**, 4711–4721.

Hibino H and Kurachi Y (2007). Distinct detergent-resistant membrane microdomains (lipid rafts) respectively harvest K⁺ and water transport systems in brain astroglia. *Eur J Neurosci* **26**, 2539–2555.

Higashi K, Fujita A, Inanobe A *et al.* (2001). An inwardly rectifying K⁺ channel, Kir4.1, expressed in astrocytes surrounds synapses and blood vessels in brain. *Am J Physiol Cell Physiol* **281**, C922–931.

Hilgemann DW and Ball R (1996). Regulation of cardiac Na⁺, Ca²⁺ exchange and KATP potassium channels by PIP$_2$. *Science* **273**, 956–959.

Hilgemann DW, Feng S and Nasuhoglu C (2001). The complex and intriguing lives of PIP$_2$ with ion channels and transporters. *Sci STKE* **111**, RE19.

Hille B and Schwarz W (1978). Potassium channels as multi-ion single-file pores. *J Gen Physiol* **72**, 409–442.

Ho K, Nichols CG, Lederer WJ *et al.* (1993). Cloning and expression of an inwardly rectifying ATP-regulated potassium channel. *Nature* **362**, 31–38.

Hodgkin A, Huxley A and Katz B (1952). Measurements of current–voltage relations in the membrane of the giant axon of Laligo. *J Physiol* **116**, 424–448.

Hoffman EP (1995). Voltage-gated ion channelopathies, inherited disorders caused by abnormal sodium, chloride and calcium regulation in skeletal muscle. *Annu Rev Med* **46**, 431–441.

Horio Y, Morishige K, Takahashi N *et al.* (1996). Differential distribution of classical inwardly rectifying potassium channel mRNAs in the brain, comparison of IRK2 with IRK1 and IRK3. *FEBS Lett* **379**, 239–243.

Hropot M, Fowler N, Karlmark B *et al.* (1985). Tubular action of diuretics, distal effects on electrolyte transport and acidification. *Kidney Int* **28**, 477–489.

Huang CL, Feng S and Hilgemann DW (1998). Direct activation of inward rectifier potassium channels by PIP$_2$ and its stabilization by Gbetagamma. *Nature* **391**, 803–806.

Hughes BA, Kumar G, Yuan Y *et al.* (2000). Cloning and functional expression of human retinal kir2.4, a pH-sensitive inwardly rectifying K⁺ channel. *Am J Physiol Cell Physiol* **279**, C771–784.

Imredy JP, Chen C and MacKinnon R (1998). A snake toxin inhibitor of inward rectifier potassium channel ROMK1. *Biochemistry* **37**, 14867–14874.

Inagaki N, Gonoi T, Clement JP *et al.* (1996). A family of sulfonylurea receptors determines the pharmacological properties of ATP-sensitive K⁺ channels. *Neuron* **16**, 1011–1017.

Inagaki N, Gonoi T, Clement JP *et al.* (1995). Reconstitution of I$_{KATP}$, an inward rectifier subunit plus the sulfonylurea receptor. *Science* **270**, 1166–1170.

Inanobe A, Fujita A, Ito M *et al.* (2002). Inward rectifier K⁺ channel Kir2.3 is localized at the postsynaptic membrane of excitatory synapses. *Am J Physiol Cell Physiol* **282**, C1396–1403.

Ishii M, Fujita A, Iwai K *et al.* (2003). Differential expression and distribution of Kir5.1 and Kir4.1 inwardly rectifying K⁺ channels in retina. *Am J Physiol Cell Physiol* **285**, C260–267.

Ishii M, Horio Y, Tada Y *et al.* (1997). Expression and clustered distribution of an inwardly rectifying potassium channel, K$_{AB}$-2/Kir4.1, on mammalian retinal Muller cell membrane, their regulation by insulin and laminin signals. *J Neurosci* **17**, 7725–7735.

Isomoto S, Kondo C, Yamada M *et al.* (1996). A novel sulfonylurea receptor forms with BIR (Kir6.2) a smooth muscle type ATP-sensitive K⁺ channel. *J Biol Chem* **271**, 24321–24324.

Ito M, Inanobe A, Horio Y *et al.* (1996). Immunolocalization of an inwardly rectifying K⁺ channel, K$_{AB}$-2 (Kir4.1), in the basolateral membrane of renal distal tubular epithelia. *FEBS Lett* **388**, 11–15.

Iwashita M, Watanabe M, Ishii M *et al.* (2006). Pigment pattern in jaguar/obelix zebrafish is caused by a Kir7.1 mutation, implications for the regulation of melanosome movement. *PLoS Genet* **2**, e197.

Jeck N, Derst C, Wischmeyer E *et al.* (2001). Functional heterogeneity of ROMK mutations linked to hyperprostaglandin E syndrome. *Kidney Int* **59**, 1803–1811.

Jin W and Lu Z (1998). A novel high-affinity inhibitor for inward-rectifier K⁺ channels. *Biochemistry* **37**, 13291–13299.

Jongsma HJ and Wilders R (2001). Channelopathies, Kir2.1 mutations jeopardize many cell functions. *Curr Biol* **11**, R747–750.

Kahle KT, Wilson FH, Leng Q *et al.* (2003). WNK4 regulates the balance between renal NaCl reabsorption and K⁺ secretion. *Nat Genet* **35**, 372–376.

Kaiser M, Maletzki I, Hulsmann S *et al.* (2006). Progressive loss of a glial potassium channel (KCNJ10) in the spinal cord of the SOD1 (G93A). transgenic mouse model of amyotrophic lateral sclerosis. *J Neurochem* **99**, 900–912.

Karolyi L, Koch MC, Grzeschik KH *et al.* (1998). The molecular genetic approach to Bartter's syndrome. *J Mol Med* **76**, 317–325.

Karschin C, Dissmann E, Stuhmer W *et al.* (1996). IRK(1–3). and GIRK(1–4) inwardly rectifying K⁺ channel mRNAs are differentially expressed in the adult rat brain. *J Neurosci* **16**, 3559–3570.

Katz B (1949). Les constantes electriques de la membrane du muscle. *Arch Sci Physiol* **3**, 285–299.

Kenna S, Roper J, Ho K *et al.* (1994). Differential expression of the inwardly-rectifying K-channel ROMK1 in rat brain. *Brain Res Mol Brain Res* **24**, 353–356.

Kim E and Sheng M (2004). PDZ domain proteins of synapses. *Nat Rev Neurosci* **5**, 771–781.

Kim JH, Liao D, Lau LF *et al.* (1998). SynGAP, a synaptic RasGAP that associates with the PSD-95/SAP90 protein family. *Neuron* **20**, 683–691.

Knot HJ and Nelson MT (1998). Regulation of arterial diameter and wall [Ca²⁺] in cerebral arteries of rat by membrane potential and intravascular pressure. *J Physiol* **508** (**Pt 1**), 199–209.

Knot HJ, Zimmermann PA and Nelson MT (1996). Extracellular K⁺-induced hyperpolarizations and dilatations of rat coronary and cerebral arteries involve inward rectifier K⁺ channels. *J Physiol* **492** (**Pt 2**), 419–430.

Kofuji P, Ceelen P, Zahs KR *et al.* (2000). Genetic inactivation of an inwardly rectifying potassium channel (Kir4.1 subunit). in mice, phenotypic impact in retina. *J Neurosci* **20**, 5733–5740.

Kohda Y, Ding W, Phan E *et al.* (1998). Localization of the ROMK potassium channel to the apical membrane of distal nephron in rat kidney. *Kidney Int* **54**, 1214–1223.

Kondo C, Isomoto S, Matsumoto S *et al.* (1996). Cloning and functional expression of a novel isoform of ROMK inwardly rectifying ATP-dependent K⁺ channel, ROMK6 (Kir1.1f). *FEBS Lett* **399**, 122–126.

Konig S, Hinard V, Arnaudeau S *et al.* (2004). Membrane hyperpolarization triggers myogenin and myocyte enhancer factor-2 expression during human myoblast differentiation. *J Biol Chem* **279**, 28187–28196.

Konstas AA, Korbmacher C and Tucker SJ (2003). Identification of domains that control the heteromeric assembly of Kir5.1/Kir4.0 potassium channels. *Am J Physiol Cell Physiol* **284**, C910–917.

Koyama H, Morishige K, Takahashi N *et al.* (1994). Molecular cloning, functional expression and localization of a novel inward rectifier potassium channel in the rat brain. *FEBS Lett* **341**, 303–307.

Krapivinsky G, Medina I, Eng L *et al.* (1998). A novel inward rectifier K⁺ channel with unique pore properties. *Neuron* **20**, 995–1005.

Kubo Y and Murata Y (2001). Control of rectification and permeation by two distinct sites after the second transmembrane region in Kir2.1 K$^+$ channel. *J Physiol* **531**, 645–660.

Kubo Y, Adelman JP, Clapham DE *et al.* (2005). International Union of Pharmacology. LIV. Nomenclature and molecular relationships of inwardly rectifying potassium channels. *Pharmacol Rev* **57**, 509–526.

Kubo Y, Baldwin TJ, Jan YN *et al.* (1993). Primary structure and functional expression of a mouse inward rectifier potassium channel. *Nature* **362**, 127–133.

Kubo Y, Miyashita T and Kubokawa K (1996). A weakly inward rectifying potassium channel of the salmon brain. Glutamate 179 in the second transmembrane domain is insufficient for strong rectification. *J Biol Chem* **271**, 15729–15735.

Kucheryavykh YV, Kucheryavykh LY, Nichols CG *et al.* (2007). Downregulation of Kir4.1 inward rectifying potassium channel subunits by RNAi impairs potassium transfer and glutamate uptake by cultured cortical astrocytes. *Glia* **55**, 274–281.

Kuffler SW and Nicholls JG (1966). The physiology of neuroglial cells. *Ergeb Physiol* **57**, 1–90.

Kuo A, Gulbis JM, Antcliff JF *et al.* (2003). Crystal structure of the potassium channel KirBac1.1 in the closed state. *Science* **300**, 1922–1926.

Kurachi Y (1985). Voltage-dependent activation of the inward-rectifier potassium channel in the ventricular cell membrane of guinea-pig heart. *J Physiol* **366**, 365–385.

Kurachi Y (1995). G protein regulation of cardiac muscarinic potassium channel. *Am J Physiol* **269**, C821–830.

Kurachi Y, Nakajima T and Sugimoto T (1986a). Acetylcholine activation of K$^+$ channels in cell-free membrane of atrial cells. *Am J Physiol* **251**, H681–684.

Kurachi Y, Nakajima T and Sugimoto T (1986b). On the mechanism of activation of muscarinic K$^+$ channels by adenosine in isolated atrial cells, involvement of GTP-binding proteins. *Pflugers Arch* **407**, 264–274.

Kurachi Y, Nakajima T and Sugimoto T (1986c). Role of intracellular Mg^{2+} in the activation of muscarinic K$^+$ channel in cardiac atrial cell membrane. *Pflugers Arch* **407**, 572–574.

Kusaka S, Inanobe A, Fujita A *et al.* (2001). Functional Kir7.1 channels localized at the root of apical processes in rat retinal pigment epithelium. *J Physiol* **531**, 27–36.

Kwan HY, Leung PC, Huang Y *et al.* (2003). Depletion of intracellular Ca^{2+} stores sensitizes the flow-induced Ca^{2+} influx in rat endothelial cells. *Circ Res* **92**, 286–292.

Lacey MG, Mercuri NB and North RA (1988). On the potassium conductance increase activated by GABA$_B$ and dopamine D$_2$ receptors in rat substantia nigra neurones. *J Physiol* **401**, 437–453.

Le Maout S, Brejon M, Olsen O *et al.* (1997). Basolateral membrane targeting of a renal-epithelial inwardly rectifying potassium channel from the cortical collecting duct, CCD-IRK3, in MDCK cells. *PNAS* **94**, 13329–13334.

Lee WS and Hebert SC (1995). ROMK inwardly rectifying ATP-sensitive K$^+$ channel. I. Expression in rat distal nephron segments. *Am J Physiol* **268**, F1124–1131.

Leichtle A, Rauch U, Albinus M *et al.* (2004). Electrophysiological and molecular characterization of the inward rectifier in juxtaglomerular cells from rat kidney. *J Physiol* **560**, 365–376.

Leng Q, MacGregor GG, Dong K *et al.* (2006). Subunit–subunit interactions are critical for proton sensitivity of ROMK, evidence in support of an intermolecular gating mechanism. *PNAS* **103**, (1982-(1987.

Leonoudakis D, Conti LR, Anderson S *et al.* (2004b). Protein trafficking and anchoring complexes revealed by proteomic analysis of inward rectifier potassium channel (Kir2.x)-associated proteins. *J Biol Chem* **279**, 22331–22346.

Leonoudakis D, Conti LR, Radeke CM *et al.* (2004a). A multiprotein trafficking complex composed of SAP97, CASK, Veli and Mint1 is associated with inward rectifier Kir2 potassium channels. *J Biol Chem* **279**, 19051–19063.

Lewis DL, Ikeda SR, Aryee D *et al.* (1991). Expression of an inwardly rectifying K$^+$ channel from rat basophilic leukemia cell mRNA in Xenopus oocytes. *FEBS Lett* **290**, 17–21.

Leyland ML and Dart C (2004). An alternatively spliced isoform of PSD-93/chapsyn 110 binds to the inwardly rectifying potassium channel, Kir2.1. *J Biol Chem* **279**, 43427–43436.

Li L, Head V and Timpe LC (2001). Identification of an inward rectifier potassium channel gene expressed in mouse cortical astrocytes. *Glia* **33**, 57–71.

Lin D, Sterling H, Lerea KM *et al.* (2002). Protein kinase C (PKC)-induced phosphorylation of ROMK1 is essential for the surface expression of ROMK1 channels. *J Biol Chem* **277**, 44278–44284.

Lin DH, Sterling H, Wang Z *et al.* (2005). ROMK1 channel activity is regulated by monoubiquitination. *PNAS* **102**, 4306–4311.

Liou HH, Zhou SS and Huang CL (1999). Regulation of ROMK1 channel by protein kinase A via a phosphatidylinositol 4,5-bisphosphate-dependent mechanism. *PNAS* **96**, 5820–5825.

Liu GX, Derst C, Schlichthorl G *et al.* (2001). Comparison of cloned Kir2 channels with native inward rectifier K$^+$ channels from guinea-pig cardiomyocytes. *J Physiol* **532**, 115–126.

Liu JH, Bijlenga P, Fischer-Lougheed J *et al.* (1998). Role of an inward rectifier K$^+$ current and of hyperpolarization in human myoblast fusion. *J Physiol* **510** (**Pt 2**), 467–476.

Lopatin AN, Makhina EN and Nichols CG (1994). Potassium channel block by cytoplasmic polyamines as the mechanism of intrinsic rectification. *Nature* **372**, 366–369.

Lopes CM, Zhang H, Rohacs T *et al.* (2002). Alterations in conserved Kir channel-PIP$_2$ interactions underlie channelopathies. *Neuron* **34**, 933–944.

Lorenz JN, Baird NR, Judd LM *et al.* (2002). Impaired renal NaCl absorption in mice lacking the ROMK potassium channel, a model for type II Bartter's syndrome. *J Biol Chem* **277**, 37871–37880.

Lourdel S, Paulais M, Cluzeaud F *et al.* (2002). An inward rectifier K$^+$ channel at the basolateral membrane of the mouse distal convoluted tubule, similarities with Kir4–Kir5.1 heteromeric channels. *J Physiol* **538**, 391–404.

Lu M, Wang T, Yan Q *et al.* (2004). ROMK is required for expression of the 70-pS K channel in the thick ascending limb. *Am J Physiol Renal Physiol* **286**, F490–495.

Lu M, Wang T, Yan Q *et al.* (2002). Absence of small conductance K$^+$ channel (SK). activity in apical membranes of thick ascending limb and cortical collecting duct in ROMK (Bartter's). knockout mice. *J Biol Chem* **277**, 37881–37887.

Lu T, Nguyen B, Zhang X *et al.* (1999). Architecture of a K$^+$ channel inner pore revealed by stoichiometric covalent modification. *Neuron* **22**, 571–580.

Lu Z and MacKinnon R (1994). Electrostatic tuning of Mg^{2+} affinity in an inward-rectifier K$^+$ channel. *Nature* **371**, 243–246.

Ma D, Zerangue N, Lin YF *et al.* (2001). Role of ER export signals in controlling surface potassium channel numbers. *Science* **291**, 316–319.

Makino Y, Matsushita K, Matsuzawa Y *et al.* (2004). Effect of tertiapin on different types of recombinant inwardly rectifying potassium channels. *Med J Osaka Univ* **47**, 1–11.

Marcus DC (1984). Characterization of potassium permeability of cochlear duct by perilymphatic perfusion of barium. *Am J Physiol* **247**, C240–246.

Marcus DC, Rokugo M and Thalmann R (1985). Effects of barium and ion substitutions in artificial blood on endocochlear potential. *Hear Res* **17**, 79–86.

Marcus DC, Wu T, Wangemann P *et al.* (2002). KCNJ10 (Kir4.1) potassium channel knockout abolishes endocochlear potential. *Am J Physiol Cell Physiol* **282**, C403–407.

Masuzawa T, Ohta T, Kawakami K *et al.* (1985). Immunocytochemical localization of Na$^+$, K$^+$-ATPase in the canine choroid plexus. *Brain* **108**(**Pt 3**), 625–646.

Masuzawa T, Saito T and Sato F (1981). Cytochemical study on enzyme activity associated with cerebrospinal fluid secretion in the choroid plexus and ventricular ependyma. *Brain Res* **222**, 309–322.

Matsuda H, Saigusa A and Irisawa H (1987). Ohmic conductance through the inwardly rectifying K channel and blocking by internal Mg^{2+}. *Nature* **325**, 156–159.

McAllister RE and Noble D (1966). The time and voltage dependence of the slow outward current in cardiac Purkinje fibres. *J Physiol* **186**, 632–662.

McCarron JG and Halpern W (1990a). Potassium dilates rat cerebral arteries by two independent mechanisms. *Am J Physiol* **259**, H902–908.

McCarron JG and Halpern W (1990b). Impaired potassium-induced dilation in hypertensive rat cerebral arteries does not reflect altered Na^+,K^+-ATPase dilation. *Circ Res* **67**, 1035–1039.

McKinney LC and Gallin EK (1988). Inwardly rectifying whole-cell and single-channel K currents in the murine macrophage cell line J774.1. *J Membr Biol* **103**, 41–53.

McLerie M and Lopatin AN (2003). Dominant-negative suppression of I_{K1} in the mouse heart leads to altered cardiac excitability. *J Mol Cell Cardiol* **35**, 367–378.

McNicholas CM, MacGregor GG, Islas LD *et al.* (1998). pH-dependent modulation of the cloned renal K^+ channel, ROMK. *Am J Physiol* **275**, F972–981.

McNicholas CM, Wang W, Ho K *et al.* (1994). Regulation of ROMK1 K^+ channel activity involves phosphorylation processes. *PNAS* **91**, 8077–8081.

Mi H, Deerinck TJ, Jones M *et al.* (1996). Inwardly rectifying K^+ channels that may participate in K^+ buffering are localized in microvilli of Schwann cells. *J Neurosci* **16**, 2421–2429.

Minor DL Jr, Masseling SJ, Jan YN *et al.* (1999). Transmembrane structure of an inwardly rectifying potassium channel. *Cell* **96**, 879–891.

Miyazaki SI, Takahashi K, Tsuda K *et al.* (1974). Analysis of non-linearity observed in the current-voltage relation of the tunicate embryo. *J Physiol* **238**, 55–77.

Morishige K, Takahashi N, Jahangir A *et al.* (1994). Molecular cloning and functional expression of a novel brain-specific inward rectifier potassium channel. *FEBS Lett* **346**, 251–256.

Nagelhus EA, Horio Y, Inanobe A *et al.* (1999). Immunogold evidence suggests that coupling of K^+ siphoning and water transport in rat retinal Muller cells is mediated by a coenrichment of Kir4.1 and AQP4 in specific membrane domains. *Glia* **26**, 47–54.

Nagelhus EA, Mathiisen TM and Ottersen OP (2004). Aquaporin-4 in the central nervous system, cellular and subcellular distribution and coexpression with KIR4.1. *Neuroscience* **129**, 905–913.

Nakamura N, Suzuki Y, Sakuta H *et al.* (1999). Inwardly rectifying K^+ channel Kir7.1 is highly expressed in thyroid follicular cells, intestinal epithelial cells and choroid plexus epithelial cells, implication for a functional coupling with Na^+,K^+-ATPase. *Biochem J* **342**(**Pt 2**), 329–336.

Nehring RB, Wischmeyer E, Doring F *et al.* (2000). Neuronal inwardly rectifying K^+ channels differentially couple to PDZ proteins of the PSD-95/SAP90 family. *J Neurosci* **20**, 156–162.

Nelson MT, Cheng H, Rubart M *et al.* (1995). Relaxation of arterial smooth muscle by calcium sparks. *Science* **270**, 633–637.

Neusch C, Papadopoulos N, Muller M *et al.* (2006). Lack of the Kir4.1 channel subunit abolishes K^+ buffering properties of astrocytes in the ventral respiratory group, impact on extracellular K^+ regulation. *J Neurophysiol* **95**, 1843–1852.

Neusch C, Rozengurt N, Jacobs RE *et al.* (2001). Kir4.1 potassium channel subunit is crucial for oligodendrocyte development and in vivo myelination. *J Neurosci* **21**, 5429–5438.

Newman EA (1984). Regional specialization of retinal glial cell membrane. *Nature* **309**, 155–157.

Nilius B and Droogmans G (2001). Ion channels and their functional role in vascular endothelium. *Physiol Rev* **81**, 1415–1459.

Nilius B, Schwarz G and Droogmans G (1993). Modulation by histamine of an inwardly rectifying potassium channel in human endothelial cells. *J Physiol* **472**, 359–371.

Nin F, Hibino H, Doi K *et al.* (2008). The endocochlear potential depends on two K^+ diffusion potentials and an electrical barrier in the stria vacularis of the inner ear. *PNAS* **105**, 1751–1756.

Nishida M and MacKinnon R (2002). Structural basis of inward rectification, cytoplasmic pore of the G protein-gated inward rectifier GIRK1 at 1.8 A resolution. *Cell* **111**, 957–965.

Noble D (1965). Electrical properties of cardiac muscle attributable to inward going (anomalous) rectification. *J Cell Comp Physiol* **66**, 127–136.

Noma A (1983). ATP-regulated K^+ channels in cardiac muscle. *Nature* **305**, 147–148.

Noma A, Nakayama T, Kurachi Y *et al.* (1984). Resting K conductances in pacemaker and non-pacemaker heart cells of the rabbit. *Jpn J Physiol* **34**, 245–254.

North RA (1989). Twelfth Gaddum memorial lecture. Drug receptors and the inhibition of nerve cells. *Br J Pharmacol* **98**, 13–28.

North RA, Williams JT, Surprenant A *et al.* (1987). Mu and delta receptors belong to a family of receptors that are coupled to potassium channels. *PNAS* **84**, 5487–5491.

O'Connell AD, Leng Q *et al.* (2005). Phosphorylation-regulated endoplasmic reticulum retention signal in the renal outer-medullary K^+ channel (ROMK). *PNAS* **102**, 9954–9959.

Oh U, Ho YK and Kim D (1995). Modulation of the serotonin-activated K^+ channel by G protein subunits and nucleotides in rat hippocampal neurons. *J Membr Biol* **147**, 241–253.

Ohno Y, Hibino H, Lossin C *et al.* (2007). Inhibition of astroglial Kir4.1 channels by selective serotonin reuptake inhibitors. *Brain Res* **1178**, 44–51.

Okami T, Yamamoto A, Omori K *et al.* (1990). Immunocytochemical localization of Na^+, K^+-ATPase in rat retinal pigment epithelial cells. *J Histochem Cytochem* **38**, 1267–1275.

Oliver D, Hahn H, Antz C *et al.* (1998). Interaction of permeant and blocking ions in cloned inward-rectifier K^+ channels. *Biophys J* **74**, 2318–2326.

Olsen ML, Higashimori H, Campbell SL *et al.* (2006). Functional expression of Kir4.1 channels in spinal cord astrocytes. *Glia* **53**, 516–528.

Ookata K, Tojo A, Suzuki Y *et al.* (2000). Localization of inward rectifier potassium channel Kir7.1 in the basolateral membrane of distal nephron and collecting duct. *J Am Soc Nephrol* **11**, 1987–1994.

Oonuma H, Iwasawa K, Iida H *et al.* (2002). Inward rectifier K^+ current in human bronchial smooth muscle cells, inhibition with antisense oligonucleotides targeted to Kir2.1 mRNA. *Am J Respir Cell Mol Biol* **26**, 371–379.

Orkand RK, Nicholls JG and Kuffler SW (1966). Effect of nerve impulses on the membrane potential of glial cells in the central nervous system of amphibia. *J Neurophysiol* **29**, 788–806.

Padmawar P, Yao X, Bloch O *et al.* (2005). K^+ waves in brain cortex visualized using a long-wavelength K^+-sensing fluorescent indicator. *Nat Methods* **2**, 825–827.

Partiseti M, Collura V, Agnel M *et al.* (1998). Cloning and characterization of a novel human inwardly rectifying potassium channel predominantly expressed in small intestine. *FEBS Lett* **434**, 171–176.

Pearson WL, Dourado M, Schreiber M *et al.* (1999). Expression of a functional Kir4 family inward rectifier K^+ channel from a gene cloned from mouse liver. *J Physiol* **514**(**Pt 3**), 639–653.

Perier F, Radeke CM and Vandenberg CA (1994). Primary structure and characterization of a small-conductance inwardly rectifying potassium channel from human hippocampus. *PNAS* **91**, 6240–6244.

Pessia M, Imbrici P, D'Adamo MC *et al.* (2001). Differential pH sensitivity of Kir4.1 and Kir4.2 potassium channels and their modulation by heteropolymerisation with Kir5.1. *J Physiol* **532**, 359–367.

Pessia M, Tucker SJ, Lee K *et al.* (1996). Subunit positional effects revealed by novel heteromeric inwardly rectifying K⁺ channels. *EMBO J* **15**, 2980–2987.

Peters M, Ermert S, Jeck N *et al.* (2003). Classification and rescue of ROMK mutations underlying hyperprostaglandin E syndrome/antenatal Bartter syndrome. *Kidney Int* **64**, 923–932.

Peters M, Jeck N, Reinalter S *et al.* (2002). Clinical presentation of genetically defined patients with hypokalemic salt-losing tubulopathies. *Am J Med* **112**, 183–190.

Plaster NM, Tawil R, Tristani-Firouzi M *et al.* (2001). Mutations in Kir2.1 cause the developmental and episodic electrical phenotypes of Andersen's syndrome. *Cell* **105**, 511–519.

Poopalasundaram S, Knott C, Shamotienko OG *et al.* (2000). Glial heterogeneity in expression of the inwardly rectifying K⁺ channel, Kir4.1, in adult rat CNS. *Glia* **30**, 362–372.

Preisig-Muller R, Schlichthorl G, Goerge T *et al.* (2002). Heteromerization of Kir2.x potassium channels contributes to the phenotype of Andersen's syndrome. *PNAS* **99**, 7774–7779.

Pruss H, Derst C, Lommel R *et al.* (2005). Differential distribution of individual subunits of strongly inwardly rectifying potassium channels (Kir2 family) in rat brain. *Brain Res Mol Brain Res* **139**, 63–79.

Qu Z, Yang Z, Cui N *et al.* (2000). Gating of inward rectifier K⁺ channels by proton-mediated interactions of N- and C-terminal domains. *J Biol Chem* **275**, 31573–31580.

Quayle JM, McCarron JG, Brayden JE *et al.* (1993). Inward rectifier K⁺ currents in smooth muscle cells from rat resistance-sized cerebral arteries. *Am J Physiol* **265**, C1363–1370.

Raab-Graham KF, Radeke CM and Vandenberg CA (1994). Molecular cloning and expression of a human heart inward rectifier potassium channel. *Neuroreport* **5**, 2501–2505.

Ramu Y, Klem AM and Lu Z (2001). Titration of tertiapin-Q inhibition of ROMK1 channels by extracellular protons. *Biochemistry* **40**, 3601–3605.

Rodriguez-Soriano J (1998). Bartter and related syndromes, the puzzle is almost solved. *Pediatr Nephrol* **12**, 315–327.

Romanenko VG, Fang Y, Byfield F *et al.* (2004). Cholesterol sensitivity and lipid raft targeting of Kir2.1 channels. *Biophys J* **87**, 3850–3861.

Romanenko VG, Rothblat GH and Levitan I (2002). Modulation of endothelial inward-rectifier K⁺ current by optical isomers of cholesterol. *Biophys J* **83**, 3211–3222.

Rougier O, Vassort G and Stampfli R (1968). Voltage clamp experiments on frog atrial heart muscle fibres with the sucrose gap technique. *Pflugers Arch Gesamte Physiol Menschen Tiere* **301**, 91–108.

Sabirov RZ, Tominaga T, Miwa A *et al.* (1997). A conserved arginine residue in the pore region of an inward rectifier K channel (IRK1) as an external barrier for cationic blockers. *J Gen Physiol* **110**, 665–677.

Sakmann B and Trube G (1984). Conductance properties of single inwardly rectifying potassium channels in ventricular cells from guinea-pig heart. *J Physiol* **347**, 641–657.

Sakmann B, Noma A and Trautwein W (1983). Acetylcholine activation of single muscarinic K⁺ channels in isolated pacemaker cells of the mammalian heart. *Nature* **303**, 250–253.

Schmitt BM, Berger UV, Douglas RM *et al.* (2000). Na/HCO3 cotransporters in rat brain, expression in glia, neurons and choroid plexus. *J Neurosci* **20**, 6839–6848.

Schram G, Melnyk P, Pourrier M *et al.* (2002). Kir2.4 and Kir2.1 K⁺ channel subunits co-assemble, a potential new contributor to inward rectifier current heterogeneity. *J Physiol* **544**, 337–349.

Schulte U, Hahn H, Konrad M *et al.* (1999). pH gating of ROMK (K_ir1.1) channels, control by an Arg-Lys-Arg triad disrupted in antenatal Bartter syndrome. *PNAS* **96**, 15298–15303.

Schwalbe RA, Bianchi L, Accili EA *et al.* (1998). Functional consequences of ROMK mutants linked to antenatal Bartter's syndrome and implications for treatment. *Hum Mol Genet* **7**, 975–980.

Shang L, Lucchese CJ, Haider S *et al.* (2005). Functional characterisation of missense variations in the Kir4.1 potassium channel (KCNJ10) associated with seizure susceptibility. *Brain Res Mol Brain Res* **139**, 178–183.

Sheng M (1996). PDZs and receptor/channel clustering, rounding up the latest suspects. *Neuron* **17**, 575–578.

Shenolikar S, Voltz JW, Cunningham R *et al.* (2004). Regulation of ion transport by the NHERF family of PDZ proteins. *Physiology (Bethesda)* **19**, 362–369.

Shimura M, Yuan Y, Chang JT *et al.* (2001). Expression and permeation properties of the K⁺ channel Kir7.1 in the retinal pigment epithelium. *J Physiol* **531**, 329–346.

Shuck ME, Bock JH, Benjamin CW *et al.* (1994). Cloning and characterization of multiple forms of the human kidney ROM-K potassium channel. *J Biol Chem* **269**, 24261–24270.

Shuck ME, Piser TM, Bock JH *et al.* (1997). Cloning and characterization of two K⁺ inward rectifier (Kir) 1.1 potassium channel homologs from human kidney (Kir1.2 and Kir1.3). *J Biol Chem* **272**, 586–593.

Silver MR and DeCoursey TE (1990). Intrinsic gating of inward rectifier in bovine pulmonary artery endothelial cells in the presence or absence of internal Mg²⁺. *J Gen Physiol* **96**, 109–133.

Simon DB, Karet FE, Rodriguez-Soriano J *et al.* (1996). Genetic heterogeneity of Bartter's syndrome revealed by mutations in the K⁺ channel, ROMK. *Nat Genet* **14**, 152–156.

Sims SM and Dixon SJ (1989). Inwardly rectifying K⁺ current in osteoclasts. *Am J Physiol* **256**, C1277–1282.

Stanfield PR, Davies NW, Shelton PA *et al.* (1994). A single aspartate residue is involved in both intrinsic gating and blockage by Mg²⁺ of the inward rectifier, IRK1. *J Physiol* **478(Pt 1)**, 1–6.

Starremans PG, van der Kemp AW, Knoers NV *et al.* (2002). Functional implications of mutations in the human renal outer medullary potassium channel (ROMK2) identified in Bartter syndrome. *Pflugers Arch* **443**, 466–472.

Su S, Ohno Y, Lossin C *et al.* (2006). Inhibition of astroglial inwardly rectifying Kir4.1 channels by a tricyclic antidepressant, nortriptyline. *JPET* **320(2)**, 573–80.

Taglialatela M, Ficker E, Wible BA *et al.* (1995). C-terminus determinants for Mg²⁺ and polyamine block of the inward rectifier K⁺ channel IRK1. *EMBO J* **14**, 5532–5541.

Taglialatela M, Wible BA, Caporaso R *et al.* (1994). Specification of pore properties by the carboxyl terminus of inwardly rectifying K⁺ channels. *Science* **264**, 844–847.

Takahashi N, Morishige K, Jahangir A *et al.* (1994). Molecular cloning and functional expression of cDNA encoding a second class of inward rectifier potassium channels in the mouse brain. *J Biol Chem* **269**, 23274–23279.

Takahashi T (1990). Inward rectification in neonatal rat spinal motoneurones. *J Physiol* **423**, 47–62.

Takano M and Kuratomi S (2003). Regulation of cardiac inwardly rectifying potassium channels by membrane lipid metabolism. *Prog Biophys Mol Biol* **81**, 67–79.

Takeuchi S, Ando M and Kakigi A (2000). Mechanism generating endocochlear potential, role played by intermediate cells in stria vascularis. *Biophys J* **79**, 2572–2582.

Takumi T, Ishii T, Horio Y *et al.* (1995). A novel ATP-dependent inward rectifier potassium channel expressed predominantly in glial cells. *J Biol Chem* **270**, 16339–16346.

Tanemoto M, Kittaka N, Inanobe A *et al.* (2000). *In vivo* formation of a proton-sensitive K⁺ channel by heteromeric subunit assembly of Kir5.1 with Kir4.1. *J Physiol* **525(Pt 3)**, 587–592.

Taniguchi J, Yoshitomi K and Imai M (1989). K⁺ channel currents in basolateral membrane of distal convoluted tubule of rabbit kidney. *Am J Physiol* **256**, F246–254.

Tawil R, Ptacek LJ, Pavlakis SG et al. (1994). Andersen's syndrome, potassium-sensitive periodic paralysis, ventricular ectopy and dysmorphic features. *Ann Neurol* **35**, 326–330.

Topert C, Doring F, Wischmeyer E et al. (1998). Kir2.4, a novel K$^+$ inward rectifier channel associated with motoneurons of cranial nerve nuclei. *J Neurosci* **18**, 4096–4105.

Tsai TD, Shuck ME, Thompson DP et al. (1995). Intracellular H$^+$ inhibits a cloned rat kidney outer medulla K$^+$ channel expressed in *Xenopus* oocytes. *Am J Physiol* **268**, C1173–1178.

Tucker SJ, Imbrici P, Salvatore L et al. (2000). pH dependence of the inwardly rectifying potassium channel, Kir5.1 and localization in renal tubular epithelia. *J Biol Chem* **275**, 16404–16407.

Vandenberg CA (1987). Inward rectification of a potassium channel in cardiac ventricular cells depends on internal magnesium ions. *PNAS* **84**, 2560–2564.

Vollmer M, Koehrer M, Topaloglu R et al. (1998). Two novel mutations of the gene for Kir 1.1 (ROMK) in neonatal Bartter syndrome. *Pediatr Nephrol* **12**, 69–71.

von Beckerath N, Dittrich M, Klieber HG et al. (1996). Inwardly rectifying K$^+$ channels in freshly dissociated coronary endothelial cells from guinea-pig heart. *J Physiol* **491**(Pt 2), 357–365.

Von Bekesy G (1951). DC potentials and energy balance of the cochlear partition. *J Acoust Soc Amer* **23**, 576–582.

Von Bekesy G (1952). Resting potentials inside the cochlear partition of the guinea pig. *Nature* **169**, 241–242.

Wade JB, Fang L, Liu J et al. (2006). WNK1 kinase isoform switch regulates renal potassium excretion. *PNAS* **103**, 8558–8563.

Wang W, Hebert SC and Giebisch G (1997). Renal K$^+$ channels, structure and function. *Annu Rev Physiol* **59**, 413–436.

Wang WH and Giebisch G (1991). Dual modulation of renal ATP-sensitive K$^+$ channel by protein kinases A and C. *PNAS* **88**, 9722–9725.

Wang Z, Yue L, White M et al. (1998). Differential distribution of inward rectifier potassium channel transcripts in human atrium versus ventricle. *Circulation* **98**, 2422–2428.

Wangemann P, Itza EM, Albrecht B et al. (2004). Loss of KCNJ10 protein expression abolishes endocochlear potential and causes deafness in Pendred syndrome mouse model. *BMC Med* **2**, 30.

Welling PA (1997). Primary structure and functional expression of a cortical collecting duct Kir channel. *Am J Physiol* **273**, F825–836.

Wellman GC and Bevan JA (1995). Barium inhibits the endothelium-dependent component of flow but not acetylcholine-induced relaxation in isolated rabbit cerebral arteries. *JPET* **274**, 47–53.

Wible BA, De Biasi M, Majumder K et al. (1995). Cloning and functional expression of an inwardly rectifying K$^+$ channel from human atrium. *Circ Res* **76**, 343–350.

Wible BA, Taglialatela M, Ficker E et al. (1994). Gating of inwardly rectifying K$^+$ channels localized to a single negatively charged residue. *Nature* **371**, 246–249.

Williams JT, Colmers WF and Pan ZZ (1988). Voltage- and ligand-activated inwardly rectifying currents in dorsal raphe neurons *in vitro*. *J Neurosci* **8**, 3499–3506.

Wilson FH, Disse-Nicodeme S, Choate KA et al. (2001). Human hypertension caused by mutations in WNK kinases. *Science* **293**, 1107–1112.

Wilson GF and Chiu SY (1990). Ion channels in axon and Schwann cell membranes at paranodes of mammalian myelinated fibers studied with patch clamp. *J Neurosci* **10**, 3263–3274.

Xu H, Yang Z, Cui N et al. (2000). A single residue contributes to the difference between Kir4.1 and Kir1.1 channels in pH sensitivity, rectification and single channel conductance. *J Physiol* **528 Pt 2**, 267–277.

Xu JZ, Hall AE, Peterson LN et al. (1997). Localization of the ROMK protein on apical membranes of rat kidney nephron segments. *Am J Physiol* **273**, F739–748.

Xu ZC, Yang Y and Hebert SC (1996). Phosphorylation of the ATP-sensitive, inwardly rectifying K$^+$ channel, ROMK, by cyclic AMP-dependent protein kinase. *J Biol Chem* **271**, 9313–9319.

Yamada M and Kurachi Y (1995). Spermine gates inward-rectifying muscarinic but not ATP-sensitive K$^+$ channels in rabbit atrial myocytes. Intracellular substance-mediated mechanism of inward rectification. *J Biol Chem* **270**, 9289–9294.

Yang J, Jan YN and Jan LY (1995a). Control of rectification and permeation by residues in two distinct domains in an inward rectifier K$^+$ channel. *Neuron* **14**, 1047–1054.

Yang J, Jan YN and Jan LY (1995b). Determination of the subunit stoichiometry of an inwardly rectifying potassium channel. *Neuron* **15**, 1441–1447.

Yang Z, Xu H, Cui N et al. (2000). Biophysical and molecular mechanisms underlying the modulation of heteromeric Kir4.1-Kir5.1 channels by CO2 and pH. *J Gen Physiol* **116**, 33–45.

Yoo D, Flagg TP, Olsen O et al. (2004). Assembly and trafficking of a multiprotein ROMK (Kir 1.1). channel complex by PDZ interactions. *J Biol Chem* **279**, 6863–6873.

Yoo D, Kim BY, Campo C et al. (2003). Cell surface expression of the ROMK (Kir 1.1). channel is regulated by the aldosterone-induced kinase, SGK-1 and protein kinase A. *J Biol Chem* **278**, 23066–23075.

Yun CC, Palmada M, Embark HM et al. (2002). The serum and glucocorticoid-inducible kinase SGK1 and the Na$^+$/H$^+$ exchange regulating factor NHERF2 synergize to stimulate the renal outer medullary K$^+$ channel ROMK1. *J Am Soc Nephrol* **13**, 2823–2830.

Zaritsky JJ, Eckman DM, Wellman GC et al. (2000). Targeted disruption of Kir2.1 and Kir2.2 genes reveals the essential role of the inwardly rectifying K$^+$ current in K$^+$-mediated vasodilation. *Circ Res* **87**, 160–166.

Zaritsky JJ, Redell JB, Tempel BL et al. (2001). The consequences of disrupting cardiac inwardly rectifying K$^+$ current (I$_{K1}$). as revealed by the targeted deletion of the murine Kir2.1 and Kir2.2 genes. *J Physiol* **533**, 697–710.

Zeng WZ, Babich V, Ortega B et al. (2002). Evidence for endocytosis of ROMK potassium channel via clathrin-coated vesicles. *Am J Physiol Renal Physiol* **283**, F630–639.

Zeng WZ, Li XJ, Hilgemann DW et al. (2003). Protein kinase C inhibits ROMK1 channel activity via a phosphatidylinositol 4,5-bisphosphate-dependent mechanism. *J Biol Chem* **278**, 16852–16856.

Zhu G, Chanchevalap S, Cui N et al. (1999). Effects of intra- and extracellular acidifications on single channel Kir2.3 currents. *J Physiol* **516** (Pt 3), 699–710.

Zobel C, Cho HC, Nguyen TT et al. (2003). Molecular dissection of the inward rectifier potassium current (I$_{K1}$) in rabbit cardiomyocytes, evidence for heteromeric co-assembly of Kir2.1 and Kir2.2. *J Physiol* **550**, 365–372.

4.4.2 **Kir3 family**

Atsushi Inanobe and Yoshihisa Kurachi

4.4.2.1 Introduction

Heterotrimeric GTP-binding proteins (G proteins) transduce the signals of heptahelical membrane proteins, so called G protein-coupled receptors (GPCR), to a number of effectors, including adenylyl cylase, phospholipase C, and ion channels. This is one of the major pathways for extracellular stimuli to control cell function. The G protein-gated K$^+$ (K$_G$) channel, consisting of various

Table 4.4.2.1 Nomenclature and chromosomal localization of K_G channel subunits

K_G Subunit		Species		
		Human	Rat	Mouse
Kir3.1 (GIRK1, KGA)	Gene	KCNJ3	Kcnj3	Kcnj3
	Locus ID	3760	50599	16519
	GenBank ID	U50964, NM_002239	Y12259, NM_031610	L25264, U01071, NM_008426
	PubMed ID	8804710	8642402	8355805, 8234283
	Chromosomal location	chr. 2q24.1	chr. 3	chr. 2c1.1
Kir3.2 (GIRK2, hiGIRK2)	Gene	KCNJ6	Kcnj6	Kcnj6
	Locus ID	3763	25743	16522
	GenBank ID	U24660, NM_002240	AB073753, NM_013192	U37253, NM_010606
	PubMed ID	7592809, 10659995	11883954	7499385
	Chromosomal location	chr. 21q22.13–q22.2	chr. 11q21	chr. 16, 68.75 cM
Kir3.3 (GIRK3)	Gene	KCNJ9	Kcnj9	Kcnj9
	Locus ID	3765	116560	16524
	GenBank ID	AF193615, NM_004983	L77929, NM_053834	AF130860, NM_008429
	PubMed ID	8575783	8670302	7426018, 10341034
	Chromosomal location	chr. 1q21–23	chr. 13q24	chr. 1, 94.2 cM
Kir3.4 (GIRK4, CIR)	Gene	KCNJ5	Kcnj5	Kcnj5
	Locus ID	3762	29713	16521
	GenBank ID	L47208, NM_000890	L35771, NM_017297	U33631, NM_010605
	PubMed ID	8558261	7877685	7499385
	Chromosomal location	chr. 11q24	chr. 8q21	chr. 11q23

chr, chromosome; cM, centimorgans.

combinations of Kir3.1~Kir3.4 subunits, is one of the effectors regulated by G protein signalling. This type of K⁺ channel underlies many physiological processes including deceleration of the heartbeat upon stimulation of vagal nerves, the formation of slow inhibitory postsynaptic potentials in the central nervous system, and inhibition of hormone release in endocrine cells. In this chapter, we summarize the current knowledge about K_G channel subunits and discuss the physiological functions of this type of potassium channel.

4.4.2.2 Molecular characterization

The K_G channel is a member of the inwardly rectifying K⁺ (Kir) channel family. Mammalian K_G channels are tetramers, and either heteromeric or homomeric assembly of four Kir3.x subunits, Kir3.1, Kir3.2, Kir3.3 and Kir3.4, and their splicing variants. Nomenclature and chromosomal localization for each K_G channel subunit are outlined in Table 4.4.2.1 (Kubo et al., 2005). Different combinations of Kir3.x subunits are found in different tissues and cells (Table 4.4.2.2) which means that native K_G channels exhibit substantial diversity in respect to their subunit composition. In this section, we will describe the characteristics of each subunit and the heterogeneity of the subunit assembly.

Kir3.1

Prior to the identification of genes encoding K_G channels, a K⁺ homeostasis-linked Kir channel, Kir1.1/ROMK1 (Ho et al., 1993), and a classical Kir channel, Kir2.1/IRK1 (Kubo et al., 1993a), had been isolated by expression cloning. PCR amplification with degenerative primers designed from homology within Kir1.1 and Kir2.1 led to the findings of a new gene named Kir3.1/GIRK1 in a rat heart cDNA library (Dascal et al., 1993; Kubo et al., 1993b). The Kir3.1 encodes the main subunit of K_G channels and shares 39 per cent

and 42 per cent amino acid identity to Kir1.1 and Kir2.1, respectively. Its primary structure predicts two transmembrane-spanning regions and one potential pore-forming region, which corresponds to the selectivity filter for K⁺ ions (Doyle et al., 1998). Whereas Kir3.1 when expressed alone appears to form a homomeric assembly in an overexpression system (Corey et al., 1998), it is actually

Table 4.4.2.2 Tissue distribution of K_G channel subunit composition

Subunit	Complex	Tissue	References
Kir3.1, Kir3.2a, Kir3.2c, Kir3.3 (Kir3.4)	Kir3.1/Kir3.2a Kir3.1/Kir3.2c Kir3.1/Kir3.3 Kir3.2/Kir3.4 Kir3.2/Kir3.3	CNS: most neurons	Liao et al (1996); Inanobe et al (1999) Liao et al. 1996; Inanobe et al (1999b) Jelacic et al. (2000) Lesage et al (1995) Inanobe et al (1999); Jelacic et al (2000)
Kir3.2a, Kir3.2c	Kir3.2a/Kir3.2c Kir3.2c	CNS: dopaminergic neurones (substantia nigra)	Inanobe et al (1999)
Kir3.1, Kir3.2, Kir3.4	Kir3.1/Kir3.4	Pituitary thyrotroph	Morishige et al (1999)
Kir3.1, Kir3.4	Kir3.1/Kir3.4 Kir3.4	Heart Atrial myocytes	Krapivinsky et al (1996) Corey and Clapham (1998)
Kir3.1, Kir3.2, Kir3.4		Pancreas a cell	Yoshimoto et al (1999)
Kir3.2d	Kir3.2d	Testis: spermatid	Inanobe et al (1999)
Kir3.1	Kir3.1	Testis: spermatogonia	Inanobe et al (1999a)

non-functional because it is retained within the cells due to the lack of a signal sequence to traffic it to the plasma membrane (Ma et al., 2002). When Kir3.1 is co-expressed with Kir3.2 and Kir3.4, the channels exhibit enhanced current with different single channel properties (Duprat et al., 1995; Kofuji et al., 1995; Slesinger et al., 1996; Velimirovic et al., 1996). In mice lacking either Kir3.2 (Signorini et al., 1997) or Kir3.4 (Wickman et al., 1998), or Kir3.2 and Kir3.3 (Koyrakh et al., 2005), protein expression levels of total and glycosylated Kir3.1 is reduced without any change in mRNA expression levels, further suggesting that the native K_G channels containing Kir3.1 behave as a heteromeric assembly with other Kir3.x subunits.

Kir3.2

From a mouse brain cDNA library, two additional homologues of Kir3.1 were isolated and designated Kir3.2/GIRK2 and Kir3.3/GIRK3 (Lesage et al., 1994). The amino acid sequence of Kir3.2 is 49 per cent identity and 60 per cent homology to that of Kir3.1. At least four different isoforms of Kir3.2 are alternatively spliced out from a transcript of a single gene which is composed of more than eight exons (Wei et al., 1998; Inanobe et al., 1999a; Wickman et al., 2002). They are designated Kir3.2a–d in the order of identification (Lesage et al., 1994; Isomoto et al., 1996; Inanobe et al., 1999a). Kir3.2a and Kir3.2c correspond to GIRK2 and GIRK2A respectively, (Lesage et al., 1995). Kir3.2c was also isolated from an insulinoma cell cDNA library and named K_{ATP}-2 (Stoffel et al., 1995). These variants share a central core which includes the transmembrane portion, but differ at either distal N- or C-terminal ends. Thus, the C-terminus of Kir3.2b is 87 amino acids shorter than that of Kir3.2a and 8 amino acids different from Kir3.2a. Kir3.2c is the longest isoform that possesses a C terminus with an extra 11 amino acids appended to the C-terminus of Kir3.2a. Kir3.2d possesses the C-terminus of Kir3.2c and its N-terminus is 14 amino acids shorter than other isoforms.

As shown in Table 4.4.2.2, Kir3.2 isoforms are distributed in various tissues such as brain, pituitary, pancreas and testis. The expression patterns of most of the Kir3.2 isoforms overlaps with those of the other Kir3.x subunits (Kobayashi et al., 1995; Karschin et al., 1996; Liao et al., 1996; Inanobe et al., 1999b). Furthermore, Kir3.2 is immunoprecipitated with Kir3.1 (Liao et al., 1996; Inanobe et al., 1999b), Kir3.3 (Inanobe et al., 1999b; Jelacic et al., 2000) and Kir3.4 (Lesage et al., 1995) from either brain or cells expressing recombinant subunits. On the other hand, Kir3.2 subunits are expressed in the absence of Kir3.1 in, for example, (1) dopaminergic neurons of the substantia nigra in rodent brain which express Kir3.2a and Kir3.2c which can be isolated as a complex (Inanobe et al., 1999b), and (2) spermatids of mouse testis which exclusively express Kir3.2d (Inanobe et al., 1999a). Kir3.2 isoforms evoke the agonist-dependent currents even in the absence of Kir3.1 or Kir3.4 (Kofuji et al., 1995; Lesage et al., 1995; Kofuji et al., 1996; Liao et al., 1996; Slesinger et al., 1996; Velimirovic et al., 1996; Inanobe et al., 1999b). Therefore, Kir3.2 behaves as a subunit of either heteromeric or homomeric K_G channels in native tissues.

Mutant mice lacking Kir3.2 have a reduced protein expression of Kir3.1 in the brain, develop spontaneous seizures and display a range of phenotypes related to sporadic seizures and increased susceptibility to convulsant agents (Signorini et al., 1997). The mice also reveal that Kir3.2 is essential for the formation of IPSPs in hippocampal and cerebellar neurons (Luscher et al., 1997; Signorini et al., 1997).

Kir3.3

Kir3.3 is expressed exclusively in the central nervous system and behaves as a subunit in heteromeric assemblies with other Kir3.x subunits (Lesage et al., 1994). Most expression experiments have shown that Kir3.3 does not increase the current amplitude of Kir3.1 and inhibits currents produced by homomeric Kir3.2 channels, and Kir3.1/Kir3.2 or Kir3.1/Kir3.4 heteromeric channels (Kofuji et al., 1995; Lesage et al., 1995; Schoots et al., 1999; Ma et al., 2002). Ma and colleagues (2002) reported a tetra-peptide sequence (Y-M-S-I) on the C-terminus of Kir3.3 which codes a lysosome-localization signal which prevents the expression of K_G channels containing Kir3.3 on the plasma membrane. Mutations at this sequence only weakly increase the surface expression of Kir3.3, suggesting that the sequence may not be sufficient to explain fully the effect of the Kir3.3 subunit in preventing surface expression of the K_G channel. Nevertheless, these results suggest that Kir3.3 is a regulatory subunit for the surface expression of the K_G channel.

On the other hand, there are several reports that Kir3.3 can enhance the expression of the K_G channel current when co-expressed with either Kir3.1 (Wischmeyer et al., 1997; Jelacic et al., 1999) or Kir3.2 (Jelacic et al., 2000). In this respect, the Kir3.3 protein has been isolated as a complex with Kir3.1 and Kir3.2 isoforms from the brain (Inanobe et al., 1999b; Jelacic et al., 2000). While Kir3.3-null mice exhibit a reduction in glycosylated Kir3.1 expression, their behavioural phenotype is relatively normal (Torrecilla et al., 2002). Both hippocampal CA1 neurons and locus coeruleus neurons of either Kir3.2-null or Kir3.3-null mouse show slightly depolarized resting membrane potentials and small K^+ currents induced by GABA or opioid receptor agonists (Torrecilla et al., 2002; Koyrakh et al., 2005). In neurons isolated from mice lacking both genes, the resting membrane potential is significantly depolarized and inhibitory neurotransmitters hardly elicit a K^+ current. These results suggest that Kir3.3 is one of the subunits forming neuronal K_G channels.

Kir3.4

Krapivinsky and colleagues (1995b) reported the purification of Kir3.1 from bovine atrial membrane by an immunoprecipitation method. They found an auxiliary protein band in fractions containing Kir3.1. After sequencing the proteolytic fragment of the protein, they identified that the protein was a new K_G channel subunit, Kir3.4/GIRK4/CIR. This was the first example of a heteromeric assembly of the K_G channel. The heteromeric assembly shows properties close to those of the cardiac K^+ current known as I_{KACh}. A homotetramer of Kir3.4 has also been identified in atrial myocytes (Corey and Clapham, 1998). A heteromeric complex of Kir3.1 and Kir3.4 has also been identified in the thyrotrophs of the rat pituitary (Morishige et al., 1999). As discussed below, the channel is localized on intracellular secretory granules, suggesting that the subcellular localization of the K_G channel is differentially controlled in different cells. Kir3.4 expresses weakly in some regions of the brain (Iizuka et al., 1997; Wickman et al., 2000) and in α cells in the pancreas (Yoshimoto et al., 1999). Kir3.4-deficient mice appear normal, but they lack I_{KACh} in the cardiac atrium, indicating that Kir3.4 is necessary for the formation of this current (Wickman et al., 1998). On the other hand, the mice exhibit impaired performance in spatial learning and memory tasks

(Wickman *et al.*, 2000) although there is little Kir3.4 expression in the hippocampus. As such, our understanding of the function of Kir3.4-containing K_G channels in the brain is still in its infancy and much remains to be determined.

Subunit number and stoichiometry

At first, the heteromeric K_G channel with Kir3.1 and Kir3.4 isolated from bovine atrium was considered to be a heteromultimer based on its migration position in a size exclusion chromatography experiment (Krapivinsky *et al.*, 1995b). Since the isolated membrane proteins behave as a detergent-protein complex, they have a molecular mass in solution that is higher than that of the native assembly. The physical properties of neuronal K_G channels containing Kir3.1 were examined and this channel was estimated to be a tetramer (Inanobe *et al.*, 1995a). However, the number of subunits in a single K_G channel was difficult to estimate, because of the heterogeneities of the assembly of neuronal K_G channels in detergent.

Electrophysiological analyses of concatenated subunits of Kir3.1 and Kir3.4 (Silverman *et al.*, 1996; Tucker *et al.*, 1996) revealed that the channel behaves as a tetramer in a 1:1 stoichiometry of the subunits arranged side by side. Subsequently, chemical cross-linking approaches were applied to assign the subunit stoichiometry of native I_{KACh} and it was concluded that the Kir3.1 to Kir3.4 stoichiometry was 1:1 (Corey *et al.*, 1998). Finally, the tetrameric assembly of Kir3.x subunits was proven to exist by the determination of the crystal structures of a cytoplasmic region of Kir3.1 (Nishida and MacKinnon, 2002), Kir2.1 (Pegan *et al.*, 2005), Kir3.2 (Inanobe *et al.*, 2007), the entire region of its bacterial homologue, KirBac1.1 (Kuo *et al.*, 2003) and the chimera between Kir3.1 and KirBac3.1 (Nishida and MacKinnon, 2007).

4.4.2.3 Biophysical characteristics

The biophysical characteristics of ion channels, such as ion conduction, ion selectivity and gating have been investigated, mainly using electrophysiological techniques (Hille, 2001). The main concepts generated by the long-accumulating endeavours of many researchers have been beautifully confirmed and structure–function relationships have been elucidated by crystallographic analyses, mainly by the MacKinnon group (Doyle *et al.*, 1998; Morais-Cabral *et al.*, 2001; Zhou *et al.*, 2001b; Jiang *et al.*, 2002, 2003).

Ion conduction

The transmembrane domain of Kir is composed of four helices: slide, M1, pore and M2 helices from the N- to C-terminus. The ion permeation pathway of the K_G channel consists of M1, pore and M2 helices; a configuration that is essentially shared by all K^+ channels (Doyle *et al.*, 1998; Jiang *et al.*, 2002, 2003; Kuo *et al.*, 2003), while the amphiphilic slide helix prior to the M1 helix is unique among the K^+ channel family. Briefly, the channel allows K^+ to permeate according to the difference between the equilibrium potential of K^+ (E_K) and the membrane potential. The current–voltage relationship of the K_G channel is such that a large inward conductance exists at membrane potentials below E_K and a small outward conductance at those potentials above E_K. This voltage-dependent change in the conductance indicates that the K_G channel is a strong inward rectifier. The block of outward ion flow is caused by the 'plug-in' action of intracellular cations such as Mg^{2+} and positively charged aliphatic amines, such as polyamines (Matsuda *et al.*, 1987;

Vandenberg 1987; Matsuda 1988; Ficker *et al.*, 1994; Lopatin *et al.*, 1994; Wible *et al.*, 1994; Yamada and Kurachi 1995). It is thought that these positively charged species interact with negatively charged residues at the M2 helix of Kir channels. In addition to the residues at the M2 helix, the residues at the cytoplasmic region have also been identified as being important in the inward rectification. This was confirmed by the crystal structure of the Kir3.1 cytoplasmic region which distributes such residues at a cytoplasmic pore in the tetrameric assembly (Nishida and MacKinnon, 2002; Pegan *et al.*, 2005). These observations indicate that K^+ ions traverse over 60 Å from the outer vestibule at extracellular side to the cytoplasm through Kir channels.

Ion selectivity

All K_G channel subunits possess a signature sequence (The-Ile-Gly-Tyr/Phe-Gly) for the K^+ channel. The corresponding region in the crystal structure of KcsA cages K^+ ions with the carbonyl oxygen atoms of the backbone and the hydroxyl groups of side chains (Zhou *et al.*, 2001b). Mutation of G156 to serine in the selectivity filter of Kir3.2 causes the loss of ion selectivity in the *weaver* mouse (Patil *et al.*, 1995). In contrast, mutation of Ser177 in M2 of the Kir3.2 channel affects ion selectivity, even though this residue is located away from the selectivity filter (Yi *et al.*, 2001). Furthermore, introduction of a pore-facing M2 negative charge (Asn184) in the central cavity below the selectivity filter restores the ion selectivity and inward rectification of the non-selective Ser177Trp mutant Kir3.2 channel (Bichet *et al.*, 2006). These results indicate that the selectivity filter of the K_G channel behaves as only a small part of the entire permeation pathway, and that the channel's central cavity also contributes to the ion selectivity through an electrostatic interaction between ions and lining residues.

Activation gating

As discussed in the previous chapter, the structure of KirBac1.1 reveals the essential architecture of K_G channels (Kuo *et al.*, 2003). Four identical subunits assemble to form a functional unit which is segregated into two domains: transmembrane and cytoplasmic domains. The M2 helix in the KcsA channel (Doyle *et al.*, 1998) creates the narrowest point at the intracellular surface of the membrane. In other K^+ channel structures (Jiang *et al.*, 2002, 2003), this portion is dilated enough to allow hydrated K^+ ions to pass through. Thus, the pore-lining helix M2 essentially functions to control ion permeation through active gating of the channel. This idea is partially supported by reports that introduction of mutations in the M2 helix of either Kir3.1 (Sadja *et al.*, 2001) or Kir3.2 (Yi *et al.*, 2001) causes the mutant channels to be constitutively active. The native K_G channel is a constitutively inactive channel and primarily requires the G protein βγ subunit (Gβγ) for its activation. The binding site for Gβγ on the channel distributes to the cytoplasmic region and, as such, this region controls conformational changes in the transmembrane domains. However, the precise mechanism by which the cytoplasmic region controls the K_G channel gating remains to be clarified.

4.4.2.4 Cellular and subcellular localization

Most neuronal K_G channels are heterogeneous complexes with Kir3.1, Kir3.2 isoforms, Kir3.3 and Kir3.4 (Table 4.4.2.2) (Lesage *et al.*, 1995; Liao *et al.*, 1996; Inanobe *et al.*, 1999; Koyrakh *et al.*, 2005).

Whilst the expression of Kir3.1, Kir3.2 and Kir3.3 protein is observed widely throughout the brain (Liao et al., 1996; Inanobe et al., 1999b; Grosse et al., 2003; Koyrakh et al., 2005), Kir3.4 is expressed in only a subset of brain regions (Iizuka et al., 1997; Wickman et al., 2000). The subcellular distributions of Kir3.1, Kir3.2 and Kir3.3 have been examined using electron microscopy. Kir3.1 and Kir3.2 are located at both postsynaptic and presynaptic regions (Liao et al., 1996; Morishige et al., 1996; Ponce et al., 1996; Inanobe et al., 1999b; Koyrakh et al., 2005; Kulik et al., 2006). Axonal sorting of Kir3.3 has been identified in GABA-positive interneurons of the hippocampal CA3 region (Grosse et al., 2003). Under light microscopy, Kir3.4-immunoreactivity is found at the axonal terminus (Iizuka et al., 1997; Wickman et al., 2000). These results indicate that K_G channels are involved not only in the formation of slow inhibitory postsynaptic potentials, but also in the presynaptic modulation of neuronal activity.

The relationship between the specific subunit combination and the subcellular localization has not yet been clarified. However, neuronal K_G channels at the postsynaptic membrane may be regulated by the Kir3.2c isoform which is found in complex with Kir3.2a at the inhibitory postsynaptic region in dopaminergic neurons of the substantia nigra (Inanobe et al., 1999b). The Kir3.2c specifically possesses a distal C-terminal region which interacts with PDZ domain-containing proteins including membrane-associated proteins which act as scaffolding and/or signalling molecules (Sheng and Sala, 2001). Thus, no K_G channel currents could be recorded in Xenopus oocytes expressing Kir3.2c and a G-protein coupled receptor (GPCR) unless SAP-97 was co-expressed (Hibino et al., 2000). The PDZ domain-containing anchoring proteins may therefore play a crucial role not only in the regulation of subcellular localization but also in the modulation of the function of K_G channels which contain the Kir3.2c isoform.

Subcellular localization of the same type of K_G channel may also be regulated in a tissue-specific manner (Table 4.4.2.2). The K_G channels of both atrial myocytes and thyrotrophs of anterior pituitary lobes are composed of Kir3.1/Kir3.4 subunits. The subcellular localization of heteromeric Kir3.1/Kir3.4 channels in these tissues are, however, completely different. Cardiac K_G channels are localized on the cell membrane (Inanobe and Kurachi, 2000), whereas K_G channels in resting thyrotrophs are found exclusively on intracellular secretory vesicles (Morishige et al., 1999). As mentioned below, vesicular K_G channels in thyrotrophs are functional, when inserted into the plasma membrane. Thus, the molecular mechanisms controlling the subcellular localization of the K_G channel may be varied in different cells.

4.4.2.5 Pharmacology

Inhibitors and activators of K_G channels are summarized in Table 4.4.2.3.

n-Alcohols such as methanol, ethanol and 1-propanol have been reported to activate, in a concentration-dependent manner, K_G channels formed from various combinations of Kir3.1, Kir3.2 and Kir3.4 subunits (Kobayashi et al., 1999; Lewohl et al., 1999). Kir3.2-containing K_G channels show the greatest enhancement by ethanol, while Kir1.1 and Kir2.1 are insensitive to n-alcohols.

Intracellular Na^+ concentration is recognized as an activator of K_G channels (Petit-Jacques et al., 1995; Sui et al., 1996). The Na^+-binding site has been identified as being at Asp228 in Kir3.2 and Asp223 in Kir3.4 (Ho and Murrell-Lagnado, 1999; Zhang et al., 1999). Mirshahi et al., (2002) found $G\beta$ mutants which fail to activate K_G channels, but do potentiate Na^+ mediated activation in inside-out patches. Since these mutants directly interact with the cytoplasmic region of Kir3.4, Na^+-dependent activation of the K_G channel seems to be distinct from that involved in direct $G\beta\gamma$-mediated activation of this K^+ current.

The honey bee venom contains tertiapin which has been identified as a potent inhibitor of Kir3.1/Kir3.4 and Kir1.1 with nanomolar affinity (Jin and Lu 1998). Kir amino acids involved in tertiapin recognition have been identified at the outer vestibule of Kir1.1 (Jin and Lu, 1999). The residues are partially conserved within tertiapin-sensitive Kir channels, indicating that the toxin may associate with the channel via multiple contacts. Jin and Lu (1999) created the oxidation-resistant toxin, tertiapinQ, to increase its stability. The modulated toxin specifically blocks I_{KACh}, but not I_{KATP} nor I_{K1} in isolated cardiac myocytes (Kitamura et al., 2000; Dobrev et al., 2005). Furthermore, this toxin has been shown to be effective in canine atrial fibrillation models (Cha et al., 2006; Hashimoto et al., 2006), leading to the idea that I_{KACh} is a potential anti-arrhythmic target. In this respect, it has been reported that I_{KACh} is inhibited directly by anti-arrhythmic agents such as quinidine (Kurachi et al., 1987) and AN-132 3-(diisopropylaminoethylamino)-2′,6′-dimethylpropionanilide (Kurachi et al., 1989).

Using K_G channels reconstituted in Xenopus oocytes it has been possible to demonstrate that the activity of this channel can be modulated by a number of compounds (Table 4.4.2.3) including antipsychotic drugs such as haloperidol, thioridazine, pimodine and clozapine (Kobayashi et al., 2000), antidepressants such as imipramine, desipramine, amitriptyline, nortriptyline, clomipramine, maprotiline, citalopram, and fluoxetine (Kobayashi et al., 2003 and 2004), volatile anaesthetics such as halothane, F3 (1-chloro-1,2,2-trifluorocyclobutane), isoflurane, and enflurane (Weigl and Schreibmayer 2001; Yamakura 2001; Milovic et al., 2004), and the local anaesthetic bupivacaine (Zhou et al., 2001a). These compounds block K_G channel activity, but show relatively broad specificity in terms of the subunit combinations that are affected. In addition to these compounds, the selective dopamine D_1 receptor antagonist R-(+)-7-chloro-8-hydroxy-3-methyl-1-phenyl-2,3,4,5-tetrahydro-1H-3-benzazepine hydrochloride (Kuzhikandathil and Oxford, 2002), the NMDA receptor antagonists MK-801 (Kofuji et al., 1996) and ifenprodil (Kobayashi et al., 2006), the inhaled drug of abuse toluene (Del Re et al., 2006), the Ca^{2+} channel blocker verapamil and the classical cation channel blocker QX-314 (Kofuji et al., 1996) have also been reported to inhibit K_G channels composed of various combinations of their subunits. Although the binding sites through which most compounds inhibit the K_G channels have not been determined, the binding sites of halothane and bupivacaine on K_G channels have been suggested to distribute at either the cytoplasmic region (Yamakura, 2001) or the distal C-terminal region of the channel (Zhou et al., 2001a; Milovic et al., 2004).

4.4.2.6 Physiology

Discovery of K_G channel

In the 1920s, Otto Loewi (1921) established the concept of chemical synaptic transmission through evaluation of the action of acetylcholine (ACh) in the heart. ACh is released from the vagal nerve terminals and decelerates the heart beat. This ACh-induced

Table 4.4.2.3 Modulators of K_G channel activity

	Complex	Efficacy	References
Activation			
n-alcohol	Kir3.1/Kir3.2; Kir3.1/Kir3.4	200 mM – 60–100% increase	Lewohl *et al.* (1999)
sodium ion	Kir3.2/Kir3.4; Kir3.2	200 mM – 60–100% increase	Kobayashi *et al.* (1999)
	Kir3.1/Kir3.2	EC_{50} = 27 mM	Ho *et al.* (1999)
	Kir3.1/Kir3.4	EC_{50} = 44 mM	
	Kir3.2	EC_{50} = 37 mM	
Inhibition			
Tertiapin	Kir3.1/Kir3.4	K_i = 9 nM	Jin and Lu (1998)
TertiapinQ	Kir3.1/Kir3.4	K_i = 9 nM	Jin and Lu (1998)
	Kir3.1/Kir3.2d	IC_{50} = 5.4 nM	Matsushita and Kurachi
	Kir3.2d	IC_{50} = 7 nM	(unpublished)
Anti-arrhythmic drugs			
Quinidine	GTP or GTPgS activated Kir3.1/Kir3.4 (I_{KACh})	IC_{50} = 10, 14 mM; 100% inhibition at 0.5 mM	Kurachi *et al.* (1987)
AN-132	GTP or GTPgS activated Kir3.1/Kir3.4 (I_{KACh})	IC_{50} = 0.3, >10 mM; 100, 48% inhibition at 0.1 mM	Kurachi *et al.* (1989)
Antipsychotic drugs			
Haloperidol	Kir3.1/Kir3.4, Kir3.1/Kir3.2	IC_{50} = 41, 76 mM; 30–35% inhibition at 0.3 mM	Kobayashi *et al.* (2000)
Thioridazine	Kir3.1/Kir3.4, Kir3.1/Kir3.2	IC_{50} = 84, 58 mM; 92% inhibition at 0.5 mM	
Pimodine	Kir3.1/Kir3.4, Kir3.1/Kir3.2	IC_{50} = 2.2, 3 mM; 40% inhibition at 30 mM	
Clozapine	Kir3.1/Kir3.4, Kir3.1/Kir3.2	IC_{50} = 110, 180 mM; 50% inhibition at 0.3 mM	
Antidepressant drugs			
Imipramine	Kir3.1/Kir3.4, Kir3.1/Kir3.2	IC_{50} = >50, 47 mM; 36, 62% inhibition at 1 mM	
Desipramine	Kir3.1/Kir3.4, Kir3.1/Kir3.2	IC_{50} = 54, 36 mM; 72, 76% inhibition at 1 mM	Kobayashi *et al.* (2004)
Amitriptyline	Kir3.1/Kir3.4, Kir3.1/Kir3.2	IC_{50} = 270, 111 mM; 54, 58% inhibition at 1 mM	
Nortriptyline	Kir3.1/Kir3.4, Kir3.1/Kir3.2	IC_{50} = 390, 130 mM; 59, 65% inhibition at 1 mM	
Clomipramine	Kir3.1/Kir3.4, Kir3.1/Kir3.2	IC_{50} = 250, 140 mM; 53, 63% inhibition at 1 mM	
Maprotiline	Kir3.1/Kir3.4, Kir3.1/Kir3.2	IC_{50} = 290, 113 mM; 58, 67% inhibition at 0.5 mM	
Citalopram	Kir3.1/Kir3.4, Kir3.1/Kir3.2	6, 22% inhibition at 0.1 mM	
Buproprion	Kir3.1/Kir3.4, Kir3.1/Kir3.2	4–5% inhibition at 0.1 mM	
Fluoxetine	Kir3.1/Kir3.4, Kir3.1/Kir3.2, Kir3.2	IC_{50} = 18,17, 90 mM; 63, 74, 58% inhibition at 0.3 mM	Kobayashi *et al...* (2003)
Volatile anaesthetic			
Halothane	Kir3.1/Kir3.2, Kir3.2	30–40% inhibition at 0.25 mM	Yamakura *et al.* (2001)
	Kir3.1/Kir3.4, Kir3.1F137S	90% activation at 0.9 mM	Weigl *et al.* (2001)
	Kir3.1/Kir3.4, Kir3.1F137S, Kir3.2	IC_{50} = 60 mM; 34% inhibition at 0.1–0.3 mM *	Milovic *et al.* (2004)
	Kir3.1/Kir3.4, Kir3.1F137S	20% activation at 0.9 mM*	
F3	Kir3.1/Kir3.4, Kir3.1/Kir3.2, Kir3.2	41, 78, 78% inhibition at 0.8 mM	Yamakura *et al.* (2001)
Isoflurane	Kir3.1/Kir3.2, Kir3.2	20–25% inhibition at 0.6 mM	
Enflurane	Kir3.1/Kir3.2, Kir3.2	30–40% inhibition at 0.6, 1.2 mM	
Local anaesthetic			
Bupivacaine	Kir3.1/Kir3.4, Kir3.1/Kir3.2		
D_1 receptor antagonist			
SCH23390		K_i = 22 mM	
NMDA receptor antagonist			
Ifenprodil	Kir3.1/Kir3.2, Kir3.2	K_i = 8, 80 mM; 85% inhibition at 0.3 mM	Zhou *et al.* 2001a
MK-801			Kuzhikandathil and Oxford 2002
Inhaled drug			
Toluene	Kir3.1/Kir3.2, Kir3.1/wvKir3.2	IC_{50} = ~200, 3.1 mM; ~40, 80% inhibition at 0.1 mM[†]	Kobayashi *et al...* (2006)
Ca^{2+} channel blocker			
Verapamil	Kir3.2	20% inhibition and 30% inhibition* at 3 mM	Kofuji *et al.* (1996)
Na^+ channel blocker			
QX-314	Kir3.1/Kir3.2, Kir3.1/wvKir3.2	IC_{50} = 120, 3.5 mM	Del Re *et al.* (2006)
	Kir3.1/Kir3.2, Kir3.1/wvKir3.2	IC_{50} = >200, 11 mM; ~80% inhibition at 0.3 mM[†]	Kofuji *et al.* (1996)

*, agonist activated current;

[†], Gbg activated current; all other efficacy measurements relate to basal currents.

AN-132, 3- (diisopropylaminoethylamino)-2′,6′-dimethylpropionanilide; SCH23390, *R*-(+)-7-chloro-8-hydroxy-3-methyl-1-phenyl-2,3,4,5-tetrahydro-1*H*-3-benzazepine hydrochloride; F3, 1-chloro-1,2,2-trifluorocyclobutane.

bradycardia was identified by measuring an increase in K^+ efflux (Hutter and Trautwein, 1955) across the cardiac cell membrane; an effect that leads to membrane potential hyperpolarization (Burgen and Terroux, 1953; del Castillo and Katz, 1955). Trautwein and his colleagues (Trautwein and Dudel, 1958; Noma and Trautwein, 1978; Osterrieder et al., 1981; Sakmann et al., 1983) analysed the ACh-induced K^+ current in the rabbit sinoatrial node which slowly increased in amplitude with a hyperpolarizing pulse, and proposed that ACh induced activation of a specific population of K^+ channels, which they called muscarinic K^+ (K_{ACh}) channels. Currently, this K_{ACh} channel is known as the cardiac K_G channel which is composed of Kir3.1 and Kir3.4 (Krapivinsky et al., 1995b).

Fundamentals of K_G channel activation

The G protein is a heterologous complex comprised of $G\alpha$ and $G\beta\gamma$ subunits (Gilman, 1987). In the absence of agonist, $G\alpha$ binds GDP and $G\beta\gamma$. Upon an agonist binding to a GPCR, $G\alpha$ releases GDP and instead binds GTP, leading to the dissociation of $G\beta\gamma$. Both $G\alpha$ and $G\beta\gamma$ subunits are able to regulate effectors (Clapham and Neer, 1993; Kurachi, 1995). $G\alpha$ has an intrinsic GTPase activity and, when the GTP bound to $G\alpha$ is hydrolysed to GDP, the GDP-bound $G\alpha$ re-assembles with $G\beta\gamma$ to sequester free $G\beta\gamma$.

The mechanism by which the K_G channel is activated by $G\beta\gamma$ has been mainly investigated in inside-out membrane patches from atrial myocytes (Kurachi et al., 1986; Logothetis et al., 1987; Ito et al., 1991, 1992; Yamada et al., 1993, 1994). Application of $G\beta\gamma$, but not GDP-bound nor GTP-bound $G\alpha$, to the intracellular side specifically acivates the K_G channel. Application of GDP-bound $G\alpha$ suppresses both GTP and GTPγS-induced K_G channel activity. $G\beta\gamma$-dependent activation of K_G channel has also been confirmed by the co-expression of $G\beta1$ and $G\gamma2$ subunits with Kir3.1 in Xenopus oocytes which yielded a current identical to that evoked by GPCR stimulation (Reuveny et al., 1994). A possible direct interaction between $G\beta\gamma$ and K_G channel subunit has been examined by biochemical pull-down, yeast two hybrid and electrophysiological studies (Huang et al., 1995; Inanobe et al., 1995b; Krapivinsky et al., 1995a; Kunkel and Peralta, 1995; Slesinger et al., 1995; Yan and Gautam, 1996; Huang et al., 1997; He et al., 1999; Ivanina et al., 2003). The results of these studies indicated that the $G\beta\gamma$-binding site distributes to both N- and C-termini. Furthermore, Huang and colleagues (1995, 1997) have shown that N- and C-terminal domains of Kir3.1 and Kir3.4 bind to each other and synergistically enhance $G\beta\gamma$-binding activity. In the crystal structure of the cytoplasmic region of Kir3.1, two neighbouring C-termini interact with each other, and an N-terminus is present between two neighbouring C-termini contributing to the subunit interface (Nishida and MacKinnon, 2002). These results indicate that fragments of K_G channel subunits have the opportunity to possess conformations different from those in native channels. Recently, the $G\beta\gamma$-binding site on Kir3.2 has been mapped to the βL–βM loop of the C-terminus which is present at the peripheral cytoplasmic region and near the N-terminus (Finley et al., 2004). Whereas this agrees with previous studies, it is not clearly understood how the binding of $G\beta\gamma$ to the K_G channel leads to the opening of the channel.

$G\beta\gamma$ association on K_G channels

As mentioned above, the K_G channel is a constitutively inactive channel. Internal perfusion of trypsin causes the activation of the K_G channel in atrial myocytes (Ito et al., 1992), suggesting that channel activity is attenuated by a protein component which is removed by the $G\beta\gamma$ association. Dascal and colleagues (1995) proposed that the C-terminus of Kir3.1 functions as the 'Shaker-ball' of the Kv channel (Dascal et al., 1995). On the other hand, Slesinger and colleagues (1995) demonstrated that in a Kir2.1 mutant in which its C-terminus is substituted by that of Kir3.1 there was a twofold increase in the current in the presence of $G\beta\gamma$ which could be accounted for by an increase in the open probability of each channel and the number of functional channels. This is consistent with the notion that $G\beta\gamma$ activates the native cardiac K_G channel by increasing the number of functional channels (Hosoya et al., 1996).

The cytoplasmic region of K_G channels receives physiological stimuli, leading to the channel gating. As the structure of Kir3.1 resembles that of constitutively active Kir2.1, the structural elements in the cytoplasmic region of K_G channel responsible for the channel gating remain to be elucidated. At present, $G\beta\gamma$-induced rotation and expansion at the N- and C-termini of the Kir3.1/Kir3.4 channel are predicted by fluorescence resonance energy transfer studies (Riven et al., 2003). On the other hand, Kir1.1 (Schulte et al., 1998; Rapedius et al., 2006) and KirBac3.1 (Kuo et al., 2005) are reported to undergo significant conformational change at the cytoplasmic region during channel gating. Thus, the functional coupling between the structural alteration at this cytoplasmic ring and channel gating needs to be clarified in the future.

The K_G channel containing Kir3.2 or Kir3.4 is thought to be activated by intracellular Na^+, as mentioned above (Petit-Jacques et al., 1995; Sui et al., 1996). A potential residue to sense the Na^+ ion is Asp228 in Kir3.2 which is located at the loop between βC- and βD-strands (CD loop) (Ho and Murrell-Lagnado, 1999; Zhang et al., 1999). Comparison of the crystal structure of the cytoplasmic region of Kir3.1 with that of Kir3.2 in the presence of Na^+ around half maximal activation concentration revealed the disposition of the side chain of Asp228 and the diversity of CD loop (Inanobe et al., 2007). The loop is located at the interface with the plasma membrane, suggesting association between transmembrane helices and involvement in the regulation of channel gating.

RGS proteins are involved in physiological control of ACh activation of K_G channels

Recently, regulators of G protein signalling (RGS) have been identified to accelerate intrinsic GTPase hydrolysis of the $G\alpha$ subunit (Hepler, 1999; Ross and Wilkie, 2000). RGS proteins are supposed to play essential roles in the negative regulation of various G protein-mediated cell-signalling systems. In reconstituted systems, using Xenopus oocytes or mammalian cell lines, RGS proteins accelerate the time course of activation and deactivation of K_G currents induced by agonists (Doupnik et al., 1997; Saito et al., 1997; Fujita et al., 2000). Since the RGS proteins behave as GTPase-activating proteins and thus reduce the quantity of GTP-bound $G\alpha$, their acceleration of the 'turn-off' response could be due to the accelerated sequestration of $G\beta\gamma$ by the increase of GDP-bound $G\alpha$. The acceleration of the 'turn-on' rate by RGS proteins is still unknown.

Voltage-dependent relaxation gating of K_G channels is caused by RGS proteins

When the membrane potential is suddenly shifted from one voltage to another, the plasma membrane conductance (g_K) of the K_G channel alters via two kinetically distinct processes. Upon hyperpolarization, g_K increases instantaneously to a certain level and then increases slowly to the steady-state level. The reverse reaction, i.e. fast and slow decrease in g_K, occurs upon membrane depolarization. The fast component is common to almost all Kir channels whereas the slow component is characteristic of K_G channels. The fast component of the voltage-dependent change in g_K of Kir channels is due to block of the channel pore by intracellular cations such as Mg^{2+} and polyamines. The slow component of the voltage-dependent change in g_K of K_G channels is called 'relaxation'. This is characterized by a slow time-dependent current increase during hyperpolarizing pulses and reflects a slow recovery from inhibition at depolarized voltages by an unknown mechanism. Interestingly, this characteristic depends on the concentration of agonist. Since it was first described in sinoatrial node cells (Noma and Trautwein, 1978), the molecular mechanism underlying this characteristic feature of the K_G current has been an enigma. K_G currents reconstituted in *Xenopus* oocytes by expressing Kir3.1/Kir3.4 and the M_2 muscarinic acetylcholine receptor do not exhibit relaxation, but co-expression of RGS proteins confers the native relaxation behaviour to the reconstituted K_G current (Fujita et al., 2000). This effect was mediated exclusively by the interaction of the RGS domain of the RGS protein with the PTX-sensitive $G\alpha$ subunit (Inanobe et al., 2001).

In cardiac myocytes, it has been shown that depolarization produces facilitation of RGS activity and the attenuation of K_G channel availability, leading to the relaxation of the native K_G channel (Ishii et al., 2001, 2002). The molecular signalling mechanisms for this effect have recently been elucidated as follows: at diastolic membrane potentials the GAP action of RGS proteins is inhibited by phosphatidylinositol-3,4,5-trisphosphate (PIP_3). Binding of Ca^{2+}–calmodulin (CaM) to RGS relieves PIP_3-mediated inhibition, restores the GAP activity of RGS proteins, decreases free $G\beta\gamma$ and thereby the number of active K_G channels (Ishii et al., 2005). The Ca^{2+}–CaM complex is formed by depolarization-induced Ca^{2+} influx across the plasma membrane. Therefore, at systolic membrane voltages the G protein cycle is negatively regulated and the number of active K_G channels is decreased. Thus 'relaxation' turns out not to be a gating process but to reflect a biochemical reaction between RGS and G proteins. In this context the K_G channel can be regarded as a G protein-effector molecule through which we can detect the G protein cycle in real time with high temporal resolution.

Requirement of PIP_2 on K_G channel function

Phosphatidylinositol-4,5-bisphosphate (PIP_2) is shown to regulate the activity of native and recombinant Kir channels, IP_3 receptor and transporters, such as the sodium-calcium exchanger (Sui et al., 1996). In Kir channels, it was known that intracellular ATP prevents run-down of the functional channel and restores the run-down channel in inside-out patches. PIP_2 could mimic these effects of intracellular ATP (Huang et al., 1998). Amino acid residues responsible for PIP_2-binding are located at the loop between βC- and βD-strands in the cytoplasmic domain of the K_G channel, where the domain faces the plasma membrane (Zhang et al., 1999). K_G channels composed of Kir3.1 and Kir3.4 show low dependence on the position of phosphates of the phospholipid head group and high dependence on the phospholipid acyl chains. The most potent PIP_2 analogue possesses arachidonyl and stearyl chains (Roháscs et al., 1999). Structural information about the complex of the K_G channel and PIP_2 will provide mechanistic insight into its role in channel regulation.

Vesicular K_G channel

The stimulation of thyrotrophs with thyrotropin-releasing hormone causes the fusion of secretory vesicles to the plasma membrane and enhancement of dopamine- or somatostatin-induced K_G currents (Morishige et al., 1999). This is a novel mechanism for the rapid insertion of functional ion channels into the plasma membrane which may contribute to forming an effective negative feedback control loop for hormone secretion by hyperpolarizing the membrane potential. A similar feedback mechanism might be present in the nervous system where some K_G channels are known to be localized on the presynaptic region of axonal termini (Morishige et al., 1996; Ponce et al., 1996; Koyrakh et al., 2005).

4.4.2.7 Relevance to disease states

No mutation in K_G channels has been found to cause any hereditary disease in humans.

Weaver

The *weaver* (*wv*) mutant mouse has severe locomotor defects, a deficiency in the migration of granule cells of the cerebellum (Rakic and Sidman, 1973), and cellular deficits in the midbrain dopaminergic system (Schmidt et al., 1982). The homozygous male *wv* mouse is essentially infertile due to the failure of sperm production, but females are fertile (Harrison and Roffler-Tarlov, 1994). Concentrations of hormone related to the hypothalamic–pituitary–gonadal axis are comparable to those in the wild-type (Schwartz et al., 1998). Patil and colleagues (1995) identified that a point mutation in the Kir3.2 gene is responsible for the abnormalities in the *wv* mouse. The *wv* mutation causes alteration of Gly-Tyr-Gly to Ser-Tyr-Gly in the signature amino acid sequence in the selectivity filter, which results in the loss of selectivity of K^+ ions over Na^+ ions in the *wv* K_G channels (Kofuji et al., 1996; Slesinger et al., 1996). Therefore, the presence of inhibitory neurotransmitters is thought to cause neurons to be depolarized, resulting in neuronal death. Kir3.2d, one of Kir3.2 isoforms, specifically expressed in testis, is present at the acrosome of spermatids (Inanobe et al., 1999a). Its distribution shows a significant agreement with the developmental stage of defects in the *wv* mouse. However, the function of Kir3.2 in spermogenesis is still unclear. In respect to regulation of the *wv* K_G channel, the channel is constitutively active and insensitive to G protein regulation (Navarro et al., 1996; Slesinger et al., 1996).

Atrial fibrillation

Atrial fibrillation (AF) is a common cardiac arrhythmia in the clinic. Electrical remodelling such as decreased L-type Ca^{2+} current density (Van Wagoner et al., 1999) and increased classical inward rectifier K^+ current density (Dobrev et al., 2002) is thought to be one of the mechanisms of AF. However, the molecular mechanisms underlying this arrhythmia are only partly understood. Recently, I_{KACh} has been suggested to play an important role in chronic AF (cAF). Dobrev et al., (2005) have suggested that the increased Kir currents in cAF patients include not only I_{K1}, but also constitutively active I_{KACh}. These channels may cause shortening of the refractory period of atrium. In the same myocytes, they also found the small muscarinic acetylcholine receptor-dependent I_{KACh} current. Whereas the molecular basis of constitutively active I_{KACh} in cAF patients is unknown, these results suggest that I_{KACh} may be a novel therapeutic target for AF prevention and treatment.

4.4.2.8 Concluding remarks

As a simplification, GPCRs couple predominantly to a subset of G-proteins, leading to the activation of specific effectors responsible for the physiological effects observed (Yamada *et al.*, 1998; Hille 2001). A classical view of the signal transduction that membrane components freely diffuse and interactions occur by collision has been shifted to a new idea that signalling cascades may be more structured and either protein or lipid may participate in this organization of the cascade (Jacobson *et al.*, 2007; Allen *et al.*, 2007). The direct interaction of inhibitory Gα subunits with the K_G channel has been reported to be involved in the mechanism of G protein-specificity (Huang *et al.*, 1995; Schreibmayer *et al.*, 1996; Peleg *et al.*, 2002). More recent works suggest that GPCRs, G proteins, RGSs and K_G channels may be pre-coupled without signals or may be gathered in close vicinity (Zhang *et al.*, 2002; Clancy *et al.*, 2005; Nobles *et al.*, 2005; Riven *et al.*, 2006). Furthermore, the cardiac K_G channel is isolated as a signalling complex with not only G proteins, but also various signalling molecules such as protein kinases and phosphatases (Nikolov and Ivanova-Nikolova, 2004). Comprehensive understanding of the regulation for the K_G channel and its structure–function relationship will undoubtedly shed light on the physiological function of this channel.

References

Allen JA, Halverson-Tamboli RA and Rasenic MM (2007). Lipid raft microdomains and neurotransmitter signaling. *Nat Rev Neurosci* **8**, 128–140.

Bichet D, Grabe M, Jan YN *et al.* (2006). Electrostatic interactions in the channel cavity as an important determinant of potassium channel selectivity. *PNAS* **103**, 14355–14360.

Burgen ASV and Terroux KG (1953). On the negative inotropic effect in the cat's auricle. *J. Physiol* **120**, 449–464.

Cha TJ, Ehrlich JR, Chartier D *et al.* (2006). Kir3-based inward rectifier potassium current: potential role in atrial tachycardia remodeling effects on atrial repolarization and arrhythmias. *Circulation* **113**, 1730–1737.

Clancy SM, Fowler CE, Finley M *et al.* (2005). Pertussis-toxin-sensitive Gα subunits selectively bind to C-terminal domain of neuronal GIRK channels: evidence for a heterotrimeric G-protein-channel complex. *Mol Cell Neurosci* **28**, 375–389.

Clapham DE and Neer EJ (1993). New roles for G-protein βγ-dimers in transmembrane signalling. *Nature* **365**, 403–406.

Corey S and Clapham DE (1998). Identification of native atrial G-protein-regulated inwardly rectifying K$^+$ (GIRK4) channel homomultimers. *J Biol Chem* **273**, 27499–27504.

Corey S, Krapivinsky G, Krapivinsky L *et al.* (1998). Number and stoichiometry of subunits in the native atrial G-protein-gated K$^+$ channel, I_{KACh}. *J Biol Chem* **273**, 5271–5278.

Dascal N, Doupnik CA, Ivanina T *et al.* (1995). Inhibition of function in *Xenopus* oocytes of the inwardly rectifying G-protein-activated atrial K channel (GIRK1) by overexpression of a membrane-attached form of the C-terminal tail. *PNAS* **92**, 6758–6762.

Dascal N, Schreibmayer W, Lim NF *et al.* (1993). Atrial G protein-activated K$^+$ channel: expression cloning and molecular properties. *PNAS* **90**, 10235–10239.

del Castillo J and Katz B (1955). Production of membrane potential changes in the frog's heart by inhibitory nerve impulses. *Nature* **175**, 1035.

Del Re AM, Dopico AM and Woodward JJ (2006). Effects of the abused inhalant toluene on ethanol-sensitive potassium channels expressed in oocytes. *Brain Res* 1087, 75–82.

Dobrev D, Friedrich A, Voigt N, *et al.* (2005). The G protein-gated potassium current $I_{K,ACh}$ is constitutively active in patients with chronic atrial fibrillation. *Circulation* **112**, 3697–3706.

Dobrev D, Wettwer E, Kortner A *et al.* (2002). Human inward rectifier potassium channels in chronic and postoperative atrial fibrillation. *Cardiovasc Res* **54**, 397–404.

Doupnik CA, Davidson N, Lester HA *et al.* (1997). RGS proteins reconstitute the rapid gating kinetics of gbetagamma-activated inwardly rectifying K$^+$ channels. *PNAS* **94**, 10461–10466.

Doyle DA, Morais Cabral J, Pfuetzner RA *et al.* (1998). The structure of the potassium channel: molecular basis of K$^+$ conduction and selectivity. *Science* **280**, 1152–1155.

Duprat FD, Lesage F, Guillemare E *et al.* (1995). Heterologous multimericassembly is essential for K$^+$ channel activity of neuronal and cardiac G-protein-activated inward rectifiers. *Biochem Biophys Res Commun* **212**, 657–663.

Ficker E, Taglialatela M, Wible BA *et al.* (1994). Spermine and spermidine as gating molecules for inward rectifier K$^+$ channels. *Science* **266**, 1068–1072.

Finley M, Arrabit C, Fowler C *et al.* (2004). βL–βM loop in the C-terminal domain of G protein-activated inwardly rectifying K$^+$ channels is important for Gβγ subunit activation. *J Physiol* **555**, 643–657.

Fujita S, Inanobe A, Chachin M *et al.* (2000). A regulator of G protein signalling (RGS) protein confers agonist-dependent relaxation gating to a G protein-gated K$^+$ channel. *J Physiol* **526**, 341–347.

Gilman AG (1987). G proteins: transducers of receptor-generated signals. *Annu Rev Biochem* **56**, 615–649.

Grosse G, Eulitz D, Thiele T *et al.* (2003). Axonal sorting of Kir3.3 defines a GABA-containing neuron in the CA3 region of rodent hippocampus. *Mol Cell Neurosci* **24**, 709–724.

Harrison SMW and Roffler-Tarlov S (1994). Male-sterile phenotype of the neurological mouse mutant *weaver*. *Dev Dyn* **200**, 26–38.

Hashimoto N, Yamashita T and Tsuruzoe N (2006). Tertiapin, a selective IK_{ACh} blocker, terminates atrial fibrillation with selective atrial effective refractory period prolongation. *Pharmacol Res* **54**, 136–141.

He C, Zhang H, Mirshahi T *et al.* (1999). Identification of a potassium channel site that interacts with G protein βγ subunits to mediate agonist-induced signaling. *J Biol Chem* **274**, 12517–12524.

Hepler JR (1999). Emerging roles of RGS proteins in cell signalling. *Trends in Pharmacol Sci* **20**, 376–382.

Hibino H, Inanobe A, Tanemoto M *et al.* (2000). Anchoring proteins confer G protein sensitivity to an inward-rectifier K$^+$ channel through the GK domain. *EMBO J* **19**, 78–83.

Hille B (2001). *Ion channels of excitable membranes*, Sinauer, Sunderland, Massachusetts, USA. 3rd edition.

Ho IHM and Murrell-Lagnado RD (1999). Molecular determinants for sodium-dependent activation of G protein-gated K$^+$ channels. *J Biol Chem* **274**, 8639–8648.

Ho K, Nichols CG, Lederer WJ *et al.* (1993). Cloning and expression of an inwardly rectifying ATP-regulated potassium channel. *Nature* **362**, 31–38.

Hosoya Y, Yamada M, Ito H *et al.* (1996). A functional model for G protein activation of the muscarinic K$^+$ channel in guinea pig atrial myocytes. Spectral analysis of the effect of GTP on single-channel kinetics. *J Gen Physiol* **108**, 485–495.

Huang C, Jan YN and Jan LY (1997). Binding of the G protein βγ subunit to multiple regions of G protein-gated inward-rectifying K$^+$ channels. *FEBS Lett* **405**, 291–298.

Huang CL, Feng S and Hilgemann DW (1998). Direct activation of inward rectifier potassium channels by PIP_2 and its stabilization by Gβγ. *Nature* **391**, 803–806.

Huang CL, Slesinger PA, Casey PJ *et al.* (1995). Evidence that direct binding of Gβγ to the GIRK1 G protein-gated inwardly rectifying K$^+$ channel is important for channel activation. *Neuron* **15**, 1133–1143.

Hutter OF and Trautwein W (1955). Vagal and sympathetic effects on the pacemaker fibers in the sinus venosus of the heart. *J Gen Physiol* **39**, 715–733.

Iizuka M, Tsunenari I, Momota Y *et al.* (1997). Localization of a G-protein-coupled inwardly rectifying K$^+$ channel, CIR, in the rat brain. *Neuroscience* **77**, 1–13.

Inanobe A and Kurachi Y (2000). G protein-gated K+ channels. In M Endo, Y Kurachi and M Mishina, eds, *Handbook of experimental pharmacology, Pharmacology of ionic channel function: activators and inhibitors*, **147**, 216–250.

Inanobe A, Fujita S, Makino Y *et al.* (2001). Interaction between the RGS domain of RGS4 with G protein a subunits mediates the voltage-dependent relaxation of the G protein-gated potassium channel. *J Physiol* **535**, 133–143.

Inanobe A, Horio Y, Fujita A *et al.* (1999a). Molecular cloning and characterization of a novel splicing variant of the Kir3.2 subunit predominantly expressed in mouse testis. *J Physiol* **521**, 19–30.

Inanobe A, Ito H, Ito M *et al.* (1995a). Immunological and physical characterization of the brain G protein-gated muscarinic potassium channel. *Biochem Biophysl Res Commun* **217**, 1238–1244.

Inanobe A, Matuura T, Nakagawa A *et al.* (2007). Structural diversity in the cytoplasmic region of G protein-gated inward rectifier K+ channels. *Channels* **1**, 39–45.

Inanobe A, Morishige KI, Takahashi N *et al.* (1995b). Gβγ directly binds to the carboxyl terminus of the G protein-gated muscarinic K+ channel, GIRK1. *Biochem Biophys Res Commun* **212**, 1022–1028.

Inanobe A, Yoshimoto Y, Horio Y *et al.* (1999b). Characterization of G-protein-gated K+ channels composed of Kir3.2 subunits in dopaminergic neurons of the substantia nigra. *J Neurosci* **19**, 1006–1017.

Ishii M, Fujita S, Yamada M *et al.* (2005). Phosphatidylinositol 3,4,5-trisphosphate and Ca^{2+}/calmodulin competitively bind to the regulators of G-protein-signalling (RGS) domain of RGS4 and reciprocally regulate its action. *Biochem J* **385**, 65–73.

Ishii M, Inanobe A and Kurachi Y (2002). PIP$_3$ inhibition of RGS protein and its reversal by Ca^{2+}/calmodulin mediate voltage-dependent control of the G protein cycle in a cardiac K+ channel. *PNAS* **99**, 4325–4330.

Ishii M, Inanobe A, Fujita S *et al.* (2001). Ca^{2+} elevation evoked by membrane depolarization regulates G protein cycle via RGS proteins in the heart. *Circ Res* **89**, 1045–1050.

Isomoto S, Kondo C, Takahashi N *et al.* (1996). A novel ubiquitously distributed isoform of GIRK2 (GIRK2B) enhances GIRK1 expression of the G-protein-gated K+ current in *Xenopus* oocytes. *Biochem Biophys Res Commun* **218**, 286–291.

Ito H, Sugimoto T, Kobayashi I *et al.* (1991). On the mechanism of basal and agonist-induced activation of the G protein-gated muscarinic K+ channel in atrial myocytes of guinea pig heart. *J Gen Physiol* **98**, 517–533.

Ito H, Tung RT, Sugimoto T *et al.* (1992). On the mechanism of G protein bg subunit activation of the muscarinic K+ channel in guinea pig atrial cell membrane: comparison with the ATP-sensitive K+ channel. *J Gen Physiol* **99**, 961–983.

Ivanina T, Rishal I, Varon D *et al.* (2003). Mapping the Gβγ-binding sites in GIRK1 and GIRK2 subunits of the G protein-activated K+ channel. *J Biol Chem* **278**, 29174–29183.

Jacobson K, Mouritsen OG and Anderson GW (2007). Lipid rafts: at a crossroad between cell biology and physics. *Nat Cell Biol* **9**, 7–14.

Jelacic TM, Kennedy ME, Wickman K *et al.* (2000). Functional and biochemical evidence for G-protein-gated inwardly rectifying K+ (GIRK) channels composed of GIRK2 and GIRK3. *J Biol Chem* **275**, 36211–36216.

Jelacic TM, Sims SM and Clapham DE (1999). Functional expression and characterization of G-protein-gated inwardly rectifying K+ channels containing GIRK3. *J Membr Biol* **169**, 123–129.

Jiang Y, Lee A, Chen J *et al.* (2002). Crystal structure and mechanism of a calcium-gated potassium channel. *Nature* **417**, 515–522.

Jiang Y, Lee A, Chen J *et al.* (2003). X-ray structure of a voltage-dependent K+ channel. *Nature* **423**, 33–41.

Jin W and Lu Z (1998). A novel high-affinity inhibitor for inward-rectifier K+ channels. *Biochemistry* **37**, 13291–13299.

Jin W and Lu Z (1999). Synthesis of a stable form of tertiapin: a high-affinity inhibitor for inward-rectifier K+ channels. *Biochemistry* **38**, 14286–14293.

Karschin C, Dißmann E, Stühmer W *et al.* (1996). IRK(1–3) and GIRK(1–4) inwardly rectifying K+ channel mRNAs are differentially expressed in the adult rat brain. *J Neurosci* **16**, 3559–3570.

Kitamura H, Yokoyama M, Akita H *et al.* (2000). Tertiapin potently and selectively blocks muscarinic K+ channels in rabbit cardiac myocytes. *JPET* **293**, 196–205.

Kobayashi T, Ikeda K and Kumanishi T (2000). Inhibition of various antipsychotic drugs of the G-protein-activated inwardly rectifying K+ (GIRK) channels expressed in *Xenopus* oocytes. *Br J Pharmacol* **129**, 1716–1722.

Kobayashi T, Ikeda K, Ichikawa T *et al.* (1995). Molecular cloning of a mouse G-protein-activated K+ channel (mGIRK1) and distinct distributions of three GIRK (GIRK1, 2 and 3) mRNAs in mouse brain. *Biochem Biophys Res Commun* **208**, 1166–1173.

Kobayashi T, Ikeda K, Kojima H *et al.* (1999). Ethanol opens G-protein-activated inwardly rectifying K+ channels. *Nat Neurosci* **2**, 1091–1097.

Kobayashi T, Washiyama K and Ikeda K (2003). Inhibition of G protein-activated inwardly rectifying K+ channels by fluoxetine (Prozac). *Br J Pharmacol* **138**, 1119–1128.

Kobayashi T, Washiyama K and Ikeda K (2004). Inhibition of G protein-activated inwardly rectifying K+ channels by various antidepressant drugs. *Neuropsychopharmacol* **29**, 1841–1851.

Kobayashi T, Washiyama K and Ikeda K (2006). Inhibition of G protein-activated inwardly rectifying K+ channels by ifenprodil. *Neuropsychopharmacol* **31**, 516–524.

Kofuji P, Davidson N and Lester HA (1995). Evidence that neuronal G-protein-gated inwardly rectifying K+ channels are activated by Gβγ subunits and function as heteromultimers. *PNAS* **92**, 6542–6546.

Kofuji P, Hofer M, Millen KJ *et al.* (1996). Functional analysis of the *weaver* mutant GIRK2 K+ channel and rescue of *weaver* granule cells. *Neuron* **16**, 941–952.

Koyrakh L, Lujan R, Colon J *et al.* (2005). Molecular and cellular diversity of neuronal G-protein-gated potassium channels. *J Neurosci* **25**, 11468–11478.

Krapivinsky G, Gordon EA, Wickman K *et al.* (1995b). The G-protein-gated atrial K+ channel I$_{KACh}$ is a heteromultimer of two inwardly rectifying K+-channel proteins. *Nature* **374**, 135–141.

Krapivinsky G, Krapivinsky L, Wickman K *et al.* (1995a). Gβγ binds directly to the G protein-gated K+ channel, I$_{KACh}$. *J Biol Chem* **270**, 29059–29062.

Kubo Y, Adelman JP, Clapham DE *et al.* (2005). International Union of Pharmacology. LIV. Nomenclature and molecular relationships of inwardly rectifying potassium channels. *Pharmacol Rev* **57**, 509–526.

Kubo Y, Baldwin TJ, Jan YN *et al.* (1993a). Primary structure and functional expression of a mouse inward rectifier potassium channel. *Nature* **362**, 127–133.

Kubo Y, Reuveny E, Slesinger PA *et al.* (1993b). Primary structure and functional expression of a rat G-protein-coupled muscarinic potassium channel. *Nature* **364**, 802–806.

Kulik A, Vida I, Fukazawa Y *et al.* (2006). Compartment-dependent colocalization of Kir3.2-containing K+ channels and GABA$_B$ receptors in hippocampal pyramidal cells. *J Neurosci* **26**, 443–449.

Kunkel MT and Peralta EG (1995). Identification of domains conferring G protein regulation on inward rectifier potassium channel. *Cell* **83**, 443–449.

Kuo A, Domene C, Johnson LN *et al.* (2005). Two different conformational states of the KirBac3.1 potassium channel revealed by electron crystallography. *Structure* **13**, 1463–1472.

Kuo A, Gulbis JM, Antcliff JF *et al.* (2003). Crystal structure of the potassium channel KirBac1.1 in the closed state. *Science* **300**, 1922–1926.

Kurachi Y (1995). G protein regulation of cardiac muscarinic potassium channel. *Am J Physiol* **269**, C821–830.

Kurachi Y, Nakajima T and Sugimoto T (1986). Acetylcholine activation of K⁺ channels in cell-free membrane of atrial cells. *Am J Physiol* **251**, H681-H684.

Kurachi Y, Nakajima T and Sugimoto T (1987). Quinidine inhibition of the muscarine receptor-activated K⁺ channel current in atrial cells of guinea pig. *Naunyn Schmiedebergs Arch Pharmacol* **335**, 216–218.

Kurachi Y, Nakajima T, Ito H et al. (1989). AN-132, a new class I anti-arrhythmic agent, depresses the acetylcholine-induced K⁺ current in atrial myocytes. *Eur J Pharmacol* **165**, 319–322.

Kuzhikandathil EV and Oxford GS (2002). Classic D1 dopamine receptor antagonist R-(+)-7-chloro-8-hydroxy-3-methyl-1-phenyl-2,3,4,5-tetrahydro-1H-3-benzazepine hydrochloride (SCH23390) directly inhibits G protein-coupled inwardly rectifying potassium channels. *Mol Pharmacol* **62**, 119–126.

Lesage F, Duprat F, Fink M et al. (1994). Cloning provides evidence for a family of inward rectifier and G-protein coupled K⁺ channels in the brain. *FEBS Lett* **353**, 37–42.

Lesage F, Guillemare E, Fink M et al. (1995). Molecular properties of neuronal G-protein-activated inwardly rectifying K⁺ channels. *J Biol Chem* **270**, 28660–28667.

Lewohl JM, Wilson WR, Mayfield RD et al. (1999). G-protein-coupled inwardly rectifying potassium channels are targets of alcohol action. *Nat Neurosci* **2**, 1084–1090.

Liao YJ, Jan YN and Jan LY (1996). Heteromultimerization of G-protein-gated inwardly rectifying K⁺ channel proteins GIRK1 and GIRK2 and their altered expression in weaver brain. *J Neurosci* **16**, 7137–7150.

Loewi O (1921). Über humorale Übertragbarkeit der Herznervenwirkung. *Pflugers Arch* **189**, 239–242.

Logothetis DE, Kurachi Y, Galper J et al. (1987). The βγ subunits of GTP-binding proteins activate the muscarinic K⁺ channel in heart. *Nature* **325**, 321–326.

Lopatin AN, Makhina EN and Nichols CG (1994). Potassium channel block by cytoplasmic polyamines as the mechanism of intrinsic rectification. *Nature* **372**, 366–369.

Luscher C, Jan LY, Stoffel M et al. (1997). G protein-coupled inwardly rectifying K⁺ channels (GIRKs) mediate postsynaptic but not presynaptic transmitter actions in hippocampal neurons. *Neuron* **19**, 687–695.

Ma D, Zerangue N, Raab-Graham K et al. (2002). Diverse trafficking patterns due to multiple traffic motifs in G protein-activated inwardly rectifying potassium channels from brain and heart. *Neuron* **33**, 715–729.

Matsuda H (1988). Open-state substructure of inwardly rectifying potassium channels revealed by magnesium block in guinea-pig heart cells. *J Physiol* **397**, 237–258.

Matsuda H, Saigusa A and Irisawa H (1987). Ohmic conductance through the inwardy rectifying K channel and blocking by intercellular Mg²⁺. *Nature* **325**, 156–159.

Milovic S, Steinecker-Frohnwieser B, Schreibmayer W et al. (2004). The sensitivity of G protein-activated K⁺ channels toward halothane is essentially determined by the C-terminus. *J Biol Chem* **279**, 34240–34249.

Mirshahi T, Mittal, V, Zhang, H et al. (2002). Distinct sites on G protein βγ subunits regulate different effector functions. *J Biol Chem* **277**, 36345–36350.

Morais-Cabral JH, Zhou Y and MacKinnon R (2001). Energetic optimization of ion conduction rate by the K⁺ selectivity filter. *Nature* **414**, 37–42.

Morishige K, Inanobe A, Yoshimoto Y et al. (1999). Secretagogue-induced exocytosis recruits G protein-gated K⁺ channels to plasma membrane in endocrine cells. *J Biol Chem* **274**, 7969–7974.

Morishige KI, Inanobe A, Takahashi N et al. (1996). G protein-gated K⁺ channel (GIRK1) protein is expressed presynaptically in the paraventricular nucleus of the hypothalamus. *Biochem Biophys Res Commun* **220**, 300–305.

Navarro B, Kennedy ME, Velimirovic B et al. (1996). Nonselective and Gβγ-insensitive weaver K⁺ channels. *Science* **272**, 1950–1953.

Nikolov EN and Ivanova-Nikolova TT (2004). Coordination of membrane excitability through a GIRK1 signaling complex in the atria. *J Biol Chem* **279**, 23630–23636.

Nishida M and MacKinnon R (2002). Structural basis of inward rectification: cytoplasmic pore of the G protein-gated inward rectifier GIRK1 at 1.8 angstrom resolution. *Cell* **111**, 957–965.

Nishida M and MacKinnon R (2007). Crystal structure of a Kir3.1-prokaryotic Kir channel chimera. *EMBO J* **26**, 4005–4015.

Nobles M, Benians A and Tinker A (2005). Heterotrimeric G proteins precouple with G protein-coupled receptors in living cells. *PNAS* **102**, 18706–18711.

Noma A and Trautwein W (1978). Relaxation of the ACh-induced potassium current in the rabbit sinoatrial node cell. *Pflugers Arch* **377**, 193-200.

Osterrieder W, Yang QF and Trautwein W (1981). The time course of the muscarinic response to ionophoretic acetylcholine application to the S-A node of the rabbit heart. *Pflugers Arch* **389**, 283–291.

Patil N, Cox DR, Bhar D et al. (1995). A potassium channel mutation in weaver mice implicates membrane excitability in granule cell differentiation. *Nat Genetics* **11**, 126–129.

Pegan S, Arrabit C, Zhou W et al. (2005). Cytoplasmic domain structures of Kir2.1 and Kir3.1 show sites for modulating gating and rectification. *Nat Neurosci* **8**, 279–287.

Peleg S, Varon D, Ivanina T et al. (2002). Gα_i controls the gating of the G protein-activated K⁺ channel, GIRK. *Neuron* **33**, 87–99.

Petit-Jacques J, Sui JL and Logothetis DE (1995). Synergistic activation of G protein-gated inwardly rectifying potassium channels by the βγ subunits of G proteins and Na⁺ and Mg²⁺ ions. *J Gen Physiol* **114**, 673–684.

Ponce A, Bueno E, Kentros C et al. (1996). G-protein-gated inward rectifier K⁺ channel proteins (GIRK1) are present in the soma and dendrites as well as in nerve terminals of specific neurons in the brain. *J Neurosci* **16**, 1990-2001.

Rakic P and Sidman RL (1973). Organization of cerebellar cortex secondary to deficit of granule cells in weaver mutant mice. *J Comp Neurol* **152**, 133–161.

Rapedius M, Haider S, Browne KF et al. (2006). Structural and functional analysis of the putative pH sensor in the Kir1.1 (ROMK) potassium channel. *EMBO Rep* **7**, 611–616.

Reuveny E, Slesinger PA, Inglese J et al. (1994). Activation of the cloned muscarinic potassium channel by G protein βγ subunits. *Nature* **370**, 143–146.

Riven I, Iwanir S and Reuveny E (2006). GIRK channel activation involves a local rearrangement of a preformed G protein channel complex. *Neuron* **51**, 561–573.

Riven I, Kalmanzon E, Segev L et al. (2003). Conformational rearrangements associated with the gating of the G protein-coupled potassium channel revealed by FRET microscopy. *Neuron* **38**, 222–235.

Rohács T, Chen J, Prestwich GD et al. (1999). Distinct specificities of inwardly rectifying K⁺ channels for phosphoinositides. *J Biol Chem* **274**, 36065–36072.

Ross EM and Wilkie TM (2000). GTPase-activating proteins for heterotrimeric G proteins: regulators of G protein signaling (RGS) and RGS-like proteins. *Annu Rev Biochem* **69**, 795–827.

Sadja R, Smadja K, Alagem N et al. (2001). Coupling Gβγ-dependent activation to channel opening via pore elements in inwardly rectifying potassium channels. *Neuron* **29**, 669–680.

Saito O, Kubo Y, Miyatani Y et al. (1997). RGS8 accelerates G-protein-mediated modulation of K⁺ currents. *Nature* **390**, 525–529.

Sakmann B, Noma A and Trautwein W (1983). Acetylcholine activation of single muscarinic K[+] channels in isolated pacemaker cells of the mammalian heart. *Nature* **303**, 250–253.

Schmidt MJ, Sawyer BD, Perry KW *et al.* (1982). Dopamine difficiency in the weaver mutant mouse. *J Neurosci* **2**, 376–380.

Schoots O, Wilson JM, Ethier N *et al.* (1999). Co-expression of human Kir3 subunits can yield channels with different functional properties. *Cell Signal* **11**, 871–883.

Schreibmayer W, Dessauer CW, Vorobiov D *et al.* (1996). Inhibition of an inwardly rectifying K[+] channel by G-protein α-subunits. *Nature* **380**, 624–627.

Schulte U, Hahn H, Wiesinger H *et al.* (1998). pH-dependent gating of ROMK (Kir1.1) channels involves conformational changes in both N and C termini. *J Biol Chem* **273**, 34575–34579.

Schwartz NB, Szabo M, Verina T *et al.* (1998). Hypothalamic–pituitary–gonadal axis in the mutant *weaver* mouse. *Neuroendocrinology* **68**, 374–385.

Sheng M and Sala C (2001). PDZ domains and the organization of supramolecular complexes. *Annu Rev Neurosci* **24**, 1–29.

Signorini S, Liao YJ, Duncan SA *et al.* (1997). Normal cerebellar development but susceptibility to seizures in mice lacking G protein-coupled, inwardly rectifying K[+] channel GIRK2. *PNAS* **94**, 923–927.

Silverman SK, Lester HA and Dougherty DA (1996). Subunit stoichiometry of a heteromultimeric G protein-coupled inward-rectifier K[+] channel. *J Biol Chem* **271**, 30524–30528.

Slesinger PA, Patil N, Liao YJ *et al.* (1996). Functional effects of the mouse *weaver* mutation on G protein-gated inwardly rectifying K[+] channels. *Neuron* **16**, 321–331.

Slesinger PA, Reuveny E, Jan YN *et al.* (1995). Identification of structural elements involved in G protein gating of the GIRK1 potassium channel. *Neuron* **15**, 1145–1156.

Stoffel M, Tokuyama Y, Trabb JB *et al.* (1995). Cloning of rat K$_{ATP-2}$ channel and decreased expression in pancreatic islets of male Zucker diabetic fatty rats. *Biochem Biophys Res Commun* **212**, 894–899.

Sui JL, Chan KW and Logothetis DE (1996). Na[+] activation of the muscarinic K[+] channel by a G-protein-independent mechanism. *J Gen Physiol* **108**, 381–391.

Torrecilla M, Marker CL, Cintora SC *et al.* (2002). G-protein-gated potassium channels containing Kir3.2 and Kir3.3 subunits mediate the acute inhibitory effects of opioids on locus ceruleus neurons. *J Neurosci* **22**, 4328–4334.

Trautwein W and Dudel J (1958). Mechanism of membrane effect of acetylcholine on myocardial fibers. *Pflugers Arch* **266**, 324–334.

Tucker SJ, Pessia M and Adelman JP (1996). Muscarine-gated K[+] channel: subunit stoichiometry and structural domains essential for G protein stimulation. *Am J Physiol* **271**, H379-H385.

Van Wagoner DR, Pond AL, Lamorgese M *et al.* (1999). Atrial L-type Ca[2+] currents and human atrial fibrillation. *Circ Res* **85**, 428–436.

Vandenberg CA (1987). Inward rectification of a potassium channel in cardiac ventricular cells depends on internal magnesium ions. *PNAS* **84**, 2560–2564.

Wei J, Hodes ME, Piva R *et al.* (1998). Characterization of murine Girk2 transcript isoforms: structure and differential expression. *Genomics* **51**, 379–390.

Weigl LG and Schreibmayer W (2001). G protein-gated inwardly rectifying potassium channesl are targets for volatile anesthetics. *Mol Pharmacol* **60**, 282–289.

Velimirovic BM, Gordon EA, Lim NF *et al.* (1996). The K[+] channel inward rectifier subunits form a channel similar to neuronal G protein-gated K[+] channel. *FEBS Lett* **379**, 31–37.

Wible BA, Taglialatela M, Ficker E *et al.* (1994). Gating of inwardly rectifying K[+] channels localized to a single negatively charged residue. *Nature* **371**, 246–249.

Wickman K, Karschin C, Karschin A *et al.* (2000). Brain localization and behavioral impact of the G-protein-gated K[+] channel subunit GIRK4. *J Neurosci* **20**, 5608–5615.

Wickman K, Nemec J, Gendler SJ *et al.* (1998). Abnormal heart rate regulation in GIRK4 knockout mice. *Neuron* **20**, 103–114.

Wickman K, Pu WT and Clapham DE (2002). Structural characterization of the mouse Girk genes. *Gene* **284**, 241–250.

Wischmeyer E, Doring F, Wischmeyer E *et al.* (1997). Subunit interactions in the assembly of neuronal Kir3.0 inwardly rectifying K[+] channels. *Mol Cell Neurosci* **9**, 194–206.

Yamada M and Kurachi Y (1995). Spermine gates inward-rectifying muscarinic but not ATP-sensitive K[+] channels in rabbit atrial myocytes. Intracellular substance-mediated mechanism of inward rectification. *J Biol Chem* **270**, 9289–9294.

Yamada M, Ho YK, Lee RH *et al.* (1994). Muscarinic K[+] channels are activated by βγ subunits and inhibited by the GDP-bound form of a subunit of transducin. *Biochem Biophys Res Commun* **200**, 1484–1490.

Yamada M, Inanobe A and Kurachi Y (1998). G protein regulation of potassium ion channels. *Pharmacol Rev* **50**, 723–760.

Yamada M, Jahangir A, Hosoya Y *et al.* (1993). GK* and brain Gβγ activate muscarinic K[+] channel through the same mechanism. *J Biol Chem* **268**, 24551–24554.

Yamakura T (2001). Differential effects of general anesthetics on G protein-coupled inwardly rectifying and other potassium channels. *Anesthesiology* **95**, 144–153.

Yan K and Gautam N (1996). A domain on the G protein β subunit interacts with both adenylyl cyclase 2 and the muscarinic atrial potassium channel. *J Biol Chem* **271**, 17597–17600.

Yi BA, Lin YF, Jan YN *et al.* (2001). Yeast screen for constitutively active mutant G protein-activated potassium channels. *Neuron* **29**, 657–667.

Yoshimoto Y, Fukuyama Y, Horio Y *et al.* (1999). Somatostatin induces hyperpolarization in pancreatic islet a cells by activating a G protein-gated K[+] channel. *FEBS Lett* **444**, 265–269.

Zhang H, He C, Yan X *et al.* (1999). Activation of inwardly rectifying K[+] channels by distinct PtdIns(4,5)P$_2$ interactions. *Nat Cell Biol* **1**, 183–188.

Zhang Q, Pacheco MA and Doupnik CA (2002). Gating properties of GIRK channels activated by Gα$_o$- and Gα$_i$-coupled muscarinic m$_2$ receptors in *Xenopus* oocytes: the role of receptor precoupling in RGS modulation. *J Physiol* **545**, 355–373.

Zhou W, Arrabit C, Choe S *et al.* (2001a). Mechanism underlying bupivacaine inhibition of G protein-gated inwardly rectifying K[+] channels. *PNAS* **98**, 6482–6487.

Zhou Y, Morais-Cabral JH, Kaufman A *et al.* (2001b). Chemistry of ion coordination andhydration revealed by a K[+] channel-Fab complex at 2.0 A resolution. *Nature* **414**, 43–48.

4.4.3 **ATP-sensitive K[+] channels**

Michel Vivaudou, Christophe Moreau and Andre Terzic

4.4.3.1 **Introduction**

The patch-clamp technique, introduced over two decades ago (Hamill *et al.*, 1981), has made possible the isolation and observation of individual ion channels. In the excised patch configuration, direct access to the cytoplasmic side of the plasma membrane has allowed probing of intracellular mechanisms of ion channel regulation.

This approach has enabled the discovery of K+-selective channels richly expressed in the cardiac sarcolemma, and that are uniquely inhibited by ATP acting from the cytosolic face (Noma, 1983). Adenine nucleotide-gated channels with similar properties were then found in pancreatic β-cells (Cook and Hales, 1984) and skeletal muscle fibres (Spruce et al., 1985). The inhibitory effect of ATP was mapped to direct binding, without hydrolysis, of the nucleotide to the channel pore as onset and offset of block was fast and did not require magnesium ions. Hence, the designation 'ATP-sensitive K+ channel' or K$_{ATP}$ channel. K$_{ATP}$ channels were further found to be the target of antidiabetic (Sturgess et al., 1985) and antihypertensive (Trube et al., 1986; Escande et al., 1988) drugs, which attracted particular attention in the medical community. Ever since, considerable effort has been made to dissect the biophysical, physiological and pharmacological properties of K$_{ATP}$ channels throughout the body. Their molecular identity was solved with the advent of molecular biology (Inagaki et al., 1995a; Aguilar-Bryan et al., 1995), opening the present era of K$_{ATP}$ channel (pharmaco)genetics and (patho)physiological genomics aimed at dissecting the most intimate mechanisms of channel function and malfunction, and the role of this class of K+ channel in both health and disease. Here, we focus on plasma membrane or molecularly defined K$_{ATP}$ channels. Other subcellular compartments, such as the mitochondrion, have also been implicated in harbouring channels with a pharmacology that overlaps with 'classical' K$_{ATP}$ channels (Terzic et al., 2000; Ardehali and O'Rourke, 2005; Hanley and Daut, 2005).

4.4.3.2 Molecular architecture

The K$_{ATP}$ channel subunit duo

K$_{ATP}$ channels are composed of two distinct proteins—the ~140–170 kDa sulfonylurea receptor SUR (Aguilar-Bryan et al., 1995), a member of the ABC (ATP-binding-cassette) transporter family, and a smaller ~40 kDa protein Kir6 which belongs to the inward rectifier K+ channel family (Inagaki et al., 1995a). SUR and Kir6 are obligatory partners and there is no evidence that either may function alone outside the K$_{ATP}$ channel complex. This symbiosis between an ABC protein and an ion channel is rather unique among members of these two large membrane protein families.

In mammals, two Kir6 isoforms have been identified: Kir6.2 (Inagaki et al., 1995c; Sakura et al., 1995) and Kir6.1 (Ämmälä et al., 1996; Yamada et al., 1997). An additional member, Kir6.3, is also present in zebrafish (Zhang et al., 2006). Human Kir6.1 and Kir6.2, as well as zebrafish Kir6.3, share ~70 per cent amino acid sequence identity. The divergences in sequences are not equally distributed but are concentrated in two regions, i.e., the ~20-residue extracellular loop following transmembrane helix M1 and the last 50 residues of the intracellular C-terminal extremity of the proteins. The human Kir6.1 gene *KCNJ8* contains three exons and maps to locus 12p11.23 (Inagaki et al., 1995b) while the Kir6.2 gene *KCNJ11* is intronless in the protein-coding region and maps to 11p15.1 (Inagaki et al., 1995a). No splice variants of Kir6.1 have been reported. Mammals possess two SUR genes, SUR1 (or *ABCC8*) at locus 11p15.1 and SUR2 (or *ABBC9*) at locus 12p12.1 (Thomas et al., 1995; Chutkow et al., 1996). Both genes have 39 exons although the last 2 exons of SUR2 are alternatively used as the terminal exon of the two main SUR2 isoforms, SUR2A and SUR2B (Aguilar-Bryan et al., 1998).

The *SUR1* gene follows immediately after the *Kir6.2* gene (Inagaki et al., 1995a), although both genes have distinct promoters (Ashfield and Ashcroft, 1998). The *SUR2* and *Kir6.1* genes are also tightly linked in the mouse (Isomoto et al., 1997) and, without being adjacent, co-localize on the short arm of chromosome 12 in humans. This suggests a correlated expression pattern that has, however, not been verified.

Studies of recombinant K$_{ATP}$ channels have focused largely on the full-length isoforms SUR1, SUR2A, and SUR2B that appear to be the predominant forms in most tissues where SUR genes are expressed. Other less ubiquitous alternatively spliced variants have been reported (reviewed by Shi et al., 2005). Mouse SUR1Δ31 (Gros et al., 2002), rat SUR1Δ19, SUR1Δ17Δ19, and a truncated SUR1Δ1–32 (Hambrock et al., 2002) remain intracellular and lose the capacity to bind sulfonylureas whereas SUR1Δ17 resembles the full-length SUR1 (Hambrock et al., 2002). SUR1Δ33, which lacks the C-terminal NBD because of an introduced frameshift, forms with Kir6.2, K$_{ATP}$ channels that are sulfonylurea-sensitive but do not respond to ADP or the opener diazoxide (Sakura et al., 1999).

SUR2 variants exist in the SUR2A and SUR2B background. The most frequent is SUR2Δ17, which does not deviate much functionally from the full-length protein (Aguilar-Bryan et al., 1998; Chutkow et al., 1999; Davis-Taber et al., 2000). The other variants, SUR2AΔ14 (Chutkow et al., 1996), SUR2AΔ17Δ18 (Aguilar-Bryan et al., 1998), and a C-terminal fragment SUR2BΔ1–12 having a 16-residue N-terminal addition (Szamosfalvi et al., 2002), have yet to be characterized.

Channel stoichiometry

Like all K+ channels, the K$_{ATP}$ channels adopt a tetrameric conformation. This has been demonstrated experimentally through co-expression of wild-type and fused or mutated subunits, functional analysis of the resulting population of channels (Clement et al., 1997; Inagaki et al., 1997; Shyng and Nichols, 1997; Zerangue et al., 1999) and more recently through direct imaging by electron microscopy (Mikhailov et al., 2005). Four Kir6.x subunits assemble to form a K+-selective pore which is constitutively associated to four regulatory SUR subunits (Figure 4.4.3.1).

This stoichiometry is strictly enforced by an intracellular quality-control checkpoint that keeps all incomplete channel complexes from reaching the plasma membrane. Each Kir6.x protein possesses a unique stretch of three residues, RKR, within the last 26 residues of its C-terminal extremity, that acts as an endoplasmic reticulum (ER) retention signal (Zerangue et al., 1999). Unless a SUR protein is bound to Kir6.x, this signal is exposed, keeping the protein from exiting the ER/golgi network. SUR1 harbours the same RKR signal located at the N-terminal end of NBD1. That signal keeps individual SUR1 proteins from reaching the plasma membrane unless masked by a neighbouring SUR1. Since a single exposed ER retention signal is sufficient to halt forward trafficking, only fully assembled SUR-Kir6 channels reach the membrane (Zerangue et al., 1999). As a RKR signal detector, the main suspect is the coat protein complex I (COPI) (Yuan et al., 2003).

Although originally postulated as self-contained, this masking process now appears to involve additional members of the 14-3-3 protein family. 14-3-3 proteins recognize the RKR signals of Kir6.2 tetramers and promote their forward transport by hiding the signals from the COPI watchdog (Yuan et al., 2003). However, in the SUR1-Kir6.2 complex, recent data suggest that Kir6.2 signals are

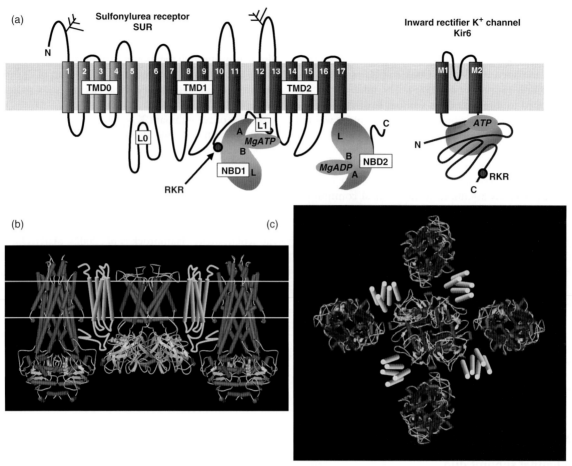

Fig. 4.4.3.1 Secondary structure and molecular models of the 2 subunits of the K_{ATP} channel. (a) Kir (~400 amino acids) and SUR (~1600 amino acids) are the constitutive subunits of the K_{ATP} channel. Four subunits of the inwardly rectifying K^+ channel, Kir6.1 or Kir6.2, associate with four ATP-binding cassette proteins, SUR, to form a functional K_{ATP} channel octamer. Kir6.x has two transmembrane helices M1 and M2, and a large cytoplasmic domain harbouring an inhibitory binding site for ATP formed by the proximal C-terminus of one subunit and N-terminus of its neighbour. SUR possesses three transmembrane domains (TMD 0, 1 and 2), and two cytoplasmic nucleotide binding domains (NBD 1 and 2) incorporating the Walker A, Walker B and Linker L consensus sequences. NBD 1 and 2 probably function as a dimer with nucleotide sites made up of Walker A and B of one NBD and linker L of the other. Cytoplasmic loops L0 and L1 are indicated. (b and c) Homology model of Kir6.2 tetramer based on the KirBac1.1 structure (centre) and model of Sav1866 together with a cartoon representation of TMD 0 and L0 in yellow. Panel b is a side view with 2 Sav1866 and TMD0 and a Kir6.2 tetramer. Panel c is viewed from the extracellular side of the full complex. The relative position of the subunits is arbitrary. α-helices are in red and β-sheets are in blue. ATP in the Kir6.2 and NBD sites is in green. See also colour plate section.

actually sterically masked by SUR1, as initially thought, and that SUR1 signals are recognized by 14-3-3 proteins (Heusser *et al.*, 2006).

It should be noted that the trafficking signal sequence RKQ of SUR2 deviates from the canonical RKR but serves the same purpose (Konstas *et al.*, 2002).

Even the most sophisticated checkpoints are meant to be bypassed. By tinkering with trafficking signal sequences, it has been shown that incomplete channels are functional although their properties deviate significantly from fully assembled channels (Tucker *et al.*, 1997; Zerangue *et al.*, 1999). However, so far there is no evidence that non-octameric plasmalemmal channels are physiologically relevant.

Mixing and matching isoforms

Even with a fixed stoichiometry, substantial diversity among K_{ATP} channels can result from the multiple ways of building an octamer with a choice of two Kir6 isoforms and three main SUR isoforms plus several variants.

It is clear that, at least in heterologous systems, each Kir6 isoform can associate with any of the SUR isoforms to form functional homotetrameric K_{ATP} channels that are endowed with unique biophysical and pharmacological properties conferred by constituent subunits. This generates six distinct combinations: SUR1+Kir6.1 (Ämmälä *et al.*, 1996), SUR1+Kir6.2 (Sakura *et al.*, 1995; Inagaki *et al.*, 1995a), SUR2A+Kir6.1 (Kono *et al.*, 2000), SUR2A+Kir6.2 (Inagaki *et al.*, 1996), SUR2B+Kir6.1 (Yamada *et al.*, 1997), and SUR2B+Kir6.2 (Isomoto *et al.*, 1996). Other combinations based on less-studied SUR splice variants yield active surface K_{ATP} channels: SUR2AΔ17 and SUR2BΔ17+Kir6.2 (Chutkow *et al.*, 1999), SUR1Δ33+Kir6.2 (Sakura *et al.*, 1999) and SUR1Δ17+Kir6.2 (Hambrock *et al.*, 2002).

A further degree of complexity arises from the possibility that distinct isoforms can cohabit within the same channel complex. In native tissues, uncertainty remains as to whether SUR1–SUR2 channels could be present. Indeed, in cardiac and neuronal tissues, some cells express both SUR1 and SUR2 isoforms (Yokoshiki *et al.*, 1999; Morrissey *et al.*, 2005a, b) and possess channels with physiological and pharmacological properties intermediate between those of SUR1-based and SUR2-based channels (Liss *et al.*, 1999).

In heterologous expression systems, SUR1 and SUR2 co-expressed with Kir6.2 do not co-assemble to form distinctive channels nor form biochemically detectable complexes (Gross *et al.*, 1999; Giblin *et al.*, 2002), although more recent works challenge such conclusion (Cheng *et al.*, 2008; Chan *et al.*, 2008). Therefore, SUR1–SUR2 co-assembly and its physiological relevance requires further experimental work. This is also the case for co-assembly of distinct SUR1 or SUR2 variants. In particular, channels incorporating both SUR2A and SUR2B are probably viable because these isoforms have almost identical sequences, although their *in vitro* and *in vivo* existence remains to be demonstrated.

Regarding Kir6 isoforms, the majority of evidence suggests that Kir6.1 and Kir6.2 have the capacity to form hybrid channels in expression systems. This evidence includes demonstration that Kir6.1 and Kir6.2 interact during trafficking (Zerangue *et al.*, 1999); viable channels with mixed conductances and regulations can be created from fused Kir6.1–Kir6.2 proteins (Kono *et al.*, 2000; Babenko *et al.*, 2000b) as well as from unfused proteins (Babenko and Bryan 2001; Cui *et al.*, 2001; Thorneloe *et al.*, 2002); coimmunoprecipation of the two isoforms (Cui *et al.*, 2001; Pountney *et al.*, 2001); and finally functional impairment of one isoform by a dominant-negative mutant of the other (Cui *et al.*, 2001; Pountney *et al.*, 2001; van Bever *et al.*, 2004). Nonetheless, the latter result conflicts with the dominant-negative demonstration of lack of Kir6.1–Kir6.2 heteromerization (Seharaseyon *et al.*, 2000).

In certain cells like cardiomyocytes, both Kir6 isoforms are expressed (Morrissey *et al.*, 2005a, b) and channels having the intermediate conductance and regulatory properties of hybrid Kir6.1–Kir6.2 channels have been identified (Baron *et al.*, 1999; Babenko and Bryan, 2001). Moreover, expression of dominant-negative Kir6.1 or Kir6.2 drastically reduced the magnitude of native K_{ATP} currents (van Bever *et al.*, 2004). These observations are in support of the presence of a sizeable population of Kir6.1–Kir6.2 hybrid channels in native tissues. Two reports are at odds with this conclusion: Seharaseyon *et al.* (2000) applied a dominant-negative assay to cardiomyocytes to show that Kir6.1 was absent from the surface membrane of heart muscle cells, and Suzuki *et al.*, (2001) did not detect K_{ATP} current in myocytes from Kir6.2 knockout mice despite maintained levels of Kir6.1 proteins. These conflicting data are puzzling and suggest that Kir6.1 and Kir6.2 are structurally fit to assemble *in vitro* but coassembly may be regulated by unidentified regulatory mechanisms *in vivo*.

Structural model

Secondary structures and molecular models of Kir6.x and SUR are presented in Figure 4.4.3.1. SUR is thought to conform to the architecture of other members of the ABCC subfamily of ABC proteins (Moreau *et al.*, 2005b), having the canonical two transmembrane domains TMD1 and TMD2 and two nucleotide-binding domains NBD1 and NBD2 of ABC transporters with the supplementary TMD0 domain of most ABCC proteins. No high-resolution structure of the K_{ATP} channel is yet available. An 18-Å resolution structure has been obtained by single-particle electron microscopy that has confirmed the octameric, compact structure of the channel complex (Mikhailov *et al.*, 2005). It is possible to model Kir6.2 using the structures of the bacterial homologues KirBac1.1 and the cytoplasmic pore of mammalian Kir3.1. This exercise has permitted

mapping of the ATP binding pocket (Antcliff *et al.*, 2005). The closest structurally defined homologue of SUR are the bacterial ABC transporters Sav1866 and MsbA (Dawson and Löcher, 2006; Ward *et al.*, 2007) which are, however, still distant in sequence and lack the domains TMD0, L0, and C-terminus that appear crucial in the interactions between SUR and Kir6.2. TMD0, which alone can assemble with Kir6.2, forms a dynamic interface between SUR and Kir6.2 (Chan *et al.*, 2003; Fang *et al.*, 2006; Hosy *et al.*, 2007). The loop L0 attaching TMD0 and TMD1 also modulates gating in response to SUR ligands by possibly interacting directly with the cytoplasmic N-terminus of Kir6.2 (Babenko *et al.*, 1999a; Babenko and Bryan, 2003). Other SUR-Kir6.2 interacting regions have been localized to the proximal C-terminal end of SUR2A between the last transmembrane helix 17 and domain NBD2 (Rainbow *et al.*, 2004; Dupuis *et al.*, 2008) and to the helix 12 of TMD2 (Mikhailov *et al.*, 2000). The 'receiving' ends of these elements in Kir6.x include the proximal C-terminal (Giblin *et al.*, 1999), the N-terminal extremity (Reimann *et al.*, 1999) and the first transmembrane helix (Schwappach *et al.*, 2000).

4.4.3.3 Molecular properties

The biophysical (Figure 4.4.3.2) and regulatory properties (Figure 4.4.3.3) of the K_{ATP} channel—permeability, selectivity, rectification, ATP inhibition, phosphatidylinositol 4,5-bisphosphate (PIP_2) activation—are imparted by the Kir6 protein. They are summarized in Table 4.4.3.1. The more complex physiological (including gating by the cellular bioenergetic state) and pharmacological regulations are conferred by the SUR subunit.

Biophysical characteristics

Ion conduction

Ion selectivity of K+ channels is determined by the filter selectivity P-loop region (Doyle *et al.*, 1998) which contains the highly conserved TxGYG sequence in most K+ channels. This sequence is TxGFG in all sequenced Kir6. The Phe residue is necessary for proper function of Kir6 as its mutation to a canonical Tyr incapacitates the channel (Proks *et al.*, 2001) but it does not affect the capacity of the channel to conduct K+ over Na+ as demonstrated with mutated Kir2.1 (So *et al.*, 2001). Accordingly, K_{ATP} channels are highly selective for K+ over Na+ (extracellular) with experimentally determined permeability ratios >50 for Kir6.2 (Gribble *et al.*, 1997a), Kir6.1 (Surah-Narwal *et al.*, 1999) as well as native pancreatic and muscle channels (Spruce *et al.*, 1987; Ciani and Ribalet, 1988; Ashcroft *et al.*, 1989). The channels are also able to conduct rubidium ions, weakly but sufficiently to justify the widespread use of radioactive Rb+ as a tracer of ionic fluxes through K_{ATP} channels (Spruce *et al.*, 1987; Ashcroft *et al.*, 1989).

In symmetrical elevated K+, the conductance of recombinant K_{ATP} channels is near 70 pS for Kir6.2-based channels although reported values range from 60 to 80 pS (Inagaki *et al.*, 1995a; Takano *et al.*, 1998; Proks and Ashcroft, 1997; Inagaki *et al.*, 1996; Okuyama *et al.*, 1998; Cui *et al.*, 2001). Kir6.1-based channels have a lower conductance of about 35 pS with a range of 29 to 37 pS (Yamada *et al.*, 1997; Takano *et al.*, 1998; Cui *et al.*, 2001; Babenko *et al.*, 2000b). This difference in conductance between Kir6.1 and Kir6.2 has been attributed to different configurations of the pore extracellular entrance and permeation pathway due to the presence

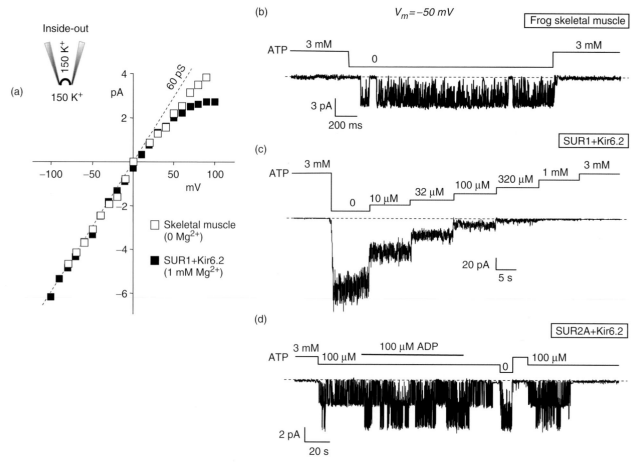

Fig. 4.4.3.2 Fundamental properties of K$_{ATP}$ channels illustrated by inside-out patch-clamp recordings from frog skeletal muscle and from *Xenopus* oocytes coexpressing Kir6.2 and SUR1 or SUR2A. (a) Unitary current vs voltage showing a conductance of ~60 pS and weak inward rectification dependent on Mg^{2+} as shown here as well as polyamines (not shown). (b) Typical bursting behaviour of K$_{ATP}$ channels with prolonged closed intervals separating bursts of rapid flickering between open and closed states. (c) ATP inhibits K$_{ATP}$ channels concentration-dependently at micromolar concentrations. (d) In the presence of Mg^{2+}, ADP upregulates gating by interacting with the SUR subunit. In the absence of nucleotide diphosphates and in submillimolar ATP, channel activity runs down spontaneously as seen at end of the record. All traces recorded at −50mV in symmerttrical high K$^+$ in inside-out patches.

of distinct residues in the extracellular loops connecting helix M1 and pore H5 (positions 113 to 115 in Kir6.2) and H5 and M2 (position 138 in Kir6.2) (Repunte *et al.*, 1999).

The conductance of hybrid Kir6.1/Kir6.2 channels varies continuously between that of Kir6.1 and Kir6.2 as a function of the proportion of each subunit in the tetrameric channel (Babenko *et al.*, 2000b) and can therefore serve as an indicator of the subunit composition of K$_{ATP}$ channels (Kono *et al.*, 2000; Cui *et al.*, 2001).

Rectification

Like all Kir channels, K$_{ATP}$ channels display inward rectification in the current–voltage relationship, which limits potassium efflux at potentials more positive than the potassium reversal potential. Rectification in K$_{ATP}$ and other Kir channels is not an intrinsic property of the channel protein, as it is in certain voltage-dependent channel counterparts (Smith *et al.*, 1996), but is mediated by interaction with cytosolic multivalent cations, such as Mg^{2+} (Horie *et al.*, 1987) and the linear polyamines putrescine, spermidine, and spermine (Lopatin *et al.*, 1994; Shyng *et al.*, 1997a). These cations bind within the channel and impede outward K$^+$ movement at

depolarized voltages, but are displaced by incoming K$^+$ ions at hyperpolarized voltages. High-affinity binding and strong voltage-dependence of these blocking molecules are made possible by an ion conduction pore that extends towards the cytosol (Nishida and Mackinnon 2002; Kuo *et al.*, 2003; Pegan *et al.*, 2005). How effectively these blockers bind, and how strong rectification is, is determined by the nature of several residues lining the cytoplasmic access pathway of the channel. These include an amino acid residue, termed the 'rectification controller' in the region of the M2 helix that delineates the proximal entrance of the channel (Lu, 2004), as well as several acidic and hydrophobic amino acids along the inner wall of the cytoplasmic pore that create a favourable docking environment for polyamines (Nishida and Mackinnon, 2002).

More precisely, mutagenesis studies have correlated the degree of rectification with the negative charges of several pore-lining residues (Wible *et al.*, 1994; Yang *et al.*, 1995; Kubo and Murata 2001; Pegan *et al.*, 2005). In the strong rectifier Kir2.1 these residues, D172, E224, E299, D255 and D259, carry five negative charges. The matching residues in Kir6.2 (N160, S212, E288, G243 and N247) and Kir6.1 (N170, S222, E297, P253 and N257) carry a single negative

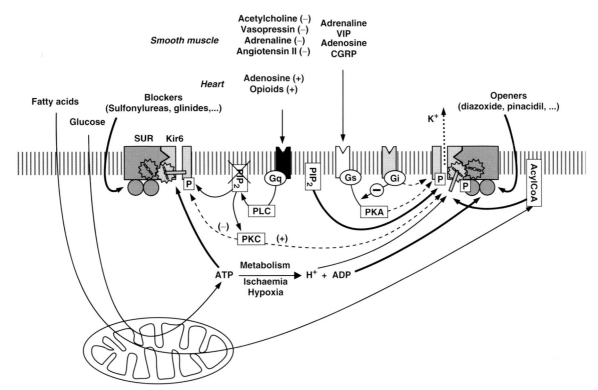

Fig. 4.4.3.3 Pathways of downregulation (left side) and upregulation (right side) of K_{ATP} channels. Thick lines indicate major regulations, inhibition by ATP and activation by MgADP, PIP_2, and acylCoA. The phosphorylation (dashed lines) as well as the receptor-mediated mechanisms are tissue-dependent and still debated. Similar PKC pathways in heart and vascular smooth muscle lead to stimulation (+) or inhibition (−). K_{ATP} channel activity is under the control of many other hormones not shown here, some through unsettled routes, some like leptin through receptors not coupled to G-proteins. See text for references. PLC, phospholipase C; PKC, protein kinase C, PKA, protein kinase A; Gx, type x heterotrimeric GTP-binding protein; AcylCoA, Long-chain acyl coenzyme A esters; CGRP, calcitonin-gene related peptide; VIP, Vasoactive intestinal peptide.

charge that confers only weak affinity for Mg^{2+} and polyamines. Thus K_{ATP} channels rectify weakly and are able to pass sizeable outward currents at depolarized potentials (Figure 4.4.3.2).

Channel gating

Gating of K_{ATP} channels is characterized by spontaneous bursts of rapid openings and closings interrupted by long closures (Figure 4.4.3.2). Channel kinetics within bursts are intrinsic to the Kir6 protein and are conserved among Kir6.1 (Babenko and Bryan, 2001) and Kir6.2 (Alekseev et al., 1997; Proks and Ashcroft, 1997; Drain et al., 1998; Babenko et al., 1999b; Fang et al., 2006). In both cases, at negative potentials mean open times equal ~2 ms (range of 0.8 to 3.2 ms) and mean fast closed times ~0.3 ms (range of 0.2 to 0.55 ms). The former increases and the latter decreases upon depolarization (Alekseev et al., 1997; Babenko et al., 1998; Okuyama et al., 1998).

Table 4.4.3.1 Basic properties of K_{ATP} channels. Biophysical characteristics and sensitivity to inhibition by nucleotides of recombinant K_{ATP} channels (Kir6.2 represents all modified versions of Kir6.2 that are functional without SUR). Indicated values are representative of those reported in the literature. No data have been reported for Kir6.1 and SUR2A+Kir6.1. Kinetic parameters were measured at −80 to −40 mV in absence of nucleotides.

Channel	g	P_O	T_O	T_{IB}	T_B	T_{C1}	T_{C2}	T_{C3}	ATP	ADP	AMP	AP4A	GDP	GTP
			Mean dwell time (ms)						IC50 (μM)					
Kir6.2	72	0.1	0.9	10	2.3	0.35	13		200 (150)	300	10000	190		(6000)
SUR1+Kir6.1	35		1.3			0.3			7.5 (16)					
SUR1+Kir6.2	75	0.6	2	4	10–80	0.3	16	800	8 (15)	120	1800	12	1400	1300
SUR2A+Kir6.2	78	0.9	2.5	4	390	0.3	40	1000	25 (25)			[17]	1800	1000
SUR2B+Kir6.1	33								(~250)					
SUR2B+Kir6.2	78		2.1			0.2			68 (200)					

Half-maximal inhibitory concentrations (IC50) were measured in absence of Mg^{2+} except for values in parenthesis obtained in the presence of 0.2–3 mM: $K^+ Mg^{2+}$. Values in brackets are from native channels. See text for references. g, conductance in pS; Po, open probability; To, open time; Tib, burst interval; Tb, burst duration; Tc_{1-2-3}, fast-medium-long closed times.

Analysis of single channel dwell-time histograms reveal additional slow closing rate constants corresponding to mean closed times of ~20 ms and >100 ms that reflect interburst kinetics (Alekseev *et al.*, 1998; Chutkow *et al.*, 1999; Proks *et al.*, 2001; Fang *et al.*, 2006).

In the absence of ligands, Kir6.2, mutated to form channels without SUR (Tucker *et al.*, 1997), has an open probability Po of ~0.1 (Drain *et al.*, 1998; Trapp *et al.*, 1998; Babenko *et al.*, 1999a). Association with SUR1 or SUR2A raises this value to ~0.6 or ~0.9, respectively (Babenko *et al.*, 1999b; Fang *et al.*, 2006), a result of physical interactions with Kir6.2 affecting mainly burst durations and interburst intervals (Lorenz *et al.*, 1998: Babenko *et al.*, 1999a). Ligands—nucleotides, blockers, and openers—also modulate channel gating by altering burst and interburst durations to a much greater extent than the rapid, intraburst kinetics (Drain *et al.*, 1998; Fan and Makielski, 1999; Li *et al.*, 2002; Ninomiya *et al.*, 2003).

These observations are consistent with the existence of two gates operating in parallel within the Kir6 protein, a fast gate responsible for intrinsic bursting and a slow gate connected to ligand binding sites. The fast gate is located in the selectivity filter region of the pore (Proks *et al.*, 2001). The location of the ligand-sensitive gate is still disputed. The slow gate of Kir6 could lie at the cytoplasmic entrance of the pore at the M2-helix bundle crossing (Loussouarn *et al.*, 2001; Phillips and Nichols 2003) as in other K⁺ channels (Webster *et al.*, 2004). It could be in or above the selectivity filter (Proks *et al.*, 2003; Xiao *et al.*, 2003) or it could be within the cytoplasmic pore (Pegan *et al.*, 2005).

Pore blockers

K_{ATP} channels are rather insensitive to tetraethylammonium (TEA) compared to other K⁺ channels (Fatherazi and Cook, 1991). Extracellularly, TEA produces a voltage-independent, incomplete block that plateaus at 40 per cent at 20 mM (Bokvist *et al.*, 1990). Intracellularly, TEA block is mainly a voltage-dependent slow block that causes full inhibition of outward currents at supramillimolar concentrations (Davies *et al.*, 1989; Kakei *et al.*, 1985). Barium ions are more effective blockers that produce voltage-dependent open-channel block with a Kd in the submillimolar range from both sides when the driving force is favourable (Quayle *et al.*, 1988; Takano and Ashcroft, 1996). Extracellular barium concentrations >1–5 mM are usually sufficient to abrogate K_{ATP} currents at any voltages (Kakei and Noma, 1984). In contrast, caesium ions (Cs⁺) are inefficient open-channel blockers from either side of the membrane although substitution of cytosolic K⁺ for Cs⁺ can serve to silence K_{ATP} currents (Quayle *et al.*, 1988; Tseng and Hoffman, 1990).

Other compounds known to target the pore structure of K⁺ channels are often used to sort out the role of various K⁺ conductances. Specifically, 4-aminopyridine, a blocker of voltage-dependent K⁺ channels, has no effect on K_{ATP} channels at external concentrations of up to 10 mM although it does cause a voltage-dependent block of outward currents at mM internal concentrations (Davies *et al.*, 1991). Charybdotoxin and iberiotoxin, subnanomolar affinity blockers of large-conductance K_{Ca} channels, do not affect K_{ATP} channels at concentrations up to 50 nM (Kovacs and Nelson, 1991; Ohtsuka *et al.*, 2006) and, neither does apamin, a blocker of small-conductance K_{Ca} channels at nM doses, at concentrations >100 nM (Castle and Haylett, 1987; Ohtsuka *et al.*, 2006).

Kir6.x effectors

Nucleotides

Inhibition by intracellular ATP is the trademark of K_{ATP} channels. When SUR was originally cloned and identified as an ABC protein (Aguilar-Bryan *et al.*, 1995), its two well-defined nucleotide-binding sites were the sole candidates to mediate ATP inhibition. This was disproved when Tucker *et al.* (1997) discovered that short truncations (26–36 residues) of the C-terminal of Kir6.2 made it competent to form channels in the absence of SUR, and that these SUR-less channels were still inhibited by ATP. Since the sequence of Kir6.2 does not reveal the signature of a classical nucleotide-binding site, extensive mutagenesis was necessary to locate key residues in the proximal N-terminal and C-terminal cytosolic regions (Tucker *et al.*, 1998; Drain *et al.*, 1998; Trapp *et al.*, 2003). This information, together with structural data on homologous channels (Nishida and Mackinnon, 2002; Kuo *et al.*, 2003), helped delineate an ATP binding pocket at the interface between Kir6.2 subunits, that incorporates residues from the N-terminal of one subunit and from the C-terminal of its neighbour (Dong *et al.*, 2005; Antcliff *et al.*, 2005). SUR-less Kir6.2 channels are blocked, non-cooperatively, by ATP with an IC_{50} of ~200 µM (see Table 4.4.3.1). Association with SUR decreases this value to ~10 µM, possibly by reshaping the ATP binding fold (Dabrowski *et al.*, 2004). Binding of ATP to a single Kir6.2 subunit appears sufficient to provoke closure of the tetramer (Markworth *et al.*, 2000). The nucleotide site binds ATP better than other nucleotides. On one hand, it fits the adenine base better than other purine and pyrimidine bases and does not tolerate even minor modifications to the adenine ring (Tucker *et al.*, 1998; Dabrowski *et al.*, 2004; Tammaro *et al.*, 2006). On the other hand, binding affinity is maximal for three or more phosphates as it decreases several-fold for each phosphate removed (Tucker *et al.*, 1998; Ribalet *et al.*, 2003) but does not increase significantly for higher order polyphosphates such as ATetraP (Koster *et al.*, 1999) and the diadenosine polyphosphates (Jovanovic *et al.*, 1997; Dabrowski *et al.*, 2004).

Lipids

Phosphatidylinositol 4,5-bisphosphate (PIP_2) is a membrane-bound signaling molecule that has emerged as a major regulator of membrane proteins, in particular inward-rectifying K⁺ channels (Suh and Hille, 2005). PIP_2 is at the intersection of multiple reactions of phosphorylation/dephosphorylation of the various PIP_x intermediates involving numerous kinases and phosphatases. It is also irreversibly hydrolyzed by receptor-activated phospholipase C (PLC) to yield the second messengers diacylglycerol and inositol 1,4,5-triphosphate.

K_{ATP} channel activity is potently upregulated by PIP_2 (Furukawa *et al.*, 1996). PIP_2 augments open probability of run-down channels in the absence of ATP (Hilgemann and Ball, 1996; Fan and Makielski, 1997) suggesting that run-down in excised patches is partly due to progressive loss of PIP_2 in the patch and that it is through replenishment of PIP_2 by phosphatidylinositol (PI) kinases that MgATP reverses run-down. PIP_2 also antagonizes ATP inhibition by greatly reducing the apparent affinity of ATP for the channel (Shyng and Nichols, 1998; Baukrowitz *et al.*, 1998; Fan and Makielski, 1999). Experiments showing that attenuation of PIP_2 effects by polycations such as polylysine and neomycin, as well as by neutralizing mutations of positively charged residues of the C-terminal extremity of Kir6.2 (Fan and Makielski, 1997; Shyng et al., 2000), demonstrate that the negatively charged PIP_2 acts through electrostatic interactions with acidic residues of the proximal C-terminal tail of Kir6.2. Mutagenesis indicates that the PIP_2 binding site is located near the inner face of the membrane in close proximity to the ATP site (Enkvetchakul and Nichols, 2003; Haider et al., 2007).

Another class of anionic lipids that strongly affect K_{ATP} channels are long-chain fatty acid acyl coenzyme A esters (LC-CoA), endogenous intermediates of fatty acid metabolism that can greatly fluctuate with metabolism. LC-CoA potently activates K_{ATP} channels through a mechanism resembling that of PIP_2. They can reverse run-down and decrease ATP inhibition by interacting with the Kir6.2 subunit (Gribble et al., 1998a; Branstrom et al., 2007) at concentrations that are physiologically relevant (Shumilina et al., 2006).

Protons

Kir6.2 channels are gated by intracellular protons in multiple ways: activation by mild acidification, inactivation by strong acidification, and increased rectification by alkalinization. As pH is made more acidic, channels are first activated reversibly (pKa ~7.2) and then inactivated persistently by a run-down-like process (pKi ~6.8) (Xu et al., 2001a). Activation results from a shortening of closed dwell times and is stronger in the presence of ATP because of an apparent competition between protons and nucleotides (Wu et al., 2002). Activation is mediated by a histidine residue at the cytoplasmic end of helix M2 (H175) while inactivation has been linked to histidines 186, 193 and 216 of the cytosolic C-terminal (Xu et al., 2001b). Deprotonation of histidine 216 is also responsible for the change from weak to strong rectification as pH is made more alkaline (pK~7.3) (Baukrowitz et al., 1999).

These stimulatory/inhibitory effects of protons have been observed for several recombinant channels, SUR1/Kir6.2, SUR1/Kir6.1, SUR2A/Kir6.2 and SUR2B/Kir6.1 (Xu et al., 2001a; Wang et al., 2003; Li et al., 2005). Although contradictory results have been reported in native cells, these results largely match observations made on native channels in several preparations (Davies 1990; Fan et al., 1994; Vivaudou and Forestier 1995; Kang et al., 2006).

Phosphorylation

Receptor-mediated activation of protein kinase A (PKA) causes activation of K_{ATP} channels whereas that of protein kinase C (PKC) generally causes activation in heart and inhibition in smooth muscle (Rodrigo and Standen, 2005). The opposite effects of PKC correlate with the presence of Kir6.2 in activated channels and of Kir6.1 in inhibited channels (Thorneloe et al., 2002; Quinn et al., 2003). Several PKC and PKA phosphorylation consensus motifs are present within the sequences of Kir6.x and SURx. Direct biochemical evidence of phosphorylation has been provided by testing protein incorporation of labelled phosphate under kinase stimulation. This was shown for Kir6.2 and SUR1 with PKA (Beguin et al., 1999), Kir6.2 with PKC (Light et al., 2000) and both Kir6.1 and SUR2B with PKA (Quinn et al., 2004).

By mutagenesis of putative sites, several residues have been identified as crucial for activation via PKA although reports have often been contradictory on the relative importance of these residues. In Kir6, these are S372 in Kir6.2, its equivalent S385 in Kir6.1, and T224 in Kir6.2. In SUR, these are S1448 in SUR1 and T633, S1351, S1387

and S1465 in SUR2B (Beguin et al., 1999; Light et al., 2000; Lin et al., 2000; Quinn et al., 2004; Shi et al., 2007). So far, only residue T180 of Kir6.2 has been implicated in the effect of PKC (Light et al., 2000).

SUR effectors

MgADP and other Mg-nucleotides

Activation by intracellular ADP is a fundamental property of K_{ATP} channels conferred by the SUR subunit and essential to their function as metabolic sensors (Nichols et al., 1996). ADP, in the presence of Mg^{2+}, stimulates channel activity several-fold at concentrations of ~10–500 μM in the presence or absence of ATP. It is thought that MgADP binds preferentially to the nucleotide NBD2 site within SUR (Ueda et al., 1997) and presumably stabilizes a posthydrolytic conformation of the SUR catalytic cycle associated with reduced ATP-induced channel inhibition promoting channel opening (Zingman et al., 2001). Activation requires Mg^{2+} and relies on the integrity of both NBD domains of SUR as it is abolished by mutations in the conserved folds of either NBD (Gribble et al., 1997b; Shyng et al., 1997b; D'hahan et al., 1999b). SUR isoforms confer distinct ADP sensitivity to Kir6.2, SUR1 and SUR2B being more potently stimulated than SUR2A (Matsuoka et al., 2000). These differences have been ascribed to a higher hydrolysis capacity of the NBD's of SUR1 over those of SUR2A (Masia et al., 2005) and to the distinct NBD2 characteristics of SUR2A and SUR2B (Reimann et al., 2000) arising from their divergent C-terminal extremities (Matsushita et al., 2002).

Other Mg-nucleotides are able to stimulate channel activity by interacting with SUR including MgATP (Gribble et al., 1998b), MgGDP and MgGTP (Trapp et al., 1997), MgUDP and MgUTP (Yamada et al., 1997; Satoh et al., 1998). These effects are particularly striking for the SUR2B/Kir6.1 channel which is inactive in excised patches unless intracellular Mg-nucleotides are present (Yamada et al., 1997)

G proteins

Heterotrimeric GTP-binding proteins (G-proteins) are activated and released as separate subunits Gα and Gβγ upon binding of ligand to membrane bound G-protein-coupled receptors. G-protein subunits act as second messengers in enzymatic cascades but they can regulate ion channels including K_{ATP} channels through direct interaction as suggested by activation with Gα subunits of native K_{ATP} channels in excised patches (Ito et al., 1992; Ribalet and Ciani 1994; Terzic et al., 1994; Ribalet and Eddlestone 1995). In reconstituted systems, SUR has been implicated as a target of G-proteins. $Gα_{i1}$ stimulates SUR1+Kir6.2 and, to a lesser extent, SUR2A+Kir6.2 while $Gα_{i1}$ stimulates only weakly SUR2A+Kir6.2 and Gβγ was ineffective (Sanchez et al., 1998). $Gβγ_2$ subunits were nonetheless found to bind to SUR1 and SUR2A and activate both SUR1+Kir6.2 and SUR2A+Kir6.2 channels by reducing ATP sensitivity while $Gα_o$ has no effect (Wada et al., 2000).

Zinc

Zinc, the most abundant heavy metal in the body, is a potent extracellular activator of SUR1+Kir6.2 channels producing half-maximal effect at ~2 μM (Bloc et al., 2000). Two extracellular histidines, H326 and H332, present in SUR1 and not in SUR2, form part of the zinc site (Bancila et al., 2005). Surprisingly, intracellular zinc also activates SUR1+Kir6.2 with a K_d of ~2 μM and SUR2A+Kir6.2 with >tenfold lower affinity (Prost et al., 2004).

4.4.3.4 Cellular and subcellular localization

K_{ATP} channels are ubiquitous in most tissues. They have been characterized most exhaustively in pancreatic β-cells, skeletal muscle, and cardiac muscle, where they are present at high density. They are also present less prominently in smooth muscle and brain. Although, for practical reasons, attention has focused on plasmalemmal channels, K_{ATP} channel subunits also reside, sometimes predominantly, in intracellular membranes where their function remains subject to speculation. This is the case for the pancreatic β-cell insulin secretory granules (Kir6.2 and SUR1) (Ozanne et al., 1995; Varadi et al., 2006) and nucleus (Quesada et al., 2002), and the pancreatic acinar cell zymogen granules (Kir6.1) (Kelly et al., 2005). The presence of SUR/Kir6 subunits in mitochondria and their contribution to mitochondrial K_{ATP} channels is still controversial (Hanley and Daut, 2005).

Since Kir6.1 and Kir6.2 confer distinct conductances and nucleotide response and because SUR1, SUR2A, and SUR2B confer distinct pharmacological properties, it is possible to infer the presence of a given subunit isoform in a native channel provided the channel can be characterized with the patch-clamp technique and does not display a mixed profile indicative of the simultaneous presence of distinct Kir6 or SUR isoforms. These functional data, together with tissue mRNA and protein expression measurements as well as analysis of transgenic animals, have identified SUR1+Kir6.2, SUR2A+Kir6.2, SUR2B+Kir6.1, and SUR2B+Kir6.2 as the major β-cell, cardiac muscle, vascular smooth muscle nucleotide-diphosphate-dependent (K_{NDP}), and non-vascular smooth muscle channels, respectively (Inagaki et al., 1995a, 1996; Yamada et al., 1997; Isomoto et al., 1996; Seino and Miki, 2003). The neuronal K_{ATP} channels are predominantly SUR1+Kir6.2 although SUR2B+Kir6.1 and SUR2B+Kir6.2 are found in some cells.

Other combinations are rare. The presence of SUR1+Kir6.1 has been postulated in glucose-receptive hypothalamic neurons (Lee et al., 1999), striatal cholinergic interneurones (Lee et al., 1998), and retinal glial cells (Eaton et al., 2002) while that of SUR2A+Kir6.1 was hypothesized in follicular cells (Fujita et al., 2007) and tight junctions of various tissues (Jons et al., 2006).

4.4.3.5 Pharmacology

Inhibitors

The hypoglycaemic properties of sulfonylureas were recognized long before (Janbon et al., 1942) they were found to block K_{ATP} channels (Sturgess et al., 1985) and subsequently used to clone the pancreatic sulfonylurea receptor SUR1 (Aguilar-Bryan et al., 1995).

The first- and second-generation sulfonylureas are exemplified by the low-affinity tolbutamide and high-affinity glibenclamide (also known as glyburide) but include numerous variants such as glimepiride, glipizide or gliclazide (see Table 4.4.3.2).

Meglitinide, representing the non-sulfonylurea part of glibenclamide, also blocks K_{ATP} channels and has given rise to the glinide class of blockers that includes mitiglinide, repaglinide, and nateglinide.

These antidiabetic molecules target the pancreatic SUR1+Kir6.2 channels although they can also block muscle SUR2-based channels (Quast et al., 2004). It appears that two blocker interaction sites exist on SUR, sites A and B, that are located on distinct,

although possibly overlapping regions. Both sites are present on SUR1 while SUR2 has only the glinide site. Site A involves the cytoplasmic loop linking helices 15 and 16 in domain TMD2 of SUR (Ashfield et al., 1999). Within this loop, the residue at position 1237 of SUR1 (= 1206 in SUR2) plays a key role since the presence of a serine, as in SUR1, favours tolbutamide block while a tyrosine, as in SUR2, hampers it (Ashfield et al., 1999; Hambrock et al., 2001). Site B involves loop L0 linking TMD0 and TMD1 (Mikhailov et al., 2001) and possibly part of the N-terminus of Kir6.2 (Vila-Carriles et al., 2007). It is thought that site A, absent or degenerate in SUR2, recognizes sulfonylurea groups while site B, conserved in SUR1 and SUR2, binds the benzamido group of glibenclamide and meglitinide.

The glibenclamide molecule, through its two halves, can bind to both sites, and therefore blocks both SUR1, with high affinity, and SUR2, with lower affinity, while tolbutamide binds to the sulfonylurea site A and is specific for SUR1 (Gribble et al., 1998c). Beyond this paradigm, other unknown structural features can affect blocker specificity. Among the glinides for instance, nateglinide and mitiglinide, but not repaglinide, display high specificity for SUR1 over SUR2 (Dabrowski et al., 2001; Reimann et al., 2001; Hansen et al., 2002; Chachin et al., 2003; Stephan et al., 2006).

All sulfonylureas and glinides have affinities for SUR1 that are higher or comparable to that for SUR2. Only a recent compound, the sulfonylthiourea HMR1098 (sodium salt of HMR1883), is cardioselective as it blocks SUR2 channels at much lower concentrations than SUR1 (Russ et al., 2001; Fox et al., 2002). HMR1098 also has the property of being selective for sarcolemmal over mitochondrial K_{ATP} channels. As such it is a widely used experimental tool, often in combination with 5-hydroxydecanoate (5-HD), a presumed mitochondrial K_{ATP} channel antagonist (Liu et al., 2001).

Another unique blocker is the guanidine U-37883A (or PNU-37883A). This molecule is the only known compound able to discriminate between Kir6.1- and Kir6.2-containing channels. Although its action depends on the SUR isoform, it inhibits Kir6.1 channels more potently (Cui et al., 2003) and therefore has the potential to selectively target the SUR2B+Kir6.1 vascular channel (Teramoto, 2006).

K_{ATP} channels are secondary targets for a number of other drugs. Imidazolines, such as the α_2-adrenoreceptor blockers phentolamine and yohimbine, block K_{ATP} channels at micromolar concentrations through direct interaction with the Kir6.2 subunit (Plant and Henquin 1990; Proks and Ashcroft 1997; Proks et al., 2002; Doyle and Egan 2003). The antimalarial quinolines, quinine, quinidine, and chloroquine act through the same mechanism with an IC_{50} of 1–100 μM (Gribble et al., 2000). Similar observations have been reported for several antiarrhythmic agents—cibenzoline, disopyramide, and flecanaide (Mukai et al., 1998; Zünkler et al., 2000; Yunoki et al., 2001).

Activators

A number of chemically diverse agents are able to bind to SUR and activate K_{ATP} channels (Table 4.4.3.3). These K_{ATP} channel openers (KCO) which do not conform to an obvious pharmacophore principle, are classified into chemical families according to their molecular structures. These include benzothiadiazines (diazoxide), benzopyrans (cromakalim), cyanoguanidines (pinacidil), thioformamides (aprikalim), nicotinamides (nicorandil), pyrimidines derivatives (minoxidil), and more recently, dihydropyridines

Table 4.4.3.2 K_{ATP} channel blockers. Concentrations in μM of pharmacological blockers causing half-maximal inhibition of recombinant K_{ATP} channels (Kir6.2 represents all modified versions of Kir6.2 that are functional without SUR). Indicated values are representative of those reported in the literature with a preference for data obtained in excised patches whenever available. The data for the postulated mitochondrial K_{ATP} channels (MitoK$_{ATP}$) are included for comparison and were obtained by measurement of K^+ flux and flavoprotein oxidation in the presence of MgATP and openers. No data have been reported for Kir6.1 and SUR2A+Kir6.1. See text for references.

Channel composition	Tolbutamide	Glibenclamide	Glimepiride	Glipizide	Gliclazide	Meglitinide	Mitiglinide	Nateglinide	Repaglinide	5-HD	U-99963	U-37883A	HMR1883
Kir6.2	1700	40	400	>1	2700	>1000		290			>10	5	
SUR1+Kir6.1		5								66		40	1000
SUR1+Kir6.2	5	0.008	0.003	0.004	0.05	0.7	0.04	0.4	0.007	81	0.07	300	1000
SUR2A+Kir6.2	1000	0.04	0.006		800	0.3	3	105	0.01	>200	0.04	>300	1
SUR2B+Kir6.1		0.04									0.01	5	5
SUR2B+Kir6.2	90	0.025	0.007	1.2	>>10	1.6	5	110	0.01	>200	0.04	15	1
MitoK$_{ATP}$		1								80		>100	

(ZM-244085), cyclobutenediones (WAY-151616) and tertiary carbinols (ZD-6169) (Mannhold, 2004; Jahangir and Terzic, 2005). Most of these KCOs activate SUR2-based channels much more potently than SUR1-type channels. This selectivity has been shown to arise from discrete differences in amino-acid sequence in the last transmembrane helix (helix 17) of SUR (D'hahan *et al.*, 1999a; Moreau *et al.*, 2000, 2005a). These observations, along with reports highlighting a role for the helix-13 to 14 loop (Uhde *et al.*, 1999), suggest that TMD2 harbors a binding site for SUR2-selective KCOs. Diazoxide, as a non-selective KCO targeting SUR1, SUR2B as well as SUR2A in the presence of MgADP (D'hahan *et al.*, 1999b), appears to operate through the TMD1 domain of SUR (Babenko *et al.*, 2000a; Dabrowski *et al.*, 2002; Moreau *et al.*, 2005b).

As the current challenge is to improve tissue and thereby SUR isoform selectivity, newer compounds are introduced such as SUR1-selective KCOs, including the nitropyrazoles (Peat *et al.*, 2004) diazoxide-derived NNC 55–0118 and NN414 (Dabrowski *et al.*, 2003).

4.4.3.6 Physiology

By virtue of their responsiveness to adenine nucleotides, K_{ATP} channels are established metabolic sensors that serve to adjust membrane electrical properties in response to the energetic status of a cell. Depletion of energetic resources opens the channels, while abundance closes them. Since the channels are weak rectifiers, they

Table 4.4.3.3 K_{ATP} channel openers. Concentrations in μM of pharmacological openers causing half-maximal activation of recombinant K_{ATP} channels (Kir6.2 represents all modified versions of Kir6.2 that are functional without SUR). Indicated values are representative of those reported in the literature with a preference for data obtained in excised patches whenever available. The data for the postulated mitochondrial K_{ATP} channels (MitoK$_{ATP}$) are included for comparison and were obtained by measurement of K^+ flux and flavoprotein oxidation. No data have been reported for Kir6.1. Italics, estimated; NE, No effect at specified test concentration. Values recorded in MgATP + ADP are in parenthesis. Values in bracket indicate conflicting reports. See text for references.

Channel	Diazoxide	Pinacidil	P1075	Cromakalim	Nicorandil	NN414
Kir6.2	NE @ 300		NE @ 100	NE @ 100		
SUR1+Kir6.1	6					
SUR1+Kir6.2	30	>1000	1000	>1000	NE @ 100	0.45
SUR2A+Kir6.1	>1000					
SUR2A+Kir6.2	>1000 (~100)	10	1	10	1000 (100)	NE @ 100
SUR2B+Kir6.1	30		0.1			
SUR2B+Kir6.2	200	17	0.05	3	20 (200)	NE @ 100
MitoKATP	1	30	0.07 [NE @ 30]	1	100	

can open at all potentials and pass sizeable outward currents at depolarized potentials. Therefore, their stimulation causes, at rest, hyperpolarization toward E_{K+} and reduced excitability, and during activity, shortening of action potentials. Conversely, downregulation will tend to depolarize the membrane, especially if few other K^+ channels are active at rest to maintain an elevated K^+ conductance, and will promote action potential firing.

Glycemic control

K_{ATP} channels are involved in the maintenance of normoglycemia at the level of the pancreas and the central nervous system through complementary mechanisms. The role of K_{ATP} channels is best understood in pancreatic β-cells (Figure 4.4.3.4). These endocrine cells release insulin in a tightly controlled manner as a function of glycaemic levels. The SUR1+Kir6.2 K_{ATP} channels provide the dominant resting K^+ conductance and therefore set the membrane potential in pancreatic β-cells. Glucose is shuttled in the cytosol by the GLUT-2 transporter, enters the glycolytic cycle, and triggers ATP production from ADP, mainly in the mitochondrion. When plasma glucose levels increase, the concomitant increase in ATP (a Kir6.2 inhibitor) and decrease in ADP (a SUR1 activator) tally up to lower K_{ATP} channel activity and depolarize the membrane potential. Depolarization triggers action potential trains during which voltage-dependent L-type Ca^{2+} channels open and the

subsequent rise in internal Ca^{2+} initiates exocytosis of insulin granules. Insulin is packed within secretory granules in the form of zinc co-crystals. Its secretion is accompanied by release of Zn^{2+}, a potent activator of SUR1+Kir6.2 channels, that could provide a negative feedback mechanism to limit excessive secretion by a given cell or its immediate neighbors. Downregulation of insulin secretion by hormones like galanin, somatostatin, or adrenaline is established but current evidence does not support receptor-mediated modulation of K_{ATP} channels in β-cells as the primary event, as confirmed in K_{ATP} channel knockout models (Dufer et al., 2004; Sieg et al., 2004). This does not appear to be the case for leptin, the product of the obese (ob) gene, that activates K_{ATP} channels (Kieffer et al., 1997) possibly through phospholipid-dependent cytoskeleton disruption (Harvey et al., 2000).

Studies of K_{ATP} channels in pancreatic α- and δ-cells have been limited because of the lower density of these cells. The role of K_{ATP} channels in somatostatin release by δ-cells resembles that in β-cells (Göpel et al., 2000a). The scheme in glucagon-secreting α-cells must be different because the glucose-secretion coupling in these cells is the reverse of that in β-cells with higher glucose causing lower, not higher, glucagon release. It has been proposed that the specificity of α-cells stems not from an altered glucose regulation of K_{ATP} channels but from a distinct set of voltage-dependent channels that reshape electrical response (Göpel et al., 2000b).

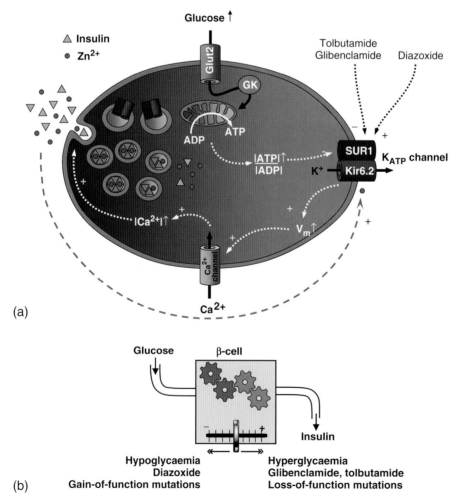

Fig. 4.4.3.4 K_{ATP} channels and insulin secretion in pancreatic β-cells. (a) In pancreatic β-cells, glucose is shuttled in the cytosol by the glucose transporter GLUT2 and metabolized by glucokinase (GK) and mitochondria to produce ATP from ADP. Cytosolic ATP raises and ADP decreases. These changes cause closure of K_{ATP} channels by the cumulative effects of increased inhibition of Kir6.2 by ATP and decreased activation of SUR1 by MgADP. The ensuing depolarization elicits opening of voltage-dependent Ca^{2+} channels, Ca^{2+} entry, and Ca^{2+}-initiated insulin granules exocytosis. Co-secreted Zn^{2+} could reactivate K_{ATP} channels in a negative feedback loop. K_{ATP} channel blockers like sulfonylureas mimic high glucose and upregulate secretion while openers like diazoxide down regulate secretion. (b) Viewed as a black box the β-cell produces insulin as a function of glucose. Output is increased by high glucose, channel blockers, or channel-disabling hyperinsulinemic mutations. It is decreased by low glucose, openers, or channel-activating diabetic mutations. See also colour plate section.

The metabolic sensing capacities of K_{ATP} channels are also used by the brain to keep glucose levels within strict limits. In severe hypoglycaemia, food intake is stimulated and secretion of counter-regulatory hormones like glucagon is augmented under autonomic input (Taborsky et al., 1998). This response is initiated in hypothalamic glucose-responsive neurons where K_{ATP} channels, as in β-cells, couple glucose to electrical activity (Miki et al., 2001). It was also recently demonstrated that, as in the hypothalamus, insulin and leptin can inhibit K_{ATP} channels in some neurons (Spanswick et al., 1997), triggering liver glucose release via the autonomic nervous system (Pocai et al., 2005).

Vascular tone control

K_{ATP} channels regulate vascular tone, and thereby the delivery of metabolic resources to match demand (Cole and Clement-Chomienne, 2003). This is due to K_{ATP} channel-dependent membrane hyperpolarization, reduction in Ca^{2+} influx through the voltage-gated Ca^{2+} channels and regulation of intracellular Ca^{2+} mobilization in smooth muscle (Quast et al., 1994). Knockout of K_{ATP} channel subunits promotes vasospasm, and hypertension (Miki et al., 2002; Chutkow et al., 2002). Conversely, activation of K_{ATP} channels controls blood pressure in the majority of cases with systemic hypertension (Jahangir and Terzic, 2005).

Metabolic sensing and protection

Integration of K_{ATP} channels with the cellular energetic network renders these heteromultimers high-fidelity metabolic sensors (Alekseev et al., 2005). This vital function is facilitated through phosphotransfer enzyme-mediated transmission of controllable energetic signals (Selivanov et al., 2004). By virtue of coupling with cellular energetic networks and the ability to decode metabolic signals, K_{ATP} channels set membrane excitability to match demand for homeostatic maintenance (Carrasco et al., 2001; Abraham et al., 2002). K_{ATP} channels are recognized for their cytoprotective role throughout the body (Zingman et al., 2003), exemplified by cardioprotection against ischaemia with channel-mediated shortening of the cardiac action potential controlling potentially deleterious calcium influx into the cytosol (Nichols 2006; Gumina et al., 2007). Sarcolemnal K_{ATP} channel activation was found to be responsible for the electrical current that underlies the characteristic ST-segment elevation of transmural ischemic injury (Li et al., 2000), and implicated in the endogenous protection mechanism of ischemic preconditioning (Suzuki et al., 2002; Gumina et al., 2003). Genetic ablation of the metabolic-sensing K_{ATP} channel disrupts an integrated homeostatic mechanism required in maintaining energetic myocardial stability under ischaemic stress (Kane et al., 2005).

More recent experimental data support a wider interpretation of this channel as a guarantor of metabolic and ionic homeostasis to diverse stressors (Zingman et al., 2003; Kane et al., 2005). K_{ATP} channels, harnessing the ability to recognize alterations in the metabolic state of the cell and translate this information into changes in membrane excitability, provide the link necessary for maintaining cellular well being in the face of stress-induced energy-demanding augmentation in performance. Conditions of sympathomimetic challenge (Zingman et al., 2002; Liu et al., 2004), physical exertion (Kane et al., 2004), mineralocorticoid-induced hypertension (Kane et al., 2006a), transverse aortic banding (Yamada et al., 2006) or septic shock (Kane et al., 2006b) have all been linked to myocardial K_{ATP} channel-mediated protection. Moreover, stress challenge has been found to be pro-arrhythmic in the K_{ATP} channel-deficient myocardium provoking early after-depolarizations, triggered activity and ventricular dysrrhythmia (Liu et al., 2004).

4.4.3.7 Pathophysiology and disease

K_{ATP} channels and insulin secretion diseases

The central role of the K_{ATP} channel in insulin secretion makes it a prime suspect as well as a prime pharmacological target for hypoglycaemic (hyperinsulinemia) and hyperglycaemic (diabetes) pathologies. As described above, insulin secretion by β-cells is inversely correlated with K_{ATP} channel activity. Conditions tending to decrease activity will boost insulin secretion above normal leading to hypoglycaemia, while an abnormally high activity will attenuate insulin secretion, leading to hyperglycaemia. This is the basis for the use of sulfonylurea blockers as antidiabetic drugs. This is also the basis for diseases arising from mutations in the genes encoding the Kir6.2 and SUR1 subunits of the pancreatic K_{ATP} channel (for a complete list of known mutations, see Gloyn et al., 2006).

Loss-of-function mutations are the most common cause of a rare disease, hyperinsulinemia of infancy (HI; OMIM #256450), also known as persistent hyperinsulinemic hypoglycemia of infancy (PHHI) or congenital hyperinsulinism (CHI). HI mutations are most frequent in the SUR1 gene. They may be separated in two classes (Ashcroft, 2005): Class I mutations reduce the number of channels at the plasma membrane because they disrupt a step (synthesis, addressing, etc.) in the biogenesis of the channel complex; Class II mutations reduce the open probability of correctly formed channels mainly by abrogating activation by MgADP (Nichols et al., 1996). HI mutations of Kir6.2 are less frequent and also result in lower channel activity (Ashcroft, 2005). In rare HI cases where channels remain functional and responsive to K_{ATP} channel openers, pharmacological treatment with diazoxide-type openers may be able to restore a reasonable level of channel activity and reduce released insulin (Dunne et al., 2004). It has also been reported that sulfonylureas can act as chemical chaperones and correct trafficking deficiencies of SUR1 mutants (Yan et al., 2004). The clinical benefit of this approach remains to be explored.

Gain-of-function mutations, discovered recently (Gloyn et al., 2004), tend to keep channels open, hyperpolarize β-cells, and reduce insulin secretion. These mutations are responsible for a rare form of diabetes melittus, neonatal diabetes (NDM; OMIM #606176 and 601410). In these conditions, channels are overactive and nominal activity can often be restored with sulfonylurea blockers (Koster et al., 2005b). NDM mutations are concentrated in the vicinity of the presumed ATP binding site of Kir6.2 and reduce the apparent blocking affinity of ATP (Ashcroft, 2005). Functionally equivalent mutations in SUR1 have also been identified (Babenko et al., 2006; Proks et al., 2006). The severity of the clinical syndromes correlate with the magnitude of the shift in ATP affinity and range from mild in the case of transient NDM (TNDM) to severe for permanent NDM (PNDM) with possible developmental and neurological complications leading to developmental delay/epileptic ND (DEND). It is of interest that clarification of the molecular etiology of these syndromes, previously diagnosed as type 1 diabetes, has led to re-evaluation of therapeutic management from insulin injections to better-suited and less painful oral sulfonylureas with refinement of pharmacogenomic approaches for individualized patient care (Pearson *et al.*, 2006; Sattiraju *et al.*, 2008).

There are indications that K_{ATP} channel gene polymorphisms are also associated with the much more widespread adult-onset type 2 diabetes (T2DM). In particular, although this has been disputed, a common Kir6.2 gene variant E23K has been linked to increased T2DM susceptibility (Gloyn *et al.*, 2003; Koster *et al.*, 2005a) and functional studies have revealed that this mutation increased channel open probability, leading to a slightly reduced sensitivity to inhibition by ATP (Schwanstecher *et al.*, 2002) and abnormal gating by long chain acyl CoA esters (Riedel *et al.*, 2005).

K_{ATP} channels and cardiac diseases

Mutations in K_{ATP} channel genes have been linked to cardiac human channelopathies, and an increased susceptibility to cardiac disease. Specifically, mutations in the regulatory SUR2A subunit of the cardiac K_{ATP} channel complex were recently demonstrated in patients with dilated cardiomyopathy and ventricular arrhythmia (Bienengraeber *et al.*, 2004), as well as in a case of adrenergic atrial fibrillation (Olson *et al.*, 2007).

Dilated cardiomyopathy with ventricular tachycardia

Dilated cardiomyopathy is an idiopathic form of heart failure characterized by ventricular dilation and reduced contractile function. This phenotype may occur as an isolated trait or in association with rhythm disorder. It has been recognized that dilated cardiomyopathy is familial in >20 per cent cases, and over 20 distinct genes have so far been linked with the heritable form of the disease (Ahmad *et al.*, 2005). The ontological spectrum of dilated cardiomyopathy-associated mutant genes has traditionally included those encoding contractile, cytoskeletal or nuclear proteins, and was more recently expanded to the distinct class of proteins that regulate ion homeostasis. This latter group includes the K_{ATP} channel, as mutations in the *ABCC9*-encoded regulatory channel subunit SUR2A have been demonstrated in a subset of patients with dilated cardiomyopathy and ventricular arrhythmia (Bienengraeber *et al.*, 2004). The identified missense and frameshift mutations were mapped to domains bordering the catalytic ATPase pocket within SUR2A. Mutant SUR2A proteins reduced the intrinsic channel ATPase activity, altering reaction kinetics, and translating into dysfunctional channel phenotype with impaired metabolic signal decoding and processing capabilities. Dilated cardiomyopathy with ventricular tachycardia due to *ABCC9* K_{ATP} channel mutations has been designated CMD10 (OMIM #608569) in the Online Mendelian Inheritance in Man database to distinguish this entity from other aetiologies causing this heterogeneous condition. Collectively, these data implicate a link between mutations in the cardioprotective K_{ATP} channel and susceptibility to cardiomyopathic disease with heart failure and electrical instability, as independently validated in knockout models lacking the Kir6.2 channel pore (Kane *et al.*, 2004; Liu *et al.*, 2004; Kane *et al.*, 2005, 2006a; Yamada *et al.*, 2006).

Atrial fibrillation

Atrial fibrillation is characterized by rapid and irregular electrical activation of the atrium, and is traditionally viewed as an acquired disorder attributable to structural heart disease in patients with comorbidities. However, recognition of familial aggregation has more recently implicated a heritable basis for atrial fibrillation (Darbar *et al.*, 2003). A primary genetic defect is particularly likely in familial cases of early onset lone atrial fibrillation. In this regard, gain-of-function mutations in the *KCNQ1* and *KCNE2* genes (Chen *et al.*, 2003; Darbar *et al.*, 2003), encoding subunits of the cardiac voltage-dependent channel I_{Ks}, were the first molecular defects identified as a cause for atrial fibrillation, based on action potential shortening and pro-arrhythmogenic reduction in refractory period. More recently, identification of a loss-of-function mutation in *KCNA5*, encoding the voltage-dependent Kv1.5 channel, has provided an alternative mechanism for atrial fibrillation (Olson *et al.*, 2006). Kv1.5 channelopathy increased propensity for prolongation of action potential duration, providing a substrate for triggered activity in the human atrium. A possibly equivalent mechanism has been reported in the case of a K_{ATP} channel mutation conferring risk for adrenergic atrial fibrillation originating from the vein of Marshall (Olson *et al.*, 2007). The vein of Marshall, a remnant of the left superior vena cava rich in sympathetic fibres, is a recognized source for adrenergic atrial fibrillation. Recent human genetic investigation has uncovered a missense mutation (T1547I) in the *ABCC9* gene, encoding the regulatory subunit of cardiac K_{ATP} channels. Structural modelling of mutant K_{ATP} channels predicted, and patch-clamp electrophysiology demonstrated, compromised function with defective stress responsiveness. Targeted knockout of the K_{ATP} channel verified the pathogenic link between channel dysfunction and predisposition to adrenergic atrial fibrillation. In this first report of genetic susceptibility to adrenergic vein of Marshall atrial fibrillation, radio-frequency ablation was curative, disrupting the gene–environment substrate for arrhythmia conferred by K_{ATP} channelopathy (Olson *et al.*, 2007).

4.4.3.8 Concluding remarks

As a unique combination of an inward rectifier K^+ channel and an ABC protein, the K_{ATP} channel is probably one of the most sophisticated channels in terms of mechanism and pharmacology. Much progress has been made in almost 25 years since its discovery. However, many questions remain at the molecular and physiological levels. We still lack high-resolution structural data on the SUR-Kir6.x complex and are left with a tenuous understanding of the interactions between SUR and Kir6.x and a vague idea of the topology of the SUR binding sites for pharmacological agents. Except in the pancreas, we are only beginning to grasp the physiological and pathophysiological roles of the K_{ATP} channels in the organs where they are present. As more advanced animal models and more specific pharmacological tools are being developed, answers should be provided and should translate into novel insights and therapies.

References

Abraham MR, Selivanov VA, Hodgson DM *et al.* (2002). Coupling of cell energetics with membrane metabolic sensing. Integrative signaling through creatine kinase phosphotransfer disrupted by M-CK gene knock-out. *J Biol Chem* **277**, 24427–24434.

Aguilar-Bryan L, Clement JP, Gonzalez G *et al.* (1998). Toward understanding the assembly and structure of K_{ATP} channels. *Physiol Rev* **78**, 227–245.

Aguilar-Bryan L, Nichols CG, Wechsler SW et al. (1995). Cloning of the β cell high-affinity sulfonylurea receptor: a regulator of insulin secretion. Science 268, 423–426.

Ahmad F, Seidman JG and Seidman CE (2005). The genetic basis for cardiac remodeling. Annu Rev Genomics Hum Genet 6, 185–216.

Alekseev AE, Brady PA and Terzic A (1998). Ligand-insensitive state of cardiac ATP-sensitive K+ channels—basis for channel opening. J Gen Physiol 111, 381–394.

Alekseev AE, Hodgson DM, Karger AB et al. (2005). ATP-sensitive K+ channel channel/enzyme multimer: metabolic gating in the heart. J Mol Cell Cardiol 38, 895–905.

Alekseev AE, Kennedy ME, Navarro B et al. (1997). Burst kinetics of co-expressed Kir6.2/SUR1 clones: Comparison of recombinant with native ATP-sensitive K+ channel behavior. J Membr Biol 159, 161–168.

Ämmälä C, Moorhouse A and Ashcroft FM (1996). The sulphonylurea receptor confers diazoxide sensitivity on the inwardly rectifying K+ channel Kir6.1 expressed in human embryonic kidney cells. J Physiol 494, 709–714.

Antcliff JF, Haider S, Proks P et al. (2005). Functional analysis of a structural model of the ATP-binding site of the KATP channel Kir6.2 subunit. EMBO J 24, 229–239.

Ardehali H and O'Rourke B (2005). Mitochondrial KATP channels in cell survival and death. J Mol Cell Cardiol 39, 7–16.

Ashcroft FM (2005). ATP-sensitive potassium channelopathies: focus on insulin secretion. J Clin Invest 115, 2047–2058.

Ashcroft FM, Kakei M and Kelly RP (1989). Rubidium and sodium permeability of the ATP-sensitive K+ channel in single rat pancreatic β-cells. J Physiol 408, 413–429.

Ashfield R and Ashcroft SJ (1998). Cloning of the promoters for the β-cell ATP-sensitive K-channel subunits Kir6.2 and SUR1. Diabetes Care 47, 1274–1280.

Ashfield R, Gribble FM, Ashcroft SJ et al. (1999). Identification of the high-affinity tolbutamide site on the SUR1 subunit of the KATP channel. Diabetes 48, 1341–1347.

Babenko AP and Bryan J (2001). A conserved inhibitory and differential stimulatory action of nucleotides on Kir6.0/SUR complexes is essential for excitation-metabolism coupling by KATP channels. J Biol Chem 276, 49083–49092.

Babenko AP and Bryan J (2003). SUR domains that associate with and gate KATP pores define a novel gatekeeper. J Biol Chem 278, 41577–41580.

Babenko AP, Gonzalez G and Bryan J (1999a). The tolbutamide site of SUR1 and a mechanism for its functional coupling to KATP channel closure. FEBS Lett 459, 367–376.

Babenko AP, Gonzalez G and Bryan J (1999b). Two regions of sulfonylurea receptor specify the spontaneous bursting and ATP inhibition of KATP channel isoforms. J Biol Chem 274, 11587–11592.

Babenko AP, Gonzalez G and Bryan J (2000a). Pharmaco-topology of sulfonylurea receptors—separate domains of the regulatory subunits of KATP channel isoforms are required for selective interaction with K+ channel openers. J Biol Chem 275, 717–720.

Babenko AP, Gonzalez G, Aguilar-Bryan L et al. (1998). Reconstituted human cardiac KATP channels—functional identity with the native channels from the sarcolemma of human ventricular cells. Circ Res 83, 1132–1143.

Babenko AP, Gonzalez GC and Bryan J (2000b). Hetero-concatemeric K(IR)6.X-4/SUR1(4) channels display distinct conductivities but uniform ATP inhibition. J Biol Chem 275, 31563–31566.

Babenko AP, Polak M, Cave H et al. (2006). Activating mutations in the ABCC8 gene in neonatal diabetes mellitus. N Engl J Med 355, 456–466.

Bancila V, Cens T, Monnier D et al. (2005). Two SUR1-specific histidine residues mandatory for zinc-induced activation of the rat KATP channel. J Biol Chem 280, 8793–8799.

Baron A, van Bever L, Monnier D et al. (1999). A novel KATP current in cultured neonatal rat atrial appendage cardiomyocytes. Circ Res 85, 707–715.

Baukrowitz T, Schulte U, Oliver D et al. (1998). PIP2 and PIP as determinants for ATP inhibition of KATP channels. Science 282, 1141–1144.

Baukrowitz T, Tucker SJ, Schulte U et al. (1999). Inward rectification in KATP channels: a pH switch in the pore. EMBO J 18, 847–853.

Beguin P, Nagashima K, Nishimura M et al. (1999). PKA-mediated phosphorylation of the human KATP channel: separate roles of Kir6.2 and SUR1 subunit phosphorylation. EMBO J 18, 4722–4732.

Bienengraeber M, Olson TM, Selivanov VA et al. (2004). ABCC9 mutations identified in human dilated cardiomyopathy disrupt catalytic KATP channel gating. Nat Genet 36, 382–387.

Bloc A, Cens T, Cruz H et al. (2000). Zinc-induced changes in ionic currents of clonal rat pancreatic β-cells: activation of ATP-sensitive K+ channels. J Physiol 529, 723–734.

Bokvist K, Rorsman P and Smith PA (1990). Block of ATP-regulated and Ca2+-activated K+ channels in mouse pancreatic β-cells by external tetraethylammonium and quinine. J Physiol 423, 327–342.

Branstrom R, Leibiger IB, Leibiger B et al. (2007). Single residue (K332A) substitution in Kir6.2 abolishes the stimulatory effect of long-chain acyl-CoA esters: indications for a long-chain acyl-CoA ester binding motif. Diabetologia 50, 1670–1677.

Carrasco AJ, Dzeja PP, Alekseev AE et al. (2001). Adenylate kinase phosphotransfer communicates cellular energetic signals to ATP-sensitive potassium channels. PNAS 98, 7623–7628.

Castle NA and Haylett DG (1987). Effect of channel blockers on potassium efflux from metabolically exhausted frog skeletal muscle. J Physiol 383, 31–43.

Chachin M, Yamada M, Fujita A et al. (2003). Nateglinide, a D-phenylalanine derivative lacking either a sulfonylurea or benzamido moiety, specifically inhibits pancreatic β-cell-type KATP channels. J Pharmacol Exp Ther 304, 1025–1032.

Chan KW, Wheeler A and Csanady L (2008). Sulfonylurea receptors type 1 and 2A randomly assemble to form heteromeric KATP channels of mixed subunit composition. J Gen Physiol 131, 43–58.

Chan KW, Zhang H and Logothetis DE (2003). N-terminal transmembrane domain of the SUR controls trafficking and gating of Kir6 channel subunits. EMBO J 22, 3833–3843.

Chen YH, Xu SJ, Bendahhou S et al. (2003). KCNQ1 gain-of-function mutation in familial atrial fibrillation. Science 299, 251–254.

Cheng WW, Tong A, Flagg TP et al. (2008). Random assembly of SUR subunits in KATP channel complexes. Channels 2, 34–36.

Chutkow WA, Makielski JC, Nelson DJ et al. (1999). Alternative splicing of sur2 exon 17 regulates nucleotide sensitivity of the ATP-sensitive potassium channel. J Biol Chem 274, 13656–13665.

Chutkow WA, Pu JL, Wheeler MT et al. (2002). Episodic coronary artery vasospasm and hypertension develop in the absence of Sur2 KATP channels. J Clin Invest 110, 203–208.

Chutkow WA, Simon MC, LeBeau MM et al. (1996). Cloning, tissue expression, and chromosomal localization of SUR2, the putative drug-binding subunit of cardiac, skeletal muscle, and vascular KATP channels. Diabetes 45, 1439–1445.

Ciani S and Ribalet B (1988). Ion permeation and rectification in ATP-sensitive channels from insulin-secreting cells (RINm5F): effects of K+, Na+ and Mg2+. J Membr Biol 103, 171–180.

Clement JP, Kunjilwar K, Gonzalez G et al. (1997). Association and stoichiometry of KATP channel subunits. Neuron 18, 827–838.

Cole WC and Clement-Chomienne O (2003). ATP-sensitive K+ channels of vascular smooth muscle cells. J Cardiovasc Electrophysiol 14, 94–103.

Cook DL and Hales CN (1984). Intracellular ATP directly blocks K+ channels in pancreatic β-cells. Nature 311, 271–273.

Cui Y, Giblin JP, Clapp LH et al. (2001). A mechanism for ATP-sensitive potassium channel diversity: functional coassembly of two pore-forming subunits. PNAS 98, 729–734.

Cui Y, Tinker A and Clapp LH (2003). Different molecular sites of action for the K$_{ATP}$ channel inhibitors, PNU-99963 and PNU-37883A. Br J Pharmacol 139, 122–128.

D'hahan N, Jacquet H, Moreau C et al. (1999a). A transmembrane domain of the sulfonylurea receptor mediates activation of ATP-sensitive K$^+$ channels by K$^+$ channel openers. Mol Pharmacol 56, 308–315.

D'hahan N, Moreau C, Prost AL et al. (1999b). Pharmacological plasticity of cardiac ATP-sensitive potassium channels toward diazoxide revealed by ADP. PNAS 96, 12162–12167.

Dabrowski M, Ashcroft FM, Ashfield R et al. (2002). The novel diazoxide analog 3-isopropylamino-7-methoxy-4H-1,2,4-benzothiadiazine 1,1-dioxide is a selective Kir6.2/SUR1 channel opener. Diabetes 51, 1896–1906.

Dabrowski M, Larsen T, Ashcroft FM et al. (2003). Potent and selective activation of the pancreatic β-cell type K$_{ATP}$ channel by two novel diazoxide analogues. Diabetologia 46, 1375–1382.

Dabrowski M, Tarasov A and Ashcroft FM (2004). Mapping the architecture of the ATP-binding site of the K$_{ATP}$ channel subunit Kir6.2. J Physiol 557, 347–354.

Dabrowski M, Wahl P, Holmes WE et al. (2001). Effect of repaglinide on cloned β cell, cardiac and smooth muscle types of ATP-sensitive potassium channels. Diabetologia 44, 747–756.

Darbar D, Herron KJ, Ballew JD et al. (2003). Familial atrial fibrillation is a genetically heterogeneous disorder. J Am Coll Cardiol 41, 2185–2192.

Davies NW (1990). Modulation of ATP-sensitive K$^+$ channels in skeletal muscle by intracellular protons. Nature 343, 375–377.

Davies NW, Pettit AI, Agarwal R et al. (1991). The flickery block of ATP-dependent potassium channels of skeletal muscle by internal 4-aminopyridine. Pflügers Arch-Eur J Physiol 419, 25–31.

Davies NW, Spruce AE, Standen NB et al. (1989). Multiple blocking mechanisms of ATP-sensitive potassium channels of frog skeletal muscle by tetraethylammonium ions. J Physiol 413, 31–48.

Davis-Taber R, Choi W, Feng JL et al. (2000). Molecular characterization of human SUR2-containing K$_{ATP}$ channels. Gene 256, 261–270.

Dawson RJ and Locher KP (2006). Structure of a bacterial multidrug ABC transporter. Nature 443, 180–185.

Dong K, Tang LQ, MacGregor GG et al. (2005). Novel nucleotide-binding sites in ATP-sensitive potassium channels formed at gating interfaces. EMBO J 24, 1318–1329.

Doyle DA, Cabral JM, Pfuetzner RA et al. (1998). The structure of the potassium channel: molecular basis of K$^+$ conduction and selectivity. Science 280, 69–77.

Doyle ME and Egan JM (2003). Pharmacological agents that directly modulate insulin secretion. Pharmacol Rev 55, 105–131.

Drain P, Li LH and Wang J (1998). K$_{ATP}$ channel inhibition by ATP requires distinct functional domains of the cytoplasmic C terminus of the pore-forming subunit. PNAS 95, 13953–13958.

Dufer M, Haspel D, Krippeit-Drews P et al. (2004). Oscillations of membrane potential and cytosolic Ca^{2+} concentration in SUR1(-/-) β cells. Diabetologia 47, 488–498.

Dunne MJ, Cosgrove KE, Shepherd RM et al. (2004). Hyperinsulinism in infancy: from basic science to clinical disease. Physiol Rev 84, 239–275.

Dupuis P, Revilloud J, Moreau CJ et al. (2008). Three C-terminal residues from the sulfonylurea receptor contribute to the functional coupling between the K$_{ATP}$ channel subunits SUR2A and Kir6.2. J Physiol 586, 3075–3085.

Eaton MJ, Skatchkov SN, Brune A et al. (2002). SUR1 and Kir6.1 subunits of K$_{ATP}$-channels are co-localized in retinal glial (Muller) cells. Neuroreport 13, 57–60.

Enkvetchakul D and Nichols CG (2003). Gating mechanism of K$_{ATP}$ channels: function fits form. J Gen Physiol 122, 471–480.

Escande D, Thuringer D, Leguern S et al. (1988). The potassium channel opener cromakalim (BRL 34915) activates ATP-dependent K$^+$ channels in isolated cardiac myocytes. Biochem Biophys Res Commun 154, 620–625.

Fan Z and Makielski JC (1997). Anionic phospholipids activate ATP-sensitive potassium channels. J Biol Chem 272, 5388–5395.

Fan Z and Makielski JC (1999). Phosphoinositides decrease ATP sensitivity of the cardiac ATP-sensitive K$^+$ channel—a molecular probe or the mechanism of ATP-sensitive inhibition. J Gen Physiol 114, 251–269.

Fan Z, Tokuyama Y and Makielski JC (1994). Modulation of ATP-sensitive K$^+$ channels by internal acidification in insulin-secreting cells. Am J Physiol 267, C1036–C1044.

Fang K, Csanady L and Chan KW (2006). The N-terminal transmembrane domain (TMD0) and a cytosolic linker (L0) of sulphonylurea receptor define the unique intrinsic gating of K$_{ATP}$ channels. J Physiol 576, 379–389.

Fatherazi S and Cook DL (1991). Specificity of tetraethylammonium and quinine for three K-channels in insulin-secreting cells. J Membr Biol 120, 105–114.

Fox JE, Kanji HD, French RJ et al. (2002). Cardioselectivity of the sulphonylurea HMR 1098: studies on native and recombinant cardiac and pancreatic K$_{ATP}$ channels. Br J Pharmacol 135, 480–488.

Fujita R, Kimura S, Kawasaki S et al. (2007). Electrophysiological and pharmacological characterization of K$_{ATP}$ channel involved in the K$^+$ current responses to FSH and adenosine in the follicular cells of Xenopus oocyte. J Physiol Sci 57(1), 51–61.

Furukawa T, Yamane T, Terai T et al. (1996). Functional linkage of the cardiac ATP sensitive K$^+$ channel to the actin cytoskeleton. Pflügers Arch-Eur J Physiol 431, 504–512.

Giblin JP, Cui Y, Clapp LH et al. (2002). Assembly limits the pharmacological complexity of ATP-sensitive potassium channels. J Biol Chem 277, 13717–13723.

Giblin JP, Leaney JL and Tinker A (1999). The molecular assembly of ATP-sensitive potassium channels—determinants on the pore-forming subunit. J Biol Chem 274, 22652–22659.

Gloyn AL, Pearson ER, Antcliff JF et al. (2004). Activating mutations in the gene encoding the ATP-sensitive potassium-channel subunit Kir6.2 and permanent neonatal diabetes. N Engl J Med 350, 1838–1849.

Gloyn AL, Siddiqui J and Ellard S (2006). Mutations in the genes encoding the pancreatic β-cell K$_{ATP}$ channel subunits Kir6.2 (KCNJ11) and SUR1 (ABCC8) in diabetes mellitus and hyperinsulinism. Hum Mutat 27, 220–231.

Gloyn AL, Weedon MN, Owen KR et al. (2003). Large-scale association studies of variants in genes encoding the pancreatic β-cell K$_{ATP}$ channel subunits Kir6.2 (KCNJ11) and SUR1 (ABCC8) confirm that the KCNJ11 E23K variant is associated with type 2 diabetes. Diabetes 52, 568–572.

Göpel SO, Kanno T, Barg S et al. (2000a). Patch-clamp characterisation of somatostatin-secreting-cells in intact mouse pancreatic islets. J Physiol 528, 497–507.

Göpel SO, Kanno T, Barg S et al. (2000b). Regulation of glucagon release in mouse α-cells by K$_{ATP}$ channels and inactivation of TTX-sensitive Na$^+$ channels. J Physiol 528, 509–520.

Gribble FM, Ashfield R, Ämmälä C et al. (1997a). Properties of cloned ATP-sensitive K$^+$ currents expressed in Xenopus oocytes. J Physiol 498, 87–98.

Gribble FM, Davis TM, Higham CE et al. (2000). The antimalarial agent mefloquine inhibits ATP-sensitive K-channels. Br J Pharmacol 131, 756–760.

Gribble FM, Proks P, Corkey BE et al. (1998a). Mechanism of cloned ATP-sensitive potassium channel activation by oleoyl-CoA. J Biol Chem 273, 26383–26387.

Gribble FM, Tucker SJ and Ashcroft FM (1997b). The essential role of the Walker A motifs of SUR1 in K$_{ATP}$ channel activation by Mg-ADP and diazoxide. EMBO J 16, 1145–1152.

Gribble FM, Tucker SJ, Haug T *et al.* (1998b). MgATP activates the β cell K$_{ATP}$ channel by interaction with its SUR1 subunit. *PNAS* **95**, 7185–7190.

Gribble FM, Tucker SJ, Seino S *et al.* (1998c). Tissue specificity of sulfonylureas: studies on cloned cardiac and β-cell K$_{ATP}$ channels. *Diabetes* **47**, 1412–1418.

Gros L, Trapp S, Dabrowski M *et al.* (2002). Characterization of two novel forms of the rat sulphonylurea receptor SUR1A2 and SUR1BD31. *Br J Pharmacol* **137**, 98–106.

Gross I, Toman A, Uhde I *et al.* (1999). Stoichiometry of potassium channel opener action. *Mol Pharmacol* **56**, 1370–1373.

Gumina RJ, O'Cochlain DF, Kurtz CE *et al.* (2007). K$_{ATP}$ channel knockout worsens myocardial calcium stress load *in vivo* and impairs recovery in stunned heart. *Am J Physiol* **292**, H1706–H1713.

Gumina RJ, Pucar D, Bast P *et al.* (2003). Knockout of Kir6.2 negates ischemic preconditioning-induced protection of myocardial energetics. *Am J Physiol* **284**, H2106–H2113.

Haider S, Tarasov AI, Craig TJ *et al.* (2007). Identification of the PIP2-binding site on Kir6.2 by molecular modelling and functional analysis. *EMBO J* **26**, 3749–3759.

Hambrock A, Löffler-Walz C, Russ U *et al.* (2001). Characterization of a mutant sulfonylurea receptor SUR2B with high affinity for sulfonylureas and openers: differences in the coupling to Kir6.x subtypes. *Mol Pharmacol* **60**, 190–199.

Hambrock A, Preisig-Müller R, Russ U *et al.* (2002). Four novel splice variants of sulfonylurea receptor 1. *Am J Physiol* **283**, C587–C598.

Hamill OP, Marty A, Neher E *et al.* (1981). Improved patch-clamp techniques for high-resolution current recording from cells and cell-free membrane patches. *Pflügers Arch-Eur J Physiol* **391**, 85–100.

Hanley PJ and Daut J (2005). K$_{ATP}$ channels and preconditioning: a re-examination of the role of mitochondrial K$_{ATP}$ channels and an overview of alternative mechanisms. *J Mol Cell Cardiol* **39**, 17–50.

Hansen AM, Christensen IT, Hansen JB *et al.* (2002). Differential interactions of nateglinide and repaglinide on the human β-cell sulphonylurea receptor 1. *Diabetes* **51**, 2789–2795.

Harvey J, Hardy SC, Irving AJ *et al.* (2000). Leptin activation of ATP-sensitive K+ (K$_{ATP}$) channels in rat CRI-G1 insulinoma cells involves disruption of the actin cytoskeleton. *J Physiol* **527**, 95–107.

Heusser K, Yuan H, Neagoe I *et al.* (2006). Scavenging of 14-3-3 proteins reveals their involvement in the cell-surface transport of ATP-sensitive K+ channels. *J Cell Sci* **119**, 4353–4363.

Hilgemann DW and Ball R (1996). Regulation of cardiac Na+,Ca^{2+} exchange and K$_{ATP}$ potassium channels by PIP2. *Science* **273**, 956–959.

Horie M, Irisawa H and Noma A (1987). Voltage-dependent magnesium block of adenosine-triphosphate-sensitive potassium channel in guinea-pig ventricular cells. *J Physiol* **387**, 251–272.

Hosy E, Derand R, Revilloud J (2007). Remodelling of the SUR-Kir6.2 interface of the K$_{ATP}$ channel upon ATP binding revealed by the conformational blocker rhodamine 123. *J Physiol* **582**, 27–39.

Inagaki N, Gonoi T and Seino S (1997). Subunit stoichiometry of the pancreatic β-cell ATP-sensitive K+ channel. *FEBS Lett* **409**, 232–236.

Inagaki N, Gonoi T, Clement JP *et al.* (1995a). Reconstitution of I-KATP: an inward rectifier subunit plus the sulfonylurea receptor. *Science* **270**, 1166–1170.

Inagaki N, Gonoi T, Clement JP *et al.* (1996). A family of sulfonylurea receptors determines the pharmacological properties of ATP-sensitive K+ channels. *Neuron* **16**, 1011–1017.

Inagaki N, Inazawa J and Seino S (1995b). cDNA sequence, gene structure, and chromosomal localization of the human ATP-sensitive potassium channel, uK$_{ATP}$-1, gene (*KCNJ8*). *Genomics* **30**, 102–104.

Inagaki N, Tsuura Y, Namba N *et al.* (1995c). Cloning and functional characterization of a novel ATP-sensitive potassium channel ubiquitously expressed in rat tissues, including pancreatic islets, pituitary, skeletal muscle, and heart. *J Biol Chem* **270**, 5691–5694.

Isomoto S, Horio Y, Matsumoto S *et al.* (1997). Sur2 and Kcnj8 genes are tightly linked on the distal region of mouse chromosome 6. *Mamm Genome* **8**, 790–791.

Isomoto S, Kondo C, Yamada M *et al.* (1996). A novel sulfonylurea receptor forms with BIR (Kir6.2) a smooth muscle type ATP-sensitive K+ channel. *J Biol Chem* **271**, 24321–24324.

Ito H, Tung RT, Sugimoto T *et al.* (1992). On the mechanism of G protein βg subunit activation of the muscarinic K+ channel in guinea pig atrial cell membrane. Comparison with the ATP-sensitive K+ channel. *J Gen Physiol* **99**, 961–983.

Jahangir A and Terzic A (2005). K$_{ATP}$ channel therapeutics at the bedside. *J Mol Cell Cardiol* **39**, 99–112.

Janbon M, Chapal J, Vedel A *et al.* (1942). Accidents hypoglycémiques graves par un sulfamidothiodiazol (le VK 57 ou 2254 RP). *Montpellier med* **441**, 21–22.

Jons T, Wittschieber D, Beyer A *et al.* (2006). K+-ATP-channel-related protein complexes: potential transducers in the regulation of epithelial tight junction permeability. *J Cell Sci* **119**, 3087–3097.

Jovanovic A, Alekseev AE and Terzic A (1997). Intracellular diadenosine polyphosphates—a novel family of inhibitory ligands of the ATP-sensitive K+ channel. *Biochem Pharmacol* **54**, 219–225.

Kakei M and Noma A (1984). Adenosine-5'-triphosphate-sensitive single potassium channel in the atrioventricular node cell of the rabbit heart. *J Physiol* **352**, 265–284.

Kakei M, Noma A and Shibasaki T (1985). Properties of adenosine-triphosphate-regulated potassium channels in guinea-pig ventricular cells. *J Physiol* **363**, 441–462.

Kane GC, Behfar A, Dyer RB *et al.* (2006a). KCNJ11 gene knockout of the Kir6.2 K$_{ATP}$ channel causes maladaptive remodeling and heart failure in hypertension. *Hum Mol Genet* **15**, 2285–2297.

Kane GC, Behfar A, Yamada S *et al.* (2004). ATP-sensitive K+ channel knockout compromises the metabolic benefit of exercise training, resulting in cardiac deficits. *Diabetes* **53**, S169–S175.

Kane GC, Lam CF, O'Cochlain F *et al.* (2006b). Gene knockout of the KCNJ8-encoded Kir6.1 K$_{ATP}$ channel imparts fatal susceptibility to endotoxemia. *FASEB J* **20**, 2271–2280.

Kane GC, Liu XK, Yamada S *et al.* (2005). Cardiac K$_{ATP}$ channels in health and disease. *J Mol Cell Cardiol* **38**, 937–943.

Kang YH, Ng B, Leung YM *et al.* (2006). Syntaxin-1A actions on sulfonylurea receptor 2A can block acidic pH-induced cardiac K$_{ATP}$ channel activation. *J Biol Chem* **281**, 19019–19028.

Kelly ML, Abu-Hamdah R, Jeremic A *et al.* (2005). Patch clamped single pancreatic zymogen granules: direct measurements of ion channel activities at the granule membrane. *Pancreatology* **5**, 443–449.

Kieffer TJ, Keller RS, Leech CA *et al.* (1997). Leptin suppression of insulin secretion by the activation of ATP-sensitive K+ channels in pancreatic β-cells. *Diabetes* **46**, 1087–1093.

Kono Y, Horie M, Takano M *et al.* (2000). The properties of the Kir6.1–6.2 tandem channel co-expressed with SUR2A. *Pflügers Arch-Eur J Physiol* **440**, 692–698.

Konstas AA, Dabrowski M, Korbmacher C *et al.* (2002). Intrinsic sensitivity of Kir1.1 (ROMK) to glibenclamide in the absence of SUR2B—implications for the identity of the renal ATP-regulated secretory K+ channel. *J Biol Chem* **277**, 21346–21351.

Koster JC, Permutt MA and Nichols CG (2005a). Diabetes and insulin secretion—the ATP-sensitive K+ channel (K$_{ATP}$) connection. *Diabetes* **54**, 3065–3072.

Koster JC, Remedi MS, Dao C *et al.* (2005b). ATP and sulfonylurea sensitivity of mutant ATP-sensitive K+ channels in neonatal diabetes—implications for pharmacogenomic therapy. *Diabetes* **54**, 2645–2654.

Koster JC, Sha Q, Shyng SL *et al.* (1999). ATP inhibition of K$_{ATP}$ channels: control of nucleotide sensitivity by the N-terminal domain of the Kir6.2 subunit. *J Physiol* **515**, 19–30.

Kovacs RJ and Nelson MT (1991). ATP-sensitive K+ channels from aortic smooth muscle incorporated into planar lipid bilayers. *Am J Physiol* **261**, H604-H609.

Kubo Y and Murata Y (2001). Control of rectification and permeation by two distinct sites after the second transmembrane region in Kir2.1 K+ channel. *J Physiol* **531**, 645–660.

Kuo AL, Gulbis JM, Antcliff JF *et al.* (2003). Crystal structure of the potassium channel KirBac1.1 in the closed state. *Science* **300**, 1922–1926.

Lee K, Dixon AK, Freeman TC *et al.* (1998). Identification of an ATP-sensitive potassium channel current in rat striatal cholinergic interneurones. *J Physiol* **510**, 441–453.

Lee K, Dixon AK, Richardson PJ *et al.* (1999). Glucose-receptive neurones in the rat ventromedial hypothalamus express K$_{ATP}$ channels composed of Kir6.1 and SUR1 subunits. *J Physiol* **515**, 439–452.

Li L, Shi Y, Wang XR *et al.* (2005). Single nucleotide polymorphisms in K$_{ATP}$ channels—muscular impact on type 2 diabetes. *Diabetes* **54**, 1592–1597.

Li LH, Geng XH and Drain P (2002). Open state destabilization by ATP occupancy is mechanism speeding burst exit underlying K$_{ATP}$ channel inhibition by ATP. *J Gen Physiol* **119**, 105–116.

Li RA, Leppo M, Miki T *et al.* (2000). Molecular basis of electrocardiographic ST-segment elevation. *Circ Res* **87**, 837–839.

Light PE, Bladen C, Winkfein RJ *et al.* (2000). Molecular basis of protein kinase C-induced activation of ATP-sensitive potassium channels. *PNAS* **97**, 9058–9063.

Light PE, Fox JE, Riedel MJ *et al.* (2002). Glucagon-like peptide-1 inhibits pancreatic ATP-sensitive potassium channels via a protein kinase A- and ADP-dependent mechanism. *Mol Endocrin* **16**, 2135–2144.

Lin YF, Jan YN and Jan LY (2000). Regulation of ATP-sensitive potassium channel function by protein kinase A-mediated phosphorylation in transfected HEK293 cells. *EMBO J* **19**, 942–955.

Liss B, Bruns R and Roeper J (1999). Alternative sulfonylurea receptor expression defines metabolic sensitivity of K$_{ATP}$ channels in dopaminergic midbrain neurons. *EMBO J* **18**, 833–846.

Liu XK, Yamada S, Kane GC *et al.* (2004). Genetic disruption of Kir6.2, the pore-forming subunit of ATP-sensitive K+ channel, predisposes to catecholamine-induced ventricular dysrhythmia. *Diabetes* **53**, S165-S168.

Liu YG, Ren GF, O'Rourke B *et al.* (2001). Pharmacological comparison of native mitochondrial K$_{ATP}$ channels with molecularly defined surface K$_{ATP}$ channels. *Mol Pharmacol* **59**, 225–230.

Lopatin AN, Makhina EN and Nichols CG (1994). Potassium channel block by cytoplasmic polyamines as the mechanism of intrinsic rectification. *Nature* **372**, 366–369.

Lorenz E, Alekseev AE, Krapivinsky GB *et al.* (1998). Evidence for direct physical association between a K+ channel (Kir6.2) and an ATP-binding cassette protein (SUR1) which affects cellular distribution and kinetic behavior of an ATP-sensitive K+ channel. *Mol Cell Biol* **18**, 1652–1659.

Loussouarn G, Phillips LR, Masia R, Rose T *et al.* (2001). Flexibility of the Kir6.2 inward rectifier K+ channel pore. *PNAS* **98**, 4227–4232.

Lu Z (2004). Mechanism of rectification in inward-rectifier K+ channels. *Annu Rev Physiol* **66**, 103–129.

Mannhold R (2004). K$_{ATP}$ channel openers: structure–activity relationships and therapeutic potential. *Med Res Rev* **24**, 213–266.

Markworth E, Schwanstecher C and Schwanstecher M (2000). ATP(4-) mediates closure of pancreatic β-cell ATP-sensitive potassium channels by interaction with 1 of 4 identical sites. *Diabetes* **49**, 1413–1418.

Masia R, Enkvetchakul D and Nichols CG (2005). Differential nucleotide regulation of K$_{ATP}$ channels by SUR1 and SUR2A. *J Mol Cell Cardiol* **39**, 491–501.

Matsuoka T, Matsushita K, Katayama Y *et al.* (2000). C-terminal tails of sulfonylurea receptors control ADP-induced activation and diazoxide modulation of ATP-sensitive K+ channels. *Circ Res* **87**, 873–880.

Matsushita K, Kinoshita K, Matsuoka T *et al.* (2002). Intramolecular interaction of SUR2 subtypes for intracellular ADP-induced differential control of K$_{ATP}$ channels. *Circ Res* **90**, 554–561.

Mikhailov MV, Campbell JD, de-Wet H *et al.* (2005). 3-D structural and functional characterization of the purified K$_{ATP}$ channel complex Kir6.2-SUR1. *EMBO J* **24**, 4166–4175.

Mikhailov MV, Mikhailova EA and Ashcroft SJ (2000). Investigation of the molecular assembly of β-cell K$_{ATP}$ channels. *FEBS Lett* **482**, 59–64.

Mikhailov MV, Mikhailova EA and Ashcroft SJ (2001). Molecular structure of the glibenclamide binding site of the β-cell K$_{ATP}$ channel. *FEBS Lett* **499**, 154–160.

Miki T, Liss B, Minami K *et al.* (2001). ATP-sensitive K+ channels in the hypothalamus are essential for the maintenance of glucose homeostasis. *Nat Neurosci* **4**, 507–512.

Miki T, Suzuki M, Shibasaki T *et al.* (2002). Mouse model of Prinzmetal angina by disruption of the inward rectifier Kir6.1. *Nat Med* **8**, 466–472.

Moreau C, Gally F, Jacquet-Bouix H *et al.* (2005a). The size of a single residue of the sulfonylurea receptor dictates the effectiveness of K$_{ATP}$ channel openers. *Mol Pharmacol* **67**, 1026–1033.

Moreau C, Jacquet H, Prost AL *et al.* (2000). The molecular basis of the specificity of action of K$_{ATP}$ channel openers. *EMBO J* **19**, 6644–6651.

Moreau C, Prost AL, Derand R *et al.* (2005b). SUR, ABC proteins targeted by K$_{ATP}$ channel openers. *J Mol Cell Cardiol* **38**, 951–963.

Morrissey A, Parachuru L, Leung M *et al.* (2005a). Expression of ATP-sensitive K+ channel subunits during perinatal maturation in the mouse heart. *Pediatr Res* **58**, 185–192.

Morrissey A, Rosner E, Lanning J *et al.* (2005b). Immunolocalization of K$_{ATP}$ channel subunits in mouse and rat cardiac myocytes and the coronary vasculature. *BMC Physiol* **5**, 1.

Mukai E, Ishida H, Horie M *et al.* (1998). The antiarrhythmic agent cibenzoline inhibits K$_{ATP}$ channels by binding to Kir6.2. *Biochem Biophys Res Commun* **251**, 477–481.

Nichols CG (2006). K$_{ATP}$ channels as molecular sensors of cellular metabolism. *Nature* **440**, 470–476.

Nichols CG, Shyng SL, Nestorowicz A *et al.* (1996). Adenosine diphosphate as an intracellular regulator of insulin secretion. *Science* **272**, 1785–1787.

Ninomiya T, Takano M, Haruna T *et al.* (2003). Verapamil, a Ca^{2+} entry blocker, targets the pore-forming subunit of cardiac type K$_{ATP}$ channel (Kir6.2). *J Cardiovasc Pharmacol* **42**, 161–168.

Nishida M and Mackinnon R (2002). Structural basis of inward rectification: cytoplasmic pore of the G protein-gated inward rectifier GIRK1 at 1.8 angstrom resolution. *Cell* **111**, 957–965.

Noma A (1983). ATP-regulated K+ channels in cardiac muscle. *Nature* **305**, 147–148.

Ohtsuka T, Ishiwa D, Kamiya Y *et al.* (2006). Effects of barbiturates on ATP-sensitive K channels in rat substantia nigra. *Neuroscience* **137**, 573–581.

Okuyama Y, Yamada M, Kondo C *et al.* (1998). The effects of nucleotides and potassium channel openers on the SUR2A/Kir6.2 complex K+ channel expressed in a mammalian cell line, HEK293T cells. *Pflügers Arch-Eur J Physiol* **435**, 595–603.

Olson TM, Alekseev AE, Liu XK *et al.* (2006). Kv1.5 channelopathy due to KCNA5 loss-of-function mutation causes human atrial fibrillation. *Hum Mol Genet* **15**, 2185–2191.

Olson TM, Alekseev AE, Moreau C *et al.* (2007). K$_{ATP}$ channel mutation confers risk for vein of Marshall adrenergic atrial fibrillation. *Nat Clin Pract Cardiovasc Med* **4**, 110–116.

Ozanne SE, Guest PC, Hutton JC *et al.* (1995). Intracellular localization and molecular heterogeneity of the sulphonylurea receptor in insulin-secreting cells. *Diabetologia* **38**, 277–282.

Pearson ER, Flechtner I, Njolstad PR *et al.* (2006). Switching from insulin to oral sulfonylureas in patients with diabetes due to Kir6.2 mutations. *N Engl J Med* **355**, 467–477.

Peat AJ, Townsend C, McKay MC *et al.* (2004). 3-trifluoromethyl-4-nitro-5-arylpyrazoles are novel K_{ATP} channel agonists. *Bioorg Med Chem Lett* **14**, 813–816.

Pegan S, Arrabit C, Zhou W *et al.* (2005). Cytoplasmic domain structures of Kir2.1 and Kir3.1 show sites for modulating gating and rectification. *Nat Neurosci* **8**, 279–287.

Phillips LR and Nichols CG (2003). Ligand-induced closure of inward rectifier Kir6.2 channels traps spermine in the pore. *J Gen Physiol* **122**, 795–804.

Plant TD and Henquin JC (1990). Phentolamine and yohimbine inhibit ATP-sensitive K+ channels in mouse pancreatic β-cells. *Br J Pharmacol* **101**, 115–120.

Pocai A, Lam TK, Gutierrez-Juarez R *et al.* (2005). Hypothalamic K_{ATP} channels control hepatic glucose production. *Nature* **434**, 1026–1031.

Pountney DJ, Sun ZQ, Porter LM *et al.* (2001). Is the molecular composition of K_{ATP} channels more complex than originally thought? *J Mol Cell Cardiol* **33**, 1541–1546.

Proks P and Ashcroft FM (1997). Phentolamine block of K_{ATP} channels is mediated by Kir6.2. *PNAS* **94**, 11716–11720.

Proks P, Antcliff JF and Ashcroft FM (2003). The ligand-sensitive gate of a potassium channel lies close to the selectivity filter. *Embo Rep* **4**, 70–75.

Proks P, Arnold AL, Bruining J *et al.* (2006). A heterozygous activating mutation in the sulphonylurea receptor SUR1 (ABCC8) causes neonatal diabetes. *Hum Mol Genet* **15**, 1793–1800.

Proks P, Capener CE, Jones P *et al.* (2001). Mutations within the P-loop of Kir6.2 modulate the intraburst kinetics of the ATP-sensitive potassium channel. *J Gen Physiol* **118**, 341–353.

Proks P, Treinies I, Mest HR *et al.* (2002). Inhibition of recombinant K_{ATP} channels by the antidiabetic agents midaglizole, LY397364 and LY389382. *Eur J Pharmacol* **452**, 11–19.

Prost AL, Bloc A, Hussy N *et al.* (2004). Zinc is both an intracellular and extracellular regulator of K_{ATP} channel function. *J Physiol* **559**, 157–167.

Quast U, Guillon JM and Cavero I (1994). Cellular pharmacology of potassium channel openers in vascular smooth muscle. *Cardiovasc Res* **28**, 805–810.

Quast U, Stephan D, Bieger S *et al.* (2004). The impact of ATP-sensitive K+ channel subtype selectivity of insulin secretagogues for the coronary vasculature and the myocardium. *Diabetes* **53**, S156-S164.

Quayle JM, Standen NB and Stanfield PR (1988). The voltage-dependent block of the ATP-sensitive potassium channels of frog skeletal muscle by caesium and barium ions. *J Physiol* **405**, 677–697.

Quesada I, Rovira JM, Martin F *et al.* (2002). Nuclear K_{ATP} channels trigger nuclear Ca^{2+} transients that modulate nuclear function. *PNAS* **99**, 9544–9549.

Quinn KV, Cui Y, Giblin JP *et al.* (2003). Do anionic phospholipids serve as cofactors or second messengers for the regulation of activity of cloned ATP-sensitive K+ channels? *Circ Res* **93**, 646–655.

Quinn KV, Giblin JP and Tinker A (2004). Multisite phosphorylation mechanism for protein kinase A activation of the smooth muscle ATP-sensitive K+ channel. *Circ Res* **94**, 1359–1366.

Rainbow RD, James M, Hudman D *et al.* (2004). Proximal C-terminal domain of sulphonylurea receptor 2A interacts with pore-forming Kir6 subunits in K_{ATP} channels. *Biochem J* **379**, 173–181.

Reimann F, Gribble FM and Ashcroft FM (2000). Differential response of K_{ATP} channels containing SUR2A or SUR2B subunits to nucleotides and pinacidil. *Mol Pharmacol* **58**, 1318–1325.

Reimann F, Proks P and Ashcroft FM (2001). Effects of mitiglinide (S 21403) on Kir6.2/SUR1, Kir6.2/SUR2A and Kir6.2/SUR2B types of ATP-sensitive potassium channel. *Br J Pharmacol* **132**, 1542–1548.

Reimann F, Tucker SJ, Proks P *et al.* (1999). Involvement of the N-terminus of Kir6.2 in coupling to the sulphonylurea receptor. *J Physiol* **518**, 325–336.

Repunte VP, Nakamura H, Fujita A *et al.* (1999). Extracellular links in Kir subunits control the unitary conductance of SUR/Kir6.0 ion channels. *EMBO J* **18**, 3317–3324.

Ribalet B and Ciani S (1994). Characterization of the G protein coupling of a glucagon receptor to the K_{ATP} channel in insulin-secreting cells. *J Membr Biol* **142**, 395–408.

Ribalet B and Eddlestone GT (1995). Characterization of the G protein coupling of a somatostatin receptor to the K_{ATP} channel in insulin-secreting mammalian HIT and RIN cell lines. *J Physiol* **485**, 73–86.

Ribalet B, John SA and Weiss JN (2003). Molecular basis for Kir6.2 channel inhibition by adenine nucleotides. *Biophys J* **84**, 266–276.

Riedel MJ, Boora P, Steckley D *et al.* (2003). Kir6.2 polymorphisms sensitize β-cell ATP-sensitive potassium channels to activation by acyl CoAs: a possible cellular mechanism for increased susceptibility to type 2 diabetes? *Diabetes* **52**, 2630–2635.

Rodrigo GC and Standen NB (2005). ATP-sensitive potassium channels. *Curr Pharm Des* **11**, 1915–1940.

Russ U, Lange U, Löffler-Walz C *et al.* (2001). Interaction of the sulfonylthiourea HMR 1833 with sulfonylurea receptors and recombinant ATP-sensitive K+ channels: comparison with glibenclamide. *J Pharmacol Exp Ther* **299**, 1049–1055.

Sakura H, Ämmälä C, Smith PA *et al.* (1995). Cloning and functional expression of the cDNA encoding a novel ATP-sensitive potassium channel subunit expressed in pancreatic β-cells, brain, heart and skeletal muscle. *FEBS Lett* **377**, 338–344.

Sakura H, Trapp S, Liss B *et al.* (1999). Altered functional properties of K_{ATP} channel conferred by a novel splice variant of SUR1. *J Physiol* **521**, 337–350.

Sanchez JA, Gonoi T, Inagaki N *et al.* (1998). Modulation of reconstituted ATP-sensitive K+-channels by GTP-binding proteins in a mammalian cell line. *J Physiol* **507**, 315–324.

Satoh E, Yamada M, Kondo C *et al.* (1998). Intracellular nucleotide-mediated gating of SUR/Kir6.0 complex potassium channels expressed in a mammalian cell line and its modification by pinacidil. *J Physiol* **511**, 663–674.

Sattiraju S, Reyes S, Kane GC *et al.* (2008). K_{ATP} channel pharmacogenomics: from bench to bedside. *Clin Pharmacol Ther* **83**, 354–357.

Schwanstecher C, Meyer U and Schwanstecher M (2002). Kir6.2 polymorphism predisposes to type 2 diabetes by inducing overactivity of pancreatic β-cell ATP-Sensitive K+ channels. *Diabetes* **51**, 875–879.

Schwappach B, Zerangue N, Jan YN (2000). Molecular basis for K_{ATP} assembly: transmembrane interactions mediate association of a K+ channel with an ABC transporter. *Neuron* **26**, 155–167.

Seharaseyon J, Sasaki N, Ohler A *et al.* (2000). Evidence against functional heteromultimerization of the K_{ATP} channel subunits Kir6.1 and Kir6.2. *J Biol Chem* **275**, 17561–17565.

Seino S and Miki T (2003). Physiological and pathophysiological roles of ATP-sensitive K+ channels. *Prog Biophys Mol Biol* **81**, 133–176.

Selivanov VA, Alekseev AE, Hodgson DM *et al.* (2004). Nucleotide-gated K_{ATP} channels integrated with creatine and adenylate kinases: Amplification, tuning and sensing of energetic signals in the compartmentalized cellular environment. *Mol Cell Biochem* **256**, 243–256.

Shi NQ, Ye B and Makielski JC (2005). Function and distribution of the SUR isoforms and splice variants. *J Mol Cell Cardiol* **39**, 51–60.

Shi Y, Wu ZY, Cui NR *et al.* (2007). PKA phosphorylation of SUR2B subunit underscores vascular K_{ATP} channel activation by beta-adrenergic receptors. *Am J Physiol* **293**, R1205-R1214.

Shumilina E, Klocker N, Korniychuk G *et al.* (2006). Cytoplasmic accumulation of long-chain coenzyme A esters activates K_{ATP} and inhibits Kir2.1 channels. *J Physiol* **575**, 433–442.

Shyng SL and Nichols CG (1997). Octameric stoichiometry of the K_{ATP} channel complex. *J Gen Physiol* **110**, 655–664.

Shyng SL and Nichols CG (1998). Membrane phospholipid control of nucleotide sensitivity of K_{ATP} channels. *Science* **282**, 1138–1141.

Shyng SL, Cukras CA, Harwood J *et al.* (2000). Structural determinants of PIP2 regulation of inward rectifier K_{ATP} channels. *J Gen Physiol* **116**, 599–607.

Shyng SL, Ferrigni T and Nichols CG (1997a). Control of rectification and gating of cloned K_{ATP} channels by the Kir6.2 subunit. *J Gen Physiol* **110**, 141–153.

Shyng SL, Ferrigni T and Nichols CG (1997b). Regulation of K_{ATP} channel activity by diazoxide and MgADP—distinct functions of the two nucleotide binding folds of the sulfonylurea receptor. *J Gen Physiol* **110**, 643–654.

Sieg A, Su JP, Munoz A, Buchenau M *et al.* (2004). Epinephrine-induced hyperpolarization of islet cells without K_{ATP} channels. *Am J Physiol* **286**, E463-E471.

Smith PL, Baukrowitz T and Yellen G (1996). The inward rectification mechanism of the HERG cardiac potassium channel. *Nature* **379**, 833–836.

So I, Ashmole I, Davies NW *et al.* (2001). The K^+ channel signature sequence of murine Kir2.1: mutations that affect microscopic gating but not ionic selectivity. *J Physiol* **531**, 37–50.

Spanswick D, Smith MA, Groppi VE *et al.* (1997). Leptin inhibits hypothalamic neurons by activation of ATP-sensitive potassium channels. *Nature* **390**, 521–525.

Spruce AE, Standen NB and Stanfield PR (1985). Voltage-dependent ATP-sensitive potassium channels of skeletal muscle membrane. *Nature* **316**, 736–738.

Spruce AE, Standen NB and Stanfield PR (1987). Studies of the unitary properties of adenosine-5'-triphosphate-regulated potassium channels of frog skeletal muscle. *J Physiol* **382**, 213–236.

Stephan D, Winkler M, Kuhner P *et al.* (2006). Selectivity of repaglinide and glibenclamide for the pancreatic over the cardiovascular K_{ATP} channels. *Diabetologia* **49**, 2039–2048.

Sturgess NC, Ashford ML, Cook DL *et al.* (1985). The sulphonylurea receptor may be an ATP-sensitive potassium channel. *Lancet* **2**, 474–475.

Suh BC and Hille B (2005). Regulation of ion channels by phosphatidylinositol 4,5-bisphosphate. *Curr Opin Neurobiol* **15**, 370–378.

Surah-Narwal S, Xu SZ, McHugh D *et al.* (1999). Block of human aorta Kir6.1 by the vascular K_{ATP} channel inhibitor U37883A. *Br J Pharmacol* **128**, 667–672.

Suzuki M, Li RA, Miki T *et al.* (2001). Functional roles of cardiac and vascular ATP-sensitive potassium channels clarified by Kir6.2-knockout mice. *Circ Res* **88**, 570–577.

Suzuki M, Sasaki N, Miki T *et al.* (2002). Role of sarcolemmal K_{ATP} channels in cardioprotection against ischemia/reperfusion injury in mice. *J Clin Invest* **109**, 509–516.

Szamosfalvi B, Cortes P, Alviani R *et al.* (2002). Putative subunits of the rat mesangial K_{ATP}: a type 2B sulfonylurea receptor and an inwardly rectifying K^+ channel. *Kidney Int* **61**, 1739–1749.

Taborsky GJ, Ahren B and Havel PJ (1998). Autonomic mediation of glucagon secretion during hypoglycemia: implications for impaired alpha-cell responses in type 1 diabetes. *Diabetes* **47**, 995–1005.

Takano M and Ashcroft FM (1996). The Ba2+ block of the ATP-sensitive K^+ current of mouse pancreatic β-cells. *Pflügers Arch-Eur J Physiol* **431**, 625–631.

Takano M, Xie LH, Otani H et al.(1998). Cytoplasmic terminus domains of Kir6.x confer different nucleotide-dependent gating on the ATP-sensitive K^+ channel. *J Physiol* **512**, 395–406.

Tammaro P, Proks P and Ashcroft FM (2006). Functional effects of naturally occurring KCNJ11 mutations causing neonatal diabetes on cloned cardiac K_{ATP} channels. *J Physiol* **571**, 3–14.

Teramoto N (2006). Pharmacological profile of U-37883A, a channel blocker of smooth muscle-type ATP sensitive K channels. *Cardiovasc Drug Rev* **24**, 25–32.

Terzic A, Dzeja PP and Holmuhamedov EL (2000). Mitochondrial K_{ATP} channels: probing molecular identity and pharmacology. *J Mol Cell Cardiol* **32**, 1911–1915.

Terzic A, Tung RT, Inanobe A *et al.* (1994). G proteins activate ATP-sensitive K^+ channels by antagonizing ATP-dependent gating. *Neuron* **12**, 885–893.

Thomas PM, Cote GJ, Wohllk N *et al.* (1995). Mutations in the sulfonylurea receptor gene in familial persistent hyperinsulinemic hypoglycemia of infancy. *Science* **268**, 426–429.

Thorneloe KS, Maruyama Y, Malcolm AT *et al.* (2002). Protein kinase C modulation of recombinant ATP-sensitive K^+ channels composed of Kir6.1 and/or Kir6.2 expressed with SUR2B. *J Physiol* **541**, 65–80.

Trapp S, Haider S, Jones P *et al.* (2003). Identification of residues contributing to the ATP binding site of Kir6.2. *EMBO J* **22**, 2903–2912.

Trapp S, Proks P, Tucker SJ *et al.* (1998). Molecular analysis of ATP-sensitive K channel gating and implications for channel inhibition by ATP. *J Gen Physiol* **112**, 333–349.

Trapp S, Tucker SJ and Ashcroft FM (1997). Activation and inhibition of K_{ATP} currents by guanine nucleotides is mediated by different channel subunits. *PNAS* **94**, 8872–8877.

Trube G, Rorsman P and Ohno-Shosaku T (1986). Opposite effects of tolbutamide and diazoxide on the ATP-dependent K^+ channel in mouse pancreatic β-cells. *Pflügers Arch-Eur J Physiol* **407**, 493–499.

Tseng GN and Hoffman BF (1990). Actions of pinacidil on membrane currents in canine ventricular myocytes and their modulation by intracellular ATP and cAMP. *Pflügers Arch-Eur J Physiol* **415**, 414–424.

Tucker SJ, Gribble FM, Proks P *et al.* (1998). Molecular determinants of K_{ATP} channel inhibition by ATP. *EMBO J* **17**, 3290–3296.

Tucker SJ, Gribble FM, Zhao C *et al.* (1997). Truncation of Kir6.2 produces ATP-sensitive K^+ channels in the absence of the sulphonylurea receptor. *Nature* **387**, 179–183.

Ueda K, Inagaki N and Seino S (1997). MgADP antagonism to Mg^{2+}-independent ATP binding of the sulfonylurea receptor SUR1. *J Biol Chem* **272**, 22983–22986.

Uhde I, Toman A, Gross I *et al.* (1999). Identification of the potassium channel opener site on sulfonylurea receptors. *J Biol Chem* **274**, 28079–28082.

van Bever L, Poitry S, Faure C *et al.* (2004). Pore loop-mutated rat Kir6.1 and Kir6.2 suppress K_{ATP} current in rat cardiomyocytes. *Am J Physiol* **287**, H850–H859.

Varadi A, Grant A, McCormack M *et al.* (2006). Intracellular ATP-sensitive K^+ channels in mouse pancreatic β cells: against a role in organelle cation homeostasis. *Diabetologia* **49**, 1567–1577.

Vila-Carriles WH, Zhao GL and Bryan J (2007). Defining a binding pocket for sulfonylureas in ATP-sensitive potassium channels. *FASEB J* **21**, 18–25.

Vivaudou M and Forestier C (1995). Modification by protons of frog skeletal muscle K_{ATP} channels: effects on ion conduction and nucleotide inhibition. *J Physiol* **486**, 629–645.

Wada Y, Yamashita T, Imai K *et al.* (2000). A region of the sulfonylurea receptor critical for a modulation of ATP-sensitive K^+ channels by G-protein βg-subunits. *EMBO J* **19**, 4915–4925.

Wang X, Wu J, Li L *et al.* (2003). Hypercapnic acidosis activates K_{ATP} channels in vascular smooth muscles. *Circ Res* **92**, 1225–1232.

Ward A, Reyes CL, Yu J *et al.* (2007). Flexibility in the ABC transporter MsbA: alternating access with a twist. *PNAS* **104**, 19005–19010.

Webster SM, del Camino D, Dekker JP *et al.* (2004). Intracellular gate opening in Shaker K+ channels defined by high-affinity metal bridges. *Nature* **428**, 864–868.

Wible BA, Taglialatela M, Ficker E *et al.* (1994). Gating of inwardly rectifying K+ channels localized to a single negatively charged residue. *Nature* **371**, 246–249.

Wu JP, Cui NG, Piao HL *et al.* (2002). Allosteric modulation of the mouse Kir6.2 channel by intracellular H+ and ATP. *J Physiol* **543**, 495–504.

Xiao J, Zhen XG and Yang J (2003). Localization of PIP_2 activation gate in inward rectifier K+ channels. *Nat Neurosci* **6**, 811–818.

Xu HX, Cui NR, Yang ZJ *et al.* (2001a). Direct activation of cloned K_{ATP} channels by intracellular acidosis. *J Biol Chem* **276**, 12898–12902.

Xu HX, Wu JP, Cui NR *et al.* (2001b). Distinct histidine residues control the acid-induced activation and inhibition of the cloned K_{ATP} channel. *J Biol Chem* **276**, 38690–38696.

Yamada M, Isomoto S, Matsumoto S *et al.* (1997). Sulphonylurea receptor 2B and Kir6.1 form a sulphonylurea-sensitive but ATP-insensitive K+ channel. *J Physiol* **499**, 715–720.

Yamada S, Kane GC, Behfar A *et al.* (2006). Protection conferred by myocardial ATP-sensitive K+ channels in pressure overload-induced congestive heart failure revealed in KCNJ11 Kir6.2-null mutant. *J Physiol* **577**, 1053–1065.

Yan FF, Lin CW, Weisiger E *et al.* (2004). Sulfonylureas correct trafficking defects of ATP-sensitive potassium channels caused by mutations in the sulfonylurea receptor. *J Biol Chem* **279**, 11096–11105.

Yang J, Jan YN and Jan LY (1995). Control of rectification and permeation by residues in two distinct domains in an inward rectifier K+ channel. *Neuron* **14**, 1047–1054.

Yokoshiki H, Sunagawa M, Seki T *et al.* (1999). Antisense oligodeoxynucleotides of sulfonylurea receptors inhibit ATP-sensitive K+ channels in cultured neonatal rat ventricular cells. *Pflügers Arch-Eur J Physiol* **437**, 400–408.

Yuan H, Michelsen K and Schwappach B (2003). 14-3-3 dimers probe the assembly status of multimeric membrane proteins. *Curr Biol* **13**, 638–646.

Yunoki T, Teramoto N, Naito S *et al.* (2001). The effects of flecainide on ATP-sensitive K+ channels in pig urethral myocytes. *Br J Pharmacol* **133**, 730–738.

Zerangue N, Schwappach B, Jan YN *et al.* (1999). A new ER trafficking signal regulates the subunit stoichiometry of plasma membrane K_{ATP} channels. *Neuron* **22**, 537–548.

Zingman LV, Alekseev AE, Bienengraeber M *et al.* (2001). Signaling in channel/enzyme multimers: ATPase transitions in SUR module gate ATP-sensitive K+ conductance. *Neuron* **31**, 233–245.

Zingman LV, Hodgson DM, Alekseev AE *et al.* (2003). Stress without distress: homeostatic role for K_{ATP} channels. *Mol Psychiatry* **8**, 253–254.

Zingman LV, Hodgson DM, Bast PH *et al.* (2002). Kir6.2 is required for adaptation to stress. *PNAS* **99**, 13278–13283.

Zünkler BJ, Kuhne S, Rustenbeck I *et al.* (2000). Disopyramide block of K_{ATP} channels is mediated by the pore-forming subunit. *Life Sci* **66**, PL245-PL252.

4.5

Ryanodine receptors

Stephan Lehnart and Andrew Marks

4.5.1 Introduction

Cytoplasmic Ca^{2+} serves as a versatile intracellular messenger controlling functions such as muscle contraction, neurotransmitter release, and cell growth. In excitable cells, plasma membrane depolarization activates voltage-dependent L-type Ca^{2+} (Ca_V) channels and the resulting Ca^{2+} influx activates intracellular Ca^{2+} release channels called ryanodine receptors (RyRs). RyRs are expressed in virtually all excitable cells. Indeed, intracellular Ca^{2+} release channels are found in all cells on the membranes of the endoplasmic reticulum (ER) or in striated muscles the specialized sarcoplasmic reticulum (SR) Ca^{2+} store. RyRs and the related inositol 1,4,5-trisphosphate receptors (IP3Rs) represent the two principle classes of intracellular Ca^{2+} release channels. Tissue-specific expression of RyR1 and RyR2 in skeletal and cardiac muscles, respectively, is essential for intracellular Ca^{2+} release during excitation–contraction (EC) coupling. RyRs are the largest known ion channels, with a relatively small pore region and a large region of the protein which functions as a cytoplasmic scaffold domain forming a macromolecular signalling complex which controls channel function. RyR1 and RyR2 interact with calstabin1 (FKBP12) and calstabin2 (FKBP12.6), respectively, subunits that stabilize the channel closed state preventing uncontrolled intracellular Ca^{2+} leak. RyR targeting proteins allow for specific interactions with kinase, phosphatase, and phosphodiesterase enzymes. Thus, the macromolecular RyR architecture allows for channel modulation within small subcellular compartments in response to specific extracellular signals. For instance activation of the sympathetic nervous system during the 'fight or flight' stress response results in phosphorylation of the cardiac RyR2 isoform by protein kinase A (PKA), increased sarcoplasmic reticulum (SR) Ca^{2+} release and increased muscle force. Importantly, a variety of genetic and acquired forms of RyR dysfunction have been linked to intracellular Ca^{2+} leak. Intracellular calcium leak from defective RyR channels contributes to contractile dysfunction during heart failure and sudden death from cardiac arrhythmias. Moreover, RyR channels are important drug targets and a novel pharmacologic strategy to prevent RyR dysfunction through Ca^{2+} release channel stabilizers has recently been discovered. Although this review is focused on RyR in muscle physiology and disease, it is likely that many important roles of RyR Ca^{2+} release in other tissues will emerge in the future.

4.5.2 Molecular characterization of ryanodine receptors

Ryanodine receptor gene family and expression

Ryanodine receptors were originally cloned from mammalian skeletal muscle (*RyR1*; human chromosome 19) and heart (*RyR2*; human chromosome 1) sharing 67 per cent homology (Table 4.5.1) (Marks *et al.*, 1989; Takeshima *et al.*, 1989; Otsu *et al.*, 1990). *RyR3* (human chromosome 15) was cloned from rabbit brain and mink lung epithelial cells, exhibiting 70 per cent homology to RyR2 (Giannini *et al.*, 1992; Hakamata *et al.*, 1992). *RyR1* is one of the largest human genes with 153,832 Kb of genomic DNA containing multiple silent DNA polymorphisms in coding regions and 106 exons 2 of which undergo alternative splicing (Gillard *et al.*, 1992; Phillips *et al.*, 1996; McCarthy *et al.*, 2000). RyRs and IP3Rs likely evolved from the same ancestral cation release channel since only one *RyR* gene exists in invertebrates (Vazquez-Martinez *et al.*, 2003; Serysheva *et al.*, 2005). Non-mammalian vertebrates express two RyRa and RyRb isoforms which are homologous to the mammalian RyR1 and RyR3 isoforms, respectively (Oyamada *et al.*, 1994; Ottini *et al.*, 1996;). A second gene duplication may have occurred in mammals with subsequent adaptation for distinct functions.

Ryanodine receptors are expressed in most tissues with highest levels found in striated muscles (Marks *et al.*, 1989; Takeshima *et al.*, 1989; Kuwajima *et al.*, 1992; Lai *et al.*, 1992; Lai and Meissner, 1992; Olivares *et al.*, 1993; Ahern *et al.*, 1994; Oyamada *et al.*, 1994; Giannini *et al.*, 1995; Ottini *et al.*, 1996; Sutko and Airey, 1996). In skeletal muscles RyR1 channels control repetitive SR Ca^{2+} release events on the millisecond timescale required for neuronal control of voluntary muscle contraction (Appelt *et al.*, 1989). RyR1 is also expressed in brain Purkinje cells and cerebellum, and in smooth muscle, adrenal glands, spleen, testes, and ovaries (Marks *et al.*, 1989; Takeshima *et al.*, 1989; Kuwajima *et al.*, 1992; Giannini *et al.*, 1995; Ottini *et al.*, 1996). RyR2 is most abundant in the heart and the brain and was found at lower levels in the stomach, lung,

Table 4.5.1 Properties of RyR channels

Channel subunit	Biophysical characteristics	Cellular and subcellular localization	Pharmacology	Physiology	Relevance to disease states
RyR1 (α-subunit)[1] (OMIM 180901) Gene: RYR1 Chromosomal location: mouse: 7A2–B3 [2] rat: 1q22 (predicted) human: 19q13.1 (OMIM 180901) [1,3]	g Ca^{2+} 791 pS [4] τ_{act} 1.2–4 ms (pH 6) τ_{inact}* 5.8–13.7 ms (pH 6) τ_{recov} 5.3–13 s (pH 7.4)	RyR1 is expressed at highest levels in skeletal muscle. RyR1 is found in most tissues including brain Purkinje cells, cerebellum, smooth muscle, adrenal glands, spleen, testes and ovaries. Many non-mammalian vertebrate skeletal muscles express α-RyR at similar amounts as the RyR1 homologue [5]	Biphasic to cytosolic $[Ca^{2+}]$: single-channel steady-state EC_{50}: $[Ca^{2+}]_{cis}$ ~1–3.1 μM (pH 7.4) [6]; single-channel steady-state IC_{50}: $[Ca^{2+}]_{cis}$ ~ 0.5 mM (pH 7.4) [6]; PKA phosphorylation: increases Po, overcomes inhibition by $[Mg^{2+}]_{cis}$, dissociates calstabin1 and induces subconductance states. CaMKII phosphorylation increases Po, overcomes inhibition by calmodulin. IC_{50}: Mg^{2+} ~1.5 μM; Ca^{2+} ~1.0 mM (pH 7.1) Ryanodine induces subconductance states Imperatoxin ($IpTx_a$) binding to cytosolic side induces subconductance states, reduces conductance, and slightly rectifies gating [4]	RyR1 is the primary isoform governing SR Ca^{2+} release in fast- and slow-twitch skeletal muscles necessary for VICR during EC coupling RyR1 channels associate in arrays in the terminal SR membranes and exhibit coupled gating under *in vitro* conditions. RyR1 knockout is lethal. α-RyR, the non-mammalian vertebrate RyR1 homologue, is exclusively expressed in rapidly contracting external eye and swimbladder muscles of fish	Linkage to malignant hyperthermia (MH) in humans (MHS1, OMIM 145600) and pigs [8,9]. MH is usually inherited in an autosomal dominant pattern, and RyR1 mutations are found in over 50% of families with MH and at a lower frequency in sporadic cases [10] Linkage to central core disease (CCD, OMIM 117000), a congenital myopathy usually inherited in an autosomal-dominant pattern (MIM 117000)[11,12], and to CCD associated with fetal akinesia syndrome [13]. CCD usually confers MH susceptibility. Most CCD patients are RyR1 mutation carriers Linkage to the moderate form of multiminicore disease (Mmd, OMIM 602271), a rare congenital myopathy usually inherited in an autosomal-recessive trait[14], and to MmD associated with ophthalmoplegia (OMIM 255320). MmD shows phenotypic overlap with CCD and MH Linkage to a autosomal-dominant form of nemaline rod myopathy. some myasthenia gravis (MG) patients have antibodies against RyR in addition to the acetylcholine receptor which correlate with disease severity. The anti-RyR antibodies recognize a region near the N-terminus important for channel regulation and inhibit SR Ca^{2+} release *in vitro*. However, evidence that anti-RyR antibodies are pathogenic in vivo is still missing Muscle fatigue in heart failure patients has been associated with RyR1 PKA hyperphosphorylation which may contribute to other fatigue syndromes
RyR2 (α-subunit) [15] (OMIM 180902) Gene: RYR2 Chromosomal location: mouse: 13A1–2 [2] rat: 17q12.2–12.3 [16] human: 1q42–q43 (OMIM 180902) [15]	pCa^{2+} / pK^+ ~6.5 pBa^{2+} / pK^+ ~6.5 pCa^{2+} / pBa^{2+} ~1.1 g Ca^{2+} 795 pS [4] bi-ionic g Ca^{2+} + K^+ 135 pS Activation: increase free $[Ca^{2+}]_{cis}$: τ_{act} ~1–2 ms (pH 7.4, caged Ca^{2+}) [7,17] deactivation: decrease free $[Ca^{2+}]_{cis}$ [18], τ_{inact} ~5.3 ms (pH 7.4, Diazo-2) [17]; adaptation: following $[Ca^{2+}]_{cis}$ decrease; $\tau_{adaptation}$ 13–15 s (pH 7.4, caged Ca^{2+}) [7,17]	RyR2 is the main isoform in heart muscle [19]. RyR2 intracellular Ca^{2+} release channels are located on the membranes of SR terminal cisternae in cardiomyocytes [20, 21]	Biphasic to cytosolic $[Ca^{2+}]$: single-channel steady-state EC_{50}: $[Ca^{2+}]_{cis}$ ~1–5 μM (pH 7.4) [6,7]; single-channel steady-state IC_{50}: $[Ca^{2+}]_{cis}$ ~ 5.3 mM (pH 7.4) [6]; PKA phosphorylation: increases Po, increases adaptation rate ~tenfold [7], overcomes inhibition by $[Mg^{2+}]_{cis}$ [22,23], dissociates calstabin2 and induces subconductance states. CaMKII phosphorylation increases Po, overcomes inhibiton by calmodulin [24]. IC_{50}: Mg^{2+} ~1.5 μM; Ca^{2+} ~1.0 mM (pH 7.1) Tetrabutyl ammonium (TBA$^+$) induces substates by partial occlusion	Major SR Ca^{2+} release channel in the heart necessary for Ca^{2+}-induced Ca^{2+} release during excitation-contraction coupling [25] RyR2 channels associate in arrays in the terminal SR membranes and exhibit coupled gating under *in vitro* conditions.	Linkage to stress-induced *Catecholaminergic Polymorphic Ventricular Tachycardia* (CPVT OMIM 604772, FPVT OMIM 192605), an autosomal-dominant electrical cardiomyopathy Linkage to stress induced CPVT and concomitant arrhythmogenic right ventricular dysplasia/cardiomyopathy type 2 (ARVD/C2 OMIM 600996), an electrical cardiomyopathy with relatively mild structural abnormalities *Heart Failure* is characterized by a chronic hyperadrenergic state with results in RyR2 PKA hyperphosphorylation and calstabin2 depletion as a possible cause of a gain-of-function defect and chronic intracellular SR Ca^{2+} leak[26–29] *Chronic Atrial Fibrillation*, the most common arrhythmia syndrome, has been associated with RyR2 PKA hyperphosphorylation, calstabin2 depletion, and a gain-of-function defect consistent with intracellular SR Ca^{2+} leak [30].

Continued

Table 4.5.1 (*Continued*) Properties of RyR channels

RyR3 (α-subunit) [31] (OMIM 180903) Gene: *RYR3* Chromosomal location: mouse: 2E5-F3 [2] rat: unknown human: 15q14–15 [31]	Does not associate with $Ca_V1.1$ in 1B5 cells and is thought to be regulated by CICR similar to RyR2 [32]. RyR3 is the least common isoform and relatively little is known about its function compared to RyR1 and RyR2. Monovalent cation conductance is significantly higher in RyR3 than in RyR1 [33, 34].	Ubiquitous expression of small amounts. Found in smooth muscle, skeletal muscle, and brain with preferential expression in the hippocampus [5]. RyR3 is detected in neonate mammalian skeletal muscles for 2 weeks after birth, followed by decrease in adult skeletal muscles to nil except for diaphragm and soleus [5, 35]. RyR3 may exist in parajunctional arrays immediately adjacent to RyR1 arrays of the junctional SR in skeletal muscle [36]. Many non-mammalian vertebrate skeletal muscles express b-RyR at similar amounts as the RyR3 homologue [5]	RyR3 is frequently compared with RyR1, however, it may have functional similarities to RyR2. RyR3 is less sensitive to Ca^{2+} inactivation than RyR1 and requires high Ca^{2+} for activation with ATP [33, 37]. Single-channel: increase in free $[Ca^{2+}]_{cis}$ from 100 nM to 1000 nM steeply increases RyR3 Po [34]. 4-chloro-m-cresol was ineffective on RyR3 [32]	RyR3 knockout mice are phenotypically normal except for reduced muscle contraction following birth consistent with peak expression levels around that time and a 2-fold higher speed of locomotion indicating an important CNS function[5]. Tissue-specific expression in the hippocampus suggests a role in behaviour [5]. Sensitivity to CICR was changed in RyR3 knockout mice. Due to the parajunctional RyR3 location it is possible that a CICR mechanism contributes to skeletal muscle function and/or that RyR3 contributes to drug-induced alterations [38]	No genetic or acquired diseases have been associated with RyR3 It was hypothesized that RyR3 may contribute to drug-induced alterations in skeletal muscle [38]
Calstabin1 (FKBP12) [39–42] (OMIM 186945) Gene: *FKBP12* Chromosomal location: human: 20p13	Associates with RyR1 and RyR3 Calstabin1 binds to a conserved motif Val-2461-Pro-2462 in RyR1 Similarly, mutation of Val-2322 in RyR3 to a leucine reduces the affinity for calstabin1 Each RyR1 tetramer binds four molecules of calstabin1 (one molecule per channel subunit) [43] The hydrophobic calstabin1 pocket involved in RyR1 interaction is formed by Tyr-26, Phe-46, Phe-48, Trp-59, Tyr-82, and Phe-99	The ubiquitous cytosolic protein calstabin1 is found at high concentrations in skeletal muscles. Calstabin1 binds preferentially to RyR1	Rapamycin binds to the hydrophobic RyR1 binding pocket on calstabin1 inducing dissociation Calstabin1 dissociation by FK506 or thapsigargin binding increases RyR1 Po and induces subconductance states[44] JTV519 increases the calstabin1 binding affinity for RyR1	Calstabin1 stabilizes the closed state of the RyR1 channel and is important for cooperative interactions among the subunits of the tetramer Calstabin1 regulates RyR1 coupled gating whereby immediately neighboring channels open simultaneously.	Calstabin1 knockout results in a *cardiac septal defect* and secondary cardiomyopathy resulting in early lethality after birth indicating an important role during embryonic heart development [45] Skeletal muscle-specific calstabin1 knockout mice display decreased VICR and decreased tetanic force production consistent with a role in antegrade and retrograde $Ca_V1.1$–RyR1 coupling [46] Abnormal fatigability in heart failure may occur from intracellular Ca^{2+} leak due to RyR1 PKA hyperphosphorylation which may also contribute to skeletal dysfunction

Calstabin2 (FKBP12.6) (OMIM 186945) Gene: FKBP12.6 Chromosomal location: human: 2p21–23	Associates with RyR2[47]. Calstabin2 binds to a conserved motif Iso-2427-Pro-2428 in RyR2 Each RyR2 tetramer binds four molecules of calstabin2 (one molecule per channel subunit)	The cytosolic protein calstabin2 is found at lower concentrations than calstabin1 in both skeletal and cardiac tissues [48]. Calstabin2 binds preferentially to RyR2 in cardiac muscle due to the much higher binding affinity	Rapamycin binds to the hydrophobic RyR2 binding pocket on calstabin2 inducing dissociation Calstabin2 dissociation by FK506 or thapsigargin increases RyR1 Po and induces subconductance states[44] JTV519 increases the calstabin2 binding affinity for RyR2	Calstabin2 stabilizes the closed state of the RyR2 channel and is important for cooperative interactions among the subunits of the tetramer Calstabin2 regulates RyR2 coupled gating whereby immediately neighboring channels open simultaneously Calstabin2 knockout results in stress-induced catecholaminergic arrhythmias and faster development of heart failure resulting from intracellular Ca²⁺ leak, suggesting an important protective role of calstabin2	In heart failure, RyR2 PKA hyperphosphorylation results in calstabin2 depletion and intracellular Ca^{2+} leak which may contribute to progression of heart failure and/or the development of ventricular arrhythmias and sudden death

* τ_{inact} of RyR2 is strongly dependent on the dynamics of Ca^{2+} buffering; CICR, Ca^{2+}-induced Ca^{2+} release; VICR, voltage-induced Ca^{2+} release; Po, open probability.

[1] Zorzato F et al. (1990) J Biol Chem 265, 2244–2256; [2] Mattei MG et al. (1994) Genomics 22, 202–204; [3] MacKenzie AE et al. (1991) Genomics 22, 202–204; [4] Tripathy A et al. (1998) J Gen Physiol 111, 679–690; [5] Sorrentino V (2003) Front Biosci 8, d176–182; [6] Copello JA et al. (1997) Biophys J 73, 141–156; [7] Valdivia HH et al. (1995) Science 267, 1997–2000; [8] Davies W et al. (1988) Anim Genet 19, 203–212; [9] Fujii J et al. (1991) Science 253, 448–51; [10] Robinson RL et al. (2003) Eur J Hum Genet 11, 342–348; [11] Schwemmle S et al. (1993) Genomics 17, 205–207; [12] Mulley JC et al. (1993) Am J Hum Genet 52, 398–405; [13] Romero NB et al. (2003) Brain 126, 2341–2349; [14] Kausch K et al. (1991) Genomics 10, 765–769; [15] Otsu K et al. (1993) Genomics 17, 507–509; [16] Zhao L et al. (2001) Circulation 103, 442–447; [17] Velez P et al. (1997) Biophys J 72, 691–697; [18] Gyorke S et al. (1993) Science 260, 807–809; [19] Sorrentino V et al. (1993) Trends Pharmacol Sci 14, 98–103; [20] Franzini-Armstrong C et al. (1998) Ann N Y Acad Sci 853, 20–30; [21] Franzini-Armstrong C et al. (1999) Biophys J 77, 1528–1539; [22] Hain J et al. (1994) Biophys J 67, 1823–1833; [23] Hain J et al. (1995) Biol Chem 270, 2074–2081; [24] Witcher DR et al. (1991) Biol Chem 266, 11144–11152; [25] Bers DM (2002) Nature 415, 198–205; [26] Reiken S et al. (2001) Circulation 104, 2843–2848; [27] Doi M et al. (2002) Circulation 105, 1374–1379; [28] Ono K et al. (2000) Cardiovasc Res 48, 323–331; [29] Yamamoto T et al. (1999) Cardiovasc Res 44, 146–155; [30] Vest JA et al. (2005) Circulation 111, 2025–2032; [31] Sorrentino V et al. (1993) Genomics 18, 163–165; [32] Fessenden JD et al. (2000) Biophys J 79, 2509–2525; [33] Chen SR et al. (1997) J Biol Chem 272, 24234–24246; [34] Murayama T et al. (1999) J Biol Chem 274, 17297–17308; [35] Ogawa Y et al. (1999) Adv Biophys 36, 27–64; [36] Felder E et al. (2002) Proc Natl Acad Sci USA 99, 1695–1700; [37] Sonnleitner A et al. (1998) EMBO J 17, 2790–2798; [38] Ogawa Y et al. (2002) Front Biosci 7, d1184–1194; [39] Waldmann R et al. (1996) J Biol Chem 271, 10433–10436; [40] Garcia-Anoveros J et al. (1997) Proc Natl Acad Sci USA 94, 1459–1464; [41] Skoyles JR et al. (1992) J Cardiothorac Vasc Anesth 6, 222–225; [42] Champigny G et al. (1998) J Biol Chem 273, 15418–15422; [43] Jayaraman T et al. (1992) Biol Chem 267, 9474–9477; [44] Mayrleitner M et al. (1994) Cell Calcium 15, 99–108; [45] Shou W et al. (1998) Nature 391, 489–492; [46] Tang W et al. (2004) Faseb J 18, 1597–1599; [47] Timerman AP et al. (1996) Biol Chem 271, 20385–20391. [48] Timerman AP et al. (1994) Biochem Biophys Res Commun 198, 701–706.

thymus, adrenal gland, and ovaries (Nakai *et al.*, 1990; Kuwajima *et al.*, 1992; Giannini *et al.*, 1995). RyR3 is found at relatively low levels in skeletal muscle, heart, specific brain areas such as the hippocampus, and also in spleen, testes, and organs containing smooth muscle (Giannini *et al.*, 1992; Hakamata *et al.*, 1992; Giannini *et al.*, 1995; Ottini *et al.*, 1996). Tissue-specific and developmentally regulated mouse RyR1 splice variants in the modulatory cytoplasmic region have been identified (Futatsugi *et al.*, 1995). A splice isoform of RyR3 has been reported to regulate Ca^{2+} release in smooth muscle cells but no functional differences between variants have been reported (Jiang *et al.*, 2003). Alternative splicing also occurs in the insect *RyR* gene potentially allowing for variant diversity (Scott-Ward *et al.*, 2001). Although alternative splicing may underlie developmental regulation of certain RyR isoforms in specific tissues, expression from multiple genes may be the major mechanism of mammalian RyR diversity (Futatsugi *et al.*, 1995).

RyR channel structure

Ryanodine receptors are homotetrameric channel complexes comprised of four identical ~565 kDa-subunits which are symmetrically arranged around a C-terminal pore region. Electron microscopy has identified a large cytosolic foot structure corresponding to an enormous N-terminal scaffold region interacting with a variety of signalling molecules important for channel regulation.

Analysis of the primary amino acid sequence shows that the transmembrane (TM) domains are clustered in the C-terminal region of the RyR channel (Otsu *et al.*, 1990; Du *et al.*, 2002). Accordingly, the RyR1 deletion mutant D183-4006 resembled a functionally active channel which lost the N-terminal regulation characteristic for wild-type RyR1 channels (Bhat *et al.*, 1997, 1999). Mutagenesis studies within the predicted pore region have identified amino acids involved in RyR cation permeation (Balshaw *et al.*, 1999; Du *et al.*, 2001, 2004). Current models predict 6 to 8 TM domains (Figure 4.5.1) (Du *et al.*, 2004), however, further characterization of the pore at the atomic level is necessary (Ludtke *et al.*, 2005). The RyR pore region contains a sequence between the last two TM domains analogous to the P loop signature sequence which defines the selectivity filter in voltage-gated ion channels (Balshaw *et al.*, 1999). Similarities with the open conformation of the Mth K^+ channel pore structure have been predicted including certain pore lining α-helices (Ludtke *et al.*, 2005) and a high-resolution cryoelectron microscopy (cryo-EM) reconstruction study showed structural homology with the inner pore region of the KcsA K^+ channel (Samso *et al.*, 2005). Kinase, phosphatase, and phosphodiesterase targeting to RyR occurs via leucine–isoleucine zipper (LIZ) motifs (Marx *et al.*, 2001b). LIZ binding motifs are also important for the formation of the IP3R macromolecular complex (Tu *et al.*, 2004) and voltage-gated ion channels of the plasma membrane (Hulme *et al.*, 2002; Marx *et al.*, 2002; Hulme *et al.*, 2003) suggesting similarities in their overall channel architecture with RyRs.

Although the three RyR isoforms are highly homologous, three 'divergent' regions (DR-I/II/III) with a high degree of variability between the isoforms may contribute to functional differences. Using GFP insertion into mouse RyR2 at Asp-4365 followed by cryo-EM three-dimensional reconstruction, DR-I has been localized to the cytoplasmic aspect of domain 3 at the junction with domain 4 in the C-terminal pore region (Figure 4.5.1) (Liu *et al.*, 2002) in agreement with a cryo-EM study using an antibody against

a DR-I epitope in rabbit RyR1 which suggested an analogous location (Benacquista *et al.*, 2000). Moreover, calmodulin binds to native RyR1 in a cleft formed between the cytoplasmic aspects of domains 3 and 7 and calstabin1 binds adjacent to region 9 and above region 3 at the corners of the tetrameric channel complex suggesting three-dimensional proximity with DR-I (Figure 4.5.1) (Wagenknecht *et al.*, 1997). DR-II is located in the cytoplasmic region to the so called 'handle' or domain 6, DR-III, to the 'clamp' or domain 9, and the central disease associated region which contains a number of RyR missense mutations, to a 'bridge' connecting regions 5 and 6 (Figure 4.5.1).

The ryanodine receptor macromolecular complex

The N-terminal domain of the ryanodine receptor is exposed to the cytosolic compartment between the junctional SR and T-tubule membranes where it serves as a scaffold for protein interaction and channel modulation (Diaz-Munoz *et al.*, 1990; Knudson *et al.*, 1993; Brillantes *et al.*, 1994; Coronado *et al.*, 1994; Jones *et al.*, 1995; Tripathy *et al.*, 1995; Marks, 1996b; Lokuta *et al.*, 1997; Wagenknecht *et al.*, 1997; Currie *et al.*, 2004; Wehrens *et al.*, 2004b). The C-terminal RyR pore region potentially interacts with and is controlled by a complex of proteins at the luminal side of the terminal SR (Rossi and Dirksen, 2006).

Calmodulin

Calmodulin (CaM) was found to modulate single RyR channel function in a Ca^{2+}-dependent manner implicating a specific cytosolic interaction (Smith *et al.*, 1989; Chu *et al.*, 1990b). A CaM binding site has been identified in the RyR1 subunit on the basis of peptide digestion and sequence prediction (Takeshima *et al.*, 1989; Porter Moore *et al.*, 1999), and cryo-EM three-dimensional reconstruction has localized CaM on the cytosolic RyR2 surface between domains 3 and 7 close to the junctional SR membrane (Figure 4.5.1a) (Wagenknecht *et al.*, 1997; Wagenknecht and Samso, 2002). Apo-CaM (Ca^{2+}-free calmodulin) activates and Ca^{2+}-CaM inhibits RyR1 function (Tripathy et al., 1995; Fruen et al., 2000; Balshaw et al., 2001). Moreover, CaM protects RyR1 from oxidation and possibly nitrosylation at Cys-3653 (Figure 4.5.1b) (Porter Moore *et al.*, 1999). However, the complex role of CaM in RyR modulation during EC coupling and repetitive channel activation requires further investigation.

Calstabin

Calstabin (for Ca^{2+} release channel-stabilizing protein, also known as FKBP for FK506 binding protein) binds with a stoichiometry of one calstabin molecule per RyR monomer or maximally four calstabin molecules per channel tetramer. The two isoforms calstabin1 and calstabin2 associate with RyR1 and RyR2, respectively (Timerman *et al.*, 1993; Marks, 1996a; Timerman *et al.*, 1996; Wehrens *et al.*, 2004a). RyR1 and RyR3 channels bind calstabin1 and calstabin2 *in vitro* (Timerman *et al.*, 1993, 1996; Van Acker *et al.*, 2004). However, in skeletal muscles RyR1 and RyR3 will preferentially bind calstabin1 due to a higher affinity and a higher prevalent cytosolic concentration (Timerman *et al.*, 1993, 1996). RyR2 channels exhibit a higher affinity for calstabin2 which is the predominant isoform bound to the cardiac channel complex (Timerman *et al.*, 1996; Jeyakumar *et al.*, 2001).

Cryo-EM reconstruction of the RyR1 channel has revealed that calstabin1 binds along the cytoplasmic sides adjacent to

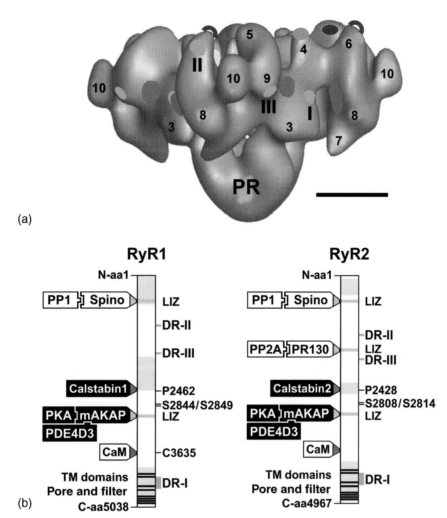

Fig. 4.5.1 The homotetrameric RyR channel complex. (a) Side view of the RyR2 channel comprised of four identical subunits by cryo-EM surface reconstruction shows large cytoplasmic 'foot' structure on top and relatively small transmembrane pore region (PR). Difference mapping has resolved the relative surface positions of calmodulin (pink) and calstabin2 (orange) relative to the RyR2 surface map. Moreover, GFP insertion mapping has revealed the relative surface localization of the three divergent regions (green, DR-I/II/III) the central CPVT-ARVD2 associated mutation region (yellow), Tyr2801 near the PKA phosphorylation site Ser2808 between domains 5 and 6 (purple), Thr2023 near the hypothetical PKA phosphorylation site Ser2030 at domain 4 (faint purple), and mAb34C Fab fragment binding corresponding to residues 2722–2769 of RyR2 (indicated by C-shaped blue symbol at the periphery of domain 6). DR-I overlaps to some extent with the putative pore region and the N-terminal CPVT-ARVD2 associated mutation region (for overview see (b)). Surface domain numbers according to Radermacher *et al.* (1994); structure possibly represents unknown 'subconductance' gating state according to Sharma *et al.* (2006). Scale bar 10 nm. As four identical subunits form the tetrameric RyR complex, regions and domains may occur more than once in the surface rendering. Adapted from Sharma *et al.* (2006)[1] with permission from the *Biophysical Journal*. (b) Schematic of the peptide sequence of RyR1 and RyR2 showing linear organization and putative C-terminal transmembrane (8-TM) pore domain. Calmodulin (CaM) and calstabin binding involves specific amino acid residues as discussed in the text. Leucine-isoleucine zipper (LIZ) motifs (light blue) bind to specific targeting proteins (Spino, spinophilin; PR130; mAKAP) and thereby associate phosphatases and a PKA-mAKAP-PDE4D3 signaling module with the RyR. Three different MH-CCD regions in RyR1 and CPVT-ARVD2 regions in RyR2 are indicated, respectively (yellow). RyR PKA and CaMKII phosphorylation sites are indicated as discussed in the text. Figure not drawn to scale. See also colour plate section.

[1] Sharma MR, Jeyakumar LH, Fleischer S and Wagenknecht T (2006). Three-dimensional visualization of FKBP12.6 binding to an open conformation of cardiac ryanodine receptor. *Biophysical Journal* **90**, 164–172.

domain 9 close to the corners of the square-shaped channel complex (Figure 4.5.1) (Wagenknecht *et al.*, 1997). Accordingly, the three-dimensional location of calstabin2 on RyR2 appears to be similar to calstabin1 on RyR1 albeit at a resolution below that required to determine the atomic structure (Sharma *et al.*, 2002, 2006). The high-affinity calstabin1 binding site on RyR1 contains Val-2461-Pro-2462 (or Ile-2427-Pro-2428 in RyR2) corresponding to the twisted-amide transition state intermediate of a peptidyl-prolyl bond (Figure 4.5.1b) (Gaburjakova *et al.*, 2001). Mutagenesis of Val-2461 to glycine abolishes calstabin1 binding

to RyR1 potentially by increased thermodynamic mobility of the peptidyl-prolyl bond from unrestrained isomerization movement at the binding site (Gaburjakova *et al.*, 2001). It has been proposed that the conserved prolyl residue introduces a break between two adjacent α-helical structures allowing calstabin to bind to a stabilized peptidyl-proline bond corresponding to a twisted amide transition state (Bultynck *et al.*, 2001). Other studies have suggested multiple N-terminal regions required for calstabin2-GST binding to RyR2 (Masumiya *et al.*, 2003) and therefore additional domains may contribute to calstabin binding.

PKA and CamKII

PKA and Ca^{2+}-calmodulin-dependent kinase II (CaMKII) are important regulatory enzymes which associate with the RyR macromolecular complex (Figure 4.5.1b). The PKA holoenzyme is comprised of catalytic and regulatory subunits which bind to the anchoring protein mAKAP (AKAP6) which targets PKA to RyR2 via binding to the respective LIZ motif (Marx *et al.*, 2000; Kapiloff *et al.*, 2001). The protein phosphatases 1 and 2A (PP1 and PP2A) are bound to RyR2 via LIZ motifs to distinct targeting proteins, spinophilin and PR130, respectively (Allen *et al.*, 1997; Marx *et al.*, 2001b). Thus, evolutionary conserved LIZ motifs in RyR1 or RyR2 form the basis for specific binding by cognate LIZ motifs in the respective kinase and phosphatase targeting proteins to the channel complex (Figure 4.5.1b) (Marx *et al.*, 2001b). We and others have shown that CaMKII associates with the RyR2 complex, although the molecular binding mechanism has not been defined (Currie *et al.*, 2004; Wehrens *et al.*, 2004b).

Sorcin

Sorcin is a cytosolic 22 kDa Ca^{2+}-binding protein which was found to associate with both RyR2 and the L-type Ca^{2+} channel (Meyers *et al.*, 1995, 1998). One model predicts that sorcin functions to reduce RyR2 open probability which can be relieved by PKA phosphorylation of sorcin (Lokuta *et al.*, 1997). Therefore, sorcin like calstabin, may stabilize the RyR closed state and prevent aberrant SR Ca^{2+} release, which can be dynamically reversed by PKA-dependent phosphorylation (Valdivia, 1998; Marks *et al.*, 2002a). However, the physiological role of sorcin during repetitive RyR channel activation requires additional characterization.

Calsequestrin, triadin and junctin

Calsequestrin, triadin and junctin are thought to be part of a SR luminal protein complex that binds to and regulates RyR activity. Junctin (Zhang *et al.*, 1997) and triadin (Flucher *et al.*, 1993) may anchor the RyR channel within specific areas of the SR membrane. Calsequestrin (CSQ) represents the major Ca^{2+} binding protein of the terminal SR Ca^{2+} store potentially providing localized high-capacity, low-affinity SR Ca^{2+} buffering (Collins *et al.*, 1990; Viatchenko-Karpinski *et al.*, 2004). Evidence suggests that the dynamic Ca^{2+}-dependent conformational changes in CSQ may ultimately modulate RyR channel activity (Howarth *et al.*, 2002). The molecular mechanisms of CSQ modulation of the SR luminal RyR domain are an area of ongoing investigation (Ohkura *et al.*, 1998; Szegedi *et al.*, 1999).

4.5.3 RyR channel biophysical characteristics

The release of Ca^{2+} from SR storage organelles greatly depends on the structures and mechanisms involved in RyR cation selection and membrane translocation. Single channel experiments have characterized RyR ion transport (Figure 4.5.2). The RyR is impermeable to anions but is permeable to a wide range of inorganic monovalent and divalent cations and to a number of monovalent organic cations (Lindsay *et al.*, 1991; Williams, 1992). All inorganic divalent cations are transported with a similarly high conductance (Table 4.5.1). However, the relative permeability of divalent cations is approximately 6.5-fold greater than that of group 1a monovalent cations indicating limited discrimination by the RyR pore (Tinker and Williams, 1992). Maximal unitary conductance of RyR is very high, reaching 200 pS at saturating Ba^{2+} concentrations (Tinker and Williams, 1992) which is approximately 10 times higher compared to L-type Ca^{2+} channels of the plasma membrane.

The properties of the RyR channel pore have been investigated with organic cations and the minimum radius of the RyR pore has been estimated as 3.5 Å based on the relative permeability of organic monovalent cations (Tinker and Williams, 1993). Block by bis-quaternary ammonium ions of varying length indicates that the voltage drop across the channel is likely to occur over a distance of approximately 10 Å (Tinker and Williams, 1995). Therefore the RyR pore appears to be a relatively short and wide structure that allows high rates of Ca^{2+} permeation across the SR membrane (Williams *et al.*, 2001). Determination of the bacterial K^+ channel pore structure at atomic resolution has greatly increased our understanding of mechanisms involved in selective cation transport (Doyle *et al.*, 1998). Similar to the RyR, the K^+ channel pore is formed by the C-termini of four identical monomers. The resulting pore selectivity filter in KcsA is approximately 12 Å long and 3 Å in diameter which is significantly different from the dimensions predicted for the RyR.

However, there are also important structural similarities between K^+ and RyR2 channels. In RyR, an amino acid sequence, GGIG, analogous to the pore filter signature sequence of K^+ channels was identified in the hypothetical luminal loop connecting the last two TM helices of the RyR (Balshaw *et al.*, 1999). Secondary structure predictions for the putative pore-forming sequence of RyR indicate that this region contains structural elements similar to K^+ channels (Welch *et al.*, 2004), some of which were modelled into the RyR1 cryo-EM pore structure (Ludtke *et al.*, 2005). Analogous to K^+ channels, negative charges located in the orifices of the RyR channel pore have been probed with the positively charged drug neomycin (Mead and Williams, 2004). Moreover, mutations within the pore sequence result in profound alterations in the rate of ion permeation while N-terminal pharmacologic channel regulation is preserved (Zhao *et al.*, 1999). Accordingly, a naturally occurring I4868T mutation in the RyR1 consensus selectivity pore region underlies an unusually severe form of central core disease (Table 4.5.1) (Balshaw *et al.*, 1999).

RyR channels represent important effector proteins for regulation by extracellular signals. An important example is β-adrenergic stimulation of cardiomyocytes which results in increased intracellular cAMP concentrations and activation of PKA. PKA phosphorylation of RyR2 depletes calstabin2 from the channel complex which increases channel activity and open probability (Figure 4.5.2) (Wehrens *et al.*, 2004b) consistent with increased intracellular SR Ca^{2+} release (Lindegger and Niggli, 2005). RyR2 PKA phosphorylation shifts the sensitivity of Ca^{2+}-dependent channel activation to the left such that the channel is more active at any given Ca^{2+} concentration (Marx *et al.*, 2000; Wehrens *et al.*, 2003; Lehnart *et al.*, 2004). RyR dephosphorylation by phosphatase activity allows for calstabin binding which stabilizes the channel closed state (Reiken *et al.*, 2003a). Moreover, phosphodiesterase PDE4D3 activity protects RyR from excess cAMP stimulation and chronic PKA phosphorylation (Lehnart *et al.*, 2005b) (Figure 4.5.2).

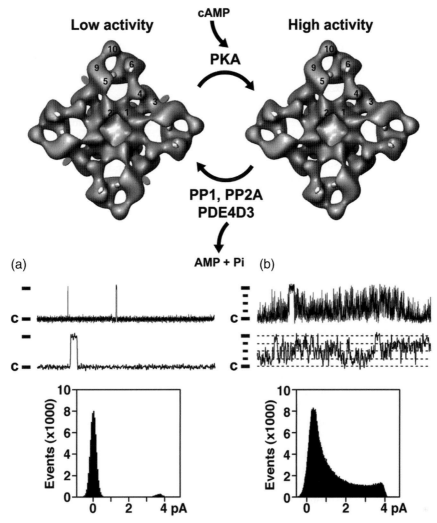

Fig. 4.5.2 RyR2 function is regulated by cAMP-dependent PKA phosphorylation. (a) Top: view of RyR2 complex from the cytosolic side. Calstabin2 (orange) bound to RyR2 stabilizes the channel closed state corresponding with low single channel activity and low open probability (middle, representative recordings, c = closed state). Bottom: RyR2 current amplitude histogram shows that the channel is predominantly in the closed state (0 pA). (b) An increase in intracellular cAMP concentration results in RyR2 PKA phosphorylation of maximally four subunits at Ser-2808 and transient calstabin2 depletion resulting in increased channel activity and open probability as well as the emergence of subconductance states (middle, representative recordings). Bottom: Current amplitude histogram reflects increased activity of maximally PKA phosphorylated channel including higher current amplitude events. Phosphatase (PP1 and PP2A) activity decreases RyR2 PKA phosphorylation resulting in calstabin2 re-binding and stabilization of the channel closed state as seen in (a). Phosphodiesterase (PDE4D3) activity shuts off cAMP-dependent PKA activity in the RyR2 complex as part of a negative feedback loop. Surface domain numbers are indicated according to Radermacher *et al.* (1994). Adapted from Sharma *et al.* (2006)[2] with permission from the *Biophysical Journal*. See also colour plate section.

[2] Sharma MR, Jeyakumar LH, Fleischer S and Wagenknecht T (2006). Three-dimensional visualization of FKBP12.6 binding to an open conformation of cardiac ryanodine receptor. *Biophysical Journal* **90**, 164–172.

4.5.4 RyR cellular and subcellular localization

Intracellular Ca²⁺ release during muscle contraction

Contractile force development in striated muscles is controlled by an increase in cytoplasmic Ca^{2+} concentrations due to SR Ca^{2+} release via RyR1 or RyR2 in skeletal muscles or heart, respectively (Fill and Copello, 2002; Wehrens and Marks, 2003). Depolarization of the plasma membrane by an incoming action potential activates voltage-gated L-type Ca^{2+} channels (skeletal isoform: Ca$_V$1.1; cardiac isoform Ca$_V$1.2) in the plasma membrane T-tubules. In skeletal muscles, Ca$_V$1.1 activation results in conformation-dependent protein–protein interaction with RyR1 which activates

intracellular SR Ca^{2+} release, referred to as voltage-induced Ca^{2+} release (VICR) (Schneider, 1981). In the heart, Ca$_V$1.2 activation results in a plasma membrane Ca^{2+} influx current which triggers Ca^{2+} release from the SR via ligand-induced RyR2 activation, referred to as Ca^{2+}-induced Ca^{2+} release (CICR) (Fabiato, 1983). The approximately tenfold increase in cytoplasmic Ca^{2+} concentrations can be visualized as the intracellular Ca^{2+} transient as a determinant of myofilament activation and ultimately muscle force production (EC coupling).

Subcellular organization: calcium release units

RyR and Ca$_V$ channels are spatially organized with respect to each other in cellular compartments called Ca^{2+}-release units (CRUs). Combined immunolabelling and EM studies have established that

RyR and Ca$_V$ channels are localized in close proximity to each other at intracellular membrane junctions of the SR terminal cisternae and plasmalemmal T-tubules, respectively (Flucher and Franzini-Armstrong, 1996; Gomez *et al.*, 1997). In cardiac muscle CRUs, approximately one Ca$_V$1.2 exists for every five to ten RyR2s channels, and random Ca$_V$1.2 distribution excludes specific structural interaction with RyR2 (Lai *et al.*, 1988; Carl *et al.*, 1995; Sham *et al.*, 1995; Sun *et al.*, 1995; Flucher and Franzini-Armstrong, 1996; Franzini-Armstrong, 1996; Protasi *et al.*, 1996). Accordingly, Ca$_V$1.2 opening allows for Ca^{2+} to diffuse through the junctional space which activates RyR2 via CICR and subsequently cardiac contraction (Fabiato, 1983; Bers, 1991; Gomez *et al.*, 1997). In skeletal muscle, direct protein–protein interactions between Ca$_V$1.1 and RyR1 allows for significantly faster signal transduction and Ca^{2+} release on a millisecond timescale allowing for rapid repetitive muscle contractions (Schneider, 1981). Accordingly, Ca$_V$1.1 are distributed in a regular pattern throughout the T-tubule membrane such that every other RyR1 channel comes into contact with four adjacent Ca$_V$1.1 voltage sensors (Paolini *et al.*, 2004).

RyR membrane organization

Analysis of the *in situ* organization of RyR channels in the terminal SR membranes by electron microscopy (EM) has revealed densely packed RyR arrays occurring in a chequerboard pattern with the four corners of each receptor contacting the corners of each of four neighbouring RyRs (Ferguson *et al.*, 1984). Moreover, EM reconstruction has shown that purified RyRs assemble spontaneously into 2D chequerboard arrays (Yin and Lai, 2000). The intrinsic structural affinity between neighbouring RyR channels may have functional implications. Accordingly, coupled gating has been found in RyR1 and RyR2 channel preparations suggesting that channel opening in native membranes may occur in concert (Marx *et al.*, 1998, 2001a). Coupled gating allows for coordinated SR Ca^{2+} release under physiological conditions and RyR uncoupling due to disease conditions may contribute to a reduced efficiency of SR Ca^{2+} release (Lehnart *et al.*, 2003).

4.5.5 Pharmacology of RyRs

The influence of physiologically occurring intracellular RyR modulators has been investigated extensively and the interested reader is therefore referred to recent reviews (Stamler and Meissner, 2001; Fill and Copello, 2002; Wehrens *et al.*, 2005a). However, much less is known about the molecular mechanisms of channel modulation by a variety of powerful pharmacologic compounds, of which only the compounds relevant to this review are discussed here. For a more comprehensive list of pharmacologic RyR modulators please refer to Zucchi and Ronca-Testoni (1997).

Ryanodine

Ryanodine is a plant alkaloid which binds with high affinity to the RyR and alters channel function (Sutko *et al.*, 1997). Ryanodine increases channel open probability, decreases the rate of ion transport (Rousseau *et al.*, 1987; Tinker *et al.*, 1996), and induces characteristic subconductance states (Tinker *et al.*, 1996). Ryanodine substrates are thought to induce and lock the RyR pore within a specific conformation of the channel pore (Tanna *et al.*, 2005). High- and low-affinity binding sites have been described at the carboxyl terminus of the receptor (Callaway *et al.*, 1994). Studies involving proteolytic degradation and photo-affinity labelling have

demonstrated that the ryanodine binding site is localized within a 76-kD region of the RyR1 C-terminus (Callaway *et al.*, 1994). Ryanodine binds preferentially to the open channel conformation suggesting modulation of allosteric gating mechanisms (Chu *et al.*, 1990a; Tanna *et al.*, 1998).

Caffeine

Caffeine is a methylxanthine that activates RyRs resulting in significantly increased open probability at low millimolar concentrations. Caffeine increases the affinity of the endogenous RyR modulators Ca^{2+} and ATP contributing to increased channel activity even at low Ca^{2+} concentrations (Rousseau *et al.*, 1988). Caffeine application to cells causes mass intracellular SR Ca^{2+} release as a measure of SR Ca^{2+} store content. The RyR1 deletion mutant D4274–4535 exhibited a higher caffeine sensitivity indicating that the binding site is located outside the C-terminal divergent region I (Du *et al.*, 2000). Theophylline and other methylxanthines show RyR actions similar to caffeine.

4-chloro-m-cresol

4-chloro-*m*-cresol (4-C*m*C) is a phenol-based compound thought to directly activate RyR1 by increasing the affinity of Ca^{2+} binding to its activator site. 4-C*m*C is 25 times more potent in activating RyR1 compared with caffeine and has distinct RyR isoform-dependent activation profiles: caffeine is more potent in activating RyR2 and RyR3 compared with RyR1, whereas 4-C*m*C is a much more effective activator of RyR1 and RyR2 and a poor activator of RyR3. Using RyR1 chimeric proteins, the 4-C*m*C activation site has been mapped to amino acids 4007 to 4180 of the RyR1 sequence (Fessenden *et al.*, 2003). Further it was suggested that the isoform-specific RyR1 sensitivity to 4-C*m*C is determined by residues Gln-4020 and Lys-4021 (Fessenden *et al.*, 2006). Like caffeine, 4-C*m*C is an important compound used to test for abnormal intracellular Ca^{2+} leak sensitivity in muscle biopsies from patients susceptible to malignant hyperthermia or other myopathies (Table 4.5.1).

FK506 and rapamycin

FK506 and rapamycin are immunosuppressive macrolide drugs which among other targets bind to the family of FK506 binding proteins (FKBPs) (Deivanayagam *et al.*, 2000). The immunophilin FKBP family has at least eight mammalian members which are named according their molecular mass (e.g., FKBP12). FKBP12 (calstabin1) is a 108-aa protein that shares 85 per cent sequence identity with FKBP12.6 (calstabin2) which differs by only 18 residues. Both calstabins are amphiphilic β-sheet proteins containing five antiparallel strands which form a hydrophobic binding pocket (Deivanayagam *et al.*, 2000). FK506 or rapamycin binding to calstabins dissociates the stabilizing subunit from RyR (Samso *et al.*, 2006) which increases the open probability and results in intracellular Ca^{2+} leak (Kaftan *et al.*, 1996; Ahern *et al.*, 1997; Xiao *et al.*, 1997). Apart from the fact that FK506 and rapamycin are useful compounds to investigate the RyR-calstabin complex, immunosuppressant intoxication may elicit intracellular Ca^{2+} leak resulting in diverse pathological consequences.

JTV-519

JTV-519, a 1,4-benzothiazepine (also known as K201), is a member of a class of drugs known as Ca^{2+} release channel stabilizers which have been shown to increase calstabin binding to RyR and to prevent ventricular arrhythmias (Yano *et al.*, 2003; Wehrens *et al.*, 2004a, 2005b). JTV519 inhibits diastolic SR Ca^{2+} leak, delayed afterdepolarizations and triggered arrhythmias as well as heart failure progression by increasing the binding affinity of calstabin2 for RyR2 in the heart (Kohno *et al.*, 2003; Yano *et al.*, 2003; Lehnart *et al.*, 2005b, 2006). However, in a R4496C knockin mouse model of catecholaminergic polymorphic ventricular tachycardia (CPVT), JTV519 showed no anti-arrhythmogenic effects (Liu *et al.*, 2006). Additionally, JTV519 was shown to prevent abnormal fatigability of skeletal muscles in a heart failure model (Wehrens *et al.*, 2005b). For a review of JTV519 and related pharmacological strategies please refer to Lehnart *et al.* (2005a).

Neomycin

Neomycin, an amino glycoside antibiotic, can be used to probe for negative charges of the cytosolic and luminal RyR pore orifices (Mead and Williams, 2002a). Neomycin functions as a polycationic RyR blocker by interacting with a site mediating a voltage drop in the conduction pathway (Mead and Williams, 2002a, b). Thus, the mechanism underlying neomycin block involves interactions between the positive charges of the neomycin with negative charges in the channel pore (Mead and Williams, 2004). This is in contrast to other channel blockers such as tetraethylamine which involve a specific binding site as opposed to the luminal block mechanism observed with neomycin (Tinker *et al.*, 1992). Negative charges in the luminal pore mouth may be important for cation conduction as indicated by the effects of neutralization of carboxyl groups in the luminal mouth of the pore (Tu *et al.*, 1994).

Dantrolene

Dantrolene is a hydantoin derivative which acts like a postsynaptic muscle relaxant and is used to treat or prevent a malignant hyperthermia crisis (Table 4.5.1) (Parness and Palnitkar, 1995). In skeletal muscle SR preparations, 10 to 90 µM dantrolene inhibited Ca^{2+} release indicating that effective therapeutic dantrolene concentrations in the range of 10 µM may indeed prevent intracellular SR Ca^{2+} leak (Flewellen *et al.*, 1983). This has been confirmed by single-channel studies, where dantrolene at micromolar concentrations decreased the open probability of skeletal muscle porcine and human RyR1 (Nelson *et al.*, 1996).

4.5.6 Physiology and cellular biophysics of RyRs

Cytoplasmic Ca^{2+} controls important cellular functions including muscle contraction, neurotransmitter release and cell growth and death. In excitable cells, plasma membrane depolarization activates voltage-dependent Ca$_V$ channels and the resulting Ca^{2+} influx activates CICR via RyR channels as part of an intracellular signal translation and amplification mechanism. RyRs which are expressed throughout excitable cells are therefore important for a variety of physiological functions.

Cardiac muscle contraction is controlled by RyR2

RyR2 is predominantly expressed in the junctional SR membranes of cardiac but not skeletal muscle. Voltage-dependent activation of Ca$_V$1.2 in the T-tubule plasma membrane activates SR Ca^{2+} release via RyR2 by CICR in the dyadic space resulting in cardiomyocyte contraction (Fabiato and Fabiato, 1975; Nabauer *et al.*, 1989). Since

cardiomyocytes are electrochemically coupled in the heart, electrical excitation can rapidly travel throughout the entire myocardium and activate EC coupling within a relatively short time. In rat cardiomyocytes, the Ca$_V$1.2 Ca^{2+} influx is amplified by a factor of ~20 by RyR2-dependent CICR (Adachi-Akahane *et al.*, 1999). In contrast to adult myocardium, embryonic cardiomyocytes rely significantly more on plasma membrane Ca^{2+} influx for EC coupling during fetal stages. RyR2 knockout results in morphologic abnormalities of the heart tube, cardiac arrest, and death at embryonic day E10 (Takeshima *et al.*, 1998). Prior to embryonic death at days E8 to E9, structurally abnormal mitochondria and abnormally enlarged SR vacuoles develop which contain high Ca^{2+} concentrations while caffeine- or ryanodine-induced Ca^{2+} release is abolished in RyR2-deficient cardiomyocytes (Takeshima *et al.*, 1998; Uehara *et al.*, 2002). RyR2 deficiency also results in intracellular Ca^{2+} toxicity due to a lack of controlled SR Ca^{2+} release during early embryonic development of EC coupling mechanisms. Therefore, RyR2 is the essential SR Ca^{2+} release channel in the heart required for EC coupling.

Skeletal muscle contraction is controlled by RyR1

RyR1 is predominantly expressed in the junctional SR membranes of skeletal muscle (Takeshima *et al.*, 1989). In contrast to cardiac muscle, membrane depolarization is transmitted into RyR1 activation by protein–protein interaction referred to as VICR (Schneider and Chandler, 1973). As a prerequisite, Ca$_V$1.1 and RyR1 are co-localized in a specific spatial pattern in the triad junction to enable conformational coupling (Paolini *et al.*, 2004). RyR1 knockouts survive fetal development but die immediately following birth, potentially from respiratory failure (Takeshima *et al.*, 1994). Moreover, skeletal abnormalities in RyR1 deficient mice are consistent with deformations resulting from a myopathy due to myofibre degeneration (Takeshima *et al.*, 1994). Voltage-induced EC coupling is consistently abolished in RyR1 deficient mice (Takeshima *et al.*, 1994; Takekura *et al.*, 1995). However, RyR1 knockout muscle retains some caffeine-induced SR Ca^{2+} release which is only abolished in double RyR1–RyR3 knockout mice, who also develop abnormally enlarged SR vacuoles that do not occur in single RyR1 knockout mice. These data confirm that RyR3 does not couple to Ca$_V$1.1 in skeletal muscle but functions as an intracellular, ligand-dependent channel (Takeshima *et al.*, 1995; Ikemoto *et al.*, 1997). In summary, RyR1 functions as the major SR Ca^{2+} release channel during skeletal muscle EC coupling.

Smooth muscle relaxation and RyRs

All RyR isoforms are expressed together with IP3Rs in the SR membranes of smooth muscle. While the role of the IP3R in agonist-induced smooth muscle contraction is well established, the role of RyR has only recently been elucidated. In smooth muscle cells of arteries RyR-dependent Ca^{2+} sparks lead to activation of neighbouring plasma membrane large-conductance (BK) K$^+$ channels causing hyperpolarization and inactivation of L-type Ca^{2+} channels which contributes to smooth muscle relaxation (Nelson *et al.*, 1995). Arterial smooth muscle from the RyR3 knockout mouse showed a normal contractile response to pharmacologic modulators (Takeshima *et al.*, 1996). However, the Ca^{2+} spark frequency of cerebral artery smooth muscle cells was significantly higher in RyR3 deficient mice, implicating a role during smooth muscle relaxation (Lohn *et al.*, 2001). In brief, as opposed to cardiac and skeletal EC coupling, RyR in smooth muscles are thought to contribute to relaxation through activation of BK channels. While more studies are needed to elucidate RyR physiology in smooth muscle, it is possible that Ca^{2+} signalling abnormalities play a major role in smooth muscle cells during development of high blood pressure.

4.5.7 Relevance of RyR dysfunction to disease states

There is evidence for genetic and acquired forms of striated muscle disease both in skeletal and cardiac muscles. Accordingly, we discuss skeletal and cardiac myopathies which involve RyR Ca^{2+} leak separately.

Skeletal myopathy

Central core disease

Central core disease (CCD) is a congenital myopathy usually inherited in an autosomal-dominant pattern. Most patients with CCD have congenital hypotonia, delayed motor development, and proximal muscle weakness (Shuaib et al., 1987). Classical skeletal muscle histology shows demarcated central cores along the myofibre length devoid of mitochondria and oxidative enzyme activity as well as sarcomere disruptions (Magee and Shy, 1956). Most CCD patients have mutations in the *RyR1* gene a majority of which cluster in the channel C-terminus comprising the transmembrane, luminal and pore-forming channel regions (Curran et al., 1999; Robinson et al., 2002; Rueffert et al., 2004; Wu et al., 2006). Notably, CCD patients with a consistent muscle pathology and early onset show a significantly higher association with C-terminal RyR1 mutations in the channel pore region (Monnier et al., 2001; Davis et al., 2003; Wu et al., 2006). RyR1 mutations linked to CCD alter RyR1-dependent Ca^{2+} release (Tilgen et al., 2001), cause SR Ca^{2+} leak (Tong et al., 1999), or result in EC uncoupling (Avila et al., 2001). In the case of RyR1 mutations in the pore region, muscle weakness in CCD patients may result from a reduced sensitivity to voltage-dependent activation of SR Ca^{2+} release (Avila et al., 2001; Dirksen and Avila, 2004). On the other hand, some CCD patients with non-C-terminal mutations display histological changes resembling multiple cores and non-C-terminal mutations were also found to result in SR Ca^{2+} leak (Avila and Dirksen, 2001; Wu et al., 2006). Moreover, a subset of patients with a C-terminal RyR1 mutation exhibit nemaline rod myopathy including both rods and cores and, similar to CCD, a type 1 fibre predominance (Scacheri et al., 2000).

Multiminicore disease

RyR1 is essential for skeletal muscle development as demonstrated by lethal birth defects and skeletal muscle abnormalities resulting from RyR1 gene inactivation (Takeshima et al., 1994). Accordingly, a severe form of multiminicore disease (MmD) is caused by a homozygous *RyR1* mutation resulting in significantly reduced RyR1 protein levels in skeletal muscles leading to respiratory insufficiency, skeletal deformities and ophthalmological paresis (Monnier et al., 2003). A distinct, moderate form of MmD has been characterized by a phenotype resembling CCD and and has been linked to a homozygous mutation in the *RyR1* gene (Ferreiro et al., 2002). Histological analysis of muscle biopsies from affected individuals of this moderate form of MmD reveals multiple minicores disrupting sarcomeres in all myofibre types and a lack of

mitochondria (Ferreiro et al., 2002). The involvement of eye muscles, histological, clinical, and genetic changes in MmD and CCD are similar, suggesting phenotypic overlap between these diseases (Monnier et al., 2003; Guis et al., 2004). However, it was only recently recognized that CCD mutations can occur outside the proposed hotspot clusters and that compound heterozygous mutations exist, indicating that previous genotype-phenotype studies may have to be re-evaluated on the basis of a more comprehensive genetic screen (Wu et al., 2006).

Malignant hyperthermia

Malignant hyperthermia (MH) represents a subclinical myopathy which is occasionally associated with myofibre cores or multiminicores (Quane et al., 1993; Zhang et al., 1993; McCarthy et al., 2000; Guis et al., 2004; Ibarra et al., 2006). MH susceptibility is a genetically heterogeneous predisposition to a pharmaco-genetic shock syndrome which is triggered by exposure to certain medications (inhalation anesthetics, depolarizing muscle relaxants) (Jurkat-Rott et al., 2000). Susceptible individuals are asymptomatic unless drug exposure provokes a life-threatening syndrome characterized by rapidly rising body temperatures, generalized muscle contracture, and rhabdomyolysis, and a hypermetabolic state. MH susceptibility has been linked to the *RyR1* gene (MacLennan et al., 1990; McCarthy et al., 1990) and it has been shown that RyR1 mutations facilitate uncontrolled intracellular Ca^{2+} leak during different forms of channel activation in heterologous expression systems and by other techniques (MacLennan and Phillips, 1992; Tong et al., 1997, 1999; Balog et al., 2001). Muscle biopsies from MH susceptible individuals reveal significantly increased contractile sensitivity to caffeine, *4-CmC* and halothane (MacLennan and Phillips, 1992) and skinned myofibres show an increased rate of SR Ca^{2+} release (Ibarra et al., 2006). MH and CCD missense mutations have been found in three tentative 'hotspot' regions at a higher frequency corresponding to the RyR1 peptide N-terminus (1–614), the center (2129–2458), or the C-terminus (3916–4973) (Figure 4.5.1b), however, more comprehensive screening methods have detected mutations throughout the *RyR1* gene (Sambuughin et al., 2005; Ibarra et al., 2006; Wu et al., 2006). Mutations in the cytosolic RyR1 region may disrupt a hypothetical domain interaction site which stabilizes the channel closed state and/or alter cytoplasmic regulation of RyR1 (Wu et al., 1997; Ikemoto and Yamamoto, 2002) whereas mutations in the transmembrane pore region may alter gating and ion conduction by the channel (Balshaw et al., 1999). Dantrolene application reduces mortality during an MH crisis by preventing intracellular Ca^{2+} leak in MH susceptible patients (Parness and Palnitkar, 1995). RyR1 mutations occur in the majority of MH cases, while other gene loci exist including $Ca_V1.1$ missense mutations in the conserved III–IV linker region (Monnier et al., 1997; Weiss et al., 2004). Moreover, certain myopathies including CCD confer MH susceptibility (Monnier et al., 2001) potentially through modifier genes (Guis et al., 2004; Ibarra et al., 2006).

Skeletal muscle fatigue

A reduced capacity for muscle force production after prolonged activity from exercise is commonly referred to as muscle fatigue and has been attributed to accumulation of lactic acid (Hill and Kupalov, 1929). Studies performed under more physiologic conditions have shown that repeated tetanic contractions do not result in significant intracellular pH changes and that acidosis may protect from fatigue (Pedersen *et al.*, 2004). Acute fatigue development has been linked to a reversible depression of intracellular Ca^{2+} transients (Westerblad and Allen, 1991) and to specific intracellular Ca^{2+} transport mechanisms (Zhao *et al.*, 2005). RyR1-dependent SR Ca^{2+} leak mechanisms have recently been demonstrated during sustained exercise and implicated in dystrophic muscle remodelling (Wang *et al.*, 2005). Chronic fatigue is a common symptom in a variety of disease forms. For instance the clinical status of heart failure patients is classified according to fatigue severity. We have recently documented intracellular Ca^{2+} leak in skeletal muscles from animals with heart failure which was attributed to chronic RyR1 PKA-mediated hyperphosphorylation resulting in a gain-of-function defect (Reiken *et al.*, 2003b). Moreover, we have shown that a drug that fixes the RyR1 Ca^{2+} leak in heart failure also improves fatigability in skeletal muscle (Wehrens *et al.*, 2005b). Thus, a chronic hyperadrenergic state during HF may cause intrinsic skeletal muscle fatigue and evolve into a debilitating symptom in patients.

Cardiac myopathy

Catecholaminergic polymorphic ventricular tachycardia (CPVT) and arrhythmogenic right ventricular cardiomyopathy Type 2 (ARVC2)

CPVT is a heterogeneous form of electrical heart disease which has been linked to RyR2 and calsequestrin mutations (Swan *et al.*, 1999; Lahat *et al.*, 2003). Although generally considered a rare genetic disease, recent post-mortem studies have found RyR2 mutations in 14 per cent of sudden death victims (Tester *et al.*, 2004). In the majority of cases RyR2 missense mutations have been linked to CPVT (Laitinen *et al.*, 2001; Priori *et al.*, 2001; Tester *et al.*, 2004) and a relatively mild form of ARVC2 (Tiso *et al.*, 2001). However, the linkage of RyR2 mutations to ARVC2 has recently been called into question since the phenotype of patients with RyR2 mutations likely does not match the diagnostic criteria of ARVC. By analogy to the MH-CCD mutations in RyR1, approximately 40 RyR2 mutations were assigned to three 'hotspot' regions: the N-terminus (176–420), the central region (2246–2534) and the C-terminal pore region (3778–4950) (Marks *et al.*, 2002b). CPVT has a very high risk of stress-induced juvenile sudden death and no specific therapeutic treatment exists (Laitinen *et al.*, 2001; Lehnart *et al.*, 2004). Unlike ARVC2, patients with CPVT have normal hearts, apart from minor structural variations reported in a few patients (Swan *et al.*, 1999; Laitinen *et al.*, 2001). We have characterized a gain-of-function defect in six different CPVT-mutant RyR2 channels which coincides with PKA phosphorylation-induced calstabin2 depletion due to a reduced binding affinity (Wehrens *et al.*, 2003; Lehnart *et al.*, 2004). Consistent with an RyR2 gain-of-function defect, intracellular Ca^{2+} leak has been confirmed in atrial tumour cells overexpressing CPVT-mutant RyR2 (George *et al.*, 2003). GFP insertion tagging at RyR2-Ser-2367 has localized the central CPVT-ARVC2 region between domains 5 and 6 of RyR2 by cryo-EM reconstruction suggesting that CPVT mutations may affect cytosolic channel regulation (Figure 4.5.1) (Liu *et al.*, 2004).

Importantly, neutralizing a charge in calstabin2-D37S increases binding to CPVT-mutant RyR2 and normalizes channel function (Wehrens *et al.*, 2003). Increased calstabin2 binding leading to normalized CPVT-mutant RyR2 single-channel function has also been achieved with the 1,4-benzothiazepine drug, JTV519, (Lehnart *et al.*,

2004; Wehrens *et al.*, 2005b). A phenotype resembling CPVT can be provoked by stress-testing of *calstabin2*-deficient mice (Wehrens *et al.*, 2003; Lehnart *et al.*, 2006;) which was prevented by treatment with the RyR2 channel stabilizer JTV519 in heterozygous *calstabin2*$^{+/-}$ but not in homozygous *calstabin2*$^{-/-}$ knockout mice (Wehrens *et al.*, 2004a; Lehnart *et al.*, 2006;). Calstabin2 deficient cardiomyocytes develop intracellular Ca^{2+} leak, a concomitant transient inward current (I_{ti}), and delayed afterdepolarizations (DADs) consistent with triggered arrhythmias (Wehrens *et al.*, 2003; Lehnart *et al.*, 2006). Bidirectional ventricular arrhythmias, I_{ti} and DADs have been reported recently in a RyR2-R4496C knockin mouse model (Liu *et al.*, 2006). However, following combined epinephrine and caffeine arrhythmia provocation in RyR2-R4496C mice JTV519 treatment did not prevent DADs or arrhythmias, the significance of which for stress-induced CPVT is currently unclear (Liu *et al.*, 2006). Moreover, CPVT has been linked to missense and nonsense mutations in cardiac calsequestrin2 (Lahat *et al.*, 2001; Postma *et al.*, 2002). One mutation results in substitution of an aspartic acid for a histidine at position 307 in the negatively charged C-terminal region involved in Ca^{2+} binding. This may cause increased SR Ca^{2+} leak by as yet undefined molecular mechanisms (Lahat *et al.*, 2004). In agreement, overexpression or knockout studies using CPVT-mutant calsequestrin2 in heterologous cells or cardiomyocytes have confirmed luminal SR Ca^{2+} leak (Terentyev *et al.*, 2006).

Heart failure

Heart failure (HF) is a leading cause of mortality, and is characterized by secondary activation of neuroendocrine pathways in a futile attempt to overcome depressed cardiac function typically following myocardial infarction, viral myocarditis, toxic cardiomyopathy, or other insults. Over weeks to months, a chronic hyperadrenergic state ensues with elevated plasma catecholamine levels which contribute to a progressive, maladaptive response including cardiac chamber remodelling, progressive pump dysfunction, and deadly arrhythmias. In HF, RyR2 is hyperphosphorylated by PKA, contributing to intracellular SR Ca^{2+} leak (Marx *et al.*, 2000; Yano *et al.*, 2000, 2003; Obayashi *et al.*, 2006). PKA hyperphosphorylation has been reported for other proteins including the plasma membrane Na^+/Ca^{2+} exchanger (Wei *et al.*, 2003), the L-type Ca^{2+} channel (Chen *et al.*, 2002), and sorcin (Matsumoto *et al.*, 2005). Importantly, while HF results in decreased SR Ca^{2+} load, RyR2-dependent SR Ca^{2+} leak can be maintained despite a reduced Ca^{2+} gradient (Maier *et al.*, 2003) potentially contributing to maladaptive remodeling and triggered arrhythmias.

We have investigated the molecular mechanisms contributing to RyR2 PKA hyperphosphorylation. During HF, PKA levels are unchanged (Marx *et al.*, 2000; Reiken *et al.*, 2003b) and phosphatase (PP1, PP2A) levels in the RyR2 protein complex are decreased contributing to a reduced rate of Ser-2808 de-phosphorylation (Marx *et al.*, 2000; Reiken *et al.*, 2003b). In parallel, due to the chronic hyperadrenergic state, desensitization of β1-adrenoceptors (β1-ARs) and reduced global intracellular cAMP synthesis occur (Cohn *et al.*, 1984; Bristow *et al.*, 1986; Feldman *et al.*, 1987). Decreased phosphodiesterase activity in the RyR2 complex could result in increased local cAMP concentration directly contributing to chronic PKA hyperphosphorylation of RyR2 and intracellular Ca^{2+} leak. A splice variant of the PDE4 family, PDE4D3, contains an N-terminal targeting motif for mAKAP, forming a PKA-mAKAP-PDE4D3 signalling module (Dodge *et al.*, 2001). Accordingly, we demonstrated a specific association of PDE4D3 with the cardiac RyR2 complex (Lehnart *et al.*, 2005b). In human HF, PDE4D3 levels in the RyR2 complex were decreased by 43 per cent and the cAMP hydrolyzing activity of RyR2-bound PDE4D3 was decreased by 42 per cent (Lehnart *et al.*, 2005b). Thus, RyR2 PKA hyperphosphorylation and calstabin2 depletion from the RyR2 complex may directly result from reduced PDE4D3 activity in the RyR2 complex (Lehnart *et al.*, 2005b). We have investigated whether PDE4D3 deficiency in the RyR2 complex contributes to the development of HF using a mouse model of PDE4D gene inactivation (Jin *et al.*, 1999). PDE4D deficient mice showed an age-dependent increase in left ventricular dimensions and a concomitant decrease in cardiac function that became significant by 9 months of age, consistent with a slowly progressive form of heart failure (Lehnart *et al.*, 2005b). To demonstrate that the cardiomyopathy and arrhythmias in PDE4D deficient mice are due to PKA phosphorylation of RyR2 we crossed these mice with RyR2-S2808A knockin mice that carry RyR2 channels that cannot be PKA phosphorylated and showed that these mice are protected against HF progression and arrhythmias (Lehnart *et al.*, 2005b). Thus, the RyR2 channel is the critical mediator of cardiovascular pathology in PDE4D deficient mice.

4.5.8 Concluding remarks

RyR Ca^{2+} release channels have been extensively characterized in skeletal and cardiac muscle where they play key roles in EC coupling. Other organ functions, e.g. blood vessel constriction or higher brain functions, may also involve RyR dependent intracellular Ca^{2+} release mechanisms. RyR protein complexes are targets of many pharmacological compounds, the mechanisms of which are increasingly investigated. Moreover, the RyR is now recognized as a potential therapeutic target for diverse human disorders including cardiac and skeletal myopathies. RyR dysfunction has been associated with an increasing number of complex genetic and acquired diseases. Therefore greater understanding of RyR structure–function relationships is critical to advance both our understanding of physiologic mechanisms and to develop specific therapeutic modulators for this target.Acknowledgements

ARM is a scientific advisor for a start-up company based on developing novel RyR-specific therapeutics.

References

Adachi-Akahane S, Cleemann L *et al.* (1999). BAY K 8644 modifies Ca2+ cross signaling between DHP and ryanodine receptors in rat ventricular myocytes. *Am J Physiol* **276**, H1178–1189.

Ahern GP, Junankar PR and Dulhunty AF (1994). Single channel activity of the ryanodine receptor calcium release channel is modulated by FK-506. *FEBS Letters* **352**, 369–374.

Ahern GP, Junankar PR and Dulhunty AF (1997). Subconductance states in single-channel activity of skeletal muscle ryanodine receptors after removal of FKBP12. *Biophys J* **72**, 146–162.

Allen P, Ouimet C and Greengard P (1997). Spinophilin, a novel protein phosphatase 1 binding protein localized to dendritic spines. *Proc Natl Acad Sci USA* **94**, 9956–9961.

Appelt D, Buenviaje B, Champ C *et al.* (1989). Quantitation of 'junctional feet' content in two types of muscle fiber from hind limb muscles of the rat. *Tissue Cell* **21**, 783–794.

Avila G and Dirksen RT (2001). Functional effects of central core disease mutations in the cytoplasmic region of the skeletal muscle ryanodine receptor. *J Gen Physiol* **118**, 277–290.

Avila G, O'Brien JJ and Dirksen RT (2001). Excitation–contraction uncoupling by a human central core disease mutation in the ryanodine receptor. *Proc Natl Acad Sci USA* **98**, 4215–4220.

Balog EM, Fruen BR, Shomer NH *et al.* (2001). Divergent effects of the malignant hyperthermia-susceptible Arg(615)–>Cys mutation on the Ca(2+) and Mg(2+) dependence of the RyR1. *Biophys J* **81**, 2050–2058.

Balshaw D, Gao L and Meissner G (1999). Luminal loop of the ryanodine receptor: a pore-forming segment? *Proc Natl Acad Sci USA* **96**, 3345–3347.

Balshaw DM, Xu L, Yamaguchi N *et al.* (2001). Calmodulin binding and inhibition of cardiac muscle calcium release channel (ryanodine receptor). *J Biol Chem* **276**, 20144–20153.

Benacquista BL, Sharma MR, Samso M *et al.* (2000). Amino acid residues 4425–4621 localized on the three-dimensional structure of the skeletal muscle ryanodine receptor. *Biophys J* **78**, 1349–1358.

Bers DM (1991). *Excitation–contraction coupling and cardiac contractile force*, vol. *122*. Boston, Kluwer Academic Publishers.

Bhat MB, Hayek SM, Zhao J *et al.* (1999). Expression and functional characterization of the cardiac muscle ryanodine receptor Ca(2+) release channel in Chinese hamster ovary cells. *Biophys J* **77**, 808–816.

Bhat MB, Zhao J, Takeshima H *et al.* (1997). Functional calcium release channel formed by the carboxyl-terminal portion of ryanodine receptor. *Biophysical J* **73**, 1329–1336.

Brillantes AB, Ondrias K, Scott A *et al.* (1994). Stabilization of calcium release channel (ryanodine receptor) function by FK506-binding protein. *Cell* **77**, 513–523.

Bristow MR, Ginsburg R, Umans V *et al.* (1986). Beta 1- and beta 2-adrenergic-receptor subpopulations in nonfailing and failing human ventricular myocardium: coupling of both receptor subtypes to muscle contraction and selective beta 1-receptor down-regulation in heart failure. *Circ Res* **59**, 297–309.

Bultynck G, Rossi D, Callewaert G *et al.* (2001). The conserved sites for the FK506-binding proteins in ryanodine receptors and inositol 1,4,5-trisphosphate receptors are structurally and functionally different. *J Biol Chem* **276**, 47715–47724.

Callaway C, Seryshev A, Wang JP *et al.* (1994). Localization of the high- and low-affinity [3H]ryanodine binding sites on the skeletal muscle Ca2+ release channel. *J Biol Chem* **269**, 15876–15884.

Carl SL, Felix K, Caswell AH *et al.* (1995). Immunolocalization of sarcolemmal dihydropyridine receptor and sarcoplasmic reticular triadin and ryanodine receptor in rabbit ventricle and atrium. *J Cell Biol* **129**, 672–682.

Chen X, Piacentino V 3rd, Furukawa S *et al.* (2002). L-type Ca2+ channel density and regulation are altered in failing human ventricular myocytes and recover after support with mechanical assist devices. *Circ Res* **91**, 517–524.

Chu A, Diaz-Munoz M, Hawkes MJ *et al.* (1990a). Ryanodine as a probe for the functional state of the skeletal muscle sarcoplasmic reticulum calcium release channel. *Mol Pharmacol* **37**, 735–741.

Chu A, Sumbilla C, Inesi G *et al.* (1990b). Specific association of calmodulin-dependent protein kinase and related substrates with the junctional sarcoplasmic reticulum of skeletal muscle. *Biochemistry* **29**, 5899–5905.

Cohn JN, Levine TB, Olivari MT *et al.* (1984). Plasma norepinephrine as a guide to prognosis in patients with chronic congestive heart failure. *N Engl J Med* **311**, 819–823.

Collins J, Tarcsafalvi A and Ikemoto N (1990). Identification of a region of calsequestrin that binds to the junctional face membrane of sarcoplasmic reticulum. *Biochem Biophys Res Commun* **167**, 189–193.

Coronado R, Morrissette J, Sukhareva M *et al.* (1994). Structure and function of ryanodine receptors. *Am J Physiol* **266**, C1485–1504.

Curran JL, Hall WJ, Halsall PJ *et al.* (1999). Segregation of malignant hyperthermia, central core disease and chromosome 19 markers. *Br J Anaesth* **83**, 217–222.

Currie S, Loughrey CM, Craig MA *et al.* (2004). Calcium/calmodulin-dependent protein kinase II delta associates with the ryanodine receptor complex and regulates channel function in rabbit heart. *Biochem J* **377**, 357–366.

Davis MR, Haan E, Jungbluth H *et al.* (2003). Principal mutation hotspot for central core disease and related myopathies in the C-terminal transmembrane region of the *RYR1* gene. *Neuromuscul Disord* **13**, 151–157.

Deivanayagam CC, Carson M, Thotakura A *et al.* (2000). Structure of FKBP12.6 in complex with rapamycin. *Acta Crystallogr D Biol Crystallogr* **56**, 266–271.

Diaz-Munoz M, Hamilton S, Kaetzel M *et al.* (1990). Modulation of Ca2+ release channel activity from sarcoplasmic reticulum by annexin V1 (67-kDa Calcimedin). *J Biol Chem* **265**, 15894–15899.

Dirksen RT and Avila G (2004). Distinct effects on Ca2+ handling caused by malignant hyperthermia and central core disease mutations in RyR1. *Biophys J* **87**, 3193–3204.

Dodge KL, Khouangsathiene S, Kapiloff MS *et al.* (2001). mAKAP assembles a protein kinase A/PDE4 phosphodiesterase cAMP signaling module. *EMBO J* **20**, 1921–1930.

Doyle DA, Morais Cabral J, Pfuetzner RA *et al.* (1998). The structure of the potassium channel: molecular basis of K+ conduction and selectivity. *Science* **280**, 69–77.

Du GG, Avila G, Sharma P *et al.* (2004). Role of the sequence surrounding predicted transmembrane helix M4 in membrane association and function of the Ca(2+) release channel of skeletal muscle sarcoplasmic reticulum (ryanodine receptor isoform 1). *J Biol Chem* **279**, 37566–37574.

Du GG, Guo X, Khanna VK *et al.* (2001). Functional characterization of mutants in the predicted pore region of the rabbit cardiac muscle Ca(2+) release channel (ryanodine receptor isoform 2). *J Biol Chem* **276**, 31760–31771.

Du GG, Khanna VK and MacLennan DH (2000). Mutation of divergent region 1 alters caffeine and Ca(2+) sensitivity of the skeletal muscle Ca(2+) release channel (ryanodine receptor). *J Biol Chem* **275**, 11778–11783.

Du GG, Sandhu B, Khanna VK *et al.* (2002). Topology of the Ca2+ release channel of skeletal muscle sarcoplasmic reticulum (RyR1). *Proc Natl Acad Sci USA* **99**, 16725–16730.

Fabiato A (1983). Calcium-induced release of calcium from the cardiac sarcoplasmic reticulum. *Am J Physiol* **245**, C1–C14.

Fabiato A and Fabiato F (1975). Contractions induced by a calcium-triggered release of calcium from the sarcoplasmic reticulum of single skinned cardiac cells. *J Physiol* **249**, 469–495.

Feldman MD, Copelas L, Gwathmey JK *et al.* (1987). Deficient production of cyclic AMP: pharmacologic evidence of an important cause of contractile dysfunction in patients with end-stage heart failure. *Circulation* **75**, 331–339.

Ferguson DG, Schwartz HW and Franzini-Armstrong C (1984). Subunit structure of junctional feet in triads of skeletal muscle: a freeze-drying, rotary-shadowing study. *J Cell Biol* **99**, 1735–1742.

Ferreiro A, Monnier N, Romero NB *et al.* (2002). A recessive form of central core disease, transiently presenting as multi-minicore disease, is associated with a homozygous mutation in the ryanodine receptor type 1 gene. *Ann Neurol* **51**, 750–759.

Fessenden JD, Feng W, Pessah IN *et al.* (2006). Amino acid residues Gln4020 and Lys4021 of the ryanodine receptor type 1 are required for activation by 4-chloro-m-cresol. *J Biol Chem* **281**, 21022–21031.

Fessenden JD, Perez CF, Goth S *et al.* (2003). Identification of a key determinant of ryanodine receptor type 1 required for activation by 4-chloro-m-cresol. *J Biol Chem* **278**, 28727–28735.

Fill M and Copello JA (2002). Ryanodine receptor calcium release channels. *Physiol Rev* **82**, 893–922.

Flewellen EH, Nelson TE, Jones WP *et al.* (1983). Dantrolene dose response in awake man: implications for management of malignant hyperthermia. *Anesthesiology* **59**, 275–280.

Flucher BE and Franzini-Armstrong C (1996). Formation of junctions involved in excitation–contraction coupling in skeletal and cardiac muscle. *Proc Natl Acad Sci USA* **93**, 8101–8106.

Flucher BE andrews SB, Fleischer S *et al.* (1993). Triad formation: organization and function of the sarcoplasmic reticulum calcium release channel and triadin in normal and dysgenic muscle *in vitro*. *J Cell Biol* **123**, 1161–1174.

Franzini-Armstrong C (1996). Functional significance of membrane architecture in skeletal and cardiac muscle. *Society of General Physiologists Series* **51**, 3–18.

Fruen BR, Bardy JM, Byrem TM *et al.* (2000). Differential Ca(2+) sensitivity of skeletal and cardiac muscle ryanodine receptors in the presence of calmodulin. *Am J Physiol Cell Physiol* **279**, C724–733.

Futatsugi A, Kuwajima G and Mikoshiba K (1995). Tissue-specific and developmentally regulated alternative splicing in mouse skeletal muscle ryanodine receptor mRNA. *Biochem J* **305**, 373–378.

Gaburjakova M, Gaburjakova J, Reiken S *et al.* (2001). FKBP12 binding modulates ryanodine receptor channel gating. *J Biol Chem* **276**, 16931–16935.

George CH, Higgs GV and Lai FA (2003). Ryanodine receptor mutations associated with stress-induced ventricular tachycardia mediate increased calcium release in stimulated cardiomyocytes. *Circ Res* **93**, 531–540.

Giannini G, Clementi E, Ceci R *et al.* (1992). Expression of a ryanodine receptor-Ca2+ channel that is regulated by TGF-beta. *Science* **257**, 91–94.

Giannini G, Conti A, Mammarella S *et al.* (1995). The ryanodine receptor/calcium channel genes are widely and differentially expressed in murine brain and peripheral tissues. *J Cell Biol* **128**, 893–904.

Gillard EF, Otsu K, Fujii J *et al.* (1992). Polymorphisms and deduced amino acid substitutions in the coding sequence of the ryanodine receptor (RYR1) gene in individuals with malignant hyperthermia. *Genomics* **13**, 1247–1254.

Gomez AM, Valdivia HH, Cheng H *et al.* (1997). Defective excitation–contraction coupling in experimental cardiac hypertrophy and heart failure. *Science* **276**, 800–806.

Guis S, Figarella-Branger D, Monnier N *et al.* (2004). Multiminicore disease in a family susceptible to malignant hyperthermia: histology, *in vitro* contracture tests and genetic characterization. *Arch Neurol* **61**, 106–113.

Hakamata Y, Nakai J, Takeshima H *et al.* (1992). Primary structure and distribution of a novel ryanodine receptor/calcium release channel from rabbit brain. *FEBS Lett* **312**, 229–235.

Hill AV and Kupalov P (1929). Anaerobic and aerobic activity in isolated muscle. *Proc R Soc Lond Series B* **105**, 313–322.

Howarth FC, Glover L, Culligan K *et al.* (2002). Calsequestrin expression and calcium binding is increased in streptozotocin-induced diabetic rat skeletal muscle though not in cardiac muscle. *Pflugers Arch* **444**, 52–58.

Hulme JT, Ahn M, Hauschka SD *et al.* (2002). A novel leucine zipper targets AKAP15 and cyclic AMP-dependent protein kinase to the C terminus of the skeletal muscle Ca2+ channel and modulates its function. *J Biol Chem* **277**, 4079–4087.

Hulme JT, Lin TW, Westenbroek RE *et al.* (2003). Beta-adrenergic regulation requires direct anchoring of PKA to cardiac CaV1.2 channels via a leucine zipper interaction with A kinase-anchoring protein 15. *Proc Natl Acad Sci USA* **100**, 13093–13098.

Ibarra MC, Wu S, Murayama K *et al.* (2006). Malignant hyperthermia in Japan: mutation screening of the entire ryanodine receptor type 1 gene coding region by direct sequencing. *Anesthesiology* **104**, 1146–1154.

Ikemoto N and Yamamoto T (2002). Regulation of calcium release by interdomain interaction within ryanodine receptors. *Front Biosci* **7**, d671–683.

Ikemoto T, Komazaki S, Takeshima H *et al.* (1997). Functional and morphological features of skeletal muscle from mutant mice lacking both type 1 and type 3 ryanodine receptors. *J Physiol* **501**, 305–312.

Jeyakumar LH, Ballester L, Cheng DS *et al.* (2001). FKBP binding characteristics of cardiac microsomes from diverse vertebrates. *Biochem Biophys Res Commun* **281**, 979–986.

Jiang D, Xiao B, Li X *et al.* (2003). Smooth muscle tissues express a major dominant negative splice variant of the type 3 Ca2+ release channel (ryanodine receptor). *J Biol Chem* **278**, 4763–4769.

Jin SL, Richard FJ, Kuo WP *et al.* (1999). Impaired growth and fertility of cAMP-specific phosphodiesterase PDE4D-deficient mice. *Proc Natl Acad Sci USA* **96**, 11998–12003.

Jones LR, Zhang L, Sanborn K *et al.* (1995). Purification, primary structure and immunological characterization of the 26-kDa calsequestrin binding protein (junctin) from cardiac junctional sarcoplasmic reticulum. *J Biol Chem* **270**, 30787–30796.

Jurkat-Rott K, McCarthy T and Lehmann-Horn F (2000). Genetics and pathogenesis of malignant hyperthermia. *Muscle Nerve* **23**, 4–17.

Kaftan E, Marks AR and Ehrlich BE (1996). Effects of rapamycin on ryanodine receptor/Ca(2+)-release channels from cardiac muscle. *Circ Res* **78**, 990–997.

Kapiloff MS, Jackson N and Airhart N (2001). mAKAP and the ryanodine receptor are part of a multi-component signaling complex on the cardiomyocyte nuclear envelope. *J Cell Sci* **114**, 3167–3176.

Knudson CM, Stang KK, Jorgenson AO *et al.* (1993). Biochemical characterization and ultrastructural localization of a major junctional sarcoplasmic reticulum glycoprotein (triadin). *J Biol Chem* **268**, 12646–12645.

Kohno M, Yano M, Kobayashi S *et al.* (2003). A new cardioprotective agent, JTV519, improves defective channel gating of ryanodine receptor in heart failure. *Am J Physiol Heart Circ Physiol* **284**, H1035–1042.

Kuwajima G, Futatsugi A, Niinobe M *et al.* (1992). Two types of ryanodine receptors in mouse brain: skeletal muscle type exclusively in Purkinje cells and cardiac muscle type in various neurons. *Neuron* **9**, 1133–1142.

Lahat H, Pras E and Eldar M (2003). RYR2 and CASQ2 mutations in patients suffering from catecholaminergic polymorphic ventricular tachycardia. *Circulation* **107**, e29.

Lahat H, Pras E and Eldar M (2004). A missense mutation in CASQ2 is associated with autosomal recessive catecholamine-induced polymorphic ventricular tachycardia in Bedouin families from Israel. *Ann Med* **36**, 87–91.

Lahat H, Pras E, Olender T *et al.* (2001). A missense mutation in a highly conserved region of CASQ2 is associated with autosomal recessive catecholamine-induced polymorphic ventricular tachycardia in Bedouin families from Israel. *Am J Hum Genet* **69**, 1378–1384.

Lai F, Erickson H, Block B *et al.* (1988). Evidence for a Ca^{2+} channel within the ryanodine receptor complex from cardiac sarcoplasmic reticulum. *Biochem Biophys Res Commun* **151**, 441–449.

Lai F, Liu Q, Xu L *et al.* (1992). Amphibian ryanodine receptor isoforms are related to those of mammalian skeletal or cardiac muscle. *Am J Physiol* **263**, C365–372.

Lai FA and Meissner G (1992). Purification and reconstitution of the ryanodine-sensitive Ca^{2+} release channel complex from muscle sarcoplasmic reticulum. In A Longstaff and P Revest, eds, *Protocols in molecular neurobiology*. Totowa, NJ, Humana Press, pp. 287–305.

Laitinen PJ, Brown KM, Piippo K *et al.* (2001). Mutations of the cardiac ryanodine receptor (RyR2) gene in familial polymorphic ventricular tachycardia. *Circulation* **103**, 485–490.

Lehnart SE, Huang F, Marx SO *et al.* (2003). Immunophilins and coupled gating of ryanodine receptors. *Curr Top Med Chem* **3**, 1383–1391.

Lehnart SE, Terrenoire C, Reiken S *et al.* (2006). Stabilization of cardiac ryanodine receptor prevents intracellular calcium leak and arrhythmias. *Proc Natl Acad Sci USA* **103**, 7906–7910.

Lehnart SE, Wehrens XH and Marks AR (2005a). Ryanodine receptors as drug targets for heart failure and cardiac arrhythmias. *Drug Discov Today (Therapeutic Strategies)* **2**, 259–269.

Lehnart SE, Wehrens XH, Laitinen PJ *et al.* (2004). Sudden death in familial polymorphic ventricular tachycardia associated with calcium release channel (ryanodine receptor) leak. *Circulation* **109**, 3208–3214.

Lehnart SE, Wehrens XH, Reiken S *et al.* (2005b). Phosphodiesterase 4D deficiency in the ryanodine-receptor complex promotes heart failure and arrhythmias. *Cell* **123**, 25–35.

Lindegger N and Niggli E (2005). Paradoxical SR Ca2+ release in guinea-pig cardiac myocytes after beta-adrenergic stimulation revealed by two-photon photolysis of caged Ca2+. *J Physiol* **565**, 801–813.

Lindsay AR, Manning SD and Williams AJ (1991). Monovalent cation conductance in the ryanodine receptor-channel of sheep cardiac muscle sarcoplasmic reticulum. *J Physiol* **439**, 463–480.

Liu N, Colombi B, Memmi M *et al.* (2006). Arrhythmogenesis in catecholaminergic polymorphic ventricular tachycardia. insights from a RyR2 R4496C knock-in mouse model. *Circ Res* **99**, 292–298.

Liu Z, Zhang J, Li P *et al.* (2002). Three-dimensional reconstruction of the recombinant type 2 ryanodine receptor and localization of its divergent region 1. *J Biol Chem* **277**, 46712–46719.

Liu Z, Zhang J, Wang R *et al.* (2004). Location of divergent region 2 on the three-dimensional structure of cardiac muscle ryanodine receptor/calcium release channel. *J Mol Biol* **338**, 533–545.

Lohn M, Jessner W, Furstenau M *et al.* (2001). Regulation of calcium sparks and spontaneous transient outward currents by RyR3 in arterial vascular smooth muscle cells. *Circ Res* **89**, 1051–1057.

Lokuta AJ, Meyers MB, Sander PR *et al.* (1997). Modulation of cardiac ryanodine receptors by sorcin. *J Biol Chem* **272**, 25333–25338.

Ludtke SJ, Serysheva II, Hamilton SL *et al.* (2005). The pore structure of the closed RyR1 channel. *Structure* **13**, 1203–1211.

MacLennan DH and Phillips MS (1992). Malignant hyperthermia. *Science* **256**, 789–794.

MacLennan DH, Duff C, Zorzato F *et al.* (1990). Ryanodine receptor gene is a candidate for predisposition to malignant hyperthermia. *Nature* **343**, 559–561.

Magee KR and Shy GM (1956). A new congenital non-progressive myopathy. *Brain* **79**, 610–621.

Maier LS, Zhang T, Chen L *et al.* (2003). Transgenic CaMKIIdeltaC overexpression uniquely alters cardiac myocyte Ca2+ handling: reduced SR Ca2+ load and activated SR Ca2+ release. *Circ Res* **92**, 904–911.

Marks AR (1996a). Cellular functions of immunophilins. *Physiol Rev* **76**, 631–649.

Marks AR (1996b). Immunophilin modulation of calcium channel gating. *Methods: A Companion to Methods in Enzymology* **9**, 177–187.

Marks AR, Marx SO and Reiken S (2002a). Regulation of ryanodine receptors via macromolecular complexes: a novel role for leucine/isoleucine zippers. *Trends Cardiovasc Med* **12**, 166–170.

Marks AR, Priori S, Memmi M *et al.* (2002b). Involvement of the cardiac ryanodine receptor/calcium release channel in catecholaminergic polymorphic ventricular tachycardia. *J Cell Physiol* **190**, 1–6.

Marks AR, Tempst P, Hwang KS *et al.* (1989). Molecular cloning and characterization of the ryanodine receptor/junctional channel complex cDNA from skeletal muscle sarcoplasmic reticulum. *Proc Natl Acad Sci USA* **86**, 8683–8687.

Marx SO, Gaburjakova J, Gaburjakova M *et al.* (2001a). Coupled gating between cardiac calcium release channels (ryanodine receptors). *Circ Res* **88**, 1151–1158.

Marx SO, Kurokawa J, Reiken S *et al.* (2002). Requirement of a macromolecular signaling complex for beta adrenergic receptor modulation of the KCNQ1–KCNE1 potassium channel. *Science* **295**, 496–499.

Marx SO, Ondrias K and Marks AR (1998). Coupled gating between individual skeletal muscle Ca2+ release channels (ryanodine receptors). *Science* **281**, 818–821.

Marx SO, Reiken S, Hisamatsu Y *et al.* (2001b). Phosphorylation-dependent regulation of ryanodine receptors. A novel role for leucine/isoleucine zippers. *J Cell Biol* **153**, 699–708.

Marx SO, Reiken S, Hisamatsu Y *et al.* (2000). PKA phosphorylation dissociates FKBP12.6 from the calcium release channel (ryanodine receptor): defective regulation in failing hearts. *Cell* **101**, 365–376.

Masumiya H, Wang R, Zhang J *et al.* (2003). Localization of the 12.6-kDa FK506-binding protein (FKBP12.6) binding site to the NH2-terminal domain of the cardiac Ca2+ release channel (ryanodine receptor). *J Biol Chem* **278**, 3786–3792.

Matsumoto T, Hisamatsu Y, Ohkusa T *et al.* (2005). Sorcin interacts with sarcoplasmic reticulum Ca(2+)-ATPase and modulates excitation–contraction coupling in the heart. *Basic Res Cardiol* **100**, 250–262.

McCarthy TV, Healy JM, Heffron JJ *et al.* (1990). Localization of the malignant hyperthermia susceptibility locus to human chromosome 19q12–13.2. *Nature* **343**, 562–564.

McCarthy TV, Quane KA and Lynch PJ (2000). Ryanodine receptor mutations in malignant hyperthermia and central core disease. *Hum Mutat* **15**, 410–417.

Mead F and Williams AJ (2002a). Block of the ryanodine receptor channel by neomycin is relieved at high holding potentials. *Biophys J* **82**, 1953–1963.

Mead F and Williams AJ (2002b). Ryanodine-induced structural alterations in the RyR channel suggested by neomycin block. *Biophys J* **82**, 1964–1974.

Mead FC and Williams AJ (2004). Electrostatic mechanisms underlie neomycin block of the cardiac ryanodine receptor channel (RyR2). *Biophys J* **87**, 3814–3825.

Meyers MB, Pickel VM, Sheu SS *et al.* (1995). Association of sorcin with the cardiac ryanodine receptor. *J Biol Chem* **270**, 26411–26418.

Meyers MB, Puri TS, Chien AJ *et al.* (1998). Sorcin associates with the pore-forming subunit of voltage-dependent L-type Ca2+ channels. *J Biol Chem* **273**, 18930–18935.

Monnier N, Ferreiro A, Marty I *et al.* (2003). A homozygous splicing mutation causing a depletion of skeletal muscle RYR1 is associated with multi-minicore disease congenital myopathy with ophthalmoplegia. *Hum Mol Genet* **12**, 1171–1178.

Monnier N, Procaccio V, Stieglitz P *et al.* (1997). Malignant-hyperthermia susceptibility is associated with a mutation of the alpha 1-subunit of the human dihydropyridine-sensitive L-type voltage-dependent calcium-channel receptor in skeletal muscle. *Am J Hum Genet* **60**, 1316–1325.

Monnier N, Romero NB, Lerale J *et al.* (2001). Familial and sporadic forms of central core disease are associated with mutations in the C-terminal domain of the skeletal muscle ryanodine receptor. *Hum Mol Genet* **10**, 2581–2592.

Nabauer M, Callewaert G, Cleemann L *et al.* (1989). Regulation of calcium release is gated by calcium current, not gating charge, in cardiac myocytes. *Science* **244**, 800–803.

Nakai J, Imagawa T, Hakamat Y *et al.* (1990). Primary structure and functional expression from cDNA of the cardiac ryanodine receptor/calcium release channel. *FEBS Lett* **271**, 169–177.

Nelson MT, Cheng H, Rubart M *et al.* (1995). Relaxation of arterial smooth muscle by calcium sparks. *Science* **270**, 633–637.

Nelson TE, Lin M, Zapata-Sudo G *et al.* (1996). Dantrolene sodium can increase or attenuate activity of skeletal muscle ryanodine receptor calcium release channel. Clinical implications. *Anesthesiology* **84**, 1368–1379.

Obayashi M, Xiao B, Stuyvers BD *et al.* (2006). Spontaneous diastolic contractions and phosphorylation of the cardiac ryanodine receptor at serine-2808 in congestive heart failure in rat. *Cardiovasc Res* **69**, 140–151.

Ohkura M, Furukawa K, Fujimori H *et al.* (1998). Dual regulation of the skeletal muscle ryanodine receptor by triadin and calsequestrin. *Biochemistry* **37**, 12987–12993.

Olivares E, Arispe N and Rojas E (1993). Properties of the ryanodine receptor present in the sarcoplasmic reticulum from lobster skeletal muscle. *Membrane Biochemistry* **10**, 221–235.

Otsu K, Willard HF, Khanna VK *et al.* (1990). Molecular cloning of cDNA encoding the Ca^{2+} release channel (ryanodine receptor) of rabbit cardiac muscle sarcoplasmic reticulum. *J Biol Chem* **265**, 13472–13483.

Ottini L, Marziali G, Conti A *et al.* (1996). Alpha and beta isoforms of ryanodine receptor from chicken skeletal muscle are the homologues of mammalian RyR1 and RyR3. *Biochem J* **315**, 207–216.

Oyamada H, Murayama T, Takagi T *et al.* (1994). Primary structure and distribution of ryanodine-binding protein isoforms of the bullfrog skeletal muscle. *J Biol Chem* **269**, 17206–17214.

Paolini C, Protasi F and Franzini-Armstrong C (2004). The relative position of RyR feet and DHPR tetrads in skeletal muscle. *J Mol Biol* **342**, 145–153.

Parness J and Palnitkar SS (1995). Identification of dantrolene binding sites in porcine skeletal muscle sarcoplasmic reticulum. *J Biol Chem* **270**, 18465–18472.

Pedersen TH, Nielsen OB, Lamb GD *et al.* (2004). Intracellular acidosis enhances the excitability of working muscle. *Science* **305**, 1144–1147.

Phillips MS, Fujii J, Khanna VK *et al.* (1996). The structural organization of the human skeletal muscle ryanodine receptor (RYR1) gene. *Genomics* **34**, 24–41.

Porter Moore C, Zhang JZ *et al.* (1999). A role for cysteine 3635 of RYR1 in redox modulation and calmodulin binding. *J Biol Chem* **274**, 36831–36834.

Postma AV, Denjoy I, Hoorntje TM *et al.* (2002). Absence of calsequestrin 2 causes severe forms of catecholaminergic polymorphic ventricular tachycardia. *Circ Res* **91**, e21–26.

Priori SG, Napolitano C, Tiso N *et al.* (2001). Mutations in the cardiac ryanodine receptor gene (hRyR2) underlie catecholaminergic polymorphic ventricular tachycardia. *Circulation* **103**, 196–200.

Protasi F, Sun XH and Franzini-Armstrong C (1996). Formation and maturation of the calcium release apparatus in developing and adult avian myocardium. *Dev Biol* **173**, 265–278.

Quane KA, Healy JM, Keating KE *et al.* (1993). Mutations in the ryanodine receptor gene in central core disease and malignant hyperthermia. *Nat Genet* **5**, 51–55.

Radermacher M, Rao V, Grassucci R *et al.* (1994). Cryo-electron microscopy and three dimensional reconstruction of the calcium release channel/ryanodine receptor from skeletal muscle. *J Cell Biol* **127**, 411–423.

Reiken S, Gaburjakova M, Guatimosim S *et al.* (2003a). Protein kinase A phosphorylation of the cardiac calcium release channel (ryanodine receptor) in normal and failing hearts. Role of phosphatases and response to isoproterenol. *J Biol Chem* **278**, 444–453.

Reiken S, Lacampagne A, Zhou H *et al.* (2003b). PKA phosphorylation activates the calcium release channel (ryanodine receptor) in skeletal muscle: defective regulation in heart failure. *J Cell Biol* **160**, 919–928.

Robinson RL, Brooks C, Brown SL *et al.* (2002). RYR1 mutations causing central core disease are associated with more severe malignant hyperthermia *in vitro* contracture test phenotypes. *Hum Mutat* **20**, 88–97.

Rossi AE and Dirksen RT (2006). Sarcoplasmic reticulum: the dynamic calcium governor of muscle. *Muscle Nerve* **33**, 715–731.

Rousseau E, Ladine J, Liu QY *et al.* (1988). Activation of the Ca2+ release channel of skeletal muscle sarcoplasmic reticulum by caffeine and related compounds. *Arch Biochem Biophys* **267**, 75–86.

Rousseau E, Smith JS and Meissner G (1987). Ryanodine modifies conductance and gating behavior of single Ca2+ release channel. *Am J Physiol* **253**, C364–368.

Rueffert H, Olthoff D, Deutrich C et al. (2004). A new mutation in the skeletal ryanodine receptor gene (RYR1) is potentially causative of malignant hyperthermia, central core disease and severe skeletal malformation. *Am J Med Genet* **124**, 248–254.

Sambuughin N, Holley H, Muldoon S et al. (2005). Screening of the entire ryanodine receptor type 1 coding region for sequence variants associated with malignant hyperthermia susceptibility in the North American population. *Anesthesiology* **102**, 515–521.

Samso M, Shen X and Allen PD (2006). Structural characterization of the RyR1-FKBP12 interaction. *J Mol Biol* **356**, 917–927.

Samso M, Wagenknecht T and Allen PD (2005). Internal structure and visualization of transmembrane domains of the RyR1 calcium release channel by cryo-EM. *Nat Struct Mol Biol* **12**, 539–544.

Scacheri PC, Hoffman EP, Fratkin JD et al. (2000). A novel ryanodine receptor gene mutation causing both cores and rods in congenital myopathy. *Neurology* **55**, 1689–1696.

Schneider MF (1981). Membrane charge movement and depolarization–contraction coupling. *Annu Rev Physiol* **43**, 507–517.

Schneider MF and Chandler WK (1973). Voltage-dependent charge movement of skeletal muscle: a possible step in excitation–contraction coupling. *Nature* **242**, 244–246.

Scott-Ward TS, Dunbar SJ, Windass JD et al. (2001). Characterization of the ryanodine receptor-Ca2+ release channel from the thoracic tissues of the lepidopteran insect *Heliothis virescens*. *J Membr Biol* **179**, 127–141.

Serysheva II, Hamilton SL, Chiu W et al. (2005). Structure of Ca2+ release channel at 14 A resolution. *J Mol Biol* **345**, 427–431.

Sham JS, Cleemann L and Morad M (1995). Functional coupling of Ca2+ channels and ryanodine receptors in cardiac myocytes. *Proc Nat Acad Sci USA* **92**, 121–125.

Sharma MR, Jeyakuma LH, Fleischer S et al. (2002). Three-dimensional visualisation of FKBP12.6 binding to cardiac ryanodine receptor (RyR2) in open buffer conditions. *Biophys J* **82**, 644a.

Sharma MR, Jeyakumar LH, Fleischer S et al. (2006). Three-dimensional visualization of FKBP12.6 binding to an open conformation of cardiac ryanodine receptor. *Biophys J* **90**, 164–172.

Shuaib A, Paasuke RT and Brownell KW (1987). Central core disease. Clinical features in 13 patients. *Medicine* **66**, 389–396.

Smith J, Rousseau E and Meissner G (1989). Calmodulin modulation of single sarcoplasmic reticulum Ca2+ channels from cardiac and skeletal muscle. *Circ Res* **64**, 352–359.

Stamler JS and Meissner G (2001). Physiology of nitric oxide in skeletal muscle. *Physiol Rev* **81**, 209–237.

Sun XH, Protasi F, Takahashi M et al. (1995). Molecular architecture of membranes involved in excitation–contraction coupling of cardiac muscle. *J Cell Biol* **129**, 659–671.

Sutko JL, Airey JA, Welch W et al. (1997). The pharmacology of ryanodine and related compounds. *Pharmacol Rev* **49**, 53–98.

Sutko JL and Airey JA (1996). Ryanodine receptor Ca^{2+} release channels: does diversity in form equal diversity in function? *Physiol Rev* **76**, 1027–1071.

Swan H, Piippo K, Viitasalo M et al. (1999). Arrhythmic disorder mapped to chromosome 1q42–q43 causes malignant polymorphic ventricular tachycardia in structurally normal hearts. *J Am Coll Cardiol* **34**, 2035–2042.

Szegedi C, Sarkozi S, Herzog A et al. (1999). Calsequestrin: more than 'only' a luminal Ca2+ buffer inside the sarcoplasmic reticulum. *Biochem J* **337**, 19–22.

Takekura H, Nishi M, Noda T et al. (1995). Abnormal junctions between surface membrane and sarcoplasmic reticulum in skeletal muscle with a mutation targeted to the ryanodine receptor. *Proc Natl Acad Sci USA* **92**, 3381–3385.

Takeshima H, Iino M, Takekura H et al. (1994). Excitation-contraction uncoupling and muscular degeneration in mice lacking functional skeletal muscle ryanodine-receptor gene. *Nature* **369**, 556–559.

Takeshima H, Ikemoto T, Nishi M et al. (1996). Generation and characterization of mutant mice lacking ryanodine receptor type 3. *J Biol Chem* **271**, 19649–19652.

Takeshima H, Komazaki S, Hirose K et al. (1998). Embryonic lethality and abnormal cardiac myocytes in mice lacking ryanodine receptor type 2. *Embo J* **17**, 3309–3316.

Takeshima H, Nishimura S, Matsumoto T et al. (1989). Primary structure and expression from complementary DNA of skeletal muscle ryanodine receptor. *Nature* **339**, 439–445.

Takeshima H, Yamazawa T, Ikemoto T et al. (1995). Ca(2+)-induced Ca2+ release in myocytes from dyspedic mice lacking the type-1 ryanodine receptor. *Embo J* **14**, 2999–3006.

Tanna B, Welch W, Ruest L et al. (1998). Interactions of a reversible ryanoid (21-amino-9alpha-hydroxy-ryanodine) with single sheep cardiac ryanodine receptor channels. *J Gen Physiol* **121**, 55–69.

Tanna B, Welch W, Ruest L et al. (2005). Voltage-sensitive equilibrium between two states within a ryanoid-modified conductance state of the ryanodine receptor channel. *Biophys J* **88**, 2585–2596.

Terentyev D, Nori A, Santoro M et al. (2006). Abnormal interactions of calsequestrin with the ryanodine receptor calcium release channel complex linked to exercise-induced sudden cardiac death. *Circ Res* **98**, 1151–1158.

Tester DJ, Spoon DB, Valdivia HH et al. (2004). Targeted mutational analysis of the RyR2-encoded cardiac ryanodine receptor in sudden unexplained death: a molecular autopsy of 49 medical examiner/coroner's cases. *Mayo Clin Proc* **79**, 1380–1384.

Tilgen N, Zorzato F, Halliger-Keller B et al. (2001). Identification of four novel mutations in the C-terminal membrane spanning domain of the ryanodine receptor 1: association with central core disease and alteration of calcium homeostasis. *Hum Mol Genet* **10**, 2879–2887.

Timerman AP, Ogunbumni E, Freund E et al. (1993). The calcium release channel of sarcoplasmic reticulum is modulated by FK-506-binding protein. Dissociation and reconstitution of FKBP-12 to the calcium release channel of skeletal muscle sarcoplasmic reticulum. *J Biol Chem* **268**, 22992–22999.

Timerman AP, Onoue H, Xin HB et al. (1996). Selective binding of FKBP12.6 by the cardiac ryanodine receptor. *J Biol Chem* **271**, 20385–20391.

Tinker A and Williams AJ (1992). Divalent cation conduction in the ryanodine receptor channel of sheep cardiac muscle sarcoplasmic reticulum. *J Gen Physiol* **100**, 479–493.

Tinker A and Williams AJ (1993). Using large organic cations to probe the nature of ryanodine modification in the sheep cardiac sarcoplasmic reticulum calcium release channel. *Biophys J* **65**, 1678–1683.

Tinker A and Williams AJ (1995). Measuring the length of the pore of the sheep cardiac sarcoplasmic reticulum calcium-release channel using related trimethylammonium ions as molecular calipers. *Biophys J* **68**, 111–120.

Tinker A, Lindsay AR and Williams AJ (1992). Block of the sheep cardiac sarcoplasmic reticulum Ca(2+)-release channel by tetra-alkyl ammonium cations. *J Membr Biol* **127**, 149–159.

Tinker A, Sutko JL, Ruest L et al. (1996). Electrophysiological effects of ryanodine derivatives on the sheep cardiac sarcoplasmic reticulum calcium-release channel. *Biophys J* **70**, 2110–2119.

Tiso N, Stephan DA, Nava A et al. (2001). Identification of mutations in the cardiac ryanodine receptor gene in families affected with arrhythmogenic right ventricular cardiomyopathy type 2 (ARVD2). *Hum Mol Genet* **10**, 189–194.

Tong J, McCarthy TV and MacLennan DH (1999). Measurement of resting cytosolic Ca2+ concentrations and Ca2+ store size in HEK-293 cells transfected with malignant hyperthermia or central core disease mutant Ca2+ release channels. *J Biol Chem* **274**, 693–702.

Tong J, Oyamada H, Demaurex N et al. (1997). Caffeine and halothane sensitivity of intracellular Ca2+ release is altered by 15 calcium release channel (ryanodine receptor) mutations associated with malignant hyperthermia and/or central core disease. *J Biol Chem* **272**, 26332–26339.

Tripathy A, Xu L, Mann G *et al.* (1995). Calmodulin activation and inhibition of skeletal muscle Ca2+ release channel (ryanodine receptor). *Biophys J* **69**, 106–119.

Tu H, Tang TS, Wang Z *et al.* (2004). Association of type 1 inositol 1,4,5-trisphosphate receptor with AKAP9 (Yotiao) and protein kinase A. *J Biol Chem* **279**, 19375–19382.

Tu Q, Velez P, Cortes-Gutierrez M *et al.* (1994). Surface charge potentiates conduction through the cardiac ryanodine receptor channel. *J Gen Physiol* **103**, 853–867.

Uehara A, Yasukochi M, Imanaga I *et al.* (2002). Store-operated Ca2+ entry uncoupled with ryanodine receptor and junctional membrane complex in heart muscle cells. *Cell Calcium* **31**, 89–96.

Valdivia HH (1998). Modulation of intracellular Ca2+ levels in the heart by sorcin and FKBP12, two accessory proteins of ryanodine receptors. *Trends Pharmacol Sci* **19**, 479–482.

Van Acker K, Bultynck G, Rossi D *et al.* (2004). The 12 kDa FK506-binding protein, FKBP12, modulates the Ca(2+)-flux properties of the type-3 ryanodine receptor. *J Cell Sci* **117**, 1129–1137.

Vazquez-Martinez O, Canedo-Merino R, Diaz-Munoz M *et al.* (2003). Biochemical characterization, distribution and phylogenetic analysis of *Drosophila melanogaster* ryanodine and IP3 receptors and thapsigargin-sensitive Ca2+ ATPase. *J Cell Sci* **116**, 2483–2494.

Viatchenko-Karpinski S, Terentyev D, Gyorke I *et al.* (2004). Abnormal calcium signaling and sudden cardiac death associated with mutation of calsequestrin. *Circ Res* **94**, 471–477.

Wagenknecht T and Samso M (2002). Three-imensional reconstruction of ryanodine receptors. *Front Biosci* **7**, d1464–1474.

Wagenknecht T, Radermacher M, Grassucci R *et al.* (1997). Locations of calmodulin and FK506-binding protein on the three-dimensional architecture of the skeletal muscle ryanodine receptor. *J Biol Chem* **272**, 32463–32471.

Wang X, Weisleder N, Collet C *et al.* (2005). Uncontrolled calcium sparks act as a dystrophic signal for mammalian skeletal muscle. *Nat Cell Biol* **7**, 525–530.

Wehrens XH and Marks AR (2003). Altered function and regulation of cardiac ryanodine receptors in cardiac disease. *Trends Biochem Sci* **28**, 671–678.

Wehrens XH, Lehnart SE and Marks AR (2005a). Intracellular calcium release and cardiac disease. *Annu Rev Physiol* **67**, 69–98.

Wehrens XH, Lehnart SE, Huang F *et al.* (2003). FKBP12.6 deficiency and defective calcium release channel (ryanodine receptor) function linked to exercise-induced sudden cardiac death. *Cell* **113**, 829–840.

Wehrens XH, Lehnart SE, Reiken S *et al.* (2005b). Enhancing calstabin binding to ryanodine receptors improves cardiac and skeletal muscle function in heart failure. *Proc Natl Acad Sci USA* **102**, 9607–9612.

Wehrens XH, Lehnart SE, Reiken SR *et al.* (2004b). Ca2+/calmodulin-dependent protein kinase II phosphorylation regulates the cardiac ryanodine receptor. *Circ Res* **94**, e61–70.

Wehrens XH, Lehnart SE, Reiken SR *et al.* (2004a). Protection from cardiac arrhythmia through ryanodine receptor-stabilizing protein calstabin2. *Science* **304**, 292–296.

Wei SK, Ruknudin A, Hanlon SU *et al.* (2003). Protein kinase A hyperphosphorylation increases basal current but decreases beta-adrenergic responsiveness of the sarcolemmal Na+-Ca2+ exchanger in failing pig myocytes. *Circ Res* **92**, 897–903.

Weiss RG, O'Connell KM, Flucher BE *et al.* (2004). Functional analysis of the R1086H malignant hyperthermia mutation in the DHPR reveals an unexpected influence of the III-IV loop on skeletal muscle EC coupling. *Am J Physiol Cell Physiol* **287**, C1094–1102.

Welch W, Rheault S, West DJ *et al.* (2004). A model of the putative pore region of the cardiac ryanodine receptor channel. *Biophys J* **87**, 2335–2351.

Westerblad H and Allen DG (1991). Changes of myoplasmic calcium concentration during fatigue in single mouse muscle fibers. *J Gen Physiol* **98**, 615–635.

Williams AJ (1992). Ion conduction and discrimination in the sarcoplasmic reticulum ryanodine receptor/calcium-release channel. *J Muscle Res Cell Motil* **13**, 7–26.

Williams AJ, West DJ and Sitsapesan R (2001). Light at the end of the Ca(2+)-release channel tunnel: structures and mechanisms involved in ion translocation in ryanodine receptor channels. *Q Rev Biophys* **34**, 61–104.

Wu S, Ibarra MC, Malicdan MC *et al.* (2006). Central core disease is due to RYR1 mutations in more than 90% of patients. *Brain* **129**, 1470–1480.

Wu Y, Aghdasi B, Dou SJ *et al.* (1997). Functional interactions between cytoplasmic domains of the skeletal muscle Ca2+ release channel. *J Biol Chem* **272**, 25051–25061.

Xiao RP, Valdivia HH, Bogdanov K *et al.* (1997). The immunophilin FK506-binding protein modulates Ca2+ release channel closure in rat heart. *J Physiol* **500**, 343–354.

Yano M, Kobayashi S, Kohno M *et al.* (2003). FKBP12.6-mediated stabilization of calcium-release channel (ryanodine receptor) as a novel therapeutic strategy against heart failure. *Circulation* **107**, 477–484.

Yano M, Ono K, Ohkusa T *et al.* (2000). Altered stoichiometry of FKBP12.6 versus ryanodine receptor as a cause of abnormal Ca(2+) leak through ryanodine receptor in heart failure. *Circulation* **102**, 2131–2136.

Yin CC and Lai FA (2000). Intrinsic lattice formation by the ryanodine receptor calcium-release channel. *Nat Cell Biol* **2**, 669–671.

Zhang L, Kelley J, Schmeisser G *et al.* (1997). Complex formation between junctin, triadin, calsequestrin and the ryanodine receptor. Proteins of the cardiac junctional sarcoplasmic reticulum membrane. *J Biol Chem* **272**, 23389–23397.

Zhang Y, Chen HS, Khanna VK *et al.* (1993). A mutation in the human ryanodine receptor gene associated with central core disease. *Nat Genet* **5**, 46–50.

Zhao M, Li P, Li X *et al.* (1999). Molecular identification of the ryanodine receptor pore-forming segment. *J Biol Chem* **274**, 25971–25974.

Zhao X, Yoshida M, Brotto L *et al.*(2005). Enhanced resistance to fatigue and altered calcium handling properties of sarcalumenin knockout mice. *Physiol Genomics* **23**, 72–78.

Zucchi R and Ronca-Testoni S (1997). The sarcoplasmic reticulum Ca2+ channel/ryanodine receptor: modulation by endogenous effectors, drugs and disease states. *Pharmacol Rev* **49**, 1–51.

4.6

IP$_3$ receptors

Randen Patterson and Damian van Rossum

4.6.1 Introduction

The 1,4,5-trisphosphate receptor (IP$_3$R) was initially identified due to its ability to bind the second messenger molecule inositol 1,4,5-trisphosphate (InsP$_3$) (Supattapone *et al.*, 1988; Worley *et al.*, 1987; Furuichi *et al.*, 1989). InsP$_3$ and diacylglycerol (DAG) are generated via phospholipase C (PLC) metabolism of phosphatidylinositol-4,5-bisphosphate (PIP$_2$) in response to the stimulation of G-protein-coupled receptors (GPCRs) or receptor tyrosine kinases (RTKs) (Berridge *et al.*, 2000) by various hormones and neurotransmitters. InsP$_3$ rapidly releases Ca^{2+} from intracellular Ca^{2+} stores contained in the endoplasmic reticulum (ER) and other cellular membranes by binding to InsP$_3$-Receptor (IP$_3$R). This process transmits numerous cellular signals. The IP$_3$R can impact a vast number of signalling processes, and a significant portion of this chapter is devoted to demonstrating the nodal importance of the IP$_3$R. Furthermore, as the IP$_3$R has multiple sites for direct and allosteric regulation, it is an excellent target for therapeutic reagent development as well as pharmacologies specific to a variety of cellular pathways.

4.6.2 Molecular characterization

To understand the importance of the IP$_3$R in animal physiology, it is important to understand the evolutionary pathway of this gene family. To date, the first instance of a gene encoding the IP$_3$R, *itr-1*, occurred in *C. elegans*. All invertebrates express only one known copy of the IP$_3$R, with deletion of the IP$_3$R disrupting proper development. When the IP$_3$R is deleted from developing *C. elegans* embryos, epithelial cell migration is disrupted, thus altering the developmental axis of the embryo (Thomas-Virnig *et al.*, 2004). Therefore, even at the earliest point in evolution where expression of the IP$_3$R can be identified, its function is critical to animal physiology. This is further supported by the fact that type 1 IP$_3$R knockout mice have a high incidence of death directly after birth, and surviving mice are ataxic and epileptic (Matsumoto and Nagata, 1999; Matsumoto *et al.*, 1996).

At the evolutionary split between invertebrates and vertebrates, the IP$_3$R gene was duplicated twice, giving rise to three IP$_3$R genes, *iptr-1*, *iptr-2*, and *iptr-3*. Although the duplication of genes at this point in evolution is common, this suggests an increased need for IP$_3$R function in vertebrates and/or new functions for the additional IP$_3$R genes, as they have been retained. Indeed, the type 1 IP$_3$R is ubiquitously expressed, while the type 2 and 3 IP$_3$Rs have more specific expression patterns (Table 4.6.1) (Nakagawa *et al.*, 1991).

In addition to the three IP$_3$R genes, there are also a number of splice variants (Figure 4.6.1a, b), which, in theory, give rise to ~23,001 variants of the tetrameric type 1 IP$_3$R (Regan *et al.*, 2005). In this seminal study, these authors determined that within the Purkinje neurons, where IP$_3$R expression is the highest in vertebrates, the number of type 1 IP$_3$R isoforms change drastically during mouse development. At p6 only 28 isoforms can be detected, rising to a high of 44 isoforms at p22, and falling to 37 isoforms in the adult animal. These data suggest that the different splice variants of the IP$_3$R have different functional roles during development, and that the function of the IP$_3$R is 'fine-tuned' through the expression of various splice variants. In cells where multiple isoforms of the IP$_3$R are expressed and heteromultimerization can occur (Monkawa *et al.*, 1995; Onoue *et al.*, 2000; Joseph *et al.*, 1995), the combinations and possible specificity of function is further augmented. It is likely that the expression of various IP$_3$R isoforms through development are cell-type specific. Taken together, the evolution of the IP$_3$R outlines a crucial and perhaps nodal role in animal physiology. This 'gradient' of isoforms could allow for extremely specific Ca^{2+} signal generation in response to multiple stimuli in numerous cell types, a topic we will elaborate on at the conclusion of this chapter.

The expression pattern of the *iptr* gene family is extremely diverse. Type 1 IP$_3$R is expressed in all tissues, with most cell types expressing some level of IP$_3$R. IP$_3$R has a unique and selective enrichment in Purkinje neurons, which enabled its initial purification and cloning (Worley *et al.*, 1989; Furuichi *et al.*, 1989; Danoff *et al.*, 1988; Ferris *et al.*, 1989). Type 1 IP$_3$R localizes to numerous membranes within the cell. These include the golgi, nucleus, numerous vesicles, plasma membrane, and its primary localization in the ER (Otsu *et al.*, 1990; Lin *et al.*, 1999; Huh and Yoo, 2003; Huh *et al.*, 2005). Type 2 IP$_3$R is expressed at high levels in cardiac muscle, but has relatively scarce expression in other tissues, although it can be seen in neurons and neuroendocrine cells (Perez *et al.*, 1997; Huh and Yoo, 2003). Type 1 IP$_3$R localizes to numerous membranes in the

Table 4.6.1 Properties of IP$_3$-receptor channels

Channel subunit	Biophysical characteristics	Cellular and subcellular localization	Pharmacology	Physiology	Relevance to disease states
Type 1 IP$_3$R (Itr1, Itpr1) Splice variants S1 (exon 12, 45bp), SII A, B, C (exons 40–42, 69, 3, 48bp respectively), and SIII (exon 23) Chromosomal location: 3p26–p25 [1]	Ca2+ dependent: splice variant dependent, in general at 100 nM InsP3: 30 nM–800 nM stimulatory, >800 nM inhibitory (max 300 nM) [4, 5] Ligand dependent-InsP3 binding is required for channel opening, EC50~500 nM [4, 6]	Present in all tissues, and almost all cells. Localizes primarily to the ER, but also found in the plasma membrane, golgi, vesicles, and nucleus [4, 7, 8]	"Specific" Pharmacology **Inhibitors:** Heparin, 2-APB (IC50 ~42 μM), Xestospongin C (IC50 ~358 nM), TMB-8 (IC50 ~50 μM). **Agonists:** Adenophostin A and B (Ka ~1-10 nM), D-3-C-trifluoro-methyl-myo 1,4,5 InsP3 (Ka ~100 nM), D-6-CH2OH-scyllo-1,2,4-InsP3 (Ka ~100 nM) **Analogues-** 2,4,5 InsP3, 1,3,4,5 InsP4 For a complete list see ref's [2,3]	Primary ER Ca^{2+} release channel in response to GPCR and RTK pathway activation. Physically couples to plasma-membrane Ca^{2+} channels gating their activation. Coupled to mitochondrial Ca^{2+} homeostasis. Scaffolds numerous proteins including dependent upon its sub-cellular localization. [9,10]	Ataxia, Duchenne muscular dystrophy (mitochondrial overload), Huntington's disease, polycystic kidney disease, hypoxia, bile secretion, Alzheimers. [11–15]
Type 2 IP$_3$R (Itpr2) Splice variants: TIPR (short form a.a. 1–179) Chromosomal location: 12p11 [1]	Ca^{2+} dependent: at 100nM InsP$_3$, 30 nM–300 nM stimulatory, >300 nm inhibitory (max ~100 nm), although may be insensitive to inhibition via Ca^{2+} [4, 5] Ligand dependent-InsP$_3$ binding is required for channel opening, EC$_{50}$~40 nM [4–6]	Present primarily in cardiac muscle, but also observed in some neurons and neuroendocrine cells. Localizes primarily to the nucleus and ER [4, 7, 8, 16]	Heparin, 2-APB, Xestospongin C, TMB-8	Nucleoplasm Ca^{2+} release channel in response to GPCR and RTK pathway activation. [9,10]	Cardiac Hypertrophy [17]
Type 3 IP$_3$R (Itpr3) Splice variants: none known Chromosomal location: 6p21 [1]	Ca^{2+} dependent: at 33 nM InsP$_3$ 10 nM–1000 nM stimulatory, >1000 nM inhibitory (max ~800 nM) [4,18] Ligand dependent-InsP$_3$ binding is required for channel opening, EC$_{50}$~3200 nM [4, 6, 18]	Present in all tissues, and almost all cells. Localizes primarily to the ER, but also found in the plasma membrane, golgi, vesicles, and nucleus [4, 7, 8]	Heparin, 2-APB, Xestospongin C, TMB-8	ER Ca^{2+} release channel in response to GPCR and RTK pathway activation(ref). Physically couples to plasma-membrane Ca^{2+} channels gating their activation. [9,10]	Bile secretion [12]

[1] Regan MR et al. (2005) Proteins 59, 312–331; [2] Bultynck G et al. (2003) Pflugers Arch 445, 629–642; [3] Wilcox RA et al. (1998) TIPS 19, 467–475; [4] Patel S et al. (1999) Cell Calcium 25, 247–264; [5] Ramos-Franco J et al. (1998) Biophys J 75, 834–839; [6] Thrower E et al. (2001) TIPS 22, 580–586; [7] Huh YH et al. (2005) FEBS Lett 579, 2597–2603; [8] Nakagawa T et al. (1991) PNAS 88, 6244–6248; [9] Berridge MJ et al. (2000) Nat Rev Mol Cell Bio. 1, 11–21; [10] Patterson RL et al. (2004) Annu Rev Biochem 73, 437–465; [11] Ramsey IS et al. (2006) Annu Rev Physiol 68, 619–47; [12] Pusl T et al. (2004) Biochem Biophys Res Commun 322, 1318–1325; [13] Tang TS et al. (2004) Eur J Neurosci 20, 1779–1787; [14] Matsumoto M et al. (1999) J Mol Med 77, 406–411; [15] Basset O et al. (2004) J Biol Chem 279, 47092–47100; [16] Huh YH et al. (2003) FEBS Lett 555, 411–418; [17] Mohler PJ et al. (2003) Nature 421, 634–639; [18] Mak DO et al. (2001) J Gen Physiol 117, 435–446.

cell whilst type 2 IP$_3$R tends to localize to the nucleus in most cell types (Echevarria et al., 2003; Huh and Yoo, 2003). Type 3 IP$_3$R is widely expressed in most tissues and localizes to membranes similar to that of type 1 (Huh et al., 2005; Joseph et al., 1995; Missiaen et al., 1998; Vermassen et al., 2004).

In accordance with the ideas presented in Regan et al. (2005), the subcellular localization of each IP$_3$R isoform appears to be specific. For example, in intestinal epithelial cells type 3 IP$_3$R localizes to tight junctions, while type 1 IP$_3$R localizes to the basolateral membrane (Colosetti et al., 2003). An excellent cryoelectron microscopy (cryo-EM) study from Yoo and colleagues (2000) examined the distribution of the IP$_3$R in chromaffin cells (Huh et al., 2005). They observed that the IP$_3$R is widely distributed, with staining appearing in all major organelles, including the ER, nucleus, golgi, plasma membrane, secretory vesicles, and even low levels appearing in the mitochondria. When these data were quantified by IP$_3$R subtype, they determined that type 3 IP$_3$R expression is markedly enhanced in the nucleus, whilst type 2 IP$_3$R is increased in the ER, and type 1 IP$_3$R is highest in the secretory granules. Overall, the secretory granules in these cells contain the highest levels of IP$_3$R, which is distinct from most cells. This distribution exemplifies the idea that, although the IP$_3$R has classically been defined as an ER resident Ca^{2+} channel, the ER need not be the predominant organelle for IP$_3$R expression. This refined localization could indeed be controlled by the myriad of splice variant/isoform combinations possible between the IP$_3$R subunits.

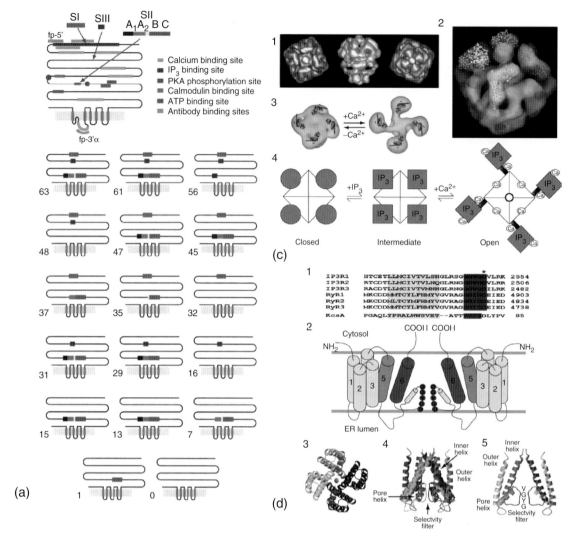

Table I. Distribution of itpr 1 splice variants in rat cerebral cortex and cerebellum

Variant[a]	No. in P6 cerebellum (freq.)	No. in P12 cerebellum (freq.)	No. in P22 cerebellum (freq.)	No. in Adult1 cerebellum (freq.)	No. in Adult2 cerebellum (freq.)	No. in Adult cerebral cortex (freq.)	No. in mixed RNA (freq.)[c]	32[b] Exon 12 (SI)	16 Exon 23B (SII)	8 Exon 40 (SIIA₁)	4 Exon 41 (SIIA₂)	2 Exon 42A (SIIB)	1 Exon 42B (SIIC)
63/61	114 (0.43)	29 (0.18)	28 (010)	7 (0.04)	13 (0.05)	145 (0.815)	48 (0.255)	+[d]	+	+	+	±	+
47/45	107 (0.40)	85 (0.53)	73 (0.26)	28 (0.16)	34 (0.13)	18 (0.101)	20 (0.106)	+		+	+	±	+
31/29	8 (0.03)	5 (0.03)	34 (013)	39 (0.22)	62 (0.24)	5 (0.028)	29 (0.154)		+	+	+	±	+
15/13	11 (0.04)	32 (0.20)	138 (0.50)	101 (0.56)	145 (0.56)	0	90 (0.479)			+	+	±	+
0	3 (0.01)	2 (0.01)	2 (0.007)	3 (0.02)	3 (0.01)	4 (0.022)	0						
48	0	1 (0.006)	1 (0.004)	1 (0.006)	0	3 (0.017)	0	+	+				
32	16 (0.06)	6 (0.04)	0	0	1 (0.004)	0	0	+					
16	1 (0.004)	0	1 (0.004)	0	0	3 (0.017)	1 (0.005)		+				
35	1 (0.004)	0	0	1 (0.006)	0	0	0	+				+	+
1	0	0	0	0	0	1 (0.006)	0						+
56	0	1 (0.006)	0	0	0	0	0	+	+	+			
7	0	1 (0.006)	0	0	0	0	0				+	+	+
37	2 (0.008)	0	0	0	0	0	0	+			+		+
Total clones	263	162	277	180	258	179	188						

Fig. 4.6.1 (a) Gallery of naturally occurring rat brain *itpr1* splice variants. Schematic representation of peptide topology, variable domains, regulatory sites, and antibody binding sites. Natural variants are identified by number as specified in the legend to Table 4.6.1. The grey region specifies the membrane bilayer, cytoplasmic side up (reproduced from (Regan *et al.*, 2005). (b) Table depicting the expression of the IP₃R splice variants in (a) over various developmental periods (reproduced from Regan *et al.*, 2005). (c) EM derived three-dimensional reconstructions of the type 1 IP₃R (reproduced from Bosanac *et al.*, 2004). (1) Highest resolution structure of type 1 IP₃R, determined in the absence of InsP₃ and Ca²⁺ (adapted from (Sato *et al.*, 2004), Figures 2A, 3A and 2F, respectively). (2) Modelling of IP₃R core, in the InsP₃-free state (ribbon and space-filled model), into the density map of the whole receptor (adapted from Sato *et al.*, 2004; Figure 4F). (3) Global structural change of type 1 IP₃R resulting from the presence of Ca²⁺. The structure of the IP₃R core (blue) in complex with InsP₃ (red) is modeled in both conformational states (adapted from Hamada *et al.*, 2002; Fig. 9B). (4) A possible three-state model for receptor activation. (d) Structure of the IP₃R channel domain and its pore (reproduced from Bosanac *et al.*, 2004). (1) Sequence alignment of three types of IP₃R and RyR at the region of the pore helix (highlighted in yellow) and the selectivity filter (highlighted in red). Sequence of the potassium channel KcsA for the same structural segments is also shown. (2) A model of the channel domain for the two IP₃R monomers based on the KcsA structure. The pore helix is yellow and residues comprising the selectivity filter are highlighted in red. Helix 5 and 6 correspond to outer and inner helices of the KcsA structure, respectively. (3) Structure of the KcsA potassium channel comprising four monomers (green, blue, red and yellow) viewed from the extracellular side. (4) Side view of KcsA channel. (5) The side view of the two monomers of KcsA with outer helix (light grey), inner helix (dark grey), pore helix (yellow) and the selectivity filter (red) shown. GenBank accession numbers mouse IP₃R1, X15373; human IP₃R2, D26350; human IP₃R3, D26351; rabbit RyR1, X15209; rabbit RyR2, U50465; rabbit RyR3, X68650; *S. lividans* KcsA, AL939132. See also colour plate section.

Structure

The structure of the IP$_3$R has been well reviewed (Bosanac *et al.*, 2004; Taylor *et al.*, 2004). In brief, each IP$_3$R isoform is more than 2700 amino acids (aa) in length and forms a homo or heterotetramer channel when functional as determined by cryo-EM structures (Figure 4.6.1c). Hamada *et al.* (2002) demonstrated two distinct conformational states of the IP$_3$R: windmill- and square-shaped. They identified the InsP$_3$ binding regions at the tips of the windmill by heparin-gold labelling and propose that the interconversion between these two structural states is regulated by Ca^{2+} (Figure 4.6.1c, d). A similar cryo-electron microscopic structure was elucidated to 30Å, which also demonstrated a windmill shape (Serysheva *et al.*, 2003). This study assigned the putative domain organization on the basis of the crystal structure of the InsP$_3$-binding domain, placing the InsP$_3$-binding domain within a spoke of the windmill. They also elucidated at least three structures corresponding to the open, closed, and an as yet undefined intermediate state (Figure 4.6.1c). This was an excellent study in which samples were prepared at physiological pH, and a large number of particles were analysed.

Building on the cryo-EM structures, Bosanac and colleagues (2002, 2004, 2005) have performed further analysis using X-ray crystallography and NMR, enabling them to propose new models on how the structure of the IP$_3$R is coordinated during tetramerization and activation. Based on their Nuclear Magnetic Resonance (NMR) data, they hypothesize that the N-terminus of the IP$_3$R has two hinge regions which coordinate InsP$_3$ binding, and that these hinges are coordinated by Ca^{2+}. This model is supported by biochemical evidence which demonstrates that InsP$_3$ affinity is indeed regulated by Ca^{2+} (Patel *et al.*, 1999b; Patterson *et al.*, 2004a).

Each IP$_3$R subunit has five domains which are structurally stable (Yoshikawa *et al.*, 1999a). The first two comprise the suppressor region and the InsP$_3$ ligand binding core; the following two encompass the regulatory region of the IP$_3$R. The last domain contains the channel domain (Figure 4.6.1d), as well as the C-terminus which is required for tetramerization and is also modified via regulatory proteins (Boehning and Joseph, 2000a).

IP$_3$R activity has been shown to be strongly regulated by both InsP$_3$ and Ca^{2+}. The InsP$_3$ binding core comprises aa 225–578 in the rat type 1 IP$_3$R. Within this region R265, K508 and R511 bind to InsP$_3$ on the position 4 and 5 phosphates and R568 is implicated in providing specificity of binding (i.e. InsP$_3$ vs InsP$_4$) (Yoshikawa *et al.*, 1996). There are eight Ca^{2+} binding sites on the IP$_3$R, most of which do not have an identified function, although it is believed that these binding sites could function in the Ca^{2+} feedback inhibition of the channel (Patel *et al.*, 1999a). One Ca^{2+} binding site which has been characterized is Asp-2100, which is required for the Ca^{2+} responsiveness of IP$_3$R because its replacement with Glu causes a tenfold decrease in Ca^{2+} stimulation (Boehning *et al.*, 2001b) (Figure 4.6.1d). The homologous residue in the ryanodine receptor also mediates Ca^{2+} sensitivity of the channel (Du *et al.*, 1998).

4.6.3 Biophysical characteristics

IP$_3$Rs are non-selective cation channels; however, due to their primary localization in Ca^{2+} containing membranes, they flux Ca^{2+} under physiological conditions (Boehning *et al.*, 2001a; Mak *et al.*, 2000). Conceivably, if they were contained in other membranes where Ca^{2+} levels were not enriched, they would flux other cations. Importantly, IP$_3$Rs are biphasically controlled via free levels of InsP$_3$ and Ca^{2+} in the cytosol (Bezprozvanny *et al.*, 1991; Bezprozvanny and Ehrlich, 1995). In membranes isolated from tissues, Ca^{2+} stimulates the type 1 IP$_3$R Ca^{2+} conductance with maximal effects at ~100–300 nM Ca^{2+} in the presence of 100 nM InsP$_3$. The maximal stimulatory effects of Ca^{2+} on the IP$_3$R require less Ca^{2+} in the presence of lower InsP$_3$ levels. Electrophysiological single channel studies of purified type 1 IP$_3$R in planar lipid bilayers and nuclear membrane patches reveal a positive cooperative increase in open probability Po at physiologic concentrations of Ca^{2+} and inhibition at low micromolar concentrations (Bezprozvanny *et al.*, 1991). This is similar to what is observed using microsomal ^{45}Ca^{2+} flux and in intact cells (Boehning *et al.*, 2001a; Ionescu *et al.*, 2006; Boehning and Joseph, 2000b). With the exception of experiments performed in planar lipid bilayers (Hagar *et al.*, 1998; Ramos-Franco *et al.*, 1998), the type 2 and type 3 IP$_3$R are also sensitive to Ca^{2+}, although the stimulation of these channels via Ca^{2+} differs from that of the type 1 IP$_3$R (Boehning and Joseph, 2000b; Boehning *et al.*, 2001a; Missiaen *et al.*, 1998).

Studies have been performed with purified type 1 IP$_3$R which did not reveal inhibition by Ca^{2+}, suggesting that a protein important in mediating the Ca^{2+} inhibition of the IP$_3$R may be lost during purification (Danoff *et al.*, 1988). Mikoshiba and colleagues demonstrated that calmodulin mediates Ca^{2+}-dependent inhibition of IP$_3$R function and may be the missing constituent after purification (Hirota *et al.*, 1999; Michikawa *et al.*, 1999). Whereas there are a number of calmodulin binding sites identified on the IP$_3$R (discussed below), the site which exerts the Ca^{2+}-dependence effect is still controversial. Nevertheless, a new study by Kasri *et al.* (2006) clearly demonstrate that calmodulin has the ability to regulate the Ca^{2+}-dependent inhibition of IP$_3$R. When high-affinity peptides which bind calmodulin are used to compete away endogenous calmodulin from the IP$_3$R, channel activity is substantially altered.

The mechanism of IP$_3$R Ca^{2+} regulation is still largely unknown. However, it is generally accepted that Ca^{2+} initially released by IP$_3$R feeds back to augment further Ca^{2+} release in a positively cooperative fashion. At high Ca^{2+} concentrations channel function is inhibited, thus creating the classical oscillatory pattern of Ca^{2+} release which occurs in response to most hormonal/ligand regulated pathways at physiological agonist concentrations.

4.6.4 Pharmacology

InsP$_3$ analogues

A great deal has been elucidated about the nature of ligand binding to the IP$_3$R through the use of InsP$_3$ analogues as pharmacological agents (Bultynck *et al.*, 2003; Wilcox *et al.*, 1998) (Table 4.6.1). For example, 2,4,5 InsP$_3$ is only 10 per cent as effective as 1,4,5 InsP$_3$ at activating the IP$_3$R. This indicates that the residues binding the 1-phosphate are crucial for transmitting the ligand binding signal to the channel pore (Wilcox *et al.*, 1998; Willcocks *et al.*, 1989). Conversely, D-3-*C*-trifluoro-methyl-*myo*inositol 1,4,5-trisphosphate is equipotent to 1,4,5 InsP$_3$ even though this molecule contains a bulky axial 3-CF3-group, demonstrating that the InsP$_3$ binding pocket is quite selective yet still flexible (Wilcox *et al.*, 1994). The 1,4,5 InsP$_3$ analogue D-6-CH$_2$OH-*scyllo*-inositol 1,2,4-trisphosphate and adenophostins also contain bulky groups which correspond to the inositol 3-OH position binding domain within the IP$_3$R

(Murphy *et al.*, 1997; Marchant *et al.*, 1997). The former is equipotent to 1,4,5 InsP$_3$ while the latter is ~10–100 times more potent than InsP$_3$ in activating the IP$_3$R. These data suggest that modification of inositol 3-OH position can be used to alter specificity as well as affinity of IP$_3$R ligands. Support for this idea comes from experiments whereby IP$_3$R in plasmalemmal membranes, which are sialic acid-rich, have limited selectivity between 1,4,5 InsP$_3$ and 1,3,4,5 InsP$_4$ (Khan *et al.*, 1992). Furthermore, in pancreatic acinar cells, 1,3,4,5 InsP$_4$ can promote Ca^{2+} release from the IP$_3$R when phospholipase A2 activity is inhibited (Rowles and Gallacher, 1996). It is interesting to contemplate that lipid, sugar, and/or protein modification of the IP$_3$R binding pocket could change the preference of the IP$_3$R from 1,4,5-InsP$_3$ to other inositol phosphates, thus promoting increased spatial regulation of distinct pockets of IP$_3$R within the cell. In-depth analysis describing the significance of InsP$_3$ analogues and their effects on IP$_3$R function is well reviewed by Wilcox and colleagues (1998).

Agonists and inhibitors

One of the major challenges of performing IP$_3$R biology is the lack of specific cell-permeant activators and inhibitors which rapidly cross the plasma membrane and alter IP$_3$R function. This is a difficult task as InsP$_3$ and its analogues are charged molecules and therefore cannot pass the plasma membrane.

The development of useful IP$_3$R agonists has been a challenging task. For example, even if the potent metabolically stable agonist adenophostin A were conjugated to an ester group making it cell permeant, its ability to activate the IP$_3$R would be limited by the rate of esterase activity within the cell. This would create a graded activation of the IP$_3$R which would not emulate cellular InsP$_3$ generation, making data collected using this methodology difficult to interpret within cellular contexts. Caged 1,4,5 InsP$_3$, which once in the cell can be un-caged using UV light, provides rapid spatial InsP$_3$ levels which emulate receptor stimulation (Walker *et al.*, 1987, 1989). Unfortunately, even when caged this compound cannot cross the plasma membrane. Therefore microinjection or other processes which are traumatic to the cell are needed to introduce this compound to the cytosol, making caged InsP$_3$ far from ideal for *in vivo* experimentation. Cell-permeant esters of InsP$_3$ have also been developed, but they suffer from their activity being dependent upon cellular esterase activity as was discussed for adenophostin (Walker *et al.*, 1989). Since the crystal structure of the IP$_3$R binding pocket has been elucidated, rational drug design techniques may lead to high-affinity cell-permeant IP$_3$R agonists.

Slightly more successful has been the identification of IP$_3$R inhibitors. The first inhibitor which was widely used to study IP$_3$R function was high-molecular-weight heparin (Joseph and Rice, 1989). Heparin has a high binding affinity (low micromolar) for the IP$_3$R which allowed for the first purification of IP$_3$R from rat cerebellar membranes (Ferris *et al.*, 1989), and has been commonly used in microinjection studies as an IP$_3$R antagonist. It is believed that heparin works by directly blocking the pore of the IP$_3$R (Patel *et al.*, 1999a). Within cellular contexts though, heparin has low specificity as an inhibitor of the IP$_3$R, due to cross reactivity with the ryanodine receptor, as well as inhibitory effects on G-protein coupled pathways (Dasso and Taylor, 1991; Bezprozvanny *et al.*, 1993). This, in addition to it being cell impermeant, makes this compound less than ideal for *in vivo* IP$_3$R studies. Similar cross-reactivity problems are present with many other IP$_3$R inhibitors

such as cyclic-ADP ribose, caffeine, Cyclosporin A, FK506, and rapamycin (ryanodine receptor modulators), chloroquine (lysozome acidification), and thimerosal (general reduction of cystine containing proteins) (Bultynck *et al.*, 2003).

Recently, two cell permeant inhibitors of the IP$_3$R have been described, 2-aminoethoxydiphenyl borate (2-APB), and xestospongins (Maruyama *et al.*, 1997b; Gafni *et al.*, 1997). Both of these compounds work to inhibit IP$_3$R, with IC$_{50}$ values of 42 mM and 358 nM, respectively. These compounds received a great deal of attention shortly after their discovery, but as with the aforementioned compounds, these agents possess non-specific activities. Xestospongin C is known to block sarcoplasmic/endoplasmic reticulum Ca^{2+} ATPases, and can release Ca^{2+} from intracellular stores at concentrations above 1 mM (De *et al.*, 1999; Broad *et al.*, 1999). 2-APB is fairly specific to the IP$_3$R as it does not block a number of other Ca^{2+} channels including the ryanodine receptor, voltage gated Ca^{2+} channels, arachidonic acid Ca^{2+} channels, or polycystic kidney disease channels (Luo *et al.*, 2001; Koulen *et al.*, 2002; Gregory *et al.*, 2001; Maruyama *et al.*, 1997a). However, at the concentrations used to inhibit the IP$_3$R *in vivo*, transient receptor potential channels (TRPC) and store-operated Ca^{2+} channels (SOCs) activities are affected (Ma *et al.*, 2000). 2-APB can also be stimulatory to both the IP$_3$R and SOCs at low concentrations (1–30 mM) (Ma *et al.*, 2002). Neither of these compounds provides exceptional specificity. However, xestospongin C and 2-APB have differing non-specific effects at concentrations near their IC$_{50}$ values. Therefore, using both of these agents in comparative studies is likely to provide results whereby specific alterations of IP$_3$R function are interpretable.

4.6.5 Physiology

As previously stated, the IP$_3$R is a ubiquitously expressed protein which resides within numerous cellular compartments. We will now address the physiological role of the IP$_3$R within each of these cellular compartments and describe proteins which bind to and/or regulate the function of the IP$_3$R in each of these spatially distinct organelles.

Endoplasmic reticulum and secretory vesicles

Luminal face

Two proteins are known to interact with the luminal face of the IP$_3$R in the ER: ankyrin and Erp44. Ankyrins are well known for their role in the structural formation of sarcomeres in skeletal muscle where they 'anchor' proteins to the cytoskeletal network (Tuvia *et al.*, 1999). Ankyrin was first reported to bind to the luminal face of the IP$_3$R, inhibiting its ability to bind the ligand InsP$_3$ (Bourguignon and Jin, 1995; Bourguignon *et al.*, 1993). A recent study by Mohler *et al.* (2003) in cardiac myocytes, where the role of the IP$_3$R has been linked to hypertrophy, demonstrates that in addition to regulating IP$_3$R activity, ankyrin can also alter channel stability. In ankyrin knockout mice, the half-life of the IP$_3$R is drastically reduced and specific and multiple ankyrin repeats (repeats 22–24 in mouse ankyrin) are needed to bind the IP$_3$R. As ankyrin domains have a high degree of identity between each repeat, these data elegantly demonstrate the concept that a subset of highly similar domains within proteins can provide specific functions.

The interaction of Erp44 with the IP$_3$R is a very recent discovery (Higo *et al.*, 2005). Erp44 is a member of the thioredoxin family, and can act as an intracellular sensor. It can regulate the IP$_3$R in response to changes in pH, Ca^{2+} concentration, and redox state. This is likely due to Erp44 binding to cystines on the luminal face of the IP$_3$R. In lipid bilayer experiments with purified type 1 IP$_3$R and Erp44, Erp44 binds to the IP$_3$R and significantly decreases the open probability of the channel. As Erp44 is related to stress pathways, this interaction is likely to be important in apoptotic as well as cytoprotective pathways.

Chromogranins provide a novel example of direct coupling between an ion channel and the storage protein for the same ion. Chromogranins are Ca^{2+}-binding proteins which are highly enriched in secretory granules of neurons, neuroendocrine, and endocrine cells, where they sequester approximately 40 mM Ca^{2+}(Yoo and Jeon, 2000; Yoo et al., 2000; Thrower et al., 2002). The IP$_3$R binds to chromogranins through the luminal face of the channel which enhances InsP$_3$-induced Ca^{2+} release (Yoo and Lewis, 2000; Yoo et al., 2000; Thrower et al., 2002). This effect was demonstrated in both IP$_3$R reconstituted into liposomes as well as in planar lipid bilayers. Under both experimental paradigms, a shift in the InsP$_3$ concentration response relationship was observed. Chromogranins bind IP$_3$R on its intraluminal loops rendering the IP$_3$R in secretory granules more sensitive to InsP$_3$ than other cellular organelles, such as the ER and nucleus.

A physiological link between IP$_3$R and chromogranins exists. Mice heterozygous for the Anx7 gene, a Ca^{2+} regulated GTPase associated with Ca^{2+}-dependent exocytosis, have defective IP$_3$R expression and exhibit chromaffin cell hyperplasia (Srivastava et al., 2002). These cells also have constitutively high levels of chromogranin A expression, presumably as a compensation mechanism.

More recent work by Erlich and colleagues demonstrates that when an IP$_3$R binding peptide from chromagranin A is expressed, it is an extremely effective inhibitor of IP$_3$R function (Choe et al., 2004). Using fluorescent Ca^{2+} imaging in PC12 cells transfected with an active or scrambled peptide, these authors monitored Ca^{2+} signals in response to muscarinic acetylcholine receptor signalling. They observed that chromagranin peptides inhibit most Ca^{2+} signalling after receptor activation. Upon quantification of these data, a 50 per cent inhibition of function was observed. This peptide may represent an excellent target to design drugs which will specifically inhibit the IP$_3$R.

What can be concluded is that the luminal region of the IP$_3$R is an active site of channel regulation in response to a variety of stimuli and environmental changes. In addition this region can control channel stability. The interaction sites identified on the IP$_3$R for these proteins certainly contain sites for pharmacological modification.

Cytosolic face (N-terminal region)

Specific proteins also interact with the N-terminal region of the IP$_3$R which accounts for ligand binding (Yoshikawa et al., 1999b). Each of these proteins has been identified to modulate either the Ca^{2+} dependence or ligand binding capability of the IP$_3$R.

Calmodulin was one of the first proteins identified to interact with the IP$_3$R in a Ca^{2+} dependent manner (Yamada et al., 1995). The IP$_3$R has two high-affinity binding domains for calmodulin, one which is Ca^{2+} dependent in the regulatory domain of IP$_3$R-1, and one in the extreme N-terminus which is Ca^{2+} independent (Yamada et al., 1995; Patel et al., 1997). The two binding sites have been mapped: the Ca^{2+} dependent binding sequence occurs at residues 1564–1589 in the rat type 1 IP$_3$R SII isoform, and the Ca^{2+} independent site resides between residues 6–159. The N-terminal site was further mapped to two adjacent regions, residues 49–81 and 106–128 (Sienaert et al., 1997). Calmodulin binding appears to be required for Ca^{2+} dependent inhibition of IP$_3$R (Michikawa et al., 1999; Hirota et al., 1999). However, in other studies a single point mutation within the Ca^{2+} dependent calmodulin-binding site eliminated binding of IP$_3$R to calmodulin in vitro without influencing Ca^{2+}-dependent inhibition of the IP$_3$R in microsomal flux or lipid bilayer assays (Zhang and Joseph, 2001; Nosyreva et al., 2002). The current data suggest that the N-terminal Ca^{2+}-independent binding site is most likely the site relevant to the Ca^{2+}-dependent inhibition of the IP$_3$R (Patel et al., 1999b; Patterson et al., 2004a). Since multiple studies have demonstrated influences of calmodulin on InsP$_3$ binding to IP$_3$R in vitro and in vivo, calmodulin is likely a key physiological regulator of IP$_3$R.

In the recent report by De Smedt and colleagues various peptides which bind to calmodulin were used to compete calmodulin from the IP$_3$R (Kasri et al., 2006). In IP$_3$R containing crude microsomes loaded with ^{45}Ca^{2+}, calmodulin peptides were applied with InsP$_3$, and calcium release was measured. A myosin light chain kinase peptide, which has a high-affinity calmodulin binding site, was observed to drastically alter IP$_3$R function by competing away calmodulin from the IP$_3$R. As predicted, lower affinity peptides from other proteins were also effective but at higher concentrations. What is clear is that calmodulin can be competed away from the IP$_3$R by calmodulin-binding peptides, although the mechanism of regulation by calmodulin has yet to be identified.

Caldendrin (CaBP) also interacts with the IP$_3$R at the N-terminal calmodulin binding sites (Yang et al., 2002; Kasri et al., 2004; Haynes et al., 2003). CaBPs are neuronal EF-hand containing calcium-binding proteins which resemble calmodulin in that they possess four EF-hand Ca^{2+}-binding motifs, as well as one or more non-functional EF-hands. The N-terminal portion of IP$_3$R binds to CaBPs and is reported to gate the IP$_3$R channel in the near absence of InsP$_3$, as demonstrated by patch clamp recordings (Yang et al., 2002). CaBP has an extremely high affinity for binding the IP$_3$R with a K$_a$ = 25 nM. This binding is enhanced by Ca^{2+} reducing the K$_a$ to 1 nM. This suggests that CaBPs may provide a mechanism for opening IP$_3$R channels in response to increased Ca^{2+} concentration without augmentation of InsP$_3$ levels. However, two other groups reported essentially opposite results with CaBP inhibiting InsP$_3$ binding and reducing channel activity (Kasri et al., 2004; Haynes et al., 2003). These differences may result from the methodologies employed: patch-clamping of Xenopus laevis nuclei versus Ca^{2+} imaging in intact cells. Regardless, CaBP regulates the IP$_3$R though an unknown yet likely similar mechanism to calmodulin.

The InsP$_3$ dependence of the IP$_3$R is also modified by the WD40-containing proteins RACK1 and the heterotrimeric G protein subunits (Patterson et al., 2004b). Using yeast-2-hybrid, these proteins were identified to interact with the N-terminal domains of the IP$_3$R. Purified RACK1 is capable of increasing the binding affinity of the IP$_3$R for InsP$_3$ in in vitro microsomal assays, thus sensitizing the receptor. Initial studies with RACK 1 reveal two binding sites, aa 90–110 and 580–600 in the rat type 1 IP$_3$R. The two binding sites for RACK1 are also critical for the proper functioning of the channel, as deletion of aa 90–110 decreases the activity of the IP$_3$R two orders of magnitude, and the 580–600 deletion mutant is completely inactive. In vivo Ca^{2+} imaging studies demonstrate that

RACK1 can alter the sensitivity of the IP_3R to $InsP_3$ in response to G-protein-coupled receptor activation in cultured cells, whereas Ca^{2+} entry is unaffected.

$G\beta\gamma$ has also been linked to IP_3R function. Unlike RACK1, $G\beta\gamma$ complexes were demonstrated to inhibit $InsP_3$ binding to IP_3R, and have the ability to activate the IP_3R directly as demonstrated by nuclear patch-clamping of the IP_3R and $^{45}Ca^{2+}$ microsomal assays (Zeng et al., 2003). The results from this study remain unclear as our own studies confirm the binding of $G\beta\gamma$ binding to the IP_3R, but do not confirm the $G\beta\gamma$-mediated activation of IP_3R in the absence of $InsP_3$.

Mikoshiba and colleagues identified IRBIT interacting with the IP_3R in a yeast-2-hybrid study, and determined that it bound to the IP_3R near the $InsP_3$ binding pocket (Ando et al., 2003). Subsequently, they have determined that IRBIT alters IP_3R function by competing with $InsP_3$ for the ligand binding pocket. When $InsP_3$ levels are high, IRBIT is released from the IP_3R (Ando et al., 2006). They have also determined that IRBIT is expressed in the pancreas and when released from the IP_3R specifically binds to, and activates, the pancreas-type Na^+/HCO_3^- cotransporter 1 (Shirakabe et al., 2006). This protein represents a unique way of amplifying the $InsP_3$ signalling cascade to other protein families without their direct contact with $InsP_3$.

Taken together, it is reasonable to consider that of the many proteins which bind to the extreme N-terminus of the IP_3R, most are likely to affect ligand binding and/or sensitivity and are themselves likely to be regulated by ligand binding.

Cytosolic face (regulatory region)

Kinases

The IP_3R is phosphorylated by multiple kinases including cyclic-AMP dependent protein kinase (PKA), cyclic-GMP dependent protein (PKG), protein kinase C (PKC), the tyrosine kinases Fyn and Lyn, and calmodulin dependent kinase II (CamKII) (Patterson et al., 2004a). The regulation of the IP_3R by these kinases appears to be important for providing cell-type and subcellular specificity of IP_3R activity.

PKA stoichiometrically phosphorylates purified Type 1 IP_3R at two known sites, serine 1589 and 1755 (S1589 and S1755) (Ferris et al., 1991a) in response to muscarinic acetylcholine receptor pathway activation (Bruce et al., 2002), ERK kinase activation (Kovalovsky et al., 2002), Drosophila larval development (Venkatesh et al., 2001), neuroprotection (Pieper et al., 2001), neostriatal signalling (Tang et al., 2003b), and PLC activation (Otani et al., 2002). PKA phosphorylation of the IP_3R can vary within the cell. In neurons, phosphorylated and non-phosphorylated-specific antibodies for type 1 IP_3R S1755 demonstrate phosphorylated IP_3Rs in dendrites, whereas non-phosphorylated forms predominate in cell bodies (Pieper et al., 2001). This suggests that IP_3R activity can be varied within subcellular microdomains to provide specific signals. Both S1589 and S1755 PKA phosphorylation sequences in the type 1 IP_3R are absent in type 2 and 3 IP_3Rs, although these channels are also phosphorylated by PKA demonstrating the global nature of this regulatory pathway (Wojcikiewicz and Luo, 1998). The PKA phosphorylation states of IP_3R are controlled via protein phosphatases 1 and 2A which form a complex with IP_3R analogous to the complex formed with RyR (Tang et al., 2003b; DeSouza et al., 2002). This complex facilitates rapid changes in IP_3R phosphorylation state and likely represents the functional unit by which cAMP regulates IP_3R function.

The IP_3R is also phosphorylated by cyclic GMP-dependent protein kinase (PKG) at the same sites as PKA (Komalavilas and Lincoln, 1994; Haug et al., 1999). PKG phosphorylation generates Ca^{2+} oscillations as does phosphorylation by PKA (Rooney et al., 1996). PKG works in concert with IP_3R associated cGMP kinase (IRAG) substrate to modulate IP_3R activity in response to nitric oxide (described later).

PKC phosphorylation of the IP_3R results in increased $InsP_3$-mediated Ca^{2+} release in vitro (Ferris et al., 1991b) and in vivo (Matter et al., 1993). As Ca^{2+} can directly activate PKC, this may provide a positive feedback loop which could specifically regulate processes such as long-term depression, neuronal exocytosis, and apoptosis (Otani et al., 2002; Zhu et al., 1999; Belmeguenai et al., 2002).

CamK-II has been linked to modification of Ca^{2+} oscillations via IP_3R in neuronal systems (Zhu et al., 1996; Bagni et al., 2000) where it can influence neurotransmitter release. IP_3R modulation by CAMK-II appears to create a positive feedback loop that influences neurotransmitter release at the neuromuscular junction (He et al., 2000). A consistent pattern emerges with respect to serine/threonine phosphorylation of the IP_3R in that these modifications alter the spatiotemporal Ca^{2+}-release properties of the channel. This is likely important for local regulation of Ca^{2+} in spatially restricted areas, providing signal specificity.

IP_3R can also be phosphorylated at tyrosine residues by the tyrosine kinase Fyn, as demonstrated in activated T cells (Jayaraman et al., 1996). Fyn directly binds the IP_3R, and in Fyn $-/-$ mice, which manifest reduced intracellular Ca^{2+} release and defective T-cell signalling, tyrosine phosphorylation of the IP_3R is significantly reduced. Lyn also tyrosine phosphorylates IP_3R in response to B-cell receptor stimulation, a process facilitated by the B-cell scaffold protein with ankyrin repeats (BANK) (Yokoyama et al., 2002). Due to the specificity of the tyrosine kinase receptor signals in lymphatic cells, tyrosine phosphorylation of IP_3R may be selective for immune responses and not mitogenic signals.

IP_3R is also autophosphorylated at the same sites used by PKA and PKC (Ferris et al., 1992). This phosphorylation in response to incubation with ATP was determined to be autophosphorylation due to the persistence of the IP_3R phosphorylation even after extensive purification, denaturation, and renaturation. Autophosphorylation is well known for many kinases, and is common in tyrosine kinase receptors, but not for ion channel receptors. An exception is the plasma membrane ion channel TRPM7 (Runnels et al., 2002). Therefore, IP_3R autophosphorylation may provide a mechanism by which sensitization/desensitization of the channel can be regulated by other IP_3R interacting proteins in the absence of protein kinase cascades.

Nucleotides

The nucleotides ATP and NADH both bind to and influence IP_3R function independently of phosphorylation (Ferris et al., 1990; Kaplin et al., 1996). The binding site of ATP to the IP_3R has been mapped to aa 1769–1781 in the rat type 1 IP_3R sequence. 100 mM ATP enhances Ca^{2+} release, while at higher concentrations ATP can compete with $InsP_3$ for binding to the ligand binding region of the channel. The effects of altering ATP concentrations on the IP_3R are believed to be physiologically relevant as the open probability of $InsP_3$-regulated Ca^{2+} channels in muscle membranes is augmented by ATP (Mak et al., 1999). Since resting levels of ATP in the cell are ~1 mM, physiological regulation of the IP$_3$R may occur when ATP levels are depleted in a spatially restricted environment near the IP$_3$R. When Ca^{2+} is released via the IP$_3$R, it is immediately followed by activation of SERCA to refill stores, and thus could deplete ATP in the vicinity of IP$_3$R, further activating Ca^{2+} release via a feedforward mechanism.

The adenine nucleotide NADH also regulates IP$_3$R in spatially restricted areas. NADH augments IP$_3$R-mediated Ca^{2+} release at physiological concentrations as demonstrated via ^{45}Ca^{2+} flux assays performed in lipid vesicles containing purified IP$_3$R (Kaplin et al., 1996). This augmentation of IP$_3$R activity is selective for NADH as its analogues do not activate the receptor. Physiological relevance has been demonstrated as hypoxia of PC12 cells or cerebellar Purkinje cells raises NADH levels and generates rapid Ca^{2+} signals from InsP$_3$-sensitive stores.

The NADH producing enzyme glyceraldehyde-3-phosphate dehydrogenase (GAPDH) binds to the N-terminus of the IP$_3$R, generating NADH in close proximity to the IP$_3$R (Patterson et al., 2005). Congruent with the results observed in PC12 cells and Purkinje neurons, GAPDH generation of NADH may provide a rapid adaptation to hypoxia. When Ca^{2+} is released from IP$_3$Rs in close proximity to the mitochondria, this would stimulate mitochondrial production of ATP, which could provide a mechanism for enhancing energy production during hypoxic inhibition of mitochondrial respiration (Szalai et al., 1999).

Proteins

Outside of the kinases previously discussed, only two additional proteins have been determined, definitively, to interact with regulatory regions 1 and 2 of the IP$_3$R in the ER. These are myosin and carbonic anhydrase related protein (CARP) (Walker et al., 2002; Hirota et al., 2003). IRAG, which promotes serine/threonine phosphorylation of the IP$_3$R in response to NO (Schlossmann et al., 2000; Ammendola et al., 2001), is also believed to bind in this region although its exact binding pocket is unknown. This chapter will not discuss the association of FKBP12 with the regulatory region of the IP$_3$R due to the extreme controversial nature of this interaction (Bultynck et al., 2003).

Myosin was identified in C. elegans to interact with the C. elegans IP$_3$R between aa 852–1193 (Walker et al., 2002). In C. elegans wherein the IP$_3$R/myosin interaction has been ablated, pharyngeal pumping is inhibited to the same degree as in the myosin knockout. Therefore the interaction of myosin with the IP$_3$R must be crucial for providing the rhythmic pharyngeal pumping action. The myosin/IP$_3$R interaction may also be important for ER localization as myosin Va knockout mice demonstrate a loss of ER innervations within the spines of Purkinje neurons, leading to a loss of long-term depression in these mice (Miyata et al., 2000).

CARP is also highly concentrated in cerebellar Purkinje cells. CARP binds to aa 1387–1647 within the rat IP$_3$R (Hirota et al., 2003). In a carefully constructed study it was determined that CARP interaction within the regulatory domain of the IP$_3$R results in inhibited InsP$_3$ binding. This interaction may account for the extremely low sensitivity of Purkinje cells to InsP$_3$-modulated Ca^{2+} release (Khodakhah and Ogden, 1993).

IRAG was discovered in a phosphorylation screen to identify substrates of cyclic-GMP kinase I. It is a widely expressed protein, with notable enrichment in smooth muscle where cGMP levels are regulated by NO signalling. Cyclic GMP Kinase 1 exists in a ternary complex with IP$_3$R and IRAG, phosphorylating both proteins (Schlossmann et al., 2000; Ammendola et al., 2001). Overexpression of cGK1 and IRAG in COS-7 cells does not alter agonist-mediated Ca^{2+} release via IP$_3$R unless cGMP is present. Inhibition of IP$_3$R function occurs when cGK1 phosphorylates IRAG at serine 696.

A recent study by Fritsch et al. (2004) demonstrates that phosphorylation of the IP$_3$R through this pathway is critical for nitric oxide regulation of Ca^{2+} signalling in the human colon. When the NO donor sodium nitroprusside (SNP) is applied to human kidney, it profoundly attenuates Ca^{2+} release through the IP$_3$R in response to bradykinin receptor stimulation due to hyperphosphorylation of the IP$_3$R. When IRAG is transiently deleted using antisense RNA, the NO donor is no longer capable of attenuating InsP$_3$ mediated Ca^{2+} release. These data provide a physiological relationship of NO signalling to IP$_3$R function.

Cytosolic face (C-terminal region)

The C-terminus of the IP$_3$R is critical for tetramerization, as well as Ca^{2+} sensitivity and channel function. Huntington (Htt), the phosphatase PP1, and band 4.1N which associate with the C-terminus of IP$_3$Rs located in the ER, and perhaps in a yet unidentified organelle.

Seminal studies by Bezprozvanni and colleagues (2004) have linked the IP$_3$R into the Huntington's disease pathway (Figure 4.6.2a). They found that Htt expression facilitates the activity of the NR2B subtype of NMDA receptors and the type 1 IP$_3$R (Tang et al., 2003a). Specifically, Htt binds to residues 2627–2736 of the rat IP$_3$R. This interaction is facilitated by Htt-associated protein which sensitizes the IP$_3$R to InsP$_3$, thus increasing the IP$_3$R open probability and channel activity.

A link between disturbances in Ca^{2+} signalling and neuronal apoptosis is well established. Therefore, this group investigated whether Htt expression induced Ca^{2+} overload and subsequent degeneration of medium spiny neurons, a group of key neurons which are lost in Huntington's disease (Tang et al., 2004). Indeed, they observed that Ca^{2+} induced apoptosis occurs in medium spiny neurons expressing the disease form of Htt. In an elegant experiment, cell death was inhibited using the cell-permeant low-molecular-weight heparin drug enoxparin. As previously mentioned, heparin has long been known to be a high-affinity specific inhibitor of the IP$_3$R. Perhaps this class of compounds needs to be revisited for its possible therapeutic effects in neuronal diseases involving Ca^{2+} overload.

Bezprozvanni and colleagues also demonstrated that the phosphatase PP1a binds to the IP$_3$R, regulating its phosphorylation by PKA (Tang et al., 2003b). PKA phosphorylation of IP$_3$R varies throughout the brain, and is important for Ca^{2+} signalling for neuronal processes like memory and synaptic transmission. Antibodies that distinguish between IP$_3$R S1755 phosphorylated and unphosphorylated forms of type 1 IP$_3$R reveal that phosphorylated IP$_3$R is enriched within dendrites, whereas non-phosphorylated forms predominate in cell bodies. Within areas where rapid signalling and Ca^{2+} release are required, the IP$_3$R is phosphorylated to increase its responsivity. Importantly, the phosphorylation state of the IP$_3$R can be altered rapidly because PKA and PP1 and 2A form a complex with IP$_3$R, analogous to the complex formed with RyR. This protein complex facilitates rapid changes in IP$_3$R phosphorylation state and likely represents the functional unit by which cAMP

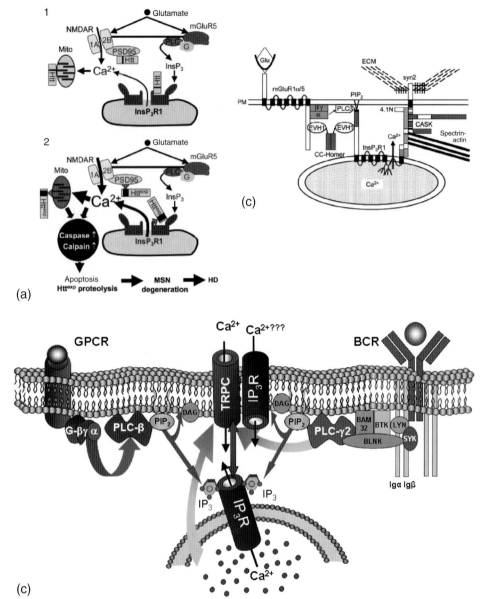

Fig. 4.6.2 (a) Hypothetical marcomolecular signalling complex at the postsynaptic density. 4.1N-FERM domain forms a ternary complex with the carboxy-terminal tail of syndecan-2 (syn2) and PDZ domain of CASK. Ectodomain of syndecans binds to the extracellular matrix. The spectrin/actin binding domain of Band4.1N binds to the spectrin–actin lattice in the postsynaptic terminal. The Band4.1N-CTD domain binds to the carboxy-terminal region of the type 1 IP_3R. The N-terminus of the IP_3R then binds to Homer and PIP_2 at the plasma-membrane. Homer links the IP_3R and mGluR1/5. This cascade of molecular interactions may help to assemble upstream (mGluR1/5) and downstream (IP_3R) signalling components of phosphoinositide signalling pathway at the postsynaptic density (reproduced from Maximov et al., 2003). (b) Huntington expression causes cytosolic and mitochondrial Ca^{2+} overload and apoptosis of striatal medium spiny neurons. Glutamate released from corticostriatal projection neurons stimulates NR1A/NR2B NMDA and mGluR5 receptors in striatal medium spiny neurons. (1) In normal medium spiny neurons, NMDAR activation triggers Ca^{2+} influx from extracellular sources. Activation of mGlu5 stimulates PLC via a heterotrimeric G protein pathway to generate DAG and $InsP_3$ second messengers. The $InsP_3$ generated downstream of mGluR5 does not trigger robust Ca^{2+} release from ER in response to low levels of glutamate, as type $1IP_3R$ sensitivity to $InsP_3$ is low. The association of Huntington with the cytosolic carboxy-terminal tail of the IP_3R is weak while the PSD95-NR1A/NR2B interaction is strong. (2) In medium spiny neurons of Huntington's disease patients, Huntington expression perturbs Ca^{2+} signalling via three synergistic mechanisms. Huntington enhances NMDAR function, possibly through decreased interaction with the PSD95-NR1A/NR2B complex. In addition, Huntington strongly binds to IP_3R carboxy-terminus and sensitizes the IP_3R to activation by $InsP_3$. As a result, low levels of glutamate released from corticostriatal projection neurons lead to supranormal Ca^{2+} influx via NMDAR and Ca^{2+} release via the IP_3R. The net result is elevated intracellular Ca^{2+} levels in medium spiny neurons that causes abnormal Ca^{2+} accumulation in mitochondria. Direct effect of Huntington expression on mitochondrial membrane ion permeability may be a contributing factor in deranged Ca^{2+} mitochondrial signalling. Eventual mitochondrial Ca^{2+} overload causes opening of the permeability transition pore, release of cytochrome C and Ca^{2+}, activation of caspases and calpain, Huntington proteolysis, apoptosis, and degeneration of medium spiny neurons. (reproduced from Bezprozvanny and Hayden, 2004). (c) Mediation of Ca^{2+} release and Ca^{2+} entry signals in response to G protein-coupled and tyrosine kinase-coupled receptors. The model depicts the dual signalling system in B cells involving either GPCRs coupling to the activation of PLC-β, or the B cell receptor which through the complex series of kinases and adaptor proteins shown, results in tyrosine phosphorylation and activation of PLC-$\gamma2$. The IP_3R plays a dual role of (a) mediating Ca^{2+} release from ER stores in response to $InsP_3$ production, and (b) modulating the ability of GPCRs to activate Ca^{2+} entry channels. The physical coupling of IP_3R to TRPC channels has been established, but whether this coupling occurs from IP_3R in the ER, vesicles or plasma-membrane has yet to be determined. Furthermore, the IP_3R itself may function as a Ca^{2+} entry channel in response to $InsP_3$ in B-cells. (Adapted from Spassova et al., 2004). See also colour plate section.

regulates IP$_3$R function. Isoform-specific modulation of IP$_3$R function may also depend on PKA actions on serines 1589 and 1755. These PKA phosphorylation sequences are present within the type 1 IP$_3$R, but are absent in type 2 and 3 IP$_3$R, although these channels are also phosphorylated by PKA (Wojcikiewicz and Luo, 1998). Further studies of this complex revealed that PP1A directly associates with the C-terminus of the IP$_3$R, and indirectly associates with the regulatory region 1 of the IP$_3$R thorough its interaction with AKAP79, which also binds to the IP$_3$R (Tu *et al.*, 2004). This complex provides an elegant mechanism for rapid alterations of the IP$_3$R phosphorylation states.

Band4.1 is an important cytoskeletal protein for cellular structure as well as protein targeting. In neurons, Maximov *et al.* (2003) described a ternary complex existing between IP$_3$R, 4.1N, CASK, and syndecan 2. This protein complex likely helps create the ultrastructure at the postsynaptic density (Figure 4.6.2B). This is an example of where the IP$_3$R may be used more as a cytoskeletal element for protein complexation than for its channel activity.

More recently Band4.1 was demonstrated to be critical for the basolateral targeting of the IP$_3$R in a kidney cell culture line (Zhang *et al.*, 2003). When fluorescently tagged full-length Band4.1 is co-expressed with the IP$_3$R, they colocalize to the basolateral membrane. Expression of a 14 a.a. fluorescent fragment of Band4.1, which binds IP$_3$R, causes neither protein to co-localize to the basolateral membrane. Instead they are found in the cytosol and the nucleus. Thus, this fragment competes away the IP$_3$R from endogenous Band 4.1, disrupting the basolateral targeting.

Reviewing even a subset of the numerous proteins which interact with the IP$_3$R in the ER solidifies the concept that these complexes perform a multitude of functions from regulating channel activity, initiating InsP$_3$ dependent signalling cascades, and cellular targeting of IP$_3$R.

Golgi apparatus

Much less is known about the protein/protein interactions of the IP$_3$R in the golgi, but its presence in Ca^{2+}-containing vesicles in this organelle has been confirmed by numerous groups (Sharp *et al.*, 1992; Otsu *et al.*, 1990; Huh *et al.*, 2005). The role of these Ca^{2+}-containing vesicles is unknown although the presence of chromagranins, calmodulin, and CaBPs has been observed in complex with IP$_3$R in the golgi. To date, only the Sec 6/8 complex (Shin *et al.*, 2000), which is an important protein for vesicle trafficking in the golgi, and nucleobindin (Lin *et al.*, 1999) have been isolated as golgi-specific protein interactors of the IP$_3$R. Nucelobindin participates in Ca^{2+} storage in the golgi, as well as in other biological processes that involve DNA-binding and protein–protein interactions. The direct binding of nucleobindin with the IP$_3$R has been observed, but as yet, no function has been associated with this interaction.

The Sec 6/8 complex has been elucidated further. Muallem and colleagues (Shin *et al.*, 2000) describe a large 'near plasma membrane' multimeric complex of sec6/8 coupling to a large Ca^{2+} signalling complex which contains actin, the plasma membrane Ca^{2+} ATP-ase, the G protein subunits Gαq and G$\beta\gamma$, the b1 isoform of phospholipase C, and the type 1 and 3 IP$_3$R (Shin *et al.*, 2000). Although direct binding is not established, these results were confirmed in pancreatic acinar cells as well as brain lysates. In patch clamp experiments of acinar cells, perfusion with Sec6 or Sec8 antibody significantly reduces InsP$_3$-mediated Ca^{2+} release in response to muscarinic receptor stimulation. This demonstrates that both Sec6 and 8 are active upstream of InsP$_3$ binding to the IP$_3$R. Furthermore, this complex can be dissociated by actin depolymerization. It stands to reason that the formation of these large Ca^{2+} signalling complexes, containing the connections for proteins important for both Ca^{2+} release and entry, would in part be established before the complex is delivered to the plasma membrane. Further study of sec6/8 may reveal which pieces of the Ca^{2+} signalling machinery are pre-assembled before reaching the plasma membrane.

Plasma membrane

It has been realized for a number of years that the IP$_3$R interacts with plasma membrane (PM)-resident proteins and is itself a component of this membrane. Interactors include the canonical transient receptor potential channels (TRPC) channels, Homer, B-cell scaffold protein with ankyrin repeats (BANK), and Junctate. Overall, our understanding of IP$_3$R function in this organelle is rudimentary at best, and there exists many conflicting reports in the literature. That said, we will provide our perspective of how the IP$_3$R works in the PM, and why this may be a crucial site for pharmacological intervention (Figure 4.6.2c).

TRPC channels are ubiquitously expressed throughout the human body (Ramsey *et al.*, 2005). These ion channels tend to flux Ca^{2+}, although some have been identified to flux large heavy metals or small ions such as protons. The TRP channel family has been linked to a number of diseases, including polycystic kidney disease which affects 1 in 1000 people and has an extremely high morbidity rate (Ramsey *et al.*, 2005). InsP$_3$-bound IP$_3$R has been reported to interact with both N and C-termini of TRPC channels (Boulay *et al.*, 1999; Zhang *et al.*, 2001), thus activating Ca^{2+} entry. This ER/PM interaction has been proposed to occur in an InsP$_3$-dependent but channel-independent manner (i.e. the IP$_3$R does not flux Ca^{2+}) (van Rossum *et al.*, 2004; Kiselyov *et al.*, 1999). If these two channels are physically coupled, the Ca^{2+} entering through the TRPC channel may create a microdomain of Ca^{2+} which could be high enough to inhibit the IP$_3$R through its Ca^{2+}-dependent feedback loop. This model is supported by the studies in which a pore dead IP$_3$R mutant has the capacity to rescue Ca^{2+} entry in cells devoid of IP$_3$R (van Rossum *et al.*, 2004). Additionally, TRPC3 channels can also be activated by a soluable N-terminal IP$_3$R fragment containing the InsP$_3$-binding site (Kiselyov *et al.*, 1999).

Alternatively, TRPC channels might couple to PM-resident IP$_3$R (Figure 4.6.2c). This mechanism would provide a rapid way of coupling these two channels, and may also allow the IP$_3$R to function independently of its channel activity. Furthermore, this organization could provide a large surface area for scaffolding other proteins involved in the regulation of TRPC channels, such as calmodulin or Homer. Recent work by Taylor and colleagues (Dellis *et al.*, 2006) presents evidence supporting IP$_3$Rs as the major Ca^{2+} entry channel in DT40 B-cells. They propose that 1–2 IP$_3$Rs, active in the PM, generate all of the Ca^{2+} entry observed in response to B-cell receptor stimulation. This work is supported by genetic engineering of the IP$_3$R with an extracellular bungarotoxin site on the N-terminus of the IP$_3$R making the channel sensitive to bungarotoxin, an effect which promoted Ca^{2+} entry in the presence of InsP$_3$. Studies in other IP$_3$R knockout models, including the type 1 IP$_3$R knockout mice, are needed to resolve the diametrically opposed results obtained from these groups.

Studies by Yuan *et al.* (2003), determined that full-length Homer binds to many of the canonical TRPC channels, in particular the TRPC1,4,5 subfamily. Analogous to the IP$_3$R, they found that Homer also binds both the N- and C-termini of TRPC channels. In this configuration, these channels are inactivated. Upon receptor stimulation and InsP$_3$ production, Homer is released from the TRPC channel, allowing for Ca^{2+} entry. Homer also binds to the IP$_3$R influencing IP$_3$R function by inhibiting Ca^{2+} release (Tu *et al.*, 1998), a process which would correlate with models supporting a channel-independent role of the IP$_3$R in TRPC activation.

A recently discovered integral membrane Ca^{2+} binding protein, Junctate, also associates with the TRPC/IP$_3$R complex. Zorozato and collegues discovered this new family of proteins during a cDNA muscle screen (Treves *et al.*, 2000). Junctate is ubiquitously expressed in many tissues, and binds directly to the IP$_3$R. Interaction with the IP$_3$R is required for proper stimulation by InsP$_3$ (Treves *et al.*, 2004). Furthermore, Junctate is able to increase coupling between the IP$_3$R and TRPC3 (Stamboulian *et al.*, 2005). When the N-terminus of Junctate is deleted, it can no longer potentiate agonist mediated Ca^{2+} entry. Studies by Stamboulian et al (2005) obtained similar results in a rat sperm model. They established specificity of TRPC-binding, in that Junctate interacts directly with TRPC2 and TRPC5 but not TRPC1. In their proposed mechanism, Junctate stimulates Ca^{2+} release from IP$_3$R stores, and induces calmodulin binding to the TRPC channels, thus regulating Ca^{2+} influx.

Whereas both Homer and Junctate are involved in the coupling of IP$_3$R to TRPC channels, they appear to have opposing roles in the regulation of the TRPC/IP$_3$R interaction. These two molecules may be important to impart specificity between alternate IP$_3$R and TRPC channel isoforms. What is clear from these studies is that elucidating the intricate mechanism(s) by which IP$_3$R and TRPC channels are regulated will be important to our global understanding of ion channel physiology. Furthermore, the sites of interaction for Homer and Junctate may provide pharmacological targets for the specific regulation of distinct IP$_3$R/TRPC complexes.

The B-cell scaffold protein with ankyrin repeats (BANK) is another protein which regulates IP$_3$R function (Yokoyama *et al.*, 2002). BANK is a substrate of tyrosine kinases which are activated via B-cell antigen receptor (BCR) stimulation. After BCR stimulation, BANK is predominantly phosphorylated by Syk. Upon phosphorylation BANK binds the IP$_3$R. This promotes Lyn tyrosine phosphorylation of the IP$_3$R, increasing InsP$_3$-mediated Ca^{2+} release. It is proposed that activation of the B-cell receptor stimulates PLCg to produce InsP$_3$ and Syk phosphorylation of BANK. Phospho-BANK then binds the IP$_3$R and scaffolds Lyn near the receptor to phosphorylate Y355 (in the rat type 1 IP$_3$R), increasing the open probability of the channel. This may provide B-cells a method for rapidly changing the Ca^{2+} output from the IP$_3$R during B-cell receptor stimulation, creating the proper Ca^{2+} waves needed for NFAT translocation and gene transcription.

Although we do not know the exact nature of IP$_3$R function in and near the PM, the functions of the IP$_3$R are required in this organelle for proper Ca^{2+} signalling, and should be an area of future work. Many disease states within cells are caused, or influenced, by increased Ca^{2+} levels: these signalling pathways are prime targets for the development of therapeutic treatments.

Mitochondria

To date, the IP$_3$R has not been demonstrated to contain IP$_3$Rs, at least not in detectable levels. Nevertheless, the IP$_3$R can have drastic effects on mitochondrial Ca^{2+} homeostasis.

It has been demonstrated that a number of mitochondrial proteins can impinge upon IP$_3$R function in apoptotic and necrotic pathways. These include proteases caspase-3 and calpain, anti-apoptotic proteins B-cell Lymphoma-2 (Bcl-2) and Bcl-XL, and the small mitochondrial signalling molecule cytochrome C.

The excellent works of Rizzuto and Csordas were the first to demonstrate that the IP$_3$R was directly juxtaposed to the mitochondria, and that the exchange of Ca^{2+} from ER to the mitochondria was mediated through the IP$_3$R. This exchange between the ER and mitochondria is so intimate that cytosolic Ca^{2+} indicators cannot detect it (Rizzuto *et al.*, 1993; Csordas *et al.*, 1999). Since then it has been determined that protein interactions with the C-terminus of the IP$_3$R are critical for Ca^{2+} signalling during apoptosis. We determined that cytochrome C binds the C-terminus of the IP$_3$R inhibiting Ca^{2+} dependent inhibition of the channel which results in Ca^{2+} poisoning of the mitochondria (Boehning *et al.*, 2003).

We confirmed the effects of cytochrome C on the IP$_3$R using a peptide (aa 2621–2636) of the IP$_3$R which binds cytochrome c, and determined that this peptide is cytoprotective in that it inhibits Fas ligand induced apoptosis in Jurkat cells (Boehning *et al.*, 2005). The Ca^{2+} overload induced by IP$_3$R activity appears to occur prior to cytochrome C binding to apoptotic protease-activating factor 1 (APAF1), and is likely a prerequisite step for the large release of cytochrome C necessary for the global activation of APAF1 and subsequently caspase 3, which can then cleave the IP$_3$R. Under high Ca^{2+} conditions, calpain is involved in the N-terminal degradation of the IP$_3$R (Wojcikiewicz and Oberdorf, 1996). It may be that when the N-terminus of the IP$_3$R is cleaved, this channel is functionally misregulated and can contribute to the ER Ca^{2+} depletion which is observed during apoptosis.

Bcl and Bcl-XL are anti-apoptotic proteins which are involved in modulating outer mitochondrial membrane permeability (Thomenius and Distelhorst, 2003). These proteins also localize to the ER, where their functional significance has been controversial (Distelhorst and Shore, 2004). The recent studies of the Distelhorst and Foskett groups have dispelled much of this controversy (Chen *et al.*, 2004; Zhong *et al.*, 2006; White *et al.*, 2005). They have clearly shown that Bcl and Bcl-XL directly interact with the C-terminus of the IP$_3$R. Upon binding, they facilitate the Ca^{2+} cross-talk between the ER and the mitochondria and increase the bioenergetics of the mitochondria to promote cell survival. Overexpression of Bcl-XL increases InsP$_3$-mediated Ca^{2+} release and increases NADPH levels in the mitochondria (White *et al.*, 2005). The interaction of Bcl molecules with the IP$_3$R is regulated by Bcl-XL, Bcl-2 associated kinase, and BH3-interacting domain death agonist, all of which can bind to Bcl-2 and sequester it from the IP$_3$R (Thomenius *et al.*, 2003; Chen *et al.*, 2004). Interactions of these pro-apoptotic proteins with Bcl-2 increases the leakiness of the IP$_3$R, depletes ER Ca^{2+} stores, inhibits Ca^{2+} crosstalk with the mitochondria, and ultimately promotes apoptosis. We hypothesize that the Bcl binding site may overlap with the cytochrome C pocket, which would provide an elegant mechanism for control of IP$_3$R/mitochondrial Ca^{2+} signalling.

In analysing the apoptotic signalling cascade associated with the IP$_3$R, it is clear that this is an extremely complex process with multiple places for regulation and checkpoints. No other process so clearly demonstrates why the possible 23,001 IP$_3$R combinations may exist: i.e. specific and graded regulation. Furthermore, as many proteins interact at regions outside the pore region of the channel, development of allosteric regulators of the IP$_3$R may be guided by the interaction of these apoptotic proteins. One could speculate that these types of compounds could be quite useful in neurodegenerative diseases such as Huntingdon's and Alzheimer's disease.

Nucleus

The discovery of the nucleoplasmic reticulum (NR) is a recent advance (Echevarria *et al.*, 2003). Eschevarra and colleagues demonstrated that the NR exists and contains functional IP$_3$R. When Hep2G cells were subjected to IP$_3$R staining, a clear reticular structure could be observed in the nucleus. This group demonstrated that the IP$_3$R localizes within the NR. Finally using fluorescent Ca^{2+} dyes they were able to demonstrate that clear nuclear Ca^{2+} signals could be induced by hepatocyte growth factor. This work has been confirmed and added to wherein the IP$_3$R has been demonstrated to function in the nuclei of neurons, and be sensitive to NO

(Sharp *et al.*, 1999; Mishra *et al.*, 2003). In skeletal muscle cells, nuclear Ca^{2+} events are also observed that can be blocked by inhibitors of the IP$_3$R (Molgo *et al.*, 2004). Whereas skeletal muscle Ca^{2+} dynamics are almost completely controlled by RyR, perhaps in these cells the IP$_3$R regulates gene transcription. This could be mediated through alteration of Ca^{2+} dynamics at a local scale much smaller than previously hypothesized.

4.6.6 Concluding remarks

After touring the cell and discussing the location and function of the IP$_3$R throughout, we are now still left with the question: why would we need 23,001 IP$_3$R isoforms? A biological example which illustrates this concept is the metabotropic glutamate receptor cascade (Figure 4.6.3).

(1) Receptor binding (cell membrane) Glutamate binds to the metabotropic glutamate receptor (mGluR) activating PLC to generate InsP$_3$. This InsP$_3$ can activate IP$_3$R in the ER and IP$_3$R which is physically coupled to the mGluR receptor by Homer.	(2) Channel activation (cell membrane) InsP$_3$ generated by PLC can bind to IP$_3$R either resident or juxtaposed to the plasma membrane. This could activate Ca^{2+} influx through cell membrane IP$_3$R, or via physical coupling to TRP channels.	(3) Enzyme activation (cell membrane) Increased Ca^{2+} levels can activate PLCδ to generate more InsP$_3$. Ca^{2+} can also activate Ca^{2+} sensitive adenylyl cyclases which generate cAMP. cAMP stimulates PKA phosphorylation of the IP3R, leading to more Ca^{2+} release.

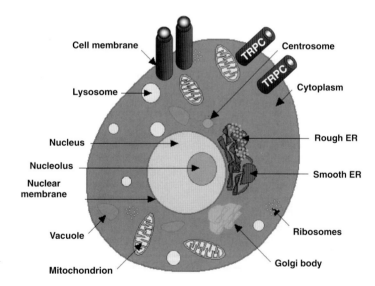

(4) Transcription (cytosol/nucleus) Ca^{2+} generated in response to mGluR stimulation activates transcription factors which translocate to the nucleus initiating gene transcription. Gene transcription can be further regulated by nuclear IP$_3$Rs.	(5) Translation (cytosol/ER/golgi/ secretory vesicles) After transcription, mRNA is translated into proteins which are trafficked through the ER, golgi, and secretory vesicles, all of which can contain and are regulated by IP$_3$R.	(6) Excitotoxicity (mitochondria /ER/ cytosol) If the glutamate levels remain high, excitotoxicity occurs inducing apoptosis via the IP$_3$R overloading the mitochondria with Ca^{2+} This causes the release of cytochrome C from the mitochondrion which locks the ER resident IP$_3$Rs in an open conformation which poisons the mitochondria with Ca^{2+}, depletes ER Ca^{2+}, and creates high cytosolic Ca^{2+} which increases apoptotic protein activity.

Fig. 4.6.3 IP$_3$R activities within the mGluR signalling pathway. The IP$_3$R is well established to release Ca^{2+} from ER stores during metabotropic glutamate receptor activation. This figure demonstrates the many other areas of the cell where IP$_3$R activity can impinge upon this signalling network. See also colour plate section.

With the IP_3R impinging upon this signalling pathway at so many places, it is easy to imagine that 23,001 isoforms of the IP_3R would be necessary to provide the specificity of function we know exists for this pathway. We hypothesize that through this gradient of receptors a multitude of cellular processes can be specifically regulated by Ca^{2+} transients derived from IP_3R signalling cascades.

In conclusion, we would like to impress that the scope of function and functional interactions of the IP_3R are only now starting to be uncovered. It is indeed possible that the predicted 23,001 isoforms of the IP_3R all have important roles in cellular signalling pathways, many of which may not be related to the channel function of this dynamic receptor.

References

Ammendola A, Geiselhoringer A, Hofmann F et al. (2001). Molecular determinants of the interaction between the inositol 1,4,5-trisphosphate receptor-associated cGMP kinase substrate (IRAG) and cGMP kinase Ibeta. *J Biol Chem* **276**, 24153–24159.

Ando H, Mizutani A, Kiefer H et al. (2006). IRBIT suppresses IP3 receptor activity by competing with IP3 for the common binding site on the IP3 receptor. *Mol Cell* **22**, 795–806.

Ando H, Mizutani A, Matsu-ura T et al. (2003). IRBIT, a novel inositol 1,4,5-trisphosphate (IP3) receptor-binding protein, is released from the IP3 receptor upon IP3 binding to the receptor. *J Biol Chem* **278**, 10602–10612.

Bagni C, Mannucci L, Dotti CG and Amaldi F (2000). Chemical stimulation of synaptosomes modulates alpha-Ca^{2+}/calmodulin-dependent protein kinase II mRNA association to polysomes. *J Neurosci* **20**, RC76.

Belmeguenai A, Leprince J, Tonon MC et al. (2002). Neurotensin modulates the amplitude and frequency of voltage-activated Ca^{2+} currents in frog pituitary melanotrophs: implication of the inositol triphosphate/protein kinase C pathway. *Eur J Neurosci* **16**, 1907–1916.

Berridge MJ, Lipp P and Bootman MD (2000). The versatility and universality of calcium signalling. *Nat Rev Mol Cell Biol* **1**, 11–21.

Bezprozvanny I and Ehrlich BE (1995). The inositol 1,4,5-trisphosphate (InsP3) receptor. *J Membr Biol* **145**, 205–216.

Bezprozvanny I and Hayden MR (2004). Deranged neuronal calcium signaling and Huntington disease. *Biochem Biophys Res Commun* **322**, 1310–1317.

Bezprozvanny I, Watras J and Ehrlich BE (1991). Bell-shaped calcium-response curves of Ins(1,4,5)P3- and calcium-gated channels from endoplasmic reticulum of cerebellum. *Nature* **351**, 751–754.

Bezprozvanny IB, Ondrias K, Kaftan E et al. (1993). Activation of the calcium release channel (ryanodine receptor) by heparin and other polyanions is calcium dependent. *Mol Biol Cell* **4**, 347–352.

Boehning D, Mak DO, Foskett JK et al. (2001b). Molecular determinants of ion permeation and selectivity in inositol 1,4,5-trisphosphate receptor Ca^{2+} channels. *J Biol Chem* **276**, 13509–13512.

Boehning D, Patterson RL, Sedaghat L et al. (2003). Cytochrome c binds to inositol (1,4,5) trisphosphate receptors, amplifying calcium-dependent apoptosis. *Nat Cell Biol* **5(12)**, 1051–1061.

Boehning D, van Rossum DB, Patterson RL et al. (2005). A peptide inhibitor of cytochrome c/inositol 1,4,5-trisphosphate receptor binding blocks intrinsic and extrinsic cell death pathways. *PNAS* **102**, 1466–1471.

Boehning D, Joseph SK, Mak DO et al. (2001a). Single-channel recordings of recombinant inositol trisphosphate receptors in mammalian nuclear envelope. *Biophys J* **81**, 117–124.

Boehning D and Joseph SK (2000a). Direct association of ligand-binding and pore domains in homo- and heterotetrameric inositol 1,4,5-trisphosphate receptors. *EMBO J* **19**, 5450–5459.

Boehning D and Joseph SK (2000b). Functional properties of recombinant type I and type III inositol 1, 4,5-trisphosphate receptor isoforms expressed in COS-7 cells. *J Biol Chem* **275**, 21492–21499.

Bosanac I, Alattia JR, Mal TK et al. (2002). Structure of the inositol 1,4,5-trisphosphate receptor binding core in complex with its ligand. *Nature* **420**, 696–700.

Bosanac I, Michikawa T, Mikoshiba K et al. (2004). Structural insights into the regulatory mechanism of IP3 receptor. *Biochim Biophys Acta* **1742**, 89–102.

Bosanac I, Yamazaki H, Matsu-ura T et al. (2005). Crystal structure of the ligand binding suppressor domain of type 1 inositol 1,4,5-trisphosphate receptor. *Mol Cell* **17**, 193–203.

Boulay G, Brown DM, Qin N et al. (1999). Modulation of Ca^{2+} entry by polypeptides of the inositol 1,4,5-trisphosphate receptor (IP_3R) that bind transient receptor potential (TRP): evidence for roles of TRP and IP_3R in store depletion-activated Ca^{2+} entry. *PNAS* **96**, 14955–14960.

Bourguignon LY and Jin H (1995). Identification of the ankyrin-binding domain of the mouse T-lymphoma cell inositol 1,4,5-trisphosphate (IP3) receptor and its role in the regulation of IP3-mediated internal Ca^{2+} release. *J Biol Chem* **270**, 7257–7260.

Bourguignon LY, Jin H, Iida N et al. (1993). The involvement of ankyrin in the regulation of inositol 1,4,5-trisphosphate receptor-mediated internal Ca^{2+} release from Ca^{2+} storage vesicles in mouse T-lymphoma cells. *J Biol Chem* **268**, 7290–7297.

Broad LM, Cannon TR and Taylor CW (1999). A non-capacitative pathway activated by arachidonic acid is the major Ca^{2+} entry mechanism in rat A7r5 smooth muscle cells stimulated with low concentrations of vasopressin. *J Physiol* **517**, 121–134.

Bruce JI, Shuttleworth TJ, Giovannucci DR et al. (2002). Phosphorylation of inositol 1,4,5-trisphosphate receptors in parotid acinar cells. A mechanism for the synergistic effects of cAMP on Ca^{2+} signaling. *J Biol Chem* **277**, 1340–1348.

Bultynck G, Sienaert I, Parys JB et al. (2003). Pharmacology of inositol trisphosphate receptors. *Pflugers Arch* **445**, 629–642.

Chen R, Valencia I, Zhong F et al. (2004). Bcl-2 functionally interacts with inositol 1,4,5-trisphosphate receptors to regulate calcium release from the ER in response to inositol 1,4,5-trisphosphate. *J Cell Biol* **166**, 193–203.

Choe CU, Harrison KD, Grant W et al. (2004). Functional coupling of chromogranin with the inositol 1,4,5-trisphosphate receptor shapes calcium signalling. *J Biol Chem* **279**, 35551–35556.

Colosetti P, Tunwell RE, Cruttwell C et al. (2003). The type 3 inositol 1,4,5-trisphosphate receptor is concentrated at the tight junction level in polarized MDCK cells. *J Cell Sci* **116**, 2791–2803.

Csordas G, Thomas AP and Hajnoczky G (1999). Quasi-synaptic calcium signal transmission between endoplasmic reticulum and mitochondria. *EMBO J* **18**, 96–108.

Danoff SK, Supattapone S and Snyder SH (1988). Characterization of a membrane protein from brain mediating the inhibition of inositol 1,4,5-trisphosphate receptor binding by calcium. *Biochem J* **254**, 701–705.

Dasso LL and Taylor CW (1991). Heparin and other polyanions uncouple alpha 1-adrenoceptors from G-proteins. *Biochem J* **280**(Pt 3), 791–795.

De SP, Parys JB, Callewaert G et al. (1999). Xestospongin C is an equally potent inhibitor of the inositol 1,4,5-trisphosphate receptor and the endoplasmic-reticulum $Ca^{(2+)}$ pumps. *Cell Calcium* **26**, 9–13.

Dellis O, Dedos SG, Tovey SC et al. (2006). Ca^{2+} entry through plasma membrane IP3 receptors. *Science* **313**, 229–233.

DeSouza N, Reiken S, Ondrias K et al. (2002). Protein kinase A and two phosphatases are components of the inositol 1,4,5-trisphosphate receptor macromolecular signaling complex. *J Biol Chem* **277(42)**, 39397–400.

Distelhorst CW and Shore GC (2004). Bcl-2 and calcium: controversy beneath the surface. *Oncogene* **23**, 2875–2880.

Du GG, Imredy JP and MacLennan DH (1998). Characterization of recombinant rabbit cardiac and skeletal muscle Ca^{2+} release channels (ryanodine receptors) with a novel [^{3}H]ryanodine binding assay. *J Biol Chem* **273**, 33259–33266.

Echevarria W, Leite MF, Guerra MT et al. (2003). Regulation of calcium signals in the nucleus by a nucleoplasmic reticulum. *Nat Cell Biol* **5**, 440–446.

Ferris CD, Cameron AM, Bredt DS et al. (1991a). Inositol 1,4,5-trisphosphate receptor is phosphorylated by cyclic AMP-dependent protein kinase at serines 1755 and 1589. *Biochem Biophym Res Commun* **175**, 192–198.

Ferris CD, Cameron AM, Bredt DS et al. (1992). Autophosphorylation of inositol 1,4,5-trisphosphate receptors. *J Biol Chem* **267**, 7036–7041.

Ferris CD, Huganir RL and Snyder SH (1990). Calcium flux mediated by purified inositol 1,4,5-trisphosphate receptor in reconstituted lipid vesicles is allosterically regulated by adenine nucleotides. *PNAS* **87**, 2147–2151.

Ferris CD, Huganir RL, Bredt DS et al. (1991b). Inositol trisphosphate receptor: phosphorylation by protein kinase C and calcium calmodulin-dependent protein kinases in reconstituted lipid vesicles. *PNAS* **88**, 2232–2235.

Ferris CD, Huganir RL, Supattapone S et al. (1989). Purified inositol 1,4,5-trisphosphate receptor mediates calcium flux in reconstituted lipid vesicles. *Nature* **342**, 87–89.

Fritsch RM, Saur D, Kurjak M et al. (2004). InsP3R-associated cGMP kinase substrate (IRAG) is essential for nitric oxide-induced inhibition of calcium signaling in human colonic smooth muscle. *J Biol Chem* **279**, 12551–12559.

Furuichi T, Yoshikawa S, Miyawaki A et al. (1989). Primary structure and functional expression of the inositol 1,4,5-trisphosphate-binding protein P400. *Nature* **342**, 32–38.

Gafni J, Munsch JA, Lam TH et al. (1997). Xestospongins: potent membrane permeable blockers of the inositol 1,4,5-trisphosphate receptor. *Neuron* **19**, 723–733.

Gregory RB, Rychkov G and Barritt GJ (2001). Evidence that 2-aminoethyl diphenylborate is a novel inhibitor of store-operated Ca^{2+} channels in liver cells and acts through a mechanism which does not involve inositol trisphosphate receptors. *Biochem J* **354**, 285–290.

Hagar RE, Burgstahler AD, Nathanson MH et al. (1998). Type III InsP3 receptor channel stays open in the presence of increased calcium. *Nature* **396**, 81–84.

Hamada K, Miyata T, Mayanagi K et al. (2002). Two-state conformational changes in inositol 1,4,5-trisphosphate receptor regulated by calcium. *J Biol Chem* **277**, 21115–21118.

Haug LS, Jensen V, Hvalby O et al. (1999). Phosphorylation of the inositol 1,4,5-trisphosphate receptor by cyclic nucleotide-dependent kinases *in vitro* and in rat cerebellar slices *in situ*. *J Biol Chem* **274**, 7467–7473.

Haynes LP, Tepikin AV and Burgoyne RD (2003). Calcium binding protein 1 is an inhibitor of agonist-evoked, inositol 1,4,5-trisphophate-mediated calcium signalling. *J Biol Chem* **279(1)**, 547–55.

He X, Yang F, Xie Z et al. (2000). Intracellular Ca(2+) and Ca(2+)/calmodulin-dependent kinase II mediate acute potentiation of neurotransmitter release by neurotrophin-3. *J Cell Biol* **149**, 783–792.

Higo T, Hattori, M Nakamura et al. (2005). Subtype-specific and ER lumenal environment-dependent regulation of inositol 1,4,5-trisphosphate receptor type 1 by ERp44. *Cell* **120**, 85–98.

Hirota J and H, Hamada K et al. (2003). Carbonic anhydrase-related protein is a novel binding protein for inositol 1,4,5-trisphosphate receptor type 1. *Biochem J* **372**, 435–441.

Hirota J, Michikawa T, Natsume T et al. (1999). Calmodulin inhibits inositol 1,4,5-trisphosphate-induced calcium release through the purified and reconstituted inositol 1,4,5- trisphosphate receptor type 1. *FEBS Lett* **456**, 322–326.

Huh YH and Yoo SH (2003). Presence of the inositol 1,4,5-triphosphate receptor isoforms in the nucleoplasm. *FEBS Lett* **555**, 411–418.

Huh YH, Yoo JA, Bahk SJ et al. (2005). Distribution profile of inositol 1,4,5-trisphosphate receptor isoforms in adrenal chromaffin cells. *FEBS Lett* **579**, 2597–2603.

Ionescu L, Cheung KH, Vais H et al. (2006). Graded recruitment and inactivation of single InsP3 receptor Ca^{2+}-release channels: implications for quantal Ca2+ release. *J Physiol* **573**, 645–662.

Jayaraman T, Ondrias K, Ondriasova E et al. (1996). Regulation of the inositol 1,4,5-trisphosphate receptor by tyrosine phosphorylation. *Science* **272**, 1492–1494.

Joseph SK and Rice HL (1989). The relationship between inositol trisphosphate receptor density and calcium release in brain microsomes. *Mol Pharmacol* **35**, 355–359.

Joseph SK, Lin C, Pierson S et al. (1995). Heteroligomers of type-I and type-III inositol trisphosphate receptors in WB rat liver epithelial cells. *J BioL Chem* **270**, 23310–23316.

Kaplin AI, Snyder SH and Linden DJ (1996). Reduced nicotinamide adenine dinucleotide-selective stimulation of inositol 1,4,5-trisphosphate receptors mediates hypoxic mobilization of calcium. *J Neurosci* **16**, 2002–2011.

Kasri NN, Holmes AM, Bultynck G et al. (2004). Regulation of InsP(3) receptor activity by neuronal Ca(2+)-binding proteins. *EMBO J* **23**, 312–321.

Kasri NN, Torok K, Galione A et al. (2006). Endogenously bound calmodulin is essential for the function of the inositol 1,4,5-trisphosphate receptor. *J Biol Chem* **281**, 8332–8338.

Khan AA, Steiner JP and Snyder SH (1992). Plasma membrane inositol 1,4,5-trisphosphate receptor of lymphocytes: selective enrichment in sialic acid and unique binding specificity. *PNAS* **89**, 2849–2853.

Khodakhah K and Ogden D (1993). Functional heterogeneity of calcium release by inositol trisphosphate in single Purkinje neurones, cultured cerebellar astrocytes and peripheral tissues. *PNAS* **90**, 4976–4980.

Kiselyov KI, Mignery GA, Zhu MX et al. (1999). The N-terminal domain of the IP$_3$ receptor gates store-operated hTrp3 channels. *Mol Cell* **4**, 423–429.

Komalavilas P and Lincoln TM (1994). Phosphorylation of the inositol 1,4,5-trisphosphate receptor by cyclic GMP-dependent protein kinase. *J Biol Chem* **269**, 8701–8707.

Koulen P, Cai Y, Geng L et al. (2002). Polycystin-2 is an intracellular calcium release channel. *Nat Cell Biol* **4**, 191–197.

Kovalovsky D, Refojo D, Liberman AC et al. (2002). Activation and induction of NUR77/NURR1 in corticotrophs by CRH/cAMP: involvement of calcium, protein kinase A and MAPK pathways. *Mol Endocrinol* **16**, 1638–1651.

Lin P, Yao Y, Hofmeister R et al. (1999). Overexpression of CALNUC (nucleobindin) increases agonist and thapsigargin releasable Ca^{2+} storage in the golgi. *J Cell Biol* **145**, 279–289.

Luo D, Broad LM, Bird GS et al. (2001). Mutual antagonism of calcium entry by capacitative and arachidonic acid-mediated calcium entry pathways. *J Biol Chem* **276**, 20186–20189.

Ma H-T, Patterson RL, van Rossum DB et al. (2000). Requirement of the inositol trisphosphate receptor for activation of store-operated Ca^{2+} channels. *Science* **287**, 1647–1651.

Ma HT, Venkatachalam K, Parys JB et al. (2002). Modification of store-operated channel coupling and inositol trisphosphate receptor function by 2-aminoethoxydiphenyl borate in DT40 lymphocytes. *J Biol Chem* **277**, 6915–6922.

Mak DO, McBride S and Foskett JK (1999). ATP regulation of type 1 inositol 1,4,5-trisphosphate receptor channel gating by allosteric tuning of Ca(2+) activation. *J Biol Chem* **274**, 22231–22237.

Mak DO, McBride S, Raghuram V et al. (2000). Single-channel properties in endoplasmic reticulum membrane of recombinant type 3 inositol trisphosphate receptor. *J Gen Physiol* **115**, 241–256.

Marchant JS, Beecroft MD, Riley AM *et al.* (1997). Disaccharide polyphosphates based upon adenophostin A activate hepatic D-*myo*-inositol 1,4,5-trisphosphate receptors. *Biochem* **36**, 12780–12790.

Maruyama T, Cui Z, Kanaji T *et al.* (1997a). Attenuation of intracellular Ca^{2+} and secretory responses by $Ins(1,4,5)P_3$-induced Ca^{2+} release modulator, 2APB, in rat pancreatic acinar cells. *Biomed Res* **18**, 297–302.

Maruyama T, Kanaji T, Nakade S *et al.* (1997b). 2APB, 2-aminoethoxydiphenyl borate, a membrane-penetrable modulator of $Ins(1,4,5)P3$-induced Ca^{2+} release. *J Biochem* **122**, 498–505.

Matsumoto M and Nagata E (1999). Type 1 inositol 1,4,5-trisphosphate receptor knock-out mice: their phenotypes and their meaning in neuroscience and clinical practice. *J Mol Med* **77**, 406–411.

Matsumoto M, Nakagawa T, Inoue T *et al.* (1996). Ataxia and epileptic seizures in mice lacking type 1 inositol 1,4,5-trisphosphate receptor. *Nature* **379**, 168–171.

Matter N, Ritz MF, Freyermuth S *et al.* (1993). Stimulation of nuclear protein kinase C leads to phosphorylation of nuclear inositol 1,4,5-trisphosphate receptor and accelerated calcium release by inositol 1,4,5-trisphosphate from isolated rat liver nuclei. *J Biol Chem* **268**, 732–736.

Maximov A, Tang TS and Bezprozvanny I (2003). Association of the type 1 inositol (1,4,5)-trisphosphate receptor with 4.1N protein in neurons. *Mol Cell Neurosci* **22**, 271–283.

Michikawa T, Hirota J, Kawano S *et al.* (1999). Calmodulin mediates calcium-dependent inactivation of the cerebellar type 1 inositol 1,4,5-trisphosphate receptor. *Neuron* **23**, 799–808.

Mishra OP, Qayyum I and ivoria-Papadopoulos M (2003). Hypoxia-induced modification of the inositol triphosphate receptor in neuronal nuclei of newborn piglets: role of nitric oxide. *J Neurosci Res* **74**, 333–338.

Missiaen L, Parys JB, Sienaert I *et al.* (1998). Functional properties of the type-3 $InsP_3$ receptor in 16HBE14o bronchial mucosal cells. *J Biol Chem* **273**, 8983–8986.

Miyata M, Finch EA, Khiroug L *et al.* (2000). Local calcium release in dendritic spines required for long-term synaptic depression. *Neuron* **28**, 233–244.

Mohler PJ, Schott JJ, Gramolini AO *et al.* (2003). Ankyrin-B mutation causes type 4 long-QT cardiac arrhythmia and sudden cardiac death. *Nature* **421**, 634–639.

Molgo J, Colasantei C, Adams DS *et al.* (2004). IP3 receptors and Ca^{2+} signals in adult skeletal muscle satellite cells *in situ*. *Biol Res* **37**, 635–639.

Monkawa T, Miyawaki A, Sugiyama T *et al.* (1995). Heterotetrameric complex formation of inositol 1,4,5-trisphosphate receptor subunits. *J Biol Chem* **270**, 14700–14704.

Murphy CT, Riley AM, Lindley CJ *et al.* (1997). Structural analogues of D-*myo*-inositol-1,4,5-triphosphate and adenophostin A: recognition by cerebellar and platelet inositol-1,4,5-triphosphate receptors. *Mol Pharmacol* **52**, 741.

Nakagawa T, Okano H, Furuichi T *et al.* (1991). The subtypes of the mouse inositol 1,4,5-trisphosphate receptor are expressed in a tissue-specific and developmentally specific manner. *PNAS* **88**, 6244–6248.

Nosyreva E, Miyakawa T, Wang Z *et al.* (2002). The high-affinity calcium[bond]calmodulin-binding site does not play a role in the modulation of type 1 inositol 1,4,5-trisphosphate receptor function by calcium and calmodulin. *Biochem J* **365**, 659–667.

Onoue H, Tanaka H, Tanaka K *et al.* (2000). Heterooligomer of type 1 and type 2 inositol 1,4,5-trisphosphate receptor expressed in rat liver membrane fraction exists as tetrameric complex. *Biochem Biophys Res Commun* **267**, 928–933.

Otani S, Daniel H, Takita M *et al.* (2002). Long-term depression induced by postsynaptic group II metabotropic glutamate receptors linked to phospholipase C and intracellular calcium rises in rat prefrontal cortex. *J Neurosci* **22**, 3434–3444.

Otsu H, Yamamoto A, Maeda N *et al.* (1990). Immunogold localization of inositol 1,4,5-trisphosphate (InsP3) receptor in mouse cerebellar Purkinje cells using three monoclonal antibodies. *Cell Struct Funct* **15**, 163–173.

Patel S, Joseph SK and Thomas AP (1999a). Molecular properties of inositol 1,4,5-trisphosphate receptors. *Cell Calcium* **25**, 247–264.

Patel S, Morris SA, Adkins CE *et al.* (1997). Ca^{2+}-independent inhibition of inositol trisphosphate receptors by calmodulin: redistribution of calmodulin as a possible means of regulating Ca^{2+} mobilization. *PNAS* **94**, 11627–11632.

Patel S, Robb-Gaspers LD, Stellato KA *et al.* (1999b). Coordination of calcium signalling by endothelial-derived nitric oxide in the intact liver. *Nat Cell Biol* **1**, 467–471.

Patterson RL, Boehning D and Snyder SH (2004a). Inositol 1,4,5-trisphosphate receptors as signal integrators. *Annu Rev Biochem* **73**, 437–465.

Patterson RL, van Rossum DB, Barrow RK *et al.* (2004b). RACK1 binds to inositol 1,4,5-trisphosphate receptors and mediates Ca^{2+} release. *PNAS* **101**, 2328–2332.

Patterson RL, van Rossum DB, Kaplin AI *et al.* (2005). Inositol 1,4,5-trisphosphate receptor/GAPDH complex augments Ca^{2+} release via locally derived NADH. *PNAS* **102**, 1357–1359.

Perez PJ, Ramos-Franco J, Fill M *et al.* (1997). Identification and functional reconstitution of the type 2 inositol 1,4,5-trisphosphate receptor from ventricular cardiac myocytes. *J Biol Chem* **272**, 23961–23969.

Pieper, AA, Brat DJ, O'Hearn E *et al.* (2001). Differential neuronal localizations and dynamics of phosphorylated and unphosphorylated type 1 inositol 1,4,5-trisphosphate receptors. *Neurosci* **102**, 433–444.

Ramos-Franco J, Fill M and Mignery GA (1998). Isoform-specific function of single inositol 1,4,5-trisphosphate receptor channels. *Biophys J* **75**, 834–839.

Ramsey IS, Delling M and Clapham DE (2005). An introduction to TRP channels. *Annu Rev Physiol* **68**, 619–47.

Regan MR, Lin DD, Emerick MC *et al.* (2005). The effect of higher order RNA processes on changing patterns of protein domain selection: a developmentally regulated transcriptome of type 1 inositol 1,4,5-trisphosphate receptors. *Proteins* **59**, 312–331.

Rizzuto R, Brini M, Murgia M *et al.* (1993). Microdomains with high Ca^{2+} close to IP3-sensitive channels that are sensed by neighboring mitochondria. *Science* **262**, 744–747.

Rooney TA, Joseph SK, Queen C *et al.* (1996). Cyclic GMP induces oscillatory calcium signals in rat hepatocytes. *J Biol Chem* **271**, 19817–19825.

Rowles SJ and Gallacher DV (1996). Ins(1, 3, 4, 5)P4 is effective in mobilizing Ca^{2+} in mouse exocrine pancreatic acinar cells if phospholipase A2 is inhibited. *Biochem J* **319**(Pt 3), 913–918.

Runnels LW, Yue L and Clapham DE (2002). The TRPM7 channel is inactivated by PIP(2) hydrolysis. *Nat Cell Biol* **4**, 329–336.

Sato, C, Hamada K, Ogura T *et al.* (2004). Inositol 1,4,5-trisphosphate receptor contains multiple cavities and L-shaped ligand-binding domains. *J Mol Biol* **336**, 155–164.

Schlossmann J, Ammendola A, Ashman K *et al.* (2000). Regulation of intracellular calcium by a signalling complex of IRAG, IP3 receptor and cGMP kinase Ibeta. *Nature* **404**, 197–201.

Serysheva II, Bare DJ, Ludtke SJ *et al.* (2003). Structure of the type 1 inositol 1,4,5-trisphosphate receptor revealed by electron cryomicroscopy. *J Biol Chem* **278**, 21319–21322.

Sharp AH, Nucifora FC Jr, Blondel O *et al.* (1999). Differential cellular expression of isoforms of inositol 1,4,5-triphosphate receptors in neurons and glia in brain. *J Comp Neurol* **406**, 207–220.

Sharp AH, Snyder SH and Nigam SK (1992). Inositol 1,4,5-trisphosphate receptors localization in epithelial tissue. *J Biol Chem* **267**, 7444–7449.

Shin DM, Zhao XS, Zeng W *et al.* (2000). The mammalian Sec6/8 complex interacts with Ca(2+) signaling complexes and regulates their activity. *J Cell Biol* **150**, 1101–1112.

Shirakabe K, Priori G, Yamada H *et al.* (2006). IRBIT, an inositol 1,4,5-trisphosphate receptor-binding protein, specifically binds to and activates pancreas-type Na$^+$/HCO3$^-$ cotransporter 1 (pNBC1). *PNAS* **103**, 9542–9547.

Sienaert I, Missiaen L, De Smedt H *et al.* (1997). Molecular and functional evidence for multiple Ca^{2+}-binding domains in the type 1 inositol 1,4,5, trisphosphate receptor. *J Biol Chem* **272**, 25899–25906.

Spassova MA, Soboloff J, He LP *et al.* (2004). Calcium entry mediated by SOCs and TRP channels: variations and enigma. *Biochim Biophys Acta* **1742**, 9–20.

Srivastava M, Kumar P, Leighton X *et al.* (2002). Influence of the Anx7 (+/−) knockout mutation and fasting stress on the genomics of the mouse adrenal gland. *Ann N Y Acad Sci* **971**, 53–60.

Stamboulian S, Moutin MJ, Treves S *et al.* (2005). Junctate, an inositol 1,4,5-triphosphate receptor associated protein, is present in rodent sperm and binds TRPC2 and TRPC5 but not TRPC1 channels. *Dev Biol* **286**, 326–337.

Supattapone S, Worley PF, Baraban JM *et al.* (1988). Solubilization, purification and characterization of an inositol trisphosphate receptor. *J Biol Chem* **263**, 1530–1534.

Szalai G, Krishnamurthy R and Hajnoczky G (1999). Apoptosis driven by IP(3)-linked mitochondrial calcium signals. *EMBO J* **18**, 6349–6361.

Tang TS, Tu H, Chan EY *et al.* (2003a). *Huntingtin* and huntingtin-associated protein 1 influence neuronal calcium signaling mediated by inositol-(1,4,5) triphosphate receptor type 1. *Neuron* **39**, 227–239.

Tang TS, Tu H, Orban PC *et al.* (2004). HAP1 facilitates effects of mutant huntingtin on inositol 1,4,5-trisphosphate-induced Ca release in primary culture of striatal medium spiny neurons. *Eur J Neurosci* **20**, 1779–1787.

Tang TS, Tu H, Wang Z *et al.* (2003b). Modulation of type 1 inositol (1,4,5)-trisphosphate receptor function by protein kinase a and protein phosphatase 1alpha. *J Neurosci* **23**, 403–415.

Taylor CW, da Fonseca PC and Morris EP (2004). IP(3) receptors: the search for structure. *Trends Biochem Sci* **29**, 210–219.

Thomas-Virnig CL, Sims PA *et al.* (2004). The inositol 1,4,5-trisphosphate receptor regulates epidermal cell migration in *Caenorhabditis elegans*. *Curr Biol* **14**, 1882–1887.

Thomenius MJ and Distelhorst CW (2003). Bcl-2 on the endoplasmic reticulum: protecting the mitochondria from a distance. *J Cell Sci* **116**, 4493–4499.

Thomenius MJ, Wang NS, Reineks EZ *et al.* (2003). Bcl-2 on the endoplasmic reticulum regulates Bax activity by binding to BH3-only proteins. *J Biol Chem* **278**, 6243–6250.

Thrower EC, Park HY, So SH *et al.* (2002). Activation of the inositol 1,4,5-trisphosphate receptor by the calcium storage protein chromogranin A. *J Biol Chem* **277**, 15801–15806.

Treves S, Feriotto G, Moccagatta L *et al.* (2000). Molecular cloning, expression, functional characterization, chromosomal localization and gene structure of junctate, a novel integral calcium binding protein of sarco(endo) plasmic reticulum membrane. *J Biol Chem* **275**, 39555–39568.

Treves S, Franzini-Armstrong C, Moccagatta L *et al.* (2004). Junctate is a key element in calcium entry induced by activation of InsP3 receptors and/or calcium store depletion. *J Cell Biol* **166**, 537–548.

Tu H, Tang TS, Wang Z *et al.* (2004). Association of type 1 inositol 1,4,5-trisphosphate receptor with AKAP9 (Yotiao) and protein kinase A. *J Biol Chem* **279**, 19375–19382.

Tu JC, Xiao B, Yuan JP *et al.* (1998). Homer binds a novel proline-rich motif and links group 1 metabotropic glutamate receptors with IP3 receptors. *Neuron* **21**, 717–726.

Tuvia S, Buhusi M, Davis L *et al.* (1999). Ankyrin-B is required for intracellular sorting of structurally diverse Ca^{2+} homeostasis proteins. *J Cell Biol* **147**, 995–1008.

van Rossum DB, Patterson RL, Kiselyov K *et al.* (2004). Agonist-induced Ca^{2+} entry determined by inositol 1,4,5-trisphosphate recognition. *PNAS* **101**, 2323–2327.

Venkatesh K, Siddhartha G, Joshi R *et al.* (2001). Interactions between the inositol 1,4,5-trisphosphate and cyclic AMP signaling pathways regulate larval molting in *Drosophila*. *Genetics* **158**, 309–318.

Vermassen E, Parys JB and Mauger JP (2004). Subcellular distribution of the inositol 1,4,5-trisphosphate receptors: functional relevance and molecular determinants. *Biol Cell* **96**, 3–17.

Walker DS, Ly S, Lockwood KC *et al.* (2002). A direct interaction between IP(3) receptors and myosin II regulates IP(3) signaling in *C. elegans*. *Curr Biol* **12**, 951–956.

Walker JW, Feeney J and Trentham DR (1989). Photolabile precursors of inositol phosphates preparation and properties of 1-(2-nitrophenyl) ethyl esters of myo-inositol 1,4,5-trisphosphate. *Biochem* **28**, 3272–3280.

Walker JW, Somlyo AV, Goldman YE *et al.* (1987). Kinetics of smooth and skeletal muscle activation by laser pulse photolysis of caged inositol 1,4,5-trisphosphate. *Nature* **327**, 249–252.

White C, Li C, Yang J, Petrenko NB *et al.* (2005). The endoplasmic reticulum gateway to apoptosis by Bcl-X(L) modulation of the InsP3R. *Nat Cell Biol* **7**, 1021–1028.

Wilcox RA, Challiss RA, Traynor JR *et al.* (1994). Molecular recognition at the myo-inositol 1,4,5-trisphosphate receptoR 3-position substituted myo-inositol 1,4,5-trisphosphate analogues reveal the binding and Ca^{2+} release requirements for high affinity interaction with the myo-inositol 1,4,5-trisphosphate receptor. *J Biol Chem* **269**, 26815–26821.

Wilcox RA, Primrose WU, Nahorski SR *et al.* (1998). New developments in the molecular pharmacology of the myo-inositol 1,4,5-trisphosphate receptor. *Trends Pharmacol Sci* **19**, 467–475.

Willcocks AL, Strupish J, Irvine RF *et al.* (1989). Inositol 1:2-cyclic, 4,5-trisphosphate is only a weak agonist at inositol 1,4,5-trisphosphate receptors. *Biochem J* **257**, 297–300.

Wojcikiewicz RJ and Oberdorf JA (1996). Degradation of inositol 1,4,5-trisphosphate receptors during cell stimulation is a specific process mediated by cysteine protease activity. *J Biol Chem* **271**, 16652–16655.

Wojcikiewicz RJH and Luo SG (1998). Phosphorylation of inositol 1,4,5-trisphosphate receptors by cAMP-dependent protein kinase—Type I, II and III receptors are differentially susceptible to phosphorylation and are phosphorylated in intact cells. *J Biol Chem* **273**, 5670–5677.

Worley PF, Baraban JM and Snyder SH (1989). Inositol 1,4,5-trisphosphate receptor binding: autoradiographic localization in rat brain. *J Neurosci* **9**, 339–346.

Worley PF, Baraban JM, Colvin JS *et al.* (1987). Inositol trisphosphate receptor localization in brain: variable stoichiometry with protein kinase C. *Nature* **325**, 159–161.

Yamada M, Miyawaki A, Saito K *et al.* (1995). The calmodulin-binding domain in the mouse type 1 inositol 1,4,5-trisphosphate receptor. *Biochem J* **308** (**Pt 1**), 83–88.

Yang J, McBride S, Mak DO *et al.* (2002). Identification of a family of calcium sensors as protein ligands of inositol trisphosphate receptor Ca(2+) release channels. *PNAS* **99**, 7711–7716.

Yokoyama K, Su Ih IH, Tezuka T *et al.* (2002). BANK regulates BCR-induced calcium mobilization by promoting tyrosine phosphorylation of IP(3) receptor. *EMBO J* **21**, 83–92.

Yoo SH and Jeon CJ (2000). Inositol 1,4,5-trisphosphate receptor/Ca^{2+} channel modulatory role of chromogranin A, a Ca^{2+} storage protein of secretory granules. *J Biol Chem* **275**, 15067–15073.

Yoo SH and Lewis MS (2000). Interaction of chromogranin B and the near N-terminal region of chromogranin B with an intraluminal loop peptide of the inositol 1,4,5-trisphosphate receptor. *J Biol Chem* **275**, 30293–30300.

Yoo SH, Kang MK, Kwon HS *et al.* (2000). Inositol 1,4,5-trisphosphate receptor and chromogranins A and B in secretory granules co-localization and functional coupling. *Adv Exp Med Biol* **482**, 83–94.

Yoshikawa F, Iwasaki H, Michikawa T *et al.* (1999a). Trypsinized cerebellar inositol 1,4,5-trisphosphate receptor structural and functional coupling of cleaved ligand binding and channel domains. *J Biol Chem* **274**, 316–327.

Yoshikawa F, Iwasaki H, Michikawa T *et al.* (1999b). Cooperative formation of the ligand-binding site of the inositol 1, 4, 5-trisphosphate receptor by two separable domains. *J Biol Chem* **274**, 328–334.

Yoshikawa F, Morita M, Monkawa T *et al.* (1996). Mutational analysis of the ligand binding site of the inositol 1,4,5-trisphosphate receptor. *J Biol Chem* **271**, 18277–18284.

Yuan JP, Kiselyov KI, Shin DM *et al.* (2003). Homer binds TRPC family channels and is required for gating of TRPC1 by IP_3-receptors. *Cell* **114**, 777–789.

Zeng W, Mak DO, Li Q *et al.* (2003). A New Mode of Ca(2+) Signaling by G protein-coupled receptors gating of IP(3) receptor Ca(2+) release channels by gbetagamma. *Curr Biol* **13**, 872–876.

Zhang S, Mizutani A, Hisatsune C *et al.* (2003). Protein 4.1N is required for translocation of inositol 1,4,5-trisphosphate receptor type 1 to the basolateral membrane domain in polarized Madin–Darby canine kidney cells. *J Biol Chem* **278**, 4048–4056.

Zhang X and Joseph SK (2001). Effect of mutation of a calmodulin binding site on Ca^{2+} regulation of inositol trisphosphate receptors. *Biochem J* **360**, 395–400.

Zhang Z, Tang J, Tikunova S *et al.* (2001). Activation of Trp3 by inositol 1,4,5-trisphosphate receptors through displacement of inhibitory calmodulin from a common binding domain. *PNAS* **98**, 3168–3173.

Zhong F, Davis MC, McColl KS *et al.* (2006). Bcl-2 differentially regulates Ca^{2+} signals according to the strength of T cell receptor activation. *J Cell Biol* **172**, 127–137.

Zhu DM, Fang WH, Narla RK *et al.* (1999). A requirement for protein kinase C inhibition for calcium-triggered apoptosis in acute lymphoblastic leukemia cells. *Clin Cancer Res* **5**, 355–360.

Zhu DM, Tekle E, Chock PB *et al.* (1996). Reversible phosphorylation as a controlling factor for sustaining calcium oscillations in HeLa cells: involvement of calmodulin-dependent kinase II and a calyculin A-inhibitable phosphatase. *Biochem* **35**, 7214–7223.

PART 5

Polymodal gated ion channels

Transient receptor potential channels

Grzegorz Owsianik, Thomas Voets and
Bernd Nilius

5.1.1 Introduction

The transient receptor potential (TRP) channels form a large superfamily of versatile channels that are likely expressed in every cell type in both invertebrates and vertebrates. They are involved in the perception of a diverse variety of different physical and chemical stimuli and in the initiation of cellular responses thereupon. These stimulation specificities as well as selectivity of TRP channels make them crucial players in the influx and/or transepithelial machinery that transports Ca^{2+}, Mg^{2+}, trace metal ions or modulate the driving force for cation entry. In this chapter we summarize our current knowledge concerning mammalian channels of the TRP channel superfamily, their activation mechanisms, biophysical characteristics, pharmacology, physiology, and relevance to disease states.

5.1.2 Molecular characteristics

All TRP members are homologues of the first characterized *Drosophila* transient receptor potential (TRP) channel, which is involved in the response to bright light (Cosens and Manning, 1969). Analysis of the photoreceptor cells of *Drosophila* mutants revealed that sustained light induced a transient rather than the normal sustained, plateau-like receptor potential, and was described as *trp* mutation. Cloning of the *trp* gene showed that it encodes a Ca^{2+}-permeable cation channel, TRP (Montell and Rubin, 1989; Hardie and Minke, 1992). Subsequently, two other close *Drosophila* homologues of TRP, named TRPL (or TRP-like) (Phillips *et al.*, 1992) and TRPg (Xu *et al.*, 2000) were identified and found to contribute to the light-induced currents in the photoreceptor cells (Phillips *et al.*, 1992; Reuss *et al.*, 1997; Xu *et al.*, 2000). *Drosophila* TRP channels function as receptor-operated channels that are activated downstream of the light-induced, phospholipase C (PLC)-mediated hydrolysis of phosphatidylinositol 4,5-bisphosphate (PIP_2) (Hardie *et al.*, 2001). However, it is still unclear whether TRPs open in response to reduced PIP_2 levels, or whether they are activated by diacyl glycerol (DAG) or by polyunsatured fatty acids derived from DAG (Chyb *et al.*, 1999).

Based on structural homology TRP channels are classified into seven main subfamilies (see Figure 5.1.1a) (Montell, 2005; Nilius and Voets, 2005; Pedersen *et al.*, 2005; Voets *et al.*, 2005). Currently, there are more than 50 TRP channels that have been identified in

ycast, worms, insects, fish, and mammals and, thanks to genome sequencing projects, it can be stated with certainty that there are 28 *trp*-related genes in mice, 27 *trp*-related genes in humans, 17 *trp*-related genes in the worm *C. elegans* and 13 *trp*-related genes in *Drosophila* (Clapham, 2003; Vriens *et al.*, 2004; Nilius and Voets, 2005; Pedersen *et al.*, 2005; Voets *et al.*, 2005; Owsianik *et al.*, 2006a). The TRPC ('canonical' or 'classical') subfamily members exhibit the highest homology to *Drosophila* TRP channels. Nomenclature of other subfamilies originate from their first identified members: the TRPV subfamily after the vanilloid receptor 1 (VR-1, now TRPV1), the TRPM subfamily after the tumour suppressor melastatin (TRPM1), the TRPA subfamily after the protein ankyrin-like with transmembrane domains 1 (ANKTM1, now TRPA1), the TRPN subfamily after the *no mechanoreceptor potential C* gene (*nompC*) from *Drosophila*, the TRPP subfamily after the polycystic kidney disease-related protein 2 (PKD2, now TRPP2), and the TRPML subfamily after mucolipin 1 (TRPML1).

TRP channels are intrinsic membrane proteins with six putative transmembrane spans (TM) and a cation-permeable pore region between TM5 and TM6 (Figure 5.1.1b). Similar to other 6 TM pore-forming proteins, TRP channels function as either homo- or heteromultimers composed of four TRP subunits (Kuzhikandathil *et al.*, 2001; Hoenderop *et al.*, 2003). The intracellular amino (N) and carboxy (C) termini are variable in their lengths. The composition of structural domains they contain varies significantly between members of respective TRP channel subfamilies (Figure 5.1.1b). The cytoplasmic domains of TRP channels play important roles in the regulation and modulation of channel function and trafficking. However, in most cases, the relevance of these different cytoplasmic domains to channel function is poorly understood.

TRPCs

The mammalian TRPC subfamily comprises seven channels that share the highest homology to classical *Drosophila* TRP channels. In general, they are PLC-dependent Ca^{2+} permeable cation channels but it is still controversial as to whether they are also regulated by the depletion of intracellular Ca^{2+} stores (Harteneck *et al.*, 2000; Clapham *et al.*, 2001; Clapham 2003; Nilius *et al.*, 2003; Nilius 2003, 2004; Freichel *et al.*, 2004; Vazquez *et al.*, 2004). TRPCs are formed by tetramerization of either identical or different TRPC channel subunits and depending on the combination of TRPCs in

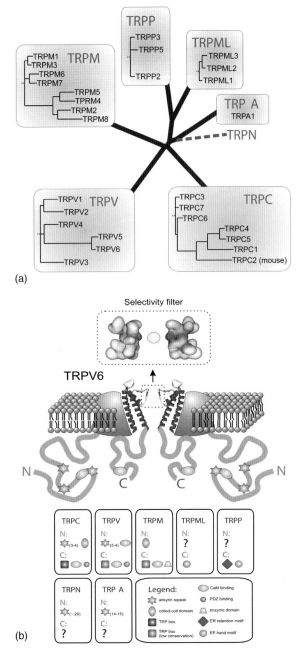

(a)

(b)

Fig. 5.1.1 Phylogeny and structural properties of TRP channels. (a) Phylogenetic relationships between human members of the TRP channel superfamily. The multiple alignment phylogenetic tree illustrates the relationship between the different TRP subfamilies. Phenograms were generated independently for each subfamily. Note that TRPC2 is a pseudogene in primates and that TRPN channels have not been identified in mammals. (b) Predicted structural topology of TRPV6 (Top). Four identical or similar subunits form a functional channel (for clarity only two subunits are shown). The pore region is formed by the loop between TM5 and TM6, which forms the pore helix and selectivity filter similar to that of K$^+$ channels (Voets et al., 2004b). The structure of TM1–4 is currently unknown. (Bottom) Overview of structural motifs found in the N- and C-terminal tails of the different TRP subfamilies. The number of motif repeats is given in brackets. The structural motifs in the cytosolic tails of members of the TRPML, TRPP, TRPN and TRPA subfamilies are not always known, hence the question marks. Adapted by permission from Macmillan Publishers Ltd *Nature chemical biology* (Voets et al., 2005), copyright 2005. See also colour plate section.

the functional tetrameric channel they play important roles in different physiological processes (Lintschinger *et al.*, 2000; Schaefer *et al.*, 2000; Wu *et al.*, 2000; Strubing *et al.*, 2001; Maruyama *et al.*, 2006; Poteser *et al.*, 2006). TRPC1 may function as a stretch-activated channel involved in cellular mechanosensitivity, TRPC2 is involved in pheromone sensing (note that the human TRPC2 gene is a pseudogene), TRPC3/4/5 take part in vasoregulation, TRPC1/4/5 are components of a signalling machinery in the central nervous system whereas TRPC3/6/7 are involved in the function of smooth muscle cells (Sakura and Ashcroft 1997; Freichel *et al.*, 2001; Strubing *et al.*, 2001; Stowers *et al.*, 2002; Tiruppathi *et al.*, 2002; Kim *et al.*, 2003; Lucas *et al.*, 2003; Maroto *et al.*, 2005; Maruyama *et al.*, 2006).

In the N-terminal region, all TRPC channels contain three to four ankyrin (ANK) repeats (Philipp *et al.*, 2000b), 33-residue motifs that are involved in specific protein–protein interactions and can interconnect membrane proteins with the spectrin-actin based membrane skeleton (Sedgwick and Smerdon, 1999; Denker and Barber, 2002). Additional possible interactions may occur via putative coiled-coil domains, protein oligomerization motifs, which are localized in N and/or C termini. In the C terminus close to TM6, all TRPCs possess a so-called TRP domain, which consists of 25 amino acids, 6 of which are invariant (EWKFAR) and referred as a TRP-box (Clapham *et al.*, 2001; Minke and Cook, 2002; Montell *et al.*, 2002; Prescott and Julius, 2003). The TRP-box was postulated to serve as a putative signature of the TRP superfamily. However, in the light of the latest TRP channels classification (Montell *et al.*, 2002; Clapham, 2003), this idea has not been substantiated since the TRP-box sequence is not conserved in TRPP, TRPML, TRPA and TRPN subfamilies. Surprisingly, until now, there is no evidence for the functional importance of the TRP-box of TRPC channels. Other structural features that have been identified by sequence homology in all TRPC channels are C-terminal binding sites for calmodulin (CaM) and the inositol (1,4,5) trisphosphate (IP$_3$)-receptor (IP$_3$R), called CIBR (CaM- and IP$_3$R-binding site), which are important for agonist-dependent activation and regulation of channel function (Trost *et al.*, 2001; Zhu and Tang, 2004; Zhu 2005).

TRPC1 is a 793 amino acid (aa) protein encoded by a 13-exon gene located on chromosome 3 (ENSG00000144935)[1] (Wes *et al.*, 1995; Zhu *et al.*, 1995). No alternatively spliced variants have been identified so far. Deletion of all three ANK repeats had no effect on dimerization and correct insertion of the channel in the membrane but results in loss of function, suggesting that ANK repeats may be involved in the correct assembly of the quaternary channel structure or interaction with regulatory proteins (Engelke *et al.*, 2002). The N terminus of TRPC1 also contains a functional coiled-coil domain which like the ANK repeats may be involved in tetramerization an interaction with other protein partners (Engelke *et al.*, 2002). It is also very likely that TRPC1 can only function as a heteromeric complex with other TRPC isoforms, such as TRPC3/4/5/7 (Wu *et al.*, 2000; Strubing *et al.*, 2001, 2003, Goel *et al.*, 2002; Hofmann *et al.*, 2002; Schilling and Goel 2004; Zagranichnaya *et al.*, 2005), and the formation of such a complex seems to be necessary to translocate TRPC1 into the plasma membrane.

1 All gene references in the text are related to human genes and are from http://www.ensmbl.org/.

The human gene encoding TRPC2 (ENSG00000182048) is a pseudogene located on chromosome 11. Its straight homologue from mouse, mTRPC2, is an 1172 aa-long protein and shares all structural features of TRPC channels (three N-terminal ANK repeats and an invariant TRP-box) (Vannier et al., 1999). There are three alternative splice variants of mTRPC2, which differ in the length of the N-terminus (Hofmann et al., 2000a).[2]

TRPC3 is an 848 aa product of a 12-exon gene (ENSG00000138741) from chromosome 4. The only alternative variant found recently, TRPC3a, contains a 73 aa N-terminal extension and, like its shorter typical variant, functions as a capacitative Ca^{2+} entry channel (Yildirim et al., 2005). The N terminus of TRPC3 displays four ANK repeats, however in contrast to TRPC1, a complete deletion of this region results in retention of the truncated channel in intracellular compartments (Wedel et al., 2003). Interestingly, similar loss of plasma membrane localization is also evoked by deletion of the CIBR domain, indicating that this domain, in addition to conferring agonist-dependent activation of the channel, is involved in targeting of TRPC3 to plasma membrane (Wedel et al., 2003). TRPC3 can form homo- as well as hetero-tetramers with TRPC1/4/6/7 (Wu et al., 2000; Strubing et al., 2003; Zagranichnaya et al., 2005; Poteser et al., 2006).

Another member of the TRPC subfamily is TRPC4, which is expressed from an 11-exon gene (ENSG00000133107) located on chromosome 13. Four different isoforms have been identified (Zhu et al., 1996; McKay et al., 2000; Mery et al., 2001; Schaefer et al., 2002). The TRPC4a variant, composed of 977 aa, is the predominant isoform of this protein. The other isoforms are products of alternative splicing in which certain C-terminal regions are missing: in TRPC4b residues between aa 785–868, in TRPC4d between 730–870, and in TRPC4g between 629–693 and 785–868 (Mery et al., 2001; Schaefer et al., 2002). In contrast to other TRPCs, TRPC4 contains only two ANK repeats. Interaction with CaM depends on the C-terminal CIBR, which binds CaM in Ca^{2+}-dependent manner (Trost et al., 2001). The extreme C terminus of TRPC4, as well as its closest homologue TRPC5, contains the PDZ-binding motif, VTTRL, which binds to the PDZ domain (PDZ is named after the three proteins in which this motif was first described: the postsynaptic density protein PSD, disc-large tumour suppressor, and the tight junction protein ZO-1) found in multidomain scaffolding proteins. PDZ-containing scaffolds assemble specific proteins into large molecular complexes at defined locations in the cell (Kim and Sheng, 2004). The PDZ-binding motifs of TRPC4 and TRPC5 are essential for interaction with PDZ domain-containing proteins such as hydrogen exchanger regulating factor (NHERF) or ezrin/moesin/radixin-binding phosphoprotein 50 (EBP50) (Tang et al., 2000; Mery et al., 2002; Lee-Kwon et al., 2004; Obukhov and Nowycky, 2004). TRPC4 and TRPC5 are able to co-assemble with NHERF as well as with the NHERF-interacting partner, $PLC_{\beta 1}$, suggesting an important mechanism for allocation and regulation of the channels (Tang et al., 2000). Deletion of the PDZ-binding motif in TRPC4 strongly impairs expression of the channel at the cell surface and also changes its general distribution in cell membranes (Mery et al., 2002). TRPC4 forms heterotetramers with TRPC1/3/5 (Schaefer et al., 2000; Hofmann et al., 2002; Strubing et al., 2003).

The TRPC5 gene (ENSG00000072315) is located on chromosome 10 and contains 11 exons. The length of TRPC5 is 973 aa (Sossey-Alaoui et al., 1999) and like TRPC4 contains two ANK repeats, the CIBR domain and the PDZ-binding motif through which it interacts with EBP50 (Obukhov and Nowycky, 2004). TRPC5 forms heterotetramers with TRPC1/4 (Schaefer et al., 2000; Strubing et al., 2001).

TRPC6 and TRPC7 are close homologues of TRPC3 and are expressed from 13 and 11 exon genes located on chromosomes 11 and 5, respectively (ENSG00000137672 and ENSG00000069018). The longest TRPC6 isoform comprises 931 aa whereas two other splice variants, TRPC6–2 and TRPC6–3, are missing the N-terminal regions between residues 316–431 and 377–431, respectively (D'Esposito et al., 1998; Hofmann et al., 1999). TRPC7 seems to exist only as a main transcription product of 862 aa (Yoon et al., 2001). Both channels contain 4 ANK repeats and C-terminal CIBR-homology regions involved in CaM and/or Ca^{2+}-dependent regulation (Boulay, 2002; Shi et al., 2004; Zhu and Tang, 2004; Zhu, 2005). TRPC6 and TRPC7 can couple together and form channels (Shi et al., 2004; Maruyama et al., 2006) but also may form heterotetramers with TRPC1 and TRPC3 (Strubing et al., 2003; Zagranichnaya et al., 2005).

TRPVs

The vanilloid subfamily of TRP channels, TRPV, comprises six mammalian homologues. TRPV1 is involved in nociception and detection/integration of thermal and diverse chemical stimuli (e.g. vanilloids, endovanilloids, and anandamide) (Caterina et al., 1997; Jordt and Julius, 2002). TRPV2 and TRPV3 are activated by noxious and warm heat ranges, respectively (Kanzaki et al., 1999a; Peier et al., 2002; Smith et al., 2002; Xu et al., 2002). TRPV4 is involved in nociception and osmo- and warmth sensation (Liedtke et al., 2000; Watanabe et al., 2002; Liedtke and Friedman, 2003; Liedtke et al., 2003; Watanabe et al., 2003; Nilius et al., 2004d; Vriens et al., 2004). TRPV5 and TRPV6, the only highly Ca^{2+}-selective TRP channels, play an important role in Ca^{2+} reabsorption in kidney and intestine (Hoenderop et al., 1999, 2000, 2001; Vennekens et al., 2000, 2001a, b; Den Dekker et al., 2003; Nijenhuis et al., 2003).

As with TRPCs, all TRPV channels contain from 3–5 N-terminal ankyrin repeats, which seem to be involved in oligomerization of the channels (Erler et al., 2004). Some TRPVs display CaM binding domains (CaMBD) that may be localized in either N or C termini. The conservation of the C-terminal TRP box motif is rather low and varies form IWxLQx (with x = K, R or W) for TRPV1/4 and LWRAQx (with x = V or I) for TRPV5/6. Interestingly, for TRPV1 and TRPV4, residues responsible for interaction with ligands are located in the transmembrane domain region between TM2 and TM4 (Jordt and Julius, 2002; Gavva et al., 2004; Vriens et al., 2004). Except for TRPV5 and TRPV6, which can form a heterotetrameric channel (Hoenderop et al., 2003), all TRPVs seem to form and function as homotetramers.

The mouse vanilloid receptor, TRPV1, is the founding member of the TRPV family (Caterina et al., 1997). The 17 exon TRPV1 gene is located on chromosome 17 (ENSG00000196689) and encodes an 839 aa long protein (Hayes et al., 2000). As a result of an alternative splicing, TRPV1 also exists as a shorter isoform that ends at the residue 511 (Cortright et al., 2001). TRPV1 functions mainly as a homotetramer and its assembly seems to require the putative

---

2   If not otherwise indicated, all functional data described in the text concern the most abundant typical isoform of the particular channel.

TRP-box (Garcia-Sanz *et al.*, 2004). Two CaMBDs determine the CaM-dependent regulation of TRPV1 activity. Binding of CaM to the N-terminal CaMBD reduces capsaicin-activated currents (Rosenbaum *et al.*, 2004), whereas binding to CaMBD in the C terminus leads to TRPV1 desensitization of these currents (Numazaki *et al.*, 2003). The C-terminal extremity of TRPV1 also contains the molecular determinants responsible for PIP_2-dependent inhibition of channel function (Chuang *et al.*, 2001; Prescott and Julius, 2003).

The structural features of TRPV2 and TRPV3 are rather weakly characterized. As with TRPV1, TRPV2 (ENSG00000187688) and TRPV3 (ENSG00000167723) genes are located on chromosome 17 and are composed of 16 or 18 exons, respectively. TRPV2 is expressed as a single transcript of 764 aa (Caterina *et al.*, 1999), whereas TRPV3 exists as three isofoms. The most abundant transcript is a 790 aa protein, whereas two other splice variants, TRPV3–2 and TRPV3–3, either have an insertion of an extra alanine after residue 759 or lack the 25 aa from the C-terminus, respectively (Smith *et al.*, 2002; Xu *et al.*, 2002). It is still unclear whether TRPV3 can co-assemble with TRPV1 and form a functional heterotetramer (Smith *et al.*, 2002; Hellwig *et al.*, 2005).

The main 871 aa isoform of TRPV4 is expressed from a 16-exon gene on chromosome 12 (ENSG00000111199). Two other isoforms, TRPV4-2 and TRPV4-3, exhibit a deletion of 59 aa in the N-terminal cytoplasmic region (residues 385–444) or modifications and a truncation within the last 27 aa of the C terminus, respectively (Liedtke *et al.*, 2000; Strotmann *et al.*, 2000; Suzuki *et al.*, 2003). The function of three N-terminal ANK repeats is still elusive. Thus, the truncation of the TRPV4 N terminus, including three ANK repeats, has no effect on the targeting to the plasma membrane or the hypotonic stimulation of the channel (Liedtke *et al.*, 2000). It has recently been shown that a proline-rich region in the N terminus of TRPV4 functionally interacts with pacsin3 (Cuajungco *et al.*, 2006), a protein implicated in synaptic vesicular membrane trafficking and regulation of dynamin-mediated endocytotic processes (Plomann *et al.*, 1998; Modregger *et al.*, 2000). This interaction affects endocytosis of TRPV4, thereby stabilizing the channel in the plasma membrane (Cuajungco *et al.*, 2006). In TRPV4, Ca^{2+}-dependent potentiation depends on a CaMBD that is formed by a stretch of basic aa in the C terminus of the channel, starting at position 814. Neutralization of these positively charged residues results in the loss of Ca^{2+}-dependent potentiation and of the spontaneous opening of TRPV4 in the absence of an agonist (Strotmann *et al.*, 2003).

TRPV5 (729 aa) and TRPV6 (725 aa) are highly homologous channels, which are encoded by 15-exon genes situated on chromosome 7 (ENSG00000127412 and ENSG00000165125, respectively). In contrast to TRPV5, the TRPV6 gene can be alternatively spliced, resulting in a shorter isoform that lacks 67 aa between residues 25–192 (Peng *et al.*, 1999; Muller *et al.*, 2000; Peng *et al.*, 2001a). Both channels have either five (TRPV5) or six (TRPV6) ANK repeats, and at least for TRPV6, the third ANK repeats seem to be involved in channel tetramerization, initiating the molecular zippering process that creates an intracellular anchor for functional assembly of TRPV6 subunits (Erler *et al.*, 2004). As with TRPV1, conserved CaMBDs in TRPV5/6 are present in both the N and C termini and are responsible for Ca^{2+}-dependent inactivation of these channels (Niemeyer *et al.*, 2001; Nilius *et al.*, 2002, 2003; Lambers *et al.*, 2004). Interestingly, for

TRPV6, an additional CaMBD is located in the intracellular loop located between TM2 and TM3 (Nilius *et al.*, 2003; Lambers *et al.*, 2004). Although no apparent PDZ-binding domains are present in TRPVs, the C-terminus of TRPV5 can interact with the scaffold protein NHERF2 (Embark *et al.*, 2004). This interaction is Ca^{2+}-independent and is required for stabilization of the channel at, or targeting to, the plasma membrane (Palmada *et al.*, 2005). More recently, it has been shown that the C termini of TRPV5/6 also interact with NHERF4, which functions as a putative plasma membrane scaffold for these channels (van de Graaf *et al.*, 2006). In the case of TRPV5, the binding site for NHERF4 is located in a region that is distinct from the binding site of NHERF2 (van de Graaf *et al.*, 2006). In contrast to TRPV1, which is inhibited PIP_2 via its interaction at its extreme C-terminal region, the positive residues in the TRP box of TRPV5 seem to be involved in PIP_2-dependent activation of the channel and might serve as a putative PIP_2 binding site (Rohacs *et al.*, 2005).

TRPMs

The TRPM subfamily comprises close homologues of melastatin, which was originally identified based on its higher expression in non-metastatic compared to highly metastatic melanomas (Duncan *et al.*, 1998). The mammalian TRPM subfamily is composed of eight members, which take part in a broad range of different processes such as Mg^{2+} homeostasis (TRPM6, TRPM7) (Nadler *et al.*, 2001; Schlingmann *et al.*, 2002; Walder *et al.*, 2002; Voets *et al.*, 2004), taste detection (TRPM5) (Perez *et al.*, 2002; Y Zhang *et al.*, 2003; Talavera *et al.*, 2005), cell proliferation (TRPM7) (Nadler *et al.*, 2001) as well as noxious cold (TRPM8) (McKemy *et al.*, 2002; Peier *et al.*, 2002a; Voets *et al.*, 2004a) and warm temperature sensing (TRPM5) (Talavera *et al.*, 2005). Interestingly, so far the Ca^{2+} activated TRPM4 and TRPM5 are the only channels of the entire TRP superfamily that are Ca^{2+}-impermeable (Launay *et al.*, 2002; Hofmann *et al.*, 2003; Nilius *et al.*, 2003; Prawitt *et al.*, 2003).

The N termini of TRPMs are rather long without ANK repeats and show rather high homology over their aa sequences. As with TRPVs, the TRP box of TRPMs exhibits a certain degree of sequence degeneration, varying from xWKFQR (with x = I, V or F) for TRPM1–3/5/7/8, YWKAQR for TRPM4 and LWKYNR for TRPM6. TRPMs sequences contain putative coiled-coil domains, whose functional roles are still very elusive. Some TRPMs contain CaMDBs, which may play important roles in Ca^{2+}-dependent regulation of channel activity. A unique feature within the TRPM subfamily is the existence of so-called chanzymes, which in addition to the 'channel' part also comprise entire functional enzyme domains in their C terminus (Fleig and Penner, 2004).

TRPM1 (melastatin) is the most poorly characterized TRPM channel. The TRPM1 gene (ENSG00000134160) is located on chromosome 15 and its 27 codons code for a1533 aa protein (Hunter *et al.*, 1998).

The first of the chanzymes is TRPM2 (1503), which is encoded by a 33 exon gene (ENSG00000142185) on chromosome 21 (Nagamine *et al.*, 1998). Two splice variants have been identified: TRPM2–2 lacks residues 538–557 in the N terminus and a part of the C-terminal 'enzyme' domain (residues 1291–1325), whereas TRPM2–3 terminated at residue 847 (Wehage *et al.*, 2002; W Zhang *et al.*, 2003). The C terminus of TRPM2 contains a so-called Nudix hydrolase domain, which functions as an ADP-ribose (ADPR) pyrophosphatase (Perraud *et al.*, 2001; Kuhn and Luckhoff, 2004).

A characteristic feature of Nudix family enzymes is the presence of the conserved Nudix box, GX5EX7REuXEEXu (X any amino acid residue and u a large hydrophobic residue), which is altered in TRPM2. Introduction of such a TRPM2-like Nudix box into the human ADPR pyrophosphatase NUDT9 causes a strong decrease in the ADPR activity similar to that which is observed for TRPM2. The crystal structure of the Nudix domain has been obtained and revealed that it functions as a monomer with the substrate binding site located in a cleft between the N-terminal and the C-terminal catalytic domain (Shen et al., 2003).

The most distinctive feature of the TRPM3 gene (ENSG00000083067; 26 exons on chromosome 9) is its ability to undergo alterative splicing creating more than 10 transcripts that encode proteins containing from 1732 to 230 aa (Grimm et al., 2003; Lee et al., 2003; Oberwinkler et al., 2005).

TRPM4 (ENSG00000130529; 25 exons) and TRPM5 (ENSG00000070985; 25 exons) genes code for closely homologous channels and are located on chromosome 19 and 11, respectively (Prawitt et al., 2000; Launay et al., 2002). The typical TRPM4b variant is composed of 1214 aa whereas its shorter versions, TRPM4a and TRPM4c, lack either the first 174 aa from its N terminus or replacement of residues between G734-L882 by a P735VG triplet (Xu et al., 2001; Hofmann et al., 2003). For TRPM5, which is a 1165 aa-long channel, alternative splice variants have not yet been identified (Prawitt et al., 2000; Hofmann et al., 2003). As with TRPV5 as well as another member of the TRPM subfamily, TRPM8, residues in the putative TRP box are involved in PIP_2-dependent regulation of channel activity (Liu and Liman, 2003; Rohacs et al., 2005). In contrast, mutations within the TRP domain (including the TRP box) of TRPM4 do not affect PIP_2 sensitivity (Nilius et al., 2006). The PIP_2-dependent regulation of TRPM4 depends on the C terminal putative plekstrin homology (PH) domain, which might function as a PIP2 binding site (Nilius et al., 2006). CaM/Ca^{2+}-dependent activation of TRPM4 is linked with CaMBDs, which have been identified in N- (two sites) and C-terminal (three sites) domains using an in vitro CaM binding assay (Nilius et al., 2005c). In vitro experiments show that only the C-terminal CaMBDs are vital for Ca^{2+} sensitivity of TRPM4 in the physiological range of intracellular Ca^{2+} concentrations (Nilius et al., 2005c). The N-terminus as well as the TM2-TM3 linker of TRPM4 contains putative ATP-binding sites, composed of Walker B and ABC signature motifs, which contribute to ATP-dependent reversal of Ca^{2+} sensitivity after desensitization of the channel (Nilius et al., 2005c).

The other chanzymes in the TRPM subfamily are TRPM6 and TRPM7, encoded by genes found on chromosome 9 (ENSG00000119121; 39 exons) and 15 (ENSG00000092439; 39 exons), respectively (Nadler et al., 2001; Riazanova et al., 2001; Schlingmann et al., 2002). TRPM7 is 1865 aa long, whereas for TRPM6 six differently spliced isoforms ranging from 2022 to 569 aa have been characterized (Schlingmann et al., 2002; Chubanov et al., 2004). An interesting feature of these two chanzymes is the presence in their C terminus of an atypical protein-kinase domain, the so-called phospholipase C interacting kinase (PLIK) domain (Runnels et al., 2001), whose crystal structure shows unexpected similarity to eukaryotic α-kinases (Yamaguchi et al., 2001). The importance of the PLIK domain for TRPM7 channel function is still controversial. TRPM7 can be regulated via the PLIK domain in a cAMP- and protein kinase A (PKA)-dependent manner

(Takezawa et al., 2004) and this activity is essential for channel function (Runnels et al., 2001). It has been shown that deletion of PLIK results in functional TRPM7 channels with increased sensitivity to Mg^{2+} and MgATP (Schmitz et al., 2003). In contrast, in a more recent study describing the autophosphorylation capabilities of the TRPM7 PLIK domain, it has been shown that a loss of kinase activity ('kinase-dead' mutants) does not affect channel function or its inhibition by internal Mg^{2+} (Matsushita et al., 2005). The authors of this latest study suggest that the PLIK domain is involved in channel assembly or subcellular localization (Matsushita et al., 2005).

The last member of TRPMs is the cold/menthol receptor TRPM8 (1104 aa), which is encoded by a 26 exon gene (ENSG00000144481) that lies on chromosome 2 (Tsavaler et al., 2001). As already mentioned, the TRP box of TRPM8 is involved in PIP_2 sensitivity (Rohacs et al., 2005).

TRPMLs

The 'mucolipin' subfamily of TRP channels, TRPML, consists of 3 members TRPML1–3. Mutations in TRPML1 (mucolipin-1; MCOLN1) are responsible for mucolipidosis type IV (MLIV), an autosomal recessive, neurodegenerative, lysosomal storage disorder characterized by psychomotor retardation and opthalmological abnormalities including corneal opacities, retinal degeneration, and strabismus (Berman et al., 1974). The TRPML proteins are relatively small and share low sequence homology to other TRP channels. Available data concerning structural domains within these channels are very limited. TRPMLs can form homo- and heteromultimers, which determine the subcellular localization of the channels (Venkatachalam et al., 2006) (see also the section Cellular and subcellular distribution).

The founding member of this subfamily, TRPML1, is expressed from a 14-exon gene (ENSG00000090674) located on chromosome 19 (Bargal et al., 2000; Bassi et al., 2000). The main transcript is 580 aa-long whereas the only identified splice variant, TRPML1–2 (545 aa), has a 35 aa deletion and modifications in the first 200 aa of the N terminus (Bargal et al., 2000; Bassi et al., 2000). TRPML1 contains a putative lipase domain in the intracellular loop between TM1 and TM2, a putative N-terminal nuclear localization signal, and a putative late endosomal–lysosomal targeting signal in the C teminus (Sun et al., 2000).

Two other members, TRPML2 (538 aa) and TRPML3 (553 aa), are found on chromosome 1 (13 exons; ENSG00000153898 and ENSG00000055732, respectively) (Di Palma et al., 2002).

TRPPs

The TRPP family is very heterogeneous and, based on structural features, can be subdivided into TRPP1 (former PDK1)-like and TRPP2 (former PKD2)-like groups.

The inclusion of the TRPP1-like proteins, TRPP1, PKDREJ, PKD1L1, PKD1L2 and PKD1L3, into the TRP superfamily is at present somewhat tentative and rests upon a degree of structural similarity between TRP channels and the six distal transmembrane domains of at least some PKD1-like members (e.g. PKD1L2, PKD1L3 and PKDREJ) (Delmas, 2005; Qian et al., 2005). TRPP1 consists of 11 TMs, a very long (~3000 aa) and complex extracellular N-terminal domain and an intracellular C-terminal domain that contains a coiled-coil domain responsible for interaction with the C-terminus of TRPP2 (Qian et al., 1997; Tsiokas et al., 1997;

Wu and Somlo, 2000). The N-terminal part of TRPP1 contains numerous structural motifs, which might participate in cell–cell and cell–matrix interactions.

The TRPP2-like channels are homologous to other TRP channels, having predicted intracellular N- and C-termini and 6TMs with a pore region between TM5 and TM6. All three members of this group, TRPP2/3/5, contain a C-terminal coiled-coil structure and form polymodal multiprotein/ion channel complexes (Delmas, 2005). Additionally, TRPP2 an TRPP3 have a putative Ca^{2+} binding EF hand motif in the C-terminus, which might be involved in Ca^{2+}-dependent regulation of the channel function (Koulen et al., 2002). In heterologous expression systems TRPP2 and TRPP3 can form functional heteromers (Somlo and Markowitz, 2000; Delmas, 2004; Delmas et al., 2004).

The TRPP2 gene (ENSG00000118762) from chromosome 4 has 15 exons and encodes a 968 aa protein (Mochizuki et al., 1996; Wu et al., 1997). Two other members, TRPP3 (805 aa) and TRPP5 (624 aa), are encoded by genes from chromosomes 10 (18 exons; ENSG00000107593) and 5 (15 exons; ENSG00000078795), respectively (Nomura et al., 1998; Veldhuisen et al., 1999). As a result of alternative splicing, both genes generate multiple isoforms that possess deletions and modifications within N- and C-terminal tails (Guo et al., 2000a, b).

TRPAs

The mammalian TRPA subfamily is represented by only one channel, TRPA1. The TRPA1 gene (ENSG00000104321) contains 27 exons and is expressed from chromosome 8 (Jaquemar et al., 1999). TRPA1 (1119 aa) exhibits 14 ANK repeats in its N terminus, which might confer a mechanosensory function of the channel (Story et al., 2003; Nagata et al., 2005; Lee et al., 2006). Crystallographic data suggest that multiple ANK repeats can form a helical structure, which may act as a gating spring that transduces mechanical stress from the N-terminally connected cytoskeletal elements to the channels (Corey, 2003; Howard and Bechstedt, 2004; Lin and Corey, 2005; Sotomayor et al., 2005). The N-terminal cysteine residues, C415, C422, and C622, are involved in activation of the TRPA1 channel by reversible covalent modification in the presence of membrane-permeable electrophilic agonists (Hinman et al., 2006; Macpherson et al., 2007).

TRPNs

The TRPN subfamily is named after its first founding member, the *no mechanoreceptor potential C* (NOMP-C) channel from *Drosophila*, which is involved in mechanosensitive processes including hearing, balance, proprioception and touch (Kernan et al., 1994; Walker et al., 2000). There are no obvious homologues of NOMP-C within the human genome and, to date, the only vertebrate TRPN family channel has been identified in zebrafish (Sidi et al., 2003). The characteristic feature of TRPN channels is the presence of multiple (~29!) ANK repeats in their N terminus.

5.1.3 Biophysical characteristics

Ion channels are pore-forming integral membrane proteins that allow permeation of ions through biological lipid bilayers. The ion permeability and selectivity properties of particular channels depend on the pore region, which like other six TMs channels, in TRP channels is located between TM5 and TM6. Despite great

progress in characterizing TRP channels, functional characterization of pore properties, biophysical aspects of cation permeation or description of pore structures of TRP channels is still rather limited (see the most recent review: Owsianik et al., 2006b).

Permeability and selectivity

TRPCs

TRPC channels were the first characterized mammalian counterparts of *Drosophila* TRPs and were identified as Ca^{2+} permeable cation channels activated by a rise in intracellular calcium and/or products of PLC-dependent pathways (Clapham 1995; Harteneck et al., 2000; Hofmann et al., 2002; Nilius 2004a; Gudermann et al., 2004; Schilling and Goel 2004; Vazquez et al., 2004; Parekh and Putney, 2005; Putney, 2005). Surprisingly, data regarding permeation properties of certain members are still very controversial. A classical example is that of TRPC4 and TRPC5, which have been described as being either Ca^{2+} selective (Philipp et al., 1996, 1998; Warnat et al., 1999) or non-selective between mono- and divalent cations (Okada et al., 1998; Plant and Schaefer, 2003). Such discrepancy, especially in heterologous expression systems, might result from interpretation of functional data that are often contaminated by the 'noisy background' caused by endogenous cation-selective, Ca^{2+}-permeable channels, which are frequently also regulated by store-depletion and/or products of PLC-dependent pathways (Clapham, 1995; Parekh and Putney, 2005; Putney, 2005).

TRPVs

Members of the TRPV subfamily are the only TRP channels for which permeation and selectivity properties are described in detail. TRPV1/2/3/4 are permeable to Ca^{2+} with a rather low discrimination between mono- and divalent cations (P_{Ca}/P_{Na} between 1 and 10) (Benham et al., 2002; Gunthorpe et al., 2002; Voets et al., 2002; Voets and Nilius 2003). For TRPV4, the relative permeability for monovalent cations corresponds to Eisenman sequence IV ($K^+ > Rb^+ > Cs^+ > Na^+ > Li^+$), indicating a relatively weak field-strength binding site. Importantly, TRPV1 is also permeable to H^+, which may trigger the intracellular acidification of nociceptive neurons after channel activation (Hellwig et al., 2004). TRPV5 and TRPV6 are unique within the TRP superfamily due to their high Ca^{2+} selectivity (P_{Ca}/P_{Na} ~100). Both TRPV5 and TRPV6 display selectivity for monovalent cations corresponding to an Eisenman sequence X ($Na^+ > Li^+ > K^+ > Rb^+ > Cs^+$) or XI ($Li^+ > Na^+ > K^+ > Rb^+ > Cs^+$) whereas for divalents, a $Ca^{2+} > Ba^{2+} > Sr^{2+} > Mn^{2+}$ selectivity sequence has been reported (Nilius et al., 2000, 2001b; Vennekens et al., 2000, 2001a, b).

TRPMs

Two members of TRPMs, TRPM4 and TRPM5, are the only Ca^{2+}-impermeable TRP channels identified so far (Launay et al., 2002; Hofmann et al., 2003; Nilius et al., 2003; Prawitt et al., 2003). TRPM2 TRPM3 and TRPM8 are Ca^{2+} permeable cation channels with rather low Ca^{2+} selectivity (Perraud et al., 2001; Sano et al., 2001; Hara et al., 2002; McKemy et al., 2002; Peier et al., 2002a; Grimm et al., 2003; Lee et al., 2003; Perraud et al., 2003) whereas TRPM6 and TRPM7 are relatively highly permeable to divalent cations (Nadler et al., 2001; Monteilh-Zoller et al., 2003; Voets et al., 2004c). So far, the permeation properties of TRPM1 have not been reliably characterized.

TRPML, TRPP, TRPA and TRPNs

In the case of TRPML, TRPP, TRPA and TRPN subfamilies, our knowledge about channel properties is rather poor. TRPML1 is a lysosomal monovalent cation channel with a prominent permeability to protons as well as very likely to Ca^{2+} (Soyombo et al., 2006). It seems to be constitutively active and probably modulated by increases in $[Ca^{2+}]_i$ (LaPlante et al., 2002, 2004; Bach, 2005). TRPP2/3/5 are permeable cation channels, usually with P_{Ca}/P_{Na} above 1 (Keller et al., 1994; Nomura et al., 1998; Guo et al., 2000a, b). Likewise, TRPA1 is a Ca^{2+}-permeable, non-selective cation channel (Story et al., 2003; Bandell et al., 2004; Jordt et al., 2004; Macpherson et al., 2005).

Pore structure

Based on available functional and theoretical data it is generally believed that the loop between TM5 and TM6 is a pore-forming element in all TRP channels (for a detailed review see Owsianik et al., 2006b).

TRPVs

The biggest insights into pore structure have been obtained for members of the TRPV subfamily. The TM5–6 linker region of all TRPVs show significant sequence homology to the selectivity filter of the prokaryotic potassium channel KcsA (Doyle et al., 1998; Zhou et al., 2001). Mutations of homologous negatively charged residues in TRPV1 and TRPV4 (D546 and D682, respectively) strongly diminish the Ca^{2+} and Mg^{2+} permeabilities of these channels and reduce their affinity for ruthenium red (Garcia-Martinez et al., 2000; Voets et al., 2002). Additional mutations at neighbouring residues in TRPV4, D672 and M680, further impair, or abolish, the selectivity for divalent cations and the relative permeability for monovalent cations. Surprisingly, mutation of the only basic pore residue in TRPV4, Lys675, has no significant effects on permeation properties of the channel (Voets et al., 2002). Thus, these data point out that the GM(L)GD motif determines permeation properties of TRPV1/2/3/4 channels and is a functional homologue of the GYGD sequence in the selectivity filter of K^+ channels (Figure 5.1.2b).

In TRPV5 (TRPV6) neutralization of D542 (D541) abolishes Ca^{2+} permeation, Ca^{2+}-dependent current decay and block by intra- or extracellular Mg^{2+} or Cd^{2+}, without changing the permeability sequence for monovalent cations (Nilius et al., 2001b; Voets et al., 2001; Hoenderop et al., 2003; Voets et al., 2003). Other negatively charged residues in this region of TRPV5, E535 and D550, have no significant impact on permeation properties, whereas E522 functions as a putative extracellular 'pH sensor', regulating pH-dependent permeation properties of TRPV5/6 (Vennekens et al., 2001a; Yeh et al., 2003). Intracellular acidification of TRPV5 evokes a conformational change of the pore helix consistent with clockwise rotation along its long axis facilitating a closure of TRPV5 by external protons (Yeh et al., 2005). Using the substituted cysteine accessibility method (SCAM) a more comprehensive insight into the structure of TRPV5/6 pore region has been obtained (Dodier et al., 2004; Voets et al., 2004b). Similar to the KcsA crystal structure (Doyle et al., 1998; Zhou et al., 2001), residues preceding D542/541 show a cyclic pattern of reactivity and very likely form a pore helix. In TRPV6, the pore helix is followed by the selectivity filter with a diameter of approximately 5.4 Å (Voets et al., 2004b) (Figure 5.1.2a). The narrowest point of the pore is shaped by D541,

which contributes to the sieving properties of the pore (Voets et al., 2004b). Thus, by analogy to the pore of voltage-gated Ca^{2+} channels (Heinemann et al., 1992; Yang et al., 1993; Talavera et al., 2001), it is very likely that a ring of four aspartate residues in the channel pore determines selectivity and permeation properties of TRPV5/6.

TRPMs

TM5-TM6 regions of TRPM channels are highly conserved and to some extent homologous to pore regions of KcsA and TRPVs channels. A putative hydrophobic pore helix is followed by an invariant aspartate, which might be located in the selectivity filter. This invariant aspartate together with other negatively charged residues from the TM5-TM6 region might form a negatively charged cluster that determines the pore properties of TRPMs. So far, the most comprehensive data about TRPM pores have been obtained for TRPM4 (Nilius et al., 2004b, c, 2005b). In TRPM4, swap of E981DMDVA986 residues with a putative selectivity filter of TRPV6 (T538IIDGP543) generates a functional channel that combines the gating properties of TRPM4 (activation by Ca^{2+} and voltage dependence) with TRPV6-like sensitivity to block by extracellular Ca^{2+} and Mg^{2+} (Nilius et al., 2005b). The glutamate E981 seems to be placed in the inner part of the pore, since its neutralization removes TRPM4 susceptibility to block by spermine (Nilius et al., 2004c, 2005b). Furthermore, mutations of D982 and D984 strongly impair the rundown and voltage dependence of the channel. On the other hand, substitution of Gln977 to a glutamate, which restores a conserved negatively charged residue in divalent cation-permeable TRPMs, alters the monovalent cation permeability and confers the channel with moderate Ca^{2+} permeability (Nilius et al., 2005b) (Figure 5.1.2b).

Additional insight into the TRPM pore region comes from functional analysis of TRPM3 splice variants, TRPM3α1–5, which differ in the length of the putative pore region as one splice site is located in the TM5–TM6 linker (Oberwinkler et al., 2005). TRPM3a1 contains an extra stretch of 12 amino acids in a putative selectivity filter that results in a channel with rather low permeability for divalent cations, whereas another isoform lacking this additional residue, TRPM3α2, shows significantly higher selectivity to Ca^{2+} and Mg^{2+} (> tenfold) and insensitivity to block by extracellular monovalent cations (Oberwinkler et al., 2005).

TRPCs

Unlike TRPVs and, to a certain extent, TRPMs, the TM5–TM6 region of TRPCs is relatively long and displays no significant sequence homology to the pore of K^+ channels. In TRPC1, one of the non-membrane spanning α-helices is located in the region between TM5 and TM6 and is probably a counterpart of the pore helix in other TRPs (Dohke et al., 2004). Neutralization of all negatively charged residues between TM5 and TM6 of TRPC1 results in reduced Ca^{2+} but not Na^+ currents and a leftward shift in the reversal potential (Liu et al., 2003). Surprisingly, the key residues E576, D581 and E615 are placed in the distal parts of the putative pore mouth. By analogy to TRPC1, neutralizations of three of the five glutamates from the TM5–TM6 loop, E543, E595 and E598, abolish La^{3+} potentiation of the channels and, in the case of the double mutant in E595/E598AA, alter single channel properties (Jung et al., 2003). Similar to TRPC1, mutations of the central glutamates of TRPC5, E559 or E570, have no effect permeation properties. Intriguingly, the aspartate D633 of TRPC5, which is located in the intracellular region between TM6 and the TRP box,

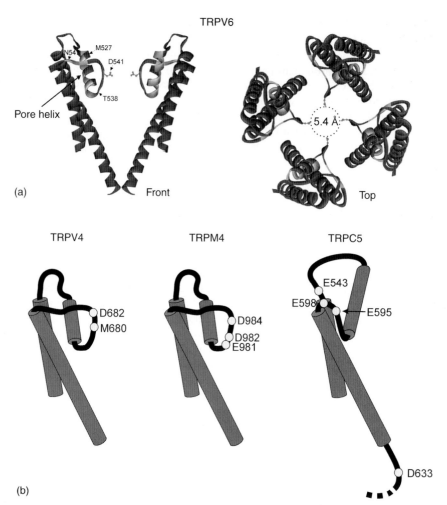

Fig. 5.1.2 Predicted topologies of the pore region of TRPV, TRPM and TRPC channels. (a) Structural model of the TRPV6 pore region, looking sideways at two opposite subunits (left) or looking down from the external solution to the complete homotetrameric channel. At the narrowest point, the pore is formed by the acidic side chain of Asp541 (orange) and has a diameter of 5.4 Å. Blue residues correspond to the residues in TM5 and TM6 and amino acids that were subjected to SCAM analysis (residues P526 to N547) are coloured in green, yellow, red, or grey. Residues in red reacted rapidly to Ag^+ (reaction rate > 5.10^6 $M^{-1}s^{-1}$), residues in yellow reacted with Ag^+ at a rate < 5.10^6 $M^{-1}s^{-1}$, and residues in green did not show significant reactivity to Ag^+. Residues where cysteine substitution resulted in non-functional channels are coloured in grey. Adapted from Voets *et al.* (2004b) with copyright permission from The American Society for Biochemistry and Molecular Biology. (b) Schematic representation of crucial residues in the putative selectivity filters of TRPV4, TRPM4 and TRPC5 channels (see text for details). Adapted from Owsianik *et al.* (2006b) with kind permission from Springer Science and Business Media (copyright 2006). See colour plate section.

is a key residue responsible for block by intracellular Mg^{2+} (Obukhov and Nowycky, 2005). Taken together, all these finding strongly predict that negatively charged residues located close to but exterior to the pore region in TRPC1 and TRPC5 determine permeation properties of these channels (Figure 5.1.2b).

Gating mechanisms

One of the amazing features of TRPs is the huge variability in gating mechanisms, ranging from PLC activation, to Ca^{2+} store depletion, membrane potential (voltage), temperature, mechanical stimuli, and ligand binding. Very often, TRP channels are activated by different types of activating stimuli, or even require the simultaneous presence of two distinct stimuli to open.

PLC- and/or Ca^{2+} store-dependent gating

A large number of studies indicate that TRPC channels constitute the machinery for the slower, sustained Ca^{2+} entry from the extracellular medium that follows PLC activation (Petersen *et al.*, 1995; Wes

et al., 1995; Zhu *et al.*, 1995; Birnbaumer *et al.*, 1996; Zitt *et al.*, 1996). TRPC3, TRPC6 and TRPC7 are activated by DAG and DAG-analogues in a membrane-delimited manner, suggesting a direct interaction between DAG and the channel protein (Hofmann *et al.*, 1999). Alternatively, all TRPCs may be activated by depletion of intracellular Ca^{2+} stores (store-dependent gating). Unfortunately, this very attractive mechanism is still under debate since several reports indicate that the level of expression may determine whether a specific TRP channel is activated in a store-dependent or store-independent manner (for review see Nilius *et al.*, 2005d; Putney, 2005).

Voltage-gating

Due to the paucity of positively charged residues in S4, TRP channels were originally considered as channels with little or no voltage dependence (Clapham, 2003). In addition, the first descriptions of voltage dependence in the TRP superfamily reported activation curves in the non-physiological positive voltage range and very low gating valencies (Gunthorpe et al., 2000; Nilius et al., 2003c, 2005d). However, more recent studies have shown that certain chemical ligands or changes in temperature can shift the voltage-dependent activation curves towards the physiological voltage range, constituting an important mechanism for TRP channel gating (see below) (Voets et al., 2004a; Nilius et al., 2005d).

Ligand-gating

For several TRP channels endogenous activators have been identified. The classical example is activation of certain TRPCs by DAG (see above). Another example is activation of TRPV1 and TRPV4 by a number of arachidonic acid (AA) derivatives. Anandamide (arachidonoylethanolamine) (Zygmunt et al., 1999) and lipoxygenase metabolites of AA such as 12,15-(S)-hydroperoxyeicosatetraenoic acid and leukotriene B_4 (Hwang et al., 2000) activate TRPV1 at low micromolar concentrations. TRPV4 is directly activated by submicromolar concentrations of 5′,6′–epoxyeicosatrienoic acid (5′,6′-EET), a cytochrome P450 epoxygenase-dependent metabolite of AA (Watanabe et al., 2003b). In the TRPM subfamily, TRPM3 is activated by D-erythro-sphingosine, a derivative of cellular sphingolipids (Grimm et al., 2005), whereas TRPM4 and TRPM5 are directly gated by rises in intracellular Ca^{2+} (Launay et al., 2002; Liu and Liman, 2003; Hofmann et al., 2003; Prawitt et al., 2003). Additionally, a broad range of exogenous ligands for different members of the TRP superfamily has also been identified (see section 5.1.5). Unfortunately, structural information about the binding site for all these ligands or insights into gating mechanisms is either very limited or completely lacking.

Modulation by temperature

TRPV1 was the first characterized heat-activated TRP channel (Caterina et al., 1999). So far, seven additional mammalian temperature-sensitive TRP channels or thermoTRPs (Patapoutian et al., 2003) have been described. The first group of channels is activated upon heating and includes TRPV1, TRPV2 (Caterina et al., 1999), TRPV3 (Peier et al., 2002b; Smith et al., 2002; Xu et al., 2002), TRPV4 (Güler et al., 2002; Watanabe et al., 2002), TRPM4 and TRPM5 (Talavera et al., 2005). The second group comprises TRPM8 (McKemy et al., 2002; Peier et al., 2002) and TRPA1 (Story et al., 2003), which open upon cooling (Story et al., 2003; Bandell et al., 2004; Kwan et al., 2006) (note that cold activation of TRPA1 could not be reproduced by some groups: Jordt et al., [2004]; Nagata et al., [2005]; Bautista et al., [2006]). All these channels together sense changes in temperature from < 10 to > 50°C.

There are several hypotheses to explain temperature sensitivity of TRP channels (Clapham, 2003). Changes in temperature might stimulate generation of an endogenous ligand that activates a particular TRP channel. A weak point of such a model is that the ligand-producing step rather than the TRP channel itself would be temperature-dependent. This mechanism is very unlikely since most thermoTRPs, except TRPV4 (Watanabe et al., 2002), display temperature sensitivity in cell-free membranes. Another possibility of a thermo-gating mechanism is related to a temperature-dependent phase transition of the lipid membrane or a conformational transition/denaturation of the channel protein, which usually happens within a narrow temperature range and may correlate with the steep temperature dependence of thermoTRP activation. More recently, a completely new and distinct fundamental thermodynamic principle to explain cold activation of TRPM8 and

heat activation of TRPV1 has been proposed (Voets et al., 2004a). Both, TRPM8 and TRPV1 are voltage-gated channels that are activated upon membrane depolarization. Changes in temperature shift the voltage dependence of activation of these two channels from strongly depolarized potentials towards the physiological potential range. Thus, temperature-dependent activation is rather gradual, which argues against a single sharp thermal threshold predicted from temperature-dependent phase transition of the lipid membrane or conformational transitions of the channel protein (Voets et al., 2004a).

Mechano-gating

The first information about mechanosensitivity of TRPs came from the TRPV subfamily. In C. elegans, mutations in the gene encoding a homologue of TRPV4, OSM-9, lead to strong impairments of avoidance reaction to noxious odours, high osmolarity and nose touch (Colbert et al., 1997). These deficiencies can be rescued by heterologous expression of mouse TRPV4, which is gated by hypotonic cell swelling (Liedtke et al., 2000; Strotmann et al., 2000; Wissenbach et al., 2000; Nilius et al., 2001a). Importantly, TRPV4 can only restore the response to hypertonicity and nose touch, but not that to noxious odours, suggesting that both channels have similar mechanosensitive properties whereas chemosensitivity is distinct (Liedtke et al., 2003). Recently, it has been shown that hypotonic stimulation of TRPV4 requires the presence of CFTR (Arniges et al., 2004) and aquaporin 5 (Liu et al., 2006; Sidhaye et al., 2006). Except TRPV4, activation by cell swelling has also been described for TRPV2 (Muraki et al., 2003), TRPM3 (Grimm et al., 2003), and Nanchung, the Drosophila TRPV channel required for hearing (Kim et al., 2003). However, it is still unclear whether they function as mechano- or osmosensors in vivo. More recently, mechanosensitivity has been suggested as a hallmark of channels from the TRPA and TRPN subfamilies. Mutations of TRPN channels in Drosophila and zebrafish led to the loss of mechanotransduction (Walker et al., 2000) or larval deafness and imbalance (Sidi et al., 2003), respectively. Subsequently, downregulation of TRPA1 expression in zebrafish or in mice strongly impairs mechanotransduction in hair cells (Corey et al., 2004), suggesting that TRPA1 can be a molecular candidate for the mechanotransduction channel upon hearing (Corey et al., 2004). Surprisingly, two independently generated TRPA1 knockout models show no obvious hearing deficiencies (Bautista et al., 2006; Kwan et al., 2006). Recently, TRPC1 has been identified as an important component and regulator of the mechanosensitive cation channel (MscCa) from Xenopus laevis oocytes (Maroto et al., 2005). These observations suggest possible involvement of TRPC1 in the formation of vertebrate MscCa (Maroto et al., 2005), challenging the general view of TRPC channels as PLC- and/or Ca^{2+}store-dependent channels (Clapham, 2003).

It is still unclear how a mechanical stimulation gates a channel. By analogy to the bacterial large conductance mechanosensitive cation channel MscL (Perozo et al., 2002a, b), mechanical stimuli may provoke alteration of the lipid bilayer tension that can be detected by TMs of the channel and result in gating of the pore. Another mechanism has been proposed for TRPN and TRPA channels that contain multiple ANK repeats in their N terminus (see section 5.1.2). Here, mechanical signal may be transduced to the channel via a direct connection between the N-terminal ANK spring and cytoskeletal elements (Corey et al., 2004; Howard and

Bechstedt, 2004; Sotomayor *et al.*, 2005). Importantly, theoretically predicted properties of the ANK spring such as its extension and stiffness match with those of a putative gating spring in vertebrate hair cells (Howard and Bechstedt, 2004; Sotomayor *et al.*, 2005). The third possible mechanism may involve auxiliary activity of mechanosensitive enzymes that induces channel gating. This hypothesis may explain the swelling-dependent activation of TRPV4 (Vriens *et al.*, 2004b). Cell swelling is known to directly activate phospholipase A2 (PLA_2) and subsequent production of AA. Inhibitiors of PLA_2 as well as cytochrome P450 epoxygenases, which metabolize AA to EETs (Vriens *et al.*, 2004b), abolish activation of TRPV4 by cell swelling without affecting direct activation by 4a-PDD. Thus, AA derivatives produced by the PLA_2-dependent pathway are agonists that activate the channel in a hypotonic environment (Watanabe *et al.*, 2003b).

Constitutively open trps

TRPV5 and TRPV6, Ca^{2+}-selective channels for Ca^{2+} entry in kidney and intestine (Hoenderop *et al.*, 2005), and TRPM6 and TRPM7, Mg^{2+} influx channels involved in cellular Mg^{2+} homeostasis reabsorption in the kidney (Schlingmann *et al.*, 2002; Walder *et al.*, 2002; Schmitz *et al.*, 2003; Voets *et al.*, 2004c), display significant open probability in the absence of any of the above-described activation mechanisms. Both channel groups undergo prominent Ca^{2+}- (TRPV5 and TRPV6) or Mg^{2+}-dependent (TRPM6 and TRPM7) inhibition, which act as negative feedback mechanisms preventing Ca^{2+} or Mg^{2+} overloads, respectively (Vennekens *et al.*, 2000; Hoenderop *et al.*, 2001; Nadler *et al.*, 2001; Yue *et al.*, 2001; Walder *et al.*, 2002).

5.1.4 Cellular and subcellular distribution

The members of the TRP superfamily are very ubiquitously expressed, and probably every cell type in every tissue expresses one or more TRP family members. Although TRP channels are mainly plasma membrane proteins, there are several examples of concomitant or exclusive residence of certain TRPs in membranes of intracellular compartments. In the following section we give short overview of the main cellular and subcellular localization of mammalian TRPs.

TRPCs

TRPC1 is one of the most ubiquitous TRPs (Zhu *et al.*, 1995). The major location of TRPC1 is intracellular, most likely in the ER, and its interactions with other TRPCs (see section 5.1.2) seems to be crucial for translocation of the heteromeric channel to the plasma membrane (Hofmann *et al.*, 2002).

The mouse TRPC2 is highly present in the dendritic tip of the vomeronasal sensory neurons, where it plays an essential role in pheromone sensing (Lucas *et al.*, 2003). It has also been detected in spermatozoa, where it may contribute to Ca^{2+} influx stimulated by glycoproteins of the ovum's extracellular matrix (Jungnickel *et al.*, 2001).

TRPC3 is highly expressed in neurons of the central nervous system (CNS) and at a much lower level in smooth and cardiac muscle cells (Li *et al.*, 1999; Hofmann *et al.*, 2002). At rest, it is predominantly restricted to intracellular compartments, and shuttles to the plasma membrane in a process that involves its CIBR domain (Wedel *et al.*, 2003).

TRPC4 is very abundant in the plasma membrane of cells of the placenta (McKay *et al.*, 2000), adrenal gland (McKay *et al.*, 2000; Philipp *et al.*, 2000a) and neurons of the CNS (Hofmann *et al.*, 2000b; Munsch *et al.*, 2003) as well as in the endothelium (McKay *et al.*, 2000; Freichel *et al.*, 2001; Tiruppathi *et al.*, 2002), smooth muscle (McKay *et al.*, 2000; Beech *et al.*, 2004), kidney (McKay *et al.*, 2000; Lee-Kwon *et al.*, 2004), and very likely in the intestinal cells of Cajal (ICC) (McKay *et al.*, 2000; Torihashi *et al.*, 2002; Walker *et al.*, 2002). In contrast, its closest homologue, TRPC5, is mainly found in brain, especially in fetal brain, with only weak expression in other tissues (Sossey-Alaoui *et al.*, 1999; Hofmann *et al.*, 2000b).

TRPC6 is an essential component of the vascular α_1-adrenoreceptor-activated Ca^{2+}-permeable plasma membrane cation channel in smooth muscle (Inoue *et al.*, 2001; Jung *et al.*, 2002). Muscarinic acetylcholine receptor activation induces translocation of TRPC6 to the plasma membrane, where it resides until the presence of a stimulus (Cayouette *et al.*, 2004). It is expressed in lung, brain, placenta, kidney (podocyte foot processes), spleen, ovary, and small intestine (Hofmann *et al.*, 2000b; Montell 2005).

The human TRPC7 appears to be primarily in the pituitary gland and kidney as well as in tissues of the CNS, with a more restricted pattern of expression in peripheral tissues (Riccio *et al.*, 2002). In contrast, the mouse orthologue is highly expressed in heart and lung, and surprisingly, almost no TRPC7 mRNAs have been detected in the CNS and kidney (Okada *et al.*, 1998), suggesting quite distinct physiological roles in these two species.

TRPVs

TRPV1 is most abundant in dorsal root ganglion (DRG) and trigeminal ganglion (TG) neurons, as well as in spinal and peripheral nerve terminals (Caterina *et al.*, 1997; Hayes *et al.*, 2000; Cortright *et al.*, 2001; McIntyre *et al.*, 2001; Planells-Cases *et al.*, 2005) and brain (Caterina *et al.*, 1997; Cortright *et al.*, 2001). In skin, it is expressed in cutaneous sensory nerve fibres, mast cells, epidermal keratinocytes, dermal blood vessels, the inner root sheet and the infundibulum of hair follicles, differentiated sebocytes, sweat gland ducts, and the secretory portion of eccrine sweat glands (Denda *et al.*, 2001; Stander *et al.*, 2004). In the pancreas its expression is necessary for substance P release (Nathan *et al.*, 2001). In bladder, it has been detected in urothelium, smooth muscle, blood vessels, and neurons (Birder *et al.*, 2001; Yiangou *et al.*, 2001; Stein *et al.*, 2004). In most cases, TRPV1 is located in the plasma membrane as well as in intracellular membranes (e.g. ER) where it functions as an intracellular Ca^{2+} release channel (Turner *et al.*, 2003).

As with TRPV1, TRPV2 has been found in DRG and CNS neurons, as well as in the gastro-intestinal tract, spleen, mast cells and smooth, cardiac and skeletal muscle cells (Kanzaki *et al.*, 1999a, b; Iwata *et al.*, 2003; Muraki *et al.*, 2003; Beech *et al.*, 2004; Stokes *et al.*, 2004). TRPV2 is located intracellularly. Insulin growth factor 1 (IGF-1) and the neuropeptide head activator (HA) promote vesicular insertion of the functional channel into the plasma membrane where it becomes active (Kanzaki *et al.*, 1999b; Boels *et al.*, 2001; Iwata *et al.*, 2003). It has been shown that maturation and surface expression of TRPV2 requires association with the recombinase gene activator (RGA) (Barnhill *et al.*, 2004; Stokes *et al.*, 2005).

TRPV3 is highly expressed in DRG, TG, brain neurons, keratinocytes, hair follicles, and at lower level in cells of tongue and testis (Peier *et al.*, 2002b; Smith *et al.*, 2002; Xu *et al.*, 2002; Moqrich *et al.*, 2005).

TRPV4 displays a high expression level in the CNS (large neurons), TG, heart, liver, kidney, skin (keratinocytes), blood vessels (endothelium), bladder (urothelium) and testis (Liedtke et al., 2000; Strotmann et al., 2000; Wissenbach et al., 2000; Suzuki et al., 2003; Vriens et al., 2005; Yang et al., 2006; Gevaert et al., 2007). In the cochlea, it is expressed in both inner and outer hair cells, and in marginal cells of the cochlear stria vascularis (Kitahara et al., 2005; Tabuchi et al., 2005; Takumida et al., 2005). In the kidney cortex, TRPV4 is strongly expressed by epithelial cells of tubules and glomeruli (Wissenbach et al., 2000; Suzuki et al., 2003).

The calcium-selective (re)absorption channels, TRPV5 and TRPV6, are highly expressed in the kidney (mostly TRPV5) and the gastrointestinal tract (mostly TRPV6) (Den Dekker et al., 2003; Nijenhuis et al., 2003a, b). Additionally, they are rather abundantly expressed in pancreas, testis, prostate, placenta, brain, and salivary gland (Muller et al., 2000; Peng et al., 2000a, b, 2001a, b; Nijenhuis et al., 2003a, b).

TRPMs

TRPM1 is highly expressed in skin melanocytes (Duncan et al., 1998, 2001) as well as in the eye cells (Lis et al., 2005).

TRPM2 mainly resides in brain, peripheral blood cells (neutrophils), lung, spleen, eye, bone marrow, heart, and liver (Nagamine et al., 1998; Perraud et al., 2001; Sano et al., 2001; Hara et al., 2002; Zhang et al., 2003).

TRPM3 is primarily expressed in the kidney and, at lower levels, in brain, testis, ovary, pancreas, and spinal cord (Grimm et al., 2003; Lee et al., 2003; Oberwinkler et al., 2005).

TRPM4 displays high expression in heart, exo- and endocrine pancreas, smooth muscle, macula densa, lung, and placenta (Colquhoun et al., 1981; Launay et al., 2002; Ullrich et al., 2005; Vennekens and Nilius, 2007) and its closest homologue, TRPM5, is found in the tongue (taste bud cells), lungs, testis, digestive system, as well as in the brain (Perez et al., 2002; Hofmann et al., 2003; Zhang et al., 2003).

High TRPM6 expression has been demonstrated in kidney, colon and intestine (Riazanova et al., 2001; Schlingmann et al., 2002) whereas TRPM7 appears to be ubiquitously expressed (Nadler et al., 2001).

The cold/menthol receptor TRPM8 was originally identified in prostate cancer and then in a number of non-prostatic primary tumours of breast, colon, lung and skin origin (Tsavaler et al., 2001). It is also found in sensory neurons of DRG and TG where it participates in pain- and thermo-sensation (McKemy et al., 2002; Peier et al., 2002; Nealen et al., 2003), in nodose ganglion cells innervating the upper gut (Zhang and Barritt, 2004), vascular smooth muscle (Yang et al., 2006), liver (Henshall et al., 2003), gastric fundus (Mustafa and Oriowo, 2005) and in urothelium of bladder and different tissues of the male genital tract (Stein et al., 2004).

TRPMLs

TRPML1 is rather widely expressed in adult and fetal tissues, and localizes in the membrane of late endosomes/lysosomes where it controls Ca^{2+} levels important for formation and recycling of these organelles (LaPlante et al., 2002, 2004; Piper and Luzio, 2004). In contrast, TRPML3 seems to be confined to hair cells where it localizes to a vesicular compartment and in the stereocilia, as well as in cells of the stria vascularis (Di Palma et al., 2002). It has recently

been shown that the subcellular distribution of TRPMLs depends on multimerization of channels (Venkatachalam et al., 2006). TRPML1 and TRPML2 homotetramers are lysosomal proteins, whereas TRPML3 homotetramers reside in the ER. Co-expression of TRPML3 with either TRPML1 or TRPML2 localizes the channels to lysosomes, suggesting that TRPML1 and TRPML2 dictate the localization of TRPML3 (Venkatachalam et al., 2006).

TRPPs

TRPP2 is strongly expressed in ovary, fetal and adult kidney, testis, and small intestine in both motile and primary cilia (Mochizuki et al., 1996; Delmas, 2004, 2005). Formation of the TRPP2-TRPP1 complex is indispensable for insertion of functional polycystin complex into the plasma membrane (Hanaoka et al., 2000). TRPP3 is expressed in adult heart, skeletal muscle, brain, spleen, testis, retina, and liver (Nomura et al., 1998; Wu et al., 1998; Veldhuisen et al., 1999; Guo et al., 2000a, b). The expression in fetal tissues seems to be higher than in adult tissues (Nomura et al., 1998). The localization of TRPP5 is still unclear and has been described as either limited to testis (Veldhuisen et al., 1999) or additionally expressed in brain and kidney (Guo et al., 2000b).

TRPAs

TRPA1 is expressed in DRG and TG neurons (Story et al., 2003) as well as at rather low levels in hair cells (Corey 2003; Corey et al., 2004) and fibroblasts (Jaquemar et al., 1999).

TRPNs

TRPN channels are expressed in mechanosensitive cells, including ciliated mechanosensory organs in *Drosophila*, mechanosensory neurons in *C. Elegans*, and the hair cells of zebrafish ear (Walker et al., 2000; Sidi et al., 2003).

5.1.5 Pharmacology

Since TRP channels are involved in several pathological conditions, identification of selective antagonists, agonists or modulators of their functions would be highly instrumental in understanding their physiological function and for the eventual treatments of TRP related diseases (see section 5.1.7). So far, only a few endogenous TRP agonists have been identified (see section 5.1.3) and very often their molecular characterization is complicated due to low intracellular concentrations or multiple intermittent stages that mask the final molecule that directly interacts with the channel. The role of endogenous agonists such as DAG, AA metabolites, hormones, D-erythro-sphingosine, PIP_2 or Ca^{2+} have already been described in the previous sections. In this section we will briefly review a growing list of exogenous natural, as well as synthetic, compounds that in the submicromolar range directly modulate the TRP channel function.

The search for the molecular targets that determine responses for natural substances, especially from plants, led to the characterization of many TRP channels. The classical example is TRPV1, which was identified as a receptor for capsaicin, the pungent compound of hot chilli peppers (Caterina et al., 1997). Other plant-derived agonists of TRPV1 are resiniferatoxin (RTX) from the cactus *Euphorbia resinifera* (Szallasi et al., 1999) and piperine from black pepper (McNamara et al., 2005). TRPV3 as well as TRPV1 are activated by camphor, the waxy substance with penetrating odour extracted from *Cinnamomum camphora* (Moqrich *et al.*, 2005; Xu *et al.*, 2005). TRPM8 has been identified as a receptor for menthol and eucalyptol that are cooling agents from the mint plant *Mentha piperita* and the tree *Eucalyptus globulus*, respectively (McKemy *et al.*, 2002; Peier *et al.*, 2002a). TRPA1 is activated by isothiocyanates from mustard oil, horseradish and wasabi, cinnamaldehyde from cinnamon oil, methyl salicylate from winter green oil and D^9-tetrahydrocannabinol from *Cannabis sativa* (Bandell *et al.*, 2004; Jordt *et al.*, 2004; Bautista *et al.*, 2006). Allicin, an unstable component of fresh garlic *Allium sativum*, activates both TRPA1 (Bautista *et al.*, 2005; Macpherson *et al.*, 2005) and TRPV1 (Macpherson *et al.*, 2005).

In addition to these plant-derived substances, the list of synthetic TRP channel ligands is still growing and includes compounds that display relatively high selectivity for distinct TRP channels, such as olvanil for TRPV1 (Appendino *et al.*, 2005) and 4a–phorbol-12,13–didecanoate (4a-PDD) for TRPV4 (Watanabe *et al.*, 2002). Acrolein, an irritant in vehicle exhaust fumes and a tear gas activates TRPA1 (Bautista *et al.*, 2006). Many others activate more than one TRP channel: 2-aminoethyl diphenylborinate (2-APB) and its close analogue diphenylboronic anhydride (DPBA) act as agonists of TRPV1/2/3 and TRPM6 (Chung *et al.*, 2004; Hu *et al.*, 2004; Li *et al.*, 2006) but on the other hand 2-APB is an antagonist of TRPC1/3/4/6 and TRPM7 (Ma *et al.*, 2001; Trebak *et al.*, 2002; Vanden Abeele *et al.*, 2003; Hu *et al.*, 2004; Li *et al.*, 2006). Icilin (AG-3–5), a synthetic cooling agent, is the most potent activator of TRPM8 (McKemy *et al.*, 2002) but also activates TRPA1 albeit at 10–100-fold higher concentrations (Story *et al.*, 2003).

The availability of antagonists for TRPs is rather limited. *N*-(4-tert. butyl-phenyl)-4-(3-chloropyridin-2-yl) tetrahydropyrazine-1-(2*H*)-carboxamide (BCTC) and thio-BCTC are the most specific blockers of TRPV1 (Behrendt *et al.*, 2004). Recently, two acyl-polyamines, AG489 and AG505, from the venom of the funnel web spider *Agelenopsis aperta* were shown to inhibit TRPV1 (IC$_{50}$ ~33 nM at −40 mV) (Kitaguchi and Swartz 2005). This inhibition is strongly voltage-dependent and occurs via block of the pore at residues E636, D646, E651 and N628 in the channel pore (Kitaguchi and Swartz, 2005). Capsazepine, which was described as a very specific competitive capsaicin antagonist for TRPV1 (Urban and Dray, 1991), can also inhibit TRPM8 at higher concentrations (Behrendt *et al.*, 2004). Ruthenium red is a very effective blocker of all TRPVs and also TRPC1/3, TRPM6 and TRPA1 (Garcia-Martinez *et al.*, 2000; Watanabe *et al.*, 2002; Flemming *et al.*, 2003; Muraki *et al.*, 2003; Embark *et al.*, 2004; Voets *et al.*, 2004c; Chung *et al.*, 2005; Nagata *et al.*, 2005). Amiloride can inhibit TRPC3/6, TRPP2/3 and TRPA1 (Inoue *et al.*, 2001; Volk *et al.*, 2003; Nagata *et al.*, 2005). Thus, the lack of specific antagonists or agonists for most of TRP channels is an obstacle that hampers their functional characterization in cells, tissues, and living animals.

5.1.6 Physiology

As described in the previous sections, the TRP superfamily channels are important components of a cation influx machinery that integrates multiple intra- and extracellular stimuli in a broad range of physiological processes. Although functions of some of the TRP members are well described at the cellular level, the systemic physiology of most TRPs is rather poorly characterized (for the most

detailed review see the special issue about functional role of TRP channels edited B. Nilius in *Pflügers Archiv – European Journal of Physiology* and available online at http://www.springer.com/00424).

TRPCs

In the TRPC subfamily, TRPC1 plays an important role in the brain where it is responsible for generation of the excitatory postsynaptic potential (EPSP) (Clapham, 2003; Kim *et al.*, 2003; Beech *et al.*, 2004) and for netrin-1 and brain-derived neurotrophic factor (BDNF)-mediated growth cone guidance (Shim *et al.*, 2005; Wang and Poo, 2005). TRPC3 may also participate in BDNF-mediated growth cone guidance, but in a TRPC1-independent manner (Li *et al.*, 1999). Together with TRPC5, TRPC1 seems to be involved in brain development processes (Strubing *et al.*, 2001, 2003). Activation of TRPC1 by the GPCR/orexin A-dependent pathway can link this channel to sleep/wakefulness states, alertness and appetite (Larsson *et al.*, 2005). As already mentioned, mammalian TRPC2 plays an essential role in pheromone detection and thereby regulates sexual and social behaviours such as gender recognition and male-male aggression in mice (Stowers *et al.*, 2002; Lucas *et al.*, 2003; Zufall *et al.*, 2005). Results obtained in *trpc4*$^{-/-}$ knockout mice demonstrate that TRPC4 is an essential component of endothelium-dependent vasorelaxation and regulation of transcellular permeation of the endothelial layer *in vivo* (Freichel *et al.*, 2001; Tiruppathi *et al.*, 2002). TRPC6 seems to be involved in vasoregulation (Inoue *et al.*, 2001) and pathogenesis of FSGS in kidney (see next section).

TRPVs

TRPV1–4 channels are involved in warmth and heat sensation (see section 5.1.3). Additionally, TRPV1 is involved in nociception, contributing to the detection and integration of (painful) chemical and thermal stimuli (Caterina *et al.*, 2000). TRPV2 likely participates to skeletal muscle and cardiac muscle degeneration (Kanzaki *et al.*, 1999a, b; Iwata *et al.*, 2003). TRPV4, a mechano- and osmosensor, is very likely involved in functions of (cardio)vascular endothelium and bladder urothelium (Nilius *et al.*, 2004d; O'Neil and Heller 2005; De Ridder and Nilius, personal communication). TRPV5/6 channels are involved in Ca^{2+} (re)absorption in the kidney and intestine (Den Dekker *et al.*, 2003; Nijenhuis *et al.*, 2003a, b; Nilius *et al.*, 2004; O'Neil and Heller, 2005).

TRPMs

TRPM8 is a cold receptor that participates in cold sensation (McKemy *et al.*, 2002; Peier *et al.*, 2002a). Two other TRPMs, TRPM4 and TRPM5, have also been classified as thermoTRPs activated above 14 °C (Talavera *et al.*, 2005). Additionally, TRPM5 is involved in transduction of sweet, amino acid (umami), and bitter taste in taste bud cells of the tongue (Zhang *et al.*, 2003). TRPM6/7 are crucial for Mg^{2+} homeostasis and reabsorption in the kidney and intestine (Schlingmann *et al.*, 2002; Chubanov *et al.*, 2004; Voets *et al.*, 2004c). Interestingly, TRPM7 also appears to be involved in skeletogenesis in zebrafish (Elizondo *et al.*, 2005). TRPM1 seems to play an important role in melanoma invasiveness (Duncan *et al.*, 1998, 2001) whereas TRPM2 might be involved in oxidative stress response (Zhang *et al.*, 2003).

TRPAs

As with TRPM8, TRPA1 is a noxious cold sensor, which is additionally involved in pain perception and mechanosensation (Bandell *et al.*, 2004; Corey *et al.*, 2004; Nagata *et al.*, 2005). As already mention, recently published data obtained in *trpa1*$^{-/-}$ knockout mice did not confirm the expected role of TRPA1 in mechanical signals upon hearing (Bautista *et al.*, 2006; Kwan *et al.*, 2006).

TRPML and TRPPs

TRPML and TRPP channels are connected with pathophysiological processes that lead to diseases that are described in the next section.

5.1.7 Disease relevance

The number of reports describing TRP-related diseases has been steadily growing during the last few years. In this section we will briefly summarize available information about known and potential relationship of TRP channel physiology to human channelopathies. Channelopathies are conventionally defined as diseases that are linked to mutations in the gene encoding the channel. So far, based on this definition, five TRP channel-related channelopathies have been identified (Nilius *et al.*, 2005e).

Mutations of the *TRPC6* gene are associated with the human proteinuric kidney disease focal and segmental glomerulosclerosis (FSGS) (Reiser *et al.*, 2005; Winn *et al.*, 2005). In this disease, the initially well-developed podocyte foot processes and the glomerular slit diaphragms loose their functionality and integrity between childhood and late adulthood (Kriz, 2005; Reiser *et al.*, 2005; Winn *et al.*, 2005). It is not clear how mutations in TRPC6 are linked to development of FSGS. Lack of nephrin, a central component of the slit diaphragm, induces upregulation of TRPC6 expression in podocytes, resulting in the augmented presence of TRPC6 in the podocyte plasma membrane (Reiser *et al.*, 2005). Enhanced TRPC6 activity leads to a Ca^{2+} overload in the podocyte that initiates apoptosis, or causes dysregulation of the permeability barrier (Reiser *et al.*, 2005). Alternatively, mutations in TRPC6 may affect adaptation of the podocyte to changes in glomerular filtration pressure. Another possibility includes a role for TRPC6 in the guidance of nephrin and podocin, which are required to maintain the filtration barrier (Winn *et al.*, 2006). However, it has to be mentioned that not all TRPC6 mutations found in FSGS patients cause a TRPC6 gain of function.

The *trpm6* gene locus has been associated with the disease hypomagnesaemia with secondary hypocalcaemia (HSH/HOMG) (Schlingmann *et al.*, 2002; Walder *et al.*, 2002). HSH/HOMG is an autosomal-recessive disorder, which is characterized by very low serum levels of Mg^{2+} and Ca^{2+}, resulting from impaired intestinal Mg^{2+} absorption in the presence of an additional renal Mg^{2+} leak. HSH/HOMG patients display a large number of neurological symptoms, including seizures and muscle spasms during infancy. There are two independent pathways for intestinal Mg^{2+} uptake in the brush-border epithelia. A passive paracellular absorption that increases linearly with increasing luminal Mg^{2+} concentrations, and an active transcellular transport, reaching saturation at high luminal Mg^{2+} concentrations (Konrad *et al.*, 2004). TRPM6 is essential for the active transcellular Mg^{2+} uptake at the

apical membrane whereas at the basolateral membrane an as yet unidentified Na^+/Mg^{2+} exchanger is responsible for onward transport of Mg^{2+} into the interstitial space and blood. Mg^{2+} overload is also prevented by TRPM6, which is highly regulated by the intracellular concentration of Mg^{2+} (Voets *et al.*, 2004c). In the absence of TRPM6 the paracellular pathway is used to allow Mg^{2+} absorption. However, in order to generate a stronger driving force for passive Mg^{2+} uptake, the luminal Mg^{2+} concentration has to be increased by a high dietary Mg^{2+} intake (Cole *et al.*, 2000). Unfortunately for HSH/HOMG patients, a high lumenal Mg^{2+} level frequently leads to severe diarrhoea, acting as an osmotic laxative. In nephrons, most Mg^{2+} is reabsorbed by the paracellular pathway in the thick ascending limb of the loop of Henle. The final reabsorption of filtered Mg^{2+} from the lumen to the blood, and thus determination of urinary Mg^{2+} excretion, takes place in the distal convoluted tubule (DCT) in an active manner involving TRPM6 located on the apical membrane of the epithelial cells. In the absence of TRPM6 function Mg^{2+} can no further be reabsorbed in DCT, resulting in the urinary leak of $Mg^{?+}$ that has often been observed in HSH patients.

The mutation T1482I of TRPM7, which is located between the channel and the kinase domain, has been described in a subgroup of both Guamanian amyotrophic lateral sclerosis (ALS-G) and Parkinsonism dementia (PD-G) patients (Hermosura *et al.*, 2005). ALS-G and PD-G are related neurodegenerative disorders with an etiology, that very likely depends on a complex interplay of genetic and environmental factors (Plato *et al.*, 2002, 2003). The T1482I TRPM7 mutant has an increased sensitivity to inhibition by intracellular Mg^{2+} within a physiologically relevant range without detectable alteration in α-kinase activity. Thus, increased sensitivity of TRPM7 to inhibition by Mg^{2+} leads to a reduced intracellular Mg^{2+} concentration (Schmitz *et al.*, 2003; Hermosura *et al.*, 2005). Importantly, a high frequency of these neurodegenerative disorders has been found in the Pacific Islands Guam and Rota where people live in a Ca^{2+}- and Mg^{2+}-deficient environment (Plato *et al.*, 2002).

Mutations in TRPP genes cause polycystic kidney disease (PKD), the most common inherited form of kidney failure. The major common characteristic of PKD is the progressive development of large epithelial-lined cysts that are filled with fluid and can occupy much of the mass of the abnormally enlarged kidneys, thereby compressing and destroying normal renal tissue and impairing kidney function (Grantham, 1993). One possible mechanism of cyst formation has recently been proposed (Benezra, 2005; Li *et al.*, 2005) i.e. membrane bound TRPP1 and TRPP2 inhibit the nuclear transfer of the helix-loop-helix (HLH) protein Id2, a crucial regulator of cell proliferation and differentiation. Formation of the TRPP1/2 complex is necessary for Id2 binding. In PKD patients, which lack this receptor-ion channel complex, Id2 can enter the nucleus and activate G1-S progression. This hypothetical mechanism appears quite attractive, as a clearly enhanced nuclear localization of Id2 has been observed in renal epithelial cells from PKD patients (Li *et al.*, 2005).

Mucolipidosis type IV (MLIV) is caused by mutations in the *TRPML1* gene. MLIV is an autosomal-recessive, neurodegenerative, lysosomal storage disorder that is clinically characterized by severe psychomotor retardation, ophthalmologic abnormalities, agenesis of the corpus callosum, blood iron deficiency and achlorohydria

(Bach, 2005). The pathophysiology of MLIV is not completely understood. TRPML1 seems to play a crucial role in Ca^{2+} release from the endosome/lysosome hybrid, which triggers, in a Ca^{2+}-dependent manner, the fusion and trafficking of these organelles. Thus, defective TRPML1 channels might block the endocytotic route to the final lysosome (LaPlante *et al.*, 2002, 2004). On the other hand, the lysosomal pH in TRPML1-deficient cells from MLIV patients is highly acidic. Since TRPML1 functions as a H^+ channel, the over-acidification of lysosomes in TRPML1 deficient cells can be explained by reduced H^+ leakage (Kiselyov *et al.*, 2005). In addition, TRPML1 deficiency causes a marked reduction in lipid hydrolysis, which may explain accumulation of lipids and membraneous material in intracellular organelles such as lysosomes of MLIV patients (Kiselyov *et al.*, 2005; Soyombo *et al.*, 2006).

In addition to inherited channelopathies, channel-related diseases can be also caused by changes in channel abundance or channel sensitization or desensitization, leading to exaggerated or diminished responses to various pathological stimuli. Moreover, altered production of various endogenous agents during early disease stages (e.g. in inflammation) can impair channel function, leading to the progression of the disease. All these pathogenic factors may cause abnormal regulation of TRP channels, which

are generally activated by diverse gating stimuli and function as molecular integrators of external and/or internal signals. Thus, alterations of TRP channel function are possibly connected with many systemic diseases, e.g. hypertension (TRPC3/6, TRPV1/4), migraine (TRPV1), asthma (TRPC1/6, TRPV1/4), allergy (TRPV1, TRPM4), acute and chronic pain (TRPV1/2/4, TRPA1, TRPM8), cancer (TRPC1, TRPV5/6, TRPM1/8) and ageing (TRPV5/6). For a more detailed account of TRP channel-related diseases, we refer to the most recent reviews focusing on this aspect (Nilius *et al.*, 2005e, 2006).

5.1.8 Concluding remarks

This review strongly emphasizes the important role of the TRP superfamily channels in many fundamental cellular functions in excitable and non-excitable cells. Understanding the variety and complexity of permeation and gating mechanisms in this superfamily forms an important challenge to many ion channel scientists. Furthermore, the vital role of TRPs in human pathopysiology and disease should be given urgent priority in biomedical sciences. Indeed, despite increasing scientific efforts, we are still quite far from fully understanding how these fascinating channels work.

Table 5.1.1 Properties of TRP channels. For details and references, please refer to the relevant sections in the text

Channel subunit	Biophysical characteristics	Cellular and subcellular localization	Pharmacology	Physiology	Relevance to disease states*
TRPC1 (TRP-1) Chromosomal location: mouse: 9 E4 human: 3q22–q24	Nonselective γ ~ 16 pS	Ubiquitous	Antagonists and blockers: 2-APB Gd La Ca²⁺–CaM	Generation of the excitatory postsynaptic potential (EPSP) in brain, and netrin-1 and brain-derived neurotrophic factor (BDNF)-mediated growth cone guidance. Activation by the GPCR/orexin A-dependent pathway can link this channel to sleep/wakefulness states, alertness and appetite. Brain development (together with TRPC5). Mechano-sensation.	Possible connections: Asthma, bronchial hyperresponsiveness, defective immunoresponse, heart hypertrophy, hypertension, Duchenne muscular dystrophy, myotonic dystrophy type 2, Seckel syndrome, neurodegenerative disorders
TRPC2 (Trp2) Chromosomal location: mouse: 7 F1 human: 11p15.4–p15.3	$P_{Ca}/P_{Na} \sim 2.7$ γ ~ 42 pS	Dendritic tips of the vomeronasal sensory neurons and spermatozoa	Agonists and activators: DAG	Pheromone detection that regulates sexual and social behaviours such as gender recognition and male–male aggression	Gender recognition and behavioral defects in mouse
TRPC3 (Trp3) Chromosomal location: mouse: 3 B human: 4q25–q27	$P_{Ca}/P_{Na} \sim 1.6$ γ ~ 60–66 pS	CNS and smooth and cardiac muscle cells	Agonists and activators: DAG intracellular Ca²⁺? Antagonists and blockers: La Gd ruthenium red 2-APB	BDNF-mediated growth cone guidance (TRPC1-independent manner)	Possible connections: Arterial hypertension, heart hypertrophy, essential hypertension, pulmonary diseases
TRPC4 (Trp4) Chromosomal location: mouse: 3 D human: 13q13.1–q13.2	$P_{Ca}/P_{Na} \sim 1.1–7.7$ γ ~ 30–42 pS	Placenta, adrenal gland, CNS, endothelium, smooth muscle cells, kidney, intestinal cells of Cajal (ICC)	Agonists and activators: calmidazolium Antagonists and blockers: 2-APB La niflumic acid DIDS high intracellular Ca²⁺	Endothelium-dependent vasorelaxation and regulation of transcellular permeation of the endothelial layer	Possible connections: Pulmonary diseases, impairments of endothelium-dependent vasorelaxation and endothelial barrier function, breast cancer, atopic dermatitis
TRPC5 (Trp5) Chromosomal location: mouse: X F2 human: Xq23–q24	$P_{Ca}/P_{Na} \sim 1.8–9.0$ γ ~ 38–64 pS	Brain, especially in fetal brain and very weak expression in other tissues	Antagonists and blockers: 2-APB La	Brain development (together with TRPC1)	Possible connections: Pulmonary diseases, coronary heart disease, migraine, X-linked mental retardation.
TRPC6 (Trp6) Chromosomal location: mouse: 9 A1 human: 11q21–q22	$P_{Ca}/P_{Na} \sim 5$ γ ~ 28–37 pS	Smooth muscle cells, lung, brain, placenta, kidney (podocyte foot processes), spleen, ovary and small intestine	Agonists and activators: DAG Antagonists and blockers: La Gd Cd amiloride 2-APB	Vasorelaxation	Channelopathy: Proteinuric kidney disease: focal and segmental glomerulosclerosis (FSGS) Possible connections: Heart hypertrophy, hypertension, mucus hypersecretion in pulmonary diseases, Duchenne muscular dystrophy, neurodegenerative disorders, lung cancer

(continued)

Table 5.1.1 (*Continued*) Properties of TRP channels. For details and references, please refer to the relevant sections in the text

Channel subunit	Biophysical characteristics	Cellular and subcellular localization	Pharmacology	Physiology	Relevance to disease states*
TRPC7 (TRP7) Chromosomal location: mouse: 13 B2 human: 5q31.2	$P_{Ca}/P_{Na} \sim 0.5–5.4$ $\gamma \sim 24–50$ pS	Pituitary glands, kidney and CNS (human) Heart and lung: weak expression in CNS and kidney (mouse)	Agonists and activators: DAG Antagonists and blockers: La	ND	ND
TRPV1 (OTRPC1, VR1) Chromosomal location: mouse: 11 B3 human: 17p13.3	$P_{Ca}/P_{Na} \sim 10$ (capsaicin activated current) $P_{Ca}/P_{Na} \sim 5$ (heat activated current) $\gamma \sim 35–80$ pS	Dorsal root and trigeminal ganglia; spinal and peripheral nerve terminals, brain, skin (cutaneous sensory nerve fibers, mast cells, epidermal keratinocytes, dermal blood vessels, the inner root sheet and the infundibulum of hair follicles, differentiated sebocytes, sweat gland ducts, and the secretory portion of eccrine sweat glands), pancreas, bladder (urothelium, smooth muscle, blood vessels and neurons)	Agonists and activators: capsaicin resiniferatoxin piperine camphor allicin arachidonic acid derivatives (anandamide, 12,15-(S)-hydroperoxyeicosatetraenoic acid, leukotriene B$_4$) 2-APB DPBA olvanil ethanol H$^+$ Antagonists and blockers: capsazepine ruthenium red PIP$_2$ BCTC thiol-BCTC acylpolyamines	Thermo-sensation (moderate heat) Nociception	Possible connections: Thermal hyperalgesia, allodynia, functional bowel disease, inflammatory bowel disease, vulvodynia, osteoarthritis, pancreatitis, gastro-oesophagal reflux disease, bladder disease, cystitis, asthma, migraine, schizophrenia, general pain, tooth pain, breast cancer, non-insulin-dependent diabetes mellitus, myasthaenic syndrome
TRPV2 (OTRPC2, VRL-1) Chromosomal location: mouse: 11 B2 human: 17p11.2	$P_{Ca}/P_{Na} \sim 1–3$	Dorsal root ganglia and CNS neurons, gastrointestinal tract, spleen, mast cells, smooth, cardiac and skeletal muscle cells	Agonists and activators: IGF-1 neuropeptide HA 2-APB DPBA Antagonists and blockers: La ruthenium red	Thermo-sensation (noxious heat) Nociception	Possible connections: Muscular dystrophy, cardiomyopathy, cardiac hypertrophy, central areolar choroidal dystrophy, myasthenic syndrome, prostate cancer
TRPV3 (VRL-3) Chromosomal location: mouse: 11 B4 human: 17p13.3	$P_{Ca}/P_{Na} \sim 2.6–12$ $\gamma \sim 172–190$ pS	Dorsal root and trigeminal ganglia neurons, brain, keratinocytes, hair follicles, tongue and testis	Agonists and activators: 2-APB DPBA camphor Antagonists and blockers: ruthenium red La	Thermo-sensation (moderate heat) Nociception	Possible connections: Breast cancer, myasthenic syndrome, non-insulin-dependent diabetes mellitus

Channel	Permeability / conductance	Expression	Agonists/activators and antagonists/blockers	Function	Possible connections
TRPV4 (OTRPC4, VRL-2, VR-OAC, Trp12) Chromosomal location: mouse: 5 F human: 12q24.1	P_{Ca}/P_{Na} ~ 6–10 γ ~ 90 pS	CNS (large neurons), trigeminal ganglia, heart, liver, kidney, skin (keratinocytes), blood vessels (endothelium), bladder (urothelium) and testis, cochlea (inner and outer hair cells, marginal cells of the cochlear stria vascularis), kidney (epithelial cells of tubules and glomeruli)	Agonists and activators: phorbol esters arachidonic acid derivatives (anandamide, epoxieicosatrienoic acids) Antagonists and blockers: ruthenium red La Gd	Thermo-sensation (moderate heat) Mechano-sensation Osmo-sensation Nociception.	Possible connections: Hypotonic hyperalgesia, allodynia, thermal hyperalgesia, asthma, bronchial hyperresponsiveness, neuropathic pain, autosomal non-syndromic hearing loss, impairment osmoregulation, hypertension, cardiopathy
TRPV5 (OTRPC3, ECaC1, CaT2) Chromosomal location: mouse: 6 B2 human: 7q35	P_{Ca}/P_{Na} > 100 γ ~ 75 pS (for monovalent cations)	Highly expressed in the kidney and less abundant in the gastrointestinal tract, pancreas, testis, prostate, placenta, brain and salivary gland	Antagonists and blockers: ruthenium red Cd Gd La Mg Cu Pb	Ca^{2+} (re)absorption in kidney and intestines	Possible connections: Hypercalciuria, osteoporosis, renal tubular acidosis, cancer
TRPV6 (ECaC2, CaT1, CaT-L) Chromosomal location: mouse: 6 B2 human: 7q33–q34	P_{Ca}/P_{Na} > 100 γ ~ 40–70 pS (for monovalent cation)	Highly expressed in the gastrointestinal tract and less abundant in the kidney, pancreas, testis, prostate, placenta, brain and salivary gland	Antagonists and blockers: ruthenium red Cd Gd La Mg	Ca^{2+} (re)absorption in intestines and kidney	Possible connections: Decreased fertility, alopecia, Ca^{2+} wasting, reduced bone density, prostate cancer, renal tubular acidosis, cancer
TRPM1 (melastatin-1, MLSN1) Chromosomal location: mouse: 7 C human: 15q13–q14	ND	Skin melanocytes, eye cells	ND	ND	Suppressor of metastatic melanomas Possible connections: Lymphoma, hypertention
TRPM2 (LTrpC2, TRPC7) Chromosomal location: mouse: 10 C1 human: 21q22.3	P_{Ca}/P_{Na} ~ 0.5–1.6 γ ~ 52–80 pS	Brain, bone marrow, peripheral blood cells (neutrophils), lung, spleen, eye, heart and liver	Agonists and activators: ADP-ribose	Oxidative stress response?	Possible connections: Bipolar disorder, non-syndromic hereditary deafness, holoprosencephaly, Knobloch syndrome, neuronal cell death, epilepsy, leukemia
TRPM3 (melastatin-2, LTrpC3, MLSN2) Chromosomal location: mouse: 19 C1 human: 9q21.13	P_{Ca}/P_{Na} ~ 1–10 (depends on splice variant) γ ~ 65 (Ca^{2+})- 130 pS	Primary in kidney and, at lower levels, in brain, testis, ovary, pancreas and spinal cord	Agonists and activators: D-erythro-sphingosine Antagonists and blockers: Gd	ND	Possible connections: Amyotrophic lateral sclerosis with frontotemporal dementia, early-onset pulverulent cataract, hemophagocytic lymphohistiocytosis, infantile nephronophthisis

(continued)

Table 5.1.1 (*Continued*) Properties of TRP channels. For details and references, please refer to the relevant sections in the text

Channel subunit	Biophysical characteristics	Cellular and subcellular localization	Pharmacology	Physiology	Relevance to disease states*
TRPM4 Chromosomal location: mouse: 7 B4 human: 19q13.32	Selective for monovalent cations $\gamma \sim 25$ pS	Heart, exo- and endocrine pancreas, smooth muscle, macula densa, lung, and placenta	Agonists and activators: intracellular Ca^{2+} Antagonists and blockers: nucleotides polyamines La Gd	ND	Possible connections: Allergy, hyperresponsiveness in immune cells, Bayliss effect, stroke, diabetes mellitus, prostate cancer, leukemia, lymphoma
TRPM5 Chromosomal location: mouse: 7 F5 human: 11p15.5	Selective for monovalent cations $\gamma \sim 16{-}25$ pS	Tongue (taste bud cells), lungs, testis, digestive system, as well as in the brain	Agonists and activators: intracellular Ca^{2+}	Taste (sweet, bitter, umami)	Possible connections: Beckwith–Wiedemann syndrome, diabetes mellitus, breast bladder and lung cancer
TRPM6 (CHAK2) Chromosomal location: mouse: 19 B human: 9q21.13	$P_{Mg}/P_{Na} \sim 6$	Kidney, colon and intestine	Agonists and activators: 2-APB Antagonists and blockers: ruthenium red	Mg^{2+} homeostasis and reabsorption in kidney and intestine	Channelopathy: Autosomal-recessive, hypomagnesaemia with secondary hypocalcaemia. Possible connections: Spastic paraplegia, deafness, amyotrophic lateral sclerosis with frontotemporal dementia
TRPM7 (CHAK1, LTrpC7, TRP-PLIK) Chromosomal location: mouse: 2 F2 human: 15q21	$P_{Ca}/P_{Na} \sim 3$ $\gamma \sim 40{-}105$ pS	Ubiquitous	Antagonists and blockers: intracellular Mg^{2+} 2-APB	Mg^{2+} homeostasis and reabsorption in kidney and intestine Skeletogenesis?	Channelopathy: Amyotrophic lateral sclerosis-Parkinsonism/dementia complex Possible connections: Neuronal cell death, defective vascular remodelling, hypertension, stroke, defective ossification, spinocerebellar ataxia, colorectal adenoma and carcinoma, susceptibility to dyslexia
TRPM8 (LTrpC6, Trp-p8) Chromosomal location: mouse: 1 C5 human: 2q37.1	$P_{Ca}/P_{Na} \sim 1{-}3$ $\gamma \sim 83$ pS	Sensory dorsal root and trigeminal ganglia neurons, nodose ganglion cells innervating the upper gut, vascular smooth muscle cells, liver, gastric fundus, bladder (urothelium) and different tissues of the male genital tract Highly expressed in prostate, breast, colon, lung and skin origin tumors	Agonists and activators: menthol icilin eucalyptol Antagonists and blockers: capsazepine	Thermo-sensation (cold)	Possible connections: Cold hyperalgesia, tumours (prostate, breast, colon, lung, skin), painful bladder syndrome, pulmonary diseases, Parkinson disease
TRPML1 (mucolipin-1, mucolipidin, MG-2, MCOLN1) Chromosomal location: mouse: 8 A1.1 human: 19p13.3–13.2	H+ permeable $P_{Ca}/P_{Na} \sim 1$ $\gamma \sim 46{-}83$ pS	Ubiquitous	ND	ND	Channelopathy: Mucolipidosis IV Possible connections: Alzheimer disease, liposarcoma, mental retardation

TRPML2 (MCOLN2) Chromosomal location: mouse: 3 H3 human: 1p22	ND	Ubiquitous?	ND	ND	Possible connections: Neurosensory disorders?
TRPML3 (MCOLN3) Chromosomal location: mouse: 3 H3 human: 1p22.3	ND	Hair cells (stria vascularis, stereocilia)	ND	Hair cell maturation?	Possible connections: Neurosensory disorders?
TRPP2 (PKD2, Polycystwin, Polycystin-2) Chromosomal location: mouse: 5 E4 human: 4q22	$P_{Ca}/P_{Na} \sim 1-5$ $\gamma \sim 40-177$ pS	Ovary, fetal and adult kidney, testis, and small intestine in both motile and primary cilia	Antagonists and blockers: amiloride Gd La	Cardiac, skeletal and renal development	Channelopathy: Autosomal dominant polycystic kidney disease Possible connections: Cardiac septal defects, distrurbances left-right axis, Bardet–Biedl syndrome, autosomal recessive polycystic kidney disease, neutropenia, susceptibility to psoriasis
TRPP3 (PKD2L1, Polycystin-L) Chromosomal location: mouse: 19 D1 human: 10q24–q25	$P_{Ca}/P_{Na} \sim 4$ $\gamma \sim 137$ pS	Adult heart, skeletal muscle, brain, spleen, testis, retina and liver	ND	Renal development	Possible connections: Autosomal recessive polycystic kidney disease, kidney-retina defects, brain defects, renal hypoplasia, optic nerve coloboma with renal disease, epilepsy, Alzhe mer desease
TRPP5 (PKD2L2, Polycystin-L2) Chromosomal location: mouse: 18 B3 human: 5q31	$P_{Ca}/P_{Na} \sim 1-5$ $\gamma \sim 300$ pS	Testis, brain and kidney	Antagonists and blockers: amiloride Gd La Ni	ND	Possible connections: Autosomal recessive polycystic kidney disease, epilepsy, allergy and asthma susceptibility, muscular dystrophy, inflammatory bowel disease
TRPA1 (ANKTM1) Chromosomal location: mouse: 1 A3 human: 8q13	$P_{Ca}/P_{Na} \sim 0.8-1.4$ $\gamma \sim 40-105$ pS	Hair cells, sensory dorsal root and trigeminal ganglia neurons, fibroblasts	Agonists and activators: icilin isothiocyanates allicin cinnamaldehyde methyl salicylate D^9-tetrahydrocannabinol acrolein Antagonists and blockers: ruthenium red amiloride	Thermo-sensation (noxious cold) Mechano-sensation Nociception	Possible connections: Inflammatory pain, cold hyperalgesia, mechanical pain, mechanical hypersensitivity, inflammatory pain, spastic paraplegia sensorineural deafness, convulsions

* For more details see the most comprehensive review see Nilius et al. (2006).

ND, not determined.

References

Appendino G, De Petrocellis L, Trevisani M *et al.* (2005). Development of the first ultra-potent capsaicinoid agonist at transient receptor potential type V1 (TRPV1) channels and its therapeutic potential. *JPET* **312**, 561–570.

Arniges M, Vazquez E, Fernandez-Fernandez JM *et al.* (2004). Swelling-activated Ca^{2+} entry via TRPV4 channel is defective in cystic fibrosis airway epithelia. *J Biol Chem* **279**, 54062–54068.

Bach G (2005). Mucolipin 1: endocytosis and cation channel-a review. *Pflugers Arch* **451**, 313–317.

Bandell M, Story GM, Hwang SW *et al.* (2004). Noxious cold ion channel TRPA1 is activated by pungent compounds and bradykinin. *Neuron* **41**, 849–857.

Bargal R, Avidan N, Ben-Asher E *et al.* (2000). Identification of the gene causing mucolipidosis type IV. *Nat Genet* **26**, 118–123.

Barnhill JC, Stokes AJ, Koblan-Huberson M *et al.* (2004). RGA protein associates with a TRPV ion channel during biosynthesis and trafficking. *J Cell Biochem* **91**, 808–820.

Bassi MT, Manzoni M, Monti E *et al.* (2000). Cloning of the gene encoding a novel integral membrane protein, mucolipidin, and identification of the two major founder mutations causing mucolipidosis type IV. *Am J Hum Genet* **67**, 1110–1120.

Bautista DM, Jordt SE, Nikai T *et al.* (2006). TRPA1 mediates the inflammatory actions of environmental irritants and proalgesic agents. *Cell* **124**, 1269–1282.

Bautista DM, Movahed P, Hinman A *et al.* (2005). Pungent products from garlic activate the sensory ion channel TRPA1. *PNAS* **102**, 12248–12252.

Beech DJ, Muraki K and Flemming R (2004). Non-selective cationic channels of smooth muscle and the mammalian homologues of *Drosophila* TRP. *J Physiol* **559**, 685–706.

Behrendt HJ, Germann T, Gillen C *et al.* (2004). Characterization of the mouse cold-menthol receptor TRPM8 and vanilloid receptor type-1 VR1 using a fluorometric imaging plate reader (FLIPR) assay. *Br J Pharmacol* **141**, 737–745.

Benezra R (2005). Polycystins: inhibiting the inhibitors. *Nat Cell Biol* **7**, 1064–1065.

Benham CD, Davis JB and Randall AD (2002). Vanilloid and TRP channels: a family of lipid-gated cation channels. *Neuropharmacology* **42**, 873–888.

Berman ER, Livni N, Shapira E *et al.* (1974). Congenital corneal clouding with abnormal systemic storage bodies: a new variant of mucolipidosis. *J Pediatr* **84**, 519–526.

Birder LA, Kanai AJ, de Groat WC *et al.* (2001). Vanilloid receptor expression suggests a sensory role for urinary bladder epithelial cells. *PNAS* **98**, 13396–13401.

Birnbaumer L, Zhu X, Jiang M *et al.* (1996). On the molecular basis and regulation of cellular capacitative calcium entry: roles for Trp proteins. *PNAS* **93**, 15195–15202.

Boels K, Glassmeier G, Herrmann D *et al.* (2001). The neuropeptide head activator induces activation and translocation of the growth-factor-regulated Ca(2+)-permeable channel GRC. *J Cell Sci* **114**, 3599–3606.

Boulay G (2002). Ca(2+).-calmodulin regulates receptor-operated Ca(2+) entry activity of TRPC6 in HEK-293 cells. *Cell Calcium* **32**, 201–207.

Caterina MJ, Leffler A, Malmberg AB *et al.* (2000). Impaired nociception and pain sensation in mice lacking the capsaicin receptor. *Science* **288**, 306–313.

Caterina MJ, Rosen TA, Tominaga M *et al.* (1999). A capsaicin-receptor homologue with a high threshold for noxious heat. *Nature* **398**, 436–441.

Caterina MJ, Schumacher MA, Tominaga M *et al.* (1997). The capsaicin receptor: a heat-activated ion channel in the pain pathway. *Nature* **389**, 816–824.

Cayouette S, Lussier MP, Mathieu EL *et al.* (2004). Exocytotic insertion of TRPC6 channel into the plasma membrane upon Gq protein-coupled receptor activation. *J Biol Chem* **279**, 7241–7246.

Chuang HH, Prescott ED, Kong H *et al.* (2001). Bradykinin and nerve growth factor release the capsaicin receptor from PtdIns(4,5). P2-mediated inhibition. *Nature* **411**, 957–962.

Chubanov V, Waldegger S, Mederos Y *et al.* (2004). Disruption of TRPM6/TRPM7 complex formation by a mutation in the TRPM6 gene causes hypomagnesemia with secondary hypocalcemia. *PNAS* **101**, 2894–2899.

Chung MK, Guler AD and Caterina MJ (2005). Biphasic currents evoked by chemical or thermal activation of the heat-gated ion channel, TRPV3. *J Biol Chem* **280**, 15928–15941.

Chung MK, Lee H, Mizuno A *et al.* (2004). 2-aminoethoxydiphenyl borate activates and sensitizes the heat-gated ion channel TRPV3. *J Neurosci* **24**, 5177–5182.

Chyb S, Raghu P and Hardie RC (1999). Polyunsaturated fatty acids activate the *Drosophila* light-sensitive channels TRP and TRPL. *Nature* **397**, 255–259.

Clapham DE (1995). Calcium signaling. *Cell* **80**, 259–68.

Clapham DE (2003). TRP channels as cellular sensors. *Nature* **426**, 517–524.

Clapham DE, Runnels LW and Strubing C (2001). The trp ion channel family. *Nat Rev Neurosci* **2**, 387–396.

Colbert HA, Smith TL and Bargmann CI (1997). OSM-9, a novel protein with structural similarity to channels, is required for olfaction, mechanosensation and olfactory adaptation in *Caenorhabditis elegans*. *J Neurosci* **17**, 8259–8269.

Cole DE, Kooh SW and Vieth R (2000). Primary infantile hypomagnesaemia: outcome after 21 years and treatment with continuous nocturnal nasogastric magnesium infusion. *Eur J Pediatr* **159**, 38–43.

Colquhoun D, Neher E, Reuter H *et al.* (1981). Inward current channels activated by intracellular Ca in cultured cardiac cells. *Nature* **294**, 752–754.

Corey DP (2003). New TRP channels in hearing and mechanosensation. *Neuron* **39**, 585–588.

Corey DP, Garcia-Anoveros J, Holt JR *et al.* (2004). TRPA1 is a candidate for the mechanosensitive transduction channel of vertebrate hair cells. *Nature* **432**, 723–730.

Cortright DN, Crandall M, Sanchez JF *et al.* (2001). The tissue distribution and functional characterization of human VR1. *Biochem Biophys Res Commun* **281**, 1183–1189.

Cosens DJ and Manning A (1969). Abnormal electroretinogram from a *Drosophila* mutant. *Nature* **224**, 285–287.

Cuajungco MP, Grimm C, Oshima K *et al.* (2006). PACSINs bind to the TRPV4 cation channel: PACSIN 3 modulates the subcellular localization of TRPV4. *J Biol Chem* **281**, 18753–18762.

D'Esposito M, Strazzullo M, Cuccurese M *et al.* (1998). Identification and assignment of the human transient receptor potential channel 6 gene TRPC6 to chromosome 11q21–>q22. *Cytogenet Cell Genet* **83**, 46–47.

Delmas P (2004). Polycystins: from mechanosensation to gene regulation. *Cell* **118**, 145–148.

Delmas P (2005). Polycystins: polymodal receptor/ion-channel cellular sensors. *Pflugers Arch* **451**, 264–276.

Delmas P, Nauli S, Li X *et al.* (2004). Gating of the polycystin ion channel signaling complex in neurons and kidney cells. *FASEB J* **18**, 740–742.

Den Dekker E, Hoenderop JG, Nilius B *et al.* (2003). The epithelial calcium channels, TRPV5 and TRPV6: from identification towards regulation. *Cell Calcium* **33**, 497–507.

Denda M, Fuziwara S, Inoue K *et al.* (2001). Immunoreactivity of VR1 on epidermal keratinocyte of human skin. *Biochem Biophys Res Commun* **285**, 1250–1252.

Denker SP and Barber DL (2002). Ion transport proteins anchor and regulate the cytoskeleton. *Curr Opin Cell Biol* **14**, 214–220.

Di Palma F, Belyantseva I, Kim H *et al.* (2002). Mutations in Mcoln3 associated with deafness and pigmentation defects in varitint-waddler (Va) mice. *PNAS* **99**, 14994–14999.

Dodier Y, Banderali U, Klein H *et al.* (2004). Outer pore topology of the ECaC-TRPV5 channel by cysteine scan mutagenesis. *J Biol Chem* **279**, 6853–6862.

Dohke Y, Oh YS, Ambudkar IS *et al.* (2004). Biogenesis and topology of the transient receptor potential Ca^{2+} channel TRPC1. *J Biol Chem* **279**, 12242–12248.

Doyle DA, Morais Cabral J, Pfuetzner RA *et al.* (1998). The structure of the potassium channel: molecular basis of K$^+$ conduction and selectivity. *Science* **280**, 69–77.

Duncan LM, Deeds J, Cronin FE *et al.* (2001). Melastatin expression and prognosis in cutaneous malignant melanoma. *J Clin Oncol* **19**, 568–576.

Duncan LM, Deeds J, Hunter J *et al.* (1998). Down-regulation of the novel gene melastatin correlates with potential for melanoma metastasis. *Cancer Res* **58**, 1515–1520.

Elizondo MR, Arduini BL, Paulsen J *et al.* (2005). Defective skeletogenesis and kidney stone formation in dwarf zebrafish with mutations in trpm7. *Curr Biol* **15**, 667–671.

Embark HM, Setiawan I, Poppendieck S *et al.* (2004). Regulation of the epithelial Ca^{2+} channel TRPV5 by the NHE regulating factor NHERF2 and the serum and glucocorticoid inducible kinase isoforms SGK1 and SGK3 expressed in *Xenopus* oocytes. *Cell Physiol Biochem* **14**, 203–212.

Engelke M, Friedrich O, Budde P *et al.* (2002). Structural domains required for channel function of the mouse transient receptor potential protein homologue TRP1beta. *FEBS Lett* **523**, 193–199.

Erler I, Hirnet D, Wissenbach U *et al.* (2004). Ca^{2+}-selective transient receptor potential V channel architecture and function require a specific ankyrin repeat. *J Biol Chem* **279**, 34456–34463.

Fleig A and Penner R (2004). The TRPM ion channel subfamily: molecular, biophysical and functional features. *Trends Pharmacol Sci* **25**, 633–639.

Flemming R, Xu SZ and Beech DJ (2003). Pharmacological profile of store-operated channels in cerebral arteriolar smooth muscle cells. *Br J Pharmacol* **139**, 955–965.

Freichel M, Suh SH, Pfeifer A *et al.* (2001). Lack of an endothelial store-operated Ca^{2+} current impairs agonist-dependent vasorelaxation in TRP4–/– mice. *Nat Cell Biol* **3**, 121–127.

Freichel M, Vennekens R, Olausson J *et al.* (2004). Functional role of TRPC proteins *in vivo*: lessons from TRPC-deficient mouse models. *Biochem Biophys Res Commun* **322**, 1352–1358.

Garcia-Martinez C, Morenilla-Palao C, Planells-Cases R *et al.* (2000). Identification of an aspartic residue in the P-loop of the vanilloid receptor that modulates pore properties. *J Biol Chem* **275**, 32552–32558.

Garcia-Sanz N, Fernandez-Carvajal A, Morenilla-Palao C *et al.* (2004). Identification of a tetramerization domain in the C terminus of the vanilloid receptor. *J Neurosci* **24**, 5307–5314.

Gavva NR, Klionsky L, Qu Y *et al.* (2004). Molecular determinants of vanilloid sensitivity in TRPV1. *J Biol Chem* **279**, 20283–20295.

Gevaert T, Vriens J, Segal A *et al.* (2007). Deletion of the transient receptor potential cation channel TRPV4 impairs murine bladder voiding. *J Clin Invest* **117**, 3453–3462.

Goel M, Sinkins WG and Schilling WP (2002). Selective association of TRPC channel subunits in rat brain synaptosomes. *J Biol Chem* **277**, 48303–48310.

Grantham JJ (1993). Polycystic kidney disease: hereditary and acquired. *Adv Intern Med* **38**, 409–420.

Grimm C, Kraft R, Sauerbruch S *et al.* (2003). Molecular and functional characterization of the melastatin-related cation channel TRPM3. *J Biol Chem* **278**, 21493–21501.

Grimm C, Kraft R, Schultz G *et al.* (2005). Activation of the melastatin-related cation channel TRPM3 by D-erythro-sphingosine. *Mol Pharmacol* **67**, 798–805.

Gudermann T, Hofmann T, Mederos Y *et al.* (2004). Activation, subunit composition and physiological relevance of DAG-sensitive TRPC proteins. *Novartis Found Symp*, **258**, 103–118; discussion 18–22, 55–59, 263–266.

Güler A, Lee H, Shimizu I *et al.* (2002). Heat-evoked activation of TRPV4 (VR-OAC). *J Neurosci* **22**, 6408–6414.

Gunthorpe MJ, Benham CD, Randall A *et al.* (2002). The diversity in the vanilloid (TRPV). receptor family of ion channels. *TIPS* **23**, 183–191.

Gunthorpe MJ, Harries MH, Prinjha RK *et al.* (2000). Voltage- and time-dependent properties of the recombinant rat vanilloid receptor (rVR1). *J Physiol* **525(Pt 3)**, 747–759.

Guo L, Chen M, Basora N *et al.* (2000). The human polycystic kidney disease 2-like (PKDL). gene. exon/intron structure and evidence for a novel splicing mechanism. *Mamm Genome* **11**, 46–50.

Guo L, Schreiber TH, Weremowicz S *et al.* (2000). Identification and characterization of a novel polycystin family member, polycystin-L2, in mouse and human: sequence, expression, alternative splicing and chromosomal localization. *Genomics* **64**, 241–251.

Hanaoka K, Qian F, Boletta A *et al.* (2000). Co-assembly of polycystin-1 and -2 produces unique cation-permeable currents. *Nature* **408**, 990–994.

Hara Y, Wakamori M, Ishii M *et al.* (2002). LTRPC2 Ca^{2+}-permeable channel activated by changes in redox status confers susceptibility to cell death. *Mol Cell* **9**, 163–173.

Hardie RC and Minke B (1992). The trp gene is essential for a light-activated Ca^{2+} channel in *Drosophila* photoreceptors. *Neuron* **8**, 643–651.

Hardie RC, Raghu P, Moore S *et al.* (2001). Calcium influx via TRP channels is required to maintain PIP2 levels in *Drosophila* photoreceptors. *Neuron* **30**, 149–159.

Harteneck C, Plant TD and Schultz G (2000). From worm to man: three subfamilies of TRP channels. *TINS* **23**, 159–166.

Hayes P, Meadows HJ, Gunthorpe MJ *et al.* (2000). Cloning and functional expression of a human orthologue of rat vanilloid receptor-1. *Pain* **88**, 205–215.

Heinemann SH, Terlau H, Stuhmer W *et al.* (1992). Calcium channel characteristics conferred on the sodium channel by single mutations. *Nature* **356**, 441–443.

Hellwig N, Albrecht N, Harteneck C *et al.* (2005). Homo- and heteromeric assembly of TRPV channel subunits. *J Cell Sci* **118**, 917–928.

Hellwig N, Plant TD, Janson W *et al.* (2004). TRPV1 acts as proton channel to induce acidification in nociceptive neurons. *J Biol Chem* **279**, 34553–34561.

Henshall SM, Afar DE, Hiller J *et al.* (2003). Survival analysis of genome-wide gene expression profiles of prostate cancers identifies new prognostic targets of disease relapse. *Cancer Res* **63**, 4196–4203.

Hermosura MC, Nayakanti H, Dorovkov MV *et al.* (2005). A TRPM7 variant shows altered sensitivity to magnesium that may contribute to the pathogenesis of two Guamanian neurodegenerative disorders. *PNAS* **102**, 11510–11515.

Hinman A, Chuang HH, Bautista DM *et al.* (2006). TRP channel activation by reversible covalent modification. *PNAS* **103**, 19564–19568.

Hoenderop JG, Nilius B and Bindels RJ (2005). Calcium absorption across epithelia. *Physiol Rev* **85**, 373–422.

Hoenderop JG, van der Kemp AW, Hartog A *et al.* (1999). The epithelial calcium channel, ECaC, is activated by hyperpolarization and regulated by cytosolic calcium. *Biochem Biophys Res Commun* **261**, 488–492.

Hoenderop JGJ, Vennekens R, Müller D *et al.* (2001). Function and expression of the epithelial Ca^{2+} channel family: comparison of the epithelial Ca^{2+} channel 1 and 2. *J Physiol* **537**, 747–761.

Hoenderop JGJ, Voets T, Hoefs S *et al.* (2003). Homo- and heterotetrameric architecture of the epithelial Ca^{2+} channels TRPV5 and TRPV6. *EMBO J* **22**, 776–785.

Hofmann T, Chubanov V, Gudermann T *et al.* (2003). TRPM5 is a voltage-modulated and Ca(2+)-activated monovalent selective cation channel. *Curr Biol* **13**, 1153–1158.

Hofmann T, Obukhov AG, Schaefer M et al. (1999). Direct activation of human TRPC6 and TRPC3 channels by diacylglycerol. *Nature* **397**, 259–263.

Hofmann T, Schaefer M, Schultz G et al. (2000a). Cloning, expression and subcellular localization of two novel splice variants of mouse transient receptor potential channel 2. *Biochem J* **351**, 115–122.

Hofmann T, Schaefer M, Schultz G et al. (2000b). Transient receptor potential channels as molecular substrates of receptor-mediated cation entry. *J Mol Med* **78**, 14–25.

Hofmann T, Schaefer M, Schultz G et al. (2002). Subunit composition of mammalian transient receptor potential channels in living cells. *PNAS* **99**, 7461–7466.

Howard J and Bechstedt S (2004). Hypothesis: a helix of ankyrin repeats of the NOMPC-TRP ion channel is the gating spring of mechanoreceptors. *Curr Biol* **14**, R224–226.

Hu HZ, Gu Q, Wang C et al. (2004). 2-aminoethoxydiphenyl borate is a common activator of TRPV1, TRPV2 and TRPV3. *J Biol Chem* **279**, 35741–35748.

Hunter JJ, Shao J, Smutko JS et al. (1998). Chromosomal localization and genomic characterization of the mouse melastatin gene (Mlsn1). *Genomics* **54**, 116–123.

Hwang SW, Cho H, Kwak J et al. (2000). Direct activation of capsaicin receptors by products of lipoxygenases: endogenous capsaicin-like substances. *PNAS* **97**, 6155–6160.

Inoue R, Okada T, Onoue H et al. (2001). The transient receptor potential protein homologue TRP6 is the essential component of vascular α_1-adrenoceptor-activated Ca^{2+}- permeable cation channel. *Circ Res* **88**, 325–332.

Iwata Y, Katanosaka Y, Arai Y et al. (2003). A novel mechanism of myocyte degeneration involving the Ca^{2+}-permeable growth factor-regulated channel. *J Cell Biol* **161**, 957–967.

Jaquemar D, Schenker T and Trueb B (1999). An ankyrin-like protein with transmembrane domains is specifically lost after oncogenic transformation of human fibroblasts. *J Biol Chem* **274**, 7325–7333.

Jordt SE and Julius D (2002). Molecular basis for species-specific sensitivity to hot chili peppers. *Cell* **108**, 421–430.

Jordt SE, Bautista DM, Chuang HH et al. (2004). Mustard oils and cannabinoids excite sensory nerve fibres through the TRP channel ANKTM1. *Nature* **427**, 260–265.

Jung S, Muhle A, Schaefer M et al. (2003). Lanthanides potentiate TRPC5 currents by an action at extracellular sites close to the pore mouth. *J Biol Chem* **278**, 3562–3571.

Jung S, Strotmann R, Schultz G et al. (2002). TRPC6 is a candidate channel involved in receptor-stimulated cation currents in A7r5 smooth muscle cells. *Am J Physiol* **282**, C347–359.

Jungnickel MK, Marrero H, Birnbaumer L et al. (2001). Trp2 regulates entry of Ca^{2+} into mouse sperm triggered by egg ZP3. *Nat Cell Biol* **3**, 499–502.

Kanzaki M, Nagasawa M, Kojima I et al. (1999a). Molecular identification of a eukaryotic, stretch-activated nonselective cation channel. *Science* **285**, 882–886.

Kanzaki M, Zhang YQ, Mashima H et al. (1999b). Translocation of a calcium-permeable cation channel induced by insulin-like growth factor-I. *Nat Cell Biol* **1**, 165–170.

Keller SA, Jones JM, Boyle A et al. (1994). Kidney and retinal defects (Krd), a transgene-induced mutation with a deletion of mouse chromosome 19 that includes the Pax2 locus. *Genomics* **23**, 309–320.

Kernan M, Cowan D and Zuker C (1994). Genetic dissection of mechanosensory transduction: mechanoreception-defective mutations of *Drosophila*. *Neuron* **12**, 1195–1206.

Kim E and Sheng M (2004). PDZ domain proteins of synapses. *Nat Rev Neurosci* **5**, 771–781.

Kim J, Chung DY, Park D et al. (2003). A TRPV family ion channel required for hearing in *Drosophila*. *Nature* **424**, 81–84.

Kim SJ, Kim YS, Yuan JP et al. (2003). Activation of the TRPC1 cation channel by metabotropic glutamate receptor mGluR1. *Nature* **426**, 285–291.

Kiselyov K, Chen J, Rbaibi Y et al. (2005). TRP-ML1 is a lysosomal monovalent cation channel that undergoes proteolytic cleavage. *J Biol Chem* **280**, 43218–43223.

Kitaguchi T and Swartz KJ (2005). An inhibitor of TRPV1 channels isolated from funnel Web spider venom. *Biochemistry* **44**, 15544–15549.

Kitahara T, Li HS and Balaban CD (2005). Changes in transient receptor potential cation channel superfamily V (TRPV). mRNA expression in the mouse inner ear ganglia after kanamycin challenge. *Hear Res* **201**, 132–144.

Konrad M, Schlingmann KP and Gudermann T (2004). Insights into the molecular nature of magnesium homeostasis. *Am J Physiol Renal Physiol* **286**, F599–605.

Koulen P, Cai Y, Geng L et al. (2002). Polycystin-2 is an intracellular calcium release channel. *Nat Cell Biol* **4**, 191–197.

Kriz W (2005). TRPC6—a new podocyte gene involved in focal segmental glomerulosclerosis. *Trends Mol Med* **11**, 527–530.

Kuhn FJ and Luckhoff A (2004). Sites of the NUDT9-H domain critical for ADP-ribose activation of the cation channel TRPM2. *J Biol Chem* **279**, 46431–46437.

Kuzhikandathil EV, Wang H, Szabo T et al. (2001). Functional analysis of capsaicin receptor (vanilloid receptor subtype 1) multimerization and agonist responsiveness using a dominant negative mutation. *J Neurosci* **21**, 8697–86706.

Kwan KY, Allchorne AJ, Vollrath MA et al. (2006). TRPA1 contributes to cold, mechanical and chemical nociception but is not essential for hair-cell transduction. *Neuron* **50**, 277–289.

Lambers TT, Weidema AF, Nilius B et al. (2004). Regulation of the mouse epithelial Ca2(+) channel TRPV6 by the Ca(2+)-sensor calmodulin. *J Biol Chem* **279**, 28855–28861.

LaPlante JM, Falardeau J, Sun M et al. (2002). Identification and characterization of the single channel function of human mucolipin-1 implicated in mucolipidosis type IV, a disorder affecting the lysosomal pathway. *FEBS Lett* **532**, 183–187.

LaPlante JM, Ye CP, Quinn SJ et al. (2004). Functional links between mucolipin-1 and Ca^{2+}-dependent membrane trafficking in mucolipidosis IV. *Biochem Biophys Res Commun* **322**, 1384–1391.

Larsson KP, Peltonen HM, Bart G et al. (2005). Orexin-A induced Ca^{2+} entry: evidence for involvement of TRPc channels and protein kinase C regulation. *J Biol Chem* **280**, 1771–1781.

Launay P, Fleig A, Perraud AL et al. (2002). TRPM4 is a Ca^{2+}-activated nonselective cation channel mediating cell membrane depolarization. *Cell* **109**, 397–407.

Lee G, Abdi K, Jiang Y et al. (2006). Nanospring behaviour of ankyrin repeats. *Nature* **440**, 246–249.

Lee N, Chen J, Sun L et al. (2003). Expression and characterization of human transient receptor potential melastatin 3 (hTRPM3). *J Biol Chem* **278**, 20890–20897.

Lee-Kwon W, Wade JB, Zhang Z et al. (2004). Expression of TRPC4 channel protein that interacts with NHERF-2 in rat descending vasa recta. *Am J Physiol* **288**, C942–949.

Li HS, Xu XZ and Montell C (1999). Activation of a TRPC3-dependent cation current through the neurotrophin BDNF. *Neuron* **24**, 261–273.

Li M, Jiang J and Yue L (2006). Functional characterization of homo- and heteromeric channel kinases TRPM6 and TRPM7. *J Gen Physiol* **127**, 525–537.

Li X, Luo Y, Starremans PG et al. (2005). Polycystin-1 and polycystin-2 regulate the cell cycle through the helix-loop-helix inhibitor Id2. *Nat Cell Biol* **7**, 1102–1112.

Liedtke W and Friedman JM (2003). Abnormal osmotic regulation in trpv4 −/− mice. *PNAS* **100**, 13698–13703.

Liedtke W, Choe Y, Marti-Renom MA *et al.* (2000). Vanilloid receptor-related osmotically activated channel (VR-OAC), a candidate vertebrate osmoreceptor. *Cell* **103**, 525–535.

Liedtke W, Tobin DM, Bargmann CI *et al.* (2003). Mammalian TRPV4 (VR-OAC). directs behavioral responses to osmotic and mechanical stimuli in Caenorhabditis elegans. *PNAS* **100**(**Suppl 2**), 14531–14536.

Lin S-Y and Corey DP (2005). TRP channels in mechanosensation. *Curr Opin Neurobiol* **15**, 350–357.

Lintschinger B, Balzer-Geldsetzer M, Baskaran T *et al.* (2000). Coassembly of Trp1 and Trp3 proteins generates diacylglycerol- and Ca^{2+}-sensitive cation channels. *J Biol Chem* **275**, 27799–27805.

Lis A, Wissenbach U and Philipp SE (2005). Transcriptional regulation and processing increase the functional variability of TRPM channels. *Naunyn Schmiedebergs Arch Pharmacol* **371**, 315–324.

Liu D and Liman ER (2003). Intracellular Ca^{2+} and the phospholipid PIP2 regulate the taste transduction ion channel TRPM5. *PNAS* **100**, 15160–15165.

Liu X, Bandyopadhyay B, Nakamoto T *et al.* (2006). A role for AQP5 in activation of TRPV4 by hypotonicity: concerted involvement of AQP5 and TRPV4 in regulation of cell volume recovery. *J Biol Chem* **281**, 15485–15495.

Liu X, Singh BB and Ambudkar IS (2003). TRPC1 is required for functional store-operated Ca^{2+} channels. Role of acidic amino acid residues in the S5–S6 region. *J Biol Chem* **278**, 11337–11343.

Lucas P, Ukhanov K, Leinders-Zufall T *et al.* (2003). A diacylglycerol-gated cation channel in vomeronasal neuron dendrites is impaired in TRPC2 mutant mice: mechanism of pheromone transduction. *Neuron* **40**, 551–561.

Ma HT, Venkatachalam K, Li HS *et al.* (2001). Assessment of the role of the inositol 1,4,5-trisphosphate receptor in the activation of transient receptor potential channels and store- operated Ca^{2+} entry channels. *J Biol Chem* **276**, 18888–18896.

Macpherson L, Geierstanger BH, Viswanath V *et al.* (2005). The pungency of garlic: activation of TRPA1 and TRPV1 in response to allicin. *Curr Biol* **15**, 929–934.

Macpherson LJ, Dubin AE, Evans MJ *et al.* (2007). Noxious compounds activate TRPA1 ion channels through covalent modification of cysteines. *Nature* **445**, 541–545.

Maroto R, Raso A, Wood TG *et al.* (2005). TRPC1 forms the stretch-activated cation channel in vertebrate cells. *Nat Cell Biol* **7**, 179–185.

Maruyama Y, Nakanishi Y, Walsh EJ *et al.* (2006). Heteromultimeric TRPC6–TRPC7 channels contribute to arginine vasopressin-induced cation current of A7r5 vascular smooth muscle cells. *Circ Res* **98**, 1520–1527.

Matsushita M, Kozak JA, Shimizu Y *et al.* (2005). Channel function is dissociated from the intrinsic kinase activity and autophosphorylation of TRPM7/ChaK1. *J Biol Chem* **280**, 20793–20803.

McIntyre P, McLatchie LM, Chambers A *et al.* (2001). Pharmacological differences between the human and rat vanilloid receptor 1 (VR1). *Br J Pharmacol* **132**, 1084–1094.

McKay RR, Szymeczek-Seay CL, Lievremont JP *et al.* (2000). Cloning and expression of the human transient receptor potential 4 (TRP4). gene: localization and functional expression of human TRP4 and TRP3. *Biochem J* **351**, 735–746.

McKemy DD, Neuhausser WM and Julius D (2002). Identification of a cold receptor reveals a general role for TRP channels in thermosensation. *Nature* **416**, 52–58.

McNamara FN, Randall A and Gunthorpe MJ (2005). Effects of piperine, the pungent component of black pepper, at the human vanilloid receptor (TRPV1). *Br J Pharmacol* **144**, 781–790.

Mery L, Magnino F, Schmidt K *et al.* (2001). Alternative splice variants of hTrp4 differentially interact with the C-terminal portion of the inositol 1,4,5-trisphosphate receptors. *FEBS Lett* **487**, 377–383.

Mery L, Strauss B, Dufour JF *et al.* (2002). The PDZ-interacting domain of TRPC4 controls its localization and surface expression in HEK293 cells. *J Cell Sci* **115**, 3497–3508.

Minke B and Cook B (2002). TRP channel proteins and signal transduction. *Physiol Rev* **82**, 429–472.

Mochizuki T, Wu G, Hayashi T *et al.* (1996). PKD2, a gene for polycystic kidney disease that encodes an integral membrane protein. *Science* **272**, 1339–1342.

Modregger J, Ritter B, Witter B *et al.* (2000). All three PACSIN isoforms bind to endocytic proteins and inhibit endocytosis. *J Cell Sci* **113**(**Pt 24**), 4511–4521.

Monteilh-Zoller MK, Hermosura MC, Nadler MJ *et al.* (2003). TRPM7 provides an ion channel mechanism for cellular entry of trace metal ions. *J Gen Physiol* **121**, 49–60.

Montell C (2005). The TRP superfamily of cation channels. *Sci STKE* **272**, re3.

Montell C and Rubin GM (1989). Molecular characterization of the *Drosophila trp* locus: a putative integral membrane protein required for phototransduction. *Neuron* **2**, 1313–1323.

Montell C, Birnbaumer L and Flockerzi V (2002). The TRP channels, a remarkable functional family. *Cell* **108**, 595–598.

Montell C, Birnbaumer L, Flockerzi V *et al.* (2002). A unified nomenclature for the superfamily of TRP cation channels. *Mol Cell* **9**, 229–231.

Moqrich A, Hwang SW, Earley TJ *et al.* (2005). Impaired thermosensation in mice lacking TRPV3, a heat and camphor sensor in the skin. *Science* **307**, 1468–1472.

Muller D, Hoenderop JG, Meij IC *et al.* (2000). Molecular cloning, tissue distribution and chromosomal mapping of the human epithelial Ca^{2+} channel (ECAC1). *Genomics* **67**, 48–53.

Munsch T, Freichel M, Flockerzi V *et al.* (2003). Contribution of transient receptor potential channels to the control of GABA release from dendrites. *PNAS* **100**, 16065–16070.

Muraki K, Iwata Y, Katanosaka Y *et al.* (2003). TRPV2 is a component of osmotically sensitive cation channels in murine aortic myocytes. *Circ Res* **93**, 829–838.

Mustafa S and Oriowo M (2005). Cooling-induced contraction of the rat gastric fundus: mediation via transient receptor potential (TRP) cation channel TRPM8 receptor and Rho-kinase activation. *Clin Exp Pharmacol Physiol* **32**, 832–838.

Nadler MJ, Hermosura MC, Inabe K *et al.* (2001). LTRPC7 is a Mg.ATP-regulated divalent cation channel required for cell viability. *Nature* **411**, 590–595.

Nagamine K, Kudoh J, Minoshima S *et al.* (1998). Molecular cloning of a novel putative Ca^{2+} channel protein (TRPC7), highly expressed in brain. *Genomics* **54**, 124–131.

Nagata K, Duggan A, Kumar G *et al.* (2005). Nociceptor and hair cell transducer properties of TRPA1, a channel for pain and hearing. *J Neurosci* **25**, 4052–4061.

Nathan JD, Patel AA, McVey DC *et al.* (2001). Capsaicin vanilloid receptor-1 mediates substance P release in experimental pancreatitis. *Am J Physiol* **281**, G1322–1328.

Nealen ML, Gold MS, Thut PD *et al.* (2003). TRPM8 mRNA is expressed in a subset of cold-responsive trigeminal neurons from rat. *J Neurophysiol* **90**, 515–520.

Niemeyer BA, Bergs C, Wissenbach U *et al.* (2001). Competitive regulation of CaT-like-mediated Ca^{2+} entry by protein kinase C and calmodulin. *PNAS* **98**, 3600–3605.

Nijenhuis T, Hoenderop JG, Nilius B *et al.* (2003). (Patho)physiological implications of the novel epithelial Ca^{2+} channels TRPV5 and TRPV6. *Pflugers Arch* **446**, 401–409.

Nijenhuis T, Hoenderop JG, van der Kemp AW *et al.* (2003). Localization and regulation of the epithelial Ca^{2+} channel TRPV6 in the kidney. *J Am Soc Nephrol* **14**, 2731–2740.

Nilius B (2003a). From TRPs to SOCs, CCEs and CRACs: consensus and controversies. *Cell Calcium* **33**, 293–298.

Nilius B (2004a). Store-operated Ca^{2+} entry channels: still elusive! *Sci STKE* **243**, pe36.

Nilius B and Voets T (2005a). Trp channels: a TR(I)P through a world of multifunctional cation channels. *Pflügers Arch Eur J Physiol* **451**, 1–10.

Nilius B, Droogmans G and Wondergem R (2003b). Transient receptor potential channels in endothelium: solving the calcium entry puzzle? *Endothelium* **10**, 5–15.

Nilius B, Mahieu F, Prenen J et al. (2006). The Ca^{2+}-activated cation channel TRPM4 is regulated by phosphatidylinositol 4,5-biphosphate. *EMBO J* **25**, 467–478.

Nilius B, Owsianik G, Voets T et al. (2007). Transient receptor potential (TRP). cation channels in disease. *Physiol Rev* **87(1)**, 165–217.

Nilius B, Prenen J, Droogmans G et al. (2003c). Voltage Dependence of the Ca^{2+}-activated cation channel TRPM4. *J Biol Chem* **278**, 30813–30820.

Nilius B, Prenen J, Hoenderop JG et al. (2002). Fast and slow inactivation kinetics of the Ca^{2+} channels ECaC1 and ECaC2 (TRPV5 and TRPV6). Role of the intracellular loop located between transmembrane segments 2 and 3. *J Biol Chem* **277**, 30852–30858.

Nilius B, Prenen J, Janssens A et al. (2005b). The selectivity filter of the cation channel TRPM4. *J Biol Chem* **280**, 22899–22906.

Nilius B, Prenen J, Janssens A et al. (2004b). Decavanadate modulates gating of TRPM4 cation channels. *J Physiol* **560**, 753–765.

Nilius B, Prenen J, Tang J et al. (2005c). Regulation of the Ca^{2+} sensitivity of the nonselective cation channel TRPM4. *J Biol Chem* **280**, 6423–6433.

Nilius B, Prenen J, Voets T et al. (2004c). Intracellular nucleotides and polyamines inhibit the Ca(2+)-activated cation channel TRPM4b. *Pflugers Arch* **448**, 70–75.

Nilius B, Prenen J, Wissenbach U et al. (2001a). Differential activation of the volume-sensitive cation channel TRP12 (OTRPC4) and volume-regulated anion currents in HEK-293 cells. *Pflügers Archiv Eur J Physiol* **443**, 227–233.

Nilius B, Talavera K, Owsianik G et al. (2005d). Gating of TRP channels: a voltage connection?. *J Physiol* **567**, 33–44.

Nilius B, Vennekens R, Prenen J et al. (2000). Whole-cell and single channel monovalent cation currents through the novel rabbit epithelial Ca^{2+} channel ECaC. *J Physiol* **527**, 239–248.

Nilius B, Vennekens R, Prenen J et al. (2001b). The single pore residue Asp542 determines Ca^{2+} permeation and Mg^{2+} block of the epithelial Ca^{2+} channel. *J Biol Chem* **276**, 1020–1025.

Nilius B, Voets T and Peters J (2005e). TRP channels in disease. *Sci STKE* **295**, re8.

Nilius B, Vriens J, Prenen J et al. (2004d). TRPV4 calcium entry channel: a paradigm for gating diversity. *Am J Physiol* **286**, C195–205.

Nilius B, Weidema F, Prenen J et al. (2003d). The carboxyl-terminus of the epithelial Ca^{2+} channel EcaC1 is involved in Ca^{2+} dependent inactivation. *Pflügers Arch Eur J Physiol* **445**, 584–588.

Nomura H, Turco AE, Pei Y et al. (1998). Identification of PKDL, a novel polycystic kidney disease 2-like gene whose murine homologue is deleted in mice with kidney and retinal defects. *J Biol Chem* **273**, 25967–25973.

Numazaki M, Tominaga T, Takeuchi K et al. (2003). Structural determinant of TRPV1 desensitization interacts with calmodulin. *PNAS* **100**, 8002–8006.

O'Neil RG and Heller S (2005). Mechanosensitive nature of TRPV channels. *Pflügers Arch Eur J Physiol* **451**, 193–203.

Oberwinkler J, Lis A, Giehl KM et al. (2005). Alternative splicing switches the divalent cation selectivity of TRPM3 channels. *J Biol Chem* **280**, 22540–22548.

Obukhov AG and Nowycky MC (2004). TRPC5 activation kinetics are modulated by the scaffolding protein ezrin/radixin/moesin-binding phosphoprotein-50 (EBP50). *J Cell Physiol* **201**, 227–235.

Obukhov AG and Nowycky MC (2005). A cytosolic residue mediates Mg^{2+} block and regulates inward current amplitude of a transient receptor potential channel. *J Neurosci* **25**, 1234–9.

Okada T, Shimizu S, Wakamori M et al. (1998). Molecular cloning and functional characterization of a novel receptor-activated TRP Ca^{2+} channel from mouse brain. *J Biol Chem* **273**, 10279–10287.

Owsianik G, D'Hoedt D, Voets T et al. (2006a). Structure-function relationship of the TRP channel superfamily. *Rev Physiol Biochem Pharmacol* **156**, 61–90.

Owsianik G, Talavera K, Voets T et al. (2006b). Permeation and selectivity of TRP channels. *Ann Rev Physiol* **68**, 685–717.

Palmada M, Poppendieck S, Embark HM et al. (2005). Requirement of PDZ domains for the stimulation of the epithelial Ca^{2+} channel TRPV5 by the NHE regulating factor NHERF2 and the serum and glucocorticoid inducible kinase SGK1. *Cell Physiol Biochem* **15**, 175–182.

Parekh AB and Putney JW Jr (2005). Store-operated calcium channels. *Physiol Rev* **85**, 757–810.

Patapoutian A, Peier AM, Story GM et al. (2003). ThermoTRP channels and beyond: mechanisms of temperature sensation. *Nat Rev Neurosci* **4**, 529–539.

Pedersen SF, Owsianik G and Nilius B (2005). TRP channels: an overview. *Cell Calcium* **38**, 233–252.

Peier AM, Moqrich A, Hergarden AC et al. (2002a). A TRP channel that senses cold stimuli and menthol. *Cell* **108**, 705–715.

Peier AM, Reeve AJ, Andersson DA et al. (2002b). A heat-sensitive TRP channel expressed in keratinocytes. *Science* **296**, 2046–2049.

Peng JB, Brown EM and Hediger MA (2001a). Structural conservation of the genes encoding CaT1, CaT2 and related cation channels. *Genomics* **76**, 99–109.

Peng JB, Chen XZ, Berger UV et al. (1999). Molecular cloning and characterization of a channel-like transporter mediating intestinal calcium absorption. *J Biol Chem* **274**, 22739–22746.

Peng JB, Chen XZ, Berger UV et al. (2000). Human calcium transport protein CaT1. *Biochem Biophys Res Commun* **278**, 326–332.

Peng JB, Chen XZ, Berger UV et al. (2000). A rat kidney-specific calcium transporter in the distal nephron. *J Biol Chem* **275**, 28186–28194.

Peng JB, Zhuang L, Berger UV et al. (2001b). CaT1 expression correlates with tumor grade in prostate cancer. *Biochem Biophys Res Commun* **282**, 729–734.

Perez CA, Huang L, Rong M et al. (2002). A transient receptor potential channel expressed in taste receptor cells. *Nat Neurosci* **5**, 1169–1176.

Perozo E, Cortes DM, Sompornpisut P et al. (2002). Open channel structure of MscL and the gating mechanism of mechanosensitive channels. *Nature* **418**, 942–948.

Perozo E, Kloda A, Cortes DM et al. (2002). Physical principles underlying the transduction of bilayer deformation forces during mechanosensitive channel gating. *Nat Struct Biol* **9**, 696–703.

Perraud AL, Fleig A, Dunn CA et al. (2001). ADP-ribose gating of the calcium-permeable LTRPC2 channel revealed by Nudix motif homology. *Nature* **411**, 595–599.

Perraud AL, Schmitz C and Scharenberg AM (2003). TRPM2 Ca^{2+} permeable cation channels: from gene to biological function. *Cell Calcium* **33**, 519–531.

Petersen CCH, Berridge MJ, Borgese MF et al. (1995). Putative capacitative calcium entry channels: expression of *Drosophila* trp and evidence for the existence of vertebrate homologues. *Biochem J* **311**, 41–44.

Philipp S, Cavalie A, Freichel M et al. (1996). A mammalian capacitative calcium entry channel homologous to *Drosophila* TRP and TRPL. *EMBO J* **15**, 6166–6171.

Philipp S, Hambrecht J, Braslavski L et al. (1998). A novel capacitative calcium entry channel expressed in excitable cells. *EMBO J* **17**, 4274–4282.

Philipp S, Trost C, Warnat J et al. (2000a). TRP4 (CCE1) protein is part of native calcium release-activated Ca^{2+}-like channels in adrenal cells. *J Biol Chem* **275**, 23965–23972.

Philipp S, Wissenbach U and Flockerzi V (2000b). Molecular biology of calcium channels. In JWJ Putney, ed., *Calcium signaling*. Boca Raton, FL, CRC Press, pp. 321–342.

Phillips AM, Bull A and Kelly LE (1992). Identification of a *Drosophila* gene encoding a calmodulin-binding protein with homology to the *trp* phototransduction gene. *Neuron* **8**, 631–642.

Piper RC and Luzio JP (2004). CUPpling calcium to lysosomal biogenesis. *Trends Cell Biol* **14**, 471–473.

Planells-Cases R, Garcia-Sanz N, Morenilla-Palao C et al. (2005). Functional aspects and mechanisms of TRPV1 involvement in neurogenic inflammation that leads to thermal hyperalgesia. *Pflugers Arch* **451**, 151–159.

Plant TD and Schaefer M (2003). TRPC4 and TRPC5: receptor-operated Ca^{2+}-permeable nonselective cation channels. *Cell Calcium* **33**, 441–450.

Plato CC, Galasko D, Garruto RM et al. (2002). ALS and PDC of Guam: forty-year follow-up. *Neurology* **58**, 765–773.

Plato CC, Garruto RM, Galasko D et al. (2003). Amyotrophic lateral sclerosis and parkinsonism-dementia complex of Guam: changing incidence rates during the past 60 years. *Am J Epidemiol* **157**, 149–157.

Plomann M, Lange R, Vopper G et al. (1998). PACSIN, a brain protein that is upregulated upon differentiation into neuronal cells. *Eur J Biochem* **256**, 201–211.

Poteser M, Graziani A, Rosker C et al. (2006). TRPC3 and TRPC4 associate to form a redox-sensitive cation channel. evidence for expression of native TRPC3-TRPC4 heteromeric channels in endothelial cells. *J Biol Chem* **281**, 13588–13595.

Prawitt D, Enklaar T, Klemm G et al. (2000). Identification and characterization of *MTR1*, a novel gene with homology to melastatin (MLSN1). and the trp gene family located in the BWS-WT2 critical region on chromosome 11p15.5 and showing allele-specific expression. *Hum Mol Genet* **9**, 203–216.

Prawitt D, Monteilh-Zoller MK, Brixel L et al. (2003). TRPM5 is a transient Ca^{2+}-activated cation channel responding to rapid changes in $[Ca^{2+}]i$. *PNAS* **100**, 15166–15171.

Prescott ED and Julius D (2003). A modular PIP_2 binding site as a determinant of capsaicin receptor sensitivity. *Science* **300**, 1284–1288.

Putney JW Jr (2005). Capacitative calcium entry: sensing the calcium stores. *J Cell Biol* **169**, 381–382.

Qian F, Germino FJ, Cai Y et al. (1997). PKD1 interacts with PKD2 through a probable coiled-coil domain. *Nat Genet* **16**, 179–183.

Qian F, Wei W, Germino G et al. (2005). The nanomechanics of polycystin-1 extracellular region. *J Biol Chem* **280**, 40723–40730.

Reiser J, Polu KR, Moller CC et al. (2005). TRPC6 is a glomerular slit diaphragm-associated channel required for normal renal function. *Nat Genet* **37**, 739–744.

Reuss H, Mojet MH, Chyb S et al. (1997). *In vivo* analysis of the drosophila light-sensitive channels, TRP and TRPL. *Neuron* **19**, 1249–59.

Riazanova LV, Pavur KS, Petrov AN et al. (2001). Novel type of signaling molecules: protein kinases covalently linked to ion channels. *Mol Biol* **35**, 321–332.

Riccio A, Mattei C, Kelsell RE et al. (2002). Cloning and functional expression of human short TRP7, a candidate protein for store-operated Ca^{2+} influx. *J Biol Chem* **277**, 12302–12309.

Rohacs T, Lopes CM, Michailidis I et al. (2005). PI(4,5).P(2). regulates the activation and desensitization of TRPM8 channels through the TRP domain. *Nat Neurosci* **8**, 626–634.

Rosenbaum T, Gordon-Shaag A, Munari M et al. (2004). Ca^{2+}/calmodulin modulates TRPV1 activation by capsaicin. *J Gen Physiol* **123**, 53–62.

Runnels LW, Yue L and Clapham DE (2001). TRP-PLIK, a bifunctional protein with kinase and ion channel activities. *Science* **291**, 1043–1047.

Sakura H and Ashcroft FM (1997). Identification of four trp1 gene variants murine pancreatic beta-cells. *Diabetologia* **40**, 528–532.

Sano Y, Inamura K, Miyake A et al. (2001). Immunocyte Ca^{2+} influx system mediated by LTRPC2. *Science* **293**, 1327–1330.

Schaefer M, Plant TD, Obukhov AG et al. (2000). Receptor-mediated regulation of the nonselective cation channels TRPC4 and TRPC5. *J Biol Chem* **275**, 17517–17526.

Schaefer M, Plant TD, Stresow N et al. (2002). Functional differences between TRPC4 splice variants. *J Biol Chem* **277**, 3752–3759.

Schilling WP and Goel M (2004). Mammalian TRPC channel subunit assembly. *Novartis Found Symp*, **258**, 18–30; discussion 30–43, 98–102, 263–266.

Schlingmann KP, Weber S, Peters M et al. (2002). Hypomagnesemia with secondary hypocalcemia is caused by mutations in TRPM6, a new member of the TRPM gene family. *Nat Genet* **31**, 166–170.

Schmitz C, Perraud AL, Johnson CO et al. (2003). Regulation of vertebrate cellular Mg^{2+} homeostasis by TRPM7. *Cell* **114**, 191–200.

Sedgwick SG and Smerdon SJ (1999). The ankyrin repeat: a diversity of interactions on a common structural framework. *Trends Biochem Sci* **24**, 311–316.

Shen BW, Perraud AL, Scharenberg A et al. (2003). The crystal structure and mutational analysis of human NUDT9. *J Mol Biol* **332**, 385–398.

Shi J, Mori E, Mori Y et al. (2004). Multiple regulation by calcium of murine homologues of transient receptor potential proteins TRPC6 and TRPC7 expressed in HEK293 cells. *J Physiol* **561**, 415–432.

Shim S, Goh EL, Ge S et al. (2005). XTRPC1-dependent chemotropic guidance of neuronal growth cones. *Nat Neurosci* **8**, 730–735.

Sidhaye VK, Guler AD, Schweitzer KS et al. (2006). Transient receptor potential vanilloid 4 regulates aquaporin-5 abundance under hypotonic conditions. *PNAS* **103**, 4747–4752.

Sidi S, Friedrich RW and Nicolson T (2003). NompC TRP channel required for vertebrate sensory hair cell mechanotransduction. *Science* **301**, 96–99.

Smith GD, Gunthorpe J, Kelsell RE et al. (2002). TRPV3 is a temperature-sensitive vanilloid receptor-like protein. *Nature* **418**, 186–190.

Somlo S and Markowitz GS (2000). The pathogenesis of autosomal dominant polycystic kidney disease: an update. *Curr Opin Nephrol Hypertens* **9**, 385–394.

Sossey-Alaoui K, Lyon JA, Jones L et al. (1999). Molecular cloning and characterization of TRPC5 (HTRP5), the human homologue of a mouse brain receptor-activated capacitative Ca^{2+} entry channel. *Genomics* **60**, 330–340.

Sotomayor M, Corey DP and Schulten K (2005). In search of the hair-cell gating spring elastic properties of ankyrin and cadherin repeats. *Structure* **13**, 669–682.

Soyombo AA, Tjon-Kon-Sang S, Rbaibi Y et al. (2006). TRP-ML1 regulates lysosomal pH and acidic lysosomal lipid hydrolytic activity. *J Biol Chem* **281**, 7294–7301.

Stander S, Moormann C, Schumacher M et al. (2004). Expression of vanilloid receptor subtype 1 in cutaneous sensory nerve fibers, mast cells and epithelial cells of appendage structures. *Exp Dermatol* **13**, 129–139.

Stein RJ, Santos S, Nagatomi J et al. (2004). Cool (TRPM8) and hot (TRPV1) receptors in the bladder and male genital tract. *J Urol* **172**, 1175–1178.

Stokes AJ, Shimoda LM, Koblan-Huberson M et al. (2004). A TRPV2-PKA signaling module for transduction of physical stimuli in mast cells. *J Exp Med* **200**, 137–147.

Stokes AJ, Wakano C, Del Carmen KA et al. (2005). Formation of a physiological complex between TRPV2 and RGA protein promotes cell surface expression of TRPV2. *J Cell Biochem* **94**, 669–683.

Story GM, Peier AM, Reeve AJ et al. (2003). ANKTM1, a TRP-like channel expressed in nociceptive neurons, is activated by cold temperatures. *Cell* **112**, 819–829.

Stowers L, Holy TE, Meister M et al. (2002). Loss of sex discrimination and male-male aggression in mice deficient for TRP2. *Science* **295**, 1493–1500.

Strotmann R, Harteneck C, Nunnenmacher K et al. (2000). OTRPC4, a nonselective cation channel that confers sensitivity to extracellular osmolarity. *Nat Cell Biol* **2**, 695–702.

Strotmann R, Schultz G and Plant TD (2003). Ca^{2+}-dependent potentiation of the nonselective cation channel TRPV4 is mediated by a carboxy terminal calmodulin binding site. *J Biol Chem* **278**, 26541–26549.

Strubing C, Krapivinsky G, Krapivinsky L et al. (2001). TRPC1 and TRPC5 form a novel cation channel in mammalian brain. *Neuron* **29**, 645–655.

Strubing C, Krapivinsky G, Krapivinsky L *et al.* (2003). Formation of novel TRPC channels by complex subunit interactions in embryonic brain. *J Biol Chem* **278**, 39014–39019.

Sun M, Goldin E, Stahl S *et al.* (2000). Mucolipidosis type IV is caused by mutations in a gene encoding a novel transient receptor potential channel. *Hum Mol Genet* **9**, 2471–2478.

Suzuki M, Mizuno A, Kodaira K *et al.* (2003). Impaired pressure sensation with mice lacking TRPV4. *J Biol Chem* **270**, 22664–22668.

Szallasi A, Szabo T, Biro T *et al.* (1999). Resiniferatoxin-type phorboid vanilloids display capsaicin-like selectivity at native vanilloid receptors on rat DRG neurons and at the cloned vanilloid receptor VR1. *Br J Pharmacol* **128**, 428–434.

Tabuchi K, Suzuki M, Mizuno A *et al.* (2005). Hearing impairment in TRPV4 knockout mice. *Neurosci Lett* **382**, 304–308.

Takezawa R, Schmitz C, Demeuse P *et al.* (2004). Receptor-mediated regulation of the TRPM7 channel through its endogenous protein kinase domain. *PNAS* **101**, 6009–6014.

Takumida M, Kubo N, Ohtani M *et al.* (2005). Transient receptor potential channels in the inner ear: presence of transient receptor potential channel subfamily 1 and 4 in the guinea pig inner ear. *Acta Otolaryngol* **125**, 929–934.

Talavera K, Staes M, Janssens A *et al.* (2001). Aspartate residues of the Glu-Glu-Asp-Asp (EEDD). pore locus control selectivity and permeation of the T-type Ca(2+). channel alpha(1G). *J Biol Chem* **276**, 45628–45635.

Talavera K, Yasumatsu K, Voets T *et al.* (2005). Heat activation of TRPM5 underlies thermal sensitivity of sweet taste. *Nature* **438**, 1022–1025.

Tang Y, Tang J, Chen Z *et al.* (2000). Association of mammalian trp4 and phospholipase C isozymes with a PDZ domain-containing protein, NHERF. *J Biol Chem* **275**, 37559–37564.

Tiruppathi C, Freichel M, Vogel SM *et al.* (2002). Impairment of store-operated Ca²⁺ entry in TRPC4(–/–). mice interferes with increase in lung microvascular permeability. *Circ Res* **91**, 70–76.

Torihashi S, Fujimoto T, Trost C *et al.* (2002). Calcium oscillation linked to pacemaking of interstitial cells of Cajal: requirement of calcium influx and localization of TRP4 in caveolae. *J Biol Chem* **277**, 19191–19197.

Trebak M St J, Bird G, McKay RR *et al.* (2002). Comparison of human TRPC3 channels in receptor-activated and store-operated modes. Differential sensitivity to channel blockers suggests fundamental differences in channel composition. *J Biol Chem* **277**, 21617–21623.

Trost C, Bergs C, Himmerkus N *et al.* (2001). The transient receptor potential, TRP4, cation channel is a novel member of the family of calmodulin binding proteins. *Biochem J* **355**, 663–670.

Tsavaler L, Shapero MH, Morkowski S *et al.* (2001). Trp-p8, a novel prostate-specific gene, is up-regulated in prostate cancer and other malignancies and shares high homology with transient receptor potential calcium channel proteins. *Cancer Res* **61**, 3760–3769.

Tsiokas L, Kim E, Arnould T *et al.* (1997). Homo- and heterodimeric interactions between the gene products of PKD1 and PKD2. *PNAS* **94**, 6965–6970.

Turner H, Fleig A, Stokes A *et al.* (2003). Discrimination of intracellular calcium store subcompartments using TRPV1 (transient receptor potential channel, vanilloid subfamily member 1). release channel activity. *Biochem J* **371**, 341–350.

Ullrich ND, Voets T, Prenen J *et al.* (2005). Comparison of functional properties of the Ca²⁺-activated cation channels TRPM4 and TRPM5 from mice. *Cell Calcium* **37**, 267–278.

Urban L and Dray A (1991). Capsazepine, a novel capsaicin antagonist, selectively antagonises the effects of capsaicin in the mouse spinal cord *in vitro*. *Neurosci Lett* **134**, 9–11.

van de Graaf SF, Hoenderop JG, van der Kemp AW *et al.* (2006). Interaction of the epithelial Ca(2+). channels TRPV5 and TRPV6 with the intestine- and kidney-enriched PDZ protein NHERF4. *Pflugers Arch* **452**(**4**), 407–417.

Vanden Abeele F, Shuba Y, Roudbaraki M *et al.* (2003). Store-operated Ca²⁺ channels in prostate cancer epithelial cells: function, regulation and role in carcinogenesis. *Cell Calcium* **33**, 357–373.

Vannier B, Peyton M, Boulay G *et al.* (1999). Mouse trp2, the homologue of the human trpc2 pseudogene, encodes mTrp2, a store depletion-activated capacitative Ca²⁺ entry channel. *PNAS* **96**, 2060–2064.

Vazquez G, Wedel BJ, Aziz O *et al.* (2004). The mammalian TRPC cation channels. *Biochim Biophys Acta* **1742**, 21–36.

Veldhuisen B, Spruit L, Dauwerse HG *et al.* (1999). Genes homologous to the autosomal dominant polycystic kidney disease genes (PKD1 and PKD2). *Eur J Hum Genet* **7**, 860–872.

Venkatachalam K, Hofmann T and Montell C (2006). Lysosomal localization of TRPML3 depends on TRPML2 and the mucolipidosis-associated protein TRPML1. *J Biol Chem* **281**, 17517–17527.

Vennekens R and Nilius B (2007). Insights into TRPM4 function, regulation and physiological role. *Handb Exp Pharmacol* **179**, 269–285.

Vennekens R, Hoenderop JG, Prenen J *et al.* (2000). Permeation and gating properties of the novel epithelial Ca²⁺ channel. *J Biol Chem* **275**, 3963–3969.

Vennekens R, Prenen J, Hoenderop JG *et al.* (2001a). Modulation of the epithelial Ca²⁺ channel ECaC by extracellular pH. *Pflügers Archiv Eur Jf Physiol* **442**, 237–242.

Vennekens R, Prenen J, Hoenderop JG *et al.* (2001b). Pore properties and ionic block of the rabbit epithelial calcium channel expressed in HEK 293 cells. *J Physiol* **530**, 183–191.

Voets T and Nilius B (2003). The pore of TRP channels: trivial or neglected?. *Cell Calcium* **33**, 299–302.

Voets T, Droogmans G, Wissenbach U *et al.* (2004a). The principle of temperature-dependent gating in cold- and heat-sensitive TRP channels. *Nature* **430**, 748–754.

Voets T, Janssens A, Droogmans G *et al.* (2004b). Outer pore architecture of a Ca²⁺-selective TRP channel. *J Biol Chem* **279**, 15223–15230.

Voets T, Janssens A, Prenen J *et al.* (2003). Mg²⁺-dependent gating and strong inward rectification of the cation channel TRPV6. *J Gen Physiol* **121**, 245–260.

Voets T, Nilius B, Hoefs S *et al.* (2004c). TRPM6 forms the Mg²⁺ influx channel involved in intestinal and renal Mg²⁺ absorption. *J Biol Chem* **279**, 19–25.

Voets T, Prenen J, Fleig A *et al.* (2001). CaT1 and the calcium release-activated calcium channel manifest distinct pore properties. *J Biol Chem* **276**, 47767–47770.

Voets T, Prenen J, Vriens J *et al.* (2002). Molecular determinants of permeation through the cation channel TRPV4. *J Biol Chem* **277**, 33704–33710.

Voets T, Talavera K, Owsianik G *et al.* (2005). Sensing with TRP channels. *Nat Chem Biol* **1**, 85–92.

Volk T, Schwoerer AP, Thiessen S *et al.* (2003). A polycystin-2-like large conductance cation channel in rat left ventricular myocytes. *Cardiovasc Res* **58**, 76–88.

Vriens J, Owsianik G, Fisslthaler B *et al.* (2005). Modulation of the Ca2 permeable cation channel TRPV4 by cytochrome P450 epoxygenases in vascular endothelium. *Circ Res* **97**, 908–915.

Vriens J, Owsianik G, Voets T *et al.* (2004a). Invertebrate TRP proteins as functional models for mammalian channels. *Pflugers Arch* **449**, 213–226.

Vriens J, Watanabe H, Janssens A *et al.* (2004b). Cell swelling, heat and chemical agonists use distinct pathways for the activation of the cation channel TRPV4. *PNAS* **101**, 396–401.

Walder RY, Landau D, Meyer P *et al.* (2002). Mutation of TRPM6 causes familial hypomagnesemia with secondary hypocalcemia. *Nature Genetics* **31**, 171–174.

Walker RG, Willingham AT and Zuker CS (2000). A *Drosophila* mechanosensory transduction channel. *Science* **287**, 2229–2234.

Walker RL, Koh SD, Sergeant GP *et al.* (2002). TRPC4 currents have properties similar to the pacemaker current in interstitial cells of Cajal. *Am J Physiol* **283**, C1637–1645.

Wang GX and Poo MM (2005). Requirement of TRPC channels in netrin-1-induced chemotropic turning of nerve growth cones. *Nature* **434**, 898–904.

Warnat J, Philipp S, Zimmer S *et al.* (1999). Phenotype of a recombinant store-operated channel: highly selective permeation of Ca^{2+}. *J Physiol* **518**, 631–638.

Watanabe H, Davis JB, Smart D *et al.* (2002). Activation of TRPV4 channels (hVRL-2/mTRP12). by phorbol derivatives. *J Biol Chem* **277**, 13569–13577.

Watanabe H, Vriens J, Janssens A *et al.* (2003a). Modulation of TRPV4 gating by intra- and extracellular Ca^{2+}. *Cell Calcium* **33**, 489–495.

Watanabe H, Vriens J, Prenen J *et al.* (2003b). Anandamide and arachidonic acid use epoxyeicosatrienoic acids to activate TRPV4 channels. *Nature* **424**, 434–438.

Watanabe H, Vriens J, Suh SH *et al.* (2002). Heat-evoked activation of TRPV4 channels in a HEK293 cell expression system and in native mouse aorta endothelial cells. *J Biol Chem* **277**, 47044–47051.

Wedel BJ, Vazquez G, McKay RR *et al.* (2003). A calmodulin/inositol 1,4,5-trisphosphate (IP3) receptor-binding region targets TRPC3 to the plasma membrane in a calmodulin/IP3 receptor-independent process. *J Biol Chem* **278**, 25758–25765.

Wehage E, Eisfeld J, Heiner I *et al.* (2002). Activation of the cation channel long transient receptor potential channel 2 (LTRPC2). by hydrogen peroxide. A splice variant reveals a mode of activation independent of ADP-ribose. *J Biol Chem* **277**, 23150–23156.

Wes PD, Chevesich J, Jeromin A *et al.* (1995). TRPC1, a human homolog of a *Drosophila* store-operated channel. *PNAS* **92**, 9652–9656.

Winn MP, Conlon PJ, Lynn KL *et al.* (2005). A mutation in the TRPC6 cation channel causes familial focal segmental glomerulosclerosis. *Science* **308**, 1801–1804.

Winn MP, Daskalakis N, Spurney RF *et al.* (2006). Unexpected role of TRPC6 channel in familial nephrotic syndrome: does it have clinical implications?. *J Am Soc Nephrol* **17**, 378–387.

Wissenbach U, Bodding M, Freichel M *et al.* (2000). Trp12, a novel Trp related protein from kidney. *FEBS Lett* **485**, 127–134.

Wu G and Somlo S (2000). Molecular genetics and mechanism of autosomal dominant polycystic kidney disease. *Mol Genet Metab* **69**, 1–15.

Wu G, Hayashi T, Park JH *et al.* (1998). Identification of PKD2L, a human PKD2-related gene: tissue-specific expression and mapping to chromosome 10q25. *Genomics* **54**, 564–568.

Wu G, Mochizuki T, Le TC *et al.* (1997). Molecular cloning, cDNA sequence analysis and chromosomal localization of mouse Pkd2. *Genomics* **45**, 220–223.

Wu X, Babnigg G and Villereal ML (2000). Functional significance of human trp1 and trp3 in store-operated Ca^{2+} entry in HEK-293 cells. *Am J Physiol* **278**, C526–536.

Xu H, Blair NT and Clapham DE (2005). Camphor activates and strongly desensitizes the transient receptor potential vanilloid subtype 1 channel in a vanilloid-independent mechanism. *J Neurosci* **25**, 8924–8937.

Xu HX, Ramsey IS, Kotecha SA *et al.* (2002). TRPV3 is a calcium-permeable temperature-sensitive cation channel. *Nature* **418**, 181–186.

Xu XZ, Chien F, Butler A *et al.* (2000). TRPg, a drosophila TRP-related subunit, forms a regulated cation channel with TRPL. *Neuron* **26**, 647–657.

Xu XZ, Moebius F, Gill DL *et al.* (2001). Regulation of melastatin, a TRP-related protein, through interaction with a cytoplasmic isoform. *PNAS* **98**, 10692–10697.

Yamaguchi H, Matsushita M, Nairn AC *et al.* (2001). Crystal structure of the atypical protein kinase domain of a TRP channel with phosphotransferase activity. *Mol Cell* **7**, 1047–1057.

Yang J, Ellinor PT, Sather WA *et al.* (1993). Molecular determinants of Ca^{2+} selectivity and ion permeation in L-type Ca^{2+} channels. *Nature* **366**, 158–161.

Yang XR, Lin MJ, McIntosh LS *et al.* (2006). Functional expression of transient receptor potential melastatin- and vanilloid-related channels in pulmonary arterial and aortic smooth muscle. *Am J Physiol* **290**, L1267–1276.

Yeh BI, Kim YK, Jabbar W *et al.* (2005). Conformational changes of pore helix coupled to gating of TRPV5 by protons. *EMBO J* **24**, 3224–3234.

Yeh BI, Sun TJ, Lee JZ *et al.* (2003). Mechanism and molecular determinant for regulation of rabbit transient receptor potential type 5 (TRPV5) channel by extracellular pH. *J Biol Chem* **278**, 51044–51052.

Yiangou Y, Facer P, Ford A *et al.* (2001). Capsaicin receptor VR1 and ATP-gated ion channel P2X3 in human urinary bladder. *BJU Int* **87**, 774–779.

Yildirim E, Kawasaki BT and Birnbaumer L (2005). Molecular cloning of TRPC3a, an N-terminally extended, store-operated variant of the human C3 transient receptor potential channel. *PNAS* **102**, 3307–3311.

Yoon IS, Li PP, Siu KP *et al.* (2001). Altered *TRPC7* gene expression in bipolar disorder. *Biol Psych* **50**, 620–626.

Yue L, Peng JB, Hediger MA *et al.* (2001). CaT1 manifests the pore properties of the calcium-release-activated calcium channel. *Nature* **410**, 705–709.

Zagranichnaya TK, Wu X and Villereal ML (2005). Endogenous TRPC1, TRPC3 and TRPC7 proteins combine to form native store-operated channels in HEK-293 cells. *J Biol Chem* **280**, 29559–29569.

Zhang L and Barritt GJ (2004). Evidence that TRPM8 is an androgen-dependent Ca^{2+} channel required for the survival of prostate cancer cells. *Cancer Res* **64**, 8365–8373.

Zhang W, Chu X, Tong Q *et al.* (2003). A novel TRPM2 isoform inhibits calcium influx and susceptibility to cell death. *J Biol Chem* **278**, 16222–16229.

Zhang Y, Hoon MA, Chandrashekar J *et al.* (2003). Coding of sweet, bitter and umami tastes: different receptor cells sharing similar signaling pathways. *Cell* **112**, 293–301.

Zhou Y, Morais-Cabral JH, Kaufman A *et al.* (2001). Chemistry of ion coordination and hydration revealed by a K^+ channel-Fab complex at 2.0 Å resolution. *Nature* **414**, 43–48.

Zhu MX (2005). Multiple roles of calmodulin and other Ca(2+).-binding proteins in the functional regulation of TRP channels. *Pflugers Arch* **451**, 105–115.

Zhu MX and Tang J (2004). TRPC channel interactions with calmodulin and IP3 receptors. *Novartis Found Symp*, **258**, 44–58; discussion 58–62, 98–102, 263–266.

Zhu X, Chu PB, Peyton M *et al.* (1995). Molecular cloning of a widely expressed human homologue for the *Drosophila trp* gene. *FEBS Lett* **373**, 193–198.

Zhu X, Jiang M, Peyton M *et al.* (1996). trp, a novel mammalian gene family essential for agonist-activated capacitative Ca^{2+} entry. *Cell* **85**, 661–671.

Zitt C, Zobel A, Obukhov AG *et al.* (1996). Cloning and functional expression of a human Ca^{2+}-permeable cation channel activated by calcium store depletion. *Neuron* **16**, 1189–1196.

Zufall F, Ukhanov K, Lucas P *et al.* (2005). Neurobiology of TRPC2: from gene to behaviour. *Pflugers Arch* **451**, 61–71.

Zygmunt PM, Petersson J, Andersson DA *et al.* (1999). Vanilloid receptors on sensory nerves mediate the vasodilator action of anandamide. *Nature* **400**, 452–457.

5.2

Connexins and pannexins

Elizabeth Hartfield, Annette Weil,
James Uney and Eric Southam

5.2.1 Introduction

The generally accepted view that the only route of communication between cells is via chemical mediators was dispelled in the 1960s by the first descriptions of direct charge and small molecule transfer between adjacent cells (Dewey and Barr, 1962; Kanno and Loewenstein, 1964) through channels that became known as gap junctions (Brightman and Reese, 1969). Subsequently, freeze fracture electron microscopy revealed that gap junctions are actually plaques, each consisting of up to several hundreds of individual channel units (Revel et al., 1971). In excitable cells, such as neurons or heart cells, gap junction-mediated electrical coupling enables direct fast transmission of electrical signals and enables the generation of synchronized and rapid responses. In non-excitable cells, such as hepatocytes, intercellular molecular signalling is supported by the movement of cytoplasmic molecules including ions, small metabolites, nucleotides and second messengers of molecular mass up to 1000 Da which may play a role in coordinating responses between cells.

It is now known that the building blocks of gap junctions are connexins (and the recently identified, functionally equivalent, pannexins), of which at least 20 have been identified in humans (Sohl and Willecke, 2003). Connexins are arranged in hexagonal arrays to form a connexon which, when docked with a second connexon in an apposing cell, permits gap junctional intercellular communication (GJIC) (Sohl and Willecke, 2003) or, when unpaired, may form a functional connexon hemichannel that directly links the cytoplasm with the extracellular milieu (Goodenough and Paul, 2003). This chapter begins with a brief description of the molecular biology of the connexin family and their aggregation into functional channels. The permeability, gating, and pharmacological properties of junctional channels formed by different connexins are then outlined. Finally, in order to gain insights into the principles of connexin biology, the roles of connexins in health and disease are explored, focusing on the normal and arrhythmic heart, physiological and seizure-related electrical activity in the brain, tumour suppression and cancer, and on human diseases resulting from connexin gene mutations. It should be emphasized, however, that with a bewildering array of connexin channel form and function, this chapter is necessarily illustrative rather than comprehensive.

5.2.2 Molecular characterization

Connexins are classified according to their predicted molecular weights which range from 20–62 kDa such that the 36 kDa human neuronal connexin, for example, is termed Cx36 (Sohl and Willecke, 2003). Since this nomenclature can cause some confusion when homologous proteins from different species are compared (e.g., the human Cx36 is homologous to the skate Cx35 protein), a second system of nomenclature was devised in which connexins are grouped according to sequence similarity and numbered according to order of discovery (Table 5.2.1). However, since the molecular weight classification is most widely used, it has been adopted throughout this chapter.

The connexin gene structure is relatively simple (Figure 5.2.1a), consisting of just two exons. Generally, the first exon contains the non-coding 5′ untranslated region (UTR) and the second contains the complete connexin coding sequence and the 3′ UTR. There are, of course, exceptions whereby the coding region (N-terminal, Cx36; C-terminal, Cx57; and splicing of multiple reading frames, Cx31.3 and Cx23) or the 5′ UTR (Cx30, Cx45, Cx32 and Cx29) is spliced (Sohl and Willecke, 2003). The connexin protein structure is highly conserved between family members (Figure 5.2.1a). Each connexin has four hydrophobic transmembrane domains (TM1–4) connected by loops, two of which are extracellular and one cytoplasmic. The N- and C-termini are both cytoplasmic; the N-terminal is important during protein synthesis and contains the signal peptide sequence (Falk et al., 1997) and the C-terminal plays a role in post-translational regulation of channel activity, for example by phosphorylation or pH (Trexler et al., 1999; Herve and Sarrouilhe, 2002; Berthoud et al., 2004; Lampe and Lau, 2004).

Connexons are oligomers of six connexins surrounding a central pore which, based on electron crystallography, appears to be lined by transmembrane α-helices (Yeager and Nicholson, 1996; Yeager and Harris, 2007). Gap junctions are formed when connexons situated in closely opposing plasma membranes dock together. The extracellular loops of connexins in opposing membranes associate non-covalently via conserved cysteine residues that form disulphide bridges during hemichannel docking. The association of two connexons and, therefore, the formation of the gap junction channel,

Table 5.2.1 Summary of human connexin genes, the biophysical, permeability, and pharmacological properties of their channels, and disease associations

Human connexin	HUGO name	Chromo-some	Primary expression	Unitaty conductance (pS)	Half-inactivation voltage (mV)	Gmin (mV)	Permeability	Pharmacology	Disease association
hCx25	GJβ7	6							
hCx26	GJβ2	13	Breast, skin, cochlea, liver, placenta	140	95	0.13	ATP, cAMP, IP3		Non-syndromal deafness; hyperkeratosis; Vohnwinkel syndrome
hCx30	GJβ6	13	Skin, brain, cochlea	180	51	0.17			Non-syndromic deafness; hydrotic ectodermal dysplasia
hCx30.2	GJγ3	7	Schwann cells, oligodendrocytes						
hCx30.3	GJβ4	1	Skin						Erythrokeratodermia variabilis
hCx31	GJβ3	1	Skin, cochlea, placenta, uterus	100	44	0.21			Erythrokeratodermia variabilis; non-syndromal deafness
hCx31.1	GJβ5	1	Skin						
hCx31.9	GJδ3	17	Heart, colon, vascular smooth muscle cells	15					
hCx32	GJβ1	X	Liver, Schwann cells	60	60	0.25	Ca++, AMP, cAMP, ADP, ATP, IP3, adenosine, glucose, glutathione, glutamate		X-linked Charcot–Marie–Tooth disease; tumour suppression
hCx36	GJδ2	15	Brain (neurons), pancreas	15			cAMP	Quinine, mefloquin ↓ Carbenoxalone ↓	Seizure
hCx37	GJα4	1	Endothelium, heart, uterus, ovary	310	28	0.27	Ca++		Association with myocardial infarction and artherosclerosis
hCx40	GJα5	1	Heart	175				Forskolin, isoprenaline ↑ Anaesthetics ↓	Atrial fibrillation
hCx40.1	GJδ4	10							
hCx43	GJα1	6	Heart, liver, brain	100	61	0.29	Ca++, AMP, cAMP, ADP, ATP, IP3, adenosine, glucose, glutathione, glutamate	Antiarrhythmic peptides, carotenoids, TPA ↑ Fenamates, thromboxane, long chain alcohols ↓	Oculodentodigital dysplasia; cardiac arrythmias; hypoplastic left heart syndrome; tumour supression
hCx45	GJγ1	17	Heart, brain, smooth muscle	30	23	0.08	cAMP	Forskolin, isoprenaline ↑ Quinine, mefloquin ↓	
hCx46	GJα3	13	Lens	140	30	0.16	cAMP		Cataract
hCx47	GJγ2	1	Brain (oligodenrocytes)				cAMP		
hCx50	GJα8	1	Lens	220	44	0.23		Quinine, mefloquin ↓	Cataract
hCx59	GJα9	1							
hCx62	GJα10	6	Retina						

↑ and ↓ refer to up and downregulation of channel activity, respectively.

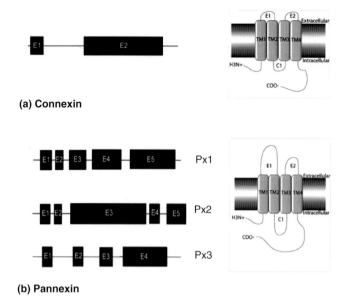

(a) Connexin

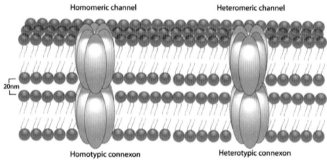

(b) Pannexin

(c) Gap junction

Fig. 5.2.1 Schematic representation of connexin and pannexin genes and proteins and connexin assembly into gap junctions. (a) The generalized connexin gene structure consists of 2 exons. The first exon (E1) is usually non-coding and the second exon (E2) contains the connexin coding sequence. The connexin protein consists of four α-helical transmembrane domains that span the plasma membrane (TM1–4) connected by two extracellular (E1 and E2) and one cytoplasmic loop (C1). Extracellular loops in apposing membranes associate via non-covalent disulphide bridges between conserved cysteine residues. Both the N- and C-termini are intracellular. (b) The pannexin gene consists of five (Px1, Px2) or four (Px3) exons but the protein maintains topography that is analogous to that of connexins with four transmembrane domains. (c) Representation of connexin assembly into gap junctions. Six connexin subunits form a connexon. Connexon hexamers may consist of a single connexin isoform (homomer) or different isoforms (heteromer). Two apposing connexons form a gap junction which may be composed of two identical connexons (homotypic) or two connexons of differing compositions (heteromeric). Figures adapted from Willecke *et al.* (2002), Baranova *et al.* (2004), Panchin (2005) and Sohl *et al.* (2005). See aslo colour plate section.

operates under a 'lock and key' mechanism whereby a 30° rotation ensures an ionically tight fit (Perkins *et al.*, 1997, 1998).

Connexons can be assembled from one connexin isoform (homomeric) or combinations of different isoforms (heteromeric). Furthermore, gap junction channels can be composed of two identical connexons (homotypic junction) or connexons of differing

compositions (heterotypic junction) (Sosinsky and Nicholson, 2005) (Figure 5.2.1c). Although there does appear to be some selectivity in connexin oligomerization (Falk *et al.*, 1997), the capacity of many connexins to form heteromers means that gap junctions may theoretically be constituted of a vast array of individual connexins: in cells expressing just two connexin isoforms, there would be 14 possible connexon configurations and 196 possible combinations of connexins to form a gap junction (Cottrell and Burt, 2005). With their properties dependent on their constituent connexins, gap junctions are functionally heterogenous. Furthermore, variability in the numbers of gap junction channels in a gap junction plaque, the dynamics of connexin synthesis, degradation, and assembly, plus the potential for post-translational modifications of connexins, all add to the potential for diversity in expression and control of gap junction function.

Biosynthesis

Due to the complexity of gap junction composition, it is essential that the biosynthesis, trafficking and degradation of their constituent connexins are tightly regulated. Connexin polypeptides are co-translationally inserted into the rough endoplasmic reticulum (ER), guided by the N-terminal signalling peptide (Falk *et al.*, 1997). Oligomerization of connexins occurs within the ER lumen, although *in vitro* models also suggest connexon assembly may occur after transport from the ER to the trans-golgi network (TGN) (Musil and Goodenough, 1993). Immunohistochemical labelling suggests that assembled connexons are retained within the TGN until required (Laird *et al.*, 1995) when they are trafficked to the plasma membrane in budding transport vesicles via microtubules (Segretain and Falk, 2004). Fusion of Cx32 with enhanced green fluorescent protein has revealed that connexins are highly dynamic: lateral movement, segregation, fusion with other fluorescent plaques, and internalization and transport away from the cell surface have all been observed (Windoffer *et al.*, 2000). The internalization and degradation of connexins is a rapid process that is reflected in the short half-life of connexins, ranging from 2 (Cx43) to 10 (Cx49) hours (Laird, 2006). Time lapse microscopy has shown that gap junction plaques are internalized by the budding of clathrin-coated double-membraned vesicles (often referred to as annular gap junctions) which are then translocated towards the centre of the cell where they are fragmented into smaller vesicles more suitable for degradation via the endosomal/lysosomal pathways (Piehl *et al.*, 2007).

Hemichannels

The fusion of apposing connexons to form an intercellular channel is not inevitable and hemichannels do exist in isolation providing a means of communication between the cell cytoplasm and the extracellular environment (Windoffer *et al.*, 2000; Paul *et al.*, 1991). These hemichannels are functional, allowing the passage of low molecular weight molecules into the cell (Paul *et al.*, 1991) and are gated by the cations Na^+, K^+ and Ca^{2+} (Srinivas *et al.*, 2006; De Vuyst *et al.*, 2006). Hemichannels are thought to play a role in the generation and spread of Ca^{2+} waves via ATP release (Gomes *et al.*, 2005; Stout *et al.*, 2002) and have also been implicated in other cellular activities, including ERK/MAPK signalling, promotion of cell survival and in electrical synapses (Goodenough and Paul, 2003).

Connexin binding partners

Most connexins contain potential phosphorylation sites within their C-terminal tails and a large number of kinases, including PKA, PKC, PKG, MAP kinase, Src tyrosine kinase, and Ca^{2+}-calmodulin dependent protein kinase, all modulate GJIC by altering the phosphorylation state of constitutive connexins (Cruciani and Mikalsen, 2002). In addition to these transient associations, more permanent binding partners have been identified. Zonula occludens -1 and -2 (ZO-1 and ZO-2) have been shown to interact with the C-terminal of Cx43 via a PDZ domain (Giepmans and Moolenaar, 1998; Nielsen et al., 2001). This interaction has since been shown to be cell-cycle specific, with ZO-1 associating with Cx43 during cellular quiescence whereas ZO-2 associations exist during both G0 and S phase (Singh et al., 2005). Immunoprecipitation experiments also indicate an association with α- and β-tubulin and Cx43 sediments with microtubules (Giepmans et al., 2001). These interactions may stabilize Cx43-containing gap junctions during the cell cycle, as suggested by dominant-negative experiments in which overexpression of the ZO-1 interacting domain in osteoblastic cells disrupted wild-type (WT) Cx43/ZO-1 interactions and decreased dye transfer. WT Cx43 remained in the cell membrane but had moved out of a junctional plaque and into the lipid raft of the plasma membrane, suggesting a role for ZO-1 in positioning Cx43 into gap junction plaques (Laing et al., 2005). So far, this interaction of ZO-1 and ZO-2 has only been demonstrated with Cx43 and since the C-terminals of connexins are not well conserved amongst family members it remains to be seen whether this interaction applies to other connexins.

Cx43 and β-catenin co-localize extensively in junctional membranes. β-catenin is normally associated with the plasma membrane as part of adhesive junctions and its association with Cx43 suggests the regulation of GJIC via the Wnt signalling pathway (Ai et al., 2000). Other protein-binding parters have been identified, including debrin, caveolin-1, NOV and CIP85 (Laird, 2006). These interactions may contribute to regulation and stabilization of gap junctions but may also provide docking sites for signalling molecules, creating a 'signalling hub' at contact points between cells.

Pannexins

Although the focus of this chapter is the connexins, the pannexins, a family of three vertebrate proteins (Px1–3) that share a similar membrane topology and function but not sequence homology to connexins, should not be overlooked (Baranova et al., 2004) (Figure 5.2.1b). Injection of rat pannexin RNAs into Xenopus oocytes revealed the formation of functional, electrically coupled channels. Pannexin 1 (Px1) can form a homomeric channel and can also oligomerize with Px2-forming heteromeric channels with different properties whereas Px2 cannot form homomeric channels (Bruzzone et al., 2003). Pannexin channels are similar to connexin gap junctions in that they are permeable to metabolites and to ions including calcium, leading to the idea that pannexin channels, particularly hemichannels, may contribute to the generation and synchronization of intercellular calcium waves (Vanden Abeele et al., 2006). Importantly, whereas connexins are sensitive to extracellular calcium concentrations, pannexins are not (Valiunas, 2002; Bruzzone et al., 2005). One recently described significant function of Px1 is its contribution to the pore-forming unit of the P2X7 receptor death complex (Locovei et al., 2007).

5.2.3 Biophysics

Permeability

Gap junctions were initially thought to be non-specific molecular sieves with their pore diameter assumed to be the key determinant of molecular permeability. However, it has since become evident that different connexins impart distinct permeabilities with pore size, charge selectivities, conductance, and gating properties all governing GJIC (Goldberg et al., 2004; Weber et al., 2004; Bedner et al., 2006; Harris, 2007).

Fluorescent dye molecules of various charges and size have been employed to investigate molecular permeability and charge selectivity (Elfgang et al., 1995; Cao et al., 1998). Pore diameter limits molecular permeability and the use of homologous uncharged molecules suggests the following rank order of maximum pore size: Cx43 > Cx32 > Cx26 = Cx26/Cx32 > Cx37 > Cx46 and Cx43 > Cx40 (Harris, 2007). It is more difficult to establish a quantitative comparison of charge selectivities between different gap junctions, due to the highly variable shape, and charge distribution of common fluorescent tracers. Qualitative comparisons indicate that gap junctions exhibit a wide range of charge selectivity, ranging from little charge selectivity (Cx43) to preferentially permeable to negatively charged dyes (e.g. Cx32) to preferentially permeable to positively charged dyes (e.g., Cx46) (Steinberg et al., 1994; Verselis et al., 2000).

While the differential permeabilities to synthetic probes have been instructive, progress in the understanding of how endogenous molecules permeate gap junctions and their underlying physiological role as chemical signals is less well advanced. This is partly due to the large number of different connexin isoforms and the even larger number of heteromeric and heterotypic connexin combinations, each with a potentially distinct set of permeability properties. So far, it is well established that cytoplasmic molecules like cAMP (Lawrence et al., 1978), nucleotides (Goldberg et al., 1999), IP_3 and Ca^{2+} (Charles et al., 1992; Niessen et al., 2000), and metabolites such as glucose (Tabernero et al., 1996) can pass through gap junction channels (Harris, 2007) (Table 5.2.1). Different connexin channels can exhibit highly differential permeabilities for cytoplasmic molecules, without obvious correlation to the permeabilities found for non-biological fluorescent dyes or to their unitary conductance. For example, a more than 100-fold difference in the rate of transfer of ATP through Cx43 and Cx32 channels, despite the similar permeabilities of these channels for some dyes (Goldberg et al., 2002) and a 33-fold difference of cAMP transfer between Cx43 and Cx36 channels (Bedner et al., 2006), are reported. An illustration of the importance of appropriate gap junction permeability is that a mutation of Cx26 that causes a selective reduction in IP_3 permeability results in deafness in humans (Beltramello et al., 2005).

The ionic conductances of gap junctions have been widely investigated using single channel electrophysiology. These studies indicate that unitary conductances of gap junctions are broadly distributed ranging from 10–15 pS for the neuronal Cx36 to 300 pS for Cx37 and, in the case of Px1, 500 pS (Harris, 2001) (Table 5.2.1). Similarly, ionic charge selectivities range from slightly anion selective (1.1:1 for Cx32) (Suchyna et al., 1999) to highly cation selective (10:1 for Cx45); (Veenstra et al., 1995). These vastly different unitary conductances and charge selectivities reflect the diversity of structural and/or electrostatic properties of connexin channels.

Connexin channel gating mechanisms

Charge conductance through homotypic gap junctions (G_j) is dependent on the transjunctional potential (V_j) and, in some cases, on the transmembrane potential (V_m). Maximal conductance (G_{jmax}) typically occurs when $V_j = 0$ and decreases symmetrically with V_j to a residual level (G_{jmin}) (Moreno et al., 1994a). Depending on the connexin, G_{jmin} varies from <5 to >50 per cent of G_{jmax} (Gonzalez et al., 2007). A quantitative comparison of the voltage gating properties of different homotypic connexin gap junctions reveals voltage sensitivities that range from highly sensitive (characterized by low G_{jmin} and/or small half inactivation voltage (V_o), e.g., Cx45) to low sensitivity (high V_o, e.g., Cx26). For heterotypic channels, the steady-state G_j/V_j relationship is asymmetric with regards to the two polarities (Gonzalez et al., 2007). For example, at a junction of apposing Cx32 and Cx26 homomeric connexons, the steady state conductance only decreases at one polarity of V_j (positive on Cx26 side), whereas, at reversed polarity no single channel transitions could be detected (Barrio et al., 1991).

Based on single channel studies, there appear to be several transitions between the open and subconductance states. The V_j-dependent reduction in macroscopic conductance is largely accounted for by channels transitioning from the main open state to a residual open state. The slower decline in junctional conductances, observed at higher V_j and voltage pulses of longer duration, is associated with transitions from a long-lived conductance substate to a closed state (Moreno et al., 1994a; Bukauskas et al., 1995, 2001). Investigations into the time dependence of gating transitions indicate that transitions between the main open and the residual state are fast (1–2 ms) whereas transitions between the residual state and the closed state are slower (>10 ms) (Bukauskas et al., 1995). Thus, a model has been proposed in which V_j gradients activate both fast and slow gating (Moreno et al., 1994a; Bukauskas et al., 2001) and the macroscopic V_j dependence of conductances is likely to include contributions from both. However, given the diversity of the connexin family, other gating processes are also likely to apply (Bukauskas and Verselis, 2004; Gonzalez et al., 2007).

Several regions of the connexin molecule are implicated in voltage gating including the carboxy and amino terminals, the first and second transmembrane segments, and the first extracellular loop (Suchyna et al., 1993; Verselis et al., 1994; Gonzalez et al., 2007). Of these, the carboxy terminal (CT) domain appears to be particularly important with regard to fast V_j gating in many connexins. Deletion of the CT domain in Cx43, for example, selectively abolished fast V_j gating, leaving the slow gating mechanism intact (Elenes et al., 2001) with voltage-gating sensitivity restored when truncated Cx43 is expressed together with a separate Cx43 CT domain (Moreno et al., 2002). This, and other studies, resulted in the formulation of a ball and chain model in which the CT domain is displaced towards the channel pore causing its occlusion.

It has long been recognized that gap junctions can also be gated by chemical parameters including intracellular calcium ion concentration (Loewenstein and Kanno, 1966) and pH (Turin and Warner, 1977). This may constitute an effective cellular protective mechanism in the case of, for example, a drop in pH in cardiac cells caused by ischaemia limiting gap junctional mediated spread of injury from damaged to normal tissue. Exogenous agents, including long-chain alkanols, local anaesthetics, acidifiers, glycyrrhetinic acid derivatives such as carbenoxolone, and quinine, may also uncouple cells (Harris, 2001) via the slow transition between conductance states ending with the full closure of the channel (Bukauskas and Peracchia, 1997). The molecular mechanism(s) of chemical gating are still poorly defined. The CT domain is again likely to be involved in the pH sensitivity of some gap junctions, as indicated by the abolition of pH sensitivity in Cx43 channels when the CT domain is absent (Morley et al., 1996). However, to explain the maintenance of chemical gating sensitivity of other connexin channels despite CT truncation (e.g., Cx45), other mechanisms must be considered (Stergiopoulos et al., 1999). Evidence that inhibition or down regulation of calmodulin prevents uncoupling induced by CO_2 led to the proposal of a 'cork-type' gating model in which chemical gating results from an interaction between calmodulin and the cytoplasmic mouth of the channel (Peracchia et al., 2000).

Connexins contain several residues within the cytoplasmic loop and the C-terminus that are susceptible to phosphorylation (Lampe and Lau, 2004; Laird, 2005; Moreno and Lau, 2007). Single channel recordings reveal that Cx43 phosphorylation, mediated by PKG, PKC, PKA and MAP kinases, enhances the prevalence of the smallest conductance state, whereas agents that promote dephosphorylation shift the preference to the largest state (Moreno et al., 1992; Takens-Kwak and Jongsma, 1992; Moreno et al., 1994b). Interestingly, phosphorylation of Cx43 by PKC produces an increase in macroscopic conductance but a decrease in unitary conductance (Moreno et al., 1992), a discrepancy that may be explained by a differential effect on open probability, assuming no substantial change in channel number. Whilst serine/threonine phosphorylation appears to change the distribution of the unitary conductance states, tyrosine protein kinase-induced phosphorylation has been shown to disrupt gap junction function (Lampe and Lau, 2004; Warn-Cramer and Lau, 2004). Phosphorylation of Cx43 on two separate tyrosine residues by v-Src reduces macroscopic junctional conductances measured using changes in dye transfer or junctional conductance (Filson et al., 1990; Cottrell et al., 2003). However, the reduction in macroscopic conductance in this case was not mediated by a shift in the main conductance states but perhaps by a reduction in open probability. As with voltage and chemical gating, the mechanism by which phosphorylation modifies the unitary conductance or open probablility of the channels is not entirely clear. Phosphorylation of the CT region appears to be required, suggesting that the resulting charge alterations may induce changes in the affinity of the carboxyl-terminus tail to the cytoplasmic loop which may shift the channel to the partially open state (Moreno, 2004).

5.2.4 Distribution

Table 5.2.1 provides an overview of connexin distribution. It should, however, be noted that this is a gross simplification of the widespread and overlapping distribution patterns of the 21 members of the human connexin family. Cx43 alone has, for example, been described in 34 different tissues (Laird, 2006). Whilst the details of this distribution are beyond the scope of this review, the widespread and overlapping distribution of connexins does provide several important insights into their function. First, the ubiquitous nature of connexin expression suggests that connexin-mediated electrotonic coupling or molecular transfer between cells, or between cells and the external milieu, is fundamental to the physiology of most tissues. Secondly, the co-expression of more than one connexin in many cell types raises the possibility of compensatory or transdominant inhibitory effects of one connexin on another. Finally, the complexity of connexin distribution patterns emphasizes the importance of the regulation of connexin expression. In this respect, numerous global and cell specific transcription factors and epigenetic gene silencing mechanisms have been described (Oyamada *et al.*, 2005).

5.2.5 Pharmacology

Although many endogenous signalling molecules and exogenous pharmacological agents are reported to modulate gap junction communication (Salameh and Dhein, 2005), several factors have impeded progress in gap junction pharmacology including the number of connexins and the diversity of assembled channels, the limited selectivity of existing pharmacological tools, and the multiple potential mechanisms of gap junction modulation ranging from a direct effect on channel gating to altered connexin expression and assembly. Of course, whilst the above factors present difficult challenges, they may ultimately increase the scope for achieving selectivity within the connexin family.

Gap junction openers

Agents reported to enhance GJIC generally appear to stimulate protein kinase activity including cAMP promoting agents (forskolin, isoprenaline, cAMP), the anti-arrythmic drug Tedisamil and eicosanoids that act through PKA, and various anti-arrythmic peptides that act through PKC (Salameh and Dhein, 2005). Of these, it is anti-arrythmic peptides such as rotigaptide that, through a combination of potency, selectivity, and pharmacokinetic properties, currently offer the greatest (only) prospect of therapeutically utility in the near term. Rotigigaptide was developed from a horseshoe-shaped hexapeptide initially derived from bovine atria and shown to synchronize chick cardiomyocytes maintained *in vitro*. Based on *in vitro* dye-coupling experiments, rotigaptide increases gap junction communication without affecting membrane conductance (Xing *et al.*, 2003) and is selective for Cx43 over Cx26 or Cx32 (Clarke *et al.*, 2006). Rotigaptide modulates gap junctions but not ion channels (Kjolbye *et al.*, 2007) and is selective for Cx43 versus a panel of 80 ion channels and receptors (Haugan *et al.*, 2005). The precise mechanism(s) of action of anti-arrhythmic peptides such as rotigaptide in modulating gap junction communication is not entirely clear but in the acute phase may involve the modulation phosphorylation of the long intracellular C-terminal tail of Cx43, probably via activation of PKC-a (Weng *et al.*, 2002; Dhein *et al.*, 2003), and in the longer-term by increasing expression of Cx43 (Stahlhut *et al.*, 2006).

Gap junction closers

Carbenoxalone, a succinyl ester of glycerrhetinic acid, has become the most widely used inhibitor of gap junctions, particularly neuronal gap junctions composed of Cx36. There are numerous reports of carbenoxalone inhibiting seizures (Gareri *et al.*, 2004; Bagirici and Bostanci, 2006; Gajda *et al.*, 2006; Nilsen *et al.*, 2006; Proulx *et al.*, 2006; Bostanci and Bagirici, 2007; Medina-Ceja *et al.*, 2008) and more physiological synchronous oscillatory activity in inhibitory networks (Sinfield and Collins, 2006; Zsiros *et al.*, 2007), including human neocortical synchronization (Gigout *et al.*, 2006), with a resultant increase in spontaneous excitatory and inhibitory

synaptic neurotransmission (Yang and Ling, 2007). Effects of carbenoxalone on memory and cognitive function have also been studied with mixed results: spatial learning in rats was impaired (Hosseinzadeh *et al.*, 2005) whereas verbal fluency in humans was enhanced (Sandeep *et al.*, 2004). The mechanism of gap junction inhibition by carbenoxalone is not well characterized but, based on the limited available information, appears more likely to be an indirect effect on phosphorylation or trafficking (Goldberg *et al.*, 1996; Guan *et al.*, 1996) of connexins rather than a direct effect on gap junction conductance. Interestingly, hemichannels comprised of pannexins are sensitive to blockade by carbenoxalone (IC_{50} ~5 mM) (Bruzzone *et al.*, 2005) raising the intriguing possibility that some of the actions of carbenoxalone attributed to connexin blockade may in fact be due to block of pannexin channels.

The gap junction blocking properties of the anti-malarial drug quinine and its analogues have been the focus of particular attention because of evidence of selectivity between connexins. Quinine itself was found to block Cx36- and Cx50-mediated gap junction communication with selectivity over Cx26, Cx32, Cx40, Cx43, and Cx45 channels (Srinivas *et al.*, 2001). The related compound mefloquine is about 100-fold more potent in blocking Cx36 ($IC_{50} = 300$ nM) and Cx50 ($IC_{50} = 1.1$ mM) channels than quinine (Cruikshank *et al.*, 2004). Mechanistic studies indicate that quinine binds to an intracellular, possibly intra-pore, site resulting in decreased open probability of the gap junction channel. However, activities at other molecular targets including voltage- and ligand-gated ion channels limit the value of quinines as experimental tools, especially *in vivo*. Thus, although quinine has been used to demonstrate that intercellular communication via gap junctions may contribute to the manifestation and propagation of seizures in adult rat neocortex (Gajda *et al.*, 2005) and even to synchronicity in excised human neocortical slices (Gigout *et al.*, 2006), anti-malarial drugs such as mefloquine are also reported to cause CNS side effects including seizures (Wooltorton, 2002) which may reflect the off-target (possibly GABA) activities of these compounds (Amabeoku and Farmer, 2005).

Other groups of compounds used experimentally to block GJIC include the fenamates and long chain alcohols. In addition to modulating a wide range of ion channels, fenamates block GJIC either via a direct interaction with channels or hemichannels or via alterations in membrane fluidity (Harks *et al.*, 2001; Srinivas and Spray, 2003). The perturbation of bulk membrane fluidity is also assumed to be the mechanism of the rapid and reversible inhibition of GJIC by long chain alcohols such as heptanol and octanol as well as general anaesthetics (e.g., halothane and isoflurane) (Bastiaanse *et al.*, 1993).

5.2.6 Physiology

Cardiac physiology

Five connexins (Cx30.2, Cx37, Cx40, Cx43, Cx45) are expressed in the heart (Bastiaanse *et al.*, 1993). Whilst the cardiac distributions of Cx37, Cx30.2 and Cx45 are relatively limited, Cx40 is found in atrial myocytes and Cx43, the most widespread of the cardiac connexins, is the primary connexin of the cardiac ventricles. The location of connexins in the cardiac conduction system and at the intercalated discs facilitates a role in propagating the cardiac action potential and aiding synchronization of electrical activity in, and hence contraction of, the myocardium (Bernstein and Morley, 2006;

Duffy *et al.*, 2006). There is considerable evidence of the interactions of connexins with associated binding proteins in cardiac myocytes, particularly at the intercalated disc, including regulatory protein kinases (e.g., c-Src) and phosphatases, structural proteins (e.g., microtubules and ZO-1), and signalling proteins (e.g., cadherins) (Giepmans, 2004). Furthermore, in the intercalated disc, gap junctions comprising Cx43 are closely associated with desmosome and adherens junctions, raising the possibility of cooperativity between different junctional types to strengthen the disc complex (Duffy *et al.*, 2006).

Neuronal gap junctions

At least six connexins have been identified at gap junctions within the nervous system (Nagy *et al.*, 2004). Whilst Cx26, Cx29, Cx30, Cx32, and Cx43 play vital roles in mediating intercellular communication within the glial syncytium, only Cx36 has been unequivocally demonstrated to mediate interneuronal gap junction, or electrical synapse, communication (Connors and Long, 2004; Nagy *et al.*, 2004). Cx36 is widespread throughout the brain but appears to be confined to parvalbumin-positive GABA-ergic interneurons (Belluardo *et al.*, 2000). Deletion of Cx36 results in mice that exhibit no obvious abnormalities of CNS structure, development or neuronal properties (Hormuzdi *et al.*, 2001). Closer examination of Cx36 knockout animals, however, provides two clues to the importance of Cx36 in facilitating neuronal network activity within the CNS. First, Cx36 appears to be essential for the propagation of rod signals across the cells of the mammalian retina and for visual acuity in low ambient light levels (Guldenagel *et al.*, 2001; Deans *et al.*, 2002; Demb and Pugh 2002). Secondly, neuronal oscillatory activity in the gamma (30–80 Hz) frequency range that is entrained by inhibitory GABA-ergic interneurons (Hormuzdi *et al.*, 2001; Deans *et al.*, 2002; Buhl *et al.*, 2003), is drastically reduced in Cx36 null mice. Gamma oscillations are associated with attention and memory (Jensen *et al.*, 2007) and Cx36 knockout mice exhibited subtle impairments in cognition and memory performance that are dependent on the complexity of the task (Frisch *et al.*, 2005). Interestingly, unlike Cx36, pannexins are prominently expressed in pyramidal neurons as well as interneurons and are therefore attractive candidates as mediators of fast ripple activity that is spared in Cx36 knockout animals. If this proves to be the case, it suggests an important role for pannexins in memory consolidation (Litvin *et al.*, 2006). Whilst consideration of the clinical significance of brain connexins and pannexins remains speculative and largely focussed on the contribution of glial gap junctions to homeostasis within the CNS (Nakase and Naus, 2004), as understanding of the importance of synchronized high-frequency neuronal oscillatory activity to normal brain function grows, so too does the recognition that a range of psychiatric and neurological disorders, including schizophrenia, autism, epilepsy, Alzheimer's disease, and Parkinson's disease, are associated with abnormal neural synchrony (Uhlhaas and Singer, 2006) and that targeting neuronal connexins and pannexins may be a valid therapeutic strategy to address them.

Tumour suppression

From the initial observation that GJIC normally exerts a profound control of fibroblast proliferation that is attenuated in malignant fibroblasts (Loewenstein and Kanno, 1966), several strands of evidence point to connexins functioning as tumour suppressor genes.

First, a consistent finding in many tumours, including carcinomas of the lung, kidney, breast, glia, cervix, and prostate, is that expression of endogenous connexins, most commonly Cx26, 32, and 43, is reduced (Vine and Bertram, 2002; Pointis *et al.*, 2007). Secondly, a role in tumour suppression may also be deduced from mouse strains bearing connexin mutations. Compared with wild-types, Cx32 knockout mice exhibit an increased incidence of chemically and radiation-induced tumours in both liver (Temme *et al.*, 1997; King and Lampe, 2004a) and lung (King and Lampe, 2004b). Furthermore, the hepatocarcinogenic potential of the gap junction blocker phenobarbital is attenuated in Cx32 knockout mice (Moennikes *et al.*, 2000). Interestingly, there does not seem to be a marked increase in spontaneous tumour formation in Cx32 null mutant animals, suggesting that this connexin is more likely to suppress tumour progression than initiation. Enhanced tumour progression is also observed in lungs of Cx43 heterozygous mice (homozygous Cx43 null mutations are lethal) (Avanzo *et al.*, 2004). Thirdly, restoration of connexin expression using gene transfer technology into tumour cell lines reinstates control of cell proliferation (Mesnil, 2002). Finally, tumour preventative agents such as retinoids and carotenoids normalize aberrant gap junction communication both *in vivo* and in cell-based systems (King and Bertram, 2005).

5.2.7 Disease

Given their importance in cardiac and CNS physiology and in tumour suppression, connexins may be involved in, and may represent pharmacological targets for the treatment of, cardiac arryhthmias, seizures, and cancer (Salameh and Dhein, 2005).

Cardiac arrythmias

The extensive literature describing altered gap junctions and connexin expression in human heart disease (Severs *et al.*, 2004) together with the cardiac abnormalities exhibited by connexin mutant mice (particularly Cx43 deficient) (Reaume *et al.*, 1995; Vaidya *et al.*, 2001) and cardiac myocyte-specific Cx43 deleted mice (Gutstein *et al.*, 2001) strongly implicate connexins in cardiac pathophysiology. It should be noted that it can be difficult to establish cause and effect relationships between cardiac connexins and their binding partners. Cardiac-specific loss of N-cadherin, for example, leads to altered connexin expression and arrhythmias (Li *et al.*, 2005). Nevertheless, interest is growing in gap junctions as potential therapeutic targets for heart disease (van der Velden and Jongsma, 2002; Salameh and Dhein, 2005). The efficacy of rotigaptide in preventing arrythmias has been demonstrated in numerous pre-clinical models including, most recently, in dogs (Guerra *et al.*, 2006; Shiroshita-Takeshita *et al.*, 2007). With adequate (intravenous) pharmacokinetics and a favourable pre-clinical safety profile, rotigaptide represents an attractive drug-like candidate for the treatment of certain cardiac arrythmias (Kjolbye *et al.*, 2007).

Cancer

Although evidence that tumour cells must isolate themselves from direct intercellular communication with unaffected cells in order to progress is compelling, the mechanism by which gap junctions are downregulated is unclear. With rare exceptions (Dubina *et al.*, 2002), cancers do not appear to be associated with mutations of connexin genes. Decreased expression of connexins in tumours may be a consequence of epigenetic silencing of connexin gene transcription via DNA methylation or histone deacetylation (Mesnil, 2002) and the reversal of this epigenetic suppression of gene transcription is a potential anti-cancer strategy. Certainly, expression of Cx43 is reported to be reinstated by 5-aza-2′-deoxycytidine, an inhibitor of DNA methylation (King et al., 2000), as is Cx26 in a breast cancer cell line (Tan et al., 2002 but see Singal et al., 2000). GJIC may be reduced by exogenous agents and is thought to be the mechanism of action of several chemical tumour promoters such as phenobarbital and 12-O-tetradecanoylphorbol-13-acetate (TPA; Matesic et al., 1994). Alternatively, altered trafficking of connexins or post-translational modification may contribute to the downregulation of GJIC in tumour cells. The observation that inhibitors of cAMP phosphodiesterase, resulting in elevated intracellular cAMP concentrations, limit the growth of micrometastases both in Cx43 expressing cell lines in vitro (Bertram, 1979) and in mouse lungs in vivo (Janik et al., 1980) raises the prospect of tumour suppression via PKA-mediated phosphorylation of C-terminal tails of connexin molecules.

A contribution of deficient GJIC to tumourogenesis is further implied by the cancer chemopreventative properties of analogues of carotenoids such as lycopene (responsible for the red colour of tomatoes) (Seren et al., 2008). Increasing evidence suggests that the anti-neoplastic effects of carotenoids are mediated, at least in part, by mechanisms independent of antioxidant or pro-vitamin A properties of caroteinoids including the upregulation of Cx43 and enhanced GJIC, thereby allowing the transfer of growth inhibitory signals from normal to neoplastic cells (Bertram, 2004).

Seizures

As understanding of the importance of synchronized high-frequency neuronal oscillatory activity to normal brain function grows, so too does the recognition that a range of psychiatric and neurological disorders, including epilepsy, schizophrenia, autism, Alzheimer's disease, and Parkinson's disease, may all be associated with abnormal neural synchrony (Uhlhaas and Singer, 2006). However, the observation that classical chemical neurotransmission is not essential for seizure activity (Taylor and Dudek, 1982) together with the recognition that gap junctions contribute to very fast oscillations in the brain that precede and possibly initiate seizures (Traub et al., 2001) and may facilitate the propagation of seizure activity (Perez Velazquez and Carlen, 2000) meant that the role of electrotonic coupling via gap junctions in the pathological electrical synchrony underlying epilepsy became the focus of particular scrutiny. Increased connexin expression in the temporal cortex of epilepsy patients (Naus et al., 1991; Fonseca et al., 2002) and in pre-clinical seizure models (Li et al., 2001; Samoilova et al., 2003; Gajda et al., 2006) is supportive of a role for connexins in epileptogenesis. Perhaps the most telling indication that gap junctions are involved in epilepsy is that gap junction blockers such as quinine and carbenoxalone, although admittedly far from ideal pharmacological tools, are consistently reported to be effective in a number of rat seizure models (Gareri et al., 2004; Gajda et al., 2005; Nilsen et al., 2006; Proulx et al., 2006; Bostanci and Bagirici, 2007; Medina-Ceja et al., 2008) whilst the putative positive modulator of GJIC trimethylamine promotes seizure activity (Gajda et al., 2003). Although far from ideal, these pharmacological agents provide a strong indication of a role for GJIC in seizure generation and propagation and suggest a potentially attractive therapeutic strategy for the future.

Human diseases associated with connexin mutations

The first human inherited disease recognized to be caused by mutations of a gap junction gene was the progressive demyelinating neuropathy X-linked Charcot Marie Tooth disease caused by mutations of Cx32 (Bergoffen et al., 1993; Nelis et al., 1999). Cx32 is expressed in Schwann cells, particularly at the Schmidt–Lantermann incisures adjacent to the nodes of Ranvier where it is speculated that Cx32 channels mediate the diffusion of small molecules, thereby circumventing the much lengthier cytoplasmic diffusion throughout the restricted Schwann cell cytoplasm of the myelin sheath (Goodenough and Paul, 2003).

Mutations in Cx30, Cx31, and particularly Cx 26 are implicated in non-syndromal sensorineural hearing loss (Petersen and Willems, 2006) and syndromes in which hearing loss is accompanied by hyperproliferation of the skin such as Vohwinkel's syndrome, palmoplanter keratoderma, keratitis–ichthyosis, and hystrix-like ichthyosis-deafness syndrome (Kelsell et al., 2001; Gerido and White 2004). In the inner ear, Cx26 is expressed in the sensory epithelium where it is important in controlling ionic homeostasis (Rabionet et al., 2000). Interestingly, Cx30 is co-expressed with Cx26 in the inner ear (Lautermann et al., 1998) suggesting that these connexins cannot compensate for each other with regards to inner ear function, perhaps because the ion flux properties of Cx26 and Cx30 heteromeric gap junctions differ from homomeric assemblies (Ahmad et al., 2003; Sun et al., 2005). Since many Cx26 mutations produce non-syndromic hearing loss, skin pathology appears unlikely to be the direct result of altered homomeric Cx26 channel gating but may impact other connexins co-expressed with Cx26 in the skin. Cx26 and Cx43 do not normally form heteromeric channels (Gemel et al., 2004) but, following the deletion of a glutamic acid residue from Cx26 (ΔE42), these connexins do co-locate and gap junction function is impaired (Rouan et al., 2001) which is indicative of the potential for Cx26 to exert transdominant inhibitory effects on another connexin.

The vertebrate lens consists of epithelial cells that line the anterior surface together with fibre cells which form the bulk of the lens structure. Lens fibre cells are dependent upon direct gap junction-mediated communication with their metabolically active epithelial cells because they lose their complement of intracellular organelles as concentrations of soluble crystallin proteins rise in their cytoplasm (White and Bruzzone, 2000; Gerido and White, 2004). Three connexins are expressed in the lens (Cx43, Cx46, and Cx50) and their importance in lens transparency is illustrated by the development of cataracts resulting from mutations of human Cx46 and Cx50 (White, 2002) and in mice in which either Cx46 or Cx50 is deleted (Cx43 knockouts are not viable) (Reaume et al., 1995; White and Bruzzone, 2000; Gerido and White, 2004).

Finally, oculodentodigital dysplasia (ODDD) is a rare inherited disorder caused by autosomal dominant mutations of Cx43 and is characterized by abnormalities of the face, eyes, hearing, limbs, teeth, and nails (Paznekas et al., 2003). Based on a mouse Cx43 mutant that exhibited a phenotype comparable with human ODDD, Cx43 mutations appear to dominantly inhibit wild-type gap junction coupling, possibly via the destabilization of wild-type

Cx43 (Flenniken *et al.*, 2005). Whilst the widespread distribution of Cx43 probably accounts for the pleiotropic nature of ODDD, given the extent of the disruption of function caused by mutations of Cx43, and the lethal nature of Cx43 deletion (Reaume *et al.*, 1995), it is perhaps surprising that the consequences of Cx43 mutations are not more severe and may point to compensatory effects of other connexins (Laird, 2006).

5.2.8 Concluding remarks

Great strides have been made in elucidating the molecular biology of the connexin family of genes and proteins, and their importance in many physiological and pathophysiological processes has become recognized. Much has been also been learned about the biophysics and gating mechanisms of connexin channels although, since this knowledge is based largely on artificial *in vitro* systems, often utilizing homomeric assemblies of connexins, the subtleties of native connexin configurations are probably not fully appreciated. However, perhaps the greatest challenge in the connexin arena is the development of potent and selective pharmacological agents that target both the opening and closure of connexin channels and which are needed to facilitate investigations into, especially native, connexin channel physiology and eventually to enable the exploitation of the therapeutic potential of connexin channels. The irony is, of course, that the enormous diversity of connexin channel form and function that has hindered pharmacological development is also the greatest hope that selective channel modulators are achievable.

References

Ahmad S, Chen S, Sun J *et al.* (2003). Connexins 26 and 30 are co-assembled to form gap junctions in the cochlea of mice. *Biochem Biophys Res Commun* **307**, 362–368.

Ai Z, Fischer A, Spray DC *et al.* (2000). Wnt-1 regulation of connexin 43 in cardiac myocytes. *J Clin Invest* **105**, 161–171.

Amabeoku GJ and Farmer CC (2005). Gamma-aminobutyric acid and mefloquine-induced seizures in mice. *Prog Neuropsychopharmacol Biol Psychiatry* **29**, 917–921.

Avanzo JL, Mesnil M, Hernandez-Blazquez FJ *et al.* (2004). Increased susceptibility to urethane-induced lung tumors in mice with decreased expression of connexin 43. *Carcinogenesis* **25**, 1973–1982.

Bagirici F and Bostanci MO (2006). Anticonvulsive effects of nimodipine on penicillin-induced epileptiform activity. *Acta Neurobiol Exp (Wars)* **66**, 123–128.

Baranova A, Ivanov D, Petrash N *et al.* (2004). The mammalian pannexin family is homologous to the invertebrate innexin gap junction proteins. *Genomics* **83**, 706–716.

Barrio LC, Suchyna T, Bargiello T *et al.* (1991). Gap junctions formed by connexins 26 and 32 alone and in combination are differently affected by applied voltage. *PNAS* **88**, 8410–8414.

Bastiaanse EM, Jongsma HJ, Van der Laarse A *et al.* (1993). Heptanol-induced decrease in cardiac gap junctional conductance is mediated by a decrease in the fluidity of membranous cholesterol-rich domains. *J Membr Biol* **136**, 135–145.

Bedner P, Niessen H, Odermatt B *et al.* (2006). Selective permeability of different connexin channels to the second messenger cyclic AMP. *J Biol Chem* **281**, 6673–6681.

Belluardo N, Mudo G, Trovato-Salinaro A *et al.* (2000). Expression of connexin 36 in the adult and developing rat brain. *Brain Res* **865**, 121–138.

Beltramello M, Piazza V, Bukauskas FF *et al.* (2005). Impaired permeability to Ins(1,4,5)P3 in a mutant connexin underlies recessive hereditary deafness. *Nat Cell Biol* **7**, 63–69.

Bergoffen J, Scherer SS, Wang S *et al.* (1993). Connexin mutations in X-linked Charcot–Marie–Tooth disease. *Science* **262**, 2039–2042.

Bernstein SA and Morley GE (2006). Gap junctions and propagation of the cardiac action potential. *Adv Cardiol* **42**, 71–85.

Berthoud VM, Minogue PJ, Laing JG *et al.* (2004). Pathways for degradation of connexins and gap junctions. *Cardiovasc Res* **62**, 256–267.

Bertram JS (1979). Modulation of cellular interactions between C3H/10T1/2 cells and their transformed counterparts by phosphodiesterase inhibitors. *Cancer Res* **39**, 3502–3508.

Bertram JS (2004). Dietary carotenoids, connexins and cancer: what is the connection? *Biochem Soc Trans* **32**, 985–989.

Bostanci MO and Bagirici F (2007). Anticonvulsive effects of quinine on penicillin-induced epileptiform activity: an *in vivo* study. *Seizure* **16**, 166–172.

Brightman MW and Reese TS (1969). Junctions between intimately apposed cell membranes in the vertebrate brain. *J Cell Biol* **40**, 648–677.

Bruzzone R, Barbe MT, Jakob NJ *et al.* (2005). Pharmacological properties of homomeric and heteromeric pannexin hemichannels expressed in *Xenopus* oocytes. *J Neurochem* **92**, 1033–1043.

Bruzzone R, Hormuzdi SG, Barbe MT *et al.* (2003). Pannexins, a family of gap junction proteins expressed in brain. *PNAS* **100**, 13644–13649.

Buhl DL, Harris KD, Hormuzdi SG *et al.* (2003). Selective impairment of hippocampal gamma oscillations in connexin-36 knock-out mouse *in vivo*. *J Neurosci* **23**, 1013–1018.

Bukauskas FF and Peracchia C (1997). Two distinct gating mechanisms in gap junction channels: CO2-sensitive and voltage-sensitive. *Biophys J* **72**, 2137–2142.

Bukauskas FF and Verselis VK (2004). Gap junction channel gating. *Biochim Biophys Acta* **1662**, 42–60.

Bukauskas FF, Bukauskiene A, Bennett MV *et al.* (2001). Gating properties of gap junction channels assembled from connexin 43 and connexin 43 fused with green fluorescent protein. *Biophys J* **81**, 137–152.

Bukauskas FF, Elfgang C, Willecke K *et al.* (1995). Biophysical properties of gap junction channels formed by mouse connexin 40 in induced pairs of transfected human HeLa cells. *Biophys J* **68**, 2289–2298.

Cao F, Eckert R, Elfgang C *et al.* (1998). A quantitative analysis of connexin-specific permeability differences of gap junctions expressed in HeLa transfectants and *Xenopus* oocytes. *J Cell Sci* **111**(Pt 1), 31–43.

Charles AC, Naus CC, Zhu D *et al.* (1992). Intercellular calcium signaling via gap junctions in glioma cells. *J Cell Biol* **118**, 195–201.

Clarke TC, Thomas D, Petersen JS *et al.* (2006). The antiarrhythmic peptide rotigaptide (ZP123) increases gap junction intercellular communication in cardiac myocytes and HeLa cells expressing connexin 43. *Br J Pharmacol* **147**, 486–495.

Connors BW and Long MA (2004). Electrical synapses in the mammalian brain. *Annu Rev Neurosci* **27**, 393–418.

Cottrell GT and Burt JM (2005). Functional consequences of heterogeneous gap junction channel formation and its influence in health and disease. *Biochim Biophys Acta* **1711**, 126–141.

Cottrell GT, Lin R, Warn-Cramer BJ *et al.* (2003). Mechanism of v-Src- and mitogen-activated protein kinase-induced reduction of gap junction communication. *Am J Physiol Cell Physiol* **284**, C511–C520.

Cruciani V and Mikalsen SO (2002). Connexins, gap junctional intercellular communication and kinases. *Biol Cell* **94**, 433–443.

Cruikshank SJ, Hopperstad M, Younger M *et al.* (2004). Potent block of Cx36 and Cx50 gap junction channels by mefloquine. *PNAS* **101**, 12364–12369.

De Vuyst E, Decrock E, Cabooter L *et al.* (2006). Intracellular calcium changes trigger connexin 32 hemichannel opening. *EMBO J* **25**, 34–44.

Deans MR, Volgyi B, Goodenough DA *et al.* (2002). Connexin 36 is essential for transmission of rod-mediated visual signals in the mammalian retina. *Neuron* **36**, 703–712.

Demb JB and Pugh EN (2002). Connexin 36 forms synapses essential for night vision. *Neuron* **36**, 551–553.

Dewey MM and Barr L (1962). Intercellular connection between smooth muscle cells: the nexus. *Science* **137**, 670–672.

Dhein S, Larsen BD, Petersen JS *et al.* (2003). Effects of the new antiarrhythmic peptide ZP123 on epicardial activation and repolarization pattern. *Cell Commun Adhes* **10**, 371–378.

Dubina MV, Iatckii NA, Popov DE *et al.* (2002). Connexin 43, but not connexin 32, is mutated at advanced stages of human sporadic colon cancer. *Oncogene* **21**, 4992–4996.

Duffy HS, Fort AG and Spray DC (2006). Cardiac connexins: genes to nexus. *Adv Cardiol* **42**, 1–17.

Elenes S, Martinez AD, Delmar M *et al.* (2001). Heterotypic docking of Cx43 and Cx45 connexons blocks fast voltage gating of Cx43. *Biophys J* **81**, 1406–1418.

Elfgang C, Eckert R, Lichtenberg-Frate H *et al.* (1995). Specific permeability and selective formation of gap junction channels in connexin-transfected HeLa cells. *J Cell Biol* **129**, 805–817.

Falk MM, Buehler LK, Kumar NM *et al.* (1997). Cell-free synthesis and assembly of connexins into functional gap junction membrane channels. *EMBO J* **16**, 2703–2716.

Filson AJ, Azarnia R, Beyer EC *et al.* (1990). Tyrosine phosphorylation of a gap junction protein correlates with inhibition of cell-to-cell communication. *Cell Growth Differ* **1**, 661–668.

Flenniken AM, Osborne LR, Anderson N *et al.* (2005). A Gja1 missense mutation in a mouse model of oculodentodigital dysplasia. *Development* **132**, 4375–4386.

Fonseca CG, Green CR and Nicholson LF (2002). Upregulation in astrocytic connexin 43 gap junction levels may exacerbate generalized seizures in mesial temporal lobe epilepsy. *Brain Res* **929**, 105–116.

Frisch C, Souza-Silva MA, Sohl G *et al.* (2005). Stimulus complexity-dependent memory impairment and changes in motor performance after deletion of the neuronal gap junction protein connexin 36 in mice. *Behav Brain Res* **157**, 177–185.

Gajda Z, Gyengesi E, Hermesz E *et al.* (2003). Involvement of gap junctions in the manifestation and control of the duration of seizures in rats *in vivo*. *Epilepsia* **44**, 1596–1600.

Gajda Z, Hermesz E, Gyengesi E *et al.* (2006). The functional significance of gap junction channels in the epileptogenicity and seizure susceptibility of juvenile rats. *Epilepsia* **47**, 1009–1022.

Gajda Z, Szupera Z, Blazso G *et al.* (2005). Quinine, a blocker of neuronal cx36 channels, suppresses seizure activity in rat neocortex *in vivo*. *Epilepsia* **46**, 1581–1591.

Gareri P, Condorelli D, Belluardo N *et al.* (2004). Anticonvulsant effects of carbenoxolone in genetically epilepsy-prone rats (GEPRs). *Neuropharmacology* **47**, 1205–1216.

Gemel J, Valiunas V, Brink PR *et al.* (2004). Connexin 43 and connexin 26 form gap junctions, but not heteromeric channels in co-expressing cells. *J Cell Sci* **117**, 2469–2480.

Gerido DA and White TW (2004). Connexin disorders of the ear, skin, and lens. *Biochim Biophys Acta* **1662**, 159–170.

Giepmans BN (2004). Gap junctions and connexin-interacting proteins. *Cardiovasc Res* **62**, 233–245.

Giepmans BN and Moolenaar WH (1998). The gap junction protein connexin 43 interacts with the second PDZ domain of the zona occludens-1 protein. *Curr Biol* **8**, 931–934.

Giepmans BN, Verlaan I and Moolenaar WH (2001). Connexin-43 interactions with ZO-1 and alpha- and beta-tubulin. *Cell Commun Adhes* **8**, 219–223.

Gigout S, Louvel J, Kawasaki H *et al.* (2006). Effects of gap junction blockers on human neocortical synchronization. *Neurobiol Dis* **22**, 496–508.

Goldberg GS, Lampe PD and Nicholson BJ (1999). Selective transfer of endogenous metabolites through gap junctions composed of different connexins. *Nat Cell Biol* **1**, 457–459.

Goldberg GS, Moreno AP and Lampe PD (2002). Gap junctions between cells expressing connexin 43 or 32 show inverse permselectivity to adenosine and ATP. *J Biol Chem* **277**, 36725–36730.

Goldberg GS, Moreno AP, Bechberger JF *et al.* (1996). Evidence that disruption of connexon particle arrangements in gap junction plaques is associated with inhibition of gap junctional communication by a glycyrrhetinic acid derivative. *Exp Cell Res* **222**, 48–53.

Goldberg GS, Valiunas V and Brink PR (2004). Selective permeability of gap junction channels. *Biochim Biophys Acta* **1662**, 96–101.

Gomes P, Srinivas SP, Van Driessche W *et al.* (2005). ATP release through connexin hemichannels in corneal endothelial cells. *Invest Ophthalmol Vis Sci* **46**, 1208–1218.

Gonzalez D, Gomez-Hernandez JM and Barrio LC (2007). Molecular basis of voltage dependence of connexin channels: an integrative appraisal. *Prog Biophys Mol Biol* **94**, 66–106.

Goodenough DA and Paul DL (2003). Beyond the gap: functions of unpaired connexon channels. *Nat Rev Mol Cell Biol* **4**, 285–294.

Guan X, Wilson S, Schlender KK *et al.* (1996). Gap-junction disassembly and connexin 43 dephosphorylation induced by 18 beta-glycyrrhetinic acid. *Mol Carcinog* **16**, 157–164.

Guerra JM, Everett TH, Lee KW *et al.* (2006). Effects of the gap junction modifier rotigaptide (ZP123) on atrial conduction and vulnerability to atrial fibrillation. *Circulation* **114**, 110–118.

Guldenagel M, Ammermuller J, Feigenspan A *et al.* (2001). Visual transmission deficits in mice with targeted disruption of the gap junction gene connexin 36. *J Neurosci* **21**, 6036–6044.

Gutstein DE, Morley GE, Tamaddon H *et al.* (2001). Conduction slowing and sudden arrhythmic death in mice with cardiac-restricted inactivation of connexin 43. *Circ Res* **88**, 333–339.

Harks EG, de Roos AD, Peters PH *et al.* (2001). Fenamates: a novel class of reversible gap junction blockers. *J Pharmacol Exp Ther* **298**, 1033–1041.

Harris AL (2001). Emerging issues of connexin channels: biophysics fills the gap. *Q Rev Biophys* **34**, 325–472.

Harris AL (2007). Connexin channel permeability to cytoplasmic molecules. *Prog Biophys Mol Biol* **94**, 120–143.

Haugan K, Olsen KB, Hartvig L *et al.* (2005). The antiarrhythmic peptide analog ZP123 prevents atrial conduction slowing during metabolic stress. *J Cardiovasc Electrophysiol* **16**, 537–545.

Herve JC and Sarrouilhe D (2002). Modulation of junctional communication by phosphorylation: protein phosphatases, the missing link in the chain. *Biol Cell* **94**, 423–432.

Hormuzdi SG, Pais I, LeBeau FE *et al.* (2001). Impaired electrical signaling disrupts gamma frequency oscillations in connexin 36-deficient mice. *Neuron* **31**, 487–495.

Hosseinzadeh H, Asl MN, Parvardeh S *et al.* (2005). The effects of carbenoxolone on spatial learning in the Morris water maze task in rats. *Med Sci Monit* **11**, BR88–BR94.

Janik P, Assaf A and Bertram JS (1980). Inhibition of growth of primary and metastatic Lewis lung carcinoma cells by the phosphodiesterase inhibitor isobutylmethylxanthine. *Cancer Res* **40**, 1950–1954.

Jensen O, Kaiser J and Lachaux JP (2007). Human gamma-frequency oscillations associated with attention and memory. *TINS* **30**, 317–324.

Kanno Y and Loewenstein WR (1964). Low-resistance coupling between gland cells. Some observations on intercellular contact membranes and intercellular space. *Nature* **201**, 194–195.

Kelsell DP, Di WL and Houseman MJ (2001). Connexin mutations in skin disease and hearing loss. *Am J Hum Genet* **68**, 559–568.

King TJ and Bertram JS (2005). Connexins as targets for cancer chemoprevention and chemotherapy. *Biochim Biophys Acta* **1719**, 146–160.

King TJ and Lampe PD (2004a). Mice deficient for the gap junction protein connexin 32 exhibit increased radiation-induced tumorigenesis associated with elevated mitogen-activated protein kinase (p44/Erk1, p42/Erk2) activation. *Carcinogenesis* **25**, 669–680.

King TJ and Lampe PD (2004b). The gap junction protein connexin 32 is a mouse lung tumor suppressor. *Cancer Res* **64**, 7191–7196.

King TJ, Fukushima LH, Donlon TA *et al.* (2000). Correlation between growth control, neoplastic potential and endogenous connexin43 expression in HeLa cell lines: implications for tumor progression. *Carcinogenesis* **21**, 311–315.

Kjolbye AL, Haugan K, Hennan JK *et al.* (2007). Pharmacological modulation of gap junction function with the novel compound rotigaptide: a promising new principle for prevention of arrhythmias. *Basic Clin Pharmacol Toxicol* **101**, 215–230.

Laing JG, Chou BC and Steinberg TH (2005). ZO-1 alters the plasma membrane localization and function of Cx43 in osteoblastic cells. *J Cell Sci* **118**, 2167–2176.

Laird DW (2005). Connexin phosphorylation as a regulatory event linked to gap junction internalization and degradation. *Biochim Biophys Acta* **1711**, 172–182.

Laird DW (2006). Life cycle of connexins in health and disease. *Biochem J* **394**, 527–543.

Laird DW, Castillo M and Kasprzak L (1995). Gap junction turnover, intracellular trafficking, and phosphorylation of connexin 43 in brefeldin A-treated rat mammary tumor cells. *J Cell Biol* **131**, 1193–1203.

Lampe PD and Lau AF (2004). The effects of connexin phosphorylation on gap junctional communication. *Int J Biochem Cell Biol* **36**, 1171–1186.

Lautermann J, ten Cate WJ, Altenhoff P *et al.* (1998). Expression of the gap-junction connexins 26 and 30 in the rat cochlea. *Cell Tissue Res* **294**, 415–420.

Lawrence TS, Beers WH and Gilula NB (1978). Transmission of hormonal stimulation by cell-to-cell communication. *Nature* **272**, 501–506.

Li J, Patel VV, Kostetskii I *et al.* (2005). Cardiac-specific loss of N-cadherin leads to alteration in connexins with conduction slowing and arrhythmogenesis. *Circ Res* **97**, 474–481.

Li J, Shen H, Naus CC *et al.* (2001). Upregulation of gap junction connexin 32 with epileptiform activity in the isolated mouse hippocampus. *Neuroscience* **105**, 589–598.

Litvin O, Tiunova A, Connell-Alberts Y *et al.* (2006). What is hidden in the pannexin treasure trove: the sneak peek and the guesswork. *J Cell Mol Med* **10**, 613–634.

Locovei S, Scemes E, Qiu F *et al.* (2007). Pannexin1 is part of the pore forming unit of the P2X(7) receptor death complex. *FEBS Lett* **581**, 483–488.

Loewenstein WR and Kanno Y (1966). Intercellular communication and the control of tissue growth: lack of communication between cancer cells. *Nature* **209**, 1248–1249.

Matesic DF, Rupp HL, Bonney WJ *et al.* (1994). Changes in gap-junction permeability, phosphorylation, and number mediated by phorbol ester and non-phorbol-ester tumor promoters in rat liver epithelial cells. *Mol Carcinog* **10**, 226–236.

Medina-Ceja L, Cordero-Romero A and Morales-Villagran A (2008). Antiepileptic effect of carbenoxolone on seizures induced by 4-aminopyridine: a study in the rat hippocampus and entorhinal cortex. *Brain Res* **1187**, 74–81.

Mesnil M (2002). Connexins and cancer. *Biol Cell* **94**, 493–500.

Moennikes O, Buchmann A, Romualdi A *et al.* (2000). Lack of phenobarbital-mediated promotion of hepatocarcinogenesis in connexin 32-null mice. *Cancer Res* **60**, 5087–5091.

Moreno AP (2004). Biophysical properties of homomeric and heteromultimeric channels formed by cardiac connexins. *Cardiovasc Res* **62**, 276–286.

Moreno AP and Lau AF (2007). Gap junction channel gating modulated through protein phosphorylation. *Prog Biophys Mol Biol* **94**, 107–119.

Moreno AP, Chanson M, Elenes S *et al.* (2002). Role of the carboxyl terminal of connexin 43 in transjunctional fast voltage gating. *Circ Res* **90**, 450–457.

Moreno AP, Fishman GI and Spray DC (1992). Phosphorylation shifts unitary conductance and modifies voltage dependent kinetics of human connexin 43 gap junction channels. *Biophys J* **62**, 51–53.

Moreno AP, Rook MB, Fishman GI *et al.* (1994a). Gap junction channels: distinct voltage-sensitive and -insensitive conductance states. *Biophys J* **67**, 113–119.

Moreno AP, Saez JC, Fishman GI *et al.* (1994b). Human connexin 43 gap junction channels. Regulation of unitary conductances by phosphorylation. *Circ Res* **74**, 1050–1057.

Morley GE, Taffet SM and Delmar M (1996). Intramolecular interactions mediate pH regulation of connexin 43 channels. *Biophys J* **70**, 1294–1302.

Musil LS and Goodenough DA (1993). Multisubunit assembly of an integral plasma membrane channel protein, gap junction connexin 43, occurs after exit from the ER. *Cell* **74**, 1065–1077.

Nagy JI, Dudek FE and Rash JE (2004). Update on connexins and gap junctions in neurons and glia in the mammalian nervous system. *Brain Res Brain Res Rev* **47**, 191–215.

Nakase T and Naus CC (2004). Gap junctions and neurological disorders of the central nervous system. *Biochim Biophys Acta* **1662**, 149–158.

Naus CC, Bechberger JF and Paul DL (1991). Gap junction gene expression in human seizure disorder. *Exp Neurol* **111**, 198–203.

Nelis E, Haites N and Van Broeckhoven C (1999). Mutations in the peripheral myelin genes and associated genes in inherited peripheral neuropathies. *Hum Mutat* **13**, 11–28.

Nielsen PA, Baruch A, Giepmans BN *et al.* (2001). Characterization of the association of connexins and ZO-1 in the lens. *Cell Commun Adhes* **8**, 213–217.

Niessen H, Harz H, Bedner P *et al.* (2000). Selective permeability of different connexin channels to the second messenger inositol 1,4,5-trisphosphate. *J Cell Sci* **113**(Pt 8), 1365–1372.

Nilsen KE, Kelso AR and Cock HR (2006). Antiepileptic effect of gap-junction blockers in a rat model of refractory focal cortical epilepsy. *Epilepsia* **47**, 1169–1175.

Oyamada M, Oyamada Y and Takamatsu T (2005). Regulation of connexin expression. *Biochim Biophys Acta* **1719**, 6–23.

Panchin YV (2005). Evolution of gap junction proteins - the pannexin alternative. *J Exp Biol* **208**, 1415–1419.

Paul DL, Ebihara L, Takemoto LJ *et al.* (1991). Connexin 46, a novel lens gap junction protein, induces voltage-gated currents in nonjunctional plasma membrane of *Xenopus* oocytes. *J Cell Biol* **115**, 1077–1089.

Paznekas WA, Boyadjiev SA, Shapiro RE *et al.* (2003). Connexin 43 (GJA1) mutations cause the pleiotropic phenotype of oculodentodigital dysplasia. *Am J Hum Genet* **72**, 408–418.

Peracchia C, Wang XG and Peracchia LL (2000). Slow gating of gap junction channels and calmodulin. *J Membr Biol* **178**, 55–70.

Perez Velazquez JL and Carlen PL (2000). Gap junctions, synchrony and seizures. *TINS* **23**, 68–74.

Perkins G, Goodenough D and Sosinsky G (1997). Three-dimensional structure of the gap junction connexon. *Biophys J* **72**, 533–544.

Perkins GA, Goodenough DA and Sosinsky GE (1998). Formation of the gap junction intercellular channel requires a 30 degree rotation for interdigitating two apposing connexons. *J Mol Biol* **277**, 171–177.

Petersen MB and Willems PJ (2006). Non-syndromic, autosomal-recessive deafness. *Clin Genet* **69**, 371–392.

Piehl M, Lehmann C, Gumpert A *et al.* (2007). Internalization of large double-membrane intercellular vesicles by a clathrin-dependent endocytic process. *Mol Biol Cell* **18**, 337–347.

Pointis G, Fiorini C, Gilleron J *et al.* (2007). Connexins as precocious markers and molecular targets for chemical and pharmacological agents in carcinogenesis. *Curr Med Chem* **14**, 2288–2303.

Proulx E, Leshchenko Y, Kokarovtseva L et al. (2006). Functional contribution of specific brain areas to absence seizures: role of thalamic gap-junctional coupling. Eur J Neurosci 23, 489–496.

Rabionet R, Gasparini P and Estivill X (2000). Molecular genetics of hearing impairment due to mutations in gap junction genes encoding beta connexins. Hum Mutat 16, 190–202.

Reaume AG, de Sousa PA, Kulkarni S et al. (1995). Cardiac malformation in neonatal mice lacking connexin 43. Science 267, 1831–1834.

Revel JP, Yee AG and Hudspeth AJ (1971). Gap junctions between electrotonically coupled cells in tissue culture and in brown fat. PNAS 68, 2924–2927.

Rouan F, White TW, Brown N et al. (2001). Trans-dominant inhibition of connexin-43 by mutant connexin-26: implications for dominant connexin disorders affecting epidermal differentiation. J Cell Sci 114, 2105–2113.

Salameh A and Dhein S (2005). Pharmacology of gap junctions. New pharmacological targets for treatment of arrhythmia, seizure and cancer? Biochim Biophys Acta 1719, 36–58.

Samoilova M, Li J, Pelletier MR et al. (2003). Epileptiform activity in hippocampal slice cultures exposed chronically to bicuculline: increased gap junctional function and expression. J Neurochem 86, 687–699.

Sandeep TC, Yau JL, MacLullich AM et al. (2004). 11Beta-hydroxysteroid dehydrogenase inhibition improves cognitive function in healthy elderly men and type 2 diabetics. PNAS 101, 6734–6739.

Segretain D and Falk MM (2004). Regulation of connexin biosynthesis, assembly, gap junction formation, and removal. Biochim Biophys Acta 1662, 3–21.

Seren S, Lieberman R, Bayraktar UD et al. (2008). Lycopene in cancer prevention and treatment. Am J Ther 15, 66–81.

Severs NJ, Coppen SR, Dupont E et al. (2004). Gap junction alterations in human cardiac disease. Cardiovasc Res 62, 368–377.

Shiroshita-Takeshita A, Sakabe M, Haugan K et al. (2007). Model-dependent effects of the gap junction conduction-enhancing antiarrhythmic peptide rotigaptide (ZP123) on experimental atrial fibrillation in dogs. Circulation 115, 310–318.

Sinfield JL and Collins DR (2006). Induction of synchronous oscillatory activity in the rat lateral amygdala in vitro is dependent on gap junction activity. Eur J Neurosci 24, 3091–3095.

Singal R, Tu ZJ, Vanwert JM et al. (2000). Modulation of the connexin 26 tumor suppressor gene expression through methylation in human mammary epithelial cell lines. Anticancer Res 20, 59–64.

Singh D, Solan JL, Taffet SM et al. (2005). Connexin 43 interacts with zona occludens-1 and -2 proteins in a cell cycle stage-specific manner. J Biol Chem 280, 30416–30421.

Sohl G, Maxeiner S and Willecke K (2005). Expression and functions of neuronal gap junction. Nat Rev Neurosci 6, 191–200.

Sohl G and Willecke K (2003). An update on connexin genes and their nomenclature in mouse and man. Cell Commun Adhes 10, 173–180.

Sosinsky GE and Nicholson BJ (2005). Structural organization of gap junction channels. Biochim Biophys Acta 1711, 99–125.

Srinivas M and Spray DC (2003). Closure of gap junction channels by arylaminobenzoates. Mol Pharmacol 63, 1389–1397.

Srinivas M, Calderon DP, Kronengold J et al. (2006). Regulation of connexin hemichannels by monovalent cations. J Gen Physiol 127, 67–75.

Srinivas M, Hopperstad MG and Spray DC (2001). Quinine blocks specific gap junction channel subtypes. PNAS 98, 10942–10947.

Stahlhut M, Petersen JS, Hennan JK et al. (2006). The antiarrhythmic peptide rotigaptide (ZP123) increases connexin 43 protein expression in neonatal rat ventricular cardiomyocytes. Cell Commun Adhes 13, 21–27.

Steinberg TH, Civitelli R, Geist ST et al. (1994). Connexin 43 and connexin 45 form gap junctions with different molecular permeabilities in osteoblastic cells. EMBO J 13, 744–750.

Stergiopoulos K, Alvarado JL, Mastroianni M et al. (1999). Hetero-domain interactions as a mechanism for the regulation of connexin channels. Circ Res 84, 1144–1155.

Stout CE, Costantin JL, Naus CC et al. (2002). Intercellular calcium signaling in astrocytes via ATP release through connexin hemichannels. J Biol Chem 277, 10482–10488.

Suchyna TM, Nitsche JM, Chilton M et al. (1999). Different ionic selectivities for connexins 26 and 32 produce rectifying gap junction channels. Biophys J 77, 2968–2987.

Suchyna TM, Xu LX, Gao F et al. (1993). Identification of a proline residue as a transduction element involved in voltage gating of gap junctions. Nature 365, 847–849.

Sun J, Ahmad S, Chen S et al. (2005). Cochlear gap junctions coassembled from Cx26 and 30 show faster intercellular Ca2+ signaling than homomeric counterparts. Am J Physiol Cell Physiol 288, C613–C623.

Tabernero A, Giaume C and Medina JM (1996). Endothelin-1 regulates glucose utilization in cultured astrocytes by controlling intercellular communication through gap junctions. Glia 16, 187–195.

Takens-Kwak BR and Jongsma HJ (1992). Cardiac gap junctions: three distinct single channel conductances and their modulation by phosphorylating treatments. Pflugers Arch 422, 198–200.

Tan LW, Bianco T and Dobrovic A (2002). Variable promoter region CpG island methylation of the putative tumor suppressor gene connexin 26 in breast cancer. Carcinogenesis 23, 231–236.

Taylor CP and Dudek FE (1982). A physiological test for electrotonic coupling between CA1 pyramidal cells in rat hippocampal slices. Brain Res 235, 351–357.

Temme A, Buchmann A, Gabriel HD et al. (1997). High incidence of spontaneous and chemically induced liver tumors in mice deficient for connexin 32. Curr Biol 7, 713–716.

Traub RD, Whittington MA, Buhl EH et al. (2001). A possible role for gap junctions in generation of very fast EEG oscillations preceding the onset of, and perhaps initiating, seizures. Epilepsia 42, 153–170.

Trexler EB, Bukauskas FF, Bennett MV et al. (1999). Rapid and direct effects of pH on connexins revealed by the connexin 46 hemichannel preparation. J Gen Physiol 113, 721–742.

Turin L and Warner A (1977). Carbon dioxide reversibly abolishes ionic communication between cells of early amphibian embryo. Nature 270, 56–57.

Uhlhaas PJ and Singer W (2006). Neural synchrony in brain disorders: relevance for cognitive dysfunctions and pathophysiology. Neuron 52, 155–168.

Vaidya D, Tamaddon HS, Lo CW et al. (2001). Null mutation of connexin 43 causes slow propagation of ventricular activation in the late stages of mouse embryonic development. Circ Res 88, 1196–1202.

Valiunas V (2002). Biophysical properties of connexin-45 gap junction hemichannels studied in vertebrate cells. J Gen Physiol 119, 147–164.

van der Velden HM and Jongsma HJ (2002). Cardiac gap junctions and connexins: their role in atrial fibrillation and potential as therapeutic targets. Cardiovasc Res 54, 270–279.

Vanden Abeele F, Bidaux G, Gordienko D et al. (2006). Functional implications of calcium permeability of the channel formed by pannexin 1. J Cell Biol 174, 535–546.

Veenstra RD, Wang HZ, Beblo DA et al. (1995). Selectivity of connexin-specific gap junctions does not correlate with channel conductance. Circ Res 77, 1156–1165.

Verselis VK, Ginter CS and Bargiello TA (1994). Opposite voltage gating polarities of two closely related connexins. Nature 368, 348–351.

Verselis VK, Trexler EB and Bukauskas FF (2000). Connexin hemichannels and cell–cell channels: comparison of properties. Braz J Med Biol Res 33, 379–389.

Vine AL and Bertram JS (2002). Cancer chemoprevention by connexins. Cancer Metastasis Rev 21, 199–216.

Warn-Cramer BJ and Lau AF (2004). Regulation of gap junctions by tyrosine protein kinases. *Biochim Biophys Acta* **1662**, 81–95.

Weber PA, Chang HC, Spaeth KE *et al.* (2004). The permeability of gap junction channels to probes of different size is dependent on connexin composition and permeant-pore affinities. *Biophys J* **87**, 958–973.

Weng S, Lauven M, Schaefer T *et al.* (2002). Pharmacological modification of gap junction coupling by an antiarrhythmic peptide via protein kinase C activation. *FASEB J* **16**, 1114–1116.

White TW (2002). Unique and redundant connexin contributions to lens development. *Science* **295**, 319–320.

White TW and Bruzzone R (2000). Intercellular communication in the eye: clarifying the need for connexin diversity. *Brain Res Brain Res Rev* **32**, 130–137.

Willeche K, Eiberger J, Degen J *et al.* (2002). Structural and functional diversity of connexin genes in the mouse and human genome. *Biol Chem* **383**, 725–737.

Windoffer R, Beile B, Leibold A *et al.* (2000). Visualization of gap junction mobility in living cells. *Cell Tissue Res* **299**, 347–362.

Wooltorton E (2002). Mefloquine: contraindicated in patients with mood, psychotic or seizure disorders. *CMAJ* **167**, 1147.

Xing D, Kjolbye AL, Nielsen MS *et al.* (2003). ZP123 increases gap junctional conductance and prevents reentrant ventricular tachycardia during myocardial ischemia in open chest dogs. *J Cardiovasc Electrophysiol* **14**, 510–520.

Yang L and Ling DS (2007). Carbenoxolone modifies spontaneous inhibitory and excitatory synaptic transmission in rat somatosensory cortex. *Neurosci Lett* **416**, 221–226.

Yeager M and Harris AL (2007). Gap junction channel structure in the early 21st century: facts and fantasies. *Curr Opin Cell Biol* **19**, 521–528.

Yeager M and Nicholson BJ (1996). Structure of gap junction intercellular channels. *Curr Opin Struct Biol* **6**, 183–192.

Zsiros V, Aradi I and Maccaferri G (2007). Propagation of postsynaptic currents and potentials via gap junctions in GABAergic networks of the rat hippocampus. *J Physiol* **578**, 527–544.

Index

The index entries appear in word-by-word alphabetical order.
Entries in italics indicate information contained in tables and figures.